AF575090

HANDBOOK OF OPTICS

Other McGraw-Hill Books of Interest

Hecht • THE LASER GUIDEBOOK
Manning • STOCHASTIC ELECTROMAGNETIC IMAGE PROPAGATION
Nishihara, Haruna, Suhara • OPTICAL INTEGRATED CIRCUITS
Rancourt • OPTICAL THIN FILMS USERS' HANDBOOK
Sibley • OPTICAL COMMUNICATIONS
Smith • MODERN OPTICAL ENGINEERING
Smith • MODERN LENS DESIGN
Waynant, Ediger • ELECTRO-OPTICS HANDBOOK
Wyatt • ELECTRO-OPTICAL SYSTEM DESIGN

HANDBOOK OF OPTICS

Volume I
Fundamentals, Techniques, and Design

Second Edition

Sponsored by the
OPTICAL SOCIETY OF AMERICA

Michael Bass Editor in Chief

The Center for Research and
Education in Optics and Lasers (CREOL)
University of Central Florida
Orlando, Florida

Eric W. Van Stryland Associate Editor

The Center for Research and Education
in Optics and Lasers (CREOL)
University of Central Florida
Orlando, Florida

David R. Williams Associate Editor

Center for Visual Science
University of Rochester
Rochester, New York

William L. Wolfe Associate Editor

Optical Sciences Center
University of Arizona
Tucson, Arizona

McGRAW-HILL, INC.

New York San Francisco Washington, D.C. Auckland Bogotá
Caracas Lisbon London Madrid Mexico City Milan
Montreal New Delhi San Juan Singapore
Sydney Tokyo Toronto

Library of Congress Cataloging-in-Publication Data

Handbook of optics / sponsored by the Optical Society of America ;
Michael Bass, editor in chief. — 2nd ed.
p. cm.
Includes bibliographical references and index.
Contents: 1. Fundamentals, techniques, and design — 2. Devices, measurement, and properties.
ISBN 0-07-047740-X
1. Optics—Handbooks, manuals, etc. 2. Optical instruments—Handbooks, manuals, etc. I. Bass, Michael. II. Optical Society of America.
QC369.H35 1995
535—dc20 94-19339
CIP

1 2 3 4 5 6 7 8 9 DOC/DOC 9 0 9 8 7 6 5 4

ISBN 0-07-047740-X

The sponsoring editor for this book was Stephen S. Chapman, the editing supervisor was Peggy Lamb, and the production supervisor was Pamela A. Pelton. It was set in Times Roman by The Universities Press (Belfast) Ltd.

Printed and bound by R.R. Donnelly & Sons Company.

This book was printed on acid-free paper.

CONTENTS

Part 6. Imaging Detectors 20.1

Part 10. Optical Fabrication 40.1

Chapter 40. Optical Fabrication *Robert Parks* 40.3

Chapter 41. Fabrication of Optics by Diamond Turning *Richard L. Rhorer and Chris J. Evans* 41.1

Part 11. Optical Properties of Films and Coatings 42.1

Chapter 42. Optical Properties of Films and Coatings *John A. Dobrowolski* 42.3

CONTRIBUTORS

Joseph H. Altman *Institute of Optics, University of Rochester, Rochester, New York* (CHAP. 20).

Martin S. Banks *School of Optometry, University of California, Berkeley, Berkeley, California* (CHAP. 25).

Jean M. Bennett *Research Department, Michelson Laboratory, Naval Air Warfare Center, China Lake, California* (CHAP. 5).

Craig F. Bohren *Meteorology Department, Pennsylvania State University, University Park, Pennsylvania* (CHAP. 6).

John E. Bowers *Department of Electrical and Computer Engineering, University of California, Santa Barbara, Santa Barbara, California* (CHAP. 17).

David Brainard *Department of Psychology, University of California, Santa Barbara, Santa Barbara, California* (CHAP. 26).

Robert P. Breault *Breault Research Organization, Inc., Tucson, Arizona* (CHAP. 38).

Stephen A. Burns *The Schepens Eye Research Institute, Boston, Massachusetts* (CHAP. 28).

William H. Carter *Naval Research Laboratory, Washington, D.C.* (CHAP. 4).

W. N. Charman *Institute of Science and Technology, Department of Ophthalmic Optics, University of Manchester, Manchester, United Kingdom* (CHAP. 24).

Eugene L. Church *Brookhaven National Laboratory, Upton, New York* (CHAP. 7).

James Churnside *National Oceanic and Atmospheric Administration, Wave Propagation Laboratory, Boulder, Colorado* (CHAP. 44).

William Cowan *Department of Comptuer Science, University of Waterloo, Waterloo, Ontario, Canada* (CHAP. 27).

M. George Craford *Hewlett-Packard Company, San Jose, California* (CHAP. 12).

Pamela L. Derry *Boeing Defense and Space Group, Aerospace and Electronics Division, Seattle, Washington* (CHAP. 13).

Jean-Claude Diels *Department of Physics and Astronomy, University of New Mexico, Albuquerque, New Mexico* (CHAP. 14).

John A. Dobrowolski *Institute for Microstructural Sciences, National Research Council of Canada, Ottawa, Ontario, Canada* (CHAP. 42).

Chris J. Evans *National Institute of Standards and Technology, Gaithersburg, Maryland* (CHAP. 41).

Bart Farrell *Institue for Sensory Research, Syracuse University, Syracuse, New York* (CHAP. 29).

Luis Figueroa *Boeing Defense and Space Group, Aerospace and Electronics Division, Seattle, Washington* (CHAP. 13).

Wilson S. Geisler *Department of Psychology, University of Texas, Austin, Texas* (CHAP. 25).

Douglas S. Goodman *IBM Research, Yorktown Heights, New York* (CHAP. 1).

Joseph W. Goodman *Department of Electrical Engineering, Stanford University, Stanford, California* (CHAP. 30).

John E. Greivenkamp, Jr. *Optical Sciences Center, Tucson, Arizona* (CHAP. 2).

Roland H. Haitz *Hewlett-Packard Company, San Jose, California* (CHAP. 12).

Michael E. Harrigan *Eastman Kodak Company, Electronic Imaging Research Laboratory, Rochester, New York* (CHAP. 33).

Brian Henderson *Department of Physics and Applied Physics, University of Strathclyde, Glasgow, United Kingdom* (CHAP. 8).

Chi-Shain Hong *Boeing Defense and Space Group, Aerospace and Electronics Division, Seattle, Washington* (CHAP. 13).

C. Bruce Johnson *Litton Electron Devices, Tempe, Arizona* (CHAP. 21).

Abhay M. Joshi *Discovery Semiconductors, Inc., Cranbury, New Jersey* (CHAP. 16).

Dennis Killinger *Department of Physics, University of South Florida, Tampa, Florida* (CHAP. 44).

Walter F. Kosonocky *Electrical and Computer Engineering, New Jersey Institute of Technology, University Heights, Newark, New Jersey* (CHAP. 23).

Lester J. Kozlowski *Rockwell International Science Center, Thousand Oaks, California* (CHAP. 23).

Paul W. Kruse *Consultant, Edina, Minnesota* (CHAP. 19).

Anthony LaRocca *ERIM, Ann Arbor, Michigan* (CHAP. 10).

Masud Mansuripur *Optical Sciences Center, University of Arizona, Tucson, Arizona* (CHAP. 31).

Arvind S. Marathay *Optical Sciences Center, University of Arizona, Tucson, Arizona* (CHAP. 3).

Alan Miller *Department of Physics and Astronomy, University of St. Andrews, St. Andrews, Fife, United Kingdom, and Center for Research and Education in Optics and Lasers (CREOL), University of Central Florida, Orlando, Florida* (CHAP. 9).

Curtis D. Mobley *Senior Research Engineer, Applied Electromagnetics and Optics Laboratory, SRI Industries, Menlo Park, California* (CHAP. 43).

Paul Norton *Santa Barbara Research Center, Goleta, California* (CHAP. 15).

Gregory H. Olsen *Sensors Unlimited, Inc., Princeton, New Jersey* (CHAP. 16).

Donald C. O'Shea *Georgia Institute of Technology, Center for Optical Science and Engineering and School of Physics, Atlanta, Georgia* (CHAP. 33).

Larry D. Owen *Litton Electron Devices, Tempe, Arizona* (CHAP. 21).

Robert Parks *Optical Sciences Center, University of Arizona, Tucson, Arizona* (CHAP. 40).

Denis G. Pelli *Institute for Sensory Research, Syracuse University, Syracuse, New York* (CHAP. 29).

Richard L. Rhorer *Group Leader, Fabrication Development, Los Alamos National Laboratory, Los Alamos, New Mexico* (CHAP. 41).

Michael Roberts *Pilkington Optronics, St. Asaph, Clwyd, Wales, United Kingdom* (CHAP. 39).

Phillip Rodgers *Pilkington Optronics, St. Asaph, Clwyd, Wales, United Kingdom* (CHAP. 39).

Laurence S. Rothman *Air Force Geophysics Directorate/Phillips Laboratory, Optical Environment Division, Hanscom Air Force Base, Massachusetts* (CHAP. 44).

Robert R. Shannon *Optical Sciences Center, University of Arizona, Tucson, Arizona* (CHAPS. 35 AND 36).

William T. Silfvast *Center for Research and Education in Optics and Lasers (CREOL), University of Central Florida, Orlando, Florida* (CHAP. 11).

Douglas C. Sinclair *Sinclair Optics, Fairport, New York* (CHAP. 34).

Warren J. Smith *Kaiser Electro-Optics Inc., Carlsbad, California* (CHAP. 32).

Peter Z. Takacs *Brookhaven National Laboratory, Upton, New York* (CHAP. 7).

Timothy J. Tredwell *Eastman Kodak Company, Sensor Systems Division, Imager Systems Development Laboratory, Rochester, New York* (CHAP. 22).

Robert H. Webb *The Schepens Eye Research Institute, Boston, Massachusetts* (CHAP. 28).

Robert H. Weissman *Hewlett-Packard Company, San Jose, California* (CHAP. 12).

Y. G. Wey *Department of Electrical and Computer Engineering, University of California, Santa Barbara, Santa Barbara, California* (CHAP. 17).

John R. Willison *Stanford Research Systems, Sunnyvale, California* (CHAP. 18).

William L. Wolfe *Professor, Optical Sciences Center, University of Arizona, Tucson, Arizona* (CHAP. 19).

Paul R. Yoder, Jr. *Consultant in Optical Engineering, Norwalk, Connecticut* (CHAP. 37).

Xin Miao Zhao *Department of Physics and Astronomy, University of New Mexico, Albuquerque, New Mexico* (CHAP. 14).

PREFACE

The *Handbook of Optics,* Second Edition, is designed to serve as a general purpose desktop reference for the field of Optics yet stay within the confines of two books of finite length. Our purpose is to cover as much of optics as possible in a manner enabling the reader to deal with both basic and applied problems. To this end, we present articles about basic concepts, techniques, devices, instruments, measurements, and optical properties. In selecting subjects to include, we also had to select which subjects to leave out. The criteria we applied when excluding a subject were: (1) was it a specific application of optics rather than a core science or technology and (2) was it a subject in which the role of optics was peripheral to the central issue addressed. Thus, such topics as medical optics, laser surgery, and laser materials processing were not included. The resulting *Handbook of Optics,* Second Edition, serves the long-term information needs of those working in optics rather than presenting highly specific papers of current interest.

The authors were asked to prepare archival, tutorial articles which contain not only useful data but also descriptive material and references. Such articles were designed to enable the reader to understand a topic sufficiently well to get started using that knowledge. They also supply guidance as to where to find more in-depth material. Most include cross references to related articles within the Handbook. While applications of optics are mentioned, there is not space in the Handbook to include articles devoted to all of the myriad uses of optics in today's world. If we had, the Handbook would have been many volumes long and would have been too soon outdated.

The *Handbook of Optics,* Second Edition, contains 83 chapters organized into 17 broad categories or parts. The categorization enables the reader to find articles on a specific subject, say Vision, more easily and to find related articles within the Handbook. Within the categories the articles are grouped to make it simpler to find related material.

Volume I presents tutorial articles in the categories of Geometric Optics, Physical Optics, Quantum Optics, Optical Sources, Optical Detectors, Imaging Detectors, Vision, Optical Information and Image Processing, Optical Design Techniques, Optical Fabrication, Optical Properties of Films and Coatings, and Terrestrial Optics. This material is, for the most part, in a form which could serve to teach the underlying concepts of optics and its implementation. In fact, by careful selection of what to present and how to present it, the contents of Volume I could be used as a text for a comprehensive course in Optics.

The subjects covered in Volume II are Optical Elements, Optical Instruments, Optical Measurements, Optical and Physical Properties of Materials, and Nonlinear and Photorefractive Optics. As can be seen from these titles, Volume II concerns the specific devices, instruments, and techniques which are needed to employ optics in a wide variety of problems. It also provides data and discussion to assist one in the choice of optical materials.

The *Handbook of Optics*, Second Edition, would not have been possible without the support of the staff of the Optical Society of America and in particular Mr. Alan N. Tourtlotte and Ms. Kelly Furr.

For his pivotal roles in the development of the Optical Society of America, in the development of the profession of Optics, and for his encouragement to us in the task of preparing this Handbook, the editors dedicate this edition to Dr. Jarus Quinn.

Michael Bass, Editor-in-Chief
Eric W. Van Stryland, Associate Editor
David R. Williams, Associate Editor
William L. Wolfe, Associate Editor

GLOSSARY AND FUNDAMENTAL CONSTANTS

Introduction

This glossary of the terms used in the Handbook represents to a large extent the language of optics. The symbols are representations of numbers, variables, and concepts. Although the basic list was compiled by the author of this section, all the editors have contributed and agreed to this set of symbols and definitions. Every attempt has been made to use the same symbols for the same concepts throughout the entire Handbook, although there are exceptions. Some symbols seem to be used for many concepts. The symbol α is a prime example, as it is used for absorptivity, absorption coefficient, coefficient of linear thermal expansion, and more. Although we have tried to limit this kind of redundancy, we have also bowed deeply to custom.

Units

The abbreviations for the most common units are given first. They are consistent with most of the established lists of symbols, such as given by the International Standards Organization ISO[1] and the International Union of Pure and Applied Physics, IUPAP.[2]

Prefixes

Similarly, a list of the numerical prefixes[1] that are most frequently used is given, along with both the common names (where they exist) and the multiples of ten that they represent.

Fundamental Constants

The values of the fundamental constants[3] are listed following the sections on SI units.

Symbols

The most commonly used symbols are then given. Most chapters of the Handbook also have a glossary of the terms and symbols specific to them for the convenience of the reader. In the following list, the symbol is given, its meaning is next, and the most customary unit of measure for the quantity is presented in brackets. A bracket with a dash in it indicates that the quantity is unitless. Note that there is a difference between units and dimensions. An angle has units of degrees or radians and a solid angle square degrees or steradians, but both are pure ratios and are dimensionless. The unit symbols as recommended in the SI system are used, but decimal multiples of some of the dimensions are sometimes given. The symbols chosen, with some cited exceptions, are also those of the first two references.

RATIONALE FOR SOME DISPUTED SYMBOLS

The choice of symbols is a personal decision, but commonality improves communication. This section explains why the editors have chosen the preferred symbols for the Handbook. We hope that this will encourage more agreement.

Fundamental Constants

It is encouraging that there is almost universal agreement for the symbols for the fundamental constants. We have taken one small exception by adding a subscript B to the k for Boltzmann's constant.

Mathematics

We have chosen i as the imaginary almost arbitrarily. IUPAP lists both i and j, while ISO does not report on these.

Spectral Variables

These include expressions for the wavelength, λ, frequency, v, wave number, σ, ω for circular or radian frequency, k for circular or radian wave number and dimensionless frequency x. Although some use f for frequency, it can be easily confused with electronic or spatial frequency. Some use $\tilde{v}$ for wave number, but, because of typography problems and agreement with ISO and IUPAP, we have chosen σ; it should not be confused with the Stefan-Boltzmann constant. For spatial frequencies we have chosen ξ and η, although f_x and f_y are sometimes used. ISO and IUPAP do not report on these.

Radiometry

Radiometric terms are contentious. The most recent set of recommendations by ISO and IUPAP are L for radiance [$\mathrm{Wcm^{-2}\,sr^{-1}}$], M for radiant emittance or exitance [$\mathrm{Wcm^{-2}}$], E for irradiance or incidance [$\mathrm{Wcm^{-2}}$], and I for intensity [$\mathrm{Wsr^{-2}}$]. The previous terms, W, H, N, and J, respectively, are still in many texts, notably Smith and Lloyd,[4] but we have used the revised set, although there are still shortcomings. We have tried to deal with the vexatious term *intensity* by using *specific intensity* when the units are $\mathrm{Wcm^{-2}\,sr^{-1}}$, *field intensity* when they are $\mathrm{Wcm^{-2}}$, and *radiometric intensity* when they are $\mathrm{Wsr^{-1}}$.

There are two sets of terms for these radiometric quantities, which arise in part from the terms for different types of reflection, transmission, absorption, and emission. It has been proposed that the *ion* ending indicate a process, that the *ance* ending indicate a value associated with a particular sample, and that the *ivity* ending indicate a generic value for a "pure" substance. Then one also has reflectance, transmittance, absorptance, and emittance as well as reflectivity, transmissibity, absorptivity, and emissivity. There are now two different uses of the word emissivity. Thus the words *exitance, incidance,* and *sterance* were coined to be used in place of emittance, irradiance, and radiance. It is interesting that ISO uses radiance, exitance, and irradiance whereas IUPAP uses radiance, excitance [*sic*], and irradiance. We have chosen to use them both, i.e., emittance, irradiance, and radiance will be followed in square brackets by exitance, incidance, and sterance (or vice versa). Individual authors will use the different endings for transmission, reflection, absorption, and emission as they see fit.

We are still troubled by the use of the symbol E for irradiance, as it is so close in meaning to electric field, but we have maintained that accepted use. The spectral concentrations of these quantities, indicated by a wavelength, wave number, or frequency subscript (e.g., L_λ) represent partial differentiations; a subscript q represents a photon quantity; and a subscript v indicates a quantity normalized to the response of the eye. Thereby, L_v is luminance, E_v illuminance, and M_v and I_v luminous emittance and luminous intensity. The symbols we have chosen are consistent with ISO and IUPAP.

The refractive index may be considered a radiometric quantity. It is generally complex and is indicated by $\tilde{n} = n - ik$. The real part is the relative refractive index and k is the extinction coefficient. These are consistent with ISO and IUPAP, but they do not address the complex index or extinction coefficient.

Optical Design

For the most part ISO and IUPAP do not address the symbols that are important in this area.

There were at least 20 different ways to indicate focal ratio; we have chosen FN as symmetrical with NA; we chose f and efl to indicate the effective focal length. Object and image distance, although given many different symbols, were finally called s_o and s_i since s is an almost universal symbol for distance. Field angles are θ and ϕ; angles that measure the slope of a ray to the optical axis are u; u can also be $\sin u$. Wave aberrations are indicated by W_{ijk}, while third order ray aberrations are indicated by σ_i and more mnemonic symbols.

Electromagnetic Fields

There is no argument about **E** and **H** for the electric and magnetic field strengths, Q for quantity of charge, ρ for volume charge density, σ for surface charge density, etc. There is no guidance from References 1 and 2 on polarization indication. We chose $\perp$ and $\parallel$ rather than p and s, partly because s is sometimes also used to indicate scattered light.

There are several sets of symbols used for reflection, transmission, and (sometimes) absorption, each with good logic. The versions of these quantities dealing with field amplitudes are usually specified with lower case symbols: r, t, and a. The versions dealing with power are alternately given by the uppercase symbols or the corresponding Greek symbols: R and T versus ρ and τ. We have chosen to use the Greek, mainly because these quantities are also closely associated with Kirchhoff's law that is usually stated symbolically as $\alpha = \epsilon$. The law of conservation of energy for light on a surface is also usually written as $\alpha + \rho + \tau = 1$.

Base SI Quantities

length	m	meter
time	s	second
mass	kg	kilogram
electric current	A	ampere
Temperature	K	kelvin
Amount of substance	mol	mole
Luminous intensity	cd	candela

Derived SI Quantities

energy	J	joule
electric charge	C	coulomb
electric potential	V	volt
electric capacitance	F	farad
electric resistance	Ω	ohm
electric conductance	S	siemens

magnetic flux	Wb	weber
inductance	H	henry
pressure	Pa	pascal
magnetic flux density	T	tesla
frequency	Hz	hertz
power	W	watt
force	N	newton
angle	rad	radian
angle	sr	steradian

Prefixes

Symbol	Name	Common name	Exponent of ten
E	exa		18
P	peta		15
T	tera	trillion	12
G	giga	billion	9
M	mega	million	6
k	kilo	thousand	3
h	hecto	hundred	2
da	deca	ten	1
d	deci	tenth	−1
c	centi	hundredth	−2
m	milli	thousandth	−3
μ	micro	millionth	−6
n	nano	billionth	−9
p	pico	trillionth	−12
f	femto		−15
a	atto		−18

Constants

c	speed of light in vacuo [299792458 ms^{-1}]
c_1	first radiation constant = $2\pi c^2 h = 3.7417749 \times 10^{-16}$ [Wm^2]
c_2	second radiation constant = $hc/k = 0.01438769$ [mK]
e	elementary charge [$1.60217733 \times 10^{-19}$ C]
g_n	free fall constant [9.80665 ms^{-2}]
h	Planck's constant [$6.6260755 \times 10^{-34}$ Ws]
k_B	Boltzmann constant [1.380658×10^{-23} JK^{-1}]
m_e	mass of the electron [$9.1093897 \times 10^{-31}$ kg]
N_A	Avogadro constant [6.0221367×10^{23} mol^{-1}]
R_∞	Rydberg constant [10973731.534 m^{-1}]
ϵ_o	vacuum permittivity [$\mu_o^{-1} c^{-2}$]
σ	Stefan-Boltzmann constant [5.67051×10^{-8} $Wm^{-1} K^{-4}$]
μ_o	vacuum permeability [$4\pi \times 10^{-7}$ NA^{-2}]
μ_B	Bohr magneton [$9.2740154 \times 10^{-24}$ JT^{-1}]

General

B	magnetic induction [Wbm^{-2}, $kgs^{-1} C^{-1}$]
C	capacitance [f, $C^2 s^2 m^{-2} kg^{-1}$]
C	curvature [m^{-1}]

c	speed of light in vacuo [ms^{-1}]
c_1	first radiation constant [Wm^2]
c_2	second radiation constant [mK]
$\mathbf{D}$	electric displacement [Cm^{-2}]
E	incidance [irradiance] [Wm^{-2}]
e	electronic charge [coulomb]
E_v	illuminance [lux, lmm^{-2}]
$\mathbf{E}$	electrical field strength [Vm^{-1}]
E	transition energy [J]
E_g	band-gap energy [eV]
f	focal length [m]
f_c	Fermi occupation function, conduction band
f_v	Fermi occupation function, valence band
FN	focal ratio (f/number) [—]
g	gain per unit length [m^{-1}]
g_{th}	gain threshold per unit length [m^1]
$\mathbf{H}$	magnetic field strength [Am^{-1}, $Cs^{-1}\,m^{-1}$]
h	height [m]
I	irradiance (see also E) [Wm^{-2}]
I	radiant intensity [Wsr^{-1}]
I	nuclear spin quantum number [—]
I	current [A]
i	$\sqrt{-1}$
Im()	Imaginary part of
J	current density [Am^{-2}]
$\mathbf{j}$	total angular momentum [$kg\,m^2\,sec^{-1}$]
$J_1()$	Bessel function of the first kind [—]
k	radian wave number $= 2\pi/\lambda$ [$rad\,cm^{-1}$]
$\mathbf{k}$	wave vector [$rad\,cm^{-1}$]
k	extinction coefficient [—]
L	sterance [radiance] [$Wm^{-2}\,sr^{-1}$]
L_v	luminance [$cd\,m^{-2}$]
L	inductance [h, $m^2\,kg\,C^{-2}$]
L	laser cavity length
L, M, N	direction cosines [—]
M	angular magnification [—]
M	radiant exitance [radiant emittance] [Wm^{-2}]
m	linear magnification [—]
m	effective mass [kg]
MTF	modulation transfer function [—]
N	photon flux [s^{-1}]
N	carrier (number) density [m^{-3}]
n	real part of the relative refractive index [—]
$\tilde{n}$	complex index of refraction [—]
NA	numerical aperture [—]
OPD	optical path difference [m]
P	macroscopic polarization [$C\,m^{-2}$]
Re()	real part of [—]
R	resistance [Ω]
$\mathbf{r}$	position vector [m]
r	(amplitude) reflectivity
S	Seebeck coefficient [VK^{-1}]
s	spin quantum number [—]
s	path length [m]

s_o	object distance [m]
s_i	image distance [m]
T	temperature [K, C]
t	time [s]
t	thickness [m]
u	slope of ray with the optical axis [rad]
V	Abbé reciprocal dispersion [—]
V	voltage [V, $m^2\,kgs^{-2}\,C^{-1}$]
x, y, z	rectangular coordinates [m]
Z	atomic number [—]

Greek Symbols

α	absorption coefficient [cm^{-1}]
α	(power) absorptance (absorptivity)
ϵ	dielectric coefficient (constant) [—]
ϵ	emittance (emissivity) [—]
ϵ	eccentricity [—]
ϵ_1	Re (ϵ)
ϵ_2	Im (ϵ)
τ	(power) transmittance (transmissivity) [—]
ν	radiation frequency [Hz]
ω	circular frequency $= 2\pi\nu$ [$rads^{-1}$]
ω_p	plasma frequency [H_2]
λ	wavelength [μm, nm]
σ	wave number $= 1/\lambda$ [cm^{-1}]
σ	Stefan Boltzmann constant [$Wm^{-2}\,K^{-1}$]
ρ	reflectance (reflectivity) [—]
θ, ϕ	angular coordinates [rad,°]
ξ, η	rectangular spatial frequencies [m^{-1}, r^{-1}]
ϕ	phase [rad, °]
ϕ	lens power [m^{-1}]
Φ	flux [W]
χ	electric susceptibility tensor [—]
Ω	solid angle [sr]

Other

$\Re$	responsivity
$\exp(x)$	e^x
$\log_a(x)$	log to the base a of x
$\ln(x)$	natural log of x
$\log(x)$	standard log of x: $\log_{10}(x)$
Σ	summation
Π	product
Δ	finite difference
δx	variation in x
dx	total differential
∂x	partial derivative of x
$\delta(x)$	Dirac delta function of x
δ_{ij}	Kronecker delta

REFERENCES

1. Anonymous, *ISO Standards Handbook 2: Units of Measurement,* 2d ed., International Organization for Standardization, 1982.
2. Anonymous, *Symbols, Units and Nomenclature in Physics,* Document U.I.P. 20, International Union of Pure and Applied Physics, 1978.
3. E. Cohen and B. Taylor, "The Fundamental Physical Constants," *Physics Today,* 9, August 1990.
4. W. J. Smith, *Modern Optical Engineering,* 2d ed., McGraw-Hill, 1990; J. M. Lloyd, *Thermal Imaging Systems,* Plenum Press, 1972.

William L. Wolfe
Optical Science Center
University of Rochester
Rochester, New York

CONTENTS IN BRIEF: VOLUME I

CONTENTS IN BRIEF: VOLUME II

Part 3. Optical Measurements 24.1

Part 4. Optical and Physical Properties of Materials 33.1

HANDBOOK OF OPTICS

P · A · R · T · 1

GEOMETRIC OPTICS

CHAPTER 1
GENERAL PRINCIPLES OF GEOMETRIC OPTICS

Douglas S. Goodman
Polaroid
Cambridge, Massachusetts

1.1 GLOSSARY

(NS)	indicates nonstandard terminology
italics	definition or first usage
∇	gradient $(\partial/\partial x, \partial/\partial y, \partial/\partial z)$
prime, unprime	before and after, object and image space (not derivatives)
A	auxiliary function for ray tracing
A, A'	area, total field areas, object and image points
AB	directed distance from A to B
$\mathbf{a}$	unit axis vector, vectors
a_O, a_B, a_I	coefficients in characteristic function expansion
B	matrix element for symmetrical systems
B	auxiliary function for ray tracing
B, B'	arbitrary object and image points
$\mathbf{b}$	binormal unit vector of a ray path
$\mathcal{B}$	interspace (between) term in expansion
C	matrix element for conjugacy
$C(\mathcal{O}, \mathcal{B}, \mathcal{I})$	characteristic function
c	speed of light in vacuum
c	surface vertex curvature, spherical surface curvature
c_s	sagittal curvature
c_t	tangential curvature
D	auxiliary distance function for ray tracing
d	distance from origin to mirror
d	nominal focal distance
d, d'	arbitrary point to conjugate object, image points $d = AO$, $d' = A'O'$
d, d'	axial distances, distances along rays
d_H	hyperfocal distance

d_N	near focal distance
d_F	far focal distance
dA	differential area
ds	differential geometric path length
E	image irradiance
E_0	axial image irradiance
E, E'	entrance and exit pupil locations
e	eccentricity
e_x, e_y, e_z	coefficients for collineation
F	matrix element for front side
F, F'	front and rear focal points
FN	F-number
FN_m	F-number for magnification m
$F(\ )$	general function
$F(x, y, z)$	general surface function
f, f'	front and rear focal lengths $f = PF$, $f' = P'F'$
G	diffraction order
g, g'	focal lengths in tilted planes
h, h'	ray heights at objects and images, field heights, $\sqrt{x^2 + y^2}$
$\mathcal{H}$	hamiltonian
I, I'	incidence angles
$\mathbf{I}$	unit matrix
i, i'	paraxial incidence angles
$\mathcal{I}$	image space term in characteristic function expansion
L	surface x-direction cosine
L	paraxial invariant
l, l'	principal points to object and image axial points $l = PO$, $l' = P'O'$ axial distances from vertices of refracting surface $l = VO$, $l' = V'O'$
$\mathcal{L}$	lagrangian for heterogeneous media
M	lambertian emittance
M	surface z-direction cosine
m	transverse magnification
m_L	longitudinal magnification
m_α	angular magnification
m_E	paraxial pupil magnification
m_N	nodal point magnification $= n/n'$
m_P	pupil magnification in direction cosines
m_O	magnification at axial point
m_x, m_y, m_z	magnifications in the x, y, and z directions
N	surface z-direction cosine
N, N'	nodal points
NA, NA'	numerical aperture
n	refractive index

$\mathbf{n}$	normal unit vector of a ray path
O, O'	axial object and image points
$\mathcal{O}$	object space term in expansion
P	power (radiometric)
P, P'	principal points
$P(\alpha, \beta; x, y)$ $P'(\alpha', \beta'; x', y')$	pupil shape functions
p	period of grating
$\mathbf{p}$	ray vector, optical direction cosine $\mathbf{p} = n\,\mathbf{r} = (p_x, p_y, p_z)$
p	pupil radius
p_x, p_y, p_z	optical direction cosines
$Q(\alpha, \beta; x, y)$ $Q'(\alpha', \beta'; x', y')$	pupil shape functions relative to principal direction cosines
q	resolution parameter
q_i	coordinate for Lagrange equations
$\dot{q}_i$	derivative with respect to parameter
q, q'	auxiliary functions for collineation
$\mathbf{q}$	unit vector along grating lines
R	matrix element for rear side
r	radius of curvature, vertex radius of curvature
$\mathbf{r}$	ray unit direction vector $\mathbf{r} = (\alpha, \beta, \gamma)$
$\mathbf{S}$	surface normal $\mathbf{S} = (L, M, N)$
$S(x, y, x', y')$	point eikonal $V(x, y, z_0; x', y', z_{0'})$
s	geometric length
s	axial length
s, s'	distances associated with sagittal foci
$\mathcal{S}$	skew invariant
$T(\alpha, \beta; \alpha', \beta')$	angle characteristic function
t	thickness, vertex-to-vertex distance
t, t'	distances associated with tangential foci
t	time
$\mathbf{t}$	tangent unit vector of a ray path
U, U'	meridional ray angles relative to axis
u, u'	paraxial ray angles relative to axis
u_M	paraxial marginal ray angle
u_C	paraxial chief ray angle
u_1, u_2, u_3, u_4	homogeneous coordinates for collineation
V	optical path length
$V(\mathbf{x}; \mathbf{x}')$	point characteristic function
V, V'	vertex points

v	speed of light in medium
W_{LMN}	wavefront aberration term
W_x, W_y, W_z	wavefront aberration terms for reference shift
$W(\xi, \eta; x, y, z)$	wavefront aberration function
$W'(\alpha, \beta; x', y')$	angle-point characteristic function
$W(x, y; \alpha', \beta')$	point-angle characteristic function
$\mathbf{x} = (x, y, z)$	position vector
$\mathbf{x}(\sigma)$	parametric description of ray path
$\dot{\mathbf{x}}(\sigma)$	derivative with respect to parameter
$\ddot{\mathbf{x}}(\sigma)$	second derivative with respect to parameter
y	meridional ray height, paraxial ray height
y_M	paraxial marginal ray height
y_C	paraxial chief ray height
y_P, y'_P	paraxial ray height at the principal planes
z	axis of revolution
$z(\rho)$	surface sag
z_{sphere}	sag of a sphere
z_{conic}	sag of a conic
z, z'	focal point to object and image distances $z = FO$, $z' = F'O'$
α, β, γ	ray direction cosines
α, β, γ	entrance pupil directions
α', β', γ'	exit pupil direction cosines
α_0, β_0	principal direction of entrance pupil
α'_0, β'_0	principal direction of exit pupil
$\alpha_{max}, \alpha_{min}$	extreme pupil directions
β_{max}, β_{min}	extreme pupil directions
Γ	$n' \cos I' - n \cos I$
$\delta x, \delta y, \delta z$	reference point shifts
$\Delta\alpha, \Delta\beta$	angular aberrations
$\Delta x, \Delta y, \Delta z$	shifts
ε	surface shape parameter
$\varepsilon_x, \varepsilon_y$	transverse ray aberrations
ξ, η	pupil coordinates—not specific
θ	ray angle to surface normal marginal ray angle plane tilt angle

κ	conic parameter
κ	curvature of a ray path
λ	wavelength
ψ	aximuth angle field angle
ϕ	power, surface power azimuth
ρ	radius of curvature of a ray path distance from axis radial pupil coordinate
σ	ray path parameter general parameter for a curve
τ	reduced axial distances torsion of a ray path
$\tau(\alpha', \beta'; x', y')$	pupil transmittance function
ω, ω'	reduced angle $\omega = nu$, $\omega' = n'u'$
$d\omega$	differential solid angle

1.2 INTRODUCTION

The Subject

Geometrical optics is both the object of abstract study and a body of knowledge necessary for design and engineering. The subject of geometric optics is small, since so much can be derived from a single principle, that of Fermat, and large since the consequences are infinite and far from obvious. Geometric optics is deceptive in that much that seems simple is loaded with content and implications, as might be suggested by the fact that some of the most basic results required the likes of Newton and Gauss to discover them. Most of what appears complicated seems so because of obscuration with mathematical terminology and excessive abstraction. Since it is so old, geometric optics tends to be taken for granted and treated too casually by those who consider it to be "understood." One consequence is that what has been long known can be lost if it is not recirculated by successive generations of textbook authors, who are pressed to fit newer material in a fairly constant number of pages.

The Contents

The material in this chapter is intended to be that which is most fundamental, most general, and most useful to the greatest number of people. Some of this material is often thought to be more esoteric than practical, but this opinion is less related to its essence than to its typical presentation. There are no applications per se here, but everything is applicable, at least to understanding. An effort has been made to compensate here for what is lacking elsewhere and to correct some common errors. Many basic ideas and useful results have not found their way into textbooks, so are little known. Moreover, some basic principles are rarely stated explicitly. The contents are weighted toward the most common type of optical system, that with rotational symmetry consisting of mirrors and/or lens elements of homogeneous materials. There is a section on heterogeneous media, an application of which is gradient index optics discussed in another chapter. The treatment here is mostly monochromatic. The topics of caustics and anisotropic media are omitted, and there is little specifically about systems that are not figures of revolution. The section on aberrations is short and mostly descriptive, with no discussion of lens design, a vast field concerned with the practice of aberration control. Because of space limitations, there are too few diagrams.

Terminology

Because of the complicated history of geometric optics, its terminology is far from standardized. Geometric optics developed over centuries in many countries, and much of it has been rediscovered and renamed. Moreover, concepts have come into use without being named, and important terms are often used without formal definitions. This lack of standardization complicates communication between workers at different organizations, each of which tends to develop its own optical dialect. Accordingly, an attempt has been made here to provide precise definitions. Terms are italicized where defined or first used. Some needed nonstandard terms have been introduced, and these are likewise italicized, as well as indicated by "NS" for "nonstandard."

Notation

As with terminology, there is little standardization. And, as usual, the alphabet has too few letters to represent all the needed quantities. The choice here has been to use some of the same symbols more than once, rather than to encumber them with superscripts and subscripts. No symbol is used in a given section with more than one meaning. As a general practice nonprimed and primed quantities are used to indicate before and after, input and output, and object and image space.

References

No effort has been made to provide complete references, either technical or historical. (Such a list would fill the entire section.) The references were not chosen for priority, but for elucidation or interest, or because of their own references. Newer papers can be found by computer searches, so the older ones have been emphasized, especially since older work is receding from view beneath the current flood of papers. In geometric optics, nothing goes out of date, and much of what is included here has been known for a century or so—even if it has been subsequently forgotten.

Communication

Because of the confusion in terminology and notation, it is recommended that communication involving geometric optics be augmented with diagrams, graphs, equations, and numeric results, as appropriate. It also helps to provide diagrams showing both first order properties of systems, with object and image positions, pupil positions, and principal planes, as well as direction cosine space diagrams, as required, to show angular subtenses of pupils.

1.3 FUNDAMENTALS

What Is a Ray?

Geometric optics, which might better be called *ray optics,* is concerned with the light ray, an entity that does not exist. It is customary, therefore, to begin discussions of geometric optics with a theoretical justification for the use of the ray. The real justification is that, like other successful models in physics, rays are indispensable to our thinking, notwithstanding their shortcomings. The ray is a model that works well in some cases and not at all in others, and light is necessarily thought about in terms of rays, scalar waves, electromagnetic waves, and with quantum physics—depending on the class of phenomena under consideration.

Rays have been defined with both corpuscular and wave theory. In corpuscular theory, some definitions are (1) the path of a corpuscle and (2) the path of a photon. A difficulty here is that energy densities can become infinite. Other efforts have been made to define rays as quantities related to the wave theory, both scalar and electromagnetic. Some are (1) wavefront normals, (2) the Poynting vector, (3) a discontinuity in the electromagnetic field (Luneburg 1964,[1] Kline & Kay 1965[2]), (4) a descriptor of wave behavior in short wavelength or high frequency limit, (Born & Wolf 1980[3]) (5) quantum mechanically (Marcuse 1989[4]). One problem with these definitions is that there are many ordinary and simple cases where wavefronts and Poynting vectors become complicated and/or meaningless. For example, in the simple case of two coherent plane waves interfering, there is no well-defined wavefront in the overlap region. In addition, rays defined in what seems to be a reasonble way can have undesirable properties. For example, if rays are defined as normals to wavefronts, then, in the case of gaussian beams, rays bend in a vacuum.

An approach that avoids the difficulties of a physical definition is that of treating rays as mathematical entities. From definitions and postulates, a variety of results is found, which may be more or less useful and valid for light. Even with this approach, it is virtually impossible to think "purely geometrically"—unless rays are treated as objects of geometry, rather than optics. In fact, we often switch between ray thinking and wave thinking without noticing it, for instance in considering the dependence of refractive index on wavelength. Moreover, geometric optics makes use of quantities that must be calculated from other models, for example, the index of refraction. As usual, Rayleigh (Rayleigh 1884[5]) has put it well: "We shall, however, find it advisable not to exclude altogether the conceptions of the wave theory, for on certain most important and practical questions no conclusion can be drawn without the use of facts which are scarcely otherwise interpretable. Indeed it is not to be denied that the too rigid separation of optics into geometrical and physical has done a good deal of harm, much that is essential to a proper comprehension of the subject having fallen between the two stools."

The ray is inherently ill-defined, and attempts to refine a definition always break down. A definition that seems better in some ways is worse in others. Each definition provides some insight into the behavior of light, but does not give the full picture. There seems to be a problem associated with the uncertainty principle involved with attempts at definition, since what is really wanted from a ray is a specification of both position and direction, which is impossible by virtue of both classical wave properties and quantum behavior. So the approach taken here is to treat rays without precisely defining them, and there are few reminders hereafter that the predictions of ray optics are imperfect.

Refractive Index

For the purposes of this chapter, the optical characteristics of matter are completely specified by its refractive index. The *index of refraction* of a medium is defined in geometrical optics as

$$n = \frac{\text{speed of light in vacuum}}{\text{speed of light in medium}} = \frac{\mathrm{c}}{\mathrm{v}} \tag{1}$$

A *homogeneous medium* is one in which n is everywhere the same. In an *inhomogeneous* or *heterogeneous medium* the index varies with position. In an *isotropic medium* n is the same for light traveling in all directions and with all polarizations, so the index is described by a scalar function of position. Anisotropic media are not treated here.

Care must be taken with equations using the symbol n, since it sometimes denotes the ratio of indices, sometimes with the implication that one of the two is unity. In many cases, the difference from unity of the index of air ($\simeq$1.0003) is important. Index varies with wavelength, but this dependence is not made explicit in this section, most of which is implicitly limited to monochromatic light. The output of a system in polychromatic light is the sum of outputs at the constituent wavelengths.

Systems Considered

The optical systems considered here are those in which spatial variations of surface features or refractive indices are large compared to the wavelength. In such systems ray identity is preserved; there is no "splitting" of one ray into many as occurs at a grating or scattering surface.

The term *lens* is used here to include a variety of systems. *Dioptric* or *refractive* systems employ only refraction. *Catoptric* or *reflective* systems employ only reflection. *Catadioptric* systems employ both refraction and reflection. No distinction is made here insofar as refraction and reflection can be treated in a common way. And the term lens may refer here to anything from a single surface to a system of arbitrary complexity.

Summary of the Behavior and Attributes of Rays

Rays propagate in straight lines in homogeneous media and have curved paths in heterogeneous media. Rays have positions, directions, and speeds. Between any pair of points on a given ray there is a geometrical path length and an optical path length. At smooth interfaces between media with different indices rays refract and reflect. Ray paths are reversible. Rays carry energy, and power per area is approximated by ray density.

Reversibility

Rays are reversible; a path can be taken in either direction, and reflection and refraction angles are the same in either direction. However, it is usually easier to think of light as traveling along rays in a particular direction, and, of course, in cases of real instruments there usually is such a direction. The solutions to some equations may have directional ambiguity.

Groups of Rays

Certain types of groups of rays are of particular importance. Rays that originate at a single point are called a *normal congruence* or *orthotomic system,* since as they propagate in isotropic media they are associated with perpendicular wavefronts. Such groups are also of interest in image formation, where their reconvergence to a point is important, as is the path length of the rays to a reference surface used for diffraction calculations. Important in radiometric considerations are groups of rays emanating from regions of a source over a range of angles. The changes of such groups as they propagate are constrained by conservation of brightness. Another group is that of two paraxial rays, related by the two-ray invariant.

Invariance Properties

Individual rays and groups of rays may have *invariance properties*—relationships between the positions, directions, and path lengths—that remain constant as a ray or group of rays passes through an optical system (Welford 1986, chap. 6[6]). Some of these properties are completely general, e.g., the conservation of etendue and the perpendicularity of rays to wavefronts in isotropic media. Others arise from symmetries of the system, e.g., the skew invariant for rotationally symmetric systems. Other invariances hold in the paraxial limit. There are also differential invariance properties (Herzberger 1935,[7] Stavroudis 1972, chap. 13[8]). Some ray properties not ordinarily thought of in this way can be thought of as invariances. For example, Snell's law can be thought of as a refraction invariant $n \sin I$.

Description of Ray Paths

A *ray path* can be described parametrically as a locus of points $\mathbf{x}(\sigma)$, where σ is any monotonic parameter that labels points along the ray. The description of curved rays is elaborated in the section on heterogeneous media.

Real Rays and Virtual Rays

Since rays in homogeneous media are straight, they can be extrapolated infinitely from a given region. The term *real* refers to the portion of the ray that "really" exists, or the accessible part, and the term *virtual* refers to the extrapolated, or inaccessible, part.

Direction

At each position where the refractive index is continuous a ray has a unique direction. The direction is given by that of its unit *direction vector* $\mathbf{r}$, whose cartesian components are direction cosines (α, β, γ), i.e.,

$$\mathbf{r} = (\alpha, \beta, \gamma)$$

where $|\mathbf{r}|^2 = \alpha^2 + \beta^2 + \gamma^2 = 1.$ (2)

The three direction cosines are not independent, and one is often taken to depend implicitly on the other two. In this chapter it is usually γ, which is

$$\gamma(\alpha, \beta) = \sqrt{1 - \alpha^2 - \beta^2} \tag{3}$$

Another vector with the same direction as $\mathbf{r}$ is

$$\mathbf{p} = n\mathbf{r} = (n\alpha, n\beta, n\gamma) = (p_x, p_y, p_z)$$

where $|\mathbf{p}|^2 = n^2$. (4)

Several names are used for this vector, including the *optical direction cosine* and the *ray vector*.

Geometric Path Length

Geometric path length is geometric distance measured along a ray between any two points. The differential unit of length is

$$ds = \sqrt{dx^2 + dy^2 + dz^2} \tag{5}$$

The path length between points $\mathbf{x}_1$ and $\mathbf{x}_2$ on a ray described parametrically by $\mathbf{x}(\sigma)$, with derivative $\dot{\mathbf{x}}(\sigma) = d\mathbf{x}(\sigma)/d\sigma$ is

$$s(\mathbf{x}_1; \mathbf{x}_2) = \int_{\mathbf{x}_1}^{\mathbf{x}_2} ds = \int_{\mathbf{x}_1}^{\mathbf{x}_2} \frac{ds}{d\sigma} d\sigma = \int_{\mathbf{x}_1}^{\mathbf{x}_2} \sqrt{|\dot{\mathbf{x}}(\sigma)|^2}\, d\sigma \tag{6}$$

Optical Path Length

The *optical path length* between two points $\mathbf{x}_1$ and $\mathbf{x}_2$ through which a ray passes is

$$\text{Optical path length} = V(\mathbf{x}_1; \mathbf{x}_2) = \int_{\mathbf{x}_1}^{\mathbf{x}_2} n(\mathbf{x})\, ds = \mathrm{c} \int \frac{ds}{\mathrm{v}} = \mathrm{c} \int dt \tag{7}$$

The integral is taken along the ray path, which may traverse homogeneous and inhomogeneous media, and include any number of reflections and refractions. Path length can be defined for virtual rays. In some cases, path length should be considered positive definite, but in others it can be either positive or negative, depending on direction (Forbes & Stone 1993[9]). If $\mathbf{x}_0$, $\mathbf{x}_1$, and $\mathbf{x}_2$ are three points on the same ray, then

$$V(\mathbf{x}_0; \mathbf{x}_2) = V(\mathbf{x}_0; \mathbf{x}_1) + V(\mathbf{x}_1; \mathbf{x}_2) \tag{8}$$

Equivalently, the time required for light to travel between the two points is

$$\text{Time} = \frac{\text{optical path length}}{\mathrm{c}} = \frac{V}{\mathrm{c}} = \frac{1}{\mathrm{c}} \int_{\mathbf{x}_1}^{\mathbf{x}_2} n(\mathbf{x})\, ds = \int_{\mathbf{x}_1}^{\mathbf{x}_2} \frac{ds}{\mathrm{v}} \tag{9}$$

In homogeneous media, rays are straight lines, and the optical path length is $V = n \int ds =$ (index) $\times$ (distance between the points).

The optical path length integral has several interpretations, and much of geometrical optics involves the examination of its meanings. (1) With both points fixed, it is simply a scalar, the optical path length from one point to another. (2) With one point fixed, say $\mathbf{x}_0$, then treated as a function of $\mathbf{x}$, the surfaces $V(\mathbf{x}_0; \mathbf{x}) =$ constant are geometric wavefronts for light originating at $\mathbf{x}_0$. (3) Most generally, as a function of both arguments $V(\mathbf{x}_1; \mathbf{x}_2)$ is the *point characteristic function,* which contains all the information about the rays between the region containing $\mathbf{x}_1$ and that containing $\mathbf{x}_2$. There may not be a ray between all pairs of points.

Fermat's Principle

According to Fermat's principle (Magie 1963,[10] Fermat 1891,[11,12] Feynman 1963,[13] Rossi 1956,[14] Hecht 1987[15]) the optical path between two points through which a ray passes is an extremum. Light passing through these points along any other nearby path would take either more or less time. The principle applies to different *neighboring* paths. The optical path length of a ray may not be a global extremum. For example, the path lengths of rays through different facets of a Fresnel lens have no particular relationship. Fermat's principle applies to entire systems, as well as to any portion of a system, for example to any section of a ray. In a homogeneous medium, the extremum is a straight line or, if there are reflections, a series of straight line segments.

The extremum principle can be described mathematically as follows (Klein 1986[16]). With the end points fixed, if a nonphysical path differs from a physical one by an amount proportional to δ, the nonphysical optical path length differs from the actual one by a quantity proportional to δ^2 or to a higher order. If the order is three or higher, the first point is imaged at the second-to-first order. Roughly speaking, the higher the order, the better the image. A point is imaged stigmatically when a continuum of neighboring paths have the same length, so the equality holds to all orders. If they are sufficiently close, but vary slightly, the deviation from equality is a measure of the aberration of the imaging. An extension of Fermat's principle is given by Hopkins (H. Hopkins 1970[17]).

Ray and wave optics are related by the importance of path length in both (Walther 1967,[18] Walther 1969[19]). In wave optics, optical path length is proportional to phase change, and the extremum principle is associated with constructive interference. The more alike the path lengths are from an object point to its image, the less the differences in phase of the wave contributions, and the greater the magnitude of the net field. In imaging this connection is manifested in the relationship of the wavefront aberration and the eikonal.

Fermat's principle is a unifying principle of geometric optics that can be used to derive laws of reflection and refraction, and to find the equations that describe ray paths and geometric wavefronts in heterogeneous and homogeneous media. Fermat's is one of a number of variational principles based historically on the idea that nature is economical, a unifying principle of physics. The idea that the path length is an extremum could be used mathematically without interpreting the refractive index in terms of the speed of light.

Geometric Wavefronts

For rays originating at a single point, a *geometric wavefront* is a surface that is a locus of constant optical path length from the source. If the source point is located at $\mathbf{x}_0$ and light leaves at time t_0, then the wavefront at time t is given by

$$V(\mathbf{x}_0; \mathbf{x}) = c(t - t_0) \tag{10}$$

The function $V(\mathbf{x}; \mathbf{x}_0)$, as a function of $\mathbf{x}$, satisfies the *eikonal equation*

$$\begin{aligned} n(\mathbf{x})^2 &= \left(\frac{\partial V}{\partial x}\right)^2 + \left(\frac{\partial V}{\partial y}\right)^2 + \left(\frac{\partial V}{\partial z}\right)^2 \\ &= |\nabla V(\mathbf{x}; \mathbf{x}_0)|^2 \end{aligned} \tag{11}$$

This equation can also be written in relativistic form, with a four-dimensional gradient as $0 = \sum (\partial V/\partial x_i)^2$ (Landau & Lifshitz 1951, sec. 7.1[20]).

For constant refractive index, the eikonal equation has some simple solutions, one of which is $V = n[\alpha(x - x_0) + \beta(y - y_0) + \gamma(z - z_0)]$, corresponding to a parallel bundle of rays with directions (α, β, γ). Another is $V = n[(x - x_0)^2 + (y - y_0)^2 + (z - z_0)^2]^{1/2}$, describing rays traveling radially from a point (x_0, y_0, z_0).

In isotropic media, rays and wavefronts are everywhere perpendicular, a condition referred to as *orthotomic.* According to the *Malus-Dupin principle*, if a group of rays emanating fron a single point is reflected and/or refracted any number of times, the perpendicularity of rays to wavefronts is maintained. The direction of a ray from $\mathbf{x}_0$ at $\mathbf{x}$ is that of the gradient of $V(\bar{x}_0; \bar{x})$

$$\mathbf{p} = n\mathbf{r} = \nabla V$$

or

$$n\alpha = \frac{\partial V}{\partial x} \qquad n\beta = \frac{\partial V}{\partial y} \qquad n\gamma = \frac{\partial V}{\partial z} \tag{12}$$

In a homogeneous medium, all wavefronts can be found from any one wavefront by a construction. Wavefront normals, i.e., rays, are projected from the known wavefront, and loci of points equidistant therefrom are other wavefronts. This gives wavefronts in both directions, that is, both subsequent and previous wavefronts. (A single wavefront contains no directional information.) The construction also gives virtual wavefronts, those which would occur or would have occurred if the medium extended infinitely. This construction is related to that of Huygens for wave optics. At each point on a wavefront there are two principal curvatures, so there are two foci along each ray and two caustic surfaces (Stavroudis 1972,[8] Kneisly 1964[21]).

The geometric wavefront is analogous to the surface of constant phase in wave optics, and the eikonal equation can be obtained from the wave equation in the limit of small wavelength (Born & Wolf 1980,[3] Marcuse 1989[4]). A way in which wave optics differs from ray optics is that the phase fronts can be modified by phase changes that occur on reflection, transmission, or in passing through foci.

Fields of Rays

In many cases the optical direction cosine vectors $\mathbf{p}$ form a field, where the optical path length is the potential, and the geometric wavefronts are equipotential surfaces. The potential changes with position according to

$$dV = n\alpha\, dx + n\beta\, dy + n\gamma\, dz = n\mathbf{r} \cdot d\mathbf{x} = \mathbf{p} \cdot d\mathbf{x} \tag{13}$$

If $d\mathbf{x}$ is in the direction of a ray, then $dV/dx = n$, the maximum rate of change. If $d\mathbf{x}$ is perpendicular to a ray, then $dV/dx = 0$. The potential difference between any two wavefronts is

$$V_2 - V_1 = \int_{\mathbf{x}_1}^{\mathbf{x}_2} dV \tag{14}$$

where $\mathbf{x}_1$ and $\mathbf{x}_2$ are any two points on the respective wavefronts, and the integrand is independent of the path. Other relationships for rays originating at a single point are

$$0 = \nabla \times \mathbf{p} = \nabla \times (n\mathbf{r}) \qquad \text{and} \qquad 0 = \oint \mathbf{p} \cdot d\mathbf{x} \tag{15}$$

where the integral is about a closed path (Born & Wolf 1980[3]). These follow since $\mathbf{p}$ is a gradient, Eq. (13). In regions where rays are folded onto themselves by refraction or reflections, **p** and V are not single-valued, so there is not a field.

1.4 CHARACTERISTIC FUNCTIONS

Introduction

Characteristic functions contain all the information about the path lengths between pairs of points, which may either be in a contiguous region or physically separated, e.g., on the two sides of a lens. These functions were first considered by Hamilton (Hamilton 1931[22]), so their study is referred to as *hamiltonian optics.* They were rediscovered in somewhat different form by Bruns (Bruns 1895,[23] Schwarzschild 1905[24]) and referred to as eikonals, leading to a confusing set of names for the various functions. The subject is discussed in a number of books (Czapski-Eppenstein 1924,[25] Steward 1928,[26] Herzberger 1931,[27] Synge 1937,[28] Caratheodory 1937,[29] Rayleigh 1908,[30] Pegis 1961,[31] Luneburg 1964,[32] Brouwer and Walther 1967,[33] Buchdahl 1970,[34] Born and Wolf 1980,[35] Herzberger 1958[36]).

Four parameters are required to specify a ray. For example, an input ray is defined in the $z = 0$ plane by coordinates (x, y) and direction (α, β). So four functions of four variables specify how an incident ray emerges from a system. In an output plane $z' = 0$, the ray has coordinates $x' = x'(x, y, \alpha, \beta)$, $y' = y'(x, y, \alpha, \beta)$, and directions $\alpha' = \alpha'(x, y, \alpha, \beta)$, $\beta' = \beta'(x, y, \alpha, \beta)$. Because of Fermat's principle, these four functions are not independent, and the geometrical optics properties of a system can be fully characterized by a single function (Luneburg 1964, sec. 19[32]).

For any given system, there is a variety of characteristic functions related by Legendre transformations, with different combinations of spatial and angular variables (Buchdahl 1970[34]). The different functions are suited for different types of analysis. *Mixed* characteristic functions have both spatial and angular arguments. Those functions that are of most general use are discussed below. The others may be useful in special circumstances. If the regions have constant refractive indices, the volumes over which the characteristic functions are defined can be extended virtually from physically accessible to inaccessible regions.

From any of its characteristic functions, all the properties of a system involving ray paths can be found, for example, ray positions, directions, and geometric wavefronts. An important use of characteristic functions is demonstrating general principles and fundamental limitations. Much of this can be done by using the general properties, e.g., symmetry under rotation. (Unfortunately, it is not always known how closely the impossible can be approached.)

Point Characteristic Function

The *point characteristic function* is the optical path integral $V(\mathbf{x}; \mathbf{x}') = V(x, y, z; x', y', z')$ taken as a function of both points $\mathbf{x}$ and $\mathbf{x}'$. At point $\mathbf{x}$ where the index is n,

$$-n\alpha = \frac{\partial V}{\partial x} \qquad -n\beta = \frac{\partial V}{\partial y} \qquad -n\gamma = \frac{\partial V}{\partial z} \qquad \text{or} \qquad -\mathbf{p} = \nabla V \tag{16}$$

Similarly, at $\mathbf{x}'$, where the index is n',

$$n'\alpha' = \frac{\partial V}{\partial x'} \qquad n'\beta' = \frac{\partial V}{\partial y'} \qquad n'\gamma' = \frac{\partial V}{\partial z'} \qquad \text{or} \qquad \mathbf{p}' = \nabla' V \tag{17}$$

It follows from the above equations and Eq. (4) that the point characteristic satisfies two conditions:

$$n^2 = |\nabla V|^2 \qquad \text{and} \qquad n'^2 = |\nabla' V|^2 \tag{18}$$

Therefore, the point characteristic is not an arbitrary function of six variables. The total differential of V is

$$dV(\mathbf{x}; \mathbf{x}') = \mathbf{p}' \cdot d\mathbf{x}' - \mathbf{p} \cdot d\mathbf{x} \tag{19}$$

"This expression can be said to contain all the basic laws of optics" (Herzberger 1958[36]).

Point Eikonal

If reference planes in object and image space are fixed, for which we use z_0 and z_0', then the *point eikonal* is $S(x, y; x', y') = V(x, y, z_0; x', y', z_0')$. This is the optical path length between pairs of points on the two planes. The function is not useful if the planes are conjugate, since more than one ray through a pair of points can have the same path length. The function is arbitrary, except for the requirement (Herzberger 1936[38]) that

$$\frac{\partial^2 S}{\partial x\, \partial x'}\frac{\partial^2 S}{\partial y\, \partial y'} - \frac{\partial^2 S}{\partial x\, \partial y'}\frac{\partial^2 S}{\partial x'\, \partial y} \neq 0 \tag{20}$$

The partial derivatives of the point eikonal are

$$-n\alpha = \frac{\partial S}{\partial x} \qquad -n\beta = \frac{\partial S}{\partial y} \qquad \text{and} \qquad n'\alpha' = \frac{\partial S}{\partial x'} \qquad n'\beta' = \frac{\partial S}{\partial y'} \tag{21}$$

The relative merits of the point characteristic function and point eikonal have been debated. (Herzberger 1936,[38] Herzberger 1937,[39] Synge 1937[40]).

Angle Characteristic

The *angle characteristic function* $T(\alpha, \beta; \alpha', \beta')$, also called the *eikonal,* is related to the point characteristic by

$$\begin{aligned} T(\alpha, \beta; \alpha', \beta') = V(x, y, z; x', y', z') &+ n(\alpha x + \beta y + \gamma z) \\ &- n'(\alpha' x' + \beta' y' + \gamma' z') \end{aligned} \tag{22}$$

Here the input plane z and output plane z' are fixed and are implicit parameters of T.

FIGURE 1 Geometrical interpretation of the angle characteristic function for constant object and image space indices. There is, in general, a single ray with directions (α, β, γ) in object space and $(\alpha', \beta', \gamma')$ in image space. Point O is the coordinate origin in object space, and O' is that in image space. From the origins, perpendiculars to the ray are constructed, which intersect the ray at Q and Q'. The angle characteristic function $T(\alpha, \beta; \alpha', \beta')$ is the path length from Q to Q'.

This equation is really shorthand for a Legendre transformation to coordinates $p_x = \partial V / \partial x$, etc. In principle, the expressions of Eq. (16) are used to solve for x and y in terms of α and β, and likewise Eq. (17) gives x' and y' in terms of α' and β', so

$$\begin{aligned} T(\alpha, \beta; \alpha', \beta') = V(x(\alpha, \beta), y(\alpha, \beta), z; x'(\alpha', \beta'), y'(\alpha', \beta'), z') \\ + n[\alpha x(\alpha, \beta) + \beta y(\alpha, \beta) + \sqrt{1 - \alpha^2 - \beta^2}\, z] \\ - n'[\alpha' x'(\alpha', \beta') + \beta' y'(\alpha', \beta') + \sqrt{1 - \alpha'^2 - \beta'^2}\, z'] \end{aligned} \tag{23}$$

The angle characteristic is an arbitrary function of four variables that completely specify the directions of rays in two regions. This function is not useful if parallel incoming rays give rise to parallel outgoing rays, as is the case with afocal systems, since the relationship between incoming and outgoing directions is not unique. The partial derivatives of the angular characteristic function are

$$\frac{\partial T}{\partial \alpha} = n\left(x - \frac{\alpha}{\gamma} z\right) \qquad \frac{\partial T}{\partial \beta} = n\left(y - \frac{\beta}{\gamma} z\right) \tag{24}$$

$$\frac{\partial T}{\partial \alpha'} = -n'\left(x' - \frac{\alpha'}{\gamma'} z'\right) \qquad \frac{\partial T}{\partial \beta'} = -n'\left(y' - \frac{\beta'}{\gamma'} z'\right) \tag{25}$$

These expressions are simplified if the reference planes are taken to be $z = 0$ and $z' = 0$. The geometrical interpretation of T is that it is the path length between the intersection point of rays with perpendicular planes through the coordinate origins in the two spaces, as shown in Fig. 1 for the case of constant n and n'. If the indices are heterogeneous, the construction applies to the tangents to the rays. Of all the characteristic functions, T is most easily found for single surfaces and most easily concatenated for series of surfaces.

Point-Angle Characteristic

The *point-angle characteristic function* is a mixed function defined by

$$\begin{aligned} W(x, y, z; \alpha', \beta') &= V(x, y, z; x', y', z') - n'(\alpha' x' + \beta' y' + \gamma' z') \\ &= T(\alpha, \beta; \alpha', \beta') - n(\alpha x + \beta y + \gamma z) \end{aligned} \tag{26}$$

As with Eq. (22), this equation is to be understood as shorthand for a Legendre transformation. The partial derivatives with respect to the spatial variables are related by equations like those of Eq. (16), so $n^2 = |\nabla W|^2$, and the derivatives with respect to the angular variables are like those of Eq. (25). This function is useful for examining transverse ray aberrations for a given object point, since $\partial W/\partial \alpha'$, $\partial W/\partial \beta'$ give the intersection points (x', y') for rays originating at (x, y).

Angle-Point Characteristic

The *angle-point characteristic function* is

$$\begin{aligned} W'(\alpha, \beta; x', y', z') &= V(x, y, z; x', y', z') + n(\alpha x + \beta y + \gamma z) \\ &= T(\alpha, \beta; \alpha', \beta') - n'(\alpha' x' + \beta' y' + \gamma' z) \end{aligned} \tag{27}$$

Again, this is shorthand for the Legendre transformation. This function satisfies relationships like those of Eq. (17) and satisfies $n'^2 = |\nabla' W'|^2$. Derivatives with respect to angular variables are like those of Eq. (24). It is useful when input angles are given, and output intersection points are to be found.

Expansions About an Arbitrary Ray

If two points on a ray that are not conjugate are taken as coordinate origins, and the z axes of the coordinate systems are taken to lie along the rays, then the expansion to second order of the point eikonal about these points is

$$\begin{aligned} S(x_1, y_1; x_2, y_2) = v &+ a_1 x_1^2 + b_1 x_1 y_1 + c_1 y_1^2 + a_2 x_2^2 + b_2 x_2 y_2 + c_2 y_2^2 \\ &+ d x_1 x_2 + e y_1 y_2 + f x_1 x_2 + g y_1 x_2 \end{aligned} \tag{28}$$

The other characteristic functions have similar expansions. These expansions have three types of terms, those associated with the input space, the output space, and "interspace" terms. From the coefficients, information about imaging along a known ray is obtained. This subject is treated in the references for the section "Images About Known Rays."

Expansions About the Axis

For rotationally symmetric systems, the building blocks for an expansion about the axis are

$$\text{Object space term:} \quad \mathcal{O} = x^2 + y^2 \quad \text{or} \quad \alpha^2 + \beta^2 \tag{29}$$

$$\text{Image space term:} \quad \mathcal{I} = x'^2 + y'^2 \quad \text{or} \quad \alpha'^2 + \beta'^2 \tag{30}$$

$$\begin{aligned} \text{Interspace term:} \quad \mathcal{B} = xx' + yy' \quad &\text{or} \quad \alpha\alpha' + \beta\beta' \quad \text{or} \quad x\alpha' + y\beta' \\ &\text{or} \quad \alpha x' + \beta y' \end{aligned} \tag{31}$$

(Here $\mathcal{B}$ = "between.") The interspace term combines the variables included in $\mathcal{O}$ and $\mathcal{I}$. The general form can be written as a series

$$C(\mathcal{O}, \mathcal{B}, \mathcal{I}) = \sum_{L,M,N} a_{LMN} \mathcal{O}^L \mathcal{B}^M \mathcal{I}^N \tag{32}$$

To second order, the expansion is

$$\begin{aligned} C(\mathcal{O}, \mathcal{B}, \mathcal{I}) = a_0 &+ a_{100}\mathcal{O} + a_{010}\mathcal{B} + a_{001}\mathcal{I} + a_{200}\mathcal{O}^2 + a_{020}\mathcal{B}^2 + a_{002}\mathcal{I}^2 \\ &+ a_{110}\mathcal{O}\mathcal{B} + a_{101}\mathcal{O}\mathcal{I} + a_{011}\mathcal{B}\mathcal{I} + \cdots \end{aligned} \tag{33}$$

The constant term is the optical path length between coordinate origins in the two spaces. It is often unimportant, but it does matter if two systems are used in parallel, as in an interferometer. The three first-order terms give the paraxial approximation. For imaging systems, the second-order terms are associated with third-order ray aberrations, and so on (Rayleigh 1908[30]). It is also possible to expand the characteristic functions in terms of three linear combinations of $\mathcal{O}$, $\mathcal{B}$, and $\mathcal{I}$. These combinations can be chosen so that the characteristic function of an aberration-free system depends on only one of the three terms, and the other two describe the aberrations (Steward 1928,[26] Smith 1945,[37] Pegis 1961[31]).

Paraxial Forms for Rotationally Symmetric Systems

These functions contain one each of the object space, image space, and interspace terms, with coefficients a_O, a_I, and a_B. The coefficients of the object and image space terms depend on the input and output plane locations. That of the interspace term depends on the system power. Point eikonal:

$$S(x', y'; x, y) = a + a_O(x^2 + y^2) + a_B(xx' + yy') + a_I(x'^2 + y'^2) \tag{34}$$

Angle characteristic:

$$T(\alpha', \beta'; \alpha, \beta) = a + a_O(\alpha^2 + \beta^2) + a_B(\alpha\alpha' + \beta\beta') + a_I(\alpha'^2 + \beta'^2) \tag{35}$$

Point-angle characteristic:

$$W(x, y; \alpha', \beta') = a + a_O(x^2 + y^2) + a_B(x\alpha' + y\beta') + a_I(\alpha'^2 + \beta'^2) \tag{36}$$

Angle-point characteristic:

$$W'(\alpha, \beta, x', y') = a + a_O(\alpha^2 + \beta^2) + a_B(\alpha x' + \beta y') + a_I(x'^2 + y'^2) \tag{37}$$

The coefficients in these expressions are different. The familiar properties of paraxial and gaussian optics can be found from these functions by taking the appropriate partial derivatives.

Some Ideal Characteristic Functions

For a system that satisfies certain conditions, the form of a characteristic function can sometimes be found. Thereafter, all of its properties can be determined. Some examples of characteristic functions follow, in each of which expression the function F is arbitrary.

For maxwellian perfect imaging (defined below) by a rotationally symmetric system between planes at $z = 0$ and $z' = 0$ related by transverse magnification m, the point characteristic function, defined for $z' \neq 0$, is

$$V(x', y', z'; x, y) = F(x^2 + y^2) + [(x' - mx)^2 + (y' - my)^2 + z'^2]^{1/2} \tag{38}$$

Expanding the expression above for small x, x', y, y' give the paraxial form, Eq. (34). The form of the point-angle characteristic is

$$W(x, y; \alpha', \beta') = F(x^2 + y^2) - m(n'\alpha' x + n'\beta' y) \tag{39}$$

The form of the angle-point characteristic is

$$W'(\alpha, \beta; x', y') = F(x'^2 + y'^2) + \frac{1}{m}(n\alpha x' + n\beta y') \tag{40}$$

The functions F are determined if the imaging is also stigmatic at one additional point, for example, at the center of the pupil (Steward 1928,[26] Smith 1945,[37] Buchdahl 1970,[34] Velzel 1991[41]). The angular characteristic function has the form

$$T(\alpha, \beta; \alpha', \beta') = F((n\alpha - mn'\alpha')^2 + (n\beta - mn'\beta')^2) \tag{41}$$

where F is any function.

For a lens that stigmatically images objects at infinity in a plane, and does so in either direction,

$$S(x, y; x', y') = -\phi(xx' + yy') \qquad \text{and} \qquad T(\alpha, \beta; \alpha', \beta') = \frac{nn'}{\phi}(\alpha\alpha' + \beta\beta') \tag{42}$$

Partially differentiating with respect to the appropriate variables shows that for such a system, the heights of point images in the rear focal plane are proportional to the sines of the incident angles, rather than the tangents.

1.5 RAYS IN HETEROGENEOUS MEDIA

Introduction

This section provides equations for describing and determining the curved ray paths in a heterogeneous or inhomogeneous medium, one whose refractive index varies with position. It is assumed here that $n(\mathbf{x})$ and the other relevant functions are continuous and have continuous derivatives to whatever order is needed. Various aspects of this subject are discussed in a number of books and papers (Heath 1895,[42] Herman 1900,[43] Synge 1937,[44] Luneburg 1964,[45] Stavroudis 1972,[46] Ghatak 1978,[47] Born & Wolf 1980,[48] Marcuse 1989[49]). This material is often discussed in the literature on gradient index lenses (Marchand 1973,[50] Marchand 1978,[51] Sharma, Kumar, & Ghatak 1982,[52] Moore 1992,[53] Moore 1994[54]) and in discussions of microwave lenses (Brown 1953,[55] Cornbleet 1976,[56] Cornbleet 1983,[57] Cornbleet 1984[58]).

Differential Geometry of Space Curves

A curved ray path is a space curve, which can be described by a standard parametric description, $\mathbf{x}(\sigma) = (x(\sigma), y(\sigma), z(\sigma))$, where σ is an arbitrary parameter (Blaschke 1945,[59] Kreyszig 1991,[60] Stoker 1969,[61] Struik 1990,[62] Stavroudis 1972[46]).

Different parameters may be used according to the situation. The path length s along the ray is sometimes used, as is the axial position z. Some equations change form according to the parameter, and those involving derivatives are simplest when the parameter is s. Derivatives with respect to the parameter are denoted by dots, so $\dot{\mathbf{x}}(\sigma) = d\mathbf{x}(\sigma)/d\sigma = (\dot{x}(\sigma), \dot{y}(\sigma), \dot{z}(\sigma))$. A parameter other than s will be a function of s, so $d\mathbf{x}(\sigma)/ds = (d\mathbf{x}/d\sigma)(d\sigma/ds)$.

Associated with space curves are three mutually perpendicular unit vectors, the tangent vector **t**, the principal normal **n**, and the binormal **b**, as well as two scalars, the curvature and the torsion. The direction of a ray is that of its unit *tangent vector*

$$\mathbf{t} = \frac{\dot{\mathbf{x}}(\sigma)}{|\dot{\mathbf{x}}(\sigma)|} = \dot{\mathbf{x}}(s) = (\alpha, \beta, \gamma) \tag{43}$$

The tangent vector **t** is the same as the direction vector **r** used elsewhere in this chapter. The rate of change of the tangent vector with respect to path length is

$$\kappa\mathbf{n} = \dot{\mathbf{t}}(s) = \ddot{\mathbf{x}}(s) = \left(\frac{d\alpha}{dx}, \frac{d\beta}{ds}, \frac{d\gamma}{ds}\right) \tag{44}$$

The *normal vector* is the unit vector in this direction

$$\mathbf{n} = \frac{\ddot{\mathbf{x}}(s)}{|\ddot{\mathbf{x}}(s)|} \tag{45}$$

The vectors **t** and **n** define the *osculating plane.* The *curvature* $\kappa = |\ddot{\mathbf{x}}(s)|$ is the rate of change of direction of **t** in the osculating plane.

$$\kappa^2 = \frac{|\dot{\mathbf{x}}(\sigma) \times \ddot{\mathbf{x}}(\sigma)|^2}{|\dot{\mathbf{x}}(\sigma)|^6} = |\ddot{\mathbf{x}}(s)|^2 = \left(\frac{d\alpha}{ds}\right)^2 + \left(\frac{d\beta}{ds}\right)^2 + \left(\frac{d\gamma}{ds}\right)^2 \tag{46}$$

The radius of curvature is $\rho = 1/\kappa$. Perpendicular to the osculating plane is the unit *binormal vector*

$$\mathbf{b} = \mathbf{t} \times \mathbf{n} = \frac{\dot{\mathbf{x}}(s) \times \ddot{\mathbf{x}}(s)}{|\ddot{\mathbf{x}}(s)|} \tag{47}$$

The *torsion* is the rate of change of the normal to the osculating plane

$$\tau = \mathbf{b}(s) \cdot \frac{d\mathbf{n}(s)}{ds} = \frac{(\dot{\mathbf{x}}(\sigma) \times \ddot{\mathbf{x}}(\sigma)) \cdot \dddot{\mathbf{x}}(\sigma)}{|\dot{\mathbf{x}}(\sigma) \times \ddot{\mathbf{x}}(\sigma)|^2} = \frac{(\dot{\mathbf{x}}(s) \times \ddot{\mathbf{x}}(s)) \cdot \dddot{\mathbf{x}}(s)}{|\ddot{\mathbf{x}}(s)|^2} \tag{48}$$

The quantity $1/\tau$ is the *radius of torsion.* For a plane curve, $\tau = 0$ and **b** is constant. The rates of change of **t**, **n**, and **b** are given by the Frenet equations:

$$\dot{\mathbf{t}}(s) = \kappa\mathbf{n} \qquad \dot{\mathbf{n}}(s) = -\kappa\mathbf{t} + \tau\mathbf{b} \qquad \dot{\mathbf{b}}(s) = -\tau\mathbf{n} \tag{49}$$

In some books, $1/\kappa$ and $1/\tau$ are used for what are denoted here by κ and τ.

Differential Geometry Equations Specific to Rays

From the general space curve equations above and the differential equations below specific to rays, the following equations for rays are obtained. Note that n here is the refractive index, unrelated to **n**. The tangent and normal vectors are related by Eq. (59), which can be written

$$\nabla \log n = \kappa\mathbf{n} + (\nabla \log n \cdot \mathbf{t})\mathbf{t} \tag{50}$$

The osculating plane always contains the vector ∇n. Taking the dot product with **n** in the above equation gives

$$\kappa = \frac{\partial \log n}{\partial N} = \mathbf{n} \cdot \nabla \log n = \mathbf{b} \cdot (\dot{\mathbf{x}} \times \nabla \log n) \tag{51}$$

The partial derivative $\partial/\partial N$ is in the direction of the principal normal, so rays bend toward regions of higher refractive index. Other relations (Stavroudis 1972[46]) are

$$\mathbf{n} = \rho\dot{\mathbf{x}}(s) \times (\nabla \log n \times \dot{\mathbf{x}}(s)) \tag{52}$$

$$\mathbf{b} = \rho\dot{\mathbf{x}}(s) \times \nabla \log n \qquad \text{and} \qquad 0 = \mathbf{b} \cdot \nabla n \tag{53}$$

$$\tau = \frac{(\dot{\mathbf{x}}(s) \times \nabla n) \cdot \nabla \dot{n}}{|\nabla n \times \dot{\mathbf{x}}(s)|^2} \tag{54}$$

Variational Integral

Written in terms of parameter σ, the optical path length integral, Eq. (7) is

$$V = \int n\,ds = \int \left(n \frac{ds}{d\sigma}\right) d\sigma = \int \mathscr{L}\,d\sigma \tag{55}$$

The solution for ray paths involves the calculus of variations in a way analogous to that used in classical mechanics, where the time integral of the lagrangian $\mathscr{L}$ is an extremum (Goldstein 1980[63]). If $\mathscr{L}$ has no explicit dependence on σ, the mechanical analogue to the optics case is that of no explicit time dependence.

Differential Equations for Rays

General Differential Equations. Because the optical path length integral is an extremum, the integrand $\mathscr{L}$ satisfies the Euler equations (Stavroudis 1972[46]). For an arbitrary coordinate system, with coordinates q_1, q_2, q_3 and the derivatives with respect to the parameter $\dot{q}_i = dq_i/d\sigma$, the differential equations for the path are

$$0 = \frac{d}{d\sigma}\frac{\partial \mathscr{L}}{\partial \dot{q}_i} - \frac{\partial \mathscr{L}}{\partial q_i} = \frac{d}{d\sigma}\left(n \frac{\partial}{\partial \dot{q}_i}\frac{ds}{d\sigma}\right) - \frac{\partial}{\partial q_i}\left(n \frac{ds}{d\sigma}\right) \qquad i = 1, 2, 3 \tag{56}$$

Cartesian Coordinates with Unspecified Parameter. In cartesian coordinates $ds/d\sigma = (\dot{x}^2 + \dot{y}^2 + \dot{z}^2)^{1/2}$, so the x equation is

$$0 = \frac{d}{d\sigma}\left(n \frac{\partial}{\partial \dot{x}}\frac{ds}{d\sigma}\right) - \frac{ds}{d\sigma}\frac{\partial n}{\partial x} = \frac{d}{d\sigma}\left[\frac{n\dot{x}}{(\dot{x}^2 + \dot{y}^2 + \dot{z}^2)^{1/2}}\right] - (\dot{x}^2 + \dot{y}^2 + \dot{z}^2)^{1/2}\frac{\partial n}{\partial x} \tag{57}$$

Similar equations hold for y and z.

Cartesian Coordinates with Parameter $\sigma = s$. With $\sigma = s$, so $ds/d\sigma = 1$, an expression, sometimes called the *ray equation,* is obtained (Synge 1937[28]).

$$\nabla n = \frac{d}{ds}\left(n \frac{d\mathbf{x}(s)}{ds}\right) = n \frac{d^2\mathbf{x}(s)}{ds^2} + \frac{dn(\mathbf{x}(s))}{ds}\frac{d\mathbf{x}(s)}{ds} \tag{58}$$

Using $dn/ds = \nabla n \cdot \dot{\mathbf{x}}$, the ray equation can also be written

$$\nabla n = n\ddot{\mathbf{x}} + (\nabla n \cdot \dot{\mathbf{x}})\dot{\mathbf{x}} \qquad \text{or} \qquad \nabla \log n = \ddot{\mathbf{x}} + (\nabla \log n \cdot \dot{\mathbf{x}})\dot{\mathbf{x}} \tag{59}$$

Only two of the component equations are independent, since $|\dot{\mathbf{x}}| = 1$.

Cartesian Coordinates with Parameter $\sigma = \int ds/n$. The parameter $\sigma = \int ds/n$, for which $ds/d\sigma = n$ and $n^2 = \dot{x}^2 + \dot{y}^2 + \dot{z}^2$, gives (Synge 1937[44])

$$\frac{d^2\mathbf{x}}{d\sigma^2} = \nabla(\tfrac{1}{2}n^2) \tag{60}$$

This equation is analogous to Newton's law of motion for a particle, $\mathbf{F} = m\, d^2\mathbf{x}/dt^2$, so the ray paths are like the paths of particles in a field with a potential proportional to $n^2(\mathbf{x})$. This analogy describes paths, but not speeds, since light travels slower where n is greater, whereas the particles would have greater speeds (Arnaud 1979,[64] Evans & Rosenquist 1986[65]).

Euler Equations for Parameter $\sigma = z$. If $\sigma = z$, then $ds/d\sigma = (\dot{x}^2 + \dot{y}^2 + 1)^{1/2}$ and $\mathcal{L} = \mathcal{L}(x, y; \dot{x}, \dot{y}; z)$. This gives (Luneburg 1964,[45] Marcuse 1989[49])

$$0 = \frac{d}{dz}\left(n \frac{\partial}{\partial \dot{x}} \frac{ds}{dz}\right) - \frac{ds}{dz}\frac{\partial n}{\partial x} = \frac{d}{dz}\left[\frac{n\dot{x}}{(1 + \dot{x}^2 + \dot{y}^2)^{1/2}}\right] - (1 + \dot{x}^2 + \dot{y}^2)^{1/2}\frac{\partial n}{\partial x} \tag{61}$$

with a similar equation for y. The equations can also be written (Moore 1975,[66] Marchand 1978, app. A[51]) as

$$n\ddot{x} = (1 + \dot{x}^2 + \dot{y}^2)\left(\frac{\partial n}{\partial x} - \frac{\partial n}{\partial z}\dot{x}\right) \qquad n\ddot{y} = (1 + \dot{x}^2 + \dot{y}^2)\left(\frac{\partial n}{\partial y} - \frac{\partial n}{\partial z}\dot{y}\right) \tag{62}$$

This parameter is particularly useful when n is rotationally symmetric about the z axis.

Hamilton's Equations with cartesian Coordinates for Parameter $\sigma = z$. A set of Hamilton's equations can also be written in cartesian coordinates using z as the parameter. (Luneburg 1964,[45] Marcuse 1989[49]) The canonical momenta in cartesian coordinates are the optical direction cosines

$$p_x = \frac{\partial \mathcal{L}}{\partial \dot{x}} = n\alpha \qquad p_y = \frac{\partial \mathcal{L}}{\partial \dot{y}} = n\beta \tag{63}$$

The hamiltonian is

$$\mathcal{H}(x, y, ; p_x, p_y; z) = \dot{x}p_x + \dot{y}p_y - \mathcal{L} = -\sqrt{n^2(x, y, z) - (p_x^2 + p_y^2)} \tag{64}$$

Hamilton's equations are

$$\frac{dx}{dz} = \frac{\partial \mathcal{H}}{\partial p_x} \qquad \frac{dy}{dz} = \frac{\partial \mathcal{H}}{\partial p_y} \qquad \frac{dp_x}{dz} = -\frac{\partial \mathcal{H}}{\partial x} \qquad \frac{dp_y}{dz} = -\frac{\partial \mathcal{H}}{\partial y} \tag{65}$$

It is not possible to write a set of Hamilton's equations using an arbitrary parameter and three canonical momenta, since they are not independent (Forbes 1991[67]). Another equation is

$$\frac{\partial \mathcal{H}}{\partial z} = \frac{d\mathcal{H}}{dz} = -\frac{1}{\gamma}\frac{\partial n}{\partial z} \tag{66}$$

Paraxial Form of Hamilton's Equations for $\sigma = z$. In the paraxial limit, if n_0 is the average index, the above set of equations gives (Marcuse 1989[49])

$$\frac{d^2x(z)}{dz^2} = \frac{1}{n_0}\frac{\partial n}{\partial x} \qquad \frac{d^2y(z)}{dz^2} = \frac{1}{n_0}\frac{\partial n}{\partial y} \tag{67}$$

Other Forms. A variety of additional differential equations can be obtained with various parameters (Forbes 1991[67]). Time cannot be used as a parameter (Landau & Lifshitz 1951[68]). The equations can also be expressed in a variety of coordinate systems (Buchdahl 1973,[69] Cornbleet 1976,[56] Cornbleet 1978,[70] Cornbleet 1979,[71] Cornbleet 1984[58]).

Refractive Index Symmetries

When the refractive index has symmetry or does not vary with one or more of the spatial variables, the above equations may simplify and take special forms. If, in some coordinate system, n does not vary with a coordinate q_i, so $\partial n/\partial q_i = 0$, and if, in addition, $\partial/\partial q_i(ds/d\sigma) = 0$, then

$$\frac{\partial \mathscr{L}}{\partial q_i} = 0 \qquad \text{and} \qquad \frac{\partial \mathscr{L}}{\partial \dot{q}} = n \frac{\partial}{\partial \dot{q}}\left(\frac{ds}{d\sigma}\right) = \text{constant} \tag{68}$$

There is an associated invariance of the ray path (Synge 1937,[44] Cornbleet 1976,[56] 1984,[58] Marcuse 1989[49]). (This is analogous to the case in mechanics where a potential does not vary with some coordinate.) There are a number of special cases. A more esoteric approach to symmetries involves Noether's theorem (Blaker 1974,[72] Joyce 1975[73]).

If the index is rotationally symmetric about the z axis, $n = n(x^2 + y^2, z)$, then $\partial \mathscr{L}/\partial \phi = 0$, where ϕ is the azimuth angle, and the constant of motion is analogous to that of the z component of angular momentum in mechanics for a potential with rotational symmetry. The constant quantity is the *skew invariant,* discussed elsewhere.

If the refractive index is a function of radius, $n = n(r)$, there are two constants of motion. The ray paths lie in planes through the center ($r = 0$) and have constant angular motion about an axis through the center that is perpendicular to this plane, so $\mathbf{x} \times \mathbf{p}$ is constant. If the plane is in the x-y plane, then $n(\alpha y - \beta x)$ is constant. This is analogous to motion of a particle in a central force field. Two of the best-known examples are the Maxwell fisheye (Maxwell 1854,[74] Born and Wolf 1980[48]) for which $n(r) \propto (1 + r^2)^{-1}$, and the Luneburg lens (Luneburg 1964,[45] Morgan 1958[75]), for which $n(r) = \sqrt{2 - r^2}$ for $r \le 1$ and $n = 1$ for $r > 1$.

If n does not vary with z, then $\mathscr{H} = n\gamma$ is constant for a ray as a function of z, according to Eq. (66).

If the medium is layered, so the index varies in only the z direction, then $n\alpha$ and $n\beta$ are constant. If θ is the angle relative to the z axis, then $n(z) \sin \theta(z)$ is constant, giving Snell's law as a special case.

The homogeneous medium, where $\partial n/\partial x = \partial n/\partial y = \partial n/\partial z = 0$, is a special case in which there are three constants of motion, $n\alpha$, $n\beta$, and $n\gamma$, so rays travel in straight lines.

1.6 CONSERVATION OF ÉTENDUE

If a bundle of rays intersects a constant z plane in a small region of size $dx\,dy$ and has a small range of angles $d\alpha\,d\beta$, then as the light propagates through a lossless system, the following quantity remains constant:

$$n^2\,dx\,dy\,d\alpha\,d\beta = n^2\,dA\,d\alpha\,d\beta = n^2\,dA \cos\theta\,d\omega = dx\,dy\,dp_x\,dp_y \tag{69}$$

Here $dA = dx\,dy$ is the differential area, $d\omega$ is the solid angle, and θ is measured relative to the normal to the plane. The integral of this quantity

$$\int n^2\,dx\,dy\,d\alpha\,d\beta = \int n^2\,dA\,d\alpha\,d\beta = \int n^2\,dA\cos\theta\,d\omega = \int dx\,dy\,dp_x\,dp_y \tag{70}$$

is the *étendue,* and is also conserved. For lambertian radiation of radiance L_e, the total power transferred is $P = \int L_e n^2\,d\alpha\,d\beta\,dx\,dy$. The étendue and related quantities are known by a variety of names (Steel 1974[76]), including *generalized Lagrange invariant, luminosity, light-gathering power, light grasp, throughput, acceptance, optical extent,* and *area-solid-angle-product.* The angle term is not actually a solid angle, but is weighted. It does approach a solid angle in the limit of small extent. In addition, the integrations can be over area, giving $n^2\,d\alpha\,d\beta \int dA$, or over angle, giving $n^2\,dA \int d\alpha\,d\beta$. A related quantity is the geometrical vector flux (Winston 1979[77]), with components $(\int dp_y\,dp_z, \int dp_x\,dp_z, \int dp_x\,dp_y)$. In some cases these quantities include a brightness factor, and in others they are purely geometrical. The étendue is related to the information capacity of a system (Gabor 1961[78]).

As special case, if the initial and final planes are conjugate with transverse magnification $m = dx'/dx = dy'/dy$, then

$$n^2\,d\alpha\,d\beta = n'^2 m^2\,d\alpha'\,d\beta' \tag{71}$$

Consequently, the angular extents of the entrance and exit pupil in direction cosine coordinates are related by

$$n^2 \int_{\text{entrance pupil}} d\alpha\,d\beta = n'^2 m^2 \int_{\text{exit pupil}} d\alpha'\,d\beta' \tag{72}$$

See also the discussion of image irradiance in the section on apertures and pupils.

This conservation law is general; it does not depend on index homogeneity or on axial symmetry. It can be proven in a variety of ways, one of which is with characteristic functions (Welford & Winston 1978,[79] Welford 1986,[80] Welford & Winston 1989[81]). Phase space arguments involving Liouville's theorem can also be applied (di Francia 1950,[82] Winston 1970,[83] Jannson & Winston 1986,[84] Marcuse 1989[85]). Another type of proof involves thermodynamics, using conservation of radiance (or brightness) or the principal of detailed balance (Clausius 1864,[86] Clausius 1879,[87] Helmholtz 1874,[88] Liebes 1969[89]). Conversely, the thermodynamic principle can be proven from the geometric optics one (Nicodemus 1963,[90] Boyd 1983,[91] Klein 1986[92]). In the paraxial limit for systems of revolution the conservation of etendue between object and image planes is related to the two-ray paraxial invariant, Eq. (152). Some historical aspects are discussed by Rayleigh (Rayleigh 1886[93]) and Southall (Southall 1910[94]).

1.7 SKEW INVARIANT

In a rotationally symmetric system, whose indices may be constant or varying, a *skew ray* is one that does not lie in a plane containing the axis. The *skewness* of such a ray is

$$\mathcal{S} = n(\alpha y - \beta x) = n\alpha y - n\beta x = p_x y - p_y x \tag{73}$$

As a skew ray propagates through the system, this quantity, known as the *skew invariant,* does not change (T. Smith 1921,[95] H. Hopkins 1947,[96] Marshall 1952,[97] Buchdahl 1954, sec. 4,[98] M. Herzberger 1958,[99] Welford 1968,[100] Stavroudis 1972, p. 208,[101] Welford 1974, sec. 5.4,[102] Welford 1986, sec. 6.4[103] Welford & Winston 1989, p. 228[104]). For a meridional ray, one lying in a plane containing the axis, $\mathcal{S} = 0$. The skewness can be written in vector form as

$$\mathcal{S} = \mathbf{a} \cdot (\mathbf{x} \times \mathbf{p}) \tag{74}$$

where **a** is a unit vector along the axis, **x** is the position along a ray, **p** is the optical cosine and vector at that position.

This invariance is analogous to the conservation of the axial component of angular momentum in a cylindrical force field, and it can be proven in several ways. One is by performing the rotation operations on α, β, x, and y (as discussed in the section on heterogeneous media). Another is by means of characteristic functions. It can also be demonstrated that $\mathcal{S}$ is not changed by refraction or reflection by surfaces with radial gradients. The invariance holds also for diffractive optics that are figures of rotation.

A special case of the invariant relates the intersection points of a skew ray with a given meridian. If a ray with directions (α, β) in a space of index n intersects the $x = 0$ meridian with height y, then at another intersection with this meridian in a space with index n', its height y' and direction cosine α' are related by

$$n\alpha y = n'\alpha' y' \tag{75}$$

The points where rays intersect the same meridian are known as *diapoints* and the ratio y'/y as the *diamagnification* (Herzberger 1958[99]).

1.8 REFRACTION AND REFLECTION AT INTERFACES BETWEEN HOMOGENEOUS MEDIA

Introduction

The initial ray direction is specified by the unit vector $\mathbf{r} = (\alpha, \beta, \gamma)$. After refraction or reflection the direction is $\mathbf{r}' = (\alpha', \beta', \gamma')$. At the point where the ray intersects the surface, its normal has direction $\mathbf{S} = (L, M, N)$.

The *angle of incidence* I is the angle between a ray and the surface normal at the intersection point. This angle and the corresponding outgoing angle I' are given by

$$\begin{aligned} |\cos I| &= |\mathbf{r} \cdot \mathbf{S}| = |\alpha L + \beta M + \gamma N| \\ |\cos I'| &= |\mathbf{r}' \cdot \mathbf{S}| = |\alpha' L + \beta' M + \gamma' N| \end{aligned} \tag{76}$$

In addition

$$|\sin I| = |\mathbf{r} \times \mathbf{S}| \qquad |\sin I'| = |\mathbf{r}' \times \mathbf{S}| \tag{77}$$

The signs of these expressions depend on which way the surface normal vector is directed. The surface normal and the ray direction define the *plane of incidence,* which is perpendicular to the vector cross product $\mathbf{S} \times \mathbf{r} = (M\gamma - N\beta, N\alpha - L\gamma, L\beta - M\alpha)$. After refraction or reflection, the outgoing ray is in the same plane. This symmetry is related to the fact that optical path length is an extremum.

The laws of reflection and refraction can be derived from Fermat's principle, as is done in many books. At a planar interface, the reflection and refraction directions are derived from Maxwell's equations using the boundary conditions. For scalar waves at a plane interface, the directions are related to the fact that the number of oscillation cycles is the same for incident and outgoing waves.

Refraction

At an interface between two homogeneous and isotropic media, described by indices n and n', the incidence angle I, and the outgoing angle I' are related by *Snell's law* (Magie 1963[105]):

$$n' \sin I' = n \sin I \tag{78}$$

If $\sin I' > 1$, there is total internal reflection. Another relationship is

$$n' \cos I' = \sqrt{n'^2 - n^2 \sin^2 I} = \sqrt{n'^2 - n^2 - n^2 \cos^2 I} \tag{79}$$

Snell's law can be expressed in a number of ways, one of which is

$$n[\mathbf{r} + (\mathbf{r} \cdot \mathbf{S})\mathbf{S}] = n'[\mathbf{r}' + (\mathbf{r}' \cdot \mathbf{S})\mathbf{S}] \tag{80}$$

Taking the cross product of both sides with **S** gives another form

$$n'(\mathbf{S} \times \mathbf{r}') = n(\mathbf{S} \times \mathbf{r}) \tag{81}$$

A quantity that appears frequently in geometrical optics (but which has no common name or symbol) is

$$\Gamma = n' \cos I' - n \cos I \tag{82}$$

It can be written several ways

$$\Gamma = (n\mathbf{r} - n'\mathbf{r}') \cdot \mathbf{S} = -n \cos I + \sqrt{n'^2 - n^2 \sin^2 I} = -n \cos I + \sqrt{n'^2 - n^2 + n^2 \cos^2 I} \tag{83}$$

In terms of Γ, Snell's law is

$$n'\mathbf{r}' = n\mathbf{r} + \Gamma\mathbf{S} \tag{84}$$

or

$$n'\alpha' = n\alpha + L\Gamma \qquad n'\beta' = n\beta + M\Gamma \qquad n'\gamma' = n\gamma + N\Gamma \tag{85}$$

The outgoing direction is expressed explicitly as a function of incident direction by

$$n'\mathbf{r}' = n\mathbf{r} + \mathbf{S}[n\mathbf{r} \cdot \mathbf{S} - \sqrt{n'^2 - n^2 + (n\mathbf{r} \cdot \mathbf{S})^2}] \tag{86}$$

If the surface normal is in the z direction, the equations simplify to

$$n'\alpha' = n\alpha \qquad n'\beta' = n\beta \qquad n'\gamma' = \sqrt{n'^2 - n^2 + n^2\gamma^2} \tag{87}$$

If $\beta = 0$, this reduces to $n'\alpha' = n\alpha$, the familiar form of Snell's law, written with direction cosines, with $n'\gamma' = (n'^2 - n^2\alpha^2)^{1/2}$, corresponding to Eq. (79). Another relation is

$$\frac{n'\alpha' - n\alpha}{L} = \frac{n'\beta' - n\beta}{M} = \frac{n'\gamma' - n\gamma}{N} = \Gamma \tag{88}$$

All of the above expressions can be formally simplified by using $\mathbf{p} = n\mathbf{r}$ and $\mathbf{p}' = n'\mathbf{r}'$. For a succession of refractions by parallel surfaces,

$$n_1 \sin I_1 = n_2 \sin I_2 = n_3 \sin I_3 = \cdots \tag{89}$$

so the angles within any two media are related by their indices alone, regardless of the intervening layers. Refractive indices vary with wavelength, so the angles of refraction do likewise.

Reflection

The reflection equations can be derived from those for refraction by setting the index of the final medium equal to the negative of that of the incident medium, i.e., $n' = -n$, which gives $\Gamma = -2n \cos I$. The angles of incidence and reflection are equal

$$I' = I \tag{90}$$

The incident and reflected ray directions are related by

$$\mathbf{S} \times \mathbf{r}' = -\mathbf{S} \times \mathbf{r} \tag{91}$$

Another expression is

$$\mathbf{r}' = \mathbf{r} - (2\mathbf{S} \cdot \mathbf{r})\mathbf{S} = \mathbf{r} - (2 \cos I)\mathbf{S} \tag{92}$$

The components are

$$\alpha' = \alpha - 2L \cos I \qquad \beta' = \beta - 2M \cos I \qquad \gamma' = \gamma - 2N \cos I \tag{93}$$

This relationship can be written in terms of dyadics (Silberstein 1918[106]) as $\mathbf{r}' = (\mathbf{I} - \mathbf{SS}) \cdot \mathbf{r}$. This is equivalent to the matrix form (T. Smith 1928,[107] Luneburg 1964,[108] R. E. Hopkins 1965,[109] Levi 1968,[110] Tongshu 1991[111])

$$\begin{pmatrix} \alpha' \\ \beta' \\ \gamma' \end{pmatrix} = \begin{pmatrix} 1 - 2L^2 & -2LM & -2LN \\ -2LM & 1 - 2M^2 & -2MN \\ -2LN & -2MN & 1 - 2N^2 \end{pmatrix} \begin{pmatrix} \alpha \\ \beta \\ \gamma \end{pmatrix} \tag{94}$$

Each column of this matrix is a set of direction cosines and is orthogonal to the others, and likewise for the rows. The matrix associated with a sequence of reflections from plane surfaces is calculated by multiplying the matrices for each reflection. Another relationship is

$$\frac{\alpha' - \alpha}{L} = \frac{\beta' - \beta}{M} = \frac{\gamma' - \gamma}{N} \tag{95}$$

If the surface normal is in the z direction, so $(L, M, N) = (0, 0, 1)$, then

$$\alpha' = \alpha \qquad \beta' = \beta \qquad \gamma' = -\gamma \tag{96}$$

Reflection by a Plane Mirror—Positions of Image Points

If light from a point (x, y, z) is reflected by a plane mirror whose distance from the coordinate origin is d, and whose surface normal has direction (L, M, N), the image point coordinates (x', y', z') are given by

$$\begin{pmatrix} x' \\ y' \\ z' \\ 1 \end{pmatrix} = \begin{pmatrix} 1 - 2L^2 & -2LM & -2LN & 2dL \\ -2LM & 1 - 2M^2 & -2MN & 2dM \\ -2LN & -2MN & 1 - 2N^2 & 2dN \\ 0 & 0 & 0 & 1 \end{pmatrix} \begin{pmatrix} x \\ y \\ z \\ 1 \end{pmatrix} \tag{97}$$

This transformation involves both rotation and translation, with only the former effect applying if $d = 0$. It is an affine type of collinear transformation, discussed in the section on collineation. The effect of a series of reflections by plane mirrors is found by a product of such matrices. The transformation can also be formulated in terms of quaternions (Wagner 1951,[112] Levi 1968, p. 367[110]).

Diffractive Elements

The changes in directions produced by gratings or diffractive elements can be handled in an ad hoc geometrical way (Spencer & Murty 1962,[113] di Francia 1950[114])

$$n'\mathbf{r}'_G \times \mathbf{S} = n\mathbf{r} \times \mathbf{S} + G\frac{\lambda}{\mathrm{p}}\mathbf{q} \tag{98}$$

Here λ is the vacuum wavelength, p is the grating period, $\mathbf{q}$ is a unit vector tangent to the surface and parallel to the rulings, and G is the diffraction order. Equations (81) and (91) are special cases of this equation for the 0th order.

1.9 IMAGING

Introduction

Image formation is the principal use of lenses. Moreover, lenses form images even if this is not their intended purpose. This section provides definitions, and discusses basic concepts and limitations. The purposes of the geometric analysis of imaging include the following: (1) discovering the nominal relationship between an object and its image, principally the size, shape, and location of the image, which is usually done with paraxial optics; (2) determining the deviations from the nominal image, i.e., the aberrations; (3) estimating image irradiance; (4) understanding fundamental limitations—what is inherently possible and impossible; (5) supplying information for diffraction calculations, usually optical path lengths.

Images and Types of Images

A definition of image (Webster 1934[115]) is: "The optical counterpart of an object produced by a lens, mirror, or other optical system. It is a geometrical system made up of foci corresponding to the parts of the object." The point-by-point correspondence is the key, since a given object can have a variety of different images.

Image irradiance can be found only approximately from geometric optics, the degree of accuracy of the predictions varying from case to case. In many instances wave optics is required, and for objects that are not self-luminous, an analysis involving partial coherence is also needed.

The term *image* is used in a variety of ways, so clarification is useful. The light from an object produces a three-dimensional distribution in image space. The *aerial image* is the distribution on a mathematical surface, often that of best focus, the locus of points of the images of object points. An aerial image is never the final goal; ultimately, the light is to be captured. The *receiving surface* (NS) is that on which the light falls, the distribution of which there can be called the *received image* (NS). This distinction is important in considerations of defocus, which is a relationship, not an absolute. The record thereby produced is the *recorded image* (NS). The recorded image varies with the position of the receiving surface, which is usually intended to correspond with the aerial image surface. In this section, "image" means aerial image, unless otherwise stated.

Object Space and Image Space

The object is said to exist in *object space*; the image, in *image space.* Each space is infinite, with a physically accessible region called *real,* and an inaccessible region, referred to as *virtual.* The two spaces may overlap physically, as with reflective systems. Corresponding quantities and locations associated with the object and image spaces are typically denoted by the same symbol, with a prime indicating image space. Positions are specified by a coordinate system (x, y, z) in object space and (x', y', z') in image space. The refractive indices of the object and image spaces are n and n'.

Image of a Point

An *object point* is thought of as emitting rays in all directions, some of which are captured by the lens, whose internal action converges the rays, more or less, to an *image point,* the term "point" being used even if the ray convergence is imperfect. Object and image points are said to be *conjugate.* Since geometric optics is reversible, if A' is the image of A, then A is the image of A'.

Mapping Object Space to Image Space

If every point were imaged stigmatically, then the entire object space would be mapped into the image space according to a transformation

$$x' = x'(x, y, z) \qquad y' = y'(x, y, z) \qquad z' = z'(x, y, z) \tag{99}$$

The mapping is reciprocal, so the equations can be inverted. If n and n' are constant, then the mapping is a collinear transformation, discussed below.

Images of Extended Objects

An *extended object* can be thought of as a collection of points, a subset of the entire space, and its stigmatic image is the set of conjugate image points. A surface described by $0 = F(x, y, z)$ has an image surface

$$0 = F'(x', y', z') = F(x(x', y', z'), y(x', y', z'), z(x', y', z')) \tag{100}$$

A curve described parametrically by $\mathbf{x}(\sigma) = (x(\sigma), y(\sigma), z(\sigma))$ has an image curve

$$\mathbf{x}'(\sigma) = (x'(x(\sigma), y(\sigma), z(\sigma)), y'(x(\sigma), y(\sigma), z(\sigma)), z'(x(\sigma), y(\sigma), z(\sigma))) \tag{101}$$

Rotationally Symmetric Lenses

Rotationally symmetric lenses have an *axis,* which is a ray path (unless there is an obstruction). All planes through the axis, the *meridians* or *meridional planes,* are planes with respect to which there is bilateral symmetry. An *axial object point* is conjugate to an *axial image point.* An axial image point is located with a single ray in addition to the axial one. Off-axis object and image points are in the same meridian, and may be on the same or opposite sides of the axis. The *object height* is the distance of a point from the axis, and the *image height* is that for its image. It is possible to have rotational symmetry without bilateral symmetry, as in a system made of crystalline quartz (Buchdahl 1970[116]), but such systems are not discussed here. For stigmatic imaging by a lens rotationally symmetric about the z axis, Eq. (99) gives

$$x' = x'(x, z) \qquad y' = y'(y, z) \qquad z' = z'(z) \tag{102}$$

Planes Perpendicular to the Axis

The arrangement most often of interest is that of planar object and receiving surfaces, both perpendicular to the lens axis. When the terms *object plane* and *image plane* are used here without further elaboration, this is the meaning. (This arrangement is more common for manufactured systems with flat detectors, than for natural systems, for instance, eyes, with their curved retinas.)

Magnifications

The term *magnification* is used in a general way to denote the ratio of conjugate object and image dimensions, for example, heights, distances, areas, volumes, and angles. A single number is inadequate when object and image shapes are not geometrically similar. The term magnification implies an increase, but this is not the case in general.

Transverse Magnification

With object and image planes perpendicular to the axis, the relative scale factor of length is the *transverse magnification* or *lateral magnification,* denoted by m, and usually referred to simply as "the magnification." The transverse magnification is the ratio of image height to object height, $m = h'/h$. It can also be written in differential form, e.g., $m = dx'/dx$ or $m = \Delta x'/\Delta x$. The transverse magnification is signed, and it can have any value from $-\infty$ to $+\infty$. Areas in such planes are scaled by m^2. A lens may contain plane mirrors that affect the image parity or it may be accompanied by external plane mirrors that reorient images and change their parity, but these changes are independent of the magnification at which the lens works.

Longitudinal Magnification

Along the rotational axis, the *longitudinal magnification,* m_L, also called *axial magnification,* is the ratio of image length to object length in the limit of small lengths, i.e., $m_L = dz'/dz$.

Visual Magnification

With visual instruments, the perceived size of the image depends on its angular subtense. *Visual magnification* is the ratio of the angular subtense of an image relative to that of the object viewed directly. Other terms are used for this quantity, including "magnification," "power," and "magnifying power." For objects whose positions can be controlled, there is arbitrariness in the subtense without the instrument, which is greatest when the object is located at the near-point of the observer. This distance varies from person to person, but for purposes of standardization the distance is taken to be 250 mm. For instruments that view distant objects the question of subtense with direct viewing does not arise.

Ideal Imaging and Disappointments in Imaging

Terms such as *perfect imaging* and *ideal imaging* are used in various ways. The ideal varies with the circumstances, and there are applications in which imaging is not the purpose, for instance, energy collection and Fourier transformation. The term *desired imaging* might be more appropriate in cases where that which is desired is fundamentally impossible. Some deviations from what is desired are called *aberrations,* whether their avoidance is possible or not. Any ideal that can be approximated must agree in its paraxial limit ideal with what a lens actually does in its paraxial limit.

Maxwellian Ideal for Single-Plane Imaging

The most common meaning of perfect imaging is that elucidated by Maxwell (Maxwell 1858[117]), and referred to here as *maxwellian ideal* or *maxwellian perfection.* This ideal is fundamentally possible. The three conditions for such imaging of points in a plane perpendicular to the lens axis are: (1) Each point is imaged stigmatically. (2) The images of all the points in a plane lie on a plane that is likewise perpendicular to the axis, so the field is flat, or free from field curvature. (3) The ratio of image heights to object heights is the same for all points in the plane. That is, transverse magnification is constant, or there is no distortion.

The Volume Imaging Ideal

A more demanding ideal is that points everywhere in regions of constant index be imaged stigmatically and that the imaging of every plane be flat and free from distortion. Such imaging is described mathematically by the collinear transformation, discussed below. It is inherently impossible for a lens to function in this way, but the mathematical apparatus of collineation is useful in obtaining approximate results.

Paraxial, First-Order, and Gaussian Optics

The terms "paraxial," "first order," and "gaussian" are often used interchangeably, and their consideration is merged with that of collineation. The distinction is often not made, probably because these descriptions agree in result, although differing in approach. One of the few discussions is that of Southall (Southall 1910[118]). A *paraxial* analysis has to do with the limiting case in which the distances of rays from the axis approach zero, as do the angles of the rays relative to the axis. The term *first order* refers to the associated mathematical approximation in which the positions and directions of such rays are computed with terms to the first order only in height and angle. *Gaussian* refers to certain results of the paraxial optics, where lenses are black boxes whose properties are summarized by the existence and locations of cardinal points. In the limit of small heights and angles, the equations of collineation are identical to those of paraxial optics. Each of these is discussed in greater detail below.

Fundamental Limitations

There are fundamental geometrical limitations on optical systems, relating to the fact that a given ray passes through many points and a given point lies on many rays. So the images of points on the same line or plane, or on different planes, are not independent. A set of rays intersecting at several points in object space cannot be made to satisfy arbitrary requirements in image space. Such limitations are best studied by the methods of hamiltonian optics.

Stigmatic Imaging

If all rays from an object point converge precisely to an image point, the imaging of this point is said to be *stigmatic*. The optical path lengths of all rays between two such points is identical. A stigmatic image point is located by the intersection of any two rays that pass through the points. An *absolute instrument* is one which images all points stigmatically (Born & Wolf 1980[119]). For such imaging

$$n\,\delta x = n'\,\delta x' \qquad n\,\delta y = n'\,\delta y' \qquad n\,\delta z = n'\,\delta z' \tag{103}$$

where conjugate length elements are δx and $\delta x'$, δy and $\delta y'$, δz and $\delta z'$.

Path Lengths and Conjugate Points

All the rays from an object point to its stigmatic image point have the same optical path length. For focal lenses, the paths lengths for different pairs of conjugate points in a plane perpendicular to the axis are different, except for points on circles centered on the axis. For afocal lenses path lengths are normally the same for all points in planes perpendicular to the axis. For afocal lenses with transverse magnification $\pm n/n'$, path lengths can be the same for all points. In general, the path lengths between different points on an object and image surface are equal only if the shape of the image surface is that of a wavefront that has propagated from a wavefront whose shape is that of the object surface.

The Cosine Condition

The *cosine condition* relates object space and image space ray angles, if the imaging is stigmatic over some area (T. Smith 1922,[120] Steward 1928,[121] Buchdahl 1970[116]). Let the x-y plane lie in the object surface and the x'-y' plane in the conjugate surface (Fig. 2). Two

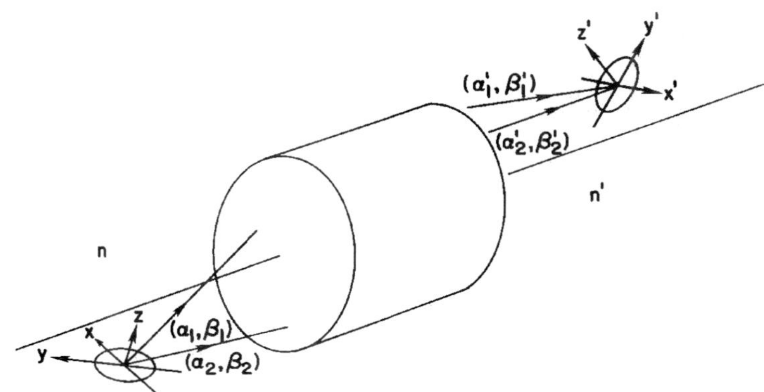

FIGURE 2 The cosine condition. A small area in object space about the origin in the x-y plane is imaged to the region around the origin of the x'-y' plane in image space. A pair of rays from the origin with direction cosines (α_1, β_1) and (α_2, β_2) arrive with (α_1', β_1') and (α_2', β_2'). The direction cosines and the transverse magnification in the planes are related by Eq. (104).

rays leaving a point in the object region have direction cosines (α_1, β_1) and (α_2, β_2), and the rays on the image side have (α_1', β_1') and (α_2', β_2'). If the imaging is stigmatic, with local transverse magnification m on the surface, then

$$m = \frac{n(\alpha_1 - \alpha_2)}{n'(\alpha_1' - \alpha_2')} = \frac{n(\beta_1 - \beta_2)}{n'(\beta_1' - \beta_2')} \tag{104}$$

In the limit as $\alpha_1 \rightarrow \alpha_2$ and $\beta_1 \rightarrow \beta_2$, the cosine condition gives

$$m = \frac{n\, d\alpha}{n'\, d\alpha'} = \frac{n\, d\beta}{n'\, d\beta'} \tag{105}$$

This condition also holds in the more general case of *isoplanatic* imaging, where there is aberration that is locally constant across the region in question (Welford 1976,[122] Welford 1986, sec. 94[123]).

The Abbe Sine Condition

The *sine condition* or *Abbe sine condition* (Abbe 1879,[124] Born and Wolf 1980[119]) is a special case of the cosine condition for object and image planes perpendicular to the axis in regions about the axis. For a plane with transverse magnification m, let θ be the angle relative to the axis made by a ray from an axial object point, and θ' be that in image space. If the lens is free of coma

$$m = \frac{n \sin\theta}{n' \sin\theta'} = \frac{n\alpha}{n'\alpha'} = \frac{n\beta}{n'\beta'} \tag{106}$$

for all θ and θ'. There are signs associated with θ and θ', so that $m > 0$ if they have the same sign, and $m < 0$ if the signs are opposite. This equation is sometimes written with m replaced by the ratio of paraxial angles. There is sometimes the implication that θ and θ' refer only to the extreme angles passing through the lens, when in fact the sine condition dictates that the ratio of the sines is the constant for all angles. For an object at infinity, the sine condition is

$$\sin\theta' = -\frac{y}{f'} \qquad \text{or} \qquad n'\beta' = -y\phi \tag{107}$$

where y is the height of a ray parallel to the axis, ϕ is the power of the lens, and f' is the rear focal length. These relationships hold to a good approximation in most lenses, since small deviations are associated with large aberrations. A deviation from this relationship is called *offense against the sine condition,* and is associated with coma (Conrady 1992,[125] H. Hopkins 1946,[126] T. Smith 1948,[127] H. Hopkins 1950,[128] Welford 1986[123]). The sine condition does not apply where there are discontinuities in ray behavior, for example, in devices like Fresnel lenses, or to diffraction-based devices like zone plates.

The Herschel Condition

The *Herschel condition* is a relationship that holds if the imaging is stigmatic for nearby points along the axis (Herschel 1821,[129] H. Hopkins 1946,[130] Born and Wolf 1980[119]). The two equivalent relations are

$$m = \frac{n \sin\left(\frac{1}{2}\theta\right)}{n' \sin\left(\frac{1}{2}\theta'\right)} \qquad \text{and} \qquad m_L = \frac{n \sin^2\left(\frac{1}{2}\theta\right)}{n' \sin^2\left(\frac{1}{2}\theta'\right)} = \frac{n(1-\gamma)}{n'(1-\gamma')} \tag{108}$$

The Herschel condition is inconsistent with the sine condition unless $m \pm n/n'$. So, in general, stigmatic imaging in one plane precludes that in others.

Sine and Herschel Conditions for Afocal Systems

For afocal systems the sine condition and Herschel condition are identical. For rays entering parallel to the axis at y and leaving at y' they are

$$m = \frac{y'}{y} \tag{109}$$

That is, the ratio of incoming and outgoing heights is independent of the incoming height. (H. Hopkins,[128] Chap. III "The Sine Condition and Herschel's Condition.")

Stigmatic Imaging Possibilities

For object and image spaces with constant refractive indices, stigmatic imaging is only possible for the entire spaces for afocal lenses with identical transverse and longitudinal magnifications $m = \pm n/n'$ and $|m_L| = |m|$. Such lenses re-create not only the intersection points, but the wavefronts, since the corresponding optical path lengths are the same in both spaces, Eq. (103). For other lenses with constant object and image space indices, the maxwellian ideal can be met for only a single surface. In addition, a single point elsewhere can be imaged stigmatically (T. Smith 1948,[127] Velzel 1991[131]). Nonplanar surfaces can be imaged stigmatically, a well-known example being the imaging of spherical surfaces by a spherical refracting surface, for a particular magnification (Born & Wolf 1980, sec. 4.2.3[119]). For systems with spherical symmetry, it is possible that two nonplanar surfaces be stigmatically imaged (T. Smith 1927[132]). In addition, various systems with heterogeneous indices can image stigmatically over a volume.

1.10 DESCRIPTION OF SYSTEMS OF REVOLUTION

Introduction

This section is concerned with the optical description of lens and mirror systems that are figures of revolution.[133–145] From a mechanical viewpoint, optical systems are comprised of lenses and mirrors. From the point of view of the light, the system is regions of media with different indices, separated by interfaces of various shapes. This section is limited to homogeneous isotropic media. It is further restricted to reflecting and refracting surfaces that are nominally smooth, and to surfaces that are figures of revolution arranged so their axes are collinear, so the entire system is a figure of revolution about the *lens axis*. (The often-used term "optical axis" is also used in crystallography. Moreover, the axis is often mechanical as well as "optical.") The lens axis is the z axis of an orthogonal coordinate system, with the x-y plane perpendicular. The distance from a point to the axis is $\rho = \sqrt{x^2 + y^2}$. Along the axis, the positive direction is from left to right.

Terminology

A *meridian* or *meridional plane* contains the axis, all such planes being equivalent. Meridional planes are planes of bilateral symmetry if the indices are homogeneous and isotropic. Some optical systems are comprised of pieces of surfaces of revolution, in which case it is still useful to discuss the behavior about the axis.

Reflection, Unfolded Diagrams

Light passes through refractive systems more or less in the same direction relative to the axis. In reflective and catadioptric systems, the light may change directions. (It may not, in the case of grazing incidence optics.) In order to consider all types of systems in a unified way, pictorially and algebraically, reflections can often be "unfolded," i.e., represented pictorially as transmission, with mirrors replaced by hypothetical lenses with the same properties, Figs. 3 and 18. Some structures must be taken into account several times in unfolding. For example, a hole may block light at one point along a ray and transmit it at another. (In some considerations, unfolding can be misleading—for instance, those involving stray light.)

Description of Surfaces

A *surface* is an interface between media with different refractive indices—either refracting or reflecting. The surface is the optical element, produced by a lens, which is a mechanical element. Surfaces can be described mathematically in many ways. (For example, conics can be described as loci of points with certain relationships.) In optical instruments, the entire surface is rarely used, and the axial region is usually special, so the description usually begins there and works out. The *vertex* of a figure of revolution intersects with

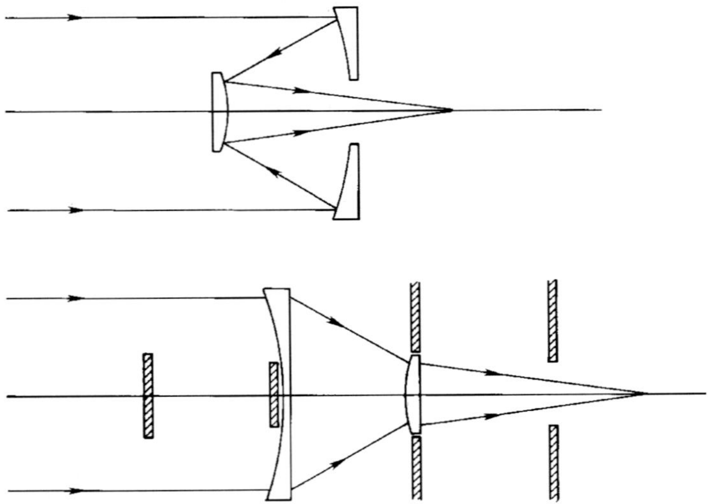

FIGURE 3 Example of an unfolded diagram. The two-mirror system above has an unfolded representation below. The reflective surfaces are replaced by thin lens equivalents. Their obstructions and the finite openings are accounted for by dummy elements.

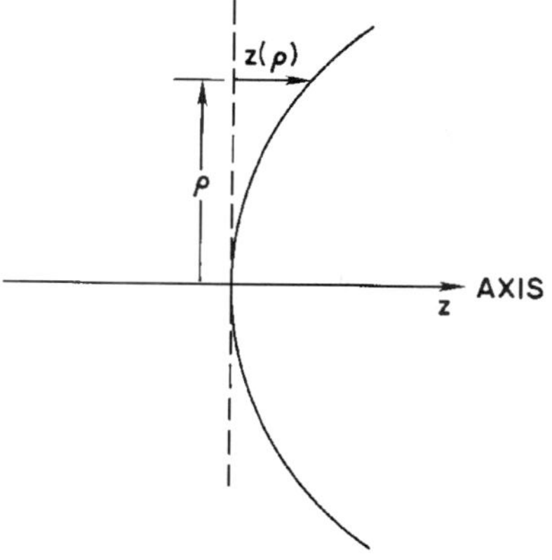

FIGURE 4 Description of a surface of revolution. The distance from the axis is ρ, and the sag $z(\rho)$ is the distance from the vertex tangent plane to the surface.

the axis, and is a local extremum. The plane perpendicular to the axis and tangent to the vertex will be referred to as the *vertex plane* (NS). A surface can be described by its *sag*, the directed distance $z(\rho)$ from the vertex plane to the surface, Fig. 4. The vertex is usually taken to have $z(0) = 0$. The *vertex curvature* or *paraxial curvature* c and radius of curvature r are given by

$$c = \frac{1}{r} = \left.\frac{\partial^2 z(\rho)}{\partial \rho^2}\right|_{\rho=0} \tag{110}$$

For an arbitrary surface, this curvature is identical to that of the sphere which is a best fit on axis. The sign convention for curvature and radius is that c and r are positive if the center of curvature is to the right of the vertex, as in the case shown in Fig. 4. In general, the curvature is mathematically more foolproof than radius, since curvature can be zero, but it is never infinite, whereas radius is never zero, but may be infinite.

Spherical Surfaces

The spherical surface is the most common in optics, since it is most naturally produced. Spherical is the default, and is assumed when no other mention is made. Aspheres are thought of as deviating from spheres, rather than spheres as special cases of more general forms. The equation for a sphere with radius r, curvature c, and a vertex at $z = 0$ is

$$\rho^2 + (z - r)^2 = r^2 \tag{111}$$

The sag is given by

$$z(\rho) = r - \sqrt{r^2 - \rho^2} = r(1 - \sqrt{1 - c^2\rho^2}) = \frac{c\rho^2}{1 + \sqrt{1 - c^2\rho^2}} \tag{112}$$

The Taylor expansion is

$$z(\rho) = \tfrac{1}{2}c\rho^2 + \tfrac{1}{8}c^3\rho^4 + \tfrac{1}{16}c^5\rho^6 + \tfrac{5}{128}c^7\rho^8 + \tfrac{7}{256}c^9\rho^{10} + \cdots \tag{113}$$

At the point (x, y, z) on the surface of the sphere, the surface normal has direction cosines

$$(L, M, N) = \left(\frac{x}{r}, \frac{y}{r}, \frac{z-r}{r}\right) = (cx, cy, cz - 1) \tag{114}$$

Conics of Rotation

The general form of a conic of rotation about the z axis is

$$z(\rho) = \frac{r}{\varepsilon}\left(1 - \sqrt{1 - \varepsilon c^2 \rho^2}\right) = \frac{c\rho^2}{1 + \sqrt{1 - \varepsilon c^2 \rho^2}} \tag{115}$$

The value of ε determines the type of conic, as given in the table below. It is common in optics to use κ, the *conic parameter* or *conic constant,* related by

$$\kappa = \varepsilon - 1 \qquad \text{or} \qquad \varepsilon = 1 + \kappa \tag{116}$$

Another parameter used to describe conics is the *eccentricity* e, used in the polar coordinate form for conics about their focus: $r(\theta) = a/(1 + e\cos\theta)$ where $e^2 = -\kappa$. In the case of paraboloids, the first form of Eq. (115) breaks down. A cone can be approximated by a hyperbola with $\kappa = -\sec^2\theta$, where θ is the cone half angle.

Conic Type and Value of Parameter

Parameter	ε	κ	e
Oblate ellipsoid	$\varepsilon > 1$	$\kappa > 0$	$e^2 < 0$
Sphere	$\varepsilon = 1$	$\kappa = 0$	0
Prolate ellipsoid	$0 < \varepsilon < 1$	$-1 < \kappa < 0$	$0 < e < 1$
Paraboloid	$\varepsilon = 0$	$\kappa = -1$	$e = 1$
Hyperboloid	$\varepsilon < 0$	$\kappa < -1$	$e > 1$

The Taylor expansion for the sag is

$$z(\rho) = \tfrac{1}{2}c\rho^2 + \tfrac{1}{8}\varepsilon c^3\rho^4 + \tfrac{1}{16}\varepsilon^2 c^5\rho^6 + \tfrac{5}{128}\varepsilon^3 c^7\rho^8 + \tfrac{7}{256}\varepsilon^5 c^9\rho^{10} + \cdots \tag{117}$$

The surface normals are

$$(L, M, N) = [1 - 2c(\varepsilon - 1)z + c^2\varepsilon(\varepsilon - 1)z^2]^{-1/2}(cx, cy, cz - 1) \tag{118}$$

The sagittal and tangential curvatures are

$$c_s = \frac{c}{[1 + (1 - \varepsilon)c^2\rho^2]^{1/2}} \qquad c_t = \frac{c}{[1 + (1 - \varepsilon)c^2\rho^2]^{3/2}} \tag{119}$$

General Asphere of Revolution

For an arbitrary figure of revolution all of whose derivatives are continuous, the Taylor expansion is

$$z(\rho) = \tfrac{1}{2}c\rho^2 + q_4\rho^4 + q_6\rho^6 + \cdots \tag{120}$$

Aspheres are often treated as a sphere that matches at the vertex and a deviation therefrom:

$$z(\rho) = z_{\text{sphere}}(\rho) + a_4\rho^4 + a_6\rho^6 + \cdots \tag{121}$$

Alternatively, nonconic aspheres can be treated as conics and a deviation therefrom:

$$z(\rho) = z_{\text{conic}}(\rho) + b_4\rho^4 + b_6\rho^6 + \cdots \tag{122}$$

The expansion coefficients are different in each case. Additional information on the coefficients is given by Malacara (Malacara 1992[144]) and Brueggemann (Brueggemann 1968[135]). The sagittal and tangential curvatures are given in general by

$$c_s = \frac{\dot{z}(\rho)}{\rho[1 + \dot{z}(\rho)^2]^{1/2}} \qquad c_t = \frac{\ddot{z}(\rho)}{[1 + \dot{z}(\rho)^2]^{3/2}} \tag{123}$$

Here $\dot{z}(\rho) = dz(\rho)/d\rho$ and $\ddot{z}(\rho) = d^2z(\rho)/d\rho^2$.

1.11 TRACING RAYS IN CENTERED SYSTEMS OF SPHERICAL SURFACES

Introduction

Ray tracing is the process of calculating the paths of rays through optical systems. Two operations are involved, propagation from one surface to the next and refraction or reflection at the surfaces. Exact equations can be written for spherical surfaces and conics of revolution with homogeneous media (Spencer & Murty 1962,[146] Welford 1974, chap. 4,[147] Kingslake 1978,[148] Kinglake 1983,[149] Slyusarev 1984,[150] Klein and Furtak 1986, sec. 3.1,[151] Welford 1986, chap. 4,[152] W. J. Smith 1992, chap. 10[153]). Conics are discussed by Welford (Welford 1986, sec. 4.7[152]). For general aspheres, the intersection position is found by iterating (Nussbaum & Phillips, p. 95,[154] W. J. Smith 1992, sec. 10.5[153]). Nonsymmetric systems are discussed by Welford (Welford 1986, chap. 5[152]).

Description and Classification of Rays in a Lens

For optical systems with rotational symmetry, rays are typically described in terms of the axial parameter z. A ray crosses each constant z plane at a point (x, y) with direction cosines (α, β, γ), where γ is not independent. Thus a ray is described by $(x(z), y(z))$ and $(\alpha(z), \beta(z))$.

For systems that are figures of revolution, *meridional rays* are those lying in a meridional plane, a plane that includes the axis, and other rays are *skew rays.* The *axial ray* corresponds to the axis of revolution. Rays are also classified according to their proximity to the axis. *Paraxial rays* are those in the limit of having small angles and small distances from the axis. Nonparaxial rays are sometimes referred to as *finite rays* or *real rays. Ray fans* are groups of rays lying in a plane. A *tangential fan* lies in a meridian, and intersects at a *tangential focus*. A *sagittal fan* lies in a plane perpendicular to a meridian, and intersects at a *sagittal focus*.

Transfer

In propagating through a homogeneous medium, a ray originating at (x_1, y_1, z_1) with directions (α, β, γ) intersects a z_2 plane at

$$x_2 = x_1 + \frac{\alpha}{\gamma}(z_2 - z_1) \qquad \text{and} \qquad y_2 = y_1 + \frac{\beta}{\gamma}(z_2 - z_1) \tag{124}$$

Intersection Points

Let the intersection points of the ray with the vertex plane at $z = 0$ be $(x_0, y_0, 0)$, Fig. 5. Define auxiliary functions

$$A(x_0, y_0; \alpha, \beta; c) = \gamma - c(\alpha x_0 + \beta y_0) \qquad \text{and} \qquad B(x_0, y_0, c) = c^2(x_0^2 + y_0^2) \tag{125}$$

The distance D along the ray from this point to the surface is given by

$$cD = A - \sqrt{A^2 - B} = \frac{B}{A + \sqrt{A^2 - B}} \tag{126}$$

The intersection point has the coordinates

$$x = x_0 + \alpha D \qquad y = y_0 + \beta D \qquad z = \gamma D \tag{127}$$

The incidence angle I at the intersection point is given by

$$\cos I = \sqrt{A^2 - B} \tag{128}$$

so

$$\Gamma = -n\sqrt{A^2 - B} + \sqrt{n'^2 - n^2 + n^2(A^2 - B)} \tag{129}$$

Mathematically, double intersection points are possible, so they must be checked for. If

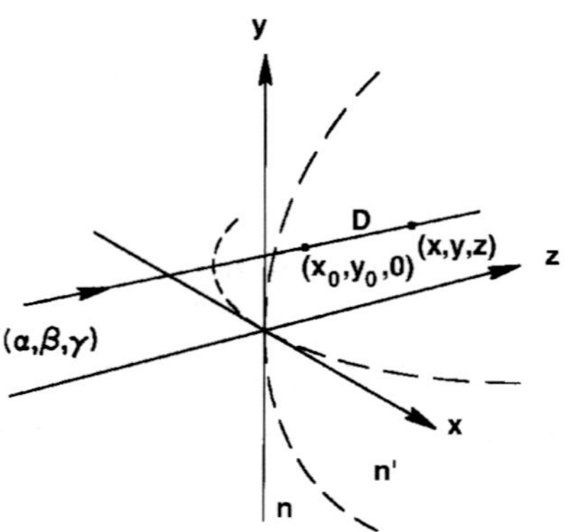

FIGURE 5 Intersection points. A ray with direction cosines (α, β, γ) intersects the vertex tangent plane at $(x_0, y_0, 0)$ and the optical surface at (x, y, z). The distance between these points is D, given by Eq. (126).

the ray misses the surface, then $A^2 < B$. If there is total internal reflection, the second square root in Eq. (129) is imaginary.

Refraction and Reflection by Spherical Surfaces

Rays refract or reflect at surfaces with reference to the local normal at the intersection point. The surface normal given by Eq. (114) is substituted in the general form for refraction, Eq. (85), to give

$$n'\alpha' = n\alpha - \Gamma cx \qquad n'\beta' = n\beta - \Gamma cy \qquad n'\gamma' = n\gamma - \Gamma(1 - cz) \tag{130}$$

For reflection, the above equations are used, with $n' = -n$, so $\Gamma = 2n \cos I = 2n\sqrt{A^2 - B}$.

Meridianal Rays

The meridian is customarily taken to be that for which $x = 0$, so the direction cosines are $(0, \beta, \gamma)$. Let U be the angle between the axis and the ray, so $\beta = \sin U$ and $\gamma = \cos U$. The transfer equation, Eq. (124) becomes

$$y_2 = y_1 + \tan U(z_2 - z_1) \tag{131}$$

The second equation of Eq. (130) can be written

$$n' \sin U' - n \sin U = -cy(n' \cos I' - n \cos I) \tag{132}$$

If the directed distance from the vertex to the intersection point of the incident ray with the axis is l, the outgoing angle is

$$U' = U + \arcsin\left[(lc - 1)\sin U\right] - \arcsin\left[\frac{n}{n'}(lc - 1)\sin U\right] \tag{133}$$

The directed distance l' from the vertex to the axial intersection of the refracted ray is given by

$$cl' = 1 + (cl - 1)\frac{n \sin U}{n' \sin U'} \tag{134}$$

For reflection, setting $n' = -n$ gives

$$U' = U + 2 \arcsin\left[(lc - 1)\sin U\right] \tag{135}$$

1.12 PARAXIAL OPTICS OF SYSTEMS OF REVOLUTION

Introduction

The term *paraxial* is used in different ways. In one, paraxial rays are those whose distances from the axis and whose angles relative to the axis are small. This leaves questions of how small is small enough and how this varies from system to system. The other interpretation of the term, which is used here, is that paraxial rays represent a limiting case in which the distances from the axis and angles relative to the axis vanish. *Paraxial optics* then describes the behavior of systems in this limit. The ray-tracing equations in the paraxial limit are linear in angle and in distance from the axis, hence the term *first-order optics,* which is often considered equivalent to paraxial. (There are no 0th-order terms since the expansion is taken about the axis, so a ray with an initial height and angle of zero, i.e., a ray along the axis, has the same outgoing height and angle.) The linearity of the paraxial equations makes them simple and understandable, as well as expressible in matrix form. Paraxial ray tracing is discussed to some extent by almost every book that treats geometrical optics.

Paraxial ray tracing is done to determine the gaussian properties of lenses, to locate image positions and magnifications, and to locate pupils and determine their sizes. Another use of paraxial ray tracing, not discussed here, is the computation of third-order aberrations (W. J. Smith 1992[155]).

Paraxial imaging is perfect in the sense that it agrees with the Maxwell ideal and with that of collineation. Point images everywhere are stigmatic, fields are flat, and there is no distortion. Aberration is often thought of as the difference between the behavior of finite rays and that of paraxial rays. If this approach is taken, then in the process of lens design, finite rays are made to agree, insofar as possible, with the paraxial ones, which cannot be similarly changed. In the paraxial limit, surfaces are described by their vertex curvatures, so conics, aspheres, and spheres are indistinguishable, the difference being in the fourth power and above. Consequently, aberrations can be altered by changing the surface asphericity without changing paraxial properties. A paraxial treatment can be done even if a system is missing the axial region, as in the case with central obstructions and off-axis sections of figures of revolution.

This section is concerned with systems of mirrors and lenses with rotational symmetry and homogeneous refractive indices. In this case, it suffices to work in a single meridian. Generalizations are found in the sections in this chapter on images about known rays and rays in heterogeneous media. Other generalizations involve expansions about given rays in systems that are not rotationally symmetric.

The Paraxial Limit

The lens axis is the z axis, and rays in the $x = 0$ meridian are considered. Ray heights are y, and angles relative to the axis are u. In the paraxial limit, the quantities u, $\tan u$, and $\sin u = \beta$ are indistinguishable. The z-direction cosine is $\gamma = \cos u \simeq 1$. Since the ray angles and heights are small, incidence angles are likewise, so $i \simeq \sin i$, $\cos I \simeq 1$, $\cos I' \simeq 1$, and $\Gamma = n' \cos I' - n \cos I \simeq n' - n$.

Transfer

In traversing a distance t between two planes, the height of a meridional ray changes from y to y' according to Eq. (124), $y' = y + t\beta/\gamma$. In the paraxial limit, this equation becomes

$$y' = y + tu \tag{136}$$

If a ray travels from one curved surface to the next, the distance t equals the vertex separation to first order, since the correction for the surface sag is of second order in height and angle. This term is given above in Eq. (127).

Refraction

The paraxial form of Snell's law, Eq. (78), is

$$n'i' = ni \tag{137}$$

Reflection

The law of reflection is the same for paraxial as for finite rays,

$$i' = -i \tag{138}$$

Angle of Incidence at a Surface

A ray with an angle u, which intersects a surface of curvature c at height y, makes an angle i with the local surface normal of the surface given by

$$i = u + yc \tag{139}$$

This equation is easily remembered from two special cases. When $y = 0$, the intersection is at the vertex, so $i = u$. When $u = -cy$, the ray is directed through the center of curvature, so $i = 0$.

Refraction at a Surface

The above equation combined with that for Snell's law gives

$$n'u' = nu - yc(n' - n) \tag{140}$$

This equation can also be obtained from the exact equation, $n'\beta' = n\beta - \Gamma cy$, Eq. (125). In the paraxial limit, $\Gamma = n' - n$, and the intersection height y is that in the vertex plane.

Reflection at a Surface

The relationship between incident and outgoing angles at a reflecting surface is found by combining Eqs. (138) and (139), to be

$$u' = -u - 2cy \tag{141}$$

Refraction and Reflection United—Surface Power

Reflection and refraction can be treated the same way mathematically by thinking of reflection as refraction with $n' = -n$, in which case Eq. (140) gives Eq. (141). A reflecting surface can be represented graphically as a thin convex-plano or concave-plano thin lens with index $-n$, where n is the index of the medium, Fig. 18. For both refraction and reflection,

$$n'u' = nu - y\phi \tag{142}$$

where the surface *power* ϕ is

$$\phi = c(n' - n) \tag{143}$$

If the surface is approached from the opposite direction, then n and n' are switched, as is the sign of c, so ϕ is the same in both directions. Thus ϕ is a scalar property of the interface, which can be positive, negative, or zero. The power is zero if $n' = n$ or $c = 0$. If $n' = n$, the surface is "invisible," and the rays are not bent. If $c = 0$, the rays are bent. For a planar refracting surface $n'u' = nu$, and a planar reflecting surface gives $u' = -u$.

Principal Focal Lengths of a Surface

A ray approaching a surface parallel to the axis ($u = 0$) with a height y has an outgoing angle given by

$$n'u' = -y\phi \tag{144}$$

This ray intercepts the axis at the *rear focal point,* whose directed distance from the vertex is $f' = y/u' = n'/\phi$. This directed distance is the *rear focal length.* Similarly, a ray entering from the right with $u' = 0$ intercepts the axis at the *front focal point,* a directed distance from the vertex of $f = y/u = -n/\phi$, the *front focal length.* Thus, a surface has a single power and two focal lengths, among which the following relationships hold:

$$f' = \frac{n}{\phi} \qquad f = -\frac{n}{\phi} \qquad \phi = -\frac{n}{f} = \frac{n'}{f'} \qquad \frac{f'}{f} = -\frac{n'}{n} \tag{145}$$

For a refracting surface, the signs of f' and f are opposite. For a reflecting surface $f' = f$.

Axial Object and Image Locations for a Single Surface

A ray from an axial point a directed distance l from the vertex of a surface that makes an angle u with the axis intersects the surface at height $y = -l/u$. After refraction or reflection, the ray angle is u', and the ray intersects the axis at a distance $l' = -y/u'$ from the vertex, Fig. 6. Substituting for u and u' in Eq. (142), the relationship between axial object and image distances is

$$\frac{n'}{l'} = \frac{n}{l} + \phi \tag{146}$$

This can also be written

$$n\left(\frac{1}{r} - \frac{1}{l}\right) = n'\left(\frac{1}{r} - \frac{1}{l'}\right) \tag{147}$$

This is a special case of the equations above for imaging about a given ray. The transverse magnification is $m = l'/l$.

Paraxial Ray Tracing

Paraxial rays are traced through an arbitrary lens by a sequence of transfers between surfaces and power operations at surfaces. Each transfer changes height but not angle, and each power operation changes angle but not height. An image can be found by applying Eq. (136) and Eq. (142) successively. Alternatively, matrix methods described below can be used.

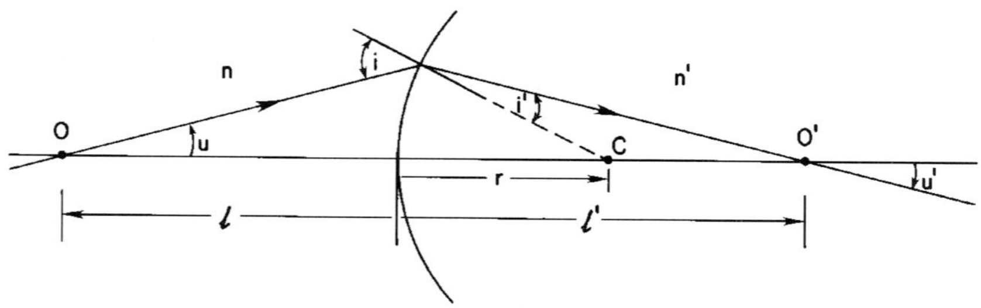

FIGURE 6 Refraction at a single spherical surface with center C and radius r. Axial object point O is imaged at O'.

Linearity of Paraxial Optics

For both the transfer and power operations, the outgoing heights and angles depend linearly on the incoming heights and angles. So a system described by a sequence of such operations is also linear. Therefore, a ray that enters with height y and angle u leaves with $y'(y, u)$ and $u'(y, u)$ given by

$$y' = \left(\frac{\partial y'}{\partial y}\right)y + \left(\frac{\partial y'}{\partial u}\right)u \qquad \text{and} \qquad u' = \left(\frac{\partial u'}{\partial y}\right)y + \left(\frac{\partial u'}{\partial u}\right)u \tag{148}$$

These equations can also be thought of as the first terms of Taylor expansions of exact expressions for $y'(y, u)$ and $u'(y, u)$. These partial derivatives depend on the structure of the system, and they can be determined by tracing two rays through the system. The partial derivatives, other than $\partial u'/\partial y$, also depend on the axial locations of the input and output surfaces. The changes with respect to these locations are treated easily by matrix methods.

The Two-Ray Paraxial Invariant

The various rays that pass through a lens are not acted upon independently, so there are several invariants that involve groups of rays. Consider two meridional paraxial rays that pass through a lens. At a given plane, where the medium has an index n, one ray has height y_1 and angle u_1, and the other has y_2 and u_2. The quantity

$$L = ny_1u_2 - ny_2u_1 = n(y_1u_2 - y_2u_1) \tag{149}$$

which we refer to as the *paraxial invariant* (NS), is unchanged as the rays pass through the system. Applying Eq. (136) and Eq. (142) to the above expression shows that this quantity does not change upon transfer or upon refraction and reflection. The invariant is also related to the general skew invariant, Eq. (73), since a paraxial skew ray can be decomposed into two meridional rays.

Another version of the invariance relationship is as follows. Two objects with heights y_1 and y_2 are separated axially by d_{12}. If their image heights are y_1' and y_2', and the image separation is d_{12}', then

$$n\frac{y_1 y_2}{d_{12}} = n'\frac{y_1' y_2'}{d_{12}'} \tag{150}$$

An additional version of the invariance relationship is

$$\left(\frac{\partial y'}{\partial y}\right)\left(\frac{\partial u'}{\partial u}\right) - \left(\frac{\partial y'}{\partial u}\right)\left(\frac{\partial u'}{\partial y}\right) = \frac{n}{n'} \tag{151}$$

where the partial derivatives, Eq. (148), describe the action of any system.

The invariant applies *regardless* of the system. Thus, for example, if the lens changes, as with a zoom system, so that both of the outgoing rays change, their invariant remains. The invariant arises from basic physical principles that are manifested in a variety of ways, for example, as conservation of brightness and Liouville's theorem, discussed above in the section on conservation of etendue. This invariance shows that there are fundamental limits on what optical systems can do. Given the paraxial heights and angles of two input rays, only three of the four output heights and angles can be chosen arbitrarily. Likewise, only three of the four partial derivatives above can be freely chosen. The invariant is not useful if it vanishes identically. This occurs if the two rays are scaled versions of one another, which happens if both $u_1 = 0$ and $u_2 = 0$ for some z, or if both rays pass through the same axial object point, in which case $y_1 = 0$ and $y_2 = 0$. The invariant also vanishes if one of the rays lies along the axis, so that $y_1 = 0$ and $u_1 = 0$.

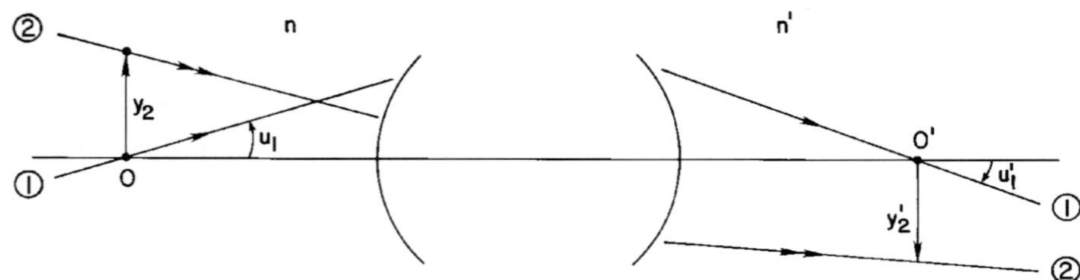

FIGURE 7 An object and image plane with ray 1 through the axial points and ray 2 through off-axis points. The location and magnification of an image plane can be found by tracing a ray from the axial object point O to axial image point O'. The magnification is given by Eq. (153). In the case pictured, u_1 and u_1' have opposite signs, so the transverse magnification is negative.

Image Location and Magnification

To locate an image plane, any ray originating at the axial object point can be traced through the system to determine where it again intersects the axis, Fig. 7. The magnification for these conjugates can be found in two ways. One is to trace an arbitrary ray from any off-axis point in the object plane. The ratio of its height in the image plane to that in the object plane is the transverse magnification.

Alternately, the magnification can be found from the initial and final angles of the ray through the axial points. Let ray 1 leave the axial object point, so $y_1 = 0$. Let ray 2 originate in the object plane some distance from the axis. At the object plane $L = ny_2u_1$, and at the image plane $y_2' = 0$, so $L = n'y_2'u_1'$. Therefore,

$$L = ny_2u_1 = n'y_2'u_1' \tag{152}$$

So the magnification is

$$m = \frac{y_2'}{y_2} = \frac{nu_1}{n'u_1'} \tag{153}$$

The relative signs of u and u' determine that of the magnification. Equation (153) is a paraxial form of the sine condition Eq. (106). Squaring this equation gives $L^2 = n^2y_2^2u_1^2$, which is proportional to a paraxial form of the etendue. These matters are discussed further in the sections on conservation of etendue and on apertures. The quantity ny_2u_1 is sometimes referred to as *the* invariant, but it is not the most general form.

Three-Ray Rule

A further consequence of the paraxial invariant and of the linearity of paraxial optics, is that once the paths of two paraxial meridional rays has been found, that of any third ray is determined. Its heights and angles are a linear combination of those of the other two rays. Given three rays, each pair has an invariant: $L_{12} = n(y_1u_2 - y_2u_1)$, $L_{23} = n(y_2u_3 - y_3u_2)$, and $L_{31} = n(y_3u_1 - y_1u_3)$. Therefore, in every plane

$$y_3 = -\frac{L_{23}}{L_{12}}y_1 + \frac{L_{31}}{L_{12}}y_2 \qquad \text{and} \qquad u_3 = -\frac{L_{23}}{L_{12}}u_1 + \frac{L_{31}}{L_{12}}u_2 \tag{154}$$

This assumes that no pair of the three rays are simply scaled versions of one another, i.e. that both $L_{23} \neq 0$ and $L_{31} \neq 0$.

Switching Axial Object and Viewing Positions

If an axial object and axial viewing position are switched, the apparent size of the image is unchanged. Put more precisely, let an object lie in a plane intersecting the axial point A and let its image be viewed from an axial point B' in image space that is not conjugate to

A. If the object and viewing positions are switched, so the eye is at *A* and the object plane is at *B′*, the subtense of the object as seen by the eye is unchanged. (Rayleigh 1886,[156] Southall 1910,[157] Herzberg 1935,[158] Brouwer 1967[159]).

1.13 IMAGES ABOUT KNOWN RAYS

Given a ray, referred to here as the *central ray* (also "base ray"), other rays from a point on the central ray making a small angle with respect to it are focused at or near other points on the central ray. These foci can be determined if the path of a central ray is known, as well as the indices of the media through which it passes, and the principal curvatures at the surfaces where it intersects. Here indices are constant. At each intersection point with an optical surface, the wavefront has two principal curvatures, as does the surface. After refraction or reflection, the wavefront has two different principal curvatures. Accordingly, if a single point is imaged, there are two astigmatic focal lines at some orientation. These foci are perpendicular, but they do not necessarily lie in planes perpendicular to that of the central ray. The imaging of a small extended region is generally skewed, so, for example, a small square in a plane perpendicular to the central ray can be imaged as a rectangle, parallelogram, or trapezoid.

This is a generalization of paraxial optics, in which the central ray is the axis of a system of revolution. While not difficult conceptually, the general case of an arbitrary central ray and an arbitrary optical system is algebraically complicated. This case can also be analyzed with a hamiltonian optics approach, using an expansion of a characteristic function about the central ray, like that of Eq. (28). The subject is sometimes referred to as *parabasal optics,* and the central ray as the *base ray*. This subject has been discussed by numerous authors[160–186] under various names, e.g., "narrow beams," "narrow pencils," "first order."

The following is limited to the case of meridional central rays and surfaces that are figures of revolution. The surface, at the point of intersection, has two principal curvatures c_s and c_t. [See Eqs. (119), (123).] For spherical surfaces, $c_s = c_t = c$, and for planar surfaces $c = 0$. There is a focus for the sagittal fan and one for the tangential one, Fig. 8, the two

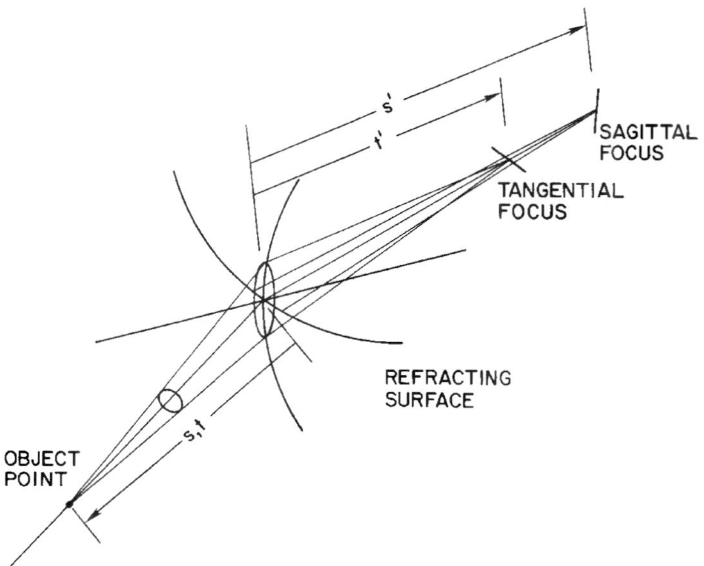

FIGURE 8 Astigmatic imaging of a point by a single refracting surface. The distance from the surface to the object point along the central ray of the bundle is $s = t$. The distances from the surface to the sagittal focus is s', and that to the tangential focus is t', as given by Eq. (155).

foci coinciding if the imaging is stigmatic. After one or more surfaces are encountered, the separated foci are the sources for subsequent imaging. Let s and t be the directed distances from the intersection point of the central ray and the surface to the object point, and s' and t' be the distances from intersection point to the foci. The separation $|s' - t'|$ is known as the *astigmatic difference.*

For refraction

$$\frac{n'}{s'} = \frac{n}{s} + c_s\Gamma \quad \text{and} \quad \frac{n' \cos^2 I'}{t'} = \frac{n \cos^2 I}{t} + c_t\Gamma \tag{155}$$

where $\Gamma = n' \cos I' - n \cos I$, Eq. (82). The sagittal equation is simpler, providing a mnemonic for remembering which equation is which: "S" = sagittal = simpler. If the surface is spherical, and the ray fan makes an arbitrary angle of ψ with the meridian, then (H. Hopkins 1950[175])

$$\frac{n'}{d'}(1 - \cos^2 \psi \sin^2 I') = \frac{n}{d}(1 - \cos^2 \psi \sin^2 I) + c\Gamma \tag{156}$$

where d and d' are the distances along the central ray from the surface to the object and image points. For normal incidence at a spherical surface $\Gamma = n' - n$, so both equations become

$$\frac{n'}{d'} = \frac{n}{d} + c(n' - n) \tag{157}$$

This also applies to surfaces of revolution if the central ray lies along the axis. This equation is identical to the paraxial equation, Eq. (146).

The corresponding relations for reflection are obtained by setting $n' = -n$ and $I' = I$ in the refraction equations, giving

$$\frac{1}{s'} = -\frac{1}{s} + 2c_s \cos I \quad \text{and} \quad \frac{1}{t'} = -\frac{1}{t} + \frac{2c_t}{\cos I} \tag{158}$$

For stigmatic imaging between the foci of reflective conics, $s = t$ is the distance from one focus to a point on the surface, and $s' = t'$ is that from the surface to the other focus. Therefore, $c_t = c_s \cos^2 I$. The reflection analogue to Eq. (156), for a spherical surface is

$$\frac{1}{d'} = -\frac{1}{d} + \frac{2c \cos I}{1 - \cos^2 \psi \sin^2 I} \tag{159}$$

These equations are known by several names, including *Coddington's equations, Young's astigmatic formulae,* and the s- and *t-trace formulae.*

1.14 GAUSSIAN LENS PROPERTIES

Introduction

The meaning of the term *gaussian optics* is not universally agreed upon, and it is often taken to be indistinguishable from paraxial optics or first-order optics, as well as collineation. Here the term is considered to apply to those aspects of paraxial optics discovered by Gauss (Gauss 1840[187]), who recognized that all rotationally symmetric systems of lens elements can be described paraxially by certain system properties. In particular, lenses can be treated as black boxes described by two axial length parameters and the locations of special points, called *cardinal points,* also called *Gauss points.* Once a lens is so characterized, knowledge of its actual makeup is unnecessary for many purposes, and repeated ray traces need not be performed. For example, given the object location, the image location and magnification are determined from the gaussian parameters. From the gaussian description of two or more lenses, that of a coaxial combination can be found. Another consequence of Gauss's discovery is that there is an infinity of specific embodiments for any external prescription.

The lenses considered in this section are figures of revolution with uniform object space and image space indices n and n'. All quantities discussed in this section are paraxial, so the prefix "paraxial" is not repeated. For the purposes of this section, no distinction is made between real and virtual rays. Those in each space are considered to extend infinitely, and intersection points may be either accessible or inaccessible. The quantities used in this section are found in Table 1.

Power, Focal Lenses, and Afocal Lenses

A paraxial ray entering a lens parallel to the axis height y leaves with some angle u', Fig. 9. Likewise, a ray entering from the opposite side with height y' leaves with angle u. The *power* of the lens is defined by

$$\phi = -n'\frac{u'}{y} = n\frac{u}{y'} \tag{160}$$

The outgoing ray can have any angle, and the power can be positive, negative, or zero. If $u' = 0$, then $\phi = 0$ and the lens is *afocal* or *telescopic.* Lenses for which $\phi \neq 0$ are referred to here as *focal,* although the term "nonafocal" is more common. Afocal lenses are fundamentally different from focal ones, and are treated separately below. Power is the same in both directions, i.e. whether the ray enters from left to right or from right to left. The lens in Fig. 9 has $\phi > 0$, and that in Fig. 10 has $\phi < 0$. Diagrams such as Fig. 11 show the location of the principal focal point, but not the sign of the power; two rays enter and two leave, but there is no indication of which is which. (Note that some negative lenses have accessible rear focal points.) Another expression for power involves two rays at arbitrary angles and heights. If two incident rays have (y_1, u_1) and (y_2, u_2), and a nonzero invariant $L = n(y_1u_2 - y_2u_1)$, and the outgoing ray angles are u_1' and u_2', then

$$\phi = -\frac{nn'}{L}(u_1'u_2 - u_2'u_1) \tag{161}$$

Focal Lenses

Focal lenses are those for which $\phi \neq 0$. Their cardinal points are the principal focal points, the principal points, and the nodal points. These points may be located anywhere on axis relative to the physical lens system. If they are inside a lens, then the intersection points referred to below are virtual. The cardinal points are pairs consisting of a member in object space and one in image space. The one in object space is often referred to as *front,* and the one in image space as *rear,* but this terminology may be misleading, since the points can be any sequence along the axis.

Principal Focal Points. Rays entering a lens parallel to its axis cross the axis at the *principal focal points* or *focal points.* Rays parallel to the axis in object space intersect the axis at the *rear focal point* F' in image space and those parallel in image space intersect at the *front focal point* F in object space, Fig. 9. The *principal focal planes* or *focal planes* are the planes perpendicular to the axis at the focal points. The terms *focal point* and *focal plane* are often used to refer to the images of any point or plane. In this chapter, *image point* is used for other points where rays are focused and *image plane* for other planes.

Principal Planes. The *principal planes* are the conjugate planes for which the transverse

TABLE 1 Gaussian Notation and Definitions

By convention, in the diagrams the object space is to the left of the lens, image space is to the right, and rays go left to right. Object space quantities are unprimed, and image space quantities are primed, and quantities or positions that correspond in some way have same symbol, primed and unprimed. This correspondence can have several forms, e.g., the same type of thing or conjugate. The term *front* refers to object space, or left side, and *rear* to image space, or right side. A "front" entity may actually be behind a "rear" one. For example, a negative singlet has its object space focal point behind lens.

Scalars

n and n' object and image space refractive indices
ϕ power
m transverse magnification
m_N nodal point magnification $= n/n'$
m_L longitudinal magnification
m_α angular magnification
u and u' paraxial ray angles (the positive direction is counterclockwise from the axis)
y and y' paraxial ray heights
y_P paraxial ray height at the principal planes $= y'_P$

Axial points

Cardinal points:
Focal points F and F', not conjugate
Principal points P and P', conjugate $m = +1$
Nodal points N and N', conjugate $m_N = n/n'$
Other points:
Axial object and image points O and O', conjugate
Arbitrary object and image points A and A', B and B'
Vertices V and V', not conjugate, in general

Directed axial distances

These distances here are between axial points and are directed.
Their signs are positive if from left to right and vice versa.
Types of distances: entirely in object or image space, between spaces
Principal focal lengths: $f = PF$ and $f' = P'F'$
Principal points to object and image axial points: $l = PO$ and $l' = P'O'$
Front and rear focal points to object and image axial points: $z = FO$ and $z' = F'O'$
Relations: $l = f + z$ and $l' = f' + z'$
Arbitrary point to conjugate object and image points: $d = AO$ and $d' = A'O'$

Distances between object space and image space points involve distances within both spaces, as well as a distance *between* the spaces, e.g., PP', FF', VV', and OO'. The distances between spaces depend on the particular structure of the lens. They can be found by paraxial ray tracing.

magnification is unity, Fig. 12. The intersections of the principal planes and the axis are the *principal points,* denoted by P and P'. The *rear principal plane* is the locus of intersections between $u = 0$ rays incident from the left and their outgoing portions, Fig. 9. Likewise, the *front principal plane* is the intersection so formed with the rays for which $u' = 0$. A ray intersecting the first principal plane with height y_P and angle u leaves the second principal plane with height $y' = y_P$ and an angle given by

$$n'u' = nu - y_P\phi \tag{162}$$

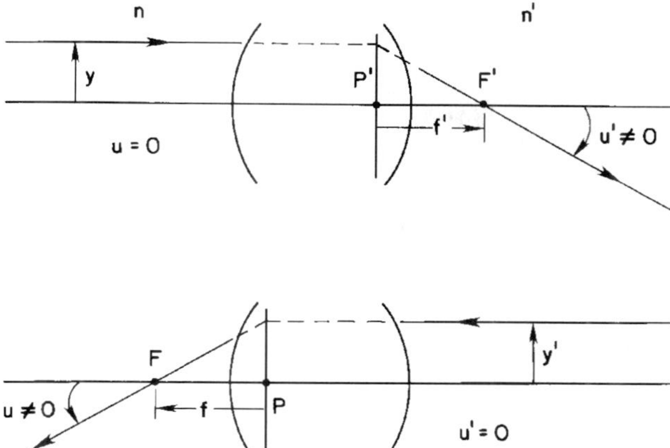

FIGURE 9 Diagrams for determining power, focal points, and focal lengths. Rays parallel to the axis in one space cross the axis in the other space at the focal points. The principal planes are at the intersections of entering and leaving rays. The power is given by Eq. (159). The lens in this diagram has positive power, a positive rear focal length, and a negative front focal length.

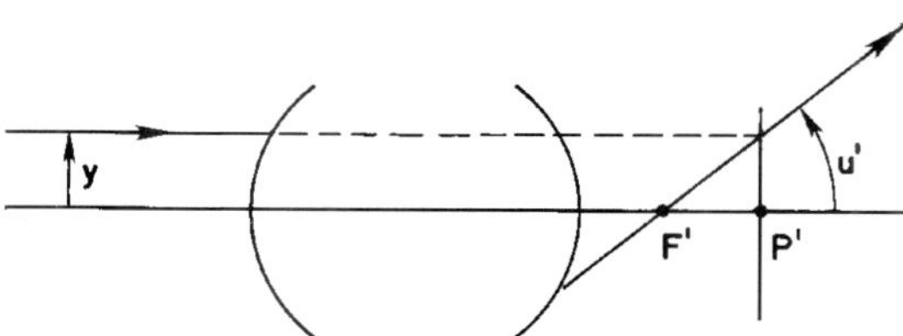

FIGURE 10 A lens with negative power and negative rear focal length. An incoming ray parallel to the axis with a positive height leaves the lens with a positive angle. The rear focal plane precedes the rear principal plane.

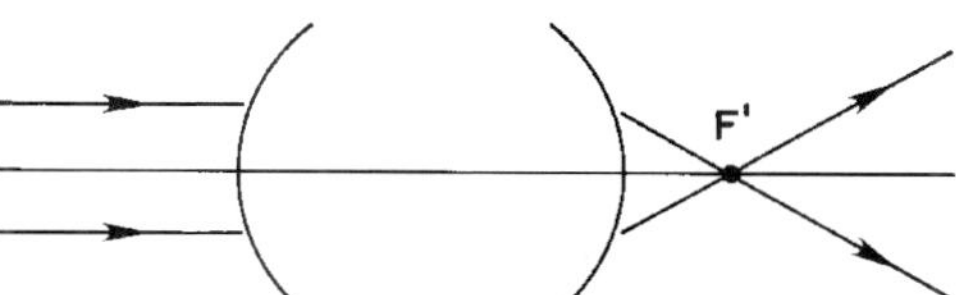

FIGURE 11 An ambiguous diagram. Two rays that enter a lens parallel to its axis converge at the rear focal point F'. Without specifying which ray is which, the sign of the power is not known.

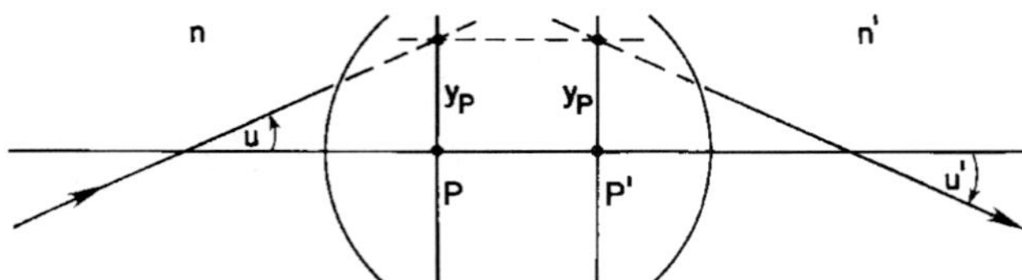

FIGURE 12 Principal planes as effective ray-bending surfaces. Incoming and outgoing paraxial rays intersect the object and image space principal planes at the same height y_P. The angles are related by Eq. (161).

The lens behaves as if the incoming ray intercepts the front principal plane, is transferred to the second with its height unchanged, and is bent at the second by an amount proportional to its height and to the power of lens. The power of the lens determines the amount of bending. For rays passing through the principal points, $y_p = 0$, so $u'/u = n/n'$.

Principal Focal Lengths. The *focal lengths,* also called *effective focal lengths,* are the directed distances from the principal points to the focal points. The front and rear focal lengths are

$$PF = f = -\frac{n}{\phi} \quad \text{and} \quad P'F' = f' = \frac{n'}{\phi} \tag{163}$$

The two focal lengths are related by

$$\phi = -\frac{n}{f} = \frac{n'}{f'} \quad \text{and} \quad \frac{f}{f'} = -\frac{n}{n'} \tag{164}$$

This ratio is required by the paraxial invariant (Kingslake 1965, p. 214[188]). If $n = n'$, then $f' = -f$. If $n = n' = 1$, then

$$f' = -f = \frac{1}{\phi} \tag{165}$$

The focal lengths are the axial scaling factors for the lens, so axial distances in all equations can be scaled to them.

Nodal Points. The *nodal points* are points of unit angular magnification. A paraxial ray entering the *object space nodal point* N leaves the *image space point* N' at the same angle, Fig. 13. The planes containing the nodal points are called *nodal planes.* A *nodal ray* is one that passes through the nodal points. Such a ray must cross the axis, and the point where it does so physically is sometimes called the *lens center.* In general, this point has no special properties. (Gauss suggested an alternate "lens center," the point midway between the

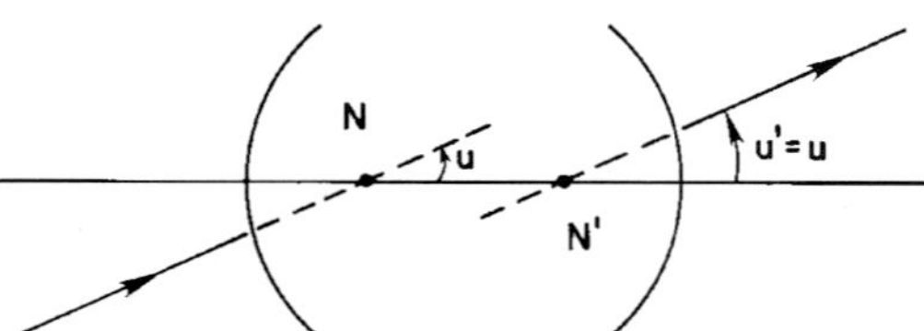

FIGURE 13 Nodal points. A paraxial ray through the object space nodal point N passes through image space nodal point N' with the same angle.

principal points. Rotating a lens front and rear about this point would leave object and image positions and magnifications unchanged.)

If the refractive indices of the object space and image space are the same, the nodal points correspond to the principal points. If not, both nodal points are shifted according to

$$PN = P'N' = \frac{n' - n}{\phi} = f + f' \tag{166}$$

The distances from the nodal points to the focal points are

$$N'F' = -f \qquad \text{and} \qquad NF = -f' \tag{167}$$

The nodal points are conjugate, and the transverse magnification of the nodal planes is

$$m_N = \frac{n}{n'} \tag{168}$$

These equations can be recalled by the simple example of the single refracting surface, for which both nodal points correspond with the center of curvature.

Conjugate Equations. For an object plane perpendicular to the axis at point O, there is an image plane perpendicular to the axis at O', in which the transverse magnification is m. Note that specifying magnification implies both object and image positions. There is a variety of *conjugate equations* (NS) that relate their positions and magnifications. The equations differ in which object and image reference points are used from which to measure the directed distances to the object and image. These equations can be written in several ways, as given below, and with reference to Fig. 14. Axial distances can be scaled to the focal lengths, or the distances can be scaled to the indices, with a common power term remaining.

The simplest conjugate equation is *Newton's equation,* for which the reference points are the focal points and the lengths therefrom are $z = FO$ and $z' = F'O'$. The equation can be written in several forms:

$$zz' = ff' \qquad \text{or} \qquad \frac{z'}{f'}\frac{z}{f} = 1 \qquad \text{or} \qquad \frac{z'}{n'}\frac{z}{n} = \frac{1}{\phi^2} \tag{169}$$

More generally, if A and A' are any pair of axial conjugate points, as are B and B', then

$$FA \times F'A' = FB \times F'B' \tag{170}$$

Another form is that for which the reference points are the principal points and the directed distances are $l = PO$ and $l' = P'O'$:

$$1 = \frac{f'}{l'} + \frac{f}{l} \qquad \text{or} \qquad \frac{n'}{l'} = \frac{n}{l} + \phi \tag{171}$$

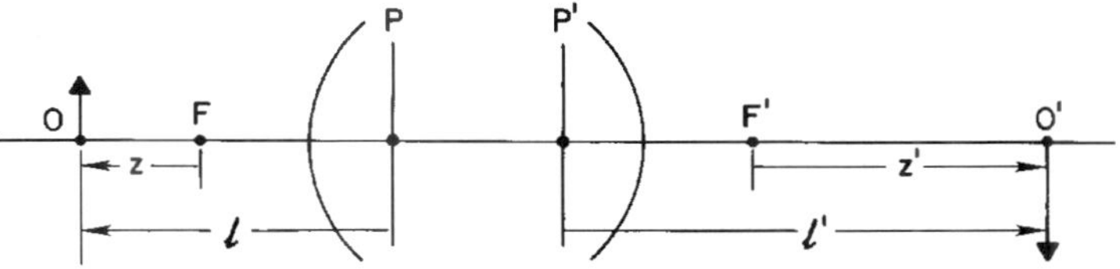

FIGURE 14 Directed distances used in common conjugate equations for focal lenses. Distances z and z' are from the focal points to axial conjugate points. Distances l and l' are from the principal points to axial conjugate points.

If the reference points are arbitrary conjugates with magnification m_A and the axial distances are $d = AO$ and $d' = A'O'$, then

$$m_A \frac{n'}{d'} = \frac{1}{m_A} \frac{n}{d} + \phi \qquad \text{or} \qquad \frac{d'}{f'} = \frac{m_A^2 \dfrac{d}{f}}{1 - m_A \dfrac{d}{f}} \tag{172}$$

This equation also relates the curvatures of a wavefront at conjugate points. For a point source at A the radius of the wavefront at O is d, so at O' the radius is d'.

If the reference points are the nodal points, $m_A = m_N = n/n'$, and the axial distances are $d = NO$ and $d' = N'O'$, then

$$1 = \frac{f}{d'} + \frac{f'}{d} \qquad \text{or} \qquad \frac{n}{d'} = \frac{n'}{d} + \phi \tag{173}$$

The most general equation relating conjugate points is obtained when both reference points are arbitrary. Let the reference point in object space be a point A, at which the magnification is m_A, and that in image space be B', associated with magnification m'_B. If $d = AO$ and $d' = B'O'$, then

$$\frac{1}{\phi}\left(1 - \frac{m'_B}{m_A}\right) = \frac{1}{m_A} d - m'_B d' + \phi\, d\, d' \qquad \text{or}$$

$$d' = \frac{\dfrac{1}{m_A} d + \left(\dfrac{m'_B}{m_A} - 1\right)\dfrac{1}{\phi}}{\phi\, d - m'_B} \tag{174}$$

All the other conjugate equations are special cases of this one with the appropriate choice of m_A and m'_B.

If the reference point in object space is the focal point, and that in image space is the principal plane, then $m_A = \infty$ and $m'_B = 1$, giving

$$\frac{n'}{z'\phi} = \frac{l\phi}{n} + 1 \qquad \text{or} \qquad \frac{f'}{z'} = \frac{l}{f} + 1 \tag{175}$$

Likewise, if the object space reference point is P and the image space reference is F', then

$$\frac{n'}{l'\phi} = \frac{z\phi}{n} + 1 \qquad \text{or} \qquad \frac{f'}{l'} = \frac{z}{f} + 1 \tag{176}$$

A relationship between distances to the object and image from the principal points and those from the focal points is

$$1 = \frac{z'}{l'} + \frac{z}{l} = \frac{F'O'}{P'O'} + \frac{FO}{PO} \tag{177}$$

Transverse Magnification. In planes perpendicular to the axis, the *transverse magnification,* usually referred to simply as the *magnification,* is

$$m = \frac{x'}{x} = \frac{y'}{y} = \frac{dx'}{dx} = \frac{dy'}{dy} \tag{178}$$

There are several equations for magnification as a function of object position or image position, or as a relationship between the two. Newton's equations are

$$m = -\frac{f}{z} = -\frac{z'}{f'} = \frac{f}{f - l} = \frac{f' - l'}{f'} \tag{179}$$

Other relationships are

$$m = \frac{n}{n'}\frac{l'}{l} = -\frac{f}{f'}\frac{l'}{l} = \frac{z'}{l'}\frac{l}{z} \tag{180}$$

If $n = n'$, then $m = l'/l$. Another form, with respect to conjugate planes of magnification m_A is

$$mm_A = \frac{n}{d}\frac{d'}{n'} = \frac{f}{d}\frac{d'}{f'} \tag{181}$$

If d and d' are distances from the nodal points, $m = d'/d$. The change of magnification with respect to object or image positions with conjugacy maintained is

$$\frac{dm}{dz'} = -\frac{1}{f'} = \frac{m}{z'} \qquad \text{and} \qquad \frac{dm}{dz} = \frac{f}{z^2} = \frac{m^2}{f} = \frac{m}{z} \tag{182}$$

Images of Distant Objects. If an object at a great distance from the lens subtends an angle ψ from the axis at the lens, then its paraxial linear extent is $y = z\psi$. The image height is

$$y' = my = \frac{-f}{z}y = -f\psi = \frac{n}{n'}f'\psi \qquad \text{and} \qquad \frac{dy'}{d\psi} = \frac{n}{n'}f' \tag{183}$$

If a distant object moves perpendicularly to the axis, then its image moves in the opposite direction if $f' > 0$ and in the same direction if $f' < 0$, so long as n and n' have the same sign.

Distance Between Object and Image. The directed distance from an axial object point to its image contains three terms, one in object space, one in image space, and one relating the two spaces. The first two depend on the magnification and focal lengths. The interspace term depends on the particular structure of the lens, and is found by paraxial ray tracing. The most commonly used interspace distance is PP', since it equals zero for a thin lens, but the equations using FF' are simpler. Newton's equations give $z = -f/m$ and $z' = -mf'$, so the object-to-image distance is

$$OO' = FF' - z + z' = FF' - f'm + \frac{f}{m} = FF' - \frac{1}{\phi}\left(n'm + \frac{n}{m}\right) \tag{184}$$

This is the basic equation from which all others are derived. If the reference points are the principal points, then

$$OO' = PP' + f'(1 - m) - f\left(1 - \frac{1}{m}\right) = PP' + \frac{1}{\phi}\left[n'(1 - m) + n\left(1 - \frac{1}{m}\right)\right] \tag{185}$$

If the object-to-image distance is given, the magnification is

$$m = \frac{1}{2n'}\left(-q \pm \sqrt{q^2 - 4nn'}\right)$$

where $q = \phi(OO' - PP') - n - n'$. (186)

There are two magnifications for which OO' is the same. The magnitude of their product is n/n'. The derivative of the object-to-image distance with respect to the magnification is

$$\frac{d}{dm}OO' = -f' - \frac{f}{m^2} = -f' - \frac{z^2}{f} = \frac{1}{\phi}\left(\frac{n}{m^2} - n'\right) \tag{187}$$

Extrema occur at $m \pm \sqrt{n/n'}$, giving $m = \pm 1$ if $n = n'$. The extrema are

$$OO' - FF' = \pm \frac{2}{\phi}\sqrt{nn'} = \pm 2\sqrt{-ff'} \tag{188}$$

or

$$OO' - PP' = \frac{1}{\phi}(n + n' \pm 2\sqrt{nn'}) = f' - f \pm 2\sqrt{-ff'} \tag{189}$$

For the common case of $n' = n$, the object-to-image distance is

$$OO' = PP' + f'\left(2 - m - \frac{1}{m}\right) \tag{190}$$

OO' is the same for magnifications m and $1/m$. For a lens with $f' > 0$, the extremum object-to-image distances are $OO' - PP' = 4f'$ with $m = -1$ and $OO' - PP' = 0$ for $m = +1$. If the object-to-image distance and the focal length are given, then the magnification is

$$m = -\tfrac{1}{2}s \pm \sqrt{\tfrac{1}{4}s^2 - 1}$$

where $s = \frac{1}{f'}(OO' - PP') - 2.$ (191)

The two values of m are reciprocal.

Axial Separations and Longitudinal Magnification. Two axial points A and B are imaged at A' and B' with magnifications m_A and m_B. Newton's equations give the object separation

$$\Delta z = z_A - z_B = \frac{m_A m_B}{m_B - m_A} f \tag{192}$$

The separation of their images is

$$\Delta z' = z'_A - z'_B = (m_B - m_A) f' \tag{193}$$

The ratio of the image and object separations is

$$\frac{\Delta z'}{\Delta z} = \frac{z'_A - z'_B}{z_A - z_B} = \frac{A'B'}{AB} = \frac{n'}{n} m_A m_B = -\frac{f'}{f} m_A m_B \tag{194}$$

If m_A and m_B have different signs, then the direction of $A'B'$ is opposite that of AB. This occurs when A and B are on opposite sides of the front focal point. In the limit as the separation between A and B vanishes, m_A and m_B both approach the same magnification m. The *longitudinal magnification* m_L is the ratio of axial separations in the limit of small separations

$$m_L = \underset{A \to B}{\text{LIMIT}} \frac{A'B'}{AB} = \frac{dz'}{dz} = \frac{n'}{n} m^2 = -\frac{z'}{z} \tag{195}$$

This quantity is also called the *axial magnification.* Since m^2 is always positive, as an object moves axially in a given direction, its image moves in a constant direction. There is a discontinuity in image position when the object crosses the focal point, but the direction of motion stays the same. At the nodal points, the transverse and longitudinal magnifications are equal.

Angular Magnification. The ratio of the outgoing to incoming ray angles, u'/u is sometimes called the *angular magnification* m_α. If the ray passes through conjugate axial points with magnification m, then the angular magnification is

$$m_\alpha = \frac{u'}{u} = \frac{n}{n'}\frac{1}{m} \tag{196}$$

If the ray leaves an object point with height y in a plane for which the magnification is m, the outgoing ray angle is given by

$$n'u' = \frac{1}{m} nu - y\phi = \frac{1}{m}(nu - y'\phi) \tag{197}$$

The ratio u'/u is not constant unless $y = 0$ or $\phi = 0$.

Relationship Between Magnifications. The transverse, angular, and longitudinal magnifications are related by

$$m_\alpha m_L = m \tag{198}$$

This relationship is connected to the paraxial invariant and also holds for afocal lenses.

Reduced Coordinates. Many relationships are formally simplified by using reduced axial distances $\tau = z/n$ and $\tau' = z'/n'$ and reduced angles $\omega = nu$, $\omega' = n'u'$, which are paraxial optical direction cosines. For example, the angular magnification is $\omega'/\omega = 1/m$, and the longitudinal magnification is $d\tau'/d\tau = m^2$.

Mechanical Distances. The cardinal points can be located anywhere on axis relative to the physical structure of the lens. The *vertex* of a lens is its extreme physical point on axis. The object space vertex is denoted by V and the image space vertex by V'. The two vertices are not, in general, conjugate. The *front focal distance* FV is that from the vertex to the front focal point, and the *rear focal distance* $V'F'$ is that from the rear vertex to the rear focal point. Likewise, the *front working distance* OV is the distance from the object to the vertex, and the *rear working distance* $V'O'$ is that from the vertex to the image. These lengths have no significance to the gaussian description of a lens. For example, a lens of a given focal length can have any focal distance and vice versa. For a *telephoto* lens the focal length is greater than the focal distance, and for a *retrofocus* lens the focal distance is greater than the focal length.

Afocal Lenses

An *afocal* or *telescopic* lens (Wetherell 1987,[189] Wetherell 1994,[190] Goodman 1988[191]) is one for which $\phi = 0$. A ray entering with $u = 0$ leaves with $u' = 0$, Fig. 15. There are no

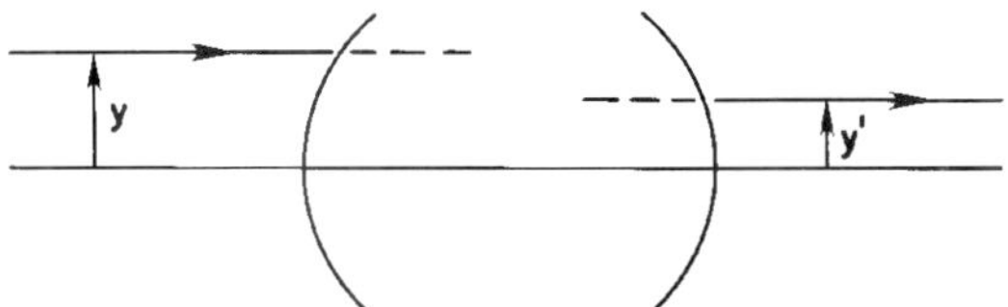

FIGURE 15 Afocal lens. Paraxial rays entering parallel to the axis leave parallel, in general at a different height. The ratio of the heights is the transverse magnification, which is constant.

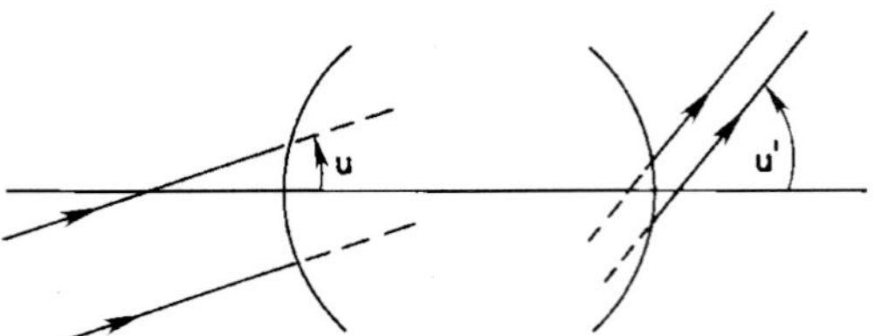

FIGURE 16 Afocal lens. Groups of paraxial rays entering parallel leave parallel, in general at a different angle. The ratio of the angles is the angular magnification, which is constant.

principal focal points or focal lengths. In general, the $u = 0$ ray leaves at a different height than that at which it enters. The ratio y'/y is the same for all such rays, so the transverse magnification m is constant. Likewise, the longitudinal magnification is constant, equaling $m_L = (n'/n)m^2$, as is the angular magnification $u'/u = m_\alpha = n/(n'm)$. A parallel bundle of rays entering at angle u leaves as a parallel bundle at $u' = m_\alpha u$, Fig. 16. Summarizing:

$$m = \text{const} \qquad m_L = \frac{n'}{n} m^2 = \text{const} \qquad m_\alpha = \frac{n}{n'}\frac{1}{m} = \text{const} \qquad m = m_L m_\alpha \tag{199}$$

Any two of these magnifications provide the two scaling factors that describe the system. If $m = n/n'$, then $m_L = m$ and $m_\alpha = 1$, so image space is a scaled version of object space.

Afocal lenses differ fundamentally from focal lenses. Objects at infinity are imaged by afocal lenses at infinity, and objects at finite distances are imaged at finite distances. An afocal lens has no cardinal points and the focal length is undefined. Afocal lenses have no principal planes. If $m \neq 1$ there are no unit magnification conjugates, and if $m = 1$ there is only unit magnification. Likewise, there are no nodal points; the angular magnification is either always unity or always differs from unity. It is sometimes stated or implied that an afocal lens is a focal one with an infinite focal length, but this description is dubious. For example, the above equations relating magnification and conjugate positions to focal length are meaningless for afocal lenses, and they cannot be made useful by substituting $f = \infty$. The equations for the afocal lenses can be obtained from those for focal lenses with a limiting process, but for most purposes this approach is not helpful.

If the positions for a single axial conjugate pair A and A' are known, other pairs are located from the property of constant longitudinal magnification. If O and O' are another pair of conjugates, then

$$A'O' = m_L AO \tag{200}$$

As a function of distance AO, the object-to-image distance OO' is

$$OO' = AA' + (m_L - 1)AO \tag{201}$$

where AA' is the separation between the initially known conjugates. If $m_L = 1$, the object-to-image distance is constant. Otherwise, it can take any value. For all afocal lenses, except those for which $m_L = 1$, there is a position, sometimes called the *center*, at which $OO' = 0$, so the object and image planes coincide.

A principal use of afocal lenses is in viewing distant objects, as with binoculars. An object of height h a great distance d from the lens subtends an angle $\psi \simeq h/d$. The image height is $h' = mh$, and the image distance is approximately $d' = m^2 d$. So the image subtends an angle $\psi' \simeq m\psi = \psi/m_\alpha$. Thus a telescope used visually produces an image which is actually smaller, but which is closer by a greater factor, so the subtense increases.

Determination of Gaussian Parameters

If a lens prescription is given, its gaussian properties can be obtained by paraxially tracing any two meridional rays whose invariant is not zero. A common choice for focal lenses is the rays with $u = 0$ and $u' = 0$, which give F, P, F', and P'. If a lens is known to be afocal, a single ray not parallel to the axis suffices, since such a ray gives a pair of conjugates and the angular magnification. If it is not known that the lens is afocal, two rays show that it is, as well as giving the required information about conjugates. Alternately, a matrix representation of the lens can be determined, from which the cardinal points are found, as described in the matrix section. The gaussian properties can also be determined experimentally in a number of ways.

Basic Systems

Single Refracting Surface. Media of indices n and n' are separated by a surface of curvature c and radius r. The power is $\phi = (n' - n)c$. The principal points coincide at the vertex. The nodal points coincide at the center of curvature. The distance from principal points to nodal points is r.

Thick Lens. The term *thick lens* usually denotes a singlet whose vertex-to-vertex distant is not negligible, where negligibility depends on the application. For a singlet of index n in vacuum with curvatures c_1 and c_2 and thickness t, measured from vertex to vertex

$$\phi = \frac{1}{f'} = (n-1)\left[c_1 - c_2 - \frac{n-1}{n} t c_1 c_2\right] \tag{202}$$

A given power may be obtained with a variety of curvatures and indices. For a given power, higher refractive index gives lower curvatures. The principal planes are located relative to the vertices by

$$VP = \frac{n-1}{n}\frac{tc_2}{\phi} \qquad \text{and} \qquad V'P' = -\frac{n-1}{n}\frac{tc_1}{\phi} \tag{203}$$

These equations can be derived by treating the lens as the combination of two refracting surfaces. Two additional relationships are

$$PP' = VV' - \frac{n-1}{n}\frac{t(c_1 + c_2)}{\phi} \qquad \text{and} \qquad \frac{V'P'}{VP} = \frac{r_2}{r_1} = \frac{c_1}{c_2} \tag{204}$$

Thin Lens. A *thin lens* is the limiting case of a refracting element whose thickness is negligible, so the principal planes coincide, and the ray bending occurs at a single surface, Fig. 17. In the limit as $t \to 0$, for a lens in vacuum the thick lens expressions give

$$\phi = \frac{1}{f'} = (n-1)(c_1 - c_2) \qquad VP = V'P' = 0 \qquad PP' = 0 \tag{205}$$

Single Reflecting Surface. A reflecting surface has power $\phi = 2n/r = 2nc$. The principal points are located at the vertex. The nodal points are at the center of curvature.

Mirror as a Thin Lens. In unfolding systems, a mirror can be thought of as a convex or

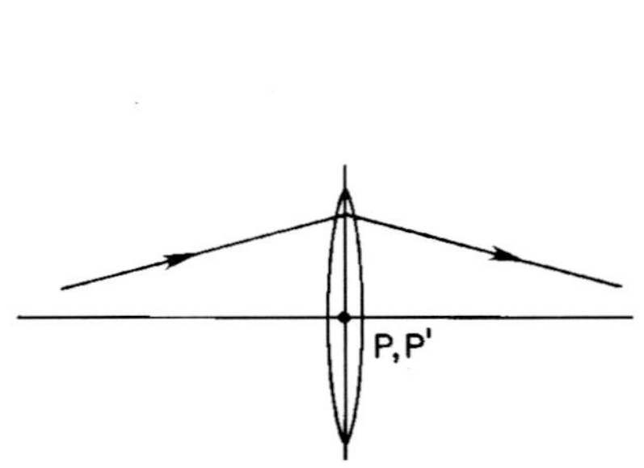

FIGURE 17 The thin lens approximation. The thickness of the lens is negligible, and the principal planes are coincident, so rays bend at the common plane.

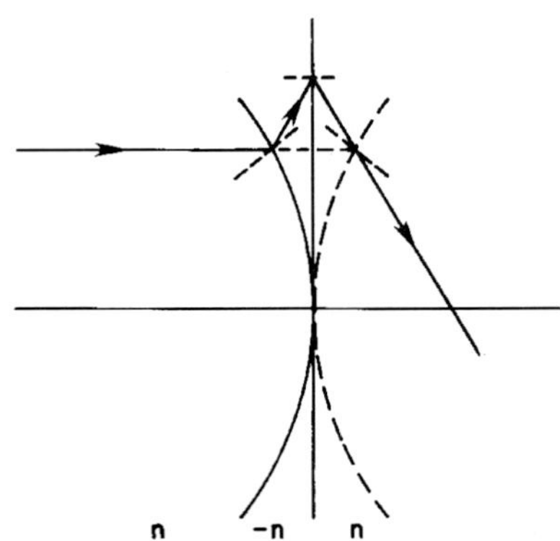

FIGURE 18 Reflecting surface represented unfolded. A convex mirror is represented as a convex-plano thin lens with index $n' = -n$, where n is the index of the medium. Snell's law gives $I' = -I$.

concave plano thin lens, with an index $-n$, where n is the index of the medium in which it works, Fig. 18. All the thin lens equations apply, as well as those for third-order aberration equations, which are not discussed here.

1.15 COLLINEATION

Introduction

Collineation is a mathematical transformation that approximates the imaging action of a lens with homogeneous refractive indices in both spaces. This transformation takes points to points, lines to lines, and planes to planes. With an actual lens, incoming rays become outgoing rays, so lines go exactly to lines. In general, however, rays that interest in object space do not intersect in image space, so points do not go to points, nor planes to planes. The collinear transformation is an approximate description of image geometry with the intervening optical system treated as a black box, not a theory that describes the process of image formation. Collineation is also referred to as *projective transformation.* The historical development of this approach, which was first applied to optics by Möbius (Möbius 1855[192]), is discussed by Southall (Southall, 1910[193]). Several authors give extensive discussions (Czapski 1893,[194] Drude 1901[195] Southall 1910,[193] Wandersleb 1920,[196] Chrétien 1980[197]). Projective transformation is used in computer graphics, and is discussed in this context in a number of recent books and papers.

The imaging described by collineation is, be definition, stigmatic everywhere, and planes are imaged without curvature. And for rotationally symmetric lenses, planes perpendicular to the axis are imaged without distortion. So the three conditions of maxwellian perfection are satisfied for all conjugates. Consequently, collineation is often taken as describing ideal imaging of the entire object space. However, it is physically impossible for a lens to image as described by collineation, except for the special case of an afocal lens with $m = m_L = n/n'$. The putative ray intersections of collineation violate the equality of optical path lengths for the rays involved in the imaging of each point. The intrinsic impossibility manifests itself in a variety of ways. As an example, for axial points in a plane with transverse magnification m and ray angles θ and θ' relative to the axis, collineation gives $m \propto \tan\theta/\tan\theta'$, but optical path length considerations require that $m \propto \sin\theta/\sin\theta'$. Another violation is that of the skew invariant $\mathscr{S} = n(\alpha y - \beta x)$. The ratio of this quantity before and after collineation is not unity, but $\mathscr{S}'/\mathscr{S} = \gamma'/\gamma$, where γ is the axial direction cosine in object space and γ' is that in image space.

The expressions for collineation do not contain refractive indices, another manifestation of their not accounting for optical path length. Rather than the refractive index ratio n'/n, which occurs in many imaging equations, the expressions of collineation involve ratios of focal lengths. For afocal lenses there are ratios of transverse and longitudinal magnifications or ratios of the focal lengths of the lenses making up the afocal system.

The expressions for actual ray behavior take the form of collineation in the paraxial, and, more generally, parabasal limits. So paraxial calculations provide the coefficients of the transformation for any particular system.

Collineation is most often treated by starting with the general form, and then reducing its complexity by applying the symmetries of a rotationally symmetric system, to give familiar simple equations such as Newton's (Born & Wolf 1980[198]). Alternatively, it is possible to begin with the simple forms and to derive the general ones therefrom with a succession of images, along with translations and rotations. However, the more important use of these properties is in treating lenses lacking rotational symmetry. This includes those comprised of elements that are arbitrarily oriented, that is, tilted or decentered—either intentionally or unintentionally. Other examples are nonplanar objects, tilted object planes, and arbitrary three-dimensional object surfaces.

Lenses, along with plane mirror systems, can form a succession of images and can produce translations and rotations. Correspondingly, a succession of collinear transformations is a collinear transformation, and these transformations form a group. It is associative, corresponding to the fact that a series of imaging operations can be associated pairwise in any way. There is a unit transformation, correspondingly physically to nothing or to a unit magnification afocal lens. There is an inverse, so an image distorted as a result of object or lens tilt can be rectified by an appropriately designed system—to the extent that collineation validly describes the effects.

General Equations

The general form of the collinear transformation is

$$x' = \frac{a_1x + b_1y + c_1z + d_1}{ax + by + cz + d} \qquad y' = \frac{a_2x + b_2y + c_2z + d_2}{ax + by + cz + d} \qquad z' = \frac{a_3x + b_3y + c_3z + d_3}{ax + by + cz + d} \tag{206}$$

At least one of the denominator coefficients, a, b, c, d, is not zero. The equations can be inverted, so there is a one-to-one correspondence between a point (x, y, z) in object space and a point (x', y', z') in image space. The inverted equations are formally identical, and can be written by replacing unprimed quantities with primed ones and vice versa in the above equation. It is seen that a plane is transformed to a plane, since $a'x' + b'y' + c'z' + d' = 0$ has the same form as a function of (x, y, z). An intersection of two planes gives a line. It can also be shown that a line transforms to a line by writing the equation for a line in parametric form, with parameter σ, $x(\sigma) = x_0 + \alpha\sigma$, $y(\sigma) = y_0 + \beta\sigma$, $z(\sigma) = z_0 + \gamma\sigma$. Substituting in the transformation equations, it is found that $dx'/dy' = (dx'/d\sigma)/(dy'/d\sigma)$ is constant, as are other such ratios.

These equations contain 16 coefficients, but it is possible to divide all three equations through by one of the coefficients, so there are 15 independent coefficients in general. Since the location of an image point is described by three coordinates, five points that are not coplanar determine the transformation.

The ratios of the coefficient dimensions are determined by the fact that x, y, z and x', y', z' are lengths. A variety of schemes can be used and, in the expressions below, a given symbol may have different dimensions.

There are two major categories of the transformation, according to whether the denominator varies or is constant. That with a varying denominator corresponds to focal lenses. For afocal lenses, the demonimator is constant, and the general form of the transformation is

$$x' = a_1 x + b_1 y + c_1 z + d_1 \qquad y' = a_2 x + b_2 y + c_2 z + d_2 \qquad z' = a_3 x + b_3 y + c_3 z + d_3 \tag{207}$$

Here coefficient d has been normalized to unity and the remaining coefficients are dimensionless. Such a transformation is called *affine* or *telescopic.*

Coordinate Systems and Degrees of Freedom

The transformation involves two coordinate systems. The origin of each is located by three parameters, as is the orientation of each. This leaves three parameters that describe the other aspects of the transformation for the most general case of no symmetry. The number is reduced to two if there is rotational symmetry.

In addition to considering the transformation of the entire space, there are other cases, especially the imaging of planes. In each situation, there are specific coordinate systems in which the aspects of the relationship, other than position and orientation, are most simply expressed. Accordingly, different coordinate systems are used in the following sections. Thus, for example, the z axis in one expression may not be the same as that for another.

Simplest Form of the General Transformation

For focal lenses, the denominators are constant for a set of parallel planes

$$ax + by + cz + d = \text{constant} \tag{208}$$

Each such plane is conjugate to one of a set of parallel planes in the other space. Within each of these planes, the quantities $\partial x'/\partial x$, $\partial x'/\partial y$, $\partial x'/\partial z$ are constant, as are the other such derivatives. Therefore, magnifications do not vary with position over these planes, although they do vary with direction. There is one line that is perpendicular to these planes in one space whose conjugate is perpendicular to the conjugate planes in the other space. It can be taken to be the z axis in one space and the z' axis in the other. The aximuths of the x-y and x'-y' axes are found by imaging a circle in each space, which gives an ellipse in the other. The directions of the major and minor axes determine the orientations of these coordinate axes. The *principal focal planes,* are the members of this family of planes for which

$$0 = ax + by + cz + d \tag{209}$$

Lines that are parallel in one space have conjugates that intersect at the principal focal plane in the other. The *principal focal points* are the intersection of the axes with the focal planes.

Using these simplifying coordinate systems, the general transformation is

$$x' = \frac{a_1 x}{cz + d} \qquad y' = \frac{b_1 y}{cz + d} \qquad z' = \frac{c_3 z + d_3}{cz + d} \tag{210}$$

One of the six coefficients can be eliminated, and two of the others are determined by the choice of origins for the z axis and z' axis. If the origins are taken to be at the principal focal points, the transformation equations are

$$x' = \frac{e_x x}{z} \qquad y' = \frac{e_y y}{z} \qquad z' = \frac{e_z}{z} \tag{211}$$

where e_x, e_y, e_z are constants. Unless $e_x = e_y$, the images of shapes in constant z planes vary with their orientations. Squares in one orientation are imaged as rectangles, and in others as parallelograms. Squares in planes not perpendicular to the axes are imaged, in general, with four unequal sides.

For afocal lenses, the simplest form is

$$x' = m_x x \qquad y' = m_y y \qquad z' = m_z z \tag{212}$$

Spheres in one space are imaged as ellipsoids in the other. The principal axes of the ellipsoids give the directions of the axes for which the imaging equations are simplest.

Conjugate Planes

A pair of conjugate planes can be taken to have $x = 0$ and $x' = 0$, so the general transformation between such planes is

$$y' = \frac{b_2 y + c_2 z + d_2}{by + cz + d} \qquad z' = \frac{b_3 y + c_3 z + d_3}{by + cz + d} \tag{213}$$

There are eight independent coefficients, so four points that are not in a line define the transformation. In each space, two parameters specify the coordinate origins and one the orientation. Two parameters describe the other aspects of the transformation.

The simplest set of coordinates is found by a process like that described above. For focal lenses, constant denominators define a line set of parallel lines

$$by + cz + d = \text{constant} \tag{214}$$

with similar conjugate lines in the other space. There is a line that is perpendicular to this family in one space, whose conjugate is perpendicular in the other, which can be taken as the z axis on one side and the z' axis on the other. There is a *principal focal line* in the plane in each space, and a *principal focal point,* at its intersection with the axis. In this coordinate system the transformation is

$$y' = \frac{b_2 y}{cz + d} \qquad z' = \frac{c_3 z + d_3}{cz + d} \tag{215}$$

Of the five coefficients, four are independent and two are fixed by the choice of origins. If $z = 0$ and $z' = 0$ are at the principal focal points, then

$$y' = \frac{e_y y}{z} \qquad z' = \frac{e_z}{z} \tag{216}$$

where e_y and e_z are constants.

For afocal lenses, the general transformation between conjugate planes is

$$y' = b_2 y + c_2 z + d_2 \qquad z' = b_3 y + c_3 z + d_3 \tag{217}$$

The simplest form of the transformation is

$$y' = m_y y \qquad z' = m_z z \tag{218}$$

where m_y and m_z are constants.

Conjugate Lines

A line can be taken to have $x = 0$, $y = 0$, $x' = 0$, $y' = 0$, so its transformation is

$$z' = \frac{c_3 z + d_3}{cz + d} \tag{219}$$

There are three independent coefficients, so three points determine them. The origins in the two spaces account for two of the parameters, leaving one to describe the relative scaling. The simplest forms are

$$\text{Focal:} \quad z' = \frac{e_z}{z} \qquad \text{Afocal:} \quad z' = m_z z \tag{220}$$

There is a relationship between distances along a line (or ray) that is unchanged in collineation (Southall 1910,[193] Southall 1933[199]). If four points on a line A, B, C, D have images A', B', C', D', the *double ratio* or *cross ratio* is invariant under projective transformation, that is,

$$\frac{AC}{BC}\frac{BD}{AD} = \frac{A'C'}{B'C'}\frac{B'D}{A'D} \tag{221}$$

where AC is the distance from A to C, and likewise for other pairs.

Matrix Representation of the Transformation

The transformation can be expressed in linear form by using the variables (u_1, u_2, u_3, u_4) and (u_1', u_2', u_3', u_4'), where $x = u_1/u_4$, $y = u_2/u_4$, $z = u_3/u_4$ and $x' = u_1'/u_4'$, $y' = u_2'/u_4'$, $z' = u_3'/u_4'$. These are referred to as *homogeneous coordinates.* The transformation can be written

$$\begin{pmatrix} u_1' \\ u_2' \\ u_3' \\ u_4' \end{pmatrix} = \begin{pmatrix} a_1 & b_1 & c_1 & d_1 \\ a_2 & b_2 & c_2 & d_2 \\ a_3 & b_3 & c_3 & d_3 \\ a & b & c & d \end{pmatrix} \begin{pmatrix} u_1 \\ u_2 \\ u_3 \\ u_4 \end{pmatrix} \tag{222}$$

In terms of the cartesian coordinates and an additional pair of terms q and q', the transformation can be expressed as

$$\begin{pmatrix} q'x' \\ q'y' \\ q'z' \\ q' \end{pmatrix} = \begin{pmatrix} a_1 & b_1 & c_1 & d_1 \\ a_2 & b_2 & c_2 & d_2 \\ a_3 & b_3 & c_3 & d_3 \\ a & b & c & d \end{pmatrix} \begin{pmatrix} qx \\ qy \\ qz \\ q \end{pmatrix} \tag{223}$$

The dimensions of q and q' depend on the choice of coefficient dimensions. Here $q'/q = ax + by + cz + d$, the equation for the special set of planes.

Certain sections of the matrix are associated with various aspects the transformation (Penna & Patterson 1986[200]). The first three elements in the rightmost column have to do with translation. This is shown by setting $(x, y, z) = (0, 0, 0)$ to locate the conjugate in the other space. The first three elements in the bottom row are related to perspective transformation. The upper left-hand 3×3 array expresses rotation, skew, and local magnification variation.

For the simple form of the transformation expressed in Eq. (211), $a_1 = e_x$, $b_2 = e_y$, $d_3 = e_z$, $c = 1$, and the rest of the coefficients vanish. The general matrix representation for the afocal transformation is

$$\begin{pmatrix} x' \\ y' \\ z' \\ 1 \end{pmatrix} = \begin{pmatrix} a_1 & b_1 & c_1 & d_1 \\ a_2 & b_2 & c_2 & d_2 \\ a_3 & b_3 & c_3 & d_3 \\ 0 & 0 & 0 & 1 \end{pmatrix} \begin{pmatrix} x \\ y \\ z \\ 1 \end{pmatrix} \tag{224}$$

The quantities q and q' can also be included, in which case $q' = q$. In the simplest afocal form, Eq. (212), the matrix is diagonal with $a_1 = m_x$, $b_2 = m_y$, $d_3 = m_z$, and the rest of the nondiagonal coefficients vanishing. A succession of collineations can be treated by multiplying the matrices that describe them (Chastang 1990[201]). To combine lenses with arbitrary orientations and to change coordinate systems, compatible rotation and translation matrices are required. The transformation for a pure rotation with direction cosines (L, M, N) is

$$\begin{pmatrix} x' \\ y' \\ z' \\ 1 \end{pmatrix} = \begin{pmatrix} 1-2L^2 & -2LM & -2LN & 0 \\ -2LM & 1-2M^2 & -2MN & 0 \\ -2LN & -2MN & 1-2N^2 & 0 \\ 0 & 0 & 0 & 1 \end{pmatrix} \begin{pmatrix} x \\ y \\ z \\ 1 \end{pmatrix} \tag{225}$$

The transformation for translation by $(\Delta x, \Delta y, \Delta z)$ is

$$\begin{pmatrix} x' \\ y' \\ z' \\ 1 \end{pmatrix} = \begin{pmatrix} 1 & 0 & 0 & \Delta x \\ 0 & 1 & 0 & \Delta y \\ 0 & 0 & 1 & \Delta z \\ 0 & 0 & 0 & 1 \end{pmatrix} \begin{pmatrix} x \\ y \\ z \\ 1 \end{pmatrix} \tag{226}$$

The quantities q and q' can be included if necessary. The transformations associated with conjugate planes can likewise be expressed with 3×3 matrices, and the transformations of lines with 2×2 matrices.

Rotationally Symmetric Lenses

For rotationally symmetric lenses, the simplest forms are obtained with the z and z' axes corresponding to the lens axis in the two spaces, and the x and x' meridians corresponding. There is one less degree of freedom than in the general case, and $a_1 = b_2$ in Eq. (210). The general transformation is thus

$$x' = \frac{a_1 x}{cz + d} \qquad y' = \frac{a_1 y}{cz + d} \qquad z' = \frac{c_3 z + d_3}{cz + d} \tag{227}$$

There are four degrees of freedom, two associated with the lens and one with the choice of coordinate origins. For focal lenses, the two axial length parameters are f and f'. If the coordinate origins are at the focal points,

$$x' = -\frac{fx}{z} \qquad y' = -\frac{fy}{z} \qquad z' = \frac{f'f}{z} \tag{228}$$

If the coordinate origins are conjugate and related by magnification m_0, then

$$x' = \frac{m_0 x}{1 + z/f} \qquad y' = \frac{m_0 y}{1 + z/f} \qquad z' = \frac{(f'/f)m_0^2 z}{1 + z/f} \tag{229}$$

The constant term in the numerator of the z' expression is the longitudinal magnification for $z = 0$, for which point $dz'/dz = (f'/f)m_0^2$. A special case of these equations is that for which the principal points are the origins, so $m_0 = 1$.

For rotationally symmetric afocal lenses, the two degrees of freedom are the transverse magnification $m_x = m_y = m$, and the longitudinal magnification $m_z = m_L$. The simplest set of transformation equations is

$$x' = mx \qquad y' = my \qquad z' = m_L z \tag{230}$$

where $z = 0$ and $z' = 0$ are conjugate. If $m = \pm 1$ and $m_L = \pm 1$ the image space replicates object space, except possibly for orientation. If $m_L = m$, the spaces are identical except for overall scaling and orientation. The m and m_L appear as functions of ratios of focal lengths of the lenses that make up the afocal system.

Rays for Rotationally Symmetric Lenses

A skew ray with direction cosines (α, β, γ) in object space is described in parametric form with parameter z as follows

$$x(z) = x_0 + \frac{\alpha}{\gamma} z \qquad y(z) = x_0 + \frac{\beta}{\gamma} z \tag{231}$$

For a focal lens, if $z = 0$ is taken to be the front focal plane, and $z' = 0$ is the rear focal plane, the parametric form of the ray in image space is

$$x'(z') = \left(-f\frac{\alpha}{\gamma}\right) + \left(-\frac{x_0}{f'}\right)z' \qquad y'(z') = \left(-f\frac{\beta}{\gamma}\right) + \left(-\frac{y_0}{f'}\right)z' \tag{232}$$

So $x_0' = -f\alpha/\gamma$, $y_0' = -f\beta/\gamma$, $\alpha'/\gamma' = -x_0/f'$, $\beta'/\gamma' = -y_0/f'$. For meridional rays with $x = 0$, if θ and θ' are the ray angles in the two spaces, then $\tan\theta = \beta/\gamma$, $\tan\theta' = -y_0/f'$, and

$$\frac{\tan\theta}{\tan\theta'} = \frac{f'}{f} m \tag{233}$$

where m is the transverse magnification in a plane where the meridional ray crosses the axis.

For afocal lenses, if $z = 0$ and $z' = 0$ are conjugate planes, the ray in image space is given by

$$x'(z') = mx_0 + \left(\frac{m}{m_L}\frac{\alpha}{\gamma}\right)z' \qquad y'(z') = my_0 + \left(\frac{m}{m_L}\frac{\beta}{\gamma}\right)z' \tag{234}$$

For meridianal rays with $x = 0$,

$$\frac{\tan\theta}{\tan\theta'} = \frac{m_L}{m} \tag{235}$$

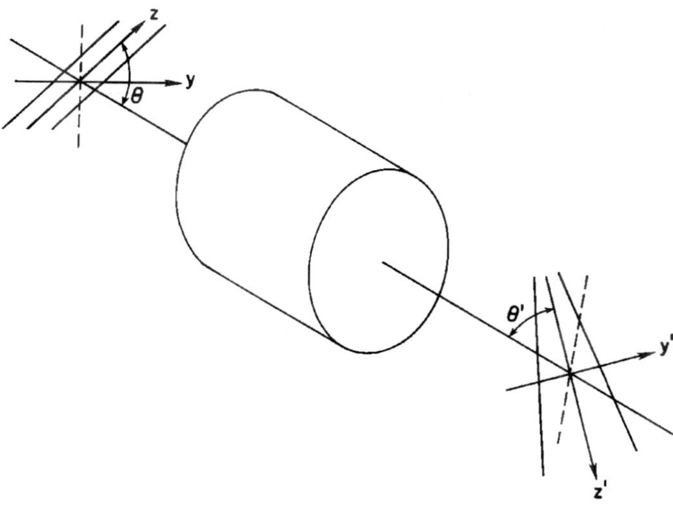

FIGURE 19 The image plane for a tilted object plane. The y-z plane is the object plane and the y'-z' plane is the image plane. The angles between the planes and the lens axis are θ and θ', which are related by Eq. (232). The conjugate points in these planes are related by Eq. (235).

Tilted Planes with Rotationally Symmetric Lenses

A plane making an angle θ with the lens axis in object space has an image plane that makes an angle θ', given by Eq. (233), the so-called *Scheimpflug condition* (Scheimpflug 1907,[202] Sasian 1992[203]). A tilted plane and its image are perpendicular to a meridian of the lens, Fig. 19. There is bilateral symmetry on these planes about the intersection line with the meridian, which is taken to be the z axis in object space and the z' axis in image space. The perpendicular, coordinates are y and y'. Letting m_0 be the transverse magnification for the axial point crossed by the planes, the transform equations are

$$y' = \frac{m_0 y}{1 + z/g} \qquad z' = \frac{(g'/g)m_0^2 z}{1 + z/g} \tag{236}$$

Here g and g' are the focal lengths in the tilted planes, the distances from the principal planes to the focal planes of the lens, measured along the symmetry line, so

$$g = \frac{f}{\cos\theta} \qquad g' = \frac{f'}{\cos\theta'} \qquad \text{and} \qquad \frac{g'}{g} = \sqrt{\left(\frac{f'}{f}\right)^2 \cos^2\theta + \frac{1}{m_0^2}\sin^2\theta} \tag{237}$$

As $\theta \to 90°$, g and g' become infinite, and $(g'/g)m_0 \to 1$, giving $y' \to m_0 y$ and $z' \to m_0 z$. (Forms like Newton's equations may be less convenient here, since the distances from the axes to the focal points may be large.)

For an afocal lens with transverse magnification m and longitudinal magnification m_L, the object and image plane angles are related by Eq. (235). The conjugate equations for points in the planes are

$$y' = my \qquad z' = (m_L^2 \cos^2\theta + m^2 \sin^2\theta)^{1/2} z \tag{238}$$

Here the origins may be the axial intersection point, or any other conjugate points.

Some General Properties

For all collinear transformations, points go to points, lines to lines, and planes to planes. In general, angles at intersections, areas, and volumes are changed. The degree of a curve is unchanged, so, for example, a conic is transformed into a conic. For focal systems, a "closed" conic, an ellipse or circle, may be imaged as either a closed or an "open" one, a parabola or hyperbola. For afocal systems, the closedness and openness are preserved. With focal systems, the imaging of a shape varies with its location, but for afocal systems it does not. For afocal systems parallelness of lines is maintained, but for focal systems the images of parallel lines intersect. For afocal systems, equal distances along lines are imaged as equal distances, but are different unless the magnification is unity.

1.16 SYSTEM COMBINATIONS—GAUSSIAN PROPERTIES

Introduction

This section deals with combinations of systems, each of which is of arbitrary complexity, ranging from a single element, thought of as a combination of two surfaces. From a gaussian description of each lens and the geometry of the combination, the gaussian description of the net system can be found. If two rotationally symmetric lenses are put in series with a common axis, the resultant system is also rotationally symmetric. Its gaussian description is found from that of the two constituent lenses and their separations. The net magnification is the product of the two contributions, i.e., $m = m_1 \times m_2$. Matrix methods are particularly convenient for handling such combinations, and the results below can be demonstrated easily thereby. If two rotationally symmetric lenses are combined so their axes do not coincide, the combination can be handled with appropriate coordinate translations and rotations in the intermediate space, or by means of collineation. In the most general case, where subsystems without rotational symmetry are combined, the general machinery of collineation can be applied. There are three classes of combinations: focal-focal, focal-afocal, and afocal-afocal.

Focal-Focal Combination—Coaxial

The first lens has power ϕ_1 and principal points at P_1 and P_1', Fig. 20. The index preceding the lens is n and that following it is n_{12}. The second lens has power ϕ_2 and principal points at P_2 and P_2', with preceding index n_{12} and following index n'. The directed distance from the rear principal point of the first lens to the first principal point of the second lens is

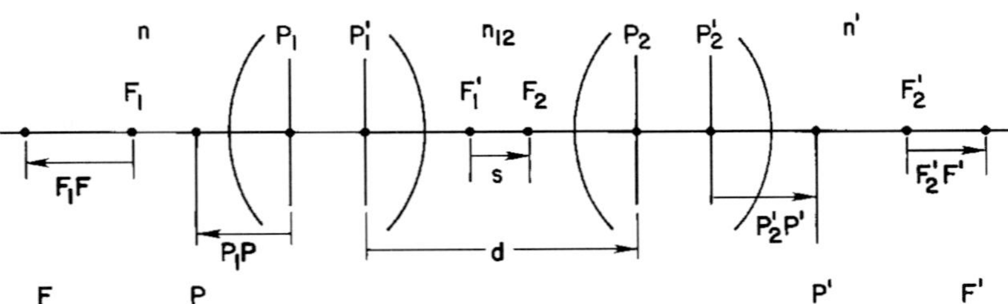

FIGURE 20 Coaxial combination of two focal lenses. The cardinal points of the two lenses are shown above the axis and those of the system below. The directions in this drawing are only one possible case.

$d = P_1'P_2'$, which may be positive or negative, since the lenses may have external principal planes. The power of the combination is

$$\phi = \phi_1 + \phi_2 - \frac{1}{n_{12}} d\phi_1\phi_2 \tag{239}$$

The two principal planes of the combination are located relative to those of the contributing lenses by directed distances

$$P_1P = +\frac{n}{n_{12}} d\frac{\phi_2}{\phi} \qquad P_2'P' = -\frac{n'}{n_{12}} d\frac{\phi_1}{\phi} \tag{240}$$

If $\phi = 0$, the combination is afocal and there are no principal planes. In applying these equations, the inner-space index n_{12} must be the same as that for which the two lenses that are characterized. For example, if two thick lenses are characterized in air and combined with water between them, these equations cannot be used by simply changing n_{12}. It would be necessary to characterize the first lens with water following it and the second lens with water preceding it.

Another set of equations involves the directed distance from the rear focal point of the first lens to the front focal point of the second, $s = F_1'F_2$. The power and focal lengths of the combination are

$$\phi = -\frac{1}{n_{12}} s\phi_1\phi_2 \qquad f = +\frac{f_1 f_2}{s} \qquad f' = -\frac{f_1' f_2'}{s} \tag{241}$$

The focal points are located with respect to those of the contributing lenses by

$$F_1F = +\frac{nn_{12}}{s\phi_1^2} = \frac{n_{12}}{n}\frac{f_1^2}{s} \qquad F_2'F' = -\frac{n'n_{12}}{s\phi_2^2} = -\frac{n_{12}}{n'}\frac{f_2'^2}{s} \tag{242}$$

Another relationship is $(F_1F)(F_2'F') = ff'$. The system is afocal if $s = 0$. There are many special cases of such combinations. Another case is that when the first principal point of the second lens is at the rear focal point of the first, in which case the system focal length is that of the first. These relationships are proven by Welford (Welford 1986, p. 35[204]).

Focal-Afocal—Coaxial

A focal lens combined with an afocal lens is focal, Fig. 21. Here we take the afocal lens to be to the left, with magnification m_1. The focal lens to the right has power ϕ_2 and rear

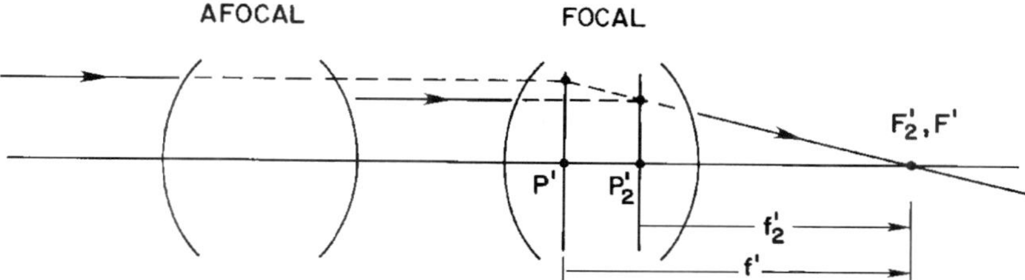

FIGURE 21 Coaxial combination of a focal lens and an afocal lens. In this drawing the afocal lens has a transverse magnification $0 < m_1 < 1$ and the focal lens has a positive power. The combination is a focal lens with focal length $f' = f_2'/m_1$. The focal point on the side of the focal lens is at the focal point of that lens alone.

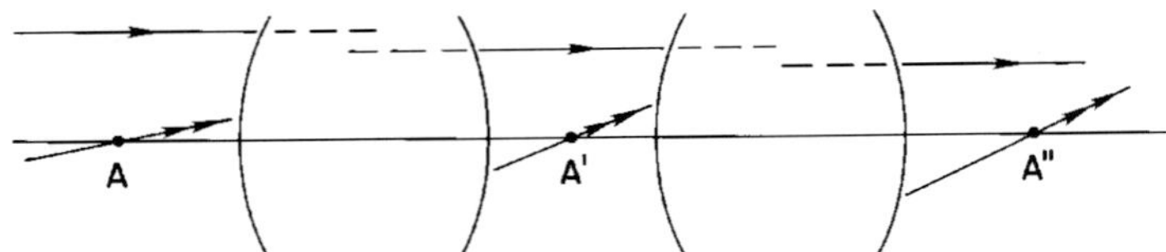

FIGURE 22 Coaxial combination of two afocal lenses. An internal point A' has an object space conjugate A and an image space conjugate A''. These two points can be used for position references in the overall object and image spaces.

focal length f_2'. The power of the combination is $\phi_2 m_1$, and the rear focal length of the combination is $f' = f_2'/m_1$. On the side of the focal lens, the location of the principal focal point is unchanged. On the side of the afocal lens, the system focal point is located at the image of the focal point of the focal lens in the space between the two. Changing the separation between the lenses does not change the power or the position of the principal focal point relative to that of the focal lens. The principal focal point on the afocal lens side does move.

Afocal-Afocal—Coaxial

The combination of two afocal lenses is itself afocal, Fig. 22. If the two lenses have transverse magnifications m_1 and m_2, the combination has $m = m_1 m_2$. A pair of conjugate reference positions is found from the conjugates in the outer regions to any axial point in the inner space. If the separation between the two lenses changes, the combination remains afocal and the magnification is fixed, but the conjugate positions change. This result extends to a combination of any number of afocal lenses.

Noncoaxial Combinations—General

The most general combinations can be handled by the machinery of collineation. The net collineation can be found by multiplying the matrices that describe the constituents, with additional rotation and translation matrices to account for their relative positions. After obtaining the overall matrix, object and image space coordinate systems can be found in which the transformation is simplest. This approach can also be used to demonstrate general properties of system combinations. For example, by multiplying matrices for afocal systems, it is seen that a succession of afocal lenses with any orientation is afocal.

1.17 PARAXIAL MATRIX METHODS

Introduction

Matrix methods provide a simple way of representing and calculating the paraxial properties of lenses and their actions on rays. These methods contain no physics beyond that contained in the paraxial power and transfer equations, Eq. (136) and Eq. (142), but they permit many useful results to be derived mechanically, and are especially useful for lens combinations. The matrix description of systems is also useful in elucidating fundamental paraxial properties. With the symbolic manipulation programs now available, matrix methods also provide a means of obtaining useful expressions.

The optical system is treated as a black box represented by a matrix. The axial positions of the input and output planes are arbitrary. The matrix describes the relationship between what enters and what leaves, but contains no information about the specifics of the system within, and there is an infinity of systems with the same matrix representation.

The origin of matrix methods in optics is not clear. Matrices were used by Samson (Samson 1897[205]) who referred to them as "schemes." Matrices appear without comment in a 1908 book (Leathem 1908[206]). Matrix methods are treated in papers (Halbach 1964,[207] Sinclair 1973[208]) and in many books (O'Neil 1963,[209] Brouwer 1964,[210] Blaker 1971,[211] Longhurst 1973,[212] Gerrard & Burch 1974,[213] Naussbaum & Phillips 1976,[214] Kogelnik 1979,[215] Klein & Furtak 1986,[216] Moller 1988,[217] Guenther 1990[218]). Notation is not standardized, and many treatments are complicated by notation that conceals the basic structures.

This section is limited to rotationally symmetric lenses with homogeneous media. References are provided for systems with cylindrical elements. This treatment is monochromatic, with the wavelength dependence of index not made explicit.

The matrices are simplified by using *reduced axial distances* $\tau = t/n$ and *reduced angles* $\omega = nu$. The paraxial angles u are equivalent to direction cosines, and the reduced angles are optical direction cosines in the paraxial limit. For brevity, ω and τ are usually referred to in this section simply as "angle" and "distance."

Basic Idea—Linearity

Paraxial optics is concerned with the paraxial heights and paraxial angles of rays. A meridional ray entering a system has a given height y and angle ω and leaves with another height y' and angle ω'. Paraxial optics is linear, as discussed above, in the sense that both the outgoing height and angle depend linearly on the incoming height and angle. Writing Eq. (148) in terms of ω's gives

$$y' = \left(\frac{\partial y'}{\partial y}\right) y + \left(\frac{\partial y'}{\partial \omega}\right)\omega \qquad \text{and} \qquad \omega' = \left(\frac{\partial \omega'}{\partial y}\right) y + \left(\frac{\partial \omega'}{\partial \omega}\right)\omega \tag{243}$$

The partial derivatives are constant for a given system. This linearity is the basis of the matrix treatment, since these equations can be written in matrix form:

$$\begin{pmatrix} y' \\ \omega' \end{pmatrix} = \begin{pmatrix} \dfrac{\partial y'}{\partial y} & \dfrac{\partial y'}{\partial \omega} \\ \dfrac{\partial \omega'}{\partial y} & \dfrac{\partial \omega'}{\partial \omega} \end{pmatrix} \begin{pmatrix} y \\ \omega \end{pmatrix} \tag{244}$$

Basic Operations

The basic operations in paraxial ray tracing are transfer, Eq. (136), between surfaces and refraction or reflection at surfaces, Eq. (142).

Transfer Matrix

Transfer changes the height of a ray, in general, leaving the angle unchanged. In terms of reduced quantities, the relationships are:

$$y' = y + tu = y + \frac{t}{n}un = y + \tau\omega \qquad \text{and} \qquad \omega' = \omega \tag{245}$$

The transfer matrix is:

$$\begin{pmatrix} 1 & \tau \\ 0 & 1 \end{pmatrix} \tag{246}$$

For left-to-right transfer, $\tau > 0$. This gives a difference in signs between some of the terms in expressions here and those in the gaussian section, where directed distances are measured from the lens to the object.

Power Matrix

Refraction or reflection changes the angle of a ray, but not its height. The equations for reduced quantities are

$$n'u' = nu - y\phi = \omega' = \omega - y\phi \qquad \text{and} \qquad y' = y \tag{247}$$

Here $\phi = c(n' - n)$ for refraction and $\phi = -2nc$ for reflection, where c is the surface curvature, Eq. (143). The power matrix is:

$$\begin{pmatrix} 1 & 0 \\ -\phi & 1 \end{pmatrix} \tag{248}$$

A planar reflecting or refracting surface has $\phi = 0$, so it is represented by the unit matrix.

Arbitrary System

A general system consists of a series of surfaces with powers $\phi_1, \phi_2, \ldots$ that are separated from one another by distances $\tau_1, \tau_2, \ldots$. Its matrix is the product

$$\begin{pmatrix} 1 & \tau_N \\ 0 & 1 \end{pmatrix} \cdots \begin{pmatrix} 1 & 0 \\ -\phi_2 & 1 \end{pmatrix} \begin{pmatrix} 1 & \tau_2 \\ 0 & 1 \end{pmatrix} \begin{pmatrix} 1 & 0 \\ -\phi_1 & 1 \end{pmatrix} \begin{pmatrix} 1 & \tau_1 \\ 0 & 1 \end{pmatrix} \tag{249}$$

By convention, the successive matrices are concatenated from right to left, whereas ray tracing is done left to right.

A special case is a succession of transfers, itself a transfer.

$$\text{Succession of transfers:} \quad \begin{pmatrix} 1 & \tau_1 + \tau_2 + \cdots \\ 0 & 1 \end{pmatrix} \tag{250}$$

Another is a series of refractions with no intervening transfer, itself a power operation.

$$\text{Succession of powers:} \quad \begin{pmatrix} 1 & 0 \\ -(\phi_1 + \phi_2 + \cdots) & 1 \end{pmatrix} \tag{251}$$

Matrix Elements

Each matrix element has a physical significance, and the terms can be given mnemonic symbols associated with the conditions under which they are zero. (This practice is not standard.) If the initial ray angle is zero, the outgoing angles depend on the incident ray heights and the power of the system, according to $\omega' = -\phi y$, so $\partial\omega'/\partial y = -\phi$. If the initial surface is at the front focal plane, the outgoing ray angles depend only on the incident height, so $\partial\omega'/\partial\omega = 0$. This term is denoted by F for "front." Similarly, if the final surface is at the real focal plane, the outgoing ray heights depend only on the incoming angles, so $\partial y'/\partial y = R$ for "rear." If the initial and final planes are conjugate, then all incoming rays at a given height y have the outgoing height $y' = my$, regardless of their angle, so $\partial y'/\partial\omega = 0$ for conjugate planes. Since this term is related to the condition of conjugacy, $\partial y'/\partial\omega = C$ for "conjugate." With this notation, the general matrix is

$$\begin{pmatrix} R & C \\ -\phi & F \end{pmatrix} \tag{252}$$

Dimensions

The terms R and F are dimensionless. C has the dimensions of length, and those of ϕ are inverse length. Dimensional analysis, as well as the consideration of Eq. (248), shows that the F and R terms will always contain products of equal numbers of ϕ_l's and τ_k's, such as $\phi_k\tau_l$. The ϕ expression contains terms like ϕ_k and $\tau_k\phi_l\phi_m$, with one more power term than distance terms. Similarly, C has terms like τ_k and $\tau_k\tau_l\phi_m$.

Determinant

Both the transfer and power matrices have unit determinants. Therefore, any product of such matrices has a unit determinant, a fact that is related to the two-ray paraxial invariant.

$$\begin{vmatrix} R & C \\ -\phi & F \end{vmatrix} = FR + C\phi = 1 \tag{253}$$

This provides an algebraic check. For afocal lenses and conjugate arrangements, $FR = 1$.

Possible Zeros

The possible arrangements of zeros in a system matrix is limited by the unit determinant restriction. There can be a single zero anywhere. In this case, either $C = 1/\phi$ or $F = 1/R$, and the remaining nonzero term can have any value. There can be two zeros on either diagonal. No row or column can contain two zeros, since a system represented by such a matrix would violate conservation of brightness. A matrix with double zeros in the bottom row would collimate all rays, regardless of their incoming position and direction. A matrix with all zeros in the top row represents a system that would bring all incoming light to a single point. A system whose matrix has double zeros in the first column would bring all incoming light to a focus on the axis. For double zeros in the second row, the system would concentrate all light diverging from an input point in a single output point with a single direction.

Operation on Two Rays

Instead of considering a single input and output ray, the matrix formalism can be used to treat a pair of rays, represented by a 2×2 matrix. In this case

$$\begin{pmatrix} y_1' & y_2' \\ \omega_1' & \omega_2' \end{pmatrix} = \begin{pmatrix} R & C \\ -\phi & F \end{pmatrix} \begin{pmatrix} y_1 & y_2 \\ \omega_1 & \omega_2 \end{pmatrix} \tag{254}$$

Since the system matrix has a unit determinant, the determinants of the incoming and outgoing ray matrices are identical:

$$L_{12} = y_1'\omega_2' - y_2'\omega_1' = y_1\omega_2 - y_2\omega_1 \tag{255}$$

This is the paraxial invariant, Eq. (149). It is possible to operate on more than two rays, but never necessary, since any third ray is a linear combination of two, Eq. (154). Operations on two rays can also be handled with a complex notation in which two ray heights and two angles are each represented by a complex number (Marechal 1956,[219] Marechal 1961[220]).

Conjugate Matrix

For conjugate planes, $y' = my$, so $C = 0$, $R = m$, and $F = 1/m$, giving

$$\begin{pmatrix} m & 0 \\ -\phi & 1/m \end{pmatrix} \tag{256}$$

The $1/m$ term gives the angular magnification, $u'/u = n/n'm$, Eq. (196). This matrix also holds for afocal lenses, in which case $\phi = 0$.

Translated Input and Output Planes

For a given system, the locations of the input and output planes are arbitrary. If the input plane is translated by τ and the output plane by τ', the resultant matrix is

$$\begin{pmatrix} R - \tau'\phi & C + \tau R + \tau' F - \tau\tau'\phi \\ -\phi & F - \tau\phi \end{pmatrix} \tag{257}$$

Note that the object-space translation term τ is grouped with F and the image-space term τ' with R. The equation $C = 0 = \tau R - \tau' F - \tau\tau'\phi$ gives all pairs of τ and τ' for which the input and output surfaces are conjugate.

Principal Plane-to-Principal Plane

If the input and output planes are the principal planes, then the matrix is a conjugate one, for which $m = +1$.

$$\begin{pmatrix} 1 & 0 \\ -\phi & 1 \end{pmatrix} \tag{258}$$

This is also the matrix representing a thin lens.

Nodal Plane-to-Nodal Plane

The nodal points are conjugate, with unit angular magnification, so $u' = u$ and $\omega' = n'\omega/n$. Thus

$$\begin{pmatrix} n/n' & 0 \\ -\phi & n'/n \end{pmatrix} \tag{259}$$

The transverse magnification $m_N = n/n'$ equals unity when $n = n'$. This matrix has no meaning for afocal lenses.

Focal Plane-to-Focal Plane

If the initial surface is at the front principal focal plane and the final surface is at the rear focal plane, the matrix is

$$\begin{pmatrix} 0 & 1/\phi \\ -\phi & 0 \end{pmatrix} \tag{260}$$

This is the "Fourier transform" arrangement, in which incident heights are mapped as angles and vice versa.

Translation from Conjugate Positions

If the input plane is translated τ from a plane associated with magnification m and the output plane is translated a distance τ' from the conjugate plane, the matrix is

$$\begin{pmatrix} m - \tau'\phi & m\tau + \tau'/m - \tau\tau'\phi \\ -\phi & 1/m - \tau\phi \end{pmatrix} \tag{261}$$

Setting $C = 0$ gives an equation that locates all other pairs of conjugate planes relative to the first one, Eq. (172).

Translation from Principal Planes

If the initial conjugate planes are the principal planes, then

$$\begin{pmatrix} 1 - \tau'\phi & \tau + \tau' - \tau\tau'\phi \\ -\phi & 1 - \tau\phi \end{pmatrix} \tag{262}$$

The equation for other conjugates is $C = 0 = \tau + \tau' - \tau\tau'\phi$, corresponding to Eq. (170). It follows that the distance from the input surface to the first principal plane is $\tau = (1 - F)/\phi$ and the distance from the output surface to the second principal plane is $\tau' = (1 - R)/\phi$.

Translation from Focal Planes

If the input plane is a distance τ from the front focal plane and the output plane a distance τ' from the rear focal plane, the matrix is

$$\begin{pmatrix} -\phi\tau' & \dfrac{1}{\phi}(1 - \phi^2\tau\tau') \\ -\phi & -\phi\tau \end{pmatrix} \tag{263}$$

Thus F and R are proportional to the distances of the input and output surfaces from the object space and image space focal planes. Using Newton's formulas, this can also be written

$$\begin{pmatrix} m' & \dfrac{1}{\phi}\left(1-\dfrac{m'}{m}\right) \\ -\phi & \dfrac{1}{m} \end{pmatrix} \tag{264}$$

Here m' is the magnification that would obtain if the image point were as located by R, and m is that if the object point were located by F. The conjugate term vanishes when $m = m'$.

Conjugate Relative to Principal Focal Planes

If Eq. (263) is a conjugate matrix, it becomes

$$\begin{pmatrix} -\phi\tau' & 0 \\ -\phi & -\phi\tau \end{pmatrix} \tag{265}$$

The vanishing C term gives $0 = 1/\phi - \phi\tau\tau'$, which is the Newton equation usually written as $zz' = ff'$. The magnification terms are the other Newton's equations, $m = -\phi\tau'$ and $1/m = -\phi\tau$, which are usually written as $m = -z'/f' = -f/z$.

Afocal Lens

For afocal lenses $\phi = 0$. Since the determinant is unity, $F = 1/R$. And since the transverse magnification is constant, $R = m$, giving

$$\begin{pmatrix} m & C \\ 0 & 1/m \end{pmatrix} \tag{266}$$

A ray with $\omega = 0$ has $y' = my$, and $\omega' = \omega/m$ for all y. At conjugate positions, an afocal lens has the matrix

$$\begin{pmatrix} m & 0 \\ 0 & 1/m \end{pmatrix} \tag{267}$$

Performing a translation in both object and images spaces from the conjugate position gives

$$\begin{pmatrix} m & m\tau + \dfrac{\tau'}{m} \\ 0 & 1/m \end{pmatrix} \tag{268}$$

Setting $C = 0$ gives $\tau' = -m^2\tau$, which relates the location of a single conjugate pair to all others, Eq. (200).

Symmetrical Lenses

For lenses with symmetry about a central plane, $F = R$, so the matrix has the form

$$\begin{pmatrix} B & C \\ -\phi & B \end{pmatrix} \tag{269}$$

where $B^2 = 1 - \phi C$. The conjugate matrix has $m = \pm 1$.

Reversing Lenses

When a lens is flipped left to right, the matrix of the reversed system is obtained from that of the original one by switching the F and R terms.

$$\begin{pmatrix} F & C \\ -\phi & R \end{pmatrix} \tag{270}$$

This reversal maintains the exterior references planes, that is, the input surface for the initial system becomes the output surface for the flipped one and vice versa.

Inverse Systems

By the "inverse" of a lens is meant a second system that undoes the effect of a given one. That is, the rays at the output surface of the second system have the same height and angle as those at the input of the first system. The combination of a system and its inverse is afocal with unit magnification. The matrix representing the inverse system is the inverse of that representing the system.

$$\begin{pmatrix} F & -C \\ \phi & R \end{pmatrix} \tag{271}$$

The matrix provides no instruction as to how such a lens is made up. Alternatively, the inverse matrix can be interpreted as that whose input is y' and ω', with outputs y and ω.

Series of Arbitrary Lenses

The matrix for two successive lenses is

$$\begin{pmatrix} R_1R_2 - C_2\phi_1 & C_1R_2 + C_2F_1 \\ -\phi_1F_2 - \phi_2R_1 & F_1F_2 - C_1\phi_2 \end{pmatrix} = \begin{pmatrix} R_2 & C_2 \\ -\phi_2 & F_2 \end{pmatrix} \begin{pmatrix} R_1 & C_1 \\ -\phi_1 & F_1 \end{pmatrix} \tag{272}$$

For example, two given lenses separated by some distance have the matrix

$$\begin{pmatrix} R_2 & C_2 \\ -\phi_2 & F_2 \end{pmatrix} \begin{pmatrix} 1 & \tau \\ 0 & 1 \end{pmatrix} \begin{pmatrix} R_1 & C_1 \\ -\phi_1 & F_1 \end{pmatrix} \tag{273}$$

Multiplying from right to left gives a running product or "cumulative matrix," that shows the effect of the system up to a given plane.

Decomposition

Matrix multiplication is associative, so the system representation can be broken up in a number of ways. For example, the portion of a lens before and after the aperture stop can be used to find the pupil locations and magnifications. An arbitrary lens matrix can be written as a product of three matrices (Macukow & Arsenault 1983[221]):

$$\begin{pmatrix} R & C \\ -\phi & F \end{pmatrix} = \begin{pmatrix} 1 & 0 \\ -\phi/R & 1 \end{pmatrix} \begin{pmatrix} R & 0 \\ 0 & 1/R \end{pmatrix} \begin{pmatrix} 1 & C/R \\ 0 & 1 \end{pmatrix} \tag{274}$$

or

$$\begin{pmatrix} R & C \\ -\phi & F \end{pmatrix} = \begin{pmatrix} 1 & C/F \\ 0 & 1 \end{pmatrix} \begin{pmatrix} 1/F & 0 \\ 0 & F \end{pmatrix} \begin{pmatrix} 1 & 0 \\ -\phi/F & 1 \end{pmatrix} \tag{275}$$

Thus a general lens is equivalent to a succession of three systems. One has power and works at unit magnification. The second is a conjugate afocal matrix. The third is a translation. Each of these systems is defined by one of the three terms, either R, ϕ/R, C/R or F, ϕ/F, C/F. This is another manifestation of the three degrees of freedom of paraxial systems.

Matrix Determination by Two-Ray Specification

If a two-ray input matrix is given along with the desired output, or the two input and output rays are measured by determining the matrix of an unknown lens, Eq. (254) gives

$$\begin{pmatrix} R & C \\ -\phi & F \end{pmatrix} = \begin{pmatrix} y_1' & y_2' \\ \omega_1' & \omega_2' \end{pmatrix} \begin{pmatrix} y_1 & y_2 \\ \omega_1 & \omega_2 \end{pmatrix}^{-1} \tag{276}$$

so

$$\begin{pmatrix} R & C \\ -\phi & F \end{pmatrix} = \frac{1}{y_1\omega_2 - y_2\omega_1} \begin{pmatrix} y_1'\omega_2 - y_2'\omega_1 & y_2'y_1 - y_1'y_2 \\ \omega_1'\omega_2 - \omega_2'\omega_1 & \omega_2'y_1 - y_2\omega_1' \end{pmatrix} \tag{277}$$

The denominator of the multiplicative factor is the paraxial invariant associated with the two rays, Eq. (149). As a special case, the two rays could be the marginal and chief rays. The input and output pairs must have the same invariant, or the matrix thus found will not have a unit determinant.

Experimental Determination of Matrix Elements

The matrix elements for an unknown lens can, in principle, be determined experimentally. One method, as mentioned in the preceding section, is to measure the heights and angles of an arbitrary pair of rays. Another method is as follows. The power term is found in the usual way by sending a ray into the lens parallel to the axis and measuring its outgoing angle. To find $C = \partial y'/\partial\omega$, the input ray angle is varied, while its height is unchanged. If the output height is graphed, its slope is C. Likewise, the other partial derivatives in Eq. (243) can be found by changing one of the input parameters while the other is fixed. The four measurements are redundant, the unit determinant providing a check of consistency.

Angle Instead of Reduced Angle

The matrices above can be modified to use the angles u and u', instead of the reduced angles. In terms of matrix theory, this amounts to a change in basis vectors, which is accomplished by multiplying by diagonal vectors with elements 1 and n or 1 and n'. The result is

$$\begin{pmatrix} y \\ u' \end{pmatrix} = \begin{pmatrix} R & nC \\ -\frac{1}{n'}\phi & \frac{n}{n'}F \end{pmatrix} \begin{pmatrix} y \\ u \end{pmatrix} \tag{278}$$

This matrix has a constant determinant n/n'. The form Eq. (252) is simpler.

Other Input-Output Combinations

Referring to Eq. (244), any pair of the four quantities y, ω, y', and ω' can be taken as inputs, with the other two as outputs, and the relationships can be expressed in matrix form. The four matrices in this section cannot be multiplied to account for the concatenation of lenses. If the angles are given, the heights are

$$\begin{pmatrix} y \\ y' \end{pmatrix} = \frac{1}{\phi} \begin{pmatrix} F & -1 \\ 1 & -R \end{pmatrix} \begin{pmatrix} \omega \\ \omega' \end{pmatrix} \tag{279}$$

The matrix is undefined for afocal lenses, for which the relationship of ω and ω' is independent of heights. Similarly, the angles can be expressed as functions of the heights by

$$\begin{pmatrix} \omega \\ \omega' \end{pmatrix} = \frac{1}{C} \begin{pmatrix} -R & 1 \\ -1 & F \end{pmatrix} \begin{pmatrix} y \\ y' \end{pmatrix} \tag{280}$$

For conjugates the expression breaks down, since there is no fixed relationship between heights and angles. If the input is a height on one side and an angle on the other, then

$$\begin{pmatrix} y' \\ \omega \end{pmatrix} = \frac{1}{F} \begin{pmatrix} 1 & C \\ \phi & 1 \end{pmatrix} \begin{pmatrix} y \\ \omega' \end{pmatrix} \tag{281}$$

For the inverse situation,

$$\begin{pmatrix} y \\ \omega' \end{pmatrix} = \frac{1}{R} \begin{pmatrix} 1 & -C \\ -\phi & 1 \end{pmatrix} \begin{pmatrix} y' \\ \omega \end{pmatrix} \tag{282}$$

The determinants of these matrices are, respectively, C, ϕ, R, and F.

Derivative Matrices

If the axial position of the input surface changes, the rate of change of the output quantities is

$$\begin{pmatrix} dy'/dz \\ d\omega'/dz \end{pmatrix} = \begin{pmatrix} 0 & R \\ 0 & -\phi \end{pmatrix} \begin{pmatrix} y \\ \omega \end{pmatrix} \tag{283}$$

If the axial position of the output surface can change, the rate of change of output quantities is

$$\begin{pmatrix} dy'/dz' \\ d\omega'/dz' \end{pmatrix} = \begin{pmatrix} -\phi & F \\ 0 & 0 \end{pmatrix} \begin{pmatrix} y \\ \omega \end{pmatrix} \tag{284}$$

Higher derivatives vanish.

Skew rays

The matrix formalism can be used to treat a paraxial skew ray, represented by a 2×2 matrix of x and y positions and directions α and β. In this case

$$\begin{pmatrix} x' & y' \\ n'\alpha' & n'\beta' \end{pmatrix} = \begin{pmatrix} R & C \\ -\phi & F \end{pmatrix} \begin{pmatrix} x & y \\ n\alpha & n\beta \end{pmatrix} \tag{285}$$

Since the lens matrix has a unit determinant, the determinants of the incoming and outgoing ray matrices are identical:

$$n'(y'\alpha' - x'\beta') = n(y\alpha - x\beta) \tag{286}$$

From Eq. (73), this is the skew invariant.

Relationship to Characteristic Functions

A lens matrix can be related to any one of the four paraxial characteristic functions, Eqs. (34) through (37), each of which has three first coefficients, associated with the three degrees of freedom of the matrix. Brouwer and Walther (Brouwer & Walther 1967[222]) derive the paraxial matrices from more general matrices based on the point angle characteristic function.

Nonrotationally Symmetric Systems

Systems comprised of cylindrical lenses can also be treated paraxially by matrices (Arsenault 1979,[223] Arsenault 1980,[224] Arsenault 1980,[225] Keating 1981,[226] Arsenault & Macukow 1983,[227] Macukow & Arsenault 1983,[221] Attard 1984[228]). The more general case of a treatment around an arbitrary ray is also represented by a 4×4 matrix (Stone & Forbes 1992[229]). This is treated by several of the references to the section "Images About Known Rays."

1.18 APERTURES, PUPILS, STOPS, FIELDS, AND RELATED MATTERS

Introduction

This section is concerned with the finite sizes of lens and their fields, as expressed in various limitations of linear dimensions and angles, and with some of the consequences of these limits. (Other consequences, for example, resolution limitations, are in the domain of wave optics.) Terminology in this area is not well defined, and the terms typically used are insufficient for all the aspects of the subject, so this section deals considerably with definitions.

Field Size and Field Stop

The *field* or *field of view* of a lens is the region of object from which light is captured or the region of image space that is used. The field size may be described in angular, linear, or area units, depending on the circumstances. (It can be described in still other ways, e.g., the number of pixels.) In and of itself, a lens does not have a definite field size, but

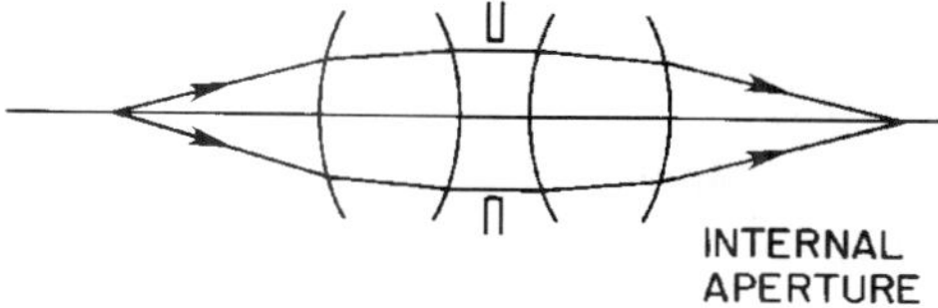

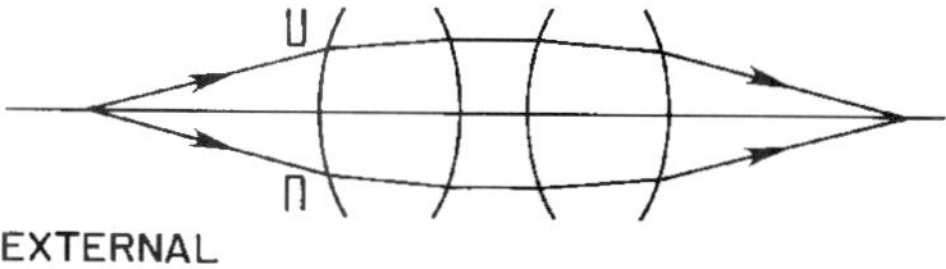

FIGURE 23 Axial ray cone and aperture stop. The upper lens has an internal aperture, and the lower one has an external aperture on the object side.

beyond a certain size, image quality diminishes, both with respect to aberration correction and to light collection. A *field stop* is a physical delimiter of the field, which may be in either object or image space. A detector may be the delimiter.

Aperture Stop

Each object point can be thought of as emitting rays in all directions. Since lenses are finite in size, only some of the rays pass through them. The rays that do pass are referred to as *image-forming rays,* the ensemble of which is the *image-forming bundle,* also called the *image-forming cone,* although the bundle may not be conical. The bundle associated with each object point is delimited by one or more physical structures of the lens. For axial object points, the delimiting structure is called the *aperture,* the *stop,* or the *aperture stop.* The aperture may be either within the lens or outside of it on either side, Fig. 23. The aperture may be a structure whose sole purpose is delimiting the bundle, or it may be the edge of an optical element or a lens mount. The aperture stop may be fixed or adjustable, for instance an iris. Which structure acts as the aperture can change with object position, Fig. 24. The size and position of the aperture do not effect the gaussian properties of the lens, i.e., the cardinal points and the conjugate locations and magnifications. They do

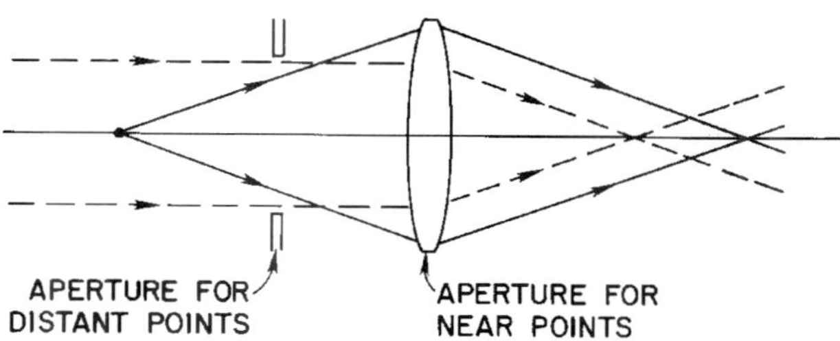

FIGURE 24 An example of change of aperture with axial object position. For distant points the aperture is the nominal stop. For near points the aperture is the rim of the lens.

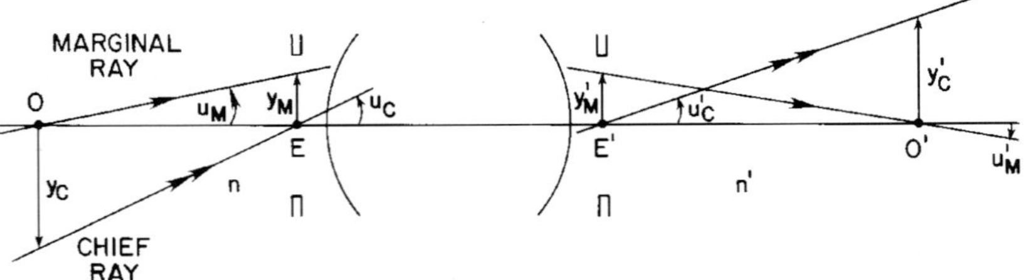

FIGURE 25 Schematic diagram of a lens with object and image planes, entrance and exit pupils, and marginal and chief rays. The entrance pupil is located at E and the exit pupil at E'. The chief ray passes through the edges of the fields and the centers of the pupils. The marginal ray passes through the axial object and image points and the edges of the pupils.

affect the image irradiance, the aberrations, and the effects of defocus. The aperture is most commonly centered on axis, but this is not always so. With visual instruments, the aperture stop for the entire system may be either an aperture in the optics or the iris of the observer's eye.

Marginal Rays and Chief Rays

Ray bundles are described to a considerable extent by specifying their central and extreme rays. For object planes perpendicular to the lens axis, there are two meridional rays of particular importance, defining the extremities of field and aperture, Fig. 25. These rays are reciprocal in that one is to the pupil what the other is to the field.

The *marginal ray* originates at the axial object point, intersects the conjugate image point, and passes through the edge of the aperture. This term is also used for rays from other field points that pass through the extremes of the aperture. The *paraxial marginal ray* is the marginal ray in the paraxial limit.

The *chief ray* or *principal ray* originates at the edge of the object field, intersects the edge of the image field, and passes approximately through the center of the aperture, and hence approximately through the center of the pupils. (Here we use "chief ray," since the prefix "principal" is so commonly used for other entities.) The term is also used for the central ray of other bundles. The *paraxial chief ray* passes exactly through the centers of the aperture and both paraxial pupils.

Field Angle

The *field angle* is that subtended by the field of view at the lens. This term is ambiguous, since several angles can be used, as well as angles in both object and image space. A nodal ray angle is the same in both spaces. If the nodal points are not at the pupils, the chief ray angle differs on the two sides. The ratio of paraxial chief ray angles is proportional to the paraxial pupil magnification, as discussed later, Eq. (289). If the lens is telecentric, the chief ray angle is zero. An afocal lens has no nodal points, and the paraxial ratio of output angles to input angles is constant. The concept of field angle is most useful with objects and/or images at large distances, in which case on the long conjugate side the various ray angles are nearly identical. On the short conjugate side, ambiguity is removed by giving the focal length, the linear size of the detector, and the principal plane and exit pupil positions. For finite conjugates, such information should be provided for both spaces.

Pupils

The term *pupil* is used in several ways, and care should be taken to distinguish between them. There are paraxial pupils, "real" pupils, pupils defined as ranges of angles, and pupil reference spheres used for aberration definition and diffraction calculations. The *entrance pupil* is the aperture as seen from object space—more precisely, as seen from a particular point in object space. If the aperture is physically located in object space, the entrance pupil is identical to the aperture. Otherwise, the entrance pupil is the image of the aperture in object space formed by the portion of the lens on the object side of the aperture. If the aperture is in image space, the entrance pupil is its image formed by the entire lens. Similarly, the *exit pupil* is the aperture as seen from image space. A *real pupil* is a physically accessible image of the aperture or the aperture itself, and a *virtual pupil* is an inaccessible image. Visual instruments often have external pupils, where the user's eye is located. The axial entrance pupil point is denoted here by E and the exit pupil by E'.

The pupils can be located anywhere on axis, except that they cannot coincide with the object or image. It is common to draw pupils as shown in Fig. 25, but they can also be on the side of the object or image away from the lens. The pupils are usually centered on axis, but not necessarily. Aberrations may shift pupils from nominal axial centration.

Both pupils are conjugate to the aperture, so they are conjugate to each other. The term *pupil imaging* refers to the relationship of the pupils with respect to each other and to the aperture. In pupil imaging, the chief ray of the lens is the marginal ray and vice versa. The *pupil magnification* m_P denotes the ratio of exit pupil size to entrance pupil size. The size may be specificed as linear or an angular extent, and the pupil magnification may be a transverse magnification, finite or paraxial, or a ratio of angular subtenses. In general, there is *pupil aberration,* so the image of the aperture in each space is aberrated, as is that of the imaging of one pupil to the other. Pupil imaging is subject to chromatic aberration, so positions, sizes, and shapes of pupils may vary with wavelength.

There is ambiguity about pupil centers and chief rays for several reasons. The center can be taken with respect to linear, angular, or direction cosine dimensions. Because of spherical pupil aberration, a ray through the center of the pupil may not also pass through the center of the aperture, and vice versa. The angular dimensions of pupils may change with field position. Pupil aberrations cause the actual pupil shape to be different from that of the paraxial pupil.

Pupils that are not apertures can have any linear size, since the aperture can be imaged at any magnification. If the aperture is within the lens, there is no particular relationship between the positions and linear sizes of the entrance and exit pupils, since the portions of the lens that precede and follow the aperture have no specific relationship. There is a relationship between the angular subtense of the pupils, as discussed below.

The angular size and shape of the pupils can vary with field position, and the pupils can change position if the aperture changes with object position. If the lens changes internally, as with a zoom, the sizes and positions of the pupils change.

Paraxial Description

The *paraxial pupils* are the paraxial images of the aperture. They are usually planar and perpendicular to the axis and are implicitly free from aberration. The paraxial chief ray passes through the center of both pupils and the aperture, and the paraxial marginal ray through the edges. The object and pupil magnifications and the distances from object to entrance pupil and from exit pupil to image are related by Eq. (194). If the object at O

is imaged at O' with magnification m, and the pupil magnification from entrance pupil at E to exit pupil at E' is m_E, then from Eq. (194)

$$O'E' = \frac{n'}{n} m m_E OE \tag{287}$$

Paraxial Invariant for Full Field and Full Aperture

Let the height of the paraxial marginal ray be y_M at the entrance pupil and y'_M at the exit pupil, and that of the paraxial chief ray by y_C, at the object plane and y'_C at the image plane, Fig. 25. Let the angles of these rays be u_M, u_C, u'_M, u'_C. The two-ray paraxial invariant, Eq. (149), is

$$L = n y_C u_M = n y_M u_C = n' y'_M u'_C = n' y'_C u'_M \tag{288}$$

This relationship was rediscovered several times, so the conserved quantity is referred to by a variety of names, including the *Lagrange invariant,* the *Helmholtz invariant,* the *Smith invariant,* and with various hyphenated combinations of the proper names (Rayleigh 1886,[230] Southall 1910[231]). Further discussions are found in the sections on paraxial optics and on the étendue. The paraxial transverse magnification and paraxial pupil magnifications are related to the paraxial marginal and chief ray angles by

$$m = \frac{y'_C}{y_C} = \frac{n u_M}{n' u'_M} \qquad \text{and} \qquad m_P = \frac{y'_M}{y_M} = \frac{n u_C}{n' u'_C} \tag{289}$$

Pupil Directions

For some purposes, pupils are best described as ranges of directions, specified in direction cosines, rather than by linear extents of aperture images. Here the term *pupil directions* (NS) is used. This is particularly the case when dealing with a given region of the object. The construction for this description is shown in Fig. 26. The x and y axes of the

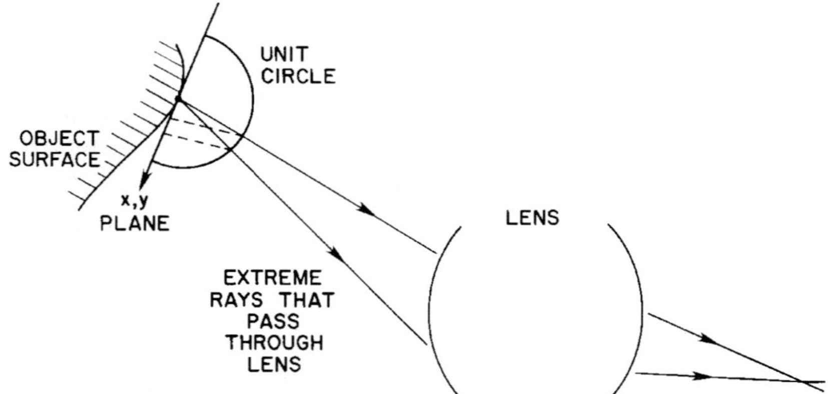

FIGURE 26 Construction for the description of the pupils with direction cosines. An x-y plane is tangent to the object surface at the object point, and a unit sphere is centered on the point. The intersections with the unit sphere of the rays are projected to the tangent plane to give the pupil direction cosines.

object-space coordinate system lie in the object surface, and the x' and y' axes. From a point on the object plane, the extreme set of rays that passes through the lens is found. Its intersection with a unit sphere about the object point is found, and perpendiculars are dropped to the unit circle on (or tangent to) the object plane, giving the extent in direction cosines.

The entrance pupil is delimited by a closed curve described by a relationship $0 = P(\alpha, \beta; x, y)$, and the exit pupil is likewise defined by $0 = P'(\alpha', \beta'; x', y')$. The spatial argument is included to indicate that the shape varies, in general, with the field position. There may be multiple regions, as in the case of central obstructions. It is usually preferable to define the pupils relative to *principal directions* (NS) (α_0, β_0) in object space and (α_0', β_0') in image space, where the two directions are those of the same ray in the two spaces, usually a meridional ray. The principal directions are analogous to the chief rays. The entrance pupil is then given by $0 = Q(\alpha - \alpha_0, \beta - \beta_0; x, y)$ and the exit pupil by $0 = Q'(\alpha' - \alpha_0', \beta' - \beta_0'; x', y')$. For example, for a field point on the $x = 0$ meridian, the expression for the pupil might be well approximated by an ellipse, $0 = a\alpha^2 + b(\beta - \beta_0)^2$, where $(0, \beta_0)$ is the chief ray direction. If the imaging is stigmatic, the relationship between entrance and exit pupil angular shapes is provided by the cosine condition, Eq. (104).

$$Q'(\alpha', \beta'; x', y') = Q(m_P\alpha' - \alpha_0', m_P\beta' - \beta_0'; x, y) \tag{290}$$

The entrance and exit pupils have the same shapes when described in direction cosine space. They are scaled according to the *pupil angular magnification* (NS) $m_P = n/n'm$. The orientations may be the same or rotated 180°. There is no particular relationship between (α_0, β_0) and (α_0', β_0'), which can, for example, be changed by field lenses. The principal directions are, however, usually in the same meridian as the object and image points, in which case $\alpha_0/\beta_0 = \alpha_0'/\beta_0'$. If the field point is in the x meridian, and the central ray is in this meridian, then $\alpha_0 = 0$ and $\alpha_0' = 0$. Even with aberrations, Eq. (290) usually holds to a good approximation. The aberration *pupil distortion* refers to a deviation from this shape constancy.

Pupil Directional Extent: Numerical Aperture and Its Generalizations

The angular extent of a pupil extent is limited by some extreme directions. In the example above of the elliptical shape, for instance, there are two half widths

$$\tfrac{1}{2}(\alpha_{\max} - \alpha_{\min}) \qquad \text{and} \qquad \tfrac{1}{2}(\beta_{\max} - \beta_{\min}) \tag{291}$$

For a rotationally symmetric lens with a circular aperture, the light from an axial object point in a medium of index n is accepted over a cone whose vertex angle is $\theta_{\max}$. The *object space numerical aperture* is defined as

$$NA = n \sin \theta_{\max} = n\sqrt{(\alpha^2 + \beta^2)_{\max}} = n\alpha_{\max} = n\beta_{\max} \tag{292}$$

Likewise, on the image side, where the index is n' and the maximum angle is $\theta_{\max}'$, the *image space numerical aperture* is

$$NA' = n' \sin \theta_{\max}' = n'\sqrt{(\alpha'^2 + \beta'^2)_{\max}} = n'\alpha_{\max}' = n'\beta_{\max}' \tag{293}$$

If the lens is free of coma, the sine condition, Eq. (106), gives for finite conjugates

$$m = \frac{n \sin \theta_{\max}}{n' \sin \theta'_{\max}} = \frac{NA}{NA'} \tag{294}$$

For infinite conjugates

$$\sin \theta'_{\max} = -\frac{y_{\max}}{f'} \quad \text{or} \quad n' \sin \theta'_{\max} = NA' = n\beta_{\max} = -y_{\max}\phi \tag{295}$$

If there is coma, these relationships are still good approximations. For a given lens and a given aperture size, the numerical aperture varies with the axial object position.

F-Number and Its Problems

The *F-number* is written in a variety of ways, including F/no. and F/#. It is denoted here by *FN*. The F-number is not a natural physical quantity, is not defined and used consistently in the literature, and is often used in ways that are both wrong and confusing (Hatch 1980,[232] Goodman 1993[233]). Moreover, there is no need to use the F-number, since everything that it purports to describe or approximately describes is treated properly with direction cosines. The most common definition for F-number, applied to the case of an object at infinity, is

$$FN = \frac{\text{focal length}}{\text{entrance pupil diameter}} = \frac{1}{2 \tan \theta'} \tag{296}$$

where θ' is the outgoing angle of the axial imaging done. In general, the F-number is associated with the tangents of collinear transformations, rather than the sines (or direction cosines) that are physically appropriate. It presumes that a nonparaxial ray entering parallel to the axis at height y leaves the rear principal plane at the same height and intersects the rear focal point, so that $\tan \theta' = y/f'$. However, this particular presumption contradicts Eq. (294), and in general, collineation does not accurately describe lens behavior, as discussed above.

Other problems with F-number, as it is used in the literature, include the following: (1) It is not defined consistently. For example, the literature also contains the definition F-number = (focal length)/(exit pupil diameter). (2) For lenses used at finite conjugates, the F-number is often stated for an object at infinity. In fact, given only the numerical aperture for an object at infinity, that for other conjugates cannot be determined. (3) There are confusing descriptions of variation of F-number with conjugates, for example, the equation $FN_m = (1 + m)FN_\infty$, where FN_m is the F-number for magnification m and FN_∞ is that for an object at infinity. In fact, numerical apertures for various magnification are not so related. (4) The object and image space numerical apertures are related by Eq. (293), but there is no such relationship for tangents of angles, except that predicted by collineation, Eq. (232), which is approximate. (5) With off-axis field points and noncircular pupils, the interpretation of F-number is more ambiguous. (6) Afocal systems have finite numerical apertures when used at finite conjugates, but they have no analogue to Eq. (295). (7) Object and image space refractive indices are not accounted for by the F-number, whereas they are by the numerical aperture. (8) The F-number is often used as a descriptor of radiometric throughput, rather than of ray angles per se.

A related quantity is the *T-number* (W. J. Smith 1992[234]), which accounts for both the convergence angle of the imaging cone and the fraction of power transmitted by the lens.

This is useful as a single-number descriptor, but it is subject to all the confusion associated with the F-number.

Image Irradiance for Lambertian Objects

If the light from a region of an object is lambertian with a power/area M, then the emitted power per angle with angle according to $(M/\pi)\cos\theta\, d\omega = (M/\pi)\, d\alpha\, d\beta$. The power captured by the entrance pupil from a small object area dA is

$$dP = \frac{1}{\pi} M\, dA \int_{\text{entrance pupil}} d\alpha\, d\beta \tag{297}$$

(For a full hemisphere $\int d\alpha\, d\beta = \pi$, giving $dP = M\, dA$.) If there are no losses within the lens, the power reaching the conjugate image region dA' is the same. Using the conservation of étendue equation, Eq. (72), the image irradiance is

$$E = \frac{dP}{dA'} = \frac{1}{\pi} M \frac{n'^2}{n^2} \int_{\text{exit pupil}} d\alpha'\, d\beta' \tag{298}$$

The image irradiance does not depend explicitly on the magnification, but magnification is included implicitly, since, for a given lens, the subtense of the exit pupil varies with conjugates.

This equation obtains everywhere in the field, and it applies to arbitrary object surface positions and orientations, so long as the direction cosines are defined with respect to the local object and image surface normals. These equations apply regardless of the chief ray angles, so they are applicable, for example, with telecentricity. In general, the pupil shape and principal direction vary with field position, so there is a gradation of irradiance in the image of a uniform lambertian object.

These equations do not account for all that influences image irradiance, for example lens absorption and reflection. These effects can be included in the above expressions by adding an appropriate weighting function of angle and field in the above integrals, giving

$$E(x', y') = \frac{dP}{dA'} = \frac{1}{\pi} M(x, y) \frac{n'^2}{n^2} \int \tau(\alpha', \beta'; x', y')\, d\alpha'\, d\beta' \tag{299}$$

where $\tau(\alpha', \beta'; x', y')$ is the lens transmittance as a function of the direction cosines for the image point (x', y'). With externally illuminated objects that are not lambertian scatterers, these relationships do not hold. For example, in optical projectors the illumination is matched to the object and imaging lens to give nominally uniform image irradiance.

Axial Image Irradiance for Lambertian Objects

In the special case of circular pupils and axial object surfaces perpendicular to the axis, the collected power and image irradiance given above are

$$dP = M\, dA \sin^2\theta \qquad \text{and} \qquad E = M \frac{n'^2}{n^2} \sin^2\theta' \tag{300}$$

Power/Pixel

From wave optics, a lens working at the "resolution limit" has an image pixel size $q\lambda/n' \sin\theta'$, where λ is the vacuum wavelength and q is a dimensionless factor, typically of the order of unity. Applying Eq. (300), this gives

$$\text{Power/pixel} = q^2 M\left(\frac{\lambda}{n}\right)^2 \tag{301}$$

$M(\lambda/n)^2$ is the energy emitted per square wavelength of object area. This is a fundamental radiometric quantity. Increasing q gives a greater numerical aperture than is nominally required for resolution, but in practice the aberration correction may be such that the actual resolution is not greater.

Cosine-to-the-Fourth Approximation

For distant, planar, uniform, lambertian objects perpendicular to the lens axis, if the entrance pupil is well approximated by a circle, then the image irradiance varies approximately with the object space field angle ψ according to the *cosine-to-the-fourth* relationship

$$E(\psi) = E_0 \cos^4 \psi \tag{302}$$

where E_0 is the axial irradiance. There are three contributions to this dependence. (1) The angular distribution of a lambertian emitter varies as $\cos\psi$. (2) The distance from the field point to the entrance pupil varies as $1/d^2 \propto \cos^2\psi$. (3) Insofar as the pupil behaves as a rigid circle, its projected solid angle varies approximately as $\cos\psi$. The cosine-to-the-fourth relationship should be used only as a guideline, since ray tracing permits more accurate calculations, and because of the ambiguities in the meaning of the field angle, as discussed above, and elsewhere (Kingslake 1945,[235] Reiss 1945,[236] Gardner 1947,[237] Reiss 1948,[238] Kingslake 1965[239]). For example, field angle is meaningless with telecentricity. Some lenses, especially wide-angle ones, are specifically designed so the pupil subtense increases with the field angle in order to compensate for effects (1) and (2) above, to produce a sufficiently uniform image (Slyusarev 1941[240]).

Total Lens Étendue

The total amount of power from a lambertian object that can be transferred through a lens is

$$\frac{1}{\pi} M \int_{\text{field}} dx\, dy \int_{\text{pupil}} d\alpha\, d\beta \tag{303}$$

The pupil integral may vary over the field. If the pupil is round and constant over the field, the étendue is proportional to $A(\text{NA})^2$, where A is the area of the field. This quantity is also related to the total number of pixels in the field, and the ability of the lens to transfer information (Gabor 1961[241]). The term "area-solid angle product" is sometimes used, but this is an approximation. The total etendue is proportional paraxially to $\sim L^2$, where L is given by Eq. (288).

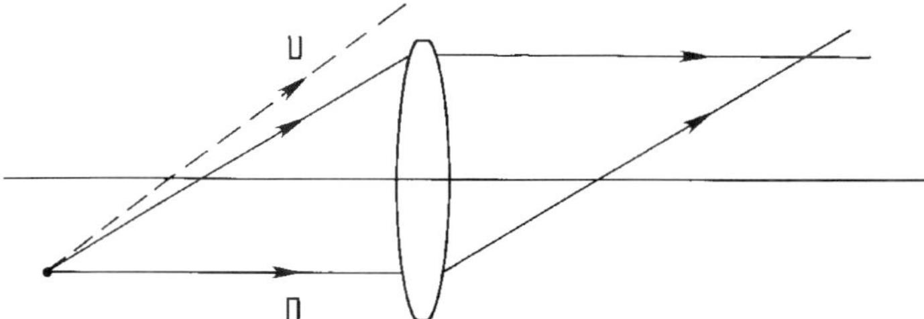

FIGURE 27 Example of vignetting. The dashed ray passes through the aperture, but misses the lens.

Vignetting

Vignetting occurs when an image-forming bundle is truncated by two or more physical structures in different planes, Fig. 27. Typically, one is the nominal aperture and another is the edge of a lens. Another case is that of central obstructions away from the aperture. When vignetting occurs, the image irradiance is changed, and its diminution with field height is faster than it otherwise would be. Aberration properties are also changed, so vignetting is sometimes used to eliminate light that would unacceptably blur the image.

Lens Combinations and Field Lenses

When lenses are used to relay images, the light is transferred without loss only if the exit pupil of one corresponds with the entrance pupil of the next. An example of the failure to meet this requirement is shown in Fig. 28. The axial point is reimaged satisfactorily, but off-axis bundles are vignetted. To transfer the light properly, a *field lens* in the vicinity

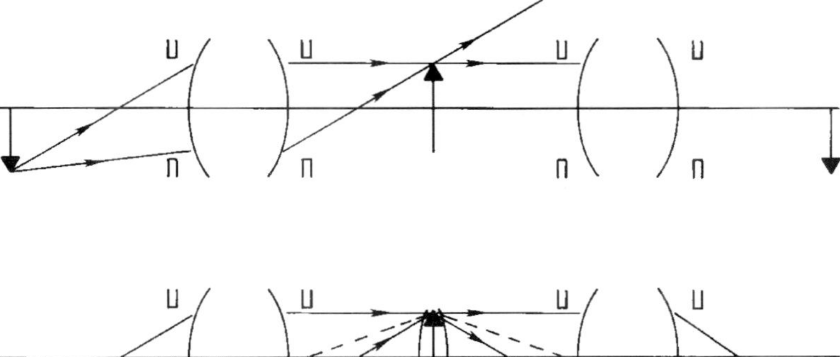

FIGURE 28 A pair of lenses relaying an image with and without a field lens. In the top figure, there is no field lens, and some of the light forming the intermediate image does not pass through the second lens. The amount lost depends on the two numerical apertures and increases with distance from the axis. In the lower figure, a field lens at the intermediate image forms an image of the exit pupil of the first lens into the entrance pupil of the next. No light is lost unless the numerical aperture of the second lens is less than that of the first.

of the intermediate image is used to image the exit pupil of the preceding lens into the entrance pupil of the next one. If the field lens is a thin lens in the image plane, then its magnification with respect to the image is unity. In practice, the field lens is usually shifted axially, so scratches or dust on its surface are out of focus. Its magnification then differs from unity. The focal length of a thin field lens in air is given by $1/f' = 1/a + 1/b$, where a is the distance from exit pupil of first lens to the field lens, and b is that from field lens to the entrance pupil of the second lens. The exit pupil is reimaged with a magnification b/a. If the sizes of the various pupils and their images are not matched, then the aperture of the combination is determined by the smallest. Field lenses affect aberrations.

Defocus

When the object and image-receiving surface are not conjugate there is *defocus*. If either the object or the receiving surface is considered to be correctly positioned, the defocus is associated with the other. Another situation is that in which the object and receiving surfaces are conjugate, but both are wrongly located, so that the image is sharp but the magnification is not what is desired, a condition that might be called *misfocus* (NS).

Defocus has two basic geometric effects, if there are no aberrations, Fig. 29. One is blurring, since the rays from an object point do not converge to a single point on the receiving surface. The blur size varies linearly with the axial defocus in image space and with the cone angle of the image-forming bundle. The shape of the blur is that of the exit pupil, projected on the receiving surface. The other effect of defocus is a lateral shift in position of the blur's centroid relative to that of the correctly focused point. The shift depends on the chief ray angle on the side of the lens where the defocus occurs. In the simplest case, the shift is approximately linear with field height, so acts as a change of magnification. If the object is tilted or is not flat, the effects of defocus vary across the field in a more complicated way. Aberrations affect the nature of the blur. With some aberrations, the blur is different on the two sides of focus. With spherical aberration, the blur changes in quality, and with astigmatism the orientation of the blur changes.

In considering the geometrical imaging of a small region of a lambertian object, there is an implict assumption that the pupil is filled uniformly with light. In imaging an extended object that is externally illuminated, the light from a given region may not fill the pupil uniformly, so the character of the blurring is affected by the angular properties of the illumination and scattering properties of the object.

The amount of defocus can be described in either object or image space, and it can be measured in a variety of ways, for example, axial displacement, displacement along a chief ray, geometrical blur size, and wavefront aberration. The axial displacements in object

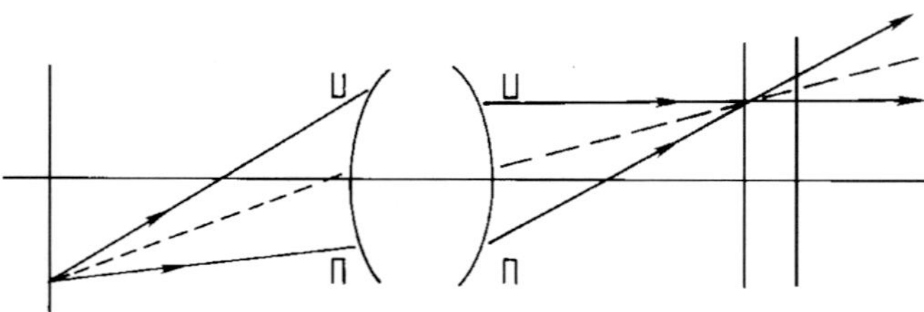

FIGURE 29 Defocus of the receiving surface. A receiving surface is shown in focus and shifted axially. The image of a point on the shifted surface is blurred, and its centroid is translated radially.

and image space differ, in general, and are related by the longitudinal magnification. As expressed in wavefront aberration, i.e., optical path length, defocus is the same in both spaces. There are also various functional measurements of defocus, for example, the sizes of recorded images through focus.

Telecentricity

A lens is *telecentric* if the chief rays are parallel to one another. Most commonly, they are also parallel to the lens axis and perpendicular to the object and/or image planes that are perpendicular to the axis, Fig. 30. Telecentricity is often described by speaking of pupils at infinity, but the consideration of ray angles is more concrete and more directly relevant. A lens is *telecentric in object space* if the chief rays in object space are parallel to the axis, $\alpha_0 = 0$ and $\beta_0 = 0$. In this case the image of the aperture formed by the portion of the lens preceding it is at infinity and the aperture is at the rear focal plane of the portion preceding it. Similarly, a lens is *telecentric in image space* if the aperture is at the front focal point of the subsequent optics, so $\alpha_0' = 0$ and $\beta_0' = 0$. More generally, but less commonly, the chief rays can be parallel to each other, but not necessarily to the axis, and not necessarily perpendicular to a (possibly tilted) object or image plane.

With tilted object and image surfaces and nonaxial pupils, the chief rays are not perpendicular to the object and/or image surfaces, but their angles are everywhere the same, so defocus can result in a rigid shift of the entire image.

A focal lens can be nontelecentric or telecentric on either side, but it cannot be doubly telecentric. An afocal lens can be nontelecentric, or doubly telecentric, but it cannot be telecentric on one side. A doubly telecentric lens must be afocal, and a singly telecentric lens cannot be afocal.

For a lens that is telecentric in image space, if the receiving surface is defocused, the image of a point is blurred, but its centroid is fixed. However, if it is not telecentric in object space, then the scale changes if the object is defocused. The converse holds for object-space telecentricity without image-space telecentricity. For a doubly telecentric lens, an axial shift of either the object or the receiving plane produces blurring without a centroid shift. Although the magnification of an afocal lens does not change with conjugates, there can be an effective change with defocus if it is not telecentric. If the pupil is not on the axis or if the object and image planes are tilted, there can be telecentricity without the chief rays being perpendicular to the object and/or image planes. In these cases, defocus results in a rigid shift of the entire image.

Nominal telecentricity can be negated in several ways. Pupil aberrations may change the chief ray angles across the field. For an extended object that is externally illuminated the pupil may not be filled uniformly by light from a given region, so defocus can product a lateral image shift.

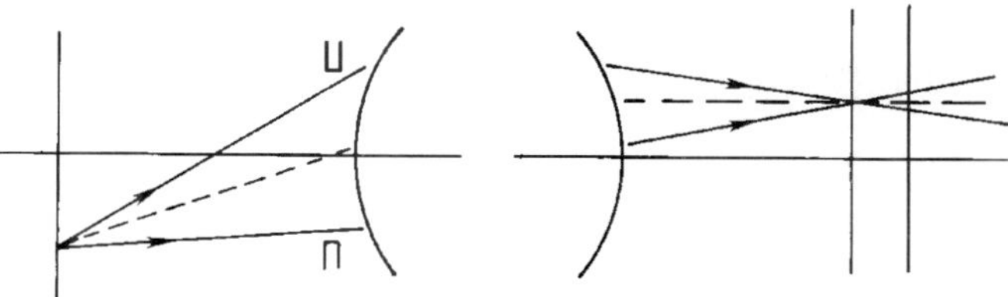

FIGURE 30 Example of telecentricity. The lens shown is telecentric in image space, in which ray bundles are parallel to the axis. An axial shift in the receiving surface results in blurring, but does not translate the centroid, so there is no change in image scale.

Depth of Focus and Depth of Field

The *depth of focus* and *depth of field* are the amounts of defocus that the receiving surface or object may undergo before the recorded image becomes unacceptable. The criterion depends on the application—the nature of the object, the method of image detection, and so on, and there are both ray and wave optics criteria for goodness of focus. For example, a field of separated point objects differs from that of extended objects. Depth of focus is usually discussed in terms of blurring, but there are cases where lateral shifts become unacceptable before blurring. For example, in nature photography blurring is more critical than geometrical deformation, while the opposite may be true in metrology.

Range of Focus and Hyperfocal Distance

In some cases, a geometrical description of defocus is applicable, and the allowable blur is specified as an angle (Ray 1988,[242] Kingslake 1992,[243] W. Smith 1992[234]). The *hyperfocal distance* is

$$\text{Hyperfocal distance} = \frac{\text{diameter of the entrance pupil}}{\text{maximum acceptable angular blur}} = d_H \tag{304}$$

Let the object distance at which the lens is focused be d, the nearest distance at which the image is acceptable be d_N, and the furthest distance be d_F. All of these quantities are positive definite. The following relations obtain:

$$d_F = \frac{d_H d}{d_H - d} \qquad \text{and} \qquad d_N = \frac{d_H d}{d_H + d} \tag{305}$$

The distances to either side of best focus are

$$d_F - d = \frac{d^2}{d_H - d} \qquad \text{and} \qquad d - d_N = \frac{d^2}{d_H + d} \tag{306}$$

The total range of focus is

$$d_F - d_N = \frac{2d^2 d_H}{d_H^2 - d^2} = \frac{2d}{(d_H/d)^2 - 1} \tag{307}$$

For $d > d_H$ the above quantities involving d_F are infinite (not negative). If the lens is focused at the hyperfocal distance or beyond, then everything more distant is adequately focused. If the lens is focused at the hyperfocal distance, i.e., $d = d_H$, the focus is adequate everywhere beyond half this distance, and this setting gives the greatest total range. If the lens is focused at infinity, then objects beyond hyperfocal distance are adequately focused. The hyperfocal distance decreases as the lens is stopped down.

1.19 GEOMETRIC ABERRATIONS OF POINT IMAGES—DESCRIPTION

Introduction

In instrumental optics, the term *aberration* refers to a departure from what is desired, whether or not it is physically possible. Terms such as "perfect system" and "ideal system" indicate what the actual is compared to, and these terms themselves are not absolute, but depend on what is wished for. The ideal may be intrinsically impossible, in which case a deviation therefrom is not a defect. A further distinction is between aberrations inherent in a design and those that result from shortcomings in fabrication.

This section considers only the description of aberrations of point images, with the lens treated as a black box, whose action with respect to aberrations is accounted for by what leaves the exit pupil. A full consideration of aberrations involves, among other things, their causes, their correction, their various manifestations, and their evaluation. Aberrated images of extended objects are formed by overlapping blurs from the individual points. The analysis of such images is object- and application-dependent, and is beyond the scope of this section. Aberrations do vary with wavelength, but most of this discussion involves monochromatic aberrations, those at a single wavelength. In addition, aberrations vary with magnification. Aberrations are discussed to some extent in many books that treat geometric optics (Conrady 1929,[244] H. Hopkins 1950,[245] Buchdahl 1954,[246] Herzberger 1958,[247] Kingslake 1978,[248] Born & Wolf 1980,[249] Slyusarev 1984,[250] Welford 1986,[251] Welford 1986,[252] W. Smith 1992[253]).

Descriptions

Aberration has many manifestations, and can be described in a variety of ways. For example, geometric wavefronts, path lengths, ray angles, and ray intersection points can all differ from the nominal (and in wave optics there are additional manifestations). Terms such as "wavefront aberration" and "ray aberration" do not refer to fundamentally different things, but to different aspects of the same thing. Often, a single manifestation of the aberration is considered, according to what is measurable, what best describes the degradation in a particular application, or what a lens designer prefers to use for optimization during the design process.

Classification

Aberrations are classified and categorized in a variety of ways. These include pupil dependence, field dependence, order, evenness and oddness, pupil and field symmetry, and the nature of change through focus—symmetrical and unsymmetrical. In addition, there are natural groupings, e.g., astigmatism and field curvature. The classification systems overlap, and the decompositions are not unique. The complete aberration is often described as a series of terms, several schemes being used, as discussed below. The names of aberrations, such as "spherical," "coma," and "astigmatism," are not standardized, and a given name may have different meanings with respect to different expansions. Furthermore, the effects of aberrations are not simply separated. For example, "pure coma" can have effects usually associated with distortion. Defocus is sometimes taken to be a type of aberration, and it is useful to think of it in this way, since it is represented by a term in the same expansion and since the effects of aberrations vary with focus. The number of terms in an expansion is infinite, and familiar names are sometimes associated with unfamiliar terms. To improve clarity, it is recommended that all the terms in an expansion be made explicit up to agreed-upon values too small to matter, and that, in addition, the net effect be shown graphically. Further, it is often helpful to show more than one of an aberration's manifestations.

Pupil and Field Coordinates

In this section, all the quantities in the equation are in image space, so primes are omitted. Field coordinates are x and y, with $h^2 = x^2 + y^2$, and (x, y) is the nominal image point in a plane $z = 0$. Direction cosines equally spaced on the exit pupil should be used for pupil coordinates but, in practice, different types of coordinates are used, including linear positions, spatial frequencies, and direction cosines. Here the pupil coordinates are ξ and

η, which are dimensionless, with $\rho^2 = \xi^2 + \eta^2$. The overall direction of the pupil may vary with field. Here the $(\xi, \eta) = (0, 0)$ is always taken at the pupil center, the meaning of which may not be simple, as discussed in the section on pupils above. The angle of a meridian in the pupil is ψ. Entrance and exit pupil coordinates must be distinguished. For diffraction calculations, the exit pupil should be sampled at equal intervals in direction cosines, but a set of rays from an object point that is equally spaced in direction cosines may leave with uneven spacing, as a result of aberrations.

Wavefront Aberration

If an object point is imaged stigmatically, then the optical path lengths of all rays from the object point to its image are identical, and the geometric wavefronts leaving the exit pupil are spherical. In the presence of aberrations, the wavefront is no longer spherical. Rather than describing the wavefront shape, it is usually preferable to consider the difference between the actual wavefront, and a nominal wavefront, often called the *reference sphere,* centered at a *reference point* that is usually the nominal image point. This reference sphere is usually taken to intersect the center of the pupil, since this gives the most accurate diffraction calculations. The *wavefront aberration* W is the optical path length from reference sphere to wavefront, or vice versa, according to the convention used, Fig. 31. Two sign conventions are in use; a positive wavefront aberration may correspond either to a wavefront which lags or leads the reference sphere. For each nominal image point (x, y, z), the wavefront aberration is a function of the pupil coordinates (ξ, η), so the functional form is $W(\xi, \eta; x, y, z)$, with the z usually suppressed, since the image plane is usually taken to be fixed. For a given lens prescription, W is found by tracing a set of rays from each object point to the reference sphere and calculating their path lengths. If the absolute path length is unimportant, the choice of the reference sphere's radius is not critical. Considered from the point of view of wave optics, the image of a point is degraded by phase differences across the reference sphere, so absolute phase is of no consequence, and the zero of the wavefront aberration can be chosen arbitrarily. By convention and convenience, the zero is usually taken at the center of the pupil, so $W(0, 0, x, y) = 0$.

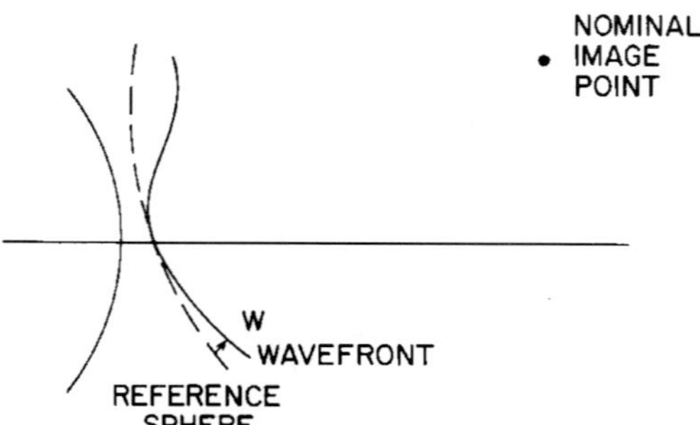

FIGURE 31 Wavefront aberration. The reference sphere is concentric with the nominal image point. The wavefront is taken that is tangent to the reference sphere in the center of the pupil. The wavefront aberration function is the distance from the reference sphere to the wavefront as a function of pupil coordiantes.

Absolute optical path lengths are significant for imaging systems with paths that separate between object and image in cases where there is coherence between the various image contributions. An error in absolute optical path length is called *piston error*. This results in no ray aberrations, so it is omitted from some discussions.

Ray Aberrations

In the presence of aberrations, the rays intersect any surface at different points than they would otherwise. The intersection of the rays with the receiving surface, usually a plane perpendicular to the axis, is most often of interest. The *transverse ray aberration* is the vectorial displacement $(\varepsilon_x, \varepsilon_y)$ between a nominal intersection point and the actual one. The displacement is a function of the position of the nominal image point (x, y) and the position in the pupil through which the ray passes (ξ, η). A complete description of transverse ray aberrations is given by

$$\varepsilon_x(\xi, \eta; x, y) \qquad \text{and} \qquad \varepsilon_y(\xi, \eta; x, y) \tag{308}$$

The *longitudinal aberration* is the axial displacement from nominal of an axial intersection point. This description is useful for points on the axis of rotationally symmetrical systems, in which case all rays intersect the axis. Such aberrations have both transverse and longitudinal aspects. The intersection with a meridian can also be used. The *diapoint* is the point where a ray intersects the same meridian as that containing the object point (Herzberger 1958[247]). For an image nominally located at infinity, aberrations can be described by the slope of the wavefront relative to that of the nominal, that is, by ray angles rather than intersection points. A hypothetical ideal focusing lens can also be imagined to convert to transverse aberrations.

A *ray intercept diagram* shows the intersection points of a group of rays with the receiving surface (O'Shea 1994[254]). The rays are usually taken to arise from a single object point and to uniformly sample the pupil, with square or hexagonal arrays commonly used. The ray intercept diagrams can suffer from artifacts of the sampling array, which can be checked for by using more than one type of array. Other pupil loci, for instance, principal meridians and annuli, can be employed to show particular aspects of the aberration. Intercept diagrams can also be produced for a series of surfaces through focus. Image quality may be better than ray diagrams suggest, since destructive interference can reduce the irradiance in a region relative to that predicted by the ray density.

Relationship of Wavefront and Ray Aberrations

Since rays are normal to geometric wavefronts, Fig. 32, transverse ray aberrations are proportional to the slope of the wavefront aberration function. For systems of rotation

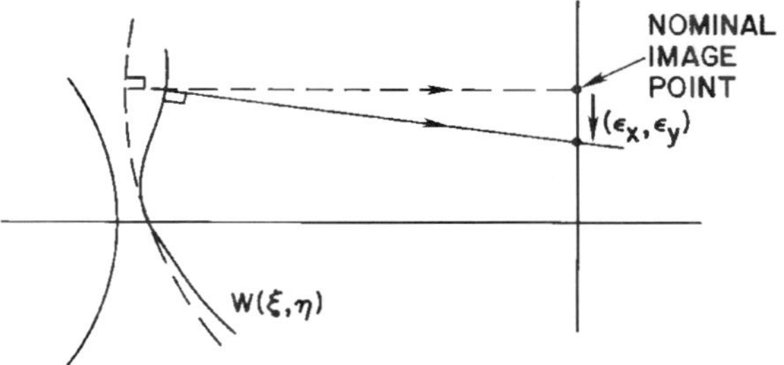

FIGURE 32 Ray aberration. Rays intersect the receiving plane at positions shifted from the nominal.

with image space index n and marginal ray angle θ, the transverse aberrations are to a good approximation (Welford 1986[251])

$$\varepsilon_x = \frac{1}{n \sin \theta} \frac{\partial W}{\partial \xi} \qquad \varepsilon_y = \frac{1}{n \sin \theta} \frac{\partial W}{\partial \eta} \tag{309}$$

The refractive index appears since W is an optical path length. If the rays are nominally parallel, then the partial derivatives give the angular ray errors

$$\Delta\alpha = \frac{1}{np} \frac{\partial W}{\partial \xi} \qquad \Delta\beta = \frac{1}{np} \frac{\partial W}{\partial \eta} \tag{310}$$

where p is the linear radius of the exit pupil, which cannot be infinite if the image is at infinity. These expressions may also have a multiplicative factor of -1, depending on the sign conventions. A sum of wavefront aberrations gives a transverse aberration that is the sum of the contributing ones.

Ray Densities

The density of rays near the nominal image point is (Welford 1986[251])

$$\frac{1}{\text{Density}} \propto \left(\frac{\partial^2 W}{\partial \xi^2}\right)\left(\frac{\partial^2 W}{\partial \eta^2}\right) - 2\left(\frac{\partial^2 W}{\partial \xi\, \partial \eta}\right)^2 \tag{311}$$

Caustics are the surfaces where ray densities are infinite. Here, geometric optics predicts infinite power/area, so the ray model is quantitatively inaccurate in this case.

Change of Reference Points

The center of the reference sphere may be displaced from the nominal image point. If the reference point is changed by linear displacement $(\delta x, \delta y, \delta z)$, then the wavefront aberration function changes from W to W' according to

$$W'(\xi, \eta; x, y; \delta x, \delta y, \delta z) = W(\xi, \eta; x, y) + W_x \xi + W_y \eta + W_z(\xi^2 + \eta^2) \tag{312}$$

where $W_x = n \sin \theta\, \delta x$,

$$\begin{aligned} W_y &= n \sin \theta\, \delta y \\ W_z &= \tfrac{1}{2} n \sin^2 \theta\, \delta z \end{aligned} \tag{313}$$

The transverse ray aberration ε'_x and ε'_y with respect to the new reference points are

$$\varepsilon'_x = \varepsilon_x + \delta x + \sin \theta\, \delta z \qquad \varepsilon'_y = \varepsilon_y + \delta y + \sin \theta\, \delta z \tag{314}$$

The change through focus is accounted for by varying δz. Setting $\varepsilon'_x = \varepsilon'_y = 0$ gives the parametric equations $x(\delta z)$ and $y(\delta z)$ for a ray with pupil coordinates (ξ, η), relative to the nominal ray near the nominal image point.

Aberration Symmetries for Systems with Rotational Symmetry

If the lens, including the aperture, is a figure of rotation, only certain aberration forms are possible. For object points on axis, the wavefront aberration and the image blur are figures of revolution. For off-axis points, both wavefront aberration and blur are bilaterally symmetrical about the meridional plane containing the object point. For object points on a circle centered on the axis, the wavefront and ray aberrations are independent of azimuth, relative to the local meridian. In practice, there is always some imperfection, so the symmetries are imperfect and additional aberration forms arise.

Wavefront Aberration Forms for Systems with Rotational Symmetry

Here the pupil is taken to be circular, with the coordinate zero taken at the center. The field coordinates are normalized so $x^2 + y^2 = h^2 = 1$ at the edge of the field. The pupil coordinates are normalized, so that $\xi^2 + \eta^2 = \rho^2 = 1$ on the rim of the pupil. The algebra is simplified by using dimensionless coordinates. To add dimensions and actual sizes, replace the ξ by ξ/ξ_{max}, and likewise for other variables. The simplest combinations of pupil and field coordinates with rotational symmetry are

$$x^2 + y^2 = h^2 \qquad \xi^2 + \eta^2 = \rho^2 \qquad \xi x + \eta y \tag{315}$$

The general wavefront aberration function can be expressed as a series of such terms raised to integral powers,

$$W(x, y; \xi, \eta) = \sum_{L,M,N=0} W_{LMN}(x^2 + y^2)^L(\xi^2 + \eta^2)^M(x\xi + y\eta)^N \tag{316}$$

where L, M, N are positive integers. The terms can be grouped in *orders* according to the sum $L + M + N$, where, by convention, the order equals $2(L + M + N) - 1$. The order number refers more directly to ray aberration forms than to wavefront forms, and it is always odd. The first-order terms are those for which $L + M + N = 1$, for the third-order terms the sum is two, and so on. The number of terms in the Qth order is $1 + (Q + 1)(Q + 7)/8$. For orders 1, 3, 5, 7, 9 the number of terms is 3, 6, 10, 15, 21. For each order, one contribution is a piston error, which is sometimes excluded from the count.

The expression of Eq. (316) is related to the characteristic function for a rotationally symmetrical system, Eq. (32). If the spatial coordinates are taken to be those of the object point, this is the point-angle characteristic function. In the hamiltonian optics viewpoint, the characteristic function is a sum of two parts. The first-order terms specify the nominal properties, and those of higher orders the deviation therefrom. This is discussed in the references given in that section. The term for which $L = M = N = 0$ has to do with absolute optical path length.

Since there is bilateral symmetry about all meridians, the expansion can be simplified by considering object points in a single meridian, customarily taken to be that for which $x = 0$. Doing so and letting the fractional field height be $y = h$ gives the wavefront aberration function

$$W(h; \rho, \eta) = \sum_{L,M,N=0} W_{LMN}h^{2L+N}\rho^{2M}\eta^N = \sum_{A,B,C} W'_{ABC}h^A\rho^B\eta^C \tag{317}$$

where $A = 2L + N$, $B = 2M$, $C = N$, and the order equals $(A + B + C) - 1$. Another form is obtained with the fractional pupil radius ρ and the pupil azimuth ψ, the angle from the $x = 0$ meridian, so $\eta = \rho \cos \psi$. With these pupil variables the wavefront aberration function is

$$W(h; \rho, \psi) = \sum_{L,M,N=0} W_{LMN} h^{2L+N} \rho^{2M+N} \cos^N \psi = \sum_{A,B,C} W''_{ABC} h^A \rho^B \cos^C \psi \tag{318}$$

where $A = 2L + N$, $B = 2M + N$, $C = N$, and the order is $A + B - 1$. For orders above the first, the W_{LMN}, W'_{ABC}, and W''_{ABC} are the *wavefront aberration coefficients.*

For a given field position, the wavefront aberration function for circular pupils can also be decomposed into the *Zernike polynomials,* also called *circle polynomials,* a set of functions complete and orthonormal on a circle (Zernike 1934,[255] Kim & Shannon 1987,[256] Malacara 1978,[257] Born & Wolf 1980[249]).

Third-Order Aberrations and Their Near Relatives

There are six third-order terms. The *Seidel aberrations* are spherical, coma, astigmatism, field curvature, distortion, and there is also a piston-error term. Expressions for these aberrations are given below, along with some higher-order ones that fall in the same classification. The terminology of higher-order aberrations is not standardized, and there are forms that do not have third-order analogues. This section uses the notation of the second expression of Eq. (318), without the primes on the coefficients.

It is useful to include *defocus* as a term in aberration expansions. Its wavefront aberration and transverse ray aberrations are

$$W = W_{020}\rho^2 \qquad \varepsilon_x \propto 2W_{020}\xi \qquad \varepsilon_y \propto 2W_{020}\eta \tag{319}$$

Coefficient W_{020} is identical to W_z, Eq. (313).

In *spherical aberration* the wavefront error is a figure of revolution in the pupil. The individual terms of the expansion have the form ρ^{2N}. The form that appears on axis, and which is independent of field position is

$$W = W_{020}\rho^2 + W_{040}\rho^4 + W_{060}\rho^6 + \cdots \tag{320}$$

where defocus has been included. The W_{040} term is the third-order term, the W_{060} is the fifth-order term, etc. The ray aberrations are

$$\begin{aligned} \varepsilon_x &\propto 2W_{020}\xi + 4W_{040}\rho^2\xi + 6W_{060}\rho^4\xi + \cdots \\ \varepsilon_y &\propto 2W_{020}\eta + 4W_{040}\rho^2\eta + 6W_{060}\rho^4\eta + \cdots \end{aligned} \tag{321}$$

There are also higher-order off-axis terms, called *oblique spherical aberration,* with forms $h^{2L}\rho^{2M}$. Spherical is an even aberration.

In *coma,* the wavefront aberration varies linearly with field height, so the general form is $h\rho^{2M}\eta = h\rho^{2M+1} \cos \psi$. Coma is an odd aberration. The wavefront expansion is

$$W = (W_{131}\rho^2 + W_{151}\rho^4 + \cdots)\eta h = (W_{131}\rho^3 + W_{151}\rho^5 + \cdots) \cos \psi h \tag{322}$$

The ray aberrations are

$$\begin{aligned} \varepsilon_x &\propto [W_{131}(2\xi\eta) + 4W_{151}(\xi^2 + \eta^2)\xi\eta + \cdots]h \\ \varepsilon_y &\propto [W_{131}(\xi^2 + 3\eta^2) + W_{151}(\xi^4 + 5\xi^2\eta^2 + 6\eta^4) + \cdots]h \end{aligned} \tag{323}$$

In *astigmatism* the wavefront aberration is cylindrical. The third-order term is

$$W = W_{222}h^2\eta^2 = W_{222}h^2\rho^2 \cos^2 \psi \tag{324}$$

with ray aberration

$$\varepsilon_x = 0 \qquad \varepsilon_y \propto 2W_{222}h^2\eta \tag{325}$$

Field curvature, also known as *Petzval curvature,* is a variation of focal position in the axial direction with field height. In its presence, the best image of a planar object lies on a nonplanar surface. Its absence is called *field flatness.* The wavefront aberration form is

$$W = (W_{220}h^2 + W_{420}h^4 + W_{620}h^6 + \cdots)\rho^2 \tag{326}$$

with symmetrical blurs given by

$$\begin{aligned}\varepsilon_x &\propto (W_{220}h^2 + W_{420}h^4 + W_{620}h^6 + \cdots)\xi \\ \varepsilon_y &\propto (W_{220}h^2 + W_{420}h^4 + W_{620}h^6 + \cdots)\eta\end{aligned} \tag{327}$$

The curvature of the best focus surface may have the same sign across the field, or there may be curvatures of both signs.

Astigmatism and field curvature are often grouped together. Combining defocus, third-order astigmatism, and third-order field curvature, the wavefront aberration can be written

$$W = W_{020}(\xi^2 + \eta^2) + [W_{220}\xi^2 + (W_{220} + W_{222})\eta^2]h^2 \tag{328}$$

The resultant ray aberration is

$$\varepsilon_x \propto [W_{020} + W_{220}h^2]\xi \qquad \varepsilon_y \propto [W_{020} + (W_{222} + W_{220})h^2]\eta \tag{329}$$

A tangential fan of rays, one that lies in the $x = 0$ meridian, has $\xi = 0$, so $\varepsilon_x = 0$. The *tangential focus* occurs where $\varepsilon_y = 0$, which occurs for a defocus of $W_{020} = -(W_{220} + W_{222})h^2$. Combining this result with Eq. (314) gives $\delta z \propto h^2$, the equation for the tangential focal surface. A sagittal fan of rays crosses the pupil in the $\eta = 0$ meridian, so $\varepsilon_y = 0$. The *sagittal focus* occurs where $\varepsilon_x = 0$, i.e., on the surface given by $W_{020} = -W_{220}h^2$.

In general, *distortion* is a deviation from geometric similarity between object and image. For rotationally symmetrical lenses and object and image planes perpendicular to the axis, the error is purely radial, and can be thought of as a variation of magnification with field height. The aberration forms are

$$W = (W_{111}h + W_{311}h^3 + W_{511}h^5 + \cdots)\eta \tag{330}$$

with

$$\varepsilon_x = 0 \qquad \varepsilon_y \propto W_{111}h + W_{311}h^3 + W_{511}h^5 + \cdots \tag{331}$$

In *pincushion* distortion the magnitude of magnification increases monotonically with field height, so the image is stretched radially. In *barrel distortion* the magnitude decreases, so the image is squeezed. In general, the aberration coefficients can be both positive and negative, so the direction of distortion can change as a function of field height and the distortion may vanish for one or more field heights.

For *piston error* the wavefront differs uniformly across the pupil from its nominal in a way that varies with field height.

$$W = W_{000} + W_{200}h^2 + W_{400}h^4 + W_{600}h^6 + \cdots \qquad \varepsilon_x = \varepsilon_y = 0 \tag{332}$$

There are no transverse ray aberrations.

Chromatic Aberrations

In general, the properties of optical systems vary with wavelength. The term *chromatic aberration* often refers to the variation in paraxial properties as a function of wavelength. Thus, *axial color* is related to differences of focal length and principal plane location with wavelength, and *lateral color* is related to variations of magnification with wavelength. Also, the monochromatic aberrations vary in magnitude with wavelength. Reflective systems have identical ray properties at all wavelengths, but their wave properties vary with color, since a given variation in path length has an effect on phase that varies with wavelength.

Stop Size and Aberration Variation

For a given system, if the size of the aperture is changed, the marginal ray is changed, but not the chief ray. If the aperture is reduced, depth of focus and depth of field increase and image irradiance decreases. The rays from axial object points are more nearly paraxial, so the imaging tends to be better corrected. For off-axis points, some aberrations are changed and others are not. Distortion, as defined with respect to the chief ray, is not changed. Field curvature per se does not change, since the aperture size does not change the location of the best image surface (if there are no other aberrations), but the depth of focus does change, so a flat detector can cover a larger field.

Stop Position and Aberration Variation

For a given system, if the aperture is moved axially, the image-forming bundle passes through different portions of the lens elements. Accordingly, some aberrations vary with the position of the stop. Lens design involves an operation called the *stop shift,* in which the aperture is moved axially while its size is adjusted to keep the numerical apertures constant. In this operation, the marginal ray is fixed, while the chief ray is changed. This does not change the aberrations on axis. Most of those for off-axis points are changed, but third-order field curvature is unchanged.

REFERENCES

1. R. K. Luneburg, *Mathematical Theory of Optics,* University of California Press, Berkeley, 1964 chap. 1.
2. M. Kline and I. W. Kay, *Electromagnetic Theory and Geometrical Optics,* Interscience Publishers, New York, 1965.
3. M. Born and E. Wolf, *Principles of Optics,* Pergamon, New York, 1980.
4. D. Marcuse, *Light Transmission Optics,* Van Nostrand Reinhold, New York, 1989 chap. 3.6.
5. L. Rayleigh, "Optics," *Encyclopedia Britannica* vol. XVII, 1884, and *Scientific Papers,* Dover, New York, 1964, sec. 119, p. 385.
6. W. T. Welford, *Aberrations of Optical Systems,* Hilger, Bristol, 1986, chap. 6.

7. M. Herzberger, "On the Fundamental Optical Invariant, the Optical Tetrality Principle, and on the New Development of Gaussian Optics Based on This Law," *J. Opt. Soc. Am.,* vol. 25, 1935, pp. 295–304.
8. O. N. Stavroudis, *The Optics of Rays, Wavefronts, and Caustics,* Academic, New York, 1972, chap. 13, "The Fundamental Optical Invariant."
9. G. Forbes and B. D. Stone, "Restricted Characteristic Functions for General Optical Configurations," *J. Opt. Soc. Am.,* vol. 10, 1993, pp. 1263–1269.
10. P. Fermat, "Fermat's principle," *A Source Book in Physics,* W. F. Magie, Harvard, Cambridge, Mass., 1963, p. 278.
11. P. de Fermat, *Oeuvres,* vol. II, P. Tomnery et C. Henry (eds.), Paris, 1891.
12. P. de Fermat, *Oeuvres de Fermat,* C. Henry, P. Tannery (eds.), vols. i–iv and supplement Gauthier-Villars, Paris, 1854–1922 vols. I–IV and supplement vol. III, French translations of Latin papers.
13. R. P. Feynman, R. B. Leighton, and M. Sands, *The Feynman Lectures on Physics,* Addison Wesley, Reading, Mass., 1963 vol 1, chap. 6, "Optics: The Principle of Least Time."
14. B. B. Rossi, *Optics,* Addison-Wesley, Reading, Mass., 1957.
15. E. Hecht, *Optics,* Addison-Wesley, Reading, Mass., 1987.
16. M. V. Klein and T. E. Furtak, *Optics,* 2d ed. Wiley, New York, 1986.
17. H. H. Hopkins, "An Extension of Fermat's Theorem," *Optica Acta,* 1970, vol. 17, pp. 223–225.
18. A. Walther, "Systematic Approach to the Teaching of Lens Theory," *Am. J. Phys.,* 1967, vol. 35, pp. 808–816.
19. A. Walther, "Lenses, Wave Optics, and Eikonal Functions," *J. Opt. Soc. Am.,* 1969, vol. 59, pp. 1325–1333.
20. L. Landau and E. Lifshitz, *Classical Theory of Fields,* Addison-Wesley, Reading, Mass., 1951, sec. 7.1, "Geometrical Optics."
21. J. A. Kneisly, "Local Curvature of Wavefronts in an Optical System," *J. Opt. Soc. Am.,* 1965, vol. 54, pp. 229–235.
22. W. R. Hamilton, *Geometrical Optics. Mathematical Papers of Sir William Rowan Hamilton,* vol. 1, A. W. Conway and J. L. Synge (eds.), Cambridge University Press, 1931.
23. H. Bruns, "Das Eikonal," Abh. Königl. sächs. Abhand. Ges. Wiss., Leipzig, Math Phys K1, 1895, vol. 21, pp. 325–436.
24. K. Schwarzschild, *Untersuchungen zür geometrischen Optik,* I–II, Abh. Königl. Ges. Wiss. Goettingen, Math Phys. K1, Neue Folge 4, 1905, pp 1–54.
25. Czapski-Eppenstein, *Grundzuge der Theorie der Optische Instrumente,* Leipzig, 1924.
26. C. G. Steward, *The Symmetrical Optical System,* Cambridge University Press., 1928.
27. M. Herzberger, *Strahlenoptik,* Springer, Berlin, 1931.
28. J. L. Synge, *Geometrical Optics. An Introduction to Hamilton's Method,* Cambridge University Press, 1937.
29. C. Caratheodory, *Geometrische Optik,* Springer-Verlag, Berlin, 1937.
30. L. Rayleigh, "Hamilton's Principle and the Five Aberrations of Von Seidel," *Phil. Mag.,* vol. XV, 1908, pp. 677–687, or *Scientific Papers,* Dover, New York, 1964.
31. R. J. Pegis, "The Modern Development of Hamiltonian Optics" *Progress in Optics,* E. Wolf (ed.), vol. 1, North Holland, 1961, pp. 1–29.
32. R. K. Luneburg, *Mathematical Theory of Optics,* University of California Press, Berkeley, 1964.
33. W. Brouwer and A. Walther, "Geometrical Optics," *Advanced Optical Techniques,* A. C. S. Van Heel (ed.), North-Holland, Amsterdam, 1967, pp. 503–570.
34. H. A. Buchdahl, *An Introduction to Hamiltonian Optics,* Cambridge U.P., 1970.
35. M. Born and E. Wolf, *Principles of Optics,* Pergamon, New York, 1980 chap. 4.
36. M. Herzberger, "Geometrical Optics," *Handbook of Physics,* E. U. Condon and H. Odishaw (ed.), McGraw-Hill, 1958, chap. 6-2.
37. T. Smith, "On Perfect Optical Instruments," *Proc. Roy. Soc.,* vol. 55, 1945, pp. 293–304.

38. M. Herzberger, "On the Characteristic Function of Hamilton, the Eiconal of Bruns, and Their Use in Optics," *J. Opt. Soc. Am.*, vol. 26, 1936, pp. 177–180.
39. M. Herzberger, "Hamilton's Characteristic Function and Bruns Eiconal," *J. Opt. Soc. Am.*, vol. 27, 1937, pp. 133–317.
40. J. L. Synge, "Hamilton's Method in Geometrical Optics," *J. Opt. Soc. Am.*, vol. 27, 1937, pp. 75–82.
41. C. H. F. Velzel, "Breaking the Boundaries of Optical System Design and Construction," *Trends in Modern Optics,* J. W. Goodman (ed.), Academic, Boston, 1991, pp. 325–338.
42. R. S. Heath, *A Treatise on Geometrical Optics,* Cambridge University Press, 1895, chap. 13, "Refraction through Media of Varying Density."
43. R. A. Herman, *A Treatise on Geometrical Optics,* Cambridge University Press, 1900, chap. 13 "Heterogeneous Media."
44. J. L. Synge, *Geometrical Optics. An Introduction to Hamilton's Method,* Cambridge University Press, 1937, chap. 5, "Heterogeneous Isotropic Media."
45. R. K. Luneburg, *Mathematical Theory of Optics,* University of California Press, Berkeley, 1964, chaps. 2, 3.
46. O. N. Stavroudis, *The Optics of Rays, Wavefronts, and Caustics,* Academic, New York, 1972, chap. 11 "The Inhomogeneous Medium."
47. A. K. Ghatak and K. Thyagarajan, *Contemporary Optics,* Plenum, New York, 1978.
48. M. Born and E. Wolf, *Principles of Optics,* Pergamon, New York, 1980, chap. 3.2.
49. D. Marcuse, *Light Transmission Optics,* Van Nostrand Reinhold, New York, 1989, chap. 3.
50. E. W. Marchand, "Gradient Index Lenses," *Progress in Optics,* vol. 11, North Holland, Amsterdam, 1973, pp. 305–337.
51. E. W. Marchand, *Gradient Index Optics,* Academic, New York, 1978.
52. A. Sharma, D. V. Kumar, and A. K. Ghatak, "Tracing Rays Through Graded-Index Media: A New Method," *Appl. Opt.* vol. 21, 1982, pp. 984–987.
53. D. T. Moore, ed., *Selected Papers on Gradient-Index Optics,* SPIE, 1992.
54. D. Moore, "Gradient Index Optics," *Handbook of Optics,* vol. II, chap. 9, McGraw-Hill, New York, 2d ed., 1994.
55. J. Brown, *Microwave Lenses,* Methuen, London, 1953.
56. S. Cornbleet, *Microwave Optics,* Academic, New York, 1976, chap. 2, "Non-Uniform Media."
57. S. Cornbleet, "Geometrical Optics Reviewed: A New Light on an Old Subject," *Proc. IEEE,* vol. 71, 1983, pp. 471–502.
58. S. Cornbleet, *Microwave and Optical Ray Geometry,* Wiley, New York, 1984.
59. W. Blaschke, *Vorlesungen über Differentialgeometrie,* Springer, 1930, Dover, 1945.
60. E. Kreyszig, *Differential Geometry,* Univ. Toronto, 1959, Dover, New York, 1991.
61. J. J. Stoker, *Differential Geometry,* Wiley, New York, 1969.
62. D. J. Struik, *Lectures in Classical Differential Geometry,* Addison-Wesley, 1961, Dover, 1990.
63. H. Goldstein, *Classical Mechanics* Addison-Wesley, Reading, Mass., 1980, 2d edition. Variational principles in physics are discussed in many books on classical mechanics, a good reference list is supplied by Goldstein.
64. J. A. Arnaud, "Analogy Between Optical Rays and Nonrelativistic Particle Trajectory: A Comment," *Am. J. Phys.*, vol. 44, 1976, pp. 1067–1069.
65. J. Evans, M. Rosenquist, "'$F = ma$' Optics," *Am. J. Phys.*, vol. 54, 1986, pp. 876–883.
66. D. T. Moore, "Ray Tracing in Gradient-Index Media," *J. Opt. Soc. Am.*, vol. 65, 1975, pp. 451–455.

67. G. W. Forbes, "On Variational Problems in Parametric Form," *Am. J. Phys.*, vol. 59, 1991, pp. 1130–1140.

68. L. Landau and E. Lifshitz, *Classical Theory of Fields,* Addison-Wesley, Reading, Mass., 1951, sec. 7.1, "Geometrical Optics."

69. H. A. Buchdahl, "Rays in Gradient Index Media: Separable Systems," *J. Opt. Soc. Am.*, vol. 63, 1973, pp. 46–49.

70. S. Cornbleet, "Ray Paths in Nonuniform Axially Symmetric Medium," *IEE Journal of Microwaves Optics and Acoustics,* vol. 2, 1978, pp. 194–200.

71. S. Cornbleet, "Ray Tracing in a General Inhomogeneous Cylindrical Medium," *IEE Journal of Microwaves Optics and Acoustics,* vol. 3, 1979, pp. 181–185.

72. J. W. Blaker and M. A. Tavel, "The Application of Noether's Theorem to Optical Systems," *Amer. Jour. Phys.*, vol. 42, 1974, pp. 857–861.

73. W. B. Joyce, "Comments on 'The Application of Noether's Theorem to Optical Systems,'" *Amer. Jour. Phys.*, vol. 43, 1975, p. 455.

74. J. C. Maxwell, "Solutions of Problems," problem no. 2, *Cambridge and Dublin Math. Journal,* vol. 8, 1854, p. 188, and *The Scientific Papers of James Clerk Maxwell,* Cambridge University Press, 1890, Dover, New York, 1965, pp. 74–79.

75. S. P. Morgan, "General Solution of the Luneberg Lens Problem," *J. Appl. Phys.*, vol. 29, 1958, pp. 1358–1368.

76. W. H. Steel, "Luminosity, Throughput, or Etendue?" *Appl. Opt.*, vol. 13, 1974, pp. 704–705.

77. R. Winston, W. T. Welford, "Geometrical Vector Flux and Some Nonimaging Concentrators," *J. Opt. Soc. Am.*, vol. 69, 1979, pp. 532–536.

78. D. Gabor, "Light and Information," *Progress in Optics,* vol. 1, North-Holland, Amsterdam, 1961, chap. 4, pp. 111–153.

79. W. T. Welford, R. Winston, *The Optics of Nonimaging Concentrators,* Academic, New York, 1978, app. A.

80. W. T. Welford, *Aberrations of Optical Systems,* Hilger, Bristol, 1986.

81. W. T. Welford and R. Winston, *High Collection Nonimaging Optics,* Academic, New York, 1989, app. A.

82. G. T. di Francia, "Parageometrical Optics," *J. Opt. Soc. Am.*, vol. 40, 1950, pp. 600–602.

83. R. Winston, "Light Collection within the Framework of Geometrical Optics," *J. Opt. Soc. Am.*, vol. 60, 1970, pp. 245–247.

84. T. Jannson and R. Winston, "Liouville's Theorem and Concentrator Optics," *J. Opt. Soc. Am.* A., vol. 3, 1986, pp. 7–8.

85. D. Marcuse, *Light Transmission Optics,* Van Nostrand Reinhold, New York, 1989, chap. 3.7, "Liouville's Theorem," pp. 112–124.

86. R. Clausius, "Die Concentration von Wärme und Lichstrahlen und die Grenzen ihrer Wirkung," *Pogg. Ann.*, cxxi, S. 1, 1864.

87. R. Clausius, *The Mechanical Theory of Heat,* Macmillan, London, 1879, "On the Concentration of Rays of Heat and Light and on the Limits of its Action." (English translation.)

88. H. Helmholtz, "Die theoretische Grenze für die Leitungsföhigkeit der Mikroskope," *Pogg. Ann. Jubelband,* 1874, pp. 556–584.

89. S. Liebes, "Brightness—On the Ray Invariance of B/n^2," *Am. J. Phys.*, vol. 37, 1969, pp. 932–934.

90. Nicodemus, "Radiance," *Am. J. Phys.*, vol. 31, 1963, pp. 368–377.

91. R. W. Boyd, *Radiometry and the Detection of Optical Radiation,* Wiley, New York, 1983, chap. 5.

92. M. V. Klein, T. E. Furtak, *Optics,* 2d ed., Wiley, New York, 1986, sec. 4.2, "Radiometry and Photometry."

93. L. Rayleigh, "Notes, Chiefly Historical, on Some Fundamental Propositions in Optics," *Phil. Mag,* vol. 21, 1886, pp. 466–476. Collected works, vol. II, Dover, New York, 1964, article 137, pp. 513–521.

94. J. P. C. Southall, *Principles and Methods of Geometric Optics,* Macmillan, London, 1910.
95. T. Smith, "On Tracing Rays through an Optical System," *Proc. Phys. Soc. London,* vol. 33, 1921, pp. 174–178.
96. H. H. Hopkins, "An Optical Magnification for Finite Objects," *Nature,* vol. 159, 1947, pp. 573–574.
97. J. Marshall, "Particle Counting by Cerenkov Radiation," *Phys. Rev.,* vol. 86, 1952, pp. 685–693.
98. H. A. Buchdahl, *Optical Aberration Coefficients,* Oxford, London, 1954, sec. 4.
99. M. Herzberger, *Modern Geometrical Optics,* Interscience Publishers, New York, 1958.
100. W. T. Welford, "A Note on the Skew Invariant of Optical Systems," *Optica Acta,* vol. 15, 1968, pp. 621–623.
101. O. N. Stavroudis, *The Optics of Rays, Wavefronts, and Caustics,* Academic, New York, 1972, p. 208.
102. W. T. Welford, *Aberrations of the Symmetrical Optical System,* Academic Press, London, 1974, sec. 5.4, p. 66.
103. W. T. Welford, *Aberrations of Optical Systems,* Hilger, Bristol, 1986, sec. 6.4, p. 84.
104. W. T. Welford and R. Winston, *High Collection Nonimaging Optics,* Academic, New York, 1989, p. 228.
105. R. Descartes, "Snell's Law," *A Source Book in Physics,* W. F. Magie, ed., Harvard, Cambridge, Mass., 1963, p. 265.
106. L. Silberstein, *Simplified Methods of Tracing Rays through any Optical System of Lenses, Prisms, and Mirrors,* Longmans, Green, and Co., London, 1918, sec. 8, "Dyadic Representing the Most General Reflector."
107. T. Smith, "On Systems of Plane Reflecting Surfaces," *Trans. Opt. Society,* vol. 30, 1928, pp. 68–78.
108. R. K. Luneburg, *Mathematical Theory of Optics,* University of California Press, Berkeley, 1964, app. II.
109. R. E. Hopkins, *Applied Optics and Optical Engineering,* R. Kingslake (ed.), vol. 3, Academic, New York, 1965, chap. 7, pp. 269–308. "Mirror and Prism Systems."
110. L. Levi, *Applied Optics: A Guide to Modern Optical System Design,* vol. 1, Wiley, New York, 1968, pp. 346–354.
111. Lian Tongshu, *Theory of Conjugation for Reflecting Prisms: Adjustment and Image Stabilization of Optical Instruments,* International Academic Publishers, Pergamon, Oxford, 1991.
112. H. Wagner, "Zur mathematischen Behandlung von Spiegelungen," *Optik,* vol. 8, 1951, pp. 456–472.
113. G. H. Spencer, M. V. R. K. Murty, "General Ray-Tracing Procedure," *J. Opt. Soc. Am.,* vol. 52, 1962, pp. 672–678.
114. G. T. di Francia, "Parageometrical Optics," *J. Opt. Soc. Am.,* vol. 40, 1950, pp. 600–602.
115. *Webster's International Unabridged Dictionary,* 2d ed., 1934.
116. H. A. Buchdahl, *An Introduction to Hamiltonian Optics,* Cambridge University Press, 1970.
117. J. C. Maxwell, "On the General Laws of Optical Instruments," *Quarterly Journal of Pure and Applied Mathematics,* Feb. 1858, and *The Scientific Papers of James Clerk Maxwell,* Cambridge UP, 1890, Dover, New York, 1965.
118. J. P. C. Southall, *The Principles and Methods of Geometrical Optics,* Macmillan, London, 1910.
119. M. Born and E. Wolf, *Principles of Optics,* Pergamon, New York, 1980.
120. T. Smith, "The Optical Cosine Law," *Trans. Opt. Soc. London,* vol. 24, 1922–23, pp. 31–39.
121. G. C. Steward, *The Symmetrical Optical Systems,* Cambridge University Press., 1928.
122. W. T. Welford, "Aplanatism and Isoplanatism," *Progress in Optics,* vol. 13, E. Wolf (ed.), North-Holland, Amsterdam, 1976, pp. 267–292.
123. W. T. Welford, *Aberrations of Optical Systems,* Hilger, Bristol, 1986.
124. E. Abbe, "Über die Bedingungen des Aplanatismus der Linsensysteme, Sitzungsber. der

Jenaischen Gesellschaft für Med. u. Naturw.," 1879, pp. 129–142; *Gessammelte Abhandlungen,* Bd. I, pp. 213–226; *Carl. Repert. Phys.,* vol. 16, 1880.

125. A. E. Conrady, *Applied Optics and Optical Design,* Oxford University Press, 1929, Dover, New York, 1992.
126. H. H. Hopkins, "The Optical Sine-Condition," *Proc. Phys. Soc. Lond.,* vol. 58, 1946, pp. 92–99.
127. T. Smith, "On Perfect Optical Instruments," *Proc. Phys. Soc.,* vol. LX 1948, pp. 293–304.
128. H. H. Hopkins, *Wave Theory of Aberrations,* Clarendon Press, Oxford, 1950.
129. J. F. W. Herschel, "On the Aberration of Compound Lenses and Object Glasses," *Phil. Trans. Roy. Soc.,* vol. 8, 1821, pp. 222–267.
130. H. H. Hopkins, "Herschel's Condition," *Proc. Phys. Soc. Lond.,* vol. 58, 1946, pp. 100–105.
131. C. H. F. Velzel, "Breaking the Boundaries of Optical System Design and Construction," *Trends in Modern Optics,* J. W. Goodman (ed.), Academic, Boston, 1991, pp. 325–338.
132. T. Smith, "The Theory of Aplanatic Surfaces," *Trans. Opt. Soc.,* vol. 29, 1927–28, pp. 179–186.
133. Military Standardization Handbook MIL-HDBK-141, *Optical Design,* Defense Supply Agency, 1962.
134. R. Kingslake, "Basic Geometrical Optics," *Applied Optics and Optical Engineering,* vol. I, R. Kingslake (ed.), Academic, New York, 1965, chap. 6.
135. H. P. Brueggemann, *Conic Mirrors,* Focal, New York, 1968.
136. R. Kingslake, *Lens Design Fundamentals,* Academic, New York, 1978.
137. R. Kingslake, *Optical System Design,* Academic, New York, 1983.
138. G. G. Slyusarev, *Aberration and Optical Design Theory,* Hilger, Bristol, 1984.
139. M. V. Klein and T. E. Furtak, *Optics,* 2d ed., Wiley, New York, 1986.
140. R. E. Hopkins, "Geometric Optics," *Geometrical and Instrumental Optics,* D. Malacara (ed.), vol. 25 of *Methods of Experimental Physics,* Academic, New York, 1988, pp. 7–58.
141. W. J. Smith, *Modern Optical Engineering,* McGraw-Hill, New York, 1990.
142. R. Shannon, "Aspheric Surfaces," *Applied Optics and Optical Engineering,* vol. 8 Academic, New York, 1980, chap. 3, pp. 55–85.
143. W. T. Welford, *Aberrations of Optical Systems,* Hilger, Bristol, 1986.
144. D. Malacara, "An Optical Surface and Its Characteristics," *Optical Shop Testing,* 2d ed., D. Malacara (ed.), Wiley, New York, 1992, app. I.
145. *ISO Fundamental Standards Optics and Optical Instruments,* "Indications in Optical Drawings Part 12: Aspheric Surfaces," April, ISO/TC 172/SC 1, 1992.
146. G. H. Spencer, M. V. R. K. Murty, "General Ray-Tracing Procedure," *J. Opt. Soc. Am.,* vol. 52, 1962, pp. 672–678.
147. W. T. Welford, *Aberrations of the Symmetrical Optical System,* Academic Press, London, 1974.
148. R. Kingslake, *Lens Design Fundamentals,* Academic, New York, 1978.
149. R. Kingslake, *Optical System Design,* Academic, New York, 1983.
150. G. G. Slyusarev, *Aberration and Optical Design Theory,* Hilger, Bristol, 1984.
151. M. V. Klein and T. E. Furtak, *Optics,* 2d ed., Wiley, New York, 1986.
152. W. T. Welford, *Aberrations of Optical Systems,* Hilger, Bristol, 1986.
153. W. J. Smith, *Modern Optical Engineering,* McGraw-Hill, New York, 1992.
154. A. Nussbaum and R. A. Phillips, *Contemporary Optics for Scientists and Engineers,* Prentice-Hall, Englewood Cliffs, N. J., 1976.
155. W. J. Smith, *Modern Optical Engineering,* McGraw-Hill, New York, 1992.
156. L. Rayleigh, "Notes, Chiefly Historical, on some Fundamental Propositions in Optics," *Phil. Mag.,* vol. 21, 1886, pp. 466–476. Collected works, Dover, New York, 1964, vol. II, article 137, pp. 513–521.
157. J. P. C. Southall, *Principles and Methods of Geometric Optics,* Macmillan, London, 1910.
158. M. Herzberger, "On the Fundamental Optical Invariant, the Optical Tetrality Principle, and on

the New Development of Gaussian Optics Based on This Law," *J. Opt. Soc. Am.,* vol. 25, 1935, pp. 295–304.

159. W. Brouwer and A. Walther, "Geometrical Optics," *Advanced Optical Techniques,* A. C. S. Van Heel (ed.), North-Holland, Amsterdam, 1967, p. 576.
160. T. Young, *Phil. Trans. Roy. Soc.,* vol. 102, 1801, pp. 123–128.
161. H. Coddington, *A Treatise on the Reflexion and Refraction of Light,* Cambridge, 1829, pp. 66*ff.*
162. J. C. Maxwell, "On the Application of Hamilton's Characteristic Function to the Theory of an Optical Instrument Symmetrical about Its Axis," *Proc. London Math Society,* vol. VI, 1875. *The Scientific Papers of James Clerk Maxwell,* Cambridge University Press, 1890, Dover, 1965.
163. J. C. Maxwell, "On Hamilton's Characteristic Function for a Narrow Beam of Light," *Proc. London Math. Society,* vol. VI, 1847–75. *The Scientific Papers of James Clerk Maxwell,* Cambridge University Press, 1890, Dover, 1965.
164. J. C. Maxwell, "On the Focal Lines of a Refracted Pencil," *Proc. London Math. Society,* vol. VI, 1874–75. *The Scientific Papers of James Clerk Maxwell,* Cambridge University Press, 1890, Dover, 1965.
165. J. Larmor, "The Characteristics of an Asymmetric Optical Computation," *Proc. London Math. Soc.,* vol. 20, 1889, pp. 181–194.
166. J. Larmor, "The Simplest Specification of a Given Optical Path, and the Observations Required to Determine It," *Proc. London Math. Soc.,* vol. 23, 1892, pp. 165–173.
167. R. S. Heath, *A Treatise on Geometrical Optics,* Cambridge University Press, 1895, chap. 8 "On the General Form and Properties of a Thin Pencil. General Refraction of Thin Pencils."
168. R. W. Sampson, "A Continuation of Gauss's 'Dioptrische Untersuchungen,'" *Proc. London. Math. Soc.,* vol. 24, 1897, pp. 33–83.
169. T. Smith, "Note on Skew Pencils Traversing a Symmetrical Instrument," *Trans. Opt. Soc.,* vol. 30, 1928–29, pp. 130–133.
170. T. Smith, "Imagery Around a Skew Ray," *Trans. Opt. Soc.,* vol. 31, 1929–30, pp. 131–156.
171. T. Smith, "Canonical Forms in the Theory of Asymmetric Optical Systems," *Trans. Opt. Soc.,* vol. 29, 1928–1929, pp. 88–98.
172. M. Herzberger, "First-Order Laws in Asymmetrical Optical Systems. Part I. The Image of a Given Congruence: Fundamental Conceptions," *J. Opt. Soc. Am.,* vol. 26, 1936, pp. 345–359.
173. J. L. Synge, *Geometrical Optics. An Introduction to Hamilton's Method,* Cambridge University Press, 1937, chap. 3, "Thin Bundles of Rays."
174. H. H. Hopkins, "A Transformation of Known Astigmatism Formulae," *Proc. Phys. Soc.,* vol. 55, 1945, pp. 663–668.
175. H. H. Hopkins, *Wave Theory of Aberrations,* Clarendon Press, Oxford, 1950, chap. 5, "Astigmatism."
176. J. A. Kneisly, "Local Curvature of Wavefronts in an Optical System," *J. Opt. Soc. Am.,* vol. 54, 1963, pp. 229–235.
177. P. J. Sands, "When is First-Order Optics Meaningful," *J. Opt. Soc. Am.,* vol. 58, 1968, pp. 1365–1368.
178. H. A. Buchdahl, *An Introduction to Hamiltonian Optics,* Cambridge University Press, 1970, sec. 10, "Parabasal Optics in General."
179. O. N. Stavroudis, *The Optics of Rays, Wavefronts, and Caustics,* Academic, New York, 1972, chap. 10, "Generalized Ray Tracing."
180. P. J. Sands, "First-Order Optics of the General Optical System," *J. Opt. Soc. Am.,* vol. 62, 1972, pp. 369–372.
181. H. H. Hopkins, "Image Formation by a General Optical System. 1: General Theory," *Appl. Opt.,* vol. 24, 1985, pp. 2491–2505.
182. H. H. Hopkins, "Image Formation by a General Optical System. 2: Computing Methods," *Appl. Opt.,* vol. 24, 1985, pp. 2506–2519.
183. W. T. Welford, *Aberrations of Optical Systems,* Hilger, Bristol, 1986, sec. 9.5, "Optics Round a Finite Principal Ray," p. 185.

184. C. H. F. Velzel, "Image Formation by a General Optical System, Using Hamilton's Method," *J. Opt. Soc. Am. A,* vol. 4, 1987, pp. 1342–1348.

185. W. J. Smith, *Modern Optical Engineering,* McGraw-Hill, New York, 1992, chap. 10.6, "Coddington's Equations."

186. B. D. Stone and G. W. Forbes, "Characterization of First-Order Optical Properties for Asymmetric Systems," *J. Opt. Soc. Am. A,* vol. 9, 1992, pp. 478–489.

187. C. F. Gauss, *Dioptrische Untersuchungen,* delivered to the Royal Society, Dec 10, 1840. *Transactions of the Royal Society of Sciences at Göttingen,* vol. I, 1843, "Optical Investigations" (translated by Bernard Rosett), July 1941. (Photocopies of a typed translation.)

188. R. Kingslake, "Basic Geometrical Optics," *Applied Optics and Optical Engineering,* vol. I, R. Kingslake (ed.), Academic, New York, 1965, chap. 6.

189. W. B. Wetherell, "Afocal Lenses," *Applied Optics and Optical Engineering,* vol. 10, R. R. Shannon and J. C. Wyant (eds.), Academic, New York, 1987, chap. 3, pp. 110–192.

190. W. B. Wetherell, "Afocal Lenses and Mirror Systems," *Handbook of Optics,* vol. II, chap. 2, McGraw-Hill, New York, 2d ed., 1994.

191. D. S. Goodman, "Afocal Systems," *Geometrical and Instrumental Optics,* vol. 25 of *Methods of Experimental Physics,* D. Malacara, ed., Academic, New York, 1988, pp. 132–142.

192. A. F. Möbius, "Entwickelung der Lehre von dioptrischen Bildern mit Huelfe der Collineations-Verwandschaft," *Leipziger Berichte,* vol. 7, 1855 pp. 8–32. (Translation: "Development of the teaching (or science) of dioptric images by means of the collineation relationship.")

193. J. P. C. Southall, *The Principles and Methods of Geometrical Optics,* Macmillan, London, 1910.

194. S. Czapski, *Theorie der Optischen Instrumente nach Abbe,* Verlag von Eduard Trewendt, Breslau, 1893.

195. P. K. L. Drude, *The Theory of Optics,* Longmans, Green, and Co., New York, 1901, Dover, New York, 1959. (1900 English translation, C. R. Mann, R. A. Millikan.)

196. E. Wandersleb, "Abbe's Geometrical Image Formation of Optical Images," *The Formation of Images in Optical Instruments,* Von Rohr, ed., London, 1920, pp. 83–124. (English translation, H. M. Stationary Office.)

197. H. Chrétien, *Calcul des Combinaison Optique,* Mason, Paris, 1980, pp. 25*ff.*

198. M. Born and E. Wolf, *Principles of Optics,* Pergamon, New York, 1980, sec. 4.3, "Projective Transformations (Collineation) with Axial Symmetry."

199. J. P. C. Southall, *Mirrors, Prisms and Lenses,* Macmillan, New York, 1933.

200. M. A. Penna, R. R. Patterson, *Projective Geometry and Its Applications to Computer Graphics,* Prentice-Hall, Englewood Cliffs, N.J., 1986.

201. J. C. Chastang, "Oblique Imaging in the Paraxial Approximation," 1990 OSA annual meeting, Boston.

202. T. Scheimpflug, "Die Herstellung von Karten und Platen auf photographischem Wege," *Journal of the Royal Academy of Sciences,* Vienna, vol. 116, 1907, pp. 235–266.

203. J. Sasian, "Image Plane Tilt in Optical Systems," *Optical Engineering,* vol. 31, 1992, pp. 527–532.

204. W. T. Welford, *Aberrations of Optical Systems,* Hilger, Bristol, 1986.

205. R. A. Samson, "A Continuation of Gauss's *Dioptrische Untersuchungen,*" *Proc. London Mathematical Society,* vol. 29, 1897, pp. 33–83.

206. G. G. Leathem, *The Elementary Theory of the Symmetrical Optical Instrument,* Hafner, New York, 1908.

207. K. Halbach, "Matrix Representation of Gaussian Optics," *Am. J. Phys.,* vol. 32, 1964, pp. 90–108.

208. D. C. Sinclair, "The Specification of Optical Systems by Paraxial Matrices," *Applications of Geometrical Optics,* W. J. Smith (ed.), *SPIE Proceedings,* vol. 39, SPIE, Bellingham, Wash., 1973, pp. 141–149.

209. E. L. O'Neil, *Introduction to Statistical Optics,* Addison-Wesley, Reading, Mass., 1963.

210. W. Brouwer, *Matrix Methods in Optical Instrument Design,* W. A. Benjamin, New York, 1964.

211. J. W. Blaker, *Geometric Optics: The Matrix Theory,* Marcel Dekker, New York, 1971.
212. R. A. Longhurst, *Geometrical and Physical Optics,* Longman, London, 1973.
213. A. Gerrard and J. M. Burch, *Introduction of Matrix Methods in Optics,* Wiley, London, New York, 1974.
214. A. Nussbaum and R. A. Phillips, *Contemporary Optics for Scientists and Engineers,* Prentice-Hall, Englewood Cliffs, N.J., 1976.
215. H. Kogelnik, "Propagation of Laser Beams," *Applied Optics and Optical Engineering,* vol. 7, R. R. Shannon and J. C. Wyant (eds.), Academic, New York, 1979, chap. 6, pp. 156–190.
216. M. V. Klein, T. E. Furtak, *Optics,* 2d ed., Wiley, New York, 1986.
217. K. D. Moller, *Optics,* University Science, 1988.
218. R. Guenther, *Modern Optics,* Wiley, New York, 1990.
219. A. Marechal, "Optique Geometrique General," *Fundamentals of Optics,* S. Fluegge (ed.), *Handbuch der Physik,* vol. 24, Springer-Verlag, Berlin, 1956, pp. 44–170.
220. A. Marechal, "Optique Geometrique et Algebre des Matrices," *Application de L'Algebre Moderne a Quelque Probleme de Physique Classique,* M. Parodi, Gauthier-Villars, Paris, 1961, chap. 15, pp. 306–346.
221. B. Macukow, H. H. Arsenault, "Matrix Decompositions for Nonsymmetrical Optical Systems," *J. Opt. Soc. Am.,* vol. 73, 1983, pp. 1360–1366.
222. W. Brouwer and A. Walther, "Geometrical Optics," *Advanced Optical Techniques,* A. C. S. Van Heel, (ed.), North-Holland, Amsterdam, 1967, chap. 16, pp. 505–570.
223. H. H. Arsenault, "The Rotation of Light Fans by Cylindrical Lenses," *Opt. Commun.,* vol. 31, 1979, pp. 275–278.
224. H. H. Arsenault, "Generalization of the Principal Plane Concept in Matrix Optics," *Am. J. Phys.,* vol. 48, 1980, pp. 397–399.
225. H. H. Arsenault, "A Matrix Representation for Non-Symmetrical Optical Systems," *J. Opt.,* vol. 11, Paris, 1980, pp. 87–91.
226. M. P. Keating, "A System Matrix for Astigmatic Optical Systems. Introduction and Dioptric Power Relations," *Am. J. Optometry and Physiological Optics,* vol. 58, 1981, pp. 810–819.
227. H. H. Arsenault and B. Macukow, "Factorization of the Transfer Matrix for Symmetrical Optical Systems," *J. Opt. Soc. Am.,* vol. 73, 1983, pp. 1350–1359.
228. A. E. Attard, "Matrix Optical Analysis of Skew Rays in Mixed Systems of Spherical and Orthogonal Cylindrical Lenses," *Appl. Opt.,* vol. 23, 1984, pp. 2706–2709.
229. B. D. Stone and G. W. Forbes, "Characterization of First-Order Optical Properties for Asymmetric Systems," *J. Opt. Soc. Am. A,* vol. 9, 1992, pp. 478–489.
230. L. Rayleigh, "Notes, Chiefly Historical, on Some Fundamental Propositions in Optics," *Phil. Mag.,* vol. 21, 1886, pp. 466–476. Collected works, vol. II, Dover, New York, 1964, article 137, pp. 513–521.
231. J. P. C. Southall, *Principles and Methods of Geometric Optics,* Macmillan, London, 1910.
232. M. R. Hatch and D. E. Stoltzman, "The F-Stops Here," *Optical Spectra,* June 1980, pp. 88–91.
233. D. S. Goodman, "The F-Word in Optics," *Optics and Photonics News,* vol. 4, April 1993, pp. 38–39.
234. W. J. Smith, *Modern Optical Engineering,* McGraw-Hill, New York, 1992.
235. R. Kingslake, "The Effective Aperture of a Photographic Objective," *J. Opt. Soc. Am.,* vol. 32, 1945, pp. 518–520.
236. M. Reiss, "The Cos^4 Law of Illumination," *J. Opt. Soc. Am.,* vol. 35, 1945, pp. 283–288.
237. J. C. Gardner, "Validity of the Cosine-Fourth-Power Law of Illumination," *Journal of Research of the National Bureau of Standards,* vol. 39, 1947, pp. 213–219.
238. M. Reiss, "Notes on the Cos^4 Law of Illumination," *J. Opt. Soc. Am.,* vol. 38, 1948, pp. 980–986.
239. R. Kingslake, "Illumination in Optical Instruments," *Applied Optics and Optical Engineering,* vol. 2, Academic, New York, 1965, chap. 5, pp. 195–228.

240. G. Slyusarev, "L'Eclairement de L'Image Formeé par les Objectifs Photographiques Grand-Angulaires," *Journal of Physics* (USSR), vol. 4, 1941, pp. 537–545.
241. D. Gabor, "Light and Information," *Progress in Optics,* E. Wolf, ed., vol. 1, 1961.
242. S. F. Ray, *Applied Photographic Optics,* Focal Press, London, 1988, chap. 22, "Depth of Field and Depth of Focus."
243. R. Kingslake, *Lenses in Photography,* Barnes, New York, 1963, SPIE, Bellingham WA, 1992.
244. A. E. Conrady, *Applied Optics and Optical Design,* vol. 1, 2, A. E. Conrady, Dover, New York, 1929.
245. H. H. Hopkins, *Wave Theory of Aberrations,* Clarendon Press, Oxford, 1950.
246. H. A. Buchdahl, *Optical Aberration Coefficients,* Oxford, London, 1954.
247. M. Herzberger, *Modern Geometrical Optics,* Interscience Publishers, New York, 1958.
248. R. Kingslake, *Lens Design Fundamentals,* Academic, New York, 1978.
249. M. Born and E. Wolf, *Principles of Optics,* Pergamon, New York, 1980.
250. G. C. Slyusarev, *Aberration and Optical Design Theory,* Hilger, Bristol, 1984.
251. W. T. Welford, *Aberrations of the Symmetrical Optical Systems,* Academic Press, London, 1974.
252. W. T. Welford, *Aberrations of Optical Systems,* Hilger, Bristol, 1986.
253. W. J. Smith, *Modern Optical Engineering,* McGraw-Hill, New York, 1992.
254. D. O'Shea, "Aberration Curves in Lens Design," *Handbook of Optics,* vol. I, chap. 33, McGraw-Hill, New York, 2d ed., 1994.
255. F. Zernike, "Beugungstheorie des Schneidenver-Eahrens und Seiner Verbesserten Form, der Phasenkontrastmethode," *Physica,* vol. 1, 1934, p. 689.
256. C.-J. Kim, R. R. Shannon, "Catalogue of Zernike Polynomials," *Applied Optics and Optical Engineering,* vol. 10, R. R. Shannon, J. C. Wyant (eds.), Academic, New York, 1987, chap. 4, pp. 193–233.
257. "Zernike Polynomials and Wavefront Fitting," D. Malacara (ed.), *Optical Shop Testing,* Wiley, New York, 1978, app. 2, pp. 489–505.

P · A · R · T · 2

PHYSICAL OPTICS

CHAPTER 2
INTERFERENCE

John E. Greivenkamp
Optical Sciences Center
University of Arizona
Tucson, Arizona

2.1 GLOSSARY

A	amplitude
$\mathbf{E}$	electric field vector
$\mathbf{r}$	position vector
x, y, z	rectangular coordinates
ϕ	phase

2.2 INTRODUCTION

Interference results from the superposition of two or more electromagnetic waves. From a classical optics perspective, interference is the mechanism by which light interacts with light. Other phenomena, such as refraction, scattering, and diffraction, describe how light interacts with its physical environment. Historically, interference was instrumental in establishing the wave nature of light. The earliest observations were of colored fringe patterns in thin films. Using the wavelength of light as a scale, interference continues to be of great practical importance in areas such as spectroscopy and metrology.

2.3 WAVES AND WAVEFRONTS

The *electric field vector* due to an electromagnetic field at a point in space is composed of an amplitude and a phase

$$\mathbf{E}(x, y, z, t) = \mathbf{A}(x, y, z, t)e^{i\phi(x,y,z,t)} \tag{1}$$

or

$$\mathbf{E}(\mathbf{r}, t) = \mathbf{A}(\mathbf{r}, t)e^{i\phi(\mathbf{r},t)} \tag{2}$$

where $\mathbf{r}$ is the position vector and both the amplitude $\mathbf{A}$ and phase ϕ are functions of the spatial coordinate and time. As described in Chap. 5, "Polarization," the polarization state of the field is contained in the temporal variations in the amplitude vector.

This expression can be simplified if a linearly polarized monochromatic wave is assumed:

$$\mathbf{E}(x, y, z, t) = \mathbf{A}(x, y, z)e^{i[\omega t - \phi(x,y,z)]} \tag{3}$$

where ω is the angular frequency in radians per second and is related to the frequency ν by

$$\omega = 2\pi\nu \tag{4}$$

Some typical values for the optical frequency are 5×10^{14} Hz for the visible, 10^{13} Hz for the infrared, and 10^{16} Hz for the ultraviolet. Note that in the expression for the electric field vector, the time dependence has been eliminated from the amplitude term to indicate a constant linear polarization. The phase term has been split into spatial and temporal terms. At all locations in space, the field varies harmonically at the frequency ω.

Plane Wave

The simplest example of an electromagnetic wave is the *plane wave*. The plane wave is produced by a monochromatic point source at infinity and is approximated by a collimated light source. The complex amplitude of a linearly polarized plane wave is

$$\mathbf{E}(x, y, z, t) = \mathbf{E}(\mathbf{r}, t) = \mathbf{A}e^{i[\omega t - \mathbf{k}\cdot\mathbf{r}]} \tag{5}$$

where $\mathbf{k}$ is the wave vector. The wave vector points in the direction of propagation, and its magnitude is the wave number $k = 2\pi/\lambda$, where λ is the wavelength. The wavelength is related to the temporal frequency by the speed of light v in the medium:

$$\lambda = \frac{v}{\nu} = 2\pi\frac{v}{\omega} = \frac{c}{n\nu} = 2\pi\frac{c}{n\omega} \tag{6}$$

where n is the index of refraction, and c is the speed of light in a vacuum. The amplitude $\mathbf{A}$ of a plane wave is a constant over all space, and the plane wave is clearly an idealization.

If the direction of propagation is parallel to the z axis, the expression for the complex amplitude of the plane wave simplifies to

$$\mathbf{E}(x, y, z, t) = \mathbf{A}e^{i[\omega t - kz]} \tag{7}$$

We see that the plane wave is periodic in both space and time. The spatial period equals the wavelength in the medium, and the temporal period equals $1/\nu$. Note that the wavelength changes with index of refraction, and the frequency is independent of the medium.

Spherical Wave

The second special case of an electromagnetic wave is the spherical wave which radiates from an isotropic point source. If the source is located at the origin, the complex amplitude is

$$E(r, t) = (A/r)e^{i[\omega t - kr]} \tag{8}$$

where $r = (x^2 + y^2 + z^2)^{1/2}$. The field is spherically symmetric and varies harmonically with time and the radial distance. The radial period is the wavelength in the medium. The amplitude of the field decreases as $1/r$ for energy conservation. At a large distance from the source, the spherical wave can be approximated by a plane wave. Note that the vector characteristics of the field (its polarization) are not considered here as it is not possible to describe a linear polarization pattern of constant amplitude that is consistent over the

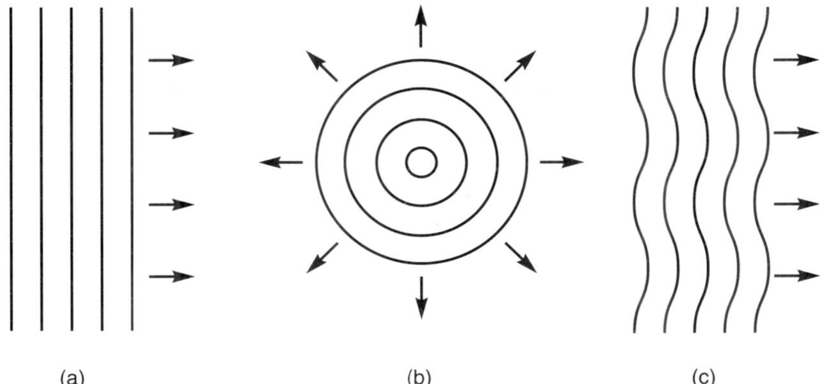

FIGURE 1 Examples of wavefronts: (*a*) plane wave; (*b*) spherical wave; and (*c*) aberrated plane wave.

entire surface of a sphere. In practice, we only need to consider an angular segment of a spherical wave, in which case this polarization concern disappears.

Wavefronts

Wavefronts represent surfaces of constant phase for the electromagnetic field. Since they are normally used to show the spatial variations of the field, they are drawn or computed at a fixed time. Wavefronts for plane and spherical waves are shown in Fig. 1*a* and *b*. The field is periodic, and a given value of phase will result in multiple surfaces. These surfaces are separated by the wavelength. A given wavefront also represents a surface of constant optical path length (OPL) from the source. The OPL is defined by the following path integral:

$$\text{OPL} = \int_S^P n(s)\, ds \tag{9}$$

where the integral goes from the source S to the observation point P, and $n(s)$ is the index of refraction along the path. Variations in the index or path can result in irregularities or aberrations in the wavefront. An aberrated plane wavefront is shown in Fig. 1*c*. Note that the wavefronts are still separated by the wavelength.

The local normal to the wavefront defines the propagation direction of the field. This fact provides the connection between wave optics and ray or geometrical optics. For a given wavefront, a set of rays can be defined using the local surface normals. In a similar manner, a set of rays can be used to construct the equivalent wavefront.

2.4 INTERFERENCE

The net complex amplitude is the sum of all of the component fields,

$$\mathbf{E}(x, y, z, t) = \sum_i \mathbf{E}_i(x, y, z, t) \tag{10}$$

and the resulting field intensity is the time average of the modulus squared of the total complex amplitude

$$I(x, y, z, t) = \langle |\mathbf{E}(x, y, z, t)|^2 \rangle \tag{11}$$

where ⟨ ⟩ indicates a time average over a period much longer than $1/\nu$. If we restrict ourselves to two interfering waves $\mathbf{E}_1$ and $\mathbf{E}_2$, this result simplifies to

$$I(x, y, z, t) = \langle |\mathbf{E}_1|^2 \rangle + \langle |\mathbf{E}_2|^2 \rangle + \langle \mathbf{E}_1 \cdot \mathbf{E}_2^* \rangle + \langle \mathbf{E}_1^* \cdot \mathbf{E}_2 \rangle \tag{12}$$

or

$$I(x, y, z, t) = I_1 + I_2 + \langle \mathbf{E}_1 \cdot \mathbf{E}_2^* \rangle + \langle \mathbf{E}_1^* \cdot \mathbf{E}_2 \rangle \tag{13}$$

where I_1 and I_2 are the intensities due to the two beams individually, and the (x, y, z, t) dependence is now implied for the various terms.

This general result can be greatly simplified if we assume linearly polarized monochromatic waves of the form in Eq. (3):

$$\mathbf{E}_i(x, y, z, t) = \mathbf{A}_i(x, y, z) e^{i[\omega_i t - \phi_i(x,y,z)]} \tag{14}$$

The resulting field intensity is

$$I(x, y, z, t) = I_1 + I_2 + 2(\mathbf{A}_1 \cdot \mathbf{A}_2) \cos [(\omega_1 - \omega_2)t - (\phi_1(x, y, z) - \phi_2(x, y, z))] \tag{15}$$

The interference effects are contained in the third term, and we can draw two important conclusions from this result. First, if the two interfering waves are orthogonally polarized, there will be no visible interference effects, as the dot product will produce a zero coefficient. Second, if the frequencies of the two waves are different, the interference effects will be modulated at a temporal beat frequency equal to the difference frequency.

Interference Fringes

We will now add the additional restrictions that the two linear polarizations are parallel and that the two waves are at the same optical frequency. The expression for the intensity pattern now becomes

$$I(x, y, z) = I_1 + I_2 + 2\sqrt{I_1 I_2} \cos [\Delta\phi(x, y, z)] \tag{16}$$

where $\Delta\phi = \phi_1 - \phi_2$ is the phase difference. This is the basic equation describing interference. The detected intensity varies cosinusoidally with the phase difference between the two waves as shown in Fig. 2. These alternating bright and dark bands in the

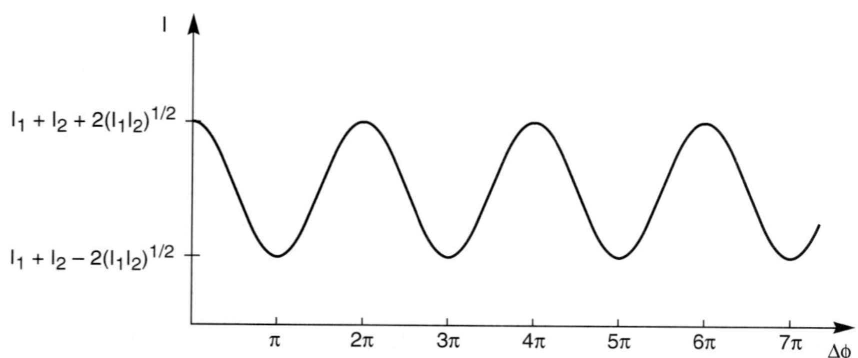

FIGURE 2 The variation in intensity as a function of the phase difference between two interfering waves.

TABLE 1 The Phase Difference and OPD for Bright and Dark Fringes (m an Integer)

	$\Delta\phi$	OPD
Bright fringe	$2m\pi$	$m\lambda$
Dark fringe	$2(m+1)\pi$	$(m+1/2)\lambda$

intensity pattern are referred to as *interference fringes,* and along a particular fringe, the phase difference is constant.

The phase difference is related to the difference in the optical path lengths between the source and the observation point for the two waves. This is the *optical path difference* (OPD):

$$\text{OPD} = \text{OPL}_1 - \text{OPL}_2 = \left(\frac{\lambda}{2\pi}\right)\Delta\phi \tag{17}$$

or

$$\Delta\phi = \left(\frac{2\pi}{\lambda}\right)\text{OPD} \tag{18}$$

The phase difference changes by 2π every time the OPD increases by a wavelength. The OPD is therefore constant along a fringe.

Constructive interference occurs when the two waves are in phase, and a bright fringe or maximum in the intensity pattern results. This corresponds to a phase difference of an integral number of 2π's or an OPD that is a multiple of the wavelength. A dark fringe or minimum in the intensity pattern results from *destructive interference* when the two waves are out of phase by π or the OPD is an odd number of half wavelengths. These results are summarized in Table 1. For conditions between these values, an intermediate value of the intensity results. Since both the OPD and the phase difference increase with the integer m, the absolute value of m is called the *order of interference.*

As we move from one bright fringe to an adjacent bright fringe, the phase difference changes by 2π. Each fringe period corresponds to a change in the OPD of a single wavelength. It is this inherent precision that makes interferometry such a valuable metrology tool. The wavelength of light is used as the unit of measurement. Interferometers can be configured to measure small variations in distance, index, or wavelength.

When two monochromatic waves are interfered, the interference fringes exist not only in the plane of observation, but throughout all space. This can easily be seen from Eq. (16) where the phase difference can be evaluated at any z position. In many cases, the observation of interference is confined to a plane, and this plane is usually assumed to be perpendicular to the z axis. The z dependence in Eq. (16) is therefore often not stated explicitly, but it is important to remember that interference effects will exist in other planes.

Fringe Visibility

It is often more convenient to rewrite Eq. (16) as

$$I(x, y) = I_0(x, y)\{1 + \gamma(x, y)\cos[\Delta\phi(x, y, z)]\} \tag{19}$$

or

$$I(x, y) = I_0(x, y)\{1 + \gamma(x, y)\cos[2\pi\text{OPD}(x, y)/\lambda]\} \tag{20}$$

where $I_0(x, y) = I_1(x, y) + I_2(x, y)$, and

$$\gamma(x, y) = \frac{2[I_1(x, y)I_2(x, y)]^{1/2}}{I_1(x, y) + I_2(x, y)} \tag{21}$$

Since the cosine averages to zero, $I_0(x, y)$ represents the average intensity, and $\gamma(x, y)$ is the local *fringe contrast* or *visibility*. The fringe visibility can also be equivalently calculated using the standard formula for modulation:

$$\gamma(x, y) = \frac{I_{max}(x, y) - I_{min}(x, y)}{I_{max}(x, y) + I_{min}(x, y)} \tag{22}$$

where I_{max} and I_{min} are the maximum and minimum intensities in the fringe pattern.

The fringe visibility will have a value between 0 and 1. The maximum visibility will occur when the two waves have equal intensity. Not surprisingly, the visibility will drop to zero when one of the waves has zero intensity. In general, the intensities of the two waves can vary with position, so that the average intensity and fringe visibility can also vary across the fringe pattern. The average intensity in the observation plane equals the sum of the individual intensities of the two interfering waves. The interference term redistributes this energy into bright and dark fringes.

Two Plane Waves

The first special case to consider is the interference of two plane waves of equal intensity, polarization and frequency. They are incident at angles θ_1 and θ_2 on the observation plane, as shown in Fig. 3. The plane of incidence is the x-z plane (the two k-vectors are contained in this plane). According to Eq. (5), the complex amplitude for each of these plane waves is

$$\mathbf{E}_i(x, y, z, t) = \mathbf{A}e^{i[\omega t - kz\cos(\theta_i) - kx\sin(\theta_i)]} \tag{23}$$

where the dot product has been evaluated. For simplicity we will place the observation plane at $z = 0$, and the phase difference between the two waves is

$$\Delta\phi(x, y) = kx(\sin\theta_1 - \sin\theta_2) = (2\pi x/\lambda)(\sin\theta_1 - \sin\theta_2) \tag{24}$$

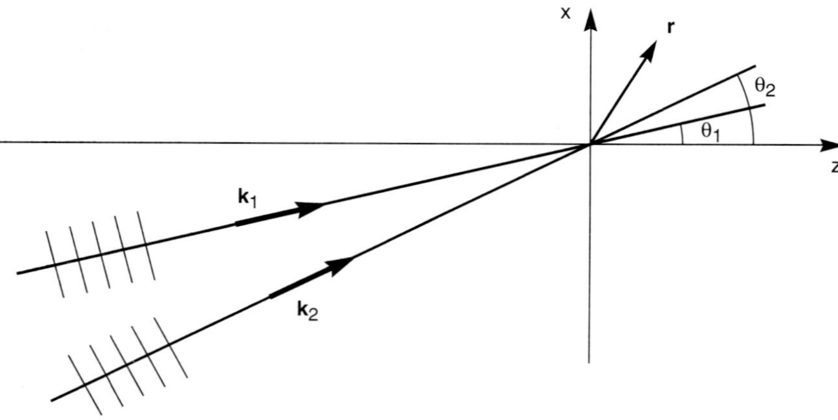

FIGURE 3 The geometry for the interference of two plane waves.

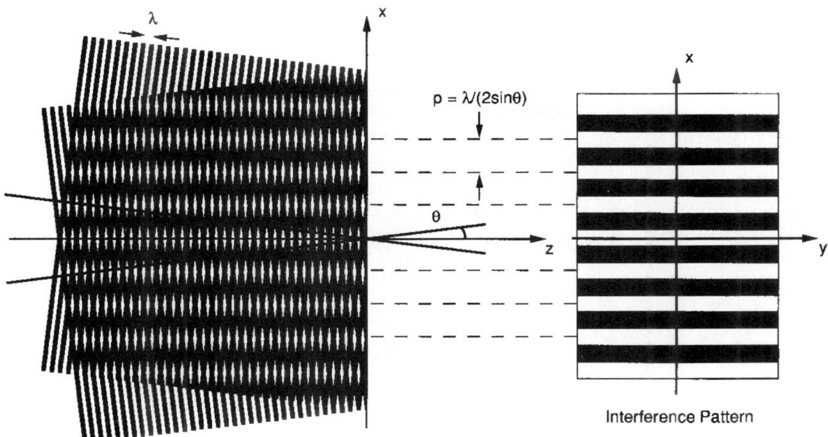

FIGURE 4 The interference of plane waves incident at $\pm\theta$ resulting in straight fringes.

The resulting intensity from Eq. (19) is

$$I(x, y) = I_0\{1 + \cos [(2\pi x/\lambda)(\sin \theta_1 - \sin \theta_2)]\} \tag{25}$$

where $I_0 = 2A^2$ is twice the intensity of each of the individual waves. Straight equispaced fringes are produced. The fringes are parallel to the y axis, and the fringe period depends on the angle between the two interfering beams.

The fringe period p is

$$p = \frac{\lambda}{\sin \theta_1 - \sin \theta_2} \tag{26}$$

and this result can also be obtained by noting that a bright fringe will occur whenever the phase difference equals a multiple of 2π. A typical situation for interference is that the two angles of incidence are equal and opposite, $\theta_1 = -\theta_2 = \theta$. The angle between the two beams is 2θ. Under this condition, the period is

$$p = \frac{\lambda}{2 \sin \theta} \approx \frac{\lambda}{2\theta} \tag{27}$$

and the small-angle approximation is given. As the angle between the beams gets larger, the period decreases. For example, the period is 3.8λ at 15° (full angle of 30°) and is λ at 30° (full angle of 60°). The interference of two plane waves can be visualized by looking at the overlap or moiré of two wavefront patterns (Fig. 4). Whenever the lines representing the wavefronts overlap, a fringe will result. This description also clearly shows that the fringes extend parallel to the z axis and exist everywhere the two beams overlap.

Plane Wave and Spherical Wave

A second useful example to consider is the interference of a plane wave and a spherical wave. Once again the two waves have the same frequency. The plane wave is at normal incidence, the spherical wave is due to a source at the origin, and the observation plane is located at $z = R$. The wavefront shape at the observation plane will be a spherical shell of radius R. Starting with Eq. (8), the complex amplitude of the spherical wave in the observation plane is

$$E(\rho, t) = (A/R)e^{i[\omega t - k(R^2+\rho^2)^{1/2}]} \approx (A/R)e^{i[\omega t - k(R+\rho^2/2R)]} \tag{28}$$

where $\rho = (x^2 + y^2)^{1/2}$, and the square root has been expanded in the second expression. This expansion approximates the spherical wave by a parabolic wave with the same vertex radius. An additional assumption is that the amplitude of the field A/R is constant over the region of interest. The field for the plane wave is found by evaluating Eq. (23) at $z = R$ and $\theta = 0$. The phase difference between the plane and the sphere is then

$$\Delta\phi(\rho) \approx \frac{\pi\rho^2}{\lambda R} \tag{29}$$

and the resulting intensity pattern is

$$I(\rho) = I_0\left[1 + \cos\left(\frac{\pi\rho^2}{\lambda R}\right)\right] \tag{30}$$

The fringe pattern is comprised of concentric circles, and the radial fringe spacing decreases as the radius ρ increases. The intensities of the two waves have been assumed to be equal at the observation plane. This result is valid only when ρ is much smaller than R.

The radius of the mth bright fringe can be found by setting $\Delta\phi = 2\pi m$:

$$\rho_m = \sqrt{2mR} \tag{31}$$

where m is an integer. The order of interference m increases with radius. Figure 5 shows a visualization of this situation using wavefronts. This fringe pattern is the Newton's ring pattern and is discussed in more detail later, under "Fizeau Interferometer." This picture also shows that the radii of the fringes increase as the square root of R.

The analysis of the spherical wave could also have been done by using the sag of a spherical wavefront to produce an OPD and then converting this value to a phase difference. The quadratic approximation for the sag of a spherical surface is $\rho^2/2R$. This

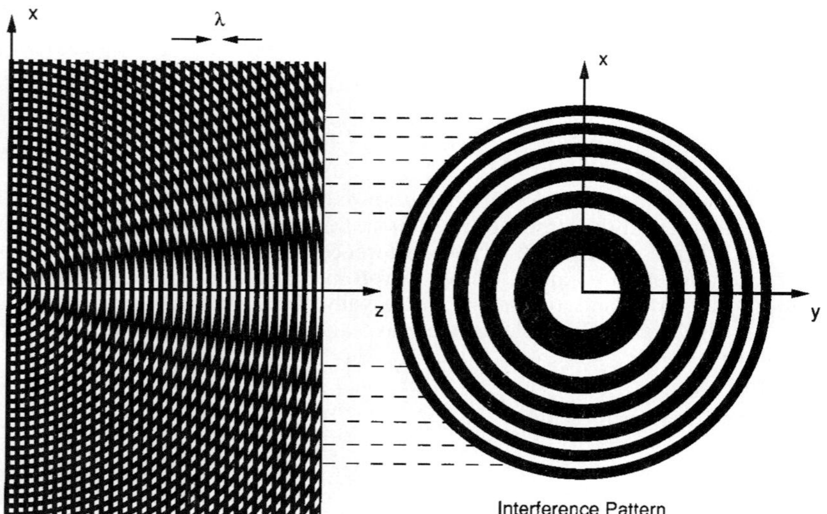

FIGURE 5 The interference of a plane wave and a spherical wave.

corresponds to the OPD between the spherical and planar wavefronts. The equivalent phase difference [Eq. (18)] is then $\pi\rho^2/\lambda R$, as before.

Two Spherical Waves

When considering two spherical waves, there are two particular geometries that we want to examine. The first places the observation plane perpendicular to a line connecting the two sources, and the second has the observation plane parallel to this line. Once again, the sources are at the same frequency.

When the observations are made on a plane perpendicular to a line connecting the two sources, we can use Eq. (28) to determine the complex amplitude of the two waves:

$$E_i(\rho, t) \approx (A/R)e^{i[\omega t - k(R_i + \rho^2/2R_i)]} \tag{32}$$

Let $d = R_1 - R_2$ be the separation of the two sources. For simplicity, we have also assumed that the amplitudes of the two waves are equal (R is an average distance). The phase difference between the two waves is

$$\Delta\phi = \left(\frac{\pi\rho^2}{\lambda}\right)\left(\frac{1}{R_1} - \frac{1}{R_2}\right) + \frac{2\pi d}{\lambda} \approx \frac{2\pi d}{\lambda} - \left(\frac{\pi\rho^2}{\lambda}\right)\left(\frac{d}{R^2}\right), \tag{33}$$

where the approximation $R_1 R_2 \approx R^2$ has been made. There are two terms to this phase difference. The second is a quadratic phase term identical in form to the result obtained from spherical and plane waves. The pattern will be symmetric around the line connecting the two sources, and its appearance will be similar to Newton's rings. The equivalent radius of the spherical wave in Eq. (29) is R^2/d. The first term is a constant phase shift related to the separation of the two sources. If this term is not a multiple of 2π, the center of the fringe pattern will not be a bright fringe; if the term is π, the center of the pattern will be dark. Except for the additional phase shift, this intensity pattern is not distinguishable from the result in the previous section. It should be noted, however, that a relative phase shift can be introduced between a spherical wave and a plane wave to obtain this same result.

An important difference between this pattern and the Newton's ring pattern is that the order of interference ($|m|$ defined by $\Delta\phi = 2\pi m$) or phase difference is a maximum at the center of the pattern and decreases with radius. The phase difference is zero at some finite radius. The Newton's ring pattern formed between a plane and a spherical wave has a minimum order of interference at the center of the pattern. This distinction is important when using polychromatic sources.

There are several ways to analyze the pattern that is produced on a plane that is parallel to a line connecting the two sources. We could evaluate the complex amplitudes by using Eq. (28) and moving the center of the spherical waves to $\pm d/2$ for the two sources. An equivalent method is to compare the wavefronts at the observation plane. This is shown in Fig. 6. The OPD between the two wavefronts is

$$\mathrm{OPD}(x, y) = \frac{[(x + d/2)^2 + y^2]}{2L} - \frac{[(x - d/2)^2 + y^2]}{2L} \tag{34}$$

where the quadratic approximation for the wavefront sag has been assumed, and L is the distance between the sources and the observation plane. After simplification, the OPD and phase differences are

$$\mathrm{OPD}(x, y) = \frac{xd}{L} \tag{35}$$

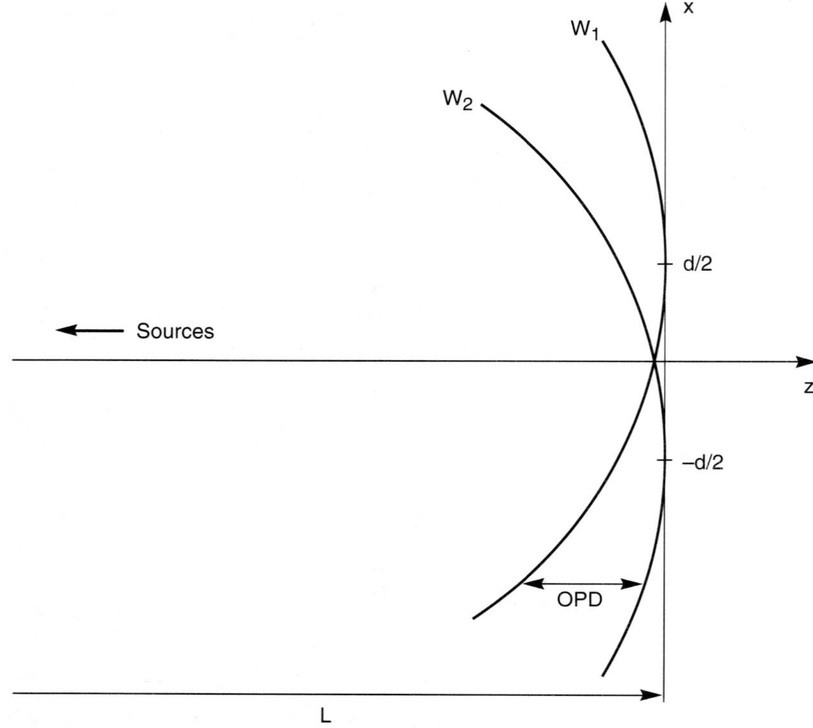

FIGURE 6 The interference of two spherical waves on a plane parallel to the sources.

and

$$\Delta\phi(x, y) = \frac{2\pi x d}{\lambda L} \tag{36}$$

Straight equispaced fringes parallel to the y axis are produced. The period of the fringes is $\lambda L/d$. This fringe pattern is the same as that produced by two plane waves. Note that these fringes increase in spacing as the distance from the sources increases. The approximations used require that L be much larger than ρ and d.

Figure 7 shows the creation of the fringe patterns for two point sources. The full three-dimensional pattern is a series of nested hyperboloids symmetric about the line connecting the sources. Above the two sources, circular fringes approximating Newton's rings are produced, and perpendicular to the sources, the fringes appear to be straight and equispaced.

Aberrated Wavefronts

When an aberrated or irregularly shaped wavefront is interfered with a reference wavefront, an irregularly shaped fringe pattern is produced. However, the rules for analyzing this pattern are the same as with any two wavefronts. A given fringe represents a contour of constant OPD or phase difference between the two wavefronts. Adjacent fringes differ in OPD by one wavelength or equivalently correspond to a phase difference

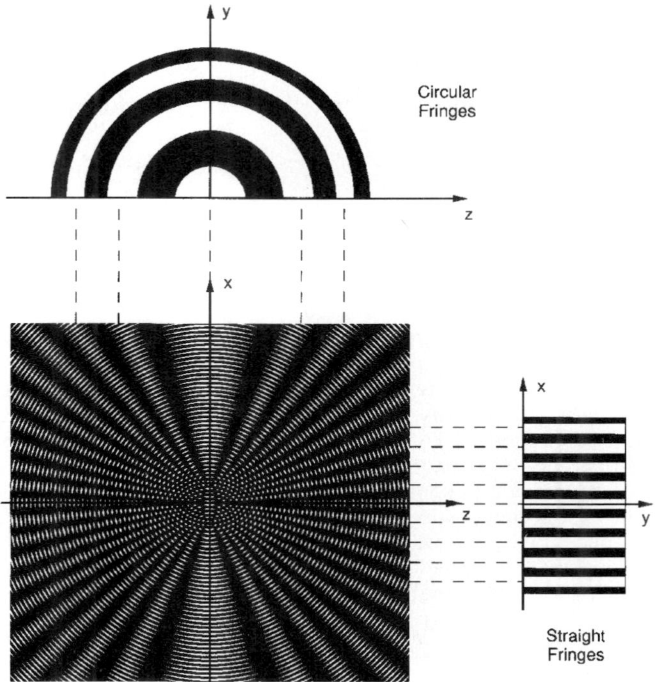

FIGURE 7 The interference of two spherical waves.

of 2π. If the reference is a plane wave, the absolute shape of the irregular wavefront is obtained. If the reference is a spherical wave, or another aberrated wave, the measured OPD or phase difference map represents the difference between the two wavefronts.

Temporal Beats

In Eq. (15) it was noted that if the waves are at different frequencies, the interference effects are modulated by a beat frequency. Rewriting this expression assuming equal-intensity parallel-polarized beams produces

$$I(x, y, t) = I_0\{1 + \cos [2\pi \, \Delta\nu t - \Delta\phi(x, y)]\} \tag{37}$$

where $\Delta\nu = \nu_1 - \nu_2$. The intensity at a given location will now vary sinusoidally with time at the beat frequency $\Delta\nu$. The phase difference $\Delta\phi$ appears as a spatially varying phase shift of the beat frequency. This is the basis of the heterodyne technique used in a number of interferometers. It is commonly used in distance-measuring interferometers.

In order for a heterodyne system to work, there must be a phase relationship between the two sources even though they are at different frequencies. One common method for obtaining this is accomplished by starting with a single source, splitting it into two beams, and frequency-shifting one beam with a known Doppler shift. The system will also work in reverse; measure the interferometric beat frequency to determine the velocity of the object producing the Doppler shift.

Coherence

Throughout this discussion of fringe patterns, we have assumed that the two sources producing the two waves have the same frequency. In practice, this requires that both sources be derived from a single source. Even when two different frequencies are used [Eq. (37)] there must be an absolute phase relation between the two sources. If the source has finite size, it is considered to be composed of a number of spatially separated, independently radiating point sources. If the source has a finite spectral bandwidth, it is considered to be composed of a number of spatially coincident point sources with different frequencies. These reductions in the spatial or temporal coherence of the source will decrease the visibility of the fringes at different locations in space. This is referred to as *fringe localization.* These effects will be discussed later in this chapter and also in Chap. 4, "Coherence Theory."

There are two general methods to produce mutually coherent waves for interference. The first is called *wavefront division,* where different points on a wavefront are sampled to produce two new wavefronts. The second is *amplitude division,* where some sort of beamsplitter is used to divide the wavefront at a given location into two separate wavefronts. These methods are discussed in the next sections.

2.5 INTERFERENCE BY WAVEFRONT DIVISION

Along a given wavefront produced by a monochromatic point source, the wavefront phase is constant. If two parts of this wavefront are selected and then redirected to a common volume in space, interference will result. This is the basis for *interference by wavefront division.*

Young's Double-Slit Experiment

In 1801, Thomas Young performed a fundamental experiment for demonstrating interference and the wave nature of light. Monochromatic light from a single pinhole illuminates an opaque screen with two additional pinholes or slits. The light diffracts from these pinholes and illuminates a viewing screen at a distance large compared to the pinhole separation. Since the light illuminating the two pinholes comes from a single source, the two diffracted wavefronts are coherent and interference fringes form where the beams overlap.

In the area where the two diffracted beams overlap, they can be modeled as two spherical waves from two point sources, and we already know the form of the solution for the interference from our earlier discussion. Equispaced straight fringes are produced, and the period of the fringes is $\lambda L/d$, where L is the distance to the screen and d is the separation of the pinholes. The fringes are oriented perpendicular to the line connecting the two pinholes.

Even though we already know the answer, there is a classic geometric construction we should consider that easily gives the OPD between the two wavefronts at the viewing

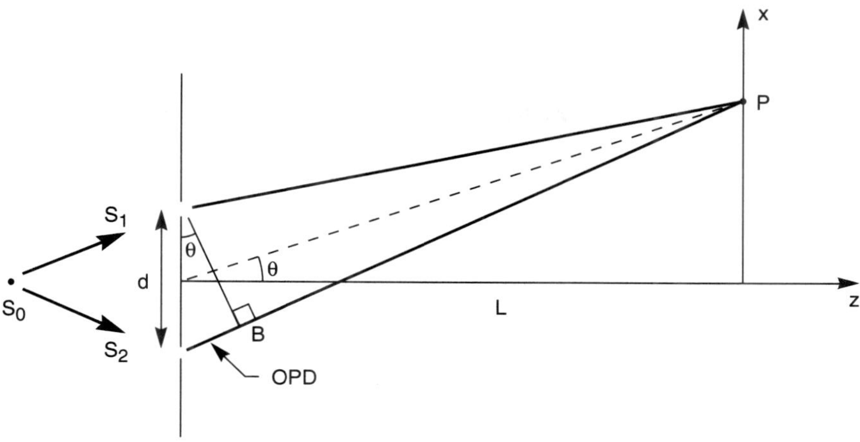

FIGURE 8 Young's double-slit experiment.

screen. This is shown in Fig. 8. S_0 illuminates both S_1 and S_2 and is equidistant from both slits. The OPD for an observation point P at an angle θ or position x is

$$\text{OPD} = \overline{S_2P} - \overline{S_1P} \tag{38}$$

We now draw a line from S_1 to B that is perpendicular to the second ray. Since L is much larger than d, the distances from B to P and S_1 to P are approximately equal. The OPD is then

$$\text{OPD} \approx \overline{S_2B} = d \sin \theta \approx d\theta \approx \frac{dx}{L} \tag{39}$$

and constructive interference or a bright fringe occurs when the OPD is a multiple of the wavelength: $\text{OPD} = m\lambda$, where m is an integer. The condition for the mth order bright fringe is

$$\text{Bright fringe:} \quad \sin(\theta) \approx \theta = \frac{m\lambda}{d} \qquad \text{or} \qquad x = \frac{m\lambda L}{d} \tag{40}$$

This construction is useful not only for interference situations, but also for diffraction analysis.

Effect of Slit Width

The light used to produce the interference pattern is diffracted by the pinholes or slits. Interference is possible only if light is directed in that direction. The overall interference intensity pattern is therefore modulated by the single-slit diffraction pattern (assuming slit apertures):

$$I(x) = I_0 \operatorname{sinc}^2\left(\frac{Dx}{\lambda L}\right)\left[1 + \gamma(x) \cos\left(\frac{2\pi x d}{\lambda L}\right)\right] \tag{41}$$

where D is the slit width, and a one-dimensional expression is shown. The definition of a sinc function is

$$\operatorname{sinc}(\alpha) = \frac{\sin(\pi\alpha)}{\pi\alpha} \tag{42}$$

where the zeros of the function occur when the argument α is an integer. The intensity variation in the y direction is due to diffraction only and is not shown. Since the two slits are assumed to be illuminated by a single source, there are no coherence effects introduced by using a pinhole or slit of finite size.

The term $\gamma(x)$ is included in Eq. (41) to account for variations in the fringe visibility. These could be due to unequal illumination of the two slits, a phase difference of the light reaching the slits, or a lack of temporal or spatial coherence of the source S_0.

Other Arrangements

Several other arrangements for producing interference by division of wavefront are shown in Fig. 9. They all use a single source and additional optical elements to produce two separate and mutually coherent sources. Fresnel's biprism and mirror produce the two

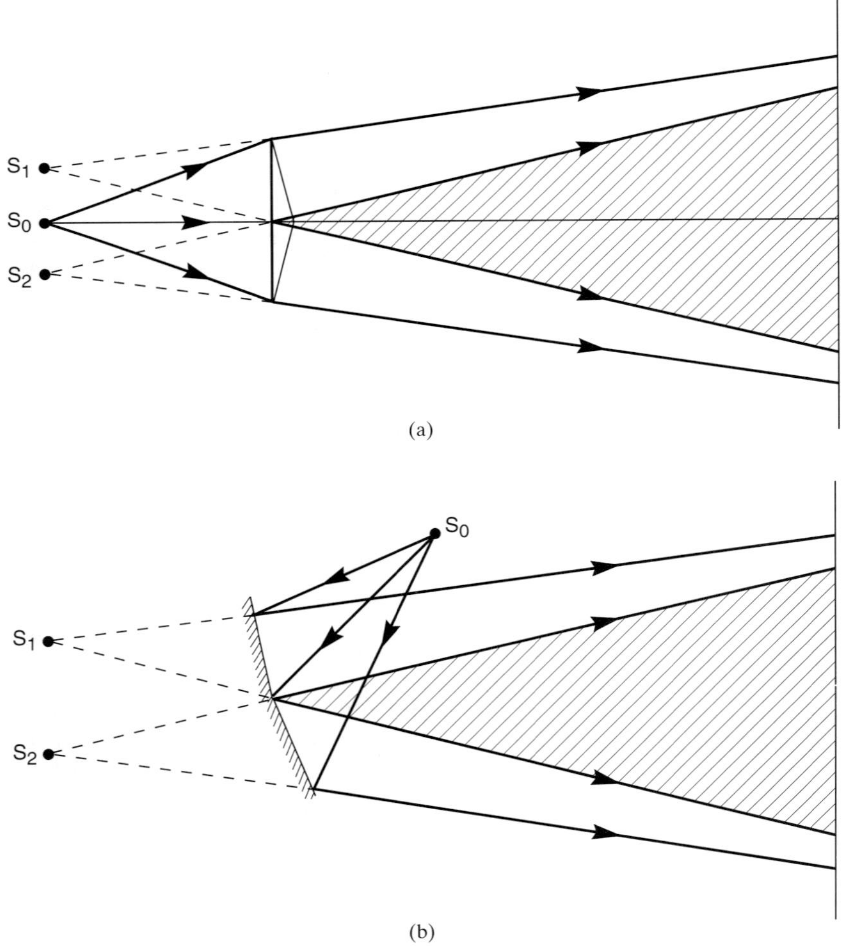

FIGURE 9 Arrangements for interference by division of wavefront: (*a*) Fresnel's biprism; (*b*) Fresnel's mirror; (*c*) Billet's split lens; and (*d*) Lloyd's mirror.

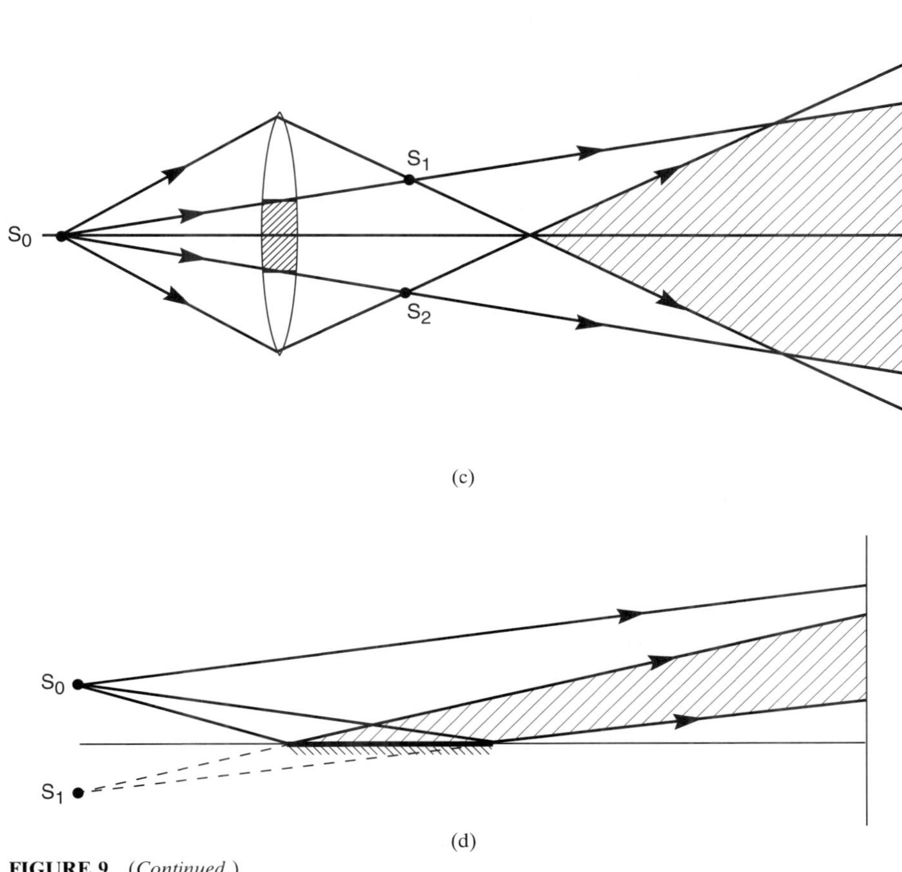

FIGURE 9 (*Continued.*)

virtual source images, Billet's split lens produces two real source images, and Lloyd's mirror produces a single virtual source image as a companion to the original source. Interference fringes form wherever the two resulting waves overlap (shaded regions). One significant difference between these arrangements and Young's two slits is that a large section of the initial wavefront is used instead of just two points. All of these systems are much more light efficient, and they do not rely on diffraction to produce the secondary wavefronts.

In the first three of these systems, a bright fringe is formed at the zero OPD point between the two sources as in the double-slit experiment. With Lloyd's mirror, however, the zero OPD point has a dark fringe. This is due to the π phase shift that is introduced into one of the beams on reflection from the mirror.

Source Spectrum

The simple fringe pattern produced by the two-slit experiment provides a good example to examine the effects of a source with a finite spectrum. In this model, the source can be considered to be a collection of sources, each radiating independently and at a different wavelength. All of these sources are colocated to produce a point source. (Note that this is an approximation, as a true point source must be monochromatic.) At each wavelength, an independent intensity pattern is produced:

$$I(x, \lambda) = I_0\left[1 + \cos\left(\frac{2\pi xd}{\lambda L}\right)\right] = I_0\left[1 + \cos\left(\frac{2\pi \text{ OPD}}{\lambda}\right)\right] \tag{43}$$

where the period of the fringes is $\lambda L/d$, and a fringe visibility of one is assumed. The total intensity pattern is the sum of the individual fringe patterns:

$$I(x) = \int_0^\infty S(\lambda)I(x, \lambda)\, d\lambda = \int_0^\infty S(\nu)I(x, \nu)\, d\nu \tag{44}$$

where $S(\lambda)$ or $S(\nu)$ is the source intensity spectrum which serves as a weighting function.

The effect of this integration can be seen by looking at a simple example where the source is composed of three different wavelengths of equal intensity. To further aid in visualization, let's use Blue (400 nm), Green (500 nm), and Red (600 nm). The result is shown in Fig. 10*a*. There are three cosine patterns, each with a period proportional to the wavelength. The total intensity is the sum of these curves. All three curves line up when the OPD is zero ($x = 0$), and the central bright fringe is now surrounded by two-colored dark fringes. These first dark fringes have a red to blue coloration with increasing OPD. As we get further away from the zero OPD condition, the three patterns get out of phase, the pattern washes out, and the color saturation decreases. This is especially true when the source is composed of more than three wavelengths.

It is common in white light interference situations for one of the two beams to undergo an additional π phase shift. This is the situation in Lloyd's mirror. In this case, there is a central dark fringe at zero OPD with colored bright fringes on both sides. This is shown in Fig. 10*b*, and the pattern is complementary to the previous pattern. In this case the first

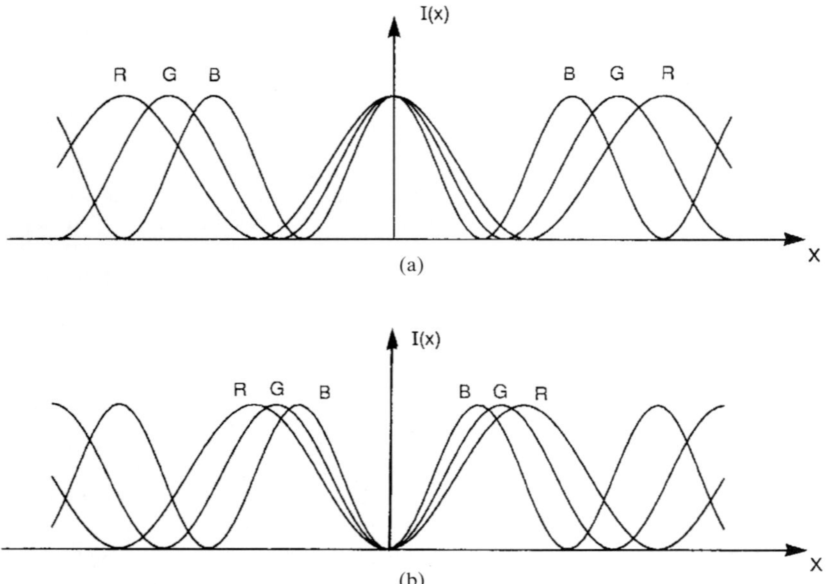

FIGURE 10 The interference pattern produced by a source with three separate wavelengths: (*a*) zero OPD produces a bright fringe; and (*b*) zero OPD produces a dark fringe.

bright fringe shows a blue to red color smear. The dark central fringe is useful in determining the location of zero OPD between the two beams.

The overall intensity pattern and resulting fringe visibility can be computed for a source with a uniform frequency distribution over a frequency range of $\Delta\nu$:

$$I(x) = \frac{1}{\Delta\nu}\int_{\nu_0-\Delta\nu/2}^{\nu_0+\Delta\nu/2} I(x, \nu)\, d\nu = \frac{1}{\Delta\nu}\int_{\nu_0-\Delta\nu/2}^{\nu_0+\Delta\nu/2} I_0\left[1+\cos\left(\frac{2\pi\nu x d}{cL}\right)\right] d\nu \tag{45}$$

where ν_0 is the central frequency, and the $1/\Delta\nu$ term is a normalization factor to assure that the average intensity is I_0. After integration and simplification, the result is

$$I(x) = I_0\left[1 + \operatorname{sinc}\left(\frac{xd\,\Delta\nu}{cL}\right)\cos\left(\frac{2\pi\nu_0 x d}{cL}\right)\right] \tag{46}$$

where the sinc function is defined in Eq. (42). A fringe pattern due to the average optical frequency results, but it is modulated by a sinc function that depends on $\Delta\nu$ and x. The absolute value of the sinc function is the fringe visibility $\gamma(x)$, and it depends on both the spectral width and position of observation. The negative portions of the sinc function correspond to a π phase shift of the fringes.

It is informative to rewrite this expression in terms of the OPD:

$$I(x) = I_0\left[1 + \operatorname{sinc}\left(\frac{\text{OPD}\,\Delta\nu}{c}\right)\cos\left(\frac{2\pi\,\text{OPD}}{\lambda_0}\right)\right] \tag{47}$$

where λ_0 is the wavelength corresponding to ν_0. Good fringe visibility is obtained only when either the spectral width is small (the source is quasi-monochromatic) or the OPD is small. The fringes are *localized* in certain areas of space. This result is consistent with the earlier graphical representations. In the area where the OPD is small, the fringes are in phase for all wavelengths. As the OPD increases, the fringes go out of phase since they all have different periods, and the intensity pattern washes out.

This result turns out to be very general: for an incoherent source, the fringes will be localized in the vicinity of zero OPD. There are two other things we should notice about this result. The first is that the first zero of the visibility function occurs when the OPD equals $c/\Delta\nu$. This distance is known as the *coherence length* as it is the path difference over which we can obtain interference. The second item is that the visibility function is a scaled version of the Fourier transform of the source frequency spectrum. It is evaluated for the OPD at the measurement location. The Fourier transform of a uniform distribution is a sinc function. We will discuss this under "Coherence and Interference" later in the chapter.

2.6 INTERFERENCE BY AMPLITUDE DIVISION

The second general method for producing interference is to use the same section of a wavefront from a single source for both resulting wavefronts. The original wavefront amplitude is split into two or more parts, and each fraction is directed along a different optical path. These waves are then recombined to produce interference. This method is called *interference by amplitude division.* There are a great many interferometer designs based on this method. A few will be examined here, and many more will be discussed in Chap. 21 of Vol. 2, "Interferometers."

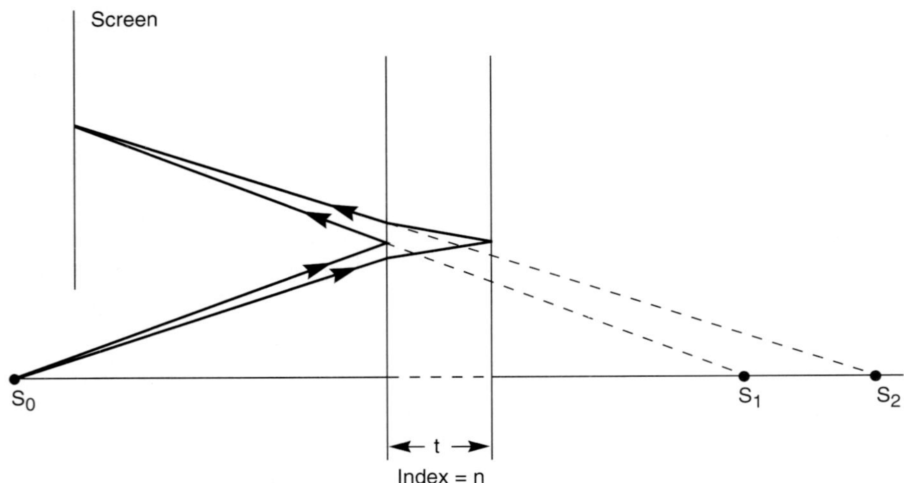

FIGURE 11 Interference from a plane-parallel plate and a point source.

Plane-Parallel Plate

A first example of interference by amplitude division is a plane-parallel plate illuminated by a monochromatic point source. Two virtual images of the point source are formed by the Fresnel reflections at the two surfaces, as shown in Fig. 11. Associated with each of the virtual images is a spherical wave, and interference fringes form wherever these two waves overlap. In this case, this is the volume of space on the source side of the plate. The pattern produced is the same as that found for the interference of two spherical waves (discussed earlier under "Two Spherical Waves"), and nonlocalized fringes are produced. The pattern is symmetric around the line perpendicular to the plate through the source. If a screen is placed along this axis, a pattern similar to circular Newton's ring fringes are produced as described by Eq. (33), where $d = 2t/n$ is now the separation of the virtual sources. The thickness of the plate is t, its index is n, and the distance R is approximately the screen-plate separation plus the source-plate separation. We have ignored multiple reflections in the plate. As with the interference of two spherical waves, the order of interference is a maximum at the center of the pattern.

The interference of two plane waves can be obtained by illuminating a wedged glass plate with a plane wavefront. If the angle of incidence on the first surface is θ and the wedge angle is α, two plane waves are produced at angles θ and $\theta + 2n\alpha$ due to reflections at the front and rear surfaces. Straight equispaced fringes will result in the volume of space where the two reflected waves overlap. The period of these fringes on a screen parallel to the plate is given by Eq. (26), where the two reflected angles are used.

Extended Source

An *extended source* is modeled as a collection of independent point sources. If the source is quasi-monochromatic, all of the point sources radiate at the same nominal frequency, but without a phase relationship. Each point source will produce its own interference pattern, and the net intensity pattern is the sum or integral of all the individual intensity patterns. This is the spatial analogy to the temporal average examined earlier under "Source Spectrum."

With an extended source, the fringes will be localized where the individual fringe position or spacing is not affected by the location of the point sources that comprise the extended source. We know from our previous examples that a bright fringe (or a dark fringe, depending on phase shifts) will occur when the OPD is zero. If there is a location where the OPD is zero independent of source location, all of the individual interference patterns will be in phase, and the net pattern will show good visibility. In fact, the three-dimensional fringe pattern due to a point source will tend to shift or pivot around this zero-OPD location as the point source location is changed. The individual patterns will therefore be out of phase in areas where the OPD is large, and the average intensity pattern will tend to wash out in these regions as the source size increases.

The general rule for fringe visibility with an extended quasi-monochromatic source is that the fringes will be localized in the region where the OPD between the two interfering wavefronts is small. For a wedged glass plate, the fringes are localized in or near the wedge, and the best visibility occurs as the wedge thickness approaches zero and is perhaps just a few wavelengths. The allowable OPD will depend on the source size and the method of viewing the fringes. This result explains why, under natural light, interference effects are seen in thin soap bubbles but not with other thicker glass objects. An important exception to this rule is the plane-parallel plate where the fringes are localized at infinity.

Fringes of Equal Inclination

There is no section of a plane-parallel plate that produces two reflected wavefronts with zero OPD. The OPD is constant, and we would expect, based on the previous section, that no high-visibility fringes would result with an extended source. If, however, a lens is used to collect the light reflected from the plate, fringes are formed in the back focal plane of the lens. This situation is shown in Fig. 12, and any ray leaving the surface at a particular angle θ is focused to the same point P. For each incident ray at this angle, there are two parallel reflected rays: one from the front surface and one from the back surface. The reflections from different locations on the plate at this angle are due to light from different points in the extended source. The OPD for any pair of these reflected rays is the same regardless of the source location. These rays will interfere at P and will all have the same phase difference. High-visibility fringes result. Different points in the image plane correspond to different angles. The formation of these fringes localized at infinity depends on the two surfaces of the plate being parallel.

The OPD between the reflected rays is a function of the angle of incidence θ, the plate index n, and thickness t:

$$\text{OPD} = 2nt \cos \theta' \tag{48}$$

where θ' is the internal angle. Taking into account the half-wave shift due to the phase change difference of π between an internal and an external reflection, a dark fringe will result for angles satisfying

$$2nt \cos \theta' = m\lambda \qquad \text{or} \qquad \cos \theta' = \frac{m\lambda}{2nt} \tag{49}$$

where m is an integer. Since only the angle of incidence determines the properties of the interference (everything else is constant), these fringes are called *fringes of equal inclination.* They appear in the back focal plane of the lens and are therefore localized at infinity since infinity is conjugate to the focal plane. As the observation plane is moved away from the focal plane, the visibility of the fringes will quickly decrease.

When the axis of the lens is normal to the surfaces of the plate, a beamsplitter arrangement is required to allow light from the extended source to be reflected into the lens as shown in Fig. 13. Along the axis, $\theta = \theta' = 90°$, and symmetry requires that the

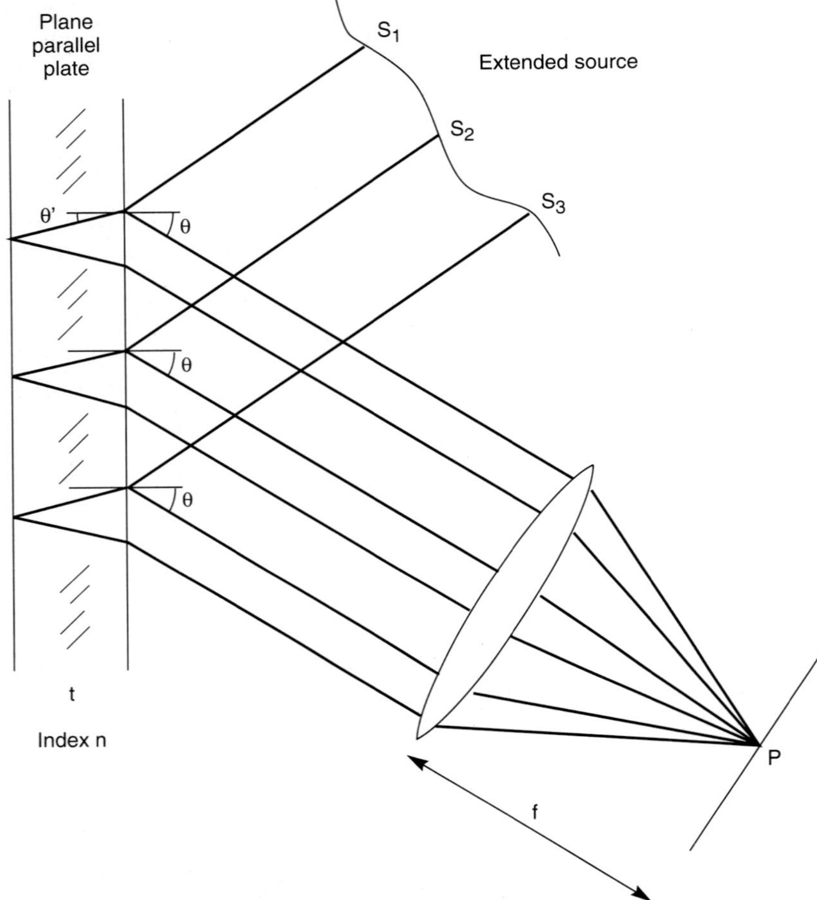

FIGURE 12 The formation of fringes of equal inclination.

fringes are concentric about the axis. In this special case, these fringes are called *Haidinger fringes,* and they are identical in appearance to Newton's rings [Eq. (30)]. If there is an intensity maximum at the center, the radii of the other bright fringes are proportional to the square roots of integers. As with other fringes formed by a plane-parallel plate (discussed earlier), the order of interference decreases with the observation radius on the screen. As θ' increases, the value of m decreases.

Fringes of Equal Thickness

The existence of fringes of equal inclination depends on the incident light being reflected by two parallel surfaces, and the angle of incidence is the mechanism which generates changes in the OPD. There are many arrangements with an extended source where the reflections are not parallel, and the resulting changes in OPD dominate the angle-of-incidence considerations. The fringes produced in this situation are called *fringes of equal*

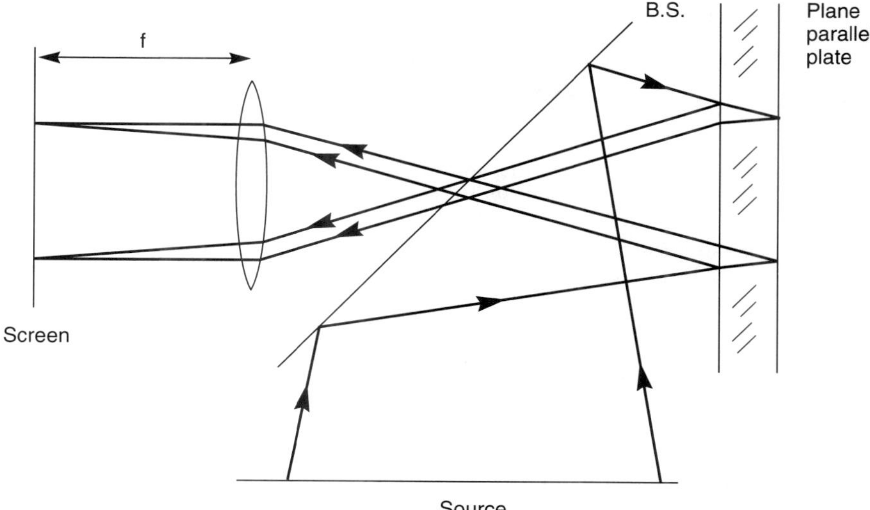

FIGURE 13 The formation of Haidinger fringes.

thickness, and we have stated earlier that they will be localized in regions where the OPD between the two reflections is small.

An example of fringes of equal thickness occurs with a wedged glass plate illuminated by a quasi-monochromatic extended source. We know that for each point in the source, a pattern comprised of equispaced parallel fringes results, and the net pattern is the sum of all of these individual patterns. However, it is easier to examine this summation by looking at the OPD between the two reflected rays reaching an observation point *P* from a source point *S*. This is shown in Fig. 14. The wedge angle is α, the thickness of the plate at this

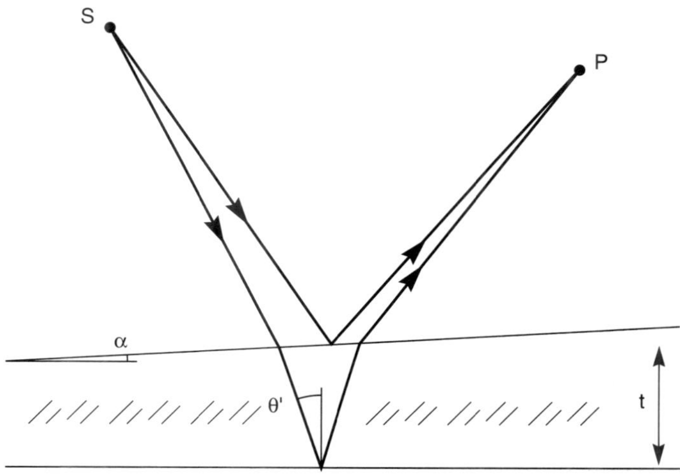

FIGURE 14 The ray path between a point source and an observation point for a wedged plate.

location is t, its index is n, and the internal ray angle is θ'. The exact OPD is difficult to calculate, but under the assumption that α is small and the wedge is sufficiently thin, the following result for the OPD is obtained:

$$\text{OPD} \approx 2nt \cos \theta' \tag{50}$$

As other points on the source are examined, the reflection needed to get light to the observation point will move to a different location on the plate, and different values of both t and θ' will result. Different source points may have greatly different OPDs, and in general the fringe pattern will wash out in the vicinity of P.

This reduction in visibility can be avoided if the observation point is placed in or near the wedge. In this case, all of the paths between S and P must reflect from approximately the same location on the wedge, and the variations in the thickness t are essentially eliminated. The point P where the two reflected rays cross may be virtual. The remaining variations in the OPD are from the different θ's associated with different source points. This variation may be limited by observing the fringe pattern with an optical system having a small entrance pupil. This essentially limits the amount of the source that is used to examine any area on the surface. A microscope or the eye focused on the wedge can be used to limit the angles. If the range of values of θ' is small, high-visibility fringes will appear to be localized at the wedge. The visibility of the fringes will decrease as the wedge thickness increases.

It is common to arrange the system so that the fringes are observed in a direction approximately normal to the surface. Taking into account the additional phase shift introduced at the reflection from one of the surfaces, the conditions for bright and dark fringes are then

$$\text{Bright:} \quad 2nt - \frac{\lambda}{2} = m\lambda \tag{51}$$

and

$$\text{Dark:} \quad 2nt = m\lambda \tag{52}$$

where m is an integer greater than or equal to zero. Since t increases linearly across the wedge, the observed pattern will be straight equispaced fringes.

These same conditions hold for any plate where the two surfaces are not parallel. The surfaces may have any shape, and as long as the surface angles are small and the plate is relatively thin, high-visibility fringes localized in the plate are observed. Along a given fringe the value of m is constant, so that a fringe represents a contour of constant optical path length nt. If the index is constant, we have fringes of equal thickness. The fringes provide a contour map of the plate thickness, and adjacent fringes correspond to a change of thickness of $\lambda/2n$. An irregularly shaped pattern will result from the examination of a plate of irregular thickness.

Thin Films

With the preceding background, we can easily explain the interference characteristics of *thin films*. There are two distinct types of films to consider. The first is a thin film of nonuniform thickness, and examples are soap bubbles and oil films on water. The second type is a uniform film, such as would be obtained by vacuum deposition and perhaps used as an antireflection coating. Both of these films share the characteristic of being extremely thin—usually not more than a few wavelengths thick and often just a fraction of a wavelength thick.

With a nonuniform film, fringes of equal thickness localized in the film are produced.

There will be a dark fringe in regions of the film where it is substantially thinner than a half wave. We are assuming that the film is surrounded by a lower-index medium such as air so that there is an extra π phase shift. If white light is used for illumination, colored bands will be produced similar to those diagramed in Fig. 10*b* (the curves would need to be modified to rescale the x axis to OPD or film thickness). Each color will produce its first maximum in intensity when the optical thickness of the film is a quarter of that wavelength. As the film thickness increases, the apparent fringe color will first be blue, then green, and finally red. These colored fringes are possible because the film is very thin, and the order of interference m is often zero or one [Eqs. (51) and (52)]. The interference patterns in the various colors are just starting to get out of phase, and interference colors are visible. As the film thickness increases, the various wavelength fringes become jumbled, and distinct fringe patterns are no longer visible.

When a uniform thin film is examined with an extended source, fringes of equal inclination localized at infinity are produced. These fringes will be very broad since the thickness of the film is very small, and large angles will be required to obtain the necessary OPD for a fringe [Eq. (49)]. A common use of this type of film is as an antireflection coating. In this application, a uniform coating that has an optical thickness of a quarter wavelength is applied to a substrate. The coating index is lower than the substrate index, so an extra phase shift is not introduced. A wave at normal incidence is reflected by both surfaces of the coating, and these reflected waves are interfered. If the incident wavelength matches the design of the film, the two reflected waves are out of phase and interfere destructively. The reflected intensity will depend on the Fresnel reflection coefficients at the two surfaces, but will be less than that of the uncoated surface. When a different wavelength is used or the angle of incidence is changed, the effectiveness of the antireflection coating is reduced. More complicated film structures comprised of many layers can be produced to modify the reflection or transmission characteristics of the film.

Fizeau Interferometer

The *Fizeau interferometer* compares one optical surface to another by placing them in close proximity. A typical arrangement is shown in Fig. 15, where the extended source is filtered to be quasi-monochromatic. A small air gap is formed between the two optical surfaces, and fringes of equal thickness are observed between the two surfaces. Equations (51) and (52) describe the location of the fringes, and the index of the thin wedge is now that of air. Along a fringe, the gap is of constant thickness, and adjacent fringes correspond to a change of thickness of a half wavelength. This interferometer is sometimes referred to as a *Newton interferometer.*

This type of interferometer is the standard test instrument in an optical fabrication shop. One of the two surfaces is a reference or known surface, and the interferometric comparison of this reference surface and the test surface shows imperfections in the test part. Differences in radii of the two surfaces are also apparent. The fringes are easy to interpret, and differences of as little as a twentieth of a wavelength can be visually measured. These patterns and this interferometer are further discussed in Chap. 30, Vol. 2, "Optical Testing." The interferometer is often used without the beamsplitter, and the fringes are observed in the direct reflection of the source from the parts.

The classic fringe pattern produced by a Fizeau interferometer is *Newton's rings.* These are obtained by comparing a convex sphere to a flat surface. The parabolic approximation for the sag of a sphere of radius R is

$$\text{sag}\,(\rho) = \frac{\rho^2}{2R} \tag{53}$$

and ρ is the radial distance from the vertex of the sphere. If we assume the two surfaces

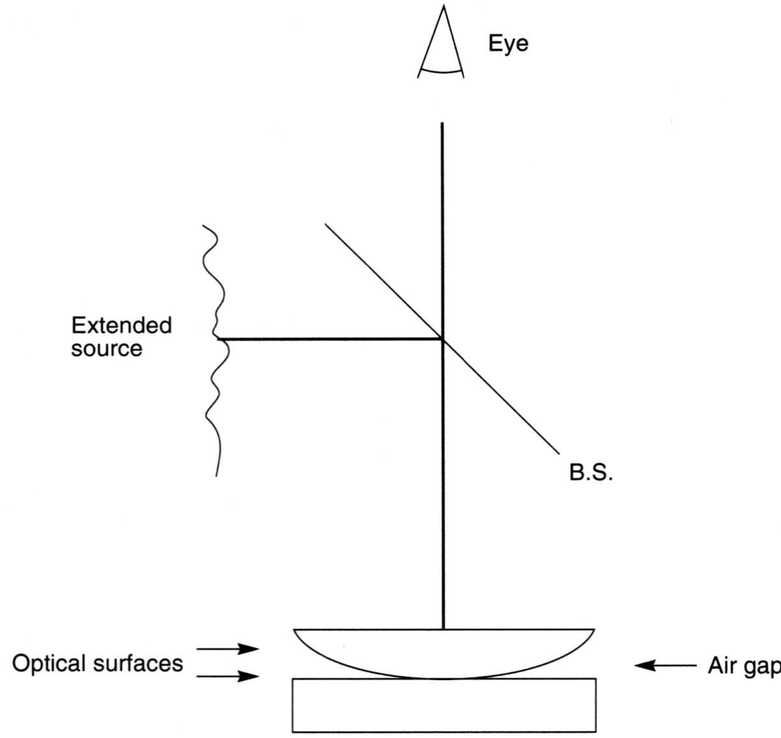

FIGURE 15 Fizeau interferometer.

are in contact at $\rho = 0$, the OPD between the reflected waves is twice the gap, and the condition for a dark fringe is

$$\rho = \sqrt{m\lambda R} \tag{54}$$

Circular fringes that increase in radius as the square root of ρ are observed. Note that a dark fringe occurs at the center of the pattern. In reflection, this point must be dark, as there is no interface at the contact point to produce a reflection.

Michelson Interferometer

There are many two-beam interferometers which allow the surfaces producing the two wavefronts to be physically separated by a large distance. These instruments allow the two wavefronts to travel along different optical paths. One of these is the *Michelson interferometer* diagramed in Fig. 16*a*. The two interfering wavefronts are produced by the reflections from the two mirrors. A plate beamsplitter with one face partially silvered is used, and an identical block of glass is placed in one of the arms of the interferometer to provide the same amount of glass path in each arm. This cancels the effects of the dispersion of the glass beamsplitter and allows the system to be used with white light since the optical path difference is the same for all wavelengths.

Figure 16*b* provides a folded view of this interferometer and shows the relative optical

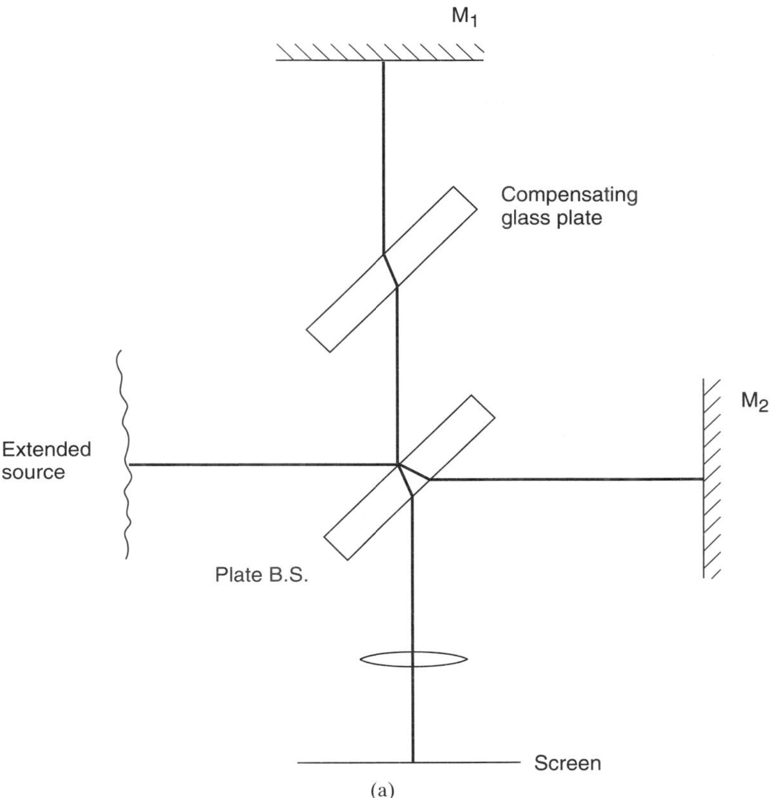

FIGURE 16 Michelson interferometer: (*a*) schematic view; and (*b*) folded view showing the relative optical position of the two mirrors.

position of the two mirrors as seen by the viewing screen. It should be obvious that the two mirrors can be thought of as the two surfaces of a "glass" plate that is illuminated by the source. In this case, the index of the fictitious plate is one, and the reflectivity at the two surfaces is that of the mirrors. Depending on the mirror orientations and shapes, the interferometer either mimics a plane-parallel plate of adjustable thickness, a wedge of arbitrary angle and thickness, or the comparison of a reference surface with an irregular or curved surface. The type of fringes that are produced will depend on this configuration, as well as on the source used for illumination.

When a monochromatic point source is used, nonlocalized fringes are produced, and the imaging lens is not needed. Two virtual-source images are produced, and the resulting fringes can be described by the interference of two spherical waves (discussed earlier). If the mirrors are parallel, circular fringes centered on the line normal to the mirrors result as with a plane-parallel plate. The source separation is given by twice the apparent mirror separation. If the mirrors have a relative tilt, the two source images appear to be laterally displaced, and hyperbolic fringes result. Along a plane bisecting the source images, straight equispaced fringes are observed.

When an extended monochromatic source is used, the interference fringes are localized. If the mirrors are parallel, fringes of equal inclination or Haidinger fringes (as described earlier) are produced. The fringes are localized at infinity and are observed in the rear

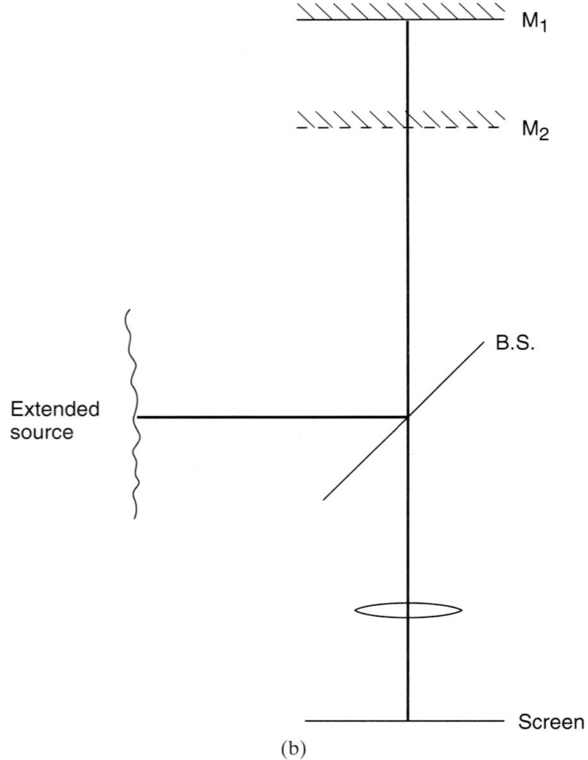

FIGURE 16 (*Continued.*)

focal plane of the imaging lens. Fringes of equal thickness localized at the mirrors are generated when the mirrors are tilted. The apparent mirror separation should be kept small, and the imaging lens should focus on the mirror surface.

If the extended source is polychromatic, colored fringes localized at the mirrors result. They are straight for tilted mirrors. The fringes will have high visibility only if the apparent mirror separation or OPD is smaller than the coherence length of the source. Another way of stating this is that the order of interference m must be small to view the colored fringes. As m increases, the fringes will wash out. The direct analogy here is a thin film. As the mirror separation is varied, the fringe visibility will vary. The fringe visibility as a function of mirror separation is related to the source frequency spectrum (see under "Source Spectrum" and "Coherence and Interference"), and this interferometer forms the basis of a number of spectrometers. This topic is further discussed in Chap. 29, Vol. 2, "Metrology." When the source spectrum is broad, chromatic fringes cannot be viewed with the mirrors parallel. This is because the order of interference for fringes of equal inclination is a maximum at the center of the pattern.

An important variation of the Michelson interferometer occurs when monochromatic collimated light is used. This is the *Twyman-Green interferometer,* and is a special case of point-source illumination with the source at infinity. Plane waves fall on both mirrors, and if the mirrors are flat, nonlocalized equispaced fringes are produced. Fringes of equal thickness can be viewed by imaging the mirrors onto the observation screen. If one of the mirrors is not flat, the fringes represent changes in the surface height. The two surfaces are compared as in the Fizeau interferometer. This interferometer is an invaluable tool for optical testing.

2.7 MULTIPLE BEAM INTERFERENCE

Throughout the preceding discussions, we have assumed that only two waves were being interfered. There are many situations where multiple beams are involved. Two examples are the diffraction grating and a plane-parallel plate. We have been ignoring multiple reflections, and in some instances these extra beams are very important. The net electric field is the sum of all of the component fields. The two examples noted above present different physical situations: all of the interfering beams have a constant intensity with a diffraction grating, and the intensity of the beams from a plane-parallel plate decreases with multiple reflections.

Diffraction Grating

A *diffraction grating* can be modeled as a series of equispaced slits, and the analysis bears a strong similarity to the Young's double slit (discussed earlier). It operates by division of wavefront, and the geometry is shown in Fig. 17. The slit separation is d, the OPD between successive beams for a given observation angle θ is $d \sin(\theta)$, and the corresponding phase difference $\Delta\phi = 2\pi d \sin(\theta)/\lambda$. The field due to the nth slit at a distant observation point is

$$E_j(\theta) = Ae^{i(j-1)\Delta\phi} \qquad j = 1, 2, \ldots, N \tag{55}$$

where all of the beams have been referenced to the first slit, and there are N total slits. The net field is

$$E(\theta) = \sum_{j=1}^{N} E_j(\theta) = A \sum_{j=1}^{N} (e^{i\Delta\phi})^{j-1} \tag{56}$$

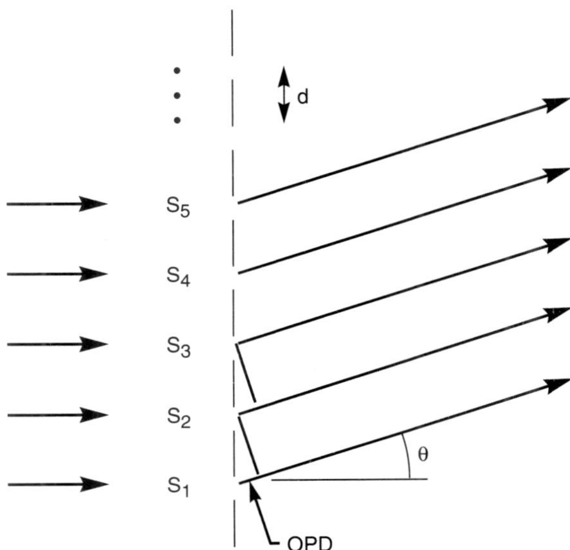

FIGURE 17 Diffraction grating: multiple-beam interference by division of wavefront.

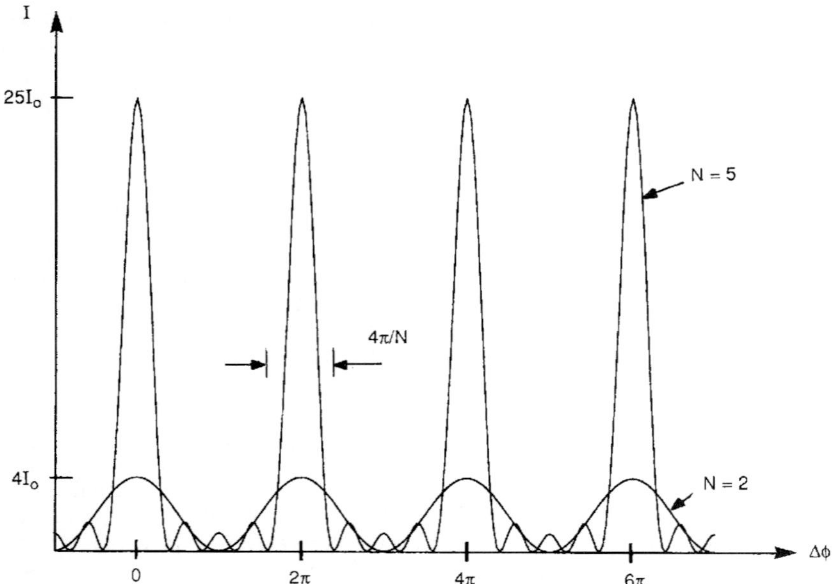

FIGURE 18 The interference patterns produced by gratings with 2 and 5 slits.

which simplifies to

$$E(\theta) = A\left(\frac{1 - e^{iN\Delta\phi}}{1 - e^{i\Delta\phi}}\right) \tag{57}$$

The resulting intensity is

$$I(\theta) = I_0\left[\frac{\sin^2\left(\frac{N\Delta\phi}{2}\right)}{\sin^2\left(\frac{\Delta\phi}{2}\right)}\right] = I_0\left[\frac{\sin^2\left(\frac{N\pi d \sin(\theta)}{\lambda}\right)}{\sin^2\left(\frac{\pi d \sin(\theta)}{\lambda}\right)}\right] \tag{58}$$

where I_0 is the intensity due to an individual slit.

This intensity pattern is plotted in Fig. 18 for $N = 5$. The result for $N = 2$, which is the double-slit experiment, is also shown. The first thing to notice is that the locations of the maxima are the same, independent of the number of slits. A maximum of intensity is obtained whenever the phase difference between adjacent slits is a multiple of 2π. These maxima occur at the diffraction angles given by

$$\sin(\theta) = \frac{m\lambda}{d} \tag{59}$$

where m is an integer. The primary difference between the two patterns is that with multiple slits, the intensity at the maximum increases to N^2 times that due to a single slit, and this energy is concentrated into a much narrower range of angles. The full width of a diffraction peak between intensity zero corresponds to a phase difference $\Delta\phi$ of $4\pi/N$.

The number of intensity zeros between peaks is $N-1$. As the number of slits increases, the angular resolution or resolving power of the grating greatly increases. The effects of a finite slit width can be added by replacing I_0 in Eq. (58) by the single-slit diffraction pattern. This intensity variation forms an envelope for the curve in Fig. 18.

Plane-Parallel Plate

The plane-parallel plate serves as a model to study the interference of multiple waves obtained by division of amplitude. As we shall see, the incremental phase difference between the interfering beams is constant but, in this case, the beams have different intensities. A plate of thickness t and index n with all of the reflected and transmitted beams is shown in Fig. 19. The amplitude reflection and transmission coefficients are ρ and ρ', and τ and τ', where the primes indicate reflection or transmission from within the plate. The first reflected beam is 180° out of phase with the other reflected beams since it is the only beam to undergo an external reflection, and $\rho = -\rho'$. Note that ρ' occurs only in odd powers for the reflected beams. Each successive reflected or transmitted beam is reduced in amplitude by ρ^2. The phase difference between successive reflected or transmitted beams is the same as we found when studying fringes of equal inclination from a plane-parallel plate:

$$\Delta\phi = \left[\frac{4\pi nt \cos(\theta')}{\lambda}\right] \tag{60}$$

where θ' is the angle inside the plate.

The transmitted intensity can be determined by first summing all of the transmitted amplitudes:

$$E(\Delta\phi) = \sum_{j=1}^{\infty} E_j = A\tau\tau' \sum_{j=1}^{\infty} (\rho^2 e^{i\Delta\phi})^{j-1} \tag{61}$$

where the phase is referenced to the first transmitted beam. The result of the summation is

$$E(\Delta\phi) = \left(\frac{A\tau\tau'}{1-\rho^2 e^{i\Delta\phi}}\right) \tag{62}$$

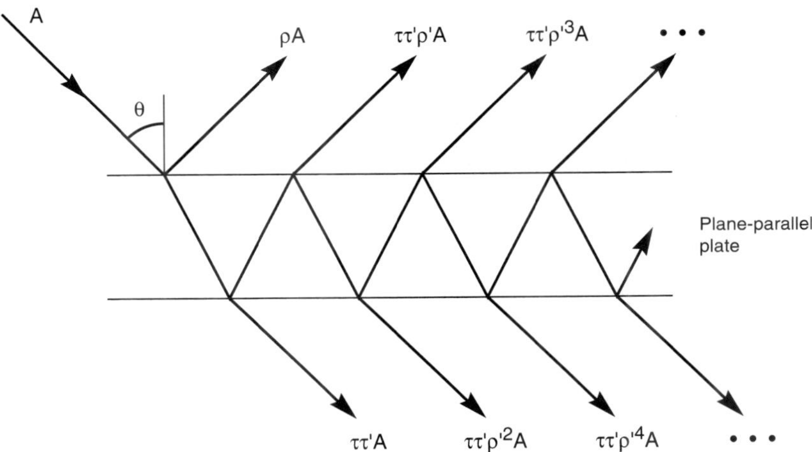

FIGURE 19 Plane-parallel plate: multiple-beam interference by division of amplitude.

The transmitted intensity I_t is the squared modulus of the amplitude which, after simplification, becomes

$$\frac{I_t}{I_0} = \frac{1}{1 + \left(\frac{2\rho}{1-\rho^2}\right)^2 \sin^2(\Delta\phi/2)} \tag{63}$$

where I_0 is the incident intensity. We have also assumed that there is no absorption in the plate, and therefore $\tau\tau' + \rho^2 = 1$. Under this condition of no absorption, the sum of the reflected and transmitted light must equal the incident light: $I_t + I_r = I_0$. The expressions for the transmitted and reflected intensities are then

$$\frac{I_t}{I_0} = \frac{1}{1 + F\sin^2(\Delta\phi/2)} \tag{64}$$

and

$$\frac{I_r}{I_0} = \frac{F\sin^2(\Delta\phi/2)}{1 + F\sin^2(\Delta\phi/2)} \tag{65}$$

and F is defined as

$$F \equiv \left(\frac{2\rho}{1-\rho^2}\right)^2 \tag{66}$$

F is the *coefficient of finesse* of the system and is a function of the surface reflectivity only. The value of F will have a large impact on the shape of the intensity pattern. Note that the reflected intensity could also have been computed by summing the reflected beams.

A maximum of transmitted intensity, or a minimum of reflected intensity, will occur when $\Delta\phi/2 = m\pi$, where m is an integer. Referring back to Eq. (60), we find that this corresponds to the angles

$$\cos\theta' = \frac{m\lambda}{2nt} \tag{67}$$

This is exactly the same condition that was found for a plane-parallel plate with two beams [Eq. (49)]. With an extended source, fringes of equal inclination are formed, and they are localized at infinity. They must be at infinity since all of the reflected or transmitted beams are parallel for a given input angle. The fringes are observed in the rear focal plane of a viewing lens. If the optical axis of this lens is normal to the surface, circular fringes about the axis are produced. The locations of the maxima and minima of the fringes are the same as were obtained with two-beam interference.

The shape of the intensity profile of these multiple beam fringes is not sinusoidal, as it was with two beams. A plot of the transmitted fringe intensity [Eq. (64)] as a function of $\Delta\phi$ is shown in Fig. 20 for several values of F. When the phase difference is a multiple of 2π, we obtain a bright fringe independent of F or ρ. When F is small, low-visibility fringes are produced. When F is large, however, the transmitted intensity is essentially zero unless the phase has the correct value. It drops off rapidly for even small changes in $\Delta\phi$. The transmitted fringes will be very narrow bright circles on an essentially black background. The reflected intensity pattern is one minus this result, and the fringe pattern will be very dark bands on a uniform bright background. The reflected intensity profile is plotted in Fig. 21 for several values of F.

The value of F is a strong function of the surface reflectivity $R = \rho^2$. We do not obtain appreciable values of F until the reflectivity is approximately one. For example, $R = 0.8$ produces $F = 80$, while $R = 0.04$ gives $F = 0.17$. This latter case is typical for uncoated glass, and dim broad fringes in reflection result, as in Fig. 21. The pattern is approximately sinusoidal, and it is clear that our earlier assumptions about ignoring multiple reflections when analyzing a plane-parallel plate are valid for many low-reflectivity situations.

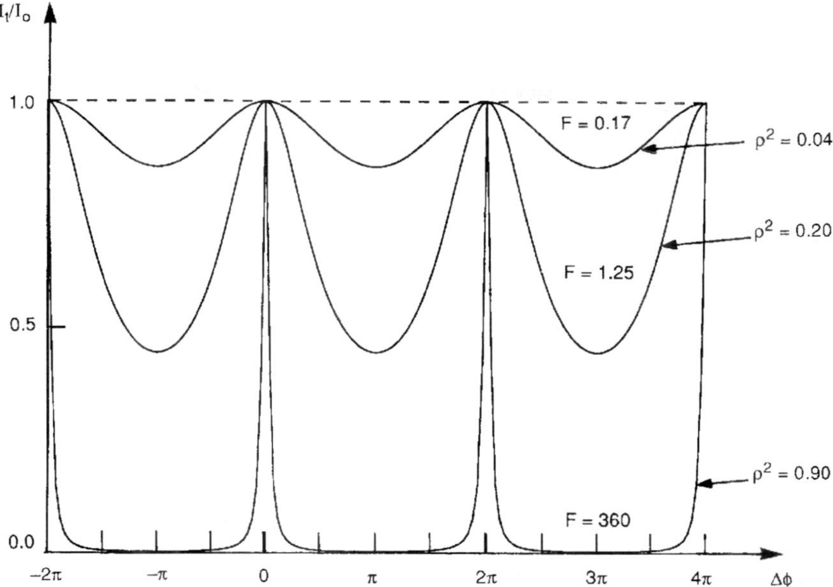

FIGURE 20 The transmitted intensity of a multiple-beam interference pattern produced by a plane-parallel plate.

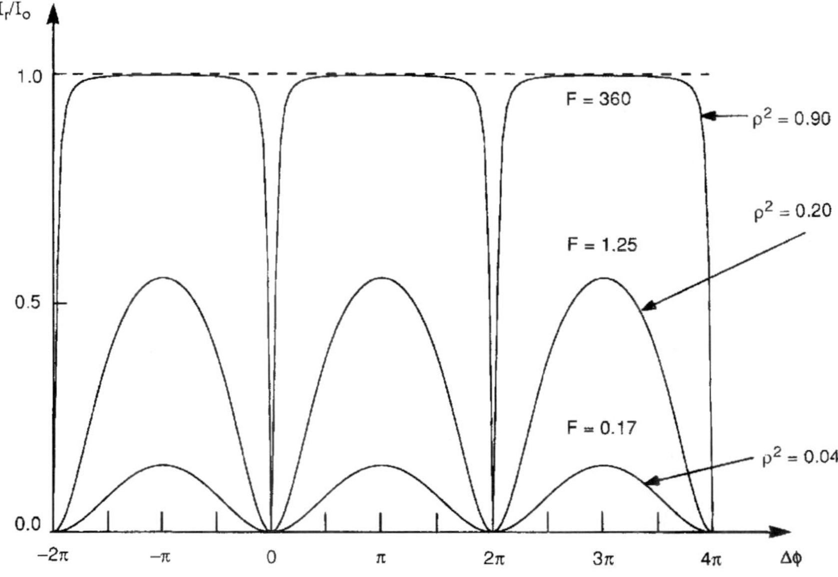

FIGURE 21 The reflected intensity of a multiple-beam interference pattern produced by a plane-parallel plate.

The multiple beam interference causes an energy redistribution much like that obtained from a diffraction grating. A strong response is obtained only when all of the reflected beams at a given angle add up in phase. The difference between this pattern and that of a diffraction pattern is that there are no oscillations or zeros between the transmitted intensity maxima. This is a result of the unequal amplitudes of the interfering beams. With a diffraction grating, all of the beams have equal amplitude, and the resultant intensity oscillates as more beams are added.

Multiple-beam fringes of equal thickness can be produced by two high-reflectivity surfaces in close proximity in a Fizeau interferometer configuration. The dark fringes will narrow to sharp lines, and each fringe will represent a contour of constant OPD between the surfaces. As before, a dark fringe corresponds to a gap of an integer number of half wavelengths. The area between the fringes will be bright. The best fringes will occur when the angle and the separation between the surfaces is kept small. This will prevent the multiple reflections from walking off or reflecting out of the gap.

Fabry-Perot Interferometer

The *Fabry-Perot interferometer* is an important example of a system which makes use of multiple-beam interference. This interferometer serves as a high-resolution spectrometer and also as an optical resonator. In this latter use, it is an essential component of a laser. The system is diagrammed in Fig. 22, and it consists of two highly reflective parallel surfaces separated by a distance t. These two separated reflective plates are referred to as a *Fabry-Perot etalon* or cavity, and an alternate arrangement has the reflected coatings applied to the two surfaces of a single glass plate. The two lenses serve to collimate the light from a point on the extended source in the region of the cavity and to then image this point onto the screen. The screen is located in the focal plane of the lens so that fringes of equal inclination localized at infinity are viewed. As we have seen, light of a fixed wavelength will traverse the etalon only at certain well-defined angles. Extremely sharp multiple-beam circular fringes in transmission are produced on the screen, and their profile is the same as that shown in Fig. 20.

If the source is not monochromatic, a separate independent circular pattern is formed for each wavelength. Equation (67) tells us that the location or scale of the fringes is dependent on the wavelength. If the source is composed of two closely spaced wavelengths, the ring structure is doubled, and the separation of the two sets of rings allows the hyperfine structure of the spectral lines to be evaluated directly. More

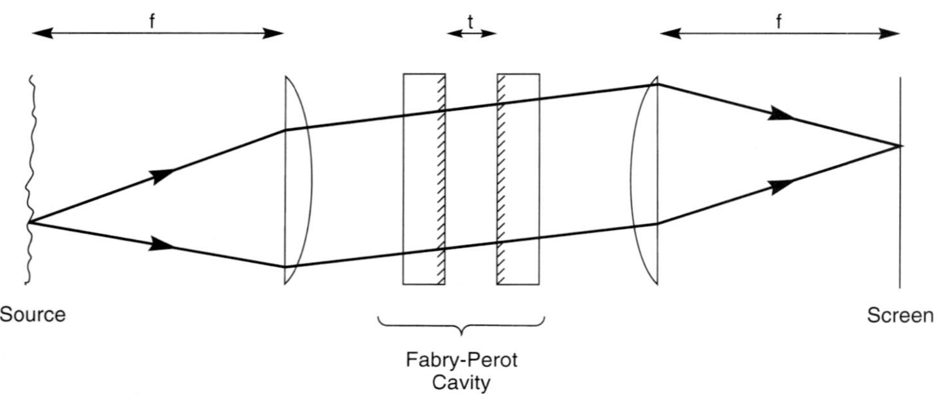

FIGURE 22 Fabry-Perot interferometer.

complicated spectra, usually composed of discrete spectral lines, can also be measured. This analysis is possible even though the order of interference is highest in the center of the pattern. If the phase change $\Delta\phi$ due to the discrete wavelengths is less than the phase change between adjacent fringes, nonoverlapping sharp fringes are seen.

A quantity that is often used to describe the performance of a Fabry-Perot cavity is the *finesse* $\mathcal{F}$. It is a measure of the number of resolvable spectral lines, and is defined as the ratio of the phase difference between adjacent fringes to the full width-half maximum FWHM of a single fringe. Since the fringe width is a function of the coefficient of finesse, the finesse itself is also a strong function of reflectivity. The phase difference between adjacent fringes is 2π, and the half width-half maximum can be found by setting Eq. (64) equal to $\frac{1}{2}$ and solving for $\Delta\phi$. The FWHM is twice this value, and under the assumption that F is large,

$$\text{FWHM} = \frac{4}{\sqrt{F}} \tag{68}$$

and the finesse is

$$\mathcal{F} \equiv \frac{2\pi}{\text{FWHM}} = \frac{\pi\sqrt{F}}{2} = \frac{\pi\rho}{1-\rho^2} = \frac{\pi\sqrt{R}}{1-R} \tag{69}$$

where ρ is the amplitude reflectivity, and R is the intensity reflectivity. Typical values for the finesse of a cavity with flat mirrors is about 30 and is limited by the flatness and parallelism of the mirrors. There are variations in $\Delta\phi$ across the cavity. Etalons consisting of two curved mirrors can be constructed with a much higher finesse, and values in excess of 10,000 are available.

Another way of using the Fabry-Perot interferometer as a spectrometer is suggested by rewriting the transmission [Eq. (64)] in terms of the frequency ν:

$$T = \frac{I_t}{I_0} = \frac{1}{1 + F\sin^2(2\pi t\nu/c)} \tag{70}$$

where Eq. (60) relates the phase difference to the wavelength, t is the mirror separation, and an index of one and normal incidence ($\theta' = 0$) have been assumed. This function is plotted in Fig. 23, and a series of transmission spikes separated in frequency by $c/2t$ are

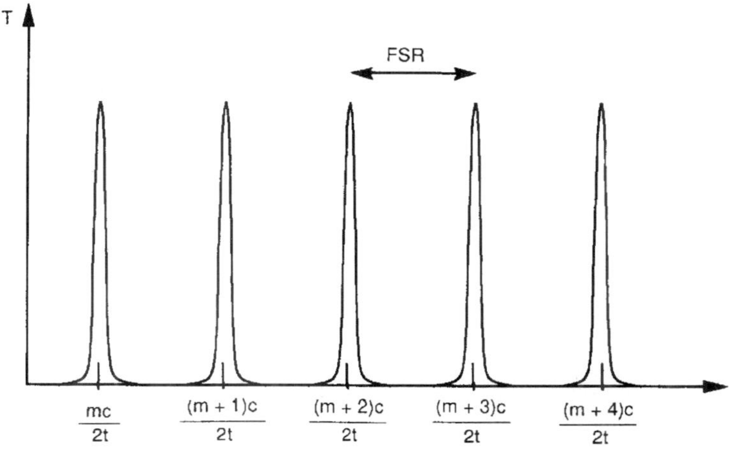

FIGURE 23 The transmission of a Fabry-Perot cavity as a function of frequency.

seen. A maximum occurs whenever the value of the sine is zero. The separation of these maxima is known as the *free spectral range,* FSR. If the separation of the mirrors is changed slightly, these transmission peaks will scan the frequency axis. Since the order of interference m is usually very large, it takes only a small change in the separation to move the peaks by one FSR. In fact, to scan one FSR, the required change in separation is approximately t/m. If the on-axis transmitted intensity is monitored while the mirror separation is varied, a high-resolution spectrum of the source is obtained. The source spectrum must be contained within one free spectral range so that the spectrum is probed by a single transmission peak at a time. If this were not the case, the temporal signal would contain simultaneous contributions from two or more frequencies resulting from different transmission peaks. Under this condition there are overlapping orders, and it is often prevented by using an auxiliary monochromator with the scanning Fabry-Perot cavity to preselect or limit the frequency range of the input spectrum. The resolution $\Delta\nu$ of the trace is limited by the finesse of the cavity.

For a specific cavity, the value of m at a particular transmission peak, and some physical insight into the operation of this spectrometer, is obtained by converting the frequency of a particular transmission mode $mc/2t$ into wavelength:

$$\lambda = \frac{2t}{m} \qquad \text{or} \qquad t = m\frac{\lambda}{2} \tag{71}$$

For the mth transmission maximum, exactly m half waves fit across the cavity. This also implies that the round-trip path within the cavity is an integer number of wavelengths. Under this condition, all of the multiply-reflected beams are in phase everywhere in the cavity, and therefore all constructively interfere. A maximum in the transmission occurs. Other maxima occur at different wavelengths, but these specific wavelengths must also satisfy the condition that the cavity spacing is an integer number of half wavelengths.

These results also allow us to determine the value of m. If a 1-cm cavity is used and the nominal wavelength is 500 nm, $m = 40{,}000$ and $\text{FSR} = 1.5 \times 10^{10}$ Hz. The wavelength interval corresponding to this FSR is 0.0125 nm. If a 1-mm cavity is used instead, the results are $m = 4000$ and $\text{FSR} = 1.5 \times 10^{11}\ \text{Hz} = 0.125$ nm. We see now that to avoid overlapping orders, the spectrum must be limited to a very narrow range, and this range is a function of the spacing. Cavities with spacings of a few tens of μm's are available to increase the FSR. Increasing the FSR does have a penalty. The finesse of a cavity depends only on the reflectivities, so as the FSR is increased by decreasing t, the FWHM of the transmission modes increases to maintain a constant ratio. The number of resolvable spectrum lines remains constant, and the absolute spectral resolution decreases.

A mirror translation of a half wavelength is sufficient to cover the FSR of the cavity. The usual scanning method is to separate the two mirrors with a piezoelectric spacer. As the applied voltage is changed, the cavity length will also change. An alternate method is to change the index of the air in the cavity by changing the pressure.

2.8 COHERENCE AND INTERFERENCE

The observed fringe visibility is a function of the spatial and temporal coherence of the source. The classical assumption for the analysis is that every point on an extended source radiates independently and therefore produces its own interference pattern. The net intensity is the sum of all of the individual intensity patterns. In a similar manner, each wavelength or frequency of a nonmonochromatic source radiates independently, and the temporal average is the sum of the individual temporal averages. *Coherence theory* allows the interference between the light from two point sources to be analyzed, and a good visual model is an extended source illuminating the two pinholes in Young's double slit. We need to determine the relationship between the light transmitted through the two pinholes. Coherence theory also accounts for the effects of the spectral bandwidth of the source.

With interference by division of amplitude using an extended source, the light from many point sources is combined at the observation point, and the geometry of the interferometer determines where the fringes are localized. Coherence theory will, however, predict the spectral bandwidth effects for division of amplitude interference. Each point on the source is interfered with an image of that same point. The temporal coherence function relates the interference of these two points independently of other points on the source. The visibility function for the individual interference pattern due to these two points is computed, and the net pattern is the sum of these patterns for the entire source. The temporal coherence effects in division of amplitude interference are handled on a point-by-point basis across the source.

In this section, the fundamentals of coherence theory as it relates to interference are introduced. Much more detail on this subject can be found in Chap. 4, "Coherence Theory."

Mutual Coherence Function

We will consider the interference of light from two point sources or pinholes. This light is derived from a common origin so that there may be some relationship between the complex fields at the two sources. We will represent these amplitudes at the pinholes as $E_1(t)$ and $E_2(t)$, as shown in Fig. 24. The propagation times between the two sources and the observation point are t_1 and t_2, where the times are related to the optical path lengths by $t_i = \mathrm{OPL}_i/c$. The two complex amplitudes at the observation point are then $E_1(t - t_1)$ and $E_2(t - t_2)$, where the amplitudes have been scaled to the observation plane. The time-average intensity at the observation point can be found by returning to Eq. (13), which is repeated here with the time dependence:

$$I = I_1 + I_2 + \langle E_1(t - t_1)E_2^*(t - t_2)\rangle + \langle E_1^*(t - t_1)E_2(t - t_2)\rangle \tag{72}$$

where I_1 and I_2 are the intensities due to the individual sources. If we now shift our time origin by t_2, we obtain

$$I = I_1 + I_2 + \langle E_1(t + \tau)E_2^*(t)\rangle + \langle E_1^*(t + \tau)E_2(t)\rangle \tag{73}$$

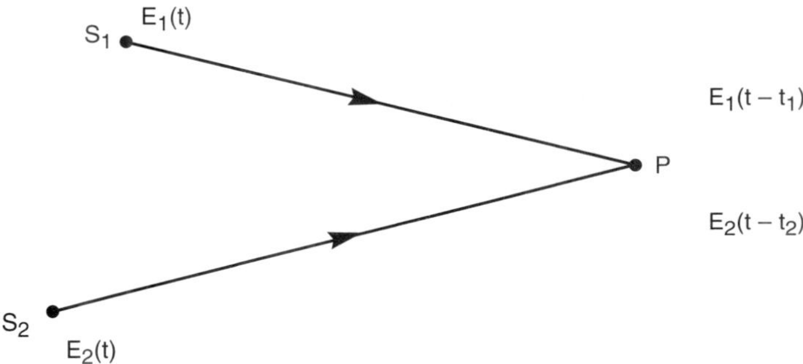

FIGURE 24 Geometry for examining the mutual coherence of two sources.

where

$$\tau = t_2 - t_1 = \frac{\text{OPL}_2 - \text{OPL}_1}{c} = \frac{\text{OPD}}{c} \tag{74}$$

The difference in transit times for the two paths is τ. The last two terms in the expression for the intensity are complex conjugates, and they contain the interference terms.

We will now define the *mutual coherence function* $\Gamma_{12}(\tau)$:

$$\Gamma_{12}(\tau) = \langle E_1(t + \tau)E_2^*(t)\rangle \tag{75}$$

which is the cross correlation of the two complex amplitudes. With this identification, the intensity of the interference pattern is

$$I = I_1 + I_2 + \Gamma_{12}(\tau) + \Gamma_{12}^*(\tau) \tag{76}$$

or, recognizing that a quantity plus its complex conjugate is twice the real part,

$$I = I_1 + I_2 + 2\,\text{Re}\,\{\Gamma_{12}(\tau)\} \tag{77}$$

It is convenient to normalize the mutual coherence function by dividing by the square root of the product of the two self-coherence functions. The result is the *complex degree of coherence*:

$$\gamma_{12}(\tau) = \frac{\Gamma_{12}(\tau)}{\sqrt{\Gamma_{11}(0)\Gamma_{22}(0)}} = \frac{\Gamma_{12}(\tau)}{\sqrt{\langle|E_1(t)|^2\rangle\langle|E_2(t)|^2\rangle}} = \frac{\Gamma_{12}(\tau)}{\sqrt{I_1 I_2}} \tag{78}$$

and the intensity can be rewritten:

$$I = I_1 + I_2 + 2\sqrt{I_1 I_2}\,\text{Re}\,\{\gamma_{12}(\tau)\} \tag{79}$$

We can further simplify the result by writing $\gamma_{12}(\tau)$ as a magnitude and a phase:

$$\gamma_{12}(\tau) = |\gamma_{12}(\tau)|\,e^{i\phi_{12}(\tau)} = |\gamma_{12}(\tau)|\,e^{i[\alpha_{12}(\tau) - \Delta\phi(\tau)]} \tag{80}$$

where $\alpha_{12}(\tau)$ is associated with the source, and $\Delta\phi(\tau)$ is the phase difference due to the OPD between the two sources and the observation point [Eq. (18)]. The quantity $|\gamma_{12}(\tau)|$ is known as the *degree of coherence.* The observed intensity is therefore

$$I = I_1 + I_2 + 2\sqrt{I_1 I_2}\,|\gamma_{12}(\tau)|\cos\,[\alpha_{12}(\tau) - \Delta\phi(\tau)] \tag{81}$$

The effect of $\alpha_{12}(\tau)$ is to add a phase shift to the intensity pattern. The fringes will be shifted. A simple example of this situation is Young's double-slit experiment illuminated by a tilted plane wave or a decentered source. With quasi-monochromatic light, the variations of both $|\gamma_{12}(\tau)|$ and $\alpha_{12}(\tau)$ with τ are slow with respect to changes of $\Delta\phi(\tau)$, so that the variations in the interference pattern in the observation plane are due primarily to changes in $\Delta\phi$ with position.

A final rewrite of Eq. (81) leads us to the intensity pattern at the observation point:

$$I = I_0\left\{1 + \frac{2\sqrt{I_1 I_2}}{I_1 + I_2}|\gamma_{12}(\tau)|\cos\,[\alpha_{12}(\tau) - \Delta\phi(\tau)]\right\} \tag{82}$$

where $I_0 = I_1 + I_2$. The fringe visibility is therefore

$$\gamma(\tau) = \frac{2\sqrt{I_1 I_2}}{I_1 + I_2}|\gamma_{12}(\tau)| \tag{83}$$

and is a function of the degree of coherence and τ. Remember that τ is just the temporal measure of the OPD between the two sources and the observation point. If the two intensities are equal, the fringe visibility is simply the degree of coherence: $\gamma(\tau) = |\gamma_{12}(\tau)|$. The degree of coherence will take on values betwen 0 and 1. The source is *coherent* when $|\gamma_{12}(\tau)| = 1$, and completely incoherent when $|\gamma_{12}(\tau)| = 0$. The source is said to be *partially coherent* for other values. No fringes are observed with an incoherent source, and the visibility is reduced with a partially coherent source.

Spatial Coherence

The spatial extent of the source and its distance from the pinholes will determine the visibility of the fringes produced by the two pinhole sources (see Fig. 25). Each point on the source will produce a set of Young's fringes, and the position of this pattern in the observation plane will shift with source position. The value of $\alpha_{12}(\tau)$ changes with source position. The existence of multiple shifted patterns will reduce the overall visibility. As an example, consider a quasi-monochromatic source that consists of a several point sources arranged in a line. Each produces a high modulation fringe pattern in the observation plane (Fig. 26*a*), but there is a lateral shift between each pattern. The net pattern shows a fringe with the same period as the individual patterns, but it has a reduced modulation due to the shifts (Fig. 26*b*). This reduction in visibility can be predicted by calculating the degree of coherence $|\gamma_{12}(\tau)|$ at the two pinholes.

Over the range of time delays between the interfering beams that are usually of interest, the degree of coherence is a slowly varying function and is approximately equal to the value at $\tau = 0$: $|\gamma_{12}(\tau)| = |\gamma_{12}(0)| = |\gamma_{12}|$. The *van Cittert–Zernike theorem* allows the degree of coherence in the geometry of Fig. 25 to be calculated. Let θ be the angular separation of the two pinholes as seen from the source. This theorem states that degree of coherence between two points is the modulus of the scaled and normalized Fourier transform of the source intensity distribution:

$$|\gamma_{12}| = \left| \frac{\iint_S I(\xi, \eta) e^{i(2\pi/\lambda)(\xi\theta_x + \eta\theta_y)} d\xi \, d\eta}{\iint_S I(\xi, \eta) \, d\xi \, d\eta} \right| \tag{84}$$

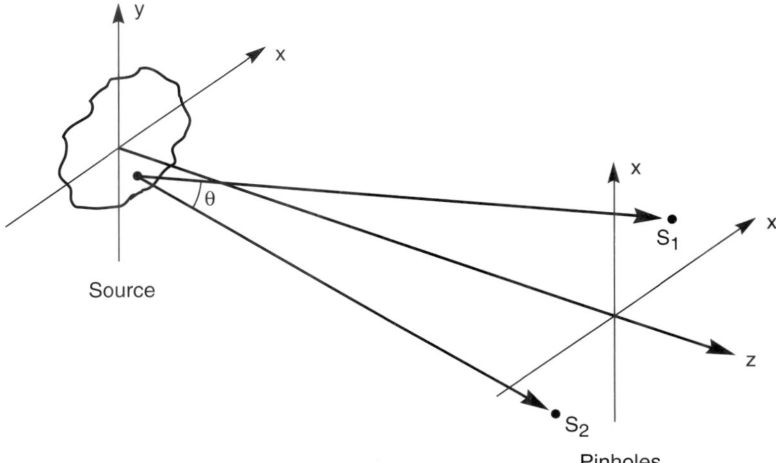

FIGURE 25 An extended source illuminating two pinholes.

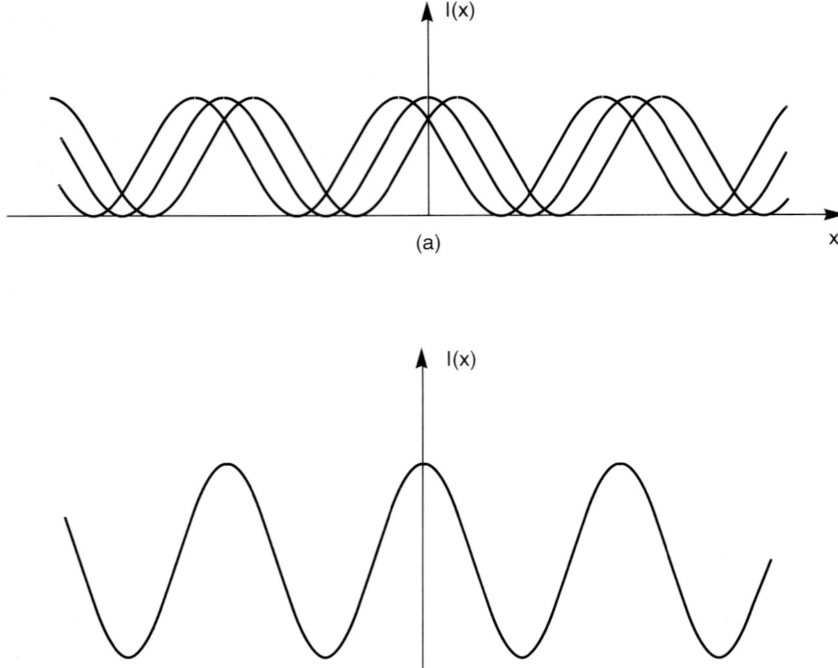

FIGURE 26 The interference pattern produced by a linear source: (*a*) the individual fringe patterns; and (*b*) the net fringe pattern with reduced visibility.

where θ_x and θ_y are the x and y components of the pinhole separation θ, and the integral is over the source.

Two cases that are of particular interest are a slit source and a circular source. The application of the van Cittert–Zernike theorem yields the two coherence functions:

$$\text{Slit source of width } w\text{:} \quad |\gamma_{12}| = \left|\text{sinc}\left(\frac{w\theta_x}{\lambda}\right)\right| = \left|\text{sinc}\left(\frac{wa}{\lambda z}\right)\right| \tag{85}$$

$$\text{Circular source of diameter } d\text{:} \quad |\gamma_{12}| = \left|\frac{2J_1\left(\frac{\pi d\theta_x}{\lambda}\right)}{\frac{\pi d\theta_x}{\lambda}}\right| = \left|\frac{2J_1\left(\frac{\pi da}{\lambda z}\right)}{\frac{\pi da}{\lambda z}}\right| \tag{86}$$

where a is the separation of the pinholes, z is the distance from the source to the pinholes, the sinc function is defined by Eq. (42), and J_1 is a first-order Bessel function. The pinholes are assumed to be located on the x-axis. These two functions share the common characteristic of a central core surrounded by low-amplitude side lobes. We can imagine these functions of pinhole spacing mapped onto the aperture plane. The coherence function is centered on one of the pinholes. If the other pinhole is then within the central core, high-visibility fringes are produced. If the pinhole spacing places the second pinhole outside the central core, low-visibility fringes result.

Michelson Stellar Interferometer

The *Michelson stellar interferometer* measures the diameter of stars by plotting out the degree of coherence due to the light from the star. The system is shown in Fig. 27. Two small mirrors separated by the distance a sample the light and serve as the pinholes. The spacing between these mirrors can be varied. This light is then directed along equal path lengths into a telescope, and the two beams interfere in the image plane. To minimize chromatic effects, the input light should be filtered to a small range of wavelengths. The modulation of the fringes is measured as a function of the mirror spacing to measure the degree of coherence in the plane of the mirrors. This result will follow Eq. (86) for a circular star, and the fringe visibility will go to zero when $a = 1.22\lambda\alpha$, where $\alpha = d/z$ is the angular diameter of the star. We measure the mirror separation that produces zero visibility to determine α. In a similar manner, this interferometer can be used to measure the spacing of two closely spaced stars.

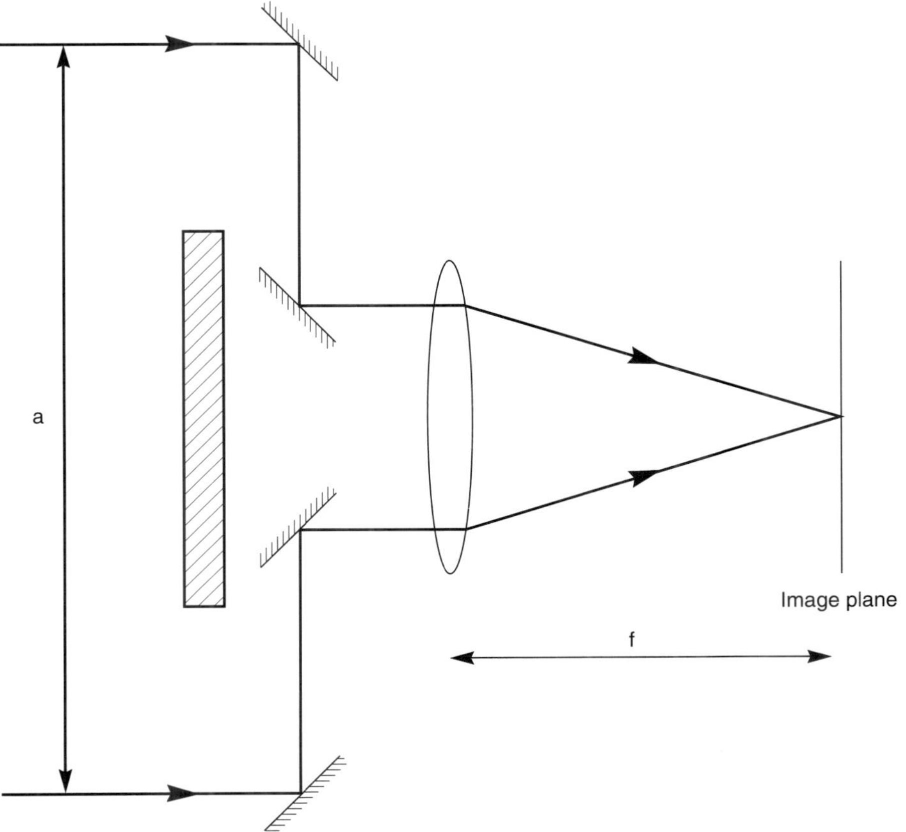

FIGURE 27 Michelson stellar interferometer.

Temporal Coherence

When examining temporal coherence effects, we use a source of small dimensions (a point source) that radiates over a range of wavelengths. The light from this source is split into two beams and allowed to interfere. One method to do this is to use an amplitude-splitting interferometer. Since the two sources are identical, the mutual coherence function becomes the *self-coherence function* $\Gamma_{11}(\tau)$. Equal-intensity beams are assumed. The complex degree of temporal coherence becomes

$$\gamma_{11}(\tau) = \frac{\Gamma_{11}(\tau)}{\Gamma_{11}(0)} = \frac{\langle E_1(t+\tau)E_1^*(t)\rangle}{\langle |E_1(t)|^2\rangle} \tag{87}$$

After manipulation, it follows from this result that $\gamma_{11}(\tau)$ is the normalized Fourier transform of the source intensity spectrum $S(\nu)$:

$$\gamma_{11}(\tau) = \frac{FT\{S(\nu)\}}{\int_0^\infty S(\nu)\,d\nu} = \frac{\int_0^\infty S(\nu)e^{i2\pi\nu\tau}\,d\nu}{\int_0^\infty S(\nu)\,d\nu} \tag{88}$$

The fringe visibility is the modulus of this result. Since $\gamma_{11}(\tau)$ has a maximum at $\tau = 0$, the maximum fringe visibility will occur when the time delay between the two beams is zero. This is consistent with our earlier observation under "Source Spectrum" that the fringes will be localized in the vicinity of zero OPD.

As an example, we will repeat the earlier problem of a uniform source spectrum:

$$S(\nu) = \text{rect}\left(\frac{\nu - \nu_0}{\Delta\nu}\right) \tag{89}$$

where ν_0 is the average frequency and $\Delta\nu$ is the bandwidth. The resulting intensity pattern is

$$I = I_0\{1 + \text{Re}\{\gamma_{12}(\tau)\}\} = I_0[1 + \text{sinc}(\tau\Delta\nu)\cos(2\pi\tau\nu_0)] \tag{90}$$

where the sinc function is the Fourier transform of the rect function. Using $\tau = \text{OPD}/c$ from Eq. (74), we can rewrite this equation in terms of the OPD to obtain the same result expressed in Eq. (47).

Laser Sources

The laser is an important source for interferometry, as it is a bright source of coherent radiation. Lasers are not necessarily monochromatic, as they may have more than one longitudinal mode, and it is important to understand the unique temporal coherence properties of a laser in order to get good fringes. The laser is a Fabry-Perot cavity that contains a gain medium. Its output spectrum is therefore a series of discrete frequencies separated by $c/2nL$, where L is the cavity length. For gas lasers, the index is approximately equal to one, and we will use this value for the analysis. If $G(\nu)$ is the gain bandwidth, the frequency spectrum is

$$S(\nu) = G(\nu)\,\text{comb}\left(\frac{2L\nu}{c}\right) \tag{91}$$

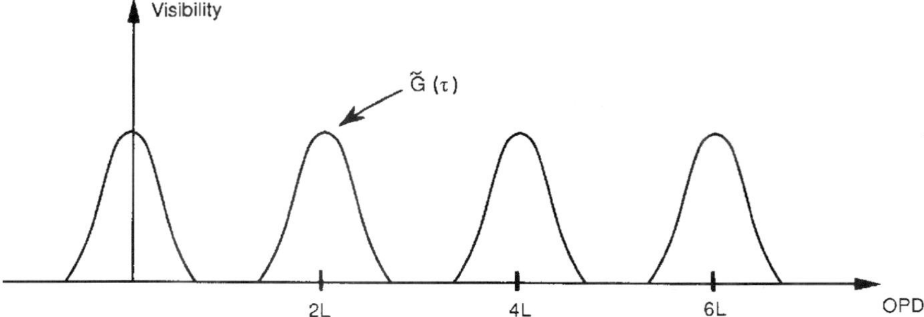

FIGURE 28 The fringe visibility versus OPD for a laser source.

where a comb function is a series of equally spaced delta functions. The number of modes contained under the gain bandwidth can vary from 1 or 2 up to several dozen. The resulting visibility function can be found by using Eq. (88):

$$\gamma(\tau) = |\gamma_{11}(\tau)| = \left|\tilde{G}(\tau) * \text{comb}\left(\frac{c\tau}{2L}\right)\right| = \left|\tilde{G}(\tau) * \text{comb}\left(\frac{\text{OPD}}{2L}\right)\right| \tag{92}$$

where $\tilde{G}(\tau)$ is the normalized Fourier transform of the gain bandwidth, and $*$ indicates convolution. This result is plotted in Fig. 28, where $\tilde{G}(\tau)$ is replicated at multples of $2L$. The width of these replicas is inversely proportional to the gain bandwidth. We see that as long as the OPD between the two optical paths is a multiple of twice the cavity length, high-visibility fringes will result. This condition is independent of the number of longitudinal modes of the laser. If the laser emits a single frequency, it is a coherent source and good visibility results for any OPD.

2.9 REFERENCES

M. Born and E. Wolf, *Principles of Optics,* Pergamon, New York, 1975.

R. W. Ditchburn, *Light,* Academic, New York, 1976.

M. Francon, *Optical Interferometry,* Academic, New York, 1966.

M. H. Freeman, *Optics,* Butterworths, London, 1990.

J. D. Gaskill, *Linear Systems, Fourier Transforms and Optics,* Wiley, New York, 1978.

P. Hariharan, *Optical Interferometry,* Academic Press, San Diego, Calif., 1985.

P. Hariharan (ed.), *Selected Papers on Interferometry,* SPIE, Bellingham, Wash., 1990.

P. Hariharan, *Basics of Interferometry,* Academic Press, San Diego, Calif., 1992.

E. Hecht, *Optics,* Addison-Wesley, Reading, Mass., 1989.

F. A. Jenkins and H. E. White, *Fundamentals of Optics,* McGraw-Hill, New York, 1976.

R. A. Longhurst, *Geometrical and Physical Optics,* Longman, London, 1973.

D. Malacara (ed.), *Selected Papers on Optical Shop Metrology,* SPIE, Bellingham, Wash., 1990.

D. Malacara (ed.), *Optical Shop Testing,* 2d edition, Wiley, New York, 1992.

A. S. Marathay, *Elements of Optical Coherence Theory,* Wiley, New York, 1982.

J. R. Meyer-Arendt, *Introduction to Classical and Modern Optics,* Prentice-Hall, Englewood Cliffs, N.J., 1989.

G. O. Reynolds, J. B. De Velis, G. B. Parrent, and B. J. Thompson, *The New Physical Optics Notebook: Tutorials in Fourier Optics,* SPIE, Bellingham, Wash., 1989.

B. E. A. Saleh and M. C. Teich, *Fundamentals of Photonics,* Wiley, New York, 1991.

W. H. Steel, *Interferometry,* Cambridge, London, 1967.

F. G. Smith and J. H. Thompson, *Optics,* Wiley, New York, 1971.

R. W. Wood, *Physical Optics,* Optical Society of America, Washington D.C., 1988.

CHAPTER 3
DIFFRACTION

A. S. Marathay
Optical Sciences Center
University of Arizona
Tucson, Arizona

3.1 GLOSSARY

A	amplitude
$\mathbf{E}$	electric field
f	focal length
G	Green function
I	irradiance
p, q	direction cosines
$\mathbf{r}$	spatial vector
$\mathbf{S}$	Poynting vector
t	time
ϵ	dielectric constant
μ	permeability
ν	frequency
ψ	wave function
$\hat{}$	Fourier transform

3.2 INTRODUCTION

Starting with waves as solutions to the wave equation obtained from Maxwell's equations, basics of diffraction of light are covered in this article. The discussion includes those applications where the geometry permits analytical solutions. At appropriate locations references are given to the literature and/or textbooks for further reading. The discussion is limited to an explanation of diffraction, and how it may be found in some simple cases with plots of fringe structure.

Selection of topics is not easy, some topics that are normally included in textbooks may not be found in this chapter. One thing that is not covered is Cornu's spiral and all associated phasor diagrams. These form convenient tools but are not absolutely necessary to understand the results. Material on these topics is in all of the textbooks listed in the article. Derivations are here excluded. Use of the two-pinhole experiments to describe the influence of partial coherence of light on the visibility of fringes is not mentioned [Thompson, B. J., and E. Wolf, "Two-Beam Interference with Partially Coherent Light," *J. Opt. Soc. Am.* **47**(10):895–902 (1957)]. So also the diffractive effect of the finite size of the apertures on this experiment is not described [Marathay, A. S., and D. B. Pollock, "Young's Interference Fringes with Finite-Sized Sampling Apertures," *J. Opt. Soc. Am.* A, **1**:1057–1059 (1984)]. Walls of mathematics and computer programs and/or plots have been avoided. These are found in the current literature, vector-diffraction calculation being a case in point.

Closely related chapters in the handbook are Chap. 1 by Douglas Goodman, Chap. 2 by J. Greivenkamp, Chap. 4, by William H. Carter, Chap. 5 by Bennett, Chap. 6 by Bohren, and Vol. II, Chap. 5 by Zissis. Appropriate chapters will be cross-referenced as needed.

3.3 LIGHT WAVES

Light waves propagate through free space or a vacuum. They exhibit the phenomenon of diffraction with every obstacle they encounter. Maxwell's equations form a theoretical basis for describing light in propagation, diffraction, scattering and, in general, its interaction with material media. Experience has shown that the electric field **E** plays a central role in detection of light and interaction of light with matter. We begin with some mathematical preliminaries.

The electric field **E** obeys the wave equation in free space or a vacuum

$$\nabla^2 \mathbf{E} - \frac{1}{c^2}\frac{\partial^2 \mathbf{E}}{\partial t^2} = 0 \tag{1}$$

where c is the velocity of light in a vacuum. Each cartesian component $\mathbf{E}_j$, $(j = x, y, z)$ obeys the equation and, as such, we use a scalar function $\psi(\mathbf{r}, t)$ to denote its solutions, where the radius vector $\mathbf{r}$ has components, $\mathbf{r} = \hat{i}x + \hat{j}y + \hat{k}z$.

The wave equation is a linear second-order partial differential equation. Linear superposition of its two linearly independent solutions offers the most general solution. It has plane waves, spherical waves and cylindrical waves as solutions. These solutions represent optical wave forms. A frequently used special case of these solutions is the time harmonic version of these waves. We start with the Fourier transform on time,

$$\psi(\mathbf{r}, t) = \int \hat{\psi}(\mathbf{r}, \nu) \exp(-i2\pi\nu t)\, d\nu \tag{2}$$

where ν is a temporal (linear) frequency in Hz. The spectrum $\hat{\psi}(\mathbf{r}, \nu)$ obeys the Helmholtz equation,

$$\nabla^2 \hat{\psi} + k^2 \hat{\psi} = 0 \tag{3}$$

with the propagation constant $k = 2\pi/\lambda = 2\pi\nu/c \equiv \omega/c$, where λ is the wavelength and ω is the circular frequency. A Fourier component traveling in a medium of refractive index

$n = \sqrt{\epsilon}$, where ϵ is the dielectric constant, is described by the Helmholtz equation with k^2 replaced by n^2k^2. The effect of each component with proper account of dispersion, n as a function of frequency, is superimposed as dictated by the Fourier integral to yield the field $\psi(\mathbf{r}, t)$.

As a further special case, a wave may be harmonic in time as well as in space (see Chap. 2).

$$\psi(\mathbf{r}, t) = A \cos(\mathbf{k} \cdot \mathbf{r} - \omega t) \tag{4}$$

where $\mathbf{k} \equiv k\hat{s}$, $\hat{s}$ is a unit vector in the direction of propagation and A is a constant. An expanding spherical wave may be written in the form,

$$\psi(r, t) = \frac{A}{r} \cos(kr - \omega t) \tag{5}$$

For convenience of operations, a complex form frequently is used. For example, we write

$$\psi(\mathbf{r}, t) = A \exp[i(\mathbf{k} \cdot \mathbf{r} - \omega t)] \tag{6}$$

in place of Eq. (4) bearing in mind that only its real part corresponds to the optical wave form. The function $\psi(\mathbf{r}, t)$ is called the optical "disturbance" while the coefficient A is the amplitude.

In the general discussion of diffraction phenomena throughout this chapter several classic source books have been used.[1–9] This discussion is a blend of ideas contained in these sources.

The mathematical solutions described heretofore, although ideal, are nevertheless often approximated. A suitable experimental arrangement with a self-luminous source and a condensing lens to feed light to a small enough pinhole fitted with a narrowband spectral filter serves as a quasi-monochromatic, or almost monochromatic, point source. In Fig. 1, light behind the pinhole S is in the form of ever-expanding spherical waves. These waves are of limited spatial extent; all are approximately contained in a cone with its apex at the pinhole. When intercepted by a converging lens, L_1, with the pinhole on its axis and at the front focal point, these spherical waves are converted to plane waves behind L_1. These plane waves also are limited spatially to the extent dictated by the aperture of the converging lens. A second converging lens, L_2, is behind the first converging lens and is oriented so that both lenses have a common optical axis and can form an image of the pinhole. The image S' is on the axis at the focal point behind the second lens and is formed by converging spherical waves. These waves, which converge toward the image, are

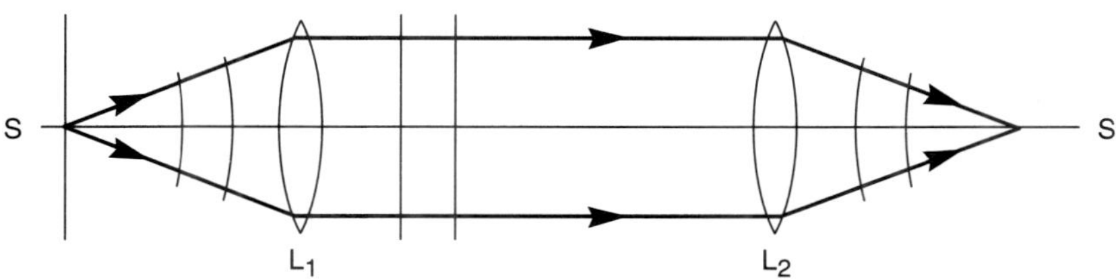

FIGURE 1 Experimental layout to describe the notation used for spherical and plane waves. S: pinhole source. L_1, L_2: lenses. S': image.

limited spatially to the extent dictated by the aperture of the second lens and are approximately contained in a cone with its apex at the image of the pinhole.

It is necessary to clarify that "a small enough pinhole" means that the optics behind the pinhole is not able to resolve its structure.[1] A "narrowband filter" means that its pass band, $\Delta\nu$, is much smaller than the mean frequency $\bar{\nu}$, that is, $\Delta\nu \ll \bar{\nu}$. In this situation, the experimental arrangement may be described by a quasi-monochromatic theory, provided that the path differences, Δl, of concern in the optics that follows the pinhole are small enough, as given by, $\Delta l \leq c/\Delta\nu$. If the path differences Δl involved are unable to obey this condition, then a full polychromatic treatment of the separate Fourier components contained within $\Delta\nu$ is necessary, even if $\Delta\nu \ll \bar{\nu}$. See, for example, Beran and Parrent[10] and Marathay.[11]

Limiting the extent of plane waves and spherical waves, as discussed before, causes diffraction, a topic of primary concern in this chapter. The simplifying conditions stated above are assumed to hold throughout the chapter, unless stated otherwise.

As remarked earlier, the electric field $\mathbf{E}$ $[V/m]$ plays a central role in optical detection (see Chap. 16). There are detectors that attain a steady state for constant incident beam power $[W]$, and there are those like the *photographic plate* that integrate the incident power over a certain time (see Chap. 20). For a constant beam power, the darkening of the photographic plate depends on the product of the power and exposure time. Since detectors inherently take the time average, the quantity of importance is the average radiant power $\Phi[W]$. Furthermore, light beams have a finite cross-sectional area, so it is meaningful to talk about the average power in the beam per unit area of its cross section measured in square meters or square centimeters. In the radiometric nomenclature this sort of measurement is called *irradiance* $[Wm^{-2}]$. For a plane wave propagating in free space, the irradiance, I, may be expressed in terms of the Poynting vector, $\mathbf{S}$ by

$$I = |\langle \mathbf{S} \rangle| = \left(\frac{1}{2}\right)\left(\frac{\epsilon_0}{\mu_0}\right)^{1/2} \langle \mathbf{E} \cdot \mathbf{E} \rangle \tag{7}$$

The constants given in Eq. (7) may not be displayed with every theoretical result. (We have used the symbol I; however, the radiometrically accepted symbol E is easily confused with the electric field E.) The Poynting vector and irradiance are discussed further in Ref. 11 (pp. 280–285).[11]

Light is properly described by a transverse vector field (see Chap. 6). Nevertheless, a scalar field is a convenient artifice to use in understanding the wave nature of light without the added complication of the vector components. The transverse nature of the field will be accounted for when the situation calls for it.

3.4 HUYGENS-FRESNEL CONSTRUCTION

Without the benefit of a fundamental theory based on Maxwell's equations and the subsequent mathematical development, Huygens sought to describe wave propagation in the days before Maxwell. Waves are characterized by constant-phase surfaces, called *wavefronts.* If the initial shape at time t of such a wavefront is known in a vacuum or in any medium, Huygens proposed a geometrical construction to obtain its shape at a later time, $t + \Delta t$. He regarded each point of the initial wavefront as the origin of a new disturbance that propagates in the form of secondary wavelets in all directions with the same speed as the speed v of propagation of the initial wave in the medium. These secondary wavelets of radii $v\,\Delta t$ are constructed at each point of the initial wavefront. A surface tangential to all these secondary wavelets, called the *envelope* of all these wavelets, is then the shape and position of the wavefront at time $t + \Delta t$. With this construct Huygens explained the phenomena of reflection and refraction of the wavefront. To explain the phenomenon of diffraction, Fresnel modified Huygens' construction by attributing the property of mutual interference to the secondary wavelets (see Chap. 2). The modified Huygens construction is called the Huygens-Fresnel construction. With further minor modifications it helps explain the phenomenon of diffraction and its various aspects, including those that are not so intuitively obvious.

We will begin with the Huygens-Fresnel construction and will show how it explains some simple optical experiments on diffraction. This will be followed by the Fresnel-Kirchhoff formula and related topics.

Fresnel Zones

Let P_0 be a point source of light that produces monochromatic spherical waves. A typical spherical wave, $A/r_0 \exp[-i(\omega t - kr_0)]$, of radius r_0 at time t is shown in Fig. 2. The coefficient A stands for the amplitude of the wave at unit distance from the source P_0. At a later time this wave will have progressed to assume a position passing through a point of observation P with radius, $r_0 + b$. Fresnel zone construction on the initial wave offers a way to obtain the wave in the future by applying the Huygens-Fresnel construction. The zone construction forms a simple basis for studying and understanding diffraction of light.

From the point of observation P, we draw spheres of radii b, $b + \lambda/2$, $b + 2\lambda/2$, $b + 3\lambda/2, \ldots$, $b + j\lambda/2, \ldots$, to mark zones on the wave in its initial position, as shown in Fig. 2. The zones are labeled $z_1, z_2, \ldots z_j \ldots$ The zone boundaries are successively half a wavelength away from the point of observation P. By the Huygens-Fresnel construction, each point of the wave forms a source of a secondary disturbance. Each secondary source produces wavelets that are propagated to the point P. A linear superposition of the contribution of all such wavelets yields the resulting amplitude at the point P. It is reasonable to expect that the contribution of the secondary wavelets is not uniform in all directions. For example, a wavelet at C is in line with the source P_0 and the point of

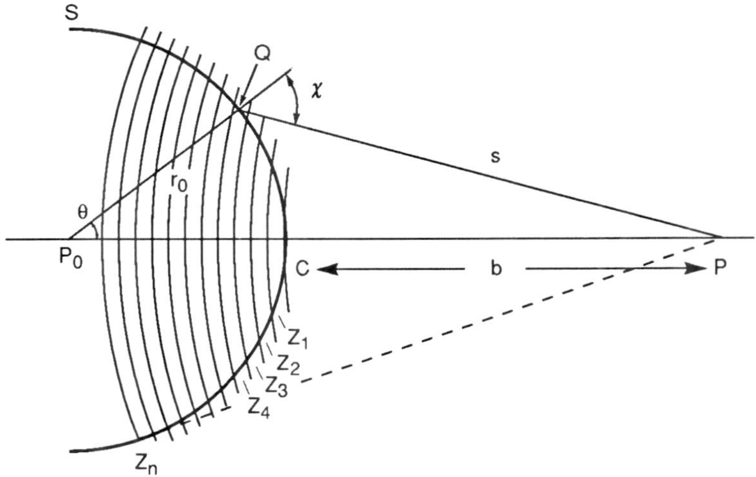

FIGURE 2 Fresnel zone construction. P_0: point source. S: wavefront. r_0: radius of the wavefront b: distance CP. s: distance QP. (*After Born and Wolf*[1].)

observation P, while a wavelet at Q sees the point P at an angle χ with respect to the radius vector from the source P_0. To account for this variation, an obliquity or inclination factor $K(\chi)$ is introduced. In the phenomenological approach developed by Fresnel, no special form of $K(\chi)$ is used. It is assumed to have the value unity at C where $\chi = 0$, and it is assumed to decrease at first slowly and then rapidly as χ increases. The obliquity factors for any two adjacent zones are nearly equal and it is assumed that it becomes negligible for zones with high enough index j.

The total contribution to the disturbance at P is expressed as an area integral over the primary wavefront,

$$\psi(P) = A \frac{\exp\left[-i(\omega t - kr_0)\right]}{r_0} \iint_S \frac{\exp(iks)}{s} K(\chi)\, dS \tag{8}$$

where dS is the area element at Q. The subscript S on the integrals denotes the region of integration on the wave surface. The integrand describes the contribution of the secondary wavelets. Fresnel-zone construction provides a convenient means of expressing the area integral as a sum over the contribution of the zones.

For optical problems, the distances involved, such as r_0 and b, are much larger than the wavelength, λ. This fact is used very effectively in approximating the integral. The phases of the wavelets within a zone will not differ by more than π. The zone boundaries are successively $\lambda/2$ further away from the point of observation P. The average distance of successive zones from P differs by $\lambda/2$; the zones, therefore, are called half-period zones. Thus, the contributions of the zones to the disturbance at P alternate in sign,

$$\psi(P) = \psi_1 - \psi_2 + \psi_3 - \psi_4 + \psi_5 - \psi_6 + \ldots \tag{9}$$

where ψ_j stands for the contribution of the jth zone, $j = 1, 2, 3, \ldots$ The contribution of each annular zone is directly proportional to the zone area and is inversely proportional to the average distance of the zone to the point of observation P. The ratio of the zone area to its average distance from P is independent of the zone index j. Thus, in summing the contributions of the zones we are left with only the variation of the obliquity factor, $K(\chi)$. To a good approximation, the obliquity factors for any two adjacent zones are nearly equal and for a large enough zone index j the obliquity factor becomes negligible. The total disturbance at the point of observation P may be approximated by

$$\psi(P) = 1/2(\psi_1 \pm \psi_n) \tag{10}$$

where the index n stands for the last zone contributing to P. The $\pm$ sign is taken according to whether n is odd or even. For an unobstructed wave, the integration is carried out over the whole spherical wave. In this case, the last term ψ_n is taken to be zero. Thus, the resulting disturbance at the point of observation P equals one-half of the contribution of the first Fresnel zone,

$$\psi(P) = 1/2\psi_1 \tag{11}$$

The contribution ψ_1 is found by performing the area integral of Eq. (8) over the area of the first zone. The procedure results in

$$\psi(P) = \frac{A}{r_0 + b} \lambda \exp\{-i[\omega t - k(r_0 + b) - \pi/2]\} \tag{12}$$

whereas a freely propagating spherical wave from the source P_0 that arrives at point P is known to have the form

$$\psi(P) = \frac{A}{r_0 + b} \exp\{-i[\omega t - k(r_0 + b)]\} \tag{12'}$$

The synthesized wave of Eq. (12) can be made to agree with this fact, if one assumes that the complex amplitude of the secondary waves, $\exp(iks)/s$ of Eq. (8) is $[1/\lambda \exp(-i\pi/2)]$ times the primary wave of unit amplitude and zero phase. With the time dependence $\exp(-i\omega t)$, the secondary wavelets are required to oscillate a quarter of a period ahead of the primary.

The synthesis of propagation of light presented above has far-reaching consequences. The phenomenon of light diffraction may be viewed as follows. Opaque objects that interrupt the free propagation of the wave block some or parts of zones. The zones, or their portions that are unobstructed, contribute to the diffraction amplitude (disturbance) at the point of observation P. The obstructed zones do not contribute.

Diffraction of Light from Circular Apertures and Disks

Some examples of unobstructed zones are shown in Fig. 3. Suppose a planar opaque screen with a circular aperture blocks the free propagation of the wave. The center C of the aperture is on the axis joining the source point S and the observation point P, as shown in

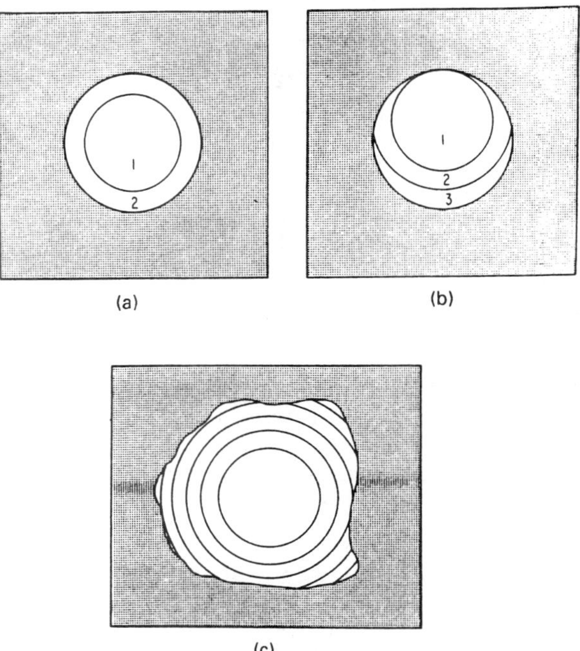

FIGURE 3 Some examples of unobstructed Fresnel zones that contribute to the amplitude at the observation point P. (*After Andrews.*[8])

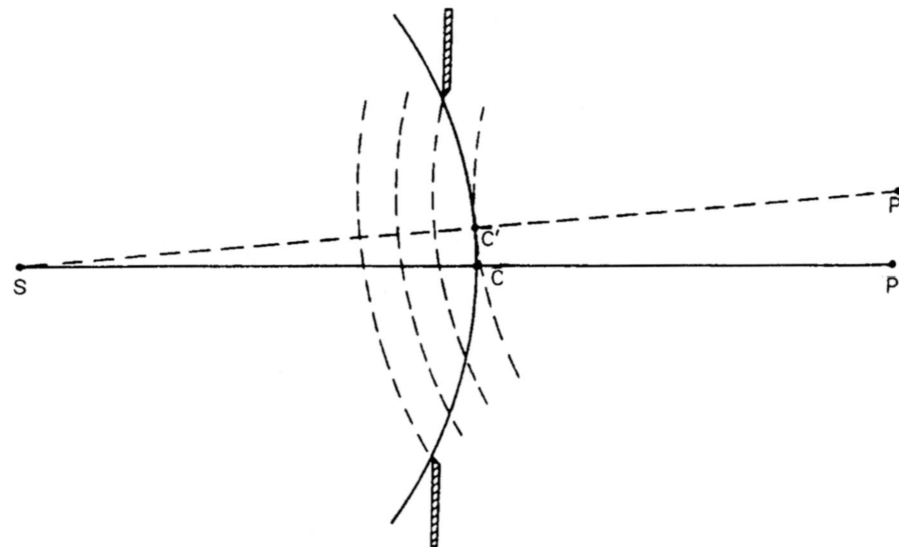

FIGURE 4 The redrawn zone structure for use with an off-axis point P'. (*After Andrews.*[8])

Fig. 4. The distance and the size of the aperture are such that, with respect to point P, only the first two zones are uncovered as in Fig. 3*a*. To obtain the diffraction amplitude for an off-axis point such as P', one has to redraw the zone structure as in Fig. 4. Figure 3*b* shows the zones and parts of zones uncovered by the circular aperture in this case. Figure 3*c* shows the uncovered zones for an irregularly shaped aperture.

In Fig. 3*a* the first two zones are uncovered. Following Eq. (9), the resulting diffraction amplitude at P for this case is

$$\psi(P) = \psi_1 - \psi_2 \tag{13}$$

but, since these two contributions are nearly equal, the resulting amplitude is $\psi(P) = 0$!

Relocating point P necessitates redrawing the zone structure. The first zone may just fill the aperture if point P is placed farther away from it. In this case the resulting amplitude is

$$\psi(P) = \psi_1 \tag{14}$$

which is twice what it was for the unobstructed wave! Therefore the irradiance is four times as large!

On the other hand, if the entire aperture screen is replaced by a small opaque disk, the irradiance at the center of the geometrical shadow is the same as that of the unobstructed wave! To verify this, suppose that the disk diameter and the distance allows only one Fresnel zone to be covered by the disk. The rest of the zones are free to contribute and *do* contribute. Per Eq. (9) we have

$$\psi(P) = -\psi_2 + \psi_3 - \psi_4 + \psi_5 - \psi_6 + \cdots$$

The discussion after Eq. (9) also applies here and the resulting amplitude on the axis behind the center of the disk is

$$\psi(P) = -\tfrac{1}{2}\psi_2 \tag{15}$$

which is the same as the amplitude of the unobstructed wave. Thus, the irradiance is the same at point P as though the wave were unobstructed. As the point P moves farther

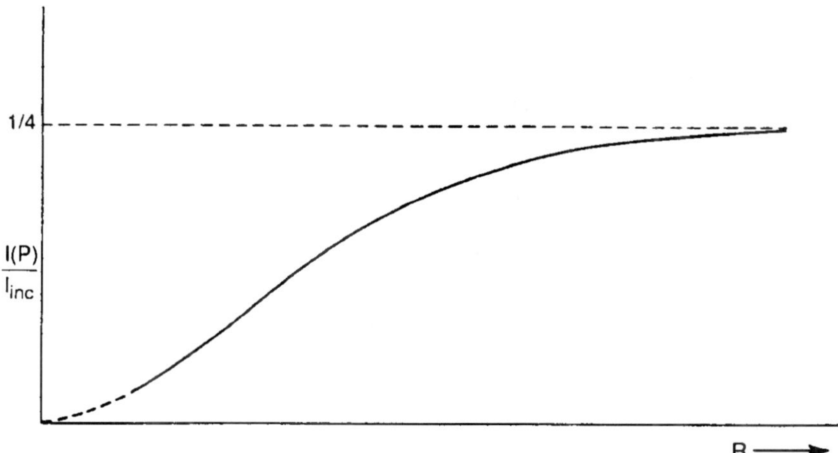

FIGURE 5 Variation of on-axis irradiance behind an opaque disk. R: distance along the axis behind the disk. (*From Marion and Heald.*[5])

away from the disk, the radius of the first zone increases and becomes larger than the disk. Alternatively one may redraw the zone structure starting from the edge of the disk. The analysis shows that the point P continues to be a bright spot of light. As the point P moves closer to the disk, more and more Fresnel zones get covered by the disk, but the analysis continues to predict a bright spot at P. There comes a point where the unblocked zone at the edge of the disk is significantly weak; the point P continues to be bright but has reduced irradiance. Still closer to the disk, the analysis ceases to apply because P enters the near-field region, where the distances are comparable to the size of the wavelength. In Fig. 5, the variation of irradiance on the axial region behind the disk is shown. It is remarkable that the axial region is nowhere dark! For an interesting historical note, see Refs. 1 and 5.

For comparison we show the variation of on-axis irradiance behind a circular opening in Fig. 6. It shows several on-axis locations where the irradiance goes to zero. These correspond to the situation where an even number of zones are exposed through the circular aperture. Only the first few zeros are shown, since the number of zeros per unit length (linear density) increases as the point P is moved closer to the aperture. The linear density increases as the square of the index j when P moves closer to the aperture. While far enough away, there comes a point where the first zone fills the aperture and, thereafter, there are no more zeros as the distance increases.

Figure 7 shows a series of diffraction patterns from a circular aperture. The pictures are taken at different distances from the aperture to expose one, two, three, etc., zones. Each time an odd number of zones is uncovered the center spot becomes bright. As we approach the pictures at the bottom right, more zones are exposed.

Babinet Principle

Irradiances for the on-axis points are quite different for the circular disk than for the screen with a circular opening. The disk and the screen with a hole form of a pair of complementary screens, that is, the open areas of one are the opaque areas of the other and vice versa. Examples of pairs of such complementary screens are shown in Fig. 8. Observe that the open areas of screen S_a taken with the open areas of the complementary screen S_b add up to no screen at all.

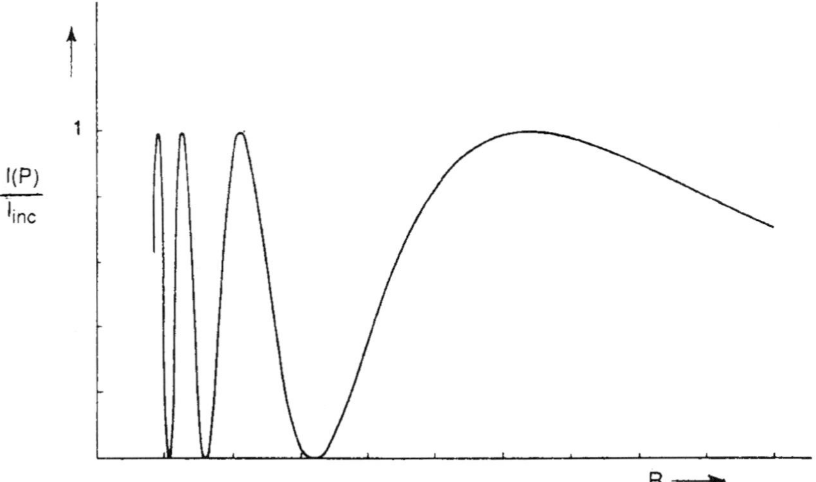

FIGURE 6 Variation of on-axis irradiance behind a circular opening. R: distance along the axis behind the opening. (*From Marion and Heald*[5])

The Babinet principle states that the wave disturbance $\psi_S(P)$ at any point of observation P due to a diffracting screen S_a added to the disturbance $\psi_{CS}(P)$ due to the complementary screen S_b at the same point P equals the disturbance at P due to the unobstructed wave. That is,

$$\psi_S(P) + \psi_{CS}(P) = \psi_{UN}(P) \tag{16}$$

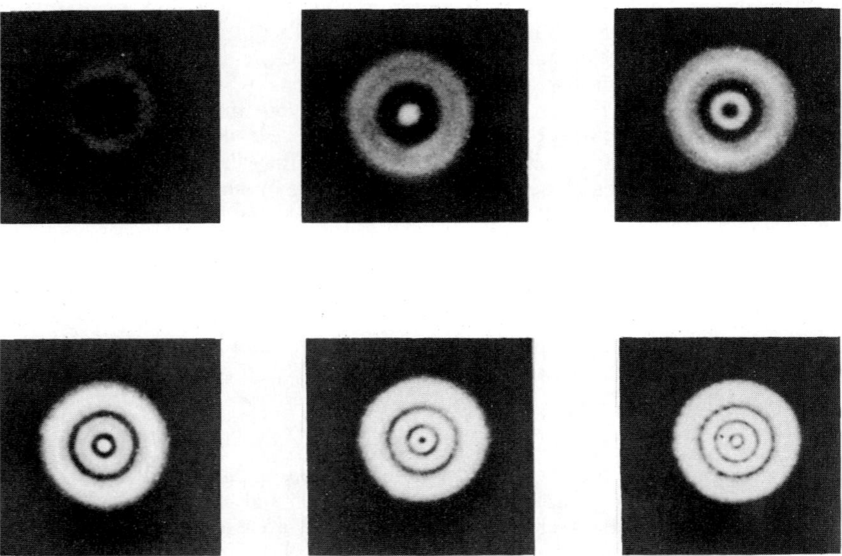

FIGURE 7 A series of pictures of diffraction patterns from circular apertures. (*After Andrews.*[8])

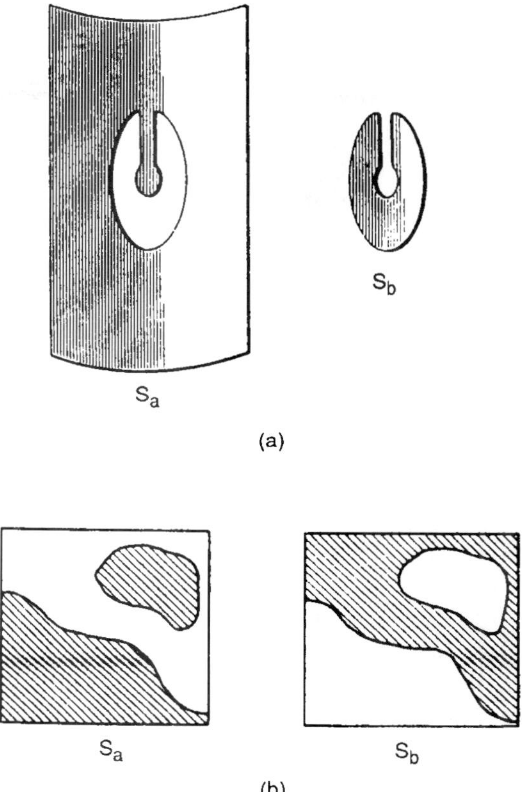

FIGURE 8 Examples of complementary screens, labeled S_a and S_b. (*After Jackson*[12] *and Andrews.*[8])

Recall that the wave disturbance at any point of observation P behind the screen is a linear superposition of the contributions of the unobstructed zones or portions thereof. This fact, with the observation that the open areas of screen S_a taken with the open areas of the complementary screen S_b add up to no screen at all, implies the equality indicated by the Babinet principle.

The behavior of the on-axis irradiance for the case of the opaque disk is quite different from that of the complementary circular aperture. There is no simple relationship between the irradiances of the two cases because they involve a squaring operation that brings in cross terms.

Zone Plate

If alternate zones are blocked the contribution of the unblocked zones will add in phase to yield a large irradiance at the point of observation. An optical device that blocks alternate zones is called a *zone plate.* Figure 9*a* and *b* shows two zone plates made up of concentric circles with opaque alternate zones. They block odd-indexed or even-indexed zones, respectively. The radii of the zone boundaries are proportional to the square root of

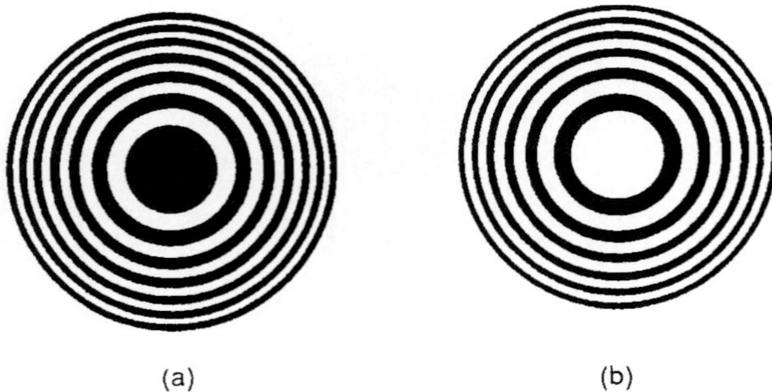

FIGURE 9 Two zone plates made up of concentric circles with alternate zones made opaque. They block odd-indexed or even-indexed zones respectively. The radii of the zone boundaries are proportional to the square root of natural numbers. (*From Hecht and Zajac.*[9])

natural numbers. As in Fig. 2, we place a point source at a distance r_0 in front of the zone plate. If R_m is the radius of the mth zone, a bright image of this source is observed at a distance b behind the plate, so that

$$\frac{1}{r_0}+\frac{1}{b}=\frac{m\lambda}{R_m^2} \tag{17}$$

where λ is the wavelength of light from the source. This equation for the condition on the distance b is like the paraxial lens formula from which the focal length of the zone plate may be identified or may be obtained by setting the source distance $r_0 \to \infty$ (see Chap. 1). The focal length f_1 so obtained is

$$f_1=\frac{R_m^2}{m\lambda} \tag{18}$$

and is called the *primary focal length*. For unlike, the case of the lens, the zone plate has several secondary focal lengths. These are given by

$$f_{2n-1}=\frac{R_m^2}{(2n-1)m\lambda} \tag{19}$$

where $n = 1, 2, 3\ldots$ In the case of the primary focal length, each opaque zone of the zone plate covers exactly one Fresnel zone. The secondary focal length f_3 is obtained when each opaque zone covers three Fresnel zones. It is a matter of regrouping the right-hand side of Eq. (9) in the form

$$\begin{aligned}\psi(P) &= (\psi_1-\psi_2+\psi_3)-\langle\psi_4-\psi_5+\psi_6\rangle+(\psi_7-\psi_8+\psi_9)\\ &\quad -\langle\psi_{10}-\psi_{11}+\psi_{12}\rangle+(\psi_{13}-\psi_{14}+\psi_{15})-\cdots\end{aligned} \tag{20}$$

The zone plate in Fig. 9*b*, for example, blocks all even-indexed zones. It corresponds to omitting the terms enclosed in the angular brackets, $\langle \cdots \rangle$, in Eq. (20). The remaining terms grouped in parentheses add in phase to form a secondary image of weaker irradiance. The higher-order images are formed successively closer to the zone plate and are successively weaker in irradiance.

Further discussion is given in several books of the bibliography, for example, Ref. 9 (p. 375). The radii of the concentric circles in a zone plate are proportional to the square root of natural numbers. For equidistant source and image locations, say 10.0 cm at a wavelength of 500 nm, $R_m = \sqrt{m}$ (0.16 mm). Due to the smallness of the radii, a photographic reduction of a large-scale drawing is used.

Incidentally, the pair of zone plates of Fig. 9*a* and *b* form a pair of complementary screens. Per the Babinet principle, the groupings are

$$\begin{aligned}\psi_{UN}(P) &= \psi_S(P) + \psi_{CS}(P) \\ &= (\psi_1 + \psi_3 + \psi_5 + \psi_7 + \psi_9 + \psi_{11} \cdots) \\ &\quad - (\psi_2 + \psi_4 + \psi_6 + \psi_8 + \psi_{10} + \psi_{12} + \cdots)\end{aligned} \tag{21}$$

The first group of terms correspond to the zone plate of Fig. 9*b* and the second group of items correspond to Fig. 9*a*.

3.5 CYLINDRICAL WAVEFRONT

A line source generates cylindrical wavefronts. It is frequently approximated in practice by a slit source, which, in turn, can illuminate straightedges, rectangular or slit apertures (see Fig. 10*a*). In this case, as we shall see, the phenomena of diffraction can be essentially reduced to a one-dimensional analysis for this source and aperture geometry.

Fresnel zones for cylindrical wavefronts take the form of rectangular strips, as shown in Fig. 10(*a*). The edges of these strip zones are $\lambda/2$ farther away from the point of observation *P*. The treatment for the cylindrical wave parallels the treatment used for the spherical wave in Section 3.4. The line M_0 on the wavefront intersects at right angles to the line joining the source *S* and the point of observation *P*. Refer to M_0 as the axis line of the wavefront with respect to the point *P*. Let *a* be the radius of the wavefront with respect to the source slit and let *b* the distance of *P* from M_0. Fresnel zones are now in the form of strips above and below M_0 and are parallel to it. The line pairs M_1M_1', M_2M_2', etc., are marked $\lambda/2$ farther away from the point of observation *P*. Fresnel zones are now half-period strips. Thus $PM_m = b + m\lambda/2$ and, to a good approximation, the arc length $(M_mM_{m+1}) = aK(\sqrt{m+1} - \sqrt{m})$, where the constant, $K = \sqrt{mb\lambda/a(a+b)}$. For small values of *m* such as 1, 2, etc., the arc widths decrease rapidly while, for large values of *m*, the neighboring strips have nearly equal widths. The lower-order strips have much larger areas compared to the ones further up from M_0. This effect is much more dominant than the variation of the obliquity factor $K(\chi)$ that is neglected in this analysis.

Consider one single strip as marked in Fig. 10*b*. Imagine that this strip is divided into half-period sections as shown. The wavefront is almost planar over the width of this strip. All the sections on either side of the arc M_1M_2 contribute to the disturbance at *P*. The boundaries of these are marked N_1, N_2, etc. The area of these sections are proportional to $K'(\sqrt{n+1} - \sqrt{n})$, where the constant $K' = \sqrt{c\lambda}$. The areas of those half-period sections decrease rapidly at first and then slowly. The contribution to the disturbance at *P* from the higher-order sections is alternately positive and negative with respect to the first section near M_1M_2. Consequently, their contribution to the total disturbance at *P* is nearly zero.

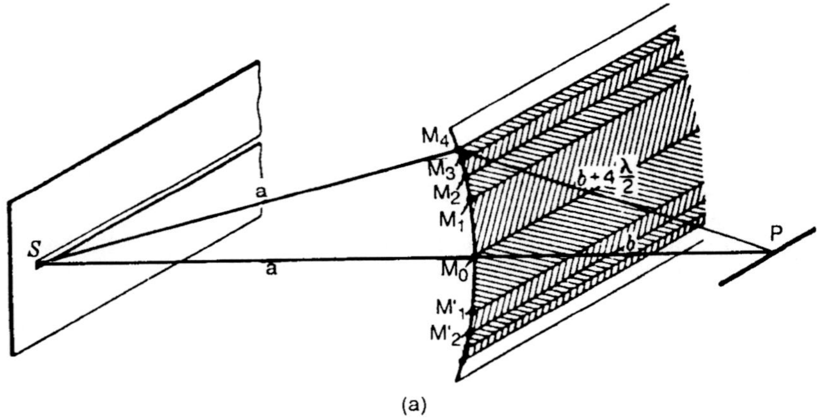

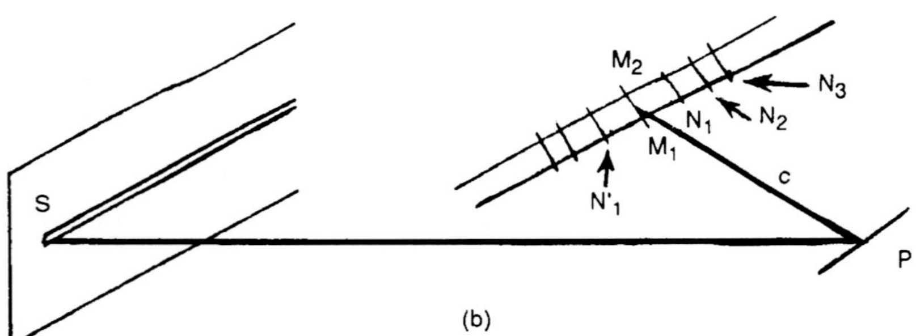

FIGURE 10 Fresnel zones for cylindrical wavefront. S: slit source; a: radius of the cylindrical wavefront; b: distance M_0P. M_1, M_1', etc.: zone boundaries. (*Adapted from Jenkins and White.*[2])

The disturbance at P due to a single strip consists of the dominant contribution of the two sections from N_1 to N_1'. This conclusion holds for all the strips of the cylindrical wave.

Following the procedure of Eq. (9) employed for the spherical wave,

$$\psi(P) = \psi_1 - \psi_2 + \psi_3 - \psi_4 + \psi_5 - \psi_6 + \cdots \tag{9'}$$

where $\psi(P)$ is the disturbance at P and ψ_m denotes the secondary wavelet contributions from strip zones of either side of the axis line M_0 of Fig. 10*a*. As in Eq. (11) the series can be summed, but here we need to account for the strip zone contribution from both sides of the axis line M_0: therefore, we have

$$\psi(P) = \psi_1 \tag{11'}$$

The first zone contributions can be computed and compared with the freely propagating cylindrical wave. Without going into the details as carried out with Eqs. (12) and (12′) for the spherical wave, we shall discuss qualitative features of some typical diffraction patterns resulting from cylindrical waves.

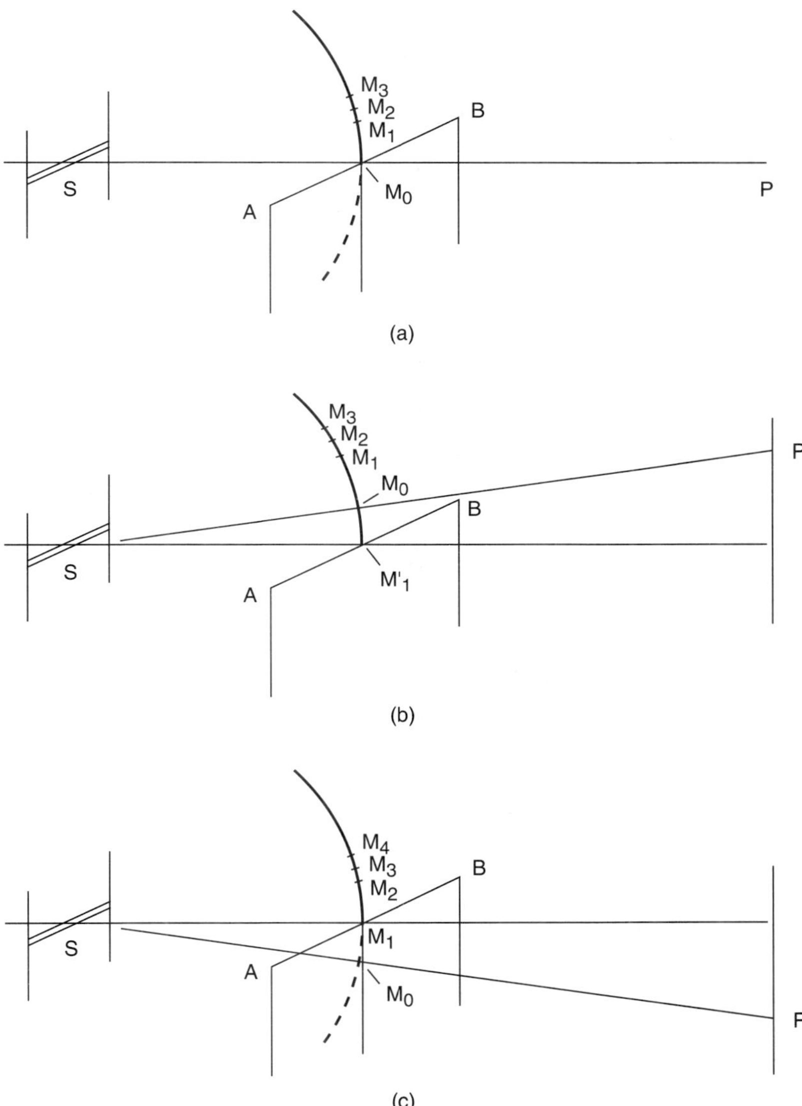

FIGURE 11 Fresnel zones for a cylindrical wavefront. The edges of these strip zones are $\lambda/2$ farther away from the point of observation P. S: slit source. P: point of observation. M_0: axis line of the cylindrical wave. AB: straight-edge opaque obstruction. (*After Jenkins and White.*[2])

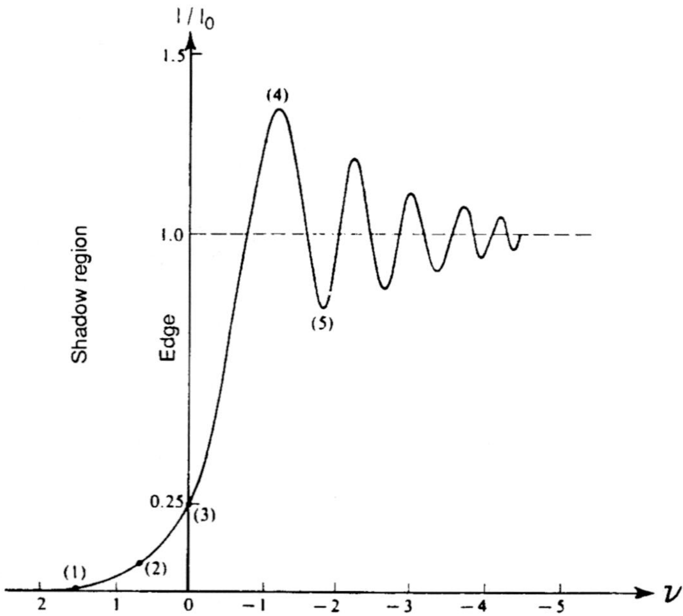

FIGURE 12 The plot of irradiance in the diffraction pattern of a straight edge *AB*. The plot is normalized to unity for the irradiance of the unobstructed wave. Labels 1 and 2 show points *P* in the geometrical shadow. Label 3 is at the edge of the geometrical shadow, while Labels 4 and 5 are in the illuminated region. v is a unitless variable to label distances along the plane of observation. (*From Hecht and Zajac.*[9])

Straightedge

A cylindrical wave from a slit source S illuminates an opaque screen with a straightedge AB oriented parallel to the slit, as shown in Fig. 11. It shows three special positions of the point of observation P. In Fig. 11*a*, P is such that all the strip zones above the axis line M_0 are exposed, while those below are blocked. The point P in Fig. 11*b* is such that the strip zones above M_0 and one zone below, marked by the edge M_1', are exposed. In Fig. 11*c*, P has moved into the geometrical shadow region. The strip M_1M_0 and all those below M_0 are blocked.

A plot of the irradiance in the diffraction pattern of a straightedge is shown in Fig. 12. Following the discussion in Section 3.4, we discuss the disturbance at P for the three cases of Fig. 11. At the edge of the geometrical shadow

$$\psi_a(P) = 1/2\psi_1 \tag{22}$$

For Fig. 11*b* we have

$$\psi_b(P) = 1/2\psi_1 + \psi_1 = 3/2\psi_1 \tag{23}$$

As P is moved further up, there comes a point for which two strip zones below M_0 are exposed, resulting in $\psi(P) = 3/2\psi_1 - \psi_2$. As P explores the upper half of the observation plane, the amplitude and, hence, the irradiance go through maxima and minima according to whether an odd or even number of (lower) strip zones is exposed. Furthermore, the maxima decrease gradually while the minima increase gradually until the fringes merge into a uniform illumination that corresponds to the unobstructed wave.

In the geometrical shadow (see Fig. 11*c*),

$$\psi_c(P) = -1/2\psi_2 \tag{24}$$

As P goes further down, we get $\psi(P) = 1/2\psi_3$; in general, the number of exposed zones decreases and the irradiance falls off monotonically.

A plot of the irradiance in the diffraction pattern of a straightedge is shown in Fig. 12. It is interesting to observe that the edge is neither at the maximum of the first fringe nor at the half-way point. It appears at one-fourth of the irradiance of the unobstructed wave.

Rectangular Aperture

Figure 13 is a series of diagrams of irradiance distributions for light diffracted by single-slit apertures. A pair of marks on the horizontal axis indicate the edges of the geometrical shadow of the slit relative to the diffraction pattern. In all cases, relatively little light falls in the geometrical shadow region. The last diagram corresponds to a rather wide slit. It appears as two opposing straightedge diffraction patterns corresponding to the two edges of the slit.

These patterns may be interpreted as obtained with the plane of observation fixed for different-size slits. Alternately, the slit size may be held fixed but move the plane of observation. For the first diagram the plane is far away. For the successive diagrams the plane is moved closer to the slit. The plane of observation is the closest for the last

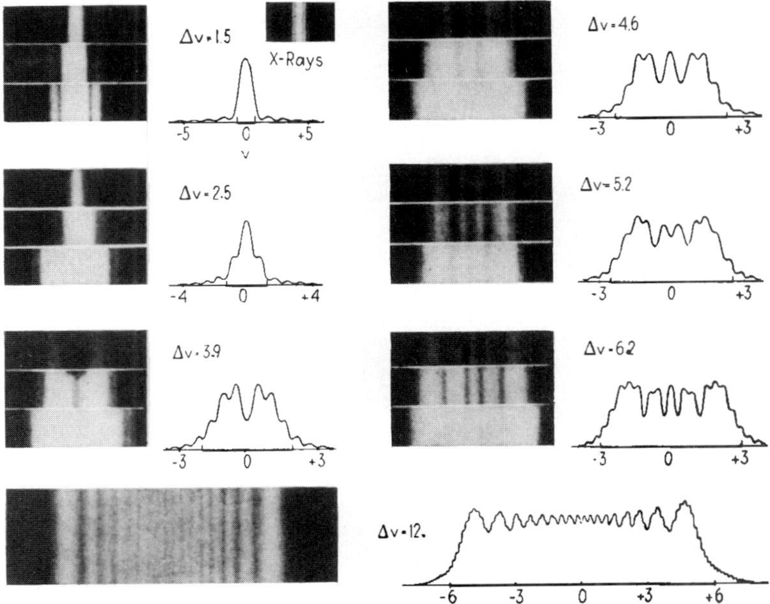

FIGURE 13 A series of diagrams of irradiance distributions for light diffracted by single-slit apertures of different widths. (*From Jenkins and White.*[2])

diagram of Fig. 13. The important parameter is the angular subtense of the slit to the observation plane. A similar comment applies to the case of the circular aperture shown in Fig. 7. See Jenkins and White[2] for further discussion.

Opaque Strip Obstruction

A slit aperture and an opaque strip or a straight wire form a pair of complementary screens. In Fig. 14 photographs of Fresnel diffraction patterns produced by narrow wires are shown with the corresponding theoretical curves. These theoretical curves show some more detail. Generally, the figures show the characteristic unequally spaced diffraction fringes of a straight edge on either side of the geometrical shadow. These fringes get closer and closer together, independently of the width of the opaque obstruction, and finally merge into a uniform illumination. Figure 14 also shows the maximum in the center and equally spaced narrow fringes within the shadow. The width of these fringes is inversely proportional to the width of the obstruction. We shall now discuss this detail.

Figure 15 shows the arrangement of the source S, opaque strip AB, and the plane of observation. A point x in the geometrical shadow receives light from Fresnel zones of both sides of the opaque strip. At each edge of the opaque strip the exposed zones add up effectively to one-half of the contribution of a single zone adjacent to that edge. Owing to the symmetry, the resulting disturbance from each edge starts out in phase. Light from the two edges adds constructively or destructively according to whether the path difference to

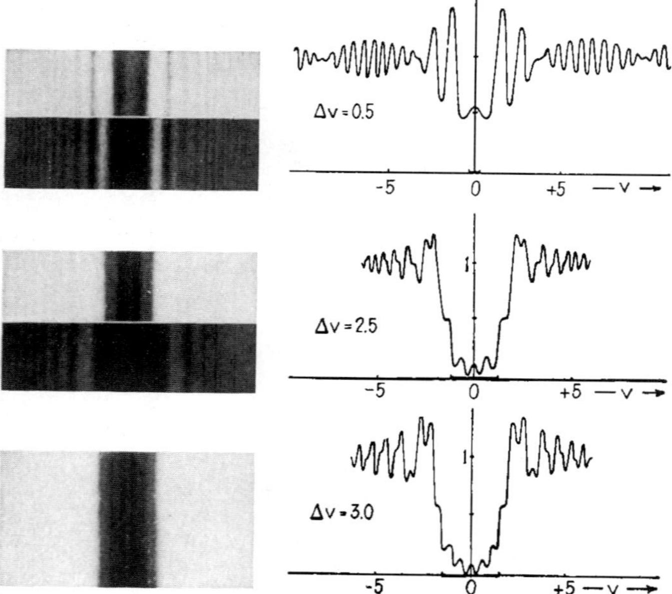

FIGURE 14 Fresnel diffraction patterns produced by narrow wires are shown with the corresponding theoretical curves. (*From Jenkins and White.*[2])

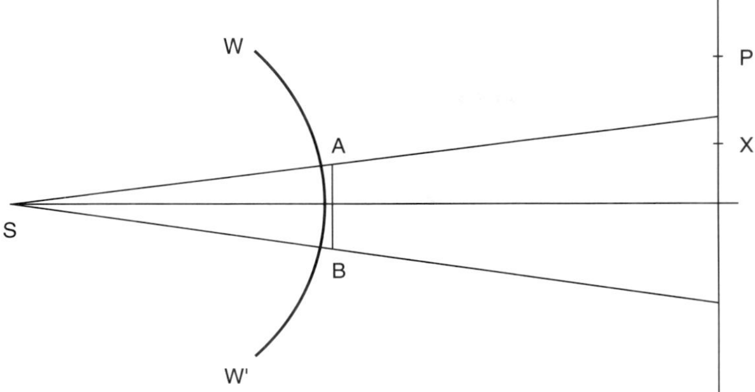

FIGURE 15 Arrangement of the source *S*, opaque strip *AB*, and the plane of observation. Point *x* is in the geometrical shadow region.

the point x in the shadow region is an even or an odd multiple of $\lambda/2$ (see Chap. 2). The situation is similar to two coherent sources separated by the width of the opaque strip. Young examined these fringes inside the geometrical shadow. In particular, he showed that if an opaque screen is introduced on one side of the opaque strip to block that part of the wave, then the straightedge diffraction fringes due to that edge, as well as the interference fringes in the shadow region, vanished.

3.6 MATHEMATICAL THEORY OF DIFFRACTION

Kirchhoff showed that the Huygens-Fresnel construction follows from an integral theorem starting from the wave equation. The resulting mathematical expression is called the Fresnel-Kirchhoff diffraction formula.[1] This theory was further modified by Rayleigh and Sommerfeld.[6,11] At each stage the modifications previously referred to were for removing certain mathematical inconsistencies. In view of the approximations applicable to practical optical situations, the mathematical details will not be discussed here. There are several other theoretical aspects and different approaches[1] to the mathematics of diffraction theory.

It is well known in wave theory that the field values inside a volume enclosed by a bounding surface are determined by the values of the field and/or its normal derivative on this bounding surface. The solution is expressed in terms of the Green's function of the problem, as in

$$\psi(P) = \left(\frac{1}{4\pi}\right)\iint_S \left\{\psi\left(\frac{\partial G}{\partial n}\right) - G\left(\frac{\partial \psi}{\partial n}\right)\right\} dS \tag{25}$$

where G is the Green's function of the problem. The integral is over the arbitrary closed surface S. The symbol $(\partial/\partial n)$ stands for the normal derivative with the normal pointing into the volume enclosed by the surface.[1] A convenient Green's function is the expanding spherical wave, $\exp(iks)/s$ from the point of observation P. The closed surface for the diffraction problem is made up of the aperture plane and a large partial sphere centered at

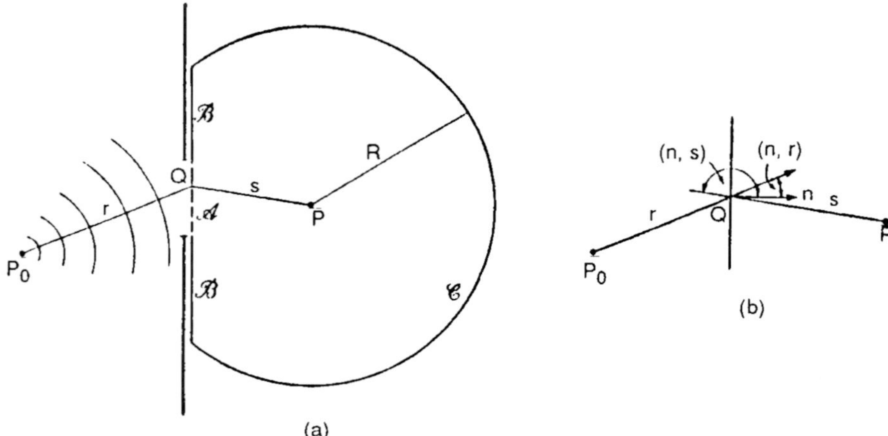

FIGURE 16 The closed surface for the diffraction problem. It is made up of the aperture plane and a large partial sphere centered at the point of observation *P*. (*From Born and Wolf.*[1])

the point of observation *P*, as shown in Fig. 16. This is the starting point of Kirchhoff theory. It requires specifying the field values and its normal derivative on the bounding surface to obtain the field $\psi(P)$ at *P* in the volume enclosed by the surface. It is possible to show that the contribution of the surface integral from the partial sphere is zero. Kirchhoff assumed that the field and its normal derivative are zero on the opaque portion of the aperture plane. On the open areas of the aperture plane he assumed the values to be the same as (unperturbed) incident values. If the incident field is an expanding spherical wave $(a/r)\exp(ikr)$, then the field $\psi(P)$ is given by

$$\psi(P) = -\left\{\frac{ia}{2\lambda}\right)\iint_A \left[\frac{\exp(ikr)}{r}\right]\left[\frac{\exp(iks)}{s}\right][\cos(n, r) - \cos(n, s)]\, dS \tag{26}$$

The area integral is over the open areas *A* of the aperture. As shown in Fig. 16, (n, s) and (n, r) are the angles made by *s* and *r*, respectively, with the normal to the aperture plane. The above equation is referred to as the Fresnel-Kirchhoff diffraction formula.

From a strictly mathematical point of view the specification of field and its normal derivative is overspecifying the boundary conditions. It is possible to modify the Green function so that either the field ψ or its normal derivative $\partial\psi/\partial n$ needs to be specified. With this modification one obtains

$$\psi(P) = -\left(\frac{ia}{\lambda}\right)\iint_A \left[\frac{\exp(ikr)}{r}\right]\left[\frac{\exp(iks)}{s}\right]\cos(n, s)\, dS \tag{27}$$

This is referred to as the Rayleigh-Sommerfeld diffraction formula. Other than mathematical consistency, both formulas yield essentially similar results when applied to practical optical situations. They both use the approximate boundary conditions, namely, that the field is undisturbed in the open areas of the aperture and zero on the opaque regions of the aperture plane. The cosine factors in the above formulas play the role of the obliquity factor of the Huygens wave used in Section 3.4 Eq. (8).

More generally the field (for a single temporal frequency) at the point of observation P, x, y, z may be expressed by[11]

$$\psi(x, y, z) = \iint_A \psi(x_s, y_s, 0)\left[\frac{i}{2\pi}\frac{z}{\rho}(1 - ik\rho)\frac{\exp(ik\rho)}{\rho^2}\right] dx_s\, dy_s \tag{28}$$

where $\psi(x_s, y_s, 0)$ are the field values in the aperture A, at $z = 0$. The expression in the square brackets is the normal derivative of the modified Green function. In this expression $\rho = [(x - x_s)^2 + (y - y_s)^2 + z^2]^{1/2}$, the distance between a point in the aperture and the point of observation P, and the ratio z/ρ is the direction cosine of the difference vector. In the far zone where $k\rho \gg 1$, Eq. (28) reduces to Eq. (27) for the case of a spherical wave illumination. Since the expression in the square brackets depends on the coordinate difference, $(x - x_s)$ and $(y - y_s)$, Eq. (28) has the form of a convolution integral. It is well known that the convolution integral has a corresponding product relationship in the Fourier-spatial-frequency domain. The two-dimensional Fourier decomposition of the field ψ is

$$\psi(x, y, z) = \iint \hat{\psi}(p/\lambda, q/\lambda, z) \exp[+i2\pi(px + qy)/\lambda]\, d(p/\lambda)\, d(q/\lambda) \tag{29}$$

where p and q are the two-direction cosines. The third-direction cosine m is defined by

$$m = +(1 - p^2 - q^2)^{1/2}, \text{ for } p^2 + q^2 \le 1$$

and

$$m = +i(p^2 + q^2 - 1)^{1/2}, \text{ for } p^2 + q^2 > 1 \tag{30}$$

A similar decomposition as in Eq. (29) is used for the field in the aperture at $z = 0$, wherein the finite area of the aperture is included in the description of the incident field. With the help of Weyl's plane-wave decomposition of a spherical wave,

$$\frac{\exp(ikr)}{r} = \frac{i}{\lambda}\iint \frac{1}{m}\exp(ikmz)\exp\left[\frac{i2\pi}{\lambda}(px + qy)\right] dp\, dq \tag{31}$$

the Fourier transform of the expression in square brackets in Eq. (28) can be found.[11] The product relationship in the Fourier domain has the form

$$\hat{\psi}(p/\lambda, q/\lambda, z) = \hat{\psi}(p/\lambda, q/\lambda, 0)\exp(ikmz) \tag{32}$$

The inverse Fourier transform yields the disturbance in x, y, z space at point P. At $z = 0$ it reproduces the assumed boundary conditions, a property not shared by the Fresnel-Kirchhoff formula.

A plane-wave decomposition describes a function in (x, y, z) space in terms of the weighted sum of plane waves, each propagating in a direction given by the direction cosines (p, q, m). Equation (32) may be referred to as the angular spectrum formulation of diffraction. For application of this formulation see Chap. 4, Coherence Theory by W. H. Carter.

Fresnel and Fraunhofer Approximations

In practical optical situations diffraction is mainly studied in the forward direction, that is, for small angles from the direction of propagation of the incident field. Furthermore, the distances involved are much larger than the wavelength λ, $r \gg \lambda$. In this situation the distance ρ of Eq. (28) may be approximated by the low-order terms in the binomial expansion of the square root,

$$\rho = [(x - x_s)^2 + (y - y_s)^2 + z^2]^{1/2}$$
$$\simeq \left(r - \frac{xx_s + yy_s}{r} + \frac{x_s^2 + y_s^2}{2r}\right) \tag{33}$$

where r is the radial distance of the observation point P, $r = \sqrt{x^2 + y^2 + z^2}$. When the terms quadratic in the aperture variables are retained, namely, $(x_s^2 + y_s^2)$, we have a description of Fresnel diffraction. Let d be the maximum dimension of the aperture. If the plane of observation is moved to distance $z \gg d^2/\lambda$ the quadratic terms are negligible and Eq. (28) is approximated by

$$\psi(x, y, z) = \left(-\frac{i}{\lambda r}\right) \exp(ikr) \iint_A \psi(x_s, y_s, 0) \exp\left[-\frac{ik(xx_s + yy_s)}{r}\right] dx_s\, dy_s \tag{34}$$

This is the formula for Fraunhofer diffraction.

Fraunhofer Diffraction

Far enough away from the aperture, $z \gg d^2/\lambda$, Fraunhofer-type diffraction is found. Eq. (34) shows that it has the form of a Fourier transform of the light distribution in the aperture. For more general conditions on the distance and angles to obtain Fraunhofer diffraction, see Born and Wolf,[1] sec. 8.3.3. Thus, instead of moving the observation plane to the far field, parallel light incident on the aperture can be brought to a focus by a converging lens as in Fig. 17, thus producing a Fraunhofer pattern of the aperture in the focal plane.

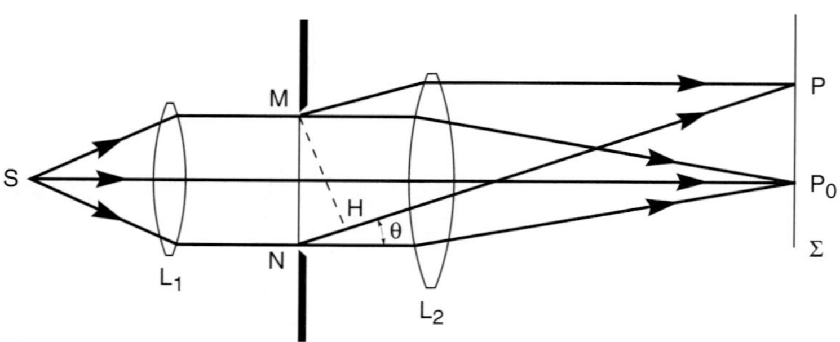

FIGURE 17 Arrangement to observe a Fraunhofer diffraction by a slit aperture. (*After Rossi.*[7])

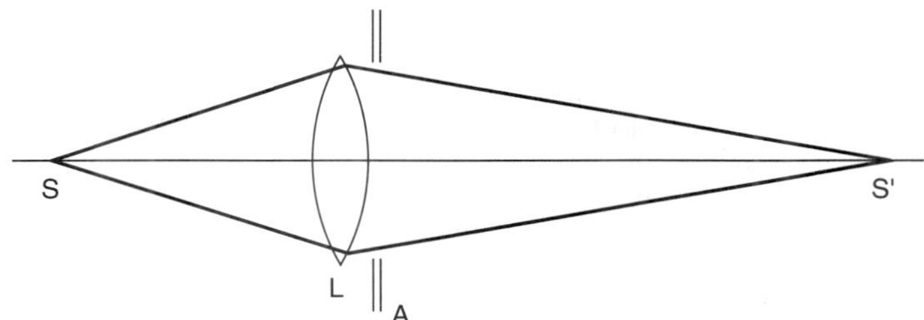

FIGURE 18 Fraunhofer diffraction of an aperture A, with a converging spherical wave. S: point source. L: converging lens. S': image.

In an imaging situation (see Fig. 18), a diverging spherical wave is brought to a focus in the image plane. This is also an example of Fraunhofer diffraction pattern of the light distribution in the aperture A by a converging spherical wave. To realize Fraunhofer diffraction, a similar situation is obtained when a narrow diffracting aperture is held next to the eye when it is focused on a distant point source. The diffraction pattern is observed in the plane of the source.

An optical processing setup is shown in Fig. 19 where collimated or parallel light is incident normally on plane 1. In this arrangement an inverted image of plane 1 is formed in plane 3 (see Chap. 1). The imaging process may be thought of as a Fourier transform (Fraunhofer diffraction) of the light distribution in plane 1 onto plane 2, followed by another Fourier transform of the light distribution in plane 2 onto plane 3.

Recall our earlier discussion in relation to Eqs. (33) and (34). When the quadratic phase factor, $\exp[+i\pi(x_s^2 + y_s^2)/\lambda r]$, may be approximated by unity, we are in the domain of Fraunhofer diffraction. From the point of view of Fresnel zone construction, the far-field condition, $z \gg d^2/\lambda$, means that for these distances z the first Fresnel zone overfills the aperture. The entire aperture contributes to the disturbance at any point in the Fraunhofer pattern. In Fresnel diffraction only relatively small portions of the aperture contribute to any one point in the pattern.

In this context the term Fresnel number is frequently used. It is defined in terms of the product of two ratios. The radius r of the aperture to the wavelength λ times the radius of the aperture to the distance b measured from the aperture to the plane of observation:

$$\text{Fresnel number} \equiv N = \frac{r}{\lambda} \cdot \frac{r}{b} = \frac{1}{4b} \frac{d^2}{\lambda} \tag{35}$$

Thus, Fresnel number can also be expressed as the ratio of the far-field distance, d^2/λ, to

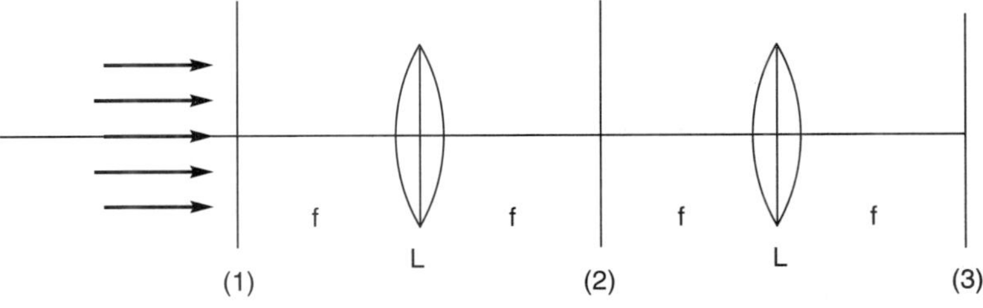

FIGURE 19 Optical processing arrangement. Collimated or parallel light is incident normally on plane 1. The system images plane 1 onto plane 3 with a Fourier transform in plane 2.

the distance, b, from the aperture. With the definition of the Fresnel zones Section 3.4, these ratios indicate that the Fresnel number equals the number of Fresnel zones that may be drawn within the aperture from a point P at a distance b from the aperture.

Thus, well within the Fresnel region, $b \ll d^2/\lambda$, where the Fresnel number is (high) large. There are many zones in the aperture. As seen in Figs. 7 and 13, very little light falls within the geometrical shadow region; most of the light is in the confines of the aperture boundary dictated by geometrical optics. In the study of cavity resonators[12,13] and modes it is found that diffraction losses are small for large Fresnel numbers, $N \gg 1$. In the Fraunhofer region $b > d^2/\lambda$, $N < 1$ where the first Fresnel zone overfills the aperture as pointed out before—well into the Fraunhofer region $N \ll 1$.

In Figs. 20 and 21 the theoretical plots of Fraunhofer patterns of a rectangular aperture and a circular aperture, respectively, are shown. In the rectangular case the central maximum has equally spaced zeros on either side, while in the circular case the central maximum is surrounded by unequally spaced concentric dark rings. In both cases the central maximum occurs at the geometrical image of the point source that produced parallel light illumination on the aperture. The unitless variable x shown in the plots is defined as follows. (1) In the case of rectangular aperture, $x = 2\pi ap/\lambda$, where $2a$ is the width of the aperture in the x_s direction. In the other dimension $y = 2\pi bq/\lambda$, and $2b$ is the dimension in the y_s direction. As before, p, q are the direction cosines of the vector joining the center of the aperture to the point of observation. (2) In the case of circular aperture, the unitless radial variable $x = 2\pi aw/\lambda$, where $2a$ is the diameter of the aperture in the x_s,y_s plane and $w = \sqrt{p^2 + q^2}$.

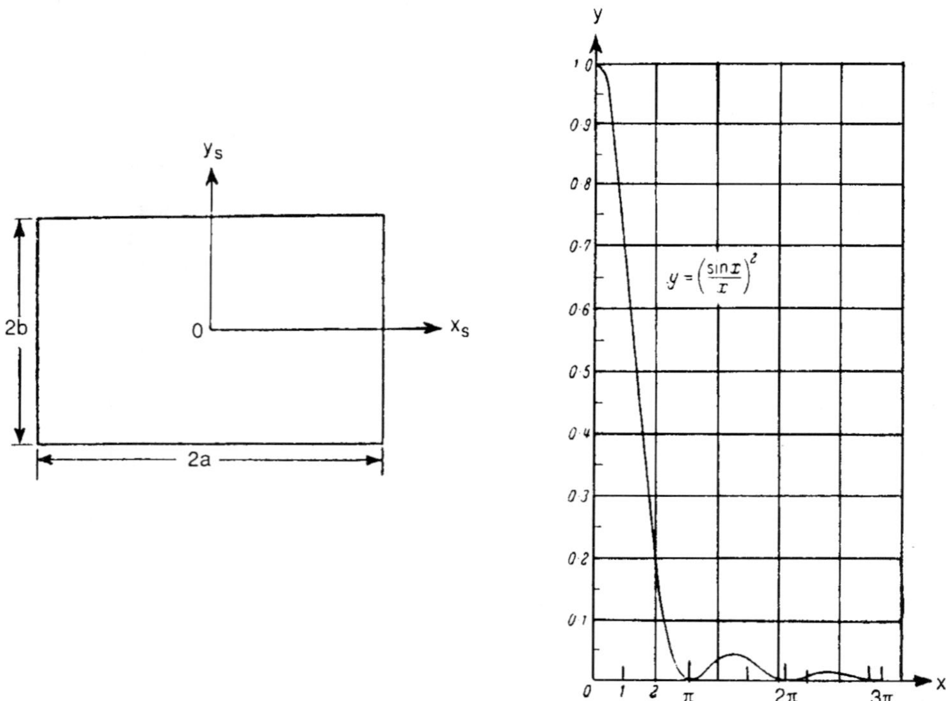

FIGURE 20 Rectangular aperture (in coordinates x_s and y_s) and one section of the diffraction pattern. Normalized irradiance y plotted against a unitless variable x as discussed in the text. (*From Born and Wolf.*[1])

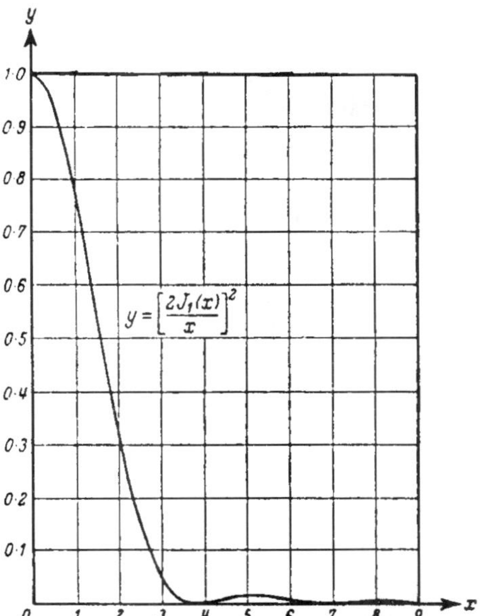

FIGURE 21 A section of the diffraction pattern of a circular aperture. The normalized irradiance y is plotted against a unitless variable x in the plane of observation as discussed in the text. (*From Born and Wolf.*[1])

In the far field the size of the diffraction pattern is very large compared to the aperture that produced it. In the focal plane of the lens, $z = f$ and the size of the diffraction pattern is much smaller than the aperture. In both cases the patterns are in a reciprocal width relationship, i.e., if the aperture is narrow in the x_s direction compared to y_s, the pattern is broader in the x direction compared to the y. A converging spherical lens illuminated by a plane wave produces in the focal plane a Fraunhofer diffraction pattern of the amplitude and phase of the aperture of a circular lens. When the lens has negligible phase errors, the diffraction pattern has a bright disk in the center surrounded by concentric dark rings. This is called an *Airy disk* and it plays an important role in the Rayleigh criterion of resolving power. This topic is discussed in Chap. 1, "Geometric Optics," by D. Goodman.

Fraunhofer Diffraction Pattern of a Double Slit

The diffraction pattern of two slits may be observed by using the optical arrangement of Fig. 22. The center-to-center separation of the two slits is h. The off-axis point P is in the direction θ from the axis as shown in the figure. The maxima and minima are determined according to whether the path difference O_1H is an even or odd multiple of a half-wave. Let I_0 be the irradiance at the center of the single-slit diffraction pattern. The irradiance distribution in the plane of observation Σ, is given by

$$I = 4I_0\left(\frac{\sin\alpha}{\alpha}\right)^2(\cos\delta)^2 \tag{36}$$

where $\delta = \pi h \sin\theta/\lambda$. The irradiance at the center of the double-slit pattern is $4I_0$. The second term, $(\sin\alpha/\alpha)^2$, describes the diffraction pattern of a single slit of width $2a$. Here

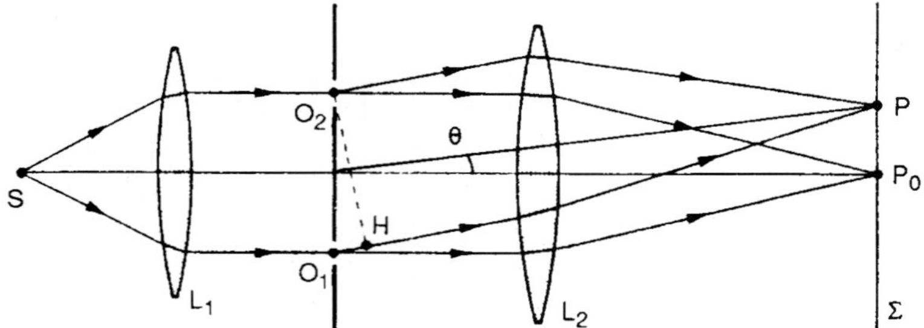

FIGURE 22 Arrangement to observe the Fraunhofer diffraction of an aperture consisting of two slits. S: point source. P: point of observation. O_1, O_2: two slit apertures with center-to-center separation h. (*From Rossi.*[7])

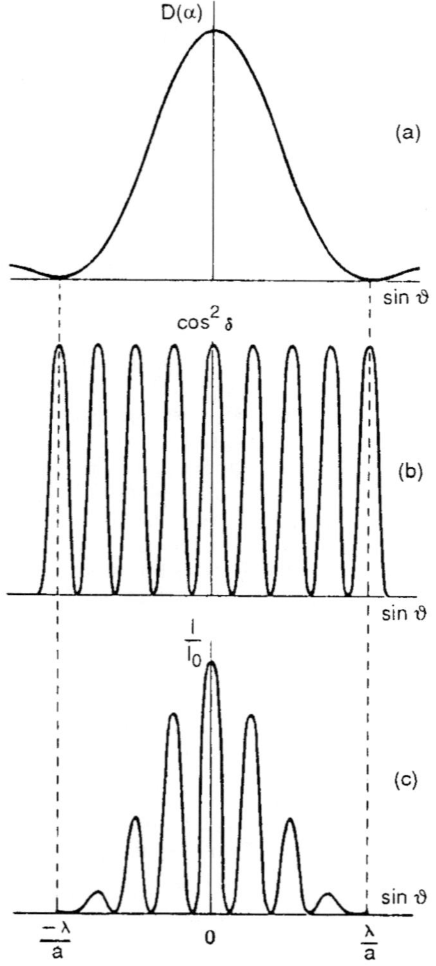

FIGURE 23 (a) Plot of a single-slit diffraction pattern $D(\alpha)$; (b) Plot of a two-slit interference pattern; and (c) their product I/I_0. (*From Rossi.*[7])

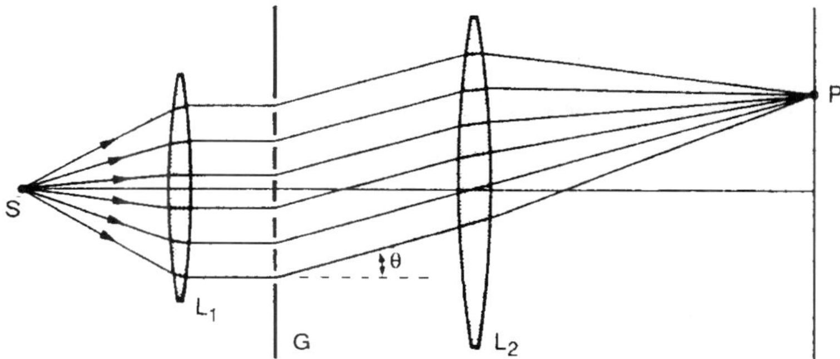

FIGURE 24 Arrangement to observe the Fraunhofer diffraction of an aperture consisting of N slits. S: slit source, P: point of observation. G: grating. (*From Rossi.*[7])

$\alpha = 2\pi ap/\lambda$, where $p = \sin\theta$. The term $(\cos\delta)^2$ is the interference pattern of two slits. These two patterns as well as their product are sketched in Fig. 23.

Diffraction Grating

In Fig. 24 an arrangement similar to Fig. 22 permits observation of the Fraunhofer diffraction pattern of a grating, of N parallel and equidistant slits. The center-to-center separation between neighboring slits is h. As in the two-slit case, the Fraunhofer pattern consists of the diffraction due to one slit times the interference pattern of N slits. The irradiance distribution in the plane of observation is given by

$$I = N^2 I_0 \left(\frac{\sin\alpha}{\alpha}\right)^2 \left(\frac{\sin N\gamma}{N\sin\gamma}\right)^2 \tag{37}$$

where $\gamma = \pi h \sin\theta/\lambda$ and $N^2 I_0$ is proportional to the irradiance at the center of the N-slit pattern. The term $(\sin\alpha/\alpha)^2$ is the single-slit pattern as used with Eq. 36. In the case of multiple slits each slit is very narrow; hence, this pattern is very broad, a characteristic of Fraunhofer diffraction. The interference term $(\sin N\gamma/N\sin\gamma)^2$ shows prominent maxima when both the numerator and denominator are simultaneously zero; this happens when $\gamma = \pi h \sin\theta/\lambda = m\pi$, where m is an integer. It leads to the grating equation, namely,

$$h \sin\theta = m\lambda \tag{38}$$

There are several, $(N-1)$, subsidiary minima in between the principal maxima. This happens when the numerator is zero but the denominator is not: $\gamma = m\pi/N$. For the case of $N = 10$, these effects are sketched in Fig. 25, which shows the effect of the product of the diffraction and interference terms.

In general, as N increases the subsidiary maxima become more nearly negligible, while the principal maxima become narrower, being proportional to $(1/N)$. The location of the principal maxima other than the zeroth order ($m = 0$) are proportional to the wavelength λ. The diffraction grating thus forms an important spectroscopic tool. Further discussion of gratings is given in Chap. 5, by Zissis, in Vol. II of this Handbook.

3.7 VECTOR DIFFRACTION

The popularity of the Fresnel-Kirchhoff diffraction formula in the scalar case stems from the fact that it is widely applicable and relatively easy to use. In the study of

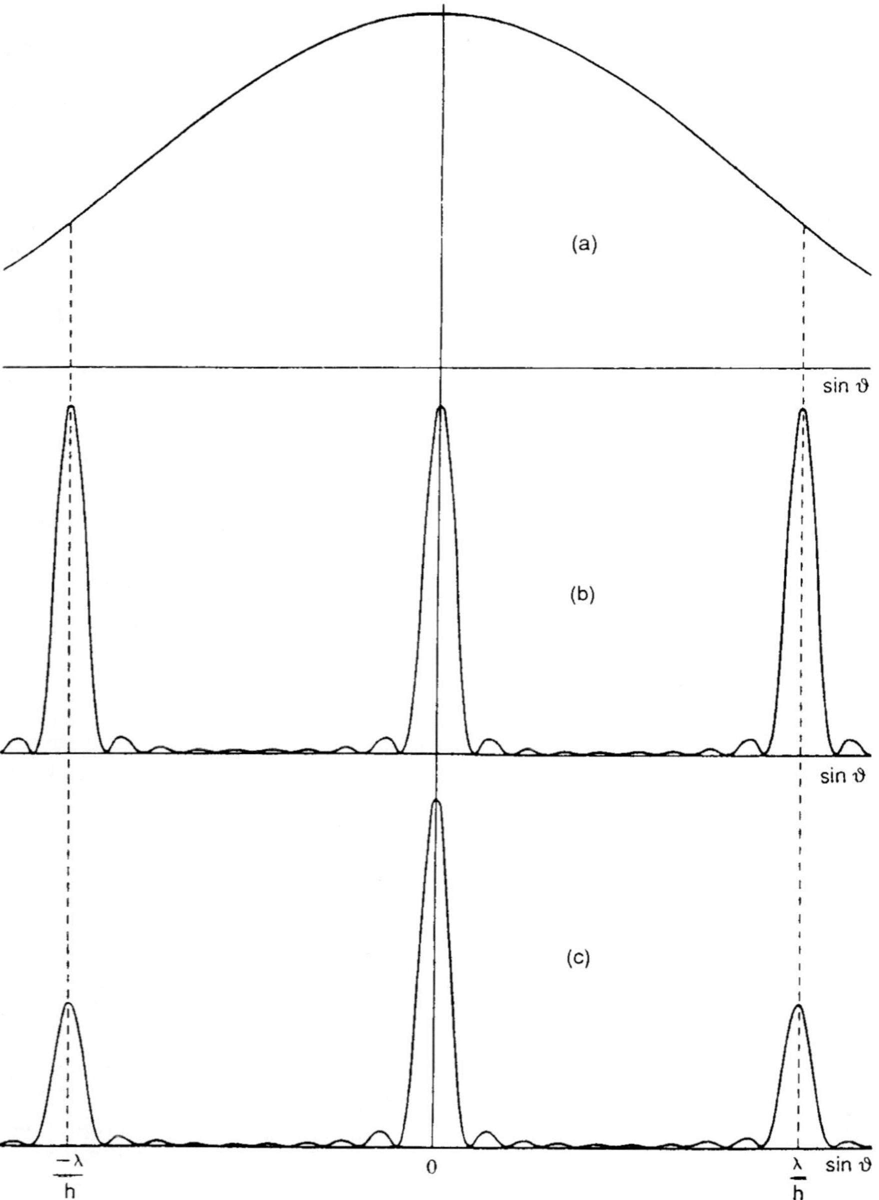

FIGURE 25 (*a*) Plot of a single-slit diffraction pattern; (*b*) partial plot of an $N = 10$ slit interference pattern; and (*c*) their product (*From Rossi.*[7])

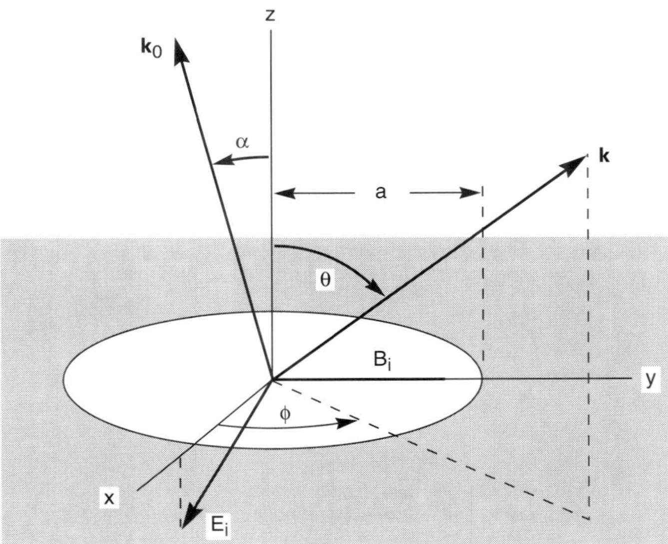

FIGURE 26 Coordinate system and aperture geometry for vector diffraction. α: angle of incidence. The *E* field is in the *xz* plane. $\mathbf{k}_0$: the wave vector of the incident wave. and **k**: the diffracted field. (*From Jackson.*[12])

electromagnetic diffraction[14,15] a similar formula can be obtained [see Eq. (9.156) in sec. 9.12 of Ref. 14] but has limited applicability because of the boundary conditions that must be satisfied. These conditions are the ones related to perfectly conducting screens. They are not adequately approximated at optical frequencies. The study with finite conductivity makes for complicated mathematical procedures. From the point of view of instrumental optics the applicability of the theory then is severely limited.

In the optical literature, periodic structures such as gratings (both shallow and deep compared to the wavelength) have been studied.[16,17] Boundary conditions are applied to perfectly conducting grating profiles. The equation of the grating dictating the angular positions of the diffraction orders such as Eq. (38) continues to apply; the amount of power found in the different orders is significantly different in the vector theory compared to the scalar theory.

Without going into the mathematical procedures and/or displaying involved formulas we discuss a special case of interest discussed in detail by Jackson.[14] We consider a plane wave incident at an angle α on a thin, perfectly conducting screen with a circular hole of radius a in the x-y plane. The polarization vector (**E** field) of the incident wave lies in the x-z plane, which is taken to be the plane of incidence. The arrangement is shown in Fig. 26 where $\mathbf{k}_0$ stands for the wave vector of the incident wave and **k** is used for the diffracted field.

The vector and scalar approximations are compared in Fig. 27. The angle of incidence is equal to 45° and the aperture is one wavelength in diameter, $ka = \pi$. The angular distribution is shown in Fig. 27 for two cases. Figure 27*a* shows the distribution of the power per unit solid angle in the plane of incidence which contains the **E** field and Fig. 27*b* the distribution for the plane perpendicular to it. Both vector and scalar theories contain the Airy-disk-type distribution; the differences show in the angular distribution. For

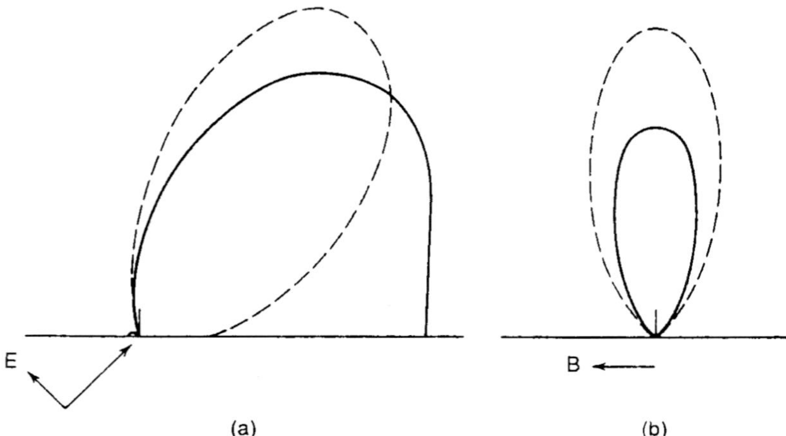

FIGURE 27 Fraunhofer diffraction pattern for a circular opening one wavelength in diameter in a thin-plane conducting screen. The angle of incidence is 45°. (*a*) power-per-unit solid angle (radiant intensity) in the plane of incidence; and (*b*) perpendicular to it. The solid (dotted) curve gives the vector (scalar) approximation in each case. (*From Jackson.*[12])

normal incidence $\alpha = 0$ and $ka \gg 1$ the polarization dependence is unimportant and the diffraction is confined to very small angles in the forward direction (Airy-disk-type distribution) as we found before in Fig. 21 under Fraunhofer diffraction.

3.8 REFERENCES

Exhaustive and/or the latest listing of references is not the intent but the following source books were used throughout the chapter.

1. M. Born and E. Wolf, *Principles of Optics,* 6th ed. Pergamon, Oxford, 1980.
2. F. A. Jenkins and H. E. White, *Fundamentals of Optics,* McGraw-Hill, New York, 1957.
3. R. S. Longhurst, *Geometrical and Physical Optics,* Wiley, New York, 1967.
4. B. K. Mathur and T. P. Pandya, *Geometrical and Physical Optics,* Gopal Printing Press, Kanpur, India, 1972.
5. J. B. Marion and M. A. Heald, *Classical Electromagnetic Radiation,* 2d ed., Academic Press, New York, 1980.
6. J. W. Goodman, *Introduction to Fourier Optics,* McGraw-Hill, New York, 1968.
7. B. Rossi, *Optics,* Addison-Wesley, Reading, Massachusetts, 1957.
8. C. L. Andrews, *Optics of the Electromagnetic Spectrum,* Prentice-Hall, Englewood Cliffs, New Jersey, 1960.
9. E. Hecht and A. Zajac, *Optics,* Addison-Wesley, Reading, Massachusetts, 1976.
10. M. J. Beran and G. B. Parrent Jr., *Theory of Partial Coherence,* Prentice-Hall, Englewood Cliffs, New Jersey, 1964.
11. A. S. Marathay, *Elements of Optical Coherence Theory,* Wiley, New York, 1985.
12. K. D. Möller, *Optics* University Science Books, California, 1988.

13. H. Kogelnik and T. Li, "Laser Beams and Resonators," *Appl. Opt.* **5,** 1966, pp. 1550–1565.
14. J. D. Jackson, *Classical Electrodynamics,* Wiley, New York, 1962.
15. J. A. Stratton, *Electromagnetic Theory,* McGraw-Hill, New York, 1941.
16. R. Petit, *Electromagnetic Theory of Gratings,* Springer-Verlag, New York, 1980.
17. T. K. Gaylord and M. G. Moharam, "Analysis and Applications of Optical Diffraction by Gratings," *Proc. IEEE* **73,** 1985, pp. 894–937.

Acknowledgments

I greatly appreciate the editorial help by Anna Elias-Cesnik in preparing this manuscript. The article is dedicated to my wife Sunita and family.

CHAPTER 4
COHERENCE THEORY

William H. Carter*
Naval Research Laboratory
Washington, D.C.

4.1 GLOSSARY

I	intensity [use irradiance or field intensity]
k	radian wave number
$\mathbf{p}$	unit propagation vector
t	time
U	field amplitude
u	Fourier transform of U
W_ω	cross-spectral density function
$\mathbf{x}$	spatial vector
$\Gamma_{12}(\tau)$	mutual coherence function
$\Delta\ell$	coherence length
$\Delta\tau$	coherence time
μ_ω	complex degree of spatial coherence
ϕ	phase
ω	radian frequency
Real ( )	real part of ( )

4.2 INTRODUCTION

Classical Coherence Theory

All light sources produce fields that vary in time with highly complicated and irregular waveforms. Because of diffraction, these waveforms are greatly modified as the fields propagate. All light detectors measure the intensity time averaged over the waveform. This measurement depends on the integration time of the detector and the waveform of the light at the detector. Generally this waveform is not precisely known. Classical coherence theory[1–8] is a mathematical model which is very successful in describing the effects of this unknown waveform on the observed measurement of time-averaged intensity. It is based on the electromagnetic wave theory of light as formulated from Maxwell's equations, and uses statistical techniques to analyze the effects due to fluctuations in the waveform of the field in both time and space.

* The author was a visiting scientist at the Johns Hopkins University Applied Physics Laboratory while this chapter was written. He is now Acting Program Director for Quantum Electronic Waves and Beams at the National Science Foundation, 4201 Wilson Blvd., Arlington, VA 22230.

Quantum Coherence Theory

Classical coherence theory can deal very well with almost all presently known optical coherence phenomena; however, a few laboratory experiments require a much more complicated mathematical model, quantum coherence theory[9–11] to explain them. This theory is not based on classical statistical theory, but is based on quantum electrodynamics.[12–14] While the mathematical model underlying classical coherence theory uses simple calculus, quantum coherence theory uses the Hilbert space formulation of abstract linear algebra, which is very awkward to apply to most engineering problems. Fortunately, quantum coherence theory appears to be essential as a model only for certain unusual (even though scientifically very interesting) phenomena such as squeezed light states and photon antibunching. All observed naturally occuring phenomena outside of the laboratory appear to be modeled properly by classical coherence theory or by an approximate semiclassical quantum theory. This article will deal only with the simple classical model.

4.3 SOME ELEMENTARY CLASSICAL CONCEPTS

Analytical Signal Representation

Solutions of the time-dependent, macroscopic Maxwell's equations yield six scalar components of the electric and the magnetic fields which are functions of both time and position in space. As in conventional diffraction theory, it is much more convenient to treat monochromatic fields than it is to deal with fields that have complicated time dependencies. Therefore, each of these scalar components is usually represented at some typical point in space (given with respect to some arbitrary origin by the radius vector $\mathbf{x} = (x, y, z)$ by a superposition of monochromatic real scalar components. Thus the field amplitude for a typical monochromatic component of the field with radial frequency ω is given by

$$U_r(\mathbf{x}, \omega) = U_0(\mathbf{x}) \cos [\phi(\mathbf{x}) - \omega t] \tag{1}$$

where $U_0(\mathbf{x})$ is the field magnitude and $\phi(\mathbf{x})$ is the phase. Trigonometric functions like that in Eq. (1) are awkward to manipulate. This is very well known in electrical circuit theory. Thus, just as in circuit theory, it is conventional to represent this field amplitude by a "phasor" defined by

$$U(\mathbf{x}, \omega) = U_0(\mathbf{x}) e^{i\phi(\mathbf{x})} \tag{2}$$

The purpose for using this complex field amplitude, just as in circuit theory, is to eliminate the need for trigonometric identities when adding or multiplying field amplitudes. A time-dependent complex analytic signal (viz., Ref. 15, sec. 10.2) is usually defined as the Fourier transform of this phasor, i.e.,

$$u(\mathbf{x}, t) = \int_0^\infty U(\mathbf{x}, \omega) e^{-i\omega t}\, d\omega \tag{3}$$

The integration in Eq. (3) is only required from zero to infinity because the phasor is defined with hermitian symmetry about the origin, i.e., $U(-\mathbf{x}, \omega) = U^*(\mathbf{x}, \omega)$. Therefore, all of the information is contained within the domain from zero to infinity. To obtain the actual field component from the analytical signal just take the real part of it. The Fourier transform in Eq. (3) is well defined if the analytical signal represents a deterministic field. However, if the light is partially coherent, then the analytic signal is usually taken to be a stationary random process. In this case the Fourier inverse of Eq. (3) does not exist. It is then possible to understand the spectral decomposition given by Eq. (3) only within the theory of generalized functions (viz., see Ref. 16, the appendix on generalized functions, and Ref. 17, pp. 25–30).

Scalar Field Amplitude

Each monochromatic component of an arbitrary deterministic light field propagating through a homogeneous, isotropic medium can always be represented using an angular spectrum of plane waves for each of the six scalar components of the vector field. The six angular spectra are coupled together by Maxwell's equations so that only two are independent.[18–20] Any two of the six angular spectra can be used to define two scalar fields from which the complete vector field can be determined. A polarized light field can be represented in this way by only one scalar field.[20,21] Thus it is often possible to represent one polarized component of a vector electromagnetic field by a single scalar field. It has also been found useful to represent completely unpolarized light by a single scalar field. In more complicated cases, where the polarization properties of the light are important, a vector theory is sometimes needed as discussed later under Explicit Vector Representations.

Temporal Coherence and Coherence Time

Within a short enough period of time, the time dependence of any light field at a point in space can be very closely approximated by a sine wave (Ref. 15, sec. 7.5.8). The length of time for which this is a good approximation is usually called the *coherence time* $\Delta\tau$. The coherence time is simply related to the spectral bandwidth for any light wave by the uncertainty principle, i.e.,

$$\Delta\tau\,\Delta\omega \geq 1 \tag{4}$$

For a light wave which is also highly directional within some region of space (like a beam) so that it propagates generally in some fixed direction (given by the unit vector $\mathbf{p}$), the field amplitude is given by

$$u(\mathbf{x}, t) = f(\mathbf{p}\cdot\mathbf{x} - ct) \tag{5}$$

Such a traveling wave will be approximately sinusoidal (and hence coherent) over some coherence length $\Delta\ell$ in the direction of $\mathbf{p}$ where from Eq. (4) we see that

$$\Delta\ell = c\,\Delta\tau \approx c/\Delta\omega \tag{6}$$

so that the coherence length varies inversely with bandwidth.

Spatial Coherence and Coherence Area

The time-dependent waveform for any light field is approximately the same at any point within a sufficiently small volume of space called the *coherence volume*. The projection of this volume onto a surface is termed a *coherence area*. If we have a field that, within some region, is roughly directional so that its field amplitude is given by Eq. (5), then the coherence length gives the dimension of the coherence volume in the direction of propagation $\mathbf{p}$, and the coherence area gives the dimensions of the coherence volume normal to this direction.

Measurements of Coherence

Coherence is usually measured by some form of interferometer that takes light from two test points in a light field, $\mathbf{x}_1$ and $\mathbf{x}_2$, and then allows them to interfere after introducing a time advance τ in the light from $\mathbf{x}_1$ relative to that from $\mathbf{x}_2$. If the field intensity of the interference pattern is measured as a function of τ, then in general it has the form (see Ref. 15, sec. 10.3.1)

$$I(\tau) = I(\mathbf{x}_1) + I(\mathbf{x}_2) + 2\ \mathrm{Real}\ (\Gamma_{12}(\tau)) \tag{7}$$

where $I(\mathbf{x}_i)$ is the intensity at the ith test point, and $\Gamma_{12}(\tau)$ is the mutual coherence function which measures the τ advanced correlation between the waveforms at the two test points (as subsequently defined under "Mutual Coherence Function"). There are many interferometers which have been developed to measure $\Gamma_{12}(\tau)$ in this way. One of the earliest techniques was developed by Thompson and Wolf.[22] They used a diffractometer to measure the coherence over a surface normal to the direction of propagation for a collimated beam from a partially coherent source. More recently, Carter[23] used an interferometer made from a grating and a microscope to similarly measure the coherence of a beam transverse to the direction of propagation.

4.4 DEFINITIONS OF COHERENCE FUNCTIONS

Mutual Coherence Function

In an older form of coherence theory[1] the principal coherence function was the mutual coherence function defined by[24]

$$\Gamma_{12}(\tau) \underset{T\to\infty}{\approx} \frac{1}{2T}\int_{-T}^{T} u(\mathbf{x}_1, t+\tau)u^*(\mathbf{x}_2, t)\,dt \tag{8}$$

where $u(\mathbf{x}, t)$ represents the complex analytic time-dependent signal at some point $\mathbf{x}$ and some time t as defined in Eq. (3). This definition was originally motivated by the fact that the intensity, as actually measured, is precisely this time averaged function with $\mathbf{x}_1 = \mathbf{x}_2$ and $\tau = 0$, and that this function is the most readily measured since it appears directly in Eq. (7). Thus it was clearly possible to measure $\Gamma_{12}(\tau)$ over some input plane, propagate it to an output plane,[25] and then find the intensity over the output plane from $\Gamma_{12}(\tau)$. It was assumed in this definition in Eq. (8) that $u(\mathbf{x}, t)$ is stationary in time so that $\Gamma_{12}(\tau)$ is only a function of τ and not of t. In most of the older literature, sharp brackets were usually used to represent this time average rather than an ensemble average (see Ref. 15, sec. 10.3.1). In the early 1960s it was found to be much more convenient to treat $u(\mathbf{x}, t)$ as an ergodic, stationary random process so that Eq. (8) could be replaced by

$$\Gamma_{12}(\tau) = \langle u(\mathbf{x}_1, t+\tau)u^*(\mathbf{x}_2, t)\rangle \tag{9}$$

where (everywhere in this article) the sharp brackets denote an ensemble average. After the change to ensemble averages the cross-spectral density function (to be defined shortly) became the most used correlation function in the coherence literature, because of the simpler and more general rules for its propagation (as discussed later under "Two Representations").

Complex Degree of Coherence

To obtain a function that depends only on the coherence properties of a light field it is often useful to normalize the mutual coherence function in the manner of

$$\gamma_{12}(\tau) = \frac{\langle u(\mathbf{x}_1, t+\tau)u^*(\mathbf{x}_2, t)\rangle}{\sqrt{\langle u(\mathbf{x}_1, t)u^*(\mathbf{x}_1, t)\rangle\langle u(\mathbf{x}_2, t)u^*(\mathbf{x}_2, t)\rangle}} \tag{10}$$

This is called the *complex degree of coherence.* It is a properly normalized correlation coefficient, so that $\gamma_{11}(0) = \gamma_{22}(0) = 1$. This indicates that the field at a point in space must always be perfectly coherent with itself. All other values of $\gamma_{12}(\tau)$ are generally complex with an amplitude less than one. This indicates that the fields at two different points, or at the same point after a time delay τ, are generally less than perfectly coherent with each other. The magnitude of the complete degree of spatial coherence (from zero to one) is a measure of the mutual coherence between the fields at the two test points and after a time delay τ.

Cross-Spectral Density Function

Just as in classical diffraction theory, it is much easier to propagate monochromatic light than light with a complicated time waveform. Thus the most frequently used coherence function is the cross-spectral density function, $W_\omega(\mathbf{x}_1, \mathbf{x}_2)$, which is the ensemble-averaged correlation function between a typical monochromatic component of the field at some point $\mathbf{x}_1$ with the complex conjugate of the same component of the field at some other point $\mathbf{x}_2$. It may be defined by

$$\delta(\omega - \omega')W_\omega(\mathbf{x}_1, \mathbf{x}_2) = \langle U(\mathbf{x}_1, \omega)U^*(\mathbf{x}_2, \omega')\rangle \tag{11}$$

The amplitude $U(\mathbf{x}, \omega)$ for a field of arbitrary coherence is taken to be random variable. Thus $U(\mathbf{x}, \omega)$ represents an ensemble of all of the possible fields, each of which is represented by a complex phasor amplitude like that defined in Eq. (2). The sharp brackets denote an ensemble average over all of these possible fields weighted by the probability for each of them to occur. The correlation functions defined by Eqs. (11) and (9) are related by the Fourier transform pairs

$$\Gamma_{12}(\tau) = \int_0^\infty W_\omega(\mathbf{x}_1, \mathbf{x}_2)e^{-i\omega\tau}\, d\omega \tag{12}$$

and

$$W_\omega(\mathbf{x}_1, \mathbf{x}_2) = \frac{1}{2\pi}\int_{-\infty}^{\infty} \Gamma_{12}(\tau)e^{i\omega\tau}\, d\tau \tag{13}$$

which is easily shown, formally, by substitution from Eq. (3) into (9) and then using (11). These relations represent a form of the generalized Wiener-Khintchine theorem (see Ref. 26, pp. 107–108).

Complex Degree of Spectral Coherence

Because the cross-spectral density function contains information about both the intensity (see "Intensity," which follows shortly) and the coherence of the field, it is useful to define another coherence function which describes the coherence properties only. This is the complex degree of spectral coherence (not to be confused with the complex degree of *spatial* coherence, which is a totally different function), which is usually defined by[27]

$$\mu_\omega(\mathbf{x}_1, \mathbf{x}_2) = \frac{W_\omega(\mathbf{x}_1, \mathbf{x}_2)}{\sqrt{W_\omega(\mathbf{x}_1, \mathbf{x}_1)W_\omega(\mathbf{x}_2, \mathbf{x}_2)}} \tag{14}$$

It is easy to show that this function is a properly normalized correlation coefficient which is always equal to unity if the field points are brought together, and is always less than or equal to unity as they are separated. If the magnitude of $\mu_\omega(\mathbf{x}_1, \mathbf{x}_2)$ is unity, it indicates that the monochromatic field component with radial frequency ω is perfectly coherent between the two points $\mathbf{x}_1$ and $\mathbf{x}_2$. If the magnitude of this function is less than unity it indicates less-than-perfect coherence. If the magnitude is zero it indicates complete incoherence between the field amplitudes at the two test points. For most partially coherent fields the cross-spectral density function has significantly large values only for point separations which keep the two field points within the same coherence volume. This function depends only on the positions of the points and the single radial frequency that the field components at the two points share. Field components of different frequency are always uncorrelated (and therefore incoherent), even at the same point.

Spectrum and Normalized Spectrum

Recently, the changes in the spectrum of light due to propagation have been studied using coherence theory. It is therefore useful to define the spectrum of light as just the monochromatic intensity (which is just the trace of the cross-spectral density function) as a function of omega, and the spectrum of a primary source as a very similar function, i.e.,

$$\begin{aligned} S_U(\mathbf{x}, \omega) &= \langle U_\omega(\mathbf{x}) U^*_\omega(\mathbf{x}) \rangle = W_U(\mathbf{x}, \mathbf{x}) \\ S_Q(\mathbf{x}, \omega) &= \langle \rho_\omega(\mathbf{x}) \rho^*_\omega(\mathbf{x}) \rangle = W_Q(\mathbf{x}, \mathbf{x}) \end{aligned} \tag{15}$$

where the subscript Q indicates that this is a primary source spectrum, and the subscript U indicates that this is a field spectrum. The spectrum for the primary source is a function of the phasor $\rho_\omega(\mathbf{x})$ which represents the currents and charges in this source as discussed under "Primary Sources" in the next section. It is also useful to normalize these spectra in the manner

$$s_A(\mathbf{x}, \omega) = \frac{S_A(\mathbf{x}, \omega)}{\displaystyle\int_0^\infty S_A(\mathbf{x}, \omega)\, d\omega} \tag{16}$$

where the subscript A can indicate either U or Q, and the normalized spectrum has the property

$$\int_0^\infty s_A(\mathbf{x}, \omega)\, d\omega = 1 \tag{17}$$

so that it is independent of the total intensity.

Angular Correlation Function

A new coherence function, introduced for use with the angular spectrum expansion of a monochromatic component of the field,[28] is the angular correlation function defined by

$$\mathcal{A}_\omega(\mathbf{p}_1, \mathbf{p}_2) = \langle A_\omega(\mathbf{p}_1), A^*_\omega(\mathbf{p}_2) \rangle \tag{18}$$

where $A_\omega(\mathbf{p}_i)$ is the angular spectrum which gives the complex amplitude of the plane wave component of the field which propagates in the direction given by the unit vector $\mathbf{p}_i$. This is related to the cross-spectral density function over the $z = 0$ plane by the equation

$$\mathcal{A}(\mathbf{p}_1, \mathbf{p}_2) = \frac{1}{\lambda^4} \int_{-\infty}^{\infty} \int_{-\infty}^{\infty} \int_{-\infty}^{\infty} \int_{-\infty}^{\infty} W^{(0)}_\omega(\mathbf{x}'_1, \mathbf{x}'_2) e^{-ik(\mathbf{p}_1 \cdot \mathbf{x}'_1 - \mathbf{p}_2 \cdot \mathbf{x}'_2)}\, d^2\mathbf{x}'_1\, d^2\mathbf{x}'_2 \tag{19}$$

This is the four-dimensional Fourier transform of the cross-spectral density function over the $z = 0$ plane. It represents a correlation function between the complex amplitudes of two plane wave components of the field propagating in the directions given by the unit vectors $\mathbf{p}_1$ and $\mathbf{p}_2$, respectively. It can be used to calculate the cross-spectral density function (as described later under "Angular Spectrum Representation") between any pair of points away from the $z = 0$ plane, assuming that the field propagates in a source-free homogeneous medium. In this article we will use single-primed vectors, as in Eq. (19), to represent radius vectors from the origin to points within the $z = 0$ plane, i.e., $\mathbf{x}' = (x', y', 0)$, as shown in Fig. 1.

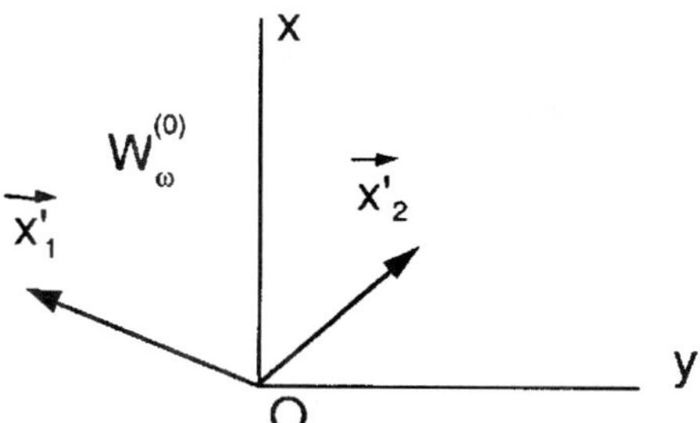

FIGURE 1 Illustrating the coordinate system and notation used for planar sources in the $z = 0$ plane.

All other vectors, such as $\mathbf{x}$ or $\mathbf{x}''$, are to be taken as three-dimensional vectors. Generally $\mathbf{s}$ and $\mathbf{p}$ are three-dimensional unit vectors indicating directions from the origin, a superscript (0) on a function indicates that it is the boundary condition for that function over the $z = 0$ plane, and a superscript (∞) on a function indicates that it is the asymptotic value for that function on a sphere of constant radius R from the origin as $R \to \infty$.

Intensity

The intensity is usually considered to be the observable quantity in coherence theory. Originally it was defined to be the trace of the mutual coherence function as defined by Eq. (8), i.e.,

$$I(\mathbf{x}_1) \triangleq \Gamma_{11}(0) \underset{T\to\infty}{\approx} \frac{1}{2T} \int_{-T}^{T} u(\mathbf{x}_1, t) u^*(\mathbf{x}_1, t)\, dt \tag{20}$$

which is always real. Thus it is the time-averaged square magnitude of the analytic signal. This represents the measurement obtained by the electromagnetic power detectors always used to detect light fields. Since the change to ensemble averages in coherence theory, it is almost always assumed that the analytic signal is an ergodic random process so that the intensity can be obtained from the equation

$$I(\mathbf{x}_1) \triangleq \Gamma_{11}(0) = \langle u(\mathbf{x}_1, t) u^*(\mathbf{x}_1, t) \rangle \tag{21}$$

where the sharp brackets indicate an ensemble average. Usually, in most recent coherence-theory papers, the intensity calculated is actually the spectrum, which is equivalent to the intensity of a single monochromatic component of the field which is defined as the trace of the cross-spectral density function as given by

$$I_\omega(\mathbf{x}_1) \triangleq W_\omega(\mathbf{x}_1, \mathbf{x}_1) = \langle U(\mathbf{x}_1, \omega) U^*(\mathbf{x}_1, \omega) \rangle \tag{22}$$

Since different monochromatic components of the field are mutually incoherent and cannot interfere, we can always calculate the intensity of the total field as the sum over the intensities of its monochromatic components in the manner

$$I(\mathbf{x}_1) \triangleq \int_0^\infty I_\omega(\mathbf{x}_1)\, d\omega \tag{23}$$

Since in most papers on coherence theory the subscript omega is usually dropped, the reader should be careful to observe whether or not the intensity calculated is for a monochromatic component of the field only. If it is, the total measurable intensity can be obtained simply by summing over all omega, as indicated in Eq. (23).

Radiant Emittance

In classical radiometry the *radiant emittance* is defined to be the power radiated into the far field by a planar source per unit area. A wave function with some of the properties of radiant emittance has been defined by Marchand and Wolf using coherence theory (see Ref. 29, eq. (32), and Ref. 30). However, because of interference effects, the far-field energy cannot be subdivided into components that can be traced back to the area of the source that produced them. The result is that the radiant emittance defined by Marchand and Wolf is not nonnegative definite (as it is in classical radiometry) except for the special case of a completely incoherent source. Since, as discussed in the next section under "Perfectly Incoherent Source," perfectly incoherent sources exist only as a limiting case, radiant emittance has not been found to be a very useful concept in coherence theory.

Radiant Intensity

In classical radiometry the *radiant intensity* is defined to be the power radiated from a planar source into a unit solid angle with respect to an origin at the center of the surface. This can be interpreted in coherence theory as the field intensity over a portion of a surface in the far field of the source, in some direction given by the unit vector **s**, which subtends a unit solid angle from the source. Thus the radiant intensity for a monochromatic component of the field can be defined in coherence theory as

$$J_\omega(\mathbf{s}) \underset{R\to\infty}{\approx} W_\omega^{(\infty)}(R\mathbf{s}, R\mathbf{s})R^2 \tag{24}$$

To obtain the total radiant intensity we need only sum this function over all omega.

Radiance

In classical coherence theory the radiance function is the power radiated from a unit area on a planar source into a unit solid angle with respect to an origin at the center of the source. In the geometrical optics model, from which this concept originally came, it is consistent to talk about particles of light leaving an area at a specified point on a surface to travel in some specified direction. However, in a wave theory, wave position and wave direction are Fourier conjugate variables. We can have a configuration space wave function (position) or a momentum space wave function (direction), but not a wave function that depends independently on both position and direction. Thus the behavior of a wave does not necessarily conform to a model which utilizes a radiance function.[31] Most naturally occurring sources are quasi-homogeneous (discussed later). For such sources, a radiance function for a typical monochromatic component of the field can be defined as the Wigner distribution function[32–35] of the cross-spectral density function over the $z = 0$ plane, which is given by

$$B_\omega(\mathbf{x}'_+, \mathbf{s}) = \frac{\cos\theta}{\lambda^2}\int_{-\infty}^{\infty}\int_{-\infty}^{\infty} W_\omega^{(0)}(\mathbf{x}'_+ + \mathbf{x}'_-/2, \mathbf{x}'_+ - \mathbf{x}'_-/2)e^{-ik\mathbf{s}\cdot\mathbf{x}'_-}\, d^2\mathbf{x}'_- \tag{25}$$

where θ is the angle that the unit vector **s** makes with the $+z$ axis. For quasi-homogeneous fields this can be associated with the energy radiated from some point $\mathbf{x}'_+$ into the far field in the direction **s**. Such a definition for radiance also works approximately for some other light fields, but for light which does not come from a quasi-homogeneous source, no such definition is either completely equivalent to a classical radiance function[31] or unique as an approximation to it. Much progress has been made toward representing a more general class of fields using a radiance function.[36] In general, waves do not have radiance functions.

Higher-Order Coherence Functions

In general, the statistical properties of a random variable are uniquely defined by the probability density function which can be expanded into a series which contains correlation functions of all orders. Thus, in general, all orders of correlation functions are necessary to completely define the statistical properties of the field. In classical coherence theory we usually assume that the partially coherent fields arise from many independent sources so that, by the central limit theorem of statistics, the probability distribution function for the real and imaginary components of the phasor field amplitude are zero-mean gaussian random variables (See Ref. 8, sec. 2.72*e*). Thus, from the gaussian moment theorem, the field correlation functions of any order can be calculated from the second-order correlation functions, for example,

$$\langle U(\mathbf{x}_1, \omega)U^*(\mathbf{x}_1, \omega)U(\mathbf{x}_2, \omega)U^*(\mathbf{x}_2, \omega)\rangle = I_\omega(\mathbf{x}_1)I_\omega(\mathbf{x}_2) + |W_\omega(\mathbf{x}_1, \mathbf{x}_2)|^2 \tag{26}$$

Thus, for gaussian fields, the second-order correlation functions used in coherence theory completely define the statistical properties of the field. Some experiments, such as those involving intensity interferometry, actually measure fourth- or higher-order correlation functions.[37]

4.5 MODEL SOURCES

Primary Sources

In coherence theory it is useful to talk about primary and secondary sources. A primary source distribution is the usual source represented by the actual charge and current distribution which give rise to the field. For propagation from a primary source through a source-free media, the field amplitude is defined by the inhomogeneous Helmholtz equation [see Ref. 38, eq. (6.57)], i.e.,

$$\left(\nabla^2 + \frac{\omega^2}{c^2}\right)U_\omega(\mathbf{x}) = -4\pi\rho_\omega(\mathbf{x}) \tag{27}$$

where $\rho_\omega(\mathbf{x})$ represents the charge-current distribution in the usual manner. A solution to this wave equation gives

$$W_U(\mathbf{x}_1, \mathbf{x}_2) = \int_{-\infty}^{\infty}\int_{-\infty}^{\infty}\int_{-\infty}^{\infty}\int_{-\infty}^{\infty}\int_{-\infty}^{\infty}\int_{-\infty}^{\infty} W_Q(\mathbf{x}_1'', \mathbf{x}_2'')K(\mathbf{x}_1, \mathbf{x}_1'')K^*(\mathbf{x}_2, \mathbf{x}_2'')\, d^3\mathbf{x}_1''\, d^3\mathbf{x}_2'' \tag{28}$$

for the cross-spectral density function, where $k = \omega/c = 2\pi/\lambda$

$$K(\mathbf{x}, \mathbf{x}'') = \frac{e^{ik|\mathbf{x}-\mathbf{x}''|}}{|\mathbf{x} - \mathbf{x}''|} \tag{29}$$

is the free-space propagator for a primary source, $W_U(\mathbf{x}_1, \mathbf{x}_2)$ is the cross-spectral density function for the fields (with suppressed ω-dependence), as defined by Eq. (11), and

$$W_Q(\mathbf{x}_1'', \mathbf{x}_2'') = \langle \rho_\omega(\mathbf{x}_1'')\rho_\omega^*(\mathbf{x}_2'')\rangle \tag{30}$$

is the cross-spectral density function for the source. This three-dimensional primary source can be easily reduced to a two-dimensional source over the $z = 0$ plane by simply defining the charge current distribution to be

$$\rho_\omega(\mathbf{x}) = \rho'_\omega(x, y)\delta(z) \tag{31}$$

where $\delta(z)$ is the Dirac delta function.

Secondary Sources

Often, however, it is more convenient to consider a field which arises from sources outside of the region in space in which the fields are of interest. Then it is sometime useful to work with boundary conditions for the field amplitude over some surface bounding the region of interest. In many coherence problems these boundary conditions are called a *planar secondary source* even though they are actually treated as boundary conditions. For example, most conventional diffraction equations assume the boundary condition over the $z = 0$ plane is known and use it to calculate the fields in the $z > 0$ half-space. Then the field obeys the homogeneous Helmholtz equation [See Ref. 38, eq. (7.8)], i.e.,

$$\left(\nabla^2 + \frac{\omega^2}{c^2}\right)U_\omega(\mathbf{x}) = 0 \tag{32}$$

which has the solution

$$W_\omega(\mathbf{x}_1, \mathbf{x}_2) = \int_{-\infty}^{\infty}\int_{-\infty}^{\infty}\int_{-\infty}^{\infty}\int_{-\infty}^{\infty} W_\omega^{(0)}(\mathbf{x}'_1, \mathbf{x}'_2)h(\mathbf{x}_1, \mathbf{x}'_1)h^*(\mathbf{x}_2, \mathbf{x}'_2)\, d^2\mathbf{x}'_1\, d^2\mathbf{x}'_2 \tag{33}$$

where $W_\omega^{(0)}(\mathbf{x}'_1, \mathbf{x}'_2)$ is the boundary condition for the cross-spectral density function of the fields over the $z = 0$ plane (as shown in Fig. 1), $W_\omega(\mathbf{x}_1, \mathbf{x}_2)$ is the cross-spectral density function anywhere in the $z > 0$ half-space, and

$$h(\mathbf{x}, \mathbf{x}') = \frac{-1}{2\pi}\frac{d}{dz}\frac{e^{ik|\mathbf{x}-\mathbf{x}'|}}{|\mathbf{x}-\mathbf{x}'|} \tag{34}$$

is the free-space propagator for the field amplitude. This is a common example of a secondary source over the $z = 0$ plane.

Perfectly Coherent Source

The definition of a perfectly coherent source, within the theory of partial coherence, is somewhat complicated. The light from an ideal monochromatic source is, of course, always perfectly coherent. Such a field produces high-contrast interference patterns when its intensity is detected by any suitable instrument. However, it is possible that light fields exist that are not monochromatic but can also produce similar interference patterns and therefore must be considered coherent. The ability of light fields to produce interference patterns at a detector is measured most directly by the complex degree of coherence

$\gamma_{12}(\tau)$, defined by Eq. (10). If a field has a complex degree of coherence that has unit magnitude for every value of τ and for every point pair throughout some domain D, then light from all points in D will combine to produce high-contrast interference fringes.[39] Such a field is defined to be perfectly coherent within D. Mandel and Wolf[27] have shown that the mutual coherence function for such a field factors within D in the manner

$$\Gamma_{12}(\tau) = \psi(\mathbf{x}_1)\psi^*(\mathbf{x}_2)e^{-i\omega\tau} \tag{35}$$

We will take Eq. (35) to be the definition of a perfectly coherent field. Coherence is not as easily defined in the space-frequency domain because it depends on the spectrum of the light as well as on the complex degree of spectral coherence. For example, consider a field for which every monochromatic component is characterized by a complex degree of spectral coherence which has unit magnitude between all point pairs within some domain D. Mandel and Wolf[40] have shown that for such a field the cross-spectral density function within D factors is

$$W_\omega(\mathbf{x}_1''\mathbf{x}_2'') = \bar{U}(\mathbf{x}_1'', \omega)\bar{U}^*(\mathbf{x}_2'', \omega) \tag{36}$$

However, even if Eq. (36) holds for this field within D, the field may not be perfectly coherent [as perfect coherence is defined by Eq. (35)] between all points within the domain. In fact it can be completely incoherent between some points within D.[41] A secondary source covering the $z = 0$ plane with a cross-spectral density function over that plane which factors as given by Eq. (36) will produce a field in the $z > 0$ half-space (filled with free space or a homogeneous, isotopic dielectric) which has a cross-spectral density function that factors in the same manner everywhere in the half-space. This can be easily shown by substitution from Eq. (36) into Eq. (28), using Eq. (29). But, even if this is true for every monochromatic component of the field, Eq. (35) may not hold within the half-space for every point pair, so we cannot say that the field is perfectly coherent there. Perfectly coherent light sources never actually occur, but sometimes the light from a laser can behave approximately in this way over some coherence volume which is usefully large. Radio waves often behave in this manner over very large coherence volumes.

Quasi-Monochromatic Source

In many problems it is more useful not to assume that a field is strictly monochromatic but instead to assume that it is only quasi-monochromatic so that the time-dependent field amplitude can be approximated by

$$u(\mathbf{x}, t) = u_0(\mathbf{x}, t)e^{-i\omega t} \tag{37}$$

where $u_0(\mathbf{x}, t)$ is a random process which varies much more slowly in time than $e^{-i\omega t}$. Then the mutual coherence function and the complex degree of spatial coherence can be usefully approximated by (see Ref. 15, sec. 10.4.1)

$$\begin{aligned} \Gamma_{12}(\tau) &= \Gamma_{12}(0)e^{-i\omega\tau} \\ \gamma_{12}(\tau) &= \gamma_{12}(0)e^{-i\omega\tau} \end{aligned} \tag{38}$$

within coherence times much less than the reciprocal bandwidth of the field, i.e., $\Delta\tau \ll 1/\Delta\omega$. In the pre-1960 coherence literature, $\Gamma_{12}(0)$ was called the *mutual intensity*. Monochromatic diffraction theory was then used to define the propagation properties of this monochromatic function. It was used instead of the cross-spectral density function to formulate the theory for the propagation of a partially coherent quasi-monochromatic field. While this earlier form of the theory was limited by the quasi-monochromatic approximation and was therefore not appropriate for very wideband light, the newer formulation (in terms of the cross-spectral density function) makes no assumptions about the spectrum of the light and can be applied generally. The quasi-monochromatic approximation is still very useful when dealing with radiation of very high coherence.[42]

Schell Model Source

Other, more general, source models have been developed. The most general of these is the Schell model source (see Ref. 43, sec. 7.5, and Ref. 44), for which we only assume that the complex degree of spectral coherence for either a primary or secondary source is stationary in space, so that from Eq. (14) we have

$$W_A(\mathbf{x}_1, \mathbf{x}_2) = \mu_A(\mathbf{x}_1 - \mathbf{x}_2)\sqrt{W_A(\mathbf{x}_1, \mathbf{x}_1)W_A(\mathbf{x}_2, \mathbf{x}_2)} \tag{39}$$

where the subscript A stands for U in the case of a Schell model secondary source and Q in the case of a Schell model primary source. The Schell model does not assume low coherence and, therefore, can be applied to spatially stationary light fields of any state of coherence. The Schell model of the form shown in Eq. (39) has been used to represent both three-dimensional primary sources[45,46] and two-dimensional secondary sources.[43,47,48]

Quasi-Homogeneous Source

If the intensity of a Schell model source is essentially constant over any coherence area, then Eq. (39) may be approximated by

$$W_A(\mathbf{x}_1, \mathbf{x}_2) = \mu_A(\mathbf{x}_1 - \mathbf{x}_2)I_A[(\mathbf{x}_1 + \mathbf{x}_1)/2] \tag{40}$$

where the subscript A can be either U, for the case of a quasi-homogeneous secondary source or Q, for the case of a quasi-homogeneous primary source. This equation is very useful in coherence theory because of the important exact mathematical identity.[49]

$$\int_{-\infty}^{\infty}\int_{-\infty}^{\infty}\int_{-\infty}^{\infty}\int_{-\infty}^{\infty} \mu_\omega^{(0)}(\mathbf{x}_1' - \mathbf{x}_2')I_\omega^{(0)}[(\mathbf{x}_1' + \mathbf{x}_2')/2]e^{-ik(\mathbf{x}_1'\cdot\mathbf{p}_1 - \mathbf{x}_2'\cdot\mathbf{p}_2)}\, dx_1'\, dy_1'\, dx_2'\, dy_2'$$

$$= \int_{-\infty}^{\infty}\int_{-\infty}^{\infty} \mu_\omega^{(0)}(\mathbf{x}_-')e^{-ik(\mathbf{x}_-'\cdot\mathbf{p}_+)}\, dx_-'\, dy_-' \int_{-\infty}^{\infty}\int_{-\infty}^{\infty} I_\omega^{(0)}(\mathbf{x}_+')e^{-ik(\mathbf{x}_+'\cdot\mathbf{p}_-)}\, dx_+'\, dy_+' \tag{41}$$

where

$$\mathbf{x}_+' = (\mathbf{x}_1' + \mathbf{x}_2')/2 \qquad \mathbf{x}_-' = \mathbf{x}_1' - \mathbf{x}_2'$$

and

$$\mathbf{p}_+ = (\mathbf{p}_1 + \mathbf{p}_2)/2 \qquad \mathbf{p}_- = \mathbf{p}_1 - \mathbf{p}_2$$

which allows the four-dimensional Fourier transforms that occur in propagating the correlation functions for secondary sources [for example, Eqs. (49) or (54)] to be factored into a product of two-dimensional Fourier transforms. An equivalent identity also holds for the six-dimensional Fourier transform of the cross-spectral density function for a primary source, reducing it to the product of two three-dimensional Fourier transforms. This is equally useful in dealing with propagation from primary sources [for example, Eq. (53)]. This model is very good for representing two-dimensional secondary sources with sufficiently low coherence that the intensity does not vary over the coherence area on the input plane.[49,50] It has also been applied to primary three-dimensional sources.[45,46] to primary and secondary two-dimensional sources,[51,52] and to three-dimensional scattering potentials.[53,54]

Perfectly Incoherent Source

If the coherence volume of a field becomes much smaller than any other dimensions of interest in the problem, then the field is said to be *incoherent.* It is believed that no field can be incoherent over dimensions smaller than the order of a light wavelength. An incoherent field can be taken as a special case of a quasi-homogeneous field for which the complex degree of spectral coherence is approximated by a Dirac delta function, i.e.,

$$\begin{aligned} W_A(\mathbf{x}_1, \mathbf{x}_2) &= I(\mathbf{x}_1)\delta^2(\mathbf{x}_1 - \mathbf{x}_2) \\ W_Q(\mathbf{x}_1, \mathbf{x}_2) &= I(\mathbf{x}_1)\delta^3(\mathbf{x}_1 - \mathbf{x}_2) \end{aligned} \tag{42}$$

where the two-dimensional Dirac delta function is used for any two-dimensional source and a three-dimensional Dirac delta function is used only for a three-dimensional primary source. Even though this approximation is widely used, it is not a good representation for the thermal sources that predominate in nature. For example, the radiant intensity from a planar, incoherent source is not in agreement with Lambert's law. For thermal sources the following model is much better.

Thermal (Lambertian) Source

For a planar, quasi-homogeneous source to have a radiant intensity in agreement with Lambert's law it is necessary for the complex degree of spectral coherence to have the form

$$\mu_A(\mathbf{x}_1 - \mathbf{x}_2) = \frac{\sin(k\,|\mathbf{x}_1 - \mathbf{x}_2|)}{k\,|\mathbf{x}_1 - \mathbf{x}_2|} \tag{43}$$

to which arbitrary spatial frequency components with periods less than a wavelength can be added since they do not affect the far field.[55] It can also be shown that, under frequently obtained conditions, blackbody radiation has such a complex degree of spectral coherence.[56] It is believed that most naturally occurring light can be modeled as quasi-homogeneous with this correlation function.

4.6 PROPAGATION

Perfectly Coherent Light

Perfectly coherent light propagates according to conventional diffraction theory. By substitution from Eq. (36) into Eq. (33) using Eq. (34) we obtain

$$U_\omega(\mathbf{x}) = \frac{1}{2\pi}\int_{-\infty}^{\infty}\int_{-\infty}^{\infty} U_\omega^{(0)}(\mathbf{x}')\,\frac{1}{2\pi}\frac{d}{dz'}\frac{e^{ik|\mathbf{x}-\mathbf{x}'|}}{|\mathbf{x}-\mathbf{x}'|}\,d^2\mathbf{x}' \tag{44}$$

which is just Rayleigh's diffraction integral of the first kind. Perfectly coherent light does not lose coherence as it propagates through any medium which is time-independent. For perfectly coherent light propagating in time-independent media, coherence theory is not needed.

Hopkin's Formula

In 1951 Hopkins[57] published a formula for the complex degree of spatial coherence for the field from a planar, secondary, incoherent, quasi-monochromatic source after propagating through a linear optical system with spread function $h(\mathbf{x}, \mathbf{x}')$, i.e.,

$$\gamma_{12}(0) = \frac{1}{\sqrt{I(\mathbf{x}_1)I(\mathbf{x}_2)}} \int_{-\infty}^{\infty}\int_{-\infty}^{\infty} I_\omega^{(0)}(\mathbf{x}_1')h(\mathbf{x}_1 - \mathbf{x}_1')h^*(\mathbf{x}_2 - \mathbf{x}_1')\, d^2\mathbf{x}_1' \tag{45}$$

where $I_\omega^{(0)}(\mathbf{x}')$ is the intensity over the source plane. This formula can be greatly generalized to give the complex degree of spectral coherence for the field from any planar, quasi-homogeneous, secondary source[58] after transmission through this linear optical system, i.e.,

$$\mu_{12}(\mathbf{x}_1, \mathbf{x}_2) = \frac{1}{\sqrt{I(\mathbf{x}_1)I(\mathbf{x}_2)}} \int_{-\infty}^{\infty}\int_{-\infty}^{\infty} I_\omega^{(0)}(\mathbf{x}')h(\mathbf{x}_1, \mathbf{x}')h^*(\mathbf{x}_2, \mathbf{x}')\, d^2\mathbf{x}' \tag{46}$$

provided that the spread function $h(\mathbf{x}, \mathbf{x}')$ can be assumed to be constant in its $\mathbf{x}'$ dependence over any coherence area in the source plane.

van Cittert–Zernike Theorem

Hopkins' formula can be specialized for free-space propagation to calculate the far-field coherence properties of planar, secondary, quasi-homogeneous sources of low coherence. In 1934 van Cittert[59] and, later, Zernike[60] derived a formula equivalent to

$$\mu_\omega(\mathbf{x}_1, \mathbf{x}_2) = \frac{1}{\sqrt{I(\mathbf{x}_1)I(\mathbf{x}_2)}} \int_{-\infty}^{\infty}\int_{-\infty}^{\infty} I_\omega^{(0)}(\mathbf{x}') \frac{e^{ik[|\mathbf{x}_1 - \mathbf{x}'| - |\mathbf{x}_2 - \mathbf{x}'|]}}{|\mathbf{x}_1 - \mathbf{x}'|\,|\mathbf{x}_2 - \mathbf{x}'|}\, d^2\mathbf{x}' \tag{47}$$

for the complex degree of spectral coherence between any pair of points in the field radiated from an incoherent planar source, assuming that the points are not within a few wavelengths of the source. We can obtain Eq. (47) by substitution from Eq. (34) into Eq. (46) and then approximating the propagator in a standard manner [see Ref. 61, eq. (7)]. Assume next, that the source area is contained within a circle of radius a about the origin in the source plane as shown in Fig. 4. Then, if the field points are both located on a sphere of radius R, which is outside of the Rayleigh range of the origin (i.e., $|\mathbf{x}_1| = |\mathbf{x}_2| = R \gg ka^2$), we can apply the Fraunhofer approximation to Eq. (47) to obtain

$$\mu_{12}(\mathbf{x}_1, \mathbf{x}_2) = \frac{1}{\sqrt{I(\mathbf{x}_1)I(\mathbf{x}_2)}} \int_{-\infty}^{\infty}\int_{-\infty}^{\infty} I_\omega^{(0)}(\mathbf{x}')e^{ik\mathbf{x}'\cdot(\mathbf{x}_1 - \mathbf{x}_2)/R}\, d^2\mathbf{x}' \tag{48}$$

This formula is very important in coherence theory. It shows that the complex degree of spectral coherence from any planar, incoherent source with an intensity distribution $I_\omega^{(0)}(\mathbf{x}')$ has the same dependence on $(\mathbf{x}_1 - \mathbf{x}_2)$, over a sphere of radius R in the far field, as the diffraction pattern from a closely related perfectly coherent planar source with a real amplitude distribution proportional to $I_{\omega}^{(0)}(\mathbf{x}')$ (see Ref. 15, sec. 10.4.2*a*). Equation (48) can also be applied to a planar, quasi-homogeneous source[49] that is not necessarily incoherent, as will be shown later under "Reciprocity Theorem."

Angular Spectrum Representation

Much more general equations for propagation of the cross-spectral density function can be obtained. Equation (33) is one such expression. Another can be found if we expand the fields in a source-free region of space into an angular spectrum of plane waves. Then we find that the cross-spectral density function over any plane can be calculated from the same function over a parallel plane using a linear systems approach. For an example, consider the two planes illustrated in Fig. 2.

We assume that the cross-spectral density function is known over the $z = 0$ plane in the figure, and we wish to calculate this function over the $z = d$ plane. To do this we first take the Fourier transform of the cross-spectral density function over the $z = 0$ plane according to Eq. (19)

$$\mathcal{A}_{\text{in}}(\mathbf{p}_1, \mathbf{p}_2) = \frac{1}{\lambda^4} \int_{-\infty}^{\infty} \int_{-\infty}^{\infty} \int_{-\infty}^{\infty} \int_{-\infty}^{\infty} W_{\omega}^{(0)}(\mathbf{x}_1', \mathbf{x}_2') e^{-ik(\mathbf{p}_1 \cdot \mathbf{x}_1' - \mathbf{p}_2 \cdot \mathbf{x}_2')} \, d^2\mathbf{x}_1' \, d^2\mathbf{x}_2' \tag{49}$$

to obtain the angular correlation function in which all of the phase differences between the plane wave amplitudes are given relative to the point at the origin. Second, we shift the phase reference from the origin to the point on the z axis in the output plane by multiplying the angular correlation function by a transfer function, i.e.,

$$\mathcal{A}_{\text{out}}(\mathbf{p}_1, \mathbf{p}_2) = \mathcal{A}_{\text{in}}(\mathbf{p}_1, \mathbf{p}_2) \exp\left[ik(m_1 - m_2)d\right] \tag{50}$$

where d is the distance from the input to the output plane along the z axis (for back propagation d will be negative) and m_i is the third component of the unit vector $\mathbf{p}_i = (p_i, q_i, m_i)$, $i = 1$ or 2, which is defined by

$$\begin{aligned} m_i &= \sqrt{1 - p_i^2 - q_i^2}, \text{ if } p_i^2 + q_i^2 \leq 1 \\ &= i\sqrt{p_i^2 + q_i^2 - 1}, \text{ if } p_i^2 + q_i^2 > 1 \end{aligned} \tag{51}$$

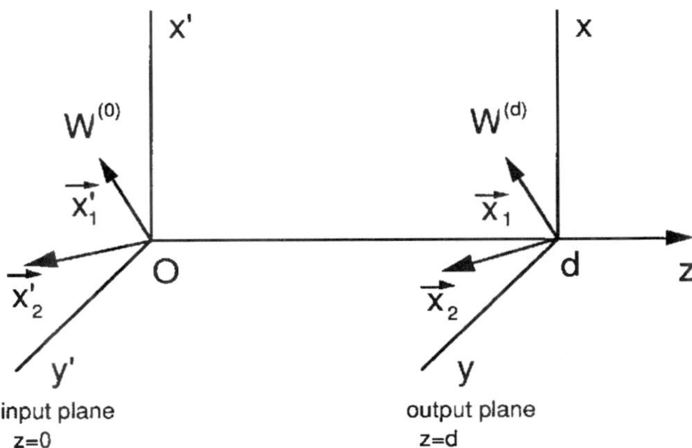

FIGURE 2 Illustrating the coordinate system for propagation of the cross-spectral density function using the angular spectrum of plane waves.

and is the cosine of the angle that $\mathbf{p}_i$ makes with the $+z$ axis for real m_i. Finally, to obtain the cross-spectral density function over the output plane, we simply take the Fourier inverse of $\mathscr{A}_{\text{out}}(\mathbf{p}_1, \mathbf{p}_2)$, i.e.,

$$W_{\omega}^{(d)}(\mathbf{x}_1\mathbf{x}_2) = \int_{-\infty}^{\infty}\int_{-\infty}^{\infty}\int_{-\infty}^{\infty}\int_{-\infty}^{\infty} \mathscr{A}_{\text{out}}(\mathbf{p}_1, \mathbf{p}_2) e^{ik(\mathbf{p}_1\cdot\mathbf{x}_1 - \mathbf{p}_2\cdot\mathbf{x}_2)}\, d^2\mathbf{p}_1\, d^2\mathbf{p}_2 \tag{52}$$

where, in this equation only, we use $\mathbf{x}_i$ to represent a two-dimensional radius vector from the point $(0, 0, d)$ to a field point in the $z = d$ plane, as shown in Fig. 2. This propagation procedure is similar to the method usually used to propagate the field amplitude in Fourier optics. In coherence theory, it is the cross-spectral density function for a field of any state of coherence that is propagated between arbitrary parallel planes using the linear systems procedure. The only condition for the validity of this procedure is that the volume between the planes must either be empty space or a uniform dielectric medium.

Radiation Field

The cross-spectral density function far away from any source, which has finite size, can be calculated using a particularly simple equation. Consider a primary, three-dimensional source located within a sphere of radius a, as shown in Fig. 3. For field points, $\mathbf{x}_1$ and $\mathbf{x}_2$ which are much farther from the origin than the Rayleigh range ($|\mathbf{x}_i| \gg ka^2$), in any direction, a form of the familiar Fraunhofer approximation can be applied to Eq. (28) to obtain [see Ref. 62, eq. (2.5)]

$$W_U^{(\infty)}(\mathbf{x}_1, \mathbf{x}_2) = \frac{e^{ik(|\mathbf{x}_1| - |\mathbf{x}_2|)}}{|\mathbf{x}_1|\,|\mathbf{x}_2|} \int_{-\infty}^{\infty}\int_{-\infty}^{\infty}\int_{-\infty}^{\infty}\int_{-\infty}^{\infty}\int_{-\infty}^{\infty}\int_{-\infty}^{\infty} W_Q(\mathbf{x}_1'', \mathbf{x}_2'')$$
$$\times\, e^{-ik[\mathbf{x}_1\cdot\mathbf{x}_1''/|\mathbf{x}_1| - \mathbf{x}_2\cdot\mathbf{x}_2''/|\mathbf{x}_2|]}\, d^3\mathbf{x}_1''\, d^3\mathbf{x}_2'' \tag{53}$$

Thus the cross-spectral density function of the far field is proportional to the six-dimensional Fourier transform of the cross-spectral density function of its sources. A very

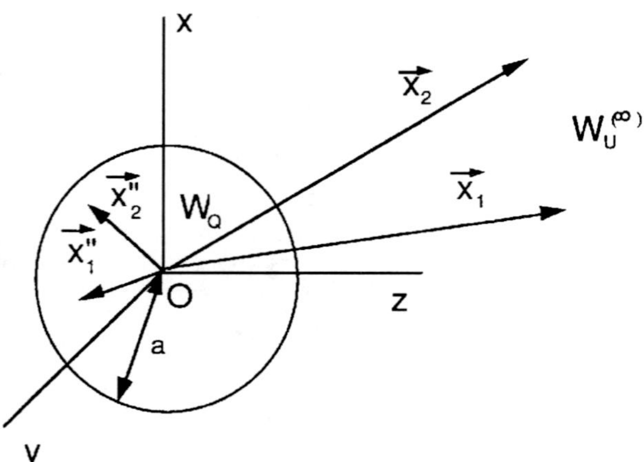

FIGURE 3 Illustrating the coordinate system used to calculate the cross-spectral density function in the far field of a primary, three-dimensional source.

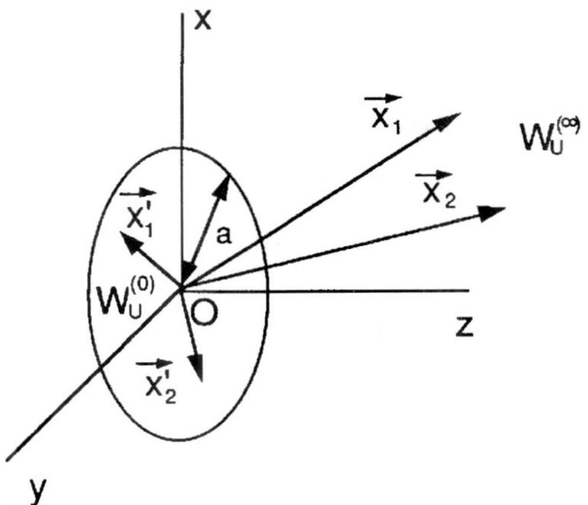

FIGURE 4 Illustrating the coordinate system for calculating the far field cross-spectral density function for a planar, secondary source distribution in the $z=0$ plane. Single-primed coordinates indicate radius vectors from the origin to points within the $z=0$ plane.

similar expression can also be found for a two-dimensional, secondary source distribution over the $z=0$ plane, as illustrated in Fig. 4.

If the sources are all restricted to the area within a circle about the origin of radius a, then the cross-spectral density function for all points which are outside of the Rayleigh range ($|\mathbf{x}_i| \gg ka^2$) from the origin and also in the $z>0$ half-space can be found by using a different form of the Fraunhofer approximation in Eq. (33) [see Ref. 62, eq. (3.3)] to get

$$W_U^{(\infty)}(\mathbf{x}_1, \mathbf{x}_2) = \frac{1}{\lambda^2}\frac{z_1}{|\mathbf{x}_1|}\frac{z_2}{|\mathbf{x}_2|}\frac{e^{ik(|\mathbf{x}_1|-|\mathbf{x}_2|)}}{|\mathbf{x}_1|\,|\mathbf{x}_2|}\int_{-\infty}^{\infty}\int_{-\infty}^{\infty}\int_{-\infty}^{\infty}\int_{-\infty}^{\infty} W_U^{(0)}(\mathbf{x}'_1, \mathbf{x}'_2) \times e^{-ik[\mathbf{x}_1\cdot\mathbf{x}'_1/|\mathbf{x}_1| - \mathbf{x}_2\cdot\mathbf{x}'_2/|\mathbf{x}_2|]}\, d^2\mathbf{x}'_1\, d^2\mathbf{x}'_2 \tag{54}$$

Because of both their relative simplicity (as Fourier transforms) and great utility, Eqs. (53) and (54) are very important in coherence theory for calculating both the radiant intensity and the far-field coherence properties of the radiation field from any source.

Representations

Several equations have been described here for propagating the cross-spectral density function. The two most general are Eq. (33), which uses an expansion of the field into spherical Huygens wavelets, and Eqs. (49), (50), and (52), which use an expansion of the field into an angular spectrum of plane waves. These two formulations are completely equivalent. Neither uses any approximation not also used by the other method. The choice of which to use for a particular calculation can be made completely on a basis of convenience. The far-field approximations given by Eqs. (53) and (54), for example, can be derived from either representation. There are two bridges between the spherical and plane wave representations, one given by Weyl's integral, i.e.,

$$\frac{e^{ik|\mathbf{x}-\mathbf{x}''|}}{|\mathbf{x}-\mathbf{x}''|} = \frac{i}{\lambda}\int_{-\infty}^{\infty}\int_{-\infty}^{\infty}\frac{1}{m}e^{ik[p(x-x'')+q(y-y'')+m|z-z''|]}\,dp\,dq \tag{55}$$

where

$$m = \sqrt{1 - p^2 - q^2} \text{ if } p^2 + q^2 \leq 1,$$
$$= i\sqrt{p^2 + q^2 - 1} \text{ if } p^2 + q^2 > 1 \tag{56}$$

and the other by a related integral

$$\frac{-1}{2\pi}\frac{d}{dz}\frac{e^{ik|\mathbf{x}-\mathbf{x}''|}}{|\mathbf{x}-\mathbf{x}''|} = \frac{\pm 1}{\lambda^2}\int_{-\infty}^{\infty}\int_{-\infty}^{\infty} e^{ik[p(x-x'')+q(y-y'')+m|z-z''|]}\,dp\,dq \tag{57}$$

which can be easily derived from Eq. (55). In Eq. (57) the $\pm$ sign holds according to whether $(z - z'') \gtrless 0$. With these two equations it is possible to transform back and forth between these two representations.

Reciprocity Theorem

The radiation pattern and the complex degree of spectral coherence obey a very useful reciprocity theorem for a quasi-homogeneous source. By substituting from Eq. (40) into Eq. (53), and using Eq. (24) and the six-dimensional form of Eq. (41), it has been shown that the radiant intensity in the direction of the unit vector **s** from any bounded, three-dimensional, quasi-homogeneous primary source distribution is given by [see Ref. 45, eq. (3.11)]

$$J_\omega(\mathbf{s}) = J_0\int_{-\infty}^{\infty}\int_{-\infty}^{\infty}\int_{-\infty}^{\infty} \mu_Q(\mathbf{x}''_-)e^{-ik\mathbf{x}''_-\cdot\mathbf{s}}\,d^3\mathbf{x}''_- \tag{58}$$

where $\mu_Q(\mathbf{x}''_-)$ is the (spatially stationary) complex degree of spectral coherence for the source distribution as defined in Eq. (40), and

$$J_0 = \int_{-\infty}^{\infty}\int_{-\infty}^{\infty}\int_{-\infty}^{\infty} I_Q(\mathbf{x}''_+)\,d^3\mathbf{x}''_+ \tag{59}$$

where $I_Q(\mathbf{x}''_+)$ is the intensity defined in Eq. (40). Note that the far-field radiation pattern depends, not on the source intensity distribution, but only on the source coherence. We also find from this calculation that the complete degree of spectral coherence between any two points in the far field of this source is given by [see Ref. 45, eq. (3.15)]

$$\mu^{(\infty)}u(R_1\mathbf{s}_1, R_2\mathbf{s}_2) = \frac{e^{ik(R_1-R_2)}}{J_0}\int_{-\infty}^{\infty}\int_{-\infty}^{\infty}\int_{-\infty}^{\infty} I_Q(\mathbf{x}''_+)e^{ik\mathbf{x}''_+\cdot(\mathbf{s}_1-\mathbf{s}_2)}\,d^3\mathbf{x}''_+ \tag{60}$$

Note that the coherence of the far field depends, not on the source coherence, but rather on the source intensity distribution, $I_Q(\mathbf{x}''_+)$. Equation (60) is a generalization of the van Cittert–Zernike theorem to three-dimensional, primary quasi-homogeneous sources, which are not necessarily incoherent. Equation (59) is a new theorem, reciprocal to the van Cittert–Zernike theorem, which was discovered by Carter and Wolf.[45,49] Equations (58) and (60), taken together, give a reciprocity relation. For quasi-homogeneous sources, far-field coherence is determined by source intensity alone and the far-field intensity pattern is determined by source coherence alone. Therefore, coherence and intensity are reciprocal when going from the source into the far field. Since most sources which occur naturally are believed to be quasi-homogeneous, this is a very useful theorem. This reciprocity theorem has been found to hold much more generally than just for three-dimensional primary sources. For a planar, secondary source, by substitution from Eq. (40) into Eq. (54) and then using Eq. (41), we obtain[49]

$$J_\omega(\mathbf{s}) = J_0' \cos^2\theta \int_{-\infty}^{\infty}\int_{-\infty}^{\infty} \mu_U^{(0)}(\mathbf{x}_-')e^{-ik\mathbf{x}_-'\cdot\mathbf{s}}\, d^2\mathbf{x}_-' \tag{61}$$

where θ is the angle that $\mathbf{s}$ makes with the $+z$ axis, and where

$$J_0' = \int_{-\infty}^{\infty}\int_{-\infty}^{\infty} I_U^{(0)}(\mathbf{x}_+')\, d^2\mathbf{x}_+' \tag{62}$$

and we also obtain

$$\mu_U^{(\infty)}(R_1\mathbf{s}_1, R_2\mathbf{s}_2) = \frac{1}{J_0'}\int_{-\infty}^{\infty}\int_{-\infty}^{\infty} I_U^{(0)}(\mathbf{x}_+')e^{-ik\mathbf{x}_+'\cdot(\mathbf{s}_1-\mathbf{s}_2)}\, d^2\mathbf{x}_+' \tag{63}$$

Very similar reciprocity relations also hold for scattering of initially coherent light from quasi-homogeneous scattering potentials,[53,54] and for the scattering of laser beams from quasi-homogeneous sea waves.[63] Reciprocity relations which apply to fields that are not necessarily quasi-homogeneous have also been obtained.[64]

Nonradiating Sources

One additional important comment must be made about Eq. (58). The integrals appearing in this equation form a three-dimensional Fourier transform of $\mu_Q(\mathbf{x}_-'')$. However, $J_\omega(\mathbf{s})$, the radiant intensity that this source radiates into the far field at radial frequency ω, is a function of only the direction of $\mathbf{s}$, which is a unit vector with a constant amplitude equal to one. It then follows that only the values of this transform over a spherical surface of unit radius from the origin affect the far field. Therefore, sources which have a complex degree of spectral coherence, $\mu_Q(\mathbf{x}_-'')$, which do not have spatial frequencies falling on this sphere, do not radiate at frequency ω. It appears possible from this fact to have sources in such a state of coherence that they do not radiate at all. Similar comments can also be made in regard to Eq. (53) which apply to all sources, even those that are not quasi-homogeneous. From Eq. (53) it is clear that only sources which have a cross-spectral density function with spatial frequencies (with respect to both $\mathbf{x}_1''$ and $\mathbf{x}_2''$) on a unit sphere can radiate. It is believed that this is closely related to the phase-matching condition in nonlinear optics.

Perfectly Incoherent Sources

For a completely incoherent primary source the far-field radiant intensity, by substitution from Eq. (42) into Eq. (58), is seen to be the same in all directions, independent of the shape of the source. By substitution from Eq. (40) into Eq. (28), and then using Eqs. (14), (29), and (42), we find that the complex degree of spectral coherence for this incoherent source is given by

$$\mu_U(\mathbf{x}_1, \mathbf{x}_2) = \frac{1}{\sqrt{I(\mathbf{x}_1)/(\mathbf{x}_2)}}\int_{-\infty}^{\infty}\int_{-\infty}^{\infty}\int_{-\infty}^{\infty} I_Q(\mathbf{x}'')\frac{e^{ik[|\mathbf{x}_1-\mathbf{x}''|-|\mathbf{x}_2-\mathbf{x}''|]}}{|\mathbf{x}_1-\mathbf{x}''|\,|\mathbf{x}_2-\mathbf{x}''|}\, d^3\mathbf{x}'' \tag{64}$$

Thus it is only the source intensity distribution that affects the field coherence. Comparison of this equation with Eq. (47) shows that this is a generalization of the van Cittert–Zernike theorem to primary sources. It is clear from this equation that the complex degree of spectral coherence depends only on the shape of the source and not the fact that the source is completely incoherent. For a completely incoherent, planar, secondary source the radiant intensity is given by

$$J_\omega(\mathbf{s}) = J_0' \cos^2 \theta \tag{65}$$

independent of the shape of the illuminated area on the source plane, where θ is the angle that $\mathbf{s}$ makes with the normal to the source plane. This can be proven by substitution from Eq. (42) into Eq. (61). Note that such sources do not obey Lambert's law. The far-field coherence, again, depends on the source intensity as given by the van Cittert–Zernike theorem [see Eq. (47)] and not on the source coherence.

Spectrum

For a quasi-homogeneous, three-dimensional primary source the spectrum of the radiation $S_U^{(\infty)}(R\mathbf{s}, \omega)$ at a point $R\mathbf{s}$ in the direction $\mathbf{s}$ (unit vector) and at a distance R from the origin, in the far field of the source, can be found, as a function of the source spectrum $S_Q^{(0)}(\omega)$, by substitution from Eqs. (24) and (15) into Eq. (58) to get

$$S_U^{(\infty)}(R\mathbf{s}, \omega) = \frac{c^3 S_Q(\omega)}{\omega^3 R^2} \int_{-\infty}^{\infty} \int_{-\infty}^{\infty} \int_{-\infty}^{\infty} \mu_Q(\mathbf{x}_-'', \omega) e^{-ik\mathbf{x}_-''\cdot\mathbf{s}} \, d^3(k\mathbf{x}_-'') \tag{66}$$

where we explicitly indicate the dependence of the complex degree of spectral coherence on frequency. Notice that the spectrum of the field is not necessarily the same as the spectrum of the source and, furthermore, that it can vary from point to point in space. The field spectrum depends on the source coherence as well as on the source spectrum. A very similar propagation relation can be found for the far-field spectrum from a planar, secondary source in the $z = 0$ plane. By substitution from Eqs. (24) and (15) into Eq. (61) we get

$$S_U^{(\infty)}(R\mathbf{s}, \omega) = \frac{S_U^{(0)}(\omega)}{R^2} \int_{-\infty}^{\infty} \int_{-\infty}^{\infty} \mu_U^{(0)}(\mathbf{x}_-', \omega) e^{-ik\mathbf{x}_-'\cdot\mathbf{s}} \, d^2(k\mathbf{x}_-') \tag{67}$$

This shows that the spectrum for the field itself is changed upon propagation from the $z = 0$ plane into the far field and is different in different directions $\mathbf{s}$ from the source.[65]

4.7 SPECTRUM OF LIGHT

Limitations

The complex analytic signal for a field that is not perfectly coherent, as defined in Eq. (3), is usually assumed to be a time-stationary random process. Therefore, the integral

$$\int_{-\infty}^{\infty} u(\mathbf{x}, t) e^{i\omega t} \, dt \tag{68}$$

does not converge, so that the analytic signal does not have a Fourier transform. Therefore, it is only possible to move freely from the space-time domain to the

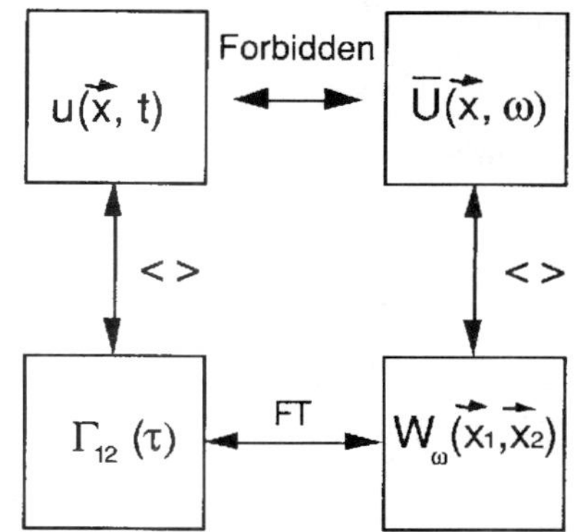

FIGURE 5 Illustrating the transformations which are possible between four functions that are used in coherence theory.

space-frequency domain along the path shown in Fig. 5. Equation (3) does not apply to time-stationary fields within the framework of ordinary function theory.

Coherent Mode Representation

Coherent mode representation (see Wolf[66,67]) has shown that any partially coherent field can be represented as the sum over component fields that are each perfectly self-coherent, but mutually incoherent with each other. Thus the cross-spectral density function for any field can be represented in the form

$$W_A(\mathbf{x}_1, \mathbf{x}_2) = \sum_n \lambda_n \phi_{A,n}(\mathbf{x}_1)\phi^*_{A,n}(\mathbf{x}_2) \tag{69}$$

where $\phi_{A,n}(\mathbf{x}_i)$ is a phasor amplitude for its nth coherent component. This representation can be used either with primary $(A = Q)$, or secondary $(A = U)$ sources. The phasor amplitudes $\phi_{A,n}(\mathbf{x}_i)$ and the complex expansion coefficients λ_n in Eq. (69) are eigenfunctions and eigenvalues of $W_A(\mathbf{x}_1, \mathbf{x}_2)$, as given by the equation

$$\int_{-\infty}^{\infty}\int_{-\infty}^{\infty}\int_{-\infty}^{\infty} W_A(\mathbf{x}_1, \mathbf{x}_2)\phi_{A,n}(\mathbf{x}_2)\, d^3\mathbf{x}_2 = \lambda_n(\omega)\phi_{A,n}(\mathbf{x}_1) \tag{70}$$

Since $W_A(\mathbf{x}_1, \mathbf{x}_2)$ is hermitian, the eigenfunctions are complete and orthogonal, i.e.,

$$\sum_n \phi_{A,n}(\mathbf{x}_1)\phi^*_{A,n}(\mathbf{x}_2) = \delta^3(\mathbf{x}_1 - \mathbf{x}_2) \tag{71}$$

and

$$\int_{-\infty}^{\infty}\int_{-\infty}^{\infty}\int_{-\infty}^{\infty} \phi_{A,n}(\mathbf{x}_1)\phi^*_{A,m}(\mathbf{x}_1)\, d^3\mathbf{x}_1 = \delta_{nm} \tag{72}$$

where δ_{nm} is the Kronecker delta function, and the eigenvalues are real and nonnegative definite, i.e.,

$$\begin{aligned} \text{Real}\,\{\lambda_n(\omega)\} &\geq 0 \\ \text{Imag}\,\{\lambda_n(\omega)\} &= 0 \end{aligned} \tag{73}$$

This is a very important concept in coherence theory. Wolf has used this representation to show that the frequency decomposition of the field can be defined in a different manner than was done in Eq. (3). A new phasor,

$$\bar{U}_A(\mathbf{x}, \omega) = \sum_n a_n(\omega)\phi_{A,n}(\mathbf{x}, \omega) \tag{74}$$

where $a_n(\omega)$ are random coefficients such that

$$\langle a_n(\omega)a_m^*(\omega)\rangle = \lambda_n(\omega)\delta_{nm} \tag{75}$$

can be introduced to represent the spectral components of the field. It then follows[66] that, if the $\phi_{A,n}(\mathbf{x}_i)$ are eigenfunctions of the cross-spectral density function, $W_A(\mathbf{x}_1, \mathbf{x}_2)$, then the cross-spectral density function can be represented as the correlation function between these phasors at the two spatial points $\mathbf{x}_1$ and $\mathbf{x}_2$, i.e.,

$$W_A(\mathbf{x}_1, \mathbf{x}_2) = \langle \bar{U}_A(\mathbf{x}_1, \omega)\bar{U}_A^*(\mathbf{x}_2, \omega)\rangle \tag{76}$$

in a manner very similar to the representation given in Eq. (11) in respect to the older phasors. Notice, by comparison of Eqs. (11) and (76), that the phasors $\bar{U}_A(\mathbf{x}_i, \omega)$ and $U_A(\mathbf{x}_i, \omega)$ are not the same. One may formulate coherence theory either by defining $u(\mathbf{x}, t)$ in Fig. 5 and then moving in a counterclockwise direction in this figure to derive the correlation functions using the Wiener-Khintchene theory or, alternatively, defining $\bar{U}_A(\mathbf{x}_i, \omega)$ and moving in a clockwise direction to define the correlation functions using the coherent mode expansion.

Wolf Shift and Scaling law

From Eqs. (66) and (67) it is clear that the spectrum of a radiation field may change as the field propagates. It is not necessarily equal to the source spectrum as it is usually assumed. This brings up an important question. Why do the spectra for light fields appear so constant, experimentally, that a change with propagation was never suspected? Wolf[65] has provided at least part of the answer to this question. He discovered a scaling law that is obeyed by most natural fields and under which the normalized spectrums for most fields do remain invariant as they propagate. We can derive this scaling law by substitution from Eqs. (66) and (67) into Eq. (16). We then find that, if the complex degree of spectral coherence for the source is a function of $k\mathbf{x}_-$ only, i.e.,

$$\mu_A(\mathbf{x}_-, \omega) = f(k\mathbf{x}_-) \tag{77}$$

(so that this function is the same for each frequency component of the field, provided that the spatial separations of the two test points are always scaled by the wavelength), then the normalized spectrum in the far field is given by

$$s_U^{(\infty)}(R\mathbf{s}, \omega) = \frac{S_Q(\omega)/\omega^3}{\int_0^\infty S_Q(\omega)/\omega^3\, d\omega} \neq f(R\mathbf{s}) \tag{78}$$

if $S_Q(\omega)$ is the complex degree of spectral coherence for a primary, quasi-homogeneous source [see Ref. 90, Eq. (65)], and

$$s_U^{(\infty)}(R\mathbf{s}, \omega) = \frac{S_U^{(0)}(\omega)}{\int_0^\infty S_U^{(0)}(\omega)\, d\omega} = s_U^{(0)}(\omega) \neq f(R\mathbf{s}) \tag{79}$$

if $s_U^{(0)}(\omega)$ is the normalized spectrum for a secondary, quasi-homogeneous source [see Ref. 90, Eq. (51)]. In each case the field spectrum does not change as the field propagates. Since the cross-spectral density function for a thermal source [see Eq. (43)] obeys this scaling law, it is not surprising that these changes in the spectrum of a propagating light field were never discovered experimentally. The fact that the spectrum can change was verified experimentally only after coherence theory pointed out the possibility.[68–72]

4.8 POLARIZATION EFFECTS

Explicit Vector Representations

As discussed earlier under "Scalar Field Amplitude", the scalar amplitude is frequently all that is required to treat the vector electromagnetic field. If polarization effects are important, it might be necessary to use two scalar field amplitudes to represent the two independent polarization components. However, in some complicated problems it is necessary to consider all six components of the vector field explicitly. For such a theory, the correlation functions between vector components of the field become tensors,[73,74] which propagate in a manner very similar to the scalar correlation functions.

4.9 APPLICATIONS

Speckle

If coherent light is scattered from a stationary, rough surface, the phase of the light field is randomized in space. The diffraction patterns observed with such light displays a complicated granular pattern usually called *speckle* (see Ref. 8, sec. 7.5). Even though the light phase can be treated as a random variable, the light is still perfectly coherent. Coherence theory deals with the effects of time fluctuations, not spatial variations in the field amplitude or phase. Despite this, the same statistical tools used in coherence theory have been usefully applied to studying speckle phenomena.[75,76] To treat speckle, the ensemble is usually redefined, not to represent the time fluctuations of the field but, rather, to represent all of the possible speckle patterns that might be observed under the conditions of a particular experiment. An observed speckle pattern is usually due to a single member of this ensemble (unless time fluctuations are also present), whereas the intensity observed in coherence theory is always the result of a weighted average over all of the ensemble. To obtain the intensity distribution over some plane, as defined in coherence theory, it would be necessary to average over all of the possible speckle patterns explicitly. If this is done, for example, by moving the scatterer while the intensity of a diffraction pattern is time-averaged, then time fluctuations are introduced into the field during the measurement; the light becomes partially coherent; and coherence theory can be properly used to model the measured intensity. One must be very careful in applying the coherence theory model to treat speckle phenomena, because coherence theory was not originally formulated to deal with speckle.

Statistical Radiometry

Classical radiometry was originally based on a mechanical treatment of light as a flux of particles. It is not totally compatible with wave theories. Coherence theory has been used to incorporate classical radiometry into electromagnetic theory as much as has been found possible. It has been found that the usual definitions for the radiance function and the radiant emittance cause problems when applied to a wave theory. Some of this work is discussed earlier (pp. 4.8 to 4.10, 4.12, 4.13, 4.15 to 4.16, 4.18). Other radiometric functions, such as the radiant intensity, have clear meaning in a wave theory.

Spectral Representation

It was discovered, using coherence theory, that the spectrum of a light field is not the same as that of its source and that it can change as the light field propagates away from its source into the radiation field. Some of this work was discussed earlier under "Perfectly Incoherent Sources" and "Coherent Mode Representation." This work has been found very useful for explaining troublesome experimental discrepancies in precise spectroradiometry.[71]

Laser Modes

Coherence theory has been usefully applied to describing the coherence properties of laser modes.[77] This theory is based on the coherent mode representation discussed under "Spectrum of Light."

Radio Astronomy

Intensity interferometry was used to apply techniques from radio astronomy to observation with light.[37] A lively debate ensued as to whether optical interference effects (which indicate partial coherence) could be observed from intensity correlations.[78] From relations like Eq. (26) and similar calculations using quantum coherence theory, it quickly became clear that they could.[79,80] More recently, coherence theory has been used to model a radio telescope[42] and to study how to focus an instrument to observe emitters that are not in the far field of the antenna array.[81] It has been shown[82] that a radio telescope and a conventional optical telescope are very similar, within a coherence theory model, even though their operation is completely different. This model makes the similarities between the two types of instruments very clear.

Noncosmological Red Shift

Cosmological theories for the structure and origin of the universe make great use of the observed red shift in the spectral lines of the light received from distant radiating objects, such as stars.[83] It is usually assumed that the spectrum of the light is constant upon propagation and that the observed red shift is the result of simple Doppler shift due to the motion of the distant objects away from the earth in all directions. If this is true, then clearly the observed universe is expanding and must have begun with some sort of explosion, called "the big bang." The size of the observable universe is estimated based on the amount of this red shift. A new theory by Wolf[84–87] shows that red shifts can occur naturally without Doppler shifts as the light propagates from the source to an observer if the source is not in thermal equilibrium, i.e., a thermal source as discussed earlier under "Thermal (Lambertian) Source." The basis of Wolf's theory was discussed on pp. 4: 32–33 and 34–36.

4.10 REFERENCES

1. L. Mandel and E. Wolf, "Coherence Properties of Optical Fields," *Rev. Mod. Phys.* **37,** 1965, pp. 231–287.
2. L. Mandel and E. Wolf, *Selected Papers on Coherence and Fluctuations of Light,* vol. I, Dover, New York, 1970. Reprinted in Milestone Series, SPIE press, Bellingham, Wash., 1990.
3. L. Mandel and E. Wolf, *Selected Papers on Coherence and Fluctuations of Light,* vol. II, Dover, New York, 1970. Reprinted in Milestone Series, SPIE press, Bellingham, Wash., 1990.
4. M. J. Beran and G. B. Parrent, *Theory of Partial Coherence,* Prentice-Hall, Engelwood Cliffs, N.J., 1964.
5. B. Crosignani and P. Di Porto, *Statistical Properties of Scattered Light,* Academic Press, New York, 1975.
6. A. S. Marathay, *Elements of Optical Coherence Theory,* Wiley, New York, 1982.
7. J. Perina, *Coherence of Light,* Reidel, Boston, 1985.
8. J. W. Goodman, *Statistical Optics,* Wiley, New York, 1985.
9. R. J. Glauber, "The Quantum Theory of Optical Coherence," *Phys. Rev.* **130,** 1963, pp. 2529–2539.
10. R. J. Glauber, "Coherent and Incoherent States of the Radiation Field," *Phys. Rev.* **131,** 1963, pp. 2766–2788.
11. C. L. Mehta and E. C. G. Sudarshan, "Relation between Quantum and Semiclassical Description of Optical Coherence," *Phys. Rev.* **138,** B274–B280, 1965.
12. W. Louisel, *Radiation and Noise in Quantum Electronics,* McGraw-Hill, New York, 1964.
13. W. H. Louisell, *Quantum Statistical Properties of Radiation,* Wiley, New York, 1973.
14. M. Sargent, M. O. Scully, and W. E. Lamb, *Laser Physics,* Addison-Wesley, Reading, MA, 1974.
15. M. Born and E. Wolf, *Principles of Optics,* Pergamon Press, New York, 1980.
16. H. M. Nussenvieg, *Causality and Dispersion Relations,* Academic Press, New York, 1972.
17. A. M. Yaglom, *An Introduction to the Theory of Stationary Random Functions,* Prentice-Hall, Engelwood Cliffs, N.J., 1962.
18. G. Borgiotti, "Fourier Transforms Method in Aperture Antennas Problems," *Alta Frequenza* **32,** 1963, pp. 196–205.
19. D. R. Rhodes, "On the Stored Energy of Planar Apertures," *IEEE Trans.* **AP14,** 1966, pp. 676–683.
20. W. H. Carter, "The Electromagnetic Field of a Gaussian Beam with an Elliptical Cross-section," *J. Opt. Soc. Am.* **62,** 1972, pp. 1195–1201.
21. W. H. Carter, "Electromagnetic Beam Fields," *Optica Acta* **21,** 1974, pp. 871–892.
22. B. J. Thompson and E. Wolf, "Two-Beam Interference with Partially Coherent Light," *J. Opt. Soc. Am.* **47,** 1957, pp. 895–902.
23. W. H. Carter, "Measurement of Second-order Coherence in a Light Beam Using a Microscope and a Grating," *Appl. Opt.* **16,** 1977, pp. 558–563.
24. E. Wolf, "A Macroscopic Theory of Interference and Diffraction of Light from Finite Sources II: Fields with Spatial Range and Arbitrary Width," *Proc. Roy. Soc.* A **230,** 1955, pp. 246–265.

25. G. B. Parrent, "On the Propagation of Mutual Coherence," *J. Opt. Soc. Am.* **49,** 1959, pp. 787–793.

26. W. B. Davenport and W. L. Root, *An Introduction to the Theory of Random Signals and Noise,* McGraw-Hill, New York, 1960.

27. L. Mandel and E. Wolf, "Spectral Coherence and the Concept of Cross-spectral Purity," *J. Opt. Soc. Am.* **66,** 1976, pp. 529–535.

28. E. W. Marchand and E. Wolf, "Angular Correlation and the Far-zone Behavior of Partially Coherent Fields," *J. Opt. Soc. Am.* **62,** 1972, pp. 379–385.

29. E. W. Marchand and E. Wolf, "Radiometry with Sources of Any State of Coherence," *J. Opt. Soc. Am.* **64,** 1974, pp. 1219–1226.

30. E. Wolf, "The Radiant Intensity from Planar Sources of any State of Coherence," *J. Opt. Soc. Am.* **68,** 1978, pp. 1597–1605.

31. A. T. Friberg, "On the Existence of a Radiance Function for Finite Planar Sources of Arbitrary States of Coherence," *J. Opt. Soc. Am.* **69,** 1979, pp. 192–198.

32. E. Wigner, "Quantum Correction for Thermodynamic Equilibrium" *Phys. Rev.* **40,** 1932, pp. 749–760.

33. K. Imre, E. Ozizmir, M. Rosenbaum, and P. F. Zweifer, "Wigner Method in Quantum Statistical Mechanics," *J. Math. Phys.* **8,** 1967, pp. 1097–1108.

34. A. Papoulis, "Ambiguity Function in Fourier Optics," *J. Opt. Soc. Am.* **64,** 1974, pp. 779–788.

35. A. Walther, "Radiometry and Coherence," *J. Opt. Soc. Am.* **63,** 1973, pp. 1622–1623.

36. J. T. Foley and E. Wolf, "Radiometry as a Short Wavelength Limit of Statistical Wave Theory with Globally Incoherent Sources," *Opt. Commun.* **55,** 1985, pp. 236–241.

37. R. H. Brown and R. Q. Twiss, "Correlation between Photons in Two Coherent Beams of Light," *Nature* **177,** 1956, pp. 27–29.

38. J. D. Jackson, *Classical Electrodynamics,* Wiley, New York, 1962.

39. C. L. Mehta and A. B. Balachandran, "Some Theorems on the Unimodular Complex Degree of Optical Coherence," *J. Math. Phys.* **7,** 1966, pp. 133–138.

40. L. Mandel and E. Wolf, "Complete Coherence in the Space Frequency Domain," *Opt. Commun.* **36,** 1981, pp. 247–249.

41. W. H. Carter, "Difference in the Definitions of Coherence in the Space-Time Domain and in the Space-Frequency Domain," *J. Modern Opt.* **39,** 1992, pp. 1461–1470.

42. W. H. Carter and L. E. Somers, "Coherence Theory of a Radio Telescope," *IEEE Trans.* **AP24,** 1976, pp. 815–819.

43. A. C. Schell, *The Multiple Plate Antenna,* doctoral dissertation, M.I.T., 1961.

44. A. C. Schell, "A Technique for the Determination of the Radiation Pattern of a Partially Coherent Aperture," *IEEE Trans.* **AP-15,** 1967, p. 187.

45. W. H. Carter and E. Wolf, "Correlation Theory of Wavefields Generated by Fluctuating Three-dimensional, Primary, Scalar Sources: II. Radiation from Isotropic Model Sources," *Optica Acta* **28,** 1981, pp. 245–259.

46. W. H. Carter and E. Wolf, "Correlation Theory of Wavefields Generated by Fluctuating Three-dimensional, Primary, Scalar Sources: I. General Theory," *Optica Acta* **28,** 1981, pp. 227–244.

47. E. Collett and E. Wolf, "Is Complete Coherence Necessary for the Generation of Highly Directional Light Beams?," *Opt. Letters* **2,** 1978, pp. 27–29.

48. W. H. Carter and Bertolotti, "An Analysis of the Far Field Coherence and Radiant Intensity of Light Scattered from Liquid Crystals," *J. Opt. Soc. Am.* **68,** 1978, pp. 329–333.

49. W. H. Carter and E. Wolf, "Coherence and Radiometry with Quasi-homogeneous Planar Sources," *J. Opt. Soc. Am.* **67,** 1977, pp. 785–796.

50. W. H. Carter and E. Wolf, "Inverse Problem with Quasi-Homogeneous Sources," *J. Opt. Soc. Am.* **2A,** 1985, pp. 1994–2000.

51. E. Wolf and W. H. Carter, "Coherence and Radiant Intensity in Scalar Wave Fields Generated by Fluctuating Primary Planar Sources," *J. Opt. Soc. Am.* **68,** 1978, pp. 953–964.
52. W. H. Carter, "Coherence Properties of Some Isotropic Sources," *J. Opt. Soc. Am.* **A1,** 1984, pp. 716–722.
53. W. H. Carter and E. Wolf, "Scattering from Quasi-homogeneous Media," *Opt. Commun.* **67,** 1988, pp. 85–90.
54. W. H. Carter, "Scattering from Quasi-homogeneous Time Fluctuating Random Media," *Opt. Commun.* **77,** 1990, pp. 121–125.
55. W. H. Carter and E. Wolf, "Coherence Properties of Lambertian and Non-Lambertian Sources," *J. Opt. Soc. Am.* **65,** 1975, pp. 1067–1071.
56. J. T. Foley, W. H. Carter, and E. Wolf, "Field Correlations within a Completely Incoherent Primary Spherical Source," *J. Opt. Soc. Am.* A **3,** 1986, pp. 1090–1096.
57. H. H. Hopkins, "The Concept of Partial Coherence in Optics," *Proc. Roy. Soc.* **A208,** 1951, pp. 263–277.
58. W. H. Carter, "Generalization of Hopkins' Formula to a Class of Sources with Arbitrary Coherence," *J. Opt. Soc. Am.* A **2,** 1985, pp. 164–166.
59. P. H. van Cittert, "Die wahrscheinliche Schwingungsverteilung in einer von einer Lichtquelle direkt oder mittels einer Linse beleuchteten Ebene," *Physica* **1,** 1934, pp. 201–210.
60. F. Zernike, "The Concept of Degree of Coherence and Its Application to Optical Problems," *Physica* **5,** 1938, pp. 785–795.
61. W. H. Carter, "Three Different Kinds of Fraunhofer Approximations. I. Propagation of the Field Amplitude," *Radio Science* **23,** 1988, pp. 1085–1093.
62. W. H. Carter, "Three Different Kinds of Fraunhofer Approximations. II. Propagation of the Cross-spectral Density Function," *J. Mod. Opt.* **37,** 1990, pp. 109–120.
63. W. H. Carter, "Scattering of a Light Beam from an Air-Sea Interface," *Applied Opt.* **32,** 3286–3294 (1993).
64. A. T. Friberg and E. Wolf, "Reciprocity Relations with Partially Coherent Sources," *Opt. Acta* **36,** 1983, pp. 417–435.
65. E. Wolf, "Invariance of the Spectrum of Light on Propagation," *Phys. Rev. Letters* **56,** 1986, pp. 1370–1372.
66. E. Wolf, "New Theory of Partial Coherence in the Space-frequency Domain. Part I: Spectra and Cross Spectra of Steady-state Sources," *J. Opt. Soc. Am.* **72,** 1982, pp. 343–351.
67. E. Wolf, "New Theory of Partial Coherence in the Space-frequency Domain. Part II: Steady-state Fields and Higher-order Correlations," *J. Opt. Soc. Am.* A **3,** 1986, pp. 76–85.
68. G. M. Morris and D. Faklis, "Effects of Source Correlation on the Spectrum of Light," *Opt. Commun.* **62,** 1987, pp. 5–11.
69. M. F. Bocko, D. H. Douglass, and R. S. Knox, "Observation of Frequency Shifts of Spectral Lines Due to Source Correlations," *Phys. Rev. Letters* **58,** 1987, pp. 2649–2651.
70. F. Gori, G. Guattari, and C. Palma, "Observation of Optical Redshift and Blueshifts Produced by Source Correlations," *Opt. Commun.* **67,** 1988, pp. 1–4.
71. H. C. Kandpal, J. S. Vashiya, and K. C. Joshi, "Wolf Shift and Its Application in Spectro-radiometry," *Opt. Commun.* **73,** 1989, pp. 169–172.
72. G. Indebedouw, "Synthesis of Polychromatic Light Sources with Arbitrary Degree of Coherence: From Experiments," *J. Mod. Opt.* **36,** 1989, pp. 251–259.
73. W. H. Carter, "Properties of Electromagnetic Radiation from a Partially Correlated Current Distribution," *J. Opt. Soc. Am.* **70,** 1980, pp. 1067–1074.
74. W. H. Carter and E. Wolf, "Far-zone Behavior of Electromagnetic Fields Generated by Fluctuating Current Distributions," *Phys. Rev.* **36A,** 1987, pp. 1258–1269.
75. J. C. Dainty, "The Statistics of Speckle Patterns," in E. E. Wolf, *Progress in Optics,* North Holland, Amsterdam, 1976.
76. J. C. Dainty, *Laser Speckle and Related Phenomena,* Springer Verlag, Berlin, 1975.

77. E. Wolf and G. S. Agarwal, "Coherence Theory of Laser Resonator Modes," *J. Opt. Soc. Am.* A **1,** 1984, pp. 541–546.

78. E. M. Purcell, "The Question of Correlation Between Photons in Coherent Light Rays," *Nature* **178,** 1956, pp. 1449–1450.

79. R. H. Brown and R. Q. Twiss, "Interferometry of the Intensity Fluctuations in Light: I. Basic Theory: the Correlation between Photons in Coherent Beams of Radiation," *Proc. Roy. Soc.* **242,** 1957, pp. 300–324.

80. R. H. Brown and R. Q. Twiss, "Interferometry of the Intensity Fluctuations in Light: II. An Experimental Test of the Theory for Partially Coherent Light," *Proc. Roy. Soc.* **243,** 1957, pp. 291–319.

81. W. H. Carter, "Refocusing a Radio Telescope to Image Sources within the Near Field of the Antenna Array," NRL Report no. 9141, August 17, 1988.

82. W. H. Carter, "A New Theory for Image Formation with Quasi-homogeneous Sources and Its Application to the Detection and Location of ICBM Launches," NRL Report no. 9336 (September 17, 1991).

83. J. V. Narliker, "Noncosmological Redshifts," *Space Sci. Revs.* **50,** 1989, pp. 523–614.

84. E. Wolf, "Non-cosmological Redshifts of Spectral Lines," *Nature* **326,** 1987, pp. 363–365.

85. E. Wolf, "Red Shifts and Blue Shifts of Spectral Lines Emitted by Two Correlated Sources," *Phys. Rev. Letters* **58,** 1987, pp. 2646–2648.

86. A. Gamliel and E. Wolf, "Spectral Modulation by Control of Source Correlations," *Opt. Commun.* **65,** 1988, pp. 91–96.

87. E. Wolf, J. T. Foley, and F. Gori, "Frequency Shifts of Spectral Lines Produced by Scattering from Spatially Random Media," *J. Opt. Soc. Am.* A **6,** 1989, pp. 1142–1149.

88. W. H. Carter, "Difference in the Definition of Coherence in the Space-Time Domain and in the Space-Frequency Domain," *J. Mod. Opt.* **39,** 1992, pp. 1461–1470.

89. W. H. Carter, "Coherence Properties of Some Broadband Highly Coherent Light Fields," *J. Op. Soc. Am.* A **10,** 1993, pp. 1570–1578.

90. W. H. Carter, "A Study of the Propagation of Optical Spectra Using a Wavelength Independent Formulation of Diffraction Theory and Coherence Theory," *J. Mod. Opt.* **40,** 1993, pp. 2433–2449.

CHAPTER 5
POLARIZATION†

Jean M. Bennett
Senior Scientist (Optics),
Research Department,
Michelson Laboratory,
Naval Air Warfare Center,
China Lake, California

GLOSSARY

c	velocity of light
d	thickness
E	electric field
$\mathbf{k}$	wave vector ($k = 2\pi/\lambda$)
k	extinction coefficient
m	number of reflections
N	retardation per wavelength
n	real refractive index
$\tilde{n}$	complex refractive index
$\hat{\mathbf{n}}$	unit normal vector
p	degree of polarization
p	parallel polarization
R	intensity reflection coefficient
r	amplitude reflection coefficient
$\mathbf{r}$	position vector
s	senkrecht or perpendicular polarization
t	amplitude transmission coefficient
t	time
z	cartesian coordinate

† This material was originally prepared under the auspices of the U.S. government and is not subject to copyright.

$\alpha, \beta, a, b, c, d$	intermediate parameters
α	absorption coefficient
γ	$2\pi nd \cos\theta/\lambda$
δ	phase angle
ε	dielectric constant
η	effective refractive index
θ_B	Brewster angle
θ	angle
κ	absorption index
λ	wavelength
ρ	extinction ratio
σ	conductivity
ω	radian or angular frequency
∇	laplacian operator
0	first medium
1	second medium

The material on polarization is abridged from the much more complete treatment by Bennett and Bennett.[1] Information on polarizers is found in Volume 2, Chapter 3, "Polarizers."

1. Basic Concepts and Conventions Optical polarization was discovered by E. L. Malus in 1808. A major triumph of nineteenth- and early twentieth-century theoretical physics was the development of electromagnetic theory and the demonstration that optical polarization is completely described by it. This theory is phenomenological in that instead of trying to explain why materials have certain fundamental characteristics, it concentrates on the resulting properties which any material with those characteristics will display. In the optical case, the polarization and all other optical properties of a material are determined by two or more phenomenological parameters called *optical constants.* Electromagnetic theory has little or nothing to say about *why* a material should have these particular optical constants or *how* they are related to its atomic character. This problem has been extensively investigated in twentieth-century solid-state physics and is still only partially understood. It is clear, however, that the optical constants are a function not only of the atomic nature of the material, i.e., its position in the periodic table, but are also quite sensitive to how it is prepared. Perhaps *optical parameters* would be a better term than optical constants. Nevertheless, the concept of optical constants is an extremely useful one and makes it possible to predict quantitatively the optical behavior of a material and, under certain conditions, to relate this behavior to nonoptical parameters.

Since the optical constants are so fundamental, differences in their definition are particularly unfortunate. The most damaging of these differences arises from an ambiguity in the initial derivation. Maxwell's equations, which form the basis of electromagnetic theory, result in the wave equation, which in mks units is

$$\nabla^2\mathbf{E} = \frac{\varepsilon}{c^2}\frac{\partial^2\mathbf{E}}{\partial t^2} + \frac{4\pi\sigma}{c^2}\frac{\partial\mathbf{E}}{\partial t} \tag{1}$$

where ∇^2 = laplacian operator
$\mathbf{E}$ = electric field vector of traveling wave
t = time
c = velocity of light
σ = conductivity of material at frequency of wave motion
ε = dielectric constant of material at frequency of wave motion

A solution to this equation is

$$\mathbf{E} = \mathbf{E}_0 \exp\left[i(\omega t + \delta)\right] \exp\left(-i\mathbf{k}\cdot\mathbf{r}\right) \exp\left(-\frac{\alpha z}{2}\right) \tag{2}$$

where $\mathbf{E}_0$ = amplitude of wave
ω = angular frequency of wave
δ = phase vector
$\mathbf{k}$ = wave vector
$\mathbf{r}$ = position vector
z = direction wave is traveling
α = absorption coefficient

The wave vector $\mathbf{k}$ is assumed to be real and equal to $(2\pi/\lambda_m)\hat{\mathbf{n}}$, where λ_m is the wavelength in the medium in which the wave is traveling and $\mathbf{n}$ is a unit vector in the k direction.† Equation (2) can also be written in terms of $\tilde{n}$, the complex index of refraction, defined as

$$\tilde{n} = n - ik \tag{3}$$

where n is the index of refraction and k the extinction coefficient. In this form, Eq. (2) is

$$E = E_0 \exp\left[i\omega\left(t - \frac{\tilde{n}z}{c}\right)\right] \tag{4}$$

when $\delta = 0$. By setting the imaginary part of the exponent equal to zero one obtains

$$z = \frac{c}{n}t \tag{5}$$

To show that Eq. (4) represents a wave traveling in the positive z direction with phase velocity c/n, we note that the phase ϕ of the wave in Eq. (4) is $\omega t - (\omega\tilde{n}z)/c = \phi$. For a wave propagating with a constant phase, $d\phi = 0$, so that $\omega\, dt - (\omega\tilde{n}/c)\, dz = d\phi = 0$, and hence the phase velocity $v_p = dz/dt = c/n$.[2] The amplitude of the wave at z is, from Eq. (4),

$$|E| = E_0 e^{-2\pi kz/\lambda} \tag{6}$$

where λ is the wavelength in vacuum. The wave is thus exponentially damped, and the amplitude penetration depth, or distance below an interface at which the *amplitude* of the wave falls to $1/e$ times its initial value, is $z = \lambda/2\pi k$. The absorption coefficient α, or

† Frequently the wave vector is taken to be complex, that is, $\tilde{\mathbf{k}} = (2\pi/\lambda_m - i\alpha/2)\mathbf{n}$, and Eq. (2) is written $\mathbf{E} = \mathbf{E}_0 \exp\left[i(\omega t + \delta)\right] \exp\left(-i\mathbf{k}'\cdot\mathbf{r}\right)$.

the reciprocal of the distance in which the *intensity* of the wave falls to $1/e$ times its initial value, is

$$\alpha = \frac{4\pi k}{\lambda} \tag{7}$$

This development follows that commonly given by those working at optical or radio frequencies. The confusion in the definition of the optical constants arises because an equally valid solution to Eq. (1) is

$$E' = E_0 \exp\left[-i\omega\left(t - \frac{\tilde{n}'z}{c}\right)\right] \tag{8}$$

which also represents an exponentially damped wave traveling in the $+z$ direction *provided that the complex index of refraction is defined to be*

$$\tilde{n}' = n + ik \tag{9}$$

where the primes indicate the alternative solution. When the wave equation arises in quantum mechanics, the solution chosen is generally the negative exponential, i.e. Eq. (8) rather than Eq. (4). Solid-state physicists working in optics thus often define the complex index of refraction as the form given in Eq. (9) rather than that in Eq. (3). Equally valid, self-consistent theories can be built up using either definition, and as long as only intensities are considered, the resulting expressions are identical. However, when phase differences are calculated, the two conventions usually lead to contradictory results. Even worse, an author who is not extremely careful may not consistently follow either convention, and the result may be pure nonsense. Some well-known books might be cited in which the authors are not even consistent from chapter to chapter.

There are several other cases in optics in which alternative conventions are possible and both are found in the literature. Among these, the most distressing are the use of a left-handed rather than a right-handed coordinate system, which makes the p and s components of polarized light have the same phase change at normal incidence (see Par. 2), and defining the optical constants so that they depend on the angle of incidence, which makes the angle of refraction given by Snell's law real for an absorbing medium. There are many advantages to be gained by using a single set of conventions in electromagnetic theory. In any event, an author should *clearly* state the conventions being used and then *stay with them.*

Finally, the complex index of refraction is sometimes written

$$\tilde{n} = n(1 - i\kappa) \tag{10}$$

In this formulation the symbol κ is almost universally used instead of k, which is reserved for the imaginary part of the refractive index. Although k is more directly related to the absorption coefficient α than κ [see Eq. (7)] and usually makes the resulting expressions slightly simpler, in areas such as attenuated total reflection the use of κ results in a simplification. To avoid confusion between k and κ, if Eq. (10) is used, κ could be called the *absorption index* to distinguish it from the extinction coefficient k, and the absorption coefficient α.

2. Fresnel Equations The Fresnel equations are expressions for the reflection and transmission coefficients of light at nonnormal incidence. In deriving these equations, the coordinate system assumed determines the signs in the equations and therefore the phase changes on reflection of the p and s components. In accordance with the Muller

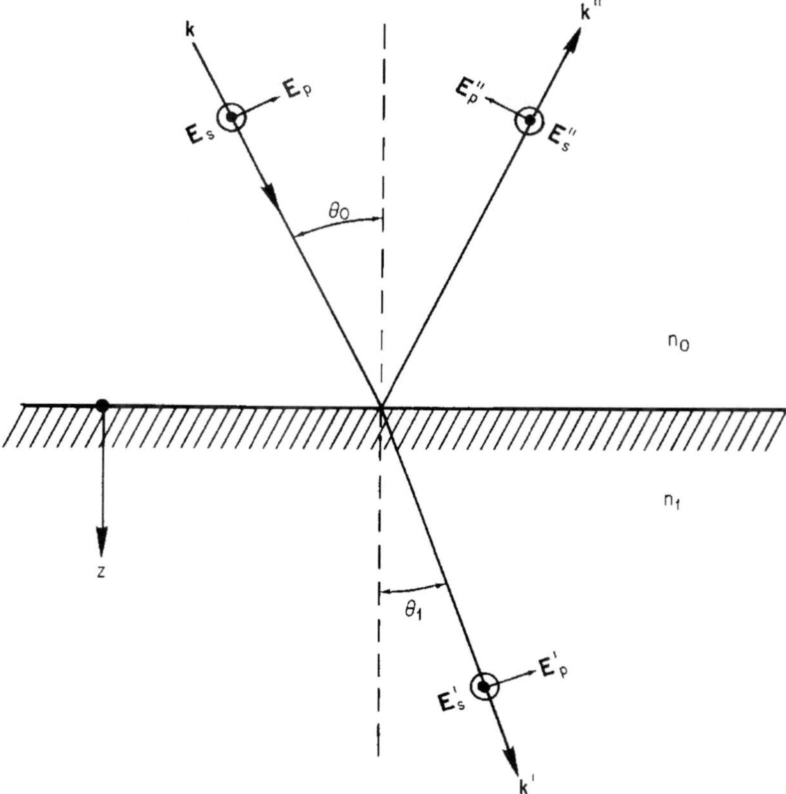

FIGURE 1 Coordinate system for measuring the E vectors of a plane wave reflected and refracted at a boundary between a medium of refractive index n_0 and a medium of refractive index n_1 (may be absorbing). The positive direction for the coordinates of the E_s, E_s', and E_s'' components is out of the paper, and that for the coordinates of the E_p components is in the plane of the paper, as indicated by the arrows. The wave vector **k**, the direction the wave is traveling z, and angles of incidence and refraction θ_0 and θ_1 are also shown. [*Modified from Muller, Ref.* 3.]

convention,[3] we shall assume that the coordinate system is as shown in Fig. 1. In this system, the angle of incidence is θ_0, and the angle of refraction is θ_1. The *s component* of polarization is the plane of vibration of the E wave which is perpendicular to the plane of the paper, and the *p component* is the plane of vibration which is in the plane of the paper.† (The plane of incidence is in the plane of the paper.) The positive directions for the vibrations are indicated in Fig. 1 by the dots for E_s, E_s', and E_s'' and by the arrows for the corresponding p components. Note that the positive direction for E_p'' is as shown in the

† Unfortunately, when Malus discovered that light reflected at a certain angle from glass is, as he said, "polarized," he defined the plane of polarization" of the reflected light as the plane of incidence. Since the reflected light in this case has its E vector perpendicular to the plane of incidence, the "plane of polarization" is perpendicular to the plane in which the E vector vibrates. This nomenclature causes considerable confusion and has been partially resolved in modern terminology by discussing the *plane of vibration* of the E vector and avoiding, insofar as possible, the term *plane of polarization.* In this chapter, when specifying the direction in which light is polarized, we shall give the direction of vibration, *not* the direction of polarization.

figure because of the *mirror-image effect.* By convention, one always looks *against the direction of propagation of the wave* so that the positive direction of E_p is to the right and the positive direction of E''_p is also to the right. The positive directions of the reflected E vectors are not the same as the actual directions of the reflected E vectors. These latter directions will depend on the refractive index of the material and may be either positive or negative. For example, if $n_1 > n_0$, at normal incidence E''_s will be in the negative direction and E''_p will be in the positive direction. Thus we say that there is a phase change on reflection of 180° for the s wave and a phase change of 0° for the p wave.

With this coordinate system, the Fresnel amplitude reflection coefficients for a single interface, obtained from Eq. (4) by setting up and solving the boundary-value problem, can be written

$$\frac{E''_s}{E_s} \equiv r_s = \frac{n_0 \cos \theta_0 - n_1 \cos \theta_1}{n_0 \cos \theta_0 + n_1 \cos \theta_1} \tag{11}$$

and

$$\frac{E''_p}{E_p} \equiv r_p = \frac{n_1 \cos \theta_0 - n_0 \cos \theta_1}{n_1 \cos \theta_0 + n_0 \cos \theta_1} \tag{12}$$

The amplitude transmission coefficients are

$$\frac{E'_s}{E_s} \equiv t_s = \frac{2n_0 \cos \theta_0}{n_0 \cos \theta_0 + n_1 \cos \theta_1} \tag{13}$$

and

$$\frac{E'_p}{E_p} \equiv t_p = \frac{2n_0 \cos \theta_0}{n_1 \cos \theta_0 + n_0 \cos \theta_1} \tag{14}$$

Other forms of the Fresnel amplitude reflection and transmission coefficients containing only the angles of incidence and refraction are somewhat more convenient. These relations can be derived using Snell's law

$$\frac{\sin \theta_0}{\sin \theta_1} = \frac{n_1}{n_0} \tag{15}$$

to eliminate n_0 and n_1 from Eqs. (1) to (14):

$$r_s = \frac{-\sin (\theta_0 - \theta_1)}{\sin (\theta_0 + \theta_1)} \tag{16}$$

$$r_p = \frac{\tan (\theta_0 - \theta_1)}{\tan (\theta_0 + \theta_1)} \tag{17}$$

$$t_s = \frac{2 \sin \theta_1 \cos \theta_0}{\sin (\theta_0 + \theta_1)} \tag{18}$$

$$t_p = \frac{2 \sin \theta_1 \cos \theta_1}{\sin (\theta_0 + \theta_1) \cos (\theta_0 - \theta_1)} \tag{19}$$

For nonabsorbing materials the intensity reflection coefficients R_s and R_p are simply the squares of Eqs. (16) and (17):

$$R_s = r_s^2 = \frac{\sin^2(\theta_0 - \theta_1)}{\sin^2(\theta_0 + \theta_1)} \tag{20}$$

$$R_p = r_p^2 = \frac{\tan^2(\theta_0 - \theta_1)}{\tan^2(\theta_0 + \theta_1)} \tag{21}$$

and, at normal incidence,

$$R_s = R_p = \frac{(n_0 - n_1)^2}{(n_0 + n_1)^2} \tag{22}$$

from Eqs. (11) and (12). In the lower part of Fig. 2, R_s and R_p are given as a function of angle of incidence for various values of the refractive-index ratio n_1/n_0 with k_1 for the material equal to zero. The curves for $n_1/n_0 = 1.3$, 1.8, and 2.3 show that the normal-incidence reflectance increases as n_1 increases. The curves for $n_1/n_0 = 0.3$ and 0.8 and $k_1 = 0$ have no physical significance as long as the incident medium is air. However,

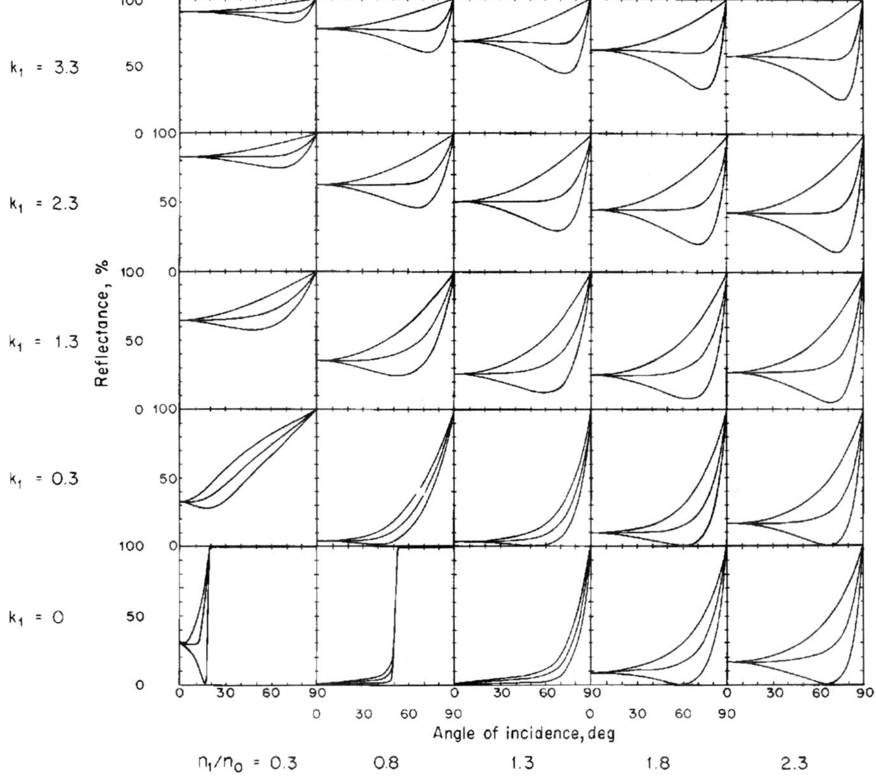

FIGURE 2 R_s (upper curves), R_p (lower curves), and $R_{av} = (R_s + R_p)/2$ as a function of angle of incidence for various values of the refractive-index ratio n_1/n_0 and k_1. The incident medium, having refractive index n_0, is assumed to be nonabsorbing. [*Modified from Hunter, Ref.* 4.]

they are representative of *intenal reflections* in materials of refractive index $n_0 = 3.33$ and 1.25, respectively, when the *other* medium is air ($n_1 = 1$).

The intensity transmission coefficients T_s and T_p are obtained from the Poynting vector and for nonabsorbing materials are

$$T_s = 1 - R_s = \frac{n_1 \cos \theta_1}{n_0 \cos \theta_0} t_s^2 = \frac{4n_0 n_1 \cos \theta_0 \cos \theta_1}{(n_0 \cos \theta_0 + n_1 \cos \theta_1)^2} = \frac{\sin 2\theta_0 \sin 2\theta_1}{\sin^2 (\theta_0 + \theta_1)} \tag{23}$$

$$T_p = 1 - R_p = \frac{n_1 \cos \theta_1}{n_0 \cos \theta_0} t_p^2 = \frac{4n_0 n_1 \cos \theta_0 \cos \theta_1}{(n_1 \cos \theta_0 + n_0 \cos \theta_1)^2}$$

$$= \frac{\sin 2\theta_0 \sin 2\theta_1}{\sin^2 (\theta_0 + \theta_1) \cos^2 (\theta_0 - \theta_1)} \tag{24}$$

These coefficients are for light passing through a single boundary and hence are of limited usefulness. In actual cases, the light is transmitted through a slab of material where there are two boundaries, generally multiple reflections within the material, and sometimes interference effects when the boundaries are smooth and plane-parallel.

The intensity transmission coefficient T_{sample} for a slab of transparent material in air is given by the well-known Airy equation[5] when the sample has smooth, plane-parallel sides and coherent multiple reflections occur within it:

$$T_{\text{sample}} = \frac{1}{1 + [4R_{s,p}/(1 - R_{s,p})^2] \sin^2 \gamma} \tag{25}$$

where

$$\gamma = \frac{2\pi n_1 d \cos \theta_1}{\lambda} \tag{26}$$

The values of R_s and R_p can be determined from Eqs. (20) to (22); d is the sample thickness, λ the wavelength, n_1 the refractive index of the material, and θ_1 the angle of refraction. Equation (25) holds for all angles of incidence including the Brewster angle, where $R_p = 0$ [see Eq. (48)]. The Airy equation predicts that at a given angle of incidence the transmission of the sample will vary from a maximum value of 1 to a minimum value of $(1 - R_{s,p})^2/(1 + R_{s,p})^2$ as the wavelength or the thickness is changed. If the sample is very thick, the oscillations in the transmittance will occur at wavelengths very close together and hence will be unresolved. A complete oscillation occurs every time γ changes by π, so that the wavelength interval $\Delta\lambda$ between oscillations is

$$\Delta\lambda \approx \frac{\lambda^2}{2n_1 d \cos \theta_1} \tag{27}$$

An an example, a sample 1 mm thick with an index of 1.5 at 5000 Å will have transmission maxima separated by 0.83 Å when measured at normal incidence ($\cos \theta_1 = 1$). These maxima would not be resolved by most commercial spectrophotometers. In such a case, one would be measuring the average transmission $T_{\text{sample,av}}$:

$$T_{\text{sample,av}} = \frac{1 - R_{s,p}}{1 + R_{s,p}} \tag{28}$$

For nonabsorbing materials, this is the same value as that which would be obtained if the multiply reflected beams did not coherently interfere within the sample. If the sample is wedge-shaped, so that no multiply reflected beams contribute to the measured transmittance, T_{sample} is simply T_s^2 or T_p^2 and can be calculated from Eq. (23) or (24).

When the material is absorbing, i.e., has a complex refractive index, it is not so easy to calculate the reflectance and transmittance since the angle of refraction is complex. However, Snell's law [Eq. (15)] and Fresnel's equations (11) and (12) are sometimes used with complex values of n_1 and θ_1. The resulting amplitude reflection coefficients are written

$$r_s = |r_s|\, e^{i\delta_s} \tag{29}$$

and

$$r_p = |r_p|\, e^{i\delta_p} \tag{30}$$

where $|r_s|$ and $|r_p|$ are the magnitudes of the reflectances and δ_s and δ_p are the phase changes on reflection. The intensity reflection coefficients are

$$R_{s,p} = r_{s,p} r^*_{s,p} \tag{31}$$

An alternative approach is to use the method of effective indexes to calculate R_s and R_p. In the medium of incidence, which is assumed to be nonabsorbing, the effective indexes η_{0s} and η_{0p} for the s and p components are

$$\eta_{0s} = n_0 \cos\theta_0 \tag{32}$$

$$\eta_{0p} = \frac{n_0}{\cos\theta_0} \tag{33}$$

where n_0 generally equals 1 for air. In the absorbing material both η's are complex and can be written, according to the Bernings,[6,7]

$$\tilde{\eta}_{1s} = \tilde{n}_1 \cos\theta_1 \tag{34}$$

$$\tilde{\eta}_{1p} = \frac{\tilde{n}_1}{\cos\theta_1} \tag{35}$$

where $\tilde{n}_1 = n_1 - ik_1$ is the complex refractive index of the material, and

$$\cos\theta_1 = \left[\frac{(\alpha_1^2 + \beta_1^2)^{1/2} + \alpha_1}{2}\right]^{1/2} - i\left[\frac{(\alpha_1^2 + \beta_1^2)^{1/2} - \alpha_1}{2}\right]^{1/2} \tag{36}$$

$$\alpha_1 = 1 + \left(\frac{n_0 \sin\theta_0}{n_1^2 + k_1^2}\right)^2 (k_1^2 - n_1^2) \tag{37}$$

and

$$\beta_1 = -2n_1 k_1 \left(\frac{n_0 \sin\theta_0}{n_1^2 + k_1^2}\right)^2 \tag{38}$$

Abelès' method[8] also uses effective indexes for the absorbing material, but they are calculated differently:

$$\tilde{\eta}_{1s} = a - ib \tag{39}$$

$$\tilde{\eta}_{1p} = c - id \tag{40}$$

where

$$a^2 - b^2 = n_1^2 - k_1^2 - n_0^2 \sin^2\theta_0 \tag{41}$$

$$ab = n_1 k_1 \tag{42}$$

$$c = a\left(1 + \frac{n_0^2 \sin^2\theta_0}{a^2 + b^2}\right) \tag{43}$$

$$d = b\left(1 - \frac{n_0^2 \sin^2\theta_0}{a^2 + b^2}\right) \tag{44}$$

In both methods employing effective indexes, the amplitude reflection coefficients are

$$r_s = \frac{\eta_{0s} - \eta_{1s}}{\eta_{0s} + \eta_{1s}} \tag{45}$$

$$r_p = \frac{\eta_{1p} - \eta_{0p}}{\eta_{1p} + \eta_{0p}} \tag{46}$$

which are equivalent to Eqs. (29) and (30) and reduce to Eqs. (11) and (12) when $k_1 = 0$. The intensity reflection coefficients are given by Eq. (31), as before. At normal incidence,

$$R_s = R_p = \frac{(n_0 - n_1)^2 + k_1^2}{(n_0 + n_1)^2 + k_1^2} \tag{47}$$

Values of R_s and R_p are plotted as a function of angle of incidence in Fig. 2 for various values of n_1 and k_1. (The incident medium is assumed to be air with $n_0 = 1$ unless otherwise noted.) As n_1 increases with $k_1 > 0$ held constant, the magnitudes of R_s and R_p at normal incidence both decrease. As k_1 increases with n_1 held constant, the magnitudes of R_s and R_p at normal incidence both increase. Tables of R_s and R_p for various values of n_1 and k_1 are given for angles of incidence from 0 to 85° by Holl.[9]

The absolute phase changes on reflection δ_s and δ_p are also of interest in problems involving polarization. When the material is nonabsorbing, the phase changes can be determined from the amplitude reflection coefficients, Eqs. (11) and (12); when $\theta_0 = 0$ and $n_1 > n_0$, $\delta_s = 180°$ and $\delta_p = 360°$.† This is an apparent contradiction since at normal incidence the s and p components should be indistinguishable. However, the problem is resolved by recalling that by convention we are always looking against the direction of propagation of the light (see Fig. 1). To avoid complications, the phase change on reflection at normal incidence (often defined as β) is identified with δ_s.

For a dielectric, if $n_0 < n_1$, δ_s remains 180° for all angles of incidence from 0 to 90°, as can be seen from the numerator of Eq. (11). However, there is an abrupt discontinuity in δ_p, as can be seen from Eq. (12). If $n_0 < n_1$, $\delta_p = 360°$† at normal incidence and at larger angles for which the numerator of Eq. (12) is positive. Since $\cos\theta_0$ becomes increasingly less than $\cos\theta_1$ as θ_0 increases, and since $n_1 > n_0$, there will be an angle for which $n_1 \cos\theta_0 = n_0 \cos\theta_1$. At this angle δ_p undergoes an abrupt change from 360 to 180°, and it remains 180° for larger angles of incidence. At the transition value of θ_0, which is called the *Brewster angle* θ_B since $R_p = 0$,

$$\tan\theta_B = \frac{n_1}{n_0} \tag{48}$$

(This angle is also called the *polarizing angle* since $\theta_0 + \theta_1 = 90°$.)

The phase changes δ_s and δ_p are not simply 360 or 180° for an absorbing material. At normal incidence it follows from Eq. (45) that

$$\tan\delta_s = \frac{2n_0 k_1}{n_0^2 - n_1^2 - k_1^2} \tag{49}$$

† Since 360° and 0° are indistinguishable, many optics books state that $\delta_p = 0°$ for dielectrics at normal incidence, but this makes the ellipsometric parameter $\Delta = \delta_p - \delta_s < 0$, which is incompatible with ellipsometric conventions—see the section on Ellipsometry.

so that $\delta_s = 180°$ only if $k_1 = 0$. As before, $\delta_p = \delta_s + 180°$, as seen by comparing Eqs. (45) and (46). At nonnormal incidence

$$\tan \delta_s = \frac{2\eta_{0s} b}{\eta_{0s}^2 - a^2 - b^2} \tag{50}$$

and

$$\tan \delta_p = \frac{-2\eta_{0p} d}{c^2 + d^2 - \eta_{0p}^2} \tag{51}$$

where the relations for a, b, c, and d have been given in Eqs. (41) to (44). The following relations between these quantities may also prove helpful:

$$a^2 + b^2 = [(n_1^2 - k_1^2 - n_0^2 \sin^2 \theta_0)^2 + 4n_1^2 k_1^2]^{1/2} \tag{52}$$

$$c^2 + d^2 = \frac{(n_1^2 + k_1^2)^2}{a^2 + b^2} \tag{53}$$

$$b^2 = \frac{n_1^2 - k_1^2 - n_0^2 \sin^2 \theta_0}{2} + \frac{a^2 + b^2}{2} \tag{54}$$

Figure 3 shows how δ_s and δ_p change as a function of angle of incidence for an absorbing material. At normal incidence they are 180° apart because of the mirror-image effect, mentioned previously. As the angle of incidence increases, δ_p approaches δ_s, and at the *principal angle* $\bar{\theta}$ the two quantities differ by only 90°. At grazing incidence they coincide.

The reflectance R_p does not reach zero for an absorbing material as it does for a dielectric, but the angle for which it is a minimum is called the *pseudo Brewster angle* θ_B'. Two other angles closely associated with the pseudo Brewster are also of interest. The angle for which the ratio R_p/R_s is a minimum is sometimes called the *second Brewster*

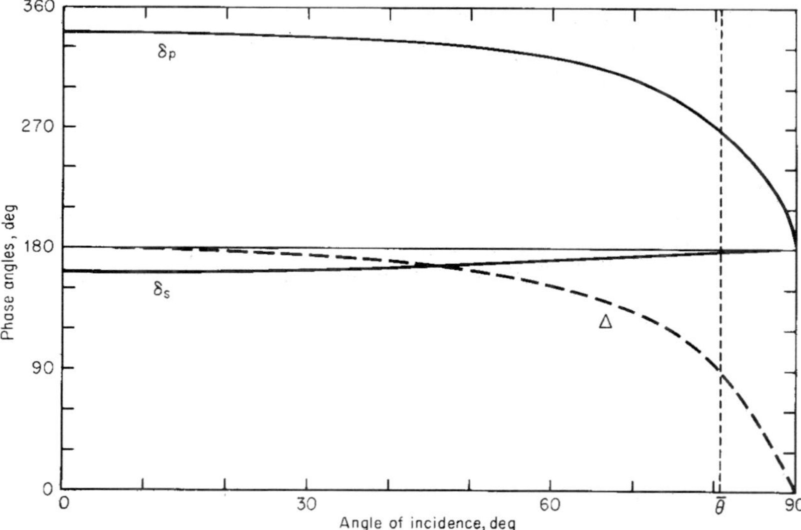

FIGURE 3 Phase changes on reflection δ_s and δ_p and phase difference $\Delta = \delta_p - \delta_s$ as a function of angle of incidence for an absorbing material. The principal angle, for which $\Delta = 90°$, is also shown. [*Bennett and Bennett, Ref.* 10.]

angle. It is generally only slightly larger than θ'_B. The *principal angle* $\bar{\theta}$, at which $\delta_p - \delta_s = 90°$, is always larger than the second Brewster angle and θ'_B. For most metals θ'_B and $\bar{\theta}$ are only a fraction of a degree apart, but it is possible for them to differ by as much as 45°.[9] There is no polarizing angle as such for an absorbing material because the angle of refraction is complex.

3. Basic Relations for Polarizers A linear† polarizer is anything which when placed in an incident unpolarized beam produces a beam of light whose electric vector is vibrating primarily in one plane, with only a small component vibrating in the plane perpendicular to it. If a polarizer is placed in a plane-polarized beam and is rotated about an axis parallel to the beam direction, the transmittance T will vary between a maximum value T_1 and a minimum value T_2 according to the law

$$T = (T_1 - T_2)\cos^2\theta + T_2 \tag{55}$$

Although the quantities T_1 and T_2 are called the *principal transmittances,* in general $T_1 \gg T_2$; θ is the angle between the plane of the principal transmittance T_1 and the plane of vibration (of the electric vector) of the incident beam. If the polarizer is placed in a beam of unpolarized light, its transmittance is

$$T = \tfrac{1}{2}(T_1 + T_2) \tag{56}$$

so that a perfect polarizer would transmit only 50 percent of an incident unpolarized beam.‡

When two identical polarizers are placed in an unpolarized beam, the resulting transmittance will be

$$T_\parallel = \tfrac{1}{2}(T_1^2 + T_2^2) \tag{57}$$

when their principal transmittance directions are parallel and will be

$$T_\perp = T_1 T_2 \tag{58}$$

when they are perpendicular. In general, if the directions of principal transmittance are inclined at an angle θ to each other, the transmittance of the pair will be

$$T_\theta = \tfrac{1}{2}(T_1^2 + T_2^2)\cos^2\theta + T_1 T_2 \sin^2\theta \tag{59}$$

The polarizing properties of a polarizer are generally defined in terms of its *degree of polarization* P§,‖

$$P = \frac{T_1 - T_2}{T_1 + T_2} \tag{60}$$

or its *extinction ratio* ρ_p

$$\rho_p = \frac{T_2}{T_1} \tag{61}$$

When one deals with nonnormal-incidence reflection polarizers, one generally writes P

† Circular polarizers are discussed in Par. 6.

‡ Jones[11] has pointed out that a perfect polarizer can transmit more than 50 percent of an incident unpolarized beam under certain conditions.

§ Bird and Shurcliff[12] distinguish between *degree of polarization,* which is a constant of the light beam, and *polarizance,* which is a constant of the polarizer. The polarizance is defined as being equal to the degree of polarization the polarizer produces in an incident monochromatic beam that is unpolarized. In practice, incident beams are often slightly polarized, so that the polarizance values differ slightly from the ideal degree of polarization. Other authors have not followed this distinction.

‖ Authors dealing with topics such as scattering from aerosols sometimes define *degree of polarization* (of the scattered light) in terms of the Stokes vectors (Par. 7) as $P = (S_1^2 + S_2^2 + S_3^2)^{1/2}/S_0$.

and ρ_p in terms of R_p and R_s, the reflectances of light polarized parallel and perpendicular to the plane of incidence, respectively. As will be shown in Par. 4, R_s can be equated to T_1 and R_p to T_2, so that Eqs. (60) and (61) become $P = (R_s - R_p)/(R_s + R_p)$ and $\rho_p = R_p/R_s$. If either ρ_p or P is known, the other can be deduced since

$$P = \frac{1 - \rho_p}{1 + \rho_p} \tag{62}$$

and

$$\rho_p = \frac{1 - P}{1 + P} \tag{63}$$

If one is determining the degree of polarization or the extinction ratio of a polarizer, the ratio of $T_\perp$ to $T_\parallel$ can be measured for two identical polarizers in unpolarized light. From Eqs. (57) and (58),

$$\frac{T_\perp}{T_\parallel} = \frac{T_1 T_2}{(T_1^2 + T_2^2)/2} \approx \frac{2T_2}{T_1} = 2\rho_p \tag{64}$$

if $T_2^2 \ll T_1^2$. If a perfect polarizer or a source of perfectly plane-polarized light is available, T_2/T_1 can be determined directly by measuring the ratio of the minimum to the maximum transmittance of the polarizer. Other relations for two identical partial polarizers are given by West and Jones,[13] as well as the transmittance $T_{\theta ab}$ of two dissimilar partial polarizers *a* and *b* whose principal axes are inclined at an angle θ with respect to each other. This latter expression is

$$T_{\theta ab} = \tfrac{1}{2}(T_{1a}T_{1b} + T_{2a}T_{2b})\cos^2\theta + \tfrac{1}{2}(T_{1a}T_{2b} + T_{1b}T_{2a})\sin^2\theta \tag{65}$$

where the subscripts 1 and 2 refer to the principal transmittances, as before.

Spectrophotometric measurements can involve polarizers and dichroic samples. Dichroic (optically anisotropic) materials are those which absorb light polarized in one direction more strongly than light polarized at right angles to that direction. (Dichroic materials are to be distinguished from birefringent materials, which may have different refractive indices for the two electric vectors vibrating at right angles to each other but similar, usually negligible, absorption coefficients.) When making spectrophotometric measurements, one should know the degree of polarization of the polarizer and how to correct for instrumental polarization. This latter quantity may arise from nonnormal-incidence reflections from a grating, dispersing prism, or mirrors. Light sources are also sometimes polarized. Simon,[14] Charney,[15] Gonatas *et al.*,[16] and Wizinowich[17] suggest methods for dealing with imperfect polarizers, dichroic samples, and instrumental polarization. In addition, when a dichroic sample is placed between a polarizer and a spectrophotometer which itself acts like an imperfect polarizer, one has effectively three polarizers in series. This situation has been treated by Jones,[18] who showed that anomalies can arise when the phase retardation of the polarizers takes on certain values. Mielenz and Eckerle[19] have discussed the accuracy of various types of polarization attenuators.

4. Polarization by Nonnormal-Incidence Reflection (Pile of Plates) Pile-of-plates polarizers make use of reflection or transmission of light at nonnormal incidence, frequently near the Brewster or polarizing angle [Eq. (48) in Par. 2]. The extinction ratio and "transmittance" of these polarizers can be calculated directly from the Fresnel equations. Some simplifications occur for nonabsorbing or slightly absorbing plates. Equations (20) and (21) give the values of the intensity reflection coefficients R_s and R_p for light vibrating perpendicular to the plane of incidence (s component) and parallel to the plane of incidence (p component). The angle of refraction θ_1 in those equations is related

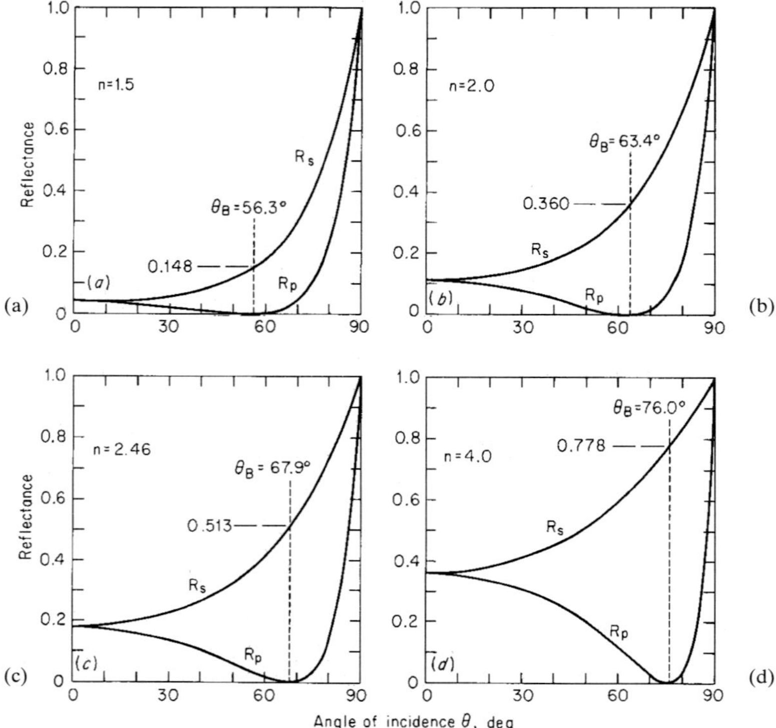

FIGURE 4 Reflectance of light polarized parallel R_p and perpendicular R_s to the plane of incidence from materials of different refractive index n as a function of angle of incidence: (*a*) $n = 1.5$ (alkali halides in ultraviolet and sheet plastics in infrared), (*b*) $n = 2.0$ (AgCl in infrared), (*c*) $n = 2.46$ (Se in infrared), and (*d*) $n = 4.0$ (Ge in infrared). The Brewster angle θ_B (at which R_p goes to 0) and the magnitude of R_s at θ_B are also indicated.

to the refractive index n of the material by Snell's law [Eq. (15)†]. At the Brewster angle $R_p = 0$, so that the reflected light is, in principle, completely plane-polarized. This is the basis for all Brewster angle reflection polarizers.

Let us now see how the characteristics of a reflection polarizer depend on its refractive index. In Fig. 4 the reflectances R_s and R_p have been plotted for different values of the refractive index, roughly representing alkali halides in the ultraviolet and sheet-plastic materials, silver chloride, selenium, and germanium in the infrared. The Brewster angle, given by Eq. (48), is also indicated, as well as the magnitude of R_s at the Brewster angle. We note from these graphs that if light is polarized by a single reflection from a nonabsorbing material, the polarizer with the highest refractive index will have the largest throughput. In reflection polarizers, the quantity R_s is essentially the principal "transmittance" of the polarizer [T_1 in Eqs. (55) to (65)] except that it must be multipled by the reflectance of any other mirrors used to return the beam to its axial position.

The reflectance R_p can be equated to T_2, the minimum "transmittance" of the polarizer, so that the extinction ratio ρ_p of a reflection polarizer [Eq. (61)] is $\rho_p = R_p/R_s$. If R_p is

† Since we are assuming that the medium of incidence is air, $n_0 = 1$ and $n_1 = n$, the refractive index of the material.

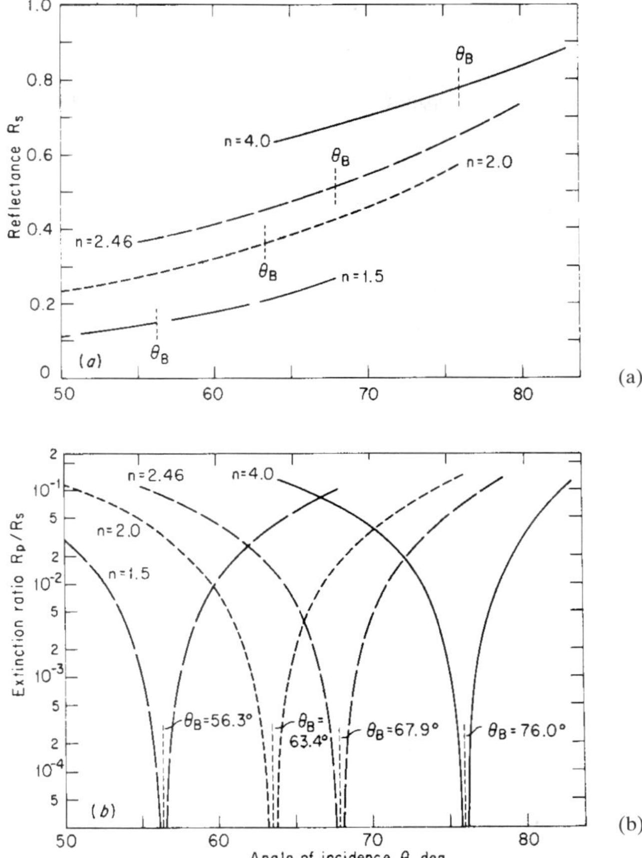

FIGURE 5 (*a*) Reflectance R_s and (*b*) extinction ratio R_p/R_s for materials of different refractive index at angles near the Brewster angle θ_B. A single surface of the material is assumed.

really zero at the Brewster angle, the extinction ratio will be zero for all materials *independent of the value of n.* If a given extinction ratio is desired, for example, 10^{-3} [corresponding to 99.8 percent polarization; see Eq. (62)], then the convergence angle of the light beam must be small so that all the angles of incidence lie within about ±1° of the Brewster angle. The convergence angle depends only weakly on the refractive index for this case, varying from ±1.2° for $n = 1.5$ to ±0.8° for $n = 4.0$.

If a good extinction ratio is required for a beam of larger convergence angle, two polarizing reflections may be used. Then all the exponents in Fig. 5*b* are doubled, and the convergence angles for a given extinction ratio are greatly increased. To obtain a value of 10^{-3} with two reflections, the angle of incidence must be within about ±6° of the Brewster angle for values of n less than 3.5; for $n = 4$ it is reduced slightly and becomes more asymmetric (+4.0 and −5.2°). A disadvantage of having two reflections from the polarizing materials is that the throughput is reduced. All the values of R_s in Fig. 5*a* are squared, so that for $n = 4$, $R_s = 0.78$ but $R_s^2 = 0.61$; for smaller refractive indexes the reduction in throughput is much greater.

The information shown graphically in Figs. 4 and 5 is given analytically in a paper by Azzam[20] who is concerned about the angular sensitivity of Brewster-angle reflection polarizers, particularly those made with silicon or germanium plates. Also, Murty and Shukla[21] show analytically that the shadowy extinction patterns sometimes seen with a crossed Brewster angle reflection polarizer and analyzer are caused by light incident on the surfaces at angles different from the Brewster angle.

Although in many cases multiple reflections within a plate degrade its polarizing properties, this is not true for Brewster angle reflection polarizers. For multiple reflections within a plane-parallel plate of material

$$(R_{s,p})_{\text{plate}} = \frac{2R_{s,p}}{1 + R_{s,p}} \tag{66}$$

assuming no interference or absorption; R_s and R_p are given by Eqs. (20) and (21). Multiple reflections have a minor effect on the extinction ratio but the increase in R_s is appreciable. To fulfill the conditions of Eq. (66), the plate must have plane-parallel sides and be unbacked. We are also assuming that the plate is thick or nonuniform enough for interference effects within it to be neglected.

All the preceding discussion applies only to nonabsorbing materials. If a small amount of absorption is present, R_p will have a minimum that is very close to zero and the material will still make a good reflection polarizer. However, if the extinction coefficient k becomes appreciable, the minimum in R_p will increase and the polarizing efficiency will be degraded. By referring to Fig. 2 one can see roughly what the ratio of R_p to R_s will be for a given set of optical constants. Exact values of R_p and R_s can be calculated from n and k using Eqs. (45), (46), (31), and the other pertinent relations in Par. 2. When choosing materials for possible use as metallic reflection polarizers, one wants the largest difference between R_s and R_p and the smallest magnitude of R_p at the minimum. Thus, ideally n should be much larger than k.

The Abelès condition[22] applies to the amplitude reflectances r_s and r_p for either dielectrics or metals at 45° angle of incidence. At this angle

$$r_s^2 = r_p \tag{67}$$

and

$$2\delta_s = \delta_p \tag{68}$$

where the δ's are the absolute phase changes on reflection for the p and s components (see Par. 2). Relation (67) is frequently applied to the intensity reflectances R_s and R_p, which are directly related to the amplitude reflectances [Eqs. (20), (21), and (31)].

5. Polarization by Nonnormal-Incidence Transmission (Pile of Plates) The theory of Brewster angle transmission polarizers follows directly from that given for reflection polarizers. Table 1 lists the relations giving the s and p transmittances of the polarizers with various assumptions about multiple reflections, interference, absorption, etc.† All these relations contain R_s and R_p, the reflectances at a single interface, which are given at the bottom of the table.

At the Brewster angle, R_p at a single interface equals zero, and the transmittances of the plates can be expressed in terms of the refractive index of the material and the number of plates. The relations for the s and p transmittances at this angle are given in Table 2. Most references that contain an expression for the degree of polarization of a pile of plates give the formula of Provostaye and Desains,[23] which assumes an infinite series of multiple reflections between all surfaces, i.e., multiple reflections within and between plates. This

† Transmission polarizers in which the multiply internally reflected beams are coherent and produce interference effects are discussed in Chap. 3, "Polarizers," in Vol. II of this Handbook.

TABLE 1 Transmittances and Degree of Polarization for a Single Plate and Multiple Plates at any Angle of Incidence in Terms of R_s and R_p for a Single Surface†

	One plate (two surfaces) $(T_{s,p})_{\text{sample}}$	*m* plates (2*m* surfaces) $(T_{s,p})_{\text{sample}}$	*m* plates (2*m* surfaces) $P = \frac{T_p - T_s}{T_p + T_s} = \frac{1 - T_s/T_p}{1 + T_s/T_p}$
Single transmitted beam, no multiple reflections, no absorption	$(1 - R_{s,p})^2$	$(1 - R_{s,p})^{2m}$ ‡	$\frac{1 - \cos^{4m}(\theta_0 - \theta_1)}{1 + \cos^{4m}(\theta_0 + \theta_1)}$ ‡
Multiple reflections within plate, no interference effects, no absorption	$\frac{1 - R_{s,p}}{1 + R_{s,p}}$ ‖	$\left(\frac{1 - R_{s,p}}{1 + R_{s,p}}\right)^{m}$ ‡, ‖	$\frac{1 - \left[\cos^2(\theta_0 - \theta_1)\frac{1 + R_p}{1 + R_s}\right]^m}{1 + \left[\cos^2(\theta_0 - \theta_1)\frac{1 + R_p}{1 + R_s}\right]^m}$ ‡, ‖
		$\frac{1 - R_{s,p}}{1 + (2m - 1)R_{s,p}}$ §	$\frac{1 - \cos^2(\theta_0 - \theta_1)\frac{1 + (2m - 1)R_p}{1 + (2m - 1)R_s}}{1 + \cos^2(\theta_0 - \theta_1)\frac{1 + (2m - 1)R_p}{1 + (2m - 1)R_s}}$ §
Multiple reflections within plate, no interference effects, small absorption	$\frac{(1 - R_{s,p})^2 e^{-\alpha d}}{1 - R_{s,p}^2 e^{-2\alpha d}}$	$\frac{(1 - R_{s,p})^{2m} e^{-m\alpha d}}{(1 - R_{s,p}^2 e^{-2\alpha d})^m}$ ‡	$\frac{1 - \cos^{4m}(\theta_0 - \theta_1)\left(\frac{1 - R_p^2 e^{-2\alpha d}}{1 - R_s^2 e^{-2\alpha d}}\right)^m}{1 + \cos^{4m}(\theta_0 - \theta_1)\left(\frac{1 - R_p^2 e^{-2\alpha d}}{1 - R_s^2 e^{-2\alpha d}}\right)^m}$ ‡
Multiple reflections within plate, interference within plate, no absorption	$\frac{1}{1 + \frac{4R_{s,p}}{(1 - R_{s,p})^2}\sin^2\gamma}$	$\left[\frac{1}{1 + \frac{4R_{s,p}}{(1 - R_{s,p})^2}\sin^2\gamma}\right]^m$ ‡ $\begin{cases} \max 1 \\ \min \left(\frac{1 - R_{s,p}}{1 + R_{s,p}}\right)^{2m} \end{cases}$	
Single surface	$R_s = \frac{\sin^2(\theta_0 - \theta_1)}{\sin^2(\theta_0 + \theta_1)}$ $T_s = 1 - R_s = \frac{\sin 2\theta_0 \sin 2\theta_1}{\sin^2(\theta_0 + \theta_1)}$	$R_p = \frac{\tan^2(\theta_0 - \theta_1)}{\tan^2(\theta_0 + \theta_1)}$ $T_p = 1 - R_p = \frac{\sin 2\theta_0 \sin 2\theta_1}{\sin^2(\theta_0 + \theta_1)\cos^2(\theta_0 - \theta_1)}$	$P = \frac{1 - \cos^2(\theta_0 - \theta_1)}{1 + \cos^2(\theta_0 - \theta_1)}$

† $\alpha = 4\pi k/(\lambda \cos\theta_1)$, $\gamma = 2\pi n d \cos\theta_1/\lambda$, θ_0 = angle of incidence, θ_1 = angle of refraction, n = refractive index = $(\sin\theta_0)/(\sin\theta_1)$, k = extinction coefficient, d = plate thickness, λ = wavelength.

‡ No multiple reflections between plates.

§ Multiple reflections between plates.

‖ Also holds for coherent multiple reflections averaged over one period of $\sin^2\gamma$.

TABLE 2 Transmittances and Degree of Polarization for a Single Plate and Multiple Plates at the Brewster Angle θ_B, where $\tan \theta_B = n$ and $\theta_0 + \theta_1 = 90°$†

	One plate (two surfaces)		m plates (2m surfaces)		
	$(T_p)_{\text{sample}}$	$(T_s)_{\text{sample}}$	$(T_p)_{\text{sample}}$	$(T_s)_{\text{sample}}$	$P = \frac{T_p - T_s}{T_p + T_s} = \frac{1 - T_s/T_p}{1 + T_s/T_p}$
Single transmitted beams, no multiple reflections, no absorption	1	$\left(\frac{2n}{n^2+1}\right)^4$	1	$\left[\frac{2n}{n^2+1}\right]^{4m}$‡	$\frac{1 - [2n/(n^2+1)]^{4m}}{1 + [2n/(n^2+1)]^{4m}}$ ‡
Multiple reflections within plate, no interference effects, no absorption	1	$\frac{2n^2}{n^4+1}$ ¶	1	$\left[\frac{2n^2}{n^4+1}\right]^{m}$‡,¶	$\frac{1 - [2n^2/(n^4+1)]^{m}}{1 + [2n^2/(n^4+1)]^{m}}$ ‡,¶
			1	$\frac{1}{1 + m(n^2-1)^2/2n^2}$ §	$\frac{m}{m + [2n/(n^2-1)]^2}$ §,‖
Multiple reflections within plate, no interference effects, small absorption	$e^{-\alpha d}$	$\frac{\left(\frac{2n}{n^2+1}\right)^4 e^{-\alpha d}}{1 - \left(\frac{n^2-1}{n^2+1}\right)^4 e^{-2\alpha d}}$	$e^{-\alpha d}$	$\frac{\left(\frac{2n}{n^2+1}\right)^{4m} e^{-m\alpha d}}{\left[1 - \left(\frac{n^2-1}{n^2+1}\right)^4 e^{-2\alpha d}\right]^m}$ ‡	$\frac{1 - \frac{[2n/(n^2+1)]^{4m}}{\{1 - [(n^2-1)/(n^2+1)]^4 e^{-2\alpha d}\}^m}}{1 + \frac{[2n/(n^2+1)]^{4m}}{\{1 - [(n^2-1)/(n^2+1)]^4 e^{-2\alpha d}\}^m}}$ ‡
Multiple reflections within plate, interference within plate, no absorption	1	$\frac{1}{1 + \frac{(n^4-1)^2}{4n^4}\sin^2\gamma}$	1	$\left[\frac{1}{1 + \frac{(n^4-1)^2}{4n^4}\sin^2\gamma}\right]^{m}$‡ $\begin{cases} \max 1 \\ \min \left(\frac{2n^2}{n^4+1}\right)^{2m} \end{cases}$	
Single surface		$R_p = 0$		$R_s = \left(\frac{n^2-1}{n^2+1}\right)^2$	
		$T_p = 1 - R_p = 1$		$T_s = 1 - R_s = \left(\frac{2n}{n^2+1}\right)^2$	$P = \frac{1 - [2n/(n^2+1)]^2}{1 + [2n/(n^2+1)]^2}$

† Where $\alpha = 4\pi k(n^2+1)^{1/2}/\lambda n$, $\gamma = 2\pi n^2 d/\lambda(n^2+1)^{1/2}$, n = refractive index, k = extinction coefficient, d = plate thickness, λ = wavelength.
‡ No multiple reflections between plates.
§ Multiple reflections between plates.
‖ Formula of Provostaye and Desains, Ref. 23.
¶ Also holds for coherent multiple reflections averaged over one period of $\sin^2 \gamma$.

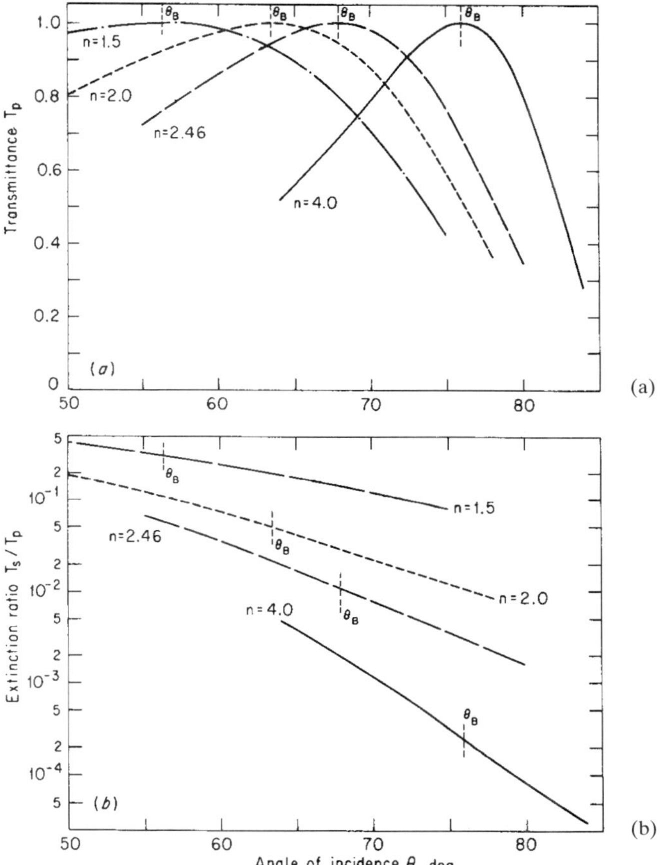

FIGURE 6 (*a*) Transmittance and (*b*) extinction ratio of four plane-parallel plates of refractive index n as a function of angle of incidence, for angles near the Brewster angle. Assumptions are multiple reflections but no interference within each plate and no reflections between plates.

assumption is not valid for most real transmission polarizers (see Chap. 3, "Polarizers," in Vol. II of this Handbook, specifically Brewster Angle Transmission Polarizers).

For most parallel-plate polarizers it is reasonable to assume incoherent multiple reflections within each plate and no reflections between plates. Figure 6 shows the principal transmittance (p component) and extinction ratio for several four-plate polarizers having the refractive indexes indicated.† The extinction ratio improves considerably with increasing refractive index. It is also improved by using the plates at an angle of incidence slightly above the Brewster angle. This procedure, which is most helpful for high refractive index plates, reduces the transmission per plate so that a trade-off is required between losses resulting from absorption or scattering when many plates are used and the

† The extinction ratio of a pile of m plates (no multiple reflections between plates) is simply the product of the extinction ratios of the individual plates.

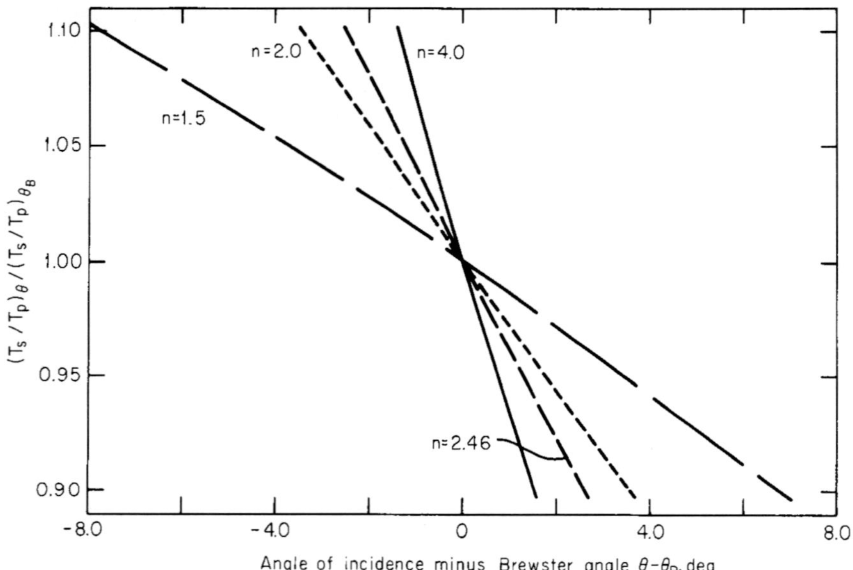

FIGURE 7 Variation of extinction ratio (per film) as a function of angle near the Brewster angle $\theta - \theta_B$. The ordinate is the extinction ratio at θ divided by the extinction ratio at θ_B.

reflectance loss per plate when only a few plates are used above the Brewster angle. In some cases significant improvements have been achieved by following the latter course.[24]

When the number of plates of a given refractive index is increased, the transmittance is unaffected (in the absence of absorption) and the extinction ratio is greatly increased, as shown in the earlier Polarization chapter.[1] In the absence of absorption, comparable transmittances and extinction ratios are obtained with a large number of low-refractive-index plates or a small number of high refractive index plates. Small amounts of absorption decrease the transmittance, but have little effect on the extinction ratio.[1] Tuckerman[25] has derived exact expressions for light reflected from or transmitted through a pile of absorbing plates. He has also noted mistakes that have been perpetuated in some of the formulas for light reflected from or transmitted through a pile of nonabsorbing plates.

A figure of merit giving the variation of the extinction ratio with angle of incidence can be defined as in Fig. 7, where the ordinate is the extinction ratio at a given angle of incidence divided by the extinction ratio at the Brewster angle. The angles of incidence are referred to the Brewster angle, and curves for different values of the refractive index are shown. These curves are calculated from the ratio

$$\frac{\left(\dfrac{T_s}{T_p}\right)_\theta}{\left(\dfrac{T_s}{T_p}\right)_{\theta_B}} = \frac{\left(\dfrac{1-R_s}{1+R_s}\,\dfrac{1+R_p}{1-R_p}\right)_\theta}{\left(\dfrac{1-R_s}{1+R_s}\right)^{\theta_B}} \tag{69}$$

and are for a single transparent film or plate having multiple incoherent internal reflections within the material. As an example of how to use the graphs, consider an optical system having a two-plate germanium polarizer with a refractive index of 4.0. If the angles of incidence vary from −1.4 to +1.5° around the Brewster angle, the ratio of the extinction ratios will vary between $1.10^2 = 1.21$ and $0.90^2 = 0.81$, respectively. (For m plates it would be 1.10^m and 0.90^m.) Thus, in order to restrict the percent variation of the extinction ratio to a given value, one must use a smaller acceptance angle when using more plates.

We have assumed that there are multiple incoherent reflections within each plate and no multiple reflections between plates. The difference in extinction ratios for a series of four plates with and without internal reflections is shown in Fig. 8. The principal transmittance is essentially the same as in Fig. 6 for values of T_p above 0.70 (and only about 0.025 lower when T_p drops to 0.30). However, the extinction ratio of high-refractive-index

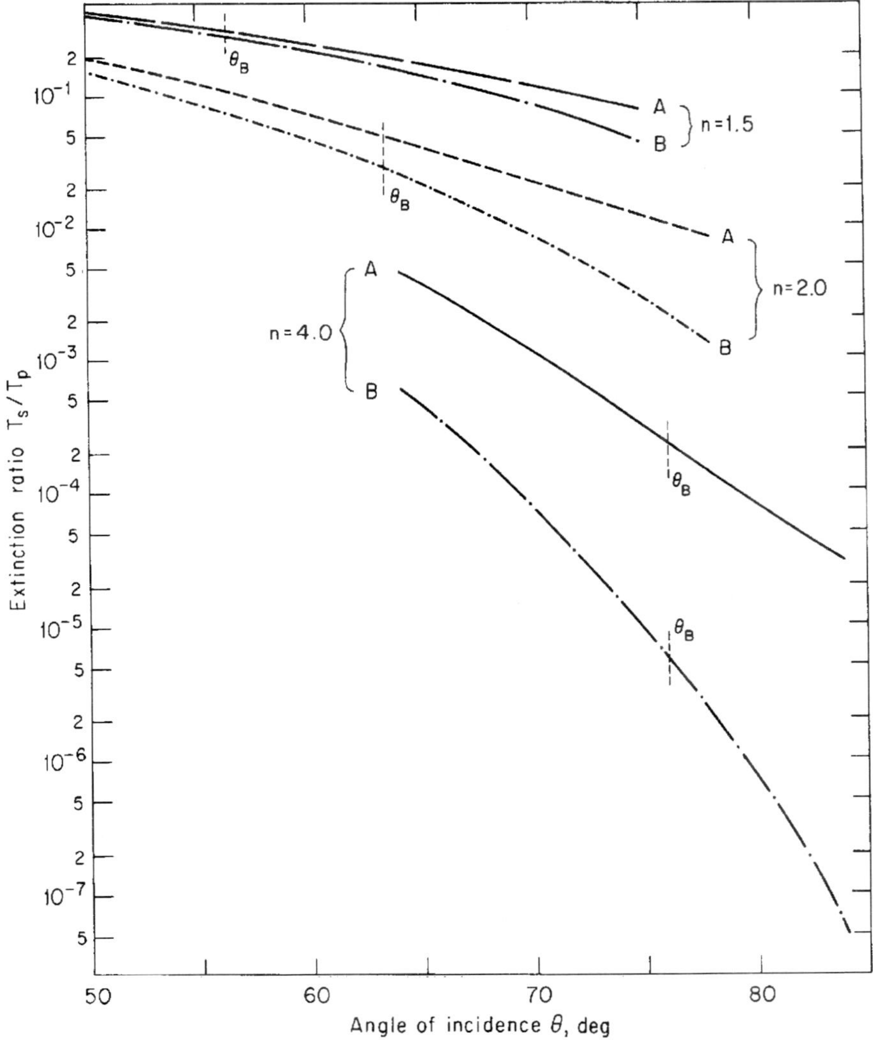

FIGURE 8 Extinction ratio of four plane-parallel plates of refractive index n as a function of angle of incidence for angles near the Brewster angle. Assumptions are A, multiple reflections but no interference within each plate and no reflections between plates; B, no multiple reflections within each plate or between plates. The transmittances for conditions A and B are essentially identical (see Fig. 6a).

materials is much better without multiple internal reflections; for low-refractive-index materials the difference in extinction ratios is small.

The effect of multiple reflections on the extinction ratio can readily be seen from the three relations for the transmittances of the p and s components:

No multiple reflections:

$$(T_{s,p})_{\text{sample}} = (1 - R_{s,p})^{2m} = 1 - 2mR_{s,p} + 2m^2R_{s,p}^2 - mR_{s,p}^2 + \cdots \tag{70}$$

Multiple reflections within plates:

$$(T_{s,p})_{\text{sample}} = \left(\frac{1 - R_{s,p}}{1 + R_{s,p}}\right)^m = 1 - 2mR_{s,p} + 2m^2R_{s,p}^2 + \cdots \tag{71}$$

Multiple reflections within and between plates:

$$(T_{s,p})_{\text{sample}} = \frac{1 - R_{s,p}}{1 + (2m-1)R_{s,p}} = 1 - 2mR_{s,p} + 4m^2R_{s,p} - 2mR_{s,p}^2 + \cdots \tag{72}$$

At the Brewster angle, $R_p = 0$, $T_p = 1$, and the extinction ratio will be smallest, i.e., highest degree of polarization, for the smallest values of the s transmittance. The first three terms in Eqs. (70) and (71) are identical, but Eq. (70) has an additional negative term in R_s^2 and so it will give a slightly smaller value of the s transmittance. Equation (72), from which the formula of Provostaye and Desains was derived, has twice as large a third term as the other two equations, and the negative fourth term is only $1/2m$ of the third term, so that it does not reduce the overall value of the expression appreciably. Thus, Eq. (72) gives an appreciably larger value of the s transmittance, but fortunately it is a limiting case and is rarely encountered experimentally.

6. Quarter-Wave Plates and Other Phase Retardation Plates A retardation plate is a piece of birefringent, uniaxial (or uniaxial-appearing) material in which the ordinary and extraordinary rays travel at different velocities. Thus, one ray is retarded relative to the other, and the path $N\lambda$ between the two rays is given by

$$N\lambda = \pm d(n_e - n_o) \tag{73}$$

where n_o = refractive index of ordinary ray
n_e = refractive index of extraordinary ray
d = physical thickness of plate
λ = wavelength

The positive sign is used when $n_e > n_o$, that is, a positive uniaxial crystal, and the negative sign is used for a negative uniaxial crystal, for which $n_e < n_o$. Since $N\lambda$ is the path difference between the two rays, N can be considered the retardation expressed in fractions of a wavelength. For example, $N = 1/4$ for a quarter-wave (or $\lambda/4$) plate, 1/2 for a half-wave (or $\lambda/2$) plate, 3/4 for a three-quarter-wave (or $3\lambda/4$) plate, etc.

The phase difference between two rays traveling through a birefringent material is $2\pi/\lambda$ times the path difference, so that the phase retardation δ is

$$\delta = 2\pi N = \pm \frac{2\pi d(n_e - n_o)}{\lambda} \tag{74}$$

Thus, phase differences of $\pi/2$, π, and $3\pi/2$ are introduced between the two beams in quarter-wave, half-wave, and three-quarter-wave plates, respectively.

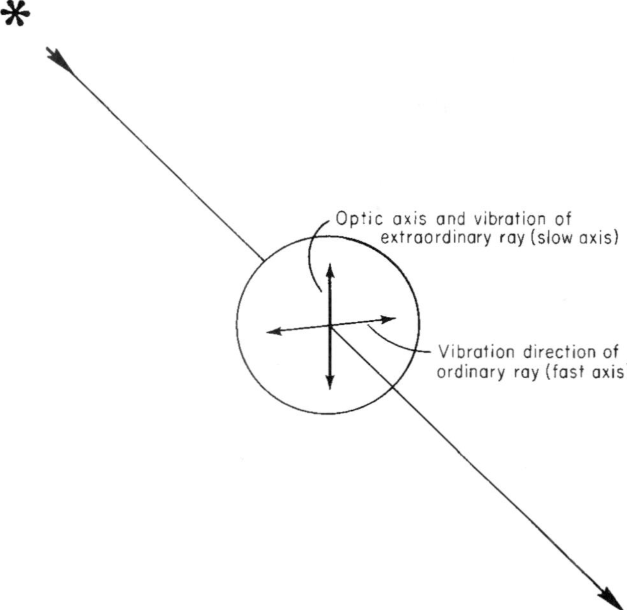

FIGURE 9 Light incident normally on the front surface of a retardation plate showing the vibration directions of the ordinary and extraordinary rays. In a positive uniaxial crystal, the fast and slow axes are as indicated in parentheses; in a negative uniaxial crystal, the two axes are interchanged.

A retardation plate can be made from a crystal which is cut so that the optic axis lies in a plane parallel to the face of the plate, as shown in Fig. 9. Consider a beam of unpolarized or plane-polarized light normally incident on the crystal. It can be resolved into two components traveling along the same path through the crystal but vibrating at right angles to each other. The ordinary ray vibrates in a direction perpendicular to the optic axis, while the extraordinary ray vibrates in a direction parallel to the optic axis. In a positive uniaxial crystal $n_e > n_o$, so that the extraordinary ray travels more slowly than the ordinary ray. The fast axis is defined as the direction in which the faster-moving ray vibrates; thus in a positive uniaxial crystal, the fast axis (ordinary ray) is perpendicular to the optic axis, while the slow axis (extraordinary ray) coincides with the optic axis. For a negative uniaxial crystal, the fast axis coincides with the optic axis.

Figure 10 shows how the state of polarization of a light wave changes after passing through retardation plates of various thicknesses when the incident light is plane-polarized at an azimuth of 45° to the fast axis of the plate. If the plate has a retardation of $\lambda/8$, which means that the ordinary and extraordinary waves are out of phase by $\pi/4$ with each other, the transmitted light will be elliptically polarized with the major axis of the ellipse coinciding with the axis of the original plane-polarized beam. As the retardation gradually increases (plate gets thicker for a given wavelength or wavelength gets shorter for a given plate thickness), the ellipse gradually turns into a circle, but its major axis remains at 45° to the fast axis of the retardation plate. For a retardation of $\lambda/4$, the emerging light is right circularly polarized. As the retardation continues to increase, the transmitted light becomes elliptically polarized with the major axis of the ellipse lying perpendicular to the plane of the incident polarized beam, and then the minor axis of the ellipse shrinks to zero

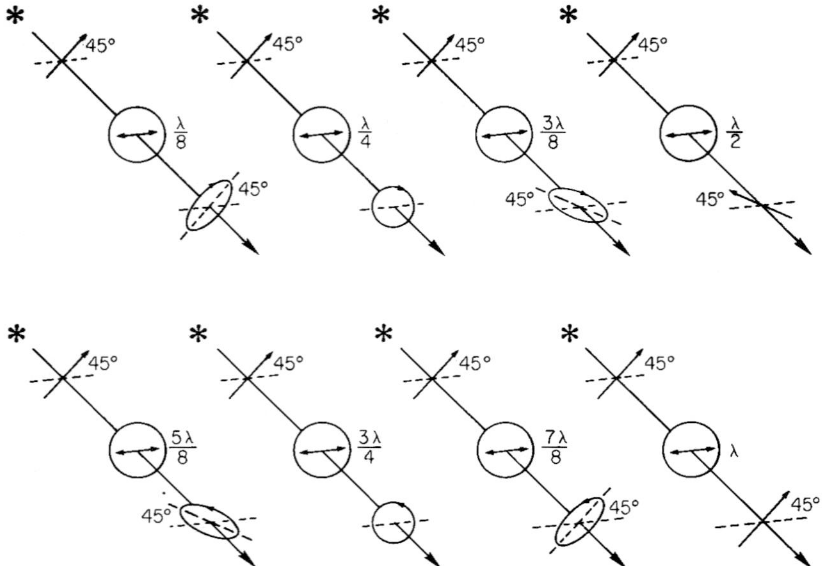

FIGURE 10 State of polarization of a light wave after passing through a crystal plate whose retardation is indicated in fractions of a wavelength (phase retardation $2\pi/\lambda$ times these values) and whose fast axis is indicated by the double arrow. In all cases the incident light is plane-polarized at an azimuth of 45° to the direction of the fast axis.

and plane-polarized light is produced when the retardation becomes $\lambda/2$. As the retardation increases further, the patterns change in opposite order and the polarized light is left circularly polarized when the retardation equals $3\lambda/4$. Finally, when the retardation is a full wave, the incident plane-polarized light is transmitted unchanged although the slow wave has now been retarded by a full wavelength relative to the fast wave.

The most common type of retardation plate is the quarter-wave plate. Figure 11 shows how this plate affects the state of polarization of light passing through it when the fast axis is positioned in the horizontal plane and the azimuth of the incident plane-polarized light is changed from $\theta = 0°$ to $\theta = 90°$. When $\theta = 0°$, only the ordinary ray (for a positive birefringent material) passes through the plate, so that the state of polarization of the beam is unchanged. When θ starts increasing, the transmitted beam is elliptically polarized with the major axis of the ellipse lying along the fast axis of the $\lambda/4$ plate; $\tan \theta = b/a$, the

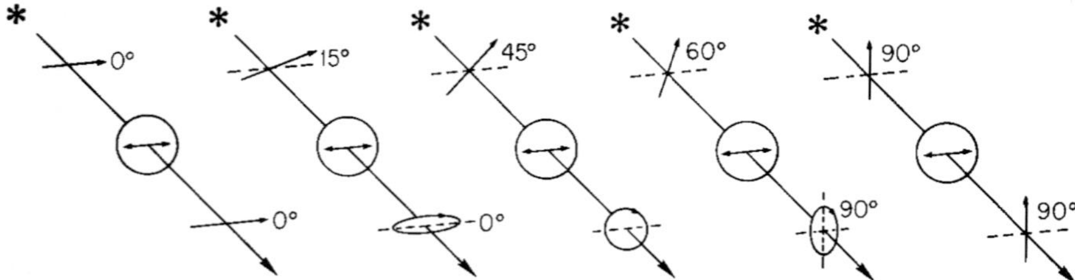

FIGURE 11 State of polarization of a light wave after passing through a $\lambda/4$ plate (whose fast axis is indicated by the double arrow) for different azimuths of the incident plane-polarized beam.

ratio of the minor to the major axis of the ellipse. In the next case, $\theta = 15°$ and $\tan\theta = 0.268$, and so the ellipse is long and narrow. When the plane of vibration has rotated to an azimuth of 45°, the emerging beam is right circularly polarized (the same situation as that shown in the second part of Fig. 10). For values of θ between 45 and 90°, the light is again elliptically polarized, this time with the major axis of the ellipse lying along the direction of the slow axis of the $\lambda/4$ plate. The angle shown in the figure is 60°, and $\tan 60° = 1.732$, so that b/a (referred to the fast axis) is greater than unity. When θ increases to 90°, the plane of vibration coincides with the slow axis and the transmitted light is again plane-polarized. As θ continues to increase, the transmitted patterns repeat those already described and are symmetric about the slow axis, but the direction of rotation in the ellipse changes from right-handed to left-handed, so that left-circularly polarized light is produced when $\theta = 135°$.

The definition of right- and left-circularly polarized light should be clear from Figs. 10 and 11. When the rotation is *clockwise* with the observer looking *opposite to the direction of propagation,* the light is called *right-circularly polarized*; if the rotation is *counterclockwise,* the light is called *left-circularly polarized.*[26] When circularly polarized light is reflected from a mirror, the direction of propagation is reversed, so that the sense of the circular polarization changes; i.e., left-circularly polarized light changes on reflection into right-circularly polarized light and vice versa. Therefore, in experiments involving magnetic fields in which the sense of the circularly polarized light is important,[27,28] it is important to know which kind one started with and how many mirror reflections occurred in the rest of the light path. Cyclotron resonance experiments can sometimes be used to determine the sense of the circular polarization.[28] Another method utilizing a polarizer and $\lambda/4$ plate has been described by Wood.[29]

The behavior of a half-wave plate in a beam of plane-polarized light is completely different from that of a quarter-wave plate; the transmitted light is always plane-polarized. If the incident plane of vibration is at an azimuth θ with respect to the fast axis of the $\lambda/2$ plate, the transmitted beam will be rotated through an angle 2θ relative to the azimuth of the incident beam. The case showing $\theta = 45°$ where the phase of vibration is rotated through 90° is illustrated in the fourth part of Fig. 10. In this situation the extraordinary beam is retarded by half a wavelength relative to the ordinary beam (for a positive birefringent material), hence the name, half-wave plate. If the polarizer is fixed and the $\lambda/2$ plate is rotated (or vice versa), the plane of vibration of the transmitted beam will rotate at twice the frequency of rotation of the $\lambda/2$ plate.

Quarter-wave plates are useful for analyzing all kinds of polarized light. In addition, they are widely employed in experiments using polarized light, e.g., measurements of the thickness and refractive index of thin films by ellipsometry or measurements of optical rotary dispersion, circular dichroism, or strain birefringence. Polarizing microscopes, interference microscopes, and petrographic microscopes are usually equipped with $\lambda/4$ plates. In some applications the $\lambda/4$ plate is needed only to produce circularly polarized light, e.g., for optical pumping in some laser experiments, or to convert a partially polarized light source into one which appears unpolarized, i.e., has equal amplitudes of vibration in all azimuths. For these and similar applications, one can sometimes use a circular polarizer which does not have all the other properties of a $\lambda/4$ plate (see Pars. 73 to 76 in Ref. 1).

The customary application for a $\lambda/2$ plate is to rotate the plane of polarization through an angle of 90°. In other applications the angle of rotation can be variable. Automatic setting ellipsometers or polarimeters sometimes employ rotating $\lambda/2$ plates in which the azimuth of the transmitted beam rotates at twice the frequency of the $\lambda/2$ plate.

7. Matrix Methods for Computing Polarization In dealing with problems involving polarized light, it is often necessary to determine the effect of various types of polarizers (linear, circular, elliptical, etc.), rotators, retardation plates, and other polarization-sensitive devices on the state of polarization of a light beam. The Poincaré sphere construction is helpful for giving a qualitative understanding of the problem; for quantitative calculations, one of several forms of matrix calculus can be used. The matrix methods are based on the fact that the effect of a polarizer or retarder is to perform a linear transformation (represented by a matrix) on the vector representation of a polarized light beam. The advantage of these methods over conventional techniques is that problems are reduced to simple matrix operations; thus since one does not have to think through the physics of every problem, the probability of making an error is greatly reduced. The most common forms of matrix calculus are the Mueller calculus and the Jones calculus, but the coherency-matrix formulation is also gaining popularity for dealing with problems involving partially polarized light. We give here a brief description of the Poincaré sphere and the different matrix methods, indicating how they are used, the different types of problems for which they are helpful, and where complete descriptions of each may be found.

The *Poincaré sphere* is a useful device for visualizing the effects of polarizers and retarders on a beam of polarized light. The various states of polarization are represented on the sphere as follows. The equator represents various forms of linear polarization, the poles represent right- and left-circular polarization,[†] and other points on the sphere represent elliptically polarized light. Every point on the sphere corresponds to a different polarization form. The radius of the sphere indicates the intensity of the light beam (which is usually assumed to be unity). The effects of polarizers and retarders are determined by appropriate displacements on the sphere. Partially polarized light or absorption may be dealt with approximately by ignoring the intensity factor, since one is generally interested only in the state of polarization; however, the construction is most useful when dealing with nonabsorbing materials. Good introductory descriptions of the Poincaré sphere, including references, can be found in *Polarized Light* by Shurcliff,[30‡] *Ellipsometry and Polarized Light* by Azzam and Bashara,[31] and *Polarized Light in Optics and Spectroscopy* by Kliger, Lewis, and Randall;[32§] illustrative examples and problems are given in Sutton and Panati.[33] More comprehensive treatments are given by Ramachandran and Ramaseshan[34§] and Jerrard[35§] and include numerous examples of applications to various types of problems. The new book *Polarized Light, Fundamentals and Applications* by Collett[36] has a comprehensive 35-page chapter on the mathematical aspects of the Poincaré sphere; this material can be best understood after reading some of the introductory descriptions of the Poincaré sphere. The main advantage of the Poincaré sphere, like other graphical methods, is to reveal by essentially a physical argument which terms in exceedingly complex equations are negligible or can be made negligible by modifying the experiment. It is characteristic of problems in polarized light that the trigonometric equations are opaque to inspection and yield useful results only after exact calculation with the aid of a computer or after complex manipulation and rather clever trigonometric identities. The Poincaré sphere thus serves as a guide to the physical intrepretation of otherwise obscure polarization phenomena. It can be used for solving problems involving retarders or combinations of retarders,[30,32,36–39] compensators, half-shade devices, and depolarizers,[34] and it has also been applied to ellipsometric problems[40] and stress-optical measurements.[41]

The Poincaré sphere is based on the Stokes vectors, which are sometimes designated S_0, S_1, S_2, and S_3. The physical interpretation of the vectors is as follows. S_0 is the intensity of the light beam, corresponding to the radius of the Poincaré sphere. S_1 is the difference in intensities between the horizontal and vertical polarization components of the beam; when S_1 is positive, the preference is for horizontal polarization, and when it is negative, the

† Right-circularly polarized light is defined as a *clockwise* rotation of the electric vector when the observer is looking *against the direction the wave is traveling.*

‡ Schurcliff and Kliger, Lewis, and Randall have the S_3 axis pointing down, so that the upper pole represents left-circular polarization. The more logical convention, followed by most others, is for the upper pole to represent right-circular polarization.

§ The notation is similar to that used by Schurcliff,[30] with the upper pole representing left-circular polarization.

preference is for vertical polarization.† S_2 indicates preference for +45° or −45° polarization, depending upon whether it is positive or negative, and S_3 gives the preference for right or left circular polarization. The Stokes vectors S_1, S_2, and S_3 are simply the three cartesian coordinates of a point on the Poincaré sphere: S_1 and S_2 are perpendicular to each other in the equatorial plane, and S_3 points toward the north pole of the sphere.‡ Thus, any state of polarization of a light beam can be specified by these three Stokes vectors. The intensity vector S_0 is related to the other three by the relation $S_0^2 = S_1^2 + S_2^2 + S_3^2$ when the beam is completely polarized. If the beam is partially polarized, $S_0^2 > S_1^2 + S_2^2 + S_3^2$. Good introductory material on Stokes vectors is given by Shurcliff,[30] Azzam and Bashara,[31] Kliger *et al.*,[32] Sutton and Panati,[33] and Walker.[42] A comprehensive discussion of the Stokes vectors has been given by Collett.[36] Rigorous definitions of the simple vectors and those for partially coherent light can be found in Born and Wolf;[43] other authors are cited by Shurcliff[30] and Collett.[36] Stokes vectors are generally used in conjunction with the Mueller calculus, and some examples of applications will be given there. We note here that Budde[44] has demonstrated a method for experimentally determining the Stokes vectors and other polarization parameters from a Fourier analysis of measured quantities. Ioshpa and Obridko[45] have proposed a photoelectric method for simultaneously and independently measuring the four Stokes parameters. Collett[46] has developed a method for measuring the four Stokes vectors using a single circular polarizer. Azzam and coworkers[47–51] have built, tested, analyzed, and calibrated a four-detector photopolarimeter for measuring normalized Stokes vectors of a large number of polarization states, and have given a physical meaning to the rows and columns in the instrument matrix. Other methods for measuring Stokes parameters are discussed by Collett.[36] Hauge[52] has surveyed different types of methods for completely determining the state of polarization of a light beam using combinations of Stokes vectors.

The matrix methods for solving problems involving polarized light have certain properties in common. All use some type of representation for the original light beam (assumed to be a plane wave traveling in a given direction) that uniquely describes its state of polarization. Generally the beam is completely polarized, but for some of the matrix methods it can also be unpolarized or partially polarized or its phase may be specified. The beam encounters one or more devices which change its state of polarization. These are called *instruments* and are represented by appropriate matrices. After the instruments operate on the light beam, it emerges as an outgoing plane wave in an altered state of polarization. The basic problem for all the methods is to find a suitable representation for the incident plane wave (usually a two- or four-component column vector), and the correct matrices (2×2 or 4×4) to represent the instruments. Once the problem is set up, one can perform the appropriate matrix operations to obtain a representation for the outgoing plane wave. Its properties are interpreted in the same way as the properties of the incident plane wave.

An introduction to the Jones and Mueller calculus is given by Shurcliff,[30] Azzam and Bashara,[31] and Kliger *et al.*,[32] and an excellent systematic and rigorous discussion of all the matrix methods has been given by O'Neill[53] and Collett.[36] All references contain tables of vectors for the various types of polarized beams and tables of instrument matrices. More complete tables are given by Sutton and Panati.[33] In the Mueller calculus the beam is represented by the four-component Stokes vector, written as a column vector. This vector has all real elements and gives information about *intensity* properties of the beam. Thus it is not able to handle problems involving phase changes or combinations of two beams that are coherent. The instrument matrix is a 4×4 matrix with all real elements. In the Jones calculus, the Jones vector is a two-component column vector that generally has complex elements. It contains information about the *amplitude* properties of the beam and hence is

† Some authors dealing with light scattering from aerosols define S_1 as positive when the preference is for vertical polarization.

‡ See Shurcliff and Kliger, Lewis, and Randall footnote, p. 5.26.

well suited for handling coherency problems. However, it cannot handle problems involving depolarization, as the Mueller calculus can. The Jones instrument matrix is a 2×2 matrix whose elements are generally complex.

Shurcliff[30] has noted some additional differences between Jones calculus and Mueller calculus. The Jones calculus is well suited to problems involving a large number of similar devices arranged in series in a regular manner and permits an investigator to arrive at an answer expressed explicitly in terms of the number of such devices. The Mueller calculus is not suited for this type of problem. The Jones instrument matrix of a train of transparent or absorbing nondepolarizing polarizers and retarders contains no redundant information. The matrix contains four elements each of which has two parts, so that there are a total of eight constants, none of which is a function of any other. The Mueller instrument matrix of such a train contains much redundancy; there are 16 constants but only 7 of them are independent.

In order to handle problems involving partially coherent polarized light, coherency-matrix formalism has been developed. In this system the beam is represented by a 4×4 matrix called a *coherency* or *density matrix,* which is the time average of the product of the Jones vector with its hermitian conjugate. The instrument matrices are the same as those used in the Jones calculus. O'Neill[53] and Born and Wolf[43] have good basic descriptions of coherency-matrix formalism; later extensions of the theory are given by Marathay.[54,55]

There have been some modifications of the various matrix methods. Priebe[56] has introduced an operational notation for the Mueller matrices that facilitates the analysis by simplifying the functional description of a train of optical components. Collins and Steele[57] have suggested a modification of the Jones calculus in which the light vector is expressed as the sum of two circularly polarized (rather than linearly polarized) components. Schmieder[58] has given a unified treatment of Jones calculus and Mueller calculus including the coherency matrix and has shown that if the Stokes parameters are ordered in a different way from that customarily used, familiar relationships are preserved and the rotation matrix looks like a rotation matrix rather than like a rearranged one. Tewarson[59] presents a generalized reciprocity equation expressing an algebraic relationship between the parameters of an optical system and its reciprocal system and has verified the equation for both plane-polarized and circularly polarized light beams. Since his equation follows from the reciprocity law in the Mueller calculus, that law is verified also. Cernosek[60] presents a simple geometric method based on the properties of quaternions to give a quick, quantitative analysis of the effect of any combination of linear retarders and rotators on the state of polarization of a system.

Among the applications of Mueller calculus and Jones calculus to problems involving polarized light, McCrackin[61] has used both matrix methods to analyze instrumental errors in ellipsometry, and Hellerstein[62] has used Mueller calculus to study the passage of linearly, circularly, and elliptically polarized light through a Sénarmont polariscope. Azzam and Bashara[63] have used Jones calculus to give a unified analysis of errors in ellipsometry, including effects of birefringence in cell windows, imperfect components, and incorrect azimuth angles. Azzam[64] also describes a simple photopolarimeter with rotating polarizer and analyzer for measuring Jones and Mueller matrices.

REFERENCES

1. H. E. Bennett and J. M. Bennett, "Polarization," in *Handbook of Optics,* W. G. Driscoll and W. Vaughan, eds. (McGraw-Hill, New York, 1978), pp. 10-1 to 10-164.
2. E. Collett, private communication, 1992.
3. R. H. Muller, Proc. Symp Recent Dev. Ellipsometry, *Surf. Sci.* **16,** 14–33 (1969).
4. W. R. Hunter, *J. Opt. Soc. Am.* **55,** 1197–1204 (1965).
5. W. E. Williams, *Applications of Interferometry,* 4th ed. (Wiley, New York, 1950), pp. 77–78.

6. J. A. Berning and P. H. Berning, *J. Opt. Soc. Am.* **50,** 813–815 (1960).
7. P. H. Berning, "Theory and Calculations of Optical Thin Films," in *Physics of Thin Films,* vol. 1, G. Hass ed. (Academic Press, New York, 1963), pp. 78–81.
8. F. Abelès, *Prog. Opt.* **2,** 251–288 (1963).
9. H. B. Holl, *The Reflection of Electromagnetic Radiation,* vols. 1 and 2 (U.S. Army Missile Command, Redstone Arsenal, Huntsville, Alabama, 1963), Report RF-63-4.
10. H. E. Bennett and J. M. Bennett, "Precision Measurements in Thin Film Optics," in *Physics of Thin Films,* vol. 4, G. Hass and R. E. Thun, eds. (Academic Press, New York, 1967), pp. 69–78.
11. R. C. Jones, *J. Opt. Soc. Am.* **52,** 747–752 (1962).
12. G. R. Bird and W. A. Shurcliff, *J. Opt. Soc. Am.* **49,** 235–237 (1959).
13. C. D. West and R. C. Jones, *J. Opt. Soc. Am.* **41,** 976–982 (1951).
14. I. Simon, *J. Opt. Soc. Am.* **41,** 336–345 (1951).
15. E. Charney, *J. Opt. Soc. Am.* **45,** 980–983 (1955).
16. D. P. Gonatas, X. D. Wu, G. Novak, and R. H. Hildebrand, *Appl. Opt.* **28,** 1000–1006 (1989).
17. P. L. Wizinowich, *Opt. Eng.* **28,** 157–159 (1989).
18. R. C. Jones, *J. Opt. Soc. Am.* **46,** 528–533 (1956).
19. K. D. Mielenz and K. L. Eckerle, *Appl. Opt.* **11,** 594–603 (1972).
20. R. M. A. Azzam, *Appl. Opt.* **26,** 2847–2850 (1987).
21. M. V. R. K. Murty and R. P. Shukla, *Appl. Opt.* **22,** 1094–1098 (1983).
22. F. Abelès, *C. R. Acad. Sci.* **230,** 1942–1943 (1950).
23. M. F. de la Provostaye and P. Desains, *Ann. Chim. Phys.,* ser. 3, **30,** 158 (1850).
24. H. E. Bennett, J. M. Bennett, and M. R. Nagel, *J. Opt. Soc. Am.* **51,** 237 (1961).
25. L. B. Tuckerman, *J. Opt. Soc. Am.* **37,** 818–825 (1947).
26. F. A. Jenkins and H. E. White, *Fundamentals of Optics,* 3d ed. (McGraw-Hill, New York, 1957), p. 229.
27. E. D. Palik, *Appl. Opt.* **2,** 527–539 (1963).
28. P. L. Richards and G. E. Smith, *Rev. Sci. Instrum.* **35,** 1535–1537 (1964).
29. R. W. Wood, *Physical Optics,* 3d ed. (Macmillan, New York, 1934), p. 360.
30. W. A. Shurcliff, *Polarized Light* (Harvard University Press, Cambridge, Mass., 1962), pp. 15–29, 95–98, 109–123.
31. R. M. A. Azzam and N. M. Bashara, *Ellipsometry and Polarized Light* (North-Holland Publishing Company, Amsterdam, 1977), Chapters 1 and 2.
32. D. S. Kliger, J. W. Lewis, and C. E. Randall, *Polarized Light in Optics and Spectroscopy* (Academic Press, Inc., San Diego, 1990), Chapters 1 to 5.
33. A. M. Sutton and C. F. Panati, Lecture notes, Modern Optics, RCA Institute Clark, N.J. (1969).
34. G. N. Ramachandran and S. Ramaseshan, "Crystal Optics," in *Handbuch der Physik,* S. Flügge, ed., vol. 25/1 (Springer, Berlin, 1961), pp. 1–54.
35. H. G. Jerrard, *J. Opt. Soc. Am.* **44,** 634–640 (1954).
36. E. Collett, *Polarized Light: Fundamentals and Applications* (Marcel Dekker, Inc., New York, 1992).
37. C. J. Koester, *J. Opt. Soc. Am.* **49,** 405–409 (1959).
38. S. Pancharatnam, *Proc. Indian Acad. Sci.* **A41,** 130–136 (1955).
39. S. Pancharatnam, *Proc. Indian Acad. Sci.* **A41,** 137–144 (1955).
40. F. L. McCrackin, E. Passaglia, R. R. Stromberg, and H. L. Steinberg, *J. Res. Natl. Bur. Stand (U.S.)* **67A,** 363–377 (1963).
41. A. J. Michael, *J. Opt. Soc. Am.* **58,** 889–894 (1968).
42. M. J. Walker, *Am. J. Phys.* **22,** 170–174 (1954).

43. M. Born and E. Wolf, *Principles of Optics,* 6th ed. (Pergamon Press, New York, 1980), pp. 30–32, 544–555.
44. W. Budde, *Appl. Opt.* **1,** 201–205 (1962).
45. B. A. Ioshpa and V. N. Obridko, *Opt. Spectrosc.* (*USSR*) **15,** 60–62 (1963).
46. E. Collett, *Opt. Commun.* **52,** 77–80 (1984).
47. R. M. A. Azzam, *Opt. Lett.* **10,** 110–112 (1985).
48. R. M. A. Azzam, I. M. Elminyawi, and A. M. El-Saba, *J. Opt. Soc. Am.* **A5,** 681–689 (1988).
49. R. M. A. Azzam, E. Masetti, I. M. Elminyawi, and F. G. Grosz, *Rev. Sci. Instr.* **59,** 84–88 (1988).
50. R. M. A. Azzam and A. G. Lopez, *J. Opt. Soc. Am.* **A6,** 1513–1521 (1989).
51. R. M. A. Azzam, *J. Opt. Soc. Am.* **A7,** 87–91 (1990).
52. P. S. Hauge, *Proc. Soc. Photo-Opt. Instrum. Eng.* **88,** 3–10 (1976).
53. E. L. O'Neill, *Introduction to Statistical Optics* (Addison-Wesley, Reading, Mass, 1963), pp. 133–156.
54. A. S. Marathay, *J. Opt. Soc. Am.* **55,** 969–980 (1965).
55. A. S. Marathay, *J. Opt. Soc. Am.* **56,** 619–623 (1966).
56. J. R. Priebe, *J. Opt. Soc. Am.* **59,** 176–180 (1969).
57. J. G. Collins and W. H. Steele, *J. Opt. Soc. Am.* **52,** 339 (1962).
58. R. W. Schmieder, *J. Opt. Soc. Am.* **59,** 297–302 (1969).
59. S. P. Tewarson, *Indian J. Phys.* **40,** 281–293, 562–566 (1966).
60. J. Cernosek, *J. Opt. Soc. Am.* **61,** 324–327 (1971).
61. F. L. McCrackin, *J. Opt. Soc. Am.* **60,** 57–63 (1970).
62. D. Hellerstein, *Appl. Opt.* **2,** 801–805 (1963).
63. R. M. A. Azzam and N. M. Bashara, *J. Opt. Soc. Am.* **61,** 600–607, 773–776, 1380–1391 (1971).
64. R. M. A. Azzam, *Opt. Commun.* **25,** 137–140 (1978).

CHAPTER 6
SCATTERING BY PARTICLES

Craig F. Bohren
Pennsylvania State University
University Park, Pennsylvania

6.1 GLOSSARY

a	radius
a_n, b_n	scattering coefficients
C	cross section
D_n	logarithmic derivative, $d/d\rho[\ln \psi_n(\rho)]$
E	Electric field strength
$\mathbf{e}_x$	unit vector in the x direction
f	Nv
G	projected particle area
h	thickness
I	irradiance
I, Q, U, V	Stokes parameters
j	running index
k	$2\pi/\lambda$
k	imaginary part of the refractive index
m	relative complex refractive index
N	number
n	running index
P_n	associated Legendre functions of the first kind
p	phase function, normalized differential scattering cross section
Q	efficiencies or efficiency factors
r	distance
S	element of the amplitude scattering matrix
v	volume

W	power
X	scattering amplitude
x	size parameter, ka
α	absorption coefficient
θ	angle
λ	wavelength
π_n	$P_n^1/\sin\theta$
τ_n	$dP_n^1/d\theta$
ψ, ξ	Riccati-Bessel functions
Ω	solid angle
ω	radian frequency
$\parallel$	parallel
$\perp$	perpendicular
Re	real part of

Subscripts

abs	absorbed
ext	extinction
sca	scattered

6.2 INTRODUCTION

Light scattering by particles plays starring and supporting roles on a variety of stages: astronomy, cell biology, colloid chemistry, combustion engineering, heat transfer, meteorology, paint technology, solid-state physics—the list is almost endless. The best evidence of the catholicity of scattering by particles is the many journals that publish papers about it.

Scattering by single particles is the subject of monographs by van de Hulst,[1] Deirmendjian,[2] Kerker,[3] Bayvel and Jones,[4] Bohren and Huffman,[5] Barber and Hill[6] and of a collection edited by Kerker.[7] Two similar collections contain papers on scattering by atmospheric particles[8] and by chiral particles[9] (ones not superposable on their mirror images); scattering by chiral particles is also treated by Lakhtakia et al.[10] Papers on scattering by particles are included in collections edited by Gouesbet and Gréhan[11] and by Barber and Chang.[12] Within this *Handbook* scattering by particles is touched upon in chaps. 44 and 45. A grand feast is available for those with the juices to digest it. What follows is a mere snack.

A particle is an aggregation of sufficiently many molecules that it can be described adequately in macroscopic terms (i.e., by constitutive parameters such as permittivity and permeability). It is a more or less well-defined entity unlike, say, a density fluctuation in a gas or a liquid. Single molecules are not particles, even though scattering by them is in some ways similar (for a clear but dated discussion of molecular scattering, see Martin[13]).

Scattering by single particles is discussed mostly in the wave language of light, although multiple scattering by incoherent arrays of many particles can be discussed intelligibly in the photon language. The distinction between single and multiple scattering is observed more readily on paper than in laboratories and in nature. Strict single scattering can exist only in a boundless void containing a lone scatterer illuminated by a remote source, although single scattering often is attained to a high degree of approximation. A distinction made less frequently is that between scattering by coherent and by incoherent arrays. In treating scattering by coherent arrays, the wave nature of light cannot be ignored: phases *must* be taken into account. But in treating scattering by incoherent arrays, phases *may* be ignored.

Pure water is a coherent array of water molecules; a cloud is an incoherent array of water droplets. In neither of these arrays is multiple scattering negligible, although the theories used to describe them may not explicitly invoke it.

The distinction between incoherent and coherent arrays is not absolute. Although a cloud of water droplets is usually considered to be an incoherent array, it is not such an array for scattering in the forward direction. And although most of the light scattered by pure water is accounted for by the laws of specular reflection and refraction, it also scatters light—weakly yet measurably—in directions not accounted for by these laws.[13]

A single particle is itself a coherent array of many molecules, but can be part of an incoherent array of many particles, scattering collectively in such a way that the phases of the waves scattered by each one individually are washed out. Although this section is devoted to single scattering, it must be kept in mind that multiple scattering is not always negligible and is not just scaled-up single scattering. Multiple scattering gives rise to phenomena inexplicable by single-scattering arguments.[14]

6.3 SCATTERING: AN OVERVIEW

Why is light scattered? No single answer will be satisfactory to everyone, yet because scattering by particles has been amenable to treatment mostly by classical electromagnetic theory, our answer lies within this theory.

Although palpable matter may appear to be continuous and is often electrically neutral, it is composed of discrete electric charges. Light is an oscillating electromagnetic field, which can excite the charges in matter to oscillate. Oscillating charges radiate electromagnetic waves, a fundamental property of such charges with its origins in the finite speed of light. These radiated electromagnetic waves are scattered waves, waves excited or driven by a source external to the scatterer: an incident wave from the source excites secondary waves from the scatterer; the superposition of all these waves is what is observed. If the frequency of the secondary waves is (approximately) that of the source, these waves are said to be *elastically* scattered (the term *coherently* scattered is also used).

Scientific knowledge grows like the accumulation of bric-a-brac in a vast and disorderly closet in a house kept by a sloven. Few are the attempts at ridding the closet of rusty or obsolete gear, at throwing out redundant equipment, at putting things in order. For example, spurious distinctions are still made between reflection, refraction, scattering, interference, and diffraction despite centuries of accumulated knowledge about the nature of light and matter.

Countless students have been told that specular reflection is localized at smooth surfaces, that photons somehow rebound from them. Yet this interpretation is shaky given that even the smoothest surface attainable is, on the scale of a photon, as wrinkled as the back of a cowboy's neck. Photons conceived of as tiny balls would be scattered in all directions by such a surface, for which it is difficult even to define what is meant by an angle of incidence.

Why do we think of reflection occurring at surfaces rather than because of them whereas we usually do not think of scattering by particles in this way? One reason is that we can see the surfaces of mirrors and ponds. Another is the dead hand of traditional approaches to the laws of specular reflection and refraction.

The empirical approach arrives at these laws as purely geometrical summaries of what is observed—and a discreet silence is maintained about underlying causes. The second approach is by way of continuum electromagnetic theory: reflected and refracted fields satisfy the Maxwell equations. Perhaps because this approach, which also yields the Fresnel formulas, entails the solution of a boundary-value problem, reflected and refracted fields are mistakenly thought to originate from boundaries rather than from all the illuminated matter they enclose. This second approach comes to grips with the nature of light but not of matter, which is treated as continuous. The third approach is to explicitly recognize that reflection and refraction are consequences of scattering by discrete matter. Although this scattering interpretation was developed by Paul Ewald and Carl Wilhelm Oseen early in this century, it has diffused with glacial slowness. According to this interpretation, when the optically smooth interface between optically homogeneous dissimilar media is illuminated, the reflected and refracted waves are superpositions of vast numbers of secondary waves excited by the incident wave. Thus reflected and refracted light is, at heart, an interference pattern of scattered light. Doyle[15] showed that although the Fresnel equations are obtained from macroscopic electromagnetic theory, they can be dissected to reveal their microscopic underpinnings.

No optics textbook would be complete without sections on interference and diffraction, a distinction without a difference: there is no diffraction without interference. Moreover, diffraction is encumbered with many meanings. Van de Hulst[1] lists several: small deviations from rectilinear propagation; wave motion in the presence of an obstacle; scattering by a flat particle such as a disk; an integral relation for a function satisfying the wave equation. To these may be added scattering near the forward direction and by a periodic array.

Van de Hulst stops short of pointing out that a term with so many meanings has no meaning. Even the etymology of diffraction is of little help: it comes from a Latin root meaning to break.

There is no fundamental distinction between diffraction and scattering. Born and Wolf[16] refer to scattering by a sphere as diffraction by a sphere. I leave it as a penance for the reader to devise an experiment to determine whether a sphere scatters light or diffracts it.

The only meaningful distinction is that between approximate theories. Diffraction theories obtain answers at the expense of obscuring the physics of the interaction of light with matter. For example, an illuminated slit in an opaque screen may be the mathematical source but it is not the physical source of a diffraction pattern. Only the screen can give rise to secondary waves that yield the observed pattern. Yet generations of students have been taught that empty space is the source of the radiation diffracted by a slit. To befuddle them even more, they also have been taught that two slits give an interference pattern whereas one slit gives a diffraction pattern.

If we can construct a mathematical theory (diffraction theory) that enables us to avoid having to explicitly consider the nature of matter, all to the good. But this mathematical theory and its quantitative successes should not blind us to the fact that we are pretending. Sometimes this pretense cannot be maintained, and when this happens a finger is mistakenly pointed at "anomalies," whereas what is truly anomalous is that a theory so devoid of physical content could ever give adequate results.

A distinction must be made between a physical process and the superficially different theories used to describe it. There is no fundamental difference between specular reflection and refraction by films, diffraction by slits, and scattering by particles. All are consequences of light interacting with matter. They differ only in their geometries and the approximate theories that are sufficient for their quantitative description. The different terms used to describe them are encrustations deposited during the slow evolution of our understanding of light and matter.

6.4 SCATTERING BY PARTICLES: BASIC CONCEPTS AND TERMINOLOGY

A single particle can be considered a collection of tiny dipolar antennas driven to radiate (scatter) by an incident oscillating electric field. Scattering by such a coherent array of antennas depends on its size and shape, the observation angle (scattering angle), the response of the individual antennas (composition), and the polarization state and frequency of the incident wave. Geometry, composition, and the properties of the illumination are the determinants of scattering by particles.

Perhaps the only real difference between optics and electrical engineering is that electrical engineers can measure amplitudes and phases of fields whereas the primary observable quantity in optics is the time-averaged Poynting vector (irradiance), an amplitude squared. Several secondary observables are inferred from measurements of this primary observable. Consider, for example, a single particle illuminated by a beam with irradiance I_i. The total power scattered by this particle is W_{sca}. Within the realm of linear optics, the scattered power is proportional to the incident irradiance. This proportionality can be transformed into an equality by means of a factor C_{sca}:

$$W_{\text{sca}} = C_{\text{sca}} I_i \tag{1}$$

For Eq. (1) to be dimensionally homogeneous C_{sca} must have the dimensions of area, hence C_{sca} has acquired the name *scattering cross section*.

Particles absorb as well as scatter electromagnetic radiation. The rate of absorption W_{abs} by an illuminated particle, like scattered power, is proportional to the incident irradiance:

$$W_{\text{abs}} = C_{\text{abs}} I_i \tag{2}$$

where C_{abs} is the *absorption cross section*. The sum of these cross sections is the *extinction cross section*:

$$C_{\text{ext}} = C_{\text{sca}} + C_{\text{abs}} \tag{3}$$

Implicit in these definitions of cross sections is the assumption that the irradiance of the incident light is constant over lateral dimensions large compared with the size of the illuminated particle. This condition is necessarily satisfied by a plane wave infinite in lateral extent, which, much more often than not, is the source of illumination in light-scattering theories.

The extinction cross section can be determined (in principle) by measuring transmission by a slab populated by N identical particles per unit volume. Provided that multiple scattering is negligible, the incident and transmitted irradiances I_i and I_t are related by

$$I_t = I_i e^{-NC_{\text{ext}}h} \tag{4}$$

where h is the thickness of the slab. Only the sum of scattering and absorption can be obtained from transmission measurements. To separate extinction into its components requires additional measurements.

Equation (4) requires that all particles in the slab be identical. They are different if they differ in size, shape, composition, or orientation (incident beams are different if they differ in wavelength or polarization state). Equation (4) is generalized to a distribution of particles by replacing NC_{ext} with

$$\sum_j N_j C_{ext,j} \tag{5}$$

where j denotes all parameters distinguishing one particle from another.

Instead of cross sections, normalized cross sections called *efficiencies* or *efficiency factors*, Q_{sca}, Q_{abs}, and Q_{ext}, often are presented. The normalizing factor is the particle's area G projected onto a plane perpendicular to the incident beam. No significance should be attached to efficiency used as shorthand for normalized cross section. The normalization factor is arbitrary. It could just as well be the total area of the particle or, to honor Lord Rayleigh, the area of his thumbnail.

Proper efficiencies ought to be less than unity, whereas efficiencies for scattering, absorption, and extinction are not so constrained. Moreover, some particles—soot aggregates, for example—do not have well-defined cross-sectional areas. Such particles have cross sections for scattering and absorption but the corresponding efficiencies are nebulous.

If any quantity deserves the designation efficiency it is the cross section per particle volume v. Equation (4) can be rewritten to display this:

$$I_t = I_i e^{-fh(C_{ext}/v)} \tag{6}$$

where $f = Nv$ is the total volume of particles per unit slab volume. For a given particle loading, specified by fh (volume of particles per unit slab area), transmission is a minimum when C_{ext}/v is a maximum.

Each way of displaying extinction (or scattering) versus particle size or wavelength of the incident beam tells a different story. This is illustrated in Fig. 1, which shows the scattering cross section, scattering efficiency, and scattering cross section per unit volume of a silicate sphere in air illuminated by visible light. These curves were obtained with van de Hulst's simple *anomalous diffraction* approximation[1] (all that is anomalous about it is that it gives such good results). Each curve yields a different answer to the question, what size particle is most efficient at scattering light? And comparison of Figs. 1*c* and 2 shows that scattering by a particle and specular reflection are similar.

At sufficiently large distances r from a scatterer of bounded extent, the scattered field $\mathbf{E}_s$ decreases inversely with distance and is transverse:

$$\mathbf{E}_s \sim \frac{e^{ik(r-z)}}{-ikr}\mathbf{X}E \qquad (kr \gg 1) \tag{7}$$

where $k = 2\pi/\lambda$ is the wave number of the incident plane harmonic wave $\mathbf{E}_i = \mathbf{e}_x E$, $E = E_0 \exp(ikz)$ propagating along the z axis. The *vector scattering amplitude* is written as $\mathbf{X}$ as a reminder that the incident wave is linearly polarized along the x axis. Here and elsewhere the time-dependent factor $\exp(-i\omega t)$ is omitted.

The extinction cross section is related in a simple way to the scattering amplitude;

$$C_{ext} = \frac{4\pi}{k^2}\operatorname{Re}\{(\mathbf{X}\cdot\mathbf{e}_x)_{\theta=0}\} \tag{8}$$

This remarkable result, often called the *optical theorem*, implies that plane-wave extinction depends only on scattering in the forward direction $\theta = 0$, which seems to contradict the

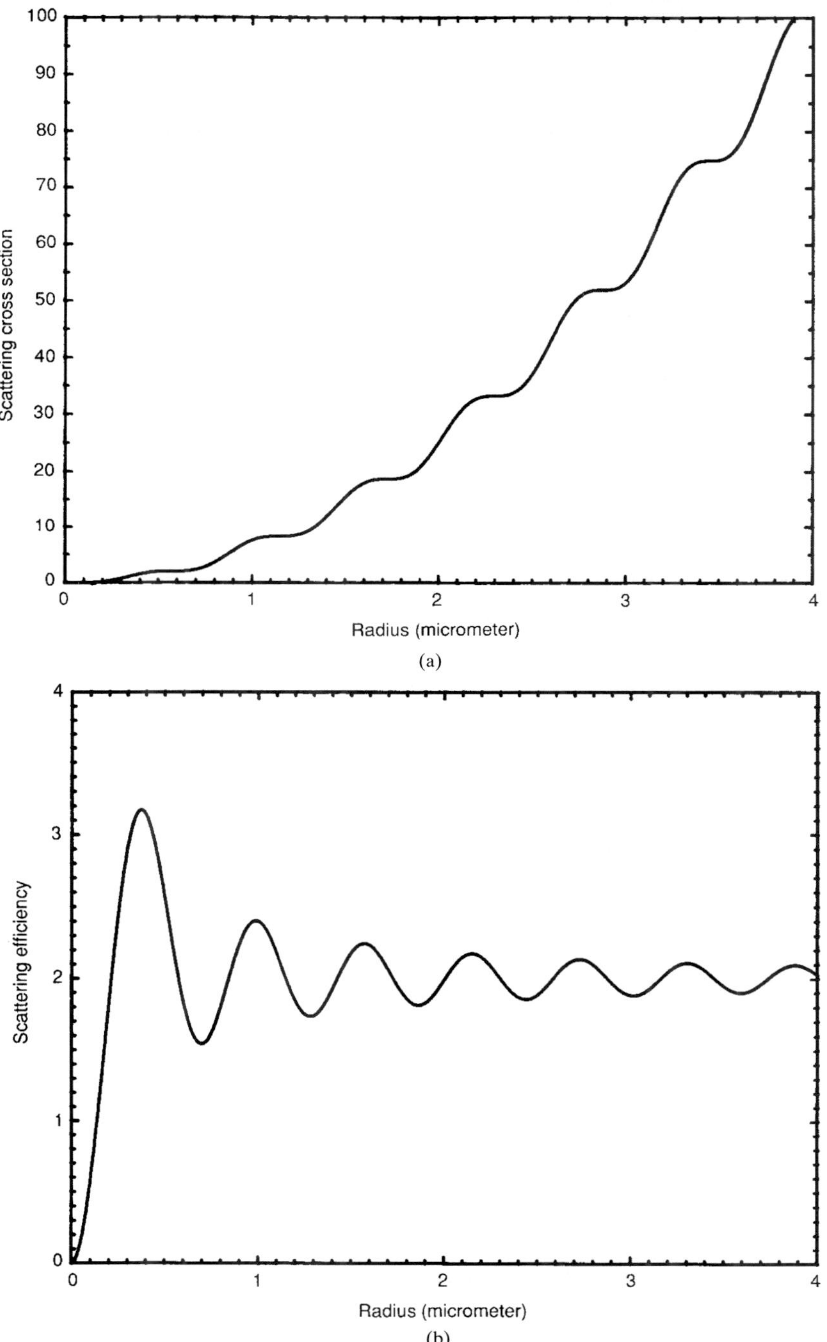

FIGURE 1 Scattering of visible light by a silicate sphere calculated using the anomalous diffraction approximation: (*a*) scattering cross section; (*b*) scattering efficiency (cross section normalized by projected area); (*c*) volumetric scattering cross section (cross section per unit particle volume).

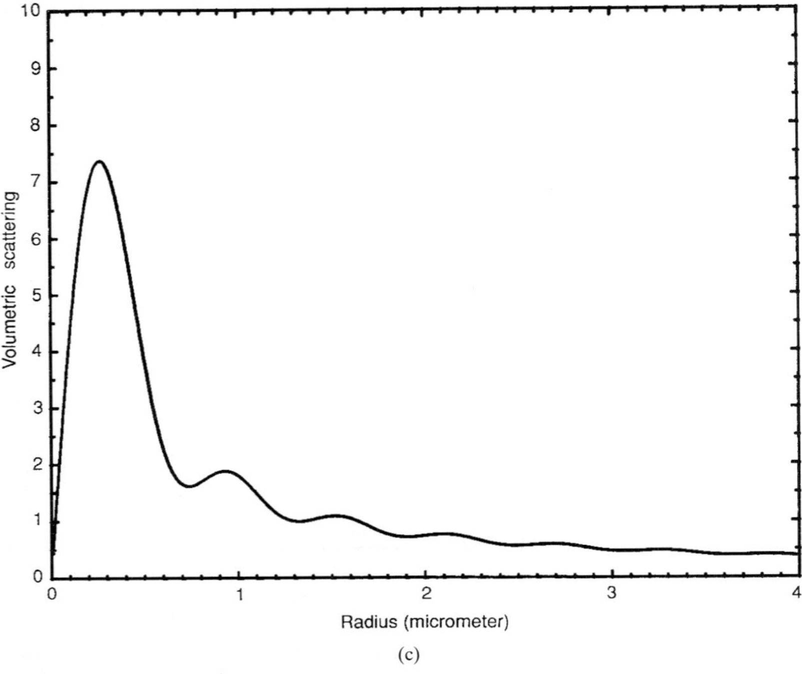

(c)

FIGURE 1 (*Continued*)

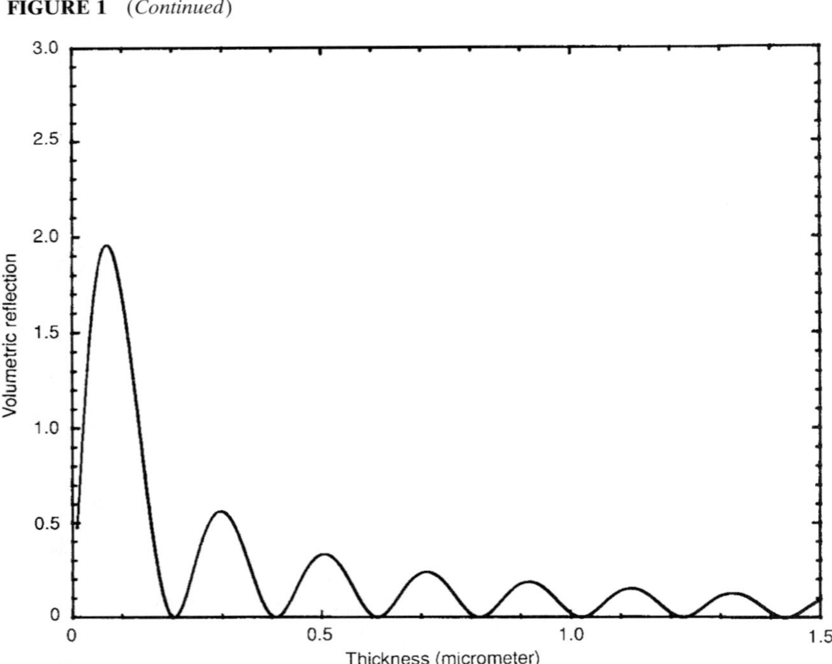

FIGURE 2 Reflected power per unit incident irradiance and unit volume of a silicate slab normally illuminated by visible light (reflectance divided by slab thickness).

interpretation of extinction as the sum of scattering in *all* directions and absorption. Yet extinction has two interpretations, the second manifest in the optical theorem: extinction is interference between incident and forward-scattered waves.

The scattering cross section is also obtained from the vector scattering amplitude by an integration over all directions:

$$C_{\mathrm{sca}} = \int_{4\pi} \frac{|\mathbf{X}|^2}{k^2}\, d\Omega \tag{9}$$

At wavelengths far from strong absorption bands, the scattering cross section of a particle small compared with the wavelength satisfies (approximately)

$$C_{\mathrm{sca}} \propto \frac{v^2}{\lambda^4} \qquad (ka \rightarrow 0) \tag{10}$$

where a is a characteristic linear dimension of the particle. This result was first obtained by Lord Rayleigh in 1871 by dimensional analysis (his paper is included in Ref. 8).

The extinction cross section of a particle large compared with the wavelength approaches the limit

$$C_{\mathrm{ext}} \rightarrow 2G \qquad (ka \rightarrow \infty) \tag{11}$$

The fact that C_{ext} approaches *twice* G instead of G is sometimes called the *extinction paradox.* This alleged paradox arises from the expectation that geometrical optics should become a better approximation as a particle becomes larger. But all particles have edges because of which extinction by them always has a component unaccounted for by geometrical optics. This additional component, however, may not be observed because it is associated with light scattered very near the forward direction and because all detectors have finite acceptance angles. Measured extinction is theoretical extinction reduced by the scattered light collected by the detector.

No particle scatters light equally in all directions; isotropic scatterers exist only in the dreams of inept theorists. The angular dependence of scattering can be specified by the *differential scattering cross section,* written symbolically as $dC_{\mathrm{sca}}/d\Omega$ as a reminder that the *total* scattering cross section is obtained from it by integrating over all directions:

$$C_{\mathrm{sca}} = \int_{4\pi} \frac{dC_{\mathrm{sca}}}{d\Omega}\, d\Omega \tag{12}$$

The normalized differential scattering cross section p

$$p = \frac{1}{C_{\mathrm{sca}}} \frac{dC_{\mathrm{sca}}}{d\Omega} \tag{13}$$

is sometimes called the *phase function.* This coinage of astronomers (after the phases of astronomical bodies) confuses those who are perplexed by phase attached to a quantity from which phase in the usual optical sense is absent. To add to the confusion, the phase function is sometimes normalized to 4π instead of to unity.

A salient characteristic of scattering by particles is strong forward-backward asymmetry. Small metallic particles at far infrared wavelengths provide one of the few examples in which backscattering is larger than forward scattering. Except for very small particles,

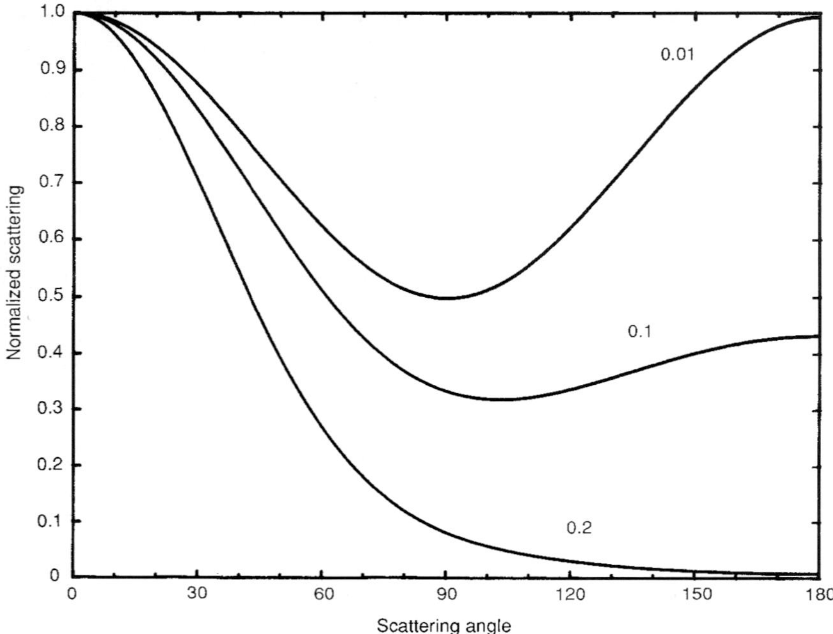

FIGURE 3 Scattering of unpolarized visible light by spheres of radii 0.01, 0.1, and 0.2 μm calculated according to the Rayleigh-Gans approximation.

scattering is peaked in the forward direction; the larger the particle, the sharper the peak. Examples are given in Fig. 3, which shows differential scattering cross sections for unpolarized visible light illuminating spheres of various radii. These curves were obtained using the Rayleigh-Gans approximation,[1,3,5] valid for particles optically similar to the surrounding medium. Forward scattering is much greater than backscattering even for a sphere as small as 0.2 μm.

A simple explanation of forward-backward asymmetry follows from the model of a scatterer as an array of N antennas. If we ignore mutual excitation (the antennas are excited solely by the external source), the total scattered field is the sum of N fields, the phases of which, in general, are different except in the forward direction. Scattering by noninteracting scatterers in this direction—and only in this direction—is in-phase regardless of their separation and the wavelength of the source. Thus as N increases, the scattered irradiance increases more rapidly in the forward direction than in any other direction.

Particles are miniature polarizers and retarders: they scatter differently the orthogonal components into which incident fields can be resolved. Similarly, an optically smooth surface can be both a polarizer and retarder. Just as polarization changes upon reflection are described by decomposing electric fields into components parallel and perpendicular to the plane of incidence, it is convenient to introduce a *scattering plane,* defined by the directions of the incident and scattered waves, for describing scattering by particles.

The incident plane wave is transverse, as is the scattered field at large distances. Thus these fields can be decomposed into two orthogonal components, one parallel, the other perpendicular to the scattering plane. The orthonormal basis vectors are denoted by $\mathbf{e}_\parallel$ and $\mathbf{e}_\perp$ and form a right-handed triad with the direction of propagation $\mathbf{e}_p$ (of either the incident or scattered waves): $\mathbf{e}_\perp \mathbf{x} \mathbf{e}_\parallel = \mathbf{e}_p$. Incident and scattered fields are specified relative

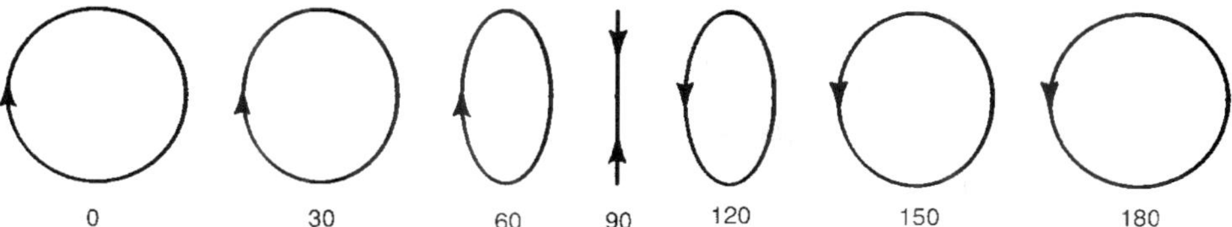

FIGURE 4 Vibration ellipses at various scattering angles for light scattered by a sphere much smaller than the wavelength of the incident right-circularly polarized light.

to different basis vectors. With this decomposition the relation between fields can be written[1,5]

$$\begin{pmatrix} E_{\parallel s} \\ E_{\perp s} \end{pmatrix} = \frac{\mathbf{e}^{ik(r-z)}}{-ikr} \begin{pmatrix} S_2 & S_3 \\ S_4 & S_1 \end{pmatrix} \begin{pmatrix} E_{\parallel i} \\ E_{\perp i} \end{pmatrix} \tag{14}$$

where i and s denote incident and scattered, respectively. The elements of this *amplitude scattering matrix* (or Jones matrix) are complex-valued functions of the scattering direction.

If a single particle is illuminated by completely polarized light, the scattered light is also completely polarized but possibly differently from the incident light, and differently in different directions. An example is given in Fig. 4, which shows vibration ellipses of light scattered by a small sphere. The polarization state of the scattered light varies from right-circular (the polarization state of the incident light) in the forward direction, to linear (perpendicular to the scattering plane) at 90°, to left-circular in the backward direction.

Just as unpolarized light can become partially polarized upon specular reflection, scattering of unpolarized light by particles can yield partially polarized light varying in degree and state of polarization in different directions. Unlike specular reflection, however, an ensemble of particles can transform completely polarized incident light into partially polarized scattered light if all the particles are not identical.

Transformations of polarized light upon scattering by particles are described most conveniently by the *scattering matrix* (or Mueller matrix) relating scattered to incident Stokes parameters:[1,5]

$$\begin{pmatrix} I_s \\ Q_s \\ U_s \\ V_s \end{pmatrix} = \frac{1}{k^2r^2} \begin{pmatrix} S_{11} & S_{12} & S_{13} & S_{14} \\ S_{21} & S_{22} & S_{23} & S_{24} \\ S_{31} & S_{32} & S_{33} & S_{34} \\ S_{41} & S_{42} & S_{43} & S_{44} \end{pmatrix} \begin{pmatrix} I_i \\ Q_i \\ U_i \\ V_i \end{pmatrix} \tag{15}$$

The scattering matrix elements S_{ij} for a single particle are functions of the amplitude scattering matrix elements. Only seven of these elements are independent, corresponding to the four amplitudes and three phase differences of the S_j.

The scattering matrix for an ensemble of particles is the sum of matrices for each of them provided they are separated by sufficiently large random distances. Although all 16 matrix elements for an ensemble can be nonzero and different, symmetry reduces the number of matrix elements. For example, the scattering matrix for a rotationally and mirror symmetric ensemble has the form

$$\begin{pmatrix} S_{11} & S_{12} & 0 & 0 \\ S_{12} & S_{22} & 0 & 0 \\ 0 & 0 & S_{33} & S_{34} \\ 0 & 0 & -S_{34} & S_{44} \end{pmatrix} \tag{16}$$

6.5 SCATTERING BY AN ISOTROPIC, HOMOGENEOUS SPHERE: THE ARCHETYPE

An isotropic, homogeneous sphere is the simplest finite particle, the theory of scattering by which is attached to the name of Gustav Mie.[17] So firm is this attachment that in defiance of logic and history every particle under the sun has been dubbed a "Mie scatterer," and Mie scattering has been promoted from a particular theory of limited applicability to the unearned rank of general scattering process.

Mie was not the first to solve the problem of scattering by an arbitrary sphere.[18] It would be more correct to say that he was the last. He gave his solution in recognizably modern notation and also addressed a real problem: the colors of colloidal gold. For these reasons, his name is attached to the sphere-scattering problem even though he had illustrious predecessors, most notably Lorenz.[19] This is an example in which eponymous recognition has gone to the last discoverer rather than to the first.

Mie scattering is not a physical process; Mie theory is one theory among many. It isn't even exact because it is based on continuum electromagnetic theory and on illumination by a plane wave infinite in lateral extent.

Scattering by a sphere can be determined using various approximations and methods bearing little resemblance to Mie theory: Fraunhofer theory, geometrical optics, anomalous diffraction, coupled-dipole method, T-matrix method, etc. Thus, is a sphere a Mie scatterer or an anomalous diffraction scatterer or a coupled-dipole scatterer? The possibilities are endless. When a physical process can be described by several different theories, it is inadvisable to attach the name of one of them to it.

There is no distinct boundary between so-called Mie and Rayleigh scatterers. Mie theory includes Rayleigh theory, which is a limiting theory strictly applicable only as the size of the particle shrinks to zero. Even for spheres uncritically labeled "Rayleigh spheres," there are always deviations between the Rayleigh and Mie theories. By hobbling one's thinking with a supposed sharp boundary between Rayleigh and Mie scattering, one risks throwing some interesting physics out the window. Whether a particle is a Mie or Rayleigh scatterer is not absolute. A particle may be graduated from Rayleigh to Mie status merely by a change of wavelength of the illumination.

One often encounters statements about Mie scattering by cylinders, spheroids, and other nonspherical particles. Judged historically, these statements are nonsense: Mie never considered any particles other than homogeneous spheres.

Logic would seem to demand that if a particle is a Mie scatterer, then Mie theory can be applied to scattering by it. This fallacious notion has caused and will continue to cause mischief, and is probably the best reason for ceasing to refer to Mie particles or Mie scatterers. Using Mie theory for particles other than spheres is risky, especially for computing scattering toward the backward direction.

More often than not, a better term than Mie or Rayleigh scattering is available. If the scatterers are molecules, molecular scattering is better than Rayleigh scattering (itself an imprecise term):[20] the former term refers to an agent, the latter to a theory. Mie scatterer is just a needlessly aristocratic name for a humble sphere. Wherever Mie scatterer is replaced with sphere, the result is clearer. If qualifications are needed, one can add small or large compared with the wavelength or comparable to the wavelength.

Briefly, the solution to the problem of scattering by an arbitrary homogeneous sphere illuminated by a plane wave can be obtained by expanding the incident, scattered, and internal fields in a series of vector spherical harmonics. The coefficients of these expansion functions are chosen so that the tangential components of the electric and magnetic fields are continuous across the surface of the sphere. Thus this scattering problem is formally identical to reflection and refraction because of interfaces, although the sphere problem is more complicated because the scattered and internal fields are not plane waves.

Observable quantities are expressed in terms of the coefficients a_n and b_n in the expansions of the scattered fields. For example, the cross sections are infinite series:

$$C_{\text{ext}} = \frac{2\pi}{k^2} \sum_{n=1}^{\infty} (2n+1) \operatorname{Re} \{a_n + b_n\} \tag{17}$$

$$C_{\text{sca}} = \frac{2\pi}{k^2} \sum_{n=1}^{\infty} (2n+1)(|a_n|^2 + |b_n|^2) \tag{18}$$

If the permeability of the sphere and its surroundings are the same, the scattering coefficients can be written

$$a_n = \frac{[D_n(mx)/m + n/x]\psi_n(x) - \psi_{n-1}(x)}{[D_n(mx)/m + n/x]\xi_n(x) - \xi_{n-1}(x)} \tag{19}$$

$$b_n = \frac{[mD_n(mx) + n/x]\psi_n(x) - \psi_{n-1}(x)}{[mD_n(mx) + n/x]\xi_n(x) - \xi_{n-1}(x)} \tag{20}$$

ψ_n and ξ_n are Riccati-Bessel functions and the logarithmic derivative

$$D_n(\rho) = \frac{d}{d\rho} \ln \psi_n(\rho) \tag{21}$$

The *size parameter* x is ka, where a is the radius of the sphere and k is the wavenumber of the incident light in the surrounding medium, and m is the complex refractive index of the sphere relative to that of this (nonabsorbing) medium. Equations (19) and (20) are one of the many ways of writing the scattering coefficients, some of which are more suited to computations than others.

During the Great Depression mathematicians were put to work computing tables of trigonometric and other functions. The results of their labors now gather dust in libraries. Today, these tables could be generated more accurately in minutes on a pocket calculator. A similar fate has befallen Mie calculations. Before fast computers were inexpensive, tables of scattering functions for limited ranges of size parameter and refractive index were published. Today, these tables could be generated in minutes on a personal computer. The moral is to give algorithms rather than only tables of results, which are mostly useless except as checks for someone developing and testing algorithms.

These days it is not necessary to reinvent the sphere: documented Mie programs are readily available. The first widely circulated program was published as an IBM report by Dave in 1968, although it no longer seems to be available. A Mie program is given in Ref. 5. Reference 6 includes a diskette containing scattering programs for spheres (and other particles). Wiscombe[21,22] suggested techniques for increasing the speed of computations, as did Lentz,[23] whose method makes use of continued fractions. Wang and van de Hulst[24] recently compared various scattering programs.

The primary tasks in Mie calculations are computing the functions in Eqs. (19) and (20) and summing series like Eqs. (17) and (18). Bessel functions are computed by recurrence. The logarithmic derivative, the argument of which can be complex, is usually computed by downward recurrence. $\psi_n(x)$ and $\xi_n(x)$ can be computed by upward recurrence if one does not generate more orders than are needed for convergence, approximately the size parameter x. When a program with no logical errors falls ill, it often can be cured by promoting variables from single to double precision.

Cross sections versus radius or wavelength convey physical information; efficiencies versus size parameter convey mathematical information. The size parameter is a variable with less physical content than its components, the whole being less than the sum of its parts. Moreover, cross section versus size parameter (or its inverse) is not equivalent to cross section versus wavelength. Except in the fantasy world of naive modelers, refractive indices vary with wavelength, and the Mie coefficients depend on x and m, wavelength being explicit in the first and implicit in the second.

The complex refractive index is written dozens of different ways, one of which is $n + ik$ (despite the risk of confusing the imaginary part with the wavenumber). The quantities n and k are called *optical constants.* But just as the Lord Privy Seal is neither a lord nor a privy nor a seal, optical constants are neither optical nor constant.

Few quantities in optics are more shrouded in myth and misconception than the complex refractive index. The real part for any medium is often defined as the ratio of the velocity of light c in free space to the phase velocity in the medium. This definition, together with notions that nothing can go faster than c, has engendered the widespread misconception that n must be greater than unity. But n can take on any value, even zero. The phase velocity is not the velocity of any palpable object or of any signal, hence is not subject to speed limits enforced by the special relativity police. The least physically relevant property of a refractive index is that it is a ratio of phase velocities. A refractive index is a response function (or better, is simply related to response functions such as permittivity and permeability): it is a macroscopic manifestation of the microscopic response of matter to a periodic driving force.

When we turn to the imaginary part of the refractive index, we enter a ballroom in which common sense is checked at the door. It has been asserted countless times that an imaginary index of, say, 0.01 corresponds to a weakly absorbing medium (at visible and near-visible wavelengths). Such assertions are best exploded by expressing k in a more physically transparent way. The absorption coefficient α is

$$\alpha = \frac{4\pi k}{\lambda} \tag{22}$$

The inverse of α is the *e-folding distance* (or skin depth), the distance over which the irradiance of light propagating in an unbounded medium decreases by a factor of e. At visible wavelengths, the e-folding distance corresponding to $k = 0.01$ is about 5 μm. A thin sliver of such an allegedly weakly absorbing material would be opaque.

When can a particle (or any object) be said to be strongly absorbing? A necessary condition is that $\alpha d \gg 1$, where d is a characteristic linear dimension of the object. But this condition is not sufficient. As k increases, absorption increases—up to a point. As k approaches infinity, the absorption cross section of a particle or the absorptance of a film approaches zero.

One of the most vexing problems in scattering calculations is finding optical constants dispersed throughout dozens of journals. Palik[25] edited a compilation of optical constants for several solids. The optical constants of liquid water over a broad range were compiled by Hale and Querry;[26] Warren[27] published a similar compilation for ice. For other materials, you are on your own. Good hunting!

For small x and $|m|x$, the extinction and scattering efficiencies of a sphere are approximately

$$Q_{\text{ext}} = 4x \,\text{Im}\left\{\frac{m^2 - 1}{m^2 + 2}\right\} \tag{23}$$

$$Q_{\text{sca}} = \frac{8}{3}x^4 \left|\frac{m^2 - 1}{m^2 + 2}\right|^2 \tag{24}$$

These equations are the source of a nameless paradox, which is disinterred from time to time, a corpse never allowed eternal peace. If the sphere is nonabsorbing (m real), Eq. (23) yields a vanishing extinction cross section, whereas Eq. (24) yields a nonvanishing scattering cross section. Yet extinction never can be less than scattering. But note that Eq. (23) is only the first term in the expansion of Q_{ext} in powers of x. The first nonvanishing term in the expansion of Q_{sca} is of order x^4. To be consistent, Q_{ext} and Q_{sca} must be expanded to the same order in x. When this is done, the paradox vanishes.

The amplitude-scattering matrix elements for a sphere are

$$S_1 = \sum_n \frac{2n+1}{n(n+1)} (a_n \pi_n + b_n \tau_n) \tag{25}$$

$$S_2 = \sum_n \frac{2n+1}{n(n+1)} (a_n \tau_n + b_n \pi_n) \tag{26}$$

where the angle-dependent functions are

$$\pi_n = \frac{P_n^1}{\sin\theta} \qquad \tau_n = \frac{dP_n^1}{d\theta} \tag{27}$$

and P_n^1 are the associated Legendre functions of the first kind. The off-diagonal elements of the amplitude-scattering matrix vanish, because of which the scattering matrix is block-diagonal and $S_{12} = S_{21}$, $S_{43} = -S_{34}$, $S_{44} = S_{33}$. Thus, when the incident light is polarized parallel (perpendicular) to the scattering plane, so is the scattered light, a consequence of the sphere's symmetry.

6.6 SCATTERING BY REGULAR PARTICLES

The field scattered by any spherically symmetric particle has the same form as that scattered by a homogeneous, isotropic sphere; only the scattering coefficients are different. One such particle is a uniformly coated sphere. Scattering by a sphere with a single layer was first treated by Aden and Kerker.[28] Extending their analysis to multilayered spheres is straightforward.[29]

New computational problems arise in going from uncoated to coated spheres. The scattering coefficients for both contain spherical Bessel functions, which are bounded only if their arguments are real (no absorption). Thus, for strongly absorbing particles, the arguments of Bessel functions can be so large that their values exceed computational bounds. This does not occur for uncoated spheres because the only quantity in the scattering coefficients with complex argument is the logarithmic derivative, a ratio of Bessel functions computed as an entity instead of by combining numerator and denominator, each of which separately can exceed computational bounds. It is not obvious how to write the scattering coefficients for a coated sphere so that only ratios of possibly large quantities are computed explicitly. For this reason the applicability of the coated-sphere program in Ref. 5 is limited. Toon and Ackerman,[30] however, cast the coated-sphere coefficients in such a way that this limitation seems to have been surmounted.

Bessel functions of large complex argument are not the only trap for the unwary. A coated sphere is two spheres. The size parameter for the outer sphere determines the number of terms required for convergence of series. If the inner sphere is much smaller than the outer, the various Bessel functions appropriate to the inner sphere are computed for indices much greater than needed. More indices are not always better. Beyond a certain number, round-off error can accumulate to yield terms that should make ever smaller contributions to sums but may not.

Scattering by spheres and by infinitely long circular cylinders illuminated normally to their axes are in some ways similar. Spherical Bessel functions in the sphere scattering coefficients correspond to cylindrical Bessel functions in the cylinder scattering coefficients. Unlike a sphere, however, an infinitely long cylinder cannot be enclosed in a finite volume. As a consequence, the field scattered by such a cylinder decreases inversely as the square root of distance r instead of inversely as r (for sufficiently large r).

Infinite particles may be mathematically tractable but they are physically unrealizable. In particular, cross sections for infinite cylinders are infinite. But cross sections per unit length of infinite cylinders are finite. Such cross sections may be applied to a finite cylindrical particle by multiplying its length by the cross section per unit length of the corresponding infinite particle. If the aspect ratio (length/diameter) of the finite particle is sufficiently large, what are vaguely called "end effects" may be negligible. Because no exact theory for a finite cylinder exists, the aspect ratio at which differences between finite and infinite cylinders become negligible is not known with certainty, although the value 10 is bruited about. Nevertheless, there always will be differences between scattering by finite and infinite particles, which may or may not be of concern depending on the application.

A physical difference between scattering by spheres and by cylinders is that cross sections for cylinders depend on the polarization state of the incident plane wave. But normally incident light illuminating an infinite cylinder and polarized perpendicular (parallel) to the plane defined by the incident wave and the cylinder axis excites only scattered light polarized perpendicular (parallel) to the plane defined by the scattered wave and the cylinder axis. Obliquely incident linearly polarized light can, however, excite scattered light having both copolarized and cross-polarized components.

Obliquely illuminated uncoated cylinders pose no special computational problems. Coated cylinders, however, pose the same kinds of problems as coated spheres and are even more difficult to solve. Toon and Ackerman's[30] algorithm for coated spheres is based on the fact that spherical Bessel functions can be expressed in a finite number of terms. Because cylindrical Bessel functions cannot be so expressed, this algorithm cannot be extended to coated cylinders, for which Bessel functions must be computed separately rather than as ratios and can have values beyond computational bounds. Even if such bounds are not exceeded, problems still can arise.

Although Barabás[31] discussed in detail scattering by coated cylinders, Salzman and Bohren[32] found that his computational scheme is unsuitable when absorption is large. They attempted, with only partial success, to write programs for *arbitrary* coated cylinders. Care must be taken in computing Bessel functions. The often-used Miller algorithm can be inadequate for large, complex arguments.

The simplest nonspherical, finite particle is the spheroid, prolate or oblate. Because the scalar wave equation is separable in spheroidal coordinates, scattering by spheroids can be solved in the same way as for spheres and cylinders. The expansion functions are based on spheroidal rather than spherical or cylindrical wave functions. Asano and Yamamoto[33] were the first to solve in this way the problem of scattering by an arbitrary spheroid. Although Asano[34] subsequently published an extensive set of computations based on this solution, it has not seen widespread use, possibly because of the intractability and exoticness of spheroidal functions.

Computational experience with spheroids and even simpler particles such as spheres and cylinders leads to the inescapable conclusion that hidden barriers lie between a mathematical solution to a scattering problem and an algorithm for reliably and quickly extracting numbers from it.

6.7 COMPUTATIONAL METHODS FOR NONSPHERICAL PARTICLES

The widespread notion that randomly oriented nonspherical particles are somehow equivalent to spheres is symptomatic of a failure to distinguish between the symmetry of an ensemble and that of its members. Considerable effort has been expended in seeking prescriptions for equivalent spheres. This search resembles that for the Holy Grail—and has been as fruitless.

From extensive studies of scattering by nonspherical particles, Mugnai and Wiscombe[35] concluded that "after examining hundreds of nonspherical results and observing that they all cluster relatively close together, relatively far from the equivolume spheres (except at forward angles), we have come to regard nonspherical particles as normal, and spheres as the most unrepresentative shape possible—almost a singularity." This serves as a warning against using Mie theory for particles of all shapes and as a spur to finding methods more faithful to reality. We now turn to some of these methods. Keep in mind that no matter how different they may appear on the surface, they are all linked by the underlying Maxwell equations.

The *T-matrix method* is based on an integral formulation of scattering by an arbitrary particle. It was developed by Waterman, first for a perfect conductor,[36] then for a particle with less restricted properties.[37] It subsequently was applied to scattering problems under the name *extended boundary condition method* (EBCM).[38] Criticism of the T-matrix method was rebutted by Varadan et al.,[39] who cite dozens of papers on this method applied to electromagnetic scattering. Another source of papers and references is the collection edited by Varadan and Varadan.[40] Reference 6 is accompanied by a diskette containing T-matrix programs.

Linearity of the field equations and boundary conditions implies that the coefficients in the spherical harmonic expansion of the field scattered by any particle are linearly related to those of the incident field. The linear transformation connecting these two sets of coefficients is called the T (for transition) matrix.

The T-matrix elements are obtained by numerical integration. Computational difficulties arise for particles with high absorption or large aspect ratios. These limitations of the original T-matrix method have been surmounted somewhat by Iskander et al.,[41] whose extension is dubbed the iterative extended boundary condition method.

Although the T-matrix method is not restricted to axisymmetric particles, it almost exclusively has been applied to spheroids and particles defined by Chebyshev polynomials.[35,42,43]

Despite its virtues, the T-matrix method is not readily grasped in one sitting. Another method, variously called the Purcell-Pennypacker,[44] coupled-dipole,[45] digitized Green's function[46] method and discrete dipole approximation,[47] is mathematically much simpler—the most complicated function entering into it is the exponential—and physically transparent. Although originally derived by heuristic arguments, the coupled-dipole method was put on firmer analytical foundations by Lakhtakia.[48]

In this method, a particle is approximated by a lattice of N dipoles small compared with the wavelength but still large enough to contain many molecules. The dipoles often are, but need not be, identical and isotropic. Each dipole is excited by the incident field and by the fields of all the other dipoles. Thus the field components at each site satisfy a set of $3N$ linear equations. These components can be calculated by iteration[44,49] or by inverting the $3N \times 3N$ coefficient matrix.[45] The coefficient matrix for only one particle orientation need be inverted. This inverse matrix then can be used to calculate scattering for other orientations.[50] A disadvantage of matrix inversion is that the number of dipoles is limited by computer storage.

Arrays of coupled dipoles were considered long before Purcell and Pennypacker entered the scene. More than half a century ago Kirkwood[51] treated a dielectric as an array of molecules, the dipole moment of each of which is determined by the external field and by the fields of all the other molecules. What Purcell and Pennypacker did was to apply the coupled-dipole method to absorption and scattering by optically homogeneous particles. They bridged the gap between discrete arrays and continuous media with the Clausius-Mosotti theory. Because this theory, like every effective-medium theory, is not exact, critics of their method have pronounced it guilty by association. But the Clausius-Mosotti theory is merely the effective-medium theory that Purcell and Pennypacker happened to use. Whatever flaws their method may have, one of them is not that it is forever chained to the ghosts of Clausius and Mosotti. Alleged violations of the optical theorem are easily remedied by using the exact expression for the polarizability of a finite sphere,[52] which in no way changes the structure of the method.

Draine[47] applied this method (under the label *discrete dipole approximation*) to extinction by interstellar grains, obtaining the field components with the conjugate gradient method. An outgrowth of his paper is that by Flatau et al.,[53] who considered scattering by rectangular particles. Goedecke and O'Brien[46] baptized their version of the digitized Green's function method and applied it to scattering of microwave radiation by snowflakes.[54] Varadan et al.[55] applied the method to scattering by particles with anisotropic optical constants. It also has been applied to scattering by helices,[56] by a cylinder on a reflecting surface,[57] and extended to intrinsically optically active particles.[58]

Although Yung's analysis[59] of a large (15,600) array of dipoles representing a sphere suggests that there are no intrinsic limitations to the coupled-dipole method, it is plagued with practical limitations, most notably its inability to treat particles (especially compact ones or ones with large complex refractive indices) much larger than the wavelength of the illumination. Chiapetta,[60] then Singham and Bohren,[61] reformulated the coupled-dipole method, expressing the total field at each dipole as the sum of the incident field and the fields scattered once, twice, and so on by all the other dipoles. Although this formulation is appealing because each term in the scattering-order series has a simple physical interpretation, the series can diverge. The greater the refractive index, the smaller the particle for which the series diverges. For a particle of given volume, fewer terms are needed for convergence the more the particle departs from sphericity. The greater the average separation between dipoles, the weaker the average interaction.

Except for improvements and refinements[52,62,63] that increase accuracy and speed but do not remove barriers imposed by particle size and composition, the coupled-dipole method has not changed much since it first was used by Purcell and Pennypacker. It is not, of course, limited to optically homogeneous particles. It can be applied readily to aggregates of small particles. Indeed, it is best suited to aggregates with low fractal dimension. Berry and Percival[64] considered scattering by fractal aggregates using what they call the *mean-field approximation,* which is essentially the Rayleigh-Gans approximation, in turn a form of the scattering-order formulation of the coupled-dipole method in which the dipoles are excited only by the incident field.

The arbitrary border separating electrical engineering from optics is never more obvious than when it comes to methods for computing scattering. The engineers have theirs, the optical scientists have theirs, and rarely do the twain meet. In the hope of promoting smuggling, even illegal immigration, I must at least mention two methods that fall almost exclusively in the domain of electrical engineering: the method of moments and the finite-difference time-domain technique (FDTD).

Anyone interested in the method of moments must begin with Harrington's book,[65] a focal point from which paths fan out in all directions.

As its name implies, the FDTD technique is applied to what electrical engineers call the *time domain* (as opposed to the *frequency domain,* which most optical scientists inhabit even though they may not know it) and is explicitly labeled a finite-difference method (all methods for particles other than those of simple shape entail discretization in one form or another). Papers by Yee,[66] Holland et al.,[67] Mur,[69] Luebbers et al.,[69] and references cited in them will get you started on the FDTD technique.

When pondering the welter of species and subspecies of methods keep in mind that the differences among them and their ranges of validity are probably smaller than their adherents think or are willing to admit. There is no method that will happily compute scattering of arbitrary waves by particles of arbitrary size and composition in a finite amount of time. Moreover, each method, whatever its merits and demerits, often requires a tedious climb up a learning curve.

6.8 REFERENCES

1. H. C. van de Hulst, *Light Scattering by Small Particles,* Wiley, New York, 1957. Reprinted by Dover, New York, 1981.
2. D. Deirmendjian, *Electromagnetic Scattering on Polydispersions,* Elsevier, New York, 1969.
3. M. Kerker, *The Scattering of Light and Other Electromagnetic Radiation,* Academic, New York, 1969.
4. L. P. Bayvel and A. R. Jones, *Electromagnetic Scattering and Its Applications,* Elsevier, London, 1981.
5. C. F. Bohren and D. R. Huffman, *Absorption and Scattering of Light by Small Particles,* Wiley-Interscience, New York, 1983.
6. P. W. Barber and S. C. Hill, *Light Scattering by Particles: Computational Methods,* World Scientific, Singapore, 1990.
7. *Selected Papers on Light Scattering,* SPIE vol. 951, pts. 1 and 2, Milton Kerker (ed.), Society of Photo-Optical Instrumentation Engineers, Bellingham, Wash., 1988.
8. *Selected Papers on Scattering in the Atmosphere,* SPIE vol. MS 7, Craig F. Bohren (ed.), SPIE Optical Engineering Press, Bellingham, Wash., 1989.
9. *Selected Papers on Natural Optical Activity,* SPIE vol. MS 15, Akhlesh Lakhtakia (ed.), SPIE Optical Engineering Press, Bellingham, Wash., 1990.
10. A. Lakhtakia, V. K. Varadan, and V. V. Varadan, *Time-Harmonic Electromagnetic Fields in Chiral Media,* Springer-Verlag, Berlin, 1989.
11. *Optical Particle Sizing: Theory and Practice,* Gérard Gouesbet and Gérard Gréhan (eds.), Plenum, New York, 1988.
12. *Optical Effects Associated with Small Particles,* P. W. Barber and R. K. Chang (eds.), World Scientific, Singapore, 1988.
13. W. H. Martin, "The Scattering of Light in One-Phase Systems," in *Colloid Chemistry: Theoretical and Applied,* vol. I, Jerome Alexander (ed.), Chemical Catalog Co., New York, 1926, pp. 340–352.
14. C. F. Bohren, "Multiple Scattering of Light and Some of Its Observable Consequences," *Am. J. Phys.* **55:**524 (1987).
15. W. T. Doyle, "Scattering Approach to Fresnel's Equations and Brewster's Law," *Am. J. Phys.* **53:**463 (1985).
16. Max Born and E. Wolf, *Principles of Optics,* 3d ed., Pergamon, Oxford, 1965.
17. P. Lilienfeld, "Gustav Mie: The Person," *Appl. Opt.* **30:**4696 (1991).
18. N. A. Logan, "Survey of Some Early Studies of the Scattering of Plane Waves by a Sphere," *Proc. IEEE* **53:**773 (1965).
19. Helge Kragh, "Ludvig Lorenz and Nineteenth Century Optical Theory: The Work of a Great Danish Scientist," *Appl. Opt.* **30:**4688 (1991).
20. A. T. Young, "Rayleigh Scattering," *Phys. Today* **35:**2 (1982).
21. W. J. Wiscombe, "Mie Scattering Calculations: Advances in Techniques and Fast, Vector-Speed Computer Codes," NCAR/TN-140 + STR, National Center for Atmospheric Research, Boulder, Colo., 1979.
22. W. J. Wiscombe, "Improved Mie Scattering Algorithms," *Appl. Opt.* **19:**1505 (1980).

23. W. J. Lentz, "Generating Bessel Functions in Mie Scattering Calculations Using Continued Fractions," *Appl. Opt.* **15:**668 (1976).

24. Ru T. Wang and H. C. van de Hulst, "Rainbows: Mie Computations and the Airy Approximation," *Appl. Opt.* **30:**106 (1991).

25. *Handbook of Optical Constants of Solids,* E. W. Palik (ed), Academic, New York, 1985.

26. G. M. Hale and M. Querry, "Optical Constants of Water in the 200-nm to 200-μm Wavelength Region," *Appl. Opt.* **12:**555 (1973).

27. S. G. Warren, "Optical Constants of Ice from the Ultraviolet to the Microwave," *Appl. Opt.* **23:**1206 (1984).

28. A. L. Aden and M. Kerker, "Scattering of Electromagnetic Waves from Two Concentric Spheres," *J. Appl. Phys.* **22:**601 (1951).

29. R. Bhandari, "Scattering Coefficients for a Multilayered Sphere: Analytic Expressions and Algorithms," *Appl. Opt.* **24:**1960 (1985).

30. O. B. Toon and T. P. Ackerman, "Algorithms for the Calculation of Scattering by Stratified Spheres," *Appl. Opt.* **20:**3657 (1981).

31. M. Barabás, "Scattering of a Plane Wave by a Radially Stratified Tilted Cylinder," *J. Opt. Soc. Am.* **A4:**2240 (1987).

32. G. C. Salzman and C. F. Bohren, "Scattering by Infinite Coated Cylinders Illuminated at Oblique Incidence," TCN 90203, U.S. Army Research Office, Research Triangle Park, N.C.

33. S. Asano and G. Yamamoto, "Light Scattering by a Spheroidal Particle," *Appl. Opt.* **14:**29 (1975).

34. S. Asano, "Light Scattering Properties of Spheroidal Particles," *Appl. Opt.* **18:**712 (1979).

35. A. Mugnai and W. Wiscombe, "Scattering from Nonspherical Chebyshev Particles, 3. Variability in Angular Scattering Patterns," *Appl. Opt.* **28:**3061 (1989).

36. P. C. Waterman, "Matrix Formulation of Electromagnetic Scattering," *Proc. IEEE* **53:**805 (1965).

37. P. C. Waterman, "Symmetry, Unitarity, and Geometry in Electromagnetic Scattering," *Phys. Rev.* **D3:**825 (1971).

38. P. Barber and C. Yeh, "Scattering of Electromagnetic Waves by Arbitrarily Shaped Dielectric Bodies," *Appl. Opt.* **14:**2864 (1975).

39. V. V. Varadan, A. Lakhtakia, and V. K. Varadan, "Comments on Recent Criticism of the T-Matrix Method," *J. Acoust. Soc. Am.* **84:**2280 (1988).

40. *Acoustic, Electromagnetic and Elastic Wave Scattering—Focus on the T-Matrix Approach,* V. K. Varadan and V. V. Varadan (eds.), Pergamon, New York, 1980.

41. M. F. Iskander, A. Lakhtakia, and C. H. Durney, "A New Procedure for Improving the Solution Stability and Extending the Frequency Range of the EBCM," *IEEE Trans. Antennas Propag.* **AP-31:**317 (1983).

42. A. Mugnai and W. Wiscombe, "Scattering from Nonspherical Chebyshev Particles, 1. Cross Sections, Single-Scattering Albedo, Asymmetry Factor, and Backscattered Fraction," *Appl. Opt.* **25:**1235 (1986).

43. W. J. Wiscombe and A. Mugnai, "Scattering from Nonspherical Chebyshev Particles, 2. Means of Angular Scattering Patterns," *Appl. Opt.* **27:**2405 (1988).

44. E. M. Purcell and C. R. Pennypacker, "Scattering and Absorption of Light by Nonspherical Dielectric Grains," *Astrophys. J.* **186:**705 (1973).

45. Shermila Brito Singham and G. C. Salzman, "Evaluation of the Scattering Matrix of an Arbitrary Particle Using the Coupled-Dipole Approximation," *J. Chem. Phys.* **84:**2658 (1986).

46. G. H. Goedecke and S. G. O'Brien, "Scattering by Irregular Inhomogeneous Particles via the Digitized Green's Function Algorithm," *Appl. Opt.* **27:**2431 (1988).

47. B. T. Draine, "The Discrete-Dipole Approximation and Its Application to Interstellar Graphite Grains," *Astrophys. J.* **333:**848 (1988).

48. Akhlesh Lakhtakia, "Macroscopic Theory of the Coupled Dipole Approximation Method," *Opt. Comm.* **79:**1 (1990).

49. S. D. Druger, M. Kerker, D.-S. Wang, and D. D. Cooke, "Light Scattering by Inhomogeneous Particles," *Appl. Opt.* **18:**3888 (1979).
50. M. K. Singham, S. B. Singham, and G. C. Salzman, "The Scattering Matrix for Randomly Oriented Particles," *J. Chem. Phys.* **85:**3807 (1986).
51. J. G. Kirkwood, "On the Theory of Dielectric Polarization," *J. Chem. Phys.* **4:**592 (1936).
52. C. E. Dungey and C. F. Bohren, "Light Scattering by Nonspherical Particles: A Refinement to the Coupled-Dipole Method," *J. Opt. Soc. Am.* **A8:**81 (1991).
53. P. J. Flatau, G. L. Stephens, and B. T. Draine, "Light Scattering by Rectangular Solids in the Discrete-Dipole approximation: A New Algorithm Exploiting the Block-Toeplitz Structure," *J. Opt. Soc. Am.* **A7:**593 (1990).
54. S. G. O'Brien and G. H. Goedecke, "Scattering of Millimeter Waves by Snow Crystals and Equivalent Homogeneous Symmetric Particles," *Appl. Opt.* **27:**2439 (1988).
55. V. V. Varadan, A. Lakhtakia, and V. K. Varadan, "Scattering by Three-Dimensional Anisotropic Scatterers," *IEEE Trans. Antennas Propag.* **37:**800 (1989).
56. S. B. Singham, C. W. Patterson, and G. C. Salzman, "Polarizabilities for Light Scattering from Chiral Particles," *J. Chem. Phys.* **85:**763 (1986).
57. M. A. Taubenblatt, "Light Scattering from Cylindrical Structures on Surfaces," *Opt. Lett.* **15:**255 (1990).
58. Shermila Brito Singham, "Intrinsic Optical Activity in Light Scattering from an Arbitrary Particle," *Chem. Phys. Lett.* **130:**139 (1986).
59. Y. L. Yung, "Variational Principle for Scattering of Light by Dielectric Particles," *Appl. Opt.* **17:**3707 (1978).
60. P. Chiapetta, "Multiple Scattering Approach to Light Scattering by Arbitrarily Shaped Particles," *J. Phys. A* **13:**2101 (1980).
61. Shermila Brito Singham and C. F. Bohren, "Light Scattering by an Arbitrary Particle: The Scattering Order Formulation of the Coupled-Dipole Method," *J. Opt. Soc. Am.* **A5:**1867 (1988).
62. Shermila Brito Singham and C. F. Bohren, "Hybrid Method in Light Scattering by an Arbitrary Particle," *Appl. Opt.* **28:**517 (1989).
63. J. J. Goodman, B. T. Draine, and P. J. Flatau, "Application of Fast-Fourier Transform Techniques to the Discrete-Dipole Approximation," *Opt. Lett.* **16:**1198 (1991).
64. M. V. Berry and I. C. Percival, "Optics of Fractal Clusters Such as Smoke," *Optica Acta* **33:**577 (1986).
65. R. F. Harrington, *Field Computation by Moment Methods,* Robert E. Krieger, Malabar, Fla., 1987. Reprint of 1968 edition.
66. K. S. Yee, "Numerical Solution of Initial Boundary Value Problems Involving Maxwell's Equations in Isotropic Media," *IEEE Trans. Antennas Propag.* **AP14:**302 (1966).
67. R. Holland, L. Simpson, and K. S. Kunz, "Finite Difference Analysis of EMP Coupling to Lossy Dielectric Structures," *IEEE Trans. Electromag. Compat.* **EMC22:**203 (1980).
68. G. Mur, "Absorbing Boundary Conditions for the Finite Difference Approximation of the Time Domain Electromagnetic Field Equations," *IEEE Trans. Electromag. Compat.* **EMC23:**377 (1981).
69. R. Luebbers, F. P. Hunsberger, K. S. Kunz, R. B. Standler, and M. Schneider, "A Frequency-Dependent Finite-Difference Time-Domain Formulation for Dispersive Materials," *IEEE Trans. Electromag. Compat.* **EMC32:**222 (1990).

CHAPTER 7
SURFACE SCATTERING

E. L. Church*

and

P. Z. Takacs
Brookhaven National Laboratory
Upton, New York

7.1 GLOSSARY

A_0	area
a, b	polarization parameters = $\perp$, $\parallel$
BRDF	bidirectional reflectance distribution function
BSDF	bidirectional scattering distribution function
f	spatial frequency; $f_x = \xi$, $f_y = \eta$
G	power spectral density function including the specular contribution
I	intensity in power per steradian
P	generalized pupil function
R	Fresnel intensity reflection coefficient
S_1	one-dimensional or profile power spectral density
S_2	two-dimensional or area power spectral density
θ, φ	angles
σ_{ba}	bistatic radar cross section
Φ_{ba}	polarization factor

7.2 INTRODUCTION

The theory of wave scattering by imperfect surfaces has been highly developed in the fields of radiophysics, acoustics, and optics. It is the subject of a number of books,[1–3] book sections,[4–9] and review articles,[10–16] and its practical application to scattering from optical

* Work performed in part while the Author was with the U.S. Army Armament Research, Development and Engineering Center, Picatinny, N.J.

surfaces has been discussed in the readable and informative texts of Bennett[17] and Stover.[18] Bennett has also published a selection of reprints from the extensive literature in the related areas of surface finish and its measurement.[19]

The purpose of this article is to provide a frame of reference for interpreting the existing literature and understanding more recent developments in the description, characterization, and specification of scattering from optical-quality surfaces.

Discussions of related subjects can be found in the articles on "Measurement of Scatter," "Control of Stray Light," "X-Ray Optics," and "Diffraction" elsewhere in this *Handbook*.

7.3 NOTATION

The most important quantity in the discussion of surface scattering is the bidirectional reflectance distribution function (BRDF), which is defined precisely in Stover's article on "Measurement of Scatter." The BRDF is a multidimensional function that describes the angular and polarization dependence of the intensity of light reflected and scattered from a surface, and is related to the better-known angular-resolved or differential scattered intensity $dI/d\Omega$ and the bistatic radar cross section σ_{ba}, according to

$$\mathrm{BRDF}_{ba}(\theta_i, \theta_s, \varphi_s) = \frac{1}{\cos\theta_s} \cdot \frac{1}{I_i}\left(\frac{dI_s}{d\Omega}\right)_{ba} = \frac{\sigma_{ba}}{4\pi A_0 \cos\theta_i \cos\theta_s} \tag{1}$$

Here I_i is the incident intensity (power per steradian), I_s is the far-field reflected and scattered intensity, A_0 is the illuminated surface area, and the subscripts a and b denote the states of linear polarization of the incident and scattered radiation (note the right-to-left ordering). When the electric vector is perpendicular (or parallel) to the plane defined by the propagation direction and the surface normal, $a, b = s = \perp$ (or $p = \|$) in optics, and TE (or TM) in radiophysics. Polarization aficionados will note that the quantities defined in Eq. (1) are the square magnitudes of the components of the 2×2 Jones scattering matrix, and correspond to the upper-left quadrant of the 4×4 Stokes scattering matrix.[20]

Acoustic (scalar) scattering is described in terms of the analogous acoustic cross section, but without the polarization notation. Instead, there are two extreme forms of the acoustic boundary conditions, which correspond to the scattering of $\perp$- and $\|$-polarized electromagnetic radiation from a perfectly reflecting (i.e., infinitely conducting) surface. The first is called a "soft" surface, which uses the Dirichlet boundary condition, and the second is a "hard" surface, which uses the Neumann boundary condition.

As explained later, the angular distribution of the scattered intensity is essentially a Fourier spectrogram of the surface errors, so that it is convenient to discuss scattering in the frequency domain, that is, in terms of the spatial frequency vector **f** lying in the surface plane. The components of this vector are related to the scattering angles by

$$\mathbf{f} = \begin{pmatrix} f_x \\ f_y \end{pmatrix} = \frac{1}{\lambda}\begin{pmatrix} \sin\theta_s \cos\varphi_s - \sin\theta_i \\ \sin\theta_s \sin\varphi_s \end{pmatrix} \tag{2}$$

which can be viewed as a generalization of the grating equation for first-order diffraction from a grating with the spatial wavelength $d = 1/f$. f_x and f_y are frequently denoted by ξ and η in the Fourier-optics literature.

7.4 SCATTERING THEORY

Surface scattering arises from any deviation from the ideal interface that upsets the delicate interference effects required for rectilinear propagation. In principle, relating the BRDF to specific scattering structures is a straightforward application of Maxwell's equations but, more often than not, approximations are necessary to get practically useful results. The best known of these approximations are the physical-optics (Kirchhoff) and the perturbation (Rayleigh-Rice) theories, which despite their advanced age are still subjects of lively investigation.[21]

In the important case of small-angle scattering from weakly scattering surfaces, however, the predictions of all such theories are the same since in those limits the scattering is determined by simple linear-momentum considerations. For the present purposes we consider the most elementary scattering theory that captures that physics—the first-order Born approximation. This approximation has the added advantages that *it includes the effects of the measurement system and surface effects on an equal footing, and can be readily generalized to include both topographic and nontopographic scattering mechanisms.*

In general, scattering problems require the solution of an integral equation that relates the incident and scattered fields. The first-order Born approximation is the first term in an iterative solution of that equation, obtained by setting the total field in the integrand equal to the incident field.[3–5,9] The key quantity in the first-order Born approximation for the BRDF is the square magnitude of an integral containing three factors—an incident plane wave, a scattered plane wave, and an interaction term that couples them—integrated over the illuminated area of the surface. In particular

$$\mathrm{BRDF}_{ba} = \frac{1}{\lambda^2}\Phi_{ba}(\varphi_s)R_a(\theta_i)G(\mathbf{f}) \tag{3}$$

where

$$G(\mathbf{f}) = \frac{1}{A_0}\left|\int_{A_0} d\mathbf{r}\, e^{i2\pi\mathbf{f}\cdot\mathbf{r}}\, P(\mathbf{r})\right|^2 \tag{4}$$

and $\mathbf{r}$ is the position vector in the surface plane. Note that the single exponential factor appearing here, which causes the integral to appear in the form of a Fourier transform, is the product of the input and scattered plane waves, and P is the interaction that couples them. The remaining factors in Eq. (3) also have a simple physical interpretation: Φ_{ba} is a polarization factor which equals $\cos^2\varphi_s$ when $a = b$ and $\sin^2\varphi_s$ when $a \neq b$, and R_a is the Fresnel intensity reflection coefficient of the smooth surface.

The interaction, P can be factored into system and surface terms

$$P(\mathbf{r}) = P_{\mathrm{syst}}(\mathbf{r}) \cdot P_{\mathrm{surf}}(\mathbf{r}) \tag{5}$$

where P_{syst} is the pupil function of the measuring system and P_{surf} accounts for surface imperfections. It is convenient to write $P_{\mathrm{surf}} = 1 + p(\mathbf{r})$, where the 1 represents the ideal surface and p is the "perturbation" that describes deviations from a perfectly smooth surface.

For a perfect surface, $p = 0$, and the BRDF given by Eqs. (3) to (5) is just the response function of the measuring system—a sharp spike in the specular direction whose shape and width are determined by the system pupil function. In the case of an imperfect surface, $p \neq 0$, Eqs. (3) to (5) predict a modification of shape and intensity of this specular core, plus scattering out of the core into larger angles. In the following discussion we concentrate on the scattering distribution and return to core effects under "Finish Specification." To make the distinction clear we call the scattering part of the BRDF the bidirectional scattering distribution function, or BSDF.

In general, $p(\mathbf{r})$, the perturbation responsible for scattering, has an irregular or random dependence on surface position, which leads to a scattered intensity distribution in the form of a highly irregular speckle pattern. In general, we are not interested in the fine-scale structure of that pattern, but only its average or smoothed value as a function of scattering angle. In order to relate individual scattering measurements to surface statistics, the locally smoothed values of the scattered intensity are taken to equal the average scattering pattern of an ensemble of statistically equivalent surfaces measured under the same conditions.

That average takes the form

$$\langle \mathrm{BSDF}_{ba} \rangle = \frac{1}{\lambda^2} \Phi_{ba}(\varphi_s) R_a(\theta_i) S_p(\mathbf{f}) \tag{6}$$

where $\langle\ \rangle$ denotes the ensemble average and

$$S_p(\mathbf{f}) = \lim_{A_0 \to \infty} \left\langle \frac{1}{A_0} \left| \int_{A_0} d\mathbf{r}\, e^{i2\pi \mathbf{f} \cdot \mathbf{r}} p(\mathbf{r}) \right|^2 \right\rangle \tag{7}$$

is the two-dimensional power spectral density (PSD) of the perturbation p. The power spectrum is the most important factor in the expression for the BSDF since it is a purely surface quantity and contains all of the information about the surface responsible for scattering.

In the case of polished surfaces the power spectrum of the surface errors is generally a smooth and broad function of spatial frequency, and the smoothed value of the BSDF is independent of the system pupil function. On the other hand, if the surface roughness contains periodic components, such as tool marks in precision-machined surfaces, the BSDF will contain sharp diffraction lines. The positions and intensities of those lines are related to the feed rates and amplitudes of the tool marks, while their widths are determined by the system pupil function.[22–24]

If the surface roughness is spatially isotropic, as is usually assumed for randomly polished surfaces, the PSD depends only on the magnitude of the surface spatial frequency and is independent of its direction in the surface plane. This means that the BSDF depends only on the magnitude of $\mathbf{f}$ in Eq. (2), so that aside from angular-limit (band-limit) effects, all of the information about the surface spectrum can be obtained from scattering measurements made in the plane of incidence ($\varphi_s = 0$).

There are two other cases where scanning in the plane of incidence is sufficient: first, when the surface is "one-dimensional"—i.e., gratinglike—with its grooves oriented perpendicularly to the plane of incidence, and second, when an arbitrary surface is illuminated at an extreme glancing angle of incidence and the scattered intensity is measured by a slit detector perpendicular to the plane of incidence. In both of these cases the Born formalism described above leads to the one-dimensional form for the BSDF:

$$\langle \mathrm{BSDF}_{ba} \rangle = \frac{1}{\lambda} \delta_{b,a} R_a(\theta_i) S_p(f_x) \tag{8}$$

This expression differs from the two-dimension form in Eq. (6) in two ways: it is proportional to $1/\lambda$ rather than $1/\lambda^2$ and involves the one-dimensional or profile PSD instead of the two-dimensional or area PSD.[24] The definition of the one-dimensional power spectrum and its connection with the two-dimensional form are given later, under "Profile Measurements."

There is a final feature of the Born-approximation expression of the BSDF that is worth noting. In the plane of incidence both the one- and two-dimensional forms of the BSDF, although different, are functions of the quantity

$$|\lambda f_x| = |\sin \theta_s - \sin \theta_i| \tag{9}$$

rather than θ_i and θ_s individually. This is sometimes written as $\beta - \beta_0$ and the reduced dependency is described as "shift invariance."[25] The Born formalism, however, shows that it is a manifestation of the grating equation or, more fundamentally, linear momentum conservation in first-order scattering.

7.5 SURFACE MODELS

Surface models introduce new features into the BSDF which can be used to identify specific scattering mechanisms and to determine values of their characteristic physical parameters. One of the nice features of the Born approximation is that it permits a variety of scattering mechanisms to be discussed in a common language. Three such scattering mechanisms are discussed below.

Topographic Scattering

Surface roughness is the principal source of scattering from most optical surfaces at visible wavelengths. It comes from the phase fluctuations impressed on the reflected wavefront by the surface height fluctuations Z:

$$P_{\text{surf}}(\mathbf{r}) = e^{i(4\pi/\lambda)\cos\theta_i \cdot Z(\mathbf{r})} \approx 1 + i\frac{4\pi}{\lambda}\cos\theta_i \cdot Z(\mathbf{r}) \tag{10}$$

Optical-quality surfaces are usually "smooth" in the sense that the exponent above is so small that the two-term expansion on the right is sufficient for describing the scattering outside the specular core. The second term on the right is then p_z, the perturbation responsible for topographic scattering. In that case Eqs. (6) and (7) give

$$\langle \text{BSDF}_{ba} \rangle_z = \frac{16\pi^2}{\lambda^4}\cos^2\theta_i \Phi_{ba}(\varphi_s) R_a(\theta_i) S_z(\mathbf{f}) \tag{11}$$

where the factor S_z is the two-dimensional PSD of the height fluctuations, Z. More refined scalar and vector scattering theories give similar results in the smooth-surface limit. For example, the widely used first-order Rayleigh-Rice vector perturbation theory[4,26] gives

$$\langle \text{BSDF}_{ss} \rangle_z = \frac{16\pi^2}{\lambda^4}\cos\theta_i \cos\theta_s \cos^2\varphi_s \sqrt{R_s(\theta_i) R_s(\theta_s)}\, S_z(\mathbf{f}) \tag{12}$$

for s to s (i.e., $\perp$ to $\perp$) scatter. The simplicity of this expression has earned it the name of "the golden rule" of smooth-surface scattering. The corresponding expressions for *p-s, s-p,* and *p-p* scattering involve more complicated polarization-reflectivity factors and are given in the literature.[4,18,23,24]

In the case of rougher surfaces, higher terms in the expansion of the exponential in Eq. (10) must be taken into account. The resulting expression for the BSDF involves a more complicated dependence on the PSD of the height fluctuations, as well as a dependence on their height distribution function, which is irrelevant in the smooth-surface limit.[1–7]

Material Scattering

A perfectly smooth surface ($Z = 0$) may still scatter because of fluctuations in the composition or density of the surface material. For the lack of a better term we call this *material scattering.* The simplest way of describing such effects is to write the Fresnel-amplitude reflection coefficient of the surface in the form

$$P_{\text{surf}}(\mathbf{r}) = 1 + [\zeta_a(\mathbf{r}) - \bar{\zeta}_a]/\bar{\zeta}_a \tag{13}$$

where $\bar{\zeta}_a$ is the average reflection coefficient. The unity on the right corresponds to a perfect surface and the second term represents the fluctuations about that average, and is the perturbation, p_m, responsible for scattering. The corresponding BSDF is then

$$\langle \text{BSDF}_{ba} \rangle_m = \frac{1}{\lambda^2} \Phi_{ba}(\varphi_s) R_a(\theta_i) S_m(\mathbf{f}) \tag{14}$$

where S_m is the PSD of p_m given by Eqs. (7) and (13). This PSD can, in principle, be related to specific models of the material inhomogeneities, such as the magnitudes and spatial distribution of variations in composition.

Elson, Bennett, and Stover have recently appplied this model to the study of nontopographic scatter from composite beryllium mirrors.[27] A more fundamental source of material scattering from such mirrors is the fact that beryllium metal is optically anisotropic, and the randomly oriented crystallites in the surface lead to significant nontopographic scattering even for pure materials.[28]

If the fluctuations in material properties that are responsible for scattering are distributed throughout the volume of the material, the expression for the BSDF involves the three-dimensional PSD of those fluctuations rather than the two-dimensional form appearing in surface scattering. However, in the limit where the skin depth of the reflecting material is less than the characteristic size of the inhomogeneities—that is, the inhomogeneities can be considered to be columnar—the three-dimensional formalism reduces to the two-dimensional form in Eq. (14).[29–31]

Defect Scattering

The surface models considered above take the perturbations responsible for scattering to be distributed broadly and continuously over the surface. An alternative is to have them localized at isolated regions of the surface, such as pits or bumps in the topographic case and patches of different reflectivity in the case of material scattering.

As an illustration, consider a sparse distribution of identical pits or bumps. In that case the surface height deviations can be written

$$Z(\mathbf{r}) = \sum_j Z_d^{(j)}(\mathbf{r} - \mathbf{r}_j) \tag{15}$$

where $\mathbf{r}_j$ is the location of the jth defect and $Z_d^{(j)}$ is its topographic shape. If the defects are randomly distributed,

$$\langle \text{BSDF}_{ba} \rangle_d = \frac{1}{\lambda^2} \Phi_{ba}(\varphi_s) R_a(\theta_i) S_d(\mathbf{f}) \tag{16}$$

where S_d is the PSD of the collection of defects.[24] For a collection of cylindrical pits or protrusions, for example,

$$S_d(\mathbf{f}) = \frac{1}{A_0} \sum_j \left[2\pi r_j^2 \cdot \frac{J_1(2\pi f r_j)}{\pi f r_j} \cdot \sin\left(\frac{2\pi}{\lambda} h_j \cos\theta_i\right) \right]^2 \tag{17}$$

where r_j and h_j are the radius and height of the jth defect, and the sum is taken over all defects within the illuminated surface area, A_0. This result is instructive but would be quantitative unrealistic for large values of h_j because of the omission of shadowing and other effects in this elementary calculation.

A number of exact calculations of the scattering from particular defects and structures can be found in the literature which may be more useful in modeling real "scratch-and-dig" defects in surfaces than the analysis given above.[4,32–35] Another useful calculation is that for the scattering of dust on smooth surfaces by Videen, Wolfe, and Bickel.[36–37]

The measured BSDFs of real surfaces may involve contributions from one or more scattering mechanisms, such as those discussed above. The next two sections discuss methods of identifying the dominant mechanisms involved through their different characteristics, such as their different dependencies on the radiation wavelength.

7.6 WAVELENGTH SCALING

All first-order expressions for the BSDF appear as the product of factors, each corresponding to a different feature of the scattering process. Equation (12), for example, involves an inverse power of the radiation wavelength, obliquity factors, a reflectivity factor, and finally, the power spectral density of the perturbation responsible for scattering.

The wavelength factor comes from two sources: an inherent factor of $1/\lambda^2$ from the dimensionality of the problem [$1/\lambda$ in the one-dimensional case, Eq. (8)], and an additional factor of $1/\lambda^2$ for topographic scatter since the radiation "sees" the height fluctuations in terms of its wavelength.[31] This additional factor means that, although the topographic component may dominate scattering at visible wavelengths, its contribution to the total BSDF can be overwhelmed by nontopographic or "anomalous" scattering mechanisms at infrared wavelengths.[38]

The reflectivity factor in the BSDF depends on the radiation wavelength through the index of refraction or permittivity of the surface material, although the reflection coefficient of highly reflective mirrors may be effectively constant over a wide range of wavelengths.

The final factor in the BSDF, the PSD, depends on the radiation wavelength principally through its argument, the spatial frequency, although its magnitude can also involve inherent wavelength dependencies. In the case of topographic scattering, the PSD is a purely geometric quantity, independent of the radiation wavelength, while the PSD of material fluctuations can depend on the wavelength.

The radiation-wavelength dependence of the argument of the power spectrum comes from purely geometric considerations. According to Eq. (2), the spatial frequency corresponding to a given scattering direction is inversely proportional to the radiation wavelength. This predictable dependence can be eliminated in data analysis by comparing BSDF measurements made at different radiation wavelengths as a function of spatial frequency rather than scattering angle.

Recent results indicate that mirrors made of some materials, such as molybdenum and silicon, behave as purely topographic scatterers from visible to 10-micron radiation wavelengths, while others show significant nontopographic contributions.[38–40]

If different scattering mechanisms are present, and they are statistically independent, their BSDFs add. If they are not independent, there may be important interference effects which have an intermediate wavelength dependence.[31,41]

7.7 PROFILE MEASUREMENTS

The models described above relate the PSD of surface imperfections to more fundamental physical quantities that are known a priori or can be measured independently. Surface topography, for example, can be measured directly using optical or mechanical profiling techniques, which raises the question of how to use profile data to determine the two-dimensional PSD appearing in Eq. (12). That is, how can BSDFs be predicted from profile measurements?

Algorithms for estimating two-dimensional spectra from two-dimensional arrays of profile data are readily available.[42] The usual procedure, though, is to use the PSDs estimated from profiles taken by "one-dimensional" profiling instruments, or profiles stripped from rectangular arrays of "two-dimensional" measurements. In either case, one gets estimates of the *one-dimensional power spectral density* of the surface:

$$S_z(f_x) = \lim_{L\to\infty} \left\langle \frac{2}{L} \left| \int_{-L/2}^{+L/2} dx e^{i2\pi f_x x} Z(x) \right|^2 \right\rangle \tag{18}$$

which contains information over the range of surface frequencies

$$\frac{1}{L} < f_x < \frac{N}{2L} \tag{19}$$

where L is the profile length and N is the number of data points in the profile. The dynamic range of such measurements—the ratio of the highest to lowest frequency included—is then $N/2$, which can be 1000 or more in practice. In principle, the estimation of power spectra from limited data sets is straightforward, but it involves some subtleties and must be done with care.[42–45] Procedures for doing this are presently being codified in an ASTM standard.[46]

It is important to recognize, though, that the profile spectrum is not just a special form of the area spectrum, but is related to it in a more complicated way than might have been supposed. In particular, they are connected by the integral transform

$$S_1(f_x) = 2 \int_{-\infty}^{+\infty} df_y\, S_2(f_x, f_y) \tag{20}$$

where $f_x > 0$. This accounts for their different units—for example, $S_1(f_x)$ for topographic errors is usually expressed in units of micrometers to the third power, while $S_2(f_x, f_y)$ is in micrometers to the fourth power.[24,45] (We have added subscripts 1 and 2 in Eqs. (20) and (21) to emphasize the distinction between one- and two-dimensional spectra.

In general, Eq. (20) cannot be solved to give $S = S_2$ (the quantity that appears in the expression for the BSDF) in terms of S_1 determined from profile measurements. However, in the special case of an isotropically rough surface, it can, and S_2 can be written as a different but related integral transform of S_1, namely:[24,45]

$$S_2(\mathbf{f}) = -\frac{1}{2\pi}\int_f^\infty \frac{df_x}{\sqrt{f_x^2 - f^2}} \cdot \frac{d}{df_x} S_1(f_x) \tag{21}$$

Unfortunately, the integral on the right cannot be evaluated directly from measured (estimated) profile spectra since real data sets involve strong speckle fluctuations and do not include the high-frequency information needed to extend the upper limit of the integration to infinity. The usual way of handling these difficulties is fit the measured profile spectrum to a physically reasonable analytic model before evaluating Eq. (21). This does three things: it smooths the fluctuations, adds a priori physical information required to extrapolate the data outside the measurement bandwidth, and condenses the data into a few finish parameters.[45]

A useful model for this purpose is the *ABC* or K-correlation model,

$$S_1(f_x) = A[1 + (Bf_x)^2]^{-C/2} \tag{22}$$

where *A*, *B*, and *C* are the adjustable model parameters. *A* is the value of the profile spectrum in the low-frequency limit, $B/2\pi$ is a "correlation length" which determines the location of the transition between the low- and high-frequency behavior, and *C* is the exponent of the power-law fall-off at high frequencies. The corresponding two-dimensional form of the *ABC* spectrum, determined from the above transformations, is

$$S_2(\mathbf{f}) = A'[1 + (Bf)^2]^{-(C+1)/2} \tag{23}$$

where $A' = AB\Gamma((C+1)/2)/(2\sqrt{\pi}\,\Gamma(C/2))$ for $C > 0$.[40,45]

The *ABC* model is useful since it contains a minimal set of physical parameters, is mathematically tractable, and reduces to a number of more familiar forms as special cases. For example, $C = 2$ gives a lorentzian spectrum, and large *B* gives the fractal-like forms:

$$S_1(f_x) = K_n f_x^{-n} \tag{24}$$

and

$$S_2(\mathbf{f}) = \frac{\Gamma((n+1)/2)}{2\sqrt{\pi}\,\Gamma(n/2)} \cdot K_n f^{-(n+1)} \tag{25}$$

Equation (25) corresponds to an inverse-power-law form for the BSDF. Isolated examples of such scattering distributions have been reported for a number of years[24,25,47] but their ubiquity and importance is only now being recognized.[48]

Equations (22) to (25) provide a mechanism for comparing profile and BSDF measurements of a given surface, and this—along with wavelength scaling—offers another way of distinguishing topographic and nontopographic scattering mechanisms in real surfaces. Such comparisons have been reported in several recent papers.[27,40,–49]

A related procedure, that does not involve scattering as such, is to make mechanical and optical profile measurements of a given surface and to compare spectra deduced from them over a common frequency range. Differences could come from the fact that mechanical measurements are sensitive to topographic effects, while optical measurements are sensitive to material effects as well. Preliminary comparisons of this type indicate that machined silicon is essentially a pure topographic scatterer at HeNe wavelengths.[50,51]

7.8 FINISH PARAMETERS

Optical finish measurements are frequently reported in terms of one or more classical topographic finish parameters—the rms (root mean square) surface roughness σ, the rms profile slope μ, and the surface correlation length l. These are appealing parameters since they are "obvious" measures of quality and numerical values can be estimated directly from raw profile data using simple, intuitive algebraic expressions (estimators).

On the other hand, they suffer from the fact that their values can depend significantly on the measuring instrument used, and do not necessarily represent intrinsic properties of the surfaces being measured. In the early days, when there were only one or two instruments that could measure optical surface finish at all, and measurements were used mainly for comparative purposes, instrumental effects were not an important concern. But nowadays, when many different measurement techniques are available and there is a need for a more quantitative understanding of the measured data, these instrumental effects must be recognized and understood.

The royal road to understanding the effects of the measurement process lies in the frequency domain, which has already appeared naturally in the discussion of surface scattering. For example, it is easy to see that for linear measurement systems the measured value of the mean-square profile roughness and slope are given by weighted integrals or moments of the profile PSD:

$$\begin{pmatrix} \sigma \\ \mu \end{pmatrix}^2_{\text{meas}} = \int_0^\infty df_x\, S_1(f_x) T(f_x) \begin{pmatrix} 1 \\ 2\pi f_x \end{pmatrix}^2 \tag{26}$$

The factor $T(f_x)$ in the integrand is the square magnitude of the transfer function of the particular measurement process involved, and accounts for the fact that real measurements are sensitive to only a limited range of surface spatial frequencies. In contrast, the intrinsic values of these parameters are defined by Eq. (26) where $T(f_x) = 1$.

In the case of mechanical profile measurements $T(f_x)$ is usually taken to be a unit rectangle; that is, a constant equal to unity within the measurement bandpass and zero outside. In the case of optical profile measurements it is usually taken to be a triangle function that falls from unity at zero frequency to zero at the incoherent cutoff of the optical system involved.[43,44,51,52] Although Eq. (26) applies to profile measurements, it is readily generalized to two-dimensional measurement processes,[44,53] where the area in the frequency plane defined by the nonvanishing transfer function is called the *frequency footprint* of the measurement process.[53]

The literature also describes topographic finish in terms of the "correlation length," a length parameter characterizing its transverse properties. Unfortunately, though, most of the values in the literature have been determined without taking the possibly large effects of the measurement bandwidth into account, and must be interpreted with considerable care.[43,48,54]

The standards literature does not yet completely address the issue of measurement bandwidths. The 1985 version of ANSI/ASME B46, which is principally directed towards mechanical surfaces, lists low-frequency limits but not high-frequency ones, although the updated version will.[55] On the other hand, ASTM F1408, for the total integrated scatter (TIS) method for measuring the effective rms surface roughness[56] specifies wavelength and angular parameters that translate to a measurement bandwidth of

$$0.07 < f < 1.48\ \mu\text{m}^{-1} \tag{27}$$

It is important to note that these limits correspond to a dynamic range of only 21.5, which is much less than that of many current profiling instruments. This means that the effective roughness value of a given surface determined by conventional TIS techniques can be significantly smaller than that calculated from profile measurements.[53,57] Noll and Glenn, for example, have reported TIS roughness values that are less than Talystep roughnesses by a factor of 3 or more.[58]

The most striking example of bandwidth effects occurs for fractal surfaces, that is, surfaces with inverse-power-law spectra. Such surfaces do not have finite intrinsic values of the roughness, slope, or correlation length, although they do have *finite measured values* of these quantities, since the low- and high-frequency divergences are cut off by the finite measurement band-pass.[24,48,59–61]

In the light of this, it is more realistic to use the classical topographic finish parameters measured under standardized conditions for descriptive and comparative purposes only, and to use the parameters of the power spectrum for a more comprehensive and precise description—for example, A, B, and C in the ABC model and n and K_n in the fractal model.

The advantages of the power-spectrum description come from the fact that the PSD is directly related to the BSDF, it is an intrinsic surface property and can be used to calculate conventional finish parameters from Eq. (26).[38,40,49]

7.9 FINISH SPECIFICATION

Imperfect surface finish is undesirable since it degrades the performance of an optical system. This raises the question of how to specify surface finish in terms of performance requirements. Although this subject is outside the scope of this article, it is useful to indicate how the ideas discussed above are involved.

The simplest performance requirement is that the distribution of the scattered light intensity is less than some prescribed function of the scattering angle away from the specular direction. This can be translated directly into limitations on error spectra using Eqs. (12) (14) and (16). This method—using an inverse-power-law spectrum—was proposed a number of years ago for x-ray mirrors.[47] and has recently been adopted in an international (ISO) standard for the designation of optical surface finish.[62]

A variant of this is the requirement that the total integrated intensity scattered out of the image core be less than some fraction of the total reflected intensity. In the case of topographic errors this appears in the form of the famous Strehl ratio:

$$\frac{\text{Core intensity, } Z \neq 0}{\text{Core intensity, } Z = 0} \approx e^{-(4\pi\sigma \cos \theta_i/\lambda)^2} \tag{28}$$

where σ is the effective rms roughness of the surface evaluated by integrating the two-dimensional power spectrum of the equivalent surface roughness using the two-dimensional analog of Eq. (26), where the frequency limits of $T(\mathbf{f})$ are determined from the angular limits through the grating equation, Eq. (2). The extreme values of these angles are determined by the core size in one limit, and scattering parallel to the surface plane in the other.[57]

The integrated scattering intensity is important since it represents a loss of the core intensity. In the case of imaging optics, the effects of surface errors on the detailed distribution of the core intensity are also of interest: for example, its on-axis image intensity, the image width, and its spatial-frequency distribution. Expressions for these effects follow from precisely the same Born-approximation result, Eqs. (3) to (5), that to this point we have used only to examine the properties of the scattered radiation, that is, the BSDF.

One interesting result of this analysis is the fact that the effects on imaging are automatically finite for surfaces with fractal surface errors, even though fractal power spectra diverge at long surface spatial wavelengths. This is because the finite size of the pupil function makes the imaging effects insensitive to spatial wavelengths longer than the size of the illuminated area of the mirror. The interested reader is referred to the literature for details of these and related issues.[59–61]

7.10 REFERENCES

1. P. Beckmann and A. Spizzichino, *The Scattering of Electromagnetic Waves from Rough Surfaces,* Pergamon Press, New York, N.Y., 1963.
2. F. G. Bass and I. M. Fuks, *Wave Scattering from Statistically Rough Surfaces,* Pergamon Press, New York, N.Y., 1979.
3. J. A. Ogilvy, *Theory of Wave Scattering from Random Rough Surfaces,* Adam Hilger, New York, N.Y., 1991.
4. G. T. Ruck, D. E. Barrick, W. D. Stuart, and C. K. Krichbaum, *Radar Cross-Section Handbook,* 2 vols., Plenum Press, New York, N.Y., 1970.
5. A. Ishimaru, *Wave Propagation and Scattering in Random Media,* 2 vols., Academic Press, New York, N.Y., 1978.
6. F. T. Ulaby, R. K. Moore, and A. K. Fung, *Microwave Remote Sensing: Active and Passive,* Addison-Wesley, Reading, Mass., 1981, vols. 1, 2, and 3, Artech House, Dedham, Mass., 1986.
7. L. Tsang, J. A. Kong, and R. T. Shin, *Theory of Microwave Remote Sensing,* John Wiley, New York, N.Y., 1985.
8. S. M. Rytov, Y. A. Kravtsov, and V. I. Tatarskii, *Principles of Statistical Radiophysics,* 4 vols., Springer-Verlag, New York, N.Y., 1987.
9. M. Nieto-Vesperinas, *Scattering and Diffraction in Physical Optics,* John Wiley, New York, N.Y., 1991.
10. V. Twersky, "On Scattering and Reflection of Electromagnetic Waves by Rough Surfaces," *IRE Transactions on Antennas and Propagation* **AP-5:**81–90 (1957).
11. V. Twersky, "Rayleigh Scattering," *Applied Optics* **3:**1150–1162 (1964).
12. A. B. Shmelev, "Wave Scattering by Statistically Uneven Surfaces," *Soviet Physics Uspekhi,* **15**(2):173–183 (1972).
13. M. V. Berry, "The Statistical Properties of Echoes Diffracted from Rough Surfaces," *Proc. Royal Soc. London* **273:**611–654 (1973).
14. G. R. Valenzuela, "Theories for the Interaction of Electromagnetic and Ocean Waves—A Review," *Boundary-Layer Metrology* **13:**61–85 (1978).
15. J. M. Elson, H. E. Bennett, and J. M. Bennett, "Scattering from Optical Surfaces," *Applied Optics and Optical Engineering,* Academic Press, New York, N.Y., **VIII:**191–224 (1979).
16. A. K. Fung, "Review of Random Surface Scatter Models," *Proc. SPIE* **385:**87–98 (1982).
17. J. M. Bennett and L. Mattsson, *Introduction to Surface Roughness and Scattering,* Optical Society of America, Washington, D.C., 1989.
18. J. C. Stover, *Optical Scattering: Measurement and Analysis,* McGraw-Hill, New York, N.Y., 1990.
19. J. M. Bennett, *Surface Finish and Its Measurement,* 2 vols., Optical Society of America, Washington, D.C., 1992.
20. H. C. van de Hulst, *Light Scattering by Small Particles,* Dover Publications, New York, N.Y., 1981.

21. E. I. Thorsos and D. R. Jackson, "Studies in Scattering Theories Using Numerical Methods," *Waves in Random Media* **3:**165–190 (1991).

22. E. L. Church and J. M. Zavada, "Residual Surface Roughness of Diamond-Turned Optics," *Applied Optics* **14:**1788–1795 (1975).

23. E. L. Church, H. A. Jenkinson, and J. M. Zavada, "Measurement of the Finish of Diamond-Turned Metal Surfaces by Differential Light Scattering," *Optical Engineering* **16:**360–374 (1977).

24. E. L. Church, H. A. Jenkinson, and J. M. Zavada, "Relationship Between Surface Scattering and Microtopographic Features," *Optical Engineering* **18:**125–136 (1979).

25. J. E. Harvey, "Light-Scattering Characteristics of Optical Surfaces," *Proc. SPIE* **645:**107–115 (1986).

26. W. H. Peake, "Theory of Radar Return from Terrain," *IRE Convention Record,* **7**(1):34–41 (1959).

27. J. M. Elson, J. M. Bennett, and J. C. Stover, "Wavelength and Angular Dependence of Light Scattering from Beryllium: Comparison of Theory and Experiment," *Applied Optics* **32:**3362–3376 (1993).

28. E. L. Church, P. Z. Takacs, and J. C. Stover, "Scattering by Anisotropic Grains in Beryllium Mirrors," *Proc. SPIE* **1331:**205–220 (1990).

29. A. Stogryn, "Electromagnetic Scattering by Random Dielectric-Constant Fluctuations in a Bounded Medium," *Radio Science* **9:**509–518 (1974).

30. J. M. Elson, "Theory of Light Scattering from a Rough Surface with an Inhomogeneous Dielectric Permittivity," *Phys. Rev.* **B30:**5460–5480 (1984).

31. E. L. Church and P. Z. Takacs, "Subsurface and Volume Scattering from Smooth Surfaces," *Proc. SPIE* **1165:**31–41 (1989).

32. V. Twersky, "Reflection Coefficients for Certain Rough Surfaces," *J. Appl. Phys.* **24:**659–660 (1953).

33. J. E. Burke and V. Twersky, "On Scattering of Waves by an Elliptic Cylinder and by a Semielliptic Cylinder on a Ground Plane," *J. Opt. Soc. Amer.* **54:**732–744 (1964).

34. D. E. Barrick, "Low-Frequency Scatter from a Semielliptic Groove in a Ground Plane," *J. Opt. Soc. Amer.* **60:**625–634 (1970).

35. T. C. Rao and R. Barakat, "Plane-Wave Scattering by a Conducting Cylinder Partially Buried in a Ground Plane, 1. TM Case," *J. Opt. Soc. Amer.* **A6:**1270–1280 (1989).

36. G. Videen, "Light Scattering from a Sphere Near a Surface," *J. Opt. Soc. Amer.* **A8:**483–489 (1991).

37. G. Videen, W. L. Wolfe, and W. S. Bickel, "Light-Scattering Mueller Matrix for a Surface Contaminated by a Single Particle in the Rayleigh Limit," *Opt. Engineering* **31:**341–349 (1992).

38. J. C. Stover, M. L. Bernt, D. E. McGary, and J. Rifkin, "Investigation of Anomalous Scatter from Beryllium Mirrors," *Proc. SPIE* **1165:**100–106 (1989).

39. Y. Wang and W. L. Wolfe, "Scattering from Microrough Surfaces: Comparison of Theory and Experiment," *J. Opt. Soc. Amer.* **73:**1596–1602 (1983), **A1:**783–784 (1984).

40. E. L. Church, T. A. Leonard, and P. Z. Takacs, "The Prediction of BRDFs from Surface-Profile Measurements," *Proc. SPIE* **1165:**136–150 (1989).

41. J. M. Elson, "Anomalous Scattering from Optical Surfaces with Roughness and Permittivity Perturbations," *Proc. SPIE* **1530:**196–207 (1991).

42. W. H. Press, B. P. Flannery, S. A. Teukolsky, and W. T. Vetterling, *Numerical Recipes,* Cambridge University Press, New York, N.Y., 1986. (Fortran, Pascal, and Basic versions available.)

43. E. L. Church and P. Z. Takacs, "Instrumental Effects in Surface-Finish Measurements," *Proc. SPIE* **1009:**46–55 (1989).

44. E. L. Church and P. Z. Takacs, "Effects of the Optical Transfer Function in Surface-Profile Measurements," *Proc. SPIE* **1164:**46–59 (1989).

45. E. L. Church and P. Z. Takacs, "The Optimal Estimation of Finish Parameters," *Proc. SPIE* **1530:**71–85 (1991).
46. "Standard Practice for Characterizing Roughness of Optical Surfaces from Surface Profile Data," ASTM Subcommittee E12.09. (This Standard was in the process of being written, 1993.)
47. E. L. Church, "Role of Surface Topography in X-ray Scattering," *Proc. SPIE* **184:**196–202 (1979).
48. E. L. Church, "Fractal Surface Finish," *Applied Optics* **27:**1518–1526 (1988).
49. W. K. Wong, D. Wang, R. T. Benoit, and P. Barthol, "A Comparison of Low-Scatter Mirror PSDs Derived from Multiple-Wavelength BRDFs and Wyko Profilometer Data," *Proc. SPIE* **1530:**86–103 (1991).
50. E. L. Church, T. V. Vorburger, and J. C. Wyant, "Direct Comparison of Mechanical and Optical Measurements of the Finish of Precision-Machined Optical Surfaces," *Optical Engineering* **24:**388–395 (1985).
51. E. L. Church, J. C. Dainty, D. M. Gale, and P. Z. Takacs, "Comparison of Optical and Mechanical Measurements of Surface Finish," *Proc. SPIE* **1531:**234–250 (1991).
52. E. L. Church and P. Z. Takacs, "Use of an Optical-Profiling Instrument for the Measurement of the Figure and Finish of Optical-Quality Surfaces," *Wear* **109:**241–257 (1986).
53. E. L. Church, "The Precision Measurement and Characterization of Surface Finish," *Proc. SPIE* **429:**86–95 (1983).
54. E. L. Church, "Comments on the Correlation Length," *Proc. SPIE* **680:**102–111 (1987).
55. ANSI/ASME B46.1, *Surface Texture (Surface Waviness and Lay),* American Society of Mechanical Engineers, New York, N.Y., 1985.
56. ASTM F1048, *Standard Test Method for Measuring the Effective Surface Roughness of Optical Components by Total Integrated Scattering,* 1987.
57. E. L. Church, G. M. Sanger, and P. Z. Takacs, "Comparison of Wyko and TIS Measurements of Surface Finish," *Proc. SPIE* **740:**65–73 (1987).
58. R. J. Noll and P. Glenn, "Mirror Surface Autocovariance Functions and Their Associated Visible Scattering," *Applied Optics* **21:**1824–1838 (1982).
59. E. L. Church and P. Z. Takacs, "Specification of the Figure and Finish of Optical Elements in Terms of System Performance," *Proc. SPIE* **1781:**118–130 (1992).
60. E. L. Church and P. Z. Takacs, "Specification of Surface Figure and Finish in Terms of System Performance," *Applied Optics* **32:**3344–3353 (1993).
61. E. L. Church and P. Z. Takacs, "Specifying the Surface Finish of X-ray Mirrors," *Proceedings on Soft X-ray Projection Lithography,* vol. 18, *OSA Proceedings Series,* Optical Society of America, Washington, D.C., 1993.
62. ISO 10110 Standard, *Indications on Optical Drawings,* "Part 8: Surface Texture," 1992.

P · A · R · T · 3

QUANTUM OPTICS

CHAPTER 8
OPTICAL SPECTROSCOPY AND SPECTROSCOPIC LINESHAPES

Brian Henderson
University of Strathclyde
Glasgow, United Kingdom

8.1 GLOSSARY

A_{ba}	Einstein coefficient for spontaneous emission
a_0	Bohr radius
B_{if}	the Einstein coefficient between initial state, $\|i\rangle$, and final state, $\|f\rangle$
E_{DC}	Dirac Coulomb term
E_{hf}	hyperfine energy
E_n	eigenvalues of quantum state, n
$E(t)$	electric field at time t
$E(\omega)$	electric field at frequency, ω
e	charge on the electron
ED	electric dipole term
EQ	electric quadrupole term
$\langle f\| \mathbf{V}' \|i\rangle$	matrix element of perturbation, V'
g_a	degeneracy of ground level
g_b	degeneracy of excited level
g_N	gyromagnetic ratio of nucleus
H_{so}	spin-orbit interaction hamiltonian
$\hbar$	Planck's constant
I	nuclear spin
$I(t)$	the emission intensity at time t
$\mathbf{j}$	total angular momentum vector given by $\mathbf{j} = 1 \pm \frac{1}{2}$
l_i	orbital state
M_N	mass of nucleus N
MD	magnetic dipole term

m	mass of the electron
$n_\omega(T)$	equilibrium number of photons in a blackbody cavity radiator at angular frequency, ω, and temperature, T
QED	quantum electrodynamics
R_∞	Rydberg constant for an infinitely heavy nucleus
$R_{nl}^{(r)}$	radial wavefunction
s	spin quantum number with the value $\frac{1}{2}$
s_i	electron spin
T	absolute temperature
W_{ab}	transition rate in absorption transition between states $\|a\rangle$ and $\|b\rangle$
W_{ba}	transition rate in emission transition from state $\|b\rangle$ to state $\|a\rangle$
Z	charge on the nucleus
$\alpha = e^2/4\pi\varepsilon_0\hbar c$	fine structure constant
ε_0	permittivity of free space
μ_B	Bohr magneton
$\rho(\omega)$	energy density at frequency, ω
$\zeta(r)$	spin-orbit parameter
τ_R	radiative lifetime
ω	angular velocity
$\Delta\omega$	natural linewidth of the transition
$\Delta\omega_D$	Doppler width of transition
ω_k	mode, k, with angular frequency, ω

Spectroscopic measurements have, throughout the century, played a key role in the development of quantum theory. This article will present a simple description of the quantum basis of spectroscopic phenomena, as a prelude to a discussion of the application of spectroscopic principles in atomic, molecular, and solid-state physics. A brief survey will be presented of the multielectron energy-level structure in the three phases of matter and of the selection rules which determine the observation of optical spectra. Examples will be given of the fine-structure, hyperfine-structure, and spin-orbit splittings in the spectra of atoms, molecules, and solids. Solid-state phenomena to be considered will include color center, transition metal, and rare earth ion spectra.

The intrinsic or homogeneous lineshapes of spectra are determined by lifetime effects. Other dephasing processes, including rotational and vibrational effects, lead to splitting and broadening of spectra. There are also sources of inhomogeneous broadening associated with Doppler effects in atomic and molecular spectra and crystal field disorder in solids. Methods of recovering the homogeneous lineshape include sub-Doppler laser spectroscopy of atoms, optical hole burning, and fluorescence line narrowing.

Finally, the relationship between linewidth and lifetime are discussed and the effects of time-decay processes outlined. The consequences of measurements in the picosecond and subpicosecond regime will be described. Examples of vibrational relaxation in molecular and solid-state spectroscopy will be reviewed.

8.2 INTRODUCTORY COMMENTS

Color has been used to enhance the human environment since the earliest civilizations. Cave artists produced spectacular colorations by mixing natural pigments. These same pigments, burned into the surfaces of clays to produce color variations in pottery, were also used to tint glass. The explanation of the coloration process in solids followed from Newton's observation that white light contains all the colors of the rainbow,[1] the observed color of a solid being complementary to that absorbed from white light by the solid. Newton measured the wavelength variation of the refractive index of solids, which is responsible for dispersion, and his corpuscular theory of light explained the laws of reflection and refraction.[1] The detailed interpretation of polarization, diffraction, and interference followed from the recognition that light is composed of transverse waves, the directions of which were related to the direction of the electric field in Maxwell's electromagnetic theory:[2,3] the electronic constituents of matter are set into transverse oscillation relative to the propagating light beam. Subsequently, Einstein introduced the photon in explaining the photoelectric effect.[4] Thus the operating principles of optical components in spectrometers, such as light sources, mirrors, lenses, prisms, polarizers, gratings, and detectors, have been with us for a long time.

Many significant early developments in quantum physics led from optical spectroscopic studies of complex atoms. After Bohr's theory of hydrogen,[5] the quantum basis of atomic processes developed apace. One of Schrödinger's first applications of wave mechanics was in calculations of atomic energy levels and the strengths of spectroscopic transitions.[6] Schrödinger also demonstrated the formal equivalence of wave mechanics and Heisenberg's matrix mechanics. Extensions of spectroscopy from atomic physics to molecular physics and solid-state physics more or less coincided with the early applications of quantum mechanics in these areas.

A survey of the whole of spectroscopy, encompassing atoms, molecules, and solids, is not the present intent. Rather it is hoped that by choice of a few critical examples the more general principles linking optical spectroscopy and the structure of matter can be demonstrated. The field is now vast: Originally the exclusive domain of physicists and chemists, optical spectroscopy is now practiced by a variety of biophysicists and biochemists, geophysicists, molecular biologists, and medical and pharmaceutical chemists with applications to proteins and membranes, gemstones, immunoassay, DNA sequencing, and environmental monitoring.

8.3 THEORETICAL PRELIMINARIES

The outstanding success of the Bohr theory was the derivation of the energy-level spectrum for hydrogenic atoms:

$$E_n = -\frac{mZ^2e^4}{2(4\pi\varepsilon_0)^2n^2\hbar^2} = -\frac{Z^2hc}{n^2}R_\infty \tag{1}$$

Here the principal quantum number n is integral; $h = 2\pi\hbar$ is Planck's constant; Z is the charge on the nucleus, m and e are, respectively, the mass and charge on the electron; and ε_0 is the permittivity of free space. The Rydberg constant for an infinitely heavy nucleus, R_∞, is regarded as a fundamental atomic constant with approximate value 10,973,731 m^{-1}. Equation (1) is exactly the relationship that follows from the boundary conditions required to obtain physically realistic solutions for Schrödinger's time-independent equation for one-electron atoms. However, the Schrödinger equation did not account for the *fine structure* in the spectra of atoms nor for the splittings of spectral lines in magnetic or electric fields.

In 1927 Dirac developed a relativistic wave equation,[7] which introduced an additional angular momentum for the spinning electron of magnitude $s^*\hbar$, where $s^* = \sqrt{s(s+1)}$ and the spin quantum number s has the value $s = \frac{1}{2}$. The orbital and spin angular momenta are coupled together to form a total angular momentum vector $\mathbf{j}$, given by $\mathbf{j} = 1 \pm \frac{1}{2}$. In the hydrogenic ground state $|nl\rangle = |10\rangle$, this spin-orbit coupling yields a value of $\mathbf{j} = \frac{1}{2}$ only, giving the $1S_{1/2}$ level. In the first excited state, for which $n = 2$ the $l = 0$, $\mathbf{j} = \frac{1}{2}$ is represented by $2S_{1/2}$, while $l = 1$ leads to $\mathbf{j} = \frac{3}{2}$ ($2P_{3/2}$) and $\mathbf{j} = \frac{1}{2}(2P_{1/2})$ levels, these two levels being separated by the fine structure interval.[8] The Dirac form of the Coulomb energy, expressed as an expansion in powers of $Z \times$ the fine structure constant, $\alpha = (e^2/4\pi\varepsilon_0\hbar c)$, is then

$$E_{DC} = -\frac{Z^2}{n^2} R_\infty hc\left[1 + \frac{(Z\alpha)^2}{n}\left(\frac{1}{\mathbf{j}+\frac{1}{2}} - \frac{3}{4n}\right) + O((Z\alpha)^4)\right] \tag{2}$$

The second term in the bracket in Eq. (2) is the spin-orbit correction to the energies which scales as $(Z^4\alpha^2)/n^3$. In the case of hydrogenic atoms this relativistic coupling removes the $nP_{1/2} - nP_{3/2}$ and $nD_{3/2} - nD_{5/2}$ degeneracy, but does not split the $nS_{1/2}$ level away from the $nP_{1/2}$ level. A further relativistic correction to Eq. (1) involves replacing the electronic mass in R_∞ by the reduced mass of the electron $\mu = mM/(M + m)$, which introduces a further shift of order $(m/M)_N(1 - (Z\alpha/2n)^2)E_{DC}$. Here M_N is the mass of the nucleus.

There are two further energy-level shifts.[9] The so-called quantum electrodynamic (QED) shifts include contributions due to finite nuclear size, relativistic recoil, and radiative corrections, collectively described as the Lamb shift, as well as terms due to electron self-energy and vacuum polarization. The Lamb shift raises the degeneracy of the $nS_{1/2} - nP_{1/2}$ levels. Overall, the QED shift scales as $\alpha(Z\alpha)^4/n^3$.[10] The interaction of the electronic and nuclear magnetic moments gives rise to hyperfine structure in spectra. The hyperfine contribution to the electronic energies for a nucleus of mass M_N, nuclear spin I, and gyromagnetic ratio g_N, is given by

$$E_{hf} = \alpha^2\left(\frac{Z^3}{n^3}\right)\left(\frac{g_N m}{M_N}\right) hcR_\infty \frac{F(F-1) - I(I+1) - j(j+1)}{j(j+2)(2l+1)} \tag{3}$$

where $\mathbf{j} = \mathbf{l} + \mathbf{s}$ is the total electronic angular momentum and $\mathbf{F} = \mathbf{I} + \mathbf{j}$ is the total atomic angular momentum. E_{hf} scales as $Z^3\alpha^2/n^3$ and is larger for S-states than for higher-orbit angular momentum states. More generally, all the correction terms scale as some power of Z/n, demonstrating that the shifts are greatest for $n = 1$ and larger nuclear charge. Experiments on atomic hydrogen are particularly important, since they give direct tests of relativistic quantum mechanics and QED.

8.4 RATES OF SPECTROSCOPIC TRANSITION

The rates of transitions may be determined using time-dependent perturbation theory. Accordingly, it is necessary to consider perturbations which mix stationary states of the atom. The perturbations are real and oscillatory in time with angular velocity, ω, and have the form

$$H_1 = V\exp(-i\omega t) + V^*\exp(\omega t) \tag{4}$$

where V is a function only of the spatial coordinates of the atom. In the presence of such a time dependent perturbation, the Schrödinger equation

$$(H_0 + H_1)\Psi = i\hbar\frac{\delta\Psi}{\delta t} \tag{5}$$

has eigenstates

$$\Psi = \sum_n c_n(t)\,|nlm\rangle\exp(-iE_n t/\hbar) \tag{6}$$

which are linear combinations of the n stationary solutions of the time independent

Schrödinger equation, which have the eigenvalues, E_n. The time-dependent coefficients, $c_m(t)$, indicate the extent of mixing between the stationary state wavefunctions, $|nlm\rangle$. The value of $|c_j(t)|^2$, the probability that the electronic system, initially in state i will be in a final state f after time, t, is given by

$$|c_f(t)|^2 = 4\left|\frac{V_{fi}}{\hbar}\right|^2 \frac{\sin^2 \frac{1}{2}(\omega_{fi} - \omega)t}{(\omega_{fi} - \omega)^2} \tag{7}$$

for an absorption process in which the final state f is higher in energy than the initial state, i. This expression defines the Bohr frequency condition,

$$\hbar\omega = E_f - E_i \tag{8}$$

and $\omega_{fi} = (E_f - E_i)/\hbar$. Obviously, $|c_f(t)|^2$ has a maximum value when $\omega_{fi} = \omega$, showing that the probability of an absorption transition is a maximum when $E_f - E_i = \hbar\omega$. The emission process comes from the $V^* \exp(\omega t)$ term in Eq. (4): the signs in the numerator and denominator of Eq. (7) are then positive rather than negative. For the probability to be significant then requires that $\omega_{fi} + \omega = 0$, so that the final state, f, is lower in energy than the initial state. If the radiation field has a density of oscillatory modes $u(\omega)$ per unit frequency range, then Eq. (7) must be integrated over the frequency distribution. The transition rate is then

$$W_{fi} = \frac{2\pi}{\hbar^2} |V_{fi}^{\omega}|^2 \, u(\omega_{fi}) \tag{9}$$

in which the V_{fi}^{ω} indicates that only a narrow band of modes close to $\omega = \omega_{fi}$ has been taken into account in the integration. This equation, which gives the probability of a transition from $|i\rangle \rightarrow |f\rangle$ per unit time, is known as Fermi's Golden Rule.

In Eq. (9) $V_{fi} = \langle f|\, V\, |i\rangle$ and $V_{fi}^* = \langle f|\, \mathbf{V}^*\, |i\rangle$ determine the transition probabilities for absorption and emission between the initial, i, and final states, f. In fact $|V_{fi}^{\omega}|$ and $|V_{if}^{\omega^*}|$ are identical and the transition probabilities for absorption and emission are equal. For the kth mode with angular frequency ω_k the perturbation takes the form

$$V_k^{\omega} \cong \sum_i \left(e r_i \cdot E_k^0 + \frac{e}{2m}(l_i + 2s_i) \cdot B_k^0 + \tfrac{1}{2} e r_i \cdot r_i \cdot k E_k^0 \right) \tag{10}$$

The first term in Eq. (10) is the electric dipole (ED) term. The second and third terms are the magnetic dipole (MD) and electric quadrupole terms (EQ), respectively. The relative strengths of these three terms are in the ratio $(ea_0)^2{:}(\mu_B/c)^2{:}ea_0^2/\lambda^2$ where a_0 and μ_B are the Bohr radius and Bohr magneton, respectively. These ratios are then approximately $1{:}10^{-5}{:}10^{-7}$. Since the electromagnetic energy per unit volume contained in each mode, including both senses of polarization, is given by $2\varepsilon_0\kappa\, |E_k^0|^2$, the energy density, $\rho(\omega)$, per unit volume per unit angular frequency is just $4\varepsilon_0\kappa\, |E_k^0|^2\, u_k(\omega)$. Hence, from Eq. (9) and using only the first term Eq. (10) the electric dipole transition rate is determined as

$$W_{if} = \frac{\pi}{2\varepsilon_0\kappa\hbar^2} \sum_{\mathbf{k}} \left| \langle f| \sum_i e r_i \cdot \hat{\varepsilon}_k \,|i\rangle \right|^2 \rho(\omega) \tag{11}$$

where the summations are over the numbers of electrons, i, and polarization vectors, $\mathbf{k}$. For randomly polarized radiation Eq. (11) becomes

$$W_{if} = \frac{2\pi}{6\varepsilon_0\kappa\hbar^2} \left| \langle f| \sum_i e r_i \,|i\rangle \right|^2 \rho(\omega) \tag{12}$$

If the radiation has all the $\mathbf{E_k}$ vectors pointing along the z-direction then only this mode is taken into account and

$$W_{if} = \frac{\pi}{2\varepsilon_0 \kappa \hbar^2} \left| \langle f | \sum_i e z_i | i \rangle \right|^2 \rho(\omega) \tag{13}$$

These relationships Eqs. (12) and (13) are used subsequently in discussing experimental techniques for measuring optical absorption and luminescence spectra. They result in the selection rules that govern both polarized and unpolarized optical transitions.

For the most part the succeeding discussion is concerned with radiative transitions between the ground level, a, and an excited level, b. These levels have degeneracies g_a and g_b with individual ground and excited states labeled by $|a_n\rangle$ and $|b_m\rangle$ respectively. The probability of exciting a transition from state $|a_n\rangle$ to state $|b_m\rangle$ is the same as that for a stimulated transition from $|b_m\rangle$ to $|a_n\rangle$. The transition rates in absorption, W_{ab}, and in emission, W_{ba}, are related through

$$g_a W_{ab} = g_b W_{ba} \tag{14}$$

assuming the same energy density for the radiation field in absorption and emission. Since the stimulated transition rate is defined by

$$W_{ab} = B_{ab} \rho(\omega) \tag{15}$$

the Einstein coefficient, B_{ab}, for stimulated absorption is directly related to the squared matrix element $|\langle b_m| \sum_i |er_i |a_n\rangle|^2$. Furthermore, the full emission rate is given by

$$W_{ba} = A_{ba}[1 + n_\omega(T)] \tag{16}$$

where $n_\omega(T)$ is the equilibrium number of photons in a blackbody cavity radiator at angular frequency, ω, and temperature, T. The first term in Eq. (16) (i.e., A_{ba}) is the purely spontaneous emission rate, related to the stimulated emission rate by

$$A_{ba} = 2B_{ba} u_k(\omega) \hbar \omega_k \tag{17}$$

Equation (17) shows that the spontaneous transition probability is numerically equal to the probability of a transition stimulated by one photon in each electromagnetic mode, **k**. Similarly the stimulated absorption rate is given by

$$W_{ab} = B_{ab} \rho(\omega) = \frac{g_b}{g_a} A_{ba} n_\omega(T) \tag{18}$$

These quantum mechanical relationships show how the experimental transition rates, for both polarized and unpolarized radiation, are determined by the mixing of the states by the perturbing oscillatory electric field. Since the radiative lifetime, τ_R, is the reciprocal of the Einstein A coefficient for spontaneous emission (i.e., $\tau_R = (A_{ba})^{-1}$) we see the relationship between luminescence decaytime and the selection rules via the matrix element $|\langle b_m| \sum_i er_i |a_n\rangle|$.

8.5 LINESHAPES OF SPECTRAL TRANSITIONS

Consider the excitation of optical transitions between two nondegenerate levels $|a\rangle$ and $|b\rangle$. The instantaneous populations of the upper level at some time t after the atomic system has been excited with a very short pulse of radiation of energy $\hbar\omega_{ab}$ is given by

$$N_b(t) = N_b(0) \exp(-A_{ba} t) \tag{19}$$

where A_{ba} is the spontaneous emission rate of photons from level $|b\rangle$ to level $|a\rangle$. Since the energy radiated per second $I(t) = A_{ba}N_b(t)\hbar\omega_{ab}$, the emission intensity at time t and frequency ω_{ba} is given by $I(t) = I(0)\exp(-t/\tau_R)$, where the radiative decaytime, τ_R, is defined as the reciprocal of the *spontaneous decay rate,* i.e., $\tau_R = (1/A_{ba})$. The expectation value of the time that the electron spends in the excited state, $\langle t\rangle$, is calculated from

$$\langle t\rangle = \frac{1}{N_b(0)}\int_{-\infty}^{+\infty} N_b(t)\,dt = (A_{ba})^{-1} = \tau_R \tag{20}$$

This is just the average time, or *lifetime,* of the electron in the excited state. In consequence, this simple argument identifies the radiative decaytime with the lifetime of the electron in the excited state. Typically for an allowed electric dipole transition $\tau_R \sim 10^{-8}$ s.

The radiation from a collection of atoms emitting radiation at frequency ω_{ba} at time $t > 0$ has an associated electric field at some nearby point given by

$$E(t) = E_0 \exp(i\omega_{ba}t)\exp(-t/2\tau_R) \tag{21}$$

(i.e., the electric field oscillates at the central frequency of the transition on the atom). The distribution of frequencies in $E(t)$ is obtained by Fourier analyzing $E(t)$ into its frequency spectrum, from which

$$E(\omega) = \frac{E_0}{\sqrt{(\omega_{ab} - \omega)^2 + (2\tau_R)^{-2}}}\exp(i\phi(\omega)) \tag{22}$$

where $\phi(\omega)$ is a constant phase factor. Since $I(t) \simeq E(t)^2$ we obtain the intensity distribution of frequencies given by

$$I(\omega) = \frac{I_0}{(\omega_{ab} - \omega)^2 + (2\tau_R)^{-2}} \tag{23}$$

This classical argument shows that the distribution of frequencies in the transition has a *lorentzian* shape with full width at half maximum (FWHM), $\Delta\omega$, given by

$$\Delta\omega = \frac{1}{\tau_R} = A_{ba} \tag{24}$$

An identical lineshape is derived from wave mechanics using time dependent perturbation theory. This relationship between the *natural linewidth* of the transition, $\Delta\omega$, and the radiative decaytime, τ_R, is related to the uncertainty principle. The time available to measure the energy of the excited state is just $\langle t\rangle$; the width in energy of the transition is $\Delta E = \hbar\Delta\omega$. Hence $\Delta E\langle\tau\rangle = \hbar\Delta\omega\tau_R = \hbar$ follows from Eq. (24). For $\tau_R \cong 10^{-8}$ s the energy width $\Delta E/c \simeq 5 \times 10^{-2}\,\mathrm{m}^{-1}$. Hence the natural linewidths of a transition in the visible spectrum is $\Delta\lambda \simeq 2 \times 10^{-3}$ nm.

The broadening associated with the excited state lifetime is referred to as *natural* or *homogeneous* broadening. There are other processes which modulate the energy levels of the atom thereby contributing to the overall decay rate, τ^{-1}. It is this overall decay rate, $\tau^{-1} > \tau_R^{-1}$, which determines the width of the transition. Examples of such additional processes include lattice vibrations in crystals and the vibrations/rotations of molecules. In gas-phase spectroscopy, random motion of atoms or molecules leads to inhomogeneous broadening via the Doppler effect. This leads to a gaussian spectral profile of FWHM given by

$$\Delta\omega_D = \frac{4\pi}{c}\left(\frac{2kT}{M}\ln 2\right)^{1/2}\omega_{ba} \tag{25}$$

showing that the Doppler width varies as the square root of temperature and is smaller in heavier atoms. In solids distortions of the crystal field by defects or growth faults lead to strain which is manifested as inhomogeneous broadening of spectra. The resulting lineshape is also a gaussian. The great power of laser spectroscopy is that spectroscopists recover the true homogeneous width of a transition against the background of quite massive inhomogeneous broadening.

8.6 SPECTROSCOPY OF 1-ELECTRON ATOMS

Figure 1*a* shows the energy level structure of atomic hydrogen for transitions between $n = 3$ and $n = 2$ states, i.e., the Balmer α transition. Electric dipole transitions are indicated by vertical lines. The relative strengths of the various lines are indicated by the lengths of the vertical lines in Fig. 1*b*. Also shown in Fig. 1*b* is a conventional spectrum obtained using a discharge tube containing deuterium atoms cooled to $T = 50$ K. This experimental arrangement reduces the Doppler width of the Balmer α transition at

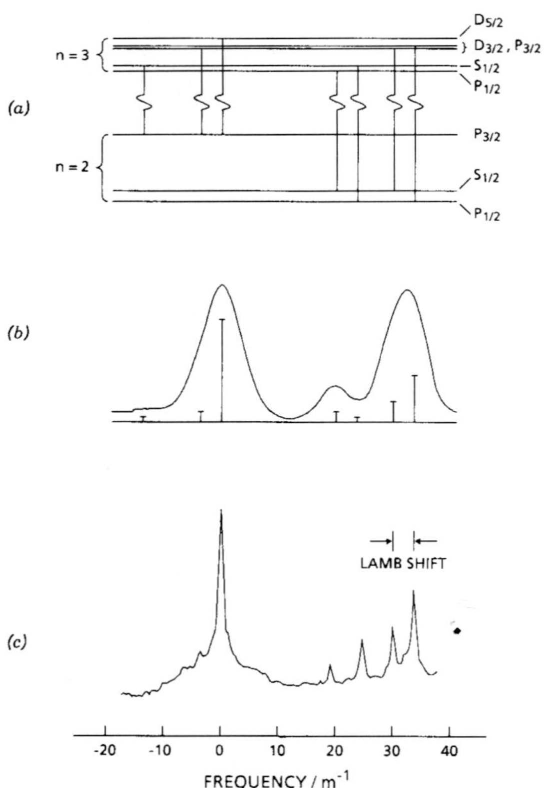

FIGURE 1 (*a*) The structure of the $n = 3$ and $n = 2$ levels of hydrogen showing the Balmer α-transitions. (*b*) A low resolution discharge spectrum of deuterium cooled to 50 K. (*c*) A Doppler-free spectrum of the Balmer α-spectrum of hydrogen. *(After Hänsch et al.*[12]*)*

656 nm to 1.7 GHz. Nevertheless, only three transitions $2P_{3/2} \rightarrow 3D_{5/2}$, $2S_{1/2} \rightarrow 3P_{1/2}$, and $2P_{1/2} \rightarrow 3D_{3/2}$ are resolved.[11] However, sub-Doppler resolution is possible using laser saturation spectroscopy.[12] Figure 1*c* shows the Doppler free Balmer α spectrum to comprise of comparatively strong lines due to $2P_{3/2} \rightarrow 3D_{5/2}$, $2S_{1/2} \rightarrow 3P_{1/2}$, $2S_{1/2} \rightarrow 3P_{3/2}$, and $2P_{1/2} \rightarrow 3D_{3/2}$ transitions, as well as a very weak $2P_{3/2} \rightarrow 3D_{3/2}$ transition. Also evident is a cross-over resonance between the two transitions involving a common lower level $2S_{1/2}$, which is midway between the $2S_{1/2} \rightarrow 3P_{1/2}$, $3P_{3/2}$ transitions. The splitting between $2P_{3/2}$, $2P_{1/2} \rightarrow 3D_{3/2}$ transitions measures the spin-orbit splitting in the n = 2 state, which from Eq. (2) is about 36.52 m^{-1}. The Lamb shift is measured from the splitting between $2S_{1/2} \rightarrow 3P_{3/2}$ and $2P_{1/2} \rightarrow 3D_{3/2}$ lines to be 3.53 m^{-1}, which compares well with the original microwave measurement (3.537 m^{-1}).[13] Subsequently, Hänsch et al.[14] made an interferometric comparison of a Balmer α-line with the 632.8 nm line from He-Ne locked to a component of $^{129}I_2$, thereby deriving a value of R_∞ of 10973731.43(10) m^{-1}, at that time an order of magnitude improvement in accuracy on previous values. Neither the $2S_{1/2}$ nor $2P_{1/2}$ hfs was resolved in this experiment, both splittings being less than the system resolution of *ca* 0.05 m^{-1}. This probably followed from the use of pulsed dye lasers where the laser linewidth exceeds by factors of ten the linewidth available from single-frequency CW dye lasers. Subsequent measurements using CW dye lasers standardized against I_2-stabilized He-Ne lasers gave further improvements in the value of R_∞.[15,16] Also using the Balmer α-transition Stacey et al[17] have studied the isotope shifts between spectra from hydrogen, deuterium, and tritium. The spectrum shown in Fig. 2 reveals the $2S_{1/2}$ hyperfine splitting on the $2S_{1/2} \rightarrow P_{3/2}$ and $2S_{1/2} \rightarrow 3D_{3/2}$ transitions. These measurements yield isotope shifts of 124259.1(1.6) MHz and 41342.7(1.6) MHz for *H-D* and *D-T* respectively, which accord well with theoretical values.

The literature on H-atom spectroscopy is vast, no doubt fueled by the unique relationship between experimental innovation and fundamental tests of relativistic quantum mechanics and QED. This article is not a comprehensive survey. However, it would be seriously remiss of the author to omit mention of three other categories of experimentation. The first experimental arrangement uses crossed atomic and laser beams: a well-collimated beam of atoms propagates perpendicular to a laser beam which, after traversing the atomic beam, is reflected to propagate through the atomic beam again in the opposite direction. The interaction between the counterpropagating beams and the atoms in the atomic beam is signaled by a change in the beam flux. The atomic beam replaces

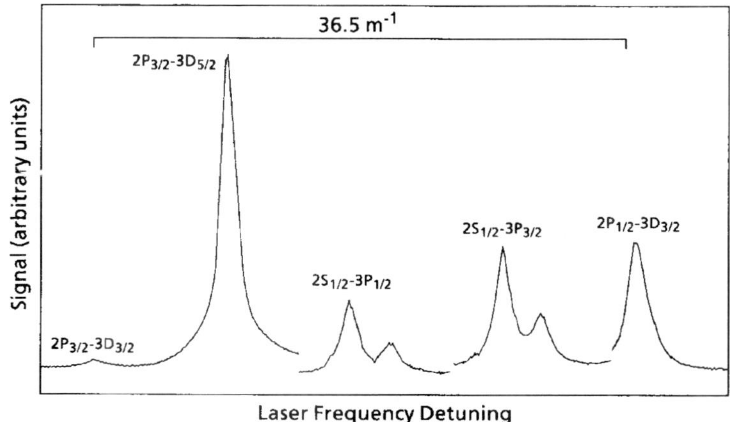

FIGURE 2 Showing hyperfine structure splittings of the $2S_{1/2} \rightarrow 3P_{3/2}$, $3D_{3/2}$ transition in hydrogen. *(After Stacey.[17])*

the discharge unit used in conventional atomic spectroscopy, thereby reducing errors due to the electric fields in discharges. This experiment is the optical analogue of the Lamb-Retherford radio-frequency experiment[13] and has been much used by workers at Yale University.[18] They reported a value of $R_\infty = 10973731.573(3)\ \text{m}^{-1}$ in experiments on the Balmer β ($n = 2$ to $n = 4$) transition.

There have been several other studies of the Balmer β transition, which has a narrower natural linewidth than the Balmer α transition. However, because it is weaker than the Balmer α transition, Wieman and Hänsch used a polarization scheme to enhance the sensitivity of the saturation absorption scheme.[19] Finally, the metastable $2S_{1/2}$ level may decay spontaneously to the $1S_{1/2}$ ground state with the emission of two photons with energies that sum to the energy separation between $1S_{1/2}$ and $2S_{1/2}$. Such a process has a radiative decaytime of 0.14 s, giving a natural linewidth for the $2S_{1/2} \rightarrow 1S_{1/2}$ transition of order 1 Hz! The probability of a two-photon absorption transition is quite low. However, as with laser absorption saturation spectroscopy, two photon absorption experiments are made feasible by Doppler-free resolution. Wieman and Hänsch[20] used an amplified CW laser beam at 243 nm to excite a two-photon absorption transition, which they detected by observing the Lyman α emission at 121 nm. In addition, part of the laser beam was split off and used to measure simultaneously the Balmer β spectrum. The coupled experiment permitted a direct measurement of the ground-state Lamb shift of 8161(29) MHz. Recently Ferguson and his colleagues have developed standard cells using $^{130}Te_2$ lines for studies of the Balmer β and $1S_{1/2} \rightarrow 2S_{1/2}$ transitions.[10,21]

8.7 MULTIELECTRON ATOMS

In order to calculate the energy level spectrum of a multielectron atom we require a suitable hamiltonian describing the interaction of all the electrons with the nucleus and with each other. A convenient starting point is the simplified Dirac equation for a one-electron atom, viz.,

$$H = H_0 + H_{so} \tag{26}$$

where H_0 is the simplified hamiltonian for the electron in the field of a nucleus of charge Ze at rest, i.e.,

$$H_0 = \frac{p^2}{2m} - \frac{Ze^2}{4\pi\varepsilon_0 r} \tag{27}$$

and $H_{so} = -\zeta(r)l \cdot s$ is the spin-orbit hamiltonian. Wavefunctions which satisfy Eq. (27) are

$$\psi_{nlm_l}(r) = R_{nl}^{(r)} Y_l^m(\theta\phi) \tag{28}$$

where the labels n, l, and m are quantum numbers which characterize the eigenstates. The eigenvalues, given in Eq. (1), depend only on the principal quantum number, n, which takes positive integral values. The quantum number, l, characterizing the orbital angular momentum also takes integral values, $l = 0, 1, 2, 3, \ldots, (n-1)$, whereas m_1 measures the z-component of the orbital angular momentum. There are $21 + 1$ integral values of l given by $m = l, (l-1), (l-2), \ldots, -(l-1), -l$, and for a given value of n there are several different orbital states with identical energy. H_{so}, among other interactions, raises this degeneracy.

It is convenient to represent the orbital wavefunction, ψ_{nlm} by the ket $|nlm\rangle$. Including spin angular momentum we represent a *spin orbital* by $|nlsmm_s\rangle$ or more simply by $|nlmm_s\rangle$. Recalling the brief discussion of the coupled representation, whereby $j = l + s$, an equally valid representation is $|nljm_j\rangle$. Indeed the new basis states $|nljm_j\rangle$ are just linear combinations of the $|nlmm_s\rangle$ basis states.[8] Each wavefunction has a definite parity. The parity of wavefunctions is important in determining the selection rules of spectra. The inversion operator P_i, defined by $P_i f(r) = f(-r)$ for any function of r, gives the following result

$$P_i \,|nlm\rangle = (-1)^l \,|\, nlm\rangle \tag{29}$$

Hence, for even values of l the wavefunctions are said to have even parity since they do not change sign under inversion of coordinates. For l odd the wavefunctions have odd parity. The strength of an optical transition is determined by a matrix element $\langle\psi_b|\, \mu \,|\psi_a\rangle$, where the integration is taken over the volume of the atom. In the electric dipole approximation $\mu = -er$ so that the matrix element is zero except that the wavefunction ψ_a and ψ_b have opposite parity. This defines the Laporte selection rule which states that the parity of a state must change from odd to even (or vice versa) in an electric dipole transition.

The hamiltonian for multielectron aotms is a sum over all N electrons of one-electron operators [see Eq. (1)] plus an electron-electron Coulomb repulsion between electrons i and j separated by a distance r_{ij}. Hence we may write this as

$$H = \sum_i \left(\frac{P_i^2}{2m} - \frac{Ze^2}{4\pi\varepsilon_0 r_i} + \zeta(r_i) l_i \cdot s_i\right) + \sum_{i \neq j} \frac{e^2}{4\pi\varepsilon_0 r_{ij}} \tag{30}$$

The computational complexity militates in favor of an approximate solution because the spin-orbit and electron-electron interactions are not spherically symmetric. In consequence, the first stage of the approximation is to Eq. (30) in the form

$$H = \sum_i \frac{p_i^2}{2m} + V_i'(r_i) + \zeta(r_i) l_i \cdot s_i. \tag{31}$$

where $V_i'(r_i)$ is a spherically symmetric one-electron operator which represents the potential energy of the ith electron in the field of the nucleus and all the electrons. The first two terms in this sum constitute the orbital hamiltonian, H_0, a sum of one-electron hydrogen-like hamiltonians (Eq. (27)), but with a more complicated radial potential energy function, $V'(r)$. The radial and angular parts of each one electron hamiltonian are separable and we write orbital functions

$$R_{nl}'(r_i) Y_l^m(\theta, \phi) = |nlm\rangle \tag{32}$$

However, $R_{nl}'(r_i)$ is the solution of the radial equation involving the central potential, $V'(r_i)$, which is characterized by the quantum numbers n and l. In consequence, the energy of the one electron state also depends on both n and l. The complete spin orbital is characterized by four quantum numbers including spin (i.e., $u = |nlmm_s\rangle$) and the many electron eigenstate of H_0 is a product of one electron states

$$U = \prod_i |nlmm_s\rangle_i \tag{33}$$

The energy E_u of this product state is

$$E_u = \sum_i E_{n_i l_i} \tag{34}$$

which depends on the set of $n_i l_i$ values. However, since E_u does not depend on m_l and m_s these eigenstates have a large degeneracy.

Experimentally the complete wavefunctions of electrons are antisymmetric under the exchange of orbital and spin coordinates of any two electrons. The product wavefunction, Eq. (33), does not conform to the requirement of interchange symmetry. Slater solved this problem by organizing the spin orbitals into an antisymmetric N-electron wavefunction in determinantal form.[22] The application of the Hartree-Fock variational approach to determine the central field potential consistent with the best Slater wavefunctions is described in detail by Tinkham.[23] There are many different sets of energy eigenfunctions that can be chosen; the net result is that the eigenstates of H_0 can be classified by a set of quantum numbers LSM_LM_S for each (n_il_i) electron configuration, where $L = \sum_i l_i$ and $S = \sum_i s_i$. That the eigenfunctions must be antisymmetric restricts the number of possible L and S values for any given configuration. Since $J = L + S$ is also a solution of H_0, we can represent the eigenstates of the configuration by the ket $|LSM_LM_S\rangle$ or alternatively by $|LSJM_J\rangle$ where the eigenstates of the latter are linear combinations of the former.

There is a particular significance to the requirement of antisymmetric wavefunctions in the Slater determinantal representation. A determinant in which any two rows or columns are identical has the value of zero. In the present context, if two one electron states are identical, then two columns of the Slater determinant are identical, and the wavefunction is identically zero. This is a statement of the Pauli Exclusion Principle: no two electrons in an atom can occupy identical states (i.e. can have the same four quantum numbers). The Slater wavefunctions indicate those one-electron states which are occupied by electrons. To see how this works consider two equivalent electrons in p-states on an atom. For $n = n'$ for both electrons, the l-values may be combined vectorially to give $L = 2$, 1, and 0. Similarly, the two electron spins, $s = \frac{1}{2}$, may be combined to give $S = 1$ or 0. The antisymmetric requirement on the total wavefunction means that the symmetric orbitals D $(L = 2)$ and S $(L = 0)$ states can only be combined with the antisymmetric spin singlet, $S = 0$. The resulting spin orbitals are represented by ^1D and ^1S. In contrast the antisymmetric P state must be combined with the spin triplet, which is a symmetric function, yielding the antisymmetric spin orbital, ^3P. However, for two inequivalent p-electrons in an excited state of the atom both spin singlet and spin triplet states are possible for the S, P, and D orbital states.

8.8 OPTICAL SPECTRA AND THE OUTER ELECTRONIC STRUCTURE

Optical spectroscopy probes those electronic transitions associated with a small number of electrons outside the closed shells of electrons. This gives further simplification to the computational problem since the multielectron hamiltonian

$$H = \sum_i \frac{p_i^2}{2m} + V'(r_i) + \zeta(r_i) l_i \cdot s_i + \sum_{i>j} \left(\frac{e^2}{4\pi\varepsilon_0 r_i} \right) \tag{35}$$

is summed only over the outer electrons where each of these electrons moves in the central field of the nucleus and the inner closed-shell electrons, $V'(r_i)$. Neglecting the smallest term due to the spin-orbit interaction, the hamiltonian in Eq. (35) takes the form $H_0 + H'$, where H_0 is a sum of one-electron hamiltonian with the radial potential functions $V'(r_i)$ and H' is the energy of the Coulomb interaction between the small number of outer electrons. The corrections to the one electron energies are then the diagonal matrix elements $\langle n_il_im_im_{si}| H' |n_il_im_im_{si}\rangle$, expressed either in terms of Racah parameters, A, B, C, or Slater parameters, F_0, F_2, F_4 . . . .[24] Transition metal ion energy levels are normally described in terms of the Racah parameters and rare earth ion energy levels in terms of the Slater functions. The effect of H' is to split each configuration (n_il_i) into a number of LS terms for each of which there are $(2L + 1)$ $(2S + 1)$ distinct energy eigenstates. We represent the energy eigenstates by the kets $|(n, l)LSJM_j\rangle$, which are defined as linear combinations of the $|(n_i l_i)LSM_L M_S\rangle$ states.

Returning briefly to the $(np)^2$ configuration (characteristic of the Group 4 elements, C, Si, Ge, etc. of the Periodic Table) it is noted that the diagonal matrix elements evaluated in terms of Slater parameters are given by $E(^1D) = F_0 + F_2$, $E(^1S) = F_0 + 10F_2$ and $E(^3P) = F_0 - 5F_2$. For the different atoms it is the relative values of F_0 and F_2 which change through the series $(2p)^2$, $(3p)^2$, $(4p)^2$, etc. Note that it is the term with maximum multiplicity, ^3P, which is lowest in energy in conformity with Hund's rule. A similar situation arises for the $(np)^4$ configuration of, for example, atomic O, S, and Se which might equally and validly be considered as deriving from two holes in the $(np)^2$ configuration. The general conclusion from this type of analysis is that the energy level structures of atoms in the same period of the Periodic Table are identical, with the principal differences being the precise energies of the eigenstates. This is evident in Fig. 3, the term scheme for atomic Li, which has the outer electron configuration $(2S)^1$; this may be looked on as a pseudo-one-electron atom. There is general spectroscopic similarity with atomic hydrogen, although the $(ns)^1 - (np)^1$ splittings are much larger than in hydrogen. The $3S_{1/2} \leftrightarrow 2S_{1/2}$ transition is forbidden in Li just as the $2S_{1/2} \leftrightarrow 1S_{1/2}$ transition is in hydrogen. However, it is unlikely to be observed as a two-photon process in emission because the $3S_{1/2} \rightarrow 2P_{3/2,1/2}$ and $2P_{3/2,1/2} \rightarrow 2S_{1/2}$ transitions provide a much more effective pathway to the ground state.

The comparison between Li and Na is much more complete as Figs. 3 and 4 show: assuming as a common zero energy the ground state, the binding energies are $E(2S_{1/2}, \text{Li}) = 43{,}300\ \text{cm}^{-1}$ and $E(3S_{1/2}, \text{Na}) = 41{,}900\ \text{cm}^{-1}$. The energies of the corresponding higher lying nS, nP, and nD levels on Li are quite similar to those of the $(n+1)S$, $(n+1)P$, and $(n+1)D$ levels on Na. This similarity in the energy level structure is reflected in the general pattern of spectral lines, although the observed wavelengths are a little different. For example, the familiar D-lines in the emission spectrum of Na occur at $\lambda \simeq 589.9$ nm whereas the corresponding transition in Li occur at 670.8 nm.

Examination of the term schemes for a large number of elements reveal striking similarities between elements in the same group of the Periodic Table, due to this structure being determined by the number of electrons outside the closed shell structure. These term diagrams also reveal a very large number of energy levels giving rise to line spectra in the ultraviolet, visible, and infrared regions of the spectrum. The term diagrams are not very accurate. Tables of accurate energy levels determined from line spectra have been compiled by C. E. Moore[25] for most neutral atoms and for a number of ionization states of the elements. This comprehensive tabulation reports energy levels to a very large principal quantum number ($n \cong 10, 11$).

The term diagram of the neutral Tl atom, $(6p)^1$ configuration (see Fig. 5) shows two interesting features in comparison with those for alkali metals (see Figs. 3 and 4). In the $(6p)^1$ state the spin-orbit splitting into $6P_{1/2}$ and $6P_{3/2}$ levels amounts to almost 8000 cm^{-1} whereas the spin-orbit splitting between the $3P_{1/2}$ and $3P_{3/2}$ levels of Na is only 17 cm^{-1}. This reflects the (Z^4/n^3) dependence of the spin-orbit coupling constant. Furthermore, when Tl is in the $7S_{1/2}$ state it can decay radiatively via transitions to either $6P_{3/2}$ or $6P_{1/2}$, each with a distinct transition probability. The relative probability of these transitions is known as the branching ratio for this mode of decay. Other examples of branching are apparent in Fig. 5: the branching ratios are intrinsic properties of the excited state.

8.9 SPECTRA OF TRI-POSITIVE RARE EARTH IONS

The rare earth elements follow lanthanum (Z = 57) in the Periodic Table from cerium (Z = 58), which has the outer electron configuration $4f^1 5d^1 6s^2$ to ytterbium (Z = 70) with electron configurations $4f^{13} 5d^1 6s^2$. In the triply charged state in ionic crystals all $5d$ and $6s$

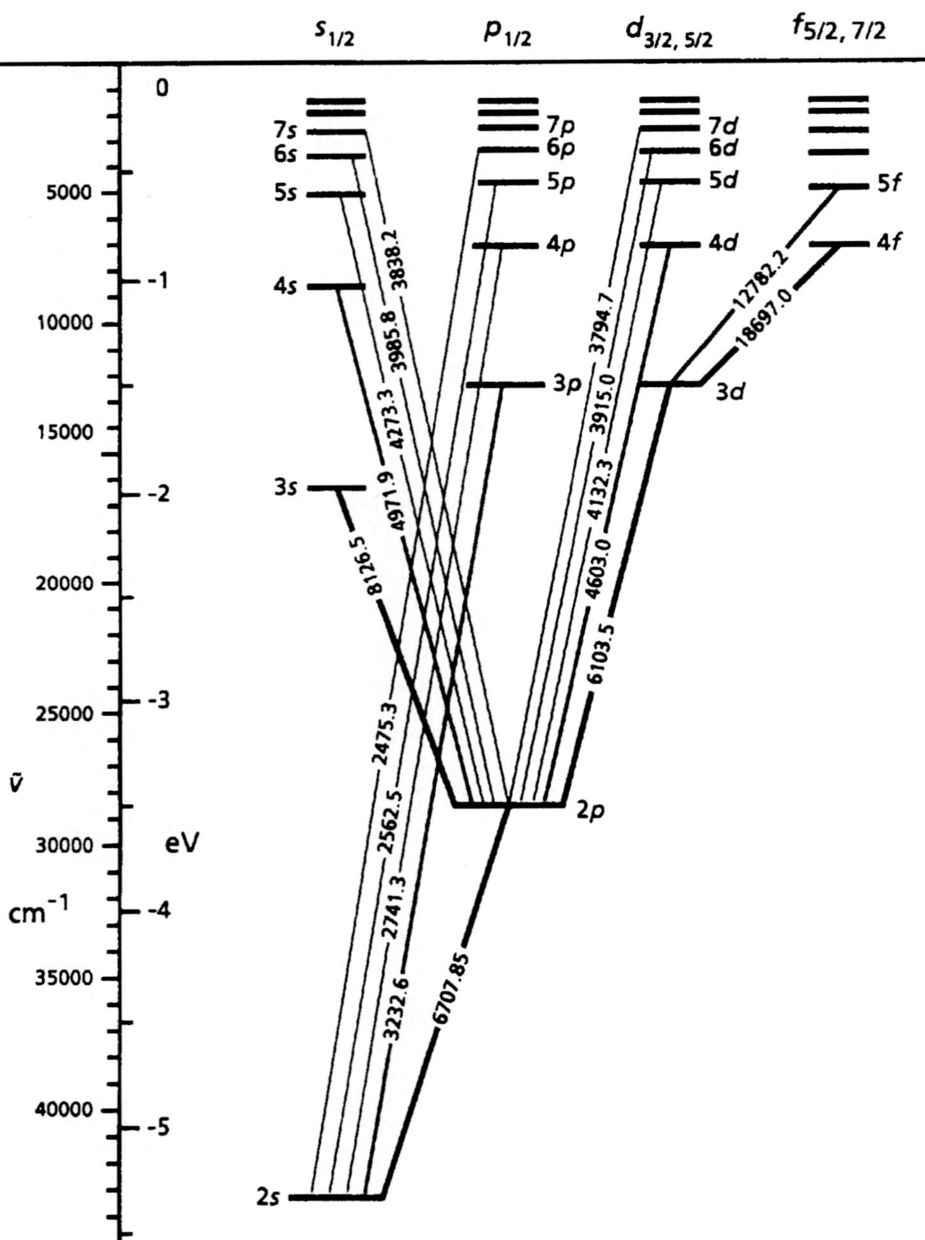

FIGURE 3 The term diagram of atomic Li, in which the slanted lines indicate the observed electric dipole transitions and the numbers on the lines are the wavelengths in Ångstrom units. *[After W. Grotian "Graphische Darstellung der Spektren von Atomen," Volume 2 (Springer Verlag, Berlin, 1928).]*

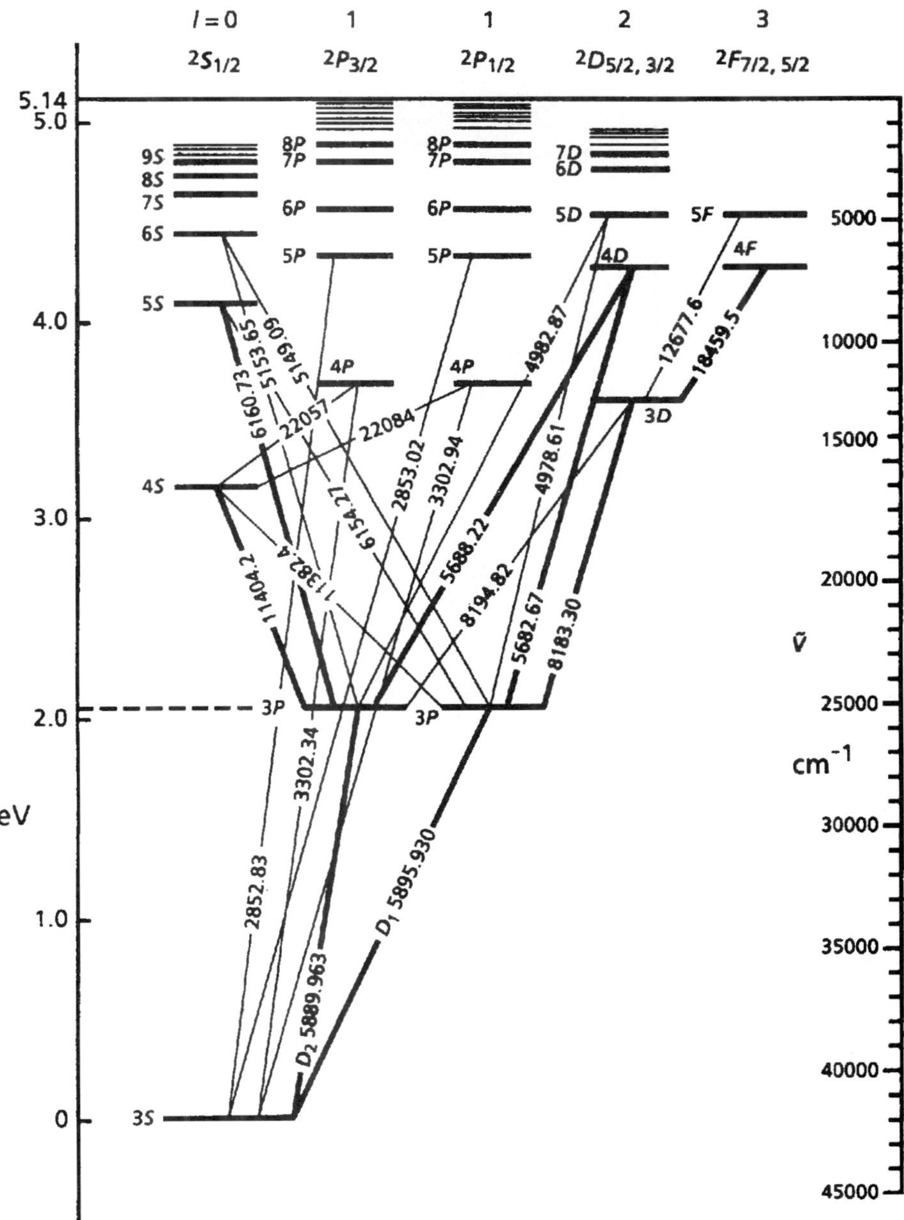

FIGURE 4 The term diagram of atomic Na. *(After Grotian, 1928 ibid.)*

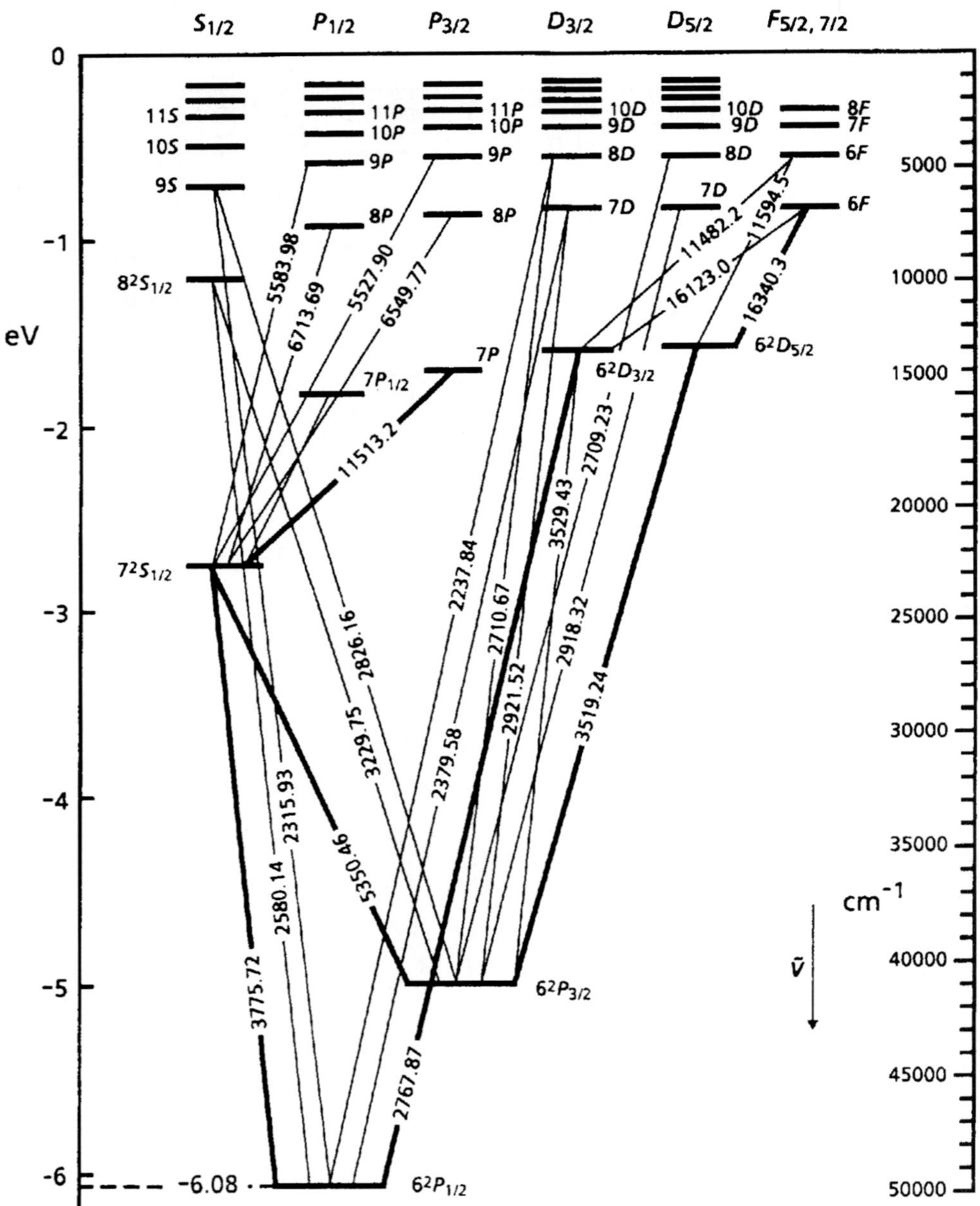

FIGURE 5 The term diagram of neutral Tl. *(After Grotian, 1928 ibid.)*

TABLE 1 The Number of Electrons (n) and Ground State of Tri-positive Rare Earth Ions

Ion	n (in ed^n)	Ground state
Ce^{3+}	1	$^2F_{5/2}$
Pr^{3+}	2	$^3H_4$
Nd^{3+}	3	$^4I_{9/2}$
Pm^{3+}	4	$^5I_4$
Sm^{3+}	5	$^6H_{5/2}$
Eu^{3+}	6	$^7F_0$
Gd^{3+}	7	^8S
Tb^{3+}	8	$^7F_6$
Dy^{3+}	9	$^6H_{15/2}$
Ho^{3+}	10	$^5I_8$
Er^{3+}	11	$^4I_{15/2}$
Tm^{3+}	12	$^3H_6$
Yb^{3+}	13	$^2F_{7/2}$

electrons are used in ionic bonding and the many energy levels of these R.E.$^{3+}$ ions are due to the partially filled $4f$ shell. The number of electrons in the $4f$ shell for each trivalent ion and the ground state configuration is indicated in Table 1. The energy levels of the unfilled $4f^n$ shells spread out over some 40,000 cm^{-1} giving rise to numerous radiative transitions with energies in the visible region. A remarkable feature of the $4f^n$ electrons is that they are shielded by the outer $5s$ and $5d$ shells of electrons, with the result that $4f$ electrons are not strongly affected by interactions with neighboring ions in crystals. In consequence, the energy levels of the $4f$ electrons in crystals are essentially the free ion levels characterized by quantum numbers L, S, and J. As with the free ions the R.E.$^{3+}$ ions in crystals have very sharp energy levels which give rise to very sharp line spectra. The crystal field interaction does split the R.E.$^{3+}$ ion levels, but this splitting is very much smaller than the splittings between the free ion levels. Hence for rare earth ions in different crystals the gross features of the optical spectra are unchanged.

As discussed in Section 8.6 the eigenstates of the $4f$ electrons are calculated using the central field approximation from which each $4f$ electron state is characterized by the ket $|n = 4,\ l = 3m_l m_s\rangle$. The effect of the Coulomb repulsion between electrons, $H' = \sum_i e^2/4\pi\varepsilon_0 r_{ij}$, is to split the energy levels of the $4f^n$ configuration into different LS terms, with wavefunctions characterized by kets $|LSM_L M_S\rangle$. The magnitudes of the electrostatic interaction for each LS-level are expressed as sums of Slater electron-electron integrals F_k, with $k = 0$, 2, 4, and 6 for $4f$ electrons. Since F_0 contributes equally to all LS states of the same $4f^n$ configuration, this term can be neglected. Generally, these Slater integrals are regarded as adjustable parameters with magnitudes determined by fitting to the measured line spectra. The values of the F_k integrals for $4f^n$ ions in many crystals vary by only about 2 percent from those obtained for free ions; they also vary slightly depending on the nature of the surrounding ions. The next largest term in the hamiltonian, after H', is spin-orbit coupling. If the spin-orbit coupling energy is much smaller than the energy separation in the LS term then the spin-orbit interaction can be written as $\zeta L \cdot S$ and the wavefunctions are characterized by $|LSJM_J\rangle$. The additional energy of the J-multiples are given by the Landé interval formula

$$E_J = \frac{\zeta}{2}[J(J+1) - L(L+1) - S(S+1)] \tag{36}$$

from which it is evident that the separation between adjacent levels is given by $J\zeta$ where J

refers to the upper J value. However, deviations from this Landé interval rule do occur because of mixing of different LS-terms when spin-orbit coupling and electron-electron interaction are of similar magnitudes. Clear examples of this are obtained from the spectra of $Pr^{3+}(4f^2)$ and $Tm^{3+}(4f^{12})$.

As representative of the type of spectra observed from R.E.$^{3+}$ ions in ionic crystals we consider just one example: $Nd^{3+}(4f^3)$ in $Y_3Al_3O_{12}$(YAG). The Nd^{3+}-YAG material is important as the gain medium in a successful commercial laser. Nd^{3+} has a multitude of levels many of which give rise to sharp line emission spectra. A partial energy level structure is shown in Fig. 6. The low temperature emission is from the $^4F_{3/2}$ level to all the ^4I_J levels of the Nd^{3+} ion. The spectra in Fig. 6 also show the splittings of the ^4I_J levels by the crystal field, corresponding to the energy level splitting patterns given in the upper

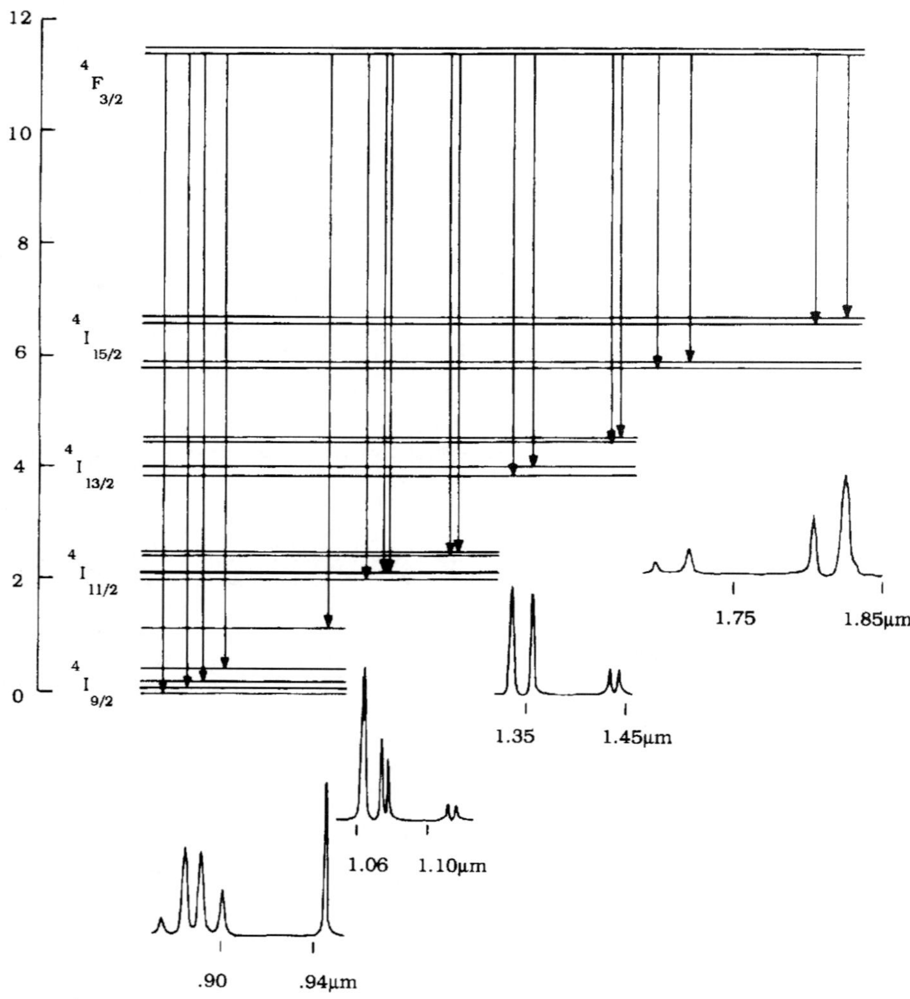

FIGURE 6 The low temperature photoluminescence spectrum from the $^4F_{3/2}$ level of Nd^{3+} in $Y_3Al_5O_{12}$ and the corresponding energy level structure. *(After Henderson and Imbusch.[8])*

portion of Fig. 6. Depending upon crystal quality these lines can be quite narrow with half-widths of order a few gigahertz. Nevertheless the low-temperature width in the crystal is determined by the distribution of internal strains and hence the lines are inhomogeneously broadened. The natural linewidth of rare earth ion spectra is of the order of a few megahertz. By using optical hole burning (OHB), which is similar to the saturated absorption spectroscopy discussed earlier for atomic hydrogen, it is possible to eliminate the inhomogeneous broadening and recover the homogeneous lineshape of the spectrum. In principle, the natural width is determined by lifetime processes of which there are more numerous sources in crystals than in atomic vapors. Indeed the width may be determined by random modulation of the optical lineshape by photons and by both nuclear and electronic spins of neighboring ions. The two most general techniques for determining the homogeneous widths of optical transitions in solids are optical holeburning (OHB) and fluorescence line narrow (FLN). Examples of these techniques are discussed elsewhere in this treatise.

8.10 VIBRATIONAL AND ROTATIONAL SPECTRA OF MOLECULES

A consultation of any one of the Tables of data in Moore's compilation[25] shows that the energy level schemes of most atoms are complex. This is confirmed by the associated atomic spectra. Considerable interpretive simplication is afforded by the construction of term diagrams (e.g., Figs. 3–5), on which a very large number of lines may be associated with a much smaller number of terms, each term corresponding to an energy level of the atom. The observed spectral lines are due to transitions between pairs of terms (not all pairs) which occur subject to an appropriate selection rule. The spectra of even simple molecules measured with low dispersion show characteristic band spectra which are even more complicated than the most complex atomic spectra. These band spectra, when studied at higher spectral resolution, are observed to consist of an enormous number of closely spaced lines. At first acquaintance, such band spectra appear to be so complex as to defy interpretation. Order can be brought to the riot of spectral components by constructing term schemes for molecules involving electronic, vibrational, and rotational energy terms, which enable the molecular spectroscopist to account for each and every line.

Molecular physics was a research topic to which quantum mechanics was applied from the very earliest times. Heitler and London developed the valence band theory of covalency in the H_2-molecule in 1927.[26] The theory shows that with both electrons in 1s states there are two solutions to the hamiltonian

$$E_{\pm} = 2E(1\text{H}) + \frac{K \pm \Im}{1 \pm \mathcal{S}} \tag{37}$$

where $E(1\text{H})$ is the energy of an electron in the ground state of atomic hydrogen, K is the Coulomb interaction due to the mutual actions of charges distributed over each atom, $\Im$ is the exchange energy, and $\mathcal{S}$ is the overlap integral. The exchange energy is a purely quantum mechanical term, representing the frequency with which the deformation of the wavefunctions by their mutual interaction oscillates from one atom to another. The positive sign refers to the symmetric combination of orbital wavefunctions for the two hydrogen atoms. Since the overall wavefunction must be antisymmetric, the combined spin states must be antisymmetric (i.e., the spin singlet $S = 0$; this state is labeled $^1\Sigma_g$. The evaluation of the integrals in Eq. (37) as a function of internuclear separation leads to Fig. 7

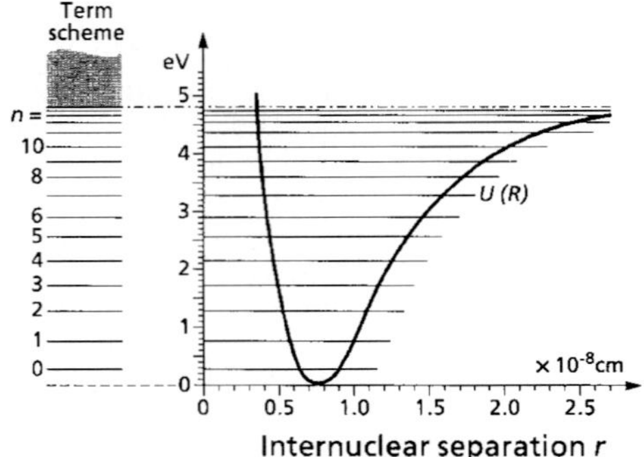

FIGURE 7 The internuclear potential, $V(R - R_0)$, in the ground state $^1\Sigma_g$ of the hydrogen molecule.

for the spin singlet state. This $^1\Sigma_g$ ground state has a potential energy minimum of about 4.8 eV (experimentally) with the nuclei separated by 0.75 nm, relative to the total energy of the two hydrogen atoms at infinity. The theoretical value of the binding energy on the valence band model is only 3.5 eV.[26] The negative sign in Eq. (37) guarantees that in the state characterized by the antisymmetric combination of orbital states and $S = 1$, i.e. $^3\Sigma_u$, the energy is monotonically ascending, corresponding to repulsion between the two hydrogen atoms for all values of R. In such a state the molecule dissociates.

The energy versus internuclear separation curve in Fig. 7 represents the effective potential well, in which the protons oscillate about their mean positions. This is the potential used in the Born-Oppenheimer treatment of the molecular vibrations in the spectra of diatomic molecules.[27] The potential function may be written as a Taylor expansion

$$V(R - R_0) = V_0 + (R - R_0)\left(\frac{dV}{d(R - R_0)}\right)_0 + \frac{(R - R_0)^2}{2}\left(\frac{d^2V}{d(R - R_0)^2}\right)_0 + \frac{(R - R_0)^3}{6}\left(\frac{d^3V}{d(R - R_0)^3}\right)_0$$

where the subscript 0 refers to values of the differentials at $R = R_0$ and higher terms have been ignored. For a molecule in stable equilibrium the potential energy is a minimum and the force at $R = R_0$ must be zero. In consequence, the potential energy function may be written as

$$V(R - R_0) = \tfrac{1}{2}(R - R_0)^2\left(\frac{d^2V}{d(R - R_0)^2}\right)_0 + \tfrac{1}{6}(R - R_0)^3\left(\frac{d^3V}{d(R - R_0)^3}\right)_0 \tag{38}$$

after setting $V_0 = 0$.

The excited states of H_2 are constructed on the assumption that only one electron is excited in the transition: the appropriate configurations may be written as (1*s*, 2*s*), (1*s*, 2*p*), (1*s*, 3*s*), etc. The electronic states of the molecule are then suitable combinations of the atomic states and categorized according to the total orbital angular momentum determined by vector addition of individual electronic orbital momenta. These orbital angular momentum states are designated as Σ, Π, Λ, etc. when the total angular momentum quantum number is 0, 1, 2, etc. Thus the electronic state in Fig. 7 is described as the $^1\Sigma_g$

(ground state) and the lowest lying excited state as $^3\Sigma_u$, where the subscripts g and u indicate even (gerade) and odd (ungerade) parity of the orbital states, respectively. The first excitation states associated with the (1*s*, 2*s*) molecular configuration, designated as $^1\Sigma_g$ and $^3\Sigma_u$, have energies

$$E_{\pm} = E(1\mathrm{H}) + E(2\mathrm{H}) + \frac{K \pm \Im}{1 \pm \mathcal{S}} \tag{39}$$

Because the charge distributions are different in the atomic 1*s* and 2*s* states, the values of K and $\Im$ are also different in the (1*s*, 2*s*) configuration relative to the ground configuration. Obviously, as the orbital degeneracy of the molecular configuration increases so does the number of molecular levels resulting from the configuration. For example, there are six molecular levels associated with (1*s*, *np*) configuration, ten associated with (1*s*, *nd*), and so on. Roughly speaking, half of these levels will have a potential minimum in a graph of electronic potential energy versus internuclear separation (e.g., Fig. 7), and are associated with "bonding" orbitals of the molecule. In the other levels the force between the constituent atoms is repulsive for all values of the internuclear separation. In such orbitals the molecules will tend to dissociate.

The potential energy curve for a molecular bonding orbital (e.g., Fig. 7) is of the form given by Eq. (38). In the lowest order approximation this relation defines a potential well which is harmonic in the displacements from the equilibrium value. Of course, not all values of the potential energy, $V(R - R_0)$, are permitted: the vibrational energy of the molecule is quantized, the quantum number n having integral values from n = 0 to infinity. The energy levels are equally spaced with separations $h\nu_v$. The second term is an anharmonic term which distorts the parabolic shape relative to the harmonic oscillator. There are two important differences between the quantized harmonic and anharmonic oscillators. First, where there is an infinite number of levels in the former, there is only a finite number of vibrational states for an anharmonic molecule. Second, the levels are not equally spaced in the anharmonic oscillator, except close to the potential minimum. For this reason the anharmonic oscillator will behave like a harmonic oscillator for relatively small values of the vibrational quantum number. It is normal to assume harmonic vibrations in the Born-Oppenheimer approximation.[27]

Molecular spectra fall into three categories, according as they occur in the far infrared (20–100 μm), near infrared (800–2000 nm), or visible/near ultraviolet (750–150 nm) region. Spectra excited by radiation in the infrared region are associated with changes in the rotational energy of the molecule. Spectra in the near-infrared region correspond to simultaneous changes in the rotational and vibrational energy of the molecule. Finally, the visible and ultraviolet spectra signal simultaneous changes in the rotational, vibrational, and electronic energies of the molecule. The latter category has the greatest potential complexity since in interpreting such visible/ultraviolet spectra we must, in principle, solve the molecular hamiltonian containing terms which represent electronic, vibrational, and rotational energies.

If we simplify the molecular problem somewhat we may represent the stationary states of a molecule by a linear sum of three energy terms: an electronic term determined by a quantum number n, a vibrational term determined by a quantum number, v, and a rotational term determined by a quantum number, r. The band spectra emitted by molecules are then characterized by three different kinds of excitations: electronic, vibrational, and rotational with frequencies in the ratio $\nu_{el}:\nu_v:\nu_r = 1:\sqrt{m/M}:m/M$, where m is the electronic mass and M is the reduced molecular mass. In an optical transition the electronic configuration (*nl*) changes, and generally we expect the vibrational/rotational state to change also. In consequence, for a particular electronic transition, we expect splittings into a number of vibrational components, each of which is split into a number of rotational components. Using the rough ratio rule given previously, and assuming the Rydberg to be a characteristic electronic transition, we find spectral splittings of 500 cm^{-1}

and 60 cm^{-1} to characterize the vibrational and rotational components, respectively. Since the quantized energy levels of the harmonic oscillator and rigid rotor are given by

$$E_v = (n + \tfrac{1}{2})h\nu_v \tag{40}$$

$$E_r = r(r+1)h\nu_r \tag{41}$$

where $\nu_v = (1/2\pi)\sqrt{k/M}$ and $\nu_r = \hbar/4\pi I$, in which k is the "spring" constant, M is the reduced molecular mass, and $I = Ml^2$ is the moment of inertia for a dumbbell shaped molecule of length l, and applying to the usual selection rules for electronic ($\Delta l = \pm 1$, with $\Delta j = 0, \pm 1$), vibrational ($\Delta v = \pm 1, \pm 2, \ldots$), and rotational ($\Delta r = 0, \pm 1$) transitions, we expect transitions at the following frequencies

$$\nu = \nu_0 + (n'' - n')/\nu_v + (r''(r''+1) - r'(r'+1))\nu_r \tag{42}$$

where the superscript primes and double primes refer to final and initial states. For simplicity we have assumed that the vibrational spring constants and rotational moments of inertia are unchanged in the transitions. The number of potential transition frequencies is obviously very large and the spectrum consists of very many closely spaced lines. Of course, the vibrational/rotational structure may be studied directly in microwave spectroscopy.[28] An early account of the interpretation of such spectra was given by Ruark and Urey.[29] For many years such complex spectra were recorded photographically; as we discuss in later sections, Fourier transform spectroscopy records these spectra electronically and in their finest detail. Subsequent detailed accounts of the energy levels of molecular species and their understanding via spectroscopic measurements have been given by Slater, Hertzberg and others.[30–32]

An example of the complexities of band spectra, for a simple diatomic molecule, is given in Fig. 8, in this case for the photographically recorded β-bands ($2\Pi \rightarrow 3\Sigma$) of the nitric oxide molecule. Under low dispersion band spectra are observed (see Fig. 8*a*) which break down into line spectra under very high resolution (see Fig. 8*b*). This band system is emitted in transitions having common initial and final electronic states, with all the electronic states corresponding to a multiplet in atomic spectra. The electronic energy difference determines the general spectral range in which the bands are observed. However, the positions of the individual bands are determined by the changes in the vibrational quantum numbers. The spectra in Fig. 8 are emission spectra: the bands identified in Fig. 8*a* by (n', n'') are so-called *progressions* in which the transition starts on particular vibrational levels ($n' = 0$) of the upper electronic state and ends on a series of

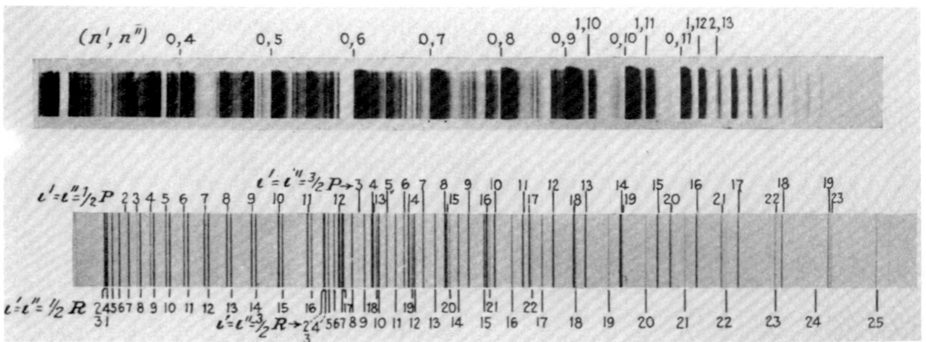

FIGURE 8 Characteristic band spectra of the diatomic NO molecule under conditions of (*a*) low resolution and (*b*) high resolution. *(After Ruark and Urey.*[29]*)*

different vibrational levels (n'') of the lower electronic level. Such n' progressions measure the vibrational energy level differences in the lower electronic levels. Specifically identified in Fig. 8*a* are band progressions $n' = 0 \rightarrow n'' = 4, 5, 6, 7 \ldots$ and $n' = 1 \rightarrow n'' = 10, 11, 12 \ldots$. Also evident, but not identified in Fig. 8*a*, are *sequences* of lines in which the difference in vibrational quantum number $n' - n''$ is a constant. For example, either side of the (0, 4) band can be seen $\Delta n = -4$ and -3 sequences, respectively.

The (0, 7) band is shown under much higher spectral resolution in Fig. 8*b*, revealing a plethora of closely spaced lines associated with the rotational structure. For transitions in which the vibrational quantum number changes by some multiple of unity, i.e., $\Delta n = 1$, 2, 3, etc., the rotational quantum number changes by $\Delta r = 0$ and ± 1. The three branches of the rotational spectrum have the following positions (measured in cm^{-1}):

$$R = A + 2B(r+1) \qquad \text{for } \Delta r = -1 \text{ and } r = 0, 1, 2 \ldots \tag{43a}$$

$$Q = A + Cr \qquad \text{for } \Delta r = 0 \text{ and } r = 0, 1, 2 \ldots \tag{43b}$$

$$P = A - 2Br \qquad \text{for } \Delta r = +1 \text{ and } r = 1, 2, 3 \ldots \tag{43c}$$

where in each case r refers to the rotational quantum number in the final electronic-vibrational state, $A = \bar{v}_e + \bar{v}_v$, and $B = \hbar/4\pi c l$. In consequence, if $C = 0$, the Q-branch consists of a single line at $Q(r) = v_e + v_v$, and the P- and R-branches shift linearly to lower and higher wavenumbers, respectively, relative to the Q-branch for increasing values of r. The spectra in Fig. 8 show that for the NO molecule $C \neq 0$ so that the lines in the P- and R-branches are shifted by $\pm 2B$ relative to the Q-branch lines. The particular Q-branches are defined by $l' = l'' = r + \frac{1}{2}$ with r being either 0 or 1, i.e., $l' = l'' = \frac{1}{2}$ or $\frac{3}{2}$. The $C \neq 0$ implies that the Eqs. (41) and (43) are inaccurate and additional rotational energy terms must be included even for vibrational states close to the bottom of the potential well (i.e., small n-values). Roughly the rotational term in Eq. (43) must be multiplied by $1 - (2\hbar^2(r+1)^2/4\pi^2 I^2)$, which has the further consequence that the separations of lines in the P- and R- branches corresponding to particular values of r increase with increasing r. A detailed analysis of Fig. 8*b* shows this to be the case.

8.11 LINESHAPES IN SOLID STATE SPECTROSCOPY

There are many modes of vibration of a crystal to which the optical center is sensitive. We will concentrate on one only, the breathing mode in which the ionic environment pulsates about the center. The variable for the lattice state, the so-called configurational coordinate, is labeled as Q. For the single mode of vibration the system oscillates about its equilibrium value Q_0^a in the ground state and Q_0^b in the excited state. The ground and excited state configurational coordinate curves are assumed to have identical harmonic shapes, and hence vibrational frequencies for the two states. This is illustrated in Fig. 9. In the Born-Oppenheimer approximation the optical center-plus-lattice system is represented in the electronic ground state by the product function[27]

$$\Phi_a(r_i, Q) = \Psi_a(r_i, Q_0^a)\chi_a(Q) \tag{44}$$

and in the electronic excited state by

$$\Phi_b(r_i, Q) = \Psi_b(r_i, Q_0^b)\chi_b(Q) \tag{45}$$

The first term in the product is the electronic wavefunction, which varies with the electronic positional coordinate r_i, and hence is an eigenstate of $H_0 = \Sigma_i(p_i^2/2m) + V'(r_i)$, and is determined at the equilibrium separation Q_0^a. The second term in the product wave function is $\chi_a(Q)$, which is a function of the configurational coordinate Q. The entire ionic potential energy in state a is then given by

$$E^a(Q) = E_0^a + V_a(Q) \tag{46}$$

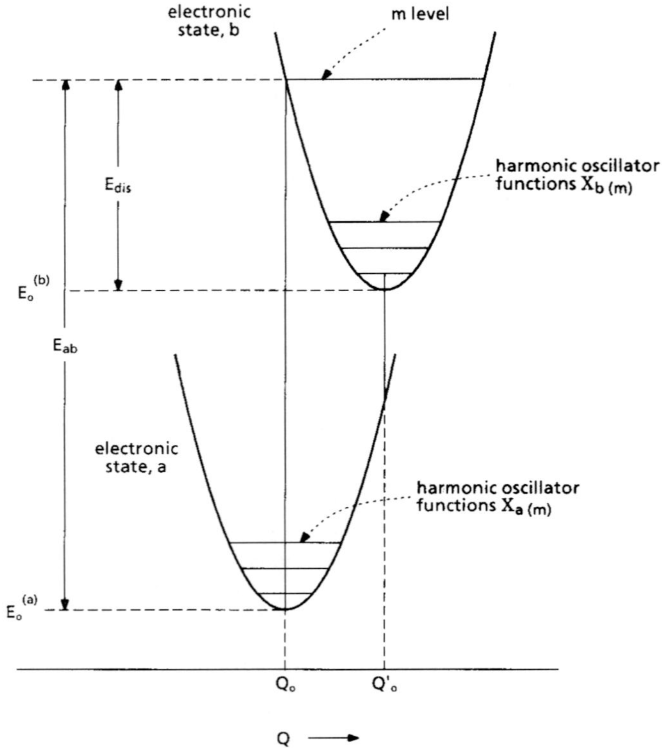

FIGURE 9 A configurational coordinate diagram showing the interionic potential in the harmonic approximation for electronic states $|a\rangle$ and $|b\rangle$. *(After Henderson and Imbusch.[8])*

The $E^a(Q)$ in Fig. 9 is an harmonic potential function. This is an approximation to the potential used in Eq. (38) and illustrated in Fig. 7. A similar expression obtains for $E^b(Q)$. This representation of the Born-Oppenheimer approximation is often referred to as the *configurational coordinate model.*[38]

Treating the electronic energy of the ground state E_0^a as the zero of energy, we write the ionic potential energy in the ground state as

$$E^a(Q) = \tfrac{1}{2}M\omega^2(Q - Q_0^a)^2 \tag{47}$$

and in the excited state as

$$E_0^b(Q) = E_{ab} + \tfrac{1}{2}M\omega^2(Q - Q_0^a)^2 - (2S)^{1/2}\hbar\omega\left(\frac{M\omega}{\hbar}\right)^{1/2}(Q - Q_0^a) \tag{48}$$

where E_{ab} is essentially the peak energy in an absorption transition between states $|a\rangle$ and $|b\rangle$ with the lattice at the coordinate, Q_0^a, where the Huang-Rhys parameter S is defined as

$$S = \frac{E_{\text{dis}}}{\hbar\omega} = \frac{1}{2}\frac{M\omega^2}{\hbar\omega}(Q_0^b - Q_0^a)^2 \tag{49}$$

If the vertical line from $Q = Q_0^a$ in Fig. 9 intersects the upper configurational coordinate curve at the vibrational level n' then

$$E_{\text{dis}} = S\hbar\omega = (n' + \tfrac{1}{2})\hbar\omega \tag{50}$$

The shapes of absorption and emission spectra are found to depend strongly on the difference in electron-lattice coupling between the states, essentially characterized by $(Q_0^b - Q_0^a)$ and by E_{dis}.

The radiative transition rate between the states $|a, n'\rangle \rightarrow |b, n''\rangle$ where n' and n'' are vibrational quantum numbers is given by,[8]

$$W(a, n' - b, n'') = |\langle \psi_b(r_i, Q_0^a)\chi_b(n'')|\ \mu\ |\psi_a(r_i, Q_0^a)\chi_a(n')\rangle|^2 \tag{51}$$

which can be written as

$$W(a, n' - b, n'') = W_{ab}\ |\langle \chi_b(n'')\ |\ \chi_a(n')\rangle|^2 \tag{52}$$

where W_{ab} is the purely electronic transition rate. The shape function of the transition is determined by the square of the vibrational overlap integrals, which are generally not zero. The absorption bandshape at $T = 0$ K where only the $n' = 0$ vibrational level is occupied, is

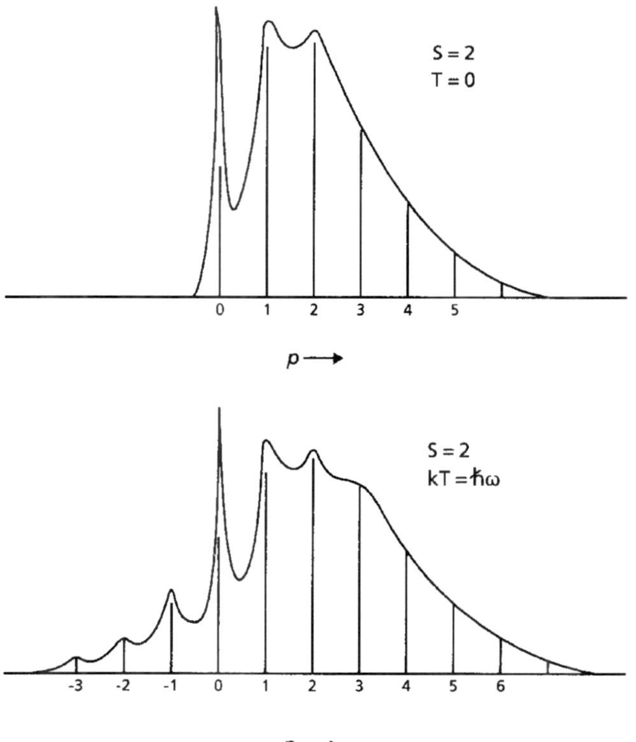

FIGURE 10 Showing the zero-phonon line and the Stokes shifted sideband on one side of the zero-phonon line at $T = 0$. When the temperature is raised the anti-Stokes sidebands appear on the other side of the zero-phonon line. *(After Henderson and Imbusch.[8])*

then given by

$$I_{ab}(E) = I_0 \Sigma_{n''} \frac{S^{n''} \exp(-S)}{n''!} \delta(E_0 + n''\hbar\omega - E) \tag{53}$$

where $E_0 = E_{b0} - E_{a0}$ is the energy of the transition between the zero vibrational levels of electronic states a and b. This is usually referred to as the *zero-phonon transition*, since $\Sigma_{n''} S^{n''} \exp(-S)/n''! = 1$, I_0 is the total intensity of the transition, which is independent of S. The intensity of the zero phonon transition I_{00} is given by

$$I_{00} = I_0 \exp(-S) \tag{54}$$

so that if $S = 0$ all the intensity is contained in the zero-phonon transition. On the other hand, when S is large, the value of I_{00} tends to zero and the intensity is concentrated in the *vibrational sidebands*. The single configurational coordinate model is relevant to the case of electron-vibrational structure in the spectra of molecules and the intensities so calculated fit observations rather well.

The net effect of this analysis is given in Fig. 10, which represents the absorption case when $S = 2$. At $T = 0$ we see a strong zero-phonon line, even stronger phonon-assisted transitions at $n'' = 1$ and 2, and then decreasing intensity in the phonon sidebands at $n'' = 3$, 4, 5, etc. These individual transitions are represented by the vertical lines in this predicted

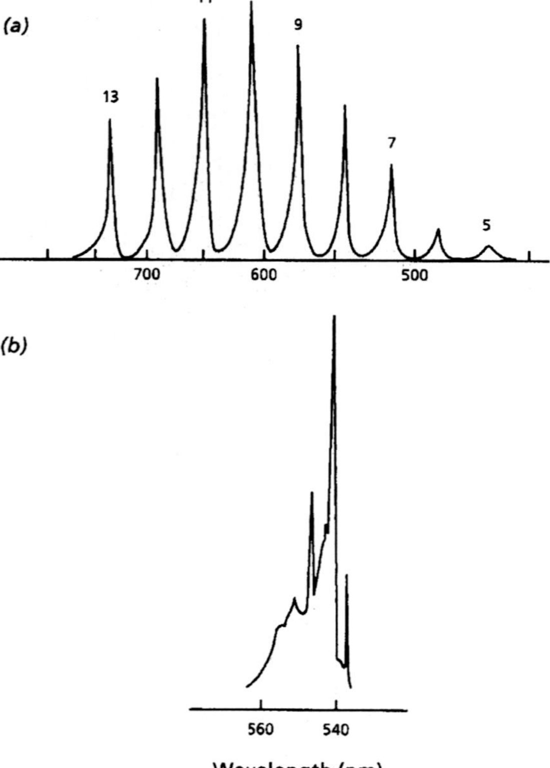

FIGURE 11 Photoluminescence spectrum of O_2^- in KBr at 77 K. *(After Rebane and Rebane.*[33]*)*

spectrum. The envelope of these sidebands, given the solid line, represents the effects of adding a finite width, $n''\hbar\omega$, to each sideband feature. The effect is to smear out the structure, especially at larger phonon numbers. Also in Fig. 10 we show the effect of temperature on the sideband shape in absorption, which is to reveal structure at lower energies than that of the zero-phonon line. Although the lineshape changes, the total intensity is independent of temperature, whereas the zero-phonon line intensity is given by

$$I_{00} = I_0 \exp\left(- S \coth\left(\frac{\hbar\omega}{2kT}\right)\right) \tag{55}$$

which decreases with increasing temperature. For small values of $S \ll 1$, the phonon sideband intensity increases according to $I_0 S \coth(\hbar\omega/2kT)$. These effects are shown in Fig. 10 also.

Three examples of the optical bandshapes in solid state spectra are considered. Figure 11 shows the luminescence spectrum of the molecular ion O_2^- in KBr measured at 77 K.[33] The O_2^- ion emits in the visible range from 400–750 nm, and the spectrum shown corresponds to the vibrational sidebands corresponding to $n''=5$ up to 13. The optical center is strongly coupled, $S=10$, to the internal vibrational mode of the O_2^- ion with energy $\hbar\omega \simeq 1000\ \text{cm}^{-1}$. However, as the detail of the $n''=8$ electronic vibrational transition is shown at $T=4.2$ K, there is also weak coupling $S \simeq 1$ to the phonon spectrum of the KBr, where the maximum vibrational frequency is only about 200 cm^{-1}.

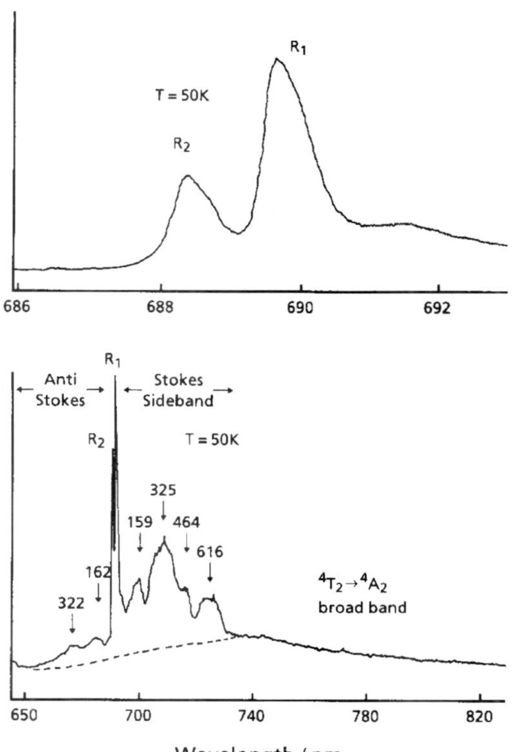

FIGURE 12 The photoluminescence spectrum of Cr^{3+} ions in $Y_3Ga_5O_{12}$.

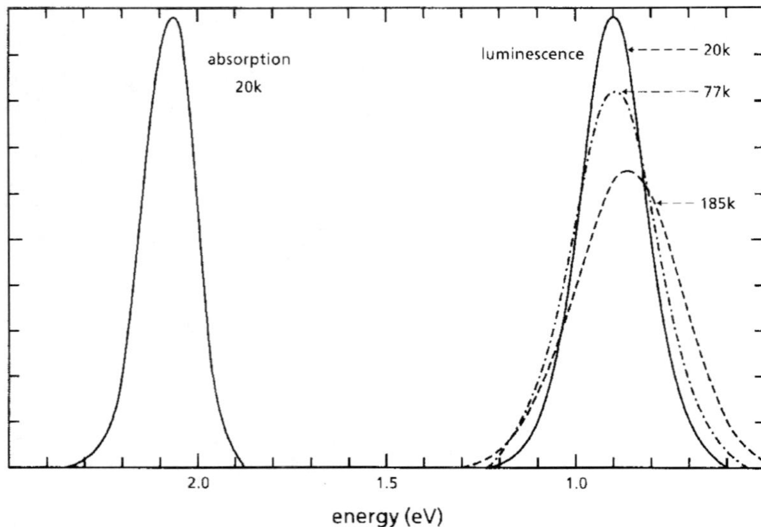

FIGURE 13 Optical absorption and photoluminescence of F-centers in KBr. *(After Gebhardt and Kuhnert.*[37]*)*

The Cr^{3+} ion occupies a central position in the folklore of solid state spectroscopy. In yttrium gallium garnet, $Y_3Ga_5O_{12}$, the Cr^{3+} ion occupies a Ga^{3+} ion site at the center of six octahedrally-disposed O^{2-} ions. The crystal field at this site is intermediate in strength[8] so that the ^2E and $^4T_2$ states are mixed by the combined effects of spin-orbit coupling and zero-point vibrations. The emission then is a composite spectrum of $^2E \rightarrow {}^4A_2$ transition and $^4T_2 \rightarrow {}^4A_2$ transition, with a common radiative lifetime.[34] As Fig. 12 shows, the composite spectrum shows a mélange of the R-line and its vibronic sideband, $^2E \rightarrow {}^4A_2$ transition with $S \sim 0.3$, and the broadband $^4T_2 \rightarrow {}^4A_2$ transition for which $S \sim 6$.[35] Understanding this particular bandshape is complex.[34–36]

The final example, in Fig. 13, shows the absorption and emission spectra of F-centers in KBr.[37] This is a strongly coupled system with $S \simeq 30$. Extension of the configurational coordinate model to the luminescence spectrum shows that the absorption and emission sidebands are mirror images of each other in the zero-phonon line. With S small (less than 6) there is structure in absorption and emission. However for S large, there is no structure, at least when a spectrum of vibrational modes interacts with the electronic states. The F-center represents strong coupling to a band of vibrational frequencies rather than to a single breathing mode of vibration. The effect of this is to broaden the spectrum to look like the envelope encompassing the spectrum of sharp sidebands shown in Fig. 10. In this case the zero-phonon line position is midway between the peaks in the absorption and emission bands of the F-center in KBr. Note also that as the temperature is raised the bands broaden and the peak shifts to longer wavelengths.

8.12 REFERENCES

1. I. Newton, *Opticks* (1704).
2. T. Preston, *The Theory of Light,* 3d ed. (Macmillan, London, 1901). Gives an elegant account of the contributions of Huyghens, Young, and Fresnel to the development of the wave theory of light.
3. J. C. Maxwell, *Treatise on Electricity and Magnetism* (1873).

4. A. Einstein, *Ann. Phys.* **17:**132 (1905).
5. N. Bohr, *The Theory of Spectra and Atomic Constitution* (Cambridge University Press, 1922).
6. E. Schrödinger, *Ann. Phys.* **79:**361 (1926); **79:**489 (1926); **79:**734 (1926); **80:**437 (1926); **81:**109 (1926).
7. P. A. M. Dirac, *Quantum Mechanics* (Oxford University Press, 1927).
8. B. Henderson and G. F. Imbusch, *Optical Spectroscopy of Inorganic Solids* (Clarendon Press, Oxford, 1989).
9. W. R. Johnson and G. Soff, *At. Data Nucl. Data Tables* **33:**406 (1985).
10. A. I. Ferguson and J. M. Tolchard, *Contemp. Phys.* **28:**383 (1987).
11. B. P. Kibble, W. R. C. Rowley, R. E. Shawyer, and G. W. Series, *J. Phys. B* **6:**1079 (1973).
12. T. W. Hänsch, I. S. Shakin, and A. L. Schawlow, *Nature* (*Lond.*) **235:**63 (1972).
13. W. E. Lamb and R. C. Retherford, *Phys. Rev.* **72:**241 (1947); **79:**549 (1950); **81:**222 (1951); **86:**1014 (1952).
14. T. W. Hänsch, M. H. Nayfeh, S. A. Lee, S. M. Curry, and I. S. Shahin, *Phys. Ref. Lett.* **32:**1336 (1974).
15. J. E. M. Goldsmith, E. W. Weber, and T. W. Hänsch, *Phys. Lett.* **41:**1525 (1978).
16. B. W. Petley, K. Morris, and R. E. Shawyer, *J. Phys. B* **13:**3099 (1980).
17. D. N. Stacey, Private communication to A. I. Ferguson, cited in Ref. 10.
18. S. R. Amin, C. D. Caldwell, and W. Lichten, *Phys. Rev. Lett.* **47:**1234 (1981); P. Zhao, W. Lichten, H. P. Layer, and J. C. Bergquist, ibid. **58:**1293 (1987) and *Phys. Rev. A.* **34:**5138 (1986).
19. C. E. Wieman and T. W. Hänsch, *Phys. Rev. Lett.* **36:**1170 (1976).
20. C. E. Wieman and T. W. Hänsch, *Phys. Rev. A* **22:**192 (1980).
21. J. R. M. Barr, J. M. Girkin, A. I. Ferguson, G. P. Barwood, P. Gill, W. R. C. Rowley, and R. C. Thompson, *Optics Commun.* **54:**217 (1985).
22. J. C. Slater, *Phys. Rev.* **35:**210 (1930).
23. M. Tinkham, *Group Theory and Quantum Mechanics* (McGraw-Hill, New York, 1964).
24. J. S. Griffith, *Theory of Transition Metal Ions* (Cambridge University Press, 1961).
25. C. E. Moore, *Atomic Energy Levels* [U.S. Nat. Bur. of Stand. Pub 467 (1950)].
26. W. Heitler and F. London, *Zeit. Phys.* **44:**455 (1927).
27. M. Born and J. P. Oppenheimer, *Ann. Phys.* **84:**457 (1927).
28. C. H. Townes and A. L. Schawlow, *Microwave Spectroscopy* (McGraw-Hill, New York, 1955). Gives an excellent account of the early days of what was then a novel sphere of study.
29. A. E. Ruark and H. C. Urey, *Atoms, Molecules and Quanta* (McGraw-Hill, New York, 1930).
30. J. C. Slater, *Quantum Theory of Molecules and Solids,* vol. 1 (McGraw-Hill, New York, 1963).
31. G. Hertzberg, *Molecular Spectra and Molecular Structure* (vols I and II, Van Nostrand, New York 1953).
32. C. N. Banwell, *Fundamentals of Molecular Spectroscopy* (McGraw-Hill, U.K., 1983).
33. K. K. Rebane and L. A. Rebane in *Optical Properties of Ions in Solids,* B. Di Bartolo (ed.) (Plenum Press, New York, 1975). See also K. K. Rebane, *Impurity Spectra of Solids* (Plenum Press, New York, 1970).
34. M. Yamaga, B. Henderson, and K. P. O'Donnell, *J. Phys.* (*Cond. Matt.*) **1:**9175 (1989).
35. M. Yamaga, B. Henderson, K. P. O'Donnell, C. Trager-Cowan, and A. Marshall, *App. Phys.* **B50:**425 (1990).
36. M. Yamaga, B. Henderson, K. P. O'Donnell, F. Rasheed, Y. Gao, and B. Cockayne, *App. Phys. B.* **52:**225 (1991).
37. W. Gebhardt and H. Kuhnert, *Phys. Lett.* **11:**15 (1964).

CHAPTER 9
FUNDAMENTAL OPTICAL PROPERTIES OF SOLIDS

Alan Miller
Department of Physics and Astronomy
University of St. Andrews
St. Andrews, Scotland
and
Center for Research and Education in Optics and Lasers (CREOL)
University of Central Florida
Orlando, Florida

9.1 GLOSSARY

$\underline{\mathbf{A}}$	vector field
$\underline{\mathbf{B}}$	magnetic induction
c	speed of light
$\underline{\mathbf{D}}$	displacement field
d_{M}	penetration depth in a metal
d_{TIR}	evanescent field depth for total internal reflection
$\underline{\mathbf{E}}$	electric field
$\mathscr{E}_{\mathrm{g}}$	band gap energy
$\underline{\mathbf{E}}_{\mathrm{loc}}$	local electric field
$-e$	electronic charge
e_{i}^{σ}	components of unit vectors of incident polarization
e_{s}^{ρ}	components of unit vectors of scattered polarization
f_{j}	oscillator strength
$\underline{\mathbf{G}}$	reciprocal lattice vector
$\mathbf{H}$	hamiltonian
$\underline{\mathbf{H}}$	magnetic field
I	irradiance
$\underline{\underline{\mathbf{I}}}$	unit tensor
$\underline{\mathbf{J}}_{\mathrm{c}}$	conduction current density
$\underline{\underline{\mathbf{K}}}$	dielectric constant or relative permittivity

$\underline{\mathbf{k}}$	electron wave vector
$\underline{\mathbf{M}}$	induced magnetization
m	electron mass
m^*	effective electron mass
$\mathbf{N}$	charge density
N_L	refractive index for left circularly polarized light
N_R	refractive index for right circularly polarized light
n	real part of refractive index
$\mathbf{P}$	Kane momentum matrix element
$\underline{\mathbf{P}}$	induced dipole moment per unit volume
$\mathbf{P_D}$	dielectric amplitude reflection polarization ratio
$\mathbf{P}_{j'j}$	momentum matrix elements
$\mathbf{P}_M$	metallic amplitude reflection polarization ratio
$P_{\mathbf{p}\sigma,\mathbf{ij}}$	elasto-optic coefficients
$\mathbf{p}$	electron momentum
$\tilde{\mathbf{q}}$	complex photon wave vector
$\hat{\mathbf{q}}$	unit photon wave vector
R	reflectance
$\mathbf{R}_{ex}$	exciton Rydberg
$\mathbf{R}_{\sigma,p}$	Raman scattering coefficient
r_p	field reflection amplitude for p-polarization
r_s	field reflection amplitude for s-polarization
$\underline{\mathbf{S}}$	Poynting vector
$\mathbf{s_{ij}}$	components of lattice strain
$\mathbf{S_R}$	Raman scattering efficiency
$\mathbf{s_j}$	phonon oscillator strength
$\mathbf{T}$	transmittance
$\mathbf{t_p}$	field transmission amplitude for p-polarization
$\mathbf{t_s}$	field transmission amplitude for s-polarization
$\mathbf{V}$	volume of unit cell
$\mathbf{V}(\underline{r})$	periodic lattice potential
$\mathbf{v}$	phase velocity
$\mathbf{W}(\omega)$	transition rate
$\mathbf{Z}$	number of electrons per atom
α	absorption coefficient
β	two-photon absorption coefficient
Γ	damping constant
Δ_D	dielectric reflection polarization phase difference
Δ_M	metallic reflection polarization phase difference
Δ_s	specific rotary power

Δ_{so}	spin-orbit energy splitting
δ	energy walk-off angle
$\underline{\underline{\varepsilon}}$	electrical permittivity
$\tilde{\varepsilon}$	complex permittivity
$\varepsilon(\infty)$	high frequency permittivity
$\varepsilon(0)$	static permittivity
ε_1	real part of permittivity
ε_2	imaginary part of permittivity
ε_0	electrical permittivity of vacuum
$\tilde{\eta}$	complex refractive index
$\bar{\theta}$	principal angle for metallic reflection
θ_{OA}	angle between optic axis and principal axis
θ_B	Brewster's angle
θ_c	total internal reflection critical angle
θ_{TIR}	phase change for total internal reflection
κ	imaginary part of refractive index
$\underline{\underline{\mu}}$	magnetic permeability
μ_0	magnetic permeability of vacuum
ξ	Hooke's Law constant
$\hat{\boldsymbol{\xi}}$	unit polarization vector
ρ_f	free charge density
ρ_p	band or polarization charge density
ρ_p	real field amplitude for p-polarization
ρ_s	real field amplitude for s-polarization
ρ_t	total charge density
$\tilde{\sigma}$	complex conductivity
$\underline{\underline{\sigma}}$	electrical conductivity
σ_1	real part of conductivity
σ_2	imaginary part of conductivity
τ	scattering or relaxation time
ϕ	scalar field
$\phi_{i'}$	ground state wave function
$\tilde{\chi}$	complex susceptibility
χ'	real part of susceptibility
χ''	imaginary part of susceptibility
$\underline{\underline{\chi}}_e$	electrical susceptibility
$\underline{\underline{\chi}}_m$	magnetic susceptibility
$\psi_{\mathbf{k}}$	Bloch solution wave function
ω	photon frequency
ω_j	resonant oscillation frequency
ω_{LO}	longitudinal optical phonon frequency

ω_p	plasma frequency
ω_s	surface plasmon frequency
ω_{TO}	transverse optical phonon frequency

9.2 INTRODUCTION

This chapter describes the fundamental interactions of light with solids. The discussion is restricted to *intrinsic* optical properties. *Extrinsic* optical properties of solids are described in Chap. 8 "Optical Spectroscopy and Spectroscopic Lineshapes." Basic formulas, definitions, and concepts are listed for reference and as a foundation for subsequent chapters of this volume of the *Handbook*. More detailed accounts of specific optical properties and particular types of solid are given in later chapters, i.e., Chap. 42 "Optical Properties of Films and Coatings," and in Vol. II, Chaps. 23 "Crystals and Glasses," 35 "Metals," 36 "Semiconductors," and 13 "Electro-Optic Modulators." The reader is referred to the many texts which provide more elaborate discussions of the optical properties of solids.[1–12]

Electrical measurements distinguish three general types of solids by their conductivities, i.e., dielectrics, semiconductors, and metals. Optically, these three groups are characterized by different fundamental band gap energies, $\mathscr{E}_g$. Although the boundaries are not sharply defined, an approximate distinction is given by, metals $\mathscr{E}_g \leq 0$, semiconductors, $0 < \mathscr{E}_g < 3$ eV, and dielectrics, $\mathscr{E}_g > 3$ eV. Solids may be found in single crystal, polycrystalline, and amorphous forms. Rudimentary theories of the optical properties of condensed matter are based on light interactions with perfect crystal lattices characterized by extended (nonlocal) electronic and vibrational energy states. These eigenstates are determined by the periodicity and symmetry of the lattice and the form of the Coulomb potential which arises from the interatomic bonding.

The principal absorption bands in condensed matter occur at photon energies corresponding to the frequencies of the lattice vibrations (phonons) in the infrared, and electronic transitions in the near infrared, visible, or ultraviolet. A quantum mechanical approach is generally required to describe electronic interactions, but classical models often suffice for lattice vibrations. Although the mechanical properties of solids can vary enormously between single crystal and polycrystalline forms, the *fundamental* optical properties are similar, even if the crystallite size is smaller than a wavelength, since the optical interaction is microscopic. However, electronic energy levels and hence optical properties are fundamentally altered when one or more dimensions of a solid are reduced to the scale of the deBroglie wavelength of the electrons. Modern crystal growth techniques allow fabrication of atomic precision epitaxial layers of different solid materials. Ultrathin layers, with dimensions comparable with or smaller than the deBroglie wavelength of an electron, may form quantum wells, quantum wires, and quantum dots in which electronic energy levels are quantized. Amorphous solids have random atomic or molecular orientation on the distance scale of several nearest neighbors, but generally have well-defined bonding and local atomic order which determine the overall optical response.

9.3 PROPAGATION OF LIGHT IN SOLIDS

Dielectrics and semiconductors provide transparent spectral regions at radiation frequencies between the phonon vibration bands and the fundamental (electronic) absorption edge. Maxwell's equations successfully describe the propagation, reflection, refraction, and scattering of harmonic electromagnetic waves.

Maxwell's Equations

Four equations relate the macroscopically averaged electric field $\mathbf{E}$ and magnetic induction $\mathbf{B}$ to the total charge density, ρ_t (sum of the bound or polarization charge, ρ_p, and the free charge, ρ_f), the conduction current density, $\mathbf{J}_c$, the induced dipole moment per unit volume, $\mathbf{P}$, and the induced magnetization of the medium, $\mathbf{M}$ (expressed in SI units):

$$\nabla \cdot \mathbf{E} = \frac{\rho_t}{\varepsilon_0} \tag{1}$$

$$\nabla \times \mathbf{E} = -\frac{\partial \mathbf{B}}{\partial t} \tag{2}$$

$$\nabla \cdot \mathbf{B} = 0 \tag{3}$$

$$\nabla \times \mathbf{B} = \mu_0\left(\varepsilon_0 \frac{\partial \mathbf{E}}{\partial t} + \frac{\partial \mathbf{P}}{\partial t} + \mathbf{J}_c + \nabla \times \mathbf{M}\right) \tag{4}$$

where $\varepsilon_0 = 8.854 \times 10^{-12}$ F/m is the permittivity of vacuum, $\mu_0 = 4\pi \times 10^{-7}\, H/m$ is the permeability of vacuum and c the speed of light in vacuum, $c = 1/\sqrt{\varepsilon_0 \mu_0}$. By defining a displacement field, $\mathbf{D}$, and magnetic field, $\mathbf{H}$, to account for the response of a medium

$$\mathbf{D} = \mathbf{P} + \varepsilon_0 \mathbf{E} \tag{5}$$

$$\mathbf{H} = \frac{\mathbf{B}}{\mu_0} - \mathbf{M} \tag{6}$$

and using the relation between polarization and bound charge density,

$$-\nabla \cdot \mathbf{P} = \rho_p \tag{7}$$

Equations (1) and (4) may also be written in the form

$$\nabla \cdot \mathbf{D} = \rho_f \tag{8}$$

and

$$\nabla \times \mathbf{H} = \frac{\partial \mathbf{D}}{\partial t} + \mathbf{J}_c \tag{9}$$

Vector $\mathbf{A}$ and scalar ϕ fields may be defined by

$$\mathbf{B} = \nabla \times \mathbf{A} \tag{10}$$

$$\mathbf{E} = -\nabla\phi - \frac{\partial \mathbf{A}}{\partial t} \tag{11}$$

A convenient choice of gauge for the optical properties of solids is the Coulomb (or transverse) gauge

$$\nabla \cdot \mathbf{A} = 0 \tag{12}$$

which ensures that the vector potential $\mathbf{A}$ is transverse for plane electromagnetic waves while the scalar potential represents any longitudinal current and satisfies Poisson's equation

$$\nabla^2 \phi = -\frac{\rho_t}{\varepsilon_0} \tag{13}$$

Three constitutive relations describe the response of conduction and bound electrons to the electric and magnetic fields:

$$\mathbf{J}_c = \underline{\underline{\sigma}}\mathbf{E} \tag{14}$$

$$\mathbf{D} = \mathbf{P} + \varepsilon_0\mathbf{E} = \underline{\underline{\varepsilon}}\mathbf{E} \tag{15}$$

$$\mathbf{B} = \mu_0(\mathbf{H} + \mathbf{M}) = \underline{\underline{\mu}}\mathbf{H} \tag{16}$$

Here $\underline{\underline{\sigma}}$ is the electrical conductivity, $\underline{\underline{\varepsilon}}$ is the electrical permittivity, and $\underline{\underline{\mu}}$ is the magnetic permeability of the medium and are in general tensor quantities which may depend on field strengths. An alternative relation often used to define a dielectric constant (or relative permittivity), $\underline{\underline{\mathbf{K}}}$, is given by $\mathbf{D} = \varepsilon_0\underline{\underline{\mathbf{K}}}\mathbf{E}$. In isotropic media using the approximation of linear responses to electric and magnetic fields, σ, ε, and μ are constant scalar quantities.

Electronic, $\underline{\underline{\chi}}_e$, and magnetic, $\underline{\underline{\chi}}_m$, susceptibilities may be defined to relate the induced dipole moment, $\mathbf{P}$, and magnetism, $\mathbf{M}$ to the field strengths, $\mathbf{E}$ and $\mathbf{H}$:

$$\mathbf{P} = \varepsilon_0\underline{\underline{\chi}}_e\mathbf{E} \tag{17}$$

$$\mathbf{M} = \varepsilon_0\underline{\underline{\chi}}_m\mathbf{H} \tag{18}$$

Thus,

$$\underline{\underline{\varepsilon}} = \varepsilon_0(\underline{\underline{\mathbf{I}}} + \underline{\underline{\chi}}_e) \tag{19}$$

and

$$\underline{\underline{\mu}} = \mu_0(\underline{\underline{\mathbf{I}}} + \underline{\underline{\chi}}_m) \tag{20}$$

where $\underline{\underline{\mathbf{I}}}$ is the unit tensor.

Wave Equations and Optical Constants

The general wave equation derived from Maxwell's equations is

$$\nabla^2\mathbf{E} - \nabla(\nabla \cdot \mathbf{E}) - \varepsilon_0\mu_0\frac{\partial^2\mathbf{E}}{\partial t^2} = \mu_0\left(\frac{\partial^2\mathbf{P}}{\partial t^2} + \frac{\partial\mathbf{J}_c}{\partial t} + \nabla \times \frac{\partial\mathbf{M}}{\partial t}\right) \tag{21}$$

For dielectric (nonconducting) solids

$$\nabla^2\mathbf{E} = \mu\varepsilon\frac{\partial^2\mathbf{E}}{\partial t^2} \tag{22}$$

The harmonic plane wave solution of the wave equation for monochromatic light at frequency, ω,

$$\mathbf{E} = \tfrac{1}{2}\mathbf{E}_0 \exp i(\mathbf{q} \cdot \mathbf{r} - \omega t) + \text{c.c.} \tag{23}$$

in homogeneous ($\nabla\varepsilon = 0$), isotropic ($\nabla \cdot \mathbf{E} = 0$, $\mathbf{q} \cdot \mathbf{E} = 0$), nonmagnetic ($\mathbf{M} = 0$) solids results in a complex wave vector, $\tilde{\mathbf{q}}$,

$$\tilde{\mathbf{q}} = \frac{\omega}{c}\sqrt{\frac{\varepsilon}{\varepsilon_0} + i\frac{\sigma}{\varepsilon_0\omega}} \tag{24}$$

A complex refractive index, $\tilde{\eta}$, may be defined by

$$\tilde{\mathbf{q}} = \frac{\omega}{c}\tilde{\eta}\hat{\mathbf{q}} \tag{25}$$

where $\hat{\mathbf{q}}$ is a unit vector and

$$\tilde{\eta} = n + i\kappa \tag{26}$$

Introducing complex notation for the permittivity, $\tilde{\varepsilon} = \varepsilon_1 + i\varepsilon_2$, conductivity, $\tilde{\sigma} = \sigma_1 + i\sigma_2$, and susceptibility, $\tilde{\chi}_e = \chi'_e + i\chi''_e$, we may relate

$$\varepsilon_1 = \varepsilon = \varepsilon_0(1 + \chi'_e) = \varepsilon_0(n^2 - \kappa^2) = -\frac{\sigma_2}{\omega} \tag{27}$$

and

$$\varepsilon_2 = \varepsilon_0\chi''_e = 2\varepsilon_0 n\kappa = \frac{\sigma}{\omega} = \frac{\sigma_1}{\omega} \tag{28}$$

Alternatively,

$$n = \left[\frac{1}{2\varepsilon_0}\left(\sqrt{\varepsilon_1^2 + \varepsilon_2^2} + \varepsilon_1\right)\right]^{1/2} \tag{29}$$

$$\kappa = \left[\frac{1}{2\varepsilon_0}\left(\sqrt{\varepsilon_1^2 + \varepsilon_2^2} - \varepsilon_1\right)\right]^{1/2} \tag{30}$$

The field will be modified locally by induced dipoles. If there is no free charge, ρ_f, the local field, $\mathbf{E}_{loc}$, may be related to the external field, $\mathbf{E}_i$, in isotropic solids using the Clausius-Mossotti equation which leads to the relation

$$\mathbf{E}_{loc} = \frac{n^2 + 2}{3}\mathbf{E}_i \tag{31}$$

Energy Flow

The direction and rate of flow of electromagnetic energy is described by the Poynting vector

$$\mathbf{S} = \frac{1}{\mu_0}\mathbf{E} \times \mathbf{H} \tag{32}$$

The average power per unit area (irradiance, I), W/m^2, carried by a uniform plane wave is given by the time averaged magnitude of the Poynting vector

$$I = \langle \mathbf{S} \rangle = \frac{cn\,|\mathbf{E}_0|^2}{2} \tag{33}$$

The plane wave field in Eq. (23) may be rewritten for absorbing media using Eqs. (25) and (26).

$$\mathbf{E}(\mathbf{r}, t) = \tfrac{1}{2}\mathbf{E}_0(\mathbf{q}, \omega)\exp\left(-\frac{\omega}{c}\kappa\hat{\mathbf{q}} \cdot \mathbf{r}\right)\exp\left[i\left(\frac{\omega}{c}n\hat{\mathbf{q}} \cdot \mathbf{r} - \omega t\right)\right] + \text{c.c.} \tag{34}$$

The decay of the propagating wave is characterized by the extinction coefficient, κ. The attenuation of the wave may also be described by Beer's law

$$I = I_0 \exp(-\alpha z) \tag{35}$$

where α is the absorption coefficient describing the attenuation of the irradiance, I, with distance, z. Thus,

$$\alpha = \frac{2\omega\kappa}{c} = \frac{4\pi\kappa}{\lambda} = \frac{\sigma_1}{\varepsilon_0 cn} = \frac{\varepsilon_2\omega}{\varepsilon_0 cn} = \frac{\chi_e''\omega}{cn} \tag{36}$$

The power absorbed per unit volume is given by

$$P_{abs} = \alpha I = \frac{\omega\chi_e''}{2}|\mathbf{E}_0|^2 \tag{37}$$

The second exponential in Eq. (34) is oscillatory and represents the phase velocity of the wave, $v = c/n$.

Anisotropic Crystals

Only amorphous solids and crystals possessing cubic symmetry are optically isotropic. In general, the speed of propagation of an electromagnetic wave in a crystal depends both on the direction of propagation and on the polarization of the light wave. The linear electronic susceptibility and dielectric constant may be represented by tensors with components of $\underline{\underline{\chi}}_e$ given by

$$P_i = \varepsilon_0 \chi_{ij} E_j \tag{38}$$

where i and j refer to coordinate axes. In an anisotropic crystal, $\mathbf{D} \perp \mathbf{B} \perp \mathbf{q}$ and $\mathbf{E} \perp \mathbf{H} \perp \mathbf{S}$, but $\mathbf{E}$ is not necessarily parallel to $\mathbf{D}$ and the direction of energy flow $\mathbf{S}$ is not necessarily in the same direction as the propagation direction $\mathbf{q}$.

From energy arguments it can be shown that the susceptibility tensor is symmetric and it therefore follows that there always exists a set of coordinate axes which diagonalize the tensor. This coordinate system defines the principal axes. The number of nonzero elements for the susceptibility (or dielectric constant) is thus reduced to a maximum of three (for any crystal system at a given wavelength). Thus, the dielectric tensor defined by the direction of the electric field vector with respect to the principal axes has the form

$$\begin{bmatrix} \varepsilon_1 & 0 & 0 \\ 0 & \varepsilon_2 & 0 \\ 0 & 0 & \varepsilon_3 \end{bmatrix}$$

The principal indices of refraction are

$$n_i = \sqrt{1 + \chi_{ij}} = \sqrt{\varepsilon_i} \tag{39}$$

with the $\mathbf{E}$-vector polarized along any principal axis, i.e. $\mathbf{E} \parallel \mathbf{D}$. This case is designated as an ordinary or **o**-ray in which the phase velocity is independent of propagation direction. An extraordinary or **e**-ray occurs when both $\mathbf{E}$ and $\mathbf{q}$ lie in a plane containing two principal axes with different n. An *optic axis* is defined by any propagation direction in which the phase velocity of the wave is independent of polarization.

Crystalline solids fall into three classes: (1) optically isotropic, (2) uniaxial, or (3) biaxial (see Table 1). All choices of orthogonal axes are principal axes and $\varepsilon_1 = \varepsilon_2 = \varepsilon_3$ in *isotropic* solids. For a *uniaxial* crystal, $\varepsilon_1 = \varepsilon_2 \neq \varepsilon_3$, a single optic axis exists for propagation in direction **3**. In this case, the ordinary refractive index, $n_0 = n_1 = n_2$, is independent of the direction of polarization in the **1-2** plane. Any two orthogonal directions in this plane can be chosen as principal axes. For any other propagation direction, the polarization can be divided into an **o**-ray component in the **1-2** plane and a perpendicular **e**-ray component

TABLE 1 Crystalographic Point Groups and Optical Properties

System		Point group symbols International	Schönflies	Optical activity	
Triclinic	biaxial	1	C_1	A	
		$\bar{1}$	S_2	-	
Monoclinic	biaxial	2	C_2	A	
		m	C_v	-	
		2/m	C_{2h}	-	
Orthorhombic	biaxial	mm	C_{2v}	-	
		222	D_2	A	
		mmm	D_{2h}	-	
Trigonal	uniaxial	3	C_3	A	
			S_6	-	
		3m	C_{3v}	-	
		32	D_3	A	
			D_{3d}	-	
Tetragonal	uiaxial	4	C_4	A	
			S_4	-	
		4/m	C_{4h}	-	
		4mm	C_{4v}	-	
			D_{2d}	A	
		42	D_4	A	
		4/mmm	D_{4h}	-	
Hexagonal	uniaxial	6	C_6	A	
			C_{3h}	-	
		6/m	C_{6h}	-	
		6mm	C_{6v}	-	
			D_{3h}	-	
		62	D_6	A	
		6/mmm	D_{6h}	-	
Cubic	isotropic	23	T	A	
		m3	T_h	-	
			T_d	-	
		432	O	A	
		m3m	O_h	-	

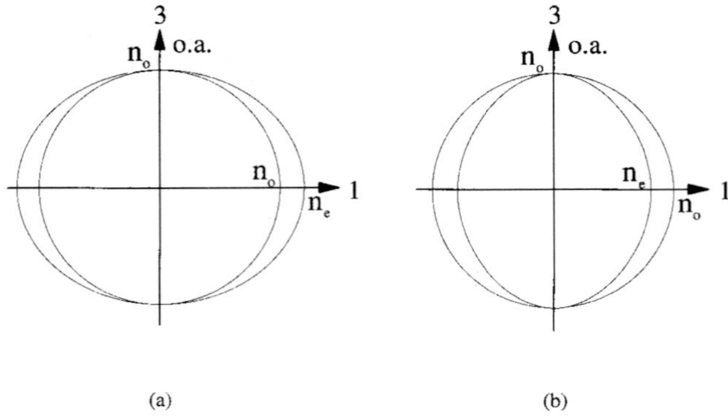

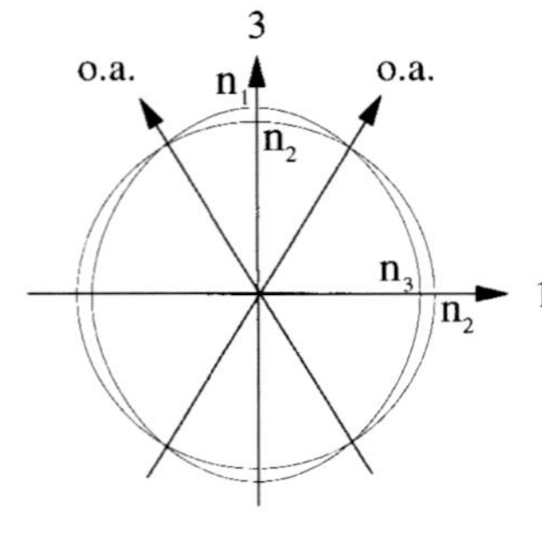

(c)

FIGURE 1 Illustration of directional dependence of refractive indices and optic axes in: (*a*) a uniaxial, positive birefringent crystal; (*b*) a uniaxial, negative birefringent crystal; and (*c*) a biaxial crystal.

(see Fig. 1). The dependence of the **e**-ray refractive index with propagation direction is given by the ellipsoid

$$n_j(\theta_i) = \frac{n_i n_j}{(n_i^2 \cos^2 \theta_i + n_j^2 \sin^2 \theta_i)^{1/2}} \tag{40}$$

where θ_i is defined with respect to optic axis, $i = 3$, and $j = 1$ or 2. $\theta_i = 90°$ gives the refractive index $n_e = n_3$ when the light is polarized along axis **3**. The difference between n_0 and n_e is the birefringence. Figure 1*a* illustrates the case of positive birefringence, $n_e > n_0$, and Fig. 1*b* negative birefringence, $n_e < n_0$. The energy walk-off angle, δ (the angle between **S** and **q** or **D** and **E**), is given by

$$\tan \delta = \frac{n^2(\theta)}{2} \left[\frac{1}{n_3^2} - \frac{1}{n_1^2} \right] \sin 2\theta \tag{41}$$

In *biaxial* crystals, diagonalization of the dielectric tensor results in three independent coefficients, $\varepsilon_1 \neq \varepsilon_2 \neq \varepsilon_3 \neq \varepsilon_1$. For orthorhombic crystals, a single set of orthogonal principal axes is fixed for all wavelengths. However, in monoclinic structures only one principal axis is fixed. The direction of the other two axes rotate in the plane perpendicular to the fixed axis as the wavelength changes (retaining orthogonality). In triclinic crystals there are no fixed axes and orientation of the set of three principal axes varies with wavelength. Equation (40) provides the **e**-ray refractive index within planes containing two principal axes. Biaxial crystals possess two optic axes. Defining principal axes such that $n_1 > n_2 > n_3$, then both optic axes lie in the **1-3** plane, at an angle θ_{OA} from axis **1**, as illustrated in Fig. 1*c*, where

$$\sin\theta_{\mathrm{OA}} = \pm\frac{n_1}{n_2}\sqrt{\frac{n_2^2 - n_3^2}{n_1^2 - n_3^2}} \tag{42}$$

Crystals with certain point group symmetries (see Table 1) also exhibit optical activity, i.e., the ability to rotate the plane of linearly polarized light. An origin for this phenomenon is the weak magnetic interaction $\nabla \times \mathbf{M}$ [see Eq. (21)], when it applies in a direction perpendicular to **P** (i.e., **M** || **P**). The specific rotary power Δ_s (angle of rotation of linearly polarized light per unit length) is given by

$$\Delta_s = \frac{\pi}{\lambda}(n_{\mathrm{L}} - n_{\mathrm{R}}) \tag{43}$$

where n_{L} and n_{R} are refractive indices for left and right circular polarization. Optical activity is often masked by birefringence, however, polarization rotation can be observed in optically active materials when the propagation is along the optic axis or when the birefringence is coincidentally zero in other directions. In the case of propagation along the optic axis of an optically active uniaxial crystal such as quartz, the susceptibility tensor may be written

$$\begin{bmatrix} \chi_{11} & i\chi_{12} & 0 \\ -i\chi_{12} & \chi_{11} & 0 \\ 0 & 0 & \chi_{33} \end{bmatrix}$$

and the rotary power is proportional to the imaginary part of the magnetic susceptibility, $\chi_m'' = \chi_{12}$:

$$\Delta_s = \frac{\pi\chi_{12}}{n\lambda} \tag{44}$$

Crystals can exist in left- or right-handed versions. Other crystal symmetries, e.g., $\bar{4}2$m, can be optically active for propagation along the **2** and **3** axes, but rotation of the polarization is normally masked by the typically larger birefringence, except at accidental degeneracies.

Interfaces

Applying boundary conditions at a plane interface between two media with different indices of refraction leads to the laws of reflection and refraction. Snell's law applies to all **o**-rays and relates the angle of incidence, θ_{A} in medium A, and the angle of refraction, θ_{B} in medium B, to the respective ordinary refractive indices n_{A} and n_{B}:

$$n_{\mathrm{A}}\sin\theta_{\mathrm{A}} = n_{\mathrm{B}}\sin\theta_{\mathrm{B}} \tag{45}$$

Extraordinary rays do not satisfy Snell's law. The propagation direction for the **e**-ray can be found graphically by equating the projections of the propagation vectors in the two media along the boundary plane. Double refraction of unpolarized light occurs in anisotropic crystals.

The field amplitude ratios of reflected and transmitted rays to the incident ray (r and t) in isotropic solids (and **o**-rays in anisotropic crystals) are given by the Fresnel relations.

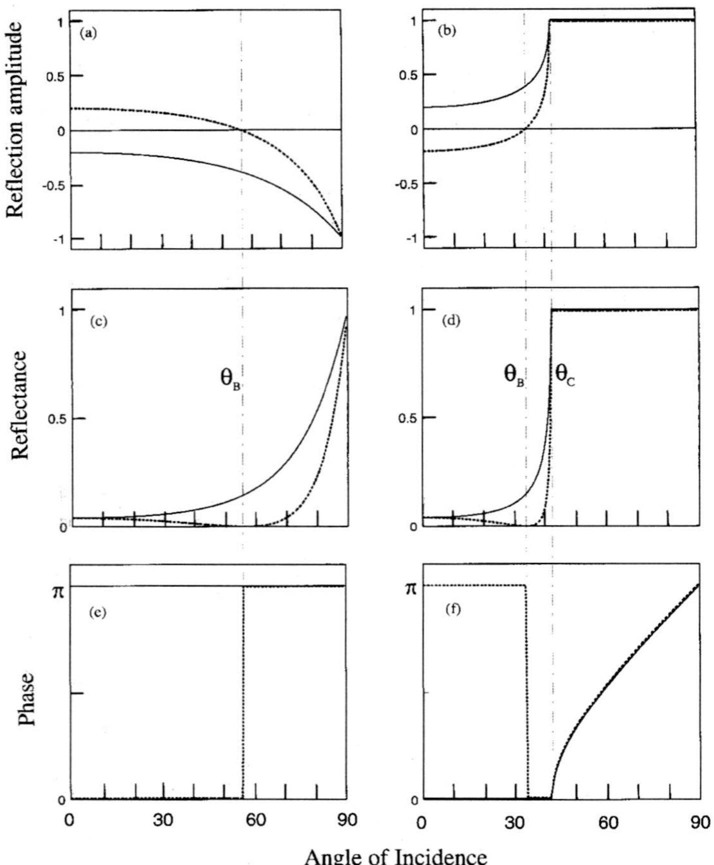

FIGURE 2 The electric field reflection amplitudes (a, b), energy reflectance (c, d), and phase change (e, f) for s- (solid lines) and p- (dashed lines) polarized light for external (a, c, e) and internal (b, d, f) reflection in the case $n_A = 1$, $n_B = 1.5$. θ_B is the polarizing or Brewster's angle and θ_c is the critical angle for total internal reflection.[2]

For s- (σ or TE) polarization (**E**-vector perpendicular to the plane of incidence) (see Fig. 2*a*) and p- (π or TM) polarization (**E**-vector parallel to the plane of incidence) (see Fig. 2*b*):

$$r_s = \frac{E_{rs}}{E_{is}} = \frac{n_A \cos\theta_A - \sqrt{n_B^2 - n_A^2 \sin^2\theta_A}}{n_A \cos\theta_A + \sqrt{n_B^2 - n_A^2 \sin^2\theta_A}} \tag{46}$$

$$r_p = \frac{E_{rp}}{E_{ip}} = \frac{n_B^2 \cos\theta_A - n_A\sqrt{n_B^2 - n_A^2 \sin^2\theta_A}}{n_B^2 \cos\theta_A + n_A\sqrt{n_B^2 - n_A^2 \sin^2\theta_A}} \tag{47}$$

$$t_s = \frac{E_{ts}}{E_{is}} = \frac{2n_A \cos\theta_A}{n_S \cos\theta_A + \sqrt{n_B^2 - n_A^2 \sin^2\theta_A}} \tag{48}$$

$$t_p = \frac{E_{tp}}{E_{ip}} = \frac{2n_A n_B \cos\theta_A}{n_B^2 \cos\theta_A + n_A\sqrt{n_B^2 - n_A^2 \sin^2\theta_A}} \tag{49}$$

At normal incidence, the energy reflectance, R (see Figs 2*c* and *d*), and transmittance, T, are

$$R = \left|\frac{E_r}{E_i}\right|^2 = \left|\frac{n_B - n_A}{n_B + n_A}\right|^2 \tag{50}$$

$$T = \left|\frac{E_t}{E_i}\right|^2 = \frac{4n_A^2}{|n_A + n_B|^2} \tag{51}$$

The p-polarized reflectivity (see Eq. (47)] goes to zero at Brewster's angle under the condition

$$\theta_B = \tan^{-1}\left(\frac{n_A}{n_B}\right) \tag{52}$$

If $n_A > n_B$, total internal reflection (TIR) occurs when the angle of incidence exceeds a critical angle:

$$\theta_c = \sin^{-1}\left(\frac{n_A}{n_B}\right) \tag{53}$$

This critical angle may be different for s- and p-polarizations in anisotropic crystals. Under conditions of TIR, the evanescent wave amplitude drops to e^{-1} in a distance

$$d_{TIR} = \frac{c}{\omega}(n_A^2 \sin^2\theta_A - n_B^2)^{-1/2} \tag{54}$$

The phase changes on reflection for external and internal reflection from an interface with $n_A = 1.5$, $n_B = 1$ are plotted in Figs. 2*e* and *f*. Except under TIR conditions, the phase change is either 0 or π. The complex values predicted by Eqs. (46) and (47) for angles of incidence greater than the critical angle for TIR imply phase changes in the reflected light which are neither 0 nor π. The phase of the reflected light changes by π at Brewster's angle in p-polarization. The ratio of s to p reflectance, P_D, is shown in Fig. 3*a* and the phase difference, $\Delta_D = \phi_p - \phi_s$, in Fig. 3*b*. Under conditions of TIR, the phase change on reflection, ϕ_{TIR}, is given by

$$\tan\frac{\phi_{TIR}}{2} = \frac{\sqrt{\sin^2\theta_A - \sin^2\theta_c}}{\cos\theta_A} \tag{55}$$

9.4 DISPERSION RELATIONS

For most purposes, a classical approach is found to provide a sufficient description of dispersion of the refractive index within the transmission window of insulators, and for optical interactions with lattice vibrations and free electrons. However the details of interband transitions in semiconductors and insulators and the effect of d-levels in transition metals require a quantum model of dispersion close to these resonances.

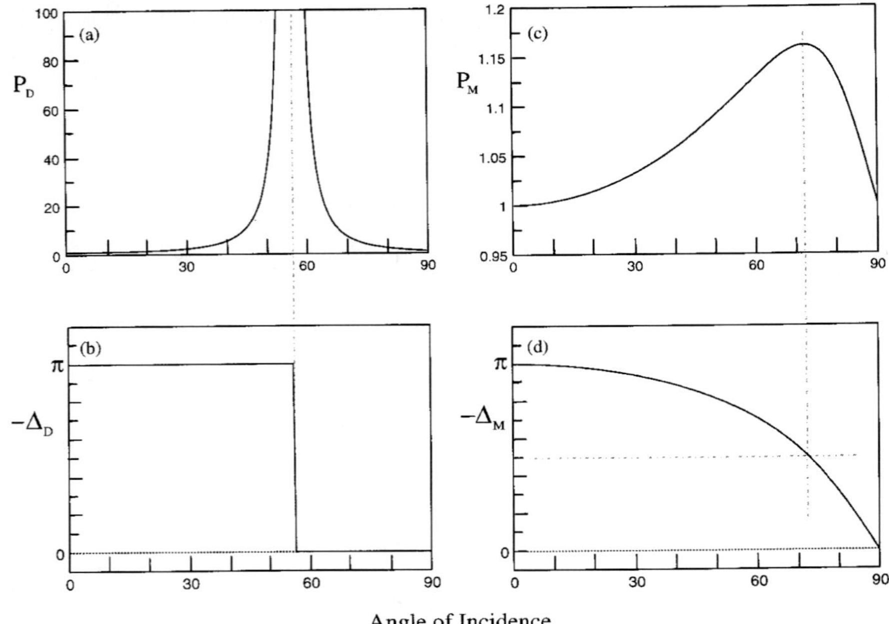

Angle of Incidence

FIGURE 3 Typical polarization ratios, P, (a, c) and phase differences, Δ, (b, d) for s- and p-polarizations at dielectric, D, and metallic, M, surfaces.

Classical Model

The Lorentz model for dispersion of the optical constants of solids assumes an optical interaction via the polarization produced by a set of damped harmonic oscillators. The polarization $\underline{\mathbf{P}}$ induced by a displacement $\mathbf{r}$ of bound electrons of density N and charge $-e$ is

$$\mathbf{P} = -Ne\mathbf{r} \tag{56}$$

Assuming the electrons to be elastically bound (Hooke's law) with a force constant, ξ,

$$-e\mathbf{E}_{\text{loc}} = \xi\mathbf{r} \tag{57}$$

the differential equation of motion is

$$m\frac{d^2\mathbf{r}}{dt^2} + m\Gamma\frac{d\mathbf{r}}{dt} + \xi\mathbf{r} = -e\mathbf{E}_{\text{loc}} \tag{58}$$

where m is the electron mass and Γ is a damping constant. Here the lattice is assumed to have infinite mass and the magnetic interaction has been neglected. Solving the equation of motion for fields of frequency ω gives a relation for the complex refractive index and dielectric constant:

$$\tilde{\eta}^2 = \frac{\tilde{\varepsilon}}{\varepsilon_0} = 1 + \frac{Ne^2}{m\varepsilon_0}\sum_{\text{j}}\frac{f_{\text{j}}}{(\omega_{\text{j}}^2 - \omega^2 - i\Gamma_{\text{j}}\omega)} \tag{59}$$

We have given the more general result for a number of resonant frequencies

$$\omega_j = \sqrt{\frac{\xi_j}{m}} \tag{60}$$

where the f_j represent the fraction of electrons which contribute to each oscillator with force constant ξ_j. f_j represent oscillator strengths.

A useful semi-empirical relation for refractive index in the transparency region of a crystal known as the Sellmeier formula follows directly from Eq. (59) under the assumption that, far from resonances, the damping constant terms $\Gamma_j\omega$ are negligible compared to $(\omega_j^2 - \omega^2)$:

$$n^2 = 1 + \sum_j \frac{A_j\lambda^2}{\lambda^2 - \lambda_j^2} \tag{61}$$

Sum Rules

The definition of oscillator strength results in the sum rule for electronic interactions

$$\sum_j f_j = Z \tag{62}$$

where Z is the number of electrons per atom. The periodicity of the lattice in solids (see Sec. 9.7) leads to the modification of this sum rule to

$$\sum_m f_{mn} = \frac{m}{\hbar^2}\frac{\partial^2 \mathscr{E}_n}{\partial k^2} - 1 = \frac{m}{m_n^*} - 1 \tag{63}$$

where m_n^* is an effective mass (see "Energy Band Structures" in Sec. 9.7).

Another sum rule for solids equivalent to Eq. (62) relates the imaginary part of the permittivity or dielectric constant and the plasma frequency, ω_p,

$$\int_0^\infty \omega\varepsilon_2(\omega)\, d\omega = \tfrac{1}{2}\pi\omega_p^2 \tag{64}$$

where $\omega_p^2 = Ne^2/\varepsilon_0 m$ (see "Drude Model" in Sec. 9.6).

Dispersion relations are integral formulas which relate refractive properties to the absorptive process. Kramers-Kronig relations are commonly used dispersion integrals based on the condition of causality which may be related to sum rules. These relations can be expressed in alternative forms. For instance, the reflectivity of a solid is often measured at normal incidence and dispersion relations used to determine the optical properties. Writing the complex reflectivity amplitude as

$$\tilde{r}(\omega) = r_r(\omega)e^{i\theta(\omega)} \tag{65}$$

the phase shift, θ, can be determined by integrating the experimental measurement of the real amplitude, r_r,

$$\theta(\omega) = -\frac{2\omega}{\pi}\mathscr{P}\int_0^\infty \frac{\ln r_r(\omega')}{\omega'^2 - \omega^2}\, d\omega' \tag{66}$$

and the optical constants determined from the complex Fresnel relation,

$$r_r(\omega)e^{i\theta} = \frac{(n-1+i\kappa)}{(n+1+i\kappa)} \tag{67}$$

Sum rules following from the Kramers-Kronig relations relate the refractive index $n(\omega)$ at a given frequency, ω, to the absorption coefficient, $\alpha(\omega')$, integrated over all frequencies, ω', according to

$$n(\omega) - 1 = \frac{c}{\omega}\mathcal{P}\int_0^\infty \frac{\alpha(\omega')\,d\omega'}{\omega'^2 - \omega^2} \tag{68}$$

Similarly, the real and imaginary parts of the dielectric constant, ε' and ε'', may be related via the integral relations,

$$\varepsilon_1(\omega) - 1 = \frac{2}{\pi}\mathcal{P}\int_0^\infty \frac{\omega'\varepsilon_2(\omega')}{\omega'^2 - \omega^2}\,d\omega' \tag{69}$$

$$\varepsilon_2(\omega) = -\frac{2\omega}{\pi}\mathcal{P}\int_0^\infty \frac{\varepsilon_1(\omega') - 1}{\omega'^2 - \omega^2}\,d\omega' \tag{70}$$

9.5 LATTICE INTERACTIONS

The adiabatic approximation is the normal starting point for a consideration of the coupling of light with lattice vibrations, i.e., it is assumed that the response of the outer shell electrons of the atoms to an electric field is much faster than the response of the core together with its inner electron shells. Further, the harmonic approximation assumes, that for small displacements, the restoring force on the ions will be proportional to the displacement. The solution of the equations of motion for the atoms within a solid under these conditions give normal modes of vibration, whose frequency eigenvalues and displacement eigenvectors depend on the crystal symmetry, atomic separation, and the detailed form of the interatomic forces. The frequency of lattice vibrations in solids are typically in the 100 to 1000 cm^{-1} range (wavelengths between 10 and 100 μm). Longitudinal and doubly degenerate transverse vibrational modes have different natural frequencies due to long range Coulomb interactions. Infrared or Raman activity can be determined for a given crystal symmetry by representing the modes of vibration as irreducible representations of the space group of the crystal lattice.

Infrared Dipole Active Modes

If the displacement of atoms in a normal mode of vibration produces an oscillating dipole moment, then the motion is dipole active. Thus, harmonic vibrations in ionic crystals contribute directly to the dielectric function, whereas higher order contributions are needed in nonpolar crystals. Since photons have small wavevectors compared to the size of the Brillouin zone in solids, only zone center lattice vibrations (i.e., long wavelength phonons) can couple to the radiation. This conservation of wavevector (or momentum) also implies that only optical phonons interact. In a dipole-active, long wavelength optical mode, oppositely charged ions within each primitive cell undergo oppositely directed displacements giving rise to a nonvanishing polarization. Group theory shows that, within the harmonic approximation, the infrared active modes have irreducible representations with the same transformation properties as x, y, or z. The strength of the light-dipole coupling will depend on the degree of charge redistribution between the ions, i.e., the relative ionicity of the solid.

Classical dispersion theory leads to a phenomenological model for the optical interaction with dipole active lattice modes. Because of the transverse nature of electromagnetic radiation, the electric field vector couples with the transverse optical (TO) phonons and the maximum absorption therefore occurs at this resonance. The resonance frequency, ω_{TO}, is inserted into the solution of the equation of motion, Eq. (59). Since electronic transitions typically occur at frequencies 10^2 to 10^3 higher than the frequency of vibration of the ions, the atomic polarizability can be represented by a single high frequency permittivity, $\varepsilon(\infty)$. The dispersion relation for a crystal with several zone center TO phonons may be written

$$\tilde{\varepsilon}(\omega) = \varepsilon(\infty) + \sum_j \frac{S_j}{(\omega_{TOj}^2 - \omega^2 - i\Gamma_j\omega)} \tag{71}$$

By defining a low frequency permittivity, $\varepsilon(0)$, the oscillator strength for a crystal possessing two atoms with opposite charge, Ze, per unit cell of volume, V, is

$$S = \frac{(\varepsilon(\infty)/\varepsilon_0 + 2)^2(Ze)^2}{9m_r V} = \omega_{TO}^2(\varepsilon(0) - \varepsilon(\infty)) \tag{72}$$

where Ze represents the "effective charge" of the ions, m_r is the reduced mass of the ions, and the local field has been included based on Eq. (31). Figure 4 shows the form of the

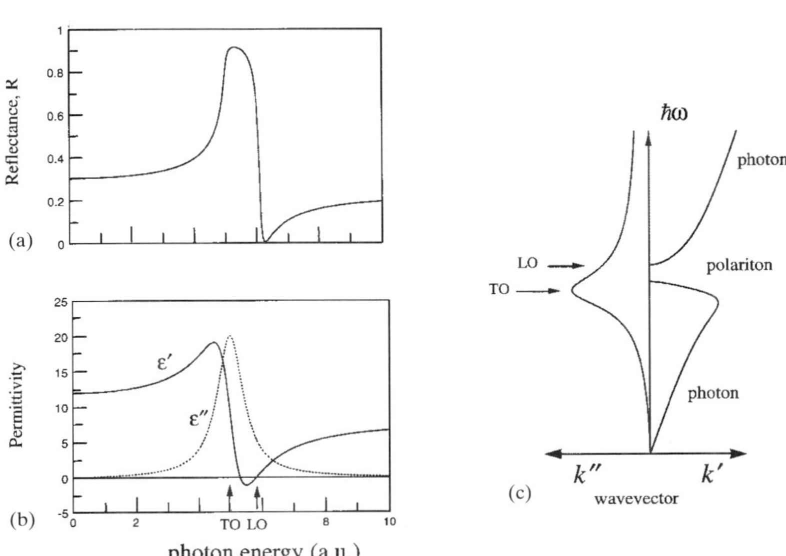

FIGURE 4 (*a*) Reflectance; and (*b*) real and imaginary parts of the permittivity of a solid with a single infrared active mode. (*c*) polariton dispersion curves (real and imaginary parts) showing the frequencies of the longitudinal and transverse optical modes.

real and imaginary parts of the dielectric constant, the reflectivity and the polariton dispersion curve. Observing that the real part of the dielectric constant is zero at longitudinal phonon frequencies, ω_{LO}, the Lyddane-Sachs-Teller relation may be derived, which in its general form for a number of dipole active phonons, is given by

$$\frac{\varepsilon(0)}{\varepsilon(\infty)} = \prod_{j} \left(\frac{\omega_{Lj}}{\omega_{Tj}}\right)^2 \tag{73}$$

These relations [Eqs. (71) to (73)] give good fits to measured reflectivities in a wide range of ionically (or partially ionically) bonded solids. The LO-TO splitting and effective charge, *Ze,* depend on the ionicity of the solid; however the magnitude of *Ze* determined from experiments does not necessarily quantify the ionicity since this "rigid ion" model does not account for the change of polarizability due to the distortion of the electron shells during the vibration.

In uniaxial and biaxial crystals, the restoring forces on the ions are anisotropic, resulting in different natural frequencies depending on the direction of light propagation as well as the transverse or longitudinal nature of the vibration. Similar to the propagation of light, "ordinary" and "extraordinary" transverse phonons may be defined with respect to the principal axes. For instance, in a uniaxial crystal under the condition that the anisotropy in phonon frequency is smaller than the LO-TO frequency splitting, infrared radiation of frequency ω propagating at an angle θ to the optic axis will couple to TO phonons according to the relation

$$\omega_T^2 = \omega_{T\parallel}^2 \sin^2\theta + \omega_{T\perp}^2 \cos^2\theta \tag{74}$$

where $\omega_{T\parallel}$ is a TO phonon propagating with atomic displacements parallel to the optic axis, and $\omega_{T\perp}$ is a TO phonon propagating with atomic displacements perpendicular to the optic axis. The corresponding expression for LO modes is

$$\omega_L^2 = \omega_{L\parallel}^2 \cos^2\theta + \omega_{L\perp}^2 \sin^2\theta \tag{75}$$

In Table 2, the irreducible representations of the infrared normal modes for the different crystal symmetries are labeled *x, y,* or *z.*

Brillouin and Raman Scattering

Inelastic scattering of radiation by acoustic phonons is known as Brillouin scattering, while the term Raman scattering is normally reserved for inelastic scattering from optic phonons in solids. In the case of Brillouin scattering, long wavelength acoustic modes produce strain and thereby modulate the dielectric constant of the medium, thus producing a frequency shift in scattered light. In a Raman active mode, an incident electric field produces a dipole by polarizing the electron cloud of each atom. If this induced dipole is modulated by a lattice vibrational mode, coupling occurs between the light and the phonon and inelastic scattering results. Each Raman or Brillouin scattering event involves the destruction of an incident photon of frequency, ω_i, the creation of a scattered photon, ω_s, and the creation or destruction of a phonon of frequency, ω_p. The frequency shift, $\omega_i \pm \omega_s = \omega_p$, is typically 100 to 1000 cm^{-1} for Raman scattering, but only a few wavenumbers for Brillouin scattering.

Atomic polarizability components have exactly the same transformation properties as the quadratic functions x^2, xy, . . . , z^2. The Raman activity of the modes of vibration of a crystal with a given point group symmetry can thus be deduced from the Raman tensors given in Table 2. The scattering efficiency, S_R, for a mode corresponding to one of the irreducible representations listed is given by

$$S_R = A\left[\sum_{\rho,\sigma} e_i^\sigma R_{\sigma,\rho} e_s^\rho\right]^2 \tag{76}$$

where A is a constant of proportionality, $R_{\sigma,\rho}$ is the Raman coefficient of the representation, and e_i^σ and e_s^ρ are components of the unit vectors of polarization of the incident, i, and scattered, s, radiation along the principal axes, where σ and $\rho = x$, y, and z.

Not all optic modes of zero wavevector are Raman active. Raman activity is often complementary to infrared activity. For instance, since the optic mode in the diamond lattice has even parity, it is Raman active, but not infrared active, whereas the zone center mode in sodium chloride is infrared active, but not Raman active because the inversion center is on the atom site and so the phonon has odd parity. In piezoelectric crystals, which lack a center of inversion, some modes can be both Raman and infrared active. In this case the Raman scattering can be anomalous due to the long-range electrostatic forces associated with the polar lattice vibrations.

The theory of Brillouin scattering is based on the elastic deformation produced in a crystal by a long wavelength acoustic phonon. The intensity of the scattering depends on the change in refractive index with the strain induced by the vibrational mode. A strain, s_{ij} in the lattice produces a change in the component of permittivity, $\varepsilon_{\mu\nu}$, given by

$$\delta\varepsilon_{\mu\nu} = -\sum_{\rho,\sigma} \varepsilon_{\mu\rho} p_{\rho\sigma,ij} \varepsilon_{\sigma\nu} s_{ij} \tag{77}$$

where $p_{\rho\sigma,ij}$ is an elasto-optical coefficient.[3] The velocity of the acoustic phonons and their anisotropy can be determined from Brillouin scattering measurements.

9.6 FREE ELECTRON PROPERTIES

Fundamental optical properties of metals and semiconductors with high densities of free carriers are well described using a classical model. Reflectivity is the primary property of interest because of the high absorption.

Drude Model

The Drude model for free electrons is a special condition of the classical Lorentz model (Sec. 9.4) with the Hooke's law force constant, $\xi = 0$, so that the resonant frequency is zero. In this case

$$\frac{\varepsilon_1}{\varepsilon_0} = n^2 - \kappa^2 = 1 - \frac{\omega_p^2}{\omega^2 + \tau^{-2}} \tag{78}$$

TABLE 2 Infrared and Raman-Active Vibrational Symmetries and Raman Tensors[13]

Monoclinic			$\begin{bmatrix} a & & d \\ & b & \\ d & & c \end{bmatrix}$	$\begin{bmatrix} & e & \\ e & & f \\ & f & \end{bmatrix}$		
	2	C_2	A(y)	B(x,z)		
	m	C_v	A'(x,z)	A"(y)		
	2/m	C_{2h}	A_g	B_g		
Orthorhombic			$\begin{bmatrix} a & & \\ & b & \\ & & c \end{bmatrix}$	$\begin{bmatrix} & d & \\ d & & \\ & & \end{bmatrix}$	$\begin{bmatrix} & & e \\ & & \\ e & & \end{bmatrix}$	$\begin{bmatrix} & & \\ & & f \\ & f & \end{bmatrix}$
	mm	C_{2v}	$A_1(z)$	A_2	$B_1(x)$	$B_2(y)$
	222	D_2	A	$B_1(z)$	$B_2(y)$	$B_3(x)$
	mmm	D_{2h}	A_g	B_{1g}	B_{2g}	B_{3g}
Trigonal			$\begin{bmatrix} a & & \\ & a & \\ & & b \end{bmatrix}$	$\begin{bmatrix} c & d & e \\ d & -c & f \\ e & f & \end{bmatrix}$	$\begin{bmatrix} d & -c & -f \\ -c & -d & e \\ -f & e & \end{bmatrix}$	
	3	C_3	A(z)	E(x)	E(y)	
	$\bar{3}$	S_6	A_g	E_g	E_g	
			$\begin{bmatrix} a & & \\ & a & \\ & & b \end{bmatrix}$	$\begin{bmatrix} c & & \\ & -c & d \\ & d & \end{bmatrix}$	$\begin{bmatrix} & -c & -d \\ -c & & \\ -d & & \end{bmatrix}$	
	3m	C_{3v}	$A_1(z)$	E(y)	E(-x)	
	32	D_3	A_1	E(x)	E(y)	
	$\bar{3}$m	D_{3d}	A_{1g}	E_g	E_g	
Tetragonal			$\begin{bmatrix} a & & \\ & a & \\ & & b \end{bmatrix}$	$\begin{bmatrix} c & d & \\ d & -c & \\ & & \end{bmatrix}$	$\begin{bmatrix} & & e \\ & & f \\ e & f & \end{bmatrix}$	$\begin{bmatrix} & & -f \\ & & e \\ -f & e & \end{bmatrix}$
	4	C_4	A(z)	B	E(x)	E(y)
	$\bar{4}$	S_4	A	B(z)	E(x)	E(-y)
	4/m	C_{4h}	A_g	B_g	E_g	E_g

			$\begin{bmatrix}a&&\\&a&\\&&b\end{bmatrix}$	$\begin{bmatrix}c&&\\&-c&\\&&\end{bmatrix}$	$\begin{bmatrix}&d&\\d&&\\&&\end{bmatrix}$	$\begin{bmatrix}&&e\\&&\\e&&\end{bmatrix}$	$\begin{bmatrix}&&\\&&e\\&e&\end{bmatrix}$	
	4mm	C_{4v}	$A_1(z)$	B_1	B_2	E(x)	E(y)	
	$\bar{4}2m$	D_{2d}	A_1	B_1	$B_2(z)$	E(y)	E(x)	
	42	D_4	A_1	B_1	B_2	E(-y)	E(x)	
	4/mmm	D_{4h}	A_{1g}	B_{1g}	B_{2g}	E_g	E_g	
Hexagonal			$\begin{bmatrix}a&&\\&a&\\&&b\end{bmatrix}$	$\begin{bmatrix}&&c\\&&d\\c&d&\end{bmatrix}$	$\begin{bmatrix}&&-d\\&&c\\-d&c&\end{bmatrix}$	$\begin{bmatrix}e&f&\\f&-e&\\&&\end{bmatrix}$	$\begin{bmatrix}f&-e&\\-e&-f&\\&&\end{bmatrix}$	
	6	C_6	A(z)	$E_1(x)$	$E_1(y)$	E_2	E_2	
	$\bar{6}$	C_{3h}	A'	E"	E"	E'(x)	E'(y)	
	6/m	C_{6h}	A_g	E_{1g}	E_{1g}	E_{2g}	E_{2g}	
			$\begin{bmatrix}a&&\\&a&\\&&b\end{bmatrix}$	$\begin{bmatrix}&&\\&&c\\&c&\end{bmatrix}$	$\begin{bmatrix}&&-c\\&&\\-c&&\end{bmatrix}$	$\begin{bmatrix}&d&\\d&&\\&&\end{bmatrix}$	$\square$	
	6mm	C_{6v}	$A_1(z)$	$E_1(y)$	$E_1(-x)$	E_2	E_2	
	$\bar{6}m2$	D_{3h}	A_1'	E"	E"	E'(x)	E'(y)	
	62	D_6	A_1	$E_1(x)$	$E_1(y)$	E_2	E_2	
	6/mmm	D_{6h}	A_{1g}	E_{1g}	E_{1g}	E_{2g}	E_{2g}	
Cubic			$\begin{bmatrix}a&&\\&a&\\&&a\end{bmatrix}$	$\begin{bmatrix}b&&\\&b&\\&&-2b\end{bmatrix}$	$\begin{bmatrix}-3^{1/2}b&&\\&3^{1/2}b&\\&&\end{bmatrix}$	$\begin{bmatrix}&&\\&&d\\&d&\end{bmatrix}$	$\begin{bmatrix}&&d\\&&\\d&&\end{bmatrix}$	$\begin{bmatrix}&d&\\d&&\\&&\end{bmatrix}$
	23	T	A	E	E	F(x)	F(y)	F(z)
	m3	T_h	A_g	E_g	E_g	F_g	F_g	F_g
	$\bar{4}3m$	T_d	A_1	E	E	$F_2(x)$	$F_2(y)$	$F_2(z)$
	432	O	A_1	E	E	F_2	F_2	F_2
	m3m	O_h	A_{1g}	E_g	E_g	F_{2g}	F_{2g}	F_{2g}

and

$$\frac{\varepsilon_2}{\varepsilon_0} = 2n\kappa = \frac{\omega_p^2}{\omega^2 + \tau^{-2}}\left(\frac{1}{\omega\tau}\right) \tag{79}$$

where ω_p is the plasma frequency,

$$\omega_p = \sqrt{\frac{Ne^2}{\varepsilon_0 m}} = \sqrt{\frac{\mu_0 \sigma c^2}{\tau}} \tag{80}$$

and τ $(=1/\Gamma)$ is the scattering or relaxation time for the electrons. In ideal metals $(\sigma \rightarrow \infty)$, $n = \kappa$. Figure 5*a* shows the form of the dispersion in the real and imaginary parts

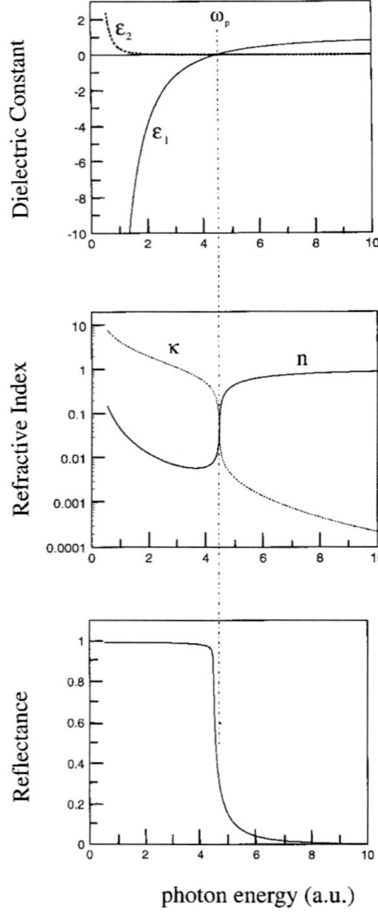

FIGURE 5 Dispersion of: (*a*) the real and imaginary parts of the dielectric constant; (*b*) real and imaginary parts of the refractive index; and (*c*) the reflectance according to the Drude model where ω_p is the plasma frequency.

of the dielectric constant for free electrons, while the real and imaginary parts of the refractive index are illustrated in Fig. 5*b*. The plasma frequency is defined by the point at which the real part changes sign. The reflectance is plotted in Fig. 5*c* and shows a magnitude close to 100 percent below the plasma frequency, but falls rapidly to a small value above ω_p. The plasma frequency determined solely by the free electron term is typically on the order of 10 eV in metals accounting for their high reflectivity in the visible.

Interband Transitions in Metals

Not all metals are highly reflective below the plasma frequency. The nobel metals possess optical properties which combine free electron intraband (Drude) and interband contributions. A typical metal has d-levels at energies a few electron volts below the electron Fermi level. Transitions can be optically induced from these d-states to empty states above the Fermi level. These transitions normally occur in the ultraviolet spectral region, but can have a significant influence on the optical properties of metals in the visible spectral region via the real part of the dielectric constant. Describing the interband effects, $\delta\varepsilon_b$, within a classical (Lorentz) model, the combined effects on the dielectric constant may be added:

$$\tilde{\varepsilon} = 1 + \delta\tilde{\varepsilon}_b + \delta\tilde{\varepsilon}_f \tag{81}$$

The interband contribution to the real part of the dielectric constant is positive and shows a resonance near the transition frequency. On the other hand, the free electron contribution is negative below the plasma frequency. The interband contribution can cause a shift to shorter wavelengths of the zero crossover in ε_1, thus causing a reduction of the reflectivity in the blue. For instance d-states in copper lie only 2 eV below the Fermi level, which results in copper's characteristic color.

Reflectivity

Absorption in metals is described by complex optical constants. The reflectivity is accompanied by a phase change and the penetration depth is

$$d_M = \frac{c}{2\omega\kappa} = \frac{\lambda}{4\pi\kappa} \tag{82}$$

At normal incidence at an air-metal interface, the reflectance is given by

$$R = \frac{(n-1)^2 + \kappa^2}{(n+1)^2 + \kappa^2} = 1 - \frac{2}{\kappa} = 1 - 2\sqrt{\frac{2\varepsilon_0}{\varepsilon_2}} \tag{83}$$

By analogy with the law of refraction (Snell's Law) a complex refractive index can be defined by the refraction equation

$$\sin\theta_t = \frac{1}{\tilde{\eta}} \sin\theta_i \tag{84}$$

Since $\tilde{\eta}$ is complex, θ_t is also complex and the phase change on reflection can take values other than 0 and π. For nonnormal incidence, it can be shown that the surfaces of constant amplitude inside the metal are parallel to the surface, while surfaces of constant phase are at an angle to the surface. The electromagnetic wave in a metal is thus inhomogeneous. The real and imaginary parts of the refractive index can be determined by measuring the amplitude and phase of the reflected light. Writing the s and p components of the complex reflected fields in the form

$$E_{rp} = \rho_p e^{i\phi_p}; \qquad E_{rs} = \rho_s e^{i\phi_s} \tag{85}$$

and defining the real amplitude ratio and phase difference as

$$P_M = \tan\psi = \frac{\rho_s}{\rho_p}; \qquad \Delta_M = \phi_p - \phi_s \tag{86}$$

then the real and imaginary parts of the refractive index are given by

$$n \approx -\frac{\sin\theta_i \tan\theta_i \cos 2\psi}{1 + \sin 2\psi \cos\Delta_M} = -\sin\bar{\theta}_i \tan\bar{\theta}_i \cos 2\bar{\psi} \tag{87}$$

$$\kappa \approx \tan 2\psi \sin\Delta_M = -\tan 2\bar{\psi} \tag{88}$$

$\bar{\theta}_i$ is the principal angle which occurs at the maximum in P_M at the condition $\Delta_M = \pi/2$ (see Fig. 3*c* and *d*), which is equivalent to Brewster's angle at an interface between two nonabsorbing dielectrics (see Fig. 3*a* and *b*).

Plasmons

Plasmons are oscillations of fluctuations in charge density. The condition for these oscillations to occur is the same as the condition for the onset of electromagnetic propagation at the plasma frequency. Volume plasmons are not excited by light at normal incidence since they are purely longitudinal. Oscillations cannot be produced by transverse electromagnetic radiation with zero divergence. However, at the surface of a solid, an oscillation in surface charge density is possible. At an interface between a metal with permittivity, ε_m, and a dielectric with permittivity, ε_d, the condition $\varepsilon_m = -\varepsilon_d$ such that (neglecting damping and assuming a free electron metal) a surface plasmon can be created with frequency

$$\omega_s = \frac{\omega_p}{(\varepsilon_d/\varepsilon_0 + 1)^{1/2}} \tag{89}$$

By altering the angle of incidence, the component of the electromagnetic wavevector can be made to match the surface plasmon mode.

9.7 BAND STRUCTURES AND INTERBAND TRANSITIONS

Advances in semiconductors for electronic and optoelectronic applications have encouraged the development of highly sophisticated theories of interband absorption in semiconductors. In addition, the development of low-dimensional structures (quantum wells, quantum wires, and quantum dots) have provided the means of "engineering" the optical properties of solids. The approach here has been to outline the basic quantum mechanical development for interband transitions in solids.

Quantum Mechanical Model

The quantum theory of absorption considers the probability of an electron being excited from a lower energy level to a higher level. For instance, an isolated atom has a characteristic set of electron levels with associated wavefunctions and energy eigenvalues. The absorption spectrum of the atom thus consists of a series of lines whose frequencies are given by

$$\hbar\omega_{fi} = \mathscr{E}_f - \mathscr{E}_i \; (\mathscr{E}_f > \mathscr{E}_i) \tag{90}$$

where $\mathscr{E}_f$ and $\mathscr{E}_i$ are a pair of energy eigenvalues. We also know that the spontaneous lifetime, τ, for transitions from any excited state to a lower state sets a natural linewidth of order $\hbar/\tau$ based on the uncertainty principle. The Schrödinger equation for the ground state with wavefunction, φ_i, in the unperturbed system,

$$\mathscr{H}_0\varphi_i = \mathscr{E}_i\varphi_i \tag{91}$$

is represented by the time-independent hamiltonian, $\mathscr{H}_0$. The optical interaction can be treated by first order perturbation theory. By introducing a perturbation term based on the classical oscillator

$$\mathscr{H}' = e\mathbf{E}\cdot\mathbf{r} \tag{92}$$

this leads to a similar expression to the Lorentz model, Eq. (59),

$$\tilde{\eta}^2 = \frac{\varepsilon}{\varepsilon_0} = 1 + \frac{Ne^2}{m\varepsilon_0}\sum_{m}\frac{f_{fi}}{(\omega_{fi}^2 - \omega^2 - i\Gamma_{fi}\omega)} \tag{93}$$

where

$$f_{j'j} = \frac{2\,|\mathbf{p}_{j'j}|^2}{m\hbar\omega_{j'j}} \tag{94}$$

and $\mathbf{p}_{j'j}$ are momentum matrix elements defined by

$$\mathbf{p}_{j'j} = \langle\varphi_{j'}\,|\,\mathbf{p}\,|\varphi_j\rangle = \int \varphi_{j'}^{*}(i\hbar\nabla)\varphi_j\,d\mathbf{r} \tag{95}$$

Perturbation theory to first order gives the probability per unit time that a perturbation of the form $\mathscr{H}(t) = \mathscr{H}_p \exp(i\omega t)$ induces a transition from the initial to final state:

$$W_{fi} = \frac{2\pi}{\hbar}|\langle\varphi_f|\,\mathscr{H}_p\,|\varphi_i\rangle|^2\delta(\mathscr{E}_f - \mathscr{E}_i - \hbar\omega) \tag{96}$$

This is known as Fermi's Golden Rule.

Energy Band Structures

If we imagine N similar atoms brought together to form a crystal, each degenerate energy level of the atoms will spread into a band of N levels. If N is large, these levels can be treated as a continuum of energy states. The wavefunctions and electron energies of these energy bands can be calculated by various approximate methods ranging from nearly free electron to tight-binding models. The choice of approach depends on the type of bonding between the atoms.

Within the one electron and adiabatic assumptions, each electron moves in the periodic potential, $V(\mathbf{r})$, of the lattice leading to the Schrödinger equation for a single particle wavefunction

$$\left[\frac{p^2}{2m} + V(\mathbf{r})\right]\psi(\mathbf{r}) = \mathscr{E}\psi(\mathbf{r}) \tag{97}$$

where the momentum operator is given by $\mathbf{p} = -i\hbar\nabla$. The simple free electron solution of the Schrödinger equation (i.e., for $V(\underline{\mathbf{r}}) = 0$) is a parabolic relationship between energy and wavevector. The solution including a periodic potential, $V(\underline{\mathbf{r}})$ has the form

$$\psi_{\mathrm{k}}(\mathbf{r}) = \exp\,(i\mathbf{k}\cdot\mathbf{r})\cdot u_{\mathrm{k}}(\mathbf{r}) \tag{98}$$

where $\underline{\mathbf{k}}$ is the electron wavevector and $u_{\mathrm{k}}(\mathbf{r})$ has the periodicity of the crystal lattice. This is known as the Bloch solution. The allowed values of k are separated by $2\pi/L$ where L is the length of the crystal. The wavevector is not uniquely defined by the wavefunction, but the energy eigenvalues are a periodic function of $\mathbf{k}$. For an arbitrarily weak periodic potential

$$\mathscr{E} = \frac{\hbar^2}{2m}\,|\mathbf{k} + \mathbf{G}|^2 \tag{99}$$

where $\mathbf{G}$ is a reciprocal lattice vector (in one dimension $G = 2\pi n/a$ where a is the lattice spacing and n is an θ integer). Thus we need only consider solutions which are restricted to a reduced zone, refered to as the first Brillouin zone, in reciprocal space (between $k = -\pi/a$ and π/a in one dimension). Higher energy states are folded into the first zone consistent with Eq. (99) to form a series of energy bands. Figure 6 shows the first Brillouin zones for face centered cubic (fcc) crystal lattices and energy levels for a weak lattice potential.

A finite periodic potential, $V(\underline{\mathbf{r}})$, alters the shape of the free electron bands. The

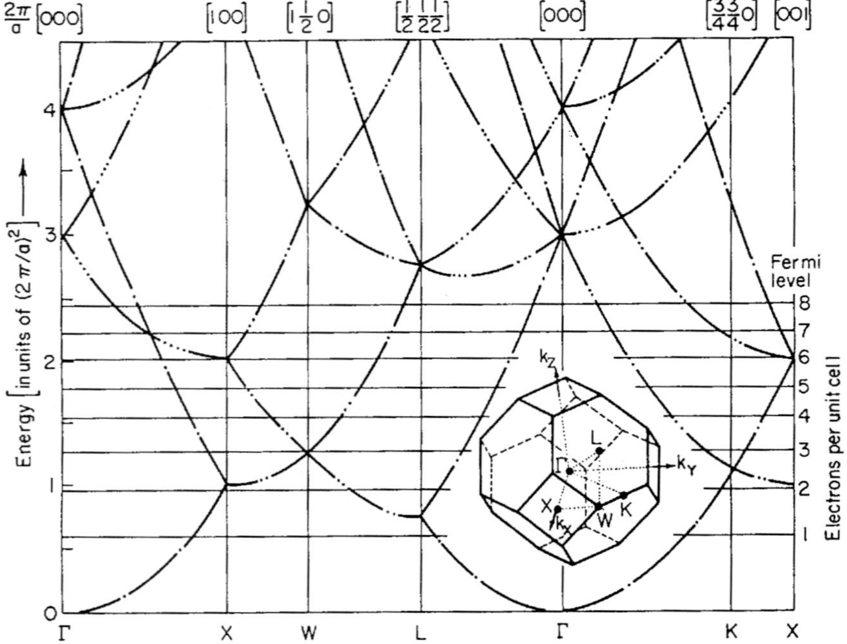

FIGURE 6 Free electron band structure in the reduced Brillouin zone for face-centered-cubic lattices. The insert shows the first Brillouin zone with principal symmetry points labeled. This applies to crystals such as Al, Cu, Ag, Si, Ge, and GaAs.[5]

curvature of the bands are described by m^*, an "effective mass," which is defined by the slope of the dispersion curve at a given **k**:

$$\frac{1}{m_k^*} = \frac{1}{\hbar^2 k}\frac{d\mathscr{E}}{dk} \tag{100}$$

At the zone center, $\mathbf{k} = 0$, this reduces to the parabolic relationship

$$\mathscr{E} = \frac{\hbar^2 k^2}{2m_0^*} \tag{101}$$

Effective masses can be related to interband momentum matrix elements and energy gaps using perturbation theory. Substituting the Bloch function [see Eq. (98)] into the Schrödinger equation [see Eq. (97)] and identifying each band by an index, j, gives

$$\left[\frac{p^2}{2m} + \frac{\hbar}{m}\mathbf{k}\cdot\mathbf{p} + \frac{\hbar^2 k^2}{2m} + V(\mathbf{r})\right]u_{j\mathbf{k}}(\mathbf{r}) = \mathscr{E}_{j\mathbf{k}}(\mathbf{r})u_{j\mathbf{k}}(\mathbf{r}) \tag{102}$$

The $\mathbf{k}\cdot\mathbf{p}$ term can be treated as a perturbation about a specific point in k-space. For any **k**, the set of all $u_{j\mathbf{k}}(r)$ (corresponding to the N energy levels) forms a complete set, i.e., the wavefunction at any value of **k** can be expressed as a linear combination of all wavefunctions at another **k**. Second order perturbation theory then predicts an effective mass given by

$$\frac{1}{m^*} = \frac{1}{m} + \frac{2}{m}\sum_{j'}\frac{|\mathbf{p}_{j'j}|^2}{\mathscr{E}_j(\mathbf{k}) - \mathscr{E}_{j'}(\mathbf{k})} \tag{103}$$

In principle the summation in Eq. (103) is over all bands, however, this can usually be reduced to a few nearest bands because of the resonant denominator. For example, in diamond and zinc blende structured semiconductors, the Kane momentum matrix element, P, defined by

$$P = -\frac{i\hbar}{m}\langle S_{CO}\,|p_x|\,X_{VO}\rangle \tag{104}$$

successfully characterizes the band structure and optical properties close to zone center. Here, S_{CO} is a spherically symmetric s-like atomic wavefunction representing the lowest zone center conduction band state and X_{VO} is a p-like function with x symmetry from the upper valence bands. In this case, including only the three highest valence bands and the lowest conduction band in the summation of Eq. (91), the conduction band effective mass is given by

$$\frac{1}{m_{CO}^*} = \frac{1}{m} + \frac{2}{3\hbar^2}\left[\frac{2P^2}{\mathscr{E}_g} + \frac{P^2}{\mathscr{E}_g + \Delta_{SO}}\right] \tag{105}$$

where $\mathscr{E}_g$ is the band gap energy and Δ_{SO} is the spin-orbit splitting. By inverting this expression, the momentum matrix element may be determined from measurements of effective mass and the band gaps. P is found to have similar magnitudes for a large number of semiconductors. Equation (105) illustrates the general rule that the effective mass of the conduction band is approximately proportional to the band gap energy.

Direct Interband Absorption

In the case of a solid, the first order perturbation of the single electron hamiltonian by electromagnetic radiation is more appropriately described by

$$\mathcal{H}'(t) = \frac{e}{mc}\mathbf{A} \cdot \mathbf{p} \tag{106}$$

rather than Eq. (92). $\mathbf{A}$ is the vector potential,

$$\mathbf{A}(\mathbf{r}, t) = A_0\hat{\boldsymbol{\xi}} \exp\left[i(\mathbf{q} \cdot \mathbf{r} - \omega t)\right] + \text{c.c.} \tag{107}$$

$\mathbf{q}$ is the wavevector, and $\hat{\xi}$ is the unit polarization vector of the electric field. (Note that this perturbation is of a similar form to the $\mathbf{k} \cdot \mathbf{p}$ perturbation described above.) Using Fermi's Golden Rule, the transition probability per unit time between a pair of bands is given by

$$W_{\text{fi}} = \frac{2\pi}{\hbar}\left(\frac{eA_0}{mc}\right)^2 |\langle\psi_{\text{f}}|\,\hat{\boldsymbol{\xi}} \cdot \mathbf{p}\,|\psi_{\text{i}}\rangle|^2\, \delta(\mathscr{E}_{\text{f}}(\mathbf{k}) - \mathscr{E}_{\text{i}}(\mathbf{k}) - \hbar\omega) \tag{108}$$

Conservation of momentum requires a change of electron momentum after the transition, however, the photon momentum is very small, so that vertical transitions in k-space can be assumed in most cases (the electric dipole approximation). The total transition rate per unit volume, $W_{\text{T}}(\omega)$, is obtained by integrating over all possible vertical transitions in the first Brillouin zone taking account of all contributing bands:

$$W_{\text{T}}(\omega) = \frac{2\pi}{\hbar}\left(\frac{2\pi e^2 I}{ncm^2\omega^2}\right)\sum_{\text{f}}\int \frac{d\mathbf{k}}{(2\pi)^3}|\hat{\boldsymbol{\xi}} \cdot \mathbf{p}_{\text{fi}}(\mathbf{k})|^2\, \delta(\mathscr{E}_{\text{f}}(\mathbf{k}) - \mathscr{E}_{\text{i}}(\mathbf{k}) - \hbar\omega) \tag{109}$$

Here the vector potential has been replaced with the irradiance, I, of the radiation through the relation

$$A_0 = \frac{2\pi c}{n\omega^2} I \tag{110}$$

Note that the momentum matrix element as defined in Eq. (95) determines the oscillator strength for the absorption. $\mathbf{p}_{\text{fi}}$ can often be assumed slowly varying in $\mathbf{k}$ so that the zone center matrix element can be employed for interband transitions and the frequency dependence of the absorption coefficient is dominated by the density of states.

Joint Density of States

The delta function in the integration of Eq. (109) represents energy conservation for the transitions between any two bands. If the momentum matrix element can be assumed slowly varying in k, then the integral can be rewritten in the form

$$J_{\text{fi}}(\omega) = \frac{1}{(2\pi)^3}\int d\mathbf{k}\, \delta(\mathscr{E}_{\text{f}}(\mathbf{k}) - \mathscr{E}_{\text{i}}(\mathbf{k}) - \hbar\omega) = \frac{1}{(2\pi)^3}\int \frac{d\mathbf{S}}{|\nabla_{\text{k}}\mathscr{E}_{\text{fi}}(\mathbf{k})|} \tag{111}$$

where $d\mathbf{S}$ is a surface element on the equal energy surface in $\mathbf{k}$-space defined by $\mathscr{E}_{\text{fi}}(\mathbf{k}) = \mathscr{E}_{\text{f}}(\mathbf{k}) - \mathscr{E}_{\text{i}}(\mathbf{k}) = \hbar\omega$. Written in this way, $J(\omega)$ is the joint density of states between the two bands (note the factor of two for spin is excluded in this definition). Points in k-space for which the condition

$$\nabla_{\text{k}}\mathscr{E}_{\text{fi}}(\mathbf{k}) = 0 \tag{112}$$

hold form critical points called van Hove singularities which lead to prominent features in the optical constants. In the neighborhood of a critical point at $\underline{\mathbf{k}}_c$, a constant energy surface may be described by the Taylor series

$$\mathscr{E}_{fi}(\mathbf{k}) = \mathscr{E}_c(\mathbf{k}_c) + \sum_{\mu=1}^{3} \beta_\mu k_\mu^2 \tag{113}$$

where μ represents directional coordinates. Minimum, maximum, and saddle points arise depending on the relative signs of the coefficients, β_μ. Table 3 gives the frequency dependence of the joint density of states in three dimensional (3-D), two-dimensional (2-D),

TABLE 3 Density of States in 3, 2, 1, and 0 Dimensions

	β_1	β_2	β_3	$E < E_c$	$E > E_c$
3-D					
M_0, min.	+	+	+	0	$C_0(E-E_c)^{\frac{1}{2}}$ 
M_1, saddle	+	+	-	$C_1 - C_1'(E_c - E)^{\frac{1}{2}}$	C_1 
M_2, saddle	+	-	-	C_2	$C_2 - C_2'(E-E_c)^{\frac{1}{2}}$ 
M_3, max.	-	-	-	$C_3(E_c - E)^{\frac{1}{2}}$	0 
2-D					
P_0, min.	+	+		0	B 
P_1, saddle	+	-		$-\frac{B}{\pi}\ln\left\lvert 1-\frac{E}{E_c}\right\rvert$	
P_2, max.	-	-		B	0
1-D					
Q_0, min.	+			0	$A(E_c - E)^{-\frac{1}{2}}$ 
Q_1, max	-			$A(E-E_c)^{-\frac{1}{2}}$	0 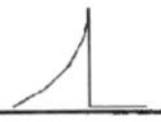
0-D				$\delta(E-E_c)$	

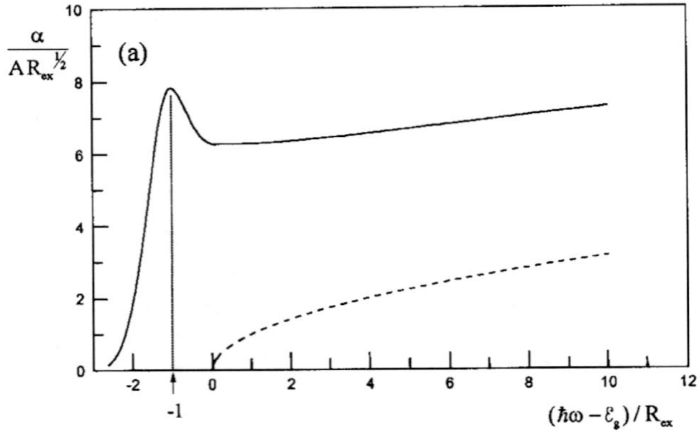

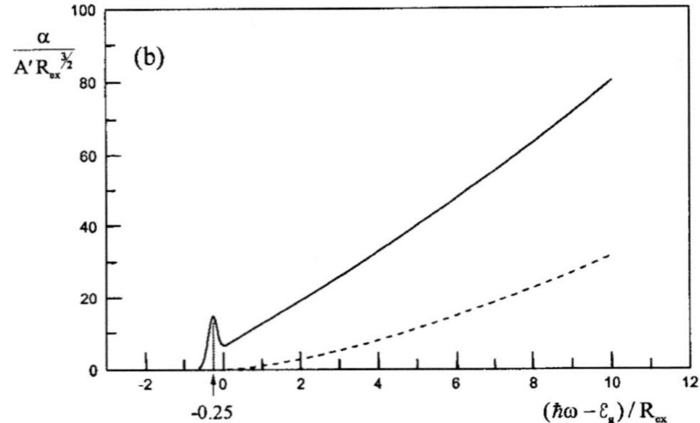

FIGURE 7 Illustration of absorption edge of crystals with: (*a*) direct allowed transitions; and (*b*) direct forbidden transitions, based on density of state (dashed lines) and excitonic enhanced absorption models (solid lines).

one-dimensional (1-D), and zero-dimensional (0-D) solids. The absorption coefficient, α, defined by Beer's Law may now be related to the transition rate by

$$\alpha(\omega) = I^{-1}\frac{dI}{dz} = \frac{\hbar\omega}{I} W_T \tag{114}$$

Thus, the minimum fundamental absorption edge of semiconductors and insulators (in the absence of excitonic effects) has the general form (see Fig. 7*a*)

$$\alpha(\omega) = C_0(\hbar\omega - \mathcal{E}_c)^{1/2} \tag{115}$$

Selection Rules and Forbidden Transitions

Direct interband absorption is allowed when the integral in Eq. (95) is nonzero. This occurs when the wavefunctions of the optically coupled states have opposite parity for single photon transitions. Transitions may be forbidden for other wavefunction symmetries. Although the precise form of the wavefunction may not be known, the selection rules can be determined by group theory from a knowledge of the space group of the crystal and symmetry of the energy band. Commonly, a single photon transition, which is not allowed at the zone center because two bands have like parity, will be allowed at finite k because wavefunction mixing will give mixed parity states. In this case, the momentum matrix element may have the form

$$p_{\mathrm{fi}}(\mathbf{k}) = (\mathbf{k} - \mathbf{k}_0) \cdot \nabla_{\mathbf{k}}[p_{\mathrm{fi}}(\mathbf{k})]_{\mathbf{k}=\mathbf{k}_0} \tag{116}$$

i.e., the matrix element is proportional to **k**. For interband transitions at an M_0 critical point, the frequency dependence of the absorption coefficient can be shown to be (see Fig. 7*b*)

$$\alpha(\omega) = C_0'(\hbar\omega - \mathscr{E}_c)^{3/2} \tag{117}$$

Indirect Transitions

Interband transitions may also take place with the assistance of a phonon. The typical situation is a semiconductor or insulator which has a lowest conduction band minimum near a Brillouin zone boundary. The phonon provides the required momentum to move the electron to this location, but supplies little energy. The phonon may be treated as an additional perturbation and therefore second order perturbation theory is needed in the analysis of this two step process. Theory predicts a frequency dependence for the absorption of the form

$$\alpha(\omega) \sim (\hbar\omega \pm \hbar\omega_{\mathrm{ph}} - \mathscr{E}_c)^2 \tag{118}$$

where ω_{ph} is the phonon frequency, absorption or emission being possible. For forbidden indirect transitions, this relationship becomes

$$\alpha(\omega) \sim (\hbar\omega \pm \hbar\omega_{\mathrm{ph}} - \mathscr{E}_c)^3 \tag{119}$$

Multiphoton Absorption

Multiphoton absorption can be treated by higher order perturbation theory. For instance, second order perturbation theory gives a transition rate between two bands

$$W_{\mathrm{fi}}^{(2)} = \frac{2\pi}{\hbar}\left(\frac{eA_0}{mc}\right)^4 \left|\sum_{\mathrm{t}} \frac{\langle\psi_{\mathrm{f}}|\hat{\xi}\cdot\mathbf{p}|\psi_{\mathrm{t}}\rangle\langle\psi_{\mathrm{t}}|\hat{\xi}\cdot\mathbf{p}|\psi_{\mathrm{i}}\rangle}{\mathscr{E}_{\mathrm{t}} - \mathscr{E}_{\mathrm{i}} - \hbar\omega}\right|^2 \delta(\mathscr{E}_{\mathrm{f}}(\mathbf{k}) - \mathscr{E}_{\mathrm{i}}(\mathbf{k}) - 2\hbar\omega) \tag{120}$$

where the summation spans all intermediate states, t. The interaction can be regarded as two successive steps. An electron first makes a transition from the initial state to an intermediate level of the system, t, by absorption of one photon. Energy is not conserved at this stage (momentum is) so that the absorption of a second photon must take the electron to its final state in a time determined by the energy mismatch and the uncertainty principle. In multiphoton absorption, one of the transitions may be an intraband self-transition. Since the probability depends on the arrival rate of the second photon, multiphoton absorption is intensity dependent. The total transition rate is given by

$$W_{\mathrm{T}}^{(2)}(\omega) = \frac{2\pi}{\hbar}\left(\frac{4\pi^2e^4I^2}{n^2c^2m^4\omega^4}\right)\sum_{\mathrm{f}}\int\frac{d\mathbf{k}}{(2\pi)^3}\left|\sum_{\mathrm{t}}\frac{\hat{\xi}\cdot\mathbf{p}_{\mathrm{ft}}(\mathbf{k})\hat{\xi}\cdot\mathbf{p}_{\mathrm{ti}}(\mathbf{k})}{\mathscr{E}_{\mathrm{t}} - \mathscr{E}_{\mathrm{i}} - \hbar\omega}\right|^2 \delta(\mathscr{E}_{\mathrm{f}}(\mathbf{k}) - \mathscr{E}_{\mathrm{i}}(\mathbf{k}) - 2\hbar\omega) \tag{121}$$

The two photon absorption coefficient is defined by the relation

$$-\frac{dI}{dz} = \alpha I + \beta I^2 \tag{122}$$

so that

$$\beta(\omega) = \frac{2\hbar\omega}{I^2} W_{\mathrm{T}}^{(2)}(\omega) \tag{123}$$

Excitons

The interband absorption processes discussed previously do not take into account Coulomb attraction between the excited electron and hole state left behind. This attraction can lead to the formation of a hydrogen-like bound electron-hole state or exciton. The binding energy of free (Wannier) excitons is typically a few meV. If not thermally washed out, excitons may be observed as a series of discrete absorption lines just below the band gap energy. The energy of formation of an exciton is

$$\mathscr{E}_{\mathrm{ex}} = \mathscr{E}_{\mathrm{g}} + \frac{\hbar^2 |\mathbf{K}|^2}{2(m_{\mathrm{e}}^* + m_{\mathrm{h}}^*)} - \frac{R_{\mathrm{ex}}}{n^2} \tag{124}$$

where R_{ex} is the exciton Rydberg

$$R_{\mathrm{ex}} = \frac{m_{\mathrm{r}}^* e^4}{2\hbar^2 \varepsilon_1^2} \tag{125}$$

m_{r}^* is the reduced effective mass, and n is a quantum number. Optically created electron-hole pairs have equal and opposite momentum which can only be satisfied if $\mathbf{K} = 0$ for the bound pair and results in discrete absorption lines. Coulomb attraction also modifies the absorption above the band gap energy. The theory of exciton absorption developed by Elliot predicts a modification to Eq. (115) for the direct allowed absorption coefficient above the band edge (see Fig. 7*a*)

$$\alpha = \frac{\pi C_0 R_{\mathrm{ex}}^{1/2} e^{\pi\gamma}}{\sinh \pi\gamma} \tag{126}$$

where

$$\gamma = \left(\frac{R_{\mathrm{ex}}}{\hbar\omega - \mathscr{E}_{\mathrm{g}}}\right)^{1/2} \tag{127}$$

Excitons associated with direct forbidden interband transitions do not show absorption at the lowest ($n = 1$) state, but transitions to excited levels are allowed. Above the band edge for direct forbidden transitions the absorption has the frequency dependence (see Fig. 7*b*)

$$\alpha = \frac{\pi C_0' R_{\mathrm{ex}}^{3/2}\left(1 + \frac{1}{\gamma^2}\right) e^{\pi\gamma}}{\sinh \pi\gamma} \tag{128}$$

Figure 7 compares the form of the absorption edge based on the density of states functions and a discrete exciton absorption line (dashed lines) with the absorption functions based on the Elliot theory and a typically broadened exciton (solid lines). Figure 7*a* is an illustration of a direct allowed gap, e.g., GaAs with the $n = 1$ exciton visible, and 7*b* shows a forbidden direct absorption edge, e.g., Cu_2O. In the latter case, optical excitation of the $n = 1$ exciton is forbidden, but the $n = 2$ and higher exciton transitions are allowed.

9.8 REFERENCES

1. M. Born and E. Wolf, *Principles of Optics,* 6th ed., Pergamon Press, Oxford, 1980.
2. E. A. Wood, *Crystals and Light: An Introduction to Optical Crystallography,* Van Nostrand, Princeton, 1964.
3. J. F. Nye, *Physical Properties of Crystals,* Oxford University Press, Oxford, 1985.
4. J. N. Hodgson, *Optical Absorption and Dispersion in Solids,* Chapman & Hall, London, 1970.
5. F. Wooten, *Optical Properties of Solids,* North-Holland, Amsterdam, 1972.
6. F. Abeles (ed.), *Optical Properties of Solids,* North-Holland, Amsterdam, 1972.
7. M. Balkanski, (ed.), *Optical Properties of Solids,* North-Holland, Amsterdam, 1972.
8. G. R. Fowles, *Introduction to Modern Optics,* rev. 2d ed., Dover, Mineola, 1989.
9. B. O. Seraphin, (ed.), *Optical Properties of Solids: New Developments,* North-Holland, Amsterdam, 1976.
10. B. E. A. Saleh and M. C. Teich, *Fundamentals of Photonics,* Wiley, New York, 1991.
11. E. Yariv and P. Yey, *Optical Waves in Crystals,* Wiley, New York, 1983.
12. P. Yey, *Optical Waves in Layered Media,* Wiley, New York, 1988.
13. R. Loudon, *Adv. Phys.* **13:**423 (1964).

P · A · R · T · 4

OPTICAL SOURCES

CHAPTER 10
ARTIFICIAL SOURCES

Anthony LaRocca
ERIM
Ann Arbor, Michigan

10.1 GLOSSARY

k	factor
R	radius of the interior
r	radius of the aperture
S	interior surface area
s	aperture area
ϵ	emissivity
ϵ_o	uncorrected emissivity

10.2 INTRODUCTION

This section deals with artificial sources of radiation as subdivided into two classes: laboratory and field sources. Much of the information on commercial sources is taken from a chapter previously written by the author.[1] Where it was feasible, similar information here has been updated. When vendors failed to comply to requests for information, the older data were retained to maintain completeness, but the reader should be aware that some sources cited here may no longer exist, or perhaps may not exist in the specification presented. Normally, laboratory sources are used in some standard capacity and field sources are used as targets. Both varieties appear to be limitless. Only laboratory sources are covered here.

The sources in this chapter were chosen arbitrarily, often depending upon manufacturer response to requests for information. The purpose of this chapter is to consolidate much of this information to assist the optical-systems designer in making reasonable choices. To attempt to include the hundreds of types of lasers, however, and the thousands of varieties, would be useless for several reasons, but particularly because they change often. A fairly comprehensive source of information on lasers can be found in the recently published CRC *Handbook of Laser Science and Technology,* Supplement I: Lasers, published by the CRC Press, Inc., 2000 Corporate Blvd., N.W., Boca Raton, FL 33431. According to the advertising, "This book updates Volumes I and II of the *Handbook of Laser Science and*

Technology by presenting the latest developments in laser sources. Together, these volumes constitute the most comprehensive and up-to-date listing available of lasers and laser transitions . . . ."

Regarding the selection of a source, Worthing[2] suggests that one ask the following questions:

1. Does it supply energy at such a rate or in such an amount as to make measurements possible?
2. Does it yield an irradiation that is generally constant or that may be varied with time as desired?
3. Is it reproducible?
4. Does it yield irradiations of the desired magnitudes over areas of the desired extent?
5. Has it the desired spectral distribution?
6. Has it the necessary operating life?
7. Has it sufficient ruggedness for the proposed problem?
8. Is it sufficiently easy to obtain and replace, or is its purchase price or its construction cost reasonable?

10.3 LABORATORY SOURCES

Standard Sources

Blackbody Cavity Theory. Radiation levels can be standardized by the use of a source that will emit a quantity of radiation that is both reproducible and predictable. Cavity configurations can be produced to yield radiation theoretically sufficiently close to Planckian that it is necessary only to determine what the imprecision is. Several theories have been expounded over the years to calculate the quality of a blackbody simulator.* Two of the older, most straightforward, and demonstrable theories are those of Gouffé and DeVos.

The Method of Gouffé.[3] For the total emissivity of the cavity forming a blackbody (disregarding temperature variations) Gouffé gives:

$$\epsilon_0 = \epsilon_0'(1 + k) \tag{1}$$

where

$$\epsilon_0' = \frac{\epsilon}{\epsilon\left(1 - \dfrac{s}{S}\right) + \dfrac{s}{S}} \tag{2}$$

and $k = (1 - \epsilon)[(s/S) - (s/S_0)]$, and is always nearly zero—it can be either positive or negative.

* Generically used to describe those sources designed to produce radiation that is nearly Planckian.

ϵ = emissivity of materials forming the blackbody surface

s = area of aperture

S = area of interior surface

S_0 = surface of a sphere of the same depth as the cavity in the direction normal to the aperture

Figure 1 is a graph for determining the emissivities of cavities with simple geometric shapes. In the lower section, the value of the ratio s/S is given as a function of the ratio $1/r$. (Note the scale change at the value for $1/r = 5$.) The values of ϵ_0' is found by reading up from this value of the intrinsic emissivity of the cavity material. The emissivity of the cavity is found by multiplying ϵ_0' by the factor $(1 + k)$.

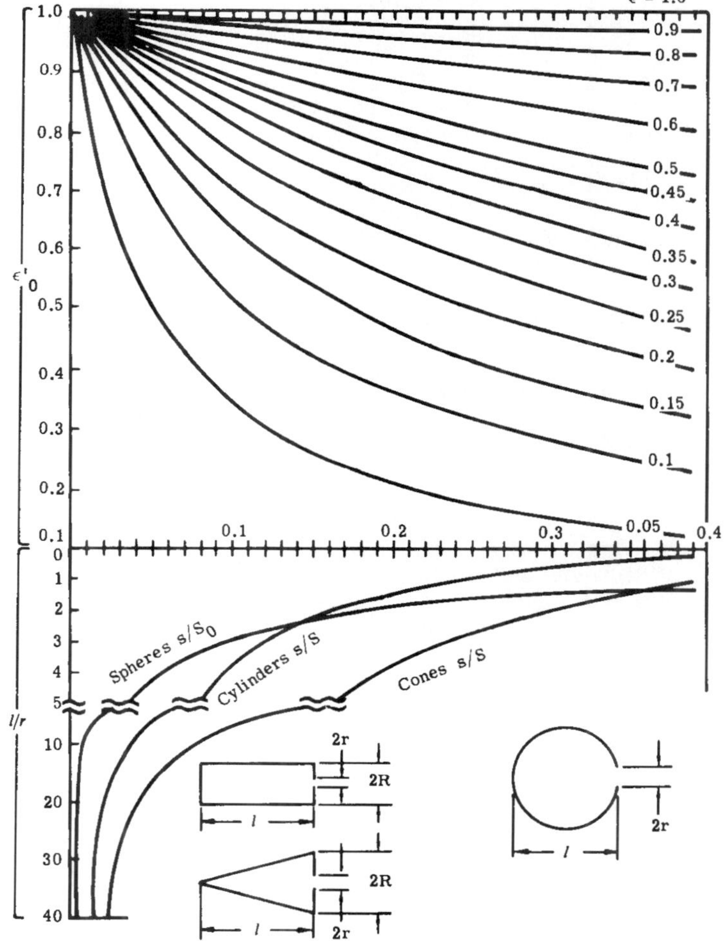

FIGURE 1 Emissivities of conical, spherical, and cylindrical cavities.

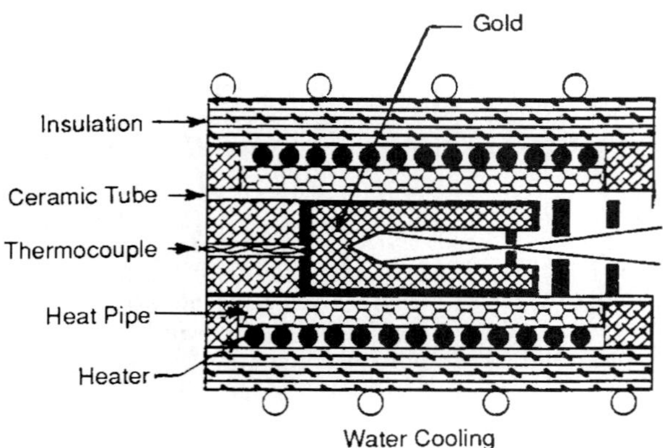

FIGURE 2a Cross section of heat-pipe blackbody furnace.

When the aperture diameter is smaller than the interior diameter of the cylindrical cavity, or the base diameter of a conical cavity, it is necessary to multiply the value of s/S determined from the graph by $(r/R)^2$, which is the ratio of the squares of the aperture and cavity radii (Fig. 1).

It is important to be aware of the effect of temperature gradients in a cavity. This factor is the most important in determining the quality of a blackbody, since it is not very difficult to achieve emissivities as near to unity as desired.

Manufacturers of blackbody simulators strive to achieve uniform heating of the cavity because it is only under this condition that the radiation is Planckian. The ultimate determination of a radiator that is to be used as the standard is the quality of the radiation that it emits.

A recent investigation on comparison of IR radiators is presented by Leupin et al.[4] There has been, incidentally, a division historically between the standards of photometry and those used to establish thermal radiation and the thermodynamic temperature scale. Thus, in photometry the standard has changed from the use of candles, the Carcel lamp, the Harcourt pentane lamp, and the Hefner lamp[5] to more modern radiators.

Baseline Standard of Radiation. Although there is no internationally accepted standard of radiation, the National Institute of Standards and Technology (NIST) uses as its substitute standard the goldpoint blackbody (see Fig. 2),[6] which fixes one point on the international temperature scale, now reported to be 1337.33 + 0.34 K. Starting from this

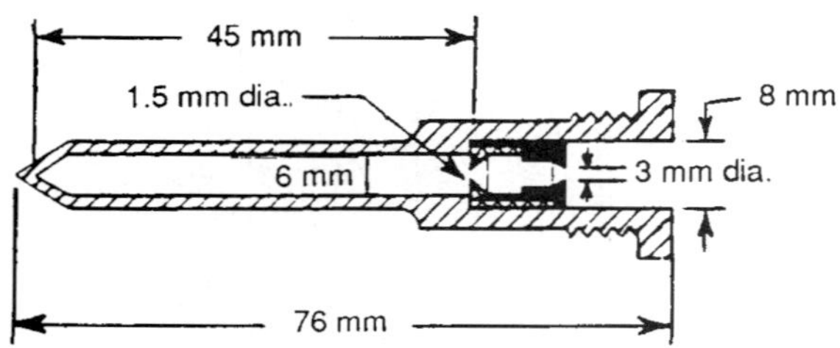

FIGURE 2b Blackbody inner cavity dimensions.

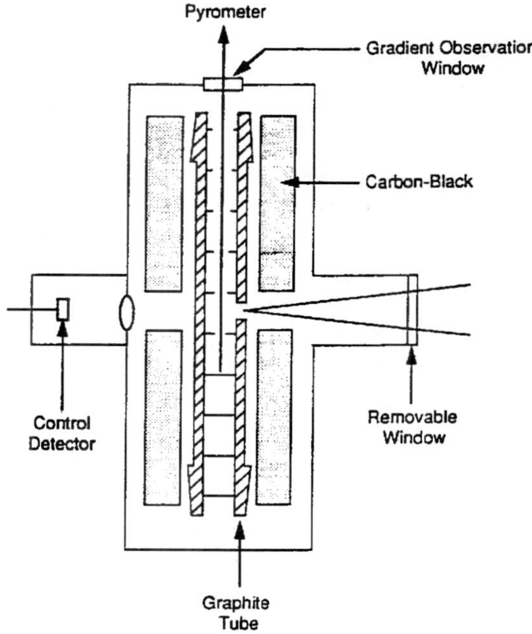

FIGURE 3*a* Variable temperature blackbody schematic.

point, NIST is able to transfer fixed radiation values to working standards of radiation through an accurately constructed variable-temperature radiator as shown in Fig. 3.[7]

The goldpoint blackbody is shown mainly for information. It is quite feasible to build a replica of the variable-temperature radiator, especially in the laboratory equipped to do fundamental radiation measurements.

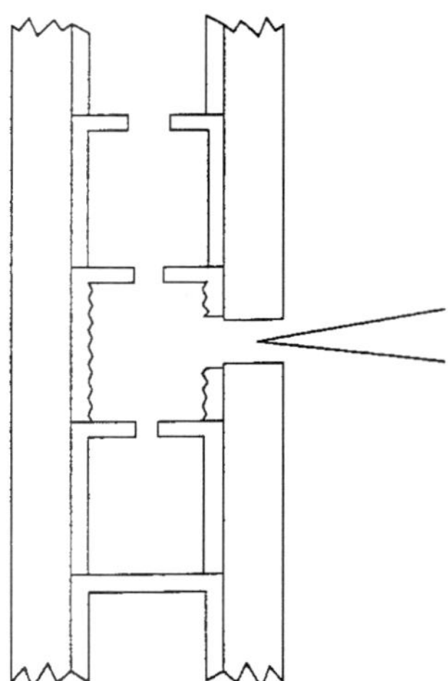

FIGURE 3*b* Central section of variable temperature blackbody.

Working Standards of Radiation. For the calibration of instruments in the ordinary laboratory, the user is likely to use a source which is traceable to NIST, and generally supplied by NIST or one of the recognized vendors of calibrated sources, mainly in the form of a heated filament, a gaseous arc enclosed in an envelope of glass or quartz (or fused silica), or in glass with a quartz or sapphire window.

Any source whose radiation deviates from that described by Planck's law is nonblackbody. Even the sources previously described are not strictly blackbodies, but can come as close as the user desires within the constraints of bulk and price. Any other source has an emissivity less than unity and can, and usually does, have a highly variable spectral emissivity. The lamps used by NIST, for example (see the following), fit into this category, but they differ in one large respect. They are transfer standards which have been carefully determined to emit specified radiation within certain specific spectral regions.

The following discussion of these types of sources is reproduced (in some cases with slight modifications), with permission, from the NIST Special Publication 250.[8] For specific details of calibration, and for the exact source designations, the user should contact NIST at:

U.S. Department of Commerce
National Institute of Standards and Technology
Office of Physical Measurement Services
Rm. B362, Physics Bldg.
Gaithersburg, MD 20899
Photometric Standards

1. Sources/Lamps

Luminous Intenity Standard (100-W Frosted Tungsten Lamp, 90 candelas)
Luminous Intensity Standard (100-W Frosted Tungsten Lamp, color temp., 2700K)
Luminous Intensity Standard (100-W Frosted Tungsten Lamp, color temp., 2856K)
Luminous Intensity Standard (500-W Frosted Tungsten Lamp, 700 candelas)
Luminous Intensity Standard (1000-W Frosted Tungsten Lamp, 1400 candelas)
Luminous Intensity Standard (1000-W Frosted Tungsten Lamp, color temp., 2856K)
Luminous Flux Standard (25-W Vacuum Lamp about 270 lumens)
Luminous Flux Standard (60-W Gas-filled Lamp about 870 lumens)
Luminous Flux Standard (100-W Gas-Filled Lamp about 1600 lumens)
Luminous Flux Standard (200-W Gas-Filled Lamp about 3300 lumens)
Luminous Flux Standard (500-W Gas-Filled Lamp about 10,000 lumens)
Luminous Flux Standard (Miniature Lamps 7 sizes 6 to 400 lumens)
Airway Beacon Lamps for Color Temperature (500-W, 1 point in range, 2000–3000K)

2. General Information

Calibration services provide access to the photometric scales realized and maintained at NIST. Lamp standards of luminous intensity, luminous flux, and color temperature, as described next, are calibrated on a routine basis.

a. Luminous Intensity Standards

Luminous intensity standard lamps supplied by NIST [100-W (90-140 cd), 500-W (approximately 700 cd), and 1000-W (approximately 1400 cd) tungsten filament lamps with C-13B filaments in inside-frosted bulbs and having medium bipost bases] are calibrated at either a set current or a specified color temperature in the range 2700 to 3000 K. Approximate 3-sigma uncertainties are 1 percent relative to the SI unit of luminous intensity and 0.8 percent relative to NIST standards.

b. Luminous Flux Standards
Vacuum tungsten lamps of 25 W and 60-, 100-, 200-, and 500-W gas-filled tungsten lamps that are submitted by customers are calibrated. Lamps must be base-up burning and rated at 120 V. Approximate 3-sigma uncertainties are 1.4 percent relative to SI units and 1.2 relative to NIST standards. Luminous flux standards for miniature lamps producing 6 to 400 lm are calibrated with uncertainties of about 2 percent.

c. Airway Beacon Lamps
Color temperature standard lamps supplied by NIST (airway beacon 500-W meduim bipost lamps) are calibrated for color temperature in the range 2000 to 3000 K with 3-sigma uncertainties ranging from 10 to 15 degrees.

IR Radiometric Standards

1. Sources/Lamps

Spectral Radiance Ribbon Filament Lamps (225 to 2400 nm)
Spectral Radiance Ribbon Filament Lamps (225 to 800 nm)
Spectral Radiance Ribbon Filament Lamps (650 to 2400 nm)
Spectral Irradiance Quartz-Halogen Lamps (250 to 1600 nm)
Spectral Irradiance Quartz-Halogen Lamps (250 to 2400 nm)
Spectral Irradiance Deuterium Lamps (200 to 350 nm)

2. General Information

a. Spectral Radiance Ribbon Filament Lamps
These spectral radiance standards are supplied by NIST. Tungsten, ribbon filament lamps (30A/T24/13) are provided as lamp standards of spectral radiance. The lamps are calibrated at 34 wavelengths from 225 to 2400 nm, with a target area 0.6 mm wide by 0.8 mm high. Radiance temperature ranges from 2650 K at 225 nm, and 2475 K at 650 nm to 1610 K at 2400 nm, with corresponding uncertainties of 2, 0.6, and 0.4 percent. For spectral radiance lamps, errors are stated as the quadrature sum of individual uncertainties at the three standard deviation level.

Figure 4 summarizes the measurement uncertainty for NIST spectral radiance calibrations.

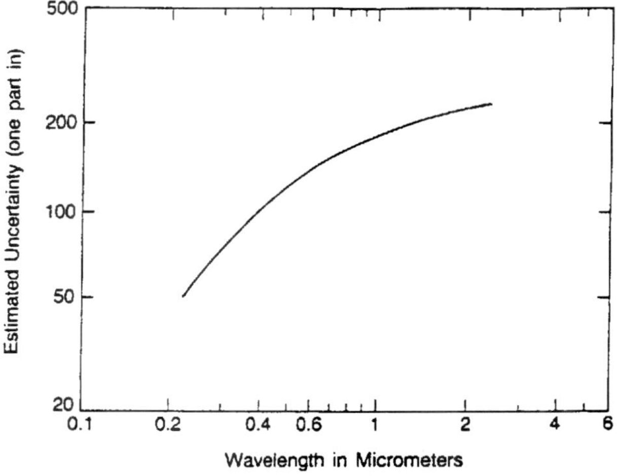

FIGURE 4 Uncertainties for NIST spectral radiance calibrations.

b. Spectral Irradiance Lamps

These spectral irradiance standards are supplied by NIST. Lamp standards of spectral irradiance are provided in two forms. Tungsten filament, 1000-watt quartz halogen type FEL lamps are calibrated at 31 wavelengths in the range 250 to 2400 nm. At the working distance of 50 cm, the lamps produce 0.2 $W/cm^2/cm$ at 250 nm, 220 $W/cm^2/cm$ at 900 nm, 115 $W/cm^2/cm$ at 1600 nm, and 40 $W/cm^2/cm$ at 2400 nm, with corresponding uncertainties of 2.2, 1.3, 1.9, and 6.5 percent. For spectral irradiance lamps, errors are stated as the quadrature sum of individual uncertainties at the three standard deviation level. Deuterium lamp standards of spectral irradiance are also provided and are calibrated at 16 wavelengths from 200 to 350 nm. At the working distance of 50 cm, the spectral irradiance produced by the lamp ranges from about 0.5 $W/cm^2/cm$ at 200 nm and 0.3 $W/cm^2/cm$ at 250 nm to 0.07 $W/cm^2/cm$ at 350 nm. The deuterium lamps are intended primarily for the spectral region 200 to 250 nm. The approximate uncertainty relative to SI units is 7.5 percent at 200 nm and 5 percent at 250 nm. The approximate uncertainty in relative spectral distribution is 3 percent. It is strongly recommended that the deuterium standards be compared to an FEL tungsten standard over the range 250 to 300 nm each time the deuterium lamp is lighted to take advantage of the accuracy of the relative spectral distribution.

Figure 5 summarizes the measurement uncertainty for NIST spectral irradiance calibrations of type FEL lamps.

Radiometric Sources in the Far Ultraviolet

1. Sources

Spectral Irradiance Standard, Argon Mini-Arc (140 to 330 nm)

Spectral Radiance Standard, Argon Mini-Arc (115 to 330 nm)

Spectral Irradiance Standard, Deuterium Arc Lamp (165 to 200 nm)

2. General Information

a. Source Calibrations in the Ultraviolet

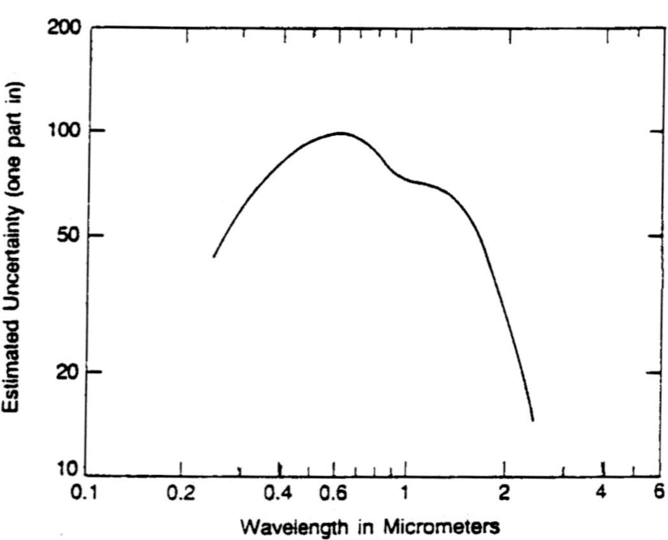

FIGURE 5 Uncertainties for NIST spectral irradiance calibrations of type FEL lamps.

NIST maintains a collection of secondary sources such as argon maxi-arcs, argon mini-arcs, and deuterium arc lamps in the near and vacuum ultraviolet radiometric standards program to provide calibrations for user-supplied sources. The calibrations of these sources are traceable to a hydrogen arc whose radiance is calculable and which NIST maintains as a primary standard. The collection also includes tungsten strip lamps and tungsten halogen lamps whose calibrations are based on a blackbody rather than a hydrogen arc. Customer-supplied sources are calibrated in both radiance and irradiance by comparing them with NIST secondary standards.

Argon arcs are used to calibrate other sources in the wavelength range 115 to 330 nm for radiance and 140 to 330 nm for irradiance. The lower wavelength limit is determined in radiance by the cutoff of the magnesium fluoride windows used in the arcs, and in irradiance by the decrease in signal produced by the addition of a diffuser. Deuterium arc lamps are used in the range 165 to 200 nm, with the low wavelength cutoff due to the onset of blended molecular lines.

The high wavelength limit is the starting point of the range for the Radiometric Standards group. The tungsten lamps are used at 250 nm and above, since their signals are too weak at shorter wavelengths. It should be noted that the wavelength range of the NIST arcs partially overlap the range of tungsten lamps, thus providing an independent check on calibrations.

An argon mini-arc lamp supplied by the customer is calibrated for spectral irradiance at 10-nm intervals in the wavelength region 140 to 300 nm. Absolute values are obtained by comparison of the radiative output with laboratory standards of both spectral irradiance and spectral radiance. The spectral irradiance measurement is made at a distance of 50 cm from the field stop. Uncertainties are estimated to be less than ±10 percent in the wavelength region 140 to 200 nm and within ±5 percent in the wavelength region 200 to 330 nm. A measurement of the spectral transmission of the lamp window is included in order that the calibration be independent of possible window deterioration or damage. The uncertainties are taken to be two standard deviations.

The spectral radiance of argon mini-arc radiation sources is determined to within an uncertainty of less than 7 percent over the wavelength range 140 to 330 nm and 20 percent over the wavelength range 115 to 140 nm. The calibrated area of the 4-mm diameter radiation source is the central 0.3-mm diameter region. Typical values of the spectral radiance are: at 250 nm, $L(\lambda) = 30\ \text{mW/cm}^2\text{/nm/sr}$; and at 150 nm, $L(\lambda) = 3\ \text{mW/cm}^2\text{/nm/sr}$. The transmission of the demountable lamp window and that of an additional MgF_2 window are determined individually so that the user may check periodically for possible long-term variations.

The deuterium arc lamp is calibrated at 10 wavelengths from 165 to 200 nm, at a distance of 50 cm, at a spectral irradiance of about 0.5 $\text{W/cm}^2\text{/cm}$ at 165 nm, 0.5 $\text{W/cm}^2\text{/cm}$ at 170 nm, and 0.5 $\text{W/cm}^2\text{/cm}$ at 200 nm. The approximate uncertainty relative to SI units is estimated to be less than 10 percent. The lamp is normally supplied by NIST and requires 300 mA at about 100 V.

10.4 COMMERCIAL SOURCES

Blackbody Simulators

Virtually any cavity can be used to produce radiation of high quality, but practicality limits the shapes to a few. The most popular shapes are cones and cylinders, the former being more popular. Spheres, combinations of shapes, and even flat-plate radiators are used occasionally. Blackbodies can be bought rather inexpensively, but there is a fairly direct correlation between cost and quality (i.e., the higher the cost the better the quality).

Few manufacturers specialize in blackbody construction. Some, whose products are

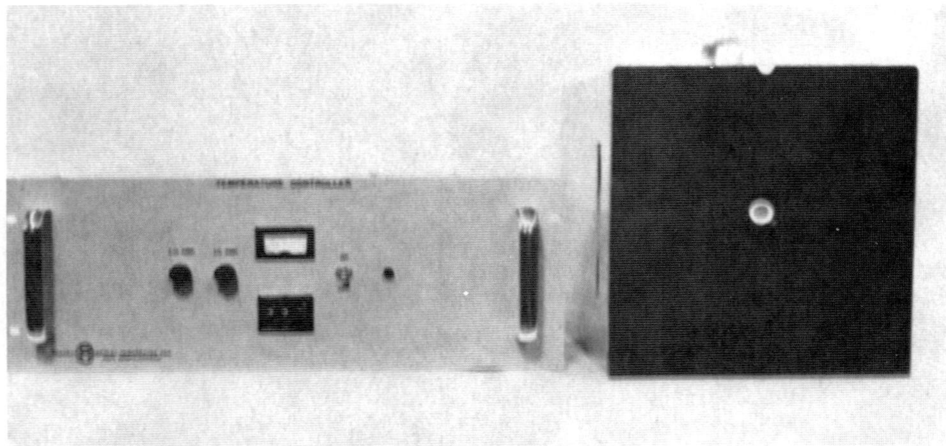

FIGURE 6 EOI blackbody.

specifically described here, have been specializing in blackbody construction for many years. Other companies of this description may be found, for example, in the latest Lasers and Optronics Buying Guide[9] or the latest Photonics Directory of Optical Industries.[10] These references are the latest as of the writing of this work. It is expected that they will continue in succeeding years.

A large selection of standard (or blackbody) radiators is offered by Electro-Optical Industries, Inc. (EOI), Santa Barbara, California.* Most blackbodies can be characterized as one of the following: primary, secondary, or working standard. The output of the primary must, of course, be checked with those standards retained at NIST. Figure 6 pictures an EOI blackbody and its controller. Figure 7 pictures a similar blackbody from Mikron, Inc. and its controller. All of the companies sell separate apertures (some of which are water cooled) for controlling the radiation output of the radiators. Another piece of auxiliary equipment which can be purchased is a multispeed chopper. It is impossible to cite all of the companies that sell these kinds of sources; therefore, the reader is referred to one of the Buyers' Guides already referenced for a relatively complete list. It is prudent to shop around for the source that suits one's own purpose.

Figure 8 demonstrates a less-conventional working standard manufactured by EOL. Its grooves-and-honeycomb structure is designed to improve the absorptance of such a large and open structure. A coating with a good absorbing paint increases its absorptance further.

Incandescent Nongaseous Sources (Exclusive of High-Temperature Blackbodies)

Nernst Glower. The Nernst glower is usually constructed in the form of a cylindrical rod or tube from refractory materials (usually zirconia, yttria, beria, and thoria) in various sizes. Platinum leads at the ends of the tube conduct power to the glower from the source.

* Many of the sources in the text are portrayed using certain specific company products, only for the sake of demonstration. This does not necessarily imply an endorsement of these products by the author. The reader is encouraged in all cases to consult the *Photonics Directory of Optical Industries*[10] or a similar directory for competitive products.

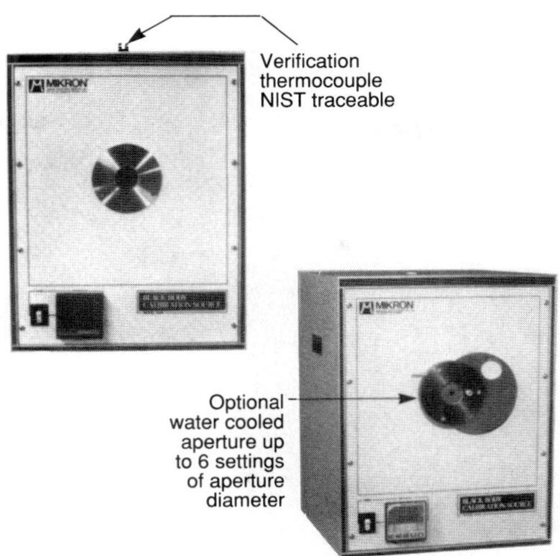

FIGURE 7 Mikron blackbody.

Since the resistivity of the material at room temperature is quite high, the working voltage is insufficient to get the glower started. Once started, its negative temperature coefficient-of-resistance tends to increase current, which would cause its destruction, so that a ballast is required in the circuit. Starting is effected by applying external heat, either with a flame or an adjacent electrically heated wire, until the glower begins to radiate.

Data from a typical glower are as follows:

1. Power requirements: 117 V, 50 to 60 A, 200 W
2. Color temperature range: 1500 to 1950 K
3. Dimension: 0.05 in. diameter by 0.3 in.

The spectral characteristics of a Nernst glower in terms of the ratio of its output to that of a 900°C blackbody are shown in Fig. 9.

The life of the Nernst glower diminishes as the operating temperature is increased. Beyond a certain point, depending on the particular glower, no great advantage is gained by increasing the current through the element. The glower is fragile, with low tensile strength, but can be maintained intact with rigid support. The life of the glower depends on the operating temperature, care in handling, and the like. Lifetimes of 200 to 1000 hours are claimed by various manufacturers.

Since the Nernst glower is made in the form of a long thin cylinder, it is particularly useful for illuminating spectrometer slits. Its useful spectral range is from the visible region to almost 30 μm, although its usefulness compared with other sources diminishes beyond about 15 μm. As a rough estimate, the radiance of a glower is nearly that of a graybody at the operating temperature with an emissivity in excess of 75 percent, especially below about 15 μm. The relatively low cost of the glower makes it a desirable source of moderate radiant power for optical uses in the laboratory. The makers of spectroscopic equipment constitute the usual source of supply of glowers (or of information about suppliers).

Globar. The globar is a rod of bonded silicon carbide usually capped with metallic caps which serve as electrodes for the conduction of current through the globar from the power source. The passage of current causes the globar to heat, yielding radiation at a temperature above 1000°C. A flow of water through the housing that contains the rod is

FIGURE 8 EOI model 1965. This model is 12 in. in diameter and 9 in. deep. The base is an array of intersecting conical cavities. The walls are hex-honeycomb and the temperature range is 175 to 340 K.

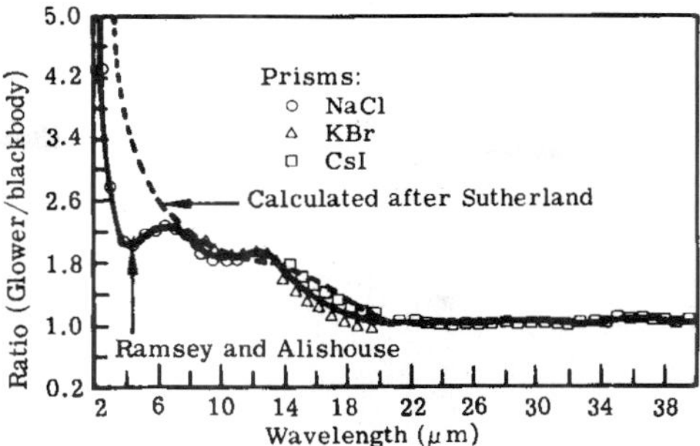

FIGURE 9 The ratio of a Nernst glower to a 900°C blackbody versus wavelength.

needed to cool the electrodes (usually silver). This complexity makes the globar less convenient to use than the Nernst glower and necessarily more expensive. This source can be obtained already mounted, from a number of manufacturers of spectroscopic equipment. Feedback in the controlled power source makes it possible to obtain high radiation output.

Ramsey and Alishouse[11] provide information on a particular sample globar as follows:

1. Power consumption: 200 W, 6 A
2. Color temperature: 1470 K

They also provide the spectral characteristics of the globar in terms of the ratio of its output to that of a 900°C blackbody. This ratio is plotted as a function of wavelength in Fig. 10. Figure 11[12] is a representation of the spectral emissivity of a globar as a function of wavelength. The emissivity values are only representative and can be expected to change considerably with use.

Gas Mantle. The Welsbach mantle is typified by the kind found in high-intensity gasoline lamps used where electricity is not available. The mantle is composed of thorium oxide with some additive to increase its efficiency in the visible region. Its near-infrared emissivity is quite small, except for regions exemplified by gaseous emission, but increases considerably beyond 10 micrometers.

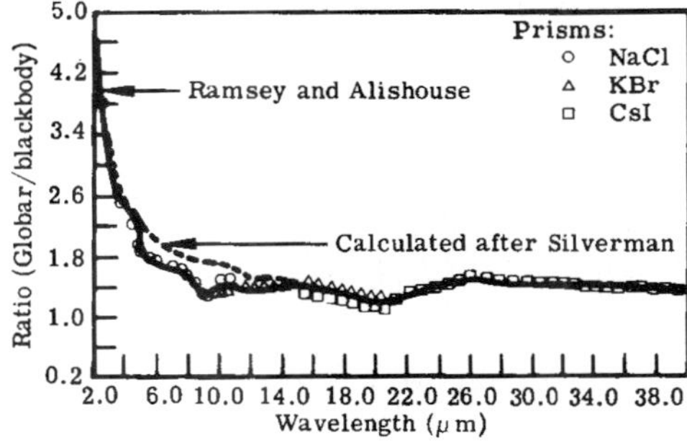

FIGURE 10 The ratio of a globar to a 900°C blackbody versus wavelength.

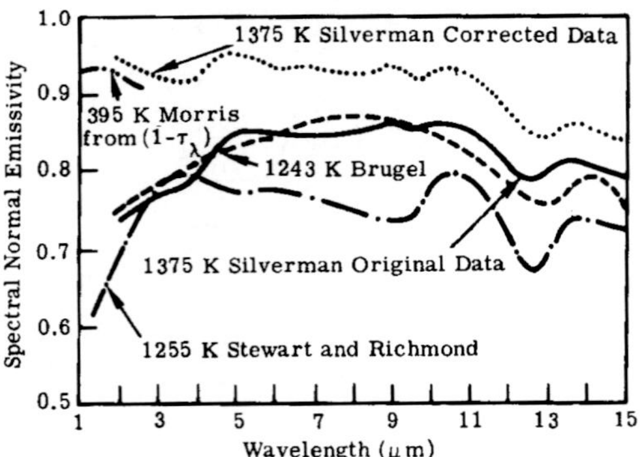

FIGURE 11 The spectral emissivity of a globar.

Ramsey and Alishouse[11] provide information on a propane-heated sample from an experiment in which a comparison of several sources is made:

1. Color temperature: 1670 K
2. Dimensions: 25.4 by 38.1 mm

The spectral characteristics of the mantle in terms of the ratio of its output to that of a 900°C blackbody are shown in Fig. 12.

Pfund modified the gas mantle so that it became more a laboratory experimental source than an ordinary radiator. By playing a gas flame on an electrically heated mantle, he was able to increase its radiation over that from the gas mantle itself.[13] Figure 13 shows a comparison of the gas mantle and the electrically heated gas mantle, with a Nernst glower. Strong[14] points out that playing a flame against the mantle at an angle produces an elongated area of intense radiation useful for illuminating the slits of a spectrometer.

Comparison of Nernst Glower, Globar, and Gas Mantle. Figure 14 compares these three types of sources, omitting a consideration of differences in the instrumentation used in making measurements of the radiation from the sources.

Availability, convenience, and cost usually influence a choice of sources. At the very long wavelength regions in the infrared, the gas mantle and the globar have a slight edge

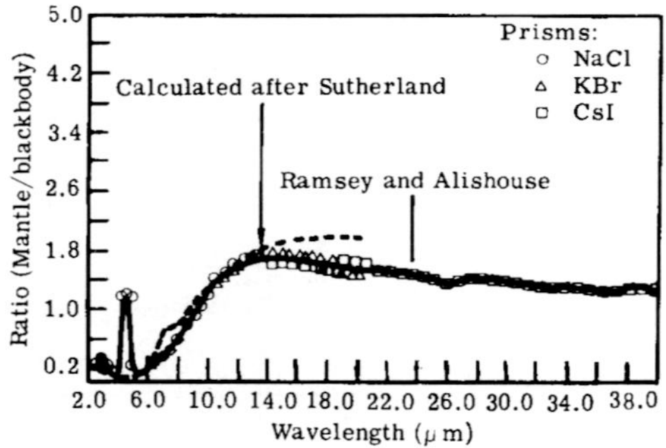

FIGURE 12 The ratio of the gas mantle to a 900°C blackbody versus wavelength.

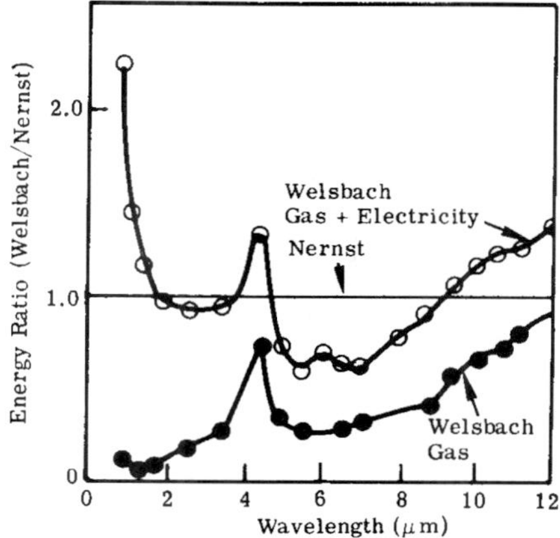

FIGURE 13 Emission relative to that of a Nernst glower (2240 K) of the gas-heated mantle (lower curve) and that of the mantle heated by gas plus electricity (upper curve).

over the Nernst glower because the Nernst glower (a convenient, small, and inexpensive source) does not have the power of the gas mantle and globar.

Tungsten-Filament Lamps. A comprehensive discussion of tungsten-filament lamps is given by Carlson and Clark.[15] Figures 15 to 17 show the configurations of lamp housings and filaments. The types and variations of lamps are too numerous to be meaningfully included in this chapter. The reader is referred to one of the Buyer's Guides for a comprehensive delineation of manufacturers from whom unlimited literature can be obtained.

Tungsten lamps have been designed for a variety of applications; few lamps are directed toward scientific research, but some bear directly or indirectly on scientific pursuits insofar as they can provide steady sources of numerous types of radiation. One set of sources cited here, particularly for what the manufacturer calls their scientific usefulness, is described in "Lamps for Scientific Purposes."[16] Their filament structures are similar to those already described, but their designs reduce extraneous radiation and ensure the quality and

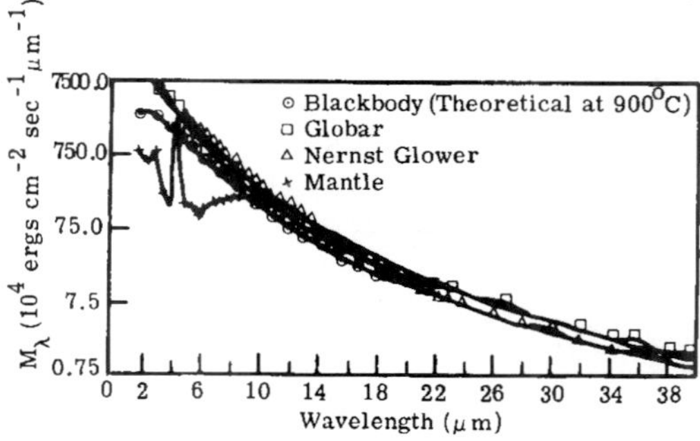

FIGURE 14 The spectral radiant emittances of a globar, Nernst glower, 900°C blackbody, and gas mantle versus wavelength.

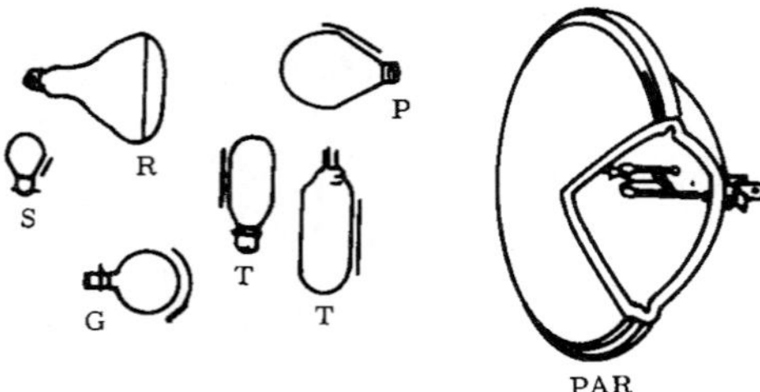

FIGURE 15 Bulk shapes most frequently used for lamps in optical devices. Letter designations are for particular shapes.

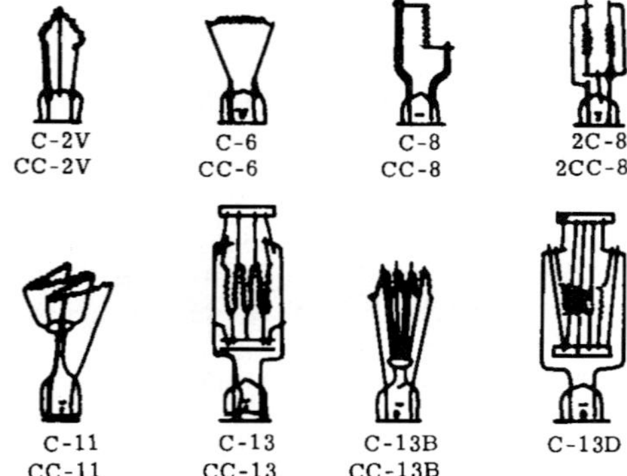

FIGURE 16 Most commonly used filament forms. Letters designate the type of filament.

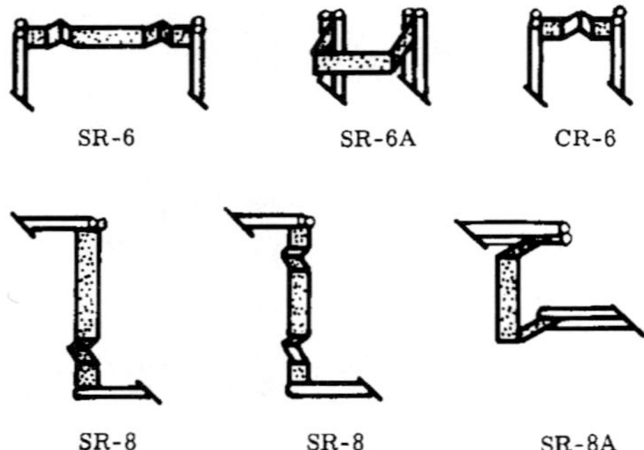

FIGURE 17 Ribbon-type tungsten filaments. Type designations are by number.

stability of the desired radiation. The lamps can be obtained with a certification of their calibration values.

The physical descriptions of some of these sources are given in Fig. 18. Applications (according to the manufacturer, Osram) are photometry, pyrometry, optical radiometry, sensitometry, spectroscopy, spectrometry, polarimetry, saccharimetry, spectrophotometry, colorimetry, microscopy, microphotography, microprojection, and stroboscopy.

Quartz Envelope Lamps. These are particularly useful as standards because they are longer lasting (due to action of iodine in the quartz-iodine series), can be heated to higher temperatures, are sturdier, and can transmit radiation to longer wavelengths in the infrared than glass-envelope lamps. Studer and Van Beers[17] have shown the spectral deviation to be expected of lamps containing no iodine. The deviation, when known, is readily acceptable in lieu of the degradation in the lamp caused by the absence of iodine. The particular tungsten-quartz-iodine lamps used in accordance with the NIST are described earlier in this chapter. Others can be obtained in a variety of sizes and wattages from General Electric, Sylvania, and a variety of other lamp manufacturers and secondary sources.

Carbon Arc

The carbon arc has been passed down from early lighting applications in three forms: low-intensity arc, flame, and high-intensity arc. The low- and high-intensity arcs are usually operated on direct current; the flame type adapts to either direct or alternating current. In all cases, a ballast must be used. "In the alternating current arc, the combined radiation from the two terminals is less than that from the positive crater of the direct-current arc of the same wattage.[2]

Spatial variation in the amount of light energy across the crater of dc arcs for different currents is shown in Fig. 19.

The carbon arc is a good example of an open arc, widely used because of its very high radiation and color temperatures (from approximately 3800 to 6500 K, or higher). The rate at which the material is consumed and expended during burning (5 to 30 cm per hour) depends on the intensity of the arc. The arc is discharged between two electrodes that are moved to compensate for the rate of consumption of the material. The anode forms a crater of decomposing material which provides a center of very high luminosity. Some electrodes are hollowed out and filled with a softer carbon material which helps keep the arc fixed in the anode and prevents it from wandering on the anode surface.

In some cored electrodes, the center is filled with whatever material is needed to produce desired spectral characteristics in the arc. In such devices, the flame between the electrodes becomes the important center of luminosity, and color temperatures reach values as high as 8000 K.[18] An example of this so-called flaming arc is shown in Fig. 20*a*. Figure 20*b* and *c* shows the low-intensity dc carbon arc and the high-intensity dc carbon arc with rotating positive electrodes. Tables 1 and 2 give characteristics of dc high intensity and flame carbon arcs.

A spectrum of low-intensity arc (Fig. 21) shows the similarity between the radiation from it and a 3800 K blackbody, except for the band structure at 0.25 and 0.39 microns. In Koller[18] an assortment of spectra are given for cored carbons containing different materials. Those for a core of soft carbon and for a polymetallic core are shown in Figs. 22 and 23. Because radiation emitted from the carbon arc is very intense, this arc supplants, for many applications, sources which radiate at lower temperatures. Among the disadvantages in using the carbon arc are its inconvenience relative to the use of other sources (e.g., lamps) and its relative instability. However, Null and Lozier[19] have studied the properties of the low-intensity carbon arc extensively and have found that under the proper operating conditions the carbon arc can be made quite stable; in fact, in their treatise they recommend its use as a standard of radiation at high temperatures.

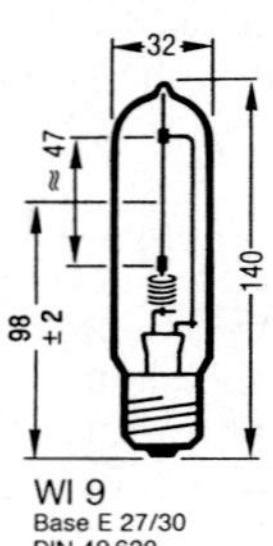

WI 9
Base E 27/30
DIN 49620

Order reference	Upper limits for electric data volt	amp.	Color temperature T_f max. [1]	Luminance T_s max.	Dimensions of luminous area width mm	height mm	Burning position [2]	Base
Lamps for scientific purposes								
WI 9	8.5	6	2856 K	–	0.2	47	s	E 27
WI 14	5	16	–	2400 K	1.6	8	s	E 27
WI 16/G	9	16	–	2600 K	21	1.6	s+h	E 27
WI 17/G	9	16	–	2600 K	1.6	20	s	E 27
WI 40/G	31	6	2856 K	–	18	18	s+h	E 27
WI 41/G	31	6	2856 K	–	18	18	s+h	E 27

Lamps for scientific purposes are gas-filled, incandescent lamps for calibration of luminous intensity, luminous flux, luminance (spectral radiant temperature), color temperature (luminance temperature and spectral radiance distribution). A test certificate can be issued for these types of lamps.

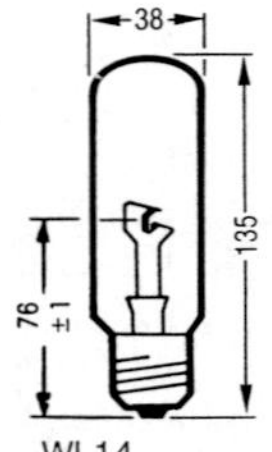

WI 14
Base E 27/30
DIN 49620

Also for other types of lamp with sufficiently constant electric and photometric data, a test certificate can be issued. To order a test certificate, the order reference of the lamp, the type of measurement, and the desired burning position have to be given. Example: Lamp 41/G, measurement of the electric data and the luminous intensity for T_f = 2856 K (light-type A), burning position vertical, base up.

Variables for which test certificates can be issued are shown in the following table by +. The sign (+) indicates that certificates can be issued for variables although the lamps were not designed for such measurements.

Type of lamp	Light intensity	Luminous flux	Luminance	Color temperature	Spectral radiance distribution [3]
WI 9	+	–	–	+	–
WI 14	(+)	–	+	+	+ 300–800 nm
WI 16/G	(+)	–	+	+	+ 300–800 nm
WI 17/G	–	–	+	(+)	+ 250–800 nm 250–2500 nm
WI 40/G	(+)	+	–	(+)	–
WI 41/G	+	–	–	+	–

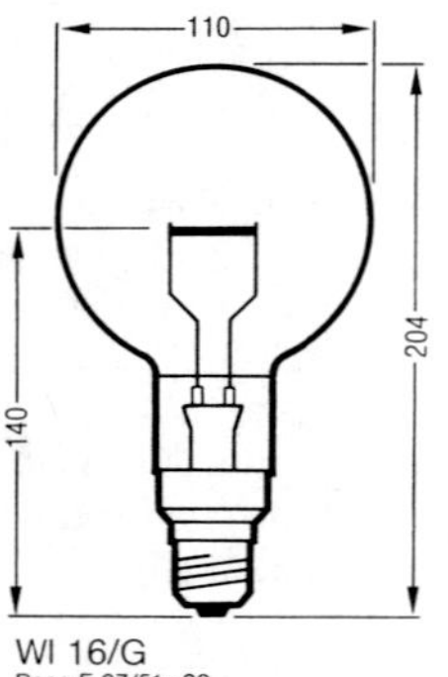

WI 16/G
Base E 27/51 x 39

Description

WI 9:
Lamp with uncoiled straight filament.

WI 14:
Tungsten ribbon lamp with tubular bulb. The portion of the tungsten ribbon to be utilized for measurement is mounted parallel to the lamp axis and positioned approx. 8 mm off-axis in the measuring direction.

[1] The color temperature of 2856 K corresponds with light-type A (DIN 5035).
[2] s = vertical (base down); h = vertical (base up)
[3] Only for additional measurement of luminance or color temperature.

FIGURE 18*a* Lamps for scientific purposes.

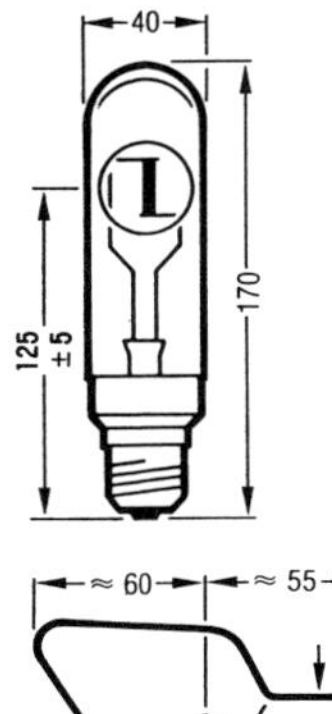

WI 16/G:
Tungsten ribbon lamp with spherical bulb. Horizontal tungsten ribbon with a small notch to indicate the measuring point. The ribbon is positioned approx. 3 mm off axis.

WI 17/G:
Tungsten ribbon lamp with horn-shaped bulb. The bulb has a tubular extension with a sealed-on quartz glass window (homogenized ultrasil). Vertical tungsten ribbon with a small notch to indicate the measuring point.

WI 40/G:
Standard lamp for total radiation, luminous flux and color temperatures with conic bulb. The bulb shape prevents reflections in the direction of the plane normal of the luminous area, which is formed by the meandrous-shaped filament.

WI 41/G:
Standard lamp for light intensity and color temperature with conic bulb. Differs from the WI 40/G lamp by a black, opaque coating which covers one side of the bulb. A window is left open in the coating opposite the filament, through which over an angle of approx. ±3° a constant light intensity is emitted. The black coating prevents stray light being reflected in the measuring direction.

≈ 60 ≈ 55 35

WI 17/G
Base E 27/51 x 39

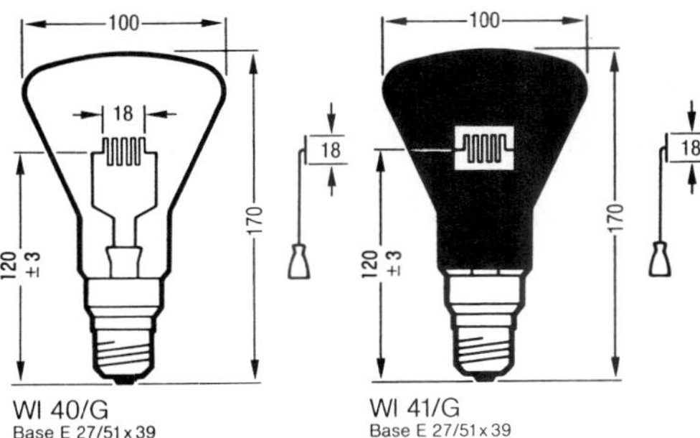

WI 40/G
Base E 27/51 x 39

WI 41/G
Base E 27/51 x 39

FIGURE 18*b* Lamps for scientific purposes.

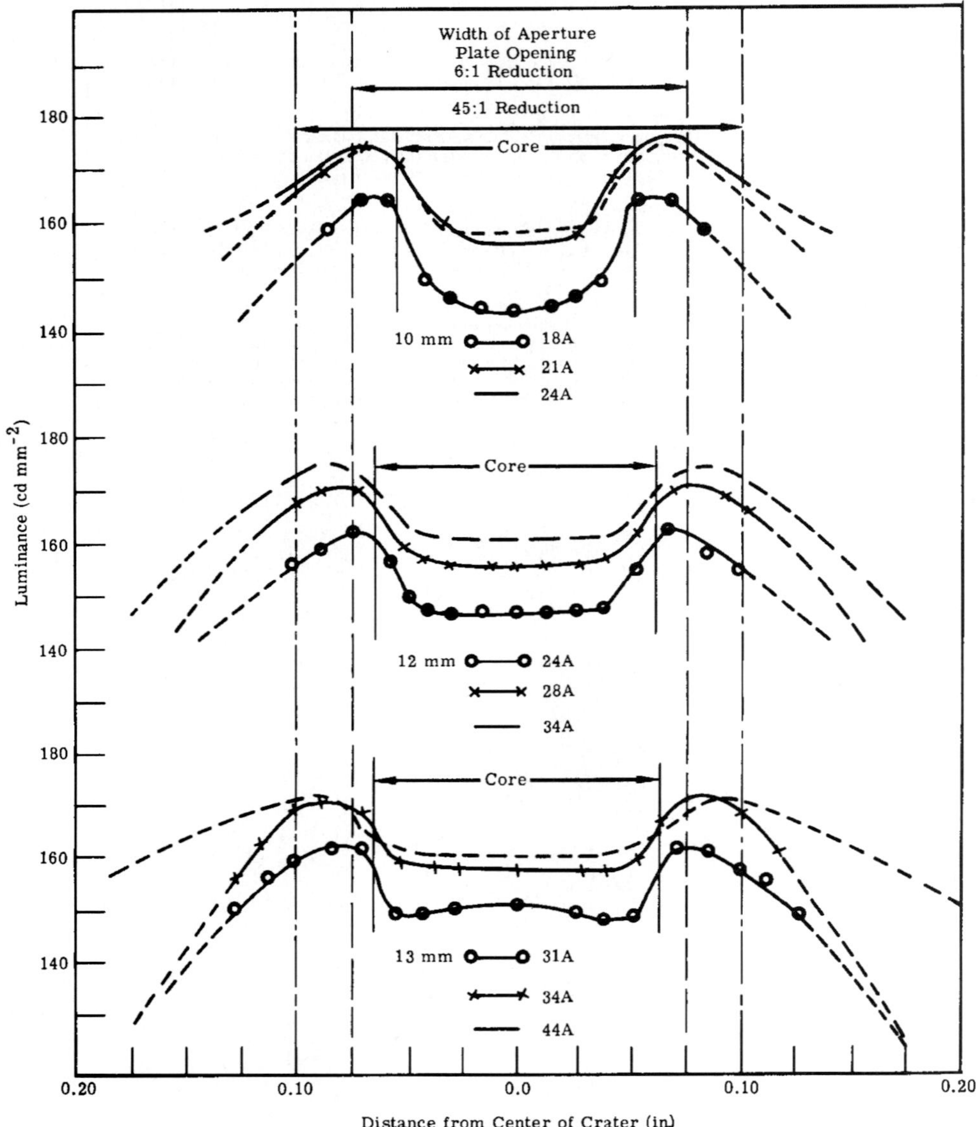

FIGURE 19 Variations in brightness across the craters of 10-, 12-, and 13-mm positive carbons of dc plain arcs operated at different currents in the regions of recommended operation.

(*a*) Flame type.

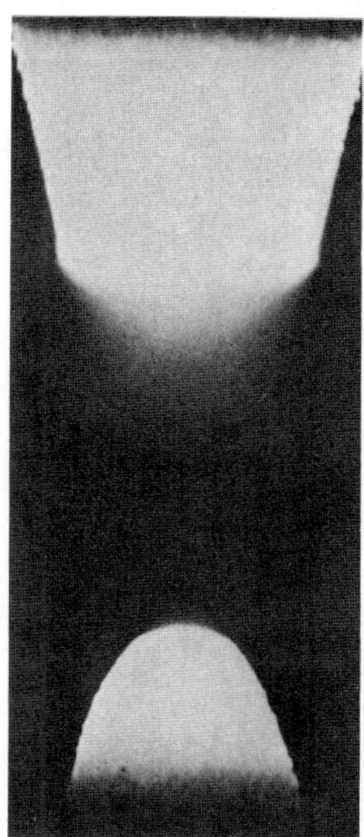

(*b*) Low-Intensity dc arc.

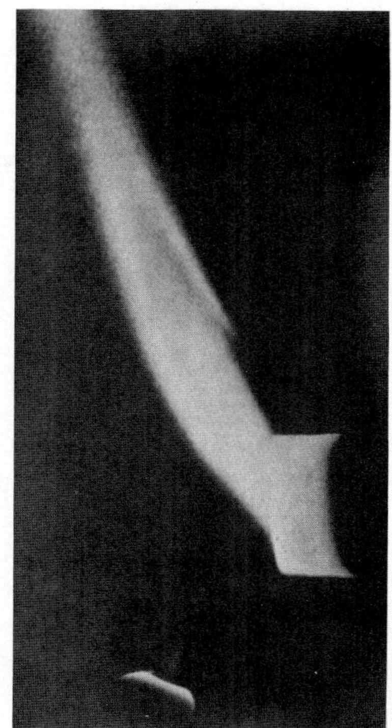

(*c*) High-intensity dc arc with rotating positive carbon.

FIGURE 20 Various types of carbon arc.

TABLE 1 dc Carbon Arcs

	Low Intensity	*Nonrotating High Intensity*		*Rotating High Intensity*						
	Application Number[a]									
	1	2	3	4	5	6	7	8	9	10
Type of carbon	Microscope	Projector	Projector	Projector	Projector	Projector	Projector	Searchlight	Studio	
Positive carbon:										
Diameter (mm)	5	7	8	10	11	13.6	13.6	16	16	16
Length (in.)	8	12-14	12-14	20	20	22	22	22	22	22-30
Negative carbon:										
Diameter	6 mm	6 mm	7 mm	11/32 in.	3/8 in.	0.5 in.	0.5 in.	11 mm	17/32 in.	7/16 in.
Length (in.)	4.5	9	9	9	9	9	9	12	9	12-48
Arc current (A)	5	50	70	105	120	160	180	150	225	400
Arc volts (dc)	59	40	42	59	57	66	74	78	70	80
Arc power (W)	295	2000	2940	6200	6840	10600	13300	11700	15800	32000
Burning rate (in. h^{-1})										
Positive carbon	4.5	11.6	13.6	21.5	16.5	17	21.5	8.9	20.2	55
Negative carbon	2.1	4.3	4.3	2.9	2.4	2.2	2.5	3.9	2.2	3.5
Approximate crater diameter (in.)	0.12	0.23	0.28	0.36	0.39	0.5	0.5	0.55	0.59	0.59
Maximum luminance of crater (cd cm^{-2})	15000	55000	83000	90000	85000	96000	95000	65000	68000	45000
Forward crater candlepower	975	10500	22000	36000	44000	63000	78000	68000	99000	185000
Crater lumens[b]	3100	36800	77000	126000	154000	221000	273000	250000	347000	660000
Total lumens[c]	3100	55000	115000	189000	231000	368000	410000	374000	521000	999000
Total lumens per arc watt	10.4	29.7	39.1	30.5	33.8	34.7	30.8	32	33	30.9
Color temperature (K)[d]	3600	5950	5500-6500	5500-6500	5500-6500	5500-6500	5500-6500	5400	4100	5800-6100

[a]Typical applications: 1, microscope illumination and projection; 2 to 7, motion-picture projection; 8, searchlight projection; 9, motion-picture-set lighting and motion-picture and television background projection.
[b]Includes light radiated in forward hemisphere.
[c]Includes light from crater and arc flame in forward hemisphere.
[d]Crater radiation only.

TABLE 2 Flame-Type Carbon Arcs

	Application Number[a]			
	1	3	3	4
Type of carbon	C	E	Sunshine	Sunshine
Flame materials	Polymetallic	Strontium	Rare earth	Rare earth
Burning position[g]	Vertical	Vertical	Vertical	Vertical
Upper carbon[d]				
Diameter	22 mm	22 mm	22 mm	22 mm
Length (in.)	12	12	12	12
Lower carbon[d]				
Diameter	13 mm	13 mm	13 mm	13 mm
Length (in.)	12	12	12	12
Arc current (A)	60	60	60	80
Arc voltage (ac)[h]	50	50	50	50
Arc power (kW)	3	3	3	4
Candlepower[i]	2100	6300	9100	10000
Lumens	23000	69000	100000	110000
Lumens per arc watt	7.6	23	33.3	27.5
Color temperature (K)			12800[j]	24000[j]
Spectral intensity ($\mu W\ cm^{-2}$) 1 m from arc axis:				
Below 270 nm	540.0	180.0	102	140
270-320 nm	540.0	150.0	186	244
320-400 nm	1800.0	1200.0	2046	2816
400-450 nm	300.0	1100.0	1704	2306
450-700 nm	600.0	4050.0	3210	3520
700-1125 nm	1580.0	2480.0	3032	3500
Above 1125 nm	9480.0	10290.0	9820	11420
Total	14930	19460	20100	24000
Spectral radiation (percent of input power):				
Below 270 nm	1.8	0.6	0.34	0.35
270-320 nm	1.8	0.5	0.62	0.61
320-400 nm	6.0	4.0	6.82	7.04
400-450 nm	1.3	3.7	5.68	5.90
450-700 nm	2.0	13.5	10.7	8.80
700-1125 nm	5.27	8.27	10.1	8.75
Above 1125 nm	31.6	34.3	32.7	28.55
Total	49.77	64.87	67.00	60.00

[a]Typical applications: 1 to 5 and 8, photochemical, therapeutic, accelerated exposure testing, or accelerated plant growth; 6, 7, and 9 blueprinting diazo printing, photo copying, and graphic arts; 10, motion-picture and television studio lighting.
[b]Photographic white-flame carbons.
[c]High intensity copper-coated sunshine carbons.
[d]Both carbons are same in horizontal, coaxial ac arcs.
[e]High-intensity photo carbons.
[f]Motion-picture-studio carbons.
[g]All combinations shown are operated coaxially.
[h]All operated on alternating current except item 10.
[i]Horizontal candlepower, transverse to arc axis.
[j]Deviates enough from blackbody colors to make color temperature of doubtful meaning.

TABLE 2 Flame-Type Carbon Arcs (*Continued*)

5	6	7[b]	8[c,d]	9[d,e]	10[f]
W	Enclosed arc	Photo	Sunshine	Photo	Studio
Polymetallic	None	Rare earth	Rare earth	Rare earth	Rare earth
Vertical	Vertical	Vertical	Horizontal	Horizontal	Vertical
22 mm	1/2 in.	1/2 in.	6 mm	9 mm	8 mm
12	3-16	12	6.5	8	12
13 mm	1/2 in.	1/2 in.	6 mm	9 mm	7 mm
12	3-16	12	6.5	8	9
80	16	38	40	95	40
50	138	50	24	30	37 dc
4	2.2	1.9	1	2.85	1.5
8400	1170	6700	4830	14200	11000
92000	13000	74000	53000	156000	110000
23	5.9	39.8	53	54.8	73.5
		7420[j]	6590	8150	4700
1020		95	11		12
1860		76	49	100	48
3120	1700	684	415	1590	464
1480	177	722	405	844	726
2600	442	2223	1602	3671	3965
3220	1681	1264	1368	5632	2123
14500	6600	5189	3290	8763	4593
27800	10600	10253	7140	20600	11930
2.55		0.5	0.11		0.08
4.65		0.4	0.49	0.35	0.32
7.80	7.7	3.6	4.15	5.59	3.09
3.70	0.8	3.8	4.05	2.96	4.84
6.50	2.0	11.7	16.02	12.86	26.43
8.05	7.6	6.7	13.68	10.75	14.15
36.25	29.9	27.3	32.90	30.60	30.62
69.50	48.0	54.0	71.40	72.20	79.53

Enclosed Arc and Discharge Sources (High-Pressure)

Koller[18] states that the carbon arc is generally desired if a high intensity is required from a single unit but that it is less efficient than the mercury arc. Other disadvantages are the short life of the carbon with respect to mercury, and combustion products which may be undesirable. Worthing[2] describes a number of the older, enclosed, metallic arc sources, many of which can be built in the laboratory for laboratory use. Today, however, it is rarely necessary to build one's own source unless it is highly specialized.

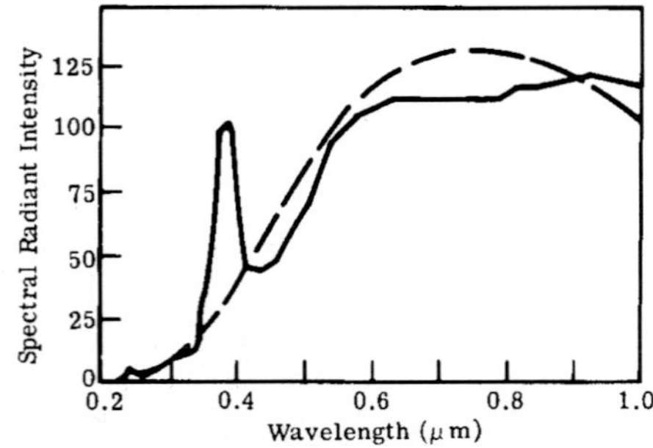

FIGURE 21 Spectral distribution of radiant flux from 30-A, 55-V dc low-intensity arc with 12-mm positive carbon (solid line) and a 3800 K blackbody radiator (broken line).

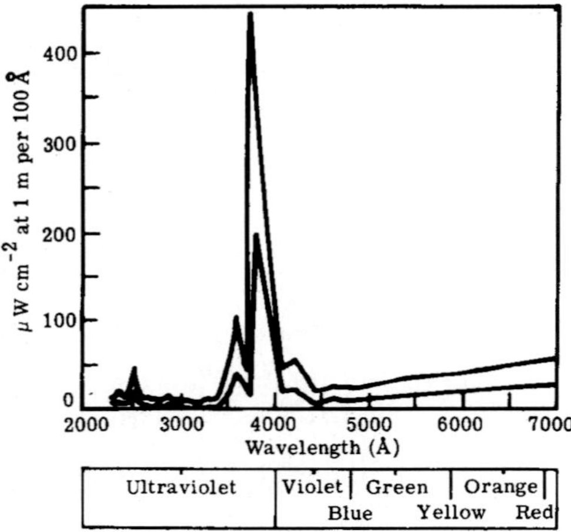

FIGURE 22 Spectral energy distribution of carbon arc with core of soft carbon. Upper curve: 60-A ac 50-V across the arc; lower curve: 30-A ac 50-V across the arc.

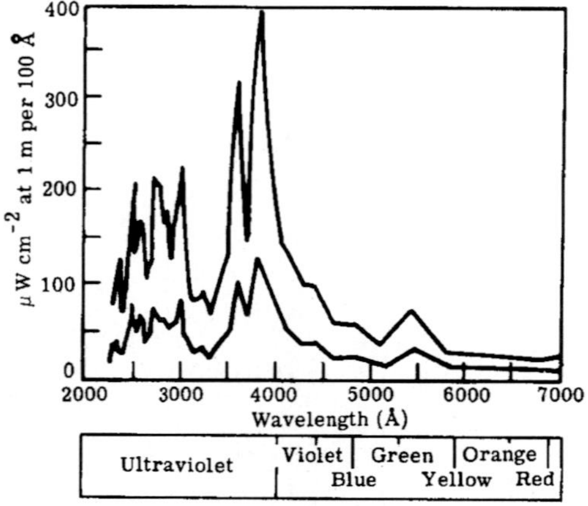

FIGURE 23 Spectral Energy distribution of carbon arc with polymetallic-cored carbons. Upper curve: 60-A ac 50-V across the arc; lower curve: 30-A ac 50-V across the arc.

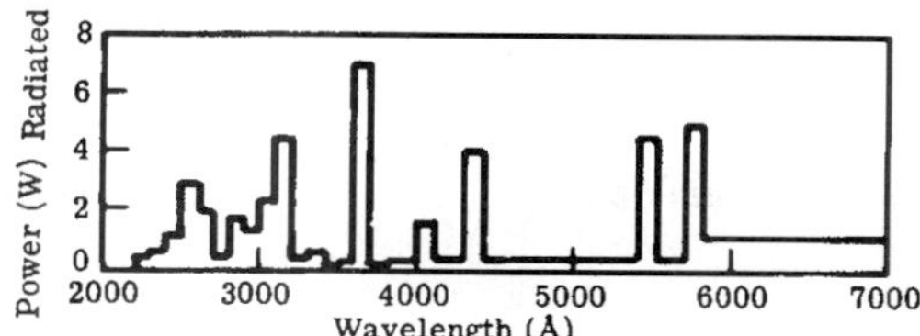

FIGURE 24 Intensity distribution of UA-2 intermediate-pressure lamp.

The Infrared Handbook[1] compiles a large number of these types of sources, some of which will be repeated here, in case that publication would not be currently available to the reader. However, the reader should take caution that many changes might have occurred in the characteristics of these sources and in the supplier whose product is preferred. Consultation with the Photonics Directory (see preceding) is usually a good procedure. In some cases a certain type of source described previously may not still exist. Thus, whereas some manufacturers were less compliant in providing data, they should be expected to respond more readily to a potential customer.

Uviarc.* This lamp is an efficient radiator of ultraviolet radiation. The energy distribution of one type is given in Fig. 24. Since the pressure of this mercury-vapor lamp is intermediate between the usual high- and the low-pressure lamps, little background (or continuum) radiation is present. In the truly high-pressure lamp, considerable continuum radiation results from greater molecular interaction. Figure 25[20] shows the dependence on pressure of the amount of continuum in mercury lamps of differing pressure. Bulb shapes and sizes are shown in Fig. 26.

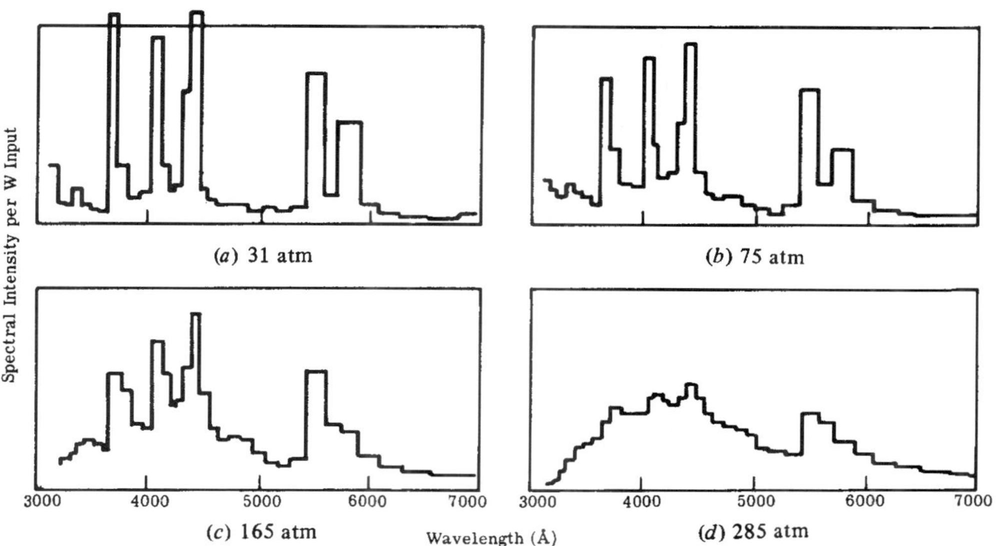

FIGURE 25 Emission spectrum of high-pressure mercury-arc lamps showing continuum background.

* Registered trademark of General Electric.

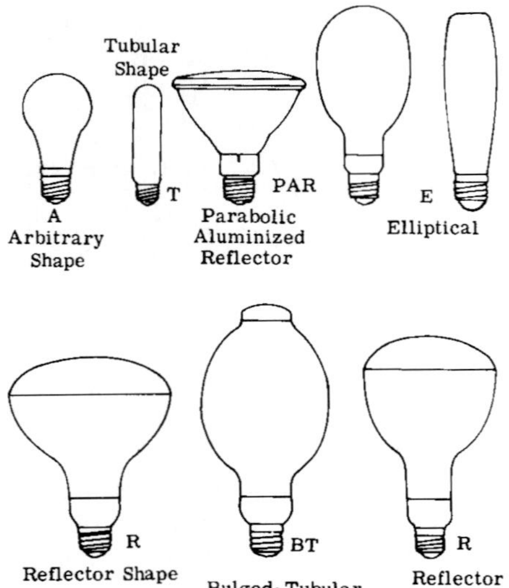

FIGURE 26 Bulb shapes and sizes (not to scale).

Mercury Arcs. A widely used type of high-pressure, mercury-arc lamp and the components necessary for its successful operation are shown in Fig. 27. The coiled tungsten cathode is coated with a rare-earth material (e.g., thorium). The auxiliary electrode is used to help in starting. A high resistance limits the starting current. Once the arc is started, the operating current is limited by ballast supplied by the high reactance of the power transformer. Spectral data for clear, 400 W mercury lamps of this type are given in Fig. 28.

Multivapor Arcs. In these lamps, argon and mercury provide the starting action. Then sodium iodide, thallium iodide, and indium iodide vaporize and dissociate to yield the bulk of the lamp radiation. The physical appearance is like that of mercury lamps of the same general nature. Ballasts are similar to their counterparts for the mercury lamp. Up-to-date

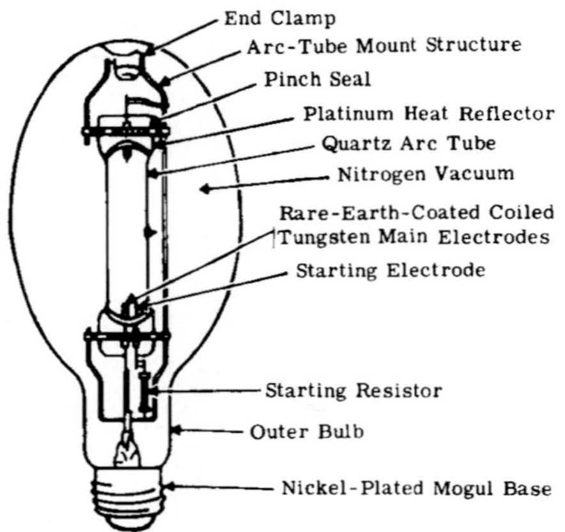

FIGURE 27 High-pressure mercury lamp showing various components.

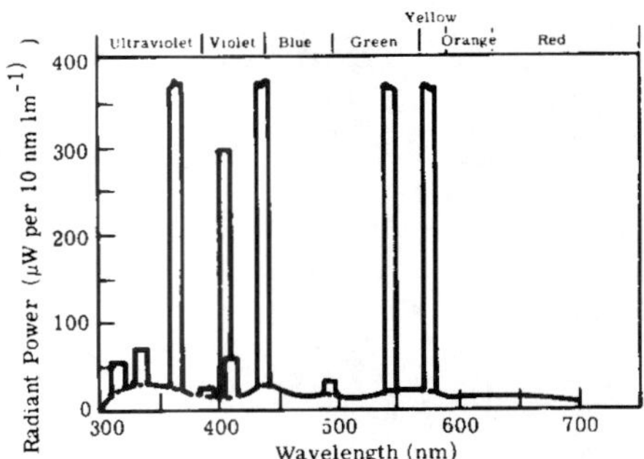

FIGURE 28 Spectral energy distribution for clear mercury-arc lamp.

information on these sources should be obtained from the General Electric Corporation Lamp Division in Nela Park near Cleveland, Ohio. Spectral features of these sources are given in Fig. 29.

Lucalox*®* *Lamps. The chief characteristics of this lamp are high-pressure sodium discharge and a high-temperature withstanding ceramic, Lucalox (translucent aluminum oxide), to yield performance typified in the spectral output of the 400-W Lucalox lamp shown in Fig. 30. Ballasts for this lamp are described in the General Electric *Bulletin TP-109R.*[21]

Capillary Mercury-Arc Lamps.[18] As the pressure of the arc increases, cooling is required to avoid catastrophic effects on the tube. The AH6 tube (Fig. 31) is constructed with a quartz bulb wall and a quartz outer jacket, to allow 2800 K radiation to pass, or a Pyrex®† outer jacket to eliminate ultraviolet. Pure water is forced through at a rapid rate, while the tube is maintained at a potential of 840 V. The spectral characteristics of certain tubes[22] are shown in Fig. 32. This company does not appear in the *Photonics Guide of 1989,* so the catalog referenced in the figure may not be current.

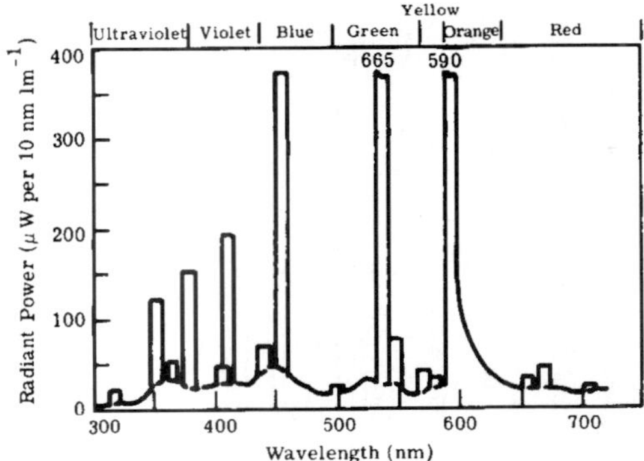

FIGURE 29 Spectral energy distribution of multivapor-arc lamp.

* Registered trademark of General Electric.
† Registered trademark of Corning Glass Works.

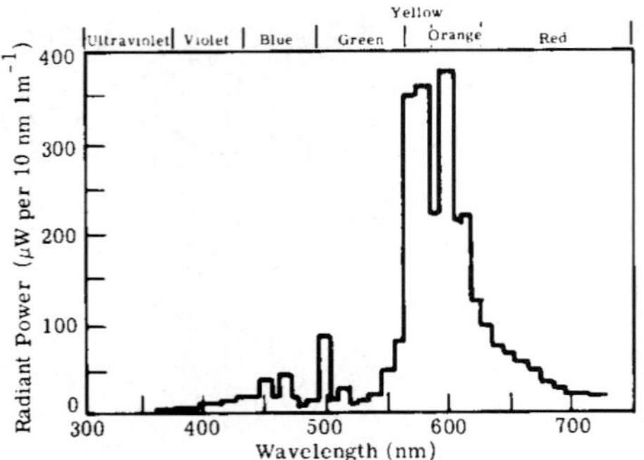

FIGURE 30 Spectral output of 400-W Lucalox lamp.

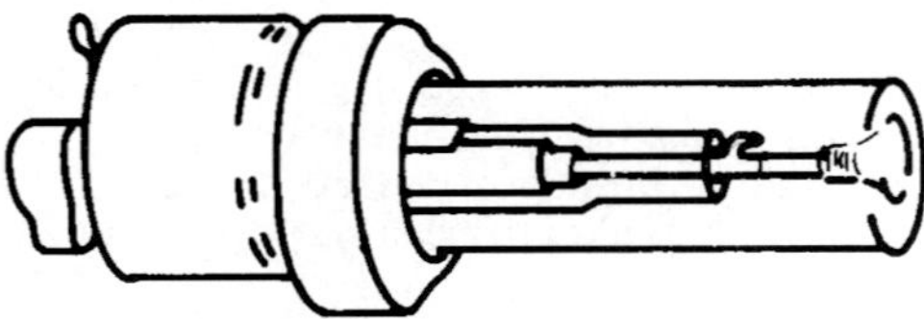

FIGURE 31 Water-cooled high-pressure (110 atm) mercury arc lamp showing lamp in water jacket.

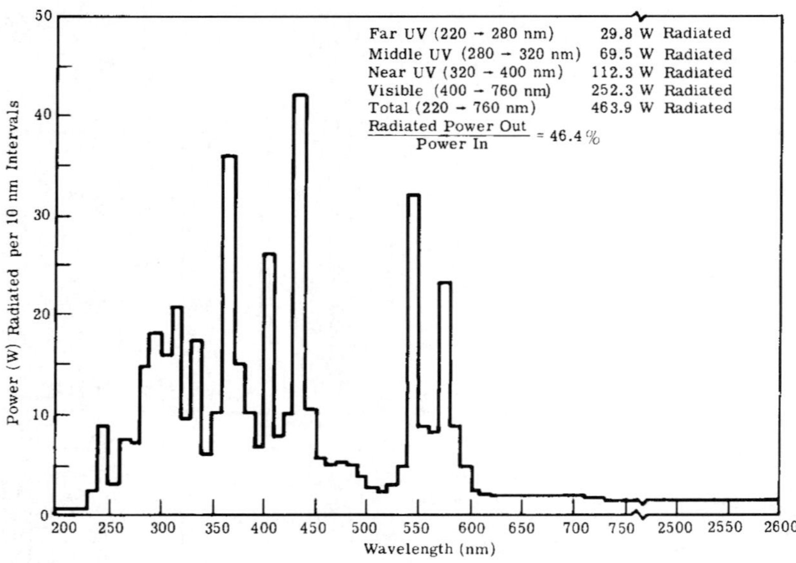

FIGURE 32 Spectral energy distribution of type BH6-1 mercury capillary lamp.

Compact-Source Arcs.[15,23] Some common characteristics of currently available compact-source arc lamps are as follows:

1. A clear quartz bulb of roughly spherical shape with extensions at opposite ends constituting the electrode terminals. In some cases, the quartz bulb is then sealed within a larger glass bulb, which is filled with an inert gas.
2. A pair of electrodes with relatively close spacing (from less than 1 mm to about 1 cm); hence the sometimes-used term short-arc lamps.
3. A filling of gas or vapor through which the arc discharge takes place.
4. Extreme electrical loading of the arc gap, which results in very high luminance, internal pressures of many atmospheres, and bulb temperatures as high as 900°C. Precautions are necessary to protect people and equipment in case the lamps should fail violently.
5. The need for a momentary high-voltage ignition pulse, and a ballast or other auxiliary equipment to limit current during operation.
6. Clean, attention-free operation for long periods of time.

These lamps are designated by the chief radiating gases enclosed as mercury, mercury xenon, and xenon lamps.

Figure 33 shows a compact-source construction for a 1000-W lamp. Since starting may be a problem, some lamps (Fig. 34) are constructed with a third (i.e., a starting) electrode, to

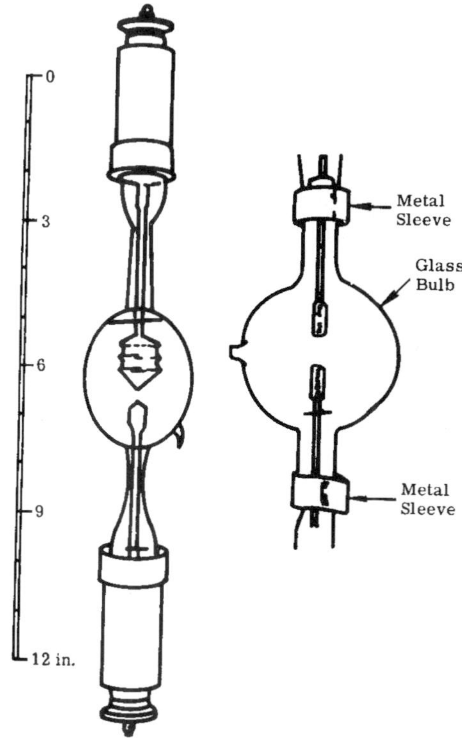

FIGURE 33 Construction of different lamps showing differences in relative sizes of electrodes for dc (*left*) and ac (*right*) operation.

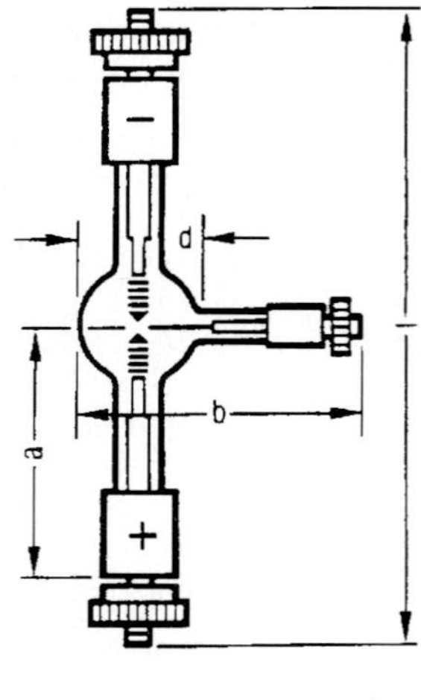

Lamp (Order Reference)		HBO 200 W¹	
Type of current		DC	AC
Lamp supply voltage	V	>105	220
Operating voltage of lamp	V L_1/L_2	65···47	61 ± 4 / 53 ± 4
Operating current at operating voltage range	A L_1/L_2	3.1···4.2	3.6 / 4.2
Rated power of lamp	W	200	
Luminous flux	lm	9500	
Luminous efficacy	lm/W	47.5	
Light intensity	cd	1000	
Average luminance	cd/cm²	40000	
Arc (width w x height h⁴)	mm	0.6 × 2.2	
Average lamp life	h	200	
Diameter d	mm	18	
Length l max.	mm	108	
Distance a	mm	41 ± 2	
Width b	mm	45	
Burning position with stamped base down		s45	

FIGURE 34 Construction of a lamp with a third, starting electrode.

which a momentary high voltage is applied for starting (and especially restarting) while hot. The usual ballast is required for compact-source arcs. For stability, these arcs, particularly mercury and mercury-xenon, should be operated near rated power on a well-regulated power supply.[23]

The spatial distribution of luminance from these lamps is reported in the literature already cited, and typical contours are shown in Fig. 35. Polar distributions are similar to those shown in Fig. 36. Spectral distributions are given in Figs. 37 through 39 for a 1000-W ac mercury lamp, a 5-kW dc xenon lamp, and 1000-W dc mercury-xenon lamp. Lamps are available at considerably less wattage.

Cann[23] reports on some interesting special lamps tested by Jet Propulsion Laboratories for the purpose of obtaining a good spectral match to the solar distribution. The types of lamps tested were: Xe; Xe-Zn; Xe-Cd; Hg-Xe-Zn; Hg-Xe-Cd; Kr; Kr-Zn; Kr-Cd;

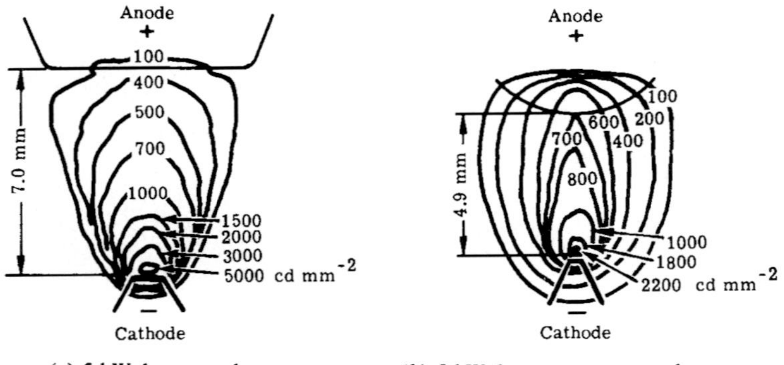

FIGURE 35 Spatial luminance distribution of compact-arc lamps.

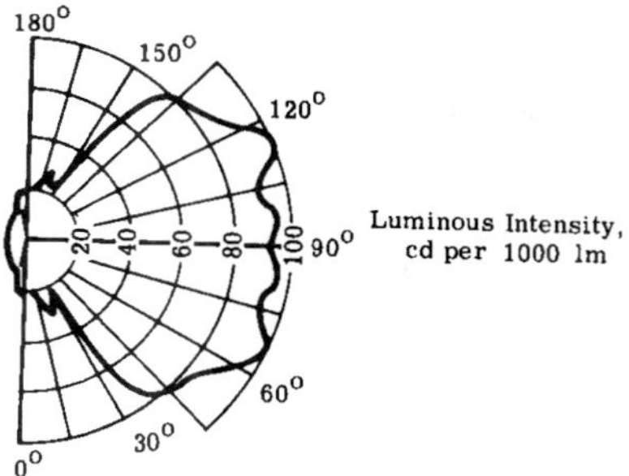

(*a*) 7.5-kW ac mercury lamp

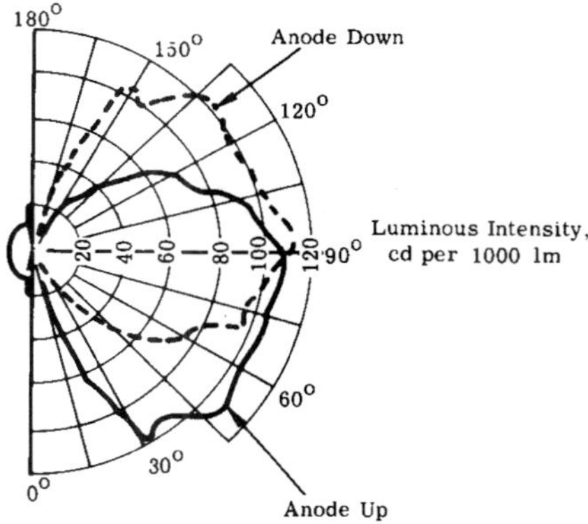

(*b*) 2.5-kW dc mercury-xenon lamp

FIGURE 36 Polar distribution of radiation in planes that include arc axis. Asymmetry in (*b*) is due to unequal size of electrodes.

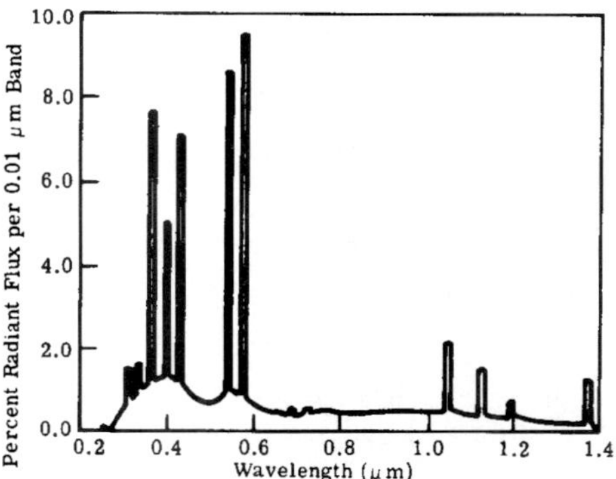

FIGURE 37 Spectral distribution of radiant intensity from a 1000-W ac mercury lamp perpendicular to the lamp axis.

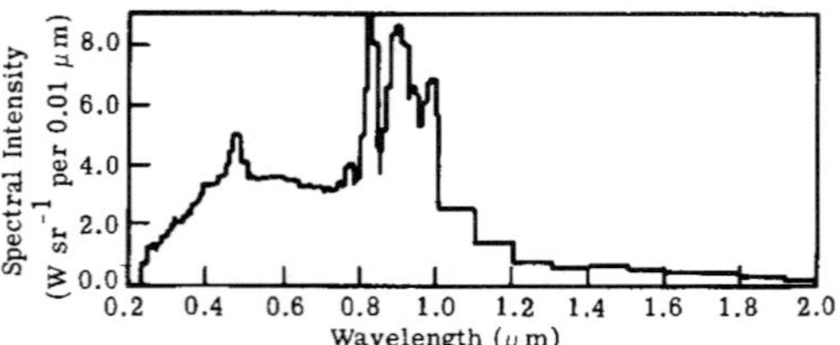

FIGURE 38 Spectral distribution of radiant intensity from a 5-kW dc xenon lamp perpendicular to the lamp axis with electrode and bulb radiation excluded.

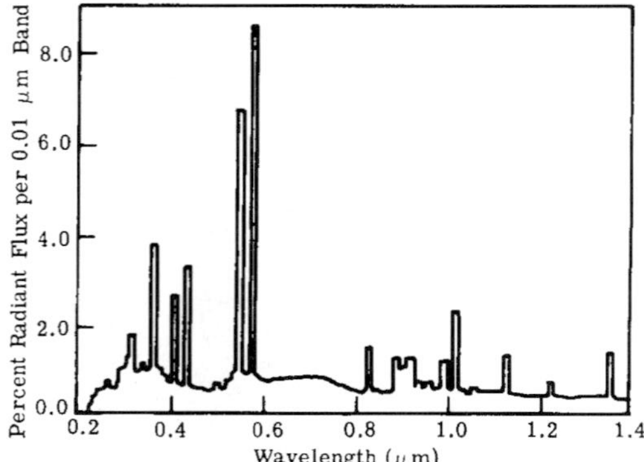

FIGURE 39 Spectral distribution of radiant flux from a 1000-W mercury-xenon lamp.

Hg-Kr-Zn; Hg-Kr-Cd; Ar; Ne; and Hg-Xe with variable mercury-vapor pressure. For details, the reader should consult the literature.

A special design of a short-arc lamp manufactured by Varian[24] is shown in Fig. 40. Aside from its compactness and parabolic sector, it has a sapphire window which allows a greater amount of IR energy to be emitted. It is operated either dc or pulsed, but the user should obtain complete specifications, because the reflector can become contaminated, with a resultant decrease in output.

FIGURE 40 High-pressure, short-arc xenon illuminators with sapphire windows. Low starting voltage, 150 through 800 watt; VIX150, VIX300, VIX500, VIX800.

Enclosed Arc and Discharge Sources (Low-Pressure)[18]

With pressure reduction in a tube filled with mercury vapor, the 2537 Å line becomes predominant so that low-pressure mercury tubes are usually selected for their ability to emit ultraviolet radiation.

Germicidal Lamps. These are hot-cathode lamps which operate at relatively low voltages. They differ from ordinary fluorescent lamps which are used in lighting in that they are designed to transmit ultraviolet, whereas the wall of the fluorescent lamp is coated with a material that absorbs ultraviolet and reemits visible light. The germicidal lamp is constructed of glass of 1-mm thickness which transmits about 65 percent of the 2537 Å radiation and virtually cuts off shorter-wavelength ultraviolet radiation.

Sterilamp®* Types. These cold cathode lamps start and operate at higher voltages than the hot-cathode type and can be obtained in relatively small sizes as shown in Fig. 41. Operating characteristics of the Sterilamps should be obtained from the manufacturer.

Black-Light Fluorescent Lamps. This fluorescent lamp is coated with a phosphor efficient in the absorption of 2537 Å radiation, emitting ultraviolet radiation in a broad

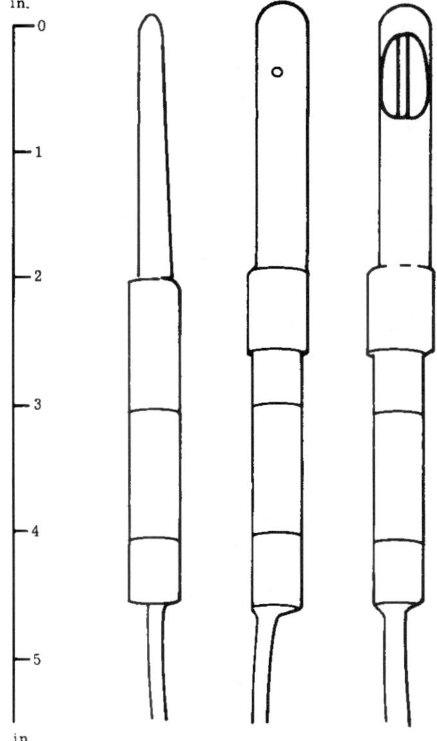

FIGURE 41 Pen-Ray low-pressure lamp. Pen-Ray is a registered trademark of Ultraviolet Products, Inc.

* Registered trademark of Westinghouse Electric.

TABLE 3 Spectral Energy Distribution for Black-Light (360 BL) Lamps

(W)	*Length* (in.)	3200-3800 Å		*Total Ultra-violet below* 3800 Å		*Total Visible* (W)	3800-7600 Å %[a]	*Erythemal Flux*
		(W)	%[a]	(W)	%[a]			
6	9	0.55	9.1	0.56	9.4	0.1	1.7	250
15	18	2.10	14.0	2.20	14.6	0.4	2.7	950
30	36	4.60	15.3	4.70	15.8	0.9	3.0	2100
40	48	6.70	16.8	6.90	17.3	1.5	3.8	3000

[a]Percentage of input power.

TABLE 4 Single-Element and Multiple-Element Hollow-Cathode Lamps

Tubes listed in this table are issued by Fisher Scientific and produced by Westinghouse Electric.

Element	*Window*[a]	*Gas Fill*[b]	*Size*[c]	*Analytical Line* (Å)	*Catalog Number*[d]
Single-Element 22000 Series					
Aluminum	Q	A	B	3092	WL22804
	P	N	A	3092	JA-45-452
	Q	A	A	3092	WL22870
	Q	N	B	3092	WL22929
	Q	N	A	3092	WL22954
Antimony	Q	A	B	2311	WL22840
	Q	A	A	2311	WL22872
	Q	N	B	2311	JA-45-461
	Q	N	A	2311	WL22956
Arsenic	Q	N	B	1937	JA-45-315
	Q	N	A	1937	WL22873
	Q	A	B	1937	JA-45-315
Barium	P	N	A	5536	JA-45-480
Beryllium	Q	N	B	2349	WL23407
Bismuth	Q	A	B	3068	WL22841
	Q	A	A	3068	WL22874
	Q	N	B	3068	JA-45-469
	Q	N	A	3068	WL22957
Boron	Q	A	B	2497	JA-45-568
	Q	A	A	2497	WL22917
Cadmium	Q	A	B	3261	WL22816
	Q	A	A	3261	WL22875
	Q	N	B	3261	JA-45-462
	Q	N	A	3261	WL22958
Calcium	P	N	A	4227	JA-45-440
Cerium	Q	N	B	–	JA-45-569
	Q	N	A	–	WL22978
Cesium	P	A	A	4556	WL22817
	P	N	A	4556	JA-45-441
Chromium	P	A	A	3579	WL22812
	Q	A	B	3579	WL22821
	Q	A	A	3579	WL22877
	Q	N	B	3579	JA-45-454
	Q	N	A	3579	WL22959
Cobalt	P	A	A	3454	WL22813
	Q	A	B	3454	WL22814
	Q	A	A	3454	WL22878
	Q	N	B	3454	JA-45-456
	Q	N	A	3454	WL22953

Element	*Window*[a]	*Gas Fill*[b]	*Size*[c]	*Analytical Line* (Å)	*Catalog Number*[d]
Copper	P	N	A	3247	JA-45-458
	Q	A	B	3247	WL22606
	Q	A	A	3247	WL22879
	Q	N	B	3247	JA-45-490
	Q	N	A	3247	WL23042
Dysprosium	Q	N	B	4212	JA-45-595
	Q	N	A	4212	WL22880
Erbium	Q	N.	B	4008	JA-45-571
	Q	N	A	4008	WL22881
Europium	Q	N	B	4594	JA-45-572
	Q	N	A	4594	WL22882
Gadolinium	P	N	A	4079	WL22975
	Q	N	B	4079	JA-45-573
	Q	N	A	4079	WL22986
Gallium	Q	N	B	4172	JA-45-470
	Q	N	A	4172	WL22884
Germanium	Q	A	B	2651	JA-45-575
	Q	N	B	2651	JA-45-313
Gold	Q	A	B	2676	WL22839
	Q	A	A	2676	WL22883
	Q	N	B	2676	JA-45-467
	Q	N	A	2676	WL22960
Hafnium	Q	N	B	3072	JA-45-303
Holmium	Q	N	A	4104	WL22885
	Q	N	B	4104	JA-45-576
Indium	Q	A	B	3040	WL22867
	Q	A	A	3040	WL22915
	Q	A	A	3040	JA-45-471
Iridium	Q	N	B	2850	JA-45-577
Iron	P	A	A	3270	WL22602
	Q	A	B	3720	WL22611
	Q	N	B	3720	JA-45-455
	P	N	A	3720	WL22820
	Q	A	A	3720	WL22886
	Q	N	A	3720	WL22887
Iron, high-purity	Q	N	B	3720	WL22837
	Q	N	A	3720	WL22888
Lanthanum	Q	A	B	5501	WL22846
	Q	A	A	5501	WL22889
	Q	N	B	5501	JA-45-495

[a]P = Pyrex, Q = quartz; [b]N = neon, A = argon; [c]A = 1½ in. diameter, B = 1 in. diameter, C = 2 in. diameter; [d]WL = Westinghouse, JA = Jarrell-Ash.

band around 3650 Å. The phosphor is a cerium-activated calcium phosphate, and the glass bulb is impervious to shorter wavelength ultraviolet radiation. Characteristics of one type are given in Table 3.

Hollow Cathode Lamps. A device described early in this century and used for many years by spectroscopists is the hollow-cathode tube. The one used by Paschen[2] consisted of a hollow metal cylinder and contained a small quantity of inert gas, yielding an intense cathode-glow characteristic of the cathode constituents. Materials that vaporize easily can be incorporated into the tube so that their spectral characteristics predominate.[2]

Several companies sell hollow-cathode lamps which do not differ significantly from those constructed in early laboratories. The external appearance of these modern tubes shows the marks of mass production and emphasis on convenience. They come with a large number of vaporizable elements, singly or in multiples, and with Pyrex® or quartz windows. A partial list of the characteristics or the lamps available from two manufacturers is given in Table 4. Their physical appearance is shown in Fig. 42*a*. A schematic of the different elements obtainable in various lamps is shown in Fig. 42*b*.[25]

Electrodeless Discharge Lamps.[26,27,28] The electrodeless lamp gained popularity when Meggers used it in his attempt to produce a highly precise standard of radiation. Simplicity of design makes laboratory construction of this type of lamp easy. Some of the simplest lamps consist of a tube, containing the radiation-producing element, and a microwave generator, for producing the electric field (within the tube) which in turn excites the

TABLE 4 Single-Element and Multiple-Element Hollow-Cathode Lamps (*Continued*)

Element	*Window*[a]	*Gas Fill*[b]	*Size*[c]	*Analytical Line* (Å)	*Catalog Number*[d]	*Element*	*Window*[a]	*Gas Fill*[b]	*Size*[c]	*Analytical Line* (Å)	*Catalog Number*[d]
Lead	Q	A	B	2833	WL22838	Palladium	Q	A	B	3404	WL22857
	Q	A	A	2833	WL22890		Q	A	A	3404	WL22911
	Q	N	B	2833	JA-45-468		Q	N	B	3404	JA-45-475
	Q	N	A	2833	WL22952		Q	N	A	3404	WL22970
Lithium 6	P	N	A	6708	JA-45-579	Phosphorus	Q	N	B	2136	JA-45-449
	P	A	A	6708	WL22925		Q	N	A	2136	WL22990
Lithium 7	P	N	A	6708	JA-45-580	Platinum	Q	A	B	2659	WL22851
	P	A	A	6708	WL22926		Q	A	A	2659	WL22896
Lithium, natural	P	A	A	6708	WL22825		Q	N	B	2659	JA-45-466
	P	N	A	6708	JA-45-444	Potassium	P	N	A	4044	JA-45-484
	Q	A	B	6708	WL23115	Praseodymium	Q	N	B	4951	JA-45-585
Lutetium	Q	N	B	3282	JA-45-581		Q	N	A	4951	WL22982
	Q	N	A	3282	WL23010	Rhenium	Q	N	B	3460	JA-45-489
Magnesium	Q	A	. B	2852	WL22609		Q	N	A	3460	WL22967
	Q	A	A	2852	WL22891	Rhodium	Q	A	B	3435	WL22850
	Q	N	A	2852	WL22951		Q	A	A	3435	WL22897
	Q	N	B	2852	JA-45-451		Q	N	B	3435	JA-45-476
Manganese	Q	A	B	2795	WL22608	Rubidium	P	N	A	7800	JA-45-443
	P	A	A	2795	WL22815		Q	N	B	7800	WL23046
	Q	N	B	2795	JA-45-472	Ruthenium	Q	A	B	3499	JA-45-586
	Q	N	A	2795	WL22961	Samarium	Q	N	B	4760	JA-45-587
	Q	A	A	2795	WL22876		Q	N	A	4760	WL22899
Mercury	Q	A	B	2537	JA-45-493	Scandium	Q	N	B	3912	JA-45-309
	Q	A	A	2537	WL22892	Selenium	Q	A	B	1960	WL22843
Molybdenum	Q	A	B	3133	WL22805		Q	A	A	1960	WL22898
	Q	A	A	3133	WL22893		Q	N	B	1960	JA-45-477
	Q	N	B	3133	JA-45-460		Q	N	A	1960	WL22963
	Q	N	A	3133	WL22962	Silicon	Q	A	B	2516	WL22832
Neodymium	Q	N	B	4925	JA-45-582		Q	A	A	2516	WL22900
	Q	N	A	4925	WL22980		Q	N	B	2516	JA-45-479
Nickel	P	A	A	3415	WL22605		Q	N	A	2516	WL22964
	Q	A	B	3415	WL22663	Silver	Q	A	B	3281	JA-45-483
	Q	N	B	3415	JA-45-457		Q	A	A	3281	WL22901
	Q	A	A	3415	WL22894	Sodium	P	A	A	5890	WL22864
	Q	N	A	3415	WL22895		P	N	A	5890	JA-45-485
Niobium	Q	N	B	4059	JA-45-486	Strontium	P	N	A	4607	JA-45-481
	Q	N	A	4059	WL22912	Sulphur	Q	N	B	–	JA-45-588
Osmium	Q	A	B	2909	JA-45-584						

TABLE 4 Single-Element and Multiple-Element Hollow-Cathode Lamps (*Continued*)

Element	*Window*[a]	*Gas Fill*[b]	*Size*[c]	*Analytical Line* (Å)	*Catalog Number*[d]
Tantalum	Q	A	B	2714	JA-45-488
	Q	A	A	2714	WL22913
	Q	N	B	2714	WL22971
	Q	N	A	2714	WL22972
Tellurium	Q	A	B	2143	WL22842
	Q	A	A	2143	WL22902
	Q	N	B	2143	JA-45-473
	Q	N	A	2143	WL22965
Terbium	Q	N	B	4326	JA-45-589
	Q	N	A	4326	WL22903
Thallium	Q	N	B	3776	WL23408
Thorium	Q	N	A	3245	WL23028
	Q	N	B	3245	JA-45-590
Thulium	Q	N	B	4105	JA-45-591
	Q	N	A	4105	WL23008
Tin	Q	A	B	2863	WL22822
	Q	A	A	2863	WL22904
	Q	N	B	2863	JA-45-463
	Q	N	A	2863	WL22966
Titanium	Q	N	B	3643	JA-45-592
	Q	N	A	3643	WL22992
Tungsten	Q	N	B	4009	JA-45-465
	Q	A	B	4009	WL22849
	Q	N	A	4009	WL22905
	Q	A	A	4009	WL22906
Uranium	Q	N	B	5027	JA-45-447
	Q	N	A	5027	WL22907
Vanadium	Q	A	B	3184	WL22856
	Q	A	A	3184	WL22910
	Q	N	B	3184	JA-45-453
	Q	N	A	3184	WL22974
Ytterbium	Q	A	B	3988	JA-45-593
	Q	A	A	3988	WL22984
Yttrium	P	N	A	4102	WL22976
	Q	N	B	4102	JA-45-594
	Q	N	A	4102	WL22988
Zinc	Q	A	B	2139	WL22607
	Q	N	B	2139	JA-45-459
	Q	A	A	2139	WL22908
	Q	N	A	2139	WL22909

Element	*Window*[a]	*Gas Fill*[b]	*Size*[c]	*Analytical Line* (Å)	*Catalog Number*[d]
Zirconium	Q	A	B	3601	JA-45-482
	Q	A	A	3601	WL22914
	Q	N	B	3601	WL22998
Single-Element 36000 Series					
Aluminum	P	N	C	3092	JA-45-36009
Antimony	Q	N	C	2311	JA-45-36010
Arsenic	Q	A	C	1937	JA-45-36011
Barium	P	N	C	5536	JA-45-36012
Beryllium	Q	N	C	2349	JA-45-36013
Bismuth	Q	N	C	3068	JA-45-36014
Boron	Q	A	C	2497	JA-45-36015
Cadmium	Q	N	C	3261	JA-45-36016
Calcium	P	N	C	4227	JA-45-36017
Cerium	Q	N	C	–	JA-45-36019
Cesium	P	N	C	4556	JA-45-36020
Chromium	P	N	C	3579	JA-45-36021
Cobalt	Q	N	C	3454	JA-45-36022
Copper	P	N	C	3247	JA-45-36024
Dysprosium	P	N	C	4212	JA-45-36025
Erbium	P	N	C	4008	JA-45-36026
Europium	P	N	C	4594	JA-45-36027
Gadolinium	P	N	C	4079	JA-45-36028
Gallium	Q	N	C	4172	JA-45-36029
Germanium	Q	N	C	2651	JA-45-36030
Gold	Q	N	C	2676	JA-45-36031
Hafnium	Q	N	C	3072	JA-45-36032
Holmium	P	N	C	4104	JA-45-36033
Indium	Q	N	C	3040	JA-45-36034
Iridium	Q	N	C	2850	JA-45-36036
Iron	Q	N	C	3720	JA-45-36037
Lanthanum	P	N	C	5501	JA-45-36038
Lead	Q	N	C	2833	JA-45-36039
Lithium 6	P	N	C	6708	JA-45-36090
Lithium 7	P	N	C	6708	JA-45-36091
Lithium, natural	P	N	C	6708	JA-45-36040
Lutetium	P	N	C	3282	JA-45-36041
Magnesium	Q	N	C	2852	JA-45-36042
Manganese	Q	N	C	2795	JA-45-36043

[a]P = Pyrex, Q = quartz; [b]N = neon, A = argon; [c]A = 1½ in. diameter, B = 1 in. diameter, C = 2 in. diameter; [d]WL = Westinghouse, JA = Jarrell-Ash.

elemental spectra. Lamps of this type can be purchased with specially designed microwave cavities for greater efficiency in coupling. Those made of fused quartz can transmit from ultraviolet to near infrared. The electrodeless lamp is better able than the arc lamp to produce stable radiation of sharp spectral lines; this makes it useful in spectroscopy and interferometry. The Hg 198 lamp makes a suitable secondary standard of radiation.

Spectral Lamps.[27] Some manufacturers produce groups of arc sources, which are similar in construction and filled with different elements and rare gases, and which yield discontinuous or monochromatic radiation throughout most of the ultraviolet and visible spectrum. They are called spectral lamps. The envelopes of these lamps are constructed of glass or quartz, depending on the part of the spectrum desired. Thus, discrete radiation can be obtained from around 2300 Å into the near infrared. Figure 43[27] represents the various atomic lines observable from Osram spectral lamps. Figure 44[29] gives a physical description of various spectral lamps obtainable from Philips. Table 5 lists the characteristics of the various types of lamps obtainable from Philips.

Pluecker Spectrum Tubes.[30] These are inexpensive tubes made of glass (Fig. 45) with an overall length of 25 cm and capillary portion of 8.5 to 10 cm long. They operate from an ordinary supply with a special transformer which supports the tubes in a vertical position

TABLE 4 Single-Element and Multiple-Element Hollow-Cathode Lamps (*Continued*)

Element	Window[a]	Gas Fill[b]	Size[c]	Analytical Line (Å)	Catalog Number[d]
Mercury	Q	A	C	2537	JA-45-36044
Molybdenum	Q	N	C	3133	JA-45-36045
Neodymium	P	N	C	4925	JA-45-36046
Nickel	Q	N	C	3415	JA-45-36047
Niobium	P	N	C	4059	JA-45-36023
Osmium	Q	A	C	2909	JA-45-36048
Palladium	Q	N	C	3404	JA-45-36049
Phosphorus	Q	N	C	2136	JA-45-36050
Platinum	Q	N	C	2659	JA-45-36051
Potassium	P	N	C	4044	JA-45-36052
Praseodymium	P	N	C	4951	JA-45-36053
Rhenium	P	N	C	3460	JA-45-36056
Rhodium	P	N	C	3435	JA-45-36057
Rubidium	P	N	C	7800	JA-45-36058
Ruthenium	P	A	C	3499	JA-45-36059
Samarium	P	N	C	4760	JA-45-36060
Scandium	P	N	C	3912	JA-45-36061
Selenium	Q	N	C	1960	JA-45-36062
Silicon	Q	N	C	2516	JA-45-36063
Silver	P	A	C	3281	JA-45-36064
Sodium	P	N	C	5890	JA-45-36065
Strontium	P	N	C	4607	JA-45-36066
Sulphur	Q	N	C	–	JA-45-36067
Tantalum	Q	A	C	2714	JA-45-36068
Tellurium	Q	N	C	2143	JA-45-36069
Terbium	P	N	C	4326	JA-45-36070
Thallium	Q	N	C	3776	JA-45-36071
Thorium	Q	N	C	3245	JA-45-36072
Thulium	P	N	C	4105	JA-45-36073
Tin	Q	N	C	2863	JA-45-36074
Titanium	P	N	C	3643	JA-45-36075
Tungsten	Q	N	C	4009	JA-45-36076
Uranium	P	N	C	5027	JA-45-36077
Vanadium	Q	N	C	3184	JA-45-36078
Ytterbium	P	A	C	3988	JA-45-36079
Yttrium	P	N	C	4102	JA-45-36080
Zinc	Q	N	C	2139	JA-45-36081
Zirconium	P	A	C	3601	JA-45-36082

Element	Window[a]	Gas Fill[b]	Size[c]	Analytical Line (Å)	Catalog Number[d]
Multiple-Element 22000 Series					
Aluminum, calcium	Q	N	B	–	WL23246
Aluminum, calcium, magnesium	Q	A	B	–	WL22604
Aluminum, calcium, magnesium	Q	A	A	–	WL22871
Aluminum, calcium magnesium	Q	N	B	–	JA-45-450
Aluminum, calcium, magnesium	Q	N	A	–	WL22955
Aluminum, calcium, magnesium, iron	Q	N	B	–	JA-45-310
Aluminum, calcium, magnesium, lithium	Q	N	B	–	JA-45-436
Aluminum, calcium, magnesium, lithium	Q	A	A	–	WL23036
Aluminum, calcium, strontium	P	N	A	–	WL23403
Antimony, arsenic, bismuth	Q	N	B	–	WL23147
Arsenic, nickel	Q	N	B	–	JA-45-434
Arsenic, selenium, tellurium	Q	N	B	–	JA-45-598
Barium, calcium, strontium	P	N	A	–	JA-45-437
Barium, calcium, silicon, magnesium	Q	N	B	–	JA-45-478
Cadmium, copper, zinc, lead	Q	N	B	–	JA-45-597
Cadmium , silver, zinc, lead	Q	N	B	–	JA-45-308
Calcium, magnesium, strontium	Q	N	B	–	WL23605
Calcium, magnesium, zinc	Q	N	B	–	JA-45-311
Calcium, magnesium, aluminum, lithium	Q	A	B	–	WL23158
Calcium, zinc	Q	N	B		JA-45-304
Chromium, iron, manganese, nickel	Q	N	B	–	JA-45-442
Chromium, cobalt, nickel	Q	N	B	–	WL23174
Chromium, copper	Q	N	B	–	JA-45-306

and maintains the voltage and current values adequate to operate the discharge and regulate the spectral intensity. Table 6 lists the various gases in available tubes.

Concentrated Arc Lamps

Zirconium Arc.[21] The cathodes of these lamps are made of a hollow refractory metal containing zirconium oxide. The anode, a disk of metal with an aperture, resides directly above the cathode with the normal to the aperture coincident with the longitudinal axis of the cathode. Argon gas fills the tube. The arc discharge causes the zirconium to heat (to about 3000 K) and produce an intense, very small source of light. These lamps have been demonstrated in older catalogs from the Cenco Company in a number of wattages (from 2 to 300). The end of the bulk through which the radiation passes comes with ordinary curvature or (for a slight increase in price) flat. Examples are shown in Fig. 46.

Tungsten-Arc (Photomicrographic) Lamp.[21] The essential elements of this discharge type lamp (see Fig. 47) are a ring electrode and a pellet electrode, both made of tungsten. The arc forms between these electrodes, causing the pellet to heat incandescently. The ring also incandesces, but to a lesser extent. Thus, the hot pellet (approximately 3100 K) provides an intense source of small-area radiation. A plot of the spectral variation of this radiation is given in Fig. 48. As with all tungsten sources, evaporation causes a steady

TABLE 4 Single-Element and Multiple-Element Hollow-Cathode Lamps (*Continued*)

Element	Window[a]	Gas Fill[b]	Size[c]	Analytical Line (Å)	Catalog Number[d]
Chromium, manganese	Q	N	B	–	WL23499
Chromium, cobalt, copper, manganese, nickel	Q	N	B	-	WL23601
Chromium, cobalt, copper, iron, manganese, nickel	Q	N	B	–	JA-45-599
Cobalt, copper	Q	N	B	–	JA-45-305
Cobalt, copper, gold, nickel	Q	N	B	–	WL23295
Cobalt, copper, zinc, molybdenum	Q	N	B	–	JA-45-596
Cobalt, iron	Q	N	B	–	WL23291
Cobalt, nickel	Q	N	B	–	WL23426
Copper, gallium	Q	N	B	–	JA-45-431
Copper, iron	Q	N	B	–	JA-45-312
Copper, iron, manganese	Q	N	B	–	JA-45-435
Copper, iron, molybdenum	Q	N	B	–	JA-45-301
Copper, iron, gold, nickel	Q	N	B	–	JA-45-307
Copper, iron, manganese, zinc	Q	N	B	–	JA-45-492
Copper, manganese	Q	N	B	–	JA-45-491
Copper, nickel	Q	N	B	–	WL23441A
Copper, nickel, zinc	Q	N	B	–	WL23405
Copper, zinc, molybdenum	Q	N	B	–	JA-45-496
Copper, zinc, lead, silver	Q	N	B	–	JA-45-448
Copper, zinc, lead, tin	Q	N	B	–	JA-45-438
Gold, nickel	Q	N	B	–	JA-45-433
Gold, silver	Q	N	B	–	WL23269
Indium, silver	Q	N	B	–	WL23294
Lead, silver, zinc	Q	N	B	–	WL23171
Mangesium, zinc	Q	N	B	–	WL23455
Sodium, potassium	P	N	A	–	JA-45-439
Sodium, potassium	P	A	A	–	WL23230
Zinc, lead, tin	Q	N	B	–	WL23404
Multiple-Element 36000 Series					
Aluminum, calcium, magnesium	Q	N	C	–	JA-45-36099
Aluminum, calcium, magnesium lithium	Q	N	C	–	JA-45-36250
Antimony, arsenic, bismuth	Q	N	C	–	JA-45-36203
Barium, calcium, strontium, magnesium	Q	N	C	–	JA-45-36228
Cadmium, silver, zinc, lead	Q	N	C	–	JA-45-36205
Cadmium, copper, zinc, lead	Q	N	C	–	JA-45-36227
Calcium, magnesium	Q	N	C	–	JA-45-36092
Calcium, magnesium, zinc	Q	N	C	–	JA-45-36097
Calcium, zinc	Q	N	C	–	JA-45-36093
Chromium, iron, manganese, nickel	Q	N	C	–	JA-45-36201
Chromium, cobalt, copper, manganese, nickel	Q	N	C	–	JA-45-36094
Chromium, cobalt, copper, manganese, nickel	Q	N	C	–	JA-45-36103
Chromium, copper, nickel, silver	Q	N	C	–	JA-45-36096
Chromium, copper, iron, nickel, silver	Q	N	C	–	JA-45-36108
Cobalt, copper, iron, manganese, molybdenum	Q	N	C	–	JA-45-36102
Copper, zinc, lead, tin	Q	N	C	–	JA-45-36202
Copper, iron	Q	N	C	–	JA-45-36200
Copper, iron, nickel	Q	N	C	–	JA-45-36101
Copper, iron, lead, nickel, zinc	Q	N	C	–	JA-45-36204
Copper, iron, manganese, zinc	Q	N	C	–	JA-45-36105
Sodium, potassium	P	A	C	–	JA-45-36095

[a]P = Pyrex, Q = quartz; [b]N = neon, A = argon; [c]A = 1½ in. diameter, B = 1 in. diameter, C = 2 in. diameter; [d]WL = Westinghouse, JA = Jarrell-Ash.

erosion of the pellet surface with the introduction of gradients, which is not serious if the pellet is used as a point source.

General Electric, manufacturer of the 30A/PS22 photomicrographic lamp, which uses a 30 Å operating current, states that this lamp requires a special heavy-duty socket obtainable through certain manufacturers suggested in its brochure, which may now be out of print in the original, but obtainable presumably as a copy from GE.

Glow Modulator Tubes[31]

According to technical data supplied by Sylvania, these are cold-cathode light sources uniquely adaptable to high-frequency modulation. (These tubes are now manufactured by The English Electric Valve Company, Elmsford, New York.) Pictures of two types are shown in Fig. 49. The cathode is a small hollow cylinder, and the high ionization density in the region of the cathode provides an intense source of radiation. Figure 50 is a graph of the light output as a function of tube current. Figure 51 is a graph depicting the response of the tube to a modulating input. The spectral outputs of a variety of tubes are shown in Fig. 52. Table 7 gives some of the glow-modulator specifications.

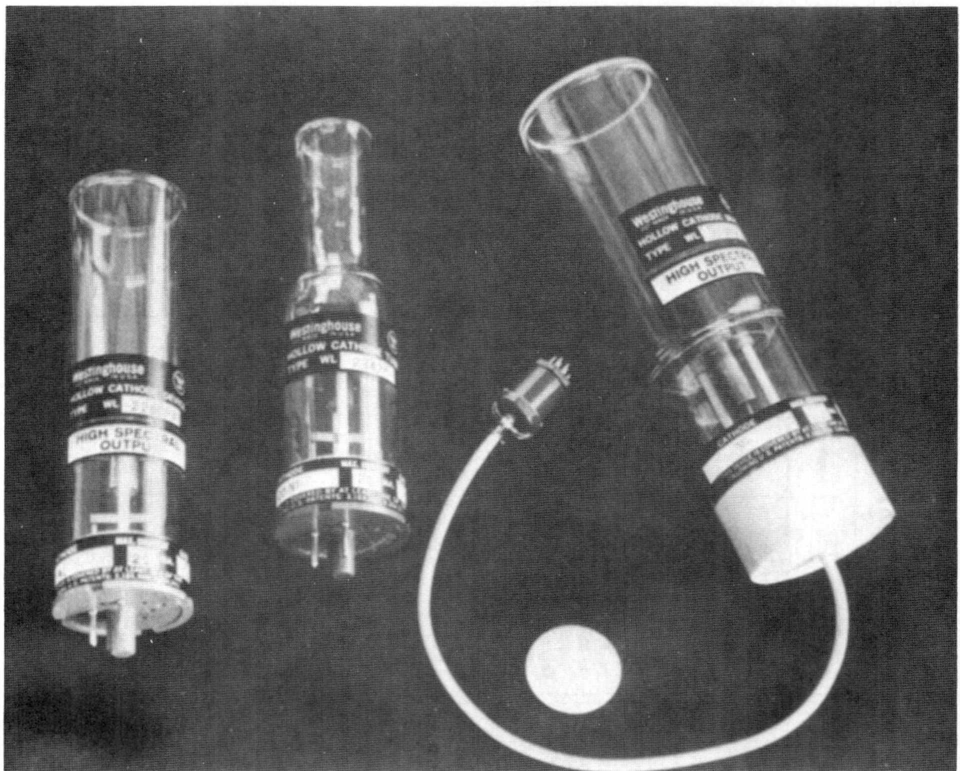

FIGURE 42*a* Hollow-cathode spectral tubes described in Table 8.

Transition Elements

Group 8

Li	Be											B					
Na	Mg											Al	Si	P	S		
K	Ca	Sc	Ti	V	Cr	Mn	Fe	Co	Ni	Cu	Zn	Ga	Ge	As	Se		
Rb	Sr	Y	Zr	Nb	Mo		Ru	Rh	Pd	Ag	Cd	In	Sn	Sb	Te		
Cs	Ba	La	Hf	Ta	W	Re	Os	Ir	Pt	Au	Hg	Tl	Pb	Bi			

Lanthanides	Ce	Pr	Nd		Sm	Eu	Gd	Tb	Dy	Ho	Er	Tm	Yb	Lu
Actinides	Th		U											

FIGURE 42*b* Periodic table showing the prevalence of elements obtainable in hollow-cathode tubes.

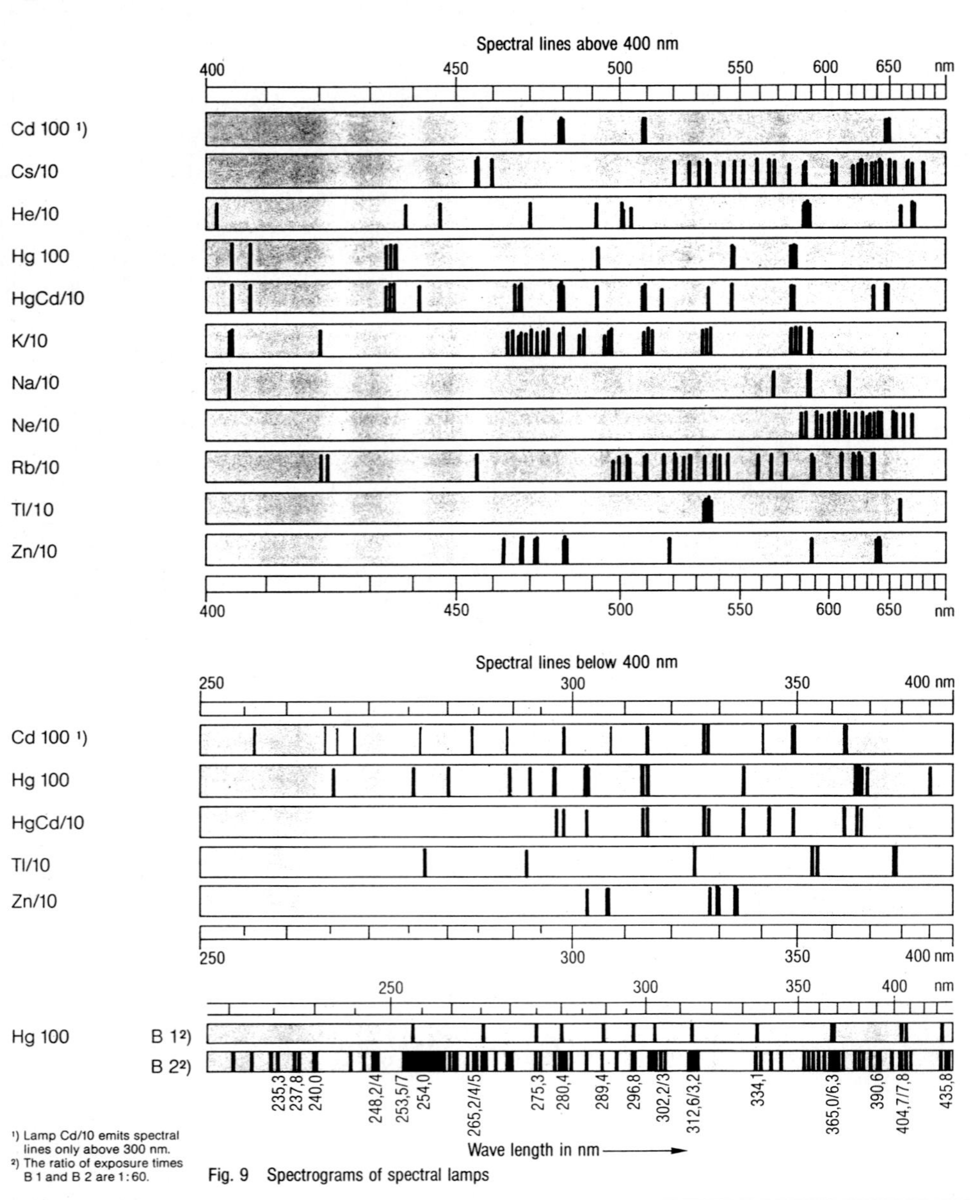

FIGURE 43 Spectral lamps.

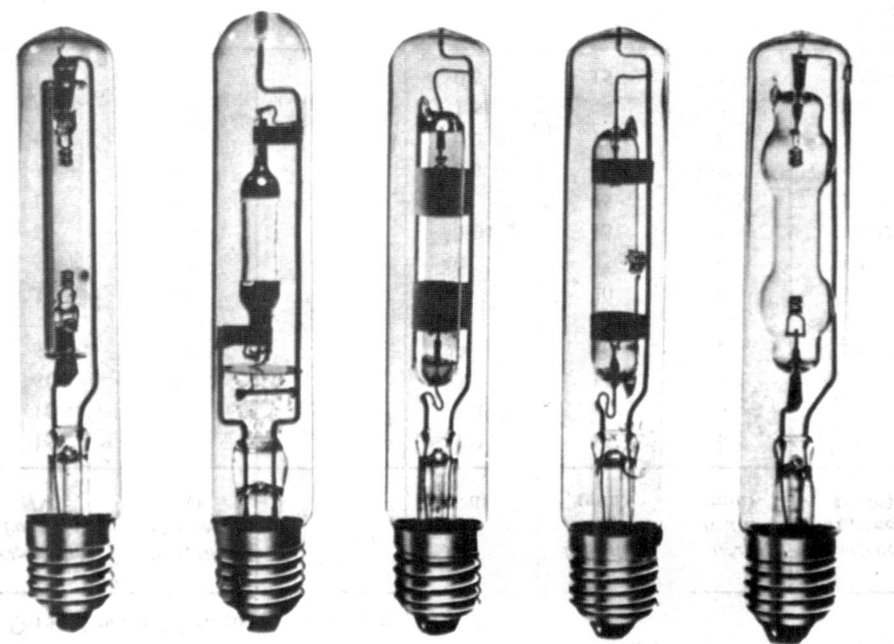

FIGURE 44 Examples of Philips spectral lamps.

TABLE 5 Specifications of Philips Spectral Lamps

Catalog Number	Symbols	Type	Material		Operating Current (A)	Wattage	Arc Length (mm)
			Burner	Envelope			
26-2709	Hg	Mercury (low-pressure)	Quartz	Glass	0.9	15	40
26-2717	Hg	Mercury (high-pressure)	Quartz	Glass	0.9	90	30
26-2725	Cd	Cadmium	Quartz	Glass	0.9	25	30
26-2733	Zn	Zinc	Quartz	Glass	0.9	25	30
26-2741	Hg, Cd, Zn	Mercury, cadmium, and zinc	Quartz	Glass	0.9	90	30
26-2758	He	Helium	Glass	Glass	0.9	45	32
26-2766	Ne	Neon	Glass	Glass	0.9	25	40
26-2774	A	Argon	Glass	Glass	0.9	15	40
26-2782	Kr	Krypton	Glass	Glass	0.9	15	40
26-2790	Xe	Xenon	Glass	Glass	0.9	10	40
26-2808	Na	Sodium	Glass	Glass	0.9	15	40
26-2816	Rb	Rubidium	Glass	Glass	0.9	15	40
26-2824	Cs	Caesium	Glass	Glass	0.9	10	40
26-2832	K	Potassium	Glass	Glass	0.9	10	40
26-2857	Hg	Mercury (low-pressure)	Quartz	Quartz	0.9	15	40
26-2865	Hg	Mercury (high-pressure)	Quartz	Quartz	0.9	90	30
26-2873	Cd	Cadmium	Quartz	Quartz	0.9	25	30
26-2881	Zn	Zinc	Quartz	Quartz	0.9	25	30
26-2899	Hg, Cd, Zn	Mercury, cadmium, and zinc	Quartz	Quartz	0.9	90	30
26-2907	In	Indium[a]	Quartz	Quartz	0.9	25	25
26-2915	Tl	Thallium	Quartz	Quartz	0.9	20	30
26-2923	Ga	Gallium	Quartz	Quartz	0.9	20	30

[a]Requires a Tesla coil to cause it to strike initially

FIGURE 45 Physical construction of Pluecker spectrum tubes.

TABLE 6 Gas Fills in Plueker Tubes[a]

Cenco Number	*Type*
87210	Argon Gas
87215	Helium Gas
87220	Neon Gas
87225	Carbonic Acid Gas
87230	Chlorine Gas
87235	Hydrogen Gas
87240	Nitrogen Gas
87242	Air
87245	Oxygen Gas
87255	Iodine Vapor
87256	Krypton Gas
87258	Xenon
87260	Mercury Vapor
87265	Water Vapor

[a]Consists of glass tube with overall length of 25 cm with capillary portion about 8.5 to 10 cm long. Glass-to-metal seal wires are welded in metal caps with loops for wire connection are firmly sealed to the ends. Power supply no. 87208 is recommended as a source of excitation.

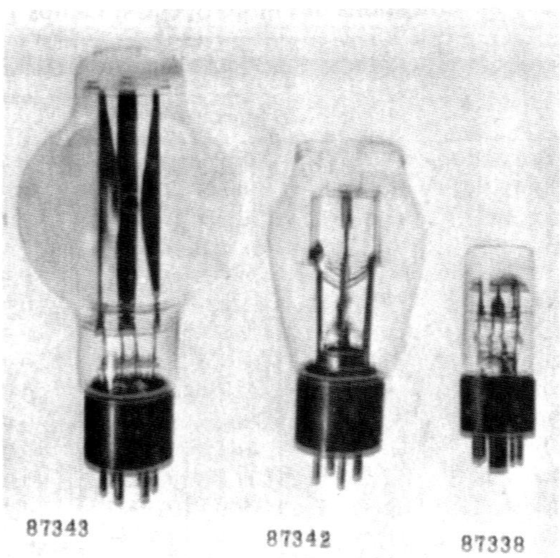

FIGURE 46 Physical construction of some zirconium arc lamps. Two 2-W lamps are available but not shown here.

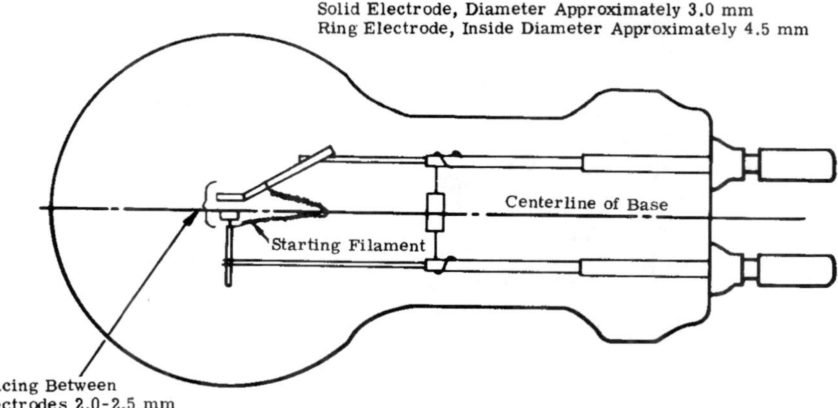

FIGURE 47 Construction of tungsten arc lamp. The lamp must be operated base-up on a well-ventilated housing and using a special high-current socket which does not distort the position of the posts.

Hydrogen and Deuterium Arcs

For applications requiring a strong continuum in the ultraviolet region, the hydrogen arc at a few millimeters pressure provides a useful source. It can be operated with a cold or hot cathode. One hot-cathode type is shown in Fig. 53. Koller[18] plots a distribution for this lamp down to about 200 Å.

Deuterium lamps (Fig. 54) provide a continuum in the ultraviolet with increased intensity over the hydrogen arc. Both lamps have quartz envelopes. The one on the left is designed for operation down to 2000 Å; the one on the right is provided with a Suprasil®* window to increase the ultraviolet range down to 1650 Å. NIST is offering a deuterium lamp standard of spectral irradiance between 200 and 350 nm. The lamp output at 50 cm from its medium bipost base is about 0.7 W cm^{-3} at 200 nm and drops off smoothly to 0.3 W cm^{-3} at 250 nm and 0.07 W cm^{-3} at 350 nm. A working standard of the deuterium lamp can be obtained also, for example, from Optronic Laboratories, Incorporated, Orlando, Florida.

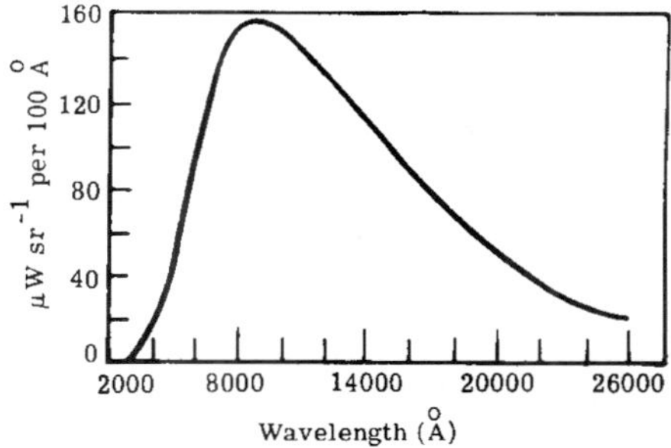

FIGURE 48 Spectral distribution of a 30 PS 22 photomicrographic lamp.

* Registered trademark of Heraeus-Amersil.

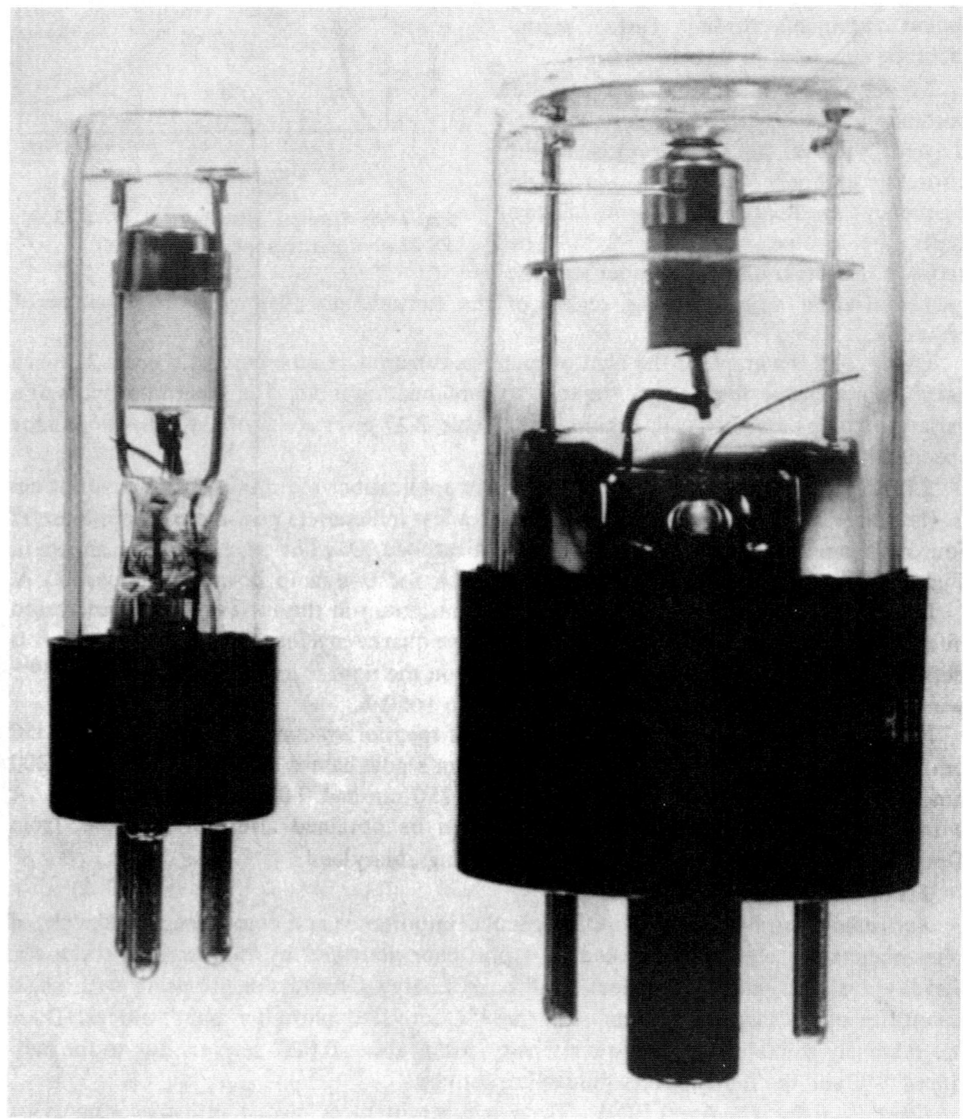

FIGURE 49 Construction of two glow modulator tubes.

Other Commercial Sources

Activated-Phosphor Sources. Of particular importance and convenience in the use of photometers are sources composed of a phosphor activated by radioactive substances. Readily available, and not subject to licensing with small quantities of radioactive material, are the $_{14}C$-activated phosphor light sources. These are relatively stable sources of low intensity, losing about 0.02 percent per year due to the half-life of $_{14}C$ and the destruction of phosphor centers.

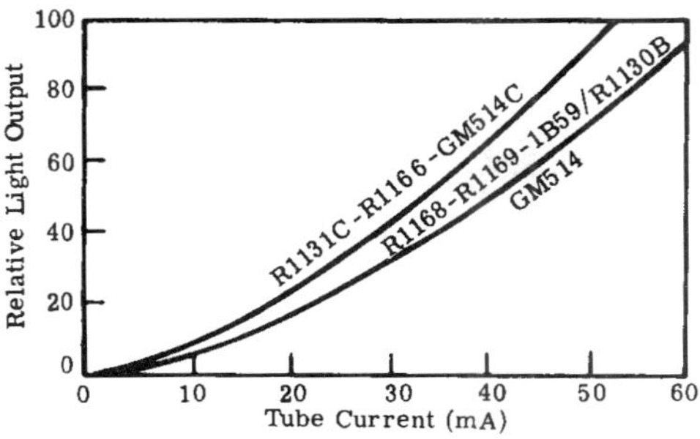

FIGURE 50 Variations of the light output from a glow modulator tube as a function of tube current.

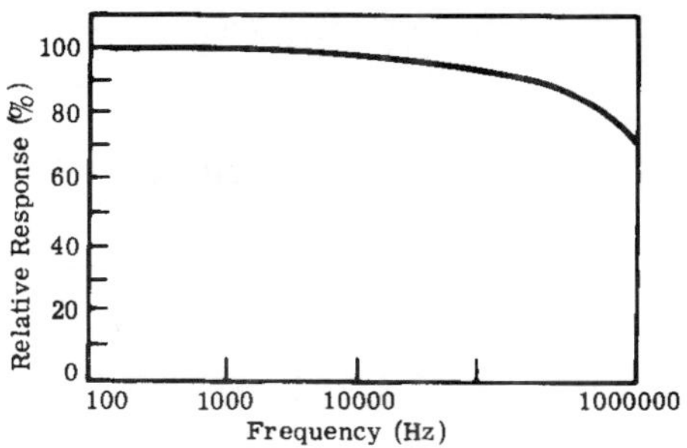

FIGURE 51 Response of the glow modulator tube to a modulating input.

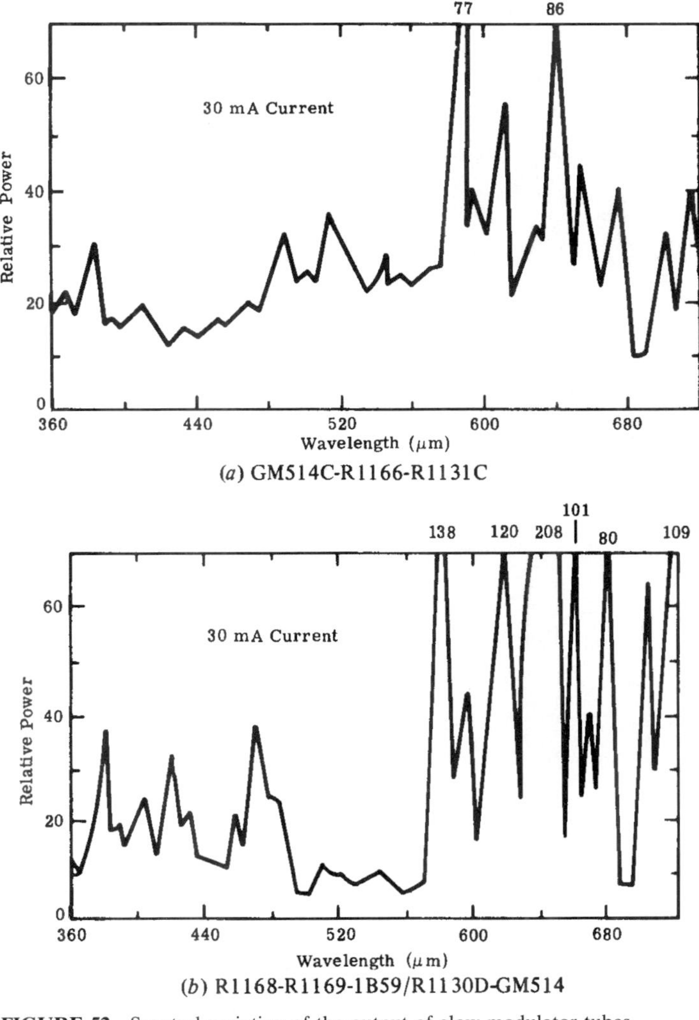

(*a*) GM514C-R1166-R1131C

(*b*) R1168-R1169-1B59/R1130D-GM514

FIGURE 52 Spectral variation of the output of glow modulator tubes.

TABLE 7 Glow-Modulator Specifications

No.[a]	*Maximum Operating Voltage*	*Current* (mA)		*Minimum Starting Voltage* (V)	*Crater Diameter* (in.)	*Approximate Light Center Length* (in.)	*Light Output* (cd)	*Brightness* (cd in.$^{-2}$)	*Rated Life* (h)	*Base Type*	*Bulb Type*	*Maximum Overall Length* (in.)	*Maximum Diameter* (in.)	*Color of Discharge*
		Average	*Peak*											
GM-514	160	5–25	55	240	0.056	1-3/4	0.1 at 25 mA	41 at 25 mA	100 at 15 mA	3-pin miniature[b]	T-4½	2-5/8	41/64	Blue-red
GM-514C	160	5–15	35	240	0.093	1-3/4	0.1 at 15 mA	15 at 15 mA	25 at 10 mA	3-pin miniature[b]	T-4½	2-5/8	41/64	White
1B59/ R-1130B	150	5–35	75	225	0.056	2	0.13 at 30 mA	43 at 30 mA	250 at 20 mA	Intermediate shell oct.[c]	T-9	3-1/16	1-9/32	Blue-red
R-1131C	150	3–25	55	225	0.093	2	0.2 at 25 mA	29 at 25 mA	150 at 15 mA	Intermediate shell oct.[c]	T-9	3-1/16	1-9/32	White
R-1166	150	3–25	55	225	0.093	2	0.2 at 25 mA	29 at 25 mA	150 at 15 mA	Intermediate shell oct.[c]	T-9	3-1/16	1-9/32	White
R-1168	150	5–15	30	225	0.015	2	0.023 at 15 mA	132 at 15 mA	150 at 15 mA	Intermediate shell oct.[c]	T-9	3-1/16	1-9/32	Blue-red
R-1169	150	5–25	45	225	0.025	2	0.036 at 15 mA	72 at 15 mA	250 at 15 mA	Intermediate shell oct.[c]	T-9	3-1/16	1-9/32	Blue-red

[a]Type R-1166 is opaque-coated with the exception of a circle 3/8 inch in diameter at end of lamp. All other types have clear-finish bulb.
[b]Pins 1 and 3 are anode; pin 2 cathode.
[c]Pin 7 anode; pin 3 cathode.

Other (High-Energy) Sources. Radiation at very high powers can be produced. Sources are synchrotrons, plasmatrons, arcs, sparks, exploding wires, shock tubes, and atomic and molecular beams, to name but a few. Among these, one can purchase in convenient, usable form precisely controlled spark-sources for yielding many joules of energy in a time interval of the order of microseconds. The number of vendors will be few, but check the directories.

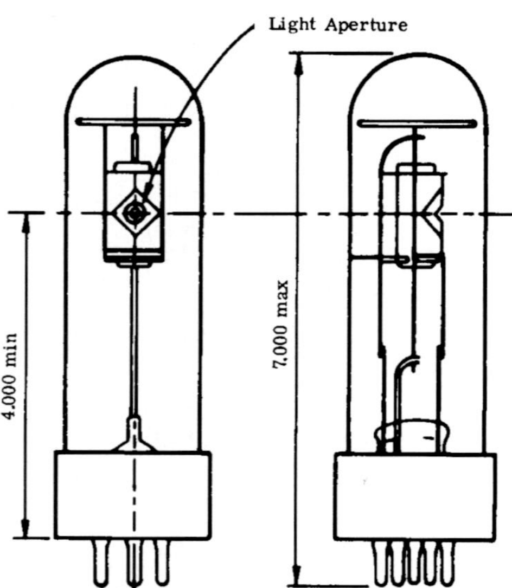

FIGURE 53 Hydrogen arc lamp.

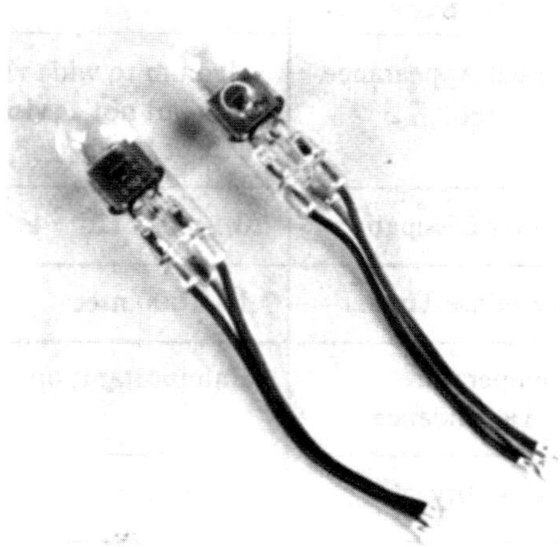

FIGURE 54 Two types of deuterium arc lamps.

Other Special Sources. An enormous number of special-purpose sources are obtainable from manufacturers and scientific instrument suppliers. One source that remains to be mentioned is the so-called miniature, sub-, and microminiature lamps. These are small, even tiny, incandescent bulbs of glass or quartz, containing tungsten filaments. They serve excellently in certain applications where small, intense radiators of visible and near-infrared radiation are needed. Second-source vendors advertise in the trade magazines.

10.5 REFERENCES

1. A. J. LaRocca, "Artificial Sources," in *The Infrared Handbook,* rev. ed., ONR, Washington, DC, 1985, Chap. 2.
2. A. G. Worthing, "Sources of Radiant Energy," in W. E. Forsythe (ed.), *Measurement of Radiant Energy,* McGraw-Hill, New York, 1937, chap. 2.
3. Andre Gouffé, "Corrections d'Ouverture des Corps-noir Artificiels Compte Tenu des Diffusions Multiples Internes," in *Revue d'Optique,* vol. 24, Masson, Paris, 1945, p. 1.
4. A. Leupin et al., "Investigation, Comparison, and Improvement of Technical Infrared Radiators," *Infrared Physics,* vol. 30, no. 3, 1990, pp. 199–258.
5. J. W. T. Walsh, *Photometry,* 3d ed., Dover, New York, 1965.
6. K. D. Mielenz, et al., "Spectroradiometric Determination of the Freezing Temperature of Gold," *Journal of Research of the National Institute of Standards and Technology,* vol. 95, no. 1, Jan.-Feb. 1990, pp. 49–67.
7. *Spectral Radiance Calibrations.* NBS Special Publication, Jan. 1987, p. 250–1.
8. *NIST Calibration Services Users Guide 1989,* NIST Special Publication, 1989, p. 250.
9. *Lasers and Optronics 1990 Buying Guide,* Elsevier Communications, Morris Plains, NJ, 1990.*
10. *Photonics Directory of Optical Industries,* Laurin Publishing Co., Pittsfield, MA, 1994.

---

* Please note that, although reference is made only to the few sources from which information was derived (in this and the succeeding reference), there are many other "so-called" trade journals which should be consulted. The best way to be aware of many of them is to access one of the on-line information services.

11. W. Y. Ramsey and J. C. Alishouse, "A Comparison of Infrared Sources," in *Infrared Physics,* vol. 8, Pergamon Publishing, Elmsford, N.Y., 1968, p. 143.
12. J. C. Morris, "Comments on the Measurements of Emittance of the Globar Radiation Source," in *Journal of the Optical Society of America,* vol. 51, Optical Society of America, Washington, DC, July 1961, p. 798.
13. A. H. Pfund, "The Electric Welsbach Lamp," in *Journal of the Optical Society of America,* vol. 26, Optical Society of America, Washington, DC, Dec. 1936, p. 439.
14. J. Strong, *Procedures in Experimental Physics,* Prentice-Hall, New York, 1938, p. 346.
15. F. E. Carlson and C. N. Clark, "Light Sources for Optical Devices," in R. Kingslake (ed.), *Applied Optics and Optical Engineering,* vol. 1, Academic Press, New York, 1965, p. 80.
16. Osram GMBH, *Lamps for Scientific Purposes,* Munchen, West Germany, 1966. (*See also* the later Osram brochure, *Light for cine projection, technology and science,* 1987.)
17. F. J. Studer and R. F. Van Beers, "Modification of Spectrum of Tungsten Filament Quartz-Iodine Lamps Due to Iodine Vapor," in *Journal of the Optical Society of America,* vol. 54, no. 7, Optical Society of America, Washington, DC, July 1964, p. 945.
18. L. R. Koller, *Ultraviolet Radiation,* 2d ed., John Wiley and Sons, New York, 1965.
19. M. R. Null and W. W. Lozier, "Carbon Arc as a Radiation Standard," in *Journal of the Optical Society of America,* vol. 52, no. 10, Optical Society of America, Washington, DC, Oct. 1962, pp. 1156–62.
20. E. B. Noel, "Radiation from High Pressure Mercury Arcs," *Illuminating Engineering,* vol. 36, 1941, p. 243.
21. General Electric, *Bulletin TP-109R,* Cleveland, 1975.
22. Illumination Industries, *Catalog No. 108–672–3M,* Sunnyvale, Calif., 1972.
23. M. W. P. Cann, *Light Sources for Remote Sensing Systems,* in NASA–CR–854, Aug. 1967.
24. Varian Associates, Palo Alto, Calif., 1969.
25. Fisher Scientific, *Special Catalog to Spectrophotometer Users,* Pittsburgh, 1972.
26. E. F. Worden, R. G. Gutmacher, and J. F. Conway, "Use of Electrodeless Discharge Lamps in the Analysis of Atomic Spectra," in *Applied Optics,* vol. 2, no. 7, Optical Society of America, Washington, DC, July 1963, pp. 707–713.
27. W. F. Meggers and F. O. Westfall, "Lamps and Wavelengths of Mercury 198," in *NBS Journal of Research,* vol. 44, National Bureau of Standards, Washington, DC, 1950, pp. 447–55.
28. W. F. Meggers, "Present Experimental Status of Rare Earth Spectra," in *Journal of the Optical Society of America,* vol. 50, Optical Society of America, Washington, DC, 1960, p. 405.
29. Ealing Corporation, *Ealing Catalog, Optical Components Section,* South Natick, Mass., 1976–77. (*See also Optical Services Supplement,* no. 1, 1969–70, p. 26; *See also Ealing Electro-optics Product Guide,* 1990.)
30. Central Scientific, *Cenco Scientific Education Catalog,* Physics-Light Section, Chicago, 1975, p. 602.
31. GTE Sylvania, *Special Purpose Lamps,* Lighting Products Group, Danvers, Mass., in TR-29R, May 1966.

CHAPTER 11
LASERS

William T. Silfvast
Center for Research and Education in Optics and Lasers (CREOL)
University of Central Florida
Orlando, Florida

11.1 GLOSSARY

A_2	radiative transition probability from level 2 to all other possible lower-lying levels
A_{21}	radiative transition probability from level 2 to level 1
a_L	scattering losses within a laser cavity for a single pass through the cavity
B_{12}	Einstein B coefficient associated with absorption
B_{21}	Einstein B coefficient associated with stimulated emission
E_1, E_2	energies of levels 1 and 2 above the ground state energy for that species
g_1, g_2	stability parameters for laser modes when describing the laser optical cavity
g_1, g_2	statistical weights of energy levels 1 and 2 that indicate the degeneracy of the levels
g_{21}	gain coefficient for amplification of radiation within a medium at a wavelength of λ_{21}
I_{sat}	saturation intensity of a beam in a medium; intensity at which exponential growth will cease to occur even though the medium has uniform gain (energy/time-area)
N_1, N_2	population densities (number of species per unit volume) in energy levels 1 and 2
r_c	radius of curvature of the expanding wavefront of a gaussian beam
R_1, R_2	reflectivities of mirrors 1 and 2 at the desired wavelength
T_1	lifetime of a level when dominated by collisional decay
T_2	average time between phase-interrupting collisions of a species in a specific excited state
t_{opt}	optimum mirror transmission for a laser of a given gain and loss

$w(z)$	beam waist at a distance z from the minimum beam waist for a gaussian beam
w_o	minimum beam waist radius for a gaussian mode
α_{12}	absorption coefficient for absorption of radiation within a medium at wavelength λ_{21}
γ_{21}	angular frequency bandwidth of an emission or absorption line
Δt_p	pulse duration of a mode-locked laser pulse
$\Delta \nu$	frequency bandwidth over which emission, absorption, or amplification can occur
$\Delta \nu_D$	frequency bandwidth (FWHM) when the dominant broadening process is Doppler or motional broadening
η	index of refraction of the laser medium at the desired wavelength
λ_{21}	wavelength of a radiative transition occurring between energy levels 2 and 1
ν_{21}	frequency of a radiative transition occurring between energy levels 2 and 1
σ_{21}	stimulated emission cross section (area)
σ_{21}^D	stimulated emission cross section at line center when Doppler broadening dominates (area)
σ_{21}^H	stimulated emission cross section at line center when homogeneous broadening dominates (area)
τ_2	lifetime of energy level 2
τ_{21}	lifetime of energy level 2 if it can only decay to level 1

11.2 INTRODUCTION

A laser is a device that amplifies light and produces a highly directional, high-intensity beam that typically has a very pure frequency or wavelength. It comes in sizes ranging from approximately one-tenth the diameter of a human hair to the size of a very large building, in powers ranging from 10^{-9} to 10^{20} watts, and in wavelengths ranging from the microwave to the soft-x-ray spectral regions with corresponding frequencies from 10^{11} to 10^{17} Hz. Lasers have pulse energies as high as 10^4 joules and pulse durations as short as 6×10^{-15} seconds. They can easily drill holes in the most durable of materials and can weld detached retinas within the human eye.

Lasers are a key component of some of our most modern communication systems and are the "phonograph needle" of compact disc players. They are used for heat treatment of high-strength materials, such as the pistons of automobile engines, and provide a special surgical knife for many types of medical procedures. They act as target designators for military weapons and are used in the checkout scanners we see everyday at the supermarket.

The word *laser* is an acronym for *L*ight *A*mplification by *S*timulated *E*mission of *R*adiation. The laser makes use of processes that increase or amplify light signals after those signals have been generated by other means. These processes include (1) stimulated emission, a natural effect that arises out of considerations relating to thermodynamic

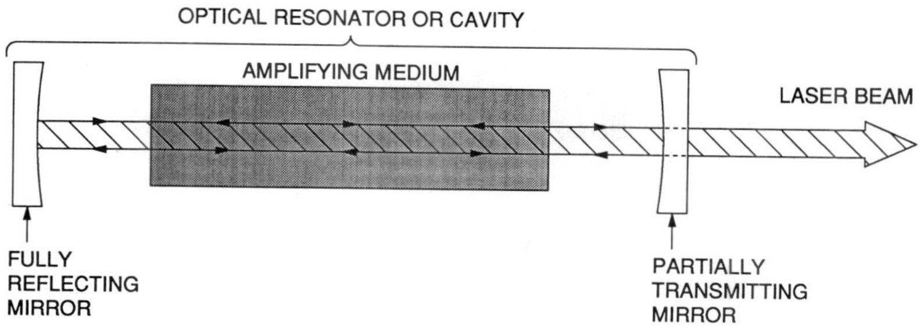

FIGURE 1 Simplified diagram of a laser, including the amplifying medium and the optical resonator.

equilibrium, and (2) optical feedback (present in most lasers) that is usually provided by mirrors. Thus, in its simplest form, a laser consists of a gain or amplifying medium (where stimulated emission occurs) and a set of mirrors to feed the light back into the amplifier for continued growth of the developing beam (Fig. 1).

The entire spectrum of electromagnetic radiation is shown in Fig. 2, along with the region covered by currently existing lasers. Such lasers span the wavelength range from the far infrared part of the spectrum ($\lambda = 1000\ \mu\text{m}$) to the soft-x-ray region ($\lambda = 3\ \text{nm}$), thereby covering a range of almost six orders of magnitude! There are several types of units that are used to define laser wavelengths. These range from micrometers (μm) in the infrared to nanometers (nm) and angstroms (Å) in the visible, ultraviolet (UV), vacuum ultraviolet (VUV), extreme ultraviolet (EUV or XUV), and soft-x-ray (SXR) spectral regions.

This chapter provides a brief overview of how a laser operates. It considers a laser as having two primary components: (1) a region where light amplification occurs which is referred to as a *gain medium* or an *amplifier,* and (2) a *cavity,* which generally consists of two mirrors placed at either end of the amplifier.

Properties of the amplifier include the concept of discrete excited energy levels and their associated finite lifetimes. The broadening of these energy levels will be associated with the emission linewidth which is related to decay of the population in these levels.

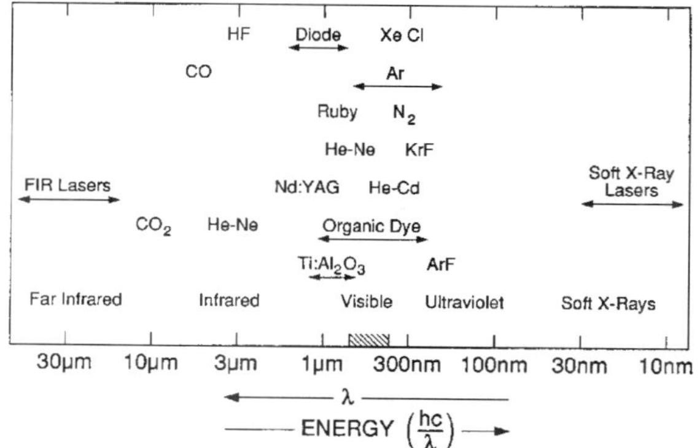

FIGURE 2 The portion of the electromagnetic spectrum that involves lasers, along with the general wavelengths of operation of most of the common lasers.

Stimulated emission will be described, and the formulas for calculating the amount of gain that can occur via stimulated emission will be given in terms of the radiative properties of the medium. The concept of the saturation intensity will be introduced and related to the amount of gain that is necessary for laser output. The addition of mirrors at the ends of the amplifier will be used to increase the gain length and to reduce the divergence of the amplified beam. The threshold conditions for laser output will be described in terms of the amplifier properties and the mirror reflectivities. This section will conclude with a review of excitation or pumping processes that are used to produce the necessary population density in the upper laser level.

Cavity properties will begin with a discussion of both longitudinal and transverse cavity modes which provide the laser beam with a gaussian-shaped transverse profile. The properties of those gaussian beams will be reviewed. The types of optical cavities that allow stable operation of laser modes will then be described. A number of special types of laser cavity arrangements and techniques will be reviewed, including unstable resonators, Q-switching, mode-locking, and ring lasers. A brief review will then be given of the various common types of gaseous, liquid, and solid-state lasers.

Additional information related to spectral lineshape and the mechanisms of spectral broadening can be found in Chap. 8 (Vol. I), "Optical Spectroscopy and Spectroscopic Lineshapes." Other related material can be found in Chap. 9 (Vol. I), "Fundamental Optical Properties of Solids," and Chap. 23 (Vol. II) "Holography and Holographic Instruments." As lasers are widely used in many of the devices and techniques discussed in other chapters in this *Handbook,* the reader is directed to those topics for information on specific lasers.

11.3 LASER PROPERTIES ASSOCIATED WITH THE LASER GAIN MEDIUM

Energy Levels and Radiation[1,2]

Nearly all lasers involve electronic charge distributions of atoms, molecules, organic dye solutions, or solids that make transitions from one energy state or level E_2 to another lower-lying level E_1. The loss of energy resulting from this transition is given off in the form of electromagnetic radiation. The relationship between the energy difference between the levels, $E_2 - E_1$ or ΔE_{21}, and the frequency ν_{21} of radiation occurring as a result of the transition, is determined by the Einstein relationship $E_{21} = h\nu_{21}$ where h is Planck's constant. It was first shown by Bohr in 1913 that the discrete set of emission wavelengths from a hydrogen discharge could be explained by the occurrence of discrete energy levels in the hydrogen atom that have a fixed relationship. This discrete arrangement of energy levels was later shown to occur in other atoms, in molecules, and also in liquids and solids. In atoms these energy levels are very precisely defined and narrow in width ($\approx 10^9$ Hz) and can be accurately calculated with sophisticated atomic physics codes. In molecules and high-density materials the locations of the levels are more difficult to calculate and they tend to be much broader in width, the largest widths occurring in liquids and solids (up to 5×10^{13} Hz).

The lowest energy level of a species is referred to as the *ground state* and is usually the most stable state of the species. There are some exceptions to this, for example, ground states of ionized species or unstable ground states of some molecular species such as excimer molecules. Energy levels above the ground state are inherently unstable and have lifetimes that are precisely determined by the arrangement of the atoms and electrons associated with any particular level as well as to the particular species or material. Thus, when an excited state is produced by applying energy to the system, that state will eventually decay by emitting radiation over a time period ranging from 10^{-15} seconds or less to times as long as seconds or more, depending upon the particular state or level involved. For *strongly allowed* transitions that involve the electron charge cloud changing from an atomic energy level of energy E_2 to a lower-lying level of energy E_1, the radiative decay time τ_{21} can be approximated by $\tau_{21} \approx 10^4 \lambda_{21}^2$ where τ_{21} is in seconds and λ_{21} is the wavelength of the emitted radiation in meters. λ_{21} is related to ν_{21} by the relationship $\lambda_{21}\nu_{21} = c/\eta$ where c is the velocity of light (3×10^8 m/sec) and η is the index of refraction of the material. For most gases, η is near unity and for solids and liquids it ranges between 1 and 10, with most values ranging from $\approx$1.3–2.0.

Using the expression suggested above for the approximate value of the lifetime of an excited energy level, one obtains a decay time of several nsec for green light ($\lambda_{21} = 5 \times 10^{-7}$ m). This represents a minimum radiative lifetime since most excited energy levels have a weaker radiative decay probability than mentioned above and would therefore have radiative lifetimes one or two orders of magnitude longer. Other laser materials such as molecules, organic dye solutions, and semiconductor lasers have similar radiative lifetimes. The one exception is the class of dielectric solid-state laser materials (both crystalline and glass) such as ruby and Nd:YAG, in which the lifetimes are of the order of 1–300 μsec. This much longer radiative lifetime in solid-state laser materials is due to the nature of the particular state of the laser species and to the crystal matrix in which it is contained. This is a very desirable property for a laser medium since it allows excitation and energy storage within the laser medium over a relatively long period of time.

Emission Linewidth and Line Broadening of Radiating Species[1,3]

Assume that population in energy level 2 decays to energy level 1 with an exponential decay time of τ_{21} and emits radiation at frequency ν_{21} during that decay. It can be shown by Fourier analysis that the exponential decay of that radiation requires the frequency width of the emission to be of the order of $\Delta\nu \cong 1/2\pi\tau_{21}$. This suggests that the energy width ΔE_2 of level 2 is of the order of $\Delta E_2 = h/2\pi\tau_{21}$. If the energy level 2 can decay to more levels than level 1, with a corresponding decay time of τ_2, then its energy is broadened by an amount $\Delta E_2 = h/2\pi\tau_2$. If the decay is due primarily to radiation at a rate A_{2i} to one or more individual lower-lying levels i, then $1/\tau_2 = A_2 = \sum_i A_{2i}$. A_2 represents the total radiative decay rate of level 2, whereas A_{2i} is the specific radiative decay rate from level 2 to a lower-lying level i.

If population in level 2 decays radiatively at a radiative rate A_2 and population in level 1 decays radiatively at a rate A_1, then the emission linewidth of radiation from level 2 to 1 is given by

$$\Delta\nu_{21} = \frac{\sum_i A_{2i} + \sum_j A_{1j}}{2\pi} \tag{1}$$

which is referred to as the *natural linewidth* of the transition and represents the sum of the widths of levels 2 and 1 in frequency units. If, in the above example, level 1 is a ground state with infinite lifetime or a long-lived metastable level, then the natural linewidth of the emission from level 2 to level 1 would be represented by

$$\Delta\nu_{21} = \frac{\sum_i A_{2i}}{2\pi} \tag{2}$$

since the ground state would have an infinite lifetime and would therefore not contribute to the broadening. This type of linewidth or broadening is known as *natural broadening* since it results specifically from the radiative decay of a species. Thus the natural linewidth associated with a specific transition between two levels has an inherent value determined only by the factors associated with specific atomic and electronic characteristics of those levels.

The emission-line broadening or natural broadening described above is the minimum line broadening that can occur for a specific radiative transition. There are a number of mechanisms that can increase the emission linewidth. These include collisional broadening, phase-interruption broadening, Doppler broadening, and isotope broadening. The first two of these, along with natural broadening, are all referred to as *homogeneous broadening.* Homogeneous broadening is a type of emission broadening in which all of the atoms radiating from the specific level under consideration participate in the same way. In other words, all of the atoms have the identical opportunity to radiate with equal probability.

The type of broadening associated with either Doppler or isotope broadening is referred to as *inhomogeneous broadening.* For this type of broadening, only certain atoms radiating from that level that have a specific property such as a specific velocity, or are of a specific isotope, participate in radiation at a certain frequency within the emission bandwidth.

Collisional broadening is a type of broadening that is produced when surrounding atoms, molecules, solvents (in the case of dye lasers), or crystal structures interact with the radiating level and cause the population to decay before it has a chance to decay by its normal radiative processes. The emission broadening is then associated with the faster decay time T_1, or $\Delta\nu = 1/2\pi T_1$.

Phase-interruption broadening or *phonon broadening* is a type of broadening that does not increase the decay rate of the level, but it does interrupt the phase of the rotating electron cloud on average over a time interval T_2 which is much shorter than the radiative decay time τ_2 which includes all possible radiative decay channels from level 2. The result of this phase interruption is to increase the emission linewidth beyond that of both natural broadening and T_1 broadening (if it exists) to an amount $\Delta\nu = 1/2\pi T_2$.

Doppler broadening is a type of inhomogeneous broadening in which the Doppler effect shifts the frequencies of atoms moving toward the observer to a higher value and the frequencies of atoms moving away from the observer to a lower value. This effect occurs only in gases since they are the only species that are moving fast enough to produce such broadening. Doppler broadening is the dominant broadening process in most visible gas lasers. The expression for the Doppler linewidth (FWHM) is given by

$$\Delta\nu_D = 7.16 \cdot 10^{-7} \nu_o \sqrt{\frac{T}{M}} \tag{3}$$

in which ν_o is the center frequency associated with atoms that are not moving either toward or away from the observer, T is the gas temperature in kelvin and M is the atomic or molecular weight (number of nucleons/atom or molecule) of the gas atoms or molecules.

Isotope broadening also occurs in some gas lasers. It becomes the dominant broadening process if the specific gas consists of several isotopes of the species and if the isotope shifts for the specific radiative transition are broader than the Doppler width of the transition. The helium-cadmium laser is dominated more by this effect than any other laser since the naturally occurring cadmium isotopic mixture contains eight different isotopes and the isotope shift between adjacent isotopes (adjacent neutron numbers) is approximately equal to the Doppler width of the individual radiating isotopes. This broadening effect can be eliminated by the use of isotopically pure individual isotopes, but the cost for such isotopes is often prohibitive.

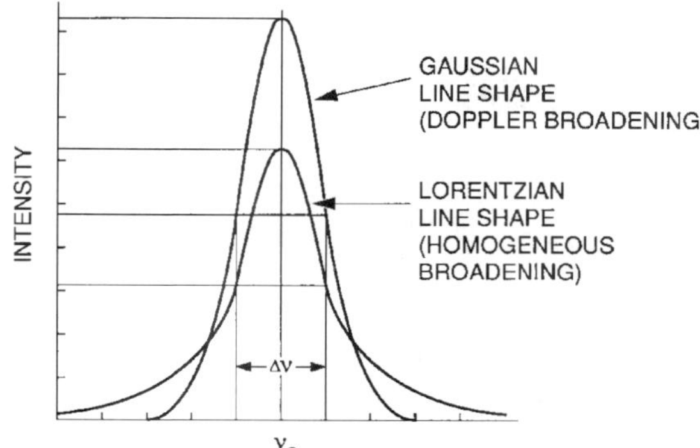

FIGURE 3 Lineshape functions for both homogeneous broadening, with a lorentzian shape, and Doppler broadening (inhomogeneous), with a gaussian shape. Both lines are arranged with equal linewidths (FWHM) and equal total intensities.

All homogeneous broadening processes have a frequency distribution that is described by a lorentzian mathematical function

$$I_{21}(\nu) = I_o \frac{\gamma_{21}/4\pi^2}{(\nu - \nu_0)^2 + (\gamma_{21}/4\pi)^2} \tag{4}$$

in which γ_{21} represents the decay rate of level 2, I_o is the total emission intensity of the transition over the entire linewidth, and ν_o is the center frequency of the emission line. In Eq. (4), γ_{21} is determined by the relationship $\gamma_{21} = 2\pi\Delta\nu_{21}$. For natural broadening, $\Delta\nu_{21}$ is given by either Eq. (1) or Eq. (2), whichever is applicable. For T_1 dominated broadening, $\Delta\nu_{21} = 1/2\pi T_1$, and for T_2 dominated broadening, $\gamma_{21} = 1/2\pi T_2$.

The frequency distribution for Doppler broadening is described by a gaussian function

$$I(\nu) = \frac{2(\ln 2)^{1/2}}{\pi^{1/2}\,\Delta\nu_D} I_o \exp - \left[\frac{4 \ln 2(\nu - \nu_o)^2}{\Delta\nu_D^2}\right] \tag{5}$$

Both of these lineshape functions are indicated in Fig. 3. In this figure, both the total emission intensity integrated over all frequencies and the emission linewidth (full width at half maximum or FWHM) for both functions are identical.

Isotope broadening involves the superposition of a series of either lorentzian shapes or gaussian shapes for each isotope of the species, separated by the frequencies associated with the isotope shifts of that particular transition.

Table 1 gives examples of the dominant broadening process and the value of the broadening for most of the common commercial lasers.

Stimulated Radiative Processes—Absorption and Emission[1,2]

Two types of stimulated radiative processes, absorption and stimulated emission, occur between energy levels 1 and 2 of a gain medium when light of frequency ν_{21} corresponding to an energy difference $\Delta E_{21} = (E_2 - E_1) = h\nu_{21}$ passes through the medium. These processes are proportional to the light intensity I as indicated in Fig. 4 for a two-level system as well as to the stimulated absorption and emission coefficients B_{12} and B_{21},

TABLE 1 Amplifier Parameters for a Wide Range of Lasers

Type of Laser	λ_{21}(nm)	τ_2(sec)	$\Delta\nu_{21}$(Hz)	$\Delta\lambda_{21}$(nm)	σ_{21}(m^2)	ΔN_{21}(m^{-3})	L(m)	g_{21}(m^{-1})
Helium-Neon	632.8	$3x10^{-7}$	$2x10^9$	$2.7x10^{-3}$	$3x10^{-17}$	$5x10^{15}$	0.2	0.15
Argon	488.0	$1x10^{-8}$	$2x10^9$	$1.6x10^{-3}$	$5x10^{-16}$	$1x10^{15}$	0.2-1.0	0.5
He-Cadmium	441.6	$7x10^{-7}$	$2x10^9$	$1.3x10^{-3}$	$8x10^{-18}$	$4x10^{16}$	0.2-1.0	0.3
Copper Vapor	510.5	$5x10^{-7}$	$2x10^9$	$1.3x10^{-3}$	$8x10^{-18}$	$6x10^{17}$	1.0-2.0	5
CO_2	10,600	4	$6x10^7$	$2.2x10^{-2}$	$1.6x10^{-20}$	$5x10^{19}$	0.2-2.0	0.8
Excimer	248.0	$9x10^{-9}$	$1x10^{13}$	2	$2.6x10^{-20}$	$1x10^{20}$	0.5-1.0	2.6
Dye (Rh6G)	577.0	$5x10^{-9}$	$5x10^{13}$	60	$1.2x10^{-20}$	$2x10^{22}$	0.01	240
Semiconductor	800	$1x10^{-9}$	$1x10^{13}$	20	$1x10^{-19}$	10^{24}	0.00025	100,000
Nd:YAG	1064.1	$2.3x10^{-4}$	$1.2x10^{11}$	0.4	$6.5x10^{-23}$	$1.6x10^{23}$	0.1	10
Nd:Glass	1054	$3.0x10^{-4}$	$7.5x10^{12}$	26	$4.0x10^{-24}$	$8x10^{23}$	0.1	3
Cr:LiSAF	840	$6.7x10^{-5}$	$9.0x10^{13}$	250	$5.0x10^{-24}$	$2x10^{24}$	0.1	10
Ti:Al_2O_3	760	$3.2x10^{-6}$	$1.5x10^{14}$	400	$4.1x10^{-23}$	$5x10^{23}$	0.1	20

respectively. These coefficients are related to the frequency ν_{21} and the spontaneous emission probability A_{21} associated with the two levels. A_{21} has units of (1/sec).

Absorption results in the loss of light of intensity I when the light interacts with the medium. The energy is transferred from the beam to the medium by raising population from level 1 to the higher-energy level 2. In this situation, the species within the medium can either reradiate the energy and return to its initial level 1, it can reradiate a different energy and decay to a different level, or it can lose the energy to the surrounding medium via collisions, which results in the heating of the medium, and return to the lower level. The absorption probability is proportional to the intensity I which has units of energy/sec-m^2 times B_{12}, which is the absorption probability coefficient for that transition. B_{12} is one of the Einstein B coefficients and has the units of m^3/energy-sec^2.

Stimulated emission results in the increase in the light intensity I when light of the appropriate frequency ν_{21} interacts with population occupying level 2 of the gain medium. The energy is given up by the species to the radiation field. In the case of stimulated

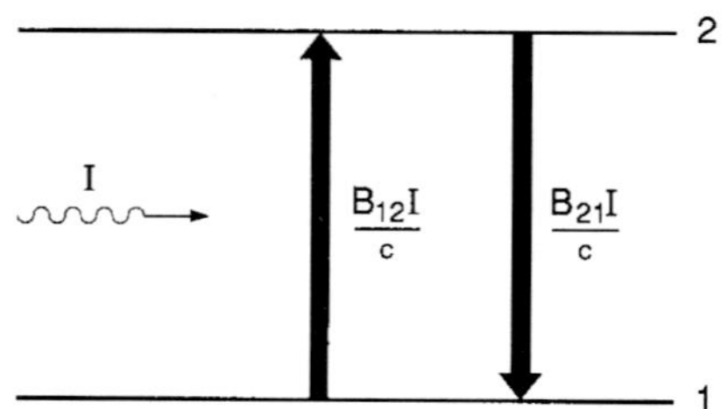

FIGURE 4 Stimulated emission and absorption processes that can occur between two energy levels, 1 and 2, and can significantly alter the population densities of the levels compared to when a beam of intensity I is not present.

emission, the emitted photons or bundles of light have exactly the same frequency ν_{21} and direction as the incident photons of intensity I that produce the stimulation. B_{21} is the associated stimulated emission coefficient and has the same units as B_{12}. It is known as the other Einstein B coefficient.

Einstein showed the relationship between B_{12}, B_{21}, and A_{21} as

$$g_2 B_{21} = g_1 B_{12} \tag{6}$$

and

$$A_{21} = \frac{8\pi h \nu_{21}^3}{c^3} B_{21} \tag{7}$$

where g_2 and g_1 are the statistical weights of levels 2 and 1 and h is Planck's constant. Since $A_{21} = 1/\tau_{21}$ for the case where radiative decay dominates and where there is only one decay path from level 2, B_{21} can be determined from lifetime measurements or from absorption measurements on the transition at frequency ν_{21}.

Population Inversions[1,2]

The two processes of absorption and stimulated emission are the principal interactions involved in a laser amplifier. Assume that a collection of atoms of a particular species is energized to populate two excited states 1 and 2 with population densities N_1 and N_2 (number of species/m^3) and state 2 is at a higher energy than state 1 by an amount ΔE_{21} as described in the previous section. If a photon beam of energy ΔE_{21}, with an intensity I_o and a corresponding wavelength $\lambda_{21} = c/\nu_{21} = hc/\Delta E_{21}$, passes through this collection of atoms, then the intensity I after the beam emerges from the medium can be expressed as

$$I = I_o e^{\sigma_{21}(N_2 - (g_2/g_1)N_1)L} = I_o e^{\sigma_{21}\Delta N_{21} L} \tag{8}$$

where σ_{21} is referred to as the *stimulated emission cross section* with dimensions of m^2, L is the thickness of the medium (in meters) through which the beam passes, and $N_2 - (g_2/g_1)N_1 = \Delta N_{21}$ is known as the *population inversion density*. The exponents in Eq. (8) are dimensionless quantities than can be either greater or less than unity, depending upon whether N_2 is greater than or less than $(g_2/g_1)N_1$.

The general form of the stimulated emission cross section per unit frequency is given as

$$\sigma_{21} = \frac{\lambda_{21}^2 A_{21}}{8\pi\,\Delta\nu} \tag{9}$$

in which $\Delta\nu$ represents the linewidth over which the stimulated emission or absorption occurs.

For the case of homogeneous broadening, at the *center* of the emission line, σ_{21} is expressed as

$$\sigma_{21}^H = \frac{\lambda_{21}^2 A_{21}}{4\pi^2\,\Delta\nu_{21}^H} \tag{10}$$

where $\Delta\nu_{21}^H$ is the homogeneous emission linewidth (FWHM) which was described earlier for several different situations.

For the case of Doppler broadening, σ_{21}^{D} can be expressed as

$$\sigma_{21}^{D} = \sqrt{\frac{\ln 2}{16\pi^3}} \frac{\lambda_{21}^2 A_{21}}{\Delta\nu_{21}^{D}} \tag{11}$$

at the *center* of the emission line and $\Delta\nu_{21}^{D}$ is the Doppler emission linewidth expressed earlier in Eq. (3).

For all types of matter, the population density ratio of levels 1 and 2 would normally be such that $N_2 \ll N_1$. This can be shown by the Boltzmann relationship for the population ratio in thermal equilibrium which provides the ratio of N_2/N_1 to be

$$\frac{N_2}{N_1} = \frac{g_2}{g_1} e^{-\Delta E_{21}/kT} \tag{12}$$

in which T is the temperature and k is Boltzmann's constant. Thus, for energy levels separated by energies corresponding to visible transitions, in a medium at or near room temperature, the ratio of $N_2/N_1 \cong e^{-100} = 10^{-44}$. For most situations in everyday life we can ignore the population N_2 in the upper level and rewrite Eq. (8) as

$$I = I_o e^{-\sigma_{12} N_1 L} = I_o e^{-\alpha_{12} L} \tag{13}$$

where $\sigma_{21} = (g_2/g_1)\sigma_{12}$ and $\alpha_{12} = \sigma_{12} N_1$. Equation (13) is known as Beer's law, which is used to describe the absorption of light within a medium. α_{12} is referred to as the absorption coefficient with units of m^{-1}.

In laser amplifiers we cannot ignore the population in level 2. In fact, the condition for amplification and laser action is

$$N_2 - (g_2/g_1)N_1 > 0 \qquad \text{or} \qquad g_1 N_2 / g_2 N_1 > 1 \tag{14}$$

since, if Eq. (14) is satisfied, the exponent of Eq. (8) will be greater than 1 and I will emerge from the medium with a greater value than I_o, or amplification will occur. The condition of Eq. (14) is a necessary condition for laser amplification and is known as a population inversion since N_2 is greater than $(g_2/g_1)N_1$. In most cases, (g_2/g_1) is either unity or close to unity.

Considering that the ratio of $g_1 N_2 / g_2 N_1$ [Eq. (14)] could be greater than unity does not follow from normal thermodynamic equilibrium considerations [Eq. (12)] since it would represent a population ratio that could never exist in thermal equilibrium. When Eq. (14) is satisfied and amplification of the beam occurs, the medium is said to have *gain* or *amplification*. The factor $\sigma_{21}(N_2 - (g_2/g_1)N_1)$ or $\sigma_{21}\,\Delta N_{21}$ is often referred to as the *gain coefficient* g_{21} and is given in units of m^{-1} such that $g_{21} = \sigma_{21}\,\Delta N_{21}$. Typical values of σ_{21}, ΔN_{21}, and g_{21} are given in Table 1 for a variety of lasers. The term g_{21} is also referred to as the *small-signal gain coefficient* since it is the gain coefficient determined when the laser beam intensity within the laser gain medium is small enough that stimulated emission does not significantly alter the populations in the laser levels.

Gain Saturation[2]

It was stated in the previous section that Eq. (14) is a necessary condition for making a laser but it is not a sufficient condition. For example, a medium might satisfy Eq. (14) by having a gain of $e^{g_{21}L} \approx 10^{-10}$, but this would not be sufficient to allow any reasonable beam to develop. Lasers generally start by having a pumping process that produces enough population in level 2 to create a population inversion with respect to level 1. As the population decays from level 2, radiation occurs spontaneously on the transition from level 2 to level 1 equally in all directions within the gain medium. In most of the directions very little gain or enhancement of the spontaneous emission occurs, since the length is not sufficient to cause significant growth according to Eq. (8). It is only in the elongated direction of the amplifier, with a much greater length, that significant gain exists and, consequently, the spontaneous emission is significantly enhanced. The requirement for a laser beam to develop in the elongated direction is that the exponent of Eq. (8) be large enough for the beam to grow to the point where it begins to significantly reduce the population in level 2 by stimulated emission. The beam will eventually grow, according to Eq. (8), to an intensity such that the stimulated emission rate is equal to the spontaneous emission rate. At that point the beam is said to reach its saturation intensity I_{sat}, which is given by

$$I_{sat} = \frac{h\nu_{21}}{\sigma_{21}^{H}\tau_{21}} \tag{15}$$

The saturation intensity is that value at which the beam can no longer grow exponentially according to Eq. (8) because there are no longer enough atoms in level 2 to provide the additional gain. When the beam grows above I_{sat} it begins to extract significant energy since at this point the stimulated emission rate exceeds the spontaneous emission rate for that transition. The beam essentially takes energy that would normally be radiated in all directions spontaneously and redirects it via stimulated emission, thereby increasing the beam intensity.

Threshold Conditions with No Mirrors

I_{sat} can be achieved by having any combination of values of the three parameters σ_{21}^{H}, ΔN_{21}, and L large enough that their product provides sufficient gain. It turns out that the requirement to reach I_{sat} can be given by making the exponent of Eq. (8) have the following range of values:

$$\sigma_{21}^{H}\Delta N_{21}L \approx 10-20 \qquad \text{or} \qquad g_{21}^{H}L \approx 10-20 \tag{16}$$

where the specific value between 10 and 20 is determined by the geometry of the laser cavity. Equation (16) suggests that the beam grows to a value of $I/I_o = e^{10-20} = 2 \times 10^4$ -5×10^8, which is a very large amplification.

Threshold Conditions with Mirrors

One could conceivably make L sufficiently long to always satisfy Eq. (16), but this is not practical. Some lasers can reach the saturation intensity over a length L of a few centimeters, but most require much longer lengths. Since one cannot readily extend the lasing medium to be long enough to achieve I_{sat}, the same result is obtained by putting mirrors around the gain medium. This effectively increases the path length by having the beam pass many times through the amplifier.

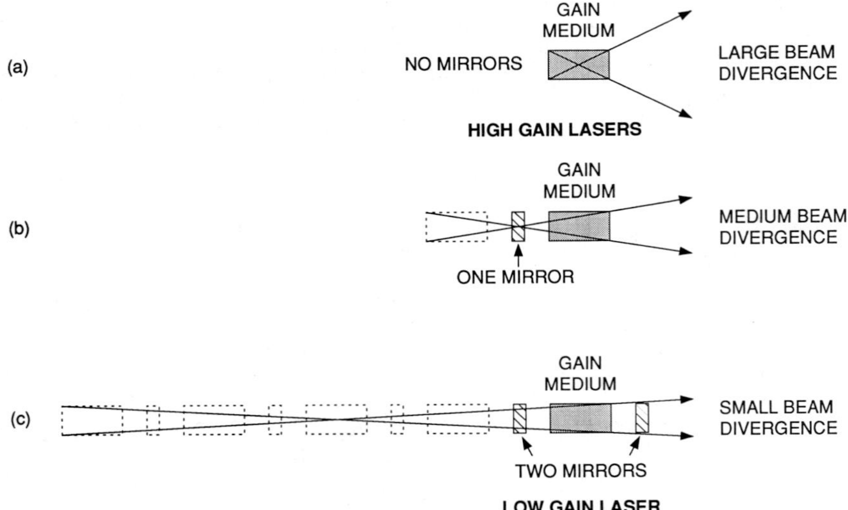

FIGURE 5 Laser-beam divergence for amplifier configurations having high gain and (*a*) no mirrors or (*b*) one mirror, and high or low gain and (*c*) two mirrors.

A simple understanding of how the mirrors affect the beam is shown in Fig. 5. The diagram effectively shows how multiple passes through the amplifier can be considered for the situation where flat mirrors are used at the ends of the amplifier. The dashed outlines that are the same size as the gain media represent the images of those gain media produced by their reflections in the mirrors. It can be seen that as the beam makes multiple passes, its divergence narrows up significantly in addition to the large increase in intensity that occurs due to amplification. For high-gain lasers, such as excimer or organic dye lasers, the beam only need pass through the amplifier a few times to reach saturation. For low-gain lasers, such as the helium-neon laser, it might take 500 passes through the amplifier in order to reach I_{sat}.

The laser beam that emerges from the laser is usually coupled out of the amplifier by having a partially transmitting mirror at one end of the amplifier which typically reflects most of the beam back into the medium for more growth. To ensure that the beam develops, the transmission of the output coupling mirror (as it is usually referred to) must be lower than the gain incurred by the beam during a round-trip pass through the amplifier. If the transmission is higher than the round-trip gain, the beam undergoes no net amplification. It simply never develops. Thus a relationship that describes the threshold for laser oscillation balances the laser gain with the cavity losses. In the most simplified form, those losses are due to the mirrors having a reflectivity less than unity. Thus, for a round-trip pass through the laser cavity, the threshold for inversion can be expressed as

$$R_1 R_2 e^{\sigma_{21} \Delta N_{21} 2L} > 1 \qquad \text{or} \qquad R_1 R_2 e^{g_{21} 2L} > 1 \tag{17a}$$

which is a similar requirement to that of Eq. (16) since it defines the minimum gain requirements for a laser.

A more general version of Eq. (17*a*) can be expressed as

$$R_1 R_2 (1 - a_L)^2 e^{(\sigma_{21} \Delta N_{21} - \alpha) 2L} = 1 \qquad \text{or} \qquad R_1 R_2 (1 - a_L)^2 e^{(g_{21} - \alpha) 2L} = 1 \tag{17b}$$

in which we have included a distributed absorption α throughout the length of the gain medium at the laser wavelength, as well as the total scattering losses a_L per pass through the cavity (excluding the gain medium and the mirror surfaces). The absorption loss α is essentially a separate absorbing transition within the gain medium that could be a separate molecule as in an excimer laser, absorption from either the ground state or from the triplet state in a dye laser, or absorption from the ground state in the broadband tunable solid-state lasers and in semiconductor lasers or from the upper laser state in most solid-state lasers. The scattering losses a_L, per pass, would include scattering at the windows of the gain medium, such as Brewster angle windows, or scattering losses from other elements that are inserted within the cavity. These are typically of the order of one or two percent or less.

Laser Operation Above Threshold

Significant power output is achieved by operating the laser at a gain greater than the threshold value defined above in Eqs. (16) and (17). For such a situation, the higher gain that would normally be produced by increased pumping is reduced to the threshold value by stimulated emission. The additional energy obtained from the reduced population inversion is transferred to the laser beam in the form of increased laser power. If the laser has low gain, as most cw (continuously operating) gas lasers do, the gain and also the power output tend to stabilize rather readily.

For solid-state lasers, which tend to have higher gain and also a long upper-laser-level lifetime, a phenomenon known as *relaxation oscillations*[7] occurs in the laser output. For pulsed (non-Q-switched) lasers in which the gain lasts for many microseconds, these oscillations occur in the form of a regularly repeated spiked laser output superimposed on a lower steady-state value. For cw lasers it takes the form of a sinusoidal oscillation of the output. The phenomenon is caused by an oscillation of the gain due to the interchange of pumped energy between the upper laser level and the laser field in the cavity. This effect can be controlled by using an active feedback mechanism, in the form of an intensity-dependent loss, in the laser cavity.

Laser mirrors not only provide the additional length required for the laser beam to reach I_{sat}, but they also provide very important resonant cavity effects that will be discussed in a later section. Using mirrors at the ends of the laser gain medium (or amplifier) is referred to as having the gain medium located within an optical cavity.

How Population Inversions Are Achieved[4]

It was mentioned earlier that population inversions are not easily achieved in normal situations. All types of matter tend to be driven towards thermal equilibrium. From an energy-level standpoint, to be in thermal equilibrium implies that the ratio of the populations of two excited states of a particular material, whether it be a gaseous, liquid, or solid material, is described by Eq. (12). For any finite value of the temperature this leads to a value of $N_2/(g_2/g_1)N_1$ that is always less than unity and therefore Eq. (14) can never be satisfied under conditions of thermal equilibrium. Population inversions are therefore produced in either one of two ways: (1) selective excitation (pumping) of the upper-laser-level 2, or (2) more rapid decay of the population of the lower-laser-level 1 than of the upper-laser-level 2, even if they are both populated by the same pumping process.

The first requirement mentioned above was met in producing the very first laser, the

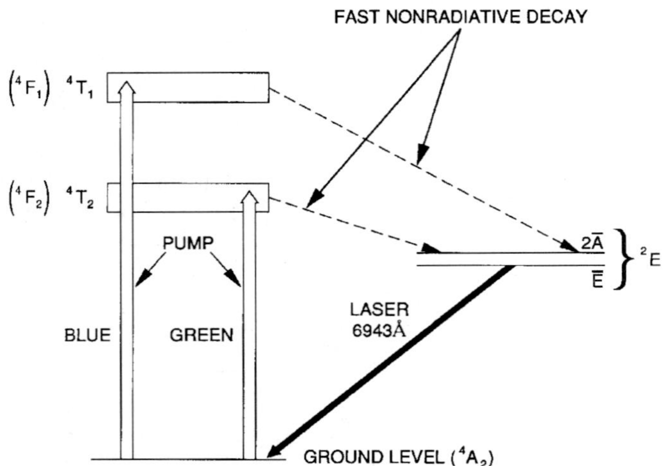

FIGURE 6 Energy-level diagram for a ruby laser showing the pump wavelength bands and the laser transition. The symbols $^4T_1$ and $^4T_2$ are shown as the appropriate designations for the pumping levels in ruby along with the more traditional designations in parenthesis.

ruby laser.[5] In this laser the flash lamp selectively pumped chromium atoms to the upper laser level (through an intermediate level) until the ground state (lower laser level) was depleted enough to produce the inversion (Fig. 6). Another laser that uses this selective pumping process is the copper vapor laser[6] (CVL). In this case, electrons in a gaseous discharge containing the copper vapor have a much preferred probability of pumping the upper laser level than the lower laser level (Fig. 7). Both of these lasers involve essentially three levels.

The second type of excitation is used for most solid-state lasers, such as the Nd^{3+} doped yttrium aluminum garnet laser[7] (commonly referred to as the Nd:YAG laser), for organic dye lasers,[8] and many others. It is probably the most common mechanism used to achieve

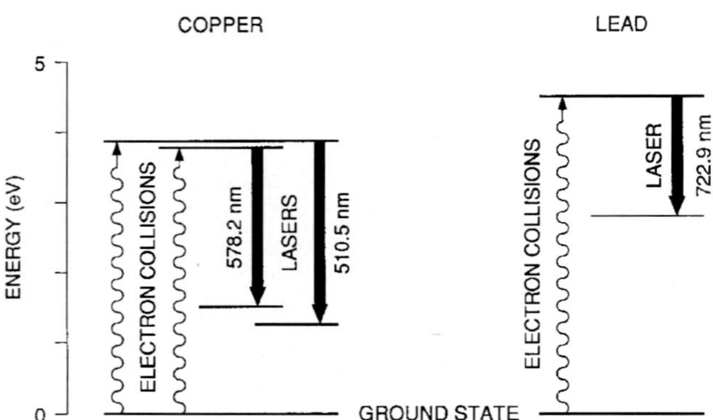

FIGURE 7 Energy-level diagrams of the transient three-level copper and lead vapor lasers showing the pump processes as well as the laser transitions.

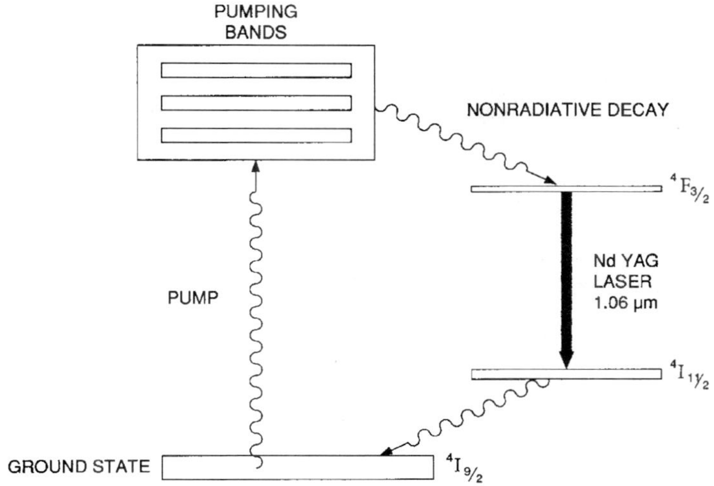

FIGURE 8 Energy-level diagram of the Nd:YAG laser indicating the four-level laser excitation process.

the necessary population inversion. This process involves four levels[2] (although it can include more) and generally occurs via excitation from the ground state 0 to an excited state 3 which energetically lies above the upper-laser-level 2. The population then decays from level 3 to level 2 by nonradiative processes (such as collisional processes in gases), but can also decay radiatively to level 2. If the proper choice of materials has been made, the lower-laser-level 1 in some systems will decay rapidly to the ground state (level 0) which allows the condition $N_2 > (g_2/g_1)N_1$ to be satisfied. This situation is shown in Fig. 8 for a Nd:YAG laser crystal.

Optimization of the Output Coupling from a Laser Cavity[9]

A laser will operate with any combination of mirror reflectivities subject to the constraints of the threshold condition of Eq. (17*a* or *b*). However, since lasers are devices that are designed to use the laser power in various applications, it is desirable to extract the most power from the laser in the most efficient manner. A simple expression for the optimum laser output coupling is given as

$$t_{\text{opt}} = (a_L g_{21} L)^{1/2} - a_L \tag{18}$$

in which a_L is the absorption and scattering losses per pass through the amplifier (the same as in Eq. (17*a* and *b*), g_{21} is the small signal gain per pass through the amplifier, and L is the length of the gain medium. This value of t_{opt} is obtained by assuming that equal output couplings are used for both mirrors at the ends of the cavity. To obtain all of the power from one end of the laser, the output transmission must be doubled for one mirror and the other mirror is made to be a high reflector.

The intensity of the beam that would be emitted from the output mirror can be estimated to be

$$I_{t_{\max}} = \frac{I_{\text{sat}} t_{\text{opt}}^2}{2a_L} \tag{19}$$

in which I_{sat} is the saturation intensity as obtained from Eq. (15). If all of the power is desired from one end of the laser, as discussed above, then $I_{t_{\max}}$ would be doubled in the above expression.

Pumping Techniques to Produce Inversions

Excitation or pumping of the upper laser level generally occurs by two techniques: (1) particle pumping and (2) optical or photon pumping. No matter which process is used, the goal is to achieve sufficient pumping flux and, consequently enough, population in the upper-laser-level 2 to satisfy or exceed the requirements of either Eq. (16) or Eq. (17).

Particle Pumping.[10] Particle pumping occurs when a high-speed particle collides with a laser species and converts its kinetic energy to internal energy of the laser species. Particle pumping occurs mostly with electrons as the pumping particle. This is especially common in a gas discharge where a voltage is applied across a low-pressure gas and the electrons flow through the tube in the form of a discharge current that can range from a few milliamps to tens of amperes, depending upon the particular laser and the power level desired. This type of excitation process is used for lasers such as the argon (Fig. 9) and krypton ion lasers, the copper vapor laser, excimer lasers, and the molecular nitrogen laser.

Two other well-known gas lasers, the helium-neon laser and the helium-cadmium laser, operate in a gas discharge containing a mixture of helium gas and the laser species (neon gas or cadmium vapor). When an electric current is produced within the discharge, the high-speed electrons first pump an excited metastable state in helium (essentially a storage reservoir). The energy is then transferred from this reservoir to the upper laser levels of

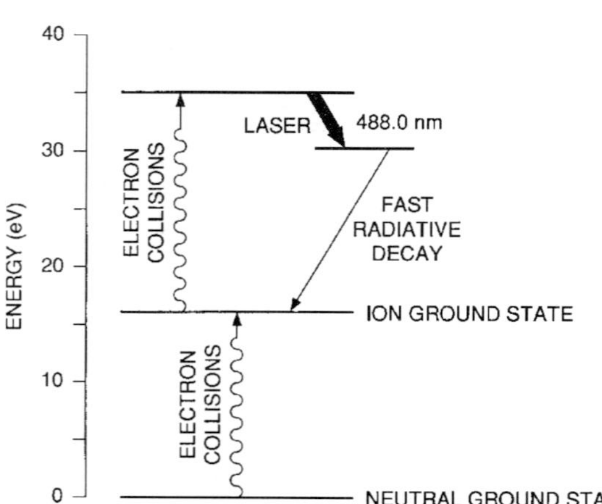

FIGURE 9 Energy-level diagram of the argon ion laser indicating the two-step excitation process.

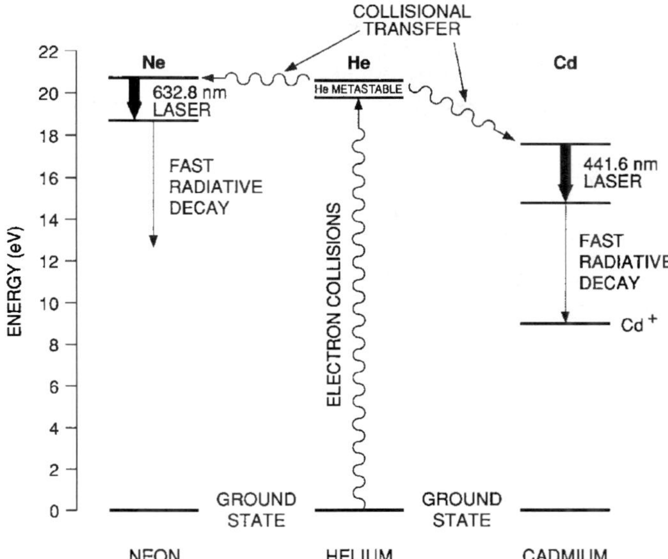

FIGURE 10 Energy-level diagrams of the helium-neon (He-Ne) laser and the helium-cadmium (He-Cd) laser that also include the helium metastable energy levels that transfer their energy to the upper laser levels by collisions.

neon or cadmium by collisions of the helium metastable level with the neon or cadmium ground-state atoms as shown in Fig. 10. Electron collisions with the cadmium ion ground state have also been shown to produce excitation in the case of the helium-cadmium laser.

In an energy-transfer process similar to that of helium with neon or cadmium, the CO_2 laser operates by using electrons of the gas discharge to produce excitation of molecular nitrogen vibrational levels that subsequently transfer their energy to the vibrational upper laser levels of CO_2 as indicated in Fig. 11. Helium is used in the CO_2 laser to control the

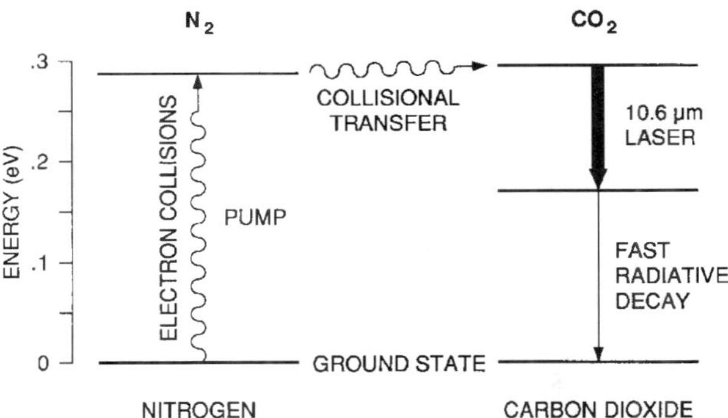

FIGURE 11 Energy-level diagram of the carbon dioxide (CO_2) laser along with the energy level of molecular nitrogen that collisionally transfers its energy to the CO_2 upper laser level.

electron temperature and also to cool (reduce) the population of the lower laser level via collisions of helium atoms with CO_2 atoms in the lower laser level.

High-energy electron beams and even nuclear reactor particles have also been used for particle pumping of lasers, but such techniques are not normally used in commercial laser devices.

Optical Pumping.[7] Optical pumping involves the process of focusing light into the gain medium at the appropriate wavelength such that the gain medium will absorb most (or all) of the light and thereby pump that energy into the upper laser level as shown in Fig. 12. The selectivity in pumping the laser level with an optical pumping process is determined by choosing a gain medium having significant absorption at a wavelength at which a suitable pump light source is available. This of course implies that the absorbing wavelengths provide efficient pumping pathways to the upper laser level. Optical pumping requires very intense pumping light sources, including flash lamps and other lasers. Lasers that are produced by optical pumping include organic dye lasers and solid-state lasers.

Flash lamps used in optically pumped laser systems are typically long, cylindrically shaped, fused quartz structures of a few millimeters to a few centimeters in diameter and 10 to 50 centimeters in length. The lamps are filled with gases such as xenon and are initiated by running an electrical current through the gas. The light is concentrated into the lasing medium by using elliptically shaped reflecting cavities that surround both the laser medium and the flash lamps as shown in Fig. 13. These cavities efficiently collect and transfer to the laser rod most of the lamp energy in wavelengths within the pump absorbing band of the rod. The most common flash-lamp-pumped lasers are Nd:YAG, Nd:glass, and organic dye lasers. New crystals such as Cr:LiSAF and HoTm:YAG are also amenable to flash-lamp pumping.

Solid-state lasers also use energy-transfer processes as part of the pumping sequence in a way similar to that of the He-Ne and He-Cd gas lasers. For example, Cr^{3+} ions are added into neodymium-doped crystals to improve the absorption of the pumping light. The energy is subsequently transferred to the Nd^{3+} laser species. Such a process of adding desirable impurities is known as *sensitizing.*

Lasers are used as optical pumping sources in situations where (1) it is desirable to be able to concentrate the pump energy into a small-gain region or (2) it is useful to have a narrow spectral output of the pump source in contrast with the broadband spectral distribution of a flash lamp.

Laser pumping is achieved by either transverse pumping (a direction perpendicular to the direction of the laser beam) or longitudinal pumping (a direction in the same direction as the emerging laser beam). Frequency doubled and tripled pulsed Nd:YAG lasers are

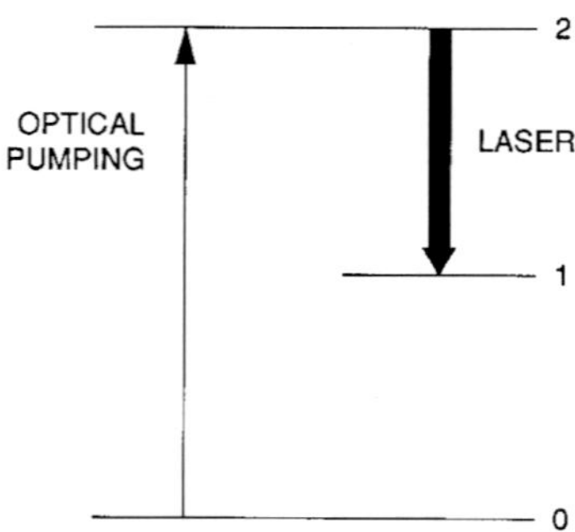

FIGURE 12 A general diagram showing optical pumping of the upper laser level.

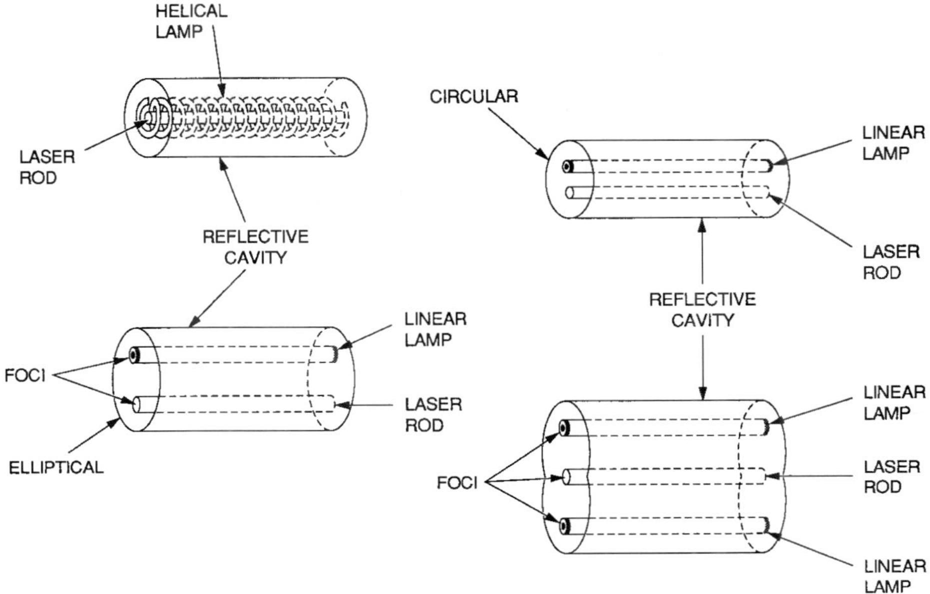

FIGURE 13 Flash-lamp pumping arrangements for solid-state laser rods showing the use of helical lamps as well as linear lamps in a circular cavity, a single elliptical cavity, and a double elliptical cavity.

used to transversely pump organic dye lasers[8] that provide continuously tunable laser radiation over the near-ultraviolet, visible, and near-infrared spectral regions (by changing dyes at appropriate wavelength intervals). For transverse pumping, the pump lasers are typically focused into the dye medium with a cylindrical lens to provide a 1–2-cm-long (but very narrow) gain medium in the liquid dye solution. The dye concentration is adjusted to absorb the pump light within a millimeter or so into the dye cell to provide the very high concentration of gain near the surface of the cell.

Both cw and mode-locked argon ion lasers and Nd:YAG lasers are typically used for longitudinal or end pumping of cw and/or mode-locked organic dye lasers and also of solid-state gain media. In this pumping arrangement the pump laser is focused into a very thin gain region, which is provided by either a thin jet stream of flowing dye solution (Fig. 14) or a solid-state crystal such as Ti:Al_2O_3. The thin gain medium is used, in the case of the generation of ultrashort mode-locked pulses, so as to allow precise timing of the

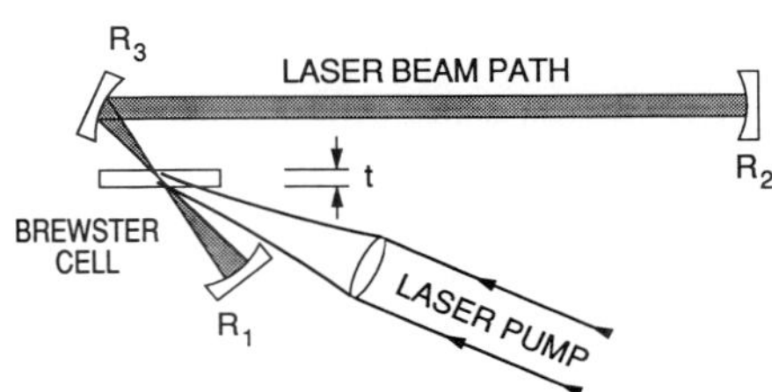

FIGURE 14 End-pumping cavity arrangement for either organic dye lasers or solid-state lasers.

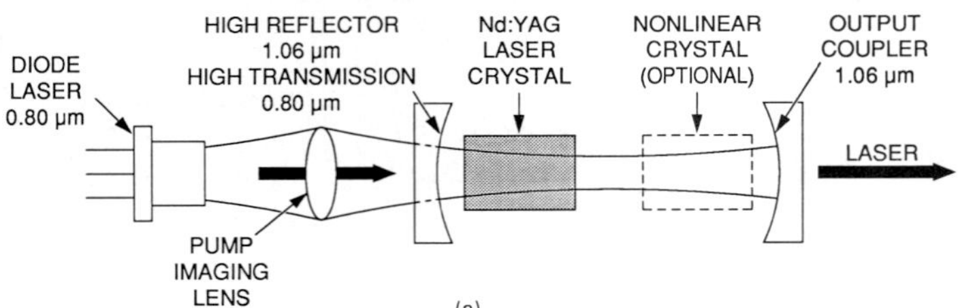

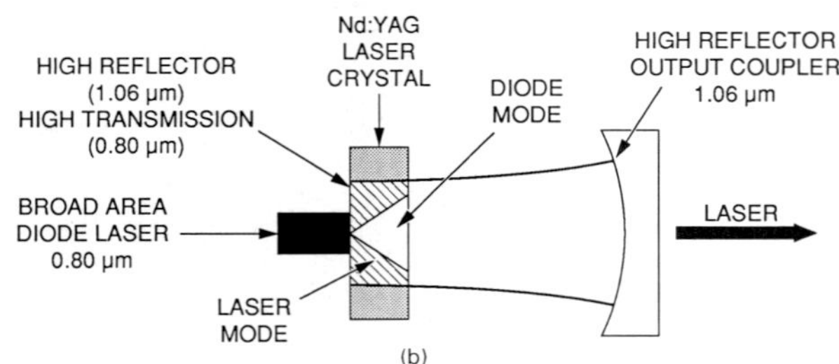

FIGURE 15 Pumping arrangements for diode-pumped Nd:YAG lasers showing both (*a*) standard pumping using an imaging lens and (*b*) close-coupled pumping in which the diode is located adjacent to the laser crystal.

short-duration pump pulses with the ultrashort laser pulses that develop within the gain medium as they travel within the optical cavity.

Gallium arsenide semiconductor diode lasers, operating at wavelengths around 0.8 μm, can be effectively used to pump Nd : YAG lasers because the laser wavelength is near that of the strongest absorption feature of the pump band of the Nd:YAG laser crystal, thereby minimizing excess heating of the laser medium. Also, the diode lasers are very efficient light sources that can be precisely focused into the desired mode volume of a small Nd:YAG crystal (Fig. 15*a*) which results in minimal waste of the pump light. Figure 15*b* shows how close-coupling of the pump laser and the Nd:YAG crystal can be used to provide compact efficient diode laser pumping. The infrared output of the Nd:YAG laser can then also be frequency doubled as indicated in Fig. 15*a*, using nonlinear optical techniques to produce a green laser beam in a relatively compact package.

11.4 LASER PROPERTIES ASSOCIATED WITH OPTICAL CAVITIES OR RESONATORS

Longitudinal Laser Modes[1,2]

When a collimated optical beam of infinite lateral extent (a plane wave) passes through two reflecting surfaces of reflectivity R and also of infinite extent that are placed normal (or nearly normal) to the beam and separated by a distance d, as shown in Fig. 16*a*, the

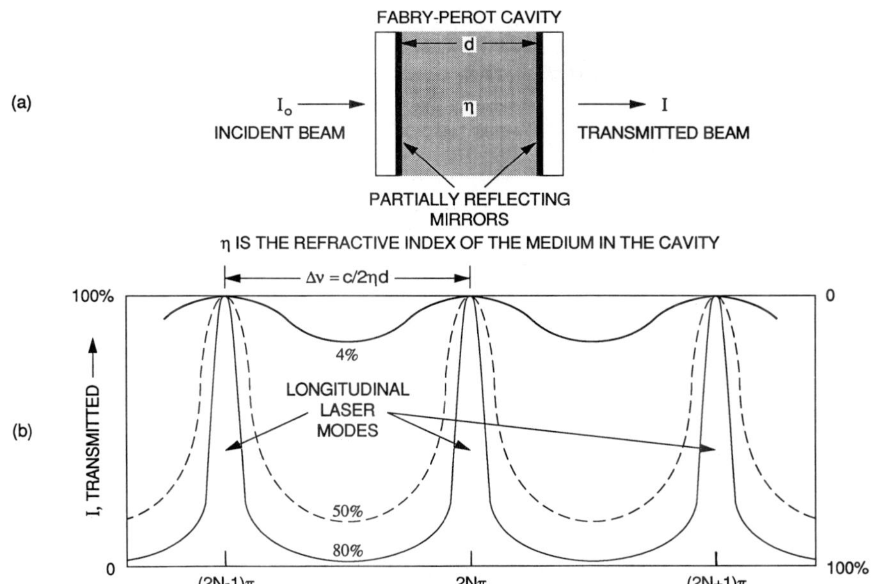

FIGURE 16 Fabry-Perot optical cavity consisting of two plane-parallel mirrors with a specific reflectivity separated by a distance d indicating the frequency spacing of longitudinal modes.

plot of transmission versus wavelength for the light as it emerges from the second reflecting surface is shown in Fig. 16*b*. The transmission reaches a maximum of 100 percent (if there are no absorption losses at the reflecting surfaces) at frequency spacings of $\Delta \nu = c/2\eta d$ where c is the speed of light in a vacuum, and η is the index of refraction of the medium between the mirrors. This frequency-selective optical device is known as a Fabry-Perot interferometer and has many useful applications in optics.

The transmission through this device, as shown in Fig. 16*b* is enhanced at regular frequency or wavelength intervals due to the development of standing waves that resonate within the optical cavity. The enhancement occurs at frequencies (or wavelengths) at which complete sinusoidal half-cycles of the electromagnetic wave exactly "fit" between the mirrors such that the value of the electric field of the wave is zero at the mirror surfaces.

If a laser amplifier is placed between two mirrors in the same arrangement as described above, the same standing waves tend to be enhanced at frequency intervals of

$$\nu = n(c/2\eta d) \tag{20}$$

where n is an integer that indicates the number of half wavelengths of the laser light that fit within the spacing d of the two mirrors. In a typical laser operating in the visible spectral region, n would be of the order of 30,000 to 40,000. In such a laser, the output of the laser beam emerging from the cavity is very strongly enhanced at the resonant wavelengths as shown in Fig. 17, since these are the wavelengths that have the lowest loss within the cavity. The widths of the resonances shown in Fig. 17 are those of a passive Fabry-Perot cavity. When an active gain medium is placed within the cavity, the linewidth of the beam that is continually amplified as it reflects back and forth between the mirrors is narrowed even further.

These enhanced regions of very narrow frequency output are known as *longitudinal*

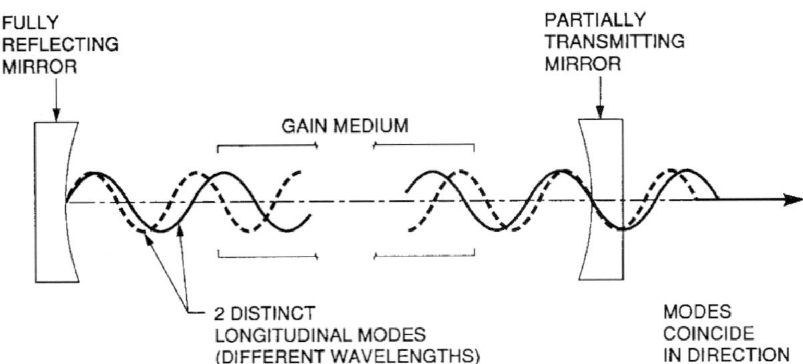

FIGURE 17 Laser resonator showing two distinct longitudinal modes traveling in the same direction but with slightly different frequencies.

modes of the laser. They are referred to as modes since they represent discrete values of frequency associated with the integral values of n at which laser output occurs. Lasers operating on a single longitudinal mode with ultrastable cavities and ultrahigh reflectivity mirrors have generated linewidths as narrow as a few hundred hertz or less. Since the longitudinal or temporal coherence length of a beam of light is determined by $c/\Delta\nu$, a very narrow laser linewidth can provide an extremely long coherence length and thus a very coherent beam of light.

For a typical gas laser (not including excimer lasers), the laser gain bandwidth is of the order of 10^9–10^{10} Hz. Thus, for a laser mirror cavity length of 0.5 m, the mode spacing would be of the order of 300 MHz and there would be anywhere from 3 to 30 longitudinal modes operating within the laser cavity. For an organic dye laser or a broadband solid-state laser, such as a $Ti:Al_2O_3$ laser, there could be as many as one million distinct longitudinal modes, each of a slightly different frequency than the next one, oscillating at the same time. However, if mode-locking is not present, typically only one or a few modes will dominate the laser output of a homogeneously broadened laser gain medium.

Transverse Laser Modes[11–14]

The previous section considered the implications of having a collimated or parallel beam of light of infinite lateral extent pass through two infinite reflecting surfaces that are arranged normal to the direction of propagation of the beam and separated by a specific distance d. We must now consider the consequences of having the light originate within the space between the two mirrors in an amplifier that has only a very narrow lateral extent limited by either the diameter of the mirrors or by the diameter of the amplifying medium. The beam evolves from the spontaneous emission within the gain medium and eventually becomes a nearly collimated beam when it reaches I_{sat} since only rays traveling in a very limited range of directions normal to the laser mirrors will experience enough reflections to reach I_{sat}. The fact that the beam has a restricted aperture in the direction transverse to the direction of propagation causes it to evolve with a slight transverse component due to diffraction, which effectively causes the beam to diverge. This slight divergence actually consists of one or more distinctly separate beams that can operate individually or in combination.

These separate beams that propagate in the z direction are referred to as *transverse*

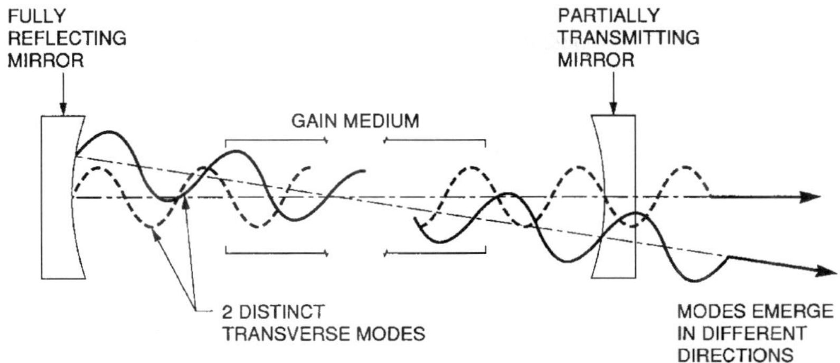

FIGURE 18 Laser resonator showing two distinct transverse modes traveling in different directions with slightly different frequencies.

modes as shown in Fig. 18. They are characterized by the various lateral spatial distributions of the electric field vector $\mathbf{E}(x, y)$ in the x-y directions as they emerge from the laser. These transverse amplitude distributions for the waves can be described by the relationship.

$$E_{pq}(x, y) = H_p\left(\frac{\sqrt{2}x}{w}\right)H_q\left(\frac{\sqrt{2}y}{w}\right)e^{-(x^2+y^2)/w^2} \tag{21}$$

In this solution, p and q are positive integers ranging from zero to infinity that designate the different modes which are associated with the order of the Hermite polynomials. Thus every set of p, q represents a specific distribution of wave amplitude at one of the mirrors, or a specific transverse mode of the open-walled cavity. We can list several Hermite polynomials as follows:

$$\begin{gathered} H_0(u) = 1 \qquad H_1(u) = 2u \\ H_2(u) = 2(2u^2 - 1) \\ H_m(u) = (-1)^m e^{u^2} \frac{d^m(e^{-u^2})}{du^m} \end{gathered} \tag{22}$$

The spatial intensity distribution would be obtained by squaring the amplitude distribution function of Eq. (21). The transverse modes are designated TEM for transverse electromagnetic. The lowest-order mode is given by TEM_{00}. It could also be written as TEM_{n00} in which n would designate the longitudinal mode number [Equation (20)]. Since this number is generally very large for optical frequencies, it is not normally given.

The lowest order TEM_{00} mode, has a circular distribution with a gaussian shape (often referred to as the *gaussian mode*) and has the smallest divergence of any of the transverse modes. Such a mode can be focused to a spot size with dimensions of the order of the wavelength of the beam. It has a minimum width or waist $2w_o$ that is typically located between the laser mirrors (determined by the mirror curvatures and separation) and expands symmetrically in opposite directions from that minimum waist according to the following equation:

$$w(z) = w_o\left[1 + \left(\frac{\lambda z}{\eta \pi w_o^2}\right)^2\right]^{1/2} = w_o\left[1 + \left(\frac{z}{z_o}\right)^2\right]^{1/2} \tag{23}$$

where $w(z)$ is the beam waist at any location z measured from w_o, η is the index of

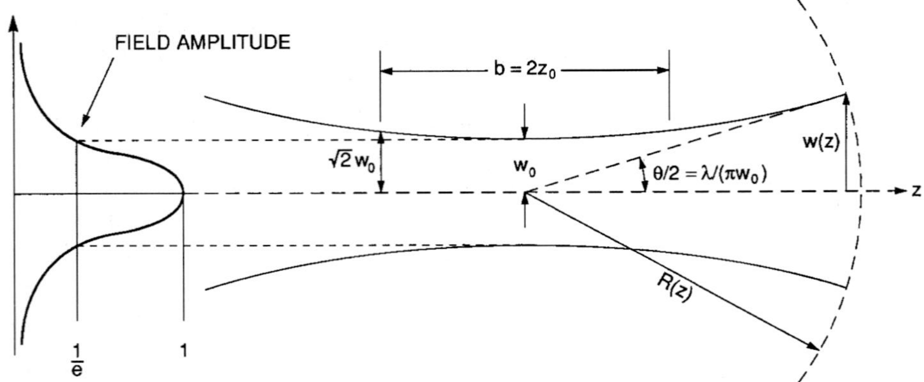

FIGURE 19 Parameters of a gaussian-shaped beam which are the features of a TEM_{00} transverse laser mode.

refraction of the medium, and $z_o = \eta \pi w_o^2/\lambda$ is the distance over which the beam waist expands to a value of $\sqrt{2}w_o$.

The waist $w(z)$ at any location, as shown in Fig. 19, describes the transverse dimension within which the electric field distribution of the beam decreases to a value of 37 percent $(1/e)$ of its maximum on the beam axis and within which 86.5 percent of the beam energy is contained. The TEM_{00} mode would have an intensity distribution that is proportional to the square of Eq. (21) for $p = q = 0$ and, since it is symmetrical around the axis of propagation, it would have a cylindrically symmetric distribution of the form

$$I(r, z) = I_o e^{-2r^2/w^2(z)} \tag{24}$$

where I_o is the intensity on the beam axis.

The beam also has a wavefront curvature given by

$$R(z) = z\left[1 + \left(\frac{\eta \pi w_o^2}{\lambda z}\right)^2\right] \tag{25}$$

which is indicated in Fig. 19, and a far-field angular divergence given by

$$\theta = \lim_{z \to \infty} \frac{2w(z)}{z} = \frac{2\lambda}{\pi w_o} = 0.64 \frac{\lambda}{w_o} \tag{26}$$

For a symmetrical cavity formed by two mirrors, each of radius of curvature R, separated by a distance d and in a medium in which $\eta = 1$ the minimum beam waist w_o is given by

$$w_o^2 = \frac{\lambda}{2\pi}[d(2R - d)]^{1/2} \tag{27}$$

and the radius of curvature r_c of the wavefront is

$$r_c = z + \frac{d(2R - d)}{4z} \tag{28}$$

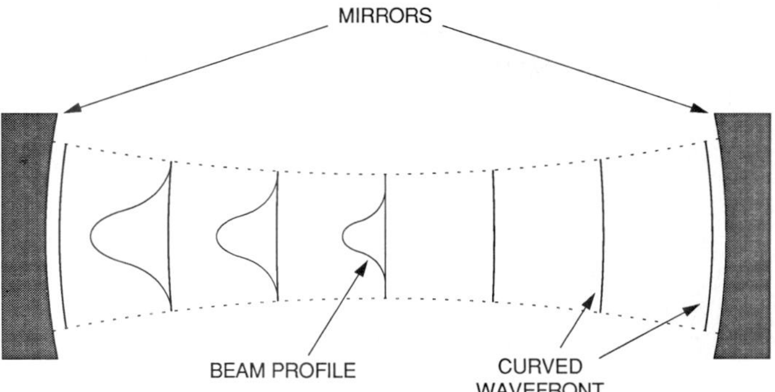

FIGURE 20 A stable laser resonator indicating the beam profile at various locations along the beam axis as well as the wavefront at the mirrors that matches the curvature of the mirrors.

For a confocal resonator in which $R = d$, w_o is given by

$$w_o = \sqrt{\frac{\lambda d}{2\pi}} \tag{29}$$

and the beam waist (spot size) at each mirror located a distance $d/2$ from the minimum is

$$w = \sqrt{\frac{\lambda d}{\pi}} \tag{30}$$

Thus, for a confocal resonator, $w(d/2)$ at each of the mirrors is equal to $\sqrt{2}w_o$ and thus at that location, $z = z_o$. The distance between mirrors for such a cavity configuration is referred to as the *confocal parameter b* such that $b = 2z_o$. In a stable resonator, the curvature of the wavefront at the mirrors, according to Eqs. (25) and (28), exactly matches the curvature of the mirrors as shown in Fig. 20.

Each individual transverse mode of the beam is produced by traveling a specific path between the laser mirrors such that, as it passes from one mirror to the other and returns, the gain it receives from the amplifier is at least as great as the total losses of the mirror, as indicated from Eq. (17), plus the additional diffraction losses produced by either the finite lateral extent of the laser mirrors or the finite diameter of the laser amplifier or some other optical aperture placed in the system, whichever is smaller. Thus the TEM_{00} mode is produced by a beam passing straight down the axis of the resonator as indicated in Fig. 19.

Laser Resonator Configurations and Cavity Stability[4,14]

There are a variety of resonator configurations that can be used for lasers. The use of slightly curved mirrors leads to much lower diffraction losses of the transverse modes than do plane parallel mirrors, and they also have much less stringent alignment tolerances. Therefore, most lasers use curved mirrors for the optical cavity. For a cavity with two

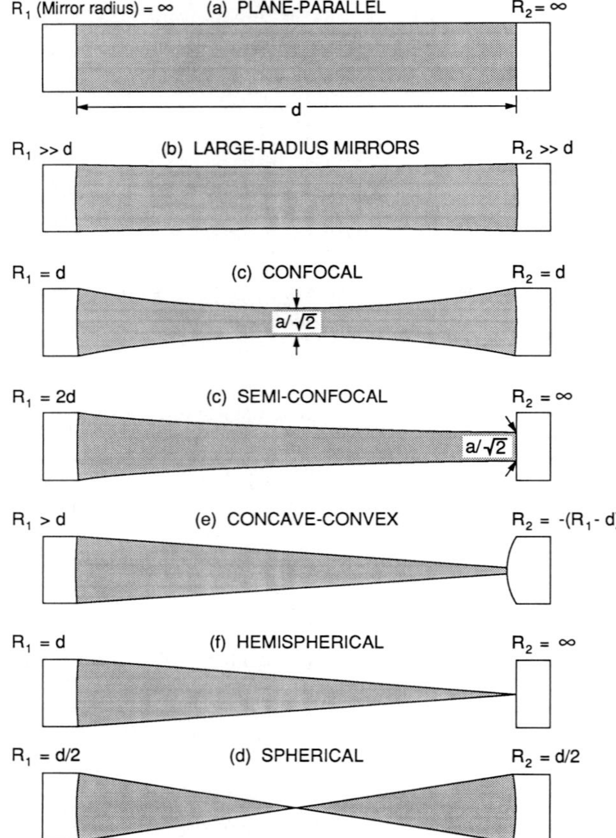

FIGURE 21 Possible two-mirror laser cavity configurations indicating the relationship of the radii of curvature of the mirrors with respect to the separation between mirrors.

mirrors of curvature R_1 and R_2, and a separation distance d, a number of possible cavity configurations are shown in Fig. 21.

A relationship between the radii of curvature and the separation between mirrors can be defined as

$$g_1 = 1 - \frac{d}{R_1} \quad \text{and} \quad g_2 = 1 - \frac{d}{R_2} \tag{31}$$

such that the condition for stable transverse modes is given by

$$0 < g_1 g_2 < 1 \tag{32}$$

A stable mode is a beam that can be maintained with a steady output and profile over a relatively long period of time. It results from a cavity configuration that concentrates the beam toward the resonator axis in a regular pattern as it traverses back and forth within the cavity, rather than allowing it to diverge and escape from the resonator. In considering the various possible combinations of curved mirror cavities, one must keep the relation

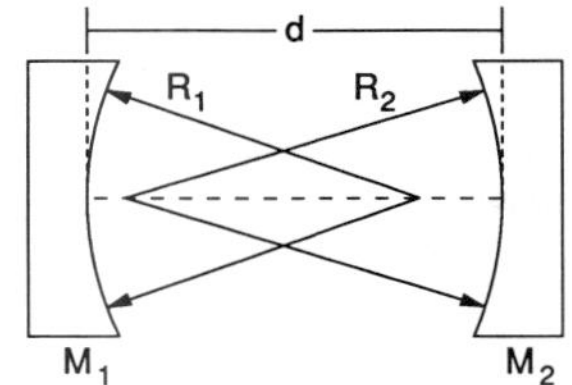

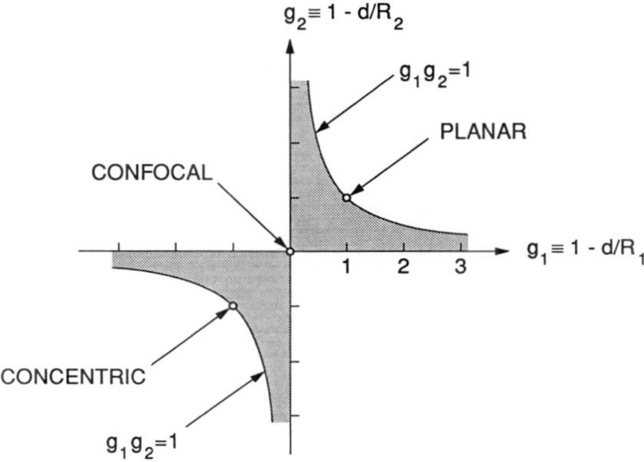

FIGURE 22 Stability diagram for two-mirror laser cavities indicating the shaded regions where stable cavities exist.

between the curvatures and mirror separation d within the stable regions of the graph, as shown in Fig. 22, in order to produce stable modes. Thus it can be seen from Fig. 22 that not all configurations shown in Fig. 21 are stable. For example, the planar, confocal, and concentric arrangements are just on the edge of stability according to Fig. 22.

There may be several transverse modes oscillating simultaneously "within" a single longitudinal mode. Each transverse mode can have the same value of n [Eq. (20)] but will have a slightly different value of d as it travels a different optical path between the resonator mirrors, thereby generating a slight frequency shift from that of an adjacent transverse mode. For most optical cavities, the mode that operates most easily is the TEM_{00} mode, since it travels a direct path along the axis of the gain medium.

11.5 SPECIAL LASER CAVITIES

Unstable Resonators[14]

A laser that is operating in a TEM_{00} mode, as outlined above, typically has a beam within the laser cavity that is relatively narrow in width compared to the cavity length. Thus, if a laser with a relatively wide gain region is used, to obtain more energy in the output beam, it is not possible to extract the energy from that entire region from a typically very narrow low-order gaussian mode. A class of resonators has been developed that can extract the energy from such wide laser volumes and also produce a beam with a nearly gaussian

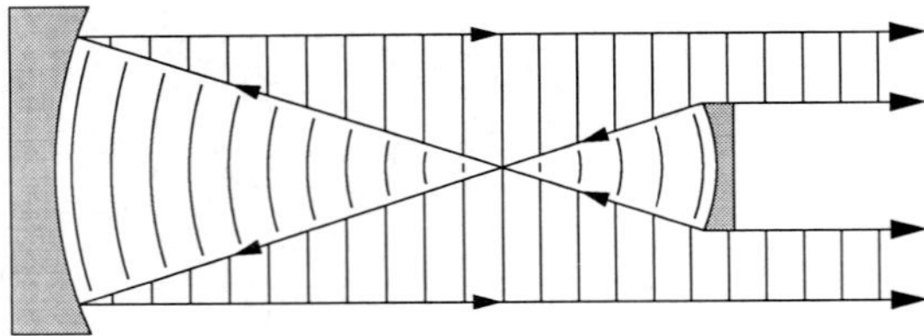

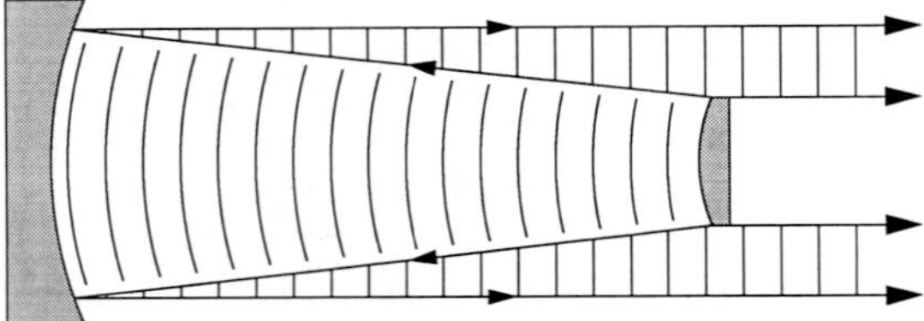

FIGURE 23 Unstable resonator cavity configurations showing both the negative and positive branch confocal cavities.

profile that makes it easily focusable. These resonators do not meet the criteria for stability, as outlined above, but still provide a good beam quality for some types of lasers. This class of resonators is referred to as *unstable resonators.*

Unstable resonators are typically used with high-gain laser media under conditions such that only a few passes through the amplifier will allow the beam to reach the saturation intensity and thus extract useful energy. A diagram of two unstable resonator cavity configurations is shown in Fig. 23. Figure 23*b* is the positive branch confocal geometry and is one of the most common unstable resonator configurations. In this arrangement, the small mirror has a convex shape and the large mirror a concave shape with a separation of length d such that $R_2 - R_1 = 2d$. With this configuration, any ray traveling parallel to the axis from left to right that intercepts the small convex mirror will diverge toward the large convex mirror as though it came from the focus of that mirror. The beam then reflects off of the larger mirror and continues to the right as a beam parallel to the axis. It emerges as a reasonably well-collimated beam with a hole in the center (due to the obscuration of the small mirror). The beam is designed to reach the saturation intensity when it arrives at the large mirror and will therefore proceed to extract energy as it makes its final pass through the amplifier. In the far field, the beam is near gaussian in shape, which allows it to be propagated and focused according to the equations described in the previous section. A number of different unstable resonator configurations can be found in the literature for specialized applications.

Q-Switching[15]

A typical laser, after the pumping or excitation is first applied, will reach the saturation intensity in a time period ranging from approximately 10 nsec to 1 μsec, depending upon the value of the gain in the medium. For lasers such as solid-state lasers the upper-laser-level lifetime is considerably longer than this time (typically 50–200 μ sec). It is possible to store and accumulate energy in the form of population in the upper laser level for a time duration of the order of that upper-level lifetime. If the laser cavity could be obscured during this pumping time and then suddenly switched into the system at a time order of the upper-laser-level lifetime, it would be possible for the gain, as well as the laser energy, to reach a much larger value than it normally would under steady-state conditions. This would produce a "giant" laser pulse of energy from the system.

Such a technique can in fact be realized and is referred to as *Q-switching* to suggest that the cavity Q is changed or switched into place. The cavity is switched on by using either a rapidly rotating mirror or an electro-optic shutter such as a Pockel cell or a Kerr cell. Nd:YAG and Nd:glass lasers are the most common Q-switched lasers. A diagram of the sequence of events involved in Q-switching is shown in Fig. 24.

Another technique that is similar to Q-switching is referred to as *cavity dumping.* With this technique, the intense laser beam inside of a normal laser cavity is rapidly switched out of the cavity by a device such as an acousto-optic modulator. Such a device is inserted at Brewster's angle inside the cavity and is normally transparent to the laser beam. When the device is activated, it rapidly inserts a high-reflecting surface into the cavity and reflects the beam out of the cavity. Since the beam can be as much as two orders of magnitude higher in intensity within the cavity than that which leaks through an output mirror, it is possible to extract high power on a pulsed basis with such a technique.

Mode-Locking[14]

In the discussion of longitudinal modes it was indicated that such modes of intense laser output occur at regularly spaced frequency intervals of $\Delta\nu = c/2\eta d$ over the gain bandwidth of the laser medium. For laser cavity lengths of the order of 10–100 cm these frequency intervals range from approximately 10^8 to 10^9 Hz. Under normal laser operation, specific modes with the highest gain tend to dominate and quench other modes (especially if the gain medium is homogeneously broadened). However, under certain conditions it is possible to obtain all of the longitudinal modes lasing simultaneously, as shown in Fig. 25*a*. If this occurs, and the modes are all phased together so that they can act in concert by constructively and destructively interfering with each other, it is possible to produce a series of giant pulses separated in a time Δt of

$$\Delta t = 2\eta d/c \tag{33}$$

or approximately 1–10 nsec for the cavity lengths mentioned above (see Fig. 25*b*). The pulse duration is approximately the reciprocal of the separation between the two extreme longitudinal laser modes or

$$\Delta t_p = \frac{1}{n\,\Delta\nu} \tag{34}$$

as can also be seen in Fig. 25*b*. This pulse duration is approximately the reciprocal of the laser gain bandwidth. However, if the index of refraction varies significantly over the gain bandwidth, then all of the frequencies are not equally spaced and the mode-locked pulse duration will occur only within the frequency width over which the frequency separations are approximately the same.

The narrowest pulses are produced from lasers having the largest gain bandwidth such as organic dye lasers, and solid-state lasers including Ti:Al_2O_3, Cr:Al_2O_3, and Cr:LiSAlF. The shortest pulses to date, of the order of 6×10^{-15} sec, are produced in

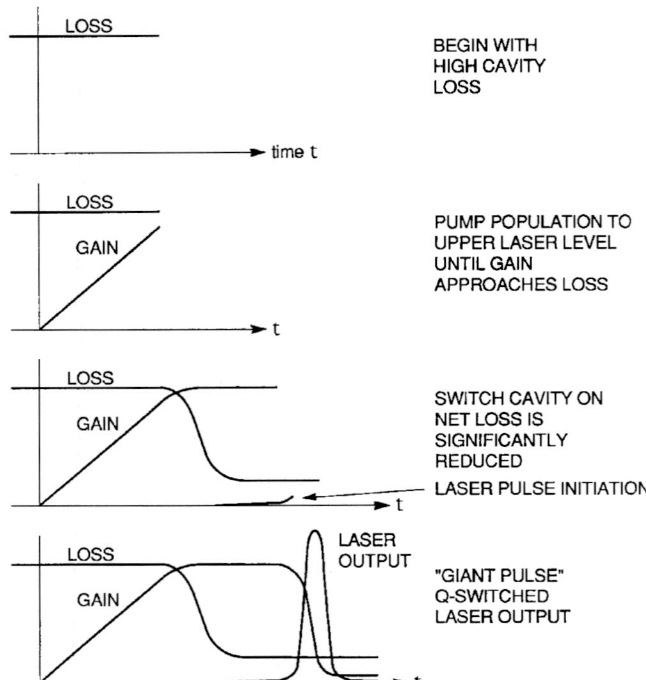

FIGURE 24 Schematic diagram of a Q-switched laser indicating how the loss is switched out of the cavity and a gaint laser pulse is produced as the gain builds up.

organic dyes where the gain bandwidth is approximately 50 nm at a center wavelength of 600 nm.

Such short-pulse operation is referred to as *mode-locking* and is achieved by inserting a very fast shutter within the cavity which is opened and closed at the intervals of the round-trip time of the short laser pulse within the cavity. This shutter coordinates the time at which all of the modes arrive at the mirror and thus brings them all into phase at that location. Electro-optic shutters, short duration gain pumping by another mode-locked laser, or passive saturable absorbers are techniques that can serve as the fast shutter. The second technique, short-duration gain pumping, is referred to as *synchronous pumping.* Three fast saturable absorber shutter techniques for solid-state lasers include colliding pulse mode-locking,[16] additive pulse mode-locking,[17] and Kerr lens mode-locking.[18]

Distributed Feedback Lasers[19]

The typical method of obtaining feedback into the laser gain medium is to have the mirrors located at the ends of the amplifier as discussed previously. It is also possible, however, to provide the reflectivity within the amplifying medium in the form of a periodic variation of either the gain or the index of refraction throughout the medium. This process is referred to as *distributed feedback* or DFB. Such feedback methods are particularly effective in

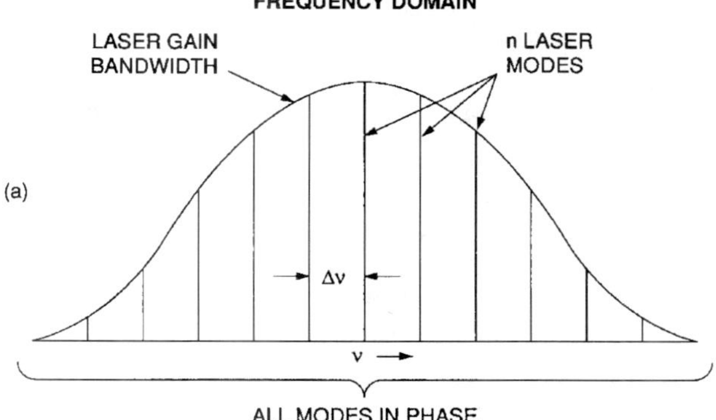

FIGURE 25 Diagrams in both the frequency and time domains of how mode-locking is produced by phasing n longitudinal modes together to produce an ultrashort laser pulse.

semiconductor lasers in which the gain is high and the fabrication of periodic variations is not difficult. The reader is referred to the reference section at the end of this chapter for further information concerning this type of feedback.

Ring Lasers[14]

Ring lasers are lasers that have an optical path within the cavity that involves the beam circulating in a loop rather than passing back and forth over the same path. This requires optical cavities that have more than two mirrors. The laser beam within the cavity consists of two waves traveling in opposite directions with separate and independent resonances within the cavity. In some instances an optical device is placed within the cavity that provides a unidirectional loss. This loss suppresses one of the beams, allowing the beam propagating in the other direction to become dominant. The laser output then consists of a traveling wave instead of a standing wave and therefore there are no longitudinal modes. Such an arrangement also eliminates the variation of the gain due to the standing waves in the cavity (spatial hole burning), and thus the beam tends to be more homogeneous than that of a normal standing-wave cavity. Ring lasers are useful for producing ultrashort mode-locked pulses and also for use in laser gyroscopes as stable reference sources.

11.6 SPECIFIC TYPES OF LASERS

Lasers can be categorized in several different ways including wavelength, material type, and applications. In this section we will summarize them by material type such as gas, liquid, solid-state, and semiconductor lasers. We will include only lasers that are available commercially since such lasers now provide a very wide range of available wavelengths and powers without having to consider special laboratory lasers.

Gaseous Laser Gain Media

Helium-Neon Laser.[10] The helium-neon laser was the first gas laser. The most widely used laser output wavelength is a red beam at 632.8 nm with a cw output ranging from 1 to 100 mW and sizes ranging from 10 to 100 cm in length. It can also be operated at a wavelength of 543.5 nm for some specialized applications. The gain medium is produced by passing a relatively low electrical current (10 mA) through a low pressure gaseous discharge tube containing a mixture of helium and neon. In this mixture, helium metastable atoms are first excited by electron collisions as shown in Fig. 10. The energy is then collisionally transferred to neon atom excited states which serve as upper laser levels.

Argon Ion Laser.[10] The argon ion laser and the similar krypton ion laser operate over a wide range of wavelengths in the visible and near-ultraviolet regions of the spectrum. The wavelengths in argon that have the highest power are at 488.0 and 514.5 nm. Power outputs on these laser transitions are available up to 20 W cw in sizes ranging from 50–200 cm in length. The gain medium is produced by running a high electric current (many amperes) through a very low pressure argon or krypton gas. The argon atoms must be ionized to the second and third ionization stages (Fig. 9) in order to produce the population inversions. As a result, these lasers are inherently inefficient devices.

Helium-Cadmium Laser.[10] The helium-cadmium laser operates cw in the blue at 441.6 nm, and in the ultraviolet at 353.6 and 325.0 nm with powers ranging from 20–200 mW in lasers ranging from 40–100 cm in length. The gain medium is produced by heating cadmium metal and evaporating it into a gaseous discharge of helium where the laser gain is produced. The excitation mechanisms include Penning ionization (helium metastables collide with cadmium atoms) and electron collisional ionization within the discharge as indicated in Fig. 10. The laser uses an effect known as *cataphoresis* to transport the cadmium through the discharge and provide the uniform gain medium.

Copper Vapor Laser.[10] This pulsed laser provides high-average powers of up to 100 W at wavelengths of 510.5 and 578.2 nm. The copper laser and other metal vapor lasers of this class, including gold and lead lasers, typically operate at a repetition rate of up to 20 kHz with a current pulse duration of 10–50 nsec and a laser output of 1–10 mJ/pulse. The copper lasers operate at temperatures in the range of 1600°C in 2–10-cm-diameter temperature-resistant tubes typically 100–150 cm in length. The lasers are self-heated such that all of the energy losses from the discharge current provide heat to bring the plasma tube to the required operating temperature. Excitation occurs by electron collisions with copper atoms vaporized in the plasma tube as indicated in Fig. 7.

Carbon-Dioxide Laser.[10] The CO_2 laser, operating primarily at a wavelength of 10.6 μm, is one of the most powerful lasers in the world, producing cw powers of over 100 kW and pulsed energies of up to 10 kJ. It is also available in small versions with powers of up to 100 W from a laser the size of a shoe box. CO_2 lasers typically operate in a mixture of carbon dioxide, nitrogen, and helium gases. Electron collisions excite the metastable levels in nitrogen molecules with subsequent transfer of that energy to carbon dioxide laser levels as shown in Fig. 11. The helium gas acts to keep the average electron energy high in the gas discharge region and to cool or depopulate the lower laser level. This laser is one of the most efficient lasers, with conversion from electrical energy to laser energy of up to 30 percent.

Excimer Laser.[20] The rare gas-halide excimer lasers operate with a pulsed output primarily in the ultraviolet spectral region at 351 nm in xenon fluoride, 308 nm in xenon chloride, 248 nm in krypton fluoride, and 193 nm in argon fluoride. The laser output, with pulse durations of 10–50 nsec, is typically of the order of 0.2–1.0 J/pulse at repetition rates up to several hundred Hz. The lasers are relatively efficient (1 to 5 percent) and are of a size that would fit on a desktop. The excitation occurs via electrons within the discharge colliding with and ionizing the rare gas molecules and at the same time disassociating the halogen molecules to form negative halogen ions. These two species then combine to form an excited molecular state of the rare gas-halogen molecule which serves as the upper laser state. The molecule then radiates at the laser transition and the lower level advantageously disassociates since it is unstable, as shown in Fig. 26, for the ArF excimer molecule. The excited laser state is an excited state dimer which is referred to as an *excimer state.*

X-Ray Laser.[24] Laser output in the soft-x-ray spectral region has been produced in plasmas of highly ionized ions of a number of atomic species. The highly ionized ions are

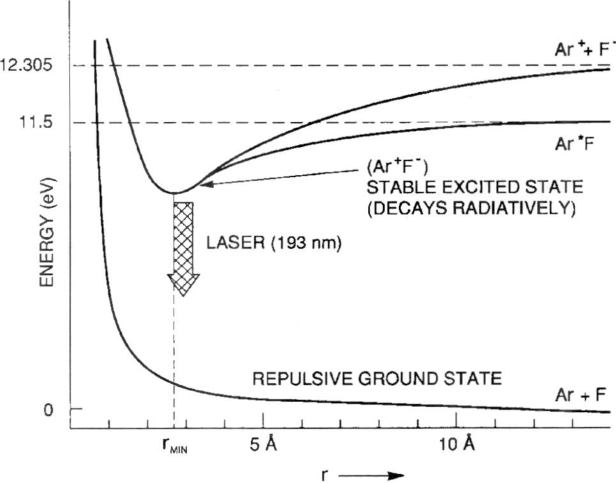

FIGURE 26 Energy-level diagram of an argon fluoride (ArF) excimer laser showing the stable excited state (upper laser level) and the unstable or repulsive ground state.

produced by the absorption of powerful solid-state lasers focused onto solid material of the desired atomic species. Since mirrors are not available for most of the soft-x-ray laser wavelengths (4–30 nm), the gain has to be high enough to obtain laser output in a single pass through the laser amplifier [Eq. (16)]. The lasers with the highest gain are the selenium laser (Se^{24+}) at 20.6 and 20.9 nm and the germanium laser (Ge^{22+}) at 23.2, 23.6, and 28.6 nm.

Liquid Laser Gain Media

Organic Dye Lasers.[8] A dye laser consists of a host or solvent material, such as alcohol or water, into which is mixed a laser species in the form of an organic dye molecule, typically in the proportion of one part in ten thousand. A large number of different dye molecules are used to make lasers covering a wavelength range of from 320 to 1500 nm with each dye having a laser bandwidth of the order of 30–50 nm. The wide, homogeneously broadened gain spectrum for each dye allows laser tunability over a wide spectrum in the ultraviolet, visible, and near-infrared. Combining the broad gain spectrum (Fig. 27) with a diffraction-grating or prism-tuning element allows tunable laser output to be obtained over the entire dye emission spectrum with a laser linewidth of 10 GHz or less. Dye lasers are available in either pulsed (up to 50–100 mJ/pulse) or continuous output (up to a few watts) in tabletop systems that are pumped by either flash lamps or by other lasers such as frequency doubled or tripled YAG lasers or argon ion lasers. Most dye lasers are arranged to have the dye mixture circulated by a pump into the gain region from a much larger reservoir since the dyes degrade at a slow rate during the excitation process.

(a)

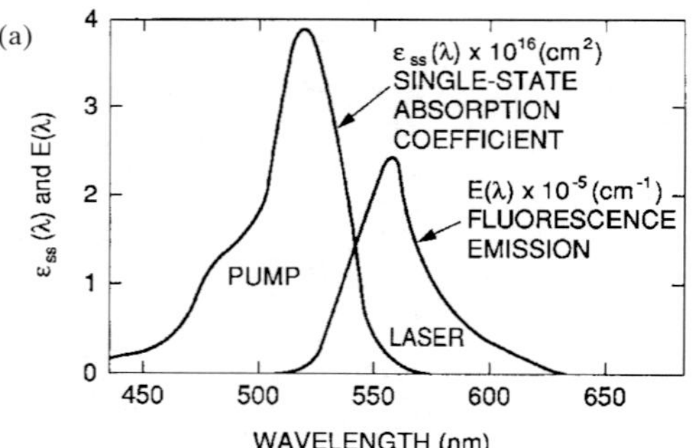

(b)

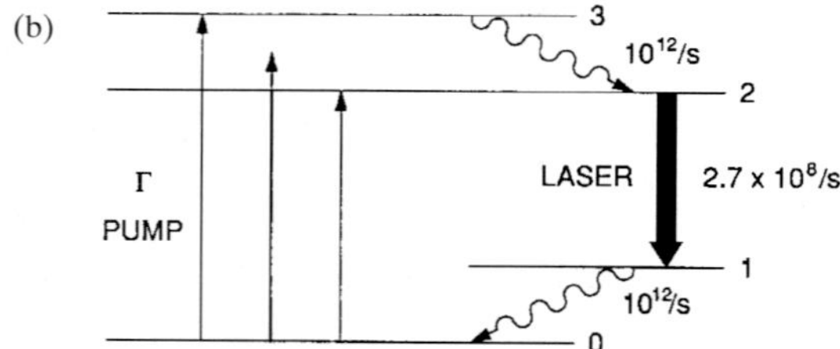

FIGURE 27 Absorption and emission spectra along with the energy-level diagram of an Rh6G organic dye laser showing both (*a*) the broad pump and emission bandwidths and (*b*) the fast decay of the lower laser level.

Dye lasers, with their broad gain spectrum, are particularly attractive for producing ultrashort mode-locked pulses. The shortest light pulses ever generated, of the order of 6×10^{-15} sec, were produced with mode-locked dye lasers. A mode-locked dye laser cavity is shown in Fig. 14 with a thin dye gain region located within an astigmatically compensated laser cavity.

Dielectric Solid-State Laser Gain Media

Ruby Laser.[5,7] The ruby laser, with an output at 694.3 nm, was the first laser ever developed. It consisted of a sapphire (Al_2O_3) host material into which was implanted a chromium laser species in the form of Cr^{3+} ions at a concentration of 0.05 percent as the amplifying medium. The ruby laser involves a three-level optical pumping scheme, with the excitation provided by flash lamps, and operates either in a pulsed or cw mode. The three-level scheme for the ruby laser (Fig. 6) requires a large fraction of the population to be pumped out of the ground state before an inversion occurs. Therefore, the ruby laser is not as efficient as other solid-state lasers such as the Nd:YAG laser which employs a four-level pumping scheme (Fig. 8). It is therefore no longer used as much as it was in the early days.

Nd:YAG and Nd:Glass Lasers.[7] Neodymium atoms, in the form of Nd^{3+} ions, are doped into host materials, including crystals such as yttrium-aluminum-garnet (YAG) and various forms of glass, in concentrations of approximately one part per hundred. The pumping scheme is a four-level system as shown in Fig. 8. When implanted in YAG crystals to produce what is referred to as an Nd:YAG laser, the laser emits primarily at 1.06 μm with continuous powers of up to 250 W and with pulsed powers as high as several MW. Difficulties in growing Nd:YAG crystals limit the size of commonly used laser rods to a maximum of 1 cm in diameter and 10 cm in length. Although this size limitation is somewhat restrictive, a YAG crystal has the advantage of high thermal conductivity allowing the rapid removal of wasted heat due to inefficient excitation. Slab geometries have recently been developed to compensate for focusing due to thermal gradients and gradient-induced stresses within the amplifier medium, thereby allowing higher average powers to be achieved. Efficient GaAs laser diodes are also being used to pump Nd:YAG amplifiers (see Fig. 15) since the diode pump wavelength matches a very strong pump absorption wavelength for the Nd^{3+} ion. This absorption wavelength is near the threshold pump energy, thereby minimizing the generation of wasted heat. Also, the diode pump laser can be made to more efficiently match the Nd laser mode volume than can a flash lamp.

Under long-pulse, flash-lamp-pumped operation, Nd:YAG lasers can produce 5 J/pulse at 100 Hz from a single rod. The pulses are trains of relaxation oscillation spikes lasting 3–4 msec which is obtained by powerful, long-pulse flash-lamp pumping. Q-switched Nd:YAG lasers, using a single Nd:YAG laser rod, can provide pulse energies of up to 1.5 J/pulse at repetition rates up to 50 Hz (75 W of average power).

Nd:glass lasers have several advantages over Nd:YAG lasers. They can be made in much larger sizes, allowing the construction of very large amplifiers. Nd:glass has a wider gain bandwidth which makes possible the production of shorter mode-locked pulses and a lower stimulated emission cross section. The latter property allows a larger population inversion to be created and thus more energy to be stored before energy is extracted from the laser. Nd:glass amplifiers operate at a wavelength near that of Nd:YAG, at 1.054 μm, for example, with phosphate glass, with a gain bandwidth 10–15 times broader than that of Nd:YAG. This larger gain bandwidth lowers the stimulated emission cross section compared to that of Nd:YAG, which allows increased energy storage when used as an amplifier, as mentioned above.

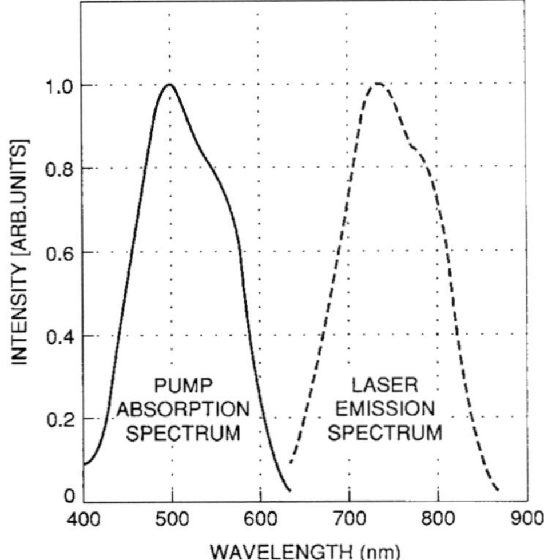

FIGURE 28 Absorption and emission spectra of a $Ti:Al_2O_3$ laser crystal showing the extremely broad emission (gain) spectrum for this material.

The large Nd:glass lasers system at Lawrence Livermore National Laboratory, referred to as NOVA, has produced a pulse energy of 100 KJ at the fundamental frequency and 40 KJ at the third harmonic frequency at 351 nm. This system was developed for studies associated with the inertial confinement laser fusion program.

$Ti:Al_2O_3$ Laser and Other Broad Bandwidth Solid-State Lasers.[7]

Another class of solid-state lasers provides emission and gain over a bandwidth of the order of 100–400 nm in the near-infrared, as indicated in Fig. 28. The pump absorption band is in the visible spectrum, thus allowing such pump sources as flash lamps and other lasers. The $Ti:AL_2O_3$ laser is perhaps the most well-known laser of this category in that it has the widest bandwidth, covering a wavelength range from 0.67 μm to greater than 1.07 μm. Other lasers of this type include alexandrite ($Cr:BeAl_2O_4$), lasing from 0.7 to 0.8 μm, and Cr:LiSAF which lases from 0.8 to 1.05 μm. These lasers are used in applications where either wide tunability or short-pulse production are desired. Pulses as short as 40 fsec have been produced with mode-locked versions of these lasers. $Ti:Al_2O_3$ lasers, although offering very wide gain bandwidth, have relatively short upper-laser-level lifetimes (3 μsec), thereby making them less efficient when pumping with conventional flash lamps. Cr:LiSAF lasers have longer upper-level lifetimes (67 μsec), closer to that of Nd:YAG or Nd:glass and have demonstrated efficient laser operation with pumping technologies developed for Nd:YAG lasers.

Color-Center Laser.[21] Color-center laser gain media are produced by a different form of impurity species than most solid-state lasers. Special defect centers (F-centers) are produced within alkali-halide crystals at a density of 1 part in 10,000 by irradiation with x rays. These defect centers have absorbing regions in the visible portion of the spectrum and emission (and gain) in the near-infrared. A variety of crystals are used to span the laser wavelength spectrum from 0.8 to 4.0 μm. Disadvantages of color-center lasers include operation at temperatures well below room temperature and the necessity to re-form the color centers at intervals of weeks or months in most cases.

Semiconductor Laser Gain Media

Semiconductor Lasers.[22] Semiconductor or diode lasers, typically about the size of a grain of salt, are the smallest lasers yet devised. They consist of a *p-n* junction formed in semiconductor crystal such as GaAs or InP in which the *p*-type material has an excess of holes (vacancies due to missing electrons) and the *n*-type material has an excess of electrons. When these two types of materials are brought together to form a junction, and an electric field in the form of a voltage is applied across the junction in the appropriate direction, the electrons and holes are brought together and recombine to produce recombination radiation at or near the wavelength associated with the bandgap energy of the material. The population of electrons and holes within the junction provides the upper-laser-level population, and the recombination radiation spectrum is the gain bandwidth $\Delta\nu$ of the laser, typically of the order of 0.5 to 1.0 nm.

The extended gain length required for these lasers is generally provided by partially reflecting cleaved parallel faces at the ends of the crystals which serve as an optical cavity. Because the cavity is so short, the longitudinal modes are spaced far apart in frequency ($\Delta\nu \cong 1\text{–}5 \times 10^{11}$ Hz or several tenths of a nanometer), and thus it is possible to obtain single longitudinal mode operation in such lasers. They require a few volts to operate with milliamperes of current.

Heterostructure semiconductor lasers include additional layers of different materials of similar electronic configurations, such as aluminum, indium, and phosphorous, grown adjacent to the junction to help confine the electron current to the junction region in order to minimize current and heat dissipation requirements (see Fig. 29). The laser mode in the transverse direction is either controlled by gain guiding, in which the gain is produced over a specific narrow lateral extent determined by fabrication techniques, or by index guiding, in which the index of refraction in the transverse direction is varied to provide total internal reflection of a guided mode. Quantum-well lasers have a smaller gain region (cross section), which confines the excitation current and thus the laser mode to an even smaller lateral region, thereby significantly reducing the threshold current and also the heat

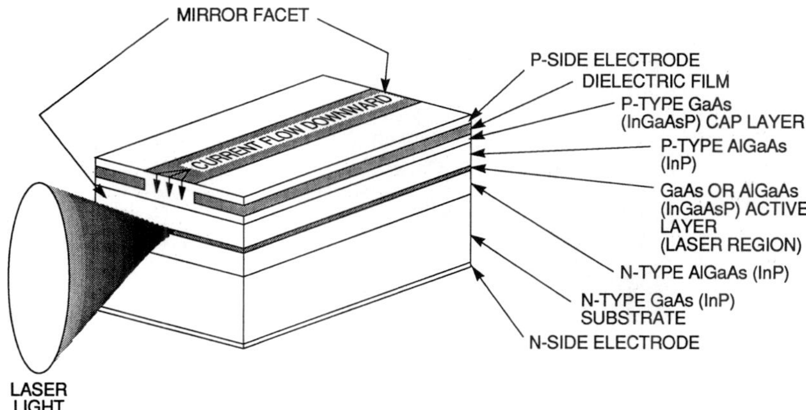

FIGURE 29 A typical heterostructure semiconductor laser showing the various layers of differential materials and the narrow region where current flows to produce the gain region in the active layer.

dissipation requirements. Because of these low threshold requirements, quantum-well semiconductor lasers are used almost exclusively for most semiconductor laser applications.

Multiple semiconductor lasers fabricated within the same bulk structure, known as *semiconductor arrays,* can be operated simultaneously to produce combined cw power outputs of up to 10 W from a laser crystal of dimensions of the order of 1 mm or less. Semiconductor lasers have also been fabricated in multiple arrays mounted vertically on a chip with the mirrors in the plane of the chip. They are known as *vertical cavity semiconductor lasers.* Each of the individual lasers has dimensions of the order of 5 μm and can be separately accessed and excited.

Semiconductor lasers operate over wavelengths ranging from 0.63 to 1.65 μm by using special doping materials to provide the expanded or contracted bandgap energies that provide the varied wavelengths. Also, new semiconductor lasers in the blue-green spectral region in crystals of ZnSe have recently been developed, but presently operate only at temperatures well below room temperature.

Laser Gain Media in Vacuum

Free Electron Laser.[23] Free-electron lasers are significantly different than any other type of laser in that the laser output does not result from transitions between discrete energy levels in specific materials. Instead, a high-energy beam of electrons, such as that produced by a synchrotron, traveling in a vacuum with kinetic energies of the order of 1 MeV, are directed to pass through a spatially varying magnetic field produced by two regular arrays of alternating magnet poles located on opposite sides of the beam as shown in Fig. 30. The alternating magnetic field causes the electrons to oscillate back and forth in a direction transverse to the beam direction. The frequency of oscillation is determined by the electron beam energy, the longitudinal spacing of the alternating poles (the magnet period), and the separation between the magnet arrays on opposite sides of the beam. The transverse oscillation of the electrons causes them to radiate at the oscillation frequency and to stimulate other electrons to oscillate and thereby radiate at the same frequency, in phase with the originally oscillating electrons. The result is an intense tunable beam of light emerging from the end of the device. Mirrors can be placed at the ends of the radiated beam to produce feedback and extra amplification.

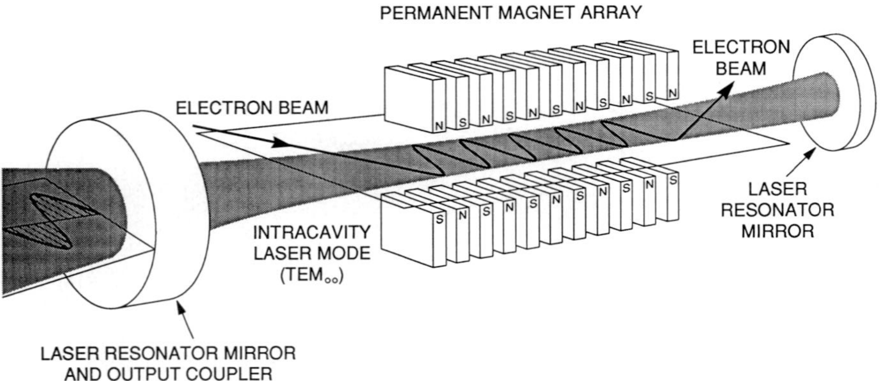

FIGURE 30 A general diagram of a free-electron laser showing how the electron beam is introduced into the cavity and how the alternating magnets cause the beam to oscillate to produce laser radiation.

Free-electron lasers have operated at wavelengths ranging from the near-ultraviolet (0.25 μm) to the far-infrared (6 mm) spectral regions. They are efficient devices and also offer the potential of very high average output powers.

11.7 REFERENCES

1. A. Corney, *Atomic and Laser Spectroscopy,* Clarendon Press, Oxford, 1977.
2. P. W. Milonni and J. H. Everly, *Lasers,* John Wiley & Sons, New York, 1988.
3. H. G. Kuhn, *Atomic Spectra,* 2d ed. Academic Press, New York, 1969.
4. J. T. Verdeyen, *Laser Electronics,* 2d ed., Prentice-Hall, Englewood Cliffs, New Jersey, 1989.
5. T. H. Maiman, *Nature* **187:**493–494 (1960).
6. W. T. Walter, N. Solimene, M. Piltch, and G. Gould, *IEEE J. of Quant. Elect.* **QE-2:**474–479 (1966).
7. W. Koechner, *Solid-State Laser Engineering,* 3d ed., Springer-Verlag, New York, 1992.
8. F. J. Duarte and L. W. Hillman (eds.), *Dye Laser Principles,* Academic Press, New York, 1990.
9. W. W. Rigrod, *J. Appl. Phys.* **34:**2602–2609 (1963); **36:**2487–2490 (1965).
10. C. S. Willett, *An Introduction to Gas Lasers: Population Inversion Mechanisms,* Pergamon Press, Oxford, 1974.
11. A. G. Fox and T. Li, *Bell Syst. Tech. J.* **40:**453–488 (1961).
12. G. D. Boyd and J. P. Gordon, *Bell Syst. Tech. J.* **40:**489–508 (1961).
13. H. Kogelnik and T. Li, *Proc. IEEE* **54:**1312–1329 (1966).
14. A. E. Siegman, *Lasers,* University Science Books, Mill Valley, Calif., 1986.
15. W. Wagner and B. Lengyel, *J. Appl. Phys.* **34:**2040–2046 (1963).
16. R. L. Fork, B. I. Greene, and C. V. Shank, *Appl. Phys. Lett.* **38:**671–672 (1981).
17. J. Mark, L. Y. Liu, K. L. Hall, H. A. Haus, and E. P. Ippen, *Opt. Lett.* **14:**48–50 (1989).
18. D. K. Negus, L. Spinelli, N. Goldblatt, and G. Feuget, *Proc. on Advanced Solid State Lasers* (Optical Soc. of Am.) **10:**120–124 (1991).
19. H. Kogelnik and C. V. Shank, *J. Appl. Phys.* **43:**2327–2335 (1972).
20. M. Rokni, J. A. Mangano, J. H. Jacobs, and J. C. Hsia, *IEEE Journ. of Quant. Elect.* **QE-14:**464–481 (1978).
21. L. F. Mollenauer and D. H. Olson, *J. Appl. Phys.* **46:**3109–3118 (1975).
22. A. Yariv, *Quantum Electronics,* John Wiley & Sons, New York, 1989.
23. L. R. Elias, W. M. Fairbank, J. M. Madey, H. A. Schewttman, and T. J. Smith, *Phys. Rev. Lett.* **36:**717–720 (1976).
24. R. C. Elton, *X-Ray Lasers,* Academic Press, New York, 1990.

CHAPTER 12
LIGHT-EMITTING DIODES

Roland H. Haitz
M. George Craford
Robert H. Weissman
Hewlett-Packard Co.,
San Jose, California

12.1 GLOSSARY

c	velocity of light
E_g	semiconductor energy bandgap
h	Planck's constant
I_T	total LED current
J	LED current density
k	Boltzmann's constant
M	magnification
n_0	low index of refraction medium
n_1	high index of refraction medium
q	electron charge
T	temperature
V	applied voltage
η_i	internal quantum efficiency
θ_c	critical angle
λ	emission wavelength
τ	total minority carrier lifetime
τ_n	nonradiative minority carrier lifetime
τ_r	radiative minority carrier lifetime

12.2 INTRODUCTION

Over the past 25 years the light-emitting diode (LED) has grown from a laboratory curiosity to a broadly used light source for signaling applications. In 1992 LED production reached a level of approximately 25 billion chips, and $2.5 billion worth of LED-based components were shipped to original equipment manufacturers.

This article covers light-emitting diodes from the basic light-generation processes to descriptions of LED products. First, we will deal with light-generation mechanisms and light extraction. Four major types of device structures—from simple grown or diffused homojunctions to complex double heterojunction devices are discussed next, followed by a description of the commercially important semiconductors used for LEDs, from the pioneering GaAsP system to the AlGaInP system that is currently revolutionizing LED technology. Then processes used to fabricate LED chips are explained—the growth of GaAs and GaP substrates; the major techniques used for growing the epitixal material in which the light-generation processes occur; and the steps required to create LED chips up to the point of assembly. Next the important topics of quality and reliability—in particular, chip degradation and package-related failure mechanisms—will be addressed. Finally, LED-based products, such as indicator lamps, numeric and alphanumeric displays, optocouplers, fiber-optic transmitters, and sensors, are described.

This article covers the mainstream structures, materials, processes, and applications in use today. It does not cover certain advanced structures, such as quantum well or strained layer devices, a discussion of which can be found in Chap. 13, "Semiconductor Lasers." The reader is also referred to Chap. 13 for current information on edge-emitting LEDs, whose fabrication and use are similar to lasers.

For further information on the physics of light generation, the reader should consult Refs. 1–11. Semiconductor material systems for LEDs are discussed in Refs. 13–24. Crystal growth, epitaxial, and wafer fabrication processes are discussed in detail in Refs. 25–29.

12.3 LIGHT-GENERATION PROCESSES

When a p-n junction is biased in the forward direction, the resulting current flow across the boundary layer between the p and n regions has two components: holes are injected from the p region into the n region and electrons are injected from the n region into the p region. This so-called minority-carrier injection disturbs the carrier distribution from its equilibrium condition. The injected minority carriers recombine with majority carriers until thermal equilibrium is reestablished. As long as the current continues to flow, minority-carrier injection continues. On both sides of the junction, a new steady-state carrier distribution is established such that the recombination rate equals the injection rate.[1,2]

Minority-carrier recombination is not instantaneous. The injected minority carriers have to find proper conditions before the recombination process can take place. Both energy and momentum conservation have to be met. Energy conservation can be readily met since a photon can take up the energy of the electron-hole pair, but the photon doesn't contribute much to the conservation of momentum. Therefore, an electron can only combine with a hole of practically identical and opposite momentum. Such proper conditions are not readily met, resulting in a delay. In other words, the injected minority carrier has a finite lifetime τ_r before it combines radiatively through the emission of a photon.[2] This average time to recombine radiatively through the emission of light can be visualized as the average time it takes an injected minority carrier to find a majority carrier with the right momentum to allow radiative recombination without violating momentum conservation.

Unfortunately, radiative recombination is not the only recombination path. There are also crystalline defects, such as impurities, dislocations, surfaces, etc., that can trap the injected minority carriers. This type of recombination process may or may not generate light. Energy and momentum conservation are met through the successive emission of phonons. Again, the recombination process is not instantaneous because the minority carrier first has to diffuse to a recombination site. This nonradiative recombination process is characterized by a lifetime τ_n.[2]

Of primary interest in design of light-emitting diodes is the maximization of the radiative recombination relative to the nonradiative recombination. In other words, it is of interest to develop conditions where radiative recombination occurs fairly rapidly compared with nonradiative recombination. The effectiveness of the light-generation process is described by the fraction of the injected minority carriers that recombine radiatively compared to the total injection. The internal quantum efficiency, η_i can be calculated from τ_r and τ_n. The combined recombination processes lead to a total minority-carrier lifetime τ given by Eq. (1):

$$\frac{1}{\tau} = \frac{1}{\tau_r} + \frac{1}{\tau_n} \tag{1}$$

η_i is simply computed from Eq. (1) as the fraction of carriers recombining radiatively:[2]

$$\eta_i = \frac{\tau_n}{\tau_r + \tau_n} \tag{2}$$

Of interest are two simple cases: in the case of excellent material quality (large τ_n) or efficient radiative recombination conditions (small τ_r), the internal quantum efficiency approaches 100 percent. For the opposite case ($\tau_n \ll \tau_r$), we find $\eta_i \approx \tau_n / \tau_r \ll 1$. As discussed under "Material Systems," there are several families of III-V compounds with internal quantum efficiencies approaching 100 percent. There are also other useful semiconductor materials with internal quantum efficiencies in the 1 to 10 percent range.

To find material systems for LEDs with a high quantum efficiency, one has to understand the band structure of semiconductors. The band structure describes the allowed distribution of energy and momentum states for electrons and holes (see Fig. 1

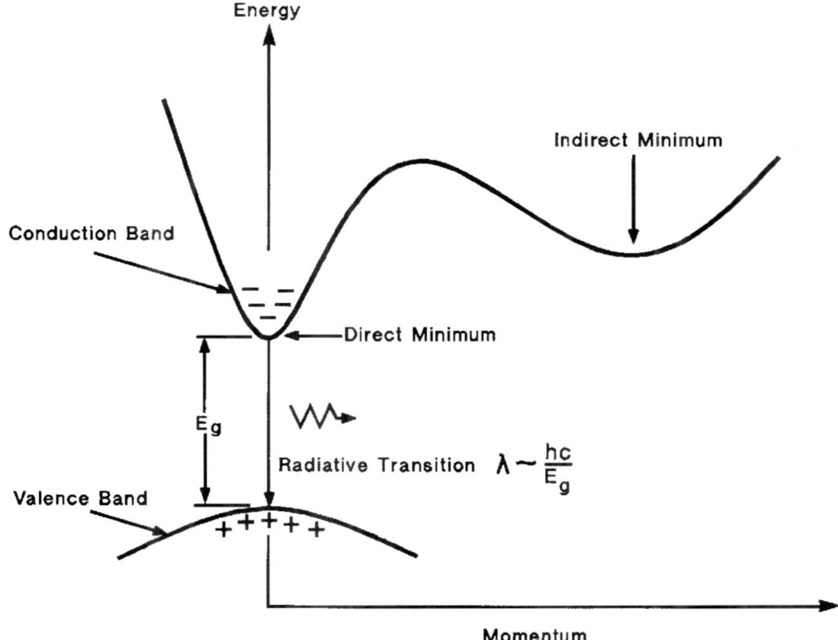

FIGURE 1 Energy band structure of a direct semiconductor showing radiative recombination of electrons in the conduction band with holes in the valence band.

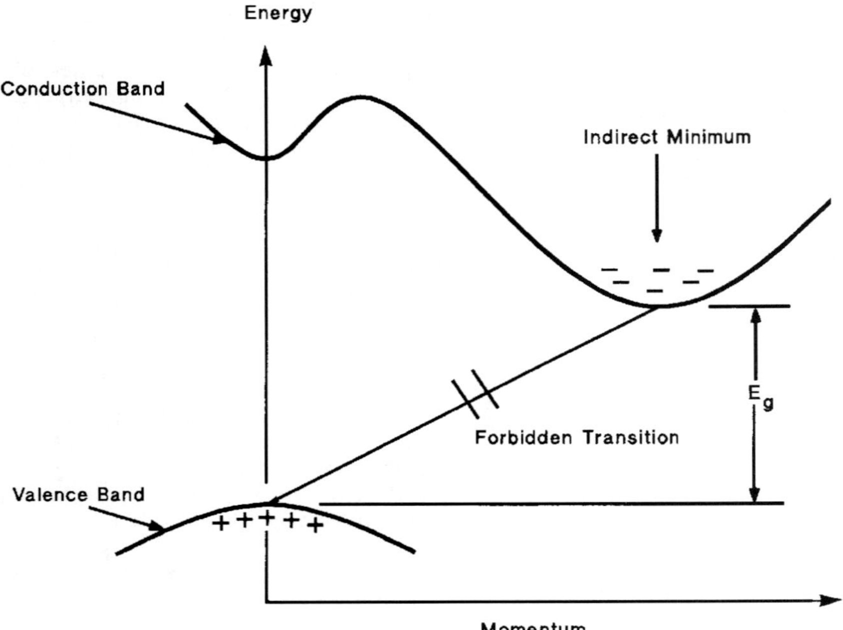

FIGURE 2 Energy band structure of an indirect semiconductor showing the conduction-band minima and valence-band maximum at different positions in momentum space. Radiative recombination of conduction-band electrons and valence-band holes is generally forbidden.

and Ref. 2). In practically all semiconductors the lower band, also known as the *valence band,* has a fairly simple structure, a paraboloid around the $\langle 0, 0, 0 \rangle$ crystalline direction. Holes will take up a position near the apex of the paraboloid and have very small momentum. The upper band, also known as the *conduction band,* is different for various semiconductor materials. All semiconductors have multiple valleys in the conduction band. Of practical interest are the valleys with the lowest energy. Semiconductor materials are classified as either *direct* or *indirect.*[1–3] In a direct semiconductor, the lowest valley in the conduction band is directly above the apex of the valence-band paraboloid. In an indirect semiconductor, the lowest valleys are not at $\langle 0, 0, 0 \rangle$, but at different positions in momentum/energy space (see Fig. 2). Majority or minority carriers mostly occupy the lowest energy states, i.e., holes near the top of the valence band paraboloid and electrons near the bottom of the lowest conduction-band valley.

In the case of a direct semiconductor the electrons are positioned directly above the holes at the same momentum coordinates. It is relatively easy to match up electrons and holes with proper momentum-conserving conditions. Thus, the resulting radiative lifetime τ_r is short. On the other hand, electrons in an indirect valley will find it practically impossible to find momentum-matching holes and the resulting radiative lifetime will be long. Injected carriers in indirect material generally recombine nonradiatively through defects.

In a direct semiconductor, such as GaAs, the radiative lifetime τ_r is in the range of 1–100 ns, depending on doping, temperature, and other factors. It is relatively easy to grow crystals with sufficiently low defect density such that τ_n is in the same range as τ_r.

For indirect semiconductors, such as germanium or silicon, the radiative recombination process is extremely unlikely, and τ_r is in the range of seconds.[1] In this case, $\tau_r \gg \tau_n$, and practically all injected carriers recombine nonradiatively.

The wavelength of the photons emitted in a radiative recombination event is determined by the energy difference between the recombining electron-hole pair. Since carriers relax quickly to an energy level near the top of the valence band (holes), or the bottom of the conduction band (electrons), we have the following approximation for the wavelength λ of the emitted photon (see Fig. 1):

$$\lambda \approx hc/E_g \tag{3}$$

where h = Planck's constant, c = velocity of light, and E_g = bandgap energy.

This relation is only an approximation since holes and electrons are thermally distributed at levels slightly below the valence-band maximum and above the conduction-band minimum, resulting in a finite linewidth in the energy or wavelength of the emitted light. Another modification results from a recombination between a free electron and a hole trapped in a deep acceptor state. See Ref. 2 for a discussion of various recombination processes.

To change the wavelength or energy of the emitted light, one has to change the bandgap of the semiconductor material. For example, GaAs with a bandgap of 1.4 eV has an infrared emission wavelength of 900 nm. To achieve emission in the visible red region, the bandgap has to be raised to around 1.9 eV. This increase in E_g can be achieved by mixing GaAs with another material with a wider bandgap, for instance GaP with $E_g = 2.3$ eV. By adjusting the ratio of arsenic to phosphorous the bandgap of the resulting ternary compound, GaAsP, can be tailored to any value between 1.4 and 2.3 eV.[3]

The resulting band structure with varying As to P ratio is illustrated in Fig. 3. Note, the

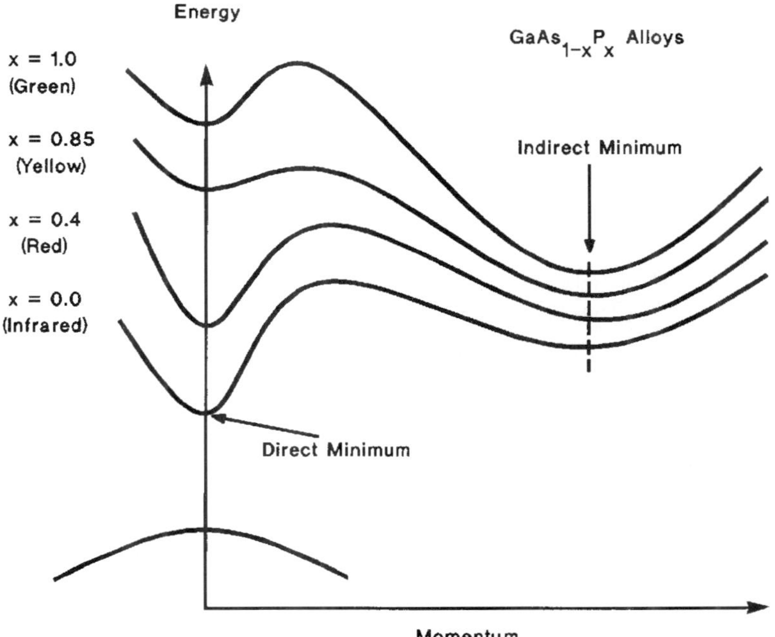

FIGURE 3 Energy band diagram for various alloys of the $GaAs_{1-x}P_x$ material system showing the direct and indirect conduction-band minima for various alloy compositions.

two conduction-band valleys do not move upward in energy space at the same rate. The direct valley moves up faster than the indirect valley with increasing phosphorous composition. At a composition of around 40 percent GaP and 60 percent GaAs, the direct and indirect valleys are about equal in energy. When the valleys are approximately equal in energy, electrons in the conduction band can scatter from the direct valley into the indirect valley. While the direct valley electrons still undergo rapid radiative recombination, the indirect valley electrons have a long radiative lifetime and either have to be scattered back to the direct valley or they will recombine nonradiatively. In other words, near this crossover between direct and indirect valleys, the radiative efficiency drops off dramatically, and for compositions with greater than 40 percent phosphorous the direct radiative recombination is practically nonexistent.[3,4]

The above discussion indicates that indirect semiconductors are not suitable for efficient generation of light through minority-carrier recombination. Fortunately, the introduction of so-called isoelectronic impurities can circumvent this limitation and introduces a new radiative recombination process.[5] A commonly used isoelectronic trap is generated by substituting a nitrogen atom for phosphorous in the GaAsP system[6–11] (see Fig. 4). Since N and P both have five electrons in their outer shell, the trap is electrically neutral. However, the stronger electronegativity of N relative to P can result in capture of an electron from the conduction band. Since the electron is very tightly bound to the impurity atom, its wave function in momentum space is spread out and has reasonable magnitude at $\langle k = 0 \rangle$ in momentum space.[10] The negatively charged defect can attract a free hole to form a loosely bound electron-hole pair or "exciton." This electron-hole pair has a high probability to recombine radiatively. The energy of the emitted light is less than E_g. Another isoelectronic trap in GaP is formed by ZnO pairs (Zn on a Ga site and O on a P

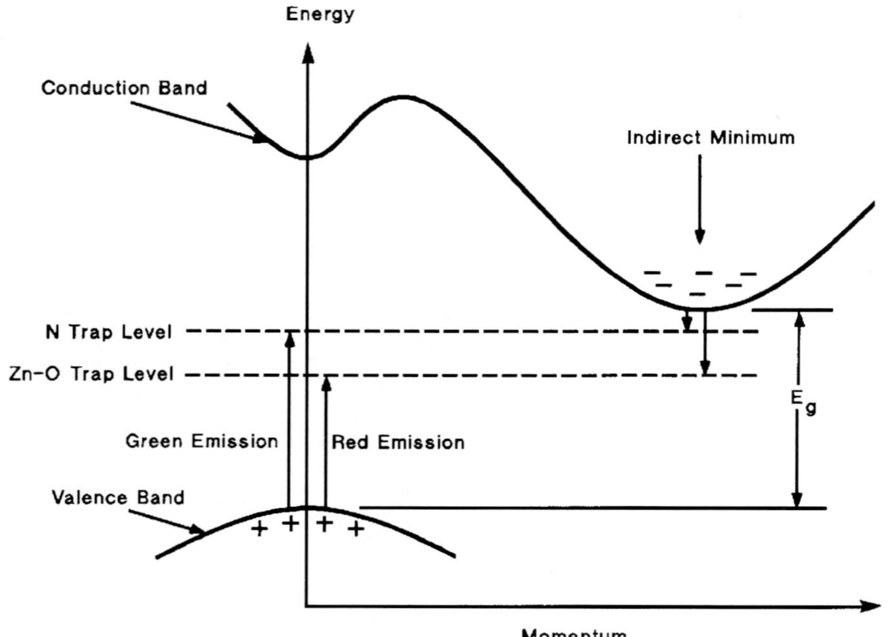

FIGURE 4 Formation of excitons (electron-hole pairs) by the addition of isoelectronic dopants N and ZnO to an indirect semiconductor. The excitons have a high probability to recombine radiatively.

site) (see Fig. 4). The ZnO trap is deeper than the N trap, resulting in longer wavelength emission in the red region of the spectrum.[1]

The recombination process for exciton recombination is quite complex. For a detailed analysis, the reader is referred to Ref. 10. One result of this analysis is the recognition that the bound exciton has a relatively long lifetime in the range of 100–1000 ns. Light emission by exciton recombination is generally slower than emission due to direct band-to-band recombination.

12.4 LIGHT EXTRACTION

Generating light efficiently within a semiconductor material is only one part of the problem to build an efficient light source. The next challenge is the extraction of light from within the LED chip to the outside. The designer must consider total internal reflection.[1] According to Snell's law, light can escape from a medium of high index of refraction n_1 into a medium of low index refraction n_0 only if it intersects the surface between the two media at an angle from normal less than the critical angle θ_c with θ_c being defined by Eq. (4):

$$\theta_c = \arcsin n_0/n_1 \tag{4}$$

Most semiconductor LEDs have an isotropic emission pattern as seen from within the light-generating material. Assuming a cubic shape for the LED chip, because of internal reflections, only a small fraction of the isotropically emitted light can escape any of the six surfaces. As a case in point, let us calculate the emission through the top surface. For typical light-emitting semiconductors, n_1 is in the range of 2.9 to 3.6. If $n_1 = 3.3$ and $n_0 = 1.0$ (air), we find $\theta_c = 17.6°$. The emission from an isotropic source into a cone with a half angle of θ_c is given by $(1 - \cos\theta_c)/2$. After correcting for Fresnel reflections, only 1.6 percent of the light generated escapes through the LED top surface into air. Depending on chip and *p-n* junction geometry, virtually all of the remaining light (98.4 percent) is reflected and absorbed within the LED chip.

The fraction of light coupled from chip to air is a function of the number of surfaces through which the chip can transmit light effectively. Most LED chips are called "absorbing substrate" (AS) chips. In such a chip, the starting substrate material (discussed later under "Substrate Technology") has a narrow bandgap and absorbs all the light with energy greater than the bandgap of the substrate. Consider the case of a GaAsP LED grown on a GaAs substrate. The emitted light ($E_g > 1.9$ eV) is absorbed by the GaAs substrate ($E_g = 1.4$ eV). Thus, a GaAsP-emitting layer on a GaAs substrate can transmit only through its top surface. Light transmitted toward the side surfaces or downward is absorbed.

To increase light extraction, the substrate or part of the epitaxial layers near the top of the chip has to be made of a material transparent to the emitted light. The "transparent substrate" (TS) chip is designed such that light transmitted towards the side surfaces within θ_c half-angle cones can escape. Assuming that there is negligible absorption between the point of light generation and the side walls, this increases the extraction efficiency by a factor of five (five instead of one escape cones).

In a TS chip, additional light can be extracted if the side walls are nonplanar, i.e., if light from outside an escape cone can be scattered into an escape cone. This process increases the optical path within the chip and is very dependent on residual absorption. In a chip with low absorption and randomizing side surfaces, most of the light should escape. Unfortunately, in practical LED structures, there are several absorption mechanisms left such as front and back contacts, crystal defects, and absorption in areas where secondary radiative recombination is inefficient.

TABLE 1 Extraction Efficiency into Air or Plastic for Three Types of Commonly Used LED Chips

Chip type	No. cones	Typical extraction efficiency	
		Air, %	Plastic, %
AS	1	1.5	4
Thick window	3	4.5	12
TS	5	7.5	18

A common approach is to use a hybrid chip with properties between AS and TS chips. These chips utilize a thick, transparent window layer above the light-emitting layer. If this layer is sufficiently thick, then most of the light in the top half of the cones transmitted toward the side surfaces will reach the side of the chip before hitting the substrate. In this case of hybrid chips, the efficiency is between that of AS and TS chips as shown in Table 1.

Another important way to increase extraction efficiency is derived from a stepwise reduction in the index of refraction from chip to air. If the chip is first imbedded in a material with an intermediate index, i.e., plastic with $n_2 = 1.5$, then the critical angle θ_c between chip and plastic is increased to 27°. The extraction efficiency relative to air increases by the ratio of $(n_2/n_0)^2$ plus some additional correction for Fresnel-reflection losses. The gain from plastic encapsulation is usually around 2.7× compared to air. Chips with multipath internal reflection will result in lower gains. It is important to note that this gain can be achieved only if the plastic/air interface can accommodate the increased angular distribution through proper lenslike surface shaping or efficient scattering optics. Table 1 illustrates the approximate extraction efficiencies achieved by the three dominant chip structures in air and in plastic. The numbers assume only first-pass extraction, limited absorption, and no multiple reflections within the chip.

12.5 DEVICE STRUCTURES

LED devices come in a broad range of structures. Each material system (see following section) requires a different optimization. The only common feature for all LED structures is the placement of the *p-n* junction where the light is generated. The *p-n* junction is practically never placed in the bulk-grown substrate material for the following reasons:

- The bulk-grown materials such as GaAs, GaP, and InP usually do not have the right energy gap for the desired wavelength of the emitted light.
- The light-generating region requires moderately low doping that is inconsistent with the need for a low series resistance.
- Bulk-grown material often has a relatively high defect density, making it difficult to achieve high efficiency.

Because of these reasons, practically all commercially important LED structures utilize a secondary growth step on top of a single-crystal bulk-grown substrate material. The secondary growth step consists of a single-crystal layer lattice matched to the substrate.

This growth process is known as *epitaxial growth* and is described in a later section of this chapter.

The commonly used epitaxial structures can be classified into the following categories:

- Homojunctions
  - grown
  - diffused
- Heterojunctions
  - single confinement
  - double confinement

Grown Homojunctions

Figure 5 illustrates one of the simplest design approaches to an LED chip. An n-type GaAs layer with low to moderate doping density is grown on top of a highly doped n-type substrate by a vapor or liquid-phase epitaxial process (see "Epitaxial Technology"). After a growth of 5–10 μm, the doping is changed to p-type for another 5–10 μm. A critical dimension is the thickness of the epitaxial p-layer. The thickness should be larger than the diffusion length of electrons. In other words, the electrons should recombine radiatively in the epitaxially grown p layer before reaching the surface. The p-layer should be of sufficiently high quality to meet the condition for efficient recombination, i.e., $\tau_n \gg \tau_r$. In addition, the side surfaces may have to be etched to remove damage. Damage and other defects where the p-n junction intercepts the chip surface can lead to a substantial leakage current that reduces efficiency, especially at low drive levels.

The structure of Fig. 5 was used in some of the earliest infrared emitters (wavelength 900 nm). Efficiency was low, typically 1 percent. Modern infrared emitters use Si-doped

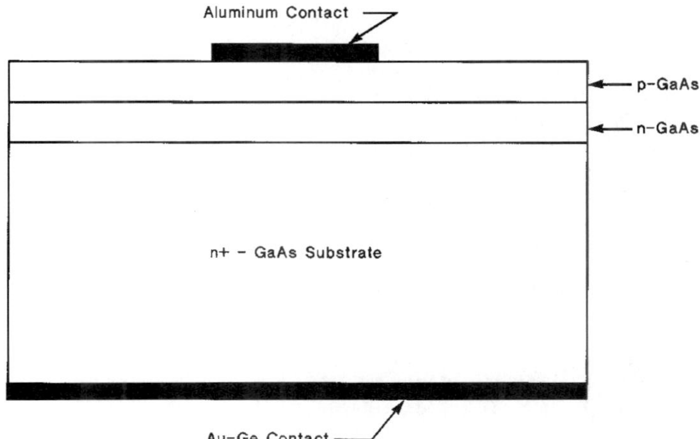

FIGURE 5 Cross section of an infrared LED chip. A p-n junction is formed by epitaxially growing n- and p-doped GaAs onto an $n+$-doped GaAs substrate.

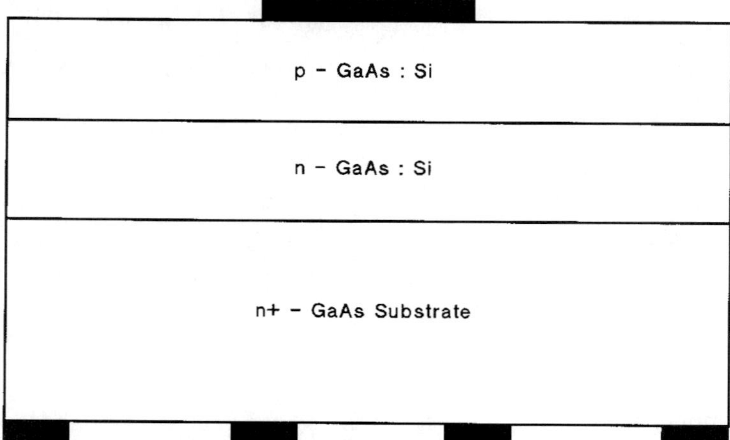

FIGURE 6 A high-efficiency IR LED made by LPE growth of GaAs that is doped with silicon on both the *p* and *n* sides of the junction. To increase external quantum efficiency, a partly reflective back contact is employed.

GaAs (see Fig. 6). The detailed recombination mechanism in GaAs:Si even today is quite controversial and goes beyond the scope of this publication. The recombination process has two important characteristics: (1) the radiative lifetime τ_r is relatively slow, i.e., in the range of 1 μs; and (2) the wavelength is shifted to 940 nm. At this wavelength the GaAs substrate is partly transparent, making this device a quasi-TS structure with efficiencies into plastic of 5 to 10 percent.

Diffused Homojunction

The chip structure of Fig. 5 can also be produced by a zinc (Zn) diffusion into a thick *n*-layer. The commercially most significant structure of this type is shown in Fig. 7. By replacing 40 percent of the As atoms with P atoms, the bandgap is increased to 1.92 eV to make a GaAsP LED that emits visible red light. In this case, the *p*-*n* junction is diffused selectively by using a deposited layer of silicon-nitride as a diffusion mask. This structure of Fig. 7 has several advantages over the structure of Fig. 5. Lateral diffusion of Zn moves the intersection of the junction with the chip surface underneath the protecting Si_3N_4 layer. This layer protects the junction from contamination and adds to the long-term stability of the device. In addition, it is important for applications requiring more than one clearly separated light-emitting area. For instance, seven-segment displays, such as those used in the early hand-held calculators, are made by diffusing seven long and narrow stripes into a single chip of GaAsP material. (See Fig. 8.) This chip consists of eight (seven segments plus decimal point) individually addressable *p*-regions (anodes) with a common *n*-type cathode. Such a chip is feasible only in an AS-type structure because the individual segments have to be optically isolated from each other. A TS-type structure results in unacceptable levels of crosstalk.

Figure 7 shows another feature of practical LED devices. The composition with 40 percent P and 60 percent As has a lattice constant (atomic spacing) that is different from the GaAs substrate. Such a lattice mismatch between adjacent layers would result in a very high density of dislocations. To reduce this problem to an acceptable level, one has to slowly increase the phosphorous composition from 0 percent at the GaAs interface to 40

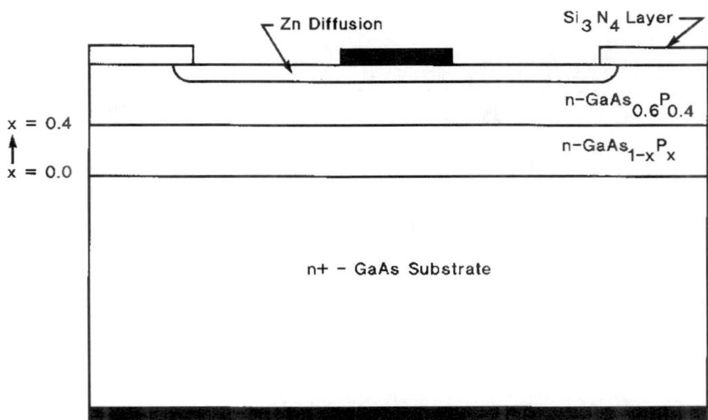

FIGURE 7 GaAsP LED which emits at 650 nm. On a GaAs substrate, a layer is grown whose composition varies linearly from GaAs to $GaAs_{0.6}P_{0.4}$, followed by a layer of constant composition. Zinc is selectively diffused using an Si_3N_4 mask to form the light-emitting junction.

percent over a 10–20-μm-thick buffer layer. Typically, the buffer layer is graded linearly. The phosphorous composition is increased linearly from bottom to top. The thicker the buffer layer, the lower is the resulting dislocation density. Cost constraints keep the layer in the 10–15-μm range. The layer of constant composition (40 percent P) has to be thick enough to accommodate the Zn diffusion plus the diffusion length of minority carriers. A thickness range of 5–10 μm is typical.

Another variation of a homojunction is shown in the TS-chip structure of Fig. 9. Instead of an absorbing GaAs substrate, one starts with a transparent GaP substrate. The graded layer has an inverse gradient relative to the chip shown in Fig. 7. The initial growth is 100 percent GaP phasing in As linearly over 10–15 μm. At 15 percent As, the emission is in the yellow range (585 nm), at 25 percent in the orange range (605 nm), and at 35 percent in the red range (635 nm). Figure 3 shows the approximate band structure of this material system. The composition range mentioned above has an indirect band structure. To obtain efficient light emission, the region of minority-carrier injection is doped with nitrogen forming an isoelectronic recombination center (see exciton recombination in the first section).

Figure 9 shows an important technique to increase extraction efficiency. In a TS-chip, the major light loss is due to free carrier absorption at the alloyed contacts. Rather than covering the entire bottom surface of the chip with contact metal, one can reduce the contact area either by depositing small contact islands (see Fig. 6) or by placing a dielectric mirror (deposited SiO_2) between the substrate and the unused areas of the back contact (Fig. 9). This dielectric mirror increases the efficiency by 20 to 50 percent at the expense of higher manufacturing cost.

Single Heterojunctions

Heterojunctions introduce a new variable: local variation of the energy bandgap resulting in carrier confinement. Figure 10 shows a popular structure for a red LED emitter chip. A *p*-type layer of GaAlAs with 38 percent Al is grown on a GaAs substrate. The GaAlAs alloy system can be lattice-matched to GaAs; therefore, no graded layer is required as in the GaAsP system of Fig. 7. Next, an *n*-type layer is grown with 75 percent Al. The variation of E_g from substrate to top is illustrated in Fig. 11. Holes accumulate in the

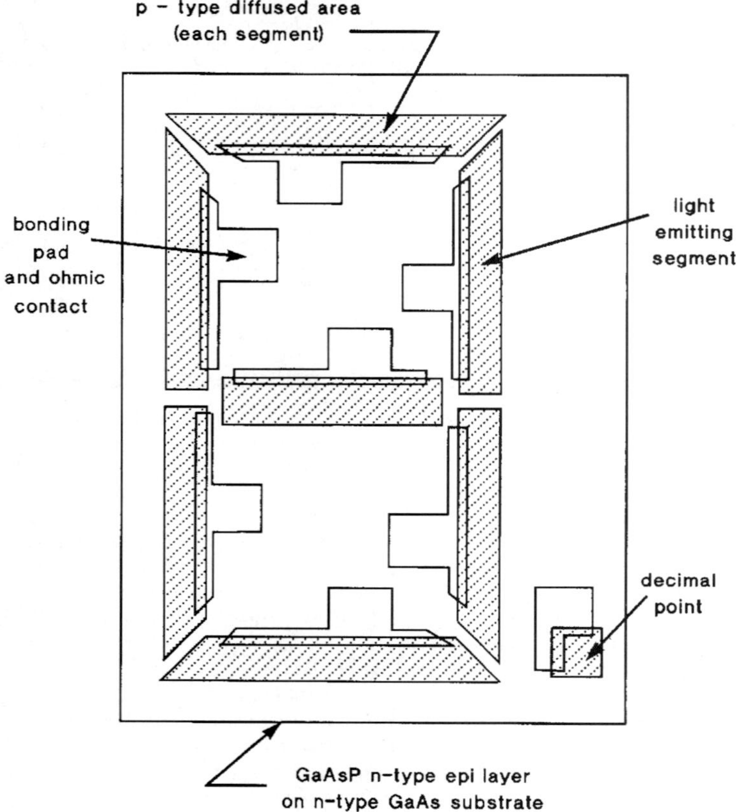

FIGURE 8 Monolithic seven-segment display chip with eight separate diffused regions (anodes) and a common cathode (the GaAs substrate).

GaAlAs p-layer with the narrower bandgap. Electrons are injected from the n-layer into this p-layer. The holes have insufficient energy to climb the potential barrier into the wide-bandgap material. Holes are confined to the p layer. In the p layer, the radiative recombination time is very short because of the high concentration of holes. As a result, the internal quantum efficiency is quite high. A variation of this structure is a widely used infrared emitter that emits at 880 nm.

Double Heterostructures

The double heterostructure shown in Fig. 12 repeats the p-side confinement of Fig. 10 on the n side. An n-type buffer layer is grown on the GaAs substrate to create a high-quality surface onto which the first n-type GaAlAs confinement layer with 75 percent Al is grown. The active or light-generation layer is a 3-μm-thick p-type layer with 38 percent Al. The top p-type confinement layer again uses 75 percent Al. This structure with the energy-band diagram of Fig. 13 has two advantages: (1) There is no hole injection into the n-type layer with reduced efficiency and a slow hole recombination in the lowly doped n layer. (2) The high electron and hole density in the active layer reduces τ_r, thus increasing device speed

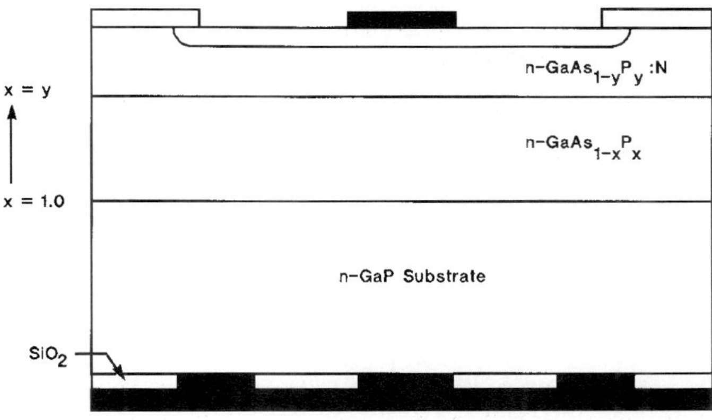

FIGURE 9 Cross section of a $GaAs_{1-x}P_x$ LED which, by changing the composition "x," produces red, orange, or yellow light. The top layer is doped with nitrogen to increase the quantum efficiency. The GaP substrate is transparent to the emitted light. Also shown is a reflective back contact made by depositing the contact metallization on top of SiO_2.

and efficiency. The increased speed is quite important for LED sources in fiber-optic communication applications. (See "Fiber Optics" subsection later in this chapter.)

The double heterostructure of Fig. 12 represents a one-dimensional containment of injected carriers. Injection and light emission occurs across the entire lateral dimension of the chip. For fiber-optic applications, the light generated over such a large area cannot be effectively coupled into small-diameter fiber. A rule of thumb for fiber coupling requires that the light-emitting area be equal to or, preferably, smaller than the fiber core diameter. This rule requires lateral constraint of carrier injection. The localized diffusion of Fig. 7 is not applicable to grown structures such as those in Fig. 12. The preferred solution inserts

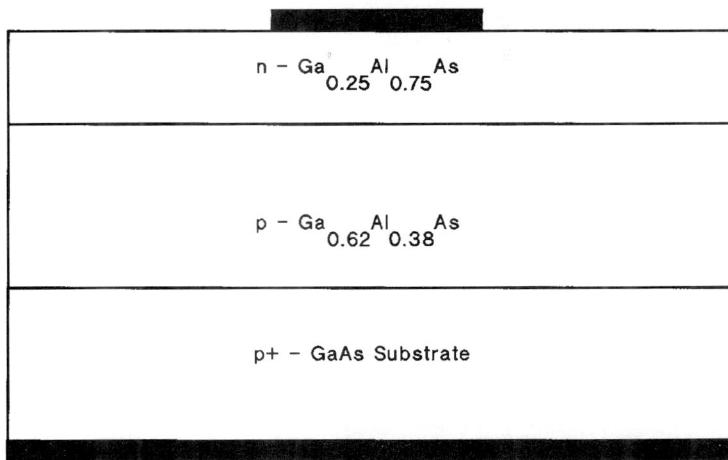

FIGURE 10 Cross section of a single heterostructure LED.

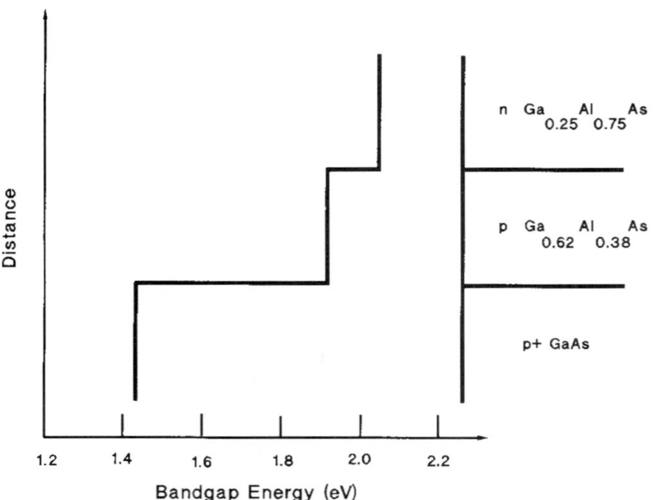

FIGURE 11 Variation in energy bandgap for the various layers in the GaAlAs LED shown in Fig. 10.

an *n* layer between buffer and lower confinement layer (see Fig. 14). A hole etched into the *n* layer allows current flow. Outside of this hole, the *p*-*n* junction between *n* layer and lower confinement layer is reverse-biased, thus blocking any current flow. The disadvantage of this approach is a complication of the growth process. In a first growth process, the *n* layer is grown. Then a hole is etched into the *n* layer using standard photolithographic etching techniques. Finally, a second epitaxial growth is used for the remaining layers.

Another technique to constrain current injection utilizes a small ohmic contact.[12] It is used frequently in conjunction with InP-based fiber-optic emitters (see Fig. 15). A SiO_2 layer limits contact to a small-diameter (typically 25 μm) hole that results in a relatively

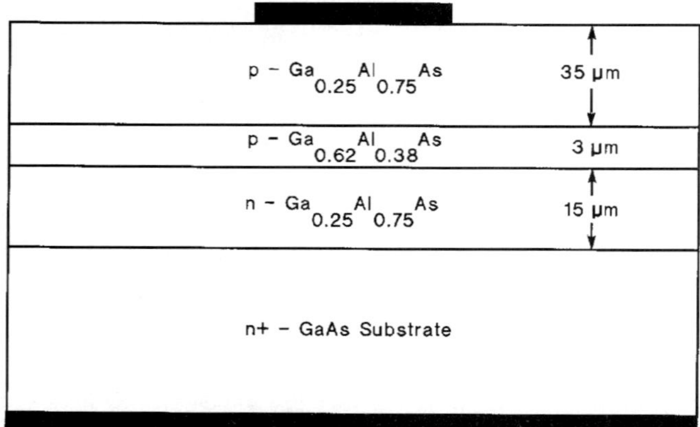

FIGURE 12 Structure of a double heterostructure (DH) GaAlAs LED. The DH is composed of a $Ga_{0.62}Al_{0.38}As$ layer surrounded on either side by a $Ga_{0.25}Al_{0.75}As$ layer. The thick top layer acts as window to increase light extraction through the side walls of the chip.

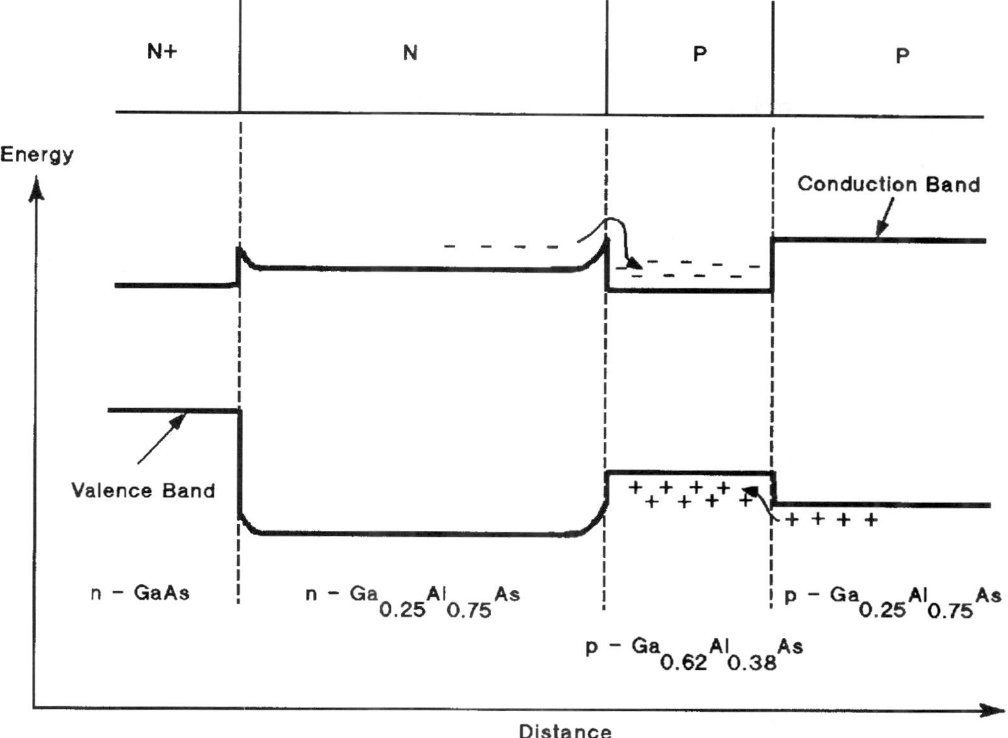

FIGURE 13 Energy band diagram of the GaAlAs LED shown in Fig. 12. The LED is forward-biased. Electrons and holes are confined in the *p*-doped $Ga_{0.62}Al_{0.38}As$ layer, which increases the radiative recombination efficiency.

small light-emitting area. The etched lens shown for this structure helps to collimate the light for more efficient coupling onto the fiber.[12]

12.6 MATERIALS SYSTEMS

The $GaAs_{1-x}P_x$ System

The most widely used alloy for LEDs is the ternary $GaAs_{1-x}P_x$ system, including its two binary components GaAs and GaP. This system is best described by the composition parameter "x" with $0 \leq x \leq 1$. For $x = 0$, we have GaAs and for $x = 1$ the composition is GaP. For $x \leq 0.4$, the alloy has a direct bandgap. GaAs was developed in the early 1960s as an infrared emitter with a wavelength of 910 nm and an efficiency in the range of 1 percent. This emitter was soon followed by a Si-doped variety. As discussed earlier, this leads to an emission wavelength of 940 nm, a wavelength at which the GaAs substrate is partly transparent. The resulting efficiency is increased substantially and, depending on configuration, is in the 5 to 10 percent range. However, the recombination process is quite slow, resulting in rise and fall times in the 0.5–1.0 μs range (see Table 2). One other drawback is caused by the low absorption coefficient of Si detectors at 940 nm. To absorb 90 percent of the light requires a detector thickness of 60–70 μm. Conventional photo

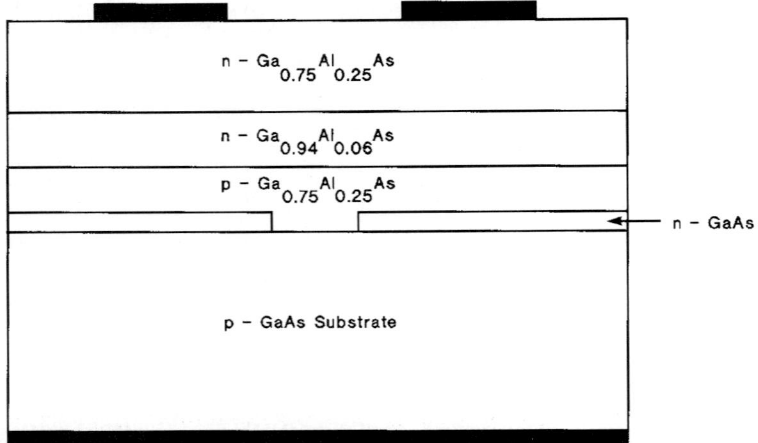

FIGURE 14 Cross section of an LED with three-dimensional carrier confinement. A DH structure is used to confine injected carriers in the $Ga_{0.94}Al_{0.06}As$ layer (direction perpendicular to the junction). The patterned, *n*-type GaAs layer is used to limit current flow in the lateral direction. The small emitting area and the 820–nm emission of this LED makes it ideal for fiber-optic applications.

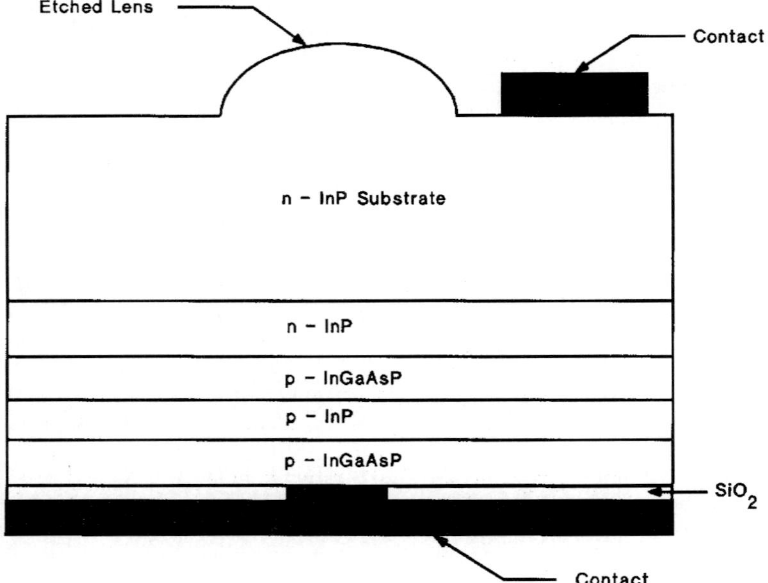

FIGURE 15 Structure of a 1300-nm LED used for optical fiber communications. The cross section shows the DH-layer configuration, limited area back contact for emission-size control, and an etched lens at the top of the chip which magnifies ($M = 2$) the source area and collimates the light for effective coupling into a fiber.

TABLE 2 Performance Summary of LED Chips in the $GaAs_{1-x}P_x$ System

x	Growth process*	Isoel. dopant	Substrate†	Dominant wavelength (nm)	Color‡	Luminous efficacy§ (lm/A)	Quantum efficiency§ (%)	Speed (ns)
0	LPE	—	AS	910	IR		1	50
0	LPE	—	TS	940	IR		10	1000
0.3	VPE	—	AS	700	IR		0.5	50
0.4	VPE	—	AS	650	Red	0.2		50
0.65	VPE	N	TS	635	Red	2.5		300
0.75	VPE	N	TS	605	Orange	2.5		300
0.85	VPE	N	TS	585	Yellow	2.5		300
1.0	LPE	N	TS	572	Y/green	6		300
1.0	LPE	—	TS	565	Green	1		
1.0	LPE	ZnO	TS	640	Red	1		

* LPE = liquid phase epitaxy; VPE = vapor phase epitaxy.
† AS = absorbing substrate; TS = transparent substrate.
‡ IR = infrared.
§ Into plastic (index = 1.5).

transistors are quite suitable as detectors. Integrated photo ICs with their 5–7-μm-thick epitaxial layers are very inefficient as detectors at 940 nm.

To shift the wavelength toward the near-infrared or into the visible spectrum, one has to grow a ternary alloy, a mixture between GaAs and GaP. Of commercial interest are two alloys with $x = 0.3$ and $x = 0.4$ grown on a GaAs substrate (see Table 2). The $x = 0.4$ alloy was the first commercially produced material with a wavelength in the visible range of the spectrum. Grown on an absorbing substrate, it has a modest luminous efficacy of around 0.2 lm/A. [Luminous efficacy is the luminous (visible) flux output measured in lumens divided by the electrical current input.] The absorbing substrate allows the integration of multiple light sources into a single chip without crosstalk. Such monolithic seven-segment displays became the workhorse display technology for hand-held calculators from 1972 to 1976.[9,13] Today this alloy is used in LED arrays for printers.[14]

The $x = 0.3$ alloy with a wavelength of 700 nm became important in the mid 1970s as a light source in applications using integrated photodetectors. It has three to five times the quantum efficiency of the $x = 0.4$ alloy (see Table 2), but has a lower luminous efficacy because of the much-reduced eye sensitivity at 700 nm.

For $x > 0.4$, the GaAsP material system becomes indirect (see earlier under "Light-Generation Process" and Fig. 3). The quantum efficiency decreases faster than the increase in eye sensitivity.[4] The only way to achieve a meaningful efficiency is through the use of isoelectronic dopants as described earlier. The choices for isoelectronic dopants that have been successful are N for GaAsP[11] and N or ZnO for GaP. Nitrogen doping is used widely for alloys with $x = 0.65$ to $x = 1.0$.[6–10] The resulting light sources cover the wavelength range from 635 nm to approximately 565 nm (see Table 2). Since these alloys are either GaP or very close in composition to GaP, they are all grown on GaP substrates. The resulting transparent-substrate chip structure increases the luminous efficacy.

In the case of the binary GaP compound, the dominant wavelength depends on N concentration. With low concentrations, practically all N atoms are isolated in single sites. With increasing concentration, N atoms can arrange themselves as pairs or triplets. The resulting electron traps have lower energy states which shift the emitted light toward longer wavelength. Phonon coupling can also reduce the emission energy. Commercially significant are two compositions: (1) undoped GaP which emits at 565 nm (dominant wavelength) with a reasonably green appearance and a low efficiency of around 1 lm/A, and (2) GaP with a nitrogen concentration in the range of 10^{19} cm^{-3} with a substantially higher luminous efficiency of around 6 lm/A at 572 nm. At this wavelength, the color appearance is yellow-green, often described as chartreuse.

Three nitrogen-doped ternary alloys of GaAsP are commercially important for red, orange, and yellow. The red source with $x = 0.65$ has an efficiency in the range of 2–3 lm/A. With increasing bandgap or decreasing wavelength, the drop in quantum efficiency is compensated by an increase in eye sensitivity, resulting in a practically wavelength-independent luminous efficiency for the range of 635 to 585 nm.[8,15]

ZnO-doped GaP is an interesting material. The quantum efficiency of such chips is relatively high, around 3 percent. However, the linewidth is quite broad. The quantum efficiency peaks at 700 nm, but the luminous efficiency peaks at 640 nm (dominant wavelength). In other words, most of the photons are emitted at wavelengths with low eye sensitivity. Another problem of GaP:ZnO is saturation. The deep ZnO electron trap causes very slow exciton recombination. At high injection currents, all traps are saturated and most of the injected carriers recombine nonradiatively. At low injection levels ($\leq$1 A/cm^2), the efficacy is relatively high, 3–5 lm/A. At a more useful density of 10–30 A/cm^2, the emission saturates, resulting in an efficacy of around 1 lm/A.

The $Al_xGa_{1-x}As$ System

The $Al_xGa_{1-x}As$ material system has a direct bandgap for $0 \leq x \leq 0.38$. This system has one very significant advantage over the GaAsP system described above, the entire alloy range from $x = 0$ to $x = 1$ can be lattice-matched to GaAs. In other words, every alloy composition can be directly grown on any other alloy composition without the need for transition layers. This feature allows the growth of very abrupt heterojunctions, i.e., abrupt transition in composition and bandgap. These heterojunctions add one important property not available in the GaAsP system: carrier containment (see earlier under "Device Structures"). Carrier containment reduces the movement of injected carriers in a direction perpendicular to the junction. Thus, carrier density can be increased beyond the diffusion-limited levels. This results in increased internal quantum efficiency and higher speed. Another benefit is reduced absorption and improved extraction efficiency (under "Light Extraction").

Of practical significance are two compositions: $x = 0.06$ and $x = 0.38$ (see Table 3). Both compositions exist in single and double heterojunction variations (see under "Device Structures"). The double heterojunctions usually have a 1.5–2.0-times advantage in efficiency and speed. In all cases, the efficiency strongly depends on the thickness of the window layer and, to a lesser degree, on the thickness of the transparent layer between active layer and absorbing substrate (see Fig. 12). Chips with a transparent substrate have an additional efficiency improvement of 1.5–3.0 times again, depending on layer thickness

TABLE 3 Performance Summary of LED Chips in the $Al_xGa_{1-x}As$ System

x	λ (nm)	Substrate*	Structure†	Efficiency or efficacy	Speed (ns)
0.06	820	AS, TW	DH	8%	30
0.06	820	TS	DH	15%	30
0.38	650	AS, TW	SH	4 lm/A	
0.38	650	AS, TW	DH	8 lm/A	
0.38	650	TS	DH	16 lm/A	

* AS = absorbing substrate; TW = thick window layer; TS = transport substrate.
† DH = double heterostructure; SH = single heterostructure.

and contact area. The efficiency variation is best understood by counting exit cones as described in the text in conjunction with Table 1. For $x = 0.06$, the internal quantum efficiency of a double heterojunction approaches 100 percent. For $x = 0.38$, the direct and indirect valleys are practically at the same level and the internal quantum efficiency is reduced to the range of 50 percent, again dependent upon the quality of the manufacturing process.

The best compromise for efficiency and speed is the $x = 0.06$ alloy as a double heterostructure. Depending on layer thickness, substrate, and contact area, these devices have efficiencies of 5 to 20 percent and rise/fall times of 20–50 ns. This alloy is becoming the workhorse for all infrared applications demanding power and speed. A structural variation as shown in Fig. 14 is an important light source for fiber-optic communication.

The $x = 0.38$ alloy is optimized for applications in the visible spectrum. The highest product of quantum efficiency and eye response is achieved at $x = 0.38$ and $\lambda = 650$ nm. The single heterostructure on an absorbing substrate has an efficacy of around 4 lm/A. The equivalent double heterostructure is in the 6–8 lm/A range. On a transparent GaAlAs substrate, the efficacy is typically in the 15–20 lm/A range and results of as high as 30 lm/A have been reported in the literature.[16] The major application for these red LEDs is in light-flux-intensive applications, such as message panels and automotive stoplights. A variation optimized for speed is widely used for optical communication using plastic fiber.

The AlInGaP System

The AlInFaP system has most of the advantages of the AlGaAs system with the additional advantage that it has a higher-energy direct energy gap of 2.3 eV that corresponds to green emission at 540 nm. AlInGaP can be lattice-matched to GaAs substrates. Indium occupies about half of the Group III atomic sites. The ratio of aluminum to gallium can be changed without affecting the lattice match, just as it can in the AlGaAs material system, since AlP and GaP have nearly the same lattice spacing. This enables the growth of heterostructures that have the efficiency advantages described in the previous section.

Various AlInGaP device structures have been grown. A simple DH structure with an AlInGaP active layer surrounded by higher bandgap AlInGaP confining layers has been effective for injection lasers, but has not produced efficient surface-emitting LEDs.[17] The main problem has been that AlInGaP is relatively resistive and the top AlInGaP layer is not effective in distributing the current uniformly over the chip. This is not a problem with lasers since the top surface is covered with metal and the light is emitted from the edge of the chip.

The top layer must also be transparent to the light that is generated. Two window layers used are AlGaAs or GaP on top of the AlInGaP heterostructure.[18,19] AlGaAs has the advantage that it is lattice-matched and introduces a minimum number of defects at the interface, but it has the disadvantage that it is somewhat absorptive to the yellow and green light which is generated for high-aluminum compositions. The highest AlInGaP device efficiencies have been achieved using GaP window layers.[18,20] GaP has the advantage that it is transparent to shorter wavelengths than AlGaAs and that it is easy to grow thick GaP layers, using either VPE or LPE, on top of the AlInGaP DH that was grown by MOVPE. Both VPE and LPE have substantially higher growth rates than MOCVD. The various growth techniques are discussed later under "Epitaxial Technology."

AlInGaP devices with 45-μm-thick GaP window layers have achieved external quantum efficiencies exceeding 5 percent in the red and yellow regions of the emission spectrum.[21] This is more than twice as bright as devices that have thinner AlGaAs window layers.

Green-emitting AlInGaP devices have also been grown which are brighter than the conventional GaP and GaP:N green emitters.[18,20,21] Substantial further improvement in green is expected since the quantum efficiency is not as high as would be expected based on the energy position of the transition from a direct to an indirect semiconductor.

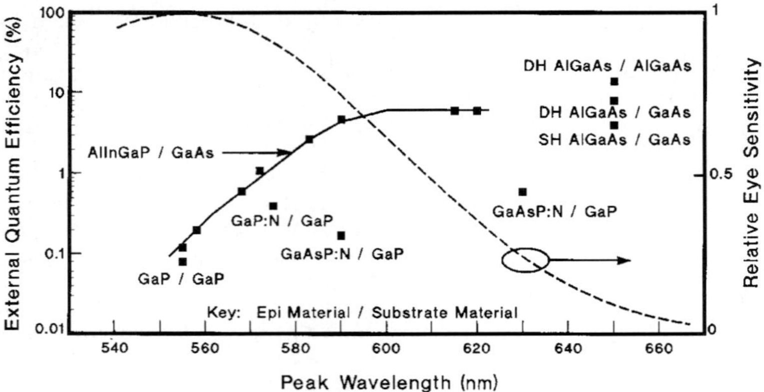

FIGURE 16 External quantum efficiency as a function of peak wavelength for various types of visible LEDs. Below 590 nm the efficiency of AlInGaP LEDs decreases due to the approaching transition from direct to indirect semiconductor. The human eye sensitivity curve is also shown. Since the eye response increases sharply from 660 to 540 nm, it partially makes up for the drop in AlInGaP LED efficiency. The resulting luminous performance is shown in Fig. 17.

The performance of AlInGaP LEDs compared to the most important other types of visible emitters is shown in Figs. 16 and 17. GaAsP on a GaAs substrate and GaP:ZnO are not shown since their luminous efficacy are off the chart at the bottom and the lower-right-hand corner, respectively. It is clear from Fig. 17 that the luminous efficacy of AlInGaP is substantially higher than the other technologies in all color regions except for red beyond 640 nm. Since forward voltage is typically about 2V, the lumen/watt value for a given device is about one-half the lumen/ampere value that is given in Tables 2 and 3. The quantum efficiency of AlInGaP is also better than all of the other technologies except for the highest-performance AlGaAs devices operating at about 650 nm. Because of the eye

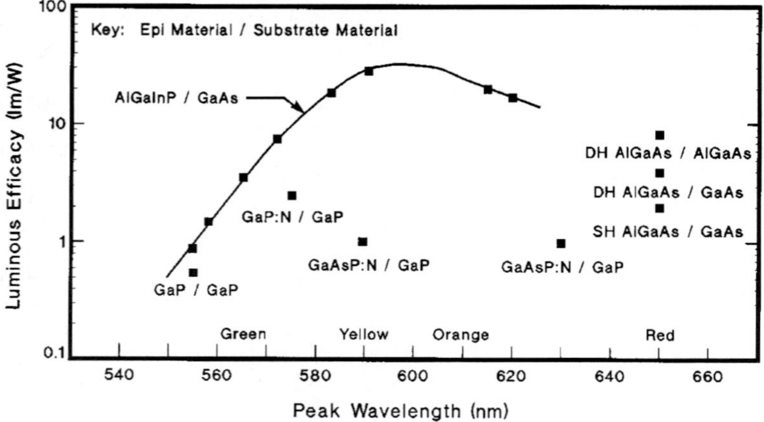

FIGURE 17 Luminous efficacy for AlInGaP LEDs versus wavelength compared to other LED technologies. AlInGaP LEDs are more than an order of magnitude brighter in the orange and yellow regions than other LEDs. AlInGaP LEDs compare favorably to the best AlGaAs red LEDs.

sensitivity variations (see C.I.E. curve in Fig. 16), the 620 nm (red/orange) AlInGaP devices have a higher luminous efficiacy than 650 nm AlGaAs LEDs (see Fig. 17).

Blue LED Technology

Blue emitters have been commercially available for more than a decade, but have only begun to have a significant impact on the market in the last few years. SiC is the leading technology for blue emitters with a quantum efficiency of about 0.02 percent and 0.04 lm/A luminous performance. SiC devices are not much used due to their high price and relatively low performance efficiency.

Other approaches for making blue LEDs are the use of II-VI compounds such as ZnSe or the nitride system GaN, AlGaN, or AlGaInN. It has been difficult to make good *p-n* junctions in these materials. Recently improved *p-n* junctions have been demonstrated in both ZnSe[22,23] and GaN.[24] Device performance is still in the 0.1 lm/A range and reliability is unproven. However, this recent progress is very encouraging for blue-emission technology and could lead to a high-performance device in the next few years. Both ZnSe and the nitride system have a major advantage over SiC because they are direct bandgap semiconductors, so a much higher internal quantum efficiency is possible. However, it is difficult to find suitable lattice-matched substrates for these materials.

12.7 SUBSTRATE TECHNOLOGY

Substrate Criteria

There are several requirements for substrates for LEDs. The substrate must be as conductive as possible both thermally and electrically to minimize power loss. In order to minimize defects it should match the epitaxial layers as closely as possible in atomic lattice spacing, and in the coefficient of thermal expansion. The substrate should also have a low defect density itself. Finally, the substrate should, if possible, be transparent to the light generated by the LED structure since this will enhance the external quantum efficiency.

Substrate Choices

The substrates used for nearly all visible LEDs are GaAs and GaP. GaAs or InP is used for infrared devices, depending upon the device structure required. Substrate parameters are summarized in Table 4, along with Si and Ge for comparison. GaAs is used for the AlGaAs and AlInGaP material systems since they can be lattice-matched to it. GaAs is

TABLE 4 Properties of Common Semiconductor Substrates

Substrate	Lattice parameter	Energy gap @ 300 K (eV)		Melting point (°C)
GaAs	5.653	1.428	Direct	1238
GaP	5.451	2.268	Indirect	1467
InP	5.868	1.34	Direct	1062
Si	5.431	1.11	Indirect	1415
Ge	5.646	0.664	Indirect	937

also used for the $GaAs_{1-x}P_x$ system for $x \leq 0.4$ because it is more nearly latticed-matched and, since it is absorbing, it is useful in multiple-junction devices where optical crosstalk must be minimized (see under "Diffused Homojunction" and Fig. 8). GaP is used for compositions of $x > 0.6$ due to its transparency and closer lattice match. However, neither GaP nor GaAs are well matched to GaAsP, and grading layers are required to grow epitaxial layers with decent quality. InP is the choice for long wavelength emitters made using the InGaAs or InGaAsP materials systems, due to the better lattice match.

Substrate Doping

Generally, substrates are n-type and are doped with Te, S, or Si, although sometimes Se and Sn are also used. In some cases, particularly for some AlGaAs LEDs and laser structures, p-type substrates are required and Zn is nearly always the dopant. The doping levels are typically in the 10^{18}-cm^{-3} range. Basically, the substrates are doped as heavily as possible to maximize conductivity. However, the doping must be below the solubility limit to eliminate precipitates and other defects. In the case of substrates that are transparent to the light which is generated, such as GaP, with a $GaAs_{1-x}P_x$ epitaxial layer, the doping should also remain below the level at which substantial free carrier absorption occurs.

Growth Techniques

Substrates can be grown by either the Bridgeman or Liquid Encapsulated Czochralski (LEC) technique. The LEC technique is the most widely used. Both techniques are described in detail elsewhere and will be only briefly summarized here.[25]

The LEC technique for GaAs consists of a crucible containing a molten GaAs solution, into which a single crystal "seed" is dipped. The temperature is carefully controlled so that the molten GaAs slowly freezes on the seed as the seed is rotated and raised out of the molten solution. By properly controlling the temperature, rotation rate, seed lift rate, etc., the seed can be grown into a single crystal weighing several kilograms and having a diameter of typically 2–4 inches for GaAs and for 2–2.5 inches for GaP. At the GaAs and GaP melting points As and P would rapidly evaporate from the growth crucible if they were not contained with a molten boric oxide layer covering the growth solution. This layer is the reason for the name "liquid encapsulated." The seed is dipped through the boric oxide to grow the crystal. The growth chamber must be pressurized to keep the phosphorus and arsenic from bubbling through the boric oxide. The growth pressure for GaP is 80 atm, and for GaAs is 20 atm or less, depending upon the approach for synthesis and growth. The LEC technique, used for GaAs, GaP, and InP, is similar to the Czochralski technique used for silicon, but the silicon process is much simpler since encapsulation is not required and the growth can be done at atmospheric pressure.

The Bridgeman and the gradient-freeze technique, which is a variation of the Bridgeman technique, can also be used to grow compound semiconductors. In this technique the growth solution and seed are contained in a sealed chamber so liquid encapsulation is not required. Growth is accomplished by having a temperature gradient in the solution, with the lowest temperature at the melting point in the vicinity of the seed. Growth can be accomplished by lowering the temperature of the entire chamber (gradient freeze), or by physically moving the growth chamber relative to the furnace (Bridgeman technique) to sweep the temperature gradient through the molten solution. The growing crystal can be in either a vertical or horizontal position.

GaAs for LEDs is commonly grown using either LEC or horizontal gradient freeze, also called "boat grown," but sometimes a vertical Bridgeman approach is used. GaP and InP are almost always grown using LEC but sometimes a vertical Bridgeman approach has also been used.

12.8 EPITAXIAL TECHNOLOGY

Growth Techniques Available

Epitaxial layers are grown using one of several techniques, depending upon the material system. The most common techniques are liquid phase epitaxy (LPE), which is primary used to grow GaAs, GaP, and AlGaAs; and vapor phase epitaxy (VPE), which is used to grow GaAsP. Metalorganic vapor phase epitaxy (MOCVD) is also used to grow AlGaAs, GaInAsP, and AlInGaP. Molecular beam epitaxy (MBE) is used for lasers and high-speed devices but is not used for high-volume commercial LEDs at this time. It has been used to grow blue ZnSe-based lasers and LEDs and could be important in the future. All of these epitaxial techniques have been discussed extensively elsewhere and will be only briefly described here.

LPE

LPE growth consists of a liquid growth solution, generally gallium, which is saturated with the compound to be grown.[26] The saturated solution is placed in contact with the substrate at the desired growth temperature, and cooled. As the solution cools, an epitaxial film is grown on the substrate. The technique has the advantage that it is relatively easy to grow high-quality epitaxial layers, and materials containing aluminum (such as AlGaAs) can be readily grown. The disadvantages are that composition control can be difficult. Also, the growth of epitaxial structures involving more than two or three layers, particularly thin layers, can be mechanically complicated since each layer requires a separate growth solution that must be carefully saturated and sequentially brought into contact with the substrate.

One important use of the LPE technique has been for the growth of GaP:ZnO and GaP:N for red and green LEDs, respectively. These devices each consist of two relatively thick layers: an *n*-type layer, followed by the growth of a *p*-type layer. While other growth techniques can be used to grow GaP LEDs, the best results have been obtained using LPE. As a result of a high-volume, low-cost production technology has evolved, which produces more visible LED chips than any other technique. Another major use of LPE is for the growth of GaAs:Si for infrared emitters. LPE is the only technique with which it has been possible to produce the recombination center that gives rise to the 940 nm emission characteristic of this material. The GaAs:Si structures are generally grown from a single growth solution. At high temperatures the silicon is incorporated on Ga sites and the layers are *n*-type. As the solution cools the silicon becomes preferentially incorporated on the As sites and the layer becomes *p*-type.

AlGaAs devices for both visible (red) and IR devices are also generally grown by LPE. AlGaAs devices can also be grown by MOCVD, but LPE has the advantage that thick layers can be more easily grown. This is important for high extraction efficiency (see under "Device Structures" and Fig. 12]. In the case of visible devices at 650 nm, the internal quantum efficiency is also higher using LPE than MOCVD. This is not understood, but the result is that virtually all of the visible AlGaAs LEDs are produced using LPE.

VPE

VPE is the other major commercial epitaxial technology for LEDs.[9,27] VPE consists of a quartz chamber containing the substrate wafers at the appropriate growth temperature. The reactants are transported to the substrates in gaseous form. The technique is mainly used for the growth of GaAsP which, along with GaP, dominates the high-volume visible LED market. In this case HCl is passed over Ga metal to form gallium chlorides, and AsH_3 and PH_3 are used to provide the As and P compounds. Appropriate dopant gases are added to achieve the *n*- and *p*-type doping. NH_3 is used to achieve nitrogen doping for the growth of GaAsP:N. The VPE technique has the advantages that it is relatively easy to scale up the growth chamber size so large quantities of material can be grown and layer composition and thickness can be easily controlled by adjusting the flow conditions. A limitation of VPE is that it has not been possible to grow high-quality compounds containing aluminum because the aluminum-bearing reactants attack the quartz chamber resulting in contaminated films. Thus, AlGaAs and AlInGaP, the emerging high-brightness technologies, cannot be grown using this technique.

MOCVD

MOCVD growth, like conventional VPE, uses gases to transpose the reactants to the substrates in a growth chamber.[28] However, in this case metallorganic compounds such as trimethylgallium (TMG) are used for one or more of the reactants. A major difference between VPE and MOCVD is that in the case of MOCVD the decomposition of the source gas (e.g., TMG) occurs as a reaction at or near the substrate surface, and the substrate is in the hottest area of the reactor such that the decomposition occurs on the substrate instead of the walls of the growth chamber. The walls of the growth chamber remain relatively cool. This is the key factor that makes MOCVD suitable for the growth of aluminum-bearing compounds which, unlike the VPE situation, do not react significantly with the cooler reactor walls. Thus AlGaAs and AlInGaP can be readily grown with MOCVD, and this technology is widely used for infrared AlGaAs LEDs and lasers, and for the emerging visible AlInGaP laser and LED technology.

MOCVD is also used for the growth of GaN and AlGaN that are candidates for blue emission, and for the growth of II–VI compounds, such as ZnSe, that is also a potential blue emitter. However, at this time the key limitation in obtaining blue ZnSe emitters is the growth of low-resistivity *p*-type ZnSe. For reasons that are not yet understood, low-resistivity *p*-type ZnSe has been grown using MBE only.

MBE

MBE is a high vacuum growth technique in which the reactants are essentially evaporated onto the substrates under very controlled conditions.[29] MBE, like MOCVD, can be used to deposit compounds containing aluminum. The growth rates using MBE are generally slower than the other epitaxial techniques, so MBE is most suitable for structures requiring thin layers and precise control of layer thickness. MBE equipment is somewhat more expensive than the equipment used for the other types of epitaxial growth, so it has not been suitable for the high-volume, low-cost production that is required for most types of LEDs. MBE has generally been utilized for lasers and high-speed devices where control of complicated epitaxial structures is critical and where relatively low volumes of devices are required. One advantage that MBE has over the other growth technologies is that the reactants utilized are generally less hazardous. Consequently, MBE equipment is often cheaper and easier to install since there are less safety issues and safety-code restrictions to deal with.

12.9 WAFER PROCESSING

Wafer Processing Overview

Wafer processing of compound semiconductors for LED applications has many of the same general steps used to process silicon integrated-circuit wafers, namely passivation, diffusion, metallization, testing, and die fabrication. The LED device structures are much simpler, so fewer steps are required; but, due to the materials involved, the individual steps are generally different and sometimes more complicated.

Compound semiconductor processing has been described in detail elsewhere, so only a brief summary is discussed here.[30] Some types of LED structures require all of the processing steps listed here, but in many cases fewer process steps are required. An example is a GaP or AlGaAs device with a grown *p-n* junction. For these devices no passivation or diffusion is required.

Passivation

Some types of LED structures, particularly multijunction structures, require a passivation layer prior to diffusion, as shown in Fig. 7. This layer must be deposited relatively free of pinholes, be patternable with standard photolithographic techniques, and must block the diffusing element, generally zinc. In the case of silicon, a native oxide is grown which is suitable for most diffusions. Unfortunately, the compound semiconductors do not form a coherent native oxide as readily as silicon. Silicon nitride (Si_3N_4) is the most widely used passivation layer for LEDs. Si_3N_4 is grown by reacting silane (SiH_4) and ammonia at high temperature in a furnace. S_3N_4 blocks zinc very effectively, and is easily grown, patterned, and removed. Sometimes an SiO_2 layer is used in conjunction with Si_3N_4 for applications such as protecting the surface of the compound semiconductor during high-temperature processing. Silicon oxynitride can also be used instead of or in addition to pure Si_3N_4. Silicon oxynitride is somewhat more complicated to deposit and control, but can have superior properties, such as a better match of coefficient of expansion, resulting in lower stress at the interface.

Diffusion

Generally, only *p*-type impurities, usually Zn, are diffused in compound semiconductors. *N*-type impurities have prohibitively small diffusion coefficients for most applications. Zn is commonly used because it diffuses rapidly in most materials and because it is nontoxic in contrast to Be, which also diffuses rapidly. Mg is another reasonable *p*-type dopant, but it diffuses more slowly than zinc. Diffusions are generally done in evacuated and sealed ampoules using metallic zinc as source material. A column V element such as As is also generally added to the ampoule to provide an overpressure that helps to prevent decomposition of the semiconductor surface during diffusion. Diffusion conditions typically range from 600–900°C for times ranging from minutes to days, depending upon the material and device involved. Junction depths can range from a fraction of a μm to a more than 10 μm.

Open-tube diffusions have also been employed but have generally been harder to control than the sealed ampoule approach, often because of surface decomposition problems. Open-tube diffusions have the advantage that one does not have to deal with the expense and hazard of sealing, breaking, and replacing quartz ampoules.

A third type of diffusion that has been used is a "semisealed" ampoule approach in which the ampoule can be opened and reused. The diffusion is carried out at atmospheric pressure and the pressure is controlled by having a one-way pressure relief valve on the ampoule.

Contacting

The contacts must make good ohmic contact to both the *p*- and *n*-type semiconductor, and the top surface of the top contact must be well suited for high-speed wire bonding. Generally, multilayer contacts are required to meet these conditions. Evaporation, sputtering, and e-beam deposition are all employed in LED fabrication. The *p*-type contact generally uses an alloy of either Zn or Be to make the ohmic contact. An Au-Zn alloy is the most common due to the toxicity of Be. The Au-Zn can be covered with a layer of Al or Au to enable high-yield wire bonding. A refractory metal barrier layer may be included between the Au-Zn and top Al or Au layer to prevent intermixing of the two layers and the out-diffusion of Ga, both of which can have a deleterious effect on the bondability of the Al or Au top layer. The *n*-type contact can be similar to the *p*-type contact except that an element which acts as an *n*-type dopant, commonly Ge, is used instead of Zn. An Au-Ge alloy is probably most frequently used to form the *n*-type ohmic contact since it has a suitably low melting point. If the *n*-type contact is the top, or bonded, contact it will be covered by one or more metallic layers to enhance bondability.

Testing

The key parameters that need to be tested are light output, optical rise and fall times, emission wavelength, forward voltage, and leakage current. The equipment used is similar to that used to test other semiconductor devices except that a detector must be added to measure light output. Rise and fall times and wavelength are generally measured on only a sample basis and not for each device on a wafer. In order to test the individual LEDs, the devices must be isolated on the wafer. This occurs automatically for LEDs that are masked and diffused, but if the LEDs are sheet diffused or have a grown junction the top layer must be processed to isolate individual junctions. This can be accomplished by etching or sawing with a dicing saw. Generally, sawing is used, followed by an etch to remove saw damage, because the layers are so thick that etching deep ($>10\ \mu m$) grooves is required. It is advisable to avoid deep groove etching because undercutting and lateral etching often occur and the process becomes hard to control. In many cases LED junctions are not 100 percent tested. This is particularly true of GaP and AlGaAs red-emitting devices in which the top layer may be 30 μm thick. Wafers of this type can be sampled by "coring" through the top layer of the wafer in one or more places with an ultrasonic tool in order to verify that the wafer is generally satisfactory. Later, when chips are fully processed, chips can be selected from several regions of the wafer and fully tested to determine if the wafer should be used or rejected.

Die Fab

Die fab is the process of separating the wafer into individual dice so they are ready for assembly. Generally, the wafer is first mounted on a piece of expandable tape. Next the wafer is either scribed or sawed to form individual dice. Mechanical diamond scribing or laser scribing were the preferred technologies in the past. Mechanical scribing has zero kerf loss, but the chips tend to have jagged edges and visual inspection is required. Laser scribing provides uniform chips but the molten waste material from between the chips damages neighboring chips, and in the case of full function chips the edges of the junction can be damaged by the laser. As a result of the limitations of scribing, sawing (using a dicing saw with a thin diamond impregnated blade) has become the technology of choice for most LEDs.

The kerf loss for sawing has been reduced to about 40 μm and the chip uniformity is excellent such that a minimum of inspection and testing is required. For most materials a "cleanup" etch is required after sawing to remove work damage at the edges of the chips which can both affect the electrical performance and absorb light. The wafer remains on the expandable tape during the sawing process. After sawing, the tape is expanded so that the chips are separated. The tape is clamped in a ring that keeps it expanded and the chips aligned. In this form the chips are easily individually picked off the tape by the die-attach machine that places the chips in the LED package.

12.10 LED QUALITY AND RELIABILITY

LEDs offer many advantages over other types of light sources. They have long operating life, they operate over a wide temperature range, and they are unaffected by many adverse environmental conditions. LED devices also are mechanically robust, making them suitable for applications where there is high vibration, shock, or acceleration. Excellent quality and reliability are obtainable when an LED product is properly designed, fabricated, packaged, tested, and operated.

Product quality is defined as "fitness for use" in a customer's application. Quality is measured in units of the average number of defective parts per million shipped (i.e., ppm), and is inferred from product sampling and testing. LED product quality is assured by: (1) robust chip and product design, (2) high-quality piece parts, (3) well-controlled fabrication processes, (4) use of statistical process control during manufacturing, (5) careful product testing, and (6) proper handling and storage. Most III-V LEDS are comparable in quality to the best silicon devices manufactured today. Well-designed LED products have total defect levels well below 100 ppm.

Reliability measures the probability that a product will perform its intended function under defined use conditions over the useful life of the product. Probability of survival is characterized by a failure rate, which is calculated by dividing the number of failures by the total number of operating hours (number of products tested per x hours operated). Common measures for reliability are percent failures per 1000 hours (percent/khr) and number failures per 10^9 hours (FITS). LED failure rates typically are better than 0.01 percent per khr at 50°C.

The reliability of an LED product is dependent on the reliability of the LED semiconductor chip and on the robustness of the package into which the chip is placed. Interactions between the chip and package can affect product reliability as well. Aspects of LED packaging and LED chip reliability are discussed in the following paragraphs.

LED Package Reliability

The package into which the LED chip is assembled should provide mechanical stability, electrical connection, and environmental protection. To evaluate package integrity, stress tests such as temperature cycling, thermal and mechanical shock, moisture resistance, and vibration are used to establish the worst-case conditions under which a product can survive. Generally, product data sheets contain relevant information about safe conditions for product application and operation.

Plastic materials are commonly used to package LEDs (see under "LED-Based Products"). Thermal fatigue is a limiting factor in plastic-packaged LEDs. Take the case of the plastic LED lamp shown later in Fig. 21. Because of the different materials used (epoxy plastic, copper lead frame, gold wire, III-V LED, etc.) and the different coefficients of expansion of these materials, temperature changes cause internal stresses. If the package is not well designed and properly assembled, thermal changes can cause cracking, chip-attach failure, or failure of the wire bond (open circuits). Careful design can reduce these problems to negligible levels over wide temperature ranges. Today's high-quality plastic lamps are capable of being cycled from −55 to +100°C for 100 cycles without failure.

Long-term exposure to water vapor can lead to moisture penetration through the plastic, subjecting the chips to humidity. High humidity can cause chip corrosion, plastic delamination, or surface leakage problems. Plastic-packaged LEDs are typically not harmed when used under normal use conditions. Accelerated moisture resistance testing can be used to test the limits of LED packages. Plastic materials have been improved to the point that LED products can withstand 1000 hr of environmental testing at elevated temperatures and high humidity (i.e., 85°C and 85 percent RH).

The thermal stability of plastic packaging materials is another important parameter. Over normal service conditions, the expansion coefficient of plastic is relatively constant. Above the so-called glass transition temperature *Tg* the coefficient increases rapidly. Reliable operation of plastic-packaged LEDs generally requires operation at ambient temperatures below *Tg*. Failures associated with improper soldering operations, such as too high a soldering temperature or for too long a time can cause the package to fail. Excessive storage temperatures also should be avoided. When an LED product is operated, internal ohmic heating occurs; hence, the safe operating maximum temperature is generally somewhat lower than the safe storage temperature.

LED Chip Reliability

The reliability characteristics of the LED chip determine the safe limits of operation of the product. When operated at a given temperature and drive current there is some probability that the LED will fail. In general, LED failure rates can be separated into three time periods: (1) infantile failure, (2) useful life, and (3) the wearout period (see Fig. 18). During the infant mortality period, failures occur due to weak or substandard units. Typically, the failure rate decreases during the infantile period until no weak units remain. During the "useful life" period, the failure rate is relatively low and constant. The number of failures that do occur are random in nature and cannot be eliminated by more testing. The useful life of an LED is a function of the operating temperature and drive conditions. Under normal use conditions, LEDs have useful lives exceeding 100,000 hours. The wearout period is characterized by a rapidly rising failure rate. Generally, wearout for LEDs is not a concern, as the useful life far exceeds the useful life of the product that the LEDs are designed into.

The principle failure mode for LED chips is light-output degradation. In the case of a

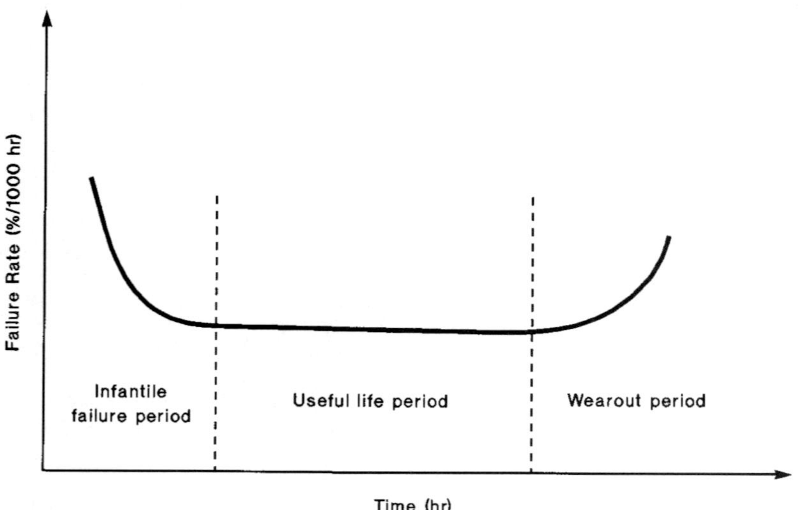

FIGURE 18 Plot of LED failure rate versus time, showing the infantile failure period (decreasing rate), the useful life period (constant, low rate) and the wear-out period (rapidly rising rate).

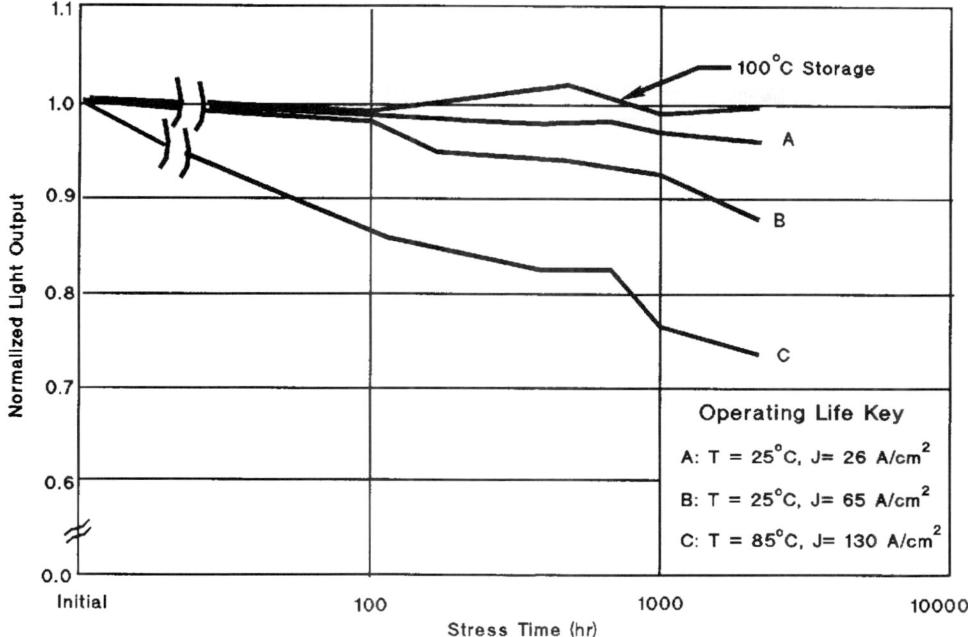

FIGURE 19 Curves of light output versus time for GaAsP indicator lamps stressed under various conditions. Light output is normalized to the initial value. Each curve shows the average degradation of 20 lamps.

visible lamp or display, failure is typically defined as a 50 percent decrease in light output from its initial value, since that is the level where the human eye begins to observe a noticeable change. For infrared emitters or visible LEDs where flux is sensed by a semiconductor detector, failure is commonly defined as a 20 to 30 percent decrease in flux output.

Figure 19 shows degradation curves for a direct bandgap GaAsP LED packaged in a 5-mm plastic lamp.[31] Current must flow for degradation to occur, as negligible change is observed after 1 khr of 100°C storage. Degradation is a function of the temperature at the *p*-*n* junction and the junction current density. As shown in Fig. 19, a larger decline in light output is observed as junction current density and/or temperature at the junction increases. The dependence of degradation on current density J is superlinear, varying as J^x with $1.5 < x < 2.5$. Hence, accelerated-aging tests typically use high currents and temperatures to shorten the time needed to observe LED degradation. The maximum stress level shown in Fig. 19 is 200 percent of the maximum allowable drive current specified in the data sheet.

Light-output degradation in GaAsP LEDs is due to an increase in the nonradiative space-charge recombination current. Total current flowing through the LED is made up of the sum of diffusion current and space-charge recombination current, as shown in the following equations:

$$I_T(V, t) = A(t)e^{qV/kT} + B(t)e^{qV/2kT} \tag{5}$$

where q is electron charge, k is Boltzmann's constant, and T is temperature.[32] The first term is diffusion current that produces light output, while the second term is space-charge

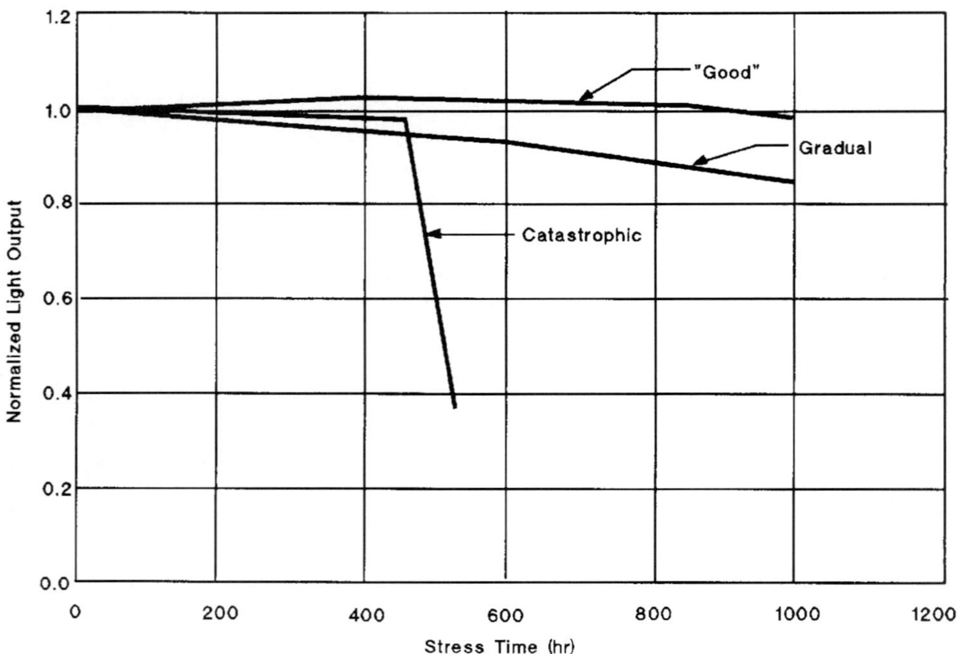

FIGURE 20 Light output degradation of AlGaAs LEDs. Three modes are shown: "good" devices with negligible light-output decrease over time, devices which degrade gradually over time, and "catastrophic" degradation devices where the flux rapidly decreases over a short time period.

recombination that is nonradiative. At fixed I_T, if $B(t)$ increases, then the diffusion term must decrease and, hence, the light output decreases. The reason for the increase in space-charge recombination in GaAsP LEDs is not fully understood.

The degradation characteristics of GaAlAs LEDs differ from those of GaAsP LEDs. Typical curves of normalized light output versus time for GaAlAs LEDs are shown in Fig. 20. "Good" units show negligible decrease in light output when operated under normal service conditions. Gradual degradation may occur, but it is relatively uncommon. The predominate failure mode in GaAlAs LEDs is catastrophic degradation. The light emission decreases very rapidly over a period of less than 100 hr and simultaneously nonradiative regions ("dark-spot" or "dark-line defects") are observed to form. The catastrophic degradation mechanism is described in detail in Refs. 33 and 34. In brief, the dark regions are caused by nonradiative recombination at dislocation networks that grow rapidly from a crystal dislocation located in the light-emitting region of the LED. Network formation depends on carrier recombination, both nonradiative, which creates mobile point defects, and radiative recombination, which enhances the movement of the point defects to the growing network.

Formation of dark-line defects is enhanced by mechanical stress either present in the LED chip or occurring during assembly. Properly designed products reduce such stress by minimizing bending caused by the different coefficients of expansion in the LED, and by stress-free die attach, wire bond, and encapsulation of the LED during assembly.

Failures due to dark-spot or dark-line defects can be effectively screened out by operating the LED at high current and temperature. Units with defects typically fail within the first few hundred hours. GaAlAs LEDs with dark-spot or dark-line defects are screened out by means of visual inspection and/or by eliminating units with large decreases in light output. GaAlAs LEDs used for fiber-optic applications have small emitting areas (see under "Fiber Optics" discussion and Fig. 14). Due to the high current densities present in such devices, high temperature and current burn-in is used extensively for these types of LEDs.

Another degradation mode in LEDs is change in the reverse breakdown characteristics over time. The reverse characteristics become soft and the breakdown voltage may decrease to a very low value. Several mechanisms have been observed in LEDs. One cause is localized avalanche breakdown due to microplasma formation at points where electric fields are high. Microplasmas have been observed in GaAs and GaAsP LEDs.

Damage or contamination of the edges of an LED chip can cause increased surface leakage and reduced reverse breakdown. Incomplete removal of damage during die separation operations and damage induced during handling and assembly are known to cause reverse breakdown changes. Die-attach materials also can unintentionally contaminate the edges of LEDs. Copper, frequently found in LED packaging materials, can diffuse into the exposed surfaces of the LED, causing excess leakage and, in some cases, light-output degradation.[35] Chips whose *p-n* junction extends to the edges (i.e., Fig. 6) are very susceptible to damage and/or contamination.

12.11 LED BASED PRODUCTS

Indicator Lamps

The simplest LED product is an indicator lamp or its infrared equivalent. The most frequently used lamp is shown in Fig. 21. A LED chip with a typical dimension of

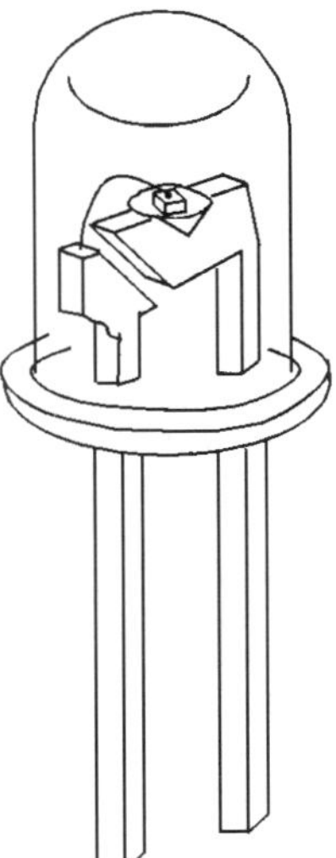

FIGURE 21 Plastic indicator lamp. The LED chip is placed in a reflector coined into the end of one electrode lead. The top of the chip is connected with a gold wire to the second electrode. The electrodes are encapsulated in plastic to form a mechanically robust package.

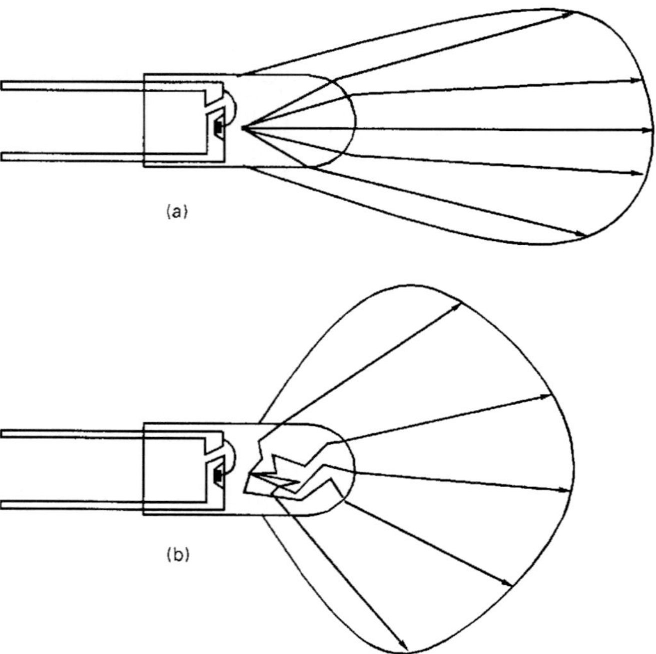

FIGURE 22 Radiation pattern of two types of LED indicator lamps: (*a*) lamp with clear plastic package with a narrow beam; (*b*) lamp with a diffusing plastic package (glass powder added) with a broader radiation pattern.

$250 \times 250\ \mu m$ is attached with conductive silver-loaded epoxy into a reflective cavity coined into the end of a silver-plated copper or steel lead frame. The top of the LED chip is connected with a thin 25-μm gold wire to the second terminal of the lead frame. The lead frame subassembly is then embedded in epoxy. The epoxy serves several functions: (1) it holds the assembly together and protects the delicate chip and bond wire; (2) it increases the light extraction from the chip (see under "Light Extraction," discussed earlier); and (3) it determines the spatial light distribution.

There are a large number of variations of the lamp shown in Fig. 21. Besides the obvious variation of source wavelength, there are variations of size, shape, radiation pattern, etc. The cross section of the plastic body ranges from 2–10 mm. The radiation pattern is affected by three factors: the shape of the dome, the relative position of the chip/reflector combination, and by the presence of a diffusant in the epoxy. Figure 22*a* shows the radiation pattern of a lamp using clear plastic as an encapsulant. The rays emanating from chip and reflector are collimated into a narrow beam. For many indicator lamps a broader viewing angle is desired such as that shown in the radiation pattern of Fig. 22*b*. This effect is achieved by adding a diffusant, such as glass powder, to the clear plastic. Another variation is shape. Common shapes are round, square, rectangular, or triangular.

Another variation is achieved by placing two different chips into the reflector cup. For instance, a lamp with green and red chips connected to the second post in an antiparallel fashon can operate as a red indicator if the reflector post is biased positive, and as a green indicator if it is biased negative. Rapidly switching between the two polarities one can achieve any color between the two basic colors, i.e., yellow or orange, depending on current and duty cycle.

A number of chips can be combined in a single package to illuminate a rectangular area. These so-called annunciator assemblies range in size from one to several cm. They typically use 4 chips per cm^2. By placing an aperture-limiting symbol or telltale in front of the lit area, these structures are cost-effective means to display a fixed message, such as warning lights in an automotive dashboard.

Numeric Displays

Numeric displays are usually made up of a nearly rectangular arrangement of seven elongated segments in a figure-eight pattern. Selectively switching these segments generates all ten digits from 0–9. Often decimal point, colon, comma, and other symbols are added.

There are two main types of LED numeric displays: (1) monolithic displays and (2) stretched segment displays. All monolithic displays are based on GaAs-GaAsP technology. Seven elongated p-doped regions and a decimal point are diffused into an n-type epitaxial layer of a single or monolithic chip (see Fig. 8). Electrically, this is a structure with eight anodes and one common cathode. This monolithic approach is relatively expensive. For arms-length viewing, a character height of 3–5 mm is required. Adding space for bonding pads, decimal point, and edge separation, such a display consumes around $10\,\mathrm{mm}^2$ of expensive semiconductor material per digit. One way to reduce material and power consumption is optical magnification. Viewing-angle limitation and distortion limit the magnification M to $M \leq 2.0$. Power consumption is reduced by M^2—an important feature for battery-powered applications.

For digits >5 mm, a stretched segment display is most cost effective. The design of Fig. 23 utilizes a $250 \times 250\,\mu\mathrm{m}$ chip to generate a segment with dimensions of up to 8×2 mm for a 20-mm digit height. This corresponds to a real magnification of $M^2 = 256$, or an equivalent linear magnification of $M = 16$. This magnification is achieved without a reduction in viewing angle by using scattering optics. An LED chip is placed at the bottom

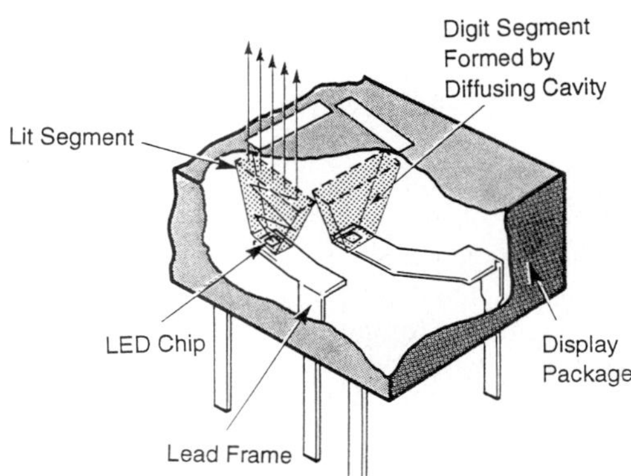

FIGURE 23 Cutaway of a seven-segment numeric LED display, showing how light from a small LED chip is stretched to a large character segment using a diffusing cavity. Characters 0–9 are created by turning on appropriate combinations of segments.

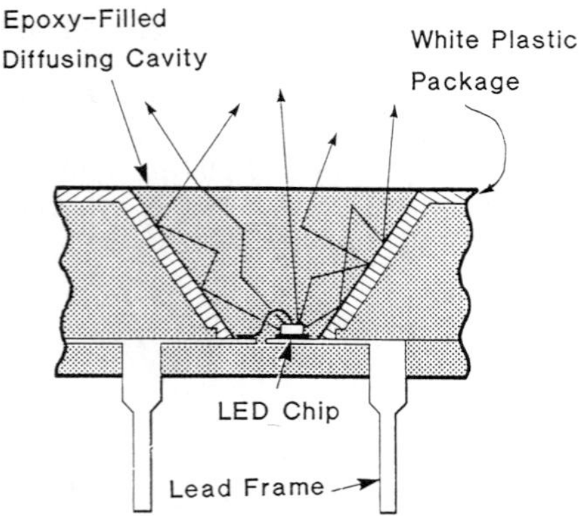

FIGURE 24 Cross section through one segment of the seven-segment numeric display shown in Fig. 23. A LED chip is placed at the bottom of a highly reflective, white plastic cavity which is filled with clear plastic containing a diffusant. Light is scattered within the cavity to produce uniform emission at the segment surface.

of a cavity having the desired rectangular exit shape and the cavity is filled with a diffusing plastic material (see Fig. 24). The side and bottom surfaces of the cavity are made as reflective as possible. Highly reflective white surfaces are typically used. A good white plastic surface measured in air may have a reflectivity of 94 percent compared with 98 percent for Ag and 91 percent for Al. Ag and Al achieve this reflectivity only if evaporated on a specularly smooth surface. Measured in plastic, the reflectivity of the white surface increases from 94 to 98 percent, the Ag surface remains at 98 percent, while the Al surface decreases to 86 percent. Both metallic surfaces have substantially lower reflectivities if they are evaporated onto a nonspecular surface, or if they are deposited by plating. Practically all numeric LED displays above 5-mm character heights are made using white cavity walls and diffusing epoxy within the cavity.

This case of magification by scattering does not result in a power saving as in the case of magnification of monolithic displays. Since there is practically no reduction in emission angle, the law of energy conservation requires an increased light flux from the chip that equals the area magnification plus reflection losses in the cavity.

Alphanumeric Displays

There are two ways LEDs are used to display alphanumeric information: either by using more than seven elongated segments, i.e., 14, 16, or 24; or by using an array of LED chips in a 5×7 dot matrix format. The multiple segment products are similar in design to the monolithic or stretched segment numeric displays described above.

In the case of small monolithic characters, the number of input terminals quickly exceeds conventional pin spacing. These products are usually clusters of 4 to 16 characters combined with a decoder/driver integrated circuit within the same package. To reduce cost and power, some modest optical magnification is usually used. The segmented displays are usually larger, i.e., 12–25 mm and limited to 14 segments per character. At this size, there is no pin density constraint and the decoder is usually placed outside the package.

The most frequently used alphanumeric LED display is based on a 5×7 matrix per character. For small characters in the range of 5–8 mm, the LED chips are directly viewed and pin density limitations require an on-board decoder/driver IC. Products are offered as end-stackable clusters of 4, 8, and 16 characters.

For larger displays, the LED chip is magnified by the same optical scattering technique described above for numerics. Exit apertures per pixel have a diameter of 2–5 mm. Products are offered as 5×7 single characters or end-stackable 20×20 tiles for large message- or graphics-display panels. At this size, pin density is not a limitation.

Optocouplers

An optocoupler is a device where signal input and signal output have no galvanic connection. It is mainly used in applications as the interface between the line voltage side of a system and the low-voltage circuit functions, or in systems where the separate ground connection of interconnected subsystems causes magnetic coupling in the galvanic loop between signal and ground connections. By interrupting the galvanic loop with an optical signal path, many sources of signal interference are eliminated.

The oldest optocouplers consist of an IR LED and a photodetector facing each other in an insulating tube. The second generation utilized the so-called dual-in-line package widely used by logic ICs. In this package an IR emitter and a phototransistor are mounted face-to-face on two separate lead frames. The center of the package between emitter and detector is filled with a clear insulating material. The subassembly is then molded in opaque plastic to shield external light and to mechanically stabilize the assembly (Fig 25). The second generation optocouplers have limited speed performance for two reasons: (1) the slow response time of the GaAs:Si LED and (2) the slow response of the phototransistor detector because of the high collector-base capacitance.

The third generation of optocouplers overcomes the speed limitation. It uses an integrated photodetector and a decoupled gain element. Integration limits the thickness of the effective detection region in silicon to 5–7 μm. This thickness range forces a shift of the source to wavelengths shorter than the 940 nm sources used in the second-generation couplers. Third-generation couplers use GaAsP (700 nm) or AlGaAs emitters (880 nm).

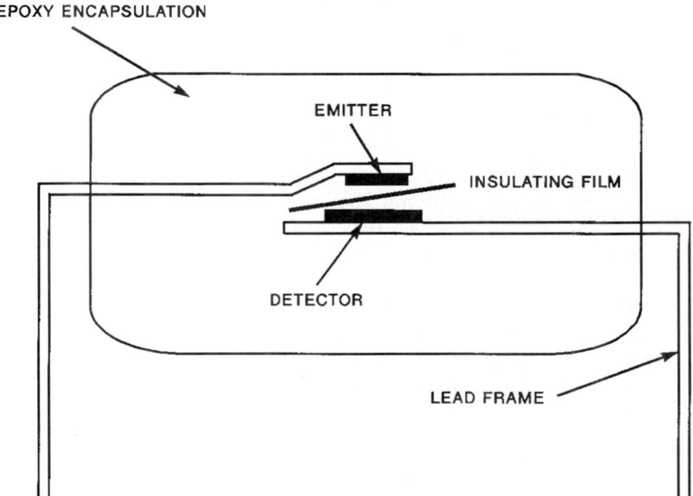

FIGURE 25 Optocoupler consisting of face-to-face emitter and detector chips. An insulating film is placed between the chips to increase the ability of the optocoupler to withstand high voltages between input and output electrodes.

Within the last decade, the optocoupler product family has seen further proliferation by adding features on the input or output side of the coupler. One proliferation resulted in couplers behaving like logic gates. Another variation used MOS FET devices on the output side, eliminating the offset voltages of bipolar devices. These couplers are comparable to the performance of conventional relays and are classified as solid-state relays. Other types use a CMOS input driver and CMOS output circuitry to achieve data transfer rates of 50 Mb/s and CMOS interface compatibility.

Fiber Optics

LEDs are the primary light source used in fiber-optic links for speeds up to 200 Mb/s and distances up to 2 km. For higher-speed and longer distances, diode lasers are the preferred source.

For fiber-optic applications, LEDs have to meet a number of requirements that go well beyond the requirements for lamps and displays. The major issues are: minimum and maximum flux coupled into the fiber, optical rise and fall times, source diameter, and wavelength. Analysis of flux budget, speed, fiber dispersion, wavelength, and maximum distance is quite complex and goes far beyond the scope of this work. A simplified discussion for the popular standard, Fiber Distributed Data Interface (FDDI) will highlight the issues. For a detailed discussion, the reader is referred to Ref. 36.

Flux budget. The minimum flux that has to be coupled into the fiber is determined by receiver sensitivity (−31 dBm), fiber attenuation over the 2-km maximum distance (3 dB), connector and coupler losses (5 dB), and miscellaneous penalties for detector response variations, bandwidth limitations, jitter, etc. (3 dB). With this flux budget of 11 dB, the minimum coupled power has to be −20 dBm. Another flux constraint arises from the fact that the receiver can only handle a maximum level of power before saturation (−14 dBm). These two specifications bracket the power level coupled into the fiber at a minimum of −20 dBm (maximum fiber and connector losses) and at a maximum of −14 dBm (no fiber or connector losses for the case of very short fiber).

Speed. The transmitter speed or baud rate directly translates into a maximum rise and fall time of the LED. The 125 Mdb FDDI specifications call for a maximum rise and fall time of 3.5 ns. As a rule of thumb, the sum of rise and fall time should be a little shorter than the inverse baud period (8 ns for 125 Mbd).

Source Diameter and Fiber Alignment. Efficient coupling of the LED to the fiber requires a source diameter that is equal to or preferably smaller than the diameter of the fiber core. For sources smaller than the core diameter, a lens between source and fiber can magnify the source to a diameter equal to or larger than the core diameter. The magnification has two benefits: (1) It increases the coupling efficiency between source and fiber. The improvement is limited to the ratio of source area to core cross-sectional area. (2) It increases the apparent spot size to a diameter larger than the fiber core. This effect relaxes the alignment tolerance between source and core and results in substantially reduced assembly and connector costs.

Wavelength. The LEDs used in fiber-optic applications operate at three narrowly defined wavelength bands 650, 820–870, and 1300 nm, as determined by optical fiber transmission characteristics.

650 nm. This band is defined by an absorption window in acrylic plastic fiber. It is a very narrow window between two C-H resonances of the polymer material. The bottom of the window has an absorption of approximately 0.17 dB/m. However, the effective absorption is in the 0.3–0.4 dB/m range because the LED linewidth is comparable to the width of the absorption window and the LED wavelength changes with temperature. The 650-nm LEDs use either $GaAs_{1-x}P_x$ with $x = 0.4$ or $GaAl_xAs_{1-x}$ with $x = 0.38$. Quantum efficiencies are at 0.2 and 1.5 percent, respectively. Maximum link length is in the range of 20–100 m, depending on source efficiency, detector sensitivity, speed, and temperature range.

820–870 nm. This window was chosen for several reasons. GaAlAs emitters (see Fig. 14) and Si detectors are readily available at this wavelength. Early fibers had an absorption peak from water contamination at approximately 870 nm. As fiber technology improved, the absorption peak was eliminated and the wavelength of choice moved from 820- to the 850–870-nm range. Fiber attenuation at 850 nm is typically 3 dB/km. Maximum link length in the 500–2000 m range, depending on data rate. In the 850–870-nm window the maximum link length is limited by chromatic dispersion. GaAlAs emitters have a half-power linewidth of approximately 35 nm. The velocity of light in the fiber is determined by the index of refraction of the fiber core. The index is wavelength-dependent resulting in dispersion of the light pulse. This dispersion grows with distance. The compounded effect of LED linewidth and fiber dispersion is a constant distance-speed product. For a typical multimode fiber and GaAlAs LED combination, this product is in the range of 100 Mbd-km.[36]

1300 nm. At this wavelength, the index of refraction as a function of wavelength reaches a minimum. At this minimum, the velocity of light is practically independent of wavelength, and chromatic dispersion is nearly eliminated. The distance-speed limitation is caused by modal dispersion. Modal dispersion can be envisioned as a different path length for rays of different entrance angles into the fiber. A ray going down the middle of the core will have a shorter path than a ray entering the fiber at the maximum acceptance angle undergoing many bends as it travels down the fiber. The resulting modal dispersion limits multimode fibers to distance bandwidth products of approximately 500 Mbd-km. LED sources used at this wavelength are GaInAsP emitters, as shown in Fig. 15. At this wavelength, fiber attenuation is typically <1.0 dB/km and maximum link length is in the range of 500–5000 m, depending on data rate and flux budget.

Sensors

LED/detector combinations are used in a wide range of sensor applications. They can be grouped into three classes: transmissive, reflective, and scattering sensors.

Transmissive Sensors. The most widely used transmissive sensor is the slot interrupter. A U-shaped plastic holder aligns an emitter and detector face-to-face. It is used widely for such applications as sensing the presence or absence of paper in printers, end-of-tape in tape-recorders, erase/overwrite protection on floppy disks and many other applications where the presence or absence of an opaque obstruction in the light path determines a system response.

A widely used slot interrupter is a two- or three-channel optical encoder. A pattern of opaque and transmissive sections moves in front of a fixed pattern with the same spatial frequency. Two optical channels positioned such that they are 90° out-of-phase to each other with regard to the pattern allow the measurement of both distance (number of transmissive/opaque sequences) and direction (phase of channel A with regard to channel B). Such encoders are widely used in industrial control applications, paper motion in printers, pen movement in plotters, scales, motor rotation, etc. A third channel is often used to detect an index pulse per revolution to obtain a quasi-absolute reference.

Reflective Sensors. In a reflective sensor an LED, detector, and associated optical elements are positioned such that the detector senses a reflection when a reflective surface (specular or diffuse) is positioned within a narrow sensing range. A black surface or nonaligned specular surface or the absence of any reflective surface can be discriminated from a white surface or properly aligned specular surface. Applications include bar-code reading (black or white surface), object-counting on a conveyer belt (presence or no presence of a reflecting surface), and many others. Many transmissive sensor applications can be replaced by using reflective sensors and visa versa. The choice is usually determined by the optical properties of the sensing media or by cost. An emerging application for reflective sensors is blood gas analysis. The concentration of O_2 or CO_2 in blood can be determined by absorption at two different LED wavelengths, i.e., red and infrared.

Scattering Sensors. One design of smoke detectors is based on light scattering. The LED light beam and the detector path are crossed. In the absence of smoke, no light from the LED can reach the detector. In the presence of smoke, light is scattered into the detector.

12.12 REFERENCES

1. A. A. Bergh and P. J. Dean, "Light-Emitting Diodes," *Proc. IEEE,* vol. 60, 1972, pp. 156–224.
2. K. Gillessen and W. Shairer, *LEDs—An Introduction,* Prentice-Hall, 1987.
3. M. G. Craford, "Properties and Electroluminescence of the GaAsP Ternary System," *Progress in Solid State Chemistry,* vol. 8, 1973, pp. 127–165.
4. A. H. Herzog, W. O. Groves, and M. G. Craford, "Electroluminescence of Diffused $GaAs_{1-x}P_x$ Diodes with Low Donor Concentrations," *J. Appl. Phys.,* vol. 40, 1969, pp. 1830–1838.
5. R. A. Faulkner, "Toward a Theory of Isoelectronic Impurities in Semiconductors," *Phys. Rev.,* vol. 175, 1968, pp. 991–1009.
6. W. O. Groves, A. J. Herzog, and M. G. Craford, "The Effect of Nitrogen Doping on $GaAs_{1-x}P_x$ Electroluminescent Diodes," *Appl. Phys. Lett.,* vol. 19, 1971, pp. 184–186.
7. M. G. Craford, R. W. Shaw, W. O. Groves, and A. H. Herzog, "Radiative Recombination Mechanisms in GaAsP Diodes with and without Nitrogen Doping," *J. Appl. Physics,* vol. 43, 1972, pp. 4075–4083.
8. M. G. Craford, D. L. Keune, W. O. Groves, and A. H. Herzog, "The Luminescent Properties of Nitrogen Doped GaAsP Light Emitting Diodes," *J. Electron. Matls.,* vol. 2, 1973, pp. 137–158.
9. M. G. Craford and W. O. Groves, "Vapor Phase Epitaxial Materials for LED Applications," *Proc. IEEE,* vol. 61, 1973, pp. 862–880.
10. J. C. Campbell, N. Holonyak Jr., M. G. Craford, and D. L. Keune, "Band Structure Enhancement and Optimization of Radiative Recombination in $GaAs_{1-x}P_x$:N (and $In_{1-x}Ga_xP$:N)," *J. Appl. Phys.,* vol. 45, 1974, pp. 4543–4553.
11. R. A. Logan, H. G. White, and W. Wiegmann, "Efficient Green Electroluminescence in Nitrogen-Doped GaP *p-n* Junctions," *Appl. Phys. Lett.,* vol. 13, 1968, p. 139.
12. A. A. Bergh and J. A. Copeland, "Optical Sources for Fiber Transmission Systems," *Proc. IEEE,* vol. 68, 1980, pp. 1240–1247.
13. M. G. Craford, "Recent Developments in Light-Emitting Diode Technology," *IEEE Trans. Electron Devices,* vol. 24, 1977, pp. 935–943.
14. H. Nather, V. Nitsche, and W. Schairer, "High Resolution Printing Capability of LED-Based Print Heads," *Proceedings SPIE,* 1988, pp. 396–404.
15. M. G. Craford, "Light-Emitting Diode Displays," in *Flat-Panel Displays and CRTs,* L. E. Tannas Jr. (ed.), Van Nostrand Reinhold, 1985, pp. 289–331.
16. L. W. Cook, M. D. Camras, S. L. Rudaz, and F. M. Steranka, "High Efficiency 650 nm Aluminum Gallium Arsenide Light Emitting Diodes," *Proceedings of the 14th International Symposium on GaAs and Related Compounds,* Institute of Physics, Bristol, 1988, pp. 777–780.

17. J. M. Dallesasse, D. W. Nam, D. G. Deppe, N. Holonyak Jr., R. M. Fletcher, C. P. Kuo, T. D. Osentowski, and M. G. Craford, "Short-Wavelength (<6400 Å) Room Temperature Continuous Operation of *p-n* InAlGaP Quantum Well Lasers," *Appl. Phys. Lett.,* vol. 19, 1988, pp. 1826–1828.

18. C. P. Kuo, R. M. Fletcher, T. D. Osentowski, M. C. Lardizabel, and M. G. Craford, "High Performance AlGaInP Visible Light-Emitting Diodes," *Appl. Phys. Lett.,* vol. 57, 1990, pp. 2937–2939.

19. H. Sugawara, M. Ishikawa, and G. Hatakoshi, "High-efficiency InGaAlP/GaAs Visible Light-Emitting Diodes," *Appl. Phys. Lett.,* vol. 58, 1991, 1010–1012.

20. R. M. Fletcher, C. P. Kuo, T. D. Osentowski, K. H. Huang, and M. G. Craford, "The Growth and Properties of High Performance AlGaInP Emitters Using a Lattice Mismatched GaP Window Layer," *Jour. Elec. Matls.,* vol. 20, 1991, pp. 1125–1130.

21. K. H. Huang, J. G. Yu, C. P. Kuo, R. M. Fletcher, T. D. Osentowski, L. J. Stinson, A. S. H. Liao, and M. G. Craford, "Twofold Efficiency Improvement in High Performance AlGaInP Light-Emitting Diodes in the 555–620 nm Spectral Region Using a Thick GaP Window Layer," *Appl. Phys. Lett.,* vol. 61, 1992, pp. 1045–1047.

22. M. A. Haase, J. Qiv, J. M. DePoydt, and H. Cheng, "Blue-green Laser Diodes," *Appl. Phys. Lett.,* vol. 58, 1991, pp. 1272–1275.

23. J. Jeon, J. Ding, A. V. Normikko, W. Xie, M. Kobayashi, and R. L. Gunshore, "ZnSe Based Multilayer *p/n* Junctions as Efficient Light Emitting Diodes for Display Applications," *Appl. Phys. Lett.,* vol. 60, 1992, pp. 892–894.

24. S. Nakamura, M. Senoh, and T. Mukai, "Highly *P*-typed Mg-doped GaN Films Grown with GaN Buffer Layers," *Jap Jour. of Appl. Phys.,* vol. 30, 1991, pp. L1701–L1711.

25. A. G. Fischer, "Methods of Growing Crystals under Pressure," in *Crystal Growth,* B. R. Pamplin (ed.), Pergamon Press, 1975, pp. 521–555.

26. R. L. Moon, "Liquid Phase Epitaxy," in *Crystal Growth,* 2d ed., B. R. Pamplin (ed.), Pergamon Press, 1980.

27. J. W. Burd, "A Multi-Wafer Growth System for the Epitaxial Deposition of GaAs and $GaAs_{1-x}P_x$," *Trans. Met. Soc. AIME,* 1969, pp. 571–576.

28. G. B. Stringfellow, *Organometallic Vapor Phase Epitaxial Growth of III-V Semiconductor: Theory and Practice,* Academic Press, 1989.

29. E. C. H. Parker (ed.), *Technology and Physics of Molecular Beam Epitaxy,* Plenum Press, 1985.

30. S. K. Ghandhi, *VLSI Fabrication Principles,* John Wiley and Sons, 1982.

31. Hewlett-Packard, *Optoelectronics/Fiber-Optics Application Manual,* 2d ed., McGraw-Hill, 1981, p. 82.

32. A. S. Grove, *Physics and Technology of Semiconductor Devices,* sec. 6.6, John Wiley, 1967.

33. O. Veda, *Material Research Society Symposium Proc.,* vol. 184, 1991, p. 125.

34. M. Fukuda, *Reliability and Degradation of Semiconductor Lasers and LEDs,* Artech House, 1991.

35. A. A. Bergh, "Bulk Degradation of GaP Red LEDs," *IEEE Trans. Electron Devices,* vol. 18, 1971, pp. 166–170.

36. D. C. Hanson, "Progress in Fiber Optic LAN and MAN Standards," *IEEE LCS Magazine,* vol. 1, 1990, pp. 17–25.

CHAPTER 13
SEMICONDUCTOR LASERS

Pamela L. Derry
Luis Figueroa
Chi-Shain Hong
Boeing Defense & Space Group
Seattle, Washington

13.1 GLOSSARY

A	Constant approximating the slope of gain versus current or carrier density
C	Capacitance
c	Speed of light
D	Density of states for a transition
D_c	Density of states for the conduction band
D_v	Density of states for the valence band
d	Active layer thickness
d_{eff}	Effective beam width in the transverse direction
d_G	Guide layer thickness
dg/dN	Differential gain
E	Energy of a transition
E_c	Total energy of an electron in the conduction band
E_g	Bandgap energy
E_n	The nth quantized energy level in a quantum well
E_n^c	The nth quantized energy level in the conduction band
E_n^v	The nth quantized energy level in the valence band
E_v	Total energy of a hole in a valence band
e	Electronic charge
F_c	Quasi-Fermi level in the conduction band
F_v	Quasi-Fermi level in the valence band
f_c	Fermi occupation function for the conduction band
f_d	Damping frequency
f_o	Resonant frequency of an LRC circuit
f_p	Peak frequency

f_r	Resonance frequency
f_v	Fermi occupation function for the valence band
g	Model gain per unit length
g_{th}	Threshold modal gain per unit length
H	Heavyside function
h	Refers to heavy holes
$\hbar$	Plank's constant divided by 2π
I	Current
I_{off}	DC bias current before a modulation pulse
I_{on}	Bias current during a modulation pulse
I_{th}	Threshold current
J	Current density
J_o	Transparency current density
J_{th}	Threshold current density
K	Constant dependent on the distribution of spectral output function
$\mathbf{k}$	Wavevector
k	Boltzmann constant
L	Inductance
L	Laser cavity length
L_c	Coherence length
L_z	Quantum well thickness
l	Refers to light holes
$\|M\|^2$	Matrix element for a transition
m	Effective mass of a particle
m_c	Conduction band mass
m_r	Effective mass of a transition
m_v	Valence band mass
N	Carrier density
N_0	Transparency carrier density
n_{eff}	Effective index of refraction
n_r	Index of refraction
n_{sp}	Spontaneous emission factor
P	Photon density
P_{off}	Photon density before a modulation pulse
P_{on}	Photon density during a modulation pulse
R	Resistance
R_F	Front facet reflectivity
R_R	Rear facet reflectivity
T	Temperature

w	Laser stripe width
α	Absorption coefficient
α	Linewidth enhancement factor
α_i	Internal loss per unit length
β	Spontaneous emission factor
Γ	Optical confinement factor
$\Delta f_{1/2}$	Frequency spectral linewidth
$\Delta\lambda_L$	Longitudinal mode spacing
$\Delta\lambda_{1/2}$	Half-width of the spectral emission in terms of wavelength
λ	Wavelength
λ_o	Wavelength of the stimulated emission peak
τ_d	Turn-on time delay
τ_p	Photon lifetime
τ_s	Carrier lifetime

13.2 INTRODUCTION

This chapter is devoted to the performance characteristics of semiconductor lasers. In addition, some discussion is provided on fabrication and applications. In the first section we describe some of the applications being considered for semiconductor lasers. The following several sections describe the basic physics, fabrication, and operation of a variety of semiconductor laser types, including quantum well and strained layer lasers. Then we describe the operation of high-power laser diodes, including single element and arrays. A number of tables are presented which summarize the characteristics of a variety of lasers. Next we discuss the high-speed operation and provide the latest results, after which we summarize the important characteristics dealing with the spectral properties of semiconductor lasers. Finally, we discuss the properties of surface emitting lasers and summarize the latest results in this rapidly evolving field.

More than 260 references are provided for the interested reader who requires more information. In this *Handbook,* Chap. 12 (LEDs) also contains related information. For further in-depth reviews of semiconductor lasers we refer the reader to the several excellent books which have been written on the subject.[1–5]

13.3 APPLICATIONS FOR SEMICONDUCTOR LASERS

The best-known application of diode lasers is in optical communication systems. However, there are many other potential applications. In particular, semiconductor lasers are being considered for high-speed optical recording,[6] high-speed printing,[7] single- and multimode database distribution systems,[8] long-distance transmission,[9] submarine cable transmission,[10] free-space communications,[11] local area networks,[12] Doppler optical radar,[13] optical signal processing,[14] high-speed optical microwave sources,[15] pump sources for other solid-state lasers,[16] fiber amplifiers,[17] and medical applications.[18]

For very high speed optical recording systems (>100 MB/s), laser diodes operating at relatively short wavelengths ($\lambda < 0.75$ μm) are required. In the past few years, much progress has been made in developing short-wavelength semiconductor lasers, although the output powers are not yet as high as those of more standard semiconductor lasers.

One of the major applications for lasers with higher power and wide temperature of operation is in local area networks. Such networks will be widely used in high-speed computer networks, avionic systems, satellite networks, and high-definition TV. These systems have a large number of couplers, switches, and other lossy interfaces that determine the total system loss. In order to maximize the number of terminals, a higher-power laser diode will be required.

Wide temperature operation and high reliability are required for aerospace applications in flight control and avionics. One such application involves the use of fiber optics to directly link the flight control computer to the flight control surfaces, and is referred to as fly-by-light (FBL). A second application involves the use of a fiber-optic data network for distributing sensor and video information.

Finally, with the advent of efficient high-power laser diodes, it has become practical to replace flash lamps for the pumping of solid-state lasers such as Nd:YAG. Such an approach has the advantages of compactness and high efficiency. In addition, the use of strained quantum well lasers operating at 0.98 μm has opened significant applications for high-gain fiber amplifiers for communications operating in the 1.55-μm wavelength region.

13.4 BASIC OPERATION

Lasing in a semiconductor laser, as in all lasers, is made possible by the existence of a gain mechanism plus a resonant cavity. In a semiconductor laser the gain mechanism is provided by light generation from the recombination of holes and electrons (see Fig. 1). The wavelength of the light is determined by the energy bandgap of the lasing material. The recombining holes and electrons are injected, respectively, from the *p* and *n* sides of a *p*-*n* junction. The recombining carriers can be generated by optical pumping or, more commonly, by electrical pumping, i.e. forward-biasing the *p*-*n* junction. In order for the light generation to be efficient enough to result in lasing, the active region of a semiconductor laser, where the carrier recombination occurs, must be a direct bandgap semiconductor. The surrounding carrier injection layers, which are called *cladding layers,*

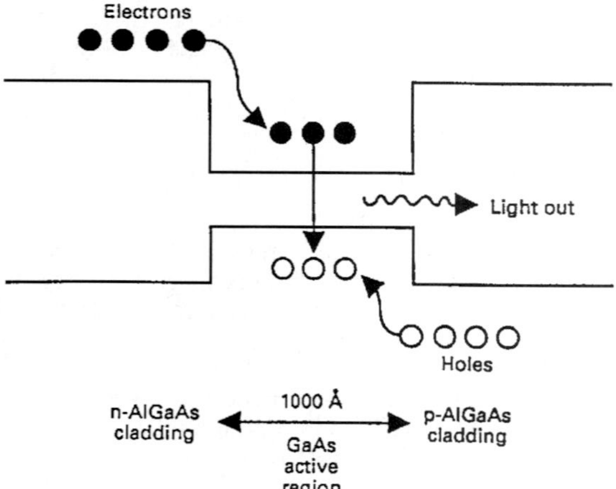

FIGURE 1 Schematic diagram of the recombination of electrons and holes in a semiconductor laser.

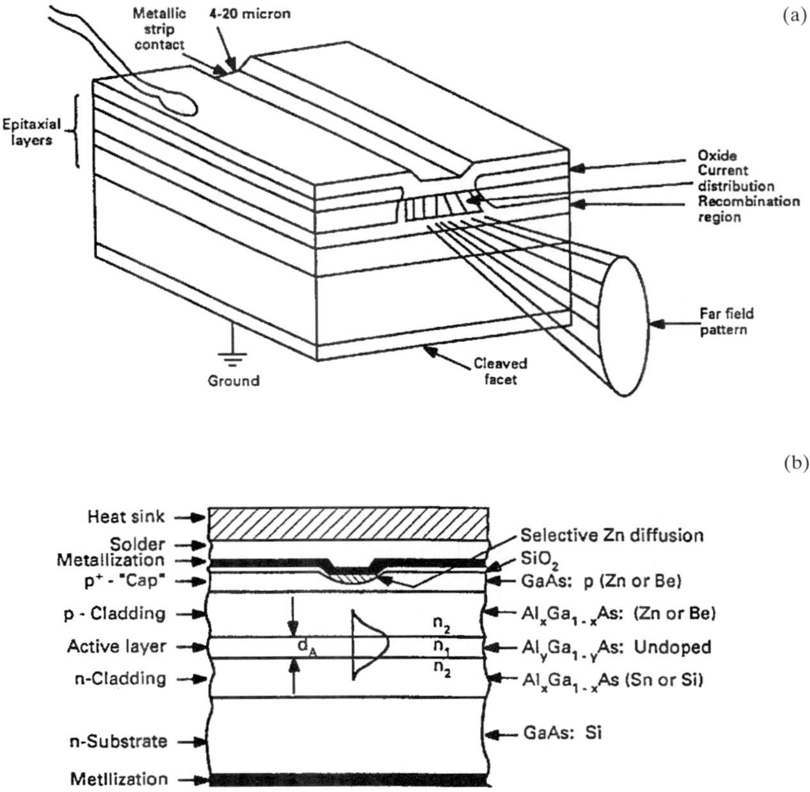

FIGURE 2 (*a*) Schematic diagram of a simple double heterostructure laser; (*b*) cross-sectional view showing the various epitaxial layers. (*After Ref. 20.*)

can be indirect bandgap semiconductors. For a discussion of semiconductor band levels see any solid-state physics textbook such as that by Kittel.[19] For a more detailed discussion of carrier recombination see Chap. 12 in this *Handbook*.

For a practical laser, the cladding layers have a wider bandgap and a lower index of refraction than the active region. This type of semiconductor laser is called a *double heterostructure* (DH) laser, since both cladding layers are made of a different material than the active region (see Fig. 2). The first semiconductor lasers were homojunction lasers,[21–24] which did not operate at room temperature; it is much easier to achieve lasing in semiconductors at low temperatures. Today, however, all semiconductor lasers contain heterojunctions. The narrower bandgap of the active region confines carrier recombination to a narrow optical gain region. The sandwich of the larger refractive index active region surrounded by cladding layers forms a waveguide, which concentrates the optical modes generated by lasing in the active region. For efficient carrier recombination the active region must be fairly thin, typically on the order of 1000 Å, so a significant fraction of the optical mode spreads into the cladding layers. In order to completely confine the optical mode in the semiconductor structure, the cladding layers must be fairly thick, usually about 1 μm.

The resonant cavity of a simple semiconductor laser is formed by cleaving the ends of the structure. Lasers are fabricated with their lasing cavity oriented perpendicular to a natural cleavage plane. For typical semiconductor materials this results in mirror reflectivities of about 30 percent. If necessary, the reflectivities of the end facets can be modified by applying dielectric coatings to them.[25] For applications where it is not possible to cleave the laser facets, it is also possible to etch them,[26–28] although this is much more difficult and usually does not work as well. Laser cavity lengths can be anywhere from about 50 to 2000 μm, although commercially available lasers are typically 200 to 1000 μm long.

Unpumped semiconductor material absorbs light of energy greater than or equal to its bandgap. When the semiconductor material is pumped optically or electrically, it reaches a point at which it stops being absorbing. This point is called *transparency.* If it is pumped beyond this point it will have optical gain, which is the opposite of absorption. A semiconductor laser is subject to both internal and external losses. For lasing to begin, i.e., to reach threshold, the gain must be equal to these optical losses. The threshold gain per unit length is given by:

$$g_{th} = \alpha_i + \frac{1}{2L} \ln \left(\frac{1}{R_F R_R} \right) \tag{1}$$

where α_i is the internal loss per unit length, L is the laser cavity length, and R_F and R_R are the front and rear facet reflectivities. (For semiconductor lasers, gain is normally quoted as gain per unit length in cm^{-1}. This turns out to be very convenient, but unfortunately is confusing for people in other fields, who are used to gain being unitless.)

The internal loss is a material parameter determined by the quality of the semiconductor layers. Mechanisms such as free-carrier absorption and scattering contribute to α_i.[1] The second term in Eq. (1) is the end loss. A long laser cavity will have reduced end loss, since the laser light reaches the cavity ends less frequently. Similarly, high facet reflectivities also decrease the end loss, since less light leaves the laser through them.

For biases below threshold, a semiconductor laser emits a small amount of incoherent light spontaneously (see Fig. 3). This is the same type of light emitted by an LEF (see Chap. 12). Above threshold, stimulated emission results in lasing. The relationship between lasing emission and the bias current of a healthy semiconductor laser is linear. To

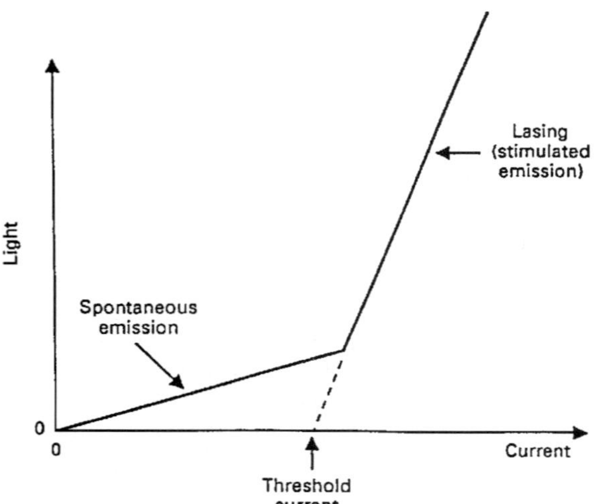

FIGURE 3 An example of the light-versus-current relationship of a semiconductor laser, illustrating the definition of threshold current.

find the threshold current of a laser this line is extrapolated to the point at which the stimulated emission is zero (see Fig. 3).

For further discussion of optical gain in semiconductors, see under "Quantum Well Lasers" later in this chapter. For more detail, see one of the books that has been written about semiconductor lasers.[1–5]

13.5 FABRICATION AND CONFIGURATIONS

In order to fabricate a heterostructure laser, thin semiconductor layers of varying composition must be grown on a semiconductor substrate (normally GaAs or InP). There are three primary epitaxial methods for growing these layers: liquid phase epitaxy (LPE), molecular beam epitaxy (MBE), and metalorganic chemical vapor deposition (MOCVD), which is also called organometallic vapor phase deposition (OMVPE).

Most of the laser diode structures which have been developed were first grown by LPE.[29] For a description of LPE see Chap. 12. Most commerically available lasers are grown by LPE; however, it is not well suited for growing thin structures such as quantum well lasers, because of lack of control and uniformity, especially over large substrates.[30] MBE and MOCVD are better suited for growths of thin, uniform structures.

MOCVD[31,32] is basically a specialized form of chemical vapor deposition. In MOCVD, gases reacting over the surface of a substrate form epitaxial layers; some of the gases are metalorganics. MOCVD is well suited for production environments, since epitaxial layers can be grown simultaneously on multiple large-area substrates and quickly, compared to MBE. It is expected that more commercial laser diodes will be grown by MOCVD in the future.

In the simplest terms, MBE[30,33,34,35] is a form of vacuum evaporation. In MBE growth occurs through the thermal reaction of thermal beams of atoms and molecules with the substrate, which is held at an appropriate temperature in an ultrahigh vacuum. MBE is different from simple vacuum evaporation for several reasons: the growth is single crystalline; the growth is much more controlled; and the vacuum system, evaporation materials, and substrate are cleaner.

With MOCVD the sources are gases, while with MBE they are solids. There are advantages and disadvantages to both types of sources. With gaseous sources the operator must work with arsine and/or phosphine, which are extremely hazardous gases. Solid-source phosphorus, however, is very flammable. Also, with MBE balancing the ratios of arsenic and phosphorus is extremely difficult; therefore, MOCVD is the preferred method for growth of GaInAsP and InP. MBE is a slower growth process (on the order of 1–2 μm per hour) than MOCVD. MBE therefore has the control necessary to grow very thin structures (10 Å), but MOCVD is more efficient for production. MBE has a cleaner background environment, which tends to make it better suited for growths at which background impurities must be kept at a minimum. Newer growth techniques,[36,37] which combine some of the advantages of both MBE and MOCVD, are gas source MBE, metalorganic MBE (MOMBE), and chemical beam epitaxy (CBE). In these growth techniques the background environment is that of MBE, but some or all of the sources are gases, which makes them more practical for growth of phosphorus-based materials.

Double heterostructure (DH) semiconductor lasers can be fabricated from a variety of lattice-matched semiconductor materials. The two material systems most frequently used are $GaAs/Al_xGa_{1-x}As$ and $In_{1-x}Ga_xAs_yP_{1-y}/InP$. All of these semiconductors are III-V alloys. The $GaAs/Al_xGa_{1-x}As$ material system has the advantage that all compositions of $Al_xGa_{1-x}As$ are closely lattice-matched to GaAs, which is the substrate. For GaAs-based lasers, the active region is usually GaAs or low Al concentration $Al_xGa_{1-x}As$ ($x < 0.15$), which results in lasing wavelengths of 0.78–0.87 μm. $Al_xGa_{1-x}As$ quantum well lasers with wavelengths as low as 0.68 μm have been fabricated,[38] but the performance is reduced.

In the $In_{1-x}Ga_xAs_yP_{1-y}$/InP material system, the active region is $In_{1-x}Ga_xAs_yP_{1-y}$ and the cladding layers and substrate are InP. Not all compositions of $In_{1-x}Ga_xAs_yP_{1-y}$ are lattice-matched to InP; x and y must be chosen appropriately to achieve both lattice match and the desired lasing wavelength.[4] The lasing wavelength range of InP-based lasers, 1.1–1.65 μm, includes the wavelengths at which optical fibers have the lowest loss (1.55 μm) and material dispersion (1.3 μm). (For more information on fibers see Vol. II, Chap. 10) This match with fiber characteristics makes $In_{1-x}Ga_xAs_yP_{1-y}$/InP lasers the preferred laser for long-distance communication applications. InP-based lasers can also include lattice-matched $In_{1-x-y}Al_xGa_yAs$ layers,[39–42] but the performance is reduced.

There is a great deal of interest in developing true visible lasers for optical data storage applications. $(Al_xGa_{1-x})_{0.5}In_{0.5}P$ lasers[43–46] lattice-matched to GaAs have proven superior to very short wavelength GaAs/$Al_xGa_{1-x}As$ lasers. Higher Al concentration layers are cladding layers and a low Al concentration layer or $Ga_{0.5}In_{0.5}P$ is the active region. In this material, system lasers with a lasing wavelength as low as 0.63 μm which operate continuously at room temperature have been fabricated.[46]

In order to fabricate a blue semiconductor laser, other material systems will be required. Recently, lasing at 0.49 μm at a temperature of 77 K was demonstrated in a ZnSe- (II-VI semiconductor) based laser.[47]

Very long wavelength (>2 μm) semiconductor lasers are of interest for optical communication and molecular spectroscopy. The most promising results so far have been achieved with GaInAsSb/AlGaAsSb lattice-matched to a GaSb substrate. These lasers have been demonstrated to operate continuously at 30°C with a wavelength of 2.2 μm.[48]

Lead salt lasers ($Pb_{1-x}Eu_xSe_yTe_{1-y}$, $Pb_{1-x}Sn_xSe$, $PbS_{1-x}Se_x$, $Pb_{1-x}Sn_xTe$, $Pb_{1-x}Sr_xS$) can be fabricated for operation at even longer wavelengths,[4,49–52] but they have not been demonstrated at room temperature. Progress has been made, however, increasing the operating temperature; currently $Pb_{1-x}Eu_xSe_yTe_{1-y}$/PbTe lasers operating continuously at 203 K with a lasing wavelength of 4.2 μm have been demonstrated.[53] Other very long wavelength lasers are possible; recently, HgCdTe lasers with pulsed operation at 90 K and a lasing wavelength of 3.4 μm have been fabricated.[54]

Laser Stripe Structures

We have discussed the optical and electrical confinement provided by a double heterostructure parallel to the direction of epitaxial growth; practical laser structures also require a confinement structure in the direction parallel to the substrate.

The simplest semiconductor laser stripe structure is called an *oxide stripe laser* (see Fig. 4*a*). The metallic contact on the *n*-doped side of a semiconductor laser is normally applied with no definition for current confinement; current confinement is introduced on the *p* side of the device. For a wide-stripe laser, a dielectric coating (usually SiO_2 or Si_3N_4) is evaporated on the *p* side of the laser. Contact openings in the dielectric are made through photolithography combined with etching of the dielectric. The *p* metallic contact is then applied across the whole device, but makes electrical contact only at the dielectric openings. A contact stripe works very well for wide stripes, but in narrow stripes current spreading is a very significant drawback, because there is no mechanism to prevent current spreading after the current is injected. Also, since the active region extends outside of the stripe, there is no mechanism to prevent optical leakage in a contact-stripe laser. Lasers like this, which provide electrical confinement, but no optical confinement are called *gain-guided lasers.*

Another type of gain-guide laser is an ion bombardment stripe (see Fig. 4*b*). The material outside the stripe is made highly resistive by ion bombardment or implantation which produces lattice defects.[55] Implantation causes optical damage,[56] so implantation should not be heavy enough to reach the active region.

A more complicated stripe structure with electrical and optical confinement is required

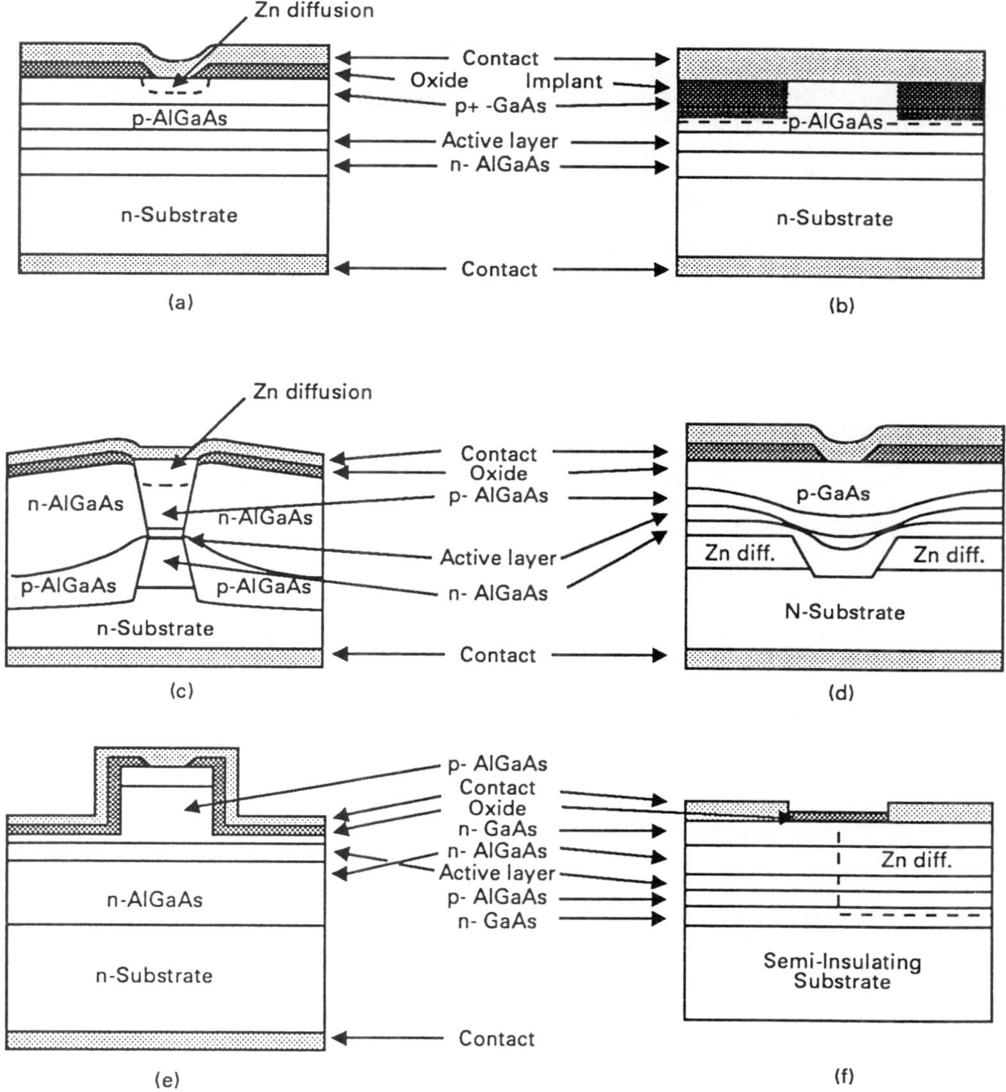

FIGURE 4 Schematic diagrams of GaAs/AlGaAs stripe laser structures. (*a*) Oxide stripe laser; (*b*) ion bombardment laser, (*c*) buried heterostructure (BH) laser; (*d*) variation on buried heterostructure laser; (*e*) ridge waveguide (RWG) laser; and (*f*) transverse junction stripe (TJS) laser.

for an efficient narrow-stripe laser. A number of structures which accomplish the necessary confinement have been developed. These structures are called *index-guided lasers,* since optical confinement is achieved through a change in refractive index.

The buried heterostructure laser (BH) was first developed by Tsukada.[57] To form a BH stripe, a planar laser structure is first grown. Stripe mesas of the laser structure are formed by photolithography combined with etching. For a GaAs-based BH laser, AlGaAs is then regrown around the lasing stripe. Figure 4*c* is a schematic diagram of a buried heterostructure. Since the active region is completely surrounded by AlGaAs, a BH has tight optical confinement. If the regrown layers are doped to produce a reverse-biased junction or are semi-insulating, a BH laser can also provide good current confinement. There are many variations on the BH structure. In some cases the active region is grown in the second growth step (see Fig. 4*d*). The tight optical confinement of BH lasers allows practical fabrication of very narrow stripes, on the order of 1 to 2 μm.

There are many other stripe structures that provide weaker optical confinement than a buried heterostructure. One of the simplest and most widely used of these is the ridge waveguide laser (RWG) (Fig. 4*e*). After epitaxial growth, most of the *p*-cladding layer is etched away, leaving a mesa where the lasing stripe will be. Only this mesa is contacted, which provides electrical confinement. The change in surrounding refractive index produces an effective change in refractive index in the active region beyond the mesa and provides optical confinement. Other stripe structures are described later in this chapter under "High-Power Semiconductor Lasers."

Another type of laser stripe is one in which confinement is provided by the *p*-*n* junction. The best-known laser of this type is the transverse junction stripe[58,59,60] (see Fig. 4*f*). In order to fabricate a TJS laser, both cladding layers are grown as *n*-AlGaAs. Zn diffusion is then used to create a *p*-*n* junction and contacts are applied on either side of the junction. In this laser the current flows parallel to the substrate rather than perpendicular to it. In a TJS laser the active region is limited to the small region of GaAs in which the Zn diffusion front ends.

The examples of laser stripe structures described here are GaAs/AlGaAs lasers. Long-wavelength laser structures (InP-based) are very similar,[4] but the active region is InGaAsP and the cladding layers are InP. With an *n*-InP substrate the substrate can be used as the *n*-type cladding, which allows greater flexibility in designing structures such as that illustrated in Fig. 4*d*. For a more detailed discussion of GaAs-based laser stripe structures see Casey and Panish[2] or Thompson.[3]

13.6 QUANTUM WELL LASERS

The active region in a conventional DH semiconductor laser is wide enough (~1000 Å) that it acts as bulk material and no quantum effects are apparent. In such a laser the conduction band and valence band are continuous (Fig. 5*a*). In bulk material the density of states, $D(E)$, for a transition energy E per unit volume per unit energy is:[26]

$$D(E) = \sum_{i=l,h} \frac{m_r^i}{\pi^2 \hbar^3} \sqrt{2m_r^i(E - E_g)} \qquad E > E_g \tag{2}$$

where E_g is the bandgap energy, $\hbar$ is Plank's constant divided by 2π, l and h refer to light and heavy holes, and m_r is the effective mass of the transition which is defined as:

$$\frac{1}{m_r} = \frac{1}{m_c} + \frac{1}{m_v} \tag{3}$$

where m_c is the conduction band mass and m_v is the valence band mass. (The split-off hole band and the indirect conduction bands are neglected here and have a negligible effect on most semiconductor laser calculations.)

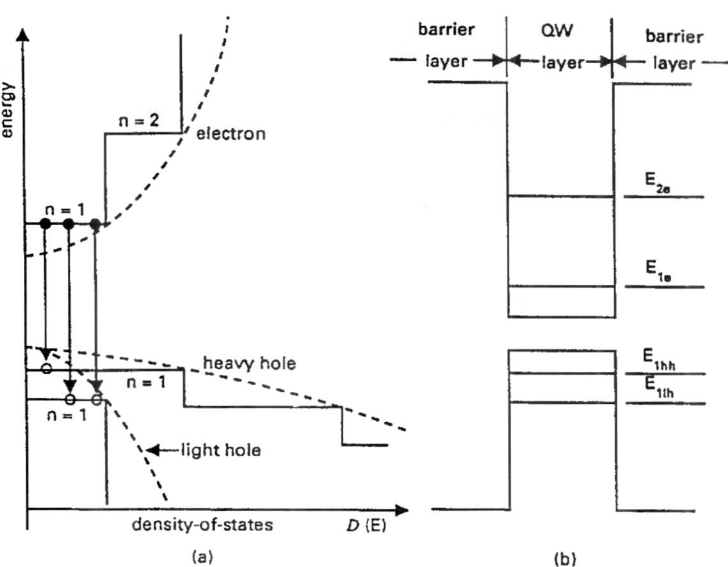

FIGURE 5 Schematic diagrams of (*a*) the density of states for a quantum well (solid line) and for a bulk DH (dotted line) and (*b*) quantized energy levels in a quantum well for $n = 1$ and 2 for the conduction band and for the light and heavy hole bands. (*After Ref. 63.*)

If the active region of a semiconductor laser is very thin (on the order of the DeBroglie wavelength of an electron) quantum effects become important. When the active region is this narrow (less than ~200 Å) the structure is called a *quantum well* (QW). (For a review of QWs see Dingle,[61] Holonyak et al.,[62] Okamoto,[63] or the book edited by P. Zory.[64]) Since the quantum effects in a QW are occurring in only one dimension they can be described by the elementary quantum mechanical problem of a particle in a one-dimensional quantum box.[65] In such a well, solution of Schrödinger's equation shows that a series of discrete energy levels (Fig. 5*b*) are formed instead of the continuous energy bands of the bulk material. With the approximation that the well is infinitely deep, the allowed energy levels are given by:

$$E_n = \frac{(n\pi\hbar)^2}{2mL_z^2} \tag{4}$$

where $n = 1, 2, 3, \ldots, m$ is the effective mass of the particle in the well, and L_z is the quantum well thickness. Setting the energy at the top of the valence band equal to zero, the allowed energies for an electron in the conduction band of a semiconductor QW become $E = E_g + E_n^c$, where E_n^c is E_n with m equal to m_c. The allowed energies for a hole in the valence band are then $E = -E_n^v$, where E_n^v is E_n with m equal to m_v. The allowed transition energies E, are limited to:

$$E = E_g + E_n^c + E_n^v + \frac{\hbar^2\mathbf{k}^2}{2m_r} \tag{5}$$

where **k** is the wavevector, rigorous **k**-selection is assumed, and transitions are limited to those with $\Delta n = 0$.

This quantization of energy levels will, of course, change the density of states. For a QW the density of states is given by:

$$D(E) = \sum_{i=l,h} \sum_{n=1}^{\infty} \frac{m_r^i}{L_z \pi \hbar^2} H(E - E_g - E_n^c - E_{n,i}^v) \tag{6}$$

where H is the Heavyside function. The difference in the density of states directly affects the modal optical gain generated by the injection of carriers. The modal gain is proportional to the stimulated emission rate:[1,66,67]

$$g(E, N) \alpha \frac{\Gamma D(E) |M|^2 (f_c(E, N) - f_v(E, N))}{E} \tag{7}$$

where Γ is the optical confinement factor, $|M|^2$ is the matrix element for the transition, N is the carrier density of either electrons or holes (the active region is undoped so they have equal densities), and $f_c(E, N)$ and $f_v(E, N)$ are the Fermi occupation factors for the conduction and valence bands. (For a detailed review of gain in semiconductor lasers see Ref. 67.)

The optical confinement factor Γ is defined as the ratio of the light intensity of the lasing mode within the active region to the total intensity over all space. Since a QW is very thin, Γ_{QW} will be much smaller than Γ_{DH}. Γ_{DH} is typically around 0.5 whereas for a single QW, Γ_{QW} is around 0.03.

The Fermi occupation functions describe the probability that the carriers necessary for stimulated emission have been excited to the states required. They are given by:[1,19]

$$f_c(E_c, N) = \frac{1}{1 + \exp((E_g + E_c - F_c)/kT)} \tag{8}$$

and

$$f_v(E_v, N) = \frac{1}{1 + \exp(-(E_v + F_v)/kT)} \tag{9}$$

where k is the Boltzmann constant, T is temperature, E_c is the energy level of the electron in the conduction band relative to the bottom of the band (including both the quantized energy level and kinetic energy), E_v is the absolute value of the energy level of the hole in a valence band, and F_c and F_v are the quasi-Fermi levels in the conduction and valence bands. Note that E_c and E_v are dependent on E, so f_c and f_v are functions of E. f_c and f_v are also functions of N through F_c and F_v. F_c and F_v are obtained by evaluating the expressions for the electron and hole densities:

$$N = \int D_c(E_c) f_c(E_c)\, dE_c \tag{10}$$

and

$$N = \int D_v(E_v) f_v(E_v)\, dE_v \tag{11}$$

where $D_c(E_c)$ and $D_v(E_v)$ are the densities of states for the conduction and valence bands and follow the same form as $D(E)$ for a transition.

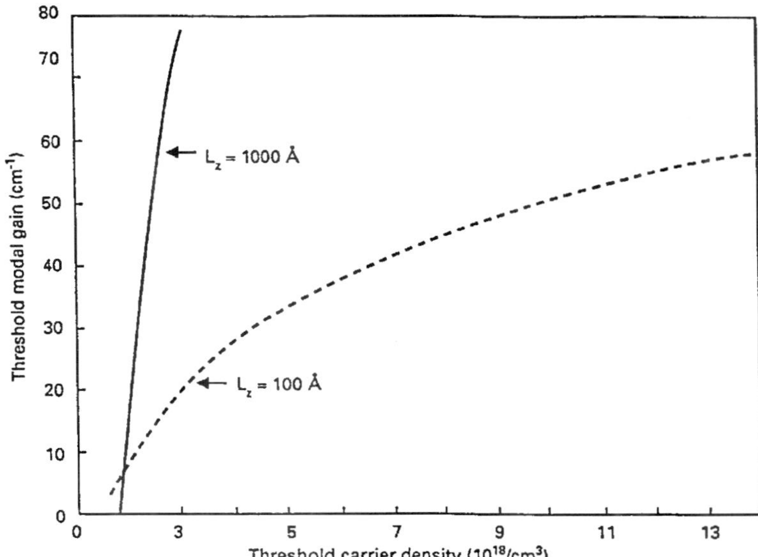

FIGURE 6 Threshold modal gain as a function of threshold carrier density for a conventional (Al, Ga)As double heterostructure with an active region thickness of 1000 Å and for a 100-Å single quantum well. (*From Ref. 68.*)

In Fig. 6, the results of a detailed calculation[68] based on Eq. (7) for the threshold modal gain as a function of threshold carrier density are plotted for a 100-Å single QW. The corresponding curve for a DH laser with an active region thickness of 1000 Å is also shown. The gain curves for the QWs are very nonlinear because of saturation of the first quantized state as the carrier density increases. The transparency carrier density N_o is the carrier density at which the gain is zero. From Fig. 6 it is clear that the transparency carrier densities for QW and DH lasers are very similar and are on the order of 2×10^{18} cm^{-3}.

The advantage of a QW over a DH laser is not immediately apparent. Consider, however, the transparency current density J_o. At transparency[28]

$$J_o = \frac{N_o L_z e}{\tau_s} \tag{12}$$

where L_z is the active layer thickness, e is the charge of an electron and τ_s is the carrier lifetime near transparency. τ_s is approximately 2 to 4 ns for either a QW or a DH laser. Since N_o is about the same for either structure, any difference in J_o will be directly proportional to L_z. But L_z is approximately ten times smaller for a QW; therefore, J_o will be approximately ten times lower for a QW than for a conventional DH laser. (A lower J_o will result in a lower threshold current density since the threshold current density is equal to J_o plus a term proportional to the threshold gain.) Note that this result is not determined by the quantization of energy levels; it occurs because fewer carriers are needed to reach the same carrier density in a QW as in a DH laser. In other words, this result is achieved because the QW is thin!

In this discussion we are considering current density instead of current. The threshold current density (current divided by the length and width of the stripe) is a more meaningful measure of the relative quality of the lasing material than is current. Current depends very strongly on the geometry and stripe fabrication of the device. In order to

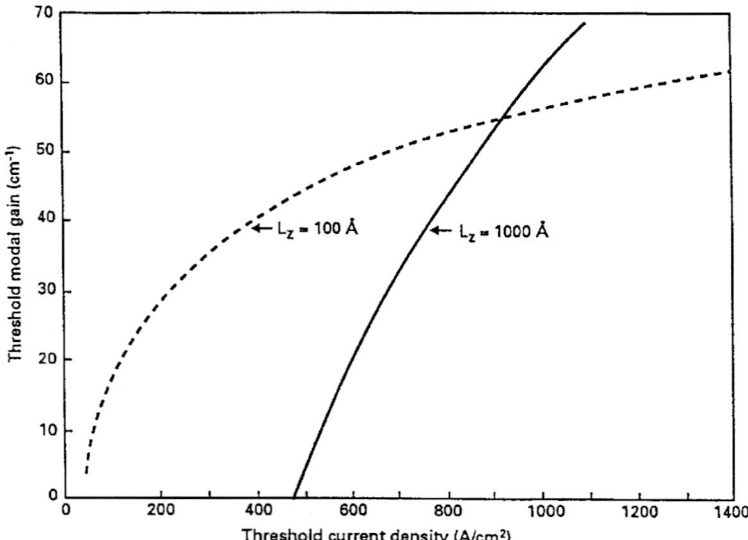

FIGURE 7 Threshold modal gain as a function of threshold current density for a conventional (Al, Ga)As double heterostructure with an active region thickness of 1000 Å and for a 100-Å single quantum well. (*From Ref. 68.*)

eliminate geometry-induced variations, current density is normally measured on broad-area (50–150 μm wide) oxide stripe lasers (see earlier under "Fabrication and Configurations"). With a narrow stripe the current spreads beyond the intended stripe width, so it is difficult to accurately measure the current density.

Figure 7 shows the results of a detailed calculation[68] of the threshold modal gain versus the threshold current density for a DH laser with an active region thickness of 1000 Å and for a 100-Å single QW. The potential for lower threshold current densities for QW lasers is clear for threshold gains less than that where the curve for the DH laser intercepts those of the QWs. With low losses, the threshold current of a QW laser will be substantially lower than that of a DH laser, since the threshold gain will be below the interception point.

To get an appreciation for how the threshold current density of a single QW will compare to that of a DH laser, consider that near transparency, the modal gain is approximately linearly dependent on the current density:

$$g(J) = A(J - J_o) \tag{13}$$

where A is a constant which should have a similar value for either a QW or a DH laser (this can be seen visually on Fig. 7). Taking Eq. (13) at threshold we can equate it to Eq. (1) and solve for J_{th} (the threshold current density):

$$J_{th} = J_o + \frac{\alpha_i}{A} + \frac{1}{2AL}\ln\left(\frac{1}{R_F R_R}\right) \tag{14}$$

α_i is related primarily to losses occurring through the interaction of the optical mode with the active region. In a QW, the optical confinement is lower, which means that the optical mode interacts less with the active region and α_i tends to be smaller. Let's substitute in the numbers in order to get an idea for the difference between a QW and a DH laser. Reasonable values are:[69] $A_{\text{QW}} \sim 0.7\ \text{A}^{-1}\,\text{cm}$, $A_{\text{DH}} \sim 0.4\ \text{A}^{-1}\,\text{cm}$, $J_o^{\text{QW}} \sim 50\ \text{A/cm}^2$, $J_o^{\text{DH}} \sim 500\ \text{A/cm}^2$, $\alpha_i^{\text{QW}} \sim 2\ \text{cm}^{-1}$, $\alpha_i^{\text{DH}} \sim 15\ \text{cm}^{-1}$, $L \sim 400\ \mu\text{m}$, and for uncoated facets $R_F = R_R = 0.32$. Substituting in we get: $J_{th}^{\text{QW}} \sim 95\ \text{A/cm}^2$ and $J_{th}^{\text{DH}} \sim 610\ \text{A/cm}^2$.

It is clear that changes in the losses will have a more noticeable effect on threshold current for a QW than for a DH laser since losses are responsible for a more significant portion of the threshold current of a QW laser. The gain curve of a QW laser saturates due to the filling of the first quantized energy level, so operating with low losses is even more important for a QW than is illustrated by the above calculation. When the gain saturates, the simple approximation of Eq. (13) is invalid. Operating with low end losses is also important for a QW, since they are a large fraction of the total losses. This explains why threshold current density results for QW lasers are typcially quoted for long laser cavity lengths (greater than 400 μm), while DH lasers are normally cleaved to lengths on the order of 250 μm. High-quality broad-area single QW lasers (without strain) have threshold current densities lower than 200 A/cm^2 (threshold current densities as low as 93 A/cm^2 have been achieved[69–71]), while the very best DH lasers have threshold current densities around 600 A/cm^2.[2,72] The end loss can also be reduced by the use of high-reflectivity coatings.[25] The combination of a single QW active region with a narrow stripe and high-reflectivity coatings has allowed the realization of submilliampere threshold current semiconductor lasers.[68,69,73] and high-temperature operation.[74,75,76]

A disadvantage of QW lasers compared to DH lasers is the loss of optical confinement. One of the advantages of a DH laser is that the active region acts as a waveguide, but in a QW the active region is too thin to make a reasonable waveguide. Guiding layers are needed between the QW and the (Al, Ga)As cladding layers. As the bandgap diagram of Fig. 8 illustrates, a graded layer of intermediate aluminum content can be inserted between the QW and each cladding layer. The advantage of this structure, which is called a *graded-index separate-confinement heterostructure* (GRIN SCH),[77,78] is separate optical and electrical confinement. The carriers are confined in the QW, but the optical mode is confined in the surrounding layers. The grading can be either parabolic (as illustrated in Fig. 8) or linear. Experimentally it has been found that the optimum AlAs mole fraction x for layers around a GaAs QW is approximately 0.2.[79] Typically, each additional layer is on the order of 2000 Å thick. In order to confine the optical model, the cladding layers need a low index of refraction compared to that of an $x = 0.2$ layer. In a simple DH laser, the cladding layers typically have x between 0.3 and 0.4, but for good confinement in an

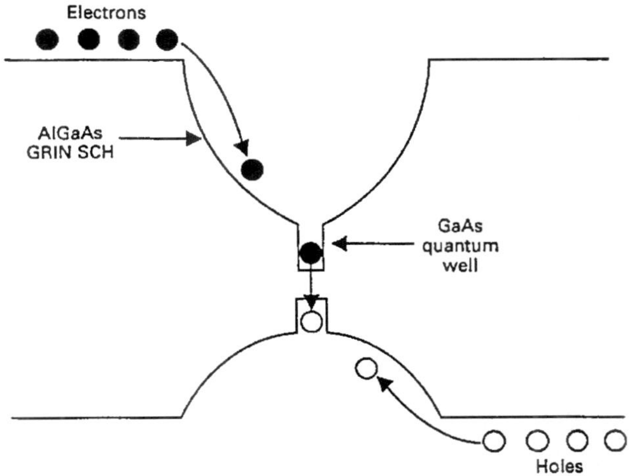

FIGURE 8 Schematic energy-band diagram for a GRIN SCH single QW.

$x = 0.2$ layer, more aluminum should be incorporated into the cladding layers; x should be between 0.5 and 0.7.

In the discussion so far we have considered only single QWs. Structures in which several quantum wells are separated by thin AlGaAs barriers are called *multiquantum wells* (MQWs) and also have useful properties. For a given carrier density, a MQW with n QWs of equal thickness, L_z has gain which is approximately n times the gain for a single QW of the same thickness L_z, but the current density is also approximately increased by a factor of n.[80] The transparency current density will be larger for the MQW than for the single QW since the total active region thickness is larger. Figure 9 shows that, as a function of current density, the gain in the single QW will start out higher than that in the MQW because of the lower transparency current, but the gain in the MQW increases more quickly so the MQW gain curve crosses that of the single QW at some point.[81]

Which QW structure has a lower threshold current will depend on how large the losses are for a particular device structure and on where the gain curves cross. The best structure for low threshold current in a GaAs-based laser is normally a single QW, but for some applications involving very large losses and requiring high gain a MQW is superior. For applications in which high output power is more important than low current a MQW is appropriate. A MQW is also preferred for high modulation bandwidth (see "Spectral Properties" later in this chapter).

An advantage of a QW structure over a bulk DH laser is that the lasing wavelength, which is determined by the bulk bandgap plus the first quantized energy levels, can be changed by changing the quantum well thickness [see Eq. (4)]. A bulk GaAs laser has a lasing wavelength of about 0.87 μm, while a GaAs QW laser of normal thickness (60–120 Å) has a lasing wavelength of 0.83–0.86 μm. Further bandgap engineering can be introduced with a strained QW.[67,81,82]

Strained Quantum Well Lasers

Normally, if a semiconductor layer of significantly different lattice constant is grown in an epitaxial structure, it will maintain its own lattice constant and generate misfit dislocations. If this layer is very thin, below a certain critical thickness,[81,83–85] it will be distorted to match the lattice constant (perpendicular to the substrate) of the surrounding layers and will not generate misfit locations. A layer with thickness above the critical thickness is called "relaxed," one below is called "strained."

Straining a semiconductor layer changes the valence-band structure. Figure 10*a* shows the band structure of an unstrained III-V semiconductor, while Fig. 10*b* and *c* show the band structure under biaxial tension and compression, respectively.[81,82] For the unstrained semiconductor, the light and heavy hole bands are degenerate at $\mathbf{k} = 0$ (Γ). Strain lifts this degeneracy and changes the effective masses of the light and heavy holes. In the direction parallel to the substrate the heavy hole band becomes light and the light hole band becomes heavy. Under biaxial tension (Fig. 10*b*) the bandgap decreases and the "heavy" hole band lies below the "light" hole band. Under biaxial compression the bandgap increases and the "heavy" hole band lies above the "light" hole band.

Figure 10 is a simplification of the true band structure.[81,82] The bands are not really exactly parabolic, especially the hole bands, and strain increases the nonparabolicity of the hole bands. The details of the band structure can be derived using $\mathbf{k} \cdot \mathbf{p}$ theory.[86–90]

For a GaAs based QW, strain can be introduced by adding In to the QW. Since InAs has a larger lattice constant than GaAs, this is a compressively strained QW. In the direction of quantum confinement the highest quantized hole level is the first heavy hole level. This hole level will, therefore, have the largest influence on the density of states and on the optical gain. The effective mass of holes in this level confined in the QW is that parallel to the substrate and is reduced by strain. The reduction of the hole mass within the QW results in a reduced density of hole states [see Eq. (6)].

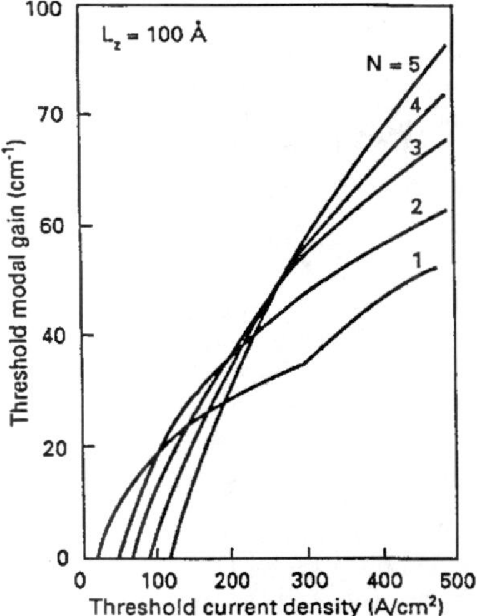

FIGURE 9 Threshold modal gain as a function of threshold current density for (Al, Ga)As single QW and MQWs with 2, 3, 4, and 5 QWs. Each QW has a thickness of 100 Å. (*From Ref. 80*.)

The reduction in density of hole states is a significant improvement. In order for a semiconductor laser to have optical gain (and lase), $f_c(E, N) - f_v(E, N)$ must be greater than zero [see Eq. (7)]. In an unstrained semiconductor the electrons have a much lighter mass than the heavy holes; the holes therefore have a higher density of states than the electrons. F_c and $f_c(E, N)$ change much more quickly with the injection of carriers than F_v and $f_v(E, N)$. Since approximately equal numbers of holes and electrons are injected into the undoped active layer, reducing the mass of the holes by introducing compressive strain reduces the carrier density required to reach transparency and therefore reduces the threshold current of a semiconductor laser.[91]

This theoretical prediction is well supported by experimental results. Strained InGaAs

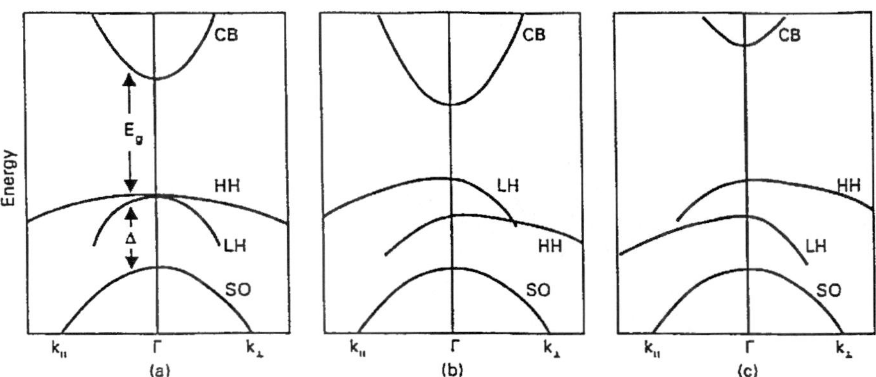

FIGURE 10 Schematic diagrams of the band structure of a III-V semiconductor: (*a*) unstrained; (*b*) under tensile strain; and (*c*) compressively strained. (*After Ref. 81*.)

single-QW lasers with record-low threshold current densities of 45 to 65 A/cm^2 have been demonstrated.[92–95] These very high quality strained QW lasers typically have lasing wavelengths from 0.98 to 1.02 μm, QW widths of 60 to 70 Å, and In concentrations of 20 to 25 percent. InGaAs QWs with wavelengths as long as 1.1 μm have been successfully fabricated,[96,97] but staying below the critical thickness of the InGaAs layers becomes a problem since the wavelength is increased by increasing the In concentration. (With higher In concentration the amount of strain is increased and the critical thickness is reduced.)

Strained InGaAs QWs have another advantage over GaAs QWs. Strained QW lasers are more reliable than GaAs lasers, i.e., they have longer lifetimes. Even at high temperatures (70–100°C), they are very reliable.[75,76] The reasons for this are not well understood, but it has been suggested that the strain inhibits the growth of defects in the active region.[98–100] Improving the reliability of GaAs-based lasers is of great practical significance since GaAs lasers are generally less reliable than InP-based lasers.[4,101,102]

Up to this point our discussion of QW lasers has been limited to GaAs-based QW lasers. QW lasers can also be fabricated in other material systems. GaInP/AlGaInP visible lasers have been improved significantly with the use of a single strained QW active region.[103–105] These are also compressively strained QWs formed by adding excess In to the active region. This is a much less developed material system than GaAs, so recent results such as 215 A/cm^2 for a single strained $Ga_{0.43}In_{0.57}P$ QW[103] are very impressive.

Long Wavelength (1.3 and 1.55 μm) Quantum Well Lasers

Long wavelength (InGaAsP/InP) QWs generally do not perform as well as GaAs-based QWs; however, with the advent of strained QW lasers significant progress has been made. Narrow bandgap lasers are believed to be significantly effected by nonradiative recombination processes such as Auger recombination[4,106–108] and intervalence band absorption.[109] In Auger recombination (illustrated in Fig. 11) the energy from the recombination of an electron and a hole is transferred to another carrier (either an electron or a hole). This newly created carrier relaxes by emitting a phonon; therefore, no photons are created. In intervalence band absorption (IVBA) a photon is emitted, but is reabsorbed to excite a hole from the split-off band to the heavy hole band. These processes reduce the performance of long wavelength QW lasers enough to make an MQW a lower threshold

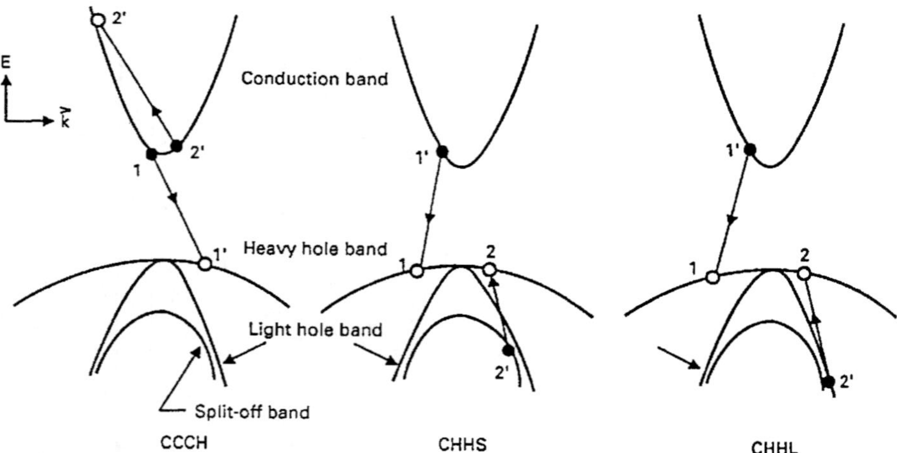

FIGURE 11 Schematic diagrams of band-to-band Auger recombination processes. (*After Ref. 4.*)

device than a single QW. As illustrated by Fig. 9, this means that the threshold gain is above the point where the gain versus current density curve of a single QW crosses that of an MQW. Good threshold current density results for lasers operating at 1.5 μm are 750 A/cm^2 for a single QW and 450 A/cm^2 for an MQW.[110]

Long-wavelength QW lasers can be improved by use of a strained QW. For these narrow bandgap lasers strain has the additional benefits of suppressing Auger recombination[111,112] and intervalence band absorption.[111] Several groups have demonstrated excellent results with compressively strained InGaAsP/InP QW lasers.[113–115] Compressively strained single QW lasers operating at 1.5 μm have been demonstrated with threshold current densities as low as 160 A/cm^2.[115]

Surprisingly, tensile strained InGaAsP/InP QW lasers also show improved characteristics.[115–117] Tensile strained QW lasers operating at 1.5 μm have been fabricated with threshold current densities as low as 197 A/cm^2.[116] These results had not been expected (although some benefit could be expected through suppression of Auger recombination), but have since been explained in terms of TM-mode lasing[118,119] (normally, semiconductor lasers lase in the TE mode) and suppression of spontaneous emission.[118]

Long-wavelength semiconductor lasers are more sensitive to increases in operating temperature than GaAs-based semiconductor lasers. This temperature sensitivity has been attributed to the strong temperature dependence of Auger recombination[106–108] and intervalence-band absorption.[109] The use of a strained QW should therefore improve the high-temperature operation of these lasers. This is in fact the case; the best results are reported for tensile strained QW lasers with continuous operation at 140°C.[115,117]

In summary, the use of QW active regions has significantly improved the performance of semiconductor lasers. In this section we have emphasized the dramatic reductions in threshold current density. Improvements have also been realized in quantum efficiency,[119] high-temperature operation,[74–76,115,117] modulation speed (discussed later in this chapter), and spectral linewidth (discussed later). We have limited our discussion to quantum wells. It is also possible to provide quantum confinement in two directions, which is called a *quantum wire,* or three directions, which is called a *quantum dot* or *quantum box.* It is much more difficult to fabricate a quantum wire than a QW, but quantum wire lasers have been successfully demonstrated.[121] For a review of these novel structures we refer the reader to Ref. 122.

13.7 HIGH-POWER SEMICONDUCTOR LASERS

There are several useful methods for stabilizing the lateral modes of an injection laser.[123–129] In this section, we will discuss techniques for the achievement of high-power operation in a single spatial and spectral mode. There are several physical mechanisms that limit the output power of the injection laser: spatial hole-burning effects lead to multispatial mode operation and are intimately related to multispectral mode operation, temperature increases in the active layer will eventually cause the output power to reach a maximum, and catastrophic facet damage will limit the ultimate power of the laser diode (GaAlAs/GaAs). Thus, the high-power laser designer must optimize these three physical mechanisms to achieve maximum power. In this section, we discuss the design criteria for optimizing the laser power.

High-Power Mode-Stabilized Lasers with Reduced Facet Intensity

One of the most significant concerns for achieving high-power operation and high reliability is to reduce the facet intensity while, at the same time, providing a method for stabilizing the laser lateral mode. Over the years, researchers have developed four approaches for performing this task: (1) increasing the lasing spot size, both perpendicular

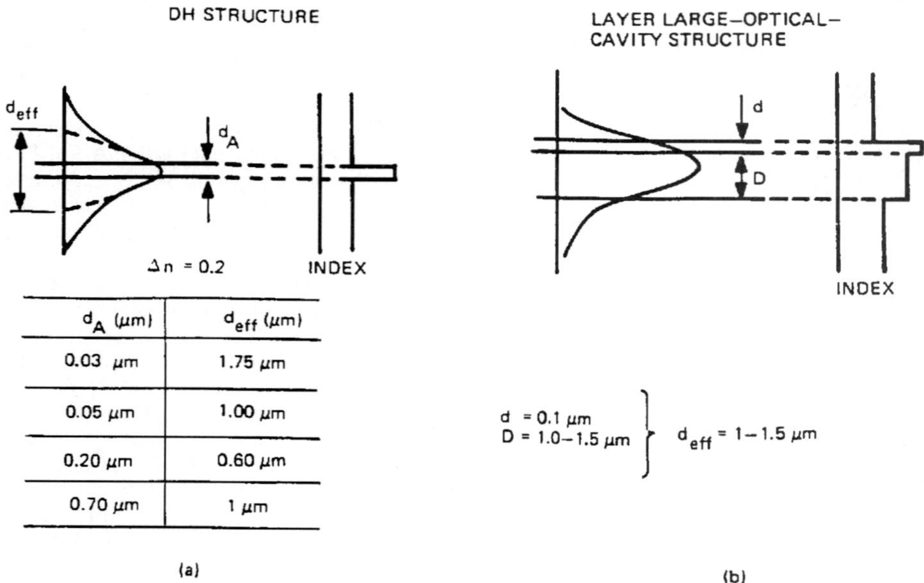

d_A (μm)	d_{eff} (μm)
0.03 μm	1.75 μm
0.05 μm	1.00 μm
0.20 μm	0.60 μm
0.70 μm	1 μm

FIGURE 12 Schematic diagrams of the two most commonly used heterostructure configurations for fabricating high-power laser diodes. (*a*) DH structure; (*b*) layer large-optical-cavity structure. The d_{eff} calculations are after Botez.[130]

to and in the plane of the junction, and at the same time introducing a mechanism for providing lateral mode-dependent absorption loss to discriminate against higher-order modes; (2) modifying the facet reflectivities by providing a combination of high-reflectivity and low-reflectivity dielectric coatings; (3) eliminating or reducing the facet absorption by using structures with nonabsorbing mirrors (NAM); (4) using laser arrays and unstable resonator configurations to increase the mode volume. Techniques 1 and 2 are the commonly used techniques and will be further discussed in this section. Techniques 3 and 4 (laser arrays) will be discussed shortly.

Given the proper heat sinking, in order to increase the output power of a semiconductor GaAlAs/GaAs laser, we must increase the size of the beam and thus reduce the power density at the facets for a given power level. The first step in increasing the spot size involves the transverse direction (perpendicular to the junction). There are two approaches for accomplishing this, with the constraint of keeping threshold current low: (1) thinning the active layer in a conventional double-heterostructure (DH) laser (Fig. 12*a*) below 1000 Å and (2) creating a large optical cavity structure (Fig. 12*b*).

Thinning the active layer from a conventional value of 0.2 to 0.03 μm causes the transverse-mode spot size to triple for a constant index of refraction step, Δn_r.[130] The catastrophic power level is proportional to the effective beam width in the transverse direction, d_{eff}. the asymmetric large optical cavity (A-LOC) concept[3,131] involves the epitaxial growth of an additional cladding layer referred to as the *guide layer* (d_G), with index of refraction intermediate between the n-AlGaAs cladding layer and the active layer. By using a relatively small index of refraction step ($\Delta n_r = 0.1$) versus 0.20–0.30 for DH lasers, it is possible to force the optical mode to propagate with most of its energy in the guide layer. The effective beam width for the A-LOC can be approximately expressed as

$$d_{eff} = d_A + d_G \tag{15}$$

where d_A is the active layer thickness. Mode spot sizes in the transverse direction of approximately 1.5 μm can be achieved.

Important Commercial High-Power Diode Lasers

In the last few years, several important high-power laser geometries have either become commercially available or have demonstrated impressive laboratory results. Table 1 summarizes the characteristics of the more important structures. It is evident that the structures that emit the highest cw power (>100 mW) (QW ridge waveguide (QWR), twin-ridge structure (TRS), buried TRS (BTRS), current-confined constricted double-heterostructure large optical cavity (CC-CDH-LOC), and buried V-groove-substrate inner stripe (BVSIS)) have several common features: (1) large spot size (CDH-LOC, TRS, BTRS, QW ridge); (2) low threshold current and high quantum efficiency; and (3) a combination of low- and high-reflectivity coatings. All the lasers with the highest powers, except for the CDH-LOC, use the thin active laser design. A recent trend has been the use of quantum well active layers.

Figure 13 contains schematic diagrams for five of the more common DH laser designs for high cw power operation, and Fig. 14 shows plots of output power versus current for various important geometries listed in Table 1.

The CC-CDH-LOC device with improved current confinement[136] (Fig. 13*a*) is fabricated by one-step liquid-phase epitaxy (LPE) above a mesa separating two dovetail channels. Current confinement is provided by a deep zinc diffusion and a narrow oxide stripe geometry. The final cross-sectional geometry of the device is very dependent on the exact substrate orientation and LPE growth conditions.[145] By properly choosing these conditions, it is possible to grow a convex-lens-shaped active layer ($Al_{0.07}Ga_{0.93}As$) on top of a concave-lens-shaped guide layer ($Al_{0.21}Ga_{0.79}As$). The combination of the two leads to a structure with antiwaveguiding properties and a large spot size. Discrimination against higher-order modes is provided by a leaky-mode waveguide. The cw threshold current is in the range of 50–70 mA. Single-mode operation has been obtained to 60 mW under 50 percent duty cycle, and the maximum cw power from the device is 165 mW. The power conversion efficiency at this power level is 35 percent considering only the front facet.

The CSP laser[127] (Fig. 13*b*) is fabricated by one-step LPE above a substrate channel. The current stripe is purposely made larger than the channel to ensure uniform current flow across the channel. However, this leads to some waste of current and thus a lower differential efficiency than other similar high-power laser structures (BVSIS, BTRS). Lateral mode control is very effectively obtained by the large difference in the absorption coefficient α between the center and edges of the channel and by changes in the index of refraction that result from changes in the geometry. By proper control of the active and n-cladding layer thicknesses, it is possible to obtain $\Delta\alpha \cong 1000\ \mathrm{cm}^{-1}$ (see Ref. 146) and $\Delta n_r \cong 10^{-2}$. Threshold currents are in the range of 55–70 mA. The transverse far field is relatively narrow due to the very thin active layer. Researchers from RCA have obtained power levels in excess of 150 mW (cw) with a CSP-type laser.[137] A detailed study of the CSP laser has been presented by Lee et al. in a recent publication.[147]

Matsushita[133] has also developed a CSP-like structure called the twin-ridge structure (TRS) that uses a 400-Å active layer thickness (Fig. 13*c*). The structure has demonstrated fundamental-mode cw power to 200 mW and single-longitudinal-mode cw power to 100 mW. The maximum available power for the TRS laser is 115 mW, and threshold currents are in the range of 80–120 mA. It appears that even though their geometry is similar to that of the CSP, lasers with ultrathin and planar-active layers have been fabricated. It should be further pointed out that one of the keys to achieving ultrahigh power from CSP-like structures is the achievement of ultrathin (<1000 Å) active layers that are highly uniform in thickness. Small nonuniformities in the active layer thickness

TABLE 1 Summary of Mode-Stabilized High-Power Laser Characteristics (GaAlAs/GaAs)*†

Manufacturer [reference]	Geometry	Construction	Rated power (mW)	Max. cw power (mW)	Spectral qual (cw)	Spatial qual (cw)	*I*th (m/A)	Slope EEF (mW/mA)	Far field
General Optronics [132]	CNS-LOC	Two-step LPE	—	60	SLM (50)	SSM (50)	50	0.67	12° × 26°
Hitachi [127]	CSP	One-step LPE (TA)	30	100	SLM (40)	SSM (40)	75	0.5	(10–12) × 27°
MATS. [133]	TRS	One-step LPE	25	115	SLM (50)	SSM (80)	90	0.43	6° × 20°
MATS. [134]	BTRS	Two-step LPE	40	200	SLM (50)	SSM (100)	50	0.8	6° × 16°
NEC [135]	BCM	Two-step LPE	—	80	SLM (80)	SSM (80)	40	0.78	7° × 20°
RCA [136]	CC-CDH	One-step LPE	—	165	SLM (50)	SSM (50)	50	0.77	6° × 30°
RCA [137]	CSP	One-step LPE	—	190	SLM (70)	SSM (70)	50	—	6.5° × 30°
Sharp [138]	VSIS	Two-step LPE	30	100	SLM (50)	SSM (50)	50	0.74	12° × 25°
Sharp [139]	BVSIS	Two-step LPE	—	100	—	SSM (70)	50	0.80	12° × 25°
HP [140]	TCSM	One-step MOCVD	—	65	SLM (65)	SSM (40)	60	0.4	—
TRW [141]	ICSP	Two-step MOCVD (AH/HR)	—	100	SLM (30)	150 (50% duty cycle)	75	0.86	(8–11°) × 35°
Ortel [142]	BH/LOC (NAM)	Two-step LPE (AH/HR)	30	90	—	90	30–50	0.85	—
Spectra Diode [143]	QWR	MOCVD	—	500	SLM (100)	SSM (180)	16	1.3	8° × 22°
BN(STC) [144]	QWR	MOCVD	—	300	SLM (150)	SSM (175)	—	0.8	—

* BH Buried heterostructure
BTRS Buried TRS
BVSIS Buried VSIS
CC-CDH Current-confined constricted double heterostructure
CNS Channeled narrow stripe
CSP Channeled-substrate planar
ISCP Inverted CSP
LOC Large optical cavity
NAM Nonabsorbing mirror
QWR Quantum-well ridge
SLM Single longitudinal mode
SSM Single spatial mode
TCSM Twin-channel-substrate mesa
TRS Twin-ridge substrate
VSIS V-groove-substrate inner stripe

† Approaches for achieving high-power GaAlAs lasers:
- Thin active (TA) or A-LOC layer to decrease facet power density
- Tight current confinement to produce high current utilization
- Combination of low/high-reflectivity facet coatings (AR/HR) to produced high differential efficiency and lower facet intensity
- Quantum well design with long cavity

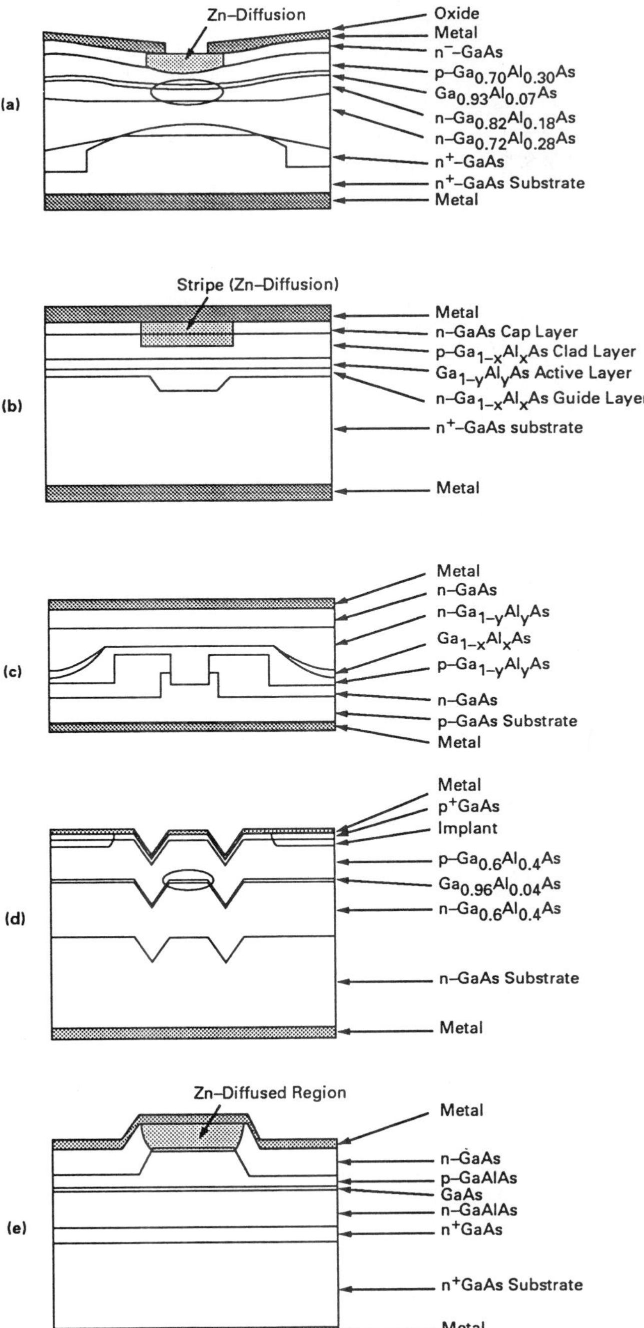

FIGURE 13 Geometries for several important high-power diode lasers. (*a*) Constricted double-heterostructure large-optical-cavity laser (CDH-LOC);[136] (*b*) channel substrate planar laser (CSP);[127] (*c*) broad-area twin-ridge structure (BTRS);[134] (*d*) twin-channel substrate mesa; (*e*) inverted CSP.

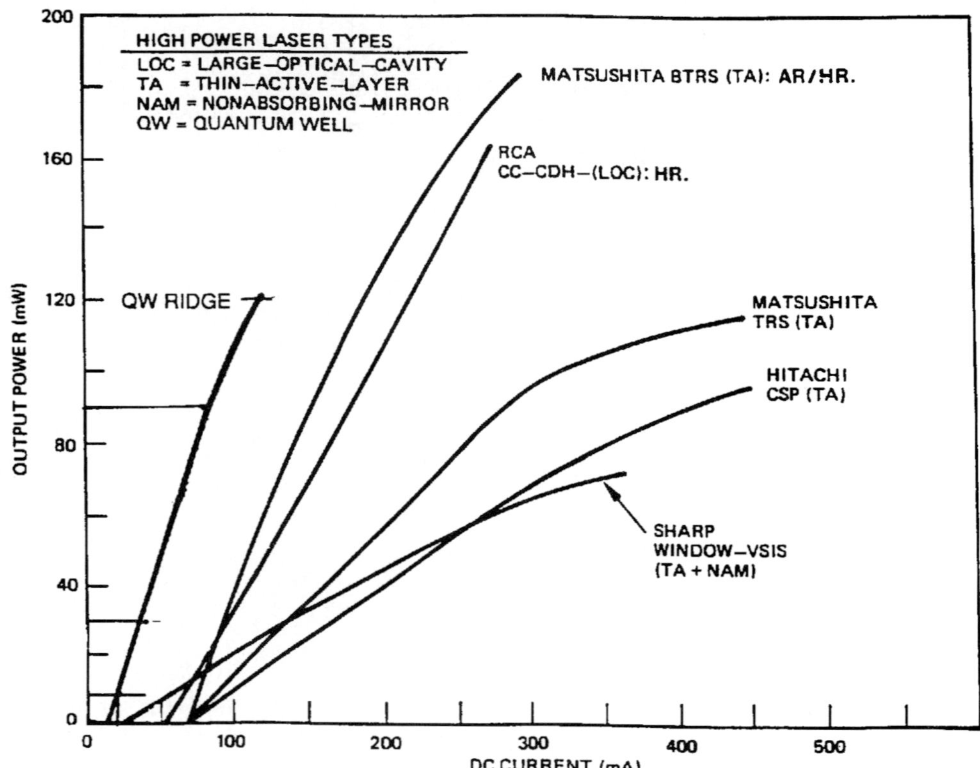

FIGURE 14 Plots showing output power versus cw current for the major high-power laser diodes. The maximum power in a single spatial mode is in the range of 100–150 mW and total cw power can approach 200 mW.

lead to a larger Δn_r difference, and thus a smaller lateral spot size, which will lead to lower power levels and reduced lateral mode stability.

Metalorganic chemical vapor deposition (MOCVD, discussed earlier) has been used to fabricate lasers with higher layer uniformity, which leads to a reduced spectral width and more uniform threshold characteristics. Several MOCVD laser structures with demonstrated high-power capability are schematically shown in Fig. 13, and their characteristics are summarized in Table 1. Figure 13*d* shows the twin-channel substrate mesa (TCSM) laser.[140] The fabrication consists of growing a DH laser structure over a chemically etched twin-channel structure using MOCVD. Optical guiding is provided by the curvature of the active layer. The TCSM laser has achieved cw powers of 40 mW in a single spatial mode and 65 mW in a single longitudinal mode. The inverted channel substrate planar (ICSP) laser[135] is schematically shown in Fig. 13*e*. This structure is one MOCVD version of the very successful CSP structure (Fig. 13*b*).[141] The ICSP laser has achieved powers in excess of 150 mW (50 percent duty cycle) in a single spatial mode and a 100-mW (cw) catastrophic power level.

More recently, quantum well lasers using the separate carrier and optical confinement (see previous section, "Quantum Well Lasers") and ridge waveguide geometries have been used for producing power levels in excess of 150 mW (cw) in a single spatial mode.[143,144] The QW ridge resembles a standard RWG (Fig. 4*e*), but with a QW active region. Such laser structures have low threshold current density and low internal absorption losses, thus permitting higher-power operation.

Future Directions for High-Power Lasers

Nonabsorbing Mirror Technology. The catastrophic facet damage is the ultimate limit to the power from a semiconductor laser. In order to prevent catastrophic damage, one has to create a region of higher-energy bandgap and low surface recombination at the laser facets. Thus, the concept of a laser with a nonabsorbing mirror (NAM) was developed. The first NAM structure was demonstrated by Yonezu et al.[148] by selectively diffusing zinc along the length of the stripe, except near the facets. This created a bandgap difference between the facet and bulk regions and permitted a three- to fourfold increase in the cw facet damage threshold and a four- to fivefold increase in pulse power operation.[149] More recent structures have involved several steps of liquid-phase epitaxy.[150,151]

The incorporation of the NAM structure is strongly device-dependent. For example, in the diffused device structures, such as deep-diffused stripe (DDS)[148] and transverse junction stripe (TJS) lasers, NAM structures have been formed by selective diffusion of zinc in the cavity direction.[149] The n-type region will have a wider bandgap than the diffused region, and thus there will be little absorption near the facets. However, most index-guided structures require an additional growth step for forming the NAM region.[150,151] the NAM structures in the past have suffered from several problems: (1) Due to their complex fabrication, they tend to have low yields. Furthermore, cw operation has been difficult to obtain. (2) Cleaving must be carefully controlled for NAM structures having no lateral confinement, in order to avoid excessive radiation losses in the NAM region. The NAM length is a function of the spot size. (3) The effect of the NAM structure on lateral mode control has not been documented, but could lead to excessive scattering and a rough far-field pattern.

It is now becoming more clear that the use of a NAM structure will be required for the reliable operation of high-power GaAlAs/GaAs laser diodes. Experimental results[152] appear to indicate that laser structures without a NAM region show a decrease in the catastrophic power level as the device degrades. However, most of the approaches currently being implemented require elaborate processing steps. A potentially more fundamental approach would involve the deposition of a coating that would reduce the surface recombination velocity and thus enhance the catastrophic intensity level.[153,154] Such coatings have been recently used by researchers from Sharp and the University of Florida to increase the uncoated facet catastrophic power level by a factor of 2.[155,156]

Recently, the use of NAM technology has been appearing in commercial products. The crank transverse junction stripe (TJS) laser (a TJS laser with NAM) can operate reliably at an output power of 15 mW (cw), while the TJS laser without the NAM can operate only at 3 mW (cw).[147] The Ortel Corporation has developed a buried heterostructure (BH) laser with significantly improved output power characteristics compared to conventional BH lasers.[142] The NAM BH laser is rated at 30 mW (cw)[142] compared to 3–10 mW for the conventional BH/LOC device.

Last, the use of alloy disordering, whereby the bandgap of a quantum well laser can be increased by diffusion of various types of impurities (for example, Zn and Si),[157] can lead to a very effective technique for the fabrication of a NAM structure. Such structures have produced an enhancement of the maximum pulsed power by a factor of 3–4.

High-Power 1.3/1.48/1.55-μm Lasers. Previous sections have discussed high cw power operation from (GaAl)As/GaAs laser devices. In the past several years there have been reports of the increasing power levels achieved with GaInAsP/InP lasers operating at $\lambda = 1.3\ \mu\text{m}$. The physical mechanisms limiting high-power operation in this material system are quite different than those for GaAlAs/GaAs lasers. The surface recombination at the laser facets is significantly lower than in GaAlAs/GaAs, and thus catastrophic damage has not been observed. Maximum output power is limited by either heating or carrier leakage effects. With the advent of structures having low threshold current density and high quantum efficiency, it was just a matter of time before high-power results would become

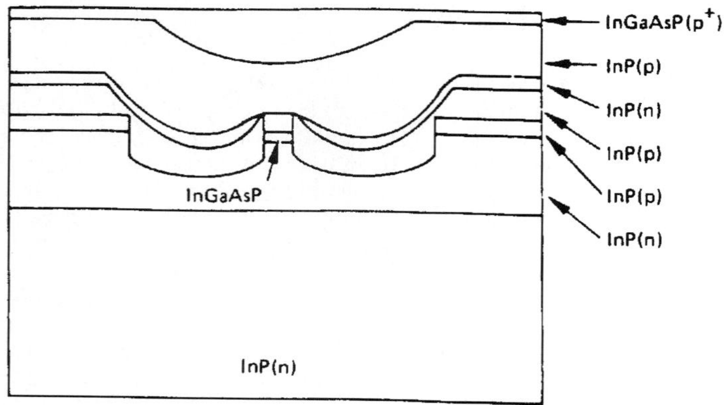

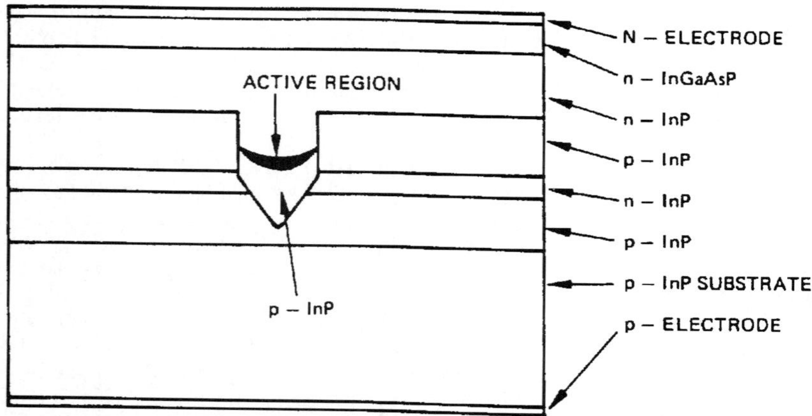

FIGURE 15 Schematic diagrams of the most prominent LPE-based 1.3-μm laser structures. (*a*) Double-channel planar-buried heterostructure (DC-PBH);[158,159] (*b*) buried crescent.[160,161]

available. Furthermore, since facet damage is not a problem, the only real need for facet coatings is for improving the output power from one facet and sealing the device for improved reliability.

In Fig. 15, we schematically show the two most common long-wavelength laser structures that have demonstrated high cw power operation. In Fig. 15*a,* the double-channel planar buried heterostructure (DC-PBH) is systematically shown.[158,159] The structure requires a two-step LPE growth process. The first step is the growth of the first and top cladding layers in addition to the active layer. This is followed by the etching of the structure, which is followed by a regrowth to form the blocking and contact layers. LPE growth of this material system is such that if the mesa region is narrow enough, no growth occurs on top of it during the deposition of the blocking layer, and this occurs for mesa widths of less than ~5 μm. Low threshold current is achieved due to the narrow mesa geometry and the good carrier and current confinement.

The DC-PBH has proved to be a laser structure with excellent output characteristics

TABLE 2 Summary of Mode-Stabilized High-Power Laser Characteristics (GaInAsP/InP)*†

Manufacturer [reference]	Geometry	Construction	Max. power (mW)	Spatial quality	I_{th} (mA)
Mitsubishi (160)	PBC	Two-step LPE (p-subst.)	140	SSM (70)	10–30
NEC [157, 158]	DC-PBH	Two-step LPE	140	SSM (140)	10–30
OKI [161, 162]	VIPS	Two-step LPE (p-subst.)	200	SSM (200)	10–30
TRW/EORC [159]	DC-PBH type	Two-step LPE	100	SSM (70)	10–30
TRW/EORC [159]	PBC	Two-step LPE (p-subst.)	107	SSM (78)	10–30
STC [164]	MQW	MOCVD	170	—	—
ATT [165]	MQW BH	MBE	200	—	—

* DC-PBH Double-channel planar buried heterostructure
MQW Multiquantum well
PBC Planar buried cresent
SSM Single spatial mode

† Approaches for high-power GalnAsP/lnP lasers:
- Tight current confinement to reduce the threshold current
- Facet coatings (reflector/low reflecting front facet)
- Diamond heat sinking
- Long cavity length

and high reliability. NEC has been able to obtain thresholds as low as 10 mA with 70 percent quantum efficiency. Degradation rates of the order of 10^{-6}/hr for an output power of 5 mW at a temperature of 70°C have also been obtained. More recently, NEC has obtained 140-mW power in a single spatial mode.[158] Lasers at 50 mW and 25°C have been placed on lifetest and show relatively low degradation rates after several hundred hours. Degradation rates at 20 and 30 mW (50°C) are 1.3×10^{-5}/hr and 2.22×10^{-5}/hr, respectively. TRW has also worked with DC-PBH/PBC-(planar buried cresent) type laser diodes and has obtained 100 mW cw.[159] A summary of the various high-power $\lambda = 1.3\ \mu m$ laser diode structures and characteristics is given in Table 2.

The other structure that has demonstrated high cw power is the buried crescent laser first investigated by Mitsubishi (Fig. 15*b*).[160] The structure is grown using a two-step LPE process and a *p* substrate. The final structure resembles a channel laser with an active layer that tapers to zero near the edges of the channel. The tapering provides good carrier and optical confinement. Researchers from Oki with a structure similar to the Mitsubishi structure have demonstrated maximum power levels of 200 and 140 mW in a single spatial mode.[162] Lifetests[163] on these lasers demonstrated a mean time to failure of $\sim 7 \times 10^5$/hr (at 20°C) at 75 percent of the maximum cw output power (maximum = 25–85 mW). These results appear to indicate that 1.3-μm lasers are reliable for high-power applications.

A more recent development has been the use of multiquantum well (MQW) high-power lasers in the 1.5–1.55-μm wavelength band. The use of a MQW ridge waveguide structure has produced power levels of ~170 mW (cw).[164] The MQW structure consists of five wells of InGaAs, 60 Å thick, separated by four GaInAsP barriers of 100-Å thickness. The two thicker, outermost barriers of GaInAsP provide a separate confinement heterostructure (SCH) waveguide. In addition, buried heterostructure (BH) lasers[165] with power levels in excess of 200 mW have been achieved by incorporating strain into MQW structures. A recent review article by Henshall describes the state of the art in more detail.[166]

TABLE 3 High-Power GaInAs Strained Layer Quantum-Well Lasers

Laser group [reference]	Ridge width (μm)	Wavelength (μm)	Threshold current (mA)	Max. power in single spatial mode (mW)	Max. cw power (mW)
JPL [168]	6	0.984–0.989	13	—	24
JPL [168]	3	0.978	8	116	400
NTT [169]	3	0.973–0.983	9	115	500
Spectra Diode [170]	4	0.9–0.91	~20	180	350
Boeing [76]	4	0.98	10–15	150	440

High-Power Strained Quantum Well Lasers. Over the last several years, there has been extensive research in the area of strained layer quantum well high-power lasers. As with GaAs QW high-power lasers, the geometry is typically a QW ridge. Table 3 summarizes some of the latest single-spatial-mode high-power results.

Thermal Properties. An important parameter in the operation of high-power laser diodes is the optimization of thermal properties of the device. In particular, optimizing the laser geometry for achieving high-power operation is an important design criterion. Arvind et al.[167] used a simple one-dimensional thermal model for estimating the maximum output power as a function of laser geometry (cavity length, active layer thickness substrate type, etc.). The results obtained for GaInAsP/InP narrow stripe PH lasers were as follows:

- Maximum output power is achieved for an optimum active layer thickness in the 0.15-μm region. This result applies only to nonquantum well lasers.
- Significantly higher output powers (25 to 60 percent) are obtained for lasers fabricated on *p* substrates compared to those on *n* substrates. The result is based on the lower electrical resistance of the top epitaxial layers in the *p* substrate compared to *n* substrate.
- Significantly higher output powers (~60 percent) are obtained for lasers mounted on diamond rather than silicon heat sinks as a result of the higher thermal conductivity of diamond compared to silicon, 22 versus 1.3 W/(°C-cm).
- Significantly higher output powers (~100 percent) are obtained for lasers having a length of 700 μm compared to the conventional 300 μm. The higher power results from the reduced threshold current density and thermal resistance for the longer laser devices.

A plot of the calculation and experimental data from Oki[162] is given in Fig. 16. Note that there are no adjustable parameters in the calculation.

An important conclusion from the thermal modeling is that longer cavity semiconductor lasers (700–1000 μm) will be able to operate at higher heat sink temperatures when the power level is nominal (~5 mW) compared to shorter cavity devices (~100–300 μm). In addition, the reliability of the longer cavity devices is also expected to be better. More recent calculations and experimental results using strained layer lasers have verified this.[95]

Semiconductor Laser Arrays

One of the most common methods used for increasing the power from a semiconductor laser is to increase the width of the emitting region. However, as the width is increased, the occurrence of multilateral modes, filaments, and lateral-mode instabilities becomes more significant. A far-field pattern is produced that is not diffraction-limited and has reduced brightness. The most practical method to overcome this problem is to use a monolithic

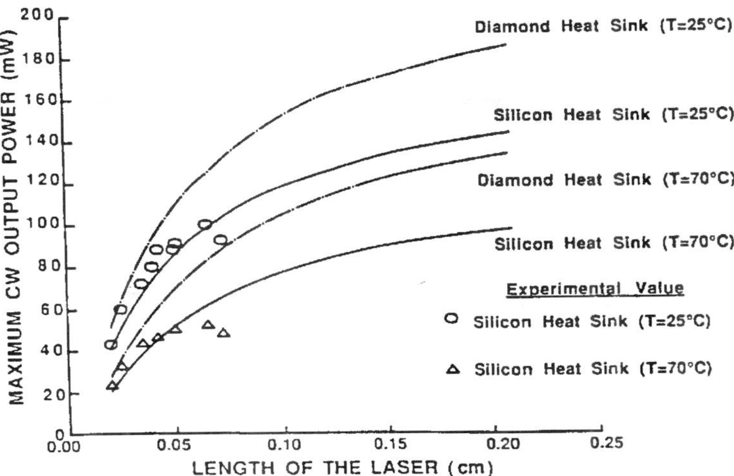

FIGURE 16 Calculation of maximum output power per facet as a function of device length for *n-p*-substrate-type lasers and different heat sinks.[167] Note the increased power level achieved for longer lasers and *p*-type substrates.

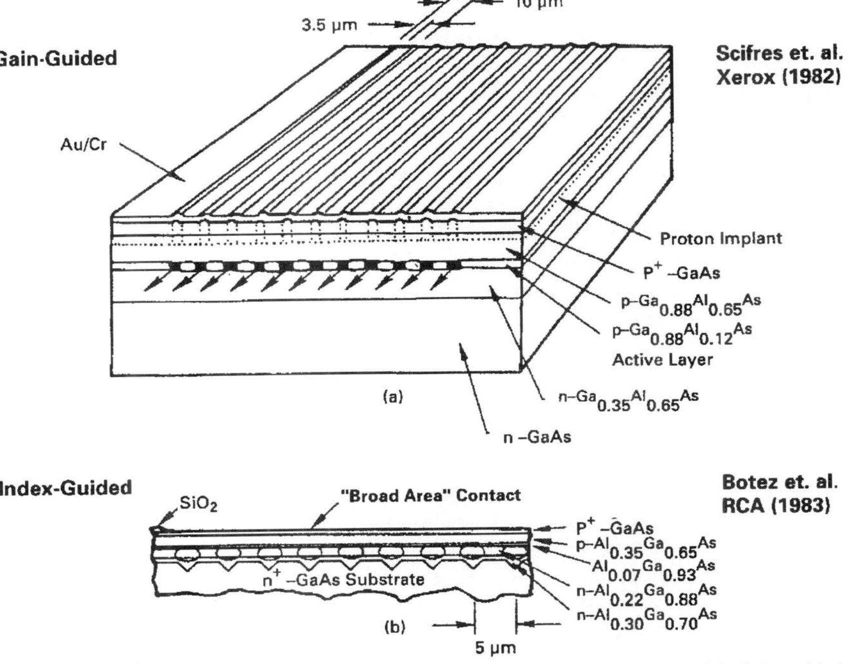

FIGURE 17 Schematic diagrams showing two types of laser array structures. (*a*) Gain-guided phased array using quantum well active layers and grown by MOCVD:[175] (*b*) index-guided phased array using CSP-LOC structures grown by LPE.[173]

TABLE 4 Summary of High-Power Phase-Locked Laser Arrays*

Laser group [reference]	No. of elements	Material system	Type of array *	Max. power (mW)	Max. power (mW)	Far field
Xerox, 1978 [172]	5	GaAlAs-GaAs	GG	60 (P)	130 (P)	SL (2°)
HP, 1981 [174]	10	GaAlAs-GaAs	IG	1W (P)	1400 (P)	DL
Xerox, 1982 [175]	10	GaAlAs-GaAs	GG	200 (P)	270 (cw)	SL (1°)
Xerox, 1983 [176]	40	GaAlAs-GaAs	GG	800 (P)	2600 (cw)	DL
RCA, 1983 [173]	10	GaAlAs-GaAs	IG	400 (P)	1000 (P)	DL
Siemens, 1984 [177]	40	GaAlAs-GaAs	GG	—	1600 (cw)	DL
Bell Labs, 1984 [178]	10	GaAlAs-GaAs	IG	—	—	DL
TRW, 1984 [179–181]	2	GaAlAs-GaAs	IG	75 (cw)	115 (cw)	SL (4–6°)
UC Berkeley, 1982 [182]	10	GaAlAs-GaAs	IG	—	200	SL (2–7°)
Cal-Tech, 1984 [183]	5	GaAlAs-GaAs	IG	—	—	SL (3°)
Xerox/Spectra Diode 1985 [184]	10	GaAlAs-GaAs	Offset stripe GG	575 (P)	—	SL (1.9°)
Bell Labs, 1985 [185]	10	InGaAsP-InP	GG	100 (cw)	600	SL (4°)
Sharp, 1985 [186]	2	GaAlAs-GaAs	IG; Y-C	65 (cw)	90 (cw)	SL (4.22°)
Mitsubishi, 1985 [187]	3	GaAlAs-GaAs	IG	100 (cw)	150 (cw)	SL (3.6°)
Xerox/Spectra Diode 1986 [188]	10	GaAlAs-GaAs	IG(Y-C) stripe GG	200 (cw)	575 (P)	SL (3°)
TRW [189]	10	GaAlAs-GaAs	ROW	380 (cw) 1500 (P)	—	SL (0.7°)

* GG Gain guided
IC Index guided
Y-C Y-coupled
ROW Resonant optical waveguide
P pulsed
DL Double lobe
SL Single lobe

array of phase-locked semiconductor lasers. Such lasers have been used to generate powers in excess of 10 W (cw)[170] and over 200 W[171] from a single laser bar.

It was not until 1978 that Scifres and coworkers[172] first reported on the phase-locked operation of a monolithic array consisting of five closely coupled proton-bombarded lasers. The original coupling scheme involved branched waveguides, but this was quickly abandoned in favor of evanescent field coupling by placing the individual elements of the array in close proximity (Fig. 17*a*). More recently, arrays of index-guided lasers[173] have been fabricated; one example is shown in Fig. 17*b*.

Recent emphasis has been on achieving higher cw power and controlling the output far-field distribution. Some of the more significant events in the development of practical semiconductor laser arrays are summarized in Table 4.

The subject of array-mode stability has become of great interest. In a series of significant papers, Butler et al.[190] and Kapon et al.[191] recognized that to a first approximation, an array can be modeled as a system of n weakly coupled waveguides. The results indicate that the general solution for the field amplitudes will consist of a superposition of these array modes. The analytic results permitted, for the first time, a simple explanation for the observed far-field patterns and provided a means for designing device structures that would operate in the fundamental array mode (i.e., all elements in phase). This particular mode will provide the greatest brightness.

Many techniques have been used for improving array-mode selection[179,188,192,197] and thus achieving a well-controlled spatial mode. Two of the more successful earlier techniques involve (1) incorporation of optical gain in the interelement regions (gain coupling of the laser array[179–184,193]) and (2) use of interferometric techniques that involves *Y*-coupled junctions.[186,188,192]

The gain-coupled arrays achieve mode selectivity by introducing optical gain in the interelement regions and thus increasing the gain of the fundamental array mode since this mode has a significant portion of its energy in the interelement regions. The first demonstration of this approach was the twin-channel laser (TCL) developed by researchers from TRW;[180–181] since then, there have been other demonstrations.[182,184,193]

The theoretical foundations of the *Y*-coupled junction were first described in a paper by Chen and Wang.[192] Mode-selectivity is accomplished because the in-phase mode adds coherently at each *Y* junction, while the out-of-phase mode has destructive interference, since the single waveguides after the *Y* junction can support only the fundamental mode. Similar interferometric and mode-selective techniques have been used in the development of optical modulators.[198]

Finally, the most recent mode control mechanism for laser arrays involves the resonant phase-locking of leaky-mode elements.[189] With a properly optimized geometry, the fundamental array mode has significant mode discrimination analogous to the high discrimination found in single-element leaky-mode devices. Recent results indicate power levels in excess of 360 mW (cw) in the fundamental array mode, and the beam broadens to ~2 times diffraction limited for output powers of ~500 mW (cw).

At the present time, it is not clear which technique will be most useful for achieving stable, fundamental array mode operation. The gain-coupling concept works well for two or three elements.[199] However, array-mode selection described by the difference in gain between the first and second array modes rapidly decreases as the number of array elements increases beyond two or three.[199] The *Y*-junction and leaky-mode approaches do not appear to have the same limitations. The resonant leaky-mode arrays appear to have the most promising performance at high power levels; however, the structures are complex and thus yield and reliability need to be more fully addressed.

Two-Dimensional, High-Power Laser Arrays

There has been a significant amount of research activity in the past few years in the area of very high power diode lasers.[200–205] The activity has been driven by the significant reductions in threshold current density of GaAlAs/GaAs lasers that can be achieved with metalorganic chemical vapor deposition (MOCVD), utilizing a quantum well design. Threshold current densities as low as 200–300 A/cm^2 with external efficiencies exceeding 80 percent have been achieved using GRIN-SCH quantum well lasers.[200,201] CW powers of ~6–9 W have been achieved from single laser bars. Such power levels correspond to a maximum of 11 W/cm from a single laser bar. Table 5 lists some of the more recent results on very high power diode laser arrays.

In order to increase the output power from laser array structures, researchers have investigated the use of a two-dimensional laser array. One particular configuration, referred to as the "rack-stack approach," is schematically shown in Fig. 18. In essence, the approach involves stacking a linear array of edge emitters into a two-dimensional array. The two-dimensional arrays are fabricated[203] by (1) cleaving linear arrays of laser diodes from a processed wafer, (2) mounting the bars on heat sinks, and (3) stacking the heat sinks into a two-dimensional array.

As shown in Table 5, the main players in this business are McDonnell-Douglas and Spectra Diode Labs. The largest stacked[204] two-dimensional array has been manufactured by McDonnell-Douglas and has an active area with five laser bars 8 mm in length. The array was operated with 150-μs pulses to the limit of the driver[204] at pulse repetition rates of 20–666 Hz. Approximately 2.5 kW/cm^2 was obtained at 20 Hz (average power ~300 W) and 0.9 kW/cm^2 at 666 Hz (average power ~92 W). Higher output powers will be obtained

TABLE 5 Summary of High-Power Laser Array Results*

Laboratory [reference]	Array type	Maximum output power	Power efficiency (%)	Slope efficiency (W/A)	Power density (W/cm^2)
General Electric [201]	1D array	80 W (200-μs pulse; 10–100 Hz)	20	0.9	80
McDonnell-Douglas [203, 204]	Broad stripe (L = 1200 μm)	6 W (cw) (W = 300 μm)	38	0.91	200
	2D: 4 bars, 8 mm	15 W (cw)	15	—	50
	2D: 5 bars, 8 mm	320W (0.3% DF†)	—	—	2560
Spectra Diode [170]	1D array	8 W (cw)	—	—	—
	1D array	134 W (150-μs pulse)	49	1.26	134

* All high-power laser structures are fabricated using MOCVD and quantum-well design.
† DF = Duty factor.

as a result of achieving the ultimate limits in threshold current density and optical losses in the individual cavities with advances in nonabsorbing mirror and active cooling techniques. Some recent progress has been seen in the latter with the use of etched silicon grooves for fluid flow, which function as radiator elements to remove the heat.

13.8 HIGH-SPEED MODULATION

In many applications semiconductors lasers are modulated in order to carry information. Semiconductor laser dynamics are usually described by the rate equations for the photon and carrier densities:[3–5,205,206]

$$\frac{dN}{dt} = \frac{I}{edLw} - \frac{cg}{n_r\Gamma}P - \frac{N}{\tau_s} \tag{16}$$

$$\frac{dP}{dt} = \frac{c}{n_r}gP - \frac{P}{\tau_p} + \Gamma\beta\frac{N}{\tau_s} \tag{17}$$

where N is the carrier density, P is the photon density, I is the current, e is the charge of an electron, d is the active layer thickness, L is the laser cavity length, w is the laser stripe width, c is the speed of light, n_r is the refractive index of the active region, g is the threshold modal gain, Γ is the optical confinement factor, τ_s is the carrier lifetime, β is the spontaneous emission factor, and τ_p is the photon lifetime of the cavity. (When these equations are written in terms of total gain instead of modal gain, Γ does not appear in Eq. (16), but multiplies the gain in Eq. (17).)

$$\tau_p = \frac{n_r}{c(\alpha_i + (1/2L)\ln(1/R_F R_R))} \tag{18}$$

where α_i is the internal loss and R_F and R_R are the front- and rear-facet reflectivities. β is the ratio of spontaneous emission power into the lasing mode to the total spontaneous emission rate.[207] (Do not confuse β with the other spontaneous emission factor, which is

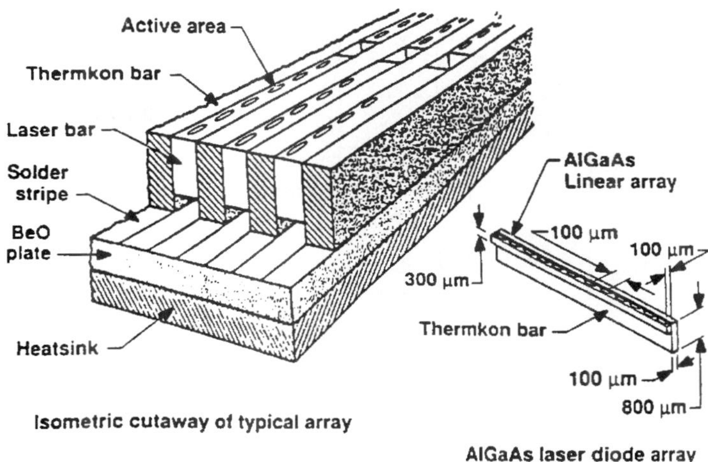

FIGURE 18 Schematic diagram of the two-dimensional rack-stack laser array architecture. (*Courtesy of R. Solarz, Lawrence Livermore.*)

used in linewidth theory and is defined as the ratio of the spontaneous emission power into the lasing mode to the stimulated emission power of the mode.)

When a semiconductor laser is modulated there is some delay before it reaches a steady state. Because it takes time for a carrier population to build up, there will be a time delay τ_d before the final photon density P_{on} is reached (see Fig. 19). Once P_{on} is reached, additional time is required for the carrier and photon populations to come into equilibrium. The output power therefore goes through relaxation oscillations before finally reaching a steady state. This type of oscillation has many parallels in other second-order systems,[208] such as the vibration of a damped spring or an *RLC* circuit.

The frequency of these relaxation oscillations, f_r is called the relaxation, resonance, or corner frequency. By considering small deviations from the steady state where $N = N_{th} + \Delta N$ and $P = P_{on} + \Delta P$, we can solve Eqs. (16) and (17) for f_r with the result:[5,205]

$$f_r \cong \frac{1}{2\pi}\sqrt{\frac{c}{n_r\Gamma}\frac{dg}{dN}\frac{P_{\text{on}}}{\tau_p}} = \frac{1}{2\pi}\sqrt{\frac{c}{n_r}\frac{dg}{dN}\frac{(I - I_{th})}{edLw}} \tag{19}$$

where dg/dN is the differential modal gain. For bulk double-heterostructure lasers the gain is linearly dependent on the carrier density and $(c/n_r)dg/dN$ is replaced by A, where A is a constant. As discussed earlier in the chapter, the gain-versus-carrier-density relationship of a quantum well is nonlinear, so A is not a constant for a QW laser.

Let us return our attention to the turn on delay τ_d between a current increase and the beginning of relaxation oscillations as illustrated in Fig. 19. If the initial current I_{off} is below I_{th}, the initial photon density P_{off} can be neglected in Eq. (16). Assuming an exponential increase in carrier density we can derive[206]

$$\tau_d = \tau_s \ln\left(\frac{I_{on} - I_{off}}{I_{on} - I_{th}}\right), \qquad I_{off} < I_{th} < I_{on} \tag{20}$$

Since τ_s is on the order of several ns, τ_d is usually very large for semiconductor lasers with I_{off} below I_{th}. For example, with $\tau_s = 4$ ns, $I_{on} = 20$ mA, $I_{th} = 10$ mA, and $I_{off} = 5$ mA, τ_d will be 1.6 ns. If a short current pulse is applied to a laser biased below threshold, τ_d may be so long that the laser barely responds (see Fig. 20). When a current pulse ends, it takes

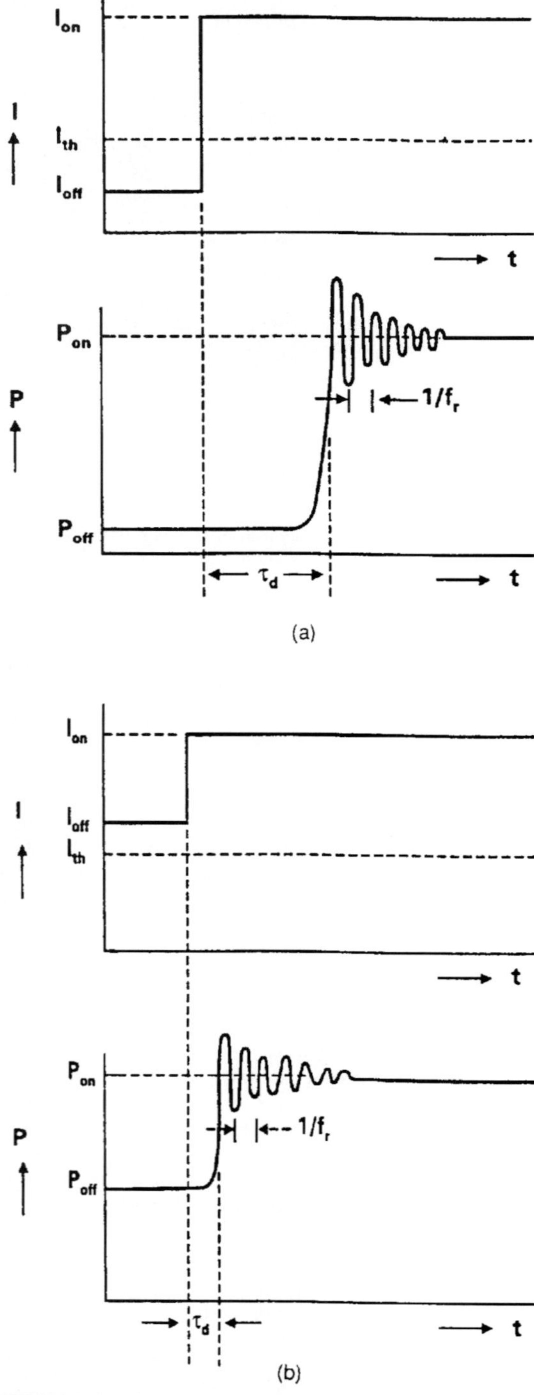

FIGURE 19 Schematic diagrams of the turn on delay and relaxation oscillations for a semiconductor laser (*a*) prebiased below threshold and (*b*) prebiased above threshold. (*After Ref. 206.*)

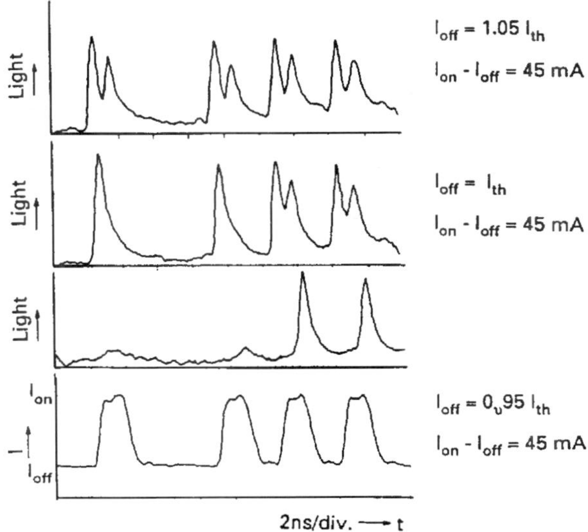

FIGURE 20 An illustration of the pattern effect for a AlGaAs laser diode with a 280 Mbit/s 10111 return to zero pattern. (*From Ref. 210*.)

time for the carrier population to decay. If another identical pulse is applied before the carrier population decays fully, it will produce a larger light pulse than the first current pulse did. This phenomenon, which is illustrated in Fig. 20, is called the *pattern effect.*[209] The pattern effect is clearly undesirable as it will distort information carried by the laser modulation. The pattern effect can be eliminated by prebiasing the laser at a current sufficient to maintain a carrier population; for most semiconductor lasers this will mean prebiasing at or above threshold. In order to modulate with a prebias below threshold, I_{on} must be much greater than I_{th}. For most lasers this will require an impractically large I_{on}, but if I_{th} is very low it may be possible.[211,212]

Even a semiconductor biased above threshold will have a nonzero τ_d before it reaches its final photon density. Equations (16) and (17) may be solved for τ_d above threshold:[206]

$$\tau_d = \frac{1}{2\pi f_r}\sqrt{2\ln\left(\frac{P_{on}}{P_{off}}\right)} = \frac{1}{2\pi f_r}\sqrt{2\ln\left(\frac{I_{on}-I_{th}}{I_{off}-I_{th}}\right)} \qquad I_{th} < I_{off} < I_{on} \tag{21}$$

τ_d will be much shorter for a prebias above threshold. For example if $f_r = 5$ GHz, $I_{on} = 40$ mA, $I_{off} = 15$ mA, and $I_{th} = 10$ mA, $\tau_d = 60$ ps. τ_d will be shortest for large P_{on} and P_{off}, so the shortest time delays will be achieved for small-scale modulation at high photon density. For digital applications in which fairly large-scale modulation is required, the maximum modulation speed of a semiconductor laser is to a large extent determined by τ_d.

For very high speed microwave applications, lasers are prebiased at a current greater than the threshold current and modulated at high frequencies through small amplitudes about the continuous current prebias. The frequency response of a semiconductor laser has the typical shape expected from a second-order system. (For a discussion of the frequency response of a second-order system see Ref. 208.) The laser amplitude response is fairly uniform at frequencies less than the relaxation oscillation frequency. At f_r the response goes through a resonance and then drops off sharply. The relaxation oscillation frequency is therefore the primary intrinsic parameter determining the modulation bandwidth. The

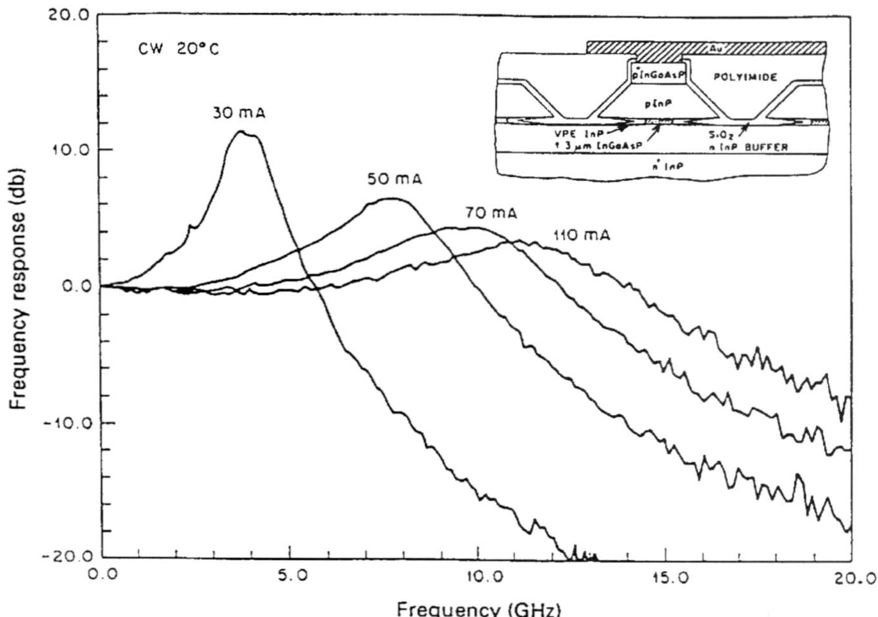

FIGURE 21 The small-signal modulation response of a 1.3-μm InGaAsP-constricted mesa laser for different bias levels. The cavity length is 170 μm and the stripe width is 1 μm. Inset: Schematic diagram of a 1.3-μm InGaAsP-constricted mesa laser (*From Refs. 213 and 214*.)

actual useful bandwidth is generally considered to be the frequency at which the response of the laser drops by 3 dB. Figure 21 is an example of the frequency response of a semiconductor laser under amplitude modulation.[213,214] In this example, the maximum 3-dB bandwidth is 16 GHz. If the 3-dB bandwidth is measured in electrical dB, as in our example, it is located at approximately $1.55f_r$. Sometimes the 3-dB frequency is quoted as that at which the optical power is reduced by a factor of 2; this actually corresponds to 6 dB in electrical power and occurs at approximately $1.73f_r$. The 0-dB frequency occurs at approximately $1.41f_r$.[213–215]

The description of the relaxation oscillation frequency given here is rather simplistic since it is based on rate equations, which consider only one type of carrier, neglect the spatial dependances of the carrier and photon distributions, and neglect the effects of carrier diffusion and nonlinear gain. In addition, the spontaneous emission term of Eq. (17) was neglected in the derivation. The neglected effects are particularly important when considering damping of the relaxation oscillations.[204–206] With significant damping[215] the measured peak frequency f_p will be more accurately determined by:

$$f_p^2 = f_r^2 - \frac{f_d^2}{4} \tag{22}$$

where f_d is the damping frequency.

We have also neglected, however, the electrical parasitics of the laser and its operating circuit (bonding wires, etc.). Figure 22 is a simple equivalent circuit, which describes the parasitic elements influencing a semiconductor laser. Here L is the inductance of bond wire, R is the laser resistance including contact resistance, and C is capacitance primarily due to bonding-pad capacitance and capacitance of the current-confining structure of the

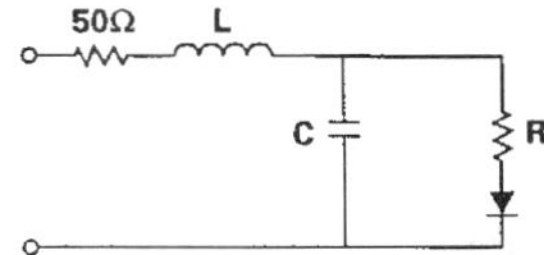

FIGURE 22 Simple equivalent circuit of the parasitics affecting the modulation of a semiconductor laser.

laser stripe.[213,214] The 50-Ω resistance is included to represent a 50-Ω drive. The resonant frequency of this circuit is

$$f_o = \frac{1}{2\pi}\sqrt{\frac{R+50}{LRC}} \tag{23}$$

The circuit is strongly damped so no resonance peak occurs; instead, the response simply drops off.[205,218] The amplitude response of a modulated semiconductor laser will begin to drop off at frequencies at which the response of the parasitics drops off even if the intrinsic peak frequency of the laser is higher. Therefore, the maximum practical modulation bandwidth may be determined by f_o instead of f_r. Figure 23 shows the modulation response of a semiconductor laser strongly affected by parasitics.[219]

The most significant parasitic limiting the performance of a semiconductor laser is normally the capacitance.[205] In order to achieve high speeds, the laser stripe must be very narrow [see Eq. (18)]; practical narrow stripe lasers are often some form of buried heterostructure (see Fig. 4*c*). The substrate doping and confinement-layer doping of buried heterostructure form a parallel plate capacitor. In order to reduce the capacitance the laser can be fabricated on a semi-insulating substrate,[205,218] the confinement layers can be semi-insulating,[124] the active area of the device can be isolated from the confinement layers by etching trenches on either side of it,[124,125] or the confinement layers can be replaced by a thick dielectric layer such as polyimide.[213,214] The inset on Fig. 21 is a schematic diagram of the high-speed laser stripe whose frequency response is shown in Fig. 21.

Assuming that the parasitics have been minimized, consider how a semiconductor laser

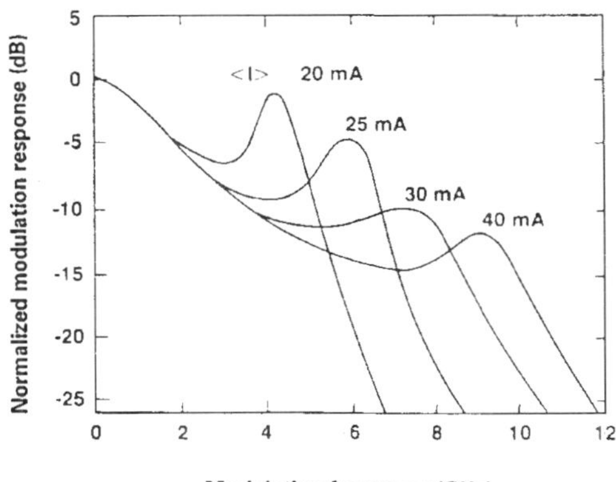

FIGURE 23 The modulation response of a 1.3-μm double-channel, planar-buried heterostructure laser with a cavity length of 80 μm and a threshold current of 18 mA. The effect of parasitics is apparent. (*From Ref. 219.*)

can be optimized for high-speed operation. As already mentioned, minimizing the stripe width is desirable. Figure 21 illustrates that increasing the photon density (or equivalently the current) increases the speed. Of course there will be a limit as to how much the photon density can be increased; when the photon density of a semiconductor laser is increased, eventually a maximum power is reached at which the laser fails due to catastrophic facet damage. InGaAsP/InP lasers have a higher threshold for catastrophic facet damage than AlGaAs/GaAs lasers; InGaAsP/InP lasers, therefore, tend to have higher bandwidths.[214]

Consideration of Eq. (19) shows that decreasing the cavity length will also increase the speed.[205,213,214,218,219] (Decreasing the length reduces τ_p.) Increasing dg/dN will also increase the speed. With a bulk active region dg/dN is approximately constant, so use of a short cavity length will not affect it. As illustrated in Fig. 6, however, dg/dN of a QW laser will decrease with increasing threshold gain, and therefore with decreasing cavity length. A single QW laser will, therefore, make a relatively poor high-speed laser. An MQW laser, however, will have higher dg/dN than a single QW. In the past, the highest modulation bandwidths were achieved with InGaAsP/InP bulk active region InGaAsP/InP lasers[213,214,222,223] with the best results on the order of 24 GHz[223] for room-temperature CW measurements. The advent of strained MQW lasers has, however, recently resulted in higher bandwidths because strain increases dg/dN.[224–226] Strained GaAs-based $In_{0.3}Ga_{0.7}As$ MQW lasers with bandwidths as high as 28 GHz have been demonstrated.[226]

So far our discussion has dealt only with the amplitude response to modulation. The phase and lasing wavelength (or optical frequency) are also affected by modulation. Figure 24 is an example of the phase response which accompanies the amplitude response of a semiconductor laser under modulation.

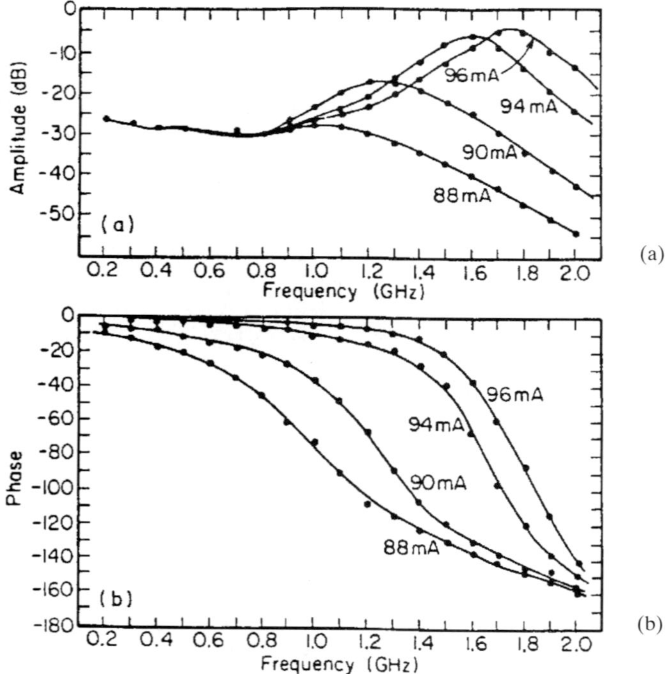

FIGURE 24 The modulation response of a proton stripe laser at various bias currents: (*a*) amplitude response and (*b*) phase. The threshold current is approximately 80 mA. (*From Ref. 205*.)

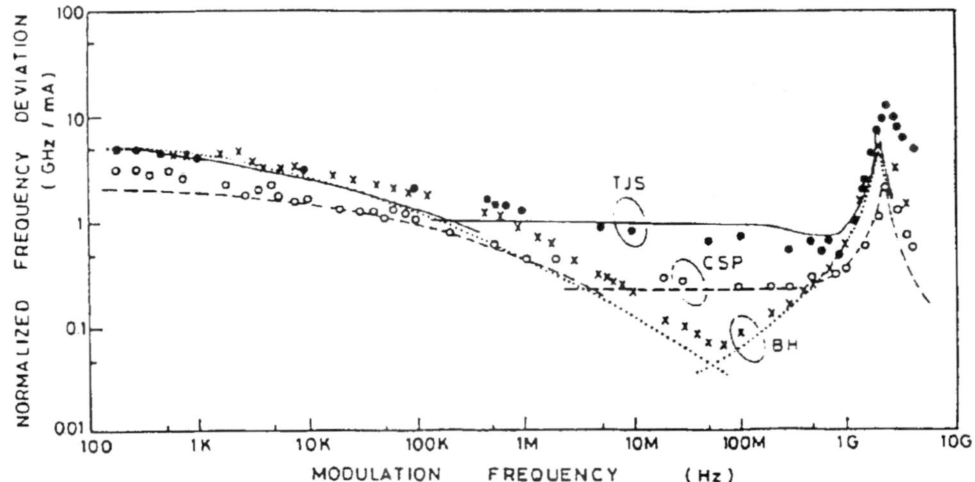

FIGURE 25 The FM response of TJS, BH, and channeled substrate planar (CSP) lasers. All the lasers are biased at 1.2 times threshold. (*From Ref. 229.*)

Assuming that a semiconductor laser lases in a single longitudinal mode under static conditions, high-frequency modulation can cause it to shift to another mode or to become multimode.[227] The tendency to become multimode increases with the depth of modulation. For many applications single-mode operation under modulation is required. In this case a laser with built-in frequency selectivity such as a distributed feedback laser[228] (DFB) (see "Spectral Properties," following) can be used to maintain single-mode operation. Even with single-mode operation under modulation, the linewidth of the lasing mode will be broadened.[206,227] This broadening, which is often called *chirp*, is discussed in more detail in the next section.

While the frequency changes associated with small-scale modulation are generally undesirable in amplitude modulation (AM) they can be utilized for frequency modulation[206,229,230] (FM). In digital systems, FM is often called frequency shift keying (FSK). FM requires only a very small amplitude modulation, so it normally refers to the effect of modulation on a single mode. In FM modulation, very fine shifts in optical frequency are detected; frequency stabilized lasers such as DFB lasers[230] as well as standard semiconductor lasers will show an FM response. Typically, FM response shows a low frequency decay below f_r and a resonance at f_r[206,229] (see Fig. 25).

If the reader requires more in-depth information on high-speed modulation of semiconductor lasers, the book by Petermann[206] or the review by Lau and Yariv[205] will be particularly helpful. For a recent review of the state of the art see the tutorial by Bowers.[215] References 3, 4, and 5 also contain chapters on high-speed modulation.

13.9 SPECTRAL PROPERTIES

One of the most important features of a semiconductor laser is its high degree of spectral coherence. There are several aspects to laser coherence. First, the laser must have spatial coherence in the various transverse directions. This is usually accomplished by controlling

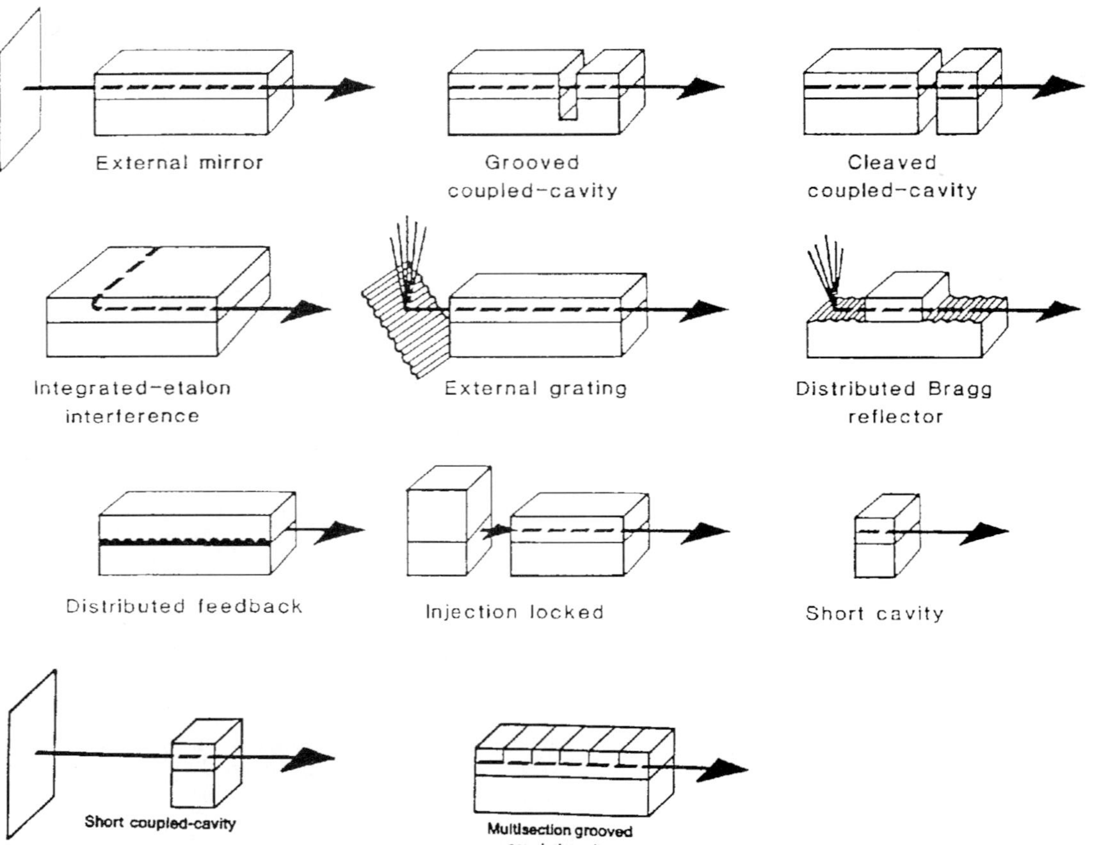

FIGURE 26 Eleven major designs for single-frequency lasers. The three in the top row and the first in the second row are coupled cavity lasers; the next three are frequency-selective-feedback lasers; the next is an injection-locked laser; the last one in the third row is a geometry-controlled laser. The left two are hybrid designs. (*From Ref. 232.*)

both the geometry and the lateral-mode geometry using a structure with a built-in index as discussed earlier under "Fabrication and Configurations." In order to achieve high spectral coherence, the semiconductor laser must operate in a single longitudinal mode. There are four technical approaches for accomplishing this:[231,232] (1) coupled cavity, (2) frequency selective feedback, (3) injection locking, and (4) geometry control. The various techniques for achieving spectral control are described in Fig. 26.

The temporal coherence of the laser is related to the spectral width of the stimulated emission spectrum by:

$$L_c = K \frac{\lambda_o^2}{\Delta\lambda_{1/2}} \tag{24}$$

where K is a constant dependent on the distribution of spectral output function, L_c is coherence length, λ_o is the wavelength of the stimulated emission peak, and $\Delta\lambda_{1/2}$ is the halfwidth of the spectral emission. $K = 1$ for rectangular; $K = 0.32$ for a lorentzian; and $K = 0.66$ for a gaussian.

The spectral linewidth, $\Delta f_{1/2}$, for a single longitudinal mode can be expressed as:[233]

$$\Delta f_{1/2} = \frac{n_{sp}}{4\pi\tau_p}\left(\frac{J}{J_{th}} - 1\right)^{-1}(1 + \alpha^2) \tag{25}$$

where n_{sp} is the spontaneous emission factor, defined as the ratio of spontaneous to stimulated emission in the lasing mode, τ_p is the cavity lifetime, and α is the linewidth enhancement factor.

Typical linewidths for a solitary single-longitudinal-mode laser are in the range of 5–20 MHz. Narrower linewidths can be achieved by using some of the techniques described in Fig. 26. More recently, the use of QW lasers, as described earlier, has led to a significant reduction in the linewidth enhancement factor and the corresponding laser linewidths.[234] Typical linewidths in the range of 0.9 to 1.3 MHz have been achieved.

Coupled cavity lasers make up a family of devices whereby spectral control is achieved by reinforcing certain wavelengths which resonate in several cavities.[235] A typical configuration is shown in Fig. 26, whereby the long cavity is cleaved into two smaller cavities. By properly controlling the length ratios and the gap width, good longitudinal mode discrimination (better than 20-dB side-mode suppression) can be obtained.

Another important technique is the use of an external resonant optical cavity (frequency-selective feedback) as shown in Fig. 27. This technique has been used by researchers at Boeing to achieve extremely narrow linewidth ($\Delta f_{1/2} \sim 1$–2 KHz) single-longitudinal-mode operation.[236]

Frequency-selective feedback can also be achieved by using either a distributed feedback (DFB) or distributed Bragg reflector (DBR) laser. As shown in Fig. 26, it differs from other types of lasers in that the feedback is provided by a grating internal to the diode laser. By using a DFB/DBR in combination with a long external cavity, it is possible to achieve linewidths below 1 MHz in a monolithic diode.[237]

Injection-locked lasers have also been under investigation at several research labs.[238] In

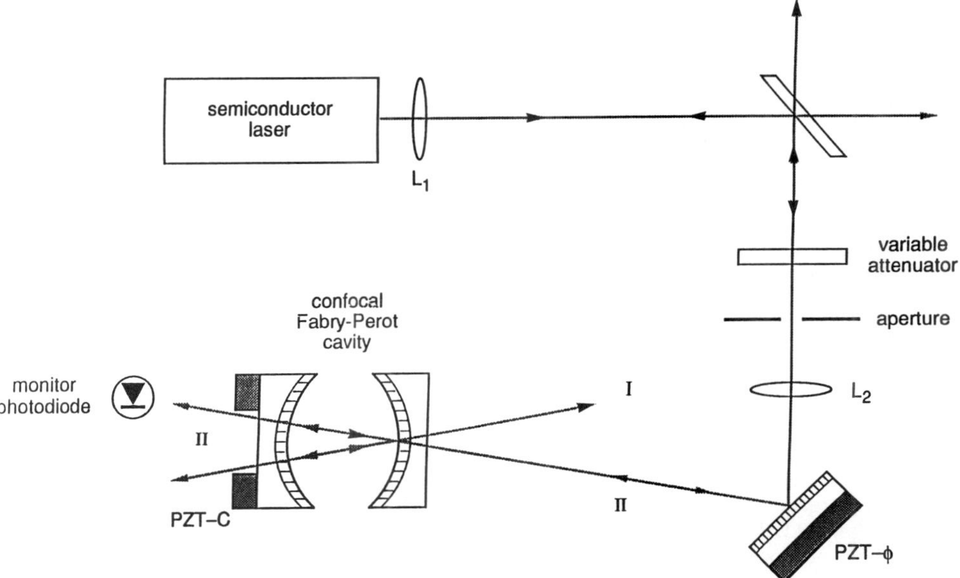

FIGURE 27 Semiconductor laser using resonant optical feedback.

this technique, a low-power, single-frequency laser, which does not have to be a semiconductor laser, is coupled to a single-mode semiconductor laser by injecting the continuous wave emission of a single wavelength of radiation into the laser's cavity.

The last technique for achieving single-longitudinal-mode operation involves the geometry-controlled cavity. Basically, this involves a short cavity 50 μm or less in length, since the longitudinal-mode spacing $\Delta\lambda_L$ in a semiconductor laser is given by:[128]

$$\Delta\lambda_L = \frac{\lambda_o^2}{2n_{eff}L} \tag{26}$$

where n_{eff} is the effective index of refraction and L is the cavity length. Then if $L < 50$ mm, $\Delta\lambda_L > 20$ Å, and the gain available to modes away from the gain maximum falls rapidly, the laser operates in a single longitudinal mode. However, the width of the spectral model can still be rather large as dictated by Eq. (25), unless special precautions are made (e.g. ultrahigh mirror reflectivities).

SURFACE-EMITTING LASERS

Monolithic two-dimensional (2-D) laser arrays are key to many important applications such as massive parallel data transfer, interconnect, processing, computing, and high-power, diode-pumped, solid-state lasers. Conventional lasers, as described in previous sections, require a pair of parallel crystalline facets (by cleaving) for delineating the laser cavity, thus limiting laser emission parallel to the junction plane. In this section, we describe laser structures and fabrication techniques which allow light to emit perpendicular to the junction plane, namely, surface-emitting lasers (SEL). SEL structures are compatible with monolithic 2-D laser array integration and requirements.

There are three designs for the fabrication of surface-emitting lasers and arrays: (1) in-plane laser with a 45° mirror, (2) in-plane laser with a distributed grating coupler, and (3) vertical cavity laser. The main body of the first two structures is very similar to the conventional cavity design with the axis parallel to the junction plane (in-plane). Light is coupled out from the surface via an integrated mirror or grating coupler. The third structure is an ultrashort cavity (10 μm) "microlaser" requiring no cleaving and compatible with photodiode and integrated-circuit processing techniques. High-density, surface-emitting laser arrays of this type have been demonstrated jointly by AT&T and Bellcore.[239] The following subsections will summarize each of the three structures.

Integrated Laser with a 45° Mirror

The development of this SEL structure requires the wafer processing of two 90° laser mirrors as well as a 45° mirror for deflecting the laser output from the junction plane as shown in Fig. 28. Dry etching techniques such as reactive ion (beam) etching (RIE), chemical assisted ion beam etching (CAIBE), and ion beam milling are usually used for the fabrication. In combination with a mass-transport process,[240] a smooth parabolic sidewall has been demonstrated for the 45° mirror of InGaAsP/InP lasers.

This approach for an SEL takes advantage of well-established layer structure growth for typical lasers. The laser performance relies on the optical quality and reliability of the facet mirrors formed. The etched-mirror lasers have been improved over the past decade to the stage very comparable with the cleaved lasers. Two-dimensional, high-power (over 1 W) laser arrays have been demonstrated by both TRW[241] and MIT Lincoln Laboratory.[240] These structures would require injection-locking or other external optical techniques in

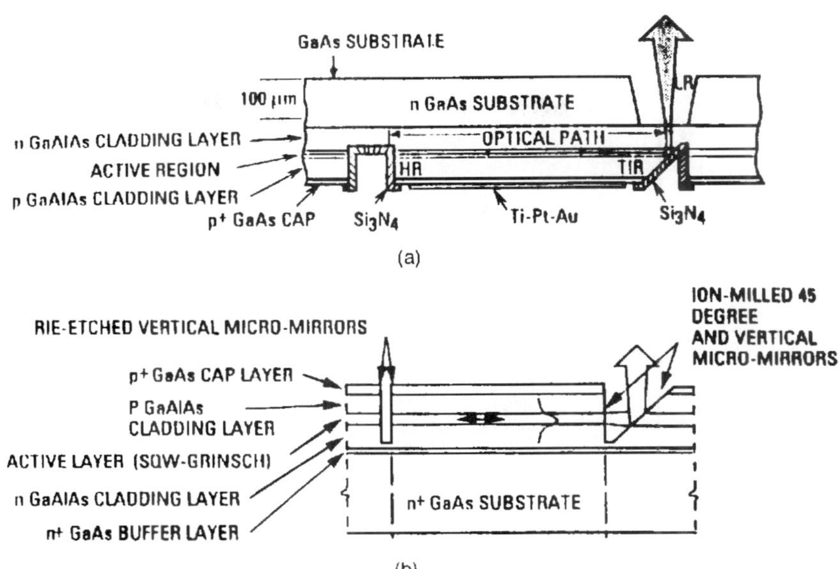

FIGURE 28 Monolithic in-plane-cavity surface-emitting lasers with 45° mirrors: (*a*) Junction-down and (*b*) junction-up configurations. (*From Ref. 241.*)

order to achieve coherent phased array operation as mentioned in the high-power laser section.

Distributed Grating Surface-Emitting Lasers

Distributed feedback (DFB) (see under "Spectral Properties," discussed earlier) and distributed Bragg reflector (DBR) lasers were proposed and demonstrated in the early 1970s. It is well known that for the second-order gratings fabricated in the laser, the first-order diffraction will be normal to the grating surface as shown in Fig. 29. Since early demonstrations, it has taken over 10 years for both the applications and processing techniques to become mature. Low-threshold, high-reliability DFB lasers with true single-mode characteristics are readily fabricated. The critical issue involved in the fabrication of the laser structure is the fabrication of the gratings with a period on the order of 2000 Å. Holographic interference techniques with an Ar^+ or He-Cd laser are generally used in many laboratories. The fabrication of large-area gratings with good throughput can be easily achieved by this technique. Another technique involves direct electron-beam writing which is effective for design iterations.

The development of DBR structure with second-order gratings for surface-emitting lasers did not occur until it was funded by the U.S. Air Force pilot program. This type of laser does not require discrete mirrors for the laser action, so that one could link an array of the lasers with residual in-plane light injection (leaking) across neighboring lasers for coherent operation. A near-diffraction-limited array operation has been demonstrated with this type of SEL. The concept was recently used for a high-power MOPA (master oscillator power amplifier) laser amplifier demonstration with the slave lasers' grating slightly off the resonant second-order diffraction condition. High-power laser arrays of this type have been demonstrated by SRI DSRC[242] and Spectra Diode Laboratories.[243] Coherent output powers of 3 to 5 W with an incremental quantum efficiency in excess of 30 percent have been obtained with the array.

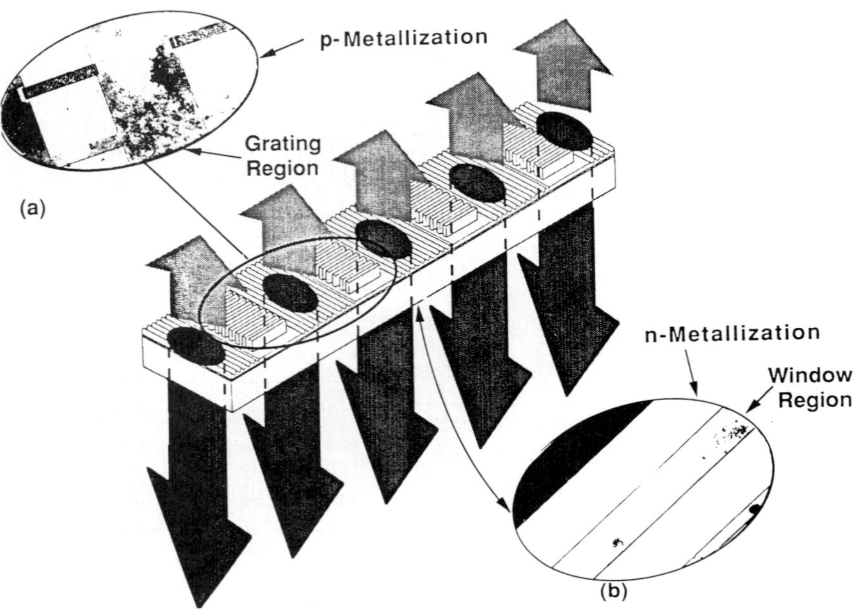

FIGURE 29 Grating surface-emitting laser array. Groups of 10 parallel ridge-guide lasers are laterally coupled in each gain section. (*From Ref. 242.*)

Vertical Cavity Lasers

The term "vertical" refers to the laser axis (or cavity) perpendicular to the wafer surface when it is fabricated. Conventional lasers have a relatively long cavity, on the order of 250 μm. It is not practical to grow such a thick layer for the laser. From the analysis, if we reduce the cavity length down to 10 μm, one needs to have a pair of very high reflectivity mirrors to make it lase at room temperature. To satisfy these conditions, researchers at Tokyo Institute of Technology[244] have used a metal thin-film or a quarter-wavelength stack of dielectric layers (Bragg reflectors) of high- and low-index material for the mirror post to the growth of laser layers. The advances in epitaxial growth techniques allow an accurate control of semiconductor layer compositions and thicknesses such that Bragg reflectors with 99.9 percent reflectivity can be attained. Therefore, a complete vertical cavity laser structure as shown in Fig. 30 consisting of a gain medium and high-reflectivity (more than 10 periods of alternate layers due to incremental index difference) mirrors can be grown successfully in one step by MBE or MOCVD techniques.

It is important to optimize the structure for optical gain. To maximize the modal gain, one can locate the standing wave field peak at the thin quantum well active layer(s) (quantum well lasers were discussed earlier in this chapter) to form a resonant periodic gain structure.[245] The issue associated with the semiconductor superlattice Bragg reflectors is the built-in carrier resistance across the abrupt heterojunction discontinuity. Without modifying the structure, the series resistance is on the order of several hundred to a thousand ohms. There have been two techniques applied to lower the resistance, namely, the use of graded junctions[246] and peripheral Zn diffusion[247] for conducting current to the active region. Both have demonstrated improvement over the original design.

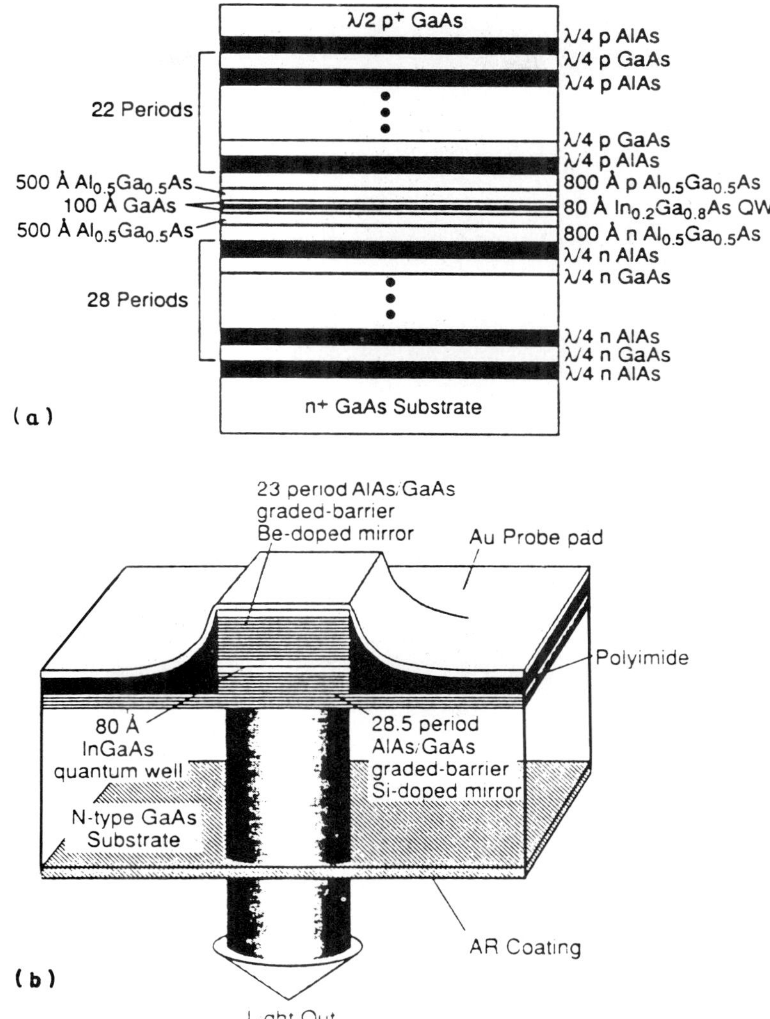

FIGURE 30 Vertical cavity surface-emitting laser with (*a*) layer structure and (*b*) device geometry. (*From Ref. 249.*)

The laser size is defined by etching into a circular column that can be mode-matched to a single-mode fiber for high coupling efficiency. It is desirable that the lasers can be planarized. Proton-bombardment-defined lasers[248] with good performance and high yield have been obtained. Meanwhile, the small size of the laser has resulted in low threshold currents close to 1 mA.[249] The differential quantum efficiency has been improved from a few percent to more than 30 percent; the output power level, modulation frequency, and maximum operating temperature have also increased over the past several years. As mentioned previously, the advantage of this SEL structure is the potential of high packing density. Bellcore researchers[250] have demonstrated a novel WDM (wavelength division multiplexing) laser source with a good histogram of wavelength distribution. The grading

TABLE 6 Performance Characteristics of Vertical Cavity Surface-Emitting Lasers Developed in Various Research Laboratories*

J_{th} (kA/cm^2)	I_{th} (mA)	V_{th} (V)	T_0 (K)	P_{max} (mW)	η (%)	Size (μm)	Structure	Reference
22.6	40		115	1.2	1.2	15ϕ	0.5 mm, DH-DBR	Ref. 246, Tai, AT&T
6.6	2.5			0.3	3.9	7ϕ	3 × 8 nm MQW strained columnar μlaser	
6.0	1.5					5 × 5	10 nm SQW strained columnar μlaser	Ref. 240 Jewell, AT&T
4.1	0.8					5ϕ	Passivated 3 × 8 nm MQS strained columnar μlaser	
3.6	3.6	3.7		0.7	4.7	10 × 10	3 × 8 nm MQW strained ion-implanted μlaser	Ref. 252 Lee 1990, AT&T
2.8	2.2	7.5		0.6	7.4	10ϕ	4 × 10 nm MQW proton-implanted μlaser	
1.4	0.7	4.0				7 × 7	8 nm SQW strained columnar μlaser	Ref. 249 Geels, UCSB
1.2	4.8			3.0*	12*	20 × 20	Strained columnar μlaser	Ref. 253 Clausen 1990 Bellcore
1.1	7.5	4.0		3.2	8.3	30ϕ	4 × 10 nm MQW proton-implanted μlaser	Ref. 254 Tell 1990, AT&T
0.8	1.1	4.0				12 × 12	8 nm SQW strained columnar μlaser	Ref. 249 Geels, UCSB
				1.5	14.5	10ϕ	4 × 10 nm GRIN-SCH proton-implanted μlaser	Ref. 254 Tell 1990, AT&T

* J_{th} = threshold current density; I_{th} = threshold current; V_{th} = threshold voltage; P_{max} = maximum power; η = overall efficiency; T_0 = characteristic temperature; * = pulsed.

of layer thickness across the wafer during a portion of growth translates into different lasing wavelengths. Two-dimensional, individually addressable lasers in a matrix form have also been demonstrated.[251] In the future, 2-D laser arrays operating at a visible wavelength will be very useful for display and optical recording/reading applications. The performance characteristics of vertical cavity SELs reported are shown in Table 6.

13.11 CONCLUSION

In this chapter we have introduced the basic properties of semiconductor lasers and reviewed some areas of the field, including high-power operation, high-speed operation, coherence, and surface-emitting lasers. We have particularly emphasized the advantages of quantum well lasers and strained quantum well lasers. Up until very recently, all the major laser diodes were fabricated using GaAs/GaAlAs and GaInAsP/InP heterostructures. However, there have been such significant advances in the use of strained quantum wells that these lasers have performance levels which exceed, in many cases, that found for the unstrained lasers. Coupled with the measured excellent reliability results and better output power/temperature performance, these types of lasers will experience a high demand in the future. High-power strained InGaAs/GaAs quantum well lasers are also of interest because their lasing wavelength range includes 0.98 μm,which makes them useful for pumping erbium-doped fiber amplifiers.[17]

Another trend in the future will be the extension of commercial semiconductor lasers to a wider variety of lasing wavelengths. Just as with the standard lasers, strained quantum wells result in significant performance improvements in these more novel laser systems. Shorter wavelength ($\lambda < 0.75\ \mu m$) lasers with high output powers are of interest for high-density optical recording and printing. In the last few years, much progress has been made in developing short wavelength semiconductor lasers. In the near future, practical visible wavelength lasers will be in the red-to-yellow range, but progress has begun on even shorter wavelengths in the blue. Recent work on very long wavelength ($\lambda > 2.0\ \mu m$) GaInAsSb/AlGaAsSb lasers is also very promising. Long wavelengths ($>1.55\ \mu m$) are of interest for eye-safe laser radar, metrology, and medical instrumentation.

Currently, commercial lasers are of the edge-emitting variety, but two-dimensional surface-emitting laser arrays have advanced considerably in the past few years. When they reach maturity, they will be used for pixel interconnect and display applications.

In a limited chapter such as this it is impossible to cover all areas of the field of semiconductor lasers in depth. One of the most important areas neglected is that of tunable lasers.[255–257] Single-mode tunable DFB and DBR lasers are of great interest for future coherent optical transmission systems. These lasers lase in a single longitudinal mode, but that mode can be tuned to a range of frequencies.

Another area not discussed in detail here is amplifiers. Amplifiers are of great interest for long-haul communication systems, for example, submarine cable systems. Amplifiers can be laser pumped fiber amplifiers[17] or laser amplifiers.[258–262] A laser amplifier has a structure similar to that of a semiconductor laser and has some optical gain, but only enough to amplify an existing signal, not enough to lase on its own.

It is hoped that the information presented in this chapter will satisfy readers who are interested in the basics of the subject and will give readers interested in greater depth the understanding necessary to probe further in order to satisfy their specific requirements.

13.12 REFERENCES

1. H. C. Casey, Jr., and M. B. Panish, *Heterostructure Lasers, Part A: Fundamental Principles,* Academic Press, Orlando, 1978.
2. H. C. Casey, Jr., and M. B. Panish, *Heterostructure Lasers, Part B: Materials and Operating Characteristics,* Academic Press, Orlando, 1978.
3. G. H. B. Thompson, *Physics of Semiconductor Laser Devices,* John Wiley & Sons, New York, 1980.
4. G. P. Agrawal and N. K. Dutta, *Long-Wavelength Semiconductor Lasers,* Van Nostrand Reinhold, New York, 1986.
5. H. Kressel and J. K. Butler, *Semiconductor Lasers and Heterojuction LEDs,* Academic Press, New York, 1977.
6. R. A. Bartolini, A. E. Bell, and F. W. Spong, *IEEE J. Quantum Electron.* **QE-17:**69 (1981).
7. R. N. Bhargava, *J. Cryst. Growth* **117:**894 (1992).
8. C. Lin (ed.), *Optoelectronic Technology and Lightwave Communication Systems,* Van Nostrand, Reinhold, New York, 1989.
9. S. E. Miller and I. Kaminow (eds.), *Optical Fiber Telecomunications,* Academic Press, Orlando, 1988.
10. M. Katzman (ed.), *Laser Satellite Communications,* Prentice-Hall, Englewood Cliffs, N.J., 1987.

11. M. Ross, *Proc. SPIE* **885:**2 (1988).
12. J. D. McClure, *Proc. SPIE* **1219:**446 (1990).
13. G. Abbas, W. R. Babbitt, M. de La Chappelle, M. L. Fleshner, and J. D. McClure, *Proc. SPIE* **1219:**468 (1990).
14. J. W. Goodman, *International Trends in Optica,* Academic Press, Orlando, 1991.
15. R. Olshansky, V. Lanzisera, and P. Hill, *J. Lightwave Technology* **7:**1329 (1989).
16. R. L. Byer, *Proc. of the CLEO/IQEC Conf.,* Plenary Session, Baltimore, Md., 1987.
17. K. Nakagawa, S. Nishi, K. Aida, and E. Yonoda, *J. Lightwave Technology* **9:**198 (1991).
18. T. F. Deutch, J. Boll, C. A. Poliafito, K. To, *Proc. of the CLEO Conf.,* San Francisco, Calif., 1986.
19. C. Kittel, *Introduction to Solid State Physics,* John Wiley & Sons, New York, 1976.
20. L. Figueroa, "Semiconductor Lasers," *Handbook of Microwave and Optical Components,* K. Chang (ed.), J. Wiley & Sons, New York, 1990.
21. R. N. Hall, G. E. Fenner, J. D. Kingsley, T. J. Soltys, and R. O. Carlson, *Phys. Rev. Lett.* **9:**366 (1962).
22. M. I. Nathan, W. P. Dumke, G. Burns, F. H. Dill, Jr., and G. Lasher, *Appl. Phys. Lett.* **1:**62 (1962).
23. N. Holonyak, Jr. and S. F. Bevacqua, *Appl. Phys. Lett.* **1:**82 (1962).
24. T. M. Quist, R. H. Rediker, R. J. Keyes, W. E. Krag, B. Lax, A. L. McWhorter, and H. J. Zeigler, *Appl. Phys. Lett.* **1:**91 (1962).
25. T. R. Chen, Y. Zhuang, Y. J. Xu, P. Derry, N. Bar-Chaim, A. Yariv, B. Yu, Q. Z. Wang, and Y. Q. Zhou, *Optics & Laser Tech.* **22:**245 (1990).
26. G. L. Bona, P. Buchmann, R. Clauberg, H. Jaeckel, P. Vettiger, O. Voegeli, and D. J. Webb, *IEEE Photon. Tech. Lett.* **3:**412 (1991).
27. A Behfar-Rad, S. S. Wong, J. M. Ballantyne, B. A. Stolz, and C. M. Harding, *Appl. Phys. Lett.* **54:**493 (1989).
28. N. Bouadma, J. F. Hogrel, J. Charil, and M. Carre, *IEEE J. Quantum Electron.* **QE-23:**909 (1987).
29. M. B. Panish, J. Sumski, and I. Hayashi, *Met. Trans.* **2:**795 (1971).
30. W. T. Tsang (ed.), *Semiconductors and Semimetals,* vol. 22, part A, Academic Press, New York, 1971, pp. 95–207.
31. R. D. Dupuis and P. D. Dapkus, *IEEE J. Quantum Electron.* **QE-15:**128 (1979).
32. R. D. Burnham, W. Streifer, T. L. Paoli, and N. Holonyak, Jr., *J. Cryst. Growth* **68:**370 (1984).
33. A. Y. Cho, *Thin Solid Films* **100:**291 (1983).
34. K. Ploog, *Crystal Growth, Properties and Applications,* vol. 3, H. C. Freyhardt (ed.), Springer-Verlag, Berlin, 1980, pp. 73–162.
35. B. A. Joyce, *Rep. Prog. Phys.* **48:**1637 (1985).
36. W. T. Tsang, *J. Cryst. Growth* **105:**1 (1990).
37. W. T. Tsang, *J. Cryst. Growth* **95:**121 (1989).
38. T. Hayakawa, T. Suyama, K. Takahashi, M. Kondo, S. Yamamoto, and T. Hijikata, *Appl. Phys. Lett.* **51:**707 (1987).
39. A. Kasukawa, R. Bhat, C. E. Zah, S. A. Schwarz, D. M. Hwang, M. A. Koza, and T. P. Lee, *Electron. Lett.* **27:**1063 (1991).
40. J. I. Davies, A. C. Marchall, P. J. Williams, M. D. Scott, and A. C. Carter, *Electron. Lett.* **24:**732 (1988).
41. W. T. Tsang and N. A. Olsson, *Appl. Phys. Lett.* **42:**922 (1983).
42. H. Temkin, K. Alavi, W. R. Wagner, T. P. Pearsall, and A. Y. Cho, *Appl. Phys. Lett.* **42:**845 (1983).
43. D. P. Bour, *Proc. SPIE* **1078:**60 (1989).

44. M. Ishikawa, K. Itaya, M. Okajima, and G. Hatakoshi, *Proc. SPIE* **1418:**344 (1991).
45. G. Hatakoshi, K. Itaya, M. Ishikawa, M. Okajima, and Y. Uematsu, *IEEE J. Quantum Electron.* **QE-27:**1476 (1991).
46. H. Hamada, M. Shono, S. Honda, R. Hiroyama, K. Yodoshi, and T. Yamaguchi, *IEEE J. Quantum Electron.* **QE-27:**1483 (1991).
47. M. A. Haase, J. Qiu, J. M. DePuydt, and H. Cheng, *Appl. Phys. Lett.* **59:**1272 (1991).
48. H. K. Choi and S. J. Eglash, *Appl. Phys. Lett.* **59:**1165 (1991).
49. D. L. Partin, *IEEE J. Quantum Electron.* **QE-24:** 1716 (1988).
50. Z. Feit, D. Kostyk, R. J. Woods, and P. Mak, *J. Vac. Sci. Technol.* **B8:**200 (1990).
51. Y. Nishijima, *J. Appl. Phys.* **65:**935 (1989).
52. A. Ishida, K. Muramatsu, H. Takashiba, and H. Fujiasu, *Appl. Phys. Lett.* **55:**430 (1989).
53. Z. Feit, D. Kostyk, R. J. Woods, and P. Mak, *Appl. Phys. Lett.* **58:**343 (1991).
54. R. Zucca, M. Zandian, J. M. Arias, and R. V. Gill, *Proc. SPIE* **1634:**161 (1992).
55. K. Wohlleben and W. Beck, *Z. Naturforsch.* **A21:**1057 (1966).
56. J. C. Dyment, J. C. North, and L. A. D'Asaro, *J. Appl. Phys.* **44:**207 (1973).
57. T. Tsukada, *J. Appl. Phys.* **45:**4899 (1974).
58. H. Namizaki, H. Kan, M. Ishii, and A. Ito, *J. Appl. Phys.* **45:**2785 (1974).
59. H. Namizaki, *IEEE J. Quantum Electron.* **QE-11:**427 (1975).
60. C. P. Lee, S. Margalit, I. Ury, and A. Yariv, *Appl. Phys. Lett.* **32:**410 (1978).
61. R. Dingle, *Festkörper Probleme XV (Advances in Solid State Physics),* H. Queisser (ed.), Pergamon, New York, 1975, pp. 21–48.
62. N. Holonyak, Jr., R. M. Kolbas, R. D. Dupuis, and P. D. Dapkus, *IEEE J. Quantum Electron.* **QE-16:**170 (1980).
63. N. Okamoto, *Jpn. J. Appl. Phys.* **26:**315 (1987).
64. *Quantum Well Lasers,* P. Zory (ed.), Academic Press, Orlando, 1993.
65. C. Cohen-Tannoudji, B. Diu, and F. Lalöe, *Quantum Mechanics,* vol. 1, John Wiley & Sons, New York, 1977.
66. G. Lasher and F. Stern, *Phys. Rev.* **133:**A553 (1964).
67. S. W. Corzine, R. H. Yan, and L. A. Coldren, *Quantum Well Lasers,* P. Zory (ed.), Academic Press, Orlando, 1993.
68. P. L. Derry, *Properties of Buried Heterostructure Single Quantum Well (Al, Ga)As Lasers,* thesis, Calif. Inst. of Tech., Pasadena, Calif., 1989.
69. P. L. Derry, A. Yariv, K. Y. Lau, N. Bar-Chaim, K. Lee, and J. Rosenberg, *Appl. Phys. Lett.* **50:**1773 (1987).
70. H. Z. Chen, A. Ghaffari, H. Morkoç, and A. Yariv, *Appl. Phys. Lett.* **51:**2094 (1987).
71. H. Chen, A. Ghaffari, H. Morkoç, and A. Yariv, *Electron. Lett.* **23:**1334 (1987).
72. R. Fischer, J. Klem, T. J. Drummond, W. Kopp, H. Morkoç, E. Anderson, and M. Pion, *Appl. Phys. Lett.* **44:**1 (1984).
73. P. L. Derry, T. R. Chen, Y. Zhuang, J. Paslaski, M. Mittlestein, K. Vahala, A. Yariv, K. Y. Lau, and N. Bar-Chaim, *Optoelectronics—Dev. and Tech.* **3:**117 (1988).
74. P. L. Derry, R. J. Fu, C. S. Hong, E. Y. Chan, K. Chiu, H. E. Hager, and L. Figueroa, *Proc. SPIE* **1634:**374 (1992).
75. R. J. Fu, C. S. Hong, E. Y. Chan, D. J. Booher, and L. Figueroa, *IEEE Photon. Tech. Lett.* **3:**308 (1991).
76. R. J. Fu, C. S. Hong, E. Y. Chan, D. J. Booher, and L. Figueroa, *Proc. SPIE* **1418:**108 (1991).
77. W. T. Tsang, *Appl. Phys. Lett.* **40:**217 (1982).
78. W. T. Tsang, *Appl. Phys. Lett.* **39:**134 (1981).
79. W. T. Tsang, *Appl. Phys. Lett.* **39:**786 (1981).

80. Y. Arakawa and A. Yariv, *IEEE J. Quantum Electron.* **QE-21:** 1666 (1985).
81. E. P. O'Reilly, *Semicond. Sci. Technol.* **4:**121 (1989).
82. E. P. O'Reilly and A. Ghiti, *Quantum Well Lasers,* P. Zory (ed.), Academic Press, Orlando, 1993.
83. J. W. Matthews and A. E. Blakeslee, *J. Cryst. Growth* **27:**118 (1974).
84. I. J. Fritz, S. T. Picraux, L. R. Dawson, T. J. Drummon, W. D. Laidig, and N. G. Anderson, *Appl. Phys. Lett.* **46:**967 (1985).
85. T. G. Andersson, Z. G. Chen, V. D. Kulakovskii, A. Uddin, and J. T. Vallin, *Appl. Phys. Lett.* **51:**752 (1987).
86. M. Altarelli, *Heterojunctions and Semiconductor Superlattices,* G. Allan, G. Bastard, N. Boccara, M. Lannoo, and M. Voos (eds.), Springer-Verlag, Berlin, 1985, p. 12.
87. J. M. Luttinger and W. Kohn, *Phys. Rev.* **97:**869 (1955).
88. P. Lawaetz, *Phys. Rev.* **B4:**3640 (1971).
89. D. Ahn and S. L. Chuang, *IEEE J. Quantum Electron.* **QE-24:**2400 (1988).
90. S. L. Chuang, *Phys. Rev.* **B43:**9649 (1991).
91. E. Yablonovitch and E.O. Kane, *J. Lightwave Tech.* **LT-4:**504 (1986).
92. H. K. Choi and C. A. Wang, *Appl. Phys. Lett.* **57:**321 (1990).
93. N. Chand, E. E. Becker, J. P. van der Ziel, S. N. G. Chu, and N. K. Dutta, *Appl. Phys. Lett.* **58:**1704 (1991).
94. C. A. Wang and H. K. Choi, *IEEE J. Quantum Electron.* **QE-27:**681 (1991).
95. R. L. Williams, M. Dion, F. Chatenoud, and K. Dzurko, *Appl. Phys. Lett.* **58:**1816 (1991).
96. S. L. Yellen, R. G. Waters, P. K. York, K. J. Beerink, and J. J. Coleman, *Electron Lett.* **27:**552 (1991).
97. P. K. York, K. J. Beerink, J. Kim, J. J. Alwan, J. J. Coleman, and C. M. Wayman, *J. Cryst. Growth* **107:**741 (1991).
98. R. G. Waters, D. P. Bour, S. L. Yellen, and N. F. Ruggieri, *IEEE Photon. Tech. Lett.* **2:**531 (1990).
99. K. Fukagai, S. Ishikawa, K. Endo, and T. Yuasa, *Japan J. Appl. Phys.* **30:**L371 (1991).
100. J. J. Coleman, R. G. Waters, and D. P. Bour, *Proc. SPIE* **1418:**318 (1991).
101. S. Tsuji, K. Mizuishi, H. Hirao, and M. Nakamura, *Links for the Future: Science, Systems & Services for Communications,* P. Dewilde and C. A. May (eds.), IEEE/Elsevier Science, North Holland, 1984, p. 1123.
102. H. D. Wolf and K. Mettler, *Proc. SPIE* **717:**46 (1986).
103. J. Hashimoto, T. Katsyama, J. Shinkai, I. Yoshida, and H. Hayashi, *Appl. Phys. Lett.* **58:**879 (1991).
104. H. B. Serreze, Y. C. Chen, and R. G. Waters, *Appl. Phys. Lett.* **58:**2464 (1991).
105. D. F. Welch and D. R. Scifres, *Electron. Lett.* **27:**1915 (1991).
106. J. I. Pankove, *Optical Processes in Semiconductors,* Dover Publ. Inc., New York, 1971.
107. N. K. Dutta, *J. Appl. Phys.* **54:**1236 (1983).
108. N. K. Dutta and R. J. Nelson, *J. Appl. Phys.* **53:**74 (1982).
109. A. R. Adams, M. Asada, Y. Suematsu, and S. Arai, *Jpn. J. Appl. Phys.* **19:**L621 (1980).
110. T. Tanbun-Ek, R. A. Logan, H. Temkin, K. Berthold, A. F. J. Levi, and S. N. G. Chu, *Appl. Phys. Lett.* **55:**2283 (1989).
111. A. R. Adams, *Electron. Lett.* **22:**249 (1986).
112. Y. Jiang, M. C. Teich, and W. I. Wang, *Appl. Phys. Lett.* **57:**2922 (1990).
113. H. Temkin, R. A. Logan, and T. Tanbun-Ek, *Proc. SPIE* **1418:**88 (1991).
114. C. E. Zah, R. Bhat, R. J. Favire, Jr., S. G. Menocal, N. C. Andreadakis, K. W. Cheung, D. M. Hwang, M. A. Koza, and T. P. Lee, *IEEE J. Quantum Electron.* **27:**1440 (1991).

115. P. J. A. Thijs, L. F. Tiemeijer, P. I. Kuindersma, J. J. M. Binsma, and T. Van Dongen, *IEEE J. Quantum Electron.* **27:**1426 (1991).

116. C. E. Zah, R. Bhat, B. Pathak, C. Caneau, F. J. Favire, Jr., N. C. Andreadakis, D. M. Hwang, M. A. Koza, C. Y. Chen, and T. P. Lee, *Electron. Lett.* **27:**1414 (1991).

117. P. J. A. Thijs, J. J. M. Binsma, E. W. A. Young, and W. M. E. Van Gils, *Electron. Lett.* **27:**791 (1991).

118. E. P. O'Reilly, G. Jones, A. Ghiti, and A. R. Adams, *Electron. Lett.* **27:**1417 (1991).

119. S. W. Corzine and L. A. Coldren, *Appl. Phys. Lett.* **59:**588 (1991).

120. A. Larsson, M. Mittelstein, Y. Arakawa, and A. Yariv, *Electron. Lett.* **22:**79 (1986).

121. S. Simhony, E. Kapon, E. Colas, R. Bhat, N. G. Stoffel, and D. M. Hwang, *IEEE Photon. Tech. Lett.* **2:**305 (1990).

122. K. J. Vahala, J. A. Lebens, C. S. Tsai, T. F. Kuech, P. C. Sercel, M. E. Hoenk, and H. Zarem, *Proc. SPIE* **1216:**120 (1990).

123. N. Chinone, *J. Appl. Phys.* **48:**3237 (1978).

124. P. A. Kirby, A. R. Goodwin, G. H. B. Thompson, D. F. Lovelace, and S. E. Turley, *IEEE J. Quantum Electron.* **QE-13:**720 (1977).

125. R. Lang, *IEEE J. Quantum Electron.* **QE-15:** 718 (1979).

126. S. Wang, C. Y. Chen, A. S. Liao, and L. Figueroa, *IEEE J. Quantum Electron.* **QE-17:**453 (1981).

127. K. Aiki, N. Nakamura, T. Kurada, and J. Umeda, *Appl. Phys. Lett.* **30:**649 (1977).

128. M. Nakamura, *IEEE Trans. Circuits Syst.* **26:**1055 (1979).

129. D. Botez, *IEEE Spectrum* **22:**43 (1985).

130. D. Botez, *RCA Rev.* **39:**577 (1978).

131. H. C. Casey, M. B. Panish, W. O. Schlosser, and T. L. Paoli, *J. Appl. Phys.* **45:**322 (1974).

132. R. J. Fu, C. J. Hwang, C. S. Wang, and B. Lolevic, *Appl. Phys. Lett.* **45:**716 (1984).

133. M. Wada, K. Hamada, H. Himuza, T. Sugino, F. Tujiri, K. Itoh, G. Kano, and I. Teramoto, *Appl. Phys. Lett.* **42:**853 (1983).

134. K. Hamada, M. Wada, H. Shimuzu, M. Kume, A. Yoshikawa, F. Tajiri, K. Itoh, and G. Kano, *Proc. IEEE Int. Semicond. Lasers Conf.,* Rio de Janeiro, Brazil, 1984, p. 34.

135. K. Endo, H. Kawamo, M. Ueno, N. Nido, Y. Kuwamura, T. Furese, and I. Sukuma, *Proc. IEEE Int. Semicond. Laser Conf.,* Rio de Janeiro, Brazil, 1984, p. 38.

136. D. Botez, J. C. Connolly, M. Ettenberg, and D. B. Gilbert, *Electron. Lett.* **19:**882 (1983).

137. B. Goldstein, J. K. Butler, and M. Ettenberg, *Proc. CLEO Conf.,* Baltimore, Md., 1985, p. 180.

138. Y. Yamamoto, N. Miyauchi, S. Maci, T. Morimoto, O. Yammamoto, S. Yomo, and T. Hijikata, *Appl. Phys. Lett.* **46:**319 (1985).

139. S. Yamamoto, H. Hayashi, T. Hayashi, T. Hayakawa, N. Miyauchi, S. Yomo, and T. Hijikata, *Appl. Phys. Lett.* **42:**406 (1983).

140. D. Ackley, *Electron. Lett.* **20:**509 (1984).

141. J. Yang, C. S. Hong, L. Zinkiewicz, and L. Figueroa, *Electron. Lett.* **21:**751 (1985).

142. J. Ungar, N. Bar-Chaim, and I. Ury, *Electron. Lett.* **22:**280 (1986).

143. D. F. Welch, W. Streifer, D. R. Scifres, *Proc. SPIE* **1043:**54 (1989).

144. D. R. Daniel, D. Buckley, B. Garrett, *Proc. SPIE* **1043:**61 (1989).

145. D. Botez, *IEEE J. Quantum Electron.* **QE-17:**2290 (1981).

146. T. Kuroda, M. Nakamura, K. Aiki, and J. Umeda, *Appl. Opt.* **17:**3264 (1978).

147. S. J. Lee, L. Figueroa, and R. Rammaswamy, *IEEE J. Quant. Electron.* **25:**1632 (1989).

148. H. Yonezu, M. Ueno, T. Kamejima, and I. Hayashi, *IEEE J. Quantum Electron.* **15:**775 (1979).

149. H. Kumabe, T. Tumuka, S. Nita, Y. Seiwa, T. Sugo, and S. Takamija, *Jpn. J. Appl. Phys.* **21:**347 (1982).

150. H. Blauvelt, S. Margalit, and A. Yariv, *Appl. Phys. Lett.* **40:**1029 (1982).
151. D. Botez and J. C. Connally, *Proc. IEEE Int. Semicond. Laser Conf.,* Rio de Janeiro, Brazil, 1984, p. 36.
152. H. Matsubara, K. Ishiki, H. Kumabe, H. Namazaki, and W. Susaki, *Proc. CLEO.,* Baltimore, Md., 1985, p. 180.
153. F. Capasso, and G. F. Williams, *J. Electrochem. Soc.* **129:**821 (1982).
154. H. H. Lee and L. Figueroa, *J. Electrochem. Soc.* **135:**496 (1988).
155. H. Kawanishi, H. Ohno, T. Morimoto, S. Kaneiwa, N. Miyauchi, H. Hayashi, Y. Akagi, Y. Nakajima, *Proc. SPIE* **1219:**309 (1990).
156. J. Yoo, H. Lee, and P. Zory, *IEEE Photonics Lett.* **3:**594 (1991).
157. Y. Suzuki, Y. Horikoshi, M. Kobayashi, and H. Okamoto, *Electron. Lett.* **20:**384 (1984).
158. M. Yamaguchi, H. Nishimoto, M. Kitumara, S. Yamazaki, I. Moto, and K. Kobayashi, *Proc. CLEO,* Baltimore, 1988, p. 180.
159. C. B. Morrison, D. Botez, L. M. Zinkiewicz, D. Tran, E. A. Rezek, and E. R. Anderson, *Proc. SPIE* **893:**84 (1988).
160. Y. Sakakibara, E. Oomura, H. Higuchi, H. Namazaki, K. Ikeda, and W. Susaki, *Electron. Lett.* **20:**762 (1984).
161. K. Imanaka, H. Horikawa, A. Matoba, Y. Kawai, and M. Sakuta, *Appl. Phys. Lett.* **45:**282 (1984).
162. M. Kawahara, S. Shiba, A. Matoba, Y. Kawai, and Y. Tamara, *Proc. Opt. Fiber Commun.* (OFC 1987), paper ME1, 1987.
163. S. Oshiba, A. Matoba, H. Hori Kawa, Y. Kawai, and M. Sakuta, *Electron. Lett.* **22:**429 (1986).
164. B. S. Bhumbra, R. W. Glew, P. D. Greene, G. D. Henshall, C. M. Lowney, and J. E. A. Whiteaway, *Electron. Lett.* **26:**1755 (1990).
165. T. Tanbun-Ek, R. A. Logan, N. A. Olsson, H. Temkin, A. M. Sergent, and K. W. Wecht, *International Semiconductor Laser Conference,* paper D-3, Davos, 1990.
166. G. D. Henshall, A. Hadjifotiou, R. A. Baker, and K. J. Warwick, *Proc. SPIE* **1418:**286 (1991).
167. M. Arvind, H. Hsing, and L. Figueroa, *J. Appl. Phys.* **63:**1009 (1988).
168. A. Larsson, S. Forouher, J. Cody, and R. J. Lang, *Proc. SPIE* **1418:**292 (1991).
169. M. Okayasu, M. Fukuda, T. Takeshita, O. Kogure, T. Hirone, and S. Uehara, *Proc. of Optical Fiber Communications Conf.,* **29,** 1990.
170. D. F. Welch, C. F. Schaus, S. Sun, M. Cardinal, W. Streifer, and D. R. Scifres, *Proc. SPIE* **1219:**186 (1990).
171. G. L. Harnagel, J. M. Haden, G. S. Browder, Jr., M. Cardinal, J. G. Endriz, and D. R. Scifres, *Proc. SPIE* **1219:**186 (1990).
172. D. R. Scifres, R. D. Burnham, and W. Steifer, *Appl. Phys. Lett.* **33:**1015 (1978).
173. D. Botez and J. C. Connally, *Appl. Phys. Lett.* **43:**1096 (1983).
174. D. E. Ackley and R. G. Engelmann, *Appl. Phys. Lett.* **39:**27 (1981).
175. D. R. Scifres, R. D. Burnham, W. Streifer, and M. Bernstein, *Appl. Phys. Lett.* **41:**614 (1982).
176. D. R. Scifres, C. Lindstrom, R. D. Burnman, W. Streifer, and T. L. Paoli, *Appl. Phys. Lett.* **19:**160 (1983).
177. F. Kappeler, H. Westmeier, R. Gessner, M. Druminski, and K. H. Zschauer, *Proc. IEEE Int. Semicond. Laser Conf.,* Rio de Janeiro, Brazil, 1984, p. 90.
178. J. P. Van der Ziel, H. Temkin, and R. D. Dupuis, *Proc. IEEE Int. Semicond. Laser Conf.,* Rio de Janeiro, Brazil, 1984, p. 92.
179. L. Figueroa, C. Morrison, H. D. Law, and F. Goodwin, *Proc. Int. Electron Devices Meeting,* 1983, p. 760.
180. L. Figueroa, C. Morrison, H. D. Law, and F. Goodwin, *J. Appl. Phys.* **56:**3357 (1984).
181. C. Morrison, L. Zinkiewicz, A. Burghard, and L. Figueroa, *Electron. Lett.* **21:**337 (1985).

182. Y. Twu, A. Dienes, S. Wang, and J. R. Whinnery, *Appl. Phys. Lett.* **45:**709 (1984).
183. S. Mukai, C. Lindsey, J. Katz, E. Kapon, Z. Rav-Noy, S. Margalit, and A. Yariv, *Appl. Phys. Lett.* **45:**834 (1984).
184. D. F. Welch, D. Scifres, P. Cross, H. Kung, W. Streifer, R. D. Burnham, and J. Yaeli, *Electron. Lett.* **21:**603 (1985).
185. N. Dutta, L. A. Kozzi, S. G. Napholtz, and B. P. Seger, *Proc. Conf. Lasers Electro-Optics (CLEO),* Baltimore, Md., 1985, p. 44.
186. M. Taneya, M. Matsumoto, S. Matsui, Y. Yano, and T. Hijikata, *Appl. Phys. Lett.* **47:**341 (1985).
187. J. Ohsawa, S. Himota, T. Aoyagi, T. Kadowaki, N. Kaneno, K. Ikeda, and W. Susaki, *Electron. Lett.* **21:**779 (1985).
188. D. F. Welch, P. S. Cross, D. R. Scifres, W. Streifer, and R. D. Burnham, *Proc. CLEO,* San Francisco, Calif., 1986, p. 66.
189. L. Mawst, D. Botez, E. R. Anderson, M. Jansen S. Ou, M. Sargent, G. L. Peterson, and T. J. Roth, *Proc. SPIE* **1418:**353 (1991).
190. J. K. Butler, D. E. Ackley, and D. Botez, *Appl. Phys. Lett.* **44:**293 (1984).
191. E. Kapon, J. Katz, and A. Yariv, *Opt. Lett.* **10:**125 (1984).
192. K. L. Chen and S. Wang, *Electron. Lett.* **21:**347 (1985).
193. W. Streifer, A. Hardy, R. D. Burnham, and D. R. Scifres, *Electron. Lett.* **21:**118 (1985).
194. S. Chinn and R. J. Spier, *IEEE J. Quantum Electron.* **20:**358 (1985).
195. J. Katz, E. Kapon, C. Lindsey, S. Margalit, U. Shreter, and A. Yariv, *Appl. Phys. Lett.* **42:**521 (1983).
196. E. Kapon, C. P. Lindsey, J. S. Smith, S. Margalit, and A. Yariv, *Appl. Phys. Lett.* **45:**1257 (1984).
197. D. Ackley, *Electron. Lett.* **20:**695 (1984).
198. T. R. Ranganath and S. Wang, *IEEE J. Quantum Electron.* **13:**290 (1977).
199. L. Figueroa, T. Holcomb, K. Burghard, D. Bullock, C. Morrison, L. Zinkiewicz, and G. Evans, *IEEE J. Quantum Electron.* **22:**241 (1986).
200. L. J. Mawst, M. E. Givens, C. A. Zmudzinski, M. A. Emanuel, and J. J. Coleman, *IEEE J. Quantum Electron.* **QE-23:**696 (1987).
201. P. S. Zory, A. R. Reisinger, R. G. Walters, L. J. Mawst, C. A. Zmudzinski, M. A. Emanuel, M. E. Givens, and J. J. Coleman, *Appl. Phys. Lett.* **49:**16 (1986).
202. R. G. Walters, P. L. Tihanyi, D. S. Hill, and B. A. Soltz, *Proc. SPIE* **893:**103 (1988).
203. M. S. Zediker, D. J. Krebs, J. L. Levy, R. R. Rice, G. M. Bender, and D. L. Begley, *Proc. SPIE* **893:**21 (1988).
204. C. Krebs and B. Vivian, *Proc. SPIE* **893:**38 (1988).
205. K. Y. Lau and A. Yariv, *Semiconductors and Semimetals Volume 22: Lightwave Communications Technology,* W. T. Tsang (ed.), Academic Press, New York, 1985, pp. 69–151.
206. K. Petermann, *Laser Diode Modulation and Noise,* Kluwer Academic Publ., Dordrecht, The Netherlands, 1988.
207. K. Petermann, *IEEE J. Quantum Electron.* **QE-15:**566 (1979).
208. G. F. Franklin, J. D. Powell, and A. Emami-Naeini, *Feedback Control of Dynamic Systems,* Addison-Wesley Publishing Company, Reading, 1986.
209. T. P. Lee and R. M. Derosier, *Proc. IEEE* **62:**1176 (1974).
210. G. Arnold, P. Russer, and K. Petermann, *Topics in Applied Physics, vol. 39: Semiconductor Devices for Optical Communication,* H. Kressel (ed.), Springer-Verlag, Berlin, 1982, p. 213.
211. K. Y. Lau, N. Bar-Chaim, P. L. Derry, and A. Yariv, *Appl. Phys. Lett.* **51:**69 (1987).
212. K. Y. Lau, P. L. Derry, and A. Yariv, *Appl. Phys. Lett.* **52:**88 (1988).
213. J. E. Bowers, B. R. Hemenway, A. H. Gnauck, and D. P. Wilt, *IEEE J. Quantum Electron.* **QE-22:**833 (1986).
214. J. E. Bowers, *Solid-State Electron.* **30:**1 (1987).

215. J. Bowers, *Conference on Optical Fiber Communication: Tutorial Sessions,* San Jose, Calif. 1992, p. 233.
216. K. Furuya, Y. Suematsu, and T. Hong, *Appl. Optics* **17:**1949 (1978).
217. D. Wilt, K. Y. Lau, and A. Yariv, *J. Appl. Phys.* **52:**4970 (1981).
218. K. Y. Lau and A. Yariv, *IEEE J. Quantum Electron.* **QE-21:**121 (1985).
219. R. S. Tucker, C. Lin, C. A. Burrus, P. Besomi, and R. J. Nelson, *Electron. Lett.* **20:**393 (1984).
220. W. H. Cheng, A. Appelbaum, R. T. Huang, D. Renner, and K. R. Cioffi, *Proc. SPIE* **1418:**279 (1991).
221. C. B. Su and V. A. Lanzisera, *IEEE J. Quantum Electron.* **QE-22:**1568 (1986).
222. R. Olshansky, W. Powazink, P. Hill, V. Lanzisera, and R. B. Lauer, *Electron. Lett.* **23:**839 (1987).
223. E. Meland, R. Holmstrom, J. Schlafer, R. B. Lauer, and W. Powazinik, *Electron. Lett.* **26:**1827 (1990).
224. S. D. Offsey, W. J. Schaff, L. F. Lester, L. F. Eastman, and S. K. McKernan, *IEEE J. Quantum Electron.* **27:**1455 (1991).
225. R. Nagarajan, T. Fukushima, J. E. Bowers, R. S. Geels, and L. A. Coldren, *Appl. Phys. Lett.* **58:**2326 (1991).
226. L. F. Lester, W. J. Schaff, X. J. Song, and L. F. Eastman, *Proc. SPIE* **1634:**127 (1992).
227. K. Y. Lau, C. Harder, and A. Yariv, *IEEE J. Quantum Electron.* **QE-20:**71 (1984).
228. Y. Sakakibara, K. Furuya, K. Utaka, and Y. Suematsu, *Electron. Lett.* **16:**456 (1980).
229. S. Kobayashi, Y. Yamamoto, M. Ito, and T. Kimura, *IEEE J. Quantum Electron.* **QE-18:**582 (1982).
230. P. Vankwikelberge, F. Buytaert, A. Franchois, R. Baets, P. I. Kuindersma, and C. W. Fredrksz, *IEEE J. Quantum Electron.* **25:**2239 (1989).
231. T. Ikegami, "Longitudinal Mode Control in Laser Diodes," *Opto-Electronics Technology and Lightwave Communication Systems,* Van Nostrand Reinhold, New York, 1989, p. 264.
232. T. E. Bell, *IEEE Spectrum* **20**(12):38 (December 1983).
233. M. Osinsky and J. Boos, *IEEE J. Quantum Electron.* **QE-23:**9 (1987).
234. S. Takano, T. Sasaki, H. Yamada, M. Kitomura, and I. Mito, *Electron Lett.* **25:**356 (1989).
235. W. T. Tsang, N. A. Olson, R. A. Linke, and R. A. Logan, *Electron Lett.* **19:**415 (1983).
236. R. Beausoleil, J. A. McGarvey, R. L. Hagman, and C. S. Hong, *Proc. of the CLEO Conference,* Baltimore, Md., 1989.
237. S. Murata, S. Yamazaki, I. Mito, and K. Koboyashi, *Electron Lett.* **22:**1197 (1986).
238. L. Goldberg and J. F. Weller, *Electron Lett.* **22:**858 (1986).
239. J. L. Jewell, A. Scherer, S. L. McCall, Y. H. Lee, S. J. Walker, J. P. Harbison, and L. T. Florez, *Electron. Lett.* **25:**1123 (1989).
240. Z. L. Liau and J. N. Walpole, *Appl. Phys. Lett.* **50:**528 (1987).
241. M. Jansen, J. J. Yang, S. S. Ou, M. Sergant, L. Mawst, T. J. Roth, D. Botez, and J. Wilcox, *Proc. SPIE* **1582:**94 (1991).
242. G. A. Evans, D. P. Bour, N. W. Carlson, R. Amantea, J. M. Hammer, H. Lee, M. Lurie, R. C. Lai, P. F. Pelka, R. E. Farkas, J. B. Kirk, S. K. Liew, W. F. Reichert, C. A. Wang, H. K. Choi, J. N. Walpole, J. K. Butler, W. F. Ferguson, Jr., R. K. DeFreez, and M. Felisky, *IEEE J. Quantum Electron.* **27:**1594 (1991).
243. D. Mehuys, D. Welch, R. Parke, R. Waarts, A. Hardy, and D. Scifres, *Proc. SPIE* **1418:**57 (1991).
244. K. Iga, F. Koyama, and S. Kinoshita, *IEEE J. Quantum Electron.* **24:**1845 (1988).
245. M. Y. A. Raja, S. R. J. Brueck, M. Osinski, C. F. Schaus, J. G. McInerney, T. M. Brennan, and B. E. Hammons, *IEEE J. Quantum Electron.* **25:**1500 (1989).
246. K. Tai, L. Yang, Y. H. Wang, J. D. Wynn, and A. Y. Cho, *Appl. Phys. Lett.* **56:**2496 (1990).

247. Y. J. Yang, T. G. Dziura, R. Fernandez, S. C. Wang, G. Du, and S. Wang, *Appl. Phys. Lett.* **58:**1780 (1991).

248. M. Orenstein, A. C. Von Lehmen, C. Chang-Hasnain, N. G. Stoffel, J. P. Harbison, L. T. Florez, E. Clausen, and J. E. Jewell, *Appl. Phys. Lett.* **56:**2384 (1990).

249. R. S. Geels and L. A. Coldren, *Appl. Phys. Lett.* **57:**1605 (1990).

250. C. J. Chang-Hasnain, J. R. Wullert, J. P. Harbison, L. T. Florez, N. G. Stoffel, and M. W. Maeda, *Appl. Phys. Lett.* **58:**31 (1991).

251. A. Von Lehmen, M. Orenstein, W. Chan, C. Chang-Hasnain, J. Wullert, L. Florez, J. Harbison, and N. Stoffel, *Proc. of the CLEO Conference,* Baltimore, Md., 1991, p. 46.

252. Y. H. Lee, J. L. Jewell, B. Tell, K. F. Brown-Goebeler, A. Scherer, J. P. Harbison, and L. T. Florez, *Electron. Lett.* **26:**225 (1990).

253. E. M. Clausen, Jr., A. Von Lehmen, C. Chang-Hasnain, J. P. Harbison, and L. T. Florez, *Techn. Digest Postdeadline Papers, OSA 1990 Annual Meeting,* Boston, Mass., 1990, p. 52.

254. B. Tell, Y. H. Lee, K. F. Brown-Goebeler, J. L. Jewell, R. E. Leibenguth, M. T. Asom, G. Livescu, L. Luther, and V. D. Mattera, *Appl. Phys. Lett.* **57:**1855 (1990).

255. T. P. Lee, *IEEE Proceedings* **79:**253 (1991).

256. M.-C. Amann and W. Thulke, *IEEE J. Selected Areas Comm.* **8:**1169 (1990).

257. K. Kobayashi and I. Mito, *J. Lightwave Tech.* **6:**1623 (1988).

258. T. Saitoh and T. Mukai, *IEEE Global Telecommunications Conference and Exhibition,* San Diego, Calif., 1990, p. 1274.

259. N. A. Olsson, *J. Lightwave Tech.* **7:**1071 (1989).

260. A. F. Mitchell and W. A. Stallard, *IEEE Int. Conference on Communications,* Boston, Mass., 1989, p. 1546.

261. T. Saitoh and T. Mukai, *J. Lightwave Tech.* **6:**1656 (1988).

262. M. J. O'Mahony, *J. Lightwave Tech.* **6:**531 (1988).

CHAPTER 14
ULTRASHORT LASER SOURCES

Xin Miao Zhao
Jean-Claude Diels
Department of Physics and Astronomy
The University of New Mexico
Albuquerque, New Mexico

14.1 GLOSSARY

$A_c(\tau)$	autocorrelation
a	linear small-signal attenuation
a_g	average (saturated) gain prior to the arrival of perturbation
d	crystal thickness
d_g	thickness of gain medium
$\mathscr{E}_s(t)$	pulse amplitude
g	linear small-signal amplification
I	pulse intensity
l	attenuation inside nonlinear medium
n	number of round-trips
P_p	amplitude of a fluctuation
P_{av}	noise of average power
p	group velocity mismatch
r	amplitude reflectivity
r	reflectivity
T_{1a}	energy relaxation time of absorbing transition
t_e	pulse rise time to reach saturation energy density
t_g	gain existing time
t_{RT}	cavity round-trip time
W	pulse energy
W_{sa}	saturation energy
W_{sg}	gain saturation energy

α	absorption coefficient
α_g	gain absorption coefficient
ΔP	power fluctuation
δ	mismatch between cavity round-trip time and the pump pulse spacing
$\epsilon(t)$	small noise burst
κ	nonlinear parameter proportional to the length and nonlinear index of medium
σ	gain cross section
σ_a	growth rate
τ_c	decay time for fluctuation
τ_p	duration of fluctuation
Φ	intensity-dependent phase
$\chi^{(2)}$	second-order susceptibility
ω_p	pump radiation frequency
ω_s	pulse frequency

14.2 INTRODUCTION

Ultrashort pulses are often generated by locking in phase a large comb of modes sustained by the laser gain bandwidth. Mode locking has been studied and used in many different lasers. A typical laser consists of a resonator formed of two mirrors and a gain medium. The simplest mode-locked laser is a resonator with an intracavity modulator. This modulator, whether it is a saturable absorber, an acousto-optic (A-O) or electro-optic (E-O) modulator, or a self-phase modulating component, modulates the amplitude or the phase of the field inside the cavity to generate short pulses. We will devote this chapter to the different aspects and variations of the mode-locking mechanism.

There are two broad categories of mode-locking mechanisms: passive and active. In the former, some intensity-dependent loss or dispersion mechanism is used to favor operation of pulsed over continuous radiation. In the latter, a coupling is introduced between cavity modes, "locking" them in phase. Between these two classes are "hybrid" and "doubly mode-locked" lasers in which both mechanisms of mode locking are used. In parallel to this categorization in "active" and "passive" lasers, one can also classify lasers as being modulated inside (the most common approach) or outside (usually in a coupled cavity) the lasers.

14.3 PASSIVELY MODE-LOCKED LASERS

Evolution of the Pulse Train

In a laser, many modes are allowed along the resonator axis with frequency separation of $c/2L$, where L is the optical length of the cavity and c is the speed of light. These modes usually oscillate with random phases and irregular amplitudes resulting in a randomly time-varying amplitude within the round-trip period $t_{RT} = 2L/c$. However, if these modes have the same phase ϕ, they will constructively interfere at the same instant $t_c = (\phi/2\pi)t_{RT}$ of the round-trip time. The output will thus consist of a series of pulses centered at $t_c, t_c + t_{RT}, t_c + 2t_{RT}, \ldots$, and the laser is said to be "mode locked." The element to fix these relative phases is referred to as the *mode locker*. The width of each of the successive pulses is approximately equal to the inverse of the frequency range spanned by the modes being locked in phase. In the laser resonator, mode locking in the time domain corresponds to a single pulse traveling back and forth between mirrors with loss on the output mirror being compensated by the gain from pumping at each round-trip.

A passive mode-locking element can be a saturable absorber, a nonlinear lens, a nonlinear mirror, or a combination of dispersion and phase modulation. Before looking at individual mode-locking elements, let us review the general mechanism of passive mode locking.

The passive mode-locking element can most often be represented by intensity-dependent intracavity loss. Larger losses at low intensity imply that the laser has less gain—and may be below threshold—for low-intensity continuous wave (cw) radiation than for pulses with higher peak intensity. This leads to the emergence of a pulse out of the amplified spontaneous emission noise of the laser. Rather than concern ourselves with the primary process of formation of a precursor of a pulse from random noise, let us follow the evolution of the pulse from its birth from noise until it has blossomed into a fully shaped stable laser pulse. We will look for simple evolution equations for the pulse energy $W = \int_{-\infty}^{\infty} I(t)\,dt$, with $I(t)$ being the pulse intensity.

The element responsible for saturable losses (gain) should have typically a *linear* loss (gain) factor at low energies, and a *constant* loss (gain) at higher energies. We thus have, at low energies: $dW/dz = \mp\alpha W$. At large energies $W \gg W_s$: $dW/dz = \mp\alpha W_s$ where W_s is the saturation energy for the chosen geometry. The simplest differential equation to combine these two limits is

$$\frac{dW}{dz} = \alpha_g W_{sg}[1 - e^{-W/W_{sg}}] \tag{1}$$

Equation (1) is written for a medium with a linear gain α_g and a saturation energy W_{sg}. It can be integrated to yield the energy W_2 at the end of the amplifier of thickness d_g, as a function of the input energy W_1:

$$W_2 = G(W_2, W_1)W_1 = W_{sg} \ln\{1 - e^{\alpha_g d_g}(1 - e^{W_1/W_{sg}})\} \tag{2}$$

A similar equation applies to the saturable absorber, with a negative absorption coefficient $-\alpha_a$ and a smaller saturation energy W_{sa}:

$$W_2 = A(W_2, W_1)W_1 = W_{sa} \ln\{1 - e^{-\alpha_a d_a}(1 - e^{W_1/W_{sa}})\} \tag{3}$$

The dominant linear loss element is the output coupler, with reflectivity r. The transfer function for that element is simply

$$W_2 = L(W_2, W_1)W_1 = rW_1 \tag{4}$$

and the energy of the output pulse is $(1 - r)W_1$. The evolution of the pulse energy in a single round-trip can be simply calculated from the product of all three transfer functions given by Eqs. (2) (3), and (4). For instance, if we consider a ring laser with the sequence: mirror, gain, and absorber, the pulse energy W_4 after the absorber is given by the product $A(W_4, W_3)G(W_3, W_2)L(W_2, W_1)$. One can also express the relation between the energy W_4 and the pulse energy W_1 before the output mirror by the algebraic relation:[1]

$$1 + a[e^{W_4/W_{sa}} - 1] = \{1 + g[e^{rW_1/W_{sg}} - 1]\}^{W_{sg}/W_{sa}} \tag{5}$$

where $a = \exp\{-\alpha_a d_a\}$ is the *linear* small-signal attenuation of the passive element and $g = \exp\{\alpha_g d_g\}$ is the *linear* small-signal amplification.

The recent trend has been to develop high-power mode-locked oscillators, for which both numbers a and g can be large, and the reflectivity of the output coupler r can be as low as 50 percent. As a result, the *order* of the elements matters in the design of the laser, and in its performances. To illustrate this point, let us assume that the passive element is totally saturated in normal operation. In full saturation, the input energy $W(0)$ is related to the output energy $W(d)$ by

$$W(d) = W(0) - \alpha_a d_a W_{sa} \tag{6}$$

Given an initial energy W_1, the energies W_4 after single passage through three different sequences of the same elements are given below. For the sequence mirror-absorber-gain

$$e^{W_4/W_{sg}} = 1 - g[1 - e^{(rW_1 - \alpha_a d_a W_{sa})/W_{sg}}] \tag{7}$$

For the sequence absorber-mirror-gain

$$e^{W_4/W_{sg}} = 1 - g[1 - e^{r(W_1 - \alpha_a d_a W_{sa})/W_{sg}}] \tag{8}$$

For the sequence absorber-gain-mirror

$$e^{W_4/W_{sg}} = \{1 - g[1 - e^{(W_1 - \alpha_a d_a W_{sa})/W_{sg}}]\}^{1/r} \tag{9}$$

Let us take a numerical example for a high-gain system such as the flash-lamp pumped Nd:glass laser, with $W_{sa}/W_{sg} = 0.1$, $\alpha_a d_a = 1$, $\alpha_g d_g = 1.5$ and output coupling of $r = 0.5$. For an initial energy $W_1/W_{sg} = 0.5$, we find for the sequence mirror-absorber-gain $W_4/W_{sg} = 0.545$, for the sequence absorber-mirror-gain $W_4/W_{sg} = 0.689$, and for the sequence absorber-gain-mirror $W_4/W_{sg} = 0.582$.

The order of the elements, the relative saturation of the gain to that of the passive element, as well as the output coupling influence the stability and output power of the laser, as shown in Ref. 1. The evolution of the pulse in the cavity can be calculated by repeated applications of products of operations such as $A(W_4, W_3)G(W_3, W_2)L(W_2, W_1)$ for the sequence (passive element, gain, output coupler), starting from a minimum value of W_1 above threshold for pulsed operation,[1] and recycling at each step the value of W_4 as the new input energy W_1. Figure 1 shows the growth of intracavity pulse energy, as a function of the round-trip index j, for different orders of the elements. The initial pulse energy is 1 percent of the saturation energy W_{sg} in the gain medium. The saturation energy and optical thickness of the absorbing medium are, respectively, $W_{sa} = 0.8W_{sg}$ and $\alpha_a d_a = 1.2$. The linear gain is $\alpha_g d_g = 1.5$, and the output coupling $r = 0.8$.

The main point of this exercise is that the order of the elements *is* important, a fact confirmed by measurement on high-gain lasers such as Ti:Al_2O_3. Such dependence on the order of the elements proves the inadequacy of analytical theories based on the approximation of infinitesimal change per element and per cavity round-trip.

Numerical codes have been developed that attempt to include all physical phenomena affecting pulse shape and duration.[2,3] Unfortunately, the shear number of these mechanisms makes it difficult to reach a physical understanding of the pulse-generation process, or even identify the essential parameters. Therefore, the most popular approach is to construct a simplified analytical model on a selected mechanism. The next section gives a brief summary of the various mode-locking methods.

Pulse Shaping in Passive Mode Locking

It is beyond the scope of this chapter to review the numerous theories of passive mode locking that have emerged since the early 1970s. An overview of these theories can be found in Refs. 1 and 4. We will limit ourselves here to a qualitative description of pulse shaping.

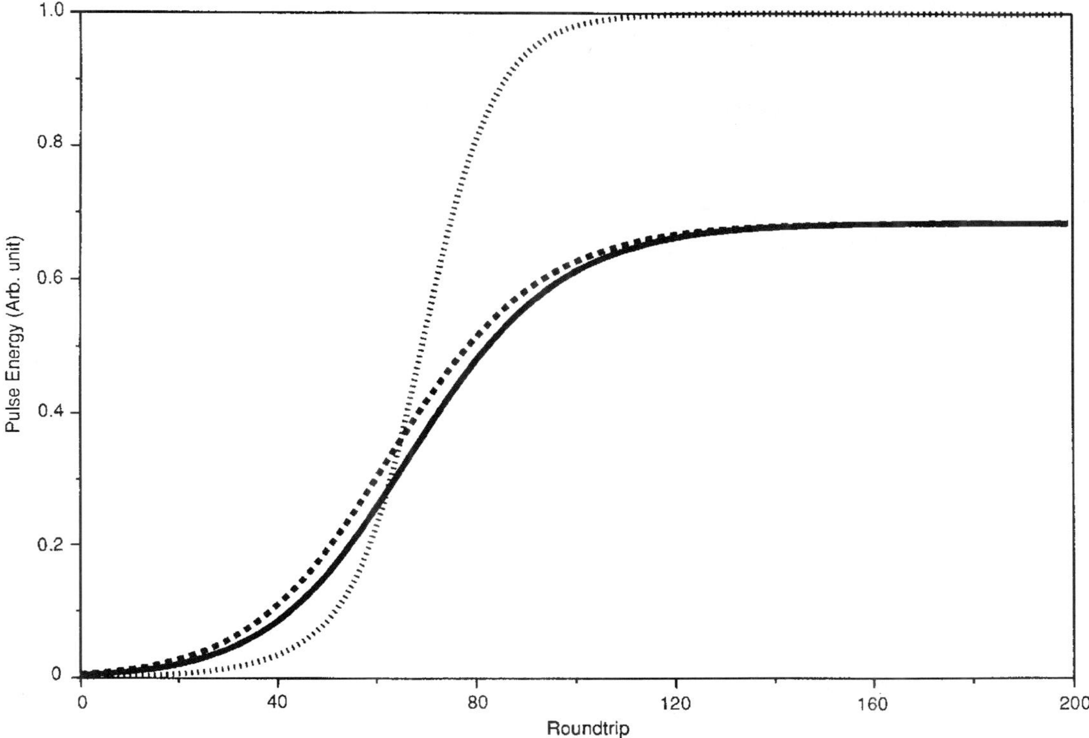

FIGURE 1 Intracavity pulse energy versus round-trip index number j. The solid, dashed, and dotted lines correspond to the sequence m(irror)-a(bsorber)-g(ain), a-g-m and a-m-g, respectively.

The main influence of the saturable absorber is to steepen the leading edge of the pulses circulating in the cavity. At the early stage of the pulse formation, the portion of the pulse prior to $t = t_s$ (t_s is the time to reach saturation) is absorbed. It is assumed at this early stage that $I(t_s) = I_{sa}$ is reached at a time $t_s \ll T_{1a}$, where T_{1a} is the energy relaxation time of the absorbing transition. As the pulse intensity increases, the pulse duration shortens because the leading edge of the pulse from $t = 0$ to $t = t_s$ is eroded by absorption, while there is much less attenuation at the trailing edge of the pulse for $t > t_s$. However, as the pulse is amplified in the cavity, the onset of saturation moves toward the earlier part of the pulse. As a result of the decreasing t_s, absorption saturation becomes less and less effective as a compression mechanism because it affects a smaller and smaller fraction of the pulse leading edge. As the pulse duration becomes much smaller than T_{1a}, the absorber affects the pulse only up to the time t_e for which the accumulated energy density reaches the saturation energy density W_{sa}:

$$\int_0^{t_e} I(t)\, dt = I_{sa} T_{1a} = W_{sa} \tag{10}$$

Trailing edge pulse shaping generally is the result of the gain depletion by stimulated emission. This mechanism becomes significant as the pulse energy approaches a steady state.

The above considerations are sufficient to establish broad criteria for the gain and absorber in a passively mode-locked laser. For the gain medium, a larger inversion at minimum cost of pump power calls for a long energy relaxation time T_{1g}. At moderate pumping, the gain recovery time (between pulses separated by the cavity round-trip time t_{RT}) is equal to—or less than—T_{1g}. Stable single-pulse operation of a passively mode-locked laser requires that the cavity round-trip time does not exceed the gain lifetime T_{1G} by more than a factor of 4 or 5.

The saturable absorber should have a larger or equal saturation cross section than the gain medium, and a relaxation time T_{1a} shorter than the cavity round-trip time. The requirement for a large saturation cross section can be relaxed by using standing wave saturation (factor of 3 effective reduction of the saturation intensity) as in the ring laser or an antiresonant ring laser,[1] or by focusing more tightly in the absorber than in the gain.

Optimal pulse compression through saturable absorption requires that the saturation energy be as large as is compatible with the self-starting requirement (buildup of the oscillation from noise).

Phase Modulation and Dispersion

In most practical situations involving passive mode locking with a saturable absorber dye, the laser radiation is applied *below resonance* with the absorber. Consequently, the index of refraction of the dye solution decreases as the absorber saturates, leading to a *downchirp* of the pulse and *self-defocusing* of the beam. Both processes can simultaneously affect the mode-locking process.[1]

A downchirped pulse has higher frequency components at its leading edge. Therefore, the pulse front will propagate slower than the pulse tail in a normally dispersive medium such as glass, and the pulse will be compressed.[5] Depending on the particular cavity configuration, the self-defocusing may result in a significant loss modulation. The induced lens, which may have a focal distance of several centimeters, will nearly always modify the beam geometry because of astigmatism.[6] The z-scan technique[7] was used to measure the lensing effect in the absorber jet of diethyloxadicarbocyanine iodide (DODCI) mixed in ethylene glycol. When a dye laser at 620 nm was used as pumping source, a focal length of -3.1 cm was measured for the induced lens in the dye jet, which corresponds to a $\lambda/100$ phase front distortion[6] at the beam waist. This lensing effect results in a total change of intensity on axis of 2.6 percent when the absorber jet is translated along the z direction[6] away from the focal point of the mirrors.

Both the self-defocusing and the phase modulation are maximum if the pulse-energy density in the dye amounts to approximately 6 times the saturation-energy density. This is the condition for which the fastest change in index takes place near the peak of the pulse. If this mode of operation is sought, the spot size in the absorber should be large enough to avoid excessive saturation of the dye. Positive dispersion—such as provided by an adjustable thickness of glass—is sufficient to compensate the downchirp and achieve pulse compression.

Positive phase modulation or "upchirp" can be obtained in a cavity where the Kerr effect dominates. This is the case in laser cavities with tight focusing in the absorber jet,[8,9] or in a solid-state gain medium such as the Ti:Al_2O_3.

The amount of phase modulation can be made very large either by using additives with large Kerr constant such as 2-methyl-4-nitroaniline (MNA) to the saturable absorber solution,[9,10] or using an external fiber in additive-pulse mode locking as discussed later, or using amplifier crystals (such as is the case for Ti:Al_2O_3). When self-phase modulation is combined with an appropriate amount of group velocity dispersion, some important effects occur. One of them is the formation of "soliton." Solitons in fibers were first observed in 1980.[11] Pulse narrowing was demonstrated associated with soliton formation by Mollenauer et al.[12] This work led subsequently to the design of a soliton laser.[13] If the modulation due to the Kerr effect dominates—such as in the Ti:sapphire laser—the steady-state pulse can also be calculated from the nonlinear Schrödinger equation.[1,14,15] The approximations made in modeling the soliton laser by the nonlinear Schrödinger equation neglect loss mechanisms. In particular, the modulation of the beam parameters caused by transient self-focusing associated with the Kerr effect is generally ignored. Recently, the importance of the spatial effect—in particular for Ti:sapphire lasers—was recognized and included in the laser analysis.[16]

Passive Pulse-Shaping Elements

The most common passive mode-locking element is the saturable absorber dye with a saturation energy of the order of 1 mJ/cm^2, and an energy relaxation time ranging from picoseconds to nanoseconds. Its functions include pulse shaping by saturation, phase modulation leading to downchirp, saturation-induced self-focusing, Kerr-effect-induced upchirp, and self-defocusing.

In addition, the absorbing dye jet provides a common time reference for the counterpropagating pulses of a ring cavity. Since the pulses generate a transient standing wave pattern in the absorber jet, a complex single- and two-photon resonant degenerate four-wave mixing coupling takes place between the two pulses.[17]

Multiple quantum wells provide a substitute saturable absorber with a smaller saturation energy required to mode-lock semiconductor lasers.[18] Also, "artificial saturable absorbers" can be designed with nonlinear optics. A detailed analysis of this concept is given in a later section, "Nonlinear Optical Sources," under the subheading "Mode Locking with Intracavity Second Harmonic Generation" and the section following it.

A last and rather complex pulse-shaping element is a two-photon absorbing semiconductor associated with a limiting aperture. The function of that element is to limit the pulse intensity to roughly the saturation intensity, and provide pulse compression through self-defocusing by the two-photon-induced carriers. A detailed description of this laser operation is given in the section "Passive Negative Feedback."

14.4 SYNCHRONOUS, HYBRID, AND DOUBLE MODE LOCKING

Synchronous Mode Locking

Short pulses can be created by exciting the gain medium at a repetition rate synchronized with the cavity-mode spacing. This simple method is called *synchronously pumped mode locking.* As compared to passive mode locking, fewer internal elements are needed in this type of laser. In its simplest form, synchronous mode locking has the advantage of a higher pumping efficiency. Indeed, no energy is wasted in trying to maintain a gain between pulses. A larger number of dyes is available for synchronously pumped operation, including dyes with a very short energy relaxation time which would be too difficult to be pumped continuously. Another important advantage over passively mode-locked lasers is its higher average output power. The reason is the low duty cycle of the pump, enabling more pump power to be concentrated in the gain jet, resulting in a larger inversion and gain than in a continuously pumped laser.

Unfortunately, the same degree of stability cannot be achieved for purely synchronously pumped lasers as for passively mode-locked lasers. Typically, the autocorrelation

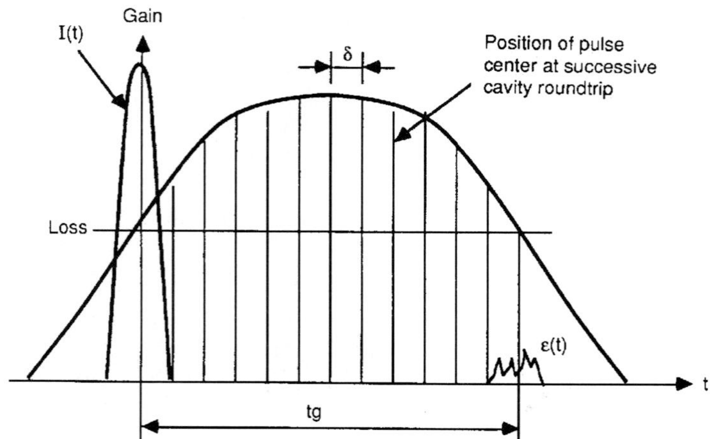

FIGURE 2 Net gain development inside a laser cavity.

trace of such a laser has exponential wings, indicating a random distribution of pulse duration along the train.[19] If the laser cavity were slightly longer than required for exact synchronization, amplified spontaneous emission would constantly accumulate at the leading edge of the pulses, resulting in long pulse durations. For best stability, the cavity length should be slightly shorter than that required for exact synchronism with the pump radiation. In fact, if δ is the mismatch between cavity round-trip time and the pump pulse spacing (synchronous with the pulsed gain $\alpha(t)$), the intracavity intensity of the jth round-trip is related to the previous one by

$$I_j(t) = I_{j-1}(t-\delta)[e^{\alpha(t-\delta)d_g} - l] \tag{11}$$

In Eq. (11), d_g is the thickness of the gain medium and l is the loss per cavity round-trip.

The net gain $(e^{\alpha(t)d_g} - l)$ exists in the cavity for a time t_g (Fig. 2). Any pulse will experience gain for $n = t_g/\delta$ round-trips. The laser oscillation will start from a small noise burst $\epsilon(t)$. The fully developed intracavity pulse is simply

$$I(t) = \epsilon(t - tg)[e^{\alpha_{\text{av}}dg - l}]^n \tag{12}$$

where we have defined an average gain $\alpha(1/tg) \int \alpha(t'/dt')$. Equation (12) indicates that in the absence of any spectral filtering mechanism, and neglecting the distortion of the gain curve $\alpha(t)$ by saturation, the pulse should be roughly n times shorter than the duration of the gain.

The somewhat oversimplified representation of Fig. 2 includes nevertheless an important noise feature of the laser. The seed $\epsilon(t)$ has random phase. As pointed out by Peter et al.,[20] the aleatory phase of the spontaneous emission source is at the origin of the noise of the laser.

The phase randomization of the "starting" pulse can be reduced by adding to $\epsilon(t)$ a minimum fraction $\eta E(t)$ of the laser output field, just large enough so that the phase of $\eta E(t) + \epsilon(t)$ comes nearly equal to the phase of the output fields $E(t)$. Both calculation and experiment have demonstrated a dramatic noise reduction by seeding the cavity with a small fraction of the pulse *in advance* of the main pulse.[20] The practical implementation is to provide a weak reflection from a microscope slide ahead of the output mirror. Since a

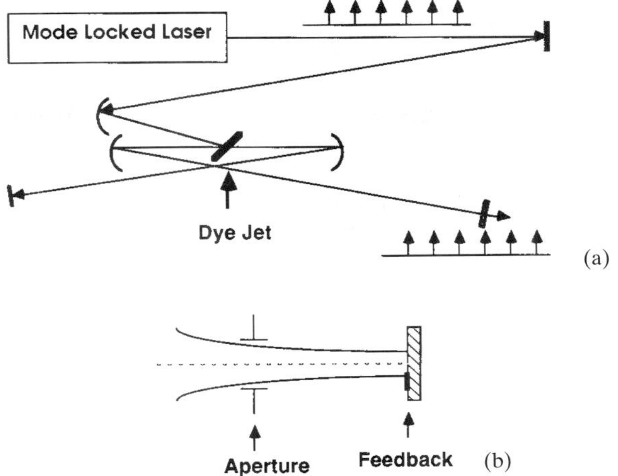

FIGURE 3 (*a*) Synchronously pumped laser, (*b*) modified end mirror.

feedback of less than 10^{-5} is required, the microscope slide needs only to interact with the edge of the beam, as shown in Fig. 3.

Ideally, the gain medium in a synchronously pumped laser should have a short lifetime, so that the duration of the laser pulse is not longer than that of the pump pulse. If the gain medium has a longer lifetime than the cavity round-trip time, there is hardly a modulation of the gain, and we would expect mode locking to be impossible. However, even a small modulation of the gain may grow because of gain saturation by the modulated intracavity radiation, resulting in a shortening of the function $\alpha(t)$, and, ultimately, ultrashort pulses. Phase modulation and dispersion may also lead to the growth of a steady-state pulse out of a small modulation. We will discuss the synchronous pumping of a very long lifetime gain medium in the section on Ti:sapphire lasers.

Hybrid and Double Mode Locking

As mentioned above, synchronously pumped mode-locked lasers have the disadvantage that their pulse (usually longer than those from passive mode-locked lasers) are unstable in duration[19] and in frequency.[21]

To improve the laser characteristics, a saturable absorber can be inserted in the cavity.[22,23] Such a laser is called *hybrid mode-locked laser.* Depending on the concentration of the absorber dye, the hybrid mode-locked laser is either a synchronously mode-locked laser perturbed by the addition of saturable absorption or a passively mode-locked laser pumped synchronously. Detailed discussion of the distinctions can be found in Ref. 1.

Double Mode Locking

Numerous applications require multiple-wavelength femtosecond sources. One possibility is to cascade several lasers synchronously pumping each other.[24] This technique allows us to obtain various synchronized sources of increasing wavelength, but such a complex setup has a low efficiency. More sophisticated and efficient is the method of *double mode*

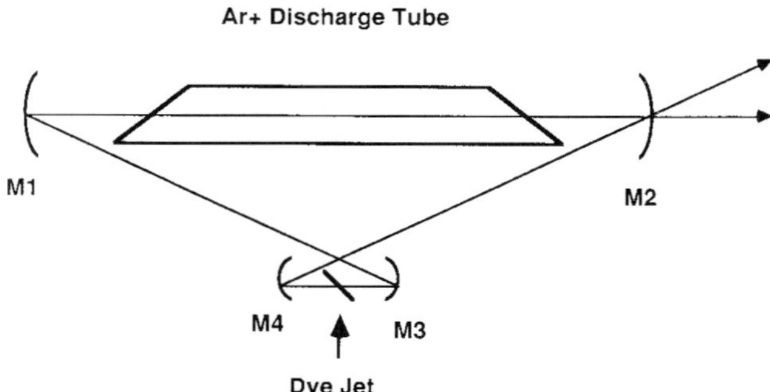

FIGURE 4 Double colliding pulse mode locking, M1, M3, and M4 are high reflectors and M2 has partial transmission at both laser wavelengths.

locking,[25,26] used in a passively mode-locked laser to serve essentially the same purposes. In the latter technique, the saturable absorber of the primary laser is the gain medium for the secondary laser. Although one would have expected pulses shorter than with simple passive mode locking because the lifetime of the common dye—absorber for the primary laser, gain for the secondary—is shortened by the secondary laser action, the improvement was not achieved. Combining double mode locking with pulse collision in the absorber (standing wave saturation), Wang[27] et al. succeeded in double mode locking an Ar^+ and dye laser in a ring cavity. A simple configuration of this double colliding mode-locked laser is shown in Fig. 4. This technique eliminates the need for active mode locking of the Ar^+ laser for synchronous or hybrid mode locking. It also has a high overall efficiency: the average output power was 50 mW for an Ar^+ pump power of 1 W. Another example of cascade operation involves a Ti:sapphire laser synchronously pumping a dye laser.[28]

Multiple wavelength operation is easily achievable within a laser cavity, provided the bandwidth of the gain medium is sufficiently large. This is in particular the case of the Ti:sapphire laser, for which various configurations for two-color femtosecond (fs) operation have been demonstrated. Simultaneous operation of two *synchronized* pulse trains is generally desired. One synchronization mechanism exploited by Dykaar and Darack[29] is cross-phase modulation via the Kerr effect, in a region of the gain medium where the pulses are spatially and temporally close enough to interact. The pump beam (an Ar^+ laser) is split to excite two adjacent volumes of the same piece Ti:sapphire crystal. These two adjacent gain volumes are associated with two cavities tuned to different wavelengths. Up to the largest wavelength offset observed of about 20 nm, the two pulse trains appear perfectly synchronous, as indicated by a cross-correlation width of 70 fs.

Another technique is to use the traditional prism sequence needed to adjust the group velocity dispersion as a spectral device to select two spectral regions in the cavity. Barros and Becker[30] inserted a double slit between the prism pair and the end cavity mirror to select two center frequencies. Dual wavelength spacing as large as 60–75 nm with less than 20 fs time jittering was achieved. A third technique applies to even larger wavelength spacing. Evans et al.[31] used two prism pairs with three prisms (sharing of the first one) and two output couplers. They obtained wavelength range as large as 90 nm and less than 70 fs time jittering between two pulse sequences.

As noted by Kafka[32] and Luo,[33] *unsynchronized* dual wavelength pulse trains can be quite useful for pump-probe spectroscopy. Let us consider, for instance, a transient transmission experiment, with a pump laser operating at a repetition rate of ν_{cp} and the probe at a repetition rate of ν_{ct}. On the screen of an oscilloscope triggered on the pump pulses, transmitted probe pulses will appear at time intervals $\Delta t = (1/v_{cp}) - (1/v_{ct})$, providing a direct display of the transient phenomena being observed, with a temporal resolution equal to Δt.

14.5 ACTIVE AND PASSIVE NEGATIVE FEEDBACK

A typical flash-lamp pumped mode-locked Nd laser generates a train of only 5 to 10 pulses of all different intensity. "Negative feedback" is a technique to modulate cavity losses and stretch the pulse train. It increases cavity losses when the peak intensity of the pulses tends to increase beyond a well-defined value. By limiting the intracavity pulse energy, the rate of depletion of the population inversion is reduced, resulting in an extended pulse train. As told by their names, active (passive) negative feedback is the implementation of this technique actively (passively) using an intracavity energy limiter. The intracavity pulse compression can work more efficiently on a pulse of constant energy that circulates for at least 100 round-trips in the cavity. Pulse duration as short as the bandwidth limit of the gain medium is achieved.[34]

The first report on active feedback is from Martinez and Spinelli,[35] who proposed to use an electro-optic (E-O) modulator to actively limit the intracavity energy in a passively mode-locked glass laser. They demonstrated that the pulse train could be extended. A fast high-voltage switching of the E-O modulator led to the generation of microsecond pulse trains in a passively mode-locked glass laser[36] and in *hybrid* mode-locked Nd:glass lasers.[37] These systems provided an alternative to cw subpicosecond[38] and femtosecond[39] lasers. If we substitute a piece of glass of high thermal conductivity for the laser rod, a microsecond long train of pulses could be generated at a repetition rate of 10 Hz. Used as a source for synchronously pumping a ring dye laser, pulses of less than 50 fs and 25 nJ are routinely extracted from the oscillator.

These electronic feedback systems are expensive and sophisticated. There is a minimum response time of one cavity round-trip before the feedback can react.[37] As a result, the pulse train envelope still exhibits some modulation. On the contrary, a passive feedback system can provide immediate response, i.e., on the time scale of the pulse duration rather than on the time scale of the cavity round-trip. The mechanism of passive negative feedback is as follows. Self-defocusing in a semiconducting two-photon absorber (typically a millimeter-thin piece of polished GaAs) sets in at a power level close to the saturation intensity in the saturable absorber dye. Because the pulse intensity is close to the pulse saturation intensity, there is an optimal pulse compression at the pulse leading edge by saturable absorption. On the other hand, because of self-defocusing by the two-photon generated carriers, the pulse trailing edge is clipped off, resulting in further pulse compression.

Stretching and stabilization of Q-switched pulses with intracavity two-photon absorbers was first demonstrated in 1967.[40,41] Simultaneous extension of the pulse train and compression of the individual pulses was only observed two decades later, with the insertion of a GaAs sample in an actively and passively mode-locked Q-switched laser.[42,43,44] A detailed experimental and theoretical analysis of the mechanism by which GaAs affects the pulse evolution in the mode-locked train is given in Ref. 34. The essential components are (1) a high-gain solid-state laser, (2) a pulse-shaping element (A-O modulator and saturable absorber), (3) an intracavity GaAs sample, and (4) a beam-limiting aperture. Typical results[45] are a *rectangular* pulse train of 100 to 200 *identical* pulses, with pulse durations of 5 to 10 ps in an Nd:YAG laser. The pulse energy is about 10 μJ pulse, and a shot-to-shot stability better than 0.5 percent, measured over 10,000 pulses.

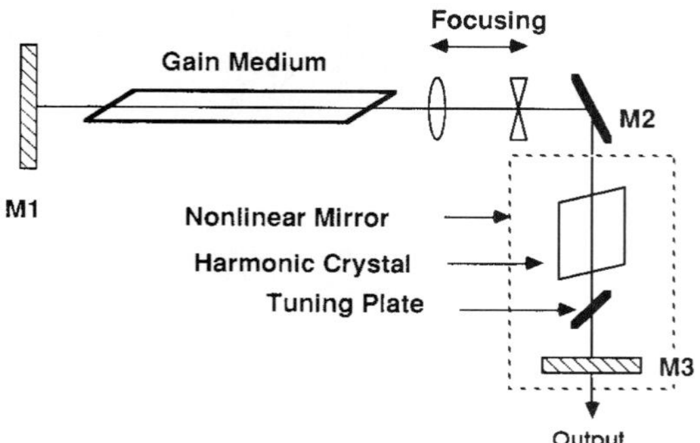

FIGURE 5 Model of laser mode locked by harmonic generation inside cavity.

14.6 NONLINEAR OPTICAL SOURCES

Mode Locking with Intracavity Second Harmonic

A new method of passive mode locking has been investigated, which uses the intracavity harmonic generation as a mode-locking mechanism. In both active and passive mode-locked lasers, harmonic generation has been observed and predicted to have a pulse-broadening effect.[46,47] Stankov proposed that intracavity harmonic generation could be used for pulse narrowing,[48] and subsequently demonstrated passive mode locking in a Q-switched laser by means of a nonlinear mirror consisting of a nonlinear crystal with a second-order nonlinearity and a dichroic mirror.[48,49] A simple cavity model for mode locking through harmonic generation (second harmonic or sum frequency) inside a laser cavity is shown in Fig. 5. The end mirror can be translated to adjust the relative phases (through the dispersion of air) between the fundamental and second harmonic wavelengths as they reenter the crystal. The phases are adjusted for reconversion of the fundamental from the second harmonic on the second passage. The whole assembly of crystal-air gap-end mirror has then a reflectivity *increasing* with intensity.

The mode-locking mechanism with intracavity harmonic generation has been studied and analyzed in the time domain[49] and the frequency domain.[50] The analysis shows that the nonlinear reflection from the nonlinear mirror behaves like a saturable absorber. This mechanism allows one to obtain two synchronized coherent wavelengths from one laser.

Intracavity Parametric Mode Locking

Another mode-locking mechanism called *parametric mode locking* was demonstrated by Zhao and McGraw.[51] This laser can be viewed as a hybridly mode-locked laser using a nonlinear process as the passive pulse-shaping element. Here, the third-order nonlinearity of a crystal applied to sum and difference frequency generation is used in the mode-locking process. Similar to the previous second harmonic case, a nonlinear mirror is used in the cavity. Instead of air, a glass plate for tuning the relative phases of the three wavelengths is inserted between the crystal and the trichroic end mirror. This nonlinear mirror generates a reflectivity at one wavelength as an increasing function of the intensity at another wavelength and therefore modulates the intracavity loss at the first wavelength inside the laser. In the particular experiment of Zhao and McGraw,[51] an infrared dye laser is pumped synchronously by a frequency doubled Nd:YAG laser. Sum and difference parametric generation synchronized with the 1.06 μm radiation of the pump source is obtained intracavity. The crystal is located near the end mirror which reflects both the generated blue and infrared dye laser radiation. Optimal pulse compression is obtained when, at the second passage through the crystal, the phase of the blue radiation, relative to the infrared radiation, is such that the blue radiation is reconverted back to the infrared radiation oscillating in the cavity. It has been shown[51] that both parameteric amplitude modulation (AM) and frequency modulation (FM) contribute to the mode locking.

This technique is useful for pulse narrowing, mode locking, and frequency upconverting of lasers. In addition, because the reflectivity of the parameteric mirror is intensity-dependent, the mechanism can be used to adjust the intracavity chirp. It provides us a unique opportunity to generate ultrashort pulses in various wavelength ranges, such as in the blue, where direct short-pulse generation is difficult due to the lack of suitable pump sources and mode-locking materials.

Optical Parametric Oscillator

Optical Parametric Oscillation. Optical parametric amplification (OPA) is a second-order nonlinear optical process. A nonlinear crystal is excited by a pump radiation of frequency ω_p. A "signal" of amplitude $\mathscr{E}_p(t)$ and frequency ω_s will grow exponentially in a crystal of thickness d, if pump depletion can be neglected:

$$\mathscr{E}_s(t) = \mathscr{E}_{s0} e^{\sigma_a d \mathscr{E}_p(t)} \tag{13}$$

with the growth rate σ_a proportional to the second-order susceptibility:

$$\sigma_a \propto \chi^{(2)} \tag{14}$$

An "idler" photon of frequency $\omega_i = \omega_p - \omega_s$ is generated with the emission of each signal photon. Two types of phase matching conditions can be achieved, either type I:

$$\mathbf{k}^{e(o)}(\omega_p) = \mathbf{k}^{o(e)}(\omega_s) + \mathbf{k}^{o(e)}(\omega_i) \tag{15}$$

or type II:

$$\mathbf{k}^{e(o)}(\omega_p) = \begin{cases} \mathbf{k}^{e}(\omega_s) + \mathbf{k}^{o}(\omega_i) \\ \mathbf{k}^{o}(\omega_s) + \mathbf{k}^{e}(\omega_i) \end{cases} \tag{16}$$

where $\mathbf{k}^{o(e)}$ is the wave vector for ordinary (extraordinary) beam.

Optical parameteric generation is achieved by single-passage amplification of vacuum fluctuation (superfluorescence OPA).[52] Optical parametric oscillation (OPO) can be obtained from an optical parametric amplifier with a cavity feedback. The cavity can be made resonant with the idler or the signal (singly resonant OPO) or with both (only practical for degenerate OPO).[53] For synchronous pumping by mode-locked solid-state lasers, the length of the OPO cavity should match that of the pump laser. A unidirectional ring cavity for the OPO is preferable because of lower loses.[54] The mechanisms of pulse compression are similar to those of other synchronously pumped lasers, except that there is no pulse compression through gain saturation as observed in dye lasers. The most important are the pulse compression through phase modulation in the gain medium and intracavity group velocity dispersion. Pulses as short as 65 fs over a broad tuning range were obtained by synchronously pumping an OPO resonant or with an Nd:glass laser with negative feedback.[55] The same approach leads to fs pulse generation, using as pump the intracavity fs pulses of a dye laser,[56] or of a Ti:sapphire laser.[57] Since the method of parameteric generation is intracavity to a mode-locked laser, there is little possibility for scaling up the energy of the pulses (typically 1 pJ). The main advantage of this method is the high repetition rate. Such parametric generators pumped by mode-locked Ti:sapphire lasers has extended the wavelength range of high-repetition-rate ps and fs sources from the near UV to the middle IR.[58,59,60,61,62] For complete recent trends in OPO, refer to Diels and Rudolph (Ref. 4).

Parametric Chirp Enhancement and Pulse Compression. Since the pump frequency is the sum of the idler and signal frequencies, the idler and signal are chirp-reversed of each other.[62] This has an important consequence for seeded-pulse parametric generation. If a seed signal (idler) pulse is sent (before passing through the nonlinear crystal) through a dispersion line, it will be broadened and chirped. The same dispersion line will act as a perfect pulse compressor for the generated idler (signal).

In some circumstances, the chirp of a pump pulse may be amplified in the parametric process. In the case of near degenerate parametric generation in type I cut KDP, a small change in the wavelength of the pump (visible) results in a larger change in the wavelength of the signal and idler. As a result, a slightly chirped pump pulse generates signal and idler with enhanced and opposite chirps.[63] The instantaneous departure from average carrier frequency of the signal ($\delta\omega_s$) and idler ($\delta\omega_i$) can be related to the frequency deviation of the pump ($\delta\omega_p$):

$$\begin{aligned} \delta\omega_s &= p\delta\omega_p \\ \delta\omega_i &= (1-p)\delta\omega_p \end{aligned} \tag{17}$$

The chirp enhancement factor p is related to the group velocity mismatch:

$$p = \frac{1/v_{gp} - 1/v_{gi}}{1/v_{gs} - 1/v_{gi}} \tag{18}$$

where the subscripts p, i, and s denote pump, idler, and signal. The signal pulses with enhanced chirp can be compressed in a grating pair compressor. Pulses of 50 femtosecond at 920 nm were obtained by this technique[63] with the natural chirp of frequency-doubled pulses from an Nd:glass laser.

14.7 ADDITIVE AND SELF-MODE-LOCKING

Mode locking through phase modulation by a Kerr liquid has been observed in a Ruby laser[65] in 1968. The presence of Kerr liquids in the Q-switched resonator gave rise to ultrashort pulses. Other means to introduce a modulation in the cavity include self-phase-modulation inside the laser material,[66,67,68,69] spatial soliton propagation,[13,70] and coherent addition of pulses from coupled cavities.[71,72] Some aspects of these various mechanisms will be discussed below.

Additive Pulse Mode Locking

There has been a large effort devoted to the development of additive pulse mode locking (APML) involving coupled cavities. One of the basic ideas—to establish the mode coupling outside the main laser resonator—was suggested in 1965[73] and later applied to mode-lock a He-Ne laser.[74] In that earlier implementation, an A-O modulator, operating at half the intermode spacing of the laser, coupled adjacent modes (outside the cavity) before reinjecting them into the cavity through the output mirror. The output mirror of the laser, together with the mirror used to reinject the modulated radiation, formed a cavity of a mode spacing equal to that of the main laser cavity. Each upshifted mode, reinjected in the cavity, will force laser oscillation in phase with the mode ("injection locking"). Since all adjacent pairs of modes are coupled in phase, the laser is mode locked.

The recent APML methods are passive implementation of this older technique. In the purely dispersive version, a pulse from the coupled cavity is given some phase modulation such that the first half of the pulse fed back into the laser adds in phase with the intracavity pulse, while the second half has opposite phase. At each round-trip, the externally injected pulse compresses the intracavity pulse, by adding a contribution to the leading edge and subtracting the same amount from the trailing edge, as sketched in Fig. 6. This technique has first been applied to shortening pulses generated through other mode-locking

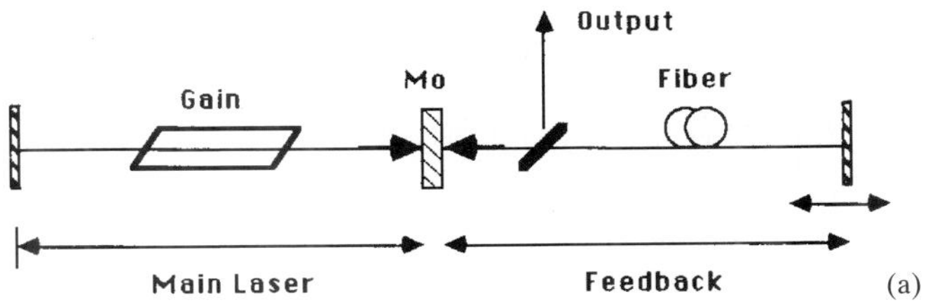

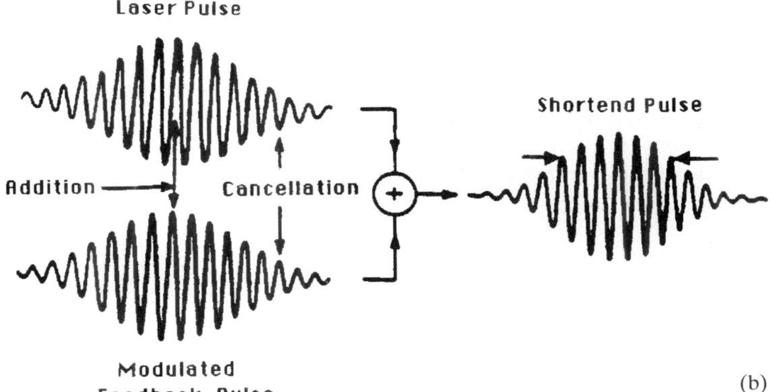

FIGURE 6 (*a*) A typical additive-pulse mode locked laser, and (*b*) modeling of its coherent field addition.

mechanisms. A reduction in pulse duration by as much as two orders of magnitudes was demonstrated with color-center lasers[72,75,76] and with Ti:Al_2O_3 lasers.[77]

It was subsequently realized that the mechanism of pulse addition through a nonlinearly coupled cavity is sufficient to passively mode lock a laser. This technique has been successfully demonstrated in a Ti:sapphire laser,[78] Nd:YAG,[79,80] Nd:YLF,[81,82] Nd:glass,[83] and KCl color-center lasers.[84] Ippen et al.[85] presented a detailed description of the coherent addition of pulses from the main laser and the extended cavity which takes place in additive-pulse mode locking.

Coherent field addition is only one aspect of the coupled-cavity mode-locked laser. The nonlinearity from the coupled cavity can be an amplitude modulation, as in the soliton laser,[13] or a resonant nonlinear reflectivity via a quantum well material.[18]

Some analyses of APML[76,85] have shown that the coupling between a laser and an external nonlinear cavity can be modeled as an intensity-dependent reflectivity of the laser end mirror. In the round-trip from the output mirror through the auxiliary cavity, the pulse accumulates a total phase shift $\phi + \Phi$ consisting of a constant (linear) part ϕ and a nonlinear part (intensity-dependent) Φ. Assuming a small nonlinear shift, we can replace the phase factor $\exp(i\Phi)$ by its first-order expansion which leads to the equation governing the reflection coefficient:[76]

$$\Gamma = r + l(1 - r^2)e^{-i\phi}(1 - i\Phi) \tag{19}$$

Here, r is the amplitude reflectivity of the output coupler; l is the attenuation inside the nonlinear medium, for instance, a fiber. In the latter case, the intensity-dependent phase is $\Phi = \kappa[|\mathscr{E}_2(t)|^2 - |\mathscr{E}_2(0)|^2]$ with $\mathscr{E}_2(t)$ ($\mathscr{E}_2(0)$) being the amplitude (peak) of the waves incident on the fiber and κ is proportional to the length and nonlinear index of the fiber. There is a differential reflectivity for different parts of the pulses. If one sets $\phi = -\pi/2$, then Γ has a maximum value at the pulse center where $\Phi = 0$, and smaller values at the wings. Adjustment of this dynamic reflectivity produces pulse shortening at each reflection until a steady-state balance is achieved between the pulse shortening and pulse spreading due to bandwidth limitation and dispersion.

Self-Starting Mode Locking

Self-Starting Mechanism. In the case of a laser mode locked passively with a saturable absorber, the oscillation builds up gradually from noise fluctuations saturating the absorber. The starting mechanism is not obvious in the case of the laser mode locked by a dispersive process, particularly when the gain medium has a long lifetime. Some lasers require an external starting mechanism; some others, such as the Ti:sapphire lasers, are "self-starting."

In a self-mode-locked laser, like in any other mode-locked system, there are intensity-dependent loss mechanisms. Temporal phase modulation implies necessarily a spatial modulation of the wavefront, hence, self-lensing. This effect is not unique to the Ti:sapphire laser: it also plays an important role in the operation of dye lasers.[6] It has been suggested[86] that mode locking can be achieved through a Kerr-induced self-focusing effect in the gain medium—an effect thereafter referred to as Kerr lens mode locking. The decreased loss with intensity for all these nonlinear effects can also be interpreted as a *net gain.* There is another effect that *opposes* the nonlinear compression mechanism, often referred to as *dynamic gain saturation.*[87,88] Laser mode locking will be self-starting if the intracavity nonlinearity—responsible for the growth of a noise spike—dominates the dynamic gain saturation.

If we define P_p as the amplitude of one fluctuation rising above the noise of average power P_{av}, the response of the nonlinear compression mechanism can be seen as a *net gain* for a power fluctuation $\Delta P = P_p - P_{av}$. To the first-order approximation the gain increase is proportional to the fluctuation

$$\Delta g(t) = b[P_p(t) - P_{av}] = b\,\Delta P(t) \tag{20}$$

where b is a characteristic of the nonlinearity and proportional to κ. The equation for the growth of the perturbation $P_p(t)$ is then

$$\frac{\partial P_p}{\partial t} = \frac{b}{t_{RT}} P_p^2 - f(P_p) \tag{21}$$

In Eq. (21), t_{RT} is the cavity round-trip time. The various theories of starting mechanism differ as to the origin of the fluctuation damping term $f(P_p)$. One school[87,88] associates the function $f(P_p)$ with the decrease in gain related to a positive power fluctuation P_p. That the derivative of Eq. (21) should be positive leads to a self-starting condition:

$$\frac{b}{a_g} \gg \sigma\tau_p \tag{22}$$

where τ_p is the duration of the fluctuation, σ is the gain cross section, and a_g is the average (saturated) gain prior to the arrival of the perturbation. Another approach is that of Krausz et al.,[89] who uses a simple damping term $-P_p/\tau_c$ for $f(P_p)$. The decay time τ_c for the fluctuation is the inverse linewidth of the first beat note. Assuming that unequal axial-mode spacing due to spatial hole burning is the major contribution to the beat note linewidth $1/\tau_c$, Krausz et al.[89] predict the following self-starting condition:

$$bP_{av} > \frac{1}{\ln(m_i)}\frac{t_{RT}}{\tau_c} \tag{23}$$

where m_i is the number of initially oscillating modes.

Kerr Lens Mode Locking. The use of the nonlinearity of Kerr effect for pulse shortening and mode locking has been investigated theoretically and experimentally for more than two decades.[66] In 1972, Dahlström succeeded in mode locking an Nd:glass laser by means of the optical Kerr effect. However, due to the weak nonlinear response of the devices, this technique was constrained to pulsed systems[66,71,90] because of the high intensities needed to produce a sufficient nonlinearity to couple the axial modes.

A breakthrough for cw pumped passive mode locking of solid lasers was achieved when strong self-phase-modulation[91] was observed in fibers and, subsequently, in Ti:sapphire crystals. The self-focusing effect in the laser rod which leads to a modification of the cavity parameters was analyzed in the Gaussian beam approximation by Salin et al.[16] In some configurations, the self-focusing results in a higher gain for pulsed than for cw operation because of an increasing overlap of the pump gain volume with the cavity mode of the circulating pulse. When observing the beam structure of a Ti:sapphire, we find a high-order transverse mode for cw operation, and a smaller one when the laser is mode locked. If an aperture is inserted to increase loss for cw operation, the laser will eliminate the cw mode and be self-mode-locked. The combination of self-lensing with the aperture acts as intensity-dependent losses, just as does a saturable absorber.

Ti:Sapphire Lasers. The large bandwidth of the Ti:sapphire laser has made it one of the most attractive candidates for ultrashort pulse generation. All mode-locking techniques developed for dye lasers have been attempted with this laser: passive mode locking,[92] acousto-optic mode locking,[93,94] injection seeding,[95] additive-pulse mode locking,[78] moving-mirror mode locking,[77] hybrid mode locking,[84] and synchronous pumping.[96,97] Naganuma et al.[98] obtained pulses as short as 50 fs, using the dye HITCI as a saturable absorber in an antiresonant ring. Unidirectional ring operation was also demonstrated.[99]

The simplest configuration, however, to produce a self-mode-locking or Kerr lens mode locking is unique to Ti:sapphire lasers. Spence et al. produced 60 fs pulses with a self-mode-locking laser without any internal mode-locking elements.[100] The label "Kerr lens mode locking" was first applied to a similar cavity configuration,[91] except that an adjustable slit inserted in front of the output coupler ensured additional loss for the cw mode. In this type of laser, the sapphire rod itself is an effective phase modulator through the nonlinearity of its index at high-peak laser intensities. Balance of phase modulation and dispersion led to the generation of pulses shorter than 35 fs by several groups.[101,102,103] The shortest pulse duration obtained, so far, from a Ti:sapphire laser is 11 fs by the group at Washington State University.[103]

The fascinating aspects of self-mode-locking or Kerr lens mode locking are still being actively investigated.[104] Whatever the details of the mechanisms involved, this type of "self-mode-locked laser," avoiding the need for active or passive mode lockers, has reduced the number of optical elements inside the cavity. Thereby, it reduced the sources of instability, increased the available power, simplified the design and operation, and, most important, reduced the cost.

14.8 OTHER ULTRASHORT PULSE SOURCES

Soliton Laser

We have seen in preceding sections that there are loss mechanisms (saturation, induced lensing) and dispersive mechanisms (phase modulation followed by dispersion) that can contribute to pulse shaping in a laser. If the mechanism of pulse compression by self-phase-modulation and dispersion dominates, the equation of the pulse evolution in the cavity can be shown to reduce to the nonlinear Schrödinger equation. The physical model for the laser is reduced to a succession of infinitesimal phase modulation followed by an infinitesimal dispersion per cavity round-trip.

Under appropriate assumptions, manipulations, and renormalizations, the nonlinear Schrödinger equation takes the form

$$i\frac{\partial \tilde{E}}{\partial z} + \frac{\partial^2 \tilde{E}}{\partial t^2} + |\tilde{E}|^2\,\tilde{E} = 0 \tag{24}$$

where z, t, and $\tilde{E}$ are normalized distance, time, and electric field, respectively. The solutions of this equation are described in an article by Zakharov and Shabat.[105]

Numerically, a soliton is an eigenfunction $\tilde{E}$ associated with eigenvalues ζ_i of the coupled differential equations. The order of the soliton is the number of poles associated with that solution. Solitons of order one are stationary solutions, which have a stable secant hyperbolic shape and propagate without distortion. Solitons of order $n > 1$, however, are periodic solutions with n characteristic frequencies.

In a soliton laser, at each round-trip, there is an infinitesimal amount of self-phase-modulation succeeded by an infinitesimal gain. In the development leading to Eq. (24), the gain and losses are assumed to be linear and nullify each other. The same mathematic development leading to the Schrödinger equation also applies to a single mode fiber, at a wavelength of negative dispersion. The physics governing the propagation equation of a short pulse also includes a combination of self-phase-modulation, negative dispersion, and negligible losses. The solitons, solutions of Eq. (24), represent steady-state pulses either circulating in some type of mode-locked laser cavities, or propagating in single-mode fibers.

In the case of fibers, because of their inherent stability, solitons can propagate without distortion over very long distance. If information can be stored and maintained in a particular time sequence of solitons, then solitons may be opening a new era of optical communications.

It is possible to combine the soliton-shaping characteristics of a fiber and a mode-locked laser cavity to produce a soliton laser. The fiber generates a soliton which is fed back into the cavity. Such a soliton laser was invented by Mollenauer in 1984.[13] The laser produced $n = 2$ soliton subpicosecond pulses at infrared wavelength (1.5 μm) for the first time. A polarization preserving fiber with anomalous dispersion is inserted into the feedback loop of a mode-locked laser. This is accomplished by directing part of the intracavity power into the fiber with a beamsplitter and a lens. A lens and a mirror at the other end of the fiber reflect the pulse back toward the cavity. A variable path portion of the feedback loop is required for adjusting the fiber arm to an integral multiple of the main cavity length. The soliton pulse duration can be adjusted from several picoseconds[13] to 50 fs[106] by selecting the length of the generating fiber.

Miniature Dye Lasers

It seems somewhat paradoxical to be using large (of the order of meters) resonators to generate pulses of spatial extension in the micron range. Two methods of short-pulse generation use either an ultrashort cavity (Fabry-Perot dye cells of thickness in the micron range) or no cavity at all (distributed feedback lasers).

Distributed Feedback Lasers. In a linear resonator, the counterpropagating pulses create a standing wave pattern in the gain medium. Stimulated emission can take place only where a stimulating field is present. There is neither emission nor gain depletion at the nodes of the standing wave pattern of the laser. Gain occurs mainly at the antinodes of this pattern. Conversely, if the radiation is produced from a spatial pattern corresponding to such standing waves, it will have the same characteristics as if a cavity were present. In the distributed feedback laser, the inversion (hence the emission) is only *produced* at the antinodes of a standing wave pattern. The wavelength selectivity $\Delta\lambda/\lambda$ is simply the inverse of the number of gain lines in the pattern. The gain pattern is produced by interfering two pump beams in the gain medium. If 1 mm-wide pump beams are used to make a pattern for 600 nm, $\Delta\lambda/\lambda = \Delta\nu/\nu \approx 1/1700$. The bandwidth of this grating is in the THz range, broad enough to support the generation of picosecond pulses.

It seems at first that a coherent source is required to pump such a laser. In principle,

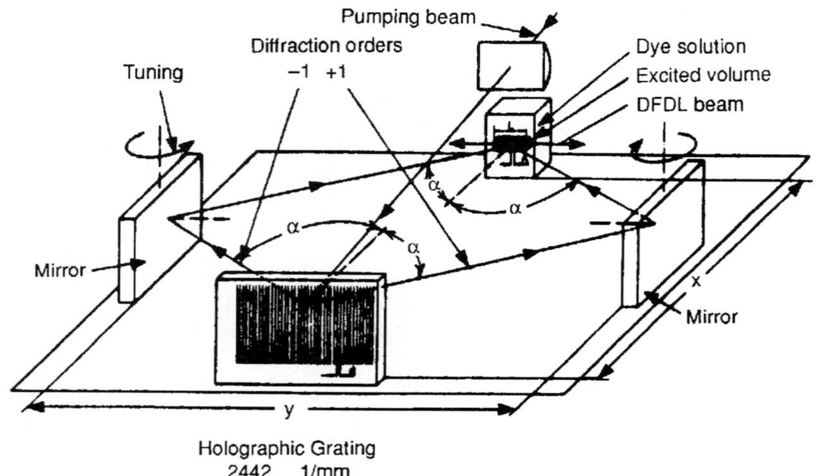

FIGURE 7 Creating an inversion grating with a source of low coherence, using a holographic grating [*courtesy of F. P. Shaefer*].

even white light can be used, if the grating is formed by imaging a holographic grating into the gain medium with the pump radiation. As shown in Fig. 7, in a perfectly symmetric configuration, the pump beam diffracted into the +1 order is made to interfere with the −1 order in a dye cell.[107]

As that laser is being pumped, a series of pulses is generated—each pulse being terminated by gain depletion. There is, however, a range of pump pulse power leading to the generation of only a single pulse, with a duration about 50 times shorter than the pump pulse. Quenching of the emission can eventually be achieved by an additional laser pulse which terminates the gain.[108] A detailed review of various experimental arrangements and applications of picosecond distributed feedback lasers can be found in a publication by Bor and Müller.[109]

Short-Cavity Lasers. Another approach to generate single short pulses is the short-cavity laser. The laser cavity has to be short enough to have a free spectral range exceeding the gain bandwidth. In a typical "short-cavity" laser,[110] the wavelength is tuned by adjusting the thickness of the dye cell within a 3- to 5-μm range with a transducer bending slightly the back mirror of the cavity. With a round-trip time only of the order of 10 femtosecond, it is obvious that the pulse duration will not be longer than that of a picosecond pump pulse. As with the distributed feedback laser, the dynamics of pump depletion can result in pulses shorter than the pump pulses. The basic operational principles of this laser can be found in Ref. 111. Technical details are given in Ref. 110.

Other Techniques

For completeness, we will list several other methods being used for pulse shortening or mode locking.

In 1990, Krausz et al.[112] reported active mode locking of lasers by piezoelectrically induced diffraction modulation. They used a piezoelectric transducer to create standing surface waves on a photo-elastic crystal ($LiNbO_3$). The standing wave causes a diffraction modulation of the light and therefore mode locking of the laser. A detailed discussion of the principles of this technique, which led to the generation of 3.8-ps pulses, can be found in Ref. 113.

In 1991, Turi and Krausz reported a mode-locked laser by regenerative feedback. The conditions to be satisfied can be found in a paper by Hugett,[114] where he referred the feedback scheme as a *regenerative feedback loop*. The natural longitudinal mode beating detected by a photodiode is amplified and used to drive an A-O modulator. In the standard active mode-locking technique, the cavity (round-trip) frequency and the electronic driver for the mode locker are independently stabilized with an accuracy of one part in 10^6. Both frequencies are simply coupled together in the regenerative feedback laser, without the need for thermal stabilization of the cavity or the stabilization of an oscillator. This technique has been successfully implemented to the mode locking of a Ti:sapphire laser.[115]

Another example, which was published in 1992 by Zhang et al.,[116] is to mode lock an Nd:YAG laser with a GaAs crystal. They used a 0.5-mm thick undoped GaAs crystal to mode lock the Nd:YAG laser and produced pulses of 10 ps. The physics is the same as for passive negative feedback discussed earlier, except that the cavity configuration is such that self-defocusing in GaAs leads to a reduction rather than an increase in losses.

14.9 AMPLIFICATION

Amplification of femtosecond pulses to higher energies is traditionally done by using relatively short dye cells transversally pumped by the second or third harmonic of a Q-switched Nd:YAG laser (see, for instance, Fork et al.[117]). A review of dye amplifiers pumped by Nd:YAG, Nd:glass, and excimer lasers can be found in a paper by Knox.[118] Sometimes, the pump module itself can be substituted to the gain dye cell, as for a 248-nm KrF amplifier[119] or the 308-nm XeCl laser.[120,121]

A trend of the last decade is to use amplifiers of higher repetition rate. From the detection point of view, the laser oscillator itself is an ideal stable clock; the pulse train has small amplitude noise at the cavity round-trip frequency, and offers a good reference signal for heterodyne detection. Often, the peak power of the laser is not sufficient for the nonlinear experiments to be performed. One solution is the CW amplification of the laser in a similar gain medium. However, because of the large pump energy required and possible resultant thermal problems, a pulsed source is desired. A cavity-dumped argon laser provides an ideal source to amplify cavity-dumped pulses from a femtosecond dye laser.[122] An elegant structure with four passes through two gain jets was used, resulting in amplification to pulse energies of roughly 10 nJ at 3 MHz.[122]

The next step in compromising repetition rate for gain is made with the introduction of copper vapor laser amplifiers, providing typically microjoule pulses at 10 to 20 kHz. Various types of configurations have been proposed and implemented, starting with an efficient but complex nine-mirror structure,[123] to a series of four-mirror amplifiers in a confocal structure,[124] to a two-mirror 13-pass structure.[125] The first challenge associated with the design of such an amplifier is to maintain a maximum gain without saturating the amplifying transition. In the 13-pass structure just mentioned, the first 11 paths are focused to a small beam waist in the gain medium: this amplifier is intentionally operating at saturation for the last few passes in order to reduce its sensitivity to fluctuations of the input. The beam is sent back for two more passes through the center of the 2-mm-diameter amplifying cell, providing unsaturated amplification to 15 μJ. In amplifiers, thermal fluctuations associated with flow turbulence cause random beam deflection and lensing. This technical difficulty, which is particularly severe with high-repetition-rate amplifiers, can be solved in the case of water soluble amplifying dyes. No observable fluctuation of the pulse output power is observed with an aqueous gain dye solution cooled down to 4°C (or 11.7°C if the solvent is heavy water) at which the thermal index change dn/dT reaches zero.[125]

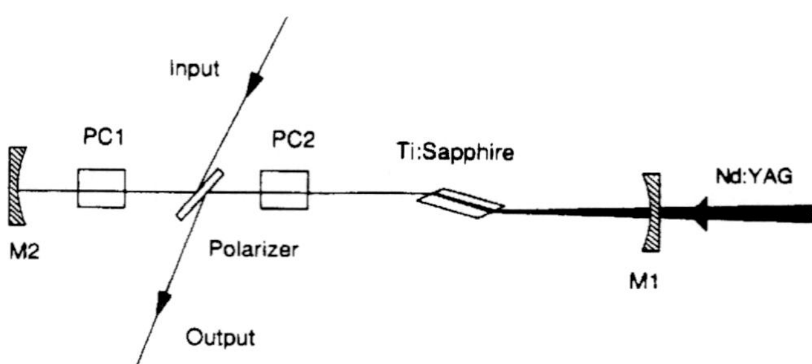

FIGURE 8 Regenerative Ti:sapphire amplifier pumped by a Nd:YAG laser, *Pc*s are Pockel cells and *P* is a polarizer beam splitter (*from Ref. 113*).

The low-repetition-rate amplifier chain has seen dramatic reduction in size and cost and increase in output energy, thanks to the method of chirped pulse amplification.[126] The principle of this method is to stretch the pulse to be amplified by a *linear, reversible* method in order to prevent catastrophic nonlinear effects due to the high peak powers. Martinez[127] has shown that a combination of gratings can chirp and broaden a pulse by as much as 3000 times. A pair of lenses added to the pair of gratings can have the effect of a *negative* distance l, hence have the opposite group velocity dispersion, and restore the original pulse duration.[128] Contrary to the particular high-repetition-rate amplifiers cited above, it is essential in this technique to have *linear* amplification in order to be able to reconstruct the original pulse duration. The method has been applied to stretch 50-ps pulses to 1 ns, amplify to 2 mJ, and recompress to 1 ps.[129] More recently, peak powers of 20 TW have been achieved through the chirped pulse amplification.[130]

The method of chirped pulse amplification is also being used to amplify femtosecond pulses from Ti:sapphire lasers. A particularly efficient system is the regenerative amplifier configuration,[131] in which a Q-switched Nd:YAG pumps a Ti:sapphire rod. The femtosecond pulse to be amplified is successively injected and extracted after several tens of roundtrips through the control of the two Pockel cells (Fig. 8).

14.10 DIAGNOSTIC TECHNIQUES

The femtosecond time scale is beyond the range of standard electronic display instruments. Many new methods have been designed to measure the duration, shape, and modulation of repetitive pulses as well as a single pulse. The subject of ultrashort pulse-shape measurement is far beyond the scope of this chapter. Only a brief overview of some of the techniques is presented in this section.

Correlations

Intensity Autocorrelation. The intensity autocorrelation is generally used to quote a "pulse duration." The standard procedure is to *assume* a pulse shape (generally a secant hyperbolic square or a gaussian shape), and to "determine" the pulse duration from the known ratio between the full width at half maximum (FWHM) of the autocorrelation and that of the pulse. A table in Ref. 21 lists these ratios for a variety of simple pulse shapes.

If a shorter pulse of known shape $I_r(t)$ is available, the temporal profile $I_s(t)$ of an optical signal can be easily determined. The method is to measure the intensity cross-correlation of the two signals:

$$A_c(\tau) = \int_{-\infty}^{+\infty} I_s(t) I_r(t-\tau)\, dt \tag{25}$$

In the ideal limit where the reference can be assimilated to a delta function, the shape of the correlation $A_c(\tau)$ is identical to that of the signal $I_s(t)$. In general, as long as the shorter pulse shape is known, the shape of the signal $I_s(t)$ can be extracted from the measurement of $A_c(\tau)$ through manipulations of Fourier transforms. Even in that ideal case of $I_r(t) \approx \delta(t)$, the intensity cross-correlation has a grave limitation: it does not provide any information on the phase content (frequency or phase modulation) of the pulse being analyzed.

A reference pulse much shorter than the signal cannot always be generated. In the ideal cases where such a pulse is available, a technique to determine the shape of that reference signal is still needed. It is therefore important to consider the limit where the signal itself has to be used as a reference. Equation (25) with $I_s(t) = I_r(t) = I(t)$ is called an *intensity autocorrelation.* An autocorrelation is always a symmetric function—this property can be understood from a comparison of the overlap integral for positive and negative arguments τ. The Fourier transform of the autocorrelation is a real function, consistent with a symmetric function in the time domain. As a result, the intensity autocorrelation provides only very little information on the pulse shape, since an infinite number of asymmetric pulse shapes can have the same autocorrelation. Nevertheless, the intensity autocorrelation is the most widely used diagnostic technique, because it can be easily implemented, and is the first tool used to determine whether a laser is producing short pulses rather than intensity fluctuations of a continuous background. A detailed discussion of the various signal "signatures" provided by the autocorrelations is given in Refs. 21 and 132.

An example of a simple second-order autocorrelator is sketched in Fig. 9. The beams to be correlated are cross-polarized, and combined with a polarizing beamsplitter. An optical delay is used to adjust the delay of the reference signal $I_r(t-\tau)$. The cross-polarized beams (polarization directions x and y) are sent into a nonlinear crystal which is cut for type II phase matched second harmonic generation. The second harmonic field is proportional to the product of the fundamental fields along each axis x and y. Design parameters for optical correlators (crystal thickness and optical components) can be found in Ref. 4.

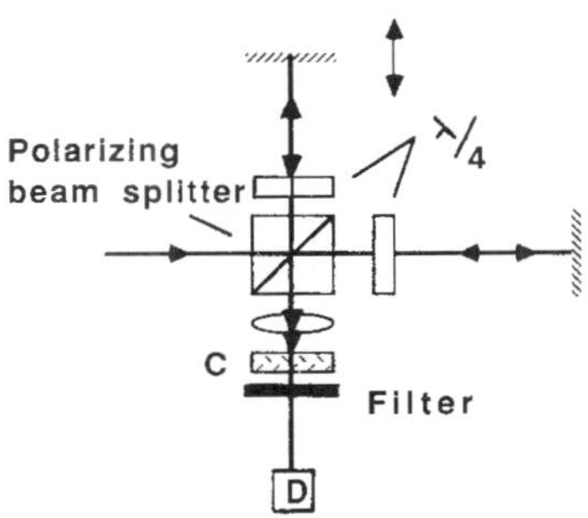

FIGURE 9 Basic intensity autocorrelator, using second harmonic type II detection.

Interferometric Correlations. With the polarizing beamsplitter replaced by a standard beamsplitter, the quarter-wave plates eliminated, and a type I phase matching crystal instead of the type II, the instrument sketched in Fig. 9 records an interferometric autocorrelation. It is essentially a Michelson interferometer in which second harmonic detection has been substituted to the standard detection. The main difference with the intensity autocorrelation is that the interferometric correlation is sensitive to phase modulation.

Instead of Eq. (25), the detected signal is now a second-order interferometric correlation, proportional to the function:

$$G_2(\tau) = \int_{-\infty}^{\infty} \{|E(t-\tau) + E(t)|^2\}^2 \, dt$$
$$= A(\tau) + B(\tau)\cos(\omega\tau) + C(\tau)\cos(2\omega\tau) \tag{26}$$

If we average out the fast oscillation terms, we measure only the d.c. term $A(\tau)$. At $\tau = 0$, the peak value of the function $A(\tau)$ is $6\int \mathscr{E}^4(t)\,dt$, where $\mathscr{E}(t)$ is the amplitude of $E(t)$. For delays larger than the pulse duration, the cross-product vanishes, leaving a background of $2\int \mathscr{E}^4\,dt$. The $A(\tau)$ term which is in fact an intensity autocorrelation has thus a peak-to-background ratio of 3 to 1.

When all terms of the autocorrelation are recorded, the constructive interference terms at zero delay add up to $16\int \mathscr{E}^4(t)\,dt$. The peak-to-background ratio for the interferometric autocorrelation is thus 8 to 1.

As is any autocorrelation, the interferometric autocorrelation is a symmetric function. However, as opposed to the intensity autocorrelation, it contains phase information. There are three different ways to exploit the shape and phase sensitivity of the interferometric autocorrelation:

1. As a *qualitative* test of the absence or presence of phase modulation and, eventually, a determination of the type of modulation
2. As a quantitative measure of *linear* chirp
3. In combination with the pulse spectrum, to determine the pulse shape and phase by fitting procedures

There are various patterns or "signatures" for different types of chirp. For instance, the lower and upper envelopes of the interference pattern split evenly from the background level in the case of an unchirped pulse. In the case of a linearly chirped pulse, the interference pattern is much narrower than the pulse-intensity autocorrelation.[21] With increasing delay, the upper and lower envelopes of the interferometric autocorrelation merge to coincide with the intensity autocorrelation (for delays exceeding the coherence time of the pulse, the terms B and C of Eq. (26) average to zero, leaving only the term A, which is the intensity autocorrelation). The level at which the interference pattern starts can be used as a measure of the chirp.[21]

Single-Shot Technique. A prism can tilt the energy front of a pulse with respect to the wavefront. Szabo et al.[133] have exploited this property to provide a variable delay along a transverse coordinate of the beam. The single-shot interferometric autocorrelator has an advantage over the scanning version used in regular correlators, because the spacing of the fringes is controlled by the tilt of one of the mirrors of the interferometer. It is therefore possible to compromise accuracy and volume of data acquisition.

Iterative Fitting. Since the autocorrelations are symmetric and do not provide any information about the pulse asymmetry, an additional measurement is required to determine the signal shape. The pulse spectrum can provide the additional information. Simple pulse shapes can be fitted to the autocorrelations and pulse spectra. As a guide to such fitting procedures, analytical expressions have been derived for three functions for standard symmetric and nonsymmetric pulse shapes.[21] These functions have been used to tabulate characteristic quantities such as the duration-bandwidth product, and the ratio of the FWHM of the intensity autocorrelation to the pulse duration.

For the cases of complex pulse shapes, a systematic fitting algorithm has been developed by Naganuma et al.[34] which converges to the final signal amplitude and phase function without any a priori knowledge of the pulse parameters.

Pulse Amplitude and Phase Reconstruction

We have already introduced in the previous sections the method of iterative fitting of the interferometric autocorrelations and pulse spectra.[21,134] Numerous other methods have been proposed and implemented. For the sake of conciseness and clarity, we will present only a few examples in this section.

As mentioned earlier, a cross-correlation between a short pulse and a longer one is an approximation of the shape of the latter. Several techniques provide the pulse shape and phase by stretching or compressing the pulse of one arm of the autocorrelator, hence transforming it into a cross-correlator.

For instance, the symmetry of an autocorrelator is broken by insertion of a block of glass in one of its arms. A downchirped pulse is compressed by propagation through glass.[21] As the delay is being scanned, the measured-intensity cross-correlation is an approximation of the pulse-intensity profile. From the interferometric cross-correlation, we can easily extract phase information. Indeed, the depth of modulation of the interference pattern is a measure of the frequency offset of the probing (compressed) femtosecond pulse, as compared to the portion of the pulse being probed.

The pulse is completely determined by the measurement of the intensity and interferometric cross-correlation. The same technique applied to unchirped or upchirped pulses will determine the amplitude and phase of the pulse broadened through quadratic dispersion in glass. But, since the dispersion properties of the glass are known, the shape of the original shorter pulse can be calculated from the amplitude and phase of the broadened one. A detailed analysis of this method can be found in Ref. 135.

Methods of higher nonlinearity have been developed recently such as fringe resolved optical gating,[136] where the spectrum of a gated signal is recorded. Research is still in progress on methods of low nonlinearity, applicable to weak fs pulses.[137,138]

14.11 REFERENCES

1. J.-C. Diels, "Femtosecond Dye Lasers," *Dye Laser Principles,* D. Hillman (ed.), Academic Press, Boston, 1990, chap. 3, pp. 41–132.
2. V. Petrov, W. Rudolph, U. Stamm, and B. Wilhelmi, *Physical Review* **A40:**1474 (1989).
3. H. Avramopoulos, P. M. W. French, J. A. R. Williams, G. H. C. New, and J. R. Taylor, *IEEE J. of Quan. Elec.* **QE-24:**1884 (1988).
4. J.-C. Diels and W. Rudolph, *Femtosecond Physics,* Academic Press, Boston, 1994.
5. W. Dietel, J. J. Fontaine, and J.-C. Diels, *Opt. Lett.* **8:**4 (1983).
6. Xin Miao Zhao and J.-C. Diels, *J. Optical Society of America* **B10:**1159 (1993).
7. M. Sheik-Bahae, A. A. Said, and E. W. Van Stryland, *Opt. Lett.* **14:**955 (1989).
8. J. A. Valdmanis, R. L. Fork, and J. P. Gordon, *Opt. Lett.* **10:**131 (1985).

9. N. Jamasbi, J.-C. Diels, and L. Sarger, *J. of Modern Opt.* **35:**1891 (1988).
10. M. Yamashita, K. Torizuka, and T. Sato, *Opt. Lett.* **13:**24 (1988).
11. L. F. Mollenauer, R. H. Stolen, and J. P. Gordon, *Phys. Rev. Lett.* **45:**1095 (1980).
12. L. F. Mollenauer, R. H. Stolen, J. P. Gordon, and W. J. Tomlinson, *Opt. Lett.* **8:**289 (1983).
13. L. F. Mollenauer and R. H. Stolen, *Opt. Lett.* **9:**13 (1984).
14. J.-C. Diels, W. Dietel, J. J. Fontaine, W. Rudolph, and B. Wilhelmi, *J. Opt. Soc. Am.* **B2:**680 (1985).
15. F. Stalin, P. Grangier, G. Roger, and A. Brun, *Phys. Rev. Lett.* **56:**1132 (1986).
16. F. Salin, J. Squier, and M. Piché, *Opt. Lett.* **16:**1674 (1991).
17. J.-C. Diels and I.C. McMichael, *J. Opt. Soc. of Am.* **B3:**535 (1986).
18. U. Keller, G. W. 'tHooft, W. H. Knox, and J. E. Cuninham, *Opt. Lett.* **16:**1022 (1991).
19. E. W. Van Stryland, *Opt. Commu.* **31:**93 (1979).
20. D. S. Peter, P. Beaud, W. Hodel, and H. P. Weber, *Opt. Lett.* **16:**405 (1991).
21. J.-C. Diels, J. J. Fontaine, I. C. McMichael, and F. Simoni, *Appl. Opt.* **24:**1270 (1985).
22. J. P. Ryan, L. S. Goldberg, and D. J. Bradley, *Opt. Commu.* **27:**127 (1978).
23. B. Couillaud, V. Fossati-Bellani, and G. Mitchel, *SPIE Proceedings,* vol. 533, Springer-Verlag, 1985, p. 46.
24. J. P. Heritage and R. K. Jain, *Appl. Phys. Lett.* **32:**101 (1978).
25. Z. A. Yasa, A. Dienes, and J. R. Whinnery, *Appl. Phys. Lett.* **30:**24 (1977).
26. G. Arjavalingam, A. Dienes, and J. R. Whinnery, *Opt. Lett.* **7:**193 (1982).
27. C. Y. Wang, Q.-R. Xing, and Xin Miao Zhao, *Opt. Commu.* **55:**135 (1985).
28. W. H. Knox and F. A. Beisser, *Opt. Lett.* **17:**1012 (1992).
29. D. R. Dykaar and S. B. Darack, *CLEO, CFG1,* vol. 11, Optical Society of America, Baltimore, Md., 1993, p. 600.
30. M. R. X. de Barros and P. C. Becker, *CLEO, CFG2,* vol. 11, Optical Society of America, Baltimore, Md., 1993, p. 600.
31. J. M. Evans, D. E. Spence, D. Burns, and W. Sibbett, *CLEO, CFG3,* vol. 11, Optical Society of America, Baltimore, Md., 1993, p. 600.
32. J. D. Kafka, J. W. Pieterse, and M. L. Watts, *Opt. Lett.* **17:**1286 (1992).
33. N. Luo, F. Ratsavong, and G. W. Robinson, OSA annual meeting, Paper ThY5, Albuquerque, N.M., 1992.
34. A. Agnesi, A. Del Corno, J.-C. Diels, P. Di Trapani, M. Fogliani, V. Kubecek, G. C. Reali, C.-Y. Yeh, and Xin Miao Zhao, *IEEE J. of Quan. Elec.* **QE-28:**710 (1992).
35. O. E. Martinez and L. A. Spinelli, *Appl. Phys. Lett.* **39:**875 (1981).
36. K. Burneika, R. Grigonis, A. Piskarskas, G. Sinkyavichyus, and V. Sirutkaĭtis, *Kvantovaya Elektron.* **15:**1658 1988.
37. P. Heinz, W. Kriegleder, and A. Laubereau, *Appl. Phys.* **A43:**209 (1987).
38. G. Angel, R. Gagel, and A. Laubereau, *Opt. Commu.* **63:**259 (1987).
39. P. Heinz and A. Laubereau, *J. Opt. Soc. Am.* **B6:**1574 (1989).
40. J. Schwartz, C. S. Naiman, and R. K. Chang, *Appl. Phys. Lett.* **11:**242 (1967).
41. A. Hordvik, *IEEE J. of Quan. Elec.* **QE-6:**199 (1970).
42. A. V. Babushkin, N. S. Vorob'ev, A. M. Prokhorov, and M. Ya. Shchelev, *Sov. J. Quan. Elec.* **19:**1310 (1989).
43. A. Del Corno, G. Gabetta, G. C. Reali, V. Kubecek, and J. Marek, *Opt. Lett.* **15:**734 (1990).
44. A. Agnesi, J.-C. Diels, P. Di Trapani, V. Kubecek, J. Marek, G. Reali, and C. Y. Yeh, *Picosecond Phenomena VII,* Mourou Harris, Ippen, and Zewail (eds.), Springer-Verlag, Berlin, 1990, pp. 38–40.

45. A. V. Babushkin, N. S. Vorob'ev, A. M. Prokhorov, and M. Ya. Shchelev, *Sov. J. Quan. Elec.* **19:**1310 (1989).
46. J. Falk, *IEEE J. of Quan. Elec.* **QE-11:**21 (1975).
47. T. R. Zhong, G. Focht, P. E. Williams, and M. C. Downer, *IEEE J. of Quan. Elec.* **QE-24:**1877 (1988).
48. K. A. Stankov, *Appl. Phys.* **B45:**191 (1988).
49. K. A. Stankov, *Appl. Opt.* **28:**942 (1989).
50. K. A. Stankov, *Opt. Lett.* **16:**639 (1991).
51. Xin Miao Zhao and D. McGraw, *IEEE J. of Quan. Elec.* **QE-28:**930 (1992).
52. T. Nishikawa, N. Uesugi, and J. Yumoto, *Appl. Phys. Lett.* **58:**1943 (1991).
53. A. S. Piskarskas, V. J. Smilgevicius, A. P. Umbrasas, J. P. Juodisius, A. S. L. Gomes, and J. R. Taylor, *Opt. Lett.* **14:**557 (1989).
54. A. S. Piskarskas, V. J. Smilgevicius, A. P. Umbrasas, A. Fix, and R. Wallenstein, *Opt. Commu.* **77:**335 (1990).
55. R. Laenen, H. Graener, and A. Laubereau, *Opt. Lett.* **15:**971 (1990).
56. E. S. Wachmann, D. C. Edelstein, and C. L. Tang, *Opt. Lett.* **15:**136 (1990).
57. C. L. Tang, R. J. Elligson, P. E. Powers, S. Ramakrishna, and K. L. Cheng, *CLEO'93, JWB6,* vol. 11, Optical Society of America, Baltimore, Md., 1993, p. 283.
58. W. S. Pelouch, P. E. Powers, and C. L. Tang, *Opt. Lett.* **17:**1070 (1992).
59. Q. Fu, G. Mark, and H. M. van Driel, *Opt. Lett.* **17:**1006 (1992).
60. A. Nebel, C. Fallnich, and R. Beigang, *CLEO, JWB3,* vol. 11, Optical Society of America, Baltimore, Md., 1993, p. 282.
61. P. E. Powers, W. S. Pelouch, S. Ramakrishna, C. L. Tang, and K. L. Cheng, *CLEO'93, CThK2,* vol. 11, Optical Society of America, Baltimore, Md., 1993, p. 432.
62. M. Itoh, T. Kishimoto, and K. Kato, *CLEO, CThK5,* vol. 11, Optical Society of America, Baltimore, Md., 1993, p. 434.
63. R. Danielius, A. Piskarskas, V. Sirutkaitis, A. Stabinis, and A. Yankauskas, *JETP Lett.* **42:**122 (1985).
64. A. Jankauskas, A. Piskarskas, D. Podenas, A. Stabinis, and A. Umbrasas, *Proceedings of the V International Symposium on Ultrafast Phenomena in Spectroscopy,* Z. Rudzikas, A. Piskarskas, and R. Baltramiejunas (eds.), World Scientific, Vilnius, 1987, pp. 22–32.
65. J. Comly, E. Garmire, J. P. Laussade, and A. Yariv, *Appl. Phys. Lett.* **13:**176 (1968).
66. L. Dahlstróm, *Opt. Commu.* **5:**157 (1972).
67. T. F. Carruthers and I. N. Duling, *Opt. Lett.* **15:**804 (1990).
68. J. Squier, F. Salin, J. S. Coe, P. Bado, and G. Mourou, *Opt. Lett.* **16:**85 (1991).
69. M. Hofer, M. E. Fermann, F. Haberl, M. H. Ober, and A. J. Schmidt, *Opt. Lett.* **16:**502 (1991).
70. A. Barthèlèmy, C. Froehle, S. Maneuf, and G. Sheng, *Proc. Soc. Photo-Opt. Instrum. Eng.,* vol. 813, Springer-Verlag, 1987, p. 387.
71. F. Ouellette and M. Piche, *Opt. Commu.* **60:**99 (1986).
72. K. J. Blow and D. Wood, *J. Opt. Soc. Am.* **B5:**629 (1988).
73. L. C. Foster, M. D. Ewy, and C. B. Crumly, *Appl. Phys. Lett.* **6,** 1965.
74. R. W. Dunn, A. T. Hendow, J. G. Small, and E. Stijns, *Appl. Opt.* **21:**3984 (1982).
75. P. N. Kean, X. Zhu, D. W. Crust, R. S. Grant, N. Langford, and W. Sibbett, *Opt. Lett.* **14:**39 (1989).
76. J. Mark, L. Y. Liu, K. L. Hall, H. A. Haus, and E. P. Ippen, *Opt. Lett.* **14:**48 (1989).
77. P. M. French, J. A. R. Williams, and R. Taylor, *Opt. Lett.* **14:**686 (1989).
78. J. Goodberlet, J. Wang, J. G. Fujimoto, and P. A. Shulz, *Opt. Lett.* **14:**1125 (1989).
79. J. Goodberlet, J. Jacobson, J. G. Fujimoto, P. A. Schulz, and T. Y. Fan, *Opt. Lett.* **15:**504 (1990).

80. L. Y. Liu, J. M. Huxley, E. P. Ippen, and H. A. Haus, *Opt. Lett.* **15:**553 (1990).
81. J. M. Liu and J. K. Chee, *Opt. Lett.* **15:**685 (1990).
82. J. Goodberlet, J. Jackson, J. G. Fujimoto, and P. A. Schultz, *Ultrafast Phenomena VII,* C. B. Harris, E. P. Ippen, G. A. Mourou, and A. H. Zewail, Springer-Verlag, Berlin, 1990, p. 11.
83. F. Krausz, Ch. Spielmann, T. Brabec, E. Wintner, and A. J. Schmidt, *Opt. Lett.* **15:**737 (1990).
84. W. Sibbett, *Ultrafast Phenomena VII,* E. P. Ippen C. B. Harris, G. A. Mourou, and A. H. Zewail (eds.), Springer-Verlag, Berlin, 1990, p. 2.
85. E. P. Ippen, H. A. Haus, and L. Y. Liu, *J. Opt. Soc. Am.* **B6:**1736 (1989).
86. M. Piche, N. McCarthy, and F. Salin, *OSA Annual Meeting,* Paper MB8, vol. 15, 1990.
87. E. P. Ippen, L. Y. Liu, and H. A. Haus, *Opt. Lett.* **15:**183 (1990).
88. J. Goodberlet, J. Wang, J. G. Fujimoto, and P. A. Schulz, *Opt. Lett.* **15:**1300 (1990).
89. F. Krausz, T. Brabec, and Ch. Spielmann, *Opt. Lett.* **16:**235 (1991).
90. K. Sala, M. C. Richardson, and N. R. Isenor, *IEEE J. Quan. Elec.* **QE-13:**915 (1977).
91. L. Spinelli, B. Couillaud, N. Goldblatt, and D. K. Negus, *CLEO,* Postdeadline Paper CPDP7, June 1991.
92. N. Sarukura, Y. Ishida, H. Nakano, and T. Yamamoto, *Appl. Phys. Lett.* **56:**814 (1989).
93. R. Roy, P. A. Schulz, and A. Walter, *Opt. Lett.* **12:**672 (1987).
94. J. D. Kafka, A. J. Alfrey, and T. Baer, *Picosecond Phenomena VI,* Springer-Verlag, 1988, p. 64.
95. P. A. Schulz, M. J. Lagasse, R. W. Schoenlein, and J. G. Fujimoto, *OSA Annual Meeting,* Paper MEE2, vol. 11, 1988.
96. G. B. Al'tshuler, V. B. Karasev, N. V. Kondratyuk, G. S. Kruglik, A. V. Okishev, G. A. Shirpko, V. S. Urbanovich, and A. P. Shkadarevich, *Sov. Tech. Phys. Lett.* **13:**324 (1987).
97. Ch. Spielmann, F. Krausz, T. Brabec, E. Wintner, and A. J. Schmidt, *Opt. Lett.* **16:**1180 (1991).
98. K. Naganuma and K. Mogi, *Opt. Lett.* **16:**738 (1991).
99. K. Tamura, J. Jacobson, E. P. Ippen, H. A. Haus, and J. G. Fujimoto, *Opt. Lett.* **18:**220 (1993).
100. D. E. Spence, P. N. Kean, and W. Sibbett, *Opt. Lett.* **16:**42 (1991).
101. F. Krausz, Ch. Spielman, T. Brabec, E. Wintner, and A. J. Schmidt, *Opt. Lett.* **17:**1 (1992).
102. C. P. Huang, H. C. Kapteyn, J. McIntosh, and M. M. Murnane, *Opt. Lett.* **17:**139 (1992).
103. M. T. Asaki, C. P. Huang, D. Garvey, J. Zhu, H. Nathel, H. Kapteyn, and M. M. Murnane, *Proceedings of the Annual Meeting of OSA 92,* vol. 23, Postdeadline Paper PD17, Optical Society of America, Albuquerque, New Mexico, 1992.
104. S. Chen and J. Wang, *Opt. Lett.* **16:**1689 (1991).
105. V. E. Zakharov and A. B. Shabat, *Trans. Soviet Phys.* **JETP34:**62 (1972).
106. F. M. Mitschke and L. F. Mollenauer, *Opt. Lett.* **12:**407 (1987).
107. Zs. Bor, *Opt. Commu.* **29:**103 (1979).
108. Zs. Bor, S. Szatmari, and A. Müller, *Appl. Phys. B,* **B32:**101 (1983).
109. Zs. Bor and A. Müller, *IEEE J. of Quan. Elec.* **QE-22:**1524 (1986).
110. P. H. Chin, P. Pex, L. Marshall, F. Wilson, and R. Aubert, *Opt. and Lasers in Eng.* **10:**55 (1989).
111. H. P. Kurz, A. J. Cox, G. W. Scott, D. M. Guthals, H. Nathel, S. W. Yeh, S. P. Webb, and J. H. Clark, *IEEE J. of Quan. Elec.* **QE-21:**1795 (1985).
112. F. Krausz, Ch. Spielman, T. Brabec, E. Wintner, and A. J. Schmidt, *Opt. Lett.* **15:**1082 (1990).
113. L. Turi, Cs. Kuti, and F. Krausz, *IEEE J. Quan. Elec.* **QE-26:**1234 (1990).
114. G. R. Hugett, *Appl. Phys. Lett.* **13:**186 (1968).
115. B. E. Lernoff and C. P. J. Borrty, *Opt. Lett.* **17:**1367 (1992).
116. Z. Zhang, L. Qian, D. Fan, and X. Deng, *Appl. Phys. Lett.* **60:**419 (1992).
117. R. L. Fork, C. V. Shank, and R. Yen, *Appl. Phys. Lett.* **41:**223 (1982).
118. W. H. Knox, *IEEE J. of Quan. Elec.* **QE-24:**388 (1988).

119. S. Szatmari and F. P. Schaefer, *Ultrafast Phenomena VI,* T. Yajima, K. Yoshihara, C. B. Harris, and S. Shionoya (eds.), Springer-Verlag, Berlin, Heidelberg, 1988, pp. 82–86.

120. J. H. Glownia, *J. Opt. Soc. Amer.* **B4:**1061 (1987).

121. S. Szatmari, B. Racz, and F. P. Schaefer, *Opt. Commu.* **62:**271 (1987).

122. M. C. Downer, R. L. Fork, and M. Islam, *Ultrafast Phenomena IV,* D. H. Auston and K. B. Eisenthal (eds.), Springer-Verlag, Berlin, Heidelberg, 1984, pp. 27–29.

123. W. H. Knox, M. C. Sowner, R. L. Fork, and C. V. Shank, *Opt. Lett.* **9:**552 (1984).

124. R. L. Fork, H. Avramopoulos, H. L. Fragnito, P. C. Becker, K. L. Scheher, and C. Hirlimann, *Opt. Lett.* **14:**1068 (1989).

125. M. Lai, J.-C. Diels, and C. Yan, *Appl. Opt.* **30:**4365 (1991).

126. M. Pessot, P. Maine, and G. Mourou, *Opt. Commu.* **62:**419 (1987).

127. O. E. Martinez, *IEEE J. of Quan. Elec.* **QE-23:**1385 (1987).

128. O. E. Martinez, *IEEE J. of Quan. Elec.* **QE-23:**59 (1987).

129. P. Maine and G. Mourou, *Opt. Lett.* **13:**467 (1988).

130. C. Sautered, D. Husson, G. Thiell, S. Seznec, S. Gary, and A. Migus, *Opt. Lett.* **16:**238 (1991).

131. P. Georges, F. Estable, F. Salin, J. P. Poizat, P. Graier, and A. Brun, *Opt. Lett.* **16:**144 (1991).

132. G. H. C. New, *IEEE J. of Quan. Elec.* **QE-10:**115 (1974).

133. G. Szabo, Z. Bor, and A. Müller, *Opt. Lett.* **13:**746 (1988).

134. K. Naganuma, K. Mogi, and H. Yamada, *IEEE J. of Quan. Elec.* **QE-25:** 1225 (1989).

135. Chi Yan and J.-C. Diels, *J. of the Opt. Soc. Am. B,* **8:**1259 (1991).

136. D. K. Kane and Tribino, *Opt. Lett.,* **18:**823 (1993).

137. V. Wong and I. A. Walmsley, *Opt. Lett.,* **19:**287 (1994).

138. S. Diddams, Xin Miao Zhao, and J.-C. Diels, SPIE Proceedings, paper 2116-36, Los Angeles, SPIE, SPS Press (1994).

P · A · R · T · 5

OPTICAL DETECTORS

CHAPTER 15
PHOTODETECTORS

Paul R. Norton
Santa Barbara Research Center
Santa Barbara, California

Revised and updated from article by **Stephen F. Jacobs***

15.1 SCOPE

The purpose of this chapter is to describe the range of detectors commercially available for sensing optical radiation. Optical radiation over the range from vacuum ultraviolet to the far-infrared or submillimeter wavelength (25 nm to 1000 μm) is considered. We will refer to the following spectral ranges:

25–200 nm	vacuum ultraviolet	VUV
200–400 nm	ultraviolet	UV
400–700 nm	visible	VIS
700–1000 nm	near-infrared	NIR
1–3 μm	short-wavelength infrared	SWIR
3–5 μm	medium-wavelength infrared	MWIR
5–14 μm	long-wavelength infrared	LWIR
14–30 μm	very long wavelength infrared	VLWIR
30–100 μm	far-infrared	FIR
100–1000 μm	submillimeter	SubMM

We begin by giving a brief description of the photoesensitive mechanism for each type of detector. The usefulness and limitations of each detector type are also briefly described. Definitions of the technical terms associated with the detection process are listed. The concept of sensitivity is defined, and *D*-star (D^*) is presented as a measure of ideal performance. Examples are then given of the limiting cases for D^* under several conditions. In addition, other detector performance parameters are described which may be of critical interest in a specific application, including spectral response, responsivity, quantum efficiency, noise, uniformity, speed, and stability. Finally, manufacturers' specifications for a range of available detectors are compiled and a list of manufacturers is included for each type of detector.

The sensitivity of many detectors has reached the limit set by room-temperature background photon fluctuations (radiation shot noise). For these detectors, sensitivity may be enhanced by providing additional cooling, while restricting their spatial field of view and/or spectral bandwidth. At some point, other factors such as amplifier noise may limit the improvement.

* In *Handbook of Optics,* first edition, McGraw-Hill, 1978. Section 4, "Nonimaging Detectors," by Stephen F. Jacobs, Optical Sciences Center, University of Arizona, Tucson, Arizona.

Techniques for evaluating detector performance are not covered in this treatment but can be found in references (Wolfe, 1965; IRE, 1962; RCA, 1963; Vincent, 1989; Dereniak, 1984; *The Infrared and Electro-Optical Systems Handbook,* vol. 3, 1993).

15.2 THERMAL DETECTORS

Thermal detectors sense the change in temperature produced by absorption of incident radiation. Their spectral response can therefore be as flat as the absorption spectrum of their blackened coating and window* will allow. This makes them useful for spectroscopy and radiometry. These detectors are generally operated at room temperature, where their sensitivity is limited by thermodynamic considerations (Jones, 1949; Bauer, 1943) to 3 pW for 1-s measurement time and 1-mm^2 sensitive area. This limit has been very nearly reached in practice, whereas cooled bolometers have been made to reach the background photon noise limit.

Construction of the detector seeks to minimize both the thermal mass of the sensitive element and the heat loss from either conductive or convective mechanisms. Heat loss may ideally be dominated by radiation. This allows the incident photon flux to give a maximum temperature rise (maximum signal), but results in a correspondingly slow response time for this class of detectors. The response time τ of thermal detectors is generally slower than 1 ms, depending on thermal capacity K and heat loss per second per degree V, through the relation

$$\tau = \frac{K}{V}$$

A short time constant requires a small K. However, for room-temperature operation, ultimate sensitivity is limited by the mean spontaneous temperature fluctuation

$$\Delta T = T\sqrt{\frac{k}{K}}$$

where k is Boltzmann's constant. There is thus a tradeoff between time constant and ultimate sensitivity.

Thermocouple/Thermopile

The thermocouple receiver is a thin, blackened flake connected thermally to the junction of two dissimilar metals or semiconductors. Heat absorbed by the flake causes a temperature rise of the junction, and hence a thermoelectric emf is developed which can be measured, for example, with a voltmeter.

Thermocouples are limited in sensitivity by thermal (Johnson-Nyquist) noise but are nevertheless respectably sensitive. Their usefulness lies in the convenience of room-temperature operation, wide spectral response, and their rugged construction. Thermocouples are widely used in spectroscopy.

Thermopiles consist of thin-film arrays of thermocouples in series. This device multiplies the thermocouple signal corresponding to the number of junctions in series. The device may be constructed with half the thermocouples acting as reference detectors attached to a heat sink.

* No windows exist without absorption bands somewhere between the visible and millimeter region. Some useful window materials for the far-infrared are diamond, silicon, polyethylene, quartz, and CsI.

Bolometer/Thermistor

The receiver is a thin, blackened slab whose impedance is highly temperature dependent. The impedance change may be sensed using a bridge circuit with a reference element in the series or parallel arm. Alternatively, a single bolometer element in series with a load and voltage source may be used.

Most bolometers in use today are of the thermistor type made from oxides of manganese, cobalt, or nickel. Their sensitivity closely approaches that of the thermocouple for frequencies higher than 25 Hz. At lower frequencies there may be excess or $1/f$ noise. Construction can be very rugged for systems applications. Some extremely sensitive low-temperature semiconductor bolometers are available commercially.

Pyroelectric

Ferroelectric material exhibits a residual polarization in the absence of any electric field. Dipole moments, initially aligned by applying an external field, result in a surface charge which is normally slowly neutralized by leakage. This polarization is temperature-dependent (pyroelectric effect), and when a pulse of incident radiation heats a portion of an electroded sample, there is a change in surface charge (open-circuit voltage) which is proportional to the incident radiation power. Electrically, the device behaves like a capacitor, requiring no bias and therefore exhibiting no current noise. The signal, however, must be chopped or modulated. Sensitivity is limited either by amplifier noise or by loss-tangent noise. Response speed can be engineered, with a proportional decrease in sensitivity, making pyroelectric detectors useful for moderately fast laser pulse detection. Other common applications include power meters. Microphonic noise in applications associated with vibrations can be a problem with some of these devices.

15.3 QUANTUM DETECTORS

Photon detectors respond in proportion to incident photon rates rather than to photon energies (heat). Thus, the spectral response of an ideal photon detector is flat on an incident-photon-rate basis but linearly rising with wavelength on an incident-power (per watt) basis. The sensitivity of efficient quantum detectors can approach the limits of photon noise fluctuations provided that the detector temperature is sufficiently low for photon-induced mechanisms to dominate thermally induced mechanisms in the detector. Quantum detectors generally have submicrosecond time constants. Their main disadvantage is the associated cooling required for optimum sensitivity. (These remarks do not apply to photographic detection, which measures cumulative photon numbers.)

Photoemissive

The radiation is absorbed by a photosensitive surface which usually contains alkali metals (cesium, sodium, or potassium). Incident quanta release photoelectrons, via the photoelectric effect, which are collected by a positively biased anode. This is called a diode phototube; it can be made the basis for the multiplier phototube (photomultiplier phototube, or photomultiplier) by the addition of a series of biased dynodes which serve as secondary emission multipliers.

In spectral regions where quantum efficiency is high ($\lambda < 550$ nm), the photoemissive detector is very nearly ideal. Sensitivity is high enough to count individual photons. Amplification does not degrade the signal-to-noise ratio. The sensitive area is conveniently large. Photomultiplier signal response time (transit spread time) can be made as short as 0.1 ns. Since the sensitivity in red-sensitive tubes is limited by thermally generated electrons, sensitivity can be improved by cooling.

Photoconductive

The radiation is absorbed by a photoconductive material, generally a semiconductor, either in thin-film or bulk form. Each incident quantum may release an electron-hole pair or an impurity-bound charge carrier, thereby increasing the electrical conductivity. The devices are operated in series with a bias voltage and load resistor. Very low impedance photoconductors may be operated with a transformer as the load. Since the impedance of photoconductors varies with device type and operating conditions from less than 50 ohms to more than 10^{14} ohms, the load resistor and preamplifier must be chosen appropriately. Photoconductors which utilize excitation of an electron from the valence to conduction band are called *intrinsic* detectors. Those which operate by exciting electrons into the conduction band or holes into the valence band from an impurity-bound state are called *extrinsic* detectors. Intrinsic detectors are most common at the short wavelengths, out to about 20 μm. Extrinsic detectors are most common at longer wavelengths. A key difference between intrinsic and extrinsic detectors is that intrinsic detectors do not require as much cooling to achieve high sensitivity at a given spectral response cutoff as extrinsic detectors. Thus, intrinsic photoconductors such as HgCdTe will operate out to 15 to 20 μm at 77 K, while comparable extrinsic detectors with similar cutoff must be cooled below 30 to 40 K.

A further distinction may be made by whether the semiconductor material has a direct or indirect bandgap. This difference shows up near the long-wavelength limit of the spectral response where detectors made from direct bandgap materials such as InGaAs, InSb, or HgCdTe have a sharper spectral cutoff than indirect bandgap materials such as silicon and germanium.

Photoconductors can have high quantum efficiency from the visible region out to the far-infrared but lack the nearly ideal high amplification of photomultipliers. They are therefore most commonly used in the spectral region beyond 1 μm, where efficient photoemitters are unavailable. Photoconductors do, however, provide current gain which is equal to the recombination time divided by the majority-carrier transit time. This current gain leads to higher responsivity than is possible with (nonavalanching) photovoltaic detectors. For applications where photovoltaic detection would be amplifier-noise-limited, the larger photoconductive responsivity makes it possible to realize greater sensitivity with the photoconductor. In general, lower-temperature operation is associated with longer-wavelength sensitivity in order to suppress noise due to thermally induced transitions between close-lying energy levels. Ideally, photoconductors are limited by generation-recombination noise in the photon-generated carriers. Response time can be shorter than 1 μs and in some cases response times can be shorter than 1 ns for small elements. Response across a photoconductive element can be nonuniform due to recombination mechanisms at the electrical contacts, and this effect may vary with electrical bias.

Photovoltaic*

The most widely used photovoltaic detector is the *pn* junction type, where a strong internal electric field exists across the junction even in the absence of radiation. Photons incident

* The use of a photovoltaic detector at other than zero bias is often referred to as its *photoconductive mode* of operation because the circuit then is similar to the standard photoconductor circuit. This terminology is confusing with regard to detection mechanism and will not be used here.

on the junction of this film or bulk material produce free hole-electron pairs which are separated by the internal electric field, causing a change in voltage across the open-circuited cell or a current to flow in the short-circuited case.

As with the photoconductor, quantum efficiency can be high from the visible to the very long wavelength infrared, generally about 20 μm. The limiting noise level can ideally be $\sqrt{2}$ times lower than that of the photoconductor, thanks to the absence of recombination noise. Lower temperatures are associated with longer-wavelength operation. Response times can be less than a nanosecond, and are generally limited by device capacitance. The *pin* diode has been developed to minimize capacitance for high-bandwidth applications. The advantages of nearly ideal internal amplification have now become available in avalanche photodiodes sensitive out to 1.55 μm. This internal gain is most important for high-frequency operation, where external load resistance must be kept small and would otherwise introduce limiting thermal noise.

Photoelectromagnetic

A thin slab of photoconductive material is oriented with radiation incident on a large face and a magnetic field perpendicular to it. Electron-hole pairs generated by the incident photons diffuse through the material and are separated by the magnetic field, causing a potential difference at opposite ends of the detector.

These detectors require no cooling or biasing electric field but do require a (permanent) magnetic field. Photoelectromagnetic InSb at room temperature has response out to 7.5 μm, where it is as sensitive as a thermocouple of equal size, and has a response time less than 1 μs. Another competing uncooled detector is InAs, which is far more sensitive out to 3.5 μm. Cooled infrared detectors are one to two orders of magnitude more sensitive.

Photographic

The receiver is an emulsion containing silver halide crystals. Incident photons are absorbed by the halide ion, which subsequently loses its electron. This electron eventually recombines with a silver ion and reduces it to a neutral silver atom. As more photons are absorbed, this process is repeated until a small but stable cluster of reduced silver atoms is formed within the crystal (latent image). Internal amplification is provided by introduction of an electron donor (photographic) developer, which, using the latent image as a catalytic center, reduces to silver all the remaining silver ions within the exposed crystal. The density of reduced crystals is a measure of the total radiation exposure.

The spectral region of sensitivity for photographic detection coincides rather closely with that of the photoemissive detector. For $\lambda > 1.2$ μm there is too little energy in each photon to form a stable latent image. The basic detection process for both detectors operates well for higher frequency radiation. The problem in ultraviolet and x-ray operation is one of eliminating nonessential materials, e.g., the emulsion which absorbs these wavelengths.

The photographic process is an integrating one in that the output (emulsion density) measures the cumulative effect of all the radiation incident during the exposure time. The efficiency of the photographic process can be very high, but it depends upon photon energy; e.g., in the visible region it takes only 10 to 100 photons to form a stable latent image (developable grain). The photographic process enjoys a large and efficient internal amplification ability (development) wherein the very small energy of the photons' interaction is converted into readily observed macroscopic changes. An extensive discussion of photographic detection is found in Chapter 20, Photographic Films, in Vol. I of this Handbook.

Photoionization

The radiation is absorbed by a gas. If the photon energy exceeds the gas-ionization threshold, ion pairs can be produced with very high efficiency. They are collected by means of an applied voltage. Operating in a dc mode, these detectors are known as *ionization* and *gas-gain chambers*. When a pulse mode is used, the detectors are known as *proportional* and *photon* (*Geiger*) *counters*.

Photoionization detectors have a high sensitivity and a low noise level. They may also be quite selective spectrally since the choice of window and gas independently set upper and lower limits on detectable photon energies. Manufacturers' specifications are not discussed for these detectors as applications are still few enough to be treated as individual cases (Samson, 1967).

15.4 DEFINITIONS

The following definitions will be used:

Background temperature The effective temperature of all radiation sources viewed by the detector exclusive of the signal source.

Blackbody D star, D^*_{BB} (cm $\mathrm{Hz}^{1/2}$/W) Similar to $D^*(\lambda)$ or $D^*(T_B f)$ except that the source is a blackbody whose temperature must be specified.

Blackbody detectivity D_{BB} (W^{-1}) A measure of detector sensitivity, defined as $D_{\mathrm{BB}} = (\mathrm{NEP}_{\mathrm{BB}})^{-1}$.

Blackbody noise-equivalent power $\mathrm{NEP}_{\mathrm{BB}}$ Same as spectral NEP, except that the source has blackbody spectral character whose temperature must be specified, for example, NEP (500 K, 1, 800) means 500-K blackbody radiation incident, 1-Hz electrical bandwidth, and 800-Hz chopping frequency.

Blackbody responsivity R_{BB} Same as spectral sensitivity except that the incident signal radiation has a blackbody spectrum (whose temperature must be specified).

Blip detector or *blip condition* Originally meaning background-limited impurity photoconductor, this term has come to mean performance of any detector where the limiting noise is due to fluctuations in the arrival rate of background photons.

Cutoff wavelength λ_c The wavelength at which the detectivity has degraded to one-half its peak value.

Dark current The output current which flows even when input radiation is absent or negligible. (Note that although this current can be subtracted out for the dc measurements, the shot noise on the dark current can become the limiting noise.)

Detective quantum efficiency The square of the ratio of measured detectivity to the theoretical limit of detectivity.

Detective time constant τ_d $\tau_d = 1/2\pi f_d$ where f_d is the frequency at which D^* drops to 0.707 times its maximum value.

Dewar A container (cryostat) for holding detector coolant.

Equivalent noise input (ENI) A term meaning nearly the same thing as $\mathrm{NEP}_{\mathrm{BB}}$ (287 K, $1, f$). The difference is that the peak-to-peak value of square-wave chopped input flux is used, rather than the rms value of the sinusoidally chopped input flux. (See recommendation IRE, 1962.)

Excess noise A term usually referring to noises other than generation-recombination, shot, or thermal.

Extrinsic semiconductor transition Incident photons produce a free electron in the conduction band and bound hole at a donor impurity site or a bound electron at an acceptor impurity site and a free hole in the valence band by excitation of an impurity level.

Field of view (FOV) The solid angle from which the detector receives radiation.

Flicker noise See *Modulation noise.*

Generation noise Noise produced by the statistical fluctuation in the rate of production of photoelectrons.

Generation-recombination (GR) *noise* Charge carriers are generated both by (optical) photons and (thermal) photons. Fluctuations in these generation rates cause noise; fluctuations in carrier-recombination times cause recombination noise. The phonon contribution can be removed by cooling. The remaining photon contribution is indistinguishable from radiation shot noise (photon noise). With photovoltaic *pn* junctions, carriers are swept away before recombination, so that recombination noise is absent.

Guard ring An electrically biased field plate or surrounding diode used in some photodiodes, usually used to control surface recombination effects and thus reduce the leakage current in the detection circuit.

Intrinsic semiconductor transition Incident photons produce a free electron-free hole pair by direct excitation across the forbidden energy gap (valence to conduction band).

Johnson noise Same as thermal noise.

Jones Unit of measure for D^* cm $\mathrm{Hz}^{1/2}/\mathrm{W}$.

Maximized D star, $D^*(\lambda_{pk} f_o)$, cm $\mathrm{Hz}^{1/2}/\mathrm{W}$ The value of $D^*(\lambda_{pk}, f)$ corresponding to wavelength λ_{pk} and chopping frequency of maximum D^*.

Modulation (or $1/f$) *noise* A consensus regarding the origin(s) of the mechanism has not been established, and although a quantum theory has been proposed, other mechanisms may dominate. As its name implies, it is characterized by a $1/f^n$ noise power spectrum, where $0.8 < n < 2$. This type of noise is prominent in thermal detectors and may dominate the low-frequency noise characteristics of photoconductive and photovoltaic quantum detectors as well as other electronic devices such as transistors and resistors.

Multiplier phototube or *multiplier photodiode* Phototube with built-in amplification via secondary emission from electrically biased dynodes.

NEI photons/cm^2sec noise equivalent irradiance is the signal flux level at which the signal produces the same output as the noise present in the detector. This unit is useful because it directly gives the photon flux above which the detector will be photon-noise-limited. See also *Spectral noise equivalent power* (NEP_λ).

Noise spectrum The electrical power spectral density of the noise.

Nyquist noise Same as thermal noise.

Photo cell See *Photodiode.*

Photoconductive gain The ratio between carrier lifetime and carrier transit time in a biased photoconductor.

Photodiode The term photodiode has been applied both to vacuum- or gas-filled photoemissive detectors (diode phototubes, or photo cells) and to photovoltaic detectors (semiconductor *pn* junction devices).

Photomultiplier Same as multiplier phototube.

Photon counting Digital counting of individual photoelectrons in contrast to averaging of the photocurrent. This technique leads to very great sensitivity but can be used only for extremely low light levels.

Quantum efficiency The ratio of the number of countable output events to the number of incident photons, e.g., photoelectrons per photon, usually referred to as a percentage value.

Rms noise $V_{n,\text{rms}}$ That component of the electrical output which is not coherent with the radiation signal. (Generally measured with the signal radiation removed.)

Rms signal $V_{s,\text{rms}}$ That component of the electrical output which is coherent with the input signal radiation.

Response time τ Same as *Time constant.*

Responsive quantum efficiency See *Quantum efficiency.*

Sensitivity Degree to which detector can sense small amounts of radiation.

Shot noise This current fluctuation results from the random arrival of charge carriers, as in a photodiode. Its magnitude is set by the size of the unit charge.

$$i_{n,\text{ms}} = (2ei_{\text{dc}}\Delta f)^{1/2}$$

Spectral D double star $D^{**}(\lambda, f)$ A normalization of D^* to account for detector field of view. It is used only when the detector is background-noise-limited. If the FOV is 2π sr, $D^{**} = D^*$.

Spectral D star $D^*(\lambda, f)$ (cm Hz$^{1/2}$/W) A normalization of spectral detectivity to take into account the area and electrical bandwidth dependence, e.g., D^*(1 μm, 800 Hz) means D^* at $\lambda = 1$ μm and chopping frequency 800 Hz; unit area and electrical bandwidth are implied. For background-noise-limited detectors the FOV and the background characteristics must be specified. For many types of detectors this normalization is not valid, so that care should be exercised in using D^*.

Spectral detectivity $D(\lambda)$ (W^{-1}) A measure of detector sensitivity, defined as $D(\lambda) = (\text{NEP}_\lambda)^{-1}$. As with NEP, the chopping frequency electrical bandwidth, sensitive area, and, sometimes, background characteristics should be specified.

Spectral noise equivalent power NEP_λ The rms value of sinusoidally modulated monochromatic radiant power incident upon a detector which gives rise to an rms signal voltage equal to the rms noise voltage from the detector in a 1-Hz bandwidth. The chopping frequency, electrical bandwidth, detector area, and, sometimes, the background for characteristics should be specified. NEP (1 μm, 800 Hz) means noise equivalent power at 1 μm wavelength, 1 Hz electrical bandwidth, and 800 Hz chopping rate. Specification of electrical bandwidth is often simplified by expressing NEP in units of W/Hz$^{1/2}$.

Spectral responsivity $R(\lambda)$ The ratio between rms signal output (voltage or current) and the rms value of the monochromatic incident signal power. This is usually determined by taking the ratio between a sample detector and a thermocouple detector.

Temperature noise Fluctuations in the temperature of the sensitive element, due either to radiative exchange with the background or conductive exchange with a heat sink, produce a fluctuation in signal voltage. For thermal detectors, if the temperature noise is due to the former, the detector is said to be at its theoretical limit. For thermal detectors

$$\overline{(\Delta T)^2} = \frac{4kT^2K\Delta f}{K^2 + 4\pi^2 f^2 C^2}$$

where $\overline{(\Delta T)^2}$ = mean square temperature fluctuations
K = thermal conductance
C = heat capacity

Thermal noise (also known as *Johnson* or *Nyquist noise*) Noise due to the random motion of charge carriers in a resistive element

$$V_{n,\text{rms}} = (4kTR\Delta f)^{1/2} \qquad k = \text{Boltzmann's constant}$$

Thermopile A number of thermocouples mounted in series in such a way that their thermojunctions lie adjacent to one another in the plane of irradiation.

Time constant τ A measure of the detector's speed of response. $\tau = 1/2\pi f_c$, where f_c is that chopping frequency at which the responsivity has fallen to 0.707 of its maximum value:

$$\mathcal{R}(f) = \frac{\mathcal{R}_0}{(1 + 4\pi^2 f^2 r^2)^{1/2}}$$

15.5 DETECTOR PERFORMANCE AND SENSITIVITY

D*

A figure of merit defined by Jones in 1958 is used to compare the sensitivity of detectors. It is called D^*. Although the units of measure are cm $\text{Hz}^{1/2}/\text{W}$, this unit is now referred to as a *Jones*. D^* is the signal-to-noise (S/N) ratio of a detector measured under specified test conditions, referenced to a 1-Hz bandwidth and normalized by the square root of the detector area in centimeters. Specified test conditions usually consist of the blackbody signal source temperature, often 500 K for infrared detectors, and the signal chopping frequency. If the background temperature is other than room temperature (295 or 300 K in round numbers), then that should be noted.

By normalizing the measured S/N ratio by the square root of the detector area, the D^* figure of merit recognizes that the statistical fluctuations of the background photon flux incident on the detector (photon noise) are dependent upon the square root of the number of photons and thus increase as the square root of the detector area, while the signal will increase in proportion to the detector area itself. This figure of merit therefore provides a valid comparison of detector types which may have been made and tested in different sizes.

The ultimate limit in S/N ratio for any radiation power detector is set by the statistical fluctuation in photon arrival times. For ideal detectors which are photon-noise-limited, we shall discuss limiting detectivity for three cases:

1. Photon detector where arrival rate of signal photons far exceeds that of background photons (all other noise being negligible)
2. Photon detector where background photon arrival rate exceeds signal photon rate (all other noise being negligible)
3. Thermal detector, background limited

The rate of signal-carrier generation is

$$n = \eta \text{AN}_s \tag{1}$$

where η = detector quantum efficiency, and AN_s = average rate of arrival of signal photons.

It can be shown (Fink, 1940) that in a bandwidth Δf, the rms fluctuation in carrier-generation rate is

$$\delta n_{\text{rms}} = (2P_N \Delta f)^{1/2} \tag{2}$$

where P_N is the frequency dependence of the mean square fluctuations in the rate of carrier generation, i.e.,

$$P_N = A \int_{\nu 0}^{\infty} \eta(\nu)(\Delta N)^2 \, d\nu \tag{3}$$

where $(\Delta N)^2$ is the mean square deviation in the total rate of photon arrivals per unit area and frequency interval including signal and background photons. For thermally produced photons of frequency ν (see Smith et al., 1957)

$$(\Delta N)^2 = \bar{N} \frac{e^{h\nu/kT}}{e^{h\nu/kT} - 1} = \frac{2\pi\nu^2}{c^2} \frac{e^{h\nu/kT}}{(e^{h\nu/kT} - 1)^2} \tag{4}$$

where $\bar{N}$ is the average rate of photon arrivals per unit area and frequency interval. Then, for the special case $h\nu \gg kT$

$$\delta n_{\text{rms}} = (2A\eta\bar{N}\Delta f)^{1/2} \tag{5}$$

This is also the case for a laser well above threshold. Here the photon statistics become Poisson, and $(\Delta N)^2 = \bar{N}$ even when $h\nu$ is not greater than kT.

Photon Detector, Strong-signal Case. This is generally a good approximation for visible and higher photon energy detectors since the background radiation is often weak or negligible. When signal photons arrive at a much faster rate than background photons

$$\delta n_{\text{rms}} \approx (2A\eta\bar{N}_s\Delta f)^{1/2} \tag{6}$$

then

$$\text{NEP} = \frac{N_s A h\nu}{(n/\delta n_{\text{rms}})(\Delta f)^{1/2}} = \left(\frac{2N_s A}{\eta}\right)^{1/2} h\nu \tag{7}$$

or the noise-equivalent quantum rate is

$$\text{NEQ} = \left(\frac{2 \times \text{incident photon rate}}{\text{quantum efficiency}}\right)^{1/2} \tag{8}$$

Photon Detector, Background-limited Case. This is usually a good approximation for detecting low signal levels in the infrared where background flux levels exceed signal flux levels in many applications. When the background photon rate N_B, exceeds the signal photon rate $(N_B \gg N_s)$

$$\delta n_{\text{rms}} \approx (2A\eta\bar{N}_B\,\Delta f)^{1/2} \tag{9}$$

the noise-equivalent power is

$$\text{NEP} = \frac{N_s A h\nu}{(n/\delta n_{\text{rms}})(\Delta f)^{1/2}} = \left(\frac{2\bar{N}_B A}{\eta}\right)^{1/2} h\nu \tag{10}$$

The noise-equivalent quantum rate is

$$\text{NEQ} = \left(\frac{2 \times \text{incident background photon rate}}{\text{quantum efficiency}}\right)^{1/2} \tag{11}$$

or

$$D^* = \frac{A^{1/2}}{\text{NEP}} = \left(\frac{\eta}{2N_B}\right)^{1/2} \frac{1}{h\nu} \tag{12}$$

or

$$\text{Area-normalized quantum detectivity} = \frac{A^{1/2}}{\text{NEQ}} = \left(\frac{\eta}{2\bar{N}_B}\right)^{1/2} \tag{13}$$

For the general case of a detector with area A seeing 2π sr of blackbody background at temperature T, $(\Delta N)^2$ is that in Eq. (4)

$$\overline{(\delta n)^2} = 2A\,\Delta f \int_0^\infty \eta(\nu)(\Delta N)^2 d\nu = \frac{4\pi A\,\Delta f}{c^2}\int_0^\infty \eta(\nu)\nu^2 \frac{e^{h\nu/kT}}{(e^{h\nu/kT}-1)^2}\,d\nu \tag{14}$$

Then for $\Delta f = 1$ Hz,

$$\text{NEP} = \frac{h\nu}{c\eta}\left[4\pi A \int_0^\infty \eta(\nu)\frac{\nu^2 e^{h\nu/kT}}{(e^{h\nu/kT}-1)^2}\,d\nu\right]^{1/2} \tag{15}$$

or

$$D^*(T, \lambda) = \frac{c\eta}{2\pi^{1/2}h\nu}\left[\int_0^{\infty} \eta(\nu)\frac{\nu^2 e^{h\nu/kT}}{(e^{h\nu/kT}-1)^2}\,d\nu\right]^{-1/2} \tag{16}$$

Assuming $\eta(\nu)$ is independent of frequency but falls back to zero for $\nu < \nu_c$

$$D^*(T, \lambda_c) = \frac{c\eta^{1/2}}{2\pi^{1/2}h\nu}\left[\int_{\nu_c}^{\infty} \frac{\nu^2 e^{h\nu/kT}}{(e^{h\nu/kT}-1)^2}\,d\nu\right]^{-1/2} \tag{17}$$

Figure 1 shows photon-noise-limited D^* vs. cutoff wavelength λ_c for various thermal-background temperatures (Jacobs and Sargent, 1970). Note that these curves are not independent. $D^*(T, \lambda_c)$ is related to $D^*(T, \lambda_c')$ by the formula

$$D^*(T, \lambda_c) = \left(\frac{T'}{T}\right)^{5/2} D^*(T, \lambda_c') \qquad \text{where } \lambda_c' = \frac{T}{T'}\lambda_c \tag{18}$$

This relation is useful for determining values of $D^*(T, \lambda_c)$, which do not appear in Fig. 1, in terms of a value $D^*(T', \lambda_c')$, which does appear. For example, to find $D^*(1000\text{ K}, 4\ \mu\text{m})$ from the 500-K curve

$$D^*(1000, 4) = \left(\frac{50}{1000}\right)^{5/2} D^*\left(500, 4 \times \frac{1000}{500}\right) = 2.3 \times 10^9 \text{ Jones} \tag{19}$$

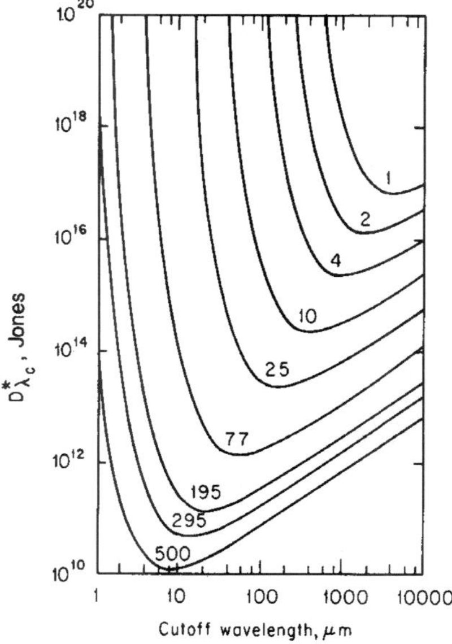

FIGURE 1 Photon-noise-limited D^* at peak wavelength assumed to be cut-off wavelength, for background temperatures 1, 2, 4, 10, 25, 77, 195, 295, and 500 K (assumes 2π FOV and $\eta = 1$). *[Reprinted from* Infrared Physics, *Vol. 10, S. F. Jacobs and M. Sargent, "Photon Noise Limited D* for Low Temperature Backgrounds and Long Wavelengths," pp. 233–235. Copyright 1990, with kind permission from Elsevier Science Ltd., The Boulevard, Langford Lane, Kidlington OX516 UK.]*

TABLE 1 D^* vs. λ_c for $T = 295$ K

λ_c, μm	$D^*(\lambda_c)$	λ_c, μm	$D^*(\lambda_c)$	λ_c, μm	$D^*(\lambda_c)$	λc, μm	$D^*(\lambda_c)$
1	2.19×10^{13}	10	5.35×10^{10}	100	1.67×10^{11}	1000	1.55×10^{12}
2	4.34×10^{13}	20	5.12×10^{10}	200	3.20×10^{11}	2000	3.10×10^{12}
3	1.64×10^{12}	30	6.29×10^{10}	300	4.74×10^{11}	3000	4.64×10^{12}
4	3.75×10^{11}	40	7.68×10^{10}	400	6.28×10^{11}	4000	6.19×10^{12}
5	1.70×10^{11}	50	9.13×10^{10}	500	7.82×10^{11}	5000	7.73×10^{12}
6	1.06×10^{11}	60	1.06×10^{11}	600	9.36×10^{11}	6000	9.28×10^{12}
7	7.93×10^{10}	70	1.21×10^{11}	700	1.09×10^{12}	7000	1.08×10^{13}
8	6.57×10^{10}	80	1.36×10^{11}	800	1.24×10^{12}	8000	1.24×10^{13}
9	5.80×10^{10}	90	1.52×10^{11}	900	1.40×10^{12}	9000	1.39×10^{13}

If higher accuracy is desired than can be determined from Fig. 1, one can use the preceding formula in combination with Table 1, which gives explicit values of $D^*(\lambda_c)$ vs. λ_c for $T = 295$ K.

The photon-noise-limited sensitivity shown in Figs. 1 and 2 applies to photovoltaic and photoemissive detectors. For photoconductors, recombination noise results in a $\sqrt{2}$ reduction in D^* at all wavelengths. Figure 3 shows the relative increase in photon-noise-limited D^* achievable by limiting the FOV through use of a cooled aperture.

Thermal Detectors. Limiting sensitivity of an ideal thermal detector has been discussed

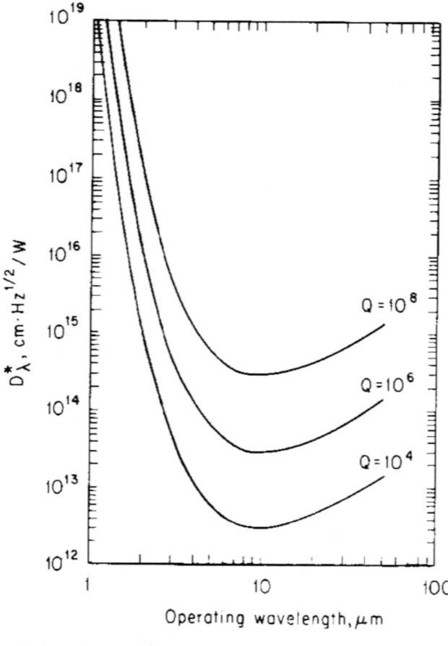

FIGURE 2 Photon noise limit of a narrowband quantum counter as a function of operating wavelength for a 290-K background, 2π FOV, and $\eta = 1$. *(From Kruse et al., 1962)*

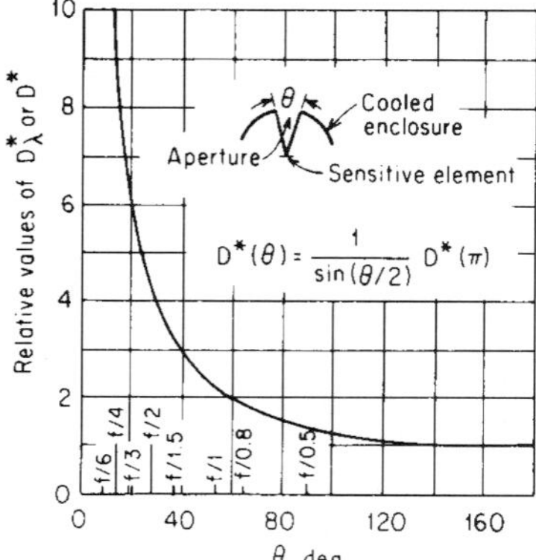

FIGURE 3 Relative increase in photon-noise-limited $D^*(\lambda_{pk})$ or D^* achieved by using a cooled aperture in front of lambertian detector. *(From Kruse et al., 1962)*

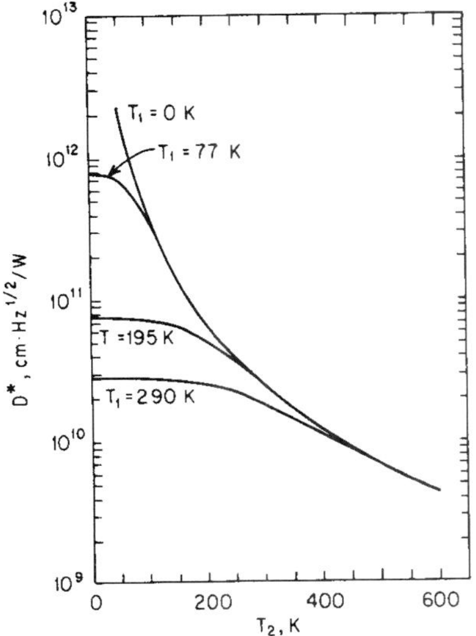

FIGURE 4 Photon-noise-limited D^* for thermal detectors as a function of detector temperature T_1 and background temperature T_2 (2π FOV: $\eta = 1$). *(From Kruse et al., 1962)*

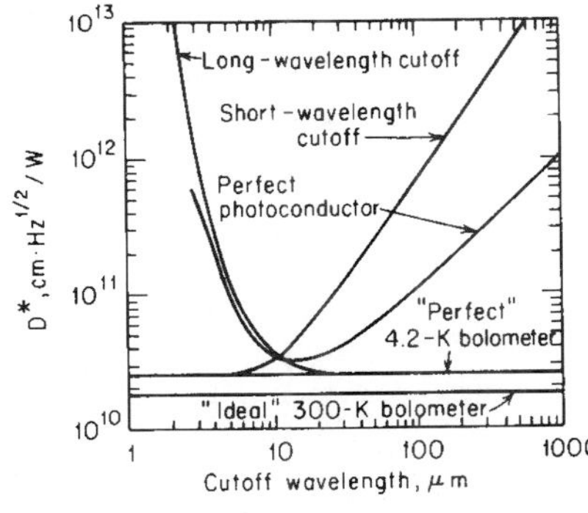

FIGURE 5 The detectivity of a "perfect" bolometer plotted as a function of both short- and long-wavelength cutoffs. Plots for a perfect photoconductor and two other cases are included for comparison. The background temperature is 300 K *(From Low and Hoffman, 1963.)*

previously (Smith et al., 1957; Felgett, 1949; Kruse et al., 1962). Assuming no shortwave or long-wavelength cutoffs exist,

$$D^* = \frac{\varepsilon}{[8\varepsilon\sigma k(T_1^5 + T_2^5)]^{1/2}} = \frac{4.0 \times 10^{16}\varepsilon^{1/2}}{(T_1^5 + T_2^5)^{1/2}}\,\text{Jones} \tag{20}$$

where T_1 = detector temperature
T_2 = background temperature
ε = detector emissivity
σ = Stefan-Boltzmann constant
k = Boltzmann constant

D^* vs. T_2 is plotted for various T_1 in Fig. 4. Figure 5 shows the effect of both short- and long-wavelength cutoffs on bolometer sensitivity (Low and Hoffman, 1963), with the ideal photoconductor curve for reference.

15.6 OTHER PERFORMANCE PARAMETERS

Spectral Response

Spectral response provides key information regarding how the detector will respond as a function of wavelength or photon energy. Spectral response may be limited by the intrinsic detector material properties, a coating on the detector, or by a window through which the radiation must pass. Relative response is the spectral response ratioed against a detector

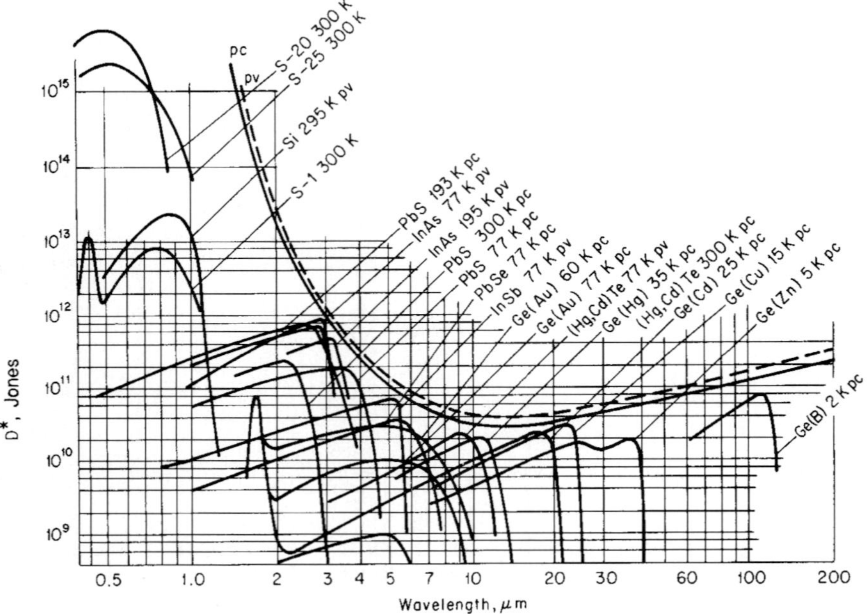

FIGURE 6 D^* vs. λ for selected detectors.

with a nominally wavelength independent response, such as a thermocouple having a spectrally broad black coating. Relative response is plotted as a function of wavelength with either a vertical scale of W^{-1} or photon^{-1}. Thermal detectors tend to be spectrally flat in the first case while quantum detectors are generally flat in the second case. The curves are typically shown with the peak value of the spectral response normalized to a value of 1. The spectral response curve can be used together with the blackbody D^* to calculate D^* as a function of wavelength, which is shown in Fig. 6 for selected detectors.

Responsivity and Quantum Efficiency

Responsivity and quantum efficiency are important figures of merit relating to the detector signal output. Responsivity is a measure of the transfer function between the input signal photon power or flux and the detector electrical signal output. Thermal detectors will typically give this responsivity in volts/watt. Photoconductors will usually quote the same units, but will also frequently reference the value to the peak value of relative response per watt from the spectral response curve. This value is actually realized if the detector bias circuit load resistor is significantly larger than the detector resistance. Photoconductor responsivity is given by:

$$\text{Responsivity}_{\text{peak}} = \frac{\eta \mathbf{q} \mathbf{R_d} \mathbf{E} \tau (\mu_n + \mu_p)}{h \nu \ell} \tag{21}$$

where η is the quantum efficiency, $\mathbf{q}$ is the electronic charge, $\mathbf{R_d}$ is the detector resistance, $\mathbf{E}$ is the electric field, τ is time constant, μ are the mobilities for electrons (n) and holes

(p), $h\nu$ is the photon energy, and ℓ is the device length. Photomultiplier tubes and photovoltaic detectors will usually reference the responsivity in amps per watt, again referenced to peak spectral response.

Detector response performance is also conveyed from the detector quantum efficiency. In the case of photovoltaic detectors which, in the absence of avalanche operation have a gain of unity, quantum efficiency is essentially the current per photon. For a blip photovoltaic detector, the quantum efficiency also determines the D^*. Quantum efficiency is not readily measured for photoconductors and photomultiplier tubes unless the internal gain is carefully calibrated. It is sometimes inferred from the measured D^* for photoconductive devices which are blip (see definition of detective quantum efficiency).

Noise, Impedance, Dark and Leakage Current

Noise has a number of potential origins. Background photon flux limited detectors have noise dominated by the square root of the number of background photons striking the detector per second [see Eq. (9)]. Other noise sources may contribute or dominate. Among these are:

- Johnson, Nyquist, or thermal noise which is defined by the detector temperature and impedance
- modulation or $1/f$ noise which may dominate at lower frequencies
- amplifier noise
- shot noise from dark or leakage current

Thus, the impedance of a photodiode may limit performance, depending upon the detector operating conditions. Fig. 7 illustrates the diode impedance per unit area ($\mathbf{R}_0A$)

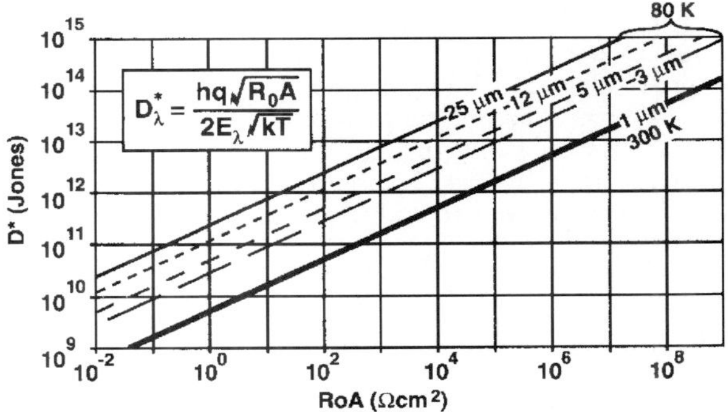

FIGURE 7 Zero-bias impedance-area product ($\mathbf{R}_0A$ or shunt resistance per unit area) of a photodiode can limit the $\mathbf{D}^*$ as shown. The limiting D^* depends on $\mathbf{R}_0A$, temperature and photon energy or wavelength. Examples are illustrated for a 1-μm diode at 300 K, and for 3-, 5-, 12-, and 25-μm devices at 80 K. Other noise mechanisms, such as photon noise, typically limit D^* to lower values than shown here.

limiting value of D^* for silicon detectors at room temperature and longer-wavelength infrared photodiodes at 80 K.

Measurement of the noise as a function of frequency can be valuable for characterizing the relevant noise sources. Selection of an appropriate preamplifier is also critical, particularly for detectors having very low or very high impedance. Integration of preamplifiers together with detectors has significantly improved the overall performance of many detectors. The use of phase-sensitive lock-in amplifiers in combination with a modulated signal can also improve the signal-to-noise ratio.

Uniformity

One cannot assume that the response of a detector will be uniform across its sensitive area. Material inhomogeneity and defects and/or fabrication variables can give rise to nonuniformity. Lateral collection from near the perimeter of a photodiode may give a gradual response decrease away from the edge (this effect will typically be accompanied by a change in response speed as well). Recombination at the electrical contacts to a photoconductor can limit the lifetime, and hence the photoconductive gain, for carriers generated near the contact, a phenomenon called sweep-out. Recombination may be enhanced at surfaces and edges also. Laser spot scanning is useful to check the detector spatial uniformity, although laser sources may not be readily available at all the wavelengths of interest. An alternative method is to move the detector around under a fixed small aperture in conjunction with a light source.

Speed

Detector response speed is often related inversely to detector sensitivity. Thermal detectors often show this characteristic because the signal is proportional to the inverse of response speed, while the noise is amplifier- or Johnson-limited. Excluding detectors with internal carrier multiplication mechanisms, the best detectors from broad experience seem limited to a D^*f^* product of a few times 10^{17} Jones Hz. D^*f^* may be proportionally higher for devices with gain, since speed can be increased to a greater extent by using a lower value of load resistance without becoming Johnson-noise-limited. The user should be aware that with many detectors it is possible to operate them in a circuit to maximize sensitivity or speed, but not both at the same time. Speed may vary across the sensitive area of the detector and with temperature, wavelength, and electrical bias.

Stability

Detector performance may change or drift with time. Changes in operating temperature, humidity, and exposure to elevated temperatures as well as to visible, ultraviolet, and high-energy radiation can affect device operation. These effects arise from the temperature dependence of electronic properties in solids, as well as from the critical role played by electrical charge conditions near the surface of many device types. Sensitivity changes in a sample of silicon detectors from four vendors illustrate this point. Wide variations in

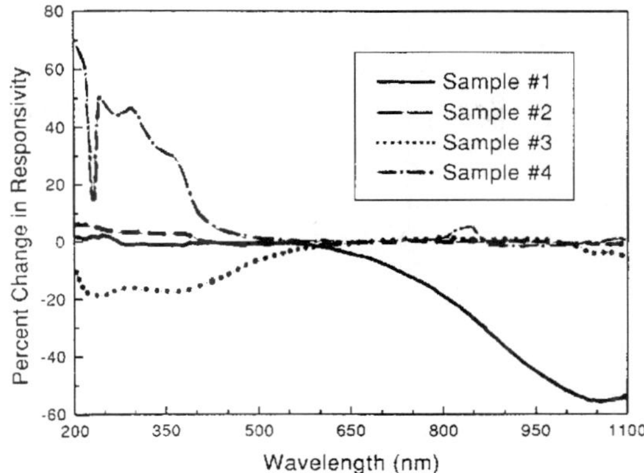

FIGURE 8 After only three hours of UV irradiation, detectors showed great variations in responsivity. *(Reprinted from the September 1993 issue of* Photonics Spectra, *© Laurin Publishing Co., Inc.)*

responsivity change after UV exposure, as shown in Fig. 8. In applications where stability is of significant concern, these effects must be carefully reviewed along with the detector supplier's experience in these matters.

15.7 DETECTOR PERFORMANCE

Manufacturers' Specifications

Detector sensitivity can be the determining factor in the design and performance of a sensor system. Detector performance is subject to the development of improved materials, fabrication techniques and the ingenuity of device engineers and inventors. The descriptions given here may improve with time, and consultation with manufacturers and users is recommended.

Thermocouple. The thermocouple offers broad uniform spectral response, a high degree of stability, and moderate sensitivity. Its slow response and relative fragility have limited its use to laboratory instruments, such as spectrometers.

Compared with thermistors, thermocouples are slower, require no bias, and have higher stability but much lower impedance and responsivity. This increases the amplification required for the thermocouple; however, the only voltage appearing is the signal voltage, so that the serious thermistor problem of bridge-circuit bias fluctuations is avoided. With proper design, performance should not be amplifier-limited but limited instead by the Johnson noise of the thermocouple. Thermocouples perform stably in dc operation, although the instability of dc amplifiers usually favors ac operation.

The inherent dc stability of thermocouples is attractive for applications requiring no

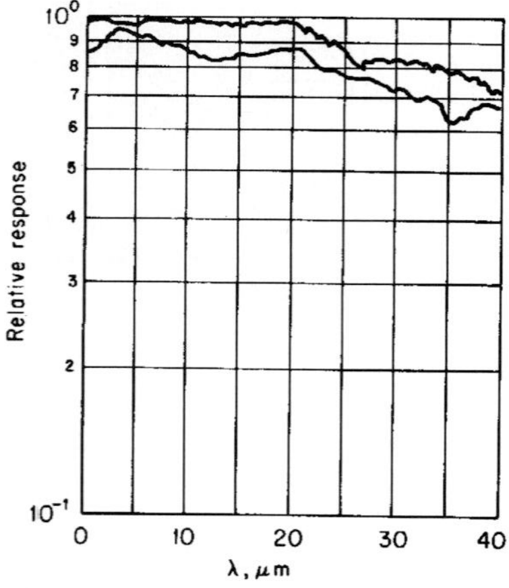

FIGURE 9 Typical thermocouple spectral-response curves (CsI window) (two different manufacturers).

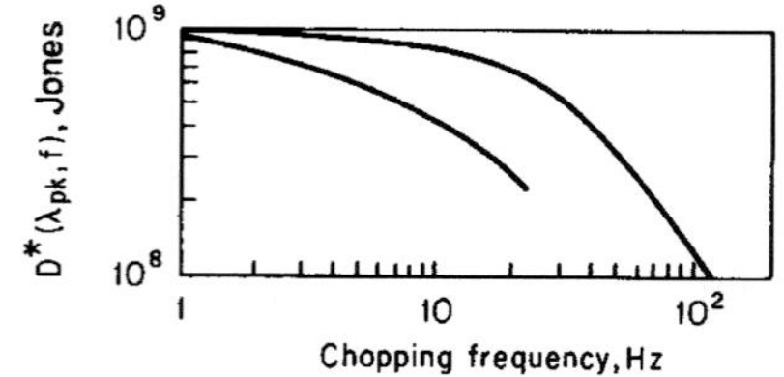

FIGURE 10 Typical thermocouple D^* (noise) frequency spectra (two manufacturers). *(From Wolfe, 1965.)*

moving parts, and recently a relatively rugged solid backed evaporated thermocouple has been developed whose sensitivity approaches that of the thermistor bolometer.

Sensitivity: $D^* = 1 \times 10^9$ Jones for 20-ms response time; spectral response depends on black coating (usually gold black) (see Fig. 9).

Noise: White Johnson noise, falling off with responsivity (see Fig. 10).

Resistance: 5–15 Ω typical.

Responsivity: 5 V/W (typical), 20–25 V/W (selected).

Time constant: 10–20 ms (typical).

Operating temperature: Normally ambient.

Sensitive area: 0.1×1–0.3×3 or 0.6×2 mm (typical).

Linearity: <0.1 percent in region investigated (6×10^{-10}–6×10^{-8} W incident).

Recommended circuit: Transformer coupled into low-noise (FET) amplifier with good low-frequency noise characteristics.

Manufacturers: Perkin-Elmer, Charles Reeder, Beckman Instruments, Farrand, Eppley Laboratory, Barnes Engineering.

Thermopile. Thermopiles are made by evaporating an array of metal junctions, such as chromel-constantan or manganin-constantan, onto a substrate. The thin-film construction is rugged, but the Coblentz-type may be quite delicate. Wire-wound thermopile arrays are also available which are very robust. Devices with arrays of semiconductor silicon junctions are also available. The array may typically be round, square, or rectangular (for matching a spectrometer slit) and consist of 10 to 100 junctions. Configuration options include matched pairs of junction arrays or compensated arrays to provide an unilluminated reference element. A black coating, such as 3M black or lampblack, is used to provide

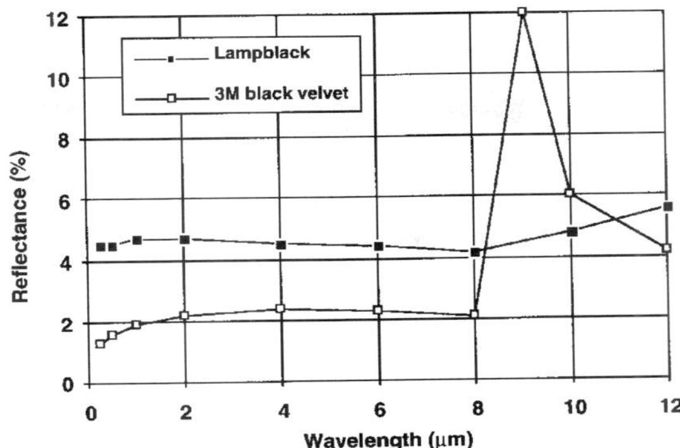

FIGURE 11 Spectral reflectance of two black coatings used in the construction of thermopile detectors. *(From Eppley Laboratory studies.)*

high absorption over a broad spectral range, as illustrated in Figs. 11 and 12. The housing window may limit the spectral range of sensitivity. Typical applications include power meters and radiometers.

Sensitivity: $D^* \approx 0.5$ to 4×10^8 Jones for 30-ms typical response time. D^* may be dependent upon sensitive area; spectral response depends on black coating and window (see Fig. 11 and Fig. 12).

Noise: White Johnson noise, falling off with responsivity, typical range is 5–30 nV/Hz$^{1/2}$.

Resistance: 2 Ω–60 kΩ typical.

Responsivity: 4–250 V/W (typical) depends on the number of junctions and time constant.

Time constant: 10 ms–2 s (typical).

Operating temperature: Normally ambient.

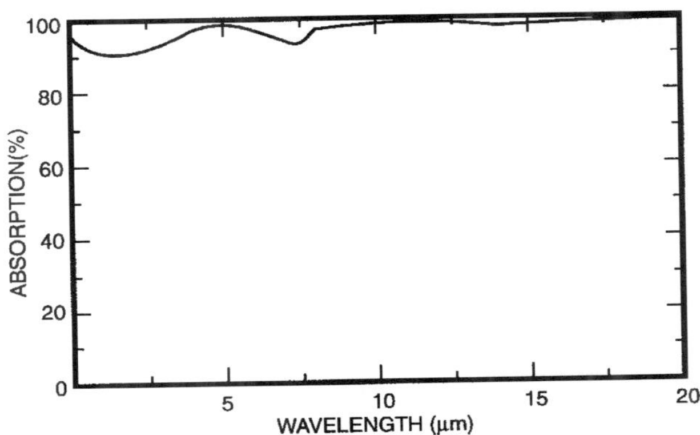

FIGURE 12 Spectral absorption of a thermopile detector coating made from a black metal oxide. *(Oriel Corporation.)*

Sensitive area: 0.5–6-mm dia., 0.025×0.025–3×3 mm, various rectangular, 0.4×3, 0.6×2, 0.6×4 mm (typical).

Recommended circuit: Low noise (0.5 μV p-p, dc to 1 Hz), low drift with voltage gain of ≈1000 and input impedance of 1 MΩ.

Manufacturers: Armtech, Beckman Instruments, Concept Engineering, Dexter Research Center, Edinburgh Instruments, EDO/Barnes Engineering, Eppley Laboratory, Farrand, Gentech, Molectron Detector, Ophir Optronics, Oriel, Scientech, Scitec, Swan Associates.

Thermistor Bolometer. Thermistors offer reliability, moderate sensitivity, and broad spectral coverage without cooling. Construction is rugged and highly resistant to vibration, shock, and other extreme environments. Response is slower than 1 ms, and tradeoff exists between speed and sensitivity.

Thermistor elements are made of polycrystalline Mn, Ni, and Co oxides. In their final form they are semiconductor flakes 10 μm thick, which undergo a temperature resistance change of ~4 percent/K. Since thermistor resistance changes with ambient temperature enough to alter the biasing significantly, it is usually operated in a bridge circuit, with a nearly identical thermistor shielded from signal radiation and used for a balance resistor.

Sensitivity:

$$\text{NEP} = 8.9 \times 10^{-10} \sqrt{\frac{A(\text{mm}^2)}{\tau_{\text{rms}}}}\ \text{W}$$

$$D^* = 1.1 \times 10^9 \sqrt{\tau_{\text{rms}}}\ \text{Jones}$$

Spectral response: Depends on coating (usually Zapon lacquer); see Fig. 13.

Quantum efficiency: Depends on blackening coating, typically 80 percent.

Noise: Thermal-noise-limited above ~20 Hz ($V_{\text{noise}} = \sqrt{4kTR\Delta f}$); below that, $1/f$ type noise (see Fig. 14). Used in balanced-bridge circuit (two flakes in parallel); limiting noise due to thermal noise in *both* flakes.

Resistance: For standard 10-μm-thick flakes, two different resistivities are available: 2.5 MΩ/sq or 250 kΩ/sq. Note that in a bolometer bridge, the resistance between the output connection and ground is half that of single flake.

Time constant: τ is 1–20-ms standard for nonimmersed detectors and 2–10-ms standard for immersed detectors.

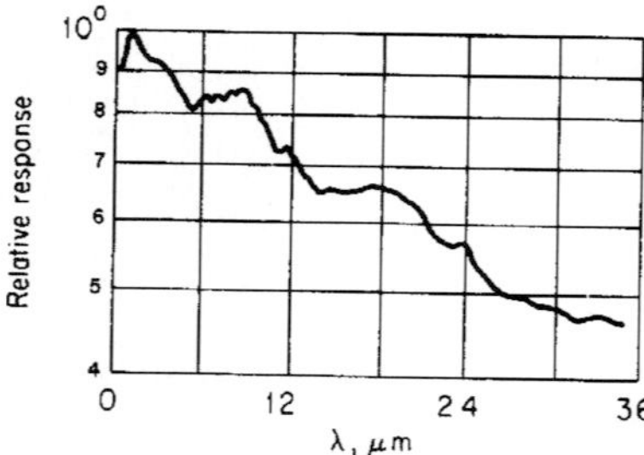

FIGURE 13 Typical thermistor spectral response (no window).

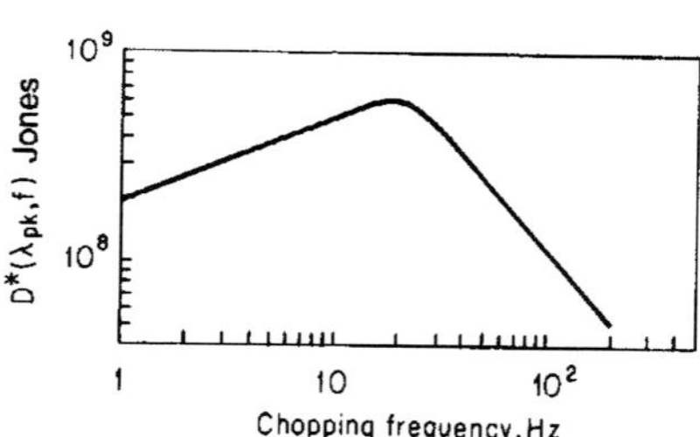

FIGURE 14 Typical thermistor D^* (noise) frequency spectrum. *(From Wolfe, 1965.)*

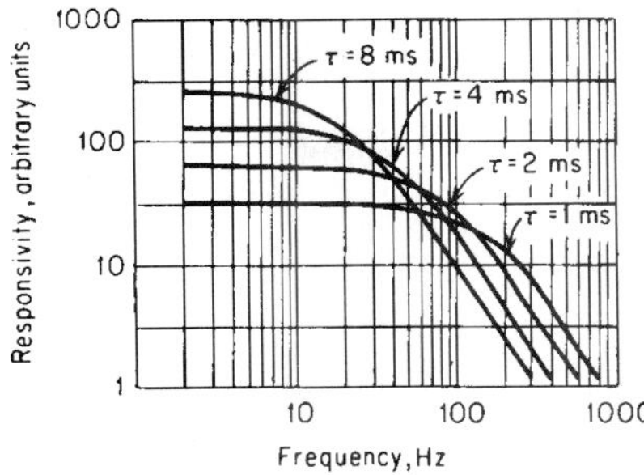

FIGURE 15 Relative responsivity vs. frequency for various time-constant thermistor detectors ($A = 1 \times 1$ mm). *(From Barnes Engineering, Bull. 2-100.)*

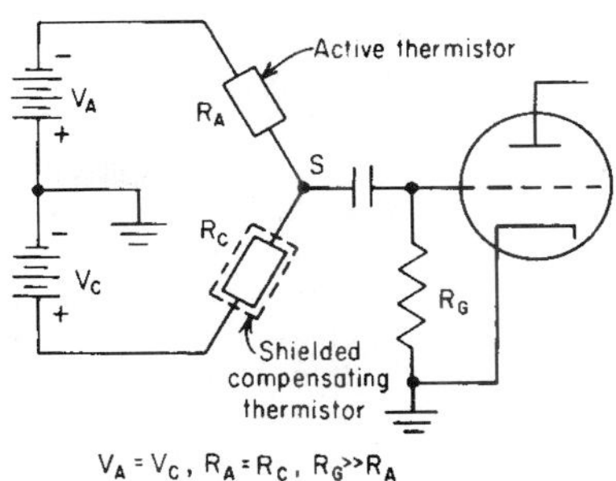

FIGURE 16 Bolometer electrical circuit. *(From Barnes Engineering, Bull. 2-100.)*

Sensitive area: 0.1×0.1 mm, 5×5-mm standard.

Operating temperature: Normally ambient, 285–370 K.

Responsivity: Depends on bias, resistance, area, and time constant $\mathcal{R} \propto \sqrt{R\tau/A} \approx 10^3$ V/W for 0.1×0.1-mm area, 250-kΩ resistance, and $\tau = 4$ ms (see Fig. 15 for frequency-time-constant dependence with given area). Output voltage (responsivity) can be increased to a limited degree by raising bias voltage. Figure 16 shows the deviation from Ohm's law due to heating. Bias should be held below 60 percent of peak voltage. Listed responsivity is that of active flake. In the bridge circuit, responsivity is half this value.

Sensitivity profile: Approximately 10 percent for 10-μm scan diameter over a 1×1-mm cell.

Linearity: ±5 percent 10^{-6}–10^{-1} W/cm^2.

Recommended circuit: See Figs. 16 and 17.

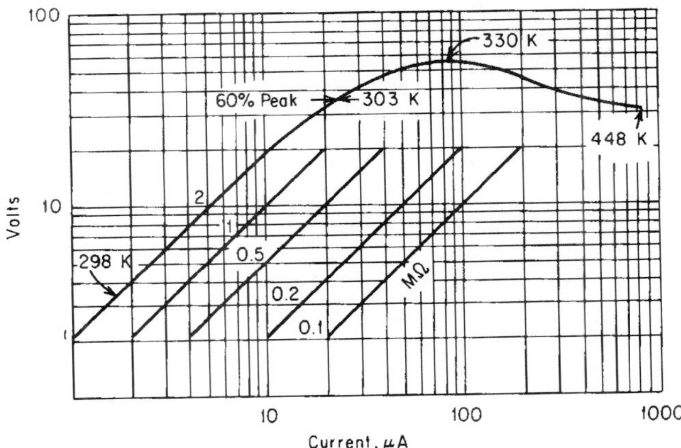

FIGURE 17 Thermistor voltage-current characteristics, showing typical flake temperatures under different conditions. *(From Barnes Engineering, Bull. 2-100.)*

Manufacturers: EDO/Barnes Engineering, Servo Corporation of America, Thermometrics.

Pyroelectric. Lithium tantalate ($LiTaO_3$), triglycine sulfate (TGS), and other pyroelectric materials provide an uncooled thermal detector with good sensitivity. The devices are capable of fast response, limited inversely by the preamplifier feedback resistance, but with responsivity and D^* traded for speed. Its principle of operation is the pyroelectric effect, which is the change of electric polarization with temperature. It offers rugged construction and absence of $1/f$ noise because no bias is involved.

Lack of $1/f$ noise, combined with the ability to easily trade off speed and sensitivity, makes pyroelectric detectors useful for scanning applications and energy measurement of pulsed optical sources. In addition, the NEP is independent of area at low frequencies (<10 Hz), so that these detectors are useful for large-area applications (preamplifier $1/f$ noise may limit, however). Pyroelectric detectors are useful for calorimetry since the pyroelectric effect is an integrated volume effect and the output signal is unaffected by spatial or temporal distribution of the radiation, up to damage threshold or depolarizing temperature. For higher damage thresholds, lead zirconate titanate ceramic (Clevite PZT-5) exhibits a much smaller pyroelectric effect than TGS, but its high Curie temperature of 638 K makes it more useful than TGS for high-energy applications.

Sensitivity: Sensitive from ultraviolet to millimeter wavelengths. For $\lambda < 2\ \mu m$, TGS must be blackened, which slows response. Normally ($\lambda > 2\ \mu m$) a transparent electrode is used, since TGS absorption is high from 2 to 300 μm. Beyond 300 μm, poor absorption and increased reflectivity reduce sensitivity. Spectral response depends largely on coating. See Fig. 18 for spectral response with modified 3M black. Fig. 19 illustrates the relative spectral response for a $LiTaO_3$ device. D^* is independent of A at low frequencies (<10 Hz) (see Figs. 20 and 21). Figure 22 shows NEP vs. A for various frequencies.

Quantum efficiency: Depends on coating absorptivity (for 3M black typically $\eta >$ 75 percent).

Noise: (See Figs. 20 and 21). Limited by loss-tangent noise up to frequencies where limited by amplifier short-circuit noise (see Fig. 23).

Operating temperature: Ambient, up to 315 K. Can be repolarized if $T > T_{curie} = 322$ K for TGS. Irreversible damage at $T = 394$ K (see Fig. 24). Other pyroelectric materials have significantly higher Curie temperatures (398–883 K).

Output impedance: 50 Ω–10 KΩ, set by built-in amplifier (see Fig. 25).

Responsivity: See Figs. 20–22, 24, and 26.

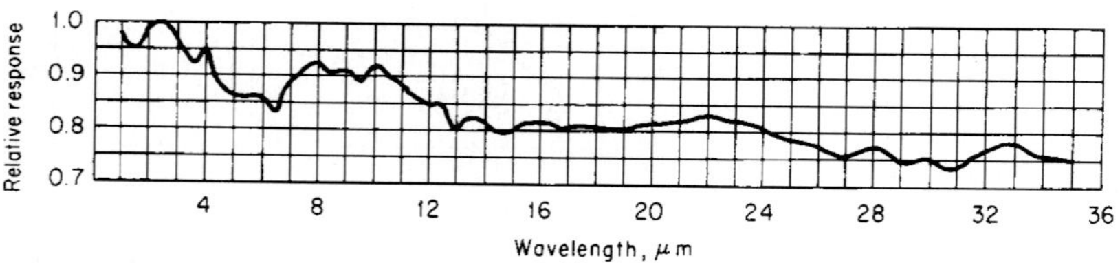

FIGURE 18 Relative spectral response of TGS detectors with modified 3M black. *(From Barnes Engineering, Bull. 2-200A.)*

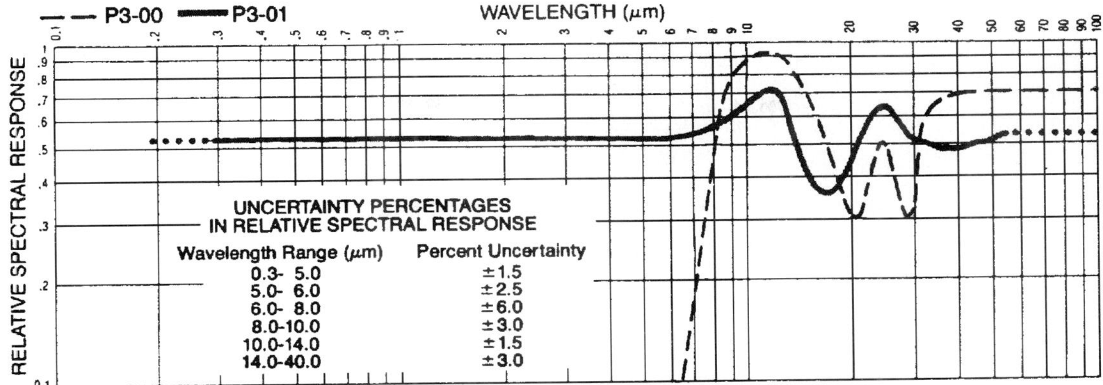

FIGURE 19 Relative spectral response of $LiTaO_3$ pyroelectric detectors showing both a black spectral coating and an optional coating tuned to the 8–14-µm LWIR band. *(From Molectron Detector, Inc.)*

Capacitance: 5 pF for 0.5 × 0.5 mm; 20 pF for 1 × 1 mm; 100 pF for 5 × 5 mm.

Sensitive area: 2–50-mm dia. round, 0.5 × 0.5-mm–10 × 10-mm square, typical.

Time constant: Not pertinent, response speed set by the preamplifier feedback resistor (see Fig. 26).

Linearity: 5 percent between 10^{-6} and 10^{-1} W/cm^2.

Sensitivity profile: Depends on coating or transparent electrode; 5–7 percent across 12 × 12-mm; spot size < 250 µm.

Recommended circuit: See Figs. 25 and 27. FET amplification stage usually built in.

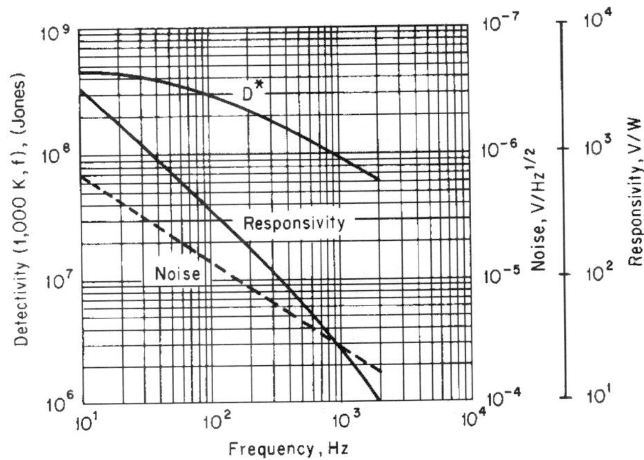

FIGURE 20 Typical D^*, responsivity, and noise vs. frequency for TGS ($A = 1 \times 1$ mm; $T = 296$ K.) *(From Barnes Engineering, Bull. 2-220A.)*

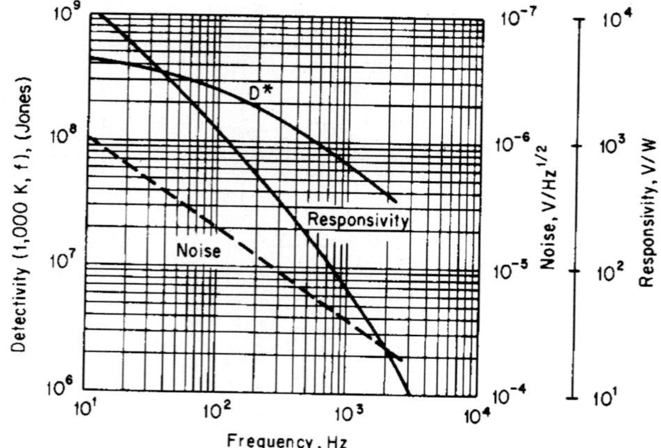

FIGURE 21 Typical D^*, responsivity, and noise vs. frequency ($A = 0.4 \times 0.4$ mm; $T = 296$ K). *(From Barnes Engineering, Bull. 2-220A.)*

Since $\mathcal{R} \propto 1/f$, use of an amplifier with $1/f$ noise and gain $\propto f$ is recommended. Then output signal and signal-to-noise ratio are independent of frequency.

Manufacturers: Alrad Instruments, Belov Technology, CSK Optronics, Delta Developments, EDO/Barnes Engineering, EG&G Heimann, Electro-Optical Systems, Eltec, Gentec, Graseby, International Light, Laser Precision, Molectron Detector, Oriel, Phillips Infrared Defence Components, Sensor-Physics, Servo Corporation of America, Spiricon, Thermometrics.

InSb Hot-Electron Bolometer At temperatures of liquid helium and lower, free carriers in indium antimonide (InSb) can absorb radiation in the far-infrared and submillimeter

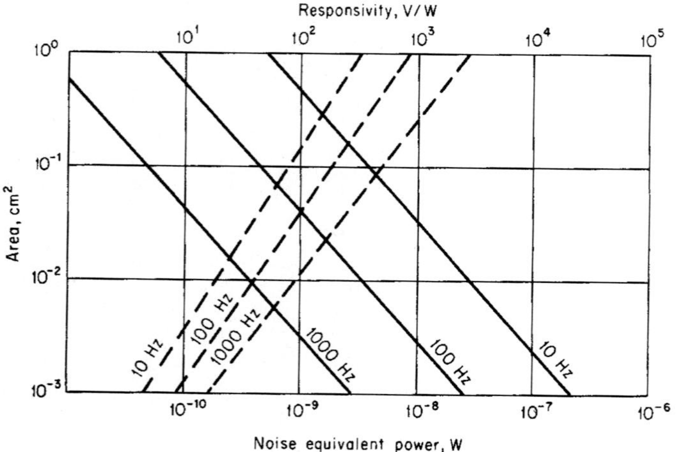

FIGURE 22 Noise equivalent power in watts (broken lines) and responsivity (heavy lines) vs. TGS detector area for various frequencies. *(From Barnes Engineering, Bull. 2-220B.)*

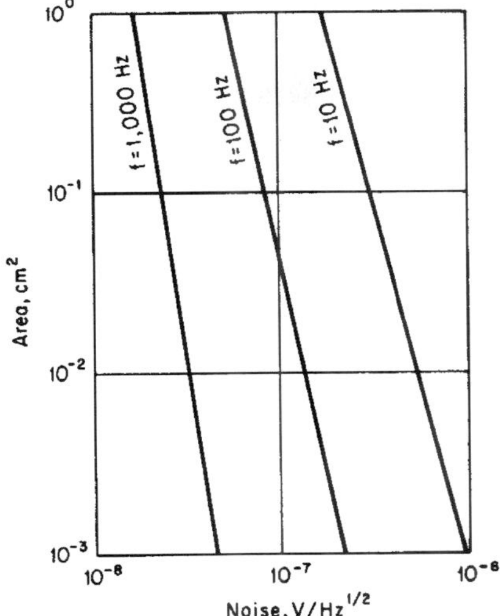

FIGURE 23 TGS noise vs. detector area for various operating frequencies. *(From Barnes Engineering, Bull. 2-220B.)*

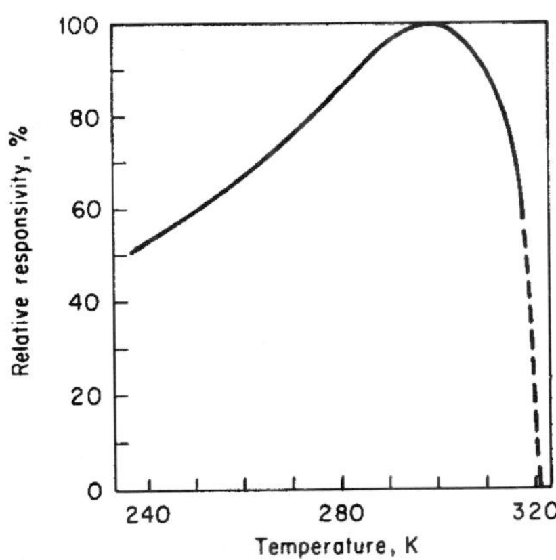

FIGURE 24 Relative responsivity vs. temperature for TGS. *(From Barnes Engineering, Bull. 2-220B.)*

spectral region. Because the mobility of the electrons varies as $T_e^{3/2}$ under these conditions, the conductivity of the material is modulated. This mechanism offers submicrosecond response and broad far-infrared coverage out to millimeter wavelengths but requires liquid-helium cooling and very sophisticated receiver design.

Technically, these devices may be classed as bolometers, since incident radiation power produces a heating effect which causes a change in free-charge mobility. In the normal bolometer, the crystal lattice absorbs energy and transfers it to the free carriers via collisions. However, in InSb bolometers incident radiation power is absorbed directly by free carriers, the crystal lattice temperature remaining essentially constant. Hence the name *electron bolometer.* Note that this mechanism differs from photoconductivity in that free-electron mobility rather than electron number is altered by incident light (hence there is no photoconductive gain).

Sensitivity: $D^*(2\text{ mm}, 900) = 4 \times 10^{11}$ Jones (see Fig. 28).

Noise: See Figs. 29 and 30.

Responsivity: ~1000 V/W.

Time constant: $\tau = 250$ ns.

Sensitive area: 5×5 mm typical.

Operating temperature: ~1.5–4.2 K.

Impedance: Without bias, 200 Ω; optimum bias, 150 Ω, depends on bias (see Fig. 29).

Recommended circuit: Optimum bias 0.5 mA (see Fig. 31).

Manufacturer: Infrared Laboratories.

Ge(Ga) Low-temperature Bolometer. The Ge(Ga) bolometer offers very high sensitivity and broad spectral coverage in the region 1.7 to 2000 μm. Liquid-helium cooling is

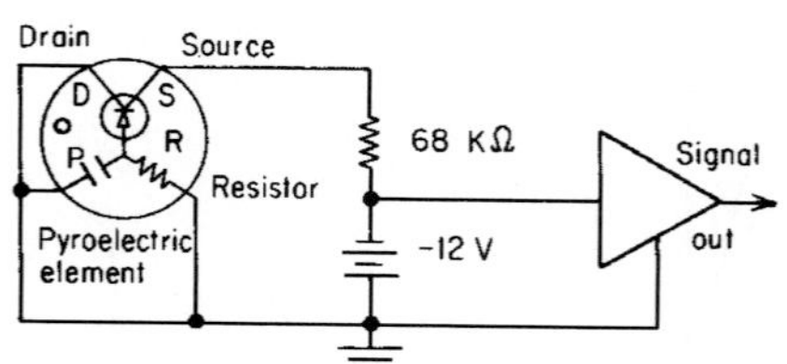

FIGURE 25 Pyroelectric detector-amplifier circuit. *(From Barnes Engineering, Bull. 2-220A.)*

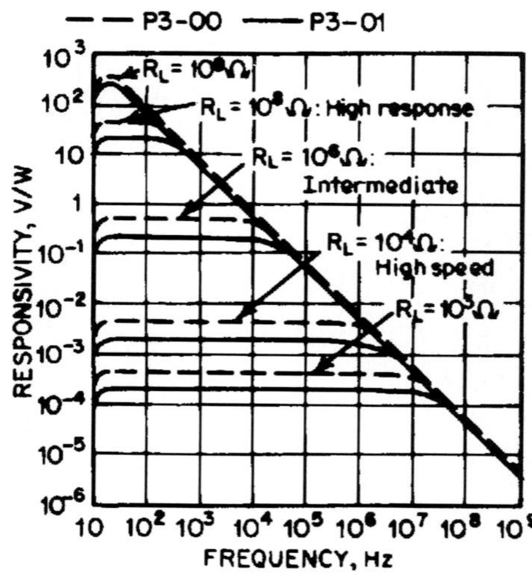

FIGURE 26 Responsivity vs. frequency for two pyroelectric detector models using various feedback resistors. *(From Molectron Detector, Inc.)*

required. A tradeoff exists between response time (seconds) and sensitivity (10^{-14} W NEP). Operation at 1000 Hz can be achieved still maintaining 2×10^{-13} W NEP.

Sensitivity: Depends on thermal conductance G (see Figs. 32 and 33). NEP(λ, 10 Hz) $= V_n/S = 3 \times 10^{-14}$ W for $G = 1\ \mu$W/K; $A = 1$ mm^2, $D^*(\lambda, 10\text{ Hz}) = 3 \times 10^{13}$ Jones ($\mathbf{Q} < 0.2\ \mu$W). For this detector, NEP $\approx 4T(kG)^{1/2}$ and does not vary with $A^{1/2}$(T is heat-sink temperature, and k is Boltzmann's constant). Thus D^* cannot be used as a valid means of comparison with other detectors; 300-K background-limited performance is achievable for 2π FOV when the bolometer is operated at 4.25 K with $G = 10^{-3}$ W/K. (For $A = 0.1$ cm^2, the time constant is 50 s.)

Responsivity: Typically, responsivity $= 2.5 \times 10^5$ V/W $= 0.7(\mathbf{R}/TG)^{1/2}$, where $\mathbf{R} =$ resistance. Responsivity, and hence NEP, depends on thermal conductance G, which in turn is set by background power. G ranges from 0.4 to 1000 μW/K.

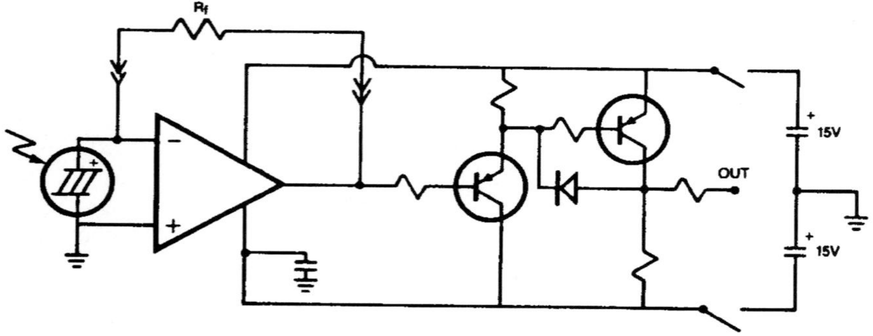

FIGURE 27 Resistive feedback circuit used with $LiTaO_3$ detectors. *(From Molectron Detector, Inc.)*

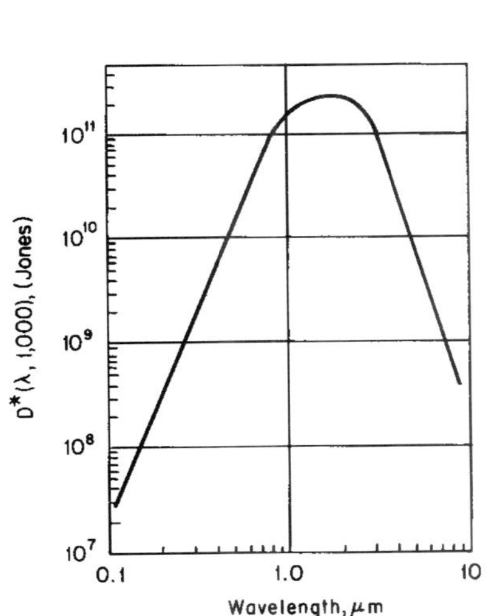

FIGURE 28 D^* vs. λ for InSb electron bolometer ($H = 0$). *(From Raytheon, IR Millimeter Wave Detector, 1967.)*

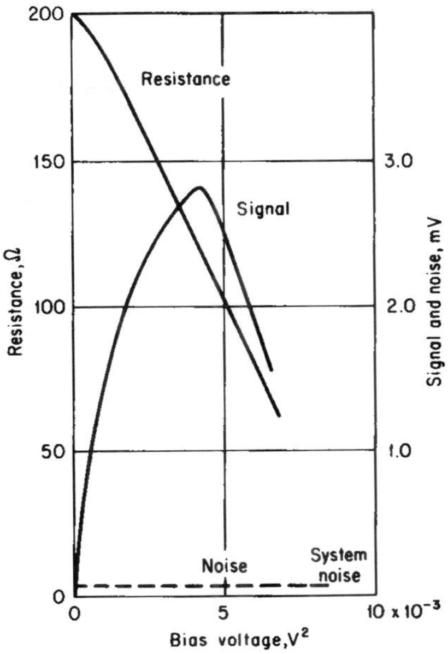

FIGURE 29 InSb electron bolometer, typical resistance, signal, and noise vs. bias voltage squared ($T = 5$ K; $\mathbf{R}_L = 200\ \Omega$; gain $= 2.4 \times 10^4$; $F = 1100$ Hz). *(From Santa Barbara Research Center, Prelim. Res. Rep., 1967.)*

Thermal conductance: Typically $G = 1\ \mu\text{W/K}$ for background $\mathbf{Q} < 0.2\ \mu\text{W}$; note that $\mathbf{Q} < 1/2$ (optimum bias power P).

Sensitive area: 0.25×0.25–10×10 mm.

Resistance: 0.5 MΩ.

Operating temperature: 2 K (see Fig. 32). In applications where radiation noise can be

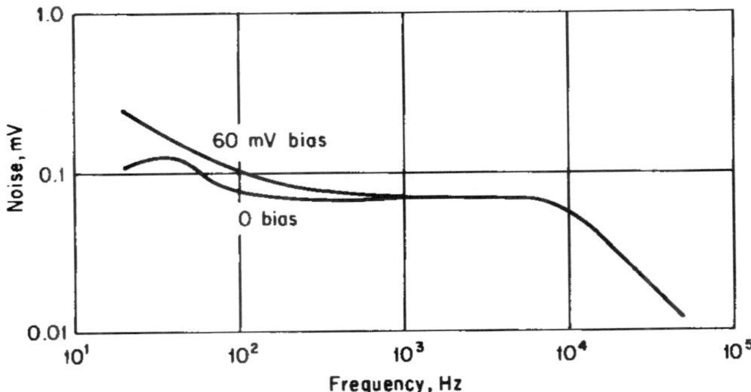

FIGURE 30 InSb electron bolometer, typical noise spectrum ($T = 5$ K, $\mathbf{R}_L = 200\ \Omega$; gain $= 2.4 \times 10^4$, $\Delta f = 5.6$ Hz). *(From Santa Barbara Research Center, Prelim. Res. Rep., 1967.)*

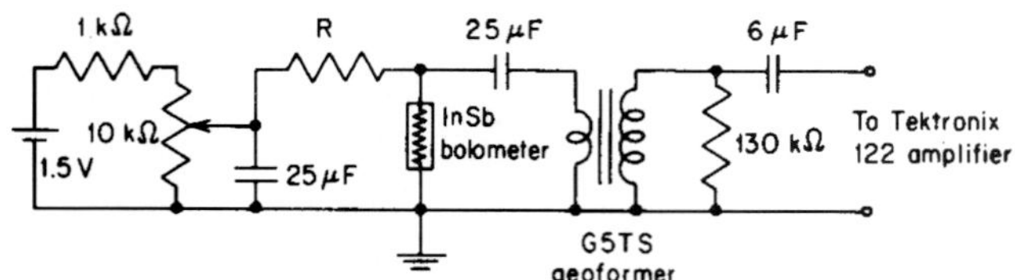

FIGURE 31 Electron-bolometer biasing circuit. *(From Santa Barbara Research Center, Prelim. Res. Rep., 1967.)*

eliminated there is much to be gained by operating at the lowest possible temperature. Figure 33 shows the theoretical NEP and time constant at 0.5 K, assuming that current noise remains unimportant.

Quantum efficiency: Depends on blackened coating and window. For $\lambda < 100\ \mu\text{m}$, absorptivity exceeds 95 percent. For $\lambda > 100\ \mu\text{m}$, efficiency varies with geometry.

Time constant: Response time constant is proportional to G^{-1}. Therefore, if G must be increased to accomodate larger background, the time constant is decreased proportionally. Responsivity and NEP, however, are degraded as $G^{-1/2}$.

Noise: $v_n = 1 \times 10^{-8}\ \text{V/Hz}^{1/2}$; thermal noise is due to $\mathbf{R}$ and $\mathbf{R}_L$.

Recommended circuit: Standard photoconductive circuit, with load resistor, grid resistor, and blocking capacitor at low temperature (see Fig. 34). See Fig. 35 for typical electrical characteristics. Bias power $P = 0.1TG$.

Manufacturer: Infrared Laboratories, Inc.

Photoemissive Detectors. Photoemissive detectors are generally the detector of choice in the UV, visible, and near-IR where high quantum efficiency is available. In the spectral region $\lambda < 600$ nm, the photomultiplier, or multiplier phototube, has close to ideal

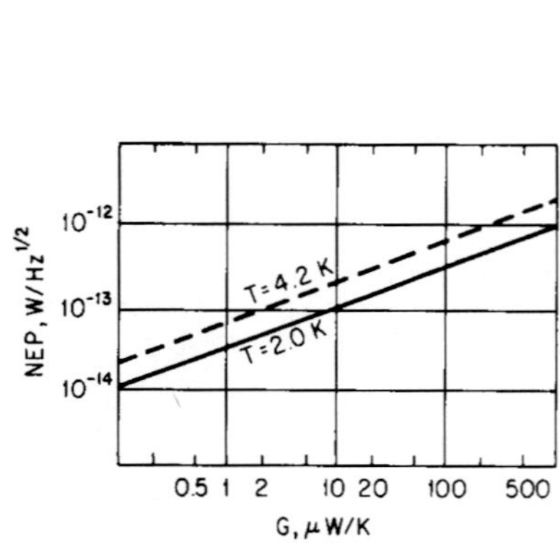

FIGURE 32 Germanium bolometer NEP vs. conductance. *(IR Labs, Inc.)*

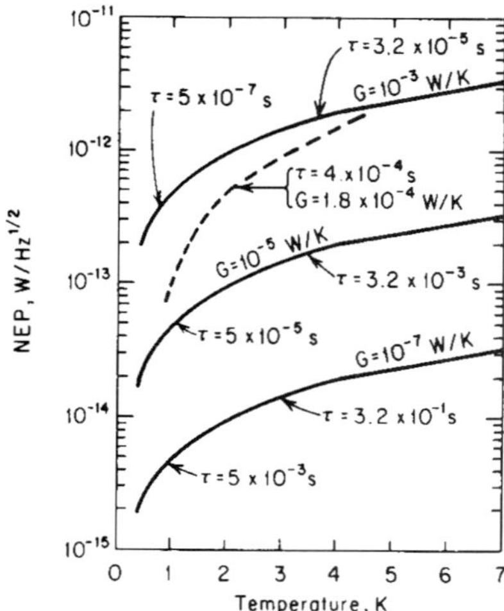

FIGURE 33 Germanium bolometer NEP vs. temperature. Solid curves are theoretical: NEP $\approx 4T(kG)^{1/2}$; dashed curve is measured value for a typical bolometer. *(From Low, 1961.)*

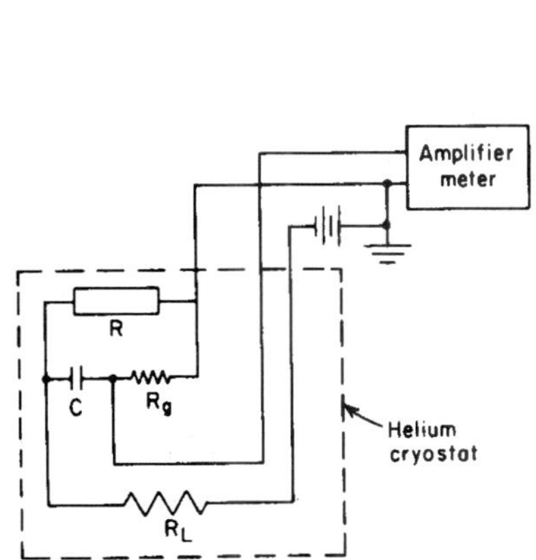

FIGURE 34 Germanium bolometer circuit and cryostat.

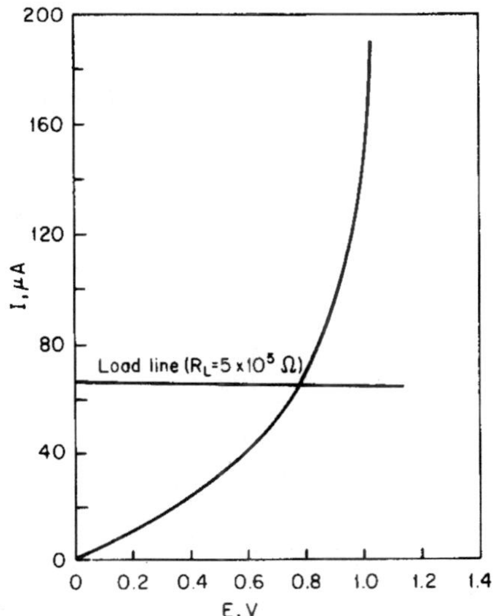

FIGURE 35 Load curve for a typical germanium bolometer ($T = 2.15$ K) with load line, showing optimum operating point. *(From Low, 1961.)*

sensitivity; i.e., selected photomultiplier tubes (PMT) are capable of detecting single photon arrivals (but at best only with about 30 percent quantum efficiency) and amplifying the photocurrent (pulse) enormously without seriously degrading the signal-to-noise ratio. Time resolution can be as short as 0.1 ns. Only very specialized limitations have precluded their use for $\lambda < 800$ nm, for example, cost, ruggedness, uniformity of manufacture, or need for still faster response. Recently these limitations have all been met individually but generally not collectively. Where adequate light is available, the simple phototube has advantages over the multiplier phototube in that high voltages are not required, the output level is not sensitive to applied voltage, and dynode fatigue is eliminated.

Microchannel plate tubes (MCPT) are a variant of the photomultiplier tube where the current amplifying dynode structure is replaced by an array of miniature tubes in which the photocathode current is amplified. MCP tubes are more compact than PMTs and are reliable in operating conditions of high environmental stress. The same range of photocathode materials is available in MCPTs as PMTs. MCPTs can provide a wide range of electron gain as available depending upon whether a single MCP or a stack of MCPs is used. The structure of a PMT and MCPT are compared in Fig. 36.

Sensitivity. In modern phototubes, shot noise due to the cathode dark current is by far the most important noise source. The most common descriptions of phototube sensitivity list both current responsivity (amperes per watt), dark current, and dark noise. Several useful measures of sensitivity are noise equivalent input (NEI) (see Sec. 15.4 "Definitions"), noise equivalent power (NEP), or its reciprocal $D \equiv 1/\text{NEP}$. NEP and NEI in the range 10^{-14} to 10^{-17} W/Hz$^{1/2}$ are not uncommon.

Detectivity is generally limited by dark-current shot noise. Dark current depends on photocathode material, area, and temperature. Thus the best detectivity is obtained with small effective sensitive area. Cooling is especially useful for red-sensitive and near-IR tubes and is generally not worthwhile for others (see "Operating Temperature"). Special tube housings which can provide thermoelectric cooling are available.

The spectral response curves shown in Fig. 37 are for the combination of photocathode

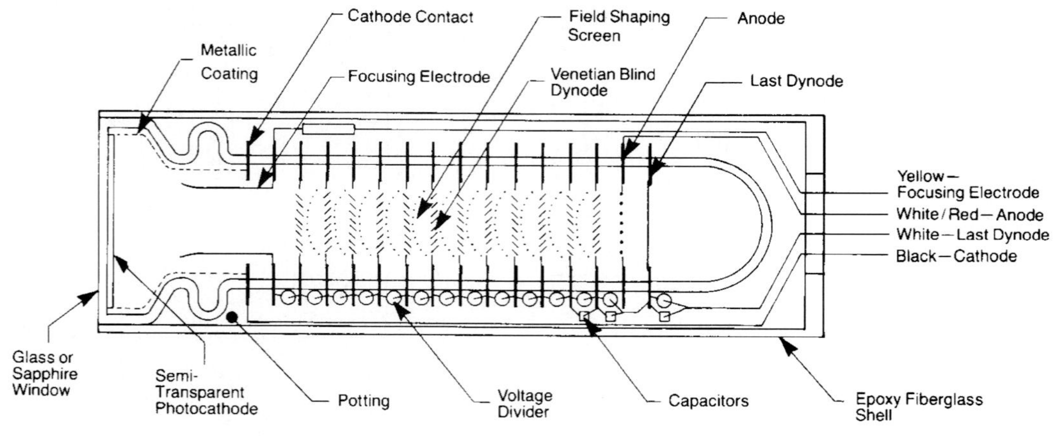

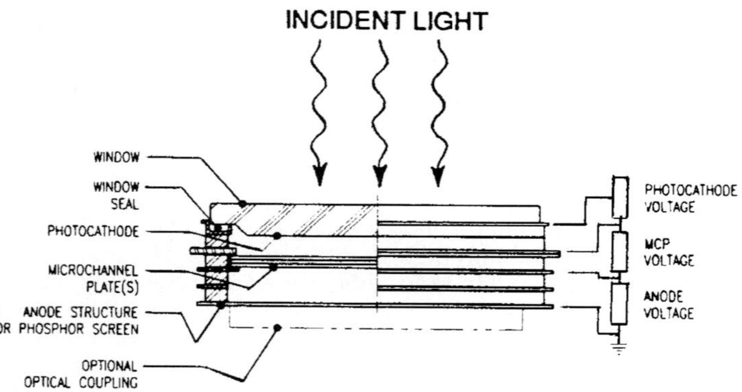

FIGURE 36 Comparison of photomultiplier tube (PMT) and microchannel plate tube (MCPT) construction. *(EMR Photoelectric.)*

and window. Historically, this method of description gave rise to the S-response designations, most of which are now obsolete. It is often desirable to separate photocathode response from window transmission. Thus, Fig. 38 shows the quantum efficiency (electrons per incident photon) of a number of photocathodes without window losses. For $\lambda < 400$ nm, each photocathode should maintain its peak quantum efficiency, up to photon energies where multiple photoemissions take place. In Fig. 39, the spectral dependence of quantum efficiency for a variety of modern photocathode/window combinations is illustrated.

D^* is a meaningful figure of merit for phototubes whose sensitivity is limited by dark noise (shot noise on the dark current) and whose emitting photocathode area is clearly defined, but D^* must be used with caution because, although modern phototubes are generally dark-noise-limited devices, they are often limited by noise in signal, i.e., the noise content of the signal itself (Eberhardt, 1967). Serious errors in predicting the detection capability of phototubes will arise if noise in signal is ignored and D^* is presumed to be the important limiting parameter (see Eqs. 6, 7, and 8). Very little reliable data are presently available on D^* for photoemitters. However D^* curves for S-1, S-20, and S-25 are shown in Figs. 40 (300 K) and 41 (PMT cooled to 200 K).

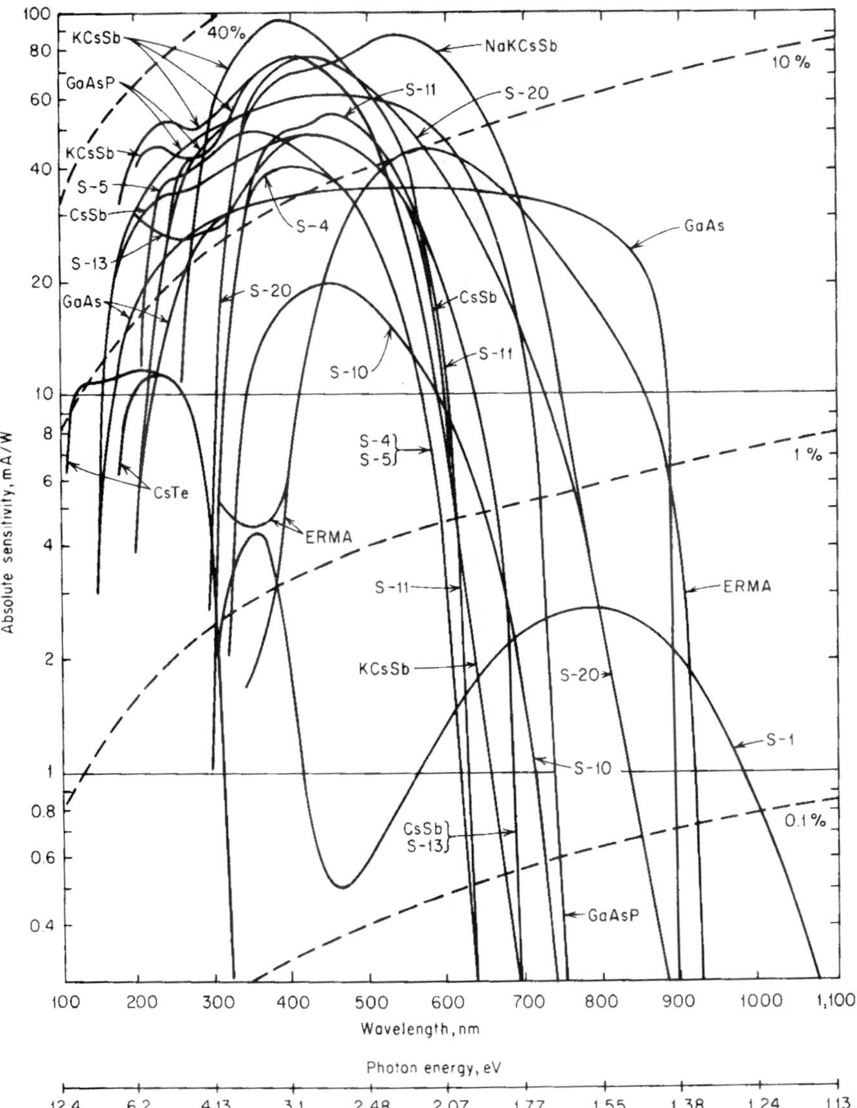

FIGURE 37 Spectral sensitivity of various photoemitters. Dotted lines indicate photocathode quantum efficiency. Chemical formulas are abbreviated to conserve space. S-1 = AgOCs with lime or borosilicate crown-glass window; S-4 = Cs_3Sb with lime or borosilicate crown-glass window (opaque photocathode); S-5 = Ss_3Sb with ultraviolet-transmitting glass window; S-8 = Cs_3Bi with lime or borosilicate crown-glass window; S-10 = AgBiOCs with lime or borosilicate crown-glass window; S-11 = Cs_3Sb with lime or borosilicate crown-glass window (semitransparent photocathode); S-13 = Cs_3Sb with fused-silica window (semitransparent photocathode); S-19 = Cs_3Sb with fused-silica window (opaque semicathode); S-20 = Na_2KCsSb with lime or borosilicate glass window. ERMA = extended red multialkali (RCA; ITT uses MA for multialkali). This curve is representative of several manufacturers' products. Many variations of this response are available, e.g., tradeoffs between short- and long-wavelength response. *(From RCA Electronic Components, chart. PIT-701B.)*

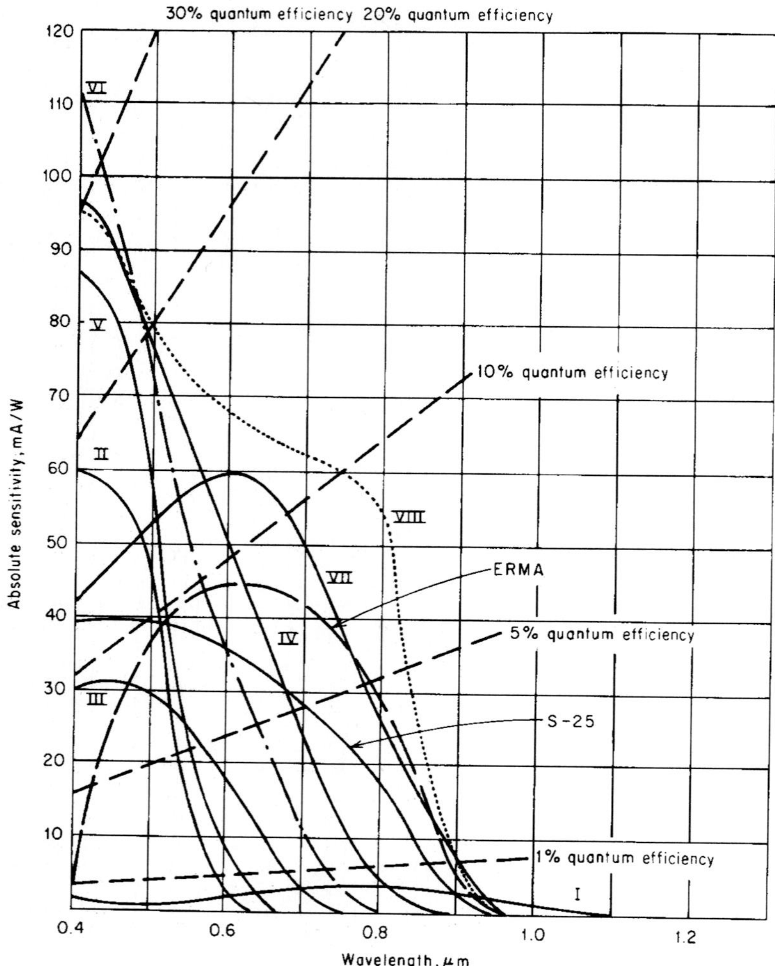

FIGURE 38 Photocathode responsivity and quantum efficiency vs. wavelength (no windows). For abbreviations, see Fig. 33. I = S-1, II = S-11, III = S-10, IV = S-20, V = K_2CsSb, VI = $K_2CsSb(O)$, VII = $NaKCsSb_3$, VIII = GaAs(Cs). *(Based on material from RCA.)*

Short-wavelength Considerations. Window considerations are as follows:

For $\lambda > 200$ nm: Windows are essential, as all useful photocathode materials are oxidized and performance would otherwise be destroyed.

200 nm $> \lambda >$ 105 nm: Photocathode materials are not oxidized by dry air (moisture degrades performance). Windows are optional. LiF windows have shortest known cutoff, 105 nm. For $\lambda < 180$ nm, it is generally advisable to flush with dry nitrogen.

$\lambda < 105$ nm: No windows are available.

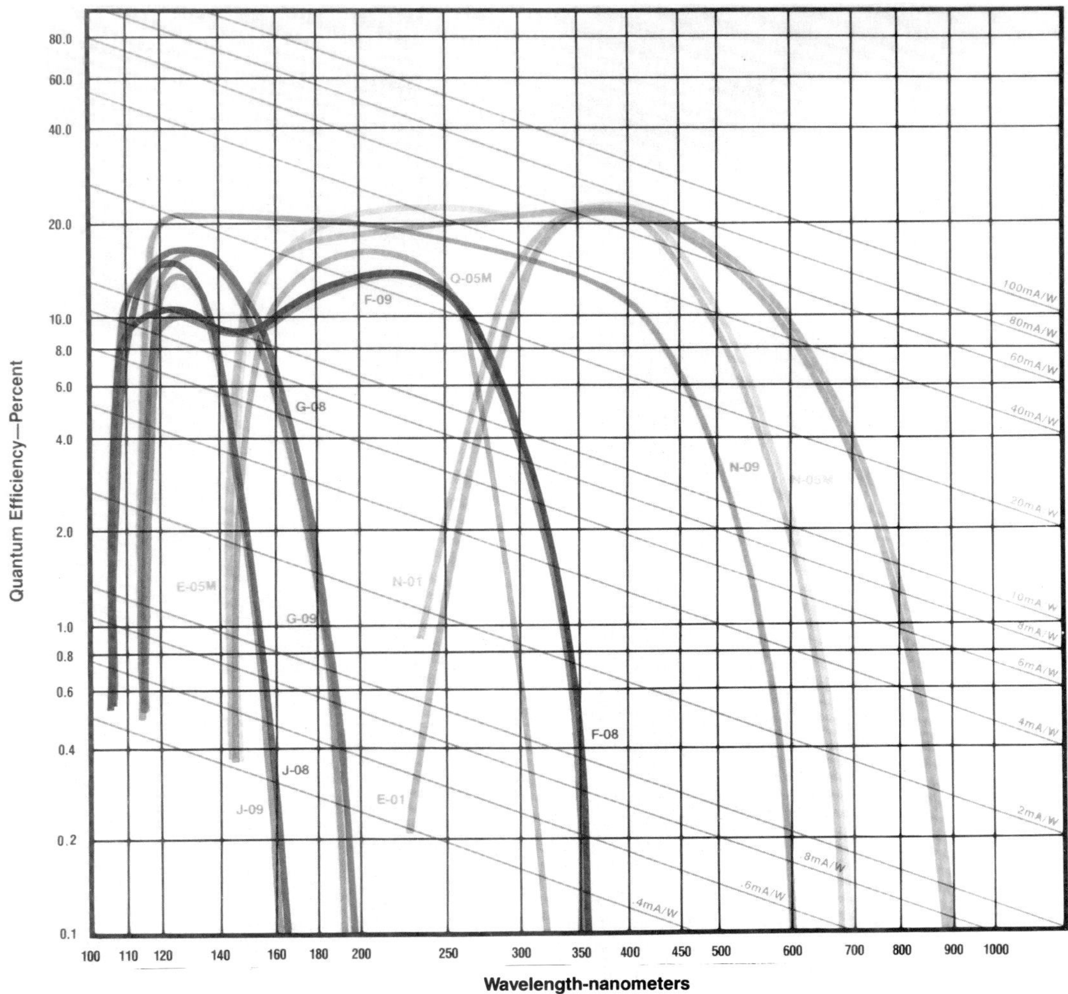

PHOTOCATHODE KEY

Key letter	Description	Long Wavelength Cutoff (Note 1)	Long Wavelength Sensitivity (Note 2)
E	Tri-Alkali (S-20)	850 nm	780 nm
F	Cesium Telluride	355 nm	340 nm
G	Cesium Iodide	195 nm	185 nm
J	Potassium Bromide	165 nm	150 nm
N	High Temperature Bi-Alkali	690 nm	640 nm
Q	Rubidium Telluride	320 nm	300 nm

Note 1 — Point at which Q.E. becomes 1% (Typical) of peak Q.E.
Note 2 — Point at which Q.E. is 1% (Typical)

WINDOW MATERIAL KEY

Key No.	Description	Short Wavelength Cutoff*
01	Borosilicate Glass	270 nm
05	UV Grade Sapphire	145 nm
08	UV Grade Lithium Fluoride	105 nm
09	Magnesium Fluoride	115 nm

*10% Energy Transmission

FIGURE 39 Quantum efficiency of photocathode/window combinations as a function of wavelength. *(EMR Photoelectric.)*

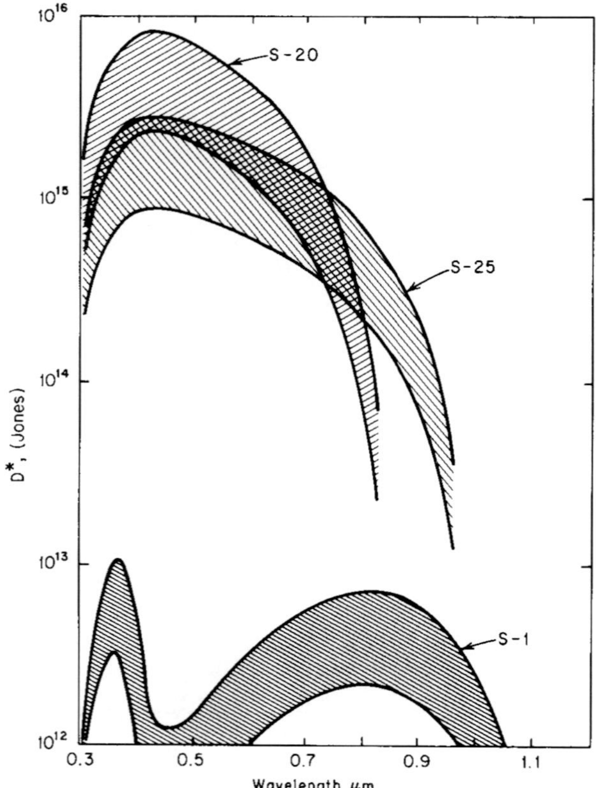

FIGURE 40 Range of D^* for uncooled photomultiplier tubes ($T = 300\ K$.) For abbreviations see Fig. 33. S-25 = same as S-20 but different physical processing. *(Based on material from RCA.)*

Since air absorbs radiation in the region 0.2 to 200 nm (ozone absorbs 200 to 300 nm), it is necessary to include the (windowless) detector in the vacuum with the source.

A useful technique for avoiding the far-ultraviolet window-absorption problem (provided $\lambda >$ vacuum ultraviolet) is to coat the outside of the window of a conventional PMT with an efficient fluorescent material, e.g., sodium salicylate, which absorbs in the ultraviolet and reemits in the blue, and is efficiently detected by most photocathodes (Samson, 1967).

Solar-blind considerations are as follows. Although most photocathodes have high quantum efficiency at short wavelengths, background-noise considerations often preclude their use at very short wavelengths, and very wide bandgap semiconductor photocathodes such as CsI, KBr, Cs_2Te, and Rb_2Te (having peak quantum efficiency a little greater than 10 percent) often give better signal-to-noise ratio. This sacrifice in quantum efficiency to obtain insensitivity to wavelengths greater than those of interest would not be necessary if suitable short-wavelength pass filters were readily available.

For applications where it is desirable that the detector not see much solar radiation, one can use photocathodes whose high work function precludes photoemission for photons of too low an energy. Figure 42 shows quantum efficiency vs. λ for three such photocathodes, tungsten, CsI, and Cs_2Te, compared with Cs_3Sb, and GaAs(Cs), which are not solar-blind.

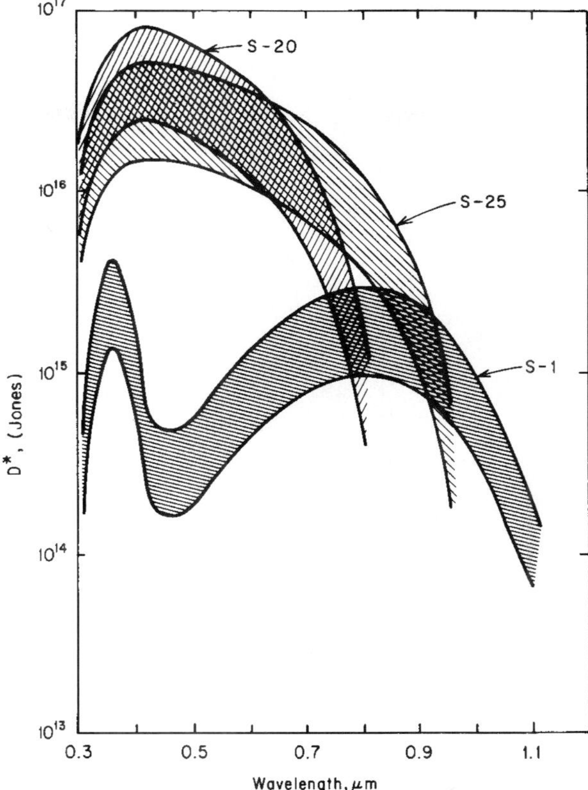

FIGURE 41 Range of D^* for cooled photomultiplier tubes ($T = 200$ K). S-25 = same as S-20 but different physical processing. *(Based on E. H. Eberhardt, "D* of Photomultiplier Tubes and Image Detectors," ITT Industrial Labs, 1969.)*

Quantum Efficiency. Figures 38, 39, and 42 show photocathode spectral quantum efficiency (probability that one photoelectron is emitted when a single photon is incident). Note that there are fairly few basic photocathode materials and that the window often determines effective quantum efficiency at short wavelengths.

For $\lambda < 40$ nm, a wide variety of photocathode materials are available at high quantum efficiency. Many of these materials, such as tungsten, are not destroyed by being subjected to air, so that open structures can be used, consisting of a photocathode multiplier chain without window. The complete windowless structure is then placed in the vacuum with the source of radiation.

The quantum efficiency at any wavelength can be calculated from the formula

$$\eta = \frac{\mathcal{R} \times 1239.5}{\lambda} \tag{22}$$

where $\mathcal{R}$ = photocathode response, A/W
λ = wavelength, nm

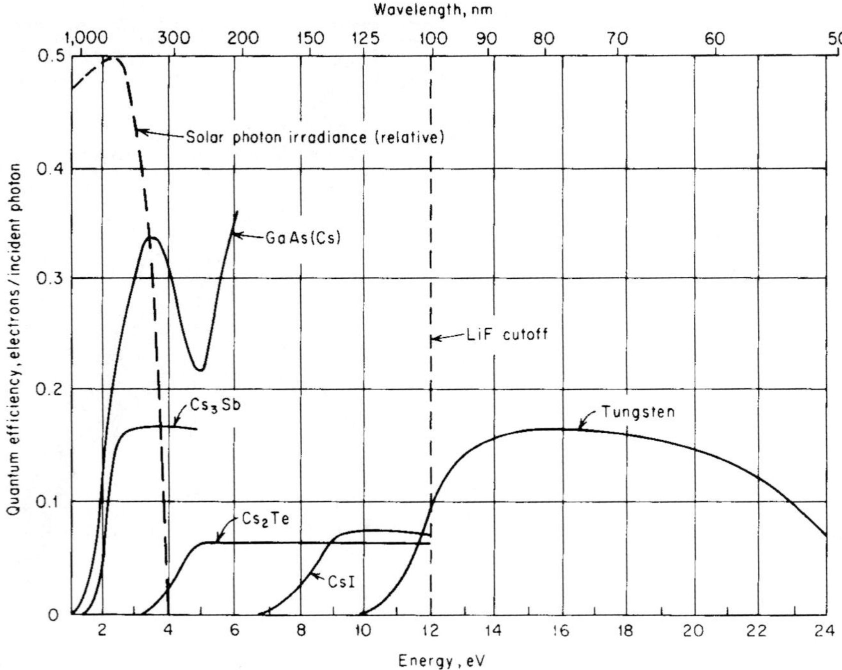

FIGURE 42 Quantum efficiency vs. wavelength (photon energy) for several photoemitters.

A useful technique for improving quantum efficiency, reported by Livingston (1966) and Gunter (1970), involves multipassing the photocathode by trapping the light inside the photocathode using a prism.

Responsivity. PMT responsivity depends upon photocathode quantum efficiency and subsequent dynode gain. For most purposes, the dynode gain in a well-designed PMT introduces no significant degradation in the photocathode signal-to-noise ratio. Figure 37 shows photocathode response expressed in photocurrent (amperes) per incident radiation power (watts).

Noise. The limiting noise in a PMT depends on the level of illumination. For low-level detection, limiting noise is the shot noise on the dark current,

$$i_n = (2ei_{\text{dark}}\Delta f)^{1/2} \tag{23}$$

For high illumination levels the shot noise on the signal photocurrent

$$i_n = (2ei_{\text{signal}}\Delta f)^{1/2} \tag{24}$$

far exceeds that on the dark current. Manufacturers usually express noise as photocathode dark current or anode dark current for given gain, which is therefore traceable to photocathode dark current.

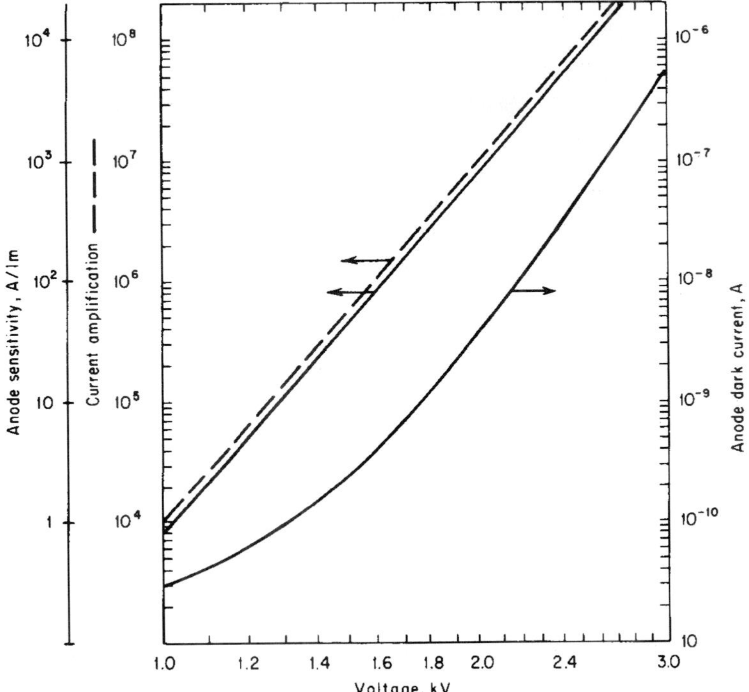

FIGURE 43 Typical current amplification and anode dark current as a function of applied voltage. *(Based on E. H. Eberhardt, "D* of Photomultiplier Tubes and Image Detectors," ITT Industrial Labs, 1969.)*

Photocathode dark current is approximately proportional to photocathode area so that small photocathode effective areas can be expected to have reduced noise. Figure 43 shows how anode dark current and gain increase with applied voltage for a typical PMT.

Minimum detectable power is related to limiting noise through responsivity via

$$\mathrm{NEP} = \frac{i_n}{\mathcal{R}} \tag{25}$$

where $\mathcal{R}$ is in amperes per watt.

Operating Temperature. Dark current due to thermionic emission, usually greater in red-sensitive tubes, can be reduced by cooling (see Cole and Ryer (1972). The trialkali (S-20) performance does not benefit from cooling below 255 K. Maximum beneficial cooling (three to four dark counts per second) for AgOCs (S-1), (Cs)Na_2KSb (S-20), and Cs_3Sb (S-11) is 195, 255, and 239 K, respectively. Most photocathodes become noisier as temperature rises above ambient because of increased thermionic emission. Because its thermionic emission starts at a very low value, (Cs)Na_2KSb is a useful photocathode up to temperatures of approximately 373 K.

Response Time. The rise time of photomultiplier tubes depends chiefly on the spread in transit time during the multiplication process. For photomultiplier tubes, this spread is about 10 ns. Some tubes with specially designed electron optics can give spread as low as

1 ns. The crossed-field PMT makes possible a spread as small as 0.1 ns. Microchannel plate tubes have response times of a nanosecond or less.

For high-speed work (<1 ns rise time), good transmission-line technique must be used to obtain impedance match and to avoid reflection. The bandwidth of the output circuit will depend upon the total capacitance (PMT circuit plus stray capacitances) and the value of the load resistance.

Linearity. Photomultiplier tubes are nearly all linear to about 1 percent for cathode currents of 0.1 μA or less. Some tubes may be linear to better than 0.1 percent but must be individually selected (Bennett, 1966). Probably most of the nonlinearity results from the dynode structure.

Sensitive Area. No fundamental limitation. Only recently available with very small effective areas for extremely low dark current. Magnetic focusing has been used so that only a small fraction of the photocathode is used electron-optically.

Sensitivity Profile. Usually uniform within 20 to 50 percent. Microchannel plate detectors may have uniformity of ±5 percent.

Stability. PMTs are subject to short- and long-term drift which can depend upon anode current, changes in anode current, storage times, and aging or anode life. They are also subject to change if exposed to magnetic fields or changes in temperature. Vibration of the tube may modulate the signal (microphonic effect).

Recommended Circuit. See Fig. 44.

1. Since a PMT is a current generator, increasing output resistance $\mathbf{R}_1$ increases output voltage. An upper limit to $\mathbf{R}_1$ may be imposed either by the time-constant limitation or by nonlinearity, which results from a space charge produced near the anode when the anode is left nearly floating electrically.
2. The rated photocathode current (referred to anode current through gain) should not be exceeded.
3. Care should be taken not to destroy the photocathode with light (heating).
4. When large currents are drawn, it may affect later dynode interstage voltage and hence gain, causing nonlinearity; e.g., in Fig. 44, if the photocurrent from DY 10 to anode becomes comparable to the biasing current, through $\mathbf{R}_{11}$, the gain of the final stage is reduced. This can be avoided by biasing the dynodes with constant-voltage sources.
5. To avoid dynode damage, final dynode current must not exceed the value suggested by the manufacturer.

Photon Counting. At the photocathode, the shot-noise-limited signal-to-noise ratio (with negligible dark current) is

$$\frac{i_s}{i_n} = \frac{i_s}{(2ei_s\Delta f)^{1/2}} \left(\frac{i_s}{2e\Delta f}\right)^{1/2} = \left(\frac{N_s}{2\Delta f}\right)^{1/2} \tag{26}$$

where N_s is the photoelectron rate at the photocathode. Thus, for extremely low levels of illumination, the ideal signal-to-noise ratio becomes very poor. At this point there is much to be gained by abandoning attempts to measure the height of the fluctuating signal (Fig. 45*a*) in favor of digitally recording the presence or absence of individual pulses (Fig. 45*e*).

Single photoelectron counting can be achieved by using a pulse amplifier (see Fig. 46), which suppresses spurious dark-noise pulses not identical in amplitude and shape to those produced by photoelectrons.

An upper practical limit for (random) photon counting is set by convenient amplifier bandwidths at about $10^5\ \mathrm{s}^{-1}$. For reasonable (1 percent) statistical accuracy, this implies a 10-MHz bandwidth.

Gallium Phosphide Dynodes. The development of GaP dynodes for increased

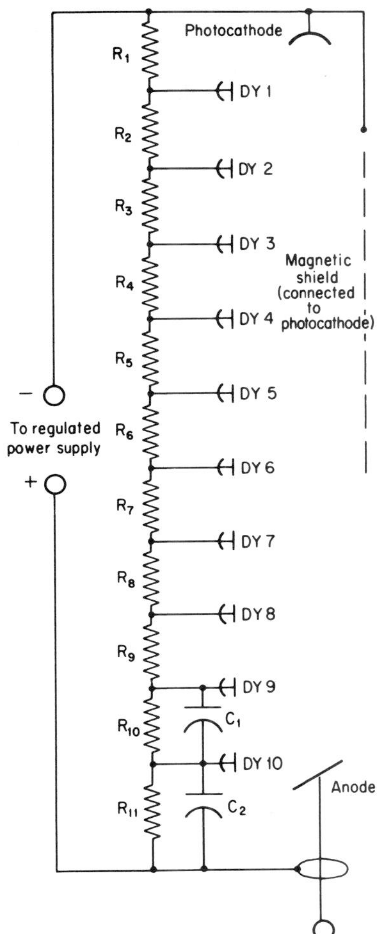

FIGURE 44 $C_1 = 68$ pF ± 10 percent, 500 V (dc working); $C_2 = 270$ pF ± 10 percent, 500 V (dc working); $\mathbf{R}_1 = 220\ \text{k}\Omega \pm 5$ percent, 1/4 W; $\mathbf{R}_2 = 240\ \text{k}\Omega \pm 5$ percent, 1/4 W; $\mathbf{R}_3 = 330\ \text{k}\Omega \pm 5$ percent, 1/4 W; $\mathbf{R}_4$ to $\mathbf{R}_{11} = 200\ \text{k}\Omega \pm 5$ percent, 1/4 W. *(Based on E. H. Eberhardt, "D* of Photomultiplier Tubes and Image Detectors," ITT Industrial Labs, 1969.)*

FIGURE 45 Oscilloscope presentation of PMT output when reviewing square-wave chopped light pulse. In (*a* to *e*) intensity is reduced and gain increased commensurately. *(Courtesy of E. H. Eberhardt.)*

secondary-electron production (Simon et al., 1968; Morton et al., 1968) makes possible unambiguous discrimination of small numbers of individual photoelectron counts which was not previously possible with lower dynode gains. This is shown in Fig. 47, where the spread in number of secondary electrons ($N \times$ gain) is just $(N \times \text{gain})^{1/2}$.

In addition to the aforementioned fundamental advantage of high dynode gain, the large gain per stage in the first dynodes also helps discriminate against noise introduced by later stages of amplification. Also, fewer stages of amplification are required.

Manufacturers. ADIT, EMR Photoelectric, Bicron, Burle, Edinburgh Instruments,

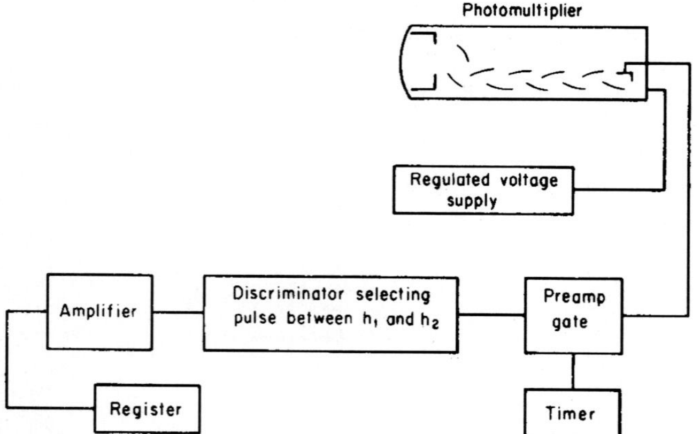

FIGURE 46 Photomultiplier and associated circuits for photon counting. *(ITT Rep. E5.)*

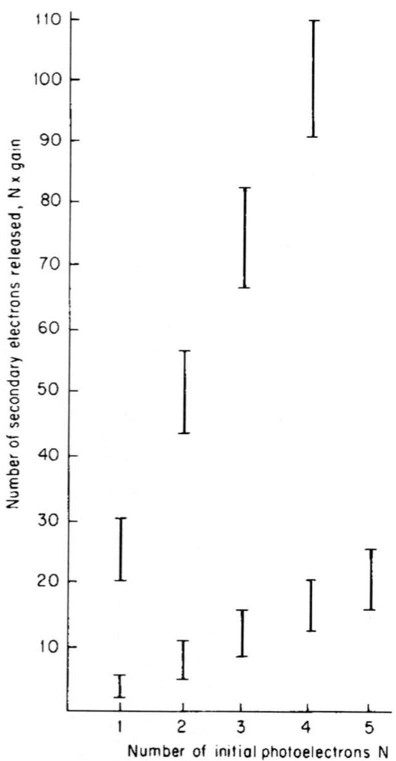

FIGURE 47 Spread in number of secondary electrons for various phototube gains.

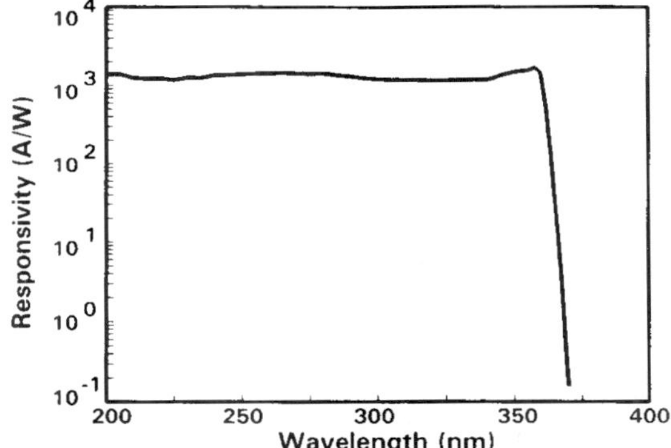

FIGURE 48 Response in amps per watt for a GaN detector. *(Reprinted from App. Phys. Lett., vol. 60, no. 23, 1992, p. 2918.)*

Galileo Electro-Optics, Hamamatsu, K and M Electronics, International Light, Optometrics USA, Oriel, Phillips Components, Photek, Photon Technology, Penta Laboratories, Thorn EMI, Varo.

GaN. Gallium nitride photoconductive and photovoltaic detectors have spectral response in the ultraviolet (UV) from ≈200 to 365 nm as illustrated in Fig. 48. Response at visible wavelengths is absent, so that no special filtering is required to detect UV in the presence of visible lighting. These solid-state devices are potentially useful for operation at elevated temperatures, in high-vibration environments, and in other environments unsuitable for photomultiplier tubes. The photoconductive version of this device uses interdigitated contact electrodes because of the very high impedance of the GaN films.

Sensitivity: $D^* = 10^{13}$ Jones.

Response: Photoconductive 100–1000 A/W; photovoltaic 0.2 A/W.

Resistance: Photoconductor $1 \times 10^{11} \Omega$.

Dynamic range: Linear over five to six orders of magnitude.

Time constant: Photoconductive 0.2–1 ms; photovoltaic 0.16 μs.

Size: 1-mm (1 × 1 mm).

Manufacturer: APA Optics.

CdS and CdSe. Cadmium sulfide and cadmium selenide photoconductors are available for detection of visible light out to 700 to 800 nm. CdS and CdSe films have sheet resistivity in the range of 20 mΩ per square at an illumination level of 2 footcandles. The devices are typically made in a linear or serpentine configuration consisting of 2 to 500 squares to maximize the length-to-width ratio. A variety of material "types" are available, offering unique spectral curves for various applications, depending upon the source color. CdS and CdSe are typically slow detectors, with response times of 5 to 100 ms, with speed improving at higher light levels. These devices exhibit "memory" or "history" effects, where the response is dependent upon the storage condition preceding use—the length of storage and time in use, and differences between the storage light level and the light level during use. These history effects may amount to changes in resistance from less than 10 percent to over 500 percent. CdSe has comparably greater memory effect than CdS.

CdS and CdSe are useful for a variety of commercial applications, both analog and

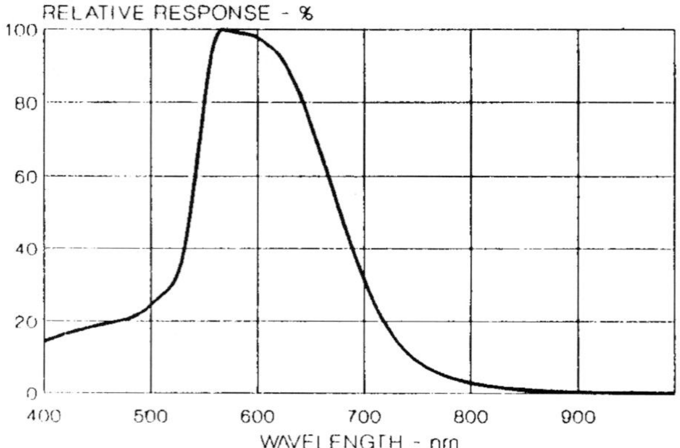

FIGURE 49 Relative spectral response for a "Type 0" CdS photoconductive cell. *(EG&G VACTEC,* Optoelectronics Data Book, *CdS Photoconductive Cells.*)

digital, such as camera exposure control, automatic focus and brightness controls, densitometers, night light controls, etc. They are comparatively inexpensive and are available in a wide range of packages and resistance values, including dual cell configurations.

Spectral response: See Figs. 49–54.

Resistance and sensitivity: See Figs. 55–60.

Temperature coefficient of resistance: See Figs. 61–66.

Light history effects: See Fig. 67.

Detector size: 4×4 mm to 12×12 mm approximate, dual elements available.

Manufacturers: EG&G VACTEK.

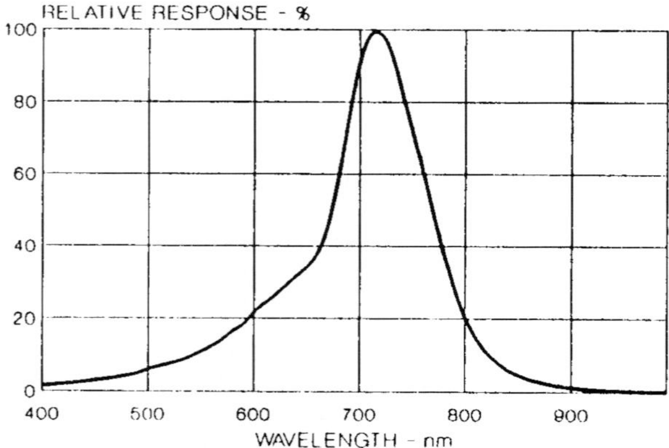

FIGURE 50 Relative spectral response for a "Type 1" CdSe photoconductive cell. *(EG&G VACTEC,* Optoelectronics Data Book, *CdS Photoconductive Cells.*)

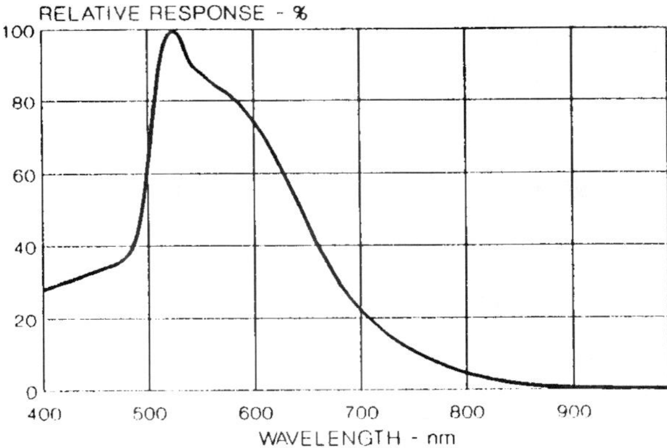

FIGURE 51 Relative spectral response for a "Type 2" CdS photoconductive cell. *(EG&G VACTEC,* Optoelectronics Data Book, *CdS Photoconductive Cells.)*

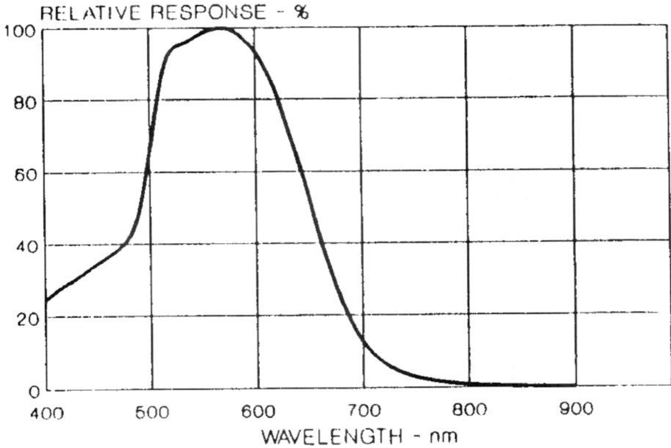

FIGURE 52 Relative spectral response for a "Type 3" CdS photoconductive cell. *(EG&G VACTEC,* Optoelectronics Data Book, *CdS Photoconductive Cells.)*

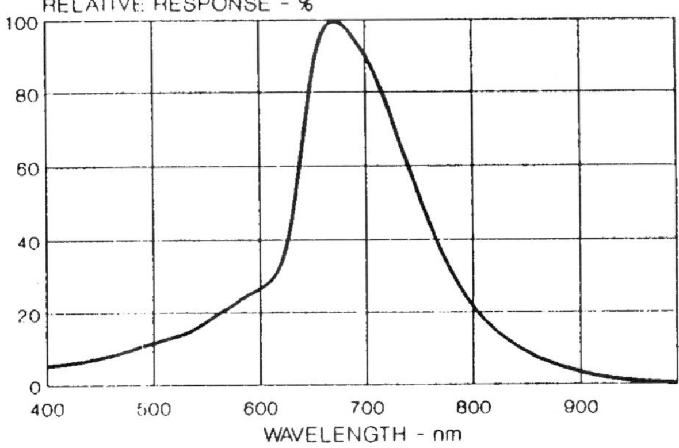

FIGURE 53 Relative spectral response for a "Type 4" CdSe photoconductive cell. *(EG&G VACTEC,* Optoelectronics Data Book, *CdS Photoconductive Cells.)*

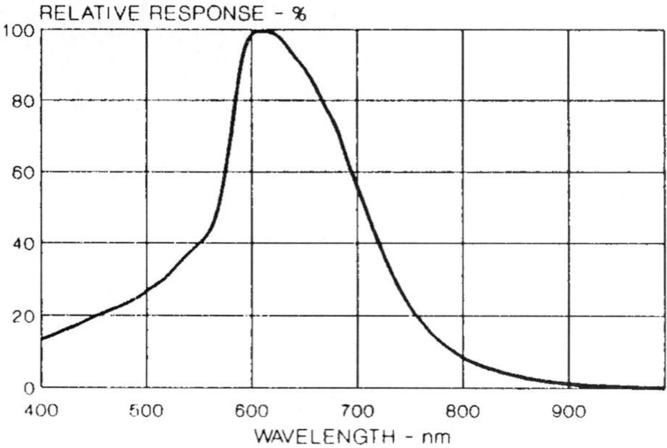

FIGURE 54 Relative spectral response for a "Type 7" CdSSe photoconductive cell. *(EG&G VACTEC,* Optoelectronics Data book, *CdS Photoconductive Cells.)*

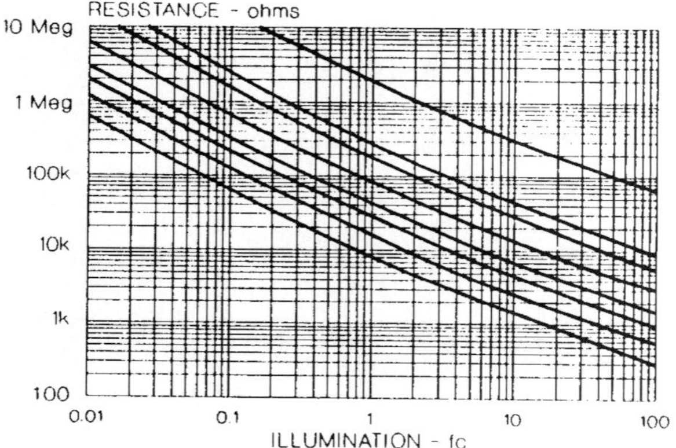

FIGURE 55 Resistance as a function of illumination for a "Type 0" CdS photoconductive cell. *(EG&G VACTEC,* Optoelectronics Data Book, *CdS Photoconductive Cells.*)

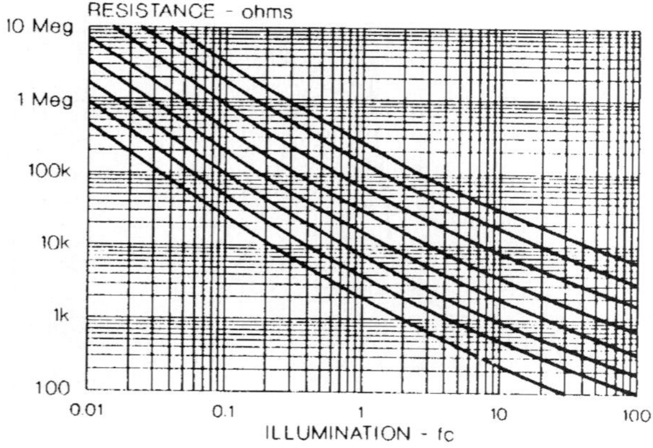

FIGURE 56 Resistance as a function of illumination for a "Type 1" CdSe photoconductive cell. *(EG&G VACTEC,* Optoelectronics Data Book, *CdS Photoconductive Cells.*)

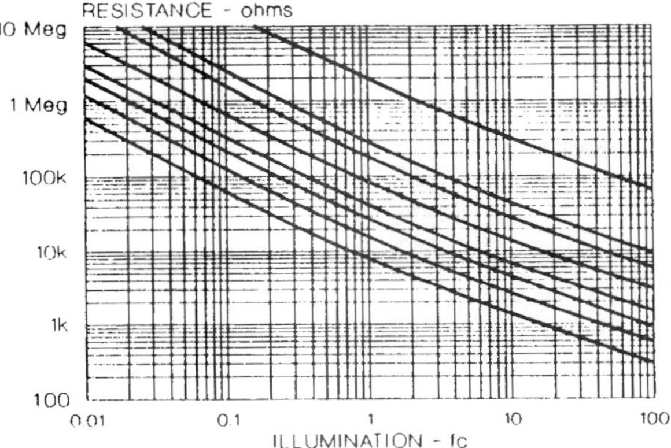

FIGURE 57 Resistance as a function of illumination for a "Type 2" CdS photoconductive cell. *(EG&G VACTEC,* Optoelectronics Data Book, *CdS Photoconductive Cells.)*

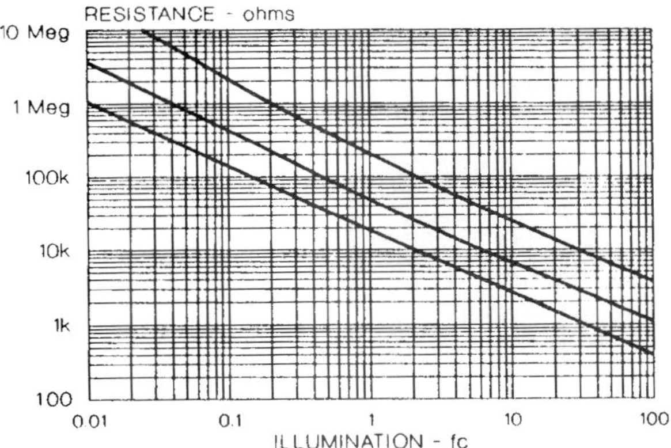

FIGURE 58 Resistance as a function of illumination for a "Type 3" CdS photoconductive cell. *(EG&G VACTEC,* Optoelectronics Data Book, *CdS Photoconductive Cells.)*

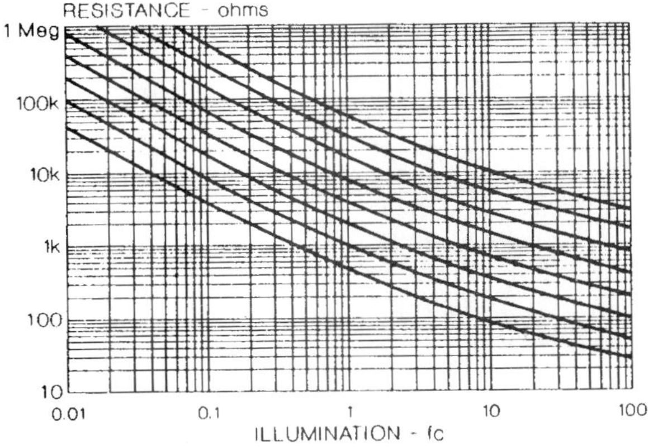

FIGURE 59 Resistance as a function of illumination for a "Type 4" CdSe photoconductive cell. *(EG&G VACTEC,* Optoelectronics Data Book, *CdS Photoconductive Cells.)*

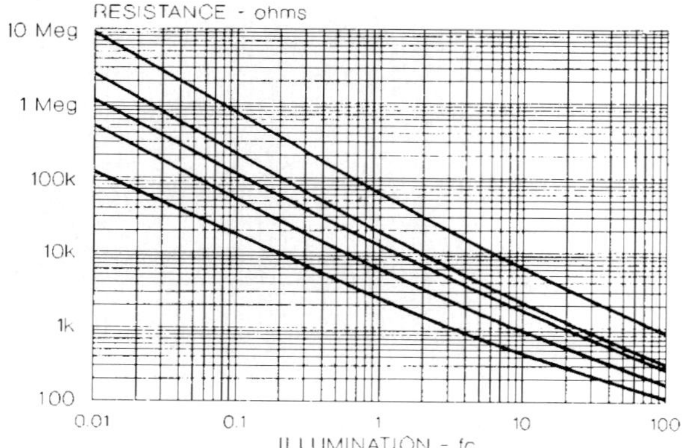

FIGURE 60 Resistance as a function of illumination for a "Type 7" CdSSe photoconductive cell. *(EG&G VACTEC,* Optoelectronics Data Book, *CdS Photoconductive Cells.)*

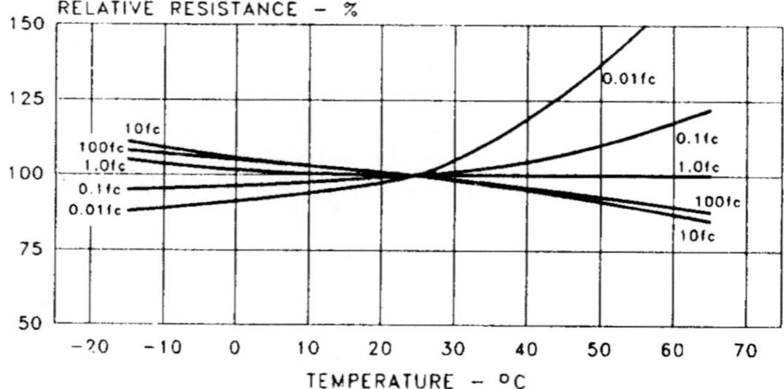

FIGURE 61 Relative resistance as a function of temperature for a "Type 0" CdS photoconductive cell. *(EG&G VACTEC,* Optoelectronics Data Book, *CdS Photoconductive Cells.)*

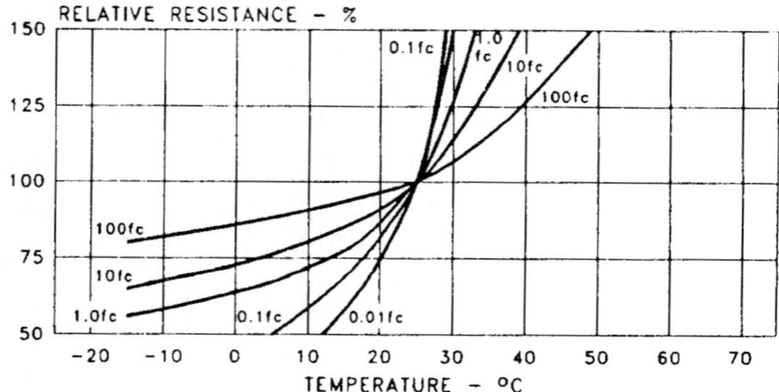

FIGURE 62 Relative resistance as a function of temperature for a "Type 1" CdSe photoconductive cell. *(EG&G VACTEC,* Optoelectronics Data Book, *CdS Photoconductive Cells.)*

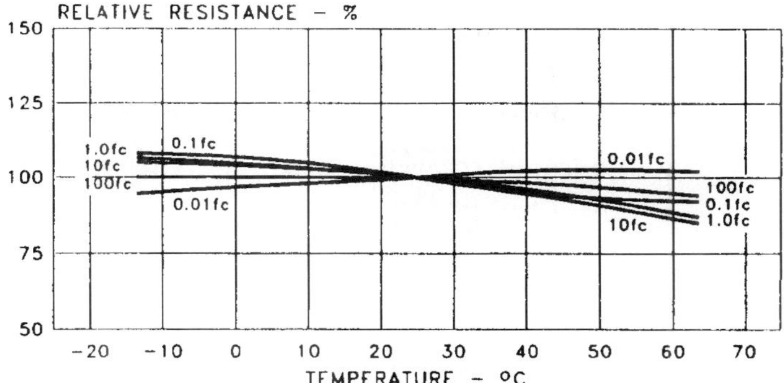

FIGURE 63 Relative resistance as a function of temperature for a "Type 2" CdS photoconductive cell. *(EG&G VACTEC,* Optoelectronics Data Book, *CdS Photoconductive Cells.)*

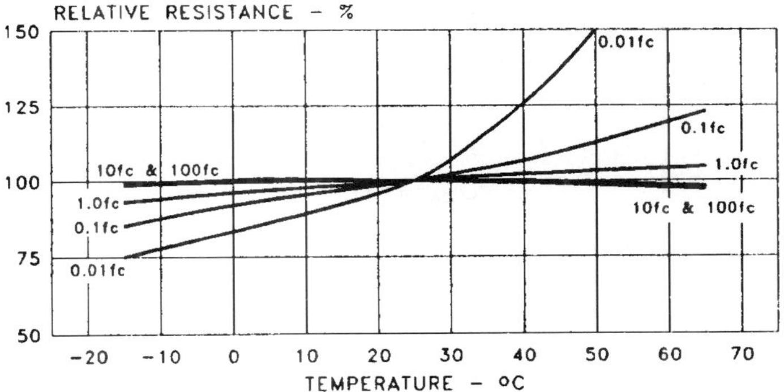

FIGURE 64 Relative resistance as a function of temperature for a "Type 3" CdS photoconductive cell. *(EG&G VACTEC,* Optoelectronics Data Book, *CdS Photoconductive Cells.)*

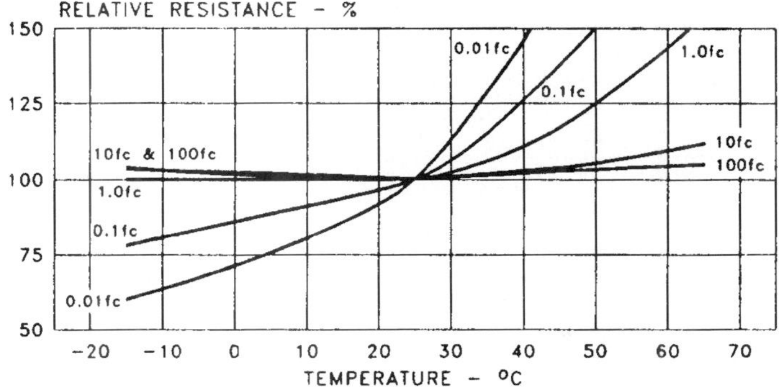

FIGURE 65 Relative resistance as a function of temperature for a "Type 4" CdSe photoconductive cell. *(EG&G VACTEC,* Optoelectronics Data Book, *CdS Photoconductive Cells.)*

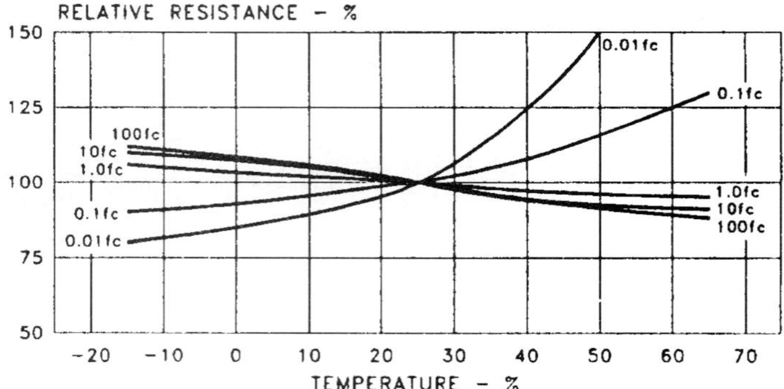

FIGURE 66 Relative resistance as a function of temperature for a "Type 7" CdSSe photoconductive cell. *(EG&G VACTEC, Optoelectronics Data Book, CdS Photoconductive Cells.)*

CdTe. Cadmium telluride and cadmium zinc telluride detectors are chemical group II-VI materials having an energy bandgap of about 1.6 eV, corresponding to a spectral cutoff in the vicinity of 775 nm. These devices, however, are principally used for gamma ray detection because of their high z number which translates into a high absorption coefficient for gamma rays. The principal advantage of CdTe in this application is its ability to operate at room temperature, in comparison with Ge gamma ray detectors which must typically be cooled to 77 K. Figure 68 illustrates the absorption of CdTe as a function of gamma ray energy out to 300 keV.

Sensitivity: See Fig. 68.

Standard sizes: Wafers in 10- and 16-mm dia.; rods 7 × 2 × 2 mm; cubes 2 × 2 × 2.

Standard thickness: 1 and 2 mm.

Bias voltage: 150–300 V/cm.

Operating temperature range: −10 + 55°C

Leakage current: 10–300 nA

Capacitance: 10 pF

VARIATION OF RESISTANCE WITH LIGHT HISTORY EXPRESSED AS A RATIO R_{LH} / R_{DH} AT VARIOUS TEST ILLUMINATION LEVELS.

MATERIAL TYPE	ILLUMINATION (fc)				
	.01	.1	1.0	10	100
TYPE 0	1.60	1.40	1.20	1.10	1.10
TYPE 1	5.50	3.10	1.50	1.10	1.05
TYPE 2	1.50	1.30	1.20	1.10	1.10
TYPE 3	1.50	1.30	1.20	1.10	1.10
TYPE 4	4.50	3.00	1.70	1.10	1.10
TYPE 7	1.87	1.50	1.25	1.15	1.08

FIGURE 67 Variation of resistance with light history expressed as a ratio of infinite light history $\mathbf{R}_{LH}$ to infinite dark history $\mathbf{R}_{DH}$ at various illumination levels. *(EG&G VACTEC,* Optoelectronics Data Book, *CdS Photoconductive Cells.)*

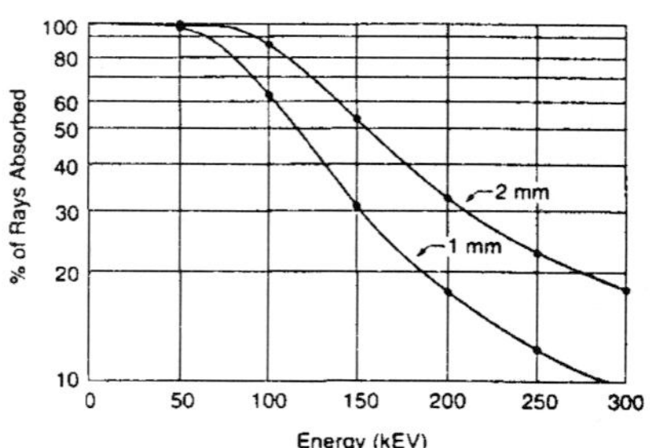

FIGURE 68 The high percentage of rays absorbed by CdTe makes these detectors highly sensitive. At 100 keV, a 2-mm-thick detector absorbs 85 percent of the rays. *(Radiation Monitoring Devices, Cadmium Telluride brochure.)*

Response time: <1 μs.

Manufacturers: Aurora, II-VI eV Products, Radiation Monitoring Devices, Santa Barbara Research Center.

Si. Silicon photovoltaic detectors are widely available. They are useful at wavelengths shorter than about 1.1 μm and can even be used for x-ray and gamma ray detection. There are four main silicon detector types:

- *pn* junction photodiodes, generally formed by diffusion, but ion implantation can also be used
- *pin* junctions, which have lower capacitance and hence higher speed, and because of a thicker active region have enhanced near-IR spectral response
- UV- and blue-enhanced photodiodes
- avalanche photodiodes with significant internal gain, combining high speed and sensitivity

The main parameters of interest are spectral response (see Fig. 69), time constant, and zero-bias resistance or reverse-bias leakage current. Silicon material has an indirect bandgap and hence the spectral cutoff is not very sharp near its long-wavelength limit as shown in Fig. 69. The effective time constant of *pn* junction silicon detectors is generally limited by RC considerations rather than by the inherent speed of the detection mechanism (drift and/or diffusion). High reverse bias may or may not shorten charge-collection time, but it generally reduces cell capacitance. Both result in faster response.

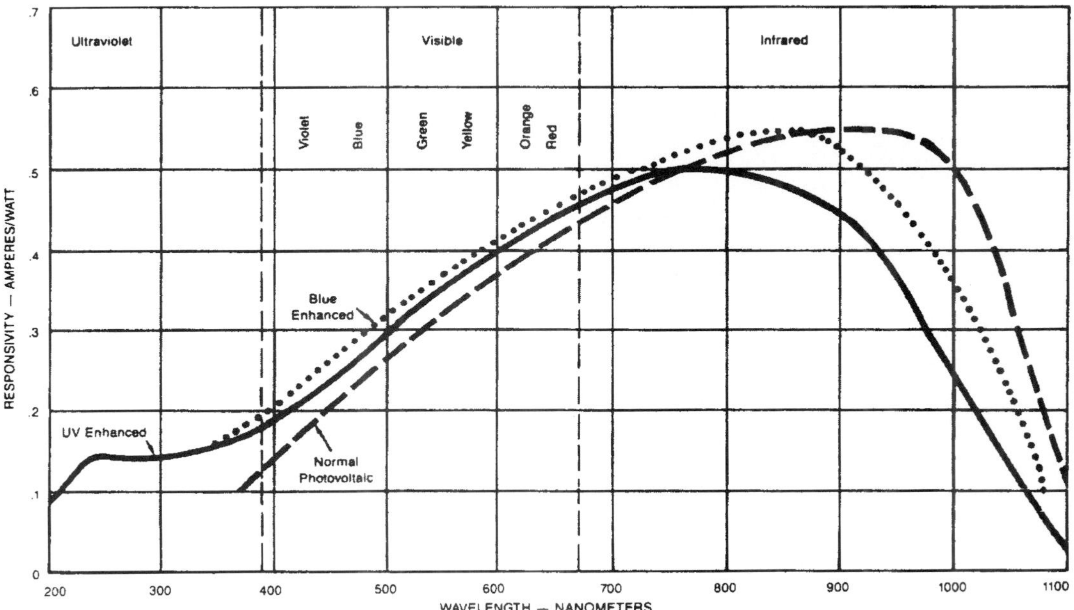

FIGURE 69 Typical spectral response for *pn* junction, blue-enhanced and UV-enhanced silicon photodiodes. *(UDT Sensors, Inc., Optoelectronic Components Catalog.)*

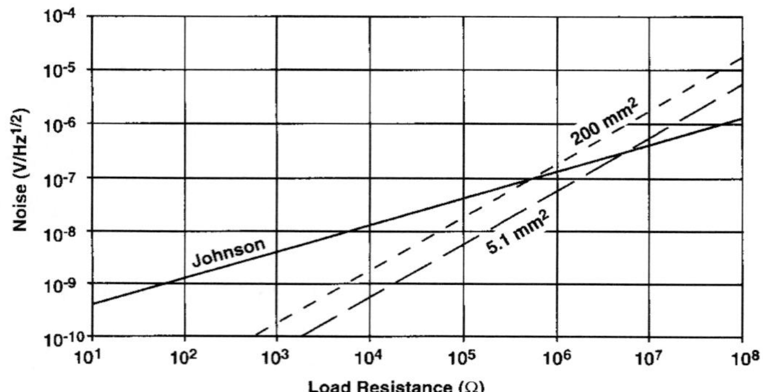

FIGURE 70 Output noise as a function of circuit load resistance for *pin* silicon photodiodes with areas of 5.1 and 200 mm^2, compared with the Johnson noise of the load resistor. Dark current measured at 10-V reverse bias for the detector with area of 5.1 mm^2 is 10 nA, and 100 nA for the detector with an area of 200 mm^2. Note that good preamplifiers have a noise level of about 1 nV/Hz$^{1/2}$, depending upon the bandwidth. *(Detector data from UDT Sensors, Optoelectronic Components Catalog.)*

On the other hand, increased reverse bias causes increased noise, so that a tradeoff exists between speed and sensitivity. For high-frequency applications, load resistance should be made small, although this makes Johnson (thermal) noise comparatively larger, which limits sensitivity (see Fig. 70). In order to keep sensitivity high when using these devices at high frequency, operational (current-mode) amplifiers, which can be built into the detector package, and avalanche photodiodes, which incorporate built-in gain before the load resistor is encountered, have been developed. Very careful regulation of the detector bias is required for stable operation of avalanche photodiodes.

Silicon pn Junction Photodiodes. These are general purpose when high sensitivity is required and time constants on the order of microsecond are permissible. The device construction is illustrated in Fig. 71. These devices are typically operated in a *photovoltaic mode* at zero bias, but can be used in a *photoconductive mode* in which the device is reversed biased.

Sensitivity: $D^*(\lambda_{pk}) \approx$ mid-10^{12} to 10^{13} Jones, $D^*(2800\text{ K}) \approx 2 \times 10^4$ Jones, becoming

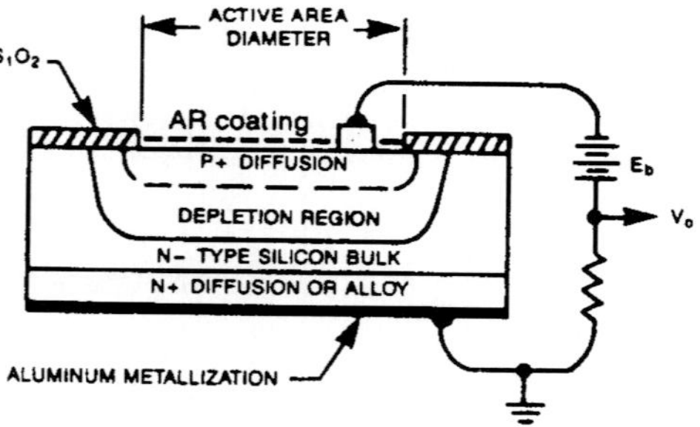

FIGURE 71 Planar diffused silicon photodiode construction. *(UDT Sensors, Optoelectronic Components Catalog.)*

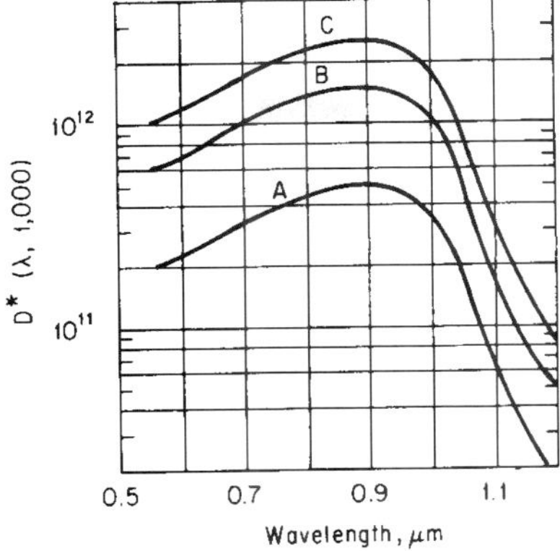

FIGURE 72 D^* vs. λ for small-area *pn* junction silicon photodiodes: curves *A*, *B*, and *C* correspond to areas of 0.02, 0.2, and 1 cm^2. The lower D^* for smaller-area detector performance is due to amplifier limitations rather than intrinsically poorer D^*, for small-area detectors. *(Texas Instruments, Infrared Devices, SC-8385-366. Reprinted by permission of Texas Instruments.)*

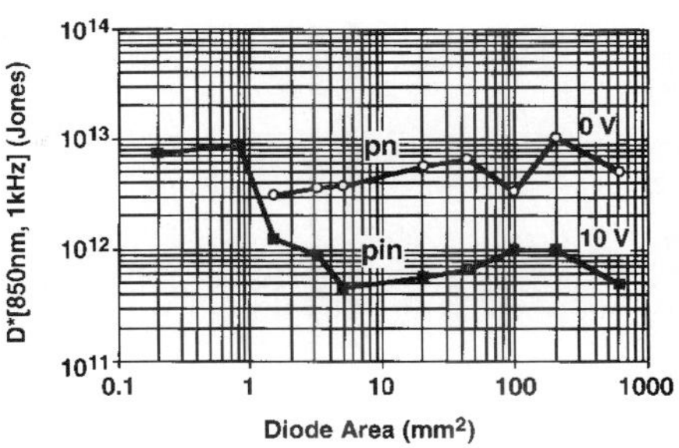

FIGURE 73 D^* as a function of diode area for *pn* junction silicon photodiodes operated in the photovoltaic mode (0 V) and *pin* junction diodes operated in the photoconductive or reverse-bias mode (10 V). *(UDT Sensors, Optoelectronic Components Catalog.)*

amplifier-limited for small-area detectors (see Figs 72 and 73). D^* can also be estimated from the $\mathbf{R}_0A$ product (detector zero-bias resistance or shunt resistance × diode area), which is illustrated in Fig. 74, in combination with Fig. 7, which illustrates the dependence of D^* on $\mathbf{R}_0A$ product.

Noise: See Figs. 75 (noise vs. bias) and 76 (noise vs. temperature); as T drops, impedance rises, so that decreasing noise current produces increasing noise voltage. However, the signal increases even faster, yielding an improved signal-to-noise ratio with cooling. Figure 77 (noise vs. frequency) shows the dependence on bias.

Capacitance: Capacitance is proportional to area and increases slightly with temperature (see Fig. 78).

Responsivity: See Figs. 69 and 75.

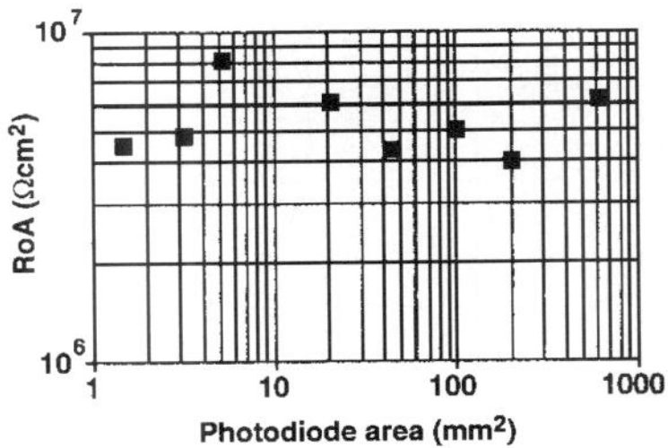

FIGURE 74 The resistance-area product at zero bias ($\mathbf{R}_0A$) and 295 K of silicon *pn* junction photodiodes. The lack of area dependence is evidence that the intrinsic properties of the junction are dominant in these devices.

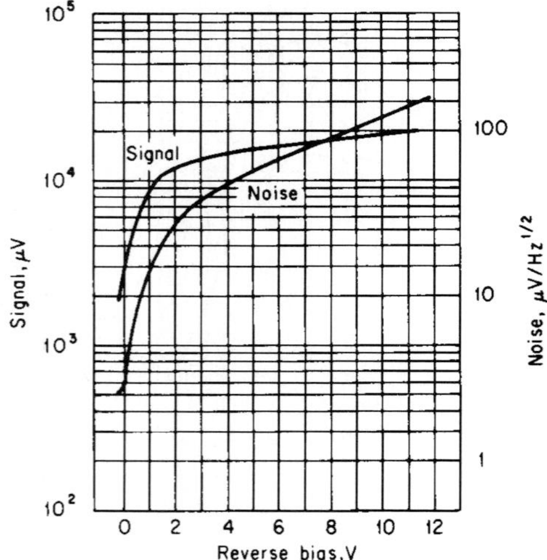

FIGURE 75 Typical *pn* junction signal and noise vs. reverse bias ($\mathbf{R}_L = 10\ \text{M}\Omega$). *(Electronuclear Laboratories, Bull. 1053, 1966.)*

Quantum efficiency: >90 percent quantum efficiency achievable with antireflection coating.

Sensitive area: 0.2–600 mm^2 areas are readily available.

Time constant: Inherently slow for high-sensitivity applications, generally limited by

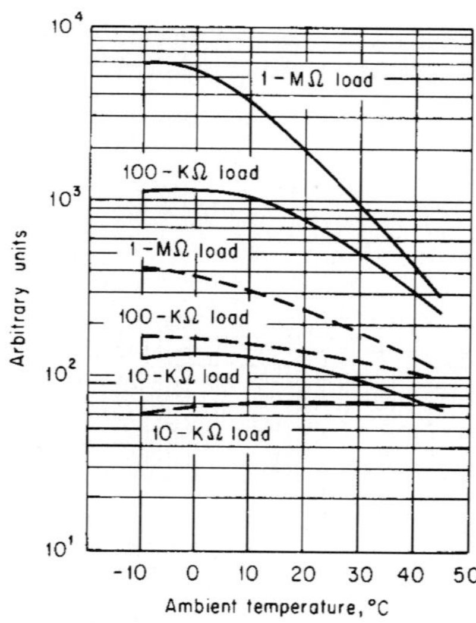

FIGURE 76 Relative signal and noise vs. temperature for *pn* junction silicon photodiode (zero bias); —— = signal; – – – – = noise. *(Electronuclear Laboratories, Bull. 1052, 1966.)*

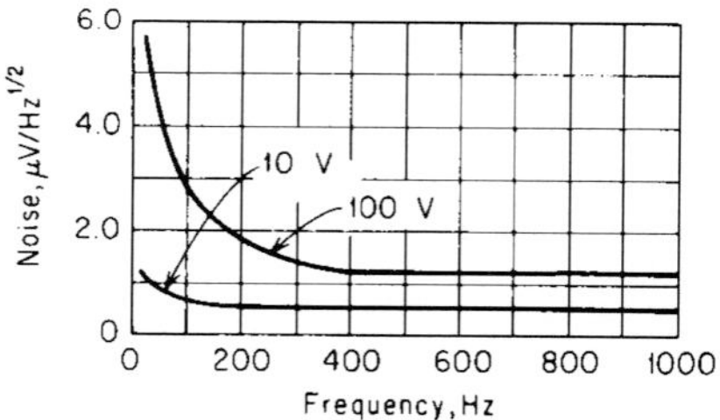

FIGURE 77 Typical *pn* junction and *pin* junction noise-frequency spectrum for different reverse bias ($A = 1 \times 1$ mm; $\mathbf{R}_L = 1\ \text{M}\Omega$). *(Electronuclear Laboratories, Bull. 1078, 1968.)*

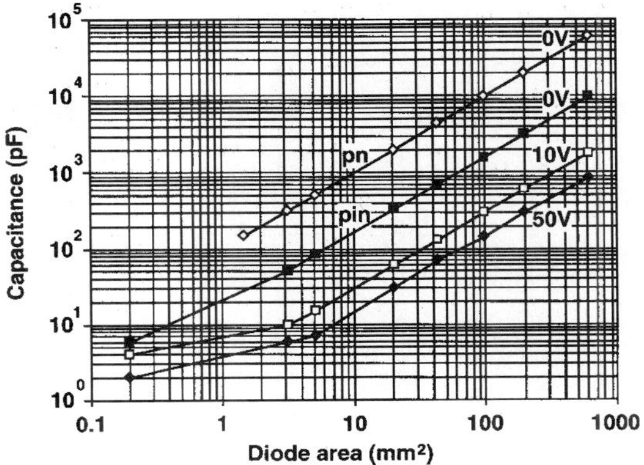

FIGURE 78 Capacitance as a function of detector area for *pn* junction silicon photodiodes operated in the photovoltaic mode (0 V) and *pin* junction photodiodes at 0-, 10-, and 50-V reverse bias. The larger depletion width, which is a consequence of the lightly doped "i" region in the *pin* device, gives *pin* diodes lower capacitance for the same device area. *(UDT Sensors, Optoelectronic Components Catalog.)*

RC (depends directly on device area), but can be limited by carrier diffusion outside the depletion region or by trapping of carriers in deep impurity centers. Typical data for a circuit using a 50-Ω load resistor is illustrated in Fig. 79.

Operating temperature: Ambient, but noise (leakage current) can be reduced by operating at lower temperatures (see Fig. 76 for typical signal and noise vs. temperature).

Uniformity: Typically ±8 percent across a diode area with a 40-μm focused light spot.

Linearity: 5 percent or better over 10 orders of magnitude flux from 10^{-13}–10^{-3} W/cm^2.

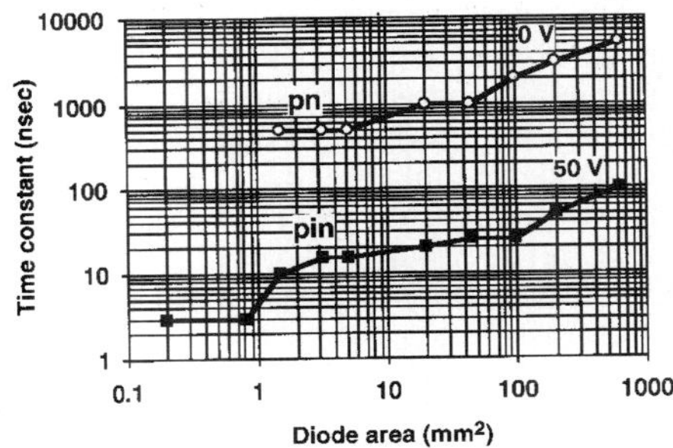

FIGURE 79 Time constant for *pn* junction silicon photodiodes operated in the photovoltaic mode (0 V) and a *pin* junction detector in the photoconductive or reverse-bias mode (50 V). A 50-Ω load was used in both cases, which limits sensitivity because of Johnson noise (see also Fig. 70). *(UDT Sensors, Optoelectronic Components Catalog.)*

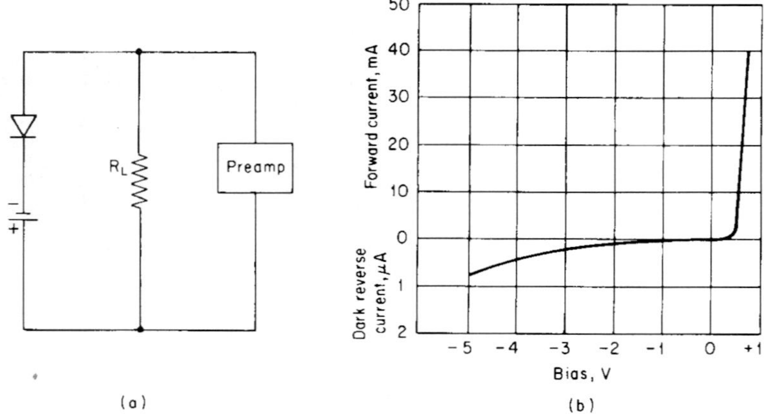

FIGURE 80 *pn* junction silicon photodiode: (*a*) recommended circuit; (*b*) typical electrical characteristics. *(Texas Instruments, Bull. SC-8385-366. Reprinted by permission of Texas Instruments.)*

Recommended circuit: See Fig. 80. High-impedance FET current-mode amplifier to supply fixed bias voltage, regardless of current.

Stability: See Fig. 8 and section relating to stability. Check with manufacturer.

Manufacturers: Advanced Photonix, EG&G Canada, EG&G Heimann, Electro-Optical Systems, Electro-Optics Technology, International Radiation Detectors, Janos Technology, Laser Precision Corp, Laser Systems Devices, Melles Griot, Newport/Klinger, Ophir Optronics, Optical Signature, Opto-Electronics, Optometrics, Oriel, Photonic Detectors, RMD, Sapidyne, Scientific Instruments, SEMICOA, Silonex, Spire, UDT Sensors.

Silicon pin Junction Photodiodes. The *pin* junction detector is faster but is also somewhat less sensitive than conventional *pn* junction detectors. *pin* photodiodes have slightly extended red response. In the normal *pn* junction, charge-collection time has a slow and a fast component. The fast component is due to photons absorbed in the depletion layer of the *pn* junction. Since the electric field in the depletion region is strong, carriers are quickly separated by drift. However, photons absorbed deeper in the material, beyond the depletion region, produce carriers which must diffuse to the junction before they are collected, and diffusion times are on the order of a microsecond. This component becomes more significant near the long-wavelength limit of the spectral response. Application of reverse bias in an ordinary *pn* junction detector reduces the capacitance, shortening the RC time constant, and increases the width of the depletion layer thereby increasing the fraction of photons absorbed within the high field region and proportionally increasing the fraction of the fast component of the response.

However, the doping level of the ordinary *pn* junction limits the extent of the depletion layer increase to only 5–10 μm at a reverse bias of 50 V (this assumes an abrupt junction with a concentration of 1×10^{15} cm^{-3}). *pin* detectors incorporate a very lightly doped region between the *p* and *n* regions which allows a modest reverse bias to form a depletion region the full thickness of the material (~500 μm for a typical silicon wafer). Extended red response in a *pin* device is a consequence of the extended depletion layer width, since longer wavelength photons will be absorbed in the active device region. Unfortunately, the higher dark current collected from generation within the wider depletion layer results in

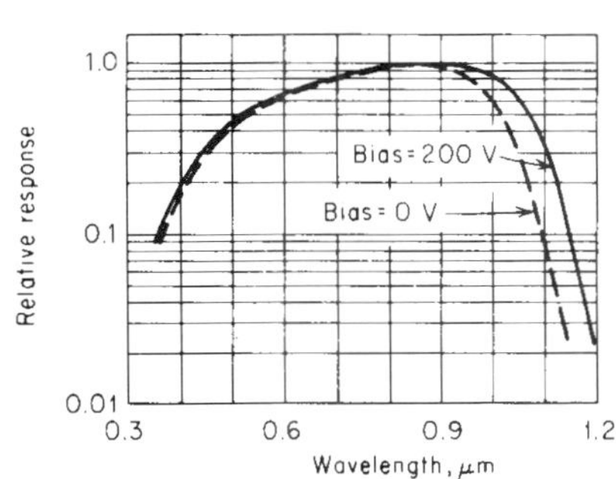

FIGURE 81 Dependence of spectral response on bias for silicon photodiodes. *(Electronuclear Laboratories, Bull. 1076, 1968.)*

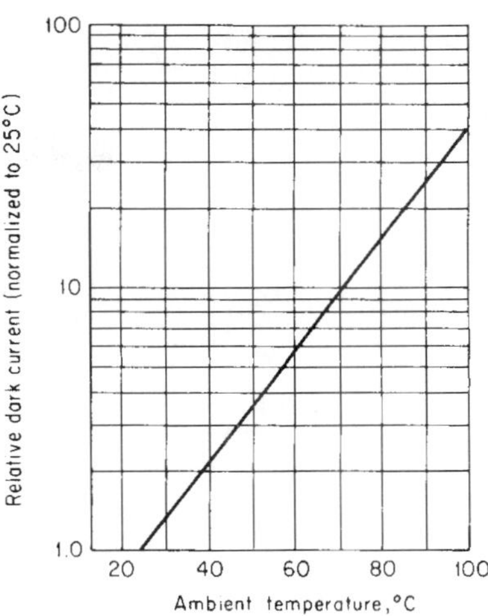

FIGURE 82 Relative dark current vs. temperature for *pin* junction silicon photodiode (bias—100 V). *(Electronuclear Laboratories, Bull. 1076, 1969.)*

lower sensitivity. Generation of carriers can be minimized by minimizing the concentration of deep-level impurity centers in the detector with careful manufacturing. Operation at lower temperature will also reduce the dark current.

Sensitivity: $D^*(\lambda_{pk}) = 1 \times 10^{12}$ Jones for 2 mm^2 area (depends slightly on bias, see Figs. 73 and 81). For high-speed operation, detectivity is lower (see Fig. 70).

Noise: Depends upon diode area and circuit load resistance. Johnson noise will dominate at low values of load resistance when circuit is optimized for fast response. Preamp noise may also limit. See Fig. 70.

Responsivity: Similar to *pn* junction. See Fig. 69.

Quantum efficiency: >90 percent quantum efficiency achievable with antireflection coating.

Capacitance: Proportional to detector area. See Fig. 78.

Operating temperature: See Fig. 82.

Time constant: Varies with capacitance (device area); see Fig. 79.

Sensitive area: 0.2–600 mm^2 readily available.

Recommended circuit: Same as for *pn* junction photodiodes. See Figs. 80 and 83.

Stability: See Fig. 8 and section relating to stability. Check with manufacturer.

Manufacturers: Same as for *pn* junction photodiodes.

UV- and Blue-enhanced Photodiodes. Blue- and UV-enhanced photodiodes may improve the quantum efficiency by 50 to 100 percent over standard photodiodes in the blue and UV spectral region. The quantum efficiency of ordinary *pn* and *pin* junction photodiodes degrades rapidly in the blue and UV spectral regions. This is because the high absorption coefficient of silicon at these wavelengths causes the photocarriers to be generated within the heavily doped $p+$ (or $n+$) contact surface where the lifetime is short due to the high doping and/or surface recombination. Blue- and UV-enhanced photodiodes optimize the response at short wavelengths by minimizing near-surface carrier

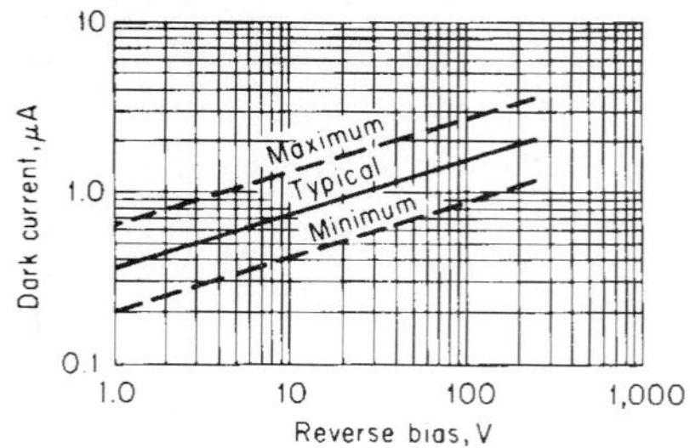

FIGURE 83 Typical dark current vs. bias for pin silicon photodiode ($A = 1 \times 1$ mm). *(Electronuclear Laboratories, Bull. 1078, 1968.)*

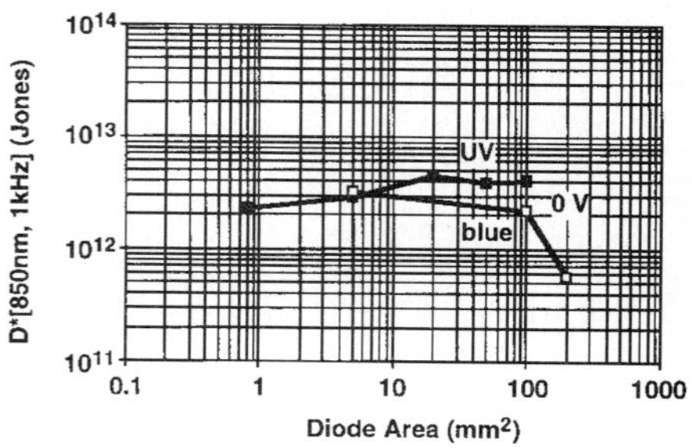

FIGURE 84 D^* as a function of diode area for blue- and UV-enhanced silicon photodiodes operated in the photovoltaic mode (0-V bias). *(UDT Sensors, Optoelectronic Components Catalog.)*

recombination. This can be achieved by using very thin and highly graded $p+$ (or $n+$ or metal Schottky) contacts, by using lateral collection to minimize the percentage of the surface area which is heavily doped, and/or passivating the surface with a fixed surface charge to repel minority carriers from the surface. These devices typically have quartz windows or UV-transmissive glass, compatible with good transmission into the UV spectrum. The user should be aware that UV and higher energy radiation in particular can alter the fixed charge conditions in the surface region of silicon and other detectors (typically in the surface oxide) which can cause the detector performance to drift and/or be unstable (see Fig. 8).

Sensitivity: See Figs. 69 and 84; $D^*(\lambda_{pk}) = 3\text{–}5 \times 10^{12}$ Jones for diodes with areas of 1–100 mm^2 at $V_R = 0$, $R_L = 40$ MΩ.

Quantum efficiency: Same as *pn* junction photodiodes, but enhanced in the UV and blue regions by 50 percent or more (see Fig. 69).

Responsivity: see Fig. 69.

Capacitance: Comparable to *pn* junction photodiodes (See Fig. 85).

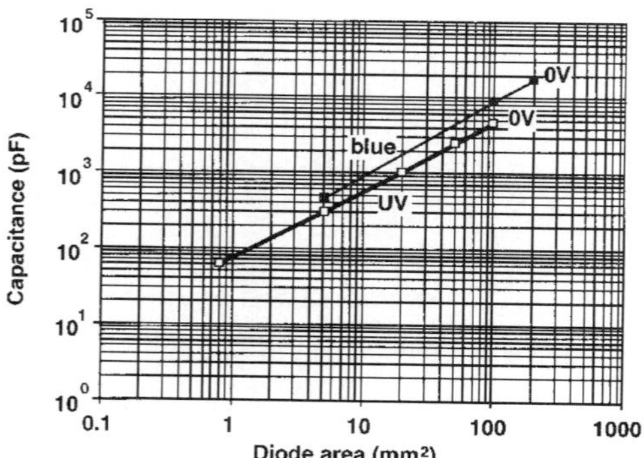

FIGURE 85 Capacitance as a function of detector area for blue- and UV-enhanced silicon photodiodes operated in the photovoltaic mode (0 V). The capacitance per unit area is close to that of *pn* junction photodiodes shown in Fig. 78. *(UDT Sensors, Optoelectronic Components Catalog.)*

Operating temperature: Ambient.

Time constant: Dependent upon device type. Increases with device area; 200 ns–6 μs for areas 1–100 mm^2 (see Fig. 86).

Sensitive area: 1–200 mm^2 readily available.

Recommended circuit: Same as *pn* junction photodiodes.

Stability: See Fig. 8 and section relating to stability. Check with manufacturer.

Manufacturers: See list for silicon *pn* junction photodiodes.

Silicon Avalanche Photodiodes (APDs). The avalanche photodiode, or APD, is especially useful where both fast response and high sensitivity are required. Whereas normal photodiodes become Johnson- or thermal-noise-limited when used with a low-impedance load resistor for fast response, avalanche photodiodes make use of internal multiplication, associated with reverse breakdown in the *pn* junction in order to keep the detector noise above the Johnson noise level. [Because the response time is usually RC-limited, small load resistors (often 50 Ω) are used to achieve fast signal response. However, as the load resistor is decreased, the detector noise voltage decreases in direct proportion, whereas the Johnson noise decreases only as the square root of the load resistor. Thus, the detector noise voltage can become lower than the Johnson noise for load resistance values smaller than a critical value. With an APD device, lower values of load resistors can be used without reaching the critical value because the internal gain boosts the detector noise voltage.]

Stable avalanche or multiplication is made possible by a guard-ring construction using n^+pp^+, Schottky-nn^+, or $n^+p\pi p^+$ structure; beveled *pin* structure (see Fig. 87); mesa structures; or other structures which prevent surface breakdown (Sze, 1981). However, very careful bias control is essential for stable performance. An optimum gain exists below which the system is limited by receiver noise and above which shot noise dominates receiver noise and the overall noise increases faster than the signal (Fig. 88).

In addition to fast-response applications, avalanche photodiodes are useful whenever amplifier noise is limiting, e.g., small-area devices. Signal-to-noise-ratio improvements of one to two orders of magnitude over a nonavalanche detector can be achieved.

APDs are sometimes used in combination with scintillator crystals such as CsI to detect high-energy radiation in the range of 10 to 1000 keV.

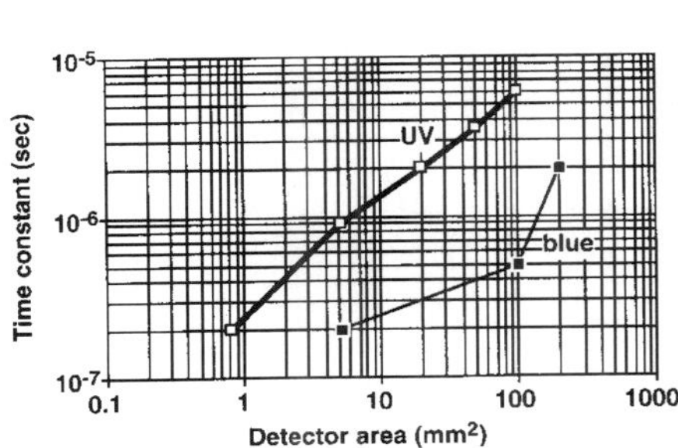

FIGURE 86 Time constant for UV- and blue-enhanced silicon photodiodes as a function of detector area at zero-volts bias. A 50-Ω load was used in both cases. *(UDT Sensors, Optoelectronic Components Catalog.)*

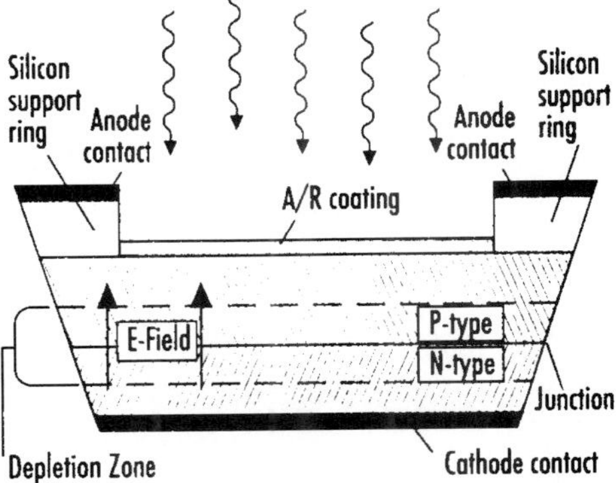

FIGURE 87 Cross section of a beveled-edge silicon avalanche photodiode. The beveled edge prevents early breakdown. *(Advanced Photonix, Avalanche Photodiode Catalog.)*

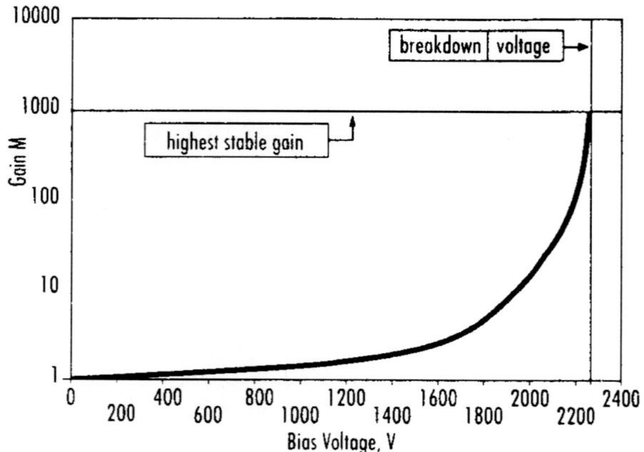

FIGURE 88 Gain as a function of reverse bias can reach 1000. This operating point is very close to breakdown and requires careful bias control. *(Advanced Photonix, Avalanche Photodiode Catalog.)*

Sensitivity: D^* 3–5 × 10^{13} Jones (see Fig. 89).

Noise: Function of detector area. As gain increases, noise increases (see Fig. 90). Optimum gain is where avalanche noise equals system noise. Thus, optimum gain is a function of system noise.

Responsivity: Photocurrent is the product of the incident optical power in watts, wavelength in micrometers, and quantum efficiency (η) divided by 1.24 and multiplied by the avalanche gain M. $I_{\text{photo}} = M(P\lambda\eta/1.24)$ (See Figs. 88 and 91).

Quantum efficiency: Typically 85 percent peak (see Fig. 91).

Capacitance: Depends on bias and area (see Fig. 92).

Sensitive area: 20–200 mm^2.

Series resistance: Depends on area; typical values are 40 Ω for 5-mm dia. to 5 Ω for 16-mm dia.

Time constant: See Fig. 93.

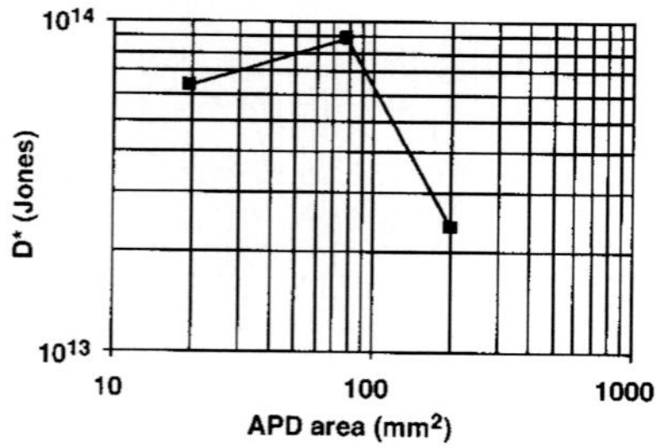

FIGURE 89 D^* for three silicon avalanche photodiodes (APDs) shown as a function of diode area. *(Advanced Photonix, Avalanche Photodiode Catalog.)*

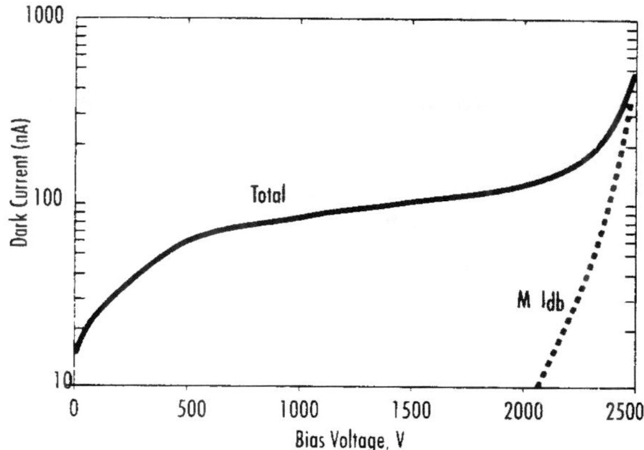

FIGURE 90 Dark current as a function of reverse bias for a 16-mm-diameter APD. At low-bias surface dark current dominates, but avalanche-multiplied bulk dark current increases rapidly as the gain increases. *(Advanced Photonix, Avalanche Photodiode Catalog.)*

Recommended circuit: Requires a filtered high-voltage dc supply that itself must have very low noise and a load resistor. The output may be ac- or dc-coupled. (See Fig. 94.)

Operating temperature: −40–+45°C.

Stability: Exposure to UV or high-energy radiation may affect dark current. See Fig. 8 and section relating to stability. Check with manufacturer.

Manufacturers: Advanced Photonix, BSA Technology, Devar, EG&G Judson, EG&G Vactec, EG&G Canada, Electro-Optical Systems, Hamamatsu, Janos Technology, Newport/Klinger, Opto-Electronics (Ontario), Oriel, Photonic Detectors, Photonic Packaging Technologies, RMD, Texas Optoelectronics, Thorn EMI Electron Tubes.

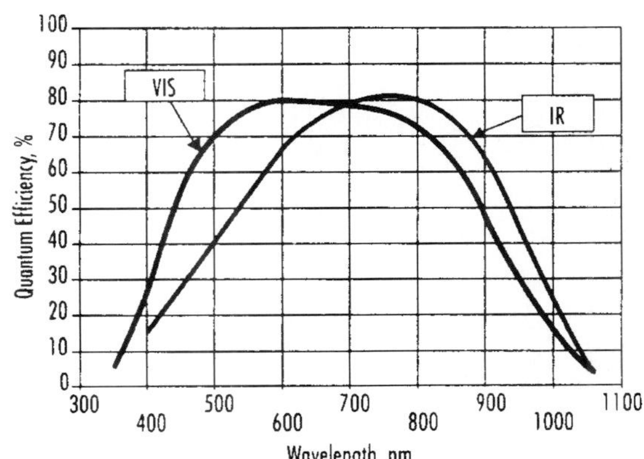

FIGURE 91 APD quantum efficiency at high gain. Adjustment of the oxide deposited on the surface produces two different curves. *(Advanced Photonix, Avalanche Photodiode Catalog.)*

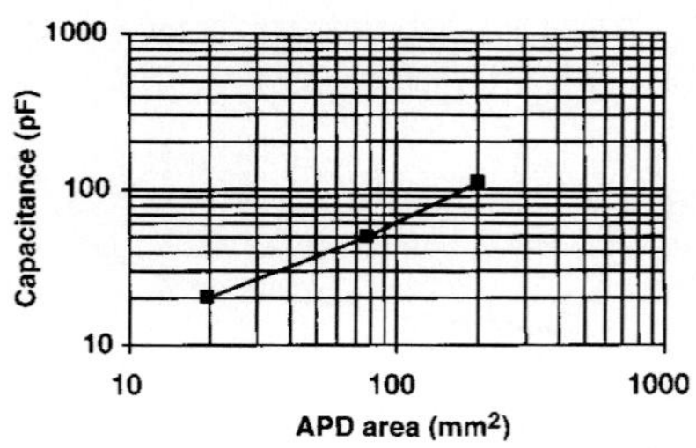

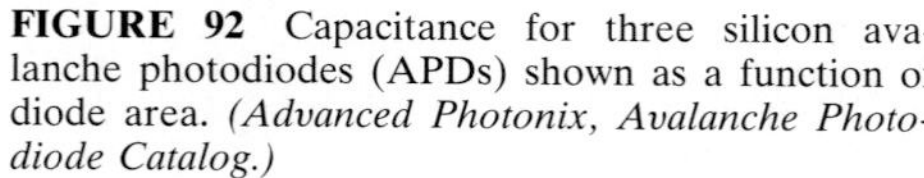

FIGURE 92 Capacitance for three silicon avalanche photodiodes (APDs) shown as a function of diode area. *(Advanced Photonix, Avalanche Photodiode Catalog.)*

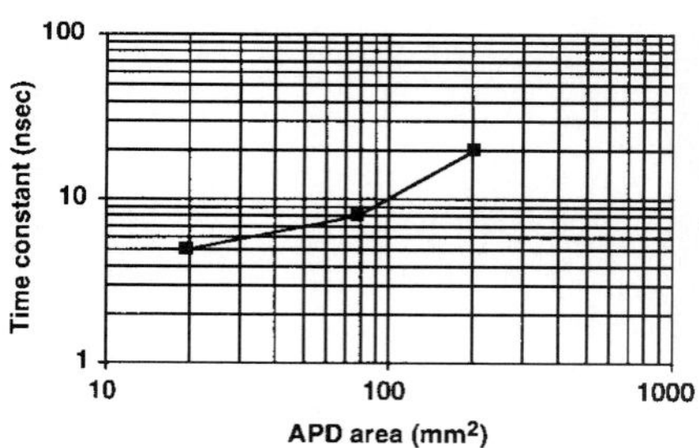

FIGURE 93 Time constant for three silicon avalanche photodiodes (APDs) shown as a function of diode area. *(Advanced Photonix, Avalanche Photodiode Catalog.)*

InGaAs. Indium gallium arsenide detectors have been developed for optimum performance with fiber-optic communications at 1.3 and 1.55 μm. This detector material has a direct bandgap and represents one of several compound semiconductor alloy systems specially developed for photodetectors. In the case of this alloy of two group III-V chemical compound semiconductors, the ratio of InAs to GaAs controls the spectral cutoff, allowing the detector to be optimized for a particular wavelength. InGaAs detectors have generally been specialized for high-speed applications with optimum sensitivity since these performance factors can drive a fiber-optic system throughput and cost. For this reason, available devices include:

- *pin* photodiodes
- avalanche photodiodes (APDs)

The significance of these devices to fiber-optic applications is reflected in the number of vendors who sell integrated packages of InGaAs photodetectors combined with preamplifiers and fiber-optic pigtails.

The InGaAs alloy system allows the spectral response to be tailored to longer wavelengths than the quartz fiber-optic bands and devices with cutoffs of 2.2 and 2.6 μm are also available.

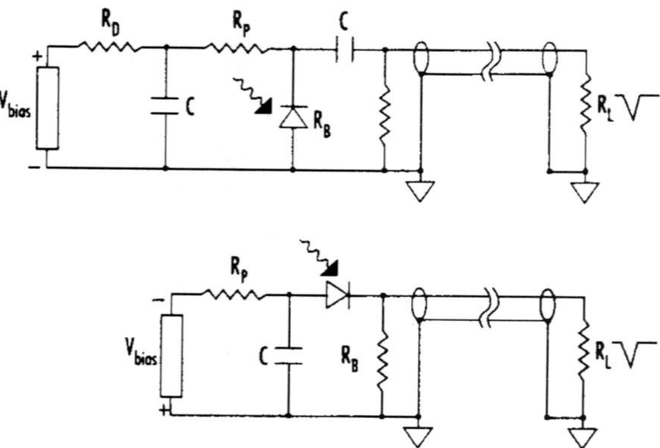

FIGURE 94 ac- and dc-coupled APD circuits with a filter following the power supply and a coaxial cable on the output. *(Advanced Photonix, Avalanche Photodiode Catalog.)*

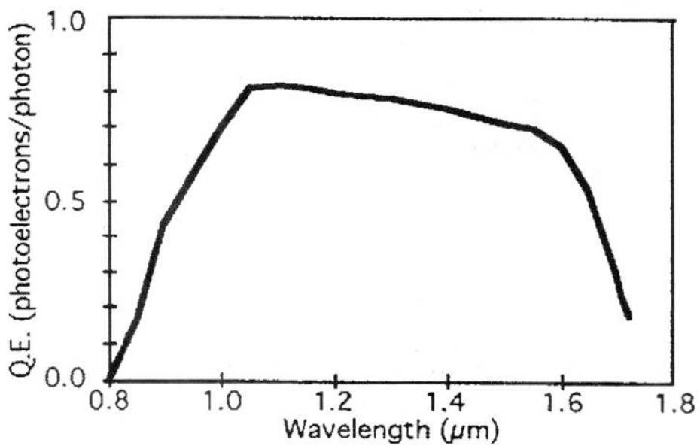

FIGURE 95 Spectral dependence of quantum efficiency for an InGaAs detector having a cutoff of 1.67 μm. *(Sensors Unlimited, data sheet.)*

InGaAs pin Photodiode

Sensitivity: D^* mid-10^{12} Jones for 1.67-μm cutoff; $D^* \approx 1 \times 10^{12}$ Jones for 2.2-μm cutoff, and $D^* \approx 5 \times 10^{11}$ Jones for 2.6-μm cutoff.

Responsivity: 0.85–0.95 A/W in the range of 1.3–1.55 μm.

Quantum efficiency: For 1.67-μm cutoff, see Fig. 95.

Dark current: See Fig. 96.

Capacitance: 0.7–1.2 × 10^4 pF/cm^2 for 1.7-μm cutoff; 2.5 × 10^4 pF/cm^2 for 1.85-μm cutoff; 3 × 10^4 pF/cm^2 for 2.15-μm cutoff; 5 × 10^4 pF/cm^2 for 2.65-μm cutoff. See also Fig. 97 for bias dependence.

Time constant: Varies with resistance-capacitance (RC) time (see Fig. 98). Since capacitance depends upon reverse bias, the time constant varies proportionally (see Fig. 97 for dependence of capacitance on bias).

Size: 0.05–3-mm dia.

Recommended circuit: Standard photodiode options; zero bias for best sensitivity, reverse bias for maximum speed.

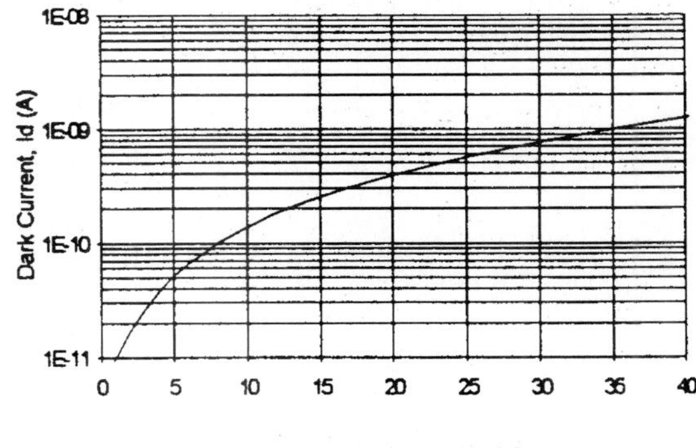

FIGURE 96 Dark current as a function of reverse-bias voltage for a 60-μm-diameter InGaAs detector having a cutoff of 1.67 μm. *(Fermionics, InGaAs Photodiodes.)*

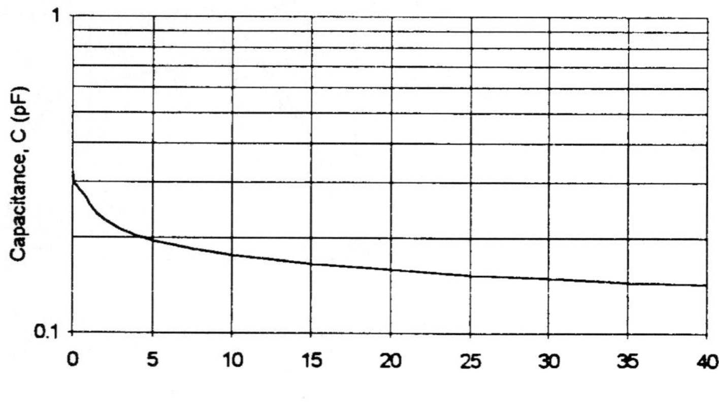

FIGURE 97 Capacitance as a function of reverse-bias voltage for a 60-μm-diameter InGaAs detector having a cutoff of 1.67 μm. *(Fermionics, InGaAs Photodiodes.)*

InGaAs Avalanche Photodiode

Sensitivity: $D^* \approx 5 \times 10^{11}$ Jones for 1.7-μm cutoff. In the fiber-optics industry, the sensitivity is also given in power units of dBm. Figure 99 compares InGaAs *pin*, APD, and Ge APD sensitivities.

Spectral response: 1.0–1.65 μm; see Fig. 100.

Responsivity: 8–10 A/W typical.

Avalanche gain: Critically depends upon reverse bias, see Fig. 101.

Capacitance: At gain of 12, 7.5×10^4 pF/cm^2.

Bandwidth: Up to 3 GHz; see Figs. 102 and 103.

Size: 0.04–0.5-mm dia.

Dark current: Dependent upon reverse bias, device structure, and temperature. See Figs. 104, 105, and 106.

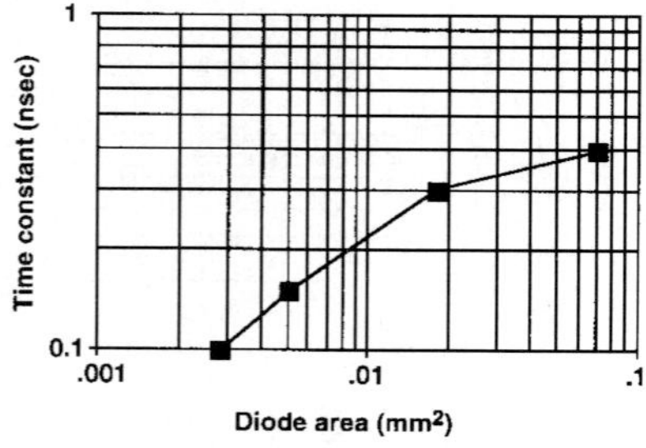

FIGURE 98 Time constant for small InGaAs *pin* photodiodes as a function of diode area measured with a 50-Ω load.

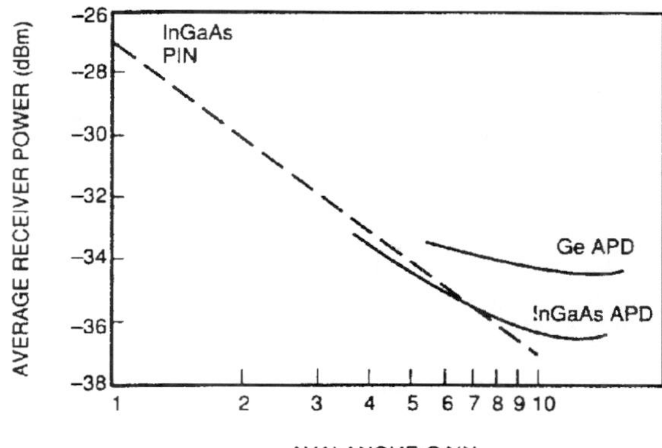

FIGURE 99 APD receiver sensitivity. Typical receiver sensitivity at a receiver rate of 1.7 Gbit/s and $\lambda = 1.3\ \mu m$ for an InGaAs *pin*, Ge APD, and InGaAs APD. *(AT&T, 126A/B,C ASTROTEC InGaAs.)*

Recommended circuit: See Fig. 107.

Manufacturers: Advanced Photonix, Atomergic Chemetals, AT&T, BSA, EG&G Canada, Electro-Optical Systems, Electro-Optics Technology, Emcore, Epitaxx, Fermionics, GCA Electronics, Germanium Power Devices, Hamamatsu, New Focus, North Coast Scientific, Opto-Electronics, Ortel, Photonic Detectors, Sensors Unlimited, Spire, Swan Associates, Telcom Devices, UDT Sensors.

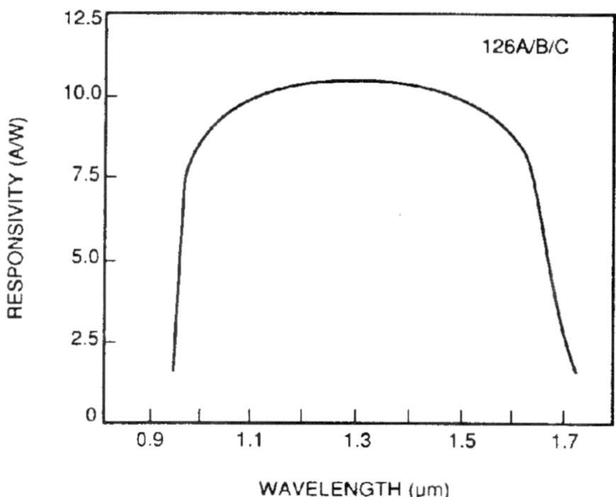

Note: Responsivity = (chip quantum efficiency) x gain x λ (μm)/1.24. The minimum chip quantum efficiency is 80%, and the minimum pigtail coupling efficiency is 90%.

FIGURE 100 Responsivity for an InGaAs avalanche photodiode vs. wavelength for avalanche gain of 12. *(AT&T, 126A/B,C, ASTROTEC InGaAs.)*

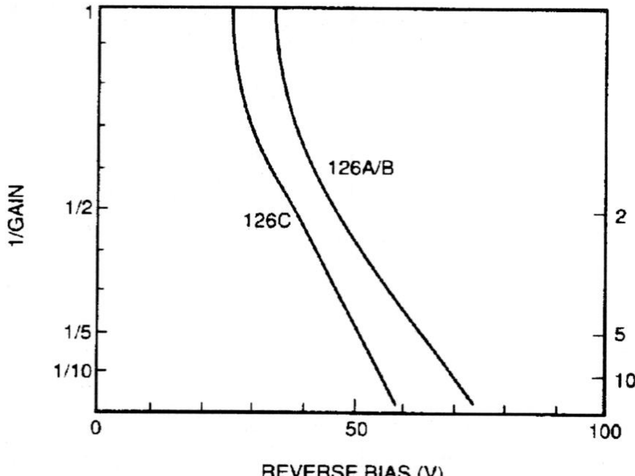

FIGURE 101 Inverse of avalanche gain for an InGaAs avalanche photodiode vs. reverse bias. *(AT&T, 126A/B,C ASTROTEC InGaAs.)*

Ge. Germanium intrinsic photodetectors are similar to intrinsic silicon detectors but offer spectral response out to 1.5 to 1.8 μm. *pn* junction photodiodes offer submicrosecond response or high sensitivity from the visible region to 1.8 μm. Zero bias is generally used for high sensitivity and large reverse bias for high speed. As in the case of silicon, germanium has an indirect bandgap and soft spectral cutoff. The previous discussion on silicon detectors applies in general, with the exception that blue- and UV-enhanced devices are not relevant to germanium detectors. Germanium detectors, because of their narrower bandgap, have higher leakage currents at room temperature, compared to silicon detectors. Detector impedance increases about an order of magnitude by cooling 20°C

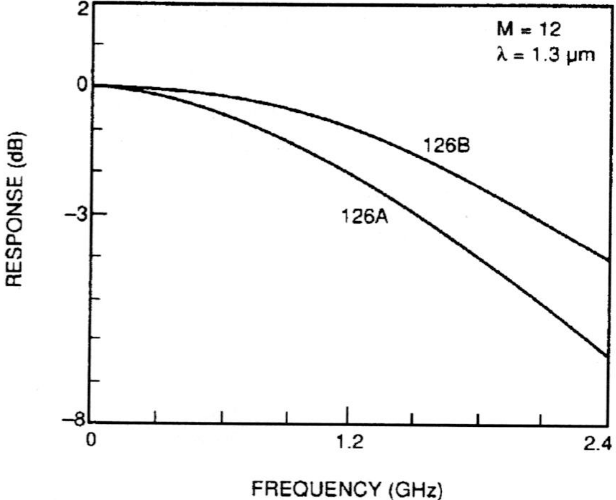

FIGURE 102 Frequency response of InGaAs APD (126A/B). *(AT&T, 126A/B,C ASTROTEC InGaAs.)*

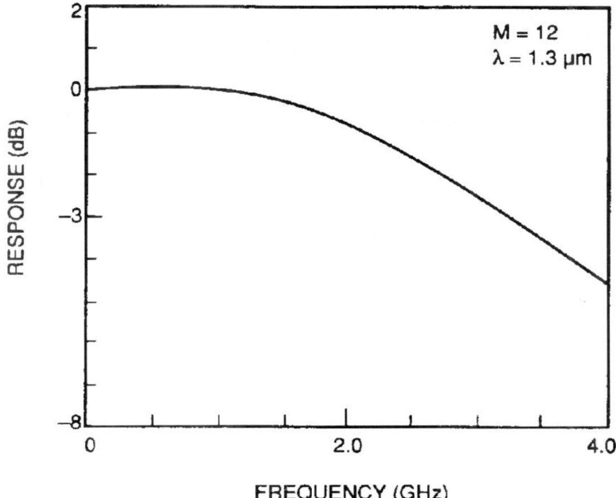

FIGURE 103 Frequency response of InGaAs APD (126C). *(AT&T, 126A/B,C ASTROTEC InGaAs.)*

below room temperature. Thus, performance can improve significantly with thermoelectric cooling or cooling to liquid nitrogen temperature.

As with silicon, the device structure and bias configuration can affect spectral response and rise time. Three detector types are available:

- *pn* junction
- *pin* junction
- avalanche photodiode (APD)

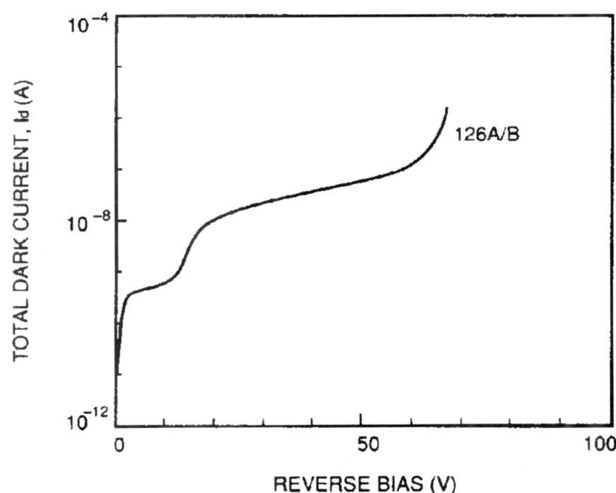

FIGURE 104 Dark current vs. reverse bias of InGaAs APD (126A/B). *(AT&T, 126A/B,C ASTROTEC InGaAs.)*

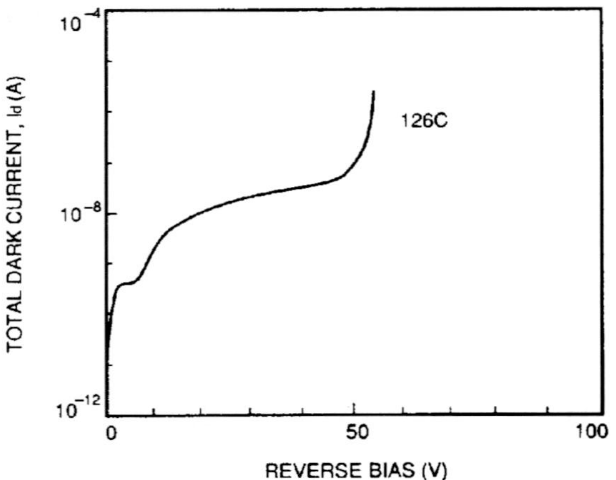

FIGURE 105 Dark current vs. reverse bias of InGaAs APD (126C). *(AT&T, 126A/B,C ASTROTEC InGaAs.)*

Germanium pn and pin

Sensitivity: $D^*(\lambda\text{peak}, 300\,\text{Hz}, \text{room temperature}) > 2 \times 10^{11}$ Jones, increases significantly with cooling by thermoelectric cooler or liquid nitrogen. (See Figs. 108, 109 and 110).

Quantum efficiency: >50 percent with antireflection coating.

Noise: See Figs. 111 and 112.

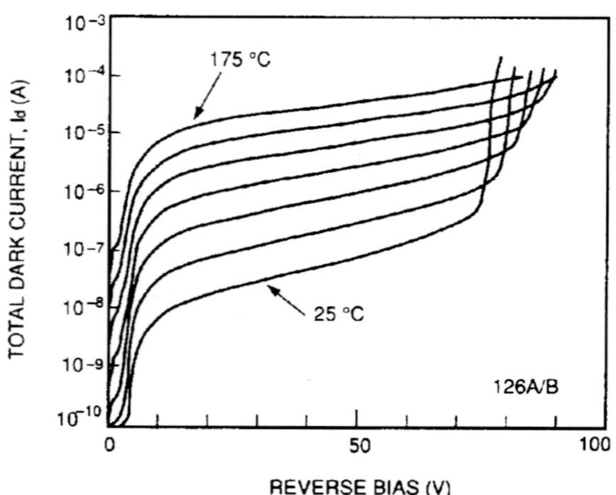

Note: The temperature dependence of the 126C dark current is the same as the 126A/B.

FIGURE 106 Dark current vs. voltage of InGaAs APD as a function of temperature at 25°C increments. *(AT&T, 126A/B,C ASTROTEC InGaAs.)*

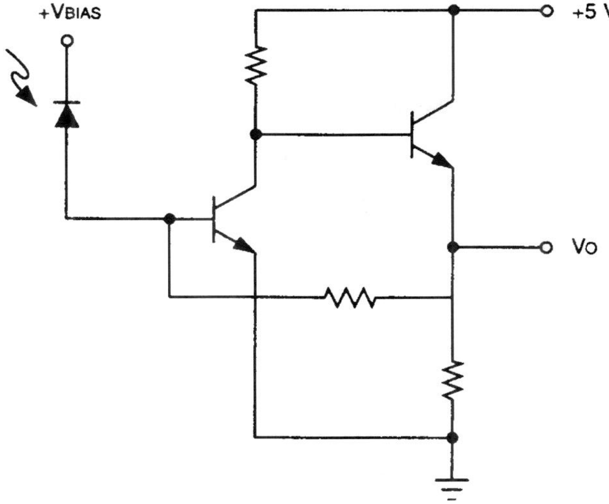

* Reference: Optical Fiber Communications, John M. Senior, © 1985, Prentice-Hall; ISBN-0-13-638248-7.

FIGURE 107 Bipolar transimpedance amplifier for InGaAs avalanche photodiode. *(AT&T, 126A/B,C ASTROTEC InGaAs.)*

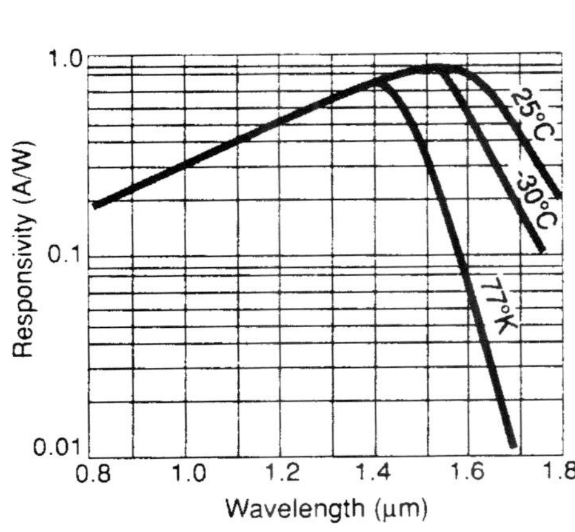

FIGURE 108 Spectral response for a germanium *pn* junction photodiode at three temperatures. *(EG&G Judson, Infrared Detectors, 1994.)*

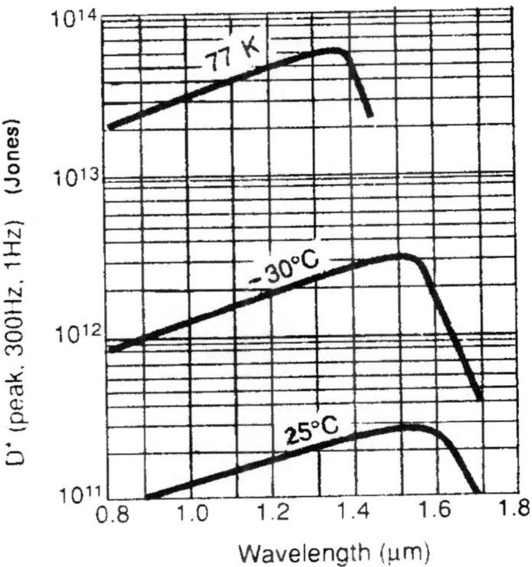

FIGURE 109 D^* as a function of wavelength for a germanium *pn* junction photodiode at three temperatures. *(EG&G Judson, Infrared Detectors, 1994.)*

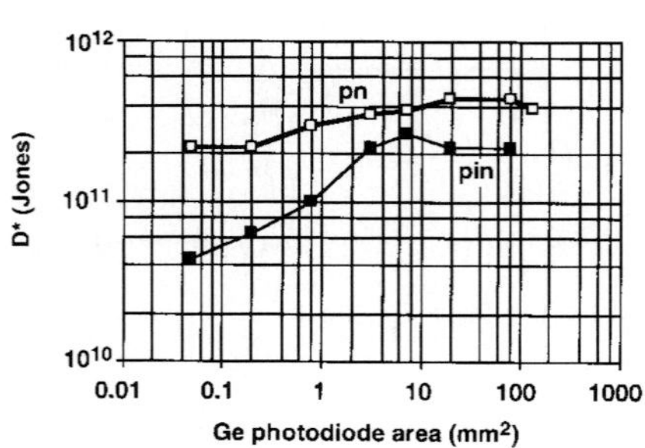

FIGURE 110 D^* for germanium *pn* and *pin* photodiodes shown as a function of diode area. *(EG&G Judson, Infrared Detectors, 1994.)*

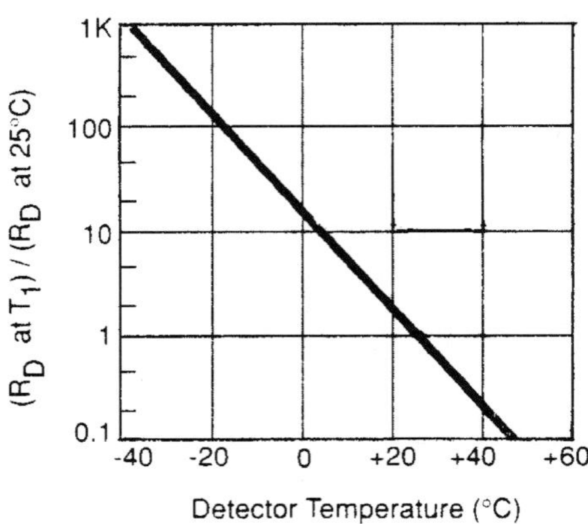

FIGURE 111 Ratio of resistance at temperature T to the resistance at 25°C for a germanium *pn* junction photodiode. *(EG&G Judson, Infrared Detectors, 1994.)*

Responsivity: 0.9 A/W at peak wavelength. see Fig. 108.

Capacitance: Lower for *pin* structure compared with *pn* diode. See Fig. 113.

Time constant: *pin* diodes provide faster response. See Fig. 114.

Sensitive area: 0.25–13-mm dia. standard.

Operating temperature: Ambient, TE-cooled, or liquid nitrogen.

Profile: ±2 percent across active area at 1.3 μm.

Linearity: Excellent over 10 orders of magnitude. See Fig. 115.

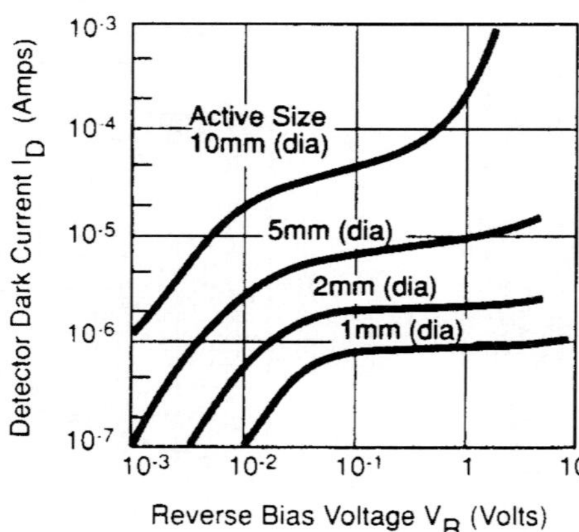

FIGURE 112 Dark current as a function of reverse bias for germanium *pn* junction photodiodes of different diameters at 25°C. *(EG&G Judson, Infrared Detectors, 1994.)*

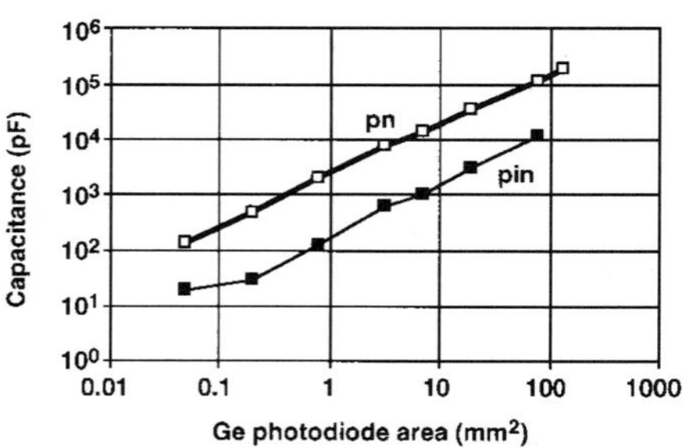

FIGURE 113 Capacitance for germanium *pn* and *pin* photodiodes shown as a function of diode area. *(EG&G Judson, Infrared Detectors, 1994.)*

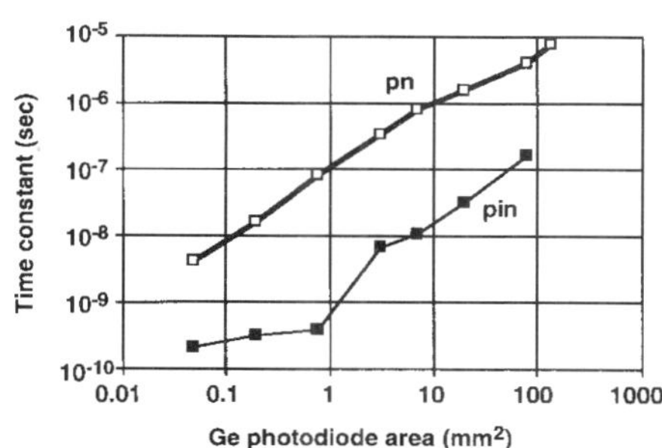

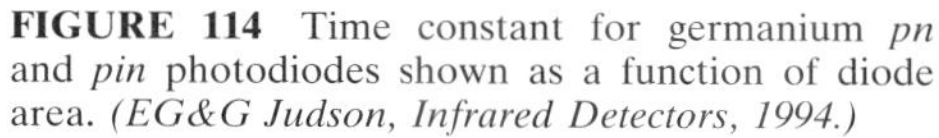

FIGURE 114 Time constant for germanium *pn* and *pin* photodiodes shown as a function of diode area. *(EG&G Judson, Infrared Detectors, 1994.)*

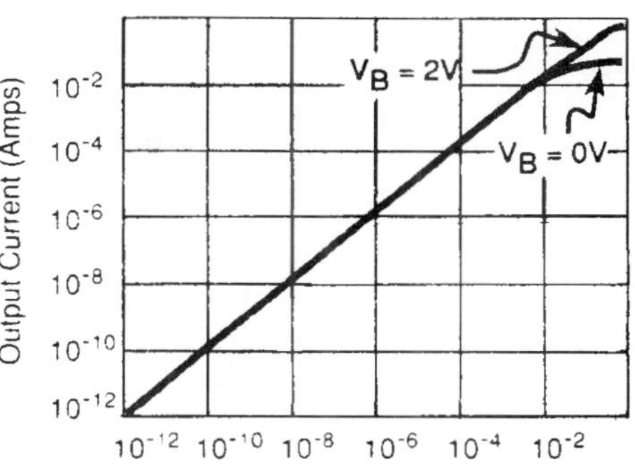

FIGURE 115 Linearity of a germanium *pn* junction photodiode. *(EG&G Judson, Infrared Detectors, 1994.)*

Recommended circuit: See previous section on silicon photodiodes.

Manufacturers: Atomergic Chemetals, Cincinnati Electronics, Edinburgh Instruments, EG&G Judson, Electro-Optical Systems, Fastpulse Technology, Germanium Power Devices, Graseby Infrared, Newport/Klinger, North Coast Scientific, Opto-Electronics, Oriel, Oxford Instruments, Scientific Instruments.

Germanium Avalanche Photodiode. The germanium avalanche photodiode (APD) is similar to the silicon APD but has lower optimum gain, longer cut-off wavelength (1.7 μm), and higher leakage current. Germanium APDs combine the sensitivity of a Ge *pn* photodiode and the speed of a *pin* Ge photodiode.

Sensitivity: D^* 2×10^{11} Jones at 30 MHz for a diode with area of $5 \times 10^{-2}\,\text{mm}^2$ or about the same as for a *pn* Ge diode with the same area, and about a factor of 4 higher than a *pin* Ge photodiode (compare with Fig. 109). D^* depends on gain.

Gain: See Fig. 116.

Dark current: See Fig. 117.

Capacitance: 2 pF at 20 V reverse bias for 100-μm diameter, 8 pF at 20 V reverse bias for 300 μm dia.

Quantum efficiency: 60–70 percent at 1.3 μm.

Responsivity: Photocurrent is the product of the incident optical power in watts, wavelength in micrometers, and quantum efficiency (η) divided by 1.24 and multiplied by the avalanche gain M. $I_{\text{photo}} = M(P\lambda\eta/1.24)$ (See Figs. 108 and 116).

Operating temperature: Ambient or TE-cooled.

Time constant: 0.2 ns for 100-μm diameter; 0.3 ns for 300-μm diameter; both at 1.3 μm, $M = 10$ and with $R_L = 50\ \Omega$. See Fig. 118.

Sensitive area: 100- and 300-μm dia. standard

Recommended circuit: See circuit recommended for Si APD.

Manufacturers: EG&G Judson, North Coast Scientific.

PbS. Photoconductive lead sulfide was one of the earliest and most successful infrared detectors. Even today it is one of the most sensitive uncooled detectors in the 1.3- to 3-μm

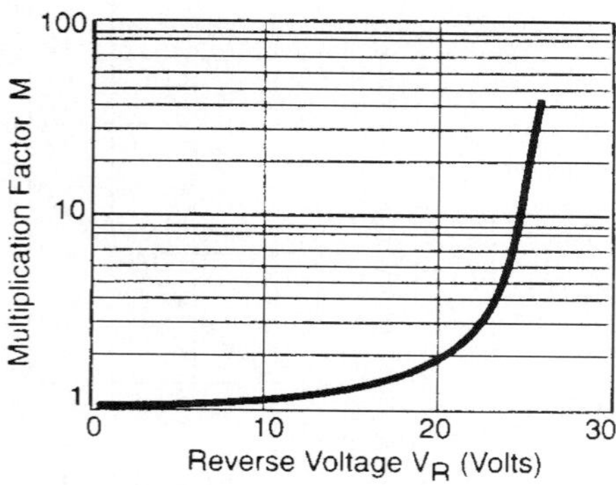

FIGURE 116 Gain as a function of reverse bias for a germanium avalanche photodiode. *(EG&G Judson, Infrared Detectors, 1994.)*

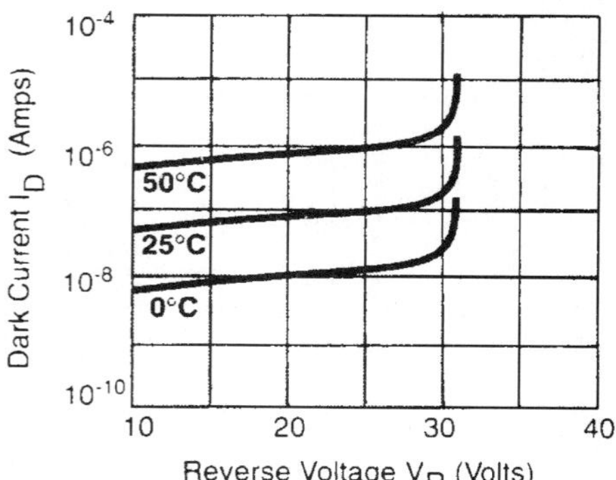

FIGURE 117 Dark current as a function of reverse bias for a germanium avalanche photodiode. *(EG&G Judson, Infrared Detectors, 1994.)*

spectral region. With cooling, PbS sensitivity is competitive with other detectors out to about 4.2 μm; however, its response time is slow.

Many PbS characteristics can be varied by adjusting the chemistry of the deposition process and/or the postdeposition heat treatment. These characteristics include spectral detectivity, time constant, resistance, and upper limit of operating temperature (Johnson, 1984). PbS is generally made by chemical reaction of Pb acetate and thiourea, except for high-temperature (373 K) applications, where evaporation is used. The material is deposited as a thin film (1 to 2 μm thick) on a variety of substrates, such as sapphire. With photolithographic processing, small sensitive areas can be made with comparatively high D^* values.

PbS may be tailored for ambient or room-temperature operation (ATO), intermediate or thermoelectrically cooled operation (ITO), and low-temperature or nitrogen-cooled operation (LTO). They are manufactured differently for particular temperature ranges, as shown in Table 2.

Sensitivity: $D^* 1.5 \times 10^{11}$ Jones at 295 K. See Figs. 119–124.

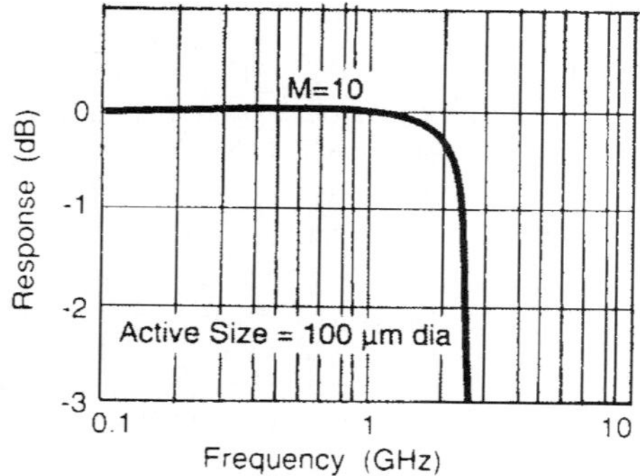

FIGURE 118 Frequency response of a germanium avalanche photodiode at two gain operating points. *(EG&G Judson, Infrared Detectors, 1994.)*

TABLE 2

	Typical operating temperature, K			
	350	273 (ATO)	193 (ITO)	77 (LTO)
Sensitivity	†	Figs. 119, 125	Figs. 120, 123, 125	Figs. 121, 125
$\dfrac{D^*(\lambda_{max})}{D^*(500\text{ K})}$		105	55	17
Noise, cm·Hz$^{1/2}$/W		Fig. 127	Fig. 127	Fig. 127
Dark resistance, MΩ/sq	<0.3	<2	<10	<20
Time constant, μs‡	50	100–500	5000	3000

† At 350 K, cutoff wavelength moves into ~2.4 μm, with $D^*(\lambda_{max}) \approx 10^{10}$ Jones.
‡ These are typical values; the time constant can be adjusted over two orders of magnitude in fabrication, but D^* is affected.

Responsivity: Depends on detector area, bias, resistance, and operating temperature (see Figs. 125 and 126).

Quantum efficiency: Generally limited by incomplete absorption in the thin film to ≈30 percent as estimated from blip D^* values.

Noise: Dominated by $1/f$ noise at low frequencies. See Figs. 122, 126, and 127.

Time constant: Can be varied in manufacturing. Typical values are 0.2 ms at 295 K, 2–3 ms at 193 and 77 K. See Fig. 126.

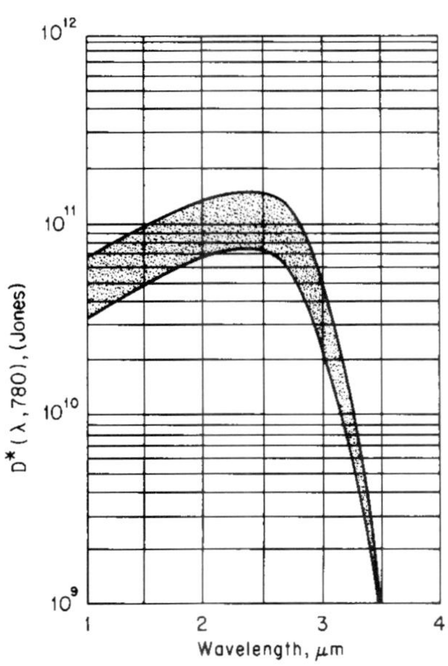

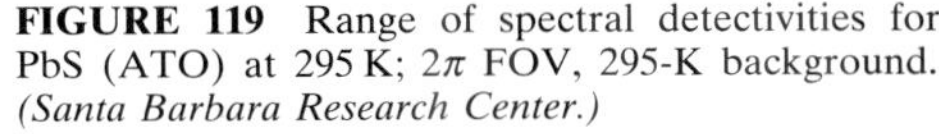
FIGURE 119 Range of spectral detectivities for PbS (ATO) at 295 K; 2π FOV, 295-K background. *(Santa Barbara Research Center.)*

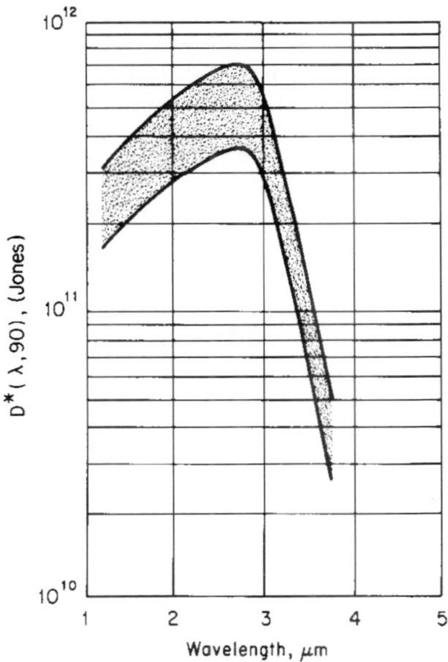

FIGURE 120 Range of spectral detectivities for PbS (ITO) at 193 K; 2π FOV, 295-K background. *(Santa Barbara Research Center.)*

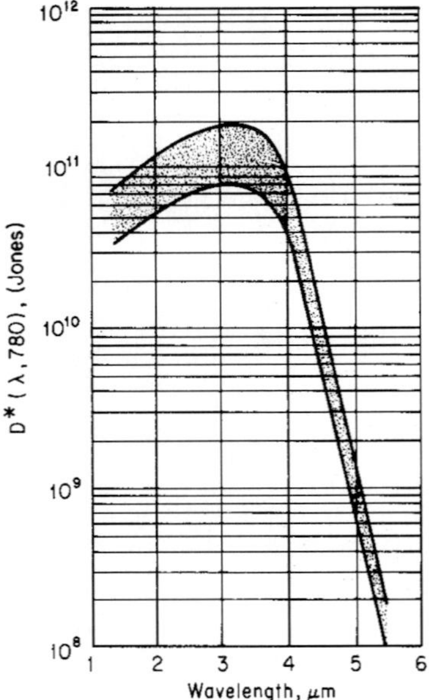

FIGURE 121 Range of spectral detectivities for PbS (LTO) at 77 K; 2π FOV, 295-K background. *(Santa Barbara Research Center.)*

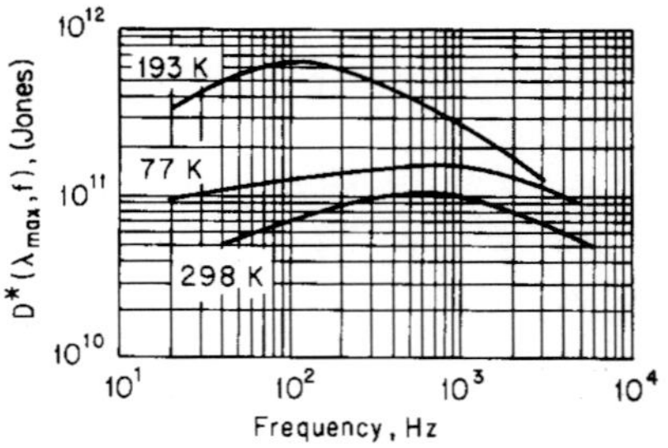

FIGURE 122 Example of detectivity vs. frequency for PbS detectors at various operating temperatures; 2π FOV, 296-K background. *(Santa Barbara Research Center.)*

Sensitive area: Typical commercial sizes are square elements with dimensions of 0.5, 1, 2, and 5 mm on a side.

Capacitance: <1 pF (limited by mounting configuration).

Recommended circuit: Standard photoconductor.

Stability: Exposure to visible and/or UV radiation can induce instability and drift. Stability will recover with storage in the dark, or by baking.

Sensitivity profile: Uniform within 10 percent.

Linearity: Excellent over broad range 10^{-8}–10^{-3} W.

Manufacturers: Alpha Omega Instruments, Atomergic Chemetals, Cal-Sensors,

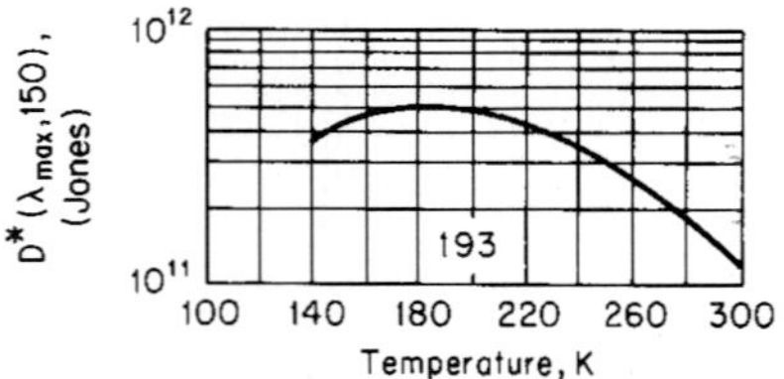

FIGURE 123 Example of detectivity vs. temperature for PbS (ITO) detectors; 2π FOV, 295-K background. *(Santa Barbara Research Center.)*

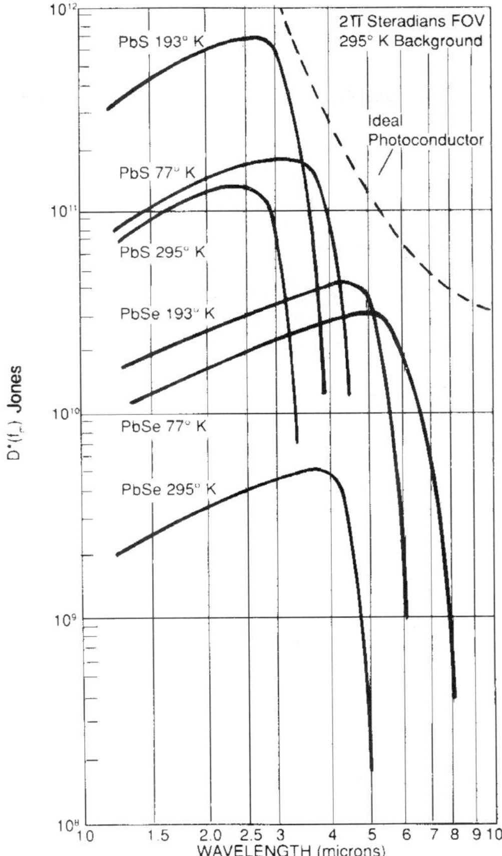

FIGURE 124 D^* vs. wavelength for PbS and PbSe detectors operating at temperatures ranging between 77 K and 295 K. *(CAL-SENSORS, Infrared Detectors.)*

Electro-Optical Systems, Graseby Infrared, Hamamatsu, Litton, OptoElectronics, Oriel, SAT.

InAs (Photovoltaic). InAs detectors are single-crystal, intrinsic, direct-bandgap photovoltaic devices for use in the 1- to 4-μm region (spectral cutoff varies with temperature). At room temperautre, InAs provides good sensitivity and submicrosecond response times. At 195 K, InAs performance equals or betters the sensitivity of any other detector in the 1- to 3.5-μm region. Devices with sapphire immersion lenses are available to increase signal responsivity for operation at higher temperatures where the detector is thermal-noise-limited.

Sensitivity: $D^*(\lambda\text{peak})$ varies from 1.2×10^9 Jones at 295 K to 6×10^{11} Jones at 77 K. See Fig. 128.

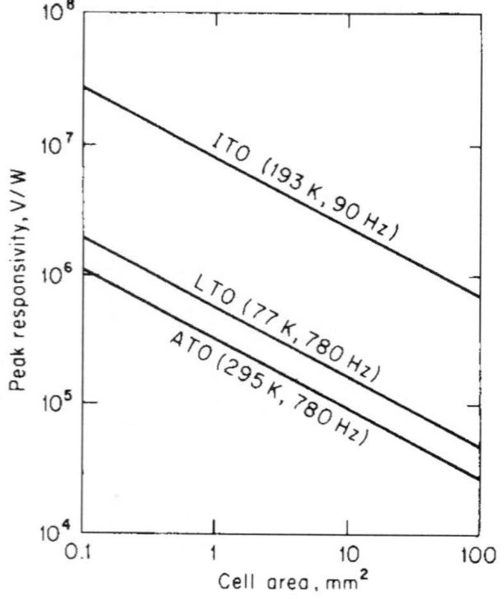

FIGURE 125 PbS typical peak responsivity vs. cell area (actual values range within a factor of two of these shown.)

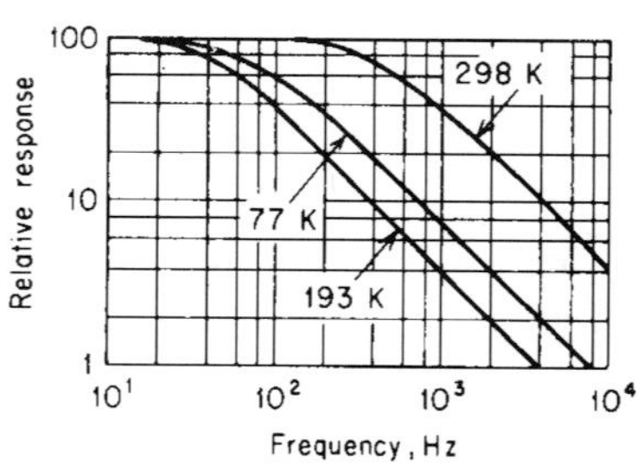

FIGURE 126 Example of signal vs. frequency for PbS detectors. *(Santa Barbara Research Center.)*

Quantum efficiency: Maximum of about 75 percent without antireflection coating.

Noise: Low impedance tends to make preamplifier noise dominate at room temperature; background limited (for 300 K background) at operating temperatures below ≈ 200 K.

Time constant: Less than 0.5 μs at all temperatures when low values of load resistor are used to lower the RC time constant.

Responsivity: 0.5–1.25 A/W at peak.

Dynamic resistance: See Fig. 129.

Diode capacitance: See Fig. 130.

Sensitive area: Standard sizes 0.25–2-mm dia.

Operating temperatures: 77–300 K.

Linearity: Anticipated to be very good over many decades.

Sensitivity profile: ±15 percent.

Recommended Circuits:

Open circuit: PV InAs detectors with areas less than $2 \times 10^{-2}\,\text{cm}^2$ require no bias when operated and can be connected directly into the input stage of amplifier (capacitor ensures elimination of dc bias from amplifier) (Fig. 131*a*).

Transformer: useful when using InAs at zero bias, particularly at room temperature where diode impedance is low (Fig. 131*b*).

Reversed bias: At temperatures greater than 225 K considerable gain in impedance and responsivity is achieved by reverse-biasing (Fig. 131*c*).

Fast response: To utilize the short intrinsic time constant, it is sometimes necessary to load the detector to lower the RC of the overall circuit (reverse bias will also lower detector capacitance) (Fig. 131*d*).

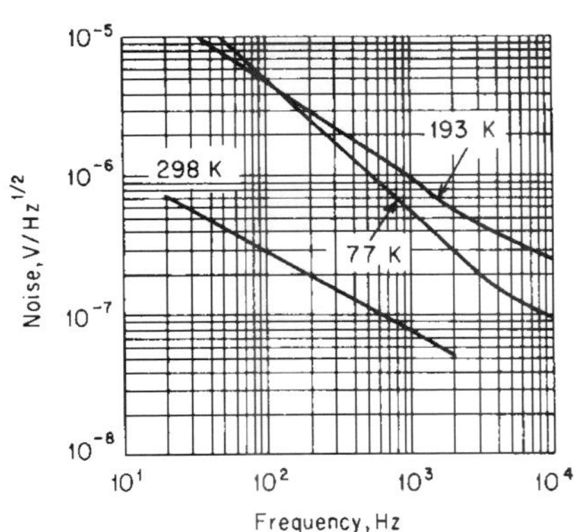

FIGURE 127 Example of noise vs. frequency for PbS detectors (1 × 1-mm area). *(Santa Barbara Research Center.)*

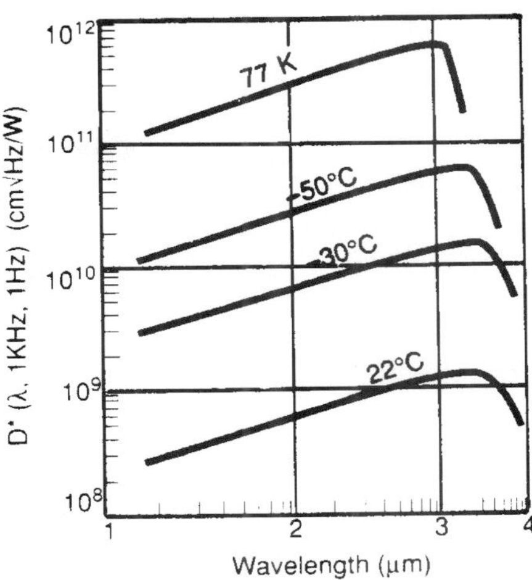

FIGURE 128 D^* vs. wavelength for InAs detector operating at temperatures ranging between 77 K and 295 K. *(EG&G Judson, Infrared Detectors, 1994.)*

Manufacturers: Atomergic Chemetals, Boston Electronics, BSA Technology, Cincinnati Electronics, EG&G Judson, MCP Wafer Technology.

PbSe. Lead selenide is an intrinsic, thin-film photoconductor, whose long-wavelength spectral response and speed of response exceeds that of PbS. At room temperature, PbSe has peak D^* which can exceed 1×10^{10} Jones with a spectral cutoff out to 4.4 μm. At liquid-nitrogen temperature, InSb offers twice the D^*, largely because PbSe offers response out to 7 μm at 77 K, considerably longer than InSb. However, for intermediate temperatures, from 180 K to room temperature, PbSe offers competitive D^* combined with moderately fast response (Johnson, 1984). PbSe technology has made a significant

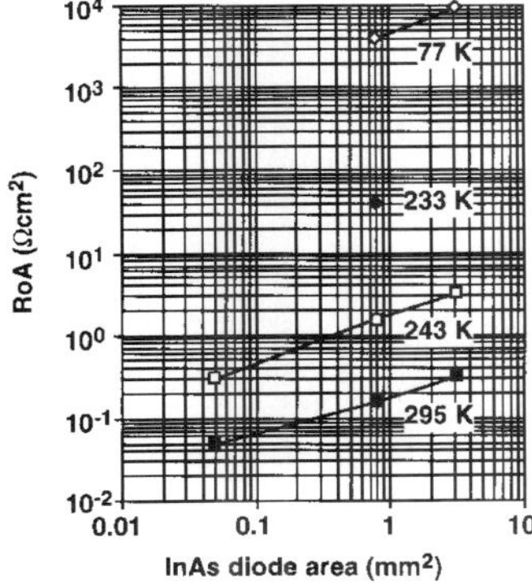

FIGURE 129 RoA of InAs photodiodes shown as a function of diode area. Lower impedance per unit area for smaller devices indicates that these devices are surface-leakage-limited. *(EG&G Judson, Infrared Detectors, 1994.)*

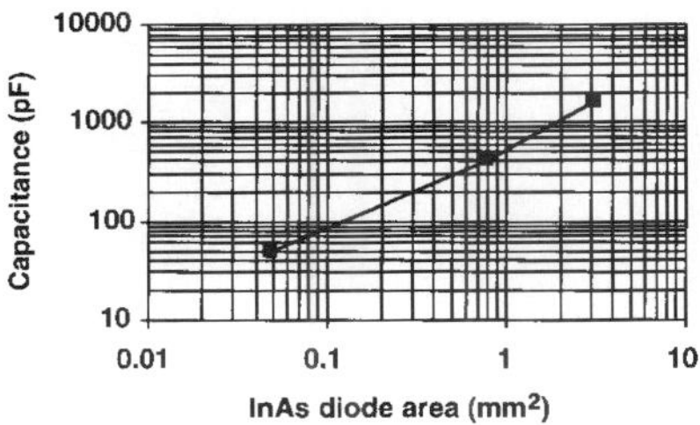

FIGURE 130 Capacitance of InAs photodiodes shown as a function of diode area. Capacitance will not change appreciably with temperature. *(EG&G Judson, Infrared Detectors, 1994.)*

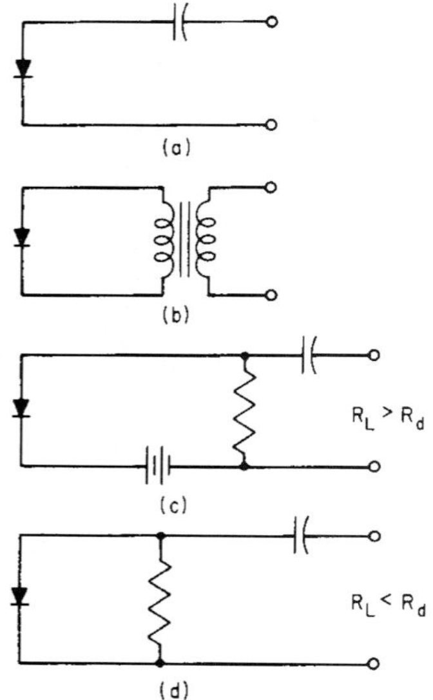

FIGURE 131 Recommended circuits for InAs detectors: (*a*) open circuit; (*b*) transformer; (*c*) reversed bias; (*d*) fast response.

advance in the past decade in some vendors being able to reproducibly make high-performance detectors.

Sensitivity: $D^* \approx 1 \times 10^{10}$ Jones at 300 K, increases with cooling (see Figs. 124 and 132). D^* is limited by $1/f$ noise at low frequencies (see Fig. 133).

Response: Figure 134 shows responsivity in amps per watt for a high-quality detector

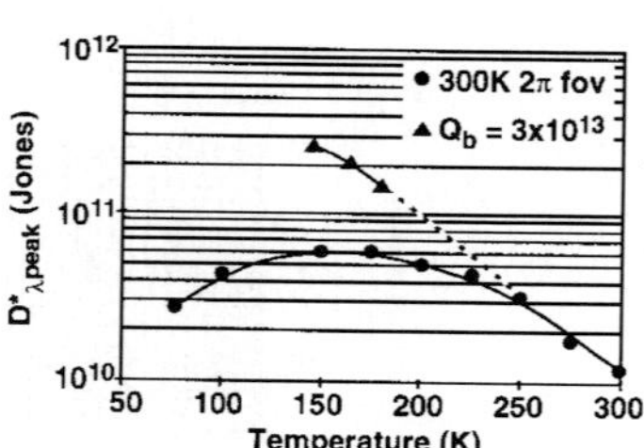

FIGURE 132 D^* of PbSe as a function of temperature for two background flux conditions: high background of 2π field of view and reduced background of 3×10^{13} photons/cm^2/sec. D^* at the higher background flux reaches a maximum around 160 K because the background noise increases at lower temperatures due to the increase in long-wavelength spectral response of the detector.

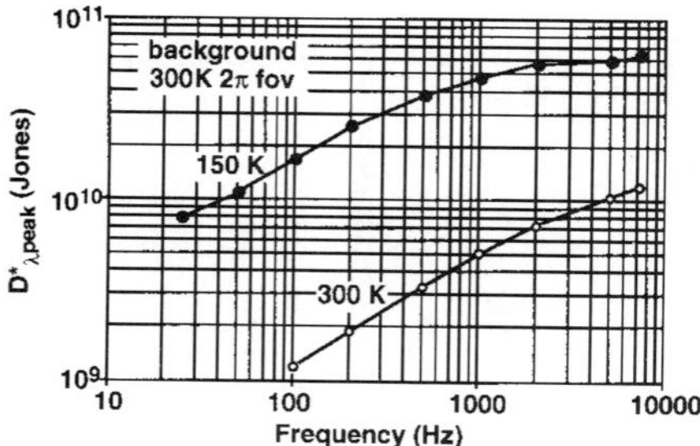

FIGURE 133 D^* of PbSe as a function of frequency for two temperatures. PbSe has considerable $1/f$ noise which reduces D^* at lower frequencies, especially at high temperatures.

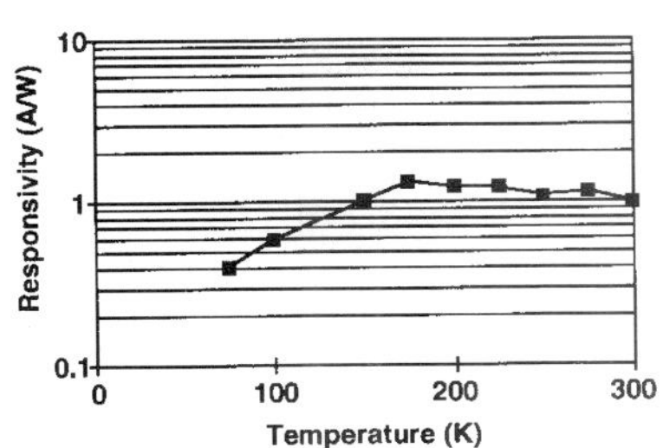

FIGURE 134 Responsivity in A/W of PbSe thin-film photoconductive detectors as a function of temperature for a high background flux level. Multiply by detector resistance to get V/W.

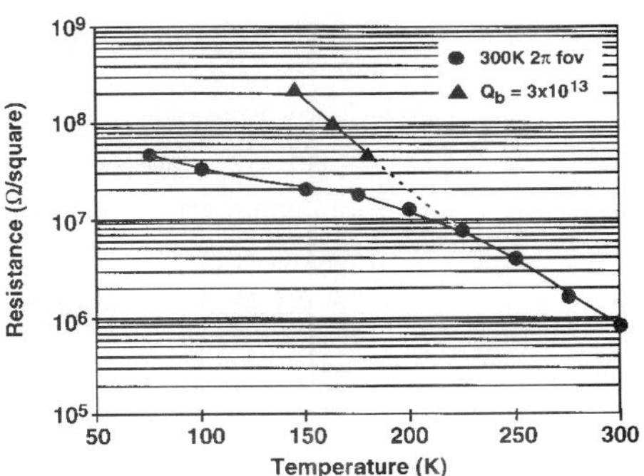

FIGURE 135 Resistance (Ω/square) of PbSe thin films as a function of temperature for two background flux levels. At any temperature, the absolute value can be varied by altering the manufacturing process in chemical deposition and/or heat treatment.

with a length of 0.016 cm and width of 0.024 cm. Responsivity in volts per watt is obtained by multiplying A/W data by resistance (see Fig. 135). Responsivity will vary inversely with detector length (see Figs. 136 and 137).

Noise: Figure 138 shows the noise as a function of temperature for a detector with a length of 0.016 cm and width of 0.024 cm. Noise as a function of frequency is shown in Fig. 139.

Resistance: Figure 135 shows the resistance as a function of temperature for a detector with a length of 0.016 cm and width of 0.024 cm.

Capacitance: 1 pF (limited by mounting configuration).

Time constant: See Figs. 140 and 141. Time constant will be longer when detector is operated in reduced background flux condition.

Stability: Exposure to visible and/or UV radiation can induce instability and drift. Stability will recover with storage in the dark at room temperature.

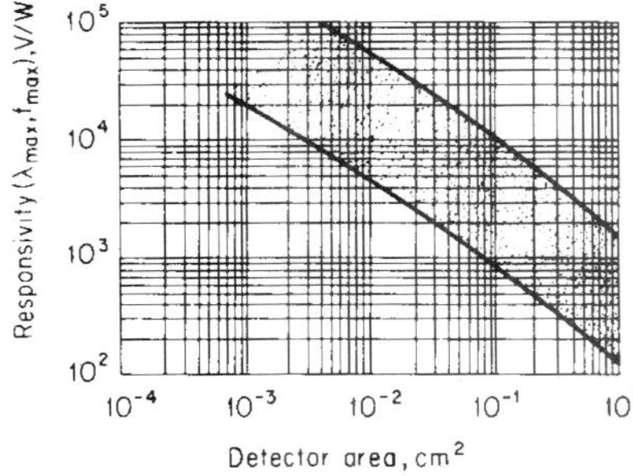

FIGURE 136 Expected range of peak responsivities vs. detector size, typical PbSe (ATO) infrared detectors. *(Santa Barbara Research Center.)*

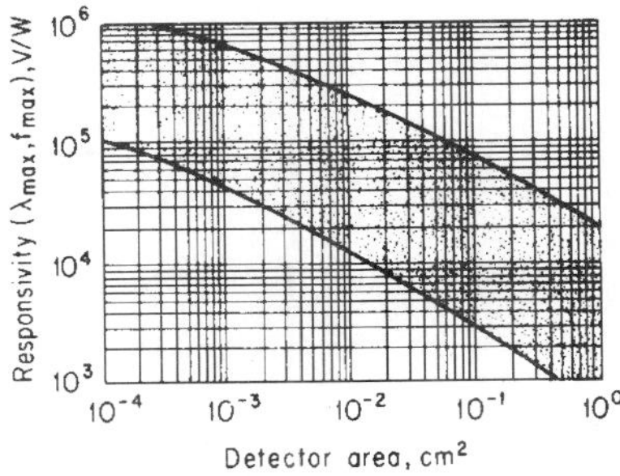

FIGURE 137 Expected range of peak responsivities vs. detector size, PbSe (ITO and LTO) infrared detectors. *(Santa Barbara Research Center.)*

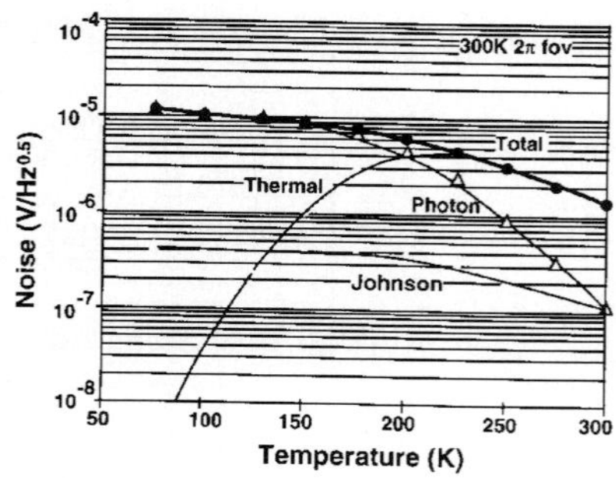

FIGURE 138 Noise voltage (per square root of bandwidth) of PbSe thin-film photoconductive detectors as a function of temperature for a high background flux level. Photon noise is dominant below 200 K. Thermal noise is dominant at higher temperatures. Total noise levels are well above typical preamplifier noise.

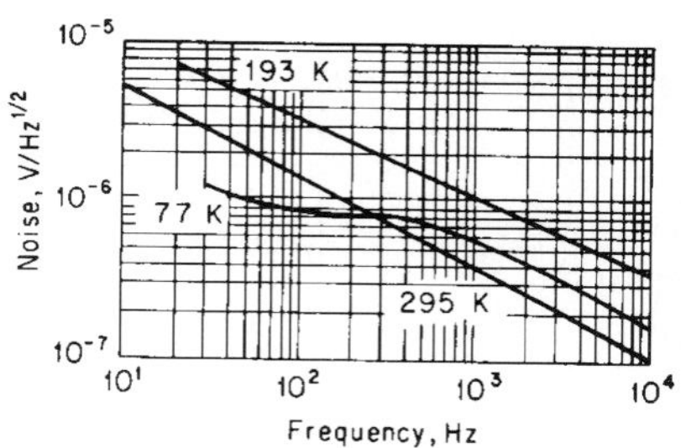

FIGURE 139 Example of noise vs. frequency for PbSe cooled detectors (ATO, ITO, and LTO types) ($A = 1 \times 1$ mm). *(Santa Barbara Research Center.)*

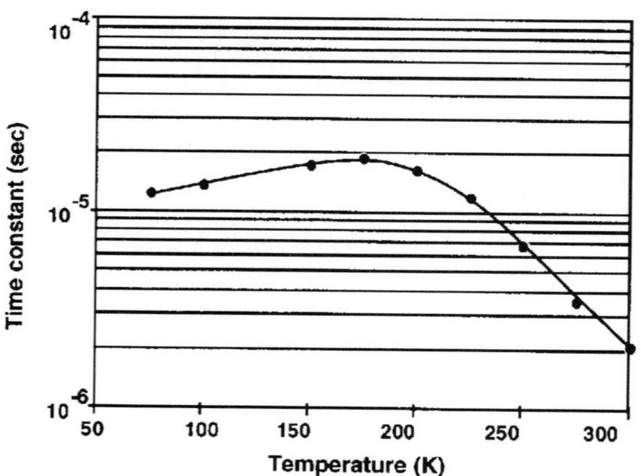

FIGURE 140 Time constant of PbSe thin films as a function of temperature.

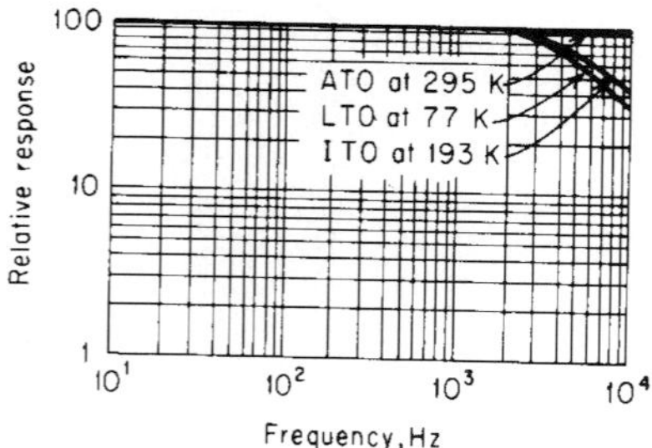

FIGURE 141 Example of signal vs. frequency for PbSe (ATO, ITO, and LTO). *(Santa Barbara Research Center.)*

Recommended circuit: Standard photoconductor.

Operating temperature: 77–300 K.

Manufacturers: Alpha Omega, Atomergic Chemetals, BSA Technology, Cal-Sensors, EDO/Barnes Engineering, Electro-Optical Systems, Graseby Infrared, Hamamatsu, Lambda Electronics, Oriel, Infrared Industries, OptoElectronics.

InSb. Historically, indium antimonide material has been used for at least four different radiation detectors, two of which, the photoconductive and photoelectromagnetic types, are no longer widely used. We discuss here the intrinsic photovoltaic device. [The very far infrared bolometer (InSb bolometer) was previously discussed in Sec. 15.7 on page 15.26.]

At 77 K, InSb photodiodes offer background limited sensitivity at medium to high background flux conditions in the 1- to 5.5-μm spectral range. At lower temperatures, they provide sensitive detectors at low background flux levels such as in astronomy applications, but with a slightly shortened long-wavelength cutoff. Operation is possible up to as much as 145 K, but because the spectral response increases with increasing temperature, the detector impedance drops rapidly leading to significant thermal noise.

Sensitivity: Spectral response out to 5.5 μm at 77 K (see Fig. 142). $D^* \approx 1 \times 10^{11}$ Jones, increases with reduced background flux (narrow field of view and/or cold filtering) as illustrated in Fig. 143.

Quantum efficiency: ~60–70 percent without antireflection coating. >90 percent with antireflection coating.

Noise: Background current limited over wide range of background flux at 77 K (see Fig. 144).

Time constant: <1 μs.

Responsivity: 3 A/W at 5 μm without antireflection coating.

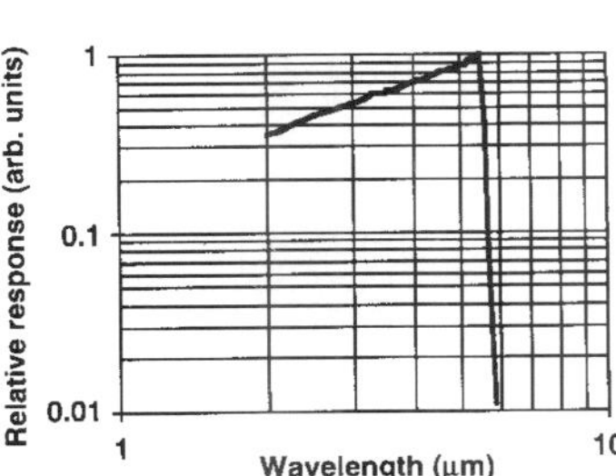

FIGURE 142 Relative spectral response per watt of an InSb photodiode without antireflection coating. The direct bandgap results in a sharp spectral cutoff.

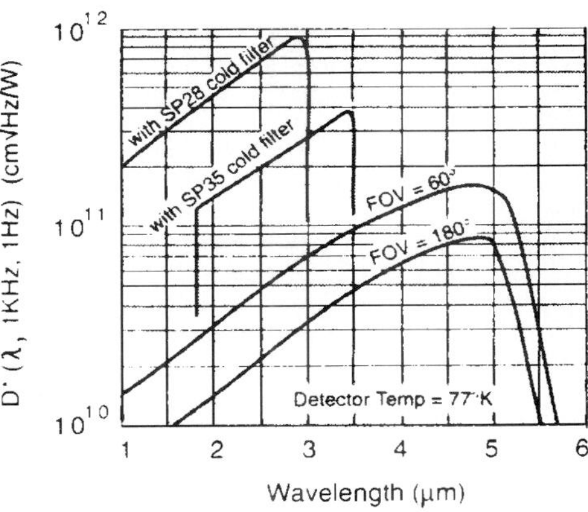

FIGURE 143 D^* as a function of wavelength for an InSb detector operating at 77 K. *(EG&G Judson, Infrared Detectors, 1994.)*

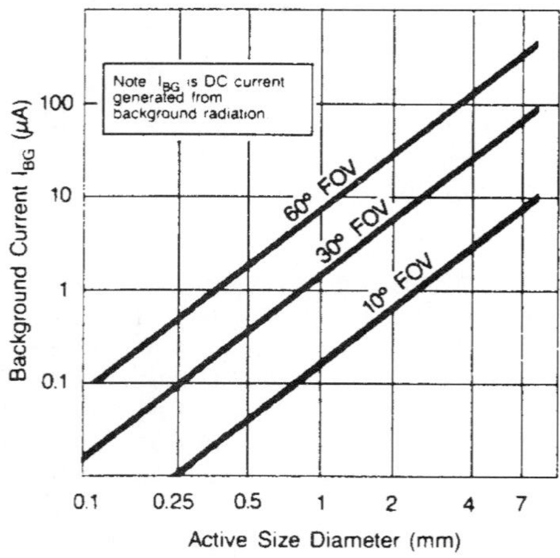

FIGURE 144 Background current as a function of photodiode area for InSb detectors operating at 77 K, shown at three values of the detector field of view. *(EG&G Judson, Infrared Detectors, 1994.)*

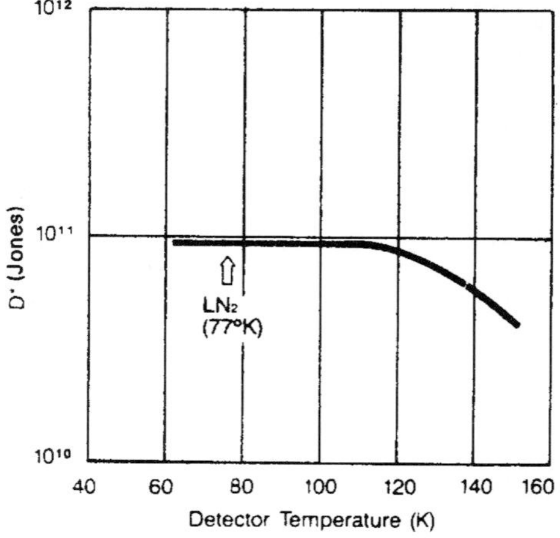

FIGURE 145 InSb photodiode D^* as a function of operating temperature between 77 K and 150 K, for a 2π (180° FOV). *(EG&G Judson, Infrared Detectors, 1994.)*

Capacitance: Typically 0.05 μF/cm^2.

Impedance: Top-grade detectors have 1–5 × 10^6 Ωcm^2 $\mathbf{R}_0 A$ product at 77 K, at zero bias, and without background flux.

Sensitive area: 0.04 × 0.04-mm square to 1 × 5-mm rectangle; 0.25–10-mm dia.

Operating temperature: Normally 77 K; InSb can be used up to approximately 145 K (see Fig. 145).

Linearity: Linear to ~1 mW/cm^2.

Sensitivity profile: ±15 percent or better.

Stability: Devices from some vendors are subject to "flashing," where exposure to visible or UV flux causes a change in the insulating surface charge thereby causing a change in the diode impedance. The detector typically recovers at room temperature.

Recommended circuit: Same as for Si and Ge photodiodes; zero or reverse bias in combination with a load resistor and low noise preamplifier. Low-impedance load resistor can be used for obtaining fast response, with consequences of reduced sensitivity.

Manufacturers: Atomergic Chemetals, BSA Technology, Cincinnati Electronics, Edinburgh Instruments, EG&G Judson, Electro-Optical Systems, Graseby Infrared, Hamamatsu, MCP Wafer Technology, SAT, Vigo Sensor.

Ge:Au. Gold-doped germanium detectors are relatively fast single-crystal p-type impurity-doped photoconductors for the 2- to 9-μm region. Although not the most sensitive detector anywhere in its range of spectral sensitivity, Ge:Au offers respectable sensitivity over a broad spectral region using liquid nitrogen cooling. Sensitivity can be improved by a factor of 2.5 by operating at $T < 65$ K (pumped liquid nitrogen or other cryogen). At these temperatures, Ge:Au becomes background-limited.

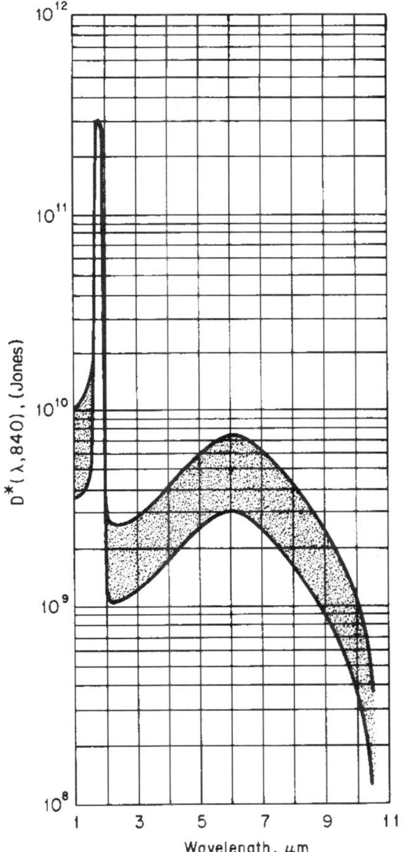

FIGURE 146 D^* vs. λ for Ge:Au; $T = 77$ K, 2π FOV; 295-K background. *(Santa Barbara Research Center.)*

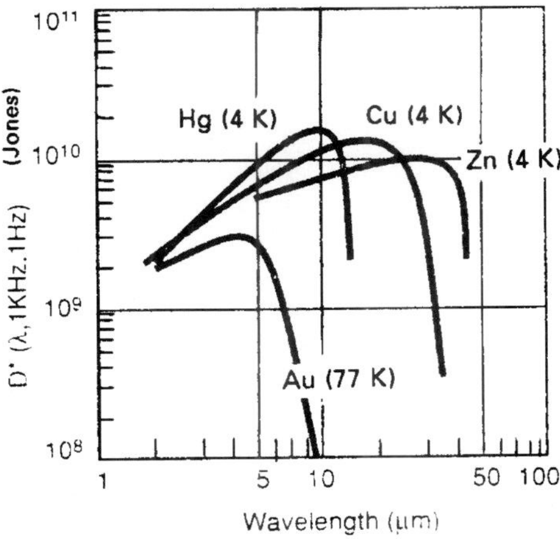

FIGURE 147 D^* as a function of wavelength for extrinsic germanium detectors doped with Au, Hg, Cu, and Zn, for a 300-K 2π (180° FOV) background flux. *(EG&G Judson, Infrared Detectors, 1994.)*

Sensitivity: See Figs. 146–149.

Quantum efficiency: Dependent on wavelength, detector geometry (absorption thickness), antireflection coating, and enclosure (integration chamber can increase absorption). $D^*(\lambda_{pk})/D^*(500\text{ K}) = 2.7$ (see Fig. 150).

Noise: See Fig. 151.

Time constant: <50 ns with full D^* [shorter response times (<2 ns) can be tailored by heavy concentration of compensating (n-type) dopant and suitable bias circuit (see circuit discussion to follow). Heavy compensation increases resistance, and hence the incoherent signal-to-noise ratio becomes limited by thermal noise of load (typically a factor of 2 degradation in the signal-to-noise ratio). Quantum efficiency, however, is not significantly altered, so that concentration does not hurt coherent-detection signal-to-noise ratio].

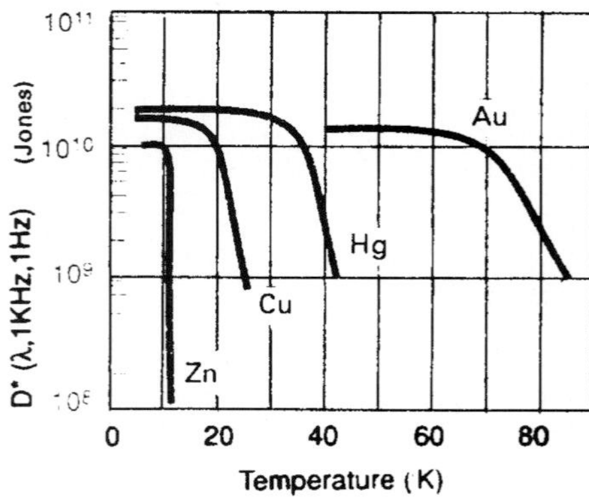

FIGURE 148 D^* as a function of operating temperature for extrinsic germanium detectors doped with Au, Hg, Cu, and Zn, for a 300-K 2π (180° FOV) background flux. *(EG&G Judson, Infrared Detectors, 1994.)*

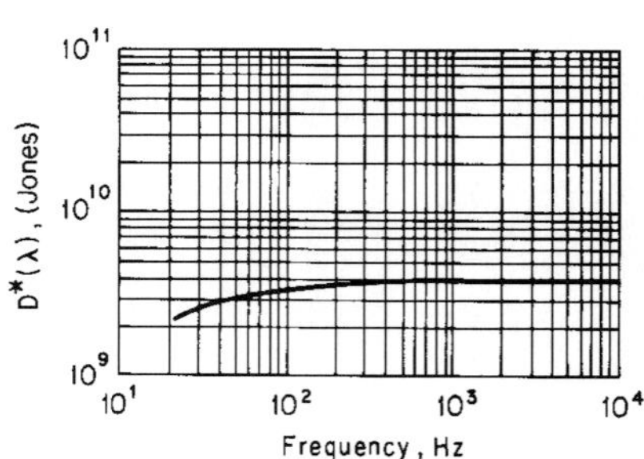

FIGURE 149 Typical D^* vs. frequency (T = 77 K). *(Santa Barbara Research Center.)*

Responsivity: Dependent upon bias and geometry, typical values are 0.1 to 0.2 A/W at 77 K. Multiply by detector resistance to get V/W.

Dark resistance: Varies with background flux and effective quantum efficiency (see previous quantum efficiency discussion), range may be 20 kΩ to 5 MΩ, or much greater under very low background flux conditions if adequately cooled to limit thermally activated conductivity. (Also see previous time-constant discussion.)

Capacitance: Depends on device geometry and mounting, typically <1 pF.

Sensitive area: 1–5-mm dia. standard.

Operating temperature: <85 K (normally 77 K, but see Fig. 148).

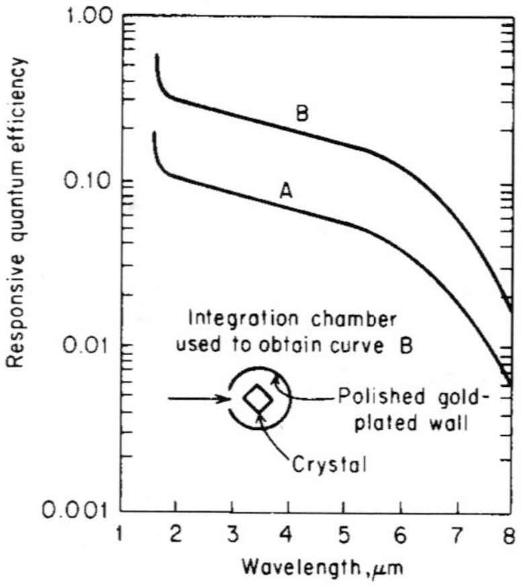

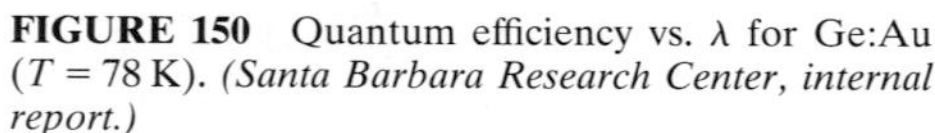

FIGURE 150 Quantum efficiency vs. λ for Ge:Au (T = 78 K). *(Santa Barbara Research Center, internal report.)*

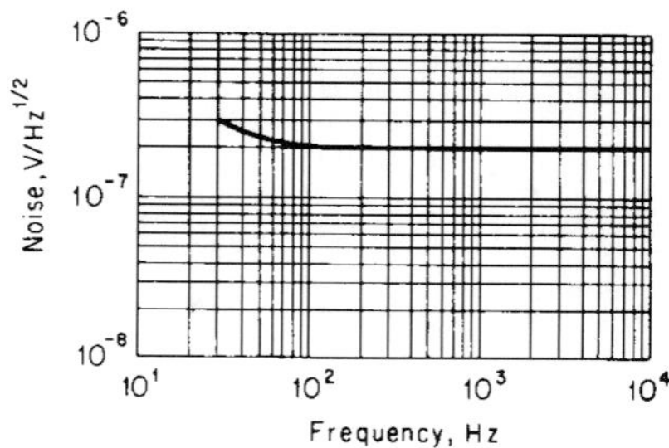

FIGURE 151 Typical noise spectrum for Ge:Au (T = 77 K; $A = 1 \times 1$ mm). *(Santa Barbara Research Center.)*

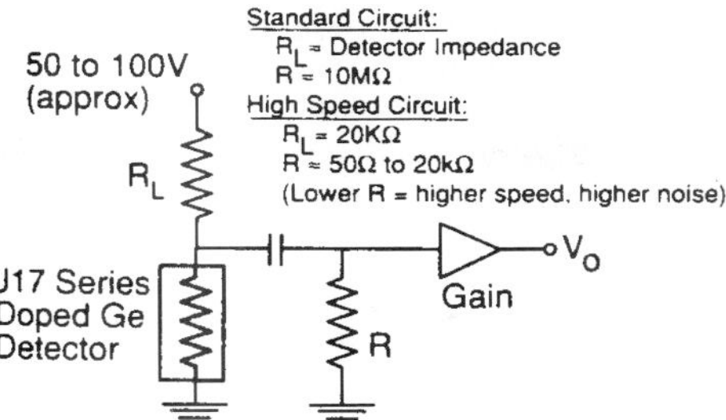

FIGURE 152 Basic operating circuit for extrinsic germanium detectors doped with Au, Hg, Cu, and Zn, for a 300-K 2π (180° FOV) high background flux. If the detector is operated in very low background flux conditions, the detector impedance can become very high. Cooled JFET ($T > \approx 50$ K) or PMOS buffer amplifiers can be helpful in impedance matching under these conditions. *(EG&G Judson, Infrared Detectors, 1994.)*

Recommended circuit: Standard photoconductive. See Fig. 152.

Manufacturers: EG&G Judson.

HgCdTe, HgZnTe, HgMnTe, etc. Mercury cadmium telluride is a direct-bandgap compound alloy semiconductor, made of chemical group II and VI elements, whose peak sensitivity at a particular temperature can be adjusted from 1 to 30 μm by varying the ratio of HgTe to CdTe (see Figs. 153 and 154). In addition to HgCdTe, other combinations of chemical groups II and VI elements can be used to produce similar variable spectral cutoff compound alloys, including HgZnTe, HgMnTe, HgCdZnTe, etc. For almost all purposes, HgCdTe will be as good as any other II-VI alloy detector, so we will speak here about it exclusively, but provide some data from the other alloys mentioned here. Both photoconductive (PC) and photovoltaic (PV) HgCdTe detectors are available for background-limited, high-speed, intrinsic photon detection in the SWIR, MWIR, and LWIR regions. Photoconductive devices with VLWIR response out to 25 μm are also available. HgCdTe detectors can be used at room temperature, with TE cooling, and at 77 K and lower temperatures. Sensitivity generally increases with cooling, depending upon

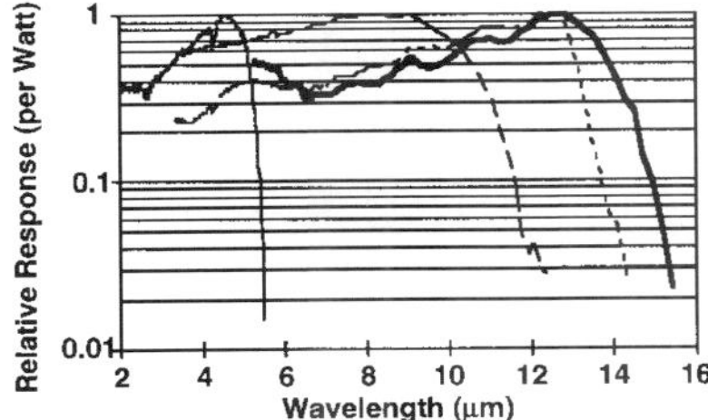

FIGURE 153 Relative spectral response per watt at 80 K for photoconductive HgCdTe detectors with antireflection coating. The curves are normalized to unity at peak value. The spectral cutoff can be adjusted by varying the ratio of HgTe to CdTe in the alloy.

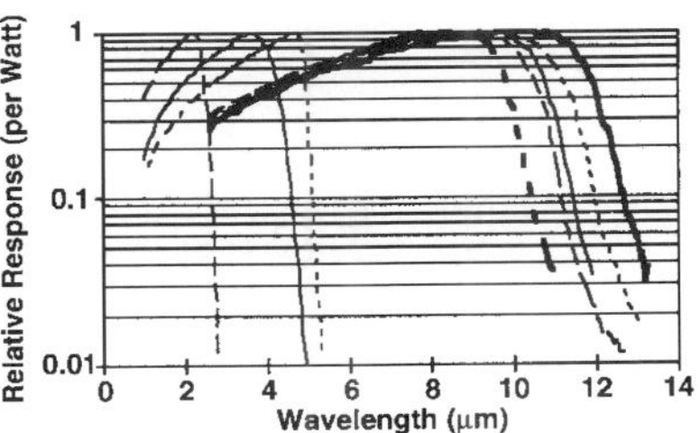

FIGURE 154 Relative spectral response per watt at 80 K for photovoltaic HgCdTe detectors without antireflection coating. The curves are normalized to unity at peak value. The spectral cutoff can be adjusted by varying the ratio of HgTe to CdTe in the alloy.

the spectral cutoff and background flux. MWIR, LWIR, and VLWIR spectral range devices are generally operated at 77 K or lower temperature for maximum sensitivity, depending upon the background flux. The photoconductive mode is advantageous when cooling is limited, since the thermal noise of a photoconductor increases less rapidly than a photodiode as the temperature is raised. Photoconductive devices with response out to 25 μm can be usefully operated at liquid nitrogen temperature and are popular for IR spectroscopy for this reason.

HgCdTe photoconductors are fabricated from thin (≈10-μm) single crystal-layers with metal contacts at each end of the element. They are low-impedance devices with 15 to 2000 Ω/square, depending upon the alloy composition, carrier concentration, operating temperature, background flux, and surface treatment. Photoconductor time constants at 77 K may be ≈2 μs for devices having a 12-μm cutoff, with longer time constants for shorter cutoffs, and shorter time constants for higher operating temperatures. In the case of small detector elements, the time constant may be reduced with increasing bias voltage because photoexcited carriers will be transported to the electrical contacts where they recombine. The spectral noise characteristics of PC HgCdTe typically exhibit $1/f$ noise out to a range of 50 Hz to 1 kHz or more, the value depending upon the detector quality, long-wavelength response, operating temperature, and background flux. The white noise levels range from less than 1 nV/$\sqrt{\text{Hz}}$ (where preamplifier noise may then dominate), up to 20 nV/$\sqrt{\text{Hz}}$, depending upon detector quality, size, geometry, applied bias, temperature, and alloy composition. Photoconductive HgCdTe detectors are typically antireflection coated with a quarter-wave ZnS film, giving a peak quantum efficiency in the range of 85 to 90 percent, although this figure is only indirectly measured because the photoconductive gain can be much greater than unity. Without antireflection coating, the quantum efficiency is typically 70 percent, limited by the optical index of ≈4.

A class of HgCdTe detectors (mostly photoconductors) is offered for detection at TE-cooled and room temperature which are optically "immersed" with a hemispherical lens of Ge, CdTe, or other high-index material (see Fig. 155). The lens increases the effective area of the detector without increasing the detector noise, provided the noise is

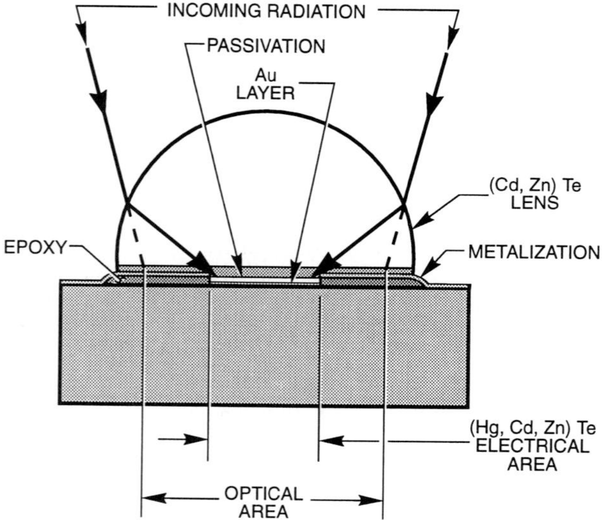

FIGURE 155 schematic of an optically immersed HgCdZnTe detector. *(Oriel Catalog Vol. II.)*

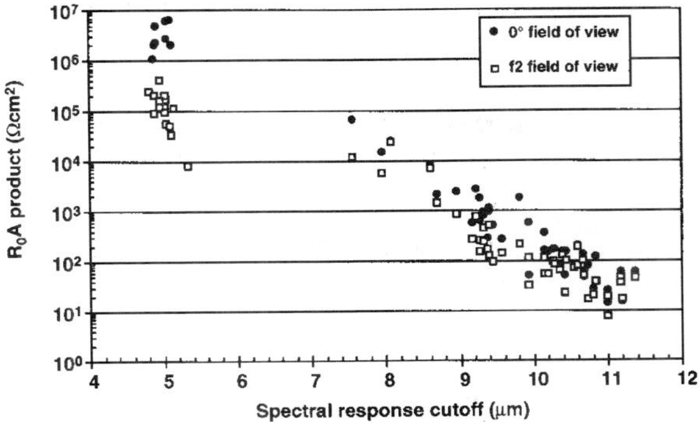

FIGURE 156 $\mathbf{R}_0 A$ product "trendline" for small (25 × 25 to 100 × 100 μm) HgCdTe photodiodes at 77 K. Data is shown for devices with zero background flux (0° FOV) and with an F2 field of view (29°) of a 300-K background. For 5-μm spectral cutoff material, the $\mathbf{R}_0 A$ product is higher at 0° FOV by about an order of magnitude, compared with an F2 background. At 10 μm, there is less of a difference between the two background conditions. Note that the $\mathbf{R}_0 A$ product will generally be somewhat lower for larger area diodes.

dominated by thermal rather than photon noise as is the case for minimal cooling. The lens must be in intimate contact with the detector surface (<1 μm) to avoid total internal reflection of off-axis rays at the lens-detector interface. Immersed detectors offer up to a factor of 16 or greater increased detector signal, which can mean an increase in D^* by the same factor for a thermal-noise-limited device. Operation of LWIR PC detectors at TE-cooled and room temperature is generally accompanied by increased $1/f$ noise which dominates out to higher frequencies.

Photovoltaic HgCdTe detectors ideally offer $\sqrt{2}$ higher D^* than the photoconductive mode. Diodes are made in both n^+p and p^+n polarities, depending upon the manufacturer's capabilities. The $\mathbf{R}_0 A$ product of HgCdTe photodiodes varies significantly with temperature, spectral cutoff, and device quality. It also varies with the amount of background flux incident on the device. The $\mathbf{R}_0 A$ product defines the maximum D^* in the limit of reduced background flux (see Fig. 156 and Fig. 7). In addition to theoretically higher D^*, high-quality PV HgCdTe detectors have lower $1/f$ noise than PC HgCdTe devices, with $1/f$ knee frequencies as low as 1 Hz or less. However, the noise of PV detectors increases more rapidly with increasing temperature than for PC detectors, making photodiodes less attractive for applications where cooling is limited. Photodiodes of high quality are more difficult to make than good photoconductors and can be expected to warrant a premium price. Antireflection coating is available from some diode producers, but is not routinely offered.

Both PC and PV HgCdTe detectors are useful for infrared heterodyne detection. When sufficient local oscillator power is available, detector cooling becomes less important, since photon noise can dominate thermal noise at comparatively higher temperatures. Other things being equal, the photovoltaic detector has $\sqrt{2}$ sensitivity (signal-to-noise voltage) advantage over the photoconductor. For 10.6-μm heterodyne detection, 0.1 × 0.1-mm HgCdTe *pin* photodiodes with sensitivity near the quantum limit of $\approx 2 \times 10^{-20}$ W/Hz are available with bandwidths up to several GHz. Ordinary photodiodes of the same area have bandwidths of several hundred MHz. Photoconductors make better 10.6-μm heterodyne detectors when cooling is limited to TE-cooled temperatures of 180 K up to room temperature. At 180 K, TE-cooled photoconductors offer bandwidths of 50 to 100 MHz

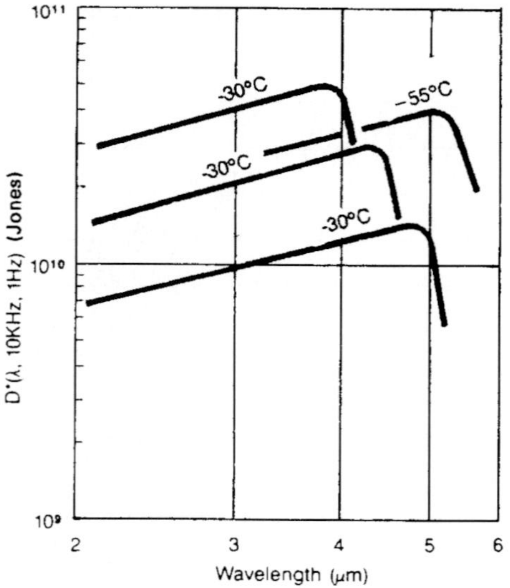

FIGURE 157 Typical D^* as a function of wavelength for a variety of MWIR HgCdTe photoconductors with thermoelectric cooling. *(EG&G Judson, Infrared Detectors, 1994.)*

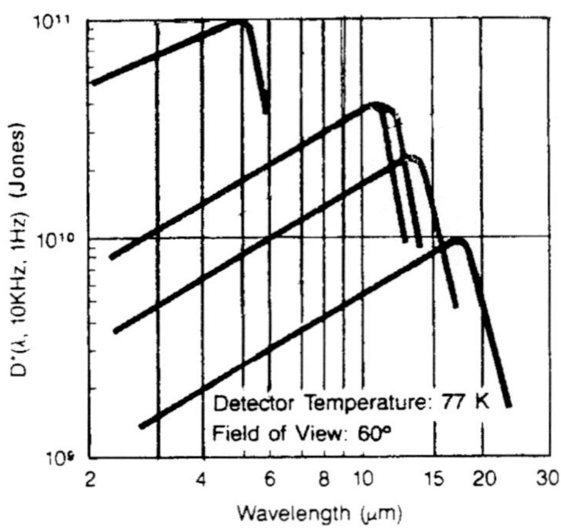

FIGURE 158 Typical D^* as a function of wavelength for a variety of LWIR and VLWIR HgCdTe photoconductors at 77 K. *(EG&G Judson, Infrared Detectors, 1994.)*

and heterodyne NEPs of 1 to 2×10^{-19} W/Hz. At room temperature, the NEP at 10.6 μm is limited to about 1×10^{-16} W/Hz. Immersion does not improve the performance of minimally cooled heterodyne detectors, since optical gain is already provided by the local oscillator.

Photoconductive HgCdTe

Sensitivity: Adjustable by varying alloy composition (see Figs. 153, 157–159).

Dark resistance: 15–2000 Ω/sq depending upon temperature, spectral cutoff, and surface passivation.

Responsivity: Varies with spectral cutoff, temperature, detector resistance, element length, and bias voltage or power. See Eq. (21) and Fig. 160 for detector elements with 50 × 50-μm dimensions.

Noise: $1/f$ noise is dominant at frequencies below 50–1000 Hz for LWIR detectors at 77 K (greater for LWIR at room temperature or TE-cooled). Generation-recombination (thermal or photon) white noise is present beyond the $1/f$ region at a level of less than 10^{-9} nV/$\sqrt{\text{Hz}}$ to 2×10^{-8} nV/$\sqrt{\text{Hz}}$, depending upon spectral cutoff, background flux, responsivity, and operating temperature. Noise and signal rolloff at high frequency determined by the time constant. See Fig. 161 for an example of the noise spectrum of an LWIR at 77 K.

Operating temperature: 77 K and below to 300 K and above for short spectral cutoffs and/or with significant D^* reduction for operation at higher temperatures.

Sensitive area: 0.025–4-mm linear dimensions.

Quantum efficiency: Typically > 70 percent, 85–90 percent with antireflection coating.

Capacitance: Low, limited by mounting configuration.

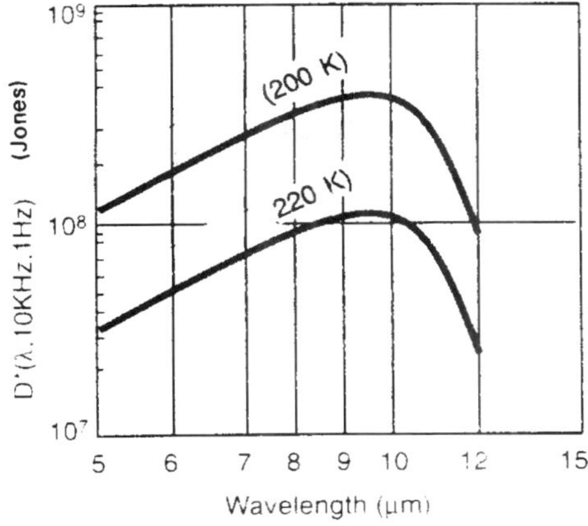

FIGURE 159 Typical D^* as a function of wavelength for LWIR HgCdTe photoconductors at 220 and 200 K. These units are cooled with three- or four-stage thermoelectric coolers. *(EG&G Judson, Infrared Detectors, 1994.)*

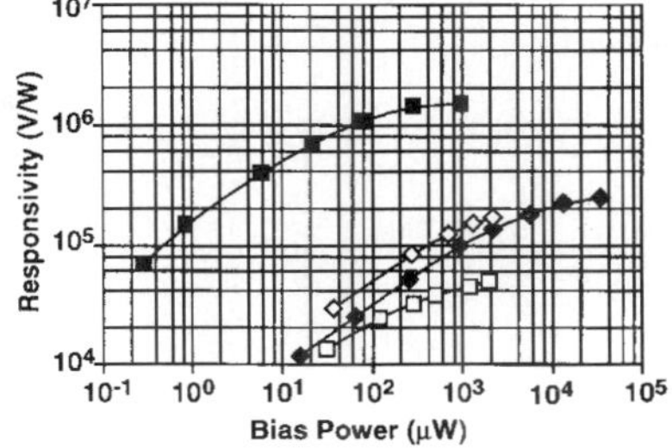

FIGURE 160 Range of peak responsivities for 12-μm cutoff HgCdTe photoconductors at 80 K. These devices have nominal dimensions of 50 × 50 μm, and resistance of 50 to 150 Ω/square.

Time constant: ~1–2 μs for LWIR at 77 K, depends on spectral cutoff, temperature, doping, and bias.

Circuit: Standard photoconductive.

Manufacturers: Belov Technology, Boston Electronics, Brimrose, Cincinnati Electronics, EG&G Judson, Graseby Infrared, Oriel, Phillips Infrared Defence Components, SAT, Spire.

Photovoltaic HgCdTe

Sensitivity: Adjustable by varying alloy composition (see Figs. 154, 156 and 162). Also compare Fig. 156 with Fig. 7 for an estimate of the extent to which D^* may increase (up to the $\mathbf{R}_0A$ or shunt resistance limit) as the background flux is reduced.

Time constant: Depends on diode capacitance (area); 10–20 ns without bias; ~0.5–

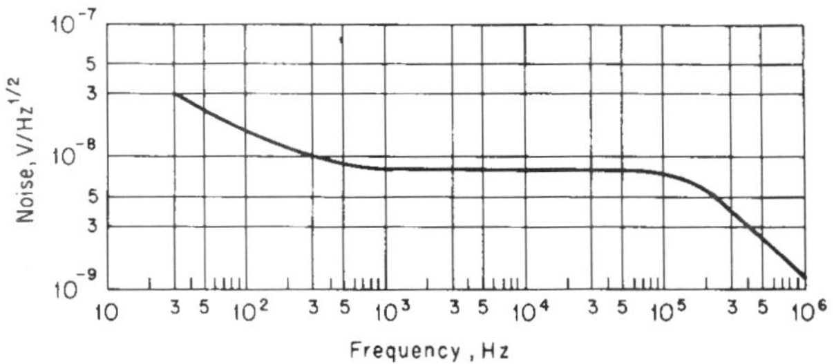

FIGURE 161 Variation of noise with frequencies for photoconductive HgCdTe at 77 K. *(GEC Marconi Infrared Ltd.)*

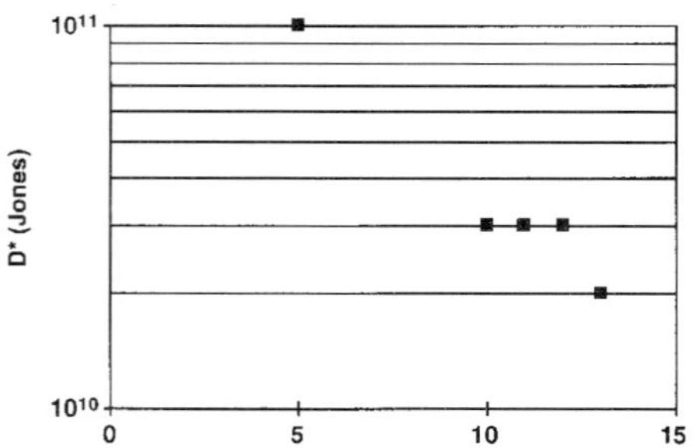

FIGURE 162 D^* specifications for small (50×50 to 250×250 μm) HgCdTe photodiodes at 77 K as a function of spectral cutoff. Data is shown for devices with 60° FOV background flux. *(Fermionics, Mercury Cadmium Telluride MWIR and LWIR Detector Series.)*

3 ns with reverse bias (some tradeoff of sensitivity). Low-capacitance *pin* devices with response out to several GHz (0.05–0.2-ns time constant) are also available for 10.6-μm CO_2 laser heterodyne detection.

Resistance: Refer to Fig. 156 for the $\mathbf{R}_0A$ product at zero bias corresponding to the detector cutoff wavelength (this figure shows very high quality diode impedances) and divide by the diode area. Large-area diodes will have somewhat lower $\mathbf{R}_0A$ product than shown in this figure. $\mathbf{R}_0A$ varies somewhat with background flux as can be noted from Fig. 156.

Operating temperature: Depends on spectral cutoff; 77 K and lower for LWIR and VLWIR detectors, up to room temperature for SWIR devices.

Noise: High-quality devices may have flat noise response from ~1 Hz out to the high-frequency limit of the time constant. $1/f$ noise may be present in lower quality devices and will vary with reverse bias.

Quantum efficiency: >50 percent (60–75 percent typical) without antireflection coating. Higher with antireflection coating.

Sensitive area: 0.01–0.25-mm square, 0.5- and 1-mm dia.

Capacitance: Depends on junction doping, area, and applied bias (very slightly dependent on spectral cutoff). For standard *pn* junction devices at zero bias and 10^{15} cm^{-3} doping, capacitance is approximately 3×10^4 pF/cm^2. Significantly lower for *pin* junction devices.

Circuit: Standard photovoltaic circuits, reverse-bias operation to enhance speed and zero bias to maximize D^* (see discussion under Si photodiodes).

Manufacturers: BSA, Fermionics, Kolmar, Oriel, Santa Barbara Research Center, SAT.

PbSnTe. PbSnTe offers an alternative semiconductor alloy system based upon the IV-VI chemical families to the II-VI (HgCdMnTeSe) families previously described for fabricating variable spectral cut-off detectors. Only photovoltaic detectors are available in PbSnTe. This technology has an advantage in the ease of material growth and in the fabrication of good quality photodiode junctions. It has a disadvantage in the very high dielectric constant of the material, combined with relatively high doping concentrations giving high-capacitance (comparatively slow) detectors. For low-frequency applications this is not a disadvantage.

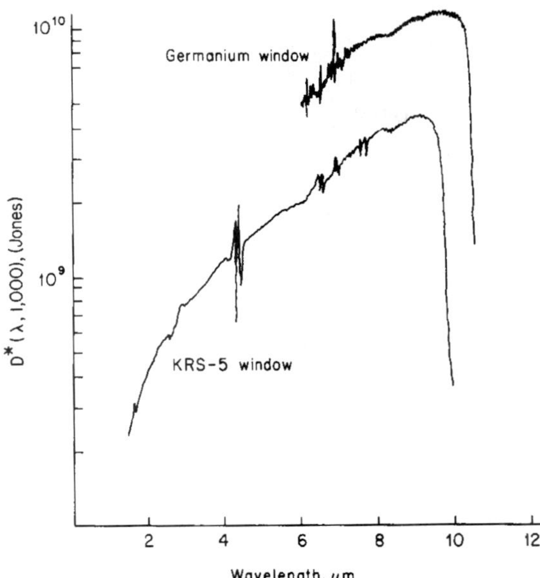

FIGURE 163 Photovoltaic (Pb,Sn)Te; 77 K, 60° FOV. *(Barnes Engineering Co.)*

Sensitivity: $D^*_{peak} > 10^{10}$ Jones (see Fig. 163).

Time constant: >50 ns.

Sensitive area: 1 × 1 mm.

Operating temperature: 77 K.

Circuit: Standard photovoltaic.

Manufacturers: Atomergic Chemetals, BSA Technology.

Ge:Hg. Mercury-doped germanium detectors are fast single-crystal impurity-doped photoconductors, sensitive out to 14 μm. Ge:Hg is especially well suited for detection through the 8- to 13-μm atmospheric window and for detection of near-ambient sources. Unfortunately, its operating temperature must be kept less than 40 K, where it becomes 300-K background-limited.

Sensitivity: See Figs. 147, 148, 164, and 165.

Quantum efficiency: 25–30 percent.

Noise: See Figs. 166 and 167.

Time constant: 100 ns with 50-Ω load for T < 28 K and electric fields < 30 V/cm. (Compensated material is available with ~5-ns time constant with a 50-Ω load. Responsivity then is reduced by 5–10× and detectivity is reduced by 2.)

Responsivity: Depends on concentration of compensating impurities, bias, area, and background flux. See Figs. 167–169, ~10^5 V/W.

Dark resistance: Depends on area and FOV: ~100 kΩ for 180° FOV (see Fig. 170).

Capacitance: <1 pF.

Sensitive area: 1–5-mm dia.

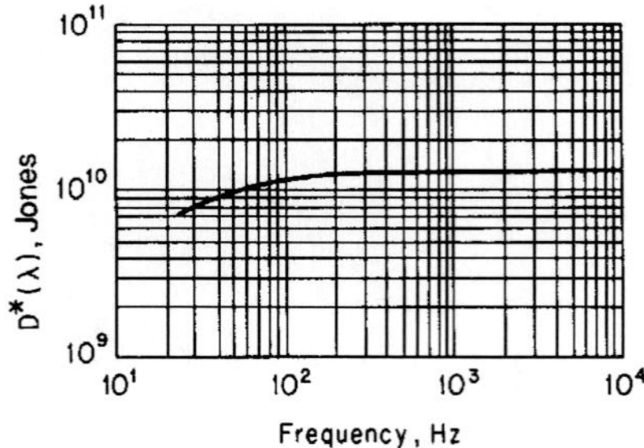

FIGURE 164 Typical D^* vs. frequency at 30 K for Ge:Hg; 1×1-mm area; essentially constant with temperature. *(Santa Barbara Research Center.)*

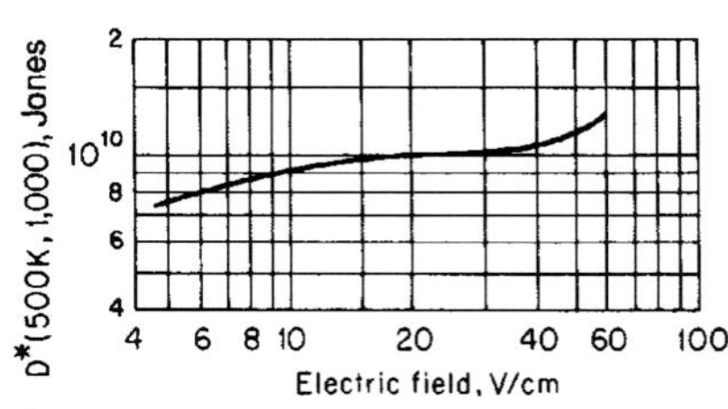

FIGURE 165 D^* vs. electric field for Ge:Hg; $T = 5$ K, 90° FOV; 300-K background; 6×10^{-4} cm^2 area; Irtran II window. *(Reprinted by permission of Texas Instruments.)*

Operating temperature: See Fig. 148.

Linearity: 10^{-3} to 10^{-8} W (size-dependent).

Sensitivity profile: ± 15 percent.

Recommended circuit: Standard photoconductive (see Ge:Au). See Fig. 171 for current-voltage characteristics.

Manufacturer: EG&G Judson.

Si:Ga. (Gallium in silicon forms an acceptor level with a binding energy of ~72 meV which is the basis of an infrared detector with spectral response out to approximately 17 μm, as shown in Fig. 172. The exact spectral cutoff and quantum efficiency will vary slightly with the gallium doping concentration. Gallium-doped silicon requires cooling to 20 K or lower for optimum performance. Background-limited performance associated with

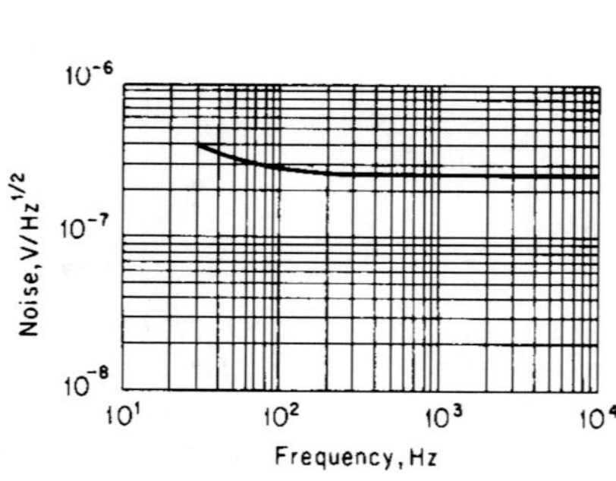

FIGURE 166 Typical noise-frequency spectrum for Ge:Hg; 1×1-mm area; essentially constant with temperature. *(Santa Barbara Research Center.)*

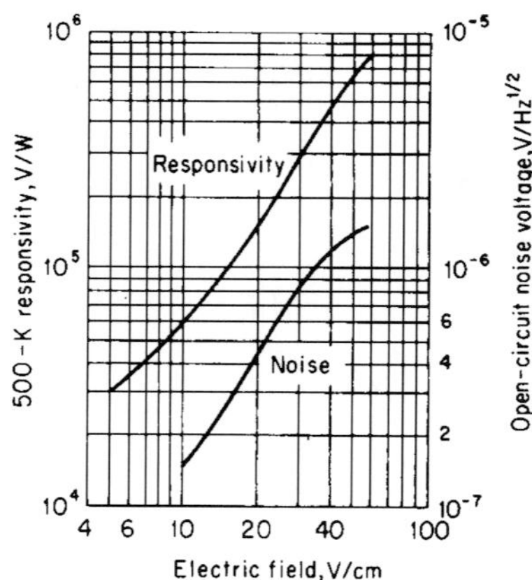

FIGURE 167 Open-circuit responsivity and noise voltage vs. electric field for Ge:Hg; $T = 5$ K; 90° FOV; 300-K background; $A = 6 \times 10^{-4}$ cm^2, Irtran II window. *(Reprinted by permission of Texas Instruments.)*

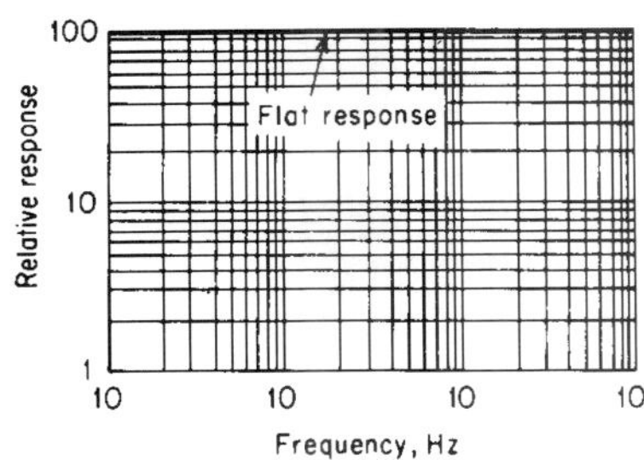

FIGURE 168 Ge:Hg typical relative response; $T = 30$ K. *(Santa Barbara Research Center.)*

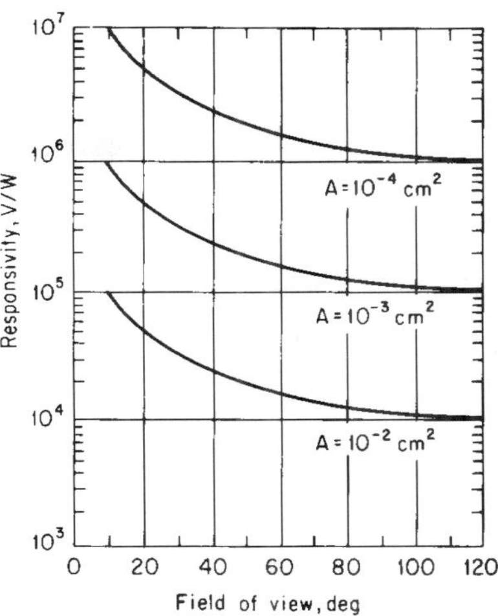

FIGURE 169 Open-circuit responsivity vs. FOV for Ge:Hg at 5 K for various detector areas; 300-K background; 500-K blackbody source. *(Reprinted by permission of Texas Instruments.)*

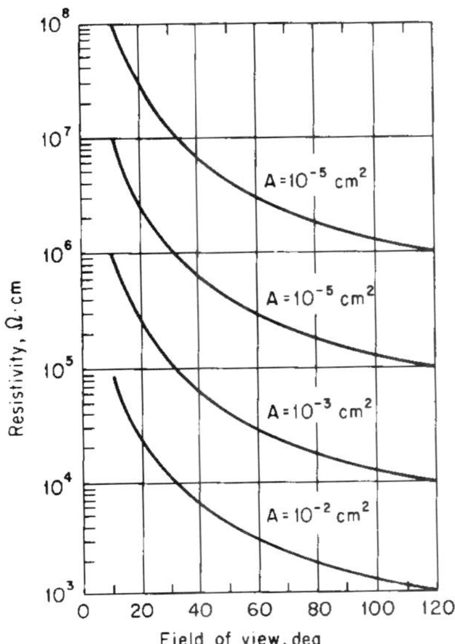

FIGURE 170 Ge:Hg resistivity vs. FOV for several detector areas; $T = 5$ K; 300-K background, optimum bias. *(Reprinted by permission of Texas Instruments.)*

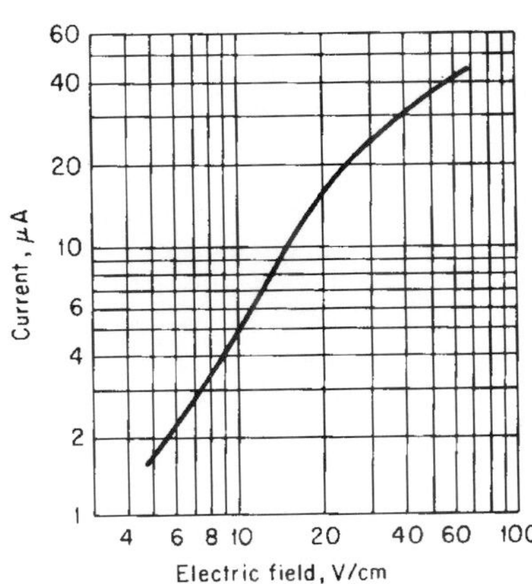

FIGURE 171 Ge:Hg bias current vs. electric field; 90° FOV; 300-K background; $T = 5$ K; $A = 6 \times 10^{-4}$ cm^2; Irtran II window. *(Reprinted by permission of Texas Instruments.)*

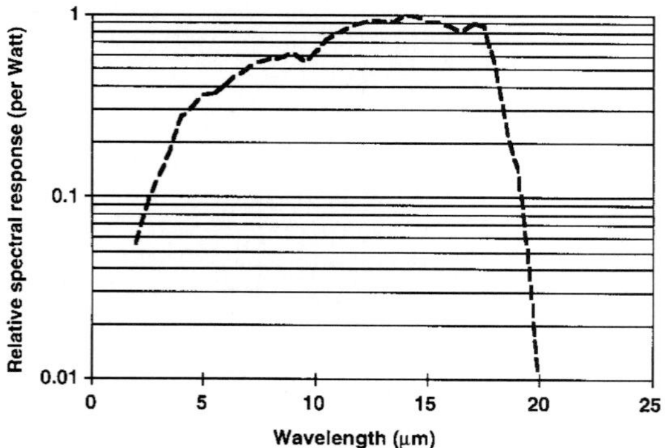

FIGURE 172 Relative spectral response per watt of Si:Ga as a function of wavelength. Data is normalized to unity at peak spectral response.

a quantum efficiency of about 15 percent is achievable over a wide range of background flux levels, provided the operating temperature is low enough to reduce thermal noise below the photon noise level.

Sensitivity: D^* is given by $D^* = 1.1 \times 10^{10} \times \sqrt{A\lambda\eta}/\sqrt{\mathbf{Q}}$ (Jones); where A is detector area, λ is the wavelength in μm, η is the quantum efficiency, and $\mathbf{Q}$ is the background flux in watts.

Responsivity: 0.9 A/W.

Time constant: <1 μs.

Resistance: Depends on background flux and detector bias (similar to Ge:Hg, see Fig. 170).

Capacitance: <1 pF (limited by mounting configuration).

Sensitive area: 0.2–2-mm square.

Operating temperature: <20 K.

Recommended circuit: Standard photoconductive.

Manufacturer: Infrared Laboratories.

Si:B. Boron in silicon forms an acceptor level with a binding energy of ~45 meV which is the basis of this infrared detector with spectral response out to approximately 30 μm. The exact spectral cutoff and quantum efficiency will vary slightly with the boron doping concentration. Boron-doped silicon requires cooling to about 15 K or lower for optimum performance. Background-limited performance associated with a quantum efficiency of about 10 percent is achievable over a wide range of background flux levels, provided the operating temperature is low enough to reduce thermal noise below the photon noise level.

Sensitivity: D^* is given by $D^* = 1.1 \times 10^{10} \times \sqrt{A\lambda\eta}/\sqrt{\mathbf{Q}}$ (Jones); where A is detector area, λ is the wavelength in μm, η is the quantum efficiency, and $\mathbf{Q}$ is the background flux in watts.

Responsivity: 2 A/W.

Time constant: <1 μs.

Resistance: Depends on background flux and detector bias (similar to Ge:Hg, see Fig. 170).

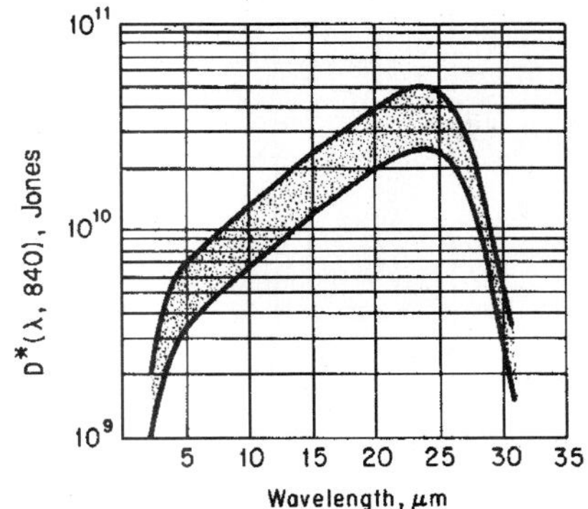

FIGURE 173 Range of spectral detectivities for Ge:Cu; 60° FOV. *(Santa Barbara Research Center.)*

Capacitance: <1 pF (limited by mounting configuration).

Sensitive area: 0.2–2-mm square.

Operating temperature: <15 K.

Recommended circuit: Standard photoconductive (see Ge:Au).

Manufacturer: Infrared Laboratories.

Ge:Cu. Copper-doped germanium detectors are fast, single-crystal, impurity-doped photoconductors, with high sensitivity in the broad region 2 to 30 μm. Operating temperature must be maintained below 20 K (ideally <14 K). Ge:Cu is then 300-K background-limited, and response time <50 ns.

Sensitivity: See Figs 147, 148, 173, and 174.

Noise: See Figs. 175 and 176.

Time constant: ~100 ns (can be doped to be faster, ~5 ns). (See discussion of time constant under Ge:Au).

Responsivity: 10^5 V/W (see Figs. 176–178).

Dark resistance: Depends on FOV (~100 kΩ for 180° FOV).

Capacitance: <1 pF.

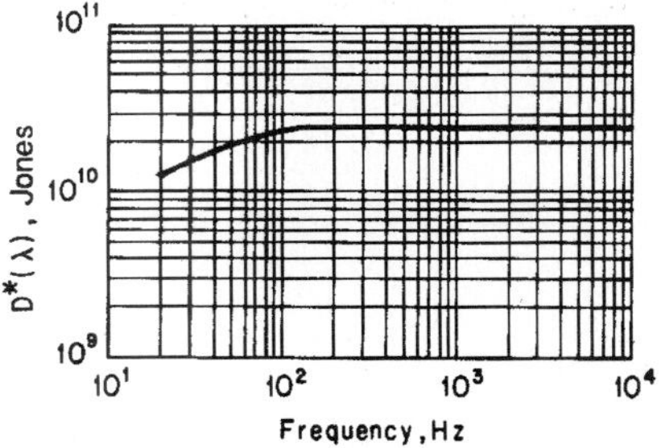

FIGURE 174 Typical D^* vs. frequency for Ge:Cu at 4.2 K. *(Santa Barbara Research Center.)*

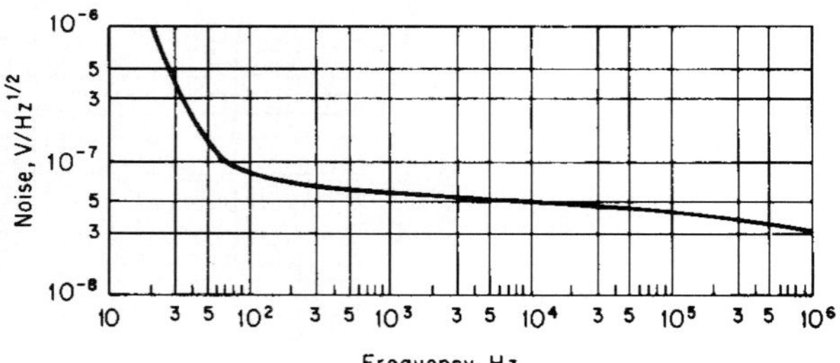

FIGURE 175 typical noise frequency spectrum for Ge:Cu. *(GEC Marconi Infra Red Ltd.)*

Sensitive area: 1–5-mm dia.

Operating temperature: See Figs. 148 and 178.

Linearity: 10^{-3}–10^{-8} W/cm^2 (depends on size).

Sensitivity profile: ±15 percent.

Stability: Stable in all ambient storage environments tested.

Recommended circuit: See Figs. 152 and 179.

Manufacturer: EG&G Judson.

Ge:Zn. Very similar to Ge:Cu except that cutoff wavelength moves out to 42 μm and operating temperature should be <10 K. A relatively low field breakdown limits the responsivity.

Sensitivity: See Figs 147, 148, and 180.

Noise: See Fig. 181.

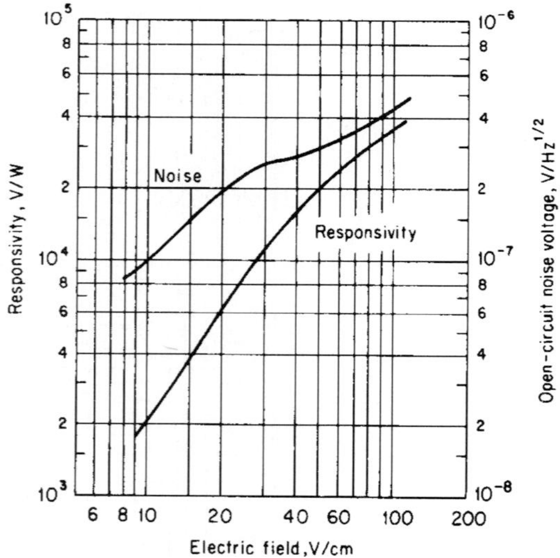

FIGURE 176 Typical noise and responsivity vs. biasing field for Ge:Cu; 5 K, 60° FOV; 300-K background, $A = 10^{-2}$ cm^2, 500-K blackbody. *(Reprinted by permission of Texas Instruments.)*

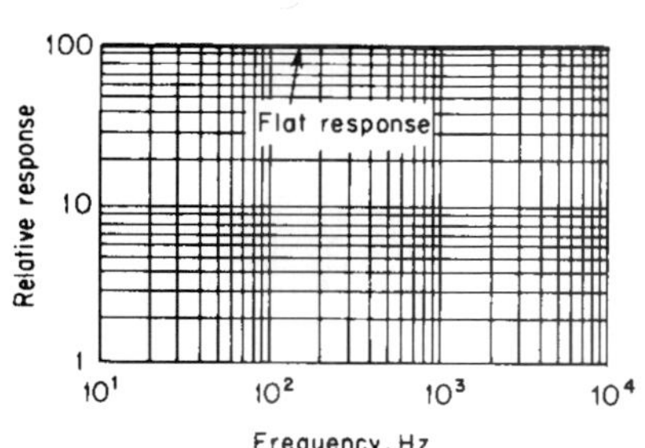

FIGURE 177 Relative response vs. frequency for Ge:Cu at 4.2 K. *(Santa Barbara Research Center.)*

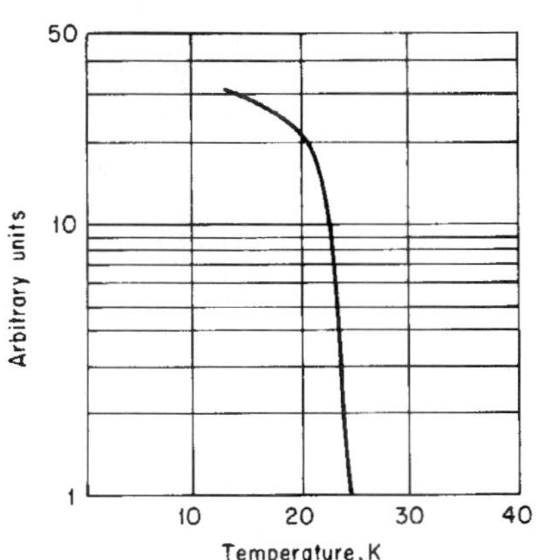

FIGURE 178 Relative responsivity vs. temperature for Ge:Cu.

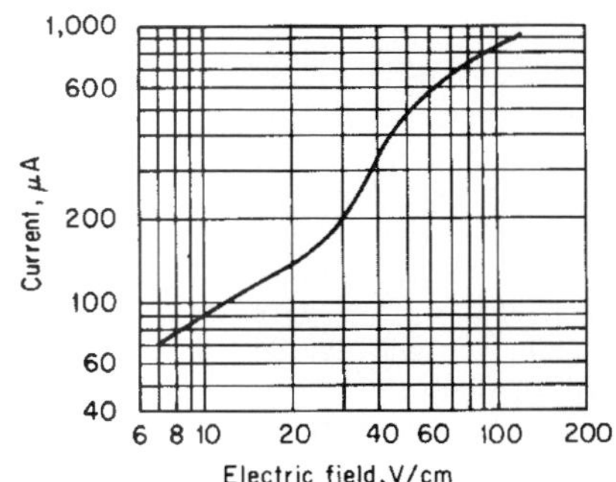

FIGURE 179 Typical current-voltage curve for Ge:Cu, 60° FOV; 300-K background: $A = 10^{-2}\,\text{cm}^2$. *(Reprinted by permission of Texas Instruments.)*

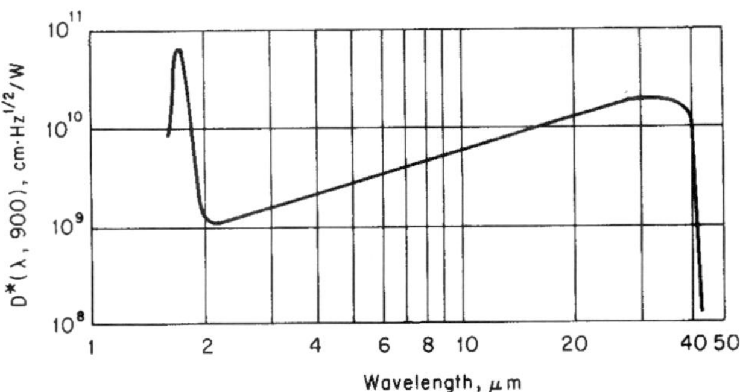

FIGURE 180 D^* vs. wavelength for Ge:Zn.

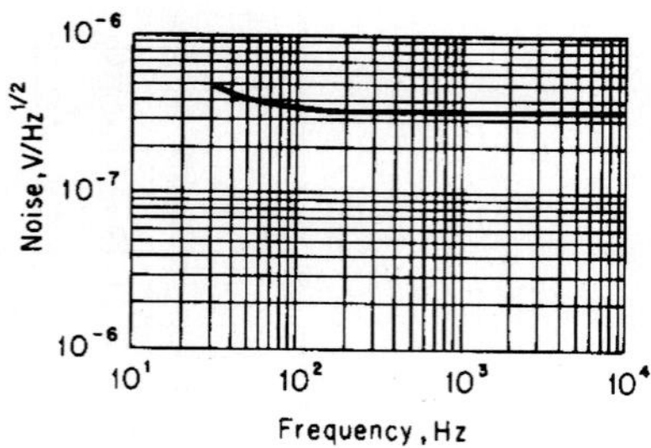

FIGURE 181 Typical noise-frequency spectrum at $T = 4.2$ K for Ge:Zn; $A = 1 \times 1$ mm. *(Santa Barbara Research Center.)*

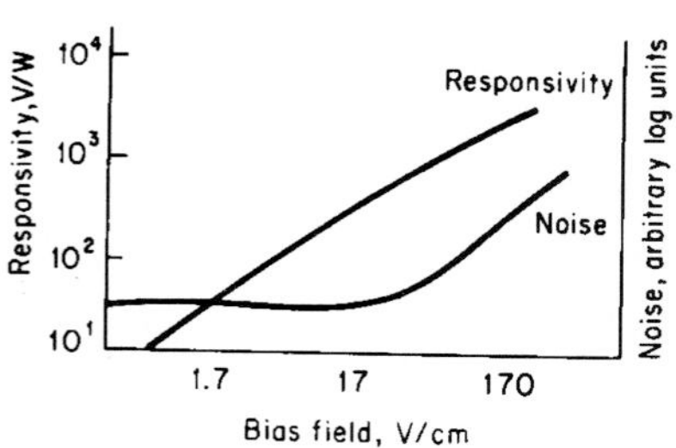

FIGURE 182 Signal and noise for Ge:Zn.

Time constant: <50 ns. (See discussion of time constant under Ge:Au.)

Responsivity: 10^3 V/W (see Fig. 182).

Dark resistance: 0.5–5 MΩ/sq (60°-FOV ambient background).

Capacitance: <1 pF (limited by mounting configuration).

Sensitive area: 1-, 2-, 3-, and 5-mm dia.

Operating temperature: <10 K.

Recommended circuit: Standard photoconductive.

Manufacturer: EG&G Judson.

Ge:Ga. The elements of B, Al, Ga, In, and Tl from chemical group III form shallow acceptor states (~10 meV) in germanium which are the basis of infrared detectors with spectral response out to approximately 120 μm. Currently, gallium-doped germanium is commercially available, but germanium doped with other group III elements (Ge:B, Ge:Al, Ge:In, Ge:Tl) will give similar detector performance. The small binding energies associated with this detector require cooling to liquid helium temperatures (4.2 K) or lower for optimum performance. Background-limited performance associated with a quantum efficiency of about 7 percent is achievable over a wide range of background flux levels, provided the operating temperature is low enough to reduce thermal noise below the photon noise level.

Sensitivity: D^* is given by $D^* = 1.1 \times 10^{10} \times \sqrt{A\lambda\eta}/\sqrt{\mathbf{Q}}$ (Jones); where A is detector area, λ is the wavelength in μm, η is the quantum efficiency, and $\mathbf{Q}$ is the background flux in watts.

Responsivity: 4 A/W.

Time constant: <1 μs.

Resistance: Depends on background flux and detector bias.

Capacitance: <1 pF (limited by mounting configuration).

Sensitive area: 0.5-, 1-, and 2-mm square.

Operating temperature: <4.2 K, best below 3 K.

Recommended circuit: Standard photoconductive.

Manufacturer: Infrared Laboratories.

Photographic. In this paragraph we present only the spectral sensitivity of some typical photographic emulsions. See Chap. 20, Photographic Films, in Vol. I of this Handbook, for a more extensive coverage.

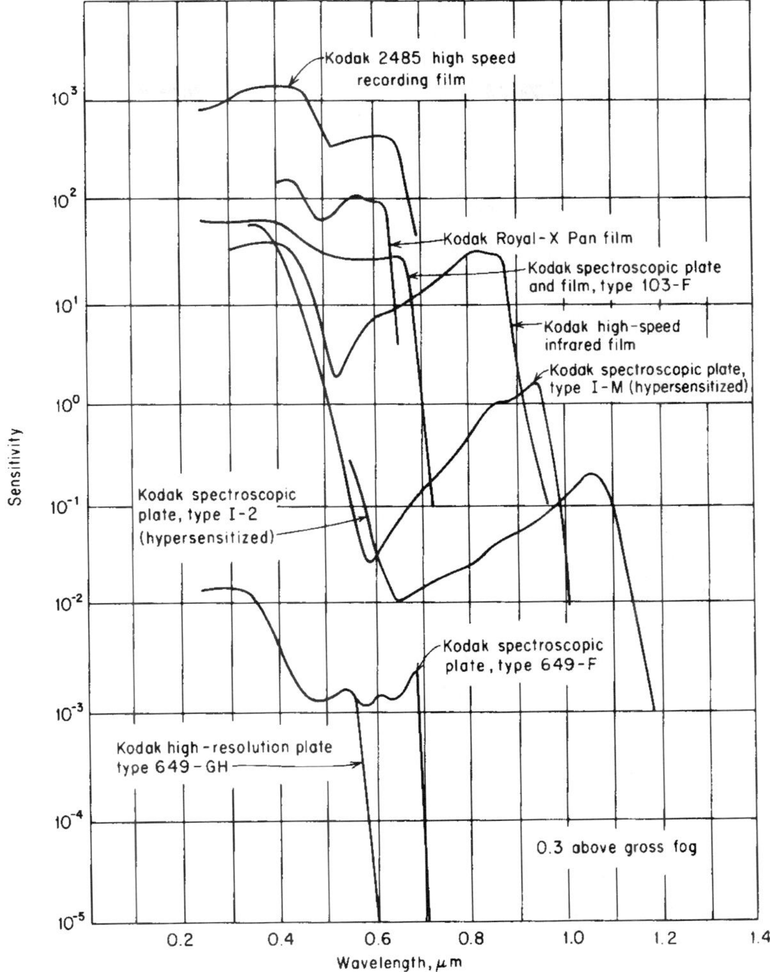

FIGURE 183 Sensitivity vs. λ for typical photographic emulsions. *(Eastman Kodak.)*

The term spectral sensitivity generally has a different meaning when applied to photographic detectors than it does when applied to the other detectors described in this chapter. It comes closer to responsivity than to minimum detectable power or energy.* In Fig. 183 sensitivity is the reciprocal of exposure, expressed in ergs/cm^2, required to produce

$$\text{Density} \equiv \log \frac{1}{\text{transmittance}} = 0.3$$

above gross fog in the emulsion when processed as recommended.

Manufacturers: AGFA, Eastman Kodak, Fuji, Polaroid.

* Work is in progress to evaluate photographic materials in terms of minimum detectable energy, a concept involving the average number of photons necessary to produce a change in density (signal) equal to that of the fog-density fluctuations (noise); see Marchant (1964), Jones (1958), Bird et al. (1969).

15.8 REFERENCES

Bauer, G., "Ein halbleiter Hochohmbolometer mit Tafel, IV," *Phys. Z.* **44:**53 (1943).

Bennett, H. E., "Accurate Method for Determining Photometric Linearity," *Appl. Opt.* **5:**1265 (1966).

Bird, G. R., R. C. Jones, and A. E. Ames, "The Efficiency of Radiation Detection by Photographic Films: State-of-the-Art and Methods of Improvement," *Appl. Opt.* **8:**2389 (1969).

Cole, M., and D. Ryer, "Cooling PM Tubes for Best Spectral Response," *Electro Optical Systems Design,* June 1972, p. 16.

Dereniak, E. L. and D. Crowe, *Optical Radiation Detectors,* Wiley, New York, 1984.

Eberhardt, E. H., "Threshold Sensitivity and Noise Ratings of Multiplier Phototubes," *Appl. Opt.* **6:**251 (1967).

Fellgett, P. B. "On the Ultimate Sensitivity and Practical Performance of Radiation Detectors," *J. Opt. Soc. Am.* **39:**970 (1949).

Fink, D. J. "Principles of Television Engineering," McGraw-Hill, New York, 1940.

Gunter, W. D., G. R. Grant, and S. A. Shaw, "Optical Devices to Increase Photocathode Quantum Efficiency," *Appl. Opt.* **9:**251 (1970).

The Infrared and Electro-Optical Systems Handbook, vol. 3, William Rogatto, ed., SPIE Press, 1993.

Institute of Radio Engineers, *IRE Standards on Electron Tubes: Methods of Testing,* publ. 62, pp. 7–51.

Jacobs, S. F. and M. Sargent III, "Photon Noise Limited D^* for Low Temperature Backgrounds and Long Wavelengths," *Infrared Phys.* **10**(4)**:**233–235 (1970).

Johnson, T. H., "Lead Salt Detectors and Arrays: PbS and PbSe," *Proc. SPIE* **443:**60–94, 1984.

Jones, R. C. "Factors of Merit for Radiation Detectors," *J. Opt. Soc. Am.* **39:**344 (1949).

———, "On the Quantum Efficiency of Photographic Negatives: On the Minimum Energy Detectable by Photographic Materials," *Photogr. Sci. Eng.* **2:**57–65j, 191–204 (1958).

Kruse, P. W., L. D. McGlaughlin, and R. B. McQuistan, *Elements of Infrared Technology,* Wiley, New York, 1962.

Livingston, W. C., "Enhancement of a Photocathode Sensitivity by Total Internal Rejection as Applied to an Image Tube," *Appl. Opt.* **5:**1335 (1966).

Low, F. J., "Low-Temperature Germanium Bolometer," *J. Opt. Soc. Am.* **51:**1300 (1961).

———, and A. R. Hoffman, "The Detectivity of Cryogenic Bolometers," *Appl. Opt.* **2:**649 (1963).

Marchant, J. C., "Exposure Criteria for the Photographic Detection of Threshold Signals," *J. Opt. Soc. Am.* **54:**79 (1964).

Morton, G. A., H. M. Smith, and H. R. Krall, "Pulse Height Resolution of High Gain First Dynode Photomultipliers," *Appl. Phys. Lett.* **13:**356 (1968).

Radio Corporation of America, "Phototubes and Photocells," *Tech. Man. PT-60,* 1963.

Samson, J. A. R., *Techniques of Ultraviolet Spectroscopy,* Wiley, New York, 1967.

Simon, R. E., A. H. Sommer, J. J. Tietjen, and B. F. Williams, "New High-Gain Dynode for Photomultipliers," *Appl. Phys. Lett.* **13:**355 (1968).

Smith, R. A., F. E. Jones, and R. P. Chasmar, *Detection and Measurement of Infrared Radiation,* Oxford University, London (1957).

Sommer, A. H., *Photoemissive Materials,* Wiley, New York, 1968.

Sommers, H. S., Jr. and E. K. Gritchell, "Demodulation of Low-Level Broadband Optical Signals with Semiconductors," *Proc. IEEE* **54:**1553 (1966).

———, and W. B. Teusch, "Demodulation of Low-Level Broadband Optical Signals with Semiconductors, II: Analysis of the Photoconductive Detector," *Proc. IEEE* **52:**144 (1964).

Sun, C. and T. E. Walsh, "Performance of Broadband Microwave-biased Extrinsic Photoconductive Detectors at 10.6 μm," *IEEE J. Quantum Electron.* **6:**450 (1970).

Sze, S. M., *Physics of Semiconductor Devices,* 2d ed., Wiley, New York, 1981, p. 773.

Vincent, D., *Fundamentals of Infrared Detector Operation and Testing,* Wiley, New York, 1989.

Wolfe, W. L. (ed.), *Handbook of Military and Infrared Technology,* Office of Naval Research, Washington, D.C., 1965.

CHAPTER 16
PHOTODETECTION

Abhay M. Joshi
Discovery Semiconductors, Inc.
Cranbury, New Jersey

Gregory H. Olsen
Sensors Unlimited, Inc.
Princeton, New Jersey

16.1 GLOSSARY

A	photodetector active area
A_0	incident photon flux
B	bandwidth of the photodetector
C	capacitance of the photodetector
D^*	detectivity
E	applied electric field
E_a	activation energy
E_g	bandgap of the semiconductor
E_i	impurity energy state
f	frequency
I_{diff}	diffusion current
$I_{g\text{-}r}$	generation-recombination current
IR_0	unity gain current
IR_1	reverse current generated by avalanche action
I_{tun}	tunneling current
k	Boltzmann's constant
L	distance traveled by a charge carrier
M	photocurrent gain
m	effective mass of electrons
N_A	acceptor impurity concentration on p side
N_D	donor impurity concentration on n side
n	refractive index of the AR coating
q	electron charge

R	sum of the detector series resistance and load resistance
R_o	detector shunt impedance
T	temperature in degrees Kelvin
t_n	transit time of electrons
t_p	transit time of holes
t_r	transit time of charge carriers (holes or electrons)
V	applied reverse bias in volts
V_B	breakdown voltage
V_{bi}	built-in potential of a *p-n* junction
W	depletion width of a *p-n* junction
α	absorption coefficient of the photodetector's absorption layer
ε_s	semiconductor permittivity
η	quantum efficiency of photodetector
θ	tunneling constant
λ	wavelength of incident photons (nm)
λ_{co}	detector cutoff wavelength (10 percent of peak response, nm)
μ	mobility of charge carriers (holes or electrons)
μ_n	mobility of electrons
μ_p	mobility of holes

16.2 INTRODUCTION

The approach of this chapter will be descriptive and tutorial rather than encyclopedic. It is assumed that the reader is primarily interested in an overview of how things work. Among the many excellent references to be consulted for further details are Sze's book,[1] and the article by Forrest.[2] For the latest in photodetector developments, consult recent proceedings of the Society of Photo-optical and Instrumentation Engineers (SPIE) conference or the IEEE Optical Fiber Conference.

A photodetector is a solid-state sensor that converts light energy into electrical energy. According to Isaac Newton, light energy consists of small packets or bundles of particles called *photons*. Albert Einstein, who won a Nobel prize for the discovery of the photoelectric effect, showed that when these photons hit a metal they could excite electrons in it. The minimum photo energy required to generate (excite) an electron is defined as the *work function* and the number of electrons generated is proportional to the intensity of the light. The semiconductor photodetectors are made from different semiconductor materials such as silicon, germanium, indium gallium arsenide, indium antimonide, and mercury cadmium telluride, to name a few. Each material has a characteristic bandgap energy (E_g) which determines its light-absorbing capabilities. Light is a form of electromagnetic radiation comprised of different wavelengths (λ). The range of light spectrum is split approximately as: ultraviolet (0–400 nm); visible (400–1000 nm); near infrared (1000–3000 nm); medium infrared (3000–6000 nm); far infrared (6000–40,000 nm); extreme infrared (40,000–100,000 nm). The equation between bandgap energy (E_g) and cutoff wavelength (λ_c) is

$$\lambda_c = \frac{1.24 \times 10^3 \text{ nm}}{E_g(\text{eV})} \tag{1}$$

TABLE 1 Important Photodetector Materials

Type	E_g (eV)	λ_c (nm)	Band
Silicon	1.12	1100	Visible
Gallium arsenide	1.42	875	Visible
Germanium	0.66	1800	Near-infrared
Indium gallium arsenide*	0.73–0.47	1700–2600	Near-infrared
Indium arsenide	0.36	3400	Near-infrared
Indium antimonide	0.17	5700	Medium-infrared
Mercury cadmium	0.7–0.1	1700–12500	Near-to-far-infrared

* The alloy composition of indium gallium arsenide and mercury cadmium telluride can be changed to alter the bandgap (E_g).

The smaller the bandgap (eV), the farther the photodetector "sees" into the infrared. Table 1 lists some prominent photodetector materials, their bandgaps, and cutoff wavelengths (λ_c) at room temperature (300 K).

Photodetectors find various applications in fiber-optic communications (800–1600 nm), spectroscopy (400 nm–6000 nm), laser range finding (400 nm–10,600 nm), photon counting (400 nm–1800 nm), and satellite imaging (200 nm–1200 nm), to name only a few topics. We will discuss three kinds of photodetectors: (1) Photoconductors, (2) *p-i-n* photodetectors (including avalanche diodes), and (3) photogates. Frequently, such detectors need to have high sensitivity, low noise, and high reliability. For fiber-optic applications, the frequency response and the cost can be a critical issue, whereas for infrared applications, many times it is the area of the photodetector. Large area (1-in diameter), high-sensitivity silicon avalanche photodetectors are now competing aggressively with conventional photomultiplier tubes for low light sensing applications in the visible. They offer the advantage of compact size and more rugged construction. We will also discuss reliability issues concerning photodetectors and take notice of a few novel photodetector structures.

16.3 PRINCIPLE OF OPERATION

When an electron in the valence band receives external energy in the form of light, the electron may overcome the nuclear attraction and become a "free electron." When light energy creates this transformation, it is termed *photonic excitation* (see Fig. 1*a*). The range of energies acquired by these free electrons is termed the *conduction band.* The energy difference between the bottom of this conduction band and the top of the valence band is

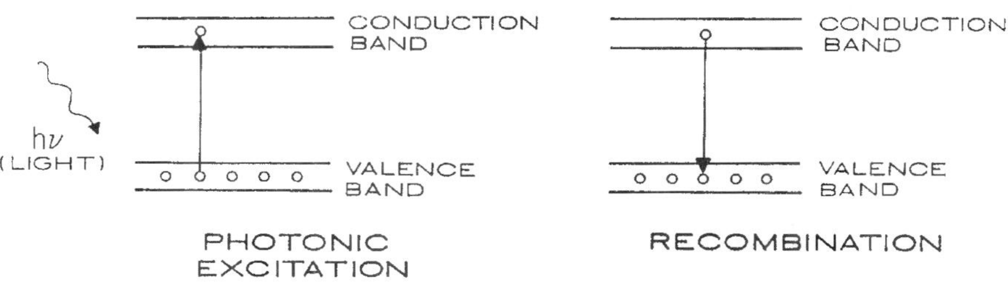

FIGURE 1 Photonic excitation and recombination in a semiconductor.

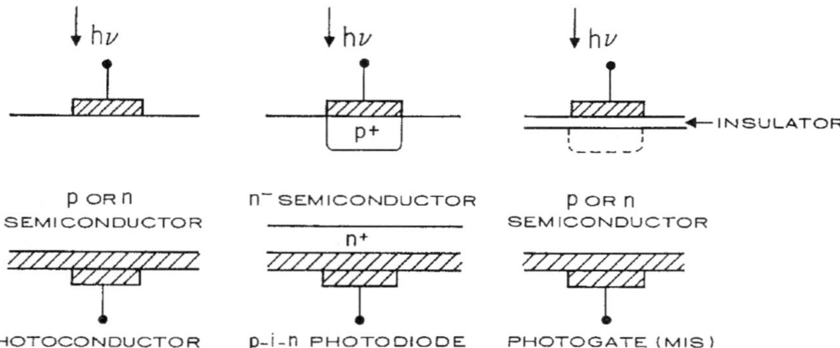

FIGURE 2 Types of photodetectors.

termed the *energy bandgap* (E_g) and represents the minimum energy of light that the material can absorb.

However, under the influence of even a small external electric field, the free electrons can "drift" in a specific direction. This is the fundamental principle of a photodetector. Figure 2 shows the three kinds of photodetectors discussed in this chapter. A brief explanation of each kind follows.

All photodetectors can be characterized by their quantum efficiency, detectivity (sensitivity), and response time.[1] Quantum efficiency (QE) is perhaps the most fundamental property, as it determines just how efficiently the device converts incoming photons into conduction electrons. Usually expressed in percentage, quantum efficiency can range from under 1 percent for PtSi Schottky barrier infrared detectors to well over 90 percent for InGaAs *p-i-n* fiber-optic photodetectors. Responsivity (R) is a related term expressed in amps/watt, which determines how much photocurrent is produced by optical power of a given wavelength. Detectivity measures how *sensitive* a detector is: that is, not only its light conversion efficiency, but also its ability to detect low-level light signals. It is limited by the various noise currents (shot, 1/f, etc.) introduced by the detector. Finally, response time describes how rapidly a detector can respond to a changing light signal. This ranges from milliseconds for certain types of PbS photoconductors to picoseconds for GaAs-like metal-insulator-semiconductor detectors.

These three parameters are frequently traded off. A large-area detector captures more light signal and thus might have greater detectivity. However, its larger capacitance would slow down the device. Similarly, response time in *p-i-n* detectors can be improved by thinning the absorbing region of the detector. However, this in turn cuts down quantum efficiency by reducing the total number of photons absorbed.

Photoconductor

A photoconductor, as the name implies, is a device whose conductivity increases with illumination. It acts as an open switch under dark (or no illumination) and as a closed switch under illumination. The simple equivalent electric circuit is shown in Fig. 3. This basic principle of a photoconductor finds numerous applications in relays and control circuits. An ideal switch should have low resistance in the closed position and, therefore, a pair of ohmic (not-rectifying) contacts are formed to the photoconductor. These ohmic contacts form the electrodes and usually have contact resistance of less than 10 ohms.

Types of Photoconductors. The two principle photoconductors are (1) intrinsic and (2)

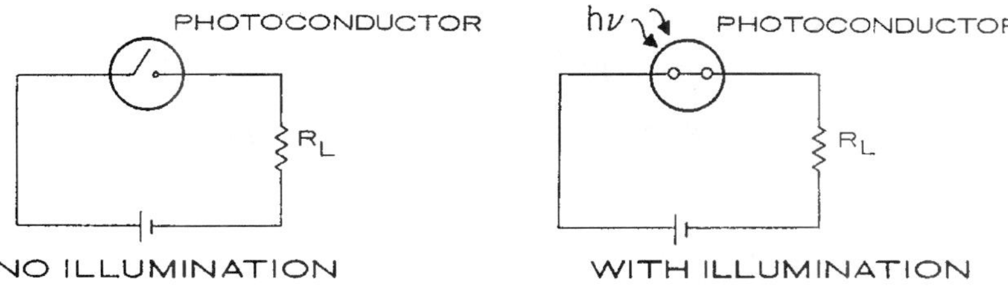

FIGURE 3 Equivalent circuit diagram of a photoconductor.

extrinsic. In an intrinsic device there is no external impurity atom. However, when an external impurity (dopant) is added to a material, it is termed *extrinsic*. This impurity atom occupies an energy state between the valence band and the conduction band. The functional difference between the intrinsic and extrinsic photoconductor is seen in Fig. 4. For an intrinsic device, the photo excitation (hv) needs to have energy greater than the bandgap energy (E_g) and its cutoff wavelength (λ_c) is given by Eq. (1). But, for an extrinsic one, the photon excitation (hv) should exceed the impurity energy state (E_i) and its cutoff wavelength λ_{co} is

$$\lambda_{co} = \frac{1.24 \times 10^3 \text{ nm}}{E_i(\text{eV})} \tag{2}$$

For intrinsic photoconductors, it is extremely difficult to achieve bandgap energies (E_g) less than 0.1 eV (refer to Table 1). This limits its capability to see in the far infrared and extreme infrared (13,000 nm–100,000 nm) and beyond. This is overcome by extrinsic devices whose E_i is less than 0.1 eV and is normally done by doping germanium or silicon. However, the extrinsic photoconductor suffers from very low absorption coefficients and, hence, poor quantum efficiencies. Also, since ambient thermal energy can excite carriers, they have to be cooled to liquid nitrogen temperature (77 K) and below, whereas most intrinsic photoconductors can operate at room temperature (300 K).

Photo Gain. The sensitivity of a photoconductor is determined by its gain. Photo gain is defined as the ratio of the output signal to the input optical signal. When photons impinge on a photoconductor, they generate electron-hole pairs and, under the influence of external fields, they are attracted toward the anode and cathode. A typical

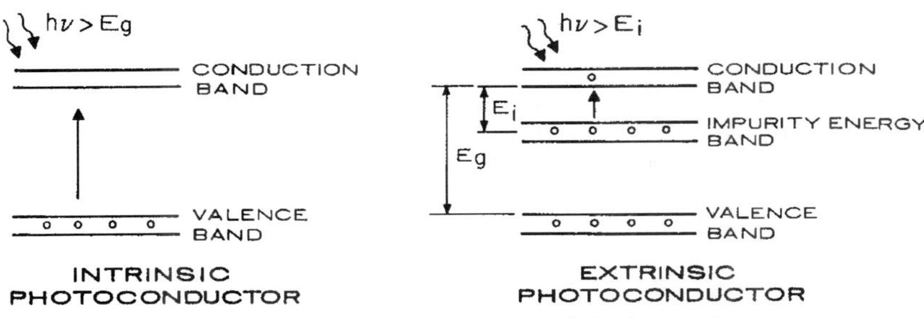

FIGURE 4 Functional diagram of an intrinsic and extrinsic photoconductor.

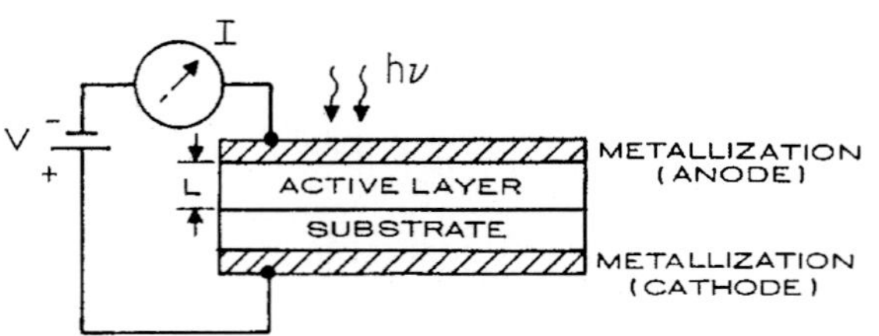

FIGURE 5 A typical photoconductor.

photoconductor is illustrated in Fig. 5 with L being the thickness of the active layer. The transit time (t_r) requires for a charge carrier to travel a distance L is given by:

$$t_r = \frac{L^2}{\mu V} \tag{3}$$

where V = applied bias and μ = mobility.

The mobility of electrons (μ_n) and that of holes (μ_p) is different, with (μ_n) being usually far higher than (μ_p). This causes a difference in the transit time of electrons and holes. Hence, photon-generated electrons are swept away more quickly than holes which result in a positive charge in the active layer. To maintain the charge neutrality, new electrons are supplied by the external voltage source. Therefore, for one incident photon, more than one electron is circulated in the electric circuit. This results in an "effective gain." Thus, photoconductor gain can be defined as the ratio of slower transit time (t_p) to faster transit time (t_n).

$$M = \frac{t_p}{t_n} \quad \text{or} \quad \frac{\mu_n}{\mu_p} \tag{4}$$

The slower the transit time, the higher the gain; however, the bandwidth of the device is reduced. Hence, high-gain photoconductors will result in slow devices and vice versa. Such "high-gain, slow devices" can be best utilized for imaging applications.[3] For high-speed optical communication applications in the 1000–1700-nm spectrum, InGaAs is the material of choice due to its high mobility. Several reports have been published on high-speed InGaAs photoconductors that find practical applications in optical receivers.[4–9] (Also see Chap. 17, "High-Speed Photodetectors," by J. Bowers.)

p-i-n Photodiode

Unlike photoconductors, a photodiode has a *p-n* junction, usually formed by diffusion or epitaxy. In a photoconductor, metal contacts are made to either *n*- or *p*-type material. However, a photodiode consists of both *n*- and *p*-type materials across which a natural electric field is generated. This field is known as the *built-in potential* (V_{bi}) and its value depends on the bandgap of its material. A silicon *p-n* junction has V_{bi} of 0.7 V whereas in germanium it is 0.3 V. The higher the bandgap (E_g), the larger the built-in potential (V_{bi}). An important physical phenomenon called *depletion* occurs when a *p*-type semiconductor is merged with an *n*-type semiconductor. After an initial exchange of charge, a potential is built up to prevent further flow of charge. This built-in field creates the depletion width (W), which is a region free of any charge carriers and is given by[10]

$$W = \sqrt{\frac{2\varepsilon_s(N_A + N_D)}{q(N_A N_D)}(V_{bi} - V)} \tag{5}$$

where N_A and N_D are impurity concentrations of p side and n side, respectively, q is the

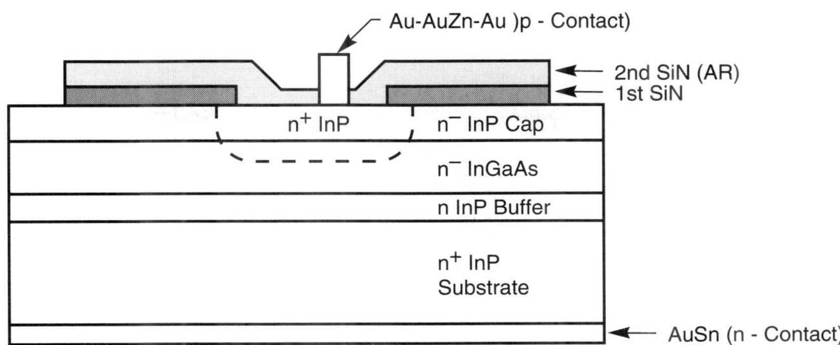

FIGURE 6 A typical InGaAs *p-i-n* photodiode.

electron charge, V_{bi} is the built-in potential and ε_s is the semiconductor permittivity, V is the applied bias and is negative for reverse-bias operation. As seen from Eq. (5), under reverse bias the depletion width (W) increases causing a decrease in the capacitance of the photodiode. A *p-i-n* photodiode is similar to a parallel plate capacitor with the anode-cathode being the two plates and the depletion width (W) being the separating medium. A typical InGaAs *p-i-n* photodiode is shown in Fig. 6 and its capacitance is termed as[1]

$$C = \frac{\varepsilon_s A}{W} \tag{6}$$

where ε_s is the semiconductor permittivity and A is the active area of the photodiode. From first principles, a decrease in capacitance improves the bandwidth (B) of the photodetector since

$$B = \frac{0.35}{2.2RC} \tag{7}$$

where, R is the sum of the detector series resistance and load resistance. For a detailed analysis on high-speed photodetectors, see Chap. 17 by John Bowers.

Dark Current. For applications ranging from optical communications (III-V compound semiconductors) to infrared sensing (Si, Ge, III-IV, and II-IV compound semiconductors), a *p-i-n* photodiode must have high sensitivity and low noise. These are largely determined by the dark currents originating in the device. Several authors have published papers on dark currents in InGaAs[11–13] and HgCdTe.[14,15] The three major components of dark current are (1) diffusion current, (2) generation-recombination current, and (3) tunneling current.

Diffusion Current. In the nondepleted region of the photodiode, electron-hole pairs are formed by the ambient temperature. These thermally generated carriers diffuse toward the depletion region and produce the diffusion current.

$$I_{\text{diff}} \propto e^{-E_g/kT} \tag{8}$$

where E_g is the bandgap of the photodiode material, k is Boltzmann's constant, and T is the ambient temperature in degrees Kelvin. From Eq. (8), clearly diffusion current is higher in low-bandgap material. Therefore, InSb ($E_g = 0.17$ eV) has far higher diffusion current than silicon ($E_g = 1.12$ eV) and in fact this makes InSb almost useless at room temperature. To overcome this excessive diffusion current, InSb photodiodes are cooled to liquid nitrogen temperature (77 K).

Generation-Recombination Current. The current generated in the depletion region of the photodiode is called the *generation-recombination current.* When impurity trap levels are present within the forbidden gap (E_g), trapped carriers can be elevated to the conduction band with less energy than for diffusion current. This "trap-assisted" current is given by

$$I_{g-r} \alpha \sqrt{(V_{bi} - V)}\, e^{-Eg/2kT} \tag{9}$$

From Eq. (5), the depletion width W is proportional to $V_{bi} - V$. Hence,

$$I_{g-r} \alpha W e^{-Eg/2kT} \tag{10}$$

Generation-recombination current is proportional to the volume of the depletion width and, hence, is reverse-bias-dependent, whereas the diffusion current in Eq. (8) is bias-independent. For high-bandgap semiconductors with bandgaps above 1.0 eV (e.g., silicon), the generation current usually dominates over the diffusion current at room temperature. However, for low-bandgap material such as indium antimonide, the diffusion current is dominant over generation current at room temperature.

Tunneling Current. When the electric field in a reverse-biased *p-n* junction exceeds 10^5 V/cm, a valence band electron can jump to the conduction band due to the quantum mechanical effect[10] called *tunneling* which occurs at high field and with geometrically narrow energy barriers. The tunneling current is given by

$$I_{\text{tun}} \alpha E V \exp\left(\frac{-\theta\sqrt{m}}{E} E_g^{3/2}\right) \tag{11}$$

where E is the applied electric field, m is the effective mass of electrons, and θ is a dimensionless constant whose value depends on the tunneling barrier height. Higher doping levels at the *p-n* junction lead to a narrower depletion width which causes higher electric fields, thus increasing the amount of tunneling current. Low-bandgap photodiodes exhibit much more tunneling than do higher-bandgap diodes. Tunneling shows a weak dependence on temperature, the only minor change being caused by the temperature dependence of the bandgap (E_g). This leads to a *decreasing* breakdown voltage with increasing temperature as opposed to an *increasing* breakdown voltage exhibited by the avalanche effect.

Quantum Efficiency, Responsivity, and Absorption Coefficient. Quantum efficiency (η) is defined as the ratio of electron-hole pairs generated for each incident photon. In a nonavalanche *p-i-n* photodiode, quantum efficiency is less than unity. Responsivity (R) is a measured quality in amps/watt or volts/watt and is related to quantum efficiency by

$$Q \cdot E(\eta) = \frac{(1240) \cdot R(A/W)}{\lambda} \tag{12}$$

where λ is the wavelength in nm of incident photons and R is the responsivity in

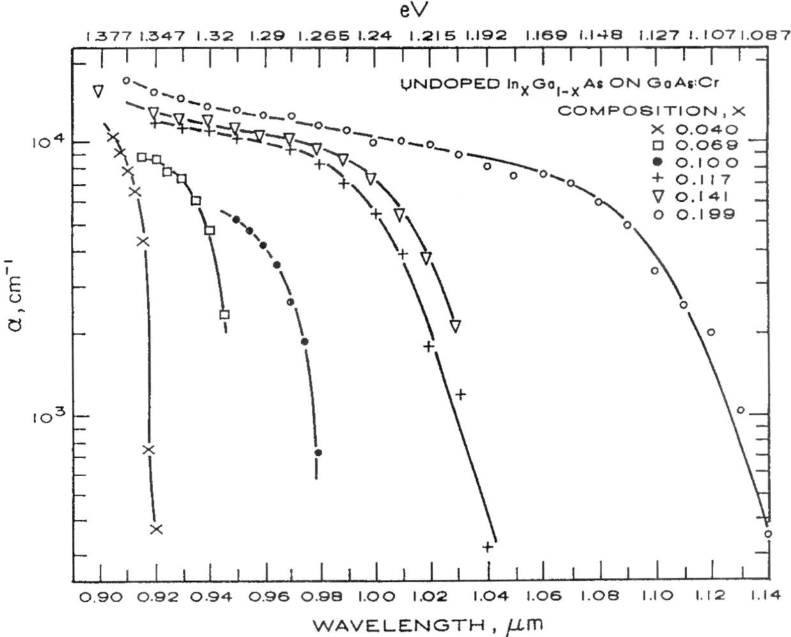

FIGURE 7 Absorption coefficients for 1-μm thick undoped $In_xGa_{1-x}As$, $0 < x < 0.25$.[16]

amps/watt. The value of η is determined by the absorption coefficient (α) of the semiconductor material and the penetration distance (x) in the absorbing layer. The light flux (A) at a distance x with the absorption layer is

$$A = A_0 e^{-\alpha x} \tag{13}$$

where, A_0 is the incident photon flux and α is a strong function of wavelength (λ). Figure 7 shows its typical values for a 1-μm thick undoped $In_xGa_{1-x}As$, $0 < x < 0.25$.[16] For optimized η, the reflectivity at the semiconductor surface has to be minimized. Hence, an antireflection (AR) coating of proper thickness is deposited on the photodiode surface. For single-layer AR coatings, the proper "quarter-wave" thickness (L) of the AR coating is:

$$L = \frac{\lambda}{4n} \tag{14}$$

where n is the refractive index of the coating. With good AR coatings, InGaAs photodiodes can achieve quantum efficiencies above 95 percent at 1300–1500 nm. For the visible region, silicon photodiodes show high η (90 percent) in the 800-nm range, and the mid-infrared InSb has a typical η of 80 percent at 5000 nm.

Avalanche Photodiodes. Avalanche photodiodes (APDs) will be briefly discussed here. For a detailed treatment, see Chap. 17 on high-speed photodetectors by John Bowers in this book. An avalanche photodiode is a *p-i-n* diode with a net efficiency or gain greater than unity. This is obtained through the process of "impact ionization" by operating the photodiode at a sufficiently high reverse bias. The typical operating voltage for an InGaAs APD is 75 V, while that for silicon can be as high as 400 V. The impact ionization process

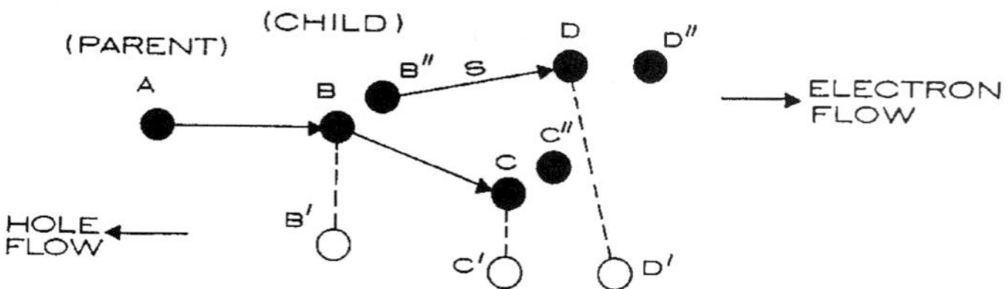

FIGURE 8 Impact ionization process.

is described in Fig. 8. Under the influence of a high electric field ($>5 \times 10^5$ V/cm), electron A gains sufficient kinetic energy to hit atom B with a tremendous force and knock out an electron hole pair B'-B''. A is called the "parent" and B'-B'' the "child" charge carriers. The child electron B'' moves through a critical distance S and acquires enough kinetic energy to create its own child particles D'-D''. The sum effect of the impact ionization of a number of electrons is termed *avalanche multiplication.* Because of this avalanche action, the gain in an APD exceeds unity, reaching useful values above 10 for InGaAs and several hundred for silicon before the multiplied noise begins to exceed the multiplied signal. A solid-state APD is a fast device with gain-bandwidth products than can exceed 20 GHz.[17,18] In spite of high operating bias, an APD can be designed for low noise operation[19] and used for numerous applications such as photon-counting, laser pulse detection,[20] and fiber-optic communication. The gain or avalanche multiplication (M) of an APD is given by

$$M = \frac{IR_1}{IR_0} \tag{15}$$

where IR_1 is the reverse current generated by avalanche action and IR_0 is the unity gain current. At voltage breakdown of the APD, the multiplication factor M tends to infinity. An empirical relation between the multiplication factor (M) and reverse bias (V) is given by[21,22]

$$M = \frac{1}{1 - (V/V_B)^n} \tag{16}$$

where V is the applied reverse bias and V_B is the breakdown voltage. The factor n varies between 3 to 6, depending on the semiconductor material and its substrate type. Typical gains are on the order of 10 to 20 for germanium and InGaAs APDs, and above 100 for silicon APDs. Due to their lower noise, InGaAs and silicon APDs have better sensitivity than their germanium counterparts.

Extended Wavelength (1000–3000 nm) Photodetectors

Detector materials used for the 1000–3000 nm spectrum include InSb, InAs, PbS, HgCdTe, and recently InGaAs. PbS is an inexpensive, reasonably sensitive detector that can operate at relatively high temperatures, even room temperature. Its major drawback is its slow (typically milliseconds) response time. InAs has higher sensitivity over the 1000–3500 nm spectrum and fast response time, but must be cooled thermoelectrically (to 230 K) or cryogenically (to 77 K). InSb has similar properties out to 5500 nm but must definitely be cryogenically cooled. HgCdTe has high sensitivity and speed and can be operated at room temperature. However, a new competitor has come on the scene recently: indium gallium arsenide. This material was originally developed for fiber-optic applications out to 1.7 μm (using $In_{0.53}Ga_{0.47}As$) and can now be used out to 2500 nm by increasing its indium content to $In_{0.8}Ga_{0.2}As$. InGaAs appears to be the *best* detector material for high-temperature

TABLE 2 Comparison of R_0A Values in HgCdTe and InGaAs (Ω-cm^2)

	$R_0A(T)$	
λ_{co} (nm)	HgCdTe	InGaAs
1400	4×10^4 (292 K)	2.5×10^5 (300 K)
	7×10^6 (230 K)	1.3×10^8 (220 K)
1700	2×10^2 (300 K)	2.5×10^5 (300 K)
	2×10^5 (220 K)	1.3×10^8 (220 K)
2100	7×10^1 (300 K)	2.5×10^3 (300 K)
	7×10^3 (220 K)	6.5×10^5 (220 K)
2500	1×10^1 (300 K)	1.3×10^2 (300 K)
	1×10^3 (210 K)	1.0×10^5 (210 K)

operation in the 1000–3000 nm spectrum. It has a 10–100× advantage in shunt resistance at room temperature compared to HgCdTe—the previously used material for this wavelength.

Table 2 contains a summary of the available data. It is extremely difficult to find data at exactly the same cutoff wavelengths and temperature with the same area device. R_0A was determined (in cases where it was given as such) by simply multiplying two numbers, where, R_0 is the shunt impedance of the detector, and A is the active area of the photodetector.

Photogate (Metal-Insulator-Semiconductor Detector)

The advent of silicon charge coupled devices (CCDs) has revolutionized the television industry and introduced one of the most popular consumer items to millions of people around the world—the CCD camcorder. From the sandy shores of Hawaii to the ski slopes of Colorado, people have captured life's best moments with a CCD camcorder. With its superior imaging quality and noise performance of a few electrons/per pixel (<20), a CCD has diverse uses from space imaging to chemical analysis spectroscopy. A CCD is a matrix of metal oxide semiconductor (MOS) devices operated in the "depletion" mode. Each individual MOS device is called a *photogate* and its schematic diagram is shown in Fig. 9. Consider an n-type semiconductor with a negative potential applied to its gate. This will repel the negatively charged electrons and create a depletion layer. As the negative potential on the gate is further increased, the volume of the depletion region increases further into the bulk. However, the surface potential at the semiconductor-insulator

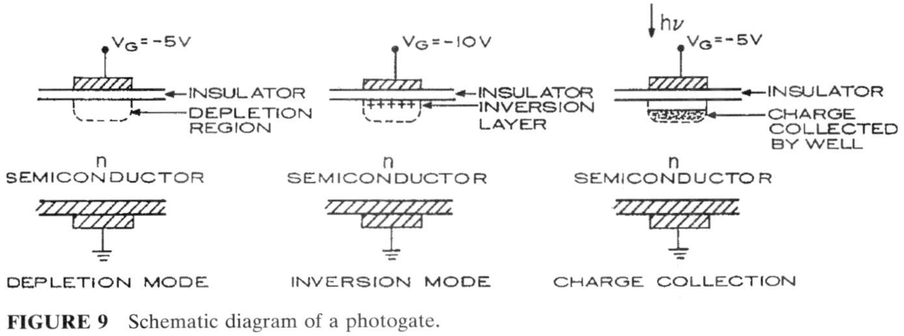

FIGURE 9 Schematic diagram of a photogate.

interface also becomes more negative. Finally, with increased gate bias, the surface potential becomes sufficiently high to attract minority carriers (holes). This creates a positive charge at the semiconductor-insulator interface and is termed the *inversion layer.* In an MOS transistor, the inversion layer forms a conducting channel between the source and the drain, and the gate bias needed to achieve inversion is termed the *threshold voltage.* Usually a photogate is operated in depletion at a gate bias lower than the threshold voltage. When incident photons create hole-electron pairs, the minority carriers drift away to the depletion region and the volume of the depletion region shrinks. The total amount of charge that a photogate can collect is defined as its *well capacity.* The total well capacity is decided by the gate bias, the insulator thickness, the area of the electrodes, and the background doping of the semiconductor. Numerous such photogates with proper clocking sequence form a CCD imaging array. For in-depth understanding of CCDs, we refer to Chap. 22 "Visible Array Detectors," by T. J. Tredwell.

16.4 APPLICATIONS

The main commercial uses for photodetectors include optical communications and infrared sensing. Although these applications often overlap, optical communication typically involves transmitting data over an optical fiber at higher rates. The format is increasingly digital (telecommunications and data links) at rates from 1 Mbit/s to over 2 Gbit/s.

However, one growing application is cable TV (CATV) where analog data rates from 1 to beyond 1000 MHz are most often found. Infrared sensing mostly involves nonfiber applications at sub-MHz analog rates. The property to be detected is usually the amplitude (in watts) and wavelength of the incoming radiation. In digital applications, the wavelength and individual pulse amplitude are relatively fixed, and successful communication occurs simply by distinguishing when the pulse is "on" or "off." Although very weak pulses must sometimes be detected, the actual amplitude of the pulse is irrelevant. The ultimate "resolving power" of the detector is when a weak pulse can no longer be distinguished from background noise, i.e., when the incoming signal strength (S) equals the background noise strength (N) or when the "signal-to-noise" ratio $S/N = 1$. Thus, the strengths of individual pulses are unimportant as long as the presence of a pulse can be detected. Information is conveyed by the timing sequence of the pulses rather than by the amplitude of the individual pulse. Analog applications, on the other hand, depend critically on the frequency content and amplitude of the transmitted signal. In an AM cable TV transmission system, the detector must be able to linearly reproduce the incoming optical signal as an electrical current of the same frequency content and amplitude, and to minimize intermodulation and harmonic distortion that is invariably produced in the detection of an AM signal.

Infrared applications often involve spectroscopy whereby the detected electrical signal depends on both the optical wavelength and strength of the incoming infrared signal. Thus, the detector must be carefully calibrated in terms of "responsivity" (electrical amps/optical watt) versus wavelength in order to accurately identify the nature of the incoming signal. Identification of gases (e.g., methane, which absorbs light near 1650 nm) depends on these properties. Other "infrared" applications include spectroscopy, remote sensing from satellite, and general laboratory detection. Not *all* infrared applications are analog, however. One notable digital application is LIDAR (LIght Detection And Ranging), which essentially is a form of laser radar. High-intensity light pulses are emitted into the atmosphere (or a gas) which absorbs, scatters, and reemits the laser pulse. The character of light pulses detected back near the source can be used to determine the nature of the gas particles that interacted with the light: the absorbing wavelength, the gas density (velocity), and the amount present. Applications include remote pollution monitoring and "windshear detection," whereby the presence of abrupt changes in wind velocity can be instantly detected at distances of several miles. This application[42] has been demonstrated with laser wavelengths of 2060 and 10,600 nm for use on aircraft and may be required on all commercial aircraft by the year 1995. The 2060-nm system works better in severe storms and does not require a cryogenically cooled detector (as does the 10,600-nm system). However, the CO_2 gas laser used for 10,600 nm, although bulky and costly, is more highly developed and more readily available than the more recent HO:YAG solid-state lasers used for 2060 nm.[42] Further developments will undoubtedly occur here over the next several years.

One important noise source in infrared applications is the so-called 1/f noise which becomes noticeable at frequencies below 10 MHz. Although poorly understood, this noise is thought to originate at heterointerfaces such as semiconductor-metal contacts and heteroepitaxial interfaces. Photodiode arrays are often required to detect low-light-level signals of a few hundred photons, and they must integrate the signal for 1 second or more. However, with longer integration times, 1/f noise may become noticeable and can degrade the S/N ratio and thus, impose an upper limit on the effectiveness of longer integration times. Limiting 1/f noise becomes critical for numerous infrared sensing applications and research indicates that surface depletion width at the semiconductor-insulator interface to be a major source of 1/f noise in the InGaAs photodiodes.[44]

One important area for detectors is the array configuration used both for spectroscopy (linear) and imaging (two-dimensional). Linear arrays are used in so-called multichannel analyzers whereby the detector is placed behind a fixed grating and the instrument functions as "motionless" or "instant" spectrometer with each pixel corresponding to a narrow band of wavelengths. The resolution of the instrument is determined by the number and spacing of pixels, so *narrow* pixel geometries are demanded along one direction whereas *tall* pixel geometries are demanded along the perpendicular direction to enhance the light collection.

One "mixed" infrared/fiber-optic application is the use of large-area (typically 3-mm diameter) detectors for optical power meters: the optical equivalent of a voltmeter which accurately measures the amount of optical power in watts or dBm (number of decibels above or below 1 mW contained in an incoming beam). The large area ensures large collection efficiency. The most important parameter here is the responsivity and the uniformity of response across the detector.

A "figure-of-merit" for infrared detectors is D^* (deestar), whereby detectors of differing area can be compared. It is related to the noise equivalent power (NEP) in watts, the lowest power a detector can detect at a signal-to-noise ratio of 1 as

$$D^*(\lambda, f, B) = (AB)^{1/2}/\mathrm{NEP} \tag{17}$$

where A is the detector area. The optical bandwidth B (often taken to be 1 Hz), frequency of signal modulation (f), and operating wavelength (λ) must be stated.

16.5 RELIABILITY

In today's global economy of severe competitiveness, new product development and innovation is incomplete without quality assurance and reliability. Reliability is the assurance that a device will perform its stated functions for a certain period of time under stated conditions, and considerable research has been done to improve the reliability of photodetectors.[23–26] The two major industrial standards for testing semiconductor device reliability are: (1) test methods and procedures for microelectronics (MIL. STD. 883C), and (2) Bellcore technical advisory (TA-TSY-00468). The former standard is generic to the semiconductor industry, while the latter is specifically developed for fiber-optic optoelectronic devices.

The tests performed under MIL. STD. 883C comprise of the following major groups:

TABLE 3 Electrical and Optical Testing of Photodetectors

Tests or measurement	Parameter	Symbol	References
Optical response	Responsivity	R	Bellcore Technology
	Gain	G	Advisory
Electrical performance	Dark current	I_d	TA-TSY-00468
	Breakdown voltage	V_{br}	Issue 2, July 1988

(1) *environmental tests,* e.g., moisture resistance, burn-in, seal, dew point, thermal shock, (2) *mechanical tests,* e.g., constant acceleration, mechanical shock, vibration, solderability, and bond strength, and (3) *electrical tests,* e.g., breakdown voltage, transition time measurements, input currents, terminal capacitance, and electrostatic discharge (ESD) sensitivity classification. Under the Bellcore Technical Advisory, each photodetector lot undergoes visual inspection, optical and electrical characterization, and screening. Visual inspection removes any photodiodes with faulty wire bonds or cracks in the glass window or in the insulating films. Table 3 lists the electrical and optical testing performed on every photodiode. After testing, all the devices are sent for screening (burn-in), e.g., some $In_{0.53}Ga_{0.47}As$ photodiodes are burned-in at 200°C for 20 hours at −20 V reverse bias to weed out any infant mortality.

Photodetector Life Test (Accelerated Aging)

To predict the lifetime or *mean-time-to-failure* (MTTF), accelerated aging tests are carried out on groups of diodes at several elevated temperatures. For example, the MTTF for 300-μm diameter InGaAs photodiodes was determined on groups of 20 screened devices at elevated temperatures of 200, 230, and 250°C. The failure criterion was a 25 percent increase in the room temperature dark current value.[26,27] The total lifetest extended over a time period of several years, and every week the samples were cooled to room temperature to check their dark current. Failed devices were removed from the sample population and the remaining good ones put back at the elevated temperature.

From the temperature-dependence of the data, it was observed that the failure mechanism is thermally activated. The Arrhenius relationship calculates the activation energy (E_a) for thermally activated failure[27] as

$$\text{MTTF}(\text{T}) = Ce^{(E_a/kT)} \tag{18}$$

where, C is a constant, k is Boltzmann's constant (8.63×10^5 eV/K), and T is the temperature in Kelvin. Figure 10 shows the MTTF for three batches of 300-μm diameter $In_{0.53}Ga_{0.47}As/InP$ photodiodes. Using least-squares fit to the data, the calculated activation energy (E_a) is 1.31 eV with a correlation coefficient (r^2) of 0.99. From Eq. (18) and an experimentally determined activation energy of 1.31 eV, the MTTF at 25°C is calculated to be 1.34×10^{14} hours. Such a "geological" lifetime may seem to be an overkill, even for the electronics industry. However, when thousands of these devices are working together in a single system (e.g., telephones), the net MTTF of all these devices chained together may be on the order of only a few years. Thus, continuing improvements in reliability must be an ongoing process. Reliability in most photodetectors is determined by a number of factors including: (1) material quality, (2) processing procedures, (3) planar technology versus mesa technology, and (4) amount of leakage current. Poor material quality can

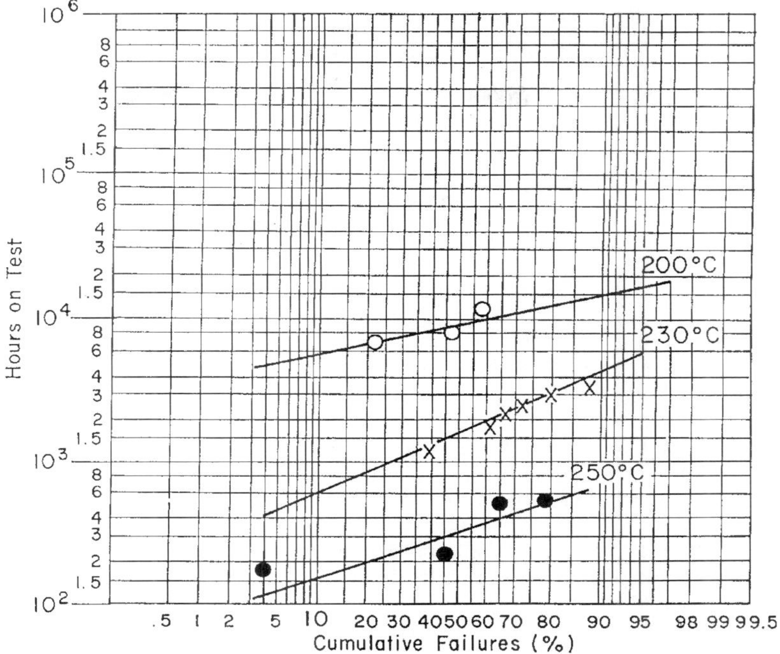

FIGURE 10 Time versus cumulative percent failure of InGaAs diodes at 200, 230, and 250°C lifetest.

introduce crystal defects such as vacancies and dislocations which can increase the dark current. Higher dark current has been directly linked to lower MTTF.[26] Device processing is probably the most crucial item in photodetector reliability. The dielectric (typically silicon nitride) used in planar detector processing serves as a diffusion mask in *p*-*n*-junction formation and a passivant (termination) for the junction so produced. Any surface states or impurities introduced here can directly increase leakage current and degrade reliability.

An important milestone in detector reliability was the changeover mesa to planar structures.[2,28] Just as the transistors in the 1950s were first made in mesa form, so were the optical photodetectors of the 1980s, due to their simplicity and ease of fabrication. However, in both cases, reliability issues forced the introduction of the more complex planar structure. A sketch of a mesa and planar photodiode is illustrated in Fig. 11. A mesa photodiode typically is formed by wet chemical etching of an epitaxially grown *p*-*n*

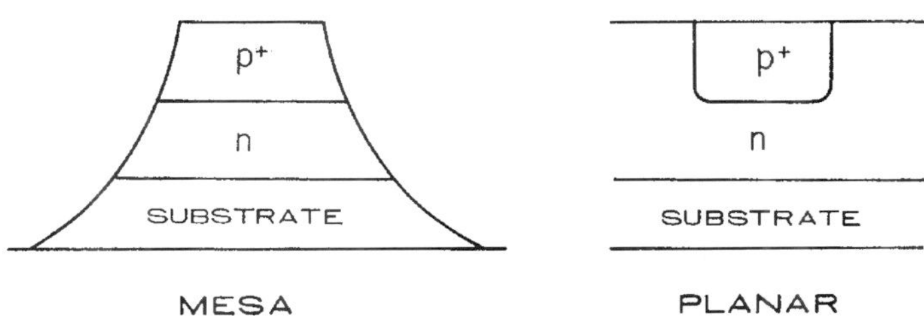

FIGURE 11 Sketch of a mesa and planar photodetector.

crystal structure, while in a planar process, a *p-n* junction is formed by diffusing a suitable *p* or *n* dopant in an *n*- or *p*-type crystal. Research shows a planar structure to be more reliable than a mesa one[28] because a *p-n* junction is never exposed to ambient conditions in a well-designed planar process. Exposure of the *p-n* junction can cause surface corrosion leading to increased leakage current and, in effect, poorer reliability.[29]

16.6 FUTURE PHOTODETECTORS

In the near future, a lateral *p-i-n* photodiode and a quantum well infrared photodetector (QWIP) may find increased commercial applications. Both these structures have already been developed in the research laboratories and show promising characteristics compared to the current photodetector structures that they may well replace. A long wavelength QWIP in the 8000–12000-nm band[30–33] has posed a severe challenge to the present favorite mercury cadmium telluride photodetectors, while a medium wavelength QWIP in the 3000–5000-nm band[34] may compete with indium antimonide and platinum silicide photodetectors. A QWIP made from GaAs/AlGaAs heterosystems promises to have higher detectivity (D^*), higher yield due to well established 3-in wafer GaAs technology, and easier monolithic integration with circuit electronics. A lateral *p-i-n* diode, as the name implies, has charge carrier flow in a lateral direction compared to the vertical direction in a conventional (vertical) photodiode structure. Because of its process compatibility and simple fabrication, a lateral *p-i-n* photodiode can be suitably integrated on an optoelectronic integrated circuit (OEIC) chip[35,36] having numerous field-effect transistors. An OEIC has a lower noise floor due to the reduced stray capacitances and inductances compared to that of hybrid detector-amplifier packages and finds applications in high-speed digital data communication.

Lateral *p-i-n* Photodetector

The vertical *p-i-n* structure in Fig. 6 has high sensitivity, low noise, low capacitance, better reliability, and an easy manufacturing process. However, such a vertical structure is nonplanar and therefore harder to integrate on an OEIC. The nonplanarity is also an issue with lasers and LEDs, and optical integration demands surface-emitting LEDs and lasers (SLEDs and SLASERs) over the conventional edge-emitting sources (ELED and ELASER). The future low-cost OEICs will probably have surface-emitting sources with lateral photodetectors in the same plane with integrating transistor amplifiers. The cross section of an AlGaAs/GaAs lateral *p-i-n* photodiode is shown in Fig. 12.

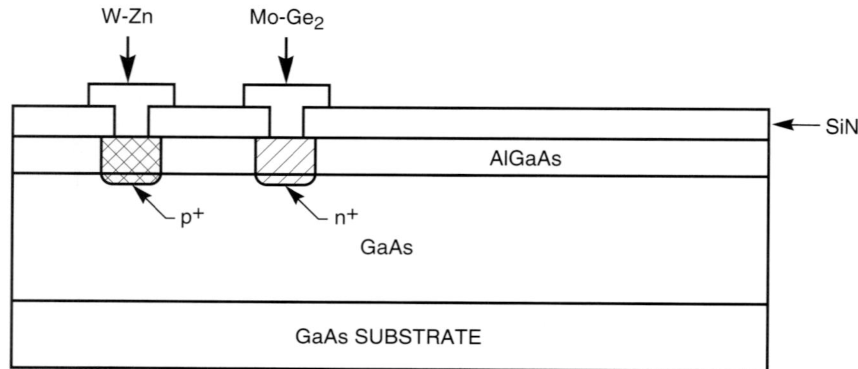

FIGURE 12 Cross section of an AlGaAs/GaAs lateral *p-i-n* photodiode.[36]

The higher-bandgap AlGaAs layer acts as a surface barrier, reducing the leakage currents. The low-bandgap GaAs layer absorbs the incoming light, and the generated carriers flow to the W-Zn and Mo-Ge_2 ohmic contacts which act as the *p* region and *n* region, respectively. The diffusion of zinc and germanium in the ohmic contacts forms a compositionally graded barrier at the AlGaAs/GaAs interface, rather than an abrupt interface. This smooth barrier helps the lateral *p-i-n* photodiode to have better speed than a lateral metal-semiconductor-metal photodetector. A comfortable spacing of 3 to 5 μm between the *p* and *n* regions gives high quantum efficiency and low capacitance, thus providing all the desirable properties of a vertical *p-i-n* structure and yet being easier to integrate.

Quantum Well Infrared Photodetector (QWIP)

QWIPs offer the possibility of achieving *long* wavelength (5000–10,000 nm) infrared detection by using materials whose bandgap normally allows them only to absorb light in the *short* wavelength (~10,000 nm) region, e.g., GaAs/AlGaAs. The use of thin (<500 Å) layers allows the absorbing wavelength to be controlled by material *geometry* rather than material *chemistry.*[37]

Before discussing the QWIPs, we take the liberty of explaining a few basic terms and concepts of quantum physics. Superlattices or quantum well structures consist of a stack of ultrathin semiconductor layers normally 50 to 500 Å in thickness. Molecular beam epitaxy (MBE) techniques are frequently employed to grow these structures because their characteristically slow growth rate of a few Angstroms/sec which helps achieve abrupt heterointerfaces. Two semiconductors of different compositions, when stacked together, form a heterointerface. Type III-V compound semiconductors such as AlGaAs/GaAs and InAlAs/InGaAs are the best candidates for growing quantum well structures, as they can be easily doped and their alloy composition readily changed to form semiconductor layers of different bandgaps. Tailoring the bandgap can alter the heterobarriers, creating exciting device results. When quantum well layers have thicknesses less than the electron mean free path (typically 50 to 100 Å), electron and holes cannot have their normal three-dimensional motion. This restricts carriers to move in two dimensions in the plane of the layer.[2,37] Because of this quantized motion, a new band of discrete energy levels is generated. Carriers no longer obey Boltzmann's statistics[1] and optical absorption becomes more complicated than the conventional band-to-band absorption given by Eq. (1). The absorption of light energy by a quantum well structure can cause an electron to jump from "multiple valence subbands" to "multiple conduction subbands," thereby enabling it to absorb light wavelengths not decided by the *material* properties (bandgap) of the semiconductor layers alone, but by its *geometrical* properties as well.

In QWIPs, the light energy transfers an electron in a bound state to an excited state in the continuum.[38] Figure 13 shows an AlGaAs/GaAs quantum well structure with L being the width of the well and V_1, V_2 being the barrier heights. The electron excited by the IR radiation is swept out of the doped GaAs well by applying an external electrical field. By controlling the barrier heights V_1, V_2, and quantum well width L, the spectral response of a QWIP can be changed for the desired IR window of 3000–5000 or 8000–12,000 nm.[38] A multiple period quantum well infrared photodetector is illustrated in Fig. 14.[32] The n^+-doped (2×10^{18} cm^{-3}) GaAs quantum wells are 40–100 Å and the undoped AlGaAs barriers are of 500-Å thickness. The multiple period stack is sandwiched between two n^+ GaAs-doped contacts. This photodetector has exhibited a blackbody D^* of 1×10^{10} cm/(Hz/W)$^{1/2}$ at 68 K for a cutoff wavelength of 10,700 nm. InGaAs/AlInAs superlattices have exhibited blackbody D^* of 2×10^{10} cm/(Hz/W)$^{1/2}$ at 120 K with peak responsivity at 4000 nm.[34]

In summary, QWIPs promise higher detectivity, good uniformity, high yield, multiple

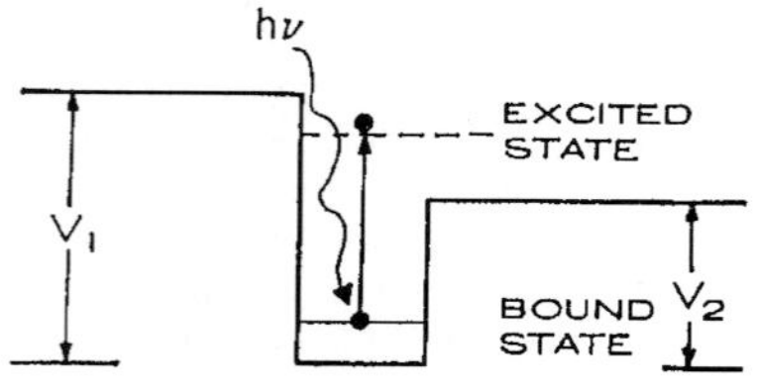

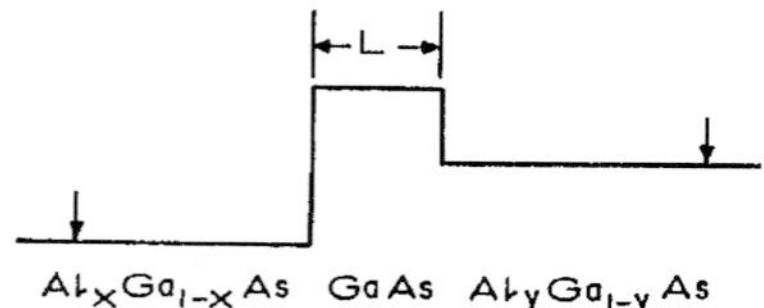

FIGURE 13 Infrared detection with an AlGaAs/GaAs quantum well.[38]

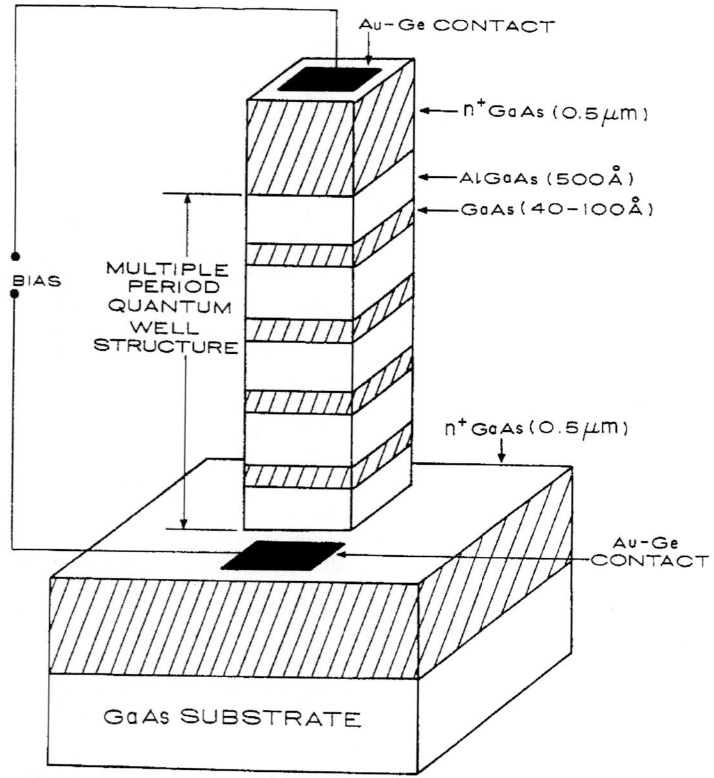

FIGURE 14 Multiple period AlGaAs/GaAs quantum well infrared photodetectors.

spectral windows, and intrinsic radiation hardness, thereby making them attractive for numerous imaging and spectroscopy applications.[39]

16.7 ACKNOWLEDGMENT

We sincerely acknowledge the support of EPITAXX, Inc., Amy Vasger, and Jennifer Romano (Sensors Unlimited) for preparing the manuscript, and Jim Rue for his technical discussions.

16.8 REFERENCES

1. S. M. Sze, *Physics of Semiconductor Devices,* Wiley, 1981.
2. S. R. Forrest, "Optical Detectors for Lightwave Communication," *Optical Fiber Telecommunications II,* 1988, pp. 569–599.
3. Z. S. Huang and T. Ando, "A Novel Amplified Image Sensor with a Si:H Photoconductor and MOS Transistors," *IEEE Trans. on Elect. Dev.,* vol. 37, no. 6, 1990, pp. 1432–1438.
4. J. C. Gammel, G. M. Metze, and J. M. Ballantyne, "A Photoconductor Detector for High Speed Fiber Communication," *IEEE Trans. on Elect. Dev.,* vol. ED-28, no. 7, 1981, pp. 841–849.
5. M. V. Rao, P. K. Bhattacharya, and C. Y. Chen, "Low Noise $In_{0.53}Ga_{0.47}As$: Fe Photoconductive Detectors for Optical Communications," *IEEE Trans. on Elect. Dev.,* vol. Ed-33, no. 1, 1986, pp. 67–71.
6. J. C. Gammel, H. Ohno, and J. M. Ballantyne, "High Speed Photoconductive Detectors Using GaInAs," *IEEE J. of Quantum Elect.,* vol. QE-17, no. 2, 1981, pp. 269–272.
7. C. Y. Chen, Y. M. Pang, K. Alavi, A. Y. Cho, and P. A. Garbinski, "Interdigitated $Al_{0.48}In_{0.52}AsGa_{0.47}In_{0.53}As$ Photoconductive Detectors," *App. Phys. Lett.* 44, 1983, pp. 99–101.
8. M. V. Rao, G. K. Chang, and W. P. Hong, "High Sensitivity, High Speed InGaAs Photoconductive Detector," *Elect. Lett.,* vol. 26, no. 11, 1990, pp. 756–757.
9. J. Degani, R. F. Leheny, R. E. Nahory, M. A. Pollack, J. P. Heritage, and J. C. DeWinter, "Fast Photoconductive Detector Using p-$In_{0.53}Ga_{0.47}As$ with Response to 1.7 μm," *Appl. Phys. Lett.* 38, 1981, pp. 27–29.
10. S. M. Sze, *Semiconductor Devices—Physics and Technology,* Wiley, 1985.
11. G. H. Olsen, "Low Leakage, High Efficiency, Reliable VPE InGaAs 1.0-1.7 μm Photodiodes," *IEEE Elect. Dev. Lett.,* vol. EDL-2, no. 9, 1981, pp. 217–219.
12. S. R. Forrest, "Performance of $In_xGa_{1-x}As_yP_{1-y}$ Photodiodes with Dark Current Limited by Diffusion, Generation Recombination and Tunneling," *IEEE J. of Quantum Elect.,* vol. QE-17, no. 2, 1981, pp. 217–226.
13. O. K. Kim, B. V. Dutt, R. J. McCoy, and J. R. Zuber, "A Low Dark-Current, Planar InGaAs *p-i-n* Photodiode with a Quaternary InGaAsP Cap Layer," *IEEE J. of Quantum Elect.,* vol. QE-21, no. 2, 1985, pp. 138–143.
14. Y. Yoshida, Y. Hisa, T. Takiguchi, and Y. Komine, "Reduction of Surface Leakage Current in $Cd_{0.2}Hg_{0.8}Te$ Photodiode," *Proc. SPIE,* vol. 972, 1988, pp. 39–43.
15. J. C. Flachet, M. Royer, Y. Carpentier, and G. Pichard, "Emission and Detection in the 1 to 3 μm Spectral Range with $Hg_{1-x}Cd_xTe$ Diodes," *Proc. SPIE,* vol. 587, 1985, pp. 149–155.
16. R. E. Enstrom, P. J. Zanucchi, and J. R. Appert, "Optical Properties of Vapor-Grown $In_xGa_{1-x}As$ Epitaxial Films on GaAs and $In_xGa_{1-x}P$ Substrates," *J. of Appl. Phys.,* vol. 45, no. 1, 1974, pp. 300–306.
17. H. W. Ruegg, "An Optimized Avalanche Photodiode," *IEEE Trans. on Elect. Dev.,* vol. Ed-14, no. 5, 1967, pp. 239–251.

18. K. Taguchi, T. Torikai, Y. Sugimoto, K. Makita, and H. Ishihara, "Planar Structure InP/InGaAsP/InGaAs Avalanche Photodiodes with Preferential Lateral Extended Guard Ring for 1.0–1.6 μm Wavelength Optical Communication Use," *J. of Lightwave Tech.,* vol. 6, no. 11, pp. 1643–1655.

19. S. R. Forrest, R. G. Smith, and O. K. Kim, "Performance of $In_{0.53}Ga_{0.47}As$/InP Avalanche Photodiodes," *IEEE J. of Quantum Elect.,* vol. QE-18, no. 12, 1982, pp. 2040–2047.

20. R. J. McIntyre, "The Distribution of Gains in Uniformly Multiplying Avalanche Photodiodes: Theory," *IEEE Trans. on Elect. Dev.,* vol. ED-19, no. 6, 1972, pp. 703–713.

21. A. S. Grove, *Physics and Technology of Semiconductor Devices,* Wiley, 1967.

22. S. L. Miller, "Avalanche Breakdown in Germanium," *Phys. Rev.,* vol. 99, 1955, p. 1234.

23. R. U. Martinelli and R. E. Enstrom, "Reliability of Planar InGaAs/InP Photodiodes Passivated with BoroPhosho-Silicate Glass," *J. of Appl. Phys.,* vol. 63, no. 1, 1988, pp. 250–252.

24. A. K. Chin, F. S. Chen, and F. Ermanis, "Failure Mode Analysis of Planar Zinc-Diffused $In_{0.53}Ga_{0.47}As$ *p-i-n* Photodiodes," *J. of Appl. Phys.,* vol. 55, no. 6, 1984, pp. 1596–1606.

25. Y. Kuhara, H. Tercuchi, and H. Nishizawa, "Reliability of InGaAs/InP Long Wavelength *p-i-n* Photodiodes Passivated with Polymide Thin Film," *IEEE J. of Lightwave Tech.,* vol. LT-4, no. 7, 1986, pp. 933–937.

26. A. M. Joshi, G. H. Olsen, and S. R. Patil, "Reliability of InGaAs Detectors and Arrays," *Proc. SPIE,* vol. 1580, 1991, pp. 34–40.

27. S. R. Forrest, V. S. Ban, G. Gasparian, D. Gay, and G. H. Olsen, "Reliability of Vapor Grown $In_{0.53}Ga_{0.47}As$/InP *p-i-n* Photodiodes With Very High Failure Activation Energy," *IEEE Elect. Dev. Lett.,* vol. 9, no. 5, 1988, pp. 217–219.

28. C. P. Skrimshire, J. R. Farr, D. F. Sloan, M. J. Robertson, P. A. Putland, J. C. D. Stokoe, and R. R. Sutherland, "Reliability of Mesa and Planar InGaAs *p-i-n* Photodiodes," *IEEE Proc.,* vol. 137, part J, no. 1, 1990, p. 7478.

29. R. R. Sutherland, J. C. D. Stokoe, C. P. Skrimshire, B. M. Macdonald, and D. F. Sloan, "The Reliability of Planar InGaAs/InP *p-i-n* Photodiodes with Organic Coatings for Use in Low Cost Receiver," *Proc. SPIE,* vol. 1174, 1989, pp. 226–232.

30. B. F. Levine, C. G. Bethea, G. Hasnain, V. O. Shen, E. Pelve, R. R. Abbott, and S. J. Hsieh, "High Sensitivity Low Dark Current 10 μm GaAs Quantum Well Infrared Photodetectors," *Appl. Physics Lett.,* vol. 56(9), 1990, pp. 851–853.

31. B. F. Levine, C. G. Bethea, V. O. Shen, and R. J. Malik, "Tunable Long-Wavelength Detectors Using Graded Quantum Wells Grown by Electron Beam Source Molecular Beam Epitaxy," *Appl. Phys. Lett.,* vol. 57(4), 1990, pp. 383–385.

32. G. Hasnain, B. F. Levine, S. Gunapala, and N. Chand, "Large Photoconductive Gain In Quantum Well Infrared Photodetectors," *Appl. Phys. Lett.,* vol. 57(6), 1990, pp. 608–610.

33. E. Pelve, F. Beltram, C. G. Bethea, B. F. Levine, V. O. Shen, S. J. Hsieh, and R. R. Abbott, "Analysis of the Dark Current in Doped Well Multiple Quantum Well AlGaAs Infrared Photodetectors," *J. of Appl. Phys.,* vol. 66(11), 1989, pp. 5656–5658.

34. G. Hasnain, B. F. Levine, D. L. Sivco, and A. Y. Cho, "Mid-Infrared Detectors in the 3–5 μm Band Using Bound to Continuum State Absorption in InGaAs/InAlAs Multiquantum Well Structures," *Appl. Phys. Lett.,* vol. 56(8), 1990, pp. 770–772.

35. S. N. Subbarao, D. W. Bechtle, R. J. Menna, J. C. Connolly, R. L. Camisa, and S. Y. Narayan, "2–4 GHz Monolithic Lateral *p-i-n* Photodetector and MESFET Amplifier on GaAs-onSi," *IEEE Trans. on Microwave Theory and Tech.,* vol. 38, no. 9, 1990, pp. 1199–1202.

36. S. Tiwari, J. Burroughs, M. S. Milshtein, M. A. Tischler, and S. L. Wright, "Lateral *p-i-n* Photodetectors with 18 GHz Bandwidth at 1.3 μm Wavelength and Small Bias Voltages," *Tech. Dig. of IEEE Int. Elect. Dev. Mtg.,* 1991, pp. 421–425.

37. D. S. Chemla, "Quantum Wells for Photonics," *Physics Today,* May 1995, pp. 56–64.

38. D. D. Coon and R. P. G. Karunasiri, "New Mode of IR Detection Using Quantum Wells," *App. Phys. Lett.,* vol. 45(6), 1984, pp. 649–651.

39. B. F. Levine, C. G. Bethea, J. W. Stayt, K. G. Glogovski, R. E. Leibenguth, S. D. Gunapala,

S. S. Pei, and J. M. Kuo, "Long Wavelength GaAs/$Al_xGa_{1-x}As$ Quantum Well Infrared Photodetectors (QWIPs)," *Proc. SPIE,* vol. 1540, 1991, pp. 232–238.

40. G. H. Olsen, "Reliable Operation of Lattice Mismatched InGaAs Detectors on Silicon," *Tech. Dig. of IEEE Int. Elect. Dev. Mtg.,* 1990, pp. 145–147.
41. E. L. Dereniak and D. G. Crowe, *Optical Radiation Detectors,* Wiley, 1984.
42. M. E. Storm, "Coherent 2 μm Sources Burst into Windshear Detection," *Laser Focus,* vol. 21, 1991, pp. 117–122.
43. G. H. Olsen, "InGaAs Fills the Near-IR Detector Array Vacuum," *Laser Focus,* vol. 21, 1991, pp. 47–50.
44. A. M. Joshi, G. H. Olsen, V. S. Ban, E. Mykietyn, M. J. Lange, and D. R. Mohr, "Reduction of $1/f$ Noise in Multiplexed Linear $In_{0.53}Ga_{0.47}As$ Detector Arrays via Epitaxial Doping," *IEEE Trans. on Elect. Dev.,* vol. 40, no. 2, 1993, pp. 303–308.

CHAPTER 17
HIGH-SPEED PHOTODETECTORS

J. E. Bowers
Y. G. Wey
Department of Electrical and Computer Engineering
University of California
Santa Barbara, California

17.1 GLOSSARY

A	area
A^{**}	modified effective Richardson constant
A_e, A_h	electron (hole) ionization parameters
B	bit rate
C_J	junction capacitance
C_P	pad capacitance
$D_e(D_h)$	diffusion coefficient for electrons (holes)
$\mathbf{E}$	electric field
$e_n(e_p)$	emission functions for electrons (holes)
F	frequency response
f	frequency
$f_{3\mathrm{dB}}$	3-dB bandwidth
G	photoconductor gain
H	transfer function
h	Plank's constant
I_d	dark current
I_{dm}	multiplied dark current
I_{du}	unmultiplied dark current
I_{ph}	photocurrent
i	current
$\langle i_{na}^2 \rangle$	amplifier noise power
J	current density

J_{DIFF}	diffusion component of current density
J_{DRIFT}	drift component of current density
$J_e(J_h)$	electron (hole) component of current density
k	ratio of electron to hole ionization coefficient
k_b	Boltzmann constant
L	absorption layer thickness
$L_e(L_h)$	diffusion length for electrons (holes)
L_s	series inductance
M	multiplication factor
$M_n(M_p)$	electron (hole) initiated multiplication factor
m	electron mass
$n(p)$	electron (hole) density
P	input optical flux
q	electron charge
R	reflectivity
R_L	load resistance
R_S	series resistance
$R_1(R_2)$	reflectivity of the surface (substrate) mirror in a resonant detector
T	temperature
t	time
t_e, t_h	transit time for electrons and holes
V_B	breakdown voltage
V_j	junction voltage
v_e, v_h	electron and hole velocities
W	thickness of the depleted region
x	position
α	absorption coefficient
α_{FC}	free carrier absorption inside the absorption layer
$\alpha_{\mathrm{FC}x}$	free carrier absorption outside the absorption layer
α_{IB}	interband absorption
α_i	electron ionization rate
α_s	scattering loss
β	propagation constant in a waveguide photodetector
β_i	hole ionization rate
Γ	confinement factor

ε	permittivity
η	quantum efficiency
κ	coupling coefficient to the waveguide of a waveguide detector
λ	wavelength
$\mu_e(\mu_h)$	mobility for electrons (holes)
ν	optical frequency
σ	charge density
σ_d	noise current spectral density
τ_e, τ_h	trapping time at a heterojunction for electrons (holes)
τ_{tr}	transit time
$\phi_{bc}(\phi_{bv})$	barrier height for the conduction band (valence band)
ω	angular frequency

17.2 INTRODUCTION

High-speed photodetectors are required for telecommunications systems, for high-capacity local area networks, and for instrumentation. Many different detector structures and materials are required to cover this range of applications. Silicon is one of the most commonly used detector materials for wavelengths from 0.4 to 1.0 μm, while Ge photodetectors are used at longer wavelengths up to 1.8 μm. Silicon and germanium have indirect bandgaps at these wavelengths, which result in relatively small bandwidth-efficiency products. Consequently, for high-speed applications, direct bandgap semiconductors such as III-V materials are more important and are the focus of this chapter. $Ga_{0.47}In_{0.53}As$ with a cutoff wavelength of 1.65 μm is especially useful for telecommunication photodetectors at 1.3 and 1.55 μm. GaAs has a cutoff wavelength around 0.9 μm and is ideal for visible and near infrared applications.

This chapter will focus on the physics and technology of high-speed photodetectors. The next section discusses the different structures that are possible. Later sections discuss some specific results and motivations for particular structures. The primary limitations to detector speed are discussed, followed by a description of specific photodetector systems. To supplement this chapter, the reader should refer to excellent chapters and articles written specifically about photodetectors,[1] photoconductors,[2] pin detectors,[3] avalanche photodetectors,[4] phototransistors,[5] and receivers.[6,7]

17.3 PHOTODETECTOR STRUCTURES

Many photodetector structures have been demonstrated and many more structures are possible. In this section, we classify the different possible structures and identify a few of the trade-offs. The optimum structure for a given application depends on the required bandwidth, efficiency, saturation power, linearity, ease of integration, and leakage current.

There are four common types of photodetectors: (1) photovoltaic detectors, (2) photoconductive detectors, (3) avalanche photodetectors (APD), and (4) phototransistors. Photovoltaic detectors have blocking contacts and operate under reverse bias. The blocking contact can be a reverse-biased *p-n* junction or a Schottky contact. The photoconductive detector has identical, nonblocking contacts such as two $n+$ regions in an undoped sample. The avalanche photodetector has a similar configuration to a photovoltaic detector except that it has a high field region that causes avalanching and results in gain in the detector. Improvements to the basic APD design include separate avalanche and gain regions (SAM APDs), and staircase APDs to increase the ratio of electron to hole (or hole to electron) multiplication rate. Phototransistors are three-terminal devices which have an integrated electronic gain region.

The second criterion is the contact type and configuration. The photogenerated carriers may be collected by means of (1) a vertical current collector, often a *p-n* or Schottky junction, (2) an interdigitated metal-semiconductor-metal (MSM) structure, or (3) a laterally grown or etched structure. These options are illustrated in Fig. 1*a*. PIN junctions are usually formed during the growth steps and tend to have low leakage current and high reliability.[8,9] Schottky junctions are simple to fabricate, but tend to have a large leakage current on narrow-gap semiconductors, such as InGaAs. MSM structures have the advantage of lower capacitance for a given cross-sectional area, but often have longer transit times, limited by the lithography capabilities possible in production. Experimental demonstrations with very fine lines (50 nm) have yielded high-speed devices with good quantum efficiencies. MSM detectors tend to be photoconductive detectors, but one could lower the capacitance for a given area of a PIN detector by using an interdigitated MSM structure with *p* and *n* regions under alternating metal fingers.

The third important aspect of photodetector design is the orientation of the light with respect to the wafer and the current collection region (Fig. 1*b*). Most commercial photodetectors are vertically illuminated and the device area is 10 μm in diameter or larger, which allows simple, high-yield packaging with single-mode optical fibers, or easy alignment to external bulk optics. The problem with this configuration is that the absorbing layer must be thin for a high-speed detector to keep the transit time of photogenerated carriers short. Consequently, the quantum efficiency is low, and single-pass vertically illuminated photodetectors tend to have bandwidth efficiency products around 30 GHz.[3] Bandwidth efficiency products are discussed in greater detail in Sec. 17.5. The bandwidth efficiency product can be increased by allowing two passes by reflecting the light off a metal layer or dielectric mirror.[10] Making a resonant cavity with multiple reflections at particular wavelengths should allow bandwidth efficiency products in excess of 100 GHz.[11–13] As will be seen below, essentially 100 percent quantum efficiency is possible with bandwidths up to 20 GHz, so there is no need for resonant detectors unless the required bandwidth is above 20 GHz or unless wavelength selectivity is needed as in a wavelength division multiplexed (WDM) system.

The other class of optical inputs are horizontally illuminated photodetectors (Fig. 1*c*). The simplest configuration is an edge-illuminated detector. The primary problem with an edge-illuminated detector is that the light is not guided. Diffraction of the incident light causes absorption to occur outside of the high-field region, and slow diffusion tails in the impulse response occur. A solution to this problem is the waveguide detector, where an optical waveguide confines the light to the high-field absorption region.[14–16] The waveguide efficiency product of this structure can be 100 to 200 GHz. However, it is limited by the capacitance of the structure, particularly if thin intrinsic layers are used for ultrahigh-speed devices. A solution to the capacitance limitation is a traveling wave photodetector where the incoming optical beam is velocity matched with the generated microwave signal.[17,18] The bandwidth efficiency product is then limited only by loss on the electrical transmission lines, and bandwidth efficiency products of hundreds of GHz are possible. Traveling wave detectors and, to a lesser extent, waveguide detectors have the important advantage that the volume of the light absorption can be quite large and, consequently, these detectors have much higher saturation powers.[19] Velocity matching in these structures requires quite narrow waveguides. Wu and Itoh[20] have suggested separating the parts of the optical waveguide with microwave delay lines to achieve velocity matching.

The fourth issue is the type of absorbing material (Fig. 1*d*) (1) bulk, (2) quantum well,

1a. Electrical Configuration

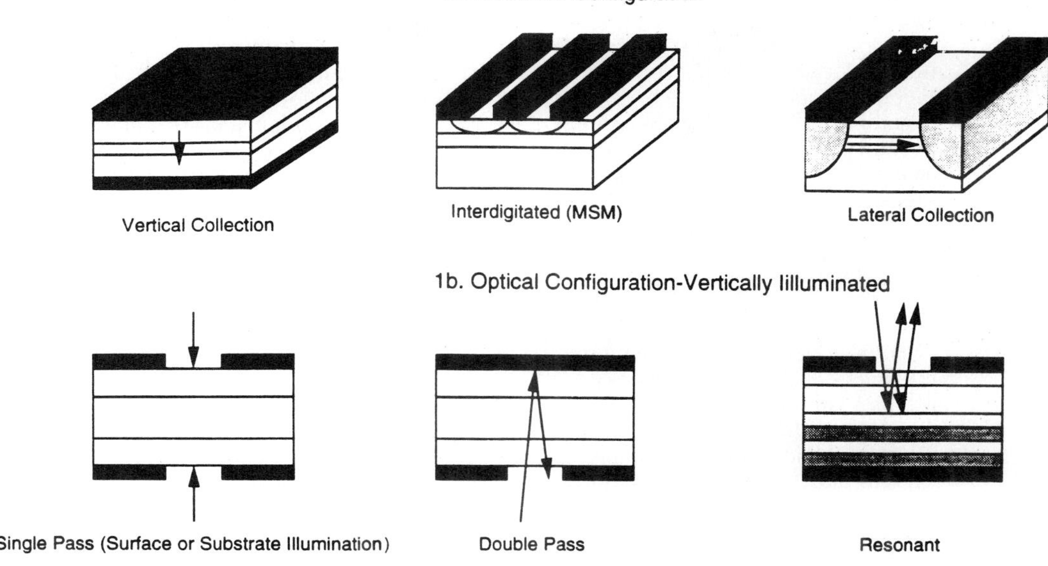

1c. Optical Configuration-Horizontally Illuminated

Edge Absorbing

Waveguide

Traveling Wave

1d. Absorbing Material

Bulk

Quantum Well

Quantum Wire

Quantum Dot

FIGURE 1 Schematic drawings of different types of photodetectors: (*a*) electrical configuration, (*b*) optical configuration-vertical illumination, (*c*) optical configuration-horizontal illumination, (*d*) absorbing material.

(3) quantum wire, (4) quantum dot, (5–7) strained quantum well, wire or dot, (8) *n-p-i-i* structure. The vast majority of commercial and experimental detectors use bulk material. However, quantum well detectors[21] are becoming increasingly important in photonic integrated circuits (PICs) because the absorbing quantum well material is also used in other parts of the PIC such as the laser. Quantum wire photodetectors[22] have potential advantages in terms of higher-bandwidth-efficiency products, but uniform quantum wires are rather difficult to fabricate. Quantum dot detectors have even higher peak absorption coefficients and more wavelength selectivity, but will probably have problems with slow impulse responses due to trapping of the carriers by the heterojunction. In other quantum-confined detectors, the carriers can be extracted along the quantum wire or well, and this problem can be avoided.[22]

The final classification is by means of the lifetime of the material. Conventional detectors have material lifetimes of typically 1 ns and achieve speed by using high field for rapid carrier collection. A second approach is to use low temperature (LT) grown material which has a very short lifetime, perhaps as low as 1 ps. A third approach is to damage the material by means of ion implantation. The final approach is to grow or diffuse in traps into the material such as iron[23] or gold.

If we combine these classifications, we find that 2600 types of photodetectors are possible, and additional subgroups such as superlattice APDs or SAGM APDs increase the total even further. In reality, about 100 types of detectors have been demonstrated. One of the points of this section is that improvements in one type of detector, such as adding a resonant cavity to a PIN detector, can be applied to other types of detectors, such as adding a resonant cavity to an APD. In the following section, we discuss in more detail some of the real limitations to the speed of a detector, and then apply this knowledge to a few important types of detectors.

17.4 SPEED LIMITATIONS

Generally speaking, the bandwidths of most photodetectors are limited by the following factors: (1) carrier transit time, (2) RC time constant, (3) diffusion current, (4) carrier trapping at heterojunctions, and (5) packaging. These limiting factors will be discussed in turn with specific application to *p-i-n* photodiodes.

Carrier Transit Time

In response to light absorbed in a material, the photogenerated carriers in the active region will travel across the high-field region and then be collected by the electrodes. As an example, Fig. 2*a* shows the *p-i-n* structure with a photogenerated electron-hole charge sheet of density σ. In response to the electric field, the electron will travel to the right and the hole to the left. This induces a displacement current and reduces the internal electric field (Fig. 2*b, c*), which is the cause of saturation in photodetectors. From Gauss's law the difference in the electric field at the position of electron or hole is

$$\Delta E = \frac{q\sigma}{\varepsilon} \tag{1}$$

where q is the electron charge, and ε is the permittivity. Due to the constant total voltage across the depletion region, the reduced electric field between the electron and hole will

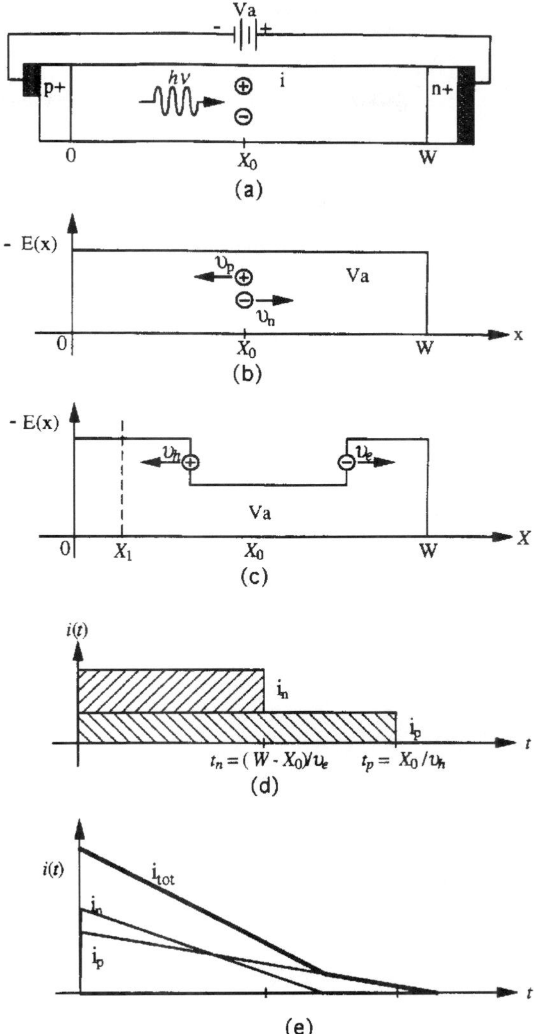

FIGURE 2 (*a*) Biased *p-i-n* structure. (*b*) Electrical field at the time when the electron-hole pairs are generated. (*c*) The perturbed electrical field due to the separated electron-hole pair. (*d*) The photocurrent due to single electron-hole pair. (*e*) The total photocurrent due to uniform illumination across the photodiode.

be compensated by the increased electric field outside. The rate of change of the electric field at the position $X = X_1$ is

$$\frac{\partial E}{\partial t} = -\frac{(v_e + v_h)\Delta E}{W} \tag{2}$$

where v_e and v_h are the saturation velocities for electrons and holes, respectively. The assumption of saturation velocities is valid at high fields. The displacement current is hence given by

$$i(t) = -\varepsilon A \frac{\partial E}{\partial t} = \frac{qAv_e\sigma}{W} + \frac{qv_h\sigma A}{W} \tag{3}$$

The current consists of two components due to the electron and hole currents. The electron current lasts for a time duration of $(W - X_0)/v_e$ and hole current of X_0/v_h. This is shown in Fig. 2*d*. Here, we note that if the fast carrier (i.e., electron) travels a longer distance, then we have a shorter pulse. The total electron and hole currents are given by

$$i_e(t) = \frac{qv_eA}{W} \int_0^W n(x, t)\, dx \tag{4}$$

$$i_h(t) = \frac{qv_hA}{W} \int_0^W p(x, t)\, dx \tag{5}$$

where $n(x, t)$ and $p(x, t)$ represent the electron and hole densities in the depletion region. The total current is the sum of Eqs. (4) and (5).

RC Time Constant

The RC time constant is determined by the equivalent circuit parameters of photodiode. For example, the intrinsic response of the *p-i-n* diode can be modeled as a current source in parallel with a junction capacitor. The diode series resistance, parasitic capacitance, and load impedance form the external circuit. Figure 3 shows the equivalent circuit of the *p-i-n* photodiode. The junction capacitance is defined by the edge of the depletion region (or space charge region). The series resistance is due to the ohmic contacts and bulk resistances. In addition, the parasitic capacitance depends on the metallization geometry. If the diode series resistance is R_S and a load resistance R_L is used to terminate the device, then the electrical 3-dB bandwidth can be approximated as

$$f_{3\text{dB}} = \frac{1}{2\pi(C_J + C_P)(R_L + R_S)} \tag{6}$$

If the photodiode is bonded by a section of gold wire, additional series inductance will be

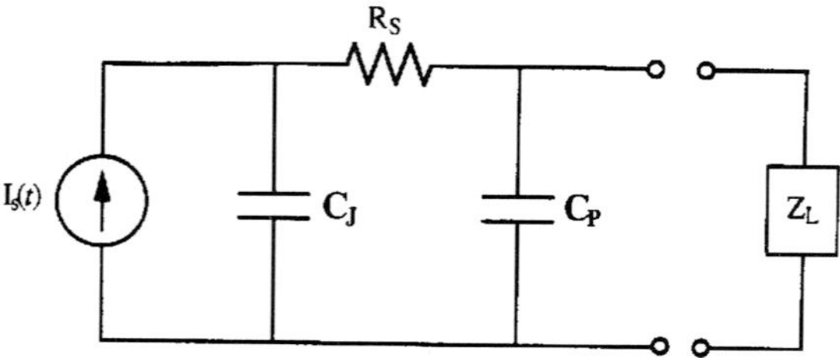

FIGURE 3 Equivalent circuit of a photodiode.

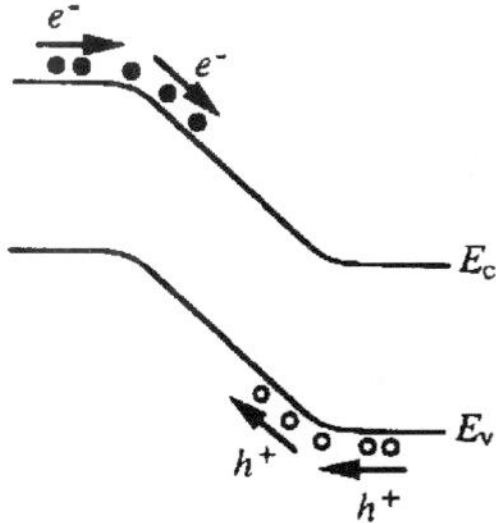

FIGURE 4 The origins of diffusion current.

included in the load impedance. The 3-dB bandwidth due to parasitics in this case is then given in Ref. 3.

Diffusion Current

Diffusion current is important in detectors in which significant absorption occurs in regions outside the high-field region. This effect is reduced to some extent by recombination in these highly doped contact layers. Those carriers within about one diffusion length of the depletion region will have a chance to diffuse into the active region. This diffusion current will contribute a slow tail to the detector impulse response (Fig. 4). The electron diffusion current at the edge of the depletion region is given by

$$J_e = qD_e \frac{\partial n}{\partial x} = qD_e \frac{\Delta n}{L_e} \tag{7a}$$

and

$$J_h = -qD_h \frac{\partial p}{\partial x} = qD_h \frac{\Delta p}{L_h} \tag{7b}$$

where $D_e(D_h)$ and $L_e(L_h)$ are the diffusion coefficient and diffusion length, respectively, for electrons (holes). The diffusion process is a relatively slow process compared with the drift process. Assuming the photocarrier density is n, with the Einstein relation and Eq. (7*a*), the electron diffusion current can be written as

$$J_{\text{DIFF}} = qn\mu_e\left(\frac{kT/q}{L_e}\right) \tag{8}$$

and the drift current term for electron can be written as

$$J_{\text{DRIF}} = qn\mu_e\mathbf{E} \tag{9}$$

where μ_e is the electron mobility and **E** is the electric field. For most devices the electric field inside the depletion region is an order of magnitude larger than $(kT/q)L_e$. For example, the hole diffusion length for GaAs is typically around 10 μm and electric field is very often over 10 kV/cm. However, the diffusion-current terms could last as long as the carrier lifetime and the charge content in the tail can be as large as the drift component due to the slow diffusion times. For high-speed detectors, the diffusion-current problem can be eliminated with a double-heterostructure design that limits the absorbing regions to the high-field intrinsic regions.[25]

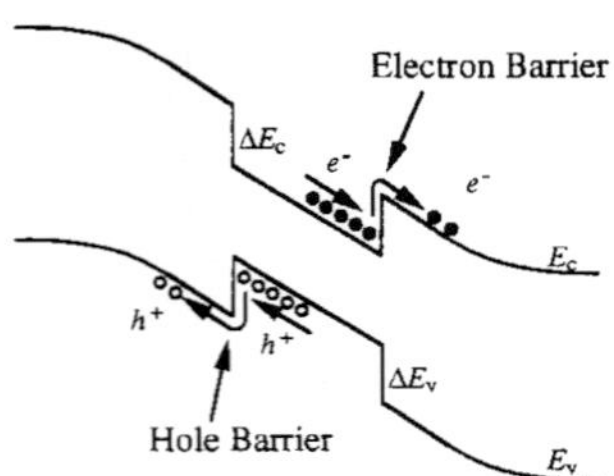

FIGURE 5 Heterostructure carrier trapping effect.

Carrier Trapping

Heterojunctions in photodetectors cause carrier trapping of electrons at conduction band discontinuities and trapping of holes at valence band discontinuities (Fig. 5). Hole trapping is a significant problem in long-wavelength photodetectors because of the large valence band discontinuity at the InP/InGaAs heterojunction. Usually, the emission rate is approximated by thermionic emission. If the interface deep-level recombination rate is significant, the total emission rate will be the sum of the two emission rates. The emission functions for electrons and holes are given by

$$e_n(t) = (1/\tau_e) \exp(-t/\tau_e)u(t) \tag{10a}$$

$$e_p(t) = (1/\tau_h) \exp(-t/\tau_h)u(t) \tag{10b}$$

where $\tau_e(\tau_h)$ represents the emission time constant for electron (hole) and $u(t)$ is the step function. The rates of thermionic emission of trapped carrier are related to the Schottky barrier height due to the bandgap discontinuity:

$$1/\tau_e = B \exp(-\phi_{be}/kT) \tag{11}$$

where B is a constant and ϕ_{bc} is the barrier height for the conduction band. The response of the carrier-trap current in time domain is often obtained by convolving an intrinsic current source with the emission function. Since the applied bias will reduce the barrier height, sufficient device bias therefore will increase the emission rate. In order to reduce the barrier height, superlattice or compositional grading is often added at the heterointerface.[25]

Packaging

The external connections to the photodetector often limit the detector performance. Another problem is that the photodiode is a high impedance load and the device has a reflection coefficient close to unity. One solution to this problem is to integrate a matching resistor with the device.[25] This can usually be added using the lower contact layer without adding any additional mask or process steps. Figure 6 shows a Smith chart plot of the impedance of a typical photodetector along with the impedance of a device with an integrated matching resistor. A good match up to 40 GHz is achieved. The disadvantage of a load resistor is the reduction in effective quantum efficiency by a factor of 2 since half of the photocurrent goes through the matching resistor. However, since the load resistance is now one-half, the RC time constant is also cut in half. Bandwidths in excess of 100 GHz have been achieved with quite large devices ($7 \times 7\ \mu m^2$) in this way. A second problem with very high speed devices is the difficulty in building external bias circuits without resonances in the millimeter range. The necessary bias capacitor and load resistor can be integrated with *p-i-n* photodetector without adding any additional mask or process steps by using a large-area *p-i-n* region as the capacitor and using the lower contact layer as the series resistor.[23] A photograph of the device is shown in Fig. 7 along with the device performance. In this case, bandwidths in excess of 100 GHz were achieved.

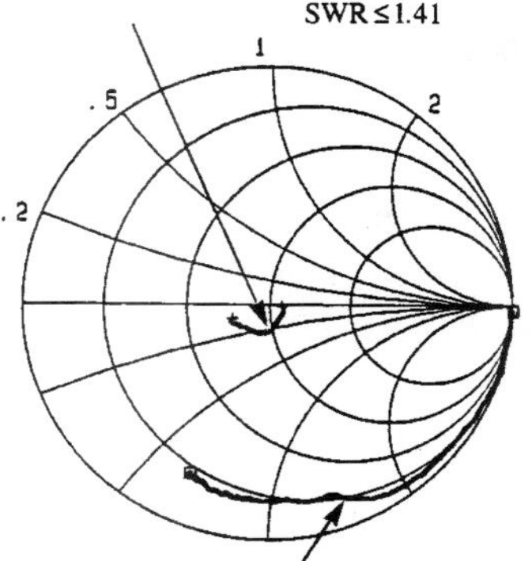

FIGURE 6 Smith chart of the impedance of typical photodiodes and photodiodes with integrated matching resistors.

Optical fiber alignment and packaging are now quite standard. Simplified alignment by means of holes etched in the substrate of back-illuminated photodiodes may allow passive alignment of optical fibers. The photodetectors must be antireflection coated to reduce the reflection to air or optical epoxy. Single dielectric layers are typically used to minimize the reflection at one wavelength. Braun et al.[26] have achieved minimum reflectivity at multiple wavelengths with one dielectric layer by using one of the semiconductor layers in a multiple antireflectivity design.

17.5 P-I-N PHOTODETECTORS

Vertically Illuminated *p-i-n* Photodiode

In order to increase the frequency response of the vertically illuminated *p-i-n* photodiode, the efficiency is always sacrificed. As the active layer thickness is reduced, the transit time decreases, and the optical absorption decreases, and there is a trade-off between the efficiency and speed. The external quantum efficiency for a surface-illuminated *p-i-n* diode is given by

$$\eta = (1 - R) \times (1 - e^{-\alpha L}) \tag{12}$$

where R is the surface reflection, α is the absorption coefficient, and L is the active layer thickness. Since the absorption coefficient is a function of wavelength $\alpha = \alpha(\lambda)$, usually α decreases as λ increases. Thus, the diode intrinsic response is wavelength-dependent. We can easily see the effect of light absorption on the transit-time-limited bandwidth by

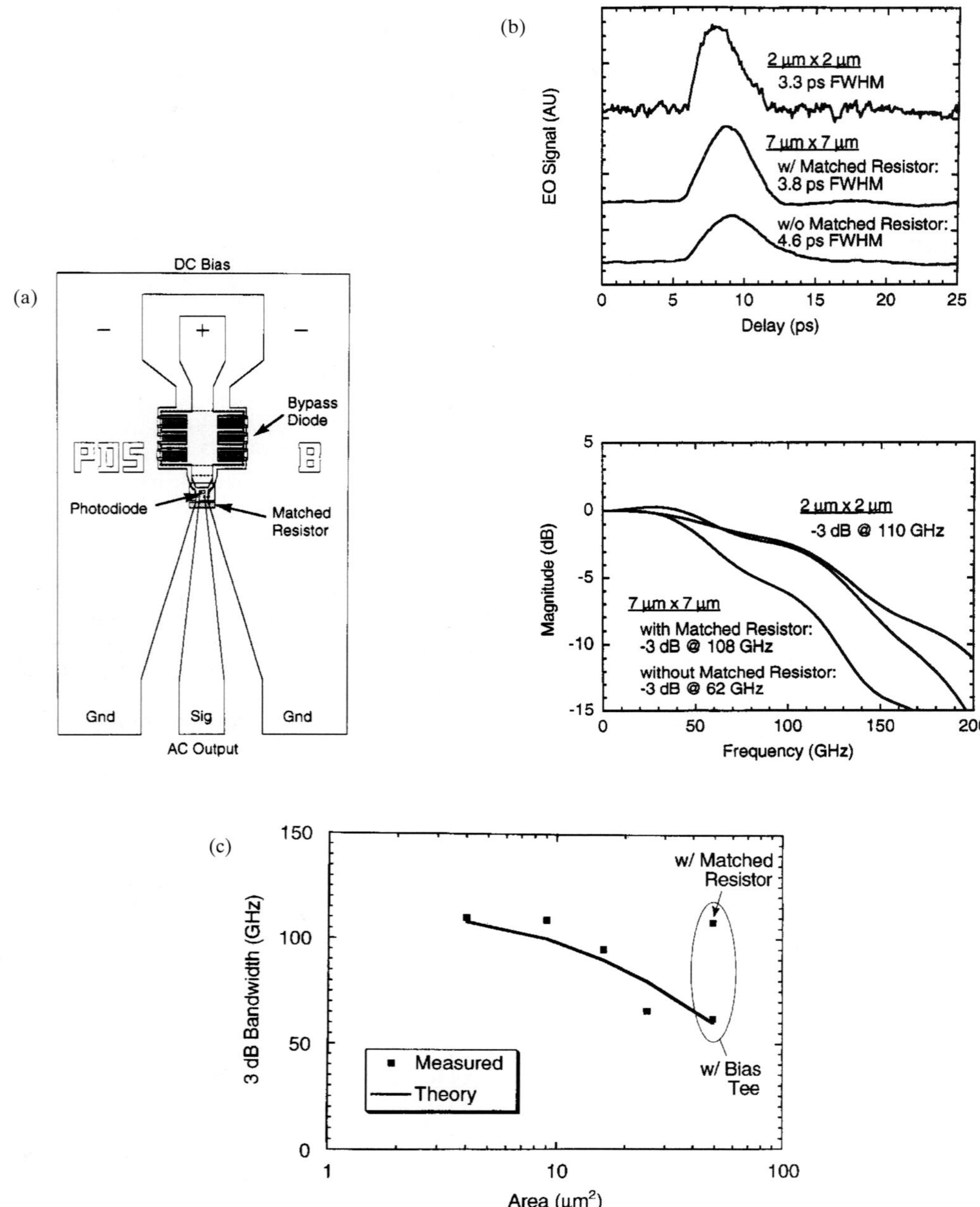

FIGURE 7 (*a*) Schematic diagram of a *p-i-n* photodiode with integrated matching resistor and bias circuit. (*b*) Impulse response of a 2-μm by 2-μm pin detector compared to 7-μm by 7-μm detectors with and without matching resistors. (*After Ref. 25.*) (*c*) Dependence of measured bandwidth on detector area and comparison to the theory presented in the text.

comparing the transit-time response[3,24] for two limiting cases for $\alpha L \rightarrow 0$ and $\alpha L \rightarrow \infty$ when $t_e = t_h = \tau_r$. The transit-time frequency response for a uniformly illuminated detector is

$$|F(\omega)_{\alpha L \rightarrow 0}| = \frac{2}{\omega \tau_{tr}} \left[1 + \frac{\sin^2 \left(\frac{\omega \tau_{tr}}{2} \right)}{\left(\frac{\omega \tau_{tr}}{2} \right)} - 2 \frac{\sin (\omega \tau_{tr})}{(\omega \tau_{tr})} \right]^{1/2} \tag{13}$$

For electron-hole pairs generated near the p side of the intrinsic region, electrons travel across the i region, and the frequency response is given by

$$|F(\omega)_{\alpha L \rightarrow \infty}| = \left| \frac{\sin \left(\frac{\omega \tau_{tr}}{2} \right)}{\left(\frac{\omega \tau_{tr}}{2} \right)} \right| \tag{14}$$

For these two limits, the transit-time-limited bandwidths are $f_{3\mathrm{dB}(\alpha L=0)} = 0.45/\tau_{tr}$ and $f_{3\mathrm{dB}(\alpha L=\infty)} = 0.55/\tau_{tr}$, respectively. For long-wavelength high-speed p-i-n diodes, the absorption layer is often very thin so that $1 - \exp(-\alpha L) \approx \alpha L$. The bandwidth efficiency product for transit-time-limited p-i-n diode is given by[3]

$$\eta \cdot f_{3\mathrm{dB}} = 0.45 \alpha v_s \tag{15}$$

Figure 8 shows the calculated 3-dB bandwidth for GaInAs/InP p-i-n diodes on the device area versus thickness plane for wavelength $\lambda = 1.3\ \mu$m. The horizontal axis is the active layer thickness (which corresponds to the quantum efficiency) and the vertical axis is the device area. As we can see, when the device active layer thickness decreases,

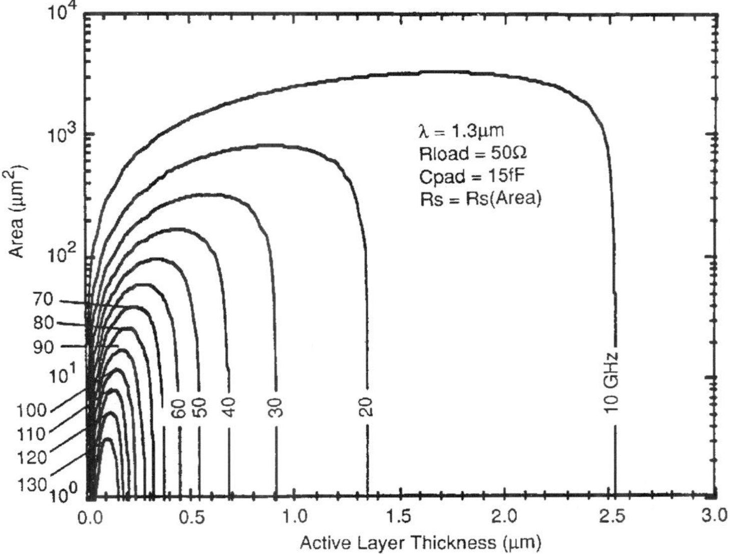

FIGURE 8 Calculated-3 dB bandwidth contours for a GaInAs pin vertically illuminated photodiode.

the quantum efficiency of *p-i-n* diode also decreases due to the insufficient light absorption in the active layer. The capacitance decreases as the device area decreases. Optimization of the device bandwidth is reached when transit-time-limited bandwidth approximately equals the RC limited bandwidth.

To minimize the bonding-pad capacitance, a semi-insulating substrate and thick polyimide layer are often used. Sometimes the series inductance of the bond wire is used to resonate the parasitic capacitance, and this results in a slightly peaked response with an increased 3-dB corner frequency. The electrical transfer function with series inductance is given by

$$H(\omega) = \frac{R_L}{[1 - \omega^2(R_S R_L C_J C_P + L_S(C_J + C_P))] - j[\omega(R_L(C_J + C_P) + R_S C_J) - \omega^3 R_S C_J C_P L_S]} \tag{16}$$

where L_S is the series inductance.

To achieve high detector bandwidth, *double heterostructure* InP/GaInAs/InP *p-i-n* photodiodes have been fabricated to reduce the diffusion-current problem. However, carrier trapping can limit the impulse response. This effect can be characterized by the emission function $e_{e,h}(t) = (1/\tau_{e,h}) \exp(-t/\tau_{e,h})$ where $\tau_{e,h}$ is emission time for electron (hole). The current-source response due to the electron and hole trapping at the heterointerfaces for *p*-side illumination is given by

$$\frac{i_s(\omega)}{i_s(0)} = \frac{1}{(1 - e^{-\alpha L})} \left\{ \left(\frac{1 - e^{-j\omega t_e}}{j\omega t_e} - e^{-\alpha L} \frac{1 - e^{-\alpha L} e^{-j\omega t_e}}{j\omega t_e - \alpha L} \right) \left(\frac{1}{1 + j\omega \tau_e} \right) + \left(\frac{1 - e^{-\alpha L} e^{-j\omega t_h}}{j\omega t_e + \alpha L} - e^{-\alpha L} \frac{1 - e^{-j\omega t_h}}{j\omega t_h} \right) \left(\frac{1}{1 + j\omega \tau_h} \right) \right\} \tag{17}$$

and for *n*-side illumination is given by

$$\frac{i_s(\omega)}{i_s(0)} = \frac{1}{(1 - e^{-\alpha L})} \left\{ \left(\frac{1 - e^{-\alpha L} e^{-j\omega t_e}}{j\omega t_e + \alpha L} - e^{-\alpha L} \frac{1 - e^{-j\omega t_e}}{j\omega t_e} \right) \left(\frac{1}{1 + j\omega \tau_e} \right) + \left(\frac{1 - e^{-j\omega t_h}}{j\omega t_h} - e^{-\alpha L} \frac{1 - e^{-\alpha L} e^{-j\omega t_h}}{j\omega t_h - \alpha L} \right) \left(\frac{1}{1 + j\omega \tau_h} \right) \right\} \tag{18}$$

where $\tau_{e,h}$ is the electron (hole) transit time. Other than the original *p-i-n* diode response, the extra terms $1/(1 + j\omega\tau_{e,h})$ are due to the trapping effect. For InGaAs/InP heterostructure *p-i-n* diodes, the valence band offset is larger than the conduction offset and the hole effective mass is much larger than the electron effective mass. Thus, hole trapping is worse than the electron trapping in an InGaAs/InP *p-i-n* diode.

Waveguide *p-i-n* Photodiode

The main advantages of waveguide detectors are the very thin depletion region resulting in a very short transit time and the long absorption region resulting in a high bandwidth photodetector with a high saturation power (Fig. 1*c*). Due to the thin intrinsic layer, it can often operate at zero bias.[27] The absorption length of a waveguide detector is usually

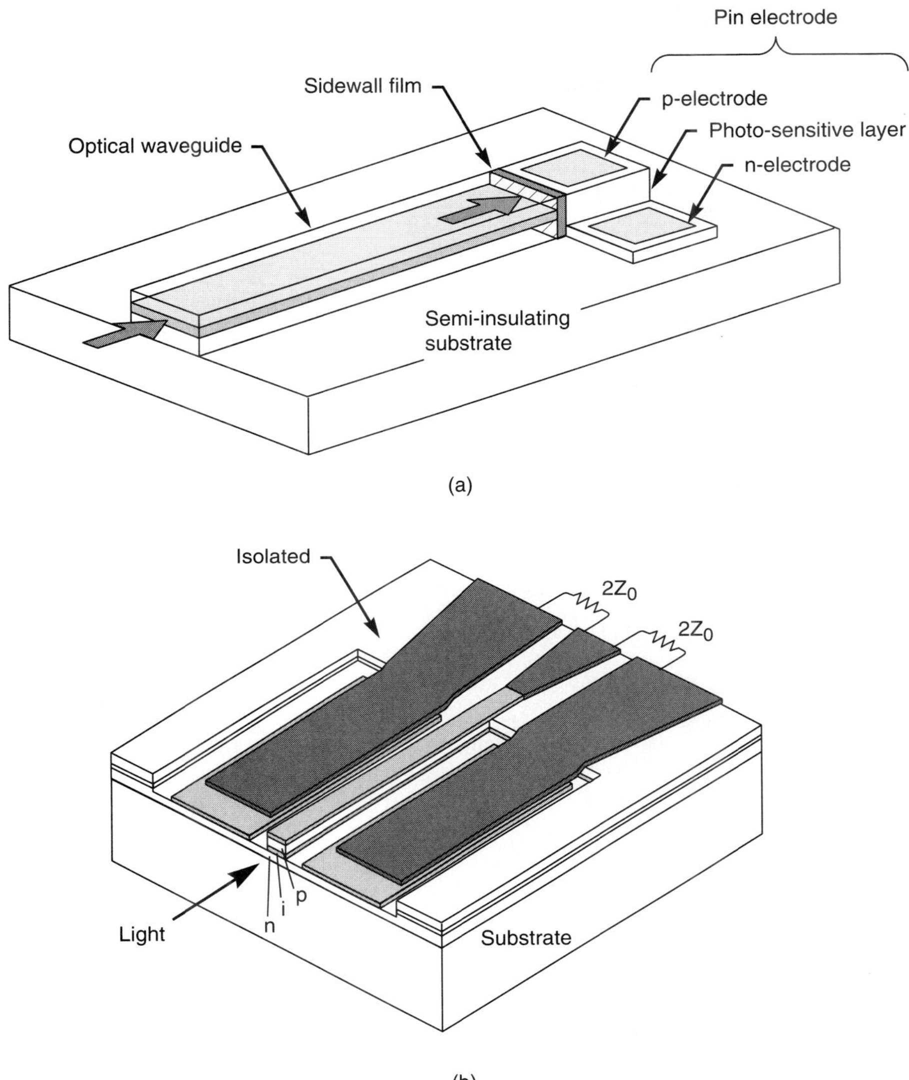

FIGURE 9 Schematic diagram of (*a*) a waveguide photodetector (*After Ref. 14*) and (*b*) a traveling wave photodetector. (*After Ref. 17.*)

designed to be long enough (>5 μm) to ensure full absorption. The waveguide structure design (Fig. 9) is often required to have low coupling loss due to modal mismatch and reasonable effective absorption coefficient. The external quantum efficiency of a waveguide *p-i-n* detector is[28,29]

$$\eta = \kappa(1 - R)\frac{\Gamma\alpha_{\mathrm{IB}}}{\alpha}(1 - e^{\alpha L}) \tag{19}$$

where κ is the coupling efficiency due to the modal mismatch, Γ is the mode confinement factor, α_{IB} is the interband absorption. The loss coefficient α is given by

$$\alpha = \Gamma\alpha_{\mathrm{IB}} + \Gamma\alpha_{\mathrm{FC}} + (1-\Gamma)\alpha_{\mathrm{FC}x} + \alpha_s \tag{20}$$

where α_{FC}, $\alpha_{\mathrm{FC}x}$ are the free carrier absorption loss inside and outside the absorption layer, α_s is the scattering loss. Kato et al.[14] reported an InGaAs waveguide *p-i-n* diode with bandwidth of 40 GHz. The detector quantum efficiency is 44 percent at 1.55-μm wavelength. The coupling loss estimated by an overlap integral was 2.1 dB. To reduce the coupling loss, it is important to have a good design of the layer structure to reduce modal mismatch and to coat the facet with an antireflecting (AR) film.

Resonant *p-i-n* Photodiode

A resonant detector utilizes the multiple passes in a Fabry-Perot resonator to achieve high quantum efficiency with thin absorbing layers (Fig. 1*b*). Since the speed of light is about three orders of magnitude faster than the carrier velocities, the quantum efficiency can be increased without significant pulse broadening due to the effective optical transit time.

The schematic diagram of a resonant cavity enhanced photodetector is shown in Fig. 10. The efficiency of the resonant detector is given by

$$\eta = \left[\frac{(1+R_2e^{-\alpha d})}{1-2\sqrt{R_1R_2}e^{-\alpha d}\cos(2\beta L+\phi_1+\phi_2)+R_1R_2e^{-2\alpha d}}\right]\times(1-R_1)\times(1-e^{-\alpha d}) \tag{21}$$

where R_1, R_2 are mirror reflectivities, φ_1, φ_2 are mirror phase shifts, $\beta = 2\pi/n\lambda$ is the propagation constant, and d is the thickness of active region. The quantum efficiency has its maximum when $2\beta L + \varphi_1 + \varphi_2 = 2m\pi$ $(m = 1, 2, 3, \ldots)$ and the quantum efficiency is then

$$\eta = \left[\frac{(1+R_2e^{-\alpha d})}{(1-\sqrt{R_1R_2}e^{-\alpha d})^2}\right]\times(1-R_1)\times(1-e^{-\alpha d}) \tag{22}$$

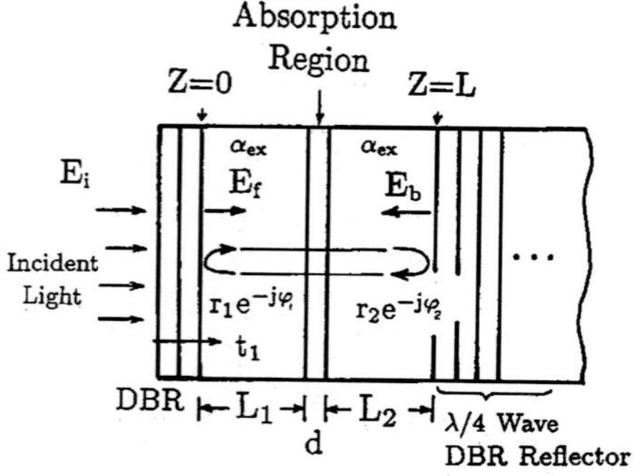

FIGURE 10 Schematic diagram of a resonant photodetector. (*After Ref. 13.*)

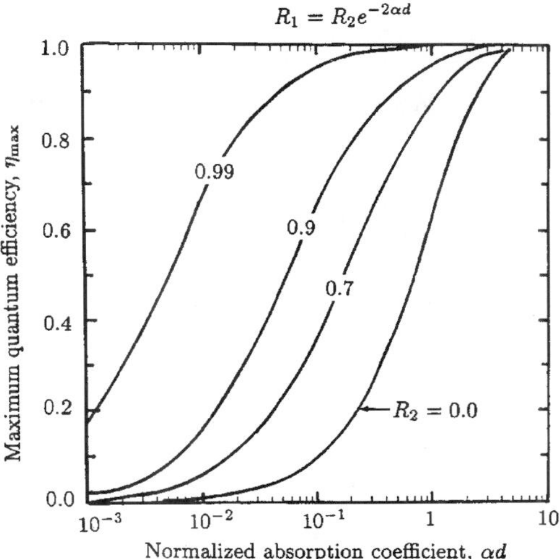

FIGURE 11 Dependence of quantum efficiency on mirror design in a resonant photodetector. (*After Ref. 13.*)

Figure 11 shows the calculated resonant quantum efficiency versus normalized absorption coefficient αd.[11] High quantum efficiency is possible even from thin absorption layers.

However, in terms of the fabrication, the material growth, and the structure design, building a high-speed resonant photodetector is not a simple task. The required low resistance and low capacitance with incorporated mirror structure is difficult to achieve due to the significant resistance of the multiple heterojunction mirror stack.

17.6 SCHOTTKY PHOTODIODE

Schottky photodiodes[30–32] are especially attractive for integration with FETs and III-V integrated circuits because of their simple material structure and easy fabrication. Figure 12 shows the Schottky barrier structure. For front-illuminated devices, the metal is very thin so that the light can penetrate the metal with very little loss. The *J-V* characteristic of a Schottky diode is given by[33]

$$J = J_0\left[\exp\left(\frac{qV}{k_B T}\right) - 1\right] \tag{23}$$

where

$$J_0 = A^{**}T^2 \exp\left(-\frac{q\Phi_b}{k_B T}\right) \tag{24}$$

and ϕ_b is the barrier height and A^{**} is the modified effective Richardson constant.[34]

The dynamics of photogenerated carriers in a Schottky diode are similar to those of a *p-i-n* diode (Fig. 2). The dynamics of both electrons and hole must be included in the analysis of a Schottky photodiode, resulting in expressions for the Schottky diode

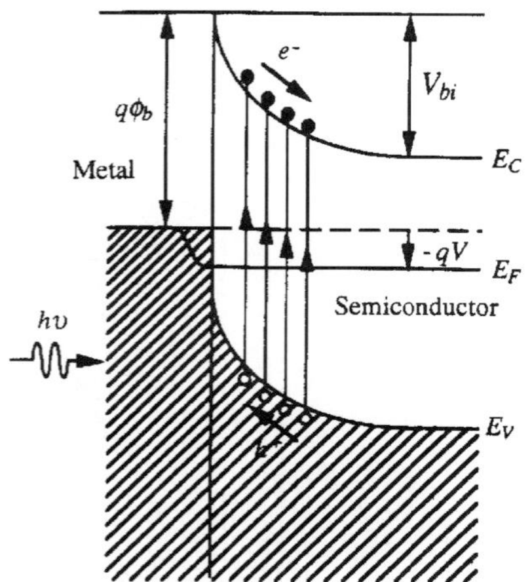

FIGURE 12 Schematic diagram of a Schottky barrier photodiode.

response given in Eqs. (13)–(18) with the exception that there is no diffusion current from the metal layer. The equivalent circuit of a Schottky diode is the same as a *p-i-n* diode.

In high-speed applications, GaAs Schottky diodes in the short-wavelength region with bandwidths over 200 GHz have been reported.[36] These devices can be combined with FETs or sampling diodes.[35,36] Figure 13 shows an integrated Schottky photodiode with a diode sampling circuit. A pair of short voltage pulses are generated by the nonlinear

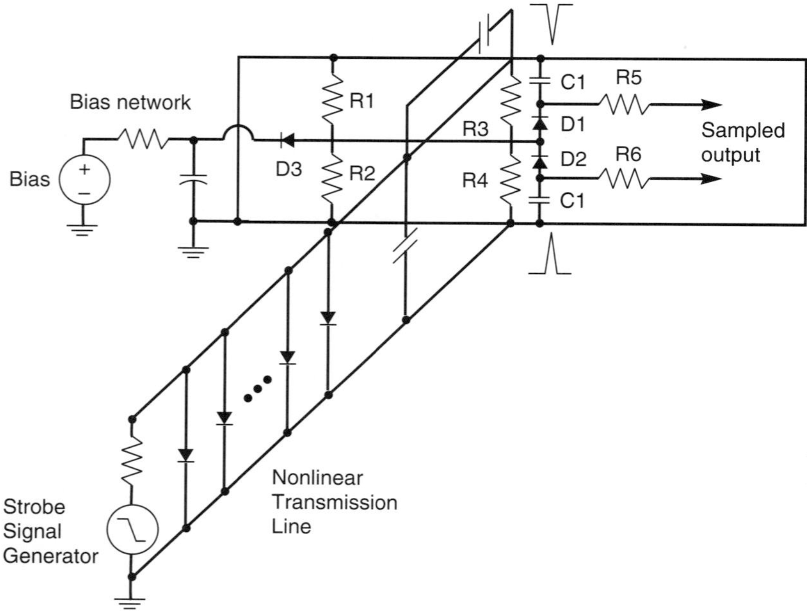

FIGURE 13 Integrated Schottky photodiode and sampling circuit. (*After Ref. 35.*)

transmission line and a differentiator. The short voltage pulses are used to control the sampling capacitors to measure the photodiode signal. The sampled signal is then passed through a low-pass filter to extract the equivalent time domain waveform. Impulse responses of under 2 ps have been demonstrated in this way.[36]

In the long-wavelength region, GaInAs Schottky diodes experience high dark current problems due to the relatively low Schottky barrier height at the metal/GaInAs interface.[37] An InP or quaternary layer is usually added at the interface in order to increase the Schottky barrier height.[38] A graded bandgap layer (e.g., GaInAsP) is then needed at the GaInAs/InP interface to reduce hole trapping.

17.7 AVALANCHE PHOTODETECTORS

High-speed avalanche photodetectors (APDs) are widely used in fiber communication. APDs with gain-bandwidth (GB) products in excess of 100 GHz have been reported.[39–43] In long-wavelength applications. InGaAs/InP APDs are better than Ge APDs due to their lower dark current and lower multiplication noise. The Ge APD also has a limited spectral response at 1.55-μm wavelength. The maximum achievable GB product of InGaAs/InP APDs is predicted to be around 140 GHz,[44] while the gain-bandwidth product of Si APDs in the near infrared region can have GB products of over 200 GHz.[45] InGaAs/InAlAs superlattice avalanche photodiodes have a lower ionization ratio ($k = 0.2$)[46] than bulk avalanche photodiodes, and lower noise and higher gain-bandwidth product can be achieved.

High-speed GaInAs/InP APDs make use of separated absorption and multiplication layers (SAM APD). Figure 14 shows the simplified one-dimensional APD structure. The narrow bandgap n-GaInAs layer absorbs the incident light and the thickness of this layer is usually thick (>1 μm) to ensure high quantum efficiency. The electric field in the absorption layer is high enough for carriers to travel at saturated velocities, yet is below the field where significant avalanching occurs and the tunneling current is negligible. The wide bandgap InP multiplication layer is thin (a few tenths of a micron) to have shorter

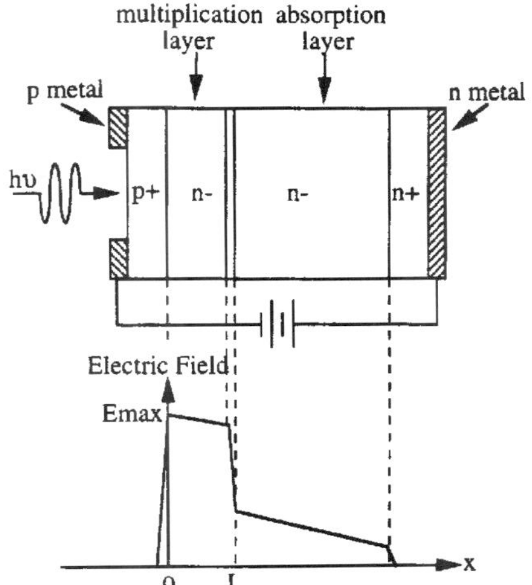

FIGURE 14 Schematic diagram of a SAGM APD.

multiplication buildup time.[47,48] The bias is applied to the fully depleted absorption layer in order to obtain effective carrier collection efficiency and, at the same time, electric field in the multiplication region must be high enough to achieve avalanche gain. A guard ring is usually added to prevent premature avalanche breakdown (or microplasma) at the corner of the diffusion edge. To reduce the hole pileup effect, a graded bandgap layer (e.g., superlattice or compositional grading) is often added at the heterointerface between the absorption layer and multiplication layer. This is the so-called separated absorption, grading, multiplication avalanche photodiode (SAGM APD).

The multiplication process in APDs can be described by the electron and hole ionization coefficients α_i and β_i. The field dependencies of ionization coefficients for electrons and holes are given by

$$\alpha_i(x) = A_e \exp\left(-B_e/E(x)\right) \tag{25a}$$

$$\beta_i(x) = A_h \exp\left(-B_h/E(x)\right) \tag{25b}$$

where $A_{e,h}$ and $B_{e,h}$ are constant parameters.[49] Since the electric field is generally position-dependent, the ionization coefficients are also position-dependent. With Eq. (25*a*, *b*) and the electric field distribution, the position-dependence of ionization coefficients can be derived. The multiplied photocurrent in the avalanche region ($0 \le x \le W$) including injected electron current density $J_n(0)$, injected hole current density $J_p(0)$, and photo-generation of electron-hole pairs $g(x)$ was derived by Lee et al.[50] The total photocurrent density is given by

$$J = \frac{J_p(w)\exp\left[-\int_0^w (\alpha_i - \beta_i)\,dx\right] + J_n(0) + q\int_0^w g(x)\exp\left[-\int_0^x (\alpha_i - \beta)\,dx'\right]dx}{1 - \int_0^w \alpha_i \exp\left[-\int_0^x (\alpha_i - \beta_i)\,dx'\right]dx} \tag{26}$$

where q is the electron charge. The electron-initiated and hole-initiated multiplication factors, M_n and M_p, can be obtained by putting $J_p(w) = g(x) = 0$ and $J_n(0) = g(x) = 0$, respectively in Eq. (26):

$$M_n = \frac{J}{J_n(0)} = \frac{1}{1 - \int_0^w \alpha_i \exp\left[-\int_0^x (\alpha_i - \beta_i)\,dx'\right]dx} \tag{27a}$$

$$M_p = \frac{J}{J_p(w)} = \frac{\exp\left[-\int_0^w (\alpha_i - \beta_i)\,dx\right]}{1 - \int_0^w \alpha_i \exp\left[-\int_0^x (\alpha_i - \beta_i)\,dx'\right]dx} \tag{27b}$$

The bandwidth of an APD is limited by the device RC time constant when the multiplication gain M is low (i.e., $M < \alpha_i/\beta_i$). As the multiplication gain increases above the ratio of the electron and hole ionization coefficients (i.e., $M > \alpha_i/\beta_i$), the avalanche buildup time becomes the dominant limitation on 3-dB bandwidth and the product of the multiplication gain and 3-dB bandwidth reaches a constant. The multiplication factor M as a function of frequency was derived by Emmons[51] and is given by

$$M(\omega) \approx \frac{M_o}{\{1 + \omega^2 M_0^2 \tau_1^2\}^{1/2}} \qquad M_o > \alpha_i/\beta_i \tag{28a}$$

$$\tau_1 \approx N(\alpha_i/\beta_i)\tau \tag{28b}$$

where τ_1 is the effective transit time, τ is the multiplication-region transit time and

$N(\beta_i/\alpha_i)$ is a number changing between 1/3 and 2 as β_i/α_i varies from 1 to 10^{-3}. The dc multiplication factor M_o is given by Miller:[52]

$$M_o = \frac{1}{1-(V_j/V_B)^n} \tag{29}$$

where V_B is the breakdown voltage, V_j is the junction voltage, and n is an empirical factor ($n < 1$).

The total APD dark current consists of two components. I_{du} is the unmultiplied current which is mainly due to the surface leakage current. I_{dm} is the bulk dark current experiencing the multiplication process. The total dark current is expressed by

$$I_d = I_{du} + MI_{dm} \tag{30}$$

where M is the avalanche gain. The noise current spectral density due to the dark current is given by

$$\sigma_d^2 = 2qI_{du} + 2qI_{dm}M^2F(M) \tag{31}$$

where $F(M)$ is the avalanche excess noise factor derived by McIntyre.[53] Excess noise factors for electron-initiated or hole-initiated multiplication are given by

$$F(M) = F_e(M) = [kM + (1-k)(2-1/M)] \tag{32a}$$

$$F(M) = F_h(M) = \left[\frac{1}{k}M + \left(1 - \frac{1}{k}\right)(2-1/M)\right] \tag{32b}$$

where k is the ratio of the ionization coefficient of holes to electrons ($k = \beta_i/\alpha_i$), and k is assumed to be a constant independent of the position. Figure 15 shows the sensitivity of an APD receiver as a function of k_{eff} which is obtained by weighting the ionization rates over the electric field profile. From Fig. 15 we can see that the smaller the k factor is, the

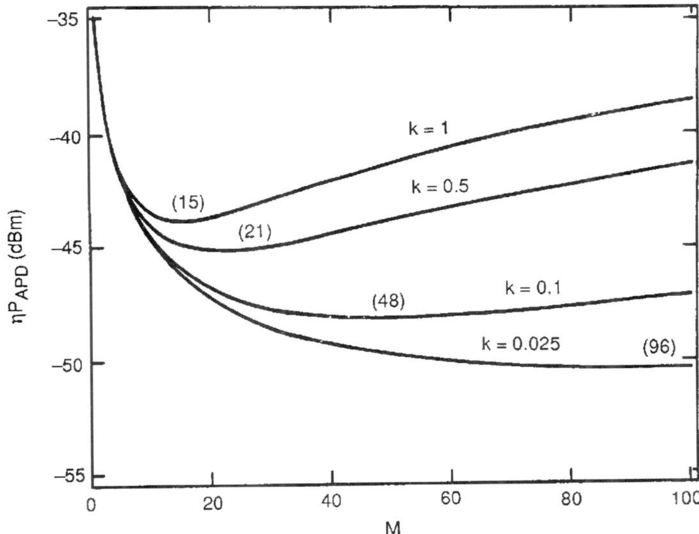

FIGURE 15 Dependence of APD receiver sensitivity on β/α in SAGM APDs (*After Ref. 7.*)

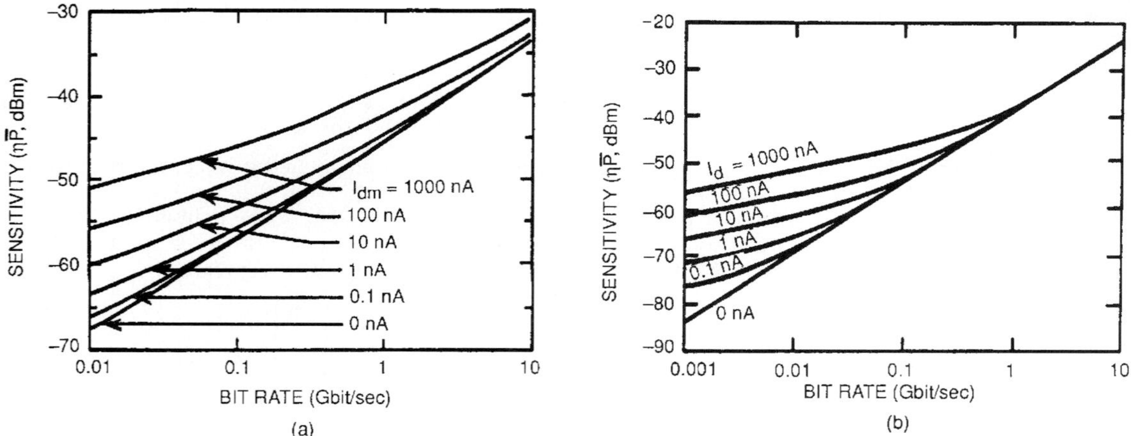

FIGURE 16 Dependence of receiver sensitivity on dark current for an (*a*) SAGM APD and (*b*) PIN detector, each with a GaAs FET preamplifier. (*After Ref. 6.*)

smaller the noise factor is and the better the receiver sensitivity is. Ge APDs have k values close to unity (0.7–1.0). GaInAs/InP APDs using an InP multiplication region have $1/k$ values from 0.3 to 0.5. Silicon is an excellent APD material since the k value for silicon is 0.02. Therefore, a Si APD has an excellent low dark current noise density and is predominantly used at short wavelengths compared with Ge APDs and GaInAs/InP APDs which are used at longer wavelengths.

In optical receiver applications,[54,55] the photodetector is used with a low-noise amplifier. The dark current noise power is given by

$$\langle i_{nd}^2 \rangle = 2qI_{du}BI_2 + 2qI_{dm}M^2F(M)BI_2 \tag{33}$$

where B is the receiver bit rate and I_2 is a parameter depending on the input optical pulse shape. The receiver sensitivity penalty[56] is given in terms of parameter ε_N.

$$\eta \bar{P} = (1 + \varepsilon_N)\eta \bar{P}_o \tag{34}$$

where $\eta \bar{P}_o$ is the sensitivity with zero dark current. For example, $\varepsilon_N = 0.023$ for a 0.1-dB penalty. The maximum allowable dark current for a given sensitivity for a *p-i-n* FET receiver is

$$I_{du} = \frac{\varepsilon_N(2 + \varepsilon_N)}{2qBI_2} \langle i_{na}^2 \rangle \tag{35}$$

where $\langle i_{na}^2 \rangle$ is the amplifier noise power and is proportional to B^3 above 100 MBits/s. Therefore, the maximum allowable dark current is proportional to B^2. For APD receivers, the maximum allowable dark current I_{dm} as a function of bit rate can be approximated by assuming sensitivity penalty is within 1 or 2 dB and optimum gain is constant. In Fig. 16, as we can see, the dark current is proportional to B at lower bit rates and $B^{1.25}$ at higher bit rates. So, as the bit rate increases, the maximum allowable dark current increases.

17.8 PHOTOCONDUCTORS

High-speed photoconductors[57–62] have become more important not only because of their simplicity in fabrication and ease of integration with MESFET amplifiers but also because of their useful applications for photodetector and photoconductor sampling gates. Usually

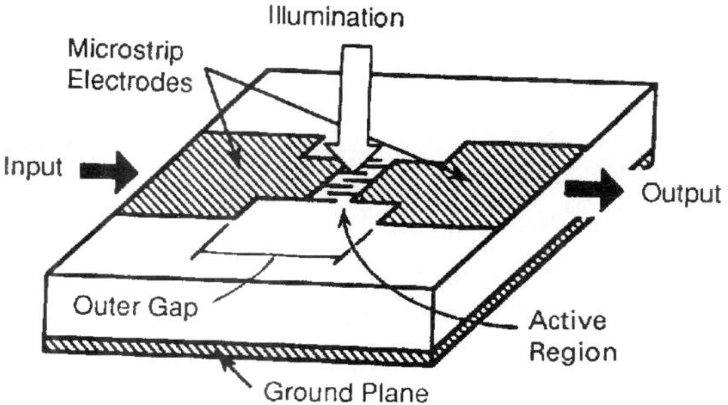

FIGURE 17 Schematic diagram of a high-speed photoconductor.

the photoconducting film has a high density of defects with the trap energy levels deep within the bandgap to shorten the material lifetime and the detector impulse response. The characteristics of the photoconductive films include: (1) high resistivity due to the fact that Fermi level is pinned at the midgap, (2) enhanced optical absorption for photon energy below the bandgap due to the introduction of new bandgap states, and (3) easy fabrication of ohmic contacts possibly due to the enhancement of tunneling through the narrow Schottky barrier with a pinned Fermi level.

Figure 17 shows a typical photoconductor on a microstrip line structure. The photoconductive film is formed on top of a semi-insulating substrate. A microstrip transmission line consists of microstrip electrodes on top and ground plane on bottom. Under a steady-state illumination, the photogenerated carrier will experience high electrical field and travel to the electrodes. The photocurrent is

$$I_{ph} = \frac{q\eta GP}{h\nu} \tag{36}$$

where q is electron charge, η is the external quantum efficiency, G is the photoconductor gain, and P is optical input flux. The photoconductor gain G is given by

$$G = \frac{\tau}{\tau_{tr}} \tag{37}$$

which is the ratio of carrier lifetime τ to the carrier transit time τ_{tr}. The frequency response of a photoconductive detector is plotted in Fig. 18 for different material lifetimes. In a detector without damage sites, the gain can be quite large at low frequencies. We can see from this figure how the increased bandwidth is achieved at the expense of quantum efficiency. Using smaller finger separations, higher quantum efficiency can be achieved for a particular bandwidth.

The standard microstrip line configuration has reflection problems in the thickness direction of the substrate and the dispersion characteristics of a microstrip line is worse than that of a coplanar stripline.[57] Coplanar striplines with "sliding-contact" excitation can have zero capacitance to first order.[58] The photoconductor using coplanar stripline has been very successful in generating short electrical pulses. To measure the short electric pulse, several techniques can be used such as photoconductor sampling or electro-optic sampling. Both of the above techniques can provide subpicosecond resolution. The coplanar strip line configuration with sliding contact and sampling gate is shown in Fig.

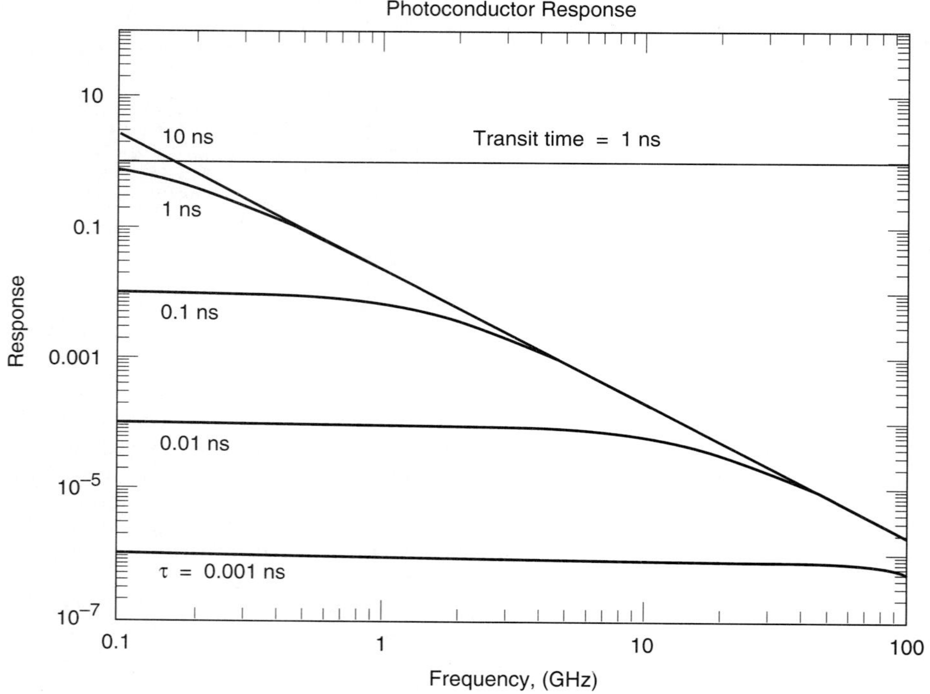

FIGURE 18 Frequency response of a photoconductor.

19*a*. The equivalent circuit is shown in Fig. 19*b*.[58] The infinite capacitances represent that the line extends without end in both directions. The generated electrical signal due to the time-varying resistance $R_s(t)$ is

$$V_{\text{out}}(t) = V_b \frac{Z_o}{Z_o + R_s(t) + R_c} \tag{38}$$

where R_c is the contact resistance. If the excitation intensity is sufficiently low to keep $R_s(t) \gg Z_o$, then

$$V_{\text{out}}(t) = V_b \frac{Z_o}{R_s(t) + R_c} \tag{39}$$

The photoconductor resistance $R_s(t)$ can be related to the photoexcited electron-hole pair density $n(t)$:[59]

$$R_s(t) = \frac{L}{qn(t)(\mu_e + \mu_e)wd_e} \tag{40}$$

where L is the gap width, w is the width of photoconductive volume and d_e is the effective absorption length. When the pulse width is of the same order of magnitude as carrier lifetime and much shorter than the carrier transit time across the switch gap, the electron-hole pair density is given by

$$n(t) = e^{-t/\tau} \int_0^t e^{t/\tau} \frac{\eta P_o(t)(1 - R)}{h\nu wd_e} dt \tag{41}$$

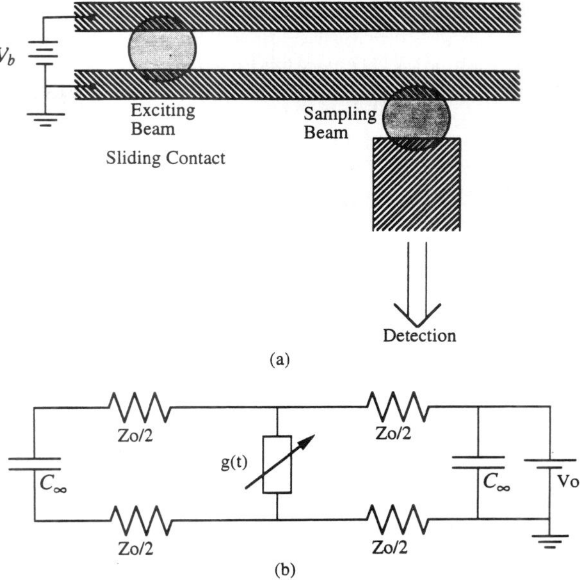

FIGURE 19 (*a*) Coplanar circuit layout of a photoconductor with sliding contact. (*b*) Equivalent circuit.

Here, we notice that the carrier density is an exponential decay function, so $G(t) = 1/R_s(t)$ is also an exponential decaying function with a time constant τ. This can be explained as a result of the convolution of the laser pulse with an exponential function with carrier life time τ.

The low temperature (LT) grown GaAs[60] can have both high carrier mobility and subpicosecond carrier lifetime when being compared with that of the ion implanted photoconductor.[61] The dislocation density in LT GaAs is about the same as that in GaAs epitaxial layer grown at normal substrate temperatures such that the LT GaAs has a mobility as high as that of the bulk material. The resistivity of the LT GaAs is greater than of semi-insulating GaAs ($>10^{17}\ \Omega$-cm) due to its high deep-level concentration. The LT GaAs photoconductive-gap switch in a coplanar strip transmission line configuration has obtained a 1.6 ps (FWHM) response with a 3-dB bandwidth of 220 GHz. Chen et al.[62] reported a high-speed photodetector utilizing LT GaAs MSM photoconductor. To achieve reasonable quantum efficiency and high-speed response, the optimum design requires carrier transit time approximately equal to carrier lifetime. With this requirement, the carriers not collected fast enough by the electrodes will be consumed by recombination. The response of a 0.2-μm finger and space MSM photodetector was measured by electro-optic sampling system. A 1.2 ps (FWHM) response with a 3-dB bandwidth of 350 GHz is obtained.

17.9 SUMMARY

Photodetector performance has steadily improved over the past decade. High-speed detectors are now available at a variety of wavelengths from 1.65 to 0.4 μm. MSM photoconductors have demonstrated the shortest impulse responses of under a picosecond. For applications that require high speed and high efficiency, the best results have been obtained using two passes through a *p-i-n* photodetector (30 percent quantum efficiency with 110-GHz bandwidth). Many applications require a high saturation power, and waveguide photodetectors have achieved the best results (20-GHz bandwidth with 0.5-A/W responsivity and 10-mW saturation power). Traveling wave photodetectors appear to offer the ultimate results in high-speed, high-responsivity, high-saturation power detectors. The combination of high-speed photodetectors with optical amplifiers is resulting in superb sensitivity of all bit rates, but requires the fabrication of high-speed photodetectors with at least 10-dBm saturation power.

An increasing amount of attention is being paid to integrating high-speed photodetectors with electronic and photonic circuits. Integration with electronic circuits increases the performance by eliminating the parasitics and limited bandwidth of bonding pads, wires, and connectors. Integration with optical waveguides decreases the optical loss associated with coupling from one device to another and reduces the packaging cost. Integration of photodetectors with optical amplifiers and wavelength tuning elements is a particularly important research direction.

17.10 REFERENCES

1. T. P. Lee and T. Li, Photodetectors, *Optical Fiber Telecommunications,* S. E. Miller and A. G. Chynoweth (eds.), Academic Press, New York, 1979).
2. D. H. Auston, *Picosecond Optoelectronic Devices,* chap. 4, C. H. Lee (ed.), Academic Press, 1984, pp. 73–117.
3. J. E. Bowers and C. A. Burrus, Jr., "Ultrawide-band Long-Wavelength *p-i-n* Photodetectors," *J. Lightwave Tech.,* vol. LT-5, no. 10, October 1987.
4. F. Capasso, "Physics of Avalanche Photodiodes," *Semiconductors and Semimetals,* 22D, W. T. Tsang (ed.), Academic Press, 1985.
5. J. Campbell, *Semiconductors and Semimetals,* 22D, W. T. Tsang (ed.), Academic Press, 1985.
6. B. Kasper, "Receiver Design," *Optical Fiber Telecommunications II,* S. E. Miller and I. P. Kaminow (eds.), Academic Press, 1988.
7. S. R. Forrest, "Avalanche Photodetector Receiver Sensitivity," *Semiconductors and Semimetals,* vol. 22D, *Lightwave Communications Technology*: *Photodetectors,* W. Tsang (ed.), Academic Press, 1985.
8. S. R. Sloan, "Processing and Passivation Techniques for Fabrication of High Speed InP/InGaAs/InP Mesa Photodetectors," *Hewlett Packard Journal,* Oct. 69, 1989.
9. K. Carey, S. Y. Wang, J. S. C. Chang and K. Nauka, "Leakage Current in GaInAs/InP Photodiodes Grown by OMVPE," *J. Crystal Growth* **98:**90 (1989).
10. Y. G. Wey, D. L. Crawford, K. Giboney, J. E. Bowers, M. J. Rodwell, P. Silvestre, M. J. Hafich, and G. Y. Robinson, "Ultrafast Graded Double-Heterostructure GaInAs/InP Photodiode," *Appl. Phys. Lett.* **58**(19):2156 (1991).
11. M. S. Unlu, K. Kishino, J. Chyi, L. Aresenault, J. Reed, and S. N. Mohammad, "Resonant Cavity Enhanced AlGaAs/GaAs Heterojunction Phototransistors with an Intermediate InGaAs Layer in the Collector," *Appl. Phys. Lett.* **57**(8), 20 August 1990, p. 750.
12. A. Chin and T. Y. Chang, "Enhancement of Quantum Efficiency in Thin Photodiodes Through Absorptive Resonance," *J. Lightwave Tech.,* vol. 9, no. 3, March 1991, p. 321.
13. K. Kishino, M. S. Unlu, J-I. Chyi, J. Reed, L. Arsenault, and H. Morkoc, "Resonant Cavity-Enhanced (RCE) Photodetectors," *IEEE J. Quantum Electron,* vol. 27, no. 8, August 1991, p. 2025.
14. K. Kato, S. Hata, A. Kozen, J. Yoshida, and K. Kawano, "High-Efficiency Waveguide InGaAs

Pin Photodiode with Bandwidth of over 40 GHz," *IEEE Photon. Tech. Lett.,* vol. 3, no. 5, May 1991, p. 473.

15. D. Wake, S. N. Judge, T. P. Spooner, M. J. Harlow, W. J. Duncan, I. D. Henning, and M. J. O'Mahony, "Monolithic Integration of 1.5 μm Optical Preamplifier and PIN Photodetector with a Gain of 20 dB and a Bandwidth of 35 GHz," *Electron. Lett.,* vol. 26, no. 15, July 19, 1990, pp. 1166–1168.
16. R. J. Deri, N. Yasuoka, M. Makiuchi, H. Hamaguchi, O. Wada, A. Kuramata, and R. J. Hawkins, "Integrated Waveguide/Photodiodes with Large Bandwidth and High External Quantum Efficiency," *IEEE Photon. Technol. Lett.,* vol. 2, 1990, pp. 496–498.
17. K. S. Giboney, M. J. W. Rodwell, and J. E. Bowers, "Travelling-Wave Photodetectors," *Photon. Tech. Lett.* **4**(12):1363–1365 (December 1992).
18. H. Taylor, O. Eknoyan, C. S. Park, K. N. Choi, and K. Chang, "Traveling Wave Photodetectors," *SPIE Proc. on Optoelectronic Signal Processing for Phased Array Antennas II,* 1990, p. 59.
19. A. R. Williams, A. L. Kellner, X. S. Jiang, and P. K. L. Yu, "InGaAs/InP Waveguide Photodetector with High Saturation Intensity," *Electron. Lett.* **28:**2258 (1992).
20. M. Wu and T. Itoh, "Ultrafast Photonic to Microwave Transformer (PMT)," *LEOS Topical Meeting on Optical Microwave Interactions,* Paper W1.2, 1993.
21. A. Larsson et al., *J. Quantum Electron.* **24:**787 (1988).
22. D. L. Crawford, R. Nagarajan, and J. E. Bowers, "Comparison of Bulk and Quantum Wire Photodetectors," *Appl. Phys. Lett.* **58**(15):1629–1631 (April 1991).
23. D. Kuhl, F. Hieronymi, E. H. Bottcher, and D. Bimberg, "High-Speed Metal-Semiconductor-Metal Photodetectors on InP: Fe," *IEEE Photon. Tech. Lett.,* vol. 2, no. 8, August 1990, p. 574.
24. G. Lucovsky, R. F. Schwarz, and R. B. Emmons, "Transit-Time Considerations in *p-i-n* Diodes," *J. of Appl. Phys.,* vol. 35, no. 3, March 1961.
25. Y. G. Wey, K. S. Giboney, J. E. Bowers, M. J. W. Rodwell, P. Silvestre, P. Thiagarajan, and G. Y. Robinson, "110 GHz GaInAs/InP *p-i-n* Photodiodes with Integrated Bias Tees and Matched Resistors," *IEEE Photonic Tech. Lett.,* August 1993.
26. D. M. Braun, "Design of Single Layer Antireflection Coatings for InP/InGaAs/InP Photodetectors for the 1200–1600 nm Wavelength Range," **27:**2006 (1988).
27. J. E. Bowers and C. A. Burrus, "High Speed Zero Bias Waveguide Photodetectors," *Electron. Lett.* **22:**905 (1986).
28. A. Alping, R. Tell and S. T. Eng, "Photodetection Properties of Semiconductor Laser Diode Detectors," *J. Lightwave Tech.,* LT-4, 1986, 1662–1668.
29. A. Alping, "Waveguide *p-i-n* Photodetectors: Theoretical Analysis and Design Criteria," *IEEE Proceedings,* vol. 136, part J, no. 3, June 1989.
30. S. Y. Wang and D. Bloom, "100 GHz Bandwidth Planar GaAs Schottky Photodiode," *Electron. Lett.,* vol. 19, no. 14, 7 July, 1983, p. 554.
31. D. G. Parker and P. G. Say, "Indium Tin Oxide/GaAs Photodiodes for Millimetric-Wave Applications," *Electron. Lett.,* vol. 22, no. 23, 6 Nov. 1986, p. 1266.
32. H. Kamiyama, Y. Kobayashi, T. Nagatsuma, and T. Kamiya, "Very Short Electrical Pulse Generation by a Composite Planar GaAs Photodetectors," *Japan. J. Appl. Phys.,* vol. 29, Sept. 1990, p. 1717.
33. S. M. Sze, *Physics of Semiconductor Devices,* 1981, pp. 255–263.
34. M. Missous and E. H. Rhoderick, "On the Richardson Constant for Aluminum/Gallium Arsenide Schottky Diodes," *J. Appl. Phys.* **69**(10):7142 (15 May 1991).
35. M. Kamegawa, K. Giboney, J. Karin, S. Allen, M. Case, R. Yu, M. J. W. Rodwell, and J. E. Bowers, "Picosecond GaAs Monolithic Optoelectronic Sampling Circuit," *Photonics Technology Lett.* **3**(6):567–569 (June 1991).
36. E. Ozbay, K. D. Li, and D. M. Bloom, "2.0 psec GaAs Monolithic Photodetector," *IEEE Photon. Tech. Lett.,* vol. 3, no. 6, June 1991, p. 570.
37. N. Emeis, H. Schumacher, and H. Beneking, "High-Speed GaInAs Schottky Photodetector," *Electron. Lett.,* vol. 21, no. 5, 28 Feb. 1985, pp. 181.
38. L. Yang, A. S. Sudbo, R. A. Logan, T. Tanbun-Ek, and W. T. Tsang, "High Performance

Fe:InP/GaAs Metal/Semiconductor/Metal Photodetectors Grown by Metalorganic Vapor Phase Epitaxy," *IEEE Photon. Tech. Lett.*, vol. 2, no. 1, January 1990, p. 56.

39. T. Mikawa, H. Kuwatsuka, Y. Kito, T. Kumai, M. Makiuchi, S. Yamazaki, O. Wada, and T. Shirai, "Flip-Chip InGaAs Avalanche Photodiode with Ultra Low Capacitance and Large Gain-Bandwidth Product," *Tech. Digest,* ThO_2, OFC 1991, p. 186.

40. H. Kuwatsuka, T. Mikawa, S. Miura, N. Yasuoka, T. Tanahashi, and O. Wada, "An AlxGa1-xSb Avalanche Photodiode with Gain Bandwidth Product of 90 GHz," *Photon. Tech. Lett.*, vol. 2, no. 1, Jan 1990, p. 54.

41. F. Capusso, H. M. Cox, A. L. Hutchinson, N. A. Olsson, and S. G. Hummel, "Pseudo-Quaternary GaInAsP Semiconductors: A New $In_{0.53}Ga_{0.47}As$/InP Graded Gap Superlattice and Its Applications to Avalanche Photodiodes," *Appl. Phys. Lett.* **45**(11):1193–11951 (1 December 1984).

42. L. E. Tarof, "Planar InP/InGaAs Avalanche Photodiodes with a Gain-Bandwidth Product Exceeding 100 GHz," *Tech. Digest,* ThO3, OFC 1991, p. 187.

43. K. Taguchi, T. Torikai, Y. Sugimoto, K. Makito, and H. Ishihara, "Planar-Structure InP/InAsAsP/InGaAs Avalanche Photodiodes with Preferential Lateral Extended Guard Ring for 1.0–1.6 μm Wavelength Optical Communication Use," *J. Lightwave Technology,* vol. 6, no. 11, Nov. 1988, p. 1643.

44. T. Shiba, E. Ishimura, K. Takahashi, H. Namizaki, and W. Susaki, "New Approach to the Frequency Response Analysis of an InGaAs Avalanche Photodiode," *J. Lightwave Technology,* vol. 6, no. 10, Oct. 1988, p. 1502.

45. K. Berchtold, O. Krumpholz, and J. Suri, "Avalanche Photodiodes with a Gain-Bandwidth Product of More Than 200 GHz," *Appl. Phys. Lett.*, vol. 26, no. 10, 15 May 1975, p. 585.

46. T. Kagawa, H. Asai, and Y. Kawamura, "An InGaAs/InAlAs Superlattice Avalanche Photodiode with a Gain Bandwidth Product of 90 GHz," *IEEE Photon. Tech. Lett.* vol. 3, no. 9, September 91, pp. 815–817.

47. H. Imai and T. Kaneda, "High-Speed Distributed Feedback Lasers and InGaAs Avalanche Photodiodes," *J. Lightwave Technology,* vol. 6, no. 11, Nov. 1988, p. 1643.

48. H. C. Hsieh and W. Sargeant, "Avalanche Buildup Time of an InP/InGaAsP/InGaAs APD at High Gain," *J. Quantum Electron.*, vol. 25, no. 9, Sept. 1989, p. 2027.

49. F. Osaka, T. Mikawa and T. Kaneda, "Impact Ionization of Electrons and Holes in (100)-Oriented Ga1 − xInxAsyP1-y," *IEEE J. Quantum Electron.*, vol. QE-21, no. 9, September 1985, pp. 1326–1338.

50. C. A. Lee, R. A. Logan, R. L. Batdorf, J. J. Kleimack, and W. Wiegmann, "Ionization Rates of Holes and Electrons in Silicon," *Phys. Rev.* vol. 134, 1964, pp. A761–A773.

51. R. B. Emmons, "Avalanche-Photodiode Frequency Response," *J. Appl. Phys.*, vol. 38, no. 9, August 1967, p. 3705.

52. S. L. Miller, "Avalanche Breakdown in Germanium," *Phys. Rev.*, vol. 99, Aug. 1955, pp. 1234–1241.

53. R. J. McIntyre, "Multiplication Noise in Uniform Avalanche Junctions," *IEEE Trans. Electron. Devices,* vol. ED-13, Jan. 1966, pp. 164–168.

54. B. L. Kasper and J. C. Campell, "Multigigabit-per-Second Avalanche Photodiode Lightwave Receivers," *J. Lightwave Tech.*, vol. LT-5, no. 10, October 1987, p. 1351.

55. M. Brain and T. P. Lee, "Optical Receiver for Lightwave Communication Systems," *J. Lightwave Tech.*, vol. LT-3, no. 6, December 1985, p. 1281.

56. T. V. Muoi, "Receiver Design for High-Speed Optical-Fiber Systems," *J. Lightwave Tech.*, vol. LT-2, no. 3, June 1984, p. 243.

57. J. A. Caldmains and G. Mourou, "Subpicosecond Electrooptic Sampling: Principles and Applications," *IEEE J. Quantum Electron.*, vol. QE-22, 1986, pp. 69–78.

58. D. R. Grischkowsky, M. B. Ketchen, C.-C. Chi, I. N. Duling, III, N. J. Halas, J-M Halbout, and P. G. May, "Capacitance Free Generation and Detection of Subpicosecond Electrical Pulses on Coplanar Transmission Lines," *IEEE J. Quantum Electron.*, vol. 24, no. 2, February 1988, pp. 221–225.

59. W. C. Nunnally and R. B. Hammond, "Optoelectronic Switch for Pulsed Powers," *Picosecond Optoelectronic Devices,* C. H. Lee (ed.), Academic Press, Orlando, Fla., 1984, pp. 373–398.

60. F. W. Smith, H. Q. Le, V. Diadiuk, M. A. Hollis, A. R. Calawa, S. Gupta, M. Frankel, D. R. Dykaar, G. A. Mourou, and T. Y. Hsiang, "Picosecond GaAs-Based Photoconductive Optoelectronic Detectors," *Appl. Phys. Lett.* **54**(10):890 (6 March 1989).

61. N. G. Paulter, A. J. Gibbs, and D. N. Sinha, "Fabrication of High-Speed GaAs Photoconductive Pulse Generators and Sampling Gates by Ion Implantation," *IEEE Trans. Electron Device,* vol. 35, no. 12, December 1988, pp. 2343–2348.

62. Y. Chen, S. Williamson, and T. Brock, "1.2 ps High Sensitivity Photodetector/Switch Based on Low-Temperature-Grown GaAs," Postdeadline papers, CPDP 10/591, CLEO 1991.

CHAPTER 18
SIGNAL DETECTION AND ANALYSIS

John R. Willison
Stanford Research Systems, Inc.
Sunnyvale, California

18.1 GLOSSARY

A	dimensionless material constant for 1/f noise
C	capacitance (farads)
I	current (amps)
$I_{\text{shot noise}}$	shot noise current (amps)
k	Boltzmann's constant
q	electron charge (coulombs)
R	resistance (ohms)
S/N	signal-to-noise ratio
T	temperature (degrees Kelvin)
$V_{\text{Johnson,rms}}$	RMS Johnson noise voltage (V)
Δf	bandwidth (Hz)

18.2 INTRODUCTION

Many optical systems require a quantitative measurement of light. Applications range from the very simple, such as a light meter using a photocell and a d'Arsenval movement, to the complex, such as the measurement of a fluorescence lifetime using time-resolved photon counting.

Often, the signal of interest is obscured by noise. The noise may be fundamental to the process: photons are discrete quanta governed by Poisson statistics which gives rise to shot noise. Or, the noise may be from more mundane sources, such as microphonics, thermal emf's, or inductive pickup.

This article will describe methods for making useful measurements of weak optical signals, even in the presence of large interfering sources. The article will emphasize the electronic aspects of the problem. Important details of optical systems and detectors used in signal recovery are covered in chapters 15–17 in this *Handbook*.

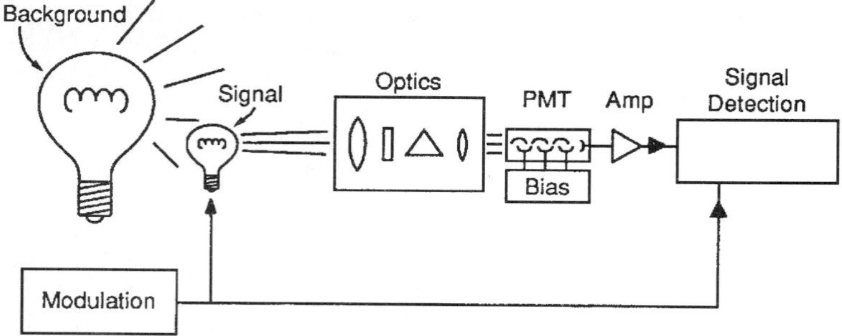

FIGURE 1 Prototypical optical measurement.

18.3 PROTOTYPE EXPERIMENT

Figure 1 details the elements of a typical measurement situation. We wish to measure light from the source of interest. This light may be obscured by light from background sources. The intensity of the source of interest, and the relative intensity of the interfering background, will determine whether some or all of the techniques shown in Fig. 1 should be used.

Optics

The optical system is designed to pass photons from the source of interest and reject photons from background sources. The optical system may use spatial focusing, wavelength, or polarization selection to preferentially deliver photons from the source of interest to the detector.

There are many trade-offs to consider when designing the optical system. For example, if the source is nearly monochromatic and the background is broadband, then a monochrometer may be used to improve the signal-to-background ratio of the light reaching the detector. However, if the source of interest is an extended isotropic emitter, then a monochrometer with narrow slits and high f number will dramatically reduce the number of signal photons from the source which can be passed to the detector. In this case, the noise of the detector and amplifiers which follow the optical system may dominate the overall signal-to-noise ratio (S/N).

Photodetectors

There are many types of nonimaging photodetectors. Key criteria to select a photodetector for a particular application include: sensitivity for the wavelength of interest, gain, noise, and speed. Important details of many detector types are given in other chapters in the *Handbook*. Operational details (such as bias circuits) of photomultipliers which are specific to boxcar integration and photon counting will be discussed in Sec. 18.5.

Amplifiers

In many applications, the output of the detector must be amplified or converted from a current to a voltage before the signal may be analyzed. Selection criteria for amplifiers include type (voltage or transconductance), gain, bandwidth, and noise.

Signal Analysis

There are two broad categories of signal analysis, depending on whether or not the source is modulated. Modulating the source allows the signal to be distinguished from the background. Often, source modulation is inherent to the measurement. For example, when a pulsed laser is used to induce a fluorescence, the signal of interest is present only after the laser fires. Other times, the modulation is "arranged," as when a cw source is chopped. Sometimes the source cannot be modulated or the source is so dominant over the background as to make modulation necessary.

18.4 NOISE SOURCES

An understanding of noise sources in a measurement is critical to achieving signal-to-noise performance near theoretical limits. The quality of a measurement may be substantially degraded by a trivial error. For example, a poor choice of termination resistance for a photodetector may increase current noise by several orders of magnitude.[1]

Shot Noise

Light and electrical charge are quantized, and so the number of photons or electrons which pass a point during a period of time are subject to statistical fluctuations. If the signal mean is M photons, the standard deviation (noise) will be $\sqrt{M}$, hence the $S/N = M/\sqrt{M} = \sqrt{M}$. The mean M may be increased if the rate is higher or the integration time is longer. Short integration times or small signal levels will yield poor S/N values. Figure 2 shows the S/N which may be expected as a function of current level and integration time for a shot-noise limited signal.

"Integration time" is a convenient parameter when using time domain signal recovery

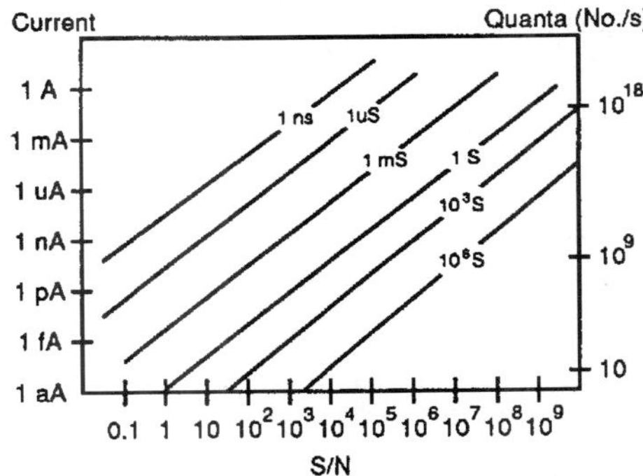

FIGURE 2 Signal to noise vs. flux and measurement time.

techniques. "Bandwidth" is a better choice when using frequency domain techniques. The rms noise current in the bandwidth Δf Hz due to a "constant" current, I amps, is given by

$$I_{\text{shot noise}} = \sqrt{(2qI\,\Delta f)} \tag{1}$$

where $q = 1.6 \times 10^{-19} C$

Johnson Noise

The electrons which allow current conduction in a resistor are subject to random motion which increases with temperature. This fluctuation of electron density will generate a noise voltage at the terminals of the resistor. The rms value of this noise voltage for a resistor of R ohms, at a temperature of T degrees Kelvin, in a bandwidth of Δf Hz is given by

$$V_{\text{Johnson,rms}} = \sqrt{(4kTR\,\Delta f)} \tag{2}$$

where k is Boltzmann's constant. The noise voltage in a 1-Hz bandwidth is given by

$$V_{\text{Johnson,rms}}(\text{per}\ \sqrt{\text{Hz}}) = 0.13\ \text{nV} \times \sqrt{(R\ (\text{ohms}))} \tag{3}$$

Since the Johnson noise voltage increases with resistance, large-value series resistors should be avoided in voltage amplifiers. For example, a 1-kΩ resistor has a Johnson voltage of about 4.1 $\text{nV}/\sqrt{\text{Hz}}$. If detected with a 100-MHz bandwidth, the resistor will show a noise of 41 μV rms, which has a peak-to-peak value of about 200 μV.

When a resistor is used to terminate a current source, or as a feedback element in a current-to-voltage converter, it will contribute a noise current equal to the Johnson noise voltage divided by the resistance. Here, the noise current in a 1-Hz bandwidth is given by

$$I_{\text{Johnson,rms}}(\text{per}\ \sqrt{\text{Hz}}) = 130\ \text{pA}/\sqrt{R(\text{ohms})} \tag{4}$$

As the Johnson noise current increases as R decreases, small-value resistors should be avoided when terminating current sources. Unfortunately, small terminating resistors are required to maintain a wide frequency response. If a 1-kΩ resistor is used to terminate a current source, the resistor will contribute a noise current of about 4.1 $\text{pA}/\sqrt{\text{Hz}}$, which is about 1000× worse than the noise current of an ordinary FET input operational amplifier.

l/f Noise

The voltage across a resistor carrying a constant current will fluctuate because the resistance of the material used in the resistor varies. The magnitude of the resistance fluctuation depends on the material used: carbon composition resistors are the worst, metal film resistors are better, and wire wound resistors provide the lowest l/f noise. The rms value of this noise source for a resistance of R ohms, at a frequency of f Hz, in a bandwidth of Δf Hz is given by

$$V_{\text{l/f,rms}} = IR \times \sqrt{(A\,\Delta f/f)} \tag{5}$$

where the dimensionless constant A has a value of about 10^{-11} for carbon. In a measurement in which the signal is the voltage across the resistor (IR), then the $S/N = 3 \times 10^5 \sqrt{(f/\Delta f)}$. Often, this noise source is a troublesome source of low-frequency noise in voltage amplifiers.

Nonessential Noise Sources

There are many discrete noise sources which must be avoided in order to make reliable low-level light measurements. Figure 3 shows a simplified noise spectrum on log-log scales. The key features in this noise spectra are frequencies worth avoiding: diurnal drifts (often seen via input offset drifts with temperature), low frequency (1/f) noise, power line frequencies and their harmonics, switching power supply and crt display frequencies, commercial broadcast stations (AM, FM, VHF, and UHF TV), special services (cellular telephones, pagers, etc.), microwave ovens and communications, to RADAR and beyond.

Your best alternatives for avoiding these noise sources are:

1. Shield to reduce pickup.
2. Use differential inputs to reject common mode noise.
3. Bandwidth limit the amplifier to match expected signal.
4. Choose a quiet frequency for signal modulation when using a frequency domain detection technique.
5. Trigger synchronously with interfering source when using a time domain detection technique.

Common ways for extraneous signals to interfere with a measurement are illustrated in Fig. 4*a–f*.

Noise may be injected via a stray capacitance as in Fig. 4*a*. The stray capacitance has an impedance of 1/jwC. Substantial currents may be injected into low-impedance systems (such as transconductance inputs), or large voltages may appear at the input to high-impedance systems.

Inductive pickup is illustrated in Fig. 4*b*. The current circulating in the loop on the left will produce a magnetic field which in turn induces an emf in the loop on the right. Inductive noise pickup may be reduced by reducing the areas of the two loops (by using twisted pairs, for example), by increasing the distance between the two loops, or by

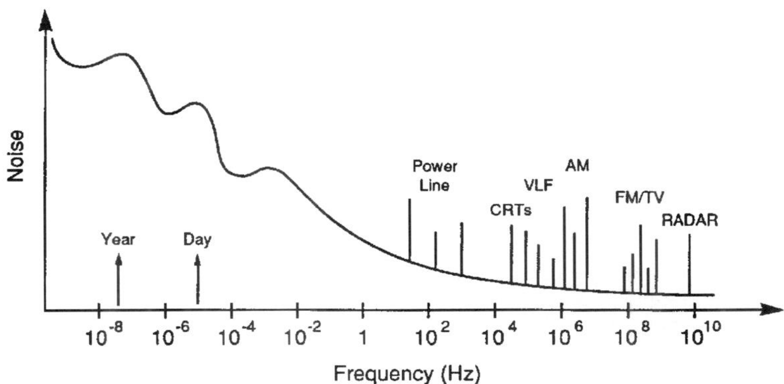

FIGURE 3 Simplified noise spectrum.

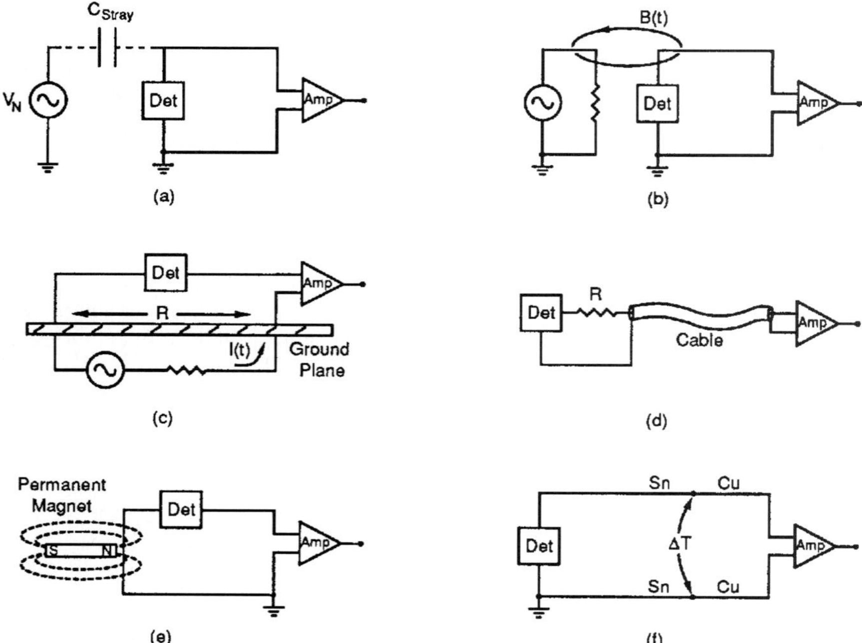

FIGURE 4 Coupling of noise sources.

shielding. Small skin depths at high frequencies allow nonmagnetic metals to be effective shields; however, high-mu materials must be used to shield from low frequency magnetic fields.

Resistive coupling, or a "ground loop," Is shown in Fig. 4*c*. Here, the detector senses the output of the experiment plus the *IR* voltage drop from another circuit which passes current through the same ground plane. Cures for ground-loop pickup include: grounding everything to the same point, using a heavier ground plane, providing separate ground return paths for large interfering currents, and using a differential connection between the signal source and amplifier.

Mechanical vibrations can create electrical signals (microphonics) as shown in Fig. 4*d*. here, a coaxial cable is charged by a battery through a large resistance. The voltage on the cable is $V = Q/C$. Any deformation of the cable will modulate the cable's capacitance. If the period of the vibration which causes the deformation is short compared to the RC time constant then the stored charge on the cable, Q, will remain constant. In this case, a 1-ppm modulation of the cable capacitance will generate an ac signal with an amplitude of 1 ppm of the dc bias on the cable, which may be larger than the signal of interest.

The case of magnetic microphonics is illustrated in Fig. 4*e*. Here, a dc magnetic field (the earth's field or the field from a permanent magnet in a latching relay, for example) induces an emf in the signal path when the magnetic flux through the detection loop is modulated by mechanical motion.

Unwanted thermocouple junctions are an important source of offset and drift. As shown in Fig. 4*f*, two thermocouple junctions are formed when a signal is connected to an amplifier. For typical interconnect materials (copper, tin) one sees about 10 μV/°C of offset. These extraneous junctions occur throughout instruments and systems: their impact may be eliminated by making ac measurements.

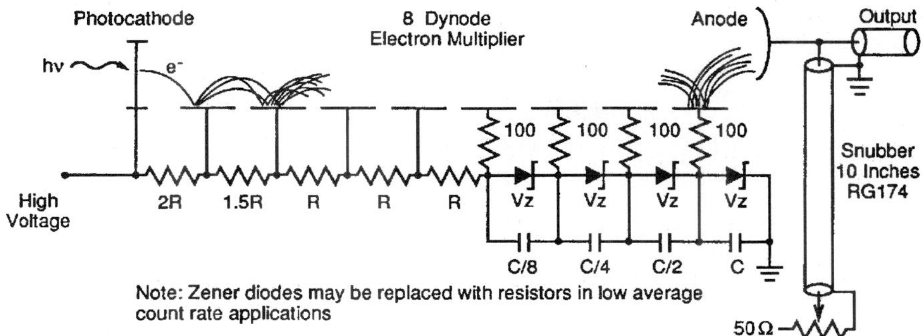

FIGURE 5 PMT base for photon counting or fast integration.

18.5 APPLICATIONS USING PHOTOMULTIPLIERS

Photomultiplier tubes (PMTs) are widely used for detection of light from about 200 to 900 nm. Windowless PMTs can be used from the near UV through the x-ray region, and may also be used as particle detectors. Their low noise, high gain, wide bandwidth, and large dynamic range have placed them in many applications. They are the only detectors which may be recommended for low-noise photon counting applications.[2,3]

In this article, we are primarily concerned with the electrical characteristics of PMTs. Understanding these characteristics is important if we are to realize the many desirable features of these devices.

A schematic representation of a PMT, together with a typical bias circuit, is shown in Fig. 5. While the concepts depicted here are common to all PMTs, the particulars of biasing and termination will change between PMT types and applications.

PMTs have a photocathode, several dynodes (6 to 14), and an anode. They are usually operated from a negative high voltage, with the cathode at the most negative potential, each successive dynode at a less negative potential, and the anode near ground. An incident photon may eject a single photoelectron from the photocathode which will strike the first dynode with an energy of a few hundred volts. A few (2–5) electrons will be ejected from the first dynode by the impact of the photoelectron: these electrons will in turn strike the second dynode, ejecting more electrons. The process continues at each dynode until all of the electrons are collected by the anode.

Quantum Efficiency

The quantum efficiency (QE) of a PMT is a measure of the probability that a photon will eject a photoelectron at the photocathode. The QE depends on the type of material used in the cathode and the wavelength of light. QEs may be as high as 10 to 30 percent at their peak wavelength. The cathode material will also affect the dark count rate from the PMT: a cathode with good red sensitivity may have a high dark count rate.

Gain

A PMT's gain depends on the number of dynodes, the dynode material, and voltage between the dynodes. PMT gains range from 10^3 to 10^7. The anode output from the PMT will typically go to an electronic amplifier. To avoid having the system noise be dominated by the amplifier's noise, the PMT should be operated with enough gain so that the dark current times the gain is larger than the amplifier's input current noise.

Bandwidth

The frequency response, speed, rise time, and pulse-pair resolution of PMTs depend on the structure of the dynode multiplier chain. The leading edges of the anode output have transition times from 2 to 20 ns. Trailing edges are usually about three times slower. Much faster PMTs, with rise times on order 100 ps, use microchannel plate multiplers.

When using gated integrators to measure PMT outputs, the pulse width of the anode signal should be less than the gate width so that timing information is not lost. For photon counting, the pulse width should be smaller than the pulse-pair resolution of the counter/discriminator to avoid saturation effects. When using lock-in amplifiers, pulse width is usually not important, since the slowest PMTs will have bandwidths well above the modulation frequency.

Pulse Height

In pulsed experiments, the criteria for a detectable signal often depends on the electrical noise environment of the laboratory and the noise of the preamplifier. In laboratories with Q-switched lasers or pulsed discharges, it is difficult to reduce the noise on any coaxial cable below a few millivolts. A good, wide bandwidth preamplifier will have about $1.5\,\text{nV}/\sqrt{\text{Hz}}$, or about 25-μV rms over a 300-MHz bandwidth. Peak noise will be about 2.5 times the rms noise, so it is important that the PMT provide pulses of greater than 1-mV amplitude.

Use manufacturer's specifications for the current gain and rise time to estimate the pulse amplitude from the PMT:

$$\text{Amplitude (mV)} = 4 \times \text{gain (in millions)/rise time (in ns)} \tag{6}$$

This formula assumes that the electrons will enter a 50-Ω load in a square pulse whose duration is twice the rise time. (Since the rise time will be limited by the bandwidth of the preamplifier, use the larger of the amplifier or PMT rise times in this formula.)

If the PMT anode is connected via a 50-Ω cable to a large load resistance, then the pulse shape may be modeled by the lumped parameters of the cable capacitance (about 100 pF/meter for RG-58) and the termination resistance. All of the charge in the pulse is deposited on the cable capacitance in a few nanoseconds. The voltage on the load will be $V = Q/C$ where C = cable capacitance. This voltage will decay exponentially with a time constant of RC where R is the load resistance in ohms. In this case, the pulse height will be

$$\text{Amplitude (mV)} = 160 \times \text{gain (in millions)/cable } C \text{ (in pF)} \tag{7}$$

The current gain of a PMT is a strong function of the high voltage applied to the PMT. Very often, PMTs will be operated well above the high voltage recommended by the manufacturer, and thus substantially higher current gains (10× to 100× above specs). There are usually no detrimental effects to the PMT as long as the anode current is kept well below the rated value.

Dark Counts

PMTs are the quietest detectors available. The primary noise source is thermionic emission of electrons from the photocathode and from the first few dynodes of the electron multiplier. PMT housings which cool the PMT to about −20°C can dramatically reduce the dark counts (from a few kHz to a few Hz). The residual counts arise from radioactive decays of materials inside the PMT and from cosmic rays.

PMTs which are specifically designed for photon counting will specify their noise in terms of the rate of output pulses whose amplitudes exceed some fraction of a pulse from a single photon. More often, the noise is specified as an anode dark current. Assuming the primary source of dark current is thermionic emission from the photocathode, the dark count rate is given by

$$\text{Dark count (kHz)} = 6 \times \text{dark current (nA)/gain (millions)} \tag{8}$$

PMT Base Design

PMT bases which are designed for general-purpose applications are not appropriate for photon counting or fast-gated integrator applications (gates <10–20 ns). General-purpose bases will not allow high count rates, and often cause problems such as double counting and poor plateau characteristics. A PMT base with the proper high-voltage taper, bypassing, snubbing, and shielding is required for good time resolution and best photon counting performance.

Dynode Biasing. A PMT base provides bias voltages to the PMTs photocathode and dynodes from a single, negative, high-voltage power supply. The simplest design consists of a resistive voltage divider. In this configuration the voltage between each dynode, and thus the current gain at each dynode, is the same. Typical current gains are three to five, so there will typically be four electrons leaving the first dynode, with a variance of about two electrons. This large relative variance (due to the small number of ejected electrons) gives rise to large variations in the pulse height of the detected signal. Since statistical fluctuations in pulse height are dominated by the low gain of the first few stages of the multiplier chain, increasing the gain of these stages will reduce pulse-height variations and so improve the pulse-height distribution. This is important for both photon counting and analog detection. To increase the gain of the first few stages, the resistor values in the bias chain are increased to increase the voltage in the front end of the multiplier chain. The resistor values are tapered slowly so that the electrostatic focusing of electrons in the multiplier chain is not adversely affected.[4]

Current for the electron multiplier is provided by the bias network. Current drawn from the bias network will cause the dynode potentials to change, thus changing the tube gain. This problem is of special concern in lifetime measurements. The shape of exponential decay curves will be changed if the tube gain varies with count rate. To be certain that this is not a problem, lifetime measurements should be repeated at reduced intensity. The problem of gain variation with count rate is avoided if the current in the bias network is about 20 times the output current from the PMT's anode.

There are a few other methods to avoid this problem which do not require high bias currents. These methods depend on the fact that the majority of the output current is drawn from the last few dynodes of the multiplier:

1. Replace the last few resistors in the bias chain with Zener diodes. As long as there is some reverse current through a Zener, the voltage across the diodes is nearly constant. This will prevent the voltage on these stages from dropping as the output current is increased.
2. Use external power supplies for the last few dynodes in the multiplier chain. This approach dissipates the least amount of electrical power since the majority of the output current comes from lower-voltage power supplies. However, it is the most difficult to implement.
3. If the average count rate is low, but the peak count rate is high, then bypass capacitors on the last few stages may be used to prevent the dynode voltage from dropping (use 20× the average output current for the chain current). For a voltage drop of less than 1 percent, the stored charge on the last bypass capacitor should be 100× the charge

output during the peak count rate. For example, the charge output during a 1-ms burst of a 100-MHz count rate, each with an amplitude of 10 mV into 50 Ω and a pulse width of 5 ns, is 0.1 uC. If the voltage on the last dynode is 200 Vdc, then the bypass capacitor for the last dynode should have a value given by

$$C = 100Q/V = 100 \times 0.15C/200V = 0.05\ \mu\text{F} \tag{9}$$

The current from higher dynodes is smaller so the capacitors bypassing these stages may be smaller. Only the final four or five dynodes need to be bypassed, usually with a capacitor which has half the capacitance of the following stage. To reduce the voltage requirement for these capacitors, they are usually connected in series.

Bypassing the dynodes of a PMT may cause high-frequency ringing of the anode output signal. This can cause multiple counts for a single photon or poor time resolution in a gated integrator. The problem is significantly reduced by using small resistors between the dynodes and the bypass capacitors.

Snubbing. Snubbing refers to the practice of adding a network to the anode of the PMT to improve the shape of the output pulse for photon counting or fast-gated integrator applications. This "network" is usually a short piece of 50-Ω coax cable which is terminated into a resistor of less than 50 Ω. The snubber will delay, invert, and sum a small portion of the anode signal to itself.

Snubbing should not be used when using a lock-in amplifier since the current conversion gain of a 50-Ω resistor is very small.

There are four important reasons for using a snubber network:

1. Without some dc resistive path between the anode and ground, anode dark current will charge the signal cable to a few hundred volts (last dynode potential). When the signal cable is connected to an amplifier, the stored charge on the cable may damage the front end of the instrument. PMT bases without a snubber network should include a 100-MΩ resistor between the anode and ground to protect the instruments.
2. The leading edge of the output current pulse is often much faster than the trailing edge. A snubber network may be used to sharply increase the speed of the trailing edge, greatly improving the pulse pair resolution of the PMT. This is especially important in photon counting applications.
3. Ringing (with a few-nanoseconds period) is very common on PMT outputs. A snubber network may be used to cancel these rings which can cause multiple counts from a single photon.
4. The snubber network will help to reverse terminate reflections from the input to the preamplifier.

The round-trip time in the snubber cable may be adjusted so that the reflected signal cancels anode signal ringing. This is done by using a cable length with a round-trip time equal to the period of the anode ringing.

Cathode Shielding. Head-on PMTs have a semitransparent photocathode which is operated at negative high voltage. Use care so that no objects near ground potential contact the PMT near the photocathode.

Magnetic Shielding. Electron trajectories inside the PMT will be affected by magnetic fields. A field strength of a few gauss can dramatically reduce the gain of a PMT. A magnetic shield made of a high permeability material should be used to shield the PMT.

PMT Base Summary

1. Taper voltage divider for higher gain in first stages.

2. Bypass last few dynodes in pulsed applications.
3. Use a snubber circuit to shape the outputs pulse for photon counting or fast-gated integration.
4. Shield the tube from electrostatic and magnetic fields.

18.6 AMPLIFIERS

Several considerations are involved in choosing the correct amplifier for a particular application. Often, these considerations are not independent, and compromises will be necessary. The best choice for an amplifier depends on the electrical characteristics of the detector, and on the desired gain, bandwidth, and noise performance of the system.

Voltage Amplifiers

High Bandwidth Photon counting and fast-gated integration require amplifiers with wide bandwidth. A 350-MHz bandwidth is required to preserve a 1-ns rise time. The input impedance to these amplifiers is usually 50 Ω in order to terminate coax cables into their characteristic impedance. When PMTs (which are current sources) are connected to those amplifiers, the 50-Ω input impedance serves as the current-to-voltage converter for the PMT anode signal. Unfortunately, the small termination resistance and wide bandwidth yield lots of current noise.[5]

High Input Impedance. It is important to choose an amplifier with a very high input impedance and low-input bias current when amplifying a signal from a source with a large equivalent resistance. Commercial amplifiers designed for such applications typically have a 100-MΩ input impedance. This large input impedance will minimize attenuation of the input signal and reduce the Johnson noise current drawn through the source resistance, which can be an important noise source. Field effect transistors (FETs) are used in these amplifiers to reduce the input bias current to the amplifiers. Shot noise on the input bias current can be an important noise component, and temperature drift of the input bias current is a source of drift in dc measurements.[6]

The bandwidth of a high-input impedance amplifier is often determined by the *RC* time constant of the source, cable, and termination resistance. For example, a PMT with 1 meter of RG-58 coax (about 100 pF) terminated into a 1-MΩ resistor will have a bandwidth of about 1600 Hz. A smaller resistance would improve the bandwidth, but increase the Johnson noise current.

Moderate Input Impedance Bipolar transistors offer an input noise voltage which may be several times smaller than the FET inputs of high-input impedance amplifiers, as low as $1\ \mathrm{nV}/\sqrt{\mathrm{Hz}}$. Bipolar transistors have larger input bias currents, hence larger shot noise current, and so should be used only with low-impedance (<1 kΩ) sources.

Transformer Inputs. When ac signals from very low source impedances are to be measured, transformer coupling offers very quiet inputs. The transformer is used to step up the input voltage by its turns-ratio. The transformer's secondary is connected to the input of a bipolar transistor amplifier.

Low Offset Drift. Conventional bipolar and FET input amplifiers exhibit input offset drifts on the order of 5 μV/C. In the case where the detector signal is a small dc voltage, such as from a bolometer, this offset drift may be the dominate noise source. A different amplifier configuration, chopper-stabilized amplifiers, essentially measure their input offsets and subtract the measured offset from the signal. A similar approach is used to "autozero" the offset on the input to sensitive voltmeters. Chopper-stabilized amplifiers exhibit very low input offsets with virtually no input offset drift.

Differential. The use of "true-differential" or "instrumentation" amplifiers is advised to provide common mode rejection to interfering noise, or to overcome the difference in grounds between the voltage source and the amplifier. This amplifier configuration amplifies the difference between two inputs, unlike a single-ended amplifier, which amplifies the difference between the signal input and the amplifier ground. In high-frequency applications, where good differential amplifiers are not available or are difficult to use, a balun or common mode choke may be used to isolate disparate grounds.

Transconductance Amplifiers

When the detector is a current source (or has a large equivalent resistance) then a transconductance amplifier should be considered. Transconductance amplifiers (current-to-voltage converters) offer the potential of lower noise and wider bandwidth than a termination resistor and a voltage amplifier; however, some care is required in their application.[7]

A typical transconductance amplifier configuration is shown in Fig. 6. An FET input op amp would be used for its low-input bias current. (Op amps with input bias currents as low as 50 fA are readily available.) The detector is a current source, I_o. Assuming an ideal op amp, the transconductance gain is $A = V_{out}/I_{in} = R_f$, and the input impedance of the circuit is R_{in} to the op amp's virtual null. (R_{in} allows negative feedback, which would have been phase shifted and attenuated by the source capacitance at high frequencies, to assure stability.) Commercial transconductance amplifiers use R's as large as 10 MΩ, with R_{in}'s which are typically $R_f/1000$. A low-input impedance will ensure that current from the source will not accumulate on the input capacitance.

This widely used configuration has several important limitations which will degrade its gain, bandwidth, and noise performance. The overall performance of the circuit depends critically on the source capacitance, including that of the cable connecting the source to the amplifier input. Limitations include:

1. The "virtual null" at the inverting input to the op amp is approximately R_f/A_v where A_v is the op amp's open loop gain at the frequency of interest. While op amps have very high gain at frequencies below 10 Hz (typically a few million), these devices have gains of only a few hundred at 1 kHz. With an R_f of 1 GΩ, the virtual null has an impedance of 5 MΩ at 1 kHz, hardly a virtual null. If the impedance of the source capacitance is less than the input impedance, then most of the ac input current will go to charging this capacitance, thereby reducing the gain.

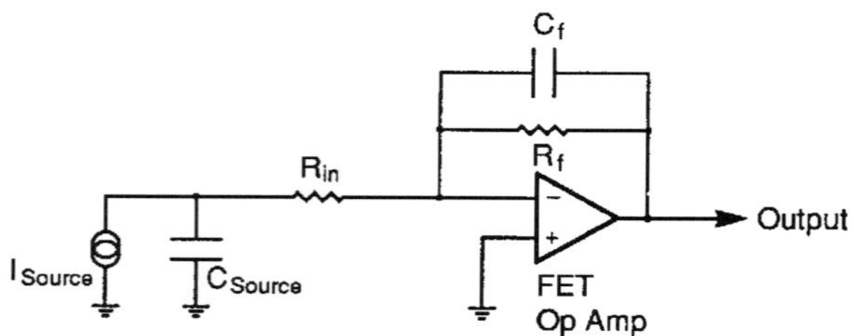

FIGURE 6 Typical transconductance amplifier.

2. The configuration provides high gain for the voltage noise at the noninverting input of the op amp. At high frequencies, where the impedance of the source capacitance is small compared to R_{in}, the voltage gain for noise at the noninverting input is R_f/R_{in}, typically about 1000. As FET input op amps with very low bias currents tend to have high-input-voltage noise, this term can dominate the noise performance of the design.
3. Large R_f's are desired to reduce the Johnson noise current; however, large R_f's degrade the bandwidth. If low values of R_f are used, the Johnson noise current can dominate the noise performance of the design.
4. To maintain a flat frequency response, the size of the feedback capacitance must be adjusted to compensate for different source capacitances.

As many undesirable characteristics of the transconductance amplifier can be traced to the source capacitance, a system may benefit from integrating the amplifier into the detector, thereby eliminating interconnect capacitance. This approach is followed in many applications, from microphones to CCD imagers.

18.7 SIGNAL ANALYSIS

Unmodulated Sources

For unmodulated sources, a strip-chart recorder, voltmeter, A/D converter, or oscilloscope may be used to measure the output of the amplifier or detector. In the case of low-light-level measurement, continuous photon counting would be the method of choice.

A variety of problems are avoided by modulating the signal source. When making dc measurements, the signal must compete with large low-frequency noise sources. However, when the source is modulated, the signal may be measured at the modulation frequency, away from these large noise sources.

Modulated Sources

When the source is modulated, one may choose from gated integration, boxcar averaging, transient digitizers, lock-in amplifiers, spectrum analyzers, gated photon counters, or multichannel scalers.

Gated Integration. A measurement of the integral of a signal during a period of time can be made with a gated integrator. Commercial devices allow gates from about 100 ps to several milliseconds. A gated integrator is typically used in a pulsed laser measurement. The device can provide shot-by-shot data which is often recorded by a computer via an A/D converter. The gated integrator is recommended in situations where the signal has a very low duty cycle, low pulse repetition rate, and high instantaneous count rates.[8]

The noise bandwidth of the gated integrator depends on the gate width: short gates will have wide bandwidths, and so will be noisy. This would suggest that longer gates would be preferred; however, the signal of interest may be very short-lived, and using a gate which is much wider than the signal will not improve the S/N.

The gated integrator also behaves as a filter: the output of the gated integrator is proportional to the average of the input signal during the gate, so frequency components of the input signal which have an integral number of cycles during the gate will average to zero. This characteristic may be used to "notch out" specific interfering signals.

It is often desirable to make gated integration measurements synchronously with an interfering source. (This is the case with time-domain signal detection techniques, and not

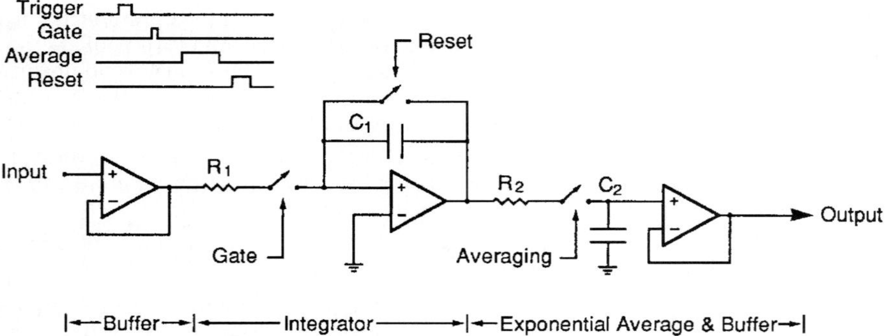

FIGURE 7 Gated integrator and exponential averager.

the case with frequency-domain techniques such as lock-in detection.) For example, by locking the pulse repetition rate to the power-line frequency (or to any submultiple of this frequency) the integral of the line interference during the short gate will be the same from shot to shot, which will appear as a fixed offset at the output of the gated integrator.

Boxcar Averaging. Shot-by-shot data from a gated integrator may be averaged to improve the S/N. Commercial boxcar averagers provide linear or exponential averaging. The averaged output from the boxcar may be recorded by a computer or used to drive a strip-chart recorder. Figure 7 shows a gated integrator with an exponential averaging circuit.

Lock-in Amplifiers. Phase-sensitive synchronous detection is a powerful technique for the recovery of small signals which may be obscured by interference which is much larger than the signal of interest. In a typical application, a cw laser which induces the signal of interest will be modulated by an optical chopper. The lock-in amplifier is used to measure the amplitude and phase of the signal of interest relative to a reference output from the chopper.[9]

Figure 8 shows a simplified block diagram for a lock-in amplifier. The input signal is ac-coupled to an amplifier whose output is mixed (multiplied by) the output of a phase-locked loop which is locked to the reference input. The operation of the mixer may be understood through the trigonometric identity

$$\cos(\omega_1 t + \Phi) * \cos(\omega_2 t) = \tfrac{1}{2}[\cos((\omega_1 + \omega_2)t + \Phi) + \cos((\omega_1 - \omega_2)t + \Phi)] \tag{10}$$

When $\omega_1 = \omega_2$ there is a dc component of the mixer output, $\cos(\Phi)$. The output of the

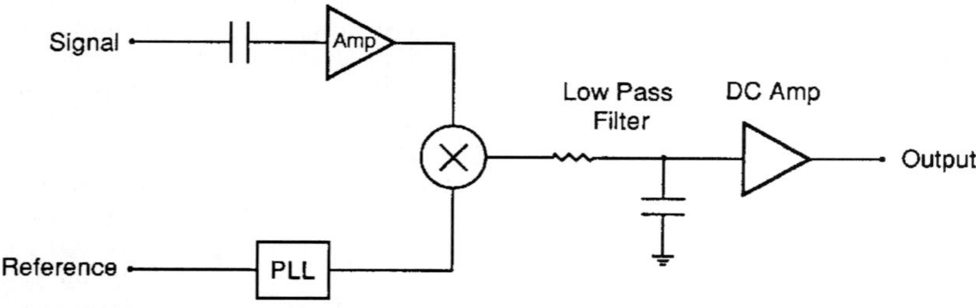

FIGURE 8 Lock-in amplifier block diagram.

mixer is passed through a low-pass filter to remove the sum frequency component. The time constant of the filter is selected to reduce the equivalent noise bandwidth: selecting longer time constants will improve the S/N at the expense of longer response times.

The simplified block diagram shown in Figure 8 is for a "single-phase" lock-in amplifier, which measures the component of the signal at one set phase with respect to the reference. A dual-phase lock-in has another channel which measures the component of the signal at 90° relative to the first channel, which allows simultaneous measurement of the amplitude and phase of the signal.

Digital signal processing (DSP) techniques are rapidly replacing the older analog techniques for the synchronous detection of the signal. In these instruments, the input signal is digitized by a fast, high-resolution A/D converter, and the signal's amplitude and phase are determined by high-speed computations in a digital signal processor. To maintain the 100-kHz bandwidth of the analog designs, the DSP designs must complete a quarter million 16-bit A/D conversions and 20 million multiply-and-accumulate operations each second. Many artifacts of the analog designs are eliminated by the DSP approach; for example, the output drift and dynamic range of the instruments are dramatically improved.[10]

Photon Counting. Photon counting techniques offer several advantages in the measurement of light: very high sensitivity (count rates as low as 1 per minute can be a usable signal level), large dynamic range (signal levels as high as 100 MHz can be counted, allowing a 195-dB dynamic range), discrimination against low-level noise (analog noise below the discriminator thresholds will not be counted), and ability to operate over widely varying duty cycles.[11]

Key elements of a photon counting system include: a high-gain PMT operated with sufficiently high voltage so that a single photoelectron will generate an anode pulse of several millivolts into a 50-Ω load, a fast discriminator to generate logic pulses from anode signals which exceed a set threshold, and fast-gated counters to integrate the counts.

Transient Photon Counting. In situations where the time evolution of a light signal must be measured (LIDAR, lifetime measurements, chemical kinetics, etc.) transient photon counters allow the entire signal to be recorded for each event. In these instruments, the discriminated photon pulses are summed into different bins depending on their timing with respect to a trigger pulse. Commercial instruments offer 5-ns resolution with zero dead-time between bins. The time records from many events may be summed together in order to improve the S/N.[12]

Choosing the "Best" Technique. Which instrument is best suited for detecting signals from a photomultiplier tube? The answer is based on many factors, including the signal intensity, the signal's time and frequency distribution, the various noise sources and their time-dependence and frequency distribution.

In general, the choice between boxcar averaging (gated integration) and lock-in detection (phase-sensitive detection) is based on the time behavior of the signal. If the signal is fixed in frequency and has a 50 percent duty cycle, lock-in detection is best suited. This type of experiment commonly uses an optical chopper to modulate the signal at some low frequency. Signal photons occur at random times during the "open" phase of the chopper. The lock-in detects the average difference between the signal during the "open" phase and the background during the "closed" phase.

To use a boxcar averager in the same experiment would require the use of very long, 50 percent duty cycle gates since the photons can arrive anywhere during the "open" phase. Since the gated integrator is collecting noise during this entire gate, the signal is easily swamped by the noise. To correct for this, baseline subtraction can be used where an equal gate is used to measure the background during the "closed" phase of the chopper and subtracted from the "open" signal. This is then identical to lock-in detection.

However, lock-in amplifiers are much better suited to this, especially at low frequencies (long gates) and low signal intensities.

If the signal is confined to a very short amount of time, then gated integration is usually the best choice for signal recovery. A typical experiment might be a pulsed laser excitation where the signal lasts for only a short time (100 ps to 1 μs) at a repetition rate up to 10 kHz. The duty cycle of the signal is much less than 50 percent. By using a narrow gate to detect signal only when it is present, noise which occurs at all other times is rejected. If a longer gate is used, no more signal is measured but the detected noise will increase. Thus, a 50 percent duty cycle gate would not recover the signal well and lock-in detection is not suitable.

Photon counting can be used in either the lock-in or the gated mode. Using a photon counter is usually required at very low signal intensities or when the use of a pulse height discriminator to reject noise results in an improved S/N. If the evolution of a weak light signal is to be measured, a transient photon counter or multichannel scaler can greatly reduce the time required to make a measurement.

18.8 REFERENCES

1. P. Horowitz and W. Hill, *The Art of Electronics,* 1989, p. 428–447.
2. Photomultiplier Tubes, Hamamatsu Company catalog, 1988.
3. Photomultipliers, Thorn EMI Company catalog, 1990.
4. G. A. Morton, et al., "Pulse Height Resolution of High Gain First Dynode Photomultipliers," *Appl. Phys. Lett.,* vol. 13, p. 356.
5. Model SR445 Fast Preamplifier, *Operation and Service Manual,* Stanford Research Systems, 1990.
6. Model SR560 Low Noise Preamplifier, *Operation and Service Manual,* Stanford Research Systems, 1990.
7. Model SR570 Low Noise Current Amplifier, *Operation and Service Manual,* Stanford Research Systems, 1992.
8. Fast Gated Integrators and Boxcar Averagers, *Operation and Service Manual,* Stanford Research Systems, 1990.
9. Model SR510 Lock-in Amplifier, *Operation and Service Manual,* Stanford Research Systems, 1987.
10. Model SR850 DSP Lock-in Amplifier, *Operation and Service Manual,* 1992.
11. Model SR400 Gated Photon Counter, *Operation and Service Manual,* 1988.
12. Model SR430 Multichannel Scaler/Averager, *Operation and Service Manual,* 1989.

CHAPTER 19
THERMAL DETECTORS

William L. Wolfe
Professor, Optical Sciences Center
The University of Arizona
Tucson, Arizona

Paul W. Kruse
Consultant
Edina, Minnesota

19.1 GLOSSARY

$DTGS$	deuterated triglycine sulfate
p	pyroelectric coefficient
R_e	electrical resistance
R_{th}	thermal resistance
$\mathfrak{R}$	responsivity
S	Seebeck coefficient
TGS	triglycine sulfate
Z	figure of merit
τ_e	electrical time constant
τ_{th}	thermal time constant

19.2 THERMAL DETECTOR ELEMENTS[1]

Introduction

Thermal detectors (transducers) of optical radiation are generally considered to be those devices that absorb the radiation, increase their own temperature, and provide a resultant electrical signal. There are several types, divided according to the physical mechanism that converts the temperature change to a resultant electrical one. The oldest are bolometers and thermocouples. The bolometer changes its electrical resistance as a result of the temperature increase; the thermocouple changes its contact potential difference. There are several different types of bolometers, including thermistor, semiconducting, superconducting, carbon, and metallic. They may also be subdivided according to whether they operate at room or cryogenic temperature. Thermocouples vary according to the materials that are joined, and are sometimes connected in series to generate thermopiles. Pyroelectric detectors make use of the property of a change in the internal polarization as a function of the change in temperature, the pyroelectric effect. Golay cells and certain variations make use of the expansion of a gas with temperature. All of these detectors are governed by the fundamental equation of heat absorption in the material. Many reviews and two books of collected reprints[2] provide additional information.

Thermal Circuit Theory

In the absence of joulean heating of the detector element, the spectrum of the temperature difference $d\,\Delta\tilde{T}$ is given in terms of the spectrum of the absorbed power $\tilde{P}$ (the power is P)

$$d\,\Delta\tilde{T} = \frac{\epsilon\tilde{P}}{G(1+i\omega\tau)} \tag{1}$$

where G is the thermal conductance, given by the product of the thermal conductivity times the cross-sectional area of the path to the heat sink and divided by the length of the path to that heat sink. The time constant τ is the product of the thermal resistance and the heat capacitance. The thermal resistance is the reciprocal of the thermal conductance, while the thermal capacitance is the thermal capacity times the mass of the detector. In the absence of Joulean heating, this is a simple, single time constant thermal circuit, for which the change in temperature is given by

$$d\tilde{T} = \frac{\epsilon\tilde{P}}{G(1+i\omega\tau)} \tag{2}$$

The absorbed power is equal to the incident power times the absorptance α of the material:

$$\tilde{P} = \epsilon\tilde{P}_i \tag{3}$$

The absorptance α is usually written as ϵ (which is legitimate according to Kirchhoff's law) since α is also used for the relative temperature coefficient of resistance (some writers use η).

$$\alpha = \frac{1}{R}\frac{dR}{dT} \tag{4}$$

As radiation is absorbed, part of the heat is conducted to the sink. Some of it gives rise to an increase in temperature. Some is reradiated, but this is usually quite small and is ignored here. The dc responsivity of a thermal detector is proportional to the emissivity and to the thermal resistance. The greater proportion of radiation that is absorbed, the greater will be the responsivity. The less heat that is conducted to the sink, the greater will be the temperature rise. The time constant is a true thermal time constant, the product of thermal resistance and capacitance. The greater the heat capacitance, the more heat necessary for a given temperature increase, and the less heat conducted to the sink, the more available for temperature increase. A high absorptance is accomplished by the use of a black coating, and a sufficient amount of it. Thus, there is a direct conflict between high speed and high responsivity.

The Ideal Thermal Detector[3–6]

The ideal thermal detector has a noise that is associated only with the thermal fluctuations of the heat loss to the heat sink, and this coupling is purely radiative. Then the noise equivalent power (NEP) is given by

$$\mathrm{NEP} = \sqrt{16A\sigma kT^5/\epsilon} \tag{5}$$

where it is assumed that the detector is irradiated by a hemisphere of blackbody radiation at the same temperature T as the detector. The corresponding specific detectivity, assuming that the signal varies as the area and the noise as its square root, is

$$D^* = \frac{\epsilon^{1/2}}{4\sqrt{\sigma kT^5}} \tag{6}$$

where the detector and background are at the same temperature.

For circumstances in which the detector is in a cooled chamber, the total radiation from the sources at various temperatures must be calculated. Figure 1 shows the specific detectivity of a background limited ideal thermal detector as a function of the temperature of the surround.

No detector is ideal, and every one will be limited by the signal loss due to incomplete absorption at the surface and any transmission losses by the optical system that puts the radiation on the detector. The detector will also have noise that arises from its conductive coupling to the heat sink, and probably Johnson noise as well. The conductive mean square power fluctuation is given by

$$\langle P^2 \rangle = 4kT^2G \tag{7}$$

The Johnson noise power density is $4kT$. Therefore, the total mean square power fluctuation is given by

$$\langle P^2 \rangle = 4kT[GT + 4\epsilon A\sigma T^4 + 1] \tag{8}$$

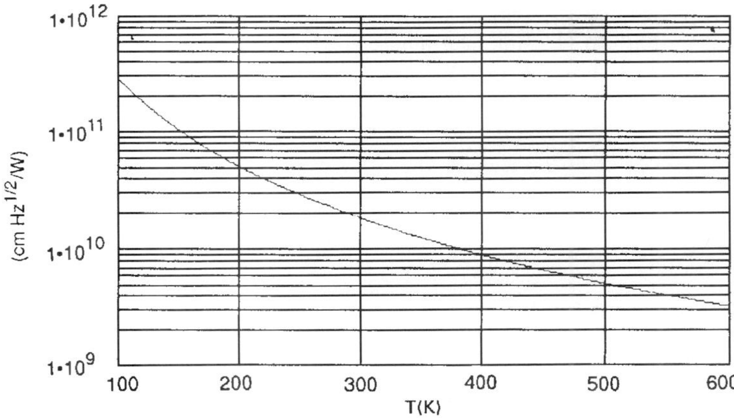

FIGURE 1 Theoretical specific detectivity for ideal thermal detectors.

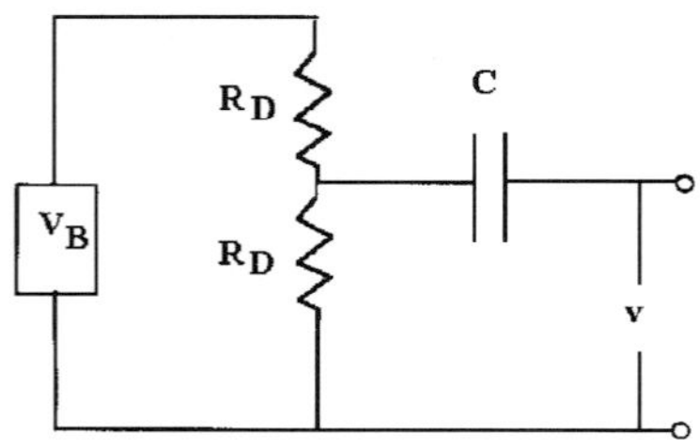

FIGURE 2 Balanced voltage divider circuit for a thermal detector.

Bolometers

Most single-element bolometers are connected in a voltage divider network, as shown in Fig. 2. A stable voltage supply is used to develop a current and consequent voltage drop across the two resistors. One is the detector, while the other should be a matching element to eliminate signals arising from a change in the ambient temperature. It should match the detector in both resistance and in the temperature coefficient of resistance. Usually another, but blinded, detector is used. The expression for power conservation is

$$C\frac{d\,\Delta T}{dt} + G\,\Delta T = \frac{d(i^2R)}{dT}\Delta T + P$$

$$C\frac{d\,\Delta T}{dt} + G\,\Delta T = \frac{V^2(R_1 - R)}{(R_1 + R)^3}\frac{dR}{dT}\Delta T + P \tag{9}$$

$$C\frac{d\,\Delta T}{dt} + \left[G - \frac{V^2R\alpha}{(R_1 + R)^2}\frac{(R_1 - R)}{(R_1 + R)}\right]\Delta T = P$$

The solution to this is a transient that has an RC time constant, where the R is the reciprocal of the bracketed term, and C is the thermal capacitance, and the same steady-state term given above. The transient decays as long as G is greater than the rest of the bracket, but the detector burns up if not. This is still another reason for matching the resistances. The dc responsivity is a function of the construction parameters, including the path to the sink, the bias voltage, and the relative change of resistance with temperature. The different bolometers are divided according to how their resistances change with temperature. (R_1 and R represent slightly different values of R_D.)

Metal bolometers. These have a linear change in resistance with temperature that may be expressed as

$$R = R_0[1 + \gamma(T - T_0)] \tag{10}$$

Therefore the thermal coefficient is

$$\alpha = \frac{\gamma}{1 + \gamma(T - T_0)} \tag{11}$$

This coefficient always decreases with temperature, and burnout does not occur. The coefficient is approximately equal to the inverse of the temperature, and is therefore never very high.

Semiconductor bolometers. These have an exponential change of resistance with temperature, given by

$$R = R_0 e^{\beta/T} \tag{12}$$

so that

$$\alpha = -\beta/T^2 \tag{13}$$

The value of β depends upon the particular material. These detectors can burn out. Two basic types exist: (1) those that are used at low temperatures and (2) those that are used at about room temperature.

The most-used low-temperature bolometer[7] is germanium in a bath of liquid helium. Pure germanium is transparent in the infrared, but with enough compensated doping it becomes a good conductor with a high-temperature coefficient of resistance.[8] Typical concentrations are about $10^{16}\,\text{cm}^{-3}$ of gallium and 10^{15} of indium. Even these are not sufficient at wavelengths shorter than 10 μm since the free-carrier absorption is proportional to wavelength. In such a case a black coating is sometimes used. Improvements have been made since Low's first work.[9–11]

Superconducting bolometers. These make use of the extremely large thermal coefficient of resistance at the transition temperature.[12–14] Originally they needed to be controlled very carefully, or a small change in ambient conditions (on the order of 0.01 K) could cause an apparent signal of appreciable magnitude. A more recent version[15] incorporates an evaporated thin film on an anodized aluminum block that is coupled to a helium bath by a brass rod. The detector has a time constant of about 3 μs due to this high thermal conductance and a good NEP of about $10^{-13}\,\text{WHz}^{-1/2}$. It still must be controlled to about 10^{-5} K, and this is accomplished with a heater current and control circuit.

Recently developed materials not only have high-temperature transition points but also have more gradual transitions, and provide a better compromise between good responsivity and the requirement for exquisite control.[16]

Carbon Bolometers. These are a form of semiconductor bolometers that have been largely superseded by germanium bolometers. They are made of small slabs of carbon resistor material, connected to a metal heat sink by way of a thin mylar film. Although their responsivities are comparable to germanium bolometers, their noise is several orders of magnitude higher.[17]

Thermocouples and Thermopiles[18,19]

A *thermocouple* is made by simply joining two dissimilar conductors. A good pair has a large relative Seebeck coefficient and gives rise to a potential difference. The materials also have large electrical conductivities and small thermal ones, so there is little voltage drop across the length and a small thermal gradient. Although there are many different couples (many are not even used for radiation detection), those most often used for this application are bismuth telluride, copper, and constantan. The expression for the responsivity is given in terms of the relative Seebeck coefficient S_{12} (the difference in the voltage change with temperature between the two materials) and the expression derived above for the thermal circuit

$$\Re = \frac{S\epsilon}{G(1 + i\omega\tau)} \tag{14}$$

Good materials are those that have a large Seebeck coefficient, a high electrical conductivity, and a small thermal conductivity, and the figure of merit is often defined as

$$Z_{12} = \frac{S_{12}^2}{[\sqrt{G_1/\sigma_1} + \sqrt{G_2/\sigma_2}]^2} \tag{15}$$

Thermopiles are arrays of thermocouples connected in series. They are manufactured in two ways. Some are carefully wound wires with junctions aligned in the desired pattern, while others are evaporated with the pattern determined by masking operations. Most of the "bulk" thermopiles are wrapped on appropriate mandrels to obtain rigidity. Both kinds are obtainable in a variety of sizes and patterns that correspond to such things as spectrometer slits, centering annuli, and staggered arrays for moderate-sensitivity thermal imaging.

The Golay Cell[20]

This detector is used mostly for laboratory operations, as it is slow and fragile, although it has high sensitivity. It is a gas-filled chamber that has a thin membrane at one end and a blackened detector area at the other. Light on the blackened surface causes the increase in temperature; this is transferred to the gas which therefore expands. The membrane bulges, and the amount of the bulge is sensed by some sort of optical lever[21] or even change in capacity of an electrical element.[22] Other versions do not use a blackened surface, but allow the radiation to interact with the gas directly, in which case they are spectral detectors that are "tuned" to the absorption spectrum of the gas.[23]

Pyroelectric Detectors[24]

Some crystals which do not have a center of symmetry experience an electric field along a crystal axis. This internal electric field results from an alignment of electric dipoles (known as polarization), and is related to the crystal temperature. In these ferroelectric crystals, this results in a charge being generated and stored on plates connected to the crystal. Polarization disappears above the so-called Curie temperature that is characteristic of each material. Thus, below the Curie temperature, a change in temperature results in a current. The equation for the response of a pyroelectric detector is

$$\Re = \frac{\omega p A_d \epsilon R_e R_{th}}{(1 + i\omega\tau_{th})(1 + i\omega\tau_e)} \tag{16}$$

where ω is the radian frequency, p is the pyroelectric coefficient, A_d is the detector area, R_e is the electrical resistance, R_{th} is the thermal resistance τ_{th} is the thermal time constant, and τ_e is the electrical time constant. The relation is shown in Fig. 3, where the responsivity is plotted as a function of frequency. In the low-frequency region the responsivity rises directly as the frequency. This is a result of the ac operation of a pyroelectric. At the (radian) frequency that is the reciprocal of the slower (often the electrical) time constant,

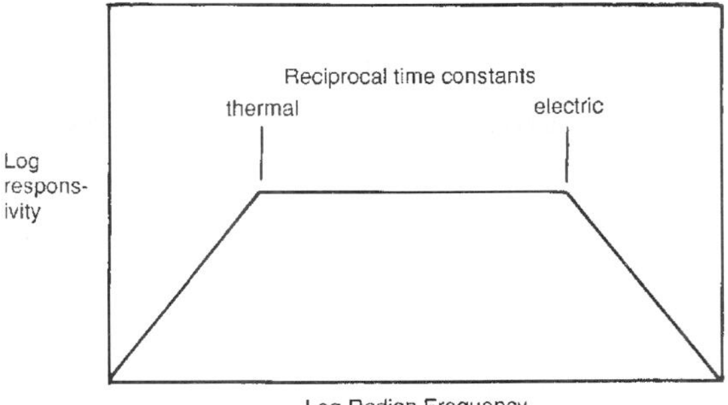

FIGURE 3 Responsivity asymptotes vs frequency.

the response levels off. This is the product of the ac rise and the thermal rolloff. Then, when the frequency corresponding to the shorter time constant is reached, the response rolls off.

Type II pyroelectric detectors work on a slightly different mechanism, which is still not fully understood. The electrodes are on the sensing surface of the detector and parallel to the polar axis. In these crystals, the temperature change is not uniform at the onset of radiation and the primary and secondary pyroelectric effects take place, thereby generating a body electric charge distribution.[25] Materials most often used for these detectors are TGS (triglycine sulfate), DTGS (deuterated TGS), Li_2SO_4, $LiNbO_3$, $LiTaO_3$, and PLZT (lead lanthanum zirconate titanate). TGS is the most used for specialized sensor systems, but has a relatively low Curie point. For higher-temperature operation, usually $LiTaO_3$ or PLZT is used in the general laboratory environment.

The two advantages of the pyreoelectric detector over the other thermal detectors, bolometers, and thermopiles, are its responsivity and its capability of rapid response. The response time and responsivity are traded by choice of the load resistor in the circuit. For instance, with a 100-MΩ load the time constant can be 1 ms and the responsivity 100 V/W, but with a 1-MΩ load the values would be 10 μs and 1.

Summary of Elemental Thermal Detector Properties

Although the user should contact suppliers for detailed information, this section provides overall property information about thermal detectors. There are several cautions about summary data. Most detectors can be tailored to have somewhat different properties. Improvements may have been made since the publication of these results. Not all parameters are available in all combinations. Table 1 does, however, give the general flavor of the performance of different thermal detectors.

19.3 ARRAYS

Introduction

As pointed out earlier, thermal detector response is governed by the thermal response time, which is the ratio of the pixel heat capacity C to the thermal conductance G of the heat leakage mechanism. High pixel responsivity is associated with high thermal isolation, i.e., low thermal conductance. Thermal detector design is driven by the thermal isolation structure. It is the structure which determines the extent to which the pixel performance

TABLE 1 General Properties of Thermal Detectors

Type	Operating Temperature (K)	D* × 10^8 ($cmHz^{1/2}W^{-1}$)	NEP × 10^{-10} ($WHz^{-1/2}$)	Time constant (ms)	Size (mm^2)
Silicon bolometer	1.6		3×10^{-5}	8	0.25–0.70
Metal bolometer	2–4	1		10	
Thermistor bolometer	300	1–6		1–8	0.01–10
Germanium bolometer	2–4		0.005	0.4	1.5
Carbon bolometer	2–4		0.03	10	20
Superconducting bolometer (NbN)	15		0.2	0.5	5 × 0.25
Thermocouples	300		2–10	10–40	0.1 × 1 to 0.3 × 3
Thermopiles	300			3.3–10	1–100
Pyroelectrics	300	2–5		10–100*	2 × 2
Golay cell	300	10	0.6	10–30	10

*Shorter values can be obtained at the expense of NEP (for laser detection).

can approach the temperature fluctuation noise limit and, ultimately, the background fluctuation noise limit. Given the value of G associated with the heat loss mechanism, the pixel heat capacity must be designed appropriately to attain the required thermal response time. Response times in the millisecond range are compatible with high thermal isolation; response times in the microsecond range are not. Thus, two-dimensional arrays of thermal detectors which operate at TV frame rates (30 Hz in the United States) are under development for applications in thermal imagers.

Noise Equivalent Temperature Difference (NETD)

Whereas elemental detectors are usually described by such figures of merit as NEP and D^*, arrays have been described by an NETD associated with their use in a camera under certain specific conditions. It is defined as the change in temperature of a blackbody which fills the field of view of a pixel of an infrared imaging system that gives rise to a change of unity in the signal-to-noise ratio at the output of the system. The measurement of the NETD should, however, be with the flooding of several pixels to avoid fringing effects and with an SNR (signal-to-noise ratio) well above 1 to obtain good accuracy. The pixel is defined as the subtense of a single element of the array. The NETD can be written in several different forms. Perhaps the simplest is

$$\text{NETD} = \frac{\sqrt{A_d B}}{D^*(dP_d/dT)} \tag{17}$$

where D^* is the specific detectivity, A_d is the area of a single pixel, B is the system bandwidth and (dP_d/dT) represents the change in power on the detector element per unit change in temperature in the spectral band under consideration. This form does not include the system noise, which is often included by the manufacturers in their calculations.

$$\frac{dP_d}{dT} = \frac{A_d \tau_o}{4FN^2} \int_{\lambda_1}^{\lambda_2} \frac{dM}{dT}\, d\lambda \tag{18}$$

where τ_o is the optics transmission, FN is the focal ratio (defined as the effective focal length divided by the entrance pupil diameter), and M is the radiant emittance of the source. This is almost the definition of the specific detectivity. The NETD can also be written in terms of the responsivity $\Re$, since the detectivity and responsivity are related in the following way:

$$D^* = \frac{\sqrt{A_d B}}{P} \frac{V_S}{V_N} = \sqrt{A_d B} \frac{\Re}{V_N} \tag{19}$$

where V_s is the signal voltage at the sensor and V_N is the rms noise voltage of a pixel in the bandwidth B. Therefore

$$\text{NETD} = \frac{V_N}{\Re(dP_d/dT)_{\lambda_1-\lambda_2}} \tag{20}$$

The power on the detector is related to the power on the aperture by the optical transmission τ_o. The expression can also be formulated in terms of the source radiance, L

$$\text{NETD} = \frac{4FN^2\sqrt{B}}{D^*\tau_a\tau_o\pi D\,\Delta\theta(dL/dT)_{\lambda_2-\lambda_1}} \tag{21}$$

where D is the diameter of the aperture, $\Delta\theta$ is the angular subtense of a pixel, L is the source radiance and τ_a is the atmospheric transmission. One last form can be generated by recognizing that, for an isotropic radiator, the radiance is the radiant emittance divided by π:

$$\text{NETD} = \frac{4FN^2V_N}{A_D\tau_a\tau_o\Re(T_S)(dM/dT)} \tag{22}$$

In this form of the expression for NETD, it is not necessary that the noise be white, nor is it necessary that the noise not include system noise. Whether or not system noise is included should be clearly stated.

Theoretical Limits

Figure 4 illustrates the theoretical limits of thermal arrays having the parameters shown and operating at 300 and 85 K as a function of thermal conductance. The performance of real thermal arrays with those parameters lies on or above the sloping line. As the conductance G is reduced (better thermal isolation), the noise equivalent temperature difference NETD is reduced (improves) until the background limit is reached, when radiant power exchange between the array and the background becomes the dominant heat transfer mechanism. Reducing the detector temperature to 85 K appropriate to a bolometer operating at the transition edge of the high-temperature superconductor $YBa_2Cu_3O_{7-x}$ (YBCO) reduces the NETD by $\sqrt{2}$ and allows the limit to be reached with less thermal isolation (higher G value).

Arrays fall into two categories: monolithic and hybrid. Monolithic arrays are prepared on a single substrate, e.g., silicon, upon which the detecting material is deposited in the form of a thin film which is subsequently processed into an array. Hybrid arrays are prepared in two parts: (1) the readout electronics array, usually in silicon, and (2) the detecting material array, usually in wafer form which is thinned by lapping, etching, and polishing. These two arrays are mated by a technique such as flip-chip bonding. Here the interconnection at each pixel must have a sufficiently high electrical conductivity, yet a sufficiently low thermal conductivity—a difficult requirement. If array cost considerations are important, then the monolithic approach, especially in silicon, is the more desirable.

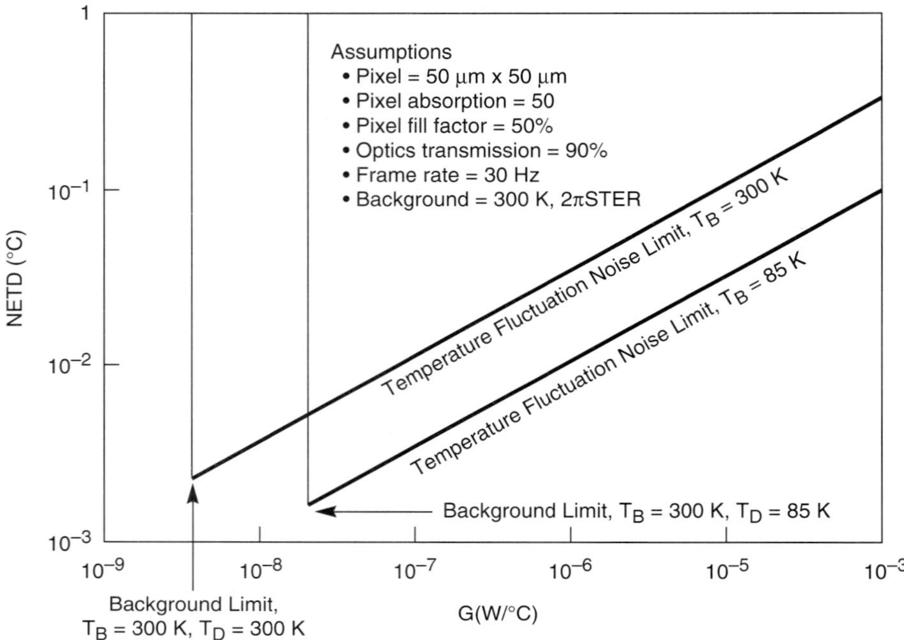

FIGURE 4 Temperature fluctuation noise limit and background fluctuation noise limit of uncooled and cryogenic thermal detector arrays.

Resistive Bolometer Arrays

The development of resistive bolometric arrays has proceeded along two paths: uncooled arrays and cryogenic arrays. Large, uncooled bolometric arrays have been developed at Honeywell by a team lead by R. A. Wood.[26,27] Silicon microstructure technology is employed to produce the arrays, a process resembling the fabrication of integrated circuits. Twelve arrays are prepared on a 4-inch-diameter silicon wafer. Each monolithic array consists of 240×336 pixels; each pixel is 50×50 μm. The detecting material is a thin film of vanadium oxide. A Si_3N_4 membrane having a thermal conductance of 1×10^{-7} WC^{-1} supports the vanadium oxide at each pixel, as shown in Fig. 5. Bipolar transistors implanted in the silicon substrate act as pixel switches for the matrix-addressed array. The response is optimized for the 8–14 μm spectral interval. The thermal response time is adjusted for a 30 Hz frame rate. Each pixel is addressed once per frame by a 5-μs pulse. A thermoelectric stabilizer maintains the array at ambient temperature. Other than a one-shot shutter, the camera has no moving parts.

The measured NETD of the camera with F/1 optics at 300 K is 0.04 K. Given the G value of 1×10^{-7} WK^{-1} it can be seen from Fig. 4 that the array is within a factor of 4 of the temperature fluctuation noise limit. Furthermore, the pixel thermal isolation is so complete that there is no measurable thermal spreading among the pixels.

Linear resistive bolometric arrays of the high-temperature superconductor YBCO on silicon microstructures have been prepared by Johnson et al.,[28] also of Honeywell. A two-dimensional array is under development.[29] The monolithic arrays operate at the transition edge from 70 to 90 K. As was true for the uncooled arrays, the superconducting ones employed a silicon nitride membrane to support the thin film and provide thermal isolation. Excess noise at the contacts limited the performance of the 12-element linear

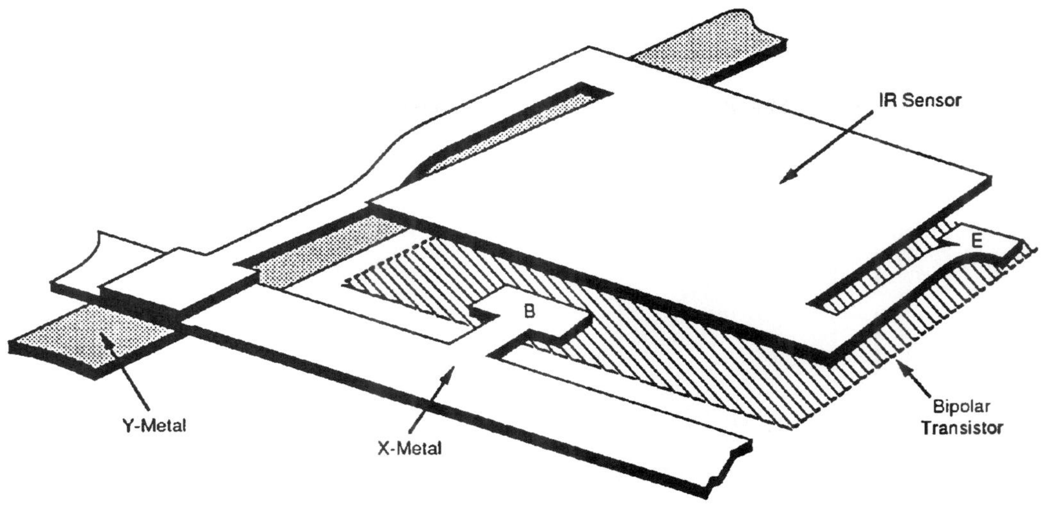

FIGURE 5 Monolithic silicon microbolometer.[26,27] (*© 1992 IEEE.*)

array. With no excess noise, the calculated NETD[29] of a 240 × 336 array with 50 μm pixels and F/1 optics would be 0.002 K, which is near the 300 K background limit, as shown in Fig. 4.

Pyroelectric Hybrid Arrays

Linear and two-dimensional pyroelectric uncooled arrays have been under development for more than two decades.[20–32] The arrays employ hybrid construction, in which a bulk pyroelectric ceramic material such as lithium tantalate or lead zirconate is mechanically thinned, etched, and polished, then bump-bonded to a silicon substrate containing readout electronics,[33] as shown in Fig. 6. Reticulation is usually employed to prevent lateral heat conduction through the pyroelectric material. The theoretical system NETD of a two-dimensional uncooled array with F/1 optics is estimated to be 0.1 K.[33] Two-dimensional uncooled arrays operating in the 8–14-μm region having 100 × 100 pixels, each 100 × 100 μm, are available commercially.[34] Their NETD (with F/1 speed) is 0.35 K. A two-dimensional pyroelectric monolithic array employing a thin film of lead titanate on a silicon microstructure is under development.[35]

Ferroelectric bolometer arrays, also known as field-enhanced pyroelectric arrays, have been developed by Texas Instruments.[36,37] Operation depends upon the temperature dependence of the spontaneous polarization and dielectric permittivity in a ferroelectric ceramic near the Curie temperature. Barium strontium titanate (BST), the selected material, has its composition (barium-to-strontium ratio) adjusted during preparation so the Curie point is 22°C. A thermoelectric stabilizer is employed to hold the BST near 22°C such that the absorbed infrared radiation changes the temperature and thus the dielectric properties. The effect is similar to the pyroelectric effect; however, a voltage is applied to enhance the signal. Construction of this array is naturally similar to that of the pyroelectric array, described above, as shown in Fig. 7. Reticulation of the ceramic is frequently applied to these arrays as well. A radiation chopper is required as both the pyroelectric

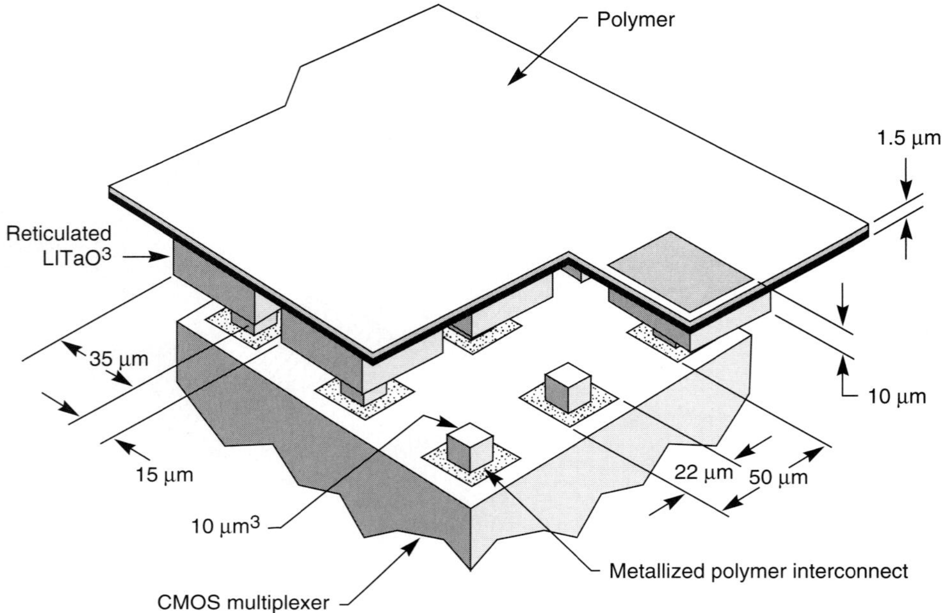

FIGURE 6 Hybrid pyroelectric array structure.[33]

and ferroelectric effects depend upon the change in temperature. The Texas Instruments BST array, incorporating 80,000 pixels, each about $50 \times 50\ \mu\text{m}$, which are matrix addressed, has an NETD of about 0.05°C (with F/1 optics).

Thermoelectric Arrays

Thermoelectric arrays prepared by silicon micromachining have been described by Choi and Wise.[38] Series-connected, thin-film thermocouples, i.e., a thermopile, are prepared on a silicon microstructure, the "hot" junctions (receiving the thermal radiation) on a silicon

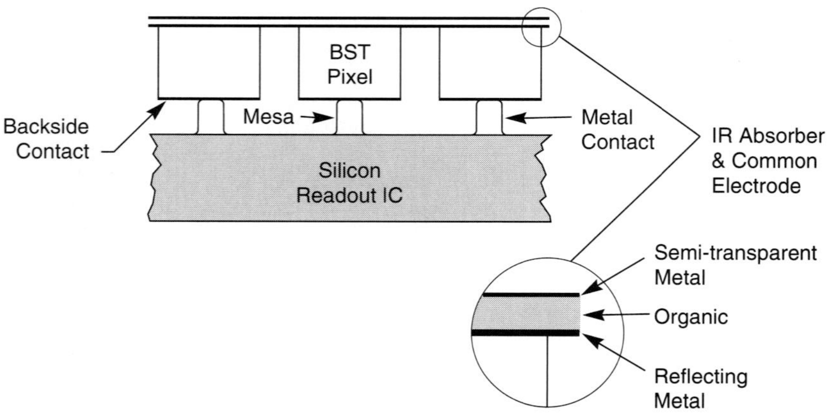

FIGURE 7 Ferroelectric bolometer hybrid array.[36,37]

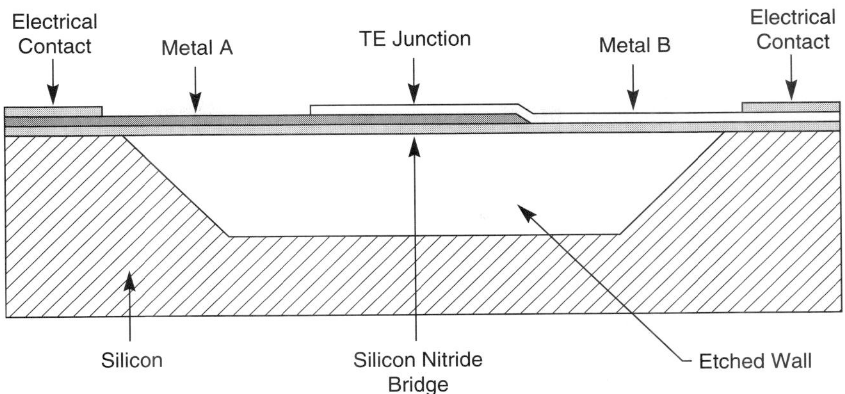

FIGURE 8 Monolithic thermoelectric array.[39] (*© 1991 Instrument Society of America. Reprinted with permission from the Symposium for Innovation in Measurement Science.*)

nitride/silicon dioxide membrane and the "cold" shielded junctions on the surrounding silicon substrate. Both 64- and 96-pixel microthermopile linear arrays in silicon microstructures have been prepared by Honeywell,[39] each microthermopile consisting of several nickel iron/chromium thermocouples connected in series, as shown in Fig. 8. The "hot" junctions are deposited on silicon nitride membranes, whereas the "cold" junctions are on the silicon substrates. A camera incorporating the linear array has been employed to image-moving targets such as automobiles. With an F/0.73 lens, the measured NETD is 0.10°C.

19.4 REFERENCES

1. P. W. Kruse, L. D. McGlauchlin, and R. B. McQuistan, *Elements of Infrared Technology,* John Wiley and Sons, New York, 1962.
2. R. D. Hudson and J. W. Hudson, *Infrared Detectors,* Halsted Press, 1975; and F. R. Arams, *Infrared to Millimeter Wave Detectors,* Artech House, 1973.
3. E. H. Putley, "Thermal Detectors," in *Optical and Infrared Detectors,* chap. 3, R. J. Keyes (ed.), Springer-Verlag, 1980.
4. R. A. Smith, F. E. Jones, and R. P. Chasmar, *The Detection and Measurement of Infrared Radiation,* Oxford University Press, 1968.
5. R. C. Jones, "The Ultimate Sensitivity of Radiation Detectors," *J. Opt. Soc. Am.* **37:**879 (1974).
6. S. Nudelman, "The Detectivity of Infrared Detectors," *Appl. Optics* **1:**627 (1962).
7. F. J. Low, "Low Temperature Germanium Bolometer," *J. Opt. Soc. Am.* **51:**1300 (1961).
8. S. Zwerdling, R. A. Smith, and J. P. Thierault, "A Fast High Responsivity Bolometer for the Very Far Infrared," *Infrared Physics* **8:**271 (1968).
9. N. Coron, "Infrared Helium Cooled Bolometers in the Presence of Background Radiation: Optimal Parameters and Ultimate Performance, *Infrared Physics* **16:**411 (1976).
10. N. Coron, G. Dambier, and J. Le Blanc, *Infrared Detector Techniques for Space Research,* V. Manno and J. Ring (eds.), Reidel, 1972.
11. N. Coron, G. Dambier, J. Le Blanc, and J. P. Moliac, "High Performance, Far Infrared Bolometer Working Directly in a Helium Bath," *Rev. Sci. Instr.* **46:**492 (1975).
12. W. H. Andrews, R. M. Milton, and W. De Sorbo, "A Fast Superconducting Bolometer," *J. Opt. Soc. Am.* **36:**518 (1946).

13. R. M. Milton, "A Superconducting Bolometer for Infrared Measurements," *Chem. Rev.* **39:**419 (1946).
14. N. J. Fuson, "The Infrared Sensitivity of Superconducting Bolometers," *J. Opt. Soc. Am.* **38:**845 (1948).
15. G. Gallinaro and R. Varone, "Construction and Calibration of a Fast Superconducting Bolometer," *Cryogenics* **15:**292 (1975).
16. K. B. Bhasin and V. O. Heinen (eds.), "Superconductivity Applications for Infrared and Microwave Devices," *Proceedings SPIE* **1292** (1990). (Includes many other references.)
17. W. S. Boyle and K. F. Rodgers, *J. Opt. Soc. Am.* **49:**66 (1959).
18. D. F. Hornig and B. J. O'Keefe, "Design of Fast Thermopiles and the Ultimate Sensitivity of Thermal Detectors, *Rev. Sci. Instr.* **18:**7 (1947).
19. P. B. Felgett, "Dynamic Impedance and the Sensitivity of Radiation Thermocouples, *Proc. Phys. Soc.* **B62:**351 (1949).
20. M. J. E. Golay, "Theoretical Considerations in Heat and Infrared Detection with Particular Reference to the Pneumatic Detector," *Rev. Sci. Instr.* **18:**347 (1947); "Pneumatic Infrared Detector," ibid., **18:**357 (1947); "Theoretical and Practical Sensitivity of the Pneumatic Detector," ibid., **20:**816 (1949).
21. J. R. Hickey and D. B. Daniels, "Modified Optical System for the Golay Detector," *Rev. Sci. Instr.* **40:**732 (1969).
22. M. Chatanier and G. Gauffre, *IEEE Transactions Instr. and Meas. IMEE* **179,** 1973.
23. A detector once made by Patterson Moos and cited by R. DeWaard and E. Wormser, "Description and Properties of Various Thermal Detectors, *Proc. IRE* **47:**1508 (1959).
24. E. H. Putley, *Semiconductors and Semimetals,* vol. 5, R. K. Willardson and A. C. Beer (eds.), Academic Press, 1970, chap. 6, "The Pyroelectric Detector"; vol. 12, 1977, chap. 7.
25. Zu-Sheng Wang and Jian-Qi Zhang, "The Mechanism of Type II Pyroelectric Detectors," *Infrared Physics,* vol. 33, no. 6, 1993, 481–486.
26. R. A. Wood, C. J. Han, and P. W. Kruse, "Integrated Uncooled Infrared Detector Imaging Array," *Proc. of the 1992 IEEE Solid State Sensor and Actuator Workshop,* Hilton Head Island, S.C., pp. 132–135.
27. R. A. Wood, "Uncooled Thermal Imaging with Monolithic Silicon Focal Plane Arrays," *Proc. SPIE* **2020:**Infrared Technology XIX (1993).
28. B. R. Johnson, T. Ohnstein, C. J. Han, R. Higashi, P. W. Kruse, R. A. Wood, H. Marsh, and S. B. Dunham, "High-T_c Superconductor Microbolometer Arrays Fabricated by Silicon Micromachining," *IEEE Trans. Appl. Superconductivity* **3:**2856 (1993).
29. B. R. Johnson and P. W. Kruse, "Silicon Microstructure Superconducting Microbolometer Infrared Arrays," *Proc SPIE* **2020:**Infrared Technology XIX (1993).
30. E. H. Putley, "The Pyroelectric Detector," *Semiconductors and Semimetals,* vol. 5, *Infrared Detectors,* R. K. Willardson and A. C. Beer (eds.), Academic Press, New York, 1970.
31. P. A. Manning, D. E. Burgess, and R. Watton, "A Linear Pyroelectric Array IR Sensor," *Proc. SPIE* **590:**2 (1985).
32. R. Watton and M. V. Mansi, "Performance of a Thermal Imager Employing a Hybrid Pyroelectric Detector Array with MOSFET Readout," *Proc. SPIE* **865:**79 (1987).
33. N. Butler and S. Iwasa, "Solid State Pyroelectric Imager," *Proc. SPIE* **1685:**146 (1992).
34. GEC-Marconi Materials Technology Ltd., 9360 Ridgehaven Court, San Diego, CA 92123.
35. B. E. Cole, R. D. Horning, and P. W. Kruse, "$PbTiO_3$ Deposited by an Alternating Dual-Target Ion-Beam Sputtering Technique," *Ferroelectric Thin Films II,* A. I. Kingon, E. R. Myers and B. Tuttle (eds.), *Materials Research Society Symposium Proceedings,* **243:**185 (1992).
36. C. Hanson, H. Beratan, R. Owen, M. Corbin, and S. McKenney "Uncooled Thermal Imaging at Texas Instruments," *Infrared Detectors: State of the Art, Proc. SPIE* **1735:**17 (1992).
37. C. M. Hanson, "Uncooled Ferroelectric Thermal Imaging," *Proc. SPIE* **2020:** Infrared Technology XIX (1993).
38. I. H. Choi and K. D. Wise, "A Silicon-Thermopile-Based Infrared Sensing Array for Use in Automated Manufacturing," *IEEE Trans. on Electron Devices* **ED-33:**72 (1986).
39. M. Listvan, M. Rhodes, and M. L. Wilson, "On-line Thermal Profiling for Industrial Process Control," *Proceedings of the Instrument Society of America, Symposium for Innovation in Measurement Science,* Geneva, N.Y., August 1991.

P · A · R · T · 6

IMAGING DETECTORS

CHAPTER 20
PHOTOGRAPHIC FILMS

Joseph H. Altman
Institute of Optics
University of Rochester
Rochester, New York

20.1 GLOSSARY

A	area of microdensitometer sampling aperture
a	projective grain area
D	optical transmission density
D_R	reflection density
DQE	detective quantum efficiency
$d(\mu)$	diameter of microdensitometer sampling aperture stated in micrometers
E	irradiance/illuminance (depending on context)
$\mathcal{G}$	Selwyn granularity coefficient
g	absorbance
H	exposure
IC	information capacity
M	modulation
M_θ	angular magnification
m	lateral magnification
NEQ	noise equivalent quanta
$P(\lambda)$	spectral power in densitometer beam
Q'	effective Callier coefficient
q	exposure stated in quanta/unit area
R	reflectance
S	photographic speed
$S(\lambda)$	spectral sensitivity
S/N	signal-to-noise ratio of the image

T	transmittance
$T(\nu)$	modulation transfer factor at spatial frequency ν
t	duration of exposure
$WS(\nu)$	value of Wiener (or power) spectrum for spatial frequency ν
γ	slope of D-log H curve
ν	spatial frequency
$\rho(\lambda)$	spectral response of densitometer
$\sigma(D)$	standard deviation of density values observed when density is measured with a suitable sampling aperture at many places on the surface
$\sigma(T)$	standard deviation of transmittance
$\phi(\tau)$	Autocorrelation function of granular structure

20.2 STRUCTURE OF SILVER HALIDE PHOTOGRAPHIC LAYERS

The purpose of this chapter is to review the operating characteristics of silver halide photographic layers. Descriptions of the properties of other light-sensitive materials, such as photoresists, can be found in Ref. 4.

Silver-halide-based photographic layers consist of a suspension of individual crystals of silver halide, called *grains,* dispersed in gelatin and coated on a suitable "support" or "base." The suspension is termed an *emulsion* in the field. The grains are composed of AgCl, AgClBr, AgBr, or AgBrI, the listing being in order of increasing sensitivity. Grain size ranges from less than 0.1 μm ("Lippmann" emulsions) to 2–3 μm, depending on the intended use of the coating. The number of grains per square centimeter of coating surface is usually very large, of the order of 10^6–10^8 grains/cm^2. The weights of silver and gelatin coated per unit area of support vary depending on intended use; usually both fall in the range 1–10 gms/m^2. The silver-to-gel ratio may also vary depending on intended use. Typically, the emulsion may be about 30 to 40 percent silver by weight, but some special-purpose materials, such as films to record Schumann-wavelength-region radiation, contain very little gelatin.

For modern materials, both the emulsion and the coating structure can be very complex. The emulsion layer is much more than silver halide in gelatin, containing additional agents such as hardeners, antifoggants, fungicides, surfactants, static control agents, etc. Likewise, the coating structure may be very complex. Even some black-and-white materials consist of layers of two different emulsions coated one over the other, and a thin, clear layer of gelatin is often coated over the emulsion(s) to provide some mechanical protection. In the case of color films, as many as 15 layers may be superimposed, some of them of the order of 1 μm thick. The thickness of the complete coating varies from about 3 μm to about 25 μm in normal films.

Commercially available emulsions are coated on a variety of glass, plastic (film), and paper supports (or "bases"). Glass is used for mechanical rigidity, spatial stability, or surface flatness.

Two different types of plastic are available commercially as film supports: cellulose acetate and polyethylene terephthalate (trade names "Cronar," "Estar," and "Mylar"). Of the two types, Mylar is superior in strength, flexibility, and spatial stability. However, the material is birefringent and its physical properties may be different in orthogonal directions. Also, these directions may not be aligned with the length or width of the sample. The anisotropic properties arise from the method of manufacture. Although not as tough as Mylar, cellulose acetate is, of course, fully adequate for most purposes. Also, this material is isotropic and easier to slit and perforate. Typical supports for roll films are around 4 mils (102 μm) thick, and for sheet films, 7 mils (178 μm), and other thicknesses are available. Most films are also coated on the back side of the support. The "backing" may be a layer of clear gelatin applied for anticurl protection, or of gelatin dyed with a dye that bleaches during processing, and provides both anticurl and antihalation protection. Lubricants and antistatic agents may also be coated, either on the front or back of the film. Properties of supports are discussed in Ref. 1.

20.3 GRAINS

The grain is the radiation-sensing element of the film or plate. It is a face-centered cubic crystal, with imperfections in the structure. For the most part, the grains act as independent receptors. In general, the larger grains are faster. The properties of the individual grains are controlled by the precipitation conditions and the after-precipitation treatment. Details of these matters are proprietary, but some discussion is given in Refs. 2 and 3. From the user's standpoint, the important fact is that when the grain is exposed to sufficient radiation it forms a "latent-image speck" and becomes *developable* by a solid-state process called the "Gurney-Mott mechanism." An excellent review of grains and their properties is given by Sturmer and Marchetti in chap. 3 of Ref. 4.

20.4 PROCESSING

The exposed halide layer is converted to a usable image by the chemical processes of development and fixation.

Development consists of reducing exposed crystals from silver halide to metallic silver, and a developing agent is an alkaline solution of mild reducer that reduces the grains having latent image specks, while not attacking the unexposed grains. Generally, once development starts the entire grain is reduced if the material is allowed to remain in the developer solution. Also, in most cases adjacent grains will not be affected, although developers can be formulated that will cause adjacent grains to be reduced ("infectious developers"). The number of quanta that must be absorbed by a grain to become developable is relatively small, of the order 4–40, while the developed grain contains on the order of 10^6 atoms. The quantum yield of the process is thus very high, accounting for the speed of silver-based materials.

The remainder of the process consists essentially of removing the undeveloped halide crystals which are still light-sensitive. The "fixer" is usually an acid solution of sodium thiosulfate $Na_2S_2O_3$, called "hypo" by photographers. The fixing bath usually serves as a gelatin hardener also. The thiosulfate reacts with the halide of the undeveloped grains to form soluble silver complexes, which can then be washed out of the emulsion layer. It is worth noting that proper washing is essential for permanent images. Additional treatments to promote permanence are available. Processing is discussed in detail in Refs. 2, 3, and 5, and image permanence in Ref. 6.

The exposed and processed silver halide layer thus consists of an array of grains of metallic silver, dispersed in a gelatin matrix. In color films, the silver is removed, and the "grains" are tiny spheres of dyed gelatin (color materials will be discussed below). Either type of grain acts as an absorber; in addition, the metallic silver grains act as scatterers.

The transmittance or reflectance of the layer is thus reduced and, from the user's standpoint, this change constitutes the response of the layer.

20.5 EXPOSURE

From fundamental considerations it is apparent that the dimensions of exposure must be energy per unit area. Exposure H is defined by

$$H = Et \tag{1}$$

where t is the time for which the radiation is allowed to act on the photosensitive layer, and therefore E must be the irradiance on the layer. The symbol H is used here for exposure in accordance with international standards, but it should be noted that in many publications, especially older ones, E is used for exposure and I for irradiance, so that the defining expression for exposure becomes $E = It$.

Strictly speaking, H and E in Eq. (1) should be in radiometric units. However, photographic exposures are customarily stated not in radiometric but in photometric units. This is done mostly for historical reasons; the English scientists Hurter and Driffield, who pioneered photographic sensitometry in 1891, measured the incident flux in their experiments in lumens/m^2, or *lux*. Their unit of exposure was thus the lux-second (old term, meter-candle-second). Strictly speaking, of course, weighting the incident flux by the relative visibility function is wrong or at least unnecessary, but in practice it works well enough because in most cases the photographer wishes to record what he or she sees, i.e., the visible spectrum. Conversion between radiometric and photometric units is discussed by Altman, Grum, and Nelson.[7]

Also, it should be noted that equal values of the exposure product (Et) may produce different outputs on the developed film because of a number of *exposure effects* which are described in the literature.[8] A complete review of radiometry and photometry is given in Chap. 24 of this *Handbook*.

20.6 OPTICAL DENSITY

As noted above, the result of exposure and processing is a change in the transmittance or reflectance of the layer. However, in photography, the response is usually measured in terms of the *optical density,* hereafter called the "density" in this chapter. For films (transmitting samples), density is defined by

$$D = -\log T = \log 1/T \tag{2}$$

where T is the transmittance. (Note: throughout this chapter, "log" indicates the base-10 logarithm.) For either silver grains or color grains, the *random dot* model of density predicts that

$$D = 0.434\, nag \tag{3}$$

where n = the number of grains per unit area of surface
a = the average projective grain area
g = the absorbance of the grain

Absorbance in turn is defined as $g = 1 - (T + R)$ where T and R are the transmittance and reflectance of the grain. For silver grains, the absorbance is taken as unity. The above expression, sometimes known as "Nutting's law," is based on a geometric approach, and

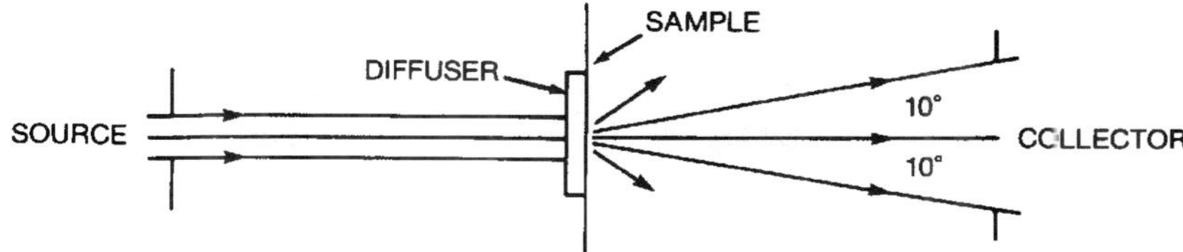

FIGURE 1 Optical system for measuring ISO/ANSI diffuse density with a 20° collection angle. (*Reprinted courtesy Eastman Kodak Co.* © *Eastman Kodak Co.*)

does not take into account any scattering by the grains. However, of course, opaque silver particles on a clear background will act as scatterers, and in fact multiple scattering usually occurs in developed silver layers. This produces an increase in the density of such layers by a factor of 2–3× from that predicted by Nutting's law. For color films, the refractive index of the gelatin in the dyed spheres is only negligibly different from that of the surround so that such layers are not scatterers. Even in the case of silver films, however, Nutting's law provides a useful model. Since for a given population of grains a and g will remain effectively constant, the law states that density should vary as the number of grains per unit area of surface. This fact is easily verified with a microscope.

Transmission Density

Transmission density is measured in a densitometer. It is worth noting that the device actually measures the transmittance of the sample and then displays the negative log of the result. In a normal or *macro*densitometer the sampling aperture area A is typically 1 mm^2 or more in size. When A is small, say, 0.1 mm^2 or less, the device becomes a *micro*densitometer. Microdensitometers present special problems and will be discussed below.

Because the scattered light may not reach the sensor of the densitometer when silver layers are measured, it is necessary to specify the angular subtenses of both the incident (influx) and emergent (efflux) beams at the sample. Clearly, if scattered light is lost to the sensor, the indicated density of the sample will *increase.* Four types of transmission density are described in an ISO standard,[9] of which two are principally important to the user. The first of these is *diffuse density,* which is diagrammed in Fig. 1. As can be seen, a collimated incident beam illuminates an opal glass diffuser. The emulsion side of the sample is placed in contact with this diffuser, and the flux contained within a cone angle of $\pm 10°$ is collected and evaluated by the sensor. The reverse of this arrangement yields the same density values and is also permitted by the standard. This is the type of density normally measured in practice. Physically, it corresponds to the conditions of contact printing.

The other case that is important in practice is projection density, which is diagrammed in Fig. 2. This case simulates the use of the layer in an optical system. As the figure shows,

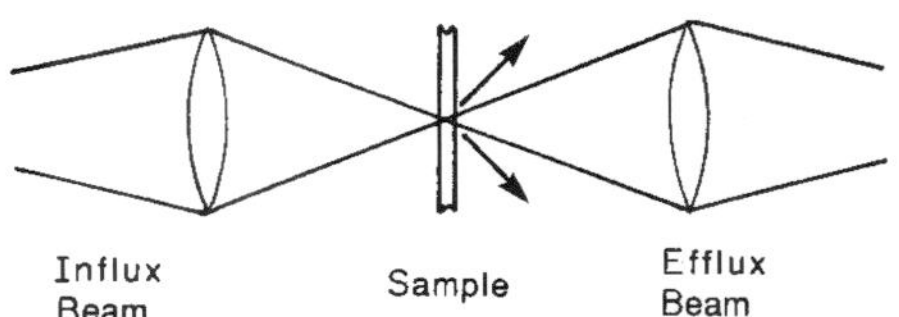

FIGURE 2 Optical conditions for projection density measurement.

light is lost to the efflux system in projection density, the exact amount depending on the numerical aperture of the optics involved and the scattering characteristics of the sample. Thus, the projection density of a silver film is usually greater than the diffuse density. The effective Callier coefficient Q' may be defined by

$$Q' = \frac{\text{projection density}}{\text{diffuse density}} \tag{4}$$

This factor can be measured experimentally. For silver films and f/2 optics, $Q' \approx 1.3$; for color films $Q' \approx 1.0$.

Nonneutral (Color) Density

In many cases, silver densities can be treated as neutrals. For dye densities, i.e., color films, measured density depends on the spectral characteristics of both the dye and the densitometer. The spectral response of the instrument is given by

$$\rho_\lambda = P_\lambda S_\lambda F_\lambda \tag{5}$$

where ρ_λ = the response at wavelength λ
P_λ = the power in the densitometer beam at λ
S_λ = the spectral sensitivity of the sensor at λ
F_λ = the transmittance of the densitometer optics at λ, specifically including any filters placed in the densitometer beam

The measured density of a nonneutral layer is then

$$D = \log \left[\frac{\int_{\lambda_1}^{\lambda_2} \rho_\lambda \, d\lambda}{\int_{\lambda_1}^{\lambda_2} \rho_\lambda T_\lambda \, d\lambda} \right] \tag{6}$$

where T_λ is the transmittance of the layer at wavelength λ, and the wavelength limits are set by the distributions. The response ρ_λ of the system is adjusted to be equal to that of the readout device with which the film is to be used.

Thus, for example, if the sample is to be viewed by an observer, ρ_λ is made equal to the visibility function, and the resulting measurement is called visual density, etc. Instrument responses have been standardized for sensitometry of color films.[10]

Reflection Density

When the emulsion is coated on paper the density is measured by reflection. Reflection density is then defined by

$$D_R = -\log R \tag{7}$$

where R is reflectance, measured under suitable geometric conditions. The measurement of reflection density is also described in the standards literature.[11]

20.7 THE D-LOG H CURVE

In routine sensitometry, samples receive a series of exposures varying by some constant factor, such as $x2$ or $x\sqrt{2}$. After processing, the measured densities are plotted against the common logarithm of the exposures that produced them. The resulting curve is known as the "D-log H curve," or the "H & D" curve (after Hurter and Driffield, the previously mentioned pioneers in the field). A typical D-log H curve is shown in Fig. 3.

As shown, the curve is divided into three regions, known as the "toe," "straight-line portion," and "shoulder," respectively. The fact that an appreciable straight-line portion is found in many cases is not an indication that the film is a linear responder in this region because, of course, both axes of the plot are logarithmic. It is worth noting here that the equation of the straight-line portion of this curve can be written

$$D = \gamma(\log H - \log C) \tag{8}$$

where γ is the slope of the straight-line portion and C is the exposure at the point where the extrapolated straight line cuts the exposure axis. Taking antilogarithms, Eq. (8) becomes

$$T = \left(\frac{H}{C}\right)^{-\gamma} \tag{8a}$$

If $\gamma = -1$, $T = (1/C)H$, and for this special case, the system becomes linear over the exposure range corresponding to the straight-line portion. A negative value of γ indicates a *positive* image.

A number of useful performance parameters for films are taken from their D-log H curves, as follows.

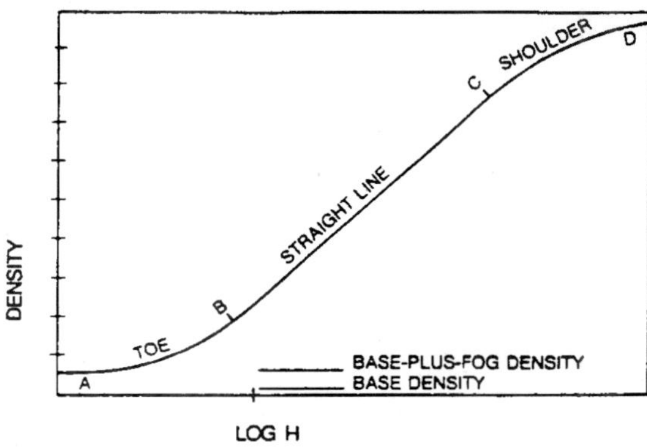

FIGURE 3 Typical D-log H curve for a negative photographic material. (*Reprinted courtesy Eastman Kodak Co. © Eastman Kodak Co.*)

1. *Fog:* For most films, a certain number of grains will be reduced even though they have received no exposure at all, or insufficient exposure to form a latent image speck. The resulting density is called *fog.* Since it is not exposure-related, and since it tends to veil any information recorded in the toe, excessive fog is very undesirable. For many purposes, the fog density plus the density of the support are subtracted from the gross density to give the value of the net density resulting from the exposure. More complicated formulas for correcting the film's response for fog grains have been proposed, but are rarely used.

2. *Gamma:* Traditionally, the slope of the straight-line portion of the D-log H curve is called the "gamma." Gamma is a crude measure of the contrast with which the original object is reproduced; it would be a good measure of this contrast if the object were in fact recorded entirely on the straight-line portion of the response curve. However, for many purposes, notably pictorial photography, an appreciable part of the toe is used. This fact led Niederpruem, Nelson, and Yule to propose the use of an average gradient that included part of the toe as a measure of the contrast of the reproduction.[12] This quantity is called the *contrast index.* Since it includes part of the toe, contrast index is less than gamma. Since information is often recorded in the toe, it is convenient to generalize the meaning of γ to refer to the gradient anywhere along the D-log H curve, and this is done in this chapter. Note that in this case, the traditional gamma is the maximum value the gradient attains.

3. *Latitude:* Latitude can be defined as the log exposure range between the point in the toe and the point in the shoulder between which the gradient is equal to or greater than the minimum value required for acceptable recording. Clearly, the latitude of the film must be at least equal to the log exposure range of the object for proper recording. In many practical cases the film's latitude easily exceeds the required range. Note that the latitude is determined in part by the maximum density that the film can produce.

4. *Speed:* Speed is defined by the general expression

$$S = \frac{K}{H_{\text{ref}}} \tag{9}$$

where K is a proportionality factor and H_{ref} is the exposure required to produce some desired effect. Since the desired effect varies depending on the type of film and the application, H_{ref} also varies. Also, H_{ref} can be stated in either radiometric or photometric units. If H_{ref} is given in radiometric units, the proportionality factor K is set to unity, and the resulting values are termed "radiometric speeds." Although radiometric speeds are the fundamental speed values, they are rarely used in practice because, as previously noted, exposures are usually given in photometric units. In this case, the factor K takes on different values that depend not only on H_{ref}, but also on the characteristics of the exposure meter, which is standardized.[13] Varying the factor K allows a single meter to be used with all kinds of films and applications, which is a practical necessity. Thus, for example, the photometric speed (usually simply the "speed") of black-and-white pictorial films is evaluated from

$$S = \frac{0.8}{H_{0.1}} \tag{9a}$$

where $H_{0.1}$ is the exposure in lux-seconds required to produce a density of 0.1 above the densities of the base plus fog in a specified process. This density level has been shown empirically to be predictive of excellent tone reproduction quality in the print. Similarly, for color slide films

$$S = \frac{10}{H_m} \tag{10}$$

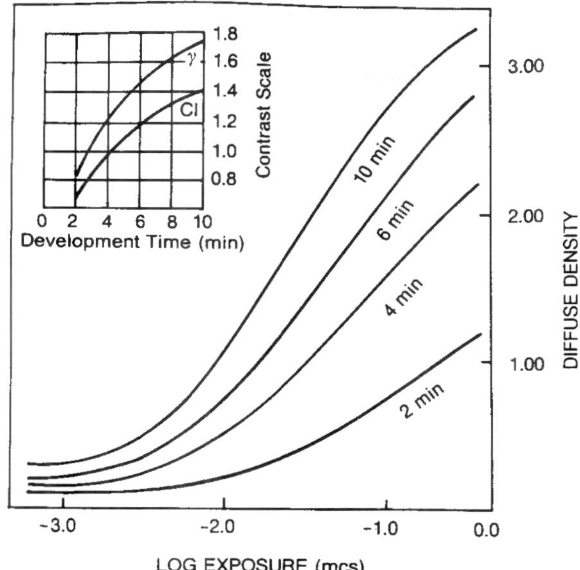

FIGURE 4 A family of characteristic curves for development times as shown, with corresponding plots of contrast index and gamma. (*Reprinted courtesy Eastman Kodak Co. © Eastman Kodak Co.*)

where H_m is the exposure to reach a specified position on the film's D-log H curve. Again, H_m was established empirically. The above two examples show how the two factors involved in determining a photometric or practical speed can vary. A number of other speeds have been defined and are described in the literature.[14]

Variation of Sensitometry with Processing

The rate of reduction of the exposed photographic grain depends on the characteristics of the grain itself, the formulation of the developer, and its temperature. In general, the reaction is allowed to continue until substantially all exposed grains have been reduced, and ended before fog becomes excessive. Many modern films are hardened and able to withstand processing temperatures up to, say, 40°C. Development times are often chosen on the basis of convenience and usually run in the order of 5–10 min in nonmachine processing. As development time and/or temperature are increased, the amount of density generated naturally increases. Thus for a given film a whole family of response curves can be produced, as shown in Fig. 4. As development time is lengthened, gamma and contrast index also increase. Typical behavior of these parameters is shown in the inset of Fig. 4.

20.8 SPECTRAL SENSITIVITY

The spectral absorption of the silver halide grain extends only to about 500 nm, and thus the inherent sensitivity of the grain is limited to regions shorter than that limit. However, it was discovered by Vogel in 1873 that sensitivity could be extended to longer wavelengths

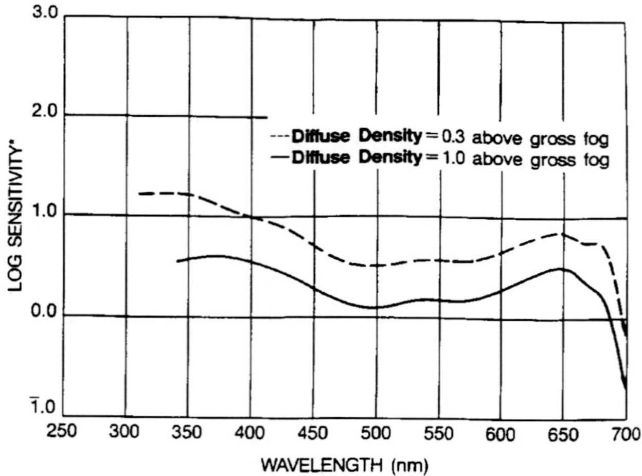

FIGURE 5 Spectral sensitivity curves for a modern negative material sensitized to about 690 nm. (*Reprinted courtesy Eastman Kodak Co. © Eastman Kodak Co.*)

by dyeing the grains, and over the years the effective sensitivity range was extended into the infrared. Presently, materials are available usefully sensitized out to about 1.2 μm, but it should be noted that IR materials tend to have poor shelf life, and may require special handling. For our purposes, we may say that the sensitizing dye absorbs the incident energy in the required spectral region and transfers the energy to the grain in a manner that produces the required latent-image speck. The mechanisms are discussed in Ref. 3, chap. 10.

The spectral sensitivity of a photographic layer is usually specified by a family of curves showing $\log(1/H_D)$ vs λ, where H_D is the exposure in ergs/cm^2 of wavelength λ required to produce some stated density. Spectral sensitivity values are thus radiometric speeds. Typical curves are shown in Fig. 5. In practical work, three broad classes of sensitization are recognized, which are called "color-blind" (or blue-sensitive), orthochromatic (additionally sensitized to green), and panchromatic (additionally sensitized to green and red). Most modern general-purpose materials are panchromatic.

In general, the shape of the spectral sensitivity curve follows that of the spectral absorption of the layer. It should also be noted that the gradient of the D-log H curve will be affected by the absorption of the layer. Gamma may therefore vary as a function of wavelength, and is generally somewhat lower in the blue and UV regions of the spectrum. This means that if the material is being used as a radiometer, it must be calibrated at the wavelength(s) of interest.

20.9 RECIPROCITY FAILURE

It was noted above that the exact response of a photographic layer may change due to exposure effects. Of these, the phenomenon of *reciprocity failure* is the most important in practical photography.

By definition [Eq. (1)] exposure is simply the product of the irradiance and time, and nothing in this definition specifies the magnitude of either factor. However, the developed *density* resulting from a given calculated exposure is often found to depend on the *rate* at which the radiation is supplied, all other factors being held constant. Broadly speaking, exposure times of about 0.01–1.0 sec are most efficient in producing density, the exact values depending on the film involved. Times much outside the above range tend to produce lower density for the same calculated exposure. The emulsion-maker has some ways of minimizing the effect, and usually attempts to optimize the response for the exposure times expected to be used with the material. Reciprocity failure is discussed in detail in Ref. 3, chap. 4, sec. II.

The loss of efficiency for short-time and correspondingly high-irradiance exposures is termed "high-intensity reciprocity failure," and that for long exposure times (low irradiances) is termed "low-intensity reciprocity failure." The Gurney-Mott mechanism explains both types of failure well. Note also that the names are misnomers; the terms should, of course, be high and low *irradiance.*

Only limited data are available, but the gradient of the D-log H curve tends to decrease as exposure times are shortened. An example is shown in data published by Hercher and Ruff.[15] Limited data also indicate that the speed loss due to high-intensity failure stabilizes for times shorter than about 10^{-5} sec.[16] Essentially, the amount of reciprocity failure is independent of exposing wavelength.[17] This is to be expected.

Reciprocity failure may be a considerable problem for photographers working in specialized time domains, such as oscilloscope photography or astronomy. Astronomers have been able to devise user treatments for minimizing low-intensity failure.[18] For the practical photographer, reciprocity failure sometimes appears as a problem in color photography. If the RF of the three sensor layers differs, the resulting picture may be "out of balance," i.e., grays may reproduce as slightly tinted. This is very undesirable, and correction filter recommendations are published for some films for various exposure times.

20.10 DEVELOPMENT EFFECTS

Beside exposure effects the final density distribution in the developed image may be affected by "development effects" arising from chemical phenomena during development. Various names such as "border effect," "fringe effect," "Eberhard effect," etc., are applied to these phenomena; what we shall here term "edge effect" may be important in practice.

Consider a sheet of black-and-white film developing in a tray, and assume for purposes of discussion that there is no motion of the developer. Since developing agent must be oxidized as halide is reduced, and since the by-products of this reaction may themselves be development inhibitors, it can be seen that local variations of developer activity will be produced in the tray, with the activity decreasing as density increases. Agitation of the solution in the tray reduces the local variations, but usually does not eliminate them entirely, because it is the developer that has diffused into the gelatin matrix that is actually reacting. Now an "edge" is a boundary between high- and low-density areas, as shown in Fig. 6. Because of the local exhaustion and the diffusion phenomena, the variation of developer activity within the layer will be as shown by the dotted line in the figure. The result is that the developed density near the edge on the low-density side tends to decrease, and on the high-density side tends to increase, as also shown in the figure. In other words, the density distribution at the edge is changed; this actually occurs to some degree in much practical work and has interesting consequences, as will be discussed below.

The local exhaustion of the developer may also be important in color films where development in one layer (see below) may affect the response of an adjacent layer. In color photography, the phenomenon is called "interimage effect."

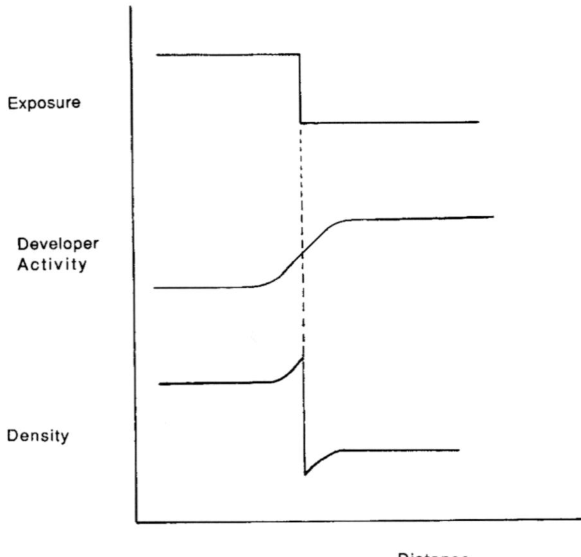

FIGURE 6 Distributions of developer activity and density resulting from a step-function input in the presence of edge effect.

20.11 COLOR PHOTOGRAPHY

Color photography has been extensively reviewed by Kapecki and Rodgers.[19] With one exception at the time of writing, all commercially available color films employ subtractive color reproduction. The exception is an instant film for color slides marketed by Polaroid Corporation, which employs additive color. The mechanism involves a reseau of very fine red, green, and blue stripes, which provide the color separation during the taking of the photograph, and also the color when the (reversed) image is projected.[20]

The basic structure of most other color films is similar to that shown in Fig. 7. The incoming light first encounters a nonspectrally-sensitized emulsion layer, which records the blue-light elements of the scene. The next layer is a yellow, or minus-blue filter, the purpose of which is to prevent any blue light from reaching the other two emulsion layers.

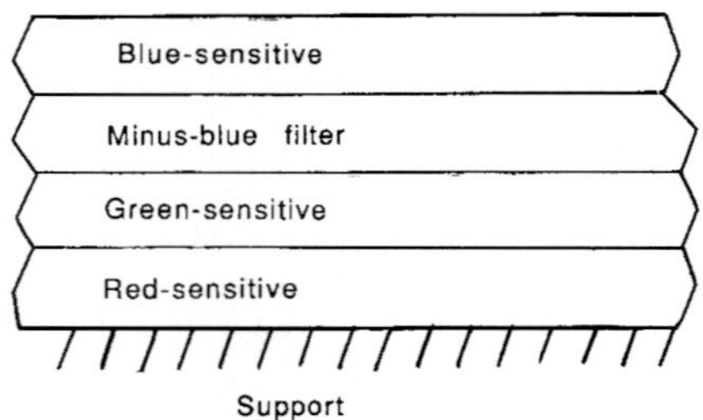

FIGURE 7 Schematic tripack structure of a color film.

This yellow filter is usually composed of "Carey Lea," or colloidal silver dispersed in gelatin. Such sols are yellow. The reason for using Carey Lea silver is that all metallic silver is removed from the film during processing anyway, and the necessary removal of the filter layer is thus accomplished automatically.

Moving downward in the stack, the next layer is an ortho-sensitized emulsion. Since any blue light has been blocked by the yellow filter, this layer records the green-light elements of the scene. The final layer in the stack is sensitized to red light but not to green light and this layer serves to record the red elements of the scene. Modern films usually contain many more than the four layers indicated here, but the operating principles of the film can be discussed in terms of such a "tripack." In accordance with the principles of subtractive color reproduction, the images in the three separation records are then converted to yellow, magenta, and cyan dyes, respectively, and the final image is composed of these dyes, without the developed silver, which is either removed chemically or left behind, depending on the exact material.

Within this broad framework, films can be separated into two classes: chromogenic and nonchromogenic. In the former class, the dyes are not coated in the film, but are formed during the processing by a reaction called "coupling." In coupling, the by-products of the halide reduction reaction serve as components for a second reaction in which dye is formed; the amount of dye thus increases as density increases. The components of the dye-forming reaction (i.e., other than the development byproducts) may be present in the developing solution (Eastman Kodak Co., "Kodachrome") or, more generally, coated within the various layers (Kodacolor, Polaroid "One Film," Agfachrome, Fujicolor, etc.). After the required dyes are formed, the developed silver is removed by a chemical process termed "bleaching."

The advantage of incorporating the couplers in the various layers of the tripack is simpler processing, but because of the additional material in the layers, such films tend to be not as sharp as the nonincorporated-coupler types. Chromogenic color films are available both as slide materials, in which the film undergoes a reversal process,[21] and as color negative—color print materials in which a dye negative is formed and then printed onto a color paper whose structure is fundamentally similar to that of the films.

The principal example of the nonchromogenic film is the Polaroid Instant Color Film. In this film the yellow, magenta, and cyan dyes are actually coated in the structure, along with the blue, green, and red-sensitive emulsions. When development takes place in a given layer, the corresponding dye is immobilized. The dye that has *not* been immobilized in the three layers migrates to a "receiver" layer, where it is mordanted. Since the amount of dye that migrates *decreases* as the original density *increases,* the result is a positive color image formed in the receiver. The material has been described in more detail in a paper by Land,[22] and also in chap. 6 of Ref. 4.

The image in a color film thus essentially consists of three superimposed dye images. Typical spectrophotometric curves for dyes formed by coupling reactions are shown in Fig. 8. The density of any one of these dye layers taken by itself is known as an *analytical* density. Note, however, that all the dyes show some "unwanted" absorption—that is, absorption in spectral regions other than the specific region that the dye is supposed to control. Thus the total density of the layer at any wavelength is the sum of the contributions of all three dyes; this type of density is known as "integral density." The integral density curve is also shown in Fig. 8.

The reproduction of color by photographic systems has been discussed by Hunt[23] and many others.[24] In general, exact colorimetric reproduction is not achieved, but for most purposes the reproduction of the hues and luminances in the original scene is satisfactory. Evans,[25] in fact, observes that under the right conditions the "magnitude of the reproduction errors that can be tolerated is astonishing." One aspect that is critical for many workers, especially expert photographers, is the ability of the system to produce good "balance," i.e., to reproduce a neutral as a neutral. This requirement is so important that one of the types of density that is measured for dye layers is the *equivalent neutral*

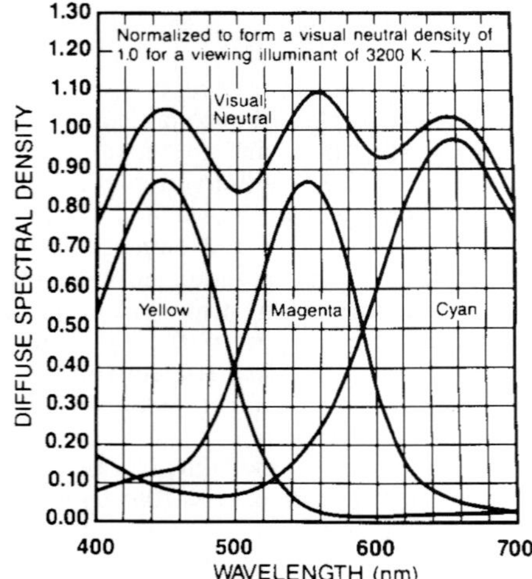

FIGURE 8 Spectral dye density curves for a color film. (*Reprinted courtesy Eastman Kodak Co.* © *Eastman Kodak Co.*)

density (END) which is defined as the visual neutral density that a dye layer would produce if combined with the correct amounts of the other two dyes (whatever those correct amounts may be). When the ENDs of the three layers are equal, the system is in balance, and color-film sensitometry is therefore often done in terms of ENDs. Further discussion of color sensitometry and densitometry may be found in Ref. 3, chap. 18.

20.12 MICRODENSITOMETERS

As indicated above, a microdensitometer is a densitometer designed to measure the density of a small area. The sampling apertures are typically slits which may be as narrow as 1–2 μm in nominal width. The sample is scanned over the aperture, creating a record of density as a function of position on the sample surface, i.e., distance.

In practical instruments, the small sampling aperture dimensions are achieved by projecting an enlarged image of the film onto a physical aperture. The optical system produces some effects not encountered in macrodensitometers, as follows.

1. In general, microdensitometers measure projection, or semispecular, density. As already noted, projection density is higher than diffuse density for silver layers, and the exact value of the effective Callier coefficient Q' depends in part on the numerical aperture of the optical system. Thus two microdensitometers fitted with optics of different NAs may give different density values for the same sample. Furthermore, macrodensity data are usually in terms of diffuse density, so that data from the microdensitometer should be corrected if intercomparisons are to be made. The effective Callier coefficient for the specific optics-sample combination is easily determined by measuring suitable areas of sample both in the microdensitometer and a macrodensitometer, and taking the ratios of the values.

2. The presence of stray light in the system tends to lower the measured density. This problem is especially troublesome in microdensitometry because of the types of images that are often encountered. Thus, for example, when an interface between clear and dense areas—that is, an edge—is scanned, stray light will distort the record in the manner shown

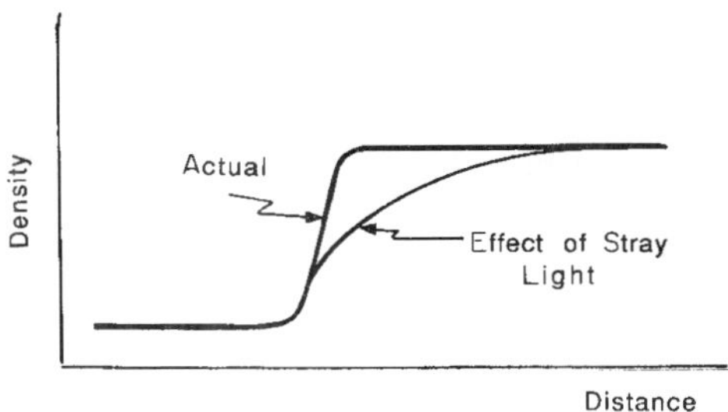

FIGURE 9 Effect of stray light in the microdensitometer on the apparent density distribution at an edge.

in Fig. 9. If the image is that of a star or spectroscopic line, this behavior results in an artifically low density reading. It is very important to control stray light as completely as possible in microdensitometry.

3. One feature of the usual microdensitometer optical system, installed for control of stray light, is the "preaperture," a field stop that limits the area of the sample that is illuminated. Because of this preaperture, and the optical system, a normal microdensitometer is a partially coherent system. This means that the instrument may respond to path-length differences in the sample as well as to density differences. This is undesirable since in practice it is the density differences that carry the information. Partial coherence in microdensitometers has been studied by Thompson and by Swing among others, and the results are summarized by Dainty and Shaw.[26] It has been shown that the coherence effects can be minimized by satisfying two conditions. In the first condition, the width W of the preslit

$$W \geq \frac{4\lambda}{NA_{\text{in}}} \tag{11a}$$

The second condition is

$$\frac{NA_{\text{in}}}{NA_{\text{eff}}} = 1 + \frac{\nu_s}{\nu_o} \tag{11b}$$

where λ = the wavelength of the light
NA_{in} = numerical aperture of the influx optics
NA_{eff} = numerical aperture of the efflux optics
ν_s = maximum spatial frequency in the sample
ν_o = spatial frequency cutoff of the scanning objective

If, for example $\nu_o = 3\nu_s$, then $NA_{\text{in}}/NA_{\text{eff}} \geq 1.3$. A microdensitiometer arranged to minimize coherence problems is called a *linear* microdensitometer. Note that the two conditions above conflict with conditions commonly adopted to control stray light.

20.13 PEFORMANCE OF PHOTOGRAPHIC SYSTEMS

The following discussion of performance is limited to those aspects which are properties of the system, and excludes such aspects as the skill of the photographer, etc. Of those aspects, which we may term "technical" quality parameters, the most important is *tone reproduction*. The subject is divided into two areas: subjective tone reproduction and objective tone reproduction. Subjective tone reproduction is concerned with the relation

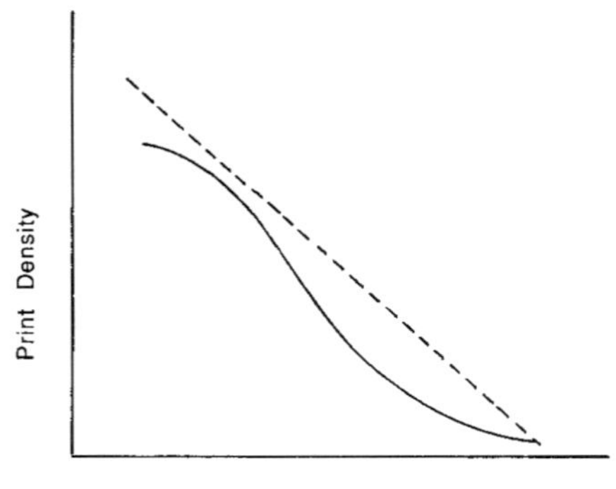

FIGURE 10 Preferred tone reproduction for viewing under typical room lighting conditions.

between the brightness sensations produced in the observer's mind when the scene is viewed and when the picture is viewed. Since the sensation of brightness depends markedly on the viewing conditions and the observer's state of adaptation, the subjective tone reproduction of a given picture is not constant, and this aspect of the general subject is not often measured in the photographic laboratory. It is discussed by Kowalski.[27] Objective tone reproduction is concerned with the reproduction of the luminance and luminance differences of the scene as luminances in the final output. Tone reproduction studies apply equally well to projected images, prints, transparencies, and video images, but a negative-print system is usually assumed for discussion. It is easy to show that the log luminance of a print area, $\log L_p = C - D_p$, where C is a constant determined by the illuminance on the print and D_p is the density of the print area. Thus tone reproduction curves are usually plotted in the form of the print density versus the log luminance of the corresponding scene element (Fig. 10). Although both the scene luminances and the print densities are fixed quantities, the viewer's reaction still depends on the illumination level at which the picture is seen.

It has been shown empirically that for paper prints viewed under typical "room lighting conditions," the preferred tone reproduction curve is the solid line in Fig. 10. Perfect objective tone reproduction, defined as the case where $\Delta D_p = -\Delta \log L_{sc}$ for all scene luminance levels, would be the dotted line in the figure. Thus under typical room lighting conditions viewers prefer a reproduction that has somewhat more contrast and less density than the "perfect" result. This preference will, however, change with illumination and stray light levels.

The exact shape of the tone reproduction curve obtained with a given system depends on the shape of the negative and positive D-log H curves, and also on the stray-light characteristic of the camera system. (Excessive stray light in the camera can be very deleterious to tone reproduction quality.) Using a graphic method devised by Jones and described by Nelson and others,[28] the effect of the D-log H curves and the stray light on the tone reproduction can be studied.

20.14 IMAGE STRUCTURE

The other two technical quality parameters of a photographic system are its sharpness and graininess, to use the most familiar terms for these properties. These properties are often lumped together under the general term "image structure." Actually, in photoscience the

TABLE 1 Performance Data for Types of Materials[1]

Type	SPD[2]	γ	Gran.[3]	ν_{50}[4]
B&W microfilm	80	3.0	6	7200
B&W very slow camera neg.	25	0.5–3.5	5–7	80–145
B&W slow camera neg.	100	0.5–1.1	8–9	65–120
B&W fast camera neg.	400	0.5–1.0	10–14	50–100
B&W very fast camera neg.	1600–3200	0.5–1.0	18	70
Color neg. very slow	25–50	0.65	4–5	40–60
Color neg. slow	100–160	0.60–0.80	4–6	30–70
Color neg. fast	400	0.65–0.80	5–7	25–40
Color neg. very fast	1000–1600	0.80	8–11	25–35
Color rev. slide very slow	25–50	1.8–2.3	9–10	30–40
Color rev. slide slow	100	2.0–3.0	10–13	25–30
Color rev. slide fast	400	2.0–2.4	15–20	20
Color rev. slide very fast	800–1600	2.2–2.8	22–28	16–20
Instant print films[5]		−1.7[6]	NA	3–4
—Black and white	3000	−1.6	NA	2.5
—Color	100			

[1] Data are as of early 1993 and were obtained from publications of the manufacturers listed in Sec. 20.21. They are presented as published. Note that products are frequently changed or improved.
[2] Speeds are calculated in different ways for various classes of product. The values given are suitable for use with standard exposure meters.
[3] Values represent 1000× the standard deviation of the diffuse density, measured at an average density of 1.0 using a 48-μm circular aperture. The exact granularity of a print depends on the characteristics of the print material and the printer as well as the granularity of the negative.
[4] Values show the spatial frequency at which the modulation transfer function is 50 percent.
[5] MTF values apply to the final print.
[6] Negative sign arises from the definition of gamma for the case of a positive image.

term sharp*ness* and graini*ness* are reserved for the subjective aspects of the phemomena, and measurement of these properties requires psychometric testing. Since such testing is expensive and time-consuming, objective correlates of both properties have been defined, and methods for measurement have been established. The objective correlate of sharpness is termed *acutance,* and of graininess is termed *granularity.* Image structure data for various kinds of materials are given in Table 1, along with speed and contrast values.

20.15 ACUTANCE

The original proposal for measuring acutance was made by Higgins and Jones[29] in 1952. It involved calculating a value from a microdensitometer trace of a test edge. However, the visual processes that occur when an observer views an edge are complex, and the straightforward calculation proposed by Higgins and Jones fails to predict the sensation of sharpness produced by some edge distributions. At about the same time, the optical transfer function (OTF) and related concepts began to be widely used in optics, and these were soon applied to photographic materials also.

The concepts of the point and line spread functions are essentially identical in optics and photoscience; in the photographic layer the smearing of the point (or line) input is

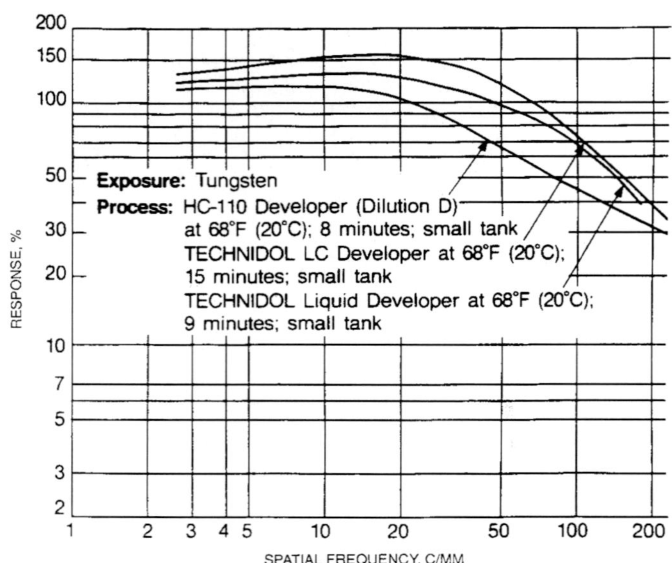

FIGURE 11 Modulation transfer functions of Kodak Technical Pan Film for three conditions of development. Note that the response at low frequencies exceeds 100%. (*Reprinted courtesy Eastman Kodak Co. © Eastman Kodak Co.*)

caused by diffraction around the grains, refraction through them, and reflection from them. These phenomena are usually lumped together and termed "scattering," and have been treated by Gasper and dePalma.[30] Likewise, the concept of the optical transfer function, or the Fourier transform of the LSF, is basically the same in optics and photography. However, three important differences should be noted for the photographic case. (1) The emulsion is isotropic, so that the PSF and LSF are always symmetrical, and the complex OTF reduces to the modulation transfer function (MTF) only. (2) Unlike lenses, photographic layers are stationary, but are generally nonlinear. Therefore, all data and calculations must be in terms of *exposure* or allied quantities. When the calculations are complete, the results are converted to density via the D-log H curve. (3) The presence of edge effects tends to raise the MTF curve, so that for low frequencies the measured response values are often found to be greater than 100 percent. The subject is treated in detail by Dainty and Shaw.[31] Typical MTF curves for a film, showing the overshoot, are given in Fig. 11. Data for MTF curves of various types of films are also given in Table 1. The value given shows the spatial frequency for which the transfer factor drops to 50 percent.

The chain relating the MTF curve to image sharpness is the same as in optics: a high MTF curve transforms to a narrow spread function, and this in turn indicates an abrupt transition of exposure—and therefore density—across the edge. Thus MTF response values greater than unity, although mathematically anomalous, indicate improved image sharpness, and this is found to be the case in practice. As a matter of fact, edge effects are often introduced deliberately to improve sharpness. This is done either by adding suitable compounds to the coating itself, or by adjusting the developer formulation. The process is similar to the electronic "crispening" often used in television.

An index of sharpness can be computed from the MTF data by a procedure first suggested by Crane and later modified by Gendron.[32] These workers were interested mainly in films, but they recognized that the film is one component of a system; e.g., a color slide system involves a camera lens, the film (and process), a projector lens, the screen, and the observer's eye. The MTF of the system is then the cascaded MTFs of these components. Gendron suggested that the area under the cascaded MTF be taken as the stimulus that produced the sensation of sharpness. A formula was proposed that produced a sharpness index scaled to 100. This index, now called CMT-Acutance (to distinguish it from the original Higgins-Jones acutance) has been found to correlate well with subjective data, and is used in the industry. Note that the treatment above has been simplified for the sake of brevity; details are available in Gendron's paper.

20.16 GRAININESS

The granular nature of the photographic image is one of its most significant characteristics. It may appear to the observer as an unpleasant roughness in what should be uniform areas, but it may also interfere with extracting information from the image. In the former case it is an aesthetic problem; in the latter case the structure is equivalent to noise in a communications channel. In either case it is desirable to measure the phenomenon objectively and in engineering terms.

The procedure now used for making these objective measurements was proposed by E. W. H. Selwyn in 1935.[33] He postulated that if the density of a uniformly exposed and processed layer were measured at many places using a suitable sampling aperture, the population of density values so obtained would be approximately gaussianly distributed around the mean. This being so, the variability for a given mean density is completely described by the standard deviation of the values. This quantity is termed the *rms-granularity*, $\sigma(D)$, and has indeed been shown to correlate with the subjective graininess.[34] It might be noted that calculating the standard deviation of the density is mathematically improper, since the underlying transmittance values are being multiplied To avoid this problem the rms-granularity is formally defined by

$$\sigma(D) = \frac{0.434\sigma(T)}{\bar{T}} \tag{12}$$

where T is transmittance. However, it can be shown that when $\sigma(T)$ is small compared to $\bar{T}$, the error involved in calculating directly in density is small, and this is often done in practice.

Selwyn also showed that the measured value of $\sigma(D)$ depended upon A, the area of the sampling aperture. The product $\sigma(D)A^{1/2}$ may be termed the *Selwyn coefficient* $\mathcal{G}$; for black-and-white films exposed to light, Selwyn showed that it should be constant, and this relation is called "Selwyn's law." Thus $\sigma(D)A^{1/2}$ is a measure of sample graininess no matter what the size of the sampling aperture. Unfortunately, the Selwyn coefficient does not remain constant with changes of aperture size for very important classes of samples. Selwyn's law may fail for prints and enlargements, black-and-white or color, for many color materials even if not enlarged, and also for radiographs, especially screened radiographs. For such materials $\sigma(D)$ still increases as A decreases, but not at a rate sufficient to keep the Selwyn coefficient constant, and it (the coefficient) is therefore not useful as an objective measure of graininess.

Stultz and Zweig[35] found that they obtained good correlation between $\sigma(D)$ and the sensation of graininess when the sampling aperture was selected in accordance with the rule

$$d(\mu) \cdot M_\theta \approx 515 \tag{13}$$

where $d(\mu)$ is the diameter of the sampling aperture in μm, and M_θ is the angular magnification[36] at which the photograph is seen by the viewer. M_θ is readily calculated from the relation

$$M_\theta = \frac{m}{4V} \tag{13a}$$

where m is the ordinary lateral magnification between the film image and the image presented to the viewer, and V is the viewing distance in meters.

An American standard[37] exists for the measurement of rms-granularity. This standard specifies that samples be scanned with a 48-μm-diameter aperture; for such an aperture, rms-granularity values for commercial films range from about 0.003 to 0.050 at an average density of 1.0. In practice, these values are often multiplied ×1000 to eliminate the decimals. It is worth noting that, experimentally, the measurement of rms-granularity is subject to many sources of error, such as sample artifacts. The standard discusses sources of error and procedures for minimizing them, and is recommended reading for those who must measure granularity.

While in practice rms-granularity serves well as an objective correlate of graininess, the situation is complicated by the fact that there are two broadly different types of granular pattern. Silver grains are small, opaque, and in nearly all cases are situated randomly and independently in the coating. The granular structure in an enlargement, however, is composed of clusters of print-stock grains that reproduce the exposure pattern coming from the enlarged negative grains. This type of granular pattern tends to be large and soft-edged compared to the pattern arising from the primary grains. The patterns found in such samples as color films and screened radiographs are generally similar to those found in enlargements. Microdensitometer traces of these two structures are illustrated in Fig. 12. A little thought will show that the two patterns shown in the figure might have the same mean and standard deviation, and yet the two patterns look entirely different. When different types of patterns are involved, the rms-granularity above is not a sufficient descriptor. The work by Bartleson which showed the correlation between graininess and rms-granularity was done with color negative films having similar granular structures.

As discussed by Dainty and Shaw,[38] further objective analysis of granular patterns may be carried out in terms of the *autocorrelation function*:

$$\phi(\tau) = \lim_{x \to \infty} \frac{1}{2x} \int_{-x}^{+x} \delta(x)\delta(x + \tau)\, dx \tag{14}$$

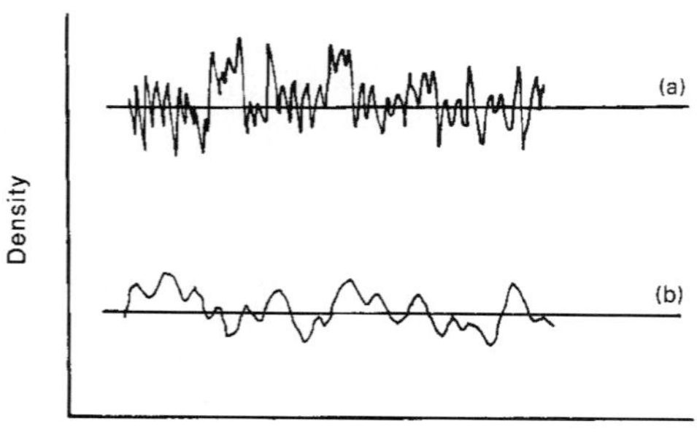

FIGURE 12 Microdensitometer traces of (*a*) primary silver grains and (*b*) granular structure of a print.

where $\delta(x) = D_x - \bar{D}$
D_x = the density reading at point x on the sample
$\bar{D}$ = mean density
$\delta(x+\tau) = D_{x+\tau} - \bar{D}$
τ = a small increment of distance

Note carefully that for the sake of simplicity it has been assumed that the sample is scanned by a very long, narrow slit, so that the autocorrelation function reduces to a one-dimensional function. If the sample is scanned by a small circular aperture it will be two-dimensional. The use of a slit is common in practice. For the case where $\tau = 0$, we have

$$\phi(0) = \lim_{x\to\infty} \frac{1}{2x} \int_{-x}^{+x} [\delta(x)]^2 \, dx = \sigma^2 \tag{15}$$

Let $\phi(\tau)$ be calculated for several different values of τ, including values smaller than the slit width. Since when $\tau < w$ the slit will contain some of the same grains at points x and $x + \tau$, correlation is observed. If there is no spatial correlation in the actual sample, $\phi(\tau) \to 0$ for $\tau > w$. Positive values of $\phi(\tau)$ for $\tau > w$ are an indication of spatial correlation in the sample; that is, large-scale grain.

In practice, it is convenient to carry out the analysis of large-scale patterns in the spatial frequency domain. The Wiener-Khintchin theorem states that what is called the "Wiener spectrum" or *power spectrum of the granularity* distribution is the Fourier transform of the autocorrelation function; that is,

$$WS(\nu) = \int_{-\infty}^{\infty} \phi(\tau) e^{i2\pi\nu\tau} \, d\tau \tag{16}$$

and also

$$\phi(\tau) = \int_{-\infty}^{\infty} WS(\nu) e^{-i2\pi\nu\tau} \, d\nu \tag{16a}$$

where ν = spatial frequency.

In practice, an approximation of the Wiener spectrum is usually obtained by a direct Fourier transform of the granularity trace itself according to the expression

$$WS(\nu) = \lim_{x\to\infty} \frac{1}{2x} \overline{\left| \int_{-x}^{x} \delta(x) e^{i2\pi\nu x} \, dx \right|^2} \tag{17}$$

where $\delta(x)$ has the same meaning as before, and the horizontal bar indicates that the average value of several different runs should be taken in order to provide a reasonable value for the approximation.

Returning to the definition of the autocorrelation function [Eq. (14)] it can be seen that the measured values of $\delta(x)$ and $\delta(x + \tau)$ are each actually the true values convolved with the slit response. By the convolution theorem of Fourier transforms, this means that in frequency space

$$WS(\nu)_{\text{measured}} = WS(\nu)_{\text{true}} x T(\nu)^2 \tag{18}$$

where $T(\nu)$ is the modulation transfer factor of the measuring system at spatial frequency ν. If $T(\nu)^2$ is divided out of $WS(\nu)_{\text{measured}}$ the underlying net value $WS(\nu)$ is obtained. When this is done for black-and-white film samples exposed to light, the underlying spectrum is found to be flat. For the kinds of samples for which Selwyn's law fails, on the other hand, the net Wiener spectrum is found to contain excess low-frequency power and lack high frequencies. Finally, by the properties of Fourier transforms, the area under the Wiener spectrum curve is the zero-value of the autocorrelation function. But by Eq. (15) this zero-value equals σ^2 for the sample, so that the area under the *WS* curve gives the rms-granularity of the sample. This can also be seen intuitively, since the value of the *WS* at special frequency ν is simply the noise power of σ^2 in a spatial frequency band $\nu \pm \Delta\nu/2$.

Doerner[39] has shown that the rms-granularity of a negative can be tracked through a printer system in terms of the Wiener spectrum of the negative and the modulation transfer function of the printer system (which includes the MTF of the print stock itself). Doerner's expression is

$$WS(\nu)_p = WS(\nu)_n \gamma_p^2 A(\nu)_{\text{syst}}^2 + WS(\nu)_{\text{ps}} \tag{19}$$

where n and p indicate negative and print respectively, "ps" the print stock itself, and the other symbols have their previous meanings. The granularity/graininess of a print thus depends on the contrast of the print stock and the spatial frequency response of the printer system as well as on the graininess of the negative.

20.17 SHARPNESS AND GRAININESS CONSIDERED TOGETHER

In the foregoing discussion, we have considered the sharpness and graininess aspects of the picture separately. But real photographs frequently suffer from less-than-optimum sharpness and graininess *both.* Bartleson[40] has studied the subjective quality of such photographs. He concluded that quality was not a linear combination of sharpness and graininess. Instead, ". . . quality tends to be determined primarily by the poorer of the two attributes. . . . If graininess is high, the print will likely be low in quality regardless of how sharp it may be or, conversely, if sharpness is low, so also will quality be low regardless of how grainy the print appears. . . ."

Bartleson's results are of interest in assessing the quality of electronic images, since, at least at the time of writing, such images exhibit low graininess, but also low sharpness. On the basis of Bartleson's work, such images would be judged to be of low subjective quality. In a comparison of electronic and film imagery published in 1990, Ikenoue and Tabei[41] rated the quality of the former as poor because of the sharpness level.

20.18 SIGNAL-TO-NOISE RATIO AND DETECTIVE QUANTUM EFFICIENCY

Since the information in a photograph is normally carried by the density variation, we may usefully define the output signal-to-noise ratio of the photography by

$$\text{S/N}_{\text{out}} = \frac{\Delta \bar{D}}{\sigma} = \frac{\gamma}{\sigma} \cdot \Delta \log H \tag{20}$$

where $\Delta\bar{D}$ is the mean density difference between an element to be detected and its surround, and σ is the rms-granularity of the surround. By Selwyn's law, $\sigma(D)$ varies as the area of the sampling aperture changes; it is convenient here to take the sampling aperture area A as equal to the area of the image element. Furthermore, for ΔH sufficiently small we may write

$$\text{S/N}_{\text{out}} = 0.434 \frac{\gamma}{\sigma} \cdot \frac{\Delta H}{H} \tag{21}$$

Note that γ/σ is a property of the film; it is termed the *detectivity.* $\Delta H/H$, on the other hand, is a property of the object; in fact, it is the object contrast. Practical tests have shown that S/N_{out} should be 4–5 if the element is to be detected against its surround, and 8–10 if it is to be recognized.[42]

In 1946, Albert Rose of RCA published a paper[43] in which he discussed the performances of the TV pickup tube, the photographic layer, and the human eye on a unified basis. His approach was to compare their performances with that of an "ideal device," that is, a radiation detector whose performance was limited only by the quantum nature of the incoming radiation. Such a perfect detector would report the arrival of every incoming quantum, and add no noise to the signal.

In 1958, R. Clark Jones of Polaroid expanded Rose's work with specific application to photographic layers.[44] Jones proposed the term *detective quantum efficiency* (DQE) for Rose's performance indicator, and defined it by the expression

$$\text{DQE} = \left[\frac{S/N_{out}}{S/N_{ideal}}\right]^2 = \left[\frac{S/N_{out}}{S/N_{max}}\right]^2 = \left[\frac{S/N_{out}}{S/N_{in}}\right]^2 \tag{22}$$

where S/N_{out} is the signal-to-noise ratio produced by the actual device, and S/N ideal is the ratio that would be produced by the ideal device, given the same input. By the definition of the ideal device, $S/N_{ideal} = S/N_{max} = S/N_{in}$, where S/N_{in} is the signal-to-noise ratio in the input, and is due to the quantum nature of the input. The ratio is squared to make the DQE compatible with various other concepts.

It is easy to derive an expression for the S/N ratio of the input; when this is combined with Eq. (21) the result is

$$\text{DQE} = \left[\frac{0.434\gamma}{\mathcal{G}}\right]^2 \cdot \frac{1}{q} \tag{23}$$

where $\mathcal{G}$ is the Selwyn coefficient and q is the average exposure received by the image in *quanta per unit area.* Since $(1/q)$ is the radiometric speed of the film, it can be seen that the DQE is a performance parameter that combines the gain (gamma), noise, and speed of the layer. As written, the DQE does not involve the sharpness aspect, but this can be included also.[45]

Since Eq. (23) involves standard photographic parameters, it is readily evaluated for a given material. When this is done, it is found that the DQE of typical materials is on the order of 1 to 3 percent, peaking sharply at low densities in the case of black-and-white films. Note that DQE can be calculated for any sensor for which S/N_{out} can be derived. It is interesting to compare the 1 to 3 percent values given above with those of other sensors. Thus, for example, Jones gives a value of 1 percent for the human eye, and 6 percent for an image orthicon tube. On the other hand, a suitable photographic layer recording electrons may approach 100 percent DQE, and a value of 30 percent has been reported for a screened x-ray film.[46] DQE has also served as a useful approach to considering silver halide mechanisms; see, for example, a paper by Bird, Jones, and Ames that appeared in *Applied Optics* in 1969.[47]

Another figure of merit allied to DQE is the *noise equivalent quanta* (NEQ) which is defined as the number of quanta that a perfect detector would need to produce a record having the same S/N ratio as the system under consideration. It can be shown that

$$q' = \text{DQE} \times q \tag{24}$$

where q is again the number of quanta/unit area used by the real system, and q' the NEQ of a unit area of image.

20.19 RESOLVING POWER

The basic procedure used to measure photographic resolving power follows that used in optics. An American Standard exists.[48] The standard provides for a suitable test object to be reduced optically onto the material to be tested. (Strictly speaking, the test thus determines the resolution of the lens-film combination, but the resolution capability of the lenses specified is high compared to that of the film.) The developed image is then studied in a microscope to determine the highest spatial frequency in which the observer is "reasonably confident" that the structure of the test pattern can still be detected. Thus, as in optics, the "last resolved" image is a threshold image. However, unlike the optical case, the limit is set not only by the progressive decrease in image modulation as spatial frequency increases, but also by the granular nature of the image.

If an exposure series of the test pattern is made, it is found that the spatial frequency of the limiting pattern goes through a maximum. It is customary to report the spatial frequency limit for the optimum exposure as the resolving power of the film.

Photographic resolving power is now not much measured, but it retains some interest as an example of a signal-to-noise ratio phenomenon. Consider the modulation in the various triplet images in the pattern. By definition,

$$M = \frac{H_{\max} - H_{\min}}{H_{\max} + H_{\min}} = \frac{\Delta H}{2\bar{H}} \tag{25}$$

and from Eq. (21),

$$\Delta D = 0.868\gamma M \tag{26}$$

so that ΔD within the pattern varies as the modulation, which in turn decreases as the spatial frequency increases. Furthermore, the area of each of the triplets in the pattern (assuming the ISO pattern configuration) $= 2.5^2\lambda^2 = 2.5^2/\nu^2$. Assuming that Selwyn's law holds, it follows that $\sigma = \mathcal{G}/A^{1/2} = 0.4\mathcal{G}\nu = C\nu$, where we have lumped the constants. As the spatial frequency increases, the S/N of the triplet decreases because ΔD decreases and the effective rms-granularity increases. The resolving power limit comes at the spatial frequency where the S/N ratio drops to the limit required for resolution. Such a system has been analyzed by Schade.[49]

20.20 INFORMATION CAPACITY

The information capacity, or number of bits per unit area that can be stored on a photographic layer, depends on the size of the point spread function, which determines the smallest element that can be recorded, and on the granularity, which determines the number of gray levels that can be reliably distinguished. The exact capacity level for a given material depends on the acceptable error rate; for one set of fairly stringent conditions, Altman and Zweig[50] reported levels up to 160×10^6 bits/cm^2. Jones[51] has published an expression giving the information capacity of films as

$$IC = \frac{1}{2} \int \int_0^\infty \log_2 \left(1 + \frac{S(\nu_x, \nu_y)}{N(\nu_x, \nu_y)}\right) d\nu_x \, d\nu_y \tag{27}$$

where S and N are the spectral distributions of power in the system's signal and noise. For a given spatial frequency, the signal power is given by

$$S = WS_i(\nu)\mathrm{MTF}^2(\nu) \tag{28}$$

where $WS_i(\nu)$ is the value of the Wiener spectrum of the input at ν and MTF(ν) is the value of the film's MTF at that frequency. The information capacity thus depends on the frequency response and noise of the system, as would be expected. The matter is discussed by Dainty and Shaw.[52]

20.21 LIST OF PHOTOGRAPHIC MANUFACTURERS

Afga Photo Division
Agfa Corporation
100 Challenger Road
Ridgefield, NJ 07660

E. I. duPont de Nemours and Co.
Imaging Systems Department
666 Driving Park Avenue
Rochester, NY 14613

Eastman Kodak Co.
343 State Street
Rochester, NY 14650
Tel. 1-800-242-2424 for product info.

Fuji Photo Film USA
555 Taxter Road
Elmsford, NY 10523

Ilford Photo Corporation
70 West Century Boulevard
Paramus, NJ 07653

3M Company
Photo Color Systems Division
3M Center
St. Paul, MN 55144-1000

Polaroid Corporation
784 Memorial Drive
Cambridge, MA 02139
Tel. 1-800-255-1618 for product info.

20.22 REFERENCES

1. W. Thomas (ed.), *SPSE Handbook of Photographic Science and Engineering,* Wiley, New York, 1973, sec. 8.
2. B. H. Carroll, G. C. Higgins, and T. H. James, *Introduction to Photographic Theory,* Wiley, New York, 1980, chaps. 6–9.
3. T. H. James (ed.), *The Theory of the Photographic Process,* 4th ed., Macmillan, New York, 1977.
4. J. Sturge, Vivian Walworth and Alan Shepp (eds.), *Imaging Processes and Materials,* Van Nostrand Reinhold, New York, 1989, chap. 3. (See also Ref. 3, chap. 4.)
5. G. Haist, *Modern Photographic Processing,* vols. I and II, Wiley, New York, 1979.
6. Klaus Hendricks, chap. 20 of Ref. 4.
7. J. H. Altman, F. Grum, and C. N. Nelson, *Phot. Sci. Eng.* **17:**513 (1973).
8. Reference 3, chaps. IV–VII.
9. *ANSI/ISO,* May 2, 1991. (See also *ANSI/ISO,* May 1, 1984, which sets forth standardized terminology.)
10. R. W. G. Hunt, *The Reproduction of Colour,* 4th ed., Fountain Press, Tolworth, England, 1987, pp. 247–257.
11. *American National Standard,* ANSI PH2.2, 1984, R1989, PH2.17 1985.
12. C. J. Niederpruem, C. N. Nelson, and J. A. C. Yule, *Phot. Sci. Eng.* **10:**35 (1966).
13. *American National Standard,* ANSI PH3.49, 1971, R1987.
14. H. N. Todd and R. D. Zakia, *Phot. Sci. Eng.* **8:**249 (1964). (See also publications of the American National Standards Institute).

15. M. Hercher and B. J. Ruff, *J. Opt. Soc. Am.* **57:**103 (1967).
16. W. F. Berg, *Proc. Roy. Soc. of London,* ser. A, **174:**5599 (1940).
17. Reference 2, p. 141.
18. *Scientific Imaging with Kodak Films and Plates,* Publication P315, Eastman Kodak Co., Rochester, N.Y., 1987, p. 63.
19. J. Kapecki and J. Rodgers, *Kirk-Othmer Encyclopedia of Chemical Technology,* 4th ed., vol. 6, Wiley, New York, 1993, p. 965.
20. S. H. Liggero, K. J. McCarthy, and J. A. Stella, *J. Imaging Technology* **10:**1 (1984).
21. Reference 5, chap. 7.
22. E. H. Land, *Phot. Sci. Eng.* **16:**247 (1972).
23. Reference 10, chap. 11 et seq.
24. See, for example, F. R. Clapper in Ref. 3, sec. II, chap. 19.
25. R. M. Evans, *Eye, Film, and Camera in Color Photography,* Wiley, New York, 1959, p. 180.
26. B. J. Thompson, *Progress in Optics VII,* E. Wolf (ed.), North-Holland Publishing Co., Amsterdam, 1969; R. E. Swing, *J. Optical Society of America* **62:**199 (1972); Dainty and Shaw, Ref. 27, chap. 9.
27. P. Kowalski, *Applied Photographic Theory,* Wiley, New York, p. 77 et seq.
28. C. N. Nelson, Ref. 3, chap. 19.
29. G. C. Higgins and L. A. Jones, *J. Soc. of Mot. Pict. & TV Engrs.* **58:**277 (1952).
30. J. Gasper and J. J. dePalma, Ref. 3, chap. 20.
31. J. C. Dainty and R. Shaw, *Image Science,* Academic Press, London, 1974, chaps. 6 and 7. (See also J. C. Dainty, *Optica Acta* **18:**795, 1971.)
32. R. M. Gendron, *J. Soc. of Mot. Pict. & TV Engrs.* **82:**1009 (1973).
33. E. W. H. Selwyn, *Photography Journal* **75:**571 (1935).
34. C. J. Bartleson, *Photography Journal* **33:**117 (1985).
35. K. F. Stultz and H. J. Zweig, *J. Opt. Soc. Am.* **49:**693 (1959).
36. F. W. Sears, *Optics,* Addison-Wesley, Reading, Mass., 1949, p. 159.
37. *American National Standard,* ANSI PH2.40, 1985, R1991.
38. Ref. 31, chap. 8.
39. E. C. Doerner, *J. Opt. Soc. Am.* **52:**669 (1962).
40. C. J. Bartleson, *J. Phot. Sci.* **30:**33 (1982).
41. S. Ikenoue and M. Tabei, *J. Imaging Sci.* **34:**187 (1990).
42. Ref. 2, p. 335.
43. A. Rose, *J. Mot. Pict. & TV Engrs.* **47:**273 (1946).
44. R. C. Jones, *Phot. Sci. Eng.* **2:**57 (1958).
45. Ref. 31, chap. 8, p. 311 et seq.
46. Ref. 4, table 3.2, p. 73.
47. G. R. Bird, R. C. Jones, and A. E. Ames, *Appl. Opt.* 1969, p. 2389.
48. *American National Standard,* ANSI PH2.33, 1983, R1990.
49. O. H. Schade, *J. Soc. Mot. Pict. & TV Engrs.* **73:**81 (1964).
50. J. H. Altman and H. J. Zweig, *Phot. Sci. Eng.* **7:**173 (1963).
51. R. Clark Jones, *J. Opt. Soc. Am.* **51:**1159 (1961).
52. Ref. 31, chap. 10.

CHAPTER 21
IMAGE TUBE INTENSIFIED ELECTRONIC IMAGING

C. B. Johnson
L. D. Owen
Litton Electron Devices
Tempe, Arizona

21.1 GLOSSARY

B_s	phosphor screen brightness, photometric units
CCDs	charge-coupled devices
CIDs	charge-injection devices
E_i	image plane illuminance, lux
E_s	scene illuminance, lux
e	charge on electron, coulombs
FO	fiberoptic
FOV	field-of-view, degrees
fc	illuminance, photometric, foot candles = lm/ft^2
f_N	spatial Nyquist frequency, cycle/mm
f_{lto}	limiting resolution at fiberoptic taper output
F_{si}	input window signal flux
ftL	luminance, photometric (brightness), foot Lamberts = lm/ft^2
G_m	VMCP electron gain, e/e
HVPS	high-voltage power supply
II	image intensifier
LLL	low-light-level
lx	illuminance, photometric, lux = lm/m^2
M_{fot}	magnification of fiberoptic taper
MCP	microchannel plate
MTF	modulation transfer function, 0 to 1.0
N_{essa}	number of stored SSA electrons per input photoelectron, e/photon
N_f	total number of frames, #
N_p	number of photoelectrons, #
$N_{\text{ps}}(\lambda)$	number of photons per second, photon/s

PDAs	photodiode arrays
P	phosphor screen efficiency, photon/eV
$P_p(\lambda)$	radiometric power spectral distribution, W
QLI	quantum limited imaging
Q_{ssa}	stored SSA charge per input photoelectron from the photocathode, C
R_s	scene reflectance, ratio
R_{sn}	signal-to-noise ratio, ratio
$S(\lambda)$	absolute spectral sensitivity, mA/W
$S(f)$	squarewave response versus frequency, cycles/mm
SIT	silicon-intensifier-target vidicon
SNR	signal-to-noise ratio
sb	luminance, photometric (brightness), stilbs = cd/cm^2
SSA	silicon self-scanned array
T_f	filter transmission, 0 to 1.0
T_{fot}	transmission of fiber-optic taper, 0 to 1.0
T_n	lens T-number = $FN/\sqrt{\tau_0}$
T_{ssa}	transmission of fiber-optic window on the SSA, 0 to 1.0
V_a	phosphor screen, actual applied voltage, V
V_d	phosphor screen, "dead-voltage," V
V_m	VMCP applied potential, V
V_s	MCP-to-screen applied potential, V
$Y(\lambda)$	quantum yield (electrons/photon), percent
Y_k	quantum yield, photoelectrons/photon
Y_{ssa}	SSA quantum yield, e/photon
τ_e	the exposure period, s
τ_i	CCD charge integration period, s
τ_o	lens transmission, 0 to 1.0
Φ_p	photon flux density, photon/m^2/s

21.2 INTRODUCTION

It is appropriate to begin our discussion of image tube intensified (II) electronic imaging with a brief review of natural illumination levels. Figure 1 illustrates several features of natural illumination in the range from full sunlight to overcast night sky conditions. Various radiometric and photometric illuminance scales are shown in this figure. Present silicon self-scanned array (SSA) TV cameras, having frame rates of 1/30 to 1/25 s, operate down to about 0.5 lx minimum illumination.

The generic term *self-scanned array* is used here to denote any one of several types of silicon solid-state sensors available today which are designed for optical input. Among these are charge-coupled devices (CCDs), charge-injection devices (CIDs), and photodiode arrays (PDAs). Vol. I, Chaps. 22 and 23 contain detailed information on these types of optical imaging detectors. Specially designed low-light-level (LLL) TV cameras making use of some type of image intensifier must be used for lower exposures, i.e., lower illumination and/or shorter exposures.

The fundamental reason for using an II SSA camera instead of a conventional SSA camera is that low-exposure applications require the low-noise optical image amplification provided by an II to produce a good signal-to-noise ratio from the SSA camera. Other

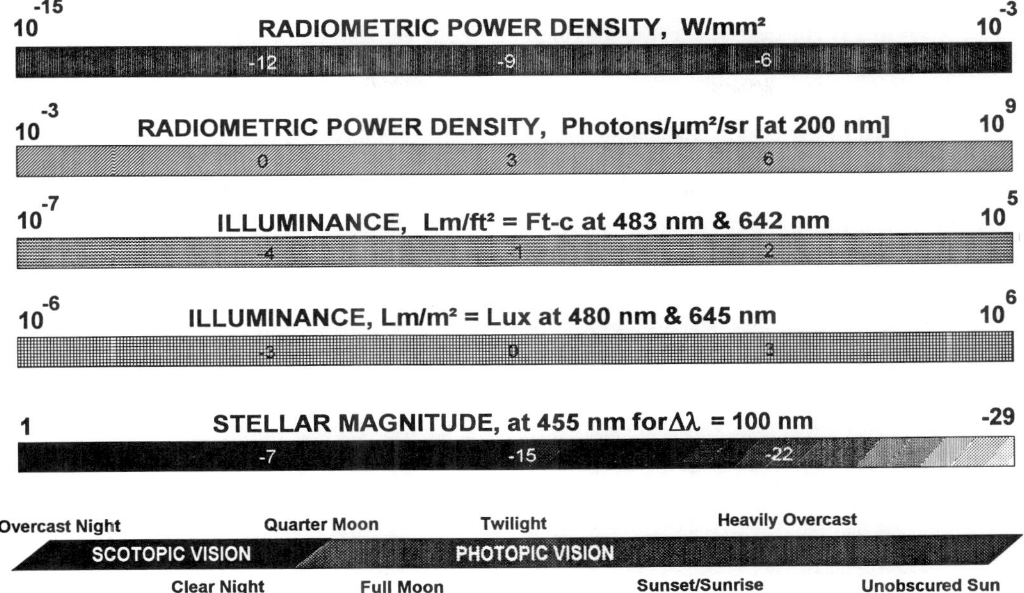

FIGURE 1 Various optical illumination ranges.

important applications arise because of the ability to electronically shutter IIs as fast as 1 ns or less and the higher sensitivity of IIs in certain spectral regions. The following sections deal with the optical interface between the object and the II SSA, microchannel plate proximity-focused IIs, and II SSA detector assemblies. By using auto-iris lenses and controlling both the electronic gain and gating conditions of the II, II SSA cameras can provide an interscene dynamic range covering the full range of twelve orders of magnitude shown in Fig. 1. Several applications for II SSAs are discussed later in the chapter under "Applications."

21.3 THE OPTICAL INTERFACE

It is necessary to begin our analysis of II SSA cameras with a brief discussion of the various ways to quantify optical input and exposure. Two fundamental systems are used to specify input illumination: radiometric and photometric. These systems are briefly described, and the fundamentals of optical image transfer are discussed. Detailed aspects of radiometry, photometry, and optical image transfer are discussed elsewhere in this *Handbook*. However, enough information is presented in this chapter to allow the reader to properly design, analyze, and apply II SSA imaging technology for a wide variety of practical applications.

Quantum Limited Imaging Conditions

Quantum limited imaging (QLI) conditions exist in a wide variety of applications. An obvious one is that of LLL TV imaging at standard frame rates, i.e., 33-ms exposure periods, under nighttime illumination conditions. For example, under full moonlight input faceplate illumination conditions, only ~1000 photons enter a $10 \times 10\,\mu m^2$ image pixel in a 33-ms frame period. Assuming a quantum yield of 10 percent, an average of only 100 electrons is generated, and the maximum SNR achievable in each pixel and each frame is only $\sqrt{100} = 10$. Alternatively, under full unobscured sunlight input faceplate illumination conditions, an electronically gated camera with gatewidth limited exposure period of 10 ns produces a total of (1E9 photons/μm^2/s)(10×10 μm^2)(10 ns) = 1000 photons, or the same SNR as for the LLL operating conditions noted above. These are both clearly QLI operating conditions. II SSA camera technology is used to obtain useful performance in both of these types of applications. Without the use of an II, a bare SSA does not meet the requirements for useful SNR under these conditions.

Radiometry

The unit of light flux in the radiometric system is the watt. The watt can be used anywhere in the optical spectrum to give the number of photons per second (N_{ps}) as a function of wavelength (λ). Since the photon energy $E_p(\lambda)$ is

$$E_p = \frac{hc}{\lambda} \tag{1}$$

where h is Planck's constant and c is the velocity of light in vacuum, the radiometric power $P_p(\lambda)$, in watts, is given by

$$P_p(\lambda) = \left(\frac{hc}{\lambda}\right) \cdot N_{ps}(\lambda) \tag{2}$$

or

$$P_p(\lambda) = (2 \cdot 10^{-25}) \cdot \frac{N_{ps}(\lambda)}{\lambda} \tag{3}$$

where N_{ps} is the number of photons per second. Alternatively, the photon rate is given by

$$N_{ps}(\lambda) = (5 \times 10^{24})\lambda P_p(\lambda), \qquad \text{photons/s} \tag{4}$$

For example, one milliwatt of 633-nm radiation from an He-Ne laser is equivalent to (5E24)(633E − 9)(1E − 3) = 3.2E15 photons/s.

Radiometric flux density, in W/m^2, represents a photon rate per unit area, and radiometric exposure per unit area is the product of the flux density times the exposure period. The active surface of a photoelectronic detector produces a current density in response to an optical flux density input, while a total signal charge is produced per unit area in the same detector during a given exposure period.

Rose[1] has shown that all types of optical detectors, e.g., photographic, electronic, or the eye, are subject to the same fundamental limits in terms of signal-to-noise ratio (R_{sn}), optical input, and exposure period. In summary, the noise in a measured signal of N_p photoelectrons during a fixed exposure period is $\sqrt{N_p}$, so that

$$R_{sn} = \sqrt{N_p} \tag{5}$$

The brightness (B_s) of a scene that produces this signal in a square pixel of dimensions ($y \cdot y$), as a result of the optical transfer and conversion from the source to the detector, possibly through a medium that absorbs, scatters, and focuses photons, is

$$B_s = \frac{C \cdot N_p}{y^2} \tag{6}$$

where C is a constant. In terms of signal-to-noise ratio,

$$B_s = \frac{C \cdot R_{sn}^2}{y^2} \tag{7}$$

Thus, for twice the signal-to-noise ratio, the scene brightness must be increased four times, or the throughput of the optical system must be doubled, etc. Also, if the pixel size is reduced by a factor of two, the same changes in scene brightness or optical throughput must be made in order to maintain the same signal-to-noise ratio. Under QLI conditions, higher resolution necessarily requires more input flux density for equal signal-to-noise ratio, and higher resolution inherently implies less sensitivity. The Rose limit should be used often as a proof check on design and performance estimates of LLL and other QLI imaging systems.

As an example, assume a simple imaging situation such as a single pixel, e.g., a star in the nighttime sky, and an II SSA camera having an objective lens of diameter D_o. Also assume that the starlight is filtered, to observe only a narrow wavelength band, and that the photon flux density from the star is Φ_p (photon/m^2/s). The number of photoelectrons produced at the photocathode of the II SSA detector (N_p) is given by

$$N_p = \Phi_p \cdot T_f \cdot \left(\frac{\pi D_o^2}{4}\right) \cdot \tau_o \cdot Y_k \cdot \tau_c \tag{8}$$

where T_f is the filter transmission, τ_o is the lens transmission, Y_k is the quantum yield of the window/photocathode assembly in the II SSA camera, and τ_e is the exposure period. Note that the II SSA camera parameters which determine the rate of production of signal photoelectrons are filter transmission, lens diameter, quantum yield, and exposure period. The key one is of course the lens diameter, and not lens f-number, for this kind of imaging; it is important, however, for extended sources such as terrestrial scenes.

Photometry and the Camera Lens

A lens on the II SSA camera is used to image a scene onto the input window/photocathode assembly of the II SSA. The relationship between the scene (E_s) and II SSA image plane (E_i) illuminances in lux (lx) is

$$E_i = \frac{\pi \cdot E_s \cdot R_s \cdot \tau_o}{(4 \cdot \mathrm{FN}^2 \cdot (m+1)^2)} \tag{9}$$

where R_s is the scene reflectance, τ_o is the optical transmission of the lens, FN is the lens f-number, and m is the scene-to-image magnification. If E_s is in foot-lamberts, then the π is dropped and E_i is in footcandles.

Alternatively, Eq. (9) becomes

$$E_i = \frac{E_s \cdot R_s}{(4 \cdot T_n^2 \cdot (m+1)^2)} \tag{10}$$

using the T-number of the lens, where

$$T_n = \frac{\mathrm{FN}}{\sqrt{\tau_o}} \tag{11}$$

The sensitivity of an II is usually given in two forms, i.e., "white-light" luminous sensitivity, in units of μA/lm, and absolute spectral sensitivity, in units of A/W as a function of wavelength, as discussed later in the section "Input Window/Photocathode Assemblies" in Sec. 21.4.

Example: A scene having an average reflectance of 50 percent receives LLL "full-moon" illumination of 1.0E–2 fc. If a lens having a T-number of 3.0 is used, and the scene is at a distance of 100 m from a lens with a focal length of 30 mm, what is the input illumination at the II SSA? Since the distance to the scene is much longer than the focal length of the lens, the magnification is much smaller than unity and m can be neglected. Thus,

$$E_i = \frac{E_s R_s}{(4 \cdot T_n^2)} \tag{12}$$

For the given values, the input illumination at the II SSA is bound to be $E_i = (1.0\text{E} - 2 \text{ fc})(0.50)/(4(3.0)^2) = 1.4\text{E} - 4 \text{ fc}$.

General Considerations

It is of prime importance in any optoelectronic system to couple the maximum amount of signal input light into the primary detector surface, e.g., the window/photocathode assembly of an II SSA. In order to achieve the maximum signal-to-noise ratio, the modulation transfer function of the input optic and the spectral sensitivity of the II SSA must be carefully chosen. As shown in Fig. 2, the spectral sensitivity of a silicon SSA is

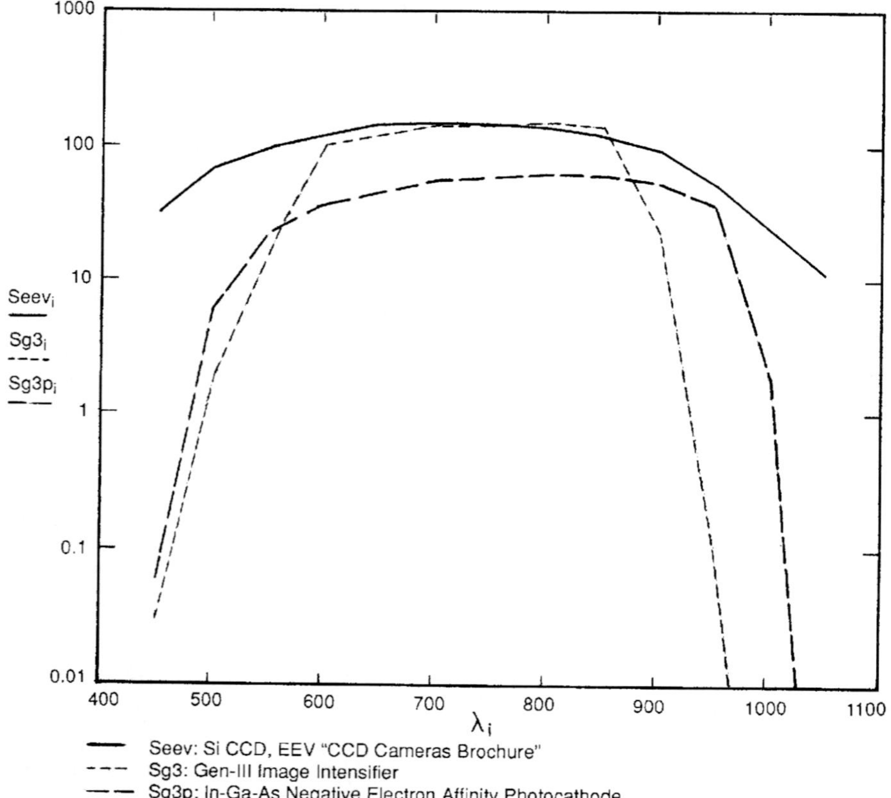

FIGURE 2 Absolute spectral sensitivity S (mA/W) versus wavelength λ (nm) of a frame-transfer type of CCD, a gen-III image intensifier, and an II having an In-Ga-As negative electron affinity photocathode.

much different than that of a Gen-3 image intensifier tube. Thus, an optimized objective lens design for a CCD will be much different than that for an II SSA. The dynamic range characteristics are also very different, since IIs will handle seven orders-of-magnitude interscene dynamic range, using a combination of II gain control and electronic duty-cycle gating, while SSAs will only provide about two orders of magnitude.[2]

Several factors must be considered if the overall system resolution and sensitivity are to be optimized. For example, the spectral responses of many optical input SSAs and/or lenses used in commercial cameras have been modified by using filters to reduce the red and near-ir responses to give more natural flesh tones. In an II SSA the filter may have little effect if the filter is on the SSA. The filter should not be used in the objective lens for the II SSA since a major portion of the signal will be filtered out. If a color SSA is to be used in an intensified system using relay lens coupling, sacrifice of both sensitivity and resolution will result. This is due to the matrix color filter used in these SSA chip designs. Most of the signal will go into green bandpass filter elements, and very little will go into the blue and red elements. The color matrix filter is usually bonded to the surface of the SSA chip; thus these SSA types are not used for fiber-optically coupled II SSAs.

The ideal objective lens design for an II SSA needs to be optically corrected over the spectral range of sensitivity of the II and the spectral range of interest. For special-purpose photosensitivity covering portions of the uv, blue, or near-ir spectral regions, appropriate adjustments must be made in the lens design. Although they may be adequate for many applications, it is very seldom that a commercial CCTV lens is optimized for nighttime illumination, or other LLL or QLI, conditions.

Another very important part of an optimized II SSA camera design is to make the proper choice of II and SSA formats. This subject is discussed in detail later under "Fiber-Optic-Coupled II/SSAs." The input of the II SSA system is the II, and the most likely choice will be one with an 18-mm active diameter, since the widest choice of II features is available in this size. Image intensifiers are also available having 25- and 12-mm active diameters, but these are generally more expensive. Regarding the SSA standard format sizes, the standard commercial TV formats are named by a longtime carryover from the days when vidicons were used extensively. Thus 2/3, 1/2, and 1/3-inch format sizes originally referred to the diameters of the vidicon envelope and not the actual image format.

21.4 IMAGE INTENSIFIERS

An image intensifier (II) module, when properly coupled to an SSA camera, produces a low-light-level electronic imaging capability that is extremely useful across a broad range of application areas, including spectral analysis, medical imaging, military cameras, nighttime surveillance, high-speed optical framing cameras, and astronomy. An immediate advantage of using an II is that its absolute spectral sensitivity can be chosen from a wide variety of window/photocathode combinations to yield higher sensitivity than that of a silicon SSA. Since recently developed IIs are very small, owing to the use of microchannel plate (MCP) electron multipliers, the small size of a solid-state SSA camera is not severely compromised. In summary, advantages of using MCP IIs are:

- Long life
- Low power consumption
- Small size and mass
- Rugged
- Very low image distortion
- Linear operation

- Wide dynamic range
- High-speed electronic gating, e.g., a few nanoseconds or less

An image intensifier can be thought of as an active optical element which transforms an optical image from one intensity level to another, amplifying the entire image at one time, i.e., all pixels are amplified in parallel and relatively independent of each other. In most cases the resultant ouptut image is more intense than that of the input image. The level of image amplification depends on the composite efficiency of all the conversion steps of the process involved in the image intensification operation and the basic definition of amplification. The term *image intensifier* is generally used to refer to a device that transforms visible and near-visible light into brighter visible images. Devices which convert nonvisible radiation, e.g., uv or ir, into visible images are generally referred to as *image converters.* For simplicity we refer to both types of image amplifiers/converters as "IIs" in this chapter.

Three general families of IIs exist, shown schematically in Fig. 3, that are based upon the three kinds of electron lenses used to extract the signal electrons from the photocathode, namely,

- Proximity focus IIs
- Electrostatic focus IIs
- Magnetic focus IIs

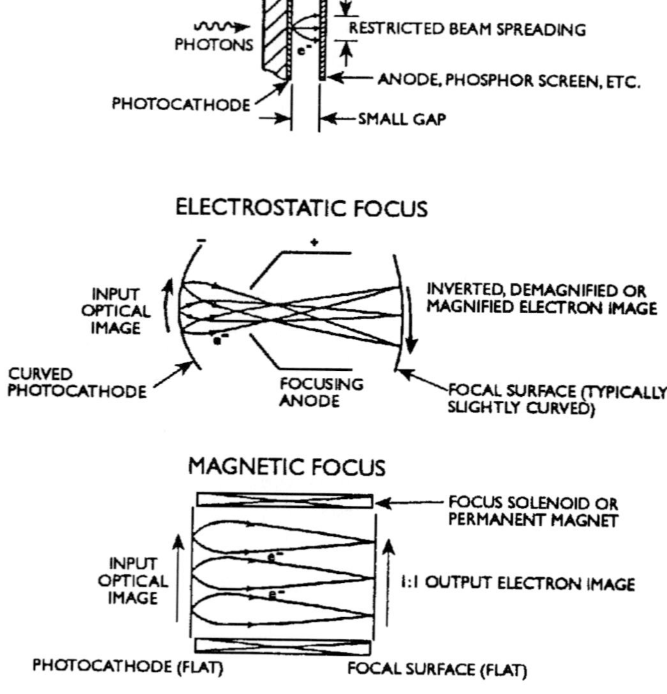

FIGURE 3 Electron lenses.

The first image tubes used a "proximity-focus" electron lens.[3] Having inherently low gain and resolution, the proximity-focus lens was dropped in favor of electrostatic focus and magnetic focus IIs. The so-called Generation-O and Generation-1 image tubes made for the U.S. Army used electrostatically focused IIs. The input end of the silicon-intensifier-target (SIT) vidicon also made use of electrostatic focusing. Magnetic focusing was used extensively in the old TV camera tubes, e.g., image orthicons, image isocons, and vidicons, and also for large-active-area and high-resolution IIs for specialized military and scientific markets.

With the development of the MCP, which was achieved for the U.S. Army's Generation-2 types of night-vision devices, it became practical to use a proximity-focused electron lens again to meet the needs for extremely small and low-mass IIs. These "Gen-2" tubes are being used extensively for military night-vision applications, e.g., night-vision goggles for helicopter pilots, individual soldier helmet mounted night-vision goggles, etc. The most recently developed "Gen-3" IIs have higher sensitivity and limiting resolution characteristics than Gen-2 IIs, and they are used in similar night-vision systems.

Both the Gen-2 and Gen-3 types of IIs are available for use as low-noise light-level amplifiers in II SSA cameras. In addition, by choosing special input window/photocathode combinations outside the military needs for Gen-2 and Gen-3 devices, a very wide range of II SSA spectral sensitivities can be achieved, well beyond silicon's range. For II SSA camera applications, we will focus our attention exclusively on the use of proximity-focused MCP IIs because of their relative advantages over other types of IIs.

The basic components of a proximity-focused MCP II are shown schematically in Fig. 4. This type of II contains an input window, a photocathode, a microchannel plate, a phosphor screen, and an output window. The *photocathode* on the vacuum side of the *input window* converts the input optical image into an electronic image at the vacuum surface of the photocathode in the II. The *microchannel plate* (MCP) is used to amplify the electron image pixel-by-pixel. The amplified electron image at the output surface of the MCP is reconverted to a visible image using the *phosphor screen* on the vacuum side of the *output window*. This complete process results in an output image which can be as much as 20,000 to 50,000 times brighter than what the unaided eye can perceive. The input window can be either plain transparent glass, e.g., Corning type 7056, fiber-optic, sapphire, fused-silica, or virtually any optical window material that is compatible with the

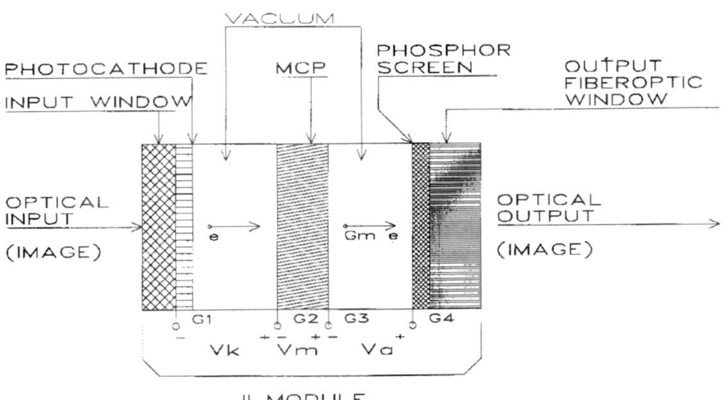

FIGURE 4 Schematic design of a proximity-focused MCP image intensifier tube module.

high-vacuum requirements of the II. The output window can be glass, but it is usually fiber-optic, with the fibers going straight through or twisted 180° for image inversion in a short distance.

A block diagram of a generalized high-voltage power supply (HVPS) used to operate the II is given in Fig. 5. For dc operation, the basic HVPS provides the following typical voltages:

$V_k = 200$ V

$V_m = 800$ V for an MCP

($V_m = 1600$ V for a VMCP)

($V_m = 2400$ V for a ZMCP)

$V_a = 600$ V

For high-speed electronic gating of the II, the photocathode is normally gated off by holding the G1 electrode a few volts positive with respect to the G2 electrode. Then, to gate the tube on and off for a short period, a pulse generator is used to control the output of the gated power supply to the normal gated on condition, i.e., $V_k = 200$ V with the polarity as shown in Fig. 5.

The dc HVPSs for IIs draw very little power, and they can be operated continuously using two AA cells, e.g., 3-V input voltage, for about two days. These dc HVPSs are available in small flat-packs or wraparound versions. Gated HVPSs, excluding the pulse generator, are generally at least two times larger than their dc counterparts.

In operation, an input image is focused onto the input window/photocathode assembly, producing a free-electron image pattern which is accelerated across the cathode-to-MCP gap by an applied bias voltage V_k. Electrons arriving at the MCP are swept into the channels, causing secondary electron emission gain due to the potential V_m applied across the MCP input and output electrodes. Finally, the amplified electron image emerging from the output end of the MCP is accelerated by the voltage V_a applied across the MCP-to-phosphor screen gap so that they strike an aluminized phosphor screen on a glass or FO output window with an energy of about 6 keV. This energy is sufficient to produce an output image which is many times brighter than the input image. The brightness gain of the MCP II is proportional to the product of the window/photocathode sensitivity to the input light, the gain of the MCP, and the conversion efficiency of the phosphor-screen/output-window assembly. Each of these key components and/or assemblies is discussed in more detail in the following sections of this section.

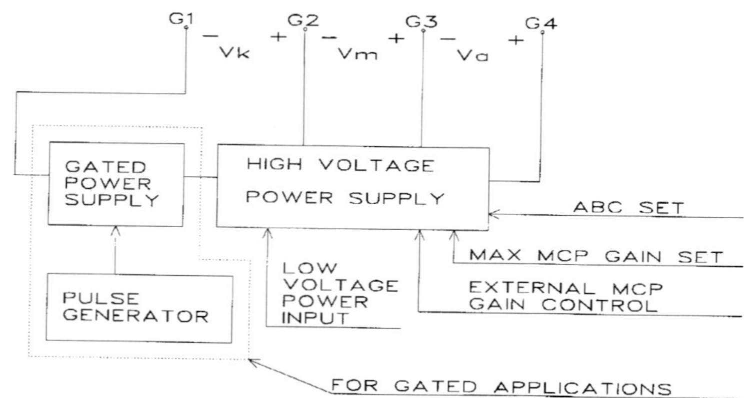

FIGURE 5 MCP image intensifier high-voltage supply.

Input Window / Photocathode Assemblies

The optical spectral range of sensitivity of an II, or the II SSA that it is used in, is determined by the combination of the optical transmission properties of the window and the spectral sensitivity of the photocathode. In practice, a photocathode is formed on the input window in a high-vacuum system to produce the window/photocathode assembly as shown in Fig. 6. This assembly is then vacuum-sealed onto the II body assembly, and the finished II is then removed from the vacuum system. This type of photocathode processing is called *remote processing* (RP), because the alkali metal generators, antimony sources, and/or other materials used to form the photocathode are located outside of the vacuum II tube. Since there is no room for these photocathode material generators, remote processing must be used for MCP IIs. Also, IIs made using remote processing are found to have significantly less spurious dark current emission than the older Gen-O and Gen-I types of IIs having internally processed photocathodes.

The short wavelength cutoff of a window/photocathode assembly is determined by the optical transmission characteristic of the wndow, i.e., its thickness and material composition. The absolute spectral sensitivity of the photocathode determines the midrange and long wavelength cutoff characteristics of the assembly. Photocathode materials having longer wavelength cutoffs also have lower bandgap energies and generally higher thermionic emission than photocathodes with shorter wavelength cutoffs.

The spectral sensitivities of various window/photocathode combinations are shown in Fig. 7 for comparison. Useful spectral bands range from the uv to the near-ir, depending upon the particular combination chosen. This figure shows the spectral sensitivity advantages that can be achieved with II SSAs. Other advantages are discussed throughout this chapter.

Note that the window/photocathode absolute spectral sensitivity $S(\lambda)$ curves given in Fig. 7 represent the ratio of photocathode output current per watt input at wavelength λ. Alternatively, window/photocathode response can be specified in terms of quantum yield $Y(\lambda)$, or quantum efficiency, defined to be the ratio of the average number of output photoelectrons produced per input photon as a function of wavelength. These two parameters are related by the convenient equation

$$Y(\lambda) = \frac{124 \cdot S(\lambda)}{\lambda} \tag{13}$$

where Y is the quantum yield in percent, S is the absolute sensitivity in mA/W, and λ is the wavelength in nm.

Microchannel Plates

The development of the microchannel plate (MCP) was a revolutionary step in the art of making IIs. Although developed for and used in modern military passive night-vision systems, MCP IIs are being used today in nearly all II SSA cameras.

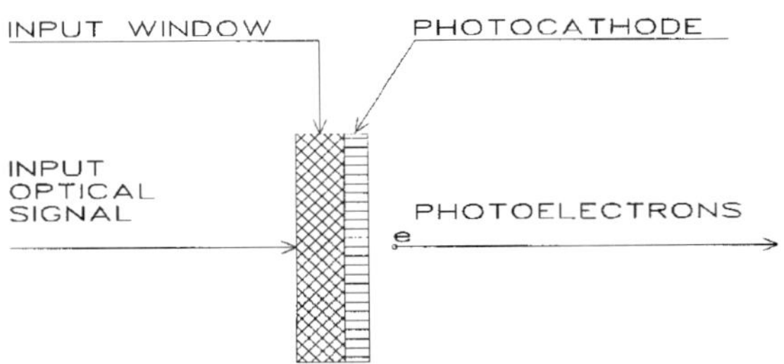

FIGURE 6 Input window/photocathode assembly.

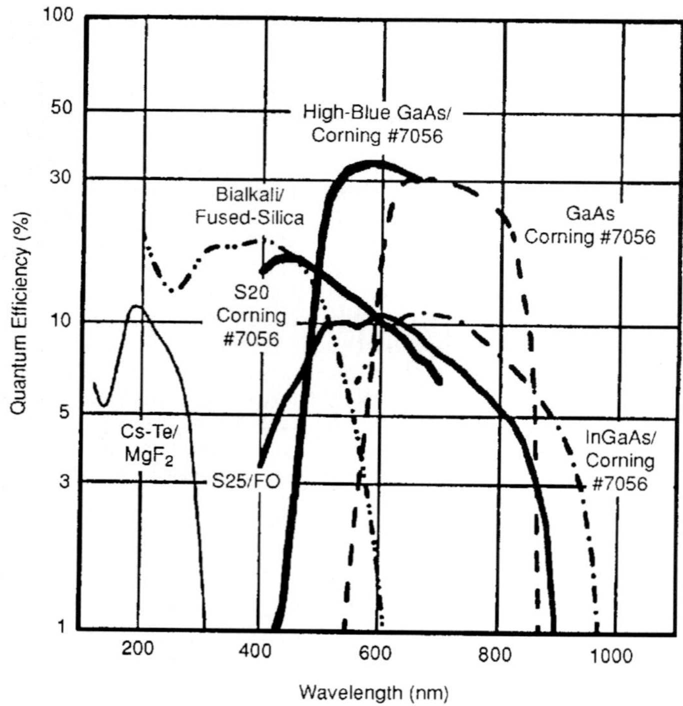

FIGURE 7 Window/cathode spectral quantum efficiencies.

An MCP is shown schematically in Fig. 8. Microchannel plates are close-packed-hexagonal arrays of channel electron multipliers. With a voltage V_m applied across its input and output electrodes, the MCP produces a low-noise gain G_m, e.g., a small electron current (I_{in}) from a photocathode produces an output current $G_m I_{in}$. In addition to its function as a low-noise current amplifier, the MCP retains the current density pattern or "electron image" from its input to output electrodes. It is also possible to operate two MCPs (VCMP) or three MCPs (ZMCP) in face-to-face contact to achieve electron gains as high as about 1E7 e/e in an II tube, as shown in Fig. 9. Other general characteristics of these types of MCP assemblies are also given in Fig. 9.

The approximate limiting spatial resolutions of MCPs depend upon the channel center-to-center spacings, as follows:

Channel diameter (μm)	Channel center-to-center spacing (μm)	Approximate limiting resolution (lp/mm)
4	6	83
6	8	63
8	10	50
10	12	42
12	15	33

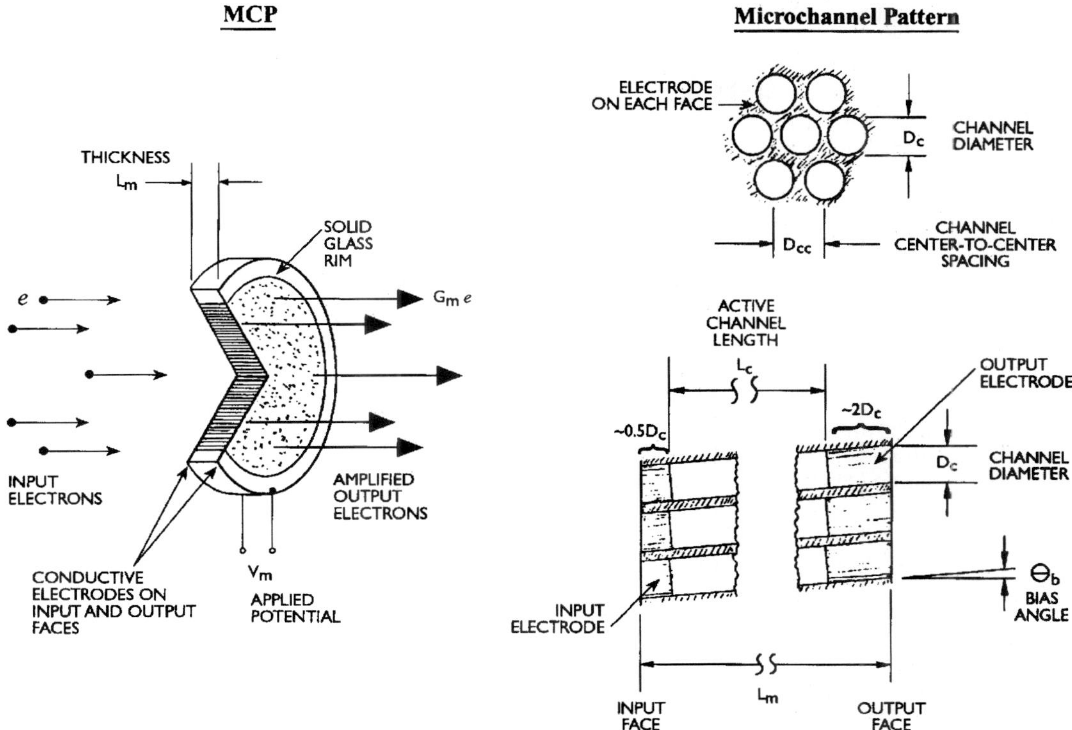

FIGURE 8 MCP parameters.

	Vm max (kV)	Gm max (e/e)	PULSE HEIGHT DISTRIBUTION (% FWHM)	RELATIVE LIMITING RESOLUTION (UNITS)
MCP:	1.0	1E3	NEGATIVE EXPONENTIAL	1.00
VMCP:	2.0	1E5	120	0.71
ZMCP:	3.0	1E7	80	0.50

FIGURE 9 General characteristics of MCPs, VMCPs, and ZMCPs.

TABLE 1 MCP Gain Equation and Gain Parameters

$$G_m(V_m) = \left(\frac{V_m}{V_c}\right)^g$$

Type	V_c (V)	g (units)	(L_m/D_c) (units)
MCP	350	8.5	40
MCP	530	13	60
VMCP	700	17	80
ZMCP	1050	25	120

As shown in Fig. 7, MCPs are made to have channel axes that make a "bias angle" (θ_b) with respect to the normal to its input and output faces. This bias angle improves electron gain and reduces noise factor by reducing "boresighting" of electrons into the channels. The MCP bias current or "strip current" (I_s) that results from the voltage applied to the MCP sets an upper limit to the maximum linear dynamic range of the MCP. Generally, when the output current density of the MCP is in excess of about 10 percent of the strip current density, the MCP ceases to remain a linear amplifier. Conventional MCPs have strip current densities of about 1 μA/cm^2, and recent high-output-technology MCPs (HOT MCPs^{TM})[4] have become available that have strip current densities as high as about 40 μA/cm^2. Electron-gain characteristics of MCP assemblies are given approximately by the equation and associated parameters shown in Table 1.

Power noise factors for conventional MCPs, used in Gen-2 IIs, and "filmed-MCPs," used in Gen-3 IIs, are approximately 2.0 and 3.5, respectively. Detailed information on MCP gain, noise factors, and other parameters are given by Eberhardt.[5] Note that MCP gain is a strong function of the channel length-to-diameter ratio. The parameter V_c in the gain equation is the "crossover" voltage for the channel, i.e., it is the MCP applied voltage at which the gain is exactly unity.

Phosphor Screens

Output spectral and temporal characteristics of a wide variety of screens are given in an Electronic Industries Association publication.[6] The phosphor materials covered in this publication are listed in Table 2. Both the old "P-type" and the new two-letter phosphor designations are given in this table. Any of these phosphor screen materials can be used in proximity-focused MCP IIs. However, one very commonly used phosphor is the type KA (P20) because it has a high conversion efficiency, its output spectral distribution matches the sensitivity of a silicon SSA reasonably well, it is fast enough for conventional 1/30-s frame times, it has high resolution, and it is commonly used in direct-view night-vision IIs.

The three main components of an aluminized phosphor-screen/output-window assembly, of the type used in a proximity focused MCP II, are shown schematically in Fig. 10. An aluminum film electrode is deposited on the electron input side of the phosphor to accelerate the MCP output to high energy, e.g., about 6 keV, and to increase the conversion efficiency of the assembly by reflecting light toward the output window. The phosphor itself is deposited on the glass or fiber-optic output window.

Decay times, or persistence, and relative output spectral distributions for a variety of phosphor types are given in Fig. 11. Key phosphor assembly parameters that should be accounted for in the design of MCP II SSAs are MCP-to-phosphor applied potential (V_a),

TABLE 2 Worldwide Phosphor-Type Designation System

Cross reference: old-to-new designations

P1	GJ	P20	KA	P38	LK
P2	GL	P21	RD	P39	GR
P3	YB	P22	X(XX)	P40	GA
P4	WW	P23	WG	P41	YD
P5	BJ	P24	GE	P42	GW
P6	WW	P25	LJ	P43	GY
P7	GM	P26	LC	P44	GX
P10	ZA	P27	RE	P45	WB
P11	BE	P28	KE	P46	KG
P12	LB	P29	SA	P47	BH
P13	RC	P31	GH	P48	KH
P14	YC	P32	GB	P49	VA
P15	GG	P33	LD	P51	VC
P16	AA	P34	ZB	P52	BL
P17	WF	P35	BG	P53	KJ
P18	WW	P36	KF	P55	BM
P19	LF	P37	BK	P56	RF
				P57	LL

Source: Adapted from Electronic Industries Association Publication, no. 116-A, 1985.

effective "dead-voltage" resulting from electron transmission losses in the aluminum film, phosphor screen energy input-to-output conversion efficiency, optical transmission of the glass or fiber-optic window, sine-wave MTF of the assembly, phosphor persistence, and output spectral distribution.

Before specifying the use of a particular phosphor, the operational requirements of the II SSA camera should be reviewed. The phosphor persistence should be short compared to the SSA frame time to minimize image smear due to rapidly moving objects. Also, the

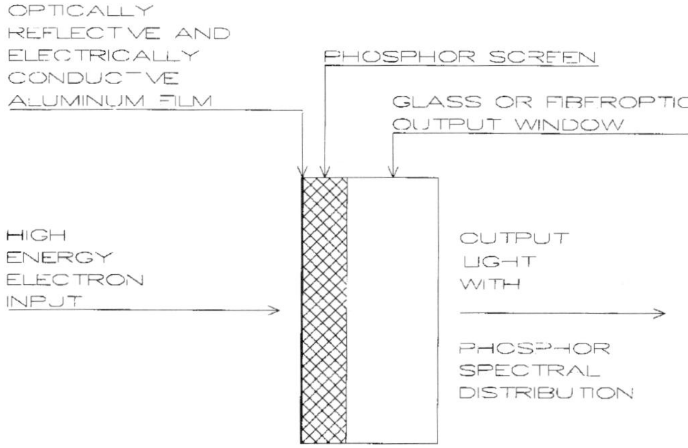

FIGURE 10 Aluminized phosphor screen and window assembly.

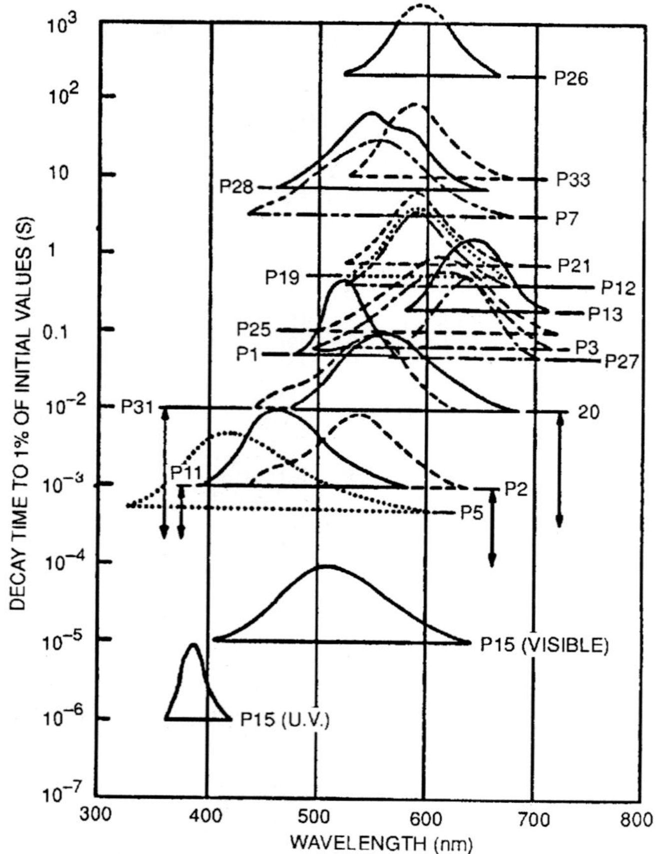

FIGURE 11 Phosphor screen decay times and spectral outputs. (*Reprinted with permission from United Mineral and Chemical Co.*)

absolute conversion efficiency of the phosphor assembly and its relative output spectral distribution whould be spectrally matched[7] to the sensitivity of the SSA for maximum coupling efficiency.

Typical absolute spectral response characteristics, i.e., the phosphor spectral efficiency (radiated watts per nanometer per watt excitation) as a function of wavelength, of aluminized phosphor screens are given in Ref. 7. The associated phosphor screen efficiencies are also given in this reference in three different ways:

- Typical quantum yield factor: photons out per eV input
- Typical absolute efficiency: radiated watts per watt excitation
- Typical luminous equivalent: radiated lumens per radiated watt.

For example, a type KA(P20) aluminized phosphor-screen/glass window assembly is found to have its peak output at 560 nm and a typical quantum yield factor of

0.063 photons/eV. Thus, an electron which leaves the MCP and strikes the assembly with 6 keV of energy, and for a "dead-voltage" of 3 kV, approximately (6 − 3) keV × 0.063 photons/eV = 190 photons will be produced at the output.

Proximity-Focused MCP IIs

By combining the image transfer and conversion properties of the three major proximity-focused MCP II assemblies discussed earlier, i.e.,

- Input window/photocathode
- Microchannel plate
- Phosphor screen/output window

the operational characteristics of the II itself, as shown in Fig. 4, can be determined.

For example, consider an II CCD application for a space-based astronomical telescope that requires more than 10 percent quantum yield in the 250–280 nm spectral range, but minimum sensitivity beyond 300 nm. It is desired that the top end of the dynamic range be at an input window signal flux (F_{si}) of 1000 photon/pixel/s at 250 nm. Let the CCD have a 1-in vidicon format, i.e., an active area of $11.9 \times 8.9\ \text{mm}^2$, with 244 vertical columns and 325 horizontal rows of pixels. The limiting resolution of even a dual-MCP (VMCP) image tube has a limiting resolution that is significantly higher than the horizontal pixel spatial Nyquist frequency (f_N) in the CCD, so that the pixel size at the input to the II will be essentially the same as that of the CCD. Let us rough-in an II design by making the following additional assumptions:

MCP-to-phosphor applied potential (V_a)	6000 V
Phosphor screen type	KA (P20)
Phosphor screen/window-quantum yield (P_q)	0.06 photon/eV
Phosphor screen dead voltage (V_d)	3000 V
CCD charge integration period (τ_i)	33 ms
CCD pixel full-well charge	1 pC = 6.3E6 e

An II with an 18-mm active diameter can be used, since the diagonal of the CCD active area is 14.9 mm. From Fig. 6, the fused-silica/Cs-Te window/photocathode assembly will be chosen, having a quantum yield (Y_k) of 0.10 at 250 nm, to meet the spectral sensitivity requirements.

Let's now proceed to estimate the required gain of the MCP structure, decide what kind of an MCP structure to use, and determine its operating point. A first-order estimate of the stored pixel charge (Q_{ccd}) for the given input signal flux density is

$$Q_{ccd} = F_{si} \cdot Y_k \cdot G_m \cdot (V_a - V_d) \cdot P_q \cdot Y_{ccd} \cdot \tau_i \tag{14}$$

Since

$$F_{si} = 1000 \text{ photon/pixel/s}$$
$$Y_k = 0.10 \text{ e/photon}$$
$$Y_{ccd} = 0.3 \text{ e/photon}$$

it is found that $Q_{ccd} = G_m$ (178 e/pixel). Setting this charge equal to the full-well pixel charge gives $G_m = 6.3\text{E}6$ e/pixel/(178 e/pixel) = 3.5E4 e/e. This MCP assembly gain is easily satisfied by using a VMCP. From Table 1, it is found that the gain of a VMCP is given approximately by $G_m = (V_m/700)^{17} = 3.5\text{E}4$ e/e. Solving for V_m gives $V_m = 1300$ V.

Thus, a first-order estimate for the general requirements to be placed in the II to do the job is as follows:

Active diameter	18 mm
Quality area	(11.9 × 8.9 mm)
Input window/photocathode	Fused-Silica/Cs-Te
MCP assembly	VMCP
Aluminized phosphor screen assembly	KA/FO window

Coupling this II to the specified FO input window CCD, e.g., by using a suitable optical cement, will meet the specified objective. Other parameters like the dark count rate per pixel as a function of temperature, the DQE of the II CCD, cosmetic, uniformity of sensitivity, and other specifications will have to be considered as well before completing the design.

Recent "Generations" of MCP IIs. The most impressive improvement in direct-view night-vision devices has come with the advent of Gen-3 technology. The improvement, which is most apparent at very low light levels, is mainly due to the use of GaAs as the photocathode material. At higher light levels, e.g., half-moon to full-moon conditions, the Gen-2+ gives somewhat better performance. Key to the detection of objects under LLL conditions is the efficiency of the photocathode; the Gen-3 typically does a factor of 3 better. Also, the spectral response of Gen-3 matches better to the night sky spectral illumination. This equates to being able to see at almost one decade lower scene illumination with Gen-3. A summary of proximity-focused MCP image intensifier general characteristics is given in Table 3.

TABLE 3 Summary of Proximity-Focused MCP Image Intensifier General Characteristics

Minimum active diameter (mm)	Input window material*	Spectral sensitivity range (nm)	MCP assembly type	Temperature rating		Output window material	Minimum limiting resolution (lp/mm)	Technology type
				Storage (°C)	Operating (°C)			
11.3	FS	160–850	MCP	−55, +65	−20, +40	FO	25	Gen-2
12.0	FS, G, FO	160–900	MCP, VMCP, ZMCP	−57, +65	−51, +45	FO, G	45, 29, 20	Gen-2
17.5	FS, G, FO	600–900	MCP, VMCP, ZMCP	−57, +65	−51, +45	FO, G	45, 25, 20	Gen-2
17.5	G, FO	500–1100	MCP, VMCP, ZMCP	−57, +95	−51, +52	FO, G	45, 25, 20	Gen-3
25.0	FS, G, FO	160–900	MCP	−57, +65	−51, +45	FO, G	40	Gen-2
25.0	G, FO	500–1100	MCP	−57, +95	−51, +52	FO, G	40	Gen-3

* FS = fused silica; G = Corning #7056 glass; FO = fiber-optic.

Technology type	Options	
	Photocathode	Phosphor
Gen-2	All but GaAs, InGaAs	Wide selection
Gen-3	GaAs, InGaAs	Wide selection

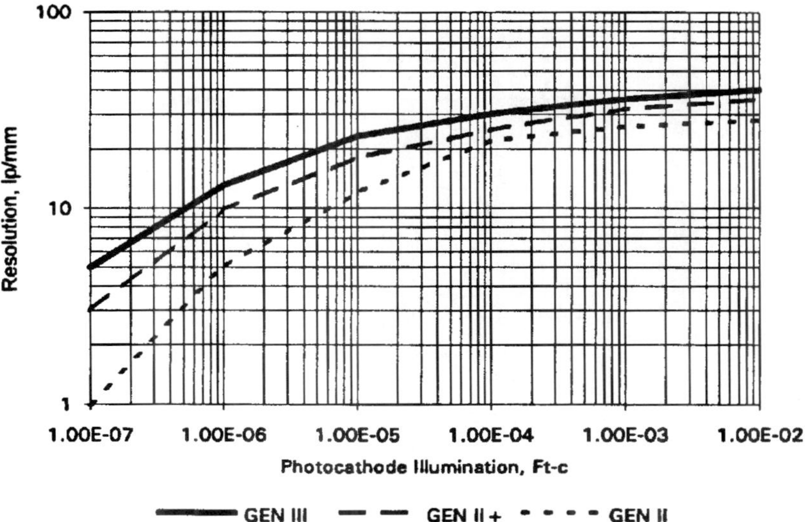

FIGURE 12 Image intensifier tube resolution curves.

For systems design work, it is useful to know the approximate characteristics of the three most recent generations in terms of II resolution versus photocathode illumination. The resolution transfer curves shown in Fig. 12 give the II resolution, observable by the eye, as a function of input illumination for Gen-2, Gen-2+, and Gen-3 IIs. These curves do not include system optics degradations, except in the sense that a human observer made the resolution measurements using a 10-power eyepiece in viewing the output image of the II.

Improved Performance Gen-2 IIs. Recent enhancements in the dynamic range performance of Gen-2 IIs for direct-view applications have been made which also benefit II SSA camera performance. Improvement goals were to increase both the usable output brightness and the LLL gain of Gen-2 IIs. Night-vision devices are normally used at light levels ranging from full moon to just below quarter-moon, or in dark city environments with ample scattered light. It is important to have good contrast over as wide a light-level range as possible. To get this extended dynamic range, the gain should be held nearly constant to as high a level as possible, for improved contrast at the high-light levels. Any gain improvement should be attained with little or no increase in noise, to ensure good performance at the minimum light levels. Reducing the objective lens f-number as low as possible also improves system performance and gain. However, f-number reduction by itself may create problems in the system dynamic range if the II and its power supply assembly is not appropriately adjusted to match the optical throughput.

Figure 13 shows the extended dynamic range of a Gen-2+ II and power supply assembly, as compared to the typical MIL-SPEC Gen-2 assembly. Increasing the gain in a standard Gen-2 assembly by increasing the gain control voltage, i.e., the MCP voltage, will not give the same benefits as the Gen-2+. Ideally, a change of one unit in input brightness should result in a proportional output brightness change. The increased near-linear gain range up to higher-output light levels in the Gen-2+ improves the contrast at the higher levels. Brightness limiting begins reducing the gain to hold the output brightness constant after the automatic brightness control (ABC) limit of the power supply is reached. The increased gain of the Gen-2+ improves the performance at the lowest-light levels as well.

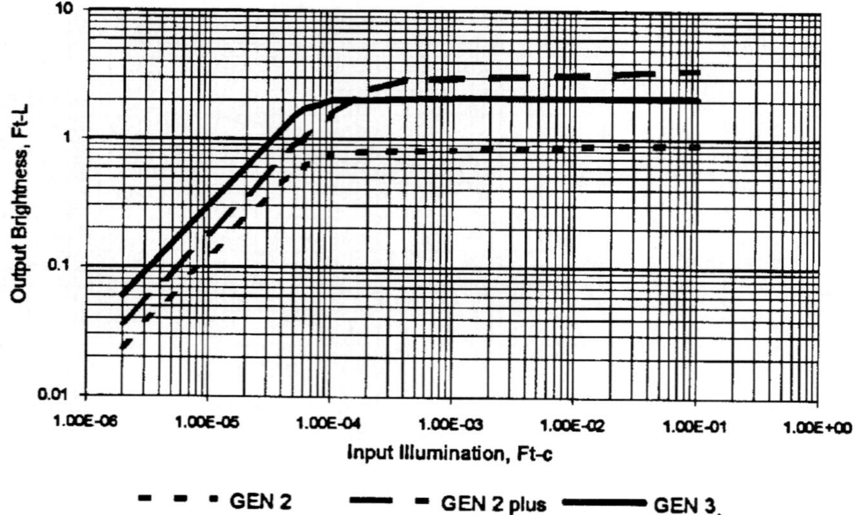

FIGURE 13 Output versus input transfer characteristics of Gen II, Gen II+, and Gen III II/power supply assemblies.

21.5 IMAGE INTENSIFIED SELF-SCANNED ARRAYS

There are several reasons to consider using an II SSA instead of an SSA alone. One obvious reason is to achieve LLL sensitivity. Figure 14 shows the limiting resolution vs.

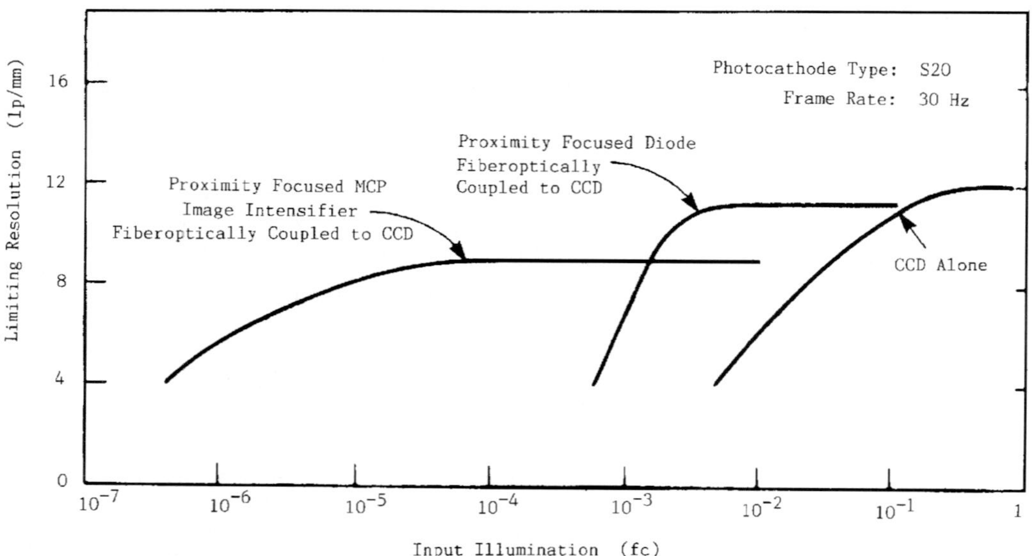

FIGURE 14 Resolution versus input illumination characteristics of a conventional optical input CCD camera and the same camera fiber-optically coupled to an MCP image intensifier tube. (*From Ref. 8.*)

faceplate illumination characteristic of CID camera operating in the unintensified and intensified modes.[8] It is seen that LLL sensitivity is achieved by coupling the CID to an image intensifier tube, albeit at the expense of reduced high-light resolution. Other reasons for using an II SSA are

- High-speed electronic gating, down to a few nanoseconds, for framing cameras, LADAR, smoke and fog penetration
- Improved spectral sensitivity
- Use in a TV camera system that operates automatically under lighting conditions ranging from nighttime to full daylight conditions.
- High-sensitivity and high-speed-gated optical multichannel analyzers (OMAs)

Fiber-Optic-Coupled II/SSAs

Figure 15 shows a schematic design of a fiber-optically (FO) coupled II SSA assembly. These designs are modular, since an II module is optically coupled to an SSA module. Virtually any type of image tube can be optically coupled to an SSA. The fiber-optically coupled design shown in Fig. 15 requires the use of an II having a fiber-optic output window and an SSA having an SSA input window. A fiber-optic taper, instead of a simple unity magnification FO window, is also generally required to efficiently couple the output of the II into the SSA, and this is shown in Fig. 15 as a separate module. The various fiber-optic modules are joined at interfaces 1, 2, and 3, using optical cement, optical grease, immersion oil, or "air." For the highest-resolution image transfer across these interfaces, it is necessary that the gap length at each interface be kept short, and the numerical aperture of the fiber-optic windows should be kept as low as possible, consistent with the SNR and gain requirements. It has been shown[9] that the first interface can be eliminated by making the fiber-optic taper part of the II and depositing the phosphor screen directly onto it, and interface 3 can also be eliminated by coupling the fiber-optic taper directly to the SSA. The properties of the image transfer and conversion components shown in Fig. 15 can be used to estimate the overall performance characteristics of the fiber-optically coupled II SSA camera.

The terminology used to define SSA image format sizes derives from the earlier

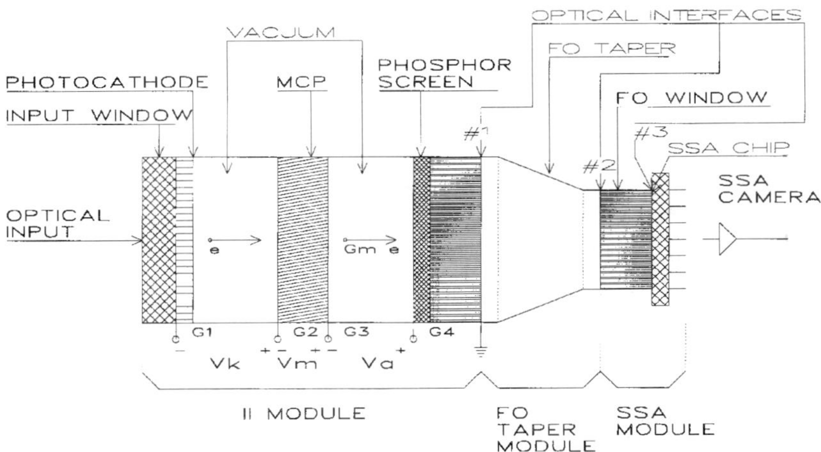

FIGURE 15 Schematic design of fiber-optically coupled IISSA assembly.

TABLE 4 Comparison of Basic Image Intensifier Diameters, SSA Format Sizes, Matching Fiber-Optic Taper Magnifications, and Limiting Resolutions at the Fiber-Optic Taper Output Surface (for 45 lp/mm Intensifier)

Image intensifier active dia. (mm)	SSA Format Vidicon (in)	SSA Format (mm)	SSA Diagonal (mm)	(M_{fot}) FOT magnification	(f_{ito}) Limiting resolution at FOT output (lp/mm)
25	1	11.9 × 8.9	14.9	0.596	76
25	2/3	8.8 × 6.6	11.0	0.440	102
18	1	11.9 × 8.9	14.9	0.828	54
18	2/3	8.8 × 6.6	11.0	0.611	74
18	1/2	6.5 × 4.85	8.1	0.451	100
12	2/3	8.8 × 6.6	11.0	0.917	49
12	1/2	6.5 × 4.85	8.1	0.676	67
12	1/3	4.8 × 3.6	6.0	0.500	90

vidicon camera tube technology. The mass, volume, and power requirements of vidicon cameras are much larger than SSA cameras. Vidicons also have image distortion and gamma characteristics which must be accounted for, whereas SSAs and II SSSAs using proximity-focused IIs are nearly distortion-free with linear, i.e., unity gamma, input/output transfer characteristics over wide intrascene dynamic ranges. Table 4 gives the basic II active diameters, SSA format sizes, SSA active-area diagonal lengths, fiber-optic taper magnifications (M_{fot}) required to couple II outputs to the SSAs, and limiting resolutions (f_{lto}) at the fiber-optic taper output. Figure 16 shows schematically the relative sizes of the standard active diameters of IIs and the standard SSA formats.

The present limiting resolution range of MCP IIs is 36 to 51 lp/mm. In an II SSA, the

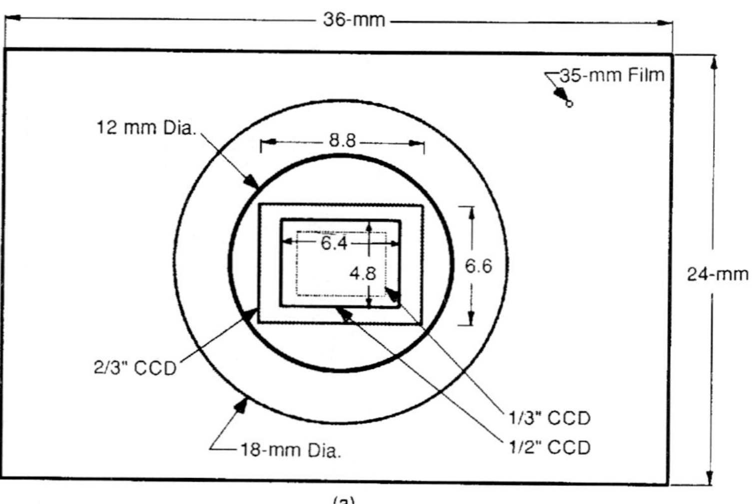

FIGURE 16*a* Typical 35-mm film, image intensifier and SSA formats.

	18-mm Φ Image Tube	12-mm Φ Image Tube	1" CCD	2/3" CCD	1/2" CCD	1/3" CCD	35-mm Film	Units
Diagonal	18.0	12.0	14.9	11.0	8.1	6.0	43.3	mm
Vertical	10.8	7.2	8.9	6.6	4.9	3.6	24.0	mm
Horizontal	14.4	9.6	11.9	8.8	6.5	4.8	36.0	mm
Area	155.5	69.1	105.9	58.1	31.5	17.3	864.0	mm^2

(b)

FIGURE 16*b* Typical dimensions for image intensifiers and SSAs using 3:4 format.

resolution of the II should be matched, in some sense, to that of the SSA. For example, it is unwise to use a low-resolution II and fiber-optic lens combination with a much higher resolution CCD.

Lens-Coupled II SSAs

Figure 17 is a schematic design for a lens-coupled II SSA assembly. The differences between this design and the fiber-optic-coupled II SSA design described earlier are that the output window of the II can be either fiber-optic or glass, and a lens is used instead of an FO taper to couple the output optical image from the II directly into a conventional optical input SSA, i.e., no FO window is required at the SSA. Although the lens-coupling efficiency is lower, its image distortion and resolution performance is superior to the FO-coupled design. Also, the chance for possible adverse rf interference at the sensitive input to the SSA camera from the II high-voltage power supply is less than for the lens-coupled design.

Parameters to Specify. Typical parameters to specify for an MCP II SSA detector assembly, using either fiber-optic or lens-coupling, are as follows:

- Sensitivity

  White-light (2856K) (μA/lm)

  Spectral sensitivity (mA/W versus nm)

  Sensitivity (mA/W at specified wavelength)

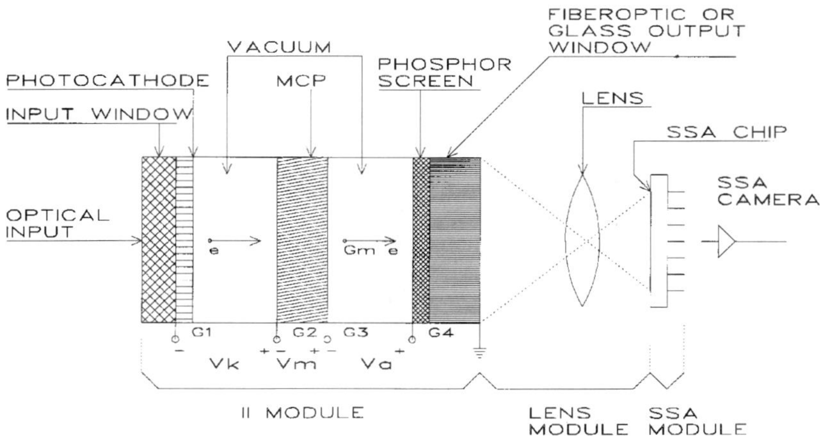

FIGURE 17 Schematic design of lens-coupled IISSA assembly.

- EBI (lm/cm^2 at 23°C)
- MCP applied potential for 10K fL/fc luminous gain (V)
- Horizontal resolution at specified input illumination (TVL)
- Shades-of-gray (units)
- Cosmetic properties
  Uniformity (percent)
  Bright spots (number allowable in format zone)
  Dark spots (number allowable in format zone)
- Burn-in (procedure)
- Mechanical specifications
- Dimensions (interface drawing)
- Mass (*g*)
- Environmental (specified)

Electron-Bombarded SSA

Since the early work by Abraham et al.[10] which showed the feasibility of achieving useful electron gain by electron bombardment (EB) of a silicon diode in a photomultiplier tube, several attempts have been made to achieve similar operation using an SSA specially designed for EB input, instead of optical input. The charge gain (G_{eb}) resulting from the electron bombardment is given by

$$G_{eb} = \frac{(V_a - V_d)}{3.6} \tag{15}$$

where V_a is the acceleration voltage and V_d is the "dead-voltage" of the EBSSA. It was quickly found that successful CCD operation could not be obtained by simply bombarding the normal optical input side of the chip with electrons, because interface states soon form which prevent readout of the chip and other problems. By thinning a CCD chip to 10–15 μm from the "backside" and operating in a backside EB-mode, useful performance is achieved. In this way, 100 percent of the silicon chip is sensitive to incident photoelectrons, and it becomes technically feasible to make EBSSA cameras.

Proximity Focused EBSSAs. A proximity-focused EBSSA is shown schematically in Fig. 18. In this design, the input light enters the window/photocathode assembly to generate

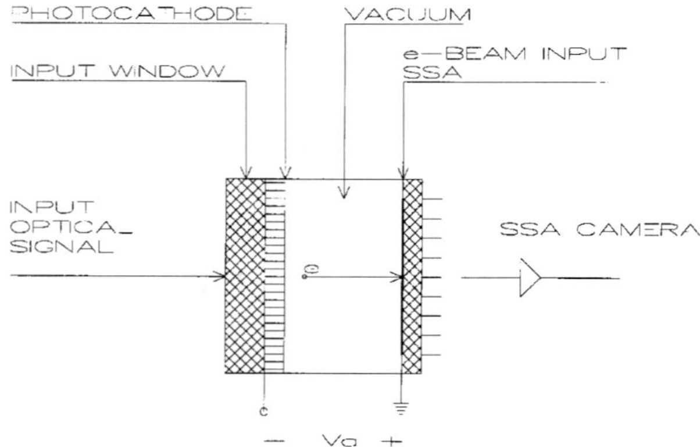

FIGURE 18 Electron bombarded SSA (EBSSA).

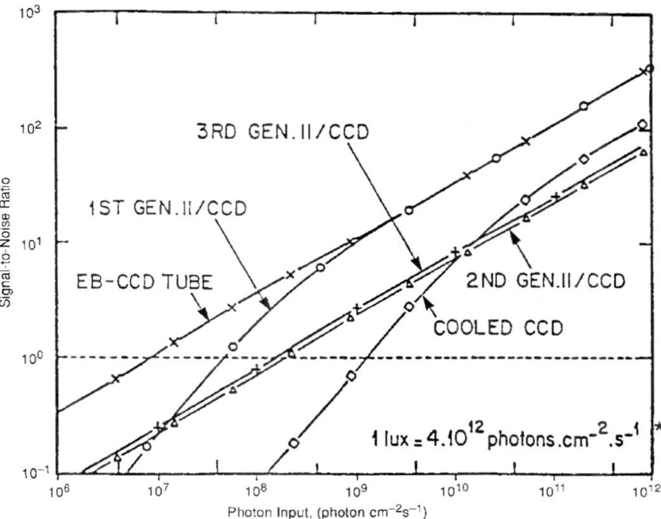

FIGURE 19 Comparison of the signal-to-noise ratio of various optoelectronic imagers versus the photon input. (*From Ref. 14.*)

the signal photoelectrons which are accelerated to about 10-keV energy and bombard the thinned backside of the EBSSA. Note that no MCP, no MCP-to-screen gap, no phosphor screen/output window assembly, and no fiber-optic or lens coupling is used to transfer the electronic image to the SSA for readout. Thus, higher limiting resolution is attainable. Also, the power noise factor associated with the EBSSA gain process is lower than that of MCP devices, and image lag is eliminated because no phosphor is used. Early work on proximity-focused EBDDs was done by Barton et al.,[11] Williams,[12] and Cuny et al.[13] By 1979, a 100×160 pixel TI CCD was used in this type of detector and put into a miniature TV camera. With an acceleration voltage of $V_a = 15\,\mathrm{kV}$, an electron gain of 2000 was achieved, along with a Nyquist limited resolution of 20 lp/mm. Recent advances have brought this technology closer to extensive usage possibilities. Richard et al.[14] have compared the SNR characteristics of an EB CCD tube, various other types of II CCDs, and bare CCDs. Their results are shown in Fig. 19.

In order to achieve its full performance capabilities, the energy of the bombarding electrons must be absorbed by the active silicon SSA material, photoelectrons must not be lost, the exposure of the EBSSA to high-energy electrons should not cause a life problem, and it must be possible to read out the stored charge pattern in the SSA. It is found that recombination phenomena at the EB-input face can be reduced with a p^+ passivation layer, e.g., by using $3E17\,\mathrm{cm}^{-3}$ boron doping, which reduces back-diffusion of signal electrons, front-diffusion of "dark" charges from the rear face, reduced diffusion length, separation of holes and electrons by the built-in electric field, and higher surface conductivity, thus better voltage stability, at the rear face.

Internally processed (IP) and remotely processed (RP) or "transfer" photocathodes have been used in EBSSAs. It is generally found that the internal processing produces consistently higher-background and spurious noise problems due to field emission from tube body parts and the photocathode. Both types of photocathode processes have yield long-life EBCCD detectors.

Proven applications to date for EBSSA detectors:

- Photon-counting wavefront sensor (adaptive optics), European Space Organization 3.6-m telescope at La Silla, Chile
- NASA, Goddard Space Flight Center, Oblique Imaging EB CCD UV sensitive camera

Advantages of EBSSA cameras over MCP II based II SSAs:

- No image lag
- Higher resolution
- Single photoelectron detection per frame per pixel
- Higher DQE

Digital II SSA Cameras. Consider a photon-counting imaging detector consisting of an MCP image intensifier tube (II) that is fiber-optically coupled to a silicon solid-state self-scanned array (SSA) chip in a TV camera. Incoming photons at wavelength λ pass through the input window of the II and produce an average quantum yield of Y_k photoelectrons per photon at the photocathode. The resulting photoelectrons (e) are accelerated into the MCP electron multiplier assembly. Amplified output electrons from this low-noise electron multiplier are accelerated into an aluminized phosphor screen on the output window of the II. The number of output photons from the II per photoelectron is proportional to the electron gain in the MCP (G_m), the effective electron bombardment energy at the phosphor screen (εV_s), and finally the electron-input to photon-output conversion efficiency (P) at the phosphor screen. As discussed earlier the optical transmission of the input window and the actual quantum yield of the photocathode are usually factored together in the average quantum yield parameter Y_k, and the optical transmission of the output window is also normally factored together with the actual conversion efficiency of the phosphor screen in the screen efficiency parameter P.

The output photon pulse from the II, resulting from the single detected input photon, is coupled into the SSA via the fiber-optic taper, which matches the output size of the II to the size of the SSA, and a fiber-optic window on the SSA. This photon pulse is then converted to an electron signal charge packet (Q_{ssa}) at the SSA. The number of electrons stored per pixel in the SSA depends upon the area of the photon pulse at the SSA, the spatial distribution of photons in this pulse, and the area per pixel in the SSA. Thus, in addition to the above II factors, the stored charge in the SSA per photoelectron is also proportional to the optical transmissions of the FO taper (T_{fot}) and SSA window (T_{ssa}), and the quantum yield of the SSA (Y_{ssa}).

By using two or three conventional MCPs in cascade, i.e., VMCPs or ZMCPs, the gain can be made so large that it completely overrides any normal room-temperature thermal dark current in an SSA at a conventional RS-170 rate. In this photon-counting mode of operation, a charge signal above a preset threshold value is looked for. When it is found in a given pixel, a "1" is stored in memory for that pixel's address, "Os" are stored in pixel addresses where this condition is not met, and the entire frame is read out. By reading out a total of N_f frames, the dynamic range can be made as high as N_f if the dark count rate is negligible. Thus, photon-counting imaging can achieve a very large dynamic range.

Another advantage of photon-counting imaging is that the image resolution can also be made very high by centroiding the detected charge packets in the SSA. Since the performance of a centroiding camera depends upon the signal-processing algorithm, this will not be analyzed here. Instead, the reader is referred to several references in which centroiding is discussed.[15]

Let us next calculate the stored charge and number of stored electrons in a photon-counting II SSA per photoelectron. Assume that a proximity-focused VMCP II is coupled to the SSA with a fiber-optic taper. For our analysis, some typical values will be used for the operating voltage and gain of a VMCP: the acceleration voltage between the VMCP and the phosphor screen, the efficiency of an aluminized type KA (P20) phosphor screen, the optical transmissions of a fiber-optic taper and an SSA fiber-optic window, and the quantum yield of an SSA.

Definitions for the parameters that will be used are summarized as follows.

Q_{ssa}	stored SSA charge per input photoelectron from the photocathode
Y_k	photocathode quantum yield, e/photon
G_m	VMCP electron gain, e/e
V_m	VMCP applied potential, V
V_s	MCP-to-screen applied potential, V
V_d	phosphor screen "dead-voltage"
P	phosphor screen efficiency, photon/eV
T_{fot}	transmission of fiber-optic taper
T_{ssa}	transmission of fiber-optic window on the SSA
Y_{ssa}	quantum yield of SSA, e/photon
e	electron charge, 1.6E − 19 C
N_{essa}	number of stored SSA electrons per input photoelectron

Using these definitions, the general equation for the charge stored in the SSA per input photoelectron is given by Eq. (16).

$$Q_{ssa} = e \cdot G_m \cdot (V_s - V_d) \cdot P \cdot T_{fot} \cdot T_{ssa} \cdot Y_{ssa} \tag{16}$$

Thus, Q_{ssa} is given by the product of the VMCP gain, the effective electron bombardment energy at the aluminized phosphor screen, the conversion efficiency of the phosphor screen assembly, the transmissions of the FO taper and the SSA's FO window, and finally the quantum yield of the SSA.

For

$$V_m = 1380\ \mathrm{V}$$

the VMCP electron gain is

$$G_m(V_m) = \left(\frac{V_m}{700V}\right)^{17}$$

$$G_m(V_m) = 1 \times 10^5$$

By using the following values for the additional parameters

$V_s = 5500$ V
$V_d = 2500$ V
$P = 0.06$ photon/eV
$T_{fot} = 0.6$
$T_{ssa} = 0.8$
$Y_{ssa} = 0.5$ e/photon

it is found that the stored charge per photoelectron is

$$Q_{ssa} = 7 \times 10^{-13}\ \mathrm{C}$$

and that the number of electrons stored in the SSA per input photoelectron is

$$N_{essa} = Q_{ssa}/\mathrm{e}$$

$$N_{essa} = 4 \times 10^6 \text{ electrons}$$

Since the full-well or saturation charge for an SSA pixel in on the order of 1 pC, this VMCP II/SSA assembly is seen to qualify as a photon-counting imaging detector.

Modulation Transfer Function and Limiting Resolution. The modulation transfer function (MTF) of an II SSA camera is determined by a convolution of the individual MTFs of the camera lens, II, II-to-SSA-coupling fiber-optic or lens, fiber-optic-to-fiber-optic interfaces, fiber-optic-to-SSA interface, SSA, etc. There are several ways to determine the MTF of an existing II SSA camera.

For example, the II SSA camera can be focused on a spatial frequency burst pattern, i.e., a periodic pattern of black and white bars in which the spatial frequency of the bars increases in one direction. Alignment of the pattern's bars with pixel columns and readout of the modulation of the spatial pattern in the pixel row direction gives the squarewave MTF of the camera S(f), where f is the spatial frequency in cycles/mm. Conversion of this square-wave MTF to a sine-wave MTF is accomplished by using the Fourier transform at a given frequency:

$$T(f) = \left(\frac{\pi}{4}\right) \cdot \left(S(f) + \frac{S(3f)}{3} - \frac{S(5f)}{5} + \frac{S(7f)}{7} + \cdots\right) \tag{17}$$

By calculating several values of $T(f)$ from the known square-wave function, the sine-wave MTF can be determined and used for camera system optical image transfer analysis. The limiting resolution is often taken to be the spatial frequency value for this sine-wave MTF at which the modulation drops to a few percent.

Also, to use the above example as an illustration, the MTF of an II SSA is a function of the direction in which the spatial frequency burst pattern is aligned. Self-scanned array pixels are not generally square, and the distances between centers of pixels in the horizontal and vertical directions are not generically the same. Thus, the corresponding MTFs in the horizontal and vertical directions are different, and the MTF is a function of the angle that the burst pattern makes with the rows and columns of pixels.

A convenient specification of the spatial frequency response of an SSA or an II SSA camera is the Nyquist frequency (f_N), defined to be the reciprocal of twice the distance between the pixels. For example, if an SSA has rows of pixels spaced on 20-μm center-to-center, then the horizontal Nyquist spatial frequency is

$$f_{N,h} = \frac{1}{(2 \cdot (0.02 \text{ mm}))} = 25 \text{ cycles/mm}$$

Assuming an II limiting resolution (f_{II}) of 32 cycle/mm, and assuming that the MTFs are gaussian, then an estimated value for the limiting resolution of the II SSA camera is

$$\left(\frac{1}{f_{\text{cam}}^2}\right) = \left(\frac{1}{f_{\text{II}}^2}\right) + \left(\frac{1}{f_{\text{ssa}}^2}\right) \tag{18}$$

or

$$f_{\text{cam}} = \frac{f_{\text{II}} \cdot f_{\text{ssa}}}{\sqrt{(f_{\text{II}}^2) + (f_{\text{ssa}}^2)}} \tag{19}$$

For the values used in this example, $f_{\text{cam}} = 20$ cycle/mm. Although this gaussian estimate is convenient to use, a more exact estimate can be made by multiplying the various component sine-wave MTFs to find the II SSA camera's MTF, and from this its limiting resolution can also be found.

For example, recent actual MTF measurements[9] on an II SSA camera, having a proximity-focused 18-mm active diameter MCP II fiber-optically coupled to a CCD with an $m = 8 \text{ mm}/18 \text{ mm} = 0.44$ magnification taper, showed that the MTF of the CCD, referred to the II input, is given by $T_{\text{CCD}}(f) = \exp - (f/9.0)^{1.4}$, where f is the spatial frequency in cycles/mm. The Nyquist frequency of the CCD was $f_N = 20$ cycle/mm, and the MTF of the complete camera was found to be $T_{\mathrm{II\,SSA}}(f) = \exp - (f/6.3)^{1.1}$, both referred to the II input.

21.6 APPLICATIONS

In time, it is expected that most of the quantum-limited and LLL TV applications will use some form of II SSA camera, instead of an intensified vidicon-based camera. A few of the major application areas for II SSA cameras are highlighted in this section.

Optical Multichannel Analyzers

Optical multichannel analyzers (OMAs) are instruments used to measure optical radiation in linear patterns, e.g., spectra or two-dimensional images. Photographic film, single-channel photomultiplier tubes, and TV camera tubes, e.g., vidicons, have been replaced by SSAs, e.g., CCDs, and II SSAs, e.g., image tube intensified CCDs or photodiode arrays (PDAs). Four distinct application areas exist for OMAs using II SSA detectors[16]:

Application	Detector*	Time
	Type	Resolution
Spectroscopy	IILPDA	50 ms
Time-resolved pulsed laser spectroscopy	GIILPDA	<5 ns
Time-resolved pulsed laser imaging spectroscopy	GIISSCCD	<5 ns/spectrum
Time-resolved imaging	GIISSCCD	<5 ns/spectrum

* IILPDA = image intensified photodiode array; GIILPDA = gated image intensified photodiode array; GIISSCCD = gated image intensified charge-coupled device.

In each of these types of systems, the incoming radiation is converted to charge packets that are stored in each pixel of the SSA. Each line and/or field of pixels is read out by a suitable camera electronics, and the pixel charge values are stored in a computer for subsequent processing and analysis. The sensitivity of the II is chosen for best performance over the range of wavelengths being investigated. The dynamic range of OMAs can be as high as 18 bits. In comparison with the older single-element scanning system, modern II SSA-based OMAs acquire spectra up to 1000 times faster and/or with higher SNR during a given measuring period. Three common OMA applications are Raman spectroscopy, multiple input spectroscopy, and small-angle light scattering.

Range Gating and LADAR

Range gating is becoming increasingly important because the required technology now exists at an affordable cost and the signal-to-noise ratio improvement is much higher than nongated conventional TV imaging. A gated laser sends out a laser pulse of only a few nanoseconds duration while the II SSA camera is gated off. No reflection from scattering or reflection in the medium between the camera and the object is allowed to be registered in the II SSA camera.[17] The II SSA camera is gated on only at the moment when the light packet from the object returns to the camera and, after exposing for the duration of the outgoing pulse, it is returned to a gated-off condition. By repeating this process and controlling the image-storing conditions, high-contrast images having high signal-to-noise ratios can be achieved. Another obvious advantage of the range-gated system is that the time-of-flight between pulse output and receipt gives the range to the object being viewed, thus leading to the realization of a *la*ser *d*etection *a*nd *r*anging (LADAR) system.

Microchannel plate image tubes offer high-speed gating and spectral response advantages to LADAR systems. They can be electronically gated to a few nanoseconds, i.e., providing distance resolutions of a few feet for LADAR systems. Present MCP IIs offer the user a broad range of spectral sensitivity, including near-ir imaging at 1060 nm, so that powerful and efficient lasers may be used for optimum LADAR system performance. Also, the ruggedness, extremely small size, and low power drain characteristics of II SSA cameras make them very attractive for LADAR applications for spacecraft and unmanned autonomous vehicles.

Day/Night Cameras

Full day-night interscene dynamic range capability, while maintaining a high signal-to-noise ratio, is achievable using an II SSA camera in conjunction with an auto-iris camera lens.[18] The principle of operation of this type of camera can be described as follows. Assume that operation begins at the lowest light level to be encountered. The MCP voltage in the II is set to operate with high SNR for the camera lens, an auto-iris lens, set to its lowest f-number. As the light level is increased, the system operates over two orders-of-magnitude dynamic range. For four orders-of-magnitude higher light level inputs, the f-setting of the auto-iris lens is increased by a feedback circuit, driven by the peak-to-peak video output signal from the SSA camera. The effective exposure of the SSA is next automatically reduced as the light level increases by four-and-one-half orders-of-magnitude, again to maintain a high SNR, by duty-cycle gating the MCP II. Finally, another two-and-one-half orders-of-magnitude in input light level are accommodated by reducing the gain of the MCP, i.e., by reducing its applied operating potential. The total interscene dynamic range achievable with this type of automatically controlled day/night camera is 13 orders-of-magnitude. In addition to its wide dynamic range capability, this type of camera is also able to make rapid narrow-band spectral samples across a wide spectral range, e.g., from the uv to the near-ir.

Mosaic II SSA Cameras

For very high amounts of image information throughput, multiple SSAs are used to read out large area IIs. For example, a 75-mm active diameter MCP II can be coupled fiber-optically to four individual SSA cameras. The fiber-optic couplers are made to butt against each other at the output of the II, and their output ends are optically coupled to the SSAs. Parallel readout of the SSAs is then accomplished, giving the advantage of high-resolution readout without the disadvantage of having to use a wide bandwidth video electronic system or lower frame rates.

One such II SSA camera is designed for x-ray radiology image input.[19] The x-ray input image is converted to a visible light image at a scintillator screen that is in optical contact with the 6-inch-diameter fiber-optic window. This scintillator/window assembly is, in turn, coupled to the input of a 6-inch-diameter proximity-focused diode II, i.e., without an MCP, for modest light gain and good image quality. Six fiber-optic tapers in a 2×3 matrix couple the images from the six adjacent output sections of the II to six CCD cameras which operate in parallel to continuously read out the converted x-ray image.

Other Applications

Other applications for II SSA cameras are the following:

- Semiconductor circuit inspection
- Astronomical observations
- X-ray imaging
- Coronary angiography
- Mammography
- Nondestructive testing
- Multispectral video systems

Active Imaging. Active imaging is becoming increasingly important because the required technology now exists at an affordable cost and the signal-to-noise ratio improvement is much higher than nonactive conventional TV imaging. Two types of active imaging presently exist, i.e., "line-scanned" and "range-gated."

In a line-scanned imaging system, a narrow beam from a cw laser is raster-scanned across the object to be viewed, and the resulting reflected light is collected by a lens and detector assembly which receives and measures light from the illuminated field-of-view (FOV). Large FOV scenes can be scanned in a short period of time, which is a major advantage of this system. The signal from the detector is finally processed by a video electronics system and displayed, for direct viewing, or image processed as required. Limiting-resolution is set by the beam diameter achievable at the object. Thus systems operating in space; atmospheric and underwater environments have significantly different limiting-resolution characteristics.

In a range-gated type of system, a gated laser sends out a laser pulse of only a few nanoseconds duration while the TV camera is gated off. No reflection from scattering or reflection in the medium between the camera and the object is allowed to be registered in the TV camera. The TV camera is gated on only at the moment when the light packet from the object returns to the camera and, after exposing for the duration of the outgoing pulse, the TV camera is returned to a gated-off condition. By repeating this process and controlling the image-storing conditions, high-contrast images having high signal-to-noise ratios can be achieved.

Another obvious advantage of the range-gated system is that the time of flight between pulse output and receipt gives the range to the object viewed, thus leading to the realization of a laser detection and ranging (LADAR) system. Microchannel plate image tubes offer high-speed gating and spectral response advantages to LADAR systems. They can be electronically gated to a few nanoseconds, i.e., distance resolutions of a few feet, and in some parts of the optical spectrum they offer high sensitivity.

21.7 REFERENCES

1. A. Rose, "A Unified Approach to the Performance of Photographic Film, Television Pickup Tubes, and the Human Eye," *J. Soc./Motion Picture Engrs.* **47:**273–294 (1946).
2. D. E. Caudle, "Dynamic Range Enhancement Techniques for Gated, Solid-State Intensified Cameras," *SPIE* **1155:**104–109 (1990).
3. G. Holst, J. H. de Boer, and C. F. Veenemans, *Physica* **1:**297 (1934).
4. HOT MCP™ is a trademark of Galileo Electro-Optics Corp.
5. E. H. Eberhardt, "An Operational Model for Microchannel Plate Devices," *IEEE Trans. Nucl. Sci.* **NS-28:**712–717 (1981).

6. "Optical Characteristics of Cathode Ray Tubes," EIA Publication, no. 116-A, 1985.
7. H. P. Westman, ed., *Reference Data for Radio Engineers,* 5th ed., Howard W. Sams & Co, Inc, Indianapolis, 1969, pp. 16-34–16-37.
8. J. J. Cuny, T. F. Lynch, and C. B. Johnson, "Proximity Focused Image Tube Intensified Charge Injection Device (CID) Camera for Low Light Level Television," *SPIE* **203:**75–79 (1979).
9. G. M. Williams, Jr., "A High Performance LLLTV CCD Camera for Nighttime Pilotage," *SPIE* **1655:**14–32 (1992).
10. J. M. Abraham, L. G. Wolfgang, and C. N. Inskeep, "Application of Solid-State Elements to Photoemission Devices," *Adv. EEP* **22B:**671 (1966).
11. J. B. Barton, J. J. Curry, and D. R. Collins, "Performance Analysis of EBS-CCD Imaging Tubes/Status of ICCD Development," *Proc. Int. Conf. Apps. of CCDs,* San Diego, Calif., 1975.
12. J. T. Williams, "Test Results on Intensified Charge Coupled Devices," *SPIE* **78:**78–82 (1976).
13. J. J. Cuny, T. F. Lynch, and C. B. Johnson, "Small Intensified Charge Injection Device Cameras for Low Light Television," *MEDE '79 Conference Proceedings,* Interavia SA, Publishers, 1979, pp. 836–845.
14. J. C. Richard, M. Vittot, and J. C. Rebuffie, "Recent Developments and Applications of Electron-Bombarded CCD in Imaging," *Proc. Conf. on Photoelectronic Image Devices,* SEP91, London, IOP Pub. Ltd., 1992.
15. A. Boksenberg and D. E. Burgess, "An Image Photon Counting System for Optical Astronomy," *Adv. EEP* **33B:**835–849 (1972).
16. H. W. Messinger, "Modern Optical Multichannel Analyzers Capture Images Without Film," *Laser Focus World,* March 1992, pp. 91–94.
17. R. A. Sturz and D. E. Caudle, "Capabilities of New Cost-Effective Near Infrared Imaging," *Advanced Imaging,* April 1993, pp. 60–62.
18. D. E. Caudle, "Low Light Level Imaging Systems Application Considerations and Calculations," *SPIE* **1346:**54–63 (1990).
19. H. Roehrig et al., "High Resolution X-Ray Imaging Device," *SPIE* **1072:**88–99 (1989).

CHAPTER 22
VISIBLE ARRAY DETECTORS

Timothy J. Tredwell
Eastman Kodak Company
Sensor Systems Division
Imager Systems Development Laboratory
Rochester, New York

22.1 GLOSSARY

A	area of the pixel
C_{FD}	total capacitance of the floating diffusion in a CCD output
C_g	gate capacitance per unit area
C_r	readout line capacitance
$\mathbf{J}_D$	total dark current per unit area
$\mathbf{J}_s$	surface generation current
L_e	diffusion length of electrons in silicon
L_p	diffusion length of holes in silicon
N_A	p-type dopant concentration in silicon
N_D	n-type dopant concentration in silicon
N_e	total number of electrons collected in a pixel
$N(0)$	number of photons entering the silicon
$N(x)$	number of photons remaining a distance x below the surface
n_i	intrinsic carrier concentration in silicon
$P(x)$	probability that an electron-hole pair generated a distance x from the surface will be collected before recombination
$\mathbf{q}$	electron charge
s_o	surface recombination velocity
T_{int}	integration time of light in an image sensor
$T(\lambda)$	transmission of light
V_{bi}	built-in voltage for a silicon pn junction
$W(V)$	width of the depletion layer at a given bias voltage in the MOS capacitor or junction-photodiode

$\alpha(\lambda)$	absorption coefficient of light
ε_s	silicon dielectric constant
μ_e	electron mobility in silicon
μ_p	hole mobility in silicon
τ_o	depletion-layer lifetime
τ_p	minority carrier hole lifetime in silicon
Φ_s	electrostatic potential at the silicon-silicon dioxide, also called surface potential

22.2 INTRODUCTION

Since the invention of the image sensor in 1964, solid state image sensors have advanced in resolution, sensitivity, and image quality to the point where they have replaced other methods of converting visible light to electronic signals in nearly all imaging applications. There are two types of image sensors: *area image sensors,* which are used in cameras, and *linear sensors,* which are used in scanning applications. Cameras using area image sensors dominate the camcorder, video and broadcast, machine vision, scientific, and medical fields. Area image sensors for camcorder applications are typically 400,000 picture elements in resolution, 60 dB in dynamic range, and have noise levels of a few tens of electrons. Area image sensors for scientific applications may have resolutions of over six million elements, dynamic ranges exceeding 80 dB, and noise levels approaching the single electron level. Production of area image sensors exceeded eight million devices in 1993. Scanners employing linear solid-state image sensors dominate facsimile, document scanner, digital copier, and film scanner applications. Linear sensors range from 2000-element monochrome arrays with 40 dB of dynamic range used in facsimile applications to 8000 or more element trilinear arrays with 80 dB of dynamic range for high-performance color scanning applications. Production of linear CCD image sensors also exceeded eight million devices in 1993.

The steps involved in image sensing consist of (1) converting the incoming photons to charge at picture element (pixel), and (2) transferring that charge to an output amplifier and converting the charge to a voltage or current signal which can be sensed by circuits external to the sensor. The image-sensing elements will be described first, followed by readout elements. Sensor architectures for area and linear sensors will then be described.

22.3 IMAGE SENSING ELEMENTS

There are four basic types of structures which are used for image sensing: the junction photodiode, the photocapacitor, the pinned (p^+np) photodiode, and the photoconductor.* The first three are generally fabricated in single-crystal silicon as part of the image sensor; the photoconductor is usually fabricated from amorphous silicon deposited over the image sensor. The photoconversion process begins with the absorption

* For some scientific applications in which high quantum efficiency and fill factor are essential, the silicon wafer will be thinned to 10 μm or less in thickness and illuminated from the backside. The frontside contains an area charge coupled device, which is used to collect the photogenerated carriers.

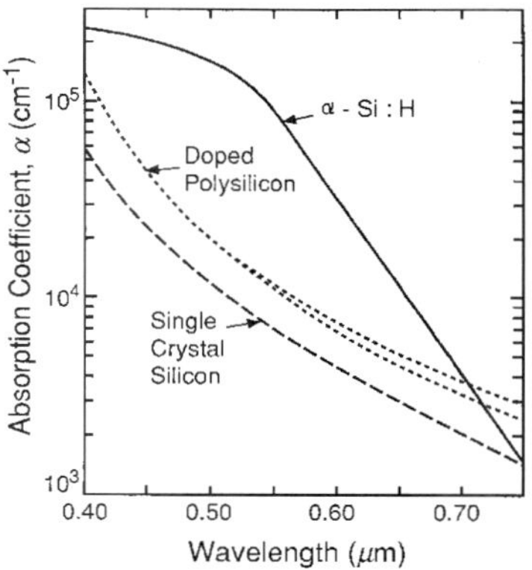

FIGURE 1 Absorption coefficient for light in single-crystal silicon, heavily doped polycrystalline silicon, and hydrogenated amorphous silicon as a function of wavelength.

of a photon in the silicon resulting in the generation of a single electron-hole pair. The absorption of light at a particular wavelength is given by:

$$N(x, \lambda) = N(0)e^{-\alpha(\lambda)x} \tag{1}$$

where $N(x)$ is the number of photons remaining a distance x below the surface, $N(0)$ is the number of photons entering the silicon, and $\alpha(\lambda)$ is the absorption coefficient.[1] The absorption coefficient is shown as a function of wavelength λ for single-crystal silicon, doped polycrystalline silicon, and hydrogenated amorphous silicon in Fig. 1. The absorption depth is defined as the inverse of the absorption coefficient $[d(\lambda) = 1/\alpha(\lambda)]$. In single-crystal silicon the absorption depth is 0.4 μm in the blue (450 nm), 1.5 μm in the green (550 nm), and 3.0 μm in the red (640 nm). In the infrared, the absorption depth increases to 10.5 μm at 800 nm. Beyond 1100 nm, the absorption is virtually zero because the photon energy is less than the 1.1-eV silicon bandgap.

Junction Photodiode

The junction photodiode is one of the most common image-sensing elements. The physical structure and band diagram of the junction photodiode are shown in Fig. 2*a* for a p-type substrate. The n-type region is formed by ion implantation or diffusion of phosphorous or arsenic to a depth of 2000 to 10,000 Å into the p-type silicon. The n-type dopant region is usually graded, with the highest concentration at the surface. The gradient in n-type dopant concentration results in a gradient in electrostatic potential which accelerates photogenerated carriers (holes in the n-type region) away from the surface. This reduces loss of photogenerated carriers to surface recombination. The photodiode is typically operated with a reverse bias V of 1 to 5 volts. A depletion layer is formed between the n- and p-type regions.* The width of the depletion layer $W(V)$ for an abrupt $n+p$ junction is given by:

$$W(V) = \sqrt{\frac{2\varepsilon_s(V + V_{\text{bi}})}{\mathbf{q}N_A}} \tag{2}$$

* For a detailed review of the device physics of junction diodes and MOS capacitors, see Sze, *Physics of Semiconductor Devices,* Wiley, New York, 1969.

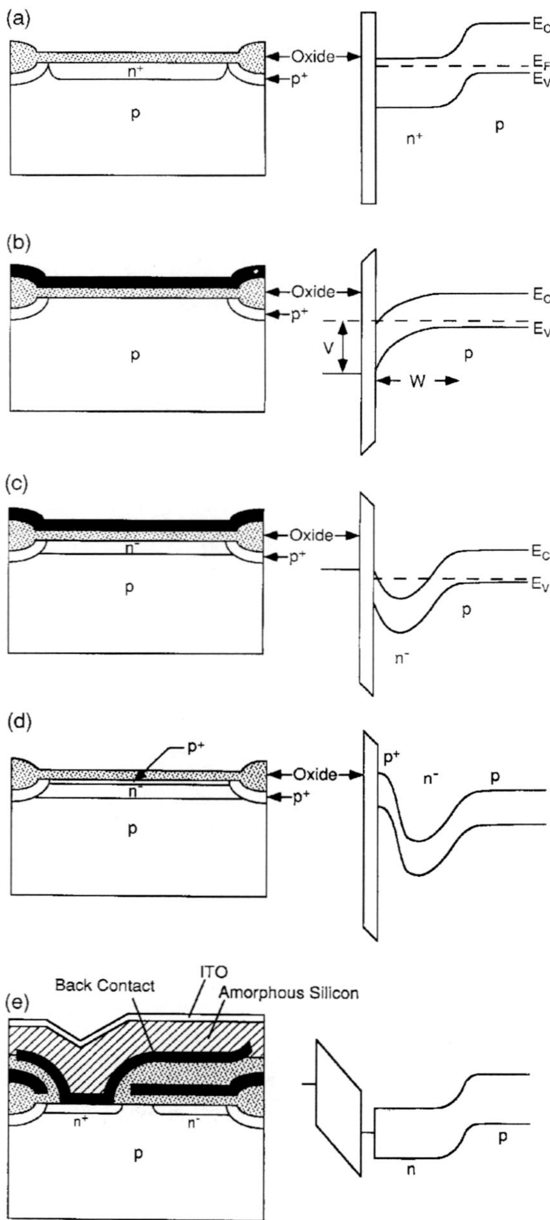

FIGURE 2 Cross-sectional diagrams and band diagrams for (*a*) junction photodiode; (*b*) surface-channel MOS capacitor; (*c*) buried-channel MOS capacitor; (*d*) pinned, or hole-accumulated, diode; (*e*) amorphous silicon photoconductor.

where $\mathbf{q}$ is the electronic charge, ε_s is the silicon dielectric constant, N_A is the p-type dopant concentration, and V_{bi} is the built-in voltage given by

$$V_{\text{bi}} = \frac{kT}{\mathbf{q}} \ln \frac{N_A}{n_i} \tag{3}$$

where n_i is the intrinsic carrier concentration. For silicon doped at $1 \times 10^{16}\,\text{cm}^{-3}$, the depletion width at 5-V reverse bias is 0.8 μm.

If the diode is illuminated, some of the photons will be absorbed in the n-region, some in the depletion layer, and the remainder in the p-type substrate. The quantum efficiency $\eta(\lambda)$ is the ratio of the charge collected to the number of photons incident on the diode (i.e., 100-percent quantum efficiency refers to one electron-hole pair collected for every incident photon). The quantum efficiency depends on three factors: transmission $T(\lambda)$ of light through the overlying layers into the silicon, absorption of light in the silicon, and the probability $P(x)$ that an electron-hole pair generated a distance x from the surface will be collected before recombination:

$$\eta(\lambda) = T(\lambda) \int_{x=0}^{x=\infty} (1 - \mathbf{e}^{-\alpha(\lambda)x}) P(x)\, dx \tag{4}$$

The transmission of light through the overlying layers into the silicon can be calculated using standard multilayer interference models.

The collection of the photogenerated charge takes place by two processes: drift and diffusion. Drift is the movement of electrons and holes due to an electric field. Even for small electric fields, transport of carriers by drift will dominate diffusion. This is the case in the depletion region. Outside the depletion region, such as in the p-type substrate, there is no electric field, and carrier transport occurs by diffusion. For example, a photon absorbed in the p-type region will excite an electron from the valence band to the conduction band. The electron will move in a three-dimensional random walk until it recombines or until it encounters the edge of the depletion region, where it is swept across the junction by the electric field. The probability that an electron a distance x' from the junction can diffuse to the junction before recombining is given by:

$$P(x') = \mathbf{e}^{-x'/L_\mathbf{e}} \qquad \text{where } L_\mathbf{e} = \sqrt{\frac{KT}{\mathbf{q}} \mu_\mathbf{e} \tau_\mathbf{e}} \tag{5}$$

in which $L_\mathbf{e}$ is the diffusion length, $\mu_\mathbf{e}$ is the electron mobility, and $\tau_\mathbf{e}$ the electron lifetime (typically $\sim$1 μs). For a 1-μs lifetime, the electron diffusion length in p-type silicon is 50 μm. Electrons generated from photons absorbed less than the diffusion length from the junction have a high probability of collection.*† Similarly, photons absorbed in the n-type

---

* In image sensors, the collection of photogenerated carriers can be complicated by a variety of factors. The doping concentration may not be uniform on either the n- or p-sides. On the n-side, the dopant concentration is designed to decrease from the surface to the junction, building in a potential gradient for holes away from the surface and toward the junction, preventing surface recombination. On the p-side, the dopant concentration may not be uniform owing to the use of epitaxial layers or wells diffused into the silicon. Additionally, the carrier lifetime may not be uniform. Impurity gettering, a process used in many image sensors to remove metallic contaminants during processing, will leave a region of crystalline defects in the silicon starting 20 to 50 μm beneath the silicon surface. The defects result in a very short electron lifetime in this region. Finally, diffusion takes place in three dimensions; a carrier absorbed beneath a given pixel in an array will diffuse laterally by the same amount it diffuses vertically. This could cause it to be collected in adjacent pixels.

† See, for example, Lavine et al., "Steady State Photocarrier Collection in Silicon Imaging Devices," *IEEE Transactions on Electron Devices* **ED-30:**1123–1133 (Sept. 1983).

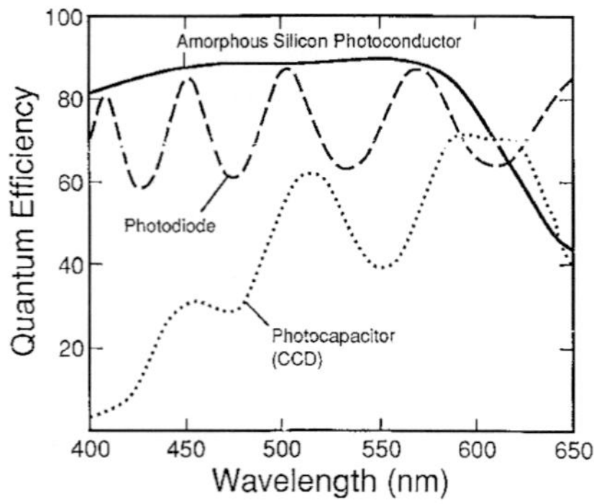

FIGURE 3 Quantum efficiency as a function of wavelength for photocapacitor, photodiode, and amorphous silicon photoconductor.

layer create electron-hole pairs. The holes must travel by diffusion to the junction to be collected. The probability that a hole a distance x' from the junction can diffuse to the junction is given by

$$P(x') = \mathbf{e}^{-x'/L_p} \qquad \text{where } L_p = \sqrt{\frac{kT}{\mathbf{q}}\mu_p \tau_p} \tag{6}$$

in which L_p is the diffusion length, μ_p is the hole mobility, and τ_p is the hole lifetime. For a 1-μs lifetime, the hole diffusion length in n-type silicon is 30 μm. Since the n-type region in an $n+p$ junction is less than 1 μm thick, the holes have no difficulty diffusing through the n-type region to the junction unless the n-type region is so heavily damaged or so heavily doped that the hole lifetime is very short. More typically, loss of quantum efficiency on the n-side results from recombination at the surface.

The quantum efficiency of a junction photodiode is illustrated as a function of wavelength in Fig. 3. In the ultraviolet, the light is absorbed very near to the surface and some of the photogenerated charge can be lost to surface recombination. In the 420- to 700-nm region, most of the light is absorbed close to the junction and is easily collected. The structure in the quantum efficiency illustrated in Fig. 3 results from the multilayer reflections of light in the oxide layer which overlies the photodiodes used in image sensors. The positions of the minima and maxima depend on the thicknesses and induces of refraction of the layers overlying the silicon. Beyond 800 nm, some of the photons are absorbed sufficiently deep in the silicon that the carriers recombine before they can diffuse to the junction; this results in the decrease in quantum efficiency at longer wavelengths.

In most image sensing applications, the junction diode at each picture element is used not only to collect the photogenerated carriers but also to store the carriers until they can be read out. In an imaging array, each photodiode would be reset to a reverse bias V, by a MOS gate. The capacitance of the diode of area A for an abrupt $n+p$ junction is given by $C(V) = \varepsilon_s A/[W(V)]$ where $W(V)$ is the depletion width.

When a photogenerated carrier is collected by the junction, it is stored on the junction until it is read out. Storage of a carrier will cause the voltage on the junction to decrease by $q/C(V)$. When the photogenerated charge is removed from the junction during readout, the junction voltage is restored to its original value.

If so much photogenerated charge is stored on the photodiode that the voltage drops to zero before the charge can be read out, additional charge cannot be collected and will diffuse into the p-type substrate. This condition is referred to as saturation. The diffusion of excess charge into neighboring photosites is called *blooming*.

One of the difficulties encountered in using the junction photodiode in image sensor applications is *image lag*. The combination of the capacitance of the photodiode and the channel resistance of the MOS transfer gate used to read out the diode give rise to a time constant for transferring the photogenerated charge from the diode to the readout structure. As a result, not all the charge can be completely drained from the diode during the short reset times typically used in imaging applications. The remaining charge is drained in successive readouts, causing an afterimage. This effect is called image lag.[3]

MOS Capacitor

The MOS capacitor consists of the silicon substrate (taken to be p-type in this section), a thin layer of silicon dioxide (typically 200 to 1000 Å thick), and an electrode (typically polycrystalline silicon doped heavily n-type with phosphorous). The physical structure and the band diagrams of the surface channel MOS capacitor are shown in Fig. 2b for a p-type substrate. If the gate of the MOS capacitor is biased positive, a depletion layer is created in the p-type silicon substrate. The depth of the depletion layer depends on the substrate doping, gate voltage, and oxide thickness. The calculation of the depth of the depletion layer is somewhat more complex than the photodiode* and depends on the electrostatic potential at the surface, called the surface potential Φ_s. For values typical of image sensors ($N_A = 1 \times 10^{15}$, 5-V gate bias, 500-Å oxide thickness) the depletion layer is 2.4 μm deep. On the edges of the photocapacitor (see Fig. 2b) is a heavily doped p-type region overlaid by a thick (2000 to 5000-Å) oxide layer. This is called the field or *channel stop* region. Because of the heavier p-type doping, the field is not depleted by the voltage on the gate. The channel stops confine the electrons to the channel region.

If the MOS capacitor is illuminated, a fraction of the light will be reflected, a fraction will be absorbed in the polysilicon, and the rest will be transmitted into the silicon substrate. The absorption coefficient of heavily doped polysilicon is shown as a function of wavelength in Fig. 1. The absorption coefficient is $4 \times 10^4\,\text{cm}^{-1}$ at 450 nm and $1.2 \times 10^4\,\text{cm}^{-1}$ at 550 nm. For a 3000-Å-thick polysilicon electrode, less than 30 percent of the blue light and less than 70 percent of the green light is transmitted through the polysilicon. The photons entering the silicon are absorbed, either in the depletion layer of the MOS capacitor or in the undepleted p-type silicon beneath the depletion layer. Those photogenerated electrons created in the undepleted p-type region move by diffusion until they are captured by the depletion layer or until they recombine. The electrons are held at the silicon-SiO_2 interface, where they remain until they are read out. The quantum efficiency as a function of wavelength is illustrated in Fig. 3. The efficiency is low in the blue owing to absorption in the polysilicon. The structure in the quantum efficiency as a function of wavelength is due to multilayer interference in the polysilicon-oxide-silicon stack.

Charge is stored in the MOS capacitor at the silicon-silicon dioxide interface as a layer of sheet-charge only a few hundred angstroms thick. As additional photogenerated charge is added, the surface potential decreases. If sufficient photogenerated charge is added, the surface potential becomes zero and no additional charge can be stored. This condition is saturation.

The capacitance per unit area on which photogenerated charge is stored is the parallel capacitance of the oxide and the depletion layer:

$$C^{-1} = \left(\frac{t_{ox}}{\varepsilon_{ox}} + \frac{W(\Phi_s)}{\varepsilon_{si}}\right) \tag{7}$$

* To determine the depletion depth, the surface potential ϕ_s must be calculated. Φ_s depends on the voltage on the gate of the MOS capacitor, the oxide thickness, substrate doping, and weakly on temperature. See Sze, *Physics of Semiconductor Devices*.

where the depletion width is $W(\Phi_s)$.

$$W(\Phi_s) = \sqrt{\frac{2\varepsilon_{\text{si}}\Phi_s}{\mathbf{q}N_A}} \tag{8}$$

in nearly all cases, the oxide capacitance is the dominant term. As a result, the storage capacity of the MOS capacitor is significantly larger than the junction diode in which the charge is stored only on the depletion capacitance.

A variant of the surface-channel photocapacitor is the buried-channel photocapacitor. The structure and band diagram are shown in Fig. 2*c*. In this device, a lightly-doped n-type region is diffused or implanted into the silicon surface early in the fabrication process. This n-type region is sufficiently lightly doped that it is fully depleted. The n-type dopant in the buried channel results in a band diagram with a potential minimum, or well, for electrons just below the surface.* This well is separated from the surface by a few thousand angstroms in distance and about 1 V in potential. When the buried-channel photocapacitor is illuminated, the electrons collect in the buried channel and do not contact the surface. The primary purpose of the buried channel is to prevent electrons from being trapped by interface states at the silicon-silicon dioxide interface.

The photocapacitor has several advantages over the junction photodiode. These include higher storage capacity per unit area, zero lag readout, and generally lower dark current. The principal disadvantage is the low quantum efficiency in the blue. In some applications, a transparent electrode, such as indium-tin-oxide (ITO) may be substituted to improve the blue response.[4] ITO has very low absorption over the visible (420 to 750 nm) and can be deposited sufficiently conductive for use as a gate electrode in an image sensor. Materials and processing complexity have prevented incorporation of ITO into commercial image sensors in volume until recently.

Another approach to achieving high quantum efficiency is the thinned backside-illuminated charge-coupled device (CCD). In this approach, the CCD (which is an array of MOS capacitors) is fabricated on the frontside of a silicon wafer. The wafer is then thinned from the backside to 10 μm or less in thickness. The backside is passivated to prevent surface recombination. Photons entering the backside are absorbed in the silicon beneath the MOS capacitors (usually buried channel). The photogenerated carriers diffuse to the capacitors, where they are held until they are read out. This device has quantum efficiency similar to the photodiode in the visible. However, because the silicon is thin, some of the photons at wavelengths beyond 700 nm will not be absorbed and so the quantum efficiency falls off beyond 700 nm. Owing to their extreme complexity in process and their extremely fragile design, backside-illuminated image sensors are limited to special scientific applications, especially astronomy.

Pinned Photodiode

The third type of photosensitive element is the pinned (p^+np) photodiode.[5] This is sometimes called the *hole accumulation diode,* or HAD. This element combines the best features of the photodiode and photocapacitor, offering the high blue response of the photodiode with the high charge capacity, zero lag, and low dark current of the buried-channel photocapacitor. The pinned photodiode consists of a very shallow (<2000 Å) P^+ layer overlying an n-type buried-channel region. The structure and band

* For a review of the calculation of electrostatic potential and charge capacity of buried-channel MOS capacitors and charge-coupled devices, see B. C. Burkey, G. Lubberts, E. A. Trabka, and T. J. Tredwell, *IEEE Transactions on Electron Devices* **ED-31**(4):423 (April 1984).

diagrams are shown in Fig. 2*d*. The p^+ surface layer, which contacts the p^+ channel stop region on the sides, holds the electrostatic potential at the surface at 0 V. When the photodiode is illuminated, the photogenerated electrons are held in the n-type buried-channel region just below the surface.

The quantum efficiency of the pinned photodiode is nearly identical to that of the photodiode shown in Fig. 2. Because there is no overlying polysilicon electrode, the blue response is very high, similar to that of the photodiode. Because the buried-channel region can be completely emptied, the pinned photodiode does not have the lag of a normal junction (n^+p or p^+n) photodiode. The pinned photodiode is becoming the most widely used image-sensing element in interline area CCDs used for camcorders and for industrial and medical cameras. The pinned photodiode is also used in some linear image sensors, particularly where low image lag is critical.

Photoconductor

The last type of photosensitive device is the photoconductor. The most common material for the photoconductor is hydrogenated amorphous silicon, although other material systems have been explored. Amorphous silicon photoconductors have been used for two types of image sensors: area image sensors, where it is deposited on top of an area array to improve fill factor (i.e., the proportion of the picture element which is photosensitive), and contact linear sensors, where it is deposited on large ceramic or glass substrates to fabricate very long line or linear arrays.

The structure of an amorphous silicon photoconductor on a CCD image sensor and the corresponding band diagrams are shown in Fig. 2*e*. The hydrogenated amorphous silicon photoconductor consists of a back electrode, an undoped amorphous silicon layer approximately 1 μm thick, and a transparent top electrode.[6] Additional doped amorphous silicon or silicon nitride layers may be added to the amorphous silicon front or back surfaces to improve performance. Photons absorbed in the amorphous silicon generate electron-hole pairs. Photogenerated electrons and holes are quickly swept to the back and front electrodes, respectively, because of the high electric field in the photoconductor. When the amorphous silicon is used as part of an area image sensor, the electrons on the back electrode can be transferred into the readout element; when used as part of a contact linear array, the voltage change can be amplified and read out through a multiplexer.

The quantum efficiency of an amorphous silicon photoconductor is shown in Fig. 3. Owing to the wider bandgap of the amorphous silicon, photons of wavelength greater than about 650 nm are not absorbed. The advantage of the amorphous silicon is the high quantum efficiency across the visible wavelengths and the ability to fabricate devices either on top of area CCDs for higher fill factor or to process on glass or ceramic for very large linear sensors. The disadvantages include charge trapping at defects in the amorphous silicon and low carrier mobility, both of which lead to field-dependent nonlinear response and image lag when used in an image sensor. Recent improvements in material and device technology have mitigated many of these disadvantages at the expense of process and device complexity.[7]

Antiblooming in Charge-sensing Elements

Blooming in image sensors occurs when the charge generated in an image-sensing element exceeds its capacity. If no method is provided to remove this excess charge, it will be injected either onto the readout element (CCD or MOS readout line) or into the substrate. If the excess charge is injected onto the readout element, it will usually appear as a bright line in the image. If it is injected into the substrate, the charge can diffuse in a circular pattern and be collected by neighboring elements.

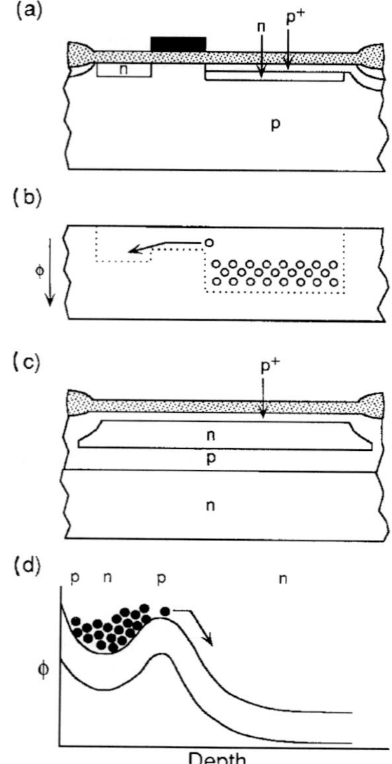

FIGURE 4 Antiblooming methods for image sensors: (*a*) cross section of lateral antiblooming structure; (*b*) illustration of electrostatic potential and charge overflow in lateral antiblooming; (*c*) cross section of vertical antiblooming structure; (*d*) band diagram and illustration of charge overflow in vertical antiblooming.

There are two basic types of antiblooming circuits: lateral and vertical.[8,*] In lateral antiblooming, illustrated in Fig. 4*a* and Fig. 4*b,* an MOS antiblooming gate and an antiblooming drain are provided adjacent to each image-sensing element. Excess charge on the sensing element overflows the antiblooming gate onto the antiblooming drain. The antiblooming drains of all elements on the array are connected and the current sunk in a bias supply.

In a vertical antiblooming structure, illustrated in Fig. 4*c* and 4*d* for a pinned photodiode sensing element, the image-sensing element is fabricated in a shallow, lightly doped *p*-well. A large (10- to 30-V) bias is applied to the *n*-type silicon substrate, causing the *p*-well underneath the photodiode to become completely depleted. Once the charge on the diode exceeds its capacity, the excess charge flows over the saddle-point in the *p*-well,

* Other types of antiblooming are occasionally used in image-sensing arrays. One of these is charge pumping, in which an MOS gate is repeatedly clocked in order to cause excess charge held underneath it to recombine at interface states at the silicon-silicon dioxide interface. In charge pumping, the MOS gate is pulsed sufficiently negative to cause accumulation, resulting in the interface stated filling with holes. When the gate is pulsed out of accumulation, excess electrons can recombine with the holes.

under the diode, and into the substrate. The substrate is connected to a bias supply which sinks the blooming current. The vertical antiblooming structure has the advantage of requiring no additional silicon area, so that antiblooming is achieved with no reduction in fill factor. Lateral antiblooming, on the other hand, requires additional silicon area in each pixel, reducing fill factor. The vertical antiblooming has the disadvantage of lower quantum efficiency at wavelengths longer than about 500 nm. Because any light absorbed below the photodiode or photocapacitor is drained into the substrate, the quantum efficiency of photodiodes or photocapacitors with vertical antiblooming is reduced.

Dark Current in Photosensing Elements

Signal in photosensing elements is the result of collection of electron-hole pairs generated by the absorption of light. However, charge is generated at each photosensing element even in the absence of light. This generation is called dark current and is a result of thermal generation of electron-hole pairs. The thermal generation occurs at defects, such as impurities or crystalline defects, in the bulk of the silicon and at surface states at the silicon-silicon dioxide interface.

There are four sources of dark current: (1) diffusion current, which is the thermal generation of carriers in the undepleted *n*- and *p*-type regions; (2) depletion layer generation current, which occurs in the depletion layer of a diode or MOS capacitor; (3) surface generation current, which is the generation of electron-hole pairs at interface states at the Si–SiO_2 interface; and (4) leakage, which refers to generation at extended defects such as impurity clusters or stacking faults, particularly in the presence of a large electric field. These sources of dark current are illustrated for a photodiode and a photocapacitor in Fig. 5. The generation of the electron-hole pairs in both diffusion current and depletion-layer generation-recombination current occurs almost exclusively at impurity sites. Impurities with energy levels near midgap, such as gold, copper, and iron, are

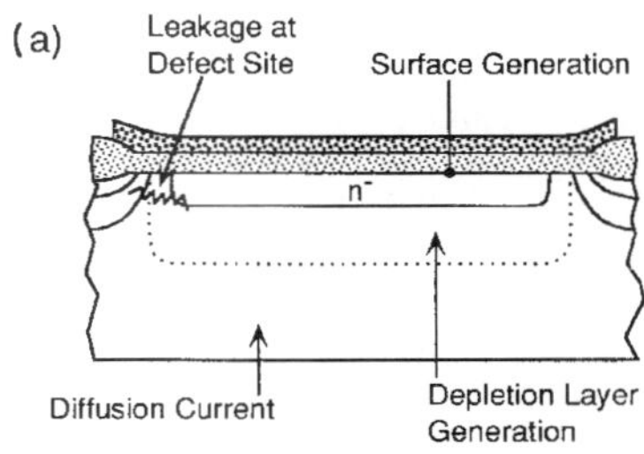

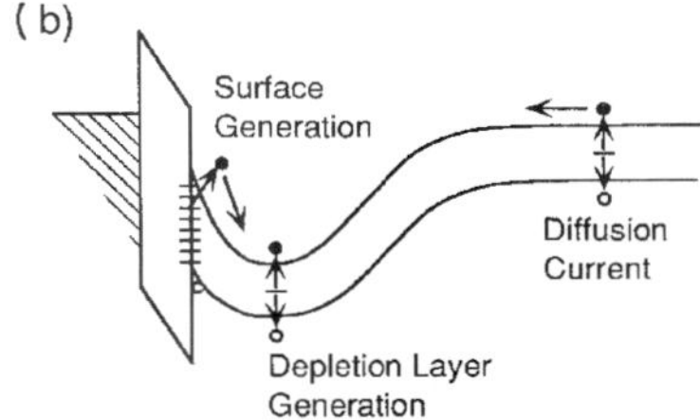

FIGURE 5 (*a*) Cross section of buried-channel MOS capacitor; (*b*) band diagram of buried-channel MOS capacitor illustrating mechanisms for dark current generation.

particularly effective in the thermal generation of charge. The depletion-layer generation current is given by:*

$$J_g(V) = \frac{\mathbf{q} n_i W(V)}{\tau_o} \tag{9}$$

where $W(V)$ is the width of the depletion layer at a given bias voltage in the MOS capacitor or junction-photodiode, n_i is the intrinsic carrier concentration, and τ_o is the depletion-layer lifetime, for carriers. Typical values of τ_o in clean MOS processes would be 1 to 10 ms and values of J_{gr} would be 30 to 300 pA/cm^2.

The diffusion current generation current for a p-type region is given by:†

$$J_{\text{diff}} = \frac{\mathbf{q} n_i^2 L_{\mathbf{e}}}{N_a \tau_e} \tag{10}$$

Since the intrinsic carrier concentration n_i depends on temperature as $\mathbf{e}^{-\mathbf{E}g/2kT}$, depletion-layer generation current and diffusion current will have different temperature dependences. Depletion-layer generation current will increase as $\mathbf{e}^{-\mathbf{E}_g/2kT}$, which corresponds to a doubling in dark current for every 9 to 11°C increase in temperature near room temperature. Diffusion current will increase as $\mathbf{e}^{-\mathbf{E}_g/kT}$ which corresponds to a doubling every 4.5 to 5.5°C increase in temperature.

The surface generation current $\mathbf{J}_s$ occurs almost exclusively at regions where the depletion layer intersects the Si-SiO$_2$ interface, such as the surface region between the n^+- and p-regions around a photodiode or in the depleted surface under a MOS capacitor. It is given by:

$$\mathbf{J}_s = \frac{\mathbf{q} n_i s_o}{2} \tag{11}$$

where s_o is the surface recombination velocity. A typical value of $\mathbf{J}_s$ for a MOS surface would be 100 pA/cm^2.[9],‡ The surface generation depends on temperature in the same manner as the depletion-layer current. In very clean MOS processes (i.e., low concentration of metallic impurities in the silicon), the surface current is often the largest component of the overall dark current.

Leakage current occurs at extended defects in the silicon, such as impurity clusters and stacking faults, particularly when these defects are in a depletion layer and so are subject to a high electric field. While there is no single analytical expression for the current generated by an extended defect, they are characterized by very high values of dark current (≫1 nA/cm^2), a very strong dependence on the electric field, and, in regions of

---

* The expressions here are for the generation current. The total current is the sum of the generation and recombination current and is given by

$$J_{gr}(V) = (\mathbf{q} n_i W(V)/\tau_o)(\mathbf{e}^{\mathbf{q}V/2kT} - 1)$$

However, for diodes under 0.2 V or more of reverse bias ($\mathbf{q}V/kT > 8$), such as would be the case in an image sensor, the recombination current is negligible.

† The full expression for the diffusion current from the undepleted p-region in a diode is:

$$J_{\text{diff}} = \frac{\mathbf{q} n_i^2 L_{\mathbf{e}}}{N_a \tau_{\mathbf{e}}}(\mathbf{e}^{\mathbf{q}V/kT} - 1)$$

However, for reverse biases more than 0.2 V ($\mathbf{q}V/kT > 8$), as would be the case in a photodiode in an image sensor, only the generation term is important.

‡ In calculating the total surface current, the current density $\mathbf{J}_s$ is multiplied by the area of depleted surface. For a pn junction diode, this would be the area of depleted surface region around the diode between the n- and p-regions (i.e., approximately the junction perimeter times the depletion-layer width); for a MOS capacitor it would be the entire area under the MOS capacitor unless sufficient sheet charge of electrons had formed at the surface to invert the surface. The surface recombination velocity is given by $s_o = N_{st} v_{th} \sqrt{\sigma_n \sigma_p}$ where N_{st} is the density of surface states near midgap, v_{th} is the thermal velocity (2.0×10^7 cm/s) and σ_n and σ_p are the electron and hole cross sections for midgap surface states. See Sze, *Physics of Semiconductor Devices,* for a complete explanation.

high electric field, only a very slight dependence on temperature. In an image sensor, they are visible as "bright spots" in a few isolated pixels against the otherwise low level of background thermal generation.[10]

Values for the dark current vary widely because of variation in the amount of impurities in the silicon; the dark current can range from 0.01 nA/cm^2 in very high quality image sensors to >10 nA/cm^2 in sensors with significant metallic contamination. The total number of electrons N_e collected in a pixel is:

$$N_e = \mathbf{J} A T_{int}/\mathbf{q} \qquad (12)$$

where $\mathbf{J}$ is the total dark current per unit area, A is the area of the pixel, and T_{int} is the integration time. For a 1/2-in format CCD such as would be used in a camcorder, typical values would be a dark current of 0.5 nA/cm^2, a pixel area of 100 μm^2, and an integration time of 1/30 s; the number of thermally generated electrons would be 105 electrons per pixel. Image sensors developed for scientific purposes might have a dark current 5 to 10 times lower at room temperature. These same devices might also be operated below room temperature to reduce the thermal generation to levels below one electron per pixel.

There are two types of noise associated with the charge generated by the dark current: shot noise and pattern noise. The shot noise due to the dark current is the square root of the number of dark electrons in a pixel. Pattern noise is due to pixel-to-pixel variations in the dark current and is often highly correlated between neighboring pixels. A numerical value for the dark pattern noise is typically obtained by using the standard deviation of the pixel values in the dark from a large region of an imager, where the values are obtained by averaging over many frames to eliminate shot noise and other temporal noise sources.*

22.4 READOUT ELEMENTS

The readout element transfers the charge from the image-sensing element (photodiode, photocapacitor, or photoconductor) to the output of the image sensor. In a linear sensor, the readout is only in one direction. In an area sensor, both x and y readout is required. There are two basic types of readout elements: charge-coupled devices, or CCDs, and x-y addressed photodiode arrays (typically called MOS arrays owing to the MOS transistors used in the addressing of the pixels). CCDs are by far the most widely used readout elements owing to their very low noise. Nearly all consumer camcorders, facsimile machines, scanners, copiers, and professional and scientific cameras utilize CCDs. Applications of MOS arrays are largely restricted to those where addressing of an individual pixel or subarray is required.

Charge-Coupled Device (CCD)

CCD Operation. The CCD works by moving packets of charge physically at or near surface of the silicon from the image-sensing element to an output, where the charge packet is converted into a voltage. The CCD is formed by an array of overlapping MOS capacitors.[11–13] There are a number of different types of CCD, including four-phase, three-phase, two-phase, virtual (single-) phase, and ripple-clocked CCDs. The number of phases refers to the number of separately clocked elements within a single stage of the

* A histogram of the dark current values of the pixels is rarely gaussian. In most cases, it exhibits an extended tail at high dark current values due to pixels with crystalline defects. The histogram may also exhibit quantization due to pixels with integral numbers of impurities (i.e., 1, 2, or 3 gold atoms); the quantization may even be used as a signature of the impurity present. See, for example, McColgin et al., "Effects of Deliberate Metal Contamination on CCD Image Sensors," *Materials Research Society Symposium Proceedings,* **262:** 769 (1992) and "Dark Current Quantization in CCD Image Sensors," *Proc. 1992 International Electron Device Meeting,* Washington, D.C., 113–116 (1992).

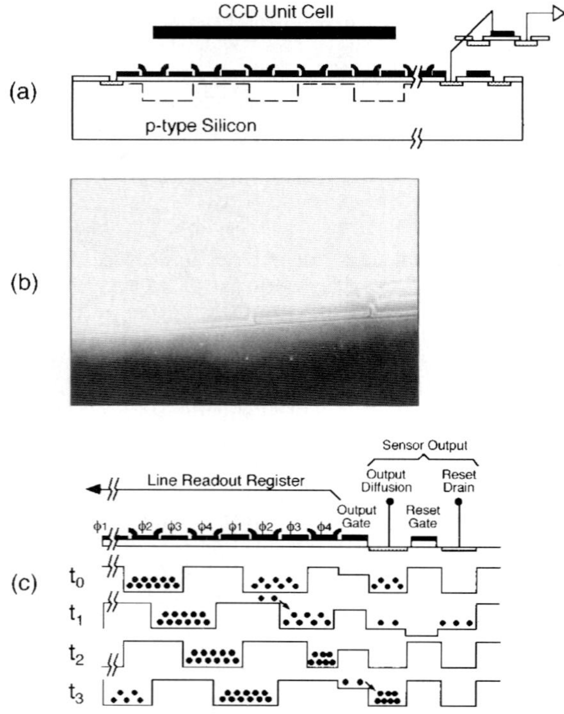

FIGURE 6 (*a*) Cross-sectional diagram of unit cell of a four-phase CCD; (*b*) scanning electron micrograph of a unit cell of a four-phase CCD; (*c*) illustration of charge transfer along a four-phase CCD, showing both transfer along the register and at the sensor output.

CCD. These will be described later. For understanding the principle of charge transfer in a CCD, consider the four-phase CCD illustrated in Fig. 6*a*.

The four-phase CCD physically comprises a silicon substrate (assumed to be *p*-type for purposes of illustration), a gate oxide 300 to 1000 Å thick, and overlapping polysilicon electrodes 1000 to 5000 Å thick which have been heavily doped with phosphorus to lower their resistance. For a four-phase CCD, two levels of electrode are required. The first level is deposited, then defined photolithographically to form two phases ($\phi 1$ and $\phi 3$). A thin (500-Å) oxide is grown over the first level of polysilicon to insulate it from the second level. The second level of polysilicon is then deposited, doped with phosphorus, and defined to form the other two phases ($\phi 2$ and $\phi 4$). An electron micrograph of a four-phase CCD with six-micron long gates is shown in Fig. 6*b*.

The process of charge transfer in a four-phase CCD is illustrated in Fig. 6*c*. In order to hold a packet of electrons, two adjacent gates ($\phi 2$ and $\phi 3$, for example) would be held at a high positive potential ($\sim$+5 V) while the other two phases would be held at a low potential ($\sim$0 V). A depletion layer, or well, is formed under $\phi 2$ and $\phi 3$, allowing electrons to be held at or below the surface. The other two phases, $\phi 1$ and $\phi 4$, serve as potential barriers, keeping the charge packet under $\phi 2$ and $\phi 3$. To transfer the electrons through the silicon, the electrode ahead of the charge packet ($\phi 4$) is clocked positive and the electrode behind ($\phi 2$) is clocked negative. The electrons move along the silicon surface following the positive potential. This is repeated through all four phases, during which the charge packet is moved forward one pixel.

The CCD may be either surface-channel or buried-channel. In a surface-channel CCD (Figs. 2*b* and 6*a*) the electrons are held at the silicon surface. Surface-channel CCDs are rarely used owing to the trapping of electrons at interface states at the silicon surface. At the silicon-SiO_2 interface there is a density of states of about $1 \times 10^{10}\ cm^{-2}\ eV^{-1}$. These states can trap electrons from one charge packet and reemit the electrons into a later charge packet. This results in transfer efficiency. In the buried-channel CCD (Figs. 2*c* and 6*d*), a lightly doped *n*-type layer is formed in the silicon. This *n*-type layer results in a potential well for electrons just below the surface rather than at the surface (Fig. 2*b*). As a result, the electrons remain separated from the interface and are not trapped by interface states. This results in the ability to transfer charge from one stage to another with very high efficiency. Virtually all CCDs for image-sensing applications utilize buried-channel CCDs. The clock voltages used to drive CCD gates typically swing by 5 to 8 V between high and low levels.

For buried-channel CCDs, transfer rates of up to 20 megapixels per second can usually be achieved without special design considerations. Above this rate, special attention must be paid to optimizing the electric field along the direction of transfer to speed up charge transfer. Transfer rates exceeding 50 MPix/s have been achieved in optimized CCD designs.[14]

CCD Output. At the output of the CCD is a circuit to convert the charge packets into a voltage signal. By far the most common type of output circuit is the floating diffusion with source follower amplifier. The floating diffusion output is shown schematically in Fig. 7*a* and a photograph of the end of a CCD shift register with the first stage of the amplifier is shown in Fig. 7*b*. The floating diffusion output consists of an output gate (OG), a floating diffusion, a reset gate (RG), and a reset drain. The floating diffusion is an n^+-type region in between the output gate and the reset gate. The floating diffusion is connected to the gate of a source-follower amplifier. The output gate is held at a fixed dc voltage, creating a barrier potential over which the packet of electrons can be transferred onto the floating diffusion when the last phase of the CCD register is clocked to its low (~0 V) voltage.

The sequence of events in the CCD output is shown in Fig. 6*c*. As the charge packet is transferred along the CCD, it arrives at the last phase ($\phi 4$) before the output gate (time t_2 in Fig. 6c). When $\phi 4$ is clocked low (time t_3), the packet of electrons is transferred over the output gate onto the floating diffusion (time t_0). The voltage of the floating diffusion changes by an amount

$$V = N\mathbf{q}/C \tag{13}$$

where N is the number of electrons and C is the total capacitance of the floating diffusion itself, the interconnect to the source-follower amplifier and the input capacitance of the amplifier. In typical CCDs, this capacitance would be in the 10-fF to 50-fF range. Because this capacitance is so small, there is a large change in voltage for a small change in charge. The charge-to-voltage conversion is an important parameter in CCD design; for a 10-fF capacitance the charge-to-voltage conversion is 16 microvolts per electron. After the change in voltage has been sensed by the amplifier, the charge packet must be removed from the floating diffusion before the next packet arrives. This is achieved by clocking the reset gate positive (time t_1 in Fig. 6*c*), allowing the charge packet to flow from the floating diffusion to the reset drain. The reset drain is held at a constant positive voltage, typically ~10 V. The reset drain is then turned off and the floating diffusion is prepared to accept the next packet of electrons.

The change in voltage from the floating diffusion is typically buffered on-chip by a source-follower (Fig. 7*c*). A two-stage source follower is most often used. The first stage utilizes very small-dimensioned FET transistors in order to minimize the input capacitance. The second uses much larger FET transistors in order to achieve sufficient drive current to

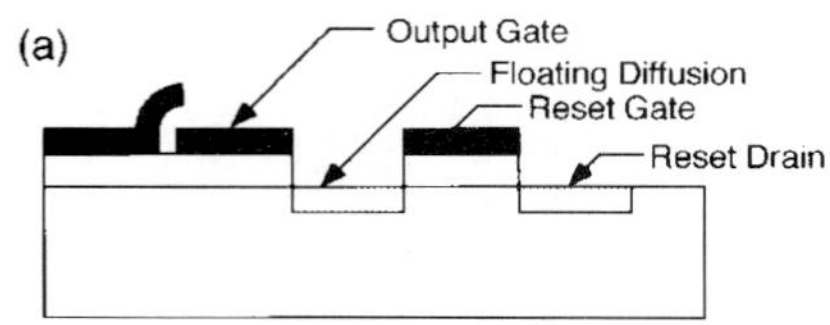

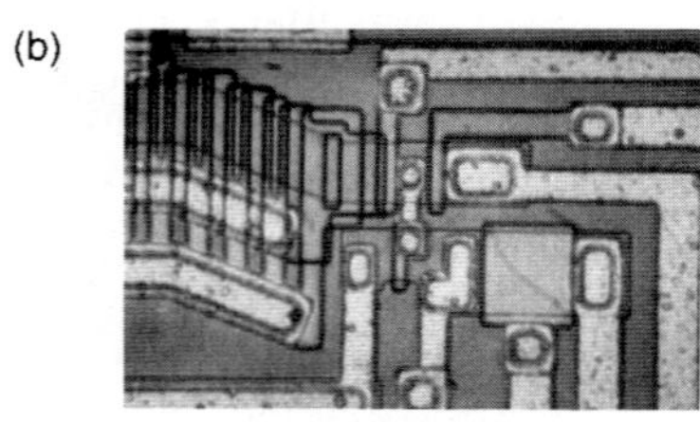

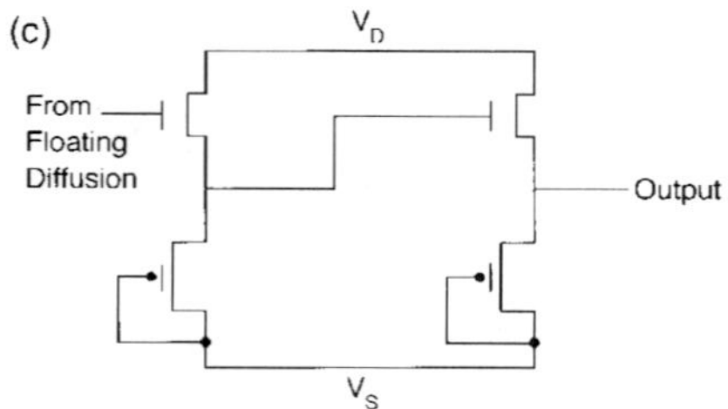

FIGURE 7 (*a*) Cross section of the floating diffusion output for a CCD; (*b*) photomicrograph of a floating diffusion output including the first stage of the source-follower amplifier; (*c*) two-stage source-follower amplifier with on-chip loads.

overcome off-chip capacitances such as package and lead capacitances on the circuit board. The source-follower amplifier is typically designed to have a bandwidth on the order of ten times the CCD pixel rate. For high-pixel-rate applications (>10 MPix/s), three-stage source-follower amplifiers are employed.

For special purpose applications, such as CCD signal processing, nondestructive readout is required. Floating gate output amplifiers are used in these applications. These amplifiers are similar to the floating diffusion except that the floating diffusion is replaced by a MOS gate which is connected to the MOS amplifier. Other outputs, such as buried-channel JFET structures, have seen limited use in very low noise applications.[15]

Types of CCDs. In addition to the four-phase CCD described here, there are a number of other types of CCD shift registers, including three-phase, two-phase, and virtual phase. The different types are illustrated in Fig. 8. The four-phase CCD (Fig. 8*a*) was described previously. The three-phase (Fig. 8*b*) consists of three different layers of polysilicon electrodes. The charge is normally held under one of the three; during transfer, the gate ahead of the charge packet is clocked positive and the gate holding the charge is clocked negative in order to transfer the charge one gate ahead. The three-phase CCD has the advantage of a shorter unit cell than the four-phase but at the expense of additional processing complexity (i.e., a third polysilicon layer).

In the two-phase CCD, each phase has a barrier and a well region. The barrier is

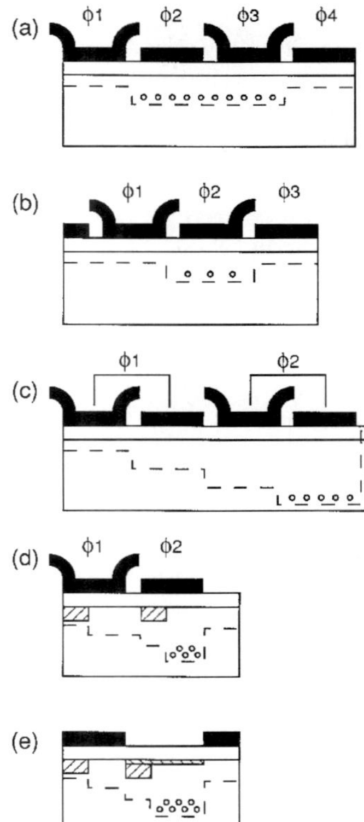

FIGURE 8 Types of CCD registers: (*a*) four-phase; (*b*) three-phase; (*c*) pseudo-two-phase; (*d*) true two-phase; (*e*) virtual phase. In the pseudo-two-phase CCD, each phase consists of two polysilicon gates, one of which is offset in potential from the other by an implanted threshold adjustment. In the true two-phase CCD, each phase consists of a single polysilicon gate in which the implanted threshold adjust region is formed underneath a portion of the gate.

formed by doping the barrier region slightly less n-type than the well region, making the electrostatic potential in the barrier region a few volts lower than the well region for the same gate voltage. Electrons will flow over the barrier region and be held in the well region. There are two methods of fabricating a two-phase CCD. In the first method (Fig. 8*c*), two separate gates are used for each phase; one gate receives a threshold adjust ion implantation in order to create the barrier region. The two gates, however, are connected and thus driven at the same voltage. In the other method (Fig. 8*d*), the barrier and well regions are created under a single polysilicon gate. The two-phase CCD is very commonly used for three reasons. First, it requires only two clocks ($\phi 1$ and $\phi 2$), simplifying drive circuitry. Second, the clocks are complementary. This reduces clock feedthrough. Third, the two-phase has a higher horizontal density, especially the two-phase with barrier and well regions under the same gate.

The virtual phase CCD[16,17] (Fig. 8*e*) consists of one gate with both barrier and well regions within it and a second region, the virtual phase, in which a shallow, high-dose ion implantation is used to create a heavily doped surface region which pins the surface potential at 0 V. The virtual phase has both barrier and well regions within it and acts the same as a polysilicon gate held at a constant voltage. Charge is first transferred from the clocked phase into the virtual phase, then the charge is transferred from the virtual phase into the next clocked phase. The virtual phase CCD has a number of advantages, including higher quantum efficiency then two-polysilicon-level area CCD sensors and simpler clocking. The disadvantages of the virtual phase include the need for larger voltage swings on the single-clock and high-clock feedthrough into the output due to the lack of complementary clocks.

CCD Characteristics. The four major performance parameters of a CCD shift register and output amplifier are charge-handling capacity, charge-transfer efficiency, charge-to-voltage conversion ratio, and noise. The charge capacity is the number of electrons which can be held and transferred in the CCD shift register. As charge is added to a shift register, a point is reached at which excess charge cannot be held; the excess charge either overflows into adjacent pixels or overflows into the bulk beneath the CCD or, in the case of a buried-channel CCD, overflows the barrier to the $Si\text{-}SiO_2$ surface. The charge capacity is a function of the device design, device layout, and CCD process. Figure 9*a* shows the electrostatic potential and charge distribution in a buried-channel CCD at three levels of charge: approximately one-quarter of saturation, at saturation, and beyond saturation. Below saturation, the electrons fill the center of the buried channel in the region of largest potential, separated from the surface by approximately 0.2 μm in distance and about 500 mV in potential. As additional charge is added, the electron distribution spreads towards the surface and the potential barrier to the surface drops. Beyond saturation, the electrons contact the surface directly, resulting in charge-transfer inefficiency and blooming to neighboring pixels. The charge capacity of most CCD shift registers is of the order 1×10^{12} electrons/cm^2 of area in which the charge is held. In most CCD cells, the charge is held in only a fraction of the total cell area both along the CCD register and across the register. Typically, area CCD image sensors are designed with CCD charge capacities in the range of 50,000 to 200,000 electrons. Linear CCD image sensors, because of the larger amount of silicon area available for the CCD shift register, often are designed for 100,000 to 1,000,000 electrons.

The second major performance parameter for CCD shift registers is charge transfer efficiency. The transfer of charge from one stage to the next is neither instantaneous nor complete, limiting both transfer rate and the total number of stages in the CCD. There are two intrinsic mechanisms governing charge transfer: drift and diffusion. Drift is the movement of charge in the presence of an electric field. There are two origins of the electric field seen by an electron during charge transfer: the self-induced field resulting from the other electrons under the gate and the externally induced, or fringing, field. During the early stages of charge transfer, the self-induced field is large and is the dominant factor. After the charge concentration under the gate has decreased to a low value, the remainder of the charge transfer will be governed either by diffusion or by drift due to externally induced, or fringing, fields.

For a surface-channel CCD, the self-induced fields can be estimated from the formula:

$$\mathbf{E} = -\frac{\mathbf{q}}{C}\frac{dN}{dx} \tag{14}$$

where C is the gate capacitance per unit area and $N(x)$ is the density of electrons as a function of distance x from the edge of the electrode. In the early stages of charge transfer, both N and dN/dx are large. As the transfer proceeds, both N and dN/dx become small and the charge transfer is governed by fringing fields and diffusion.

Fringing fields are due to the two-dimensional nature of the electrostatic potential. If

(a)

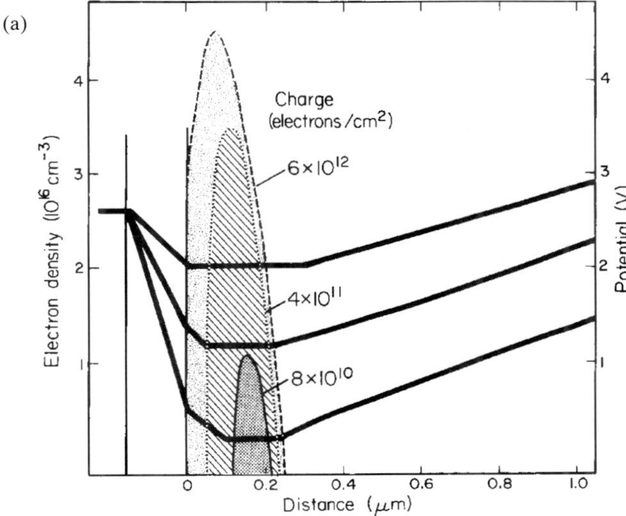

(b)

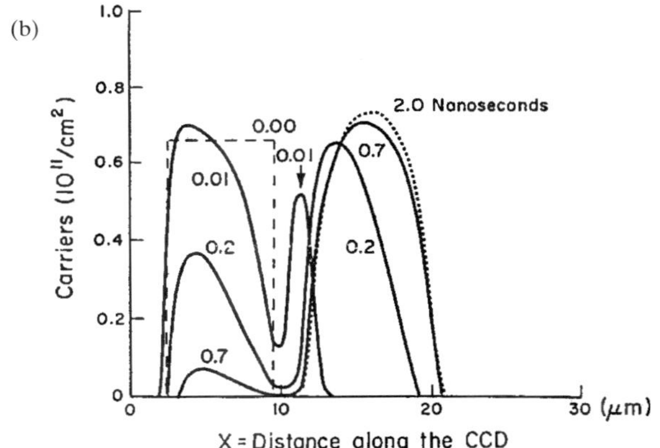

FIGURE 9 (*a*) Charge distribution and electrostatic potential as a function of distance from the surface for a buried-channel CCD for three levels of charge density; (*b*) charge density as a function of distance along a CCD for various times following the beginning of charge transfer from one 8-μm stage to another.

the charge is being transferred from gate 1 to gate 2, the potential will change smoothly between the two gates. The effect of the potential from one gate will typically extend 1 to 3 μm into a neighboring gate. The charge within the range of the fringing field will move by drift to the neighboring gate. Charge out of the range of the fringing fields will move by diffusion.

Figure 9*b* shows an example of charge transfer calculated for charge transfer from one 8-μm-long CCD gate into an adjacent gate. The charge density is shown as a function of distance along the CCD at several times after the start of charge transfer. At the start of the transfer, all the charge is under the first gate. At 0.01 ns into the transfer, the charge has moved from the edge of the first gate into the second gate, as a result of self-induced

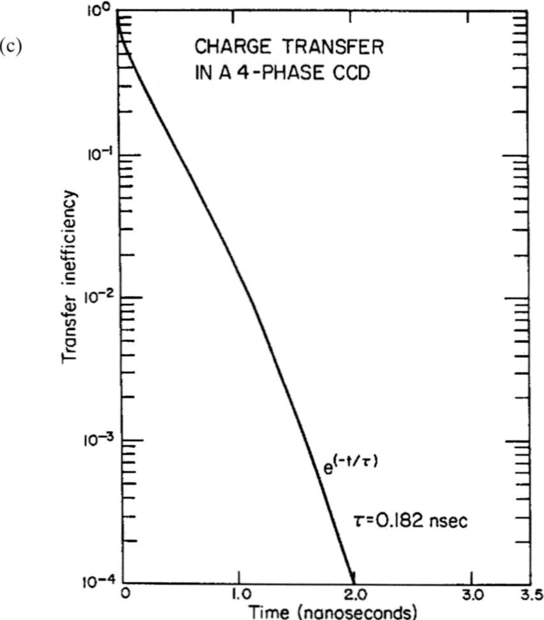

FIGURE 9 (*Continued*) (*c*) charge transfer inefficiency as a function of time from the start of charge transfer for the same example.

drift. By 0.2 ns, approximately half the charge has been transferred. By 0.7 ns, over 90 percent of the charge has moved into the second gate, leaving a residual in the first. At this point, the self-induced drift is sufficiently small that drift due to fringing fields and diffusion are the dominant mechanisms. Figure 9*c* shows the charge transfer inefficiency as a function of time for this example. Two slopes are evident in the transfer inefficiency. In the first 0.5 ns, the charge is transferred rapidly owing to the self-induced drift. For times longer than 0.7 ns, the transfer is due to fringing fields in this example.

For CCDs with longer gates or lower fringing fields than the example above, the final ~10% of the charge must transfer by diffusion. Charge transfer by diffusion follows an exponential time dependence.

$$N(t) = N(0)e^{-t/\tau_{\mathrm{diff}}} \tag{15}$$

where the time constant τ_{diff} for diffusion is

$$\tau_{\mathrm{diff}} = \frac{4L_g^2}{\pi^2 D} \tag{16}$$

where L_g is one gate length and $D = KT\mu_e/q$.

For electrons at room temperature, $D = 25.8\ \mathrm{cm^2/s}$. For an 8-μm long gate, the diffusion time constant is ~10 ns. To achieve a transfer inefficiency below 2×10^{-5} per transfer, 11 time-constants, or 110 ns in this example, are required. For this reason CCDs are typically designed with gate lengths shorter than ~8 μm and are also designed to build in fringing fields.

In the simplest case for very low levels of transfer inefficiency, the total transfer inefficiency in a CCD register is the product of the number of stages N in the register and the transfer inefficiency per stage. Each stage will require two or more transfers. Virtual phase and two-phase require two transfers per stage, three-phase requires three transfers, and four-phase requires four transfers. The inefficiency per stage, however, will likely depend in a complicated manner on the amount of charge in the charge packet, the amount of charge in preceding charge packets,* the voltages, and frequency of operation.

In addition to the intrinsic sources of transfer inefficiency, there are a variety of extrinsic sources. These include surface and bulk traps and potential wells and barriers. The traps and the potential obstacles hold back an amount of charge from a charge packet. The charge is reemitted to later charge packets. The inefficiency due to traps depends on whether the traps have been filled by preceding charge packets and the emission time constants of the traps, and so is not modeled in a simple manner.

The intrinsic sources of noise in CCD shift registers include dark current and output amplifier noise. In addition, other sources of noise not intrinsic to the CCD itself include noise due to clock feedthrough from the CCD clocks to the output signal and noise in external electronics. The generation of dark current in CCD shift registers is the same as described at the end of Sec. 22.3 for photosensing elements. Associated with the dark current are both shot noise and pattern noise. The magnitude of the pattern noise in a CCD shift register is reduced over that of a single element since the charge packet averages the dark current over many pixels as it is transferred to the output.

The noise associated with the CCD output consists of the Johnson or thermal noise, the $1/f$ noise of the output amplifier, and the kTC noise associated with resetting the floating diffusion. The kTC noise is due to uncertainty in the amount of charge remaining on the floating diffusion following reset owing to thermal fluctuations in the reset gate. The rms noise σ_n in the number of electrons caused by kTC noise is given by

$$\sigma_n = \sqrt{\frac{\mathrm{kT}}{q} C_{\mathrm{FD}}} \tag{17}$$

where C_{FD} is one total floating diffusion capacitance. For a 10-fF floating diffusion capacitance, the rms noise is ~40 electrons.

The kTC noise may be eliminated entirely by use of a signal-processing technique called correlated double sampling (CDS). In correlated double sampling, the output level is sampled before the charge packet is transferred onto the floating diffusion and again after transfer of the charge packet (times t_2 and t_0 respectively in Fig. 6*c*). The two values are subtracted either by an analog circuit or by digital subtraction. Any uncertainty in the voltage level of the floating diffusion following reset is subtracted and thus the kTC noise eliminated. The output amplifier low-frequency noise, called $1/f$ noise because of the inverse frequency $(1/f)$ shape of the noise power spectrum, is also reduced but not eliminated by correlated double sampling. The total noise of a CCD output amplifier is in the range of 7 to 40 rms electrons per pixel depending on the amplifier design and the operating speed. Values of a single rms electron or less have been obtained for very slow pixel rates under cooled conditions.[18]

MOS Readout

The other major category of readout structures is the MOS readout. The individual light-sensing elements (photodiodes, photocapacitors, or photoconductors) at each pixel are connected to a readout line by means of a transfer gate. Each pixel along the readout line is addressed separately by addressing circuitry. When a particular pixel is addressed, the transfer gate is turned on and the charge transferred from the pixel to the readout

* The charge in the preceding packets will affect the filling of both bulk and interface traps. See, for example, Sequin and Thompsett, pp. 70–108.

line. An amplifier at the end of the readout line senses the change in voltage or current resulting from the charge transfer.

Typically, the pixels would be addressed serially along the line. The first pixel would be addressed, causing the charge from the image-sensing element to be transferred onto the readout line. The voltage change or current would be sensed, the readout line reset to its original voltage if necessary, and the next pixel addressed. This is different from a CCD. In the CCD, charge from all pixels is transferred into the CCD register simultaneously. Individual pixels or groups of pixels cannot easily be addressed in a random fashion by the CCD, but this random addressing can be accomplished readily by the MOS readout.

There are several types of MOS readout devices. These include the CID,[19] the AMI,[20] and the CMD[21] in addition to the normal MOS array. The CID has no readout line. Each pixel consists of two overlapping gates, one controlled by a row address and the other by column address.[11] When neither row nor column of a particular pixel is being addressed, the photogenerated charge is held under both gates and can be transferred between them. When a row of a pixel is addressed, the charge transfers onto the column gate. Then both row and column are addressed, the charge is injected into the substrate, and the current sensed. The CID is not widely used in visible imaging applications because the charge conversion sensitivity is very poor and noise very high compared to the CCD or to the other MOS architectures.

In the amplified MOS imager (AMI), the image-sensing element at each pixel consists of a phototransistor rather than a simple photodiode. The photogenerated charge is stored on the gate of the MOS transistor.[19] When a particular pixel is addressed, the photogenerated charge modulates the transistor current. This current amplification at each pixel helps to overcome many of the noise and speed limitations of conventional MOS arrays.

MOS readout differs in an important way from CCD readout. In MOS readout, the charge is transferred from a single pixel onto a readout line and the change in voltage or current in the readout line is sensed. In a CCD, the charge packets are kept intact while being transferred physically to a low-capacitance output. The lower sensitivity of a simple MOS array can be illustrated as follows. The change in voltage on the readout line is given by $V = N\mathbf{q}/C$ where N is the number of electrons, $\mathbf{q}$ the electron charge, and C the readout line capacitance. Because the readout line covers the full length of the array, its capacitance is in the picofarad range (typically 2 to 10 pF depending on design and process). This compares to the 10-femtofarad capacitance for the CCD output. As a result, the voltage swings on the readout line are very small (16 nV/electron for a 10-pF readout line capacitance). This leads to a high sensitivity to clock noise due to capacitive feedthrough of the row and column address clocks onto the readout line. The feedthrough may be many times larger than the signal in most MOS sensors. Once the charge has been transferred onto the readout line, it is sensed either by a current-sensitive amplifier or by a voltage-sensitive amplifier, followed by a reset of the readout line to its original voltage.

CCD readout has the advantages of very high sensitivity and low noise. However, CCD readouts are limited in charge-handling capacity, while MOS readouts are capable of carrying very large amounts of charge and so are not as limited on the high end of the dynamic range. However, because the MOS readout line has much higher capacitance than the CCD, the sensitivity is lower and the noise is higher. Another difference is in the readout architecture. The CCD readout is essentially a serial readout device; it is not suited to random readout or partial-array readout. The MOS array, however, can be addressed in a manner similar to a memory, making it well-suited to pixel or partial-array addressing.

22.5 SENSOR ARCHITECTURES

Solid-state image sensors are classified into two basic groupings: linear and area. Linear sensors include single-line arrays, multilinear arrays for color scanning, and time delay and integrate (TDI) arrays for low-light-level scanning. Area sensor architectures include the frame transfer CCD, the interline transfer CCD, and various forms of MOS x–y addressed arrays.

Linear Image Sensor Arrays

Linear sensors are used almost exclusively in scanning systems for scanning of documents, film, and three-dimensional still objects. There are two basic classes of scanning systems: contact scanners and reduction scanners. These are illustrated in Figs. 10*a* and 10*b*. In reduction scanners (Fig. 10*a*), the sensor is smaller than the document to be scanned; lenses are used to image the document onto the sensor. In contact scanners (Fig. 10*b*), the sensor is the same width as the item to be scanned, usually a document. Relay optics is used between the sensor and the document. Selfoc lenses (Fig. 10*c* and 10*d*) and roof-mirror-lens arrays are the two types of relay optics used most frequently.

There are three basic architectures for linear sensing arrays: MOS line arrays, CCD linear and multilinear sensors, and time-delay and integrate (TDI) sensors. These architectures are illustrated in Fig. 11. The MOS array is used most often in contact scanning applications where material or processing problems make CCD arrays impractical. These applications include arrays fabricated from polysilicon or amorphous silicon on nonsilicon substrates, arrays covering large distances, or arrays requiring special processing (such as logarithmic amplification) at each pixel. The CCD linear and multilinear arrays are used most often in reduction scanning where wide dynamic range and small pixel size is required. However, contact arrays are also often realized by butting multiple CCD arrays end-to-end. TDI arrays are used in very low light level scanning applications where integration over many lines is required to achieve adequate signal-to-noise.

The MOS linear array (Fig. 11*a*) consists of individual photosensing elements, an

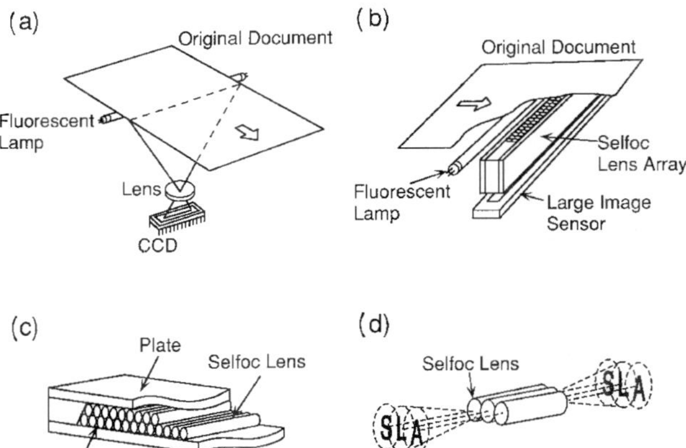

FIGURE 10 (*a*) Reduction document scanner using linear image sensor, in which the image on the page is reduced by a lens onto the line array; (*b*) contact document scanner in which the length of the image sensor is the same as the document width and one-to-one relay optics is used to transfer the image from the document to the array; (*c*) diagram of a selfoc lens array used as transfer optics between document and array in a contact scanner; (*d*) operation of a selfoc lens array.

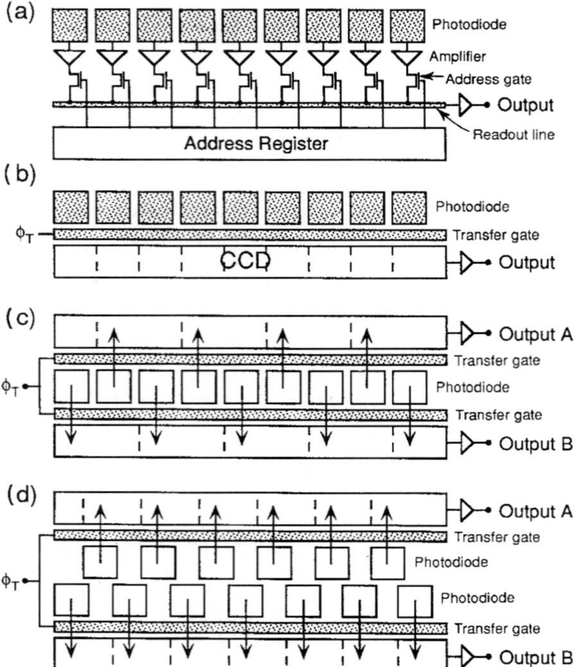

FIGURE 11 Architectures for linear image sensors: (*a*) MOS line array consisting of photodiodes, preamplifier, MOS switches addressed by an address register, and a readout line with amplifier; (*b*) linear CCD image sensor consisting of photodiodes, transfer gate, and CCD readout; (*c*) linear CCD image sensor with two CCD output registers, one for the odd diodes and the other for the even diodes, for higher horizontal pitch; (*d*) staggered linear CCD image sensor with two rows of photodiodes offset by one-half pixel to increase horizontal sampling.

amplifier, an address switch, and a readout line and amplifier. The photosensing element is usually a photodiode, although photoconductors are used in contact scanning arrays fabricated from amorphous silicon. Since the charge generated in the diode is generally very small (in the hundreds to thousands of electrons), a simple amplifier is usually placed at each pixel to drive the high-capacitance readout line. The use of amplification at each pixel can allow some signal-processing functions, such as logarithmic amplification, clipping, triggering and latching, etc., to be performed at the pixel. A MOS switch placed after the amplifier allows each pixel to be addressed in sequence; the switch is driven by an address register. At the end of the readout line is an amplifier which may buffer and/or amplify the signal. The MOS array has the advantage of process simplicity and the ability to perform signal processing at each pixel; it has the disadvantage of low signal level (because of the large readout line capacitance) and pattern noise introduced by feedthrough from the switching transistor.

The linear CCD image sensor is the most often used architecture for scanning applications owing to its low noise, high sensitivity, wide dynamic range, and small pixel pitch. Figure 11*b* shows the simplest type of linear CCD, in which a single row of photodiodes is connected to a single CCD register via a transfer gate. In operation, the

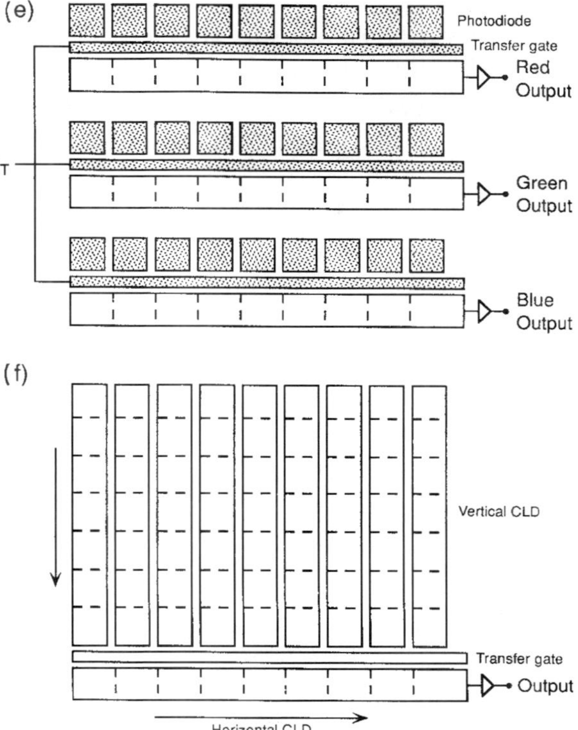

FIGURE 11 (*Continued*) (*e*) trilinear CCD in which three CCDs are fabricated on the same silicon die, each with its own color filter; (*f*) time-delay and integrate (TDI) array, in which the charge in the vertical registers is clocked in phase with the motion of the scene or document being imaged in order to increase signal-to-noise.

signal is integrated on the image-sensing element (generally a photodiode) for a line time. The horizontal CCD is then stopped, the transfer gate opened, and the charge transferred from all the photodiodes simultaneously to the CCD. The transfer time is typically a few microseconds. The transfer gate is then closed, integration resumed for the next line, and the CCD clocked to read out the charge packets. Many arrays also feature antiblooming for situations where the light level may not be controlled, as well as electronic shuttering, which allows an integration time on the photodiodes to be less than the readout time of the CCD.

For linear-sensing applications requiring a higher pixel density, a double-sided readout is often used (Fig. 11*c*). In this architecture, a CCD array is placed on either side of the line of photodiodes and charge transfer from the diodes alternates between the top and bottom CCDs. This architecture uses lower horizontal clock rates and a higher pixel pitch, since the diode pitch can usually be made smaller than the CCD pitch. The charge packets from the two arrays may be multiplexed into one output if desired. The disadvantage of this architecture is slight differences between even and odd pixels, due to slight differences in the two outputs (or slight differences due to multiplexing the two registers).

Another architecture which is used to further decrease the sampling pitch is the staggered linear array (Fig. 11*d*). In the staggered array, two rows of photodiodes are offset by a half pixel. The two rows are read out by CCD arrays. The first array is delayed (usually in a digital line store) and then combined with the second to form a double-density scan.

Multilinear arrays (Fig. 11*d*) have recently been developed for color scanning applications. In this architecture, several (usually three) linear arrays are combined on the same silicon die separated by a distance equivalent to an integral number of scan lines. Color filters (either integral or in close proximity) are aligned over the arrays. External digital line delays are used to realign the three arrays. Separate electronic shuttering may be provided for each array in order to adjust for differences in intensity in each of the bands.

The third major class of line arrays is time-delay and integrate, or TDI, arrays.[22,23] The TDI architecture is shown in Fig. 11*f*. TDI arrays are used when inadequate signal-to-noise ratio from a single-line array requires averaging over multiple lines. Applications for TDI arrays include high-speed document scanners and space-based imaging systems. Instead of a single row of photodiodes, the TDI array utilizes CCD stages in the vertical dimension which are clocked synchronously with the movement of the document to be scanned. The signal level in a TDI array increases linearly with the number of stages. The noise level, however, increases most as the square root of the number of stages.

Area Image Sensor Arrays

There are three major classes of area image sensor architectures: MOS diode arrays, frame-transfer CCDs, and interline-transfer CCDs. Within each there are a number of variations. CCDs have come to dominate the majority of applications owing to their higher sensitivity. However, MOS arrays are still used for specialized applications where addressability or high readout rate is important.

Historically, the physical dimensions of the active imaging areas of CCD arrays for consumer and commercial applications are specified by the size of the vidicon tube which it replaces. The common format sizes include $\frac{1}{4}$-in, $\frac{1}{3}$-in, $\frac{1}{2}$-in, and 1-in. In most cases, the aspect ratio is 4:3, reflecting the television standard. Table 1 lists the formats and the corresponding dimensions of the imaging area of the array.

TABLE 1 Image Area Dimensions and Pixel Dimensions for Various Format Image Sensors

The format name is given in inches based on historic image tube formats. The pixel dimensions are based on a 484 × 768 *pixel array.*

Optical format	1-in	$\frac{2}{3}$-in	$\frac{1}{2}$-in	$\frac{1}{3}$-in	$\frac{1}{4}$-in
Active area (4:3 aspect ratio)					
Height (mm)	9.6	6.6	4.8	3.3	2.4
Width (mm)	12.8	8.8	6.4	4.4	3.2
Diagonal (mm)	16	11	8	5.5	4
Pixel dimensions (484 lines × 768 pixels)					
Height (μm)	19.8	13.6	9.9	6.8	4.9
Width (μm)	16.7	11.5	8.3	5.7	4.2

The standards for the number of vertical lines in arrays for consumer and professional video are usually based on the corresponding television standards, including NTSC, PAL, and the various HDTV standards. NTSC has 484 active lines, PAL has 575, the Japanese HDTV standard has 1035, and the European HDTV standard has 1150. In all cases the lines are interlaced; i.e., odd lines are read out in one field and even lines in the next. Historically, for sensors for NTSC television, the number of pixels horizontally in a line has been associated with multiples of the color subcarrier frequency; common horizontal pixel counts include 384, 576, and 768. Sensors for the Japanese HDTV standard typically have 1920 horizontal pixels. The pixels are rectangular rather than square. Table 1 lists approximate pixel dimensions in micrometers for a few common formats for a 484 (V) × 768 (H) pixel image sensor.

For industrial, scientific, graphics electronic photography, digital television, and multimedia applications, however, the nonsquare pixels and interlaced readout of sensors based on television standards are significant disadvantages. Image sensors designed for these industrial and commercial applications typically have square pixels and progressive scan readout. In interlaced NTSC readout, for example, the first field consists of lines 1, 3, 5 · · · 483 and is read out in the first 1/60 second of a frame. The second field consists of lines 2, 4, 6 · · · 484 and is read out in the second 1/60 second of a frame. The resulting temporal and spatial displacement between the two fields is undesirable for these applications. In addition, digital compression of interlaced scan moving images, especially with motion estimation, is difficult and also introduces artifacts.

In progressive scan readout each line is read out sequentially. There is no even or odd field, only a single frame. As a result there are no temporal and spatial sampling displacement differences. However, for a given resolution and frame rate, the readout rate of a progressive scan image sensor is double that of an interlaced scan. In addition, for these applications, the number of pixels is often based on powers of 2: such as 512 × 512 or 1024 × 1024—facilitating memory mapping and image processing.

MOS Area Array Image Sensors. The architecture of MOS area arrays is illustrated in Fig. 12.[24] It consists of the imaging array, vertical and horizontal address registers, and output amplifiers. The pixel of a MOS array consists of an image-sensing element (photodiode, photocapacitor, phototransistor, or photoconductor), a row-address gate, and a vertical readout line. The row-address gate is bussed horizontally across the array and is driven from a row-address register on the side(s) of the array. At the start of a line, a single row is addressed, causing the charge from all the photodiodes in a row to be transferred onto the vertical readout line. Horizontal address gates are placed at the bottom of the vertical readout line. Buffer amplifiers to drive the horizontal readout line may also be placed at the end of the vertical readout line. The horizontal address register then serially addresses each vertical readout line, sequentially turning on the horizontal address gates. After the horizontal addressing is completed, the readout lines may be reset and precharged and the next row addressed.

There is a wide range of variations on this basic architecture. An example is the charge modulation device, or CMD,[25,26] array in which a phototransistor is placed at each pixel site in order to achieve amplification and to achieve high currents to drive the capacitance of the vertical and horizontal readout lines. Other examples include arrays with sophisticated charge collection circuits at the end of each vertical readout line.

Frame Transfer CCD Image Sensors. CCD area arrays fall into two categories: frame transfer and interline transfer. The simplest form of frame transfer CCD is the full-frame CCD, shown in Fig. 13*a*. A photograph of a few pixels of a frame transfer CCD is shown in Fig. 14*a*. The array consists of a single image area composed of vertical CCDs and a single horizontal register with an output amplifier at its end. In this architecture, the pixel consists of a single stage of a vertical CCD. This type of device requires an external shutter. When the shutter is opened, the entire surface of the sensor is exposed and the

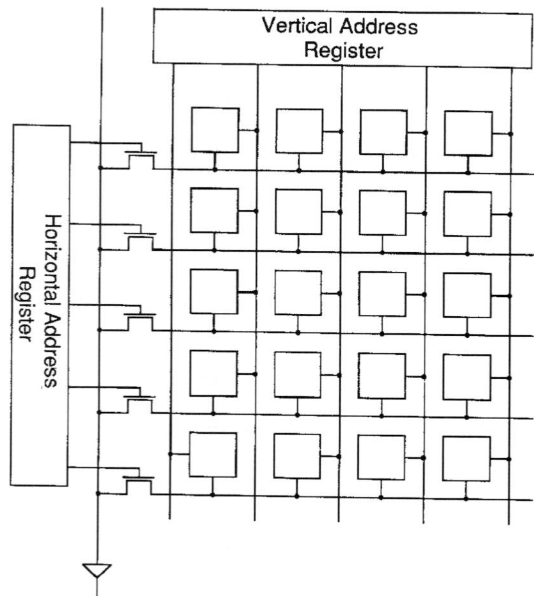

FIGURE 12 MOS photodiode array, consisting of vertical and horizontal address registers and readout line.

charge is collected in the CCD potential wells at each pixel. After the shutter is closed, the sensor is read out a row at a time by clocking a row of the vertical register into the horizontal register, then clocking the horizontal register to read out the row through the output amplifier. For higher readout rates dual horizontal CCDs are used in parallel. The full-frame CCD has the advantage of progressive scan readout high fill factor, very low noise, and wide dynamic range. However, it requires an external shutter. It is most often used in still electronic photography, scientific, industrial, and graphics applications.

For motion imaging applications, a shutter is not practical. In order to overcome the need for a shutter, frame transfer CCDs incorporate a storage area in addition to imaging area. For interlaced video applications, this storage area is sufficiently large to hold a field (242 lines in NTSC television). The image area consists of vertical CCDs. For interlaced NTSC video, there are 242 pixels vertically in the image area. Interlace is achieved by changing the gates under which integration is performed in the even and odd fields. For example, for a four-phase CCD, integration would be performed under phases 1 and 2 in one field and 3 and 4 in another, effectively shifting the sampling area by half a pixel in each field. The storage area consists also of 242 pixels vertically. The device operation is as follows. The image area integrates for a field time and the photogenerated charge held in the vertical CCDs. The vertical CCDs are then clocked in order to rapidly transfer the charge from the image area into the storage area. Because the sensor is still under illumination, this transfer time must be much shorter than the integration time. This transfer typically requires 0.2 to 0.5 ms. The storage area is then read out a row at a time by transferring a row into the horizontal register and clocking the horizontal register. While this readout is occurring, the image area is integrating the next field. The most significant disadvantage of frame transfer CCDs is the image smear caused by illumination during the transfer from the image to the storage area. This smear can be on the order of 3 percent.

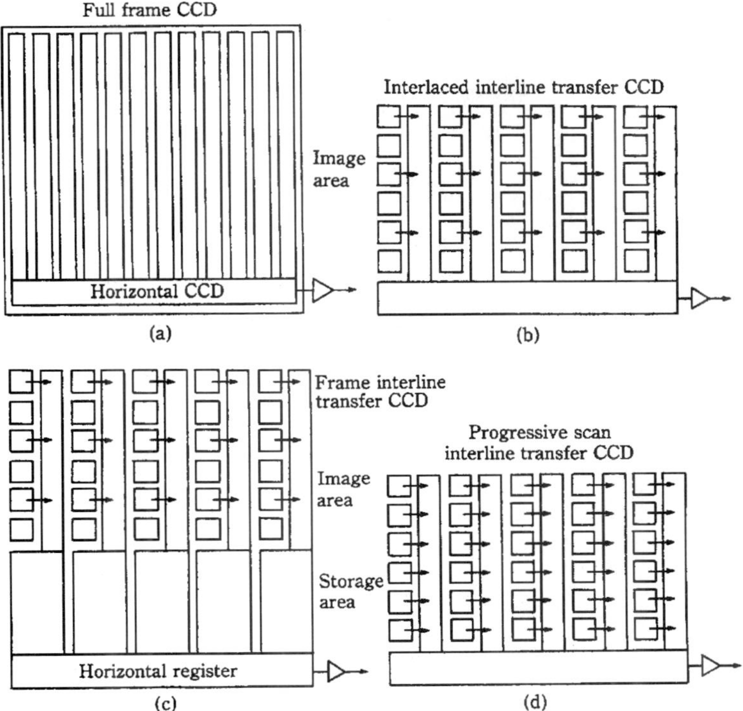

FIGURE 13 (*a*) Architecture of full-frame CCD; (*b*) architecture of interlaced interline transfer CCD; (*c*) architecture of frame interline transfer CCD, in which a storage area is provided to reduce smear during readout; (*d*) progressive scan interline transfer CCD, in which every photodiode is read out into the vertical CCD simultaneously.

In both frame transfer and full-frame devices, the light must pass through the polysilicon electrodes before being absorbed in the silicon. Owing to the high absorption of short wavelengths in the polysilicon, the quantum efficiency in the blue is only about 20 percent and the quantum efficiency in the green is about 50 percent. Figure 14*b* shows the quantum efficiency for a full-frame image sensor with polysilicon gates. Three approaches have been used to improve the efficiency: the virtual phase CCD, transparent electrodes, and backside illumination. In the virtual phase CCD (see "Types of CCDs" in Sec. 22.4), the second polysilicon electrode is replaced with a very shallow highly doped p^+ layer, very similar to the pinned photodiode (Fig. 2*d*). Since there is no electrode over this phase to absorb the light, higher quantum efficiency is achieved, particularly at wavelengths less than 500 nm. Backside illumination provides nearly unity quantum efficiency but requires that the sensor be thinned to less than 10 μm. Owing to its cost, backside thinning is employed only in sensors for very specialized scientific or aerospace applications. Indium-tin-oxide, or ITO, is the most commonly used transparent electrode.[27] Usually it is substituted for the second level of polysilicon. Figure 14*b* also shows the quantum efficiency of a full-frame device in which ITO has been substituted for one of the polysilicon gates.

Interline Transfer CCD Image Sensors. The interline transfer CCD is fundamentally

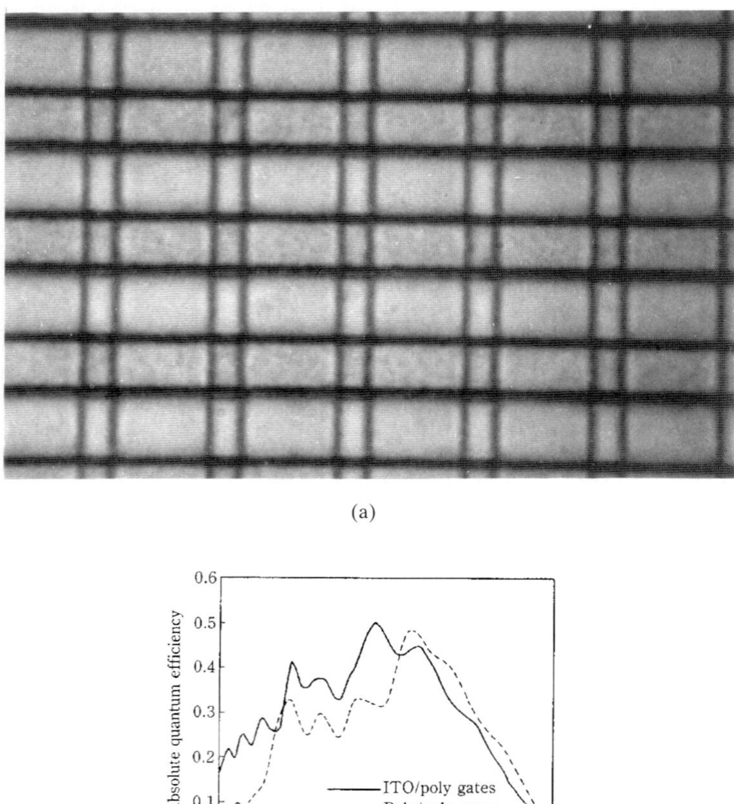

FIGURE 14 (*a*) Photograph of a few pixels in the image area of a full-frame CCD; (*b*) quantum efficiency as a function of wavelength for a full-frame CCD with polysilicon and with transparent indium-tin-oxide electrodes.

different from the frame-transfer CCD in that, in addition to the vertical CCD, the pixel also contains a separate image-sensing element (photodiode, pinned photodiode, photocapacitor, or photoconductor) and a transfer region between the photodiode and the vertical CCD. Interline CCDs with photodiodes for sensing elements are considered first. Figure 13*b* illustrates the architecture of an interline transfer CCD and Fig. 15 illustrates a pixel of the CCD. The CCD and the transfer region between the diode and CCD are covered with a light shield (typically aluminum, although metal silicides are also used). The light shield prevents any light from entering the vertical CCD registers, allowing them to be read out while the sensor is illuminated. When the sensor is illuminated, the photogenerated charge is held on the photodiode. During the vertical retrace interval at the end of a field, the photogenerated charge is transferred into the vertical CCD by clocking the CCD gate over the transfer region. Once the charge has been transferred from the diodes into the vertical CCDs, the diodes resume integrating and the vertical

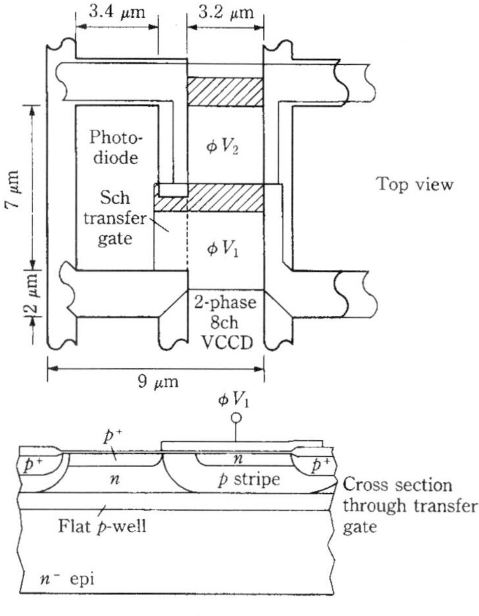

(a)

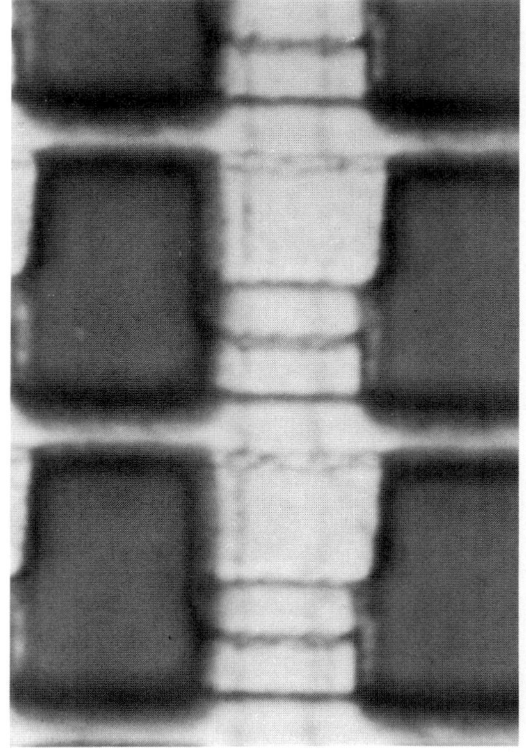

(b)

FIGURE 15 (*a*) Diagram of a pixel of an interline transfer CCD; (*b*) photograph of a pixel from an interline transfer CCD.

CCDs are clocked in order to transfer a row at a time from the image area into the horizontal CCD.

For consumer and most commercial applications, interlaced interline CCDs are used to be consistent with television standards. In interlaced interline CCDs there is one vertical CCD stage for every two photodiodes. During the retrace time before the first field the charge from the odd rows of photodiodes is transferred into the vertical CCDs; the charge is then transferred out a row at a time into the horizontal CCD. During the retrace time before the second field the charge from the even rows of photodiodes is transferred into the vertical CCDs; once again, the charge is transferred out a row at a time into the horizontal CCD. In NTSC television the fields are approximately 1/60 second long; in PAL the field time is 1/50 second.

Because the vertical CCDs in an interline CCD are covered by a light shield, very little stray light is absorbed in the CCD. However, due to light scattering under the lightshield and lateral diffusion of photogenerated electrons, some charge can reach the vertical registers as they are read out during a field. This results in smear. For consumer applications the level of smear is not noticeable. However, for especially demanding applications such as television broadcast cameras, a field storage area is added to the bottom of the image area. This architecture is called the frame interline transfer, or FIT CCD[28] (Fig. 13*c*). Following transfer of the photogenerated charge from the diodes into the vertical CCDs, the vertical CCDs are clocked to rapidly shift the charge from the image area into the storage area. This transfer typically takes less than 0.5 ms, reducing smear 30-fold. One line at a time is transferred from the storage area to the horizontal register and the horizontal register readout.

For still electronic photography, scientific, computer-related, graphics and professional applications, interlaced video is not desirable. For these applications a progressive scan architecture is utilized. In a progressive scan interline CCD there is a full CCD stage for every photodiode.[28] Following integration, the photogenerated charge from all the photodiodes is transferred into the vertical CCDs. The photodiodes resume integration and the charge packets from the vertical registers are transferred into the horizontal a row at a time. The progressive scan interline CCD requires twice the vertical CCD density and is therefore more complicated to fabricate. However, progressive scan readout provides many advantages in image quality for both motion and still imaging and is emerging as a likely standard for future digital HDTV systems as well as multimedia.

Nearly all interline CCDs used in camcorder, broadcast camera, or commercial applications utilize vertical antiblooming in order to prevent blooming when the sensor is illuminated beyond saturation. A cross-section diagram of an interline CCD with vertical antiblooming is shown in Fig. 15*b*. The device illustrated uses a pinned photodiode photosensing element. The CCD is built on an *n*-type silicon substrate. A *p*-well is formed about 2 μm deep in the *n*-type silicon. An *n*-type buried-channel is then formed followed by a $p+$ surface layer. The *n*-type substrate is reverse biased with respect to the *p*-well. When the photodiode is illuminated above saturation, the excess electrons spill out of the *n*-type buried channel and into the substrate. Owing to the vertical overflow, however, photogenerated carriers from photons absorbed below the *p*-well are drawn into the substrate and are not collected by the diode. Thus the quantum efficiency of these devices falls rapidly at wavelengths beyond 550 nm. Figure 16 shows the quantum efficiency of an interline CCD with vertical antiblooming as a function of wavelength. The internal quantum efficiency of the photodiode is nearly 100 percent to about 550 nm, after which it decreases. However, since the photodiode only occupies about 20 percent of the pixel, and this aperture is typically reduced further by the light shield in order to eliminate optical scattering into the vertical CCD, the actual efficiency of the device is only about 15 percent. The photoresponse is linear at low signal levels but becomes nonlinear at charge levels near saturation.[29]

Two major approaches are used to improve the fill factor (and therefore the quantum efficiency) of interline CCDs. These are the use of microlens arrays to focus the light incident on a pixel onto the photodiode and vertical integration of image sensors with

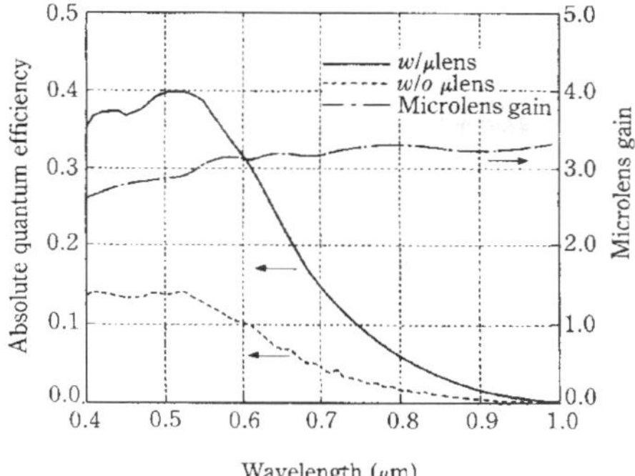

FIGURE 16 Quantum efficiency as a function of wavelength for an interline transfer CCD with and without microlens array.

amorphous silicon photoconductors to achieve a high fill factor. Figure 17 shows a microlens array on top of an interline CCD.[30] The lenses are formed by coating a spacer layer on the wafer of CCD devices followed by a lens-forming layer. The lens forming layer is patterned and then reflowed to form the lens arrays. The quantum efficiency of an interline CCD with and without a microlens array is shown in Fig. 18. A 3-fold improvement in quantum efficiency is achieved because light from nearly the entire pixel area is focused onto the photodiode.

The second approach to improving fill factor in interline CCD devices is to vertically integrate the device by use of an amorphous silicon photoconductor.[7] The structure of the photoconductor and its band diagram are illustrated in Fig. 2*e*. The amorphous silicon is

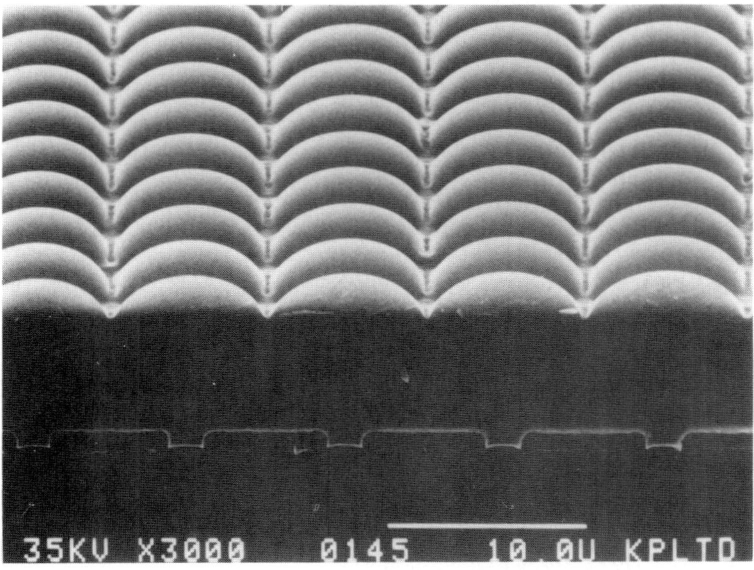

FIGURE 17 Scanning electron micrograph of microlens array on top of an interline transfer CCD.

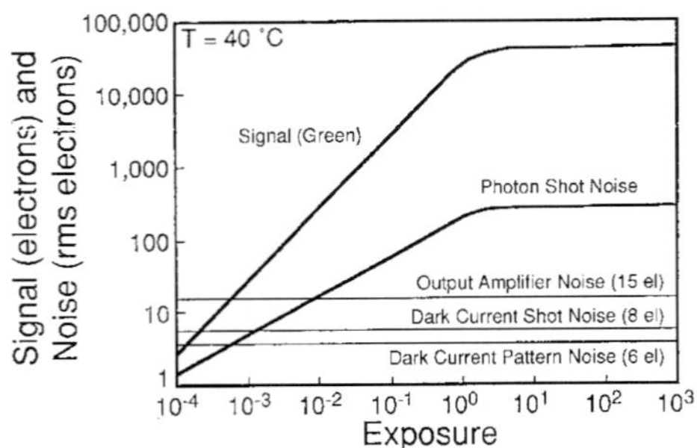

FIGURE 18 Quantum efficiency of interline CCD with and without microlens array.

about 1 μm thick; owing to its high absorption coefficient in the visible, it can achieve nearly 100 percent internal quantum efficiency. The back contact of the photoconductor contacts the diode in the interline CCD. Light absorbed in the amorphous silicon generates electron-hole pairs. The amorphous silicon is biased such as to create a high field which sweeps out the electrons to the back contact, where they can be stored on the diode until they are transferred into the CCD. Because the photoconductor is fabricated on top of the CCD pixel, it can have nearly 100 percent fill factor. In addition, because of the wide bandgap of the amorphous silicon, its dark current (due to thermal generation of carriers) is often lower than the single-crystal silicon. However, the amorphous silicon photoconductors suffer difficulties related to charge trapping. Owing to impurities and dangling or strained bonds, there is a high density of traps in amorphous silicon. These can trap charge carriers and reemit them at a later time, causing image lag. Because the trapping and detrapping is field-dependent, it can also result in nonlinear response.

CCD Performance. In all CCDs, both interline and frame transfer, the signal output is linear with the illumination except at levels approaching saturation. This linear response is in marked contrast to image tubes, which exhibit highly nonlinear response. For interline CCDs, the total charge capacity ranges from 100,000 electrons for larger cells (such as the 13.6 (V) × 11.6 (H)-μm cell typical for a $\frac{2}{3}$-in format) to 20,000 electrons or less for smaller cells (such as the 6.8 × 5.8-μm cell in a $\frac{1}{3}$-in format 484 × 768-pixel CCD).

The principal noise sources in both interline and full-frame image sensors are dark current pattern and shot noise and output amplifier noise. To illustrate the contributions, modern CCDs have dark current levels at 40°C of less than 1 nA/cm^2. For a $\frac{2}{3}$-in format sensor, this would correspond to about 320 electrons per pixel dark level. The corresponding shot noise would be 18 rms electrons and the pattern noise would typically be at a similar level. However, because of the nonrandom nature of pattern noise, its appearance is considerably more noticeable. The output amplifier noise would also be at about the 15-rms electron level. Thus, the overall noise for this example would be in the 30-electron range and the dynamic range would be over 2000 for a charge capacity of 90,000 electrons. For scientific applications where the sensor can be cooled and the readout performed at a lower frequency, noise levels less than 5 electrons can be achieved and even subelectron noise has been reported.[18]

Color Imaging

Silicon based CCDs are monochrome in nature. That is, they have no natural ability to determine the varying amounts of red, green, and blue (RGB) illumination presented to the photodetectors. There are three techniques to extract color information.

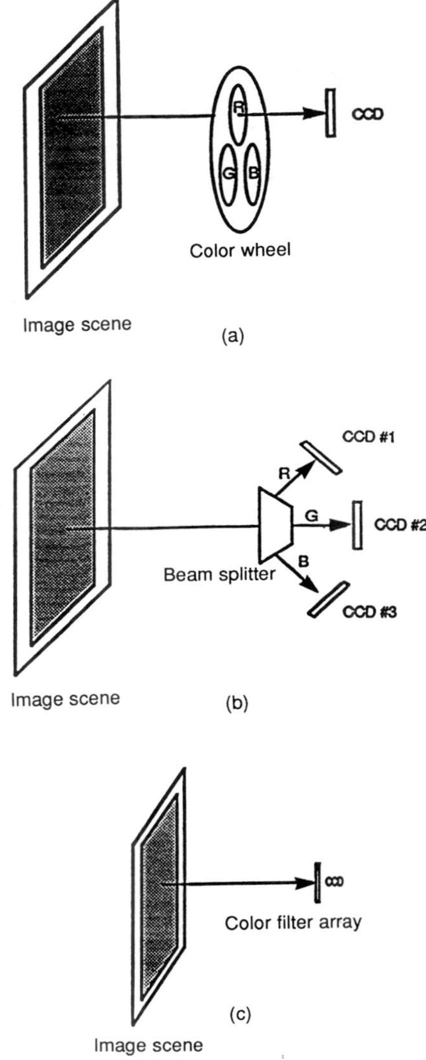

FIGURE 19 Methods of color separation in cameras with area image sensors: (*a*) color sequential using a color wheel; (*b*) a prism with dichroic beam splitters and three image sensors; (*c*) single-chip image sensor with integral color filter array.

1. *Color Sequential* (Fig. 19)—A color image can be created using a CCD by taking three successive exposures while switching in optical filters having the desired RGB transmittances. This approach is normally used only to provide still images of stationary scenes. The resulting image is then reconstructed off-chip. The advantage to this technique is that resolution of each color can remain that of the CCD itself. The disadvantage is that three exposures are required, reducing frame times by more than a factor of three. Color misregistration can also occur due to subject or camera motion. The filter switching assembly also adds to the mechanical complexity of the system.

2. *Three-chip Color* (Fig. 19)—Three-chip color systems use an optical system to split the scene into three separate color images. A dichroic prism beam splitter is normally used to provide RGB images. Color images can then be detected by synchronizing the outputs of the three CCDs. The disadvantage to such a system is that the optical complexity is very high and registration between sensors is difficult.

3. *Integral color Filter Arrays (CFA)* (Fig. 19)—Instead of performing the color filtering off-chip, filters of the appropriate characteristics can be fabricated above individual photosites.[31,32] This approach can be performed during device fabrication using dyed (e.g., cyan, magenta, yellow) photoresists in various patterns. The major problem with this approach is that each pixel is sensitive to only one color. Off-chip processing is required to "fill in" the missing color information between pixels.[33,34]

In order to minimize size, weight, and cost, most consumer color camcorders use a CCD sensor with an integral CFA. The photosites are covered with individual color filters—for example, a red, green, and blue striped filter, or a green, magenta, cyan, and yellow mosaic filter. Some popular CFA patterns are shown in Fig. 20. Because each photosite can sense only one color, the color sampling is not coincident. For example, a blue pixel might be seeing a white line, while nearby red and green pixels are seeing a dark line in the scene. As a result, high-frequency luminance edges can be aliased into bright color bands. These color bands depend not only on the color filter pattern used, but also on the optical prefilter and CFA interpolation algorithm. The color bands are caused by aliasing, which is a property of any sampled system. Aliasing occurs when the frequency of the input signal is greater than the Nyquist limit of one-half the sampling frequency. If the input frequency is well below the Nyquist limit, there are many samples per cycle. This allows the input to be reconstructed perfectly, if a proper reconstruction filter is used. When the input frequency is greater than the Nyquist limit, there are less than two samples per cycle. The sample values now define a new curve, which has a frequency lower than the input frequency. In effect, the high frequency takes on the alias of a lower frequency.

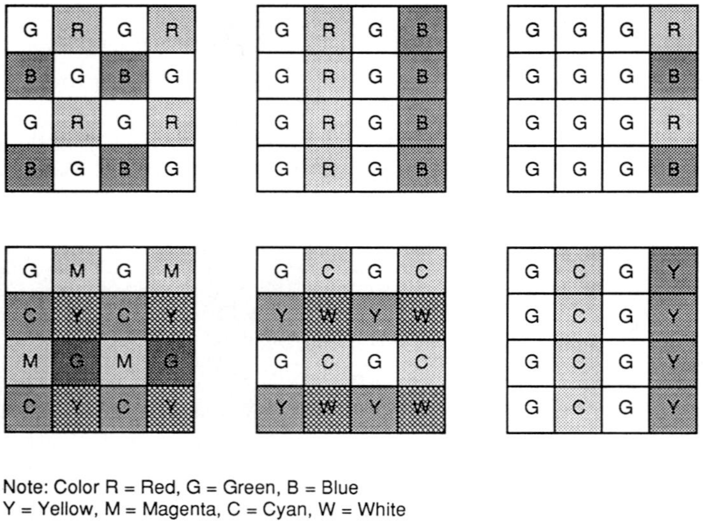

FIGURE 20 Common color filter array patterns.

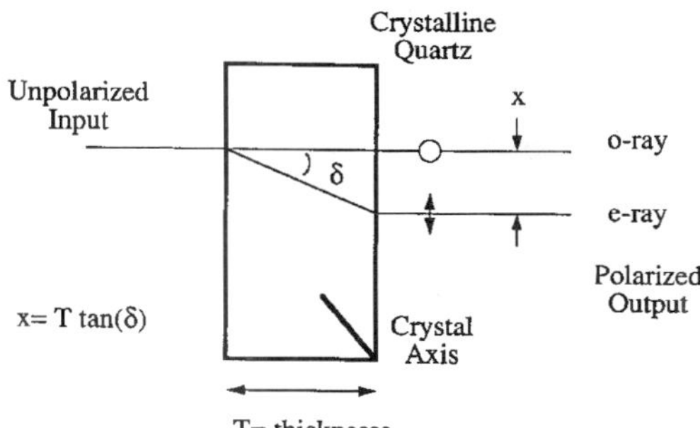

FIGURE 21 Birefringent blur filter used to reduce aliasing in single-chip color image sensors.

Aliasing is a particular problem with color sensors, since the sampling phase is different for the different color photosites. Therefore, the aliased signal has different phases for different colors. This creates the color bands.

The color aliasing is reduced by using an optical anti-aliasing or "blur" filter, positioned in front of the color CCD sensor.[35] Blur filters are typically made of birefringent quartz, with the crystal axis oriented at a 45° angle, as shown in Fig. 21. In this orientation, the birefringent quartz exhibits the double refraction effect. An unpolarized input ray emerges as two polarized output rays. The output ray separation is proportional to the filter's thickness. A 1.5-mm-thick plate will give a separation of about 9 μm. Figure 21 shows a simple "two-spot" filter. More complex filters use three or more pieces of quartz cemented in a stack.

22.6 REFERENCES

1. G. Lubberts and B. C. Burkey, "Optical and Electrical Properties of Heavily Phosphorous-doped Epitaxial Silicon Layers," *J. Appl. Phys.* **55**(3):760–763 (Feb. 1, 1984).
2. O. S. Heavens, *Optical Properties of Thin Solid Films,* Dover, New York, 1965.
3. N. Teranishi et al., "An Interline CCD Image Sensor with Reduced Image Lag," *IEEE Transactions on Electron Devices,* **ED-31**(12):1829–1833 (Dec. 1984).
4. S. Kosman et al., "A Large Area 1.3 Megapixel Full-Frame CCD Image Sensor with a Lateral Overflow Drain and a Transparent Gate Electrode," *Proc. IEEE International Electron Device Meeting,* Washington, D.C. (Dec. 1990).
5. B. Burkey et al., "The Pinned Photodiode for an Interline Transfer CCD Image Sensor," *Proc. 1984 International Electron Device Meeting,* Washington, D.C. (Dec. 1986), p. 28.
6. M. Sasaki, H. Ihara, and Y. Matsunaga, "A $\frac{2}{3}$-in 400 k-Pixel Sticking-Free Stact CCD Image Sensor," *IEEE Transactions on Electron Devices,* **ED-28**(11) (Nov. 1993).
7. H. Yamashita et al., "A $\frac{2}{3}$-inch 2-megapixel stack CCD Imager," *Proceedings of the 1994 International Solid State Circuits Conference,* p. 224.
8. E. G. Stevens et al., "A 1-Megapixel, Progressive-Scan Image Sensor with Antiblooming Control and Lag-Free Operation," *IEEE Transactions on Electron Devices,* **ED-38**(5) (May 1991).
9. D. K. Schroder, "The Concept of Generation and Recombination Lifetimes in Semiconductors," *IEEE Transactions on Electron Devices* **ED-29:**(8) (August 1982), p. 17.
10. J. Van der Spiegel and G. J. Declerck, *Solid State Electronics* **27:**147 (1984).
11. Carlo, H. Sequin and Michael F. Tompsett, *Charge Transfer Devices Advances in Electronics and Electron Physics.* suppl. 8, Academic Press, Inc., New York, 1975.

12. M. J. Howes and D. V. Morgan, (eds.), *Charge Coupled Devices and Systems,* Wiley, Chicago, 1979.
13. D. F. Barbe et al., *Charge-Coupled Devices,* Springer-Verlag, New York, 1980.
14. M. Azuma et al., "A Fixed Pattern Noise Free $\frac{2}{3}$", 1.3 M-Pixel CCD Image Sensor for HDTV Camera System," *Proc. 1991 IEEE International Solid-State Circuits Conference,* San Francisco, 212–213 (Feb. 1991).
15. Y. Matsunaga and S. Ohsawa, "$\frac{1}{3}$ Inch Interline Transfer CCD Image Sensor with Negative Feedback 94 dB Dynamic Range Charge Detector," *Proc. 1991 IEEE International Solid-State Circuits Conference,* San Francisco, 210–211 (Feb. 1991).
16. J. Janesick, "Open Pinned-Phase CCD Technology," *SPIE Proceedings,* vol. 1159, San Diego (Nov. 1989).
17. J. Hynecek, "Virtual Phase Technology: A New Approach to Fabrication of Large-Area CCDs," *IEEE Transactions on Electron Devices* **ED-28**(5):483–489 (May 1981).
18. J. Janesick et al., "Fano-Noise-Limited CCDs," *Optical and Optoelectric Applied Science and Engineering Symposium: X-Ray Instrumentation in Astronomy,* San Diego (Aug. 14–19, 1988).
19. J. Carbone et al., "New Low Noise Random Access, Radiation Resistant and Large Format Charge Injection Device (CID) Imagers," *Proc. of SPIE Conference* ***1900*** (Jan. 31–Feb. 4, 1993).
20. M. Sugawara et al., "An Amplified MOS Imager Suited for Image Processing," *Proc. 1994 IEEE International Solid-State Circuits Conference,* San Francisco, 228–229 (Feb. 1994).
21. M. Ogata et al., "A Small Pixel CMD Image Sensor," *IEEE Transactions on Electron Devices,* **ED-38**(5):1005–1010 (May 1991).
22. D. H. McCann, M. H. White, A. P. Turley, and R. A. Frosch, "Time Delay and Integration Detectors Using Charge Transfer Devices," U.S. Patent Number 4,280,141, July 1981.
23. Thompson et al., "Time-Delay-and-Integration Charge-Coupled Devices Using Tin Oxide Gate Technology," *IEEE Transactions on Electron Devices* **ED-25**(2):132–134 (Feb. 1978).
24. Ohba et al., "MOS Area Sensor: Part 2: Low-noise MOS Arèa Sensor with Antiblooming Photodiodes," *IEEE Trans. Electron Devices* **ED-27**(8):1682–1687 (Aug. 1980).
25. T. Nakamura et al., "A New MOS Image Sensor Operating in a Non-Destructive Readout Mode," *Proc. 1986 IEEE International Electron Device Meeting,* Washington, D.C. (Dec. 1986), 353–356.
26. Nomoto et al., "A $\frac{2}{3}$ Inch 2M-Pixel CMD Image Sensor with Multi-Scanning Functions," *Proc. 1993 IEEE International Solid-State Circuits Conference,* San Francisco (Feb. 1993), 196–197.
27. McCann et al., "Buried Channel CCD Imaging Arrays with Tin Oxide Transparent Gates," *IEEE International Solid State Circuits Conference,* pp. 30, 31, 261, 262 (Feb. 1978).
28. K. Harada, "A $\frac{2}{3}$-Inch, 2M-Pixel Frame Interline Transfer CCD HDTV Image Sensor,"*Proc. 1992 IEEE International Solid-State Circuits Conference,* San Francisco, 170–171 (Feb. 1992).
29. E. Stevens, "Photoresponse Nonlinearity of Solid State Image Sensors with Antiblooming Protection," *IEEE Transactions on Electron Devices,* **ED-38**(2):229–302 (Feb. 1991).
30. A. Weiss et al., *J. Electrochem. Soc.* **133:**110C (1986).
31. P. Dillion et al., "Fabrication and Performance of Color Filter Arrays for Solid-State Imagers," *IEEE Trans. Electron Devices* **ED-25:**97–101 (1978).
32. P. Dillion, D. Lewis, and F. Kaspar, "Color Imaging System Using a Single CCD Array," *IEEE Trans. Electron Devices,* **ED-25:**102–107 (Feb. 1978).
33. K. A. Parulski, et al., "A Digital Color CCD Imaging System Using Custom VLSI Circuits," *IEEE Trans. Consumer Electronics* **35**(3):382–389 (Aug. 1989).
34. K. A. Parulski, "Color Filters and Processing Alternatives for One-Chip Cameras," *IEEE Trans. Electron Devices* **ED-32**(8):1381–1389.
35. J. E. Greivenkamp, "Color Dependent Optical Prefilter for the Suppression of Aliasing Artifacts," *Appl. Opt.* **29**(5):676–684, (Feb. 1990).

CHAPTER 23
INFRARED DETECTOR ARRAYS

Lester J. Kozlowski
Rockwell International Science Center
Thousand Oaks, California

Walter F. Kosonocky
New Jersey Institute of Technology
University Heights
Newark, New Jersey

23.1 GLOSSARY

A_{det}	detector area
A_I	gate modulation current gain (ratio of integration capacitor current to load current)
A_V	amplifier voltage gain
C_{amp}	amplifier capacitance
C_{det}	detector capacitance
$C_{F/S}$	fill-and-spill gate capacitance for a Tompsett type CCD input
C_{fb}	CTIA feedback or Miller capacitance
C_{gd}	FET gate-drain overlap capacitance
C_{gs}	FET gate-source capacitance
C_{L}	CTIA band-limiting load capacitance
C_{out}	sense node capacitance at the CCD output
C_T	effective feedback (transcapacitance) or integration capacitance for a capacitive transimpedance amplifier
$\mathrm{C}_{\mathrm{T}\lambda}$	spectral photon contrast
cte	charge transfer efficiency
$D^*_{\lambda pk}$	peak detectivity (cm-Hz$^{1/2}$/W or Jones)
D^*_{bb}	blackbody detectivity (cm-Hz$^{1/2}$/W or Jones)
D^*_{th}	thermal detectivity (cm-Hz$^{1/2}$/W or Jones)
e-	electron
E_g	detector energy gap
$f/\#$	conventional shorthand for the ratio of the focal length of a lens to its diameter
f_{chop}	chopper frequency

f_{frame}	display frame rate
f_{knee}	frequency at which the 1/f noise intersects the broadband noise
f_s	spatial frequency (cycles/radian)
$g_{m,\text{LOAD}}$	gate transconductance of the load FET in the gate modulated input circuit
g_m	gate transconductance of a Field Effect Transistor
h	Planck's constant
I_D	FET drain current
I_{det}	detector current
I_{photo}	detector photocurrent
k	Boltzmann constant
K_{amp}	amplifier FET noise spectral density at 1 Hz
K_{det}	detector noise spectral density at 1 Hz
K_{FET}	FET noise spectral density at 1 Hz
L	length-to-width ratio of a bar chart (always set to 7)
MRT	minimum resolvable temperature (K)
MTF	modulation transfer function for the optics, detector, readout, the integration process, or the composite sensor
n	detector junction ideality or diffusion constant
$N_{\text{amp,1/f}}$	number of noise carriers for one integration time due to amplifier FET 1/f noise
$N_{\text{amp,white}}$	number of noise carriers for one integration time due to amplifier FET white noise
N_c	number of photo-generated carriers integrated for one integration time
n_{det}	detector junction ideality or diffusion constant
NE ΔT	noise equivalent temperature difference (K)
n_{FET}	subthreshold FET ideality
N_{FPA}	composite (total) FPA noise in carriers
$N_{kTC,\text{channel}}$	CTIA broadband channel noise in carriers
$N_{\text{load,white}}$	number of noise carriers for one integration time due to CTIA load FET white noise
N_{os}	display overscan ratio
N_{PHOTON}	shot noise of photon background in carriers
NSD	noise spectral density of a detector or field effect transistor; the 1/f noise is often specified by the NSD at a frequency of 1 Hz
N_{sf}	source follower noise
N_{ss}	serial scan ratio
q	electron charge in coulombs
Q_B	photon flux density (photons/cm^2-s) incident on a focal plane array
Q_D	charge detected in a focal plane array for one integration time

$Q_{\max}$	maximum charge signal at saturation
R_{det}	detector resistance
R_{LOAD}	gate modulation load resistance
R_0	detector resistance at zero-bias resistance
R_0A	detector resistance-are product at zero-bias voltage
R_{r}	detector resistance in reverse-bias resistance
S/N	signal-to-noise ratio
SNR_{T}	target signal-to-noise ratio
S_V	readout conversion factor describing the ratio of output voltage to detected signal carriers
T	operating temperature
tce	thermal coefficient of expansion
TCR	thermal coefficient of resistance for bolometer detectors
T_D	time constant for correlated double sampling process normally set by Nyquist rate
t_{int}	integration time
U	residual nonuniformity
V_{br}	detector reverse-bias breakdown voltage, sometimes defined as the voltage where $R_r = R_0$
V_D	FET drain voltage
V_{det}	detector bias voltage
V_{DS}	FET grain-to-source voltage
V_G	FET gate voltage
v_n	measured rms noise voltage
ΔA_I	gate modulation current gain nonuniformity
Δf	noise bandwidth (Hz)
ΔI_{photo}	differential photocurrent
ΔT	scene temperature difference creating differential photocurrent ΔI_{photo}
ΔV_S	signal voltage for differential photocurrent ΔI_{photo}
Δx	horizontal detector subtense (mradian)
Δy	vertical detector subtense (mradian)
η	detector quantum efficiency
η_{BLIP}	percentage of BLIP
$\eta_{\text{inj,DI}}$	injection efficiency of detector current into the source-modulated FET of the direct injection input circuit
η_{inj}	injection efficiency of detector current
η_{noise}	injection efficiency of DI circuit noise into integration capacitor
η_{pc}	quantum efficiency of photoconductive detector

η_{pv}	quantum efficiency of photovoltaic detector
λ_c	detector cutoff wavelength (50 percent of peak response, μm)
σ_{det}	noise spectral density of total detector noise including photon noise
$\sigma_{\text{input,ir}}$	noise spectral density of input-referred input circuit noise
σ_{LOAD}	noise spectral density of input-referred load noise
$\sigma_{\text{mux,ir}}$	noise spectral density of input-referred multiplexer noise
σ_{VT}	rms threshold voltage nonuniformity across an FPA
τ_{amp}	amplifier time constant (s)
τ_{eye}	eye integration time (s)
τ_o	optical transmission
ω	angular frequency (radians)
$\langle e_{\text{amp}} \rangle$	buffer amplifier noise for buffered direct injection circuit

23.2 INTRODUCTION

Infrared sensors have been available since the 1940s to detect, measure, and image the thermal radiation emitted by all objects. Due to advanced detector materials and microelectronics, large scanning and staring focal plane arrays (FPA) with few defects are now readily available in the short wavelength infrared (SWIR; 1 to 3 μm), medium wavelength infrared (MWIR; ≈3 to 5 μm), and long wavelength infrared (LWIR; ≈8 to 14 μm) spectral bands. We discuss in this chapter the disparate FPA technologies, including photon and thermal detectors, with emphasis on the emerging types.

IR sensor development has been driven largely by the military. Detector requirements for missile seekers and forward looking infrared (FLIR) sensors led to high-volume production of photoconductive (PC) HgCdTe arrays starting in the 1970s. Though each detector requires direct connection to external electronics for purposes of biasing, signal-to-noise ratio (SNR) enhancement via time delay integration (TDI), and signal output, the first generation FPAs displaced the incumbent Pb-salt (PbS, PbSe) and Hg-doped germanium devices, and are currently being refined using custom analog signal processing,[1] laser-trimmed solid-state preamplifiers, etc.

Size and performance limitations of first-generation FLIRs necessitated development of self-multiplexed FPAs with on-chip signal processing. Second-generation thermal imaging systems use high-density FPAs with relatively few external connections. Having many detectors that integrate longer, low-noise multiplexing and on-chip TDI (in some scanning arrays), second-generation FPAs offer higher performance and design flexibility. Video artifacts are suppressed due to the departure from ac-coupling and interlaced raster scan, and external connections are minimized. Fabricated in monolithic and hybrid methodologies, many detector and readout types are used in two basic architectures (staring and scanning). In a monolithic FPA, the detector array and the multiplexing signal processor are integrated in a single substrate. The constituents are fabricated on separate substrates and interconnected in a hybrid FPA.

FPAs use either photon or thermal detectors. Photon detection is accomplished using intrinsic or extrinsic semiconductors and either photovoltaic (PV), photoconductive (PC), or metal insulator semiconductor (MIS) technologies. Thermal detection relies on capacitive (ferro- and pyroelectric) or resistive bolometers. In all cases, the detector signal is coupled into a multiplexer and read out in a video format.

Infrared Applications

Infrared FPAs are now being applied to a rapidly growing number of civilian, military, and scientific applications such as industrial robotics and thermography (e.g., electrical and mechanical fault detection), medical diagnosis, environmental and chemical process monitoring, Fourier transform IR spectroscopy and spectroradiometry, forensic drug analysis, microscopy, and astronomy. The combination of high sensitivity and passive operation is also leading to many commercial uses. The passive monitoring provided by the addition of infrared detection to gas chromatography–mass spectroscopy (GC-MS), for example, yields positive chemical compound and isomer detection without sample alteration. Fusing IR data with standard GC-MS aids in the rapid discrimination of the closely related compounds stemming from drug synthesis. Near-IR (0.7–0.1 μm) and SWIR spectroscopy and fluorescence are very interesting near-term commercial applications since they pave the way for high-performance FPAs in the photochemical, pharmaceutical, pulp and paper, biomedical, reference quantum counter, and materials research fields. Sensitive atomic and molecular spectroscopies (luminescence, absorpion, emission, and Raman) require FPAs having high quantum efficiency, low dark current, linear transimpedance, and low read noise.

Spectral Bands

The primary spectral bands for infrared imaging are 3–5 and 8–12 μm because atmospheric transmission is highest in these bands. These two bands, however, differ dramatically with respect to contrast, background signal, scene characteristics, atmospheric transmission under diverse weather conditions, and optical aperture constraints. System performance is a complex combination of these and the ideal system requires dual band operation. Factors favoring the MWIR include its higher contrast, superior clear-weather performance, higher transmissivity in high humidity, and higher resolution due to ~3× smaller optical diffraction. Factors favoring the LWIR include much-reduced background clutter (solar glint and high-temperature countermeasures including fires and flares have much-reduced emission), better performance in fog, dust, and winter haze, and higher immunity to atmospheric turbulence. A final factor favoring the LWIR, higher S/N ratio due to the greater radiance levels, is currently moot because of technology limitations. Due to space constraints and the breadth of sensor applicability, we focus on target/background metrics in this section.

The signal collected by a visible detector has higher daytime contrast than either IR band because it is mainly radiation from high-temperature sources that is subsequently reflected off earth-based (ambient temperature; ≈290 K) objects. The high-temperature sources are both solar (including the sun, moon, and stars) and synthetic. Since the photon flux from high- and low-temperature sources differs greatly at visible wavelengths from day to night, scene contrasts of up to 100 percent ensue.

Reflected solar radiation has less influence as the wavelength increases to a few microns since the background radiation increases rapidly and the contrast decreases. In the SWIR band, for example, the photon flux density from the earth is comparable to visible room light (10^{13} photons/cm^2-s). The MWIR band (~10^{15} photon flux density) has lower, yet still dynamic, daytime contrast, and can still be photon-starved in cold weather or at night.

The net contribution from reflected solar radiation is even lower at longer wavelengths. In the LWIR band, the background flux is equivalent to bright sunlight (≈10^{17} photons/cm^2-s). This band thus has even lower contrast and much less background clutter, but the "scene" and target/background metrics are similar day and night. Clear-weather performance is relatively constant.

Depending on environmental conditions, however, IR sensors operating in either band must discern direct emission from objects having temperatures very near the average background temperature (290 K) in the presence of the large background and degraded atmospheric transmissivity. Under conditions of uniform thermal soak, such as at diurnal equilibrium, the target signal stems from minute emissivity differences.

The spectral photon incidence for a full hemispheric surround is

$$Q = \tau_{cf} \int_{\lambda_1}^{\lambda_2} Q_\lambda(\lambda)\, d\lambda \tag{1}$$

if a zero-emissivity bandpass filter having in-band transmission, τ_{cf}, cut-on wavelength λ_1, and cutoff wavelength λ_2 is used (zero emissivity obtained practically by cooling the spectral filter to a temperature where its self-radiation is negligible). The photon flux density, Q_B (photons/cm^2-s), incident on a focal plane array is

$$Q_B = \frac{1}{4(f/\#)^2 + 1} Q \tag{2}$$

where $f/\#$ is the conventional shorthand for the ratio of the focal length to the diameter (assumed circular) of the limiting aperture or lens. The cold shield $f/\#$ limits the background radiation to a field-of-view consistent with the warm optics to eliminate extraneous background flux and concomitant noise. The background flux in the LWIR band is approximately two orders of magnitude higher than in the MWIR.

The spectral photon contrast, $C_{T\lambda}$, is the ratio of the derivative of spectral photon incidence to the spectral photon incidence, has units K^{-1}, and is defined

$$C_{T\lambda} = \left(\frac{\dfrac{\partial Q}{\partial T}}{Q} \right) \tag{3}$$

Figure 1 is a plot of $C_{T\lambda}$ for several MWIR subbands (including 3.5–5, 3.5–4.1, and 4.5–5 μm) and the 8.0–12 μm LWIR spectral band. The contrast in the MWIR bands at 300 K is 3.5–4 percent compared to 1.6 percent for the LWIR band. While daytime MWIR contrast is even higher due to reflected sunlight, an LWIR FPA offers higher sensitivity if it has the larger capacity needed for storing the larger amounts of photogenerated (due to the higher background flux) and detector-generated carriers (due to the narrow bandgap). The photon contrast and the background flux are key parameters that determine thermal resolution as will be described later under "Performance Figures of Merit."

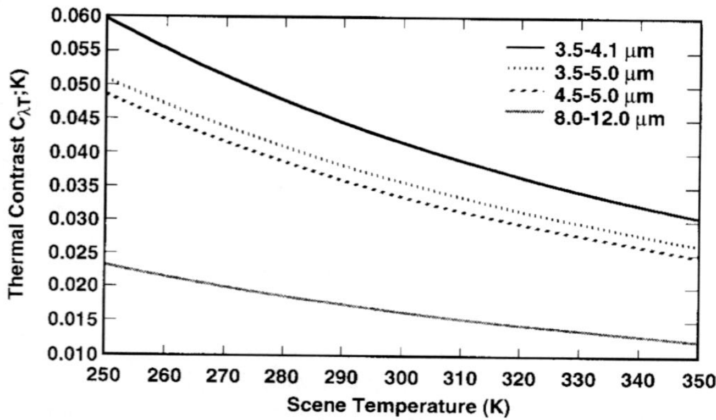

FIGURE 1 Spectral photon contrast in the MWIR and LWIR.

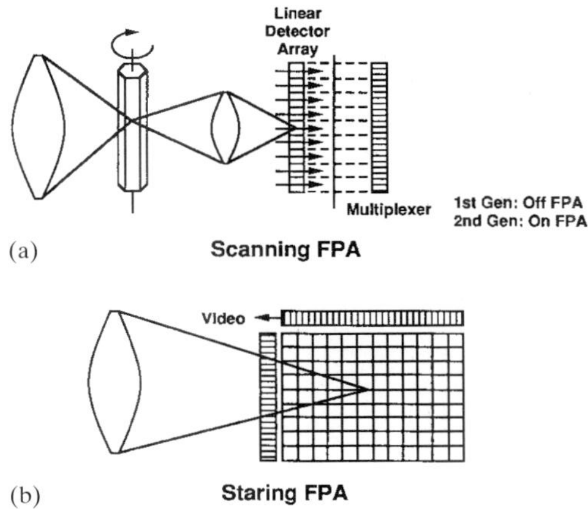

FIGURE 2 Scanning (*a*) and staring (*b*) focal plane arrays.

Scanning and Staring Arrays

The two basic types of FPA are scanning and staring. The simplest scanning device consists of a linear array as shown in Fig. 2*a*. An image is generated by scanning the scene across the strip. Since each detector scans the complete horizontal field-of-view (one video raster line) at standard video frame rates, each resolution element or pixel has a short integration time and the total detected charge can usually be accommodated.

A staring array (Fig. 2*b*) is the two-dimensional extension of a scanning array. It is self-scanned electronically, can provide enhanced sensitivity, and is suitable for lightweight cameras. Each pixel is a dedicated resolution element, but synchronized dithering of sparsely populated arrays is sometimes used to enhance the effective resolution, minimize spatial aliasing, and increase the effective number of pixels. Although theoretically charge can be integrated for the full frame time, the charge-handling capacity is inadequate at terrestrial LWIR backgrounds.

Detectors

Infrared detectors convert IR photons and energy to electrical signals. Many types are used in FPAs (as shown in Fig. 3[2]) including photon and thermal detectors that address diverse requirements spanning operating temperatures from 4 K to toom temperature. Figure 4 compares the quantum efficiencies of several detector materials.

Intrinsic detectors[3] usually operate at higher temperatures than extrinsic devices, have higher quantum efficiencies, and dissipate less power. Backside-illuminated devices, consisting of an absorbing epitaxial layer on a transparent substrate, are used in hybrid FPAs and offer the advantages of nearly 100 percent active detector area, good mechanical support, and high quantum efficiency. The most popular intrinsic photovoltaics are HgCdTe and InSb. These detectors are characterized by their quantum efficiency (η), zero-bias resistance (R_0), reverse-bias resistance (R_r), junction ideality or diffusion constant n, excess noise (if any) versus bias, and reverse-bias breakdown voltage (V_{br}), which is sometimes defined as the voltage where $R_r = R_0$.

PtSi is a photovoltaic Schottky barrier detector (SBD) which is the most mature for large monolithic FPAs. IR detection is via internal photoemission over a Schottky barrier (0.21–0.23 eV). Characteristics include low ($\approx$0.5 percent for broadband 3.5–5.0 μm) but very uniform quantum efficiency, high producibility that is limited only by the Si readout circuits, full compatibility with VLSI technology, and soft spectral response with peak

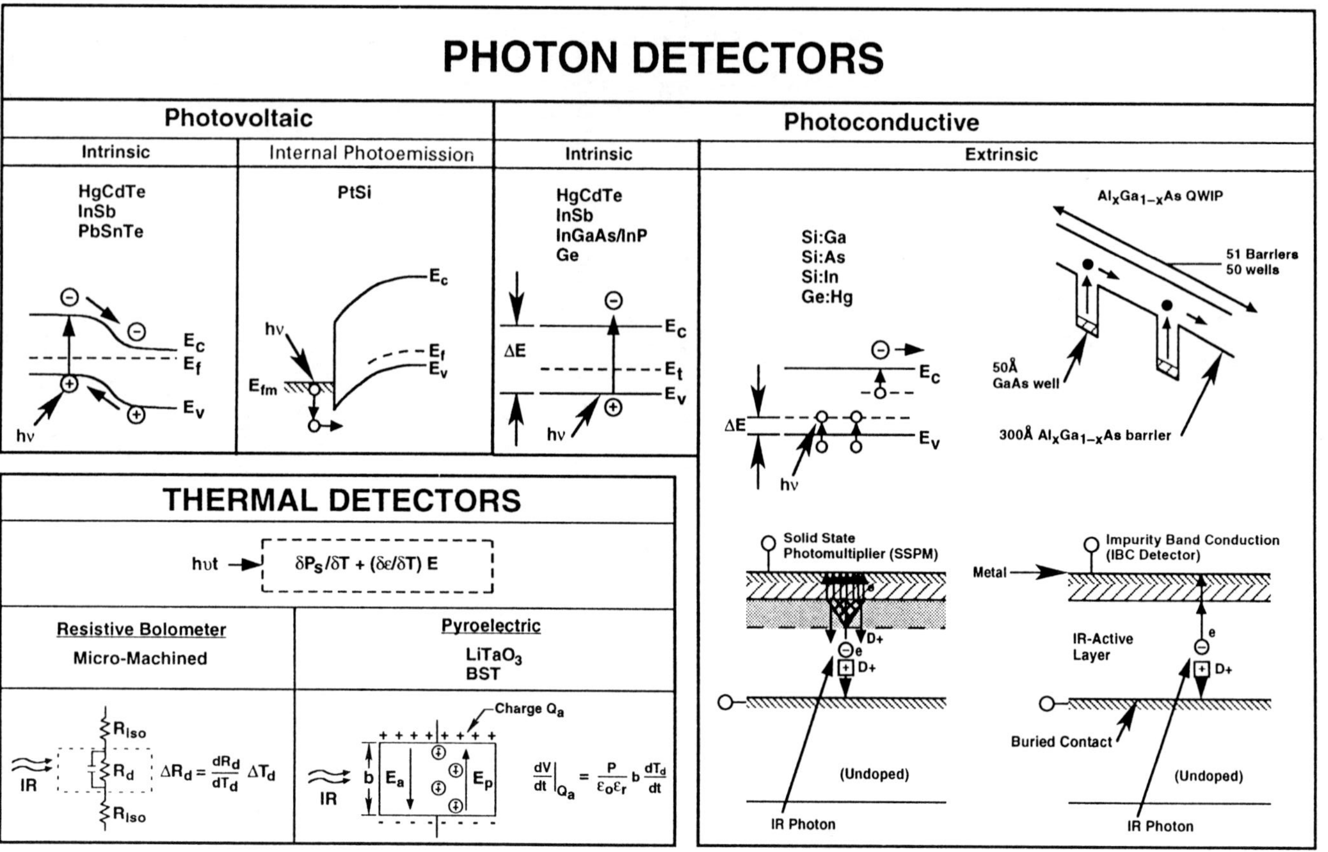

FIGURE 3 Photon and thermal detectors.

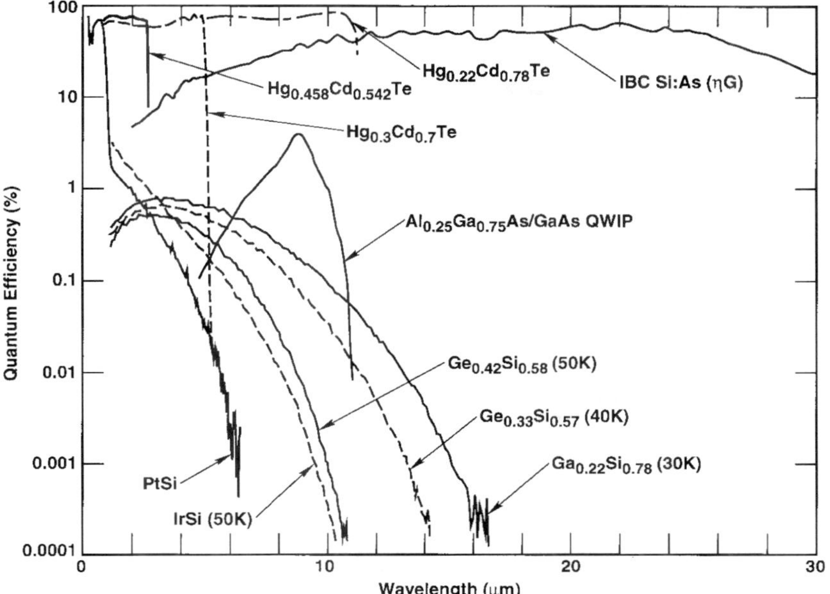

FIGURE 4 Quantum efficiency versus wavelength for several detector materials.

below 2 μm and zero response just beyond 6 μm. Internal photoemission dark current requires cooling below 77 K.

HgCdTe is the most popular intrinsic photoconductor, and various linear arrays in several scanning formats are used worldwide in first-generation FLIRs. For reasons of producibility and cost, HgCdTe photoconductors have historically enjoyed a greater utilization than PV detectors despite the latter's higher quantum efficiency, higher D^* by a factor of $(2\eta_{pv}/\eta_{pc})^{1/2}$, and superior modulation transfer function (MTF). Nevertheless, not all photoconductors are good candidates for FPAs due to their low detector impedance. This includes the intrinsic materials InSb and HgCdTe.

The most popular photoconductive material system for area arrays is doped extrinsic silicon (Si: x; where x is In, As, Ga, Sb, etc.), which is made in either conventional or impurity band conduction [IBC or blocked impurity band (BIB)] technologies. Early monolithic arrays were doped-Si devices, due primarily to compatibility with the silicon readout. Extrinsic photoconductors must be made relatively thick (up to 30 mils; doping density of IBCs, however, minimizes this thickness requirement but does not eliminate it) because they have much lower photon capture cross section than intrinsic detectors. This factor adversely affects their MTF in systems having fast optics.

Historically, Si:Ga and Si:In were the first mosaic focal plane array PC detector materials because early monolithic approaches were compatible with these dopants. Nevertheless, problems in fabricating the detector contacts, early breakdown between the epitaxial layer and the detector material (double injection), and the need for elevated operating temperatures helped force the general move to monolithic PtSi and intrinsic hybrids.

The most advanced extrinsic photoconductors are IBC detectors using Si:As and Si:Ga.[4] These have reduced recombination noise (negating the $\sqrt{2}$ superiority in S/N that PV devices normally have) and longer spectral response than standard extrinsic devices due to the higher dopant levels. IBC detectors have a unique combination of PC and PV characteristics, including extremely high impedance, PV-like noise (reduced recombination noise since IBC detectors collect carriers both from the continuum and the "hopping" impurity band), linear photoconductive gain, high uniformity, and superb stability. The photo-sensitive layer in IBCs is heavily doped to achieve hopping-type conduction. A thin, lightly doped ($10^{10}/cm^2$) silicon layer blocks the hopping current before it reaches the device electrode to reduce noise. Specially doped IBCs (see cross-sectional views in Fig. 3) operate as solid-state photomultipliers (SSPM) and visible light photon counters (VLPC) in which photoexcited carriers are amplified by impact ionization of impurity-bound carriers.[5] The amplification allows counting of individual photons at low flux levels. Standard SSPMs respond from 0.4 to 28 μm.

A rapidly developing alternative detector is the GaAs/AlGaAs quantum well infrared photodetector (QWIP). Various QWIP photoconductive[6] and photovoltaic[7] structures are being investigated as low-cost alternatives to II-VI LWIR detectors like HgCdTe. Infrared detection in the typical PC QWIP is via intersubband or bound-to-extended-state transitions within the multiple quantum well superlattice structure. Due to the polarization selection rules for transitions between the first and second quantum wells, the photon electric field must have a component parallel to the superlattice direction. Light absorption in n-type material is thus anisotropic with zero absorption at normal incidence. The QWIP detector's spectral response is narrowband, peaked about the absorption energy. The wavelength of peak response can be adjusted via quantum well parameters and can be made bias-dependent.

Various bolometers, both resistive and capacitive (pyroelectric), are also available. Bolometers sense incident radiation via energy absorption and concomitant change in device temperature in both cooled moderate-performance and uncooled lower-performance schemes. Much recent research, which was previously highly classified, has focused on both hybrid and monolithic uncooled arrays and has yielded significant improvements in the detectivity of both resistive and capacitive bolometer arrays. The resistive bolometers currently in development consist of a thin film of a temperature-sensitive resistive material film which is suspended above a silicon readout. The pixel support struts provide electrical interconnect and high thermal resistance to maximize pixel sensitivity. Recent work has focused on the micromachining necessary to fabricate mosaics with low thermal conductance using monolithic methodologies compatible with silicon.

Capacitive bolometers sense a change in elemental capacitance and require mechanical chopping to detect incident radiation. The most common are pyroelectric detectors. J. Cooper[8] suggested the use of pyroelectric detectors in 1962 as a possible solution for applications needing a low-cost IR FPA with acceptable performance. These devices have temperature-dependent spontaneous polarization. Ferroelectric detectors are pyroelectric detectors having reversible polarization. There are over a thousand pyroelectric crystals, including several popularly used in hybrid FPAs; e.g. lithium tantalate ($LiTaO_3$), triglycine sulfate (TGS), and barium strontium titanate[9] ($BaSrTiO_3$).

23.3 MONOLITHIC FPAs

A monolithic FPA consists of a detector array and the readout multiplexer integrated on the same substrate. The progress in the development of the monolithic FPAs in the last two decades has been strongly influenced by the rapid advances in the silicon VLSI technology. Therefore, the present monolithic FPAs can be divided into three categories reflecting their relationship to the silicon VLSI technology. The first category includes the "complete" monolithic FPAs in which the detector array and the readout multiplexer are integrated on the same silicon substrate using processing steps compatible with the silicon VLSI technology. They include the extrinsic Si FPAs reported initially in the 1970s,[10]

FPAs with Schottky barrier,[11] heterojunction detector FPAs, and the recently reported microbolometer FPAs.[12]

The second category will be referred to here as the "partial monolithic" FPAs. This group includes narrowband detector arrays of HgCdTe[13] and InSb[14] integrated on the same substrate only with the first level of multiplexing, such as the row and column readout from a two-dimensional detector array. In this case the multiplexing of the detected signal is completed by additional silicon IC chips usually packaged on the imager focal plane.

The third category represents "vertically integrated" photodiode (VIP) FPAs. These FPAs are functionally similar to hybrids in the sense that a silicon readout multiplexer is used with the narrow-bandgap HgCdTe detectors. However, while in the hybrid FPA the completed HgCdTe detector array is typically connected by pressure contacts via indium bumps to the silicon multiplex pads; in the case of the vertically integrated FPAs, HgCdTe chips are attached to a silicon multiplexer wafer and then the fabrication of the HgCdTe photodiodes is completed including the deposition and the definition of the metal connections to the silicon readout multiplexer.

In the following sections we will review the detector readout structures, and the main monolithic FPA technologies. It should also be noted that most of the detector readout techniques and the architectures for the monolithic FPAs were originally introduced for visible silicon imagers. This heritage is reflected in the terminology used in the section.

Architectures

The most common structures for the photon detector readout and architectures of monolithic FPAs are illustrated schematically in Fig. 5 and 6.

MIS Photogate FPAs: CCD, CID, and CIM. Most of the present monolithic FPAs use either MIS photogates or photodiodes as the photon detectors. Figure 5 illustrates a direct integration of the detected charge in the potential well of a MIS (photogate) detector for a charge coupled device (CCD) readout in (*a*) a charge injection device (CID) readout in (*b*) and a charge-integration matrix CIM) readout in (*c*). The unique characteristic of the

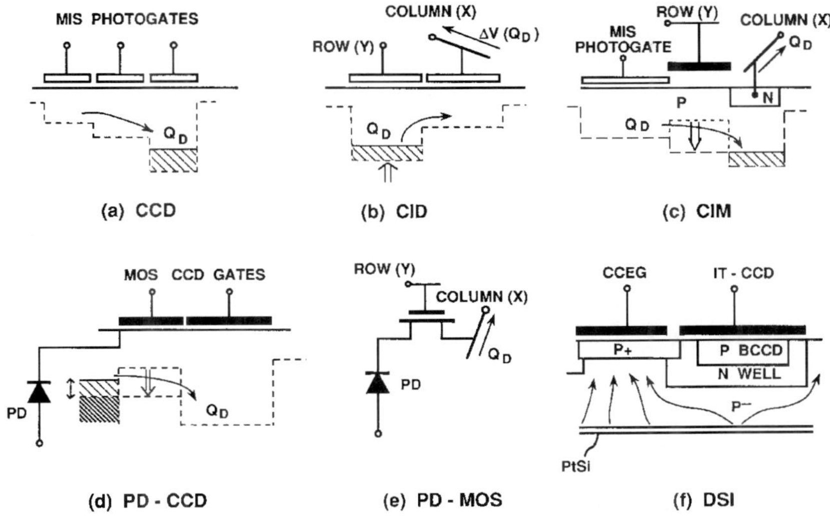

FIGURE 5 Photodetector readout structures.

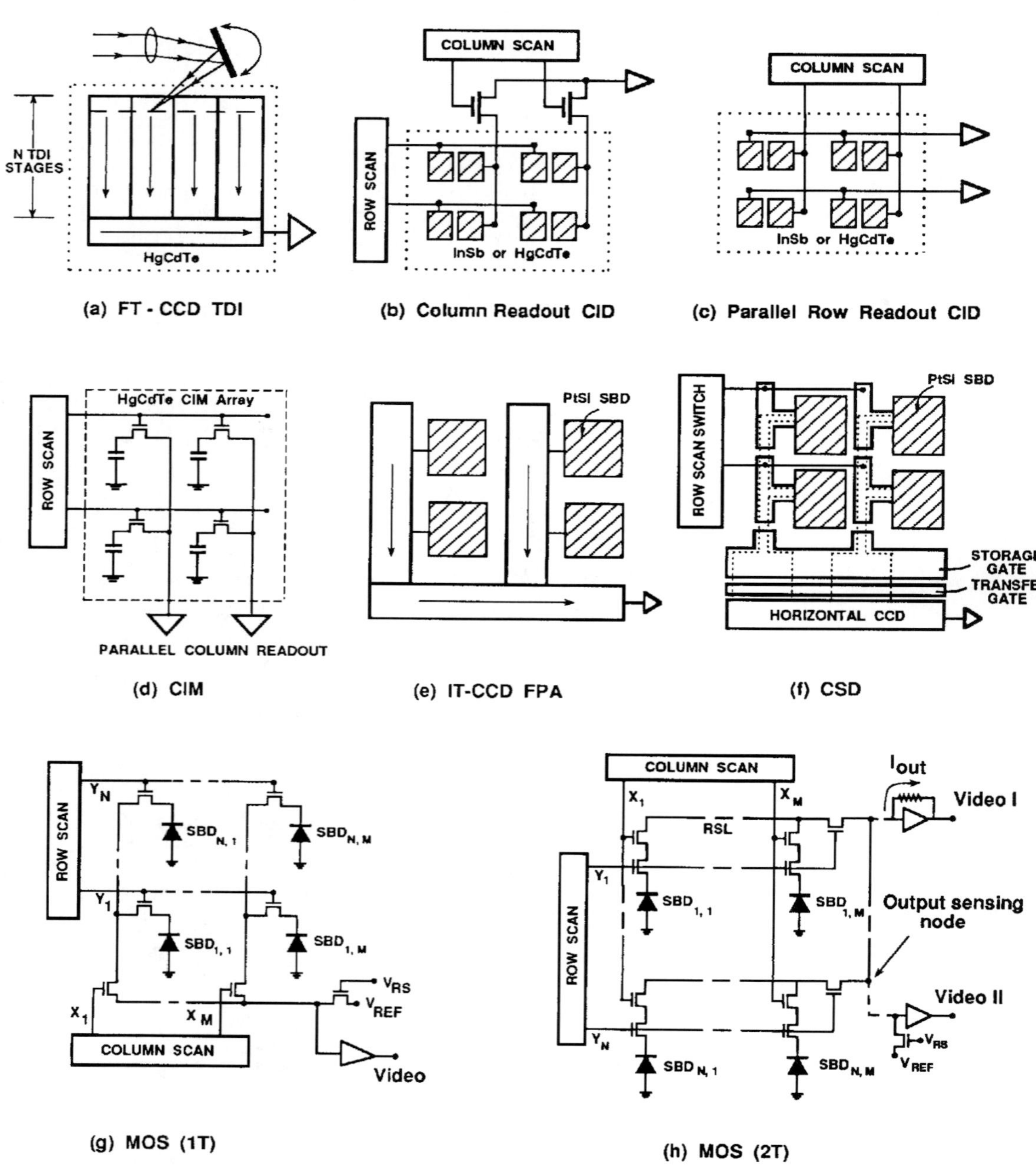

FIGURE 6 Monolithic FPA architectures.

CCD readout is the complete transfer of charge from the integration well without readout noise. Also in a CCD FPA the detected charge, Q_D, can be transferred via potential wells along the surface of the semiconductor that are induced by clock voltages but isolated from the electrical pickup until it is detected by a low-capacitance (low kTC noise) on-chip amplifier. However, because of the relatively large charge transfer losses ($\sim 10^{-3}$ per transfer) and a limited charge-handling capacity, the use of nonsilicon CCD readout has been limited mainly to HgCdTe TDI FPAs. Such TDI imager is shown in Fig. 6*a* as a frame transfer (FT) type CCD area imager performing a function of a line sensor with the effective optical integration time increased by the number of TDI elements (CCD stages) in the column CCD registers. In this imager the transfer of the detected charge signal between CCD wells of the vertical register is adjusted to coincide with the mechanical motion of the image.

In the FT-CCD TDI FPA, the vertical registers perform the functions of charge detection and integration as well as transfer. The detected image is transferred one line at a time from the parallel vertical registers to the serial output registers. From there it is transferred at high clock rate to produce the output video. Similar TDI operation can also be produced by the interline-transfer (IT) CCD architecture shown in Fig. 6*e*. However, in the case of IT-CCD readout, the conversion of infrared radiation into charge signal photodetection is performed by photodiodes.

In the CID readout, see Fig. 5*b*, the detected charge signal is transferred back and forth between the potential wells of the MIS photogates for nondestructive X-Y addressable readout, $\Delta V(Q_D)$, that is available at a column (or a row) electrode due to the displacement current induced by the transfer of the detected charge signal, Q_D. At the end of the optical integration time, the detected charge is injected into the substrate by driving both MIS capacitors into accumulation.

CID FPAs with column readout for single-output-port and parallel-row readout are illustrated schematically in Fig. 6*b* and *c*, respectively. Another example of a parallel readout is the CIM FPA shown in Fig. 6*d*. The parallel readout of CID and CIM FPAs is used to overcome the inherent limitation on charge-handling capacity of these monolithic FPAs by allowing a short optical integration time with fast frame readout and off-chip charge integration by supporting silicon ICs.

Silicon FPAs: IT-CCD, CSD, and MOS FPAs. The monolithic FPAs fabricated on silicon substrate can take advantage of the well-developed silicon VLSI process. Therefore, silicon ($E_g = 1.1$ eV), which is transparent to infrared radiation having wavelength longer than 1.0 μm, is finding an increasing use for the fabrication of monolithic CCD and MOS FPAs with infrared detectors that can be formed on silicon substrate. In the 1970s there was great interest in the development of monolithic silicon FPAs with extrinsic Si:In and Si:Ga photoconductors. However, since the early 1980s most progress was reported on monolithic FPAs with Schottly-barrier photodiodes, GeSi/Si heterojunction photodiodes, vertically integrated photodiodes, and resistive microbolometers. With the exception of the resistive microbolometers, in the form of thin-film semiconductor photoresistors formed on micromachined silicon structures, all of the above infrared detectors can be considered to be photodiodes and can be read out either by IT-CCD or MOS monolithic multiplexer.

The photodiode (PD) CCD readout, see Fig. 5*d*, is normally organized as the interline-transfer (IT) CCD staring FPA, shown in Fig. 6*e*. The IT-CCD readout has been used mostly for PtSi Schottky-barrier detectors (SBDS). The operation of this FPA consists of direct integration of the detected charge signal on the capacitance of the photodiode. At the end of the optical integration time, a frame readout is initiated by a parallel transfer of the detected charge from the photodiodes to the parallel vertical CCD registers. From there the detected image moves by parallel transfers one line at a time into the horizontal output register for a high-clock-rate serial readout.

The design of the Schottky-barrier IT-CCD FPA involves a trade-off between fill factor (representing the ratio of the active detector area to the pixel area) and maximum saturation charge signal (Q_{max}). This trade-off can be improved by the charge sweep device (CSD) architecture, shown in Fig. 6*f,* that has also been used as a monolithic readout multiplexer for PtSi and IrSi SBDs.

In CSD FPA the maximum charge signal is limited only by the SBD capacitance since its operation is based on transferring the detected charge signal from one horizontal line corresponding to one or two rows of SBDs (depending on the type of the interlacing used) into minimum geometry vertical CCD registers. During the serial readout of the previous horizontal line, the charge signal is swept into a potential well under the storage gate by low-voltage parallel clocking of the vertical registers. Then, during the horizontal blanking time, the line charge signal is transferred in parallel to the horizontal CCD register for serial readout during the next horizontal line time.

The main advantage of the silicon CCD multiplexer is a low readout noise, of 50 to 150 electrons/pixel, so that a shot-noise-limited operation can be achieved at relatively low signal levels. But as the operating temperature is lowered below 60 K the charge transfer losses of buried-channel CCDs (BCCDs) become excessive due to the freeze-out of the BCCD implant. Therefore, InSr SBD and Ge:Si detectors requiring operation at 40 K are more compatible with the MOS readout.

Photodiode (PD) MOS readout (see Fig. 5*e*) represents another approach to construction of an *X-Y* addressable silicon multiplexer. These types of monolithic MOS multiplexers used for readout of PtSi SBDs are illustrated in Fig. 6*g* and *h*.

A single-output-port FPA with one MOSFET switch per detector, MOS (1T), is shown in Fig. 6*g*. During the readout of the FPA, the vertical scan switch transfers the detected charge signal from one row of detectors to the column lines. Then the column lines are sequentially connected by MOSFET switches to the output sense line under the control of the horizontal scan switch. The main limitation of the MOS (1T) FPA is a relatively high readout noise (on the order of 1000 electrons/pixel) due to sensing of small charge signals on large-capacitance column lines. This readout noise can be decreased with a row readout MOS (2T) FPA having two MOSFET switches per detector. In this case, low readout noise can be achieved for current sensing, it is limited by the noise of the amplifier, and for voltage sensing it can be reduced by correlated double sampling (CSD).[15] A readout noise of 300 rms electrons/pixel was achieved at Sarnoff for a 640 × 480 low-noise PtSi MOS (2T) FPA designed with row buffers and 8:1 multiplexing of the output lines.[16]

An alternative form of the MOS (1T) FPA architecture is an MOS FPA with parallel column readout for fast frame operation. This silicon VIP FPA, resembling CIM architecture in Fig. 6*d,* can be used with HgCdTe vertically integrated PV detectors.

Direct-Charge-Injection Silicon FPAs. All of the silicon monolithic FPAs thus far described use separately defined detectors. A direct-charge-injection type monolithic silicon FPA with a single detector surface is a PtSi direct Schottky injection (DSI) imager that is made on thinned silicon substrate having a CCD or MOS readout on one side and PtSi SBD charge-detecting surface on the other side.[17] A cross-sectional area of one pixel of this FPA for IT-CCD readout is shown in Fig. 5*f.* In the operation of this imager, the *p*-type buried-channel CCD formed in an *n*-well removes charge from a P^+ charge-collecting electrode that in turn depletes a high resistivity *p*-type substrate. Holes injected from the PtSi SBD surface into the *p*-type substrate drift through the depleted *p*-type substrate to the P^+ charge-collecting electrode. The advantages of the DSI FPA include 100 percent fill factor, a large maximum charge due to the large capacitance between the charge collecting electrode and the overlapping gate, and that the detecting surface does not have to be defined. A 128 × 128 IT-CCD PtSi DSI FPA was demonstrated;[18] however, the same basic structure could also be used with other internal photoemission surfaces such as IrSb or Ge:Si.

Microbolometer FPAs. A microbolometer FPA for uncooled applications consists of thin-film semiconductor photoresistors micromachined on a silicon substrate. The uncooled IR FPA is fabricated as an array of microbridges with a thermoresistive element in each microbridge. The resistive microbolometers have high thermal coefficient of resistance (TCR) and low thermal conductance between the absorbing area and the readout circuit which multiplexes the IR signal. As each pixel absorbs IR radiation, the microbridge temperature changes accordingly and the elemental resistance changes.

Metal films have traditionally been used to make the best bolometer detectors because of their low 1/f noise. These latest devices use semiconductor films of 500 Å thickness having TCR of 2 percent per °C. The spacing between the microbridge and the substrate is selected to maximize the pixel absorption in the 8–14-μm wavelength range. Standard photolithographic techniques pattern the thin film to form detectors for individual pixels. The thin film TCR varies over an array by ±1 percent, and produces responsivity of 70,000 V/W in response to 300 K radiation. This has been sufficient to yield 0.1°C NEΔT with an f/1 lens. Potential low-cost arrays at prices similar to that of present large IC memories are possible with this technology.

Scanning and Staring Monolithic FPAs

In an earlier section we have reviewed the available architectures for the construction of two-dimensional scanning TDI, scanning, and staring FPAs. The same basic readout techniques, however, are also used for line-sensing imagers with photodiodes and MIS photogate detectors. For example, a line-scanning FPA corresponds to a vertical column CCD with the associated photodiodes of an IT-CCD, a column readout CID, a MOS (IT) FPA with only one row of detectors. However, since the design of a staring FPA is constrained by the size of the pixels, there is more space available for the readout multiplexer of a line of detectors. Therefore, design of the monolithic silicon multiplexer for a line detector array may also resemble the complexity of silicon multiplexer for hybrid FPAs.

23.4 HYBRID FPAS

Hybrid FPAs are made by interconnecting, via either direct or indirect means, a detector array to a multiplexing readout. Several approaches are currently pursued in both two-dimensional and three-dimensional configurations. Hybrids are typically made by either epoxying detector material to a processed silicon wafer (or readout) and subsequently forming the detectors and electrical interconnects by, for example, ion-milling; by mating a fully processed detector array to a readout to form a "two-dimensional" hybrid;[19] or by mating a fully processed detector array to a stack of signal processors to form a three-dimensional stack (*Z*-hybrid). The detector is usually mounted on top of the multiplexer and infrared radiation impinges on the backside of the detector array. Indium columns typically provide electrical and, often in conjunction with various epoxies, mechanical interconnect.

Hybrid methodology allows independent optimization of the detector array and the readout. Silicon is the preferred readout material due to performance and the leveraging of the continuous improvements funded by commercial markets. Diverse state-of-the-art processes and lithography are hence available at a fraction of their original development cost.

Thermal Expansion Match. In a hybrid FPA, the detector array is attached to a multiplexer which can be of a different material. In cooling the device from room temperature to operating temperature, mechanical strain builds up in the hybrid due to the differing coefficients of thermal expansion. Hybrid integrity requires detector material that has minimum thermal expansion mismatch with silicon. Based on this criterion, the III-V and II-VI detectors are favored over Pb-salts. Silicon-based detectors are matched perfectly to the readout; these include doped-Si and PtSi. The issue of hybrid reliability has prompted the fabrication of II-VI detectors on alternative substrates to mitigate the mismatch. HgCdTe, for example, is being grown on sapphire (PACE-I),[20] GaAs (liftoff techniques are available for substrate thinning or removal), and silicon in addition to the lattice-matched Cd(Zn)Te substrates. Detector growth techniques[21] include liquid phase epitaxy and vapor phase epitaxy (VPE). The latter includes metal organic chemical vapor deposition (MOCVD) and molecular beam epitaxy (MBE).

Hybrid Readout

Hybrid readouts perform the functions of detector interface, signal processing, and video multiplexing.[22] The hybrid FPA readout technologies include:

- Surface channel charge coupled device (SCCD)
- Buried channel charge coupled device (BCCD)
- *x-y* addressed switch-FET (SWIFET) or direct readout (DRO) FET arrays
- Combination of MOSFET and CCD (MOS/CCD)
- Charge-injection device (CI)

Early hybrid readouts were either CIDs[23] or CCDs, and the latter are still popular for silicon monolithics. However, *x-y* arrays of addressed MOSFET switches are superior for most hybrids for reasons of yield, design flexibility, and simplified interface. The move to the FET-based, direct readouts is key to the dramatic improvements in staring array producibility and is a consequence of the spin-off benefits from the silicon memory markets. DROs are fabricated with high yield and are fully compatible with advanced processes that are available at captive and commercial foundries. We will thus focus our discussion on these families. Though not extensively, CCDs are still sometimes used in hybrid FPAs.[24]

Nonsilicon Readouts. Readouts have been developed in Ge, GaAs, InSb, and HgCdTe. The readout technologies include monolithic CCD, charge-injection device (CID), charge-injection matrix (CIM), enhancement/depletion (E/D) MESFET (GaAs), complementary heterostructure FET (C-HFET; GaAs),[25] and JFET (GaAs and Ge). The CCD, CID, and CIM readout technologies generally use MIS detectors for monolithic photon detection and signal processing.

The CID and CIM devices rely on accumulation of photogenerated charge within the depletion layer of a MIS capacitor that is formed using a variety of passivants (including CVD and photo-SiO_2, anodic SiO_2, and ZnS). A single charge transfer operation then senses the accumulated charge. Device clocking and signal readout in the CIDs and CIMs relies on support chips adjacent to the monolithic IR FPA. Thus, while the FPA is monolithic, the FPA assembly is actually a multichip hybrid.

CCDs have been demonstrated in HgCdTe and GaAs.[26] The *n*-channel technology is preferred in both materials for reasons of carrier mobility and device topology. In HgCdTe, for example, *n*-MOSFETs with CVD SiO_2 gate dielectric have parametrics that are in good agreement with basic silicon MOSFET models. Fairly elaborate circuits have been demonstrated on CCD readouts, e.g., an on-chip output amplifier containing a correlated double sampler (CDS).

GaAs has emerged as a material that is very competitive for niche applications including IR FPAs. Since GaAs has very small thermal expansion mismatch with many IR detector materials including HgCdTe and InSb, large hybrids may be developed eventually, and VPE detector growth capability suggests future development of composite monolithic FPAs. The heterostructure (H-)MESFET and C-HFET technologies are particularly interesting for IR FPAs because low 1/f noise has been demonstrated; noise spectral densities at 1 Hz of as low as 0.5 $\mu V/\sqrt{Hz}$ for p-HIGFET and 2 $\mu V/\sqrt{Hz}$ for the enhancement H-MESFET[27] at 77 K have been achieved. The H-MESFET has the advantage of greater fabrication maturity (16 K SRAM and 64 × 2 readout demonstrated), but the C-HFET offers lower power dissipation.

Direct Readout Architectures. The DRO multiplexer consists of an array of FET switches. The basic multiplexer has several source follower stages that are separated at the cell, row, and column levels by MOSFET switches which are enabled and disabled to perform pixel access, reset, and multiplexing. The signal voltage from each pixel is thus direct-coupled through the cascaded source follower architecture as shown, for example, in Fig. 7. Shift registers generate the various clock signals; a minimum of externally supplied clocks is required. Since CMOS logic circuitry is used, the clock levels do not require precise adjustment for optimum performance. The simple architecture also gives high functional yield even in readout materials less mature than silicon.

Owing to the relatively low internal impedances beyond the input circuit, multiplexer noise is usually negligible. The inherent dynamic range is often >100 dB and the FPA dynamic range is limited only by the output-referred noise of the input circuit and the maximum signal excursion. The minimum read noise for DROs in imaging IR FPAs is typically capacitor reset noise. Correlated double samplers are thus used to suppress the reset noise for highest possible SNR at low integrated signal level.

In addition to excellent electrical characteristics, the DRO has excellent electro-optical properties including negligible MTF degradation and no blooming. Crosstalk in DRO-based FPAs is usually detector-limited since the readouts typically have low (<0.005 percent) electrical crosstalk. DROs also have higher immunity to clock feedthrough noise due to their smaller clock capacitances. Substrate charge pumping, which causes significant FET backgating[28] and transconductance degradation in SCCDs, is low in DROs.

X-Y Addressing and Clock Generation. Both static and dynamic shift registers are used to generate the clock signals needed for cell access, reset, and pixel multiplexing. Static registers offer robust operation and increased hardness to ionizing radiation in trade

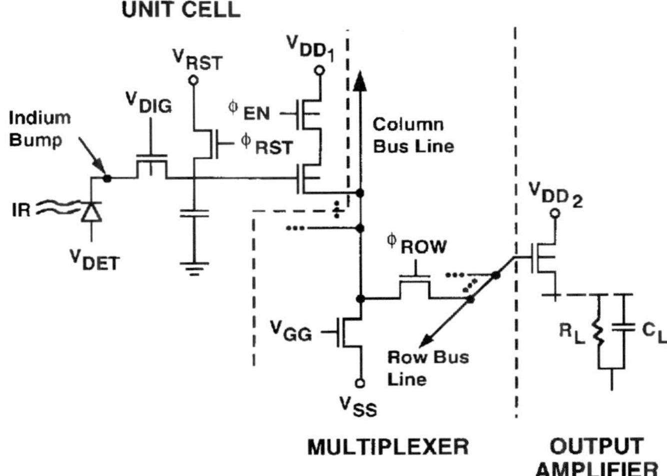

FIGURE 7 Direct readout schematic (shown with direct injection input).

for increased FET count and preference for CMOS processes. Dynamic registers use fewer transistors in NMOS or PMOS processes, but require higher voltages and have lower maximum clocking rate.

Dynamic shift registers use internal bootstrapping to regenerate the voltage at each tap. The circuit techniques limit both the lowest and highest clock rates and require fine-tuning of the MOSFET design parameters for the specific operating frequency. More importantly, the high internal voltages stress conventional CMOS processes.

Electronically Scanned Staring FPAs. The inability to integrate photogenerated charge for full staring frame times is often handled by integration time management. Since the photon background for the full 8–12-μm spectral band is over two orders of magnitude larger than the typical MWIR passband, and since the LWIR detector dark current is several orders of magnitude larger than similarly sized MWIR devices, LWIR FPA duty cycle can be quite poor. It is sometimes prudent to concede the limited duty cycle by electronically scanning the staring readout. Electronic scanning refers to a modified staring FPA architecture wherein the FPA is operated like a scanning FPA but without optomechanical means. Sensitivity is enhanced beyond that of a true scanning FPA by, for example, using multiple readout bus lines to allow integration time overlap. The sharing reduces circuit multiplicity and frees unit cell real estate to share circuitry and larger integration capacitance, and to use the otherwise parasitic bus capacitance to further increase capacity. More charge can thus be integrated even though the duty cycle is preset to $1/N$, where N is the number of elements on each common bus.

Time Delay Integration Scanning FPAs. While they are no longer extensively used for staring readouts, SCCDs are used in scanning readouts to incorporate on-focal plane TDI since they have higher dynamic range than FET bucket brigades. Dynamic range >72 dB and as high as 90 dB are typically achieved with high TDI efficacy. Two architectures dominate. In one, the CCD is integrated adjacent to the input circuit in a contiguous unit cell. In the second, the input circuit is segregated from the CCD in a sidecar configuration. The latter offers superior cell-packing density and on-chip signal processing in trade for circuit complexity. Figure 8 shows the schematic circuit for a channel of a scanning

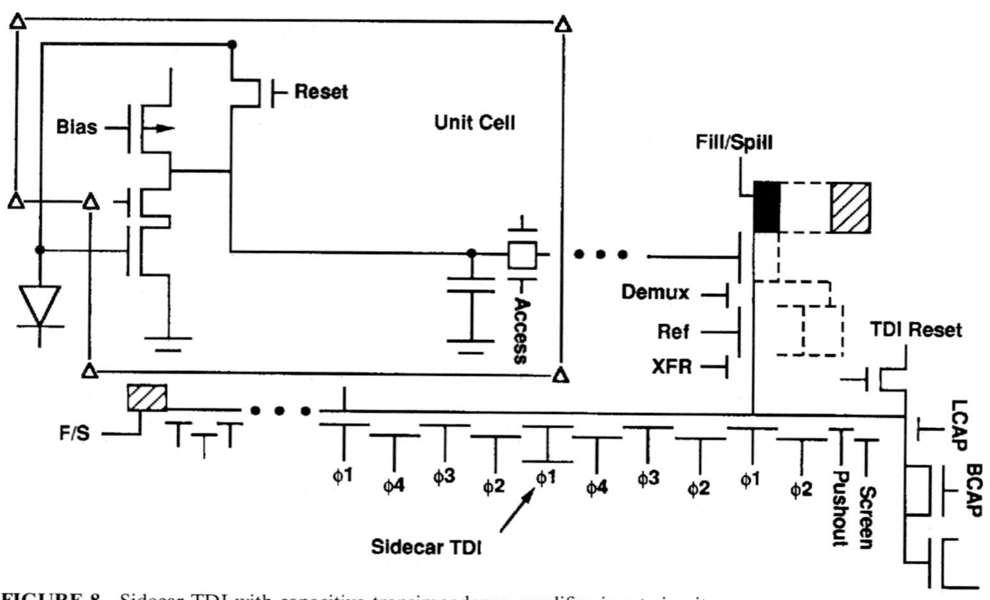

FIGURE 8 Sidecar TDI with capacitive transimpedance amplifier input circuit.

readout having capacitive transimpedance amplifier input circuit (discussed earlier), common TDI channel bus, and fill-and-spill[29] input to a sidecar SCCD TDI. This scheme integrates CMOS and CCD processes for much on-chip signal processing in very fine orthoscan pitch.[30]

The readout conversion factor, i.e., volts out per electrons in, for the sidecar CTIA scheme is

$$S_V = \frac{\Delta V}{e^-} = \frac{C_{F/S}}{C_T} A_{V_1} A_{V_2} \frac{q}{C_{\text{out}}} \tag{4}$$

where $C_{F/S}$ is the fill-and-spill gate capacitance, C_T is the integration/feedback capacitance, A_{Vx} characterizes the various source follower gains, and C_{out} is the sense node capacitance at the CCD output. The ratio of $C_{F/S}$ to C_T sets a charge gain that allows design-tailoring for dynamic range management or low input-referred noise. High charge-gain yields read noise that is limited by the input circuit and not by the transfer noise[31] of the high-carrier-capacity SCCD.

MOSFET bucket brigades are also used as TDI registers since simpler, all-MOS designs and processes can be used. Advantages include compatibility with standard MOS and CMOS, and capability for external clocking using 0–5 V, CMOS-compatible clock levels. The latter potential advantage is mitigated in the sidecar TDI scheme by appropriately sizing the SCCD registers and the charge gain to yield the desired CCD clock levels, for example. Disadvantages include higher TDI register noise due to kTC being added at each transfer and limited signal excursion.

Output Circuits. Output circuitry is usually kept to a minimum to minimize power dissipation. Circuit design thus tends to focus on the trades between voltage-mode and current-mode output amplifiers, although on-chip signal processing is increasing, including low-speed A/D conversion, switched-capacitor filtering,[32] and on-chip nonuniformity correction. Voltage-mode outputs offer better S/N performance across a wider range in backgrounds for a given readout transimpedance. Current-mode outputs offer wider bandwidth and better drive capability.

Detector Interface: Input Circuit. After the incoming photon flux is converted into a signal by the detector, it is coupled into the readout via a detector interface circuit.[33] Signal input is optical in a monolithic FPA, so signal conditioning is limited. In hybrid FPAs and some composite material monolithics, the signal is injected electrically into the readout. The simplest input schemes offering the highest mosaic densities include direct detector integration (DDI) and direct injection (DI). More complex schemes trade simplicity for input impedance reduction [buffered direct injection (BDI) and capacitive transimpedance amplification (CTIA)], background suppression (e.g., gate modulation), or ultralow read noise with high speed (CTIA). We briefly describe the more popular schemes and their performance. Listed in Table 1 are approximate performance-describing equations for comparing the circuits schematically shown in Fig. 9.

Direct Detector Integration. Direct detector integration (Figure 9*a*), also referred to as source follower per detector (SFD), is used at low backgrounds and long frame times (frame rates typically ≤15 Hz in large staring arrays). Photocurrent is stored directly on the detector capacitance, thus requiring the detector to be heavily reverse-biased for adequate dynamic range. The changing detector voltage modulates the gate of a source follower whose drive FET is in the cell and whose current source is common to all the detectors in a column or row. The limited cell constrains the source followers' drive capability and thus the bandwidth.

The DDI unit cell typically consists of the drive FET, cell enable transistor(s), and reset transistor(s). A detector site is read out by strobing the appropriate row clock, thus enabling the output source follower. The DDI circuit is capable of read noise in the range of 20 to 50 rms e-per pixel.

TABLE 1 Focal Plane Array Performance

Input Circuit	Percentage of BLIP	Detector Noise	Input-referred Circuit Noise	Input-referred MUX Noise	Transimpedance
Direct Detector Integration	$\dfrac{(\eta A_{det} Q_B \tau_{int})^{1/2}}{\left[\eta A_{det} Q_B \tau_{int} + \dfrac{2kT\tau_{int}}{qR_{det}} + N_{sf}^2 + \dfrac{2kT\tau_{int}}{qR_{int}}\right]}$	$I_{det}^2 = \left[\dfrac{4kT}{R_{det}} + 2qI_{det}\right]\Delta f + \int^{\Delta f}\left(\dfrac{K_{det}}{f^\alpha}\right)df$	$N_{sf} \approx \dfrac{\sqrt{2}}{S_V}\left[\int^{\Delta f} V_n^2(f)\dfrac{(1-\cos 2\pi f t)}{[1+(2\pi f T_D)]}df\right]^{1/2}$ $S_v = \left(\dfrac{C_{det}A_v}{q}\right)^{-1}$	$\sigma_{mux,ir}^2 = \dfrac{1}{A_v^2}kTC_{bus}\Delta f$	$\left(\dfrac{t_{int}}{C_{det}}\right)A_v$
Direct Injection	$\dfrac{(2qI_{photo}\Delta f)^{1/2}}{(\sigma_{det}^2 + \sigma_{input,ir}^2 + \sigma_{mux,ir}^2)^{1/2}}$		$\sigma_{input,ir}^2 = \int^{\Delta f}\left[\dfrac{1+\omega^2 C_{det}^2 R_{det}^2}{g_m^2 R_{det}^2}\left(\dfrac{8}{3}kTg_m + \dfrac{K_{FET}}{f^\alpha}\right)\right]df$	$\sigma_{mux,ir}^2 = \dfrac{1}{\eta_{inj}^2}kTC_{input}\Delta f$	$\left(\dfrac{t_{int}}{C_{int}}\right)A_v$
Buffered Direct Injection	$\dfrac{(2qI_{photo}\Delta f)^{1/2}}{(\sigma_{det}^2 + \sigma_{input,ir}^2 + \sigma_{mux,ir}^2)^{1/2}}$		$\sigma_{input,ir}^2 = \int^{\Delta f}\left[\eta_{noise}^2\left(\dfrac{8}{3}kTg_m + \dfrac{K_{FET}}{f^\alpha}\right) + \Lambda_{amp}^2\langle e_{amp}^2\rangle\right]df$ $\eta_{noise} = \dfrac{1 + j\omega R_{det}C_{int}}{1+(1+A_v)g_m R_{det} + j\omega R_{det}(1+A_v)C_{int}}$ $\Lambda_{amp} = \left(\dfrac{g_m}{R_{det}}\right)\dfrac{(1+j\omega R_{det}C_{det})}{1+(1+A_v)g_m R_{det} + j\omega[C_{det}+(1+A_v)C_{int}]R_{det}}$	$\sigma_{mux,ir}^2 = \dfrac{1}{\eta_{inj}^2}kTC_{input}\Delta f$	$\left(\dfrac{t_{int}}{C_{int}}\right)A_v$
Chopper Stabilized BDI	$\dfrac{(2qI_{photo}\Delta f)^{1/2}}{(\sigma_{det}^2 + \sigma_{input,ir}^2 + \sigma_{mux,ir}^2)^{1/2}}$		$\sigma_{input,ir}^2 = \int^{\Delta f}\left[\eta_{noise}^2\left(\dfrac{8}{3}kTg_m\right) + \Lambda_{amp}^2\langle e_{amp}^2\rangle\right]df$ $\eta_{noise} = \dfrac{1 + j\omega R_{det}C_{int}}{1+(1+A_v)g_m R_{det} + j\omega R_{det}(1+A_v)C_{int}}$ $\Lambda_{amp} = \left(\dfrac{g_m}{R_{det}}\right)\dfrac{(1+j\omega R_{det}C_{det})}{1+(1+A_v)g_m R_{det} + j\omega[C_{det}+(1+A_v)C_{int}]R_{det}}$	$\sigma_{mux,ir}^2 = \dfrac{1}{\eta_{inj}^2}kTC_{input}\Delta f$	$\left(\dfrac{t_{int}}{C_{int}}\right)A_v$
Gate Modulation (FET Load)	$\dfrac{(2qI_{photo}\Delta f)^{1/2}}{(\sigma_{det}^2 + \sigma_{load}^2 + \sigma_{input,ir}^2 + \sigma_{mux,ir}^2)^{1/2}}$		$\sigma_{input,ir}^2 = \int^{\Delta f}\left[\dfrac{1}{A_I^2}\left(2qI_{input} + \dfrac{K_{FET,input}}{f^\alpha}\right)\right]df$	$\sigma_{mux,ir}^2 = \dfrac{1}{A_I^2}kTC_{input}\Delta f$	$\left(\dfrac{A_I t_{int}}{C_{int}}\right)A_v$
Capacitive Trans-impedance Amplifier	$\dfrac{\eta A_{det} Q_B \tau_{int}}{\left[\eta A_{det} Q_B \tau_{int} + \dfrac{2kT\tau_{int}}{qR_{det}} + N_{amp,1/f}^2 + N_{amp,white}^2 + N_{load,white}^2\right]}$		$N_{amp,1/f} \approx \dfrac{C_{det}K_{amp}\sqrt{2}}{q}\left[\ln\dfrac{5\tau_{int}}{\tau_{amp}}\right]$ $N_{amp,white} \approx \dfrac{C_{det}}{q}\sqrt{\dfrac{8}{3}\dfrac{kT}{g_m \tau_{amp}\pi}}$ $N_{load,white} \approx \dfrac{C_{det}}{q A_{v,amp} A_{v,sf}^2}\sqrt{\dfrac{2kT}{C_L}}$	$I_{mux,ir}^2 = kTC_{input}\Delta f$	$Z_T = \dfrac{\tau_{int}}{C_T}A_{v,sf}$ $C_T = \dfrac{\left[(C_{gd}+C_{gs}+C_{det}) + A_v(C_{fb}+C_{gd})\right]}{A_v}$

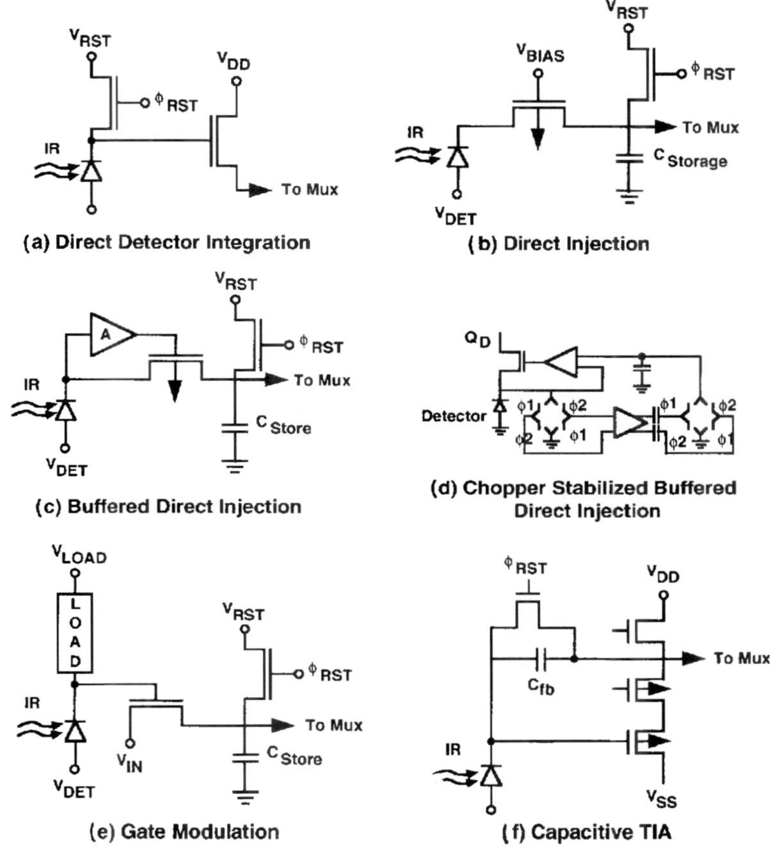

FIGURE 9 Hybrid FPA detector interface circuits.

Direct Injection. Direct injection (Fig. 9*b*) is perhaps the most widely used input circuit due to its simplicity and high performance. The detector directly modulates the source of a MOSFET. The direct coupling requires that detectors with *p*-on-*n* polarity, as is the case with InSb and most photovoltaic LWIR detectors, interface *p*-type FETs (and vice versa) for carrier collection in the integration capacitor. In surface channel CCDs, the FETs drain is virtual, as formed by a fully enhanced well, and doubles as the integration capacitor.

Practical considerations, including limited charge-handling capacity, constrain the DI input to operation with high-impedance MWIR or limited cutoff ($\lambda_c \leq 9.5\ \mu\text{m}$) LWIR detectors. The associated background photocurrent for the applications where direct injection can be used mandates that the DI FET operate subthreshold.[34] The subthreshold gate transconductance, g_m, is independent of FET geometry:[35]

$$g_m = \left(\frac{\partial I_D}{\partial V_G}\right)\Bigg|_{V_{DS}=\text{constant}} = \frac{q\left(\eta_{\text{inj}}\left\{I_{\text{photo}} + \dfrac{V_{\text{det}}}{R_{\text{det}}} - I_{\text{det}_0}\left(e^{(qV_{\text{det}}/n\text{det}kT)} - 1\right)\right\}\right)}{nkT} \approx \frac{qI_D}{nkT} \tag{5}$$

The injection efficiency, η_{inj}, of detector current into the DI FET is

$$\eta_{\text{inj,DI}} = \frac{g_m R_{\text{det}}}{1 + g_m R_{\text{det}}} \left[\frac{1}{1 + \dfrac{j\omega C_{\text{det}} R_{\text{det}}}{1 + g_m R_{\text{det}}}} \right] \tag{6}$$

where R_{det} and C_{det} are the detectors' dynamic resistance and capacitance, respectively. Poor DI circuit bandwidth occurs at low-photon backgrounds due to low g_m.

The injection efficiency varies across an FPA due to FET threshold, detector bias, and detector resistance nonuniformity. Changes in detector current create detector bias shifts since the input impedance is relatively high. In extreme cases a large offset in threshold gives rise to excess detector leakage current and 1/f noise, in addition to fixed pattern noise. The peak-to-peak threshold voltage nonuniformity spans the range from ≈1 mV for silicon *p*-MOSFETs to over 125 mV for some GaAs-based readouts.

Depending upon the interface to the multiplexing bus, the noise-limiting capacitance, C_{input}, is approximately the integration capacitance or the combined integration and bus capacitance. Some direct-injection cells are thus buffered with a source follower (see Fig. 7). Omitting the source follower reduces the readout transimpedance (due to charge splitting between the integration capacitor and thus bus line capacitance) in trade for larger integration capacitance since more unit-cell real estate is available.

Increasing the pixel density has required a continuing reduction in cell pitch. Figure 10 plots the charge-handling capacity as functions of cell pitch and minimum gate length for representative DI designs using the various minimum feature lengths. Also plotted is the maximum capacity assuming the cell is composed entirely of integration capacitor (225 Å SiO_2). A recent 27-μm DI cell, fabricated in 1.25-μm CMOS, thus has similar cell capacity as an earlier 60-μm DI cell in 3-μm CMOS. Limitations on cell real estate, operating voltage, and the available capacitor dielectrics nevertheless dictate maximum integration times that are often shorter than the total frame time. This duty cycling equates to degradation in detective quantum efficiency.

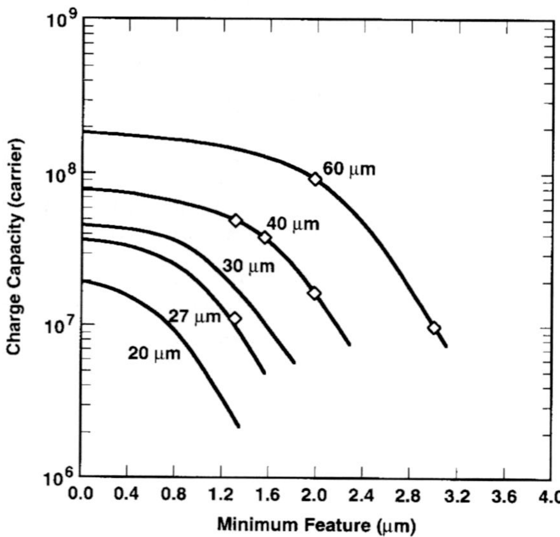

FIGURE 10 Direct injection charge-handling capacity versus cell pitch for standard CMOS processes.

Buffered Direct Injection. A significant advantage of the source-coupled input is MOSFET noise suppression. This suppression is implied in the η_{BLIP} expression shown in Table 1. When injection efficiency is poor, however, MOSFET noise becomes a serious problem along with bandwidth. These deficiencies are ameliorated via buffered direct injection (BDI),[36] wherein a feedback amplifier with open-loop gain $-A_V$ (Fig. 9*c*) is added to the DI circuit. The buffering increases injection efficiency to near-unity, increases bandwidth by orders of magnitude, and suppresses the DI FET noise.

BDI has injection efficiency

$$\eta_{\text{inj}} \cong \frac{g_m R_{\text{det}}(1 + A_V)}{1 + g_m R_{\text{det}}(1 + A_v) + j\omega R_{\text{det}}[(1 + A_v)C_{\text{amp}} + C_{\text{det}}]} \tag{7}$$

where C_{amp} is the Miller capacitance of the amplifier. Circuit bandwidth is maximized by lowering the amplifiers' Miller capacitance to provide detector-limited frequency response that is lower than that possible with DI by the factor $(1 + A_V)$.

The noise margin of the BDI circuit is superior to DI, even though two additional noise sources associated with the feedback amplifier are added. The dominant circuit noise stems from the drive FET in the amplifier. The noise power for frequencies less than $1/(2\pi R_{\text{det}} C_{\text{det}})$ is directly proportional to the detector impedance. Amplifier noise is hence a critical issue with low impedance ($<1\ \text{M}\Omega$) detectors including long wavelength photovoltaics operating at temperatures above 80 K.

Chopper-Stabilized Buffered Direct Injection. The buffered direct injection circuit has high 1/f noise with low-impedance detectors. While the MOSFET 1/f noise can be suppressed somewhat by reverse-biasing the detectors to the point of highest resistance, detector 1/f noise may then dominate. Other approaches include enlarging MOSFET gate area, using MOS input transistors in the lateral bipolar mode, or using elaborate circuit techniques such as autozeroing and chopper stabilization. Chopper stabilization is useful if circuit real estate is available, as in a scanning readout. Figure 9*d* shows a block diagram schematic circuit of chopper-stabilized BDI.

Chopper stabilization refers to the process of commutating the integrating detector node between the inverting and noninverting inputs of an operational amplifier having open-loop gain, A_V. This chopping process shifts the amplifier's operating frequency to higher frequencies where the amplifier's noise is governed by white noise, not 1/f noise. At chopping frequencies $f_{\text{chop}} \gg f_k$, the equivalent low-frequency input noise of the chopper amplifier is equal to the original amplifier white-noise component.[37] The amplifier's output signal is subsequently demodulated and filtered to remove the chopping frequency and harmonics. This scheme also reduces the input offset nonuniformity by the reciprocal of the open-loop gain, thereby generating uniform detector bias. Disadvantages include high circuit complexity and the possibility of generating excess detector noise via clock-feedthrough-induced excitation of traps, particularly with narrow bandgap photovoltaic detectors.

Gate Modulation. Signal processing can be incorporated in a small unit cell by using a gate modulated input structure (c.f. MOSFET load gate modulation in Fig. 9*e*).[38] The use of a MOSFET as an active load device, for example, provides dynamic range management via automatic gain control and user-adjustable background pedestal offset since the detector current passes through a load device with resistance R_{LOAD}. The differential gate voltage applied to the input FET varies for a change in photocurrent, ΔI_{photo}, as

$$\Delta V_G = R_{\text{LOAD}}\, \eta_{\text{inj,DI}} \Delta I_{\text{photo}} \tag{8}$$

The current injected into the integration capacitor is

$$I_{\text{input}} = g_m R_{\text{LOAD}}\, \eta_{\text{inj,DI}} \Delta I_{\text{photo}} \tag{9}$$

The ratio of I_{input} to I_{photo} is the current gain, A_I, which is

$$A_I = \frac{g_m}{g_{m,\text{LOAD}}} \eta_{\text{inj,DI}} \tag{10}$$

The current gain can self-adjust by orders of magnitude depending on the total detector current. Input-referred read noise of tens of electrons has thus been achieved with high-impedance SWIR detectors at low-photon backgrounds. The same circuit has also been used at LWIR backgrounds with LWIR detectors having adequate impedance for good injection efficiency.

The current gain expression suggests a potential shortcoming for imaging applications since the transfer characteristic is nonlinear, particularly when the currents in the load and input FETs differ drastically. In conjunction with tight specifications for threshold uniformity, pixel functionality can be decreased, dynamic range degraded, and imagery dominated by spatial noise. The rms fractional gain nonuniformity (when operating subthreshold) of the circuit is approximately

$$\frac{\Delta A_I}{A_I} = \frac{q\sigma_{VT}}{n_{\text{FET}}kT} \tag{11}$$

where σ_{VT} is the rms threshsold nonuniformity. At 80 K, state-of-the-art σ_{Vt} of 0.5 mV for a 128×128 FPA, and $n = 1$, the minimum rms nonuniformity is ≈ 7 percent.

Capacitive Transimpedance Amplifier (*CTIA*). Many CTIAs have been successfully demonstrated. The most popular approach uses a simple CMOS inverter[39] for feedback amplification (Figure 9*f*). Others use a more elaborate differential amplifier. The two schemes differ considerably with respect to open-loop voltage gain, bandwidth, power dissipation, and cell real estate. The CMOS inverter-based CTIA is more attractive for high-density arrays. The latter is sometimes preferred for scanning or *Z*-hybrid applications where real estate is available primarily to minimize power dissipation.[40]

In either case, photocurrent is integrated directly onto the feedback capacitor of the transimpedance amplifier. The minimum feedback capacitance is set by the amplifier's Miller capacitance and defines the maximum circuit transimpedance. Since the Miller capacitance can be made very small (<5 fF), the resulting high transimpedance yields excellent margin with respect to downstream system noise. The transimpedance degrades when the circuit is coupled to large detector capacitances, so reducing pixel size serves to minimize read noise and the circuits' attractiveness will continue to increase in the future.

The CTIA allows extremely small currents to be integrated with high efficiency and tightly regulated detector bias. The amplifier open-loop gain, $A_{V,\text{amp}}$, ranges from as low as fifteen to higher than several thousand for noncascoded inverters, and many thousands for some cascoded inverter and differential amplifier designs. The basic operating principle is to apply the detector output to the inverting input of a high-gain CMOS differential amplifier operated with capacitive feedback. The feedback capacitor is reset at the detector sampling rate. The noise equivalent input voltage of the amplifier is referred to the detector impedance, just as in other circuits.

The CTIA's broadband channel noise sets a lower limit on the minimum achievable read noise, is the total amplifier white noise, and can be approximated for condition of large open-loop gain by

$$N^2_{kTC,\text{channel}} = \frac{kTC_{\text{fb}}}{q^2} \left[\frac{C_{\text{det}} + C_{\text{fb}}}{C_L + \dfrac{C_{\text{fb}} C_{\text{det}}}{C_{\text{fb}} + C_{\text{det}}}} \right] \tag{12}$$

where C_{fb} is the feedback capacitance including the Miller capacitance and integration capacitance and C_L is the output load capacitance. This expression provides an intuitive formula for minimizing noise: detector capacitance must be low (i.e., minimize detector shunting capacitance which reduces closed-loop gain) and output capacitance high (i.e., limit bandwidth). Of the three amplifier noise sources listed in Table 1, amplifier 1/f noise is often largest. For this reason, the CMOS inverter-based CTIA has best performance with *p*-on-*n* detectors and *p*-MOSFET amplifier FET due to the lower 1/f noise.

23.5 PERFORMANCE: FIGURES OF MERIT

In the early days of infrared technology, detectors were characterized by the noise equivalent power (NEP) in a 1-Hz bandwidth. This was a good specification for single detectors, since their performance is usually amplifier-limited. The need to compare detector technologies for application to different geometries and the introduction of FPAs having high-performance on-board amplifiers and small parasitics necessitated normalization to the square root of the detector area for comparing S/N. R. C. Jones[41] thus introduced detectivity (D^*), which is simply the reciprocal of the normalized NEP and has units cm-$\sqrt{\text{Hz}}$/W (or Jones).

While D^* is well-suited for specifying infrared detector performance, it can be misleading to the uninitiated since the D^* is highest at low background. An LWIR FPA operating at high background with background-limited performance (BLIP) S/N can have a D^* that is much lower than for a SWIR detector having poor S/N relative to the theoretical limit. Several figures of merit that have hence proliferated include other ways of specifying detectivity: e.g., thermal D^* (D^*_{th}) blackbody D^*(D^*_{bb}), peak D^* ($D^*_{\lambda pk}$), percentage of BLIP (%BLIP or η_{BLIP}), and noise equivalent temperature difference (NE ΔT). Since the final output is an image, however, the ultimate figure of merit is how well objects of varying size are resolved in the displayed image. The minimum resolvable temperature (MRT) is thus a key benchmark. These are briefly discussed in this section.

Detectivity (D^*)

D^* is the S/N ratio normalized to the electrical bandwidth and detector area. In conjunction with the optics area and the electrical bandwidth, it facilitates system sensitivity estimation. However, D^* can be meaningless unless the test conditions, including magnitude and spectral distribution of the flux source (e.g., blackbody temperature), detector field-of-view, chopping frequency (lock-in amplifier), background temperature, and wavelength at which the measurement applies. D^* is thus often quoted as "blackbody," since the spectral responsivity is the integral of the signal and background characteristics convolved with the spectral response of the detector. D^*_{bb} specifications are often quantified for a given sensor having predefined scene temperature, filter bandpass and cold shield f/# using a generalized expression.

Peak detectivity is sometimes preferred by detector engineers specializing in photon detectors. The background-limited peak detectivity for a photovoltaic detector is

$$D^*_{\lambda pk} = \sqrt{\frac{\eta}{2Q_B}} \frac{\lambda_{pk}}{hc} \tag{13}$$

and refers to measurement at the wavelength of maximum spectral responsivity. For detector-limited scenarios, such as at higher operating temperatures or longer wavelengths (e.g., $\lambda_c > 12\ \mu$m at operating temperature less than 78 K or $\lambda_c > 4.4\ \mu$m at $\geq$195 K), the peak detectivity is limited by the detector and not the photon shot noise. In these cases the maximum detector-limited peak D^* in the absence of excess bias-induced noise (both 1/f and shot noise) is

$$D^*_{\lambda pk} = \frac{\eta q}{2}\sqrt{\frac{R_0 A}{kT}}\frac{\lambda}{hc} \tag{14}$$

The R_0A product of a detector thus describes detector quality even though other parameters may actually be more relevant for FPA operation.

Percentage of BLIP

Whereas D^* compares the performance of dissimilar detectors, FPA designers often need to quantify an FPA's performance relative to the theoretical limit at a specific operating background. Percentage of BLIP, η_{BLIP}, is one such parameter and is simply the ratio of photon noise to composite FPA noise

$$\eta_{\text{BLIP}} = \left(\frac{N^2_{\text{PHOTON}}}{N^2_{\text{PHOTON}} + N^2_{\text{FPA}}}\right)^{1/2} \tag{15}$$

NE ΔT

the NE ΔT of a detector represents the temperature change, for incident radiation, that gives an output signal equal to the rms noise level. While normally thought of as a system parameter, detector NE ΔT and system NE ΔT are the same except for system losses (conservation of radiance). NE ΔT is defined:

$$\text{NE }\Delta\text{T} = v_n\left(\frac{\partial T}{\partial Q}\right)\bigg/\left(\frac{\partial V_s}{\partial Q}\right) = v_n\frac{\Delta T}{\Delta V_s} \tag{16}$$

where v_n is the rms noise and ΔV_s is the signal measured for the temperature difference ΔT. It can be shown that

$$\text{NE }\Delta\text{T} = (\tau_o C_{T\lambda}\eta_{\text{BLIP}}\sqrt{N_c})^{-1} \tag{16a}$$

where τ_o is the optics transmission, $C_{T\lambda}$ is the thermal contrast from Fig. 1, and N_c is the number of photogenerated carriers integrated for one integration time, t_{int}:

$$N_c = \eta A_{\text{det}} t_{\text{int}} Q_B \tag{16b}$$

The distinction between an integration time, and the FPA's frame time must be noted. It is often impossible at high backgrounds to handle the large amount of carriers generated over frame times compatible with standard video frame rates. The impact on system D^* is often not included in the FPA specifications provided by FPA manufacturers. This practice is appropriate for the user to assess relative detector quality, but must be coupled with

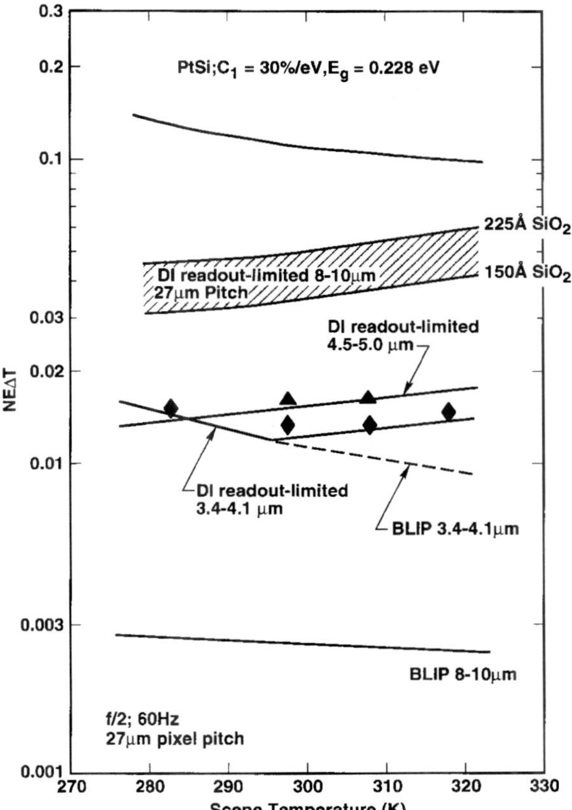

FIGURE 11 NE ΔT versus background temperature for several prominent spectral bands.

usable FPA duty cycle, read noise, and excess noise to give a clear picture of FPA utility. Off-FPA frame integration can be used to attain a level of sensor sensitivity that is commensurate with the detector-limited D^* and not the charge-handling-limited D^*.

The inability to handle a large amount of charge nevertheless is a reason why the debate as to whether LWIR or MWIR operation is superior is still heated. While the LWIR band should offer order-of-magnitude higher sensitivity, staring readout limitations often reduce LWIR imager sensitivity to below that of competing MWIR cameras. However, submicron photolithography, alternative dielectrics, and refinements in readout architectures are ameliorating this shortfall and LWIR FPAs having sensitivity superior to MWIR counterparts have now been demonstrated. Figure 11 shows the effect on high quantum efficiency FPA performance and compares the results to PtSi at TV-type frame rate. The figure illustrates BLIP and measured NEΔTs versus background temperature for several spectral bands assuming a state-of-the-art 640×480 DI readout multiplexer (27-μm pixel pitch and $\approx$1-μm-minimum feature) is hybridized to high-quantum-efficiency MWIR and LWIR detectors. Though the device has large charge-handling capacity, there is large shortfall in the predicted LWIR FPA performance relative to the BLIP limit. The measured MWIR FPA performance values, as shown by the data points, is in good agreement with the predicted trends.

Spatial Noise. Estimation of IR sensor performance must include a treatment of spatial noise that occurs when FPA nonuniformities cannot be compensated correctly. This requires consideration of cell-to-cell response variations. Mooney et al.[42] have given a comprehensive discussion of the origin of spatial noise. The total noise determining the sensitivity of a staring array is the composite of the temporal noise and the spatial noise. The spatial noise is the residual nonuniformity U after application of nonuniformity compensation, multiplied by the signal electrons N. Photon noise, equal to $\sqrt{N}$, is the dominant temporal noise source for the high infrared background signals for which spatial noise is significant (except for TE-cooled or uncooled sensors). The total noise equivalent temperature difference is

$$\text{Total NE}\,\Delta\text{T} = \frac{\sqrt{N + U^2N^2}}{\dfrac{\partial N}{\partial T}} = \frac{\sqrt{1/N + U^2}}{\dfrac{1}{N}\dfrac{\partial N}{\partial T}} \tag{17}$$

where $\partial N/\partial T$ is the signal change for a 1 K source temperature change. The denominator, $(\partial N/\partial T)/N$, is the fractional signal change for a 1 K source temperature change. This is the relative scene contrast due to $C_{T\lambda}$ and the FPA's transimpedance.

The dependence of the total NE ΔT on residual nonuniformity is plotted in Fig. 12 for 300 K scene temperature, two sets of operating conditions, and three representative detectors: LWIR HgCdTe, MWIR HgCdTe, and PtSi. Operating case A maximizes the detected signal with f/1.4 optics, 30-Hz frame rate, and 3.4–5.0-μm passband. Operating case B minimizes the solar influence by shifting the passband to 4.2–5.0 μm and trades off signal for the advantages of lighter, less expensive optics (f/2.0) at 60-Hz frame rate. Implicit in the calculations are charge-handling capacities of 30 million e- for MWIR HgCdTe, 100 million e- for LWIR HgCdTe, and 1 million for PtSi. The sensitivity at the lowest nonuniformities is independent of nonuniformity and limited by the shot noise of the detected signal. The LWIR sensitivity advantage is achieved only at nonuniformities less than 0.01 percent, which is comparable to that achieved with buffered input circuits. At the reported direct-injection MWIR HgCdTe residual nonuniformity of 0.01 to 0.02

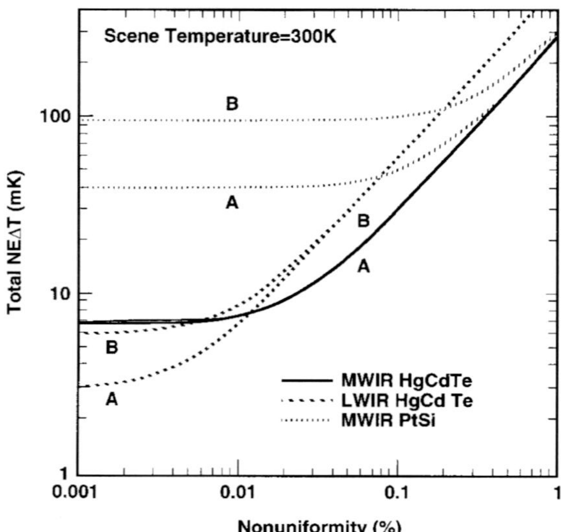

FIGURE 12 Total NE ΔT versus nonuniformity at 300 K scene temperature.

percent, the total NE ΔT is about 0.007 K, which is comparable to the MWIR BLIP limit. At the reported PtSi residual nonuniformity of 0.05 percent with direct detector integration, total NE ΔT is higher at 0.04 K, but exceeds the BLIP limit for the lower quantum efficiency detectors.

Minimum Resolvable Temperature (MRT)

The minimum resolvable temperature is often the preferred figure of merit for imaging infrared sensors. MRT is a function of spatial resolution and is defined as the signal-to-noise ratio required for an observer to resolve a series of standard four-bar targets. While many models exist due to the influence of human psycho-optic response, a representative formula[43] is

$$\mathrm{MRT}(f_s, T_{\mathrm{SCENE}}) = \frac{2\mathrm{SNR}_T \mathrm{NE}\,\Delta\mathrm{T}(T_{\mathrm{SCENE}})}{\mathrm{MTF}(f_s)} \left[\frac{f_S^2 \Delta x\, \Delta y}{L\tau_{\mathrm{eye}} f_{\mathrm{frame}} N_{\mathrm{OS}} N_{\mathrm{SS}}}\right]^{1/2} \tag{18}$$

where f_s is the spatial frequency in cycles/radian, a target signal-to-noise ratio (SNR_T) of five is usually assumed, the MTF describes the overall modulation transfer function including the optics, detector, readout, and the integration process, Δx and Δy are the respective detector subtenses in mRad, τ_{eye} is the eye integration time, f_{frame} is the display frame rate, N_{OS} is the overscan ratio, N_{SS} is the serial scan ratio, and L is the length-to-width ratio of a bar chart (always set to 7). While the MRT of systems with temporal noise-limited sensitivity can be adequately modeled using the temporal NE ΔT, scan noise in scanned system and fixed pattern noise in staring cameras requires that the MRT formulation be appropriately modified.

Shown in Fig. 13 are BLIP (for 70 percent quantum efficiency) MRT curves at 300 K background temperature for narrow-field-of-view (high-resolution) sensors in the MWIR and LWIR spectral bands. Two LWIR curves are included to show the impact of matching the diffraction-limited blur to the pixel pitch versus 2× oversampling of the blur. The latter

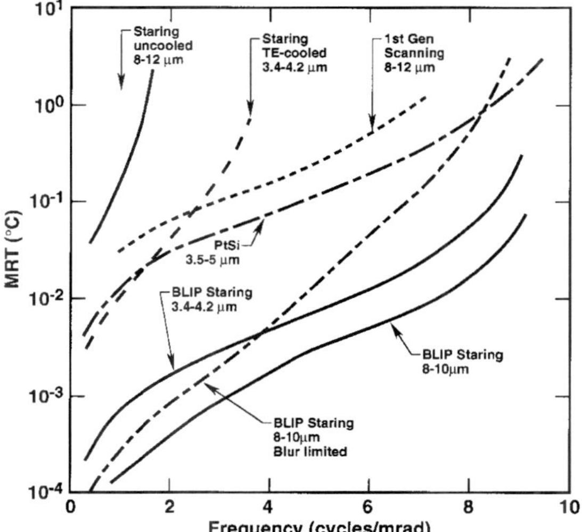

FIGURE 13 BLIP MRT for representative staring FPA configurations in the various bands.

case commonly arises, for example, when the LWIR FPA is miniaturized to minimize die size for enhancing yield and, for hybrid FPAs, alleviating thermal expansion mismatch. Also included for comparison are representative curves for first-generation scanning, staring uncooled, staring TE-cooled and staring PtSi sensors assuming 0.1, 0.1, 0.05, and 0.1 K NE ΔTs, respectively. Theoretically, the staring MWIR sensors have order of magnitude better sensitivity while the staring LWIR bands have two orders of magnitude better sensitivity than the first-generation sensor. In practice, due to charge-handling limitations, an LWIR sensor has only slightly better MRT than the MWIR sensor. The uncooled sensor is useful for short-range applications; the TE-cooled sensor provides longer range than the uncooled, but less than the high-density PtSi-based cameras.

23.6 CURRENT STATUS AND FUTURE TRENDS

Status

Over the last decade, dramatic improvements in detector and readout technology have resulted in a 200× increase in the size of the largest available FPAs. The density increase roughly parallels the transistors-per-chip trendlines[44] established by microprocessors and DRAMs, lagging the former by four years and the latter by six years. Consequently, whereas various 64×64 FPAs were available in the early 1980s, several vendors are now producing monolithic PtSi FPAs in the TV-compatible $\sim 640 \times 480$ formats, and one has demonstrated a 1040×1040 prototype.[45] Various hybrid FPA technologies closely trail the monolithics by about two years. PV MWIR HgCdTe and InSb 256×256's are in full scale "production"[46] and a 640×480[47] has been fully demonstrated in laboratories. Scanning FPAs, which require the highest possible quantum efficiencies, are thus available only in the higher-performance materials in a variety of hybrid formats including, for example, 64×2, 50×10, 96×20, 128×1, 256×1, 256×4, $480 \times N$ ($N = 1$–6-in TDI), 960×1, and 960×4.

Figure 14 summarizes the current performance of the most prominent detector technologies. The figure, a plot of the D^*_{th} (300 K, 0° field-of-view) versus operating temperature, clearly shows the performance advantage that the intrinsic photovoltaics have over the other technologies. Thermal detectivity is used here to compare the various technologies for equivalent NE ΔT irrespective of wavelength. While the extrinsic silicon detectors offer very high sensitivity, high producibility, and very long cutoff wavelengths, the very low operating temperature is often prohibitive. Also shown is the relatively low

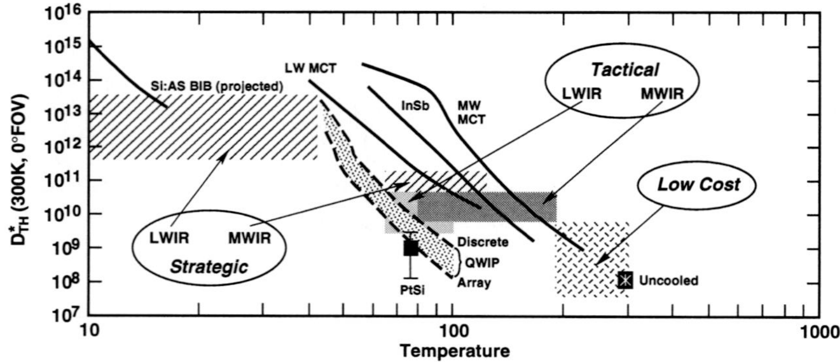

FIGURE 14 Comparison of photon detector D^*_{th} (300 K, 0°) for varous IR technologies for equivalent NE ΔT.

and slightly misleading detectivity of PtSi, which is offset by its excellent array uniformity (i.e., array performance is commensurate with best diode performance).

Similar in performance at cryogenic temperatures, InSb and HgCdTe have comparable array size and pixel yield at MWIR cutoff wavelengths. Wavelength tunability and high radiative efficiency, however, have often made HgCdTe the preferred material because the widest possible bandgap semiconductor can be configured and thus the highest possible operating temperature achieved for a given set of operating conditions. The associated cooling and system power requirements can thus be optimally distributed.

FPA costs are currently very similar for all the second-generation FPA technologies. Though it is often argued that FPA cost for IR cameras will be irrelevant once it reaches a certain minimum threshold level, nevertheless the key determinants as to which FPA technology succeeds in the coming decade are availability and cost.

SWIR. The SWIR band (0.8 to roughly 3 μm) has limited applicability to imagers, but has experienced increasing developmental interest in the last few years, particularly for astronomical and commercial laser applications. The dominant FPA detector materials include SWIR PACE-I HgCdTe, with λ_c from 1.4 to 3.5 μm, PtSi (monolithic and hybrid variants), and InSb. The latter material is operated at temperatures as low as 5 K to minimize dark current. However, constraints on material quality and performance of cryogenic electronics normally dictate best InSb FPA performance at 30–40 K. Recent additions to the list of FPA alternatives include InGaAs/InP, which can be compositionally adjusted for cutoff wavelengths from 1.7–3.0 μm, and side-illuminated SSPMs. In the former case, 1.7-μm cutoff yields the highest lattice-matched performance.

HgCdTe and InSb offer the largest hybrid production format, 256×256, with HgCdTe selected for retrofit into the Hubble Space Telescope[48] and InSb currently baselined for the Space Infrared Telescope Facility (SIRTF)[49] instrument. The various staring hybrids use DDI inputs having charge-storage capacities of roughly one million carriers and read noise ≤50 e-. The 2.5-μm HgCdTe FPA has dark current <1 e-/sec and readout noise <30 e- at 78 K. Several InSb 256×256's offer longer spectral response and up to 90 percent quantum efficiency but need cooling to 30–40 K for dark current of 10 e-/s.

CTIA-based FPAs have been demonstrated in staring formats to 48×48. Larger devices are being developed. A scanning FPA having a 10×132 with read noise as low as 10 e- has been reported.[50] This device has CTIA input in sidecar TDI architecture and internal charge gain of up to 50 to provide the transimpedance needed to attain BLIP sensitivity at short integration times and low backgrounds.

MWIR. The MWIR bands are serviced by many IR detector materials and FPA technologies. The largest MWIR FPAs have roughly 300,000 pixels (640×480, 512×512, etc.) and are available in PtSi, HgCdTe, and InSb. Though neither is yet available at the low costs needed for commercialization, many start-up companies have identified suitable niche markets and competition is increasing. The PtSi cameras typically exhibit 0.05–0.2 K NE ΔT with very good MRT limited only by temporal noise. The high quantum efficiency alternatives such as HgCdTe and InSb are in limited production, but order of magnitude higher sensitivity (NE $\Delta T \approx 0.01$ K) has been demonstrated reproducibly.

MIS HgCdTe arrays with up to 64×64 elements in the form of CCD-TDI FPAs were developed in the early 1980s for scanning and staring MWIR and CIM FPAs for LWIR. However, at the present time most of the LWIR HgCdTe FPAs are being developed for construction with silicon multiplexer either as hybrid FPAs or as VIP FPAs with vertically integrated photodiodes.

High-resolution (up to 512 elements) scanning InSb FPAs and staring InSb FPAs with up to 256×256 elements were developed in the 1980s with monolithic CID readout multiplexer. More recently, however, InSb detector arrays with silicon hybrid multiplexers are becoming more prevalent.

Monolithic silicon FPAs with PtSi Schottky-barrier detectors (SBDs) represent the

TABLE 2 State-of-the-Art Staring PtSi FPAs

Type of FPA	Pixel size (μm^2)	Fill factor	Design rules (μm)	$Q_{max} \times 10^6$ (e^-/p)	NE ΔT (K)	f/#	Year	Company	References
320 × 244 IT-CCD	40 × 40	44	2.0	1.4	0.04	1.4	1988	Sarnoff	52
680 × 480 MOS	24 × 24	38	1.5	1.5	0.06	1.0	1991		53
324 × 487 IT-CCD	42 × 21	42	1.5	0.25	0.10	1.0	1988	NEC	54
648 × 487 IT-CCD	21 × 21	40	1.3	1.5	0.10	1.0	1991		55
256 × 244 IT-CCD	31.5 × 25	36	1.8	0.55	0.07	1.8	1989	Loral	56
512 × 512 IT-CCD								Fairchild	
640 × 486 IT-CCD	25 × 25	54	1.2					Kodak	
noninterl. 4-port				0.23	0.15	2.8	1990		57
interlaced 1-port				0.55	<0.1	2.8	1991		
512 × 512 CSD	26 × 20	39	2.0	0.7	0.1	1.5	1987	Mitsubishi	58
		71	1.2	2.9	0.033	1.5	1992		59
1040 × 1040 CSD 4-port	17 × 17	53	1.5	1.6	0.1	1.2	1991		45
400 × 244 hybrid	24 × 24	84	2.0	0.75	0.08	1.8	1990	Hughes	60
640 × 488 hybrid 4-point	20 × 20	80	2.0	0.75	0.1	2.0	1991		61

most mature FPA technology for SWIR and MWIR applications that are compatible with relatively low quantum efficiency of PtSi SBDs. Since these FPAs are fabricated by a well-established silicon VLSI process, they can be fabricated with very good response uniformity and high resolution.

A scanning PtSi FPA with 4 banks of 4096-element bilinear CCD line sensors was developed by Mitsubishi and the 2048 × 16 TDI-elements FPAs were reported by Kodak and Itek[51] for space-born remote sensing applications.

Table 2 summarizes the characteristics of recently developed monolithic and hybrid PtSi FPAs with IR-CCD, CSD, and MOS silicon multiplexers. In addition to the pixel size and fill factor, this table also shows the process design rules, maximum charge signal, Q_{max}, the reported noise equivalent temperature (NE ΔT) for operation with specific f/# optics, the associated company, and the pertinent references.

The largest resolution was demonstrated with 1040 × 1040 CSD, and the largest fill factor for a monolithic 512 × 512 CSD FPA is 71 percent. This should be compared with a fill factor of 80 percent for a 640 × 480 hybrid FPA. Thermal imaging demonstrated with Sarnoff 640 × 480 PtSi low-noise MOS FPA is shown in Fig. 15.

While PtSi SBDs operating at about 77 K are useful for SWIR and MWIR applications, LWIR response can be obtained with IrSi SBDs that require operation at about 40 K for low dark current.

High quantum efficiency FPAs proliferate the 256 × 256 format since this is the minimum format needed for imagers with near-TV resolution. About a half-dozen vendors offer DI-based FPAs, but there is variability in sensor sensitivity. The best devices have yielded field-tested camera NE ΔT's approaching 10 mK. PACE-I 256 × 256 FPAs, for example, offer this sensitivity at temperatures above 100 K with residual nonuniformity of 0.012 percent after correction, which is the lowest yet reported by any FPA technology. The residual nonuniformity is nearly a factor of 10 better than that previously considered by spatial noise analysts. InSb-based imagers should perform similarly at lower temperatures, but the data is apparently unavailable at this time.

Whereas improved materials technology has already resulted in increased pixel count for unprecedented sensitivity at TV-type spatial resolution, HgCdTe advocates are also

FIGURE 15 Thermal image of a man holding a match detected by the 640×480 MOS imager. (*Courtesy of Sarnoff*).

working these improvements to increase operating temperature with moderate performance penalty. When coupled to CMOS multiplexers that dissipate little power and need minimum support circuitry, a viable thermoelectrically cooled staring FPA technology is implied in the near future. TE-cooled 128×128 FPAs having NE $\Delta T < 0.1$ K have been reported.

LWIR. LWIR FPA development has been a key military focus that will likely yield significant spin-off benefit because the improvements in materials have greatly enhanced industry capability. The primary emphasis has been on LPE HgCdTe/CdZnTe, which is the IR FPA performance leader for temperatures ≥ 40 K. BLIP detector D^* has been achieved over a wide range in operating temperatures and spectral regions. This thrust has produced high-sensitivity devices in several formats, including 64×64, 128×128, and 640×480, and has helped advance the state of the art in sensitivity at temperatures from ≥ 78 K. The highest-sensitivity (i.e., low temporal and spatial noise) devices typically use electronically scanned architectures that have demonstrated excellent linearity and low spatial noise.

The pure performance leader for cameras requiring longer wavelengths (up to 28 μm) are staring Si:X IBC FPAs, which have been demonstrated in staring formats to 128×128 and scanning formats to 10×50 at temperatures from 4–18 K.

Uncooled microbolometer and pyroelectric FPAs have shown sufficient maturity to give camera NE ΔT that is becoming highly competitive with PtSi. Both technologies have demonstrated $\sim 320 \times 240$ arrays with roughly 0.08 K NE ΔT.[62]

Development of GaAs/AlGaAs QWIPs and SiGe heterojunction internal photoemission[63] (HIP) detector arrays has been successful in quickly producing large arrays.[64] These competing detector materials are presently niche technologies that require cooling to 77 K or below. However, the "designer" nature of these detector materials suggests that operating temperature and spectral passband, for example, can be improved.

Both technologies are dramatically superior to an earlier "low-cost" alternative, IrSi, and the GaAs/AlGaAs devices are currently superior to SiGe.

SiGe heterojunction photodiodes in the form of Ge_xSi_{1-x} alloy on silicon substrate represent another internal photoemission-type detector that can be designed for LWIR operation and also require operation at 40 to 30 K, depending on cutoff wavelength. A 400 × 400 GeSi TT-CCD was reported with cutoff wavelength of 9.3 μm at 50 K.[65] To achieve lower temperature of operation, InSi and SiGe FPAs with monolithic MOS readout are under development.[66]

Future Trends and Technology Directions

The 1980s saw the maturation of PtSi and the emergence of HgCdTe and InSb as producible MWIR detector materials. Many indicators point to the 1990s being the decade that IR FPA technology reaches the consumer marketplace. Figure 16 shows the chronological development of IR FPAs including monolithic and hybrid technologies. Specifically compared is the development of various hybrids (primarily MWIR) to PtSi, the pace-setting technology with respect to array size. The hybrid FPA data includes Pb-salt, HgCdTe, InSb, and PtSi devices. This database suggests that monolithic PtSi leads all other technologies with respect to array size by about two years, indicates that HDTV-compatible IR FPAs should begin to surface by 1993, and clearly shows the thermal mismatch barrier confronted and overcome by hybrid FPA developers in the mid-1980s.

In addition to further increases in pixel density to $>10^{16}$ pixels, several trends are clear. Future arrays will have much more on-chip signal processing, need less cooling, have higher sensitivity (particularly intrinsic LWIR FPAs), and offer multispectral capability. If the full performance potential of the uncooled technologies is realized, either the microbolometer arrays or, less likely, the pyroelectric arrays will capture the low-cost markets. It is not unreasonable that the uncooled arrays may obsolete "low-cost" PtSi, QWIP, and HIP FPAs, and render the intrinsic TE-cooled developments inconsequential. To improve hybrid reliability, alternative detector substrate materials including silicon along with alternative readout materials will become sufficiently mature to begin

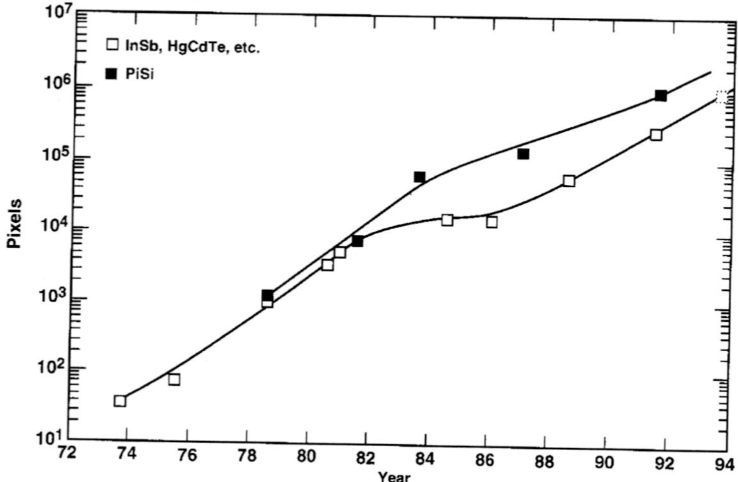

FIGURE 16 Chronological development of IR FPAs.

monolithic integration of the intrinsic materials with highest radiative performance. True optoelectronic FPAs consisting of IR sensors with optical output capability may be developed for reducing thermal loading and improving immunity to noise pickup. Availability of inexpensive, commercial devices is imminent along with the development of IR neural networks for additional signal processing capability.

23.7 REFERENCES

1. W. P. McCracken, "CCDing in the Dark," *IEEE Spectrum.* (1992).
2. P. R. Norton, "Infrared Image Sensors," *Optical Engineering,* **30:**11 (1991).
3. A. Rogalski and J. Piotrowski, "Intrinsic Infrared Detectors," *Progress in Quantum Electronics* 12, Pergamon Press, 1988, pp. 2–3.
4. S. B. Stetson, D. B. Reynolds, M. G. Stapelbroek, and R. L. Stermer, "Design and Performance of Blocked-Impurity-Band Detector Focal Plane Arrays," *SPIE Proceedings,* vol. 686 (1986).
5. M. D. Petroff, G. Stapelbroek, and W. A. Kleinhans, "Solid State Photomultiplier," U.S. Patent No. 4,586,068 (1983).
6. B. F. Levine, K. K. Choi, C. G. Bethea, J. Walker, and R. J. Malik, *Applied Phys. Lett.* **50:**1092 (1987).
7. C. S. Wu et al., "Novel GaAs/AlGaAs Multiquantum-Well Schottky Junction Device and Its Photovoltaic LWIR Detection," *IEEE Trans. Electron. Devices, ED*-3**9:**2 (1992).
8. J. Cooper, *Rev. Sci. Instrum.* **33:**92 (1962).
9. C. Hanson, H. Beratan, R. Owen, M. Corbin, and S. McKenney, "Uncooled Thermal Imaging at Texas Instruments," *SPIE Proceedings,* vol. 1735, 1992.
10. D. H. Pommerrenig, "Extrinsic Silicon Focal Plane Array," *SPIE Proceedings,* vol. 443, 1983.
11. (*a*) W. F. Kosonocky, "Review of Infrared Image Sensors with Schottky-Barrier Detectors," *Optoelectronics—Devices and Technologies,* vol. 6, no. 2, December 1991, pp. 173–203. (*b*) W. F. Kosonocky, "State-of-the-Art in Schottky-Barrier IR Image Sensors," *SPIE Proceedings,* vol. 1681, Orlando, Fla., 1992.
12. Private communications with N. A. Foss from Sensors and Systems Center, Honeywell, Inc., Bloomington, Minn.
13. C. G. Roberts, "HgCdTe Charge Transfer Focal Planes," *SPIE Proceedings,* vol. 443, 1983.
14. M. D. Gibbons et al., "Status of IrSb Charge Injection Device (CID) Detection Technology," *SPIE Proceedings,* vol. 443, 1983.
15. M. H. White et al., "Characterization of Surface CCD Image Arrays at Low Light Levels," *IEEE J. Solid-State Circuits,* vol. SC-9, 1974, pp. 1–12.
16. D. J. Sauer, F. V. Shallcross, F. L. Hsueh, G. M. Meray, P. A. Levine, H. R. Gilmartin, T. S. Villani, B. J. Esposito, and J. R. Tower, "640 × 480 MOS PtSI IR Sensor," *SPIE Proceedings,* vol. 1540, 1991, pp. 285–296.
17. W. F. Kosonocky, T. S. Villani, F. V. Shallcross, G. M. Meray, and J. J. O'Neill, III, "A Schottky-Barrier Image Sensor with 100% Fill Factor," *SPIE Proceedings,* vol. 1308, 1990, pp. 70–80.
18. M. Denda, M. Kimata, S. Iwade, N. Yutani, T. Kondo, and N. Tsubouchi, "4 × 4096-Element SWIR Multispectral Focal Plane Array," *SPIE Proceedings,* vol. 819-24, 1987, pp. 279–286.
19. K. Chow, J. P. Rode, D. H. Seib, and J. D. Blackwell, "Hybrid Infrared Focal Plane Arrays," *IEEE Trans. Electron Devices* **ED-29**(1), January, 1982.
20. E. R. Gertner, W. E. Tennant, J. D. Blackwell and J. P. Rode, "HgCdTe on Sapphire—A New Approach to Infrared Detector Arrays," *J. Cryst. Growth* **72:**465 (1985).

21. E. R. Gertner, *Ann. Rev. Mater. Sci.* **15:**303–28 (1985).
22. E. R. Fossum, "Infrared Readout Electronics," *SPIE Proceedings,* vol. 1684 (1992).
23. D. F. Barbe, "Imaging Devices Using the Charge-Coupled Concept," *Proc. IEEE* **63**(1) (1975).
24. P. Felix, M. Moulin, B. Munier, J. Portmann, and J.-P. Reboul, "CCD Readout of Infrared Hybrid Focal Plane Arrays," *IEEE Trans. Electron Devices* **ED-27**(1) (1980).
25. S. T. Baier, "Complementary Heterostructure (CHFET) Readout Technology for Infrared Focal Plane Arrays," *SPIE Proceedings,* vol. 1684 (1992).
26. R. Sahai, R. L. Pierson, R. J. Anderson, E. H. Martin, E. A. Sover, and J. Higgins, "GaAs CCD's with Transparent (ITO) Gates for Imaging and Optical Signal Processing," *IEEE Electron Device Lett.* **EDL-4** (1983).
27. L. J. Kozlowski and R. E. Kezer, "2×64 GaAs Readout for IR FPA Application," *SPIE Proceedings,* vol. 1684 (1992).
28. J. S. Bugler and P. G. A. Jespers, "Charge Pumping in MOS Devices," *IEEE Trans. on Electron Devices* **ED-16:**3 (1969).
29. M. F. Tompsett, "Surface Potential Equilibration Method of Setting Charge in Charge-coupled Devices," *IEEE Trans. Electron Devices,* **ED-22:**6 (1975).
30. L. J. Kozlowski, K. Vural, W. E. Tennant, R. E. Kezer, and W. E. Kleinhans, "10×132 CMOS/CCD Readout with 25 μm Pitch and On-Chip Signal Processing Including CDS and TDI," *SPIE Proceedings,* vol. 1684 (1992).
31. M. F. Tompsett, "The Quantitative Effects of Interface States on the Performance of Charge-Coupled Devices," *IEEE Trans. Electron Devices* **ED-20:**1 (1973).
32. P. W. Bosshart, "A Multiplexed Switched Capacitor Filter Bank," *IEEE Journal Solid-State Circuits,* **SC-15**:6 (1980).
33. W. S. Chan, "Detector-Charge-Coupled Device (CCD) Interface Methods," *SPIE Proceedings,* vol. 244 (1980).
34. Swanson and Meindl, "Ion-Implanted Complementary MOS Transistors in Low-Voltage Circuits," *IEEE Journal of Solid-State Circuits* **SC-7:**2 (1972).
35. R. Troutman and S. N. Chakravarti, "Subthreshold Characteristics of Insulated-Gate Field-Effect Transistors, *IEEE Trans. Circuit Theory,* **CT-20:**6 (1973).
36. N. Bluzer and R. Stehlik, "Buffered Direct Injection of Photocurrents into Charge-Coupled Devices," *IEEE Journal of Solid-State Circuits,* **SC-13:**1 (1978).
37. C. C. Enz, E. A. Vittoz, and F. Krummenacher, "A CMOS Chopper Amplifier," *IEEE JSSC* **SC-22:**3 (1987).
38. S. G. Chamberlain and J. P. Y. Lee, "A Novel Wide Dynamic Range Silicon Photodetector and Linear Imaging Array," *IEEE Trans. Electron Devices* **ED-31:**2 (1984).
39. F. Krummenacher, E. Vittoz, and M. DeGrauwe, "Class AB CMOS Amplifier for Micropower SC Filters," *Electronics Letters* **17:**13 (1981).
40. E. Vittoz and J. Fellrath, "CMOS Analog Integrated Circuits Based on Weak Inversion Operation," *IEEE Journal Solid State Circuits* **SC-12:**3 (1977).
41. R. D. Hudson, *Infrared System Engineering,* John Wiley and Sons, 1969.
42. J. M. Mooney et al., "Responsivity Nonuniformity Limited Performance of Infrared Staring Cameras," *Opt. Eng.* **28:**1151 (1989).
43. D. L. Shumaker, J. T. Wood, and C. R. Thacker, *FLIR Performance Handbook,* DCS Corporation, Alexandria, Va. (1988).
44. J. A. Cunningham, "The Remarkable Trend in Defect Densities and Chip Yields," Semiconductor International, 1992.
45. N. Yutani, M. Kimata, H. Yagi, J. Nakanishi, S. Nagayoshi, and N. Tubouchi, "1040×1040 Element PtSi Schottky-Barrier IR Image Sensor," IEDM, Washington, D.C., 1991.
46. Production quantities are miniscule relative to typical commercial products.
47. L. J. Kozlowski, R. B. Bailey, D. E. Cooper, K. Vural, E. R. Gertner, and W. E. Tennant, "Large Staring IRFPA's of HgCdTe on Alternative Substrates," *SPIE Proceedings,* vol. 1540 (1991).
48. L. J. Kozlowski, K. Vural, V. H. Johnson, J. K. Chen, R. B. Bailey, and D. Bui, "256×256 PACE-1 PV HgCdTe Focal Plane Arrays for Medium and Short Wavelength Infrared Applica-

tions," *SPIE Proceedings,* vol. 1308, *SPIE Technical Symposium on Optical Engineering and Photonics in Aerospace Sensing,* Orlando, Fla., 16–20 April 1990.

49. N. Lum, J. Asbrock, R. White, R. Augustine, A. Hoffman, and S. C. Nystrom, "Low-Noise, 256 × 256, 10-Kelvin Staring FPA," *SPIE Proceedings,* vol. 1684, 1992.
50. L. J. Kozlowski, S. A. Cabelli, R. Kezer, and W. E. Kleinhans, "10 × 132 Readout with 25 μm Pitch and On-Chip Signal Processing Including CDS and TDI," *SPIE Proceedings,* vol. 1684, April 1992.
51. M. T. Daigle, D. Colvin, E. T. Nelson, S. Brickman, K. Wong, S. Yoshizumi, M. Elzinga, P. Sorlie, D. Rockafellow, P. Travers, and R. Avel, "High Resolution 2048 × 16 TDI PtSi IR Imaging CCD," *SPIE Proceedings,* vol. 1308, 1990, pp. 88–98.
52. T. S. Villani, W. F. Kosonocky, F. V. Shallcross, J. V. Groppe, G. M. Meray, J. J. O'Neill, III, and B. J. Esposito, "Construction and Performance of a 320 × 244-Element IR-CCD Imager with PtSi SBDs," *SPIE Proceedings,* vol. 1107-01, 1989, pp. 9–21.
53. D. J. Sauer, F. V. Shallcross, F. L. Hsueh, G. M. Meray, P. A. Levine, H. R. Gilmartin, T. S. Villani, B. J. Esposito, and J. R. Tower, "640 × 480 MOS PtSi IR Sensor," *SPIE Proceedings,* vol. 1540, 1991, pp. 285–296.
54. K. Konuma, N. Teranishi, S. Tohyama, K. Masubuchi, S. Yamagata, T. Tanaka, E. Oda, Y. Moriyama, N. Takada, and N. Yoshioka, "324 × 487 Schottky-Barrier Infrared Imager," *IEEE Trans. Electron Devices,* vol. 37, no. 3, 1990 pp. 629–635.
55. K. Konuma, S. Tohyama, A. Tanabe, K. Masubuchi, N. Teranishi, T. Saito, and T. Muramatsu, "A 648 × 487 Pixel Schottky-Barrier Infrared CCD Image Sensor," *1991 ISSCC Digest of Tech. Papers,* 1991, pp. 156–157.
56. H. Elabd, Y. Abedini, J. Kim, M. Shih, J. Chin, K. Shah, J. Chen, F. Nicol, W. Petro, J. Lehan, M. Duron, M. Manderson, H. Balopole, P. Coyle, P. Cheng, and W. Shieh, "488 × 512 and 244 × 256-Element Monolithic PtSi Schottky IR Focal Plane Arrays," *SPIE Proceedings,* vol. 1107-29, presented at *SPIE Aerospace Sensor Symposium,* Orlando, Fla., March 1989.
57. E. T. Nelson, K. Y. Wong, S. Yoshizumi, D. Rockafellow, W. DesJardin, M. Elzinga, J. P. Lavine, T. J. Tredwell, R. P. Khosla, P. Sorlie, B. Howe, S. Brickman, and S. Refermat, "Wide Field of View PtSi Infrared Focal Plane Array," *SPIE Proceedings,* vol. 1308, 1990, pp. 36–44.
58. M. Kimata, M. Denda, N. Yutani, S. Iwade, and N. Tsubouchi, "A 512 × 512-Element PtSi Schottky-Barrier Infrared Image Sensor," *IEEE J. Solid-State Circuits,* vol. SC-22, no. 6, 1987, pp. 1124–1129.
59. H. Yagi, N. Yutani, S. Nagayoshi, J. Nakanishi, M. Kimata, and N. Tsubouchi," Improved 512 × 512 IR CSD with Large Fill Factor and Large Saturation Level," *SPIE Proceedings,* vol. 1685-04, 1992.
60. J. Edwards, J. Gates, H. Altin-Mees, W. Connelly, and A. Thompson, "244 × 400 Element Hybrid Platinum Silicide Schottky Focal Plane Array," *SPIE Proceedings,* vol. 1308, 1990, pp. 99–100.
61. J. L. Gates, W. G. Connelly, T. D. Franklin, R. E. Mills, F. W. Price, and T. Y. Wittwer, "488 × 640-Element Hybrid Platinum Silicode Schottky Focal Plane Array," *SPIE Proceedings,* vol. 1540, *Infrared Technology XVII,* 1991, pp. 262–273.
62. C. Hanson, H. Beratan, R. Owen, M. Corbin, and S. McKenney, "Uncooled Thermal Imaging at Texas Instruments," *SPIE Proceedings,* vol. 1735 (1992).
63. T. L. Lin, A. Ksendzov, T. N. Krabach, J. Maserjian, M. L. Huberman, and R. Terhune, "Novel $Si_{1-x}Ge_x$/Si Heterojunction Internal Photoemission Long-Wavelength Infrared Detectors," *Appl. Phys. Lett.* **57** (1990).
64. L. J. Kozlowski, G. M. Williams, G. J. Sullivan, C. W. Farley, R. J. Anderson, J. Chen, D. T. Cheung, W. E. Tennant, and R. E. DeWames, "LWIR 128 × 128 GaAs/AlGaAs Multiple Quantum Well Hybrid Focal Plane Array," *IEEE Trans. Electron Devices,* **ED-38:**5 (1991).
65. B.-Y. Tsaur, C. K. Chen, and S. A. Marino, "Long Wavelength Ge_xSi_{1-x}/Si Heterojunction Infrared Detectors and 400 × 400 Element Imager Arrays," *IEEE Electron Device Letters,* **12:**293–296 (1991).
66. B.-Y. Tsaur, C. K. Chen, and J.-P. Mattia, "PtSi Schottky-Barrier Focal Plane Arrays, for Multispectral Imaging in Ultraviolet, Visible, and Infrared Spectral Bands," *IEEE Electron Device Letters,* vol. 11, no. 4, 1990, pp. 162–164.

P · A · R · T · 7

VISION

CHAPTER 24
OPTICS OF THE EYE

W. N. Charman
Department of Optometry and Vision Sciences
UMIST
Manchester, United Kingdom

24.1 GLOSSARY

F, F′	focal points
N, N′	nodal points
P, P′	principal points

Equation (1)

$L_{e,\lambda}$	spectral radiance per unit wavelength interval per unit solid angle
$m(\theta, \phi, \lambda)$	areal magnification factor
$p(\theta, \phi)$	area of pupil as seen from direction (θ, ϕ)
$t(\theta, \phi, \lambda)$	fraction of incident radiant flux which is transmitted by eye
$\delta\lambda$	wavelength interval
(θ, ϕ)	angular direction coordinates

Equation (2)

r	distance from Stiles-Crawford peak
$\eta/\eta_{\max}$	relative luminous efficiency
ρ	coefficient in S-C equation

Equation (3)

d	pupil diameter
P_A	ratio of effective to true pupil area

Equation (4)

I	normalized illuminance
z	dimensionless diffraction unit

Equation (5)

d	pupil diameter
γ	angular distance from center of the Airy pattern
λ	wavelength

Equation (6)

$\theta_{\min}$	angular resolution by Rayleigh criterion

Equation (7)

R	spatial frequency
R_R	reduced spatial frequency

Equation (8)

ΔF	dioptric error of focus
g	number of Rayleigh units of defocus

Equation (9)

β	angular diameter of retinal blur circle

Equation (10)

R	spatial frequency
$T(R)$	MTF

Equation (11)

E	illuminance produced by glare source at eye
L_{eq}	equivalent veiling luminance
ω	angle between direction of glare source and fixation axis

pp. 24.21–24.22

$M_{it}(R)$	threshold modulation on retina
$M_{ot}(R)$	external threshold modulation

Equation (12)

n	exponent of expression
R_c	critical spatial frequency

Equation (13)

DOF_{go} total depth-of-focus by geometrical optics

β_{tol} tolerable angular diameter of retinal blur circle

Equation (14)

DOF_{po} total depth-of-focus by physical optics (diffraction-limited)

Equation (15)

l object distance

p interpupillary distance

δl mimimum detectable difference in distance

$\delta\theta$ stereo acuity

Equation (16)

M transverse magnification

N relative increase in effective interpupillary distance

24.2 INTRODUCTION

Although the human eye (Fig. 1) is a relatively simple optical system, it is capable of near diffraction-limited performance close to its axis under good lighting conditions when the

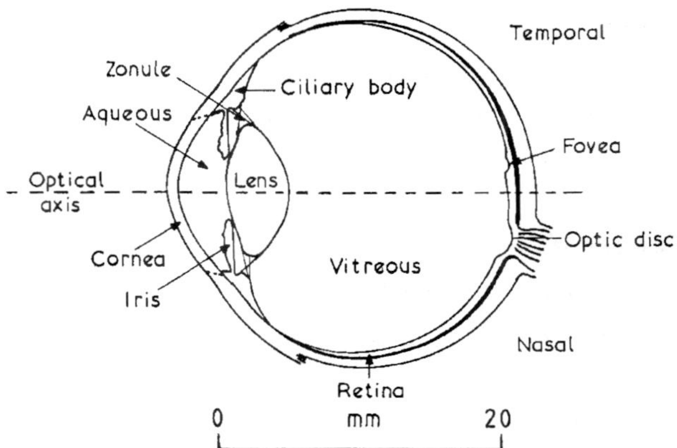

FIGURE 1 Schematic horizontal section of the eye. The bar gives the approximate scale.

pupil is small (about 2–3 mm). It also has a very wide field of view (about 65, 75, 60, and 100 deg in the superior, inferior, nasal, and temporal meridians, respectively, for a fixed frontal direction of gaze, the exact values being dependent upon the individual's facial geometry). Optical image quality, while somewhat degraded in the peripheral field is, in general, apparently adequate to meet the needs of the neural network which it serves. Control of ocular aberrations is helped by the presence of aspheric corneal and lenticular surfaces and by gradients of refractive index in the lens, the index progressively reducing from the lens center toward its outer layers.

Attempts to make general statements about the optical characteristics of the eye are complicated by the fact that there are continuous changes in the dimensions and other parameters of the eye throughout life.[1–6] Although most of these changes are largest in childhood, the lens thickness continues to increase approximately linearly throughout life,[6,7] by about 50 percent between birth and the age of 70, with accompanying changes in its index distribution.[8]

Each eye also tends to have its own idiosyncracies, involving such factors as a lack of symmetry about the anterior-posterior axis, or small tilts or sideways displacements of surfaces or the pupil. Surface curvatures, component separations and axial lengths show considerable variation (~10 percent) between individuals.[1,9,10] Thus the values that are given below (Table 1) are merely representative of the typical adult eye. In general, however, the parameters of the individual eye are correlated, so that the overall refractive power of the unaccommodated eye approximates to that required to produce a sharp retinal image of a distant object, i.e., most eyes approach *emmetropia* (see later under "Ocular Ametropia").

TABLE 1 Parameters of Some Paraxial Models of the Human Eye

		Schematic eye*	Simplified schematic eye†	Reduced eye†
Radii of surfaces (mm)	Anterior cornea	7.80	7.80	5.55
	Posterior cornea	6.50	—	—
	Anterior lens	10.20	10.00	—
	Posterior lens	−6.00	−6.00	—
Distances from anterior cornea (mm)	Posterior cornea	0.55	—	—
	Anterior lens	3.60	3.60	—
	Posterior lens	7.60	7.20	—
	Retina	24.20	23.90	—
	1st principal point P	1.59	1.55	0
	2nd principal point P′	1.91	1.85	0
	1st nodal point N	7.20	7.06	5.55
	2nd nodal point N′	7.51	7.36	5.55
	1st focal point F	−15.09	−14.99	−16.67
	2nd focal point F′	24.20	23.90	22.22
Refractive indices	Cornea	1.3771	—	4/3
	Aqueous humor	1.3374	1.333	4/3
	Lens	1.4200	1.416	4/3
	Vitreous humor	1.3360	1.333	4/3

* See Ref. 15.
† See Ref. 12.

24.3 EYE MODELS

On the basis of experimental measurements of individual ocular parameters made by a variety of techniques, many authors have attempted to produce models which allow the characteristics of the retinal image to be predicted. Most of these have been limited to the paraxial region, where the rays make small angles with the axis, the optical surfaces being assumed to be spherical.

Paraxial Models

Figure 2 shows examples of three typical types of paraxial model, the relevant parameters being given in Table 1. In constructing such models the values of the parameters are normally selected to be reasonably representative of those found in real eyes but are then adjusted slightly to make the eye emmetropic.[11–18] In *schematic eyes* the cornea and lens are each represented by a pair of surfaces (although sometimes the lens is divided into a central nuclear and outer cortical regions, these being assigned differing refractive indices): a single surface is used for the cornea in the *simplified schematic eye.* The close proximity of the two nodal and two principal points has encouraged the use of *reduced eye* models consisting of a single refracting surface.

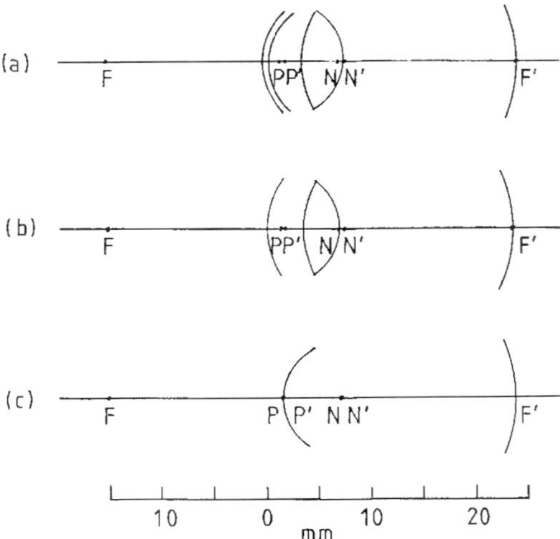

FIGURE 2 Examples of paraxial models of the human eye. In each case F, F′; P, P′; N, N′ represent the first and second focal, principal and nodal points respectively. (*a*) Unaccommodated schematic eye with four refracting surfaces (Le Grand and El Hage[15]) (*b*) Simplified schematic eye with three refracting surfaces (Emsley[12]) (*c*) Reduced eye with a single refracting surface (Emsley[12]). Note that progressive reduction in the number of surfaces used in the model produces only minor changes in the positions of the cardinal points.

These paraxial models are useful for predicting the approximate dimensions of paraxial images (1 deg in the visual field corresponds to about 0.290 mm on the retina) and their changes with accommodation and correcting lenses.[18] They are also used as an aid in determining the power of intraocular lenses required following cataract surgery[19] and in evaluating retinal radiation hazards.[20] However, they are of doubtful value in indicating retinal image quality or off-axis performance, since they cannot adequately represent the aberrations of real eyes.

Wide-Angle Models

Following gradual advances in understanding of the form of the aspheric surfaces of the cornea[21–23] and lens[24–27] and of the lenticular refractive index distribution,[8,28–30] and the availability of improved methods of ray tracing through index gradients, several authors have attempted to produce more sophisticated wide-angle eye models and to validate them by comparing their predictions of off-axis aberrations with those of real eyes.[31–36] As yet these models are only partly successful, but progressive refinement in experimental data should ultimately allow much more realistic modeling of the eye's overall optical performance.

Ocular ametropia

Although the ocular parameters are normally reasonably well correlated so that distant objects are sharply imaged on the retina, this is not always the case. In some individuals the power of the optical elements is too great for the axial length of the eye and a distant object is imaged in front of the retina (*myopia*). In others the relative power is too low and the image is formed behind the retina (*hypermetropia*). These defects can be corrected by the use of appropriately powered diverging (myopia) or converging (hypermetropia) spectacle or contact lenses to respectively reduce or supplement the spherical power of the eye (see, e.g., Refs. 37–41 for reviews). In some individuals the ocular dioptrics lack rotational symmetry, one or more optical surfaces being toroidal, tilted, or displaced from the axis. This leads to the condition of *ocular astigmatism,* in which two longitudinally separated, mutually perpendicular, line images of a point object are formed. In the vast majority of cases (*regular astigmatism*), the meridians of maximal and minimal power of the eye are perpendicular:[42] there is a strong tendency for the principal meridians to be approximately horizontal and vertical but this is not always the case. Correction can be achieved by including an appropriately oriented cylindrical component in any correcting lens.

Surgical methods of correction for both spherical and astigmatic refractive errors are currently at an experimental stage (see, e.g., Refs. 43–46) and their long-term effectiveness remains to be proven. They involve the production of a change in the corneal contour or index. Less controversial is the surgical insertion of an intraocular lens to replace the natural crystalline lens when the latter loses its transparency due to cataract: this is now a well-proven technique[19] for single-vision implants, although bifocal and multifocal implants remain at an experimental stage.[47]

Figure 3 shows representative data for the frequency of occurrence of different spherical[48] and astigmatic[49] refractive errors in adults. Note particularly the clustering of the spherical errors around emmetropia. Not all individuals with ametropia actually wear a correction; the fraction of the population that does[50] is also shown in Fig. 3. The increase

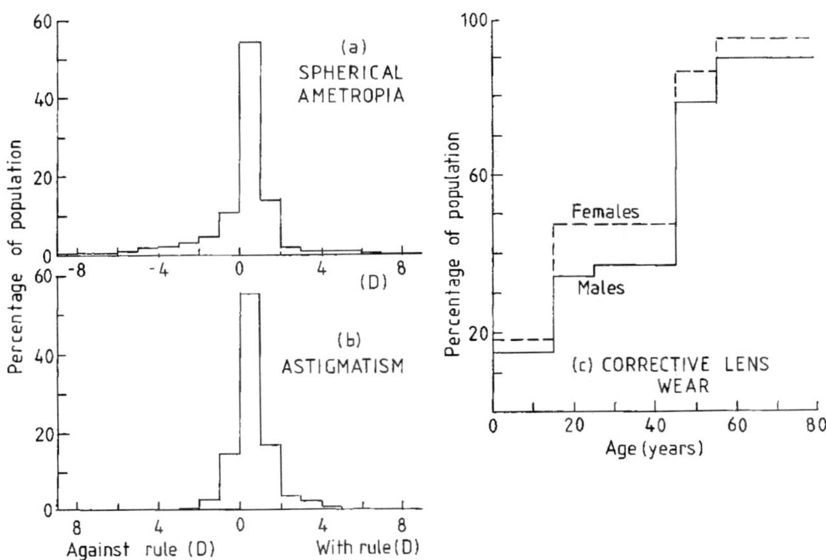

FIGURE 3 Typical data for frequency of occurrence of spherical and cylindrical refractive errors and fraction of the population wearing corrective spectacle or contact lenses. (*a*) Spherical errors (diopters) in young adult males (*after Stenstrom*[9]). (*b*) Cylindrical errors in an adult population (*based on Lyle*[49]). Cases in which the meridian of greatest power is within 30 degrees of horizontal (against-the-rule) and within 30 degrees of the vertical (with-the-rule) are shown: the remaining 4 percent of the population have axes in oblique meridians. (*c*) Proportion of population wearing corrective lenses (*after Farrell and Booth*[50]).

in lens wear beyond the age of 40 is due the need for a reading correction as a result of failing accommodation (see later section "Age-Dependent Effects in Accommodation" in Sec. 24.7).

It is desirable in many visual instruments such as telescopes and microscopes, particularly those containing a graticule, to make provision for focusing the eyepiece to compensate for any spherical refractive error of the observer. Since the refractive errors of the two eyes of an individual may not be identical, provision for differential focusing of the eyepieces should be provided in binocular instruments. Correction for cylindrical errors is inconvenient to incorporate into eyepieces and astigmatic users of instruments must usually employ their normal optical corrections. For spectacle wearers, where the distance of the lenses in front of the eyes is usually 10–18 mm, this implies that the exit pupil of the instrument must have an adequate eye clearance or eye relief (at least 20 mm and preferably 25 mm to avoid contact between the spectacle lens and the eyepiece).

24.4 OCULAR TRANSMITTANCE AND RETINAL ILLUMINANCE

The amount, spectral distribution, and polarization characteristics of the light reaching the retina are modified with respect to the original stimulus in a way that depends upon the transmittance characteristics of the eye. Light may be lost by spectrally varying reflection, scattering, and absorption in any of the optical media anterior to the retina.[51] Fresnel reflection losses are in general small, reaching a maximum of 3–4 percent at the anterior cornea. Wavelength-dependent absorption and scattering are much more important.

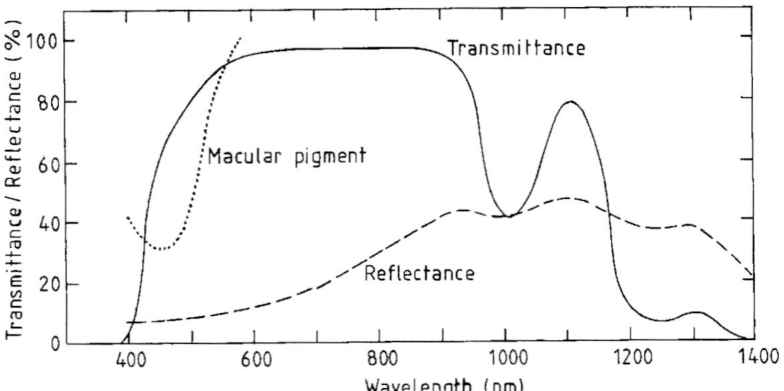

FIGURE 4 Spectral dependence of the transmittance of the ocular media and of the retinal reflectance. — Average transmittance for 7 eyes in the range 23–78 years (*after Geeraets and Berry*[53]). ···· Typical transmittance of macular pigment (*after Weale*[67]). ---- Mean reflectance from retina, pigment epithelium and choroid of 28 human eyes (*after Geeraets and Berry*[53]).

Transmittance

The measured transmittance depends to some extent upon the measuring technique, in particular upon the amount of scattered light included, but representative data[52,53] are shown in Fig. 4. The transmittance rises rapidly above about 400 nm to remain high in the visible and near infrared, falling rapidly with a series of well-marked absorption bands to zero at about 1400 nm. Retinal reflectance is quite high in the longer wavelength part of this transmission band, so that it is possible to design a variety of instruments for ocular examination (e.g., autorefractors, nonmydriatic fundus cameras) which depend upon the use of infrared light reflected from the fundus, such light being invisible to the patient. Equally, such transmittance implies the existence of potential retinal hazards from infrared sources, particularly lasers, operating in this spectral region.[20,54,55]

Although much of the absorption at short wavelengths occurs at the cornea (where it may disrupt the surface cells, leading to photokeratitis, e.g., snow blindness or welder's flash, the lowest damage thresholds[56] of about 0.4 J cm^{-2} being at 270 nm), there is also a substantial light absorption in the lens at the short wavelength end of the visible spectrum. This lenticular absorption increases markedly with age[57–59] and can adversely affect color vision: most absorption occurs in the lens nucleus.[60] There is evidence that UV absorption in the lens may be a causative factor in some types of cataracts.[61,62] Excessive visible light at the violet-blue end of the spectrum is thought to cause accelerated aging and resultant visual loss at the retinal level.[63]

In the foveal region, the macular pigment extending over the central few degrees of the retina[64,65] and lying anterior to the receptor outer segments[66] absorbs heavily at shorter wavelengths[67] (Fig. 4). It has been argued that this absorption is helpful in reducing the blurring effects of longitudinal chromatic aberration[68] and in protecting the fovea against light-induced aging processes,[69] although the amount of pigment varies widely between individuals.[70]

Since the cornea, lens, and the oriented molecules of macular pigment all show birefringence, the polarization characteristics of the light entering the eye are modified before it reaches the retinal receptors. In general, these effects are of little practical significance although they can be demonstrated by suitable methods (see, e.g., Bour[71] for review).

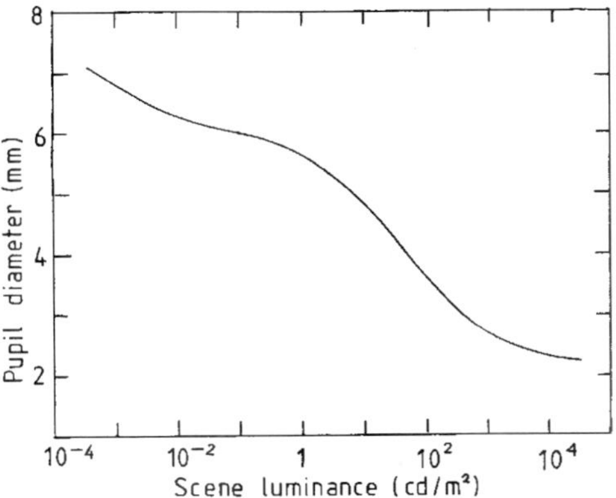

FIGURE 5 Entrance pupil diameter as a function of scene luminance for young adult observers: the curve is the weighted average of 6 studies (*after Farrell and Booth*[50]).

Pupil Diameter

The circular opening in the iris, located approximately tangential to the anterior surface of the lens, plays the important role of aperture stop of the eye. It therefore controls the amount of light flux contributing to the retinal image, as well as influencing the image quality through its effects on diffraction, aberration, and depth-of-focus (see below). What is normally observed and measured is the image of the true pupil as viewed through the cornea, i.e., the entrance pupil of the eye. This is some 13 percent larger than the true pupil. Although ambient lighting has the most important influence on entrance pupil diameter (Fig. 5) the latter is also influenced by many other factors including age, accommodation, emotion, and drugs.[72–76] For any given scene luminance, the pupils are slightly smaller under binocular conditions of observation.[77] The gradual constriction with age plays an important part in reducing the visual performance of older individuals under dim lighting conditions.[6,78] The pupil can respond to changes in light level at frequencies up to about 4 Hz.[73] Shifts in the pupil center of up to about 0.6 mm may occur when the pupil dilates, and these may be of some significance in relation to the pupil-dependence of ocular aberration and retinal image quality.[79,80] It has been suggested[81,82] that the major value of pupillary construction in light is that it reduces retinal illuminance and hence prepares the eye for a return to darkness: following a change to a dark environment the dilation of the mobile pupil allows substantially better stimulus detection during the first few minutes of dark adaptation than would be found with a fixed pupil.

Ocular Radiometry and Retinal Illuminance

If we confine ourselves to uniform object fields subtending at least 1 deg at the eye, so that diffraction and blurring due to aberration or defocus have little effect, the retinal image at moderate field angles would also be expected to be uniform across its area, except at the edges. Wyszecki and Stiles[51] show that the retinal irradiance in a wavelength interval $\delta\lambda$, corresponding to an external stimulus of spectral radiance $L_{e\lambda}(\theta, \phi)$ per unit wavelength interval per cm^2 per unit solid angle of emission, in a direction with respect to the eye given by the angular coordinates (θ, ϕ) is:

$$[L_{e,\lambda}(\theta, \phi) \cdot \delta\lambda \cdot p(\theta, \phi) \cdot t(\theta, \phi, \lambda)]/m(\theta, \phi, \lambda) \tag{1}$$

where $p(\theta, \phi)$ cm^2 is the apparent area of the pupil as seen from the direction (θ, ϕ); $t(\theta, \phi, \lambda)$ is the fraction of incident radiant flux which is transmitted through the eye; and $m(\theta, \phi, \lambda)$ is an areal magnification factor (cm^2) relating the area of the retinal image to the angular subtense of the stimulus at the eye, which will vary somewhat with the parameters of the individual eye.

Use of this equation near the visual axis is straightforward. In the peripheral field, however, complications arise. With increasing field angle, the entrance pupil appears as an ellipse of increasing eccentricity and reduced area.[83] Also, due to the retina lying on the curved surface of the eyeball, both the distance between the exit pupil of the eye and the retina and the retinal area corresponding to the image of an object area of constant angular subtense diminish with field angle. Remarkably, theoretical calculations[34,84–88] show that these pupil and retinal effects tend to compensate one another so that an extended object field of constant luminance (i.e., a *Ganzfeld*) results in a retinal illuminance which is almost constant with peripheral field angle. This theoretical result is broadly confirmed by practical measurements[89] showing that from the photometric point of view the design of the eye as a wide-angle system is remarkably effective.

A full discussion of the photometric aspects of point and extended sources in relation to the eye is given by Wright.[90]

The Stiles-Crawford Effect

One factor which limits the applicability of eq. (1) is the Stiles-Crawford effect of the first kind.[91] This results in light which enters the periphery of the pupil being less effective at stimulating the retina than light which passes through the pupil center (Fig. 6). The effect varies slightly with the individual and is not always symmetric about the center of the entrance pupil. Under photopic conditions giving cone vision there is typically a factor of

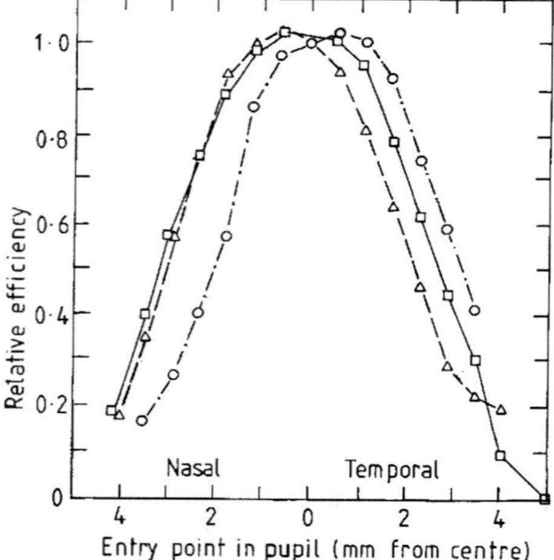

FIGURE 6 The relative luminous efficiency of a foveally viewed target as a function of horizontal pupillary position of the beam, under photopic conditions for three subjects (*after Stiles and Crawford*[91]). Note that the peak of the Stiles-Crawford function does not necessarily lie at the center of the pupil.

about 8 between the central effectiveness and that at the edge of a fully dilated 8-mm pupil; the effect is much weaker under scotopic, rod conditions.[92,93] In fact, of course, unless pupil-dilating drugs are used the natural pupil will normally be constricted at photopic levels (see Fig. 5), although the influence of the effect is still significant for photopic pupils of 3–4-mm diameter.

A variety of equations have been proposed to fit data of the type illustrated in Fig. 6. The simplest, due to Stiles,[94] can be written:

$$\log_{10}(\eta/\eta_{\max}) = -\rho r^2 \tag{2}$$

where $\eta/\eta_{\max}$ is the relative luminous efficiency and r is the distance within the entrance pupil from the function peak (in mm). Values of ρ equal to about 0.07 are typical, the value varying somewhat with wavelength.[94]

It is evident that the Stiles-Crawford effect of the first kind results in the effective photopic retinal stimulus being weaker than that predicted by Eq. (1). It can be accounted for by using an effective pupil area instead of the actual entrance pupil area. An equation giving the ratio P_A of the effective to true pupil areas of the form

$$P_A = 1 - 0.0106d^2 + 0.0000417d^4 \tag{3}$$

has been suggested,[95] where d is the pupil diameter in mm.

The Stiles-Crawford effect of the second kind involves a shift in the hue of monochromatic light as it enters different areas of the pupil.[92,94,96]

There seems little doubt that the Stiles-Crawford effects involve the waveguide properties of the outer segments of the small-diameter retinal receptors (see Refs. 92 and 93 for reviews). Effectively, the receptor can only trap light efficiently if the latter is incident in directions within a few degrees of the receptor axis. For this reason the receptor outer segments must always be aligned toward the exit pupil of the eye. An obvious advantage of such directional sensitivity is that it helps to suppress the degrading effects of intraocular stray light. Such light is ineffective at stimulating the receptors if it is incident obliquely on the retina.

24.5 FACTORS AFFECTING RETINAL IMAGE QUALITY

The Aberration-Free (Diffraction-Limited) Eye

In the absence of aberration or variation in transmittance across the pupil, the only factors influencing the retinal image quality would be the diffraction effects associated with the finite wavelength, λ, of light and the pupil diameter d. This case therefore represents the upper limit of possible optical performance for real eyes. For such an eye the point-spread function (PSF) is the well-known Airy diffraction pattern[97] whose normalized illuminance distribution takes the form:

$$I = [2J_1(z)/z]^2 \tag{4}$$

where $J_1(z)$ is the Bessel function of the first kind of order 1 of the variable z. In the case of the eye, the dimensionless distance z has the value:

$$z = [d\pi \cdot \sin\gamma]/\lambda \tag{5}$$

γ is the angular distance from the center of the pattern, measured at the second nodal

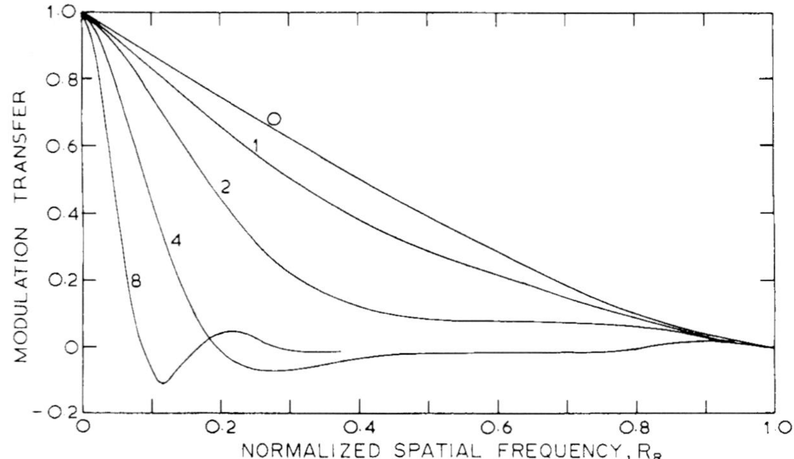

FIGURE 7 Modulation transfer functions for a diffraction-limited optical system with a circular pupil, working in monochromatic light. It suffers from the errors of focus indicated: defocus is expressed in Rayleighs, i.e., the number of quarter wavelengths of defocus wavelength aberration (*based on Levi*[106]).

point, this being equal to the corresponding angular distance in the object space, measured at the first nodal point.[98,99] The size of the PSF increases on either side of focus.[100–102] For the in-focus case, the angular resolution $\theta_{\min}$ by the Rayleigh criterion for two neighboring, equally luminous, incoherent object points is given by:

$$\theta_{\min} = 1.22\lambda/d \quad \text{radians} \tag{6}$$

$\theta_{\min}$ is about 1 minute of arc when d is 2.3 mm and λ is 555 nm. Evidently the Rayleigh criterion is somewhat arbitrary, since it depends upon the ability of the visual system to detect the 26 percent drop in irradiance between the adjacent image peaks: for small pupils actual visual performance is usually somewhat better than this.[103] Images of more complex objects can be obtained by regarding the image as the summation of an appropriate array of point-spread functions, i.e., convolving the PSF with the object radiance distribution.[98]

The in-focus line-spread function (LSF), the edge image,[104] and modulation transfer (MTF) function also take standard forms. The phase transfer function (PTF) is always zero because the diffractive blur is symmetrical. Figure 7 shows the form of the MTF as a function of focus.[105] Extensive tables of numerical values are given by Levi.[106] Relative spatial frequencies, R_R, in Fig. 7 have been normalized in terms of the cutoff value beyond which the modulation transfer is always zero. Errors of focus have been expressed in what Levi calls "Rayleigh units," i.e., the number of quarter-wavelengths of defocus wavefront aberration. To convert these dimensionless units to true spatial frequencies R c/deg and dioptric errors of focus ΔF the following relations may be used:

$$\begin{aligned} R &= [10^6 \cdot d \cdot R_R]/\lambda \quad \text{c/rad} \\ &= (1.746 \times 10^4 \cdot d \cdot R_R)/\lambda \quad \text{c/deg} \end{aligned} \tag{7}$$

$$\Delta F = (2 \times 10^{-3} \lambda g)/d^2 \quad \text{diopters} \tag{8}$$

where the entrance pupil diameter d is in mm, the wavelength λ is in nm and g is the

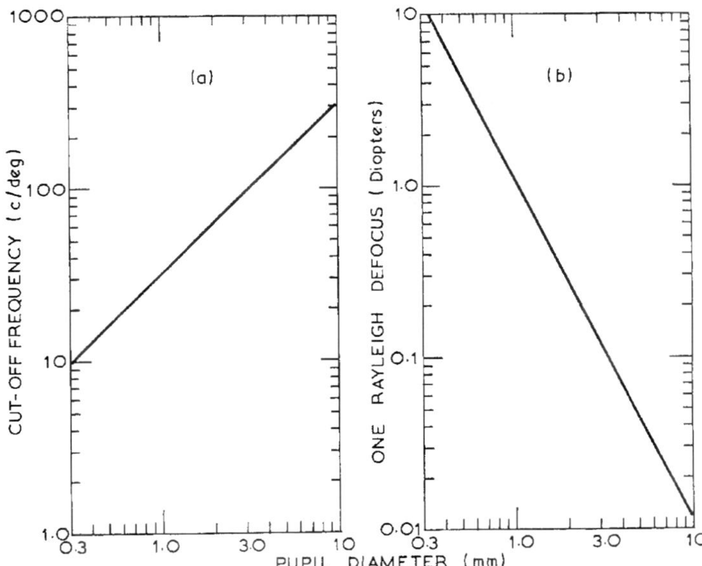

FIGURE 8 Values of (*a*) cut-off frequency $R_R = 1.0$ and (*b*) dioptric defocus equivalent to one Rayleigh, for a diffraction limited eye working in monochromatic light of wavelength 555 nm.

number of Rayleighs of defocus. Figure 8 illustrates the variation in these parameters with pupil diameter for the case where $\lambda = 555$ nm. In such green light and with the typical photopic pupil diameter of 3 mm, the cutoff frequency imposed by diffraction is about 100 c/deg: under the same conditions, one Rayleigh of defocus (i.e., a quarter-wavelength) corresponds to a focus error of about 0.12 D. Note that the modulation transfer is most sensitive to defocus at intermediate normalized spatial frequencies $R_R \sim 0.5$.[107,108]

Sets of diffraction-limited ocular MTF curves for various specific combinations of pupil diameter, wavelength, and defocus have been illustrated by various authors (e.g., Refs. 109–115).

When errors of focus become large, the geometric approximation in which the point-spread function is a uniform blur circle becomes increasingly valid.[105,114,116–118] The angular diameter β of the retinal circle for a pupil diameter d mm and error of focus ΔF diopters is:

$$\beta = (0.18d \cdot \Delta F)/\pi \quad \text{deg.} \tag{9}$$

The corresponding geometrical optical MTF is:

$$T(R) = [2J_1(\pi\beta R)]/[\pi\beta R] \tag{10}$$

where R is the spatial frequency (c/deg) and $J_1(\pi\beta R)$ is the Bessel function of the first kind of order 1 of $(\pi\beta R)$. Smith[118] gives detailed consideration to the range of ocular parameters under which the geometrical optical approximation may be applied.

Monochromatic Ocular Aberration

Traditionally it has been common to assume that the eye is symmetrical about a unique optical axis which differs only slightly from the visual axis. This angle (angle α) is about 5 degrees, although some controversy still exists as to how it should be defined and measured (e.g., Duke-Elder and Abrams[14]). In any case, since the angle is small, the eye is

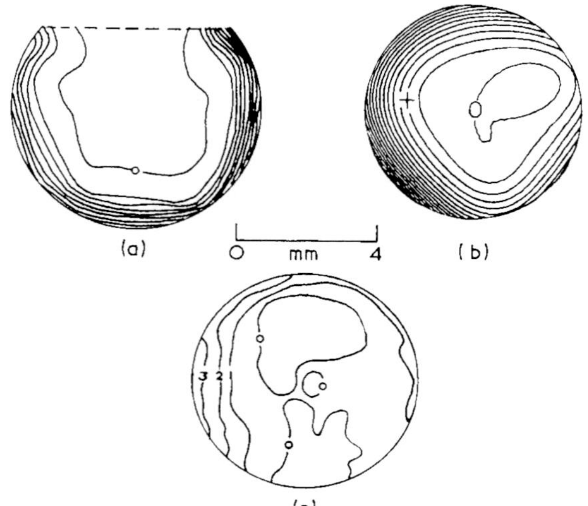

FIGURE 9 Examples of wavefront aberration data for eyes of different individual subjects. The contour lines represent the amount by which the image wavefront lies in front of the reference sphere. For clarity, the more closely spaced contours given by some authors have been omitted. (*a*) Contours are at 580 nm intervals (*after Smirnov*[119]). (*b*) Contours are at 250 nm intervals (*after Walsh and Charman*[124]). (*c*) For an eye accommodated for a target at 1 m, contours are at 546 nm intervals (*after Berny and Slansky*[120]).

often described as suffering primarily from spherical aberration close to its visual axis, with oblique astigmatism becoming more important at larger field angles.

In practice, few eyes appear to display the symmetry required for this description to be valid. This is demonstrated by measurements of the wavefront aberration on the visual axis[119–127] (Fig. 9). It is rare to find the circular contours of wavefront aberration that are characteristic of spherical aberration, although averaging data for a large number of eyes gives results which approximate more closely to pure spherical aberration.[128]

Although there are wide variations between the aberrations of individual eyes it would appear that a mixture of comatic and spherical aberrations is common on the visual axis, with third-order coma-like aberrations probably playing a surprisingly important role in degrading the optical image.[119,121–127] Fourth-order aberration becomes relatively more important as the pupil diameter increases, however.[127] Some doubt still exists as to whether aberrations higher than fourth order are of importance.[129]

The spherical aberration varies with accommodation, showing a tendency to change from undercorrected (positive) values to overcorrected (negative) values as the accommodation increases,[127,130–132] although this does not occur for all subjects. These minor changes in spherical aberration are, in any case, probably less important than the defocus blur effects produced by errors in the accommodation response (see under "The Accommodation Response").

At optimal focus, the wavefront aberration over the central 2-mm diameter of the pupil usually satisfies the Rayleigh criterion that the aberration should everywhere be less than a quarter-wavelength,[120] and the Marechal criterion,[133] that the root-mean-square deviation of the wavefront should be less than $\lambda/14$, over somewhat larger diameters (up to about 3 mm[122,123]). Hence performance for small pupil diameters approaches the diffraction limit.

In general, on axis the monochromatic aberrations are much smaller than are

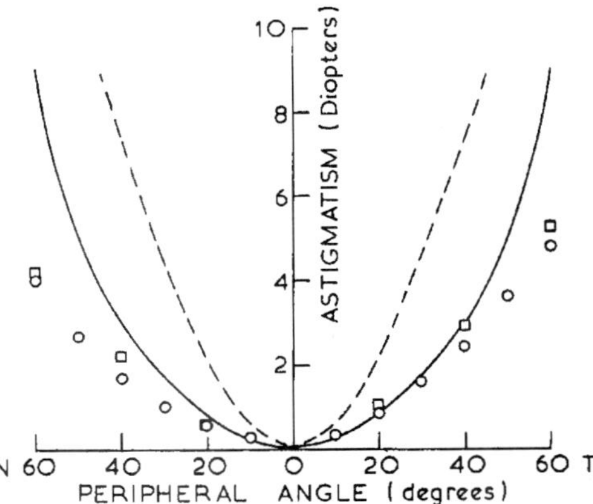

FIGURE 10 The mean oblique astigmatism of the human eye. N and T represent cases where the image falls on the nasal and temporal retina respectively. Squares are mean experimental data from Lotmar and Lotmar[141] based on the original data of Rempt et al.[136] for 726 eyes, and circles are data from Millodot[139] for 62 eyes. The broken and continuous curves are theoretical predictions from the schematic eye models of Le Grand and El Hage[15] and Lotmar[31] which use spherical and aspherical surfaces respectively.

calculated for schematic eye models with spherical surfaces and a homogeneous lens. It would appear that for photopic pupils up to about 4 mm in diameter the aspheric optical surfaces and gradient-index lens of the human eye reduce near-axial aberrations to a low basic level, onto which are superimposed small perturbations due to the idiosyncratic asymmetries of the individual eye. For larger pupils (> about 4 mm) undercorrected spherical aberration tends to be present in the unaccommodated eye, but the amounts are relatively small (about 0.5 D for a fully dilated pupil) and may be masked by individually varying irregular aberrations.

Off-axis, while on average increasing amounts of oblique astigmatism are encountered (Fig. 10), these may show substantial variations with the individual and with the meridian of the eye studied, as may also the relation of the tangential and sagittal image shells to the retinal surface.[134–139] There is little systematic change in the mean oblique astigmatism with the axial refraction of the eye.[140] The general characteristics of ocular oblique astigmatism can be well simulated by a model eye with aspheric surfaces[31,141,142] (see Fig. 10), individual differences being explicable in terms of small translations, tilts, and other asymmetries of the components.[143] The same aspheric surfaces cannot simultaneously control both spherical aberration and astigmatism, so that the index gradients of the lens must also play a role in balancing the monochromatic aberrations of the eye.

Chromatic Aberration

Chromatic aberration arises from the dispersive nature of the ocular media, the refractive index, and hence the ocular power, being higher at shorter wavelengths. Constringence values for the ocular media are generally quoted[18,144] as being around 50, although there is evidence that these may need modification.[145,146] Both longitudinal or axial chromatic

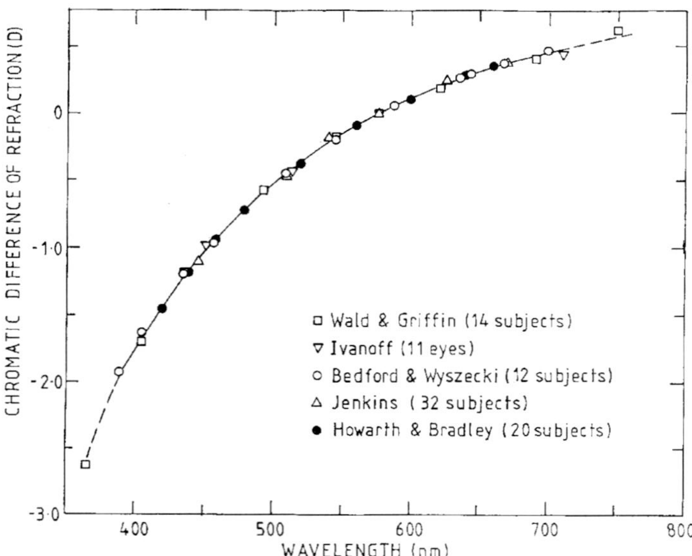

FIGURE 11 Average data for the longitudinal (axial) chromatic aberration of the eye as obtained by different authors. The data for different subjects have been displaced in power so that the chromatic aberration is always zero at 578 nm. □ Ivanoff;[150] △ Jenkins;[132] ○ Wald and Griffin;[149] ● Bedford and Wyszecki;[151] ▽ Howarth and Bradley.[152]

aberration (LCA) and transverse or lateral chromatic aberration (TCA) occur (see Refs. 147, 148 for reviews).

For LCA, what is normally measured experimentally is not the change in power of the eye across the spectrum but rather the change in its refractive error, or the *chromatic difference of refraction.*[18] All investigations[132,149–152] show only minor intersubject differences in LCA and the basic variation in ocular refraction with wavelength, equivalent to about 2 D of LCA across the visible spectrum, is well established (Fig. 11). LCA has little effect on visual acuity for high-contrast test objects in white light.[153] This appears to be because the spectral weighting introduced by the photopic luminosity curve, which is heavily biased to the central region of the spectrum, results in the primary effect of the LCA being to degrade modulation transfer at intermediate spatial frequencies rather than at the high spatial frequencies involved in the high-contrast acuity limit.[111]

TCA results in a wavelength-dependent change in the magnification of the retinal image, the red image being larger than the blue. Thus, in white light, the images of off-axis object points are drawn out into a short radial spectrum whose length in a paraxial model increases linearly with the field angle. TCA therefore affects modulation transfer for tangentially oriented gratings.

While for a centered optical system the TCA would be zero on the optical axis, in the eye the ~5 deg difference in orientation between the visual axis and the approximate optical axis (angle α) means that some TCA might be expected at the fovea.[154–156] Recent measurements suggest that the amount of foveal TCA varies widely between subjects but is of the order of 36 sec arc.[147,157,158] Although this value is less than is predicted by schematic eye models, it is large enough to cause some orientation-dependent image degradation. This may be substantially increased if an artificial pupil is used which for any reason becomes decentered: such a situation may arise when using visual instruments having exit pupils which are much smaller than the entrance pupil of the eye.

With binocular viewing, the TCA associated with small decentrations of the natural pupils and of the foveas from the optical axes leads to the phenomenon of chromostereopsis, whereby objects of different colors placed at the same distance may appear to be at different distances to an observer.[18,159,160] The exact effect varies with the individual and, when artificial pupils are used, with the separation of the pupils. It is thus of some practical significance in relation to the design of instruments such as binocular microscopes, in which interpupillary distance settings may not always be optimal for the observer.[161]

In the periphery, TCA may play a significant role in limiting the detection of tangential as opposed to radial gratings[156,162,163] but there is as yet no firm consensus as to its magnitude, although some measurements have been made.[157]

Intraocular Scattered Light and Related Effects

A variety of regular and irregular small-scale inhomogeneities exist within the eye which may serve to scatter light. Further stray light may arise from reflections from the various optical surfaces and the retina itself. Quantitative studies of the effects on vision of this light were pioneered by Holladay[164] and Stiles,[165] who expressed its effect for white light in terms of an equivalent veiling luminance, L_{eq} cd . m^{-2}, that would produce the same masking effect as a glare source giving an illuminance E at the eye, as function of the angular distance, ω deg, between the glare source and the fixation point. Vos et al.[166] have summarized more recent work by the approximate relationship:

$$L_{eq} = (29E)/(\omega + 0.13)^{2.8} \quad (0.15 \text{ deg} < \omega < 8 \text{ deg}) \tag{11}$$

This expression relates to young adult eyes. Scattering increases throughout life by a factor of at least 2–3 times[167–170] and glare formulae can be modified to take account of this (e.g., Ref. 171). Roughly a quarter of the stray light comes from the cornea[172,173] and a further quarter from the retinal reflections:[174,175] the rest comes almost entirely from the lens,[176] there being little contribution from the aqueous or vitreous humors.

Ohzu and Enoch[177] attempted to measure a retinal MTF which included the effects of forward scatter, by focusing grating images onto the anterior surface of an excised retina and measuring the image transfer to the far side. They argued that the receptor outer segments effectively act as a fiber-optics bundle and that transmission through this bundle produces image degradation which supplements that due to the main optical elements of the eye. Recent psychophysical measurements by McLeod et al.[178] suggest, however, that forward scatter in the inner retina is negligible, implying that postmortem changes in the retina may have degraded Ohzu and Enoch's MTFs.

24.6 FINAL RETINAL IMAGE QUALITY

Estimates of the final quality of the retinal image can be made in three main ways: by calculation from wavefront aberration or similar data; by a psychophysical method; and by direct measurement of the light distribution on the retina using a double-pass ophthalmoscopic technique. Although each method has its limitations, the various methods yield reasonably compatible results if allowance is made for intersubject differences, particularly at larger pupil diameters.

Image Quality on the Visual Axis

Calculation from Aberration Data. The OTF can be calculated by autocorrelation of the complex pupil function with its complex conjugate, using methods originally devised by Hopkins.[179] The pupil function gives the variation in amplitude and phase across the exit pupil of the system. The phase at each point can be deduced from the corresponding value of wavefront aberration (each wavelength of aberration corresponds to 2π of phase) and it may either be assumed that the amplitude across the pupil is uniform or, if desired, pupil weighting functions may be introduced to take account of the Stiles-Crawford effect (Stiles-Crawford apodisation[180]) or pupil-dependent lenticular absorption. The PSF and LSF can also be calculated from the wavefront aberration.[102,179,181,182]

This approach was first used with wavefront aberration data as deduced from Foucault knife-edge measurements, by Berny[183] to estimate ocular MTFs and by Berny and Slansky[120] to compare the retinal PSF with the ideal Airy diffraction pattern; these latter authors concluded that the eye was essentially diffraction-limited for pupil diameters up to about 2 mm. Van Meeteren[154] took a more indirect approach in that he estimated the typical wavefront aberration from published measurements of the individual Seidel aberrations and other sources of image degradation. He found that allowance for the Stiles-Crawford effect had little influence on the overall MTF when the pupil diameter was in the normal photopic range (<5 mm). In white light, chromatic aberration had the dominant degrading effect for photopic pupils.

The subjective aberroscope method of Howland and Howland[121,122] which involves asking subjects to sketch the appearance of a distorted grid image on the retina, allowed them to study a large sample of eyes. The resultant MTF data suggested strongly that substantial intersubject variations in MTF occurred for any pupil diameter. This was confirmed by later, objective measurements using the same basic technique.[123,124] Figure 12 shows examples of these variations in monochromatic MTFs for a 5-mm-diameter pupil: it will be observed that the optical performance of even the best eyes is well below the diffraction limit. The associated phase transfer functions[124,128] differ substantially from zero, as would be expected if asymmetric, coma-like aberrations occur on the visual axis: these phase shifts may play some role in limiting the visual system's ability to discriminate spatial phase.[128] Coma-like aberrations, which have also been found by ray-aberration methods,[184] would be expected to lead to an orientation-dependent MTF for larger pupils.[185]

Calculations of retinal image quality from wavefront aberration data fail to include the effects of scattered light and hence may give too optimistic a view of the final image quality. Van Meeteren[154] argued that it was appropriate to multiply the aberration-derived MTFs by the MTF derived by Ohzu and Enoch[177] for image transfer through the retina. When this is done the final estimated MTFs agree quite well with those found by the double-pass ophthalmoscopic technique,[123] although as noted earlier the Ohzu and Enoch MTF may overestimate the effects of retinal degradation.[178] A further problem with MTFs derived from aberroscope results is that the aberroscope method only allows the wavefront aberration to be estimated at a limited number of points across the pupil: small-scale variations in aberration are therefore not detected, with consequent uncertainties in the derived MTFs.

Psychophysical Comparison Method. This method depends upon the comparison of modulation (contrast) thresholds for a normally viewed series of gratings of differing spatial frequencies with those for gratings which are produced directly onto the retina by interference techniques.

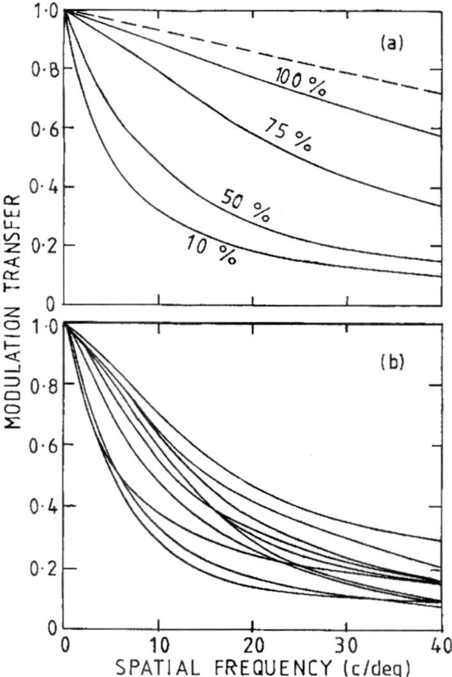

FIGURE 12 Ocular MTFs for vertical gratings and eyes with a 5-mm-diameter entrance pupil, as calculated from wavefront aberration data. (*a*) Rank ordered MTFs for a group of 55 eyes and light of wavelength 555 nm. The 100% curve represents the best eye of the set, the dashed curve being the MTF for a diffraction-limited eye (*after Howland and Howland*[122]). (b) MTFs for a sample of 10 eyes at a wavelength of 590 nm (*after Walsh and Charman*[124]). Note the substantial intersubject variations.

Suppose an observer directly views a sinusoidal grating of spatial frequency R. Then if the grating has modulation $M_o(R)$ the modulation of the retinal image will be $M_o(R) \cdot T(R)$ where $T(R)$ is the modulation transfer of the eye at this spatial frequency, under the wavelength and pupil diameter conditions in use. If now the modulation of the grating is steadily reduced until it appears to be just at threshold, the threshold modulation $M_{it}(R)$ on the retina will be given by

$$M_{it}(R) = M_{ot}(R) \cdot T(R)$$

where $M_{ot}(R)$ is the measured modulation of the external grating at threshold. The reciprocal of $M_{ot}(R)$ is the corresponding conventional contrast sensitivity and the measurement of $M_{ot}(R)$ as a function of R corresponds to the procedure used to establish the contrast sensitivity function

It is clear that $M_{it}(R)$ corresponds to the threshold for the retina/brain portion of the visual system. If it can be established independently, it will be possible to determine $T(R)$. $M_{it}(R)$ can in fact be measured by bypassing the dioptrics of the eye and forming a

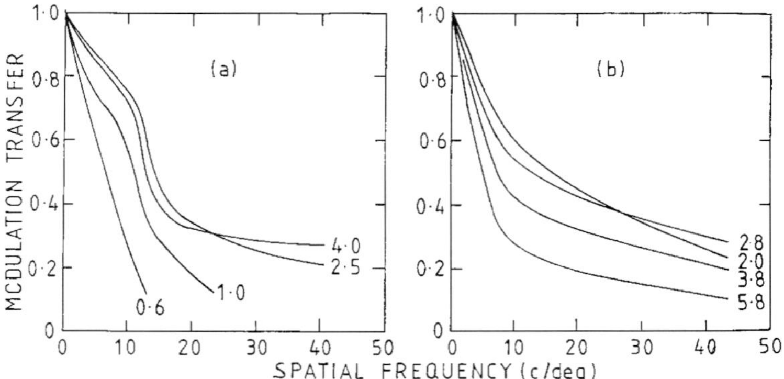

FIGURE 13 Ocular MTFs derived by the contrast threshold psychophysical method. The figures on the curves indicate the corresponding pupil diameters in mm (*a*) Mean of 3 subjects (*after Arnulf and Dupuy*[188]). (*b*) One subject (*after Campbell and Green*[110]).

system of interference fringes directly on the retina.[98,110,186–191] Two mutually coherent point sources are produced close to the nodal points of the eye and the two resultant divergent beams overlap on the retina to generate a system of Young's fringes whose angular separation, γ rads, is given by $\gamma = \lambda/a$, where λ is the wavelength and a is the source separation, both measured in air. If the sources have equal intensity, the fringes will nominally be of unit modulation; fringe modulation can be controlled either by varying the relative intensities of the two sources or by adding a controllable amount of uniform background light. The threshold contrast $M_{it}(R)$ for the retina/brain can then be determined as a function of R, allowing the modulation transfer for the ocular dioptrics to be deduced from the relation:

$$T(R) = M_{it}(R)/M_{ot}(R)$$

Figure 13 gives some typical results for the MTF as a function of pupil size, as found by this method. The results are very similar to those found from wavefront aberration data. Bour[191] has extended the method to cover through-focus effects and finds that irregular aberration may lead to more than one position of optimal focus.

Both sets of thresholds in the psychophysical method are equally affected by any retinal contribution to image degradation and hence, like the wavefront-derived results, the MTFs do not include such effects, although these may be small.[178] The other major criticism of the method is the assumption that the modulation of the interference fringes is totally unaffected by the ocular media. Ocular scatter may lower the contrast of the retinal interference fringes, particular in older eyes. This would lead to an overestimation of the MTF (but see again Ref. 178). In practice, too, subjects may find it difficult to maintain constant criteria in judging internal and external contrast thresholds since the grating systems may differ in color, field size, and appearance due to the speckle associated with the use of coherent light to generate the fringes.

Ophthalmoscopic Methods. When the image of an object is thrown onto the retina, some of the light will be reflected back out of the eye and can be collected by an appropriate observing system to form an external image. If, for example, the object is a narrow line, the external image will be the LSF for the double passage of the eye. It is usual to make the important assumption that the retina acts as a diffuse reflector,[192] coherence of the image light being lost in the reflection. In this case, the external image has suffered two identical stages of image degradation, so that the MTF deduced from the Fourier transform of the external LSF is the square of the single-pass ocular MTF. The possibility that the inward and outward MTFs might differ due to the differing scattering geometries is ignored.

Flamant's pioneering study[193] used photography to record the external LSF, but later authors have all used photoelectronic recording.[192,194–206] Although earlier work concentrated on measurement of the external LSF, more recently Santamaria and his colleagues[203–206] have used a television camera and image analysis system to record the PSF. This has the advantage that it is possible to deduce the LSF and OTF for any orientation, although recording of the low-illuminance outer parts of the PSF may be a problem, leading to truncation errors and to overestimation of modulation transfer.[207] Vos et al.[166] have attempted to overcome the truncation problem by combining ophthalmoscopic estimates of the retinal PSF with estimates of wider-angle entoptic stray light, to produce a realistic estimate of light profiles in the foveal image of a white light point source. A vexing question which has yet to be fully answered is the identity of the retinal layer or layers at which the retinal reflection occurs: it seems likely that this is wavelength-dependent.[114,208] If more than one layer is involved, the estimated MTF will be somewhat too low.

All investigators again agree that near the visual axis the eye performs close to the diffraction limit for pupil diameters up to about 2 mm. As the pupil size is increased further, aberration starts to play a more important role, overall performance usually being optimal for pupil diameters of about 2.5 to 3 mm. Thereafter the degrading effects of aberration start to dominate and modulation transfer falls. The changing balance between the effects of diffraction and aberration can be clearly seen in the pupil-dependence of the LSF (Fig. 14).

For calculation purposes, the ocular axial MTFs can be approximated reasonably well by an expression of the form

$$T(R) = \exp\left[-(R/R_c)^n\right] \tag{12}$$

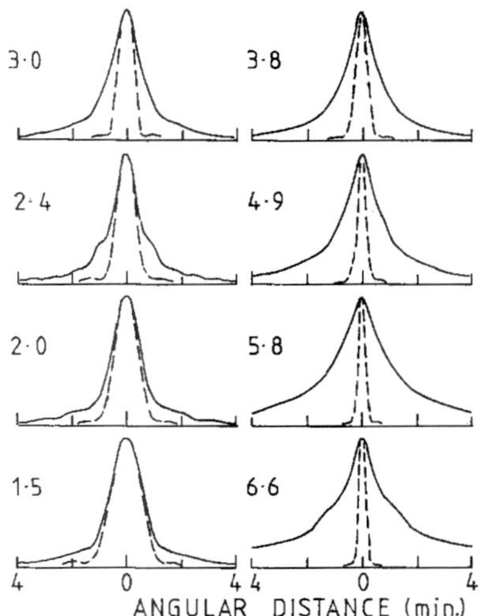

FIGURE 14 Retinal line-spread functions for the different pupil diameters (mm) indicated. In each case the dashed curve gives the corresponding LSF for a diffraction-limited eye (*after Campbell and Gubisch*[192]).

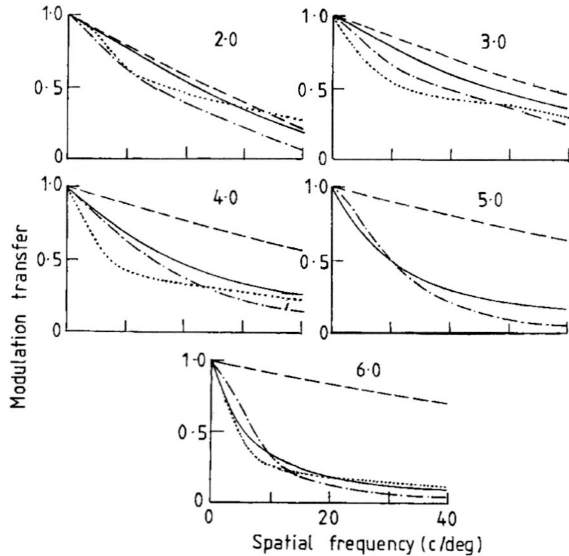

FIGURE 15 Comparison of MTF results for young adult subjects at the pupil diameters (mm) indicated, as found by three different methods: — From wave aberration,[124] mean of 10 subjects, 590 nm. ···· By psychophysical comparison,[110] one subject (pupil diameters are 2.0, 2.8, 3.8, and 5.8 mm). -·-·-· By double-pass ophthalmoscopic method[286] at 633 nm ---- Diffraction-limited 555 nm.

where, for each pupil size, R_c is some constant spatial frequency and n is a further constant.[209,210]

Comparison of MTF Results by Different Methods. Figure 15 compares typical axial MTF results for various pupil diameters as obtained from the wave aberration,[124] psychophysical comparison,[110] and double-pass ophthalmoscopic[206] methods: the corresponding theoretical diffraction-limited MTFs are also shown. When allowance is made for the slightly different wavelength and other conditions used by the different authors, and for the possibility of intersubject differences in the MTFs, there is quite good agreement between the results derived by the different methods. This suggests that the uncertainties associated with each individual method can have only minor influences on the derived MTFs.

Off-Axis Image Quality

Only a few investigators have attempted to measure retinal image quality away from the fovea, using the ophthalmoscopic method. On the visual axis it is relatively straightforward to use a cycloplegic drug to paralyze the accommodation and dilate the natural pupil, and then to use artificial pupils to study the pupil-dependence of the MTF. This is, however, difficult to carry out satisfactorily off-axis, since the approximately elliptical geometry of the natural pupil changes with field angle and it is difficult to mimic this with artificial pupils or to ensure that these are correctly located with respect to a dilated natural pupil. Some investigators have therefore simply used a cycloplegic to paralyze the

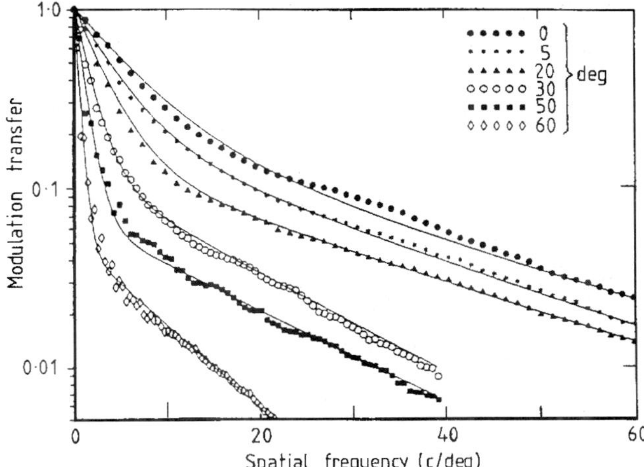

FIGURE 16 MTF, averaged over all orientations, as a function of field angle. Natural accommodation was used and the pupil diameter is approximately 4 mm (*after Navarro et al.*[211]). Note the steady decline in MTF as the field angle increases.

accommodation and keep the pupil at a constant large diameter.[199,200] This has the disadvantage that aberrations are then at their highest but, since supplementary lenses can be used to vary the retinal focus, it allows the effects of oblique astigmatism to be studied. Other authors have used a natural pupil and free accommodation and have hoped that pupil diameter and focus would remain approximately constant during the course of the experiments.[199,211] Both types of study show that optical performance tends to decline steadily with field angle, due to the effects of ocular oblique astigmatism, although it is relatively constant over the central ±15–20 deg of the visual field. Figure 16 shows data[211] for natural observing conditions at the field angles indicated.

It is of interest to know whether this falloff in optical performance with field angle is responsible for the well-known decline in acuity and other aspects of overall visual performance in the peripheral field, or whether this is largely a neural effect. Early measurements comparing cutoff frequencies for the detection of gratings when natural viewing and interference fringes which bypassed the ocular dioptrics were used gave somewhat conflicting results.[212–214] More recent work by Thibos et al.[163] suggests that peripheral grating resolution thresholds are not markedly improved when the degrading effects of the eye's optics are bypassed and that aliasing occurs under normal viewing conditions with grating objects beyond the resolution limit, so that optical image quality must be in excess of the needs of the neural sampling array. Thus optical degradation can play only a minor role in setting the limits to peripheral performance.

Retinal Image Quality with Visual Instruments

When an object (for example, a visual display unit) is observed by the naked eye with no intervening optics, the spatial frequency spectrum of the retinal image is simply the spectrum of the object multiplied by the OTF of the eye under the observing conditions in use. The situation is, however, more complex when instruments such as microscopes, telescopes, or binoculars are used. In general, it is incorrect to assume that the OTF of the instrument-eye combination is simply the product of the individual OTFs. Instead, for any field position the aberrated wavefront from the instrument is modified by the wavefront aberrations of the eye, the two wavefront aberrations being added together point by point across the common pupil. Under particularly favorable circumstances, this *coherent coupling*[215–218] can obviously result in the summed aberration being smaller than the individual wavefront aberrations and hence in performance being superior to that which would be expected on the basis of the product of the individual OTFs.

Two factors are of particular importance in this coupling: first, the size relationship between the instrument exit pupil and the ocular entrance pupil and, second, the state of focus of the eye. It is clear that if the entrance pupil of the eye is smaller than the exit pupil of the instrument (the two pupils coinciding), the eye pupil will constitute the aperture stop of the combined system. Movement of this smaller eye pupil within the instrument's exit pupil will sample different areas of the instrumental wavefront aberration and may therefore influence retinal image quality. Conversely, if the exit pupil is smaller than the eye pupil, the instrument's pupil provides the aperture stop and controls the aberrations. For example, with an exit pupil of 2-mm diameter or less centered to the eye pupil, the combined performance would be governed almost entirely by the instrumental aberrations, since the eye would make a negligible contribution to the combined wavefront aberration. Any focus error in the instrument or accommodation error in the observer will add an additional defocus term to the wavefront aberration. This may again be beneficial under some circumstances; for example, instrumental field curvature may be compensated for by changes in ocular accommodation (Refs. 219, 220, and see later under "Vergence Input").

Considerations of this type mean that the MTF of any instrument alone may not be a very good guide to the visual performance that can be achieved with it.[221] One possible solution is to test the instrument in combination with a model eye which simulates as accurately as possible the aberrations and focus of the real eye.[222] Burton and Haig[223] have explored the tolerance of the human visual system to different levels and types of wavefront aberration and conclude that instrumental assessment criteria based on the Strehl intensity ratio may ultimately be more useful than MTF tests. The role of longitudinal chromatic aberration has been considered by Mouroulis and Woo,[224] who find that under some circumstances quite substantial amounts of instrumental LCA can be tolerated.

24.7 DEPTH-OF-FOCUS AND ACCOMMODATION

In the foregoing it has been tacitly assumed that the eye is always optimally focused. In practice, since clear vision is required for objects over a range of distances, exact focus may not always be achieved. It is of interest that, even if objects are at a fixed distance, small changes in focus may be required to optimize modulation transfer at different spatial frequencies in the presence of some types of aberration (e.g., spherical aberration[114,225–227]).

Ocular Depth-of-Focus

As in any optical system, very small changes in ocular focus have little effect on the retinal image but image quality deteriorates progressively as the error of focus increases. In the geometrical optical approximation, for an aberration-free eye the angular diameter of the retinal blur circle β degrees increases linearly with the error of focus ΔF diopters and pupil diameter d mm [Eq. (9)]. Thus the limits of the depth-of-focus correspond to the blur circle diameter reaching some assumed tolerable limit β_{tol} degrees: the corresponding value of ΔF_{tol} is then determined from Eq. (9) to yield for the total geometrical depth-of-focus DOF_{go}

$$DOF_{go} = 2\,\Delta F_{tol} = (34.9\beta_{tol})/d \quad \text{diopters} \tag{13}$$

When the effects of physical optics are considered, for a diffraction-limited system, it is conventional to use the Rayleigh criterion and say that the limits of depth-of-focus are set by the requirement that the optical path difference between light from the center and edge of the pupil should not exceed a quarter-wavelength [one Rayleigh unit of defocus, Eq. (8)]. For the eye this implies that the total physical optical depth-of-focus (DOF_{po}) should correspond to 2 Rayleigh units of defocus, i.e.,

$$DOF_{po} = 4.10^{-3}\lambda/d^2 \quad \text{diopters} \tag{14}$$

where d is in mm and λ is in nm. Optimal focus will lie midway through this total depth-of-focus. Unlike the geometrical approximation of Eq. (13), Eq. (14) predicts that depth-of-focus will be inversely proportional to the square of the pupil diameter.

In reality, the ocular depth-of-focus depends upon a variety of additional factors. From the purely optical point of view, even if the eye is diffraction-limited, the loss in modulation transfer with defocus is spatial-frequency-dependent (Fig. 7), so that the detectable error of focus is likely to depend upon the spatial frequency content of the object under observation as well as the pupil diameter. Low spatial frequencies are relatively insensitive to focus change.[107,108,226] Both monochromatic and chromatic aberration will further modify the effect of any focus change.

Equally importantly, the perceptible focus changes will depend upon the neural characteristics of the visual system. Under many conditions, the limited capabilities of the retina/brain system will mean that defocus tolerance may become larger than would be expected on purely optical grounds. With large pupils and photopic conditions the Stiles-Crawford effect may play a role in reducing the effective pupil diameter and hence increasing the depth-of-focus. At low luminances, since only low spatial frequency information can be perceived,[228] there is an increase in tolerance to defocus blur[229,230]; a similarly increased tolerance is found in low-vision patients at photopic levels.[108]

Figure 17 shows a selection of experimental depth-of-focus data obtained by a variety of different techniques.[229,231–233] Although the exact results depend upon the methods and criteria used, it is clear that for larger pupils the visual depth-of-focus substantially exceeds the purely optical Rayleigh limit predictions. More importantly from the practical viewpoint, errors of focus in excess of about 0.2–0.5 diopters are likely to lead to perceptible image blur under photopic conditions with pupils of diameter 3–4 mm. Thus the eye must either change its focus (accommodate) for the range of distances that are of interest in everyday life (e.g., from infinity down to about 0.2 m or less, or vergences from 0 to −5 diopters), or use some form of optical aid such as reading glasses.

The Accommodation Response

As with any focusing system, three aspects of accommodation are of interest: its speed, its stability, and its time-averaged steady-state characteristics. All of these are age-dependent. Discussion of the response is complicated by the fact that, under normal binocular conditions of observation, accommodation (the focusing system) is intimately linked with the vergence system which ensures that the eyes converge appropriately to bring the images of any object of regard onto the foveas of both eyes (see under "Vergence Input" and "Movements of the Eyes"). Due to this linkage, accommodation can drive convergence and vice versa. There are also links to the pupil system, so that the pupil

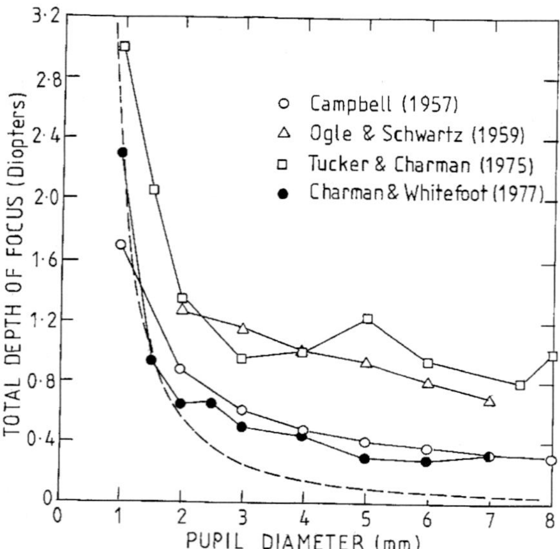

FIGURE 17 Examples of experimental measurements of photopic, total monocular depth-of-focus as a function of pupil diameter (optimal focus lies midway through the total depth-of-focus). ○ Campbell[229] based on the just perceptible blur for a small disc, one subject, white light. △ Ogle and Schwartz[231] based on a 50% probability of resolving a 20/25 checkerboard, mean of 3 subjects, white light. □ Tucker and Charman[232] based on 80% probability of achieving 90% of the optimal Snellen acuity, mean of 2 subjects, white light. ● Charman and Whitefoot[233] based on the detectable movement of laser speckles, mean of 6 subjects, 633 nm. The dashed line gives the depth-of-focus based on Rayleigh's quarter-wavelength criterion for an aberration-free eye in monochromatic light of wavelength 555 nm.

generally contracts slightly during near vision (*accommodative miosis*[234]), although the view that accommodation alone necessarily stimulates a pupil change has recently been challenged.[235] The three functions (accommodation, vergence, pupil) are sometimes known as the near triad. One important result of this linkage is that the accommodation responses in the two eyes are always essentially the same in both dynamic and static aspects, even when one eye is occluded.[236]

Although details of the physiological mechanisms responsible for accommodation and related functions are beyond the scope of this section (see, e.g., Refs. 237, 238 for reviews), it will be helpful to remember that, in simple terms, the crystalline lens is supported around its equator by a system of approximately radially oriented zonular fibers. The far ends of these fibers are anchored in the ciliary body which effectively forms a ring surrounding the lens (see Fig. 1). The lens and its enclosing capsule are elastic and, if the lens is free of the forces applied by the zonular fibers, the lens surfaces will naturally assume the relatively steep curvatures required for near vision.

Under conditions of distant viewing the diameter of the ciliary ring is relatively large, this leading to a correspondingly high tension in the zonular fibers. These forces when applied to the periphery of the elastic lens and its capsule cause the lens surfaces to flatten slightly, reducing its optical power. During active accommodation for near vision the

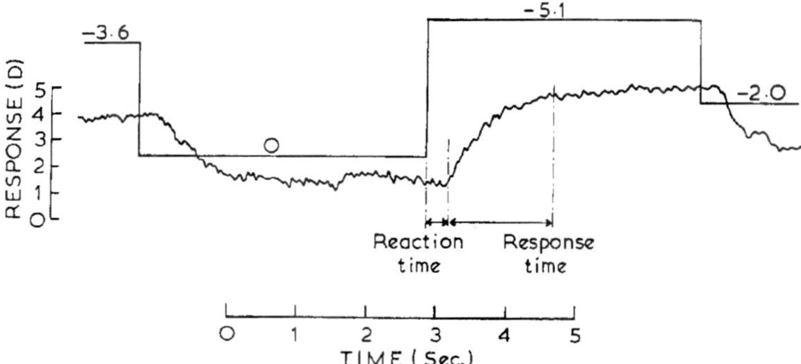

FIGURE 18 Typical dynamic accommodation response (lower trace) to a change in stimulus (upper trace), showing reaction and response times and the fluctuations in accommodation.

ciliary ring reduces in diameter, relaxing the tension in the zonular fibers and allowing the surfaces of the elastic lens to assume a steeper curvature and the power of the lens to increase. Attempts have been made to model this process.[239,240]

Dynamics of Response. A typical record showing a young adult's monocular accommodation response to a change in the position of an accommodation target is illustrated in Fig. 18. It can be seen that there is a short reaction time, or latency (about 0.4 sec), following the change in target position, during which the response does not change. The response then moves to a new level, the minimum response time typically being around 0.6 sec.[241–248] Under conditions where there are a variety of cues to target distance this response usually occurs in a single sweep, but if binocular and other cues to target distance apart from blur are eliminated, the response may become uncertain and may initially be in the wrong direction, leading to longer response times. The response times become larger for larger dioptric changes.[247,248]

One way of characterizing the dynamics of the response is in terms of its frequency response characteristics. These may be assessed by determining the gain and phase of the response obtained when a target is observed whose vergence is changing sinusoidally with time, as a function of temporal frequency.[249] As the temporal frequency increases, the gain diminishes and the phase lags increase, the system being unable to follow vergence changes above a few Hertz (Fig. 19). It should, however, be stressed that these characteristics are not the output of a simple reflex system but depend upon higher-order involvement. They are strongly influenced by training[246] and motivation and, with such

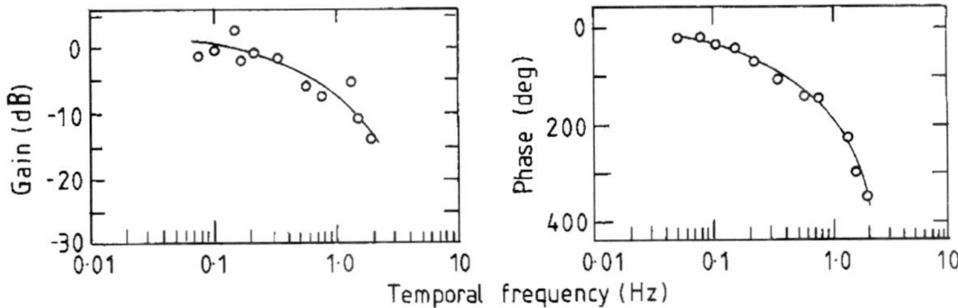

FIGURE 19 Typical gain and phase plots as a function of temporal frequency for accommodation to predictable stimuli having a sinusoidal change in vergence (*after Krishnan et al.*[250]).

repetitive stimuli, by the knowledge that the required response has a predictable periodic form. When the response to an abrupt, unexpected, step change in target vergence is analyzed in terms of the corresponding frequency response, much larger phase lags are given.[250,251]

The importance of perceptual factors in relation to the accommodation response is exemplified by studies in which the distance of the target (a Maltese cross) was kept constant but its lateral scale was varied sinusoidally. Observers tended to interpret this size variation at constant distance as a variation in distance of a target of constant size, and hence changed their accommodation accordingly, even though this blurred the retinal image.[252–254]

Stability of the Response. When a target at a fixed distance is viewed with steady fixation the accommodation response typically shows small fluctuations ~0.25 D occurring at frequencies up to about 5 Hz; the fluctuations are correlated in the two eyes (see Refs. 238, 255 for reviews). The magnitude of the fluctuations tends to increase as the target under observation approaches the eye.[256] A pronounced peak in the frequency spectrum of these fluctuations is often observed[257,258] at frequencies of 1–2 Hz; it appears that this is correlated with the arterial pulse.[259]

The possible role of these fluctuations in accommodation control remains contentious. Some have argued that they simply represent plant noise[249,260] and are of no utility. Others suggest that they could both guide the direction of the initial response and help to maintain accurate focus,[113,258,261–263] the basic hypothesis being that if a fluctuation in one direction improves the clarity of the image the control system responds by moving the mean level of accommodation in that direction. An intermediate view is that the higher-frequency components represent plant noise but the lower-frequency components (below 0.5 Hz) contribute to the control system.[255] In any case, under most circumstances the fluctuations appear to produce little degradation in visual acuity[264] although their effects can just be detected.[265,266]

Accuracy of Response. Following the pioneering work of Morgan,[267] numerous studies have shown that rather than there being one-to-one matching between the dioptric stimulus level and the corresponding accommodation response, steady-state errors are an intrinsic part of the accommodation control system. This is true under both monocular and binocular conditions, although the magnitude of the errors differs under the two states.[267,268] These focus errors are probably the major cause of foveal retinal image degradation under many conditions, rather than the aberrations discussed earlier under "Factors Affecting Retinal Image Quality." Figure 20*a* illustrates a schematic accommodation response/stimulus curve and emphasizes that the steady-state response is characterized by overaccommodation for far targets and underaccommodation for near targets. The range of stimulus vergence over which there is no noticeable image blur is termed the *subjective amplitude of accommodation*; it obviously includes depth-of-focus effects. The corresponding somewhat smaller range of actual accommodation response is termed the *objective amplitude of accommodation.*

The slope of the quasi-linear central region of the curve depends upon the observing conditions, including target form and contrast,[112,113,269–278] ocular pupil diameter,[279–281] luminance level,[230,269,282,283] and the acuity of the observer.[270,276,284–286] The common feature is that as the quality of the stimulus available to the visual system degrades, the slope of the central region of the response/stimulus curve diminishes (see Fig. 20*b* to *d* for examples). To a close approximation, for any individual as the stimulus degrades the curve pivots about the point *P* (Fig. 20*a*) at which the response and stimulus are equal, i.e., where the curve crosses the ideal one-to-one response/stimulus line.[287] It is of interest that in many studies it appears that there is a linear relationship between the slope of the central region of the response/stimulus curve and the minimum angle of resolution under the target and observing conditions in use.[288]

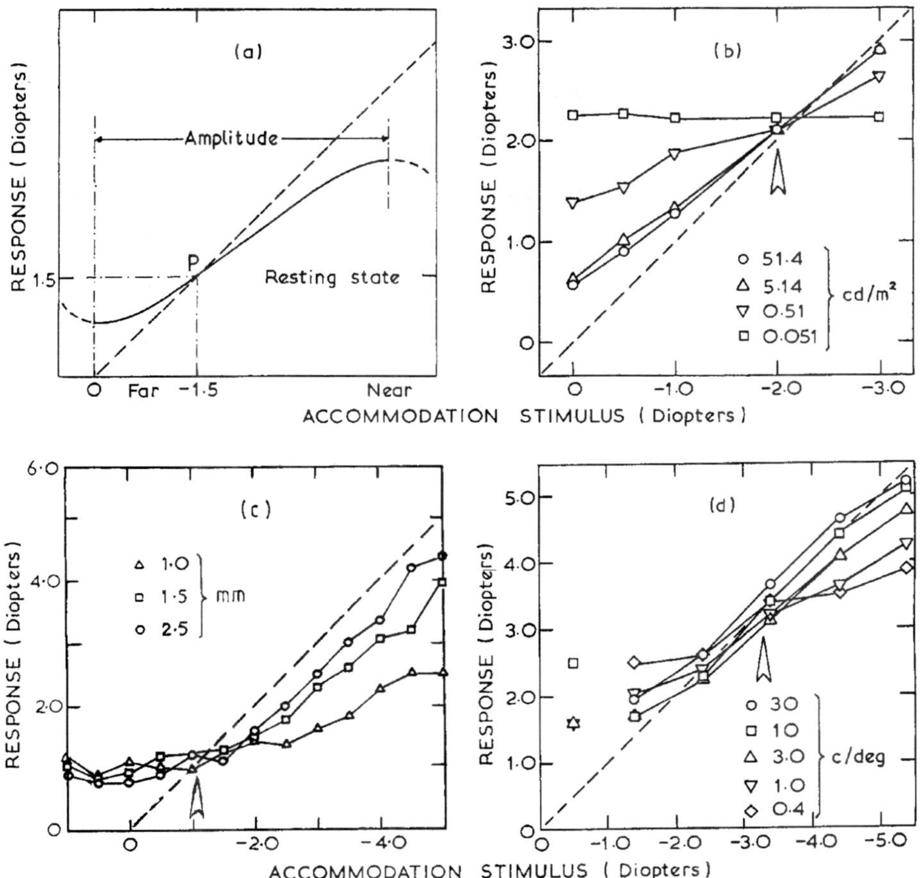

FIGURE 20 (*a*) Schematic static accommodation response/stimulus curve under photopic conditions. The dashed straight line indicates the ideal one-to-one response. The two intersect at the point P corresponding to the level of tonic accommodation (about 1.5 D). (*b*) Effect of luminance change on the lower part of the stimulus/response curve (*after Johnson*[283]): the curve collapses to the tonic level at scotopic target luminances. (*c*) Effect of change in pupil diameter at constant retinal illuminance (*after Ward and Charman*[281]). With small pupils, large depth-of-focus allows large errors in accommodation to be tolerated. (*d*) Effect of spatial frequency for sinusoidal grating targets (*after Charman and Tucker*[112]). Responses tend to be less accurate for targets of low spatial frequency. In (*b*), (*c*), and (*d*) the vertical arrow indicates the tonic accommodation level.

The extreme case in this process is where the stimulus is either a uniform photopic field, completely lacking in detail (a Ganzfeld) or the field is completely dark. In both cases no spatial information is available to the accommodation control system and the accommodation response/stimulus curve becomes completely flat. The response remains constant at the somewhat myopic value at which, under normal photopic conditions with a structured target, response equaled stimulus (i.e., P in Fig. 20*a*). The refractive states under these stimulus-free light and dark conditions are known as *empty field myopia* and *dark focus*, respectively, and for any individual these states are well correlated.[289–292]

These observations have led to the concept that this intermediate myopic level of

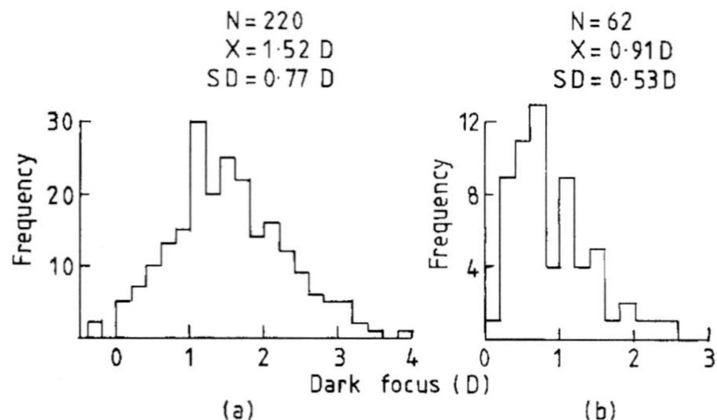

FIGURE 21 Frequency distribution of tonic accommodation as estimated by the dark focus: (*a*) laser optometer measurements for nominally emmetropic subjects in darkness (*after Leibowitz and Owens*[291]); (*b*) difference in autorefractor measurements of refraction under photopic conditions and darkness (*after McBrien and Millodot*[293]).

accommodation is, in fact, the *tonic* level of accommodation (resting state, equilibrium level) to which the system returns in the absence of an adequate stimulus, and that accommodation must actively change from this level to view both distant and near targets. This tonic level varies somewhat according to the observer (Fig. 21) but appears to have a mean value of around 1 D.[291,293]

Current theories suggest that the characteristics of the system are dictated by the balance between the sympathetic and parasympathetic innervation to the system.[238,294,295]

Vergence Input. As noted earlier, the vergence required to maintain single vision of an object during binocular observation provides an input to the control system which supplements that are due to pure defocus blur. This may be particularly useful under conditions where the accommodation stimulus alone is relatively ineffective, for example, under mesopic conditions where the onset of *night* (*twilight*) *myopia* is delayed when viewing is binocular.[296]

Control system models for the combined accommodation-vergence system involving both fast feedback loops and slow adaptive elements have been proposed by various authors (see, e.g., Refs. 249, 294, 297–301).

Application to Instrumentation. From the practical point of view, it is evident that errors in accommodation and the resultant degradation in retinal image quality will be minimized if targets such as displays which are viewed by the naked eyes are placed at vergences which approximately match the typical tonic levels of observers, i.e., at distances of about 1 m. Acuity will then be optimal[283] and the accommodative effort and, hence, potential fatigue will be minimized. With color displays, the longitudinal chromatic aberration of the eye means that more accommodative effort is required for red/black symbol combinations than for blue/black combinations.[302] Only minor variations in response have, however, been reported when two-color isoluminant symbol/background combinations are used.[303]

It is of interest that when focusing visual instruments most observers prefer to set the focus so that the light leaving the eyepiece is slightly divergent, i.e., they prefer to accommodate slightly when viewing the instrumental image. This *instrument myopia* correlates closely with the empty field myopia and the dark focus of the individual observer[289–292] suggesting that these myopic states have a common origin and that the instrument focus is selected to minimize accommodative effort. This implies that fixed-focus visual instruments should not be designed to have parallel light leaving their eyepieces but that focus should be set so that the image appears at a vergence of about −1 diopter.[304] In instruments where the imagery is at infinity and the optical axes of the eyepieces are arranged to be parallel, *proximal* (*psychic*) *convergence and accommodation* often occur, that is, most young observers tend to converge and accommodate slightly, leading to a loss in visual performance.[305,306] This is presumed to be due to the perception that the targets lie within the "black box" constituted by the instrument. Smith et al.[307] have discussed the various problems that arise when the focus of binocular instruments is not compatible with the angle between the eyepiece tubes.

Many visual instruments display field curvature and, provided that the vergences involved are negative, this can be at least partly compensated for by accommodation by the observer as the field is scanned.[216,219,220]

Age-Dependent Effects in Accommodation. As the lens ages and thickens, its elastic constants change.[308–310] This, in combination with other ocular changes,[311,312] causes the efficiency of the accommodation system to diminish. Data on dynamic changes are rather sparse, although there is no doubt that response times increase with age.[313,314] The changes in the steady-state response/stimulus curve have, however, been studied in some detail.

The major effect (Fig. 22) is that the subjective amplitude of accommodation declines steadily with age through adulthood, falling to a small constant level by the age of around 50–55:[315] this residual amplitude represents depth-of-focus rather than true accommodation.[7,316,317] It appears that the level of tonic accommodation also declines to

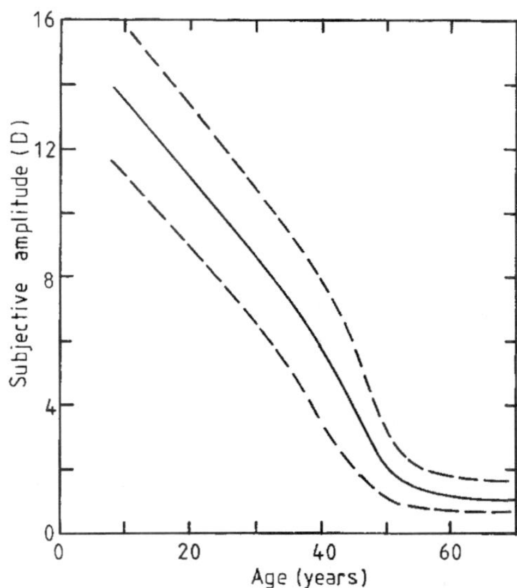

FIGURE 22 Changes in the subjective amplitude of accommodation with age (*after Duane*[315]): the dashed curves indicate the limits of the range found at each age. The data refer to 4200 eyes, amplitudes being measured at the spectacle point 14 mm anterior to the cornea.

zero with age, as would of course be expected,[318] and there is a progressive age-dependent collapse in the response-simulus curve as presbyopia is approached.[317]

The loss in amplitude becomes of major practical significance when the amplitude declines toward that required for normal reading distances (around 1/3 m). It is tiring to exercise the full amplitude of accommodation, so that problems usually start to arise in the early forties at amplitude levels of about 5 diopters. These difficulties steadily increase as the amplitude declines further. Initially, increases in reading distance may help to ease the problems by reducing the stimulus demand but by the age of about 45 all emmetropes or corrected ametropes inevitably require some form of optical correction for close work.

24.8 MOVEMENTS OF THE EYES

Movements of the eyes are of significance in relation to visual optics from two major points of view. First, since both optical and neural performances are optimal on the visual axis, the eye movement system must be capable of rapidly directing the eyes so that the images of the detail of interest fall on the central foveas of both eyes where visual acuity is highest (*gaze shifting* leading to *fixation*). A scene is explored through a series of such fixational movements for different points within the field. Second, the system must be capable of maintaining the images on the two foveas both when the object is fixed in space (*gaze holding*) and, ideally, when it is moving. Any lateral movement of the images with respect to the retina is likely to result in degraded visual performance, due to the limited temporal resolution of the visual system and the fall-off in acuity with distance from the central fovea.

These challenges to the eye movement control system are further complicated by the fact that the eyes are mounted in what is, in general, a moving rather than a stationary head. Movements of the eyes therefore need to be linked to information derived from the vestibular system or labyrinth of the inner ear, which signals rotational and translation accelerations of the head. The compensatory *vestibulo-ocular responses* take place automatically (i.e., they are reflex movements) whereas the fixational changes required to foveate a new object point are voluntary responses. Details of the subtle physiological mechanism which have evolved to meet these requirements will be found elsewhere.[319–323]

Basics

Each eye is moved in its orbit by the action of the three pairs of extraocular muscles attached to the outside of the globe. Their action rotates the eye about an approximate center of rotation lying some 13.5 mm behind the cornea, although there is in fact no single fixed point within the eye or orbit around which the eye can rotate to all the possible positions that it can assume.[319,321] In particular, the "center of rotation" for vertical movements lies about 2 mm nearer the cornea than that for horizontal movements (mean values 12.3 and 14.9 mm, respectively[324]).

Although the two eyes can scan a field extending about 45 deg in all directions from the straight-ahead or *primary* position, in practice eye movements rarely exceed about 20 deg, fixation on more peripheral objects usually being achieved by a combination of head and eye movements.

If the angle between the two visual axes does not change during the movement, the latter is described as a *version* (or conjugate) movement. However, the lateral separation of the eyes in the head implies the need for an additional class of movements to cope with the differing convergence requirements of objects at different distances. These movements involving a change in the angle between the visual axes are termed *vergence* (or disjunctive) movements. Fixational changes may in general involve both types of movement, which appear to be under independent neurological control.[325]

Characteristics of the Movements

The version movements involved in bringing a new part of the visual field onto the fovea (saccades) are very rapid, being completed in around 100 msec with angular velocities often reaching more than 700 deg/sec, depending upon the amplitude of the movement[326,327] (see Fig. 23*a*): the saccade latency is about 200 msec.[328] Interestingly, it appears that during the saccade, when the image is moving very rapidly across the retina, vision is largely, although not completely, suppressed. This *saccadic suppression* (or perhaps, more properly, saccadic attenuation) results in, for example, the retinal thresholds for brief flashes of light being elevated, the elevation commencing some 30–40 msec before the actual saccadic movement starts. The subject is normally unaware of this temporary impairment of vision and the exact mechanisms responsible for it remain controversial[321] although an explanation may lie in the masking effect of the clear images available before and after the saccade.[329]

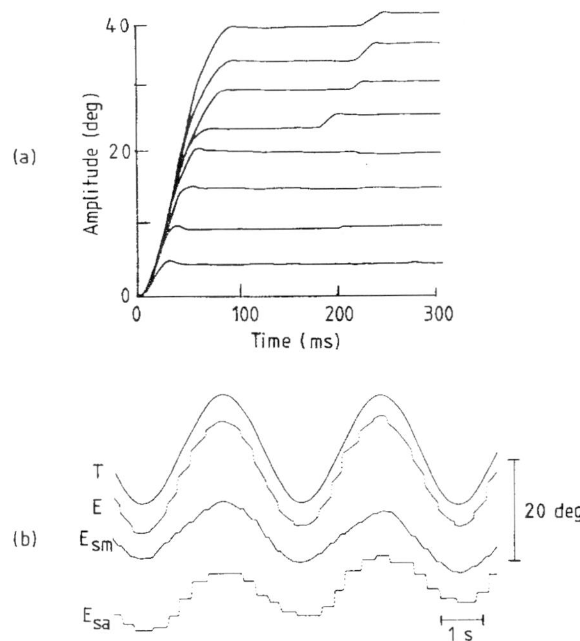

FIGURE 23 (*a*) Time course of saccades of different sizes (*after Robinson*[326]). The traces have been superimposed so that the beginning of each saccade occurs at time zero. (*b*) Separation of smooth (E_{sm}) and saccadic (E_{sa}) components of foveal tracking (*after Collewijn and Tamminga*[330]): T is the target position and E is the eye position.

Smooth voluntary pursuit movements of small targets moving sinusoidally in a horizontal direction are accurate at temporal frequencies up to a few hertz. Their peak velocities range up to about 25 deg/sec. In practice, when a small moving target is tracked, the following movements usually consist of a mixture of smooth movements and additional corrective saccades (e.g., Ref. 330; Fig. 23*b*). With repetitive or other predictable stimuli, tracking accuracy tends to improve markedly with experience, largely through the reduction in the phase lag between target and eye.

When the image of a substantial part of the visual field moves uniformly across the retina, the eyes tend to rotate to approximately stabilize the image in a following movement until eventually the gaze is carried too far from the primary position, when a rapid anticompensatory flick-back occurs. Thus a continuously moving visual scene, such as the view from a railway carriage, causes a series of slow following movements and fast recoveries, so that a record of the eye movements takes a quasi-regular sawtooth form. At slow object field speeds, the velocity of the slow phase is close to that of the field, but as the field velocity rises, greater lags in eye velocity occur until at about 100 deg/sec the following movements break down. Although this *optokinetic nystagmus* is basically a reflex, it is influenced by instructions given to the subject.

The angular speeds of vergence eye movements (about 5–10 deg/sec per deg of vergence[331]) are usually described as being much lower than those for version movements. Recent studies involving more natural 3-D objects suggest, however, that it is possible for the speeds to be much higher (up to 200 deg/sec for 35 deg vergence movements[332,333]). Vergence movements typically follow a quasi-exponential course with a time constant of about 300 msec. Their latency is about 200 msec. Although the primary stimulus for vergence movements is disparity, or the difference in position of the images in the two eyes with respect to their foveas, vergence movements can also be driven by accommodation and by perceptual factors such as perspective in line drawings.[334]

The vergence system is primarily designed to cope with the differences in the directions of the visual axes of the two eyes arising from their horizontal separation in the head. However, some vertical vergence movements and relative torsional movements about the approximate direction of the visual axis can also occur. Such fusional movements can compensate for corresponding small relative angular misalignments in the eyes or in the alignment of the optical systems of binocular or biocular instruments. The maximum amplitude of these movements is normally small (about 1 deg for vertical movements and a few deg for cyclofusional movements; see, e.g., Ref. 335) and they also take longer to complete than horizontal vergence movements (some 8–10 sec as compared to about 1 sec). Due to the limited effectiveness of vertical and torsional vergence movements, it is typically recommended that in instruments where the two eyes view an image through separate optical systems a 10 min arc tolerance be set for vertical alignment and a 30 min arc limit be set for rotational differences between the two images.

Stability of Fixation

When an observer attempts to fixate steadily on a fixed target, it is found that the eyes are not steady but that a variety of small amplitude eye movements occur. These *miniature eye movements* can be broken down into three basic components: tremor, drift, and microsaccades. The frequency spectrum of the tremor falls essentially monotically with frequency above about 10 Hz, extending up to about 200 Hz.[336] The amplitude is small (probably less than the angle subtended at the nodal point of the eye by the diameter of the smallest cones, i.e., 24 sec arc[321]). Drift movements are much larger and slower, with amplitudes of about 2–5 min arc at velocities of about 4 min/sec.[337] The errors in fixation brought about by the slow drifts (which are usually disassociated in the two eyes) are

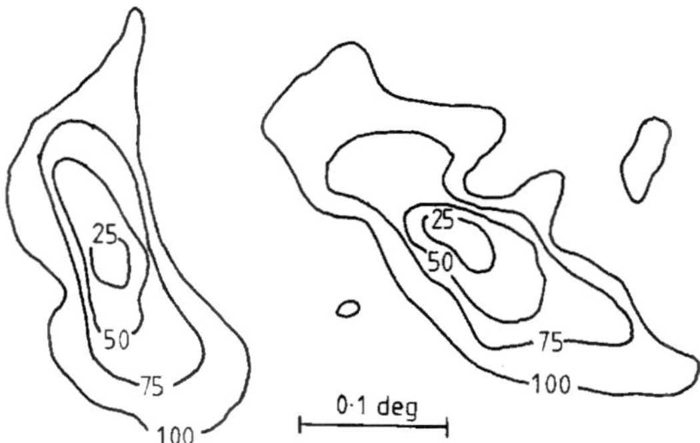

FIGURE 24 Stability of fixation for two subjects. The contours define areas within which the point of fixation was to be found 25, 50, 75, and 100% of the time (*after Bennet-Clark*[339]).

corrected by microsaccades, which are correlated in the two eyes. There are large intersubject differences in both mean microsaccade amplitude (from 1 to 23 min) and intersaccadic interval (from about 300 ms to 5 sec[338]) which probably reflect voluntary control under experimental conditions.

The overall stability of fixation can be illustrated by considering the statistical variation in the point of fixation[339] (Fig. 24). For most of the time the point of regard lies within a few minutes of the target. Although it has been suggested that these small eye movements could have some role in reducing potential aliasing problems,[340] experiments measuring contrast sensitivity for briefly presented interference fringes on the retina suggest that this is unlikely.[341] Interestingly, when a suitable optical arrangement is used to counteract these small changes in fixation and to stabilize the image on the retina, the visual image may fragment or even disappear completely,[337] so that these small movements are important for normal visual perception. Fincham's suggestion[342] that small eye movements are of importance in the ability of the eye to respond correctly to a change in accommodation stimulus has never been properly explored.

24.9 TWO EYES AND STEROPSIS

Although binocular vision confers a variety of advantages, ranging from an extension of the field of view to a lowering of contrast thresholds, attention here will be largely confined to its relevance for steropsis and stereoscopic instruments. It may, however, be remarked that little is known about the extent to which the monochromatic aberrations of the two eyes of either emmetropic or of optically corrected subjects are correlated (but see Ref. 184). It is, perhaps, possible that the brain can make selective use of the better of the two retinal images under any particular set of observing conditions due to some form of probability summation. Certainly the apparently drastic technique of monovision contact lens correction, in which one eye of a presbyopic observer receives a distance correction and the other a near correction appears to work well in yielding acceptable vision over a range of distances (e.g., Ref. 40). It may be that studies of monocular optical performance can sometimes give an unduly pessimistic view of the optical information available under binocular conditions.

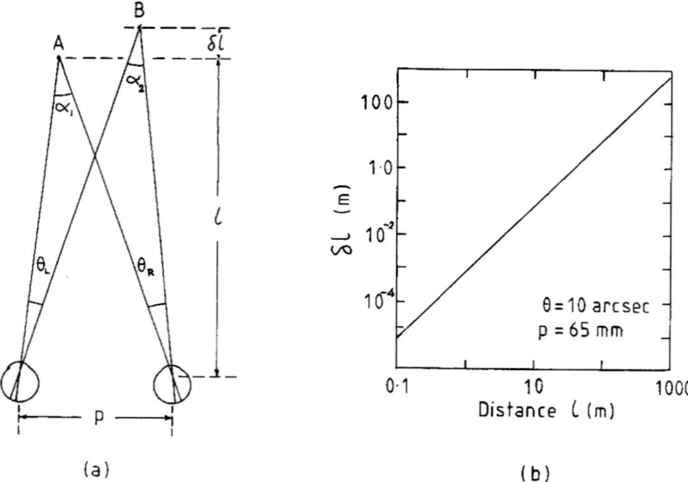

FIGURE 25 (*a*) Geometry of stereopsis. It is assumed that points A and B can just be discriminated in depth. (*b*) Theoretical just discriminable distance δl as a function of mean object distance l for the assumed values of p and $\delta\theta$ indicated.

Basics of Stereoscopic Acuity

Due to the lateral separation of the eyes in the head, the apparent angular separations of objects at differing distances are slightly different for the two eyes. The resultant disparities between the two retinal images can be used by the visual system to estimate the *relative* distances of the objects, although absolute distance judgment is usually much more dependent upon such monocular cues as perspective and size constancy.

In Fig. 25*a*, suppose that the observer can just detect that the two object points A and B lie at different distances, l and $l + \delta l$, respectively. Then the corresponding stereo acuity $\delta\theta$ is given by:

$$\delta\theta = \theta_R - \theta_L = \alpha_2 - \alpha_1$$

Approximating all angles as being small [i.e., $l \gg \delta l$, p where p is the lateral separation of the nodal points of the eyes or interpupillary distance (IPD)], and using the binomial expansion with omission of higher-order terms yields:

$$\delta\theta \approx p \cdot \delta l / l^2 \qquad \text{or} \qquad \delta l \approx l^2 \delta\theta / p \tag{15}$$

where $\delta\theta$ is in radians. Thus the minimum detectable difference in object distance is directly proportional to the square of the viewing distance and inversely proportional to the separation between the eyes (see Schor and Flom[343] for a more detailed analysis).

Figure 25*b* plots this approximate predicted value of the just-detectable difference in distance of two objects as a function of their mean distance, on the assumption that $p = 65$ mm and $\delta\theta$ is 10 sec arc. Note that stereopsis breaks down at distances in excess of about 500 m. In the real world, of course, the binocular cues to object distance are supplemented by a variety of monocular cues such as perspective, overlay (interposition), motion parallax, etc.

For any observer, $\delta\theta$ varies with such parameters as the target luminance and angular distance from fixation and the observation time allowed (e.g., Refs 344–348) being optimal at high luminance levels with extended observation times. Actual values also vary

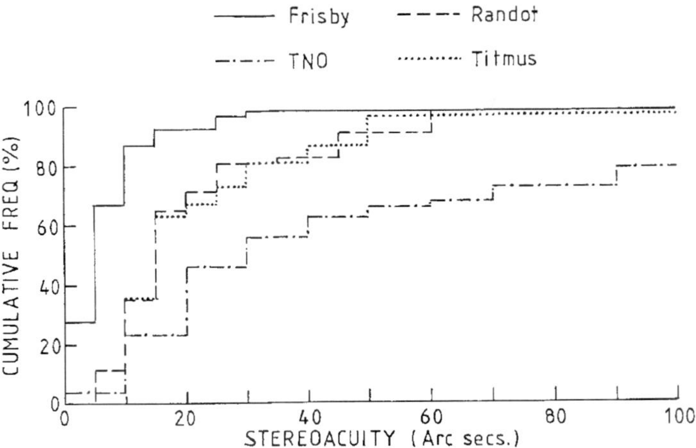

FIGURE 26 Cumulative frequency distribution for stereoscopic acuity, as measured by various clinical tests at a viewing distance of 400 mm, based on a sample of 51 adult subjects with normal vinocular vision (*after Heron et al.*[350]).

with the nature of the test used. Clinical tests of stereoacuity (e.g., Ref. 349), which are usually carried out at a distance of about 40 cm and assume a fixed value of p of 65 mm, typically yield normal stereoacuities of about 20–40 sec arc (Fig. 26, Ref. 350). Two-needle tests[351] carried out at longer distances usually give rather smaller values, of around 5–10 sec arc.[345] IPD, p, ranges between about 50 and 76 mm, with a mean value of about 65 mm.[50]

Stereoscopic Instruments

The stereoscopic acuity of natural viewing can be enhanced both by extending the effective IPD from p to Np with the aid of mirrors or prisms and by introducing transverse magnification, M, into the optical paths before each eye. This nominally has the effect of changing the just-detectable distance to

$$\delta l = (l^2 \cdot \delta\theta)/(MNp) \tag{16}$$

although in practice this improvement in performance is not always fully realized. Such changes will in general vary the spatial relationships in the perceived image, so that the object appears as having either enhanced or reduced depth in proportion to its lateral scale. If, for example, the magnification M is greater than one but N is unity, the object appears nearer but foreshortened. Simple geometrical predictions of such effects are, however, complicated by a variety of factors such as the reduced depth-of-field of magnifying systems (see, e.g., Refs. 344, 352, 353).

In any instrument involving viewing with both eyes, it is important that adequate adjustment for IPD be provided (preferably covering 46–78 mm), with a scale so that users can set their own IPD. Convergence should be appropriate to the distance at which the image is viewed, i.e., the angle between the visual axes should be approximately 3.7 degrees for each diopter of accommodation exercised.

24.10 CONCLUSION

The major uncertainties in our present understanding of the ocular dioptrics remain the refractive index distribution of the lens, aberration changes with accommodation, and the optical performance in the periphery, particularly with respect to the influence of transverse chromatic aberration. Although the axial monochromatic aberrations of the eye are quite well controlled under photopic conditions with natural pupils of the order of 2–4 mm, it is evident that under many circumstances significant defocus blur may occur due both to accommodation errors and, in the absence of optical correction, to refractive errors. Nevertheless the optical and motor characteristics of the eyes are such that, in general, the retinal image quality closely matches the needs of the neural parts of the visual system.

24.11 REFERENCES

1. A. Sorsby, B. Benjamin, M. Sheridan, and J. M. Tanner, *Emmetropia and Its Aberrations*: *Spec. Rep. Ser. Med. Res. Coun.,* no. 293, HMSO, London, 1957.
2. A. Sorsby, B. Benjamin, and M. Sheridan, *Refraction and Its Components During Growth of the Eye from the Age of Three*: *Spec. Rep. Ser. Med. Res. Coun.,* no. 301, HMSO, London, 1961.
3. A. Sorsby and G. A. Leary, *A Longitudinal Study of Refraction and Its Components During Growth*: *Spec. Rep. Ser. Med. Res. Coun.,* no. 309, HMSO, London, 1970.
4. T. Grosvenor, "Reduction in Axial Length with Age: An Emmetropizing Mechanism for the Adult Eye?," *Am. J. Optom. Physiol. Opt.* **64:**657–663 (1987).
5. F. A. Young and G. A. Leary, "Refractive Error in Relation to the Development of the Eye," *Vision and Visual Dysfunction,* vol 1, *Visual Optics and Instrumentation,* W. N. Charman (ed.), Macmillan, London, 1991, pp. 29–44.
6. R. A. Weale, *The Aging Eye,* H. K. Lewis, London, 1963, chap. 6.
7. J. F. Koretz, P. C. Kaufman, M. W. Neider, and P. A. Goeckner, "Accommodation and Presbyopia in the Human Eye—Aging of the Anterior Segment," *Vision Res.* **29:**1685–1692 (1989).
8. B. K. Pierscionek, D. Y. C. Chan, J. P. Ennis, G. Smith, and R. C. Augusteyn, "Non-Destructive Method of Constructing Three-Dimensional Gradient Index Models for Crystalline Lenses: 1 Theory and Experiment," *Am. J. Optom. Physiol. Opt,* **65:**481–491 (1988).
9. S. Stenstrom, "Untersuchengen über die Variation und Kovariation der optischen Elements des menschlichen Auges," *Acta Ophthal. Kbh,* suppl., **26** (1946).
10. A. Sorsby, B. Benjamin, and A. G. Bennett, "Steiger on Refraction: A Reappraisal," *Br. J. Ophthalmol.* **65:**805–811 (1981).
11. A. Gullstrand, appendices to *Physiological Optics,* vol. 1 by H. von Helmholtz, trans. J. P. C. Southall (trans.), Optical Society of America, Rochester, 1924.
12. H. H. Emsley, *Visual Optics,* vol. 1, 5th ed., Hatton Press, London, 1953.
13. K. N. Ogle, *Optics,* 2d ed., Thomas, Springfield, Ill., 1968.
14. S. Duke-Elder and D. Abrams, *Ophthalmic Optics and Refraction*: *System of Ophthalmology,* vol. V, Kimpton, London, 1970.
15. Y. Le Grand and S. G. El Hage, *Physiological Optics,* Springer-Verlag, Berlin, 1980, pp. 64–66.
16. L. P. O'Keefe and D. C. Coile, "A BASIC Computer Program for Schematic and Reduced Eye Construction," *Ophthal. Physiol. Opt.* **8:**97–100 (1988).
17. A. G. Bennett and R. B. Rabbetts, "Letter to the Editor," *Ophthal. Physiol. Opt.* **9:**228–230 (1989).

18. A. G. Bennett and R. B. Rabbetts, *Clinical Visual Optics,* 2d ed., Butterworths, London, 1989, pp. 249–274.
19. A. E. A. Ridgway, "Intraocular Lens Implants," *Vision and Visual Dysfunction,* vol. 1, *Visual Optics and Instrumentation,* W. N. Charman (ed.), Macmillan, London, 1991, pp. 120–137.
20. D. Sliney and M. Wolbarsht, *Safety with Lasers and Other Optical Sources,* Plenum, New York, 1980.
21. P. M. Kiely, G. Smith, and L. G. Carney, "The Mean Shape of the Human Cornea," *Optica Acta* **29:**1027–1040 (1982).
22. M. Guillon, D. P. M. Lydon, and C. Wilson, "Corneal Topography: A Clinical Model," *Ophthal. Physiol. Opt.* **6:**47–56 (1986).
23. S. A. Dingeldein and S. D. Klyce, "The Topography of Normal Corneas," *Arch. Ophthalmol.* **107:**512–518 (1989).
24. J. A. Parker, "Aspheric Optics of the Human Lens," *Can. J. Ophthalmol.* **7:**168–175 (1972).
25. M. J. Howcroft and J. A. Parker, "Aspheric Curvatures for the Human Lens," *Vision Res.* **17:**1217–1223 (1977).
26. N. Brown, "The Change in Shape and Internal Form of the Lens of the Eye on Accommodation," *Exp. Eye Res.* **15:**441–459 (1973).
27. N. Brown, "The Shape of the Lens Equator," *Exp. Eye Res.* **19:**571–576 (1974).
28. S. Nakao, K. Mine, K. Nishioka, and S. Kamiya, "The Distribution of Refractive Indices in the Human Crystalline Lens," *Jap. J. Clin. Ophthalmol.* **23:**41–44 (1969).
29. M. C. W. Campbell and A. Hughes, "The Refractive Index Distribution and Image Quality of a Human Crystalline Lens," *Advances in Diagnostic Visual Optics,* A. Fiorentini, D. Guyton, and I. M. Siegel (eds.), Springer-Verlag, Heidelberg, 1987, pp. 42–46.
30. D. Y. C. Chan, J. P. Ennis, B. K. Pierscionek, and G. Smith, "Determination and Modeling of the 3-D Gradient Refractive Indices in Crystalline Lenses," *Appl. Opt.* **27:**926–931 (1988).
31. W. Lotmar, "Theoretical Eye Model with Aspherics," *J. Opt. Soc. Am.* **61:**1522–1529 (1971).
32. J. W. Blaker, "Towards an Adaptive Model of the Human Eye," *J. Opt. Soc. Am.* **70:**220–223 (1980).
33. N. Drasdo and W. C. Peaston, "Sampling Systems for Visual Field Assessment and Computerised Perimetry," *Br. J. Ophthalmol.* **64:**705–709 (1980).
34. A. C. Kooijman, "Light Distribution on the Retina of a Wide Angle Theoretical Eye," *J. Opt. Soc. Am.* **73:**1544–1550 (1983).
35. O. Pomarentzeff, P. Dufault, and R. Goldstein, "Wide-Angle Optical Model of the Eye," *Advances in Diagnostic Visual Optics,* G. M. Breinin and I. M. Siegel (eds.), Springer-Verlag, Berlin, 1983, pp. 12–21.
36. R. Navarro, J. Santamaria, and J. Bescos, "Accommodation-Dependent Model of the Human Eye with Aspherics," *J. Opt. Soc. Am.* **A2:**1273–1281 (1985).
37. M. Jalie, *The Principles of Ophthalmic Lenses,* 4th ed., Association of Dispensing Opticians, London, 1984.
38. T. E. Fannin and T. Grosvenor, *Clinical Optics,* Butterworths, London, 1987.
39. C. Fowler, "Correction by Spectacle Lenses," *Vision and Visual Dysfunction.* vol. 1, *Visual Optics and Instrumentation,* W. N. Charman (ed.), Macmillan, London, 1991, pp. 64–79.
40. A. J. Phillips and J. Stone, *Contact Lenses,* 3d ed., Butterworths, London, 1989.
41. N. Efron, "Contact Lenses," *Vision and Visual Dysfunction,* vol. 1, *Visual Optics and Instrumentation,* W. N. Charman (ed.), Macmillan, London, 1991, pp 80–119.
42. W. F. Long, "Why Is Ocular Astigmatism Regular?," *Am. J. Optom. Physiol. Opt.* **59:**520–522 (1982).
43. G. O. Waring, "Ophthalmic Procedures Assessment: Radial Keratotomy for Myopia," *Ophthalmology* **89:**671–687 (1989).
44. B. L. Halliday, "Surgical Techniques for Refractive Correction," *Vision and Visual Dysfunction,*

vol. 1, *Visual Optics and Instrumentation,* W. N. Charman (ed.), Macmillan, London, 1991, pp. 138–162.

45. P. J. McDonnell, H. Moreira, T. N. Clapham, J. D'Arcy, and C. R. Munnerlyn, "Photorefractive Keratectomy for Astigmatism," *Arch. Ophthalmol.* **109:**1370–1373 (1991).
46. N. A. Sher, V. Chen, R. A. Bowers, J. M. Franz, D. C. Brown, R. Eiferman, S. S. Lane, P. Parker, C. Ostrov, D. S. Doughman, E. Carpel, R. Zabel, T. Gothard, and R. L. Lindstrom, "The use of the 193 nm Excimer Laser for Myopic Photorefractive Keratectomy in Sighted Eyes," *Arch. Ophthalmol.* **109:**1525–1530 (1991).
47. S. P. B. Percival and S. S. Setty, "Comparative Analysis of Three Prospective Trials of Multifocal Implants," *Eye* **5:**712–716 (1991).
48. A. Sorsby, M. Sheridan, G. A. Leary, and B. Benjamin, "Vision, Visual Acuity and Ocular Refraction of Young Men: Findings in a Sample of 1033 Subjects," *Br. Med J.* **1:**1394–1398 (1960).
49. W. M. Lyle, "Changes in Corneal Astigmatism with Age," *Am. J. Optom. Arch. Am. Acad. Optom.* **48:**467–478 (1971).
50. R. J. Farrell and J. M. Booth, *Design Handbook for Imagery Interpretation Equipment,* Boeing Aerospace Co., Seattle, Sec. 3.2, p. 8 (1984).
51. G. Wyszecki and W. S. Stiles, *Colour Science,* Wiley, New York, 1967, pp. 214–219.
52. E. A. Boerttner and J. R. Wolter, "Transmission of Ocular Media," *Invest. Ophthalmol.* **1:**776–783 (1962).
53. W. J. Geeraets and E. R. Berry, "Ocular Spectral Characteristics as Related to Hazards from Lasers and Other Sources," *Am. J. Ophthalmol.* **66:**15–20 (1968).
54. M. Waxler and V. M. Hitchins, *Optical Radiation and Visual Health,* CRC Press, Boca Raton, Fla., 1986.
55. A. F. McKinlay, F. Harlen, and M. J. Whillock, *Hazards of Optical Radiation,* Adam Hilger, Bristol, 1988.
56. D. G. Pitts, "The Ocular Effects of Ultraviolet Radiation," *Am. J. Optom. Physiol. Opt.* **55:**19–35 (1978).
57. S. Lerman, *Radiant Energy and the Eye,* Balliere Tindall, London, 1980.
58. J. Pokorny, V. C. Smith, and M. Lutze, "Aging of the Human Lens," *Appl. Opt.* **26:**1437–1440 (1987).
59. P. A. Sample, F. D. Esterson, R. N. Weinreb, and R. M. Boynton, "The Aging Lens: *In vivo* Assessment of Light Absorption in 84 Human Eyes," *Invest. Ophthalmol. Vis. Sci.* **29:**1306–1311 (1988).
60. J. Mellerio, "Yellowing of the Human Lens: Nuclear and Cortical Contributions," *Vision Res.* **27:**1581–1587 (1987).
61. R. F. Borkman, "Cataracts and Photochemical Damage in the Lens," *Human Cataract Formation, CIBA Foundation Symposium 106,* Pitmans, London, 1984, pp. 88–109.
62. C. Schmidtt, J. Schmidt, A. Wegener, and O. Hockwin, "Ultraviolet Radiation as a Risk Factor in Cataractogenesis," *Risk Factors in Cataract Development, Dev. Ophthalmol.* **17**, Karger, Basel, 1989 pp. 169–172.
63. J. Marshall, "Radiation and the Ageing Eye," *Ophthal. Physiol. Opt.* **5:**241–263 (1985).
64. A. Stanworth and E. J. Naylor, "The Measurement and Clinical Significance of the Haidinger Effect," *Trans. Ophthalmol. Soc. UK* **75:**67–79 (1955).
65. P. E. Kilbride, K. R. Alexander, M. Fishman, and G. A. Fishman, "Human Macular Pigment Assessed by Imaging Fundus Reflectometry," *Vision Res.* **26:**663–674 (1989).
66. D. M. Snodderly, J. D. Auran, and F. C. Delori, "The Macular Pigment. II Spatial Distribution in Primate Retinas," *Invest. Ophthalmol. Vis. Sci.* **25:**674–685 (1984).
67. R. A. Weale, "Natural History of Optics," *The Eye,* vol. 6., H. Davson and L. T. Graham (eds.), Academic Press, London, 1974, pp. 1–110.
68. V. M. Reading and R. A. Weale, "Macular Pigment and Chromatic Aberration," *J. Opt. Soc. Am.* **64:**231–234 (1974).

69. G. Haegerstrom-Portnoy, "Short-Wavelength Sensitive Cone Sensitivity Loss with Aging: A Protective Role for Macular Pigment?," *J. Opt. Soc. Am.* **A5:**2140 (1988).

70. M. H. Bornstein, "Developmental Pseudocyanopsia: Ontogenetic Change in Human Colour Vision," *Am. J. Optom. Physiol. Opt.* **54:**464–469 (1977).

71. L. J. Bour, "Polarized Light and the Eye," *Vision and Visual Dysfunction,* vol. 1, *Visual Optics and Instrumentation,* W. N. Charman (ed.), Macmillan, London, 1991, pp. 310–325.

72. E. H. Hess, "Attitude and Pupil Size," Sci. Am. **212**(4):46–54 (1965).

73. O. Lowenstein and I. E. Loewenfeld, "The Pupil," *The Eye,* vol. 3, H. Davson (ed.), Academic Press, New York, 1969, chap. 9, pp. 255–337.

74. K. M. Zinn, *The Pupil,* Thomas, Springfield Ill., 1972.

75. M. P. Janisse, *Pupillary Dynamics and Behaviour,* Plenum Press, New York, 1973.

76. H. S. Thompson, "The Pupil," *Adler's Physiology of the Eye*: *Clinical Application,* R. A. Moses and W. M. Hart (eds.), Mosby, St. Louis, 1987, chap. 12.

77. P. Reeves, "Rate of Pupillary Dilation and Contraction," *Psychol. Rev* **25:**330–340 (1918).

78. R. A. Weale, *Focus on Vision,* Hodder and Stoughton, London, 1982 p. 133.

79. G. Walsh, "The Effect of Mydriasis on the Pupillary Centration of the Human Eye," *Ophthal. Physiol. Opt.* **8:**178–182 (1988).

80. M. A. Wilson, M. C. W. Campbell, and P. Simonet, "Change of Pupil Centration with Change of Illumination and Pupil Size," *Optom. Vis. Sci.* **69:**129–136 (1992).

81. J. M. Woodhouse, "The Effect of Pupil Size on Grating Detection at Various Contrast Levels," *Vision Res.* **15:**645–648 (1975).

82. J. M. Woodhouse and F. W. Campbell, "The Role of the Pupil Light Reflex in Aiding Adaptation to the Dark," *Vision Res.* **15:**649–653 (1975).

83. B. S. Jay, "The Effective Pupillary Area at Varying Perimetric Angles," *Vision Res.* **1:**418–428 (1962).

84. F. W. Fitzke, "Optical Properties of the Eye" (abstract), *Invest. Ophthalmol. Vis. Sci.* **20**(suppl.):44 (1981).

85. H. E. Bedell and L. M. Katz, "On the Necessity of Correcting Peripheral Target Luminance for Pupillary Area," *Am. J. Optom. Physiol. Opt.* **59:**767–769 (1982).

86. W. N. Charman, "The Retinal Image in the Human Eye," *Progress in Retinal Research,* N. Osborne and G. Chader (eds.), vol. 2, Pergamon, Oxford, 1983, pp. 1–50.

87. W. N. Charman, "Light on the Peripheral Retina," *Ophthal. Physiol. Opt.* **9:**91–92 (1989).

88. K. P. Pflibsen, O. Pomarentzeff, and R. N. Ross, "Retinal Illuminance Using a Wide-Angle Model of the Eye," *J. Opt. Soc. Am.* **A5:**146–150 (1988).

89. A. C. Kooijman and F. K. Witmer, "Ganzfeld Light Distribution on the Retina of Human and Rabbit Eyes: Calculations and *in vitro* Measurements," *J. Opt. Soc. Am.* **A3:**2116–2120 (1986).

90. W. D. Wright, *Photometry and the Eye,* Hatton Press, London, 1949.

91. W. S. Stiles and B. H. Crawford, "The Luminous Efficiency of Rays Entering the Eye Pupil at Different Points," *Proc. R. Soc. Lond. B* **112:**428–450 (1933).

92. J. M. Enoch and H. E. Bedell, "The Stiles-Crawford Effects," *Vertebrate Photoreceptor Optics,* J. M. Enoch and F. L. Tobey (eds.) Springer-Verlag, Berlin, 1981, pp. 83–126.

93. J. M. Enoch and V. Lakshminarayanan, "Retinal Fibre Optics," *Vision and Visual Dysfunction,* vol 1, *Visual Optics and Instrumentation,* W. N. Charman (ed.), Macmillan, London, 1991, pp. 280–309.

94. W. S. Stiles, "The Luminous Efficiency of Monochromatic Rays Entering the Eye Pupil at Different Points and a New Colour Effect," *Proc. R. Soc. London* **B123:**90–118 (1937).

95. P. Moon and D. E. Spencer, "On the Stiles-Crawford Effect," *J. Opt. Soc. Am.* **34:**319–329 (1944).

96. M. Alpern, "The Stiles-Crawford Effect of the Second Kind (SCII): A Review," *Perception* **15:**785–799 (1986).

97. G. Airy, "On the Diffraction of an Object Glass with Circular Aperture," *Trans. Camb. Phil. Soc.* **5:**283 (1835).
98. G. M. Byram, "The Physical and Photochemical Basis of Resolving Power. I. The Distribution of Illumination in Retinal Images," *J. Opt. Soc. Am.* **34:**571–591 (1944).
99. U. Hallden, "Diffraction and Visual Resolution. 1. The Resolution of Two Point Sources of Light," *Acta Ophthalmol.* **51:**72–79 (1973).
100. E. Lommel, "Die Beugungerscheinungen einer Kreisrunden Oeffnung und eines runden schirmchens," abh der *K. Bayer. Akad. d. Wissenschaft.* **15:**229–328 (1884).
101. E. H. Linfoot and E. Wolf, "Phase Distribution Near Focus in an Aberration-Free Diffraction Image," *Proc. Phys. Soc.* **B69:**823–832 (1956).
102. M. Born and E. Wolf, *Principles of Optics,* 5th ed., Pergamon, Oxford, 1975.
103. L. A. Riggs, "Visual Acuity," *Vision and Visual Perception,* C. H. Graham (ed.), Wiley, New York, 1966, pp. 321–349.
104. H. Struve, "Beitrag zür der Theorie der Diffraction an Fernrohren," *Ann. Physik. Chem.* **17:**1008–1016 (1882).
105. H. H. Hopkins, "The Frequency Response of a Defocused Optical System," *Proc. R. Soc.* **A231:**91–103 (1955).
106. L. Levi, *Handbook of Tables for Applied Optics,* CRC Press, Cleveland, 1974.
107. W. N. Charman, "Effect of Refractive Error in Visual Tests with Sinusoidal Gratings," *Brit. J. Physiol. Opt.* **33**(2):10–20 (1979).
108. G. E. Legge, K. T. Mullen, G. C. Woo, and F. W. Campbell, "Tolerance to Visual Defocus," *J. Opt. Soc. Am.* **A4:**851–863 (1987).
109. G. Westheimer, "Pupil Size and Visual Resolution," *Vision Res.* **4:**39–45 (1964).
110. F. W. Campbell and D. G. Green, "Optical and Retinal Factors Affecting Visual Resolution," *J. Physiol.* (Lond.) **181:**576–593 (1965).
111. F. W. Campbell and R. W. Gubisch, "The Effect of Chromatic Aberration on Visual Acuity," *J. Physiol.* (Lond.) **192:**345–358 (1967).
112. W. N. Charman and J. Tucker, "Dependence of Accommodation Response on the Spatial Frequency Spectrum of the Observed Object," *Vision Res.* **17:**129–139 (1977).
113. W. N. Charman and J. Tucker, "Accommodation as a Function of Object Form," *Am. J. Optom. Physiol. Opt.* **55:**84–92 (1978).
114. W. N. Charman and J. A. M. Jennings, "The Optical Quality of the Retinal Image as a Function of Focus," *Br. J. Physiol. Opt.* **31:**119–134 (1976).
115. W. N. Charman and G. Heron, "Spatial Frequency and the Dynamics of the Accommodation Response," *Optica Acta* **26:**217–228 (1979).
116. H. H. Hopkins, "Geometrical Optical Treatment of Frequency Response," *Proc. Phys. Soc.* **B70:**1162–1172 (1957).
117. G. A. Fry, "Blur of the Retinal Image," *Progress in Optics,* vol. 8., E. Wolf (ed.), North-Holland, Amsterdam, 1970, pp. 53–131.
118. G. Smith, "Ocular Defocus, Spurious Resolution and Contrast Reversal," *Ophthal. Physiol. Opt.* **2:**5–23 (1982).
119. M. S. Smirnov, "Measurement of the Wave Aberration of the Human Eye," *Biofizika* **6:**687–703, and *Biophysics* **6:**766–795 (1962).
120. F. Berny and S. Slansky, "Wavefront Determination Resulting from Foucault Test as Applied to the Human Eye and Visual Instruments," *Optical Instruments and Techniques,* J. Home-Dickson (ed.), Oriel Press, London, 1969 pp. 375–385.
121. B. Howland and H. C. Howland, "Subjective Measurement of High-Order Aberrations of the Eye," *Science* **193:**580–582 (1976).
122. H. C. Howland and B. Howland, "A Subjective Method for the Measurement of the Monochromatic Aberrations of the Eye," *J. Opt. Soc. Am.* **67:**1508–1518 (1977).

123. G. Walsh, W. N. Charman, and H. C. Howland, "Objective Technique for the Determination of Monochromatic Aberrations of the Human Eye," *J. Opt. Soc. Am.* **A1:**987–992 (1984).

124. G. Walsh and W. N. Charman, "Measurement of the Axial Wavefront Aberration of the Human Eye," *Ophthal. Physiol. Opt.* **5:**23–31 (1985).

125. J. F. Bille, A. Dreher, and G. Zinser, "Scanning Laser Tomography of the Living Human Eye," *Noninvasive Diagnostic Techniques in Ophthalmology,* B. R. Masters (ed.), Springer-Verlag, Berlin, 1990, pp. 528–547.

126. W. N. Charman, "Wavefront Aberration of the Eye: A Review," *Optom. Vis. Sci.* **68:**574–583 (1991).

127. D. A. Atchison, M. J. Collins, and C. F. Wildsoet, "Ocular Aberrations and Accommodation," *Ophthalmic and Visual Optics: 1992 Technical Digest Ser.,* vol. 3, Optical Society of America, Washington, D.C., 1992, pp. 55–58.

128. W. N. Charman and G. Walsh, "The Optical Phase Transfer Function of the Eye and the Perception of Spatial Phase," *Vision Res.* **25:**619–623 (1985).

129. H. C. Howland and J. Buettner, "Computing High Order Wave Aberration Coefficients from Variations in Best Focus for Small Artificial Pupils," *Vision Res.* **29:**979–983 (1989).

130. M. Koomen, R. Tousey, and R. Skolnik, "The Spherical Aberration of the Eye," *J. Opt. Soc. Am.* **39:**370–376 (1949).

131. A. Ivanoff, "About the Spherical Aberration of the Eye," *J. Opt. Soc. Am.* **46:**901–903 (1956).

132. T. C. A. Jenkins, "Aberrations of the Eye and Their Effects on Vision. Parts I and II.," *Br. J. Physiol. Opt.* **20:**50–91 and 161–201 (1963).

133. A. Marechal, Etude des effets combinés de la diffraction et des aberrations géométriques sur l'image d'un point lumineux, *Rev. d'Optique* **26:**257–277 (1947).

134. C. E. Ferree, G. Rand, and C. Hardy, "Refraction for the Peripheral Field of Vision," *Arch. Ophthalmol.* (Chicago) **5:**717–731 (1931).

135. C. E. Ferree and G. Rand, "The Refractive Conditions for the Peripheral Field of Vision," *Report of Joint Discussion on Vision,* Physical Society, London, 1932, pp. 244–262.

136. F. Rempt, J. Hoogerheide, and W. P. H. Hoogenboom, "Peripheral Retinoscopy and the Skiagram," *Ophthalmologica* **162:**1–10 (1971).

137. J. Hoogerheide, F. Rempt, and W. P. H. Hoogenboom, "Acquired Myopia in Young Pilots," *Ophthalmologica* **163:**209–215 (1971).

138. M. Millodot and A. Lamont, "Refraction of the Periphery of the Eye," *J. Opt. Soc. Am.* **64:**110–111 (1974).

139. M. Millodot, "Effect of Ametropia on Peripheral Refraction," *Am. J. Optom. Physiol. Opt.* **58:**691–695 (1981).

140. W. N. Charman and J. A. M. Jennings, "Ametropia and Peripheral Refraction," *Am. J. Optom. Physiol. Opt.* **59:**885–889 (1982).

141. W. Lotmar and T. Lotmar, "Peripheral Astigmatism in the Human Eye: Experimental Data and Theoretical Model Predictions," *J. Opt. Soc. Am.* **64:**510–513 (1974).

142. M. C. M. Dunne, D. A. Barnes, and R. C. Clements, "A Model for Retinal Shape Changes in Ametropia," *Ophthal. Physiol. Opt.* **7:**159–160 (1987).

143. D. A. Barnes, M. C. M. Dunne, and R. A. Clement, "A Schematic Eye Model for the Effects of Translation and Rotation of Ocular Components on Peripheral Astigmatism," *Ophthal. Physiol. Opt.* **7:**153–158 (1987).

144. Y. Le Grand, *Form and Space Vision,* M. Millodot and G. C. Heath (trans.), Indiana U.P., Bloomington, 1967, pp. 5–23.

145. J. G. Sivak and T. Mandelman, "Chromatic Dispersion of the Ocular Media," *Vision Res.* **22:**997–1003 (1982).

146. T. Mandelman and J. G. Sivak, "Longitudinal Chromatic Aberration of the Vertebrate Eye," *Vision Res.* **23:**1555–1559 (1983).

147. L. N. Thibos, A. Bradley, F. D. L. Still, X. Zhang, and P. A. Howarth, "Theory and Measurement of Ocular Chromatic Aberration," *Vision Res.* **30:**33–49 (1990).

148. L. N. Thibos, A. Bradley, and X. Zhang, "Effect of Ocular Chromatic Aberration on Monocular Visual Performance," *Optom. Vis. Sci.* **68:**599–607 (1991).

149. G. Wald and D. R. Griffin, "The Change in Refractive Power of the Human Eye in Dim and Bright Light," *J. Opt. Soc. Am.* **37:**321–336 (1947).

150. A. Ivanoff, *Les Aberrations de l'Oeil,* Revue d'Optique, Paris, 1953.

151. R. E. Bedford and G. Wyszecki, "Axial Chromatic Aberration of the Human Eye," *J. Opt. Soc. Am.* **47:**564–565 (1957).

152. P. A. Howarth and A. Bradley, "The Longitudinal Chromatic Aberration of the Human Eye, and Its Correction," *Vision Res.* **26:**361–366 (1986).

153. H. Hartridge, "The Visual Perception of Fine Detail," *Phil. Trans. R. Soc. A.* **232:**519–671 (1947).

154. A. Van Meeteren, "Calculations on the Optical Modulation Transfer Function of the Human Eye for White Light," *Optica Acta* **21:**395–412 (1974).

155. P. A. Howarth, "The Lateral Chromatic Aberration of the Human Eye," *Ophthal. Physiol. Opt.* **4:**223–236 (1984).

156. L. N. Thibos, "Calculation of the Influence of Lateral Chromatic Aberration on Image Quality Across the Visual Field," *J. Opt. Soc. Am.* **A4:**1673–1680 (1987).

157. Y. U. Ogboso and H. E. Bedell, "Magnitude of Lateral Chromatic Aberration Across the Retina of the Human Eye," *J. Opt. Soc. Am.* **A4:**1666–1672 (1987).

158. P. Simonet and M. C. W. Campbell, "The Optical Transverse Chromatic Aberration on the Fovea of the Human Eye," *Vision Res.* **30:**187–206 (1990).

159. J. J. Vos, "Some New Aspects of Color Stereoscopy," *J. Opt. Soc. Am.* **50:**785–790 (1960).

160. R. C. Allen and M. L. Rubin, "Chromostereopsis," *Survey Ophthalmol.* **26:**22–27 (1981).

161. D. A. Owens and H. W. Leibowitz, "Chromostereopsis and Small Pupils," *J. Opt. Soc. Am.* **65:**358–359 (1975).

162. F. E. Cheney and L. N. Thibos, "Orientation Anisotropy for the Detection of Aliased Patterns by Peripheral Vision Is Optically Induced," *J. Opt. Soc. Am.* **A4:**P79 (1987).

163. L. N. Thibos, F. E. Cheney, and D. J. Walsh, "Retinal Limits to the Detection and Resolution of Gratings," *J. Opt. Soc. Am.* **A4:**1524–1529 (1987).

164. L. L. Holladay, "The Fundamentals of Glare and Visibility," *J. Opt. Soc. Am.* **12:**271–319 (1926).

165. W. S. Stiles, "The Effect of Glare on the Brightness Threshold," *Proc. R. Soc. Lond.* **B104:**322–351 (1929).

166. J. J. Vos, J. Walraven, and A. Van Meeteren, "Light Profiles of the Foveal Image of a Point Source," *Vision Res.* **16:**215–219 (1976).

167. M. J. Allen and J. J. Vos, "Ocular Scattered Light and Visual Performance as a Function of Age," *Am. J. Optom. Arch. Am. Acad. Optom.* **44:**717–727 (1967).

168. A. Spector, S. Li, and J. Sigelman, "Age-Dependent Changes in the Molecular Size of Human Lens Proteins and Their Relationship to Light Scatter," *Invest. Ophthalmol.* **13:**795–798 (1974).

169. R. P. Hemenger, "Intraocular Light Scatter in Normal Visual Loss with Age," *Appl. Opt.* **23:**1972–1974 (1984).

170. R. A. Weale, "Effects of Senescence," *Vision and Visual Dysfunction,* vol. 5, *Limits of Vision,* J. J. Kulikowski, V. Walsh, and I. J. Murray (eds.), Macmillans, London, 1991, pp. 277–285.

171. W. Adrian and A. Bhanji, "A Formula to Describe Straylight in the Eye as a Function of the Glare Angle and Age," *1st Int. Symp. on Glare,* Orlando Fla., Oct. 24 and 25, 1991 (The Lighting Research Institute, New York, 1991), pp. 185–193.

172. J. J. Vos and J. Boogaard, "Contribution of the Cornea to Entoptic Scatter," *J. Opt. Soc. Am.* **53:**869–873 (1963).

173. R. M. Boynton and F. J. J. Clarke, "Sources of Entoptic Scatter in the Human Eye," *J. Opt. Soc. Am.* **54:**110–119, 717–719 (1964).

174. J. J. Vos, "Contribution of the Fundus Oculi to Entoptic Scatter," *J. Opt. Soc. Am.* **53:**1449–1451 (1963).

175. J. J. Vos and M. A. Bouman, "Contribution of the Retina to Entopic Scatter," *J. Opt. Soc. Am.* **54:**95–100 (1964).

176. R. P. Hemenger, "Small-Angle Intraocular Scattered Light: A Hypothesis Concerning Its Source," *J. Opt. Soc. Am.* **A5:**577–582 (1988).

177. J. H. Ohzu and J. M. Enoch, "Optical Modulation by the Isolated Human Fovea," *Vision Res.* **12:**245–251 (1972).

178. D. I. A. MacLeod, D. R. Williams, and W. Makous, "A Visual Non-Linearity Fed by Single Cones," *Vision Res.* **32:**347–363 (1992).

179. H. H. Hopkins, "The Application of Frequency Response Techniques in Optics," *Proc. Phys. Soc.* **79:**889–919 (1962).

180. H. Metcalf, "Stiles-Crawford Apodisation," *J. Opt. Soc. Am.* **55:**72–74 (1965).

181. A. Nussbaum and R. A. Phillips, *Contemporary Optics for Scientists and Engineers,* Prentice-Hall, Englewood Cliffs, N.J., 1976.

182. W. J. Smith, "Image Formation: Geometrical and Physical Optics," *Handbook of Optics,* W. G. Driscoll and W. Vaughan (eds.), McGraw-Hill, New York, 1978, sec. 2.

183. F. Berny, "Etude de la formation des images rétiniennes et détermination de l'aberration de sphéricité de l'oeil humaine," *Vision Res.* **9:**977–990 (1969).

184. M. C. W. Campbell, E. M. Harrison, and P. Simonet, "Psychophysical Measurement of the Blur on the Retina Due to Optical Aberrations of the Eye," *Vision Res.* **30:**1587–1602 (1990).

185. W. N. Charman and G. Walsh, "Variation in Ocular Modulation and Phase Transfer Functions with Grating Orientation," *Ophthal. Physiol. Opt.* **12**, 1992.

186. Y. Le Grand, "Sur une mode de vision éliminant les défauts optiques de l'oeil," *Rev. d'Optique Theor. Instrum.* **15:**6–11 (1936).

187. G. Westheimer, "Modulation Thresholds for Sinusoidal Light Distributions on the Retina," *J. Physiol.* (London) **152:**67–74 (1960).

188. A. Arnulf and O. Dupuy, "La transmission des contrastes par la système optique de l'oeil et les seuils de contrastes rétiniens," *C. r. hebd. Seanc. Acad. Paris* **250:**2757–2759 (1960).

189. S. Berger-Lheureux, "Mesure de la fonction de transfert de modulation du système optique de l'oeil et des seuils de modulation rétiniens," *Rev. Opt. theor. instrum.* **44:**294–323 (1965).

190. F. W. Campbell, "The Eye as an Optical Filter," *Proc. IEEE* **56:**1009–1014 (1968).

191. L. J. Bour, "MTF of the Defocused Optical System of the Human Eye for Incoherent Monochromatic Light," *J. Opt. Soc. Am.* **70:**321–328 (1980).

192. F. W. Campbell and R. W. Gubisch, "Optical Quality of the Human Eye," *J. Physiol.* (London) **186:**558–578 (1966).

193. F. Flamant, "Etude de la repartition de la lumière dans l'image rétinienne d'une fente," *Revue Opt. theor. instrum.* **34:**433–459 (1955).

194. J. Krauskopf, "Light Distribution in Human Retinal Images," *J. Opt. Soc. Am.* **52:**1046–1050 (1962).

195. J. Krauskopf, "Further Measurements of Human Retinal Images," *J. Opt. Soc. Am.* **54:**715–716 (1964).

196. R. Rohler, "Die Abbildungseigenschaften der Augenmedien," *Vision Res.* **2:**391–429 (1962).

197. G. Westheimer and F. W. Campbell, "Light Distributions in the Image Formed by the Living Human Eye," *J. Opt. Soc. Am.* **52:**1040–1045 (1962).

198. R. Rohler, U. Miller, and M. Aberl, "Zür Messung der Modulationsubertragugungs-Funktion des lebenden menschlichen Augen in reflektierten Licht," *Vision Res.* **9:**407–428 (1969).

199. J. A. M. Jennings and W. N. Charman, "Optical Image Quality in the Peripheral Retina," *Am. J. Optom. Physiol. Opt.* **55:**582–590 (1978).

200. J. A. M. Jennings and W. N. Charman, "Off-Axis Image Quality in the Human Eye," *Vision Res.* **21:**445–455 (1981).

201. J. M. Gorrand, O. Dupuy, F. Farfal, M. T. Plantagenest, and S. Slansky, "Contributions de la rétine et du système optique de l'oeil a l'abaissement de la modulation de l'image par double traversée de l'oeil," *J. Optics* (Paris) **9:**359–364 (1978).
202. J. M. Gorrand, "Diffusion of the Human Retina and Quality of the Optics of the Eye on the Fovea and the Peripheral Retina," *Vision Res.* **19:**907–912 (1979).
203. J. Santamaria, P. Artal, and J. Bescos, "Determination of the Point-Spread Function of Human Eyes Using a Hybrid Optical-Digital Method," *J. Opt. Soc. Am.* **A4:**1109–1114 (1987).
204. P. Artal, J. Santamaria, and J. Bescos, "Retrieval of Wave Aberration of Human Eyes from Actual Point-Spread-Function Data," *J. Opt. Soc. Am.* **A5:**1201–1206 (1988).
205. P. Artal, J. Santamaria, and J. Bescos, "Phase-Transfer Function of the Human Eye and Its Influence on Point-Spread Function and Wave Aberration," *J. Opt. Soc. Am.* **A5:**1791–1795 (1988).
206. P. Artal, "Calculations of Two-Dimensional Foveal Retinal Images in Real Eyes," *J. Opt. Soc. Am.* **A7:**1374–1381 (1990).
207. J. F. Simon and P. M. Denieul, "Influence of the Size of Test Field Employed in Measurements of Modulation Transfer Function of the Eye," *J. Opt. Soc. Am.* **63:**894–896 (1973).
208. W. N. Charman and J. A. M. Jennings, "Objective Measurements of the Longitudinal Chromatic Aberration of the Human Eye," *Vision Res.* **16:**999–1005 (1976).
209. J. A. M. Jennings and W. N. Charman, "An Analytical Approximation for the Modulation Transfer of the Eye," *Br. J. Physiol. Opt.* **29:**64–72 (1974).
210. R. J. Deeley, N. Drasdo, and W. N. Charman, "A Simple Parametric Model of the Human Ocular Modulation Transfer Function," *Ophthal. Physiol. Opt.* **11:**91–93 (1991).
211. R. Navarro, P. Artal, and D. R. Williams, "Modulation Transfer of the Human Eye as a Function of Retinal Eccentricity," *J. Opt. Soc. Am.* **A**, 1993.
212. D. G. Green, "Regional Variation in the Visual Acuity for Interference Fringes on the Retina," *J. Physiol.* (London) **207:**351 (1970).
213. J. M. Enoch and G. M. Hope, "Interferometric Resolution Determinations in the Fovea and Parafovea," *Doc. Ophthalmol.* **34:**143–156 (1973).
214. L. Frisen and A. Glansholm, "Optical and Neural Resolution in Peripheral Vision," *Invest. Ophthalmol.* **14:**528–536 (1975).
215. P. B. DeVelis and G. B. Parrent, "Transfer Function for Cascaded Optical Systems," *J. Opt. Soc. Am.* **57:**1486–1490 (1967).
216. I. Overington, "Interaction of Vision with Optical Aids," *J. Opt. Soc. Am.* **63:**1043–1049 (1973).
217. I. Overington, "The Importance of Coherence of Coupling When Viewing Through Visual Aids," *Opt. Laser Technol.* **5:**216–220 (1973).
218. I. Overington, "Some Considerations of the Role of the Eye as a Component of an Imaging System," *Optica Acta* **22:**365–374 (1975).
219. W. N. Charman and H. Whitefoot, "Astigmatism, Accommodation and Visual Instrumentation," *Applied Opt.* **17:**3903–3910 (1978).
220. P. Mouroulis, "On the Correction of Astigmatism and Field Curvature in Telescopic Systems," *Optica Acta* **29:**1133–1159 (1982).
221. G. J. Burton and N. D. Haig, "Criteria for Testing of Afocal Instruments," *Proc. Soc. Photo-Opt. Instrum. Eng.* **274:**191–201 (1981).
222. J. Eggert and K. J. Rosenbruch, "Vergleich der visuell und der photoelektrisch gemessen Abbildungsguete von Fernrohren," *Optik* **48:**439–450 (1977).
223. G. J. Burton and N. D. Haig, "Effects of the Seidel Aberrations on Visual Target Discrimination," *J. Opt. Soc. Am.* **A1:**373–385 (1984).
224. P. Mouroulis and G. C. Woo, "Chromatic Aberration and Accommodation in Visual Instruments," *Optik* **80:**161–166 (1989).
225. M. Koomen, R. Skolnik, and R. Tousey, "A Study of Night Myopia," *J. Opt. Soc. Am.* **41:**80–90 (1951).

226. D. G. Green and F. W. Campbell, "Effect of Focus on the Visual Response to a Sinusoidally Modulated Spatial Stimulus," *J. Opt. Soc. Am.* **55:**1154–1157 (1965).

227. W. N. Charman, J. A. M. Jennings, and H. Whitefoot, "The Refraction of the Eye in Relation to Spherical Aberration and Pupil Size," *Br. J. Physiol. Opt.* **32:**78–93 (1978).

228. F. L. Van Nes and M. A. Bouman, "Spatial Modulation Transfer in the Human Eye," *J. Opt. Soc. Am.* **57:**401–406 (1967).

229. F. W. Campbell, "The Depth-of-Field of the Human Eye," *Optica Acta* **4:**157–164 (1957).

230. J. Tucker and W. N. Charman, "Depth of Focus and Accommodation for Sinusoidal Gratings as a Function of Luminance," *Am. J. Optom. Physiol. Opt.* **63:**58–70 (1986).

231. K. N. Ogle and J. T. Schwartz, "Depth-of-Focus of the Human Eye," *J. Opt. Soc. Am.* **49:**273–280 (1959).

232. J. Tucker and W. N. Charman, "The Depth-of-Focus of the Human Eye for Snellen Letters," *Am. J. Optom. Physiol. Opt.* **52:**3–21 (1975).

233. W. N. Charman and H. Whitefoot, "Pupil Diameter and the Depth-of-Field of the Human Eye as Measured by Laser Speckle," *Optica Acta* **24:**1211–1216 (1977).

234. J. Semmlow and L. Stark, "Pupil Movements to Light and Accommodative Stimulation: A Comparative Study," *Vision Res.* **13:**1087–1100 (1973).

235. M. Stakenburg, "Accommodation without Pupillary Constriction," *Vision Res.* **31:**267–273 (1991).

236. F. W. Campbell, "Correlation of Accommodation Between the Two Eyes," *J. Opt. Soc. Am.* **50:**738 (1960).

237. H. Davson, *Physiology of the Eye,* 5th ed., Macmillan, London, 1990.

238. K. J. Ciuffreda, "Accommodation and Its Anomalies," *Vision and Visual Dysfunction,* vol. 1, *Visual Optics and Instrumentation,* W. N. Charman (ed.), Macmillan, London, 1991, pp. 231–279.

239. G. H. Handelman and J. F. Koretz, "A Mathematical Representation of Lens Accommodation," *Vision Res.* **22:**924–927 (1982).

240. H. J. Wyatt, "Some Aspects of the Mechanisms of Accommodation," *Vision Res.* **28:**75–86 (1988).

241. F. W. Campbell and G. Westheimer, "Dynamics of the Accommodation Responses of the Human Eye," *J. Physiol.* **151:**285–295 (1960).

242. W. P. O'Neill and L. Stark, "Triple Function Ocular Monitor," *J. Opt. Soc. Am.* **58:**570–573 (1968).

243. S. Phillips, D. Shirachi, and L. Stark, "Analysis of Accommodation Times Using Histogram Information," *Am. J. Optom. Arch. Am. Acad. Optom.* **49:**389–401 (1972).

244. G. Heron, "A Study of Accommodation Using an Infra-Red Optometer," M.Sc. thesis, University of Manchester, Manchester, 1972.

245. L. M. Smithline, "Accommodative Response to Blur," *J. Opt. Soc. Am.* **64:**1512–1516 (1974).

246. R. J. Randle and M. R. Murphy, "The Dynamic Response of Visual Accommodation over a Seven-Day Period," *Am. J. Optom. Physiol. Opt.* **51:**530–544 (1974).

247. D. Shirachi, J. Liu, M. Lee, J. Jang, J. Wong, and L. Stark, "Accommodation Dynamics. I Range Non-Linearity," *Am. J. Optom. Physiol. Opt.* **55:**531–641 (1978).

248. J. Tucker and W. N. Charman, "Reaction and Response Times for Accommodation," *Am. J. Optom. Physiol. Opt.* **56:**490–503 (1979).

249. L. Stark, *Neurological Control Systems: Studies in Bioengineering,* Plenum Press, New York, 1968, sect. III.

250. V. V. Krishnan, S. Phillips, and L. Stark, "Frequency Analysis of Accommodation, Accommodative Vergence and Disparity Vergence," *Vision Res.* **13:**1545–1554 (1973).

251. G. J. Van der Wildt, M. A. Bouman, and J. van der Kraats, "The Effect of Anticipation on the Transfer Function of the Human Lens System," *Optica Acta* **21:**843–860 (1974).

252. P. B. Kruger and J. Pola, "Changing Target Size Is a Stimulus for Accommodation," *J. Opt. Soc. Am.* **A2:**1832–1835 (1985).

253. P. B. Kruger and J. Pola, "Stimuli for Accommodation: Blur, Chromatic Aberration and Size," *Vision Res.* **26:**957–961 (1986).

254. P. B. Kruger and J. Pola, "Dioptric and Non-Dioptric Stimuli for Accommodation: Target Size Alone and with Blur and Chromatic Aberration," *Vision Res.* **27:**555–567 (1987).

255. W. N. Charman and G. Heron, "Fluctuations in Accommodation: A Review," *Ophthal. Physiol. Opt.* **8:**153–164 (1988).

256. P. Denieul, "Effects of Stimulus Vergence on Mean Accommodation Response, Microfluctuations of Accommodation and Optical Quality of the Human Eye," *Vision Res.* **22:**561–569 (1982).

257. F. W. Campbell, J. G. Robson, and G. Westheimer, "Fluctuations of Accommodation under Steady Viewing Conditions," *J. Physiol.* (London.) **145:**579–594 (1959).

258. J. C. Kotulak and C. M. Schor, "Temporal Variations in Accommodation during Steady-State Conditions," *J. Opt. Soc. Am.* **A3:**223–227 (1986).

259. B. Winn, J. R. Pugh, B. Gilmartin, and H. Owens, "Arterial Pulse Modulates Steady State Accommodation," *Curr. Eye Res.* **9:**971–975 (1990).

260. L. Stark and Y. Takahashi, "Absence of an Odd-Error Signal Mechanism in Human Accommodation," *IEEE Trans. Biomed. Eng.* **BME-12:**138–146 (1965).

261. M. Alpern, "Variability of Accommodation During Steady Fixation at Various Levels of Illuminance," *J. Opt. Soc. Am.* **48:**193–197 (1958).

262. H. D. Crane, *A Theoretical Analysis of the Visual Accommodation System in Humans,* Report NASA CR-606, NASA, Washington, D.C., 1966.

263. G. K. Hung, J. L. Semmlow, and K. J. Ciuffreda, "Accommodative Oscillation Can Enhance Average Accommodation Response: A Simulation Study," *IEEE Trans. Syst. Man. Cybern.* **SMC-12:**594–598 (1982).

264. M. Millodot, "Effet des microfluctuations de l'accommodation sur l'acuité visuelle," *Vision Res.* **8:**73–80 (1968).

265. G. Walsh and W. N. Charman, "Visual Sensitivity to Temporal Change in Focus and Its Relevance to the Accommodation Response," *Vision Res.* **28:**1207–1221 (1988).

266. B. Winn, W. N. Charman, J. R. Pugh, G. Heron, and A. S. Eadie, "Perceptual Detectability of Ocular Accommodation Microfluctuations," *J. Opt. Soc. Am.* **A6:**459–462 (1989).

267. M. W. Morgan, "Accommodation and Its Relationship to Convergence," *Am. J. Optom. Arch. Am. Acad. Optom.* **21:**183–185 (1944).

268. H. Krueger, "Schwannkingen der Akkommodation des menschlichen Auges bei mon- und binokularer Beobachtung," *Albrecht v. Graefes Arch. Ophthal.* **205:**129–133 (1978).

269. M. C. Nadell and H. A. Knoll, "The Effect of Luminance, Target Configuration and Lenses upon the Refractive State of the Eye. Parts I and II," *Am. J. Optom. Arch. Am. Acad. Optom.* **33:**24–42 and 86–95 (1956).

270. G. G. Heath, "The Influence of Visual Acuity on Accommodative Responses of the Eye," *Am. J. Optom. Physiol. Opt.* **33:**513–534 (1956).

271. J. Tucker, W. N. Charman, and P. A. Ward, "Modulation Dependence of the Accommodation Response to Sinusoidal Gratings," *Vision Res.* **26:**1693–1707 (1986).

272. D. A. Owens, "A Comparison of Accommodative Responsiveness and Contrast Sensitivity for Sinusoidal Gratings," *Vision Res.* **20:**159–167 (1980).

273. L. J. Bour, "The Influence of the Spatial Distribution of a Target on the Dynamic Response and Fluctuations of the Accommodation of the Human Eye," *Vision Res.* **21:**1287–1296 (1981).

274. J. E. Raymond, I. M. Lindblad, and H. W. Leibowitz, "The Effect of Contrast on Sustained Detection," *Vision Res.* **24:**183–188 (1984).

275. J. Tucker and W. N. Charman, "Effect of Target Content at Higher Spatial Frequencies on the Accuracy of the Accommodation Response," *Ophthal. Physiol. Opt.* **7:**137–142 (1987).

276. K. J. Ciuffreda and D. Rumpf, "Contrast and Accommodation in Amblyopia," *Vision Res.* **25:**1445–1447 (1985).
277. K. J. Ciuffreda, M. Dul, and S. K. Fisher, "Higher-Order Spatial Frequency Contribution to Accommodative Accuracy in Normal and Amblyopic Observers," *Clin. Vis. Sci.* **1:**219–229 (1987).
278. K. J. Ciuffreda, "Accommodation to Gratings and More Naturalistic Stimuli," *Optom. Vis. Sci.* **68:**243–260 (1991).
279. H. Ripps, N. B. Chin, I. M. Siegel, and G. M. Breinin, "The Effect of Pupil Size on Accommodation, Convergence and the AC/A Ratio," *Invest. Ophthalmol.* **1:**127–135 (1962).
280. R. T. Hennessy, R. Iida, K. Shiina, and H. W. Leibowitz, "The Effect of Pupil Size on Accommodation," *Vision Res.* **16:**587–589 (1976).
281. P. A. Ward and W. N. Charman, "Effect of Pupil Size on Steady-State Accommodation," *Vision Res.* **25:**1317–1326 (1985).
282. F. W. Campbell, "The Minimum Quantity of Light Required to Elicit the Accommodation Reflex in Man," *J. Physiol.* (Lond.) **123:**357–366 (1954).
283. C. A. Johnson, "Effects of Luminance and Stimulus Distance on Accommodation and Visual Resolution," *J. Opt. Soc. Am.* **66:**138–142 (1976).
284. G. G. Heath, "Accommodative Responses of Totally Color Blind Observers," *Am. J. Optom. Arch. Am. Acad. Optom.* **33:**457–465 (1956).
285. I. C. J. Wood and A. T. Tomlinson, "The Accommodative Response in Amblyopia," *Am. J. Optom. Physiol. Opt.* **52:**243–247 (1975).
286. J. Otto and D. Safra, "Ergebnisse objectiver Akkommodationsmessungen an Augen mit organisch bedingtem Zentralskotom," *Albrecht von Graefes Arch. Klin. Ophthalmol.* **192:**49–56 (1974).
287. D. A. Owens, "The Resting State of the Eyes," *Am. Sci.* **72:**378–387 (1984).
288. W. N. Charman, "Static Accommodation and the Minimum Angle of Resolution," *Am. J. Optom. Physiol. Opt.* **63:**915–921 (1986).
289. H. W. Leibowitz and D. A. Owens, "Anomalous Myopias and the Intermediate Dark Focus of Accommodation," *Science* **189:**1121–1128 (1975).
290. H. W. Leibowitz and D. A. Owens, "Night Myopia and the Intermediate Dark Focus of Accommodation," *J. Opt. Soc. Am.* **65:**133–147 (1975).
291. H. W. Leibowitz and D. A. Owens, "New Evidence for the Intermediate Position of Relaxed Accommodation," *Doc. Ophthalmol.* **46:**1121–1128 (1978).
292. G. Smith, "The Accommodative Resting States, Instrument Accommodation and Their Measurement," *Optica Acta* **30:**347–359 (1983).
293. N. A. McBrien and M. Millodot, "The Relationship Between Tonic Accommodation and Refractive Error," *Invest. Ophthalmol. Vis. Sci.* **28:**997–1004 (1987).
294. F. M. Toates, "Accommodation Function of the Human Eye," *Physiol. Rev.* **52:**828–863 (1972).
295. G. Gilmartin, "The Role of Sympathetic Innervation of the Ciliary Muscle in Ocular Accommodation: A Review," *Ophthal. Physiol. Opt.* **6:**23–37 (1986).
296. H. W. Leibowitz, K. W. Gish, and J. B. Sheehy, "Role of Vergence Accommodation in Correcting for Night Myopia," *Am. J. Optom. Physiol. Opt.* **65:**383–386 (1988).
297. G. C. Hung, K. J. Ciuffreda, J. L. Semmlow, and S. C. Hokoda, "Model of Static Accommodative Behaviour in Human Amblyopia," *IEEE Trans. Biomed. Eng.* **BME-30:**665–672 (1983).
298. J. L. Semmlow and G. K. Hung, "The Near Response: Theories of Control," *Vergence Eye Movements: Basic and Clinical Concepts,* C. M. Schor and K. J. Ciuffreda (eds.), Butterworths, Boston, 1983, pp. 175–195.
299. C. M. Schor, "Adaptive Regulation of Accommodative Vergence and Vergence Accommodation," *Am. J. Optom. Physiol. Opty.* **63:**587–609 (1986).
300. C. M. Schor, "Influence of Accommodative and Vergence Adaptation on Binocular Motor Disorders," *Am. J. Optom. Physiol. Opt.* **65:**464–475 (1988).

301. M. Rosenfield and B. Gilmartin, "The Effect of Vergence Adaptation on Convergent Accommodation," *Ophthal. Physiol. Opt.* **8:**172–177 (1988).

302. W. N. Charman and J. Tucker, "Accommodation and Color," *J. Opt. Soc. Am.* **68:**459–471 (1978).

303. J. V. Lovasik and H. Kergoat, "Accommodative Performance for Chromatic Displays," *Ophthal. Physiol. Opt.* **8:**443–449 (1988).

304. R. Home and J. Poole, "Measurement of the Preferred Binocular Dioptric Settings at a High and Low Light Level," *Optica Acta* **24:**97 (1977).

305. G. G. Heath, "Components of Accommodation," *Am. J. Optom. Arch. Am. Acad. Optom.* **33:**569–579 (1956).

306. S. C. Hokoda and K. J. Ciuffreda, "Theoretical and Clinical Importance of Proximal Vergence and Accommodation," *Vergence Eye Movements: Basic and Clinical Concepts,* C. M. Schor and K. J. Ciuffreda (eds.), Butterworths, Boston, 1983, pp. 75–97.

307. G. Smith, K. C. Tan, and M. Letts, "Binocular Optical Instruments and Binocular Vision," *Clin. Exper. Optom.* **69:**137–144 (1986).

308. R. F. Fisher, "The Elastic Constants of the Human Lens," *J. Physiol.* (London) **212:**147–180 (1971).

309. R. F. Fisher, "Presbyopia and the Changes with Age in the Human Crystalline Lens," *J. Physiol.* (London) **228:**765–779 (1973).

310. R. F. Fisher, "The Mechanics of Accommodation in Relation to Presbyopia," *Eye* **2:**646–649 (1988).

311. D. Adler-Grinberg, "Questioning Our Classical Understanding of Accommodation and Presbyopia," *Am. J. Optom. Physiol. Opt.* **63:**571–580 (1986).

312. R. A. Weale, "Presbyopia Towards the End of the 20th Century," *Surv. Ophthalmol.* **34:**15–30 (1989).

313. M. J. Allen, "The Influence of Age on the Speed of Accommodation," *Am. J. Optom. Arch. Am. Acad. Optom.* **33:**201–208 (1956).

314. F. Sun and L. Stark, "Dynamics of Accommodation: Measurements for Clinical Application," *Exp. Neurol.* **91:**71–79 (1986).

315. A. Duane, "Studies in Monocular and Binocular Accommodation with Their Clinical Applications," *Am. J. Ophthalmol. Ser. 3* **5:**865–877 (1922).

316. D. Hamasaki, J. Ong, and E. Marg, "The Amplitude of Accommodation in Presbyopia," *Am. J. Optom. Arch. Am. Acad. Optom.* **33:**3–14 (1956).

317. C. Ramsdale and W. N. Charman, "A Longitudinal Study of the Changes in Accommodation Response," *Ophthal. Physiol. Opt.* **9:**255–263 (1989).

318. M. J. Simonelli, "The Dark Focus of the Human Eye and Its Relationship to Age and Visual Defect," *Human Factors* **25:**85–92 (1983).

319. M. A. Alpern, "Movements of the Eyes," *The Eye,* vol. 3, *Muscular Mechanisms,* H. Davson (ed.), Academic Press, New York, 1969, pp. 5–214.

320. B. L. Zuber (ed.), *Models of Oculomotor Behavior and Control,* CRC Press, Boca Raton, Fla., 1980.

321. R. H. S. Carpenter, *Movements of the Eyes*, 2d ed., Pion. Lond, 1988.

322. R. H. S. Carpenter (ed.), *Vision and Visual Dysfunction,* vol. 8, *Eye Movements,* Macmillan, London, 1991.

323. R. J. Leigh and D. S. Zee, *The Neurology of Eye Movements,* 2d ed., Davis, Philadelphia, 1991.

324. G. Fry and W. W. Hill, "The Mechanics of Elevating the Eye," *Am. J. Optom. Arch. Am. Acad. Optom.* **40:**707–716 (1963).

325. C. Rashbass and G. Westheimer, "Independence of Conjunctive and Disjunctive Eye Movements," *J. Physiol.* (London) **159:**361–364 (1961).

326. D. A. Robinson, "The Mechanics of Human Saccadic Eye Movements," *J. Physiol.* (London) **174:**245–264 (1964).

327. A. T. Bahill, A. Brockenbrough, and B. T. Troost, "Variability and Development of a Normative Data Base for Saccadic Eye Movements," *Invest. Ophthalmol.* **21:**116–125 (1981).

328. R. H. S. Carpenter, "Oculomotor Procrastination," *Eye Movements: Cognition and Visual Perception,* D. F. Fisher, R. A. Monty, and J. W. Senders (eds.), Lawrence Erlbaum, Hillsdale, N.J., 1981, pp. 237–246.

329. F. W. Campbell and R. H. Wurtz, "Saccadic Omission: Why We Do Not See a Grey Out During a Saccadic Eye Movement," *Vision Res.* **18:**1297–1303 (1978).

330. H. Collewijn and E. P. Tamminga, "Human Smooth Pursuit and Saccadic Eye Movements During Voluntary Pursuit of Different Target Motions on Different Backgrounds," *J. Physiol.* (London) **351:**217–250 (1984).

331. C. Rashbass and G. Westheimer, "Disjunctive Eye Movements," *J. Physiol.* (London) **159:**339–360 (1961).

332. C. J. Erkerlers, J. Van der Steen, R. M. Steinman, and H. Collewijn, "Ocular Vergence under Natural Conditions. I Continuous Changes of Target Distance Along the Median Plane," *Proc. R. Soc. Lond.* **B236:**417–440 (1989).

333. C. J. Erkelers, R. M. Steinman, and H. Collewijn, "Ocular Vergence under Natural Conditions. II Gaze Shifts Between Real Targets Differing in Distance and Direction," *Proc. R. Soc. Lond.* **B236:**441–465 (1989).

334. J. T. Enright, "Perspective Vergence: Oculomotor Responses to Line Drawings," *Vision Res.* **27:**1513–1526 (1987).

335. A. E. Kertesz, "Vertical and Cyclofusional Disparity Vergence," *Vergence Eye Movements: Clinical and Applied,* C. M. Schor and K. J. Ciuffreda (eds.), Butterworths, Boston, 1983, pp. 317–348.

336. J. M. Findlay, "Frequency Analysis of Human Involuntary Eye Movement," *Kybernetik* **8:**207–214 (1971).

337. R. W. Ditchburn, *Eye Movements and Visual Perception,* Clarendon Press, Oxford, 1973.

338. J. Nachmias, "Determinants of the Drift of the Eye During Monocular Fixation," *J. Opt. Soc. Am.* **51:**761–766 (1961).

339. H. C. Bennet-Clark, "The Oculomotor Response to Small Target Displacements," *Optica Acta* **11:**301–314 (1964).

340. W. H. Marshall and S. A. Talbot, "Recent Evidence for Neural Mechanisms in Vision Leading to a General Theory of Sensory Acuity," *Biol. Symp.* **7:**117–164 (1942).

341. O. Packer and D. R. Williams, "Blurring by Fixational Eye Movements," *Vision Res.* **32:**1931–1939 (1992).

342. E. F. Fincham, "The Accommodation Reflex and Its Stimulus," *Brit. J. Ophthalmol.* **35:**381–393 (1951).

343. C. M. Schor and M. C. Flom, "The Relative Value of Stereopsis as a Function of Viewing Distance," *Am. J. Optom. Arch. Am. Acad. Optom.* **46:**805–809 (1969).

344. N. A. Valyus, *Stereoscopy,* Focal Press, London, 1966.

345. C. H. Graham, "Visual Space Perception, *Vision and Visual Perception,* C. H. Graham (ed.), Wiley, London, 1966, chap. 18.

346. A. Lit, "Depth Discrimination Thresholds as a Function of Binocular Differences of Retinal Illuminance at Scotopic and Photopic Levels," *J. Opt. Soc. Am.* **49:**746–752 (1959).

347. S. C. Rawlings and T. Shipley, "Stereoscopic Acuity and Horizontal Angular Distance from Fixation," *J. Opt. Soc. Am.* **59:**991–993 (1969).

348. W. P. Dwyer and A. Lit, "Effect of Luminance-Matched Wavelength on Depth Discrimination at Scotopic and Photopic Levels of Target Illuminance," *J. Opt. Soc. Am.* **60:**127–131 (1970).

349. R. R. Fagin and J. R. Griffin, "Stereoacuity Tests: Comparison of Mathematical Equivalents," *Am. J. Optom. Physiol. Opt.* **59:**427–438 (1982).

350. G. Heron, S. Dholakia, D. E. Collins, and H. McLaughlan, "Stereoscopic Thresholds in Children and Adults," *Am. J. Optom. Physiol. Opt.* **62:**505–515 (1985).

351. H. J. Howard, "A Test for the Judgment of Distance," *Am. J. Ophthalmol.* **2:**656–675 (1919).

352. G. Westheimer, "Effect of Binocular Magnification Devices on Stereoscopic Depth Resolution," *J. Opt. Soc. Am.* **45:**278–280 (1956).

353. W. N. Charman and J. A. M. Jennings, "Binocular Vision in Relation to Stereoscopic Instruments and Three-Dimensional Displays," *Vision and Visual Dysfunction,* vol. 1, *Visual Optics and Instrumentation,* W. N. Charman (ed.), Macmillan, London, 1991, pp. 326–344.

CHAPTER 25
VISUAL PERFORMANCE

Wilson S. Geisler
Department of Psychology
University of Texas
Austin, Texas

Martin S. Banks
School of Optometry
University of California
Berkeley, California

25.1 GLOSSARY

A	amplitude
a	interpupillary distance
A_p	effective area of the entrance pupil
C	contrast
c_o	maximum concentration of photopigment
d	horizontal disparity
d_e	distance from the image plane to the exit pupil of an optical system
d_θ	average disparity between two points and the convergence point
$E(\lambda)$	photopic spectral illuminance distribution
$E_e(\lambda)$	spectral irradiance distribution
f	spatial frequency of a sinusoid
I_o	half-bleaching constant
J_o	maximum photocurrent
l	length of outer segment
$L(\lambda)$	photopic spectral luminance distribution
$L_e(\lambda)$	spectral radiance distribution
m	magnificaton of the exit pupil relative to the actual pupil
N	total effective photons absorbed per second
n_r	index of refraction of the media where the image plane is located
$n(\lambda)$	spectral photon-flux irradiance distribution
p	proportion of unbleached photopigment
t_o	time constant of photopigment regeneration
$t(\lambda)$	transmittance function of the ocular media

$V(\lambda)$	standard photopic spectral sensitivity function of the human visual system
$\alpha(\lambda)$	absorptance spectrum
ΔC	contrast increment or decrement
Δf	frequency increment or decrement
Δz	distance between any pair of points in the depth dimension
ε	retinal eccentricity
$\varepsilon(\lambda)$	extinction spectrum
θ	convergence angle of the eyes
θ	orientation of a sinusoid
κ	collection area, or aperture, of a photoreceptor
λ	wavelength of the light in a vacuum
ξ	isomerization efficiency
Σ	covariance matrix for a gaussian noise process
σ_o	half-saturation constant
τ	time interval
τ_{opt}	optimum time interval
$\Phi(\cdot)$	cumulative standard normal probability function
ϕ	phase of a sinusoid

25.2 INTRODUCTION

Physiological optics concerns the study of (1) how images are formed in biological eyes, (2) how those images are processed in the visual parts of the nervous system, and (3) how the properties of image formation and neural processing manifest themselves in the perceptual performance of the organism. The previous chapter reviewed image formation; this chapter briefly describes the neural processing of visual information in the early levels of the human visual system, and summarizes, somewhat more extensively, what is known about human visual performance.

An enormous amount of information about the physical environment is contained in the light reaching the cornea of the eye. This information is critical for many of the tasks the human observer must perform, including identification of objects and materials, determination of the three-dimensional structure of the environment, navigation through the environment, prediction of object trajectories, manipulation of objects, and communication with other individuals. The performance of a human observer in a given visual task is limited by the amount of information available in the light at the cornea and by the amount (and type) of information encoded and transmitted by the successive stages of visual processing. This chapter presents a concise description of visual performance in a number of fundamental visual tasks. It also presents a concise description of the physiological and psychological factors believed to underlie performance in those tasks. Two major criteria governed the selection of the material to be presented. First, we attempted to focus on quantitative data and theories that should prove useful for developing rigorous models or characterizations of visual performance. Second, we attempted to focus on data and theories that have a firm empirical basis, including at least some knowledge of the underlying biological mechanisms.

25.3 OPTICS, ANATOMY, PHYSIOLOGY OF THE VISUAL SYSTEM

Image Formation

The processing of visual information begins with the optics of the eye, which consists of three major components: the cornea, pupil, and lens. The optical components are designed to form a sharp image at the layer of the photoreceptors in the retina. Charman (Chap. 24) discusses many of the details concerning how these optical components affect image quality. We briefly describe a few of the most useful formulas and methods for computing the approximate size, location, quality, and intensity of images formed at the photoreceptors. Our aim is to provide descriptions of image formation that might prove useful for developing models or characterizations of performance in perceptual tasks.

The sizes and locations of retinal images can be found by projecting points on the objects along straight lines through the image (posterior) nodal point until they intersect the retinal surface. The intersections with the retinal surface give the image locations corresponding to the points on the objects. The angles between pairs of projection lines are *visual angles.* Image locations are usually described in terms of the visual angles that projection lines make with respect to a reference projection line (the *visual axis*), which passes through the nodal point and the center of the fovea. The radial visual angle between a projection line from an object and the visual axis is the object's *eccentricity.* Image sizes are often described by the visual angles between key points on the object. For purposes of computing the size and location of images, it is usually sufficient to use a simplified model of the eye's optics, such as the *reduced eye,* which consists of a single spherical refracting surface (radius of curvature = 5.5 mm) and a retinal surface located 16.7 mm behind the nodal point (see Fig. 2c, in Vol. I, Chap. 24 of this *Handbook*).

A general method for computing the quality of retinal images is by convolution with a point-spread function $h(x, y)$. Specifically, if $o(x, y)$ is the luminance (or radiance) of the object, and $i(x, y)$ is the image illuminance (or irradiance), then

$$i(x, y) = o(x, y) * * h(x, y) \tag{1}$$

where $**$ represents the two-dimensional convolution operator. The shape of the point-spread function varies with wavelength and with retinal location. The precise way to deal with wavelength is to perform a separate convolution for each wavelength in the spectral luminance distribution of the object. In practice, it often suffices to convolve with a single point-spread function, which is the weighted average, across wavelength, of the monochromatic point-spread functions, where the weights are given by the shape of the spectral luminance distribution. To deal with retinal location, one can make use of the fact that the human point-spread function changes only gradually with retinal eccentricity out to about 20 deg;[1] thus, a large proportion of the visual field can be divided into a few annular regions, each with a different point-spread function.

Calculation of the point-spread function is normally accomplished by finding the transfer function, $H(u, v)$.[2,3] (The point-spread function can be obtained, if desired, by an inverse Fourier transform.) The transfer function is given by the autocorrelation of the generalized pupil function followed by normalization to a peak value of 1.0:

$$T(u, v) = p(x, y)e^{iW(x,y,\lambda)} \otimes \otimes p(x, y)e^{-iW(x,y,\lambda)}\big|_{x=(\lambda d_e/mn_r)u,\, y=(\lambda d_e/mn_r)v} \tag{2}$$

$$H(u, v) = \frac{T(u, v)}{T(0, 0)} \tag{3}$$

where λ is the wavelength of the light in a vacuum, n_r is the index of refraction of the media where the image plane is located, d_e is the distance from the image plane to the exit pupil of the optical system, and m is the magnification of the exit pupil relative to the actual pupil. The generalized pupil function is the product of the simple pupil function, $p(x, y)$ (transmittance as a function of position within the actual pupil), and the aberration function, $e^{iW(x,t,\lambda)}$. The exit pupil is the apparent pupil, when viewed from the image plane. The size of and distance to the exit pupil can be found by ray-tracing a schematic eye. For the Le Grand eye, the relevant parameters are approximately as follows: $m = 1.03$, $d_e = 20.5$ mm, and $n = 1.336$. Average values of the monochromatic and chromatic aberrations are available (see Chap. 24) and can be used as estimates of $W(x, y, \lambda)$.

Equation (3) can be used to compute approximate point-spread functions (and hence image quality) for many stimulus conditions. However, for some conditions, direct measurements (or psychophysical measurements) of the point-spread function are also available (see Chap. 24), and are easier to deal with. For broadband (white) light and a well-accommodated eye, the axial point-spread functions directly measured by Campbell and Gubisch[4] are representative. A useful set of monochromatic point-spread functions was measured at various eccentricities by Navarro et al.[1] These point-spread functions can be used directly in Eq. (1) to compute approximate image quality.

The approximate retinal irradiance for extended objects is given by the following formula:

$$E_e(\lambda) = \frac{A_p}{278.3} L_e(\lambda)t(\lambda) \tag{4}$$

where $E_e(\lambda)$ is the retinal spectral irradiance distribution (watts $\cdot$ m^{-2} $\cdot$ nm^{-1}), $L_e(\lambda)$ is the spectral radiance distribution of the object (watts $\cdot$ m^{-2} $\cdot$ sr^{-1} $\cdot$ nm^{-1}), $t(\lambda)$ is the transmittance of the ocular media (see Chap. 24), and A_p is the effective area of the entrance pupil (mm^2). (Note, $A_p = \iint p(x/m', y/m')\, dx\, dy$, where m' is magnification of the entrance pupil relative to the actual pupil; the entrance pupil is the apparent size of the pupil when viewed from outside the eye.)

Photopic retinal illuminance, $E(\lambda)$ (candelas $\cdot$ nm^{-1}), is computed by an equivalent formula where the spectral radiance distribution, $L_e(\lambda)$, is replaced by the spectral luminance distribution, $L(\lambda)$ (candelas $\cdot$ m^{-2} $\cdot$ nm^{-1}), defined by

$$L(\lambda) = 683\, V(\lambda)L_e(\lambda) \tag{5}$$

where $V(\lambda)$ is the standard photopic spectral sensitivity function of the human visual system.[5]

In theoretical calculations, it is often useful to express light levels in terms of photon flux rather than in terms of watts (see later). The photon-flux irradiance on the retina, $n(\lambda)$ (quanta $\cdot$ sec^{-1} $\cdot$ deg^{-2} $\cdot$ nm^{-1}) is computed by multiplying the retinal irradiance, $E_e(\lambda)$, by 8.4801×10^{-8} which converts m^2 to deg^2 (based upon the *reduced eye*), and by λ/ch which converts watts into quanta/sec (where c is the speed of light in a vacuum, and h is Planck's constant). Thus,

$$n(\lambda) = 1.53 \times 10^6 A_p L_e(\lambda)t(\lambda)\lambda \tag{6}$$

and, by substitution of Eq. (5) into Eq. (6),

$$n(\lambda) = 2.24 \times 10^3 A_p \frac{L(\lambda)}{V(\lambda)} t(\lambda)\lambda \tag{7}$$

Most light-measuring devices report radiance, $L_e(\lambda)$, or luminance, $L(\lambda)$; Eqs. (6) and (7) allow conversion to retinal photon-flux irradiance, $n(\lambda)$. For more details on the calculation of retinal intensity, see Wyszecki and Stiles.[5]

Image Sampling by the Photoreceptors

The image formed at the receptor layer is described by a four-dimensional function $n(x, y, t, \lambda)$, which gives the mean photon-flux irradiance (quanta $\cdot$ sec^{-1} $\cdot$ deg^2 $\cdot$ nm^{-1}) as a function of space (x, y), time (t), and wavelength (λ). This four-dimensional function also describes the photon noise in the image. Specifically, photon noise is adequately described as an inhomogeneous Poisson process; thus, the variance in the number of photons incident in a given interval of space, time, and wavelength is equal to the mean number of photons incident in that same interval.

The photoreceptors encode the (noisy) retinal image into a discrete representation in space and wavelength, and a more continuous representation in time. The image sampling process is a crucial step in vision that can, and often does, result in significant information loss. The losses occur because physical and physiological constraints make it impossible to sample all four dimensions with sufficiently high resolution.

As shown in the schematic diagram in Fig. 1, there are two major types of photoreceptors: rods and cones. They play very different functional roles in vision; rods subserve vision at low light levels and cones at high light levels. There are three types of cones, each with a different spectral sensitivity (which is the result of having different photopigments in the outer segment). The "long" (L), "middle" (M), and "short" (S) wavelength cones have peak spectral sensitivities at wavelengths of approximately 570, 540, and 440 nm, respectively. Information about the spectral wavelength distribution of the light falling on the retina is encoded by the relative activities of the L, M, and S cones. All rods have the same spectral sensitivity (and the same photopigment), peaking at about 500 nm.

The quality of spatial, temporal, and wavelength information encoded by the photoreceptors depends upon: (1) the spatial distribution of the photoreceptors across the retina, (2) the efficiency with which individual photoreceptors absorb light at different wavelengths (the absorptance spectrum), (3) the area over which the individual photoreceptors collect light (the receptor aperture), and (4) the length of time over which the individual photoreceptors integrate light.

The spatial distribution of cones and rods is highly nonuniform. Figure 2*a* shows the typical density distribution of rod and cone photoreceptors across the retina (although there are individual differences; e.g., see Ref. 6). Cone density decreases precipitously with eccentricity; rods are absent in the fovea and reach a peak density at 20 deg. If the receptor lattice were perfectly regular, the highest unambiguously resolvable spatial frequency (the Nyquist limit), would be half the linear density (in cells $\cdot$ deg^{-1}). Under normal viewing conditions, the Nyquist limit does not affect vision in the fovea because the eye's optics eliminate spatial frequencies at and above the limit. However, the presentation of interference fringes (which avoid degradation by the eye's optics) yields visible spatial aliasing for spatial frequencies above the Nyquist limit.[7–9]

The densities and retinal distributions of the three cone types are quite different. The S cones form a rather regular lattice comprising less the 2 percent of cones in the central fovea and somewhat less than 10 percent of the cones elsewhere;[10–12] they may be absent in the central 20′–25′ of the fovea.[13] It is much more difficult to distinguish individual L and M cones anatomically, so their densities and distributions are less certain. Psychophysical evidence indicates that the ratio of L to M cones is approximately 2 : 1,[14,15] but the available physiological data in monkey suggest a ratio closer to 1 : 1.[16,17]

From the Beer-Lambert law, the absorptance spectrum of a receptor depends upon the concentration of the photopigment, $c_o p$, in the receptor outer segment, the length of the outer segment, l, and the extinction spectrum, $\varepsilon(\lambda)$, of the photopigment,

$$\alpha(\lambda) = 1 - 10^{-lc_o p\varepsilon(\lambda)} \tag{8}$$

where c_o is the concentration of photopigment in the dark-adapted eye and p is proportion of unbleached photopigment. The specific absorptance spectra for the different classes of

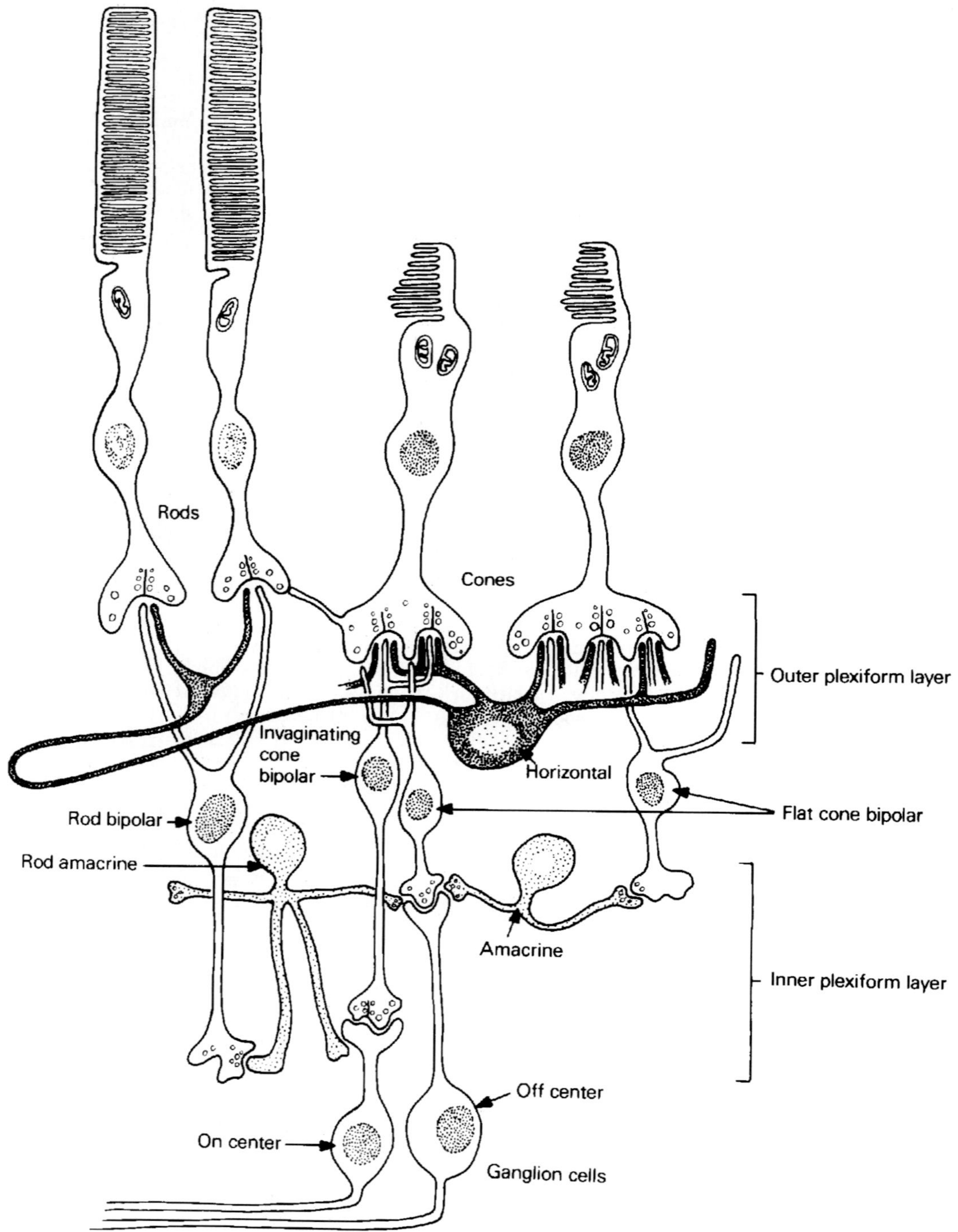

FIGURE 1 Schematic diagram of the retinal neurons and their major synaptic connections. Note that rods, rod bipolar cells, and rod amacrine cells are absent in the fovea. *(From Ref. 262.)*

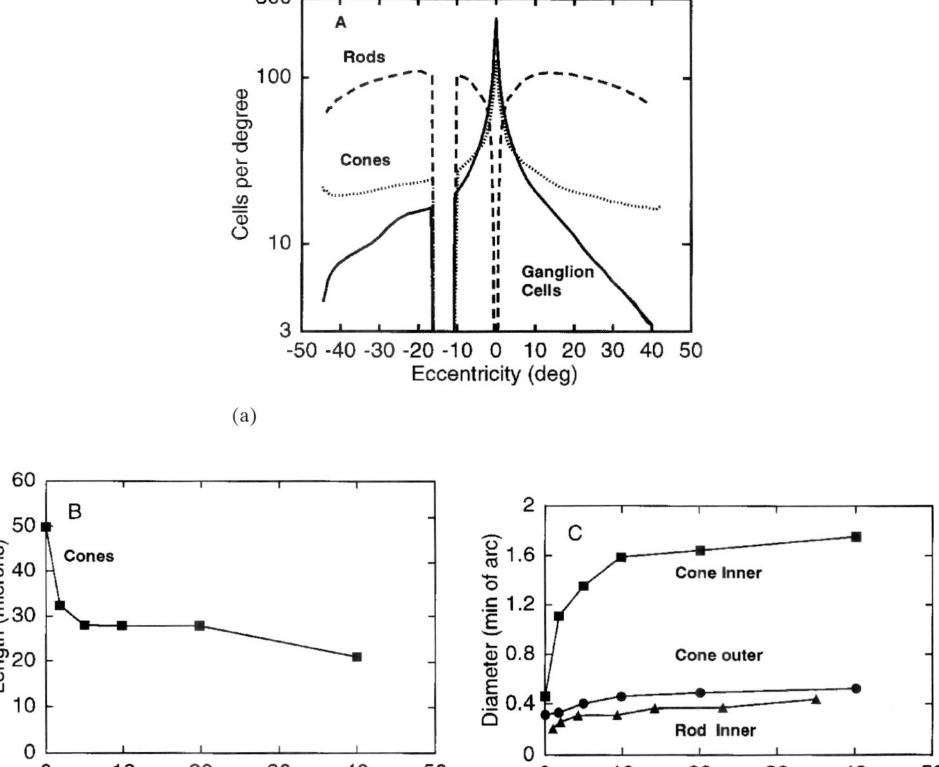

FIGURE 2 (*a*) Linear density of cones, rods, and ganglion cells as a function of eccentricity in the human retina. (The data were modified from Refs. 6 and 32.) Conversion from cells/mm^2 to cells/deg^2 was computed assuming a posterior nodal point 16.68 mm from the retina, and a retinal radius of curvature of 12.1 mm. Conversion to cells/deg was obtained by taking the square root of areal density. Ganglion cell density in the central 10 deg was derived assuming a 3:1 ratio of ganglion cells to cones in the fovea.[32] (*b*) Human cone outer segment length. (Modified from Ref. 125.) (*c*) Human cone inner segment, cone outer segment and rod diameter as a function of eccentricity. (Modified from Ref. 125.)

receptors are described in Chap. 26 (Brainard). At the peak wavelength, the photoreceptors absorb approximately 50 percent of the incident photons (although there are some variations with eccentricity).

Because of photopigment bleaching and regeneration processes within the photoreceptor, the absorption spectra of the photopigments change depending upon the history of photon absorptions. For example, prolonged exposure to high intensities depletes a significant fraction of the available photopigment, reducing the overall optical density. Reflection densitometry measurements[18–20] have shown, to first approximation, that the proportion p of available photopigment at a given point in time is described by a first-order differential equation:

$$\frac{dp}{dt} = -\frac{n(\lambda)\xi(1 - 10^{-lc_o\varepsilon(\lambda)p})}{lc_o} + \frac{1-p}{t_o} \tag{9}$$

where $n(\lambda)$ is the photon-flux irradiance distribution (quanta $\cdot$ sec^{-1} $\cdot$ deg^{-2} $\cdot$ nm^{-1}), t_o is the exponential time constant of regeneration, and ξ is the isomerization efficiency (which is believed to be near 1.0). For broadband light, Eq. (9) simplifies and can be expressed in terms of retinal illumination:[19]

$$\frac{dp}{dt} = -\frac{Ip}{Q_e} + \frac{(1-p)}{t_o} \tag{10}$$

where I is the retinal illumination (in trolands*) and Q_e is the energy of a flash (in troland $\cdot$ seconds) required to reduce the proportion of unbleached photopigment to $1/e$. For the cone photopigments, the time constant of regeneration is approximately 120 sec, and Q_e (for broadband light) is approximately 2.4×10^6 troland $\cdot$ sec. For rods, the time constant of regeneration is approximately 360 sec, and Q_e is approximately 1.0×10^7 scotopic troland $\cdot$ sec. Equation (10) implies that the steady-state proportion of available photopigment is approximately

$$p(I) = \frac{I_o}{I_o + I} \tag{11}$$

where $I_o = Q_e/t_o$. (I_o is known as the *half-bleaching constant.*) Equation (11) is useful when computing photon absorptions at high ambient light levels; however, photopigment bleaching is not significant at low to moderate light levels. Equation (7) also implies that bleaching and regeneration can produce changes in the shapes of the absorptance spectra (see Chap. 26).

The lengths of the photoreceptor outer segments (and hence their absorptances) change with eccentricity. Figure 2*b* shows that the cone outer segments are longer in the fovea. Rod outer segments are approximately constant in length. If the dark-adapted photopigment concentration, c_o, is constant in a given class of photoreceptor (which is likely), then increasing outer segment length increases the amount of light collected by the receptor (which increases signal-to-noise ratio).

The light collection area of the cones is believed to be approximately 70 to 90 percent of the cross-sectional area of the inner segment at its widest point (see Fig. 1), although direct measurements are not available (see Refs. 21, 22). The collection area of the rod receptors is equal to the cross-sectional area of the inner and outer segments (which is the same). Figure 2c plots inner segment diameter as a function of eccentricity for cones and rods. As can be seen, the cone aperture increases with eccentricity, while the rod aperture is fairly constant. Increasing the cone aperture increases the light collected (which increases signal-to-noise ratio), but slightly reduces contrast sensitivity at high spatial frequencies (see later).

The data and formulas above [Eqs. (6)–(10)] can be used to compute (approximately) the total effective photons absorbed per second, N, in any receptor, at any eccentricity:

$$N = \int \kappa \xi \alpha(\lambda) n(\lambda)\, d\lambda \tag{12}$$

where κ is the collection area, or aperture, of the receptor (in deg^2).

Photons entering through the pupillary center have a greater chance of being absorbed in the cones than photons entering through the pupillary margins. This phenomenon, known as the Stiles-Crawford effect, can be included in the above calculations by modifying the pupil function [Eqs. (3) and (4); also see Vol. I, Chap. 24]. We also note that the cones are oriented toward the exit pupil.

* The troland is defined to be the retinal illumination produced by viewing a surface with a luminance of 1 cd/m^2 through a pupil with an area of 1 mm^2.

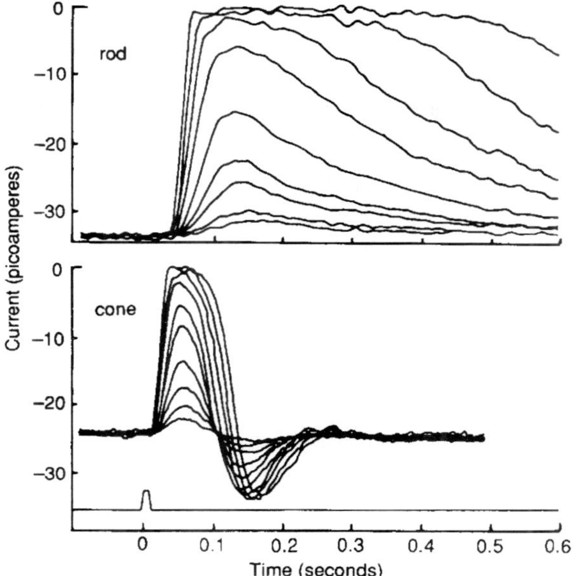

FIGURE 3 Photocurrent responses of macaque rod and cone photoreceptors to flashes of light. Each trace represents the response to a different intensity level. The flash intensities were varied by factors of two. *(From Ref. 23, "How Photoreceptor Cells Respond to Light," Julie Schnapf and Denis A. Baylor. Copyright © 1989 by Scientific American, Inc. All rights reserved.)*

The temporal response properties of photoreceptors are more difficult to measure than their spatial integration properties. Figure 3 shows some physiological measurements of the photocurrent responses of primate (macaque) rods and cones at low and moderate light levels.[23] Rods integrate photons over a substantially longer duration than do cones. In addition, rods produce reliable responses to single photon absorptions, but cones do not.

Photoreceptors, like all neurons, have a limited dynamic range. Direct recordings from receptors in macaque have shown that the peak responses of photoreceptors to brief flashes of light are adequately described by either a modified Michaelis-Menton function

$$f(z) = \frac{J_o z^n}{z^n + \sigma_o^n} \tag{13}$$

or an exponential function

$$f(z) = J_o(1 - 2^{-z^n/\sigma_o^n}) \tag{14}$$

where J_o is the maximum photocurrent, σ_o is the half-saturation constant (the value of z that produces exactly half the maximum response) and n is an exponent that has a value near 1.0.[17,24,25] As the above equations imply, the photoreceptors respond approximately linearly to transient intensity changes at low to moderate light levels, but respond nonlinearly (ultimately reaching saturation) at higher light levels. Recent electroretinogram (ERG) measurements suggest that the flash response of human photoreceptors is similar to those of the macaque.[26]

The nonlinear response saturation suggests that the receptors should not be able to encode high image intensities accurately (because any two intensities that saturate receptor response will be indistinguishable). However, the cones avoid saturation effects under many circumstances by adjusting their gain (σ_o) depending upon the ambient light level; the rods do not adjust their gain nearly as much. The gain adjustment is accomplished in a few ways. Photopigment depletion (see earlier) allows multiplicative gain changes, but only operates at very high ambient light levels. Faster-acting mechanisms in the phototransduction sequence are effective at moderate to high light levels.[27] At this time, there remains some uncertainty about how much gain adjustment (adaptation) occurs within the cones.[17,24,28]

Retinal Processing

The information encoded by the photoreceptors is processed by several layers of neurons in the retina. The major classes of retinal neuron in the primate (human) visual system and their typical interconnections are illustrated schematically in Fig. 1. Although the retina has been studied more extensively than any other part of the nervous system, its structure and function are complicated and not yet fully understood. The available evidence suggests that the primate retina is divided into three partially overlapping neural pathways: a *rod pathway,* a *cone parvo pathway,* and a *cone magno pathway* (for reviews see Refs. 29–31). These pathways, illustrated in Fig. 4, carry most of the information utilized in high-level visual processing tasks. There are other, smaller pathways (which will not be described here) that are involved in functions such as control of the pupil reflex, accommodation, and eye movements. Although Fig. 4 is based upon the available evidence, it should be kept in mind that there remains considerable uncertainty about some of the connections.

Retinal Anatomy. As indicated in Fig. 4, the photoreceptors (R, C) form electrical synapses (gap-junctions) with each other, and chemical synapses with *horizontal* cells (H) and *bipolar* cells (RB, MB +, MB −, DB +, DB −). The electrical connections between receptors are noninverting* and hence produce simple spatial and temporal summation. The dendritic processes and the axon-terminal processes of the horizontal cells provide spatially extended, negative-feedback connections from cones onto cones and from rods onto rods, respectively. It is likely that the horizontal cells play an important role in creating the surround response of ganglion cells. Rod responses probably do not influence cone responses (or vice versa) via the horizontal cells.

In the rod pathway, rod bipolar cells (RB) form noninverting chemical synapses with *amacrine* cells (AII, A). The rod amacrine cells (AII) form electrical (noninverting) synapses with the on-center bipolar cells (MB +, DB +), and form chemical (inverting) synapses with off-center bipolar cells (MB −, DB −) and with off-center ganglion cells (PG −, MG −). Other amacrine cells (A) form reciprocal negative-feedback connections to the rod bipolar cells.

In the cone pathways, on-bipolar cells (MB +, DB +) form chemical synapses with amacrine cells (A) and with on-center *ganglion* cells (PG +, MG +).† Off-bipolar cells (FMB, FDB) form chemical synapses with amacrine cells and off-center ganglion cells (PG −, MG −).

The ganglion cells are the output neurons of the retina; their mylenated axons form the optic nerve. In the fovea and parafovea, each P ganglion cell (PG +, PG −) forms

* By *noninverting* we mean that changes in the response of the presynaptic neuron, in a given direction, produce changes in the response of the postsynaptic neuron in the same direction.

† The parvo ganglion cells and magno ganglion cells are also referred to as *midget ganglion cells* and *parasol ganglion cells,* respectively.

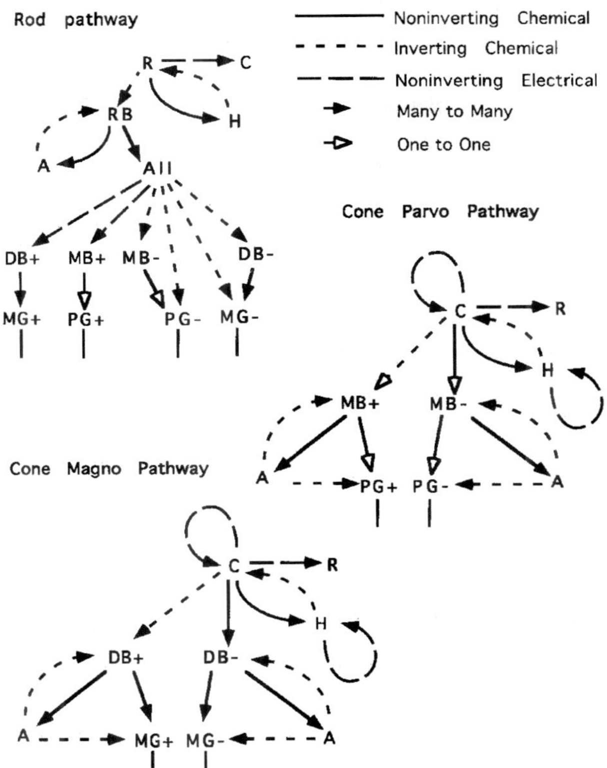

FIGURE 4 Schematic diagram of neural connections in the primate retina. R (rod), C (cone), H (horizontal cell), RB (rod bipolar), MB+ (midget on-bipolar), MB− (midget off-bipolar), DB+ (diffuse on-bipolar), DB− (diffuse off-bipolar), A (amacrine), AII (rod amacrine), PG+ (P or midget on-center ganglion cell), PG− (P or midget off-center ganglion cell), MG+ (M or parasol on-center ganglion cell), MG− (M or parasol off-center ganglion cell).

synapses with only one midget bipolar cell (MB + , MB −), and each midget bipolar cell forms synapses with only one cone; however, each cone makes contact with a midget on-bipolar and a midget-off bipolar. Thus, the P pathway (in the fovea) is able to carry fine spatial information. Current evidence also suggests that the P pathway carries most of the chromatic information (see Chap. 26). The M ganglion cells (MG+, MG−) form synapses with diffuse bipolar cells (DB+, DB−), and each diffuse bipolar cell forms synapses with many cones. Thus, the M pathway is less able to carry fine spatial information. However, the larger spatial summation (and other boosts in gain) give the M pathway greater sensitivity to changes in contrast. Furthermore, the larger sizes of neurons in the M pathway provide somewhat faster responses and hence somewhat faster transmission of information.

Figure 2*a* shows the density of ganglion cells as a function of eccentricity in the human retina;[32] for macaque retina, see Ref. 33. At each eccentricity, the P cells (PG+, PG−) comprise approximately 80 percent of the ganglion cells and the M cells (MG+, MG−)

approximately 10 percent. Half the P and M cells are on-center (PG+ and MG+) and half are off-center (PG− and MG−). Thus, the approximate densities of the four major classes of ganglion cells can be obtained by scaling the curve in Fig. 2*a*. As can be seen, ganglion cell density decreases more quickly with eccentricity than does cone density, but in the fovea there are at least 2–3 ganglion cells per cone (a reasonable assumption is that there are two P cells for every cone). The Nyquist limit of the retina would appear to be set by the cone density in the fovea and by the ganglion cell density in the periphery.

Retinal Physiology. Ganglion cells transmit information to the brain via action potentials propagating along axons in the optic nerve. The other retinal neurons transmit information as graded potentials (although amacrine cells generate occasional action potentials). Much remains to be learned about exactly how the computations evident in the responses of ganglion cells are implanted within the retinal circuitry.*

The receptive fields† of ganglion cells are approximately circular and, based upon their responses to spots of light, can be divided into a center region and an antagonistic, annular, surround region.[34] In on-center ganglion cells, light to the center increases the response, whereas light to the surround suppresses the response; the opposite occurs in off-center ganglion cells.

For small to moderate contrast modulations around a steady mean luminance, most P and M cells respond approximately linearly, and hence their spatial and temporal response properties can be usefully characterized by a spatio-temporal transfer function. The spatial components of the transfer function are adequately described as a difference of gaussian functions with separate space constants representing the center and surround;[35–37] see Fig. 5*a*. The available data suggest that the surround diameter is typically 3–6 times the center diameter.[37] The temporal components of the transfer function have been described as a cascade of simple feedback and feed-forward linear filters (e.g., Ref. 38; see Fig. 5*b*). There also appears to be a small subset of M cells that is highly nonlinear, similar to the Y cells in cat.[39]

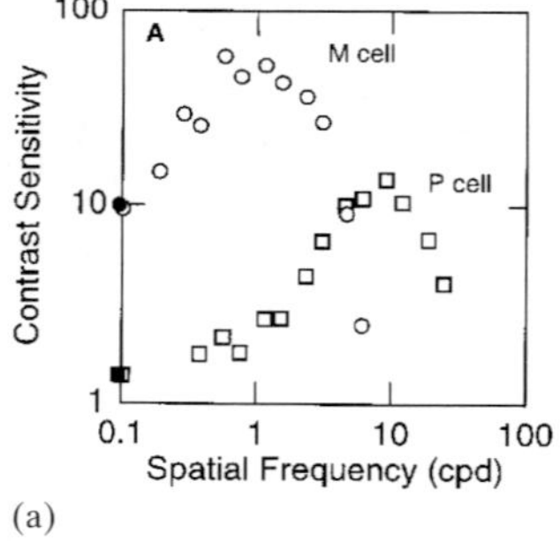

(a)

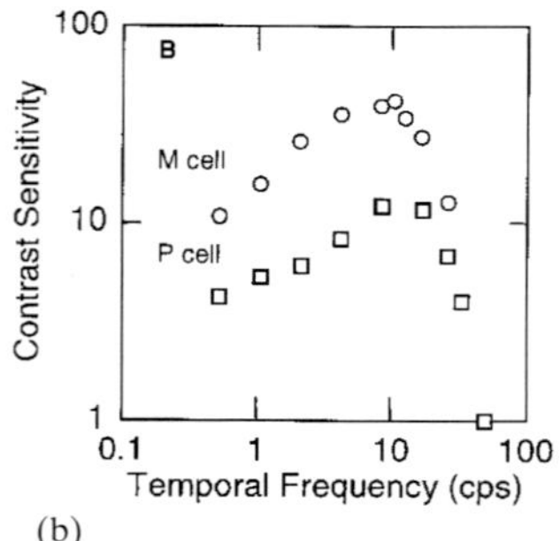

(b)

FIGURE 5 Typical spatial and temporal contrast sensitivity functions of P and M ganglion cells. (*a*) Spatial contrast sensitivity functions for sinusoidal gratings with a mean intensity of 1400 td, drifting at 5.2 cps. The solid symbols are modulation thresholds for a uniform field flickering sinusoidally at 5.2 cps. The threshold criterion was approximately 10 extra impulses per sec. (*Adapted from Ref. 37.*) (*b*) Temporal contrast sensitivity functions for drifting sinusoidal gratings with mean intensities of 4600 td (P cell) and 3600 td (M cell), and spatial frequencies of 3 cpd (P cell) and 1.6 cpd (M cell). The threshold criterion was approximately 10 extra impulses per sec. *(Adapted from Ref. 38.)*

* In this subsection we describe the electrophysiology of M and P cells. This description is based on a composite of data obtained from ganglion cells and geniculate cells. The response properties of M and P ganglion cells and M and P geniculate cells are very similar.

† The receptive field of a neuron is defined to be the region of the visual field (or, equivalently, of the receptor array) where light stimulation has an effect on the response of the neuron.

The response amplitude of M and P cells as a function of sinewave contrast is reasonably well described by a Michaelis-Menton function [i.e., by Eq. (14)], but where z and σ_o are contrasts rather than intensities. As indicated by Fig. 5, the M ganglion cells are (over most spatial and temporal frequencies) about 4–10 times more sensitive than the P ganglion cells; in other words, σ_o is 4–10 times smaller in M cells than P cells.

When mean (ambient) light level increases, the high-frequency fall-offs of the temporal transfer functions of M and P cells shift toward higher frequencies, corresponding to a decrease in the time constants of the linear filters.[38,40] The effect is reduction in gain (an increase in σ_o), and an increase in temporal resolution.

Relatively little is known about how the spatial transfer functions of primate ganglion cells change with mean light level. In cat, the relative strength of the surround grows with mean luminance, but the space constants (sizes) of the center and surround components appear to change relatively little.[41,42] The major effect of increasing adaptation luminance is an increase in the magnitude of the low-frequency fall-off of the spatial transfer function.

The center diameters of both P and M cells increase with eccentricity,[42,43] roughly in inverse proportion to the square root of ganglion-cell density (cf., Fig. 2*a*). However, the precise relationship has not been firmly established because of the difficulty in measuring center diameters of ganglion cells near the center of the fovea, where the optical point-spread function is likely to be a major component of center diameter.[37]

Central Visual Processing

Most of the information encoded in the retina is transmitted via the optic nerve to the lateral geniculate nucleus (LGN) in the thalamus. The neurons in the LGN then relay the retinal information to the primary visual cortex (V1). Neurons in V1 project to a variety of other visual areas. Less is known about the structure and function of V1 than of the retina or LGN, and still less is known about the subsequent visual areas.

Central Anatomy. The LGN is divided into six layers (see Fig. 6); the upper four layers (the parvocellular laminae) receive synaptic input from the P ganglion cells, the lower two layers (the magnocellular laminae) receive input from the M ganglion cells. Each layer of the LGN receives input from one eye only, three layers from the left eye and three from the right eye. The LGN also receives input from other brain areas including projections from the reticular formation and a massive projection (of uncertain function) from layer 6 of V1. The total number of LGN neurons projecting to the visual cortex is slightly larger than the number of ganglion cells projecting to the LGN.

The segregation of M and P ganglion cell afferents into separate processing streams at the LGN is preserved to some extent in subsequent cortical processing. The magnocellular neurons of the LGN project primarily to neurons in layer 4cα in V1, which project primarily to layer 4b. The neurons in layer 4b project to several other cortical areas including the middle-temporal area (MT) which appears to play an important role in high-level motion processing (for reviews see Refs. 44–46).

The parvocellular neurons of the LGN project to layers 4a and 4cβ in V1. The neurons in layers 4a and 4cβ project to the superficial layers of V1 (layers 1, 2, and 3), which project to areas V2, V3. Areas V2 and V3 send major projections to area V4 which sends major projections to IT (infero-temporal cortex). The superficial layers of V1 also send projections to layers 5 and 6.

It has been hypothesized that the magno stream subserves crucial aspects of motion and depth perception and that the parvo stream subserves crucial aspects of form and color perception. However, current understanding of the areas beyond V1 is very limited; it is likely that our views of their functional roles in perception will change substantially in the near future. For reviews of cortical functional anatomy see, for example, Refs. 46–49.

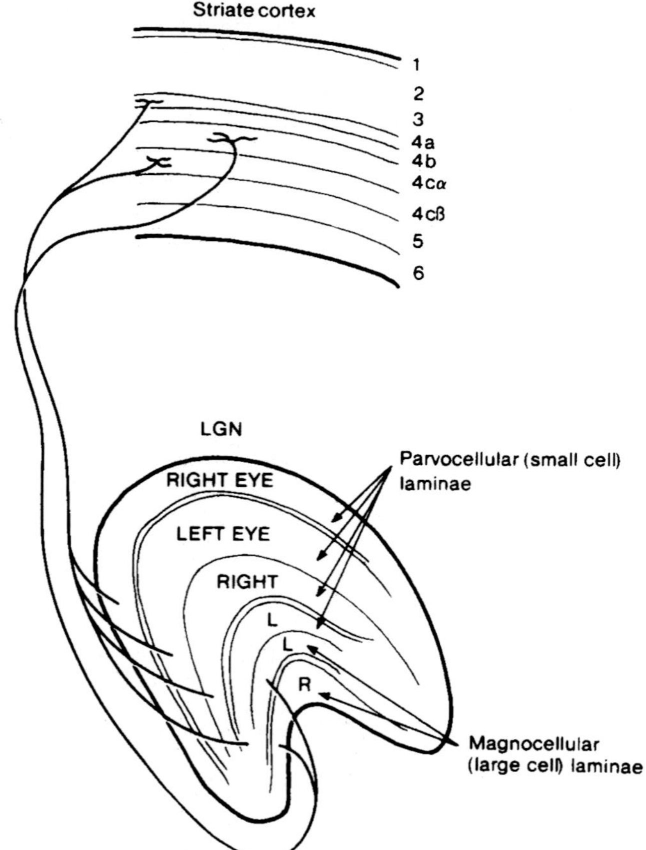

FIGURE 6 Schematic diagram of a vertical section through the lateral geniculate nucleus and through striate cortex (V1) of the primate visual system. *(Reproduced from Ref. 49.)*

Central Physiology. The receptive field properties of LGN neurons are quite similar to those of their ganglion-cell inputs.[50–52] LGN neurons have, on average, a lower spontaneous response rate than ganglion cells, and are more affected by changes in alertness and anesthetic level, but otherwise display similar center/surround organization, and similar spatial, temporal, and chromatic response properties (however, see Ref. 53).

The receptive-field properties of neurons in the primary visual cortex are substantially different from those in the retina or LGN; V1 neurons have elongated receptive fields that display a substantial degree of selectivity to the size (spatial frequency), orientation, direction of motion of retinal stimulation, and binocular disparity.[54–56] Thus, the probability that a cortical neuron will be activated by an arbitrary retinal image is much lower than for retinal and LGN neurons.

Each region of the visual field is sampled by cortical neurons selective to the full range of sizes, orientations, directions of motion, and binocular disparities. When tested with sinewave gratings, cortical neurons have a spatial-frequency bandwidth (full width at half height) of 1–1.5 octaves and an orientation bandwidth of 25–35 degrees.[57,58] The number of neurons in area V1 is more than two orders of magnitude greater than the number of LGN neurons;[59] thus, there are sufficient numbers of cortical neurons in each region of the visual field for their receptive fields to tile the whole spatial-frequency plane transmitted from retina to cortex several times. However, the number of cortical cells encoding each region of the spatial-frequency plane is still uncertain. The temporal frequency tuning of V1 neurons is considerably broader than the spatial frequency tuning with peak sensitivities mostly in the range of 5–10 cps.[60] The direction selectivity of V1 neurons varies from 0 to 100 percent with an average of 50 to 60 percent.[57]* Early measurements of disparity tuning in primate suggested that cortical neurons fall into three major categories: one selective to crossed disparities, one selective to uncrossed disparities, and one tuned to disparities near zero.[56] Recent evidence[61] in agreement with evidence in cat cortex[62,63] suggests a more continuous distribution of disparity tuning.

The chromatic response properties of V1 neurons are not yet fully understood.[49] There is some evidence that discrete regions in the superficial layers of V1, called cytochrome oxidase "blobs," contain a large proportion of neurons responsive to chromatic stimuli;[64] however, see Ref. 65.

The spatial-frequency tuning functions of cortical cells have been described by a variety of simple functions including Gabor filters, derivatives of gaussian filters, and log Gabor filters (e.g., Ref. 66). The temporal-frequency tuning functions have been described by difference of gamma filters. The full spatio-temporal tuning (including the direction-selective properties of cortical cells) has been described by quadrature (or near quadrature) pairs of separable spatio-temporal filters with spatial and temporal components drawn from the types listed above (e.g., Ref. 67)

Essentially all V1 neurons display nonlinear response characteristics, but can be divided into two classes based upon their nonlinear behavior. *Simple cells* produce approximately half-wave-rectified responses to drifting or counterphase sinewave gratings; *complex cells* produce a large unmodulated (DC) response component with a superimposed half-wave or full-wave rectified response component.[68] Simple cells are quite sensitive to the spatial phase or position within the receptor field; complex cells are relatively insensitive to position within the receptive field.[54]

Most simple and complex cells display an accelerating response nonlinearity at low contrasts and response saturation at high contrasts, which can be described by a Michaelis-Menton function [Eq. (11)] with an exponent greater than 1.0.[69] These nonlinearities impart the important property that response reaches saturation at approximately the same physical contrast independent of the spatial frequency, orientation, or direction of motion of the stimulus. The nonlinearities sharpen the spatio-temporal tuning of cortical neurons while maintaining that tuning largely independent of contrast, even though the neurons often reach response saturation of 10 to 20 percent contrast.[70,71]

Unlike retinal ganglion cells and LGN neurons, cortical neurons have little or no spontaneous response activity. When cortical cells respond, however, they display noise characteristics similar to retinal and LGN neurons; specifically, the variance of the response is approximately proportional to the mean response;[72,73] unlike Poisson noise, the proportionality constant is usually greater than 1.0.

25.4 VISUAL PERFORMANCE

A major goal in the study of human vision is to relate performance—for example, the ability to see fine detail—to the underlying anatomy and physiology. In the sections that follow, we will discuss some of the data and theories concerning the detection,

* Direction selectivity is defined as $100 \times (R_p - R_n)/R_p$, where R_p is the magnitude of response in the preferred direction and R_n is the magnitude of response in the nonpreferred direction.

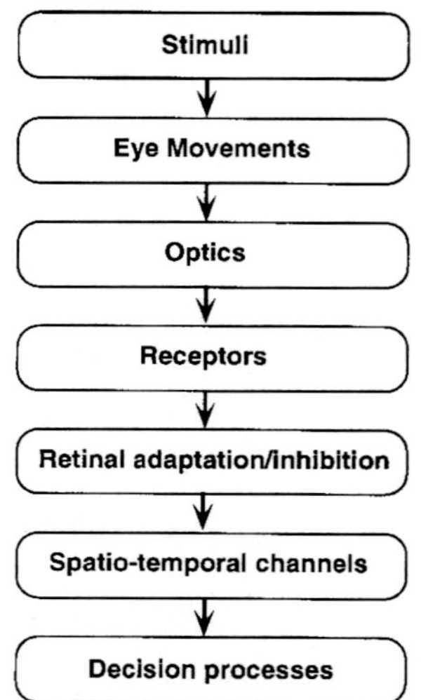

FIGURE 7 Simple information-processing model of the visual system representing the major factors affecting contrast detection and discrimination.

discrimination, and estimation of contrast, position, shape, motion, and depth by human observers. In this discussion, it will be useful to use as a theoretical framework the information-processing model depicted in Fig. 7. The visual system is represented as a cascade of processes, starting with eye movements, that direct the visual axis toward a point of interest, proceeding through a number of processing stages discussed in the previous sections and in Chaps. 24 and 26, and ending with a decision. When a human observer makes a decision about a set of visual stimuli, the accuracy of this decision depends on all the processes shown here. Psychophysical experiments measure the performance of the system as a whole, so if one manipulates a particular variable and finds it affects observers' performance in a visual task, one cannot readily pinpoint the responsible stage(s). The effect could reflect a change in the amount of information the stimulus provides, in the fidelity of the retinal image, in representation of the stimulus among spatio-temporal channels, or in the observer's decision strategy. Obviously, it is important to isolate effects due to particular processing stages as well as possible.

We begin by discussing the decisions a human observer might be asked to make concerning visual stimuli. Humans are required to perform a wide variety of visual tasks, but they can be divided into two general categories: *identification tasks* and *estimation tasks.** In an identification task, the observer is required to identify a visual image, sequence of images, or a part of an image, as belonging to one of a small number of discrete categories. An important special case is the *discrimination task* in which there are only two categories (e.g., Is the letter on the TV screen an *E* or an *F*?). When one of the two categories is physically uniform, the task is often referred to as a *detection task* (e.g., Is there a letter *E* on the screen or is the screen blank?). In the estimation task, the observer is required to estimate the value of some property of the image (e.g., What is the height of the letter?). The distinction between estimation and identification is quantitative, not

* Visual tasks can also be categorized as *objective* or *subjective.* Objective tasks are those for which there is, in principle, a precise physical standard against which performance can be evaluated (e.g., a best or correct performance). The focus here is on objective tasks because performance in those tasks is more easily related to the underlying physiology.

qualitative; strictly speaking, an estimation task can be regarded as an identification task with a large number of categories. There are two fundamental measures of performance in visual tasks: the *accuracy* with which the task is performed, and the *speed* at which the task is performed. For further discussion of methods for measuring visual performance, see Chap. 29 (Pelli and Farell).

Referring again to the information-processing model of Fig. 7, we note that it is difficult to isolate the decision strategy; the best one can do is to train observers to adhere as closely as possible to the same decision strategy across experimental conditions. So, whether the experiment involves simple identification tasks, such as discrimination or detection procedures, or estimation tasks, observers are generally trained to the point of using a consistent decision strategy.

The stimulus for vision is distributed over space, time, and wavelength. As mentioned earlier, we are primarily concerned here with spatial and temporal variations in intensity; variations in wavelength are taken up in Chap. 26.

Before considering human visual performance, we briefly describe *ideal-observer theory* (e.g., Refs 74, 75), which has proven to be a useful tool for evaluating and intepreting visual performance.

Ideal-Observer Theory

In attempting to understand human visual performance, it is critically important to quantify performance limitations imposed by the information contained in the stimulus. There is a well-accepted technique, based on the theory of ideal observers,[75] for so doing. As applied to vision (see Refs. 76–79), this approach is based on comparing human observers' performance to that of an ideal observer for the same stimuli. Ideal observers have three main elements: (1) a precise description of the stimuli, including any random variation or noise; (2) descriptions of how the stimuli are modified by the visual stages of interest (e.g., optics, receptors); and (3) an optimal decision rule. Such observers specify the best possible level of performance (e.g., the smallest detectable amount of light) given the variability in the stimulus and information losses in the incorporated visual stages. Because ideal observer performance is the best possible, the thresholds obtained are a measure of the information available in the stimulus (to perform the task) once processed by the incorporated stages. Poorer performance by the human observer must be due to information losses in the unincorporated stages. Thus, ideal-observer theory provides the appropriate physical benchmark against which to evaluate human performance.[75,80] One can, of course, use models of anatomical and physiological mechanisms to incorporate various processing stages into an ideal observer (as shown in Fig. 7); comparisons of human and ideal observer performance in such cases can allow one to compute precisely the information transmitted and lost by mechanisms of interest. This allows assessment of the contributions of anatomical and physiological mechanisms to overall visual performance.[78]

It is important to recognize that ideal-observer theory is not a theory of human performance (humans generally perform considerably below ideal); thus, the theory is not a substitute for psychophysical and physiological modeling. However, the theory is crucial for understanding the stimuli and the task. In many ways, measuring and reporting the information content of stimuli with an ideal observer is as fundamental as measuring and reporting the basic physical dimensions of stimuli.

To illustrate the concepts of ideal-observer theory, consider an identification task with response alternatives (categories) α_1 through α_m, where the probability of a stimulus from category α_j equals q_j. Suppose, further, that the stimuli are static images, with an onset at some known point in time. (By a static image, we mean there is no temporal variation within a stimulus presentation except for photon noise.) Let $\mathbf{Z}$ be the list of random values (e.g., photon counts) at each sample location (e.g., photoreceptor) in some time interval τ for a single presentation of a stimulus:

$$\mathbf{Z} = (Z_1, \ldots, Z_n) \tag{15}$$

where Z_i is the sample value at the ith location.

Overall accuracy in an identification task is always optimized (on average) by picking the most probable stimulus category given the sample data and the prior knowledge. Thus, if the goal is to maximize overall accuracy in a given time interval τ, then the maximum average percent correct (PC_{opt}) is given by the following sum:

$$PC_{opt}(\tau) = \sum_{\mathbf{z}} q_* p_\tau(\mathbf{z} \mid \alpha_*) \tag{16}$$

where $p_\tau(\mathbf{z} \mid \alpha_*)$ is the probability density function associated with $\mathbf{Z}$ for stimulus category α_j, and the subscript $_*$ represents the category j for which the product $q_* p_\tau(\mathbf{z} \mid \alpha_*)$ is maximum. (Note, $\mathbf{Z}$ is a random vector and $\mathbf{z}$ is a simple vector.)

Suppose, instead, that the goal of the task is to optimize speed. Optimizing speed at a given accuracy level is equivalent to finding the minimum stimulus duration, $\tau_{opt}(\varepsilon)$, at a criterion error rate ε. The value of $\tau_{opt}(\varepsilon)$ is obtained by setting the left side of Eq. (16) to the criterion accuracy level $(1 - \varepsilon)$ and then solving for τ.

For the detection and discrimination tasks, where there are two response categories, Eq. (16) becomes

$$PC_{opt}(\tau) = \frac{1}{2} + \frac{1}{2} \sum_{z} |q_1 p_\tau(\mathbf{z} \mid \alpha_1) - (1 - q_1) p_\tau(\mathbf{z} \mid \alpha_2)| \tag{17}$$

(The summation signs in Eqs. (16) and (17) are replaced by an integral sign if the probability density functions are continuous.)

Because the sums in Eqs. (16) and (17) are over all possible vectors $\mathbf{z} = (z_1, \ldots, z_n)$, they are often not of practical value in computing optimal performance. Indeed, in many cases there is no practical analytical solution for ideal-observer performance, and one must resort to Monte Carlo simulation.

Two special cases of ideal-observer theory have been widely applied in the analysis of psychophysical discrimination and detection tasks. One is the ideal observer for discrimination or detection tasks where the only source of stimulus variability is photon noise.[77,81,82] In this case, optimal performance is given, to close approximation, by the following formulas:[83,84]

$$PC_{opt}(\tau) = q + 1(1 - q)\Phi\left(\frac{c}{d'} + \frac{d'}{2}\right) - q\Phi\left(\frac{c}{d'} - \frac{d'}{2}\right) \tag{18}$$

where $c = \ln((1 - q)/q)$, $\Phi(\cdot)$ is the cumulative standard normal probability distribution, and

$$d' = \frac{\tau^{1/2} \sum_{i=1}^{n} (b_i - a_i) \ln (b_i/a_i)}{\left[\sum_{i=1}^{n} (b_i + a_i) \ln^2 (b_i/a_i)\right]^{1/2}} \tag{19}$$

In Eq. (19), a_i and b_i are the average numbers of photons per unit time at the ith sample

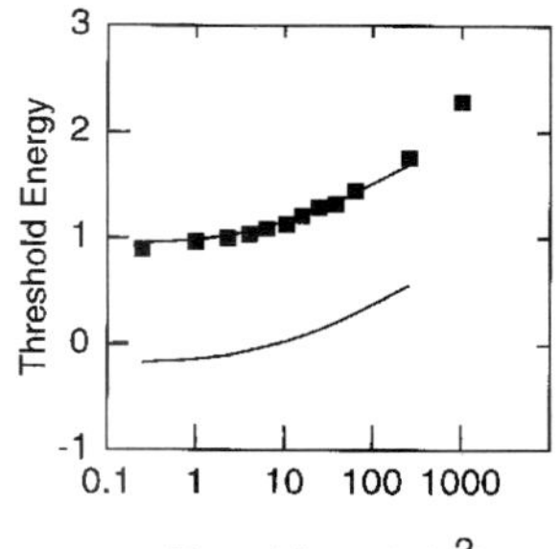

FIGURE 8 Comparison of real and ideal performance for detection of square targets on a uniform background of 10 cd/m^2 (viewed through a 3-mm artificial pupil). The symbols represent human thresholds as a function of target area. Threshold energy is in units of cd/m^2 × min^2 × sec. The lower curve is ideal performance at the level of the photoreceptors; the upper curve is the ideal performance shifted vertically to match the human data. *(Adapted from Ref. 263.)*

location, for the two alternative stimuli (α_1 and α_2). For equal presentation probabilities ($q = 0.5$), these two equations are easily solved to obtain $\tau_{opt}(\varepsilon)$.[85,86] In signal-detection theory[75] the quantity d' provides a criterion-independent measure of signal detectability or discriminability.

Figure 8 provides an example of the use of ideal-observer theory. It compares human and ideal performance for detection of square targets of different areas presented briefly on a uniform background. The symbols show the energy required for a human observer to detect the target as a function of target area. The lower solid curve shows the absolute performance for an ideal observer operating at the level of photon absorption in the receptor photopigments; the upper solid curve shows the same curve shifted vertically to match the human data. The differences between real and ideal performance represent losses of information among neural processes. The data show that neural efficiency, $[d'_{real}]^2/[d'_{ideal}]^2$, for the detection of square targets is approximately 1/2 percent for all but the largest target size.

Another frequently applied special case is the ideal observer for detection and discrimination tasks where the signal is known exactly (SKE) and white or filtered gaussian noise (i.e., image or pixel noise) has been added. Tasks employing these stimuli have been used to isolate and measure central mechanisms that limit discrimination performance[80,87,88] to evaluate internal (neural) noise levels in the visual system,[79,89] and to develop a domain of applied psychophysics relevant to the perception of the noisy images, such as those created by radiological devices and image enhancers.[90–92] For the gaussian-noise-limited ideal discriminator, Eq. (18) still applies, but d′ is given by the following:

$$d' = \frac{E(L \mid \alpha_2) - E(L \mid \alpha_1)}{\sqrt{VAR(L)}} \tag{20}$$

where

$$L = [\mathbf{m}_2 - \mathbf{m}_1]'\Sigma^{-1}\mathbf{Z} \tag{21}$$

In Eq. (21), **Z** is the column vector of random values from the sample locations (e.g., pixels), $[\mathbf{m}_2 - \mathbf{m}_1]'$ is a row vector of the differences in the mean values at each sample point (i.e., $b_1 - a_1, \ldots, b_n - a_n$), and Σ is the covariance matrix resulting from the

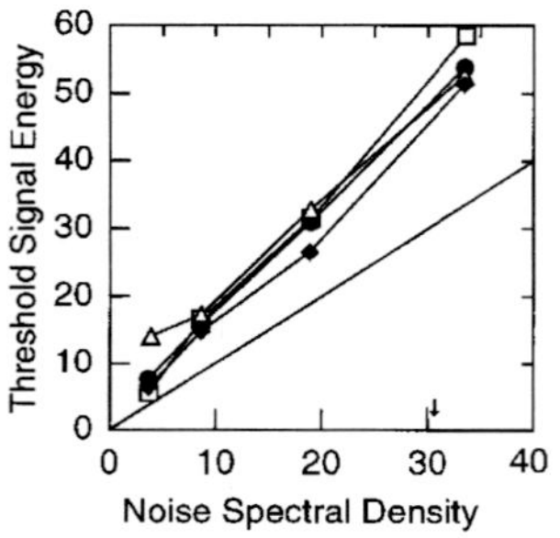

FIGURE 9 Comparison of real (symbols) and ideal (solid line) performance for contrast discrimination of spatially localized targets in white noise backgrounds with a mean luminance of 154 cd/m^2. Noise spectral density and target signal energy are in units of 10^{-7} deg^2 (the arrow indicates a noise standard deviation of 26 percent contrast per pixel). (□) Gaussian target (standard deviation = 0.054 deg); (●) Gaussian damped sinewave target (9.2 cpd, standard deviation = 0.109 deg); (△) Gaussian damped sinewave target (4.6 cpd, standard deviation = 0.217 deg); (◆) Sinewave target (4.6 cpd, 0.43 × 0.43 deg). *(Adapted from Ref. 87.)*

gaussian noise (i.e., the element $\sigma_{i,j}$ of Σ is the covariance between the ith and jth sample location). In the case of white noise, Σ is a diagonal matrix with diagonal elements equal to σ^2, and thus,*

$$d' = \frac{1}{\sigma}\sqrt{\sum (b_i - a_i)^2} \tag{22}$$

As an example, Fig. 9 compares human and ideal performance for amplitude discrimination of small targets in white noise as a function of noise spectral power density (which is proportional to σ^2). The solid line of slope 1.0 shows the absolute performance of an ideal observer operating at the level of the cornea (or display screen); thus, the difference between real and ideal performance represents all losses of information within the eye, retina, and central visual pathways. For these conditions, efficiency for amplitude discrimination of the targets ranges from about 20 to 70 percent, much higher than for detecting targets in uniform backgrounds (cf., Fig. 8). Efficiencies are lower in Fig. 8 than in Fig. 9 primarily because, with uniform backgrounds, internal (neural) noise limits human, but not ideal, performance; when sufficient image noise is added to stimulus, it begins to limit both real and ideal performance.

The following sections are a selective review of human visual performance and underlying anatomical and physiological mechanisms. The review includes spatial and temporal contrast perception, light adaptation, visual resolution, motion perception, and stereo depth perception, but due to space limitations, excludes many other active areas of research such as color vision (partially reviewed in Vol. I, Chap. 26), eye movements,[93] binocular vision,[94,95] spatial orientation and the perception of layout,[96,97] pattern and object recognition,[96–99] visual attention,[100] visual development,[101,102] and abnormal vision.[103]

* It should be noted that the performance of the photon noise limited oberver [Eq. (19)] becomes equivalent to an SKE white noise observer [Eq. (22)] for detection of targets against intense uniform backgrounds.

Contrast Detection

The study of contrast detection in the human visual system has been dominated for the last 20 years by methods derived from the concepts of linear systems analysis. The rationale for applying linear systems analysis in the study of visual sensitivity usually begins with the observation that a complete characterization of a visual system would describe the output resulting from any arbitrary input. Given that the number of possible inputs is infinite, an exhaustive search for all input-output relationships would never end. However, if the system under study is linear, then linear systems analysis provides a means for characterizing the system from the measurement of a manageably small set of input-output relationships. Of course, the visual system has important nonlinearities, so strictly speaking, the assumptions required by linear systems analysis are violated in general. Nonetheless, there have been several successful applications of linear systems analysis in vision.

In the case of spatial vision, linear systems analysis capitalizes on the fact that any spatial retinal illumination distribution, $I(x, y)$ can be described exactly by the sum of a set of basis functions, such as the set of sinusoids. Sinusoids are eigenfunctions of a linear system, which implies that the system response to a sinusoidal input can be completely characterized by just two numbers; an amplitude change and a phase change. Indeed, it is commonly assumed (usually with justification) that spatial phase is unaltered in processing, so only one number is actually required to describe the system response to a sinusoid. Linear systems analysis provides a means for predicting, from such characterizations of system responses to sinusoids, the response to any arbitrary input. For these reasons, the measurement of system responses to spatial sinusoids has played a central role in the examination of human spatial vision. We will begin our discussion with the spatial sinusoid and the system response to it.

A spatial sinusoid can be described by

$$I(x, y) = A \sin [2\pi f(x \cos (\theta) + y \sin (\theta)) + \phi] + \bar{I} \tag{23}$$

where $\bar{I}$ is the space-average intensity of the stimulus field (often expressed in trolands), A is the amplitude of the sinusoid, f is the spatial frequency (usually expressed in cycles $\cdot$ deg^{-1} or cpd), θ is the orientation of the pattern relative to axes x and y, and ϕ is the phase of the sinusoid with respect to the origin of the coordinate system. The contrast C of a spatial sinewave is defined as $(I_{max} - I_{min})/(I_{min} + I_{min})$, where I_{max} is the maximum intensity of the sinusoid and I_{min} is the minimum; thus $C = A/\bar{I}$. When the sinusoid is composed of vertical stripes, θ is zero and Eq. (23) reduces to

$$I(x, y) = A \sin [2\pi fx + \phi] + \bar{I} \tag{24}$$

When the contrast C of a sinusoidal grating (at a particular frequency) is increased from zero (while holding $\bar{I}$ fixed), there is a contrast at which it first becomes reliably detectable, and this value defines the contrast detection threshold. It is now well-established that contrast threshold varies in a characteristic fashion with spatial frequency. A plot of the reciprocal of contrast at threshold as a function of spatial frequency constitutes the *contrast sensitivity function* (CSF). The CSF for a young observer with good vision under typical indoor lighting conditions is shown in Fig. 10 (open symbols). The function is bandpass with a peak sensitivity at 3–5 cpd. At those spatial frequencies, a contrast of roughly 1/2 percent can be detected reliably (once the eye has adapted to the mean luminance). At progressively higher spatial frequencies, sensitivity falls monotonically to the so-called high-frequency cutoff at about 50 cpd; this is the finest grating an observer can detect when the contrast C is at its maximum of 1.0. At low spatial frequencies, sensitivity falls as well, although (as we will see) the steepness of this low-frequency rolloff is quite dependent

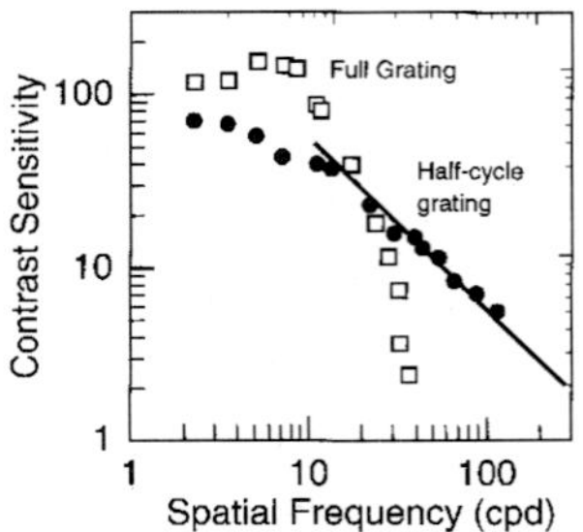

FIGURE 10 Contrast sensitivity as a function of spatial frequency for a young observer with good vision under typical indoor lighting conditions. The open symbols represent the reciprocals of contrast at detection threshold for sinusoidal grating targets. The filled symbols represent the reciprocals of contrast thresholds for half-cycle sinusoids. The solid line represents the predicted threshold function for half-cycle gratings for a linear system whose CSF is given by the open symbols. *(Adapted from Ref. 192.)*

upon the conditions under which the measurements are made. It is interesting that most manipulations affect the two sides of the function differently, a point we will expand later.

The CSF is not an invariant function; rather the position and shape of the function are subject to significant variations depending on the optical quality of the viewing situation, the average luminance of the stimulus, the time-varying and chromatic qualities of the stimulus, and the part of the retina on which the stimulus falls. We will examine all of these effects because they reveal important aspects of the visual processes that affect contrast detection.

Before turning to the visual processes that affect contrast detection, we briefly describe how CSF measurements and linear systems analysis can be used to make general statements about spatial contrast sensitivity. Figure 10 plots contrast sensitivity to targets composed of extended sinusoids and to targets composed of one half-cycle of a cosinusoid. Notice that sensitivity to high spatial frequencies is much higher with the half-cycle cosinusoids than it is with extended sinusoids.

The half-cycle wave forms can be described as cosinusoids multiplied by rectangular functions of widths equal to one-half period of the cosinusoid. The truncating rectangular function causes the frequencies in the pattern to "splatter" to higher and lower values than the nominal target frequency. Multiplication of the Fourier transforms of the half-cycle targets by the CSF obtained with extended sinusoids yields an estimate of the visual system's output response to the half-cycle targets. From there, use of a simple decision rule allows one to derive predicted half-cycle contrast sensitivities for a linear system. One expects to find greater visibility to high spatial frequencies with half-cycle gratings. The quantitative predictions are represented by the solid lines in Fig. 10 and they match the observations rather well. There are numerous cases in the literature in which the CSF yields excellent predictions of the visibility of other sorts of patterns, but there are equally many cases in which it does not. We will examine some of the differences between those two situations later.

All of the information-processing stages depicted in Fig. 7 affect the CSF, and most of those effects have now been quantified. We begin with eye movements.

Eye Movements. Even during steady fixation, the eyes are in constant motion, causing the retina to move with respect to the retinal images.[104] Such motion reduces retinal image contrast by smearing the spatial distribution of the target, but it also introduces temporal variation in the image. Thus, it is not obvious whether eye position jitter should degrade or improve contrast sensitivity. It turns out that the effect depends on the spatial frequency of the target. At low frequencies, eye movements are beneficial. When the image moves with respect to the retina, contrast sensitivity improves for spatial frequencies less than about 5 cpd.[105,106] The effect of eye movements at higher spatial frequencies is less clear, but one can measure it by presenting sinusoidal interference fringes (which bypass optical degradations due to the eye's aberrations) at durations too short to allow retinal smearing and at longer durations at which smearing might affect sensitivity.[107] There is surprisingly little attenuation due to eye position jitter; for 100-msec target presentations, sensitivity at 50 cpd decreases by only 0.2–0.3 log units relative to sensitivity at very brief durations. Interestingly, sensitivity improves at long durations of 500–2000 msec. This may be explained by noting that the eye is occasionally stationary enough to avoid smearing due to the retina's motion.

Thus, the eye movements that occur during steady fixation apparently improve contrast sensitivity at low spatial frequencies and have little effect at high frequencies. This means that with steady fixation there is little need for experimenters to monitor or eliminate the effects of eye movements in measurements of spatial contrast sensitivity at high spatial frequencies.

Optics. The optical tranfer function (OTF) of an optical system describes the attenuation in contrast (and the phase shift) that sinusoidal gratings undergo when they pass through the optical system. As shown in Chap. 24 the human OTF (when the eye is well accommodated) is a low-pass function whose shape depends on pupil size and eccentricity. The dashed curve and the solid squares in Fig. 11 show the foveal OTF and the foveal CSF, respectively, for a pupil diameter of 3.8 mm. At high spatial frequencies, the foveal

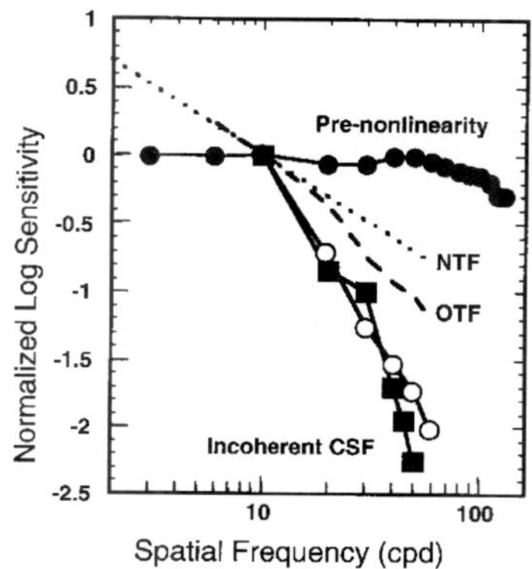

FIGURE 11 The contrast sensitivity function (CSF) and the transfer functions for various visual processing stages. All functions have been normalized to a log sensitivity of 0.0 at 10 cpd. The filled squares represent the CSF measured in the conventional way using incoherent white light.[22] The open circles represent the transfer function of the pre-nonlinearity filter measured.[22] The dotted line represents the neural transfer function that results from a bank of photon-noise-limited spatial filters of constant bandwidth.[122] The dashed line represents the optical transfer function for a 3.8-mm pupil.[4] The open squares are the product of the three transfer functions.

CSF declines at a faster rate than the foveal OTF. Thus, the OTF can only account for part of the high-frequency falloff of the CSF. Also, it is obvious that the OTF (which is a low-pass function) cannot account for the falloff in the CSF at low spatial frequencies.

Receptors. Several receptor properties affect spatial contrast sensitivity and acuity. As noted earlier, the receptors in the fovea are all cones, and they are very tightly packed and arranged in a nearly uniform hexagonal lattice. With increasing retinal eccentricity, the ratio of cones to rods decreases steadily (see Fig. 2*a*) and the regularity of the lattice diminishes.

Several investigators have noted the close correspondence between the Nyquist frequency and the highest-detectable spatial frequency[108–111] and have argued that the geometry of the foveal cone lattice sets a fundamental limit to grating acuity. However, more importance has been placed on this relationship than is warranted. This can be shown in two ways. First, grating acuity varies with several stimulus parameters including space-average luminance, contrast, and temporal duration. For example, grating acuity climbs from about 5 cpd at a luminance of 0.001 cd/m^2 to 50–60 cpd at luminances of 100 cd/m^2 or higher.[112] Obviously, the cone lattice does not change its properties in conjunction with changes in luminance, so this effect has nothing to do with the Nyquist limit. The only conditions under which the Nyquist limit might impose a fundamental limit to visibility would be at high luminances and contrasts, and relatively long target durations. Second, and more importantly, several investigators have shown that observers can actually detect targets whose frequency components are well above the Nyquist limit. For example, Williams[113] presented high-frequency sinusoidal gratings using laser interferometry to bypass the contrast reductions that normally occur in passage through the eye's optics. He found that observers could detect gratings at frequencies as high as 150 cpd, which is 2.5 times the Nyquist limit. Observers did so by detecting the gratings' "aliases," which are Moiré-like distortion products created by undersampling the stimulus.[114] The CSF under these conditions is smooth with no hint of a loss of visibility at the Nyquist limit. The observer simply needs to switch strategies above and below the Nyquist limit; when the targets are below the limit, the observer can detect the undistorted targets, and when the targets are above the limit, the observer must resort to detecting the targets' lower frequency aliases.[7] On the other hand, when the task is to resolve the stripes of a grating (as opposed to detecting the grating), performance is determined by the Nyquist limit over a fairly wide range of space-average luminances.[115]

In addition to receptor spacing, one must consider receptor size in the transmission of signals through the early stages of vision. As noted earlier, the prevailing model of cones holds that the inner segment offers an aperture to incoming photons. In the fovea, the diameter of the inner segment is roughly 0.5 min[6,21] (see Fig. 2*c*). Modeling the cone aperture by a cylinder function of this diameter, one can estimate the low-pass filtering due to such an aperture from the cylinder's Fourier transform, which is a first-order Bessel function whose first zero occurs at 146 cpd. However, the entrance to a structure only a few wavelengths of light wide cannot be described accurately by geometric optics,[116] so the receptor aperture is generally modeled by a gaussian function whose full width at half height is somewhat smaller than the anatomical estimates of the aperture diameter.[22]

The modulation transfer of the neural stages prior to a compressive nonlinearity (thought to exist in the bipolar cells of the retina or earlier) has been measured in the following way.[22] Laser interferometry was used to bypass the contrast reductions that normally occur as light passes through the eye's optics. The stimuli were two high-contrast sinusoids of the same spatial frequency but slightly different orientations; these created a spatial "beat" of a different orientation and frequency than the component gratings. Passing two sinusoids through a compressive nonlinearity, like the ones known to exist early in the visual system, creates a number of distortion products at various frequencies and orientations. MacLeod et al.[22] varied the orientations and frequencies of the components so as to create a "beat" at 10 cpd, and then measured contrast sensitivity to the "beat." Because the "beat" was always at the same spatial frequency, the filtering properties of postnonlinearity stages could not affect the measurements and, therefore, any effect of component frequency on contrast sensitivity had to be due to the filtering properties of prenonlinearity stages. Observers were able to detect the "beat" produced by component gratings of frequencies as high as 140 cpd. This implies that the prenonlinearity filter has exceptionally large bandwidth. By inverse Fourier transformation of these data, one can estimate the spatial extent of the filter; it appears to have a full width at half height of about 16 arcsec, which is a bit smaller than current estimates of the cone aperture. Thus, the early neural stages are able to pass information up to extremely high spatial frequencies once the optics are bypassed.

The solid circles in Fig. 11 display the estimate of the transfer function of the prenonlinearity neural filter for foveal viewing. Notice that it is virtually flat up to 140 cpd. Also shown (solid squares) is the CSF obtained in the conventional manner with incoherent white light from a CRT display. This function represents the product of the optics and the prenonlinearity and postnonlinearity spatial filters. As can be seen, the bandwidth of the conventional CSF is much smaller than the bandwidth of the prenonlinearity filter. Thus, optics and postnonlinearity filtering are the major constraints to contrast sensitivity at higher spatial frequencies.

As mentioned earlier, the transfer function of the optics is shown as the dashed line. The product of the optical transfer function and the prenonlinearity transfer function would be very similar to the dashed line because the prenonlinearity function is virtually flat. Thus, the product of optics and prenonlinearity filter does not match the observed CSF under conventional viewing conditions at all; the postnonlinearity filter must also contribute significantly to the shape of the CSF. The mismatch between the dashed line and the CSF is obvious at low and high spatial frequencies. We consider the high-frequency part first; the low-frequency part is considered later under "Adaptation and Inhibition."

Spatial Channels. Spatial filters with narrow tuning to spatial frequency have been demonstrated in a variety of ways.[117,118] For example, the visibility of a sinusoidal target is reduced by the presence of a narrowband noise masker whose center frequency corresponds to the spatial frequency of the sinusoid, but is virtually unaffected by the presence of a masker whose center frequency is more than 1.5 octaves from the frequency of the sinusoid (Ref. 119; see "Contrast Discrimination and Contrast Masking" later in chapter). These spatial mechanisms, which have been described mathematically by Gabor functions and other functions,[120] correspond to a first approximation with the properties of single neurons in the visual cortex.[58] They seem to have roughly constant bandwidths, expressed in octaves or log units, regardless of preferred frequency.[58,117] As a consequence, mechanisms responding to higher spatial frequencies are smaller in vertical and horizontal spatial extent than mechanisms responding to lower frequencies.[121,122] Thus, for a given stimulus luminance, a higher-frequency mechanism receives fewer photons than does a lower-frequency mechanism. The mean number of photons delivered to such a mechanism divided by the standard deviation of the number of photons—the signal-to-noise ratio—follows a square root relation,[123] so the signal-to-noise ratio is inversely proportional to the preferred frequency of the mechanism. The transfer function one expects for a set of such mechanisms should be proportional to $1/f$ where f is spatial frequency.[122] This function is represented by the dotted line in Fig. 11. If we use that relation for describing the postnonlinearity filter, and incorporate the measures of the optical and prenonlinearity filters described above, the resultant is represented by the open circles. The fit between the observed CSF and the resultant is good, so we can conclude that the filters described here are sufficient to account for the shape of the high-frequency limb of the human CSF under foveal viewing conditions.

By use of the theory of ideal observers (see earlier), one can calculate the highest possible contrast sensitivity a system could have if limited by the optics, photoreceptor properties, and constant bandwidth spatial channels described above. As can be seen (open circles), the high-frequency limb of the CSF of such an ideal observer is similar in shape to that of human observers,[122,124] although the rate of falloff at the highest spatial frequencies is not quite as steep as that of human observers.[124] More importantly, the absolute sensitivity of the ideal observer is significantly higher. Some of the low performance of the human compared to the best possible appears to be the consequence of processes that behave like noise internal to the visual system and some appears to be the consequence of employing detecting mechanisms that are not optimally suited for the target being presented.[79,89] The high detection efficiencies observed for sinewave grating patches in the presence of static white noise (e.g., Ref. 87; Fig. 9) suggest that poor spatial pooling of information is not the primary factor responsible for the low human performance (at least when the number of cycles in the grating patches are kept low).

Optical/Retinal Inhomogeneity. The fovea is undoubtedly the most critical part of the retina for numerous visual tasks such as object identification and manipulation, reading, and more. But, the fovea occupies less than 1 percent of the total retinal surface area, so it is not surprising to find that nonfoveal regions are important for many visual skills such as selecting relevant objects to fixate, maintaining posture, and determining one's direction of self-motion with respect to environmental landmarks. Thus, it is important to characterize the visual capacity of the eye for nonfoveal loci as well.

The quality of the eye's optics diminishes slightly from 0–20 deg of retinal eccentricity and substantially from 20–60 deg;[1] see Chap. 24. As described earlier and displayed in Fig. 2, the dimensions and constituents of the photoreceptor lattice and the other retinal neural elements vary significantly with retinal eccentricity. With increasing eccentricity, cone and retinal ganglion cell densities fall steadily and individual cones become broader and shorter. Given these striking variations in optical and receptoral properties across the retina, it is not surprising that different aspects of spatial vision vary greatly with retinal eccentricity. Figure 12 shows CSFs at eccentricities of 0–40 deg. With increasing eccentricity, contrast sensitivity falls off at progressively lower spatial frequencies. The high-frequency cutoffs range from approximately 2 cpd at 40 deg to 50 cpd in the fovea. Low-frequency sensitivity is similar from one eccentricity to the next. There have been many attempts to relate the properties of the eye's optics, the receptors, and postreceptoral mechanisms to contrast sensitivity and acuity. For example, models that incorporate eccentricity-dependent variations in optics and cone lattice properties, plus the assumption of fixed bandwidth filters (as in Ref. 122), have been constructed and found inadequate for explaining the observed variations in contrast sensitivity; specifically, high-frequency sensitivity declines more with eccentricity than can be explained by information losses due to receptor lattice properties and fixed bandwidth filters alone.[125] Adding a low-pass filter to the model representing the convergence of cones onto retinal ganglion cells[32] (i.e., by incorporating variation in receptive field center diameter) yields a reasonably accurate account of eccentricity-dependent variations in contrast sensitivity.[125]

Adaptation and Inhibition. Contrast sensitivity is also dependent upon the space-average intensity (i.e., the mean or background luminance) and the prior level of light adaptation. The solid symbols in Fig. 13 show how threshold amplitudes for spatial sinewaves vary as a function of background luminance for targets of 1 and 12 cpd.[126] [Recall that sinewave amplitude is contrast multiplied by background intensity; see discussion preceding Eq. (24).] The backgrounds were presented continuously and the observer was allowed to become fully light-adapted before thresholds were measured. For the low-spatial-frequency target (solid circles), threshold increases linearly with a slope of 1.0 above the low background intensities (1 log td). A slope of 1.0 in log-log coordinates implies that

$$A_T = k\bar{I} \tag{25}$$

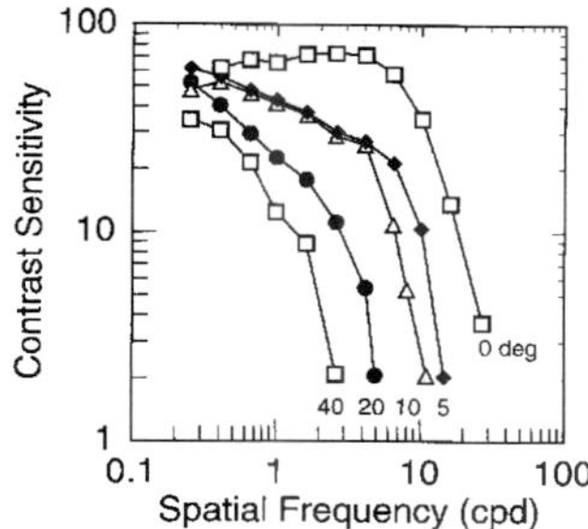

FIGURE 12 CSFs at different retinal eccentricities. Open triangles, fovea; filled triangles, 5 deg; open squares, 10 deg; filled squares, 20 deg; open circles, 40 deg. *(Adapted from Ref. 125.)*

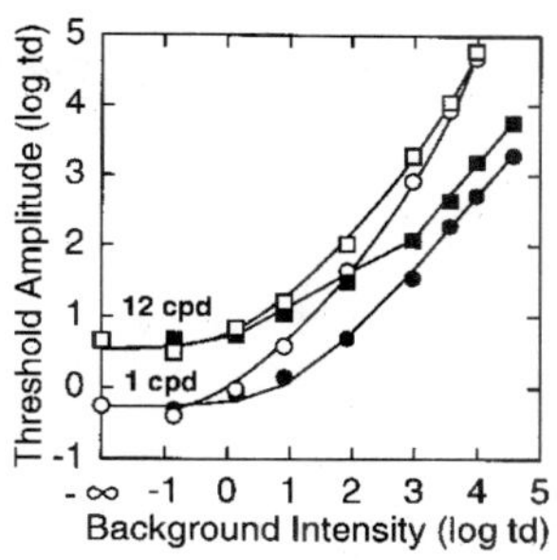

FIGURE 13 Amplitude threshold for gaussian-damped sinusoidal targets with 0.5-octave bandwidths as a function of background intensity. The targets were presented for 50 msec. The open symbols are thresholds measured at the onset of backgrounds flashed for 500 msec in the dark-adapted eye. The solid symbols are thresholds measured against steady backgrounds after complete adaptation to the background intensity. The difference between the open and solid symbols shows the effect of adaptation on contrast sensitivity. *(Adapted from Ref. 126.)*

where A_T is the amplitude threshold (or equivalently, it implies that contrast threshold, C_T, is constant). This proportional relationship between threshold and background intensity is known as *Weber's law.* For the higher spatial-frequency target (solid squares), threshold increases linearly with a slope of 0.5 at intermediate background intensities (0 to 3 log td) and converges toward a slope of 1.0 at the highest background intensities. A slope of 0.5 implies that

$$A_T = k\sqrt{I} \tag{26}$$

(or equivalently, that contrast threshold, C_T, decreases in inverse proportion to the square root of background intensity). This relationship between amplitude threshold and the square root of background intensity is known as the *square root law* or the *DeVries-Rose law.* As suggested by the data in Fig. 13, the size of the square root region gradually grows as spatial frequency is increased. The effect on the CSF is an elevation of the peak and of the high-frequency limb relative to the low-frequency limb as mean luminance is increased, thereby producing a shift of the peak toward higher spatial frequencies.[127,128] This result explains why humans prefer bright environments when performing tasks that require high spatial precision.

The visual system is impressive in its ability to maintain high sensitivity to small contrasts (as produced, for example, by small changes in surface reflectance) over the enormous range of ambient light levels that occur in the normal environment. In order to have high sensitivity to small contrasts, the response gain of visual neurons must be kept high, but high gain leaves the visual system vulnerable to the effects of response saturation (because neurons are noisy and have a limited response range). The visual system's solution to this "dynamic-range problem" is threefold: (1) use separate populations of receptors (rods and cones) to detect contrasts in the lower and upper ranges of ambient intensity, (2) adjust the retinal illumination via the pupil reflex, and (3) directly adjust the sensitivity of individual receptors and neurons (e.g., Refs. 129, 130). All three components are important, but the third is the most significant and is accomplished within the receptors and other neurons through photochemical and neural adaptation. The combined effect of the photochemical and neural adaptation mechanisms on foveal contrast detection is illustrated by the difference between the open and solid data points in Fig. 13 (all the data were obtained with a fixed pupil size). The open symbols show cone detection threshold for the 1- and 12-cpd targets on a background flashed in the dark-adapted eye (i.e., there was no chance for light adaptation) as a function of background intensity. The difference between the open and solid symbols shows the improvement in detection sensitivity due to photochemical and neural adaptation within the cone system.

The pupil reflex and photopigment depletion (see earlier under "Image Sampling by the Photoreceptors") are *multiplicative adaptation* mechanisms; they adjust the sensitivity of the eye by effectively scaling the input intensity by a multiplicative gain factor that decreases with increasing average intensity. The pupil reflex operates over most of the intensity range (see Chap. 24); photopigment depletion is only effective in the cone system above about 4 log td (see Fig. 13) and is ineffective in the rod system, at least over the range of relevant intensities (see earlier discussion). There is considerable psychophysical evidence that a substantial component of the remaining improvement in sensitivity illustrated in Fig. 12 is due to multiplicative neural adaptation mechanisms.[129,131–134] However, multiplicative adaptation alone cannot reduce threshold below a line of slope 1.0, tangent to the threshold curve in dark-adapted eye.[129,134] The remaining improvements in sensitivity (the difference between the tangent lines and the solid symbols) can be explained by neural *subtractive adaptation* mechanisms.[134,135] There is evidence that the multiplicative and subtractive adaptation mechanisms each have fast and slow components.[134–137] Threshold functions, such as those in Fig. 13, can be predicted by simple models consisting of a compressive (saturating) nonlinearity [Eq. (13)], a multiplicative adaptation mechanism, a subtractive adaptation mechanism, and either constant additive noise or multiplicative (Poisson) neural noise.

A common explanation of the low-frequency falloff of the CSF (e.g., Fig. 12) is based on center/surround mechanisms evident in the responses of retinal and LGN neurons. For example, compare the CSFs measured in individual LGN neurons in Fig. 5 with the human CSF in Fig. 12. The center and surround mechanisms are both low-pass spatial filters, but the cutoff frequency of the surround is much lower than that of the center. The center and surround responses are subtracted neurally, so at low spatial frequencies, to which center and surround are equally responsive, the neuron's response is small. With increasing spatial frequency, the higher resolution of the center mechanism yields a greater response relative to the surround, so the neuron's response increases.

While the center/surround mechanisms are undoubtedly part of the explanation, they are unlikely to be the whole story. For one thing, the time course of the development of the low-frequency falloff is slower than one expects from physiological measurements.[105] Second, the surround strength in retinal and LGN neurons is on average not much more than half the center strength;[37] this level of surround strength is consistent with the modest low-frequency falloffs observed under transient presentations, but inconsistent with the steep falloff observed for long-duration, steady-fixation conditions. Third, there is considerable evidence for slow subtractive and multiplicative adaptation mechanisms, and these ought to contribute strongly to the shape of the low-frequency limb under steady fixation conditions.[137–139] Thus, the strong low-frequency falloff of the CSF under steady viewing conditions may reflect, to some degree, slow subtractive and multiplicative adaptation mechanisms whose purpose is to prevent response saturation while maintaining high contrast sensitivity over a wide range of ambient light levels. The weaker low-frequency falloff seen for brief presentations may be the result of the fast-acting surround mechanisms typically measured in physiological studies; however, fast-acting local multiplicative and subtractive mechanisms could also play a role.

Temporal Contrast Detection. Space-time plots of most contrast-detection stimuli (plots of intensity as a function of x, y, t) show intensity variations in both space and time; in other words, most contrast detection stimuli contain both spatial and temporal contrast.

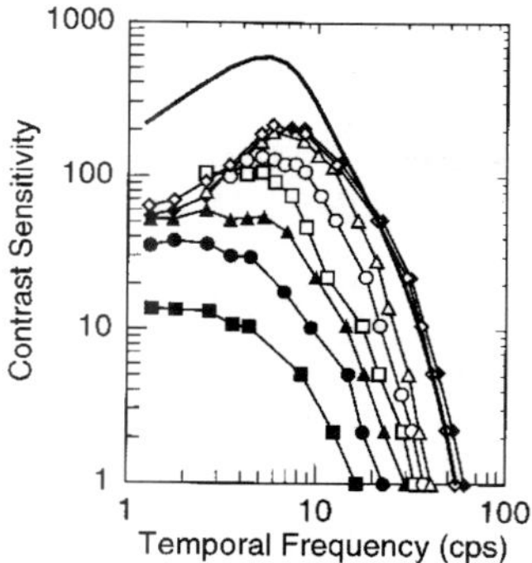

FIGURE 14 Temporal contrast sensitivity functions measured in the fovea at several mean luminances for a 2° circular test field. The test field was modulated sinusoidally and was imbedded in a large, uniform background field of the same mean luminance. Mean luminance: (■) 0.375 td, (●) 1 td, (H) 3.75 td, (□) 10 td, (○) 37.5 td, (△) 100 td, (◇) 1000 td, (◆) 10,000 td. *(Data from Ref. 141.)* The solid curve is the MTF derived from the impulse response function of a single macaque cone reported by Schnapf et al.[17] The cone MTF was shifted vertically (in log-log coordinates) for comparison with the shape of the high frequency falloff measured psychophysically.

The discussion of contrast detection has, so far, emphasized spatial dimensions (e.g., spatial frequency), but there is, not surprisingly, a parallel literature emphasizing temporal dimensions (for an recent review see Ref. 140). Figure 14 shows de Lange's[141] measurements of temporal contrast sensitivity functions at several mean luminances for uniform discs of light modulated in intensity sinusoidally over time. The major features of these results have been confirmed by other researchers.[142,143] Like spatial CSFs, temporal CSFs typically have a bandpass shape. At high mean luminances, the peak contrast sensitivity occurs at approximately 8 cycles per sec (cps) and the cutoff frequency (also known as the critical flicker frequency) at approximately 60 cps. Contrast sensitivity increases with increasing mean luminance, larger increases (in log-log coordinates) being observed at middle and high temporal frequencies. For low temporal frequencies (below 6 cps), contrast sensitivity is nearly constant (i.e., Weber's law is observed) for mean intensities greater than 10 td or so.

The entire temporal CSF can be fit by a weighted difference of linear low-pass filters, where one of the filters has a relatively lower cutoff frequency, and/or where one of the filters introduces a relatively greater time delay (see Ref. 140). However, to fit the temporal CSFs obtained under different stimulus conditions, the time constants and relative weights of the filters must be allowed to vary. The low-frequency falloff may be partly due to the biphasic response of the photoreceptors (see Fig. 3), which appears to involve a fast subtractive process.[17] Surround mechanisms, local subtractive mechanisms, and local multiplicative mechanisms also probably contribute to the low-frequency falloff.

The factors responsible for the high-frequency falloff are not fully understood, but there is some evidence that much of the high-frequency falloff is the result of temporal integration within the photoreceptors.[144,145] The solid curve in Fig. 14 is the MTF of the macaque cone photocurrent response derived from measured impulse-response functions[17] (see Fig. 3). The impulse responses were obtained in the linear response range by presenting brief, dim flashes on a dark background. Although the background was dark, there appears to be little effect of light adaptation on cone responses until background intensity exceeds at least a few hundred trolands;[17,24] thus, the cone impulse-response functions measured in the dark ought to be valid up to moderate background intensities. Following Boynton and Baron,[144] we have shifted the MTF vertically in order to compare the shape of the cone high-frequency falloff to that of the temporal CSF. The shapes are similar, lending support to the hypothesis that cone temporal integration is responsible for the shape of the high-frequency limb of the temporal CSF.

It is well established that a pulse of light briefer than some critical duration will be just visible when the product of its duration and intensity is constant. This is *Bloch's law,* which can be written:

$$I_T T = k \qquad \text{for} \qquad T < T_c \tag{27}$$

where k is constant, T_c is the critical duration and I_T is the threshold intensity at duration T. Naturally, Bloch's law can also be stated in terms of contrasts by dividing the intensities in the above equation by the background intensity. Below the critical duration, log-log plots of threshold intensity versus duration have a slope of -1. Above the critical duration, threshold intensity falls more slowly with increasing duration than predicted by Bloch's law (slope between -1 and 0) and, in some cases, it even becomes independent of duration (slope of 0).[76,146] It is sometimes difficult to determine the critical duration because the slopes of threshold-versus-duration plots can change gradually, but under most conditions Bloch's critical duration declines monotonically with increasing background intensity from about 100 msec at 0 log td to about 25 msec at 4 log td.

Bloch's law is an inevitable consequence of a linear filter that only passes temporal frequencies below some cutoff value. This can be seen by computing the Fourier transform of pulses for which the product $I_T T$ is constant. The amplitude spectrum is given by $I_T \sin(T\pi f)/\pi f$ where f is temporal frequency; this function has a value $I_T T$ at $f = 0$ and is quite similar over a range of temporal frequencies for small values of T. Indeed, to a first approximation, one can predict the critical duration T_c from the temporal CSFs shown in Fig. 14, as well as the increase in T_c with decreasing background intensity.[140]

The shape of the temporal sensitivity function depends upon the spatial frequency of the stimulus. Figure 15 shows temporal CSFs measured for spatial sinewave gratings whose contrast is modulated sinusoidally in time.[147] With increasing spatial frequency, the temporal CSF evolves from a bandpass function with high sensitivity (as in Fig. 14) to a low-pass function of lower sensitivity.[128,147] The solid curves passing through the data for the high spatial-frequency targets and the dashed curves passing through the data for the low-frequency targets are identical except for vertical shifting. Similar behavior occurs for spatial CSFs at high spatial frequencies: changes in temporal frequency cause a vertical sensitivity shift (in log coordinates). Such behavior shows that the spatio-temporal CSF is separable at high temporal and spatial frequencies; that is, sensitivity can be predicted from the product of the spatial and temporal CSFs at the appropriate frequencies. This finding suggests that the same anatomical and physiological mechanisms that determine the high-frequency slope of spatial and temporal CSFs determine sensitivity for the high-frequency regions of the spatio-temporal CSF.

The spatio-temporal CSF is, however, clearly not separable at low spatial and temporal frequencies, so an explanation of the underlying anatomical and physiological constraints is more complicated. The lack of spatio-temporal separability at low spatio-temporal frequencies has been interpreted by some as evidence for separate populations of visual neurons tuned to different temporal frequencies[143,148] and by others as evidence for different spatio-temporal properties of center and surround mechanisms within the same populations of neurons.[147,149] It is likely that a combination of the two explanations is the correct one. For a review of the evidence concerning this issue prior to 1984, see Ref. 140; for more recent evidence, see Refs. 150, 151, and 152.

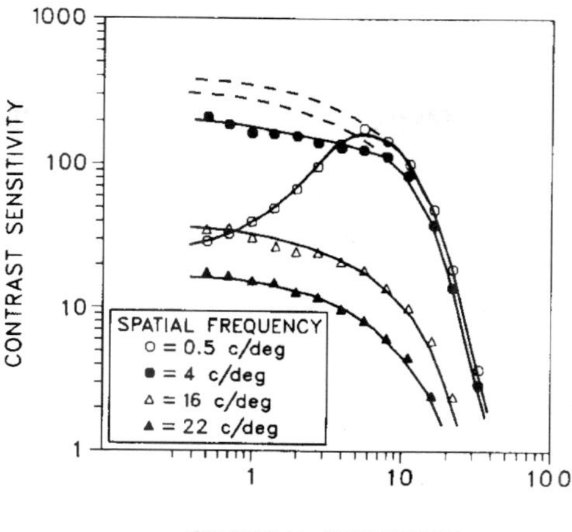

FIGURE 15 Temporal contrast sensitivity functions measured for several different spatial frequencies. The contrast of spatial sinewave gratings were modulated sinusoidally in time. (○) 0.5 cpd, (●) 4 cpd, (△) 16 cpd, (◆) 22 cpd. The solid curves passing through the data for the high spatial-frequency targets and the dashed curves passing through the data for the low-frequency targets are identical except for vertical shifting. *(Reproduced from Ref. 147.)*

The available evidence indicates that the temporal CSF varies little with retinal eccentricity once differences in spatial resolution are taken into account.[153,154]

Chromatic Contrast Detection. The discussion has, so far, concerned luminance contrast detection (contrast detection based upon changes in luminance over space or time); however, there is also a large literature concerned with chromatic contrast detection (contrast detection based upon changes in wavelength distribution over space or time). Although the focus of this chapter is on achromatic vision, it is appropriate to say a few words about chromatic contrast sensitivity functions. A chromatic spatial CSF is obtained by measuring thresholds for the detection of sinewave gratings that modulate spatially in wavelength composition, without modulating in luminance.* This is typically accomplished by adding two sinewave gratings of different wavelength distributions but the same luminance amplitude. The two gratings are added in opposite spatial phase so that the mean luminance is constant across the pattern; such a pattern is said to be *isoluminant.* Formally, an isoluminant sinewave grating can be defined as follows:

$$I(x, y) = (A \sin (2\pi fx) + \bar{I}_1) + (-A \sin (2\pi fx) + \bar{I}_2) \tag{28}$$

where the first term on the right represents the grating for one of the wavelength distributions, and the second term on the right for the other distribution. The terms $\bar{I}_1$ and

* The standard definition of luminance is based upon "the $V(\lambda)$ function" [see Eq. (5)] which is an average of spectral sensitivity measurements on young human observers. Precision work in color vision often requires taking individual differences into account. To do this a psychophysical procedure, such as flicker photometry, is used to define luminance separately for each observer in the study. For more details, see Ref. 5 and Chap. 26.

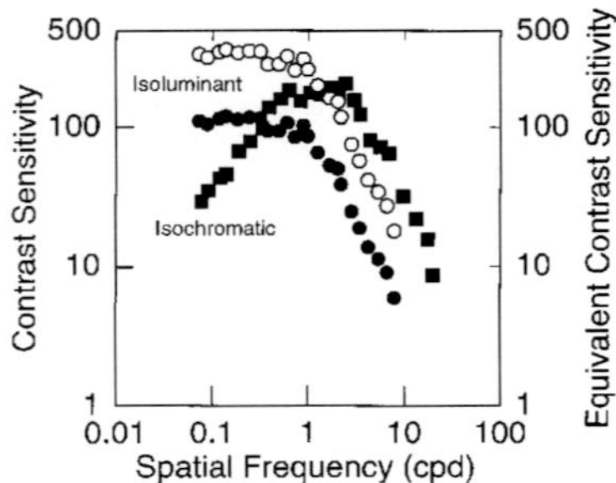

FIGURE 16 Isoluminant and isochromatic contrast sensitivity functions. The solid squares are isochromatic (luminance) CSFs for a green grating (526 nm). The solid circles are isoluminant (chromatic) CSFs for gratings modulating between red (602) and green (526 nm). The open circles are the isoluminant data replotted as equivalent luminance contrast sensitivity (right axis). The isochromatic CSF plots the same on the left and right axes. *(Data from Ref. 155.)*

$\bar{I}_2$ are equal in units of luminance. Contrast sensitivity at spatial frequency f is measured by varying the common amplitude, A. (Note that $I(x, y)$ is constant independent of A, and that when $A = 0.0$ the image is a uniform field of constant wavelength composition.) Chromatic contrast is typically defined as $C = 2A/(\bar{I}_1 + \bar{I}_2)$.

The solid circles in Fig. 16 are an isoluminant CSF for a combination of red and green gratings.[155] The squares are a luminance (isochromatic) CSF measured with the same procedure, on the same subjects. As can be seen, there are three major differences between luminance and chromatic CSFs:[156] (1) the high-frequency falloff of the chromatic CSFs occurs at lower spatial frequencies (acuity is worse); (2) the chromatic CSF has a much weaker (or absent) low-frequency falloff, even under steady viewing conditions; and (3) chromatic contrast sensitivity is greater than luminance contrast sensitivity at low spatial frequencies.

Two of the factors contributing to the lower high-frequency limb of the chromatic (red/green) CSF are the large overlap in the absorption spectra of the L (red) and M (green) cones, and the overlap (when present) of the wavelength distributions of the component gratings. Both of these factors reduce the effective contrast of the grating at the photoreceptors. A precise way to quantify these effects (as well as those of the other preneural factors) is with an ideal-observer analysis (e.g., Refs. 78, 124, 157). For example, one can physically quantify the equivalent luminance contrast of a chromatic grating by (1) computing the detection accuracy of an ideal observer, operating at the level of the photoreceptor photopigments, for the chromatic grating at the given chromatic contrast, and then (2) computing the equivalent luminance contrast required to bring the ideal observer to the same accuracy level.[124,158] The open circles in Fig. 16 are the chromatic CSF data replotted in terms of equivalent luminance contrast. (The luminance CSF data are, of course, unaffected and hence are the same on both scales.) If the differences between the high-frequency falloffs of the luminance and chromatic CSFs were due to preneural factors alone, then the high-frequency limbs (open circles and squares) would superimpose. The analysis suggests that there are also neural factors contributing to the difference between high-frequency limbs.

One difficulty in interpreting the high-frequency limb of chromatic CSFs is the potential for luminance artifacts due to chromatic aberration.[156] Recently Sekiguchi et al.[124] eliminated chromatic aberration artifacts by producing isoluminant gratings with a laser interferometer which effectively bypasses the optics of the eye. Their results are similar to those in Fig. 16.

Comparison of the open circles and squares in Fig. 16 shows that the neural mechanisms are considerably more efficient at detecting chromatic contrast than luminance contrast at spatial frequencies below 2 cpd. One possible explanation is based upon the receptive-field properties of neurons in the retina and LGN (e.g., see Refs. 48, 159). Many retinal ganglion cells (and presumably bipolar cells) are chromatically opponent in the sense that different classes of photoreceptors dominate the center and surround responses. The predicted CSFs of linear receptive fields with chromatically opponent centers and surrounds are qualitatively similar to those in Fig. 16.

Chromatic temporal CSFs have been measured in much the same fashion as chromatic spatial CSFs except that the chromatic sinewaves were modulated temporally rather than spatially (i.e., x represents time rather than space in Eq. 28). The pattern of results is also quite similar:[94,156,160] (1) the high-frequency falloff of chromatic CSFs occurs at lower temporal frequencies, (2) the chromatic CSF has a much weaker low-frequency falloff, and (3) chromatic contrast sensitivity is greater than luminance contrast sensitivity at low temporal frequencies. Again, the general pattern of results is qualitatively predicted by a combination of preneural factors and the opponent receptive field properties of retinal ganglion cells (e.g., see Ref. 48).

Contrast Discrimination and Contrast Masking

The ability to discriminate on the basis of contrast is typically measured by presenting two sinusoids of the same spatial frequency, orientation, and phase, but differing contrasts. The common contrast C is referred to as the *pedestal contrast* and the additional contrast ΔC in one of the sinusoids as the *increment contrast.* Figure 17 shows typical *contrast discrimination functions* obtained from such measurements. The increment contrast varies as a function of the pedestal contrast. At low pedestal contrasts, ΔC falls initially and then at higher contrasts becomes roughly proportional to $C^{0.6}$. Because the exponent is generally less than 1.0, Weber's law does not hold for contrast discrimination. Ideal observer calculations show that pedestal contrast has no effect on discrimination information; therefore, the variations in threshold with pedestal contrast must be due entirely to neural mechanisms. The dip at low pedestal contrasts has been modeled as a consequence of an accelerating nonlinearity in the visual pathways (e.g., Ref. 161) or the consequence of observer uncertainty (e.g., Ref. 162). The evidence currently favors the

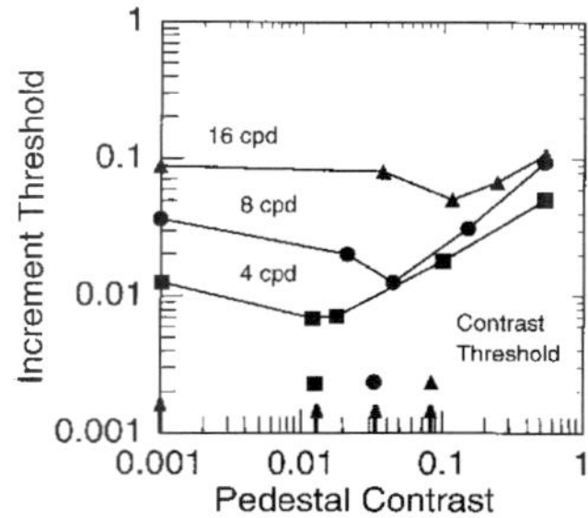

FIGURE 17 Contrast discrimination functions at different spatial frequencies. The just-discriminable contrast increment is plotted as a function of the contrast common to the two targets. The arrows indicate contrast thresholds at the indicated spatial frequencies. *(Adapted from Ref. 78.)*

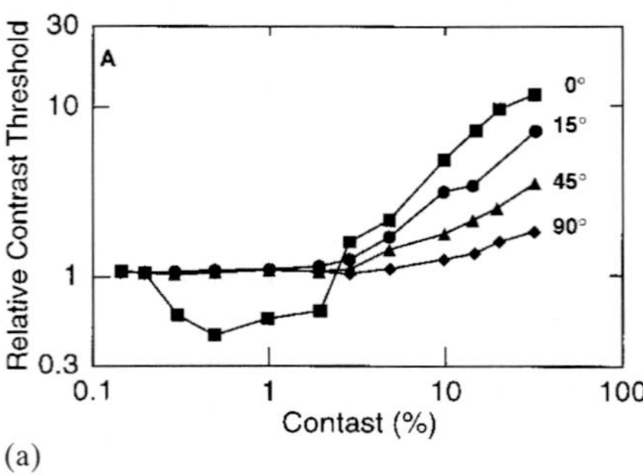

(a)

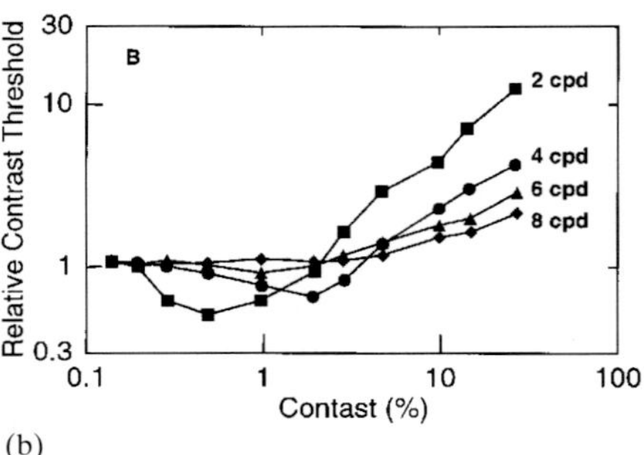

(b)

FIGURE 18 Contrast masking functions for sinewave-grating maskers of various orientations and spatial frequencies. The target was a 2 cpd sinewave grating. (*a*) Contrast-masking functions for 2 cpd maskers of various orientations. (*b*) Contrast-masking functions for 0° (vertical) maskers of various spatial frequencies. *(Adapted from Ref. 264.)*

former.[89] Contrast discrimination functions at different spatial frequencies[78,163] and at different eccentricities[163a] are, to a first approximation, similar in shape; that is to say, the functions can be superimposed by shifting the data along the log pedestal and log increment axes by the ratio of detection thresholds.

A generalization of the contrast-discrimination experiment is the contrast-masking experiment in which the increment and pedestal differ from one another in spatial frequency, orientation and/or phase. In contrast-masking experiments, the increment and pedestal are usually referred to as the *target* and *masker,* respectively. Figure 18*a* shows typical contrast-masking functions obtained for sinewave maskers of different orientations. The orientation of the sinewave target is vertical (0 deg); thus, when the orientation of the masker is also vertical (squares), the result is just a contrast discrimination function, as in Fig. 17. For suprathreshold maskers (in this case, maskers above 2 percent contrast), the threshold elevation produced by the masker decreases as the difference between target and masker orientation increases; indeed, when the target and masker are perpendicular (diamonds), there is almost no threshold elevation even at high masker contrasts. Also, the slope of the contrast-masking function (in log-log coordinates) decreases as the difference in orientation increases. Finally, notice that threshold enhancement at low masking contrasts occurs only when the masker has approximately the same orientation as the target.

Figure 18*b* shows a similar set of masking functions obtained for sinewave maskers of different spatial frequencies. The effect of varying masker spatial frequency is similar to that of varying masker orientation: with greater differences in spatial frequency between masker and target, masking is reduced, the slope of the contrast-masking function is reduced, and the threshold enhancement at low masker contrasts becomes less evident.

The substantial variations in threshold seen as a function of masker spatial frequency and orientation provide one of the major lines of evidence for the existence of multiple orientation and spatial-frequency-tuned channels in the human visual system;[119,164] for a review, see Ref. 165. Masking data (such as those in Fig. 18) have been analyzed within the framework of specific multiple-channel models in order to estimate the tuning functions of the spatial channels (e.g., Ref. 166). In one popular model, each channel consists of an initial linear filter followed by a static (zero memory) nonlinearity that is accelerating at low contrasts and compressive at high contrasts (e.g., Refs 161, 166, 167). Wilson and colleagues have shown that the spatial-channel tuning functions estimated within the framework of this model can predict discrimination performance in a number of different tasks (for a summary, see Ref. 168). However, recent physiological evidence[70,71] and psychophysical evidence[168,169] suggests that the compressive component of the nonlinearity is a broadly tuned, multiplicative gain mechanism, which is fundamentally different from the accelerating component. The full implications of these findings for the psychophysical estimation of channel-tuning functions and for the prediction of discrimination performance are not yet known.

Masking paradigms have also been used to estimate the tuning of temporal channels.[151,170]

Contrast Estimation

We have argued that the differential visibility of coarse and fine patterns can be understood from an analysis of information losses in the early stages of vision: optics, receptor sampling, the inverse relation between the preferred spatial frequency and size of tuned spatial mechanisms, and retinal adaptation mechanisms. It is important to note, however, that the differential visibility described by the CSF in Fig. 11 does not relate directly to perceptual experience. For example, if you hold this page up close and then at arm's length, the apparent contrast of the text does not change appreciably even though the spatial frequency content (in cycles per degree) does. This observation has been demonstrated experimentally by asking observers to adjust the contrast of a sinusoidal grating of one spatial frequency until it appeared to match the contrast of a sinusoid of a different frequency.[171] For example, consider two gratings, one at 5 cpd and another at 20 cpd; the former is near the peak of the CSF and the latter is well above the peak so it requires nearly a log unit more contrast to reach threshold. The 5-cpd target was set to a fixed contrast and the observer was asked to adjust the contrast of the 20-cpd target to achieve an apparent contrast match. When the contrast of the 5-cpd target was set near threshold, observers required about a log unit more contrast in the higher-frequency target before they had the same apparent contrast. The most interesting result occurred when the contrast of the 5-cpd grating was set to a value well above threshold. Observers then adjusted the contrast of the 20-cpd grating to the same physical value as the contrast of the lower-frequency target. This is surprising because, as described above, two gratings of equal contrast but different spatial frequencies produce different retinal image contrasts (see Chap. 24). In other words, when observers set 5- and 20-cpd gratings to equal physical contrasts, they were accepting as equal in apparent contrast two gratings whose retinal image contrast differed substantially. This implies that the visual system compensates at suprathreshold contrasts for the defocusing effects of the eye's optics and perhaps for the low-pass filtering effects of early stages of processing. This phenomenon has been called *contrast constancy*; the reader might recognize the similarity to deblurring techniques used in aerial and satellite photography (e.g., Ref. 172).

Visual Acuity

The ability to perceive high-contrast spatial detail is termed *visual acuity*. Measurements of visual acuity are far and away the most common means of assessing ocular health[173] and suitability for operating motor vehicles.[174] The universal use of visual acuity is well justified for clinical assessment,[175] but there is evidence that it is unjustified for automobile licensing (e.g., Ref. 176).

As implied by the discussion of contrast sensitivity, eye movements, optics, receptor properties, and postreceptoral neural mechanisms all conspire to limit acuity; one factor may dominate in a given situation, but they all contribute. To assess acuity, high-contrast patterns of various sizes are presented at a fixed distance. The smallest pattern or smallest critical pattern element that can be reliably detected or identified is taken as the threshold

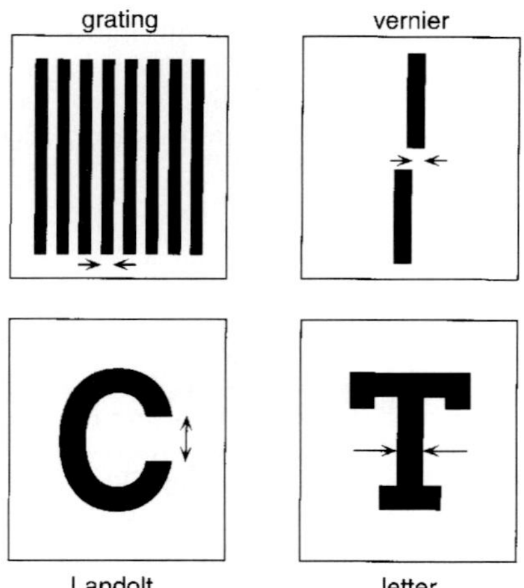

FIGURE 19 Targets commonly used to assess visual acuity. The dimension of the target that is taken as the acuity threshold is indicated by the arrows.

value and is usually expressed in minutes of arc. Countless types of stimuli have been used to measure visual acuity, but the four illustrated in Fig. 19 are most common.

Gratings are used to measure minimum angle of resolution (MAR); the spatial frequency of a high-contrast sinusoidal grating is increased until its modulation can no longer be perceived. The reader should note that this measurement is equivalent to locating the high-frequency cutoff of the spatial CSFs shown in previous sections. Under optimal conditions, the finest detectable grating bars are about 30 arcsec wide.[112]

The Landolt ring target is a high-contrast ring with a gap appearing in one of four positions. Threshold is defined as the smallest gap that can be correctly located. Under optimal conditions, threshold is again about 30 arcsec or slightly smaller.[112]

The most common test in clinical and licensing situations is the letter acuity task. The stimuli are a series of solid, high-contrast letters, varying in size. Threshold is defined by the smallest identifiable stroke width which, under optimal conditions, is nearly 30 arcsec.

Finally, the vernier acuity task involves the presentation of two nearly aligned line segments. Threshold is defined by the smallest visible offset from colinearity. Under optimal conditions, vernier acuity is about 5 arcsec.[177,178] The special label, *hyperacuity,* has been provided for spatial thresholds, like vernier acuity, that are smaller than the intercone distance.[179]

In practice, testing environments can vary widely along dimensions that affect visual acuity. These dimensions include illumination, the care chosen to correct refractive errors, and more. The National Academy of Sciences (1980) has provided a set of recommended standards for use in acuity mesurements including use of the Landolt ring.

Although these various acuity tasks are clearly related, performance on the four tasks is affected differently by a variety of factors and, for this reason, one suspects that the underlying mechanisms are not the same.[110,180] Figure 20 shows how grating and vernier acuities vary with retinal eccentricity. It is obvious from this figure that no single scaling factor can equate grating and vernier acuity across retinal eccentricities; a scaling factor that would equate grating acuity, for example, would not be large enough to offset the dependence of vernier acuity on retinal eccentricity.

Indeed, there have been several demonstrations that the two types of acuity depend

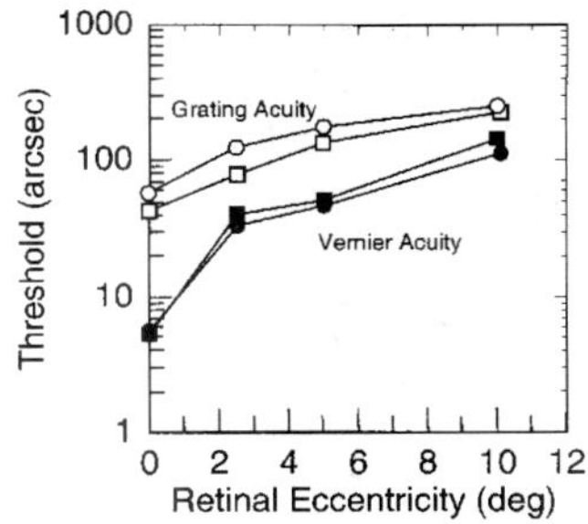

FIGURE 20 Grating and vernier acuity as a function of retinal eccentricity. The open symbols represent the smallest detectable bar widths for a high-contrast grating. The filled symbols represent the smallest detectable offset of one line segment with respect to another. Circles and squares are the data from two different observers. *(Adapted from Ref. 110.)*

differently on experimental conditions. For example, grating acuity appears more susceptible to optical defects and contrast reductions, whereas vernier, Landolt and letter acuities are more susceptible to a visual condition called *amblyopia.*[181] There are numerous models attempting to explain why different types of acuity are affected dissimilarly by experimental conditions, particularly eccentric viewing (e.g., Refs. 182, 183).

The explanation of spatial discrimination thresholds smaller than the distance between photoreceptors, such as vernier acuity, has become a central issue in contemporary models of spatial vision. One view is that the fine sensitivity exhibited by the hyperacuities[178] reveals a set of local spatial primitives or features that are encoded particularly efficiently in the early stages of vision.[184–186] The other view is that hyperacuity can be understood from an analysis of information processing among the spatially tuned mechanisms described in the earlier sections.[182,187,188] Evidence favoring the second view is manifold. It has been shown, for example, that offset thresholds smaller than the grain of the photoreceptor lattice are predicted from measurements of the linear filtering and signal-to-noise properties of retinal ganglion cells.[187] Likewise, human vernier thresholds can be predicted in many situations from measurements of contrast discrimination thresholds.[189] Also, ideal-observer analysis has demonstrated that the information available at the level of the photoreceptors predicts better performance in hyperacuity tasks than in acuity (resolution) tasks,[190] suggesting that any relatively complete set of spatially tuned mechanisms should produce better performance in the hyperacuity tasks. The remaining puzzle for proponents of the second view, however, is why vernier acuity and the other hyperacuities diminish so strikingly with increasing retinal eccentricity.

Pattern Discrimination

An important aspect of spatial vision is the ability to distinguish one pattern from another. In this section, we consider the ability to distinguish simple, suprathreshold patterns that vary along a single dimension. The dimensions considered are orientation, size or spatial frequency, and position or phase.

The detectability of a pattern varies to some degree with orientation. For most observers, for example, targets are most visible when oriented vertically or horizontally and least when oriented obliquely.[191] This *oblique effect* is demonstrable for periodic and aperiodic targets and is largest for fine-detail targets. The cause of the effect appears to be differential sensitivity among orientation-tuned mechanisms, presumably in the visual cortex.[192] The ability to discriminate gratings differing in orientation depends on several stimulus parameters including target length, contrast, and spatial frequency, but observers can generally discriminate extended, high-contrast targets that differ by 1 deg or less.

The ability to discriminate gratings differing in spatial frequency or aperiodic stimuli differing in size depends critically on the reference frequency or size. That is to say, the ratio of the frequency discrimination threshold, Δf, to the reference frequency, f, is roughly constant at about 0.03 for a wide range of reference frequencies;[193] similarly, the ratio of size discrimination threshold to reference size is roughly constant for a variety of reference sizes, except when the reference is quite small.[178]

The encoding of spatial phase is crucial to the identification of spatial patterns. Its importance is demonstrated quite compellingly by swapping the amplitude and phase spectra of two images; the appearance of such hybrid images corresponds to a much greater degree with their phase spectra than with their amplitude spectra.[194] The conventional explanation is that the phase spectrum determines the spatial structure of an image, but this is perhaps too simplistic because the amplitude spectra of most natural images are much more similar to one another than are the phase spectra (e.g., Refs. 195, 196). Also, given that the result was obtained using global Fourier transforms, it does not have direct implications about the relative importance of phase coding within spatially localized channels.[197] Nonetheless, such demonstrations illustrate the important relationship between the encoding of spatial phase and pattern identification.

In phase discrimination tasks, the observer distinguishes between patterns—usually periodic patterns—that differ only in the phase relationships of their spatial frequency components. The representation of spatial phase in the fovea is very precise: observers can, for instance, discriminate relative phase shifts as small as 2–3 deg (of phase angle) in compound gratings composed of a fundamental and third harmonic of 5 cpd;[198] this is equivalent to distinguishing a positional shift of about 5 arcsec. Phase discrimination is much less precise in extrafoveal vision.[199–201] There has been speculation that the discrimination anomalies observed in the periphery underlie the diminished ability to distinguish spatial displacements,[110] to segment textures on the basis of higher-order statistics,[202,203] and to identify complex patterns such as letters,[204] but detailed hypotheses linking phase discrimination to these abilities have not been developed.

One model of phase discrimination[205] holds that local energy computations are performed on visual inputs and that local energy peaks are singled out for further analysis. The energy computation is performed by cross-correlating the waveform with even- and odd-symmetric spatial mechanisms of various preferred spatial frequencies. The relative activities of the even- and odd-symmetric mechanisms are used to represent the sort of image feature producing the local energy peak. This account is supported by the observation that a two-channel model, composed of even- and odd-symmetric mechanisms, predicts many phase discrimination capabilities well (e.g., Refs 206–208). The model also offers an explanation for the appearance of illusory Mach bands at some types of edges and not others.[205] For this account to work, however, one must assume that odd-symmetric, and not even-symmetric, mechanisms are much less sensitive in the periphery than in the fovea because some relative phase discriminations are extremely difficult in extrafoveal vision and others are quite simple[201,208,209] (but see Ref. 210).

Motion Detection and Discrimination

The ability to discriminate the form and magnitude of retinal image motion is critical for many fundamental visual tasks, including navigation through the environment, shape estimation, and distance estimation. There is a vast literature on motion perception (some representative reviews from somewhat different perspectives can be found in Refs. 45,

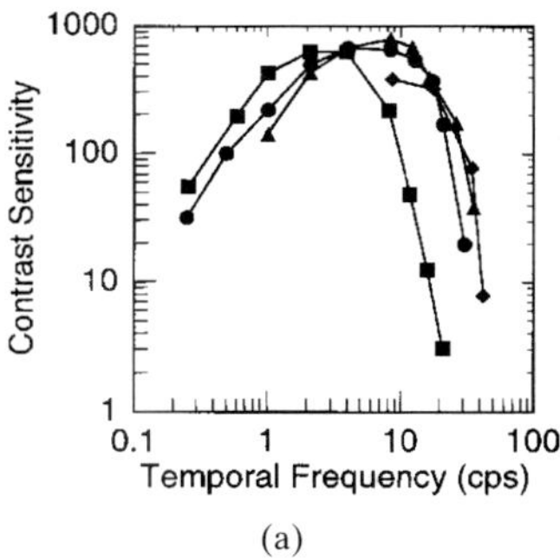

(a)

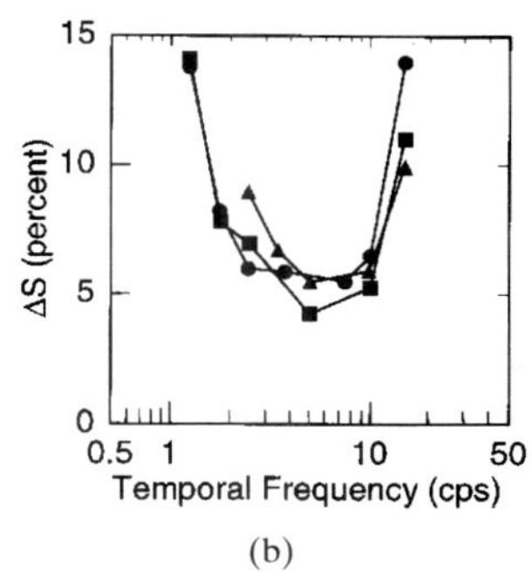

(b)

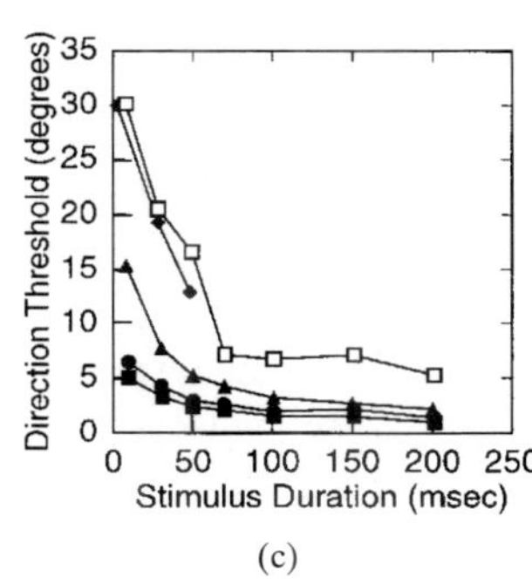

(c)

FIGURE 21 Motion detection and discrimination thresholds. (*a*) Contrast sensitivity for drifting sinewave gratings plotted as a function of temporal frequency. Each curve is for a different drift speed; (■) 1 deg/sec (●) 10 deg/sec, (▲) 100 deg/sec; (◆) 800 deg/sec. *(Adapted from Ref. 213.)* (*b*) Speed discrimination thresholds plotted as a function of temporal frequency. Each curve is for a different drift speed; (●) 5 deg/sec, (■) 10 deg/sec, (▲) 20 deg/sec. *(Adapted from Ref. 214.)* (*c*) Direction threshold for random dot patterns as a function of stimulus duration. Each curve is for a different drift speed; (□) 1 deg/sec, (▲) 4 deg/sec, (●) 16 deg/sec, (■) 64 deg/sec, (◆) 256 deg/sec. *(Adapted from Ref. 215.)*

211, 212). Here, we briefly discuss simple motion discrimination in the fronto-parallel plane, and (as an example of more complicated motion processing) observer heading discrimination.

Detection and/or discrimination of motion in the fronto-parallel plane is influenced by a number of factors, including contrast, wavelength composition, spatial-frequency content, initial speed, and eccentricity. For simplicity, and for consistency with the other sections in this chapter, we consider here only experiments employing sinewave-grating or random-dot stimuli.

Contrast sensitivity functions for detection of moving (drifting) sinewave-grating targets have the interesting property that they are nearly shape invariant on a log spatial-frequency axis; furthermore, as shown in Fig. 21*a,* they are nearly superimposed at high speeds and low spatial frequencies, when plotted as a function of temporal frequency.[106,213]* This latter result corresponds to the fact (mentioned earlier) that spatial CSFs are relatively flat at low spatial frequencies and high temporal frequencies.[128,147] In interpreting Fig. 21*a,* it is useful to note that the velocity (V) of a drifting sinewave grating is equal to the temporal frequency (f_t) divided by the spatial frequency (f_s):

$$V = \frac{f_t}{f_s} \tag{29}$$

Measurements of CSFs for moving stimuli provide relatively little insight into how the visual system extracts or represents motion information. For example, in one common paradigm[106] it is possible for subjects to perform accurately without "seeing" any motion at all. Greater insight is obtained by requiring the observer to discriminate between different aspects of motion such as speed or direction.

Representative measurements of speed discrimination for drifting sinewave grating targets are shown in Fig. 21*b*. The figure shows the just-detectable change in speed as a function of temporal frequency; each curve is for a differential initial or base speed. Similar to the motion CSFs, speed discrimination is seen to be largely dependent upon the temporal frequency and relatively independent of the speed.[214] As can be seen, the smallest detectable changes in speed are approximately 5 percent, and they occur in

* The temporal frequency of a drifting grating is the number of stripes passing a fixed reference point per unit time (sec).

the temporal frequency range of 5–10 cps, which is similar to the temporal frequency where contrast sensitivity for drifting gratings is greatest (cf., Fig. 21*a*). Interestingly, 5–10 cps is the temporal frequency range of strongest response for most neurons in the macaque's primary visual cortex.[60] For random-dot patterns, the smallest detectable change in speed is also approximately 5 percent.[215]

Representative measurements of direction discrimination for drifting random-dot patterns are shown in Fig. 21*c,* which plots direction threshold in degrees as a function of stimulus duration for a wide range of speeds. Direction discrimination improves with duration and is a U-shaped function of dot speed. Under optimal conditions, direction discrimination thresholds are in the neighborhood of 1–2°.[215,216]

The available evidence suggests that direction and speed discrimination improve quickly with contrast at low contrasts but show little improvement when contrast exceeds a few percent.[214,217,218] Motion discrimination is relatively poor at isoluminance.[219,220] The variations in motion discrimination with eccentricity can be largely accounted for by changes in spatial resolution.[221]

Psychophysical evidence suggests that there are at least two mechanisms that can mediate motion discrimination—a *short-range mechanism* which appears to be closely associated with motion-selective neurons in the early levels of cortical processing, and a *long-range mechanism* which appears to be associated with more "inferential" calculations occurring in later levels of cortical processing. Clear evidence for this view was first observed in an apparent motion paradigm in which observers were required to judge the shape of a region created by laterally displacing a subset of random dots in a larger random-dot pattern.[222] (These patterns are examples of *random-dot kinematograms,* the motion analog to random-dot stereograms.) Observers could accurately judge the shape of the region only if the displacements (1) were less than approximately 15 minarc ($D_{max} = 15$ min); (2) occurred with a delay of less than approximately 100 msec; and (3) occurred within the same eye (i.e., dichoptic presentations failed to produce reliable discrimination). These data suggest the existence of motion-sensitive mechanisms that can only detect local, monocular correlations in space-time.

The fact that motion-selective neurons in the early levels of the cortex can also only detect local spatio-temporal correlations suggests that they may be the primary source of neural information used by the subsequent mechanisms that extract shape and distance information in random-dot kinematograms. The receptive fields of motion-selective neurons decrease in size as a function of optimal spatial frequency, and they increase in size as a function of retinal eccentricity. Thus, the hypothesis that motion-selective neurons are the substrate for shape extraction in random-dot kinematograms is supported by the findings that D_{max} increases as spatial-frequency content is decreased via low-pass filtering,[223] and that D_{max} increases as total size of the kinematogram is increased.[224]

Accurate shape and motion judgments can sometimes be made in apparent-motion displays where the stimulation is dichoptic and/or the spatio-temporal displacements are large. However, in these cases the relevant shape must be clearly visible in the static frames of the kinematogram. The implication is that higher-level, inferential analyses are used to make judgments under these circumstances. A classic example is that we can infer the motion of the hour hand on a clock from its long-term changes in position even though the motion is much too slow to be directly encoded by motion-selective cortical neurons.

The evidence for short-term and long-term motion mechanisms raises the question of what mechanisms actually mediate performance in a given motion discrimination task. The studies described above (e.g., Fig. 21) either randomized stimulus parameters (such as duration and spatial frequency) and/or used random-dot patterns; hence, they most likely measured properties of the short-range mechanisms. However, even with random-dot patterns, great care must be exercised when interpreting discrimination experiments. For example, static pattern-processing mechanisms may sometimes contribute to motion discrimination performance even though performance is at chance when static views are presented at great separations of space or time. This might occur because persisting neural responses from multiple stimulus frames merge to produce "virtual" structured patterns (e.g., Glass patterns[225]). Similar caution should be applied when interpreting discrimination performance with random-dot stereograms (see below).

The discussion of motion perception to this point has focused on the estimation of velocities of various stimuli. Such estimation is importantly involved in guiding locomotion through and estimating the three-dimensional layout of cluttered environments. Motion of an observer through a rigid visual scene produces characteristic patterns of motion on the retina called the *optic flow field.*[226,227] Observers are able to judge the direction of their own motion through the environment based upon this optical flow information alone.[228,229] In fact, accurate motion perception is possible even for random-dot flow fields, suggesting that the motion-selective neurons in the early levels of the visual cortex are a main source of information in perceiving self-motion. But, how is the local motion information provided by the motion-selective neurons used by subsequent mechanisms to compute observer motion?

Gibson[226,227] proposed that people identify their direction of self-motion with respect to obstacles by locating the source of flow: the focus of expansion. Figure 22*a* depicts the flow field on the retina as the observer translates while fixating ahead. Flow is directed away from the focus of expansion and this point corresponds to the direction of translation. Not surprisingly, observers can determine their heading to within ±1 deg from such a field of motion.[229]

The situation becomes much more complicated once eye/head movements are

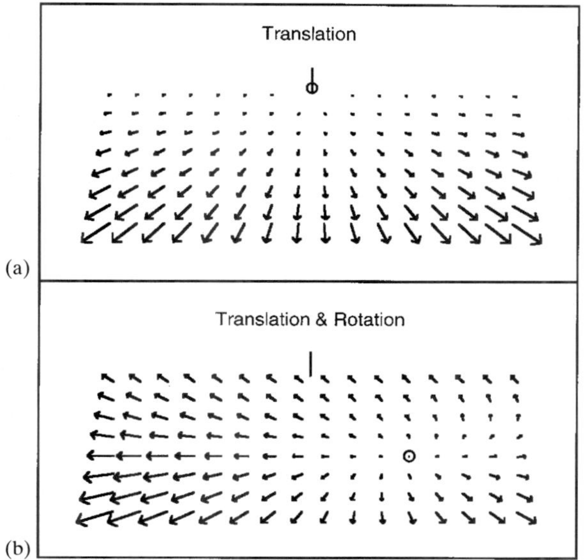

FIGURE 22 Optic flow fields resulting from forward translation across a rigid ground plane. (*a*) Flow field in the retinal image when the observer translates straight ahead while maintaining constant eye and head position; the heading is indicated by the small vertical line. (*b*) Retinal flow field when the observer translates straight ahead while making an eye movement to maintain fixation on the circle; again the heading is indicated by the small vertical line. *(Adapted from Ref. 230.)*

considered. Figure 22*b* illustrates this by portraying the flow resulting from forward translation while the observer rotates the eyes to maintain fixation on a point off to the right. This motion does not produce a focus of expansion in the image corresponding to the heading; the only focus corresponds to the point of fixation. Consequently, heading cannot be determined by locating a focus (or singularity) in the retinal flow field. Nonetheless, human observers are able to locate their heading to within ±1.5 deg.[230]

Recent theoretical efforts have concentrated on this problem of computing heading in the presence of rotations due to eye/head movements. There are two types of models. One holds that observers measure the rotational flow components due to eye/head movements by means of an extraretinal signal; that is, the velocity of rotations is signaled by proprioceptive feedback from the extraocular and/or neck muscles or by efferent information.[231–235] The rotational flow components are then subtracted from the flow field as a whole to estimate the heading. The other type of model holds that people determine heading in the presence of rotations from retinal image information alone (e.g., Refs. 232–235). These models hypothesize that the visual system decomposes the flow field into rotational and translational components by capitalizing on the fact that flows due to translation and rotation depend differently on the scene geometry. Once the decomposition is performed, the heading can be estimated from the translational components (see Fig. 22*a*).

Current experimental evidence favors the extraretinal signal model,[236,237] but there is also evidence supporting the retinal-image model when the eye-head rotations are slow.[230]

The optic flow field also contains information specifying the relative depths of objects in the visual scene: The velocities of retinal image motions are, to a first approximation, proportional to the inverse depths of the corresponding objects.[238] Human observers are quite sensitive to this depth cue as evidenced by the fact that they perceive depth variation for differential motions of as small as $\frac{1}{2}$–$\frac{1}{3}$ arcsec per sec.[239]

Binocular Stereoscopic Discrimination.

Light from the three-dimensional environment is imaged onto the two-dimensional array of photoreceptors, so the dimension associated with distance from the eye is collapsed in the two-dimensional retinal image. The images in the two eyes generally differ, however, because the eyes are separated by 6–7 cm. The differences between the images, which are called *binocular disparities,* are used by the visual system to recover information about the distance dimension. The depth percept that results from binocular disparity is called *stereopsis.* Binocular disparity (stereo) information is formally equivalent to the monocular information that results from a step translation of one eye by 6–7 cm. This observation implies a close theoretical connection between the computation of distance from motion and distance from binocular disparity. As indicated in an earlier section, the extraction of stereo and motion information presumably begins in the early stages of cortical processing (i.e., in V1 of the macaque), where many neurons are selective for disparity and direction of motion.

The geometry of stereopsis is described in Chap. 24, Fig. 25. Here, we briefly consider the nature of the information available for computing distance from disparity. The distance between any pair of points in the depth dimension (Δz) is related to the horizontal disparity (d) and the horizontal convergence angle of the eyes (θ) by the following formula:*

$$\Delta z = \frac{ad}{(\theta + d_\theta + d/2)(\theta + d_\theta - d/2)} \tag{30}$$

* This formula is an approximation based on the assumption of relatively small angles and object location near the mid-sagittal plane; it is most accurate for distances under a few meters.

where d_θ is the average disparity between the two points and the convergence point, and a is the interpupillary distance (the distance between the image nodal points of the eyes). If d_θ is zero, then the points are centered in depth about the convergence point. The locus of all points where the disparity d equals zero is known as the (Vieth-Müller) *horopter.* As Eq. (30) indicates, the computation of absolute distance between points from horizontal disparity requires knowledge of the "eye parameters" (the interpupillary distance and the angle of convergence). However, relative distance information is obtained even if the eye parameters are unknown (in other words, Δz is monotonic with d). Furthermore, absolute distance can be recovered in principle without knowing the angle of convergence from use of horizontal and vertical binocular disparities (e.g., Refs. 240, 241).

Under optimal conditions, the human visual system is capable of detecting very small binocular disparities and hence rather small changes in distance. In the fovea, disparities of roughly 10 arsec can be detected if the test objects contain sharp, high-contrast edges and if they are centered in depth about a point on or near the horopter ($d_\theta = 0.0$).[177,242] At a viewing distance of 50 cm, 10 sec corresponds to a distance change of about 0.02 cm, but at 50 m, it corresponds to a distance change of about 2 m.

In measuring stereoscopic discrimination performance, it is essential that the task cannot be performed on the basis of monocular information alone. This possibility can be assessed by measuring monocular and binocular performance with the same stimuli; if monocular performance is substantially worse than binocular performance, then binocular mechanisms per se are used in performing the stereoscopic task.[242,243] Random dot stereograms[243,244] are particularly effective at isolating binocular mechanisms because discriminations cannot be performed reliably when such stereograms are viewed monocularly.[243,245]

Stereopsis is affected by a number of factors. For example, the contrast required to produce a stereoscopic percept varies with the spatial frequency content of the stimulus. This result is quantified by the contrast sensitivity function for stereoscopic discrimination of spatially filtered random-dot stereograms; the function is similar in shape to CSFs measured for simple contrast detection,[246] but detection occurs at lower contrasts than stereopsis occurs. In other words, there is a range of contrasts that are clearly detectable, but insufficient to yield a stereoscopic percept.[246]

The fact that the CSFs for contrast detection and stereoscopic discrimination have similar shapes suggests that common spatial-frequency mechanisms are involved in the two tasks. This hypothesis receives some support from masking and adaptation experiments demonstrating spatial-frequency tuning[247–250] and orientation tuning[251] for stereo discrimination, and from electrophysiological studies demonstrating that cortical cells selective for binocular disparity are also usually selective for orientation[55,56,62] and spatial frequency.[252]

The smallest detectable disparity, which is called *stereoacuity,* improves as luminance contrast is increased.[253–255] As shown in Fig. 23*a,* stereoacuity improves approximately in inverse proportion to the square of contrast at low contrasts, and in inverse proportion to the cube root of contrast at high contrasts. Stereoacuity is also dependent upon spatial frequency; it improves in inverse proportion to target spatial frequency over the low spatial-frequency range, reaching optimum near 3 cpd (Fig. 23*b*).[254] Both the inverse-square law at low contrasts and the linear law at low spatial frequencies are predicted by signal-detection models that assume independent, additive noise and simple detection rules.[254,255]

Stereoacuity declines precipitously as a function of the distance of the test objects from the convergence plane (Fig. 23*c*).[256,257] For example, adding a disparity pedestal of 40 minarc to test objects reduces acuity by about 1 log unit. This loss of acuity is not the result of losses of information due to the geometry of stereopsis; it must reflect the properties of the underlying neural mechanisms. [Note in Eq. (30) that adding a disparity to both objects is equivalent to changing the value of d_θ.]

Much like the spatial channels described earlier under "Spatial Chanels," there are

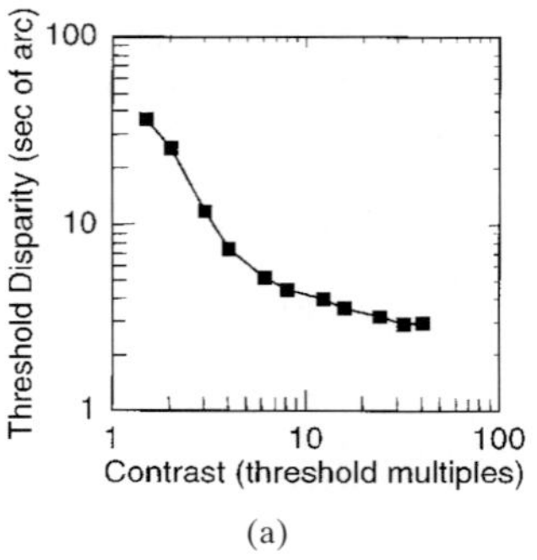

(a)

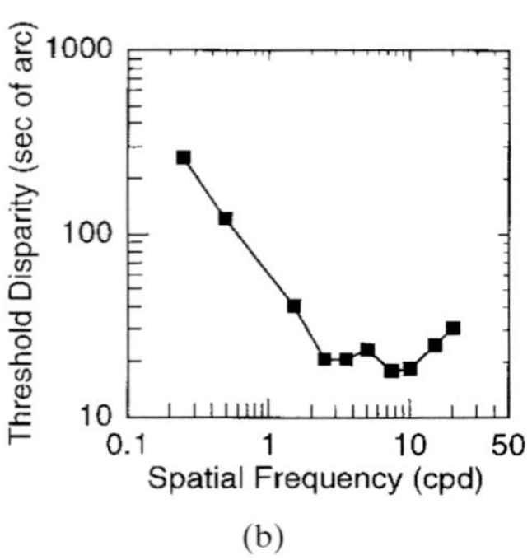

(b)

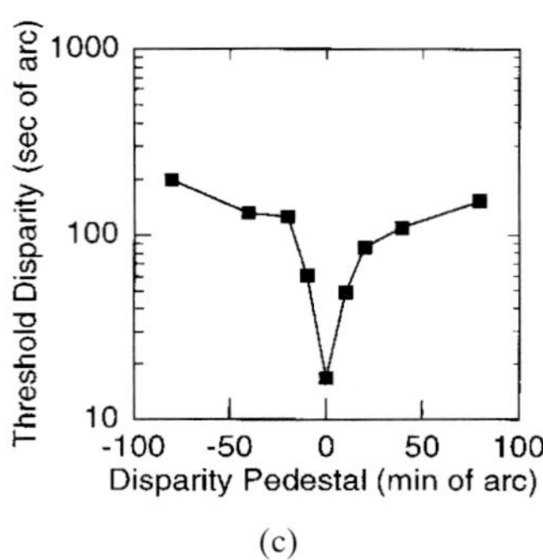

(c)

FIGURE 23 Disparity discrimination thresholds as a function of contrast, spatial frequency, and pedestal disparity. (*a*) Disparity threshold as a function of luminance contrast (in contrast threshold units) for dynamic random-dot stereograms. *(Adapted from Ref. 255.)* (*b*) Disparity threshold as a function of spatial frequency for sinewave-grating stereograms. *(Adapted from Ref. 254.* (*c*) Disparity threshold as a function of the disparity pedestal (distance from the convergence plane) for difference-of-gaussian (DOG) stereograms. *(Adapted from Ref. 257).*

multiple channels tuned to different disparities, but it is unclear whether there are a small number of such channels—"near," "far," and "tuned"[258]—or a continuum of channels.[56,63,259,260] Models that assume a continuum of channels with finer tuning near the horopter predict both the sharp decline in stereoacuity away from the horopter[260,261] and the shapes of tuning functions that have been estimated from adaptation and probability summation experiments.[259,260]

25.5 ACKNOWLEDGMENTS

This project was supported in part by NIH grant EY02688 and AFOSR grant F49620-93-1-0307 to WSG and by NIH grant HD 19927 to MSB.

25.6 REFERENCES

1. R. Navarro, P. Artal, and D. R. Williams, "Modulation Transfer Function of the Human Eye as a Function of Retinal Eccentricity," *Journal of the Optical Society of America* A, **10:**201–212 (1993).
2. J. D. Gaskill, *Linear Systems, Fourier Transforms, and Optics,* Wiley, New York, 1978.
3. J. W. Goodman, *Introduction to Fourier Optics,* Physical and Quantum Electronics Series, H. Heffner and A. E. Siegman (eds.), McGraw-Hill, New York, 1968.
4. F. W. Campbell and R. W. Gubisch, "Optical Quality of the Human Eye," *Journal of Physiology,* **186:**558–578 (London, 1966).
5. G. Wyszecki and W. S. Stiles, *Color Science: Concepts and Methods, Quantitative Data and Formulae,* 2d ed., Wiley, New York, 1982.
6. C. A. Curcio, et al., "Human Photoreceptor Topography," *Journal of Comparative Neurology* **292:**497–523 (1990).
7. D. R. Williams, "Aliasing in Human Foveal Vision," *Vision Research* **25:**195–205 (1985).
8. L. N. Thibos, F. E. Cheney, and D. J. Walsh, "Retinal Limits to the Detection and Recognition of Gratings," *Journal of the Optical Society of America* **4:**1524–1529 (1987).

9. D. R. Williams, "Topography of the Foveal Cone Mosaic in the Living Human Eye," *Vision Research* **28:**433–454 (1988).
10. R. E. Marc and H. G. Sperling, "Chromatic Organization of Primate Cones," *Science* **196:**454–456 (1977).
11. F. M. de Monasterio, et al., "Density Profile of Blue-Sensitive Cones Along the Horizontal Meridian of Macaque Retina," *Investigative Ophthalmology and Visual Science* **26:**283–288 (1985).
12. P. Ahnelt, C. Keri, and H. Kolb, "Identification of Pedicles of Putative Blue-Sensitive Cones in the Human Retina," *Journal of Comparative Neurology* **293:**39–53 (1990).
13. D. R. Williams, D. I. A. MacLeod, and M. M. Hayhoe, "Foveal Tritanopia," *Vision Research* **21:**1341–1356 (1981).
14. J. L. Nerger and C. M. Cicerone, "The Ratio of L Cones to M Cones in the Human Parafoveal Retina," *Vision Research* **32:**879–888 (1992).
15. C. M. Cicerone, "Color Appearance and the Cone Mosaic in Trichromacy and Dichromacy," *Color Vision Deficiencies,* Y. Ohta (ed.) Kugler & Ghedini, Amsterdam, 1990, p. 1–12.
16. J. K. Bowmaker, "Visual Pigments and Colour in Primates," *From Pigments to Perception: Advances in Understanding Visual Processes,* A. Valberg and B. L. Lee (eds.) Plenum Press, New York, 1991.
17. J. L. Schnapf et al., "Visual Transduction in Cones of the Monkey Macaca Fascicularis," *Journal of Physiology* **427:**681–713 (London, 1990).
18. W. A. H. Rushton, "The Difference Spectrum and the Photosensitivity of Rhodopsin in the Living Human Retina," *Journal of Physiology* **134:**11–297 (London, 1956).
19. W. A. H. Rushton and G. H. Henry, "Bleaching and Regeneration of Cone Pigments in Man," *Vision Research* **8:**617–631 (1968).
20. M. Alpern and E. N. Pugh, "The Density and Photosensitivity of Human Rhodopsin in the Living Retina," *Journal of Physiology* **237:**341–370 (London, 1974).
21. W. H. Miller and G. D. Bernard, "Averaging over the Foveal Receptor Aperture Curtails Aliasing," *Vision Research* **23:**1365–1369 (1983).
22. D. I. A. MacLeod, D. R. Williams, and W. Makous, "A Visual Nonlinearity Fed by Single Cones," *Vision Research* **32:**347–363 (1992).
23. J. L. Schnapf and D. A. Baylor, "How Photoreceptor Cells Respond to Light," *Scientific American* **256**(4):40–47 (1987).
24. J. M. Valeton and D. van Norren, "Light-Adaptation of Primate Cones: An Analysis Based on Extracellular Data," *Vision Research* **23:**1539–1547 (1982).
25. D. A. Baylor, B. J. Nunn, and J. L. Schnapf, "The Photocurrent, Noise and Spectral Sensitivity of Rods of the Monkey Macaca Fascicularis," *Journal of Physiology* **357:**576–607 (London, 1984).
26. D. C. Hood and D. G. Birch, "A Quantitative Measure of the Electrical Activity of Human Rod Photoreceptors Using Electroretinography," *Visual Neuroscience* **5:**379–387 (1990).
27. E. N. Pugh and T. D. Lamb, "Cyclic GMP and Calcium: The Internal Messengers of Excitation and Adaptation in Vertebrate Photoreceptors," *Vision Research* **30**(12):1923–1948 (1990).
28. D. C. Hood and D. G. Birch, "Human Cone Receptor Activity: The Leading Edge of the A-Wave and Models of Receptor Activity," *Visual Neuroscience* **10:**857–871 (1993).
29. R. W. Rodieck, "The Primate Retina," *Comparative Primate Biology, Neurosciences,* H. D. Steklis and J. Erwin (eds.), Liss, New York, 1988, pp. 203–278.
30. P. Sterling, "Retina," *The Synaptic Organization of the Brain,* G. M. Sherpard (ed.), Oxford University Press, New York, 1990, pp. 170–213.
31. H. Wässle and B. B. Boycott, "Functional Architecture of the Mammalian Retina," *Physiological Reviews,* **71**(2):447–480 (1991).
32. C. A. Curcio and K. A. Allen, "Topography of Ganglion Cells in the Human Retina," *Journal of Comparative Neurology* **300:**5–25 (1990).
33. H. Wässle et al., "Retinal Ganglion Cell Density and Cortical Magnification Factor in the Primate," *Vision Research* **30**(11):1897–1911 (1990).

34. S. W. Kuffler, "Discharge Patterns and Functional Organization of the Mammalian Retina," *Journal of Neurophysiology* **16:**37–68 (1953).
35. R. W. Rodieck, "Quantitative Analysis of Cat Retinal Ganglion Cell Response to Visual Stimuli," *Vision Research* **5:**583–601 (1965).
36. C. Enroth-Cugell and J. G. Robson, "The Contrast Sensitivity of Retinal Ganglion Cells of the Cat," *Journal of Physiology* **187:**517–552 (London, 1966).
37. A. M. Derrington and P. Lennie, "Spatial and Temporal Contrast Sensitivities of Neurones in Lateral Geniculate Nucleus of Macaque," *Journal of Physiology* **357:**219–240 (London, 1984).
38. K. Purpura et al., "Light Adaptation in the Primate Retina: Analysis of Changes in Gain and Dynamics of Monkey Retinal Ganglion Cells," *Visual Neuroscience* **4:**75–93 (1990).
39. R. M. Shapley and H. V. Perry, "Cat and Monkey Retinal Ganglion Cells and Their Visual Functional Roles," *Trends in Neuroscience* **9:**229–235 (1986).
40. B. B. Lee et al., "Luminance and Chromatic Modulation Sensitivity of Macaque Ganglion Cells and Human Observers," *Journal of the Optical Society of America* **7**(12):2223–2236 (1990).
41. B. G. Cleland and C. Enroth-Cugell, "Quantitative Aspects of Sensitivity and Summation in the Cat Retina," *Journal of Physiology* **198:**17–38 (London, 1968).
42. A. M. Derrington and P. Lennie, "The Influence of Temporal Frequency and Adaptation Level on Receptive Field Organization of Retinal Ganglion Cells in Cat," *Journal of Physiology* **333:**343–366 (London, 1982).
43. J. M. Cook et al., "Visual Resolution of Macaque Retinal Ganglion Cells," *Journal of Physiology* **396:**205–224 (London, 1988).
44. J. H. R. Maunsell and W. T. Newsome, "Visual Processing in Monkey Extrastriate Cortex," *Annual Review of Neuroscience* **10:**363–401 (1987).
45. K. Nakayama, "Biological Image Motion Processing: A Review," *Vision Research* **25**(5):625–660 (1985).
46. W. H. Merigan and J. H. R. Maunsell, "How Parallel are the Primate Visual Pathways," *Annual Review of Neuroscience* **16:**369–402 (1993).
47. D. C. Van Essen, "Functional Organization of Primate Visual Cortex," *Cerebral Cortex,* A. A. Peters and E. G. Jones (eds.), Plenum, New York, 1985, pp. 259–329.
48. R. L. DeValois and K. K. DeValois, *Spatial Vision,* Oxford University Press, New York, 1988.
49. P. Lennie et al., "Parallel Processing of Visual Information," *Visual Perception: The Neurophysiological Foundations,* L. Spillman and J. S. Werner (eds.), Academic Press, San Diego, 1990.
50. R. L. DeValois, I. Abramov, and G. H. Jacobs, "Analysis of Response Patterns of LGN Cells," *Journal of the Optical Society of America,* **56:**966–977 (1966).
51. T. N. Wiesel and D. H. Hubel, "Spatial and Chromatic Interactions in the Lateral Geniculate Body of the Rhesus Monkey," *Journal of Neurophysiology* **29:**1115–1156 (1966).
52. E. Kaplan and R. Shapley, "The Primate Retina Contains Two Types of Ganglion Cells, with High and Low Contrast Sensitivity," *Proceedings of the National Academy of Sciences* **83:**125–143 (U.S.A., 1986).
53. E. Kaplan, K. Purpura, and R. M. Shapley, "Contrast Affects the Transmission of Visual Information Through the Mammalian Lateral Geniculate Nucleus," *Journal of Physiology* **391:**267–288 (London, 1987).
54. D. H. Hubel and T. N. Wiesel, "Receptive Fields and Functional Architecture of Monkey Striate Cortex," *Journal of Physiology* **195:**215–243 (London, 1968).
55. D. Hubel and T. Wiesel, "Cells Sensitive to Binocular Depth in Area 18 of the Macaque Monkey Cortex," *Nature* **225:**41–42 (1970).
56. G. F. Poggio and B. Fischer, "Binocular Interaction and Depth Sensitivity in Striate and Prestriate Cortex of Behaving Rhesus Monkey," *Journal of Neurophysiology* **40:**1392–1405 (1977).
57. R. L. DeValois, E. W. Yund, and N. Hepler, "The Orientation and Direction Selectivity of Cells in Macaque Visual Cortex," *Vision Research* **22:**531–544 (1982).

58. R. L. DeValois, D. G. Albrecht, and L. G. Thorell, "Spatial Frequency Selectivity of Cells in Macaque Visual Cortex," *Vision Research* **22:**545–559 (1982).

59. H. B. Barlow, "Critical Limiting Factors in the Design of the Eye and Visual Cortex," *Proceedings of the Royal Society of London,* series B **212:**1–34 (1981).

60. K. H. Foster et al., "Spatial and Temporal Frequency Selectivity of Neurones in Visual Cortical Areas V1 and V2 of the Macaque Monkey," *Journal of Physiology* **365:**331–363 (London, 1985).

61. G. F. Poggio, F. Gonzalez, and F. Krause, "Stereoscopic Mechanisms in Monkey Visual Cortex: Binocular Correlation and Disparity Selectivity," *Journal of Neuroscience* **8:**4531–4550 (1988).

62. H. B. Barlow, C. Blakemore, and J. D. Pettigrew, "The Neural Mechanism of Binocular Depth Discrimination," *Journal of Physiology* **193:**327–342 (London, 1967).

63. S. LeVay and T. Voigt, "Ocular Dominance and Disparity Coding in Cat Visual Cortex," *Visual Neuroscience* **1:**395–414 (1988).

64. M. S. Livingstone and D. H. Hubel, "Segregation of Form, Color, Movement and Depth: Anatomy, Physiology and Perception," *Science* **240:**740–750 (1988).

65. P. Lennie, J. Krauskopf, and G. Sclar, "Chromatic Mechanisms in Striate Cortex of Macaque," *Journal of Neuroscience* **10:**649–669 (1990).

66. M. J. Hawken and A. J. Parker, "Spatial Properties of Neurons in the Monkey Striate Cortex," *Proceedings of the Royal Society of London,* series B **231:**251–288 (1987).

67. D. B. Hamilton, D. G. Albrecht, and W. S. Geisler, "Visual Cortical Receptive Fields in Monkey and Cat: Spatial and Temporal Phase Transfer Function," *Vision Research* **29:**1285–1308 (1989).

68. B. C. Skottun et al., "Classifying Simple and Complex Cells on the Basis of Response Modulation," *Vision Research* **31**(7/8):1079–1086 (1991).

69. D. G. Albrecht and D. H. Hamilton, "Striate Cortex of Monkey and Cat: Contrast Response Function," *Journal of Neurophysiology* **48**(1):217–237 (1982).

70. D. G. Albrecht and W. S. Geisler, "Motion Selectivity and the Contrast-Response Function of Simple Cells in the Visual Cortex," *Visual Neuroscience* **7:**531–546 (1991).

71. D. J. Heeger, "Nonlinear Model of Neural Responses in Cat Visual Cortex," *Computational Models of Visual Processing,* M. S. Landy and A. Movshon (eds.), MIT Press; Cambridge, Mass., 1991, pp. 119–133.

72. D. J. Tolhurst, J. A. Movshon, and A. F. Dean, "The Statistical Reliability of Signals in Single Neurons in the Cat and Monkey Visual Cortex," *Vision Research* **23:**775–785 (1983).

73. W. S. Geisler et al., "Discrimination Performance of Single Neurons: Rate and Temporal-Pattern Information," *Journal of Neurophysiology* **66:**334–361 (1991).

74. H. L. Van Trees, *Detection, Estimation, and Modulation Theory,* Wiley, New York, 1968.

75. D. M. Green and J. A. Swets, *Signal Detection Theory and Psychophysics,* Krieger, New York, 1974.

76. H. B. Barlow, "Temporal and Spatial Summation in Human Vision at Different Background Intensities," *Journal of Physiology* **141:**337–350 (London, 1958).

77. A. Rose, "The Sensitivity Performance of the Human Eye on an Absolute Scale," *Journal of the Optical Society of America* **38:**196–208 (1948).

78. W. S. Geisler, "Sequential Ideal-Observer Analysis of Visual Discrimination," *Psychological Review* **96:**267–314 (1989).

79. D. G. Pelli, "The Quantum Efficiency of Vision," *Vision: Coding and Efficiency,* C. Blakemore (ed.), Cambridge University Press, Cambridge, 1990, pp. 3–24.

80. H. B. Barlow, "The Efficiency of Detecting Changes of Density in Random Dot Patterns," *Vision Research* **18:**637–650 (1978).

81. H. de Vries, "The Quantum Character of Light and Its Bearing upon Threshold of Vision, the Differential Sensitivity and Visual Acuity of the Eye," *Physica* **10:**553–564 (1943).

82. H. B. Barlow, "Increment Thresholds at Low Intensities Considered as Signal/Noise Discriminations," *Journal of Physiology* **136:**469–488 (London, 1957).

83. H. Helstrom, "The Detection and Resolution of Optical Signals," *IEEE Transactions on Information Theory* **IT-10:**275–287 (1964).

84. W. S. Geisler, "Physical Limits of Acuity and Hyperacuity," *Journal of the Optical Society of America* A **1:**775–782 (1984).

85. W. S. Geisler and K. Chou, "Separation of Low-Level and High-Level Factors in Complex Tasks: Visual Search," *Psychological Review,* 1994 (in press).

86. M. E. Rudd, "Quantal Fluctuation Limitations on Reaction Time to Sinusoidal Gratings," *Vision Research* **28:**179–186 (1988).

87. A. E. Burgess et al., "Efficiency of Human Visual Signal Discrimination," *Science* **214:**93–94 (1981).

88. D. Kersten, "Statistical Efficiency for the Detection of Visual Noise," *Vision Research* **27:**1029–1040 (1987).

89. G. E. Legge, D. Kersten and A.E. Burgess, "Contrast Discrimination in Noise," *Journal of the Optical Society of America* A **4:**391–404 (1987).

90. A. E. Burgess, R. F. Wagner, and R. J. Jennings, "Human Signal Detection Performance for Noisy Medical Images," *Proceedings of IEEE Computer Society International Workshop on Medical Imaging,* 1982.

91. K. J. Myers et al., "Effect of Noise Correlation on Detectability of Disk Signals in Medical Imaging," *Journal of the Optical Society of America* A **2:**1752–1759 (1985).

92. H. H. Barrett et al., "Linear Discriminants and Image Quality," *Image and Vision Computing* **10:**451–460 (1992).

93. R. H. S. Carpenter, "Eye Movements," *Vision and Visual Dysfunction,* J. Cronley-Dillon (ed.), Macmillan Press, London, 1991.

94. D. Regan and C. W. Tyler, "Some Dynamic Features of Colour Vision," *Vision Research* **11:**1307–1324 (1971).

95. G. F. Poggio and T. Poggio, "The Analysis of Stereopsis," *Annual Reviews of Neuroscience* **7:**379–412 (1984).

96. I. P. Howard, *Human Visual Orientation,* Wiley, New York, 1982.

97. I. P. Howard, "The Perception of Posture, Self Motion, and the Visual Vertical," *Handbook of Perception and Human Performance,* K. R. Boff, L. Kaufman, and J. P. Thomas (eds.), Wiley, New York, 1986.

98. I. Rock, "The Description and Analysis of Object and Event Perception," *Handbook of Perception and Human Performance,* K. R. Boff, L. Kaufman, and J. P. Thomas (eds.), Wiley, New York, 1986.

99. R. J. Watt, "Pattern Recognition by Man and Machine," *Vision and Visual Dysfunction,* J. Cronly-Dillon (ed.), Macmillan Press, London, 1991.

100. A. Treisman, "Properties, Parts, and Objects," *Handbook of Perception and Human Performance,* K. R. Boff, L. Kaufman, and J. P. Thomas (eds.), Wiley, New York, 1986.

101. M. S. Banks and P. Salapatek, "Infant Visual Perception," *Handbook of Child Psychology,* M. M. Haith and J. J. Campos (eds.), Wiley, New York, 1983.

102. K. Simons, *Normal and Abnormal Visual Development,* Springer, New York, 1993.

103. J. Marshall, "The Susceptible Visual Apparatus," *Vision and Visual Dysfunction,* J. Cronly-Dillon (ed.), Macmillan Press, London, 1991.

104. F. Ratliff and L. A. Riggs, "Involuntary Motions of the Eye During Monocular Fixation," *Journal of Experimental Psychology* **46:**687–701 (1950).

105. L. E. Arend, "Response of the Human Eye to Spatially Sinusoidal Gratings at Various Exposure Durations," *Vision Research* **16:**1311–1315 (1976).

106. D. H. Kelly, "Motion and Vision II. Stabilized Spatio-Temporal Threshold Surface," *Journal of the Optical Society of America* **69:**1340–1349 (1979).

107. O. Packer and D. R. Williams, "Blurring by Fixational Eye Movements," *Vision Research* **32:**1931–1939 (1992).

108. D. G. Green, "Regional Variations in the Visual Acuity for Interference Fringes on the Retina," *Journal of Physiology* **207:**351–356 (London, 1970).

109. J. Hirsch and R. Hylton, "Quality of the Primate Photoreceptor Lattice and Limits of Spatial Vision," *Vision Research* **24:**347–355 (1984).

110. D. M. Levi, S. A. Klein, and A. P. Aitsebaomo, "Vernier Acuity, Crowding and Cortical Magnification," *Vision Research* **25:**963–977 (1985).

111. D. S. Jacobs and C. Blakemore, "Factors Limiting the Postnatal Development of Visual Acuity in Monkeys," *Vision Reseach* **28:**947–958 (1987).

112. S. Shlaer, "The Relation Between Visual Acuity and Illumination," *Journal of General Physiology* **21:**165–188 (1937).

113. D. R. Williams, "Visibility of Interference Fringes Near the Resolution Limit," *Journal of the Optical Society of America* A **2:**1087–1093 (1985).

114. R. N. Bracewell, *The Fourier Transform and Its Applications*: *Networks and Systems,* S. W. Director (ed.), McGraw-Hill, New York, 1978.

115. N. J. Coletta and H. K. Clark, "Change in Foveal Acuity with Light Level: Optical Factors," *Ophthalmic and Visual Optics,* Optical Society of America, Monterey, Calif., 1993.

116. J. M. Enoch and V. Lakshminarayanan, "Retinal Fibre Optics," *Vision and Visual Dysfunctions,* J. Cronly-Dillon (ed.), Macmillan Press, London, 1991.

117. C. B. Blakemore and F. W. Campbell, "On the Existence of Neurones in the Human Visual System Selectively Sensitive to the Orientation and Size of Retinal Images," *Journal of Physiology* **203:**237–260 (London, 1969).

118. H. R. Wilson and J. R. Bergen, "A Four Mechanism Model for Threshold Spatial Vision," *Vision Research* **19:**19–32 (1979).

119. C. F. Stromeyer and B. Julesz, "Spatial Frequency Masking in Vision: Critical Bands and the Spread of Masking," *Journal of the Optical Society of America* **62:**1221–1232 (1972).

120. J. G. Daugman, "Uncertainty Relation for Resolution in Space, Spatial Frequency, and Orientation Optimized by Two-Dimensional Visual Cortical Filters," *Journal of the Optical Society of America* **2:**1160–1169 (1985).

121. E. R. Howell and R. F. Hess, "The Functional Area for Summation to Threshold for Sinusoidal Gratings," *Vision Research* **18:**369–374 (1978).

122. M. S. Banks, W. S. Geisler, and P. J. Bennett, "The Physical Limits of Grating Visibility," *Vision Research* **27:**1915–1924 (1987).

123. A. Rose, "The Relative Sensitivities of Television Pickup Tubes, Photographic Film, and the Human Eye," *Proceedings of the Institute of Radio Engineers* **30:**293–300 (1942).

124. N. Sekiguchi, D. R. Williams, and D. H. Brainard, "Efficiency for Detecting Isoluminant and Isochromatic Interference Fringes," *Journal of the Optical Society of America* **10:**2118–2133 (1993).

125. M. S. Banks, A. B. Sekuler, and S. J. Anderson, "Peripheral Spatial Vision: Limits Imposed by Optics, Photoreceptors, and Receptor Pooling," *Journal of the Optical Society of America* **8:**1775–1787 (1991).

126. P. T. Kortum and W. S. Geisler, "Contrast Sensitivity Functions Measured on Flashed Backgrounds in the Dark-Adapted Eye," *Annual Meeting of the Optical Society of America,* Optical Society of America, Albuquerque, NM, 1992.

127. F. L. Van Nes and M. A. Bouman, "Spatial Modulation Transfer in the Human Eye," *Journal of the Optical Society of America* **57:**401–406 (1967).

128. D. H. Kelly, "Adaptation Effects on Spatio-Temporal Sine-Wave Thresholds," *Vision Research* **12:**89–101 (1972).

129. D. C. Hood and M. A. Finkelstein, "Sensitivity to Light," *Handbook of Perception and Human Performance,* K. R. Boff, L. Kaufman, and J. P. Thomas (eds.), John Wiley and Sons, New York, 1986.

130. J. Walraven et al., "The Control of Visual Sensitivity: Receptoral and Postreceptoral

Processes," *Visual Perception: The Neurophysiological Foundations,* L. Spillman and J. S. Werner (eds.), Academic Press, San Diego, 1990.

131. K. J. W. Craik, "The Effect of Adaptation on Subjective Brightness," *Proceedings of the Royal Society of London,* series B **128:**232–247, 1940.

132. D. C. Hood et al., "Human Cone Saturation as a Function of Ambient Intensity: A Test of Models of Shifts in the Dynamic Range," *Vision Research* **19:**983–993 (1978).

133. W. S. Geisler, "Adaptation, Afterimages and Cone Saturation," *Vision Research* **18:**279–289 (1978).

134. W. S. Geisler, "Effects of Bleaching and Backgrounds on the Flash Response of the Visual System," *Journal of Physiology* **312:**413–434 (London, 1981).

135. M. M. Hayhoe, N. E. Benimoff, and D. C. Hood, "The Time-Course of Multiplicative and Subtractive Adaptation Process," *Vision Research* **27:**1981–1996 (1987).

136. W. S. Geisler, "Mechanisms of Visual Sensitivity: Backgrounds and Early Dark Adaptation," *Vision Research* **23:**1423–1432 (1983).

137. M. M. Hayhoe, M. E. Levin, and R. J. Koshel, "Subtractive Processes in Light Adaptation," *Vision Research* **32:**323–333 (1992).

138. C. A. Burbeck and D. H. Kelly, "Role of Local Adaptation in the Fading of Stabilized Images," *Journal of the Optical Society of America* A **1:**216–220 (1984).

139. U. Tulunay-Keesey et al., "Apparent Phase Reversal During Stabilized Image Fading," *Journal of the Optical Society of America A* **4:**2166–2175 (1987).

140. A. B. Watson, "Temporal Sensitivity," *Handbook of Perception and Human Performance,* K. R. Boff, L. Kaufman, and J. P. Thomas (eds.), John Wiley and Sons, New York, 1986.

141. H. de Lange, "Research into the Dynamic Nature of the Human Fovea-Cortex Systems with Intermittent and Modulated Light. I. Attenuation Characteristics with White and Colored Light," *Journal of the Optical Society of America* **48:**777–784 (1958).

142. D. H. Kelly, "Visual Responses to Time-Dependent Stimuli. I. Amplitude Sensitivity Measurements," *Journal of the Optical Society of America* **51:**422–429 (1961).

143. J. A. J. Roufs, "Dynamic Properties of Vision-I. Experimental Relationships Between Flicker and Flash Thresholds," *Vision Research* **12:**261–278 (1972).

144. R. M. Boynton and W. S. Baron, "Sinusoidal Flicker Characteristics of Primate Cones in Response to Hererochromatic Stimuli, *Journal of the Optical Society of America* **65:**1091–1100 (1975).

145. D. H. Kelly, R. M. Boynton, and W. S. Baron, "Primate Flicker Sensitivity: Psychophysics and Electrophysiology," *Science* **194:**177–179 (1976).

146. G. E. Legge, "Sustained and Transient Mechanisms in Human Vision: Temporal and Spatial Properties," *Vision Research* **18:**69–81 (1978).

147. J. G. Robson, "Spatial and Temporal Contrast Sensitivity Functions of the Visual System," *Journal of the Optical Society of America* **56:**1141–1142 (1966).

148. J. J. Kulikowski and D. J. Tolhurst, "Psychophysical Evidence for Sustained and Transient Mechanisms in Human Vision," *Journal of Physiology,* **232:**149–163 (London, 1973).

149. C. Burbeck and D. H. Kelly, "Spatiotemporal Characteristics of Visual Mechanisms: Excitatory-Inhibitory Model," *Journal of the Optical Society of America* **70:**1121–1126 (1980).

150. M. B. Mandler and W. Makous, "A Three Channel Model of Temporal Frequency Perception," *Perception* **24**:1881–1887 (1984).

151. S. R. Lehky, "Temporal Properties of Visual Channels Measured by Masking," *Journal of the Optical Society of America* A **2:**1260–1272 (1985).

152. R. J. Snowden and R. F. Hess, "Temporal Properties of Human Visual Filters: Number, Shapes, and Spatial Covariation," *Vision Research* **32:**47–59 (1992).

153. J. J. Koenderink et al., "Perimetry of Contrast Detection Thresholds of Moving Spatial Sine Wave Patterns I. The Near Peripheral Field (0°–8°)," *Journal of the Optical Society of America* **68:**845–849 (1978).

154. V. Virsu et al., "Temporal Contrast Sensitivity and Cortical Magnification," *Vision Research* **22:**1211–1263 (1982).

155. K. T. Mullen, "The Contrast Sensitivity of Human Colour Vision to Red-Green and Blue-Yellow Chromatic Gratings," *Journal of Physiology* **359:**381–400 (London, 1985).

156. G. J. C. van der Horst and M. A. Bouman, "Spatiotemporal Chromaticity Discrimination," *Journal of the Optical Society of America* **59:**1482–1488 (1969).

157. M. S. Banks and P. J. Bennett, "Optical and Photoreceptor Immaturities Limit the Spatial and Chromatic Vision of Human Neonates," *Journal of the Optical Society of America A: Optics and Image Science,* 1988.

158. J. R. Jordan, W. S. Geisler, and A. C. Bovik, "Color As a Source of Information in the Stero Correspondence Process," *Vision Research* **30:**1955–1970 (1990).

159. P. Lennie and M. D'Zmura, "Mechanisms of Color Vision," *CRC Critical Reviews in Neurobiology* **3:**333–400 (1988).

160. D. H. Kelly, "Luminous and Chromatic Flickering Patterns Have Opposite Effects," *Science* **188:**371–372 (1975).

161. G. E. Legge and J. M. Foley, "Contrast Masking in Human Vision," *Journal of the Optical Society of America* **70:**1458–1470 (1980).

162. D. G. Pelli, "Uncertainty Explains Many Aspects of Visual Contrast Detection and Discrimination," *Journal of the Optical Society of America* A **2:**1508–1532 (1985).

163. A. Bradley and I. Ohzawa, "A Comparison of Contrast Detection and Discrimination," *Vision Research* **26:**991–997 (1986).

163*a*. G. E. Legge and D. Kersten, "Contrast Discrimination in Peripheral Vision," *Journal of the Optical Society of America* A **4:**1594–1598 (1987).

164. F. W. Campbell and J. J. Kulikowski, "Orientation Selectivity of the Human Visual System," *Journal of Physiology* **187:**437–445 (London, 1966).

165. N. Graham, *Visual Pattern Analyzers,* Oxford, New York, 1989.

166. H. R. Wilson, D. K. McFarlane, and G. C. Philips, "Spatial Frequency Tuning of Orientation Selective Units Estimated by Oblique Masking," *Vision Research* **23:**873–882 (1983).

167. H. R. Wilson, "A Transducer Function for Threshold and Suprathreshold Human Vision," *Biological Cybernetics* **38:**171–178 (1980).

168. H. R. Wilson, "Psychophysics of Contrast Gain," *Annual Meeting of the Association for Research in Vision and Ophthalmology,* Investigative Ophthalmology and Visual Science, Sarasota, Fla., 1990.

169. J. M. Foley, "Human Luminance Pattern Vision Mechanisms: Masking Experiments Require a New Theory," *Journal of the Optical Society of America* **A** (1994), in press.

170. R. F. Hess and R. J. Snowden, "Temporal Properties of Human Visual Filters: Number, Shapes, and Spatial Covariation," *Vision Research* **32**:47–59 (1992).

171. M. A. Georgeson and G. D. Sullivan, "Contrast Constancy: Deblurring in Human Vision by Spatial Frequency Channels," *Journal of Physiology* **252:**627–656 (London, 1975).

172. D. B. Gennery, "Determination of Optical Transfer Function by Inspection of Frequency-Domain Plot," *Journal of the Optical Society of America* **63:**1571–1577 (1973).

173. D. D. Michaels, *Visual Optics and Refraction: A Clinical Approach,* Mosby, St. Louis, 1980.

174. W. N. Charman, "Visual Standards for Driving," *Ophthalmic and Physiological Optics* **5:**211–220 (1985).

175. G. Westheimer, "Scaling of Visual Acuity Measurements," *Archives of Ophthalmology* **97:**327–330 (1979).

176. C. Owsley et al., "Visual/Cognitive Correlates of Vehicle Accidents in Older Drivers," *Psychology and Aging* **6:**403–415 (1991).

177. R. N. Berry, "Quantitative Relations Among Vernier, Real Depth and Stereoscopic Depth Acuities," *Journal of Experimental Psychology* **38:**708–721 (1948).

178. G. Westheimer and S. P. McKee, "Spatial Configurations for Visual Hyperacuity," *Vision Research* **17:**941–947 (1977).
179. G. Westheimer, "Visual Acuity and Hyperacuity," *Investigative Ophthalmology* **14:**570–572 (1975).
180. G. Westheimer, "The Spatial Grain of the Perifoveal Visual Field," *Vision Research* **22:**157–162 (1982).
181. D. M. Levi and S. A. Klein, "Vernier Acuity, Crowding and Amblyopia," *Vision Research* **25:**979–991 (1985).
182. S. A. Klein and D. M. Levi, "Hyperacuity Thresholds of 1 sec: Theoretical Predictions and Empirical Validation," *Journal of the Optical Society of America* A **2:**1170–1190 (1985).
183. H. R. Wilson, "Model of Peripheral and Amblyopic Hyperacuity, *Vision Research* **31:**967–982 (1991).
184. H. B. Barlow, "Reconstructing the Visual Image in Space and Time," *Nature* **279:**189–190 (1979).
185. F. H. C. Crick, D. C. Marr, and T. Poggio, "An Information-Processing Approach to Understanding the Visual Cortex," *The Organization of the Cerebral Cortex,* F. O. Smith (ed.), MIT Press, Cambridge, 1981.
186. R. J. Watt and M. J. Morgan, "A Theory of the Primitive Spatial Code in Human Vision," *Vision Research* **25:**1661–1674 (1985).
187. R. Shapley and J. D. Victor, "Hyperacuity in Cat Retinal Ganglion Cells," *Science* **231:**999–1002 (1986).
188. H. R. Wilson, "Responses of Spatial Mechanisms Can Explain Hyperacuity," *Vision Research* **26:**453–469 (1986).
189. Q. M. Hu, S. A. Klein, and T. Carney, "Can Sinusoidal Vernier Acuity Be Predicted by Contrast Discrimination?" *Vision Research* **33:**1241–1258 (1993).
190. W. S. Geisler and K. D. Davila, "Ideal Discriminators in Spatial Vision: Two-Point Stimuli," *Journal of the Optical Society of America* A **2:**1483–1497 (1985).
191. S. Appelle, "Perception and Discrimination As a Function of Stimulus Orientation: The 'Oblique Effect' in Man and Animals," *Psychological Bulletin* **78:**266–278 (1972).
192. F. W. Campbell, R. H. S. Carpenter, and J. Z. Levinson, "Visibility of Aperiodic Patterns Compared with That of Sinusoidal Gratings," *Journal of Physiology* **204:**283–298 (London, 1969).
193. F. W. Campbell, J. Nachmias, and J. Jukes, "Spatial Frequency Discrimination in Human Vision," *Journal of the Optical Society of America* **60:**555–559 (1970).
194. A. V. Oppenheim and J. S. Lim, "The Importance of Phase in Signals," *Proceedings of the IEEE* **69:**529–541 (1981).
195. D. J. Field, "Relations Between the Statistics of Natural Images and the Response Properties of Cortical Cells," *Journal of the Optical Society of America* A **4:**2379–2394 (1987).
196. J. Huang and D. L. Turcotte, "Fractal Image Analysis: Application to the Topography of Oregon and Synthetic Images," *Journal of the Optical Society of America* A **7:**1124–1129 (1990).
197. M. J. Morgan, J. Ross, and A. Hayes, "The Relative Importance of Local Phase and Local Amplitude in Patchwise Image Reconstruction," *Biological Cybernetics* **65:**113–119 (1991).
198. D. R. Badcock, "Spatial Phase or Luminance Profile Discrimination," *Vision Research* **24:**613–623 (1984).
199. I. Rentschler and B. Treutwein, "Loss of Spatial Phase Relationships in Extrafoveal Vision," *Nature* **313:**308–310 (1985).
200. R. F. Hess and J. S. Pointer, "Evidence for Spatially Local Computations Underlying Discrimination of Periodic Patterns in Fovea and Periphery," *Vision Research* **27:**1343–1360 (1987).
201. P. J. Bennett and M. S. Banks, "Sensitivity Loss Among Odd-Symmetric Mechanisms and Phase Anomalies in Peripheral Vision," *Nature* **326:**873–876 (1987).

202. B. Julesz, E. N. Gilbert, and J. D. Victor, "Visual Discrimination of Textures with Identical Third-Order Statistics," *Biological Cybernetics* **31:**137–147 (1978).

203. I. Rentschler, M. Hubner, and T. Caelli, "On the Discrimination of Compound Gabor Signals and Textures," *Vision Research* **28:**279–291 (1988).

204. G. E. Legge et al., "Psychophysics of Reading I. Normal Vision," *Vision Research* **25:**239–252 (1985).

205. M. C. Morrone and D. C. Burr, "Feature Detection in Human Vision: A Phase Dependent Energy Model," *Proceedings of the Royal Society of London,* series B **235:**221–245 (1988).

206. D. J. Field and J. Nachmias, "Phase Reversal Discrimination," *Vision Research* **24:**333–340 (1984).

207. D. C. Burr, M. C. Morrone, and D. Spinelli, "Evidence for Edge and Bar Detectors in Human Vision," *Vision Research* **29:**419–431 (1989).

208. P. J. Bennett and M. S. Banks, "The Effects of Contrast, Spatial Scale, and Orientation on Foveal and Peripheral Phase Discrimination," *Vision Research* **31:**1759–1786, 1991.

209. A. Toet and D. M. Levi, "The Two-Dimensional Shape of Spatial Interaction Zones in the Parafovea," *Vision Research* **32:**1349–1357 (1992).

210. M. C. Morrone, D. C. Burr, and D. Spinelli, "Discrimination of Spatial Phase in Central and Peripheral Vision," *Vision Research* **29:**433–445 (1989).

211. S. Anstis, "Motion Perception in the Frontal Plane: Sensory Aspects," *Handbook of Perception and Human Performance,* K. R. Boff, L. Kaufman, and J. P. Thomas (eds.), John Wiley and Sons, New York, 1986.

212. C. C. Hildreth and C. Koch, "The Analysis of Visual Motion: From Computational Theory to Neuronal Mechanisms," *Annual Review of Neuroscience* **10:**477–533 (1987).

213. D. C. Burr and J. Ross, "Contrast Sensitivity at High Velocities," *Vision Research* **22:**479–484 (1982).

214. S. P. McKee, G. H. Silverman, and K. Nakayama, "Precise Velocity Discrimination Despite Random Variations in Temporal Frequency and Contrast," *Vision Research* **26**(4):609–619 (1986).

215. B. De Bruyn and G. A. Orban, "Human Velocity and Direction Discrimination Measured with Random Dot Patterns," *Vision Research* **28**(12):1323–1335 (1988).

216. S. N. J. Watamaniuk, R. Sekuler, and D. W. Williams, "Direction Perception in Complex Dynamic Displays: The Integration of Direct Information," *Vision Research* **29:**47–59 (1989).

217. K. Nakayama and G. H. Silverman, "Detection and Discrimination of Sinusoidal Grating Displacements," *Optical Society of America* **2**(2):267–274 (1985).

218. A. Pantle, "Temporal Frequency Response Characteristics of Motion Channels Measured with Three Different Psychophysical Techniques," *Perception and Psychophysics* **24:**285–294 (1978).

219. P. Cavanagh and P. Anstis, "The Contribution of Color to Motion in Normal and Color-Deficient Observers," *Vision Research* **31:**2109–2148 (1991).

220. D. T. Lindsey and D. Y. Teller, "Motion at Isoluminance: Discrimination/Detection Ratios for Moving Isoluminant Gratings," *Vision Research* **30**(11):1751–1761 (1990).

221. S. P. McKee and K. Nakayama, "The Detection of Motion in the Peripheral Visual Field," *Vision Research* **24:**25–32 (1984).

222. O. J. Braddick, "Low-Level and High-Level Processes in Apparent Motion," *Philosophical Transactions of the Royal Society of London* **B290:**137–151 (1980).

223. J. J. Chang and B. Julesz, "Displacement Limits for Spatial Frequency Filtered Random-Dot Cinematograms in Apparent Motion," *Vision Research* **23**(12):1379–1385 (1983).

224. C. L. Baker and O. J. Braddick, "Does Segregation of Differently Moving Areas Depend on Relative or Absolute Displacement?" *Vision Research* **22:**851–856 (1982).

225. L. Glass, "Moire Effect from Random Dots," *Nature* **243:**578–580 (1969).

226. J. J. Gibson, *The Senses Considered As Perceptual Systems,* Houghton Mifflin, Boston, 1966.

227. J. J. Gibson, *The Perception of the Visual World,* Houghton Mifflin, Boston, 1950.

228. J. E. Cutting, *Perception with an Eye to Motion,* MIT Press, Cambridge, 1986.

229. W. H. Warren, M. W. Morris, and M. Kalish, "Perception of Translational Heading from Optical Flow," *Journal of Experimental Psychology: Human Perception and Performance* **14:**646–660 (1988).
230. W. H. Warren and D. J. Hannon, "Eye Movements and Optical Flow," *Journal of the Optical Society of America* A **7:**160–169 (1990).
231. E. von Holst, "Relations Between the Central Nervous System and the Peripheral Organs," *Animal Behavior* **2:**89–94 (1954).
232. D. J. Heeger and A. D. Jepson, "Subspace Methods for Recovering Rigid Motion. I: Algorithm and Implementation," *University of Toronto Technical Reports on Research in Biological and Computational Vision,* RBCV-TR-90-35.
233. H. C. Longuet-Higgins and K. Prazdny, "The Interpretation of a Moving Retinal Image," *Proceedings of the Royal Society of London* B **208:**385–397 (1980).
234. J. A. Perrone, "Model for Computation of Self-Motion in Biological Systems," *Journal of the Optical Society of America* A **9:**177–194 (1992).
235. J. H. Rieger and D. T. Lawton, "Processing Differential Image Motion," *Journal of the Optical Society of America* A **2:**354–360 (1985).
236. C. S. Royden, M. S. Banks, and J. A. Crowell, "The Perception of Heading During Eye Movements," *Nature* **360:**583–585 (1992).
237. A. V. van den Berg, "Robustness of Perception of Heading from Optic Flow," *Vision Research* **32:**1285–1296 (1992).
238. J. J. Gibson, P. Olum, and F. Rosenblatt, "Parallax and Perspective During Aircraft Landings," *American Journal of Psychology* **68:**373–385 (1955).
239. B. J. Rogers and M. Graham, "Similarities Between Motion Parallax and Stereopsis in Human Depth Perception," *Vision Research* **22:**261–270 (1982).
240. B. J. Rogers and M. F. Bradshaw, "Vertical Disparities, Differential Perspective and Binocular Stereopsis," *Nature* **361:**253–255 (1993).
241. J. E. W. Mayhew and H. C. Longuet-Higgins, "A Computational Model of Binocular Depth Perception," *Nature* **297:**376–378 (1982).
242. G. Westheimer and S. P. McKee, "What Prior Uniocular Processing Is Necessary for Stereopsis?" *Investigative Ophthalmology and Visual Science* **18:**614–621 (1979).
243. B. Julesz, "Binocular Depth Perception of Computer-Generated Patterns," *Bell System Technical Journal* **39:**1125–1162 (1960).
244. C. M. Aschenbrenner, "Problems in Getting Information Into and Out of Air Photographs," *Photogrammetric Engineering,* 1954.
245. B. Julesz, "Stereoscopic Vision," *Vision Research* **26:**1601–1612 (1986).
246. J. P. Frisby and J. E. W. Mayhew, "Contrast Sensitivity Function for Stereopsis," *Perception* **7:**423–429 (1978).
247. C. Blakemore and B. Hague, "Evidence for Disparity Detecting Neurones in the Human Visual System," *Journal of Physiology* **225:**437–445 (London, 1972).
248. T. B. Felton, W. Richards, and R. A. Smith, "Disparity Processing of Spatial Frequencies in Man," *Journal of Physiology* **225:**349–362 (London, 1972).
249. B. Julesz and J. E. Miller, "Independent Spatial Frequency Tuned Channels in Binocular Fusion and Rivalry," *Perception* **4:**315–322 (1975).
250. Y. Yang and R. Blake, "Spatial Frequency Tuning of Human Stereopsis," *Vision Research* **31:**1177–1189 (1991).
251. J. S. Mansfield and A. J. Parker, "An Orientation-Tuned Component in the Contrast Masking of Stereopsis," *Vision Research* **33:**1535–1544 (1993).
252. R. D. Freeman and I. Ohzawa, "On the Neurophysiological Organization of Binocular Vision," *Vision Research* **30:**1661–1676 (1990).
253. D. L. Halpern and R. R. Blake, "How Contrast Affects Stereoacuity," *Perception* **17:**483–495 (1988).

254. G. E. Legge and Y. Gu, "Stereopsis and Contrast," *Vision Research* **29:**989–1004 (1989).

255. L. K. Cormack, S. B. Stevenson, and C. M. Schor, "Interocular Correlation, Luminance Contrast, and Cyclopean Processing," *Vision Research* **31:**2195–2207 (1991).

256. C. Blakemore, "The Range and Scope of Binocular Depth Discrimination in Man," *Journal of Physiology* **211:**599–622 (London, 1970).

257. D. R. Badcock and C. M. Schor, "Depth-Increment Detection Function for Individual Spatial Channels," *Journal of the Optical Society of America* A **2:**1211–1216 (1985).

258. W. Richards, "Anomalous Stereoscopic Depth Perception," *Journal of the Optical Society of America* **61:**410–414 (1971).

259. L. K. Cormack, S. B. Stevenson, and C. M. Schor, "Disparity-Tuned Channels of the Human Visual System," *Visual Neuroscience* **10:**585–596 (1993).

260. S. B. Stevenson, L. K. Cormack, and C. M. Schor, "Disparity Tuning in Mechanisms of Human Stereopsis," *Vision Research* **32:**1685–1694 (1992).

261. S. R. Lehky and T. J. Sejnowski, "Neural Model of Stereoacuity and Depth Interpolation Based on a Distributed Representation of Stereo Disparity," *Journal of Neuroscience* **10:**2281–2299 (1990).

262. C. H. Bailey and P. Gouras, "The Retina and Phototransduction," *Principles of Neural Science,* E. R. Kandel and J. H. Schwartz (eds.), Elsevier Science Publishing Co., Inc., New York, 1985, pp. 344–355.

263. K. D. Davila and W. S. Geisler, "The Relative Contributions of Pre-Neural and Neural Factors to Areal Summation in the Fovea," *Vision Research* **31:**1369–1380 (1991).

264. J. Ross and H. D. Speed, "Contrast Adaptation and Contrast Masking in Human Vision," *Proceedings of the Royal Society of London,* series B **246:**61–69 (1991).

CHAPTER 26
COLORIMETRY

David H. Brainard
Department of Psychology
University of California, Santa Barbara
Santa Barbara, California

26.1 GLOSSARY

This glossary describes the major symbol usage for the body of the chapter. Symbols used in the appendix are generic. The following notational conventions hold throughout: (1) scalars are denoted with plain symbols, (2) vectors are denoted with lowercase bold symbols, (3) matrices are denoted with uppercase bold symbols.

$\mathbf{a}$	linear model weights
$\mathbf{B}$	linear model basis vectors
$\mathbf{b}$	spectral power distribution; basis vector
$\mathbf{M}$	color space transformation matrix
N_b	linear model dimension
N_λ	number of wavelength samples
$\mathbf{P}$	linear model for primaries
$\mathbf{p}$	primary spectral power distribution
$\mathbf{R}$	cone (or sensor) sensitivities
$\mathbf{r}$	cone (or sensor) coordinates
$\mathbf{T}$	color matching functions
$\mathbf{t}$	tristimulus coordinates
v	luminance
$\mathbf{v}$	luminous efficiency function
λ	wavelength

26.2 INTRODUCTION

Scope

The goal of colorimetry is to incorporate properties of the human color vision system into the measurement and specification of visible light. This branch of color science has been quite successful. We now have efficient quantitative representations that predict when two lights will appear identical to a human observer. Although such colorimetric representations

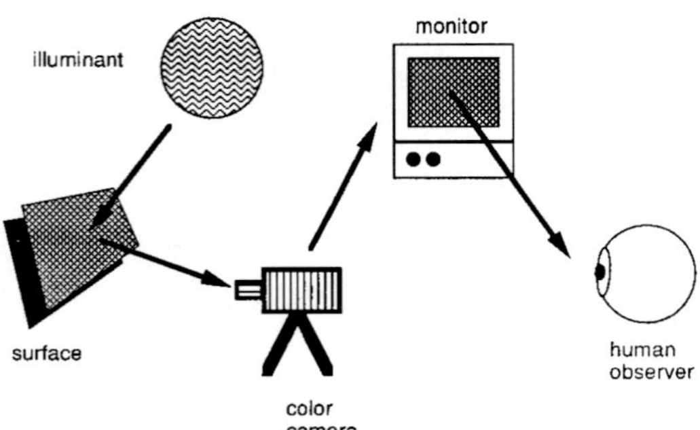

FIGURE 1 A typical image processing chain. Light reflects from a surface or collection of surfaces. This light is recorded by a color camera and stored in digital form. The digital image is processed by a computer and rendered on a color monitor. The reproduced image is viewed by a human observer.

do not directly predict the color sensation,[1–3] they do provide the foundation for the scientific study of color appearance. Moreover, colorimetry can be applied successfully in practical applications. Foremost among these is perhaps color reproduction.[4–7]

As an illustrative example, Fig. 1 shows an image processing chain. Light from an illuminant reflects from a collection of surfaces. This light is recorded by a color camera and stored in digital form. The digital image is processed by a computer and rendered on a color monitor. The reproduced image is viewed by a human observer. The goal of the image processing is to render an image with the same color appearance at each image location as the original. Although exact reproduction is not always possible with this type of system, the concepts and formulae of colorimetry do provide a reasonable solution.[4,8] To develop this solution, we will need to consider how to represent the spectral properties of light, the relation between these properties and color camera responses, the representation of the restricted set of lights that may be produced with a color monitor, and the way in which the human visual system encodes the spectral properties of light. We will treat each of these topics in this chapter, with particular emphasis on the role played by the human visual system.

Reference Sources

A number of excellent references are available that provide detailed treatments of colorimetry and its applications. Wyszecki and Stiles' comprehensive book[9] is an authoritative reference and provides numerous tables of standard colorimetric data. Pokorny and Smith[10] provide a handbook treatment complementary to the one developed here. Several publications of the Commission Internationale de l'Eclairage (International Commission on Illumination, commonly referred to as the CIE) describe current international technical standards for colorimetric measurements and calculations.[11–13] Other sources cover colorimetry's mathematical foundations,[14,15] its history,[16] its applications,[2,5,7,17] and its relation to neural mechanisms.[18,19] Chapters 27 and 28 of this volume and Vol. II, Chap. 24 are also relevant.

Chapter Overview

Colorimetry, Computers, and Linear Algebra. The personal computers that are now almost universally available in the laboratory can easily handle all standard colorimetric calculations. With this fact in mind, we have organized our treatment of colorimetry to allow direct translation between our formulation and its software implementation. In particular, we use vector and matrix representations throughout this chapter and use matrix algebra to express colorimetric formulae. Matrix algebra is being used increasingly in the colorimetric literature.[4,20–22] Appendix A reviews the elementary facts of matrix algebra required for this chapter. Numerous texts treat the subject in detail.[23–27] Various software packages provide extensive support for numerical matrix algebra.[28–31]

Chapter Organization. The rest of this chapter is organized into two main sections. The first section, "Fundamentals", reviews the empirical foundation of colorimetry and introduces basic colorimetric methods. The second section, "Topics", discusses a number of applications of colorimetry. It also includes a brief review of more advanced topics in color science.

26.3 FUNDAMENTALS

Stimulus Representation

Light at a Point. We describe the light reaching the eye from an image location by its spectral power distribution. The spectral power distribution generally specifies the radiant power density at each wavelength in the visible spectrum. For human vision, the visible spectrum extends roughly between 400 and 700 nm (but see "Sampling the Visible Spectrum" below). Depending on the viewing geometry, measures of radiation transfer other than radiant power may be used. These measures include radiance, irradiance, exitance, and intensity. The distinctions between these measures and their associated units are treated in Vol. II, Chap. 24 and are not considered here. That chapter also discusses measurement instrumentation and procedures.

Vector Representation of Spectral Functions. We will use discrete representations of spectral functions. Although a discrete representation samples the continuous functions of wavelength, the information loss caused by this sampling can be made arbitrarily small by increasing the number of sample wavelengths.

Suppose that spectral power density has been measured at N_λ discrete sample wavelengths $\lambda_1 \cdots \lambda_{N_\lambda}$, each separated by an equal wavelength step $\Delta\lambda$. As shown in Fig. 2, we can represent the measured spectral power distribution using an N_λ dimensional column vector $\mathbf{b}$. The nth entry of $\mathbf{b}$ is simply the measured power density at the nth sample wavelength multiplied by $\Delta\lambda$. Note that the values of the sample wavelengths $\lambda_1 \cdots \lambda_{N_\lambda}$ and wavelength step $\Delta\lambda$ are not explicit in the vector representation. These values must be provided as side information when they are required for a particular calculation. In colorimetric applications, sample wavelengths are typically spaced evenly throughout the visible spectrum at steps of between 1 and 10 nm. We follow the convention that the entries of $\mathbf{b}$ incorporate $\Delta\lambda$, however, so that we need not represent $\Delta\lambda$ explicitly when we approximate integrals over wavelength.

Manipulation of Light. Intensity scaling is an operation that changes the overall power of a light at each wavelength without altering the relative power between any pair of wavelengths. One way to implement intensity scaling is to place a neutral density filter in

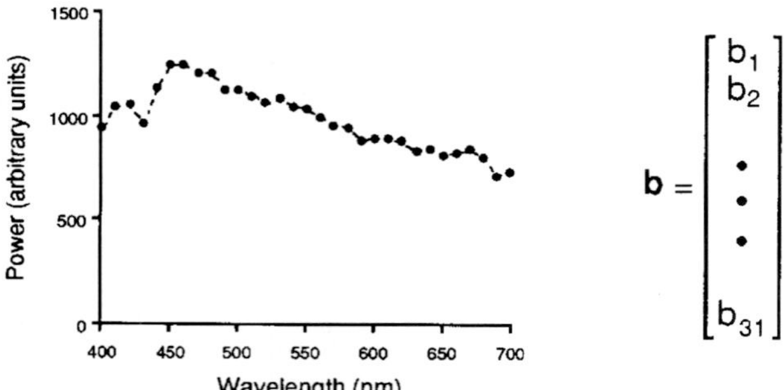

FIGURE 2 The vector representation of functions of wavelength. The plot shows a spectral power distribution measured at 10-nm intervals between 400 nm and 700 nm. Each point on the plot represents the power at a single sample wavelength. The vector **b** on the right depicts the vector representation of the same spectral power distribution. The **n**th entry of **b** is simply the measured power density at the **n**th sample wavelength times $\Delta\lambda$. Thus $\mathbf{b}_1$ is derived from the power density at 400 nm, $\mathbf{b}_2$ is derived from the power density at 410 nm, and $\mathbf{b}_{31}$ is derived from the power density at 700 nm.

the light path. The superposition of two lights is an operation that produces a new light whose power at each wavelength is the sum of the power in the original lights at the corresponding wavelength. One way to implement superposition is to use an optical beam splitter. The effects of both manipulations may be expressed using matrix algebra.

We use scalar multiplication to represent intensity scaling. If a light $\mathbf{b}_1$ is scaled by a factor a, then the result $\mathbf{b}$ is given by the equation $\mathbf{b} = \mathbf{b}_1 a$. The expression $\mathbf{b}_1 a$ represents a vector whose entries are obtained by multiplying the entries of the vector $\mathbf{b}_1 a$ by the scalar a. Similarly, we use vector addition to represent superposition. If we superimpose two lights $\mathbf{b}_1$ and $\mathbf{b}_2$, then the result $\mathbf{b}$ is given by the equation $\mathbf{b} = \mathbf{b}_1 + \mathbf{b}_2$. The expression $\mathbf{b}_1 + \mathbf{b}_2$ represents a vector whose entries are obtained by adding the entries of the vectors $\mathbf{b}_1$ and $\mathbf{b}_2$. Figures 3 and 4 depict both of these operations.

Linear Models for Spectral Functions. Intensity scaling and superposition may be used

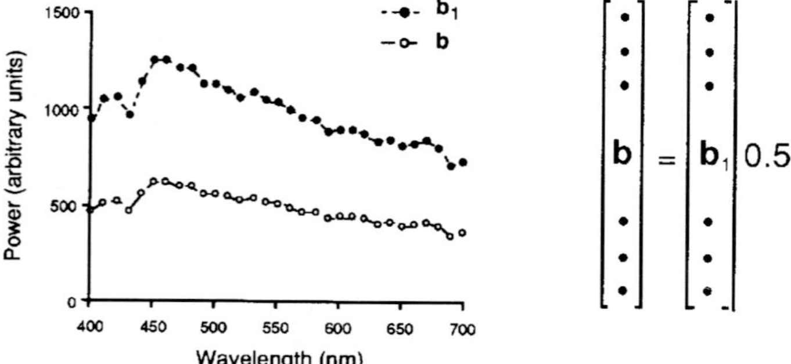

FIGURE 3 Representation of intensity scaling. Suppose that light **b** is created by reducing the power in light $\mathbf{b}_1$ by a factor of 0.5 at each wavelength. The result is shown graphically in the plot. The vector representation of the same relation is given by the equation $\mathbf{b} = \mathbf{b}_1\ 0.5$.

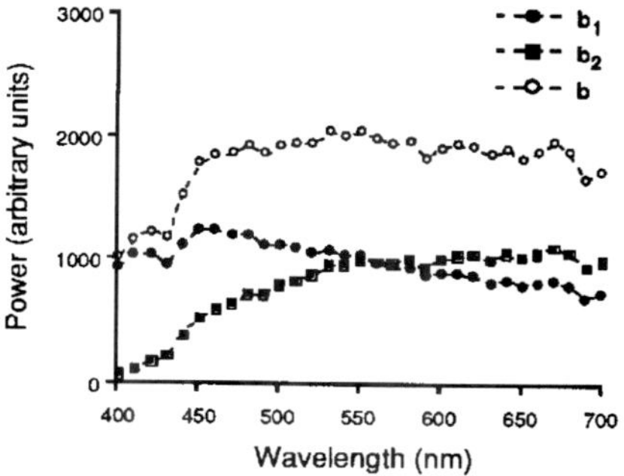

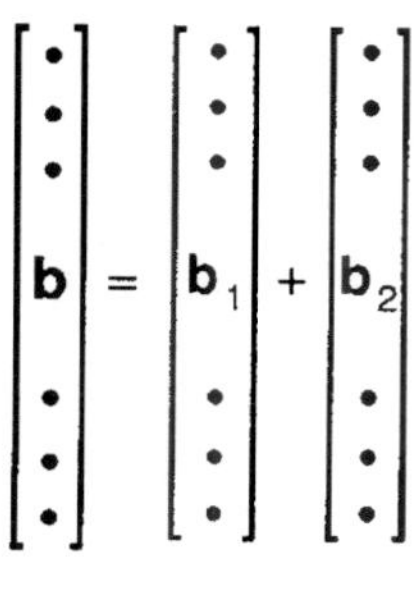

FIGURE 4 Representation of superposition. Suppose that light **b** is created by superimposing two lights **b** and $\mathbf{b}_2$. The result is shown graphically in the plot. The vector representation of the same relation is given by the equation $\mathbf{b} = \mathbf{b}_1 = \mathbf{b}_2$.

in combination to produce a wide range of spectral functions. Suppose that we have a set of N_b lights that we can individually scale and superimpose. Let the vectors $\mathbf{b}_1 \cdots \mathbf{b}_{N_b}$ represent the spectral power distributions of these lights. In this case, we can produce any spectral power distribution **b** that has the form

$$\mathbf{b} = \mathbf{b}_1 a_1 + \cdots + \mathbf{b}_{N_b} a_{N_b} \tag{1}$$

Suppose we know that a spectral function **b** is constrained to have the form of Eq. (1) where the vectors $\mathbf{b}_1 \cdots \mathbf{b}_{N_b}$ are known. Then we can specify **b** completely by providing the values of the scalars $a_1 \cdots a_{N_b}$. If the number of primaries N_b is less than the number of sample wavelengths N_λ, then this specification is more efficient (i.e., requires fewer numbers) than specifying the entries of **b** directly. We say that the spectral functions that satisfy Eq. (1) are described by (or lie within) a linear model. We call N_b the dimension of the linear model. We call the vectors $\mathbf{b}_1 \cdots \mathbf{b}_{N_b}$ the basis vectors for the model. We call the scalars $a_1 \cdots a_{N_b}$ required to construct any particular spectral function the model weights for that function.

Matrix Representation of Linear Models. Equation (1) can be written using vector and matrix notation. Let **B** be an N_λ by N_b dimensional matrix whose columns are the basis vectors $\mathbf{b}_1 \cdots \mathbf{b}_{N_b}$. We call **B** the basis matrix for the linear model. The composition of the basis matrix is shown pictorially on the left of Fig. 5. Let **a** be an N_b dimensional vector whose entries are the weights $a_1 \cdots a_{N_b}$. Figure 5 also depicts the vector **a**. Using **B** and **a** we can reexpress Eq. (1) as the matrix multiplication

$$\mathbf{b} = \mathbf{Ba} \tag{2}$$

The equivalence of Eqs. (1) and (2) may be established by direct expansion of the definition of matrix multiplication (see App. A). A useful working intuition for matrix multiplication is that the effect of multiplying a matrix times a vector (e.g., **Ba**) is to produce a new vector (e.g., **b**) that is a weighted superposition of the columns of the matrix (e.g., **B**), where the weights are given by the entries of the vector (e.g., **a**).

Use of Linear Models. When we know that a spectral function is described by a linear

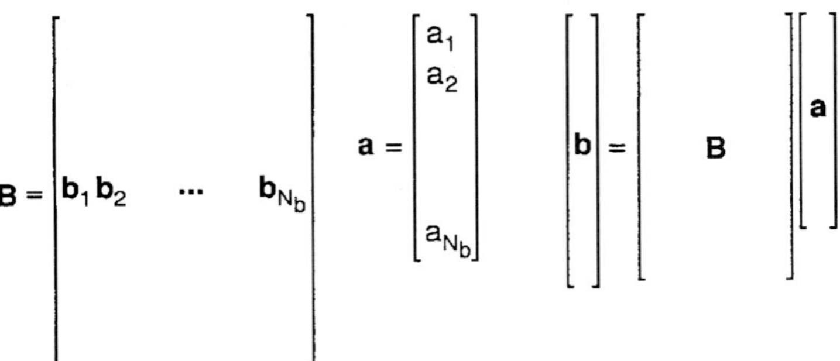

FIGURE 5 Vector representation of linear models. The matrix **B** represents the basis vectors of the linear model. The vector **a** represents the model weights required to form a particular spectral power distribution **b**. The relation between **b**, **a**, and **B** is given by Eq. (2) and is depicted on the right of the figure.

model, we can specify it by using the weight vector **a**. The matrix **B**, which is determined by the basis vectors, specifies the side information necessary to convert the vector **a** back to the discrete wavelength representation. When we represent spectral functions in this way, we say that we are representing the functions within the specified linear model.

Representing spectral functions within a small-dimensional linear model places strong constraints on the form of the functions. As the dimension of the model grows, linear models can represent progressively wider classes of functions. In many cases of interest, there is prior information that allows us to assume that spectra are indeed described by a linear model. A common example of this situation is the light emitted from a computer-controlled color monitor. Such a monitor produces different spectral power distributions by scaling the intensity of the light emitted by three different types of phosphor (see Vol. I, Chap. 27). Thus the emitted light lies within a three-dimensional linear model whose basis vectors are given by the emission spectra of the monitor's phosphors. Linear model constraints also turn out to be useful for describing naturally occurring surface and illuminant spectra (see later under "Surfaces and Illuminants").

Note that representing spectral functions within linear models is a generalization of, rather than an alternative to, the more traditional wavelength representation. To understand this, we need only note that we can choose the basis vectors of the linear model to be discrete delta functions centered at each sample wavelength. We refer to this special choice of basis vectors as the *identity basis* or *wavelength basis.* We refer to the corresponding linear model as the *identity model.* For the identity model, the basis matrix **B** is the N_λ by N_λ identity matrix, where N_λ is the number of sample wavelengths. The identity matrix contains 1's along its main diagonal and 0's elsewhere. Multiplying the identity matrix times any vector simply results in the same vector. From Eq. (2), we can see that when **B** is the identity matrix, the representation of any light **b** within the linear model is simply $\mathbf{a} = \mathbf{b}$.

Sampling the Visible Spectrum. To use a discrete representation for functions of wavelength, it is necessary to choose a sampling range and sampling increment. Standard practice varies considerably. The current recommendation of the CIE is that the visible spectrum be sampled at 5-nm increments between 380 and 780 nm.[11] Coarser sampling at 10 nm between 400 and 700 nm provides an adequate approximation for many applications. In this chapter, we provide tabulated spectral data at the latter wavelength sampling. Where possible, we provide references to where they may be obtained at finer sampling increments. In cases where a subset of the spectral data required for a calculation is not available, interpolation or extrapolation may be used to estimate the missing values.[11]

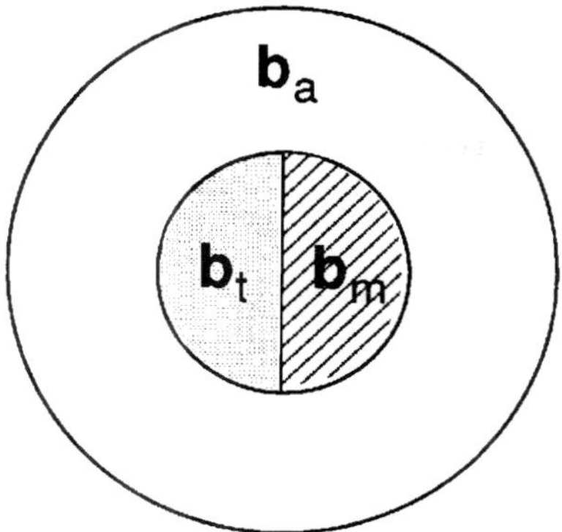

FIGURE 6 Schematic of the stimulus for the basic color-matching experiment. The observer views a bipartite field. A test is presented on one side of a bipartite field and a matching light is presented on the other side of the field. The test light's spectral power distribution $\mathbf{b}_t$ is under experimental control. The observer adjusts the spectral power distribution $\mathbf{b}_m$ of the matching light. The observer's task is to make the two halves of the bipartite field appear identical. The bipartite field is often surrounded by an annulus whose spectral power distribution $\mathbf{b}_a$ is held fixed during the matching process.

Empirical Foundations of Colorimetry

The Basic Color-Matching Experiment. In the absence of linear model constraints, the only complete representation of the spectral properties of light is the full spectral power distribution. The human visual system, however, does not encode all of the information available in the spectral power distribution. Understanding how the visual system encodes the spectral properties of light leads to an efficient representation of the information relevant to human vision. To develop this understanding, we consider the color-matching experiment.

The basic color-matching experiment is illustrated in Fig. 6. An observer views a bipartite field, as shown in the figure. Each side of the field is spatially uniform, but the two sides may differ in their spectral power distributions. The spectral power distribution $\mathbf{b}_t$ of one side of the bipartite field is under experimental control. We call this the *test light.* The spectral power distribution $\mathbf{b}_m$ of the light on the other side of the bipartite field is under the observer's control. We call this the *matching light.* The observer's task in the experiment is to adjust the matching light so that it appears identical in color to the test light. The bipartite field may be surrounded by an annulus whose spectral power distribution $\mathbf{b}_a$ is held fixed during the matching process.

Typically, the apparatus for the color-matching experiment is arranged so that the observer can adjust the matching light by controlling the intensity scaling of some number of superimposed primary lights. It is not a priori clear that it will be possible for the observer to make a match when the number of primaries is small. The results of a large number of color-matching studies, however, show that observers are able to set matches to any test light by adjusting the intensity of just three primaries. This result implies that there exist lights with different spectral power distributions that cannot be distinguished by a human observer. From the color-matching experiment, we may conclude that the human visual system does not encode all of the information available in the full spectral power distribution.

We use the symbol $\sim$ to indicate that two lights are a perceptual match. Perceptual matches are to be carefully distinguished from physical matches, which are denoted by the = symbol. Of course, when two lights are a physical match, they must be a perceptual match. Two lights that are a perceptual match but not a physical match are referred to as *metameric color stimuli* or *metamers.* The term *metamerism* is often used to refer to the fact that two physically different lights can appear identical.

Conditions for Trichromatic Color Matching. There are a number of qualifications to the empirical generalization that it is possible for observers to match any test light by adjusting the intensities of just three primaries. Some of these qualifications have to do with ancillary restrictions on the experimental conditions (e.g., the size of the bipartite field and the overall intensity of the test and matching lights). We discuss these later under "Limits of the Color-Matching Experiment." The other qualifications have to do with the choice of primaries and certain conventions about the matching procedure. First, the primaries must be chosen so that it is not possible to match any one of them with a weighted superposition of the other two. Second, the observer sometimes wishes to increase the intensity of one or more of the primaries above its maximum value. In this case, we must allow the observer to scale the intensity of the test light down. We follow the convention of saying that the match was possible, but scale up the reported primary weights by the same factor. Third, the observer sometimes wishes to decrease the intensity of one or more of the primaries below zero. In this case, we must allow the observer to superimpose each such primary on the test light rather than on the other primaries. We follow the convention of saying that the match was possible, but report with a negative sign the intensity of each transposed primary.

With these qualifications, matching with three primaries is always possible. This fact is often referred to as the *trichromacy* of normal human color vision.

Tristimulus Coordinates. The basic color-matching experiment is an empirical procedure that maps any light to three numbers: the weights on the color-matching apparatus primaries required to make a match. These three weights are often called the *tristimulus coordinates* of the test light. Given the qualifications and conventions discussed earlier, there are no restrictions on the magnitude of tristimulus coordinates. Thus, the matching light's spectral power distribution is described completely by a three-dimensional linear model whose basis vectors are the primary lights' spectral power distributions. The tristimulus coordinates of a test light are precisely the linear model weights required to form the matching light.

We denote the primary spectral power distributions by the vectors $\mathbf{p}_1 \cdots \mathbf{p}_3$. The associated linear model matrix $\mathbf{P}$ contains these vectors in its three columns. We denote the tristimulus coordinates of a light using the three dimensional vector $\mathbf{t}$. Thus we can use the tristimulus coordinates of any test light $\mathbf{b}$ to reconstruct a matching light $\mathbf{Pt}$ such that

$$\mathbf{b} \sim \mathbf{Pt} \tag{3}$$

We emphasize that, in general, $\mathbf{Pt}$ will not be equal to $\mathbf{b}$.

Critical Properties of Color Matching. Our discussion of the color-matching experiment suggests that we can represent the spectral properties of light using tristimulus coordinates. To ensure that this representation is general, however, we need to consider whether color matching exhibits a number of critical properties. We review these properties briefly below. Given that they hold, it is possible to show that tristimulus coordinates provide a representation for the spectral properties of light. Krantz provides a detailed formal treatment.[14]

Reflexivity, Symmetry, and Transitivity. Reflexivity is the requirement that a light match itself. Symmetry is the requirement that if two lights match, they will continue to match when their roles in the color-matching experiment are reversed. Transitivity is the requirement if two lights each match a common third light, then they will match each other. Apart from small failures that arise from variability in observer's judgments, these three properties do generally hold for human color matching.[14]

Uniqueness of Color Matches. The tristimulus coordinates of a light should be unique. This is equivalent to the requirement that only one weighted combination of the apparatus primaries produces a match to any given test light. The uniqueness of color matches ensures that tristimulus coordinates are well defined. In conjunction with transitivity, uniqueness also guarantees that two lights that match each other will have identical tristimulus coordinates. It is generally accepted that, apart from variability, trichromatic color matches are unique for color-normal observers.

Persistance of Color Matches. The above properties concern color matching under a single set of viewing conditions. By viewing conditions, we refer to the properties of the image surrounding the bipartite field and the sequence of images viewed by the observer before the match was made. An important property of color matching is that lights that match under one set of viewing conditions continue to match when the viewing conditions are changed. This property is referred to as the *persistence* or *stability* of color matches.[9,18] It holds to good approximation for a wide range of viewing conditions. We discuss conditions where the persistence law fails later in the chapter. The importance of the persistence law is that it allows a single set of tristimulus values to be used across viewing conditions.

Consistency Across Observers. Finally, for the use of tristimulus coordinates to have general validity, it is important that there be good agreement about matches across observers. For the majority of the population, there is good agreement about which lights match. We discuss individual differences in color matching in a later section.

Computing Tristimulus Coordinates. *Grassmann's Laws.* Tristimulus coordinates provide an efficient representation for the spectral properties of light. For this representation to be useful, we require a model of the color-matching experiment that allows us to compute tristimulus coordinates from spectral power distributions. To develop such a model, we rely on two regularities of color matching that were first characterized by Grassmann.[32] These (sometimes in conjunction with the properties discussed above) are generally referred to as Grassmann's laws. They have been tested extensively and hold well.[9,18]

The two regularities may be expressed as follows: (1) (proportionality law) if two lights match, they will continue to match if they are both scaled by the same factor, (2) (additivity law) if two pairs of lights match each other, then the superposition of the two will also match. We can express the two laws formally as follows. The proportionality law states:

$$\text{if } \mathbf{b}_1 \sim \mathbf{b}_2, \text{ then } \mathbf{b}_1 a \sim \mathbf{b}_2 a \tag{4}$$

where a is a scalar that represents any intensity scaling. The additivity law states:

$$\text{if } \mathbf{b}_1 \sim \mathbf{b}_2 \text{ and } \mathbf{b}_3 \sim \mathbf{b}_4, \text{ then } \mathbf{b}_1 + \mathbf{b}_3 \sim \mathbf{b}_2 + \mathbf{b}_4 \tag{5}$$

The proportionality law allows us to determine the relation between the tristimulus coordinates of a light and the tristimulus values of a scaled version of that light. Suppose that $\mathbf{b} \sim \mathbf{Pt}$. Applying the proportionality law, we conclude that for any scalar a, we have $\mathbf{b}a \sim (\mathbf{Pt})a$. Because matrix multiplication is associative, we can conclude that:

$$\text{if } \mathbf{b} \sim \mathbf{Pt}, \text{ then } \mathbf{b}a \sim \mathbf{P}(\mathbf{t}a) \tag{6}$$

This means that the tristimulus coordinates of a light $\mathbf{b}a$ may be obtained by scaling the tristimulus coordinates $\mathbf{t}$ of the light $\mathbf{b}$. A similar argument shows that the additivity law determines the relation between the tristimulus coordinates of two lights and the tristimulus coordinates of their superposition:

$$\text{if } \mathbf{b}_1 \sim \mathbf{Pt}_1 \text{ and } \mathbf{b}_2 \sim \mathbf{Pt}_2, \text{ then } \mathbf{b}_1 + \mathbf{b}_2 \sim \mathbf{P}(\mathbf{t}_1 + \mathbf{t}_2) \tag{7}$$

Implication of Grassmann's Laws. If the tristimulus coordinates of the basis vectors for a linear model are known, then Grassmann's laws allow us to determine the tristimulus coordinates of any light within the linear model. Let $\mathbf{t}_1 \cdots \mathbf{t}_{N_b}$ be the tristimulus coordinates corresponding to the model basis vectors and let $\mathbf{T}_B$ be the 3 by N_b matrix whose columns are $\mathbf{t}_1 \cdots \mathbf{t}_{N_b}$. For any light $\mathbf{b}$ within the linear model, we can write that $\mathbf{b} = \mathbf{Ba}$. By expanding this matrix product and applying Eqs. (6) and (7), it is possible to show that the tristimulus coordinates of $\mathbf{b}$ are given by the matrix product:

$$\mathbf{t} = \mathbf{T}_B\mathbf{a} \tag{8}$$

Equation (8) is very important. It tells how to compute the tristimulus coordinates for any light within a linear model from the tristimulus coordinates for each of the basis vectors. Thus a small number of color matches (one for each of the basis vectors) allows us to predict color matches for a large number of lights. We call the rows of the matrix $\mathbf{T}_B$ the color matching functions with respect to the linear model defined by the columns of $\mathbf{B}$. This is a generalization of the standard usage of the term *color-matching functions,* which is introduced below.

Color-Matching Functions. Let $\mathbf{T}$ be the corresponding matrix of tristimulus values for the basis vectors of the identity model. In this case, $\mathbf{T}$ has dimensions 3 by N_λ, where N_λ is the number of sample wavelengths. Each column of $\mathbf{T}$ is the tristimulus coordinates for a monochromatic light. Within the identity model, the representation of any light $\mathbf{b}$ is simply $\mathbf{b}$ itself. From Eq. (8) we conclude directly that the tristimulus values for any light are given by

$$\mathbf{t} = \mathbf{Tb} \tag{9}$$

Once we know the tristimulus coordinates for a set of monochromatic lights centered at each of the sample wavelengths, we use Eq. (9) to compute the tristimulus coordinates of any light.

We can regard each of the rows of $\mathbf{T}$ as a function of wavelength. We refer to these functions as a set of color matching functions. Color matching functions are often plotted as a function of wavelength, as illustrated in Fig. 7. It is important to note, however, they do not represent spectral power distributions.

Mechanisms of Color Matching

Cone Mechanisms. The basic color-matching experiment can be understood in terms of the action of retinal photoreceptors. Three distinct classes of cone photoreceptors are generally believed to participate in normal color vision. These are often referred to as the long (L), middle (M), and short (S) wavelength sensitive cones. The information a cone provides about a light is mediated by a single number: the rate of photopigment isomerizations caused by the absorption of light quanta.[33] If two lights produce identical absorption rates in all three classes of cones, the visual system will not be able to distinguish them. The conventional mechanistic explanation for the results of the color-matching experiment is thus that two lights match if and only if they result in the same number of photopigment absorptions in all three classes of cones.[18]

To compute a cone's quantal absorption rate from a light's spectral power distribution, we use the cone's spectral sensitivity function. For each class of cones, the function specifies the number of quanta that will be absorbed for a monochromatic light of unit power centered on each of the sample wavelengths. When the application is to predict an observer's color matches, the sensitivity function should incorporate wavelength-dependent filtering by the ocular media. To compute the quantal absorption rate, we multiply the light power by the cone's sensitivity at each wavelength and then sum the results over wavelength.

Suppose that we represent the spectral sensitivity functions of the three classes of cones by the rows of a 3 by N_λ matrix $\mathbf{R}$. Let $\mathbf{r}$ be a three-dimensional vector whose entries are the cone quantal absorption rates. For a light with spectral power distribution $\mathbf{b}$, we can compute the absorption rates through the matrix equation

$$\mathbf{r} = \mathbf{R}\mathbf{b} \tag{10}$$

This computation accomplishes the wavelength-by-wavelength multiplication and summation for each cone class.

Figure 8 shows estimates of human cone spectral sensitivity functions. The three curves are independently normalized to a maximum of one. We discuss these estimates more thoroughly later.

Cone Coordinates. We use the term *cone coordinates* to refer to the vector $\mathbf{r}$. We can relate cone coordinates to tristimulus coordinates in a straightforward manner. Suppose that in a color-matching experiment performed with primaries $\mathbf{P}$ we find that a light $\mathbf{b}$ has tristimulus coordinates $\mathbf{t}$. From our mechanistic explanation and Eq. (10) we have

$$\mathbf{r} = \mathbf{R}\mathbf{b} = \mathbf{R}\mathbf{P}\mathbf{t} \tag{11}$$

If we define the matrix $\mathbf{M}_{T,R} = (\mathbf{RP})$, we see that the tristimulus coordinates of a light are related to its cone coordinates by a linear transformation

$$\mathbf{r} = \mathbf{M}_{T,R}\mathbf{t} \tag{12}$$

By comparing Eq. (9) with Eq. (11) and noting that these equations hold for any light **b**, we derive

$$\mathbf{R} = \mathbf{M}_{R,T}\mathbf{T} \tag{13}$$

Equation (13) has the interesting implication that the color-matching experiment determines the cone sensitivities up to a free linear transformation.

Common Color Coordinate Systems

Color Coordinate Systems. For the range of conditions where the color-matching experiment obeys the properties described in the previous sections, tristimulus coordinates (or cone coordinates) provide a complete and efficient representation for human color vision. When two lights have identical tristimulus coordinates, they are indistinguishable to the visual system and may be substituted for one another. When two lights have tristimulus coordinates that differ substantially, they can be distinguished by an observer with normal color vision.

The relation between spectral power distributions and tristimulus coordinates depends on the choice of primaries used in the color-matching experiment. In this sense, the choice of primaries in colorimetry is analogous to the choice of unit (e.g., foot versus meter) in the measurement of length. We use the terms color coordinate system and color space to refer to a representation derived with respect to a particular choice of primaries. We also use the term color coordinates as a synonym for tristimulus coordinates.

Although the choice of primaries determines a color space, specifying primaries alone is not sufficient to compute tristimulus coordinates. Rather, it is the color-matching functions that characterize the properties of the human observer with respect to a particular set of primaries. Knowledge of the color-matching functions allows us to compute tristimulus coordinates through Eq. (9). As we show below, however, knowledge of a single set of color-matching functions also allows us to derive color-matching functions with respect to other sets of primaries. Thus in practice we can specify a color space either by its primaries or by its color-matching functions.

A large number of different color spaces are in common use. The choice of which color space to use in a given application is governed by a number of considerations. If all that is of interest is to use a three-dimensional representation that accurately predicts the results of the color-matching experiment, the choice revolves around the question of finding a set of color-matching functions that accurately capture color-matching performance for the set of observers and viewing conditions under consideration. From this point of view, color spaces that differ only by an invertible linear transformation are equivalent. But there are other possible uses for color representation. For example, one might wish to choose a space that makes explicit the responses of the physiological mechanisms that mediate color vision. We discuss a number of commonly used color spaces below.

Stimulus Spaces. A stimulus space is the color space determined by the primaries of a particular apparatus. For example, stimuli are often specified in terms of the excitation of three monitor phosphors. Stimulus color spaces have the advantage that they provide a direct description of the physical stimulus. On the other hand, they are nonstandard and their use hampers comparison of data collected in different laboratories. A useful compromise is to transform the data to a standard color space, but to provide enough side information to allow exact reconstruction of the stimulus. Often this side information can take the form of a linear model whose basis functions are the apparatus primaries.

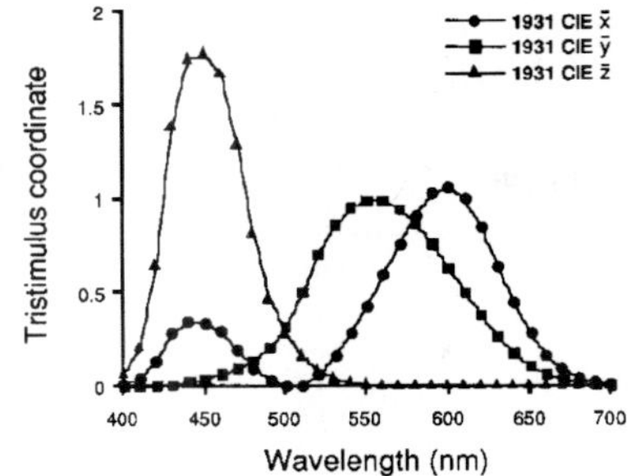

FIGURE 7 Color-matching functions. As described in the text, the rows of the matrix **T** may be viewed as a function of wavelength. Color-matching functions are often plotted as a function of wavelength. The plot shows a set of color-matching functions for normal human vision standardized by the CIE in 1931. The three individual functions are usually referred to as the CIE $\bar{x}$, $\bar{y}$, and $\bar{z}$ color-matching functions.

CIE 1931 Color-Matching Functions. In 1931, the CIE chose a standard set of primaries and integrated a large body of empirical data to determine a standard set of color-matching functions. The notion was that these functions would describe the results of a color-matching experiment performed on an "average" color-normal human observer. The result of the standardization process was a set of color-matching functions which are often called the CIE 1931 color-matching functions.[12] These functions are plotted as a function of wavelength in Fig. 7 and tabulated in Table 1. Tristimulus coordinates computed with respect to these color-matching functions are called CIE 1931 *XYZ* tristimulus coordinates. The hypothetical observer whom the CIE 1931 color-matching functions describe is often referred to as the CIE 1931 standard observer.

There is now some evidence that the color-matching functions standardized by the CIE in 1931 are slightly different from those of the average human observer.[9,34] A large body of extant data is available only in terms of the CIE 1931 system, however, and many colorimetric instruments are designed around it. Therefore, it seems likely that the CIE 1931 system will continue to be of practical importance for some time.

Judd-Vos Modified Color-Matching Functions. In 1951 Judd reconsidered the 1931 color-matching functions and came to the conclusion that they could be improved.[35] Vos[36] refined Judd's analysis. The Judd-Vos modifications lead to a set of color-matching functions that are probably more typical of the average human observer than the original CIE 1931 color-matching functions. These functions were never officially standardized. They are the basis of a number of estimates of the human cone spectral sensitivities and are thus widely used in practice, especially in vision science. We provide the Judd-Vos modified *XYZ* color-matching functions in Table 2.

1964 10° Color-Matching Functions. In 1964, the CIE standardized a second set of color-matching functions appropriate for larger field sizes. These color-matching functions take into account the fact that human color matches depend on the size of the matching fields. The CIE 1964 10° color-matching functions are an attempt to provide a standard

TABLE 1 CIE 1931 *XYZ* Color-Matching Functions

Wavelengths are in nm. The data are available at 1-nm intervals between 360 nm and 830 nm.[12]

Wavelength	X	Y	Z
400	1.43E-02	3.96E-04	6.79E-02
410	4.35E-02	1.21E-03	2.07E-01
420	1.34E-01	4.00E-03	6.46E-01
430	2.84E-01	1.16E-02	1.39E+00
440	3.48E-01	2.30E-02	1.75E+00
450	3.36E-01	3.80E-02	1.77E+00
460	2.91E-01	6.00E-02	1.67E+00
470	1.95E-01	9.10E-02	1.29E+00
480	9.56E-02	1.39E-01	8.13E-01
490	3.20E-02	2.08E-01	4.65E-01
500	4.90E-03	3.23E-01	2.72E-01
510	9.30E-03	5.03E-01	1.58E-01
520	6.33E-02	7.10E-01	7.82E-02
530	1.66E-01	8.62E-01	4.22E-02
540	2.90E-01	9.54E-01	2.03E-02
550	4.33E-01	9.95E-01	8.70E-03
560	5.95E-01	9.95E-01	3.90E-03
570	7.62E-01	9.52E-01	2.10E-03
580	9.16E-01	8.70E-01	1.65E-03
590	1.03E+00	7.57E-01	1.10E-03
600	1.06E+00	6.31E-01	8.00E-04
610	1.00E+00	5.03E-01	3.40E-04
620	8.54E-01	3.81E-01	2.00E-04
630	6.42E-01	2.65E-01	0.00E+00
640	4.48E-01	1.75E-01	0.00E+00
650	2.84E-01	1.07E-01	0.00E+00
660	1.65E-01	6.10E-02	0.00E+00
670	8.74E-02	3.20E-02	0.00E+00
680	4.68E-02	1.70E-02	0.00E+00
690	2.27E-02	8.20E-03	0.00E+00
700	1.14E-02	4.10E-03	0.00E+00

observer for these larger fields. The use of 10° color-matching functions is recommended by the CIE when the sizes of the regions under consideration are larger than 4°.[11] We provide the CIE 10° color-matching functions in Table 3.

Cone Sensitivities. For some applications it is desirable to relate the color stimulus to the responses of the cone mechanisms. A number of psychophysical and physiological techniques have been developed for determining the foveal cone spectral sensitivities.[9,34,37,38] At present, the most widely used estimates are those of Smith and Pokorny,[37,39] which are based on the Judd-Vos modified *XYZ* color-matching functions. Stockman, MacLeod, and Johnson,[34] however, argue for a set of small-field cone sensitivities derived from the CIE 1964 10° color-matching functions. We provide both the Smith-Pokorny and Stockman-MacLeod-Johnson estimates in Tables 4 and 5. Other estimates exist.[9,38]

Note that there is no universally accepted standard for scaling the sensitivities of each cone class relative to one another. One common method normalizes the sensitivities to a maximum of one for each cone class. Care should be taken when drawing conclusions that depend on the scaling chosen.

Opponent and Modulation Spaces. Cone coordinates are useful because they make explicit the responses of the initial physiological mechanisms thought to mediate color vision. A number of investigators have begun to use representations that attempt to

TABLE 2 Judd-Vos Modified *XYZ* Color-Matching Functions

Wavelengths are in nm. The data are available at 5-nm increments between 380 nm and 825 nm.[36]

Wavelength	X	Y	Z
400	3.80E-02	2.80E-03	1.74E-01
410	9.99E-02	7.40E-03	4.61E-01
420	2.29E-01	1.75E-02	1.07E+00
430	3.11E-01	2.73E-02	1.47E+00
440	3.33E-01	3.79E-02	1.62E+00
450	2.89E-01	4.68E-02	1.47E+00
460	2.33E-01	6.00E-02	1.29E+00
470	1.75E-01	9.10E-02	1.11E+00
480	9.19E-02	1.39E-01	7.56E-01
490	3.17E-02	2.08E-01	4.47E-01
500	4.85E-03	3.23E-01	2.64E-01
510	9.29E-03	5.03E-01	1.54E-01
520	6.38E-02	7.10E-01	7.66E-02
530	1.67E-01	8.62E-01	4.14E-02
540	2.93E-01	9.54E-01	2.00E-02
550	4.36E-01	9.95E-01	8.78E-03
560	5.97E-01	9.95E-01	4.05E-03
570	7.64E-01	9.52E-01	2.28E-03
580	9.16E-01	8.70E-01	1.81E-03
590	1.02E+00	7.57E-01	1.23E-03
600	1.06E+00	6.31E-01	9.06E-04
610	9.92E-01	5.03E-01	4.29E-04
620	8.43E-01	3.81E-01	2.56E-04
630	6.33E-01	2.65E-01	9.77E-05
640	4.41E-01	1.75E-01	5.12E-05
650	2.79E-01	1.07E-01	2.42E-05
660	1.62E-01	6.10E-02	1.19E-05
670	8.58E-02	3.20E-02	5.60E-06
680	4.58E-02	1.70E-02	2.79E-06
690	2.22E-02	8.21E-03	1.31E-06
700	1.11E-02	4.10E-03	6.48E-07

Source: From "Colorimetric and photometric properties of a two degree fundamental observer," J. J. Vos, *Color Research and Application,* Copyright © 1978. Reprinted by permission of John Wiley & Sons, Inc.

represent the responses of subsequent mechanisms. Two basic ideas underlie these representations. The first is the general opponent processing model described later under "Opponent Process Model." We call representations based on this idea *opponent color spaces.* The second idea is that stimulus contrast is more relevant than stimulus magnitude.[40] We call spaces that are based on this second idea *modulation color spaces.* Some color spaces are both opponent and modulation color spaces.

Cone Modulation Space. To derive coordinates in the cone modulation color space, the stimulus is first expressed in terms of its cone coordinates. The cone coordinates of a white point are then chosen. Usually these are the cone coordinates of a uniform adapting field or the spatio-temporal average of the cone coordinates of the entire image sequence. The cone coordinates of the white point are subtracted from the cone coordinates of the stimulus and the resulting differences are normalized by the corresponding cone coordinate of the white point.

The DKL Color Space. Derrington, Krauskopf, and Lennie[41] introduced an opponent modulation space that is now widely used. This space is closely related to the chromaticity diagram suggested by MacLeod and Boynton.[42] To derive coordinates in the DKL color space, the stimulus is first expressed in cone coordinates. As with cone modulation space,

TABLE 3 CIE 1964 10° *XYZ* Color-Matching Functions

Wavelengths are in nm. The data are available at 1-nm intervals between 360 nm and 830 nm.[12]

Wavelength	X	Y	Z
400	1.91E-02	2.00E-03	8.60E-02
410	8.47E-02	8.80E-03	3.89E-01
420	2.05E-01	2.14E-02	9.73E-01
430	3.15E-01	3.87E-02	1.55E+00
440	3.84E-01	6.21E-02	1.97E+00
450	3.71E-01	8.95E-02	1.99E+00
460	3.02E-01	1.28E-01	1.75E+00
470	1.96E-01	1.85E-01	1.32E+00
480	8.05E-02	2.54E-01	7.72E-01
490	1.62E-02	3.39E-01	4.15E-01
500	3.80E-03	4.61E-01	2.19E-01
510	3.75E-02	6.07E-01	1.12E-01
520	1.18E-01	7.62E-01	6.07E-02
530	2.37E-01	8.75E-01	3.05E-02
540	3.77E-01	9.62E-01	1.37E-02
550	5.30E-01	9.92E-01	4.00E-03
560	7.05E-01	9.97E-01	0.00E+00
570	8.79E-01	9.56E-01	0.00E+00
580	1.01E+00	8.69E-01	0.00E+00
590	1.12E+00	7.77E-01	0.00E+00
600	1.12E+00	6.58E-01	0.00E+00
610	1.03E+00	5.28E-01	0.00E+00
620	8.56E-01	3.98E-01	0.00E+00
630	6.48E-01	2.84E-01	0.00E+00
640	4.32E-01	1.80E-01	0.00E+00
650	2.68E-01	1.08E-01	0.00E+00
660	1.53E-01	6.03E-02	0.00E+00
670	8.13E-02	3.18E-02	0.00E+00
680	4.09E-02	1.59E-02	0.00E+00
690	1.99E-02	7.70E-03	0.00E+00
700	9.60E-03	3.70E-03	0.00E+00

the cone coordinates of a white point are then subtracted from the cone coordinates of the stimulus of interest. The next step is to reexpress the resulting difference as tristimulus coordinates with respect to a new choice of primaries that are thought to isolate the responses of postreceptoral mechanisms.[43,44] The three primaries are chosen so that modulating two of them does not change the response of the photopic luminance mechanism (discussed later). The color coordinates corresponding to these two primaries are often called the constant B and constant R and G coordinates. Modulating the constant R and G coordinate of a stimulus modulates only the S cones. Modulating the constant B coordinate modulates both the L and M cones but keeps the S cone response constant. Because the constant R and G coordinate is not allowed to change the response of the photopic luminance mechanism, the DKL color space is well defined only if the S cones do not contribute to luminance. The third primary of the space is chosen so that it has the same relative cone coordinates as the white point. The coordinate corresponding to this third primary is called the luminance coordinate. Flitcroft[45] provides a detailed treatment of the DKL color space.

Caveats. The basic ideas underlying the use of opponent and modulation color spaces seem to be valid. On the other hand, there is not general agreement about how signals from cones are combined into opponent channels, about how this combination depends on adaption, or about how adaptation affects signals originating in the cones. Since a specific

TABLE 4 Smith-Pokorny Estimates of the Cone Sensitivities

Wavelengths are in nm. The data are available at 1-nm increments between 400 nm and 700 nm.[37]

Wavelength	L	M	S
400	2.66E-03	2.82E-03	1.08E-01
410	6.89E-03	7.67E-03	2.85E-01
420	1.58E-02	1.89E-02	6.59E-01
430	2.33E-02	3.17E-02	9.08E-01
440	3.01E-02	4.77E-02	1.00E+00
450	3.43E-02	6.35E-02	9.10E-01
460	4.12E-02	8.60E-02	7.99E-01
470	6.27E-02	1.30E-01	6.89E-01
480	1.02E-01	1.89E-01	4.68E-01
490	1.62E-01	2.67E-01	2.76E-01
500	2.63E-01	3.96E-01	1.64E-01
510	4.23E-01	5.95E-01	9.56E-02
520	6.17E-01	8.08E-01	4.74E-02
530	7.73E-01	9.41E-01	2.56E-02
540	8.83E-01	9.97E-01	1.24E-02
550	9.54E-01	9.87E-01	5.45E-03
560	9.93E-01	9.22E-01	2.53E-03
570	9.97E-01	8.06E-01	1.44E-03
580	9.65E-01	6.51E-01	1.16E-03
590	8.94E-01	4.77E-01	8.12E-04
600	7.95E-01	3.18E-01	6.10E-04
610	6.70E-01	1.93E-01	3.12E-04
620	5.30E-01	1.10E-01	1.98E-04
630	3.80E-01	5.83E-02	9.03E-05
640	2.56E-01	2.96E-02	5.25E-05
650	1.59E-01	1.44E-02	2.51E-05
660	9.14E-02	6.99E-03	1.44E-05
670	4.82E-02	3.33E-03	7.56E-06
680	2.57E-02	1.64E-03	4.02E-06
690	1.24E-02	7.50E-04	1.94E-06
700	6.21E-03	3.68E-04	9.72E-07

model of these processes is implicit in any opponent or modulation color space, coordinates in these spaces must be treated carefully. This is particularly true of modulation spaces, where the relation between the physical stimulus and coordinates in the space depends on the choice of white point. As a consequence, radically different stimuli can have identical coordinates in a modulation space. For example, 100 percent contrast monochromatic intensity gratings are all represented by the same coordinates in modulation color spaces, independent of their wavelength. Nonetheless, such stimuli appear very different to human observers. Identity of coordinates in a modulation color space does not imply identity of appearance across different choices of white points.

Transformations Between Color Spaces

Because of the large number of color spaces currently in use, the ability to transform data between various color spaces is of considerable practical importance. The derivation of such transformations depends on what is known about the source and destination color spaces. Below we discuss cases where both the source and destination color space are derived from the same underlying observer (i.e., when the source and destination color

TABLE 5 Stockman-MacLeod-Johnson Estimates of Cone Sensitivities

Wavelengths are in nm. The data are available at 5-nm increments between 390 nm and 730 nm.[34]

Wavelength	L	M	S
400	2.23E-03	1.93E-03	5.74E-02
410	8.76E-03	8.46E-03	2.39E-01
420	1.75E-02	2.05E-02	5.28E-01
430	2.69E-02	3.88E-02	8.03E-01
440	3.84E-02	6.36E-02	9.87E-01
450	4.88E-02	8.79E-02	9.50E-01
460	6.52E-02	1.20E-01	8.12E-01
470	9.69E-02	1.76E-01	6.51E-01
480	1.38E-01	2.38E-01	3.94E-01
490	1.88E-01	3.04E-01	2.09E-01
500	2.97E-01	4.48E-01	1.19E-01
510	4.59E-01	6.47E-01	6.21E-02
520	6.34E-01	8.33E-01	2.94E-02
530	7.76E-01	9.46E-01	1.28E-02
540	8.83E-01	1.00E+00	5.61E-03
550	9.42E-01	9.77E-01	2.50E-03
560	9.88E-01	9.19E-01	1.15E-03
570	9.99E-01	8.03E-01	5.44E-04
580	9.68E-01	6.45E-01	2.63E-04
590	9.25E-01	4.88E-01	1.31E-04
600	8.36E-01	3.35E-01	6.65E-05
610	7.12E-01	2.10E-01	3.46E-05
620	5.64E-01	1.22E-01	1.83E-05
630	4.15E-01	6.93E-02	9.94E-06
640	2.71E-01	3.41E-02	5.49E-06
650	1.66E-01	1.57E-02	3.08E-06
660	9.40E-02	7.70E-03	1.76E-06
670	4.99E-02	3.69E-03	1.03E-06
680	2.51E-02	1.75E-03	6.07E-07
690	1.22E-02	8.33E-04	3.64E-07
700	5.87E-03	3.95E-04	2.22E-07

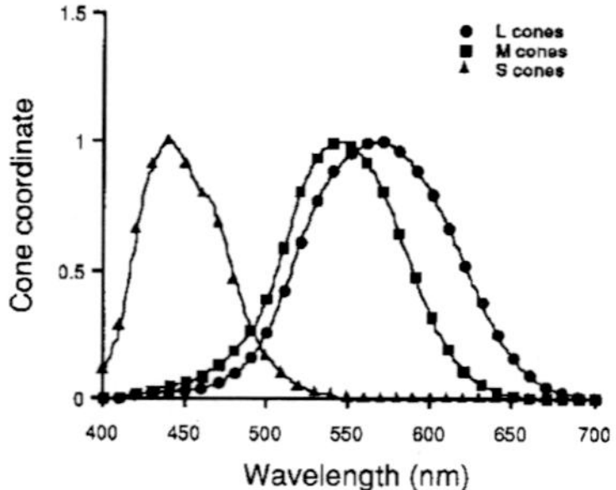

FIGURE 8 Human cone spectral responsivity functions normalized to a maximum of one. These estimates are due to Smith and Pokorny.[37,39]

TABLE 6 Color Space Transformations

The table summarizes how to form the matrix **M** that transforms color coordinates between two spaces.

Spectral functions known			
Source	Destination	M	Notes
Primaries $\mathbf{P}_1$	CMFs $\mathbf{T}_2$	$\mathbf{M} = \mathbf{T}_2\mathbf{P}_1$	
CMFs $\mathbf{T}_1$	Primaries $\mathbf{P}_2$	$\mathbf{M} = (\mathbf{T}_1\mathbf{P}_2)^{-1}$	
Primaries $\mathbf{P}_1$	Primaries $\mathbf{P}_2$	$\mathbf{M} = (\mathbf{T}\,\mathbf{P}_2)^{-1}(\mathbf{T}\,\mathbf{P}_1)$	**T** is any set of CMFs.
CMFs $\mathbf{T}_1$	CMFs $\mathbf{T}_2$	$\mathbf{T}_2 = \mathbf{M}\,\mathbf{T}_1$	Use regression to find **M**.

One space specified in terms of other	
Known tristimulus coordinates	How to construct **M**
Source primaries known in destination space.	Put them in columns of **M**.
Source CMFs known in destination space.	Put them in rows of $\mathbf{M}^{-1}$.
Destination primaries known in source space.	Put them in columns of $\mathbf{M}^{-1}$.
Destination CMFs known in source space.	Put them in rows of **M**.

* CMFs stands for Color Matching Functions

spaces both predict identical color matches). Table 6 summarizes these transformations. When the source and destination color spaces are characterized by a different underlying observer, the transformation is more difficult and often cannot be done exactly. We discuss possible approaches in a later section.

Source Primaries and Destination Color-Matching Functions Known. Let $\mathbf{P}_1$ be the matrix representing a set of known primaries. Let $\mathbf{T}_2$ be the matrix representing a known set of color-matching functions (e.g., the CIE 1931 *XYZ* color-matching functions). We would like to determine a transformation between the color coordinate system specified by $\mathbf{P}_1$ and that specified by $\mathbf{T}_2$. For example, linearized frame buffer values input to a computer-controlled color monitor may be thought of as tristimulus coordinates in a color space defined by the monitor's phosphor emission spectra. This transformation thus allows computation of the CIE 1931 *XYZ* tristimulus coordinates from the linearized frame buffer values.

We start by using Eq. (9) to compute the tristimulus coordinates, with respect to $\mathbf{T}_2$, for all three primary lights specified by $\mathbf{P}_1$. Each of these primaries is contained in a column of $\mathbf{P}_1$, so that we may perform this calculation directly through the matrix multiplication

$$\mathbf{M}_{P,T} = \mathbf{T}_2\mathbf{P}_1 \tag{14}$$

Let the matrix $\mathbf{P}_2$ represent the destination primaries. We do not need to know these explicitly, only that they exist. The meaning of Eq. (14) is that

$$\mathbf{P}_1 \sim \mathbf{P}_2\mathbf{M}_{P,T} \tag{15}$$

where we have generalized the symbol "~" to denote a column-by-column visual match for the matrices on both sides of the relation. This relation follows because the columns of $\mathbf{M}_{P,T}$ specify how the destination primaries should be mixed to match the source primaries. Equation (15) tells us that we can substitute the three lights represented by the columns of $\mathbf{P}_2\mathbf{M}_{P,T}$ for the three lights represented by the columns of $\mathbf{P}_1$ in any color-matching experiment. In particular, we may make this substitution for any light **b** with tristimulus coordinates $\mathbf{t}_1$ in the source color coordinate system. We have

$$\mathbf{b} \sim \mathbf{P}_1\mathbf{t}_1 \sim \mathbf{P}_2\mathbf{M}_{P,T}\mathbf{t}_1 \tag{16}$$

By inspection, this tells us that the three-dimensional vector

$$\mathbf{t}_2 = \mathbf{M}_{P,T}\mathbf{t}_1 \tag{17}$$

is the tristimulus coordinates of **b** in the destination color coordinate system. Equation (17) provides us with the means to transform tristimulus coordinates from a coordinate system where the primaries are known to one where the color-matching functions are known. The transformation matrix $\mathbf{M}_{P,T}$ required to perform the transformation depends only on the known primaries $\mathbf{P}_1$ and the known color-matching functions $\mathbf{T}_2$. Given these, $\mathbf{M}_{P,T}$ may be computed directly from Eq. (14).

Source Color-Matching Functions and Destination Primaries Known. A second transformation applies when the color-matching functions in the source color space and the primaries in the destination color space are known. This will be the case, for example, when we wish to render a stimulus specified in terms of CIE 1931 tristimulus coordinates on a calibrated color monitor.

Let $\mathbf{T}_1$ represent the known color-matching functions and $\mathbf{P}_2$ represent the known primaries. By applying Eq. (17) we have that the relation between source tristimulus coordinates and the destination tristimulus coordinates is given by $\mathbf{t}_1 = \mathbf{M}_{P,T}\mathbf{t}_2$. This is a system of linear equations that we may solve to find an expression for $\mathbf{t}_2$ in terms of $\mathbf{t}_1$. In particular, as long as the matrix $\mathbf{M}_{P,T}$ is nonsingular, we can convert tristimulus coordinates using the relation

$$\mathbf{t}_2 = \mathbf{M}_{T,P}\mathbf{t}_1 \tag{18}$$

where we define

$$\mathbf{M}_{T,P} = (\mathbf{M}_{P,T})^{-1} = (\mathbf{T}_1\mathbf{P}_2)^{-1} \tag{19}$$

Source and Destination Primaries Known. A third transformation applies when the primaries of both the source and destination color spaces are known. One application of this transformation is to generate matching stimuli on two different calibrated monitors.

Let $\mathbf{P}_1$ and $\mathbf{P}_2$ represent the two sets of primaries. Let $\mathbf{T}$ represent a set of color-matching functions for any human color coordinates system. (There is no requirement that the color-matching functions be related to either the source or the destination primaries. For example, the CIE 1931 *XYZ* color-matching functions might be used.) To do the conversion, we simply use Eq. (17) to transform from the color coordinate system described by $\mathbf{P}_1$ to the coordinate system described by $\mathbf{T}$. Then we use Eq. (18) to transform from the coordinates system described by $\mathbf{T}$ to the coordinate system described by $\mathbf{P}_2$. The overall transformation is given by

$$\mathbf{t}_2 = \mathbf{M}_{P,P}\mathbf{t}_1 = (\mathbf{M}_{T,P_2})(\mathbf{M}_{P_1,T})\mathbf{t}_1 = (\mathbf{T}\mathbf{P}_2)^{-1}(\mathbf{T}\mathbf{P}_1)\mathbf{t}_1 \tag{20}$$

It should not be surprising that this transformation requires the specification of a set of color-matching functions. These color-matching functions are the only source of information about the human observer in the transformation equation.

Source and Destination Color-Matching Functions Known. Finally, it is sometimes of interest to transform between two color spaces that are specified in terms of their color-matching functions. An example is transforming between the space defined by the Judd-Vos modified *XYZ* color-matching functions and the space defined by the Smith-Pokorny cone fundamentals.

Let $\mathbf{T}_1$ and $\mathbf{T}_2$ represent the source and destination color-matching functions. Our development above assures us that there is some three-by-three transformation matrix, call it $\mathbf{M}_{T,T}$, that transforms color coordinates between the two systems. Recall that the columns of $\mathbf{T}_1$ and $\mathbf{T}_2$ are themselves tristimulus coordinates for corresponding monochromatic lights. Thus $\mathbf{M}_{T,T}$ must satisfy

$$\mathbf{T}_2 = \mathbf{M}_{T,T}\mathbf{T}_1 \tag{21}$$

This is a system of linear equations where the entries of $\mathbf{M}_{T,T}$ are the unknown variables. This system may be solved using standard regression methods. Once we have solved for $\mathbf{M}_{T,T}$, we can transform tristimulus coordinates using the equation

$$\mathbf{t}_2 = \mathbf{M}_{T,T}\mathbf{t}_1 \tag{22}$$

The transformation specified by Eq. (22) will be exact as long as the two sets of color-matching functions $\mathbf{T}_1$ and $\mathbf{T}_2$ characterize the performance of the same observer. One sometimes wishes, however, to transform between two color spaces that are defined with respect to different observers. For example, one might want to convert CIE 1931 *XYZ* tristimulus values to Judd-Vos modified tristimulus values. Although the regression procedure described here will still produce a transformation matrix in this case, the result of the transformation is not guaranteed to be correct.[4] We return to this topic later.

Interpreting the Transformation Matrix. It is useful to interpret the rows and columns of the matrices derived above. Let $\mathbf{M}$ be a matrix that maps the color coordinates from a source color space to a destination color space. Both source and destination color spaces are associated with a set of primaries and a set of color-matching functions. From our derivations above, we can conclude that the columns of $\mathbf{M}$ are the coordinates of the source primaries in the destination color space [see Eq. (14)] and the rows of $\mathbf{M}$ provide the destination color-matching functions with respect to the linear model whose basis functions are the primaries of source color space. Similarly, the columns of $\mathbf{M}^{-1}$ are the coordinates of the destination primaries in the source color-matching space and the rows of $\mathbf{M}^{-1}$ are the source color-matching functions with respect to the linear model whose basis functions are the primaries of the destination color space. Thus in many cases it is possible to construct the matrix $\mathbf{M}$ without full knowledge of the spectral functions. This can be of practical importance. For example, monitor manufacturers often specify the CIE 1931 *XYZ* tristimulus coordinates of their monitors' phosphors. In additional, colorimeters that measure tristimulus coordinates directly are often more readily available than spectral radiometers.

Transforming Primaries and Color-Matching Functions. We have shown that color coordinates in any two color spaces may be related by applying a linear transformation $\mathbf{M}$. The converse is also true. If we pick any nonsingular linear transformation $\mathbf{M}$ and apply it to a set of color coordinates we have defined a new color space that will successfully predict color matches. The color-matching functions for this new space will be given by $\mathbf{T}_2 = \mathbf{M}\mathbf{T}_1$. A set of primaries for the new space will be given by $\mathbf{P}_2 = \mathbf{P}_1\mathbf{M}^{-1}$. These derived primaries are not unique. Any set of primaries that match the constructed primaries will also work.

The fact that new color spaces can be constructed by applying linear transformations has an important implication for the study of color. If we restrict attention to what we may conclude from the color-matching experiment, we can determine the psychological representation of color only up to a free linear transformation. There are two attitudes one can take toward this fact. The conservative attitude is to refrain from making any statements about the nature of color vision that depend on a particular choice of color space. The other is to appeal to experiments other than the color-matching experiment to choose a privileged representation. At present, there is not universal agreement about how to choose such a representation and we therefore advocate the conservative approach.

Visualizing Color Data

A challenge facing today's color scientist is to produce and interpret graphical representations of color data. Because the visual representation of light is three-dimensional, it is difficult to plot this representation on a two-dimensional page. Even more difficult is to represent a dependent measure of visual performance as a function of color coordinates. We discuss several approaches.

Three-Dimensional Approaches. One strategy is to plot the three-dimensional data in perspective, as shown on the top of Fig. 9. In many cases, the projection viewpoint may be chosen to provide a clear view of the regularities of interest in the data. The three-dimensional structure of the data may be emphasized by the addition of various monocular depth cues. A number of computer-graphics packages now provide facilities to aid in the preparation of three-dimensional perspective plots. Often these programs allow variation of the viewpoint and automatic inclusion of monocular depth cues.

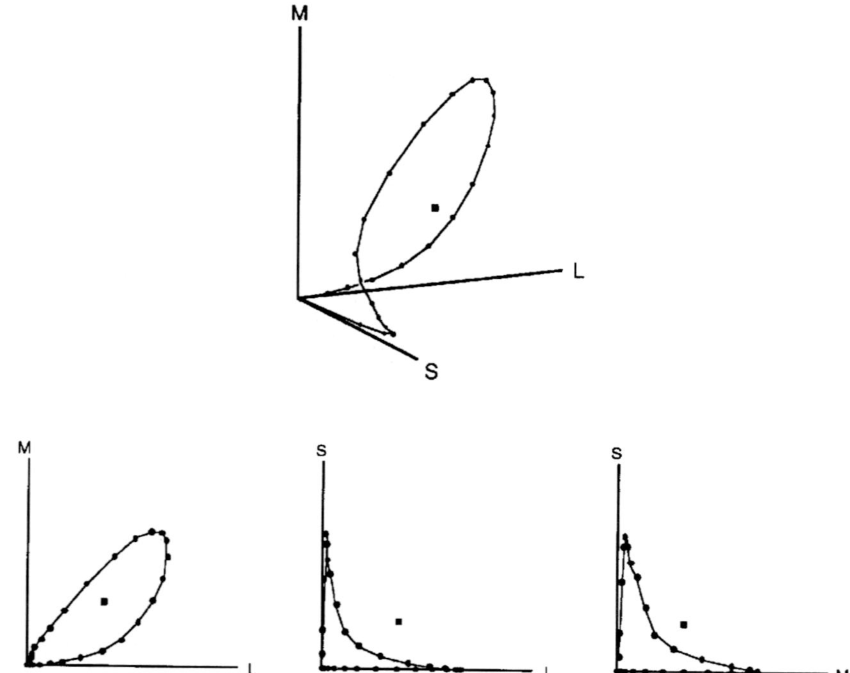

FIGURE 9 Three dimensional views of color data. The figure shows the color coordinates of an equal energy spectrum in color space defined by the human cone sensitivities (closed circles) and the color coordinates of CIE daylight D65 (closed squares). The top panel shows the data in perspective. The bottom panels show three two-dimensional views of the same data.

A second approach to showing the three-dimensional structure of color data is to provide multiple two-dimensional views, as in a drafter's sketch. This approach is shown on the bottom of Fig. 9.

Computer display technology provides promise for improved methods of viewing three-dimensional data. For example, it is now possible to produce computer animations that show plots that vary over time. Such plots have the potential for representing multidimensional data in a manner that is more comprehensible to a human viewer than a static plot. Other interesting possibilities include the use of stereo depth cues and color displays. At present, the usefulness of these techniques is largely confined to exploratory data analysis because there are no widely accepted standards for publication.

Chromaticity Diagrams. A second strategy for plotting color data is to reduce the dimensionality of the data representation. One common approach is through the use of chromaticity coordinates. Chromaticity coordinates are defined so that any two lights with the same relative color coordinates have identical chromaticity coordinates. That is, the chromaticity coordinates of a light are invariant with respect to intensity scaling. Because chromaticity coordinates have one less degree of freedom than color coordinates, they can be described by just two numbers and plotted in a plane. We call a plot of chromaticity coordinates a *chromaticity diagram.* A chromaticity diagram eliminates all information about the intensity of a stimulus.

There are many ways to normalize color coordinates to produce a set of chromaticity coordinates. When the underlying color coordinates are CIE 1931 XYZ tristimulus coordinates, it is conventional to use CIE 1931 xy chromaticity coordinates. These are given by

$$x = X/(X + Y + Z)$$
$$y = Y/(X + Y + Z) \tag{23}$$

A graphical way to understand these chromaticity coordinates is illustrated in Fig. 10.

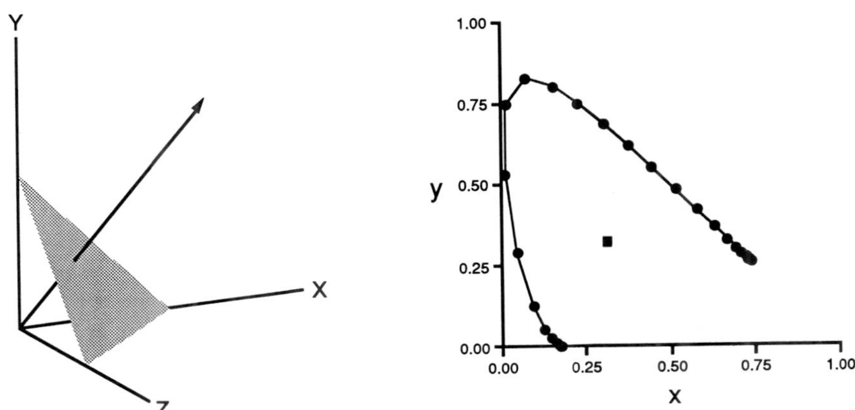

FIGURE 10 CIE 1931 xy chromaticity diagram. The left side of the figure shows a perspective view of the CIE 1931 XYZ tristimulus space. The ray shows a locus of points with constant chromaticity coordinates. The actual chromaticity coordinates for each ray are determined by where the ray intersects the plane described by the equation $X + Y + Z = 1$. This plane is indicated. The X and Y tristimulus values at the point of intersection are the x and y chromaticity coordinates for the ray. The right side of the figure shows the chromaticity coordinates for the ray. The right side of the figure shows the chromaticity coordinated of an equal energy spectrum (closed circles) and of CIE daylight D65 (closed square).

Wyszecki and Stiles[9] review a number of other standard chromaticity diagrams. MacLeod and Boynton[42] propose a chromaticity diagram that is derived from cone coordinates.

A useful property of most chromaticity diagrams is that the chromaticity coordinates of the mixture of two lights is always a weighted combination of chromaticity coordinates of the individual lights. This is easily verified for the CIE 1931 xy chromaticity diagram by algebraic manipulation. Thus the chromaticity of a mixture of lights will plot somewhere on the chord connecting the chromaticities of the individual lights.

Implicit in the use of chromaticity coordinates is the assumption that scalar multiplication of the stimuli does not affect the visual performance being plotted. If the overall intensity of the stimuli matter, then the use of chromaticity coordinates can obscure important regularities. For example, the shape of color discrimination contours (see "Color Discrimination" later in chapter) depends on how the overall intensity of the stimuli covaries with their chromaticities. Yet these contours are often plotted on a chromaticity diagram. This practice can lead to misinterpretation of the discrimination data. We recommend that plots of chromaticity coordinates be treated with some caution.

Functions of Wavelength. Color data are often represented as functions of wavelength. The wavelength spectrum parameterizes a particular path through the three-dimensional color space. The exact path depends on how overall intensity covaries with wavelength. For an equal energy spectrum, the path is illustrated in Fig. 9.

Wavelength representations are particularly useful in situations where knowing the value of a function for the set of monochromatic stimuli provides a complete characterization of performance. Color-matching functions, for example, are usefully plotted as functions of wavelength because these functions may be used to predict the tristimulus coordinates of any light. Plots of detection threshold versus wavelength, on the other hand, cannot be used to predict the detection threshold for arbitrary lights.[46] Just as the chromaticity diagram tends to obscure the potential importance of manipulating the overall intensity of light, wavelength representations tend to obscure the potential importance of considering mixtures of monochromatic lights.

Colorimetric measurements

To apply the formulae described in this chapter, it is often necessary to measure the colorimetric properties of stimuli. The most general approach is to measure the full spectral power distribution of the stimuli. Standard instrumentation and methods are discussed in Vol. II, Chap. 24. Often, however, it is not necessary to know the full spectral power distribution; knowledge of the tristimulus coordinates (in some standard color space) is sufficient. For example, the color space transformations summarized earlier in Table 6 depend on the full spectral power distributions of the primaries only through their tristimulus coordinates.

Specialized instruments, called *colorimeters,* can measure tristimulus coordinates directly. These instruments typically operate using the same principles as photometers (see Chap. 24 of Vol. II of this Handbook) with the exception that they have three calibrated filters rather than just one. Each filter mimics the spectral shape of one of the color-matching functions. Wyszecki and Stiles discuss colorimeter design.[9] Colorimeters are generally less expensive than radiometers and are thus an attractive option when full spectral data is not required.

Two caveats are worth note. First, it is technically difficult to design filters that exactly match a desired set of color-matching functions. Generally, commercial colorimeters are calibrated so that they give accurate readings for stimuli with broad spectral power distributions. For narrowband stimuli (e.g., the light emitted by the red phosphor of many color monitors) the reported readings may be quite inaccurate. Second, most colorimeters are designed to the CIE 1931 standard. This may not be an optimal choice for the purpose of predicting the matches of an average human observer.

26.4 TOPICS

Surfaces and Illuminants

As shown in Fig. 1, the light reaching the eye is often formed when light from an illuminant reflects from a surface. Illuminants and surfaces are likely to be of interest in color reproduction applications involving inks, paints, and dyes, and in lighting design applications.

Reflection Model. Illuminants are specified by their spectral power distributions. We will use the vector **e** to represent the illuminant spectral power distributions. In general, the interaction of light with matter is quite complex (see Vol. I, Chaps. 7 and 9). For many applications, however, a rather simple model is acceptable. Using this model, we describe a surface by its surface reflectance function. The surface reflectance function specifies, for each sample wavelength, the fraction of illuminant power that is reflected to the observer. We will use the vector **s** to represent surface reflectance spectra. Each entry of **s** gives the reflectance measured at a single sample wavelength. Thus the spectral power distribution **b** of the reflected light is given by the wavelength-by-wavelength product of the illuminant spectral power distribution and the surface reflectance function.

The most important consideration neglected in this formulation is viewing geometry. The relation between the radiant power emitted by a source of illumination, the material properties of a surface, and the radiant power reaching an observer can depend strongly on the viewing geometry. In our formulation, these geometrical factors must be incorporated implicitly into the specification of the illuminant and surface properties, so that any actual calculation is specific to a particular viewing geometry. Moreover, the surface reflectance must be understood as being associated with a particular image location, rather than with a particular object. A great deal of current research in photorealistic computer graphics is concerned with accurate and efficient ways to specify illuminants and surfaces for spatially complex scenes.[7,47] A second complexity that we neglect is fluorescence.

Computing the Reflected Light. The relation between the surface reflectance function and the reflected light spectral power distribution is linear if the illuminant spectral power distribution is held fixed. We form the N_λ by N_λ diagonal illuminant matrix **E** whose diagonal entries are the entries of **e**. This leads to the relation $\mathbf{b} = \mathbf{Es}$. By substituting into Eq. (9), we arrive at an expression for the tristimulus coordinates of the light reflected from a surface

$$\mathbf{t} = (\mathbf{TE})\mathbf{s} \tag{24}$$

The matrix (**TE**) in this equation plays exactly the same role as the color-matching functions do in Eq. (9). Any result that holds for spectral power distributions may thus be directly extended to a result for surface reflectance functions when the illuminant is known and held fixed.

Linear Model Representations for Surfaces and Illuminants. Judd, MacAdam, and Wyszecki[48] measured the spectral power distributions of a large number of naturally occurring daylights. They determined that a four-dimensional linear model provided a good description of their spectral measurements. Consideration of their results and other

TABLE 7 CIE Basis Vectors for Daylights

Wavelengths are in nm. The data are available at 5-nm increments from 300 nm to 830 nm.[11]

Wavelength	Vector 1	Vector 2	Vector 3
400	9.48E+01	4.34E+01	-1.10E+00
410	1.05E+02	4.63E+01	-5.00E-01
420	1.06E+02	4.39E+01	-7.00E-01
430	9.68E+01	3.71E+01	-1.20E+00
440	1.14E+02	3.67E+01	-2.60E+00
450	1.26E+02	3.59E+01	-2.90E+00
460	1.26E+02	3.26E+01	-2.80E+00
470	1.21E+02	2.79E+01	-2.60E+00
480	1.21E+02	2.43E+01	-2.60E+00
490	1.14E+02	2.01E+01	-1.80E+00
500	1.13E+02	1.62E+01	-1.50E-01
510	1.11E+02	1.32E+01	-1.30E+00
520	1.07E+02	8.60E+00	-1.20E+00
530	1.09E+02	6.10E+00	-1.00E+00
540	1.05E+02	4.20E+00	-5.00E-01
550	1.04E+02	1.90E+00	-3.00E-01
560	1.00E+02	0.00E+00	0.00E+00
570	9.60E+01	-1.60E+00	2.00E-01
580	9.51E+01	-3.50E+00	5.00E-01
590	8.91E+01	-3.50E+00	2.10E+00
600	9.05E+01	-5.80E+00	3.20E+00
610	9.03E+01	-7.20E+00	4.10E+00
620	8.84E+01	-8.60E+00	4.70E+00
630	8.40E+01	-9.50E+00	5.10E+00
640	8.51E+01	-1.09E+01	6.70E+00
650	8.19E+01	-1.07E+01	7.30E+00
660	8.26E+01	-1.20E+01	8.60E+00
670	8.49E+01	-1.40E+01	9.80E+00
680	8.13E+01	-1.36E+01	1.02E+01
690	7.19E+01	-1.20E+01	8.30E+00
700	7.43E+01	-1.33E+01	9.60E+00

daylight measurements led the CIE to standardize a three-dimensional linear model for natural daylights. The basis vectors for this model are provided in Table 7. Figure 11 depicts a daylight spectral power distribution (measured at the author's laboratory in Santa Barbara, Calif.) and its approximation using the first two basis vectors of the CIE linear model for daylight.

Cohen[49] analyzed the reflectance spectra of a large set of Munsell papers[50,51] and concluded that a four-dimensional linear model provided a good approximation to the entire data set. The basis vectors for Cohen's linear model are provided in Table 8. Maloney[52] reanalyzed these data, plus a set of natural spectra measured by Krinov[53] and confirmed Cohen's conclusion. More recently, reflectance measurements of the spectra of additional Munsell papers and of natural objects[54,55] have been described by small-dimensional linear models. Figure 12 shows a measured surface reflectance spectrum (red cloth, measured in the author's laboratory) and its approximation using Cohen's four-dimensional linear model.

It is not yet clear why natural illuminant and surface spectra are well approximated by small-dimensional linear models or how general this conclusion is. Maloney[52] provides some speculations. Nonetheless, the assumption that natural spectra do lie within small-dimensional linear models seems reasonable in light of the currently available evidence. This assumption makes possible a number of interesting practical calculations, as we illustrate in some of the following sections.

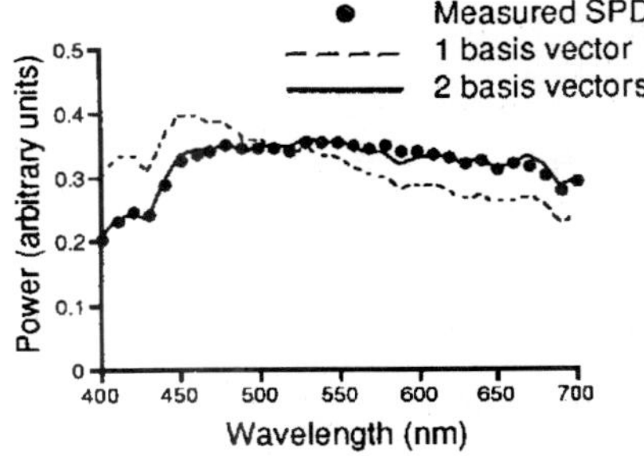

FIGURE 11 The figure shows a daylight spectral power distribution and its approximation using the CIE linear model for daylight. For this particular illuminant, only two basis functions were required to provide a very good fit.

TABLE 8 Cohen's Basis Vectors for Reflectance Functions

Wavelengths are in nm. The data are available at 10-nm increments from 380 nm to 770 nm.[49]

Wavelength	Vector 1	Vector 2	Vector 3	Vector 4
400	3.50E+00	-9.14E-01	-7.83E-01	-2.28E-01
410	3.51E+00	-9.54E-01	-7.78E-01	-2.50E-01
420	3.52E+00	-9.94E-01	-7.65E-01	-2.66E-01
430	3.54E+00	-1.05E+00	-7.39E-01	-2.64E-01
440	3.55E+00	-1.12E+00	-7.09E-01	-2.44E-01
450	3.57E+00	-1.20E+00	-6.63E-01	-2.07E-01
460	3.57E+00	-1.28E+00	-5.81E-01	-1.63E-01
470	3.56E+00	-1.35E+00	-4.62E-01	-1.13E-01
480	3.58E+00	-1.38E+00	-2.47E-01	-4.24E-02
490	3.62E+00	-1.38E+00	8.40E-03	4.18E-02
500	3.67E+00	-1.31E+00	2.79E-01	1.20E-01
510	3.80E+00	-1.13E+00	6.10E-01	2.62E-01
520	3.91E+00	-9.28E-01	8.89E-01	4.00E-01
530	3.91E+00	-7.84E-01	1.03E+00	4.07E-01
540	3.91E+00	-6.33E-01	1.10E+00	3.53E-01
550	3.94E+00	-4.66E-01	1.13E+00	2.80E-01
560	3.94E+00	-2.78E-01	1.12E+00	2.15E-01
570	4.04E+00	-7.45E-02	1.13E+00	8.13E-02
580	4.17E+00	1.71E-01	1.07E+00	-8.77E-02
590	4.42E+00	4.20E-01	1.01E+00	-1.94E-01
600	4.55E+00	6.63E-01	7.18E-01	-3.45E-01
610	4.73E+00	8.70E-01	4.75E-01	-3.74E-01
620	4.84E+00	1.00E+00	2.76E-01	-3.80E-01
630	4.90E+00	1.08E+00	1.47E-01	-3.87E-01
640	4.93E+00	1.12E+00	5.61E-02	-3.92E-01
650	4.97E+00	1.15E+00	-1.21E-02	-3.87E-01
660	5.01E+00	1.16E+00	-6.80E-02	-3.67E-01
670	5.08E+00	1.16E+00	-1.20E-01	-3.19E-01
680	5.17E+00	1.16E+00	-1.61E-01	-2.40E-01
690	5.27E+00	1.16E+00	-1.99E-01	-1.29E-01
700	5.39E+00	1.14E+00	-2.40E-01	1.28E-02

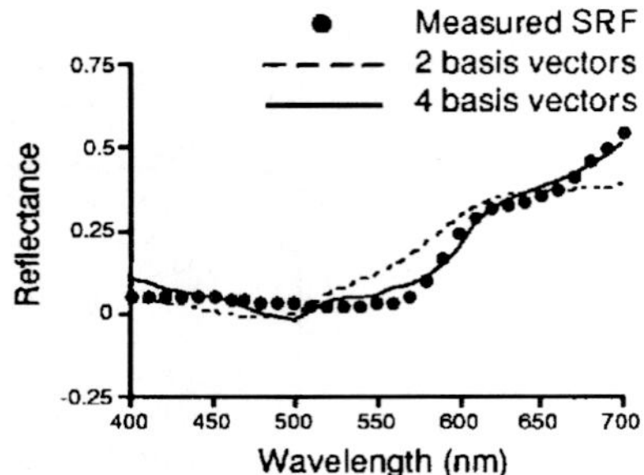

FIGURE 12 The figure shows a measured surface reflectance function and a fit to it using Cohen's 4-dimensional linear model.[49]

Determining a Linear Model from Raw Spectral Data. Given a set of spectral measurements, it is possible, for any integer N_b, to find the N_b dimensional linear model that best approximates the spectral data set (in a least-squares sense). Suppose that the data set consists of N_{meas} spectra, each of which is represented at N_λ sample wavelengths. Let $\mathbf{X}$ be an N_λ by N_{meas} data matrix whose columns represent the individual spectral measurements. The goal of the calculation is to determine an N_λ by N_b matrix $\mathbf{B}$ and an N_b by N_{meas} matrix of coefficients $\mathbf{A}$ such that the linear model approximation $\hat{\mathbf{X}} = \mathbf{BA}$ is the best least-squares approximation to the data matrix $\mathbf{X}$ over all possible choices of $\mathbf{B}$ and $\mathbf{A}$.

The process of finding the matrix $\mathbf{B}$ is called *one-mode components analysis.*[56] It is very closely related to the principle components analysis technique discussed in most multivariate statistics texts.[27,57] One-mode components analysis may be accomplished numerically through the use of the singular value decomposition.[25,29] We define the singular value decomposition in App. A. To see how the singular value decomposition is used to determine an N_b dimensional linear model for $\mathbf{X}$, consider Fig. 13. The top part of the figure depicts the singular value decomposition of an N_λ by N_{meas} matrix $\mathbf{X}$ for the case $N_{\text{meas}} > N_\lambda$, where the two matrices $\mathbf{D}$ and $\mathbf{V}^T$ have been collapsed. This form makes it clear that each column of $\mathbf{X}$ is given by a linear combination of the columns of $\mathbf{U}$. Furthermore, for each column of $\mathbf{X}$, the weights needed to combine the columns of $\mathbf{U}$ are given by the corresponding column of the matrix $\mathbf{DV}^T$. Suppose we choose an N_b

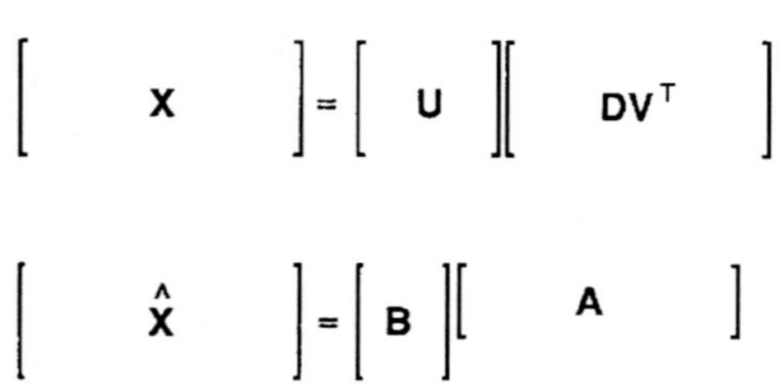

$$\left[\quad \mathbf{X} \quad\right] = \left[\ \mathbf{U}\ \right]\left[\quad \mathbf{DV}^T \quad\right]$$

$$\left[\quad \hat{\mathbf{X}} \quad\right] = \left[\ \mathbf{B}\ \right]\left[\quad \mathbf{A} \quad\right]$$

FIGURE 13 The top part of the figure depicts the singular value decomposition (SVD) of an N_λ by N_{meas} matrix $\mathbf{X}$ for the case $N_{meas} > N_\lambda$. In this view we have collapsed the two matrices $\mathbf{D}$ and $\mathbf{V}^T$. To determine an N_b dimensional linear model $\mathbf{B}$ for the data in $\mathbf{X}$ we let $\mathbf{B}$ consist of the first N_b columns of $\mathbf{U}$. As shown in bottom of the figure, the linear model approximation of the data is given by $\hat{\mathbf{X}} = \mathbf{BA}$ where $\mathbf{A}$ consists of the first N_b rows of $\mathbf{DV}^T$.

dimensional linear model **B** for the data in **X** by extracting the first N_b columns of **U**. In this case, it should be clear that we can form an approximation $\hat{\mathbf{X}}$ to the data **X** as shown in the bottom part of the figure. Because the columns of **U** are orthogonal, the matrix **A** consists of the first N_b rows of $\mathbf{DV}^T$. The accuracy of the approximation $\hat{\mathbf{X}}$ depends on how important the columns of **U** excluded from **B** were to the original expression for **X**. It can be shown that choosing **B** as above produces a linear model that minimizes the approximation error of the linear model, for any choice of N_b.[56] Thus, computing the singular value decomposition of **X** allows us to find a good linear model of any desired dimension for $N_b < N_\lambda$. Computing linear models from data is quite feasible on modern personal computers.

Although the above procedure produces the linear model that provides the best least-squares fit to a data set, there are a number of additional considerations that should go into choosing a linear model. First, we note that the choice of linear model is not unique. Any nonsingular linear combination of the columns of **B** will produce a linear model that provides an equally good account of the data. Second, the least-squares error measure gives more weight to spectra with large amplitudes. In the case of surface spectra, this means that the more reflective surfaces will tend to drive the choice of basis vectors. In the case of illuminants, the more intense illuminants will tend to drive the choice. To avoid this weighting, the measured spectra are sometimes normalized to unit length before performing the singular value decomposition. The normalization equalizes the effect of the relative shape of each spectrum in the data set.[52,58] Third, it is sometimes desired to find a linear model that best describes the variation of a data set around its mean. To do this, the mean of the data set should be subtracted before performing the singular value decomposition. When the mean of the data is subtracted, one-mode components analysis is identical to principle components analysis. Finally, there are circumstances where the linear model will be used not to approximate spectra but rather to approximate some other quantity (e.g., color coordinates) that depend on the spectra. In this case, more general techniques, closely related to those discussed here, may be used.[59]

Approximating a Spectrum with Respect to a Linear Model. Given an N_b dimensional model **B**, it is straightforward to find the representation of any spectrum with respect to the linear model. Let **X** be a matrix representing the spectra of functions to be approximated. These spectra do not need to be members of the data set that was used to determine the linear model. To find the matrix of coefficients **A** such that $\hat{\mathbf{X}} = \mathbf{BA}$ best approximates **X** we use simple linear regression. Regression routines to solve this problem are provided as part of any standard matrix algebra software package.

Digital Image Representations. If, in a given application, illuminants and surfaces may be represented with respect to small-dimensional linear models, then it becomes feasible to use point-by-point representations of these quantities in digital image processing. In typical color image processing, the image data at each point are represented by three numbers at each location. These numbers are generally tristimulus coordinates in some color space. In calibrated systems, side information about the color-matching functions or primary spectral power distributions that define the color space is available to interpret the tristimulus coordinates. It is straightforward to generalize this notion of color images by allowing the images to contain N_b numbers at each point and allowing these numbers to represent quantities other than tristimulus coordinates.[8] For example, in representing the image produced by a printer, it might be advantageous to represent the surface reflectance at each location.[60] If the gamut of printed reflectances can be represented within a small-dimensional linear model, then representing the surface reflectance functions with respect to this model would not require much more storage than a traditional color image.[8] The basis functions for the linear model need be represented only once, not at each location. But by representing reflectances rather than tristimulus values, it becomes possible to compute what the tristimulus coordinates reflected from the printed image would be under any illumination. We illustrate the calculation below. Because of the problem of metamerism, this calculation is not possible if only the tristimulus coordinates are represented in the digital image.

Simulation of Illuminated Surfaces. Consider the problem of producing a signal on a monitor that has the same tristimulus coordinates as a surface under a variety of different illuminants. The solution to this problem is straightforward and is useful in a number of applications. These include rendering digitally archived paintings,[8,61] generating stimuli for use in psychophysics,[62] and producing photorealistic computer-generated imagery.[7] We show the calculation for the data at a single image location. Let **a** be a representation of the surface reflectance with respect to an N_b dimensional linear model **B**. Let **E** represent the illuminant spectral power distribution in diagonal matrix form. Let **T** represent the color-matching functions for a human observer, and **P** represent the primary phosphor spectral power distributions for the monitor on which the surface will be rendered. We wish to determine tristimulus coordinates **t** with respect to the monitor primaries so that the light emitted from the monitor will appear identical to the light reflected from the simulated surface under the simulated illuminant. From Eqs. (2) (cast as $\mathbf{s} = \mathbf{Ba}$), (24), (18), and (19) we can write directly the desired rendering equation

$$\mathbf{t} = ((\mathbf{TP})^{-1}(\mathbf{TE})\mathbf{B})\mathbf{a} \tag{25}$$

The rendering matrix $((\mathbf{TP})^{-1}(\mathbf{TE})\mathbf{B})$ has dimensions 3 by N_b and maps the surface weights directly to monitor tristimulus coordinates. It is quite general, in that we may use it for any calibrated monitor and any choice of linear models. It does not depend on the particular surface being rendered and may be computed once for an entire image. Because the rendering matrix is of small dimension, rendering of this sort is feasible, even for very large images. As discussed earlier in the chapter, it may be possible to determine the matrix $\mathbf{M}_{T,P} = (\mathbf{TP})^{-1}$ directly. A similar shortcut is possible for the matrix $(\mathbf{TE})\mathbf{B}$. Each column of this matrix is the tristimulus coordinates of one linear model basis vector under the illuminant specified by the matrix **E**.

Color Coordinates of Surfaces. Our discussion thus far has emphasized describing the color coordinates of lights. In many applications of colorimetry, it is desirable to describe the color properties of reflective objects. One efficient way to do this, as described above, is to use linear models to describe the full surface reflectance functions. Another possibility is to specify the color coordinates of the light reflected from the surface under standard illumination. This method allows the assignment of tristimulus values to surfaces in an orderly fashion. The CIE has standardized several illuminant spectral power distributions that may be used for this purpose (see following section). Using the procedures defined above, one can begin with the spectral power distribution of the illuminant and the surface reflectance function and from there calculate the desired color-matching coordinates. The relative size of the tristimulus values assigned to a surface depends on its spectral reflectance function and on the illuminant chosen for specification. To factor the intensity of the illuminant out of the surface representation, the CIE specified a normalization of the color coordinates for use with 1931 *XYZ* tristimulus coordinates. This normalization consists of multiplying the computed tristimulus coordinates by the quantity $100/Y_0$, where Y_0 is the Y tristimulus coordinate for the illuminant.

The tristimulus coordinates of a surface provide enough information to match the surface when it is viewed under the illuminant used to compute those coordinates. It is important to bear in mind that two surfaces that have the same tristimulus coordinates under one illuminant do not necessarily share the same tristimulus coordinates under another illuminant. A more complete description can be generated using the linear model approach described above.

Standard Sources of Illumination. The CIE has standardized a number of illuminant

TABLE 9 CIE Illuminants A and D65

Wavelengths are in nm. The data are available at 1-nm increments from 300 nm to 830 nm.[13]

Wavelength	A	D65
400	1.47E+01	8.28E+01
410	1.77E+01	9.15E+01
420	2.10E+01	9.34E+01
430	2.47E+01	8.67E+01
440	2.87E+01	1.05E+02
450	3.31E+01	1.17E+02
460	3.78E+01	1.18E+02
470	4.29E+01	1.15E+02
480	4.82E+01	1.16E+02
490	5.39E+01	1.09E+02
500	5.99E+01	1.09E+02
510	6.61E+01	1.08E+02
520	7.25E+01	1.05E+02
530	7.91E+01	1.08E+02
540	8.60E+01	1.04E+02
550	9.29E+01	1.04E+02
560	1.00E+02	1.00E+02
570	1.07E+02	9.63E+01
580	1.14E+02	9.58E+01
590	1.22E+02	8.87E+01
600	1.29E+02	9.00E+01
610	1.36E+02	8.96E+01
620	1.44E+02	8.77E+01
630	1.51E+02	8.33E+01
640	1.58E+02	8.37E+01
650	1.65E+02	8.00E+01
660	1.72E+02	8.02E+01
670	1.79E+02	8.23E+01
680	1.85E+02	7.83E+01
690	1.92E+02	6.97E+01
700	1.98E+02	7.16E+01

spectral power distributions.[13] These were designed to be typical of various common viewing conditions and are useful as specific choices of illumination when the illuminant cannot be measured directly. CIE illuminant A is designed to be representative of tungsten-filament illumination. CIE illuminant D65 is designed to be representative of average daylight. The relative spectral power distributions of these two illuminants are provided in Table 9. Other CIE standard daylight illuminants may be computed using the basis vectors in Table 7 and formulae specified by the CIE.[11] Spectra representative of fluorescent lamps and other artificial sources are also available.[9,11]

Metamerism

Recovering Spectral Power Distributions from Tristimulus Coordinates. It is not possible in general to recover a spectral power distribution from its tristimulus coordinates. If some a priori information about the spectral power distribution of the color signal is available, however, then recovery may be possible. Such recovery is of most interest in applications where direct spectral measurements are not possible and where knowing the full spectrum is important. For example, the effect of lens chromatic aberrations on cone quantal absorption rates depends on the full spectral power distribution.[45,63]

Suppose the spectral power distribution of interest is known to lie within a three-dimensional linear model. We may write $\mathbf{b} = \mathbf{Ba}$, where the basis matrix $\mathbf{B}$ has dimensions N_λ by 3. Let $\mathbf{t}$ be the tristimulus coordinates of the light with respect to a set of color-matching functions $\mathbf{T}$. Following the development earlier in the chapter, we can conclude that $\mathbf{a} = (\mathbf{TB})^{-1}\mathbf{t}$, which imples

$$\mathbf{b} = \mathbf{B}(\mathbf{TB})^{-1}\mathbf{t} \tag{26}$$

When we do not have a prior constraint that the signal belongs to a three-dimensional linear model, we may still be able to place some linear model constraint, of dimension higher than 3, on the spectral power distribution. For example, when we know that the signal was produced by the reflection of daylight from a natural object, it is reasonable to assume that the color signal lies within a linear model of dimension that may be as low as 9.[64] In this case, we can still write $\mathbf{b} = \mathbf{Ba}$, but we cannot apply Eq. (26) directly because the matrix $(\mathbf{TB})$ will be singular. To deal with this problem, we can choose a reduced linear model $\hat{\mathbf{B}}$ with only three dimensions. We then proceed as outlined above, but substitute the reduced model for the true model. This will lead to an estimate $\hat{\mathbf{b}}$ for the actual spectral power distribution $\mathbf{b}$. If the reduced linear model $\hat{\mathbf{B}}$ provides a reasonable approximation to $\mathbf{b}$, the estimation error may be quite small. The estimate $\hat{\mathbf{b}}$ will have the property that it is a metamer of $\mathbf{b}$. The techniques described above for finding linear model approximations may be used to choose an appropriate reduced model.

Finding Metamers of a Light. It is often of interest to find metamers of a light. We discuss two approaches here. Wyszecki and Stiles[9] treat the problem in considerable detail.

Using a Linear Model. If we choose any three-dimensional linear model $\hat{\mathbf{B}}$ we can combine Eq. (26) with the fact the fact that $\mathbf{t} = \mathbf{Tb}$ [Eq. (9)] to compute a pair of metameric spectral power distributions $\mathbf{b}$ and $\hat{\mathbf{b}}$:

$$\hat{\mathbf{b}} = \hat{\mathbf{B}}(\mathbf{T}\hat{\mathbf{B}})^{-1}\mathbf{Tb} \tag{27}$$

Each choice of $\hat{\mathbf{B}}$ will lead to a different metamer $\hat{\mathbf{b}}$. Figure 14 shows a number of metameric spectral power distributions generated in this fashion.

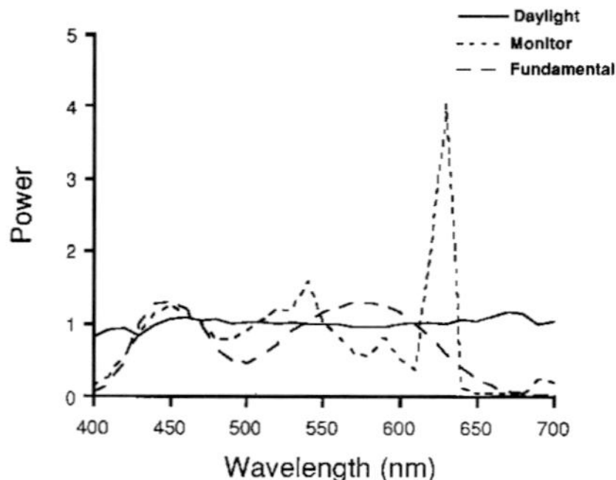

FIGURE 14 The figure shows three metameric color signals with respect to the CIE 1931 standard observer. The three metamers were computed using Eq. (27). The initial spectral power distribution **b** (not shown) was an equal energy spectrum. Three separate linear models were used: one that describes natural daylights, one typical of monitor phosphor spectral power distributions, and one that provides Cohen's "fundamental metamer."

Metameric Blacks. Another approach to generating metamers is to note that there will be some spectral power distributions $\mathbf{b}_0$ that have the property $\mathbf{T}\mathbf{b}_0 = \mathbf{0}$. Wyszecki referred to such distributions as metameric blacks, since they have the same tristimulus coordinates as no light at all.[9,65] Grassmann's laws imply that adding a metameric black $\mathbf{b}_0$ to any light $\mathbf{b}$ yields a metamer of $\mathbf{b}$. Given a linear model $\mathbf{B}$ with dimension greater than 3 it is possible to find a second linear model $\mathbf{B}_0$ such that (1) all lights that lie in $\mathbf{B}_0$ also lie in $\mathbf{B}$ and (2) all lights in $\mathbf{B}_0$ are metameric blacks. We determine $\mathbf{B}_0$ by finding a linear model for the null space of the matrix $\mathbf{TB}$. The null space of a matrix consists of all vectors that are mapped to $\mathbf{0}$ by the matrix. Finding a basis for the null space of a matrix is a standard operation in numerical matrix algebra. If we have a set of basis vectors $\mathbf{N}_0$ for the null space of $\mathbf{TB}$, we can form $\mathbf{B}_0 = \mathbf{B}\mathbf{N}_0$. This technique provides a way to generate a large list of metamers for any given light $\mathbf{b}$. We choose a set of weights $\mathbf{a}$ at random and construct $\mathbf{b}_0 = \mathbf{B}_0\mathbf{a}$. We then add $\mathbf{b}_0$ to $\mathbf{b}$ to form a metamer. To generate more metamers, we simply repeat with new choices of weight vector $\mathbf{a}$.

Surface and Illuminant Metamerism. The formal similarity between Eq. (9) (which gives the relation between spectral power distributions and tristimulus coordinates) and Eq. (24) (which gives the relation between surface reflectance functions and tristimulus coordinates when the illuminant is known) makes it clear that our discussion of metamerism can be applied to surface reflectance spectra. Two physically different surfaces will appear identical if the tristimulus coordinates of the light reflected from them is identical. This fact can be used to good purpose in some color reproduction applications. Suppose that we have a sample surface or textile whose color we wish to reproduce. It may be that we are not able to reproduce the sample's surface reflectance function exactly because of various limitations in the available color reproduction technology. If we know the illuminant under which the reproduction will be viewed, we may be able to determine a reproducible reflectance function that is metameric to that of the desired sample. This will give us a sample whose color appearance is as desired. Applications of this sort make heavy use of the methods described above to determine metamers.

But what if the illuminant is not known or if it is known to vary? In this case there is an additional richness to the topic of determining metamers. We can pose the problem of finding surface reflectance functions that will be metameric to a desired reflectance under multiple specified illuminants or under all of the illuminants within some linear model. The general methods developed here have been extended to analyze this case.[64,66] Similar issues arise in lighting design, where we desire to produce an artificial light whose color-rendering properties match those of a specified light (such as natural daylight). When wavelength-by-wavelength matching of the spectra is not feasible, it may still be possible to find a spectrum so that the light reflected from surfaces within a linear model is identical for the two light sources. Because of the symmetric role of illuminants and surfaces in reflection, this problem may be treated by the same methods used for surface reproduction.

Cohen's "Matrix R." Cohen and Kappauf[21,67,68] have proposed that a useful way to associate a spectral power distribution with a set of tristimulus coordinates is to choose as basis functions the color-matching functions themselves. That is, we choose the matrix $\hat{\mathbf{B}} = \mathbf{T}^T$. When $\hat{\mathbf{B}}$ is chosen in this way, the estimated color signal from Eq. (26) is given by $\hat{\mathbf{b}} = \mathbf{T}^T(\mathbf{T}\mathbf{T}^T)^{-1}\mathbf{t}$. From Eq. (27) we also have $\hat{\mathbf{b}} = \mathbf{T}^T(\mathbf{T}\mathbf{T}^T)^{-1}\mathbf{T}\mathbf{b}$. Cohen and Kappauf refer to this $\hat{\mathbf{b}}$ as the "fundamental metamer" of $\mathbf{b}$. The matrix $\mathbf{T}^T(\mathbf{T}\mathbf{T}^T)^{-1}\mathbf{T}$ is often referred to as "matrix R." It is easy to show that "matrix R" is invariant when a linear transformation is applied to the color-matching functions. There is no reason to believe the "fundamental metamer" will be a good estimate (in a least-squares sense) of the original spectral power distribution (see Fig. 14).

Color Cameras and Other Visual Systems

We have treated colorimetry from the point of view of specifying the spectral information available to a human observer. We have developed our treatment, however, in such a way that it may be applied to handle other visual systems. Suppose that we wish to define color coordinates with respect to some arbitrary visual system with N_{device} photosensors. This visual system might be an artificial system based on a color camera or scanner, a nonhuman biological visual system, or the visual system of a color-anomalous human observer. We assume that the sensitivities of the visual system's photosensors are known up to a linear transformation. Let $\mathbf{T}_{\text{device}}$ be an $\mathbf{N}_{\text{device}}$ by $\mathbf{N}_{\lambda}$ matrix whose entries are the sensitivities of each sensor at each sample wavelength. We can compute the responses of these sensors to any light $\mathbf{b}$. Let $\mathbf{t}_{\text{device}}$ be a vector containing the responses of each sensor type to the light. Then we have $\mathbf{t}_{\text{device}} = \mathbf{T}_{\text{device}}\mathbf{b}$. We may use $\mathbf{t}_{\text{device}}$ as the device color coordinates of $\mathbf{b}$.

Transformation Between Color Coordinates of Different Visual Systems. Suppose that we have two different visual systems and we wish to transform between the color coordinates of each. A typical example might be trying to compute the CIE 1931 *XYZ* tristimulus coordinates of a light from the responses of a color camera. Let N_s be the number of source sensors, with sensitivities specified by $\mathbf{T}_s$. Similarly, let N_d be the number of destination sensors with sensitivities specified by $\mathbf{T}_d$. For any light $\mathbf{b}$ we know that the source device color coordinates are given by $\mathbf{t}_s = \mathbf{T}_s\mathbf{b}$ and the destination device color coordinates $\mathbf{t}_d = \mathbf{T}_d\mathbf{b}$. We would like to transform between $\mathbf{t}_s$ and $\mathbf{t}_d$ without direct knowledge of $\mathbf{b}$.

If we can find an N_d by N_s matrix $\mathbf{M}$ such that $\mathbf{T}_d = \mathbf{M}\mathbf{T}_s$, then it is easy to show that the matrix $\mathbf{M}$ may be used to compute the destination device color coordinates from the source device color coordinates through $\mathbf{t}_d = \mathbf{M}\mathbf{t}_s$. We have already considered this case (in a less general form). The extension here is that we allow the possibility that the dimensions of the two color coordinate systems differ. When a linear transformation between $\mathbf{T}_s$ and $\mathbf{T}_d$ exists, it can be found by standard regression methods.

Horn demonstrated that when no exact linear transformation between $\mathbf{T}_s$ and $\mathbf{T}_d$ exists, it is not, in general, possible to transform between the two sets of color coordinates.[4] The reason for this is that there will always exist a pair of lights that have the same color coordinates for the source device but different color coordinates for the destination device. The transformation will therefore be incorrect for at least one member of this pair. When no exact linear transformation exists, it is still possible to make an approximate transformation. One approach is to use linear regression to find the best linear transformation $\mathbf{M}$ between the two sets of color-matching functions in a least-squares sense. This transformation is then applied to the source color coordinates as if it were exact.[4] Although this is an approximation, in many cases the results will be acceptable. In the absence of prior information about the spectral power distribution of the original light $\mathbf{b}$, it is a sensible approach.

A second possibility is to use prior constraints on the spectral power distribution of the light to guide the transformation.[20,69] Suppose that we know that the light is constrained to lie within an N_b dimensional linear model $\mathbf{B}$. Then we can find the best linear transformation $\mathbf{M}$ between the two matrices $\mathbf{T}_s\mathbf{B}$ and $\mathbf{T}_d\mathbf{B}$. This transformation may then be used to transform the source color coordinates to the destination color coordinates. It is easy to show that the transformation will be exact if $\mathbf{T}_d\mathbf{B} = \mathbf{M}\mathbf{T}_s\mathbf{B}$. Otherwise, it is a reasonable approximation that takes the linear model constraint into account.

Computational Color Constancy. An interesting application is the problem of estimating surface reflectance functions from color coordinates. This problem is of interest for two reasons. First, it appears that human color vision makes some attempt to perform this estimation, so that our percept of color is more closely associated with object surface properties than with the proximal properties of the light reaching the eye. Second, an artificial system that could estimate surface properties would have an important cue to aid object recognition. In the case where the illuminant is known, the problem of estimating surface reflectance properties is the same as the problem of estimating the color signal, because the illuminant spectral power distribution can simply be incorporated into the sensor sensitivities. In this case the methods outlined above for estimating color signal spectral properties can be used.

The more interesting case is where both the illuminant and the surface reflectance are unknown. In this case, the problem is more difficult. Considerable insight has been gained by applying linear model constraints to both the surface and illuminant spectral power distributions. In the past decade, a large number of approaches have been developed for recovering surface reflectance functions.[20,64,70–78] Each approach differs (1) in the additional assumptions that are made about the properties of the image and (2) in the sophistication of the model of illuminant surface interaction and scene geometry used. A thorough review of all of these methods is beyond the scope of this chapter. It is instructive, however, to review one of the simpler methods, that of Buchsbaum.[72]

Buchsbaum assumed that in any given scene, the average reflectance function of the surfaces in the scene is known. This is commonly called the "gray world" assumption. He also assumed that the illuminant was diffuse and constant across the scene and that the illuminants and surfaces in the scene are described by linear models with the same dimensionality as the number of sensors. Let $\mathbf{S}_{\text{avg}}$ be the spectral power distribution of the known average surface, represented in diagonal matrix form. Then it is possible to write the relation between the space average of the sensor responses and the illuminant as

$$\mathbf{t}_{\text{avg}} = \mathbf{T}\mathbf{S}_{\text{avg}}\mathbf{B}_e\mathbf{a}_e \tag{28}$$

where $\mathbf{a}_e$ is a vector containing the weights of the illuminant within the linear model representation $\mathbf{B}_e$. Because we assume that the dimension $N_e = N_t$, the matrix $\mathbf{T}\mathbf{S}_{\text{avg}}\mathbf{B}_e$ will be square and typically may be inverted. From this we recover the illuminant as $\mathbf{e} = \mathbf{B}_e(\mathbf{T}\mathbf{S}_{\text{avg}}\mathbf{B}_e)^{-1}\mathbf{t}_{\text{avg}}$. If we let $\mathbf{E}$ represent the recovered illuminant in matrix form, then at each image location we can write

$$\mathbf{t} = \mathbf{T}\mathbf{E}\mathbf{B}_s\mathbf{a}_s \tag{29}$$

where $\mathbf{a}_s$ is a vector containing the weights of the surface within the linear model representation $\mathbf{B}_s$. Proceeding exactly as we did for the illuminant, we may recover the surface reflectance from this equation.

Although Buchsbaum's method depends on rather strong assumptions about the nature of the scene, subsequent algorithms have shown that these assumptions can be weakened. Maloney and Wandell demonstrated that the gray world assumption can, under certain circumstances, be relaxed.[20,73] Several recent reviews emphasize the relation between computational color constancy and the study of human vision.[79–82]

Color Discrimination

Measurement of Small Color Differences. Our treatment so far has not included any discussion of the precision to which observers can judge identity of color appearance. To specify tolerances for color reproduction, it would be helpful to know how different the color coordinates of two lights must be for an observer to reliably distinguish between them. A number of techniques are available for measuring human ability to discriminate between colored lights.

One method, employed in seminal work by MacAdam,[83,84] is to examine the variability in individual color matches. That is, if we have observers set matches to the same test stimulus, we will discover that they do not always set exactly the same values. Rather, there will be some trial-to-trial variability in the settings. MacAdam and others[85,86] used the sample covariance of the individual match tristimulus coordinates as a measure of observers' color discrimination.

A second approach is to use psychophysical methods (see Vol. I, Chap. 39) to measure observers' thresholds for discriminating between pairs of colored lights. Examples include increment threshold measurements for monochromatic lights[87] and thresholds measured systematically in a three-dimensional color space.[88,89]

Measurements of small color differences are often summarized with isodiscrimination contours. An isodiscrimination contour specifies the color coordinates of lights that are equally discriminable from a common standard light. Figure 15 shows an illustrative isodiscrimination contour. Isodiscrimination contours are often modeled as ellipsoids[88,89] and the figure is drawn to the typical ellipsoidal shape. The well-known MacAdam ellipses[83] are an example of representing discrimination data using the chromaticity coordinates of a cross section of a full three-dimensional isodiscrimination contour (see the legend of Fig. 15).

CIE Uniform Color Spaces. Figure 16 shows chromaticity plots of representative isodiscrimination contours. A striking feature of the plots is that the size and shape of the contours depends on the standard stimulus. For this reason, it is not possible to predict whether two lights will be discriminable solely on the basis of the Euclidean distance between their color coordinates. The heterogeneity of the isodiscrimination contours must also be taken into account.

The CIE provides two sets of formulae that may be used to predict the discriminability of colored lights. Each set of formulae specifies a nonlinear transformation from CIE 1931 *XYZ* color coordinates to a new set of color coordinates. Specifically, the *XYZ* coordinates may be transformed either to CIE 1976 L*u*v* (CIELUV) color coordinates or to CIE

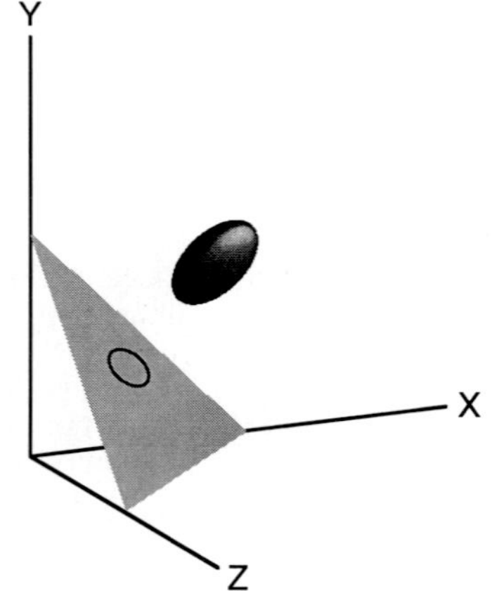

FIGURE 15 Isodiscrimination contour. The plotted ellipsoid shows a hypothetical isodiscrimination contour in the CIE *XYZ* color space. This contour represents color discrimination performance for the standard light whose color coordinates are located at the ellipsoid's center. Isodiscrimination contours such as the one shown are often summarized by a two-dimensional contour plotted on a chromaticity diagram (see Fig. 16). The two-dimensional contour is obtained form a cross section of the full contour, and its shape can depend on which cross section is used. This information is not available directly from the two-dimensional plot. A common criterion for choice of cross section is isoluminance. The ellipsoid shown in the figure is schematic and does not represent actual human performance.

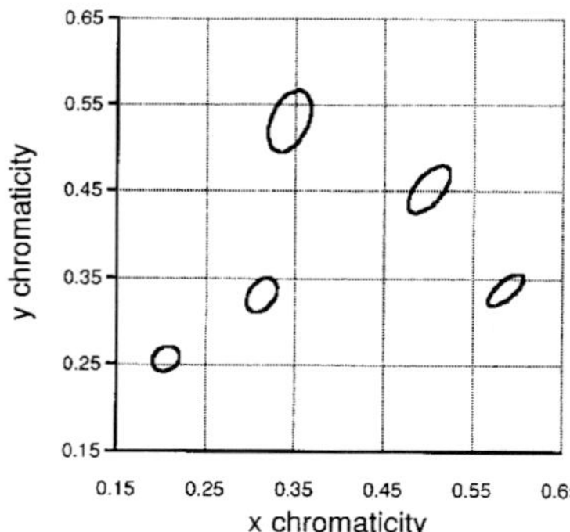

FIGURE 16 Isodiscrimination contours plotted in the chromaticity diagram. These were computed using the CIE L*u*v* uniform color space and provide an approximate representation of human performance. For each standard stimulus, the plotted contour represents the color coordinated of lights that differ from the standard by 15 ΔE^*_{uv} units but that have the same luminance as the standard. The choice of 15 ΔE^*_{uv} units magnifies the contours compared to those that would be obtained in a threshold experiment.

1976 L*a*b* (CIELAB) color coordinates. Both transformations stretch the *XYZ* color space so that the resulting Euclidean distance between color coordinates provides an approximation to the how well lights may be discriminated. Both the L*u*v* and L*a*b* systems are referred to as uniform color spaces. A more complete description of these color spaces is available elsewhere.[9,11] We provide the basic formulae and some discussion below.

Transformation to CIELUV Coordinates. The CIE 1976 L*u*v* color coordinates of a light may be obtained from its CIE *XYZ* coordinates according to the equations

$$L^* = \begin{cases} 116\left(\dfrac{Y}{Y_n}\right)^{1/3} - 16 & \dfrac{Y}{Y_n} > 0.008856 \\ 903.3\left(\dfrac{Y}{Y_n}\right) & \dfrac{Y}{Y_n} \leq 0.008856 \end{cases}$$

$$u^* = 13L^*\left(\frac{4X}{(X + 15Y + 3Z)} - \frac{4X_n}{(X_n + 15Y_n + 3Z_n)}\right)$$

$$v^* = 13L^*\left(\frac{9Y}{(X + 15Y + 3Z)} - \frac{9Y_n}{(X_n + 15Y_n + 3Z_n)}\right) \tag{30}$$

In this equation, the quantities X_n, Y_n, and Z_n are the tristimulus coordinates of a white point. Little guidance is available as to how to choose an appropriate white point. In the case where the lights being judged are formed when an illuminant reflects from surfaces, the tristimulus coordinates of the illuminant may be used. In the case where the lights being judged are on a computer-controlled color monitor, the sum of the tristimulus coordinates of the three monitor phosphors stimulated at their maximum intensity may be used.

Transformation to CIELAB Coordinates. The CIE 1976 L*a*b* color coordinates of a light may be obtained from its CIE *XYZ* coordinates according to the equations

$$L^* = \begin{cases} 116\left(\dfrac{Y}{Y_n}\right)^{1/3} - 16 & \dfrac{Y}{Y_n} > 0.008856 \\ 903.3\left(\dfrac{Y}{Y_n}\right) & \dfrac{Y}{Y_n} \leq 0.008856 \end{cases}$$

$$a^* = 500\left[f\left(\frac{X}{X_n}\right) - f\left(\frac{Y}{Y_n}\right)\right]$$

$$b^* = 200\left[f\left(\frac{Y}{Y_n}\right) - f\left(\frac{Z}{Z_n}\right)\right] \tag{31}$$

where the function $f(s)$ is defined as

$$f(s) = \begin{cases} (s)^{1/3} & s > 0.008856 \\ 7.787(s) + \frac{16}{116} & s \leq 0.008856 \end{cases} \tag{32}$$

As with the CIELUV transformation, the quantities X_n, Y_n, and Z_n are the tristimulus coordinates of a white point.

Distance in CIELUV and CIELAB Spaces. In both the CIELUV and CIELAB color spaces, the Euclidean distance between the coordinates of two lights provides a rough guide to their discriminability. The symbols ΔE^*_{uv} and ΔE^*_{ab} are used to denote distance in the two uniform color spaces and are defined as

$$\Delta E^*_{uv} = \sqrt{(\Delta L^*)^2 + (\Delta u^*)^2 + (\Delta v^*)^2}$$

$$\Delta E^*_{ab} = \sqrt{(\Delta L^*)^2 + (\Delta a^*)^2 + (\Delta b^*)^2} \tag{33}$$

where the various Δ quantities on the right represent the differences between the corresponding coordinates of the two lights. Roughly speaking, a ΔE^*_{uv} or ΔE^*_{ab} value of 1 corresponds to a color difference that can just be reliably discerned by a human observer under optimal viewing conditions. A ΔE^*_{uv} or ΔE^*_{ab} value of 3 is sometimes used as an acceptable tolerance in industrial color reproduction applications.

Limits of the CIE Uniform Color Spaces. The two CIE color difference measures ΔE^*_{uv} and ΔE^*_{ab} provide only an approximate guide to the discriminability between two lights. There are a number of reasons why this is so. One major reason is that the formulae were designed not only to predict discrimination data but also certain suprathreshold judgments of color appearance.[90] A second important reason is that color discrimination thresholds depend heavily on factors other than the tristimulus coordinates. These factors include the adapted state of the observer,[87] the spatial and temporal structure of the stimulus,[91–93] and the task demands placed on the observer.[94–97] Therefore, the complete specification of a uniform color space must incorporate these factors. At present, a model of visual performance that would allow such incorporation is not available. The transformations to CIELUV and CIELAB spaces do, however, depend on the choice of white point. This dependence is designed to provide some compensation for the adapted state of the observer.

Limits of the Color-Matching Experiment

Specifying a stimulus using tristimulus coordinates depends on having an accurate set of color-matching functions. The various standard color spaces discussed earlier under "Common Color Coordinate Systems" are designed to be representative of an "average" or standard observer under typical viewing conditions. A number of factors limit the precision to which a standard color space can predict individual color matches. We describe some of these factors below. Wyszecki and Stiles[9] provide a more detailed treatment.

For most applications, standard calculations are sufficiently precise. When high precision is required, it is necessary to tailor a set of color-matching functions to the individual and observing conditions of interest. Once such a set of color-matching functions is available, the techniques described in this chapter may be used to compute corresponding color coordinates.

Individual Differences Among Color-Normal Observers. Standard sets of color-matching functions are summaries of color-matching results for a number of color-normal observers. There is small but systematic variability between the matches set by individual observers, and this variability limits the precision to which standard color-matching functions may be taken as representative of any given color-normal observer. A number of factors underlie the variability in color matching. Stiles and Burch carefully measured color-matching functions for 49 observers using 10° fields.[98,99] Webster and MacLeod analyzed individual variation in these color-matching functions.[100] They identified six primary factors that drive the variation in individual color matches. These are macular pigment density, lens pigment density, amount of rod intrusion into the matches (discussed shortly), and variability in the absorption spectra of the L, M, and S cone photopigments.

Lens pigment density is known to increase over the life span of an individual, resulting in systematic differences in color-matching functions between populations of different ages.[101] The nature of the mechanisms underlying the variability in cone photopigment absorption spectra is a matter of considerable current interest.

Color-Deficient and Color-Anomalous Observers. Some individuals require only two (or in rare cases only one) primaries in the color-matching experiment. These individuals are referred to as color-deficient or color-blind observers. Most forms of color deficiency can be understood by assuming that the individual lacks one (or more) of the normal three types of cone photopigment. For these individuals, use of standard color coordinates will produce acceptable results, since a match for all three cone types will also be a match for any subset of these types. In very rare cases, an individual has no cones at all and the vision of such an individual is mediated entirely by rods. A second class of individuals are trichomatic but set color matches substantially different from color-normal observers. These individuals are referred to as color anomalous. The leading hypothesis about the cause of color anomaly is that the individuals have photopigments with spectral sensitivities substantially different from individuals with normal color vision.[102] Our development of colorimetry can be used to tailor color specification for color-anomalous individuals if their color-matching functions are known. Estimates of the cone sensitivities of color-anomalous observers are available.[37] Simple standard tests exist for identifying color-blind and color-anomalous individuals. These include the Ishihara pseudo-isochromatic plates[103] and the Farnsworth 100 hue test.[104]

Cone Polymorphism. Recent genetic and behavioral evidence suggests that there are multiple types of human L and possibly human M cone photopigments.[105–107] This possibility is referred to as cone polymorphism. Moreover, Neitz, Neitz, and Jacobs[108] argue that some individuals possess more than three types of cone photopigments. This claim challenges the conventional explanation for trichromacy and is controversial. The interested reader is referred directly to the current literature. Because the purported difference in spectral sensitivity of different subclasses of human L and M cone photopigments is quite small, the possibility of cone polymorphism is not of practical importance for most applications. The most notable exception is in certain psychophysical experiments where precise knowledge of the relative excitation of different cone classes is crucial.

Retinal Inhomogeneity. Most standard colorimetric systems are based on color-matching experiments where the bipartite field was small and viewed foveally. The distribution of photoreceptors and of inert visual pigment is not homogeneous across the retina. Thus color-matching functions that are accurate for the fovea do not necessarily describe color matching in the extra fovea. The CIE 1964 10° XYZ color-matching functions are designed for situations where the colors being judged subtend a large visual angle. These functions are provided in Table 3.

Rod Intrusion. Both outside the fovea and at low light levels, rods can play a role in color matching. Under conditions where rods play a role, there is a shift in the color-matching functions due to the contribution of rod signals. Wyszecki and Stiles[9] discuss approximate methods for correcting standard sets of color-matching functions when rods intrude into color vision.

Chromatic Aberrations. By some standards, even the small (roughly 2°) fields used as the basis of most color coordinate systems are rather coarse. The optics of the eye contain chromatic aberrations which cause different wavelengths of light to be focused with different accuracy. These aberrations can cause a shift in the color-matching functions if the stimuli being matched have a fine spatial structure. Two stimuli which are metameric at low spatial frequencies may no longer be so at high spatial frequencies. Such effects can be quite large.[45,63] It is possible to correct color coordinates for chromatic aberration if enough side information is available. Such correction is rare in practice but can be important for stimuli with a fine spatial structure. Some guidance is available from the literature.[63,109] Another strategy available in the laboratory is to correct the stimulus for the chromatic aberration of the eye.[110]

Pigment Self-Screening. One of the factors that determines the cone sensitivities is that the photopigment itself acts as an inert filter. In any individual cone, the spectral properties of this filter depend on the fraction of photopigment that has recently been isomerized by light quanta. As the overall intensity of the stimulus changes, this fraction changes, which changes the cones' sensitivity functions. Although such shifts may generally be neglected, they can become quite important under circumstances where very intense adapting fields are employed.[18]

Calculating the Effect of Errors in Color-Matching Functions. Given that there is some variation between different standard estimates of color-matching functions, between the color-matching functions of different individuals, and between the color-matching functions that mediate performance for different viewing conditions, it is of interest to determine whether the magnitude of this variation is of practical importance. There is probably no general method for making this determination, but here we outline one approach.

Consider the case of rendering a set of illuminated surfaces on a color monitor. If we know the spectral power distribution of the monitor's phosphors it is possible to compute the appropriate weights on the monitor phosphors to produce a light metameric to each illuminated surface. The computed weights will depend on the choice of color-matching functions. Once we know the weights, however, we can find the CIE 1976 L*u*v* (or CIE 1976 L*a*b*) coordinates of the emitted light. This suggests the following method to estimate the effect of differences in color-matching functions. First, we compute the CIE 1976 L*u*v* coordinates of surfaces rendered using the first set of color-matching functions. Then we compute the corresponding coordinates when the surfaces are rendered using the second set of color-matching functions. Finally, we compute the ΔE^*_{uv} difference between corresponding sets of coordinates. If the ΔE^*_{uv} are large, then the differences between the color-matching functions are important for the rendering application.

We have performed this calculation for a set of 462 measured surfaces[50,51] rendered under CIE Illuminant D_{65}. The two sets of color-matching functions used were the 1931 CIE *XYZ* color-matching functions and the Judd-Vos modified *XYZ* color-matching functions. The monitor phosphor spectral power distributions were measured by the author.[111] The results are shown in Fig. 17. The plot shows a histogram of the ΔE^*_{uv}

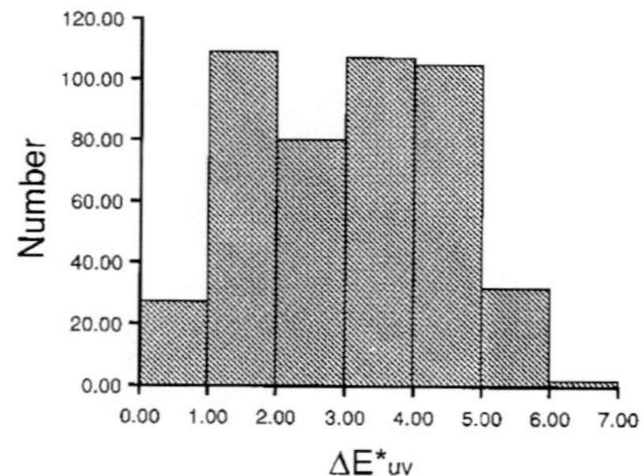

FIGURE 17 Effect of changes in color matching functions. The plot shows a histogram of the ΔE^*_{uv} differences between two sets of lights, each of which is a monitor rendering of the same set of illuminated surfaces. The two renderings were computed using different sets of color matching functions.

differences. The median difference is 3.1 ΔE^*_{uv} units. This difference is quite close to discrimination threshold and, for most applications, the differences between the two sets of color-matching functions will probably not be of great consequence.

Color Appearance

Context Effects. A naive theory of color appearance would predict that once we know the tristimulus coordinates of a light, it would be straightforward to predict its color appearance. Unfortunately, such a theory does not work. The color appearance of a light depends not only on its spectral properties but also on the context in which it is seen. By context, we refer to other attributes of the light itself, such as its size, shape, and temporal profile, to the properties of the lights that surround the light of interest, and to the adapted state of the observer. The adapted state of the observer is presumably determined by the stimuli viewed in the recent past. Some are even willing to argue that more intangible factors, such as the observer's expectations, also influence color appearance.

Wyszecki[3] and Evans[2] each review color context effects. Among the context phenomenon given the most study are color contrast, where the properties of nearby image regions affects the color appearance of a test, and observer adaptation, where the images that an observer has viewed influences his or her judgment of a subsequently viewed test. Empirically, adaptation and contrast are somewhat difficult to separate. Indeed, the two are difficult to separate conceptually, since adaptation can be viewed as a delayed contrast effect and contrast can be considered as rapidly acting adaptation. Although phenomena with names such as simultaneous color contrast, successive color contrast, color assimilation, and observer adaptation are often discussed and considered separately, there is some potential for clarity by considering the possibility that they are all manifestations of similar processes.

The fact that the color-matching experiment, from which color coordinates are derived, does not predict color appearance may seem surprising. After all, the fundamental judgment of the color-matching experiment concerns the identity of color appearance. Recall, however, that the test and matching regions are always juxtaposed, so that they are seen in identical context. As long as the context in which two lights are seen is identical, color coordinates do correctly predict color appearance.

Context effects make color reproduction difficult. If we wish to reproduce an image so that it appears identical to another image when the two will be viewed in different contexts, we must take the effect of context into account. This sort of consideration is of particular importance in color photography and television, since in these cases the reproduced image has a vastly different size from the image it depicts.[2,5,112,113] Two points are worth noting. First, if we can arrange two images so that they have identical tristimulus coordinates at each image location, then we can be assured that each presents the same context to the viewer and that the two images will match. This principle is the basis of many applied color reproduction systems and is appropriate when the source and destination images have roughly equal sizes and will be viewed under similar conditions.[6] Second, tristimulus coordinates are still of great importance when the context does vary. This is because the persistence law of color matching tells us that the context itself may be specified through the use of tristimulus coordinates. We do not need to specify the context in terms of the full spectral power distributions of the lights. Thus the basic color-matching experiment provides the foundation on which we can build theories of color appearance. A theory that successfully links tristimulus coordinates with some sort of appearance coordinates, as a function of the viewing context, would provide the quantitative framework necessary to take context into account in colorimetric applications. The search for such a theory has been at the center of color science since at least the time of von Kries.[1] Wyszecki[3] and Wyszecki and Stiles[9] review of the measurement of the effect of context on color appearance.

Color Constancy. A hypothesis about the origin of color context effects is that the visual system attempts to discount the properties of the illuminant so that color appearance is correlated with object surface reflectance. To the extent that the visual system does this, it is referred to as color constant. It has long been known that the human visual system exhibits partial color constancy.[2,114,115] The phenomenon of color constancy suggests that the spectral power distribution of the illuminant is an important parameter of the viewing context for predicting color appearance.[116] Land's retinex theory was an early account of how this parameter might affect appearance.[117–119] More recent work has examined the relation between physics-based color constancy algorithms and color appearance.[62,79–81,116,120]

Sensitivity Regulation. Another framework for thinking about color context effects has to do with the visual system's tendency to adjust its sensitivity to the ambient viewing conditions. If, as seems likely, the signals mediating the visual system's sensitivity also mediate color appearance, then we would expect sensitivity regulation to cause context effects in color appearance. Regulation of visual sensitivity is discussed in Chap. 00. It is possible that similar mechanisms could subserve both maximization of sensitivity and the achievement of color constancy.[40]

Brightness Matching and Photometry

The foundation of colorimetry is the human observer's ability to judge identity of color appearance. It is sometimes of interest to compare certain perceptual attributes of lights that do not, as a whole, appear identical. In particular, there has been a great deal of interest in developing formulae that predict when two lights with different relative spectral power distributions will appear equally bright. Colorimetry provides a partial answer to this question, since two lights that match in appearance must appear equally bright. Intuitively, however, it seems that it should be possible to set the relative intensities of any two lights so that they match in brightness.

In a heterochromatic brightness-matching experiment, observers are asked to scale the intensity of a matching light until its brightness matches that of an experimentally controlled test light. Although observers can perform the heterochromatic brightness-matching task, they often report that it is difficult and their matches tend to be highly variable.[9] For this reason, more indirect methods for equating the overall effectiveness of lights at stimulating the visual system have been developed. The most commonly used method is that of flicker photometry. In a flicker photometric experiment, two lights of different spectral power distributions are presented alternately at the same location. At moderate flicker rates (about 20 Hz), subjects are able to adjust the overall intensity of one of the lights to minimize the apparent flicker. The intensity setting that minimizes apparent flicker is taken to indicate that the two lights match in their effectiveness as visual stimuli. Two lights equated in this way are said to be equiluminant or to have equal luminance. Other methods are available for determining when two lights have the same luminance.[9]

Because experiments for determining when lights have the same luminance obey linearity properties similar to Grassmann's laws, it is possible to determine a luminous efficiency function that allows the assignment of a luminance value to any light. A luminous efficiency function specifies, for each sample wavelength, the relative contribution of that wavelength to the overall luminance. We can represent a luminous efficiency function as an N_λ dimensional row vector $\mathbf{v}$. Each entry of the matrix specifies the relative luminance of light at the corresponding sample wavelength. The luminance v of an arbitrary spectral power distribution $\mathbf{b}$ may be computed by the equation

$$v = \mathbf{v}\mathbf{b} \tag{34}$$

The CIE has standardized four luminous efficiency functions. The most commonly used of these is the standard photopic luminous efficiency function. This is identical to the 1931 XYZ $\bar{y}$ color-matching function. For lights that subtend more than 4° of visual angle, a luminous efficiency function given by the 1964 10° XYZ $\bar{y}$ color-matching function is preferred. More recently, the Judd-Vos modified $\bar{y}$ color-matching function has been made a supplemental standard.[121] A final standard luminous efficiency function is available for use at low light levels when the rods are the primary functioning photoreceptors. The symbol V_λ is often used in the literature to denote luminous efficiency functions. Note that Eq. (34) allows the computation of luminance in arbitrary units. Chapter 24, Vol. 2 of this *Handbook* discusses standard measurement units for luminance.

It is important to note that luminance is a construct derived from flicker photometric and related experiments. As such, it does not directly predict when two lights will be judged to have the same brightness. The relation between luminance and brightness is quite complicated and is the topic of active research.[9,10] It is also worth noting that there is considerable individual variation in flicker photometric judgments, even among color-normal observers. For this reason, it is common practice in psychophysical experiments to use flicker photometry to determine isoluminant stimuli for individual subjects.

Opponent Process Model

We conclude with remarks on what mechanisms might process color information after the initial transduction of light by the cone photoreceptors. Figure 18 shows a framework that in its general form underlies most current thinking. This framework is generally referred to as an opponent process model. It was first proposed in its modern form by Jameson and Hurvich.[122–124] The first stage of the model is cone photoreceptor transduction. We have already considered the implications of this stage in some detail. The L, M, and S cone responses are indicated by the triangles in the figure. After transduction, the opponent process model postulates that the outputs of the cones at each location are recombined to produce a luminance response and two chromatic responses. This recombination is often referred to as an opponent transformation, because the two chromatic responses are postulated to receive excitatory responses from some classes of cones and inhibitory responses from other classes of cones. The luminance and chromatic responses are illustrated by squares in the figure. We have denoted the luminance responses as "Lum" and the two chromatic responses as "R/G" and "B/Y," respectively. The opponent transformation itself is illustrated by a network of connections.

The opponent process framework provides a way to understand the results of flicker

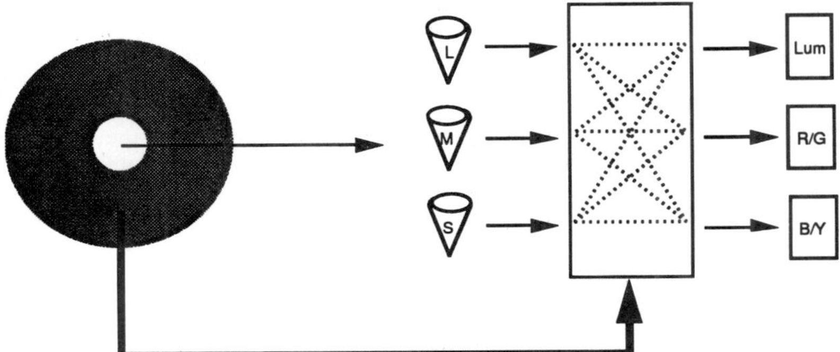

FIGURE 18 Opponent process framework for understanding color vision. See description in the text.

photometric and related experiments. Within the opponent process framework, these experiments are explained as indicating a judgment that depends solely on the luminance response, separately from the two chromatic responses. This interpretation is lent credence by the fact that observers can also make judgments that may be interpreted as tapping solely the chromatic mechanisms.[122,124–127] The opponent process framework has also been employed heavily to account for the effect of context on color appearance.[62,128–132] The idea is that the parameters of the opponent transformation depend on the viewing context. This is illustrated in the figure by the heavy arrow between the annulus that surrounds the test light and the network representing the transformation.

Although the opponent process model provides a useful framework, a number of difficult issues need to be resolved before the framework may be used to extend current colorimetric standards. One major difficulty is that the details of the opponent transformations required to explain different classes of data are not in good agreement. A second difficulty is that physiological data do not seem to support such a simple conception of biological mechanisms that mediate the processing of color information.[133,134]

26.5 APPENDIX A. MATRIX ALGEBRA

This appendix provides a brief introduction to matrix algebra. The development emphasizes the aspects of matrix algebra that are used in this chapter and is somewhat idiosyncratic. In addition, we do not prove any of the results we state. Rather, our intention is to provide the reader unfamiliar with matrix algebra with enough information to make this chapter self-contained.

Basic Notions

Vectors and matrices. A vector is a list of numbers. We use lowercase bold letters to represent vectors. We use single subscripts to identify the individual entries of a vector. The entry a_i refers to the ith number in $\mathbf{a}$. We call the total number of entries in a vector its dimension.

A matrix is an array of numbers. We use uppercase bold letters to represent matrices. We use dual subscripts to identify the individual entries of a matrix. The entry a_{ij} refers to the number in the ith row and jth column of $\mathbf{A}$. We sometimes refer to this as the ijth entry of $\mathbf{A}$. We call the number of rows in a matrix its row dimension. We call the number of columns in a matrix its column dimension. We generally use the symbol N to denote dimensions.

Vectors are a special case of matrices where either the row or the column dimension is 1. A matrix with a single column is often called a column vector. A matrix with a single row is often called a row vector. By convention, all vectors used in this chapter should be understood to be column vectors unless explicitly noted otherwise.

It is often convenient to think of a matrix as being composed of vectors. For example, if a matrix has dimensions N_r by N_c, then we may think of the matrix as consisting of N_c column vectors, each of which has dimension N_r.

Addition and Multiplication. A vector may be multiplied by a number. We call this scalar multiplication. Scalar multiplication is accomplished by multiplying each entry of the vector by the number. If $\mathbf{a}$ is a vector and b is a number, then $\mathbf{b} = \mathbf{a}b$ is a vector whose entries are given by $c_j = b_j a$.

Two vectors may be added together if they have the same dimension. We call this vector addition. Vector addition is accomplished by entry-by-entry addition. If $\mathbf{a}$ and $\mathbf{b}$ are vectors with the same dimension, the entries of $\mathbf{c} = \mathbf{a} + \mathbf{b}$ are given by $c_j = a_j + b_j$.

Two matrices may be added if they have the same row and column dimensions. We call this matrix addition. Matrix addition is also defined as entry-by-entry addition. Thus, if $\mathbf{A}$ and $\mathbf{B}$ are matrices with the same dimension, the entries of $\mathbf{C} = \mathbf{A} + \mathbf{B}$ are given by $c_{ij} = a_{ij} + b_{ij}$. Vector addition is a special case of matrix addition.

A vector may be multiplied by a matrix if the column dimension of the matrix matches the dimension of the vector. If $\mathbf{A}$ has dimensions N_r by N_c and $\mathbf{b}$ has dimension N_c, then $\mathbf{c} = \mathbf{Ab}$ is an N_r dimensional vector. The ith entry of $\mathbf{c}$ is related to the entries of $\mathbf{A}$ and $\mathbf{b}$ by the equation.

$$c_i = \sum_{j=1}^{N_c} a_{ij} b_j \tag{A.1}$$

It is also possible to multiply a matrix by another matrix if the column dimension of the first matrix matches the row dimension of the second matrix. If $\mathbf{A}$ has dimensions N_r by N and $\mathbf{B}$ has dimensions N by N_c, then $\mathbf{C} = \mathbf{AB}$ is an N_r by N_c dimensional matrix. The ik^{th} entry of $\mathbf{C}$ is related to the entries of $\mathbf{A}$ and $\mathbf{B}$ by the equation

$$c_{ij} = \sum_{j=1}^{N} a_{ij} b_{jk} \tag{A.2}$$

By comparing Eqs. (A.1) and (A.2) we see that multiplying a matrix by a matrix is a shorthand for multiplying several vectors by the same matrix. Denote the N_c columns of $\mathbf{B}$ by $\mathbf{b}_1, \ldots, \mathbf{b}_{N_c}$ and the N_c columns of $\mathbf{C}$ by $\mathbf{c}_1, \ldots, \mathbf{c}_{N_c}$. If $\mathbf{C} = \mathbf{AB}$, then $\mathbf{c}_k = \mathbf{Ab}_k$ for $k = 1, \ldots, N_c$.

It is possible to show that matrix multiplication is associative. Suppose we have three matrices $\mathbf{A}$, $\mathbf{B}$, and $\mathbf{C}$ whose dimensions are such that the matrix products $(\mathbf{AB})$ and $(\mathbf{BC})$ are both well defined. Then $(\mathbf{AB})\mathbf{C} = \mathbf{A}(\mathbf{BC})$. We often write $\mathbf{ABC}$ to denote either product. Matrix multiplication is not commutative. Even when both products are well defined, it is not in general true that $\mathbf{BA}$ is equal to $\mathbf{AB}$.

Matrix transposition. The transpose of an N_r by N_c matrix $\mathbf{A}$ is an N_c by N_r matrix $\mathbf{B}$ whose ijth entry is given by $b_{ij} = a_{ji}$. We use the superscript "T" to denote matrix transposition: $\mathbf{B} = \mathbf{A}^T$. The identity $(\mathbf{AB})^T = \mathbf{B}^T\mathbf{A}^T$ always holds.

Special Matrices and Vectors. A diagonal matrix $\mathbf{D}$ is an N_r by N_c matrix whose entries d_{ij} are zero if $i \neq j$. That is, the only nonzero entries of a diagonal matrix lie along its main diagonal. We refer to the nonzero entries of a diagonal matrix as its diagonal entries.

A square matrix is a matrix whose row and column dimensions are equal. We refer to the row and column dimensions of a square matrix as its dimension.

An identity matrix is a square diagonal matrix whose diagonal entries are all one. We use the symbol $\mathbf{I}_N$ to denote the N by N identity matrix. Using Eq. (A.2) it is possible to show that for any N_r by N_c matrix $\mathbf{A}$, $\mathbf{AI}_{N_c} = \mathbf{I}_{N_r}\mathbf{A} = \mathbf{A}$.

An orthogonal matrix $\mathbf{U}$ is a square matrix that has the property that $\mathbf{U}^T\mathbf{U} = \mathbf{UU}^T = \mathbf{I}_N$, where N is the dimension of $\mathbf{U}$.

A zero vector is a vector whose entries are all zero. We use the symbol $\mathbf{0}_N$ to denote the N dimensional zero vector.

Linear Models

Linear Combinations of Vectors. Equation (A.1) is not particularly intuitive. A more useful way to think about the effect of multiplying a vector **b** by matrix **A** is as follows. Consider the matrix **A** to consist of N_c column vectors $\mathbf{a}_1, \ldots, \mathbf{a}_{N_c}$. Then from Eq. (A.1) we have that the vector $\mathbf{c} = \mathbf{A}\mathbf{b}$ may be obtained by the operations of vector addition and scalar multiplication by

$$\mathbf{c} = \mathbf{a}_1 b_1 + \cdots + \mathbf{a}_{N_c} b_{N_c} \tag{A.3}$$

where the numbers $b_1, \ldots, b_{N_c}$ are the entries of **b**. Thus the effect of multiplying a vector by a matrix is to form a weighted sum of the columns of the matrix. The weights that go into forming the sum are the entries of the vector. We call any expression that has the form of the right-hand side of Eq. (A.3) a linear combination of the vectors $\mathbf{a}_1, \ldots, \mathbf{a}_{N_c}$.

Independence and Rank. Consider a collection of vectors $\mathbf{a}_1, \ldots, \mathbf{a}_{N_c}$. If no one of these vectors may be expressed as a linear combination of the others, then we say that the collection is independent. We define the rank of a collection of vectors $\mathbf{a}_1, \ldots, \mathbf{a}_{N_c}$ as the largest number of linearly independent vectors that may be chosen from that collection. We define the rank of a matrix **A** to be the rank of the vectors $\mathbf{a}_1, \ldots, \mathbf{a}_{N_c}$ that make up its columns. It may be proved that the rank of a matrix is always less than or equal to the minimum of its row and column dimensions. We say that a matrix has full rank if its rank is exactly equal to the minimum of its row and column dimensions.

Linear Models. When a vector **c** has the form given in Eq. (A.3), we say that **c** lies within a linear model. We call N_c the dimension of the linear model. We call the vectors $\mathbf{a}_1, \ldots, \mathbf{a}_{N_c}$ the basis vectors for the model. Thus an N_c dimensional linear model with basis vectors $\mathbf{a}_1, \ldots, \mathbf{a}_{N_c}$ contains all vectors **c** that can be expressed exactly using Eq. (A.3) for some choice of numbers $b_1, \ldots, b_{N_c}$. Equivalently, the linear model contains all vectors **c** that may be expressed as $\mathbf{c} = \mathbf{A}\mathbf{b}$ where the columns of the matrix **A** are the vectors $\mathbf{a}_1, \ldots, \mathbf{a}_{N_c}$ and **b** is an arbitrary vector. We refer to all vectors within a linear model as the subspace defined by that model.

Simultaneous Linear Equations

Matrix and Vector Form. Matrix multiplication may be used to express a system of simultaneous linear equations. Suppose we have a set of N_r simultaneous linear equations in N_c unknowns. Call the unknowns $b_1, \ldots, b_{N_c}$. Conventionally, we would write the equations in the form

$$\begin{aligned} a_{11}b_1 + \cdots + a_{1N_c}b_{N_c} &= c_1 \\ a_{21}b_1 + \cdots + a_{2N_c}b_{N_c} &= c_2 \\ \cdots \\ a_{N_r1}b_1 + \cdots + a_{N_rN_c}b_{N_c} &= c_{N_r} \end{aligned} \tag{A.4}$$

where the a_{ij} and c_i represent known numbers. From Eq. (A.1) it is easy to see that we may rewrite Eq. (A.4) as a matrix multiplication:

$$\mathbf{A}\mathbf{b} = \mathbf{c} \tag{A.5}$$

In this form, the entries of the vector **b** represent the unknowns. Solving Eq. (A.5) for **b** is equivalent to solving the system of simultaneous linear equations in Eq. (A.4).

Solving Simultaneous Linear Equations. A fundamental topic in linear algebra is finding solutions for systems of simultaneous linear equations. We will rely on several basic results in this chapter, which we state here.

When the matrix **A** is square and has full rank, it is always possible to find a unique matrix $\mathbf{A}^{-1}$ such that $\mathbf{A}\mathbf{A}^{-1} = \mathbf{A}^{-1}\mathbf{A} = \mathbf{I}_N$. We call the matrix $\mathbf{A}^{-1}$ the inverse of the matrix **A**. The matrix $\mathbf{A}^{-1}$ is also square and has full rank. Algorithms exist for determining the inverse of a matrix and are provided by software packages that support matrix algebra.

When the matrix **A** is square and has full rank, a unique solution **b** to Eq. (A.5) exists. This solution is given simply by the expression $\mathbf{b} = \mathbf{A}^{-1}\mathbf{c}$. When the rank of **A** is less than its row dimension, then there will not in general be an exact solution to Eq. (A.5). There will, however, be a unique vector **b** that is the best solution in a least-squares sense. That is, there is a unique vector **b** that minimizes the expression $\sum_{i=1}^{N_r} ((Ab)_i - c_i)^2$. We call this the least-squares solution to Eq. (A.5). Finding the least-squares solution to Eq. (A.5) is often referred to as linear regression. Algorithms exist for performing linear regression and are provided by software packages that support matrix algebra.

A generalization of Eq. (A.5) is the matrix equation

$$\mathbf{AB} = \mathbf{C} \tag{A.6}$$

where the entries of the matrix **B** are the unknowns. From our interpretation of matrix multiplication as a shorthand for multiple multiplications of a vector by a matrix, we can see immediately that this type of equation may be solved by applying the above analysis in a columnwise fashion. If **A** is square and has full rank, then we may determine **B** uniquely as $\mathbf{A}^{-1}\mathbf{C}$. When the rank of **A** is less than its row dimension, we may use regression to determine a matrix **B** that satisfies Eq. (A.6) in a least-squares sense. It is also possible to solve matrix equations of the form $\mathbf{BA} = \mathbf{C}$ where the entries of **B** are again the unknowns. An equation of this form may be converted to the form of Eq. (A.6) by applying the transpose identity (discussed earlier in this appendix). That is, we may find **B** by solving the equation $\mathbf{A}^T\mathbf{B}^T = \mathbf{C}^T$ if $\mathbf{A}^T$ meets the appropriate conditions.

Null Space. When the column dimension of a matrix **A** is greater than its row dimension N_r, it is possible to find nontrivial solutions to the equation

$$\mathbf{Ab} = \mathbf{0}_{N_r} \tag{A.7}$$

Indeed, it is possible to determine a linear model such that all vectors contained in the model satisfy Eq. (A.7). This linear model will have dimension equal to the difference between N_r and the rank of the matrix **A**. The subspace defined by this linear model is called the null space of the matrix **A**. Algorithms to find the basis vectors of a matrix's null space exist and are provided by software packages that support matrix algebra.

Singular Value Decomposition

The singular value decomposition allows us to write any N_r by N_c matrix **X** in the form

$$\mathbf{X} = \mathbf{UDV}^T \tag{A.8}$$

where **U** is an N_r by N_r orthogonal matrix, **D** is an N_r by N_c diagonal matrix, and **V** is an N_c by N_c orthogonal matrix.[25] The diagonal entries of **D** are guaranteed to be nonnegative. Some of the diagonal entries may be zero. By convention, the entries along this diagonal

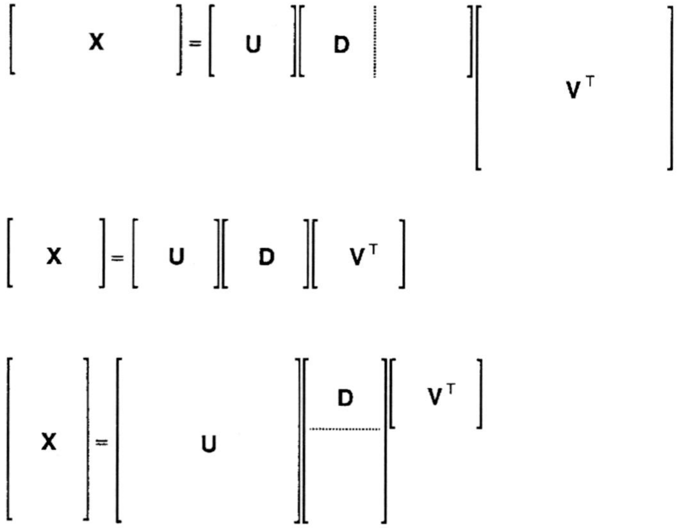

FIGURE A-1 The figure depicts the singular value decomposition (SVD) of an **N** by **M** matrix **X** for three cases $N_c > N_r$, $N_c = N_r$, and $N_c < N_r$.

are arranged in decreasing order. We illustrate the singular value decomposition in Fig. A.1. The singular value decomposition has a large number of uses in numerical matrix algebra. Routines to compute it are generally provided as part of software packages that support matrix algebra.

26.6 ACKNOWLEDGMENT

I thank J. Jacobs, F. Li, J. Loomis, P. Lennie, D. MacLeod, L. Maloney, J. Pokorny, J. Speigle, A. Stockman, J. Tietz, B. Wandell, and D. Williams.

26.7 REFERENCES

1. J. von Kries, "Chromatic Adaptation," originally published in *Festschrift der Albrecht-Ludwigs-Universitat,* 1902, pp. 145–148; D. L. MacAdam (ed.), *Sources of Color Vision,* MIT Press, Cambridge, 1970.
2. R. M. Evans, *An Introduction to Color,* Wiley, New York, 1948.
3. G. Wyszecki, "Color Appearance," *Handbook of Perception and Human Performance,* K. R. Boff, L. Kaufman, and J. P. Thomas (eds.), John Wiley & Sons, New York, 1986.
4. B. K. P. Horn, "Exact Reproduction of Colored Images," *Computer Vision, Graphics and Image Processing* **26:**135–167 (1984).
5. R. W. G. Hunt, *The Reproduction of Colour,* 4th ed., Fountain Press, Tolworth, England, 1987.
6. W. F. Schreiber, "A Color Prepress System Using Appearance Variables," *Journal of Imaging Technology* **12** (1986).
7. R. Hall, *Illumination and Color in Computer Generated Imagery,* Springer-Verlag, New York, 1989.
8. D. H. Brainard and B. A. Wandell, "Calibrated Processing of Image Color," *Color Research and Application* **15:**266–271 (1990).

9. G. Wyszecki and W. S. Stiles, *Color Science—Concepts and Methods, Quantitative Data and Formulae,* 2d ed., John Wiley & Sons, New York, 1982.
10. J. Pokorny and V. C. Smith, "Colorimetry and Color Discrimination," *Handbook of Perception and Human Performance,* K. R. Boff, L. Kaufman, and J. P. Thomas (eds.), John Wiley & Sons, 1986.
11. CIE *Colorimetry,* 2d ed., Bureau Central de la CIE, 1986.
12. ISO/CIE "CIE Standard Colorimetric Observers," 1991.
13. ISO/CIE "CIE Standard Colorimetric Illuminants," 1991.
14. D. H. Krantz, "Color Measurement and Color Theory: I. Representation Theorem for Grassmann Structures," *Journal of Mathematical Psychology* **12:**283–303 (1975).
15. P. Suppes, D. H. Krantz, R. D. Luce, and A. Tversky, *Foundations of Measurement,* Academic Press, San Diego, 1989.
16. D. L. McAdam, *Sources of Color Science,* MIT Press, Cambridge, Mass., 1970.
17. D. Judd and G. Wyszecki, *Color in Business, Science and Industry,* John Wiley & Sons, New York, 1975.
18. G. S. Brindley, *Physiology of the Retina and the Visual Pathway,* 2d ed., Williams and Wilkins, Baltimore, 1970.
19. R. M. Boynton, *Human Color Vision,* Holt, Reinhart and Winston, New York, 1979.
20. B. A. Wandell, "The Synthesis and Analysis of Color Images," *IEEE Transactions on Pattern Analysis and Machine Intelligence* **PAMI-9:**2–13 (1987).
21. J. B. Cohen, "Color and Color Mixture: Scalar and Vector Fundamentals," *Color Research and Application* **13:**5–39 (1988).
22. H. J. Trussell, "Applications of Set Theoretic Methods to Color Systems," *Color Research and Application* **16:**31–41 (1991).
23. B. Noble and J. W. Daniel, *Applied Linear Algebra,* 2d ed., Prentice-Hall, Inc., Englewood Cliffs, N.J. 1977.
24. G. H. Golub and C. F. van Loan, *Matrix Computations,* Johns Hopkins University Press, Baltimore, 1983.
25. J. M. Chambers, *Computational Methods for Data Analysis,* John Wiley & Sons, New York, 1977.
26. W. K. Pratt, *Digital Image Processing,* John Wiley & Sons, New York, 1978.
27. R. A. Johnson and D. W. Wichern, *Applied Multivariate Statistical Analysis,* Prentice-Hall, Englewood Cliffs, N.J., 1988.
28. J. Little and C. Moler, *MATLAB User's Guide.* The Matchworks, Natick, Mass., 1991.
29. W. H. Press, B. P. Flannery, S. A. Teukolsky, and W. T. Vetterling, *Numerical Recipes in C: The Art of Scientific Computing,* Cambridge University Press, Cambridge, 1988.
30. R. A. Becker, J. M. Chambers, and A. R. Wilks. *The New S Language,* Wadsworth & Brooks/Cole, Pacific Grove, Calif., 1988.
31. E. Anderson, Z. Bai, C. Bishof, J. Demmel, J. Dongarra, J. D. Croz, A. Greenbaum, S. Hammarling, A. McKenney, S. Ostrouchov, and D. Sorensen, *LAPACK User's Guide,* Society for Industrial and Applied Mathematics, Philadelphia, 1992.
32. J. G. Grassmann, "Theory of Compound Colors," originally published in *Annalen der Physik und Chemie* **89:**69–84 (1853); D. L. McAdam (ed.), *Sources of Color Vision,* MIT Press, Cambridge, 1970.
33. W. A. H. Rushton, "Visual Pigments in Man," *Handbook of Sensory Physiology,* H. J. A. Dartnall (ed.), Springer, Berlin, 1972.
34. A. Stockman, D. I. A. MacLeod, and N. E. Johnson, "The Spectral Sensitivities of Human Cones," *JOSA A* **10:**2491–2521 (1993).
35. D. B. Judd, "Report of U.S. Secretariat Committee on Colorimetry and Artificial Daylight," *CIE Proceedings* **1** (1951).

36. J. J. Vos, "Colormetric and Photometric Properties of a 2 Degree Fundamental Observer," *Color Research and Application* **3:**125–128 (1978).

37. P. DeMarco, J. Pokorny, and V. C. Smith, "Full-Spectrum Cone Sensitivity Functions for X-Chromosome-Linked Anomalous Trichromats," *Journal of the Optical Society* **A9:**1465–1476 (1992).

38. J. L. Schnapf, T. W. Kraft, and D. A. Baylor, "Spectral Sensitivity of Human Cone Photoreceptors," *Nature* **325:**439–441 (1987).

39. V. Smith and J. Pokorny, "Spectral Sensitivity of the Foveal Cone Photopigments Between 400 and 500 nm," *Vision Research* **15:**161–171 (1975).

40. J. Walraven, C. Enroth-Cugell, D. C. Hood, D. I. A. MacLeod and J. Schnapf, "The Control of Visual Sensitivity: Receptoral and Postreceptoral Processes," *Visual Perception: The Neurophysiological Foundations,* L. Spillmann and J. Werner (eds.), Academic Press, New York, 1989.

41. A. M. Derrington, J. Krauskopf, and P. Lennie, "Chromatic Mechanisms in Lateral Geniculate Nucleus of Macaque," *Journal of Physiology* **357:**241–265 (London, 1984).

42. D. I. A. MacLeod and R. M. Boynton, "Chromaticity Diagram Showing Cone Excitation by Stimuli of Equal Luminance," *Journal of the Optical Society of America* **69:**1183–1186 (1979).

43. J. Krauskopf, D. R. Williams, and D. W. Heeley, "Cardinal Directions of Color Space," *Vision Research* **22:**1123–1131 (1982).

44. J. Krauskopf, D. R. Williams, M. B. Mandler, and A. M. Brown, "Higher Order Color Mechanisms," *Vision Research* **26:**23-32 (1986).

45. D. J. Flitcroft, "The Interactions Between Chromatic Aberration, Defocus and Stimulus Chromaticity: Implications for Visual Physiology and Colorimetry," *Vision Research* **29:**349–360 (1989).

46. A. B. Poirson and B. A. Wandell, "The Ellipsoidal Representation of Spectral Sensitivity," *Vision Research* **30:**647–652 (1990).

47. J. D. Foley, A. van Dam, S. K. Feiner, and J. F. Hughes, *Computer Graphics: Principles and Practice,* 2d ed., Addison-Wesley, Reading, Mass., 1990.

48. D. B. Judd, D. L. MacAdam, and G. W. Wyszecki, "Spectral Distribution of Typical Daylight as a Function of Correlated Color Temperature," *Journal of the Optical Society of America* **54:**1031–1040 (1964).

49. J. Cohen, "Dependency of the Spectral Reflectance Curves of the Munsell Color Chips," *Psychon. Sci.* **1:**369–370 (1964).

50. K. L. Kelly, K. S. Gibson, and D. Nickerson, "Tristimulus Specification of the Munsell Book of Color from Spectrophotometric Measurements," *Journal of the Optical Society of America* **33:**355–376 (1943).

51. D. Nickerson, "Spectrophotometric Data for a Collection of Munsell Samples," U.S. Department of Agriculture, 1957.

52. L. T. Maloney, "Evaluation of Linear Models of Surface Spectral Reflectance with Small Numbers of Parameters," *Journal of the Optical Society of America* **A3:**1673–1683 (1986).

53. E. L. Krinov, "Surface Reflectance Properties of Natural Formations," *National Research Council of Canada: Technical Translation* **TT-439** (1947).

54. J. P. S. Parkkinen, J. Hallikainen, and T. Jaaskelainen, "Characteristic Spectra of Munsell Colors," *Journal of the Optical Society of America* **6:**318–322 (1989).

55. T. Jaaskelainen, J. Parkkinen, and S. Toyooka, "A Vector-Subspace Model for Color Representation," *Journal of the Optical Society of America* **A7:**725–730 (1990).

56. J. R. Magnus and H. Neudecker, *Matrix Differential Calculus with Applications in Statistics and Econometrics,* Wiley, Chichester, 1988.

57. T. W. Anderson, *An Introduction to Multivariate Statistical Analysis,* 2d ed., John Wiley & Sons, New York, 1971.

58. L. T. Maloney, "Computational Approaches to Color Constancy," unpublished Ph.D. thesis, Stanford University, 1984.

59. D. H. Marimont and B. A. Wandell, "Linear Models of Surface and Illuminant Spectra," *Journal of the Optical Society of America* **A9:**1905–1913 (1992).
60. B. A. Wandell and D. H. Brainard, "Towards Cross-Media Color Reproduction," *Proceedings OSA Applied Vision Topical Meeting,* 1989.
61. D. Saunders and A. Hamber, "From Pigments to Pixels: Measurement and Display of the Colour Gamut of Paintings," *Proceedings of the SPIE: Perceiving, Measuring, and Using Color* **1250:**90–102 (1990).
62. D. H. Brainard and B. A. Wandell, "Asymmetric Color-Matching: How Color Appearance Depends on the Illuminant," *Journal of the Optical Society of America* **A9:**1433–1448 (1992).
63. D. R. Williams, N. Sekiguchi, W. Haake, D. H. Brainard, and O. Packer. "The Cost of Trichromacy for Spatial Vision," *From Pigments to Perception,* B. B. Lee and A. Valberg (eds.), Plenum Press, New York, 1991.
64. D. H. Brainard, B. A. Wandell, and W. B. Cowan, "Black Light: How Sensors Filter Spectral Variation of the Illuminant," *IEEE Transactions on Biomedical Engineering* **36:**140–149 (1989).
65. G. Wyszecki, "Evaluation of Metameric Colors," *Journal of the Optical Society of America* **48:**451–454 (1958).
66. S. A. Burns, J. B. Cohen, and E. N. Kuznetsov, "Multiple Metamers: Preserving Color Matches Under Diverse Illuminants," *Color Research and Application* **14:** 16–22 (1989).
67. J. B. Cohen and W. E. Kappauf, "Metameric Color Stimuli, Fundamental Metamers, and Wyszecki's Metameric Blacks: Theory, Algebra, Geometry, Application," *American Journal of Psychology* **95:**537–564 (1982).
68. J. B. Cohen and W. E. Kappauf, "Color Mixture and Fundamental Metamers: Theory, Algebra, Geometry, Application," *American Journal of Psychology* **98:**171–259 (1985).
69. B. A. Wandell, "Color Rendering of Camera Data," *Color Research and Application, Supplement* **11:**S30–S33 (1986).
70. P. Sallstrom, "Color and Physics: Some Remarks Concerning the Physical Aspects of Human Color Vision," University of Stockholm, Institute of Physics, 1973.
71. M. H. Brill, "A Device Performing Illuminant-Invariant Assessment of Chromatic Relations," *J. Theor. Biol.* **71:**473–478 (1978).
72. G. Buchsbaum, "A Spatial Processor Model for Object Colour Perception," *Journal of the Franklin Institute* **310:**1–26 (1980).
73. L. T. Maloney and B. A. Wandell, "Color Constancy: A Method for Recovering Surface Spectral Reflectances," *Journal of the Optical Society of America* **A3:**29–33 (1986).
74. M. D'Zmura and P. Lennie, "Mechanisms of Color Constancy," *Journal of the Optical Society of America* **A3:**1662–1672 (1986).
75. H. Lee, "Method for Computing the Scene-Illuminant Chromaticity from Specular Highlights," *Journal of the Optical Society of America* **A3:**1694–1699 (1986).
76. B. Funt and J. Ho, "Color from Black and White," *Proceedings of the Second International Conference Comp. Vis.,* 1988.
77. B. V. Funt and M. S. Drew, "Color Constancy Computation in Near-Mondrian Scenes Using a Finite Dimensional Linear Model," *IEEE Comp. Vis & Pat. Recog.,* 1988.
78. M. D'Zmura and G. Iverson, "Color Constancy: Basic Theory of Two-Stage Linear Recovery of Spectral Descriptions for Lights and Surfaces," *Journal of the Optical Society of America,* **A10:**2148–2165.
79. M. H. Brill and G. West, "Chromatic Adaptation and Color Constancy: A Possible Dichotomy," *Color Research and Application* **11:**196–204 (1986).
80. L. T. Maloney, "Color Constancy and Color Perception: The Linear Models Framework," *Attention and Performance XIV: Synergies in Experimental Psychology, Artificial Intelligence, and Cognitive Neuroscience,* D. E. Meyer and S. E. Kornblum (eds.), MIT Press, Cambridge, Mass., 1992.
81. J. Pokorny, S. K. Shevell, and V. C. Smith, "Colour Appearance and Colour Constancy," *Vision and Visual Dysfunction,* P. Gouras (ed.), Macmillan, London, 1991.

82. D. H. Brainard, B. A. Wandell, and E. J. Chichilnisky, "Color Constancy: From Physics to Appearance," *Current Directions in Psychological Science* **2:**165–170 (1993).

83. D. L. MacAdam, "Visual Sensitivities to Color Differences in Daylight," *Journal of the Optical Society of America* **32:**247–274 (1942).

84. D. L. MacAdam, "Colour Discrimination and the Influence of Colour Contrast on Acuity," *Documenta Ophthalmologica* **3:**214–233 (1949).

85. G. Wyszecki and G. H. Fielder, "New Color-Matching Ellipses," *Journal of the Optical Society of America* **61:**1135–1152 (1971).

86. G. Wyszecki, "Matching Color Differences," *Journal of the Optical Society of America* **55:**1319–1324 (1965).

87. W. S. Stiles, "Color Vision: The Approach Through Increment Threshold Sensitivity," *Proceedings National Academy of Sciences* **45:**100–114 (U.S.A., 1959).

88. C. Noorlander and J. J. Koenderink, "Spatial and Temporal Discrimination Ellipsoids in Color Space," *Journal of the Optical Society of America* **73:**1533–1543 (1983).

89. A. B. Poirson, B. A. Wandell, D. Varner, and D. H. Brainard, "Surface Characterizations of Color Thresholds," *Journal of the Optical Society of America* **7:**783–789 (1990).

90. A. R. Robertson, "The CIE 1976 Color-Difference Formulae," *Color Research and Application* **2:**7–11 (1977).

91. H. deLange, "Research into the Dynamic Nature of the Human Fovea-Cortex Systems with Intermittent and Modulated Light. I. Attenuation Characteristics with White and Coloured Light," *Journal of the Optical Society of America* **48:**777–784 (1958).

92. H. deLange, "Research into the Dynamic Nature of the Human Fovea-Cortex Systems with Intermittent and Modulated Light. II. Phase Shift in Brightness and Delay in Color Perception," *Journal of the Optical Society of America* **48:**784–789 (1958).

93. N. Sekiguchi, D. R. Williams, and D. H. Brainard, "Efficiency for Detecting Isoluminant and Isochromatic Interference Fringes," *Journal of the Optical Society of America* **A10:**2118–2133 (1993).

94. E. C. Carter and R. C. Carter, "Color and Conspicuousness," *Journal of the Optical Society of America* **71:**723–729 (1981).

95. L. D. Silverstein and R. M. Merrifield, "The Development and Evaluation of Color Systems for Airborne Applications," *DOT/FAA/PM-85-19, U.S. Dept. of Transportation,* Federal Aviation Administration, 1985.

96. A. B. Poirson and B. A. Wandell, "Task-Dependent Color Discrimination," *Journal of the Optical Society of America* **A7:**776–782 (1990).

97. A. L. Nagy and R. R. Sanchez, "Critical Color Differences Determined with a Visual Search Task," *Journal of the Optical Society* **A7:** 1209–1217 (1990).

98. W. S. Stiles and J. M. Burch, "Interim Report to the Commission Internationale de l'Eclairage, Zurich, 1955, on the National Physical Laboratory's Investigation of Color-Matching (1955)," *Optica Acta* **2:**168–181 (1955).

99. W. S. Stiles and J. M. Burch, "NPL Colour-Matching Investigation: Final Report (1958)," *Optica Acta* **6:**1–26 (1959).

100. M. A. Webster and D. I. A. MacLeod, "Factors Underlying Individual Differences in the Color Matches of Normal Observers," *Journal of the Optical Society of America* **A5:**1722–1735 (1988).

101. J. Pokorny, V. C. Smith, and M. Lutze, "Aging of the Human Lens," *Applied Optics* **26:**1437–1440 (1987).

102. J. Pokorny, V. C. Smith, G. Verriest, and A. J. L. G. Pinckers, *Congenital and Acquired Color Vision Defects,* Grune and Stratton, New York, N.Y., 1979.

103. S. Ishihara, *Tests for Colour-Blindness,* Kanehara Shuppen Company, Ltd., Tokyo, 1977.

104. D. Farnsworth, "The Farnsworth-Munsell 100 Hue and Dichotomous Tests for Color Vision," *Journal of the Optical Society of America* **33:**568–578 (1943).

105. S. C. Merbs and J. Nathans, "Absorption Spectra of Human Cone Photopigments," *Nature* **356:**433–435 (1992).

106. J. Winderickx, D. T. Lindsey, E. Sanocki, D. Y. Teller, A. G. Motulsky, and S. S. Deeb, "Polymorphism in Red Photopigment Underlies Variation in Colour Matching," *Nature* **356:**431–433 (1992).

107. M. Neitz, J. Neitz, and G. H. Jacobs, "Spectral Tuning of Pigments Underlying Red-Green Color Vision," *Science* **252:**971–973 (1991).

108. J. Neitz, M. Neitz, and G. H. Jacobs, "More Than 3 Different Cone Pigments Among People with Normal Color Vision," *Vision Research* **33:**117–122 (1993).

109. L. N. Thibos, M. Ye, X. X. Zhang, and A. Bradley, "The Chromatic Eye—A New Reduced-Eye Model of Ocular Chromatic Aberation in Humans," *Applied Optics* **31:**3594–3667 (1992).

110. I. Powell, "Lenses for Correcting Chromatic Aberration of the Eye," *Applied Optics* **20:** 4152–4155 (1981).

111. D. H. Brainard, "Calibration of a Computer Controlled Color Monitor," *Color Research and Application* **14:**23–34 (1989).

112. R. Evans, "Visual Processes and Color Photography," *Journal of the Optical Society of America* **33:**579–614 (1943).

113. R. W. G. Hunt, "Chromatic Adaptation in Image Reproduction," *Color Research and Application* **7:**46–49 (1982).

114. H. Helmholtz, *Helmholtz's Physiological Optics,* translation from the 3d German edition, Optical Society of America, New York, 1909.

115. E. G. Boring, *Sensation and Perception in the History of Experimental Psychology,* D. Appleton Century, New York, 1942.

116. D. H. Brainard and B. A. Wandell, "A Bilinear Model of the Illuminant's Effect on Color Appearance," *Computational Models of Visual Processing,* M. S. Landy and J. A. Movshon (eds.), MIT Press, Cambridge, Mass., 1991.

117. E. H. Land and J. J. McCann, "Lightness and Retinex Theory," *Journal of the Optical Society of America* **61:**1–11 (1971).

118. E. H. Land, "Recent Advances in Retinex Theory and Some Implications for Cortical Computations: Color Vision and the Natural Image," *Proceedings of the National Academy of Sciences* **80:**5163–5169 (U.S.A., 1983).

119. D. H. Brainard and B. A. Wandell, "Analysis of the Retinex Theory of Color Vision," *Journal of the Optical Society of America* **A3:**1651–1661 (1986).

120. A. C. Hurlbert, H. Lee, and H. H. Bulthoff "Cues to the Color of the Illuminant," *Investigative Ophthalmology and Visual Science,* supplement **30:**221 (1989).

121. CIE, "CIE 1988 2° Spectral Luminous Efficiency Function for Photopic Vision," 1990.

122. D. Jameson and L. M. Hurvich, "Some Qualitative Aspects of an Opponent-Colors Theory. I. Chromatic Responses and Spectral Saturation," *Journal of the Optical Society of America* **45:**546–552 (1955).

123. L. M. Hurvich and D. Jameson, "An Opponent-Process Theory of Color Vision," *Psychological Review* **64:**384–404 (1957).

124. D. Jameson and L. M. Hurvich, "Theory of Brightness and Color Contrast in Human Vision," *Vision Research* **4:**135–154 (1964).

125. J. Larimer, D. H. Krantz, and C. M. Cicerone, "Opponent-Process Additivity—I. Red/Green Equilibria," *Vision Research* **14:**1127–1140 (1974).

126. J. Larimer, D. H. Krantz, and C. M. Cicerone, "Opponent Process Additivity—II. Yellow/Blue Equilibria and Non-Linear Models," *Vision Research* **15:**723–731 (1975).

127. D. H. Krantz, "Color Measurement and Color Theory: II. Opponent-Colors Theory," *Journal of Mathematical Psychology* **12:**304–327 (1975).

128. C. M. Cicerone, D. H. Krantz, and J. Larimer, "Opponent-Process Additivity—III. Effect of Moderate Chromatic Adaptation," *Vision Research* **15:**1125–1135 (1975).

129. J. Walraven, "Discounting the Background—The Missing Link in the Explanation of Chromatic Induction," *Vision Research* **16:**289–295 (1976).

130. S. K. Shevell, "The Dual Role of Chromatic Backgrounds in Color Perception," *Vision Research* **18:**1649–1661 (1978).

131. J. S. Werner and J. Walraven, "Effect of Chromatic Adaptation on the Achromatic Locus: The Role of Contrast, Luminance and Background Color," *Vision Research* **22:**929–944 (1982).

132. S. K. Shevell and M. F. Wesner, "Color Appearance Under Conditions of Chromatic Adaptation and Contrast," *Color Research and Application* **14** (1989).

133. P. Lennie, J. Krauskopf, and G. Scalar, "Chromatic Mechanisms in Striate Cortex of Macaque," *Journal of Neuroscience* **10:**649–669 (1990).

134. R. L. DeValois and K. K. DeValois, "A Multi-Stage Color Model," *Vision Research* **33:**1053–1065 (1993).

CHAPTER 27
DISPLAYS FOR VISION RESEARCH

William Cowan
Department of Computer Science
University of Waterloo
Waterloo, Ontario, Canada

27.1 GLOSSARY

$D(\lambda)$	spectral reflectance of the diffuser
e_P	phosphor efficiency
I_B	beam current of a CRT
M_{ij}	coefficient ij in the regression matrix linking ΔX_i to ΔR_i
m_j	vector indicating the particular sample measured
N_p	number of monochrome pixels in a color pixel
n_j	vector indicating the particular sample measured
$R(\lambda)$	spectral reflectance of the faceplate of an LCD
R_j	input coordinates of the CRT: usually R_1 is the voltage sent to the red gun; R_2 is the voltage sent to the green gun; and R_3 is the voltage sent to the blue gun
V	voltage input to a CRT
V_A	acceleration voltage of a CRT
V_0	maximum voltage input to a CRT
v_a	voltage applied to gun a of a CRT; $a = R, G, B$
v_B	maximum scanning velocity for the electron beam
v_h	horizontal velocity of the beam
v_v	vertical velocity of the beam
X_{ai}	tristimulus values of light emitted from monochrome pixel of color a

X_i	tristimulus values of light emitted from a color pixel
X_i	tristimulus value i, that is, $X = X_1$; $Y = X_2$; and $Z = X_3$
X_{0i}	tristimulus value i of a reference color
x_p	horizontal interpixel spacing, $x_p = v_h \tau_p$
y_p	vertical interpixel (interline) spacing
γ	exponent in power law expressions of gamma correction
δ	interpolation parameter for one-dimensional characterizations
ΔR_i	change in input coordinate i
ΔX_i	change in tristimulus value i
ϵ	interpolation parameter for inverting one-dimensional characterizations
$\nu_{\max}$	maximum frequency of CRT input amplifiers
μ	index of color measurements taken along a curve in color space
Φ	power of the emitted light
$\Phi_{a\lambda}^{(AMB)}(V_a)$	spectral power distribution of light emitted from monochrome pixel of color a as a result of ambiten light falling on an LCD
$\Phi_{a\lambda}^{(BL)}$	spectral power distribution of light emitted from monochrome pixel of color a as a result of the backlight
$\Phi_{a\lambda}(V_a)$	spectral power distribution of light emitted by monochrome pixel of color a, depending on the voltage with which the pixel is driven
$\Phi_{a\lambda}^{(R)}$	spectral power distribution of light reflected from the faceplate of an LCD
$\Phi(x, y)$	power of the emitted light as a function of screen position
Φ_λ	spectral power distribution of light emitted by all monochrome pixels in a color pixel
Φ_λ	spectral power of the emitted light
$\Phi_\lambda^{(AMB)}$	light output caused by ambient light falling on the faceplate of an LCD
$\Phi_\lambda^{(BL)}$	light output caused by the backlight of an LCD
Φ_0	maximum light power output from a CRT
$\Phi_{0\lambda}^{(AMB)}$	ambient light falling on the faceplate of an LCD
$\Phi_{0\lambda}^{(BL)}$	light emitted by the backlight-diffuser element of an LCD
$\Phi_\lambda(v_R, v_G, v_B)$	spectral power distribution of the light emitted when voltages v_R, v_G, v_B are applied to its inputs
$\tau_a(\lambda)$	spectral transmittance of the filter on monochrome pixel of color a
τ_d	phosphor decay time
τ_f	time spent scanning a complete frame
τ_l	time spent scanning a complete line, including horizontal flyback
τ_p	time the beam spends crossing a single pixel
$\tau(V)$	voltage-dependent transmittance of the light modulating element

27.2 INTRODUCTION

Complex images that are colorimetrically calibrated are needed for a variety of applications, from color prepress to psychophysical experimentation. Unfortunately, such images are extremely difficult to produce, especially using traditional image production technologies such as photography or printing. In the last decade technical advances in computer graphics have made digital imaging the dominant technology for all such applications, with the color television monitor the output device of choice. This chapter describes the operational characteristics and colorimetric calibration techniques for color television monitors, with its emphasis on methods that are likely to be useful for psychophysical experimentation. A short final section describes a newer display device, the color liquid crystal display, which is likely to become as important in the decade to come as the color monitor is today.

27.3 OPERATIONAL CHARACTERISTICS OF COLOR MONITORS

The color television monitor is currently the most common display device for digital imaging applications, especially when temporally varying images are required. Its advantages include good temporal stability, a large color gamut, well-defined standards, and inexpensive manufacture. A wide variety of different CRT types is now available, but all types are derived from a common technological base. This section describes the characteristics shared by most color monitors, including design, controls, and operation. Of course, a chapter such as this can only skim the surface of such a highly evolved technology. There is a vast literature available for those who wish a greater depth of technical detail. Fink et al.[1] provides a great deal of it, with a useful set of references for those who wish to dig even deeper.

Output devices that go under a variety of names, color television receivers, color video monitors, color computer displays, and so on, all use the same display component. This chapter describes only the display component. Technically, it is known as a color cathode ray tube (CRT), the term that is used exclusively in the remainder of the chapter.

In this chapter several operational characteristics are illustrated by measurements of CRT output. These measurements, which are intended only to be illustrative, were performed using a Tektronix SR690, a now obsolete CRT produced for the broadcast monitor market. While the measurements shown are typical of CRTs I have measured, they are intended to be neither predictors of CRT performance nor ideals to be pursued. In fact, color CRTs vary widely from model to model, and any CRT that is to be used in an application where colorimetry is critical should be carefully characterized before use.

Color CRT Design and Operation

The color CRT was derived from the monochrome CRT, and shares many features of its design. Thus, this section begins by describing the construction and operation of the monochrome CRT. New features that were incorporated to provide color are then discussed.

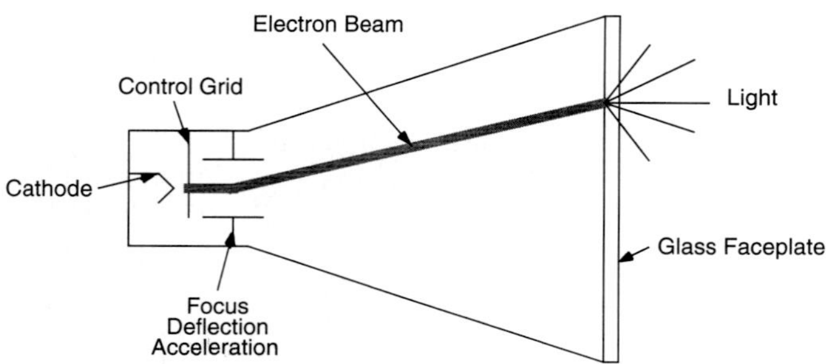

FIGURE 1 Schematic diagram of a monochrome CRT, showing the path of the electron beam and the location of the phosphor on the inside of the faceplate.

Monochrome CRTs. A schematic diagram of a monochrome CRT is shown in Fig. 1. The envelope is a sealed glass tube from which all the air has been evacuated. Electrons are emitted from the cathode, which is heated red hot. The flux of electrons in the beam, the beam current I_B, is determined by a control grid. A variety of magnetic and/or electrostatic electrodes then focus, accelerate, and deflect the beam. The beam strikes a layer of phosphor on the inside of the CRT faceplate, depositing power $I_B V_A$, where V_A is the acceleration voltage. Some of this power is converted to light by the phosphor. The power Φ in the emitted light is given by

$$\Phi = \int \Phi_\lambda \cdot \frac{hc}{\lambda} d\lambda$$

Φ_λ, the spectral power distribution of the emitted light is determined, up to a single multiplicative factor, by the chemical composition of the phosphor. The multiplicative factor is usually taken to be a linear function of the beam power. Thus the efficiency of the phosphor e_P given by

$$e_P = \Phi / I_B V_A$$

is independent of the beam current. Power not emitted as light becomes heat, with two consequences.

1. If the beam remains in the same spot on the screen long enough, the phosphor coating will heat up and boil the phosphor from the back of the faceplate leaving a hole in any image displayed on the screen.
2. Anything near the phosphor, such as the shadowmask in a color CRT, heats up.

A stationary spot on the screen produces an area of light. The intensity of the light is generally taken to have a gaussian spatial profile. That is, if the beam is centered at (x_0, y_0), the spatial profile of the light emitted is given by

$$\Phi(x, y) \propto \exp\left(\frac{1}{2\sigma_x^2}(x - x_0)^2 + \frac{1}{2\sigma_y^2}(y - y_0)^2\right)$$

The dependence of this spatial profile on the beam current is the subject of active development in the CRT industry. In most applications the beam is scanned around the screen, making a pattern of illuminated areas. The brightness of a given illuminated area depends on the beam current when the electron beam irradiates the area, with the beam current determined by the voltage applied to the control grid.

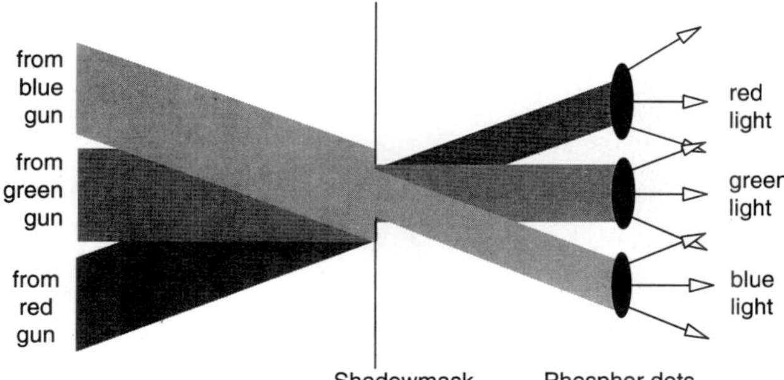

FIGURE 2 Electron beam/shadowmask/phosphor geometry in a shadowmask CRT.

Shadowmask Color CRTs. The basic technology of the monochrome CRT is extended to produce color. The standard method for producing color is to take several, usually three, monochrome images that differ in color and mix them additively to form a gamut of colors. (Even very unconventional technologies, such as the Tektronix bichromatic liquid crystal shutter technology, produce color by additive mixture.) This section describes the basic principles of shadowmask CRT technology, which is the dominant technology in color video.

Geometry of the Shadowmask CRT. A color CRT must produce three images to give a full range of color. To do so usually requires a tube with three guns. The three electron beams are scanned in exactly the way that monochrome beams are, but arrive at the screen traveling in slightly different directions. Near the faceplate is a screen called the shadowmask. It is made of metal, with a regular pattern of holes. Electrons in the beams either hit the screen to be conducted away or pass ballistically through the holes. Because the three beams are traveling in different directions, they diverge once they have passed the hole, striking the back of the faceplate in different places. The phosphor on the back of the faceplate is not uniform, but is distributed in discrete areas that radiate red, green, or blue light. The geometry is arranged so that electrons from the red gun hit red emitting phosphor, electrons from the green gun hit green emitting phosphor, and electrons from the blue gun hit blue emitting phosphor. This geometry is illustrated in Fig. 2. Several different geometries of shadowmask tube exist:

1. *Delta guns.* The three guns are arranged at the corners of an equilateral triangle, irradiating phosphor dots arranged in a triad.
2. *In-line guns.* The three guns are arranged in a line, irradiating phosphor dots side by side. The lines of phosphor dots are offset from line to line, so that the dot pattern is identical to that of the delta gun.
3. *Trinitron.* This is an in-line gun configuration, but the phosphor is arranged in vertical stripes. The holes in the shadowmask are rectangular, oriented vertically.

Other types of technology are also possible, though none is in widespread use. Beam index tubes, for example, dispense with the shadowmask, switching among the red, green, and blue signals as the electron beam falls on the different phosphor dots.

The most fundamental colorimetric property determining the colors produced by a shadowmask CRT is the light emitted by the phosphors. Only those colors that are the additive mixture of the phosphor colors (CRT primaries) can be produced. Because the color and efficiency of the phosphors are important determinants of perceived picture quality in broadcast applications phosphor chemistry undergoes continuous improvement,

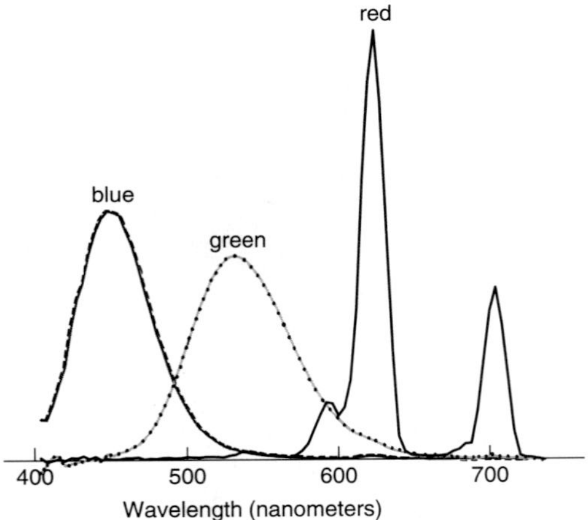

FIGURE 3 Spectral power distributions of the light output by a typical set of phosphors. The long wavelength peak of the red phosphor is sometimes missed in spectroradiometric measurements.

and varies from CRT to CRT. The emission spectra of the phosphors of a "typical" shadowmask CRT are shown in Fig. 3. The chromaticities of these phosphors are:

	x	*y*	*z*
Red phosphor	0.652	0.335	0.013
Green phosphor	0.298	0.604	0.098
Blue phosphor	0.149	0.064	0.787

Common Problems in Shadowmask CRTs. The shadowmask is a weak point in color CRT design, so that color CRTs suffer from a variety of problems that potentially degrade their performance. The three most common problems are doming, blooming, and shadowmask magnetization. Doming occurs when the shadowmask heats because of the large energy density in the electrons it stops. It then heats up and expands, often not uniformly. The result is a distortion of its shape called "doming." This geometrical distortion means that the registration of holes and dots is disturbed, and what should be uniform colors become nonuniform. Trinitron tubes have a tensioning wire to reduce this problem. It is often visible as a horizontal hairline running across the whole width of the tube about a third of the way from the top or bottom of the tube, depending on which side of the tube is installed up.

Blooming occurs when too much energy is deposited in the electron beam. Then electrons arrive at the screen in the right position but their entire kinetic energy is not immediately absorbed by the phosphor. They then can move laterally and deposit energy on nearby dots which can be both the wrong color and outside the intended boundary of the bright area. Colors become desaturated, and edges of areas become blurred.

Magnetic fields build up in the shadowmask when the electromagnetic forces produced by electrons, assisted by the heat build-up, create magnetized areas in the shadowmask. These areas are nonuniform and usually produce large regions of nonuniformity in color or brightness on the screen. CRTs usually have automatic degaussing at power-up to remove these magnetic fields. If this is insufficient to remove the field buildup, inexpensive degaussing tools are available.

CRT Electronics and Controls. The CRT receives voltage signals at its inputs, one for monochrome operation, three for color operation. The input signals must be suitably amplified to control the beam current using the grid electrode. The amplification process is described first, followed by the controls that are usually available to control the amplification.

Amplification. Two properties of the amplification are important to the quality of the displayed image: bandwidth and gamma correction.

The bandwidth of the amplifiers determines the maximum frequency at which the beam current can be modulated. In most applications the beam is moved about the screen to produce the image, so that the maximum frequency translates into a maximum spatial frequency signal. Suppose, unrealistically, that the amplifiers have a sharp cutoff frequency ν_{max} and that the beam is scanned at velocity v_B. Then the maximum spatial frequency of which the CRT is capable is ν_{max}/v_B. In some types of CRT, particularly those designed for vector use, settling time is more important than bandwidth. It can be similarly related to beam velocity to determine the sharpness of spatial transitions in the image.

For colorimetric purposes, the relationship between the voltage applied at the CRT input and the light emitted from the screen is very important. It is determined by the amplification characteristics of the input amplifiers and of the CRT itself, and creating a good relationship is part of the art of CRT design. When the relationship must be known for colorimetric purposes it is usually determined by measurement and handled tabularly. However, for explanatory purposes it is often written in the form

$$\Phi = \Phi_0(V/V_0)^\gamma$$

where V is the voltage input to the CRT, normalized to its maximum value V_0. The exponent, which is conventionally written as γ, gives this amplification process its name, gamma correction. Figure 4 shows some measured values, and a log-linear regression line drawn through them. Note the following features of this data.

1. The line is close to linear.
2. There are regions of input voltage near the top and bottom where significant deviations from linearity occur.
3. The total dynamic range (the ratio between the highest and lowest outputs) is roughly 100, a typical value. This quantity depends on the setting of brightness (black level) and contrast used during the measurement.

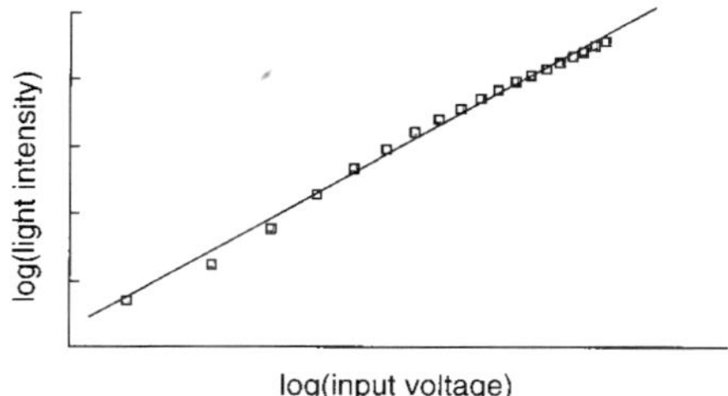

FIGURE 4 Graph of ln (light intensity) against ln (input voltage), sometimes used to determine the gamma correction exponent. Note the systematic deviation from a straight line.

There will be additional discussion of gamma correction in the section on monitor setup.

Controls that Affect Amplification in Monochrome CRTs. Monochrome monitors have several controls that adjust several properties of the electron beam. Some controls are external to the monitor, some are internal, usually in the form of small potentiometers mounted on the circuit boards. The particular configuration depends on the monitor. In terms of decreasing likelihood of being external they are: brightness (black level), contrast, focus, underscan/overscan, pedestal, gain, horizontal size, vertical size. A brief description of the action of each control follows. It is important, however, to remember that while names of controls tend to be constant from CRT to CRT, the action performed by each control often varies.

Brightness (black level). This control adjusts the background level of light on the monitor screen. It is designed for viewing conditions where ambient light is reflected from the monitor faceplate. It usually also varies the gamma exponent to a higher value when the background level (black level) is increased. A typical variation of the light intensity/input voltage relationship when brightness is varied is shown in Fig. 5.

Contrast. This control varies the ratio between the intensity of the lightest possible value and the darkest possible value. High contrast is usually regarded as a desirable attribute of a displayed image, and the contrast control is usually used to produce the highest contrast that is consistent with a sharp image. A typical variation of the light intensity/input voltage relationship when this control is varied is shown in Fig. 6.

Focus. This control varies the size of the electron beam. A more tightly focused electron beam produces sharper edges, but the beam can be focused too sharply, so that flat fields show artifactual spatial structure associated with beam motion. Focus is usually set with the beam size just large enough that no intensity minimum is visible between the raster lines on a uniform field.

Pedestal and gain. These controls, which are almost always internal, are similar to brightness and contrast, but are more directly connected to the actual amplifiers. Pedestal varies the level of light output when the input voltage is zero. Gain varies the rate at which the light output from the screen increases as the input voltage increases.

Controls Specific to Color CRTs. Color monitors have a standard set of controls similar to those of monochrome monitors. Some of these, like brightness and contrast,

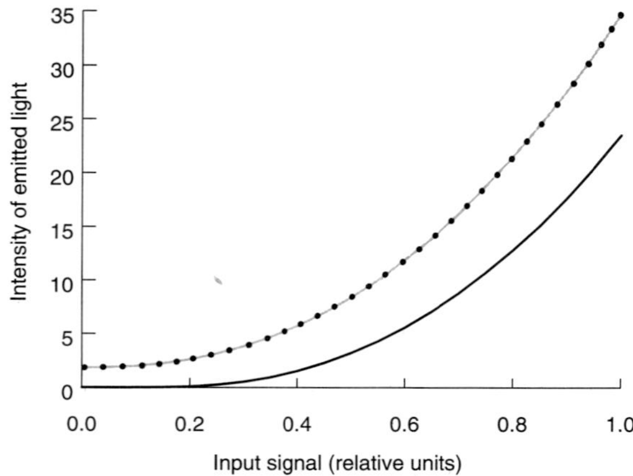

FIGURE 5 Variation of the light intensity/input voltage relationship when the brightness (black level) control is varied. The lower curve shows the relationship with brightness set near its minimum; the upper one wtih brightness set somewhat higher.

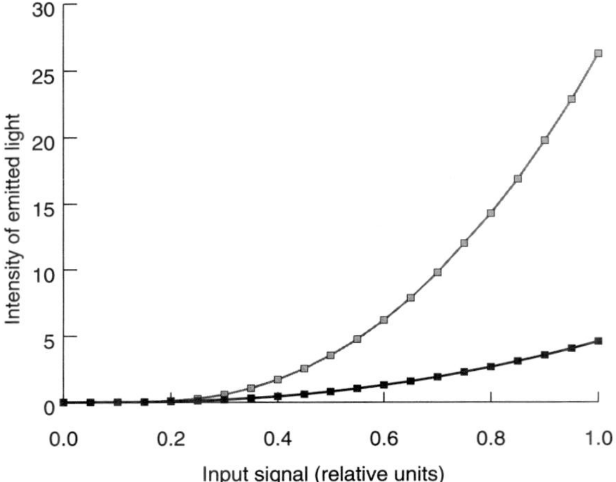

FIGURE 6 Variation of the light intensity/input voltage relationship when the contrast is varied. The lower curve shows the relationship with contrast set near its minimum; the upper one with contrast near its maximum.

have a single control applied simultaneously to each of the color components. Others, like gain and pedestal, have three controls, one for each of the color channels. There are several aspects of color that need to be controlled, however, and they are discussed in the paragraphs that follow.

Purity. Purity is an effect associated with beam/shadowmask geometry. It describes the possibility that the electron beams can cause fluorescence in inappropriate phosphors. There is no standard set of controls for adjusting purity. Generally, there are magnets in the yoke area whose position can be adjusted to control purity, but this control is very difficult to perform. There will be no need to alter them under normal conditions. Purity is most influenced by stray magnetic fields, and can often be improved by moving the CRT.

White balance. It is important for most monitor applications that when the red, green, and blue guns are turned on equally, calibration white (usually either D_{6500} or D_{9200}) appears on the screen. This should be true at all intensities. Thus, controls that alter the voltage input/light output relationship for each channel should be available. At a minimum, there will be the pedestal and gain for each channel.

Degauss. Above we mentioned magnetic fields that build up in the shadowmask. There is generally a set of wires that run around the edge of the faceplate. At power-up a degaussing signal is sent through the wires and the ensuing magnetic field degausses the shadowmask. The degauss control can produce the degauss signal at any time.

CRT Operation. The CRT forms an image on its screen by scanning the electron beam from place to place, modulating the beam current to change the brightness from one part of the image to another. A variety of scan patterns are possible, divided into two categories. Scan patterns that are determined by the context of the image (in which, for example, the beam moves following lines in the image) are used in vector displays, which were once very common but now less so. Scan patterns that cover the screen in a regular pattern independent of image content are used in raster displays: the scan pattern is called the raster. A variety of different rasters are possible; the one in most common use is a set of horizontal lines, drawn from top to bottom of the screen.

Raster Generation. Almost all raster CRTs respond to a standard input signal that

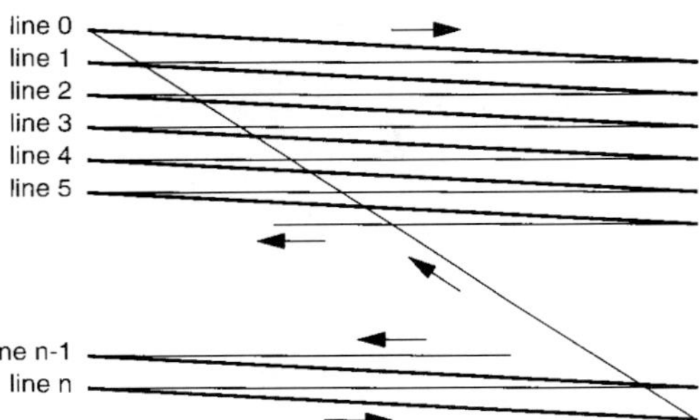

FIGURE 7 Scan pattern for a noninterlaced raster.

creates a raster consisting of a set of horizontal lines. This section describes the path of the electron beam as it traverses the CRT screen; a later section discusses the signal configurations that provide the synchronization necessary to drive it.

Frames and fields. The image on the screen is scanned out as a set of lines. Each line is scanned from left to right, as seen from a position facing the screen. Between the end of one line and the beginning of the next the beam returns very quickly to the left side of the screen. This is known as horizontal retrace or flyback. The successive lines are scanned from top to bottom of the screen. One field consists of a scan from top to bottom of the screen. Between the end of one field and the beginning of the next the beam returns very quickly to the top of the screen. This is known as vertical retrace or flyback.

One frame consists of a scan of all the lines in an image. In the simpler type of display a field is identical to a frame, and all the lines of the image are scanned out, from top to bottom of the display. The scan pattern is shown in Fig. 7. It is called noninterlaced.

A more complicated type of raster requires more than one field for each frame. The most usual case has two fields per frame, an even field and an odd field. During the even field the even-numbered lines of the image are scanned out with spaces between them. This is followed by vertical retrace. Then, during the odd field the odd-numbered lines of the image are scanned out into the spaces left during the even-field scan. A second vertical retrace completes the frame. This scan pattern, known as interlaced, is shown in Fig. 8. The purpose of interlace is to decrease the visible flicker within small regions of the screen. It works well for this purpose, provided there are no high-contrast horizontal edges in the

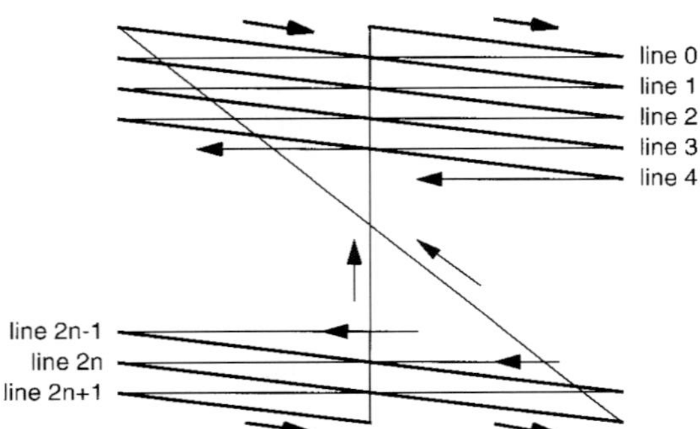

FIGURE 8 Scan pattern for an interlaced raster, two fields per frame. Line 3 eventually scans to line $2n - 1$ by way of the odd lines; line 4 scans to line $2n$.

display. If they are present they appear to oscillate up and down at the frame rate, which is usually 30 Hz. This artifact can be very visible and objectionable, particularly to peripheral vision. Interlace having more than two fields per frame is possible, but uncommon.

In viewing Figs. 7 and 8 note that the vertical scan is produced by scanning the beam down continuously. Thus the visible lines are actually sloped down from left to right while the retrace, which is much faster than the horizontal scan, is virtually unsloped. The method used to put the odd field of the interlaced scan between the lines of the even field is to scan half a line at the end of the even field followed by half a line at the beginning of the odd field. Thus an interlaced raster has an odd number of lines.

Relationship of the Raster to the CRT Input Signal. The CRT receives a serial input signal containing a voltage that controls the intensity for each location in the image. This signal must be synchronized with the raster in order to make sure that each pixel is displayed in the right location. This section describes the relationship between the raster and the CRT input.

Horizontal scanning. The input signal for one line is shown in Fig. 9. It consists of data interspersed with blank periods, called the horizontal blanking intervals. During the horizontal blanking interval the beam is stopped at the end of the line, scanned quickly back for the beginning of the next line, then accelerated before data for the next line begins. In the data portion, 0.0 volts indicates black and 1.0 volts indicates white. The signal shown would produce a line that is dim on the left, where the line starts, and bright on the right where it ends.

A second signal, the horizontal synchronization signal, has a negative-going pulse once per line, positioned during the horizontal blanking interval. This pulse is the signal for the CRT to begin the horizontal retrace.

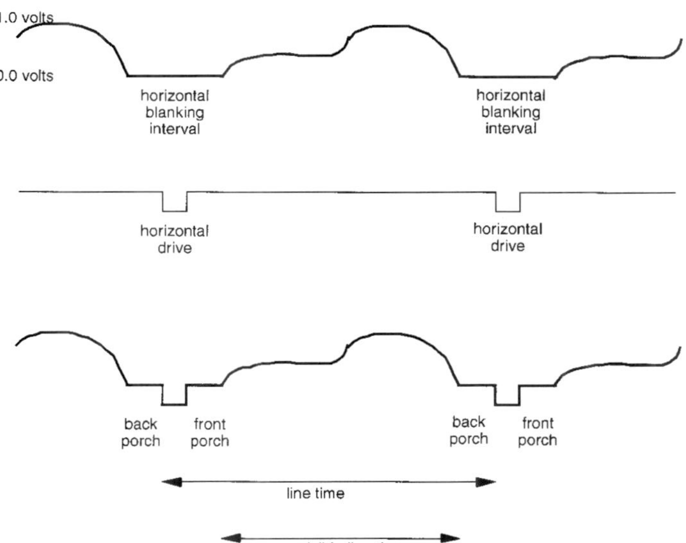

FIGURE 9 A schematic input signal that would generate a single line of raster, including the end of the preceding line and the beginning of the succeeding line. The top trace shows the signal that produces the image, including the blank between lines; the middle trace shows the synchronization signal with the horizontal drive pulse located during the blank; the bottom trace shows the synchronization signal and the picture signal in a single waveform.

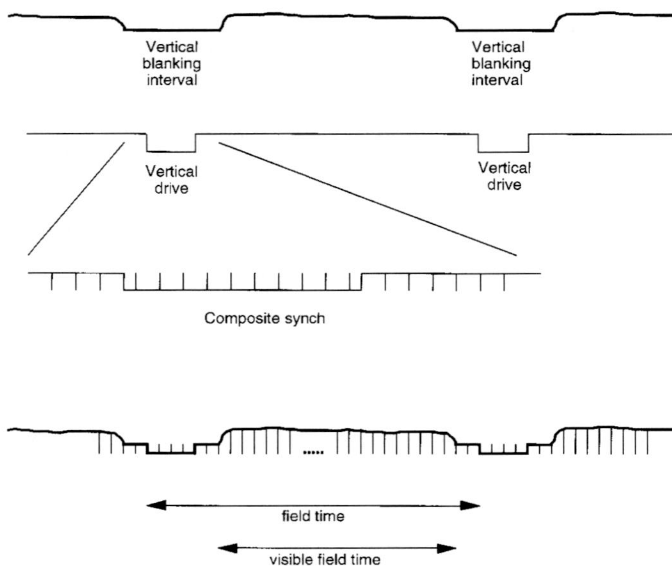

FIGURE 10 A schematic input signal that would generate a single field of a raster, including the end of the preceding field and the beginning of the succeeding field. The top trace shows the picture signal, with the vertical blank shown and the horizontal blank omitted. The second trace shows the vertical synchronization signal with the vertical drive pulse. The third trace shows the horizontal synchronization signal added to the vertical signal to give composite synch. The bottom trace shows composite synch added to the picture signal.

When the synchronization signal is combined with the data signal the third waveform in Fig. 9 is produced. The intervals in the horizontal blanking interval before and after the synchronization are known as the front and back porch. They are commonly used to set the voltage level corresponding to the black level, a process known as black clamping. Black clamping reduces the effect of low-frequency noise on the color of the image, but requires good synchronization between the precise timing of the signal and the raster production of the CRT. The ability of the monitor to hold synchronization with particular timing in this signal is an important parameter of the monitor electronics.

Vertical scanning. Figure 10 shows the input signal for one complete field. There is data for each line, separated by the horizontal blanking period, which is too short to be visible in the figure. Separating the data portion of each field is the vertical blanking interval, during which the beam is scanned back to the top of the screen.

The synchronization signal consists of a vertical drive pulse signaling the time at which the beam should be scanned to the top of the screen. This pulse is long compared to the horizontal drive pulses so that it can easily be separated from them when the synchronization signals are combined into composite synch. In composite synch positive-going pulses in positions conjugate to the positions of the horizontal drive pulses are added during the vertical drive signal. These pulses are designed to keep the phase of the horizontal oscillator from wandering during vertical drive. When they are interpreted incorrectly, as was the case in some monitors early in the history of digital electronics, the result is a small shear in the first few lines at the top of the screen as the horizontal oscillator gets back into phase.

The composite synch signal can be added to the data, as shown in the fourth illustration in Fig. 10. Most real systems have this type of synchronization, which was designed for an era when signals were to be sent long distances by broadcast or wire. Today, in computer graphics applications, we often find that electronic circuitry in the source carefully folds the two synch signals into composite synch and the composite synch signal into the data signal. This signal is carried along a short piece of wire to the receiver, where electronic circuitry strips the signals apart. Bad synchronization is often caused by the malfunction of this superfluous circuitry, but it is unlikely that this situation will change in the immediate future.

At the input to a color CRT, three or four signals like the ones described above are provided. The four signals are three input signals containing pixel intensities and blanking intervals, plus a fourth signal carrying the composite synchronization signal. To get three signals the synchonization signal is combined with one or more of the pixel signals. When only one pixel signal has synchronization information it is almost always the green one.

Controls that Affect the Raster. Several controls affect the placement and size of the raster on the CRT. They are described in this section.

Horizontal/vertical size and position. These controls provide continuous variation in the horizontal and vertical sizes of the raster, and in the position on the CRT faceplate where the origin of the raster is located.

Underscan/overscan. Most CRTs provide two standard sizes of raster, as shown in Fig. 11. This control toggles between them. In the underscan position the image is smaller than the cabinet port, so that the whole image is visible, surrounded by a black border. In the overscan position the image is slightly larger than the cabinet port, so that no edges of the image are visible. There is a standard amount of overscan, which is used in home television receivers.

Convergence. A shadowmask color CRT has, in effect, three rasters, one for each of the primary colors. It is essential that these rasters be positioned and sized exactly the same in all parts of the image; otherwise, spurious colored fringes appear at the edges in the image. To achieve this it is essential that all three electron beams should be at the same place at the same time. For example, if the green gun is slightly to the left of the red and blue guns, a white line of a black background will have a green fringe on its left side and a magenta fringe on its right side. Good convergence is relatively easy to obtain with in-line gun configurations, so the usual practice is to adjust convergence at the factory using

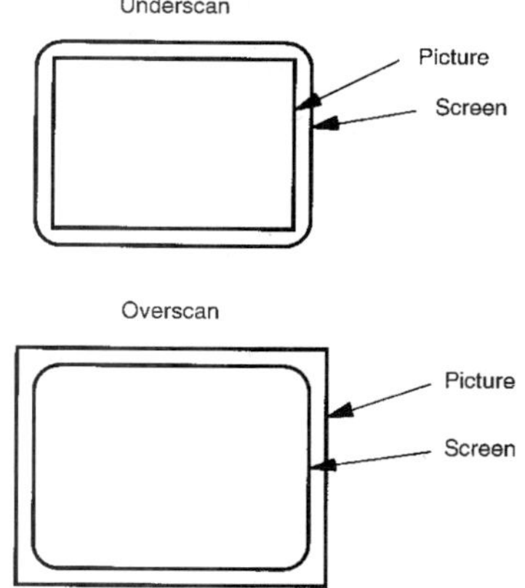

FIGURE 11 Schematic illustration of overscan and underscan. The rectangle with rounded corners is the cabinet mask that defines the viewing area; the rectangle with square corners is the displayed image.

ring-shaped magnets placed over the yoke of the CRT, then to glue them permanently in place with epoxy cement. Satisfactory readjustment is very difficult to achieve.

Delta gun configurations have, by reputation, less stable convergence. Consequently, there are often controls available to the user for adjustment of convergence. The Tektronix 690SR was an extreme case, with a pullout drawer containing 52 potentiometers for controlling convergence on different areas of the screen. However, these controls, as might be expected, are not independent, so that "fine tuning" the convergence is very difficult, even when the controls are conveniently located.

Magnetic fields are generally the main culprit when convergence is objectionable, since the paths of electrons are curved in magnetic fields. Small fields from power transformers or from other electronic equipment or even the earth's magnetic field can be the problem. Thus, before considering redoing the convergence on a new monitor try rotating it and/or moving other equipment around in the room. Another useful trick: some low-cost CRTs have poorly positioned power transformers, and convergence can be improved by moving the power transformer far outside the CRT cabinet.

Operational Characteristics of Color CRTs

Two factors influence the light output that is produced in response to input signals. One is conventional: the input signal must be sequenced and timed precisely to match what is expected by the CRT. This problem is handled by input signal standards of long standing in the television industry. Synchronization standards are described first, followed by colorimetric standards. The second is technical: the construction of CRTs produces certain variations, even when the input signal is constant. The section ends with a discussion of this variation, specifying the temporal and spatial characteristics of the light output from a CRT and describing reasonable expectations for stability and uniformity in CRT-produced images.

Timing and Synchronization Standards. Figures 9 and 10 show how the information needed to specify a CRT image are arranged in the input signal. For practical use they need to be specified numerically, thereby allowing CRT manufacturers to ensure that their products will respond appropriately to the input signals they will encounter in service. Two standards have been created by the Electronic Industries Association (EIA) for this purpose. They are actually standards for television studios, prescribing how signals should be distributed in closed-circuit applications, and specifically between television cameras and studio monitors. In practice they are more widely used. Among other applications, they specify the interface between the tuner and the CRT in television receivers, and the output of digital graphics systems allowing studio monitors to be used as display devices.

The two standards are RS-170,[2] which was designed for lower bandwidth applications, and RS-343,[3] which was designed for higher bandwidth applications. Each gives minimum and maximum timings for each part of the input signal in terms of several parameters that are allowed to vary from application to application. These parameters are then part of CRT input specification, allowing all timing characteristics to be deduced. The most important parameter is the line rate, the number of lines displayed per second. Most CRTs are intolerant of large variations in this parameter. Another important parameter is the field rate, the number of fields displayed per second. Older CRTs were quite intolerant of variations in this parameter, but most modern CRTs can handle signals with a very wide variety of field rates.

RS-170 and RS-343 are monochrome standards. When used for color they are tripled, with the input signal assumed to be in three parallel monochrome signals. As mentioned above, the synchronization signal is either placed on a fourth, or incorporated into a single color signal, usually the green one. However, although this practice is almost universal there is no official color standard for the RGB input to a monitor. This lack can have unfortunate consequences. For example, the NTSC color signal naturally decodes into three signals with peak-to-peak voltages of 0.7 volts. Thus RGB monitors were built with inputs expecting this range. RS-170 and RS-343, on the other hand, specify a peak-to-peak voltage of 1.0 volts. Early digital graphics systems were built to provide exact RS-170 and RS-343 output. These systems, naturally, overdrove standard monitors badly.

Colorimetric Standards. In broadcast applications the image transmitter should be able to specify the precise color that will be displayed on the receiver's CRT. This requirement can be supplied by a colorimetric standard. The NTSC color standard was agreed upon for use in the North American broadcast television industry. It is a complete color standard, specifying phosphor chromaticities, color representation on the carrier signal, signal bandwidth, gamma correction, color balance, and so on. Thus, if the NTSC standard were followed in both transmitter and receiver, home television would provide calibrated colors. It is not followed exactly, however, since both television manufacturers and broadcasters have discovered that there are color distortions that viewers prefer to colorimetrically precise color. Furthermore, it is not useful for high-quality imagery since the low bandwidth it allocates to the chromatic channels produces edges that are incompatible with good image quality.

Spatial and Temporal Characteristics of Emitted Light. The light emitted from a CRT is not precisely uniform, but suffers from small- and large-scale variations. The small-scale variations arise because the image is actually created by a spot that moves over the entire display surface in a time that is intended to be short compared to temporal integration times in the human visual system. In fact, a short enough glimpse of the CRT screen reveals only a single point of light; a longer one reveals a line as the point moves during the viewing time. These patterns are designed to be relatively invisible under normal viewing conditions, but often need to be considered when CRTs are used for vision experimentation, or when radiometric measurements are made. They are controlled largely by the input signal, so they are relatively constant from CRT to CRT.

The small-scale variations are described in this section; the large-scale variations, which occur as spatial nonuniformity and temporal instability in the emitted light, are discussed in the following section.

Spatial characteristics. The electron beam scans the screen horizontally, making an image that consists of a set of horizontal lines. During the scan its intensity is modulated, so that the line varies in brightness—vertical edges, for example, being created by coordinated modulation in a series of horizontal lines. The sharpness of such a line depends on two factors: the ability of the video amplifiers to produce an abrupt change in intensity and the size of the spot of light on the screen, which is essentially the same as the cross section of the electron beam. Video amplifiers vary greatly in bandwidth from one model of CRT to another, and adjusting their performance is beyond the reach of virtually all CRT users. Unfortunately, interactions between adjacent horizontal pixels are not even linear,[4–6] and the measurements needed to determine the nonlinearity of a given CRT are extremely demanding. Thus, compensating for amplifier bandwidth or even measuring its effects is beyond the scope of this chapter.

By contrast, the sharpness of a horizontal line is independent of the video amplifiers, but depends only on the spot size and its spatial profile, which is well modeled as a two-dimensional gaussian. The width of the gaussian depends on the focus electrodes, and is user-adjustable on most CRTs. The shrinking raster technique for CRT setup (discussed elsewhere in this chapter) determines a spot size that has a specific relationship to the interline spacing. Assuming this setup, it is possible to make reasonable assumptions about the contrast of images on a CRT,[7] and these assumptions can be extended to the small-scale spatial structure of arbitrary images. Note that many CRTs used primarily as display terminals are overfocused compared to the shrinking raster criterion because such overfocusing allows the display of smaller text. Overfocusing is easily detected as visible raster lines, usually in the form of closely spaced dark horizonal lines when a uniform field is displayed.

Temporal characteristics. Because of the scan pattern the light emitted from any portion of the screen has a complicated temporal dependence. The next few paragraphs describe several levels of this dependence, assuming, for simplicity, that the scan is noninterlaced. Similar results for an interlaced scan are easy to derive. The relevant variables are:

- τ_d phosphor decay time
- τ_p time the beam spends crossing a single pixel
- v_h horizontal velocity of the beam
- x_p horizontal interpixel spacing, $x_p = v_h \tau_p$
- τ_l time spent scanning a complete line, including horizontal flyback
- v_v vertical velocity of the beam
- y_p vertical interpixel (interline) spacing, $y_p = v_v \tau_l$
- τ_f time spent scanning a complete frame (identical to the time spent scanning a complete field)

These temporal factors usually change the colorimetry of the CRT. When one or more is changed, usually because the video source has been reconfigured, the color output for a given input to the CRT usually changes. Thus, a recalibration should be done after any such change. And it is probably practical to recommend that the system be used with the new video timing parameters for a while before the recalibration is done, since tuning or touch-ups will require further recalibration. The intensity of the light emitted from a vanishingly small area of the CRT screen is a function that is zero until the beam traverses the point t_s, then decays exponentially afterward,

$$\Phi(t) = \sum \Phi_0 \theta(t - t_s) \exp\left((t - t_s)/\tau_d\right)$$

τ_d ranges between $10^{-3}\tau_f$ to τ_f, and usually varies from one phosphor to another. Most often the green phosphor has the largest τ_d, the blue one the smallest. Broadcast monitors tend to have small τ_d's since they are designed to display moving imagery; data display CRTs tend to have large τ_d's since they are intended to display static imagery. Occasionally a CRT primary (most often red) is a mixture of two phosphors with different decay times. In such cases, the chromaticity of the light emitted by the primary changes over time, through the change is not usually visible.

If a second pixel is n_h pixels to the right of a given pixel and n_v lines below it, the light emitted by the second pixel lags behind the light emitted by the first pixel by

$$n_h \tau_p + n_v \tau_l \approx d_h/v_h + d_v/v_v,$$

where d_h and d_v are the horizontal and vertical screen distances.

Commonly, a detector measures a finite area of the CRT screen. Then intensity of the detected light is a closely spaced group of exponentials followed by a long gap, then another closely spaced group, and so on. Each peak in the closely spaced group is spaced about τ_l from the previous one, occurring when each line of the raster crosses the detection area. Each group is spaced about τ_f from the previous one, occurring each time a new field repaints the detected target. The time constant of this composite signal is τ_d, the decay time of the phosphor for each individual peak in the signal.

Stability and Uniformity of CRT Output. The small-scale structure of the emitted light,

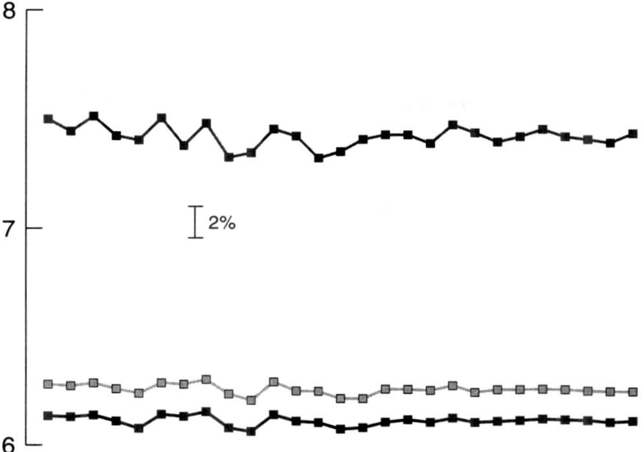

FIGURE 12 Variation of light output from a color CRT over 24 hours of continuous operation. This graph shows the three tristimulus values when a neutral color is displayed. The latter part of the graph is at night when almost all other equipment in the building is turned off.

discussed above, determines the specific form of visual stimuli produced by a CRT. The large-scale structure determines the scope of applicability of measurements made on a particular area of a CRT at a particular time.

Temporal stability. How constant is the light emitted by a CRT that receives the same input signal? Figures 12 and 13 show the results of colorimetric measurements taken over a twenty-four-hour period. In each figure the calibration bars shows a 2 percent variation compared to the average value. The variation decreases considerably in the latter part of the graph. It is the period from 5 p.m. until 9 the next morning, showing that variations in

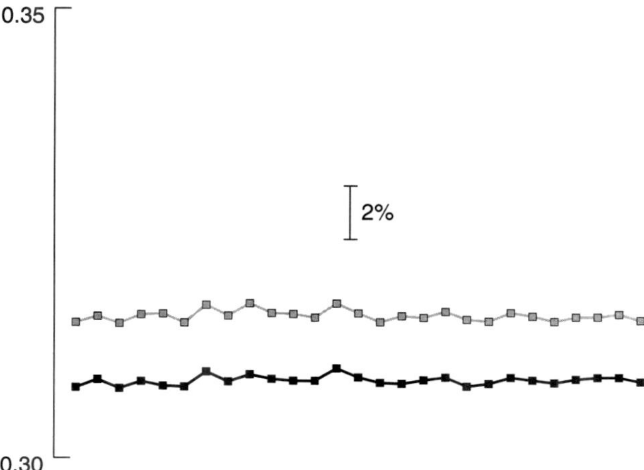

FIGURE 13 Variation of light output from a color CRT over 24 hours of continuous operation. This graph shows the chromaticity coordinates corresponding to Fig. 12. They show less variation than the tristimulus values which covary considerably.

the building power are the largest source of temporal instability. Furthermore, this variation affected all guns similarly, as is shown by the smaller variation in Fig. 13, which shows the chromaticity coordinates, thereby factoring out overall changes in intensity.

Measurements like this one are important for establishing the precision at which calibration is sensible: it isn't worthwhile to calibrate to a greater precision than the variation in the colorimetry over the period between calibrations. While this CRT has excellent temporal stability, significantly better than most other light sources, the same is not true for all CRTs. Some vary in output by as much as 20 to 30 percent over times as short as a few minutes.

The measurements shown in Figs. 12 and 13 give an idea of the variation of light output over periods of hours or days, and illustrate the precision it is possible to expect from a recently calibrated CRT. CRTs also vary on a time scale of years, but the effects are not documented. Anecdotal reports suggest the following. First, electronic components change with age, so properties that depend on the CRT electronics, such as video amplifier gain and bandwidth, change with age, almost always for the worse. Second, chemical properties do not change with age, so properties such as the spectral power of light emitted by a specific phosphor do not change. One anecdotal report[8] describes no variation in chromaticity coordinates of phosphor emission spectra over several years of CRT operation. It is possible, however, that phosphor concentrations diminish as tubes age, probably because the phosphor slowly evaporates. Such an effect would reduce the intensity of light emitted from the phosphor without changing its chromaticity. The magnitude of this effect is controversial.

Spatial Uniformity. The light emitted by a specific input signal varies a surprising amount from one area of the screen to another. Figure 14 shows the variation of luminance at constant input voltage as we measure different areas of the screen from a fixed measurement point. (Figure 15 shows the location on the screen of the measurement path.) Note that the light intensity decreases as the measurement point moves away from the center either horizontally or vertically, and is lowest in the corners. Two effects work together to create this effect. As the beam scans away from the center of the tube it meets the shadowmask at more and more oblique angles, making the holes effectively smaller. Because of the curvature of the tube and the finite distance of the observer, the edges and corners are viewed at angles off the normal to the tube. Light is emitted in a non-lambertian distribution, preferring directions closer to the normal to the tube face. The effects in Fig. 14 occur in all CRTs. How large they are, however, depends strongly on the type and setup of the monitor. Correcting for this nonuniformity is usually impractical. Doing so requires very extensive measurement.[9]

Closer examination of measured light shows, however, that many experimental

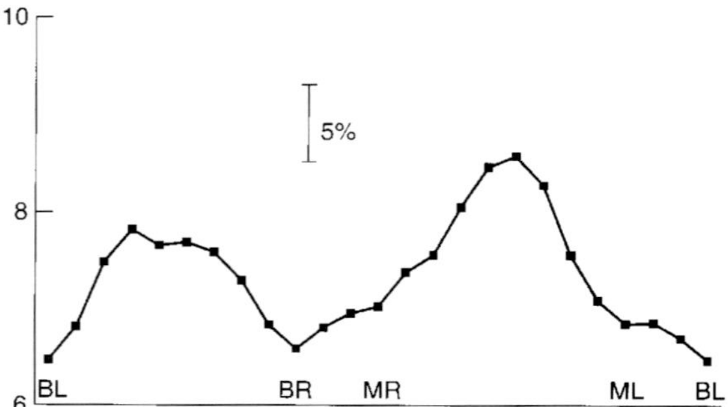

FIGURE 14 Variation of light output when different parts of a CRT screen are measured from a fixed point. Horizontal lines mark variations of about 5 percent.

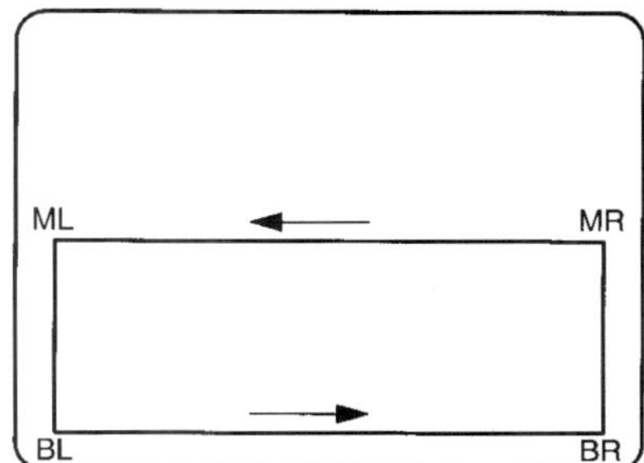

FIGURE 15 The measurement path used for the measurements shown in Fig. 14.

situations are not hampered by this nonuniformity. Usually the chromaticity coordinates are very close to being constant, even though the luminance varies greatly. General intuition about color, as well as some recent experiments,[10] shows that humans are quite insensitive to smooth luminance gradients, even when they are as large as 20 percent. This fact, combined with commonsense layout of experimental displays (making them symmetrical with respect to the center of the screen, for example) overcomes spatial nonuniformity without extensive measurement. Important to mention in this respect is the difficulty of creating good stereoscopic viewing conditions on a single monitor. First, there is only a single area of the screen that is ideal for the position of the center of an image. Second, unless one image is horizontally inverted, two images from opposite sides of the screen combined into a single image present drastically different luminance gradients to the two eyes.

Setup and Viewing Environments for Color CRTs

Many adjustments of the CRT electronics change its performance substantially. Some, such as the purity and convergence, have specific "correct" settings and are not designed to be user-adjustable. Most CRTs with in-line guns, for example, have the ring magnets that adjust the convergence glued into place at the factory. Other controls, such as contrast and brightness, are user-adjustable and should be changed when viewing conditions change if an optimal image is to be produced. These controls are provided in the expectation that the CRT will be displaying an image that is broadcast to a large number of CRTs that are viewed in very different visual environments. There is, therefore, a correct way to adjust these controls; the next few paragraphs describe the basic adjustments to be done. The procedure is based on a technical manual produced by the Canadian Broadcasting Corporation.[11] This manual also provides recommended viewing conditions for critical assessment of displayed images. The procedures and viewing conditions are expected to form the basis of an SMPTE standard for CRT image display.

When CRTs are used for visual experimentation, of course, they are often displaying specifically controlled images in unusual viewing conditions, often total darkness. For such applications the adjustment procedure described below is unlikely to be interesting, and extreme values of the controls are likely to be desired. For example, an experiment conducted in total darkness is likely to need the black level (brightness) set so that there is no background light emitted from the screen. Or an experiment to measure thresholds is likely to benefit from whatever value of the contrast control minimizes the gain of the CRT at the intensity levels where the measurement is performed. (The objective is to minimize intensity quantization of the stimulus.) In fact, modern CRTs with computer-controllable contrast and black level might be used with different control settings for different trials of a single experiment.

CRT Setup for Image Display. When a CRT is set up for image display, four adjustments are usually made: focus, brightness, contrast, and color balance and tracking. These adjustments form a rough order, with changes to one often requiring changes to succeeding ones. The following procedures are simplified from Benedikt.[11]

Focus. Focus should be adjusted by the shrinking raster method. A uniform gray field is displayed on the CRT, and the focus adjustment used to shrink the beam size until raster lines are clearly visible. The beam size is then increased until the raster lines just barely disappear. The smallest beam size for which no horizontal raster lines are visible indicates the correct setting for the focus control.

Brightness. The brightness or black-level control is set so that zero input to the CRT produces a visual impression of black. This setting *must* be performed in lighting conditions that are exactly those in which the images will be viewed, and with the observer at exactly the viewing distance at which the images will be viewed. With no signal input to the CRT, and the image set to underscan if possible, the brightness control is increased until the image area is noticeably gray. It is then reduced to the highest setting at which the image area looks black.

Contrast. An input that has all three guns fully on is used for this adjustment. With such a color displayed, the contrast control is adjusted until the luminance of the screen is the maximum luminance desired. This setting should be performed using either a luminance meter or a color comparator, a CRT adjustment device that allows an observer to view the CRT as half of a bipartite field, the other half containing a reference white at the correct luminance. In any case, it is essential that the CRT should not bloom at the contrast setting in use. Blooming occurs when a too-intense electron beam spreads after it has passed through the shadowmask, stimulating more than one phosphor. It reduces the purity of the image, with the visual consequence that bright saturated colors are washed out. Narrow lines of red, green, and blue at full intensity can be used to check for blooming.

Color balance and tracking. A white with red, green, and blue inputs equal at maximum intensity and a gray with red, green, and blue inputs equal at half intensity are used to set the color balance. In addition, colorimetric capability is needed. Visual colorimetry using a luminous white reference is usual in the broadcast industry, but this measurement can also be made instrumentally. For visual colorimetry the white reference is usually supplied by a color comparator. Adjust the red, green, and blue gain controls until the displayed white matches the reference in chromaticity. Then adjust the red, green, and blue screen controls until the displayed gray matches the reference in chromaticity. It may now be necessary to readjust the brightness and contrast. Do so, then repeat the color balance and tracking adjustment until all the three adjustments are simultaneously satisfactory.

Viewing Environments. In applications where image quality is critical it is necessary to control the viewing environment very closely. The following viewing conditions are typical of those used in the broadcast television industry.

1. The luminance of reference white is about 70 candelas per square meter.
2. The observer views the screen from a direction normal to the screen, with the screen-observer distance between four and six times the screen height.
3. The CRT should be surrounded by a neutral matte area at least eight times the screen area. The surround should have the chromaticity of the CRT reference white, and a luminance of about 10 candelas per square meter.
4. A narrow matte-black mask should frame the CRT image.

5. All room lighting should be as close as possible to the chromaticity of the CRT reference white.

27.4 COLORIMETRIC CALIBRATION OF VIDEO MONITORS

When users of color CRTs talk precisely about colorimetric calibration, a careful distinction between calibration and characterization is usually made. Measuring the input voltage/output color relationship well enough that it is possible to predict the output color from the input voltage, or to discover the input voltage needed to produce a given output color is a CRT characterization. The function mapping voltage to color characterizes the colorimetric performance of the CRT. Adjusting the CRT so that its characterization function matches the characterization function of a standard CRT is a CRT calibration. Talking loosely, however, calibration usually covers both characterization and calibration, and most CRT users are actually more interested in characterizations than in calibrations. Thus, this chapter describes several methods for creating colorimetric characterizations of CRTs. Some such characterization is an essential part of any calibration, but omits the detailed internal adjustment of a CRT, which requires significant electronic expertise if it is to be done safely.

There is no single characterization method that suits all needs. Thus, this chapter provides a variety of methods, each of which does some jobs well and others badly. They can be divided into three basic types.

1. *Exhaustive characterization methods* (*ECM*). Useful when good precision is needed over the complete monitor gamut. The same or similar algorithms can be used for the characterization of other output devices, such as printers. These methods tend to be both computationally and radiometrically expensive.
2. *Local characterization methods* (*LCM*). Useful when a precise characterization is needed for only a small portion of the monitor's output range.
3. *Model-dependent characterization methods* (*MDCM*). Useful when a characterization of moderate precision is needed for the complete monitor gamut. These methods tend to be specific to a given monitor and useful for only a small set of monitor setup methods, but they are computationally and radiometrically inexpensive. In addition they can be done so that the perceptual effects of mischaracterizations remain small, even when the colorimetric effects are large.

Here is a small set of criteria for deciding which type of method is best for a specific application:

1. High precision (better than 1 to 3 percent is impractical by any method discussed in this chapter)—ECM or LCM
2. Complete gamut characterization needed—ECM or MDCM
3. Minimal memory available for characterization tables—LCM or MDCM
4. No or slow floating point available—ECM or LCM
5. Fast inverse needed (we call the transformation from *RGB* to *XYZ* the characterization, the transformation from *XYZ* to *RGB* the inverse)—ECM, LCM, or MDCM
6. Forgiving behavior with out-of-gamut colors—LCM or MDCM
7. Photometry available but not radiometry—MDCM
8. Change of small number of parameters when monitor or monitor setup changes—LCM or MDCM

This list is not comprehensive; each type of characterization is actually a family of methods and there may be an effective method of a particular type even when general considerations seem to rule out that type. Undoubtedly more methods will be discovered as time goes on, leading to a broadening of these rules, and perhaps even a useful splitting of the characterization types into subtypes.

Important considerations when deciding on the characterization method to be used are measurement precision and monitor stability. There is no point in using a characterization method more precise than measurement precision, or more precise than the stability of the CRT over the range of conditions in which the characterization is to be used.

The characterization methods described below make use of a variety of measurements, but measurement methodology is not discussed. For reference, the methods used are (1) spectroradiometry, measurement of spectral power distributions; (2) colorimetry, measurement of tristimulus values; and (3) photometry, measurement of luminance.

The phrase "monitor coordinates" is used throughout this section. It denotes a set of controllable CRT inputs to which the experimenter has immediate access. For example, in applications where the CRT is computer controlled, it indicates the *RGB* values in the color lookup tables of the frame buffer that supplies the video signal to the CRT. The characterization techniques discussed below are independent of the monitor coordinates used. The single exception is the set of model-dependent characterization methods.

Exhaustive Characterization Methods

These are the most costly characterization methods in terms of time and computer storage. They are also the most precise and the most general. Their utility depends only on the stability of the monitor. (If the monitor doesn't have enough stability for exhaustive methods to be usable it is unlikely to have enough stability to use any other method.)

General Description. The idea behind exhaustive characterization is to measure all the colors a monitor can produce and store them in a large table. When a color of given monitor coordinates (*RGB*) is displayed, its tristimulus values are determined by looking them up in the table. When a color of given tristimulus coordinates is desired, the table is searched and that set of monitor coordinates closest to the desired value is chosen for display. (Dithering among nearby colors is also possible if more precision is desired.) This method is also useful for any kind of device, not just for monitors. Thus, software designed for monitors can be reused when, for example, printers must be characterized. The obvious drawback to this method is the number of measurements that must be made. For example, with 24-bit color (8 bits per gun), over 16 million colors can be produced. If each measurement takes 1 second, the measurement process consumes over 4500 hours, or almost 200 days, measuring around the clock. The solution to this problem is to sample the set of realizable colors, measure the samples, then interpolate between them. Thus, what we call exhaustive methods are only relatively exhaustive. Handling the practical problems is discussed in the following sections.

Sampling Densities and Interpolation Algorithms. How many measurements should be made, and which colors should be selected for measurement? The answer depends on the nature of the sampling algorithm to be used, and is an open research question for general sampling. Thus, while it is possible to discuss the issues, practical decisions depend on the experience of the user.

Sampling procedures. Most sampling algorithms sample the monitor coordinates linearly in the monitor coordinates. For example, if $512 = 2^9$ samples are to be taken for a color CRT, $8 = 2^3$ values on each of the red, green, and blue guns would be used, linearly interpolating the range of output voltages. The 512 samples would be the cartesian product of each set of gun values. Thus, if there were 256 possible values for each gun, running from 0 to 255, the 8 values chosen would be 0, 36, 72, 109, 145, 182, 218, and 255. If, as is often the case, saturation exists at the low and/or high ends of the input range, the full range is not used, and these values are scaled to the usable range. Behind this choice is the idea that a well-designed output device should have device coordinates that are close to perceptually uniform. If so, linear sampling in device coordinates roughly approximates even sampling in a perceptually uniform space. The density of the sampling is related to the interpolation algorithm, which must approximate the exact value to within the desired precision of the characterization. This necessarily entails a trade-off between measurement time and on-line computational complexity when the characterization is being used: the more dense the sampling, the simpler the interpolation algorithm, and vice versa.

Interpolation. Most often, linear interpolation is used. In this case, since we are interpolating in a cubic volume the interpolation is trilinear. Thus, given R_j (RGB), what are the tristimulus values?

1. We assume the existence of a table $X_i(R_j(n_j))$, consisting of the three tristimulus values X_i measured when the three guns are driven by the three voltages $R_j(n_j)$. Each three-vector n_j labels a different sample.
2. Find the sample m_j that has $R_j(m_j)$ less than and closest to the values to be estimated R_j.
3. Take the eight samples (m_0, m_1, m_2), $(m_0 + 1, m_1, m_2)$, $(m_0, m_1 + 1, m_2), \ldots,$ and $(m_0 + 1, m_1 + 1, m_2 + 1)$ as the vertices of a polyhedron.
4. Interpolate in R_0 on the four sides running from (m_0, m_1, m_2) to $(m_0 + 1, m_1, m_2)$, from $(m_0, m_1 + 1, m_2)$ to $(m_0 + 1, m_1 + 1, m_2)$, from $(m_0, m_1, m_2 + 1)$ to $(m_0 + 1, m_1, m_2 + 1)$, and from $(m_0, m_1 + 1, m_2 + 1)$ to $(m_0 + 1, m_1 + 1, m_2 + 1)$. The interpolation algorithm for $X_i(*, l_1, l_2)$ is given by

$$X_i(*, l_1, l_2) = \frac{R_0 - R_0(m_0)}{R_0(m_0+1) - R_0(m_0)} X_i(R_0(m_0+1), R_1(l_1), R_2(l_2)) + \frac{R_0(m_0+1) - R_0}{R_0(m_0+1) - R_0(m_0)} X_i(R_0(m_0), R_1(l_1), R_2(l_2))$$

   where l_j is either m_j or $m_j + 1$.
5. Treat the four values as the corners of a polygon. Interpolate in R_1 along the two sides running from (m_1, m_2) to $(m_1 + 1, m_2)$ and from $(m_1, m_2 + 1)$ to $(m_1 + 1, m_2 + 1)$. The interpolation algorithm for $X_i(*, *, l_2)$ is given by

$$X_i(*, *, l_2) = \frac{R_1 - R_1(m_1)}{R_1(m_1+1) - R_1(m_1)} X_i(*, R_1(m_1+1), R_2(l_2)) + \frac{R_1(m_1+1) - R_1}{R_1(m_1+1) - R_1(m_1)} X_i(*, R_1(m_1), R_2(l_2))$$

6. Treat these two values as the endpoints of a line segment, and interpolate in R_2. The final value is given by

$$X_i = \frac{R_2 - R_2(m_2)}{R_2(m_2+1) - R_2(m_2)} X_i(*, *, R_2(m_2+1)) + \frac{R_2(m_2+1) - R_2}{R_2(m_2+1) - R_2(m_2)} X_i(*, *, R_2(m_2))$$

The above equations implement trilinear interpolation within a (possibly distorted) cube. It has recently been observed that tetrahedral interpolation has some desirable properties that are lacking in trilinear interpolation. It is accomplished by subdividing the cube into five or six tetrahedra, the corners of which coincide with the corners of the cube. Barycentric coordinates are then used to determine the tetrahedron in which the displayed color lies, and to interpolate within that tetrahedron. More complex interpolation methods are also possible for nonuniform measurement sampling.[12]

Such interpolation methods implement the characterization directly in terms of the measured values. To test the adequacy of the interpolation it is important to choose regions of color space where curvature of the input/output response is expected to be high and to test the interpolation against exact measurements in that region of color space. If the interpolated values do not match the measured values there are two possible solutions: increase the measurement sampling density or improve the interpolation function. The first is expensive in terms of measurement time and computer memory; the second in terms of on-line calculation time. (There is a middle way, where a complicated algorithm is used to interpolate among coarsely sampled measurements for the purpose of generating a table large enough that linear interpolation can be done on-line. This has been used for hard-copy devices but not for monitors.) If the interpolation function is to be generalized it is possible to use higher-order interpolating functions, like splines, or to linearize the measurement space by transforming it before interpolating. For example, it is possible to build a characterization function that provides the logarithms of tristimulus values in terms of the logarithms of input coordinates. Any reasonable powerful generalization is bound to be computationally expensive. Much remains to be done to improve the sampling and interpolation process.

Inverses. Calculating inverses for multidimensional tabular functions is not straightforward. Speed of computation can be optimized by creating an inverse table, giving values of R_j for regularly sampled values of X_i. The easiest way to construct this table is not by measurement, but by calculation (off-line) from the characterization table. Once the table is available, linear interpolation can be done on-line to provide voltages corresponding to given tristimulus values. Two possible methods exist for calculating inverses from tables.

1. Newton's method of finding zeros using derivatives to calculate an error function is possible, but can have bad numerical instabilities on tabular data that contains measurement error.
2. Examine the table that drives the forward transform to find the cell that contains the desired tristimulus values. Then subdivide the cell, using trilinear interpolation to determine new values for the corners. Determine which subcell contains the desired tristimulus values, and continue subdividing until the solution is sufficiently precise. This method is robust, but too expensive computationally to use on-line.

A promising alternative, not yet in wide use, takes advantage of barycentric coordinates in a tetrahedral lattice to find the appropriate cell quickly.

Nonlinear interpolation schemes can be used in the forward transformation when producing tables for linear inverse mappings. Presumably tables for nonlinear interpolation schemes in the inverse mapping can also be produced.

Three issues about inverses stand out.

1. Nonlinear schemes have not been sufficiently explored for their potential to be clear.
2. The straightforward application of well-understood computational methodology can provide substantial improvements over present methods.
3. Whatever inversion method is employed, out-of-gamut colors remain a serious unsolved problem.

Out-of-Gamut Colors. What should be done when a program is asked to display a set of tristimulus values that are outside the color gamut of the monitor on which the color is to be displayed? The answer, of course, depends on the nature of the application. If the application demands drastic action when a request for an out-of-gamut color occurs, the solution is easy. The interpolation algorithm returns an illegal value of R_j. Then the application can display an error color in that area, or exit with an error. If, instead, the application must display a reasonable color within the monitor gamut, there is no natural solution. Solutions like projecting onto the surface of the monitor gamut have been used with success for some applications. Unfortunately, however, they are frequently computationally expensive and visually unsatisfactory.

Local Characterization Methods

In many applications only a small region of the color gamut of the monitor must be characterized, but that region must be characterized very precisely. A typical example is a threshold experiment in which there are a small number of reference stimuli. Only colors close to the reference stimuli need be characterized precisely since they are the only colors likely to arise in the experiment. Methods that are specialized for the characterization of small parts of the color gamut, such as the threshold stimulus set, are the subject of this section. They are the methods most often appropriate for vision experimentation.

General Description. Local characterization methods try to take advantage of simplifications that arise because the set of colors to be characterized is small. For example, a small set of colors all very close to a reference color can be characterized by a linear approximation to the global characterization function. Such simplifications offer many desirable properties such as high precision with minimal measurement (individual colors), linear characterizations and inverses (local regions of color), and simple inverses over extended color ranges (one-dimensional color spaces). To realize this precision, colorimetric measurements of the required precision must be available and easily usable. Detailed descriptions of three limited-gamut characterization schemes follow. Others are certainly possible, and may be worked out by analogy.

Individual colors. It is often the case that a small number of distinct colors is needed. Under such circumstances the best method available is to perform a colorimetric measurement on each color. This is easy if the colors are arbitrary but need to be known precisely. Then they can be chosen by their *RGB* coordinates, with colorimetric measurements used to establish their tristimulus values. It is less easy if a small number of colors of specified tristimulus values are needed. The best way to deal with the latter problem is to use on-line colorimetric measurement, adjusting *RGB* values until the required tristimulus values are produced. This should be repeated several times throughout the time during which the characterization is needed to establish the stability of the monitor while the characterization is in use.

This type of characterization leads most directly into the most difficult question that CRT characterization must face: how should a characterization handle spatial and angular variations of light emitted by the CRT? Suppose, for example, an experiment uses only two color stimuli, a red one and a green one. Suppose further that the red one always appears at one point on the screen, while the green one always appears at a different point. Clearly, a characterization method that performs a colorimetric measurement on each stimulus yields maximum precision for minimum effort. Now, should the green color be measured in its particular location and the red one be measured in its location, or should both be measured in a common location, probably the screen center? And how should possible variations in the color of a stimulus when the other is turned off and on be handled? There is no universal best way of resolving such questions. Each must be resolved in a way that suits the particular display application. Note that this problem does not arise only for individual color characterizations. Indeed, it arises with any characterization whatsoever. The fact that every other aspect of individual color characterizations is so simple makes it particularly noticeable in this case.

Local regions of color. Somewhat more complicated than individual color characterizations are ones in which it is necessary to characterize a small part of the color gamut surrounding a particular color. This might arise, for example, in a matching experiment, where all matches are close to a fixed reference stimulus. In such cases a small region of the color gamut must be characterized very precisely, with an emphasis on exact presentation of differences of color between the reference color and other colors in the region. The procedure is to make a precise colorimetric measurement of the reference color. Call the result $\mathbf{X}_0$ with components X_{0i}, which is the result of measuring a color having input voltages R_{0i}. Then, for small changes in input voltage ΔR_i, measure the corresponding changes in tristimulus values ΔX_i. When the results are plotted they show a region in which the changes in tristimulus value are linearly related to changes in input coordinate with nonlinear effects growing in importance near the edge of the region. The size of the nonlinear effects, when compared to the required prevision of the characterization, determines the size of the region that can be characterized linearly. Within this region the tristimulus values are given by

$$X_i = X_{0i} + \sum_{j=1}^{3} M_{ij}\,\Delta R_j.$$

The matrix M_{ij} is determined by a multilinear regression of the data, with each component of the tristimulus values regressed simultaneously against all three input coordinates. Nonlinear effects and interaction terms are not included in the regression. The matrix entry M_{ij}, which arises as a regression coefficient, is the change of X_i for a unit change in R_j. Thus, for example, M_{23} describes how much the Y tristimulus value varies with changes in the B gun of the monitor. This type of characterization is probably the most common in color research. It is most important to remember to determine the limits of such a linear characterization, or at least to determine that the limits are outside the region in which the characterization is to be used.

One-dimensional color spaces. Often it is necessary to characterize a one-dimensional set of colors, that is, a line, not necessarily straight, in color space. Here is a method for doing so easily. Make a set of colorimetric measurements spaced more or less evenly along the curve. Use the variable μ for the measurement number, numbering from one end of the line to the other. Plot the measured tristimulus values and the input voltages as functions of μ. The measurement points should be dense enough that intermediate values can be approximated by linear interpolation. Now any set of RGB values must correspond to a value of μ, not necessarily integral. How is this value determined?

1. Each measured μ has a corresponding set of RGB values.

2. Find two consecutive sets such that the desired RGB values are intermediate between the RGB values of the sets. Call the μ value of the sets μ_0 and $\mu_0 + 1$. Thus, mathematically,

$$R_i(\mu_0) \leq R_i \leq R_i(\mu_0 + 1) \quad \forall i$$

3. Now calculate "how far" between μ_0 and $\mu_0 + 1$ the desired RGB lies. The value for gun j is δ_j where

$$\delta_j = \frac{R_j - R_j(\mu_0)}{R_j(\mu_0 + 1) - R_j(\mu_0)}$$

4. If the three δ values are close together this method works. Otherwise, the line of colors must be measured more densely.

5. Use $\delta = (\delta_1 + \delta_2 + \delta_3)/3$ as the distance between μ_0 and the desired *RGB* value. This distance can then be used to interpolate in the tristimulus values.

6. The interpolated result is the characterized result.

$$X_i = X_i(\mu_0) + \delta(X_i(\mu_0 + 1) - X_i(\mu_0))$$

This method requires no special precautions if the line is relatively straight in *RGB* space. It is important to use the interpolation to calculate the tristimulus values of some measured colors to establish the precision of the measurement, or to decide if denser sampling is necessary.

Inverses. The major reason for using this family of characterization methods is the computational simplicity of the characterizations and their inverses. Neither extensive memory nor expensive computation is needed to realize them.

Individual colors. Since there is a small number of discrete colors, they can be arranged in a table. To get the characterization, you look up *RGB* in the table and read off the tristimulus values. To get the inverse, look up the tristimulus values in the output side of the table and the corresponding input gives *RGB* for the desired output. If the tristimulus values to be inverted are not in the table, then the color is not in the characterized set and it cannot be realized.

Local regions of color. To derive the inverse for local regions of color, write the characterization equation as

$$X_i - X_{0i} = \Delta X_i = \sum_{j=1}^{3} \mathbf{M}_{ij}\, \Delta R_j$$

Then, inverting the matrix $\mathbf{M}_{ij}$ to get $\mathbf{M}_{ji}^{-1}$,

$$\Delta R_j = \sum_{i=1}^{3} \mathbf{M}_{ji}^{-1}\, \Delta X_i$$

which can be written explicitly as a solution for the *RGB* values needed to generate a color of given tristimulus values. That is,

$$R_j = R_{0j} + \sum_{i=1}^{3} \mathbf{M}_{ji}^{-1}(X_i - X_{0i})$$

It is important, after doing the above calculation, to check that the *RGB* values so determined lie within the region for which the characterization offers satisfactory precision. If not, however, unlike other methods, the inverse so determined is usually a reasonable approximation to the desired color.

Instead of inverting the matrix off-line, the three linear equations in R_j can be solved on-line.

One-dimensional color spaces. For one-dimensional color sets the interpolation equations provide the inverse provided that *RGB* values and tristimulus values are reversed. To avoid confusion ϵ is used in place of δ.

1. To calculate ϵ find a consecutive pair of tristimulus values that contain the desired values. Call the first of the pair μ_0.

2. Then calculate ϵ_i using

$$\epsilon_i = \frac{X_i - X_i(\mu_0)}{X_i(\mu_0 + 1) - X_i(\mu_0)}$$

If the three ϵ_i values are not close together, then the tristimulus values designate a color that is not in the set, or the wrong interval has been found in a line of colors that must be very convoluted, and the set must be measured more densely.

3. Use $\epsilon = (\epsilon_1 + \epsilon_2 + \epsilon_3)/3$ as the distance between μ_0 and the desired tristimulus values.

4. Then the *RGB* values corresponding to the desired tristimulus values are

$$R_j = R_j(\mu_0) + \epsilon(R_j(\mu_0 + 1) - R_j(\mu_0))$$

The results of this calculation should be checked to ensure that the *RGB* values do indeed lie in the color set. This is the second check that is performed, the first being a check for a small spread in the three ϵ_i values. The checks are needed if the set is convoluted, and are always passed if the set is simple.

One way of checking whether the measurement density is adequate is to cycle an *RGB* value through the characterization and its inverse. If it fails to meet the required precision, then the measurement density is probably insufficient.

Out-of-Gamut Colors. Local gamut characterization methods are attractive because of their easy-to-use inverses. Their treatment of out-of-gamut colors is also appealing.

Individual colors. Colors not in the set of individual colors do not have inverses, and are not found in a search of tristimulus values. Because this characterization method ignores colors near to or between the measured colors, no method is possible for representing colors that are not in the measured set. However, if the search is done with finite precision, colors close to a measured color are aliased onto it in the search. The precision of the search determines the region of colors that is aliased onto any measured color.

Local regions of color. When a color is outside the local region to which the characterization applies this method returns the input coordinates that would apply to a linear characterization. The values are close to those produced by an exact characterization, but outside the bounds of the specified precision. The error increases as the color gets farther away from the characterized region, but slowly in most cases. Thus, this method clearly indicates when a color is outside the characterized region and fails gracefully outside the region, giving values that are close, if not exact.

One-dimensional color spaces. When a set of *RGB* values or tristimulus values lies outside the one-dimensional space, the fact is indicated during the calculation. The calculation can be carried through anyway; the result is a color in the space that is close to the given color. Exactly how close, and how the characterization defines the "closest" color in the one-dimensional space, depends on the details of how the measurement samples are selected, the curvature of the color set, and the appropriate experimental definition of "close." If this method comes to have wider utility, investigation of sampling methods might be warranted to find the "best" sampling algorithms. In the meantime, it is possible to say that for reasonably straight sets colors close to the set are mapped onto colors in the set that are colorimetrically near the original color.

Model-Dependent Characterization Methods

The two types of characterization methods described above are independent of any particular monitor properties. In fact, they can be used for any color-generating device. The third type of characterization makes use of a parametrized model of CRT color production. A characterization is created by measuring the parameters that are appropriate to the CRT being characterized. The next few sections describe the most common model for color CRTs, which is tried and true. Some CRTs may require it to be modified, but any modifications are likely to be small. The emphasis in this description is the set of assumptions on which the model is based, since violations of the assumptions require modifications to the model.

General Description. The standard model of a color CRT has the following parts.

1. Any displayed color is the additive mixture of three component colors. The component colors are generally taken to be the light emitted by a single phosphor.
2. The spectral power distribution of the light in each component is determined by a single input signal, *R, G,* or *B,* and is independent of the other two signals.
3. The relative spectral power distribution of the light in each component is constant. Hence, the chromaticity of the component color is constant.
4. The intensity of the light in each component is a power function of the appropriate input voltage.

Taken together, these parts form a mathematical model of CRT colorimetry.

The standard model has been described many times. For a concise presentation see Cowan;[13] for a historical one see Tannenbaum.[14]

Gun independence. The light emitted when the input coordinates are the three voltages v_R, v_G, and v_B (generically called v_a) is $\Phi_\lambda\ (v_R, v_G, v_B)$. It turns out to be convenient to form slightly different components than the standard model does, following not the physical description of how a monitor operates, but the logical description of what is done to realize a color on the monitor. The first step when generating a color is to create the *RGB* input from the separate *R, G,* and *B* inputs. Imagine turning on each input by itself, and assume that the color when all guns are turned on together is the additive mixture of the colors produced when the guns are turned on individually. This assumption is called "gun independence." In terms of tristimulus values it implies the condition

$$X_i = X_{Ri} + X_{Gi} + X_{Bi}$$

(Usually CRTs realize gun independence by exciting different phosphors independently,

$$\Phi_\lambda(v_R, v_G, v_B) = \Phi_\lambda(v_R, 0, 0) + \Phi_\lambda(0, v_G, 0) + \Phi_\lambda(0, 0, v_B)$$

This condition is, in fact, stronger than the gun-independence assumption, and only gun independence is needed for characterization.)

Gun independence was described by Cowan and Rowell,[15] along with a "shotgun" method for testing it. The tristimulus values of many colors were measured, then predictions based on the assumption of gun independence were tested. Specifically, gun independence implies consistency relationships that must hold within the set of measurements. Cowan and Rowell showed that a particular CRT had a certain level of gun independence, but didn't attempt to make measurements of many CRTs. Clearly, it is worth making measurements on a variety of monitors to determine how widely gun independence occurs, and how large violations of it are when they do occur. For CRTs without gun independence there are two ways to take corrective action:

1. Use a characterization method that is not model-dependent.
2. Modify the monitor to create gun independence. For most CRTs there are good,

inexpensive methods for improving gun independence. They will be more widely available in the future as characterization problems caused by failures of gun independence become more widely known.

Extreme settings of the monitor controls can cause violations of gun independence in otherwise sound monitors. The worst culprit is usually turning the brightness and/or contrast up so high that blooming appears at high input levels.

Phosphor constancy. When gun independence holds, it is necessary to characterize only the colors that arise when a single gun is turned on, since the rest can be derived from them. Usually, it is assumed that the colors produced when a single gun is turned on have constant chromaticity. Thus, for example, the tristimulus values of colors produced by turning on the red gun alone take the form

$$X_{Ri}(v_R) = E_R(v_R) \cdot x_{Ri}$$

where x_{Ri}, the chromaticity coordinates of the emitted light, x_R, y_R, and z_R, are independent of the voltage input. Thus the tristimulus values of the color produced depend on input voltage only through the intensity $E_R(v_R)$.

The engineering that usually lies behind phosphor constancy is the following. A CRT should be designed so that the beam current in a given electron gun is independent of the beam current in any other gun. The deflection and shadowmask geometries should be designed and adjusted so that the electrons from any gun fall only on a single phosphor type. The physical properties of the phosphor should guarantee that the phosphor emits light of which the chromaticity is independent of the intensity. Meeting these conditions ensures phosphor constancy. It is possible, but unusual, to have phosphor constancy under other conditions, if it happens that effects cancel out appropriately.

The measurement of phosphor constancy is described by Cowan and Rowell.[15] Here is a short description of how to make the required measurements for one of the guns.

1. Turn the gun on at a variety of input voltages, and make a colorimetric measurement at each input voltage.
2. Calculate chromaticity coordinates for each measurement, which are independent of input voltage if phosphor constancy is present.
3. Constancy of chromaticity should also obtain in the presence of varying input to the other guns. To consider this case, turn the two other guns on, leaving the gun to be measured off.
4. Measure the tristimulus values for a baseline value.
5. Then turn the gun to be measured on to a variety of input voltages, making a colorimetric measurement at each.
6. Subtract the baseline tristimulus values from each measurement, then calculate the chromaticity coordinates. These should be constant, independent of the input to the measured gun, and independent of the input to the other two guns as well.

One of the most common monitor characterization problems is a measured lack of phosphor constancy. This is a serious problem, since the functions derived from the CRT model cannot then be easily inverted. Sometimes the measured lack of phosphor constancy is real, in which case there is no solution but to use a different type of characterization procedure. More often, the measurement is caused by poor CRT setup or viewing conditions. This occurs as follows, first for viewing conditions. Ambient light reflected from the screen of the monitor is added to the emitted light. This light is independent of input voltage, and adds equally to all colors. Imagine a phosphor of constant chromaticity (x, y), at a variety of input voltages. The tristimulus values of light emitted from the screen are

$$X_i = X_{0i} + E(v) \cdot x_i$$

where X_{0i} are the tristimulus values of the ambient light reflected from the screen. Note that subtracting the tristimulus values measured when the input voltage is zero from each of the other measurements gives phosphor constancy. (Failure to do the subtraction gives phosphor chromaticities that tend toward white as the input voltage decreases.) This practice should be followed when ambient light is present. Psychophysical experiments usually go out of their way to exclude ambient light, so the above considerations are usually not a problem. But under more normal viewing conditions ambient light is always present.

When the black level/brightness control is set above its minimum there is a background level of light emitted from the screen even when the input voltage is zero. This light should be constant. Thus, it can be handled in exactly the same way as ambient light, and the two can be lumped together in a single normalizing measurement.

Now let's put together this gun independence and phosphor constancy. Gun independence means that the tristimulus values $X_i(v_R, v_G, v_B)$ can be treated as the sum of three tristimulus values that depend on only a single gun:

$$X_i(v_R, v_G, v_B) = \sum_{a=R,G,B} X_{ai}(v_a)$$

Phosphor constancy means that the tristimulus values from a single gun, once the background has been subtracted away, have constant chromaticity:

$$X_i(v_R, v_G, v_G) = (X_0, Y_0, Z_0) + \sum_{a=R,G,B} E_a(v_a) \cdot x_{ai}$$

To complete the characterization a model is needed for the phosphor output intensity $E_a(v_a)$.

Phosphor output models. Several different methods exist for modeling the phosphor output. The most general, and probably the best, is to build a table for each gun. To do so requires only photometric measurement, and the responsivity of the photometer need not be known. (This property of the model is very convenient. Colorimetric measurements are fairly easy to perform at light levels in the top 70 percent of the light levels produced by a CRT, but hard to perform for dark colors. In this model the top of the output range can be used when doing the colorimetric measurements required to obtain phosphor chromaticities. Then, photometric measurements can be performed easily using photodiodes to calibrate the output response at the low end of the CRT output.) Choose the largest input voltage to be used, and call it $v_{a\max}$. Measure its light output with the photometer. Then measure the light output from a range of smaller values v_a. The ratio of the light at lower input to the light at maximum input is the relative excitation $e_a(v_a)$. Store the values in a table. When excitations for intermediate voltage values are needed, find them by linear interpolation.

$$e_a(v_a) = \mu e_a(v_{a1}) + (1 - \mu) e_a(v_{a2})$$

where the interpolation parameter μ is

$$\mu = \frac{v_a - v_{a2}}{v_{a1} - v_{a2}}$$

The memory requirements of the table can be reduced by using more complicated functions, but at the price of complexity of computation in the model.

It is also possible to approximate the measurements using a parameterized function. The most commonly used form is the gamma correction equation

$$e_a(v_a) = e_{a\max}(v_a / v_{a\max})^{\gamma_a}$$

with two parameters γ_a and $e_{a\max}$. Here, $v_{a\max}$ is the maximum input voltage, $e_{a\max}$ is the maximum relative excitation, usually taken to be 1.0, and γ_a is called the gamma correction exponent. It is determined empirically by regressing the logarithm of the measured excitation against the logarithm of the input voltage.

Normalization. The excitation that enters the characterization equation $E_a(v_a)$ is the product of the voltage-dependent relative excitation $e_a(v_a)$ and a normalization coefficient N_a:

$$E_a(v_a) = N_a \cdot e_a(v_a)$$

The normalization coefficients complete the characterization model.

Summary equation. The characterization is summarized by the single equation

$$X_i = X_{0i} + \sum_{a=R,G,B} N_a \cdot e_a(v_a) \cdot x_{ai}$$

which provides the tristimulus values for any set of *RGB* input coordinates.

Conditions for Use. The assumptions discussed above are the conditions under which the model developed here can be used successfully. The assumptions should be checked carefully before this characterization method is used. The departure of the CRT from the ideal CRT, defined by perfect adherence to these conditions, is a measure of the imprecision in the characterization procedure. Note that when the CRT fails to meet the conditions, which occurs most frequently near the edges of its gamut, the erroneous colors are usually plausible. Thus, the model can be in error by a surprisingly large amount and still produce satisfactory characterizations in noncritical applications.

Partial models. It is fairly clear that there are partial models meeting some of the conditions, and that they provide useful approaches to characterization. For example, suppose that the phosphor chromaticities vary with input voltage. They can be stored in a table indexed by input voltage (exact values are produced by interpolation) and used in a variant of the characterization equation:

$$X_i = X_{0i} + \sum_{a=R,G,B} N_a \cdot e_a(v_a) \cdot x_{ai}(v_a)$$

Such a generalization works well for the transformation from input voltages to tristimulus values, but the inverse transformation is virtually unusable.

Measurement of Parameters. The characterization equation requires the measurement of a set of parameters that varies from CRT to CRT. Thus, they must be measured for the CRT to be characterized. Different types of measurements—spectral, colorimetric, and photometric—are needed, as described below.

Ambient light and black level X_{0i}. A colorimetric measurement of the screen with all input voltages set to zero produces this value. It should be measured under the exact conditions in which the characterization will be used.

Phosphor chromaticities x_{ai}. Colorimetric measurements of the screen with a single gun turned on produces these values. Careful subtraction of the ambient light and black level is important. Measuring at a variety of input voltages produces:

1. A measure of phosphor constancy
2. The range of input voltages over which phosphor constancy holds well enough for use of the characterization model
3. A value for the phosphor chromaticities, produced by averaging over the chromaticities in the characterization range

Gamma correction functions. Measurement of the relationship between the input voltages and the excitation functions requires photometric measurement at a large variety of input voltages. If interpolation into a table is the method of choice, the sampling of input voltages must be dense enough that the interpolation method chosen, usually linear, is close enough to the exact values to yield the desired accuracy. If a functional relationship, like the power function, is chosen the validity of the function chosen must be inspected very carefully. Whatever method is chosen it is essential that the ambient light/black level be subtracted from the measurements. If it is not, it is counted three times when the contributions from the three guns are added together.

Normalization coefficients. The normalization coefficients are determined through the characterization equation

$$X_i = \sum_{a=R,G,B} N_a \cdot e_a(v_a) \cdot x_{ai}$$

They may be determined in several ways, falling into two categories: measurement methods and comparison methods. A typical measurement method assumes a single color of known tristimulus values. This knowledge can be produced by a colorimetric measurement or by varying *RGB* to match a known sample such as the reference field of a color comparator. The tristimulus values are substituted into the equation above, giving three equations for the three normalization coefficients. Solving the equations gives the normalization coefficients. Note that the relative excitations and the phosphor chromaticities are both unitless. Thus, the normalization coefficients carry whatever units are used to measure the tristimulus values. The units enter naturally when the linear equations are solved.

The method for solving these linear equations utilizes the inverse transform. A matrix **M**, the entries of which are the phosphor chromaticities

$$\mathbf{M}_{ai} = x_{ai}$$

is defined, and its inverse, $\mathbf{M}_{ia}^{-1}$, determined. Multiplying it into the characterization equation gives

$$\sum_{i=1}^{3} \mathbf{M}_{ia}^{-1} X_i = N_a \cdot e_a(v_a)$$

Then N_a is calculated from

$$N_a = e_a(v_a) \cdot \sum_{i=1}^{3} \mathbf{M}_{ia}^{-1} X_i.$$

Note that $e_a(v_a)$ is known since the input voltage that matches the standard, or of the color that has been measured, is known.

The second method is used when it is possible to match a known standard. For example, it might be possible to find voltages that make colors produced by individual guns equiluminous. Call the voltages v_{YR}, v_{YG}, and v_{YB}, and substitute them into the $Y = X_2$ characterization equation

$$N_R e_R(v_{YR}) x_{R2} = N_G e_G(v_{YG}) x_{G2} = N_B e_B(v_{YB}) x_{B2}$$

This equation is easily solved for the ratios N_G/N_R and N_B/N_R, leaving a single constant, an overall normalizing coefficient, undetermined. This constant, which carries the units in which the tristimulus values are measured, can be determined only by an absolute measurement (or, equivalently, a color match to an absolute standard). However, because vision is insensitive to variations in overall intensity, the precise value of the overall normalizing coefficient is unimportant in many applications.

Inverse Transformations. One of the most attractive features of model-dependent characterizations is the simplicity of their inverses. It is simple to invert the characterization equation

$$X_i = \sum_{a=R,G,B} N_a \cdot e_a(v_a) \cdot x_{ai}$$

which determines the tristimulus values of a color in terms of the input voltages that cause it to be displayed, into the inverse equation, which specifies input voltages that create a color of given tristimulus values. Define the matrix of phosphor chromaticities

$$\mathbf{M}_{ai} = x_{ai}$$

Then determine its inverse $\mathbf{M}_{ia}^{-1}$ by conventional means. It is used to solve the characterization function

$$\sum_{i=1}^{3} \mathbf{M}_{ia}^{-1} X_i = N_a \cdot e_a(v_a)$$

Then, after the normalization coefficients are divided out

$$e_a(v_a) = \frac{1}{N_a} \sum_{i=1}^{3} \mathbf{M}_{ia}^{-1} X_i$$

It is then necessary to find the input voltages that give the appropriate excitations e_a, which requires inversion of the excitation function. Thus,

$$v_a = e_a^{-1}\left(\frac{1}{N_a} \sum_{i=1}^{3} \mathbf{M}_{ia}^{-1} X_i\right)$$

If the relative excitation function is specified in a closed form, the inverse can be calculated using only elementary algebra. Otherwise, it is necessary to develop a tabular inverse, which is most easily done by one-dimensional interpolation.

Violations of phosphor constancy. When phosphor constancy does not hold, the inverse cannot be calculated using the methods given above. Numerical inversion techniques, which are beyond the scope of this section, can be used, but it is probably a better idea to use a different calibration method.

Out-of-Gamut Colors. Most inversion methods fail internally if they are asked to produce the input coordinates needed to display an out-of-gamut color. For example, tabular inverses fail to find an interval or cell in which to interpolate. Because the functions in model-dependent characterizations are well defined beyond the gamut, these methods generally do not fail, but calculate values that lie outside the domain of input coordinates. Thus, input coordinate values calculated from inverses of model-dependent characterizations should be checked, because writing out-of-range values to color lookup tables produces unpredictable results.

Absolute Characterization versus Characterization for Interaction

The characterization methods discussed above are designed to provide absolute characterizations, which are needed for color coordinates defined in terms of an external standard. Probably just as often characterization is needed to support specific interaction techniques. Suppose, for example, an experiment calls for the observer to have three CIELAB controls, one changing L^*, one changing a^*, and one changing b^*. To make this possible a characterization of the CRT in terms of CIE tristimulus values must be done to provide a basis for the transformation from tristimulus values to $L^*a^*b^*$. This type of characterization situation has several characteristics that are quite common:

1. At any given time there is a known color on the screen and the problem is to calculate a color slightly different from it. This situation arises because the control is sampled frequently enough that large color changes are produced only cumulatively over many color updates.
2. The new color coordinates must be calculated quickly and frequently.
3. Errors in the calculated color increments are errors in quantities that are already small. Thus, they can be considerably larger (in percentage terms) than would be tolerable for an absolute characterization.
4. The inverse transformation, tristimulus values to *RGB* is usually needed.

This combination of requirements is most easily met when the characterization has a linear inverse, and as little calculation as possible in the main loop. Exhaustive characterization methods have the necessary ingredients as they stand, provided an inverse table has been built, as do local methods. But nonlinear transformations, like the one from $L^*a^*b^*$ to tristimulus values violate the conditions imposed above, and it is necessary to linearize them. They can then be combined with whatever characterization is in use to linearize the entire transformation from input coordinates to color specification.

Linearization. Consider a transformation f that takes a vector $\mathbf{x}$ into a vector $\mathbf{y}$. An example is the transformation between $L^*a^*b^*$ and XYZ. Suppose two vectors that are known to transform into each other are known: $\mathbf{x}_0$ ($L_0^*a_0^*b_0^*$) corresponds to the set $\mathbf{y}_0$ ($X_0Y_0X_0$). Now, if one vector undergoes a small change, how does the other change? If the transformation is suitably smooth (and the functions of color science are adequately smooth), small changes one variable can be expressed in terms of small changes of the other by simple matrix multiplication

$$\Delta y_i = \sum_j \mathbf{M}_{ij}\, \Delta x_j$$

The entries in the matrix **M** are the partial derivatives of f with respect to $\mathbf{x}$, evaluated at $\mathbf{x}_0$. That is,

$$\mathbf{M}_{ij} = \left.\frac{\partial f_i}{\partial x_j}\right|_{\mathbf{x}=\mathbf{x}_0}$$

Even though the computation of the derivatives may be expensive, they need to be calculated only infrequently, and the calculation can be scheduled when the system has idle resources available.

Practical Comments

CRT characterization is essentially a practical matter. It is almost always done because a superordinate goal requires it. Thus, it is appropriate to end this chapter with several practical remarks that make characterization easier and less costly.

Do as little characterization as possible! Characterization is usually done within a fixed budget of time and equipment. Identifying the least characterization that provides all the information needed for the application allows the extra resources to be used to improve precision and to check measurements.

Characterize frequently! CRTs age and drift; colleagues turn knobs unaware that they change a critical characterization. Discovering that a CRT does not match its characterization function invalidates all data collected or experiments run since the characterization was last checked. There are only two defenses. First, measure the CRT as often as is feasible. Second, devise visual checks that can be run frequently and that will catch most characterization errors. It is possible to use luminance-matching techniques like minimally distinct border or minimum motion to check characterizations in a few seconds.

Understand the principles underlying the characterization methods and alter them to fit the specific stimuli needed for the application. In particular, CRTs have many ways of being unstable. The best defense is to characterize the CRT with stimuli that are as close as possible to those that are used in the application: same size, same position on the screen, measurement apparatus at the observer's eye point, same background color and intensity, and so on. Such adherence is the best defense against inadvertently finding a new type of CRT instability that invalidates months of work.

27.5 AN INTRODUCTION TO LIQUID CRYSTAL DISPLAYS

The purpose of this section is to provide an introduction to the operational principles that provide color output from liquid crystal displays (LCDs). No actual colorimetric measurements are shown, because LCDs are evolving too fast for them to be of any lasting interest. Instead, schematic illustrations are given, showing features that the author considers likely to be present in future displays.

At the time of writing, colorimetric quality of LCDs is uneven. Nonetheless, since LCDs are semiconductor components and since they are now being produced in large quantities for low-cost computers, it is reasonable to expect a rapid decrease in cost combined with an increase in quality over the next decade. As the succeeding sections should make clear, they have a variety of interesting properties which will often make them preferred to CRTs as visual stimulators. Only by actively following their development will it be possible to use them at the earliest opportunity.

Two properties make liquid crystal devices particularly interesting to the visual scientist. First, they are usable as self-luminous, transmissive, or reflective media. They combine this flexibility with the range of temporal and spatial control possible on a CRT. This property is likely to make them particularly valuable for studies of color appearance and color constancy. Second, they have the ability to produce temporal and spatial edges that are much sharper than those producible on a CRT. This property has been called "acutance" in the display literature.[16] High acutance is potentially a serious problem for image display, since the gaussian blur that occurs at edges in CRT images helps to hide artifacts of the CRT raster. But high acutance also presents an opportunity since many small-scale irregularities that cannot be controlled on the CRT can be controlled on the LCD.

Operational Principles of Monochrome LCDs

This section describes the operation of monochrome LCDs. The components of color LCDs are so similar to those of monochrome ones that they are most easily explained based on an understanding of monochrome LCDs.

Overview. The important components of a monochrome LCD are shown schematically in Fig. 16. In the dark, the main source of light is the backlight. Its luminous flux passes through a diffuser. The uniformity of the backlight/diffuser pair determines the spatial uniformity of the display surface. Light from the diffuser then passes through a light-modulating element that consists of two crossed polarizers and an intervening space, usually a few microns thick, filled with the liquid crystal. The liquid crystal rotates the polarization of light passing through it by an amount that depends on the local electric field. Thus, the electric field can be varied to vary the amount of light passing through the light-modulating element. The electric field is produced by a capacitor that covers the area of the pixel. The different varieties of LCD differ mainly in the electronics used to drive the capacitor. At present, the most common type is bilevel, with the capacitor controlled

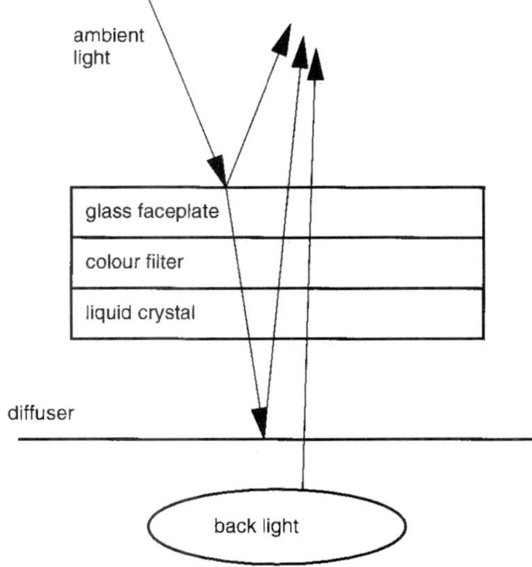

FIGURE 16 Illustration of the typical components of a single LCD pixel, seen side on. The arrows show the three light paths that contribute to the light emitted from the pixel.

by a transistor that is either on or off, thus defining two light levels, white and black. More interesting are multilevel devices, in which an analog electric field allows more or less continuous variation of light output. Currently available analog devices are capable of 16 levels of gray scale with a pixel that is about 150 microns square.

In bilevel devices the array of pixels is very similar to the layout of memory cells in a random access seimconductor memory. Thus, the display itself can be considered a write-only random-access frame buffer. Most LCDs do not, at present, make use of the random-access capability, but accept input via the RS-170 analog signal standard used for CRTs. The multilevel variant is similar, except that an analog value is stored at each location instead of 0 or 1.

An interesting feature of this design is the ability of ambient light to enter the display. It then passes through the light-modulating element and is scattered with random phase by the diffuser. From there it passes back through the light-modulating element to the viewer. Thus, under high-ambient conditions there are two contributions to light emitted from the display: one from the backlight that is linear with respect to the transmittance of the light-modulating element and one from ambient light that is quadratic with respect to the transmittance of the light-modulating element. More specifically, the contribution from the backlight, $\Phi_\lambda^{(\mathrm{BL})}$ is given by

$$\Phi_\lambda^{(\mathrm{BL})} = \tau(V)\Phi_{0\lambda}^{(BL)}$$

where $\Phi_{0\lambda}^{(BL)}$ is the light emitted from the display when the light-modulating element is in its maximally transmissive state and $\tau(V)$ is the voltage-dependent transmittance of the light-modulating element. Note that the light-modulating process is assumed to be spectrally neutral. (In fact, chemists attempt to make the liquid crystal as spectrally neutral as possible.) The contribution from ambient light $\Phi_\lambda^{(AMB)}$ is

$$\Phi_\lambda^{(AMB)} = \tau^2(V)(1 - R(\lambda))D(\lambda)\Phi_{0\lambda}^{(AMB)}$$

where $\Phi_{0\lambda}^{(AMB)}$ is the ambient light incident on the display, $R(\lambda)$ is the reflectance of the faceplate of the display, and $D(\lambda)$ is the proportion of light reflected back by the diffuser. The second contribution is particularly interesting because the display is essentially characterized by a variable reflectance $\tau^2(V)(1 - R(\lambda))D(\lambda)$. Thus it is possible to create reflective images that are computer-controlled pixel by pixel. Note the importance of the front surface reflection in this pixel model. Light that is not transmitted at the front surface produces glare; light that is transmitted creates the image. Consequently, it is doubly important that the faceplate be treated to minimize reflection.

Pixel Structure. The most interesting characteristic of the LCD is the shape and layout of its pixels, which are unlike the CRT pixel both spatially and temporally.

The time dependence of the light output form an LCD pixel is shown schematically in Fig. 17. Each refresh of the pixel produces a sharp rise followed by a slow fall. If the light output decreases from one frame to the next, the decrease occurs abruptly at the beginning of the pixel, without the exponential tail characteristic of CRTs with long persistence

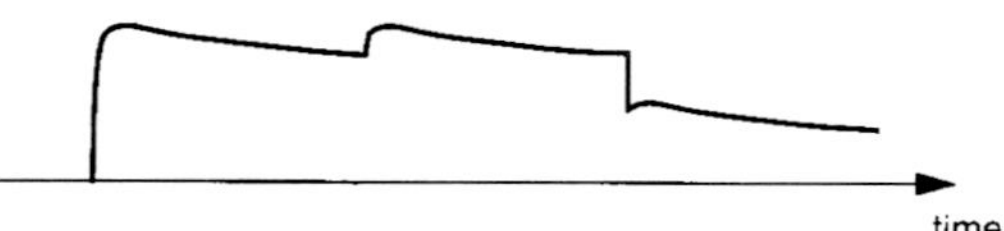

FIGURE 17 Illustration of the time dependence of the light output from an LCD pixel that is turned on for two refreshes, then turned to a lower level for the third refresh.

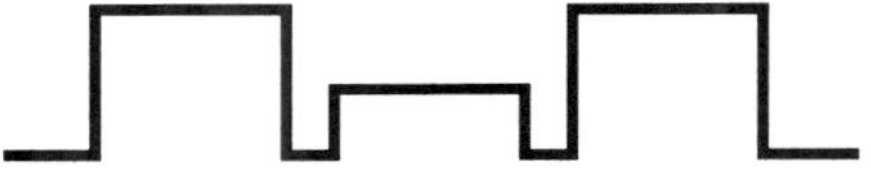

FIGURE 18 Illustration of the layout and shape of pixels on an LCD. The lower panel shows the spatial dependence of emitted light, on a cross section through the center of a row of pixels. The two middle pixels are turned on to a lower level than the outside ones.

phosphors. The turnoff is limited only by the resistance of the circuit that removes charge from the capacitor. The slow fall is produced by loss of charge from the capacitor: the circuit is essentially a sample and hold and the flatness of the pixel profile is determined by the quality of the hold.

The spatial structure of the LCD display surface is shown schematically in Fig. 18. The pixels are nonoverlapping, with dark lines between them because an interpixel mask is used to hide regions where the electric fields of neighboring capacitors overlap. The light emitted is uniform over the pixel area, so that the intensity profile is uniform. Usually there is a small amount taken off the corner of the pixel to allow space for the electronic circuits that turn the pixel off and on. The pitch of the display, the distance from one pixel to the center of the next pixel, ranges from 100 to 500 microns.

Problems. Current high resolution, analogue, color LCDs suffer from a variety of problems, which may be expected to diminish rapidly in importance as manufacturing technology improves.

Thickness variation. Uniformity of output depends critically on precise control of the thickness of the liquid crystal from pixel to pixel.

Heat. Current displays are very temperature-dependent, changing their performance as they warm up and if anything warm, like a human hand, is held in contact with them.

Gray scale. The present mass-produced LCDs have poor gray-scale capability: each pixel can be only off or on. The manufacturing processes needed to produce gray scale, which produce about 30 levels of gray scale on prototype displays, need considerable development.

Electronics. A typical display depends on thousands of address wires, connected to the display at the pitch of the pixels. These wires are delicate and subject to failure, which produces a row or column of inactive pixels, called a line-out.

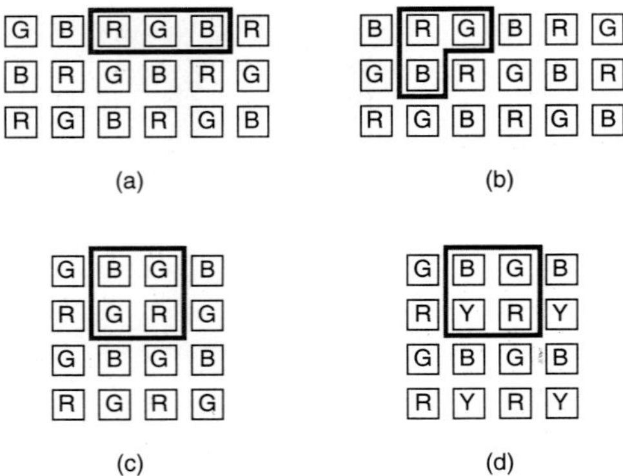

FIGURE 19 Schematic diagram of a variety of different monochrome pixel geometries; (*a*) and (*b*) show two triad geometries; (*c*) and (*d*) show two quad geometries. The color pixel is outlined in gray.

The Color LCD

The color LCD is a straightforward generalization of the monochrome LCD. Colored filters are placed in the light path to color the light emitted from each pixel. Three differently colored filters, and sometimes four, are used to create several interpenetrating colored images that combine additively to form a full-color image. The repetition pattern of the different images is regular, and it is usually possible to group a set of differently colored pixels into a single-colored pixel with three or four primaries, as shown in Fig. 19. (Terminology is important but nonstandard. In this chapter a single-colored pixel is called a monochrome pixel; a set of differently colored monochrome pixels that combines additively to produce a continuum of color at a given point is called a color pixel.)

Geometry of the Color Pixel. There are a variety of different geometrical arrangements of monochrome pixels within a color pixel. Several are illustrated in Fig. 19. It is not currently known which arrangement is best for which type of visual information. When the viewer is far enough back from the display the arrangement does not matter, only the relative number of each primary color. One strategy that is being tried is to add a fourth color, white or yellow, in a quad arrangement in order to increase brightness. Such a display offers interesting possibilities for experimentation with mesopic stimuli and for diagnosis and investigation of color-vision defects.

It is obvious that every geometry has directions in which the primaries form unwanted color patterns. Such patterns do not arise in CRT imagery because there is a random correlation between the dot triads that produce color and the pixel locations. Whether there is a regular geometry, as yet untried, that removes the patterns, whether software techniques similar to antialiasing can eliminate their visual effects, or whether a randomized primary arrangement is needed to remove the patterns is a problem as yet unresolved. (It should also be remembered that some stimuli may best be created by graphical techniques that take advantage of the existence of the patterns.) Whatever the solution, however, it must address the interrelationship between spatial and color aspects of the stimulus in the visual system of the perceiver.

Colorimetry of the Color Pixel. The dominant factor in the colorimetry of LCDs is the interaction between the spectral power distribution of the backlight and the transmittances of the filters. Typically backlight and filter design deals with a trade-off between brightness and color gamut. The higher the excitation purity of the filter the more saturated the primary but the less bright the display. An additional factor considered in this trade-off is heating of the liquid crystal, which is greater when the filters have high excitation purity. Choice of a backlight that has as much as possible of its spectral power in wavelengths passed by the filters will improve performance.

The colorimetric properties of an LCD can be summed up in a few equations. The first is that the color of a pixel is the additive mixture of the colors of its monochrome pixel components. The sum of the spectral powers is

$$\Phi_\lambda = \sum_{a=1}^{N_p} \Phi_{a\lambda}(V_a)$$

where $\Phi_{a\lambda}$ is the spectral power emitted by monochrome pixel a, which depends on the voltage V_a applied to it. Similarly, the tristimulus values are the sum of the tristimulus values of the monochrome pixels

$$X_i = \sum_{a=1}^{N_p} X_{ai}$$

where X_{ai} is the set of tristimulus values for primary a. Note that "gun independence" is assumed.

The spectral power contributed by monochrome pixel a is the sum of three components: one from the backlight, $\Phi^{(BL)}_{a\lambda}(V_a)$, one from ambient light reflected from the front surface of the LCD, $\Phi(R)_{a\lambda}$, and one from ambient light reemitted from the display, $\Phi^{(AMB)}_{a\lambda}(V_a)$.

$$\Phi_{a\lambda}(V_a) = \Phi^{(BL)}_{a\lambda}(V_a) + \Phi^{(R)}_{a\lambda} + \Phi^{(AMB)}_{a\lambda}(V_a)$$

The contribution from the backlight is

$$\Phi^{(BL)}_{a\lambda}(V_a) = \tau_a(\lambda)\tau(V_a)\Phi^{(BL)}_{0\lambda}$$

where $\Phi^{(BL)}_{0\lambda}$ is the spectral power distribution of the backlight, $\tau(V_a)$ is the voltage-dependent transmittance of the liquid crystal/polarizer sandwich, and $\tau_a(\lambda)$ is the transmittance of the color filter. The function $\tau(V)$ is independent of the primary because the display uses the same light-modulating element for each primary. The function $\tau_a(\lambda)$ is independent of the applied voltage since the light-modulating element is spectrally neutral. In the absence of ambient light, this term is the only contribution to the light emitted by the LCD. The colorimetric properties are then identical to those of a CRT with primaries of the same chromaticity, which is the chromaticity of the product $\tau_a(\lambda)\Phi^{(BL)}_{0\lambda}$, though the equivalent of the gamma function $\tau(V_a)$ is certain to have a different functional form.

The contribution from light reflected from the front surface is

$$\Phi^{(R)}_{a\lambda} = R(\lambda)\Phi^{(AMB)}_{0\lambda}$$

where $R(\lambda)$ is the reflectance of the front surface and $\Phi^{(AMB)}_{0\lambda}$ is the ambient light incident on the surface. Usually, the reflectance is spectrally neutral (independent of wavelength) and the reflected light has the same color as the incident ambient light.

The contribution from light reemitted from the display is

$$\Phi^{(R)}_{a\lambda} = (1 - R(\lambda))D(\lambda)\tau_a^2(\lambda)\tau^2(V_a)\Phi^{(AMB)}_{0\lambda}$$

where $D(\lambda)$ is the reflectance of the diffuser. This contribution can be modeled as a light of the same chromaticity as the spectral power distribution $(1 - R(\lambda))D(\lambda)\tau_a^2(\lambda)\Phi^{(AMB)}_{0\lambda}$, with its intensity modulated by the function $\tau^2(V_a)$. Note that this chromaticity is not in general the same as the chromaticity produced by light from the backlight. (It will usually be more saturated than the backlight component since it passes through the color filter twice.) Note also that the voltage dependence of the intensity of this light differs from that of the backlight component. Thus, the chromaticity of the primary changes as the voltage changes when ambient light is present, and the effect cannot be subtracted off, as can the ambient component.

Controls and Input Standards. LCDs will have standard sets of controls and input signals only when the technology is much more mature than it is at present. Currently, the most common input seems to be analog video, as used for CRTs. While this allows a one-for-one substitution of an LCD for a CRT, it seems quite inappropriate for computer applications. Specifically, information is extracted from the frame buffer, a random-access digital device, laboriously sequenced into the serial analog video signal, the reextracted for presentation for the random-access LCD. Thus, it seems likely that digital random-access input standards are likely to supersede video signals for LCD input, and that the display will be more tightly coupled to the image storage system than is common with CRTs.

Temporal and Spatial Variations in Output

LCD fabrication technology is changing too quickly for quantitative limits to spatial and temporal variability to be predictable. Nonetheless, it seems likely that certain qualitative properties are inherent in all LCD designs. They are discussed in general terms in the next few sections.

Short-Time Temporal Variation. The light output over a short time (about 100 milliseconds) is shown schematically in Fig. 17. The details shown in that figure, the rise and decay times and the turnoff time, are bound to change as fabrication techniques improve. Currently, the turnoff time is not very good for most LCDs, and images leave shadows that decay visibly for several seconds after they are removed. That defect is not inherent in LCD electronics and it should be expected to disappear as circuitry improves. The other possible defect is the ripple in the light output, which is nonetheless much smaller than the ripple of CRT light output. The size of the ripple is determined by the quality of the sample-and-hold circuit that maintains charge on the capacitor, and is likely to decrease as LCDs improve.

The interaction between ripple and turnoff time is quite different for an LCD than for a CRT. To decrease the ripple on a CRT it is necessary to increase the decay time of the phosphors; degradation of the turnoff time cannot be avoided. For an LCD, on the other hand, the turnoff is active, and independent of the ripple. Thus, it is possible to improve the ripple without degrading the turnoff. As a result, future LCDs are likely to be very useful for generating stimuli that require precise temporal control.

Long-Time Temporal Variation. Even at the current state of development, the long-time stability of LCDs is quite good, with reports showing variations of about 2 percent on a time scale of several hours. This performance is comparable to good-quality CRTs and incandescent light sources. The main contributor to instability is heat. LCDs are only stable once they are warmed up, and even small changes in cooling configuration can cause appreciable changes in light output. Thus, good stability of light output requires well-controlled temperature and a long warm-up time.

Small-Scale Spatial Variation. Small-scale variation is determined by the spatial structure of the pixel, which is illustrated in Fig. 18. The important feature is the sharp edge of the pixel: unlike CRT pixels, there is no blending at the edges of adjacent pixels. This feature makes the creation of some stimulus characteristics easy—sharp vertical and horizontal edges, for example—and makes the creation of other stimulus parameters very difficult—rounded corners, for example. Many graphical techniques and image-quality criteria have evolved to take advantage of display characteristics that are peculiar to the CRT; there is likely to be a substantial investment in devising techniques that are well suited to the very different pixel profile of the LCD.

An interesting future possibility arises because of the similarity of LCD manufacturing technology to that of random-access memories. Consequently, it is likely that future LCDs will be limited in resolution not by pixel size but by input bandwidth. If so, it would be sensible to have logical pixels within the control system that consist of many physical pixels on the display, with enough processing power on the display itself to translate commands referring to logical pixels into drive signals for physical pixels. In fact, a physical pixel could even belong to more than one logical pixel. If such a development occurs, considerable control over the pixel profile will be possible, which may greatly extend the range of spatial variation that images can possess.

Large-Scale Spatial Variation. Because LCDs have no large-scale structural features, like the beam deflection in the CRT, only manufacturing tolerances should induce spatial variation. The main source of manufacturing variability at the time of writing is the physical size of the pixels and the electronic components—capacitors, especially—that control them. Ideally, such variations would be random, so that spatial variation would add only gaussian noise to the displayed image.

Much more serious is angular variation of emitted light. The light-emitting element uses effects that have strong directional dependence, and it is sometimes even possible to find angles at which an LCD reverses contrast compared to perpendicular viewing. Although this effect is inherent in LC technology, it is reasonable to hope that display improvements will reduce the angular variability below its currently unacceptable level.

27.6 ACKNOWLEDGMENTS

The author would particularly like to express his gratitude to the late Gunter Wyszecki who provided impetus for his interest in this subject, and who provided intellectual and logistical support during the time in which the author carried out most of the work on which this chapter is based. A large number of other people have provided useful insights and technical assistance over the years, including, but not restricted to: John Beatty, Ian Bell, David Brainard, Pierre Jolicoeur, Nelson Rowell, Marueen Stone, and Brian Wandell. The author would also like to thank the editor, David Williams, for patience above and beyond the call of duty during the very difficult time when I was putting this material into the form in which it appears here.

27.7 REFERENCES

1. D. G. Fink, K. B. Benson, C. W. Rhodes, M. O. Felix, L. H. Hoke Jr, and G. M. Stamp, "Television and Facsimile Systems," *Electronic Engineers Handbook,* D. G. Fink and D. Christiansen (eds.), 3d ed., McGraw Hill, New York, 1989, pp. 20-1–20-127.
2. Electronic Industries Association, *Electrical Performance Standards—Monochrome Television Studio Facilities,* RS-170, Washington, D.C., 1957.
3. Electronic Industries Association, *Electrical Performance Standards—Monochrome Television Studio Facilities,* RS-343, Washington, D.C., 1969.
4. N. P. Lyons and J. E. Farrell, "Linear Systems Analysis of CRT Displays," *Society for Information Display Annual Symposium*: *Digest of Technical Papers* **20:**220–223 (1989).

5. J. B. Mulligan and L. S. Stone, "Halftoning Methods for the Generation of Motion Stimuli," *Journal of the Optical Society of America* **A6:**1217–1227 (1989).
6. A. C. Naiman and W. Makous, "Spatial Non-Linearities of Grayscale CRT Pixels," *Proceedings of SPIE: Human Vision, Visual Processing, and Digital Display III* **1666:**41–56 (1992).
7. Blair MacIntyre and W. B. Cowan, "A Practical Approach to Calculating Luminance Contrast on a CRT," *ACM Transaction on Graphics* **11:**336–347 (1992).
8. P. M. Tannenbaum, 1986 (personal communication).
9. D. H. Brainard, "Calibration of a Computer-Controlled Monitor," *Color Research and Application* **14:**23–34 (1989).
10. W. T. Wallace and G. R. Lockhead, "Brightness of Luminance Distributions with Gradient Changes," *Vision Research* **27:**1589–1602 (1987).
11. F. Benedikt, "A Tutorial on Color Monitor Alignment and Television Viewing Conditions," Development Report 6272, Canadian Broadcasting Corporation, Montreal, 1986.
12. I. E. Bell and W. Cowan, "Characterizing Printer Gamuts Using Tetrahedral Interpolation," *Color Imaging Conference: Transforms and Transportability of Color,* Scottsdale, Arizona, 1993.
13. W. B. Cowan, "An Inexpensive Scheme for Calibration of a Color Monitor in Terms of CIE Standard Coordinates," *Computer Graphics* **17**(3):315–321 (1983).
14. P. M. Tannenbaum, "The Colorimetry of Color Displays: 1950 to the Present," *Color Research and Application* **11:**S27–S28 (1986).
15. W. B. Cowan and N. L. Rowell, "On the Gun Independence and Phosphor Constancy of Color Video Monitors," *Color Research and Application* **11:**S34–S38 (1986).
16. L. M. Biberman, "Image Quality," *Perception of Displayed Information,* L. M. Biberman (ed.), Plenum, New York, 1973.

CHAPTER 28
OPTICAL GENERATION OF THE VISUAL STIMULUS

Stephen A. Burns
Robert H. Webb
The Schepens Eye Research Institute
Boston, Massachusetts

28.1 GLOSSARY

A	area
D	distance
E_r	illuminance at the retina (retinal illuminance)
f	focal length
f_e	focal length of the eye
L_s	intrinsic luminance of the source
td	troland (the unit of retinal illuminance)
x	position
Δ	change in position
Φ	flux (general)
Φ_p	flux at the pupil
τ	transmittance

We have also consistently used subscripted and/or subscripted versions of A and S for area, D for distance, f for focal lengths, x for positions, and Δ for changes in position.

28.2 INTRODUCTION

This chapter presents basic techniques for generating, controlling, and calibrating the spatial and temporal pattern of light on the retina (the visual stimulus). It deals with the optics of stimulus generation and the control of light sources used in the vision laboratory. Generation of stimuli by computer video displays is covered in detail in Chap. 27. Units for measuring radiation are discussed in Volume II, Chapter 24 ("Radiometry and Photometry").

28.3 THE SIZE OF THE VISUAL STIMULUS

The size of a visual stimulus can be specified either in terms of the angle which the stimulus subtends at the pupil, or in terms of the physical size of the image of the stimulus formed at the retina, for most purposes vision scientists specify stimuli in terms of the angular subtense at the pupil. If an object of size h is viewed at distance D then we express its angular extent as

$$\text{Degrees visual angle} = 2\left(\tan^{-1}\left(\frac{h}{2D}\right)\right) \tag{1}$$

or, for angles less than about 10°

$$\text{Degrees visual angle} \cong \frac{360h}{2\pi D} = \frac{57.3h}{D} \tag{2}$$

In this chapter we specify stimuli in terms of the actual retinal area, as well as the angular extent. Angular extent has the advantage that it is independent of the eye, that is, it can be specified totally in terms of externally measurable parameters. However, an understanding of the physical dimensions of the retinal image is crucial for understanding the interrelation of eye size, focal length, and light intensity, all of which are part of the design of optical systems for vision research.

28.4 FREE OR NEWTONIAN VIEWING

There are two broad classes of optical systems that are used in vision research. Free viewing, or *newtonian* viewing, forms an image of a target on the retina with minimal accessory optics (Fig. 1*a*).

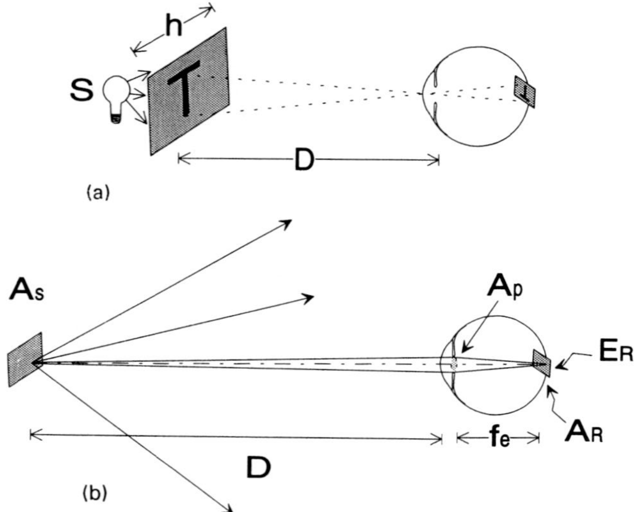

FIGURE 1 (*a*) In free viewing or newtonian viewing the eyes optics are used to image a target onto the retina. (*b*) Computation of the retinal illuminance in free-viewing. A_s the area of the source, D the distance between the source and the eye, A_p the area of the pupil, f_e the optical focal length of the eye, A_R the area of the image of the source on the retina, and E_R the retinal illuminance.

Retinal Illuminance

Photometric units incorporate the overall spectral sensitivity of the eye, but that is the only special property of this set of units. That is, there are no terms specific to color (for a discussion of colorimetry, see Chap. 26) and no allowances for the details of visual perception. The eye is treated as a linear detector which integrates across wavelengths. The retinal illuminance of an object in newtonian view is determined by the luminous intensity of the target and the size of the pupil of the eye. The luminous power at the pupil [dimensions are luminous power or energy per unit time, the SI units are lumens (lm)]

$$\Phi_P = L_s A_s \frac{A_p}{D^2} \tag{3}$$

where L_s is the luminance of the source (the SI units are lumens per meters squared per steradian (lm/m^2/sr) or candelas per meters squared (cd/m^2)), A_s is the source area, A_p is the area of the pupil, and D is the distance from the pupil to the source (so (A_p/D^2) is the solid angle the pupil subtends).

The area of the image element on the retina is

$$A'_R = A_s m^2 = A_s \left(\frac{f_e}{D}\right)^2 \tag{4}$$

where m is the magnification of the eye and f_e is the effective focal length of the eye (Fig. 1*b*). From this we compute the illuminance at the retina as

$$E_R = \frac{\Phi_P}{A'_R} = \frac{A_p L_s}{(f_e)^2} \tag{5}$$

[The SI units of illuminance are lm/m^2]. A typical value for f_e is 16.67 mm.[1,2] Note that the retinal illuminance does not depend on the distance. That is, the retinal illuminance when viewing an extended source such as a video screen is independent of the viewing distance and is dependent only on the size of the eye's pupil, the luminance of the source, and the focal length of the eye. In most cases, the focal length of the viewer's eye is not known, but it is possible to measure the size of the pupil. For this reason a standard unit for specifying retinal illuminance was developed, the *troland.*

***The Troland* (*td*).** The troland is a unit of illuminance (luminous power per unit area). The troland quantifies the luminous power per unit area at the retina (the retinal illuminance). One troland was defined as the illuminance at the retina when the eye observes a surface with luminance = 1 cd/m^2 through a pupil having an area of 1 mm^2. Using Eq. (5) and a standard f_e of 16.67 mm we find that the troland is defined by

$$1 \text{ td} = 0.0035 \text{ lumens/m}^2 \tag{6}$$

This definition ties the troland to the illuminance on the retina of a standard eye assuming no transmission loss in the ocular media at 555 nm. Thus, two observers with different-size eyes, different-size pupils, or different relative losses in the ocular media, viewing the same surface, will have different retinal illuminances. Wyszecki and Stiles[2] have recommended that the term *troland value* be used to distinguish the trolands computed for a standard eye from the actual retinal illuminance. General usage is that the retinal illuminance is determined simply by measuring the luminance of a surface in cd/m^2 and multiplying this value by the area of the pupil in mm^2.

Limitations of Free Viewing: An Example

There are two major limitations to newtonian view systems. The first is that the retinal illuminance is limited. For instance, a 60-W frosted incandescent bulb can produce a 120,000-td field, but to obtain a uniform 20° field it must be placed 17 inches from the observer's eye. This requires an accommodative effort that not all observers can make. A comparable illuminance at more realistic distances, or with variable focus, requires larger light sources or more elaborate optical systems. The second limitation of free viewing is that variations in pupil size are not readily controlled. This means that the experimenter cannot specify the retinal illuminance for different stimulus conditions or for different individuals. Maxwellian view optical systems solve these problems.

28.5 MAXWELLIAN VIEWING

Figure 2*a* shows a simple *maxwellian* view system. The key factor that distinguishes the maxwellian view system is that the illumination source is made optically conjugate to the pupil of the eye. As a result, the target which forms the stimulus is not placed at the source, but rather at a separate plane optically conjugate to the retina. In the system of Fig. 2*a* the plane conjugate to the retina, where a target should be placed, lies at the focal point of lens L_{21}, between the source and the lens. Light from the source is diverging at this point, and slight changes in target position will cause changes in both the plane of focus and the magnification of the target. For this reason most real maxwellian view systems use multiple lenses. Figure 2*b* shows such a maxwellian view system where the single lens has been replaced by two lenses. The first lens collimates light from the source and the second forms an image of the source at the pupil. This places the retinal conjugate plane at the focal plane of L_1. We label the conjugate planes starting at the eye, so R_1 is the first plane conjugate to the retina and P_1 is the first plane conjugate to the pupil.

Control of Focus and the Retinal Conjugate Plane

The focus of the maxwellian view system is controlled by varying the location of the target.[3–5] Moving the target away from R_1 allows the experimenter either to adjust for ametropia or to require the observer to accommodate. To see this we compute where lens L_1 (Fig. 2*b*) places the image of the target relative to the eye. If the target is at the focal point of L_1, the image is at infinity, which means there is a real image in focus at the retina for an emmetropic eye. If we move the target toward lens L_1, the lens of an emmetrope must shorten the focus to bring the image back into focus on the retina. We quantify this required change in focus as the change in the dioptric power of the eye (or the optical system if spectacle correction is used) where the dioptric power is simply the inverse of the focal length measured in meters. If the target is displaced by Δ from the focal point of L_1 then using the newtonian form of the lens formula ($x'x = f^2$) we find that

$$\text{Change in dioptric power} = \frac{\Delta}{(f_1)^2} \tag{7}$$

If the target move toward the lens L_1, this compensating power must be positive (the focal length of the eye or eye plus spectacle shortens). If the target moves from f_1 away from lens L_1, the required compensation is negative. If the eye accommodates to keep the target in focus on the retina, then the size of the retinal image is unchanged by the change in position.[3,4]

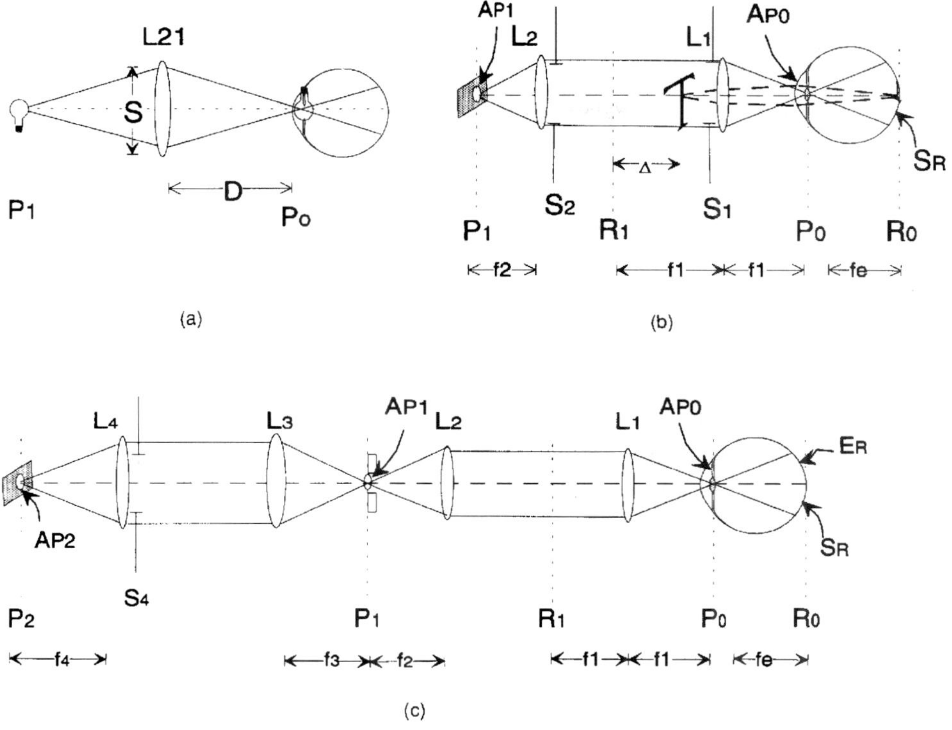

FIGURE 2 Panel (*a*): In a minimal maxwellian view system a lens L_{21} images a source in the plane of the eye's pupil. The lens is shown at a distance D from the eye equal to twice its focal length. The field stop (S) in the aperture of L_{21}. Panel (*b*): A maxwellian view optical system with an accessible retinal conjugate plane (R_1). Lens L_2 collects light from the source and collimates it. Lens L_1 images the source in the plane of the eye's pupil (P_0). Pupil conjugate planes (P_i) and retinal conjugate planes (R_i) are labeled starting at the eye. For an emmetrope lens L_1 images the retina (R_0) at R_1. This is the plane where a visual target will appear in best focus for an emmetropic observer. Moving the target away from R_1 a distance Δ changes the plane of focus of the retinal image, requiring the emmetrope to accommodate. The maximum size of the retinal image (S_R) is limited for this system to the aperture of lens $L_2(S_2)$. The pupil is imaged at position P_1, where the light source is placed. Panel (*c*) shows a more complex system where a second set of pupil conjugate and retinal conjugate planes have been added. An artificial pupil (A_{p1}) is added at P_1. It is the image of this pupil at the source (A_{p2}) that limits the retinal illuminance produced in this optical configuration. Other symbols as above.

Size

For this system the area of the illuminated retinal field is determined by the limiting aperture in a retinal conjugate plane between the light source and the retina (the field stop S_2 in Fig. 2*b*) and is

$$S_R = \left(\frac{f_e}{f_1}\right)^2 S_2 \tag{8}$$

or for the schematic eye we use:

$$S_R = \left(\frac{16.67}{f_1}\right)^2 S_2 \tag{9}$$

where f_e is the focal length of the eye (in mm) and f_1 is the focal length of lens L_1. This same formula is used to compute the physical size of an image on the retina. Note that the linear magnification is (f_e/f_1) and the areal magnification is $(f_e/f_1)^2$. The angular subtense of the field is the same as in Eq. (1), substituting the diameter of S_2 for h and f_1 for the distance D. Also note that the angular extent is independent of details of the optics of the eye. This convenience is the main reason that angular extent is the most widely used unit for specifying the stimulus in vision research.

Retinal Illuminance

One of the *principal advantages* of a maxwellian view system is that it provides a large, uniformly bright field of view. This mode of illumination is called Kohler illumination in microscopy. The light available in a maxwellian view optical system is determined by two main factors, the luminance of the light source, and the effective pupillary aperture of the optical system being used. In Fig. 2*b* we see that lens L_2 collects light from the source. The amount of light collected is the maximum amount of light that can be presented to the eye and is[1,3]

$$\Phi = A_{P1} L_S \frac{S_2}{(f_2)^2} \tag{10}$$

where A_{P1} is the area of source being considered (or, equivalently, a unit area of the source), L_S is the luminance of the source (in lm/m^2/sr), S_2 is the aperture of lens L_2, and f_2 is the focal length of lens L_2. We have used the area of lens L_2, for this example, although in actual practice this area might be reduced by later field stops. This quantity will cancel out in the subsequent calculations. Finally, if all of the light collected by lens L_2 is distributed across the retina, then the retinal illuminance is

$$E_R = \frac{\Phi}{S_R} = A_s L_S \left(\frac{f_1}{f_2 f_e}\right)^2 \tag{11}$$

where S_R is obtained from Eq. (8). Note that only the luminance of the source, the source area, and the focal lengths of the optical elements are important in setting the retinal illuminance. Using different diameters for lens L_2 will change the amount of light collected and the size of the retinal area illuminated. Such a change will be reflected by changing the area parameter in both Eqs. (8) and (10), which cancel.

In Eq. (11) the area of the source image (A_{P1}) at the pupil is

$$A_{P0} = A_{P1} \left(\frac{f_1}{f_2}\right)^2 \tag{12}$$

If the eye's pupil (A_{P0}) is smaller than this source image then the eye's pupil is limiting and

$$E_R = \frac{A_{p0} L_S}{(f_e)^2} \tag{13}$$

Equation (13) for the maxwellian view system is identical to Eq. (5) which we obtained for the newtonian view system, but now the entire field is at the retinal luminance set by the source luminance. Thus, in maxwellian view a large, high retinal illuminance field can be readily obtained.

To control the size of the entry pupil rather than allowing it to fluctuate with the natural pupil, most maxwellian view systems use a pupillary stop. Figure 2*c* shows a system where a stop has been introduced at an intermediate pupil conjugate plane A_{P1}. This has the advantage of placing the pupillary stop of the system conjugate to the eye's pupil. The projection of A_{p1} at the source is A_{p2} and it is A_{p2} that limits the available luminous area of the source. Lenses L_3 and L_4 image the source onto the artificial pupil, and lenses L_1 and L_2 image the artificial pupil in the plane of the eye's pupil (A_{p0}). The retinal illuminance of this more complex system can be computed as follows: the field stop over which light is collected by lens L_4 is S_4 and can be computed by projecting the retinal area illuminated (S_R) back to lens L_4.

$$S_4 = S_R\left(\frac{f_1 f_3}{f_2 f_e}\right)^2 \tag{14}$$

A_{p2} is the area of the source which passes through the artificial pupil:

$$A_{p2} = \left(\frac{f_4}{f_3}\right)^2 A_{p1} \tag{15}$$

Therefore the total amount of usable light collected is

$$\Phi = L_S A_{p2}\frac{S_4}{(f_4)^2} = S_R L_S A_{p1}\left(\frac{f_1}{f_2 f_e}\right)^2 \tag{16}$$

and the retinal illuminance is

$$E_R = \frac{\Phi}{S_R} = L_S A_p\left(\frac{f_1}{f_2 f_e}\right)^2 \tag{17}$$

Note that, as in Eq. (13) above, $(A_{p1}(f_1/f_2)^2)$ is the size of the image of the aperture A_{p1} when measured at the exit pupil (A_{P0}). Thus, even in this more complex case we find that the retinal illuminance is dependent only on the source luminance, the pupillary area measured at P_0, and the focal length of the eye.

Advantages of Maxwellian Viewing: Example Revisited

The strength of a maxwellian view system is that properties of the target (focus, size, shape) can be controlled independently from retinal illuminance. If the source is larger than the limiting pupil, then the maximum retinal illuminance is the luminance of the source, scaled only by the exit pupil and eye size [Eqs. (13) and (17)]. The retinal illuminance is controlled by pupillary stops, and the target size, shape, and focus are controlled by retinal stops. Additionally, focus can be controlled independently of retinal illuminance and image size.

If we use the 60-W frosted bulb that we used in the example of newtonian viewing as a source in a maxwellian view optical system we find that we can produce the same maximum retinal illuminance for a given pupil size [Eqs. (5) and (13)]. However, with the maxwellian view system relatively inexpensive, achromat lenses will allow us to generate a field size greater than 20°. In addition, we can dispense with the frosted bulb (which was needed in free viewing to give a homogeneously illuminated target) and use an unfrosted tungsten halogen bulb to produce in excess of 1 million td.

Controlling the Spatial Content of a Stimulus

The spatial frequency content of a retinal image is usually measured in cycles per degree (of visual angle). The frequencies present in the retinal image are those present in the target, cut off by the limiting aperture of the pupillary stop—either the eye's pupil or an artificial pupil in a pupillary plane *between the target and the eye.*[25] Apertures placed on the source side of the target do nothing to the spatial frequency content of the retinal image, leaving the pupil of the eye controlling the image, not the artificial pupil. (For a more detailed treatment of the effects of pupils on the spatial content of images see Refs. 6–10. For a discussion of the spatial resolving power of the eye's optics see Chap. 24 in this *Handbook*).

Positioning the Subject

One of the practical disadvantages in using maxwellian view systems is the need to position the eye's pupil at the focal point of the maxwellian lens (L_1 in Fig. 2*c*) and to maintain its position during an experimental session. One technique for stabilizing the position of the eye is to make a wax impression of the subject's teeth (a bite bar). By attaching the bite bar to a mechanism that can be accurately positioned in three dimensions (such as a milling machine stage) the eye can be aligned to the optical system and, once aligned, the eye position is maintained by having the subject bite loosely on the bite bar. Alignment of the eye can be achieved either by observing the eye, as described at the end of this chapter, or by using the subject to guide the alignment process as described in the following paragraphs.

The first step in alignment is to position the eye at the proper distance from lens L_1. This can be accomplished by noting that, when the eye's pupil is at the plane of the source image, slight movements of the head from side to side cause the target to dim uniformly. If the eye's pupil is not at the proper distance from lens L_1 then moving the head from side to side causes the target to be occluded first on one side, then the other. By systematically varying the distance of the eye from lens L_1, the proper distance of the eye can be determined. It is then necessary to precisely center the eye's pupil on the optical axis of the apparatus. There are several approaches to centering the eye.

1. The physical center of the pupil can be located by moving the translation stage such that the pupil occludes the target first on one side, then on the other. The center of these two positions is then the center (in one dimension) of the pupil. The process is then repeated for vertical translations.
2. The entry position that produces the optimum image quality of the target can be determined. One variant is to use a target that generates strong chromatic aberration (for instance, a red target on a blue background). The head can then be moved to minimize and center the chromatic fringes.[11] This process defines the achromatic axis of the eye.
3. The eye can be positioned to maximize the brightness of the target, which centers the Stiles-Crawford maximum in the pupil with respect to the exit pupil of the instrument.
4. A set of stimuli can be generated that enter the eye from different pupil positions. If the eye is centered, then all stimuli will be seen. If the eye is not centered, then the pupil occludes one of the pupil entry positions and part of the stimulus array disappears. In this case, the subject merely has to keep his or her head positioned such that all stimuli are visible.

28.6 BUILDING AN OPTICAL SYSTEM

Alternating Source Planes and Retinal Planes in a Controlled Manner

The separation of retinal and pupil conjugate planes in a maxwellian view system allows precise control over the spatial and temporal properties of the visual stimulus. By placing the light source conjugate to the pupil of the eye, every point on the source projects to every point in the retinal image and vice versa. Thus, to control the whole retinal image, such as turning a light on and off, manipulation of a pupil conjugate plane is optimal. To control the shape of the retinal image without altering the entry pupil characteristics, variation at the retinal conjugate planes is required. However, there is an exception to this rule. Light from the edges of the image traverse the pupil conjugate plane at a higher angle than light from the center of the image. For small fields (<15° diameter) the angular dependence is minimal, but for larger fields it can be significant and filters should be placed in a collimated light path.

Combining Lights in an Optical System

To control aspects of the visual stimulus independently, different light sources and targets, each with its own set of filters and shutters, can be combined. We will call these separate "channels." Three different techniques allow for combining optical channels, beamsplitters, beam separators or reflective apertures, and diffusers. All three methods are demonstrated in Fig. 3.

Beamsplitters. A beamsplitter both transmits and reflects a fraction of the incident light. Any surface with an index of refraction change can be a beamsplitter. In vision we generally use either a cube or a plate beamsplitter. By locating the beamsplitter in a collimated portion of the optical system (Fig. 3), but away from a retinal conjugate plane, two channels can be combined. For channels with different spectral compositions the beamsplitter can be dichroic, reflecting some wavelengths and transmitting others, usually by interference effects. Plate beamsplitters have the disadvantage that there are secondary

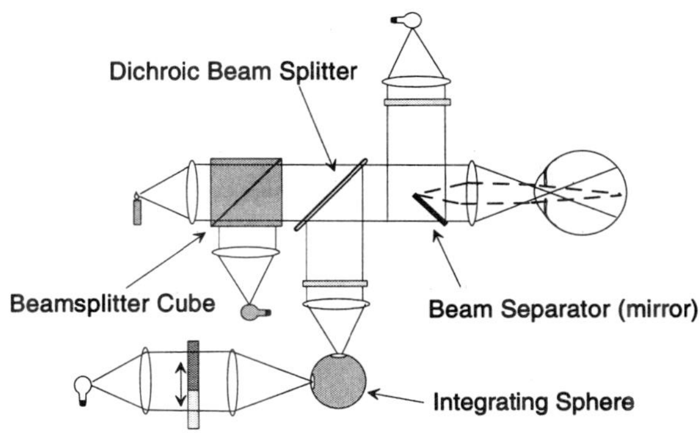

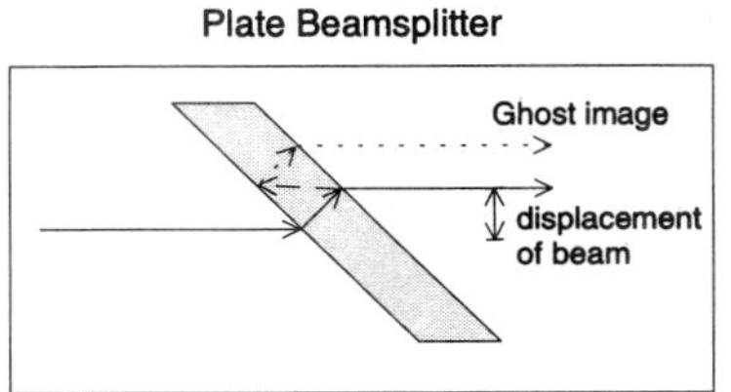

FIGURE 3 Techniques for combining light from different light sources or optical paths. The inset illustrates potential problems that need to be considered when using a plate style of beamsplitter. These problems are eliminated by the use of a cube beamsplitter or wedge beamsplitter.

reflections, slightly displaced (inset, Fig. 3) from the main beam. In an image-carrying beam this produces "ghost" images, slightly displaced from the main image. The displacement decreases with decreasing thickness. A pellicle is simply a plate beamsplitter with negligible thickness. A wedge beamsplitter displaces a ghost image a long way, and Liebman has used this to design a tapered beamsplitter with no secondary reflections.[12] Unlike plate beamsplitters, the beamsplitter cube has the advantage that there is no lateral displacement of an image.

Beam Separators (or Reflective Apertures). A beam separator combines two optical channels into one, while keeping the light spatially distinct. For instance, a mirror placed halfway across the aperture at 45 deg combines two channels at 90° to each other. Such a separator in a retinal conjugate plane is imaged at the retina to produce a bipartite field (Fig. 3). The mirror must be oriented to hide the bevel from the subject and must have a high-quality edge. A transparent plate evenly silvered in a pattern and optically sandwiched between right angle prisms is a convenient beam separator with good edges.

Diffusers, Integrating Spheres, and Optical Fibers. Diffusers also mix light. Diffusers are used when either the spatial uniformity of the final beam or thorough spatial mixing of light sources is important. The direction of light remitted from the diffuser is independent of the incident light, which simplifies the combination of sources. Both integrating spheres and integrating bars have been used to combine lights (Fig. 4*a* and *b*, respectively). Diffusers have been widely used in the construction of colorimeters.[2] In the La Jolla colorimeter[13] light is passed through a filter assembly with three monochromatic filters assembled edge to edge. Moving the filter assembly across the light path changes the relative proportions of the light impinging on each filter, which changes the average color of the light. An integrating sphere then completely mixes the colors producing a variable chromaticity light source without the spatial inhomogeneity of the filters. The major disadvantages of using diffusers are light loss, the need for careful optical baffling (since any light impinging on the diffuser gets mixed into the system) and the gradual deterioration of the diffuser due to dirt and dust. While the light loss from an integrating

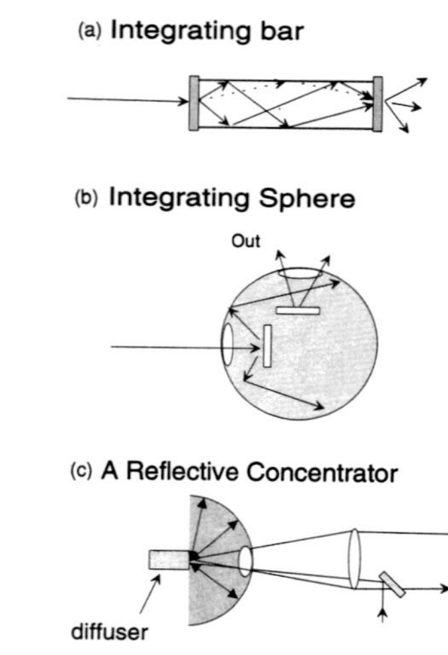

FIGURE 4 An example of different types of beam combiners based on diffusers. In general it is critical that high-efficiency diffusing materials be used in all of these techniques.

sphere is ideally very low, slight decreases in the efficiency of the diffusing surfaces cause large losses in the output. Collection of light from a simple diffuser can be increased by the use of a spherical mirror[14] (Fig. 4*c*). Similarly, a noncoherent fiber-optic bundle can be used to mix lights.

Mixing coherent light sources with diffusers requires movement to minimize interference (speckle) effects. Either liquid diffusers[15] which work by brownian motion or moving diffusers[16] can be used.

Lenses

Lens orientation affects any optical system. For instance, the most curved surface of a lens should be toward the more collimated light. This makes both sides of the lens share the work of refraction and decreases aberrations. Achromats reduce both spherical and chromatic aberrations (they have three or more refracting surfaces) and are typically designed to form a collimated beam.

Typically, the goal in aligning an optical system is to place all lenses upon a common optic axis, oriented such that a ray of light traversing that axis is not deviated. A simple prescription is

1. Define the system axis using pinholes, pointers, or a laser beam, with all lenses removed. In a multichannel optical system it is often easiest to start from the position of the eye's pupil (P_0).
2. Introduce mirrors, beamsplitters, and other nonrefractive optics in their proper positions, being careful to keep the beam at the same height. A technique for ensuring a constant height, such as a movable pinhole, also helps.
3. Starting at the alignment source, set each lens, one at a time, such that its Boys points lie on the system axis. The Boys points are the focused reflections from the various curved surfaces of lenses (see Ref. 17 for a detailed discussion of using Boys points for alignment). When using a laser they are the "bulls eye" interference patterns. Lateral translation of the lens moves the first (brightest) Boys point and rotation moves the second, so convergence is rapid.
4. To determine the position of a lens along the optical axis, a photographic loop (a magnifying lens) and a target grid (graph paper) can be used. By placing the graph paper at a known location such as at the exit pupil, and then looking with the photographic loop into the system (turn off the laser first!) the exact position of the next pupil conjugate point can be determined. To position a lens to collimate light from a source, place a mirror in the collimated beam, reflecting the light back into the lens. When the lens is placed at one focal length from the source, the mirror causes the source to be imaged back onto itself (autocollimation).

Field Quality

The uniformity of the illumination of the retinal image is controlled by the size and emission characteristics of the source and by the uniformity of illumination of the target. Ideally, a source emits uniformly into all angles subtended by the collection lens (lens L_2 in Fig. 2). LEDs do not meet this requirement and produce inhomogeneous illumination unless the light is further diffused. Tungsten-halogen sources with coiled-coil filaments uniformly illuminate the collecting lens and can produce uniform retinal images. Problems arise if a retinal conjugate is placed too close to a pupil conjugate plane. This occurs most often when the target is placed at the end of a long collimated portion of the optical path. In this case, the structure of the source becomes visible at the retinal plane. In general, collimation lengths should be kept to less than roughly twice the focal length of the collimating lens. It is also important to control the pupillary conjugate points; in some cases multiple pupil conjugate apertures can help.[18]

Controlling Wavelength

Spectral composition of a light source can be varied either by absorbing the unwanted light (in a filter), or by redirecting it (by either interference of refraction). Table 1 presents techniques for controlling wavelength. Interference filters can have blocking filters that absorb wavelengths outside the specified transmission band. However, a blocking of 10^{-3} may not be sufficient for the human visual system. For instance, a narrow 670-nm filter may transmit only 10^{-3} of the intensity at 555 nm, but the human visual system is about 30 times as sensitive to 555 nm as to 670 nm and can integrate over the entire visible spectrum. Thus, the visual effectiveness of the "blocked" light can be considerable. Monochromatic and narrowband light sources and lasers can also be used both to provide a specific wavelength and to calibrate the wavelength scale of monochromators.

Wavelength selection filters are best placed in a collimated beam to avoid changing the passband to shorter wavelengths. This shift is asymmetric, and the maximum transmission changes as the angle of incidence decreases from 90°.[2] Differently colored lights can be combined using broadband interference filters (dichroic filters) placed at an angle, but using narrowband filters designed for normal incidence at 45 deg can introduce unwanted polarization properties.[2] Filters designed for use at 45 deg are readily available.

Turning the Field On and Off

Light from an optical channel can be clocked entirely by a shutter. Shutters should generally be located at a pupil conjugate plane. At this position the source is focused to a small area and thus a smaller, faster shutter can be used. In addition, manipulation of a pupil plane changes the retinal image uniformly (see above). Parameters of interest are the speed, repetition rate (and duty cycle), and the transmission of the shutter in both its open and closed states. Table 2 summarizes details on common types of shutters. Shutters that work by means of polarization generally need to be placed in a collimated beam, and their extinction (transmission when off) and maximum transmission are wavelength-dependent.

Controlling Intensity

The visual system operates over about 11 log units of luminance (11 orders of magnitude, or 10^{11}), and within this range can make discriminations of 0.1 percent or better. This level of visual capability requires a two-stage control of retinal illuminance. The overall retinal illuminance is set by discrete neutral density filters, while smaller steps are set by either a variable density filter or an electronic system. Due to limited precision in most measuring instruments, it is hard to calibrate an individual filter with a density greater than 3.0.

Varying the Intensity of the Field. The intensity of a channel can be varied by using fixed or variable filters, or by controlling the radiance of the source. For simple, relatively slow (1/min) changes in retinal illuminance, neutral density filters are appropriate. For faster changes, modulators or direct variation of the source radiance are typically required.

TABLE 1 Controlling Wavelength

Technique	Properties	Advantages	Disadvantages/comments
Absorption filters: aka dye, film, gel, or pigment filters	Available in both gels and glass forms. Generally broadband.	Can be extremely stable over time. Wide range of colors available. Reasonably priced.	Broadband, can be damaged by too much light, as encountered with lasers and focused beams. Most fluoresce.
Narrowband interference	Half-width about 10 nm, usually bonded to an absorption filter for blocking distant wavelengths.	Convenient, readily available, and moderately priced.	If not fully sealed, humidity can cause long-term degradation. Wavelength depends on angle of incidence. Need good blockers
Broadband interference (see also dichroic beam splitters)	Half-width 40–100 nm.	Same as narrowband interference filters, less need for blocking filters.	
Grating monochromators	Tunable, bandwidth can be set by slits.	Can have low stray light.	Must match aperture of system to the interior grating to minimize stray light. Expensive.
Prism monochromators	Produces a spectrum by refraction.	High throughput.	Bulky.
Interference wedges (spatially varying interference filters)	Allow continuous change of wavelength by changing the position of light incidence.	Small, easy to use.	Some have more leakage than a monochromator, may not be adequately blocked for higher order transmission. Must be placed at a pupil conjugate point to avoid chromatic wedging on the retina.
Special sources (Na, Cd, etc.)	Discharge sources that produce a few exactly specified spectral lines.	Many lines are available. Very pure wavelengths. Larger aperture sources are available.	Mostly replaced by lasers which are stable, cheap, and easier to use but are point sources (see Table 5).

Filters. Neutral density filters either have a uniform attenuation across their spatial extent (fixed filters) or vary in their attenuation characteristics depending on the spatial location (variable filters or "wedges"). Absorbing filters should be used in a collimated beam to avoid different path lengths through the filter. Wedges are used at a pupil conjugate plane to avoid imaging the density gradient on the retina. Fixed filters are typically used to set fixed parameters (such as the average luminance), variable filters to control the brightness continuously (for instance, in an increment threshold test). Table 3 presents some types of neutral density filters.

TABLE 2 Types of Shutters

Shutter	Speed	Size of aperture (mm)	T off/T on	Comments
Mechanical shutter	ms	1 mm and up	0	Larger apertures are slower. Most designs cannot run in a continuous mode, but these are ideal for low duty-cycle use.
Galvanometers	ms	A few mm for high speeds	0	Operating at high rates requires fairly careful driver design but they can be used to excellent effect and are commercially available with drivers. Can be run continuously.
Choppers	kHz	Variable	0	For continuous on-off flicker choppers are ideal. They have no attenuation when open and no transmission when closed. Small, feedback-controlled devices are available off the shelf. Stepper motors with vanes mounted on the shaft can also be used.
Acousto-optic modulator (AOM)	μs	<2 mm	$\sim 10^{-3}$	AOMs have small aperture and high f-number and thus work best for lasers. They are inherently chromatic and orientation-sensitive. Polarization effects can be important.
LCD shutters	ms	A few cm	$\sim 10^{-2}$	Speeds are increasing rapidly, works by polarization. Maximum transmission is less than 50 percent. Use in collimated beam.
LCD displays	μs	Work as video screens; individual pixels can be μm.	$\sim 10^{-2}$	Work by polarization. Speed is increasing with development of displays for computer use.
Kerr cells	ps	cm	$\sim 10^{-2}$	Require high voltages, work by polarization.

Modulators. Passing linearly polarized light through a rotating polarizer produces sinusoidal modulation of the transmitted light. Careful optical design allows generation of an excellent variable modulation, flicker stimulus.[19,20] Acousto-optic modulators (AOM) can be used to vary a light rapidly, but have only a small aperture. Mechanical methods have also been used for temporally modulating a target.[21–23] For the most part, mechanical modulators and moving polarizers have been replaced by direct control of light sources by high-speed shutters and by video systems (see Chap. 27 of this *Handbook*).

Varying the Source. Varying the radiance of the light source is fast, and is particularly straightforward with electro-optical sources such as LEDs and diode lasers. However, control of the source is not limited to these devices and can be used with thermal sources (incandescent bulbs), xenon arcs, and fluorescent lights as well. There are four major problems to be overcome when directly varying a light source: (1) nonlinear current versus radiance characteristics, (2) changes in the current-radiance relation over time, (3)

TABLE 3 Neutral Density Filters

Method	Advantages	Disadvantages	Other comments
Metal film neutral density filters	Readily available, stable, spectral neutrality is good.	Must be cleaned with care, pinholes can cause a problem. Interreflections between filters can cause deviations from density obtained when calibrated alone.	Silver and inconel are most common. Tilt to keep reflections out of the system.
Multilayer dielectric filters	Easier to clean.	Expensive, chromatic.	
Metal film "wedges"	Same as metal film filters.	Same as metal film filters.	Tilt to aim reflections out of the system. Use in a pupil conjugate plane to avoid spatial variation of image.
Absorbing gel filters	Less expensive, easily available at good photography stores.	They become brittle with age. Somewhat chromatic. Damaged by intense light.	Keep dry and away from a hot lamp. Use in a collimated light path.
Sector disk and mechanical devices (cams, variable slits, etc.)	Spectrally flat.	Must be used in a pupil plane to get desired results.	Use in a pupil conjugate (or in a uniform beam before an integrating sphere). Variable slits can be a simple, inexpensive way to control intensity.
Photographic film or film wedges	Cheap, easily made by flashing a film and developing.	Spectral neutrality will be poor in relation to commercial filters. Must be calibrated. Stability over time will depend on the film base used.	Best used with monochromatic light.

temporal hysteresis in the current-radiance relation, and (4) changes in the wavelength distribution of the source with changes in current.

For instance, LEDs have been considered ideal light sources by some investigators, yet they show the same deviations from ideal behavior common to other sources[24–29] (see Chap. 10 in this *Handbook*). The current/radiance relation of LEDs depends on the internal temperature of the device. Thus, the relation changes with both the measurable *temperature* (due to both the environment and to the time average current) and with the immediate current history (for instance, it was just driven at a high current). Finally, the temperature also affects the probability of electronic transitions in the semiconductor, which can change the *wavelength* emitted. For LEDs used in visual work this wavelength dependence has been measured to be on the order of a 1.6-nm change in dominant wavelength with variation of the duty cycle for an ultrabright red LED.[26] Thermal sources, such as a tungsten-halogen bulb, undergo especially large changes in spectral output with changes in current. With these sources, only a narrow spectral band should be used if the source is controlled. In addition, the thermal inertia of an incandescent source precludes rapid modulation, although slow modulation (~1 Hz) can be achieved. Even fluorescent sources which have been widely used in some areas of research show significant changes in spectral output with time.[30]

While heat sinks can help to stabilize the response characteristics of many devices, linear control still requires careful driver design. Drivers can use either an analog or binary design. Analog drivers control the source radiance by varying the current, while digital drivers turn the source either on or off, with the average radiance set by the proportion of time that the source is on (the duty cycle). An advantage of the binary scheme is that transistors dissipate less power in the on and off states than in intermediate states.

There are four major approaches to linearization:

1. Calibrate the current-luminance relation and build a driver with the inverse nonlinearity. For simple applications this technique is adequate. However, for demanding applications the driver becomes complex and the demands for stability in the driver raise new problems.
2. Calibrate the current-luminance relation and use an inverse nonlinearity in computer memory (a lookup table). This technique is quick and easy to implement. A static lookup table will not compensate for dynamic nonlinearities (hysteresis) but a variation of this approach, known as *delta modulation,* will. With delta modulation a linear detector is used to precalibrate the output of a binary driver to produce the desired waveform. By later playing back the binary sequence, the waveform can be re-created exactly. Thus, delta modulation can be used to compensate for all source nonlinearities except for changes in wavelength. The disadvantage is that the waveform must be precalibrated.
3. Vary the ratio of the on and off time periods (the duty cycle) of the source using a binary driver. With a fixed cycle time, the on time can be varied (pulse-width modulation or PWM). PWM works fairly well, but is sensitive to capacitance and nonlinear switching effects that can alter the waveform for either short on or off periods. A similar approach is to use fixed (short) pulses and vary the frequency of the pulses (pulse-frequency modulation or PFM). This approach has the advantage that every pulse is identical, thus capacitance and switching effects are minimized. PFM has been used to control the luminance of LEDs[27] and AOMs[16] linearly over a 1000:1 luminance range.
4. Detect the light with a linear photodetector and use feedback within the driver to linearize the output. By using a PIN photodiode in photovoltaic mode (see Chap. 15), it is possible to construct a very linear circuit.[28,31] Light feedback can be used with either analog or binary drivers.

Generating Complex Temporal and Spatial Patterns

Any type of light source or visual stimulus, from video monitors to street scenes, can be integrated into an optical system, giving improved control of luminance and pupil position. Thus, almost anything can be used as a target. Stimuli can be moved without varying the position of light entry in the eye's pupil by (1) translating a target in the retinal plane, (2) using a cathode ray tube (CRT) or a liquid crystal display (LCD) in a retinal conjugate plane, or (3) rotating a mirror in a pupil conjugate plane.[22]

Rotating a mirror in a pupil conjugate plane changes the angle at which light enters the eye, but not the pupil entry position. This technique has been used for generating motion and counterphase gratings[21] and for decreasing the apparent inhomogeneity of a stimulus

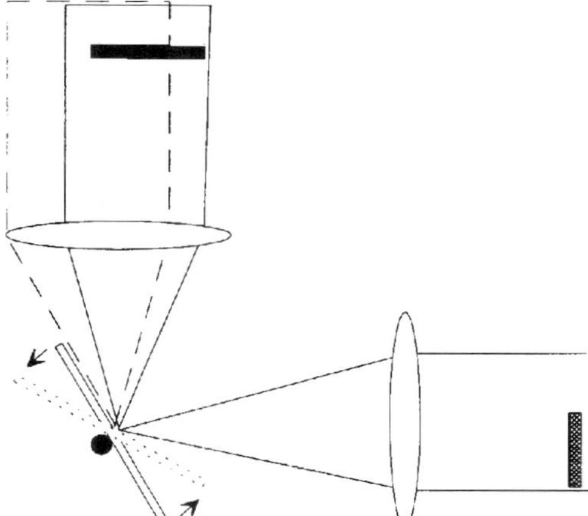

FIGURE 5 A movable mirror located in a pupil conjugate plane allows motion of the retinal image without moving the pupil conjugate images.

by moving the image at a rate above flicker fusion. Galvanometers are commonly used to generate the motion (Fig. 5).

Video Monitors and CRTs. The combination of a video display with a maxwellian view optical system allows the experimenter the advantages of precise control of the pupil, use of a broad spectral range of backgrounds (provided by the traditional optics), and the fine spatial control available in a video system. The video monitor or CRT is placed at a retinal conjugate plane and the other optical channels are added to it. By passing the combined light path through an aperture at a pupil conjugate plane it is possible to control the pupil size for the video system, insert achromatizing lenses, etc. (see Fig. 8 later in the chapter).

Liquid Crystal Displays. One type of video display that is of increasing utility is the liquid crystal display (LCD) (Chap. 27). LCDs, unlike video monitors, work by transmission or reflection; they are not self-luminous. Thus, while the wavelength composition of a monitor is determined by the characteristics of its phosphors, the LCD can be used with *any* light source. LCD displays work by changing the polarization state of local regions of the display (pixels). The pattern of pixels is controlled by a video signal or a computer. Like most polarization devices, the extinction ratio is wavelength-dependent and the proportion of light transmitted in the off state may be relatively high. With color versions of the newer active matrix LCDs it is possible to pass three wavelengths of monochromatic light through the color LCD and have a spatially controlled, high-intensity display with the maximum possible gamut. In this case the LCD is placed at a retinal conjugate plane.

Calibration

There are several approaches to calibrating the retinal illuminance of a maxwellian view system. All depend on either a standard source of known luminance (a reference source) or a calibrated, linear[31,32] photodetector. In the past, calibration typically depended on referring the illuminance of an optical system to a carefully maintained standard light or by measuring the luminance of a diffuser. Well-calibrated photometers and radiometers are

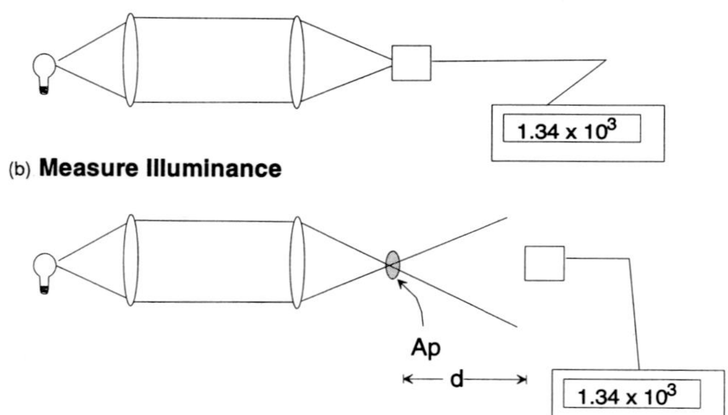

FIGURE 6 Techniques for calibrating an optical system. Panel (*a*): Measure the power of the exit pupil and assume that it is uniformly distributed in the retinal image. Panel (*b*): Measure the illuminance produced by an exit pupil of radius r at a detector located at distance d from the exit pupil.

now readily available, and we will cover techniques based on using these (the reader is referred to Westheimer[3] for other approaches). We assume that all of the light arriving at the exit pupil of the instrument enters the eye and that the exit pupil can be adequately approximated as circular with a radius of r.

Measurement of Power at Exit Pupil. With the first technique (Fig. 6*a*) a calibrated detector is placed at the exit pupil of the maxwellian view device. First, the total luminous power at the exit pupil is measured. Next, the retinal area illuminated is computed from the geometry of the stimulus [Eq. (12)]. From these two quantities the retinal illuminance can be computed. This calculation specifically assumes that all of the power is uniformly distributed across the retinal image. Conversion of retinal illuminance from $\mathrm{lm/m^2}$ to trolands is achieved using Eq. (6).

Measurement of the Illuminance on a Detector. In this technique the illuminance falling on a detector at a fixed distance from the exit pupil is measured (Fig. 6*b*; also see Nygaard and Frumkes[33]). A circular source of luminance L and radius r produces an illuminance E on a detector at distance d^2. If we assume that the source dimensions affect the calibration negligibly ($r < d/10$; $r < f_e/10$), then

$$E = \frac{L\pi r^2}{(d^2)} \tag{18}$$

We want to relate E as measured by the detector to E_R, the retinal illuminance, which is

$$E_R = \frac{L\pi r^2}{(f_e)^2} \tag{19}$$

Combining Eqs. (18) and (19) we have

$$E_R = \frac{(d)^2}{(f_e)^2} E \tag{20}$$

If the radius of the exit pupil (r_p) cannot be ignored then we need to account for its properties as a light source. If the exit pupil is circular and can be approximated as a lambertian emitter, then the illumination on the detector [Eq. (18)] is

$$E = \frac{L\pi r^2}{(r_p^2 + d^2)} \tag{21}$$

(Wyszecki and Stiles, table I(4.4); chap. 56). Likewise, the retinal illuminance [Eq. (20)] is

$$E_R = \frac{(r_p^2 + d^2)}{(r_p^2 + (f_e)^2)} E \tag{22}$$

The troland value of E_R can be computed from Eq. (16).

28.7 LIGHT EXPOSURE AND OCULAR SAFETY

There are two main mechanisms by which light can damage the eye. The first is simply by heating: too much radiation burns the retina. The second mechanism is photochemical. Light, especially short-wavelength light, causes photochemical oxidation. The byproducts of this oxidation are toxic to the retina.

Figure 7 shows the relation of the danger thresholds as defined by the ANSI 136 standard,[34] expressed in tds and lumens/m^2, for a 440-, a 550-, and a 670-nm light. These thresholds are for field sizes greater than 1°. The damage threshold for visible light is quite high, and thus lights capable of producing damage are intensely unpleasant. However, care should be taken to exclude IR and UV light. Most light sources emit considerable radiant energy in the infrared, and sources such as xenon arcs can emit ultraviolet radiation. Failures of blocking filters in these spectral regions may be visually undetectable. For safety it is advisable to have separate blocking filters to exclude unwanted radiation. It should also be noted that when using coherent light sources speckle can cause large focal variations in the retinal illuminance and the safety limits should be lowered by a factor of between 10 and 100 times.[35] Point sources and lasers have different safety standards that take into account the motion of the eye.

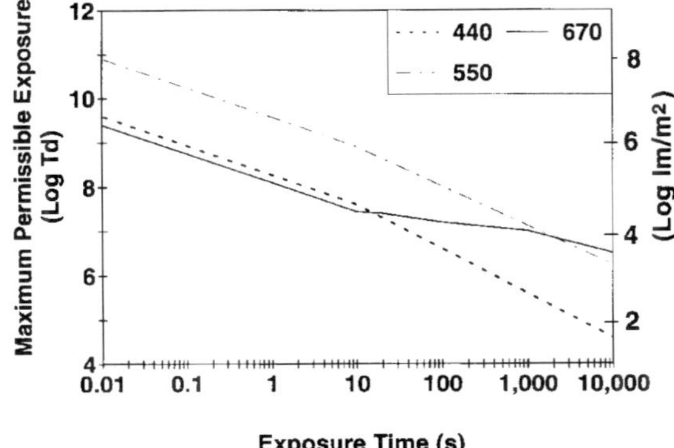

FIGURE 7 The maximum permissible retinal illuminance at different exposure times for 440 nm (dotted line), 550 nm (thin dot-dashed line), and 670 nm (solid line) lights and extended fields according to the ANSI standards.

28.8 LIGHT SOURCES

Common light sources are presented in Table 4 and lasers are presented in Table 5 and in Chap. 10. Wyszecki and Stiles[2] present detailed discussions of light sources. The considerations for deciding on a light source are stability and uniformity of the luminous area, intrinsic luminosity of the source, and the spectral distribution of light from the source.

28.9 COHERENT RADIATION

Coherent light sources can be used for generation of interference fringes,[36] a common use of lasers in vision research.[8–10] Light from the laser is directed into two similar channels. The light from the two channels arrives at the pupil of the eye separated by some distance *d*. Light spreading from the two beams then overlaps on the retina and creates an interference pattern in the intersection zone. The spacing of the interference pattern is controlled by the angle at which the beams intersect at the retina. The orientation of the pattern is controlled by the relative orientation of the two pupils. However, HeNe and easily controlled solid-state visible lasers are also excellent general-purpose light sources. The major drawback to their use is speckle.

TABLE 4 Common Light Sources

Type	Properties	Comments
Tungsten	Broadband, thermal emitter.	Available in a number of filament and power choices, hotter bulbs are short-lived.
Tungsten-halogen	Broadband, thermal emitter.	Inclusion of halogen regeneration cycle allows tungsten-halogen to run hotter (more blue light and a higher luminance) at a longer rated life than tungsten. If running at below the rated wattage the bulbs can be inverted to extend the lifetime.
Xenon arcs	Nearly flat spectrum, high luminance, some emission lines are present.	UV light can create ozone which must be vented. Tendency for the arc to move relative to the electrodes, causing movement of the exit pupil in maxwellian view systems. Luminous area is fairly small.
Fluorescent tube	Broad spectral output with superimposed emission lines. Easily obtainable, efficient.	Can be modulated rapidly. But there may be problems with compound phosphors (see Ref. 28).
Light-emitting diodes (LEDs)	Small, cheap, efficient light sources. Easily controlled.	Can be modulated rapidly, come in a variety of wavelength and power ranges (see Chap. 12).

TABLE 5 Types of Lasers

Type	Most common visible wavelengths	Comments	Typical power	Typical noise %
Argon	514, 488, (other lines are available, depending upon the design).	Power of each wavelength varies with particular laser design and cost. Most common wavelengths are 514 nm and 488 nm.	5 mW–5 W	1
HeCd	444 nm	Fairly expensive, have some high-frequency noise (>300 kHz) that may be important in scanning or short exposure applications.	0.5–50 mW	3
HeNe	543, 594, 632.8, also have orange lines.	Most common and least expensive gas laser. The 632.8-nm designs are easily available, inexpensive, and long-lived. Other wavelengths have many of the same advantages but power and beam options are limited and cost is higher.	0.1–20 mW	0.5
Krypton	588, 677.	Expensive, tend to be unstable, large.	50 mW–5 W	3
Dye lasers	Variable.	Expensive, need to be pumped by another laser. Require costly and toxic dyes to operate, not particularly stable.	50 mW–5 W	Depends on pump laser.
Slab diode lasers	~630 nm, ~670 nm, ~690 nm, some in blue.	Development is very rapid with new wavelengths and power options. The beam geometry is not ideal but they have great potential as light sources. Can be easily controlled using standard electronic techniques as are common with LEDs. Have high impedance and are easily damaged by static electricity. Rapid developments at shorter wavelengths.	0.2–100 mW	1 percent or better if temperature stabilized.
VCSELS vertical cavity lasers (microlasers)	Infrared.	These are under rapid development. They are solid-state lasers but have superior efficiency and beam properties as compared to slab lasers. Low impedance.	1 mW	

TABLE 5 Types of Lasers (*Continued*)

Type	Most common visible wavelengths	Comments	Typical power	Typical noise %
Frequency doubled YAG	532 nm.	Readily available, solid-state designs. Expensive.	5–50 mW	<1 percent.
Tunable solid-state and solid dye lasers	Variable, pumped by lasers.	An area of rapid development. By using materials that fluoresce, and tuning the laser cavity, the output wavelength can be varied while maintaining excellent beam qualities.	mW–W	Depends on pump laser.

Generation of Speckle Fields

Speckle arises when coherent light is reflected from an optically rough surface. The surface roughness causes variations in the path length between the retina and adjacent areas on the surface. This variation causes phase differences from point to point in the retinal image. The spatial frequency of speckle depends on pupil size, since the size of those adjacent areas is smaller for bigger pupils. Thus, for a very small pupil, speckles will be large. Speckle has been used to develop optometers,[37–39] and to generate a pattern on the retina in the presence of poor optic media.[40,41] However, when using lasers as general-purpose light sources, speckle needs to be minimized or eliminated. There are three ways to despeckle a source: *spatial filtering, temporal despeckling,* and *raster scanning.*

In spatial filtering[10] the light remains spatially coherent. A light source is imaged onto a pinhole which diffracts the light so that a spherical wave emerges from the pinhole. A lens is then used to collimate the light, resulting in a plane wave. Filtering should be restricted to those experiments that require the use of coherent light. The introduction of dust or even structure in the anterior segment of the eye can introduce undesirable diffraction effects.

Temporal despeckling[14,15] uses the temporal integration of the visual system to blur the speckle field. Since speckle arises as a result of surface irregularities at a very small scale, a small amount of motion can decrease speckle contrast considerably.

Scanning moves a diffraction-limited spot of light across the retina to create a visual pattern. Typically, the pattern is a raster pattern, like that of a television, and the stimulus is generated by temporal variation in the intensity of the beam.[42–44] Since at any one time only a single, diffraction-limited, retinal region is illuminated, then there is no opportunity for speckle.

28.10 DETECTORS

Detectors are used for measuring the light produced by an optical system. They can generally be characterized by their quantum sensitivity, their spectral response curve, and the temporal amplitude spectrum of their noise (see Ref. 45 and chap. 15 of this Handbook). For most purposes detectors can be treated as detecting all light power incident upon their active area, but there are limitations that make it advisable to use them for normally incident light whenever possible.[46] Table 6 presents the most common detectors (also see Chaps. 15–19). At low light levels photomultiplier tubes (PMT)[47] or an avalanche photodiodes (APD)[48] can be used in a photon-counting mode (Chap. 17).

28.11 PUTTING IT TOGETHER

This section briefly describes construction of more complex maxwellian view systems. The goal is simply to help the researcher get started building an optical device by presenting real examples.

A Two-Channel Maxwellian View Apparatus

We first consider a simple two-channel maxwellian view device (Fig. 8*a*) for measuring the detectability of a small circular target on a large circular background. The background is provided via the straight optical channel. It has a light source LS_B followed by a lens L_4 that collimates the light from the source. A field stop (target T_1) controls the size of the background field. Adjacent to the target is a beamsplitter, followed by another lens L_3. L_3 creates a source image where we place a 2.0-log-unit circular neutral density wedge, a 2-mm (diameter) stop (artificial pupil) and an achromatizing lens AL. An achromatizing lens minimizes the effects of the chromatic aberrations of the eye.[49–51] The artificial pupil will be imaged at 1:1 in the pupil of the eye, thus providing a limiting pupil. Lenses L_2 and L_1 relay the artificial pupil to the pupil of the eye. The test channel is derived from the second light source LS_T. Light is collimated by L_7, passed through a heat-rejecting filter IRF, interference filter IF_1, and a neutral density filter ND_1. An image of LS_T is then formed at P_{2T} by L_6. In this plane we set an electromechanical shutter and a neutral density wedge. We then collimate the light with lens L_5, pass it through an aperture T_2 that creates the target, and combine it with the background light at the beamsplitter. In some cases it may be desirable to obtain both the test and background light from a single light source. This is readily achieved by the addition of mirrors and avoids the possibility that slight fluctuations in the sources might produce variability in the ratio of the target and background illuminances. When appropriate, use a single light source for multiple optical channels.

A similar optical design allows full spatial control of the stimulus by incorporating a video monitor (Fig. 8*b*) or an LCD (not shown). In general, the retinal image of the monitor will need to be minified. As shown, the size of the retinal image of the monitor is set by the ratio of the focal length of lenses L_1, L_2, and L_3 [Eq. (14)]. The main advantages of the hybrid system over free-viewing the monitor are the ability to control the pupil precisely, the ability to add spectrally pure backgrounds (and, if only low contrasts are needed, the background can be quite bright), and the ability to align an achromatizing lens precisely and to monitor the eye position. We show a simple monitoring system, where the eye is monitored by a CCD camera, through the back side of a "cold mirror" (M1). A cold mirror is a dichroic beamsplitter that reflects visible light and transmits infrared radiation. IR-emitting LEDS are used to illuminate the eye diffusely. Using the full aperture of the lens L_1 means that the depth of focus will be small, and the eye can be precisely positioned (or monitored) in three dimensions. The resolution of the camera can be relatively low.

To use an LCD in transmission mode rather than a video monitor we can simply introduce the display at the retinal conjugate plane (R_2) in Fig. 8*a*. Again, it will be

TABLE 6 Detectors

Detector	Sensitivity	Speed	Problems/limitations
Photomultipliers	Best in the blue, extended multialkali cathodes extend sensitivity into the red and near IR. Many designs have very low noise.	Fast, MHz, capable of photon counting.	Low sensitivity in the red and infrared, can be damaged by high light levels, though recovery can be aided by leaving in the dark for a long time (months). Fragile. Require high voltage (kV).
Avalanche photodiodes	Have the silicon sensitivity curve, peaking in the near IR. Higher quantum efficiency than PMTs.	Fast MHz, capable of photon counting, better quantum efficiency then a PMT, but higher noise level.	Noise increases rapidly at avalanche voltage, light can affect the breakdown point. Higher noise level than PMTs, but the higher quantum sensitivity makes them better for video rate imaging. Small sensitive area.
CCDs, CIDs	Very linear, easy to use, sensitivity peaks in the near IR. These have a high quantum efficiency.	Variable, depending upon implementation, integrate between readings. kHz.	Blooming and charge spread can effect spatial properties at high radiances.[52] CID's are resistant to blooming.
PIN silicon photodiodes	Easy to use, sensitivity peaks in the near IR.	With care they can be operated at high rates (MHz); however, in most photometers the circuitry is specifically designed for low-noise, lower-frequency operation.	
Thermal (bolometers)	Measures energy, thus flat spectral responsivity.	Very slow <10 Hz.	The only real use for these devices in vision research is for calibration of the spectral sensitivity of another detector. In general, it is better to have the detector calibrated against an NIST standard by a manufacturer.
See also photoresistors, vacuum photodiodes, and phototransistors	Seldom used.		

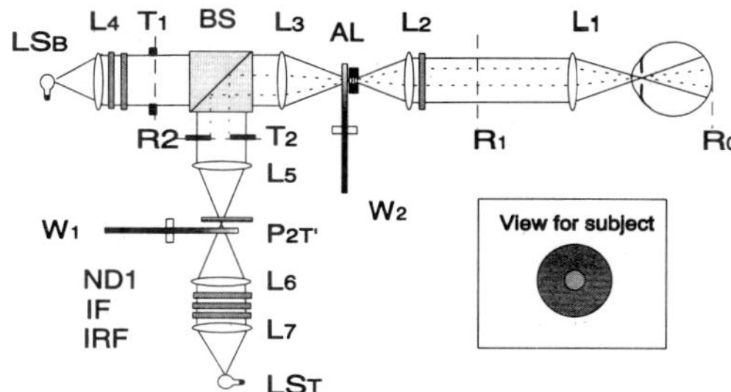

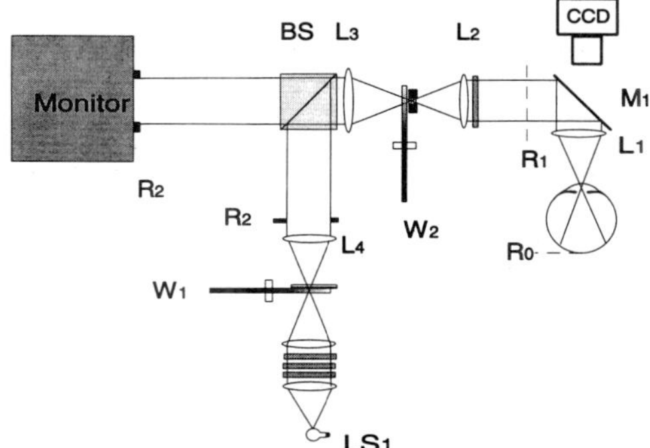

FIGURE 8 Top (a). An example of a two channel maxwellian view device. Two light sources are used (LS_B and LS_T) to form a background and a test channel respectively. The two channels are combined using a beamsplitter (BS). The intensity of the test field is controlled by a circular neutral density wedge W_1, and the intensity of both channels is controlled using wedge W_2. Additional details are provided in the text. Bottom (b). A similar apparatus, except now a video monitor is superimposed on a background derived from source LS_1. An infrared sensitive CCD camera is used to monitor the pupil through the back of a dichroic beamsplitter plate (M1) which reflects visible light and transmits infrared light (cold mirror). Infrared illumination of the eye is provided by infrared LEDs (not shown).

desirable to decrease the magnification of the LCD on the retina by choosing L_3 to have a longer focal length than lenses L_1 and L_2.

An Interferometer

We next describe a research interferometer that was built using modern electro-optical components.[10] To study the spatial sampling of the photoreceptor matrix of the eye, this instrument needed to have the following properties: (1) generation of spatial patterns on the retina with a frequency content above that imaged by the optics of the eye, (2) rapid and precise control of the modulation (contrast) and frequency (spacing) of the spatial patterns, (3) minimization of speckle, and (4) control of the two entrance pupils of the interferometer to displace them symmetrically about the center of the eye's pupil. We outline below how these objectives were achieved. The researcher should refer to the original paper for a more detailed account.

Interference can be used to generate diffraction patterns on the surface of the retina. Two coherently related sources are imaged in the plane of the eye's pupil (Fig. 9*a*). The spacing of the resulting diffraction pattern depends on the separation of the two sources in the eye's pupil, and the orientation of the pattern depends on the relative position of the two sources. Williams[10] (see also Refs. 8 and 9 for other examples) constructed a modified Mach-Zender interferometer[52] (Fig. 9*b*) that satisfies the requirements outlined above.

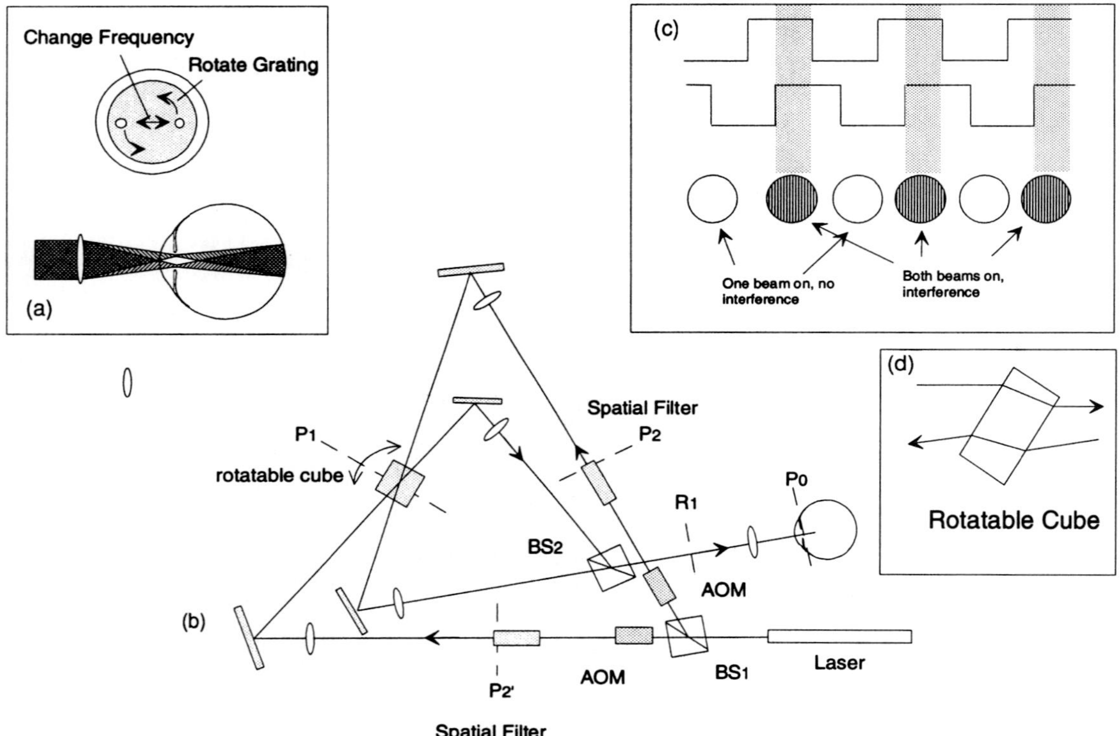

FIGURE 9 An example of a modern interferometer for vision research, modified from Ref. 10. (*a*) Factors controlling the spacing and orientation of the interference pattern on the retina. Increasing the distance between the entry pupils decreases the spacing of the pattern (increases the spatial frequency in cycles per deg). It is also possible to change the orientation of the interference pattern by rotating the two entry pupil in the plane of the eye's pupil. (*b*) A schematic of William's interferometer. This is a modified Mach-Zender interferometer and includes acousto-optic modulators (AOM) to control the modulation of the interference pattern, spatial filters to reduce speckle, a rotatable cube to control spacing and beamsplitters to separate the two beams (BS1) and to recombine them (BS2). For additional details see Ref. 10. (*c*) Contrast of the interference pattern is controlled by electronically varying the phase of the signals from the two AOMs. Each AOM is square wave modulated. The interference pattern is produced only when both beams are on at the same time. Thus, varying the relative phase of the two beams varies the proportion of the time when both are on, and thus the contrast. (*d*) The spacing of the beams in the plane of the eye's pupil is varied by rotating a deviation plate. The two beams propagate through the glass plate in opposite directions. Thus, as the plate is rotated, the beams are displaced equal but opposite amounts, resulting in symmetric deviations of the beams about the center of the eye's pupil. The actual apparatus uses an anti-reflection coated cube.

Modulation of the interference pattern is controlled using AOMs. Each of the beams is either on or off, with a 50 percent duty cycle. If both beams are on at the same time (in phase), the interference pattern is seen at full contrast. If one is off when the other is on (counterphase), then there is no interference pattern. Intermediate phases of the two beams will cause intermediate contrasts of the interference pattern (Fig. 9*c*). Thus, both beams are always operated in the same state (avoiding luminance artifacts due to temporal nonlinearities), and all control of modulation is electronic. The use of an all-electronic design allows precise control of contrast without audible cues to the subject. Spatial masking from speckle is minimized by placing spatial filters in a pupil conjugate plane. The spacing of the interference pattern is controlled by a rotable cube. Each beam traverses the cube in opposite directions (Fig. 9*d*). As the cube is rotated the beams, and thus the images in the eye's pupil, are displaced symmetrically.

28.12 CONCLUSIONS

We have covered only some of the basics of stimulus generation. We urge the interested reader to look elsewhere in this *Handbook*. In addition, the following references are good general sources for this topic.

28.13 ACKNOWLEDGMENTS

Supported by NIH EYO4395 (Burns) and DOE DE-FG02-91ER61229 (Webb). We thank Francois Delori for aid in computing the troland values for the ANSI standards and David Williams for valuable editorial help.

28.14 GENERAL REFERENCES

J. R. Meyer-Arendt, "Radiometry and Photometry: Units and Conversion Factors," *Applied Optics* **7**:2081–2084 (1968).

G. Westheimer, "The Maxwellian View," *Vision Res.* **6**:669–682 (1966).

J. W. T. Walsh, *Photometry,* Constable, London, 1953.

G. Wyszecki and W. S. Stiles, *Color Science,* Wiley, New York, 1982.

R. M. Boynton, "Vision," J. B. Sidowski (ed.), *Experimental Methods and Instrumentation in Psychology* McGraw-Hill, New York, 1966.

W. J. Smith, *Modern Optical Engineering,* 2d ed., McGraw-Hill, New York, 1990.

28.15 REFERENCES

1. Y. LeGrand, *Light Color and Vision,* Wiley, New York, 1957.
2. G. Wyszecki and W. S. Stiles, *Color Science,* Wiley, New York, 1982.
3. G. Westheimer, "The Maxwellian View," *Vision Res.* **6**:669–682 (1966).
4. F. W. Fitzke, A. L. Holden, and F. H. Sheen, "A Maxwellian-View Optometer Suitable for Electrophysiological and Psychophysical Research," *Vision Res.* **25**:871–874 (1985).

5. R. J. Jacobs, I. L. Bailey, and M. A. Bullimore, "Artificial Pupils and Maxwellian View," *Applied Optics* **31:**3668–3677 (1992).
6. G. M. Byram, "The Physical and Photochemical Basis of Visual Resolving Power Part I: The Distribution of Illumination in Retinal Images," *J. Opt. Soc. Amer.* **34:**571–591 (1944).
7. O. Dupuy, "La perception visuelle," *Vis. Res.* **8:**1507–1520 (1968).
8. F. W. Campbell and D. G. Green, "Optical and Retinal Factors Affecting Visual Resolution, *J. Physiol.* **181:**576–593 (1965).
9. F. W. Campbell and R. W. Gubisch, "Optical Quality of the Human Eye," *J. Physiol.* **186:**558–578 (1966).
10. D. R. Williams, "Aliasing in Human Foveal Vision," *Vis. Res.* **25:**195–205 (1985).
11. W. D. Wright, *Researches into Normal and Defective Colour Vision,* Kimpton, London, 1946.
12. T. W. Liebman, "Wedged Plate Beam Splitter without Ghosts Reflections," *Appl. Opt.* **31:**5905–5906 (1992).
13. R. M. Boynton and A. L. Nagy, "La Jolla Analytic Colorimeter," *J. Opt. Soc. Amer.* **72:**666–667 (1982).
14. R. H. Webb, "Concentrator for Laser Light," *Appl. Opt.* **31:**5917–5918 (1992).
15. J. R. Krauskopf, D. R. Williams, and D. H. Heeley, "Computer Controlled Color Mixer with Laser Primaries," *Vision Res.* **21:**951–953 (1981).
16. S. A. Burns, M. R. Kreitz, and A. E. Elsner, "Apparatus Note: A Computer Controlled, Two Color, Laser-Based Optical Stimulator for Vision Research," *Applied Optics* **30:**2063–2065 (1991).
17. C. A. Taylor and B. J. Thompson, "Some Improvements in the Operation of the Optical Diffractometer," *J. Sci. Instr.* **34:**439–447 (1957).
18. R. M. Boynton, M. M. Hayhoe, and D. I. A. MacLeod, "The Gap Effect: Chromatic and Achromatic Visual Discrimination as Affected by Field Separation," *Optica Acta* **24:**159–177 (1977).
19. M. C. Rynders and L. N. Thibos, "Single Channel Sinusoidally Modulated Visual Signal Generator, with Variable Temporal Contrast," *J. Opt. Soc. Amer.* **A10:**1642–1650 (1993).
20. V. V. Toi and P. A. Gronauer, "Visual Stimulator," *Rev. Sci. Instr.* **49:**1403–1406 (1978).
21. G. A. Fry, "Square Wave Grating Convoluted with a Gaussian Spread Function," *J. Opt. Soc. Amer.* **58:**1415–1416 (1968).
22. T. W. Butler and L. A. Riggs, "Color Differences Scaled by Chromatic Modulation Sensitivity Functions," *Vision Res.* **18:**1407–1416 (1978).
23. D. Vincent, "Amplitude Modulation with a Mechanical Chopper," *Applied Optics* **25:**1035–1036 (1986).
24. T. E. Cohn, "A Fast, Accurate Monochromatic Light Stimulus Generator," *Am. J. Optom. and Arch. Am. Acad. Optom.* **49:**1028–1030 (1972).
25. R. W. Nygaard and T. E. Frumkes, "LEDs: Convenient, Inexpensive Sources for Visual Experimentation," *Vis. Res.* **22:**435–440 (1982).
26. M. Yamashita and S. Takeuchi, "Temperature-Compensated Pulsed Reference Light Source Using a LED," *Review of Sci. Instr.* **54:**1795–1796 (1983).
27. W. H. Swanson, T. Ueno, V. C. Smith, and J. Pokorny, "Temporal Modulation Sensitivity and Pulse-Detection Thresholds for Chromatic and Luminance Perturbations," *J. Opt. Soc. Amer.* **A4:**1992–2005 (1987).
28. G. R. Cole, C. F. Stromeyer, III, and R. E. Kronauer, "Visual Interaction with Luminance and Chromatic Stimuli," *J. Opt. Soc. Amer.* **7:**128–140 (1990).
29. T. U. Watanabe, N. Mori, and F. Nakamura, "Technical Note: A New Suberbright LED Stimulator: Photodiode Feedback Design for Linearizing and Stabilizing Emitted Light," *Vision Res.,* 1992.
30. J. D. Mollen and P. G. Polden, "On the Time Constant of Tachistoscopes," *Quarterly J. Exp. Psychol.* **30:**555–568 (1978).

31. C. L. Sanders, "Accurate Measurements of and Corrections for Nonlinearities in Radiometers," *J. Res. Natl. Bur. Stand.,* sec. A, **76:**437–453 (1972).

32. R. G. Frehlich, "Estimation of the Nonlinearity of a Photodetector," *Appl. Opt.* **31:**5927–5929 (1992).

33. R. W. Nygaard and T. E. Frumkes, "Calibration of the Retinal Illuminance Provided by Maxwellian Views," *Vision Res.* **22:**433–434 (1982).

34. American National Standards Institute, "Safe Use of Lasers," Z-136.1 ANSI, N.Y., 1976.

35. D. L. Fried, "Laser Eye Safety: The Implication of Ordinary Speckle Statistics of Speckle-Speckle Statistics," *J. Opt. Soc. Amer.* **71:**914–916 (1981).

36. M. Francon, *Optical Interferometry,* Academic Press, N.Y., 1966.

37. R. T. Hennessy and H. W. Leibowitz, "Laser Optometer Incorporating the Badal Principle," *Behav. Res. Meth. Instr.* **4:**237–239 (1972).

38. N. W. Charman, "On the Position of the Plane of Stationarity in Laser Refraction," *Am. J. Opt. Physiol. Opt.* **51:**832–838 (1974).

39. A. Morrell and W. N. Charman, "A Bichromatic Laser Optometer," *Amer. J. Optom Physiol. Opt.* **64:**790–795 (1987).

40. W. W. Dawson and M. C. Barris, "Cortical Responses Evoked by Laser Speckle," *Invest. Ophthalmol. Vis. Sci.* **17:**1207–1212 (1978).

41. J. Fukuhara, H. Uozotot, S. Nojima, M. Saishin, and S. Nakao, Visual-Evoked Potentials Elicited by Laser Speckle Patterns," *Invest. Ophthalmol. Vis. Sci.* **24:**1400–1407 (1983).

42. R. H. Webb, G. W. Hughes, and F. C. Delori, "Confocal Scanning Laser Ophthalmoscope," *Appl. Opt.* **26:**1492–1499 (1987).

43. R. H. Webb, "Optics for Laser Rasters," *Appl. Opt.* **23:**3680–3683 (1984).

44. A. E. Elsner, S. A. Burns, R. W. Webb, and G. H. Hughes, "Reflectometry with a Scanning Laser Ophthalmoscope," *Applied Optics* **31:**3697–3710 (1992).

45. R. H. Webb and G. H. Hughes, "Detectors for Video Rate Scanning Imagers," *Appl. Optics* **32:**6227–6235 (1993).

46. J. Durnin, C. Reece, and L. Mandel, "Does a Photodetector Always Measure the Rate of Arrival of Photons?," *J. Opt. Soc. Amer.* **71:**115–117 (1981).

47. D. van Norren and J. van der Kraats, "A Continuously Recording Retinal Densitometer," *Vision Res.* **21:**897–905 (1981).

48. P. P. Webb, R. J. McIntyre, and J. Conradi, "Properties of Avalanche Photodiodes," *RCA Review* **35:**234–278 (1974).

49. R. E. Bedford and G. Wyszecki, "Axial Chromatic Aberration of the Eye," *J. Opt. Soc. Amer.* **47:**564–565 (1957).

50. A. L. Lewis, M. Katz, and C. Oehrlein, "A Modified Achromatizing Lens," *Am. J. Optom. Physiol. Opt.* **59:**909–911 (1982).

51. I. Powell, "Lenses for Correcting Chromatic Aberration of the Eye," *Appl. Opt.* **20:**4152–4155 (1981).

52. D. Malacara, *Optical Shop Testing,* chap. 4.5, Wiley Interscience, New York, 1991.

53. M. Marchywka and D. G. Socker, "Modulation Transfer Function Measurement Technique for Small-Pixel Detectors," *Applied Optics* **31:**7198–7213 (1992).

CHAPTER 29
PSYCHOPHYSICAL METHODS

Denis G. Pelli
Bart Farell
Institute for Sensory Research
Syracuse University
Syracuse, New York

29.1 INTRODUCTION

Psychophysical methods are the tools for measuring perception and performance. These tools are used to reveal basic perceptual processes, to assess observer performance, and to specify the required characteristics of a display. We are going to ignore this field's long and interesting history (Boring, 1942), and much theory as well (Gescheider, 1985; Macmillan and Creelman, 1991), and just present a user's guide. Use the supplied references for further reading.

Consider the psychophysical evaluation of the suitability of a visual display for a particular purpose. A home television to be used for entertainment is most reasonably assessed in a "beauty contest" of subjective preference (Mertz, Fowler, and Christopher, 1950), whereas a medical imaging display must lead to accurate diagnoses (Swets and Pickett, 1982; Metz, 1986) and military aerial reconnaissance must lead to accurate vehicle identifications (Scott, 1968). In our experience, the first step toward defining a psychophysically answerable question is to formulate the problem as a task that the observer must perform. One can then assess the contribution of various display parameters toward that performance. Where precise parametric assessment is desired it is often useful to substitute a simple laboratory task for the complex real-life activity, provided one can either demonstrate, or at least reasonably argue, that the laboratory results are predictive.

Psychophysical measurement is usually understood to mean measurement of behavior to reveal internal processes. The experimenter is typically not interested in the behavior itself, such as pressing a button, which merely communicates a decision by the observer about the stimulus.* This chapter reviews the various decision tasks that may be used to measure perception and performance and evaluates their strengths and weaknesses. We

* Psychophysical measurement can also be understood to include noncommunicative physiological responses such as pupil size, eye position, and electrical potentials measured on the scalp and face, which might be called "unthinking" responses. (These examples are merely suggestive, not definitive. Observers can decide to move their eyes and, with feedback, can learn to control many other physiological responses. Responses are "unthinking" only when they are not used for overt communication by the observer.) Whether these unthinking responses are called psychophysical or physiological is a matter of taste. In any case, decisions are usually easier to measure and interpret, but unthinking responses may be indicated in certain cases, as when assessing noncommunicative infants and animals.

begin with definitions and a brief review of visual stimuli. We then explain and evaluate the various psychophysical tasks, and end with some practical tips.

29.2 DEFINITIONS

At the highest level, an *experiment* answers a question about how certain "experimental conditions" affect observer performance. *Experimental conditions* include stimulus parameters, observer instruction, and anything else that may affect the observer's state. Experiments are usually made up of many individual measurements, called "trials," under each experimental condition. Each trial presents a stimulus and collects a response, a decision, from the observer.

There are two kinds of decision tasks: judgments and adjustments. It is useful to think of one as the inverse of the other. In one case the experimenter gives the observer a stimulus and asks for a classification of the stimulus or percept; in the other case the experimenter, in effect, gives the observer a classification and asks for an appropriate stimulus back. Either the experimenter controls the stimulus and the observer makes a *judgment* based on the resulting percept, or the observer *adjusts* the stimulus to satisfy a perceptual criterion specified by the experimenter (e.g., match a sample). Both techniques are powerful. Adjustments are intrinsically subjective (because they depend on the observers' understanding of the perceptual criterion), yet they can often provide good data quickly and are to be preferred when applicable. But not all questions can be formulated as adjustment tasks. Besides being more generally applicable, judgments are often easier to analyze, because the stimulus is under the experimenter's control and the task may be objectively defined. Observers typically like doing adjustments and find judgments tedious, partly because judgment experiments usually take much longer.

An obvious advantage of adjustment experiments is that they measure physical stimulus parameters, which may span an enormous dynamic range and typically have a straightforward physical interpretation. Judgment tasks measure human performance (e.g., frequency of seeing) as a function of experimental parameters (e.g., contrast). This is appropriate if the problem at hand concerns human performance per se. For other purposes, however, raw measures of judgment performance typically have a very limited useful range, and a scale that is hard to interpret. Having noted that adjustment and judgment tasks may be thought of as inverses of one another, we hasten to add that in practice they are often used in similar ways. Judgment experiments often vary a stimulus parameter on successive trials in order to find the value that yields a criterion judgment. These "sequential estimation methods" are discussed in Sec. 29.6. The functional inversion offered by sequential estimation allows judgment experiments to measure a physical parameter as a function of experimental condition, like adjustment tasks, while retaining the judgment task's more rigorous control and interpretation.

Distinguishing between judgment and adjustment tasks emphasizes the kind of response that the observer makes. It is also possible to subdivide tasks in a way that emphasizes the stimuli and the question posed. In a *detection* task there may be any number of alternative stimuli, but one is a blank, and the observer is asked only to distinguish between the blank and the other stimuli. Slightly more general, a *discrimination* task may also have any number of alternative stimuli, but one of the stimuli, which need not be blank, is designated as the reference, and the observer is asked only to distinguish between the reference and other stimuli. A decision that distinguishes among more than two categories is usually called an "identification" or "classification" (Ashby, 1992).

As normally used, the choice of term, "detection" or "discrimination," says more about the experimenter's way of thinking than it does about the actual task faced by the observer. This is because theoretical treatments of detection and discrimination usually allow for manipulation of the experimental condition by introduction of an extraneous element, often called a "mask" or "pedestal," that is added to every stimulus. Thus, one is always free to consider a discrimination task as detection in the presence of a mask. This shift in perspective can yield new insights (e.g., Campbell, Howell and Robson, 1971; Wandell, 1985; Watson, Ahumada, and Farrell, 1983—discussed by Adelson and Bergen, 1985; Klein, Casson and Carney, 1990). Since there is no fundamental difference between detection and discrimination, we have simplified the presentation below by letting detection stand in for both. The reader may freely substitute "reference" for "blank" (or suppose the presence of an extraneous mask) in order to consider the discrimination paradigm.

The idea of "threshold" plays a large role in psychophysics. Originally deterministic, *threshold* once referred to the stimulus intensity above which the stimulus was always distinguishable from blank, and below which it was indistinguishable from blank. In a discrimination task one might refer to a "discrimination threshold" or a "just-noticeable difference." Nowadays the idea is statistical; we know that the observer's probability of correct classification rises as a continuous function of stimulus intensity (see Fig. 1). Threshold is defined as the stimulus intensity (e.g., contrast) corresponding to an arbitrary level of performance (e.g., 82 percent correct). However, the old intuition, now called a "high threshold," still retains a strong hold on everyone's thinking for the good reason that the transition from invisible to visible, though continuous, is quite abrupt, less than a factor of 2 in contrast.

Most psychophysical research has concentrated on measuring thresholds. This has been motivated by a desire to isolate low-level sensory mechanisms by using operationally defined tasks that are intended to minimize the roles of perception and cognition. This program is generally regarded as successful—visual detection is well understood (e.g., Graham, 1989)—but leaves most of our visual experience and ability unexplained. This

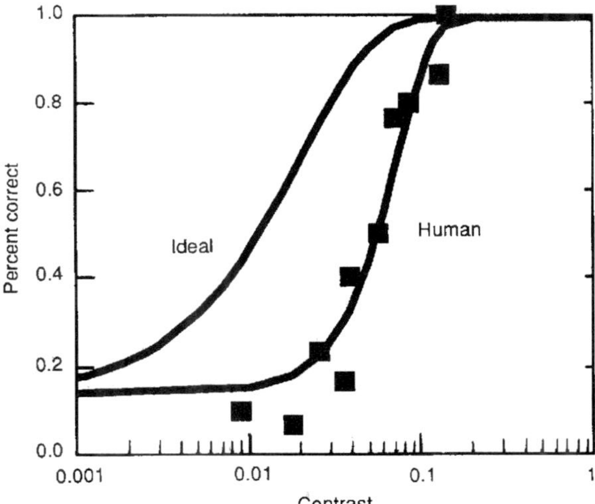

FIGURE 1 Probability of correctly identifying a letter in noise, as a function of letter contrast. The letters are bandpass filtered. Gaussian noise was added independently to each pixel. Each symbol represents the proportion correct in 30 trials. The solid curve through the points is a maximum likelihood fit of a Weibull function. The other curve represents a similar maximum likelihood fit to the performance of a computer program that implements the ideal letter classifier. Efficiency, the squared ratio of threshold contrasts, is 9 percent. (*Courtesy of Joshua A. Solomon.*)

has stimulated a great deal of experimentation with suprathreshold stimuli and nondetection tasks in recent years.

29.3 VISUAL STIMULI

Before presenting the tasks, which are general to all sense modalities, not just vision, it may be helpful to briefly review the most commonly used visual stimuli. Until the 1960s most vision research used a spot as the visual stimulus (e.g., Barlow, 1958). Then cathode ray tube displays made it easy to generate more complex stimuli, especially sinusoidal gratings, which provided the first evidence for multiple "spatial frequency channels" in vision (Campbell and Robson, 1968). Sinusoidal grating patches have two virtues. A sinusoid at the display always produces a sinusoidal image on the retina.* And most visual mechanisms are selective in space and in spatial frequency, so it is useful to have a stimulus that is restricted in both domains.

Snellen (1862), in describing his classic eye chart, noted the virtue of letters as visual stimuli—they offer a large number of stimulus alternatives that are readily identifiable (Pelli, Robson, and Wilkins, 1988; Pelli and Robson, 1991). Other stimuli include annuli, lines, arrays of such elements, and actual photographs of faces, nature, and military vehicles. There are several useful texts on image quality, emphasizing signal-to-noise ratio (Dainty and Shaw, 1974; Linfoot, 1964; Pearson, 1975; Schade, 1975). Finally, there has been some psychophysical investigation of practical tasks such as reading (Legge, Pelli, Rubin, and Schleske, 1985), flying an airplane (Rolf and Staples, 1986), or shopping in a supermarket (Pelli, 1987*b*).

The stimulus alternatives used in vision experiments are usually parametric variations along a single dimension, most commonly contrast, but frequently size and position in the visual field. *Contrast* is a dimensionless ratio: the amplitude of the luminance variation within the stimulus, normalized by the background luminance. Michelson contrast (used for gratings) is the maximum minus the minimum luminance divided by the maximum plus the minimum. Weber contrast (used for spots and letters) is the maximum deviation from the uniform background divided by the background luminance. RMS contrast is the root-mean-square deviation of the stimulus luminance from the mean luminance, divided by the mean luminance.

29.4 ADJUSTMENTS

Adjustment tasks require that the experimenter specify a perceptual criterion to the observer, who adjusts the stimulus to satisfy the criterion. Doubts about the observer's interpretation of the criterion may confound interpretation of the results. The adjustment technique is only as useful as the criterion is clear.

Threshold

Figure 2 shows contrast sensitivity (the reciprocal of the threshold contrast) for a sinusoidal grating as a function of spatial and temporal frequency (Robson, 1966). These thresholds were measured by what is probably the most common form of the adjustment task, which asks the observer to adjust the stimulus contrast up and down to the point

* This is strictly true only within an isoplanatic patch, i.e., a retinal area over which the eye's optical point spread function is unchanged.

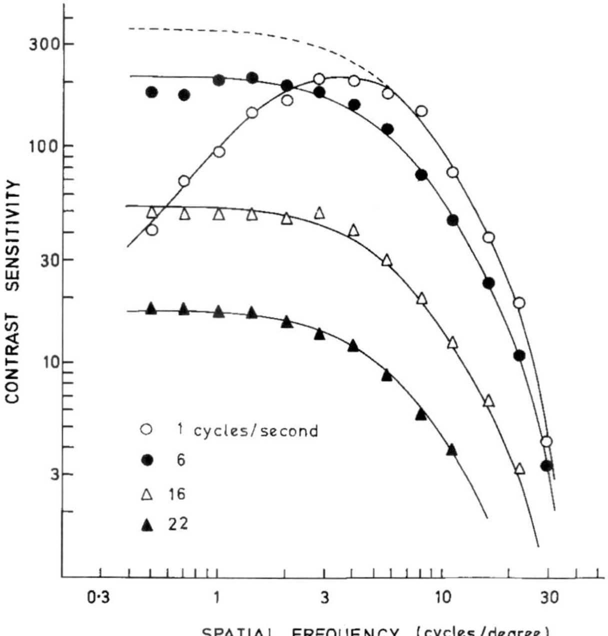

FIGURE 2 Spatial contrast sensitivity (reciprocal of threshold contrast) functions for sinusoidal gratings temporally modulated (flickered) at several temporal frequencies. The points are the means of four method-of-adjustment measurements and the curves (one with a dashed low-frequency section) differ only in their positions along the contrast-sensitivity scale. ○ 1 cycle per second, ● 6, △ 16, ▲ 22 cycles per second. [*From Robson* (1966).]

where it is "just barely detectable." While some important studies have collected their data in this way, one should bear in mind that this is a vaguely specified criterion. What should the observer understand by "barely" detectable? Seen half the time? In order to adjust to threshold the observer must form a subjective interpretation and apply it to the changing percept. It is well known that observers can be induced, e.g., by coaching, to raise or lower their criterion, and when comparing among different observers it is important to bear in mind that social and personality factors may lead to systematically different interpretations of the same vague instructions. Nevertheless, these subjective effects are relatively small (about a factor of 2 in contrast) and many questions can usefully be addressed, in at least a preliminary way, by quick method of adjustment threshold settings. Alternatively, one might ignore the mean of the settings and instead use the standard deviation to estimate the observer's discrimination threshold (Woodworth and Schlosberg, 1963).

Nulling

Of all the many kinds of adjustments, nulling is the most powerful. Typically, there is a simple basic stimulus that is distorted by some experimental manipulation, and the observer is given control over the stimulus and asked to adjust it so as to cancel the distortion (e.g., Cavanagh and Anstis, 1991). The absence of a specific kind of distortion is usually unambiguous and easy for the observer to understand, and the observer's null setting is typically very reliable.

Matching

Two stimuli are presented, and the observer is asked to adjust one to match the other. Sometimes the experiment can be designed so that the stimuli are a perfect match, utterly indistinguishable, which Brindley (1960) calls a "Class A" match. Usually, however, the stimuli are obviously different and the observer is asked to match only a particular aspect of the stimuli, which is called a "Class B" match. For example, the observer might be shown two grating patches, one fine and one coarse, and asked to adjust the contrast of one to match the contrast of the other (Georgeson and Sullivan, 1975). Or the observer might see two uniform patches of different colors and be asked to match their brightnesses (Boynton, 1979). Observers (and reviewers for publication) are usually comfortable with matching tasks, but, as Brindley points out, it is amazing that observers can seemingly abstract and compare a particular parameter of the multidimensional stimuli in order to make a Class B match. Matching tasks are extremely useful, but conclusions based on Class B matches may be less secure than those based on Class A matches because our understanding of how the observer does the task is less certain.

Magnitude Production

The observer is asked to adjust a stimulus to match a numerically specified perceptual criterion, e.g., "as bright as a 60-watt light bulb." The number may have a scale (watts in this case) or not—a pure number (Gescheider, 1985). The use of pure numbers, without any scale, to specify a perceptual criterion is obviously formally ambiguous, but in practice many experimenters report that observers seem comfortable with such instructions and produce stable results that are even reasonably consistent among different observers. Magnitude production, however, is rarely used in visual psychophysics research.

29.5 JUDGMENTS

Judgment tasks ask the observer to classify the stimulus or percept. They differ primarily in the number of alternative stimuli that may be presented on a given trial and the number of alternative responses that the observer is allowed.

The Ideal Observer

When the observer is asked to classify the stimulus (not the percept) it may be useful to consider the mathematically defined ideal classifier that would yield the most accurate performance using only the information (the stimuli and their probabilities) available to the observer (Peterson, Birdsall, and Fox, 1954; Tanner and Birdsall, 1958; Van Trees,

1968; Geisler, 1989; Pelli, 1985, 1991). Obviously this would be an empty exercise unless there is some known factor that makes the stimuli hard to distinguish. Usually this will be visual noise: random variations in the stimulus, random statistics of photon absorptions in the observer's eyes, or random variations in neural processes in the observer's visual system. If the stimuli plus noise can be defined statistically at some site—at the display, as an image at the observer's retinae, as a pattern of photon absorptions, or as a spatiotemporal pattern of neural activity—then one can solve the problem mathematically and compute the highest attainable level of performance. This ideal often provides a useful point of comparison in thinking about the actual human observer's results. A popular way of expressing such a comparison is to compute the human observer's efficiency, which will be a number between 0 and 1. For example, in Fig. 1 at threshold the observer's efficiency for letter identification is 9 percent. As a general rule, the exercise of working out the ideal and computing the human observer's efficiency is usually instructive but, obviously, low efficiencies should be interpreted as a negative result, suggesting that the ideal is not particularly relevant to understanding how the human observer does the task.

Yes-No

The best-known judgment task is yes-no. It is usually used for detection, although it is occasionally used for discrimination. The observer is either asked to classify the stimulus, "Was a nonblank stimulus present?" or classify the percept, "Did you see it?" The observer is allowed only two response alternatives: yes or no. There may be any number of alternative stimuli. If the results are to be compared with those of an ideal observer, then the kind of stimulus, blank or nonblank, must be unpredictable.

As with the method-of-adjustment thresholds discussed above, the question posed in a yes-no experiment is fundamentally ambiguous. Where is the dividing line between yes and no on the continuum of internal states between the typical percepts generated by the blank and nonblank stimuli? Theoretical considerations and available evidence suggest that observers act as if they reduced the percept to a "decision variable," a pure magnitude—a number if you like—and compared that magnitude with an internal criterion that is under their conscious control (Nachmias, 1981; Pelli, 1985). Normally we are not interested in the criterion, yet it is troublesome to remove its influence on the results, especially since the criterion may vary between experimental conditions and observers. For this reason, most investigators no longer use yes-no tasks.

As discussed next, this pesky problem of the observer's subjective criterion can be dealt with explicitly, by using "rating scale" tasks, or banished, by using unbiased "two-alternative forced choice" (2afc) tasks. Rating scale is much more work, but unless the ratings themselves are of interest, the end result of using either rating scale or 2afc is essentially the same.

Rating Scale

In a rating scale task the observer is asked to rate the likelihood that a nonblank stimulus was presented. There must be blank and nonblank stimulus alternatives, and there may be any number of alternative ratings—five is popular—but even a continuous scale may be allowed (Rockette, Gur, and Metz, 1992). The endpoints of the rating scale are "The stimulus was definitely blank" and "The stimulus was definitely nonblank," with intermediate degrees of confidence in between. The results are graphed as a receiver operating characteristic, or ROC, that plots one conditional probability against another. The observer's ratings are transformed into yes-no judgments by comparing them with an external criterion. Ratings above the criterion become "yes" and those below the criterion

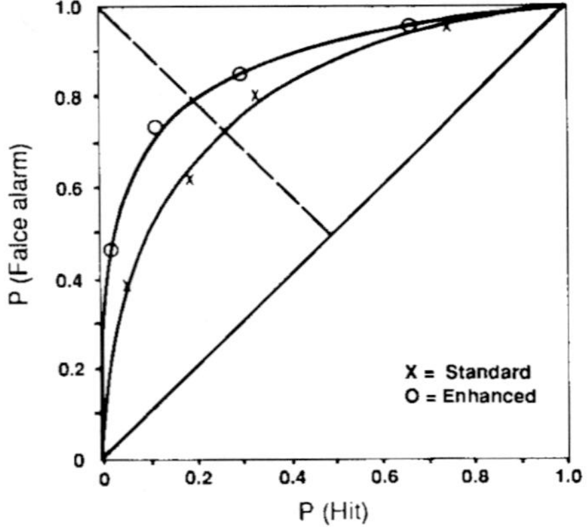

FIGURE 3 Example of an empirical ROC. Six radiologists attempted to distinguish between malignant and benign lesions in a set of 118 mammograms, 58 malignant and 60 benign, first when the mammograms were viewed in the usual manner ("standard"), and then—"enhanced"—when they were viewed with two aids, including a checklist of diagnostic features. The ratings were "very likely malignant," "probably malignant," "possibly malignant," "probably benign," and "very likely benign." The areas under the curves are 0.81 and 0.87. [*From Swets* (1988).]

become "no." This transformation is repeated for all possible values of the external criterion. Finally, the experimenter plots—for each value of the criterion—the probability of a yes when a nonblank stimulus was present (a "hit") against the probability of a yes when a blank stimulus was present (a "false alarm"). In medical contexts the hit rate is called "sensitivity" and one minus the false alarm rate is called "specificity." Figure 3 shows an ROC curve for a medical diagnosis (Swets, 1988); radiologists examined mammograms and rated the likelihood that a lesion was benign or malignant.

In real-life applications the main value of ROC curves is that they can be used to optimize yes-no decisions based on ratings, e.g., whether to refer a patient for further diagnosis or treatment. However, this requires knowledge of the prior stimulus probabilities (e.g., in Fig. 3, the incidence of disease in the patient population), the benefit of a hit, and the cost of a false alarm (Swets and Pickett, 1982; Metz, 1986). These conditions are rarely met. One usually can estimate prior probability and assess the cost of the wasted effort caused by the false alarms, but it is hard to assign a commensurate value to the hits, which may save lives through timely treatment.

The shape of the ROC curve has received a great deal of attention in the theoretical detection literature, and there are various mathematical models of the observer's detection process that can account for the shape (Nachmias and Steinman, 1965; Green and Swets, 1974; Nolte and Jaarsma, 1967). However, unless the actual situation demands rating-based decisions, the ROC shape has little or no practical significance, and the general practice is to summarize the ROC curve by the area under the curve. The area is 0.5 when the observers' ratings are independent of the stimuli. The area can be at most 1—when the observer makes no mistakes. The area can descend below 0.5 when the observer reverses the categories of blank and nonblank. We'll see in a moment that a result equivalent to ROC area can usually be obtained with much less effort by doing a two-alternative forced choice experiment instead.

Two-Alternative Forced Choice (2afc)

This task is traditionally characterized by two separate stimulus presentations, one blank and one nonblank, in random order. The two stimuli may be presented successively or side by side. The observer is asked whether the nonblank stimulus was first or second (or on the left or right). We noted above that in yes-no tasks observers seem to reduce the stimulus to a decision variable, the magnitude upon which they base their decisions. The 2afc task is said to be "unbiased" because the observer presumably chooses the presentation that generated the higher magnitude, without referring to any subjective internal criterion. At the beginning of this section we said that all judgment tasks consist of the presentation of a stimulus followed by a judgment. In this view, we might consider the two presentations in the 2afc task to be a single stimulus. The two possible composite stimuli to be discriminated are reflections of one another, either in space or time. The symmetry of the two alternatives suggests that the observer's choice between them may be unbiased.

Other related tasks are often called "two-alternative forced choice" and are similarly claimed to be unbiased. There is some confusion in the literature over which tasks should be called "2afc." In our view, the "2afc" label is of little consequence. What matters is whether the task is unbiased, i.e., are the alternative stimuli symmetric for the observer? Thus a yes-no discrimination of blank and nonblank stimuli may be biased even though there are two response alternatives and the choice is forced, whereas it may be reasonable to say that the judgment of the orientation of a grating that is either horizontal or vertical is unbiased even though there is only a single presentation. We suggest that authors wishing to claim that their task is unbiased say so explicitly and state why. This claim might be based on a priori considerations of the symmetry between the stimuli to be discriminated, or on a post hoc analysis of relative frequencies of the observer's responses.

In theory, if we accept the assumptions that each stimulus presentation produces in the observer a unidimensional magnitude (the decision variable), that the observer's ratings and 2afc decisions are based, in the proper way, on this magnitude, and that these magnitudes are stochastically independent between presentations, then the probability of a correct response on a 2afc trial must equal the area under the ROC curve (Green and Swets, 1974). Nachmias (1981) compared 2afc proportion correct and ROC area empirically, finding that ROC area is slightly smaller, which might be explained by stimulus-induced variations in the observer's rating criteria.

Magnitude Estimation

In the inverse of magnitude production, a stimulus is presented and the observer is asked to rate it numerically (Gescheider, 1985). Some practitioners provide a reference (e.g., a stimulus that rates 100), and some don't, allowing the observer to use their own scale. Magnitude estimation and rating scale are fundamentally the same. Magnitude estimation experiments typically test many different stimulus intensities a few times to plot mean magnitude versus intensity, and rating-scale experiments typically test few intensities many times to plot an ROC curve at each intensity.

Response Time

In practical situations the time taken by the observer to produce a judgment usually matters, and it will be worthwhile recording it during the course of the experiment. Some psychophysical research has emphasized response time as a primary measure of performance in an effort to reveal mental processes (Luce, 1986).

29.6 STIMULUS SEQUENCING

So far we have discussed a single trial yielding a single response from the observer. Most judgments are stochastic, so judgment experiments usually require many trials. An uninterrupted sequence of trials is called a *run.* There are two useful methods of sequencing trials within a run.

Method of Constant Stimuli

Experimenters have to worry about small, hard-to-measure variations in the observer's sensitivity that might contaminate comparisons of data collected at different times. It is therefore desirable to run the trials for the various conditions as nearly simultaneously as possible. One technique is to interleave trials for the various conditions. This is the classic "method of constant stimuli." Unpredictability of the experimental condition and equal numbers of trials for each condition are typically both desirable. These are achieved by using a randomly shuffled list of all desired trials to determine the sequence.

Sequential Estimation Methods

One can use the method of constant stimuli to measure performance as a function of a signal parameter—let us arbitrarily call it intensity—and determine, by interpolation, the threshold intensity that corresponds to a criterion level of performance.* This approach requires hundreds of trials to produce a precise threshold estimate. Various methods have been devised that obtain precise threshold estimates in fewer trials, by using the observer's previous responses to choose the stimulus intensity for the current trial. The first methods were simple enough for the experimenter to implement manually, but as computers appeared and then became faster, the algorithms have become more and more sophisticated. Even so, the requisite computer programs are very short.

In general, there are three stages to threshold estimation. First, all methods, implicitly or explicitly, require that the experimenter provide a confidence interval around a guess as to where threshold may lie. (This bounds the search. Lacking prior knowledge, we would have an infinite range of possible intensities. Without a guess, where would we place the first trial? Without a confidence interval, where would we place the second trial?) Second, one must select a test intensity for each trial based on the experimenter's guess and the responses to previous trials. Third, one must use the collected responses to estimate threshold. At the moment, the best algorithm is called ZEST (King-Smith, Grigsby, Vingrys, Benes, and Supowit, 1993), which is an improvement over the popular QUEST (Watson and Pelli, 1983). The principal virtues of QUEST are that it formalizes the three distinct stages, and implements the first two stages efficiently. The principal improvement in ZEST is an optimally efficient third stage.

29.7 CONCLUSION

This chapter has reviewed the practical considerations that should guide the choice of psychophysical methods to quickly and definitely answer practical questions related to perception and performance. Theoretical issues, such as the nature of the observer's

* The best way to interpolate frequency-of-seeing data is to make a maximum likelihood fit by an S-shaped function (Watson, 1979). Almost any S-shaped function will do, provided it has adjustable position and slope (Pelli, 1987*a*).

internal decision process have been deemphasized. The question of how well we see is answerable only after we reduce the question to measurable performance of a specific task. The task will be either an adjustment—for a quick answer when the perceptual criterion is unambiguous—or a judgment—typically to find threshold by sequential estimation.

The success of psychophysical measurements often depends on subtle details: the seemingly incidental properties of the visual display, whether the observers receive feedback about their responses, and the range of stimulus values encountered during a run. Decisions about these matters have to be taken on a case-by-case basis.

29.8 TIPS FROM THE PROS

We asked a number of colleagues for their favorite tips.

- Experiments often measure something quite different from what the experimenter intended. Talk to the observers. Be an observer yourself.
- Viewing distance is an often-neglected but powerful parameter, trivially easy to manipulate over a 100:1 range. Don't be limited by the length of your keyboard cable.
- Printed vision charts are readily available, offering objective measurement of visiblity, e.g., to characterize the performance of a night-vision system (Snellen, 1862; Ishihara, 1954; Regan and Neima, 1983; Pelli, Robson, and Wilkins, 1988).
- When generating images on a cathode ray tube, avoid generating very high video frequencies (e.g., alternating black and white pixels along a horizontal raster line) and very low video frequencies (hundreds of raster lines per cycle) since they are typically at the edges of the video amplifier's passband (Pelli and Zhang, 1991).
- Consider the possibility of aftereffects, whereby past stimuli (e.g., at high contrast or different luminance) might affect the visibility of the current stimulus (Blakemore and Campbell, 1969*a, b*; Cornsweet, 1970; Frome, MacLeod, Buck, and Williams, 1981).
- Drift of sensitivity typically is greatest at the beginning of a run. Do a few warm-up trials at the beginning of each run. Give the observer a break between runs.
- Allow the observer to see the stimulus once in a while. Sequential estimation methods tend to make all trials just barely detectable, and the observer may forget what to look for. Consider throwing in a few high-contrast trials, or defining threshold at a high level of performance.
- Calibrate your display before doing the experiment. Afterwards may be too late.

29.9 ACKNOWLEDGMENTS

Josh Solomon provided Fig. 1. Tips were contributed by Al Ahumada, Mary Hayhoe, Mary Kaiser, Gordon Legge, Walt Makous, Suzanne McKee, Eugenio Martinez-Uriega, Beau Watson, David Williams, and Hugh Wilson. Al Ahumada, Katey Burns, and Manoj Raghavan provided helpful comments on the manuscript. Supported by National Eye Institute grants EY04432 and EY06270.

29.10 REFERENCES

Adelson, E. H. and Bergen, J. R., "Spatiotemporal Energy Models for the Perception of Motion," *J. Opt. Soc. Am.* [*a*], **2:**284–299 (1985).

Ashby, F. G., "Multidimensional Models of Categorization," F. G. Ashby (ed.), *Multidimensional Models of Perception and Cognition,* Lawrence Erlbaum Associates, Hillsdale, NJ., 1992.

Barlow, H. B., "Temporal and Spatial Summation in Human Vision at Different Background Intensities," *J. of Physiol.* **141:**337–350 (1958).

Blakemore, C. and Campbell, F. W., "Adaptation to Spatial Stimuli," *J. Physiol.* **200**(1):11P–13P (1969*a*).

Blakemore, C. and Campbell, F. W., "On the Existence of Neurones in the Human Visual System Selectively Sensitive to the Orientation and Size of Retinal Images," *Journal of Physiology* **203:**237–260 (1969*b*).

Boring, E. G., *Sensation and Perception in the History of Experimental Psychology,* Irvington Publishers, New York, 1942.

Boynton, R. M., *Human Color Vision,* Holt Rinehart and Winston, New York, 1979, pp. 299–301.

Brindley, G. A., *Physiology of the Retina and the Visual Pathways,* Edward Arnold Ltd., London, 1960.

Campbell, F. W., Howell, E. R., and Robson, J. G. (1971) "The Appearance of Gratings With and Without the Fundamental Fourier Component," *J. Physiol.* **217:**17P–18P (1971).

Campbell, F. W. and Robson, J. G., "Application of Fourier Analysis to the Visibility of Gratings," *J. of Physiol.* **197:**551–566 (1968).

Cavanagh, P. and Anstis, S., "The Contribution of Color to Motion in Normal and Color-Deficient Observers," *Vision Res.* **31:**2109–2148 (1991).

Cornsweet, T. N., *Visual Perception,* Academic Press, New York, 1970.

Dainty, J. C. and Shaw, R., *Image Science,* Academic Press, New York, 1974.

Frome, F. S., MacLeod, D. I. A., Buck, S. L., and Williams, D. R., "Large Loss of Visual Sensitivity to Flashed Peripheral Targets," *Vision Res.,* **21:**1323–1328 (1981).

Geisler, W. S., "Sequential Ideal-Observer Analysis of Visual Discriminations," *Psychol Rev.* **96:**267–314 (1989).

Georgeson, M. A. and Sullivan, G. D., "Contrast Constancy: Deblurring in Human Vision by Spatial Frequency Channels," *J. of Physiol.* **252:**627–656 (1975).

Gescheider, G. A., *Psychophysics: Methods, Theory, and Application,* 2d ed., Lawrence Erlbaum and Associates, Hillsdale, N.J., 1985, pp. 174–191.

Graham, N. V. S., *Visual Pattern Analyzers,* Oxford University Press, Oxford, 1989.

Green, D. M. and Swets, J. A., *Signal Detection Theory and Psychophysics,* Krieger Press, Huntington, N.Y., 1974.

Ishihara, S., *Tests for Color Blindness,* 11th ed., Kanehara Shuppan, Tokyo, 1954.

King-Smith, P. E., Grigsby, S. S., Vingrys, A. J., Benes, S. C., and Supowit, A., "Efficient and Unbiased Modifications of the QUEST Threshold Method: Theory, Simulations, Experimental Evaluation and Practical Implementation," *Vision Res.,* **34:**885–912.

Klein, S. A., Casson, E., and Carney, T., "Vernier Acuity as Line and Dipole Detection, *Vision Res.* **30:**1703–1719 (1990).

Legge, G. E., Pelli, D. G., Rubin, G. S., and Schleske, M. M., "Psychophysics of Reading—I. Normal Vision." *Vision Res.* **25:**239–252 (1985).

Linfoot, E. H., *Fourier Methods in Optical Image Evaluation,* Focal Press, New York, 1964.

Luce, R. D., *Response Times: Their Role in Inferring Elementary Mental Organization,* Oxford University Press, New York; 1986.

Macmillan, N. A. and Creelman, C. D., *New Developments in Detection Theory,* Cambridge University Press, Cambridge, U.K., 1991.

Mertz, P., Fowler, A. D., and Christopher, H. N., "Quality Rating of Television Images," *Proceedings of the IRE* **38:**1269–1283 (1950).

Metz, C. E., "ROC Methodology in Radiologic Imaging," *Invest. Radiol.* **21:**720–733 (1986).

Nachmias, J., "On the Psychometric Function for Contrast Detection," *Vision Res.* **21:**215–223 (1981).

Nachmias, J. and Steinman, R. M., "Brightness and Discriminability of Light Flashes," *Vision Res.* **5:**545–557 (1965).

Nolte, L. W. and Jaarsma, D., "More on the Detection of One of M Orthogonal Signals," *J. Acoust. Soc. Am.* **41:**497–505 (1967).

Pearson, D. E., *Transmission and Display of Pictorial Information,* John Wiley & Sons, New York, 1975.

Pelli, D. G., "Uncertainty Explains Many Aspects of Visual Contrast Detection and Discrimination," *J. Opt. Soc. Am.* [*a*], **2:**1508–1532 (1985).

Pelli, D. G., "On the Relation Between Summation and Facilitation," *Vision Res.* **27:**119–123 (1987*a*).

Pelli, D. G., "The Visual Requirements of Mobility," G. C. Woo (ed.), *Low Vision: Principles and Application,* Springer-Verlag, N.Y., 1987*b,* pp. 134–146.

Pelli, D. G., "The Quantum Efficiency of Vision," C. Blakemore (ed.), *Vision: Coding and Efficiency,* Cambridge University Press, Cambridge, U.K., 1990, pp. 3–24.

Pelli, D. G., Robson, J. G., and Wilkins, A. J., "The Design of a New Letter Chart for Measuring Contrast Sensitivity," *Clinical Vision Sciences* **2:**187–199 (1988).

Pelli, D. G. and Robson, J. G., "Are Letters Better than Gratings?," *Clin. Vis. Sciences* **6:**409–411 (1991).

Pelli, D. G. and Zhang, L., "Accurate Control of Contrast on Microcomputer Displays," *Vision Research* **31:**1337–1350 (1991).

Peterson, W. W., Birdsall, T. G., and Fox, W. C., "Theory of Signal Detectability," *Trans. IRE PGIT* **4:**171–212 (1954).

Regan, D. and Neima, D., "Low-Contrast Letter Charts as a Test of Visual Function," *Ophthalmology* **90:**1192–1200 (1983).

Robson, J. G., "Spatial and Temporal Contrast-Sensitivity Functions of the Visual System," *J. Acoust. Soc. Am.* **56:**1141–1142 (1966).

Rockette, H. E., Gur, D., and Metz, C. E., "The Use of Continuous and Discrete Confidence Judgments in Receiver Operating Characteristic Studies of Diagnostic-Imaging Techniques," *Invest. Radiol.* **27:**169–172 (1992).

Rolf, J. M. and Staples, K. J., *Flight Simulation,* Cambridge University Press, Cambridge, U.K., 1986.

Schade, O. H., Sr., *Image Quality: A Comparison of Photographic and Television Systems,* RCA Laboratories, Princeton, N.J., 1975.

Scott, F., "The Search for a Summary Measure of Image Quality—A Progress Report," *Photographic Science and Engineering* **12:**154–164 (1968).

Snellen, H., *Scala Tipographia per Mesurare il Visus,* Utrecht, 1862.

Swets, J. A. and Pickett, R. M., *Evaluation of Diagnostic Systems: Methods from Signal Detection Theory,* Academic Press, New York, 1982.

Swets, J. A., "Measuring the Accuracy of Diagnostic Systems," *Science* **240:**1285–1293 (1988).

Tanner, W. P., Jr. and Birdsall, T. G., "Definitions of d' and η as Psychophysical Measures," *J. Acoust. Soc. Am.* **30:**922–928 (1958).

Van Trees, H. L., *Detection, Estimation, and Modulation Theory,* Wiley, New York, 1968.

Wandell, B. A., "Color Measurement and Discrimination," *J. Opt. Soc. Am.* [*a*], **2:**62–71 (1985).

Watson, A. B., "Probability Summation Over Time," *Vision Res.* **19:**515–522 (1979).

Watson, A. B., Ahumada, A., Jr., and Farrell, J. E., "The Window of Visibility: A Psychophysical Theory of Fidelity in Time-Sampled Visual Motion Displays," *NASA Technical Paper, 2211,* National Technical Information Service, Springfield, Va. 1983.

Watson, A. B. and Pelli, D. G., "QUEST: A Bayesian Adaptive Psychometric Method," *Percept Psychophys.* **33:**113–120 (1983).

Woodworth, R. S. and Schlosberg, H., *Experimental Psychology,* Holt, Rinehart, and Winston, New York: 1963, pp. 199–200.

P · A · R · T · 8

OPTICAL INFORMATION AND IMAGE PROCESSING

CHAPTER 30
ANALOG OPTICAL SIGNAL AND IMAGE PROCESSING

Joseph W. Goodman
Department of Electrical Engineering
Stanford University
Stanford, California

30.1 GLOSSARY

$C(\tau)$	cross-correlation function
d, z	distances
f	focal length
f_X, f_Y	spatial frequencies
$H(f_X, f_Y)$	transfer function
$I(x, y)$	intensity distribution
i	square root of negative one
$\mathbf{M}$	matrix
$t_A(x, y)$	amplitude transmittance of a transparency
$U(x, y)$	phasor representation of a monochromatic field
$\mathbf{u}, \mathbf{v}$	vectors
V	velocity of propagation
x, y	spatial coordinates
θ_B	Bragg angle
λ	wavelength
Λ	period of a grating
ν	optical frequency
τ	time delay

30.2 INTRODUCTION

The function of signal and image processing systems is the modification of signals and images to allow information extraction by a human observer or, alternatively, to allow fully automatic information extraction without human intervention. The origins of optical information processing are several, but certainly the invention of various techniques for visualizing the phase distribution of optical wavefronts qualifies (e.g., Ref. 1), as do also the famous Abbe-Porter experiments.[2,3] Starting in the 1950s, more general information processing tasks were undertaken with the help of optics.[4,5] This chapter presents an overview of such methods.

Optical systems are of interest both for digital processing of information and for analog processing of information. Our attention here will be restricted only to analog processing operations, which are far more mature and well developed than digital optical methods.

Certain basic assumptions will be used throughout and are detailed here. First, monochromatic optical signals will be represented by complex phasor field distributions, with the direction of phasor rotation being assumed to be clockwise. Similarly, time-varying optical fields will be represented by complex analytic signals, again with the direction of rotation in the complex plane being clockwise. In both cases the underlying real signals are recoverable as the real part of the complex representations. In all cases, a small-angle assumption will be employed, allowing paraxial approximations to be used. Polarization effects will generally be ignored, it being assumed that a scalar theory of light propagation is sufficiently accurate for our purposes.[6] The intensity of the optical waves, which is proportional to power density and is the observable quantity in an optical experiment, is defined as the squared magnitude of the complex fields.

It is very important to distinguish at the start between *coherent* and *incoherent* optical systems. For a review of optical coherence concepts, see Chap. 4 of this volume. For our purposes, we will regard an optical signal as coherent if the various optical contributions that produce an output add on an amplitude basis, with fixed and well-defined relative phases. Signals will be regarded as incoherent if the various contributions that add to produce an output at any point have time-varying phase relations, and therefore must add on an intensity or average-power basis.

30.3 FUNDAMENTAL ANALOG OPERATIONS

The fundamental components of any linear processing operation are addition and multiplication. We consider each of these operations separately.

Addition

Analog addition takes place in optical systems when light waves or wave components are superimposed. The exact nature of the addition depends on whether the optical components are mutually coherent or incoherent. In the coherent case, addition of complex phasor field components takes place. Thus if the $U_n(x, y)$ represent various optical field components that are superimposed at a given point (x, y) at the output, the resultant optical field $U(x, y)$ is given by

$$U(x, y) = \sum_n U_n(x, y) \tag{1}$$

Note that the result of such a superposition depends on the phases of the individual components.

On the other hand, if the various optical contributions at (x, y) are mutually incoherent, the addition takes place on an intensity basis. The resultant intensity $I(x, y)$ is given by

$$I(x, y) = \sum_n I_n(x, y) \tag{2}$$

where the $I_n(x, y)$ are the component intensity contributions. In this case the component intensity contributions are always positive and real, as is the resultant intensity.

In view of the above two equations, an important point can now be made. Coherent optical systems are *linear in complex amplitude,* while incoherent optical systems are *linear in intensity.* The design of an analog processing system thus depends fundamentally on whether the illumination used in the system is coherent or incoherent.

Multiplication

Analog multiplication takes place in optical systems as light passes through an absorbing or phase-shifting structure. If we define the complex amplitude transmittance $t_A(x, y)$ of a transmitting structure as the ratio of the transmitted complex field to the incident complex field, then analog multiplication in a coherent system is represented by

$$U_t(x, y) = t_A(x, y)U_i(x, y) \tag{3}$$

where $U_i(x, y)$ is the incident optical field and $U_t(x, y)$ is the transmitted optical field.

When the optical system is incoherent, then we define an intensity transmittance $t_I(x, y)$ as the ratio of the transmitted optical intensity to the incident optical intensity. Analog multiplication in such systems is represented by

$$I_t(x, y) = t_I(x, y)I_i(x, y) \tag{4}$$

Thus we have seen that the fundamental analog operations of addition and multiplication are quite naturally available in optical systems. It should be kept in mind that the operation of integration is just a generalization of addition, involving addition of an infinite number of infinitesimal components.

30.4 ANALOG OPTICAL FOURIER TRANSFORMS

Perhaps the most fundamental optical analog signal and image processing operation offered by optical systems is the Fourier transform. Such transforms occur quite simply and naturally with coherent optical systems. While Fourier sine and cosine transforms can be performed with incoherent light, the methods used are more cumbersome than in the coherent case and the numbers of resolvable spots involved in the images and transforms are more restricted. Therefore we focus here on Fourier transforms performed by coherent optical systems. The Fourier transform is normally two-dimensional in nature (image processing), although it can be restricted to a single dimension if desired (signal processing).

Focal-Plane-to-Focal-Plane Geometry

The optical system required to perform a two-dimensional Fourier transform is remarkably simple as shown in Fig. 1. We begin with a spatially coherent source of quasi-monochromatic light (a source that is both spatially and temporally coherent). The light

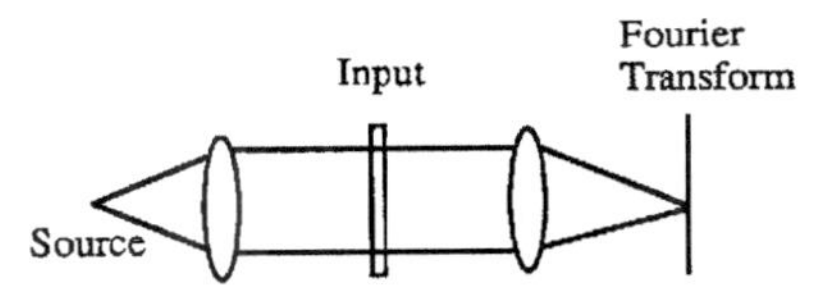

FIGURE 1 Simple optical system for performing a two-dimensional Fourier transform.

from that point source is collimated by a positive lens and a transparency of the image to be Fourier transformed is introduced in the front focal plane of a second positive lens (L_2). Under such conditions, the complex field appearing across the rear focal plane of that lens can be shown to be the two-dimensional Fourier transform of the complex field transmitted by the input transparency, as given by

$$U_f(x, y) = \frac{1}{i\lambda f} \iint_{-\infty}^{\infty} U_i(\xi, \eta) \exp\left[-i\frac{2\pi}{\lambda f}(x\xi + y\eta)\right] d\xi\, d\eta \tag{5}$$

Here λ is the optical wavelength, f is the focal length of the second lens, U_f is the field distribution across the back focal plane of lens L_2, and U_i is the field transmitted by the transparency in the front focal plane.

An intuitive explanation for the occurrence of this elegant relationship between the fields in the two focal planes can be presented as follows. If we were to mathematically Fourier transform the fields transmitted by the input transparency, each such Fourier component could be recognized as a different plane wave component of the transmitted field. Each such Fourier component represents a plane wave traveling in a unique direction with respect to the optical axis. Such representations are the basis for the so-called angular spectrum of plane waves often used in the analysis of optical wavefields (see, for example, Ref. 6, p. 48). Now consider the effect of the positive lens on a single Fourier component, i.e., a plane wave traveling at a particular angle with respect to the optical axis. As that plane wave passes through the lens L_2, it is changed into a spherical wave converging toward a focus in the rear focal plane, in a particular location determined by that plane wave's propagation direction. Thus the intensity of light at a given coordinate in the rear focal plane is proportional to the energy contained by the input wavefield at a particular Fourier spatial frequency. Hence the distribution of energy across the rear focal plane is a representation of the distribution of energy across the various spatial frequencies contained in input transparency.

Other Fourier Transform Geometries

A slightly more general configuration is one in which the input transparency is placed at an arbitrary distance d in front of the lens L_2, while the field is again considered in the rear focal plane of that lens. The relation between the input and output fields remains of the general form of a two-dimensional Fourier transform, but with the complication that a multiplicative quadratic phase factor is introduced, yielding a relation between input and focal-plane fields given by

$$U_f(x, y) = \frac{\exp\left[i\frac{k}{2f}\left(1 - \frac{d}{f}\right)(x^2 + y^2)\right]}{i\lambda f} \iint_{-\infty}^{\infty} U_i(\xi, \eta) \exp\left[-\frac{i2\pi}{\lambda f}(x\xi + y\eta)\right] d\xi\, d\eta \tag{6}$$

Three additional Fourier transform geometries should be mentioned for completeness. One is the case of an object transparency placed directly agains the lens in Fig. 1, either in front or in back of the lens. This is a special case of Eq. (6), with d set equal to 0, yielding

$$U_f(x, y) = \frac{\exp\left[i\frac{k}{2f}(x^2 + y^2)\right]}{i\lambda f} \iint_{-\infty}^{\infty} U_i(\xi, \eta) \exp\left[-\frac{i2\pi}{\lambda f}(x\xi + y\eta)\right] d\xi\, d\eta \tag{7}$$

Another situation of interest occurs when the object transparency is located behind the

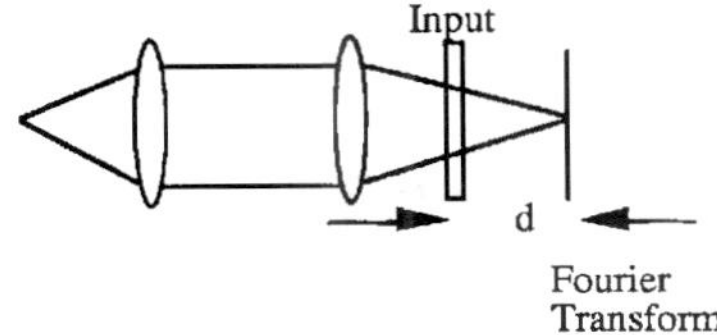

FIGURE 2 Fourier transform geometry with the object behind the lens.

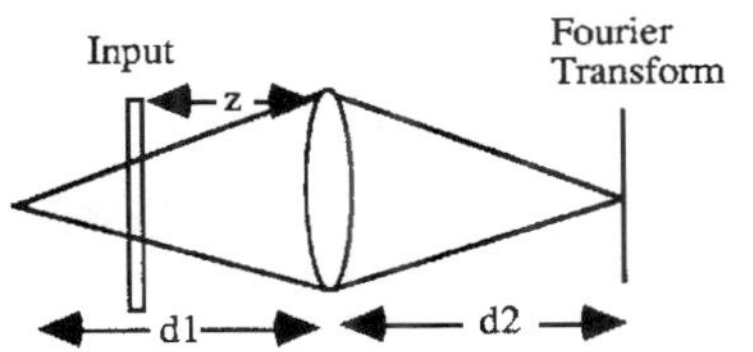

FIGURE 3 Fourier transform geometry using a single lens.

lens L_2, a distance d from the focal plane, as shown in Fig. 2. In this case the relationship between the fields transmitted by the object and incident on the focal plane becomes

$$U_f(x, y) = \frac{\exp\left[i\frac{k}{2d}(x^2 + y^2)\right]}{i\lambda d} \frac{f}{d} \iint_{-\infty}^{\infty} U_i(\xi, \eta) \exp\left[-\frac{i2\pi}{\lambda d}(x\xi + y\eta)\right] d\xi\, d\eta \tag{8}$$

Note that now the scaling distance in the Fourier kernel is d, rather than the focal length f. Therefore, by moving the object toward or away from the focal plane, the transform can be made smaller or larger, respectively.

While the Fourier transform plane in all the above examples has been the rear focal plane of the lens L_2, this is not always the case. The more general result states that the Fourier transform always appears in the plane where the original point-source of illumination is imaged by the optical system. In the previous examples, which all involved a collimating lens L_1 before the object transparency, the source was indeed imaged in the rear focal plane of L_2, where we asserted the Fourier transform lies. However, in the more general case depicted in Fig. 3, the point source of light lies in plane P_1 and its image lies in plane P_2, which for this geometry is the Fourier transform plane. A single lens L_1 performs both imaging of the source and Fourier transformation of the fields transmitted by the input transparency. If the input is placed to the right of the lens, at distance d from the image of the source, then the Fourier transform relation is identical to that presented in Eq. (8), for it does not matter what optical system illuminated the transparency with a converging spherical wave, only what distance exists between the input and the plane where the source is imaged.

If the input transparency is placed to the left of the single lens, as shown in Fig. 3, the resulting relationship between the fields transmitted by the object U_i and the fields across the plane where the source is imaged, U_f, becomes

$$U_f(x, y) = \frac{d_1}{i\lambda d_2(d_1 - z)} \exp\left\{i\frac{k}{2}\left(\frac{1}{d_2} - \frac{zd_1}{d_2^2(d_1 - z)}\right)\right\}$$
$$\times \iint_{-\infty}^{\infty} U_i(\xi, \eta) \exp\left\{-i\frac{2\pi d_1}{\lambda d_2(d_1 - z)}[x\xi + y\eta]\right\} d\xi\, d\eta \tag{9}$$

where the meanings of z, d_1 and d_2 are shown in Fig. 3, and $k = 2\pi/\lambda$. In this relation, d_1 and d_2 are connected through the lens law, $1/d_1 + 1/d_2 = 1/f$. It can be shown quite generally that the effective distance d associated with the Fourier transform kernel is $d = d_2/d_1(d_1 - z)$, while the quadratic phase factor is that associated with a diverging spherical wave in the transform plane that appears to have originated on the optical axis in the plane of the input transparency.

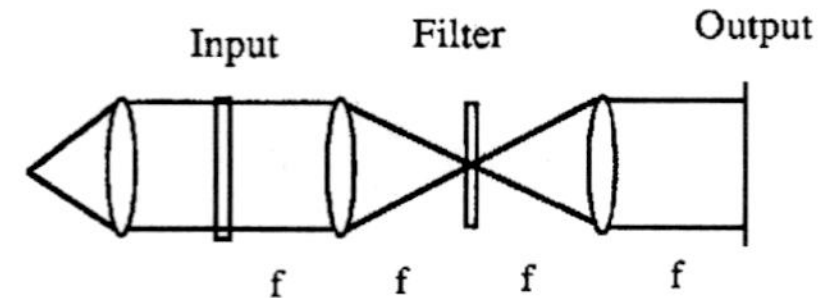

FIGURE 4 Spatial filtering system based on double Fourier transform.

30.5 SPATIAL FILTERING

Given that Fourier transforms of optical fields occur so directly in coherent optical systems, it seems natural to consider the intentional manipulation of such spectra for the purposes of signal or image processing. Given that a signal or image has been introduced into the coherent optical system, either by means of photographic film or by means of an electronically or optically controlled spatial light modulator (see Chap. 32), the idea is to insert in the plane where the Fourier transform occurs a transparency (again either film or a spatial light modulator) which intentionally alters the fields transmitted through that plane. A second Fourier transforming lens then returns the observer to an image plane, where the filtered version of the input can be measured or extracted. The simplest geometry from the conceptual point of view is that shown in Fig. 4.

The lens L_1 is again a collimating lens, the lens L_2 is a Fourier transforming lens, and the lens L_3 is a second Fourier transforming lens. The fact that a sequence of two Fourier transforms takes place, rather than a Fourier transform followed by an inverse Fourier transform, results simply in an inversion of the image at the output of the system.

Systems of this type form the basis for coherent optical spatial filtering, although the detailed geometry of the layout may vary. We will discuss several such spatial filtering problems in later sections. For the moment it suffices to say that if a filtering system is desired to have a transfer function $H(f_X, f_Y)$ then the amplitude transmittance of the transparency inserted in the Fourier plane should be

$$t_A(\xi, \eta) = H\left(\frac{\xi}{\lambda f}, \frac{\eta}{\lambda f}\right) \tag{10}$$

where λ has been defined, f is the focal length of the Fourier transforming lenses (assumed identical), and (ξ, η) represent the spatial coordinates in the filter plane.

30.6 COHERENT OPTICAL PROCESSING OF SYNTHETIC APERTURE RADAR DATA

The earliest serious application of coherent optics to signal processing was to the problem of processing data gathered by synthetic aperture radars. We explain the synthetic aperture principle, and then discuss optical signal processing architectures that have been applied to this problem.

The Synthetic Aperture Radar Principle

The synthetic radar problem is illustrated in Fig. 5. An aircraft carrying a stable local oscillator and a side-looking antenna filies a straight-line path, illuminating the terrain with microwave energy and detecting the returned energy reflected and scattered from that terrain. In the simplest case, resolution in range (i.e., perpendicular to the aircraft flight

FIGURE 5 Aircraft flight path.

FIGURE 6 Recording format for signals received.

path) is obtained by pulse echo timing, the usual radar range-measurement technique. Resolution in azimuth (the direction parallel to the flight path) is obtained by processing the Doppler-shifted returns, as will be explained. For the purpose of explaining the azimuth imaging, we neglect the pulsed nature of the radiation emitted by the aircraft, an approximation allowable because of the pulse-to-pulse coherence of the signals. The goal of the system is to obtain a two-dimensional image of the microwave reflectivity of the ground illuminated by the aircraft. The resolution of the system is not limited by the size of the antenna that is carried by the aircraft—in fact resolution increases as the size of the antenna is decreased. The system coherently combines the signals received along a portion of the flight path, thereby synthesizing the equivalent of a much longer antenna array.

If we consider the signal received at the aircraft as a function of time, originating from a single point scatterer on the ground, that signal will suffer an upward frequency shift as the aircraft approaches the scattered and a downward frequency shift as the aircraft flies away from the scatterer. This chirping signal is beat against the stable local oscillator in the aircraft, a bias is added, and the new signal is then recorded on a strip of film. Figure 6 shows the recording format. In the vertical direction, different scatterers are separated by the pulse echo timing, each being imaged on a separate horizontal line of the film. In the horizontal direction, the time histories of the chirping azimuth signals from different scatterers are recorded.

The signal recorded from a single scatterer is in fact an off-axis one-dimensional Fresnel zone plate, and as such is capable of imaging light in the horizontal direction to a focus. Such a focus constitutes the azimuthal image of the point scatterer that gave rise to this zone plate. However, the chirp rates, and therefore the focal lengths, of the zone plates produced by scatterers at different ranges are unfortunately not the same. The focal length is in fact proportional to the distance of the scatterer from the aircraft. Thus the focal points of scatterers at different ranges from the aircraft lie on a tilted plane with respect to the film plane, whereas the range images lie in the plane of the film. Thus the optical processing system must be designed to bring the two different focal planes into coincidence.

Optical Processing Systems

The earliest system used for optical processing of synthetic aperture radar data is illustrated in Fig. 7.[7] This processor uses a conical optical element, called an axicon, to change the focal lengths of all horizontal zone plates to infinity, thus moving the azimuth image to infinity. A cylindrical lens is placed one focal length from the film to likewise move the range image to infinity, and a spherical lens is placed one focal length from the final image plane to bring the infinitely distant azimuth and range planes back to a common focus.

The magnification achieved by such a system is a function of range, so the output is recorded through a vertical slit. As the input film is drawn through the system, an output

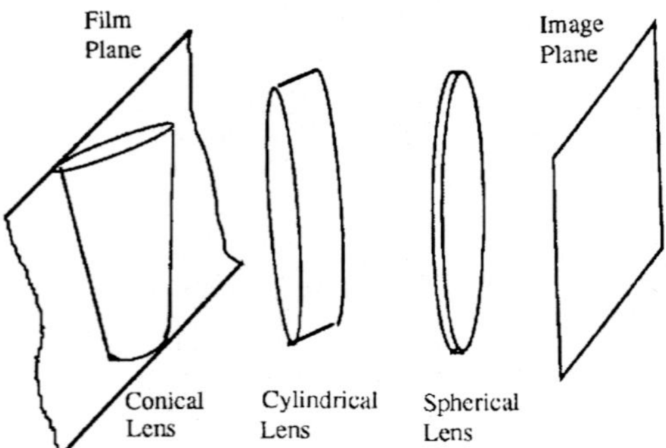

FIGURE 7 Processor using an axicon.

film strip is likewise drawn past the slit in synchronism, with the result that an image with proper magnification is recorded at the output.

Following the use of such an optical system to produce images, a far more sophisticated processing system known as the "tilted-plane processor" was developed.[8] The architecture of this system is illustrated in Fig. 8. In this case an anamorphic telescope is used to bring the range and azimuth planes into coincidence with a constant magnification, allowing a full two-dimensional image to be recorded at the output at one time. Again, motion of the input film and the output film takes place in synchronism, but the throughput of the system is much higher due to the absence of the output slit.

From the very fundamental work on processing synthetic aperture radar signals at the University of Michigan during the late 1950s and early 1960s came a multitude of extraordinary inventions, including holograms with an off-axis reference wave and the holographic matched filter, or Vander Lugt filter, discussed in Sec. 30.8.

30.7 COHERENT OPTICAL PROCESSING OF TEMPORAL SIGNALS

An important subclass of information-processing operations is those that are applied to one-dimensional signals that are functions of time. Such signals can be processed by coherent optical filtering systems once a suitable transducer is found to convert

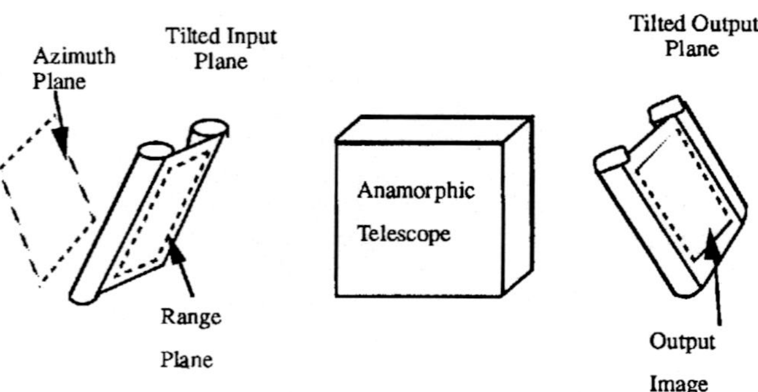

FIGURE 8 The tilted-plane processor.

time-varying voltages representing the signals into space-varying wavefields. The best developed and most common of such transducers is the acousto-optic cell.[9,10]

Acousto-Optic Cells for Inputting Signals

A time-varying electrical signal can be changed to an equivalent one-dimensional space-varying distribution of field strength by means of acousto-optic devices. In bulk form, such devices consist of a transducer, to which a time-varying voltage representing an RF signal is applied, and a transparent medium into which compressional waves are launched by the transducer. The RF signal is assumed to contain a carrier frequency, which generates a dynamic grating and, when illuminated by coherent light, produces a number of different diffraction orders, of which the +1 and −1 orders are of primary interest. Any modulation, in amplitude or phase, that may be carried by the RF signal is transferred to the spatial distributions of these diffraction orders.

Acousto-optic diffraction is characterized as either *Raman-Nath* diffraction or *Bragg* diffraction, depending on the relations that exist between the cell thickness and the period of the acoustic wave generated by the RF carrier. For cells that are sufficiently thin, Raman-Nath diffraction takes place. The acousto-optic cell then acts like a thin phase grating, generating a multitude of diffraction orders. For cells that are sufficiently thick, Bragg diffraction takes place. In this case, high diffraction efficiency into a single grating order can be achieved if the acoustic grating is illuminated at the Bragg angle θ_B which satisfies

$$\sin\frac{\theta_B}{2} = \frac{\lambda}{2\Lambda} \tag{11}$$

In this case most of the optical power is transferred to the +1 diffraction order, and other orders, including the −1 and 0 orders can be neglected.

Figure 9 illustrates Raman-Nath and Bragg diffraction from an acousto-optic cell. $v(t)$ represents the voltage driving the cell transducer. For modern-day signal processing applications, which involve very high microwave frequencies, the Bragg cell is invariably used, and the situation on the right-hand side of the figure is the one of interest.

The signal $v(t)$ is of the form (in complex notation)

$$v(t) = A(t) \exp\left[-i(2\pi \nu_0 t + \theta(t))\right] \tag{12}$$

where $A(t)$ is the amplitude modulation of the carrier, $\theta(t)$ is the phase modulation of the carrier, and ν_0 is the center frequency. If the speed of propagation of acoustic waves in the

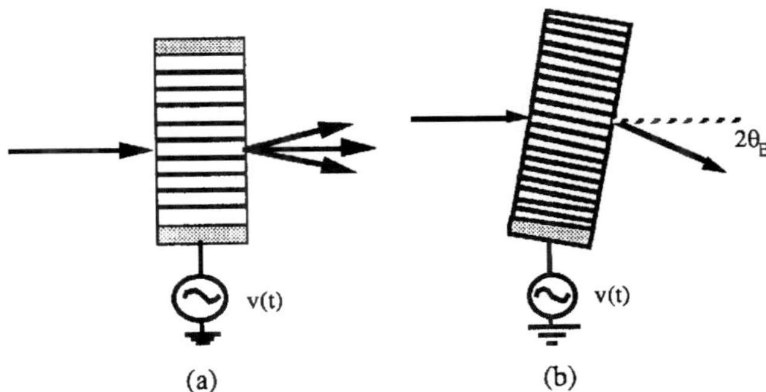

FIGURE 9 Acousto-optic diffraction in the (*a*) Raman-Nath and (*b*) Bragg regimes.

medium of the Bragg cell is V, then emerging from the right of that cell will be a spatial complex field distribution of the form

$$U(x, t) = A(x - Vt) \exp[-i\theta(x - Vt)] \tag{13}$$

where the dependence on y is suppressed due to uniformity of U in that dimension. Thus the temporal structure of the signal $v(t)$ has been changed to a spatial structure of the optical field $U(x, t)$.

The Bragg Cell Spectrum Analyzer

The most common use of coherent optics for signal processing is a method for finding and displaying the frequency (Fourier) spectrum of the electrical signal $v(t)$ applied to the cell. To construct such a spectrum analyzer, we follow the Bragg cell of the previous figure with a positive lens, which then Fourier transforms the wavefield emerging from the cell. A detector array placed in the Fourier plane then measures the amount of signal power present in each frequency bin subtended by a detector element. Note the spectrum analysis is performed over a finite sliding window, namely, the window of time stored in the Bragg cell itself. Figure 10 shows a diagram illustrating the Bragg cell spectrum analyzer.

Assuming perfect optics, the resolution of such a spectrum analyzer is determined by the diffraction limit associated with the space window that is being transformed. The spatial dimension of a resolution element is given by $\Delta x = (\lambda f / L)$ where L is the length of the cell and f is again the focal length of the lens. Given the mapping from time to space that takes place in the cell, it follows that the temporal resolution of the spectrum analyzer (in Hz) is $\Delta \nu = (V/L)$.

Bragg cell spectrum analyzers have been built with center frequencies of more than 1 GHz, with bandwidths approaching 1 GHz, and time bandwidth products (equivalent to the number of resolvable spectral elements) of the order of 1000. While the vast majority of work on this type of spectrum analyzer has used bulk devices (e.g., bulk Bragg cells, discrete lenses, etc.), work has also been carried out on integrated versions. Such devices use planar waveguides rather than free-space propagation, surface acoustic waves rather than bulk acoustic waves, integrated optic lenses, etc. Such systems are more compact than those based on bulk approaches, but their performance is so far somewhat inferior to that of the more conventional bulk systems.

The chief difficulty encountered in realizing high-performance Bragg cell spectrum analyzers is the dynamic range that can be achieved. The dynamic range refers to the ratio of the largest spectral component that can be obtained within the limit of tolerable

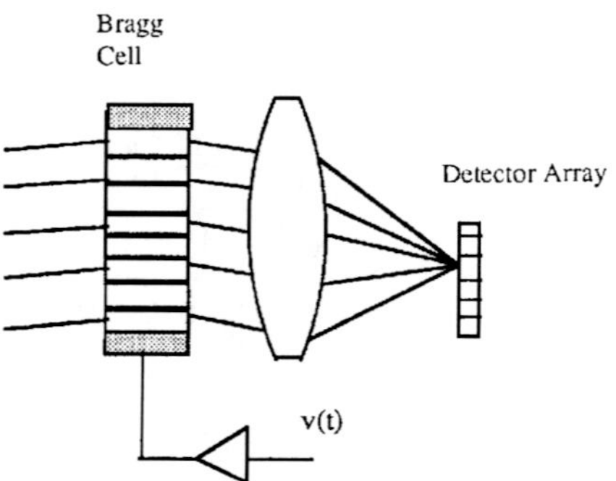

FIGURE 10 Bragg cell spectrum analyzer.

nonlinear distortion, to the smallest spectral component that can be detected above the noise floor.

Acousto-Optic Correlators

Many signal detection problems require the realization of correlators that produce cross-correlations between a local reference signal and an incoming unknown signal. A high cross-correlation between the reference and the unknown indicates a high degree of similarity between the two signals, while a low correlation indicates that the two signals are not very similar. Thus correlators play an important role in signal detection and recognition.

Given two complex-valued signals $v_1(t)$ and $v_2(t)$, the cross-correlation between those signals is defined as

$$C(\tau) = \int_{-\infty}^{\infty} v_1(t) v_2^*(t - \tau)\, dt \tag{14}$$

When v_1 and v_2 are identical, $C(\tau)$ achieves a peak value at the relative delay τ that causes the two signals to be identically aligned in time.

Two distinctly different architectures have been developed for using acousto-optic systems for cross-correlating wideband signals. We discuss each of these techniques separately.

The Space-Integrating Correlator. The older of the two architectures is known as the space-integrating correlator. As the name indicates, the integration inherent in the correlation operation is carried out over space. The variable delay τ is achieved by allowing one signal to slip past the other in time.

Figure 11 shows the structure of a time-integrating correlator. One of the signals, $v_1(t)$, is introduced by means of an input Bragg cell. Spatial filtering is used to eliminate any residual of the zeroth and unwanted first diffraction orders, retaining only a single first order. The second signal, the reference $v_2(t)$, is stored on a transparency, complete with a spatial carrier frequency representing the center frequency and acting as a high-frequency amplitude and phase modulated grating. The integration over space is provided by the final output lens. The particular diffraction order used in the final transparency is chosen to yield the conjugate of $v_2(t)$. A point detector is used at the output, and different relative delays between the two signals are achieved simply by allowing $v_1(t)$ to slide through the Bragg cell.

The Time-Integrating Correlator. A different approach to realizing the temporal cross-correlation operation is the so-called time-integrating correlator.[11,12] The

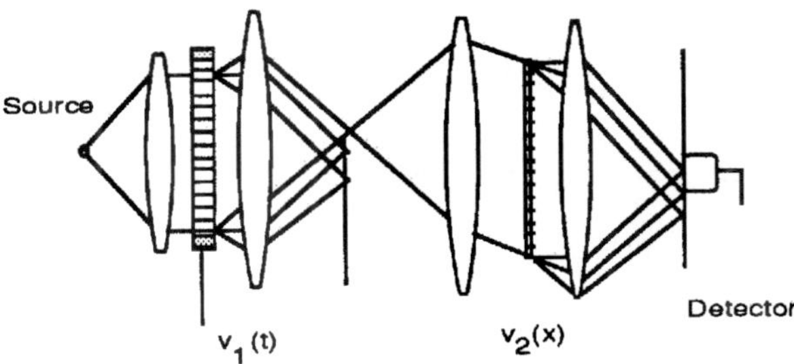

FIGURE 11 The time integrating correlator.

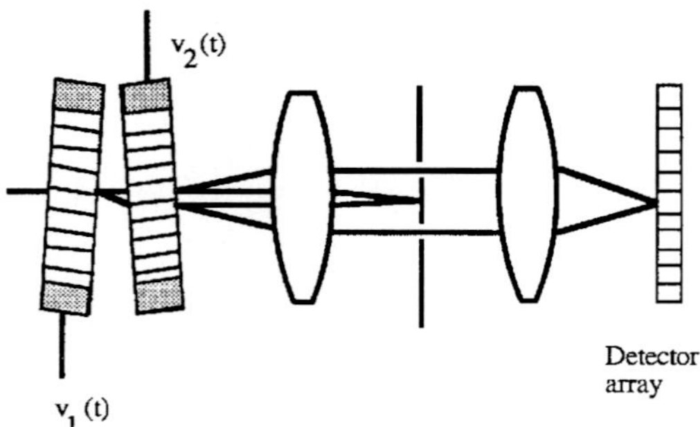

FIGURE 12 Time-integrating correlator.

architecture of a time-integrating correlator is illustrated in Fig. 12. Spatial filtering selects one component that has undergone zeroth-order diffraction by the first cell and first-order diffraction by the second, and another component that has undergone first-order diffraction by the first cell and zeroth-order by the second. These two components interfere on a time-integrating detector array.

As the name implies, the correlation integration is in this case carried out by temporal integration on an array of time-integrating detectors. Note that the two electrical signals are introduced at opposite ends of the Bragg cells, with the result that at any spatial position on the Bragg cell pair the two signals have been delayed relative to one another by different amounts, thus introducing the relative delay required in the correlation integral. The lens on the right images the pair of Bragg cells onto the detector array. Thus different detector elements measure the interference of the two signals with different relative delays, one portion of which yields

$$\mathrm{Re}\,\{C(x)\} = \mathrm{Re}\left\{\int_T v_1\left[t - \left(\frac{x + L/2}{V}\right)\right] v_2^*\left[t + \left(\frac{x - L/2}{V}\right)\right] dt\right\} \tag{15}$$

which is the real part of the correlation integral of interest. Here L represents the length of the Bragg cells, V the velocity of propagation of acoustic waves, T the total integration time of the detector, and x is the position of a particular detector on the detector array at the output. Note that for the position x on the detector array the two signals have been delayed relative to each other by the amount

$$\tau = \frac{2x}{V} \tag{16}$$

Other variants of both space-integrating and time-integrating correlators are known but will not be presented here. Likewise, many architectures for other types of acousto-optic signal processing are known. The reader may wish to consult Ref. 13 for more details.

30.8 OPTICAL PROCESSING OF TWO-DIMENSIONAL IMAGES

Because optical systems are fundamentally two-dimensional in nature, they are well suited to processing two-dimensional data. The most important type of two-dimensional data is imagery. Thus we consider now the application of optical information processing systems to image processing. The applications of optical processing in this area can be divided into two categories: (1) pattern detection and recognition, and (2) image enhancement.

Optical Matched Filtering for Pattern Recognition

By far the most well-known approach to pattern recognition is by means of the matched filter.[14] While this approach has many known defects in the pattern recognition application, it nonetheless forms the basis for many other more sophisticated approaches.

The Matched filter. A linear, invariant filter is said to be "matched" to a certain spatial image $s(x, y)$ if the impulse response (point-spread function) $h(x, y)$ of that filter is of the form

$$h(x, y) = s^*(-x, -y) \tag{17}$$

When a general signal $v(x, y)$ is applied to the input of such a filter, the output (the convolution of the input and the impulse response) is given by

$$\begin{aligned} w(x, y) &= \iint_{-\infty}^{\infty} v(\xi, \eta)h(x-\xi, y-\eta)\, d\xi\, d\eta \\ &= \iint_{-\infty}^{\infty} v(\xi, \eta)s^*(\xi - x, \eta - y)\, d\xi\, d\eta \end{aligned} \tag{18}$$

which is the *cross-correlation* between the signals $v(x, y)$ and $s(x, y)$. Thus the output of a matched filter is the cross-correlation between the input signal and the signal for which the filter is matched.

In the frequency domain, the convolution relation becomes a simple product relation. Thus the frequency domain equivalent of Eq. (18) is

$$W(f_X, f_Y) = H(f_X, f_Y)V(f_X, f_Y) = S^*(f_X, f_Y)V(f_X, f_Y) \tag{19}$$

Thus the transfer function of the matched filter is the complex conjugate of the spectrum of the signal to which the filter is matched.

The coherent optical realization of the matched filter utilizes a system identical with that shown previously in Fig. 4, where the Fourier domain transparency is one with amplitude transmittance proportional to $S^*(f_X, f_Y)$. The output of the filter, appearing at the plane on the far right in Fig. 4, consists of a bright spot at each location where the signal $s(x, y)$ is located within the input field.

Prior to 1964, key difficulty in the realization of matched filtering systems was the construction of the Fourier domain filter with the proper amplitude transmittance. To control the amplitude and phase transmittance through the Fourier plane in a relatively complicated manner was often beyond the realm of possibility. However, in 1964 Vander Lugt published his classic paper on holographically recorded matched filters, and many new applications became possible.

The Vander Lugt Filter. The method introduced by Vander Lugt[15] for recording matched filters is shown in Fig. 13. It is assumed that a mask can be constructed with

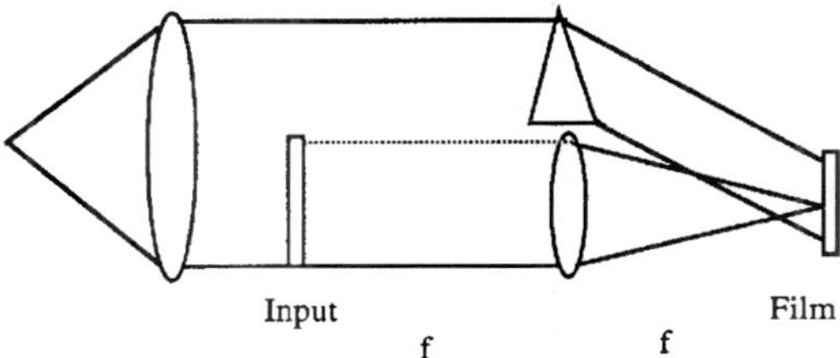

FIGURE 13 Recording a Vander Lugt filter.

amplitude transmittance proportional to the desired impulse response $s^*(-x, -y)$, which in pattern recognition problems is often real and positive. That mask is illuminated by coherent light and Fourier transformed by a positive lens. In the Fourier domain, the spectrum $S^*(f_X, f_Y)$ is allowed to interfere with an angularly inclined reference wave, often a plane wave. The result is an intensity pattern with a high spatial frequency carrier, which is amplitude modulated by the amplitude distribution associated with the incident spectrum, and phase modulated by the phase distribution of that spectrum. This recording is in fact a Fourier hologram of the desired point-spread function[16] (see also Chap. 55). The film used for recording in the Fourier domain responds to the incident optical intensity. With proper processing of the film, one of the grating orders of the resulting transparency yields a field component proportional to the desired field,

$$t_A(\xi, \eta) \approx S^*\left(\frac{\xi}{\lambda f}, \frac{\eta}{\lambda f}\right) \exp(-i2\pi\alpha\eta) \tag{20}$$

where (ξ, η) are the spatial coordinates in the filter plane, and α is the spatial frequency of the carrier. Thus the transmittance required for realizing the matched filter has been achieved, with the exception of the linear exponential term, which serves simply to shift the desired output off the optical axis.

The filter constructed as above is placed in the Fourier domain of the system in Fig. 4, and provided the correct region of the output plane in that figure is examined, the matched filter response is found. In a second region of the output plane, mirror symmetric with the matched filter region, the convolution of the signal $s(x, y)$ and the input $v(x, y)$ can be found.

Prior to Vander Lugt's invention, the only matched filters that could be realized in practice were filters with very simple transfer functions $S^*(f_X, f_Y)$. The significance of the Vander Lugt filter is that it extends the domain for which filters can be realized to those with reasonably simple impulse responses $s(-x, -y)$, a case more typically encountered in pattern recognition.

Deficiencies of the Matched Filter Concept. While the Vander Lugt filter provides an elegant solution to the problem of realizing coherent optical matched filters, nonetheless the use of coherent optics for pattern recognition has been very restricted in its applications. A major reason for this limited applicability can be traced to deficiencies of the matched filter concept itself, and is not due to the methods used for optical realization. The matched filter, in its original form, is found to be much too sensitive to parameters for which lack of sensitivity would be desired. This includes primarily rotation of the image and scale change of the image. Thus a matched filter that responds perfectly to the desired signal in its original rotation and scale size may not respond at all to that signal when it is rotated and magnified or demagnified.

Many attempts have been made to remove these undesired sensitivities of the matched filter (see for example, Refs. 17 and 18). These include the use of Mellin transforms and polar coordinate transformations to remove scale-size sensitivity and rotation sensitivity, and the use of circular harmonic decompositions to remove rotation sensitivity. These attempts have had varying degrees of success, but unfortunately they generally destroy one insensitivity that is present for a conventional matched filter, namely, insensitivity to translation of the signal. For a conventional matched filter, when an input $s(x, y)$ is translated, the resulting bright spot at the output translates in response. Realization of rotation invariance generally removes translation insensitivity, a serious loss.

Unfortunately, to date there have been no commercially successful applications of coherent optical matched filtering to pattern recognition, although several attempts to commercialize the technology have been made.

Coherent Optical Image Enhancement

Coherent optical spatial filtering systems can also be applied to the problem of image enhancement.[4] Image enhancement problems come in a wide variety of types, ranging from simple operations, such as the suppression of periodic noise in a nonperiodic image, to more complex operations, such as restoring an image that has been degraded by a known blur. We focus here on image restoration, since it is the most challenging of these problems.

The Inverse Filter. A common type of image restoration problem arises when an image produced by an incoherent imaging system has been blurred by a known, space-invariant, linear point-spread function. Let $i(x, y)$ represent the intensity of the blurred image, $o(x, y)$ represent the intensity of the object, and $b(x, y)$ represent the intensity point-spread function of the blur. These three quantities are related through a convolution equation,

$$i(x, y) = \iint_{-\infty}^{\infty} b(x - \xi, y - \eta) o(\xi, \eta)\, d\xi\, d\eta \tag{21}$$

The frequency domain equivalent is the relation

$$I(f_X, f_Y) = B(f_X, f_Y) O(f_X, f_Y) \tag{22}$$

where I, O, and B are the Fourier transforms of the lowercase quantities. The quantity B is the transfer function of the blur, and is assumed to be perfectly known.

Examination of Eq. (22) suggests an obvious approach to image restoration. Convolve the blurred image $i(x, y)$ with a kernel that provides a deblurring transfer function that is the reciprocal of the blur transfer function, i.e., having a transfer function given by $H(f_X, f_Y) = B^{-1}(f_X, f_Y)$. For obvious reasons, such a filter is referred to as an *inverse filter.* The restored image then is given, in the frequency domain, by

$$\begin{aligned} R(f_X, f_Y) &= [B(f_X, f_Y) O(f_X, f_Y)] H(f_X, f_Y) \\ &= [B(f_X, f_Y) O(f_X, f_Y)] B^{-1}(f_X, f_Y) \\ &= O(f_X, f_Y) \end{aligned} \tag{23}$$

Returning to the space domain we see that result of the image restoration operation is perfect recovery of the original object $o(x, y)$.

The inverse filter is an elegant mathematical solution to the restoration problem, but it lacks practicality. Many problems exist, both with the concept and with its implementation. The conceptual flaw, which is the most serious drawback, arises because the problem formulation completely neglected the inevitable presence of noise in the image $i(x, y)$. The inverse fllter boosts those spatial frequency components the most that were suppressed the most by the blur. In such regions of the frequency domain there is little or no image information to be boosted, but there is always noise, which then is amplified to the point that it dominates the restored image.

Other problems arise due to the very large dynamic range required of the deblurring filter transfer function in many cases. For the above reasons, the inverse filter is never used in practice, although it is an important concept to be aware of.

The Wiener Filter. The Wiener filter overcomes many of the difficulties of the inverse filter by explicitly including noise in the basic imaging model. In this case the detected image intensity is represented by

$$i(x, y) = \iint_{-\infty}^{\infty} b(x - \xi, y - \eta) o(\xi, \eta)\, d\xi\, d\eta + n(x, y) \tag{24}$$

where $o(x, y)$ and $n(x, y)$ are regarded as statistically stationary random processes. The goal of the restoration process is now to choose a restoration filter that will minimize the mean-squared error between the restored image $r(x, y)$ and the original object $o(x, y)$. The solution to this problem can be shown to be a restoration filter having a transfer function of the form

$$H(f_X, f_Y) = \frac{B^*(f_X, f_Y)}{|B(f_X, f_Y)|^2 + \dfrac{P_N(f_X, f_Y)}{P_O(f_X, f_Y)}} \tag{25}$$

where P_N and P_O represent the power spectral densities of the respective noise and object random processes.

The Wiener filter provides a balance between uncompensated blur and residual noise in just such a way as to minimize mean-squared error. Note that at spectral locations where the object power is much greater than the noise power, the Wiener filter approaches an inverse filter, while at spectral locations where the noise power dominates, the Wiener filter behaves as a matched filter with considerable attenuation.

Coherent Optical Realization of Inverse and Wiener Filters. While the inverse filter is primarily of theoretical interest, nonetheless there is much to be learned from consideration of how one might realize an approximation to such a filter. In general, the transfer function $B(f_X, f_Y)$ is complex-valued, or at least has sign reversals implying 180° phase shifts at some frequencies. This implies that the inverse filter must control both the magnitude and the phase of the transmitted fields. In most cases this implies a holographic filter and possibly a second absorbing filter.

The exact blur impulse response is assumed to be known. From a blurred image of a known point source, a photographic record of the blur impulse response can be obtained. If a filter is recorded in the geometry of Fig. 9, with the blur impulse response placed in the plane labeled "input," then an interferometrically generated transparency results, one component of amplitude transmittance being proportional to the conjugate of the blur transfer function

$$t_A(\xi, \eta) \approx B^*(f_X, f_Y) \exp(-i2\pi\alpha\eta) \tag{26}$$

where α is again a carrier frequency introduced by the offset reference wave. Passage of the blurred image through a coherent optical filtering system with this transfer function will correct any frequency-domain phase shifts associated with the blur, but will not restore the magnitude of the object spectrum correctly.

To correct the spectral magnitudes we require an additional transparency to be sandwiched with the above holographic filter. This filter can be generated in a number of ways, but the easiest to understand is a method that rests on properties of the photographic process. If a photographic emulsion is exposed to an optical intensity $I(\xi, \eta)$, then over a certain dynamic range the amplitude transmittance of the resulting negative transparency will be of the form

$$t_A(\xi, \eta) = K[I(\xi, \eta)]^{-\gamma/2} \tag{27}$$

where γ is the so-called gamma of the photographic process. If the intensity to which the emulsion is exposed is simply the intensity in the Fourier transform of the blur transfer function, as obtained by optically Fourier transforming the blur spread function (for example, as in the system of Fig. 9 but with the reference wave blocked), then if a gamma equal to 2 is achieved with the photographic processing, the second transparency will have amplitude transmittance

$$t_A(\xi, \eta) = K \left| B\left(\frac{\xi}{\lambda f}, \frac{\eta}{\lambda f}\right) \right|^{-2} \tag{28}$$

If the two transparencies discussed above are now placed in contact, the overall amplitude transmittance will be the product of the two individual transmittances, and the effective filter transfer function realized by the coherent optical processor will be

$$H(f_X, f_Y) = \frac{B^*(f_X, f_Y)}{|B(f_X, f_Y)|^2} = \frac{1}{B(f_X, f_Y)} \tag{29}$$

which is precisely the transfer function of the desired inverse filter. However, in practice there will be errors in this filter due to the limited dynamic range of the photographic media.

To realize an approximation to the Wiener filter, a different recording method can be used. In this case the full holographic recording system illustrated in Fig. 9 is used, including the reference beam. However, the intensity of the reference beam is made weak compared with the peak intensity of the $|B|^2$ component. Furthermore, the recording is arranged so that the exposure falls predominantly in a range where the amplitude transmittance of the developed transparency is proportional to the logarithm of the intensity incident during exposure. Now if amplitude transmittance is proportional to the logarithm of incident exposure, the changes of amplitude transmittance, which lead to diffraction of light by the transparency, will obey

$$\Delta t_A = \beta \Delta(\log E) \approx \beta \frac{\Delta E}{\bar{E}} \tag{30}$$

where ΔE represents changes in exposure, $\bar{E}$ represents the average exposure about which the fluctuations occur, and β is a proportionality constant. Restricting attention to the proper portion of the output plane, the following identifications can be made:

$$\begin{aligned} \Delta E &\propto \hat{B}^* \exp(-i2\pi\alpha\eta) \\ \bar{E} &\propto |\hat{B}|^2 + K \end{aligned} \tag{31}$$

where $|\hat{B}|^2$ is the squared magnitude of the blur transfer function, normalized to unity at the origin, while K is the ratio between the reference beam intensity and the maximum value of $|B|^2$. Neglecting the exponential term which leads to offset from the origin in the output plane, the amplitude transmittance of the deblurring filter becomes

$$\Delta t_A = \frac{\hat{B}^*}{|\hat{B}|^2 + K} \tag{32}$$

which is precisely the form of the Wiener filter for a constant ratio K of noise power spectral density to signal power spectral density.

Thus the Wiener filter has been achieved with a single holographic filter. If the signal-to-noise ratio in the blurred image is high, then the reference beam intensity should be much less than the object beam intensity ($K \ll 1$). Bleached filters of this kind can also be made.

30.9 INCOHERENT PROCESSING OF DISCRETE SIGNALS

The previous problems examined have all involved signals and images that are continuous functions of time or space. We turn attention now to signals that are sampled or discrete functions of time or space.

Background

A continuous signal $u(t)$ is sampled at times separated by Δt yielding a set of P samples $u(k\,\Delta t)$, which we represent by the column vector

$$\mathbf{u} = \begin{bmatrix} u_1 \\ u_2 \\ \vdots \\ u_P \end{bmatrix} \tag{33}$$

For discrete signals, the continuous operations associated with convolution and correlation become matrix-vector operations. Thus any linear transformation of an input signal $\mathbf{u}$ is represented by

$$\mathbf{v} = \mathbf{M}\mathbf{u} \tag{34}$$

where $\mathbf{v}$ is a length Q output vector containing samples of the output signal, and $\mathbf{M}$ is a $P \times Q$ matrix

$$\mathbf{M} = \begin{bmatrix} m_{11} & m_{12} & \cdots & m_{1P} \\ m_{21} & m_{22} & \cdots & m_{2P} \\ \vdots & \vdots & \vdots & \vdots \\ m_{Q1} & m_{Q2} & \cdots & m_{QP} \end{bmatrix} \tag{35}$$

In the sections that follow we examine some of the optical approaches that have been proposed and demonstrated for this kind of operation.

The Serial Incoherent Matrix-Vector Multiplier

An important starting point is provided by the serial incoherent matrix-vector multiplier invented by Bocker[19] (see also Ref. 20), and illustrated in Fig. 14. The elements of the vector $\mathbf{u}$ are applied as current pulses, with heights proportional to the u_i, to an LED. Light from the LED floods the matrix mask, which contains $Q \times P$ cells, each with an intensity transmittance proportional to a different m_{ij}. The light transmitted by the matrix

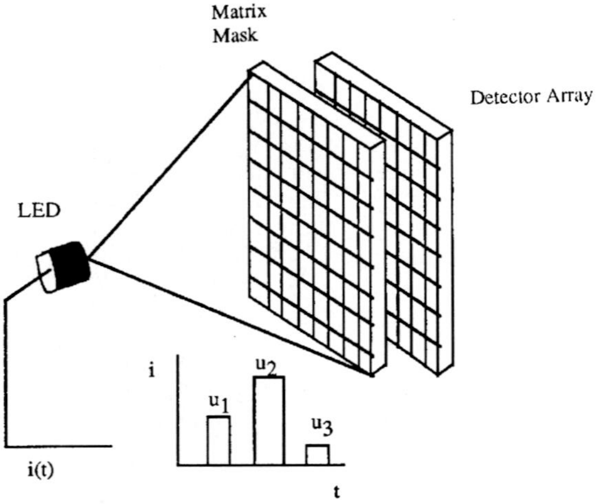

FIGURE 14 Serial matrix-vector multiplier.

mask then falls on a 2-dimensional CCD detector array, used in an unusual mode of operation. Charges are transferred horizontally along the rows of the detector array. In the first clock cycle, when the first element of the input vector is generated by the LED, the charge deposited in the first column of the detector array can be represented by a vector with elements $c_{1j} = m_{1j}u_1$. This set of charge packets is now transferred one column to the right, and the second pulse of light, proportional to u_2 is emitted. In the second column of the detector array a new charge is added to each of the existing charges, yielding a new set of charges $c_{2j} = m_{1j}u_1 + m_{2j}u_2$. After P clock cycles the column on the far right-hand side of the detector array contains a charge vector $\mathbf{c} = \mathbf{Mu}$, which within a proportionality constant is the desired output vector $\mathbf{v}$.

Thus the elements of the output vector are obtained in parallel from the right-hand column of the detector array. To compute the output vector, P cycles of the system are necessary, one for each element of the input vector. Multiplications are performed optically by passage of light through the matrix mask, while additions are performed electrically by charge addition.

The Parallel Matrix-Vector Multiplier

A fundamentally faster system for performing the matrix-vector product was discovered in 1978.[21] The architecture of this system is shown in Fig. 15.

The elements of the vector $\mathbf{u}$ are now entered in parallel as brightness values on an array of LEDs or laser diodes. The optics, not shown here, spreads the light from each source in the vertical direction to cover the height of the matrix mask, while imaging each source onto an individual column in the horizontal direction. Passage of the light through the matrix mask multiplies the input vector, element by element, by the elements of the row vectors of the mask. The second set of optics, again not shown, adds the light transmitted by each row of the mask, placing an intensity on each element of the detector array that is proportional to the sum of the products produced by one row of the mask or, equivalently, the inner product of the input vector and a unique row vector of the matrix. In this case the detectors are of the nonintegrating type, and nearly instantaneously produce an output current proportional to an element of the output vector $\mathbf{v}$. In this way a series of input vectors can be flowed through the system at high speed.

In this case both the multiplications and the additions are performed optically. A different output vector can be obtained with each cycle of the system. The result is a fundamentally faster system.

Systems of this type have had a broad impact on optical signal processing, with applications ranging from photonic switching[22] to neural networks.[23]

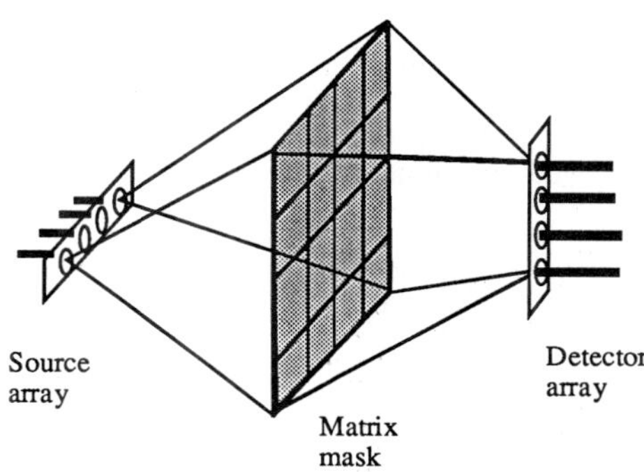

FIGURE 15 Parallel matrix-vector multiplier.

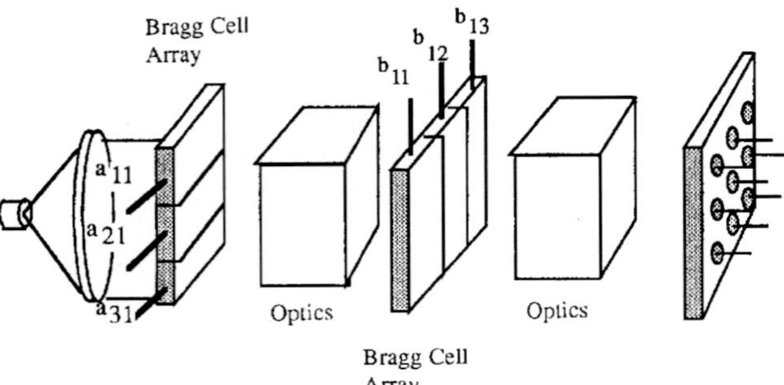

FIGURE 16 Outer product processor.

The Outer Product Processor

Another fundamentally different architecture is represented by the outer product processor,[24] shown in Fig. 16.

The goal in this case is to calculate the outer product **C** of two matrices **A** and **B**. Illustrating with a simple 3×3 example, the outer product is defined by the equation

$$\mathbf{C} = \begin{bmatrix} a_{11} \\ a_{21} \\ a_{31} \end{bmatrix} [b_{11} \quad b_{12} \quad b_{13}] + \begin{bmatrix} a_{12} \\ a_{22} \\ a_{32} \end{bmatrix} [b_{21} \quad b_{22} \quad b_{23}] + \begin{bmatrix} a_{13} \\ a_{23} \\ a_{33} \end{bmatrix} [b_{31} \quad b_{32} \quad b_{33}] \tag{36}$$

The system of Fig. 16 accomplishes this operation by use of two Bragg cell arrays oriented in orthogonal directions, and a time-integrating two-dimensional detector array. A column of **A** is entered in parallel on the first Bragg cell array, and a row of **B** on the second. The first box labeled "optics" images one array onto the other (with appropriate spatial filtering as needed to convert phase to intensity). The second box labeled "optics" images that product onto the detector array. In one cycle of the system, one of the outer products in the summation of Eq. (36) is accumulated on the elements of the detector array. In this example, three cycles are necessary, with addition of charge at the detector, to accumulate the full outer product of the matrices. More generally, if **A** is $P \times Q$ (i.e., P rows and Q columns) and **B** is $Q \times P$, then the detector array should be $P \times P$ in size, and Q cycles are required to obtain the full outer product.

Other Discrete Processing Systems

A multitude of other discrete processing systems have been proposed throughout the 1980s and 1990s. Worth special mention here are the optical systolic matrix-vector processor of Caulfield et al.[25] and the SAOBIC processor of Guilfoyle.[26] We refer the reader to the original references for details.

30.10 CONCLUDING REMARKS

Analog optical signal and image processing were strong areas of research during three decades, the 1960s through the 1980s. Many ingenious systems were devised, each motivated by one or more applications. With some exceptions, these systems seldom survived for the particular application they were conceived for, but often they led to new applications not envisioned by their inventors. The majority of applications of this technology have been to defense-related problems. Research emphasis has shifted away from analog signal processing as described above toward (1) the application of optics to digital processing, either within electronic computers or, in some cases, in all-optical digital computers, and (2) optical neural networks, which are analog optical systems. The intellectual base formed by the previous analog processing experience continues to strongly influence work on these more modern topics.

30.11 REFERENCES

1. F. Zernike, "Das Phasenkontrastverharen bei der Mikroscopischen beobachtung," *A. Tech. Phys.* **16:**454 (1935).
2. E. Abbe, "Beitrage zür theorie des mikroskops und der mikroskopischen wahrnehmung," *Archiv. Mikroskopische Anat.* **9:**413–468 (1873).
3. A. B. Porter, "On the Diffraction Theory of Microscope Vision," *Phil. Mag.* **11:**154 (1906).
4. A. Marechal and P. Croce, "Un filtre de frequences spatiales pour l'amelioration du contrast des images optiques," *C.R. Acad. Sci.* **127:**607 (1953).
5. E. L. O'Neill, "Spatial Filtering in Optics," *IRE Trans. on Info. Theory* **IT-2:**56–65 (1956).
6. J. W. Goodman, *Introduction to Fourier Optics,* McGraw-Hill, New York, 1968.
7. L. J. Cutrona, et al., "On the Application of Coherent Optical Processing Techniques to Synthetic-Aperture Radar," *Proc. IEEE* **54:**1026–1032 (1966).
8. A. Kozma, E. N. Leith, and N. G. Massey, "Tilted Plane Optical Processor," *Applied Optics* **11:**1766–1777 (1972).
9. L. Slobodin, "Optical Correlation Techniques," *Proc. IEEE* **51:**1782 (1963).
10. M. Arm, L. Lambert, and I. Weissman, "Optical Correlation Technique for Radar Pulse Compression," *Proc. IEEE* **52:**842 (1964).
11. R. M. Montgomery, "Acousto-Optical Signal Processing System," U.S. Patent 3,634,749, 1972.
12. R. A. Sprague and C. L. Koliopoulis, "Time Integrating Acousto-Optic Correlator," *Applied Optics* **15:**89–92 (1975).
13. A. Vander Lugt, *Optical Signal Processing,* John Wiley & Sons, New York, 1992.
14. G. L. Turin, "An Introduction to Matched Filters," *IRE Trans. on Info. Theory* **IT-6:**311 (1960).
15. A. B. Vander Lugt, "Signal Detection by Complex Spatial Filtering," *IEEE Trans. on Info. Theory* **IT-10:**139–145 (1964).
16. R. J. Collier, C. B. Burkhardt, and L. H. Lin, *Optical Holography,* Academic Press, New York, 1971.
17. D. Casasent and D. Psaltis, "Position, Rotation and Scale-Invariant Optical Correlation," *Applied Optics* **15:**1795–1799 (1976).
18. D. Casasent and D. Psaltis, "New Optical Transforms for Pattern Recognition," *Proc. IEEE* **65:**770–784 (1977).
19. R. P. Bocker, "Matrix Multiplication Using Incoherent Optical Techniques," *Applied Optics* **13:**1670–1676 (1974).
20. K. Bromley, "An Incoherent Optical Correlator," *Optica Acta* **21:**35–41 (1974).
21. J. W. Goodman, A. R. Dias, and L. M. Woody, "Fully Parallel, High-Speed Incoherent Optical Method for Performing the Discrete Fourier Transform," *Optics Letters* **2:**1–3 (1978).
22. A. R. Dias, R. F. Kalman, J. W. Goodman, and A. A. Sawchuk, "Fiber-Optic Crossbar Switch with Broadcast Capability," *Proceedings of the SPIE* **825:**170–177 (1987).
23. N. H. Farhat, D. Psaltris, A. Prata, and E. Paek, "Optical Implementation of the Hopfield Model," *Applied Optics* **24:**1469–1475 (1985).

24. R. A. Athale and W. C. Collins, "Optical Matrix–Matrix Multiplier Based on Outer Product Decomposition," *Applied Optics* **21:**2089–2090 (1982).
25. H. J. Caulfield, et al., "Optical Implementation of Systolic Array Processing," *Optics Communications* **40:**86–90 (1982).
26. P. S. Guilfoyle, "Systolic Acousto-Optic Binary Convolver," *Optical Engineering* **23:**20–25 (1984).

CHAPTER 31
PRINCIPLES OF OPTICAL DISK DATA STORAGE

Masud Mansuripur
Optical Sciences Center
University of Arizona
Tucson, Arizona

31.1 INTRODUCTION

Since the early 1940s magnetic recording has been the mainstay of electronic information storage worldwide. Audiotapes provided the first major application for the storage of information on magnetic media. Magnetic tape has been used extensively in consumer products such as audiotapes and videocassette recorders (VCR); it has also found application in backup/archival storage of computer files, satellite images, medical records, etc. Large volumetric capacity and low cost are the hallmarks of tape data storage, although sequential access to the recorded information is perhaps the main drawback of this technology. Magnetic hard disk drives have been used as mass-storage devices in the computer industry ever since their inception in 1957. With an areal density that has doubled roughly every two years, hard disks have been and remain the medium of choice for secondary storage in computers.* Another magnetic storage device, the floppy disk, has been successful in areas where compactness, removability, and rapid access to the recorded information have been of primary concern. In addition to providing backup and safe storage, inexpensive floppies with their moderate capacities (2 Mbytes on a 3.5-in-diameter platter is typical) and reasonable transfer rates have provided the crucial function of file/data transfer between isolated machines. All in all, it has been a great half-century of progress and market dominance for magnetic storage which is only now beginning to face a serious challenge from the technology of optical recording.

Like magnetic recording, a major application of optical data storage is the secondary storage of information for computers and computerized systems. Like the high-end magnetic media, optical disks can provide recording densities in the range of 10^7 bits/cm^2 and beyond. The added advantage of optical recording is that, like floppies, these disks can be removed from the drive and stored on the shelf. Thus the functions of the hard disk (i.e., high capacity, high data transfer rate, rapid access) may be combined with those of the floppy (i.e., backup storage, removable media) in a single optical disk drive. Applications of optical recording are not confined to computer data storage. The enormously successful compact audio disk (CD) which was introduced in 1983 and has since become the de facto

* Achievable densities on hard disks are presently in the range of 10^7 bits/cm^2; random access to arbitrary blocks of data in these devices can take on the order of 10 msec, and individual read-write heads can transfer data at the rate of several megabits per second.

standard of the music industry, is but one example of the tremendous potentials of the optical disk technology.

A strength of optical recording is that, unlike its magnetic counterpart, it can support read-only, write-once, and erasable/rewritable modes of data storage. Consider, for example, the technology of optical audio/video disks. Here the information is recorded on a master disk which is then used as a stamper to transfer the embossed patterns to a plastic substrate for rapid, accurate, and inexpensive reproduction. The same process is employed in the mass production of read-only files (CD-ROM, O-ROM) which are now being used to distribute software, catalogs, and other large databases. Or consider the write-once-read-many (WORM) technology, where one can permanently store massive amounts of information on a given medium and have rapid, random access to them afterward. The optical drive can be designed to handle read-only, WORM, and erasable media all in one unit, thus combining their useful features without sacrificing performance and ease of use. Moreover, the media can contain regions with prerecorded information as well as regions for read/write/erase operations on the same platter, thus offering opportunities for applications that have heretofore been unthinkable.

This article presents the conceptual basis for optical storage systems, with emphasis on disk technology in general and magneto-optical (MO) disk in particular. Section 31.2 is devoted to a discussion of some elementary aspects of disk data storage including the concept of track, definition of the access time, and the physical layout of data. Section 31.3 describes the function of the optical path; included are properties of the semiconductor laser diode, characteristics of the beam-shaping optics, and features of the focusing (objective) lens. The limited depth of focus of the objective lens and the eccentricity of tracks dictate that optical disk systems utilize closed-loop feedback mechanisms for maintaining the focused light spot on the right track at all times. Automatic focusing and automatic track-following schemes are described in Secs. 31.4 and 31.5. The physical process of thermomagnetic recording is the subject of Sec. 31.6, followed by a discussion of MO readout in Sec. 31.7. Certain important characteristics of MO media are summarized in Sec. 31.8. Concluding remarks and an examination of trends for future optical recording devices are the subject of Sec. 31.9.

Alternative methods of optical data storage such as reversible phase-change, photochemical spectral hole burning, three-dimensional volume holographic storage, photon echo, photon trapping, etc., will not be discussed in this article. The interested reader may consult the following references for information concerning these alternative storage schemes:

Proceedings of the International Symposium on Optical Memory, ISOM'89, published as supplement 28-3 of the *Japanese Journal of Applied Physics,* vol. 28 (1989).

Proceedings of the Optical Data Storage Conference, SPIE, vol. 1316 (1990).

Proceedings of the Optical Data Storage Conference, SPIE, vol. 1499 (1991).

Proceedings of the Optical Data Storage Conference, SPIE, vol. 1663 (1992).

R. G. Zech, "Volume Hologram Optical Memories: Mass Storage Future Perfect," *Optics and Photonics News,* vol. 3, no. 8, pp. 16–25 (1992).

31.2 PRELIMINARIES AND BASIC DEFINITIONS

The format and physical layout of recorded data on the storage medium as well as certain operational aspects of disk drive mechanism will be described in the present section.

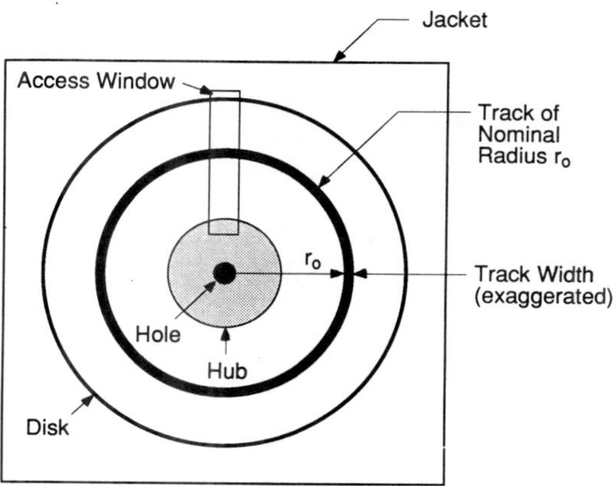

FIGURE 1 Physical appearance and general features of an optical disk. The read-write head gains access to the disk through a window in the jacket; the jacket itself is for protection purposes only. The hub is the mechanical interface with the drive for mounting and centering the disk on the spindle. The track shown here is of the concentric-ring type, with radius r_0 and width W_t.

The Concept of Track

The information on magnetic and optical disks is recorded along tracks. Typically, a track is a narrow annulus at some distance r from the disk center, as shown in Fig. 1. The width of the annulus is denoted by W_t, while the width of the guard band, if any, between adjacent tracks is denoted by W_g. The track-pitch is the center-to-center distance between neighboring tracks and is therefore equal to $W_t + W_g$. A major difference between the magnetic floppy disk, the magnetic hard disk, and the optical disk is that their respective track-pitches are presently of the order of 100, 10, and 1 μm. Tracks may be fictitious entities, in the sense that no independent existence outside the pattern of recorded marks may be ascribed to them. This is the case, for example, with the compact audio disk format where prerecorded marks simply define their own tracks and help guide the laser beam during readout. In the other extreme are tracks that are physically engraved on the disk surface before any data is ever recorded. Examples of this type of track are provided by pregrooved WORM and magneto-optical disks. Figure 2 shows micrographs from several recorded optical disk surfaces. The tracks along which data is written are clearly visible in these pictures.

It is generally desired to keep the read-write head stationery while the disk spins and a given track is being read from or written onto. Thus, in an ideal situation, not only should the track be perfectly circular, but also the disk must be precisely centered on the spindle axis. In practical systems, however, tracks are neither precisely circular, nor are they concentric with the spindle axis. These eccentricity problems are solved in low-performance floppy drives by making tracks wide enough to provide tolerance for misregistrations and misalignments. Thus the head moves blindly to a radius where the

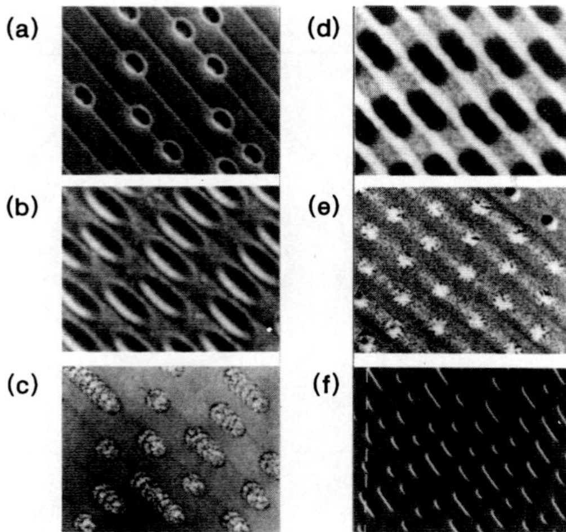

FIGURE 2 Micrographs of several types of optical storage media. The tracks are straight and narrow with a 1.6 μm pitch, and are diagonally oriented in each frame. (*a*) Ablative, write-once tellurium alloy. (*b*) Ablative, write-once organic dye. (*c*) Amorphous-to-crystalline, write-once phase-change alloy GaSb. (*d*) Erasable, amorphous magneto-optic alloy GdTbFe. (*e*) Erasable, crystalline-to-amorphous phase-change tellurium alloy. (*f*) Read-only CD-Audio, injection-molded from polycarbonate with a nickel stamper. *(From Ullmann's "Encyclopedia of Industrial Chemistry," Verlagsgesellschaft, mbH, Weinheim, 1989.)*

track center is nominally expected to be, and stays put until the reading or writing is over. By making the head narrower than the track-pitch, the track center is allowed to wobble around its nominal position without significantly degrading the performance during read-write operations. This kind of wobble, however, is unacceptable in optical disk systems which have a very narrow track, about the same size as the focused beam spot. In a typical situation arising in practice the eccentricity of a given track may be as much as 50 μm, while the track-pitch is only about 1 μm, thus requiring active track-following procedures.

A popular method of defining tracks on an optical disk is by means of pregrooves, which are either etched, stamped, or molded onto the substrate. The space between neighboring grooves is called *land* (see Fig. 3*a*). Data may be written in the groove with the land acting as a guard band. Alternatively, the land may be used for recording while the grooves separate adjacent tracks. The groove depth is optimized for generating an optical signal sensitive to the radial position of the read-write laser beam. For the push-pull method of track-error detection (described in Sec. 31.5) the groove depth is in the neighborhood of $\lambda/8$, where λ is the wavelength of the light beam.

In digital data storage each track is divided into small segments called *sectors.* A sector is intended for the storage of a single block of data which is typically either 512 or 1024 bytes. The physical length of a sector is thus several millimeters. Each sector is preceded by header information such as the identity of the sector, identity of the corresponding

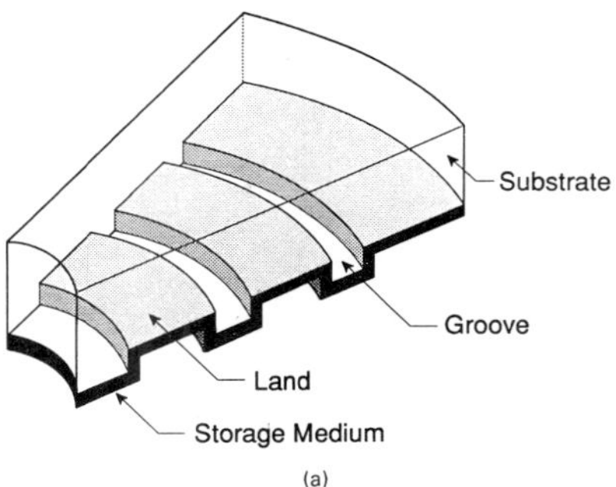

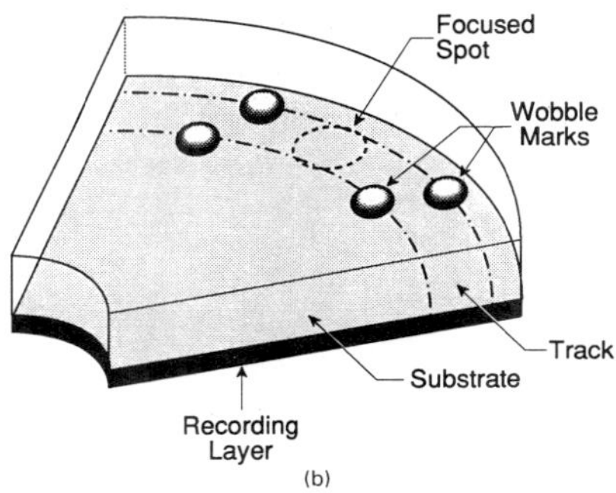

FIGURE 3 (*a*) Lands and grooves in an optical disk. The substrate is transparent, and the laser beam must pass through it before reaching the storage medium. (*b*) Sampled-servo marks in an optical disk. These marks which are offset from the track center provide information regarding the position of focused spot.

track, synchronization marks, etc. The header information may be preformatted onto the substrate, or it may be written directly on the storage layer. Pregrooved tracks may be "carved" on the optical disk either as concentric rings or as a single continuous spiral. There are certain advantages to each format. A spiral track contains a succession of sectors without interruption, whereas concentric rings may each end up with some empty space that is too small to become a sector. Also, large files may be written onto (and read from) spiral tracks without jumping to the next track, which is something that occurs when concentric tracks are used. On the other hand, multiple-path operations such as write-and-verify or erase-and-write which require two paths each for a given sector, or still-frame video are more conveniently handled on concentric-ring tracks.

Another suggested track format is based on the idea of a sampling servo. Here the tracks are identified by occasional marks placed permanently on the substrate at regular intervals, as shown in Fig. 3*b*. Details of track-following by the sampled-servo scheme will follow shortly (see Sec. 31.5), suffice it to say at this point that servo marks help the system identify the position of the focused spot relative to the track center. Once the position is determined it is fairly simple to steer the beam and adjust its position on the track.

Disk Rotation Speed

When a disk rotates at a constant angular velocity ω, a track of radius r moves with the constant linear velocity $V = r\omega$. Ideally, one would like to have the same linear velocity for all the tracks, but this is impractical except in a limited number of situations. For instance, when the desired mode of access to the various tracks is sequential, such as in audio- and video-disk applications, it is possible to place the head in the beginning at the inner radius and move outward from the center thereafter while continuously decreasing the angular velocity. By keeping the product of r and ω constant, one can achieve constant linear velocity for all tracks.* Sequential access mode, however, is the exception rather than the norm in data storage systems. In most applications, the tracks are accessed randomly with such rapidity that it becomes impossible to adjust the rotation speed for constant linear velocity. Under these circumstances the angular velocity is kept constant during normal operation. Typical rotation rates are 1200 and 1800 rpm for slower drives, and 3600 rpm for the high-end systems. Higher rotation rates (5000 rpm and beyond) are certainly feasible and will likely appear in future generations of optical storage devices.

Access Time

The direct access storage device used in computer systems for the mass storage of digital information is a disk drive capable of storing large quantities of data and accessing blocks of this data rapidly and in random order. In read-write operations it is often necessary to move the head to new locations in search of sectors containing specific data items. Such random relocations are usually time-consuming and can become the factor that limits performance in certain applications. The access time τ_a is defined as the average time spent in going from one randomly selected spot on the disk to another. τ_a can be considered the sum of a seek time τ_s, which is the average time needed to acquire the target track, and a latency τ_l, which is the average time spent on the target track waiting for the desired sector; thus $\tau_a = \tau_s + \tau_l$. The latency is half the revolution period of the disk, since a randomly selected sector is, on the average, halfway along the track from the point where the head initially lands. Thus, for a disk rotating at 1200 rpm $\tau_l = 25$ ms, while at 3600 rpm $\tau_l \simeq 8.3$ ms. The seek time, on the other hand, is independent of the rotation speed, but is determined by the travel distance of the head during an average seek, as well as by the mechanism of head actuation. (It can be shown that the average length of travel in a random seek is one-third of the full stroke.) In magnetic disk drives where the head/actuator assembly is relatively lightweight (a typical Winchester head weighs about 5 grams) the acceleration and deceleration periods are short, and seek times are typically around 10 ms. In optical disk systems, on the other hand, the head, being an assembly of discrete elements, is fairly large and heavy (typical weight $\simeq$ 50–100 grams), resulting in values of τ_s that are several times greater than those obtained in magnetic recording. The seek times reported for commercially available optical drives presently range from 20 msec in high-performance 3.5-in drives to 100 msec in larger drives. One must emphasize,

* In compact audio disk players the linear velocity is kept constant at 1.2 m/s. The starting position of the head is at the inner radius $r_{min} = 25$ mm, where the disk spins at 460 rpm. The spiral track ends at the outer radius $r_{max} = 58$ mm, where the disk's angular velocity is 200 rpm.

however, that the optical disk technology is still in its infancy; with the passage of time the integration and miniaturization of the elements within the optical head will surely produce lightweight devices capable of achieving seek times in the range of several milliseconds.

Organization of Data on Disk

For applications involving computer files and data, each track is divided into a number of sectors where each sector can store a fixed-length block of binary data. The size of the block varies among the various disk/drive manufacturers, but typically it is either 512 or 1024 bytes. As long as the disk is dedicated to a particular drive (such as in magnetic hard drives) the sector size is of little importance to the outside world. However, with removable media the sector size (among other things) must be standardized, since now various drives need to read from and write onto the same disk.

A block of user data cannot be directly recorded on a sector. First, it must be coded for protection against errors (error-correction coding) and for the satisfaction of channel requirements (modulation coding). Also, it may be necessary to add synchronization bits or other kinds of information to the data before recording. Thus a sector's capacity must be somewhat greater than the amount of raw data assigned to it. A sector also must have room for "header" information. The header is either recorded during the first use of the disk by the user, as in formatting a floppy disk, or is written by the manufacturer before shipping the disk. The header typically contains the address of the sector plus synchronization and servo bits. In magnetic disks the header is recorded magnetically, which makes it erasable and provides the option of reformatting at later times. On the negative side, formatting is time-consuming and the information is subject to accidental erasure. In contrast, the optical disk's sector headers may be mass-produced from a master at the time of manufacture, thus eliminating the slow process of soft formatting. The additional space used by the codes and by the header information constitutes the overhead. Depending on the quality of the disk, the degree of sophistication of the drive, and the particular needs of a given application, the overhead may take as little as 10 percent and as much as 30 percent of a disk's raw capacity.

31.3 THE OPTICAL PATH

The optical path begins at the light source which, in all laser disk systems in use today, is a semiconductor GaAs diode laser. Several unique features of the laser diode have made it indispensable for optical recording applications: its small size ($\approx 300 \times 50 \times 10\ \mu\text{m}$) makes possible the construction of compact head assemblies, its coherence properties allow diffraction-limited focusing to extremely small spots, and its direct modulation capability eliminates the need for external intensity modulators. The operating wavelength of the laser diode can be selected within a limited range by proper choice of material composition; presently, the shortest wavelength available from the III-V class of semiconductor materials is 670 nm.

Figure 4*a* shows a typical plot of laser power output versus input current for a GaAs-based laser diode. The lasing starts at the threshold current, and the output power rapidly increases beyond that point. Below threshold, the diode operates in the spontaneous emission mode and its output is incoherent. After threshold, stimulated emission takes place, yielding coherent radiation. Of course, the output power cannot

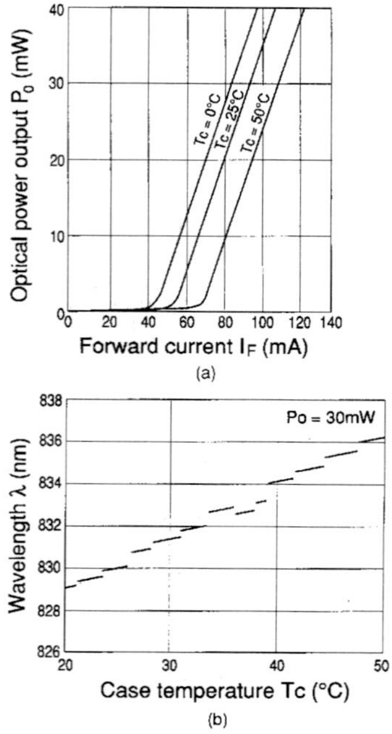

FIGURE 4 (*a*) Optical output power versus forward-bias current for a typical diode laser. Different curves were obtained at different ambient temperatures. (*b*) Variations of wavelength as function of case temperature for typical diode laser. The output power is fixed at $P_o = 30$ mW. *(From Sharp Laser Diode User's Manual.)*

increase indefinitely and beyond a certain point the laser fails catastrophically. Fortunately, the required optical power levels for the read/write/erase operations in present-day data storage systems are well below the failure levels of these lasers. Available lasers for data storage applications have threshold currents around 40 mA, maximum allowable currents of about 100 mA, and peak output powers (CW mode) around 50 mW. The relationship between the injection current and the output light power is very sensitive to the operating temperature of the laser, as evidenced by the various plots in Fig. 4*a*. Also, because the semiconductor material's bandgap is a function of the ambient temperature, there is a small shift in the operating wavelength of the device when the temperature fluctuates (see Fig. 4*b*). For best performance it is usually necessary to mount the laser on a good heat-sink, or try to steady its temperature by closed-loop feedback.

The output optical power of the laser can be modulated by controlling the injection current. One can apply pulses of variable duration to turn the laser on and off during the recording process. The pulse duration can be as short as a few nanoseconds, with rise and fall times which are typically less than 1 nanosecond. This direct-modulation capability of the laser diode is particularly welcome in optical disk systems, considering that most other sources of coherent light (such as gas lasers) require bulky and expensive devices for external modulation. Although readout of optical disks can be accomplished at constant power level in CW mode, it is customary (for noise reduction purposes) to modulate the laser at a high frequency in the range of several hundred MHz.

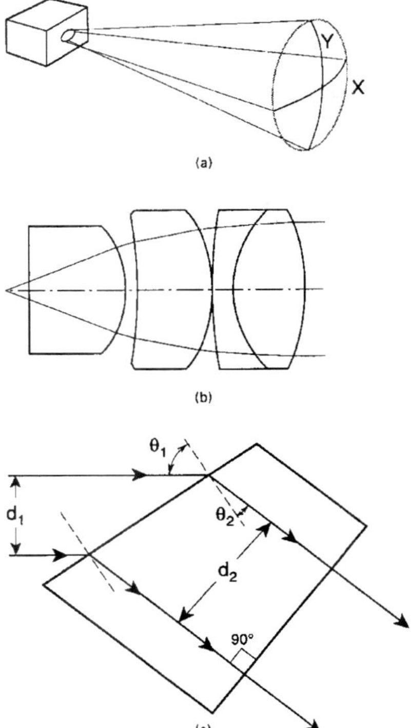

FIGURE 5 (*a*) Away from the facet, the output beam of a diode laser diverges rapidly. In general, the beam diameter along X is different from that along Y, which makes the cross section of the beam elliptical. Also, the radii of curvature R_x and R_y are not the same, thus creating a certain amount of astigmatism in the beam. (*b*) Multi-element collimator lens for laser diode applications. Aside from collimating, this lens also corrects astigmatic aberrations of the beam. (*c*) Beam-shaping by deflection at a prism surface. Θ_1 and Θ_2 are related by the Snell's law, and the ratio d_2/d_1 is the same as $\cos \Theta_2/\cos \Theta_1$. Passage through the prism circularizes the elliptical cross section of the beam.

Collimation and Beam Shaping

Since the cross-sectional area of the active region in a laser diode is only about 1 μm^2, diffraction effects cause the emerging beam to diverge rapidly. This phenomenon is depicted schematically in Fig. 5*a*. In practical applications of the laser diode, the expansion of the emerging beam is arrested by a collimating lens, such as that shown in Fig. 5*b*. If the beam happens to have aberrations (astigmatism is particularly severe in diode lasers), then the collimating lens must be designed to correct this defect as well.

In optical recording it is most desirable to have a beam with circular cross section. The need for beam shaping arises from the special geometry of the laser cavity with its rectangular cross section. Since the emerging beam has different dimensions in the

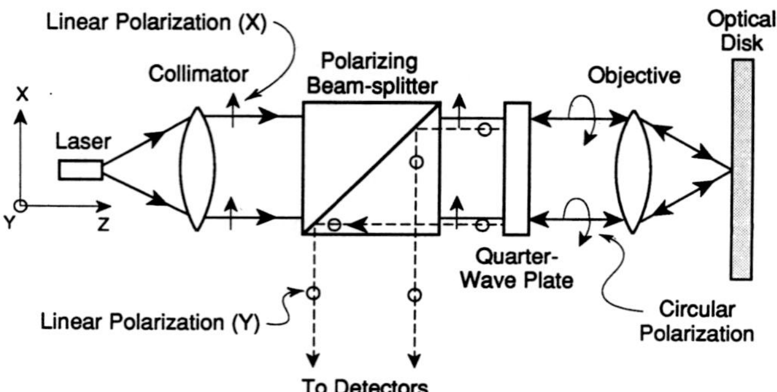

FIGURE 6 Separation of incident and reflected beams at the polarizing beam splitter (PBS). The quarter-wave plate converts the linearly polarized incident beam into one with circular polarization and converts the returning beam back to linear, but with its polarization vector orthogonal to that of the incident beam. This 90° rotation of polarization is responsible for the diversion of the reflected beam toward the detection channel.

directions parallel and perpendicular to the junction, its cross section at the collimator becomes elliptical, with the initially narrow dimension expanding more rapidly to become the major axis of the ellipse. The collimating lens thus produces a beam with elliptical cross section. Circularization may be achieved by bending various rays of the beam at a prism, as shown in Fig. 5*c*. The bending changes the beam's diameter in the plane of incidence, but leaves its diameter in the perpendicular direction intact.

The output of the laser diode is linearly polarized in the plane of the junction. In some applications (such as readout of compact disks or read-write on WORM media) the polarization state is immaterial as far as interaction with the storage medium is concerned. In such applications one usually passes the beam through a polarizing beam splitter (PBS) and a quarter-wave plate, as in Fig. 6, and converts its polarization to circular. Upon reflection from the disk, the beam passes through the quarter-wave plate once again, but this time emerges as linearly polarized in a direction perpendicular to the original direction of polarization. The returning beam is thus directed away from the laser and toward the detection module, where its data content is extracted and its phase/amplitude pattern is used to generate error signals for automatic focusing and tracking. By thoroughly separating the returning beam from the incident beam, one not only achieves efficiency in the use of the optical power, but also succeeds in preventing the beam from going back to the laser where it causes instabilities in the laser cavity and, subsequently, increases the noise level. Unfortunately, there are situations where a specific polarization state is required for interaction with the disk; magneto-optical readout which requires linear polarization is a case in point. In such instances the simple combination of PBS and quarter-wave plate becomes inadequate and one must resort to other (less efficient) means of separating the beams.

Focusing

The collimated and circularized beam of the laser is focused on the surface of the disk using an objective lens. The objective is designed to be aberration-free, so that its focused spot size is limited only by the effects of diffraction. Figure 7*a* shows the design of a typical

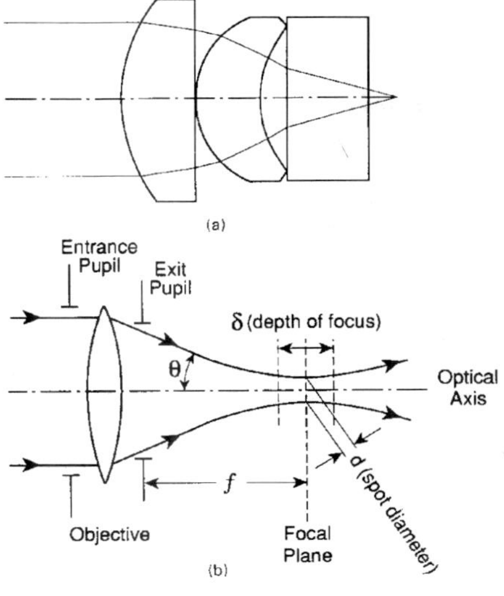

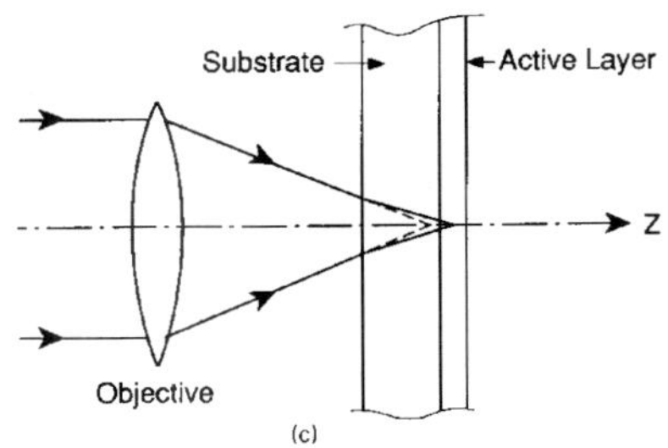

FIGURE 7 (*a*) Multi-element lens design for a high-*NA* videodisc objective. *(After D. Kuntz, "Specifying Laser Diode Optics,"* Laser Focus, *March 1984.)* (*b*) Various parameters of the objective lens. The numerical aperture is $NA = \sin \Theta$. The spot diameter d and the depth of focus δ are given by Eqs. (1) and (2), respectively. (*c*) Focusing through the substrate can cause spherical aberration at the active layer. The problem is corrected by a proper design for the objective lens, which takes the substrate into account.

objective made from spherical optics. According to the classical theory of diffraction, the diameter of the beam, d, at the objective's focal plane is,

$$d \simeq \frac{\lambda}{NA} \tag{1}$$

where λ is the wavelength of light and NA is the numerical aperture of the objective. In optical recording it is desired to achieve the smallest possible spot, since the size of the spot is directly related to the size of marks recorded on the medium. Also, in readout, the spot size determines the resolution of the system. According to Eq. (1) there are two ways to achieve a small spot: reducing the wavelength and increasing the numerical aperture. The wavelengths currently available from GaAs lasers are in the range of 670–840 nm. It is possible to use a nonlinear optical device to double the frequency of these lasers, thus achieving blue light. Good efficiencies have been demonstrated by frequency doubling. Also recent developments in II-VI materials have improved the prospects for obtaining green and blue light directly from semiconductor lasers. Consequently, there is hope that in the near future optical storage systems will operate in the wavelength range of 400–500 nm. As for the numerical aperture, current practice is to use a lens with $NA \simeq 0.5$–0.6. Although this value might increase slightly in the coming years, much higher numerical apertures are unlikely, since they put strict constraints on the other characteristics of the system and limit the tolerances. For instance, the working distance at high NA is relatively short, making access to the recording layer through the substrate more difficult. The smaller depth of focus of a high-NA lens will make attaining/maintaining proper focus more difficult, while the limited field of view might restrict automatic track-following procedures. A small field of view also places constraints on the possibility of read/write/erase operations involving multiple beams.

The depth of focus of a lens, δ, is the distance away from the focal plane over which tight focus can be maintained (see Fig. 7*b*). According to the classical theory of diffraction,

$$\delta \simeq \frac{\lambda}{NA^2} \tag{2}$$

Thus for $\lambda = 700$ nm and $NA = 0.6$ the depth of focus is about ± 1 μm. As the disk spins under the optical head at the rate of several thousand rpm, the objective must stay within a distance of $f \pm \delta$ from the active layer if proper focus is to be maintained. Given the conditions under which drives usually operate, it is impossible to make rigid enough mechanical systems to yield the required positioning tolerances. On the other hand, it is fairly simple to mount the objective lens in an actuator capable of adjusting its position with the aid of closed-loop feedback control. We emphasize that by going to shorter wavelengths and/or larger numerical apertures (as is required for attaining higher data densities) one will have to face a much stricter regime as far as automatic focusing is concerned. Increasing the numerical aperture is particularly worrisome, since δ drops with the square of NA.

A source of spherical aberrations in optical disk systems is the substrate through which the light must pass in order to reach the active layer. Figure 7*c* shows the bending of the rays at the surface of the disk, which causes the aberration. This problem can be solved by taking into account the effects of the substrate in the design of the objective, so that the lens is corrected for all aberrations, including those arising at the substrate. Recent developments in molding of aspheric glass lenses have gone a long way in simplifying the lens design problem. Figure 8 shows a pair of molded glass aspherics designed for optical storage applications; both the collimator and the objective are single-element lenses and are corrected for axial aberrations.

Laser Noise

Compared to other sources of coherent light such as gas lasers, laser diodes are noisy and unstable. Typically, within a diode laser's cavity several modes compete for dominance. Under these circumstances, small variations in the environment can cause mode-hopping which results in unpredictable power-level fluctuations and wavelength shifts. Unwanted optical feedback is specially troublesome, as even a small fraction of light returning to the cavity can cause a significant rise in the noise level. Fortunately, it has been found that

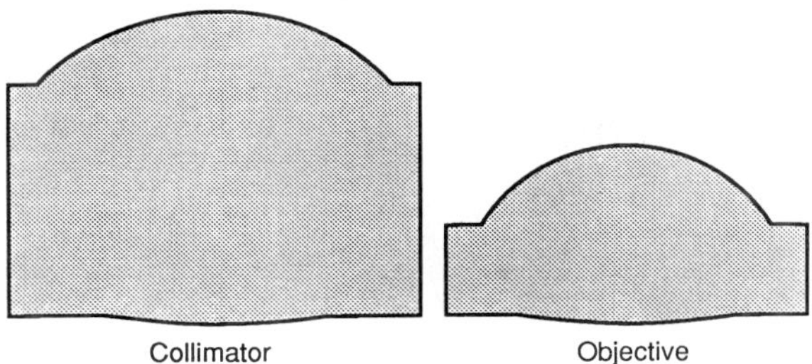

FIGURE 8 Molded glass aspheric lens pair for optical disk application. These singlets can replace the multi-element spherical lenses shown in Figs. 5(*b*) and 7(*a*).

high-frequency modulation of the injection current can be used to instigate power sharing among the modes and thereby reduce fluctuations of the output optical power. In general, a combination of efforts such as temperature stabilization of the laser, antireflection coating of the various surfaces within the system, optical isolation of the laser, and high-frequency modulation of the injection current can yield acceptable levels of noise for practical operation of the device.

31.4 AUTOMATIC FOCUSING

Since the objective lens has a large numerical aperture ($NA \geq 0.5$) its depth of focus δ is shallow ($\delta \simeq \pm 1\ \mu\text{m}$ at $\lambda = 780$ nm). During all read/write/erase operations, therefore, the disk must remain within a fraction of a micrometer from the focal plane of the objective. In practice, however, the disks are not flat and are not always mounted rigidly parallel to the focal plane, so that during any given revolution movements away from focus (by as much as $\pm 50\ \mu\text{m}$) may occur. Without automatic adjustment of the objective along the optic axis, this runout (or disk flutter) will be detrimental to the operation of the system. In practice, the objective is mounted on a small actuator (usually a voice coil) and allowed to move back and forth to keep its distance from the disk within an acceptable range. Since the spindle turns at a few thousand rpm, if the disk moves in and out of focus a few times during each revolution, then the voice coil must be fast enough to follow these movements in real time; in other words, its frequency response must extend from DC to several kHz.

The signal that controls the voice coil is obtained from the light reflected from the disk. There are several techniques for deriving the focus error signal (FES), one of which is depicted in Fig. 9*a*. In this so-called obscuration method a secondary lens with one-half of its aperture covered is placed in the path of the reflected light, and a split-detector is placed at the focal plane of this secondary lens. When the disk is in focus, the returning beam is collimated and the secondary lens will focus the beam at the center of the split-detector, giving a difference signal ΔS equal to zero. If the disk now moves away from the objective, the returning beam will become converging, as in Fig. 9*b*, sending all the light to detector 1. In this case ΔS will be positive and the voice coil will push the lens toward the disk. On the other hand, when the disk moves close to the objective, the returning beam becomes diverging and detector 2 receives the light (see Fig. 9*c*). This results in a negative ΔS which forces the voice coil to pull back and return ΔS to zero.

A given focus error detection scheme is generally characterized by the shape of its focus error signal ΔS versus the amount of defocus Δz. one such curve is shown in Fig. 9*d*. The

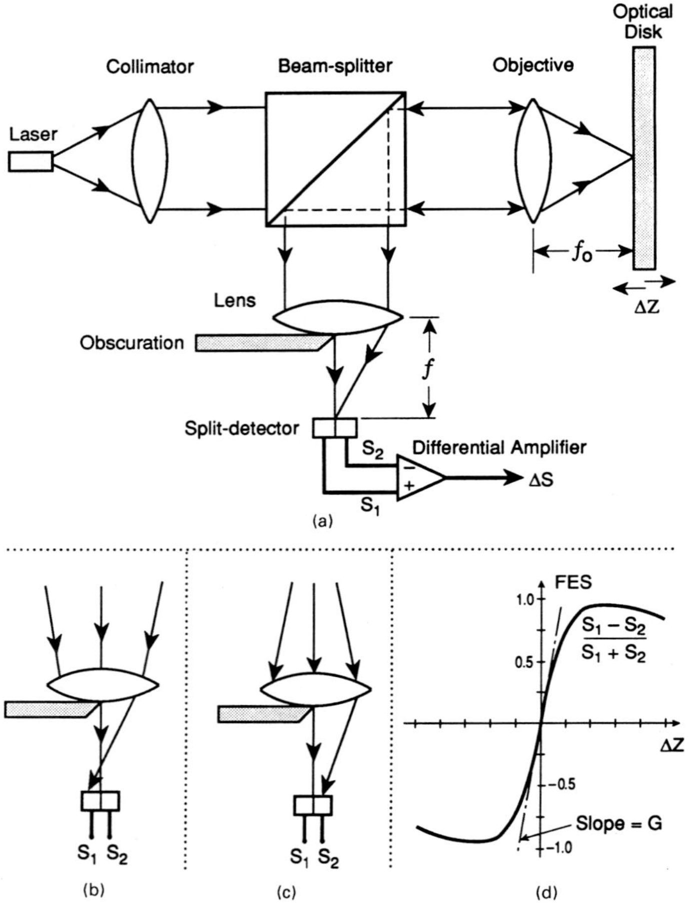

FIGURE 9 Focus error detection by the obscuration method. In (*a*) the disk is in focus, and the two halves of the split detector receive equal amounts of light. When the disk is too far from the objective (*b*) or too close to it (*c*), the balance of detector signals shifts to one side or the other. A plot of the focus error signal versus defocus is shown in (*d*), and its slope near the origin is identified as the FES gain, G.

slope of the FES curve near the origin is of particular importance, since it determines the overall performance and stability of the servo loop. In general, schemes with a large slope are preferred, although certain other aspects of system performance should also be taken into consideration. For instance, variations of the FES during seek operations (where multiple track-crossings occur) should be kept at a minimum, or else the resulting "feedthrough" might destabilize the focus servo. Also, it is important for a focus-error-detection scheme to be insensitive to slight imperfections of the optical elements, as well as to the positioning and mechanical misalignments; otherwise, the manufacturing cost of the device may become prohibitive. Finally, the focusing scheme must have a reasonable acquisition range, so that at start-up (or in those occasions where focus is lost and needs to be acquired again) the system can move in the proper direction to establish focus.

31.5 AUTOMATIC TRACKING

Consider a circular track with a certain radius, say, r_0, and imagine viewing a portion of it through the access window (see Fig. 1). It is through this window that the read-write head gains access to the disk and, by moving in the radial direction, reaches the various tracks. To a viewer looking through the window, a perfectly circular track centered on the spindle axis will look stationary, irrespective of the rotational speed of the disk. However, any track eccentricity will cause an apparent motion toward or away from the center. The peak-to-peak radial distance traveled by a track (as seen through the window) might depend on a number of factors, including centering accuracy of the hub, deformability of the disk substrate, mechanical vibrations, manufacturing tolerances, etc. For a 3.5-in plastic disk, for example, this peak-to-peak motion can be as much as 100 μm. Assuming a rotation rate of 3600 rpm, the apparent radial velocity of the track will be a few mm/sec. Now, if the focused spot (which is only about 1 μm) remains stationary while trying to read or write on this track (whose width is also about 1 μm), it is clear that the beam will miss the track for a good fraction of every revolution cycle.

Practical solutions to the above problem are provided by automatic track-following techniques. Here the objective lens is placed in a fine actuator, typically a voice coil, which is capable of moving the necessary radial distances and maintaining a lock on the desired track. The signal that controls the movement of this actuator is derived from the reflected light itself, which carries information about the position of the focused spot relative to the track. There exist several mechanisms for extracting the track-error signal (TES) from the reflected light. All these methods require some sort of structure on the disk surface to identify the position of the track. In the case of read-only disks (CD, CD-ROM, and video disk) the embossed pattern of data provides ample information for tracking purposes. In the case of write-once and erasable disks, tracking guides are impressed on the substrate during the manufacturing process. The two major formats for these tracking guides are pregrooves (for continuous tracking) and sampled-servo marks (for discrete tracking). A combination of the two schemes, known as continuous/composite format, is often used in practice. This format is depicted schematically in Fig. 10 which shows a small section containing five tracks, each consisting of the tail end of a groove, synchronization marks, a mirror area for adjusting offsets, a pair of wobble marks for sampled tracking, and header information for sector identification.

Tracking on Grooved Regions

As shown in Fig. 3*a*, grooves are continuous depressions that are embossed, etched, or molded onto the substrate prior to deposition of the storage medium. If the data is recorded on the grooves, then the lands are not used except for providing a guard band between neighboring grooves. Conversely, the land regions may be used to record the

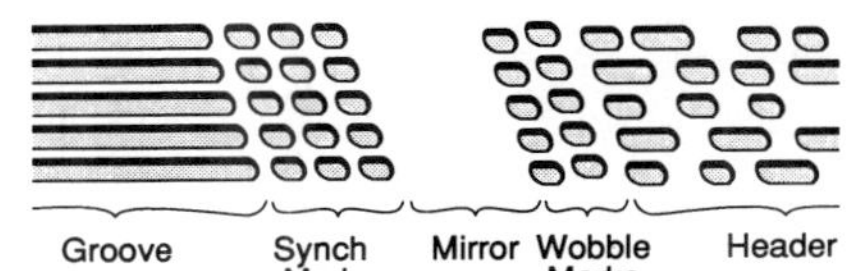

FIGURE 10 Servo offset fields in continuous/composite format contain a mirror area and offset marks for tracking. *(Marchant, 1990.)*

information, in which case grooves provide the guard band. Typical track widths are about one wavelength of the light. The guard bands are somewhat narrower, their exact shape and dimensions depending on the beam size, required track-servo accuracy, and the acceptable levels of crosstalk between adjacent tracks. The groove depth is typically around one-eighth of one wavelength ($\lambda/8$) which gives the largest TES in the push-pull method. The geometrical shape of the groove's cross section might be rectangular, trapezoidal, triangular, or some smooth version of these curves.

When the focused spot is centered on a given track, it is diffracted symmetrically from the two edges, resulting in a balanced far-field pattern. As soon as the spot moves away from the center, the symmetry breaks down and the far-field distribution tends to shift to one side or the other. A split photodetector placed in the path of the reflected light can therefore sense the relative position of the spot and provide the appropriate feedback signal (see Fig. 11). This is the essence of the push-pull method. Figure 11 also shows intensity plots at the detector plane after reflection from various locations on the grooved surface. Note how the intensity shifts to one side or the other depending on whether the spot moves to the right edge or to the left edge of the groove.

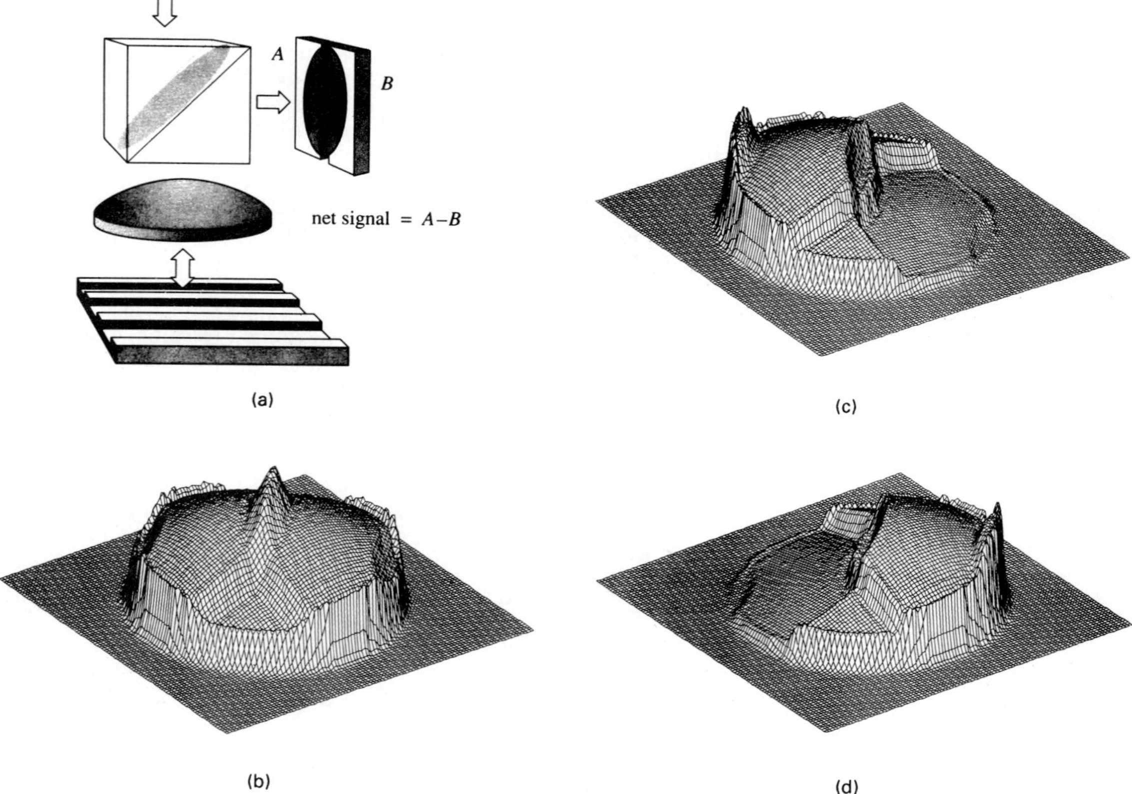

FIGURE 11 (*a*) Push-pull sensor for tracking on grooves. *(Marchant, 1990.)* (*b*) Light intensity distribution at the detector plane when the disk is in focus and the beam is centered on the track. (*c*) Light intensity distribution at the detector plane when the disk is in focus and the beam is centered on the groove edge. (*d*) Same as (*c*) except for the spot being on the opposite edge of the groove.

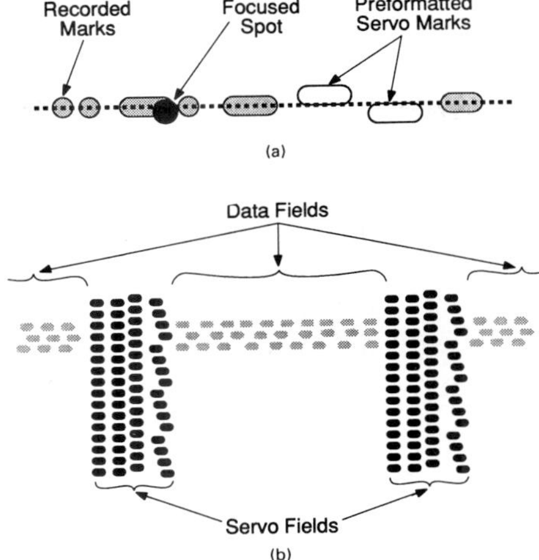

FIGURE 12 (*a*) In sampled tracking a pair of preformatted servo marks helps locate the position of the focused spot relative to the track center. (*b*) servo fields occur frequently and at regular intervals in sampled servo format. The data area shown here has data recorded on three tracks. *(Marchant, 1990.)*

Sampled Tracking

Since dynamic track runout is usually a slow and gradual process, there is actually no need for continuous tracking as done on grooved media. A pair of embedded marks, offset from the track center as in Fig. 12*a,* can provide the necessary information for correcting the relative position of the focused spot. The reflected intensity will indicate the positions of the two servo marks as two successive short pulses. If the beam happens to be on track, the two pulses will have equal magnitudes and there shall be no need for correction. If, on the other hand, the beam is off-track, one of the pulses will be stronger than the other. Depending on which pulse is the stronger, the system will recognize the direction in which it has to move and will correct the error accordingly. Sampled-servo mark pairs must be provided frequently enough to ensure proper track-following. In a typical application, the track might be divided into groups of 18 bytes, 2 bytes dedicated as servo offset areas and 16 bytes filled with other format information or left blank for user data. Figure 12*b* shows a small section from a sampled-servo disk containing a number of tracks, three of which are recorded with user data. The track-servo marks in this case are preceded by synch marks (also prerecorded on the servo offset area). Note in Fig. 12*b* that the format marks repeat a certain pattern every four tracks. This pattern is known as a "gray code," and allows the system to recognize and correct minor track-counting errors during the seek operation.

Track Counting During the Seek Operation

In the seek operation the coarse actuator moves the head assembly across the disk to a new location where the desired track is located. In order to avoid landing on a nearby track and being forced to perform a second (fine) seek, most systems in use today count the tracks as they are being crossed. In this way the head can land on the correct track and thereby minimize the overall seek time. The sampled-servo format is not suitable for this purpose, since the servo marks do not occur frequently enough to allow uninterrupted counting. In contrast, grooved media provide the necessary information for track-counting.

During a seek operation the focus servo loop remains closed, maintaining focus as the head crosses the tracks. The tracking loop, on the other hand, must be opened. The zero crossings of the TES then provide the track count. Complications may arise in this process, however, due to eccentricities of tracks. As was mentioned earlier, to an observer looking through the access window, an eccentric track moves in and out radially with a small (but not insignificant) velocity. As the head approaches the desired track and slows down to capture it, its velocity might fall just short of the apparent track velocity. Under these circumstances, a track which has already been counted may catch up with the head and be counted once again. Intelligence must be built into the system to recognize and avoid such problems. Also, through the use of gray codes and similar schemes, the system can be made to correct its occasional miscounts before finally locking onto the destination track.

31.6 THERMOMAGNETIC RECORDING PROCESS

Recording and erasure of information on a magneto-optical disk are both achieved by the thermomagnetic process. The essence of thermomagnetic recording is shown in Fig. 13. At the ambient temperature the film has a high magnetic coercivity* and therefore does not respond to the externally applied field. When a focused laser beam raises the local

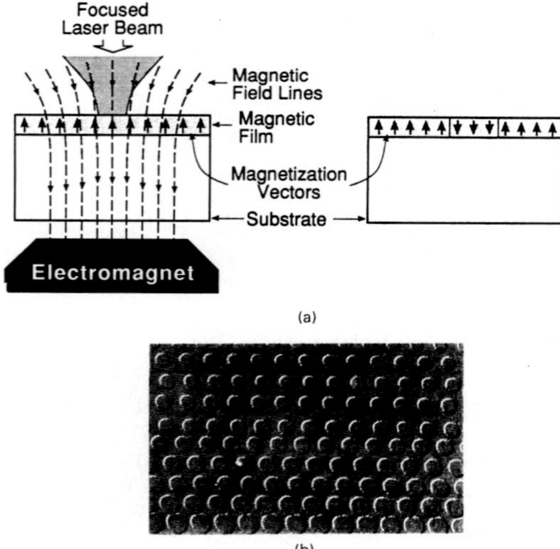

FIGURE 13 (*a*) Thermomagnetic recording process. The field of the electromagnet helps reverse the direction of magnetization in the area heated by the focused laser beam. (*b*) Lorentz micrograph of domains written thermomagnetically. The various tracks shown here were written at different laser powers, with power level decreasing from top to bottom.

* Coercivity of a magnetic medium is a measure of its resistance to magnetization reversal. For example, consider a thin film with perpendicular magnetic moment saturated in the $+Z$ direction, as in Fig. 13*a*. A magnetic field applied along $-Z$ will succeed in reversing the direction of magnetization only if the field is stronger than the coercivity of the film.

temperature of the film, the hot spot becomes magnetically soft (i.e., its coercivity drops). As the temperature rises, coercivity drops continuously until such time as the field of the electromagnet finally overcomes the material's resistance to reversal and switches its magnetization. Turning the laser off brings the temperatures back to normal, but the reverse-magnetized domain remains frozen in the film. In a typical situation in practice, the film thickness may be around 300 Å, laser power at the disk $\approx$10 mV, diameter of the focused spot $\approx$1 μm, laser pulse duration ~50 ns, linear velocity of the track $\approx$10 m/sec, and the magnetic field strength $\approx$200 gauss. The temperature may reach a peak of 500 K at the center of the spot, which is certainly sufficient for magnetization reversal, but is not nearly high enough to melt or crystalize or in any other way modify the structure of the material.

The materials of MO recording have strong perpendicular magnetic anisotropy. This type of anisotropy favors the "up" and "down" directions of magnetization over all other orientations. The disk is initialized in one of these two directions, say, up, and the recording takes place when small regions are selectively reverse-magnetized by the thermomagnetic process. The resulting magnetization distribution then represents the pattern of recorded information. For instance, binary sequences may be represented by a mapping of zeros to up-magnetized regions and ones to down-magnetized regions (NRZ scheme). Alternatively, the NRZI scheme might be used, whereby transitions (up-to-down and down-to-up) are used to represent the ones in the bit sequence.

Recording by Laser Power Modulation (LPM)

In this traditional approach to thermomagnetic recording, the electromagnet produces a constant field, while the information signal is used to modulate the power of the laser beam. As the disk rotates under the focused spot, the pulsed laser beam creates a sequence of up/down domains along the track. The Lorentz electron micrograph in Fig. 13*b* shows a number of domains recorded by LPM. The domains are highly stable and may be read over and over again without significant degradation. If, however, the user decides to discard a recorded block and to use the space for new data, the LPM scheme does not allow direct overwrite; the system must erase the old data during one revolution and record the new data in a subsequent revolution cycle.

During erasure, the direction of the external field is reversed, so that up-magnetized domains in Fig. 13*a* now become the favored ones. Whereas writing is achieved with a modulated laser beam, in erasure the laser stays on for a relatively long period of time, erasing an entire sector. Selective erasure of individual domains is not practical, nor is it desired, since mass data storage systems generally deal with data at the level of blocks, which are recorded onto and read from individual sectors. Note that at least one revolution cycle elapses between the erasure of an old block and its replacement by a new block. The electromagnet therefore need not be capable of rapid switchings. (When the disk rotates at 3600 rpm, for example, there is a period of 16 ms or so between successive switchings.) This kind of slow reversal allows the magnet to be large enough to cover all the tracks simultaneously, thereby eliminating the need for a moving magnet and an actuator. It also affords a relatively large gap between the disk and the magnet tip, which enables the use of double-sided disks and relaxes the mechanical tolerances of the system without overburdening the magnet's power supply.

The obvious disadvantage of LPM is its lack of direct overwrite capability. A more subtle concern is that it is perhaps unsuitable for the PWM (pulse width modulation) scheme of representing binary waveforms. Due to fluctuations in the laser power, spatial variations of material properties, lack of perfect focusing and track-following, etc., the length of a recorded domain along the track may fluctuate in small but unpredictable ways. If the information is to be encoded in the distance between adjacent domain walls (i.e., PWM), then the LPM scheme of thermomagnetic writing may suffer from excessive domain-wall jitter. Laser power modulation works well, however, when the information is encoded in the position of domain centers (i.e., pulse position modulation or PPM). In general, PWM is superior to PPM in terms of the recording density, and methods that allow PWM are therefore preferred.

Recording by Magnetic Field Modulation (MFM)

Another method of thermomagnetic recording is based on magnetic field modulation, and is depicted schematically in Fig. 14*a*. Here the laser power may be kept constant while the information signal is used to modulate the direction of the magnetic field. Photomicrographs of typical domain patterns recorded in the MFM scheme are shown in Fig. 14*b*. Crescent-shaped domains are the hallmark of the field modulation technique. If one assumes (using a much simplified model) that the magnetization aligns itself with the applied field within a region whose temperature has passed a certain critical value, T_{crit}, then one can explain the crescent shape of these domains in the following way: with the laser operating in the CW mode and the disk moving at constant velocity, temperature distribution in the magnetic medium assumes a steady-state profile, such as that in Fig. 14*c*. Of course, relative to the laser beam, the temperature profile is stationary, but in the frame of reference of the disk the profile moves along the track with the linear track velocity. The isotherm corresponding to T_{crit} is identified as such in the figure; within this isotherm the magnetization always aligns itself with the applied field. A succession of critical

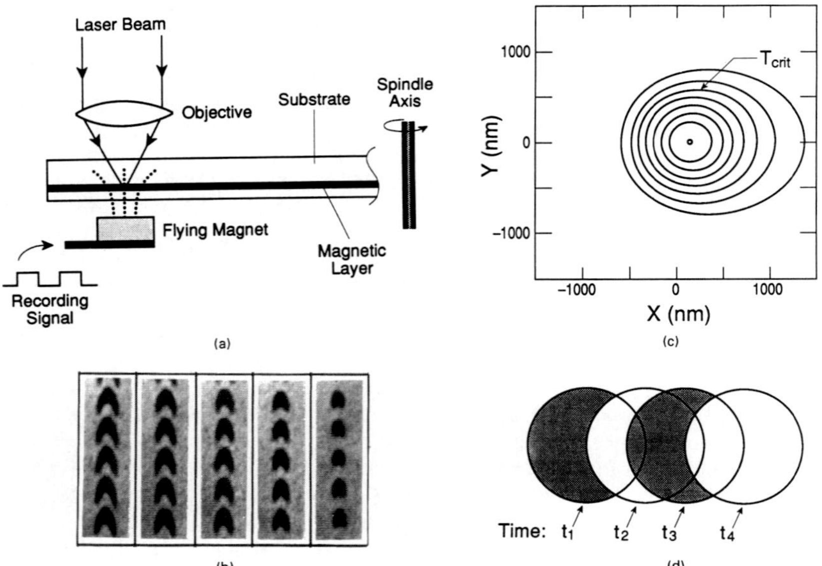

FIGURE 14 (*a*) Thermomagnetic recording by magnetic field modulation. The power of the beam is kept constant, while the magnetic field direction is switched by the data signal. (*b*) Polarized-light microphotograph of recorded domains. (*c*) Computed isotherms produced by a CW laser beam, focused on the magnetic layer of a disk. The disk moves with constant velocity under the beam. The region inside the isotherm marked as T_{crit} is above the critical temperature for writing, thus its magnetization aligns itself with the direction of the applied magnetic field. (*d*) Magnetization within the heated region (above T_{crit}) follows the direction of the applied magnetic field, whose switchings occur at times t_n. The resulting domains are crescent-shaped.

isotherms along the track, each obtained at the particular instant of time when the magnetic field switches direction, is shown in Fig. 14*d*. From this picture it is not difficult to see how the crescent-shaped domains form, and also to understand the relation between the waveform that controls the magnet and the resulting domain pattern.

The advantages of magnetic field modulation recording are that (1) direct overwriting is possible, and (2) domain wall positions along the track, being rather insensitive to defocus and laser power fluctuations, are fairly accurately controlled by the timing of the magnetic field switchings. On the negative side, the magnet must now be small and fly close to the disk surface if it is to produce rapidly switched fields with a magnitude of a few hundred gauss. Systems that utilize magnetic field modulation often fly a small electromagnet on the opposite side of the disk from the optical stylus. Since mechanical tolerances are tight, this might compromise the removability of the disk in such systems. Moreover, the requirement of close proximity between the magnet and the storage medium dictates the use of single-sided disks in practice.

Thermal Optimization of the Media—Multilayer Structures

The thermal behavior of an optical disk can be modified and improved if the active layer is incorporated into a properly designed multilayer structure, such as that shown in Fig. 15. In addition to thermal engineering, multilayers allow protective mechanisms to be built around the active layer; they also enable the enhancement of the signal-to-noise ratio in readout. (This latter feature is further explored in Sec. 31.7.) Multilayers are generally designed to optimize the absorption of light by creating an antireflection structure, whereby a good fraction of the incident optical power is absorbed in the active layer. Whereas the reflectivity of bare metal films is typically over 50 percent, a quadrilayer structure can easily reduce that to 20 percent or even less, if so desired. Multilayers can also be designed to control the flow of heat generated by the absorbed light. The aluminum reflecting layer in the quadrilayer of Fig. 15, for instance, may be used as a heat

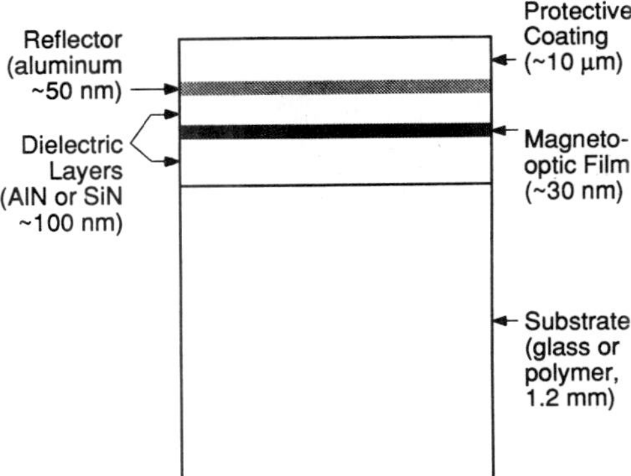

FIGURE 15 Quadrilayer magneto-optical disk structure. This particular design is for use in the substrate-incident mode, where the light goes through the substrate before reaching the MO layer. The thicknesses of the various layers can be optimized for enhancing the read signal, increasing the absorbed laser power, and controlling the thermal profile. Note in particular that the aluminum layer can play the dual roles of light reflector and heat sink.

sink for the magnetic layer, thus minimizing the undesirable effects of lateral heat diffusion within the magnetic medium.

31.7 MAGNETO-OPTICAL READOUT

The information recorded on a perpendicularly magnetized medium may be read with the aid of the polar magneto-optical Kerr effect. When linearly polarized light is normally incident on a perpendicular magnetic medium, its plane of polarization undergoes a slight rotation upon reflection. This rotation of the plane of polarization, whose sense depends on the direction of magnetization in the medium, is known as the polar Kerr effect. The schematic representation of this phenomenon in Fig. 16 shows that if the polarization vector suffers a counterclockwise rotation upon reflection from an up-magnetized region, then the same vector will rotate clockwise when the magnetization is down. A magneto-optical medium is characterized in terms of its reflectivity R and its Kerr rotation angle θ_k.* In MO readout, it is the sign of the rotation angle that carries the information about the state of magnetization of the medium, i.e., the recorded bit pattern.

The laser used for readout is usually the same as that used for recording, but its output power level is substantially reduced in order to avoid erasing (or otherwise obliterating)

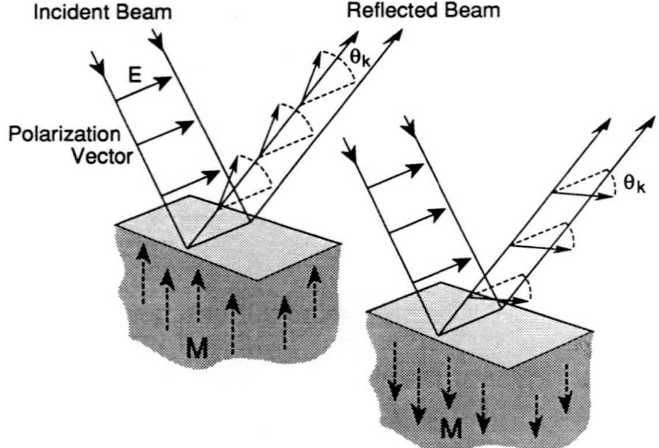

FIGURE 16 Schematic diagram describing the polar magneto-optical Kerr effect. Upon reflection from the surface of a perpendicularly magnetized medium, the polarization vector undergoes a rotation. The sense of rotation depends on the direction of magnetization **M**, and switches sign when **M** is reversed.

* In reality, the reflected state of polarization is not linear, but has a certain degree of ellipticity. One may consider the reflected polarization as consisting of two linear components: $E_{\parallel}$ which is parallel to the direction of incident polarization, and $E_{\perp}$ which is perpendicular to it. Now, if $E_{\parallel}$ is in phase with $E_{\perp}$, the net magneto-optic effect will be a pure rotation of the polarization vector. On the other hand, if $E_{\parallel}$ and $E_{\perp}$ are 90° out of phase, then the reflected polarization will be elliptical, with no rotation whatsoever. In practice, the phase difference between $E_{\parallel}$ and $E_{\perp}$ is somewhere between 0 and 90°, resulting in a reflected beam which has some degree of ellipticity ϵ_k, with the major axis of the polarization ellipse rotated by an angle θ_k (relative to the incident **E** vector). By inserting a Soleil-Babinet compensator in the reflected beam's path, one can change the phase relationship between $E_{\parallel}$ and $E_{\perp}$ in such a way as to eliminate the beam's ellipticity; the emerging polarization then will become linear with an enhanced rotation angle. In this article, reference to Kerr angle implies the effective angle which includes the above correction for ellipticity.

the previously recorded information. For instance, if the power of the write/erase beam is 20 mW, then for the read operation the beam is attenuated to about 3 or 4 mW. The same objective lens that focuses the write beam is now used to focus the read beam, creating a diffraction-limited spot for resolving the recorded marks. Whereas in writing the laser was pulsed to selectively reverse-magnetize small regions along the track, in readout it operates with constant power, i.e., in CW mode. Both up- and down-magnetized regions are read as the track passes under the focused light spot. The reflected beam, which is now polarization-modulated, goes back through the objective and becomes collimated once again; its information content is subsequently decoded by polarization-sensitive optics, and the scanned pattern of magnetization is reproduced as an electronic signal.

Differential Detection

Figure 17 shows the differential detection system that is the basis of magneto-optical readout in practically all erasable optical storage systems in use today. The beam splitter (BS) diverts half of the reflected beam away from the laser and into the detection module. The polarizing beam splitter (PBS) splits the beam into two parts, each carrying the projection of the incident polarization along one axis of the PBS, as shown in Fig. 17*b*. The component of polarization along one of the axes goes straight through, while the component along the other axis splits off to the side. If, upon reflection from the disk, the polarization did not undergo any rotations whatsoever, then the beam entering the PBS would be polarized at 45° to the PBS axes, in which case it would split equally between the two branches. Under this condition, the two detectors generate identical signals and the differential signal ΔS will be zero. Now, if the beam returns from the disk with its polarization rotated clockwise (rotation angle $= \theta_k$), then detector 1 will receive more light than detector 2, and the differential signal will be positive. Similarly, a counterclockwise rotated beam entering the PBS will generate a negative ΔS. The electronic signal ΔS thus reproduces the pattern of magnetization along the scanned track.

Enhancement of the Signal-to-Noise Ratio by Multilayering

The materials suitable for optical recording presently have very small Kerr angles (typically $\theta_k \simeq 0.5°$), with the result that the signal ΔS is correspondingly small. Multilayering schemes designed for the enhancement of the MO signal increase the interaction between the light and the magnetic medium by encapsulating a thin film of the MO material in an antireflection-type strucure. By providing a better index match between the MO film and its surroundings, and also by circulating the light through the MO film, multilayered structures manage to trap a large fraction of the incident light within the magnetized medium, and thus increase the Kerr rotation angle. These efforts inevitably result in a reduced reflectivity, but since the important parameter is the magneto-optically generated component of polarization, $E_\perp = \sqrt{R} \sin \theta_k$, it turns out that a net gain in the signal-to-noise ratio can be achieved by adopting the multilayering schemes. Reported enhancements of $E_\perp$ have been as large as a factor of 5. The popular quadrilayer structure depicted in Fig. 15 consists of a thin film of the MO material, sandwiched between two transparent dielectric layers, and capped off with a reflecting metallic layer. The entire structure, which is grown on a transparent substrate (through which light must travel to reach the MO film), is protected by a lacquer layer on the top. Numbers shown in Fig. 15 for the various layer thicknesses are representative of currently designed quadrilayers.

The advantage of sending the light through the substrate is that the front facet of the disk stays out of focus during operation. In this way, small dust particles, finger prints, and scratches will not block the passage of light, and their deteriorating effects on the quality of the focused spot (which affects the integrity of both writing and readout) will be

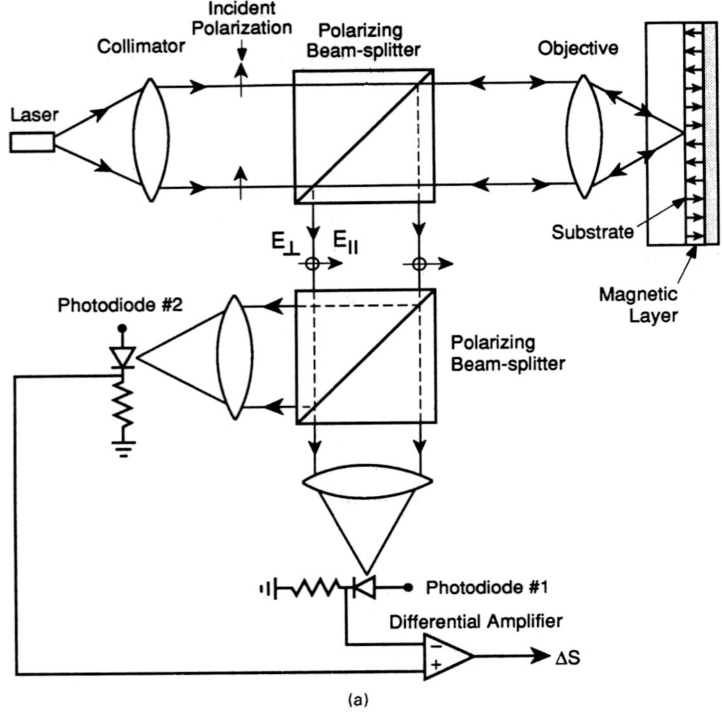

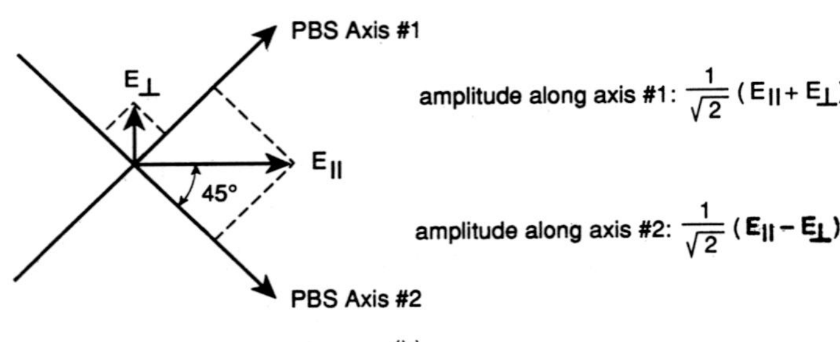

FIGURE 17 Differential detection scheme utilizes a polarizing beam splitter and two photodetectors in order to convert the rotation of polarization to an electronic signal. $E_{\parallel}$ and $E_{\perp}$ are the reflected components of polarization; they are, respectively, parallel and perpendicular to the direction of incident polarization. The diagram in (*b*) shows the orientation of the PBS axes relative to the polarization vectors.

minimized. Any optical storage medium designed for removability ought to have the kind of protection that illumination through the substrate provides. The note of caution with substrate-side illumination is that, if the objective is simply designed for focusing in the air, then the oblique rays will bend upon entering the substrate and deviate from nominal focus, causing severe aberrations (see Fig. 7*c*). Therefore, the substrate thickness and refractive index must be taken into account in the objective's design.

Sources of Noise in Readout

The read signal is always accompanied by random noise. The effective noise amplitude (relative to the strength of the signal) ultimately limits the performance of any readout system. Part of the noise is thermal in nature, arising from random fluctuations of charge carriers within the photodiodes, resistors, and amplifiers. In principle, this source of noise can be mitigated by reducing the operating temperature of the device. However, since operating below the normal room temperature is not very practical for data storage systems, one must accept some of the limitations brought about by the thermal noise.

Another source of readout noise is shot noise which, in classical language, is due to random arrival of photons at the photodetector(s). This noise is a permanent companion of the read signal and cannot be eliminated, but the system parameters may be adjusted to minimize its effect. One property of the shot noise is that its rms amplitude is proportional to the square root of the available optical power P_o. Since the signal strength is directly proportional to P_o, it is clear that by increasing the read power of the laser one can enhance the ratio of signal-to-shot noise. There is, however, an upper limit on the laser read power, since the rise in the temperature of the medium will force the decline of its magneto-optical response.

Other sources of noise in magneto-optical readout include the laser noise, the media noise, and the data noise. Laser noise is caused by amplitude/phase fluctuations of the electromagnetic radiation that comprises the optical beam. Media noise arises from variations in the reflectivity/magneto-optic activity of the medium across its surface. The presence of grooves with rough and nonuniform edges can be a source of media noise as well. The term *data noise* refers to the unpredictable variations of the read signal arising from the imperfect shape/position of the recorded marks.

Figure 18 shows the various components of noise in a typical MO readout system, as detected by a spectrum analyzer. In (*a*) the light beam is blocked and the trace on the analyzer screen is solely due to the thermal noise. The trace in (*b*) where the beam reaches the detectors but the disk is stationary shows the combined effect of thermal, shot, and laser noise. Trace (*c*) corresponds to reading an erased track on a spinning disk; the noise here includes all of the above plus the media noise. When a single-frequency tone was recorded on the track and the read-back signal was fed to the spectrum analyzer, trace (*d*)

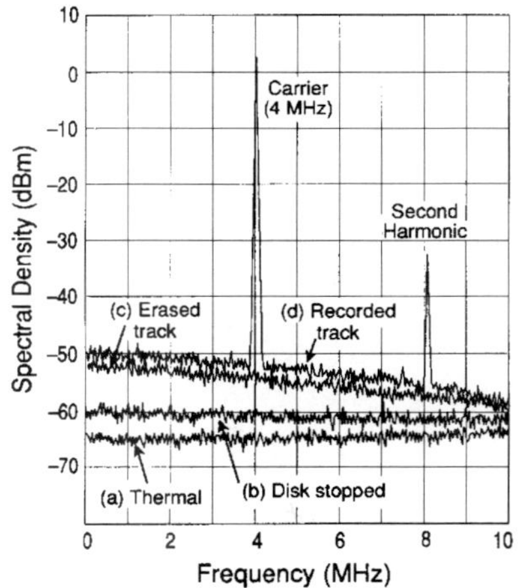

FIGURE 18 Spectra of the various noise components in magneto-optical readout.

was obtained. The narrow pulse at frequency f_0 is the first harmonic of the recorded signal; the corresponding second harmonic appears at $2f_0$. The noise level in this case is somewhat greater than that from the same track before the data was recorded. This difference is due to "data noise" and arises from jitter and nonuniformity of the recorded marks.

A commonly used measure of performance for optical recording media is the carrier-to-noise ratio or CNR. This is the ratio of the signal amplitude at the carrier frequency f_0 to the average level of noise. On a logarithmic scale the ratio is simply the difference between the two levels; in Fig. 18 the CNR is 53 decibels (dB).

31.8 MATERIALS OF MAGNETO-OPTICAL RECORDING

Amorphous rare earth transition metal alloys are presently the media of choice for erasable optical data storage applications. The general formula for the composition of the alloy may be written $(Tb_yGd_{1-y})_x(Fe_zCo_{1-z})_{1-x}$ where terbium and gadolinium are the rare earth (RE) elements, while iron and cobalt are the transition metals (TM). In practice, the transition metals constitute roughly 80 atomic percent of the alloy (i.e., $x \simeq 0.2$). In the transition metal subnetwork the fraction of cobalt is usually small, typically around 10 percent, and iron is the dominant element ($z \simeq 0.9$). Similarly, in the rare earth subnetwork Tb is the main element ($y \simeq 0.9$) while the Gd content is small or it may even be absent in some cases. Since the rare earth elements are highly reactive, RE-TM films tend to have poor corrosion resistance and, therefore, require protective coatings. In a disk structure such as that shown in Fig. 15, the dielectric layers that enable optimization of the medium for the best optical/thermal behavior also perform the crucial task of protecting the MO layer from the environment.

The amorphous nature of the material allows its composition to be continuously varied until a number of desirable properties are achieved. (In other words, the fractions x, y, z of the various elements are not constrained by the rules of stoichiometry.) Disks with large surface areas are coated uniformly with thin films of these media, and, in contrast to polycrystalline films whose grains and grain boundaries scatter the light beam and cause noise, these amorphous films are smooth and substantially noise-free. The films are deposited either by sputtering from an alloy target, or by cosputtering from multiple elemental targets. In the latter case, the substrate moves under the various targets and the fraction of a given element in the alloy film is determined by the time spent under the target as well as the power applied to that target. Substrates are usually kept at a low temperature (by water cooling, for instance) in order to reduce the mobility of deposited atoms and to inhibit crystal growth. Factors that affect the composition and short-range order of the deposited films include the type of the sputtering gas (argon, krypton, xenon, etc.) and its pressure during sputtering, the bias voltage applied to the substrate, deposition rate, nature of the substrate and its pretreatment, temperature of the substrate, etc.

Ferrimagnetism

The RE-TM alloys of interest in MO recording are ferrimagnetic, in the sense that the magnetization of the TM subnetwork is antiparallel to that of the RE subnetwork. The net magnetic moment exhibited by the material is the vector sum of the two subnetwork magnetizations. Figure 19 shows a typical temperature dependence of RE and TM magnetic moments, as well as the net saturation moment of the material. The exchange coupling between the two magnetic subnetworks is strong enough to give them the same critical temperature T_c. At $T = 0$ K the rare earth moment is stronger than that of the transition metal, giving the material a net moment along the direction of the RE magnetization. As the temperature rises, thermal disorder competes with interatomic exchange forces that tend to align the individual atomic dipole moments. The decrease of M_{RE} with the increasing temperature is faster than that of M_{TM}, and the net moment M_s begins to approach zero. At the compensation point temperature T_{comp}, the net moment vanishes. Between T_{comp} and T_c the net moment is dominated by the TM subnetwork and the material is said to exhibit TM-rich behavior (as opposed to when $T < T_{comp}$, where it exhibits RE-rich behavior). At the Curie temperature, thermal agitations finally break the hold of the exchange forces on magnetic dipoles, and the magnetic order disappears. Beyond T_c the material is in the paramagnetic state.

The composition of the materials of magneto-optical storage is chosen so that T_{comp} appears near the ambient temperature of $T_a \simeq 300$ K. Thus, under normal conditions, the net magnetization of the material is close to zero. Figure 20 shows a schematic drawing of the magnetization pattern in the cross section of a recorded track. Note that, although the net magnetization is nearly zero everywhere, the subnetwork moments have opposite orientations in adjacent domains. During readout the light from the GaAs laser interacts mainly with the transition metal subnetwork; thus, the MO Kerr signal is strong even though the net magnetization of the storage layer may be small. The magnetic electrons of iron and cobalt are in the 3*d* electronic shell, which forms the outer layer of the ion once the 4*s* electrons have escaped into the sea of conduction electrons. The magnetic electrons of Tb and Gd, in contrast, are within the 4*f* shell, concealed by the 5*s*, 5*p*, and 5*d* shells, even after the 6*s* electrons escape to the conduction band. A red or near-infrared photon is not energetic enough to penetrate the outer shell and interact with the magnetic 4*f* electrons, but it readily interacts with the exposed 3*d* electrons that constitute the magnetic moment of the TM subnetwork. It is for this reason that the MO Kerr signal in the visible and in the infrared is a probe of the state of magnetization of the TM subnetwork.

Perpendicular Magnetic Anisotropy

An important property of amorphous RE-TM alloy films is that they possess perpendicular magnetic anisotropy. The magnetization in these films favors perpendicular orientation even though there is no discernible crystallinity or microstructure that might obviously be responsible for this behavior. It is generally believed that atomic short-range order, established in the deposition process and aided by the symmetry-breaking at the surface of the film, gives preference to perpendicular orientation. Unequivocal proof of this assertion, however, is not presently available due to a lack of high-resolution observation instruments.

The perpendicular magnetization of MO media is in sharp contrast to the in-plane orientation of the magnetization vector in ordinary magnetic recording. In magnetic recording, the neighboring domains are magnetized in head-to-head fashion, which is an energetically unfavorable situation, since the domain walls are charged and highly unstable. The boundary between neighboring domains in fact breaks down into zigzags, vortices, and all manner of jagged, uneven structure in an attempt to reduce the magnetostatic energy. In contrast, adjacent domains in MO media are highly stable, since the pattern of magnetization causes flux closure, which reduces the magnetostatic energy.

Coercivity and the Hysteresis Loop

Typical hysteresis loops of an amorphous RE-TM thin film at various temperatures are shown in Fig. 21*a*. These loops, obtained with a vibrating sample magnetometer (VSM),

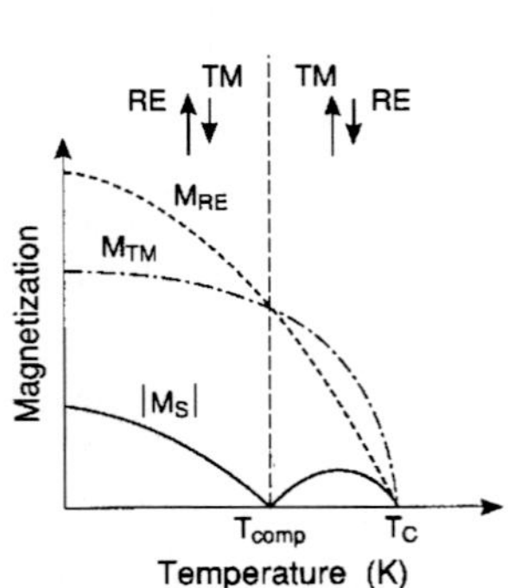

FIGURE 19 Temperature dependence of magnetization in amorphous RE-TM films. The moments of RE and TM subnetworks decrease monotonically, until they both vanish at the critical (Curie) temperature T_c. The net magnetization is the difference between the two subnetwork moments, and goes through zero at the compensation point T_{comp}.

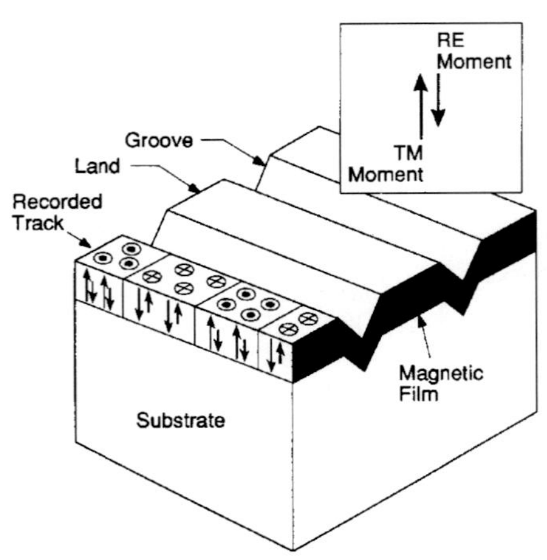

FIGURE 20 Schematic diagram showing the pattern of magnetization along a recorded track. The rare earth and the transition metal moments couple antiferromagnetically, so that the net magnetization everywhere is small. However, since the read beam interacts mainly with the TM subnetwork, the readout signal is not necessarily small.

show several characteristics of the MO media. (The VSM applies a slowly varying magnetic field to the sample and measures its net magnetic moment as a function of the field.) The horizontal axis in Fig. 21*a* is the applied field, which varies from -12 to $+12$ kOe, while the vertical axis is the measured magnetic moment per unit volume (in CGS units of emu/cm^3). The high degree of squareness of the loops signifies the following:

1. The remanent magnetization M_r is the same as the saturation magnetization M_s. Thus, once the sample is saturated with the help of an applied field, removing that field does not cause a reduction of the magnetic moment.
2. Transitions of the magnetization from up to down (or from down to up) are very sharp. The reverse field does not affect the magnetization until the critical value of H_c, the coercive field, is reached. At the coercive field the magnetization suddenly reverses direction, and saturation in the opposite direction is almost immediate.

The fact that M_r is very nearly equal to M_s in MO media is significant, since it means that the recorded domains remain fully saturated and exhibit maximum signal during readout. The coercivity H_c, in addition to being responsible for the stability of recorded domains, plays an important role in the processes of thermomagnetic recording and erasure. The coercivity at room temperature, being of the order of several thousand oersteds, prevents fields weaker than H_c from destroying (or disturbing) any recorded data. With the increasing temperature, the coercivity decreases and drops to zero at the Curie point, T_c. Figure 21*b* is the plot of H_c versus T for the same sample as in (*a*). Note that at the compensation point the coercivity goes to infinity, simply because the magnetization vanishes, and the external field does not see any magnetic moments to interact with. Above T_{comp} the coercive field decreases monotonically, which explains the process of magnetization reversal during thermomagnetic recording: **M** switches sign once the coercivity drops below the level of the applied field.

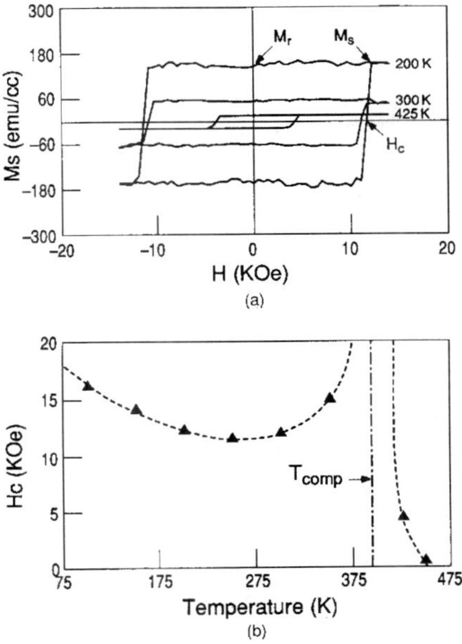

FIGURE 21 (*a*) Hysteresis loops of an amorphous $Tb_{27}(FeCo)_{73}$ film, measured by VSM at three different temperatures. The saturation moment M_s, the remanent moment M_r, and the coercive field H_c are identified for the loop measured at $T = 200$ K. (*b*) Coercivity as function of temperature for the above sample. At the compensation temperature, $T_{comp} = 400$ K, the coercivity is infinite; it drops to zero at the Curie point $T_c = 450$ K.

31.9 CONCLUDING REMARKS

In this article we have reviewed the basic characteristics of optical disk data storage systems, with emphasis on magneto-optical recording. The goal has been to convey the important concepts without getting distracted by secondary issues and less significant details. As a result, we have glossed over several interesting developments that have played a role in the technological evolution of optical data storage. In this final section some of these developments are briefly described.

Multiple-Track Read-Write with Diode Laser Arrays

It is possible in an optical disk system to use an array of lasers instead of just one, focus all the lasers simultaneously through the same objective lens, and perform parallel read/write/erase operations on multiple tracks. Since the individual lasers of an array can be modulated independently, the parallel channels thus obtained are totally independent of each other. In going from a single-channel drive to a multiple-channel one, the optics of the system (i.e., lenses, beam splitters, polarization-sensitive elements, focus and track servos, etc.) remain essentially the same; only the lasers and detectors proliferate in number. Parallel track operations boost the sustainable data rates in proportion to the number of channels used.

Diffractive Optics

The use of holographic optical elements (HOEs) to replace individual refractive optics is a promising new development. Diffractive optical elements are relatively easy to manufacture, they are lightweight and inexpensive, and can combine the functions of several elements on a single plate. These devices are therefore expected to help reduce the cost, size, and weight of optical heads, making optical drives more competitive in terms of price and performance.

An example of the application of HOEs in MO systems is shown in Fig. 22, which shows a reflection-type element consisting of four separate holograms. The light incident on the HOE at an angle of 60° has a *p* component which is the original polarization of the laser beam, and an *s* component (parallel to the hologram's grooves) which is the magneto-optically generated polarization at the disk. Nearly 90 percent of the *s* and 70 percent of the *p* polarization in this case are reflected from the HOE without suffering any diffraction (i.e., in the zero-order beam); they are captured in the differential detection module and yield the MO read signal. The four holograms deflect 20 percent of the incident *p*-polarized light in the form of first-order diffracted beams and bring them to

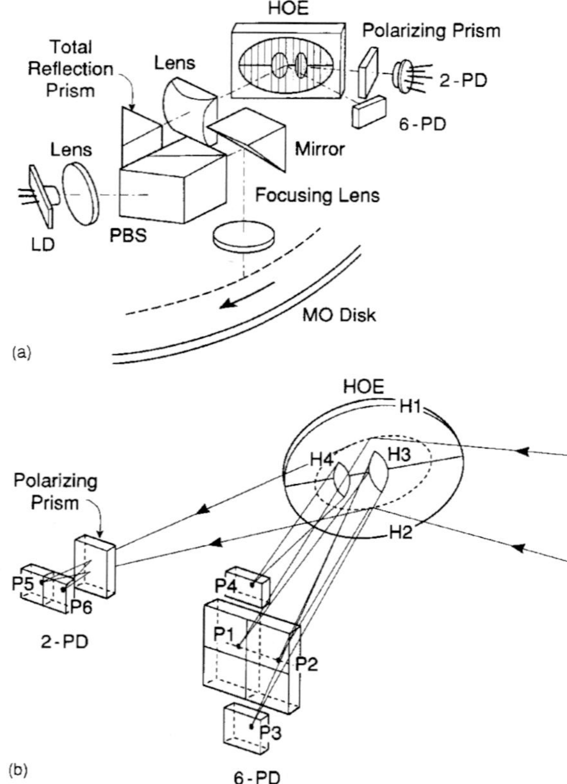

FIGURE 22 Application of holographic optical element (HOE) in optical recording. (*a*) Configuration of MO head using a polarization-sensitive reflection HOE. (*b*) Geometrical relation between holograms and detectors. *(After A. Ohba et al., SPIE Proceedings, Vol. 1078, 1989.)*

focus at four different spots on a multielement detector. The two small holograms in the middle, H_3 and H_4, focus their first-order beams on detectors P_3 and P_4, to generate the push-pull tracking-error signal. The other two holograms, H_1 and H_2, send the diffracted beams to a four-element detector in order to generate a focus-error signal based on the double knife-edge scheme. This HOE, therefore, combines the functions of beam splitting, masking, and focusing all in one compact unit.

Alternative Storage Media

The GaAlAs lasers of the present optical disk technology will likely be replaced in the future by light sources that emit in the blue end of the spectrum. Shorter wavelengths allow smaller marks to be recorded, and also enable the resolving of those marks in readout. Aside from the imposition of tighter tolerances on focusing and tracking servos, operation in the blue will require storage materials that are sensitive to the short wavelengths. The current favorites for erasable optical recording media, the amorphous RE-TM alloys, may not be suitable for readout in the blue, since their magneto-optic Kerr signal drops at short wavelengths. A new class of magnetic materials which holds promise for future-generation device applications is the class of TM/TM superlattice-type media. The best-known material in this class is the Co/Pt-layered structure which consists of very thin layers of cobalt (typically one or two atomic layers), separated by several atomic layers of platinum. These polycrystalline films which have very small crystallites (grain diameter $\simeq 200$ Å) are prepared either by electron beam evaporation or by sputtering. Co/Pt films have large perpendicular magnetic anisotropy, good signal-to-noise ratios in the blue, sufficient sensitivity for write/erase operations, and are environmentally more stable than the RE-TM media.

Direct Overwrite in Magneto-Optical Systems

The problem of direct overwrite (DOW) on MO media has been the subject of extensive research in recent years. Some of the most promising solutions have been based on exchange-coupled magnetic multilayered structures. The basic idea of recording on exchange-coupled bilayers (or trilayers) is simple and involves the writing of reverse domains that do not extend through the entire film thickness, such as those shown schematically in Fig. 23. Such domains are under pressure from their excessive wall energies to collapse and can readily be erased with a moderate-power laser pulse. DOW on exchange-coupled media is thus achieved by writing (i.e., creating reverse domains) with a high-power pulse, and erasing (i.e., eliminating domains) with a moderate-power pulse. An external magnetic field is usually required for writing on such media, but neither the direction nor the magnitude of this field needs to change during erasure.

Optical recording is an evolving technology, which will undoubtedly see many

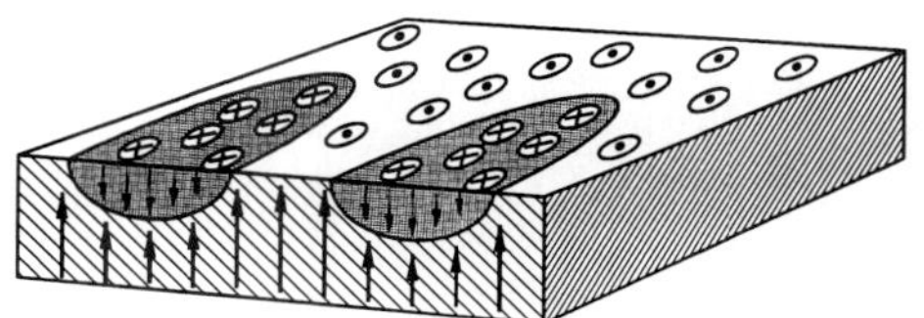

FIGURE 23 Direct overwrite in exchange-coupled magnetic multilayers involves the formation of domains that do not extend through the entire thickness of the magnetic medium.

innovations and improvements in the coming years. Some of the ideas and concepts described here will hopefully remain useful for some time to come, while others may have a shorter lifetime and limited usefulness. It is the author's hope, however, that they all serve as a stepping-stone to profound new ideas.

31.10 FURTHER INFORMATION

Proceedings of the *Optical Data Storage Conference* are published annually by SPIE, the International Society for Optical Engineering. These proceedings document each year the latest developments in the field of optical recording.

Two other conferences in this field are the *International Symposium on Optical Memory* (ISOM) whose proceedings are published as a special issue of the *Japanese Journal of Applied Physics,* and the *Magneto-Optical Recording International Symposium* (MORIS) whose proceedings appear in a special issue of the *Journal of the Magnetics Society of Japan.*

31.11 REFERENCES

1. A. B. Marchant, *Optical Recording,* Addison-Wesley, Boston, Massachusetts, 1990.
2. P. Hansen and H. Heitmann, "Media for Erasable Magneto-Optic Recording," *IEEE Trans. Mag.* **25:**4390–4404 (1989).
3. M. H. Kryder, "Data-Storage Technologies for Advanced Computing," *Scientific American* **257:**116–125 (1987).
4. M. H. Kryder, "Data Storage in 2000—Trends in Data Storage Technologies," *IEEE Trans. Mag.* **25:**4358–4363 (1989).
5. A. E. Bell, "Optical Data Storage Technology: Status and Prospects," *Computer Design,* January 1983, pp. 133–146.
6. D. S. Bloomberg and G. A. N. Connell, "Magneto-optical Recording," *Magnetic Recording,* vol. II, C. D. Mee and E. D. Daniel (eds.), McGraw-Hill, New York, 1988, chap. 6.
7. S. Miyaoka, "Digital Audio is Compact and Rugged," *IEEE Spectrum,* March 1984, pp. 35–39.
8. G. Bouwhuis and J. J. M. Braat, "Recording and Reading of Information on Optical Disks," *Applied Optics and Applied Engineering,* vol. IX, R. R. Shannon and J. C. Wyant (eds.), Academic Press, New York, 1983, chap. 3.
9. G. Bouwhuis, J. Braat, A. Huijser, J. Pasman, G. Van Rosmalen, and K. S. Immink, *Principles of Optical Disk Systems,* Adam Hilger Ltd., Bristol and Boston, 1985, chaps. 2, 3.
10. H. H. Hopkins, "Diffraction Theory of Laser Read-Out Systems for Optical Video Discs," *J. Opt. Soc. Am.* **69:**4–24 (1979).
11. Special issue of *Applied Optics* on video disks, July 1, 1978.
12. M. Mansuripur, "Certain Computational Aspects of Vector Diffraction Problems," *J. Opt. Soc. Am.* **A6:**786–805 (1989).
13. P. Sheng, "Theoretical Considerations of Optical Diffraction from RCA Video Disk Signals," *RCA Review* **39:**513–555 (1978).
14. D. O. Smith, "Magneto-Optical Scattering from Multilayer Magnetic and Dielectric Films," *Opt. Acta* **12:**13 (1985).
15. P. S. Pershan, "Magneto-Optic Effects," *J. Appl. Phys.* **38:**1482–1490 (1967).
16. K. Balasubramanian, A. S. Marathay, and H. A. Macleod, "Modeling Magneto-Optical Thin Film Media for Optical Data Storage," *Thin Solid Films* **164:**341–403 (1988).

17. Y. Tomita and T. Yoshini, "Optimum Design of Multilayer-Medium Structures in a Magneto-Optical Readout System," *J. Opt. Soc. Am.* **A1:**809–817 (1984).
18. K. Egashira and T. Yamada, "Kerr Effect Enhancement and Improvement of Readout Characteristics in MnBi Film Memory," *J. Appl. Phys.* **45:**3643–3648 (1974).
19. G. A. N. Connell, "Interference Enhanced Kerr Spectroscopy for Very Thin Absorbing Films," *Appl. Phys. Lett.* **40:**212 (1982).
20. K. Y. Ahn and G. J. Fan, "Kerr Effect Enhancement in Ferromagnetic Films," *IEEE Magnet.* **2:**678 (1966).
21. H. Wieder and R. A. Burn, "Direct Comparison of Thermal and Magnetic Profiles in Curie Point Writing on MnGaGe Films," *J. Appl. Phys.* **44:**1774 (1973).
22. P. Kivits, R. deBont, and P. Zalm, "Superheating of Thin Films for Optical Recording," *Appl. Phys.* **24:**273–278 (1981).
23. M. Mansuripur and G. A. N. Connell, "Laser-Induced Local Heating of Moving Multilayer Media," *Appl. Opt.* **22:**666 (1983).
24. J. P. J. Heemskerk, "Noise in a Video Disk System: Experiments with an (AlGa)As Laser," *Appl. Opt.* **17:**2007 (1978).
25. G. A. Acket, D. Lenstra, A. J. DenBoef, and B. H. Verbeek, "Influence of Feedback Intensity on Longitudinal Mode Properties and Optical Noise in Index-Guided Semiconductor Lasers," *IEEE J. Quant. Electron.* **QE-20:**1163 (1984).
26. A. Arimoto, M. Ojima, N. Chinone, A. Oishi, T. Gotoh, and N. Ohnuki, "Optimum Conditions for the High Frequency Noise Reduction Method in Optical Video Disk Players," *Appl. Opt.* **25:**1398 (1986).
27. M. Ojima, A. Arimoto, N. Chinone, T. Gotoh, and K. Aiki, "Diode Laser Noise at Video Frequencies in Optical Video Disk Players," *Appl. Opt.* **25:**1404 (1986).
28. J. W. Beck, "Noise Consideration of Optical Beam Recording," *Appl. Opt.* **9:**2559 (1970).
29. D. Treves and D. S. Bloomberg, "Signal, Noise, and Codes in Optical Memories," *Opt. Eng.* **25:**881 (1986).
30. D. Treves, "Magneto-Optic Detection of High-Density Recordings," *J. Appl. Phys.* **38:**1192 (1967).
31. B. R. Brown, "Readout Performance Analysis of a Cryogenic Magneto-Optical Data Storage System," *IBM J. Res. Develop,* January 1972, pp. 19–26.
32. R. L. Aagard, "Signal to Noise Ratio for Magneto-Optical Readout from MnBi Films," *IEEE Trans. Mag.* **9:**705 (1973).
33. M. Mansuripur, G. A. N. Connell, and J. W. Goodman, "Signal and Noise in Magneto-Optical Readout," *J. Appl. Phys.* **53:**4485 (1982).
34. B. G. Huth, "Calculation of Stable Domain Radii Produced by Thermomagnetic Writing," *IBM J. Res. Dev.,* 1974, pp. 100–109.
35. D. H. Howe, "Signal-to-Noise Ratio for Reliable Data Recording," *Proc. SPIE* **695:**255–261 (1986).
36. K. A. S. Immink, "Coding Methods for High-Density Optical Recording," *Philips J. Res.* **41:**410–430 (1986).
37. R. Hasegawa (ed.), *Glassy Metals: Magnetic, Chemical and Structural Properties,* CRC Press, Boca Raton, Florida, 1983.

P · A · R · T · 9

OPTICAL DESIGN TECHNIQUES

CHAPTER 32
TECHNIQUES OF FIRST-ORDER LAYOUT

Warren J. Smith
Kaiser Electro-Optics, Inc.
Carlsbad, California

32.1 GLOSSARY

A, B	scaling constants
d	distance between components
f	focal length
h	image height
I	invariant
j, k	indices
l	axis intercept distance
M	angular magnification
m	linear, lateral magnification
n	refractive index
P	partial dispersion, projection lens diameter
r	radius
S	source or detector linear dimension
SS	secondary spectrum
s	object distance
s'	image distance
t	temperature
u	ray slope
V	Abbe number
y	height above optical axis
α	radiometer field of view; projector field of view
ϕ	component power $(=1/f)$

32.2 FIRST-ORDER LAYOUT

First-order layout is the determination of the arrangement of the components of an optical system in order to satisfy the first-order requirements imposed on the system. The term "first-order" means the paraxial image properties: the size of the image, its orientations, its location, and the illumination or brightness of the image. This also implies apertures, f-numbers, fields of view, physical size limitations, and the like. It does not ordinarily include considerations of aberration correction; these are usually third- and higher-order matters, not first-order. However, ordinary chromatic aberration and secondary spectrum are first-order aberrations. Additionally, the first-order layout can have an effect on the Petzval curvature of field, the cost of the optics, the sensitivity to misalignment, and the defocusing effects of temperature changes.

The primary task of first-order layout is to determine the powers and spacings of the system components so that the image is located in the right place and has the right size and orientation. It is not necessary to deal with surface-by-surface ray-tracing here; the concern is with components. "Components" may mean single elements, cemented doublets, or even complex assemblies of many elements. The first-order properties of a component can be described by its gauss points: the focal points and principal points. For layout purposes, however, the initial work can be done assuming that each component is of zero thickness; then only the component location and its power (or focal length) need be defined.

32.3 RAY-TRACING

The most general way to determine the characteristics of an image is by ray-tracing. As shown in Fig. 1, if an "axial (marginal)" ray is started at the foot (axial intercept) of the object, then an image is located at each place that this ray crosses the axis. The size of the image can be determined by tracing a second, "principal (chief)," ray from the top of the object and passing through the center of the limiting aperture of the system, the "aperture stop"; the intersection height of this ray at the image plane indicates the image size. This size can also be determined from the ratio of the ray slopes of the axial ray at the object and at the image; this yields the magnification $m = u_0/u'_k$; object height times magnification yields the image height.

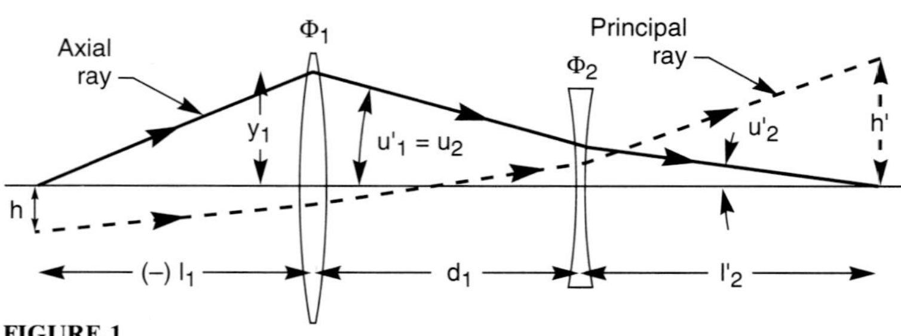

FIGURE 1

The ray-tracing equations are:

$$y_1 = -l_1 u_1 \tag{1}$$

$$u'_j = u_j - y_j \phi_j \tag{2}$$

$$y_{j+1} = y_j + d_j u'_j \tag{3}$$

$$l'_k = -y_k / u'_k \tag{4}$$

where l and l' are the axial intersection distances of the ray before and after refraction by the component, u and u' are the ray slopes before and after refraction, ϕ is the component power ($\phi = 1/f$), y_j is the height at which the ray strikes the jth component, and d_j is the distance from the jth to the $(j+1)$th component. Equations (2) and (3) are applied sequentially to the components, from object to image.

These equations can be used in two different ways. When the components and spacings are known, the image characteristics can readily be calculated. In the inverse application, the (unknown) powers and spaces can be represented by symbols, and the ray can be traced symbolically through the postulated number of components. The results of this symbolic ray-tracing can be equated to the required characteristics of the system; these equations can then be solved for the unknowns, which are the component powers and spacings.

As an example, given the starting ray data, y_1 and u_1, we get:

$$u'_1 = u_1 - y_1 \phi_1$$

$$y_2 = y_1 + d_1 u'_1 = y_1 + d_1(u_1 - y_1 \phi_1)$$

$$u'_2 = u'_1 - y_2 \phi_2$$

$$= u_1 - y_1 \phi_1 - [y_1 + d_1(u_1 - y_1 \phi_1)]\phi_2$$

$$y_3 = y_2 + d_2 u'_2 = \text{etc.}$$

Obviously the equations can become rather complex in very short order. However, because of the linear characteristics of the paraxial ray equations, they can be simplified by setting either y_1 or u_1 equal to one (1.0) without any loss of generality. But the algebra can still be daunting.

32.4 TWO-COMPONENT SYSTEMS

Many systems are either limited to two components or can be separated into two-component segments. There are relatively simple expressions for solving two-component systems.

Although the figures show thick lenses with appropriate principal planes "thin" lenses (whose thickness is zero and whose principal planes are coincident with the two coincident lens surfaces) may be used.

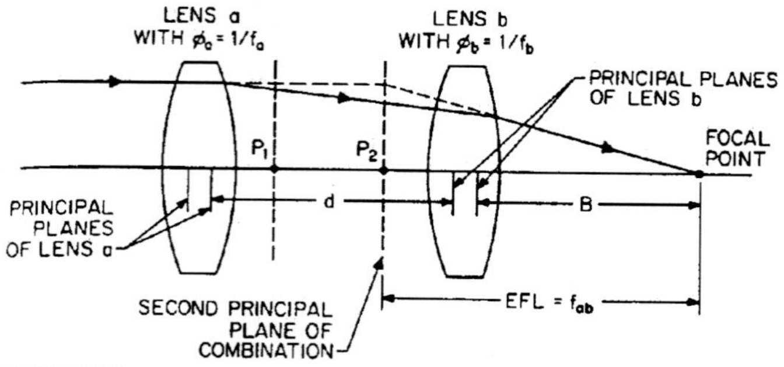

FIGURE 2

For systems with infinitely distant objects, as shown in Fig. 2, the following equations for the focal length and focus distance are useful:

$$f_{AB} = f_A f_B / (f_A + f_B - d) \tag{5}$$

$$\phi_{AB} = \phi_A + \phi_B - d\phi_A \phi_B \tag{6}$$

$$B = f_{AB}(f_A - d)/f_A \tag{7}$$

$$F = f_{AB}(f_B - d)/f_B \tag{8}$$

$$h' = f_{AB} \tan u_p \tag{9}$$

where f_{AB} is the focal length of the combination, ϕ_{AB} is its power, f_A and f_B are the focal lengths of the components, ϕ_A and ϕ_B are their powers, d is the space between the components, B is the "back focus" distance from the B component, F is the "front focus" distance, u_p is the angle subtended by the object, and h' is the image height.

If f_{AB}, d, and B (or F) are known, the component focal lengths can be found from:

$$f_A = d f_{AB} / (f_{AB} - B) \tag{10}$$

$$f_B = -dB / (f_{AB} - B - d) \tag{11}$$

These simple expressions are probably the most widely used equations in optical layout work.

If a two-component system operates at *finite* conjugates, as shown in Fig. 3, the following equations can be used to determine the layout. When the required system

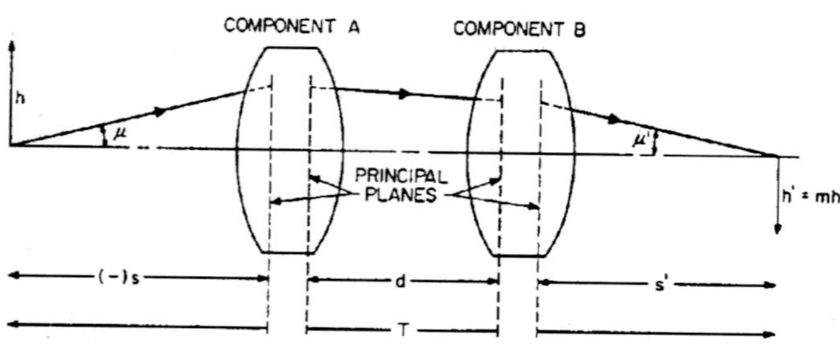

FIGURE 3

magnification and the component locations are known, the powers of the components are given by:

$$\phi_A = (ms - md - s')/msd \tag{12}$$

$$\phi_B = (d - ms + s')/ds' \tag{13}$$

where $m = h'/h$ is the magnification, s and s' are the object and image distances, and d is the spacing between components.

In different circumstances, the component powers, the object-to-image distance, and the magnification may be known and the component locations are to be determined. The following quadratic equation [Eq. (14)] in d (the spacing) is solved for d:

$$0 = d^2 - dT + T(f_A + f_B) + (m-1)^2 f_A f_B/m \tag{14}$$

and then

$$s = [(m-1)d + T]/[(m-1) - md\phi_A] \tag{15}$$

$$s' = T + s - d \tag{16}$$

32.5 AFOCAL SYSTEMS

If the system is afocal, then the following relations will apply:

$$MP = -(f_O/f_E) = (u_E/u_O) = (d_O/d_E) \tag{17}$$

and, if the components are "thin,"

$$L = f_O + f_E \tag{18}$$

$$f_O = -L \cdot MP/(1 - MP) \tag{19}$$

$$f_E = L/(1 - MP) \tag{20}$$

where MP is the angular magnification, f_O and f_E are the objective and eyepiece focal lengths, u_E and u_O are the apparent (image) and real (object) angular fields, d_O and d_E are the entrance and exit pupil diameters, and L is the length of the telescope as indicated in Fig. 4.

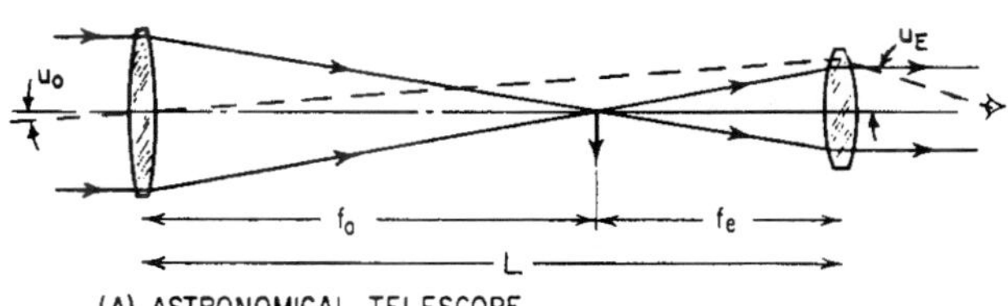

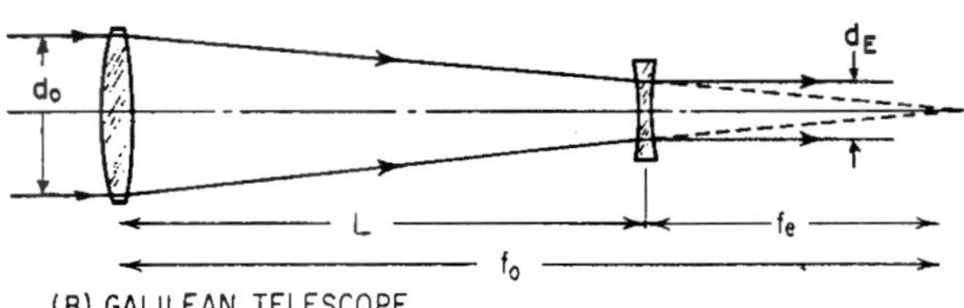

FIGURE 4

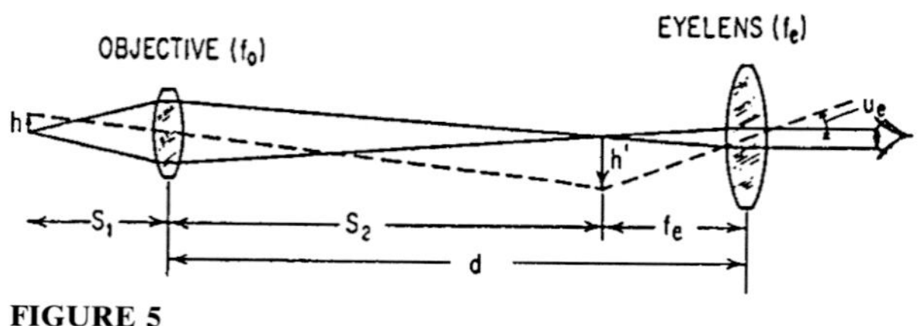

FIGURE 5

32.6 MAGNIFIERS AND MICROSCOPES

The conventional definition of magnifying power for either a magnifier or microscope compares the angular size of the image with the angular size of the object *when the object is viewed from a (conventional) distance of* 10 *inches*. Thus the magnification can be found from

$$MP = 10''/f \tag{21}$$

for either a simple microscope (i.e., magnifier) or a compound microscope, where f is the focal length of the system. Using the symbols of Fig. 5, we can also write the following for the compound microscope

$$MP = (f_E + f_O - d)10''/f_E f_O \tag{22}$$

$$MP = m_O \times m_E$$

$$= (S_2/S_1)(10''/f_E) \tag{23}$$

32.7 AFOCAL ATTACHMENTS

In addition to functioning as a telescope, beam expander, etc., an afocal system can be used to modify the characteristics of another system. It can change the focal length, power, or field of the "prime" system. Figure 6 shows several examples of an afocal device placed (in these examples) before an imaging system. The combination has a focal length equal to the focal length of the prime system multiplied by the angular magnification of the afocal device. Note that in Fig. 6*a* and *b* the same afocal attachment has been reversed to provide two different focal lengths. If the size of the film or detector is kept constant, the angular field is changed by a factor equal to the inverse of the afocal magnification.

32.8 FIELD LENSES

Figure 7 illustrates the function of the field lens in a telescope. It is placed near (but rarely exactly at) an internal image; its power is chosen so that it converges the oblique ray bundle toward the axis sufficiently so that the rays pass through the subsequent component. A field lens is useful to keep the component diameters at reasonable sizes. It acts to relay the pupil image to a more acceptable location.

The required field lens power is easily determined. In Fig. 7 the most troublesome ray is that from the bottom of the objective aperture; its slope (u) is simply the height that it climbs divided by the distance that it travels. The required slope (u') for the ray after refraction by the field lens is defined by the image height (y), the "eyelens" semidiameter, and the spacing between them. Then Eq. (2) can be solved for the field lens power,

$$\phi = (u - u')/y \tag{24}$$

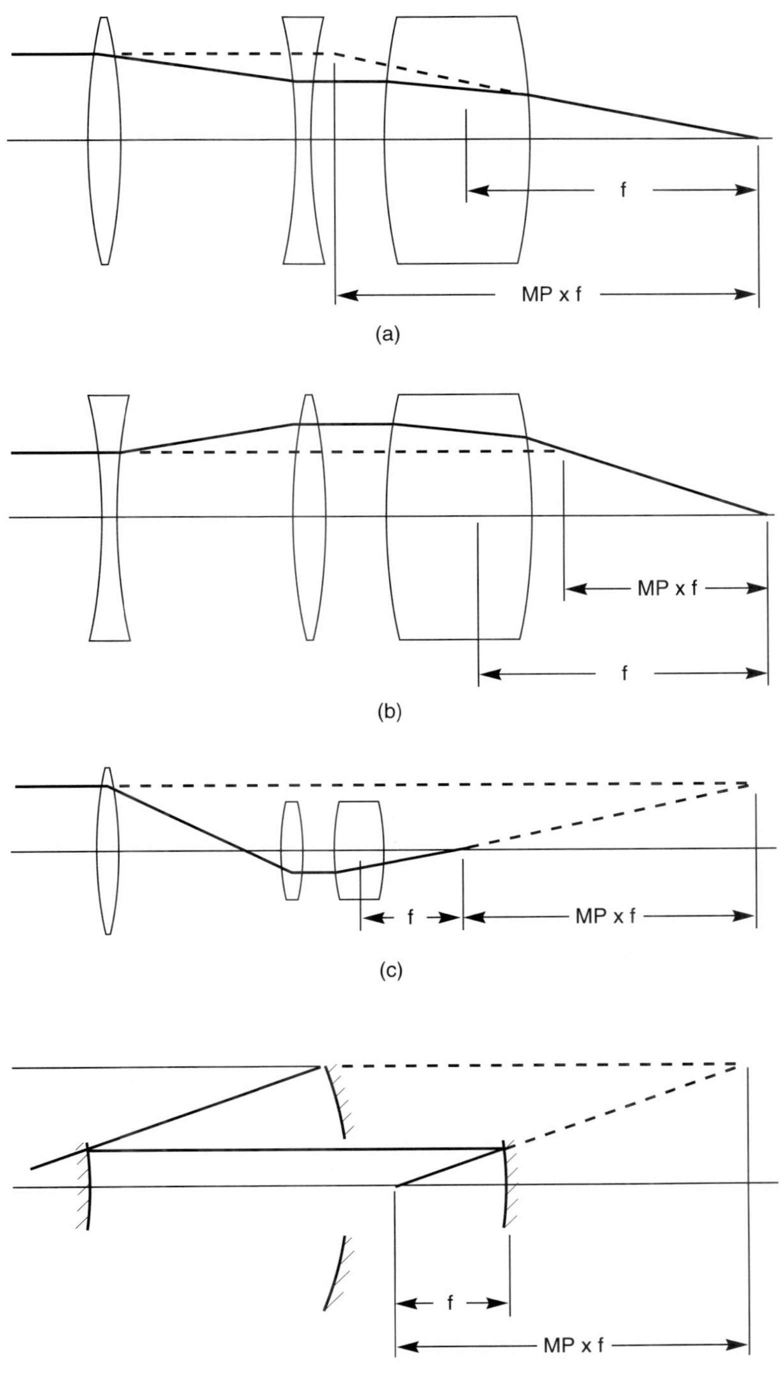

FIGURE 6

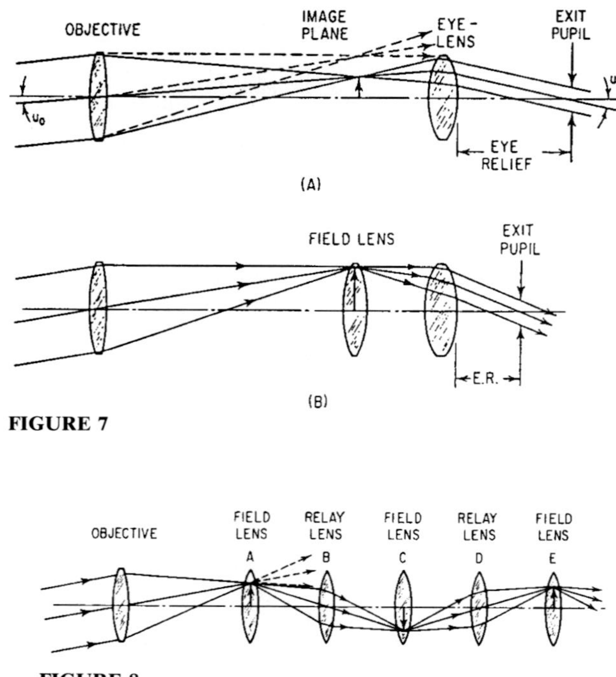

FIGURE 7

FIGURE 8

A periscope is used to carry an image through a long, small-diameter space. As shown in Fig. 8, the elements of a periscope are alternating field lenses and relay lenses. An optimum arrangement occurs when the images at the field lenses and the apertures of the relay lenses are as large as the available space allows. This arrangement has the fewest number of relay stages and the lowest power components. For a space of uniform diameter, both the field lenses and the relay lenses operate at unit magnification.

32.9 CONDENSERS

The projection/illumination condenser and the field lens of a radiation measuring system operate in exactly the same way. The condenser (Fig. 9) forms an image of the light source

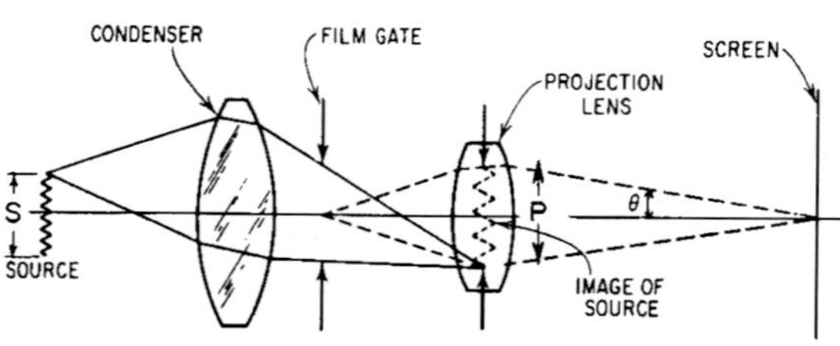

FIGURE 9

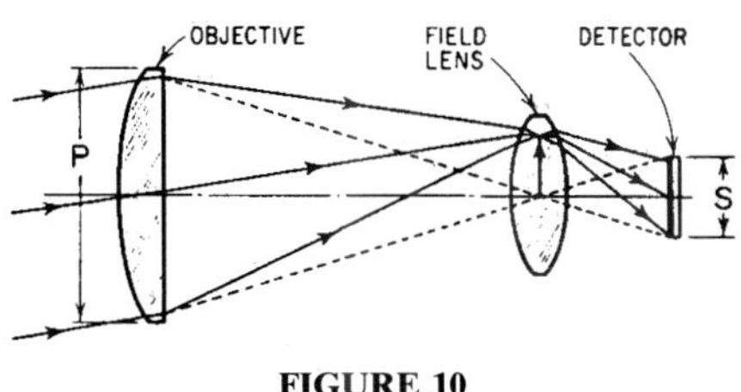

FIGURE 10

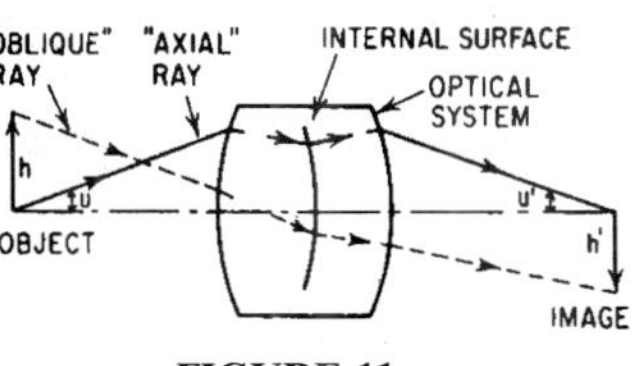

FIGURE 11

in the aperture of the projection lens, thereby producing even illumination from a nonuniform source. If the source image fills the projection lens aperture, this will produce the maximum illumination that the source brightness and the projection lens aperture diameter will allow. This is often called Köhler illumination. In a radiometer type of application (Fig. 10), the field lens images the objective lens aperture on the detector, uniformly illuminating its surface and permitting the use of a smaller detector. Often, the smallest possible source or detector is desired in order to minimize power or maximize signal-to-noise. The smallest possible size is given by

$$S = P\alpha/2n \tag{25}$$

where S is the source or detector size, P is the projection lens or objective aperture diameter, α is the field angle of projection or the radiometer field of view, and n is the index in which the source or detector is immersed. This value for S corresponds to an (impractical) system speed of F/0.5. A source or detector size twice as large is a far more realistic limit, corresponding to a speed of F/1.0.

The *invariant*, $I = n(y_2u_1 - y_1u_2)$, where y_1, u_1, y_2, and u_2 are the ray heights and slopes of two different rays, is an expression which has the same value everywhere in an optical system. If the two rays used are an axial ray and a principal (or chief) ray as shown in Fig. 11, and if the invariant is evaluated at the object and image surfaces, the result is

$$hnu = h'n'u' \tag{26}$$

32.10 ZOOM OR VARIFOCAL SYSTEMS

If the spacing between two components is changed, the effective focal length and the back focus are changed in accord with Eqs. (5) through (9). If the motions of the two components are arranged so that the image location is constant, this is a mechanically compensated zoom lens, so called because the component motions are usually effected with a mechanical cam. A zoom system may consist of just the two basic components or it may include one or more additional members. Usually the two basic components have opposite-signed powers.

If a component is working at unit magnification, it can be moved in one direction or the other to increase or decrease the magnification. There are pairs of positions where the magnifications are m and $1/m$ and for which the object-to-image distance is the same. This is the basis of what is called a "bang-bang" zoom; this is a simple way to provide two different focal lengths (or powers, or fields of view, or magnifications) for a system.

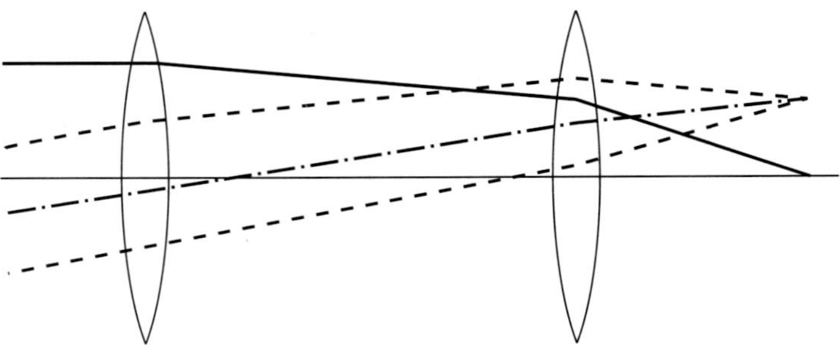

FIGURE 12

32.11 ADDITIONAL RAYS

When the system layout has been determined, an "axial" ray at full aperture and a "principal" ray at full field can be traced through the system. Because of the linearity of the paraxial equations, we can determine the ray-trace data (i.e., y and u) of *any* third ray from the data of these two traced rays by

$$y_3 = Ay_1 + By_2 \tag{27}$$

$$u_3 = Au_1 + Bu_2 \tag{28}$$

where A and B are scaling constants which can be determined from

$$A = (y_3u_1 - u_3y_1)/(u_1y_2 - y_1u_2) \tag{29}$$

$$B = (u_3y_2 - y_3u_2)/(u_1y_2 - y_1u_2) \tag{30}$$

where y_1, u_1, y_2, and u_2 are the ray heights and slopes of the axial and principal rays and y_3 and u_3 are the data of the third ray; these data are determined at any component of the system where the specifications for all three rays are known. These equations can, for example, be used to determine the necessary component diameters to pass a bundle of rays which are A times the diameter of the axial bundle at a field angle B times the full-field angle. In Fig. 12, for the dashed rays $A = +0.5$ and -0.5 and $B = 1.0$. Another application of Eqs. (27) through (30) is to locate either a pupil or an aperture stop when the other is known.

32.12 MINIMIZING COMPONENT POWER

The first-order layout may in fact determine the ultimate quality, cost, and manufacturability of the system. The residual aberrations in a system are a function of the component powers, relative apertures, and angular fields. The relationships are complex,

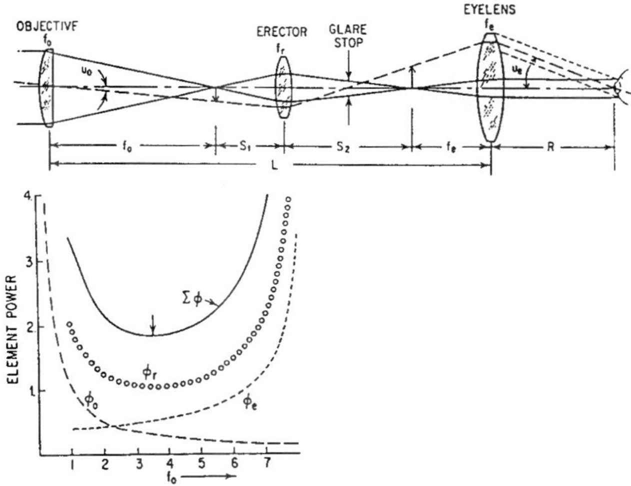

FIGURE 13

but a good choice for a system layout is one which minimizes the sum of the (absolute) component powers, or possibly the sum of the (absolute) $y\phi$ product for all the components.

For example, in Fig. 13 the length, magnification, and the eye relief of the rifle scope are specified. There are five variables: three component powers and two spaces. This is one more variable than is necessary to achieve the specified characteristics. If we take the focal length of the objective component as the free variable, the component powers which satisfy the specifications can be plotted against the objective focal length, as in Fig. 13, and the minimum power arrangement is easily determined.

Minimizing the component powers will strongly tend to minimize the aberrations and also the sensitivity of the system to fabrication errors and misalignments. The *cost* of an optical element will vary with its diameter (or perhaps the square of the diameter) and also with the product of the diameter and the power. Thus, while first-order layout deals only with components, these relationships still apply reasonably well even when applied to components rather than elements. Minimizing the component powers does tend to reduce the cost on these grounds (and also because it tends to reduce the complexity of the components).

32.13 IS IT A REASONABLE LAYOUT?

A simple way to get a feel for the reasonableness of a layout is to make a rough scale drawing showing each component as single element. An element can be drawn as an equiconvex lens with radii which are approximately $r = 2(n - 1)f$; for an element with an index of 1.5 the radii equal the focal length. The elements should be drawn to the diameter necessary to pass the (suitably vignetted) off-axis bundle of rays as well as the axial bundle. The on-axis and off-axis ray bundles should be sketched in. This will very quickly indicate which elements or components are the difficult ones. If the design is being started from scratch (as opposed to simply combining existing components), each component can be drawn as an achromat. The following section describes achromat layout, but for visual-spectrum systems it is often sufficient to assume that the positive (crown) element has twice the power of the achromat and the (negative) flint element has a power equal to that of the achromat. Thus an achromat may be sketched to the simplified, approximate prescription: $r_1 = -r_2 = f/2$ and $r_3 =$ plano.

Any elements which are too fat must then be divided or "split" until they look "reasonable." This yields a reasonable estimate of the required complexity of the system, even before the lens design process is begun.

If more or less standard design types are to be utilized for the components, it is useful to tabulate the focal lengths and diameters to get the (infinity) f-number of each component, and also its angular field coverage. The field coverage should be expressed both in terms of the angle that the object and image subtend from the component, and also the angle that the smaller of these two heights subtends as a function of the focal length (rather than as a function of that conjugate distance). This latter angle is useful because the coverage capability of a given design form is usually known in these terms, that is, h/f, rather than in finite conjugate terms. With this information at hand, a reasonable decision can be made as to the design type necessary to perform the function required of the component.

32.14 ACHROMATISM

The powers of the elements of an achromat can be determined from

$$\phi_A = \phi_{AB} V_A/(V_A - V_B) \tag{31}$$

$$\phi_B = \phi_{AB} V_B/(V_B - V_A) \tag{32}$$

$$= \phi_{AB} - \phi_A$$

where ϕ_{AB} is the power of the achromatic doublet and V_A is the Abbe V-value for the element whose power is ϕ_A, etc. For the visible spectral region $V = (n_d - 1)/(n_F - n_C)$; this can be extended to any spectral region by substituting the indices at middle, short, and long wavelengths for n_d, n_F, and n_C.

If the elements are to be spaced apart, and the back focus is B, then the powers and the spacing are given by

$$\phi_A = \phi_{AB} B V_A/(V_A B - V_B/\phi_{AB}) \tag{33}$$

$$\phi_B = -\phi_{AB} V_B/B(V_A B - V_B/\phi_{AB}) \tag{34}$$

$$D = (1 - B\phi_{AB})/\phi_A \tag{35}$$

For a complete system, the transverse *axial chromatic* aberration is the sum of $y^2\phi/Vu'_k$ for all the elements, where y is the height of the axial ray at the element and u'_k is the ray slope at the image. The *lateral color* is the sum of $yy_p\phi/Vu'_k$, where y_p is the principal ray height.

The *secondary spectrum* is the sum of $y^2\phi P/Vu'_k$, where P is the partial dispersion, $P = (n_d - n_c)/(n_F - n_c)$. Summed over two elements, this leads to an expression for the longitudinal secondary spectrum of an achromatic doublet

$$SS = f(P_B - P_A)/(V_A - V_B)$$

$$= -f(\Delta P/\Delta V) \tag{36}$$

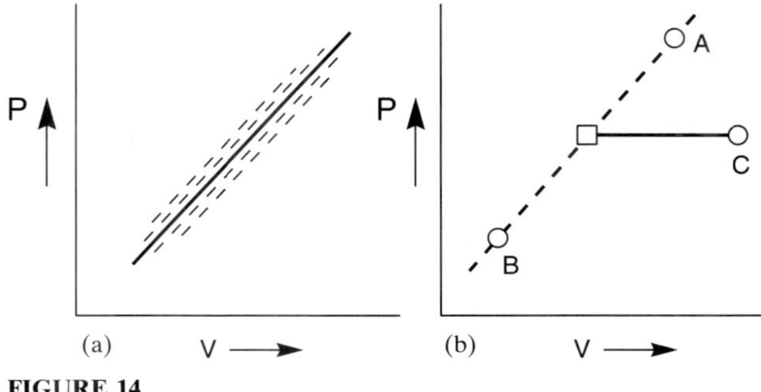

FIGURE 14

This indicates that in order to eliminate secondary spectrum for a doublet, two glasses with identical partial dispersions [so that $(P_A - P_B)$ is zero] are required. A large difference in V-value is desired so that $(V_A - V_B)$ in the denominator of Eqs. (31) and (32) will produce reasonably low element powers. As indicated in the schematic and simplified plot of P versus V in Fig. 14*a*, most glasses fall into a nearly linear array, and $(\Delta P/\Delta V)$ is nearly a constant for the vast majority of glasses. The few glasses which are away from the "normal" line can be used for apochromats, but the ΔV for glass pairs with a small ΔP tends to be quite small. In order to get an exact match for the partial dispersions so that ΔP is equal to zero, two glasses can be combined to simulate a third, as indicated in Fig. 14*b*. For a unit power ($\phi = 1$) apochromatic triplet, the element powers can be found from

$$X = [V_A(P_B - P_C) + V_B(P_C - P_A)]/(P_B - P_A) \tag{37}$$

$$\phi_C = V_C/(V_C - X) \tag{38}$$

$$\phi_B = (1 - \phi_C)(P_C - P_A)V_B/[V_B(P_C - P_A) + V_A(P_B - P_C)] \tag{39}$$

$$\phi_A = 1 - \phi_B - \phi_C \tag{40}$$

32.15 ATHERMALIZATION

When the temperature of a lens element is changed, two factors affect its focus or focal length. As the temperature rises, all dimensions of the element are increased; this, by itself, would lengthen the focal length. However, the index of refraction of the lens material also changes with temperature. For many glasses the index rises with temperature; this effect tends to shorten the focal length.

The thermal change in the power of a thin element is given by

$$d\phi/dt = -\phi[a - (dn/dt)/(n - 1)] \tag{41}$$

where dn/dt is the differential of index with temperature and a is the thermal expansion coefficient of the lens material. Then for a thin doublet

$$d\phi/dt = \phi_A T_A + \phi_B T_B \tag{42}$$

where

$$T = [-a + (dn/dt)/(n - 1)] \tag{43}$$

and ϕ is the doublet power.

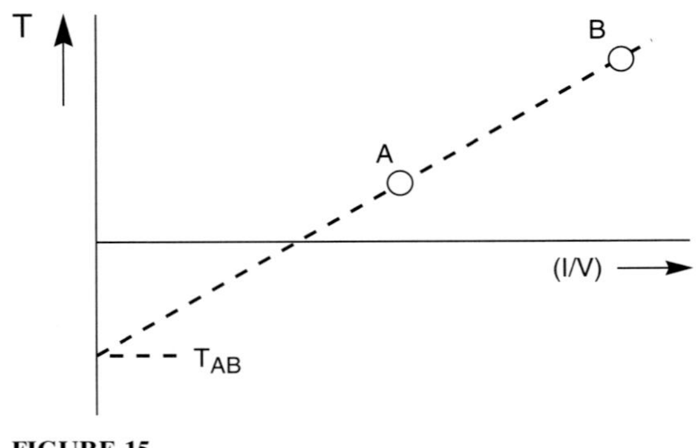

FIGURE 15

For an athermalized doublet (or one with some desired $d\phi/dt$) the element powers are given by

$$\phi_A = [(d\phi/dt) - \phi T_B]/(T_A - T_B) \tag{44}$$

$$\phi_B = \phi - \phi_A \tag{45}$$

To get an athermalized *achromatic* doublet, a plot of T against $(1/V)$ for all the glasses/materials under consideration is made. A line drawn between two glass points is extended to intersect the T axis as indicated in Fig. 15. Then the value of the $d\phi/dt$ for the achromatic doublet is equal to the doublet power times the value of T at which the line intersects the T axis. A pair of glasses with a large V-value difference and a small or zero T axis intersection is desirable.

An athermal achromatic triplet can be made with three glasses as follows:

$$\phi_A = \phi V_A(T_B V_B - T_C V_C)/D \tag{46}$$

$$\phi_B = \phi V_B(T_C V_C - T_A V_A)/D \tag{47}$$

$$\phi_C = \phi V_C(T_A V_A - T_B V_B)/D \tag{48}$$

$$D = V_A(T_B V_B - T_C V_C) + V_B(T_C V_C - T_A V_A) + V_C(T_A V_A - T_B V_B) \tag{49}$$

See also Chap. 39. "Thermal Compensation Techniques" by Rodgers.

NOTE: Figures 2, 3, 4, 5, 7, 8, 9, 10, 11, and 13 are adapted from W. Smith, *Modern Optical Engineering,* 2d ed., McGraw-Hill, New York, 1990. The remaining figures are adapted from *Critical Reviews of Lens Design,* W. Smith (Ed.), *S.P.I.E.,* vol. CR41, 1992.

CHAPTER 33

ABERRATION CURVES IN LENS DESIGN

Donald C. O'Shea
Georgia Institute of Technology
Center for Optical Science and Engineering
and School of Physics
Atlanta, Georgia

Michael E. Harrigan
Eastman Kodak Company
Electronic Imaging Research Laboratory
Rochester, N.Y.

33.1 GLOSSARY

H	ray height
NA	numerical aperture
OPD	optical path difference
P	petzval
S	sagittal
T	tangential
$\tan U$	slope

33.2 INTRODUCTION

Many optical designers use aberration curves to summarize the state of correction of an optical system, primarily because these curves give a designer important details about the relative contributions of individual aberrations to lens performance. Because a certain design technique may affect only one particular aberration type, these curves are more helpful to the lens designer than a single-value merit function. When a design is finished, the aberration curves serve as a summary of the lens performance and a record for future efforts. For applications such as photography, they are most useful because they provide a quick estimate of the effective blur circle diameter.

The aberration curves can be divided into two types: those that are expressed in terms of ray errors and those in terms of the optical path difference (OPD). OPD plots are usually plotted against the relative ray height in the entrance pupil. Ray errors can be displayed in a number of ways. Either the transverse or longitudinal error of a particular ray relative to the chief ray can be plotted as a function of the ray height in the entrance pupil. Depending upon the amount and type of aberration present, it is sometimes more appropriate to plot the longitudinal aberration as a function of field angle. For example, astigmatism or field curvature is more easily estimated from field plots, described below. Frequently, the curves are also plotted for several wavelengths to characterize chromatic performance. Because ray error plots are the most commonly used format, this entry will concentrate on them.

33.3 TRANSVERSE RAY PLOTS

These curves can take several different forms, depending on the particular application of the optical system. The most common form is the transverse ray aberration curve. It is also called lateral aberration, or ray intercept curve (also referred to by the misleading term "rim ray plots"). These plots are generated by tracing fans of rays from a specific object point for finite object distances (or a specific field angle for an object at infinity) to a linear array of points across the entrance pupil of the lens. The curves are plots of the ray error at an evaluation plane measured from the chief ray as a function of the relative ray height in the entrance pupil (Fig. 1). For afocal systems, one generally plots angular aberrations, the differences between the tangents of exiting rays and their chief ray in image space.

If the evaluation plane is in the image of a perfect image, there would be no ray error and the curve would be a straight line coincident with the abscissa of the plot. If the curve were plotted for a different evaluation plane parallel to the image plane, the curve would remain a straight line but it would be rotated about the origin. Usually the aberration is plotted along the vertical axis, although some designers plot it along the horizontal axis.

The curves in Fig. 1 indicate a lens with substantial underconnected spherical aberration as evidenced by the characteristic S-shaped curve. Since a change of the evaluation plane serves only to rotate the curve about the origin, a quick estimate of the aberrations of a lens can be made by reading the scale of the ray error axis (y axis) and mentally rotating the plot. For example, the blur spot can be estimated from the extent of a band that would enclose the curve a in Fig. 1, but a similar estimate could be made from the curves b or c, also.

The simplest form of chromatic aberration is axial color. It is shown in Fig. 2 in the presence of spherical aberration. Axial color is the variation of paraxial focus with wavelength and is seen as a difference in slope of the aberration curves at the origin as a

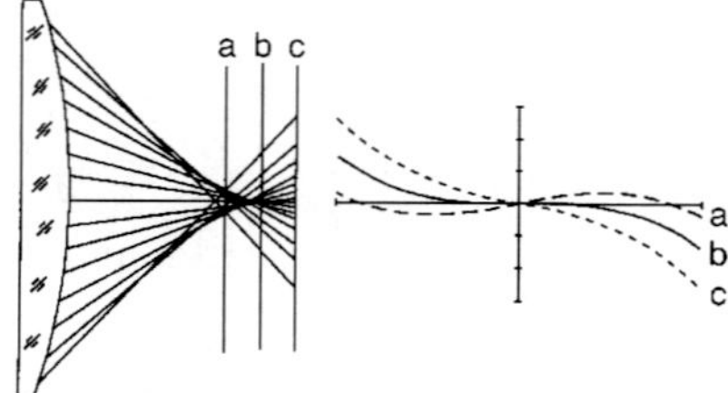

FIGURE 1 (*Left*) Rays exiting a lens are intercepted at three evaluation planes. (*Right*) Ray intercept curves plotted for the evaluation planes: (a) at the point of minimum ray error (circle of least confusion), (b) at the paraxial image plane, (c) outside the paraxial image plane.

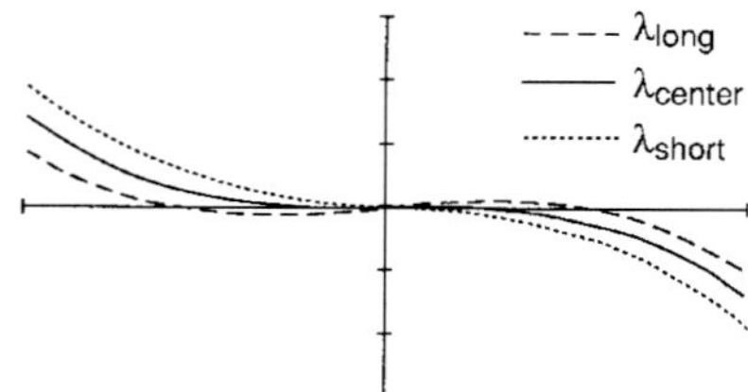

FIGURE 2 Meridional ray intercept curves of a lens with spherical aberration plotted for three colors.

function of wavelength. If the slopes of the curves at the origin for the end wavelengths are different, primary axial color is present. If primary axial color is corrected, then the curves for the end wavelengths will have the same slope at the origin. But if that slope differs from the slope of the curve for the center wavelength, then secondary axial color is present.

A more complex chromatic aberration occurs when the aberrations themselves vary with wavelength. Spherochromatism, the change of spherical aberration with wavelength, manifests itself as a difference in the shapes of the curves for different colors. Another curve that provides a measure of lateral color, an off-axis chromatic aberration, is described below.

For a point on the axis of the optical system, all ray fans lie in the meridional plane and only one plot is needed to evaluate the system. For off-axis object points, a second plot is added to evaluate a fan of skew rays traced in a sagittal plane. Because a skew ray fan is symmetrical across the meridional plane, only one side of the curve is usually plotted. For all curves the plots are departures from the chief ray location in the evaluation plane (Fig. 3). (In the case of the on-axis point, the chief ray is coincident with the optical axis.) For systems of small-field coverage only two or three object points need to be analyzed, but for wide-angle systems, four or more field points may be necessary.

What can be determined most easily from a comparison between the meridional and sagittal fans is the amount of astigmatism in the image for that field point. When astigmatism is present, the image planes for the tangential and sagittal fans are located at different distances along the chief ray. This is manifested in the ray intercept curves by different slopes at the origin for the tangential and sagittal curves. In Fig. 3 the slopes at the origins of the two curves are different at both 70 percent and full field, indicating astigmatism at both field points. The fact that the difference in the slopes of these two curves has changed sign between the two field points indicates that at some field angle between 70 percent and full field, the slopes are equal and there is no astigmatism there. In addition, the variation of slopes for each curve as a function of field angle is evidence of field curvature.

The off-axis aberration of pure primary coma would be evident on these plots as a

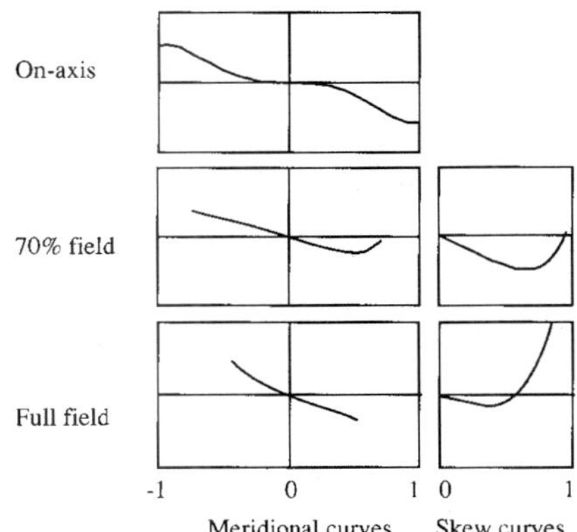

FIGURE 3 Evaluation of a lens on-axis and at two off-axis points. The reduction of the length of the curve with higher field indicates that the lens is vignetting these angles. The differences in slopes at the origin between the meridional and skew curves indicate that the lens has astigmatism at these field angles.

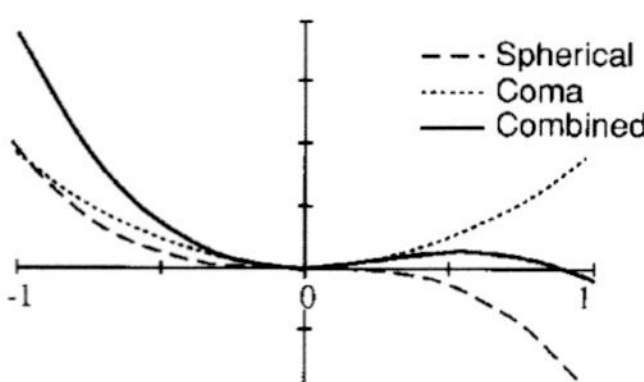

FIGURE 4 Ray intercept curve showing coma combined with spherical aberration.

U-shaped curves for the meridional fan and sagittal fans, the tangential curve being three times larger than the sagittal curve. The "U" will be either upright or upside down depending on the sign of the coma. In almost all cases coma is combined with spherical to produce an S-shaped curve that elongates one of the arms of the "S" and shortens the other (Fig. 4).

The amount of vignetting can be determined from the ray intercept curves also. When it is present, the meridional curves get progressively shorter as the field angle is increased (Fig. 3), since rays at the edges of the entrance pupil are not transmitted. Taken from another perspective, ray intercept curves can also provide the designer with an estimate of how far a system must be stopped down to provide a required degree of correction.

33.4 FIELD PLOTS

The ray intercept curves provide evaluation for a limited number of object points—usually a point on the optical axis and several field points. The field plots present information on certain aberrations across the entire field. In these plots, the independent variable is usually the field angle and is plotted vertically and the aberration is plotted horizontally. The three field plots most often used are: distortion, field curvature, and lateral color. The first of these shows percentage distortion as a function of field angle (Fig. 5).

The second type of plot, field curvature, displays the tangential and sagittal foci as a function of object point or field angle (Fig. 6*a*). In some plots the Petzval surface, the surface to which the image would collapse if there were no astigmatism, is also plotted. This plot shows the amount of curvature in the image plane and amount of astigmatism

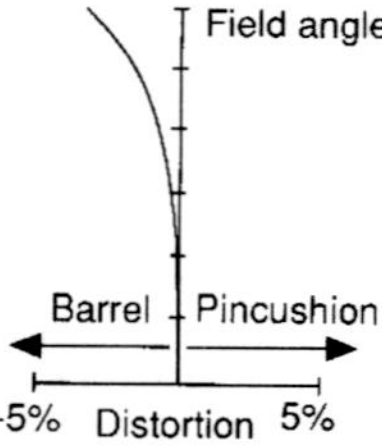

FIGURE 5 Field curve: distortion plot. The percentage distortion is plotted as a function of field angle. Note that the axis of the dependent variable is the horizontal axis.

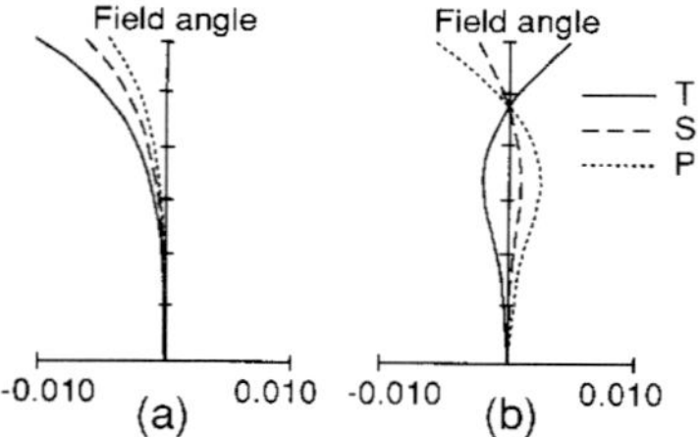

FIGURE 6 Field curve: field curvature plot. The locations of the tangential T and sagittal S foci are plotted for a full range of field angles. The Petzval surface P is also plotted. The tangential surface is always three times farther from the Petzval surface than from the sagittal surface. (*a*) An uncorrected system. (*b*) A corrected system.

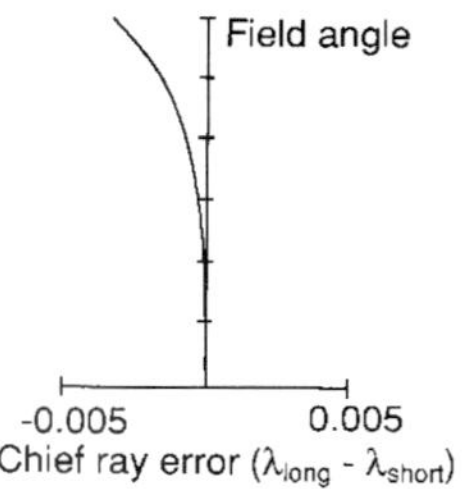

FIGURE 7 Field curve: lateral color plot. A plot of the transverse ray error between red and blue chief ray heights in the image plane for a full range of field angles. Here the distance along the horizontal axis is the color error in the image plane.

over the entire field. In cases of corrected field curvature (Fig. 6*b*), this plot provides an estimate of the residual astigmatism between the axis and the corrected zone and an estimate of the maximum field angle at which the image possesses reasonable correction.

The last of the field curves provides information on color error as a function of field angle (Fig. 7). Lateral color, the variation of magnification with wavelength, is plotted as the difference between the chief ray heights at the red and blue wavelengths as a function of field angle. This provides the designer with an estimate of the amount of color separation in the image at various points in the field. In the transverse ray error curves, lateral color is seen as a vertical displacement of the end wavelength curves from the central wavelength curve at the origin.

Although there are other plots that can describe aberrations of optical systems (e.g., plot of longitudinal error as a function of entrance pupil height), the ones described here represent the ensemble that is used in most ray evaluation presentations.

33.5 ADDITIONAL CONSIDERATIONS

In many ray intercept curves the independent variable is the relative entrance pupil coordinate of the ray. However, for systems with high NA or large field of view, where the principal surface cannot be approximated by a plane, it is better to plot the difference between the tangent of the convergence angle of the chosen ray and the tangent of the convergence angle of the chief ray. This is because the curve for a corrected image will remain a straight line in any evaluation plane.[1] When plotted this way, the curves are called H-tan U curves.

Shifting the stop of an optical system has no effect on the on-axis curves. However, it causes the origin of the meridional curves of off-axis points to be shifted along the curve. In Fig. 8, the off-axis meridional curves are plotted for three stop positions of a double Gauss lens. The center curve (Fig. 8*b*) is plotted for a symmetrically located stop; the outer curves are plots when the stop is located at lens surfaces before and after the central stop.

It is usually sufficient to make a plot of the aberrations in the meridional and sagittal sections of the beam. The meridional section, defined for an optical system with rotational symmetry, is any plane containing the optical axis. It is sometimes called the tangential section. The sagittal section is a plane perpendicular to the meridional plane containing the chief ray. There are some forms of higher-order coma that do not show in these sections.[2] In those cases where this aberration is suspected to be a problem, it may be helpful to look at a spot diagram generated from rays in all sections of the bundle.

For a rotationally symmetric system, only objects in a meridional plane need to be

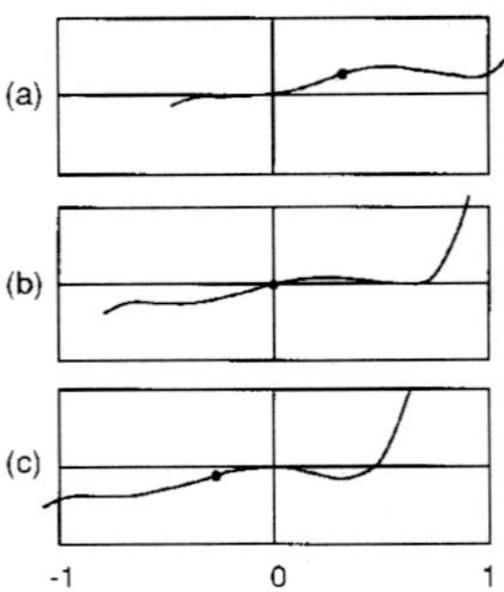

FIGURE 8 The effect of stop shifting on the meridional ray intercept curves of a double Gauss lens. (*a*) Stop located in front of the normal centrally located stop. (*b*) Stop at the normal stop position. (*c*) Stop behind the normal stop position. The dot locates the point on the curve where the origin is located for case (*b*).

analyzed. Also for such systems, only meridional ray errors are possible for purely meridional rays. To observe certain coma types, it is a good idea to plot both the meridional and sagittal ray errors for sagittal rays. It is possible for the meridional section to show no coma and have it show only in the meridional error component of the sagittal fan,[2] but this aberration is normally small.

In addition to plots of the ray error in an evaluation plane, another aberration plot is one that expresses wavefront aberrations as an optical path difference (OPD) from a spherical wavefront centered about the image point. These OPD plots are particularly useful for applications where the lens must be close to diffraction-limited.

33.6 SUMMARY

Aberration curves provide experienced designers with the information needed to enable them to correct different types of aberrations. Chromatic effects are much more easily classified from aberration curves also. In comparison to spot diagrams and modulation transfer function curves, the types of aberrations can be more easily seen and quantified. In the case of diffraction-limited systems, modulation transfer functions may provide better estimates of system performance.

33.7 REFERENCES

1. R. Kingslake, *Lens Design Fundamentals,* Academic Press, San Diego, 1978, p. 144.
2. F. D. Cruickshank and G. A. Hills, "Use of Optical Aberration Coefficients in Optical Design," *J. Opt. Soc. Am.* **50:**379 (1960).

CHAPTER 34
OPTICAL DESIGN SOFTWARE

Douglas C. Sinclair
Sinclair Optics, Inc.
Fairport, New York

34.1 GLOSSARY

a	axial ray
b	chief ray
c	curvature
d, e, f, g	aspheric coefficients
efl	effective focal length
FN	focal ratio
$f(\,)$	function
h	ray height
m	linear, lateral magnification
n	refractive index
PIV	paraxial (Lagrange) invariant
r	entrance pupil radius
t	thickness
u	ray slope
y	coordinate
z	coordinate
α	tilt about x (Euler angles)
β	tilt about y (Euler angles)
γ	tilt about z (Euler angles)
ϵ	displacement of a ray from the chief ray
μ	Buchdahl coefficients
κ	conic constant

ρ	radial coordinate
σ_1	spherical aberration
σ_2	coma
σ_3	astigmatism
σ_4	Petzval blur
σ_5	distortion

34.2 INTRODUCTION

The primary function of optical design software is to produce a mathematical description, or *prescription,* describing the shapes, locations, materials, etc., of an optical system that satisfies a given set of specifications. A typical optical design program contains three principal sections: *data entry, evaluation,* and *optimization.* The optical design programs considered here are to be distinguished from *ray-trace* programs, which are mainly concerned with evaluation, and *CAD* programs, which are mainly concerned with drawings. The essence of an optical design program is its optimization section, which takes a starting design and produces a new design that minimizes an *error function* that characterizes the system performance.

The first practical computer software for optical design was developed in the 1950s and 1960s.[1–4] Several commercially available programs were introduced during the 1970s, and development of these programs has continued through the 1980s to the present time. Despite the fact that there has now been more than 30 years' development of optical design software, there are still substantial improvements to be made in optimization algorithms, evaluation methods, and user interfaces.

This article attempts to describe a typical optical design program. It is intended for readers that have a general background in optics, but who are not familiar with the capabilities of optical design software. We present a brief description of some of the most important mathematical concepts, but make no attempt to give a detailed development. We hope that this approach will give readers enough understanding to know whether an optical design program will be a useful tool for their own work.

Of course, many different programs are available, each with its own advantages and disadvantages. Our purpose is not to review or explain specific programs, but to concentrate on the basic capabilities. Some programs work better than others, but we make no quality judgment. In fact, we avoid reference, either explicit or implicit, to any particular program. The features and benefits of particular optical design programs are more than adequately described by software vendors, who are listed in optical industry buyer's guides.[5]

Figure 1 is a flowchart of a typical optical design project. Usually, the designer not only must enter the starting design and initial optimization data, but also must continually monitor the progress of the computer, modifying either the lens data or the optimization data as the design progresses to achieve the best solution. Even when the performance requirements are tightly specified, it is often necessary to change the error function during the design process. This occurs when the error function does not correlate with the desired performance sufficiently well, either because it is ill-conceived, or because the designer has purposefully chosen a simple error function to achieve improved speed.

The fact that there are alternate choices of action to be taken when the design is not good enough has led to two schools of thought concerning the design of an optical design program. The first school tries to make the interface between the designer and the

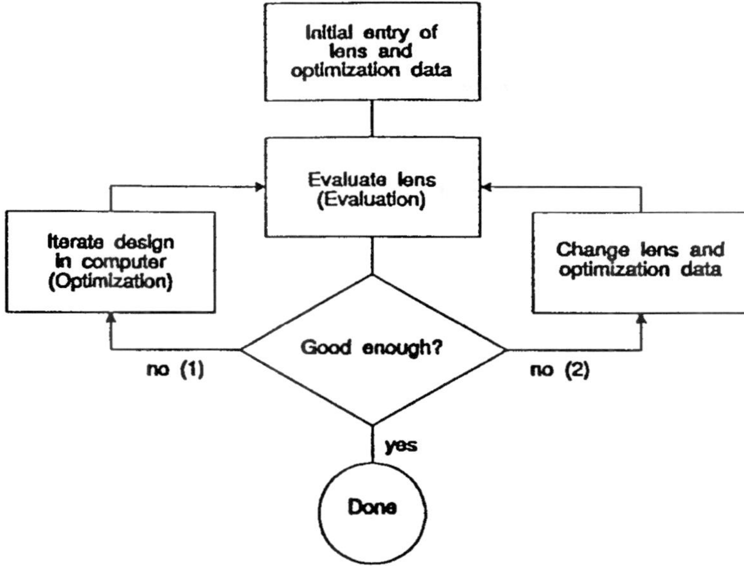

FIGURE 1 Flowchart for the lens design process. The action taken when a design is not satisfactory depends on how bad it is. The designer (or a design program) may change the lens data, or redefine the targets to ones that can be achieved.

program as smooth as possible, emphasizing the interactive side of the process. The second school tries to make the error function comprehensive, and the iteration procedure powerful, minimizing the need for the designer to intervene.

34.3 LENS ENTRY

In early lens design programs, lens entry was a "phase" in which the lens data for a starting design was read into the computer from a deck of cards. At that time, the numerical aspects of optical design on a computer were so amazing that scant attention was paid to the lens entry process. However, as the use of optical design software became more widespread, it was found that a great deal of a designer's time was spent punching cards and submitting new jobs, often to correct past mistakes. Often, it turned out that the hardest part of a design job was preparing a "correct" lens deck!

Over the years, optical design programs have been expanded to improve the lens entry process, changing the function of this part of the program from simple lens entry to what might be called lens database management. A typical contemporary program provides on-line access to a library of hundreds of lenses, interactive editing, automatic lens drawings, and many features designed to simplify this aspect of optical design.

The lens database contains all items needed to describe the optical system under study, including not only the physical data needed to construct the system (curvatures, thicknesses, etc.), but also data that describe the conditions of use (object and image location, field of view, etc.). Some programs also incorporate optimization data in the lens database, while others provide separate routines for handling such data. In any case, the lens database is often the largest part of an optical design program.

The management of lens data in an optical design program is complicated by two factors. One is that there is a tremendous range of complexity in the types of systems that can be accommodated, so there are many different data items. The other is that the data are often described indirectly. A surface curvature may be specified, for example, by the required slope of a ray that emerges from the surface, rather than the actual curvature itself. Such a specification is called a *solve,* and is based on the fact that paraxial ray tracing is incorporated in the lens entry portion of most optical design programs.

It might seem curious that paraxial ray tracing is still used in contemporary optical design programs that can trace *exact* rays in microseconds. (The term *exact* ray is used in this article to mean a real skew ray. Meridional rays are treated as a special case of skew rays in contemporary software; there is not sufficient computational advantage to warrant a separate ray trace for them.) In fact, paraxial rays have some important properties that account for their incorporation in the lens database.

First, paraxial rays provide a linear system model that leads to analysis of optical systems in terms of simple bilinear transforms. Second, paraxial ray tracing does not fail. Exact rays can miss surfaces or undergo total internal reflection. Finally, paraxial rays determine the ideal performance of a lens. In a well-corrected lens, the aberrations are balanced so that the exact rays come to the image points defined by the paraxial rays, not the other way around.

There are two basic types of data used to describe optical systems. The first are the *general* data that are used to describe the system as a whole, and the other are the *surface* data that describe the individual surfaces and their locations. Usually, an optical system is described as an ordered set of surfaces, beginning with an *object* surface and ending with an *image* surface (where there may or may not be an actual image). It is assumed that the designer knows the order in which rays strike the various surfaces. Systems for which this is not the case are said to contain *nonsequential* surfaces, which are discussed later.

General System Data

The general data used to describe a system include the aperture and field of view, the wavelengths at which the system is to be evaluated, and perhaps other data that specify evaluation modes, vignetting conditions, etc.

Aperture and Field of View. The aperture and field of view of a system determine its conditions of use. The aperture is specified by the *axial ray*, which emerges from the vertex of the object surface and passes through the edge of the entrance pupil. The field of view is specified by the *chief ray,* which emerges from the edge of the object and passes through the center of the entrance pupil.

There are various common ways to provide data for the axial and chief rays. If the object is at an infinite distance, the entrance pupil radius and semifield angle form a convenient way to specify the axial and chief rays. For finite conjugates, the numerical aperture in object space and the object height are usually more convenient.

Some programs permit the specification of paraxial ray data by image-space quantities such as the f-number and the image height, but such a specification is less desirable from a computational point of view because it requires an iterative process to determine initial ray-aiming data.

Wavelengths. It is necessary to specify the wavelengths to be used to evaluate polychromatic systems. Three wavelengths are needed to enable the calculation of primary and secondary chromatic aberrations. More than three wavelengths are required to provide an accurate evaluation of a typical system, and many programs provide additional wavelengths for this reason. There has been little standardization of wavelength specification. Some programs assume that the first wavelength is the central wavelength, while others assume that it is one of the extreme wavelengths; some require wavelengths in micrometers, while others require nanometers.

Other General Data. Several other items of general data are needed to furnish a complete lens description, but there is little consistency between programs on how these items are treated, or even what they are. The only one that warrants mention here is the aperture stop. The *aperture stop* is usually defined to be the surface whose aperture limits the angle of the axial ray. Once the aperture stop surface is given, the positions of the paraxial pupils are determined by the imaging properties of the system. Since the aperture and field of view are determined formally by the paraxial pupils, the apertures are not associated with the exact ray behavior.

The "vignetting factor" is used to account for the differences between paraxial and exact off-axis ray heights at apertures. In particular, the vignetting factor provides, in terms of fractional (paraxial) coordinates, the data for an exact ray that grazes the apertures of a system. Typically, there is an upper, lower, and skew vignetting factor. The details of how such factors are defined and handled are program-dependent.

Surface Data

Surface Location. There are two basic ways to specify the location of surfaces that make up a lens. One is to specify the position of a surface relative to the immediately preceding surface. The other is to specify its position relative to some fixed surface (for example, the first surface). The two ways lead to what are called *local* and *global* coordinates, respectively. For ordinary lenses consisting of a series of rotationally symmetric surfaces centered on an optical axis, local coordinates are more convenient, but for systems that include reflectors, tilted, and/or decentered surfaces, etc., global coordinates are simpler. Internally, optical design programs convert global surface data to local coordinates for speed in ray tracing.

Most optical design programs use a standard coordinate system and standard sign conventions, although there are exceptions.[6] Each surface defines a local right-handed coordinate system in which the z axis is the symmetry axis and the yz plane is the meridional plane. The local coordinate system, is used to describe the surface under consideration and also the origin of the next coordinate system. Tilted elements are described by an Euler-angle system in which α is a tilt around the x axis, β is a tilt around the y axis, and γ is a tilt around the z axis. Since tilting and decentering operations are not commutative; some data item must be provided to indicate which comes first.

Surface Profile. Of the various surfaces used in optical systems, the most common by far is the rotationally symmetric surface, which can be written as[7]

$$z = \frac{cr^2}{1 + \sqrt{1 - c^2(\kappa + 1)r^2}} + dr^4 + er^6 + fr^8 + gr^{10}$$

$$r = \sqrt{x^2 + y^2}$$

c is the curvature of the surface; κ is the conic constant; and d, e, f, and g are aspheric constants. The use of the above equation is almost universal in optical design programs. The description of conic surfaces in terms of a conic constant κ instead of the eccentricity e used in the standard mathematical literature allows spherical surfaces to be specified as those with no conic constant. (The conic constant is minus the square of the eccentricity.)

Although aspheric surfaces include all surfaces that are not spherical, from a design standpoint there is a demarcation between "conic" aspheres and "polynomial" aspheres described using the coefficients $d, e, f,$ and g. Rays can be traced analytically through the former, while the latter require numerical iterative methods.

Many optical design programs can handle surface profiles that are more complicated than the above, including cylinders, torics, splines, and even general aspheres of the form $z = f(x, y)$, where $f(x, y)$ is an arbitrary polynomial. The general operation of an optical design program, however, can be understood by considering only the rotationally symmetric surfaces described here.

As mentioned above, the importance of paraxial rays in optical system design has led to the indirect specification of lens data, using *solves*, as they are called, which permit a designer to incorporate the basic paraxial data describing a lens with the lens itself, rather than having to compute and optimize the paraxial performance of a lens as a separate task. Considering the jth surface of an optical system, let

$$y_j = \text{ray height on surface}$$

$$u_j = \text{ray slope on image side}$$

$$c_j = \text{curvature of surface}$$

$$n_j = \text{refractive index on image side}$$

$$t_j = \text{thickness on image side}$$

The paraxial ray trace equations can then be written as[8]

$$y_j = y_{j-1} + t_{j-1}u_{j-1}$$

$$n_j u_j = n_{j-1}u_{j-1} - y_j c_j (n_j - n_{j-1})$$

These equations can be inverted to give the curvatures and thicknesses in terms of the ray data. We have

$$c_j = \frac{n_{j-1}u_{j-1} - n_j u_j}{y_j(n_j - n_{j-1})}$$

$$t_j = \frac{y_{j+1} - y_j}{u_j}$$

The specification of curvatures and thicknesses by solves is considered to be on an equal basis with the direct specification of these items. The terminology used to specify solves is that the solves used to determine thickness are called *height solves,* and the solves used to determine curvature are called *angle solves.* Often, an axial ray height solve on the last surface is used to automatically locate the paraxial image plane, a chief ray height solve on the same surface to locate the exit pupil, and an axial ray angle solve is used to maintain a given focal length (if the entrance pupil radius is fixed). In some programs, additional types of solves are allowed, such as center of curvature solves, or aperture solves.

Of course, specifying lens data in terms of paraxial ray data means that whenever any lens data is changed, two paraxial rays must be traced through the system to resolve any following data that are determined by a solve. In an optical design program, this function is performed by a *lens setup* routine, which must be efficiently coded, since it is executed thousands of times in even a small design project.

Other functions of the lens setup routine are to precalculate values that are needed for repetitive calculations, such as refractive indices, rotation and translation matrices, etc. Many programs have the capability of specifying certain data items to be equal to (±) the value of the corresponding item on a previous surface. These are called *pickups*, and are needed for optimization of systems containing mirrors, as well as maintaining special geometrical relationships. Programs that lack pickups usually have an alternate means for maintaining the required linking between data items. Like solves, pickups are resolved by the lens setup routine, although they do not use paraxial data.

Other Surface Data. A variety of other data is required to specify surfaces. Most important are apertures, discussed below, and refractive indices. Refractive indices are usually given by specifying the name of a catalog glass. In the lens setup routine, the actual refractive indices are calculated using an index interpolation formula and coefficient supplied by the glass manufacturer, together with the design wavelengths stored with the lens data. Other surface-related items include phase data for diffractive surfaces, gradient-index data, holographic construction data, and coatings.

Apertures have a somewhat obscure status in many optical design programs. Although apertures have a major role to play in determining the performance of a typical system, they do not usually appear directly in optimization functions. Instead, apertures are usually controlled in optimization by targets on the heights of rays that define their edges. If an aperture is specified directly, it will block rays that pass outside of it and cause typical optimization procedures to become unstable. Accordingly, some programs ignore apertures during optimization. Other programs allow the apertures to be determined by a set of exact "reference rays" that graze their extremities.

Nonsequential Surfaces. In some optical systems, it is not possible to specify the order in which a ray will intersect the surfaces as it progresses through the system. The most common examples of such systems are prisms such as the corner-cube reflector, where the ordering of surfaces depends on the entering ray coordinates. Other examples of nonsequential surfaces include light pipes and a variety of nonimaging concentrators. Nonsequential surfaces can be accommodated by many optical design programs, but for the most part they are not "designed" using the program, but rather are included as a subsystem used in conjunction with another part of the system that is the actual system being designed. Data specification of nonsequential surfaces is more complicated than ordinary systems, and ray tracing is much slower, since several surfaces must be investigated to see which surface is the one actually traversed by a given ray.

Lens Setup

Whenever the lens entry process is completed, the lens must be "set up." Pickup constraints must be resolved. If the system contains an internal aperture stop, the position of the entrance pupil must be determined. Then paraxial axial and chief rays must be traced through the system so that surface data specified by solves can be computed. Depending on the program, a variety of other data may be precomputed for later use, including aperture radii, refractive indices, and various paraxial constants.

The lens setup routine must be very efficient, since it is the most heavily used code in an optical design program. In addition to running whenever explicit data entry is complete, the code is also executed whenever the lens is modified internally by the program, such as when derivatives are computed during optimization, or when configurations are changed in a multiconfiguration system. Typically, lens setup takes milliseconds (at most), so it is not noticed by the user, other than through its effects.

Programming Considerations

In writing an optical design program, the programmer must make a number of compromises between speed, size, accuracy, and ease of use. These compromises affect the usefulness of a particular program for a particular application. For example, a simple, fast, small program may be well suited to a casual user with a simple problem to solve, but this same program may not be suited for an experienced designer who routinely tackles state-of-the-art problems.

The lens entry portion of an optical design program shows, more than any other part, the difference in programming models that occurred during the 1980s. Before the 1980s, most application programs were of a type called *procedural* programs. When such a program needs data, it requests it, perhaps from a file or by issuing a prompt for keyboard input. The type of data needed is known, and the program is only prepared to accept that kind of data at any given point. Although the program may branch through different paths in response to the data it receives, the program is responsible for its own execution.

With the popularization in the 1980s of computer systems that use a mouse for input the model for an application program changed from the procedural model described above to what is called an *event-driven* model. An event-driven program has a very simple top-level structure consisting of an initialization section followed by an infinite loop usually called something like the *main event loop.* The function of the main event loop is to react to user-initiated interrupts (such as pressing a key, or clicking a mouse button), dispatching such events to appropriate processing functions. In such a program, the user controls the execution, unlike a procedural program, where the execution controls the user.

An event-driven program usually provides a better user interface than a procedural program. Unfortunately, most optical design programs were originally written as procedural programs, and it is difficult to convert a procedural program into an event-driven program by "patching" it. Usually it is easier to start over. In addition, it is harder to write an event-driven program than a procedural program, because the event-driven program must be set up to handle a wide variety of unpredictable requests received at random times. Of course, it is this very fact that makes the user interface better. There is an aphorism sometimes called the "conservation of complexity," which states that the simpler a program is to use, the more complicated the program itself must be.

The data structures used to define lens data in an optical design program may have a major impact on its capabilities. For example, for various reasons it is usually desirable to represent a lens in a computer program as an array of surfaces. If the maximum size of the array is determined at compile time, then the maximum size lens that can be accommodated is built into the program.

As additional data items are added to the surface data, the space required for storage can become unwieldy. For example, it takes about 10 items of real data to specify a holographic surface. If every surface were allowed to be a hologram, then 10 array elements would have to be reserved for each surface's holographic data. On the other hand, in most systems, the elements would never be used, so the data structure would be very inefficient. To avoid this, a more complicated data structure could be implemented in which only one element would be devoted to holograms, and this item would be used as an index into a separate array containing the actual holographic data. Such an array might have a sufficient number of elements to accommodate up to, say, five holograms, the maximum number expected in any one system.

The preceding is a simple example of how the data structure in an optical design program can grow in complexity. In fact, in a large optical design program the data structure may contain all sorts of indices, pointers, flags, etc., used to implement special data types and control their use. Managing this data while maintaining its integrity is a programming task of a magnitude often greater than the numerical processing that takes place during optical design.

Consider, for example, the task of deleting a surface from a lens. To do this, the surface data must of course be deleted, and all of the higher-numbered surfaces renumbered. But, in addition, the surface must be checked to see whether it is a hologram and, if so, the holographic data must also be deleted and that data structure "cleaned up." All other possible special data items must be tested and handled similarly. Then all the renumbered surfaces must be checked to see if any of them "pick up" data from a surface that has been renumbered, and the reference adjusted accordingly. Then other data structures such as the optimization files must be checked to see if they refer to any of the renumbered surfaces, and appropriate adjustments made. Several additional checks and adjustments must also be carried out.

Related to the lens entry process is the method used to store lens data on disc. Of course, lens data are originally provided to a program in the form of text (e.g., "TH 1.0"). The program *parses* this data to identify its type (a thickness) and value (1.0). The results of the parsing process (the binary values) are stored in appropriate memory locations (arrays). To store lens data on disc, early optical design programs used the binary data to avoid having to reparse it when it was recovered. However, the use of binary files has decreased markedly as computers have become fast enough that parsing lens input does not take long. The disadvantages of binary files are that they tend to be quite large, and usually have a structure that makes them obsolete when the internal data structure of the program is changed. The alternative is to store lens data as text files, similar in form to ordinary keyboard input files.

34.4 EVALUATION

Paraxial Analysis

Although the lens setup routine contains a paraxial ray trace, a separate paraxial ray trace routine is used to compute data for display to the user. At a minimum, the paraxial ray heights and slopes of the axial and chief ray are shown for each surface, in each color, in each configuration.

The equations used for paraxial ray tracing were described in the previous section. Although such equations become exact only for "true" paraxial rays that are infinitesimally displaced from the optical axis, it is customary to consider paraxial ray data to describe "formal" paraxial rays that refract at the tangent planes to surfaces, as shown in Fig. 2. Here, the ray ABC is a paraxial ray that provides a first-order approximation to the exact ray ADE. Not only does the paraxial ray refract at the (imaginary) tangent plane BVP, but also it bends a different amount from the exact ray.

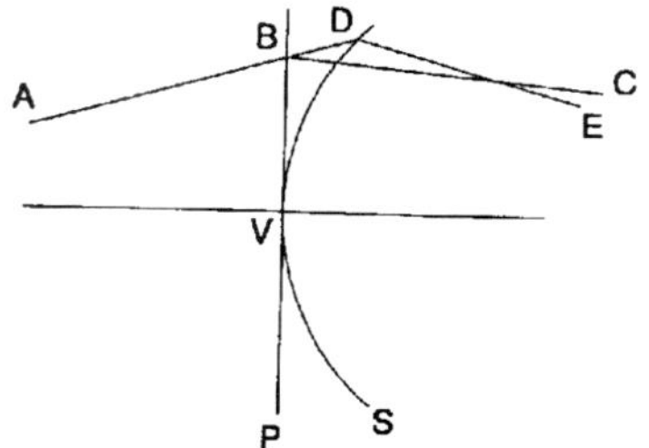

FIGURE 2 Showing the difference between a paraxial ray and a real ray. The paraxial ray propagates along ABC, while the real ray propagates along ADE.

In addition to the computation of ray heights and slopes for the axial and chief ray, various paraxial constants that characterize the overall system are computed. The particular values computed depend on whether the system is *focal* (finite image distance) or *afocal* (image at infinity). For focal systems, the quantities of interest are (at a minimum) the focal length *efl*, the f-number FN, the paraxial (Lagrange) invariant PIV, and the transverse magnification m. It is desirable to compute such quantities in a way that does not depend on the position of the final image surface. Let the object height be h, the entrance pupil radius be r, the axial ray data in object and image space be y, u and y', u', the chief ray data in object and image space be $\bar{y}, \bar{u}$ and $\bar{y}', \bar{u}'$, and the refractive indices be n and n'.

The above-mentioned paraxial constants are then given by

$$efl = \frac{-rh}{\bar{u}'r + u'h}$$

$$FN = -\frac{n}{2n'u'}$$

$$\text{PIV} = n(\bar{y}u - y\bar{u})$$

$$m = \frac{nu}{n'u'}$$

In addition to the paraxial constants, most programs display the locations of the entrance and exit pupils, which are easily determined using chief-ray data. Surprisingly, most optical design programs do not explicitly show the locations of the principal planes. In addition, although most programs have the capability to display "$y - y$bar" plots, few have integrated this method into the main lens entry routine.

Aberrations

Although most optical design is based on exact ray data, virtually all programs have the capability to compute and display first-order chromatic aberrations and third-order monochromatic (Seidel) aberrations. Many programs can compute fifth-order aberrations as well. The form in which aberrations are displayed depends on the program and the type of system under study, but as a general rule, for focal systems aberrations are displayed as equivalent ray displacements in the paraxial image plane.

In the case of the chromatic aberrations, the primary and secondary chromatic aberration of the axial and chief rays are computed. In a system for which three wavelengths are defined, the primary aberration is usually taken between the two outer wavelengths, and the secondary aberration between the central and short wavelengths.

The Seidel aberrations are computed according to the usual aberration polynomial. If we let ϵ be the displacement of a ray from the chief ray, then

$$\epsilon_y = \epsilon_{3y} + \epsilon_{5y} + \cdots$$

$$\epsilon_x = \epsilon_{3x} + \epsilon_{5x} + \cdots$$

For a relative field height h and normalized entrance pupil coordinates r and θ, the third-order terms are

$$\epsilon_{3y} = \sigma_1 \cos\theta r^3 + \sigma_2(2 + \cos 2\theta)r^2 h + (3\sigma_3 + \sigma_4)\cos\theta r h^2 + \sigma_5 h^3$$

$$\epsilon_{3x} = \sigma_1 \sin\theta r^3 + \sigma_2 \sin 2\theta r^2 h + (\sigma_3 + \sigma_4)\sin\theta r h^3$$

The interpretation of the coefficients is generally as follows, but several optical design programs display tangential coma, rather than the sagittal coma indicated in the table.

σ_1	Spherical aberration
σ_2	Coma
σ_3	Astigmatism
σ_4	Petzval blur
σ_5	Distortion

The fifth-order terms are

$$\epsilon_{5y} = \mu_1 \cos\theta r^5 + (\mu_2 + \mu_3 \cos 2\theta)r^4 h + (\mu_4 + \mu_6 \cos^2\theta)\cos\theta r^2 h^2$$
$$+ (\mu_7 + \mu_8 \cos 2\theta)r^2 h^3 + \mu_{10}\cos\theta r h^4 + \mu_{12} h^5$$

$$\epsilon_{5x} = \mu_1 \sin\theta r^5 + \mu_3 \sin 2\theta \rho^4 h + (\mu_5 + \mu_6 \cos^2\theta)\sin\theta r^3 h^2$$
$$+ \mu_9 \sin 2\theta r^2 h^3 + \mu_{11}\sin\theta r h^4$$

These equations express the fifth-order aberration in terms of the Buchdahl μ coefficients. In systems for which the third-order aberrations are corrected, the following identities exist:

$$\mu_2 = \tfrac{3}{2}\mu_3$$

$$\mu_4 = \mu_5 + \mu_6$$

$$\mu_7 = \mu_8 + \mu_9$$

μ_1	Spherical aberration
μ_3	Coma
$(\mu_{10} - \mu_{11})/4$	Astigmatism
$(5\mu_{11} - \mu_{10})/4$	Petzval blur

$\mu_4 + \mu_6$	Tangential oblique spherical aberration
μ_5	Sagittal oblique spherical aberration
$\mu_7 + \mu_8$	Tangential elliptical coma
μ_9	Sagittal elliptical coma
μ_{12}	Distortion

Some programs display only the aberrations that have corresponding third-order coefficients, omitting oblique spherical aberration and elliptical coma.

The formulas needed to calculate the chromatic and third-order aberrations are given in the *U.S. Military Handbook of Optical Design*. The formulas for calculating the fifth-order aberrations are given in Buchdahl's book.[9]

Aberration coefficients are useful in optical design because they characterize the system in terms of its symmetries, because they allow the overall performance to be expressed as a sum of surface contributions, and because they are calculated quickly. On the negative side, aberration coefficients are not valid for systems that have tilted and decentered elements for systems that cover an appreciable field of view, the accuracy of aberration coefficients in predicting performance is usually inadequate. Moreover, for systems that include unusual elements like diffractive surfaces and gradient index materials, the computation of aberration coefficients is cumbersome at best.

Ray Tracing

Exact ray tracing is the foundation of an optical design program, serving as a base for both evaluation and optimization. From the programmer's standpoint, the exact ray-trace routines must be accurate and efficient. From the user's viewpoint, the data produced by the ray-trace routines must be accurate and comprehensible. Misunderstanding the meaning of ray-trace results can be the source of costly errors in design.

To trace rays in an optical design program, it is necessary to understand how exact rays are specified. Although the details may vary from one program to the next, many programs define a ray by a two-step process. In the first step, an object point is specified. Once this has been done, all rays are assumed to originate from this point until a new object point is specified. The rays themselves are then specified by aperture coordinates and wavelength.

Exact ray starting data is usually normalized to the object and pupil coordinates specified by the axial and chief rays. That is, the aperture coordinates of a ray are specified as a fractional number, with 0.0 representing a point on the vertex of the entrance pupil, and 1.0 representing the edge of the pupil. Field angles or object heights are similarly described, with 0.0 being a point on the axis, and 1.0 being a point at the edge of the field of view.

Although the above normalization is useful when the object plane is at infinity, it is not so good when the object is at a finite distance and the numerical aperture in object space is appreciable. Then, fractional aperture coordinates should be chosen proportional to the direction cosines of rays leaving an object point. There are two reasons for this. One is that it allows an object point to be considered a point source, so that the amount of energy is proportional to the "area" on the entrance pupil. The other is that for systems without pupil aberrations, the fractional coordinates on the second principal surface should be the same as those on the first principal surface. Notwithstanding these requirements, many optical design programs do not define fractional coordinates proportional to direction cosines.

It is sometimes a point of confusion that the aperture and field of view of a system are specified by paraxial quantities, when the actual performance is determined by exact rays. In fact, the paraxial specifications merely establish a normalization for exact ray data. For example, in a real system the field of view is determined not by the angle of the paraxial chief ray, but by the angle at which exact rays blocked by actual apertures just fail to pass through the system. Using an iterative procedure, it is not to hard to find this angle, but because of the nonlinear behavior of Snell's law, it does not provide a convenient reference point.

There are two types of exact rays: *ordinary* or *lagrangian* rays, and *iterated* or *hamiltonian* rays. The designation of rays as lagrangian or hamiltonian comes from the analogy to the equations of motion of a particle in classical mechanics. Here we use the more common designation as ordinary or iterated rays. An ordinary ray is a ray that starts from a known object point in a known direction. An iterated ray also starts from a known object point, but its direction is not known at the start. Instead, it is known that the ray passes through some known (nonconjugate) point inside the system, and the initial ray direction is determined by an iterative procedure.

Iterated rays have several applications in optical design programs. For example, whenever a new object point is specified, it is common to trace an iterated ray through the center of the aperture stop (or some other point) to serve as a reference ray, or to trace several iterated rays through the edges of limiting apertures to serve as reference rays. In fact, many programs use the term *reference ray* to mean iterated ray (although in others, reference rays are ordinary rays). Iterated rays are traced using differentially displaced rays to compute corrections to the initial ray directions. Because of this, they are traced slower than ordinary rays. On the other hand, they carry more information in the form of the differentials, which is useful for computing ancillary data like field sags.

Reference rays are used as base rays in the interpretation of ordinary ray data. For example, the term *ray displacement* often refers to the difference in coordinates on the image surface between a ray and its corresponding reference ray. Similarly, the *optical path difference* of a ray may compare its phase length to that of the corresponding reference ray. The qualifications expressed in the preceding sentences indicate that the definitions are not universal. Indeed, although the terms *ray displacement* and *optical path difference* are very commonly used in optical design, they are not precisely defined, nor can they be. Let us consider, for example, the optical path difference.

Imagine a monochromatic wavefront from a specified object point that passes through an optical system. Figure 3 shows the wavefront PE emerging in image space, where it is labeled "actual wavefront." Because of aberrations, an ordinary ray perpendicular to the actual wavefront will not intersect the final image surface at the ideal image point I, but at some other point Q. The optical path difference may be defined as the optical path measured along the actual ray between the actual wavefront and a reference sphere centered on the ideal image point.

Unfortunately, the ideal image point is not precisely defined. In the figure, it is shown as the intersection of the reference ray with the image surface, but the reference ray itself

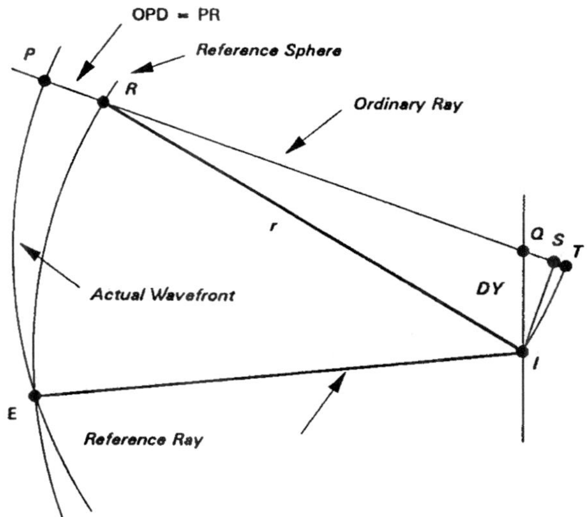

FIGURE 3 The relation between ray trajectories and optical path difference (OPD). See text.

may not be precisely defined. Is it the ray through the center of the aperture stop, or perhaps the ray through the center of the actual vignetted aperture? These two definitions will result in different reference rays, and correspondingly different values for the optical path difference. In fact, in many practical applications neither of the definitions is used, and the actual ideal image point is defined to be the one that minimizes the variance of the optical path difference (and hence maximizes the peak intensity of the diffraction image).

Moreover, the figure shows that even if the ideal image point is precisely defined, the value of the optical path difference depends on the point E where the actual wavefront interesects the reference sphere. For the particular point shown, the optical path difference is the optical length along the ordinary ray from the object point to the point T, less the optical length along the reference ray from the object point to the point I. As the radius of the reference sphere is increased, the point T merges with the point S, where a perpendicular from the ideal image point intersects the ordinary ray.

The above somewhat extended discussion is meant to demonstrate that even "well-known" optical terms are not always precisely defined. Not surprisingly, various optical design programs in common use produce different values for such quantities. There has been little effort to standardize the definitions of many terms, possibly because one cannot legislate physics. In any case, it is important for the user of an optical design program to understand precisely what the program is computing.

Virtually all optical design programs can trace single rays and display the ray heights and direction cosines on each surface. Other data, such as the path length, angles of incidence and refraction, and direction of the normal vector, are also commonly computed. Another type of ray-data display that is nearly universal is the ray-intercept curve, which shows ray displacement on the final image surface versus (fractional) pupil coordinates. A variation plots optical path difference versus pupil coordinates.

In addition to the uncertainty concerning the definition of ray displacement and optical path difference, there are different methods for handling the pupil coordinates. Some programs use entrance pupil coordinates, while others use exit pupil coordinates. In most cases, there is not a significant difference, but in the case of systems containing cylindrical lenses for example, there are major differences.

Another consideration relating to ray-intercept curves is the way in which vignetting is handled. This is coupled to the way that the program handles apertures. As mentioned before, apertures have a special status in many optical design programs. Rays can be blocked by apertures, but this must be handled as a special case by the program, because there is nothing inherent in the ray-trace equations that prevent a blocked ray from being traced, in contrast to a ray that misses a surface or undergoes total internal reflection.

Even though a surface may have a blocking aperture, it may be desirable to let the ray trace proceed anyway. As mentioned before, blocking rays in optimization can produce instabilities that prevent convergence to a solution even though all the rays in the final solution are contained within the allowed apertures. Another situation where blocking can be a problem concerns central obstructions. In such systems, the reference ray may be blocked by an obstruction, so its data are not available to compute the displacement or optical path difference of an ordinary ray (which is not blocked). The programmer must anticipate such situations and build in the proper code to handle them.

In the case of ray-intercept curves, it is not unusual for programs to display data for rays that are actually blocked by apertures. The user is expected to know which rays get through, and ignore the others, a somewhat unreasonable expectation. The justification for allowing it is that the designer can see what would happen to the rays if the apertures were increased.

In addition to ray-intercept curves, optical design programs usually display field sag plots showing the locations of the tangential and sagittal foci as a function of field angle and distortion curves. In the case of distortion, there is the question of what to choose as a reference height. It is generally easiest to refer distortion to the paraxial chief ray height in the final image surface, but in many cases it is more meaningful to refer it to the centroid height of a bundle of exact rays from the same object point. Again, it is important for the user to know what the program is computing.

Spot-Diagram Analysis

Spot diagrams provide the basis for realistic modeling of optical systems in an optical design program. In contrast to simple ray-trace evaluation, which shows data from one or a few rays, spot diagrams average data from hundreds or thousands of rays to evaluate the image of a point source. Notwithstanding this, it should be understood that the principal purpose of an optical design program is to design a system, not to simulate its performance. It is generally up to the designer to understand whether or not the evaluation model of a system is adequate to characterize its real performance, and the prudent designer will view unexpected results with suspicion.

From a programmer's point of view, the most difficult task in spot-diagram analysis is to accurately locate the aperture of the system. For systems that have rotational symmetry, this is not difficult, but for off-axis systems with vignetted apertures it can be a challenging exercise. However, the results of image evaluation routines are often critically dependent on effects that occur near the edges of apertures, so particular care must be paid to this problem in writing optical design software. Like many other aspects of an optical design program, there is a trade-off between efficiency and accuracy.

A spot diagram is an assemblage of data describing the image-space coordinates of a large number of rays traced from a single object point. The data may be either monochromatic or polychromatic. Each ray is assigned a weight proportional to the fractional energy that it carries. Usually, the data saved for each ray include its xyz coordinates on the image surface, the direction cosines klm, and the optical path length or optical path difference from the reference ray. The ray coordinates are treated statistically to calculate root-mean-square spot sizes. The optical path lengths yield a measure of the wavefront quality, expressed through its variance and peak-to-valley error.

To obtain a spot diagram, the entrance pupil must be divided into cells, usually of equal area. Although for many purposes the arrangement of the cells does not matter, for some computations (e.g., transfer functions) it is advantageous to have the cells arranged on a rectangular grid. To make the computations have the proper symmetry, the grid should be symmetrical about the x and y axis. The size of the grid cells determines the total number of rays in the spot diagram.

In computing spot diagrams, the same considerations concerning the reference point appear as for ray fans. That is, it is possible to define ray displacements with respect to the chief-ray, the paraxial ray height, or the centroid of the spot diagram. However, for spot diagrams it is most common to use the centroid as the reference point, both because many image evaluation computations require this definition, and also because the value for the centroid is readily available from the computed ray data.

$$a_i = \varepsilon_{xi} = \text{ray displacement in the } x \text{ direction}$$

$$b_i = \varepsilon_{yi} = \text{ray displacement in the } y \text{ direction}$$

$$c_i = k_i/m_i = \text{ray slope in the } x \text{ direction}$$

$$d_i = l_i/m_i = \text{ray slope in the } y \text{ direction}$$

$$w_i = \text{weight assigned to ray}$$

The displacements of rays on a plane shifted in the z direction from the nominal image plane by an amount Δz are given by

$$\delta x_i = a_i + b_i \, \Delta z$$

$$\delta y_i = c_i + d_i \, \Delta z$$

If there are n rays, the coordinates of the centroid of the spot diagram are

$$\delta \bar{x} = \frac{1}{W} \sum_{i=1}^{n} w_i \delta x_i = A + B \, \Delta z$$

$$\delta \bar{y} = \frac{1}{W} \sum_{i=1}^{n} w_i \delta y_i = C + D \, \Delta z$$

where W is a normalizing constant that ensures that the total energy in the image adds up to 100 percent, and

$$A = \frac{1}{w} \sum_{i=1}^{n} w_i a_i$$

$$B = \frac{1}{W} \sum_{i=1}^{n} w_i b_i$$

$$C = \frac{1}{W} \sum_{i=1}^{n} w_i c_i$$

$$D = \frac{1}{W} \sum_{i=1}^{n} w_i d_i$$

The mean-square spot size can then be written as

$$\text{MSS} = \frac{1}{W} \sum_{i=1}^{n} w_i \{(\delta x_i - \delta \bar{x})^2 + (\delta y_i - \delta \bar{y})^2\}$$

Usually, the root-mean-square (rms) spot size, which is the square root of this quantity, is reported. Since the MSS is a quadratic form, it can be written explicitly as a function of the focus shift by

$$\text{MSS} = P + 2Q \, \Delta z + R(\Delta z)^2$$

where

$$P = \frac{1}{W} \sum_{i=1}^{n} w_i \{(a_i^2 + c_i^2) - (A^2 + C^2)\}$$

$$Q = \frac{1}{W} \sum_{i=1}^{n} w_i \{(a_i b_i + c_i d_i) - (AB + CD)\}$$

$$R = \frac{1}{W} \sum_{i=1}^{n} w_i \{(b_i^2 d_i^2) - (B^2 + D^2)\}$$

Differentiating this expression for the MSS with respect to focus shift, then setting the derivative to zero, determines the focus shift at which the rms spot size has its minimum value:

$$\Delta z_{\text{opt}} = -Q/R$$

Although the above equations determine the rms spot size in two dimensions, similar one-dimensional equations can be written for x and y separately, allowing the ready computation of the tangential and sagittal foci from spot-diagram data. In addition, it is straightforward to carry out the preceding type of analysis using optical path data, which leads to the determination of the center of the reference sphere that minimizes the variance of the wavefront.

Beyond the computation of the statistical rms spot size and the wavefront variance, most optical design programs include a variety of image evaluation routines that are based on spot diagram data. It is useful to characterize them as belonging to geometrical optics or physical optics, according to whether they are based on ray displacements or wavefronts, although, of course, all are based on the results of geometrical ray tracing.

Geometrical Optics. Most optical design programs provide routines for computing radial diagrams and knife-edge scans. To compute a radial energy diagram, the spot-diagram data are sorted according to increasing ray displacement from the centroid of the spot. The fractional energy is then plotted as a function of spot radius. The knife-edge scan involves a similar computation, except that the spot-diagram data are sorted according to x or y coordinates, instead of total ray displacement.

Another type of geometrical image evaluation based on spot-diagram data is the so-called *geometrical optical transfer function* (GOTF). This function can be developed as the limiting case, as the wavelength approaches zero, of the actual diffraction MTF, or, alternately, in a more heuristic way as the Fourier transform of a line spread function found directly from spot-diagram ray displacements (see, for example, Smith's book[10]). From a programming standpoint, computation of the GOTF involves multiplying the ray displacements by 2π times the spatial frequency under consideration, forming cosine and sine terms, and summing over all the rays in the spot diagram. The computation is quick, flexible, and if there are more than a few waves of aberration, accurate. The results of the GOTF computation are typically shown as either plots of the magnitude of the GOTF as a function of frequency, or alternately in the form of what is called a "through-focus" MTF, in which the GOTF at a chosen frequency is plotted as a function of focus shift from the nominal image surface.

Physical Optics. The principal physical optics calculations based on spot-diagram data are the modulation transfer function, sometimes called the "diffraction" MTF, and the point spread function (PSF). Both are based on the wavefront derived from the optical path length data in the spot diagram. There are various ways to compute the MTF and PSF, and not all programs use the same method. The PSF, for example, can be computed from the pupil function using the fast Fourier transform algorithm or, alternately, using direct evaluation of the Fraunhofer diffraction integral. The MTF can be computed either as the Fourier transfer of the PSF or, alternately, using the convolution of the pupil function.[11] The decision as to which method to use involves speed, accuracy, flexibility, and ease of coding.

In physical-optics-based image evaluation, accuracy can be a problem of substantial magnitude. In many optical design programs, diffraction-based computations are only accurate for systems in which diffraction plays an important role in limiting performance. Systems that are limited primarily by geometrical aberrations are difficult to evaluate using physical optics, because the wavefront changes so much across the pupil that it is impossible to sample it sufficiently using a reasonable number of rays. If the actual wavefront in the exit pupil is compared to a reference sphere, the resultant fringe spacing defines the size required for the spot diagram grid, since there must be several sample points per fringe to obtain accurate diffraction calculations. To obtain a small grid spacing, one can either trace many rays, or trace fewer rays but interpolate the resulting data to obtain intermediate data.

Diffraction calculations are necessarily restricted to one wavelength. To obtain polychromatic diffraction results it is necessary to repeat the calculations in each color, adding the results while keeping track of the phase shifts caused by the chromatic aberration.

34.5 OPTIMIZATION

The function of the optimization part of the program is to take a *starting design* and modify its construction so that it meets a given set of specifications. The starting design may be the result of a previous design task, a lens from the library, or a new design based on general optical principles and the designer's intuition.

The performance of the design must be measured by a single number, often known in optics as the *merit function*, although the term *error function* is more descriptive and will be used here. The error function is the sum of squares of quantities called *operands* that characterize the desired attributes. Examples of typical operands include paraxial constants, aberration coefficients, and exact ray displacements. Sometimes, the operands are broken into two groups: those that must be satisfied exactly, which may be called *constraints*, and others that must be minimized. Examples of constraints might include paraxial conditions such as the focal length or numerical aperture.

The constructional parameters to be adjusted are called *variables*, which include lens curvatures, thicknesses, refractive indices, etc. Often the allowed values of the variables are restricted, either by requirements of physical reality (e.g., positive thickness) or the given specifications (e.g., lens diameters less than a prescribed value). These restrictions are called *boundary conditions,* and represent another form of constraint.

Usually, both the operands and constraints are nonlinear functions of the variables, so optical design involves nonlinear optimization with nonlinear constraints, the most difficult type of problem from a mathematical point of view. A great deal of work has been carried out to develop efficient, general methods to solve such problems. Detailed consideration of these methods is beyond the scope of this article, and the reader is referred to a paper by Hayford.[12]

In a typical optical design task, there are more operands than variables. This means that there is, in general, no solution that makes all of the operands equal to their target values. However, there is a well-defined solution called the *least-squares* solution, which is the state of the system for which the operands are collectively as close to their targets as is possible. This is the solution for which the error function is a minimum.

The Damped Least-Squares Method

Most optical design programs utilize some form of the *damped least-squares* (DLS) method, sometimes in combination with other techniques. DLS was introduced to optics in about 1960, so it has a history of more than 30 years of (usually) successful application. It is an example of what is known as a *downhill* optimizer, meaning that in a system with multiple minima, it is supposed to find the nearest local minimum. In practice, it sometimes suffers from *stagnation*, yielding slow convergence. On the other hand, many designers over the years have learned to manipulate the damping factor to overcome this deficiency, and even in some cases to find solutions beyond the local minimum.

We consider first the case of unconstrained optimization. Let the system have M operands f_i and N variables x_i. The error function ϕ is given by

$$\phi = f_1^2 + f_2^2 + \cdots + f_M^2$$

Define the following:

$$\mathbf{A} = \text{derivative matrix,} \qquad A_{ij} \equiv \frac{\partial f_i}{\partial x_j}$$

$$\mathbf{G} = \text{gradient vector,} \qquad G_k \equiv \frac{1}{2}\frac{\partial \phi}{\partial x_k}$$

$$\mathbf{x} = \text{change vector}$$

$$\mathbf{f} = \text{error vector}$$

With these definitions, we have

$$\mathbf{G} = \mathbf{A}^T\mathbf{f}$$

If we assume that the changes in the operands are linearly proportional to the changes in the variables, we have

$$\mathbf{f} = \mathbf{A}\mathbf{x} + \mathbf{f}_0$$

$$\mathbf{G} = \mathbf{A}^T\mathbf{A}\mathbf{x} + \mathbf{G}_0$$

At the solution point, the gradient vector is zero, since the error function is at a minimum. The change vector is thus

$$\mathbf{x} = -(\mathbf{A}^T\mathbf{A})^{-1}\mathbf{G}_0$$

These are called the least-squares *normal equations*, and are the basis for linear least-squares analysis. When nonlinear effects are involved, repeated use of these equations to iterate to a minimum often leads to a diverging solution. To prevent such divergence, it is common to add another term to the error function, and this limits the magnitude of the change vector **x**. In the DLS method, this is accomplished by defining a new error function

$$\varphi = \phi + p\mathbf{x}^T\mathbf{x}$$

A key property of DLS is that the minimum of φ is the same as the minimum of ϕ since, at the minimum, the change vector **x** is zero. By differentiating and setting the derivative equal to zero at the minimum, we arrive at the damped least-squares equations

$$\mathbf{x} = -(\mathbf{A}^T\mathbf{A} + p\mathbf{I})^{-1}\mathbf{G}_0$$

which look like the normal equations with terms added along the diagonal. These terms provide the damping, and the factor p is called the *damping factor*. This particular choice of damping is called *additive damping* but, more generally, it is possible to add any terms to the diagonal and still maintain the same minimum. Some optical design programs multiply the diagonal elements of the $\mathbf{A}^T\mathbf{A}$ matrix by a damping factor, while others make them proportional to second derivative terms. Although theoretical arguments are sometimes advanced to support the choice of a particular method of damping, in practice the choice of damping factor is an ad hoc way to accelerate convergence to a solution by limiting the magnitude (and changing the direction) of the change vector found from the normal equations.

In practical optical design work, it has been found that no single method for choosing the damping factor works best in all cases. In a particular problem, one method may be dramatically better than another, but in a different problem, the situation may be completely reversed. Every optical design program has its unique way of choosing the optimum damping, which makes each program different from the others, and gives it a raison d'être.

Although the principal use of the damping factor is to accelerate convergence by limiting the magnitude of the change vector, the damping factor has also been used routinely to increase the magnitude of the change vector to escape a local minimum. During the course of a minimization task, if the solution stagnates, or does not converge to what the designer believes to be an acceptable configuration, it may be possible to force the solution into another region by running one or more iterations with reduced damping in which the error function increases.

Constraints and Boundary Conditions. There are two general methods used in optical design programs for handling constraints and boundary conditions. The first is to add a term (called a *penalty* function) to the error function that targets the constraint to its desired value. In the case of boundary violations, "one-sided" terms can be added, or special weighting functions can be constructed that increase in magnitude as a violation goes farther into a forbidden region. The other method augments the number of equations by the number of constraints and solves the resulting equations using the Lagrange multiplier method. This produces a minimum that satisfies the constraints exactly and minimizes the remaining error function.

The penalty function method is more flexible and faster (since there are fewer equations) than the Lagrange multiplier method. On the other hand, the Lagrange multiplier method gives more precise control over the constraints. Both are commonly used in optical design software.

Other Methods

Although DLS is used in the vast majority of optical design applications, other methods are occasionally used,[12] and two warrant mention. These are *orthonormalization*, which has been used to overcome stagnation in some DLS problems, and *simulated annealing,* which has been used for global optimization.

Orthonormalization. The technique of orthonormalization for the solution of optical design problems was introduced by Grey.[2] Although it solves the same problem as DLS does, it proceeds in a very different fashion. Instead of forming the least-squares normal equations, Grey works directly with the operand equations

$$\mathbf{A}\mathbf{x} = -\mathbf{f}$$

To understand Grey's method, it is best to forget about optics and consider the solution of these equations strictly from a mathematical point of view. The point of view that Grey uses is that $\mathbf{f}$ represents a vector in m-dimensional space. The columns of $\mathbf{A}$ can be regarded as basis vectors in this m-dimensional space. Since there are only n columns, the basis vectors do not span the space. The change vector $\mathbf{x}$ represents a projection of $\mathbf{f}$ on the basis vectors defined by $\mathbf{A}$. At the solution point, the residual part of $\mathbf{f}$ will be orthogonal to its projection on the basis vectors.

In Grey's orthonormalization method, the solution of the equations is found by a technique similar to Gram-Schmidt orthogonalization, but during the solution process, the actual error function is evaluated several times in an effort to use the best variables to maximum advantage. Because of this, the method is computationally intensive compared to DLS. However, the extra computation is justified by a more accurate solution. The common wisdom is that orthogonalization is superior to DLS near a solution point, and inferior to DLS when the solution is far removed from the starting point.

Simulated Annealing. Simulated annealing has been applied to optical design optimization, chiefly in problems where the task is to find a global minimum. The method varies drastically from other techniques. It makes no use of derivative information, and takes random steps to form trial solutions. If a trial solution has a lower error function than the current system, the new system replaces the old. If a trial solution has a higher error function than the current system, it may be accepted, depending on how much worse it is. The probability of acceptance is taken to be $\exp(-\Delta\phi/T)$, where T is an experimentally determined. In general simulated annealing, T is provided by the user. In adaptive simulated annealing, T is reduced automatically according to algorithms that hold the system near statistical equilibrium.

Error Functions

Obviously, the choice of an error function has a major impact on the success of an optical design task. There are a number of requirements that an error function should meet. Most importantly, the error function should accurately characterize the desired properties of the system under design. There is little chance of success if the program is optimizing the wrong thing. Yet this is an area of great difficulty in computer-aided optical design, because it is at odds with efficiency. In order to obtain more accuracy, more extensive computations should be carried out, but this takes time.

There are two schools of thought concerning the implementation of error functions in optical design programs. The first holds that the designer should have complete control over the items included in the error function, while the second holds that the program itself should set up the basic error function, allowing the designer some degree of control through weighting functions. Neither school has demonstrated superiority, but the approach to error function construction taken by various optical design programs accounts for user allegiances that are sometimes remarkably strong.

The different ways that optical design programs handle error functions makes it difficult to discuss the topic here in anything other than broad detail. At one extreme are programs that provide practically no capability for the user to insert operands, displaying only the value of the overall error function, while at the other extreme are programs that make the user enter every operand individually. Regardless of the user interface, however, there are some general concepts that are universally relevant.

Error functions can be based on either aberration coefficients or exact-ray data (or both). In the early stages of design, aberration coefficients are sometimes favored because they provide insight into the nature of the design, and do not suffer ray failures. However, the accuracy of aberration coefficients for evaluating complex systems is not very good, and exact-ray data are used in virtually all final optimization work.

So far as exact-ray error functions are concerned, there is the question of whether to use ray displacements or optical path difference (or both). This is a matter of user (or programmer) preference. The use of ray displacements leads to minimizing geometrical spot sizes, while the use of optical path difference leads to minimizing the wavefront variance.

For exact-ray error functions, a suitable pattern of rays must be set up. This is often called a *ray set.* There are three common methods for setting up a ray set. The first is to allow the designer to specify the coordinates (object, pupil, wavelength, etc.) for a desired set of rays. This gives great flexibility, but demands considerable skill from the user to ensure that the resulting error function accurately characterizes performance.

The other two methods for setting up ray sets are more automatic. The first is to allow the user to specify object points, and have the program define a rectangular grid of rays in the aperture for each point. The second uses a Gaussian integration scheme proposed by Forbes to compute the rms spot size, averaged over field, aperture, and wavelength.[13] The Forbes method, which is restricted to systems having plane symmetry, leads to dividing the aperture into *rings* and *spokes.* For systems having circular pupils, the Forbes method has both superior accuracy and efficiency, but for vignetted pupils, there is little difference between the two.

Multiconfiguration Optimization

Multiconfiguration optimization refers to a process in which several systems having some common elements are optimized jointly, so that none of the individual systems are optimized, but the *ensemble* of all of the systems is optimized. The archetype of multiconfiguration systems is the *zoom* system in which the focal length is changed by changing the separation between certain elements. The system is optimized simultaneously at high, medium, and low magnifications to produce the best overall performance.

Most of the larger optical design programs have the capability to carry out multiconfiguration optimization, and this capability is probably used more for non-zoom systems than for zoom systems. A common use of this feature is to optimize a focal system for through-focus performance in order to minimize sensitivity to image plane shifts. In fact, multiconfiguration optimization is used routinely to control tolerances.

Tolerancing

Beyond the task of desensitizing a given design, considerations of manufacturing tolerances become increasingly important as the complexity of optical designs increases. It is quite easy to design optical systems that cannot be built because the fabrication tolerances are beyond the capability of optical manufacturing technology. In any case, specifying tolerances is an integral part of optical design, and a design project cannot be considered finished until appropriate tolerances are established.

Tolerancing is closely related to optimization. The basic tolerance computation is to calculate how much the error function changes for a small change in a construction parameter, which is the same type of computation carried out when computing a derivative matrix. Even more relevant, however, is the use of *compensators,* which requires reoptimization. A compensator is a construction parameter that can be adjusted to compensate for an error introduced by another construction parameter. For example, a typical compensator would be the image distance, which could be adjusted to compensate for power changes introduced by curvature errors.

There is considerable variation in how different optical design programs handle tolerancing. Some use the reoptimization method described here, while others use Monte Carlo techniques. Some stress interaction with the designer, while others use defaults for more automatic operation.

34.6 OTHER TOPICS

Of course, many other topics would be included in a full discussion of optical design software. Space limitations and our intended purpose prevents any detailed consideration, but a few of the areas where there is considerable current interest are the following.

Simulation

There is increased interest in using optical design programs to simulate the performance of actual systems. The goal is often to be able to calculate radiometric throughput of a system used in conjunction with a real extended source. It is difficult to provide software to do this with much generality, because brute force methods are very inefficient and hard to specify, while elegant methods tend to have restricted scope, and demand good judgment by the person modeling the physical situation. Nevertheless, with the speed of computers increasing as fast as it has recently, there is bound to be an increasing use of optical design software for evaluating real systems.

Global Optimization

After several years during which there was little interest in optimization methodology, the tremendous increase in the speed of new computers has spawned a renewal of efforts to find global, rather than local, solutions to optical design problems. Global optimization is a much more difficult problem than local optimization. In the absence of an analytic solution, one never knows whether a global optimum has been achieved. All solution criteria must specify a region of interest and a time limit, and the method cannot depend on the starting point. The simulated annealing method described above is one area of continuing interest. Several methods for what might be called *pseudo-global* optimization have been used in commercial optical design programs, combining DLS with algorithms that allow the solution to move away from the current local minimum.

Computing Environment

Increasingly, optical design programs are used in conjunction with other software. Drawing programs, manufacturing inventory software, and intelligent databases are all relevant to optical design. While the conventional optical design program has been a *stand-alone* application, there is increasing demand for integrating optical design into more general design tasks.

34.7 BUYING OPTICAL DESIGN SOFTWARE

The complexity of the optical design process, together with the breadth of applications of optics, has created an ongoing market for commercial optical design software. For people new to optical design, however, the abundance of advertisements, feature lists, and even technical data sheets doesn't make purchasing decisions easy. The following commentary, adapted from an article that appeared recently, may be helpful in selecting an optical design program.[14] It considers five key factors: hardware, features, user interface, cost, and support.

Hardware

It used to be that the choice of an optical design program was governed by the computer hardware available to the designer. Of course, when the hardware cost was many times higher than the software cost, this made a great deal of sense. Today, however, the software often costs more than the hardware, and many programs can be run on several different computer platforms, so the choice of computer hardware is less important. The hardware currently used for optical design is principally IBM-PC compatible, with some of the larger programs running on UNIX workstations, some of the older programs on VAXen, and a few on Macs.

To run optical design software, the fastest computer that can be obtained easily is recommended. The iterative nature of optical design makes the process interminable.

There is a rule, sometimes called the Hyde maxim, that states that an optical design is finished when the time or money runs out. Notwithstanding this, the speed of computers has ceased to be a significant impediment to ordinary optical design. Even low-cost computers now trace more than 1000 ray-surfaces/second, a speed considered the minimum for ordinary design work. The newest UNIX workstations run as much as one hundred times faster than this, and create the potential for solving new types of problems formerly beyond the range of optical design software.

Before desktop computers, optical design software was usually run on time-shared central computers accessed by terminals, and some programs are still in that mode. There seems to be general agreement, however, that the memory-mapped display found on PCs, or the high-speed bit-mapped displays found on UNIX workstations, provide a superior working environment, and dedicated desktop computer systems are currently most popular. In the future, there may be some shift back toward the shared-facility concept, but with X-windows graphics terminals substituting for the older character-based terminals.

Features

If you need a particular feature to carry out your optical design task, then it is obviously important that your optical design program have that feature. But using the number of features as a way to select an optical design program is probably a mistake. There are more important factors, such as cost, ease of use, and scope. Moreover, you might assume that all the features listed for a program work simultaneously, which may not be true. For example, if a vendor states that its program handles holograms and toric surfaces, you might assume that you can work with holographic toroids, but this may not be true.

The continuing growth of optics and the power of desktop computers has put heavy demands on software vendors to keep up with the development of new technology. Moreover, since the customer base is small and most vendors now support the same computer hardware, the market has become highly competitive. These factors have led to a "feature" contest in which software suppliers vie to outdo each other. While this is generally good for the consumer, the introduction of a highly visible new feature can overshadow an equally important but less obvious improvement (for example, fewer bugs or better documentation). In addition, the presence of a number of extra features is no guarantee that the underlying program is structurally sound.

User Interface

There is very little in common between the user interfaces used by various optical design programs. Each seems to have its own personality. The older programs, originally designed to run in batch mode on a large computer, are usually less interactive than ones that were written specifically for desktop computers. Batch programs tend to be built more around default actions than interactive programs, which require more user input. It would be hard to put any of today's major optical design programs in a box classified as either batch or interactive, but the look and feel of a program has a strong influence on its usefulness.

Many people don't realize that the most important benefit of using an optical design program is often the understanding that it provides the user about how a particular design works. It's often tempting to think that if the computer could just come up with a satisfactory solution, the design would be finished. In practice, it is important to know the trade-offs that are made during a design project. This is where the judgment of the optical engineer comes in, knowing whether to make changes in mechanical or electrical specifications to achieve the optimum balance in the overall system. Lens designers often say that the easiest lens to design is one that has to be diffraction-limited, because it is clear when to stop. If the question of how to fit the optics together with other system components is important, then the ability of the user to work interactively with the design program can be a big help.

Cost

In today's market, there is a wide range of prices for optical design software. This can be very confusing for the first-time buyer, who often can't see much difference in the specifications. The pricing of optical design software is influenced by (at least) three factors.

First, the range of tasks that can be carried out using an optical design program is enormous. The difference in complexity between the job of designing a singlet lens for a simple camera, and that of designing a contemporary objective for a microlithographic masking camera is somewhat akin to the difference between a firecracker and a hydrogen bomb.

Second, all software is governed by the factors originally studied in F. P. Brooks' famous essay *The Mythical Man-Month.*[15] Brooks was director of the group that developed the operating system for the IBM 360, a mainframe computer introduced in the 1960s. Despite its provocative title, Brooks' essay is a serious work that has become a standard reference for software developers. In it, he notes that if the task of developing a program to be used on a single computer by its author has a difficulty of 1, then the overall difficulty of producing integrated software written by a group of people and usable by anyone on a wide range of computers may be as high as 10. In recent years, the scope of the major optical design programs has grown too big for a single programmer to develop and maintain, which raises costs.

Third, there are structural differences in the way optical design software is sold. The original mainframe programs were rented, not sold. If the user did not want to continue monthly payments, the software had to be returned. PC programs, on the other hand, are usually sold with a one-time fee. In the optical design software business, several vendors offer a compromise policy, combining a permanent license with an optional ongoing support fee.

It would be nice if the buyer could feel comfortable that "you get what you pay for," but unfortunately this view is too simplistic. One program may lack essential capabilities, another may contain several unnecessary features when evaluated for a particular installation. Buying on the basis of cost, like features, is probably not a good idea.

Support

Support is an important aspect to consider in selecting an optical design program, and it is often difficult to know what is included in support. Minimal support consists of fixing outright bugs in the program. More commonly, support includes software updates and phone or fax assistance in working around problems.

Optical design programs are typically not bug-free. Unlike simple programs like word processors, optimization programs cannot be fully tested, because they generate their own data. One result of this is that software vendors are generally reluctant to offer any warranty beyond a "best-effort" attempt to fix reported problems. Unfortunately, there is no good way for buyers to know whether and when their particular problems may be fixed; the best approach is probably to assess the track record of the vendor by talking to other users.

Coupled with support is user training. Although it should be possible to use a program by studying the documentation, the major optical design software vendors offer regular seminars, often covering not only the mechanics of using their program, but also general instruction in optical design. For new users, this can be a valuable experience.

34.8 SUMMARY

As stated in the introduction, this article is intended as a survey for readers who are not regular users of optical design software. The form of an optical design program described here, consisting of lens entry, evaluation, and optimization sections, is used in many different programs. There has been little standardization in this field, so the "look and feel," performance features and extent of various programs are quite different. Nonetheless, it is hoped that with a knowledge of the basic features described here, the reader will be in a good position to judge whether an optical design program is of use, and to make an informed decision about whether one particular program is better than another.

34.9 REFERENCES

1. D. P. Feder, "Automatic Optical Design," *Appl. Opt.* **2:**1209–1226 (1963).
2. D. S. Grey, "Aberration Theories for Semiautomatic Lens Design by Electronic Computers," *J. Opt. Soc. Am.* **53:**672–680 (1963).
3. G. H. Spencer, "A Flexible Automatic Lens Correction Procedure," *Appl. Opt.* **2:**1257–1264 (1963).
4. C. G. Wynne and P. Wormell, "Lens Design by Computer," *Appl. Opt.* **2:**1233–1238 (1963).
5. See, for example, *The Photonic Industry Buyer's Guide,* Laurin Publishing, Pittsfield, MA 01202.
6. *U.S. Military Handbook for Optical Design,* republished by Sinclair Optics, Fairport, NY 14450 (1987).
7. G. H. Spencer and M. V. R. K. Murty, "Generalized Ray-Tracing Procedure," *J. Opt. Soc. Am.* **52:**672–678 (1962).
8. D. C. O'Shea, *Elements of Modern Optical Design,* John Wiley & Sons, New York (1985).
9. H. A. Buchdahl, *Optical Aberration Coefficients,* Oxford Press (1954).
10. W. J. Smith, *Modern Optical Engineering,* McGraw-Hill, New York (1990).
11. H. H. Hopkins, "Numerical Evaluation of the Frequency Response of Optical Systems," *Proc. Phys. Soc. B,* **70:**1002–1005 (1957).
12. M. J. Hayford, "Optimization Methodology," *Proc. SPIE* Vol. 531, 68–81 (1985).
13. G. W. Forbes, "Optical System Assessment for Design: Numerical Ray Tracing in the Gaussian Pupil," *J. Opt. Soc. Am. A,* **5:**1943–1956 (1988).
14. D. C. Sinclair, "Optical Design Software: What to Look For in a Program," *Photonics Spectra,* Nov. 1991.
15. F. P. Brooks, Jr., *The Mythical Man-Month,* Addison-Wesley (1975).

CHAPTER 35
OPTICAL SPECIFICATIONS

Robert R. Shannon
Optical Sciences Center
University of Arizona
Tucson, Arizona

35.1 GLOSSARY

ATF	approximate transfer factor
DTF	diffraction transfer function
MTF	modulation transfer function
W	wavefront error in units of wavelengths
W_{rms}	root-mean-square wavefront error
t-number	f-number adjusted for lens transmission
v	normalized spatial frequency

35.2 INTRODUCTION

Specifications for optical systems cover a wide range of needs. Functional specifications of the image quality or other optical characteristics are required for the satisfactory operation of a system. These functional specifications serve as the goal for the design and construction of the optical system. In addition, these specifications are a basis for tolerances placed upon the components of the optical system and lead to detailed component specifications used for procurement of the optical elements of the system. Assembly specifications and detailed specifications of optical parts to be produced by a shop can be written based upon these component specifications. The detail and extent of information required is different at each step. Over- or underspecification can contribute significantly to the cost or feasibility of design of an optical system.

Functional specifications are also used to describe the characteristics that an instrument must demonstrate in order to meet the needs of the user. This may include top-level requirements such as size, weight, image scale, image format, power levels, spectral range, and so on. The component specifications are developed after design of the system and describe the optical components, surface, and materials used in the system to the detail necessary to permit fabrication of the components. Assembly specifications are another derivative of the design and system specifications. These include the statement of tolerances upon location of the components, as well as the procedure to be used in assembling and testing the system.

The development and writing of these specifications is important both for initiating and for tracking the course of development of an optical instrument. In a business or legal sense, specifications are used to establish responsibility for a contractor or subcontractor, as well as to define the basis for bidding on the job. Thus the technical specifications can have business importance as well as engineering significance. "Meeting the customer's specifications" is an essential part of any design and fabrication task. Identifying areas where the specifications could be altered with benefit to all parties is an important business and engineering responsibility.

Specifications are usually communicated as a written document following some logical format. Although there are some international standards that may cover the details of drawings of components, there is no established uniform set of standards for stating the specifications on a system or component. The detailed or component specifications are usually added as explanatory notes to drawings of the components to be fabricated. In modern production facilities, the specifications and tolerances are often part of a digital database that is accessed as part of the production of the components of the system.

The detail and the intent of each of these classes of specifications are different. Optical specifications differ from many mechanical or other sets of specifications in that numbers are applied to surfaces and dimensions that control the cumulative effect of errors imposed on a wavefront passing through the total system. Each of the specifications must be verifiable during fabrication, and the overall result must be testable after completion.

Mechanical versus Optical Specifications

There are two types of specifications that must be applied to an optical system or assembly. One set of these includes mechanical tolerances upon the shape or location of the components that indirectly affect the optical quality of the image produced by the system. The other set consists of specialized descriptions that directly affect the image quality.

System versus Components Specifications

Some specifications have meaning only with respect to the behavior of the entire optical system. Others apply to the individual components, but may affect the ability of the entire system to function.

An example of a system specification is a set of numbers limiting the range of acceptable values of the modulation transfer function (MTF) that are required for the system. Another system specification is the total light transmission of the system.

An example of component specifications are tolerances upon surface irregularity, sphericity, and scattering. The related component specification based upon the system light transmission specification might provide detailed statements about the nature and properties of the antireflective coatings to be applied to the surface of each element.

Image Specifications

The specifications that are applied to the image usually deal with image quality. Examples are modulation transfer function, fraction of scattered light, resolution, or distortion. In some cases, these specifications can be quite general, referring to the ability of the lens to deliver an image suitable for a given purpose, such as the identification of serial numbers on specific products that are to be read by an automated scanner. In other cases, the requirements will be given in a physically meaningful manner, such as "the MTF will be greater than 40 percent at 50 lines per millimeter throughout the field of view."

Other criteria may be used for the image specifications. One example is the energy concentration. This approach specifies the concentration of light from a point object on the image surface. For example, the specification might read that "75 percent of the light shall fall within a 25-micrometer-diameter circle on the image." This quantity is obviously measurable by a photometer with appropriate-sized apertures. The function may be computed from the design data by a method of numerical integration similar to that providing the point spread function or modulation transfer function.

Wavefront Specifications

Wavefront specifications describe the extent to which the wavefront leaving the lens or components conforms to the ideal or desired shape. For example, a wavefront leaving a lens would ideally conform to a sphere centered upon the chosen focal location. The departure of the actual wavefront from this ideal, would be expressed either as a matrix or map of departure of the wavefront from the ideal sphere, as a set of functional forms representing the deviation, or as an average (usually root mean square) departure from the ideal surface. By convention, these departures are expressed in units of wavelength, although there is a growing tendency to use micrometers as the unit of measure.

The root-mean-square wavefront error, or rms wavefront error, is a specific average over the wavefront phase errors in the exit pupil. The basic definition is found by defining the nth power average of the wavefront $W(x, y)$ over the area A of the pupil and then specifically defining W_{rms} or, in words, the rms wavefront error is the square root of the mean square error minus the square of the mean wavefront error.

$$\bar{W}^n = \frac{1}{A}\int W(x, y)^n dx\, dy$$

$$W_{\text{rms}} = \sqrt{[\bar{W}^2] - [\bar{W}]^2}$$

The ability to obtain a complete specification of image quality by a single number describing the wavefront shape has been seen to be questionable in many cases. Addition of a correlation length, sometimes expressed as a phase difference between separated points, has become common. In other cases, the relative magnitude of the error when represented by various orders of Zernike polynomials is used.

There is, of course, a specific relationship between the wavefront error produced by a lens and the resulting image quality. In the lens, this is established by the process of diffraction image formation. In establishing specifications, the image quality can be determined by computation of the modulation transfer function from the known wavefront aberrations. In setting specifications, this is of some value, but an average or guide to acceptable values can be of great aid.

A perfect lens is one that produces a wavefront with no aberration, or zero rms wavefront error. By convention, any wavefront with less than 0.07 waves, rms, of aberration is considered to be essentially perfect. It is referred to as *diffraction-limited,* since the image produced by such a lens is deemed to be essentially indistinguishable from a perfect image.

The definition of image quality depends upon the intended application for the lens. In general, nearly perfect image quality is produced by lenses with wavefront errors of less than 0.15 waves, rms. Somewhat poorer image quality is found with lenses that have greater than about 0.15 waves of error. The vast majority of imaging systems operate with wavefront errors in the range of 0.1 to 0.25 waves, rms.

There are several different methods that can be used to establish this relationship. The most useful comparison is with the MTF for a lens with varying amounts of aberration. The larger the wavefront error, the lower will be the contrast at specific spatial frequencies. For rms error levels of less than 0.25 or so, the relation is generally monotonic. For larger aberrations, the MTF becomes rather complex, and the relation between rms wavefront error and MTF value can be multiple-valued. Nevertheless, an approximate relation between MTF and rms wavefront error would be useful in setting reasonable specifications for a lens.

There are several possible approximate relations, but one useful one is the empirical formula relating root-mean-square wavefront error and MTF given by

$$\text{MTF}(\nu) = \text{DTF}(\nu) \times \text{ATF}(\nu)$$

The functional forms for these values are

$$\text{DTF}(\nu) = \frac{2}{\pi}\left[\arccos(\nu) - \nu\sqrt{1-\nu^2}\right]$$

$$ATF(\nu) = \left[1 - \left(\left(\frac{W_{\text{rms}}}{0.18}\right)^2\right)(1 - 4(\nu - 0.5)^2)\right]$$

$$\nu = \frac{\text{spatial frequency}}{\text{spatial frequency cutoff}} = \frac{N}{\left(\dfrac{1}{\lambda\ f\text{-number}}\right)}$$

These look quite complicated, but are relatively simple, as is shown in Fig. 1. This is an approximation, however, and it becomes progressively less accurate as the amount of the rms wavefront error W_{rms} exceeds about 0.18 wavelength.

Figure 1 shows a plot of several values for the MTF of an optical system using this approximate method of computation. The system designer can use this information to determine the appropriate level of residual rms wavefront error that will be acceptable for the system of interest. It is important to note that this is an empirical attempt to provide a link between the wavefront error and the MTF as a single-value description of the state of correction of a system. Examination of the curves provides a method of communicating the specification to the system designer and fabricator. More detail on applying the rms wavefront error can be found in the article on tolerancing in this book.

Published Standards

There are published standards from various sources. The most frequently referred to are those from the U.S. Department of Defense, but a number of standards are being proposed by the International Standards Organization.

35.3 PREPARATION OF OPTICAL SPECIFICATIONS

Gaussian Parameters

The gaussian parameters determine the basic imaging properties of the lens. They are the starting point for setting the specifications for a lens system. In principle these numbers can be specified precisely as desired. In reality, overly tight specifications can greatly increase the cost of the lens. Some of the important parameters are shown in Table 1.

Table 1 is a sample of reasonable values that may be placed upon a lens. A specific case may vary from these nominal values. The image location, radiometry, and scale are fixed by these numbers. A specific application will require some adjustment of these nominal

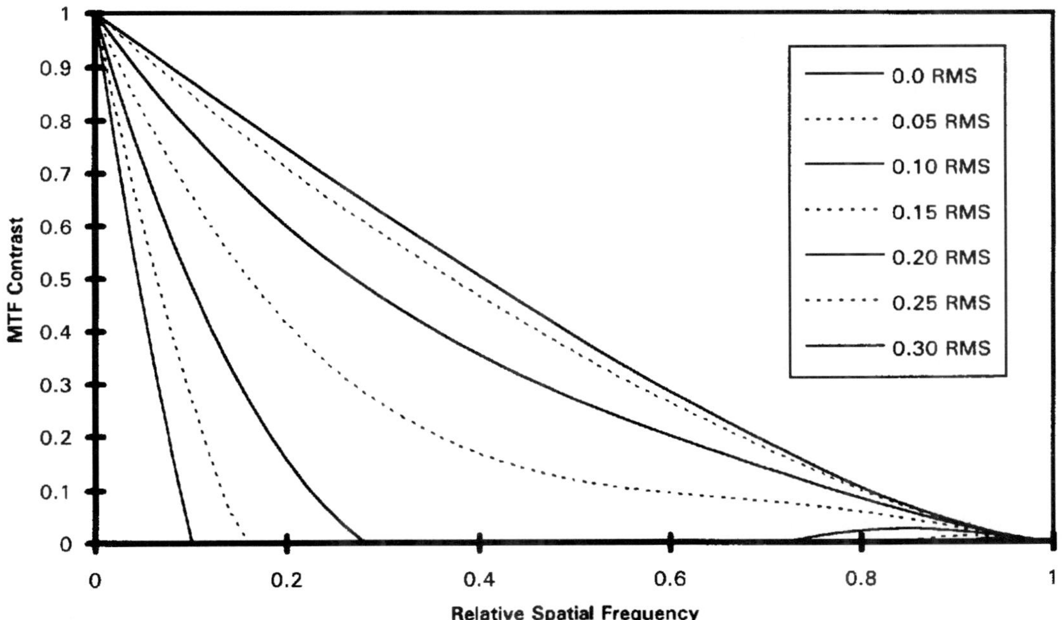

FIGURE 1 Approximate MTF curves from formula.

values. In general, specifications that are tighter than these values will likely result in increased cost and difficulty of manufacture.

There is an interaction between these numbers. For example, the tight specification of magnification and overall conjugate distance will require a very closely held specification upon the focal length. The interaction between these numbers should be considered by the user to avoid accidentally producing an undue difficulty for the fabricator. It may be appropriate to specify a looser tolerance on some of these quantities for the prototype lens, and later design a manufacturing process to bring the production values within a smaller tolerance. However, it is appropriate that this be investigated fully at the design stage.

35.4 IMAGE SPECIFICATIONS

Image Quality

The rms wavefront error and the MTF for a lens have been discussed above as useful items to specify for a lens. Frequently, the user desires to apply a detection criterion to the image. This is always related to the application for the lens.

The most familiar functional specification that is widely used for system image quality is the resolution of the system. This is usually stated as the number of line pairs per millimeter that need to be visually distinguished or recognized by the user of the system. Since this involves both the physics of image formation as well as the psychophysics of vision, this is an interesting goal, but needs to be specified more clearly to be of use to a

TABLE 1 Gaussian parameters

Parameter	Precision target	Importance	How verified
Focal length	1 to 2%	Determines focal position and image size	Lens bench
f-number	$<\pm 5\%$	Determines irradiance at image plane	Geometrical measurement
Field angle	$<\pm 2\%$	Determines extent of image	Lens bench
Magnification	$<\pm 2\%$	Determines overall conjugate distances	Trial setup of lens
Back focus	$\pm 5\%$	Image location	Lens bench
Wavelength range	As needed; set by detector and source	Describes spectral range covered by lens	Image measurement
Transmission	Usually specified as $>0.98^n$ for n surfaces	Total energy through lens	Imaging test, radiometric test of lens
Vignetting	Usually by requiring transmission to drop by less than 20% or so at the edge of the field	Uniformity of irradiance in the image	Imaging test, radiometric test of lens

designer. The reading of the resolution by a human observer is subjective, and the values obtained may differ between observers. Therefore, it is necessary to specify the conditions under which the test is to be carried out.

The type of target and its contrast need to be stated. The default standard in this case is the "standard" U.S. Air Force three-bar target, with high contrast, and a 6:1 ratio of bar length to width. This is usually selected as it will give the highest numerical values, certainly politically desirable. However, studies have shown that there is a better correlation between the resolution produced by a low-contrast target, say 2:1 contrast ratio, or 0.33 modulation contrast and the general acceptability of an image.

The resolution is, of course, related to the value of the MTF in the spatial frequency region of the resolution, as well as the threshold of detection or recognition for the observer viewing the target. If the thresholds are available, the above-described empirical relation between the rms wavefront error and the MTF can be used to estimate the allowable aberration that can be left in the system after design or fabrication.

In the case of a system not intended to produce an image to be viewed by a human, a specific definition of the required image contrast or energy concentration is usually possible. The signal-to-noise ratio of the data transmitted to some electronic device that is to make a decision can be calculated once a model for operation of the detector is assumed. The specification writer can then work backward through the required MTF to establish an acceptable level of image quality. The process is similar to that for the visual system above, except that the threshold is fully calculable.

In some cases, the fractional amount of energy collected by the aperture of the lens from a small angular source, such as a star, falling within the dimensions of a detector of a given size is desired. Such a requirement can be given directly to the designer.

Image Irradiance

The radiometry of the image is usually of importance. With an optical system containing the source, such as a viewer, projector, or printer, the usual specification is of the irradiance of the image in some appropriate units. Specifying the screen irradiance in watts per square centimeter or, more commonly, foot-lamberts, implies a number of optical properties. The radiance of the source, the transmission of the system and the apertures of the lenses are derived from this requirement.

In the more usual imaging situation, the f-number and the transmission of the lens are specified. If the lens covers a reasonable field, the allowable reduction in image irradiance over the field of view must also be specified. This leads directly to the level of vignetting that can be allowed by the designer in carrying out the setup and design of the lens system.

There is an interaction between these irradiance specifications and the image quality that can be obtained. The requirement of a large numerical aperture leads to a more difficult design problem, as the high-order aberration content is increased in lenses of high numerical aperture.

An attempt to separate the geometrical aperture effects from the transmission of the components of the lens is accomplished through the t-number specification. Since the relative amount of irradiance falling on the focal plane is inversely proportional to the square of the f/number of a lens, the effect of tranmission of the lens can be included by dividing the f-number by the transmission of the lens.

$$t/\text{number} = \frac{f\text{-number}}{\sqrt{t}}$$

where t is the transmission factor for the lens. The transmission factor for the lens is the product of the bulk transmission of the glass and the transmission factor for each of the surfaces. When this is specified, the designer must provide a combination of lens transmission and relative aperture that meets or exceeds a stated value.

Depth of Focus

The definition of the depth of focus is usually the result of a tolerance investigation. The allowable focal depth is obtained by determining when an unacceptable level of image quality is obtained. There is an obvious relation between the geometry of the lens numerical aperture and the aberrations that establishes the change of MTF with focal position. This effect can be computed for specific cases, or estimated by recognizing that the relation between rms wavefront error and focus shift is

$$W_{\text{rms}_{\text{def}}} = \frac{\delta l}{8\lambda(f\text{-number})^2}$$

which can be used in the above approximate MTF to provide an estimate of the likely MTF over a focal range.

If there is a basic amount of aberration present in the lens, then the approximation is that

$$W_{\text{rms}_{\text{total}}} = \sqrt{W^2_{\text{rms}_{\text{def}}} + W^2_{\text{rms}_{\text{lens}}}}$$

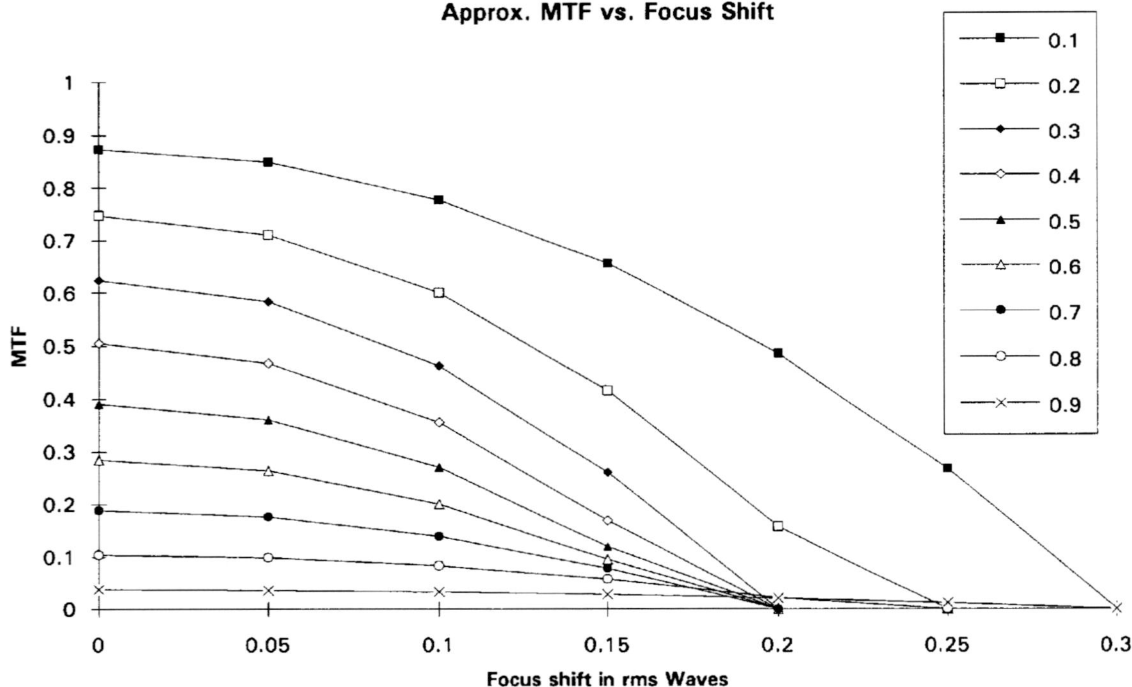

FIGURE 2 Approximate MTF as a function of focal position for various spatial frequencies.

leading to a calculation of the estimated MTF value for the given spatial frequency and focal position. As an example, Fig. 2 provides a plot of the focal position change of the MTF. The interpretation of this curve is straightforward. Each of the curves provides the MTF versus focus for a spatial frequency that is the stated fraction of the cutoff spatial frequency. The defocus is given in units of rms wavelength error, which can be obtained from reference to the appropriate formula. Using this approximate data, a specification writer can determine whether the requirements for image quality, f-number, and focal depth are realistic.

An additional consideration regarding depth of focus is that the field of the lens must be considered. The approximate model presented here is used at an individual field point. An actual lens must show the expected depth of field across the entire image surface, which places some limits upon the allowable field curvature. In general, it is the responsibility of the specification writer to establish the goal. It is the responsibility of the optical designer to determine whether the goal is realizable, and to design a system to meet the needs.

35.5 ELEMENT DESCRIPTION

Each element of the lens to be fabricated must be described in detail, usually through a drawing. All of the dimensions will require tolerances, or plus and minus values that, if met, lead to a high probability that the specified image quality goals will be met.

Mechanical Dimensions

The mechanical dimensions are specified to ensure that the element will fit into the cell sufficiently closely that the lens elements are held in alignment. This will be a result of tolerance evaluation, and must include allowances for assembly, thermal changes, and so on.

An important item for any lens is the interface specification, which describes the method of mounting the lens to the optical device used with it. For some items, such as cameras and microscopes, there are standard sizes and screw threads that should be used. In other cases, the specification needs to describe a method for coupling or mounting the optics in which there is a strain-free transfer of load between the lens and the mounting.

Optical Parameters

The optical parameters of the lens element relate to the surfaces that are part of the image-forming process. The radius of curvature of the spherical refracting (or reflecting) surfaces needs to be specified, as well as a plus or minus value providing the allowable tolerances. When tested using a test glass or an interferometer, the important radius specification is usually expressed in terms of fringes of spherical departure from the nominal radius. In addition, the shape of the surface is usually specified in terms of the fringes of irregularity that may be permitted.

When specifying a surface that will be measured on an interferometer, adjustment of focus during the test can be made. In this case, the spherical component of the surface, that is, the fringes of radius error, can be specified independently of the irregularity fringes that are applicable to the surface. When test glasses are to be used, the spherical component must be fabricated to within a small level of error to permit accurate reading of the irregularity component of the surface.

The cosmetic characteristics of the surface also need to be stated. The specification for this is as yet a bit imperfect, with the use of a scratch-and-dig number. This is actually intended to be a comparison of surface scratches with a visual standard, but is generally accepted to be in terms of a ratio, such as 20:10, which means, more or less, scratches of less than 2 μm width and digs of less than 100 μm diameter. This specification is described in MIL-O-13830, and is referred to a set of standards that are used for visual comparison to the defects on the surface. There have been several attempts to quantify this specification in detail, but no generally accepted standard has yet been achieved.

Material Specifications

The usual material for a lens is optical glass, although plastics are becoming more commonly used in optics. The specification of a material requires identification of the type, as, for example, BK7 glass from Schott. Additional data upon the homogeneity class and the birefringence needs to be stated in ordering the glass. The homogeneity is usually specified by class, currently P1 through P4 with the higher number representing the highest homogeneity, or lowest variation of index of refraction throughout the glass. The method of specifying glass varies with the manufacturer, and with the catalog date. It is necessary to refer to a current catalog to ensure that the correct specification is being used.

Similar data should be provided regarding plastics. Additional data about transmission is usually not necessary, as the type of material is selected from a catalog which provides the physical description of the material. Usually, the manufacturer of the plastic will be noted to ensure that the proper material is obtained.

Materials for reflective components similarly have catalog data describing the class and properties of the material. In specifying such materials, it is usually necessary to add a description of the form and the final shape required for the blank from which the components will be made.

Coating Specifications

The thin film coatings that are applied to the optical surfaces require some careful specification writing. In general, the spectral characteristics need to be spelled out, such as passband and maximum reflectivity for an antireflection coating. Requirements for the environmental stability also need to be described, with reference to tests for film adhesion and durability. Generally, the coating supplier will have a set of "in-house" specifications that will guarantee a specific result that can be used as the basis for the coating specification.

35.6 ENVIRONMENTAL SPECIFICATIONS

Temperature and Humidity

Specification setting should also include a description of the temperature range that will be experienced in use or storage. This greatly affects the choice of materials that can be used. The humidity and such militarily favorite specifications as salt spray tests are very important in material selection and design.

Shock and Vibration

The ruggedness of an instrument is determined by the extent to which it survives bad handling. A requirement that the lens shall survive some specified drop test can be used. In other cases, stating the audio frequency power spectrum that is likely to be encountered by the lens is a method of specifying ruggedness in environments such as spacecraft and aircraft. In most cases, the delivery and storage environment is far more stressing than the usage environment. Any specification written in this respect should be careful to state the limits under which the instrument is actually supposed to operate, and the range over which it is merely meant to survive storage.

35.7 PRESENTATION OF SPECIFICATIONS

Format for Specifications

The format used in conveying specifications for an optical system is sometimes constrained by the governmental or industrial policy of the purchaser. Most often, there is no specific format for expressing the specifications.

The best approach is to precede the specifications with a brief statement as to the goals for the use of the instrument being specified. Following this, the most important optical parameters, such as focal length, *f*-number and field size (object and image) should be stated. In some cases, magnification and overall object-to-image distance along with object dimension will be the defining quantities.

Following this, the wavelength range, detector specifications, and a statement regarding the required image quality should be given. The transmission of the lens is also important at this stage.

Following the optical specifications, the mechanical and environmental requirements should be stated. The temperature and humidity relations under which the optical system needs to operate as well as a statement of storage environment are needed. Descriptions of the mechanical environment, such as shock and vibration, are also important, even if expressed generally.

Other important pieces of information, such as a desired cost target, can then be included. Any special conditions, such as the need to be exposed to rapid temperature changes or a radiation environment, should be clearly stated. Finally, some statement of the finish quality for the optical system should be given.

In many cases, a list of applicable governmental specifications will be listed. In each case it is appropriate to ensure that these referenced documents are actually available to the individual who has to respond to the specification.

Use of International Standards

An important tool in writing specifications is the ability to refer to an established set of standards that may be applicable to the system being designed. In some cases, the development of specifications is simplified by the specification writer being able to refer to a set of codified statements about the environment or other characteristics the system must meet. In other cases, the established standards can be used as a reference to interpret parts of the specifications being written. For example, there may be a set of standards regarding interpretation of items included in a drawing.

Standards are an aid, not an end to specification. If the instrument must be interchangeable with parts from other sources, then the standards must be adhered to carefully. In other cases, the standards can serve as an indicator of accepted good practice in design or fabrication. It is the responsibility of the specification writer to ensure that the standard is applicable and meaningful in any particular situation.

At the present time there is growing activity in the preparation of standards for drawings, interfaces and dimensions, MTF, and other properties of optical and electro-optical systems. The efforts in this direction are coordinated by the International Standards organization, but there are a number of individual standards published by national standards organizations in Germany, England, Japan, and the United States. The first major publications are ISO 10110, detailing preparation of drawings for optical elements and systems and ISO 9211, on optical coatings. Other standards on optical testing and environmental requirements are in draft form.

The ISO standards are expected to provide significant detail on various standards issues, and should become the principal guiding documents. At present, the standards documents that are most used in the United States are the various military specifications, or MIL-SPEC documents, that cover many different aspects of optical systems.

Information on published standards is available from the American National Standards Institute (ANSI), 11 West 42d Street, New York, NY 11036. Information on U.S. Department of Defense standards is available from the Naval Publications and Forms Center, 5801 Tabor Ave., Philadelphia, PA 19120.

The list of standards existing or in draft is very long. There are even more standards describing the mechanical or electrical portions of an instrument. The specification writer should become familiar with those that could be applicable to his or her task.

35.8 PROBLEMS WITH SPECIFICATION WRITING

Underspecification

Failure to specify all of the conditions leaves the user vulnerable to having an instrument that will not operate properly in the real world. In many cases, the designer may not be aware of situations that may arise in operation that may affect the proper choice of design methods. Therefore, the design may not meet the actual needs.

The engineer developing a specification should examine all aspects of the problem to be solved, and carefully set the boundaries for the requirements on an optical system to meet the needs. All of the pertinent information about the image quality, environment, and relation to other systems that may interact with the lens being specified should be considered. The specifying engineer should also review the physical limits on the image quality and ensure that these are translated into realistic values.

Overspecification

Specifying image quality and focal position requirements too tightly can lead to problems. Overspecification would seem to ensure that the needs will be met, but difficulties in meeting these requirements can lead to designs that are difficult and expensive to build. Achieving the goals can be costly and may fail. In such cases, the penalty for not quite meeting very tight specifications can be serious, both economically and technically.

Boundary-Limit Specification

In most cases, the statement of goals or boundaries within which a lens must operate is better than stating specific values. This leaves the designer with some room to maneuver to find an economic solution to the design. Obviously, some fixed values are needed, such as focal length, f-number, and field angle. However, too-tight specifications upon such items as weight, space, and materials can force the design engineer into a corner where a less desirable solution is achieved.

CHAPTER 36
TOLERANCING TECHNIQUES

Robert R. Shannon
Optical Sciences Center
University of Arizona
Tucson, Arizona

36.1 GLOSSARY

a	relative tolerance error
BK7, SF2	types of optical glass
C to F	spectral region 0.486 to 0.656 micrometers
f-number	relative aperture as in F/2.8
n	refractive index
$r1, r2, r3, r4$	radii of curvature of surfaces
$t2$	airspace thickness in sample lens
V	Abbe number or reciprocal dispersion
W	wavefront or pupil function
x	change factor
Δ	finite change in a parameter
δ	small change in a parameter
λ	wavelength

36.2 INTRODUCTION

Determination of the tolerances upon an optical system is one of the most important subjects in optical design. No component can be made perfectly; thus, stating a reasonable acceptable range for the dimensions or characteristics is important to ensure that an economical, functioning instrument results. The tolerances attached to the dimensions describing the parts of the lens system are an important communication by the designer to the fabrication shop of the precision required in making the components and assembling them into a final lens.

The tolerances are not the same as the specifications. Setting specifications is discussed in Chap. 35 in this *Handbook*. The tolerances are responsive to the requested system specifications, and are intended to ensure that the final, assembled instrument meets the requested performance. The specification placed upon the individual lens elements or components are derived from the tolerances.

There are three principal issues in optical tolerancing. The first is the setting of an appropriate goal for the image quality to be expected from the system. The second is the translation of this goal into allowable changes introduced by errors occurring on each component of the system. The third is the distribution of these allowable errors against all of the components of the system, in which some components of the optical system may partially or completely compensate for errors introduced by other components.

Optical versus Mechanical Tolerances

The tolerances upon mechanical parts, in which a dimension may be stated as a specific value, plus or minus some allowable error, are familiar to any engineer. For example, the diameter of a rotating shaft may be expressed as $20.00\ \text{mm} + 0.01/-0.02\ \text{mm}$. This dimension ensures that the shaft will fit into another component, such as the inner part of a bearing, and that fabricating the shaft to within the specified range will ensure that proper operational fit occurs.

Optical tolerances are more complicated, as they are generally stated as a mechanical error in a dimension, but the allowable error is determined by the effect upon an entire set of wavefronts passing through the lens. For example, the radius of curvature of a surface may be specified as $27.00\ \text{mm} \pm 0.05\ \text{mm}$. This is interpreted that the shape of the surface should conform to a specific spherical form, but remain within a range of allowable curvatures. Meeting this criterion indicates that the surface will perform properly in producing a focused wavefront, along with other surfaces in the optical system. Verification that the tolerance is met is usually carried out by an optical test, such as examining the fit to a test plate.

Basis for Tolerances

The process involved in setting tolerances begins with setting of the minimum level of acceptable image quality. This is usually expressed as the desired level of contrast at a specific spatial frequency as expressed by the modulation transfer function. Each parameter of the system is varied to determine how large an error is allowed before the contrast is reduced to the specified level. This differential change is then used to set the allowable range of error in each component.

In most cases, direct computation of the change in the contrast is a lengthy procedure, so that a more direct function, such as the rms wavefront error, is used as the quality-defining criterion. In other cases, the quantity of importance will be the focal length, image position, or distortion.

Relating the computed individual errors in the system to the tolerances to be specified is not always a simple matter. If there are several components, some errors may compensate other errors. Thus, it would be easy to assign too tight a tolerance for each surface unless these compensating effects, as well as the probability of a specific distribution of errors, are used in assigning the final tolerances.

Tolerance Budgeting

The method of incorporating compensation of one error by another, as well as the likelihood of obtaining a certain level of error in a defined fabrication process, is called *tolerance budgeting.* As an example, in a lens system it may be found that maintaining the thickness of a component may be easier than keeping the surfaces of the component at the right spherical form. The designer may choose to allow a looser tolerance for the thickness and use some of the distributed error to tighten the tolerance on the radius of curvature. In other cases, the shop carrying out the fabrication may be known to be able to measure surfaces well, but has difficulty with the centering of the lenses. The designer may choose to trade a tight tolerance upon the irregularity of the surfaces for a loose tolerance upon the wedge in the lens components.

Finally, the effect of a plus error on one surface may be partially compensated by a minus error on another surface. If the probability of errors is considered, the designer may choose to budget a looser tolerance to both surfaces.

This budgeting of tolerances is one of the most difficult parts of a tolerancing process, since judgment, rather than hard numbers, is very much a part of the budget decisions.

Tolerance Verification

Simply stating a set of allowable errors does not complete the integrated process of design and fabrication. The errors must be measurable. Measurement of length can be gauged, but has to be within the capabilities of the shop fabricating the optics. Measurement of error in radius of curvature requires the use of an interferometer or test plates to determine the shape of the surface. Measurement of the nonspherical component of the surface, or the irregularity, requires either an estimate from the test plate, or a computation of the lack of fit to a spherical surface based upon measured fringes.

Finally, the quality of a completed lens must be measurable. Use of a criterion that cannot be measured or controlled by the shop or by the user is not acceptable. The contrast mentioned is not always obtained by the optical shop. The surface errors as measured by an interferometer or by test plates are common. Measurement of the final wavefront from an assembled optical system is most frequently calculated in a summary method by an interferometer. For this reason, wavefront tolerancing methods have become the most commonly used methods of defining and verifying tolerances.

36.3 WAVEFRONT TOLERANCES

The rms wavefront error tolerancing method will now be used as an example of the approach to evaluating the tolerances required to fabricate a lens. An example which discusses the axial image tolerances for a doublet will be used to provide insight into the tolerancing of a relatively simple system. Most optical tolerancing problems are far more complex, but this example provides an insight into the methods applied. The specific example selected for this article is an airspaced achromatic doublet using BK7 and SF2 glasses, F/2.8, 100-mm focal length. The lens design is nominally of moderately good quality, and is optimized over the usual C to F spectral range, with balanced spherical aberration, and is corrected for coma.

Figure 1 is a drawing of the sample doublet used in this article. The locations of the four radii of curvature and the airspace to be toleranced are indicated. The number of possible errors that actually can occur in such a simple lens is surprisingly large. There are four curvatures, two thicknesses, one airspace, and two materials that may have refractive index or dispersion errors. In addition to these seven quantities, there are two element wedge angles, four possible tilt, and four decenter possibilities, plus the irregularity on four surfaces and the homogeneity of two materials. So far, there are 21 possible tolerances that are required in order to completely define the lens. For interfacing to the lens mount, the element diameters, roundness after edging, and cone angle on the edges must be considered as well. More complex systems have far more possible sources of error. For the example here, only the four radii and the airspace separation will be considered.

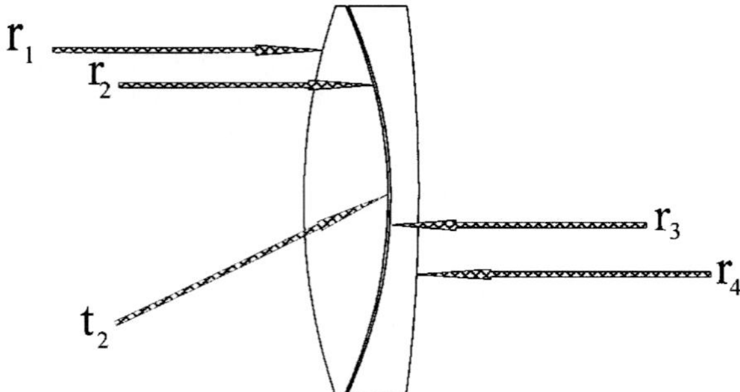

FIGURE 1 Drawing of the doublet lens used for tolerancing.

Parameter Error Quality Definitions

The starting point for a tolerance calculation is the definition of a set of levels that may be used to define the initial allowable range of variation of the parameters in the lens. The magnitude of these classes of errors is determined by experience, and usually depends upon the type of fabrication facility being used. Table 1 presents some realistic values for different levels of shop capability.

These values are based upon the type of work that can be expected from different shops, and serves as a guide for initiating the tolerancing process. It is obvious that the degree of difficulty in meeting the quality goals becomes more expensive as the required image quality increases.

Computation of Individual Tolerances

The individual tolerances to be applied to the parameters are obtained by computing the effect of parameter changes upon the image-quality function. For the example doublet, the nominal amount of rms wavefront error at the central wavelength is 0.116 waves, rms. It is determined by the user from consideration of the needs for the application that the maximum amount of error that is acceptable is 0.15 waves, rms. The tolerancing task is to

TABLE 1 Reasonable Starting Points for Tolerancing a Lens System

Parameter	Commercial	Precision	High precision
Wavefront residual	0.25 wave rms 2-wave peak	0.1 wave rms 0.5 wave peak	<0.07 wave rms <0.25 wave peak
Thickness	0.1 mm	0.01 mm	0.001 mm
Radius	1.0%	0.1%	0.01%
Index	0.001	0.0001	0.00001
V-number	1.0%	0.1%	0.01%
Homogeneity	0.0001	0.00001	0.000002
Decenter	0.1 mm	0.01 mm	0.001 mm
Tilt	1 arc min	10 arc sec	1 arc sec
Sphericity	2 rings	1 ring	0.25 ring
Irregularity	1 ring	0.25 ring	<0.1 ring

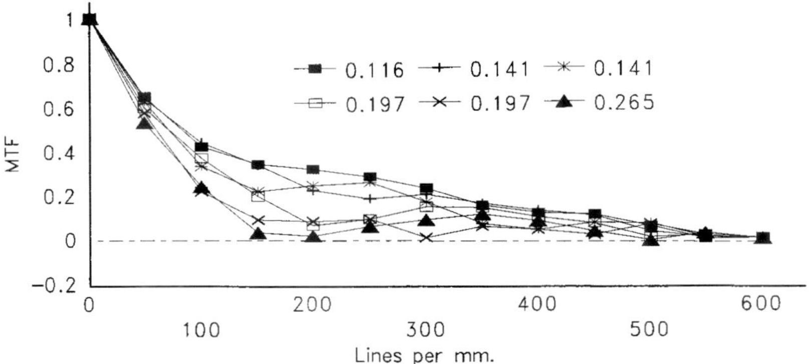

FIGURE 2 Some examples of the effect of various rms errors on the MTF of the sample doublet. (The rms error is stated in wavelengths.)

specify the tolerances upon the radii of curvature and the separation between the components such that the goal is met.

The reason for the choice of 0.15 waves rms is indicated by Fig. 2. The designer experimented with several choices of focus position to obtain a set of plots of the MTF for different amounts of error. This is not a general conclusion since the source of error follows a specified pattern, but the samples permit the intelligent selection of an upper bound to the required error. In a lens with more sources of error, and with larger amounts of aberration in the basic design, an example such as this is extremely necessary to avoid an error in the goal for the final image quality.

The first computation of the effect of nominal changes in the parameters on the rms wavefront error leads to the results in Table 2. Only the radii and thickness are considered for this example. The two columns for rms wavefront effect are first, for the aberration if no adjustment for best focus is made, and second, permitting the establishment of best focus after assembly of the lens.

Table 2 shows that the effect of a change in the parameter will have an effect proportional to the change, but that the factor relating the change to the resulting wavefront error is different for the various parameters. The amount of change if the parameter is not compensated by allowing an adjustment for best focus is quite minimal. Any one of the parameters would have to be maintained within a range far less than the delta used in computation. The allowance of a compensating focal shift does greatly loosen the tolerance.

Since the acceptable goal is 0.15 waves, rms, the amount of change of an individual parameter to attain the acceptable level is about 50 times that of $r1$ and $r4$, but the change

TABLE 2 Finite Differentials for Computing Tolerances

Parameter	Delta	Rms uncompensated	Rms compensated
$r1$	0.1%	0.740	0.117
$r2$	0.1%	1.187	0.171
$r3$	0.1%	1.456	0.157
$r4$	0.1%	0.346	0.110
$t2$	0.025 mm	1.155	0.152

of 0.1 percent would be excessive for $r2$ and $r3$. The allowable change for $t2$ alone would be just about the delta of 0.025 mm.

Combination of Tolerances

No parameter in a lens lives alone. The effect upon the image will be the result of combining the effect of all of the errors. If the errors are uncorrelated, then the usual statistical summing of errors can be used. This states that the total amount of aberration produced by the errors can be found by using

$$W_{\mathrm{rms}} = \sqrt{\sum_i W_i^2} = \sqrt{\sum_i \left(a_i x_i \frac{\partial W_{\mathrm{rms}}}{\partial x_i}\right)^2}$$

where the sum is taken over the i parameters of interest. The x factors are the amount of change used in computing the change of wavefront error and the a factors are the relative amount of tolerance error allotted to each parameter in units of the delta used in the computation.

There are implicit assumptions in the application of this method to distributing tolerances. The principal assumption is that the fabrication errors will follow normal gaussian statistics. For many fabrication processes, this is not true, and modification of the approach is required.

For the example of the doublet, Table 3 can be generated to evaluate the different possibilities in assigning tolerances. The allowable change in rms wavefront error is 0.033 waves; thus the root sum square of all of the contributors must not exceed that amount.

In Table 3 the first column identifies the parameter, the second states the delta used in the computation, the third states the amount of wavefront error caused by a delta amount. The final two columns show different budgeting of the allowable error. Distribution 1 loosens the outer radii and the thickness at the cost of maintaining the inner radii very tightly. Distribution 2 tightens the outer radii and spacing tolerances, but loosens the inner radii tolerances. Depending upon the capabilities of the shop selected to make the optics, one of these may be preferable.

The interpretation of these statistical summations is that they are the sum of a number of different random processes. Thus, if the interpretation of each of the values given is the width of a normal distribution, which implies that 67 percent of the samples lie within that value, then 67 percent of the resulting combinations will lie within that range. If the interpretation is a two- or three-sigma value, the interpretation of the result follows similarly.

TABLE 3 Two Possible Tolerance Distributions for the Doublet

Parameter	Delta	Coefficient	Distbn. 1	Distbn. 2
$r1$	0.1%	0.00071	10.0	0.1
$r2$	0.1%	0.05440	0.25	0.4
$r3$	0.1%	0.04069	0.25	0.4
$r4$	0.1%	0.00069	10.0	0.1
$t2$	0.025 mm	0.03589	0.75	0.5
		rms change	0.033	0.033

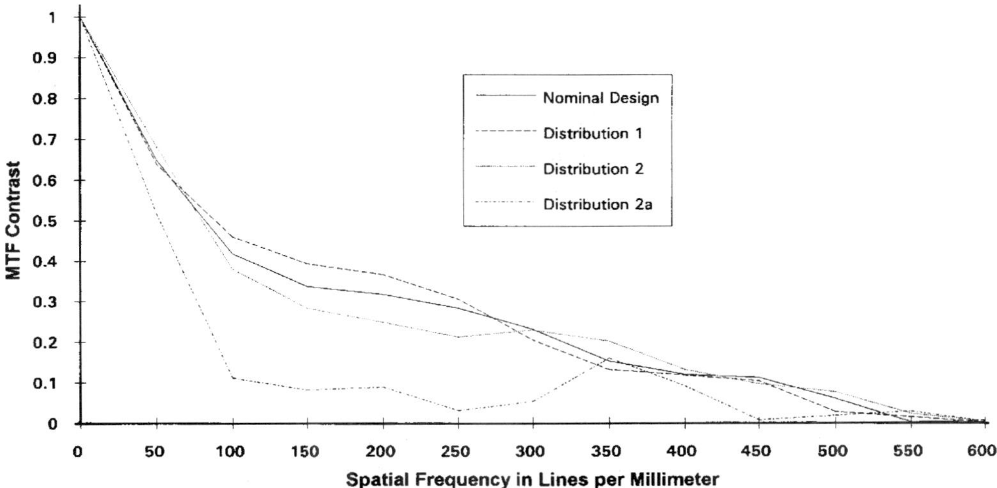

FIGURE 3 The resulting effect of different tolerance budgets for the sample doublet.

Figure 3 shows the effect of the various choices of allowable error distribution on the modulation transfer function of the lens. Either of the distributions stated in Table 3 provides an acceptable lens. For comparison, the allowable tolerances were doubled to provide distribution 2*a*, which is clearly not an acceptable lens. A spot check of some sample distributions is always relevant when doing tolerancing just to ensure that a reasonable relation between the toleranced system and acceptable image quality exists.

In any manufacturing process, the individual statistics will not necessarily follow a random rule. The interpretation is somewhat modified, but the principle still remains. In some cases, parameters may not be independent. For example, in the doublet there is a linking between the values of $r2$ and $r3$ that would loosen the tolerances if both are in error in the same direction. This could be used to advantage if the manufacturing process is carefully defined.

Use of Compensators

The use of compensators to loosen the tolerances was indicated above. An example of compensation for aberrations can be seen from a single lens element. If the first curvature is varied, both the elment power and the spherical aberration from the element will change. However, a specific change of the second radius can restore the focal position and reduce the change of spherical aberration. Thus the tolerance allowed to the first curvature needs to take into account the possibility of a correlated or deliberate change in the second curvature. It is evident that the proper use of compensators can greatly loosen the tolerances applied to a surface.

A compensation that is frequently employed is the establishment of the correct focal position after assembly of the lens. If this procedure is followed, the individual tolerances upon the surface of the elements can be loosened. It is obvious that the tolerancing and the development of a plan for fabrication and assembly must be coordinated.

36.4 OTHER TOLERANCES

Often, a particular optical parameter for the lens must be specified and maintained. Sometimes, for example, the focal length or back focus must be obtained within some tolerance. The computation of these paraxial constants for the lens can be made in the usual manner, and tolerances obtained by using differentials relating each of the parameters to the quantity, such as the focal length, and then distributing the tolerances in a manner similar to that shown above for the doublet.

Boresight

The pointing direction, or boresight, for a lens is sometimes of interest. Errors in boresight are usually due to asymmetric fabrication or mounting errors for the lens. In the simplest manner, one needs simply to trace an axial ray through the lens, and evaluate the direction of this ray as a result of introducing tilts and decenters of the surfaces, or of entire components. Tolerances upon the lens parameters can be obtained by the procedure described for the doublet above, substituting the boresight error for the wavefront error.

Distortion

Distortion is the failure of the lens to provide a constant mapping from object space to image space. There are alternate interpretations for this error, which can have radial components due to symmetric errors in the lens, as well as tangential components from tilt and decenter of the lens components. This can be toleranced in the usual manner, but may be related to some general properties of the lens, such as the overall glass thickness of the components. In the simplest case, the tolerances upon distortion may be obtained from direct aberration computation. In complex cases, it may be necessary to compute the actual location of the centroid of the image as a function of image position and in the best image location.

Assembly

Assembly tolerances are related to the tolerances on image quality. The elements must be located and held in position so that the resulting image-quality goals are met. There are additional questions of allowing sufficient clearance between the elements and the lens barrel so that the elements can be inserted into the barrel without breaking or being strained by the mountings. These must be considered in stating the allowable dimensional range in the diameter, wedge, and concentricity of the edge of the lens.

36.5 STARTING POINTS

Shop Practice

Table 3, given as part of the sample tolerancing of the doublet, provides some estimates of the accuracies to which an optical shop may operate. These generic levels of error convey what is likely to be possible. The designer carrying out a tolerance evaluation should consult with probable fabrication shops for modification to this table. The tolerances that are ultimately assigned relate errors in the system to acceptable errors in the image. However, an understanding of shop practice is of great assistance in intelligent budgeting of tolerances.

Measurement Practice

Contemporary practice in optical fabrication and testing is to use interferometry to define wavefronts and surfaces. A convention that has become common in recent years is the use of polynomials fitted to the wavefront as a method of describing the wavefront. There are several representations used, the most frequent of which is a limited set of Zernike polynomials. These ideally serve as an orthonormal set describing the wavefront or surface up to a specific order or symmetry. In tolerancing, the principal use for the coefficients of the Zernike set has been the easy computation of the rms wavefront error fitted to a given order. Thus the residual error in the system can be described after removal of low-order error such as focus or coma, which can sometimes be attributed to properties of the test setup.

36.6 MATERIAL PROPERTIES

The most important material is optical glass. Specification of the material usually includes some expected level of error in index of refraction or dispersion. In addition, glass is offered having several different levels of homogeneity of index of refraction. The usual range allows for grades of glass having index of refraction inhomogeneity ranging from ±0.00001 to less than ±0.0000001 within a single glass blank. It is usually assumed that this variation will be random, but the process of glass manufacturing does not guarantee this.

To place a tolerance upon the required glass homogeneity variation, the concept of wavefront tolerancing can be used. In general, the amount of wavefront error that can be expected along a glass path of length t through the glass is

$$\delta W = \frac{\delta N \times t}{\lambda}$$

For example, if a lens has a glass path of 5 centimeters but an error of 0.01 wavelengths is assigned to glass homogeneity, then the allowable glass homogeneity is about 0.00000013 within the glass. Thus precision-quality glass is needed for this application. For less glass path or looser tolerance assignment to glass homogeneity error, the required glass precision can be loosened. For a prism, the light path may be folded within the glass, so that an effective longer glass path occurs.

36.7 TOLERANCING PROCEDURES

The example of the doublet serves to illustrate the basic principles involved in determining the tolerances upon a lens system. Most lens design programs contain routines that carry out tolerancing to various degrees of sophistication. Some programs are capable of presenting a set of tolerances automatically with only limited input from the designer. The output is a neat table of parameters and allowable ranges in the parameters that can be handed to the shop. This appears to be a quite painless method of carrying out a complex procedure, but it must be remembered that the process is based upon application of a set of principles defined by the program writer, and the result is limited by the algorithms and specific logic used. In most cases, some trials of samples of the suggested tolerance distribution will suggest changes that can be made to simplify production of the lens system.

Direct Calculation

The preceding discussion describes methods used in calculating the tolerance distribution for a lens system. Frequently, tolerance determinations for special optical systems are required that either do not require the formal calculation described above, or may be sufficiently unusual that the use of a lens-design program is not possible. In that case, application of the principles is best accomplished directly.

The procedure is first to decide upon a meaningful measure of the image quality required. In fact, the term "image quality" may require some broader interpretation. For example, the problem may be to optimize the amount of energy that is collected by a sensor in an optical communication system; or the goal may be to scan a specific pattern with specific goals on the straightness of the projected spot or line during the scan.

The next step is to express the desired image quality in a numeric form. Usually, the rms wavefront error is the useful quantity. In some cases, other values such as the focus location of a beam waist or the size of the beam waist may be pertinent. In radiometric cases, the amount of flux within a specified area on the image surface (or within a specified angular diameter when projected to the object space) may be the pertinent value.

Once this is accomplished, the third step is to determine the relation between small changes in the parameters of the optical system and changes in the desired image quality function. This is usually accomplished by making small changes in each of the parameters, and computing the value of each differential as

$$\frac{dW}{dx_j} = \frac{\Delta W}{\Delta x_j}$$

where the right side is a finite differential. On occasion, the magnitude of this relation is nonlinear, and may require verification by using different magnitudes of change in the parameter.

This computation provides relations for independent, individual changes of the parameters. The possibility of compensation by joint changes in two or more parameters also has to be investigated. The best approach is to compute changes in the image-quality parameter in which specific coupling of parameters is included. For example, the differentials for variation of the curvatures of the two surfaces of a lens independently will be significantly different than the coupled changes of both surfaces simultaneously, either in the same or different directions.

Spreadsheet Calculation

The differentials must be combined in some manner to provide insight into the tolerance distribution. The best way to accomplish this is to develop a spreadsheet which allows simultaneous evaluation of the combination of errors using the equation for rms summation stated earlier. The use of spreadsheets for calculation is so common today that details need not be covered in this article.

Lens-Design Programs

The use of lens-design programs for tolerance calculation is becoming very widespread because of the proliferation of programs for use on the PC-level computer. The status of lens-design programs changes rapidly, so that any specific comments regarding the use of any program is sure to be out of date by the time this book appears in print. Suffice it to say that all of the principal programs have sections devoted to establishing tolerances. Usually the approach follows the procedures illustrated earlier in this chapter, with finite changes, or sometimes true computed analytical derivatives, used to establish a change table relating parameters to changes in the state of correction of the lens. In this case, the tolerances would be a listing of the allowed changes in the parameter to remain within some specified distance from the design values in aberration space. Some programs use a more complex approach where the allowable change in such quantities as the contrast value of the modulation transfer function at specified spatial frequencies is computed.

The distribution of tolerances is usually established according to the statistical addition rules given above. Some programs permit the user to specify the type of distribution of errors to be expected for various types of parameters.

As recommended above, it is strongly suggested that the user or designer not accept blindly the results of any tolerancing run but, rather, do some spot checking to verify the validity of the range of numbers computed. It is frequently found that alterations in the specified tolerances will occur as a result of such an investigation.

36.8 PROBLEMS IN TOLERANCING

Finally, it is useful to recite some of the problems remaining in establishing tolerances for a lens system. Even though the computational approach has reached a high level of sophistication for some lens-design programs, there are aspects of tolerancing that more closely approach an art than a science. The judgment of the designer or user of a tolerance program is of importance in obtaining a successful conclusion to a project.

Use of Computer Techniques

The use of a computer program mandates the application of rules that have been established by the writer of the program. These rules, of necessity, are general and designed to cover as many cases as possible. As such, they are not likely to be optimum for any specific problem. User modifications of the weighting, aberration goals, and tolerance image-quality requirements are almost always necessary.

Overtightening

The safe thing for a designer to do is to require very tight tolerances. This overtightening may ensure that the fabricated system comes close to the designed system, but the cost of production will likely be significantly higher. In some cases, the added cost generated by the overtight tolerances can raise the cost of the lens to the point where the entire project is abandoned.

The designer should consult with the fabricators of the optical system to develop an approach to assembly and testing that will allow the use of more compensating spacings or alignments to permit loosening of some of the tight tolerances.

Overloosening

A similar set of comments can be made about too-generous tolerances. In many schemes for production, these loose tolerances are justified by inclusion of an alignment step that corrects or compensates for cumulative system error. Too casual an approach to developing tolerances that require specific assembly processes, which are not fully communicated to the project, can result in a lens which is initially inexpensive to build, but becomes expensive after significant rework required to correct the errors.

Judgment factors

The last two sections really state that judgment is required. There is no completely "cookbook" approach to tolerancing any but the very simplest cases. The principles stated in this article need to be applied with a full knowledge of the relation between a change in a system parameter and the effect upon the image quality. In some cases, a completely novel relationship needs to be developed, which may include, for example, the connection between the alignment of a laser cavity and a nonlinear component included within the cavity. Finite difference calculations to obtain the output level can be developed using whatever computation techniques are appropriate. These values can be combined in a spreadsheet to examine the consequence of various distributions of the allowable errors.

CHAPTER 37
MOUNTING OPTICAL COMPONENTS

Paul R. Yoder, Jr.
Consultant in Optical Engineering
Norwalk, Connecticut

37.1 GLOSSARY

A	acceleration factor (mach number) and area (with subscript)
D	diameter
E	Young's modulus
K	special factors
m	reciprocal of v
P	total preload
p	linear preload
R	radius of curvature
S	stress
T	temperature
t	thickness
W	weight
α	thermal expansion coefficient
Δ	decentration
Δr	radial clearance
ΔT	temperature change
ϕ, φ	angle
θ_r	roll angle
v	Poisson's ratio

Subscripts

A	axial
c	other contact
E	lens edge
G	glass

M	metal
R	radial
T	toroid
t	tangent contact

37.2 INTRODUCTION AND SUMMARY

This chapter summarizes the most common techniques used to mount lenses, windows, and similar rotationally symmetric optical components as well as small mirrors and prisms to their mechanical surrounds. Typical interfaces between the optical and mechanical components are discussed. Included are considerations of mechanical (clamped) and elastomeric mounts for individual components and for assemblies of components, equations for estimating stresses within certain components due to mounting forces, and selected design methods for minimizing the adverse effects of shock, vibration, and temperature changes. Although we refer here to optics as if they are always made of glass and to mechanical parts (housings, cells, spacers, retainers, etc.) as if they are always made of metal, it should be understood that many of these mounting considerations also apply to other materials such as plastics or crystals. Examples are given to illustrate typical mounting design configurations.

Mountings for large (i.e., astronomical-telescope-sized) optics are not considered here. Readers interested in more general treatments of optical component mounting technology are referred to "Selected Papers on Optomechanical Design," *SPIE Milestone Series,* vol. 770, D. C. O'Shea, editor, (1988) and "Optomechanical Design," *SPIE Critical Reviews,* vol. CR43, Paul R. Yoder, Jr., editor, (1992).

37.3 MOUNTING INDIVIDUAL LENSES

General Considerations

Contact between a rotationally symmetric optical component and its mount can occur at the cylindrical rim, at ground bevels, or at the polished surfaces. Any force in the axial direction exerted by the mount onto the component within or outside the clear aperture is called *axial preload.* Generally, a modest axial preload is desired at assembly since this tends to prevent motion of the component relative to the mount. Forces exerted in the radial direction at the rim of the component are generally referred to as "hoop" forces. Both types of compressive forces tend to introduce stress into the glass, so should be kept within acceptable limits. Stress can introduce birefringence into the component or, in extreme cases, damage the glass. Changes in temperature, vibration, and shock are common causes of excessive applied forces which cause stress.

A glass optical component can usually withstand compressive stress as large as 3.45×10^8 N/m^2 without failure, so this value is generally accepted as a survival tolerance.[1] In some designs, component surfaces can be placed in tension as a result of applied forces. A commonly accepted tolerance for tensile stress is 6.9×10^6 N/m^2. Under operating environmental conditions, distortions of optical surfaces due to mounting stresses can degrade performance. No simple means for predicting refracting or reflecting surface deformations due to external forces are available; techniques such as finite element analysis are usually employed to evaluate such deformations. Systems using polarized light may be sensitive to stress-induced birefringence. A common tolerance for stress in such cases is 3.45×10^6 N/m^2.

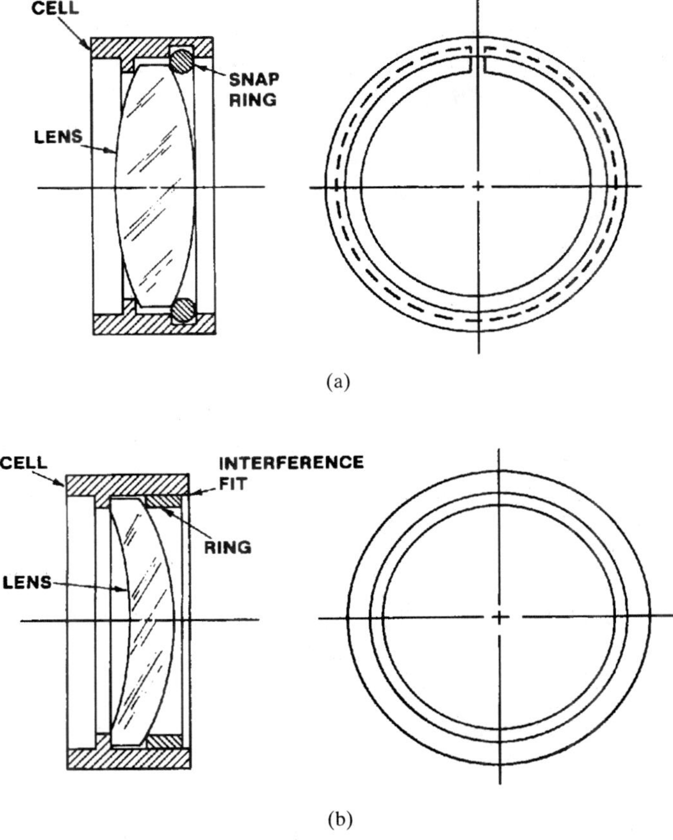

FIGURE 1 Techniques for holding a lens in a cell with (*a*) C-shaped ring snapped into a groove and (*b*) pressed-in continuous ring. *[From Ref. 7.]*

Simple, Low-Cost Designs

Two techniques sometimes used to secure lenses into housings or cells with nonthreaded retaining rings are illustrated in Fig. 1. In Fig. 1*a,* the lens is constrained between a shoulder integral to the mount and a "C"-shaped (i.e. discontinuous) ring with circular cross section that snaps into a groove machined into the inside diameter (ID) of the mount. The edge thickness of the lens, the cross-sectional diameter of the ring and the location, depth, and width of the groove all need to be controlled during manufacture if both the shoulder and ring are to touch the lens at assembly.

Another design, shown in Fig. 1*b,* uses a continuous ring of rectangular cross section and outside diameter (OD) slightly smaller (typically by 5 to 15 μm) than the ID of the mount. An interference fit will then be obtained if the ring is pressed into the cell. It is quite difficult to impose a specific axial preload on the lens if the ring must be pressed in place since it is hard to discern when contact just occurs. Assembly is facilitated if the ring is cooled and/or the mount is heated so the ring slides easily into the mount until it just touches the lens. Upon returning to thermal equilibrium, a slight axial preload will then be exerted upon the lens.

In both designs, temperature decreases will increase axial preload on the lens, while temperature increases will decrease that preload. In the design of Fig. 1*a*, the ring will tend to squeeze from the groove if the temperature drops, thereby tending to relieve axial stress. At some elevated temperature, the lens in either design may be free of axial constraint so it can move about within the mount under gravitational, shock, or vibration loadings.

In these designs, radial clearance, Δr, of the order of 0.1 to 0.3 mm is typically provided between the lens and the mount. Decentration of either lens vertex when not constrained may then be as large as Δr and the lens may tilt (or roll) within the mount through an angle θ_r as large as $2\Delta r / t_E$ in radians, where t_E is the lens edge thickness in mm.

The ring of Fig. 1*a* can be removed if necessary for maintenance purposes, but it is virtually impossible to remove that of Fig. 1*b* without damage. Neither of these designs should be used if precise control of lens position is needed under all environmental conditions or if a particular preload is desired at the time of assembly. Designs using threaded retaining rings are preferred in those cases.

Figure 2 shows a typical configuration of a lens burnished into a cell. The cell must be made of malleable material such as brass or an aluminum alloy. Usually the cell is mounted onto a spindle (note the integral chucking thread provided for this purpose); the lens is inserted and held securely against the shoulder while rotating. A burnishing tool (such as a set of three rollers inclined to the rotation axis by 45° and located at 120° intervals about that axis) is pressed against the protruding lip of the cell causing the metal to bend over the edge of the lens and clamping it in place. If the burnishing tool's forces act symmetrically, the lens may tend to center within the ID of the cell. Once again, creation of a specific axial preload is virtually impossible. Dissassembly of a lens mounted in this manner without damage is quite difficult. This mounting arrangement is most frequently used in microscope objectives, eyepieces, and small camera-lens assemblies.[2,3]

Figure 3 illustrates a simple mounting configuration described by Hopkins[4] in which a lens is appropriately centered within a cell and then secured in place with a ring of room temperature vulcanizing (RTV) adhesive contacting the polished surface. To prevent movement of the lens due to shrinkage during curing of the adhesive, temporary use of centering spacers, dabs of beeswax, or an external fixture that secures the lens to the cell is recommended. This design has the advantage of providing a relatively stress-free environment for the optical component.

Mounts Using Threaded Retaining Rings

Typical Configurations. Use of a threaded retaining ring to secure a lens against a reference surface such as a shoulder or a spacer within a cell or housing offers several advantages over the above-described lens mounts. This type mount can easily be assembled and disassembled without damage; it automatically compensates for lens edge thickness variations; it can provide a reasonably predictable axial preload; and it is compatible with sealing with an injected sealant or with an O-ring.

Figures 4 through 7 illustrate concepts for interfaces between a cell shoulder, a lens, and a threaded retainer. The simplest provides line contact at a sharp corner around the periphery of each lens surface. The corner can be a 90° intersection between a cylindrical hole and a plane (Fig. 4*a*) or an obtuse angle such as a 135° intersection between a conical surface and a plane (Fig. 4*b*). Acute corner angles should be avoided. A sharp corner interface can be used with either convex or concave lens surfaces. Accuracy of the actual intersection angle is not essential; errors of ±2° usually are tolerable. Good machining practice calls for the sharp corner to be burnished slightly to minimize burrs or nicks. Typically, a radius of the order of 0.05 mm then results.[5] A tendency for better (i.e., smoother) sharp edges to result from machining an obtuse angle intersection was reported

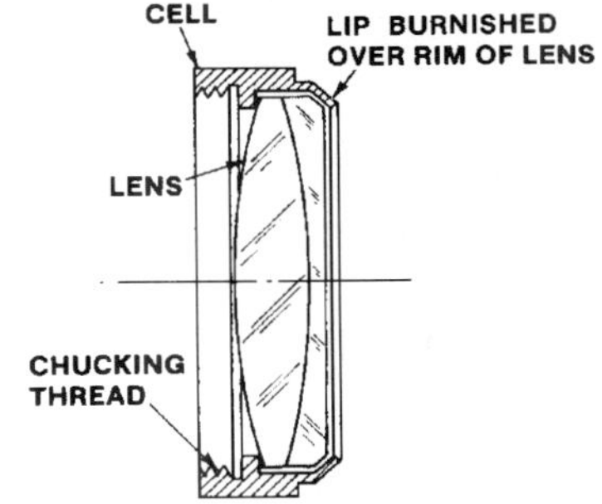

FIGURE 2 Lens burnished into a cell made of malleable metal. *[From Ref. 12.]*

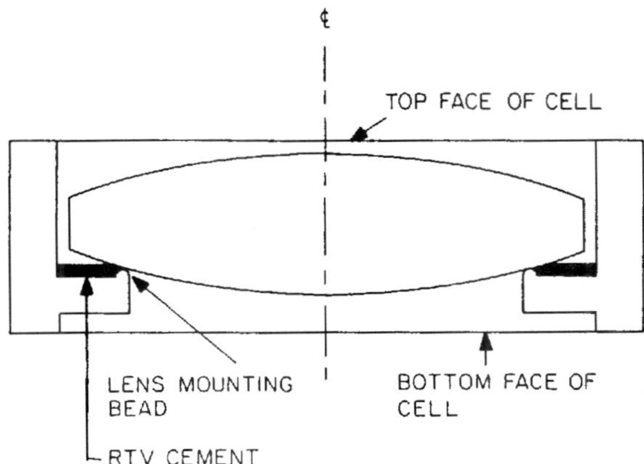

FIGURE 3 Lens secured in a cell by a ring of adhesive contacting the polished surface. *[From Ref. 4.]*

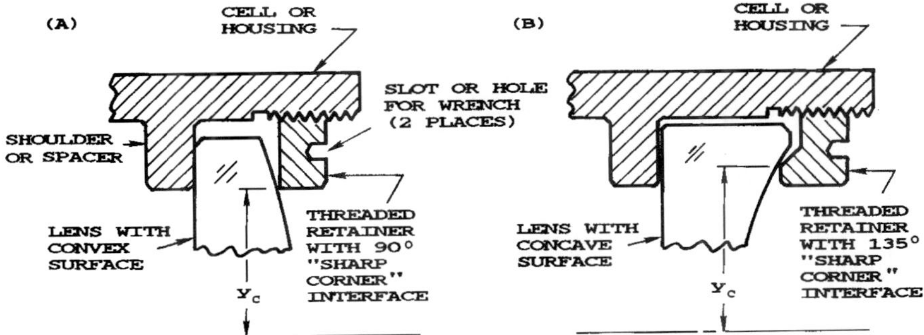

FIGURE 4 Techniques for holding a lens with sharp corner interfaces: (*a*) 90° corner on convex surface and (*b*) 135° corner on concave surface. *[From Ref. 8.]*

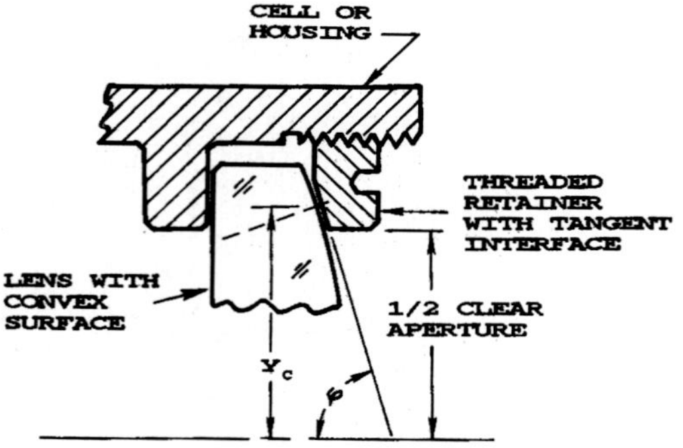

FIGURE 5 Technique for holding a lens with a conical (tangential) interface. Note this is applicable only to convex surfaces. *[From Ref. 8.]*

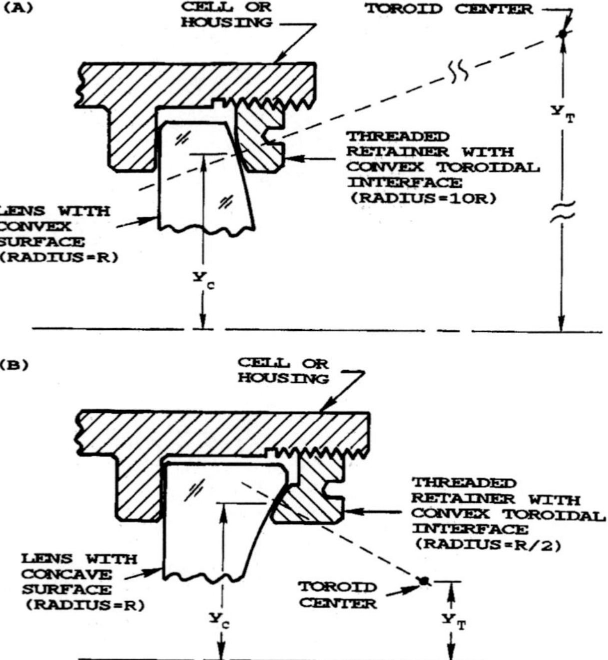

FIGURE 6 Techniques for holding a lens with toroidal interfaces: (*a*) convex surface and (*b*) concave surface. *[From Ref. 8.]*

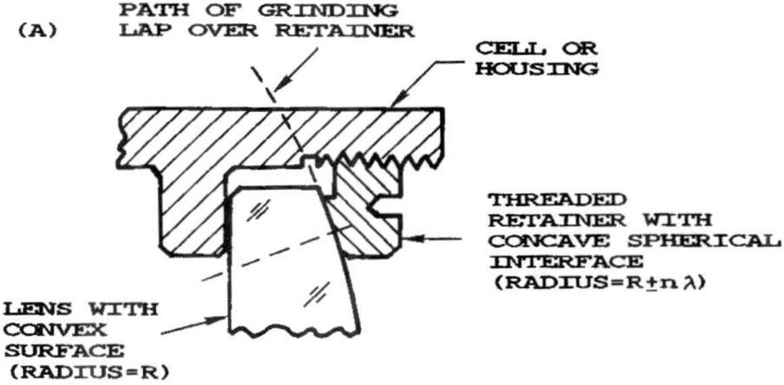

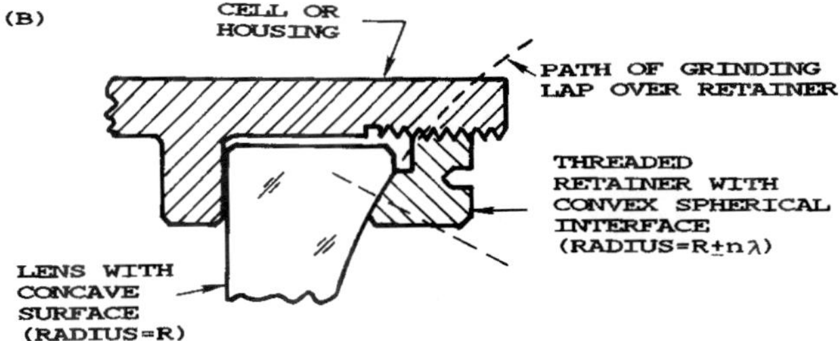

FIGURE 7 Techniques for holding a lens with spherical interfaces: (*a*) convex surface and (*b*) concave surface. *[From Ref. 8.]*

by Hopkins.[4] The radial height of line contact, y_c, usually is just slightly greater than the radius of the lens' clear aperture.

Tangential contact at a height y_c (in mm) is illustrated in Fig. 5. It can be used only with convex lens surfaces. The half-angle φ of the right-circular cone in such an interface is given by the following equation:

$$\varphi = 90°\text{-arcsin}\,(y_c/R) \tag{1}$$

where R is the lens' radius of curvature in mm and y_c in mm is typically one-half the arithmetic average of the lens' clear aperture and its OD. The typical tolerance on φ is $\pm 2°$.

Toroidal contact between the mechanical and optical components can be provided for either convex or concave lens surfaces. (See Fig. 6.) In each case, the center of curvature of the circular arc defining the doughnut-shaped toroid is located off the axis by the dimensions y_T and lies on the local normal to the lens surface at y_c where y_c is as defined for tangent contact. Preferred values for the sectional radius of the toroids are $\geq 10R$ for a convex surface and $\geq 0.5R$ for a concave surface, where R is the radius of the lens surface.[6]

If the mechanical interface with the lens is lapped to the same spherical radius as the lens surface within a few fringes of visible light, a spherical interface mount can be produced. (See Fig. 7.) If the radii of the mechanical and optical surfaces do not match adequately, the interface will degenerate into line contact at the inner or outer edge of the annular mechanical area. Manufacture of the spherical interface mount is costly due to this need for radius matching. Hence, it is generally used only in special designs.

In each of the above-described lens mounts, contact of the metal occurs on polished glass surfaces. These surfaces are usually accurately made and well aligned to each other due to the inherent nature of the optical polishing and edging processes. In most lenses,

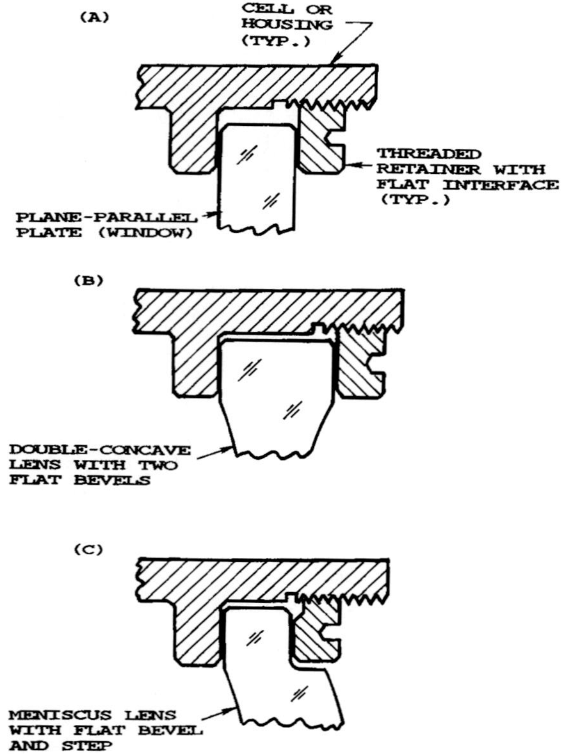

FIGURE 8 Techniques for holding plane parallel plates or lenses with flat bevels on convex and concave surfaces. *[From Ref. 8.]*

contact occurs at a zone where the optical surface is inclined with respect to the axially directed force imposed by the retainer. Radial components of that force tend to center the lens with respect to the axis. The magnitude of this centering tendency depends upon the radius of curvature of the lens surface, with sharply curved surfaces centering more easily than flatter ones.

Contact against flat bevels ground into the edges of the lens' spherical surfaces can be provided as shown in Fig. 8. Such interfaces have the advantage of distributing axial preload due to the clamping action of the retaining ring over a large area on the lens, thereby reducing stress buildup within the glass. They have the disadvantages of not being able to help self-center the lens and of referencing against secondary surfaces of potentially reduced angular accuracy as compared to the polished lens surfaces. While accurate flat bevels can be produced, they add significantly to the cost of the lens. In the sections which follow, a simplistic treatment of forces and stresses at the optomechanical interface as summarized by Yoder[6–8] is given. These analytical methods can be used to compare alternate designs or to assess the need for experimental verification or more rigorous analyses using, for example, finite element analysis methods.

Axial Preload and Axial Stress Relationships at Assembly Temperature. If we know the total preload P in newtons exerted by the retainer against the lens at the zonal contact radius y_c in mm during assembly at a temperature of (typically) 20°C, the resulting axial stress S_A in N/m^2 developed within the contact area on the glass can be estimated using the following equation:[7,8]

$$S_A = -0.798(K_1 p/K_2)^{1/2} \tag{2}$$

where p = linear preload = $P/(2\pi y_c)$ in N/mm

$K_1 = (D_1 + D_2)/D_1D_2$ for a sharp corner contacting a convex surface

$= (D_1 - D_2)/D_1D_2$ for a sharp corner contacting a concave surface

$= 1/D_1$ for a conical surface contacting a convex surface (tangent interface)

$= 1.1/D_1$ for a toroid surface of sectional radius $10D_1$ contacting a convex surface

$= 1/D_1$ for a toroid surface of sectional radius $0.5D_1$ contacting a concave surface

D_1 = twice the lens radius of curvature in mm

D_2 = 0.1 mm = twice the sharp corner radius

$K_2 = ((1 - \nu_G^2)/E_G) + ((1 - \nu_M^2)/E_M)$

ν_G = Poisson's ratio for the glass

ν_M = Poisson's ratio for the metal

E_G = Young's modulus for the glass in N/m^2

E_M = Young's modulus for the metal in N/m^2

Equations for the contact stresses introduced into a lens with flat bevel or spherical interfaces are not included here because those stresses would be relatively insignificant as compared to those with the interfaces listed. As shown by Yoder,[6] the contact stresses computed for the toroidal interface cases would be within 5 percent of those computed for the tangential case and much smaller than those for sharp corner interfaces.

Axial Preload Required to Restrain the Optical Component at Elevated Temperature. As the temperature rises, the metal of the mount expands more than the glass component within. Any axial preload existing at assembly will then be reduced. If the temperature rises sufficiently, this preload may disappear and the optic is free to move within the mount due to external forces. Ideally, unless the position and orientation of the lens are not critical in the application, the design should prevent such freedom at the highest temperature that the instrument is to survive without damage. For military applications, this highest temperature is typically 71°C. Instruments for laboratory, commercial, or consumer use may have lower survival temperature requirements.

One way of preventing this release of preload is to provide a sufficient preload at assembly, P_1, in newtons, so that the force on the lens is just reduced to zero at the highest survival temperature. Defining the change in temperature from that at assembly to the highest value as ΔT, we can use the following equation to estimate the axial preload required at assembly:[7,8]

$$P_1 = K_3(\Delta T) \tag{3}$$

where $K_3 = -E_G A_G E_M A_M(\alpha_M - \alpha_G)/(2)((E_G A_G/2) + E_M A_M)$ if $2y_c + t_E < D_G$

$K_3 = -E_G A'_G E_M A_M(\alpha_M - \alpha_G)/(2)((E_G A'_G/2) + E_M A_m)$ if $2y_c + t_E >= D_G$

$A_M = 2\pi t_c((D_M/2) + (t_c/2))$ in mm^2

t_c = cell wall thickness in mm at lens rim

$A_G = 2\pi y_c t_E$ in mm^2
t_E = lens edge thickness in mm at height y_c
$A'_G = (\pi/4)(D_G - t_E + 2y_c)(D_G + t_E - 2y_c)$ in mm^2
D_G = lens OD in mm
α_M = thermal expansion coefficient of the metal in ppm/°C
α_G = thermal expansion coefficient of the glass in ppm/°C

and all other terms are as defined earlier.

Axial Preload Required to Restrain the Optical Component Against Acceleration Forces. If the optic is expected to experience acceleration forces due to gravity, shock, or vibration and its axial location within the mount must remain unchanged during such exposure, it should be constrained by an axial preload P_2 in newtons, as given by the following equation:

$$P_2 = -9.807WA \tag{4}$$

where W = weight of optical component in kg
A = maximum imposed acceleration factor (expressed as a multiple of gravity)

Total Axial Preload to Restrain the Optic Against Acceleration and Release of Preload at the Highest Survival Temperature. By adding the preloads computed by Eqs. (3) and (4) one obtains the approximate total axial preload P_T, in newtons, required to constrain the optic against axial acceleration forces and to prevent complete release of assembly preload at the highest temperature. This preload may be very high if the lens and cell materials do not match closely in thermal expansion properties. A lower preload can be used if the motion of the lens within the clearance created by expansion of the cell away from the lens is tolerable.

Retaining Ring Torque Required to Produce a Given Axial Preload. The magnitude of the torque M, in N-cm, that should be applied to the retaining ring to introduce a given axial preload P_T, in newtons, to the edge of the lens can be approximated from the following empirically derived formula given by Kowalski:[9]

$$M = 0.2P_T D_R \tag{5}$$

where D_R is the pitch diameter in cm of the thread on the ring.

Axial Stress Introduced at Reduced Temperatures. If the temperature decreases by $-\Delta T$ in °C from that existing at assembly, the preload exerted by the retaining ring increases due to differential expansion of the mount and the optical component. The change in axial preload on the glass, P_1, in newtons, is given by Eq. (3) as defined above. The usual lower limit for survival temperature of military optical instrumentation is −62°C. It should be noted that the preload P_T at assembly should be added to that estimated by this application of Eq. (3) to give the total preload at reduced temperature. Contact stress within the glass can be computed by use of Eq. (2) and the average (or bulk) stress S_{BG} within the glass can be estimated as $2(P_1 + P_T)/A_G$ or $2(P_1 + P_T)/A'_G$ as appropriate. The average (or bulk) stress S_{BM} within the cell wall can be estimated as $(P_1 + P_T)/A_M$. These bulk stresses are generally much smaller than the corresponding contact stresses computed by Eq. (2).

Radial Stress Introduced at Reduced Temperatures. In all designs considered above, radial clearance was assumed to exist between the optic and the mount. In some designs, this clearance is the minimum allowing assembly so, at some reduced temperature, the metal touches the rim of the optic and, at still lower temperatures, a so-called "hoop" stress develops. The magnitude of this stress S_R (N/m^2) for a given temperature drop ΔT, in °C can be estimated by the following equation:[7]

$$S_R = K_4 K_5 \,\Delta T \tag{6}$$

where $K_4 = (\alpha_M - \alpha_G)/((1/E_G) + (D_G/2E_M t_C))$
$K_5 = (1 + ((2\Delta_r)/(D_g \Delta T)(\alpha_M - \alpha_G)))$
D_G = optical component OD in mm
t_C = mount wall thickness outside rim of the optic in mm
A_r = radial clearance in mm

If Δr exceeds $D_G \Delta T(\alpha_M - \alpha_G)/2$, the lens will not be constrained by the cell ID, and hoop stress will not develop within the temperature range ΔT due to rim contact.

Combined Axial/Radial Stress at Reduced Temperatures. Bayar[10] indicated that the combined effect of axial and radial stresses within the optical component can be estimated as the root sum square (rss) of the orthogonal stresses. This is most significant at worst-case reduced temperature. Hence:

$$S_{rss} = ((S_{BG})^2 + (S_R)^2)^{1/2} \tag{7}$$

where all terms are as defined above.

Growth of Radial Clearance at Increased Temperatures. The increases in radial and axial clearances, Δgap_R and Δgap_A, in mm, between the optic and the mount due to a temperature increase of ΔT in °C from that at assembly can be estimated by the following equations:

$$\Delta\text{gap}_R = K_6 \Delta T \tag{8}$$

$$\Delta\text{gap}_A = K_7 \Delta T \tag{9}$$

where $K_6 = (\alpha_M - \alpha_G)D_G/2$
$K_7 = (\alpha_M - \alpha_G)t_E$

and all terms are as defined above.

Stresses Due to Bending of the Component. If the lines of contact between the mount and optical component occur at different radii from the axis, axial forces applied through those contacts can cause bending of the surfaces. (See Fig. 9.) One of the surfaces must

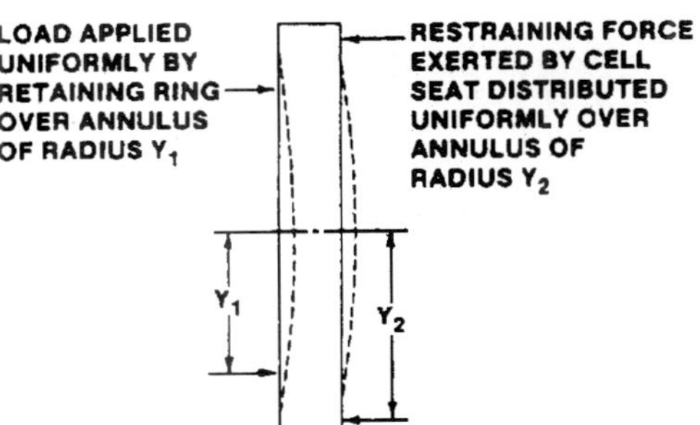

FIGURE 9 Schematic diagram of axial forces tending to bend an optical element when contact areas are not at the same radius. *[From Ref. 7.]*

necessarily be placed in tension; this is the condition in which glass is the weakest. Bayar[10] applied the following equation for the tensile stress induced by bending:

$$S_T = \frac{3P_T}{2\pi m t_E^2}\left[0.5(m-1) + (m+1)\ln\frac{Y_2}{Y_1} - (m-1)\frac{Y_1^2}{Y_2^2}\right] \tag{10}$$

where S_T = tensile stress in the component (N/m^2)
P_T = applied axial load (N)
m = 1/Poisson's ratio
t_E = edge thickness of component
Y_1 = innermost contact radius (cm)
Y_2 = outermost contact radius (cm)

As mentioned earlier, the general tolerance for tensile stress is 6.9×10^6 N/m^2.

Stresses Under Operating Conditions. Any of the above equations for stress induced into an optical component by forces exerted by the mount can be applied to operating conditions as well as worst-case conditions. In general, the stress levels will be reduced, but they still may be significant in terms of their effects upon system performance. Surface deformations due to mount-induced forces may also occur. Within limits, the effects of axial forces can be reduced by making the retainer somewhat resilient. Some techniques for doing this are illustrated schematically in Fig. 10.

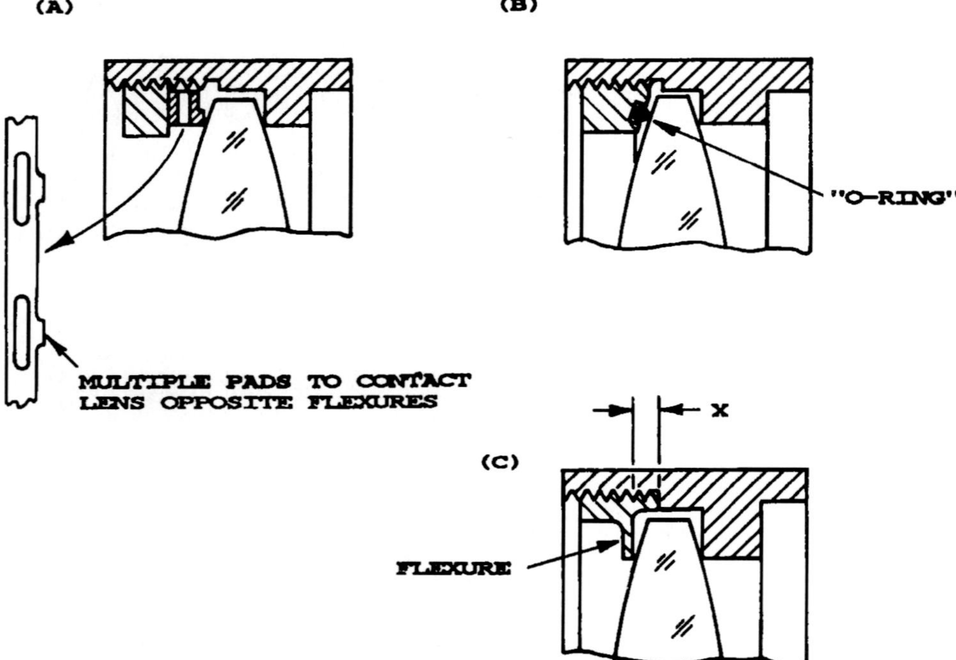

FIGURE 10 Typical resilient retainer configurations intended to more uniformly distribute axial forces and reduce surface distortions. *[From Ref. 7.]*

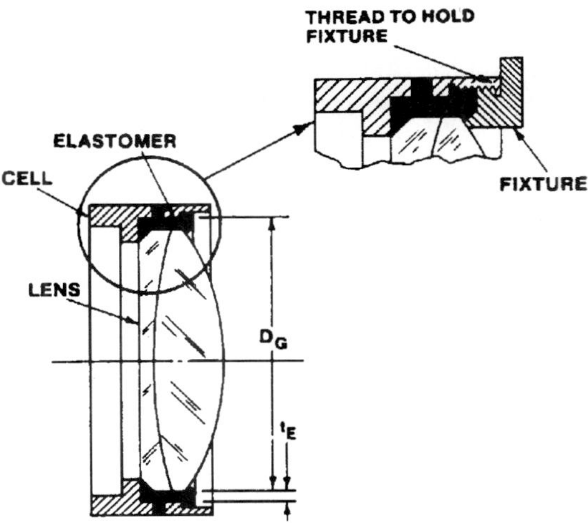

FIGURE 11 Technique for holding a lens in a cell with an annular layer of cured-in-place elastomer. *[From Ref. 7.]*

Elastomeric Mounting Designs

Resilient materials such as RTV compounds are frequently used to seal optical components into their mounts with or without mechanical constraint. Registry of at least one optical surface against a machined surface of the mount (as shown in Fig. 3) helps align the optic. The resilient layer can provide some degree of protection against shock and vibration as well as thermal isolation. Epoxy or urethane compounds are sometimes used for the same purpose. These materials are generally somewhat less resilient than RTV compounds, but they are superior in regard to structural bond strength.

Figure 11 illustrates a common technique for mounting a lens in an annular ring of elastomer. Centration can be retained by temporary shimming or external fixturing during the curing process. If the elastomer layer's radial thickness t_E (mm) is in accordance with the following equation, the design becomes relatively insensitive to temperature change:[10]

$$t_E = (D_G/2)(\alpha_M - \alpha_G)/(\alpha_E - \alpha_M) \tag{11}$$

where all terms are as defined above.

Valente and Richard[11] reported an analytical technique for estimating the decentration Δ, in m, of a lens mounted in a ring of elastomer when subjected to radial gravitational loading. Their method can be expanded to include more general radial acceleration forces resulting in the following equation:

$$\Delta = AWt_E/(\pi Rd((E_E/(1 - \nu_E^2)) + E_S)) \tag{12}$$

where A = acceleration factor

W = weight of optical component in N

t_E = thickness of elastomer layer in m

R = optical component OD/2 in m

d = optical component thickness in m

E_E = Young's modulus of elastomer in N/m^2

E_S = Shear modulus of elastomer in N/m^2

ν_E = Poisson's ratio of elastomer

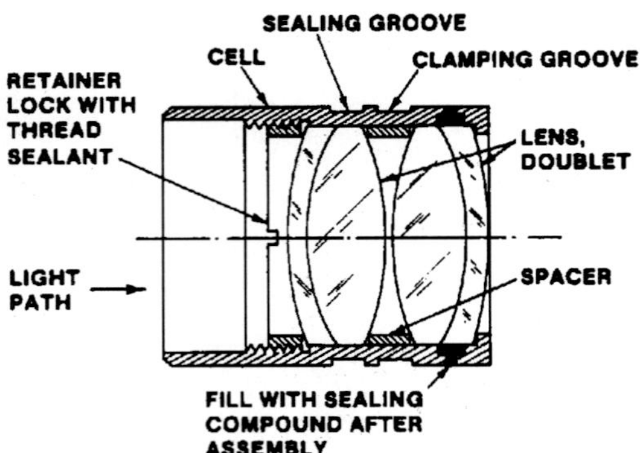

FIGURE 12 A simple eyepiece configured for assembly by dropping the lenses in place and clamping with a retainer. *[From Ref. 12.]*

The decentrations of modest-sized optics corresponding to normal gravity loading are generally quite small.

37.4 MULTICOMPONENT LENS ASSEMBLIES

Drop-In Assembly Techniques

A drop-in lens assembly is one in which the optical components are edged to diameter typically 0.075 to 0.150 mm smaller than the minimum ID of the cell or housing in which they are to be mounted. After installation of the optics and spacers as appropriate, the parts are typically held in place with a single retaining ring. The fixed-focus eyepiece for a military telescope shown in Fig. 12 typifies this type assembly. The optomechanical interfaces are all square corners. Sealing provisions are indicated. Many camera lenses and most terrestrial telescope, binocular, and microscope optics are assembled in this manner.

Figure 13 shows another design for a drop-in assembly. In this case of a fixed aperture relay lens, the lenses are individually mounted in recesses bored into the front housing and into the rear cell. Radial clearances are typically no greater than 0.013 mm and the lenses are precisely edged and beveled with respect to their optical axes to ensure alignment. Adjustment of axial spacings is accomplished by customizing thicknesses of two spacers and by machining axial dimension Y of the housing at the time of assembly.

Machine-at-Assembly Techniques

In this type assembly, the rims of optical components are precision ground to a high degree of roundness, but not held to tight diametrical tolerances. Actual ODs are measured and the IDs of the mating mechanical seats are machined to fit those lenses with radial clearances of the order of 5 μm. Axial locations of the lenses are established while the seats are being machined. This technique, sometimes referred to as "lathe assembly," is employed in high-performance applications such as aerial camera lenses or assemblies that must withstand high vibration and/or shock loads.[12] Figure 14*a* shows an airspaced

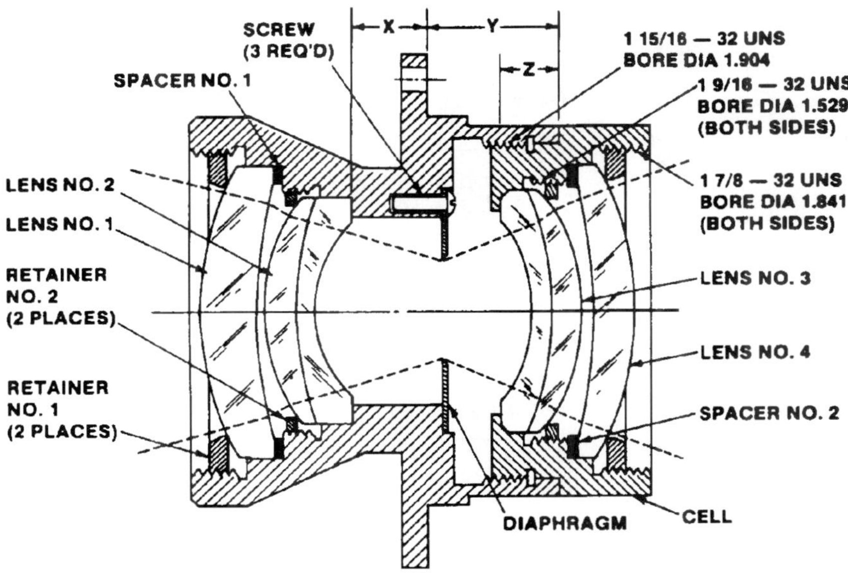

FIGURE 13 A relay lens assembly configured for dropping the lenses in place and clamping with individual retainers. *[From Ref. 7.]*

doublet mounted in this manner, while Fig. 14*b* shows the lenses. Measured values are recorded in the numbered boxes on the drawings for reference during machining. Both optical and mechanical parts are usually serialized for identification and traceability. Since the mechanical parts are customized for a specific set of lenses, replacement of a damaged lens may require considerable effort to match dimensions adequately.

A variation of the lathe-assembly principle is illustrated in Fig. 15 which shows the optomechanical configuration of a 22.8-cm focal length, f/1.5 objective.[13] Designed to image a military target at low light levels onto an image intensifier in a visual periscope and to project a coaxial laser beam for target ranging, the lenses of this objective are mounted with very small radial clearances into an aluminum barrel. Because of their large sizes, high manufacturing precision, and special coatings, the lenses were quite expensive. In order to reduce the chance of chipping the edges if inadvertently tilted as these close-fitting lenses were installed, the rims were ground spherical as shown in Fig. 16. This feature has been successfully used in several other designs involving optics with high value at the critical time of assembly.

Stacked-Cell Assembly Techniques

Figure 17 illustrates a design described by Carnel et al.[14] for a wide-field objective lens used for bubble chamber photography. A large amount of optical distortion of particular form was designed into the lens. To provide this distortion accurately, the lenses had to be precisely centered. In this assembly, the lenses were mounted in individual cells and the ODs of those cells machined to specific dimensions with each cell's mechanical axis colinear with the lens' optical axis. All the cells were then installed into the ID of the barrel with spacer rings to provide appropriate axial separations.

(a)

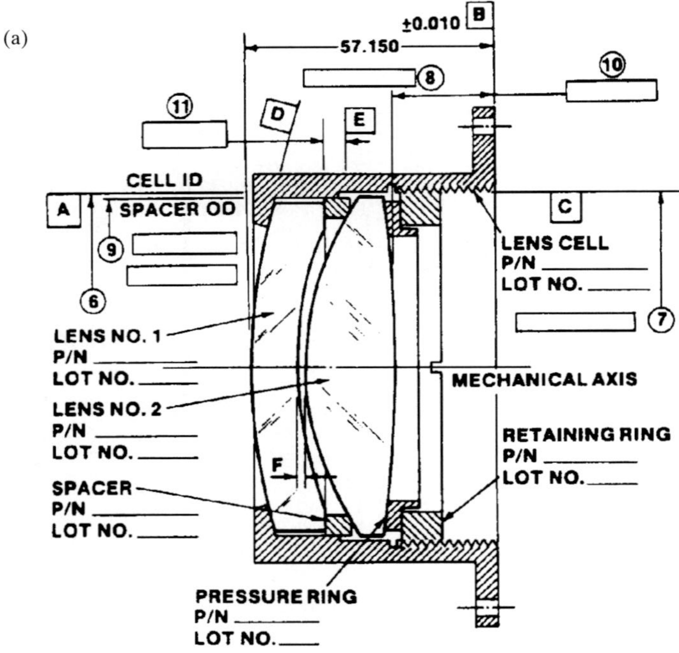

(b)

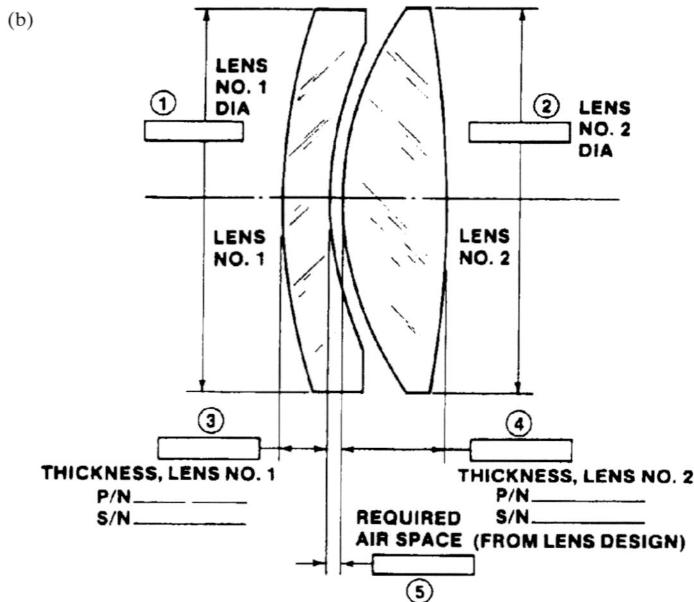

FIGURE 14 Example of the "lathe assembly" method of optical assembly: (*a*) mounted lenses and (*b*) lenses only. Measured values for serialized parts are recorded in boxes. *[From Ref. 12.]*

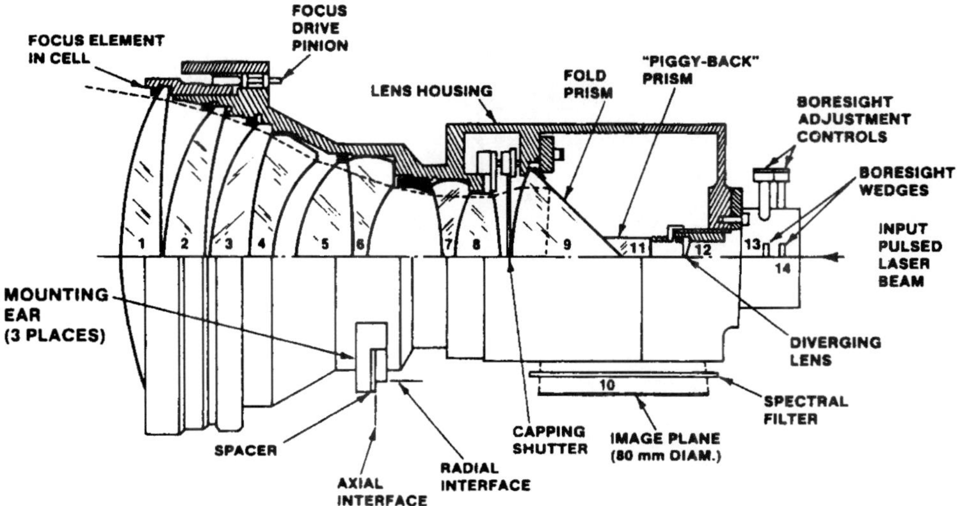

FIGURE 15 Sectional view of a low light level periscope objective. *[From Ref. 13.]*

This same principle was used by Fischer[15] in designing a low-distortion telecentric projection lens in which all lenses but one were mounted within individual stainless steel cells and the ODs of those cells machined precisely so as to just slide into the ID of a stainless steel barrel. (See Fig. 18.) The lenses were aligned to the ODs of the cells and held in place with nominally 0.38-mm-thick annular layers of epoxy. Cell 1 of this design acts as a retaining ring to hold the other cells against a shoulder in the barrel.

Figure 19 shows another design, a 2-m focal length, f/10 astrographic telescope objective, which also has its lenses mounted into cells and the cells stacked into a barrel.[16] In this case, all mechanical parts are titanium. The lenses are constrained within annular epoxy layers having nominal radial thicknesses of typically 5.08 mm. Richard and

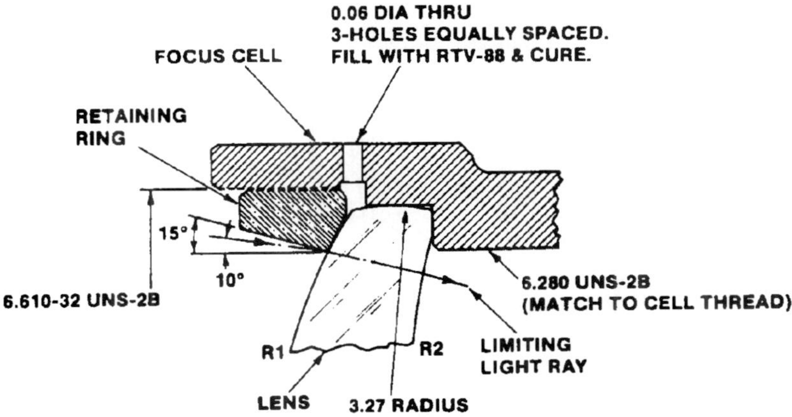

FIGURE 16 Detail view of spherical rim lenses used in the lens of Figure 15. *[From Ref. 13.]*

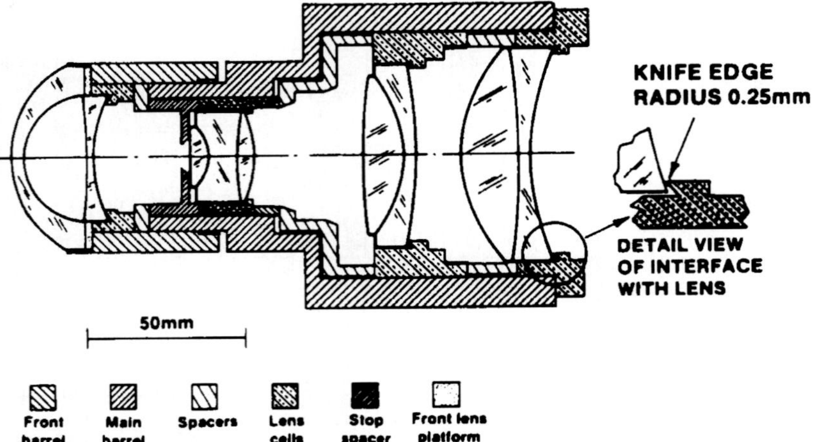

FIGURE 17 Optomechanical configuration of a bubble chamber objective with lenses individually mounted in cells and stacked in a barrel. *[From Ref. 14.]*

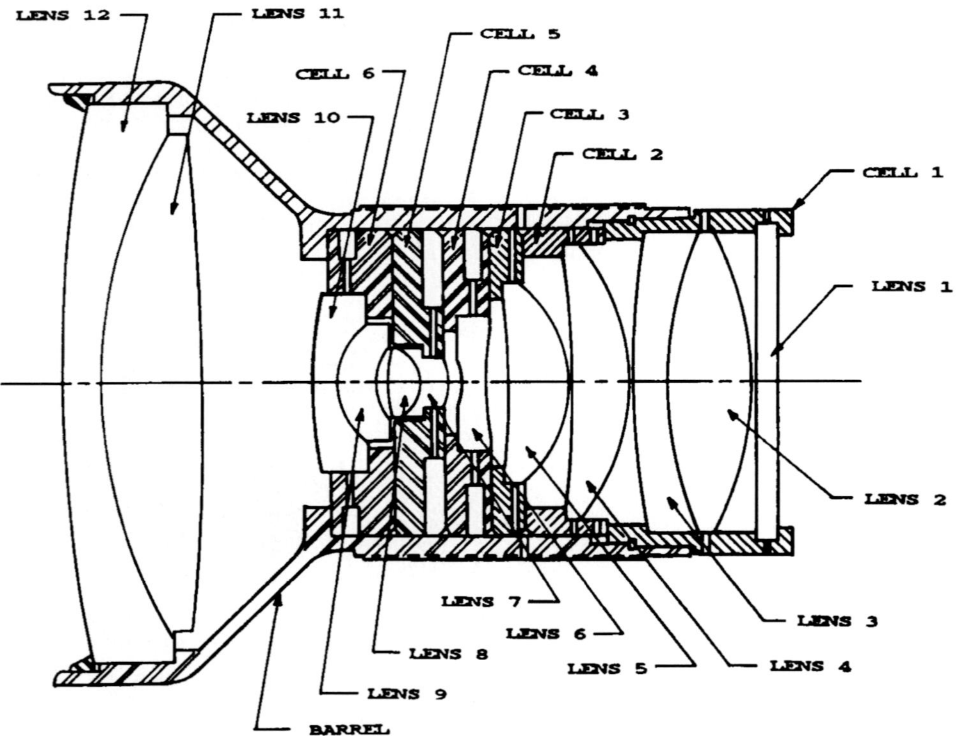

FIGURE 18 Optomechanical configuration of a low-distortion objective with stacked cell-mounted lenses. *[From Ref. 15.]*

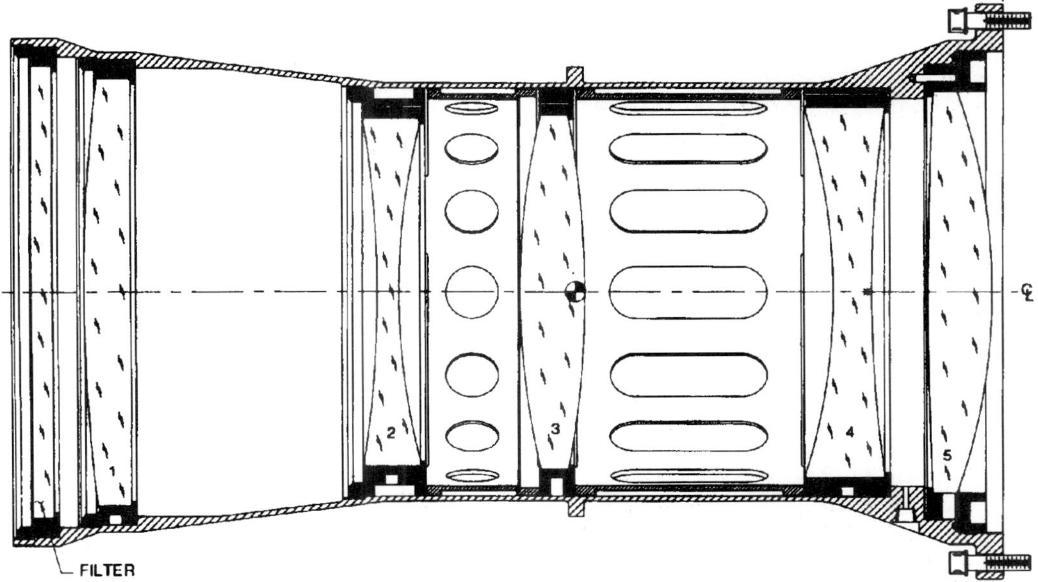

FIGURE 19 Optomechanical configuration of an astrographic telescope objective with stacked cell-mounted lenses. *[From Ref. 16.]*

Valente[17] described how the cells were designed for interference fits within the barrel and pressed in place. Valente and Richard[11] showed that the decentrations of the lenses due to radially directed gravitational force driving the lenses into the resilient layer would be only on the order of a tolerable 2.5 μm.

Assemblies Using Plastic Parts

Near the other end of the complexity spectrum from the designs so far considered here are a multitude of optomechanical devices using plastic lenses, cells, retaining means, and structures. A few examples are camera viewfinders and objectives, magnifiers, television projection systems, optics and structural parts for telescopes and binoculars, compact disc player systems, and military helmet-mounted night-vision goggles. General features of such devices are use of low-cost, lightweight materials, freedom to employ aspherics, large quantity production, consolidation of optical and mechanical functions into integrated molded components, and compatibility with automated assembly. In the following sections, we describe two optomechanical subsystem designs that illustrate some of these important features.

Figure 20 shows two configurations for an airspaced triplet consisting of plastic lenses, spacers and retainers, and a two-part molded plastic housing. In each case, the lenses are inserted axially into the housing and held by retainers that are glued or fused in place. Design a in the upper half of the figure has the joint between the mount components (i.e., the parting line of the mold) just to the right of the negative lens. The two lenses and the spacer shown on the left are inserted from that side, while the third lens is inserted from the right. In design b shown in the lower half of the figure, the housing joint is located at the extreme right end of the assembly. In this case, the IDs of the recesses for the lenses are formed by a single-stepped mold so concentration is improved over that inherent in

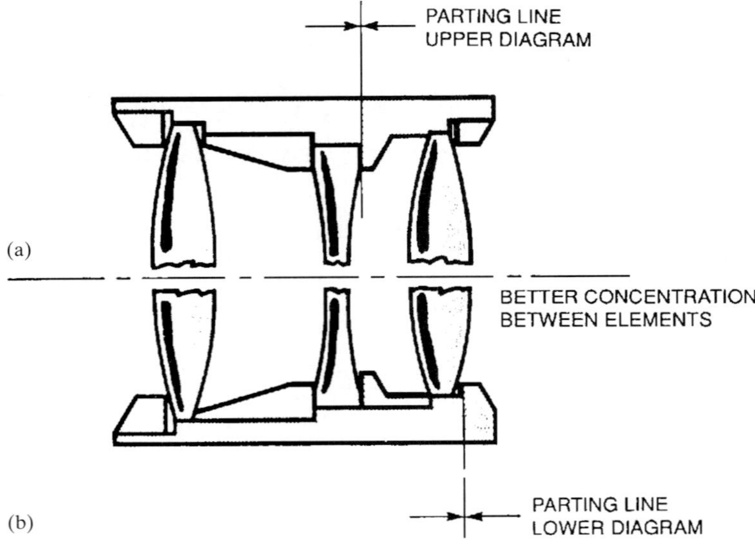

FIGURE 20 An all-plastic, three-lens optomechanical subassembly with the two-part mount designed for joining at alternate axial locations. Design *a* has lower cost but design *b* has improved centration and somewhat better performances. *[Courtesy of U.S. Precision Lens, Inc.]*

design a. Design a is less expensive to produce than design b so a trade-off between optical performance and cost must be considered in choosing which design to adopt.

A distinct advantage of plastic optics is that multiple optical elements, structural mounts, and interfaces for light sources, detectors, and mechanisms can frequently be integrated into single parts. To illustrate this, Fig. 21 shows a two-lens assembly with

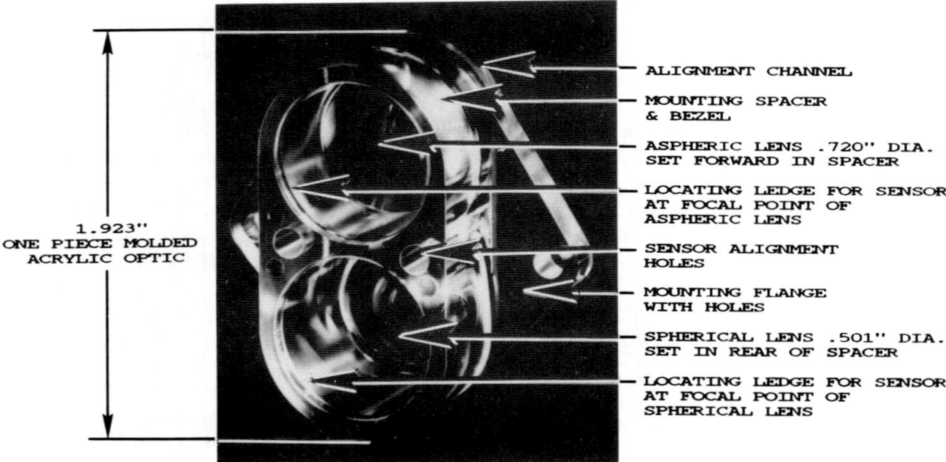

FIGURE 21 Economical construction of a one-piece optomechanical subassembly with two integral lenses, prealigned interfaces for two sensors and mounting features. *[Courtesy of U.S. Precision Lens, Inc.]*

integral mounting flange and preformed interfaces for two sensors. Used in an automatic coin changer mechanism, this assembly is prefocused and uses an aspheric element. Given large enough production quantities to amortize mold costs, this subassembly is considerably more economical than one consisting of the equivalent separate parts that would require labor to manufacture, inspect, assemble, and align.

37.5 MOUNTING SMALL MIRRORS AND PRISMS

General Considerations

The appropriateness of designs for mechanical mountings for small mirrors and prisms depends upon a variety of factors including: the tolerable movement and distortion of the refracting and/or reflecting surface(s); the magnitudes, application locations, and orientations of forces tending to move the optic with respect to its mount; steady-state and transient thermal effects (including gradients); the sizes and kinematic compatibility of interfacing optomechanical surfaces; and the rigidity and long-term stability of the structure supporting the mount. In addition, the designs must be compatible with assembly, maintenance, package size, weight, and configuration constraints, as well as be cost-effective. Examples of some very simple designs and of one more complex design are described briefly in the following sections to illustrate proven mounting techniques.

Mechanically Clamped Mountings

Figure 22 shows a simple means for attaching a first surface flat mirror to a mechanical mount such as a metal bracket or flange. The reflecting surface is pressed against three coplanar lapped pads by three leaf springs or clips. The spring contacts are directly opposite the pads to minimize bending moments. This design constrains one translation and two tilts. Translations in the plane of a circular mirror can most easily be constrained by dimensioning the spacers supporting the springs so as to just clear the rim of the

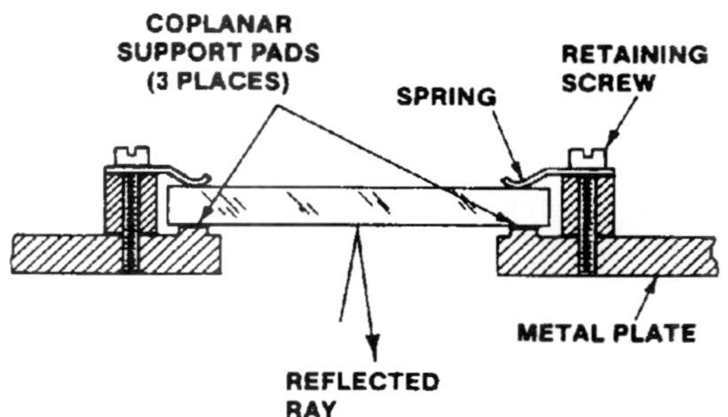

FIGURE 22 Typical construction for a spring clip-mounted flat mirror subassembly. *[From Ref. 18.]*

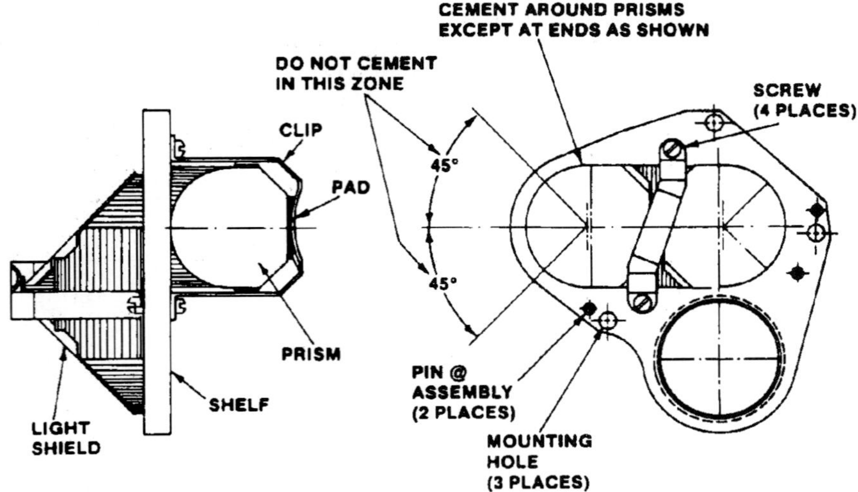

FIGURE 23 Diagram of a typical strap mounting for a Porro prism erecting subassembly. *[From Ref. 18.]*

mirror. Lengths of those spacers and the free lengths and stiffnesses of the cantilevered springs are typically chosen so as to load the mirror against the pads against expected acceleration forces with a safety factor of at least 2. Yoder[7] summarized standard techniques for design of such springs.

A common design for clamping a two-component Porro prism image erecting system to a mounting plate for use in a telescope or binocular is shown in Fig. 23.[18] The prisms fit into recesses machined into opposite sides of the plate and configured to fit the contours of the hypotenuse faces of the prisms. Straps typically made of spring-tempered phosphor bronze hold the prisms against the plate. The thin sheet metal light shields cover the reflecting surfaces to reduce stray light. Flexible adhesive (such as RTV) applied locally along the edges of the metal-to-glass interface helps prevent translation of the prisms in the recesses. Mounts for other types of prisms can be designed using the same principles described here.

Generally, clamping forces are exerted against ground surfaces of prisms rather than optically active ones to minimize surface deformations. As in the case of the mirror mount just described, spring loading against the plate should typically constrain the prisms against at least twice the expected maximum acceleration force.

Bonded Mountings

A popular technique for mounting mirrors and prisms is to bond them directly to a mounting plate or bracket. Dimensional and alignment stability are enhanced if required adjustment provisions are built into the mount rather than into the glass-to-metal interface.

Figure 24 shows a typical configuration for a flat mirror bonded to a mounting bracket. It is quite feasible to mount first surface mirrors of aperture about 15 cm or smaller directly to a mechanical support. The ratio of largest mirror dimension to mirror thickness should be at least 10:1 in order not to distort the reflecting surface. In the design shown, the mirror is crown glass with about a 6:1 diameter-to-thickness ratio. The mount is stainless steel so thermal expansion characteristics match rather well. The bonding area is circular. The adhesive layer is epoxy approximately 0.075 mm thick. Care should be exercised in

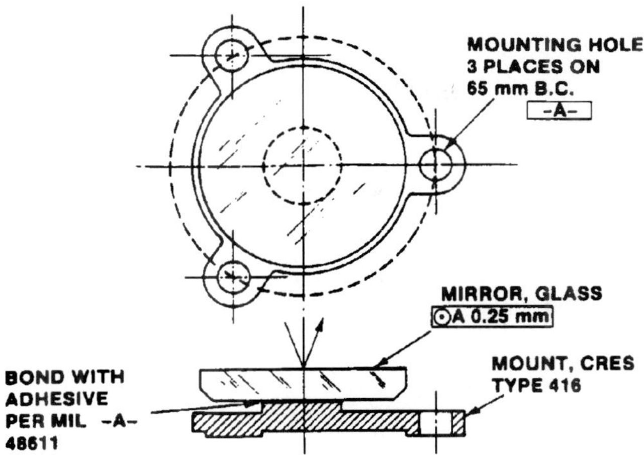

FIGURE 24 Typical construction for a first-surface mirror subassembly with glass-to-metal bond securing the mirror to its mount. *[From Ref. 18.]*

assembling such an optical component to not allow fillets of epoxy to form around the interface due to overflow, as these could lead to excess stress concentrations due to contraction during cure or at low temperatures.

A simple design for bonding a right-angle prism to a mounting bracket is illustrated in Fig. 25. The prism is cantilevered from a nominally vertical surface. Here the bond area is maximized and is adequate to maintain joint integrity under high vibration and shock loading. Yoder[19] outlined a method for determining the appropriate bonding area to hold a variety of common prism types against acceleration forces.

Flexure Mountings

Mounts for mirrors frequently consist of parts made from different types of materials so their thermal expansion properties do not match. Differential expansion as the temperature changes may cause optical components to decenter or tilt, thereby affecting critical alignment. One technique to combat such errors is to mount the optic on flexure

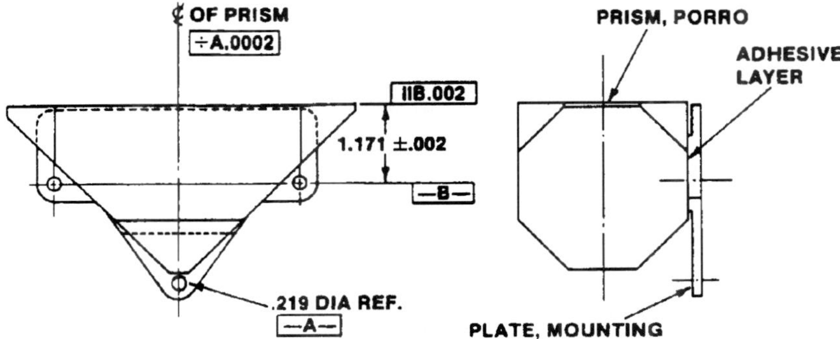

FIGURE 25 Typical configuration for a large prism bonded in cantilever fashion to a nominally vertical bracket. Dimensions are inches. *[From Ref. 18.]*

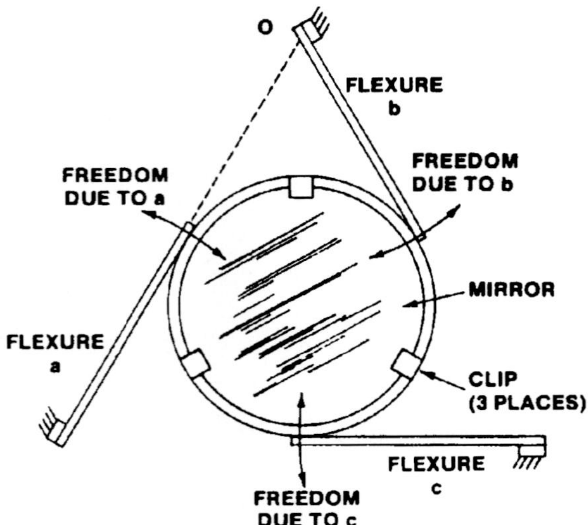

FIGURE 26 Conceptual design for mounting a circular aperture mirror on three tangent flexure blades. *[From Ref. 7.]*

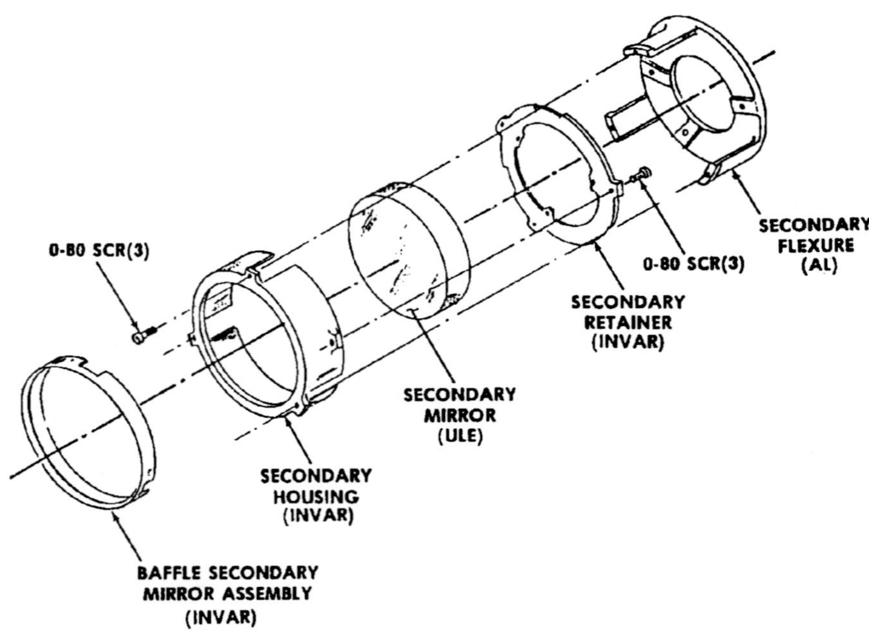

FIGURE 27 Exploded view of a flexure mounting for a small secondary mirror for a spaceborne telescope. *[From Ref. 20.]*

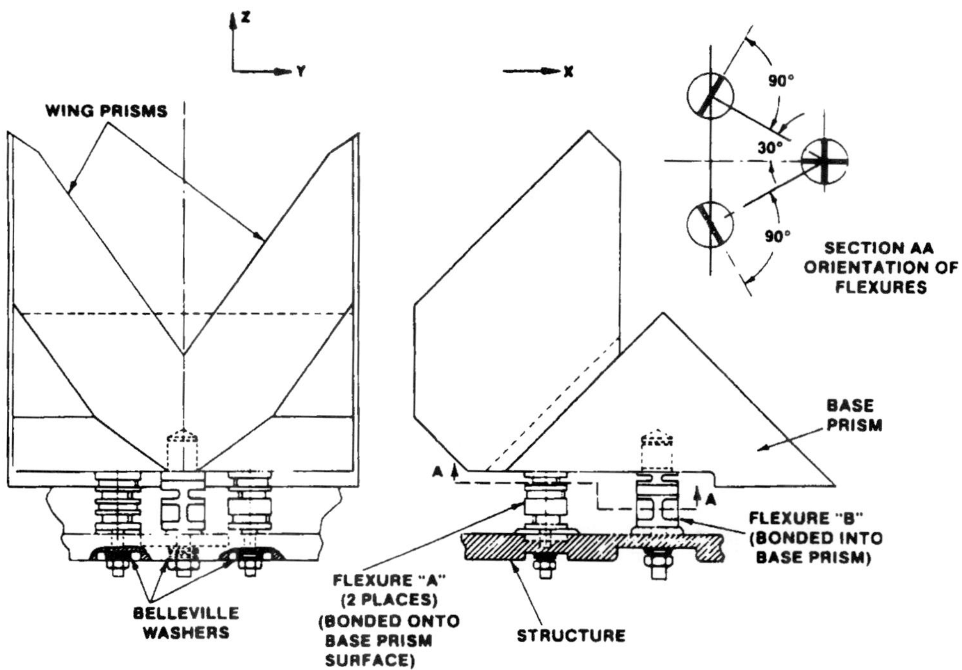

FIGURE 28 Configuration of a large prism assembly mounted on three flexure posts. *[From Ref. 7.]*

blades or posts designed to bend or twist symmetrically and maintain alignment. Three examples are considered here.

Figure 26 shows the design principle of a flexure support for a flat or curved mirror. The mirror is mounted into a cell which is connected to instrument structure by three flexure blades attached tangentially to the cell wall. The blades are stiff axially to prevent axial motion or tilt, but they bend as indicated by the curved arrows if dimensions of the mirror/cell combination or of the structure change radially. The mirror tends to remain centered, but undergoes a very small rotation about its axis as the blades flex. This mounting technique is applicable to larger mirrors as well as small ones. Examples of tangential flexure mountings for low-expansion mirrors as large as 1.5-m aperture in aluminum mounts and used in astronomical telescope or laser beam expander applications were described by Yoder.[7]

A flexure support for a telescope secondary mirror described by Hookman[20] is shown in Fig. 27. In this design, a small (3.9-cm aperture) mirror made of low-expansion ULE material is mounted in an Invar cell which is supported on three flexure blades machined integrally into an aluminum mount. Used in the telescope of a satellite-borne environmental sensor, this mirror undergoes the rigors of launch and variable temperature operation within minimum adverse effects on optical alignment or surface figure.

A flexure support for a large (14-cm aperture) Zerodur roof mirror constructed from three optically contacted prisms is illustrated in Fig. 28. This prism, part of a high-performance optical mask aligner for microlithography wafer production, is supported by three flexure posts configured with multiple orthogonal blades and cruciform sections to minimize rigid-body movements and surface distortions when mounted to a structure made of materials with significantly higher expansion properties. As indicated in the diagram at upper right, two flexures include blades allowing motion toward the third post. Two lateral translations and three rotations are controlled by this mounting arrangement.[7]

37.6 REFERENCES

1. E. B. Shand, *Glass Engineering Handbook,* 2d ed., McGraw-Hill, New York, 1958.
2. R. M. Scott, "Optical Manufacturing," chap. 2, *Applied Optics and Optical Engineering,* vol. III, Rudolf Kingslake (ed.), Academic Press, New York, 1965.
3. D. F. Horne, *Optical Production Technology,* Adam Hilger, Ltd., Bristol, 1972.
4. Robert E. Hopkins, "Lens Mounting and Centering," chap. 2, *Applied Optics and Optical Engineering,* vol. VIII, Robert R. Shannon and James C. Wyant (eds.), Academic Press, New York, 1980.
5. R. F. Delgado and M. Hallinan, "Mounting of Lens Elements," *Opt. Eng.* **14:**S-11 (1975).
6. Paul R. Yoder, Jr., "Axial Stresses with Toroidal Lens-to-Mount Interfaces," *Proc SPIE* **1533:**2 (1991).
7. Paul R. Yoder, Jr., *Opto-Mechanical Systems Design,* 2d ed., Marcel Dekker, New York, 1993.
8. Paul R. Yoder, Jr., "Advanced Considerations of the Lens to Cell Interface," *Proc. SPIE* **CR43:**305 (1992).
9. Brian J. Kowalski, "A User's Guide to Designing and Mounting Lenses and Mirrors," *Digest of Papers, OSA Workshop on Optical Fabrication and Testing,* North Falmouth (1980).
10. Mete Bayar, "Lens Barrel Optomechanical Design Principles," *Opt. Eng.* **20:**181 (1981).
11. Tina M. Valente and Ralph M. Richard, "Analysis of Elastomer Lens Mountings," *Proc. SPIE* **1533:**21 (1991).
12. Paul R. Yoder, Jr., "Lens Mounting Techniques," *Proc. SPIE* **389:**2 (1983).
13. Paul R. Yoder, Jr., "Opto-Mechanical Designs for Two Special-Purpose Objective Lens Assemblies," *Proc. SPIE* **656:**225 (1986).
14. K. H. Carnel, M. J. Kidger, A. J. Overill, R. W. Reader, F. C. Reavell, W. T. Welford, and G. C. Wynne, "Some Experiments on Precision Lens Centering and Mounting," *Optica Acta* **21:**615 (1974).
15. Robert E. Fischer, "Case Study of Elastomeric Lens Mounts," *Proc. SPIE* **1533:**27 (1991).
16. Daniel Vukobratovich, private communication (1991) Univ. of Arizona at Tucscon, Tucson, Arizona.
17. Ralph M. Richard and Tina M. Valente, "Interference Fit Equations for Lens Cell Design," *Proc. SPIE* **1533:**12 (1991).
18. Paul R. Yoder, Jr., "Non-Image-Forming Optical Components," *Proc. SPIE* **531:**206 (1985).
19. Paul R. Yoder, Jr., "Design Guidelines for Bonding Prisms to Mounts," *Proc. SPIE* **1013:**112 (1988).
20. Robert Hookman, "Design of the GOES Telescope Secondary Mirror Mounting," *Proc. SPIE* **1167:**368 (1989).

CHAPTER 38
CONTROL OF STRAY LIGHT

Robert P. Breault
Breault Research Organization
Tucson, Arizona

38.1 GLOSSARY

A	area
$BRDF$	bidirectional reflectance distribution function
GCF	geometric configuration factor
L	radiance
R	distance
θ, ϕ	angles
Φ	power
Ω	solid angle

38.2 INTRODUCTION

The analysis of stray light suppression is the study of all unwanted sources that reduce contrast or image quality. The control of stray light encompasses several very specialized fields of both experimental and theoretical research. Its basic input must consider (1) the optical design of the system; (2) the mechanical design, size, and shape of the objects in the system; (3) the thermal emittance characteristics for some systems; and (4) the scattering and reflectance characteristics of each surface for all input and output angles. It may also include spectral characteristics, spatial distribution, and polarization. Each of these areas may be concentrated on individually, but ultimately the analysis culminates in the merging of the various inputs.

Developments in detector technology, optical design software, diffraction-limited optical designs, fabrication techniques, and metrology testing have created a demand for sensors with lower levels of stray radiation. Ways to control stray light to meet these demands must be considered during the "preliminary" conceptual design. Decisions made at this time are, more often than not, irrevocable. This is because parallel studies based upon the initially accepted starting design are often very expensive. The task of minimizing the stray radiation that reaches the detector after the system has been designed by "adding on" a suppression system is very difficult. Therefore, every effort should be made to start off with a sound stray light design. To ensure a sound design, some stray light analysis should be incorporated in the earliest stages of a preliminary design study.

This chapter presents some basic concepts, tools, and methods that you, the optical or mechanical designer, can consider when creating a sensor system. You do not need to be very experienced in stray light suppression to design basic features into the system, or to consider alternative designs that may significantly enhance the sensor's performance. The concepts are applicable to all sizes of optical instrumentation and to virtually all wavelengths. In some cases, you can use the concepts to rescue a design when experimental test results indicate a major design flaw.

38.3 CONCEPTS

This section outlines some concepts that you can use to reduce stray radiation in any optical system. The section also contains some experimental and computer-calculated data as examples that should give you some idea of the magnitude of the enhancement that is possible.

The power on a collector depends on the following:

1. The power from the stray light source
2. The surface scatter characteristics of the source; these characteristics are defined by the Bidirectional Scatter Distribution Function (BSDF)
3. The geometrical relationship between the source and collector. This relationship is called the Geometrical Configuration Factor (discussed later in this section).

To reduce the power on the detector, we can try to reduce the contributions from these elements:

$$\Phi_{\text{collector power}} = \Phi_{\text{source power}} \times \text{BRDF}_{\text{source}} \times \text{GCF}_{\text{source-collector}} \times \pi \qquad (1)$$

Ways to reduce each of these factors are discussed below. The creative use of aperture stops, Lyot stops, and field stops is an important part of any attempt to reduce the GCF term of the power transfer equation.

For the discussion that follows, examples from a two-mirror Cassegrain design, with the aperture stop at the primary, will be used to stimulate thoughts about stray light reduction possibilities for other sensors.[1] The system is shown in Fig. 1.

Critical Objects

The most fundamental concept is to start the stray light analysis from the detector plane of the proposed designs. The most critical surfaces in a system are those that can be seen from the detector position or focal surface. These structures are the only ones that contribute power to the detector. For this reason, direct your initial attention toward minimizing their power contributions by removing them from the field of view of the detector.

The basic idea is to visualize what would be seen if you were to look out of the system from the image plane. Unlike most users of optical instruments, the stray light designer's primary concern is seldom the object field, but rather all the interior surfaces that scatter light. It is necessary to look beyond the radii of the imaging apertures to find the sources of unwanted energy. Removing these sources from the field of the detector is a real possibility, and will result in a significant improvement in the system.

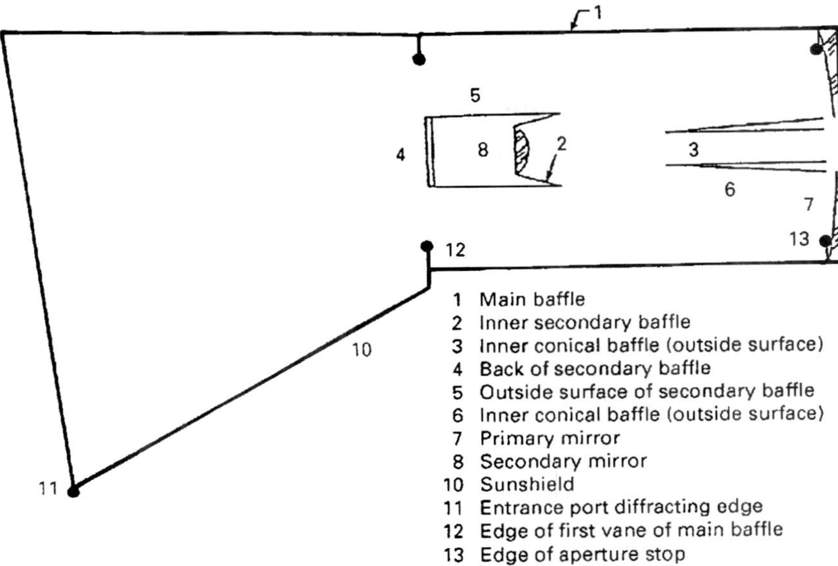

FIGURE 1 Typical Cassegrain design with the aperture stop at the primary. (*Ref. 7, p. 52.*)

Real-Space Critical Objects

I will start out by identifying a particular critical real-space object that can be seen by the detector in our example; it is the inside of the secondary baffle. The direct view discussed here is different than the image of the same baffle reflected by the secondary which is discussed in the next section.

Many Cassegrain secondary baffles have been designed to be cone-shaped (Fig. 2), usually approximating the converging cone of light from the primary. From the detector, portions of this secondary cone are seen directly as a critical surface. Since most of the unwanted energy is incident on this baffle from nearly the same direction as this surface is seen from the detector, the addition of vane structures would be of little help, assuming an

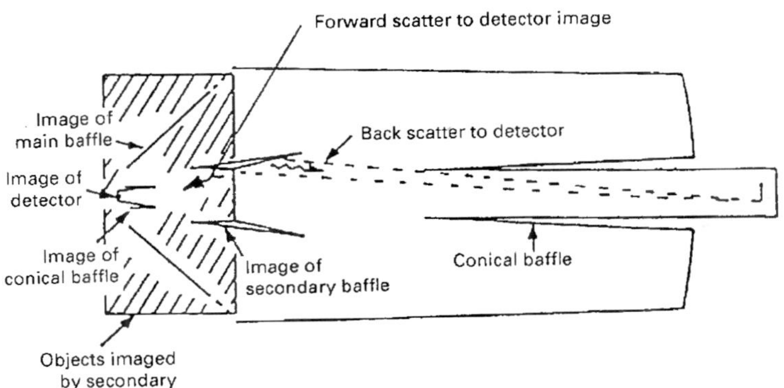

FIGURE 2 Direct and reflected scatter from the cone-shaped secondary baffle. (*Ref. 1, p. 4.*)

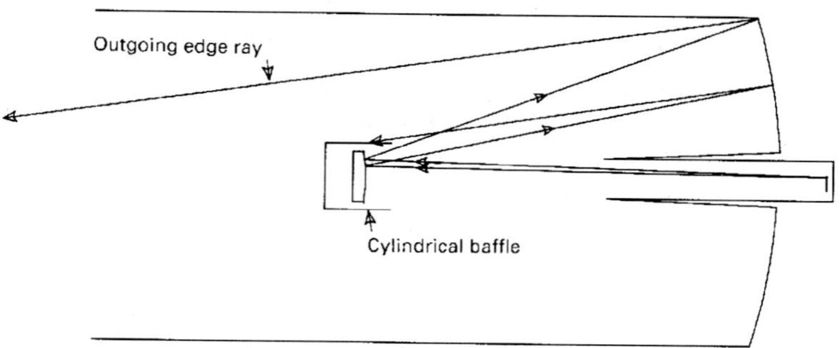

FIGURE 3 Reduced scatter from an almost cylindrical-shaped secondary baffle.

optimum coating is used on the simple baffle. If the cone is made *more* cylindrical, the amount of critical cone area is reduced, and the angle at which the surface is seen gives a smaller projected area (Fig. 3).

Avoid making the baffle cylindrical because the outside of it would be seen. Since the detector is of a finite size there is a fan of rays off the primary representing the field of view of the telescope. Although collimated for any point on the detector, any point not on axis would have its ray bundle at some angle to the optical axis, hence a cylindrical secondary baffle would be seen from off-axis positions on the detector.

Imaged Critical Objects

Imaged objects are often critical objects. They too can be seen from the detector. Determining which of the imaged objects are critical requires a bit of imagination and usually some calculations; stray light software can help you make the calculations. The Y-Ybar diagram can help you to conceptualize and determine the relative image distances and sizes with a minimum number of calculations.[2] The same could be done with other first-order imaging techniques (see Smith in this *Handbook*). Using the Cassegrain example again (Fig. 2), you can see that reflected off the secondary mirror are the images of the detector and the inside of the inner conical baffle (object 3 in Fig. 4). In some designs the outside of the conical baffle will be seen in reflection. These are *imaged critical objects.* If you wish, you can eliminate some of these images with a central

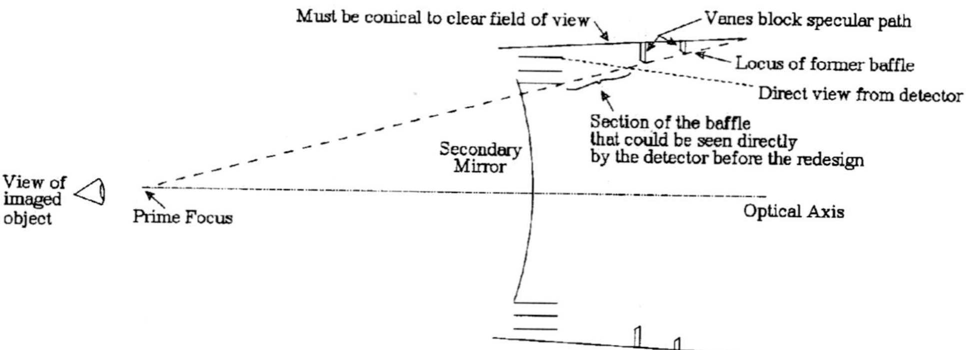

FIGURE 4 A cylindrical secondary baffle can be seen from off-axis positions on the detector.

obscuration on the secondary, or for the conical baffle, with a spherical mirror concentric about the image plane. The direct view to the inner conical baffle will remain, but the path from the image of it is removed.

The cone-shaped secondary baffle is also seen in reflection (Fig. 2). For the incident angles of radiation on this surface, the near specular (forward-scattering) characteristics will often be one of the most important stray light paths because the image of the detector is in that direction. This is an extension of *starting from the detector*. There is an image of the detector at the prime focus of the primary mirror. Often, as in this case, one location may be easier than another for you to determine what could be seen. By making the baffle more cylindrical, part of the *image* of the baffle is removed from the detector's view; as a result, the power that can scatter to the detector is reduced. Furthermore, it is sometimes possible to baffle most of this power from the field of view with one or two vanes (Fig. 4).

Continue the process of removing critical surfaces until all the critical surfaces have been considered for all points in the image plane. The power contributions from these surfaces will either go to zero, or at least be lessened after you reduce the area of the sections seen.

There is still more that can be done, since only the GCF term in the power transfer equation to the detector has been reduced. It is also possible to minimize the power onto the critical sections, which will become the source of power, Φ, at the next level of scatter. This approach can be very similar to the approach used for minimizing the power scattered to the image plane. The viewing is now forward from the critical surfaces instead of the image plane. By minimizing the BRDF and GCF factors of the surfaces scattering to the critical sections, the power incident on the ciritical surfaces will be reduced. Hence, the power to the detector is also reduced.

Illuminated Objects

Minimizing the GCFs and BRDFs for the specific input and output angle is sometimes easier if you look into the system from the position of the stray light source in object space. By doing this, you can identify the surfaces that directly receive the unwanted energy. I will call these the *illuminated objects*. If any of these illuminated surfaces contain sections that the detector can see, then you should direct your initial efforts toward eliminating these paths. These paths will usually dominate all other stray light paths because there is only a single scatter before the stray light reaches the detector. An example of such a path that is often encountered is from the source onto the inner conical baffle of multimirror systems (Fig. 1). Some of the ways that the direct radiation can be eliminated is by extending the main baffle tube, increasing the obscuration ratio by increasing the diameter of the secondary baffle (Fig. 5), or by narrowing the field of view, which will allow you to extend the secondary baffle and the inner conical baffle toward each other.

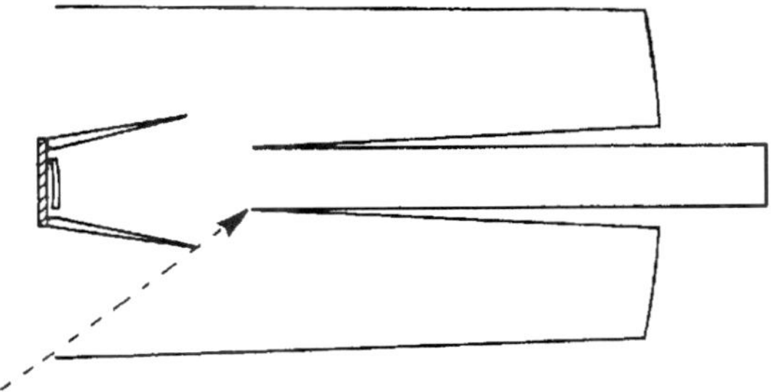

FIGURE 5 Increased obscuration ratio blocks direct path to inner conical baffle. (*Ref. 1, p. 6.*)

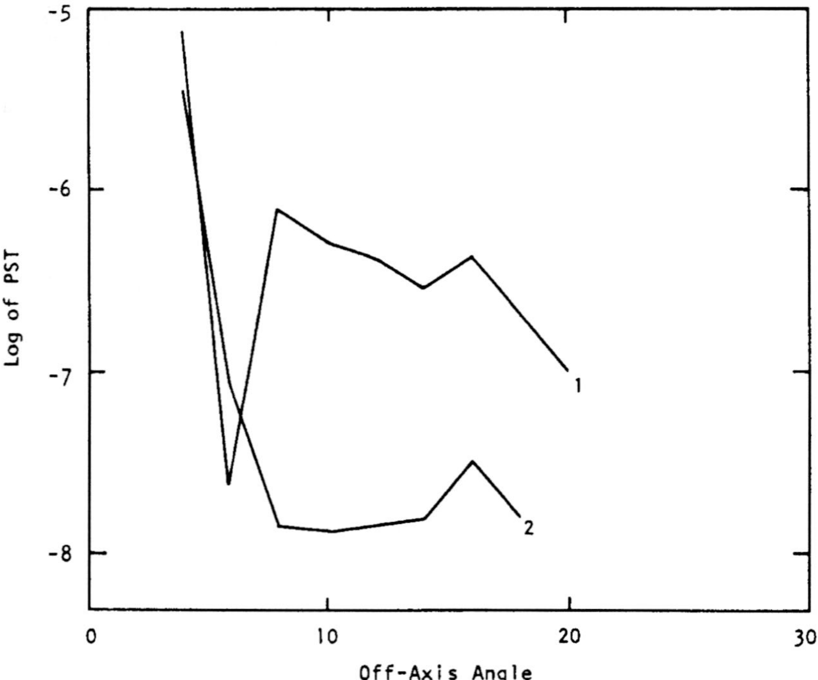

FIGURE 6 Point source transmittance with obscurations of (1) 0.333 and (2) 0.4. The 0.4 obscuration removed the direct path from the source to the inner conical baffle. (*Ref. 1, p. 6.*)

The effect of eliminating this path is shown in a composite Point Source Transmittance (PST) plot in Fig. 6.[3] The PST plot is defined as the reference plane (detector plane in most cases) irradiance divided by the input irradiance along the line of sight. (See the section called "Point Source Transmittance Definitions" for a more detailed definition of PST.) For the case shown, the unwanted irradiance on the detector is reduced by over an order of magnitude.

Aperture Placement

I will now focus on the optical design aspects of a stray light suppression system, and give a qualitative discussion of some general aspects that you might consider. All optical systems will have at least one aperture, called an *aperture stop,* that limits the size of the bundle of the incoming signal rays. Some systems will have field stops and/or Lyot stops. Each type of stop has a clearly defined role in stray radiation rejection, which is discussed in the sections below.

In many cases stop placement will have a much more noticeable effect on system performance than any vane structure, coating, or baffle redesign. Probably the only factor with more effect on the PST curve is the off-axis position of the source. Therefore, the benefits of any of the stops cannot be overemphasized.

Aperture Stops. The aperture stop is the aperture that limits the size of the cone of radiation that will reach a point on the image plane. Sometimes shifting this stop allows

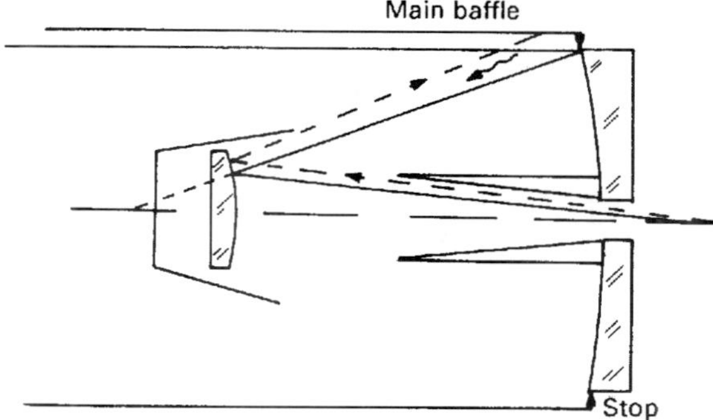

FIGURE 7 The oversized secondary allows the main baffle to be seen in reflection. (*Ref. 1, p. 8.*)

the optical designer to better balance the aberrations. In a stray radiation suppression design, it plays a similar important role. All objects in the spaces preceding the stop in the optical path will not be seen unless they are imaging elements, central obscurations, or objects that vignette the field of view. Only a limited number of critical objects is possible before the aperture stop. In the intervening spaces from the stop to the image plane it is likely that many of the baffle surfaces will be seen. Figure 7 represents a two-mirror design, and Fig. 8 represents a three-element refracting system; both have the stop at the first element. In both cases the second element is oversized to accommodate the field of view from a point in the field stop; the amount depends on the full field of view of the design. Because the elements are oversized, the main baffle following the first element will be seen. This baffle will be a critical object, a direct path of unwanted energy. The "overviewing" is characteristic of all of the optical elements past the aperture stop.

If you move the stop along the optical path toward the detector plane, its performance as a stray radiation baffle will improve. If you shift the stop to the second element, the intermediate baffle will not be seen. It is removed from the field of view of the detector, since the stop now eliminates direct paths from baffles in all spaces that precede it. Figure 9 shows the improvement in the PST curve for a two-mirror system. By moving the stop you have reduced the PST by a factor of 10. This is a desirable feature to consider for stray radiation reduction.

Direct paths from central obscurations can be blocked by a central disk located at some location deeper into the system; however, because of the parallax involved between the central obscuration disk and this central disk, the central disk obscuration will usually be a larger obstruction to imaging rays. In a reimaging design it is often possible to locate a central disk conjugate to the actual central obscuration.

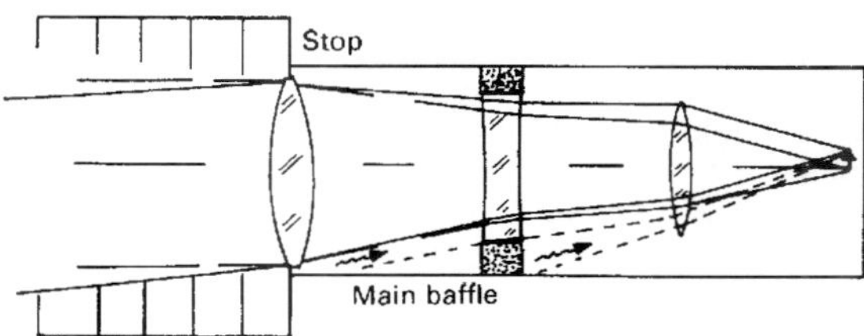

FIGURE 8 The main baffle is seen through the oversized secondary and tertiary. (*Ref. 1, p. 8.*)

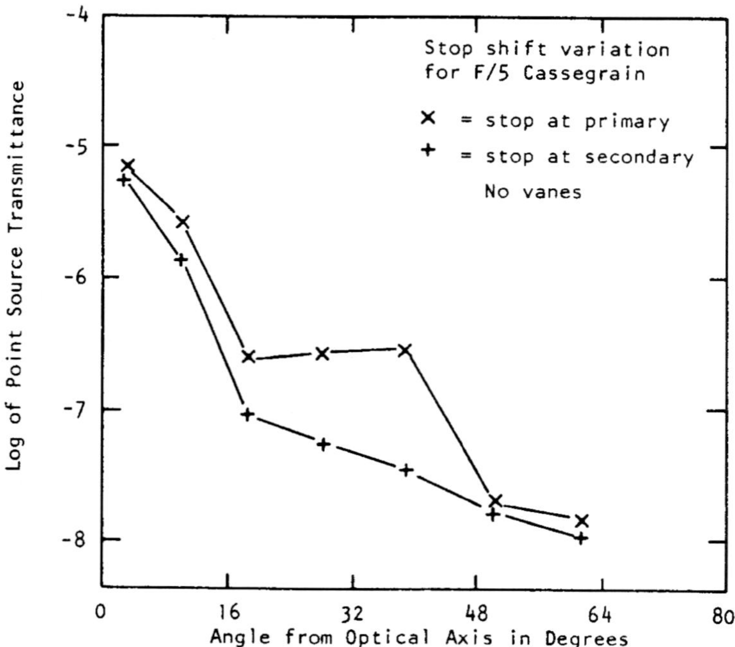

FIGURE 9 PST improvement with stop shift for the two-mirror system. (*Ref. 1, p. 8.*)

Field Stops. An aperture can be placed at intermediate *images* in a system to limit the field of view. Such an aperture will usually prevent any stray light from outside of the field of view from being directly imaged into the system beyond this field stop aperture. In a sense, its operation is just opposite that of an aperture stop. Baffle surfaces following a field stop cannot be seen from outside the field of view in the object plane, unless they are central obscurations. Note that with just a field stop, succeeding optical elements may allow out-of-field critical sections to be seen *through* the field stop, from within the field in the image plane (Fig. 11). Aperture stops are necessary to block such paths. Figures 10 and 11 show two such cases. Although for some designs the field stop is not 100 percent effective because of optical aberrations, its small size limits most of the unwanted stray light. Field stops therefore do not remove critical sections, but rather limit the propagation of power to illuminated objects. In reflecting systems, take care that the object side of the field stops does not become a critical area, which can be seen directly or in reflection from the image plane because unwanted energy is being focused onto them.[4]

Lyot Stops. A limiting aperture placed at the location of the image of an aperture stop, sometimes called a glare stop or Lyot stop, has the same property as described for aperture stops. It should be slightly smaller than the image of the aperture stop. It limits the critical sections which are out of the field of view to those objects in succeeding spaces only. Since Lyot stops are by definition further along the optical path to the detector, the number of critical surfaces seen by the detector will be reduced. Usually, these stops are incorporated into the design to block the diffracted energy from an aperture stop and field stop pair, so that only secondary or tertiary diffracted energy reaches the image. Nevertheless, both diffracted and scattered energy are removed from the direct view of the image, reimaging the largest optical element as the stop takes full advantage of both the light-gathering

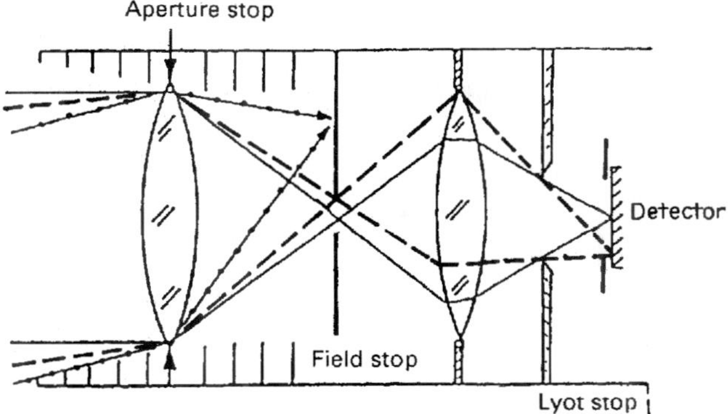

FIGURE 10 The addition of a Lyot stop prevents preceding baffles from being seen from the image.

power of the optics and the stray radiation suppression features provided by the stop. Figure 10 shows a system with a Lyot stop.

On space-based telescopes the image plane is often *shared* by one or more instruments. Each instrument reimages the telescope's image through some optical train, and eventually onto the detector. In that optical train there could be a logical place to use a Lyot stop to improve the stray light performance of the viewing instrument well beyond that of the telescope.

It is the combination of these different stops or apertures that helps minimize the propagation of unwanted energy by limiting the number of critical objects seen by the detector, and the objects illuminated by the stray light source.

When all direct paths have been eliminated, the next step is to determine the relationship between the sections that received power (illuminated objects) and the critical surfaces. This relationship takes the form of scattering *paths*; that is, stray light can scatter from the illuminated objects to the critical objects. To start reducing the stray light contributions from these paths, you can start at the critical surfaces as described above. But now you have more knowledge about where the direct incident power is being distributed throughout the system, since you can also look into the system from the source side to find the surfaces receiving direct power. With this information you can identify the possible paths between the illuminated and critical surfaces.

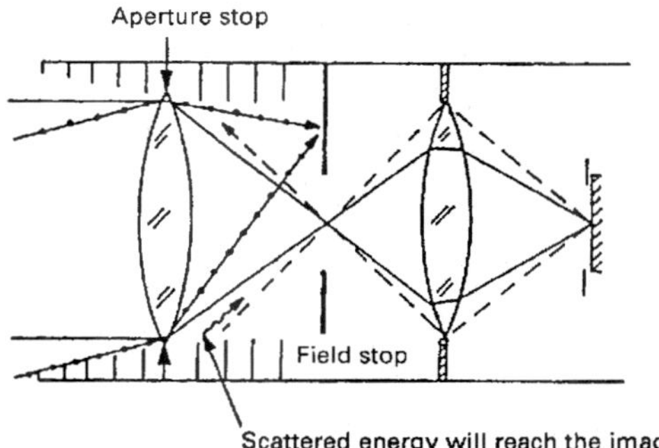

FIGURE 11 Out-of-field energy in object plane will not be imaged beyond field stop. Out-of-field energy elsewhere may be seen. (*Ref. 1, p. 9.*)

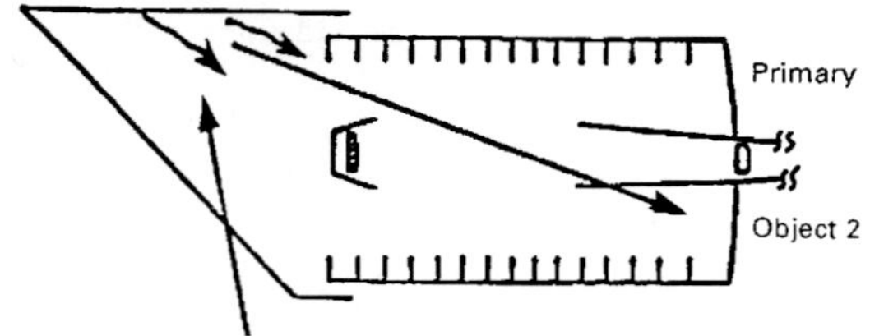

FIGURE 12 There is a direct path from the baffle to the primary mirror. (*Ref. 1, p. 7.*)

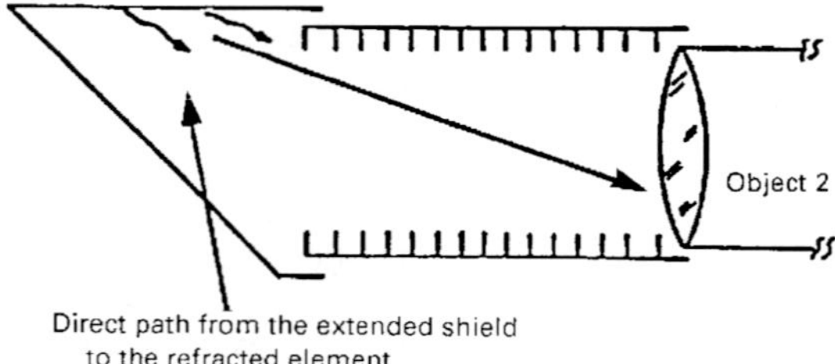

FIGURE 13 There is a direct path from the baffle to the refracting element. (*Ref. 1, p. 7.*)

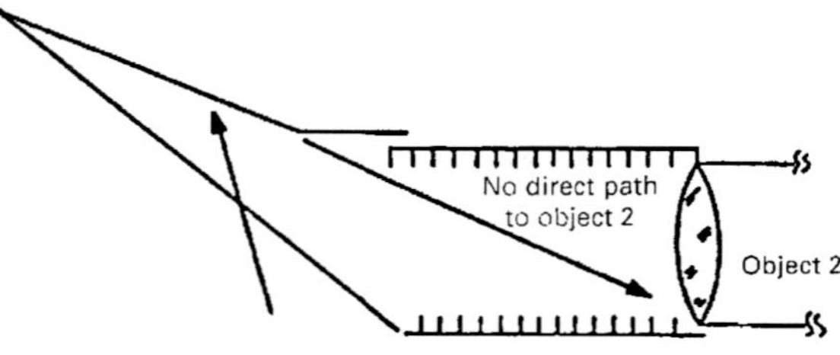

FIGURE 14 Direct paths are removed by properly tilting the shield. (*Ref. 1, p. 7.*)

Design considerations for extended baffle shields (Figs. 12 and 13) provide a good example of starting from the source side to identify possible paths. In the examples, object 2 (an optical surface) is the largest contributor of scattered radiation and is the best superpolished mirror available (Fig. 12), or if it is a lens as in Fig. 13, it has the lowest possible scattering characteristics. It cannot be removed from the view of the image plane. If the initial power incident on object 2 is only from the extended shield, then by tilting the shield (Fig. 14), the power on the shield must first scatter to the main tube and then to the optical element. The combination is then referred to as a two-stage baffle. If vanes are added to the main baffle, the scattered radiation incident on the optical element will be reduced by many orders of magnitude when all the scattering solid angles and the number of absorbing surfaces are considered. Note that without the tilt to the shield, vanes on the main baffle are worthless because there is a direct path to the objective.

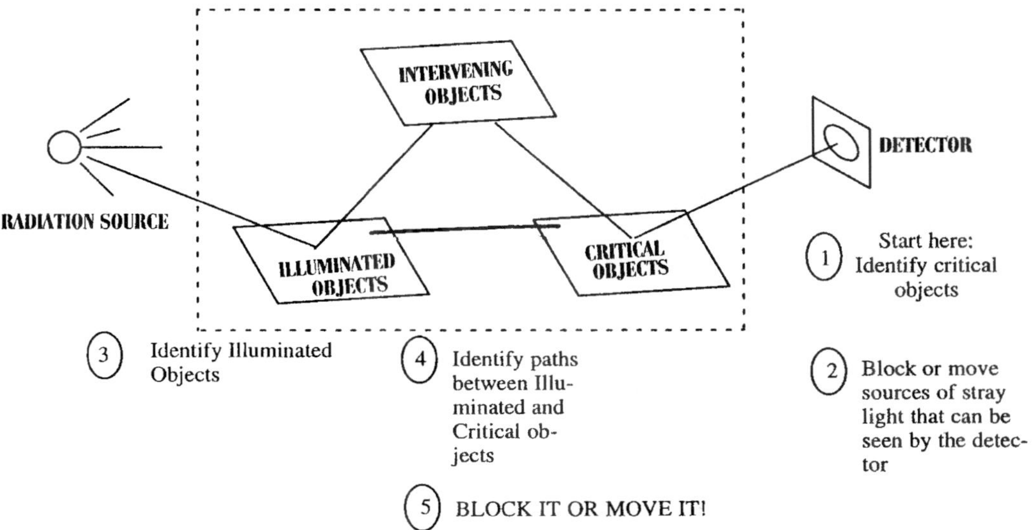

FIGURE 15 The first step in a stray light analysis begins from the detector plane, not from the source.

Figure 15 is an abstract representation of the process of reducing stray light in a sensor system. Start at the detector, then work from its conjugate image locations. Starting from the detector simplifies the analysis and directs your attention to the most productive solutions, because you can identify all the possible sources of stray light to the detector. You can then work at decreasing their number by slightly redesigning the baffles and stops. Next, identify which objects are illuminated. Discover how energy may propagate between them and you have identified the paths of stray light propagation. From then on the process of moving objects or blocking paths is quite simple, although the quantitative calculations might get difficult and may require some analysis software.

Baffles and Vanes

A few definitions are required to define baffles and vanes. Other authors have used their own different definitions. In this section the term *baffle* is used to describe conical structures (including cylindrical) that can also be described as tubelike structures. Their function is to shade, or occult, stray light from the source to one or more system components. The main baffle shields the primary mirror from direct radiation at the larger off-axis angles. *Vanes* are structures put *on* baffles to affect the scatter characteristics of the surface. Other authors have used the term "baffles," or "glare stops," to describe these vanes.

Baffles. In a well-designed system vanes play an important role only at large off-axis angles. For example, when one-tenth of the stray light falls on the primary of the Cassegrain design, then the main baffle receives the remaining 90 percent of the stray light. When the main baffle has properly designed vanes on it, light that falls on the baffles is attenuated by five orders of magnitude before it reaches the primary mirror. The resultant power on the primary mirror is then about 9.0×10^{-5} compared to the direct 10 percent that fell on the mirror. This results in less than 0.1 percent of the total on the primary. In addition, most of the subsequent scatter off the primary will be at much higher scatter

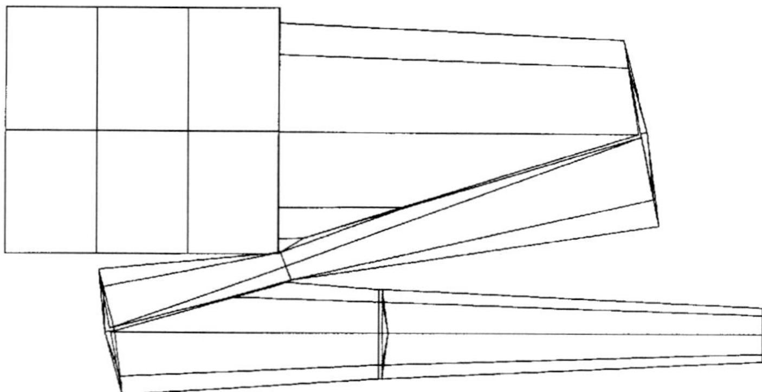

FIGURE 16 Confocal mirror system, eccentric pupil, no obscuration, low-scatter system. (*Ref. 5, p. 91.*)

angles. This will cause the scattered energy to have much lower BRDFs off the primary mirror when scattered in the direction of the detector, further reducing the scatter contribution from the baffle.

Only when no power illuminates the objective will the baffles play a significant role in the propagation paths of the stray light. Usually the system's performance merit function is then very good. Only if the stray light source has a tremendous amount of energy, like the sun, does the stray light become measurable.

Vanes. The depth, separation, angle, and bevel of vanes are variables that need to be evaluated for every design. In the following paragraphs stray light analysis results are presented for both a centrally obscured system (Cassegrain) and an unobscured eccentric pupil design (Z-system).[5] Profiles of these systems are given in Fig. 2 and 16. Of the two designs, only the eccentric pupil design has a reimager that would allow for the placement of an intermediate field stop and an accessible Lyot stop, as discussed above.

The APART stray light analysis program was used to analyze the two designs. The APART program is a substantial software package that performs deterministic calculations of stray light propagation in optical systems.[6–8]

As an example of vane design considerations, the design of vanes on a main baffle tube will be explained. With minor differences, the design steps are the same for the Cassegrain and the eccentric pupil designs. In a reimaging system, vane structure deep in the system is usually not necessary, but there are exceptions. Figure 17 shows a collecting optical element that has some small field of view (FOV). The optical element could represent a primary mirror or a refractive element. The placement of a straight, diffusely coated cylindrical tube would block the direct radiation from an external source, such as the sun, from reaching the optical element for a certain range of off-axis angles. If it were at a large off-axis angle, the forward scatter off the inside of the tube would be so high that it would normally not be acceptable. The solution is to add vanes to block this path.

Figures 18 and 19 depict the two cases that could represent the scatter from a baffle. In one case there are no vanes; in the other case there are vanes. This example shows how a propagation path is blocked by vanes. Vanes are useful, but a better approach is to make the solid angle (Ω_c) from the baffle (not the vanes) to the collectors of the scattered light go to zero, so that there is no path from the baffle and vane structure to the collecting object. By moving the baffle out of the field of view of the collector, the baffle's contribution goes to zero. There is no edge scatter, and no edge diffraction effects. That topic is in the realm of baffle design, which has already been touched on, and is well covered in the literature.[9,10]

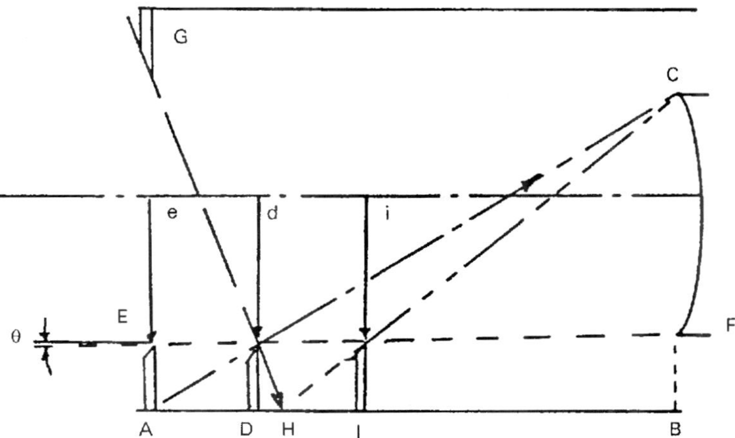

FIGURE 17 Vane placement design, lowercase letters are radii (measured from the optical axis), uppercase are z locations. (*Ref. 5, p. 94.*)

Designing the Vane Spacing and Depth[11]

A first vane is most often placed at the entrance of the baffle and an external ray is brought in from object space at a maximum off-axis position. If there is no forebaffle the angle is 90° off axis. The depth of the vane cavity is normally dictated by space and weight requirements. Too little depth will dictate the requirement for many vanes. Then vane edge scatter eventually becomes the major source of scatter instead of multiple vane scatter.

The initial ray will strike the side wall at the base of the first vane (point A in Fig. 17). From this point, a design line is drawn/calculated (AC) from the wall to the edge of the optical element on the opposite side. This line (AC) intersects the edge ray (EF), at z position D. At this point a vane could be placed. Mathematically, this assures that any point below C, including those on the optical element, would not see any directly illuminated side wall. However, practicality dictates that some offset of point D to a point D' (not shown) is required to allow for tolerance errors in fabrication of the vane, thermal effects, assembly errors, and for stray light edge scatter and diffraction effects. The tolerance allowance is company-, material-, and design-dependent. Acceptable numbers are often about 0.125 mm for fabrication and assembly tolerances. For the rest of this analysis, assume that this is accounted for.

Continue the design process by constructing another line from the edge of the entrance aperture to the tip of the second vane to the wall (line GH). Draw a new HC line to the area near the objective and determine the placement of the third vane (at I); once

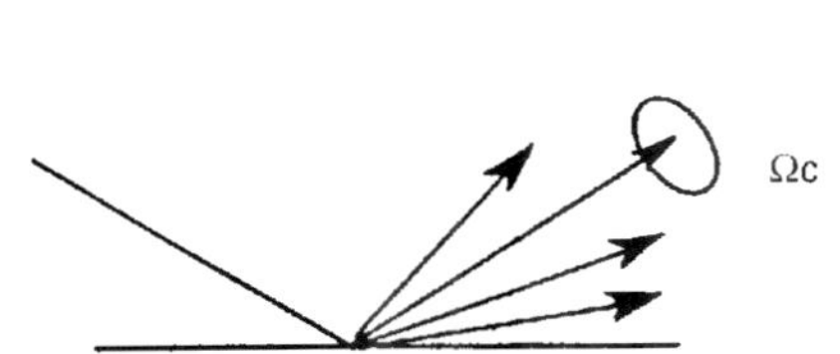

FIGURE 18 High forward scatter path. (*Ref. 5, p. 92.*)

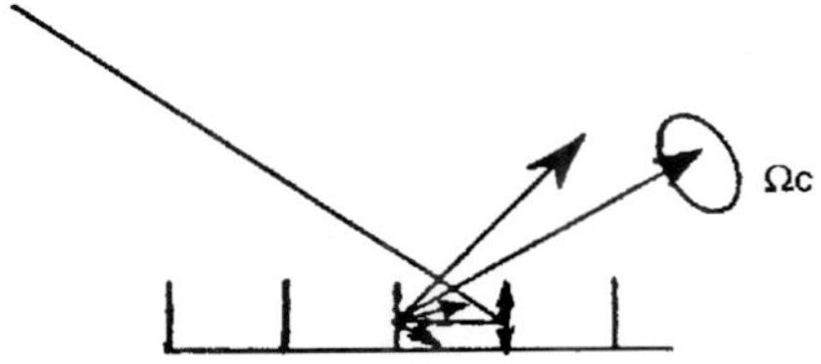

FIGURE 19 Forward scatter path highly attenuated by the vane structure. (*Ref. 5, p. 92.*)

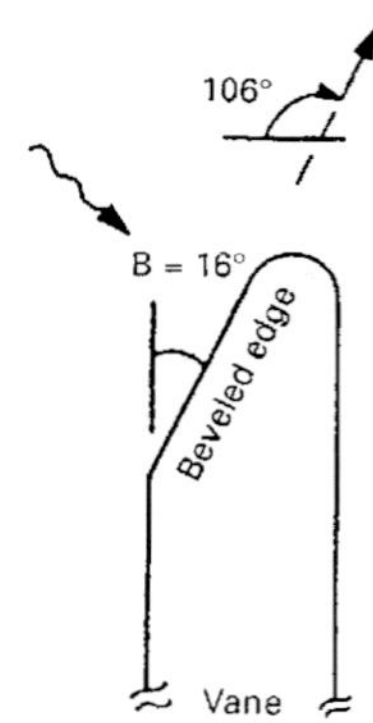

FIGURE 20 Placement of the bevel on the left side of the vane structure. (*Ref. 5, p. 96.*)

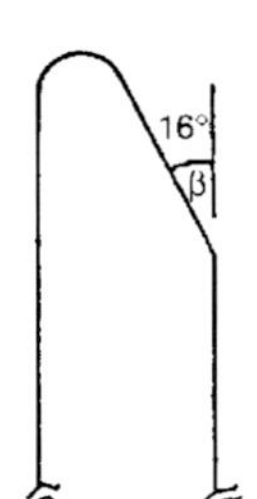

FIGURE 21 Placement of the bevel on the right side of the vane structure. (*Ref. 5, p. 96.*)

tolerances are considered, iterate the process to reach a final design. In some cases you may have to consider more than just the scatter path to the objective. In the Cassegrain design you may also have to consider the inner conical baffle opening. It is beyond the present scope to go into further detail.[12]

Bevel Placement on Vanes. In this short discussion on baffle-vane design and placement, I did not mention the placement of a slanted surface, or bevel, sometimes placed on a vane edge as shown in Fig. 20. Which side should the bevel go on? The answer is usually dictated by first-order scatter principles.[13,14] Near the front of the tube, direct radiation from a source at large off-axis angles will strike this bevel. If it is placed on the right side (Fig. 21), then the illuminated bevel will scatter its radiation all the way down into the tube to some optical surface. If placed on the left side, as depicted in Fig. 20, then it will go only 16° deeper into the system to the opposing vanes, a much better solution. For vanes deeper into the system, the bevel is placed on the right side. The point at which this is done is determined by the angle of the bevel and the diameter of the baffle tube. At some point, external radiation will not be able to directly strike the beveled edge if it is on the right-hand side of the vane. Only the nonbeveled, straight side will be illuminated. Therefore, the vane can rescatter only in the left side of the hemisphere, which is in the direction of going out of the system. If the bevel is placed on the left side, it can scatter 16° (in the example) deeper into the sensor; this is usually a needless design shortcoming that could be a significant error.

Vane Angle Considerations. Another variation on the design feature of vanes that has sometimes been incorporated onto baffles in an optical system is angled vane structure. These vanes are nonplanar objects. This makes them quite tedious to cut out of sheet metal, fabricate, and install. The next few paragraphs will present computer analysis results from two designs to show the effect of vanes on the propagated stray light. The vane angles used were 90, 70, and 45°, as depicted in Fig. 22.

FIGURE 22 Vane structure angled at 90, 70, and 45°, respectively. (*Ref. 5, p. 97.*)

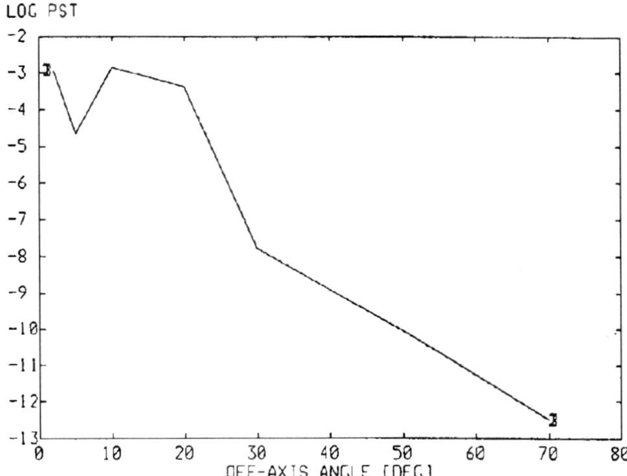

FIGURE 23 Cassegrain with Martin Black. Vane angles 1 = 90°, 2 = 70°, 3 = 45°. Log PST = detector irradiance divided by input irradiance. (*Ref. 5, p. 97.*)

The comparative stray light results for the Cassegrain system (Fig. 1) with a Martin Black coating on the vanes are shown in Fig. 23; in this system the vanes are on the main baffle, but not on the sunshade. There is no difference in the performance as the vane angle is varied from 45 to 90° (all three curves lie one on top of the other).

The comparative results for the Z-system (Fig. 16) with vanes on the sunshield are shown in Fig. 24. The results differ from the Cassegrain results for source located at

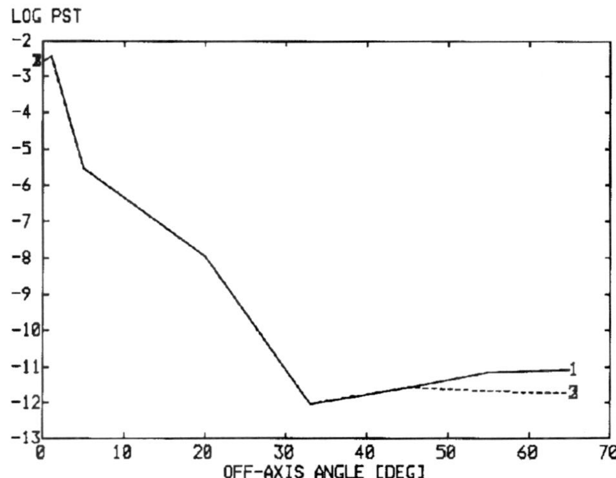

FIGURE 24 Z-system with Martin Black. Vane angles 1 = 45°, 2 = 70°, 3 = 90°. Vanes are on the sunshield. (*Ref. 5, p. 114.*)

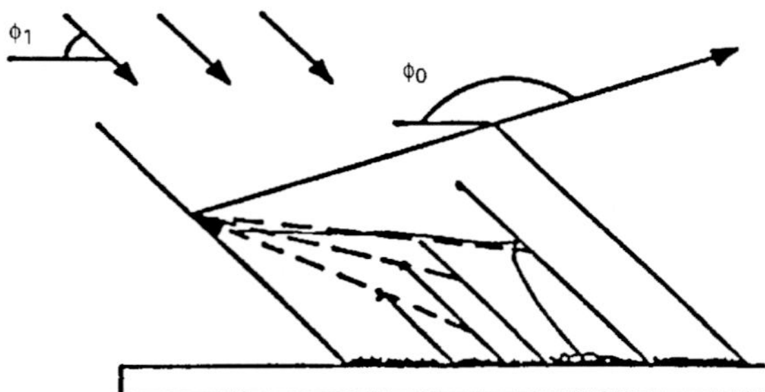

FIGURE 25 Angle-staggered vanes for fixed input angle. (*Ref. 5, p. 104.*)

angles greater than 45° off-axis. This is because the primary side of the baffle is illuminated and scatters light directly to the primary mirror. The 70° baffles would fail for sources beyond 70° off-axis. The Cassegrain system has vanes on the main baffle (not the sunshade) and the sunshade occulted the direct illumination of the primary side of the 45° vanes. This accounts for the subtle but important difference in the results.

Usually the first-order scattering properties of the vane structures are more important than whether the vanes are angled or not. The results presented above confirm this statement. There are occasions where angled vanes would be beneficial, but to fully understand those cases a much longer explanation of diffuse vane baffle scatter is necessary. These results are detailed elsewhere.[15,16]

There are special situations where angled vanes will have a significant advantage over annular vanes. One example is a bright source at a fixed offset angle. I have seen such a feature on a space-borne telescope on a platform where there was nearby a brightly sunlit rocket-thruster casing at a fixed angle outside the field of view. Figure 25 shows the design where the vanes were aimed at the thruster at an angle where the primary mirror side (right side in Fig. 25) could not be directly illuminated by the sunlight scattered off the thruster. Under those circumstances most of the stray light had to make three scatters before exiting the vane cavity. In general, as soon as the position of the bright object is moved over a range of angles, the advantage of the angled vanes is lost. Nevertheless, there are many occasions within a sensor where the relative positions of a scattering source and a collecting object are fixed along a major stray light path. The front parts of the main-barrel baffle and the opening of the inner conical baffle in the Cassegrain design is an example. Many more examples could be cited. But the point is that you, as a designer, should first consider the first-order, single scatter paths off the baffle wall, each side of the vanes, and the bevel, for the full range of input values. Based on that information you can make the decision to user planar or angular vanes.

Vane Depth Considerations. By varying the vane depth in the example analysis we can evaluate how the vane spacing-to-depth ratio affects system performance. Figure 26 gives the results of an analysis of the Cassegrain system with varying vane depths on the main baffle of 0.2, 0.4, and 0.8 inches. Figure 27 gives similar output from the Z-system analysis results. The performance of the system gets better as the vane depth increases from 0.2 to 0.4 inches, but there is little performance difference between the 0.4- and 0.8-inch baffle depths. The latter is the normal case. The 0.2-inch vane depth allows for a single path from the walls of the baffle tube, which increases the stray light propagation. Once that path is blocked by a greater vane depth, no further improvement should be expected due to further increases in vane depth.

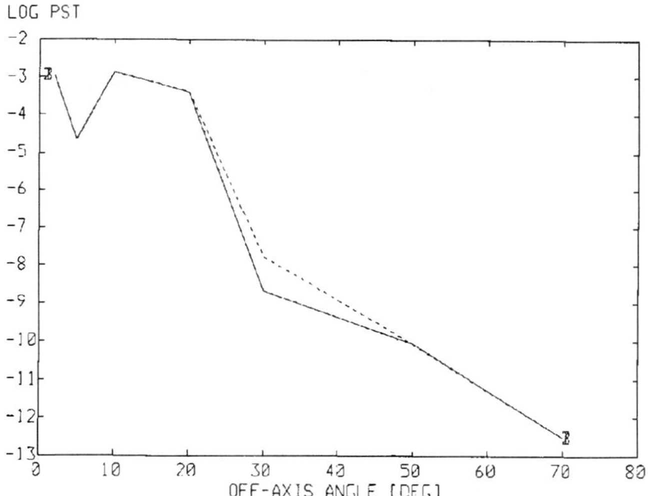

FIGURE 26 Cassegrain 90° baffles, coated with Martin Black, at varying depths; 1 = 0.2″, 2 = 0.4″, 3 = 0.8″ depth. (Scatter is dominated by baffles.) (*Ref. 5, p. 98.*)

The intent of presenting the two different optical designs was not to trade off one optical design against another. It needs to be made clear that the two optical sensors being used as examples are intentionally not equivalent from stray light design considerations. This is why the changes in performance are design-dependent. The nominal design of the eccentric pupil has a reimager, and the Cassegrain does not. The Cassegrain could have a

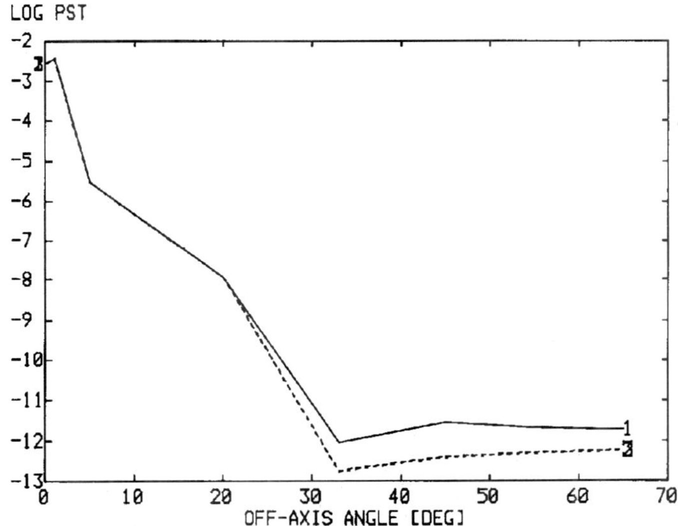

FIGURE 27 Z-system with varying vane depths. 1 = 0.2″, 2 = 0.4″, and 3 = 0.8″ depth. (*Ref. 5, p. 114.*)

reimager, in which case the stray light performance of both could be made essentially equal. It would depend on the optical design characteristics, F/#, field of view, obscuration ratio, etc. The Cassegrain design has a specular sunshield and the Z-system has a vaned diffuse baffle structure. Which would perform better could only be determined after all of these features are considered.

To summarize, the general points being made in this section are:

1. Usually, angled vane structure has little, if any, additional benefit over straight, annular vanes, and the annular vanes are much easier to fabricate and assemble.
2. Once the depth of a diffuse black vane structure is deep enough to block the single scatter path, further increases will not improve performance.

Specular Vanes. Another aspect about vane structure that has been explored, but only in a limited way, is the specular vane cavity. Previous studies indicated that specular vanes have a problem with the aberrated rays and near specular angle scatter; this problem is severe enough to degrade the performance significantly.[17,18] Another study by Ed Freniere was not always true.[19] The ASAP[20] stray light software was used to evaluate the Z-system (Fig. 28) with (1) no vane structure, but with the main barrel baffle coated with Martin Black; (2) with Martin-Black-coated vanes; and (3) with a specular vane structure. The results show a dramatic degradation in the stray light performance without the coating on the main baffle tube. A recent specular baffle design developed by Nick Stavroudis has been shown to be a major improvement over previous concepts.[21]

Contamination Levels

Light scattered from a particulate-contaminated surface can have a pronounced effect on the stray light performance of a system.[22]

I will now relate the performance of both designs (the centrally obscured Cassegrain, and the unobscured eccentric pupil) as a function of the level of scatter, per MIL-STD 1246A.[23] This analysis evaluates the sensor for different amounts of contamination on the optics only. The levels of contamination as defined in MIL-STD 1246A are for a distribution of particles with a specified range in particle sizes.

Ray Young used Mie scattering theory to predict the BRDF of a mirror covered with such MIL-STD distributions.[24] Table 1 was generated from Young's work for the 10 μm radiation. This table shows the base BRDF value and the BRDF slope that would be used in the APART stray light analysis program for input. The base value is the BRDF value at $(\beta - \beta_0) = 0.01$ and the second term is the slope of the BRDF in a (log–log) plot of BRDF versus $(\beta - \beta_0)$. β_0 is the sine of the angle of incidence and β is the sine of the observation angle.[25] The terms work equally well for out-of-plane values, but the above definitions, for simplicity, assume in-plane scattering data. See also the works of Spyak.[22]

There is a problem with specifying the optics with this standard because it is difficult to reliably relate a level of contamination by particles to a BRDF performance. Two equal sizes and distributions of particulates may not give the same BRDF, because the index of refraction, the reflectivity, and the roughness of the particulates enter the calculations. In general, few people go to the trouble to determine these other factors. These factors will vary from one distribution to another. BRDF is the most usable value when performing a stray light analysis, so it should be the stray light specification. For manufacturing specifications, other parameters may be more appropriate, but they are not as good as BRDF for a stray light specification.

The level of scatter is also given in Table 1 along with the BRDF. The BRDF data from particulate scatter for the 5-, 10-, and 20-μm wavelengths for the 100, 300, and 500 contamination level have been plotted in Fig. 29. Consensus, not factually documented, indicates that the current state of the art of contamination control is at the cleanliness

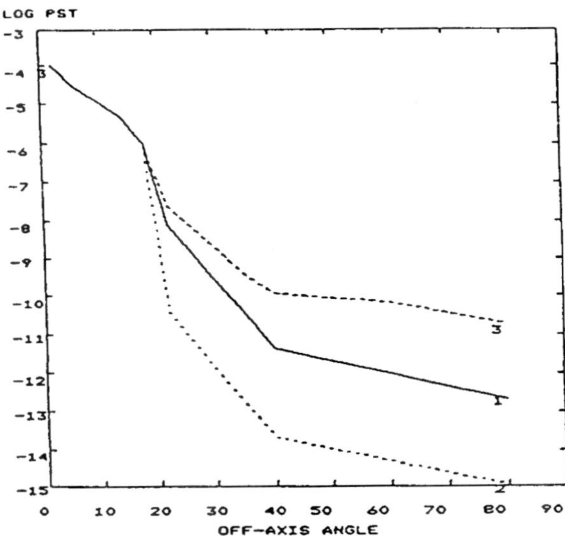

FIGURE 28 PST for unobscured pupil design without vane structure, with diffuse vane structure, and with specular vanes. 1 (solid) = no vanes, diffuse black coating; 2 (dotted) = diffuse vanes on main tube; 3 (dashed) = specular vanes on main tube. (*Ref. 5, p. 115.*)

level of 300 to 500 for the 10-μm-wavelength region. Measured BRDFs below level 200 are achievable in the lab for short periods of time. A stray light analyst is strongly advised not to predict a system's performance with values below 1.0E-3 sr^{-1} in the 10-μm region. Based on historical performance, mirrors in the IR (10-μm region) consistently degrade to this value, usually because of particulate scatter. Research work performed under Rome Air Development Center contract for the detection, prevention, and removal of contamination from the ground and in space could greatly reduce the degradation currently experienced by IR sensors.[26]

Hal Bennett presents the significance of particulate scatter, as shown in Fig. 30.[27] This figure shows an agreement between measured data and theoretical data, and illustrates why IR sensors are usually more sensitive to particulate scatter than RMS scatter; the opposite is true in the visible. Figure 30 also indicates why the wavelength scaling law does not usually relate visible BRDF measurements to BRDF measurements in the IR. The physical process is different.

TABLE 1 Mirror Scatter Relationships (Wavelength = 10 μm, BRDF slope in $\log(\beta - \beta_o)$)

BRDF at $(\beta - \beta_o) = 0.01$	BRDF slope	Cleanliness level
0.02	−1.17	500
0.01	−1.17	454
0.001	−1.17	300
0.0001	−1.17	204
0.00001	−1.17	100

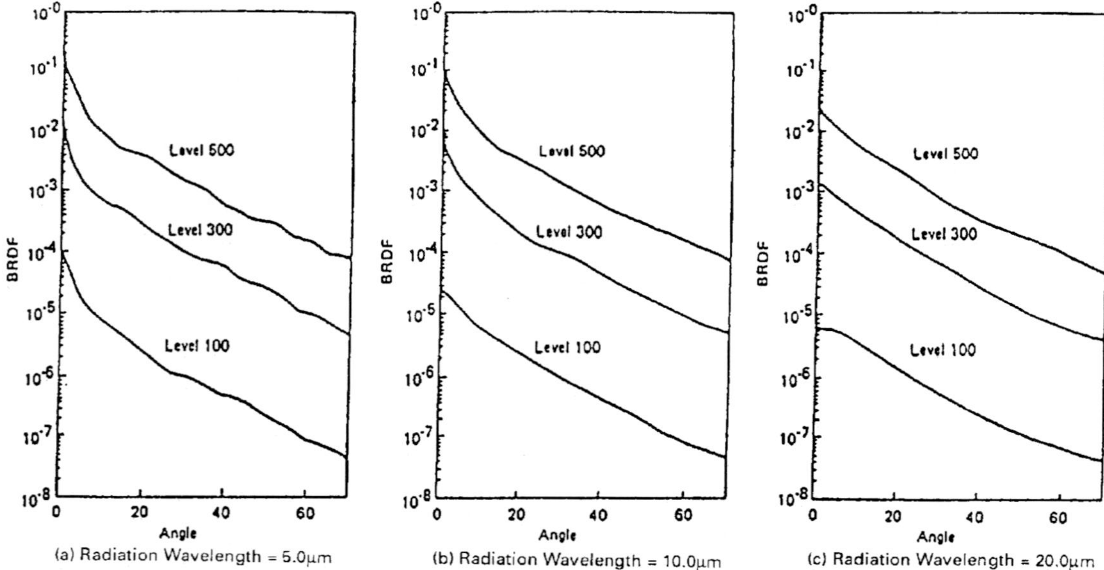

FIGURE 29 Predicted BRDFs on particles deposited on low-scatter mirror for cleanliness levels of 100, 300, and 500 at radiation wavelengths of 5.0, 10.0, and 20.0 μm (*Young, 1976*). (*Ref. 5, p. 105.*).

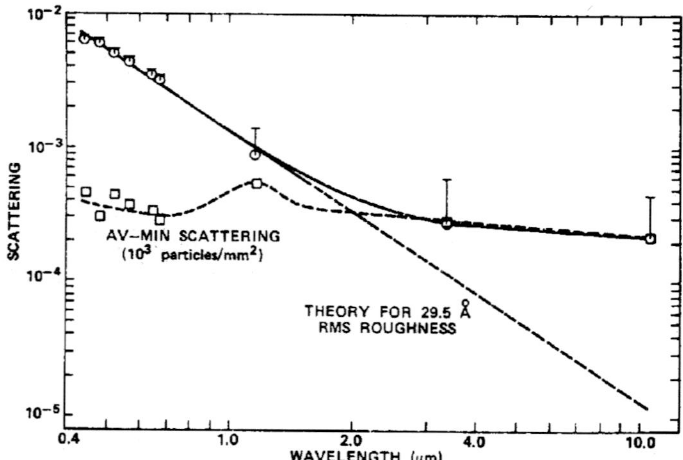

FIGURE 30 Scattering from polished dense flint glass. The diagonal line gives the contribution predicted for microirregularity scattering by a 29.5 Å rough surface. Circles indicate the minimum scattering observed, and the bars and squares the difference between the average and minimum scattering observed at several points on the surface. This difference may be related to particulate scattering. (*Ref. 27, p. 32.*)

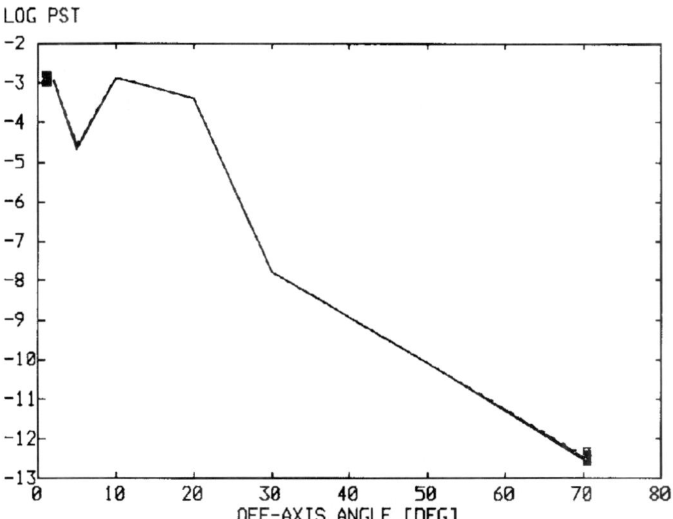

FIGURE 31 Cassegrain system with mirrors at all five contamination levels. 1 = 100, 2 = 204, 3 = 300, 4 = 454, 5 = 500. (*Ref. 5, p. 113.*)

Figures 31 and 32 are the representative Point Source Transmittances (defined as the irradiance on the detector divided by the incident irradiance) for the cleanliness levels of 100 through 500 for each design. The Cassegrain is much less affected by changes in contamination level, because the scatter from the black-coated surfaces dominates all other scatter. If the system had a reimager its performance would be better because these black surfaces would be blocked from the field of view of the detector, and the stray light

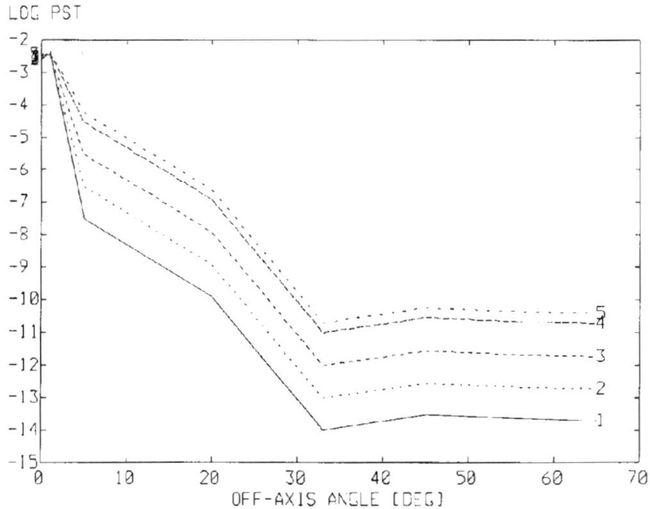

FIGURE 32 Z-system with mirrors at all five contamination levels, 1 = 100, 2 = 204, 3 = 300, 4 = 454, 5 = 500. (*Ref. 5, p. 113.*)

performance would be due to the cleanliness level of the optics. The eccentric pupil design is sensitive to changes in the mirror coatings because it does have a reimager, and the major source of scatter is from the mirror surfaces.

In summary, the impact of particle contamination on the performance of a system will depend on how well the system is designed to suppress stray light. The goal is to be limited by a single optical element, such as the collecting lens or mirror, which is the objective of the system. The eccentric pupil design (Z-system) has this design feature. The better the optical design from a stray light point of view, the more the system's performance will be degraded by particle contamination. The more the system performance is determined by the black coatings, the more it will be sensitive to degradations in the coatings on the baffles.

Strut Design

In a centrally obscured system the central obscuration must be supported. In some designs (Schmidt-Cassegrains) the obscuration can be supported by a refractive element, but in most designs some form of struts are used. The most common error in strut design is to specify manufacture from a slab or plate of coated metal. Because all detectors have some finite field of view, the scatter from the sides of the struts can be seen from the image plane. Usually the struts are out "front" and exposed to more stray light sources than the objects deeper into the system. The near off-axis angles of incidence of scattered light off of the strut sides make for very high scattering toward the detector.

The proper strut design will preclude this path by making the object end of the strut narrower than the side nearest the objective (primary). This shape, shown in profile in Fig. 33, does not allow the detector to see the sides of the struts. The angle of the taper depends upon the object space field of view of the detector. It requires only a small change in design to remove this stray light path.

Basic Equation of Radiation Transfer

This section briefly discusses the most fundamental equation needed to perform the quantitative calculations of a stray light analysis. It reenforces the concept of first identifying what the detector can see and working on the geometry of the system to limit the stray light propagation, and not the BRDF term.

The fundamental equation relating power transfer from one section to another is:

$$d\Phi_c = L_s(\theta_C, \phi_C)\, dA_s \frac{\cos(\phi_s)\, dA_c \cos(\phi_c)}{R_{sc}^2} \tag{2}$$

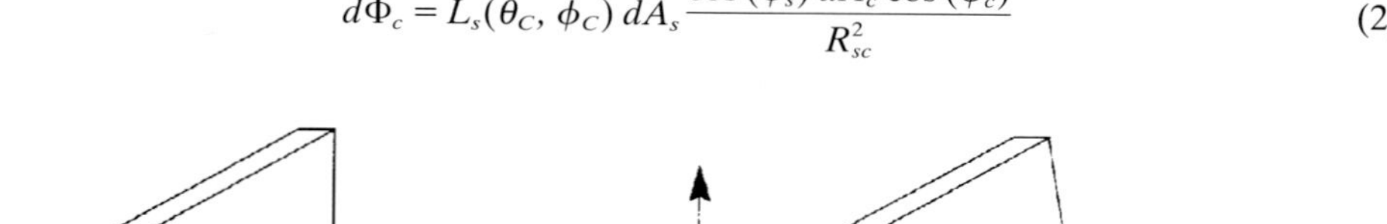
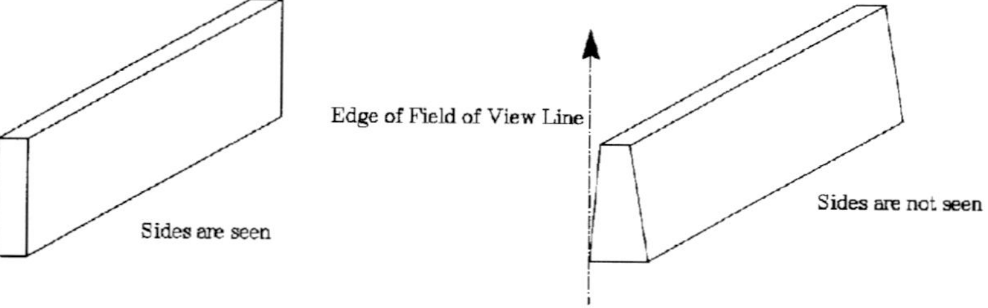

FIGURE 33 Angled strut design does not allow the detector to see the sides of the strut.

where $d\Phi_c$ is the differential power transferred, $L_s(\theta_c, \phi_c)$ is the radiance of the source section, dA_s and dA_c are the elemental areas of the source and collector, and ϕ_s and ϕ_c are the angles that the line of sight from the source to the collector makes with their respective normals. This equation can be rewritten as three factors that help clarify the reduction of scattered radiation.

$$d\Phi_c = \frac{L_s(\theta_C, \phi_C)}{E(\theta_i, \phi_i)} E(\theta_i, \phi_i)\, dA_s \frac{\cos(\phi_s)\, dA_c \cos(\phi_c)}{R_{sc}^2} \tag{3}$$

$$d\Phi_c = \text{BRDF}(\theta_i, \phi_i; \theta_C, \phi_C)\, d\Phi_s(\theta_i, \phi_i)\, d\Omega_{sc} \cos(\phi_s) \tag{4}$$

$$d\Phi_c = \text{BRDF}(\theta_i, \phi_i; \theta_c, \phi_c)\, d\Phi_s(\theta_i, \phi_i)\text{GCF}_{sc}\pi \tag{5}$$

$E(\theta_i, \phi_i)$ is the incident irradiance on the source section dA_s. GCF_{sc} is the projected solid angle from the source to the collector divided by π.

The GCF is independent of the first two terms and solely determined by the geometry of the system, including obscurations. The first term, $\text{BRDF}(\theta_i, \phi_i; \theta_c, \phi_c)$, is the bidirectional reflectance distribution function. It is usually considered independent of the second term, the incident power, and is therefore a function of the surface characteristics only. When reducing stray radiation propagation, one or more of these terms must be reduced. If any one of these terms is reduced to zero, no power will be transferred between the source and collector.

Stray Radiation Paths

Since the third term (GCF) in Eq. 4 is the *only* term that can be reduced to *zero,* it should receive attention first. This is a crucial point in a stray light analysis. Therefore, the logical starting place for stray light reduction is with the critical objects, since it is the GCF terms for these transfers which can be reduced to zero. Most novice analysts make the mistake of working on the BRDF term first.

$$\text{GCF} = \frac{\cos(\phi_s)\, dA_c \cos(\phi_c)}{\pi R_{sc}^2}$$

The apparent possibilities for decreasing the GCF are to increase R_{sc}, ϕ_s, ϕ_c or to reduce the area dA_c. Not readily apparent is that the GCF is limited by apertures and obstructions. These features will, in some cases, block out the entire view of the source section from the collector so that there is no direct path. This is the mathematical basis for the logical approach, discussed at the beginning of the chapter. First block off as many direct paths of unwanted energy to the detector as possible, and then minimize the GCF for the remaining paths.

Point Source Transmittance Definitions

There are five common ways to define the merit function of the stray light in an optical sensor. The most common and preferred method is to define it as the output irradiance divided by the input irradiance, in terms of the *Normalized Detector Irradiance* (NDI),[28] or in terms of the *Point Source Normalized Irradiance Transmittance* (PSNIT).[29] This merit function is appropriate because it describes an irradiance transmittance, and it is relatively independent of the detector size.

A term often used in the past was the *Off-Axis Rejection* (OAR), defined as the detector power divided by the input power from the same source *on axis.* The term *rejection* is a misnomer because by definition the term describes a power transmittance, which can have little correlation with the rejected stray light. The second objection is that as a merit function it varies significantly with the detector size. If you double the area of the detector, the OAR will increase by about the same factor even though the system hasn't performed significantly worse in any way.

Another term commonly used is the system's stray light *Point Source Power Transmittance* (PSPT), or its reciprocal, the *Attenuation* of the system. The PSPT is the detector power divided by the input power into the sensor from the specified *off-axis angle.* Again, this term varies with the detector size. Sometimes there is no well-defined entrance port so the denominator is impossible to define. Note that the magnitude of attenuation would normally be expressed in terms of a positive exponential. Beware that attenuations are often incorrectly called out with negative exponents.

A final PST definition that is sometimes specified is the *Point Source Irradiance Transmittance* (PSIT), defined as the output irradiance divided by the entrance port input irradiance. This definition becomes inappropriate when there is no clearly defined entrance port.

Surface Scattering Characteristics

Of the three potentially important factors in scattered radiation analysis cited above (the radiance of the undesirable source or sources, the geometry of the scattered radiation paths (GCF), and the surface scattering characteristics, (BRDF)), usually the first possibility considered is to improve the surface coatings or the addition of vane structure. In concept it *appears* to be the right place to start and that it is straightforward. Neither is the case; the BRDF never goes to zero as does the GCF, and the BRDF varies with input and output angles. However, with accurate *Bidirectional Reflectance Distribution Function* (BRDF) data and knowledge about the variations with applications, time, wavelength, and other factors, BRDF problems can be dealt with. The scattering characteristics of surfaces are discussed by Church, and the scattering characteristics of black coatings by Pompea and Breault elsewhere in this *Handbook.* The addition of vane sections on baffles can usually be considered as a specialized "coating"with its own specialized BRDF.

BRDF Characteristics

Usually, BRDF data that are presented represent only one profile of the BRDF, and many such profiles for various angles of incidence are necessary for understanding the scattering characteristics. However, studies have shown that a single profile of a mirror's surface scattering characteristics can be used, with some approximations, to define the BRDF for all angles of incidence.[30] This is a significant achievement. It reduces the amount of data that must be taken, and it makes it easier to calculate, or estimate, the BRDF value for any set of input and output angles. The BRDF can also be reconstructed for cases where only a single profile of the function has been presented, which has been the usual practice.

The approximation has its limitations, as clearly detailed by Stover.[31] The approximation is quite good for nominal angles of incidence (see Fig. 34).[32] However, it breaks down for very high θ_i and high observation angles θ_o.

It is important to understand qualitatively the scattering characteristics of diffuse black coatings. Figure 35 shows the BRDF profile of Martin Black at 10.6 μm for several angles of incidence.[33] At near-normal angles of incidence the BRDF values are bowl-shaped; the values increase at large observation angles from the normal. At high angles of incidence the BRDF vaues in the near specular direction have increased by 2.5 orders of magnitude. There is a good discussion of the qualitative characteristics of diffuse black surfaces by Pompea and Breault elsewhere in this *Handbook.*

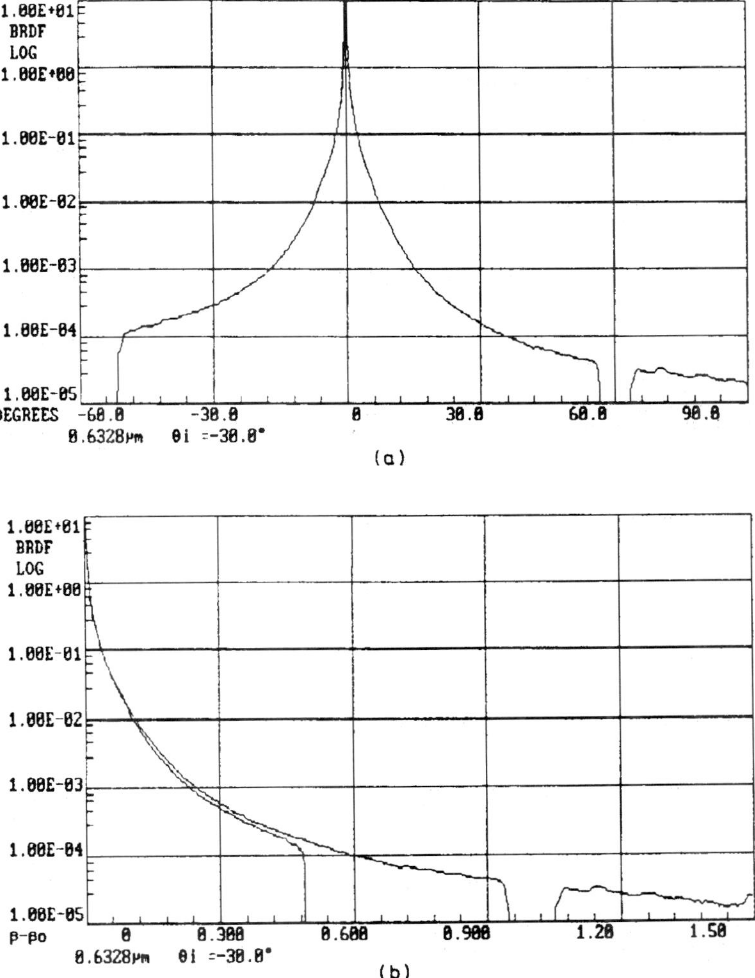

FIGURE 34 (*a*) The BRDF is asymmetrical when plotted against $\theta_s - \theta_i$; (*b*) The data in (*a*) exhibits near symmetry when plotted against $|\beta - \beta_o|$. The slight deviation from symmetry is due to the factor ($\cos \theta_s Q$), where Q is a polarization factor. (*Ref. 31, p. 69.*)

38.4 STRAY LIGHT SOFTWARE

There are several stray light software packages that should be familiar to the stray light analyst. Each uses one of two general approaches to perform stray light calculations: Monte-Carlo (ray-based) or deterministic. A Monte-Carlo technique traces real rays through the system and keeps track of the energy associated with each ray. A deterministic approach calculates power transfers between surfaces within the system; energy scatters along propagation paths to some detector, where the program sums the total incident power.

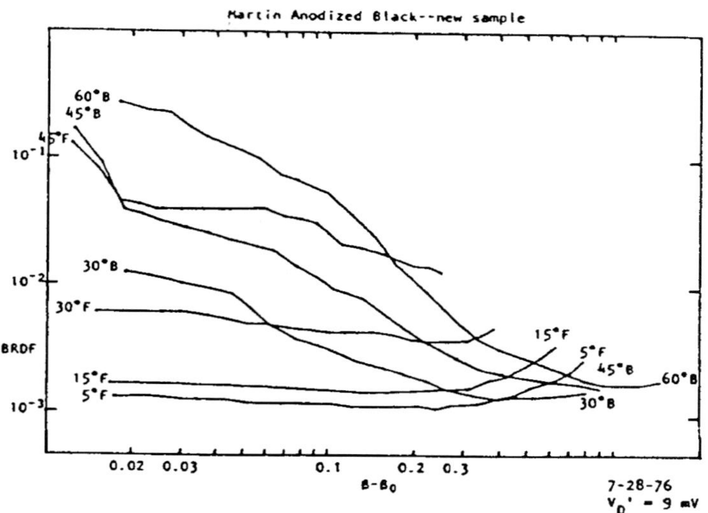

FIGURE 35 BRDF profile of Martin Black at 10 μm. (*F.O. Bartell et al., "A Study Leading to Improvements in Radiation Focusing and Control in Infrared Sensors," final report, Army Materials and Mechanics Research Center, December 1976.*)

Each of the ray-based programs share some inherent drawbacks, one of which is the large number of rays that must be traced to obtain useful calculations. In modern designs the attenuation factors exceed 10^{10}. Assuming such an attenuation, only one ray makes it to the detector for every 10^{10} input rays randomly scattered and traced. To get good statistics a very large number of rays must be traced. In the stray light analysis of an optical system with a substantial aperture size, the solid angle subtended by a scattering or diffracting edge is relatively small. It requires a very high density of rays to intercept the edge (for scatter) or even get close enough for diffraction effects.

Ray tracing can also have the inherent problem of not being smoothly distributed over the image plane. The rays create artificial noise density distribution patterns that may have no correlation with the actual irradiance pattern. Higher input ray densities can help to mitigate these problems. The increasing speed of computers may alleviate this problem in the future.

Deterministic programs are not without limitations. They can have problems with calculations involving multiple cylindrical or toroidal optical elements or specular surfaces. They also require that you indicate the propagation paths to be calculated. Since these paths are not always obvious, this can be labor intensive. In some systems with complex geometry, sampling of the surfaces can also be a problem. You may have to divide each surface in the system into many smaller sections to obtain accurate results; this increases the time needed to perform the calculations. Fast computers alleviate this problem.

New Industrial Concepts

The first program of significance was a program from New Industrial Concepts Ltd. (NIC) of Geneva, Switzerland. Its was created in the late 1960s by Dr. A. Bokenberg, and it was based upon a Monte-Carlo scattering concept. It has been enhanced to perform a modified Monte-Carlo technique, much like the GUERAP I and III programs discussed below.

GUERAP

The United States Air Force sponsored two projects to develop a more sophisticated software package. The two developers were Honeywell and Perkin-Elmer. They created two versions of the General Unwanted Energy Rejection Analysis Program, GUERAP I[34] and GUERAP II,[35] respectively. The GUERAP III version is the successor of the initial release of GUERAP I. The later version has corrections, and some additional BRDF models.

GUERAP I and III. The Honeywell program, GUERAP I, was also based on the Monte-Carlo technique, but in a more sophisticated way. Honeywell developed the principal of a modified Monte-Carlo technique that allows the analyst to direct rays toward "important" surfaces, such as the detector, optical surfaces, or specific baffle areas. This increases the probability that such a ray would eventually propagate to the detector. Without modification, this would ruin the statistical nature of the calculation, since more of the rays get through to the detector, but not in proportion to the actual performance. The solution was to reduce the energy associated with the ray to reestablish the validity of the calculation. This technique is called *importance sampling.*

The Honeywell implementation of the importance sampling included the foresight to scatter one ray outside the solid angle of the "important" direction. It is important to trace this additional ray, because it sometimes finds additional paths that you might have missed. This can be an advantage over the purely deterministic approach of a program like APART, which is discussed below.

GUERAP II. GUERAP II was conceived by Perkin Elmer as the first deterministic stray light analysis program. It has no relation to GUERAP I or III. Though never extensively used, it had some unique and promising features. Perkin Elmer created an excellent concept for incorporating diffraction effects into the analysis by extending the work of Keller,[36] Miyamoto and Wolf,[37] and others. The concept did not account for the relative phase effects of the fringes; instead, it gave the envelope of this diffracted energy. It evaluated the entire diffraction field in terms of the irradiance on an edge and associated this ray with a stationary phase point contribution. The power associated with this stationary phase point was then propagated to some surface.

APART/PADE

APART/PADE was initially developed under a NASA contract for the stray light analysis of the Hubble telescope (then called the Large Space Telescope, LST). It is a deterministic stray light analysis program. Unlike ray-based programs, it calculates the *power* propagated from a stray light source to the detector. The program calculates each of the terms in Eq. (4) for each segment of a propagation path. By directed iteration of such calculations the paths are completed.

The GCFs and BRDFs are calculated for each scattering surface encountered. The program summarizes the stray light propagation information into tables representing the power contributed by each "critical" object that can be seen from the detector, the power propagated along each stray light path to the detector, and the normalized detector irradiance (NDI), which is the focal plane irradiance for a unit input irradiance.

The Program for the Analysis of Diffracted Energy (PADE)[38] portion of the software has many of the most sophisticated diffraction analysis capabilities of the existing programs. It is based upon the stationary phase approximations, but also accounts for constant phase and other difficult diffraction effects. Like GUERAP II, it does not keep track of phase, and only calculates the envelope of the propagated diffracted energy.[39]

Since it is not a ray-based program, energy is propagated directly to edges and then scattered or diffracted; this avoids the sampling density problem of the ray-based programs.

ASAP[40]

ASAP is a very sophisticated optical analysis program that has powerful stray light capability. It is a ray-based program to some degree, but it goes well beyond the conventional concepts. By keeping track of parabasal rays it can propagate gaussian beams associated with each ray. ASAP uses gaussian beam decomposition theory to propagate wavefronts through a system. It is the only stray light program that can keep track of phase and thereby calculate interference effects of a wavefront, such as in an interferometer, or the ringing in the diffraction from an edge. It does not resort to calculating the diffraction envelope.

At any refractive boundary, the specular ray, the diffracted ray, the Fresnel reflected ray, and the scattered rays can all be propagated, while the program keeps track of their respective optical characteristics such as phase and polarization.

OARDAS

OARDAS is a proprietary ray-based program from Hughes Aircraft Co. It traces rays into and away from the focal plane of the system. By doing so it can help to establish the stray light propagation paths. At any surface encountered it can split rays into the specular ray, the reflected ray, and the scattered ray. It has no difficulty tracing rays through an all-reflective specular baffle system. It can generate ray path histories and tables of power propagated by the critical objects that can scatter or diffract energy to the focal plane.

SOAR™,[41] Mini-APART,[42] and STRAY[43]

Many small stray light analysis codes were generated following publication of an SPIE paper by Alan Greynolds[44] which detailed some of the fundamental calculations needed. None of these programs has the capability of the major software packages, but each has been useful for specific subsets of problems or quick estimates of the stray light contributions in simplified systems.

38.5 METHODS

There are two distinct methods that have been used to evaluate a system for stray radiation. You can either build the system and test it, or you can model the system and try to predict its performance. Both methods have advantages and disadvantages. Taken *together* the two methods provide the means to ensure that the system will perform as desired.

Build-and-Test Approach

A common approach is to make the system and either use it or test it for stray radiation rejection. Certainly if the system consistently performs satisfactorily *in its operational environment,* it has passed the ultimate test. But what if it does not meet the desired or expected level of performance? Making more systems to test becomes expensive rapidly. In fact, for very large systems, usually only modifications ("fixes") can be contemplated because of the high cost. This is not the only argument against the build-and-test approach. The tests are rarely designed to determine *how* the scattered radiation is propagating through the system and which surfaces contribute most of the undesired radiation.

It is this information, and a thorough knowledge of the surface scattering characteristics, that is necessary to make measurable improvements to the system. Such a test, when determining the propagation paths, should yield information about how the system is reacting to its *test* environment, including the test equipment. Unless the tests are being conducted in the environment for which the system was designed, it is imperative to determine that the *test* environment is not giving erroneous results (either better or worse). Without analyzing the test configuration, you should expect that the environment *will* affect the system stray light measurements. It is also incorrect to assume that the test environment can only add to the stray light background. It is sometimes assumed that if the system passes the stray light tests in the lab it will only perform better in space or wherever its intended environment is. This is not necessarily true.

Now that several points have been made about the difficulty of making valid experimental tests, it must be stated that valid tests can and should be made. The measurement costs need not be prohibitive. Even relatively large optical systems have been fabricated and then modestly redesigned. Changes to the system can be made until the desired information and stray radiation rejection is attained. In some cases it will be less expensive to test an existing system and modify it if necessary than to analyze the system with computer software.

The system-level test need not be extensive; it is not necessary to have an all-encompassing measurement from on-axis to 90° off-axis. Indeed, few facilities are capable of making such tests when the attenuation gets even modestly high. An important point to recognize is that the most important paths to check are those at the nearer off-axis angles where the attenuation is not so high. These can usually be measured reliably.

At small off-axis angles the stray light noise is more often much higher than the detector background noise, while at the higher off-axis angles the stray light noise is well below the electronic/detector noise. From a performance point of view, at the higher off-axis angles there is usually only one additional scattering object (scatter from the main baffle) before these same near off-axis angle paths are encountered or are reinvolved. The validation of the analysis will only be susceptible to the scatter from this one object that can't be fully tested at the system level, but most of the scatter paths, and usually all the most important ones, will have been validated by the near off-axis measurements.

This one additional surface scatter most often (especially on space-based sensors) involves the vanes on the main baffle that shields the primary objective. It will normally reduce the optical noise incident upon it by four to five orders of magnitude. That's why the optical noise goes dramatically below the electronic noise. Its most important role is to occult the direct illumination of the objective which is usually part of the most significant direct scatter path. The performance of this baffle and its vane structure could be analyzed separately and then measured independently to confirm that it too will perform as predicted.

NOTE: Contrary to some published papers you cannot, in general, multiply the stray light transmittance of two parts of a sensor and determine the system's overall performance. Although the main baffle system can be analyzed (or measured) independently from the rest of the system, it is not correct to take its performance and multiply it times the stray light performance of the rest of the system. The stray light propagation paths are far more important than the magnitudes of the two parts. In the above analysis where it was proposed that the main baffle could be measured independently it was to confirm its performance alone. A full-system stray light analysis was assumed.

Computer Analysis

As with the experimental tests, computerized analyses are also subject to errors. The three most significant ones are software limitations, scatter data of samples (not the real system), and user error. No software is capable of putting in every detail of a complex design, yet the computer model must faithfully represent the actual performance of the system. On the other hand, the software can put in "parts" with far greater mathematical precision than these parts can actually be assembled. Unless special studies are made the analyst does not usually account for assembly errors that might affect the actual system. The scatter characteristics of the surfaces, usually defined in terms of the Bi-Directional Reflectance Distribution Function (BRDF), are usually measured on sample substrates, and controls must be exercised to ensure that the samples tested represent the sensor's actual coatings, and that they do not change with time. The stray light analysis programs are also subject to errors in determining the significant paths. The experimental test is for the actual design, with real coatings, and will include any extraneous unintentional paths due to misalignment or other causes.

On the positive side, a software program can point out many flaws in the system that contribute stray radiation by considering the input BRDF characteristics of the coatings. A program can also do trade-off studies, parametric analysis, and in many other ways aid in the study of alternate designs. The analysis of the paths of scatter will suggest meaningful modifications and help to discard impossible designs. These analyses allow designers to test designs and make modifications before the design goes into production. This is very useful, since rejecting a sensor design is much easier when it is on paper than after it has already been built. It is usually much more cost-efficient than cutting new hardware, redesigning the system, or making fixes on the built system.

If you are in a field related to the optical design of a sensor, be it at the design level or the system level, you know that it would be preposterous to perform the optical design

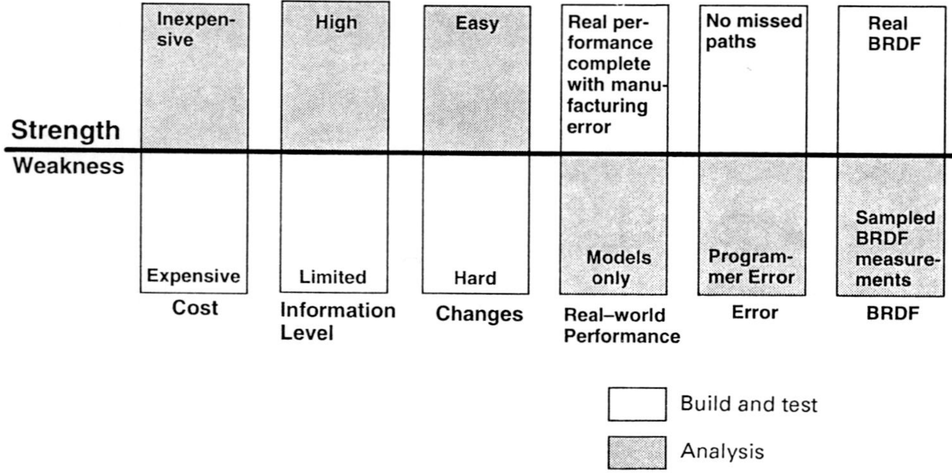

FIGURE 36 Build-and-test and analysis methods complement each other.

analysis and then put the system together without testing it for its image quality. Yet that is how far the pendulum has swung in favor of performing a stray light analysis over making a system-level stray light test. It reflects a major change in attitude since the early 1970s. It has been stated by stray light analysts that the reliability of a stray light analysis is now much higher than experimental test results, so some people avoid the latter. While there is a degree of truth in this statement, it is wrong to omit the stray light test at the system level.

The advantages and disadvantages of the two methods are summarized in Fig. 36. The disadvantages of the build-and-test approach are the strengths of the analysis method, whereas the strengths of the build-and-test approach cover the weaknesses of an analysis. Taken together these two methods give the greatest amount of reliable information which you can use to create the optimal system and have confidence in its performance. Jointly, they indicate the reliability of the analysis and test results.

38.6 CONCLUSION

In summary, the issues involved in designing a system with stray light suppression in mind are:

I. System design concepts
- **A.** Critical objects seen by the detector
- **B.** Illuminated objects
- **C.** Lyot stops
- **D.** Field stops
- **E.** Optical designs

II. Baffle and vane design
- **A.** Diffuse and specular vane cavities
- **B.** Vane edge scatter

III. Diffraction

IV. Strut design

V. Scattering theory

VI. BRDF data
- **A.** Log BRDF versus θ
- **B.** Log BRDF versus log $(\beta - \beta_o)$
- **C.** Polar plots
- **D.** Isometric projections (3-D characteristics)

VII. Coatings
- **A.** Paints and anodized surfaces
- **B.** AR coatings and other thin films
- **C.** Mirror coatings

VIII. Thermal emission

IX. Ghost images

X. Software

XI. Detection, prevention, and removal of contamination

A step by step procedure that can help you to improve your system is:

I. Start from the detector and identify what objects, called "critical objects," can be seen from various positions on the detector. Be sure to include a point near the edge of the detector.

II. Work to remove the number of critical objects that the detector can see.

III. Determine what objects the source of unwanted radiation can see, called the "illuminated objects."

IV. If possible, reduce the number of illuminated objects seen.

V. If there are illuminated objects that are also critical objects, work very hard on these paths. Orders of magnitude in improvement will be your reward.

VI. If task V is not possible, then the computations are quite easy.

- **A.** Calculate the power incident on the illuminated/critical objects.
- **B.** Use Eq. (1) to calculate the transfer of power from the critical objects to the detector. Remember to properly account for the input and output angles when calculating the BRDF. *Do not* use a straight lambertian scatter distribution; there is no such distribution in reality.

VII. Find all the paths connecting the illuminated objects to the ciritical objects.

VIII. Evaluate the corresponding input and output angles at the illuminated and critical objects.

IX. Determine if vane structure will help, or if some other redesign will effectively blocks these paths.

X. For the calculated input and output angles, evaluate which coating would be lowest.

XI. Perform the stray light calculation using Eq. (1) in an iterative fashion. This should determine the most significant stray light path and quantiify the amount of stray light on the detector.

XII. Perform the above tasks for a series of off-axis positions of the point source.

38.7 SOURCES OF INFORMATION ON STRAY LIGHT AND SCATTERED LIGHT

John D. Lytle and Howard Morrow (eds.), "Stray Light Problems in Optical Systems," *SPIE Proceedings,* vol. 107, April 18–21, 1977 (22 papers).

Mark Kahan (ed.), "Optics in Adverse Environments," *SPIE Proceedings,* vol. 216, Feb. 4–5, 1980 (30 papers).

Gary H. Hunt (ed.), "Radiation Scattering in Optical Systems," *SPIE Proceedings,* vol. 257, Sept. 30–Oct. 1, 1980 (28 papers).

Solomon Musikant (ed.), "Scattering in Optical Materials," *SPIE Proceedings,* vol. 362, Aug. 25–27, 1982 (28 papers).

Robert P. Breault (ed.), "Generation, Measurement, and Control of Stray Radiation III," *SPIE Proceedings,* vol. 384, Jan. 18–19, 1983 (15 papers).

Robert P. Breault (ed.), "Stray Radiation IV," *SPIE Proceedings,* vol. 511, Aug. 23, 1984 (14 papers).

Robert P. Breault (ed.), "Stray Radiation V," *SPIE Proceedings,* vol. 675, Aug. 18, 1986 (46 papers).

Robert P. Breault (ed.), "Stray Light and Contamination in Optical Systems," *SPIE Proceedings,* vol. 967, Aug. 17–19, 1988 (33 papers).

John C. Stover (ed.), "Scatter from Optical Components," *SPIE Proceedings,* vol. 1165, Aug. 8–10, 1989 (42 papers).

Robert P. Breault (ed.), "Stray Radiation in Optical Systems," *SPIE Proceedings,* vol. 1331, July 12–13, 1990 (29 papers).

John C. Stover (ed.), "Optical Scatter: Applications, Measurement, and Theory," *SPIE Proceedings,* vol. 1530, July 21–27, 1991.

Robert P. Breault (ed.), "Stray Light and Contamination in Optical Systems II," *SPIE Proceedings,* vol 1753, JUly 21–23, 1992.

F. O. Bartell et al., "A Study Leading to Improvements in Radiation Focusing and Control in Infrared Sensors," *Final Report Prepared for Army Materials and Mechanics Research Center,* December 1976.

J. A. Gunderson, "Goniometric Reflection Scattering Measurements and Techniques at 10.6 Micrometers," M.S. thesis, Univ. of Arizona, 1977.

Philip J. Peters, "Stray Light Control, Evaluation, and Suppression," *SPIE Proceedings,* vol. 531, January 1985.

T. W. Stuhlinger, "Bidirectional Reflectance Distribution Function (BRDF) of Gold-Plated Sandpaper," M.S. thesis, Univ. of Arizona, 1981.

A. W. Greynolds, "Computer-Assisted Design of Well-Baffled Axially Symmetric Optical Systems," M.S. thesis, Univ. of Arizona, 1981.

J. W. Figoski, "Interferometric Technique for the Reduction of Scattered Light," M.S. Thesis, Univ. of Arizona, 1977.

J. S. Fender, "An Investigation of Computer-Assisted Stray Radiation Analysis Programs," Ph.D. dissertation, Univ. of Arizona, 1981.

D. A. Thomas, "Light Scattering from Reflecting Optical Surfaces," Ph.D. dissertation, Univ. of Arizona, 1980.

Robert Paul Breault, "Suppression of Scattered Light," Ph.D. dissertation, Univ. of Arizona, 1979.

Paul Raymond Spyak, "A Cryogenic Scatterometer and Scatter from Particulate Contaminants on Mirrors," Ph.D. dissertation, University of Arizona, 1990.

Frederick Otis Bartell, "Blackbody Simulator Cavity Radiation Theory," Ph.D. dissertation, University of Arizona, 1978.

Lawrence Dean Brooks, "Microprocessor-based Instrumentation for BSDF Measurements from Visible to FIR," Ph.D. dissertation, University of Arizona, 1982.

Amy Gardner Lusk, "Measurements of the Light Scattering Profile of Small Size Parameter Fibers," M.S. thesis, University of Arizona, 1987.

Kiebong Nahm, "Light Scattering by Polystyrene Spheres on a Conducting Plane," Ph.D. dissertation, University of Arizona, 1985.

Gorden Wayne Videen, "Light Scattering from a Sphere on or Near an Interface," Ph.D. dissertation, University of Arizona, 1992.

YauJen Wang, "Comparisons of BRDF Theories with Experiment," Ph.D. dissertation, University of Arizona, 1983.

Steven Jay Wein, "Small-Angle Scatter Measurement," Ph.D. dissertation, University of Arizona, 1989.

J. M. Bennett and L. Mattsson, *Introduction to Surface Roughness and Scattering,* Optical Society of America, Washington, D.C., 1989.

John C. Stover, *Optical Scattering Measurement and Analysis,* McGraw-Hill, Inc., New York, 1990.

38.8 REFERENCES

1. R. P. Breault, "Problems and Techniques in Stray Radiation Suppression," *Stray Light Problems in Optical Systems,* John D. Lytle, Howard Morrow (eds.), *Proc. SPIE* 107, 1977, pp. 2–23.
2. R. V. Shack, "Analytic System Design with a Pencil and Ruler—The Advantages of the y–$\bar{y}$ Diagram," *Applications of Geometrical Optics, Proc. SPIE* 39, 1973.
3. S. R. Lange, R. P. Breault, and A. W. Greynolds, "APART, A First-Order Deterministic Stray Radiation Analysis Program," *Stray-Light Problems in Optical Systems,* John D. Lytle, Howard Morrow (eds.), *Proc. SPIE* 107 1977, pp. 89–97.
4. Ibid., R. P. Breault, "Problems and Techniques in Stray Radiation Suppression."
5. R. P. Breault, "Vane Structure Design Trade-Off and Performance Analysis," *Stray Light and Contamination in Optical Systems, Proc. SPIE* 967, 1988.
6. Ibid., S. R. Lange, R. P. Breault, and A. W. Greynolds, "APART, A First-Order Deterministic Stray Radiation Analysis Program."
7. R. P. Breault, A. W. Greynolds, and S. R. Lange, "APART/PADE Version 7: A Deterministic Computer Program Used to Calculate Scattered and Diffracted Energy," *Radiation Scattering in Optical Systems,* Gary Hunt (ed.), *Proc. SPIE* 257, 1980, pp. 50–63.
8. Ibid., R. P. Breault, A. W. Greynolds, and M. A. Gauvin, "Stray Light Analysis with APART/PADE, Version 8.7."
9. Ibid., R. P. Breault, "Problems and Techniques in Stray Radiation Suppression."
10. Ibid., S. R. Lange, R. P. Breault, and A. W. Greynolds, "APART, A First-Order Deterministic Stray Radiation Analysis Program."
11. This section is a part of a more complete mathematical description of the process which can be found in R. P. Breault, "Vane Structure Design Trade-Off and Performance Analysis." (Ibid.)
12. R. P. Breault, "Vane Structure Design Trade-Off and Performance Analysis" (Ibid.) contains a more detailed description of the process.
13. Ibid., R. P. Breault, "Problems and Techniques in Stray Radiation Suppression."
14. R. P. Breault, "Suppression of Scattered Light," Ph.D. dissertation, University of Arizona, 1979.
15. Ibid., R. P. Breault, "Problems and Techniques in Stray Radiation Suppression."
16. J. Gunderson, "Goniometric Reflection Scattering Measurements and Techniques at 10.6 Micrometers," M.S. thesis, University of Arizona, 1977.
17. Ibid., R. P. Breault, A. W. Greynolds, and S. R. Lange, "APART/PADE Version 7: A Deterministic Computer Program Used to Calculate Scattered and Diffracted Energy."
18. Ibid., R. P. Breault, A. W. Greynolds, and M. A. Gauvin, "Stray Light Analysis with APART/PADE, Version 8.7."
19. E. R. Freniere and D. L. Skelton, "Use of Specular Black Coatings in Well-Baffled Optical Systems," *Stray Radiation V,* Robert Breault (ed.), *Proc. SPIE* 675, 1986, pp. 126–132.
20. Alan W. Greynolds, computer code ASAP, Breault Research Organization, Inc., 1992.
21. Gary L. Peterson, Steve C. Johnston, and Jerry Thomas, "Specular Baffles," *Stray Radiation in Optical Systems II,* Robert P. Breault (ed.), *Proc. SPIE* 1753, 1992.
22. P. R. Spyak, "A Cryogenic Scatterometer and Scatter from Particulate Contaminants on Mirrors," Ph.D. dissertation, University of Arizona, 1990.
23. "Product Cleanliness Levels and Contamination Control Program," *MIL-STD-1246A,* Dept. of Defense, Global Engineering Documents, Santa Ana, Calif., 18 August 1967.
24. R. Young, "Low-Scatter Mirror Degradation by Particle Contamination," *Optical Engineering* 15, no. 6, Nov.–Dec. 1976.
25. J. E. Harvey, "Light-Scattering Characteristics of Optical Surfaces," Ph.D. dissertation, University of Arizona, 1976.
26. Contact Dierdre Dykeman at Rome Air Development Center for the best way to access the results of this work.

27. H. E. Bennett, "Reduction of Stray Light from Optical Components," *Stray Light Problems in Optical Systems,* John D. Lytle and Howard Morrow (eds.), *Proc. SPIE* 107, 1977, pp. 24–33.
28. Name coined by Dave Rock, Hughes Aircraft Co., El Segundo, Calif.
29. Name coined by Robert P. Breault; this is the name most often used in an APART stray light analysis.
30. Ibid., J. E. Harvey, "Light-Scattering Characteristics of Optical Surfaces."
31. John C. Stover, *Optical Scattering: Measurement and Analysis,* New York: McGraw-Hill, Inc., New York, 1990.
32. Ibid., John C. Stover, *Optical Scattering: Measurement and Analysis,* p. 69, reprinted with permission.
33. Ibid., J. Gunderson, "Goniometric Reflection Scattering Measurements and Techniques at 10.6 Micrometers."
34. Barry K. Likeness, *GUERAP III General Unwanted Energy Rejection Analysis Program User's Manual,* Honeywell, Inc., 1978.
35. GUERAP II User's Guide, Perkin-Elmer Corporation, Technical Report SAMSO TR. 73-309, 1974.
36. J. B. Keller, "Diffraction by an Aperture," *J. of Applied Physics,* vol. 28, no. 4, April 1957, pp. 426–44 and "Geometrical Theory of Diffraction," *J. of the Optical Society of Am.,* vol. 52, no. 6, February, 1962, pp. 116–30.
37. K. Miyamoto and E. Wolf, "Generalization of the Maggi-Rubinowicz Theory of the Boundary Diffraction Wave—Parts I and II," *J. of the Opt. Soc. of Am.,* vol. 52, no. 6, June, 1962, pp. 615–37.
38. A. W. Greynolds, "Method for Calculating Diffraction Effects in Opto-Mechanical Systems of Arbitrary Geometry," *Radiation Scattering in Optical Systems,* Gary Hunt (ed.), *Proc. SPIE* 257, 1980, pp. 64–77.
39. R. P. Breault, *APART/PADE Reference Manual,* Breault Research Organization, Inc., 1992.
40. A. W. Greynolds, *ASAP Reference Manual,* Breault Research Organization, Inc., 1992.
41. E. Freniere, Computer Code SOAR^{tm}, Telic Optics.
42. Jack A. Bamberg, "Stray Light Analysis with the HP-41C/CV Calculator," *Generation Measurement, and Control of Stray Radiation III,* Robert P. Breault (ed.), *SPIE* 384, 1983, pp. 109–116.
43. Ann S. Dinger, "STRAY: An Interactive Program for the Computation of Stray Radiation in Infrared Telescopes," *Stray Radiation V,* Robert P. Breault (ed.), *Proc. SPIE* 675, 1986, pp. 95–104.
44. A. W. Greynolds, "Formulas for Estimating Stray-Radiation Levels in Well-Baffled Optical Systems," *Radiation Scattering in Optical Systems,* Gary Hunt (ed.), *Proc. SPIE* 257, 1980, pp. 39–49.

CHAPTER 39
THERMAL COMPENSATION TECHNIQUES

P. J. Rogers
M. Roberts
Pilkington Optronics
Clwyd LL17 0LL
Wales, United Kingdom

39.1 GLOSSARY

c	surface curvature of an optical element
D	diameter
FN	F-number or focal ratio
f	paraxial focal length
G	thermo-optical constant (normalized thermal change of OPD)
h	(subscript) signifies "pertaining to the optic housing"
i	(subscript) number of a specific optical element
j	signifies a number of optical elements
K	Kelvin
k	thermal conductivity
n	refractive index
OPD	optical path difference
T	temperature
t	thickness
V	Abbe number of a refracting optical material
v	spatial frequency
α	linear coefficient of thermal expansion
γ	thermal glass constant (normalized thermal change of optical power)
Δ	small, finite change
λ	wavelength

δ infinitely small change of a parameter

ϕ optical power (reciprocal of focal length)

39.2 INTRODUCTION

In the following, the thermal effects for which compensation is required are taken to be those that affect the focus and image scale of an optical system. Methods for quantifying and offsetting these effects were described some time ago (Perry 1943[1]), similar information being provided by several other authorities (Johnson and Jeffree 1942,[2] Grey 1948,[3] Volosov 1958[4]). The thermal compensation techniques described in this section, with the exception of intrinsic athermalization, involve either mechanical movement of one or more parts of the optical system, or compensation achieved solely by choice of optical materials. Except in the section titled "Effect of Thermal Gradients," a homogeneous temperature change of all parts of the optical system is assumed.

Most optical materials undergo a change of refractive index n with temperature T, conveniently quoted as a rate of change $\delta n/\delta T$. The usual values of n and $\delta n/\delta T$ given for a material (and assumed in this section unless stated otherwise) are those relative to the surrounding air rather than the absolute values with respect to vacuum. Air has a $\delta n/\delta T$ of -1×10^{-6} at $T = 288$ K and 1 atmosphere air pressure for wavelengths between 0.25 and 20 microns (Penndorf 1957[5]): allowance for this must be made when a lens operates in a vacuum or in an enclosed space where the number of air molecules per unit volume does not change with temperature. The absolute $\delta n/\delta T$ of an optical material can be found from:

$$\left(\frac{\delta n}{\delta T}\right)_{\text{abs}} = n_{\text{air}}\left(\frac{\delta n}{\delta T}\right) + n\left(\frac{\delta n}{\delta T}\right)_{\text{air}} \tag{1}$$

where the value of n_{air} is approximately 1.0.

39.3 HOMOGENEOUS THERMAL EFFECTS

Thermal Focus Shift of a Simple Lens

The rate of change of the power ϕ (reciprocal of the focal length f) of an optical element with temperature T can be derived by differentiating the thin lens power equation $\phi = c(n-1)$, where c is the total surface curvature of the element. For a linear thermal expansion coefficient α of the material from which the element is formed this gives:

$$\text{As:}\quad f = \frac{1}{\phi} \qquad \frac{\delta f}{\delta T} = -\frac{1}{\phi^2}\cdot\frac{\delta\phi}{\delta T}$$

$$\frac{\delta\phi}{\delta T} = +\phi\left(\frac{\delta n/\delta T}{n-1} - \alpha\right) \tag{2}$$

Therefore:

$$\frac{\delta f}{\delta T} = -f\left(\frac{\delta n/\delta T}{n-1} - \alpha\right) \tag{3}$$

TABLE 1 Optical and Thermal Data for a Number of Visual Waveband Materials

Schott glass / Optical plastic Type	Refractive index, n_e*	Abbe number, V_e†	Thermal glass constant, $\gamma(\times 10^6)$‡	Thermo-optical constant, $G(\times 10^6)$‡	Thermal conductivity, $k(W \cdot m^{-1} \cdot K^{-1})$
FK52	1.487	81.4	−27	+1	0.9
FK5	1.489	70.2	−11	+4	0.9
BK7	1.519	64.0	−1	+7	1.1
PSK53A	1.622	63.2	−13	+4	—
SK5	1.591	61.0	+1	+7	1.0
BaLKN3	1.521	60.0	−3	+7	1.0
BaK2	1.542	59.4	−5	+6	—
SK4	1.615	58.4	−2	+7	0.9
LaK9	1.694	54.5	−1	+8	0.9
KzFSN4	1.617	44.1	+4	+8	0.8
LF5	1.585	40.6	−6	+7	0.9
BaSF51	1.728	37.9	+8	+14	0.7
LaFN7	1.755	34.7	+6	+12	0.8
SF5	1.678	32.0	0	+11	—
SFN64	1.711	30.1	−4	+9	—
SF6	1.813	25.2	+6	+18	0.7
Acrylic§	1.497	57	−279	−71	0.2
Polycarbonate§	1.590	30	−247	−68	0.2

* At $\lambda = 546$ nm.
† Defined as $(n_{546} - 1)/(n_{480} - n_{644})$.
‡ At $\lambda = 546$ nm and $T = 20$c.
§ Values (except conductivity) from Waxler et al. *Appl. Opt.* **18:**102 (1979).

The material-dependent factor inside the bracket in Eqs. (2) and (3) is known as the thermal "glass" constant (γ) and represents the thermal power change due to an optical material normalized to unit ϕ and unit change of T. Tables 1 to 3 give γ values for a selected number of visual and infrared materials along with the relevant V value (Abbe number) and other data. The much higher level of γ for infrared as opposed to glass

TABLE 2 Optical and Thermal Data for Selected 3- to 5-Micron Waveband Infrared Materials

Optical material	Refractive index, $n_{4\mu}$	Abbe number, $V_{3\text{–}5\mu}$	Thermal "glass" constant, γ	Thermo-optical constant, G	Thermal conductivity, $k(W \cdot m^{-1} \cdot K^{-1})$
Silicon	3.43	2.4×10^2	$+6.3 \times 10^{-5}$	$+1.7 \times 10^{-4}$	1.5×10^2
KRS5*	2.38	2.3×10^2	-2.3×10^{-4}	-1.5×10^{-4}	5.0×10^{-1}
AMTIR1†	2.51	1.9×10^2	$+3.9 \times 10^{-5}$	$+9.5 \times 10^{-5}$	3.0×10^{-1}
Zinc selenide	2.43	1.8×10^2	$+3.6 \times 10^{-5}$	$+7.3 \times 10^{-5}$	1.8×10^1
Arsenic trisulfide	2.41	1.6×10^2	-1.9×10^{-5}	$+3.4 \times 10^{-5}$	1.7×10^{-1}
Zinc sulfide	2.25	1.1×10^2	$+2.6 \times 10^{-5}$	$+5.2 \times 10^{-5}$	1.7×10^1
Germanium	4.02	1.0×10^2	$+1.3 \times 10^{-4}$	$+4.2 \times 10^{-4}$	5.9×10^1
Calcium fluoride	1.41	2.2×10^1	-5.1×10^{-5}	-1×10^{-6}	9
Magnesium oxide	1.67	1.2×10^1	$+1.9 \times 10^{-5}$	$+2.6 \times 10^{-5}$	4.4×10^1

* Thallium bromo-iodide.
† Ge/As/Ge chalcogenide from Amorphous Materials Inc.

TABLE 3 Optical and Thermal Data for Selected 8- to 12-Micron Waveband Infrared Materials

Optical material	Refractive index, $n_{10\mu}$	Abbe number, $V_{8\text{–}12\mu}$	Thermal "glass" constant, γ	Thermo-optical constant, G	Thermal conductivity, $k(W \cdot m^{-1} \cdot K^{-1})$
Germanium	4.00	8.6×10^2	$+1.2 \times 10^{-4}$	$+4.1 \times 10^{-4}$	5.9×10^1
Cesium iodide	1.74	2.3×10^2	-1.7×10^{-4}	-5.3×10^{-5}	1
Cadmium telluride	2.68	1.7×10^2	$+5.3 \times 10^{-5}$	$+1.1 \times 10^{-4}$	6
KRS5	2.37	1.7×10^2	-2.3×10^{-4}	-1.6×10^{-4}	5.0×10^{-1}
AMTIR1	2.50	1.1×10^2	$+3.6 \times 10^{-5}$	$+9.0 \times 10^{-5}$	3.0×10^{-1}
Gallium arsenide	3.28	1.1×10^2	$+7.6 \times 10^{-5}$	$+2.0 \times 10^{-4}$	4.8×10^1
Zinc selenide	2.41	5.8×10^1	$+3.6 \times 10^{-5}$	$+7.2 \times 10^{-5}$	1.8×10^1
Zinc sulfide	2.20	2.3×10^1	$+2.6 \times 10^{-5}$	$+5.0 \times 10^{-5}$	1.7×10^1
Sodium chloride	1.49	1.9×10^1	-9.5×10^{-5}	-3×10^{-6}	6

optical materials indicates that thermal defocus (focus shift) is generally a much more serious problem in the infrared wavebands. The actual value of γ varies with both wavelength and temperature range due to variations in the value of $\delta n / \delta T$ and α. In general, this is unlikely to cause major problems unless a wide wavelength or temperature range is being considered (Köhler and Strähle 1974[6]). Thermal defocus results not only from a change of optical power but also from the thermal expansion coefficient α_h of the housing. Equation (3) can be modified to allow for the effect of the latter:

$$\text{Single thin lens:} \quad \Delta f = -f \cdot (\gamma + \alpha_h) \cdot \Delta T \tag{4}$$

$$j \text{ thin lenses in contact:} \quad \Delta f = -f \cdot \left[f \sum_{i=1}^{j} (\gamma_i \phi_i) + \alpha_h \right] \cdot \Delta T \tag{5}$$

Thermal Defocus of a Compound Optical Construction

Consider a homogeneous temperature change in an optical system that comprises two thin-lens groups separated from each other, the normalized thermal power change being the same in each lens group. Taking the thermal defocus calculated from Eq. (4) as unity, then that due to a compound optic comprising two separated components and of the same overall power can be estimated from Fig. 1 (Rogers 1991[7]). The latter shows scaling of thermal defocus with respect to a simple thin lens, relative to front lens/image plane distance (overall length) for three different positions of the second lens group. The graph is divided into three basic lens constructions distinguished from each other by overall optical length and/or the sign of the power of the front lens group.

Figure 1 assumes germanium optics in an aluminum housing but change of either material, while altering values slightly, has no effect on the following two conclusions:

1. Telephoto/inverted telephoto constructions always give more (and Petzval lenses always give less) thermal defocus than an equivalent simple lens.
2. Thermal defocus reduces as the second lens group is moved toward the image plane,

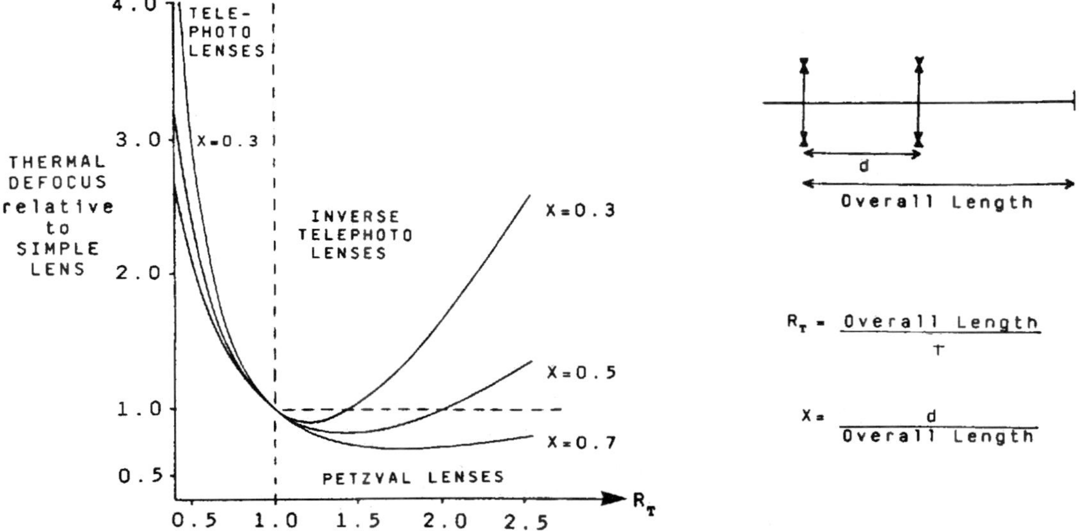

FIGURE 1 Effect of compound lens construction on thermal defocus *(from Rogers 1991)*.

irrespective of lens construction: the efficacy of this procedure is limited, however, by the increased imbalance of optical powers between the groups.

The thermal defocus scaling technique could be extended to cover an optic comprising more than two lens groups. This extension has been carried out for the Cooke triplet construction (Estelle 1980[8]) and for a series of separated thin lenses (Köhler and Strähle 1974[6]), although only for the case of a zero-expansion housing.

39.4 TOLERABLE HOMOGENEOUS TEMPERATURE CHANGE (NO COMPENSATION)

Diffraction-Limited Optic

Equation (5) can be used to establish the temperature change ΔT that will result in a quarter-wave of thermal defocus, a reasonable limit for a simple optic that is nominally diffraction-limited. Given an optic of diameter D and focal ratio FN imaging at a mean wavelength of λ:

$$\text{Diffraction-based depth of focus:} \quad \Delta f = \pm 2\lambda\,(\text{FN})^2 \tag{6}$$

$$\text{Combining Eqs. (5) and (6):} \quad \Delta T = \pm \frac{2\lambda\,(\text{FN})}{D \cdot \left[f \sum_{i=1}^{j} (\gamma_i \phi_i) + \alpha_h \right]} \tag{7}$$

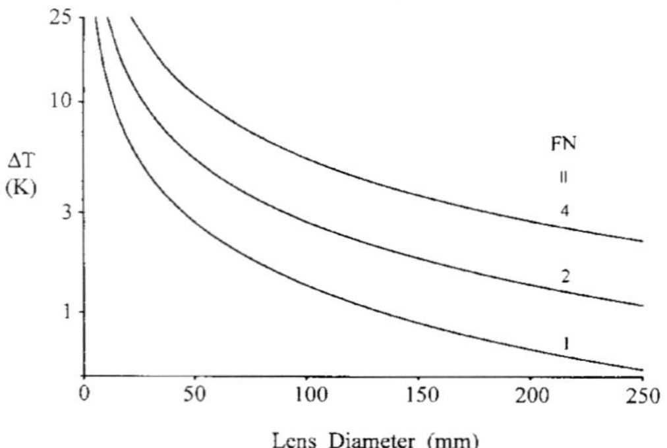

FIGURE 2 Tolerable temperature change for a simple germanium infrared lens *(from Rogers 1990).*

Figure 2 gives ΔT against D results for a simple 8–12-micron bandwidth germanium optic in an aluminum housing (Rogers 1990[9]): the curves illustrate the small temperature change that can be tolerated in germanium optics before focus compensation is required. Partial avoidance of this particular problem may be achieved by the replacement of germanium by other infrared optical materials having lower values of γ: this may also be desirable to reduce high-temperature absorption but generally leads to much greater optical complexity.

Non-Diffraction-Limited Optic

The depth of focus of an optic having a nominal performance far from the diffraction limit is a function of the residual aberration level and balance in the optic as well as its first-order parameters. An estimate related to a cutoff spatial frequency ν that gives a reasonable approximation in many cases can be obtained (Laikin 1991[10]) from:

$$\text{Approximate depth of focus:} \quad \Delta f = \pm \frac{(\text{FN})}{\nu} \tag{8}$$

$$\text{Combining Eqs. (5) and (8):} \quad \Delta T = \pm \left\{ D\nu \left[f \sum_{i=1}^{j} (\gamma_i \phi_i) + \alpha_h \right] \right\}^{-1} \tag{9}$$

Notice that, given the approximation of this method, the value of ν can be determined by extending a straight line MTF from 1.0 response at zero spatial frequency, through the MTF point of interest, to the intersection of the line with the spatial frequency axis.

39.5 EFFECT OF THERMAL GRADIENTS

The previous sections assume a homogeneous temperature change in all parts of the optical system: in situations where steady-state or transient temperature gradients exist, further consideration is required (Perry 1943[1]).

Allowance for the effect of steady-state longitudinal gradients can be made by applying a different value of T to each lens group and an average local temperature to each portion of the housing that separates two adjacent lens groups. Transient longitudinal gradients are a more difficult problem and, if severe, may require individual athermalization of each lens group in its own housing domain.

Steady-state or transient radial thermal gradients cause at least a shift of focus position, with the possible addition of a change of aberration correction. A localized radial temperature difference of ΔT through the thickness t of a plane-parallel plate will cause a deviation of a ray of light (Slyusarev 1959[11]) that can be quantified as an optical path difference (OPD):

$$\text{OPD} = t \cdot [\alpha(n-1) + \delta n/\delta T] \cdot \Delta T \tag{10}$$

The expression in the square bracket is often referred to as the thermo-optical constant G and is an approximate measure of the sensitivity of an optical material to radial gradients. More thorough analysis of the effects produced by radial thermal gradients includes computation of thermally induced stress and consequent anisotropic change of refractive index: in some cases, this may be a significant factor in image degradation (Turner 1970,[12] Mit'kin and Shchavelev 1973,[13] Reitmayer and Schroeder 1975[14]).

Tables 1 to 3 give values of G for the selected optical materials. Also tabulated is the thermal conductivity k, as in many cases G/k is a more appropriate measure of sensitivity given the greater ability of high-conductivity materials to achieve thermal equilibrium.

39.6 INTRINSIC ATHERMALIZATION

The need for athermalization can be avoided or minimized for some applications by employing optical power and mounting techniques that are inherently insensitive to temperature change. A concave spherical mirror fabricated from the same material that separates the mirror from its focal plane (for example, an aluminum mirror in an aluminum housing) is in effect "self-athermalized" for a homogeneous distribution of temperature. The optical performance of a single spherical mirror is limited, but the above principle applies for more complex all-reflective optical constructions employing conic or other aspheric surface forms. A glass spherical mirror, although not thermally matched to an aluminum mounting, may be used as part of a self-athermalized catadioptric afocal in the infrared, a germanium Mangin being used in this case as a secondary mirror lens (Rogers 1978[15]). The high thermal power change of the negatively powered lens in the germanium Mangin, used in double-pass, compensates for the thermal defocus due to the glass primary, the housing, and the remaining germanium optics in the afocal—Fig. 3 (Norrie 1986[16]).

An alternative approach to the above is to use glass-ceramic mirrors within a nickel-iron alloy housing, both materials of which can have thermal expansion coefficients approaching zero. A major advantage of this approach is its insensitivity to thermal gradients.

39.7 MECHANICAL ATHERMALIZATION

General

Mechanical athermalization essentially involves some agency moving one or more lens elements by an amount that compensates for thermal defocus—a simple manual option being to use an existing focus mechanism. Automatic methods are, however, preferable in many cases and can be divided into passive or active. Passive athermalization employs an

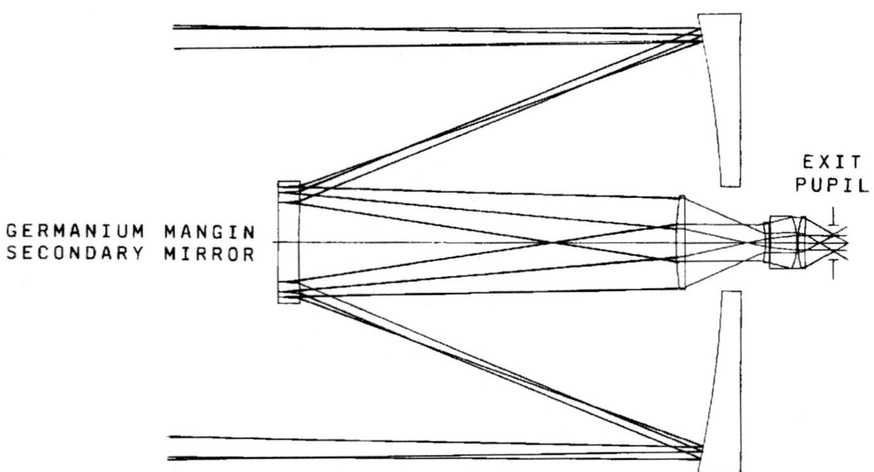

FIGURE 3 High magnification self-athermalized catadioptric afocal *(from Norrie 1986)*.

agency, often involving materials (including liquids) with abnormal thermal expansion coefficients, to maintain focus without any powered drive mechanism being required. Automatic active athermalization involves the computation of focus compensation algorithms that are stored (usually electronically) and implemented by a powered device such as an electric motor. The following sections refer to a number of passive and active athermalization methods, although the list is by no means exhaustive.

Passive Mechanical Athermalization

The principal advantages of passive thermal compensation methods are their relative simplicity and potential reliability. Disadvantages are their inadequate response to transient temperature gradients and, generally, lack of adjustment to allow for errors or unforeseen circumstances. Passive methods are ideal in glass optics (Angénieux et al. 1983)[17]) where thermal effects are low, although here it is not too difficult to achieve optical athermalization (see later in chapter under "Optical Athermalization") except where very low secondary spectrum is required. In the infrared wavebands, where thermal effects are much greater due to the nature of the optical materials, it is difficult to achieve simple passive mechanical athermalization due to the large refocusing movement required, typically 1.5×10^{-4} per unit focal length per Kelvin for an aluminum-housed germanium optic. An exception to the above is the combination of silicon and germanium in 3–5-micron optics, where thermal defocus results largely from the expansion of the housing. In this case, use of more than one nonmetallic housing material can result in an athermalized optic, even one having two fields of view (Garcia-Nuñez and Michika 1989[18])—Fig. 4.

For infrared cases other than the above, the options are either to provide a mechanism that modifies mechanical expansion effects or to reduce the required refocusing movement by optical means. Examples of the former include (Povey 1986[19]) a series of linked rods of alternatively high and low expansion coefficient—Fig. 5—and a hydraulic method where the fluid contained in a large-volume reservoir expands into a small-bore cylinder—Fig. 6.

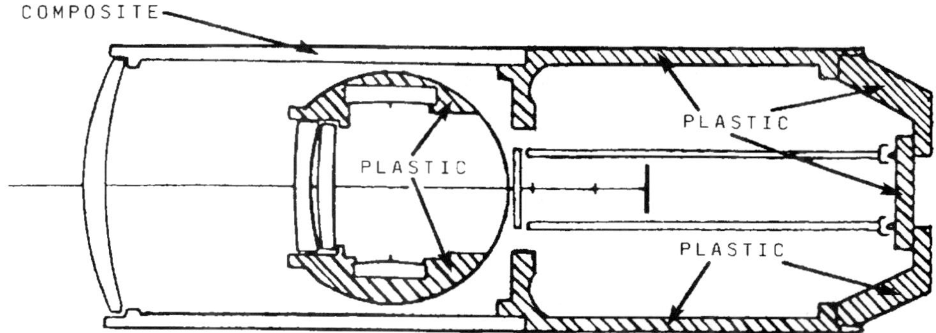

FIGURE 4 Part composite/part plastic mounting structure used for athermalization of a 3–5-micron IR optic *(from Garcia-Nuñez and Michika 1989)*.

An interesting alternative employs shape-memory metal (Michael and Hart 1980[20]) to provide a large movement over a relatively small temperature range (Povey 1986[19]). Another alternative is to employ a geodetic arrangement: in this method (Povey 1986[19]) an athermalizing adjustment of, for example, the separation between primary and secondary mirrors in a catadioptric is produced by differing expansion coefficients in the primary mirror mount and the secondary mirror struts—Fig. 7. Where none of the above methods are desirable, the option to reduce the necessary movement may be the only alternative. This may be achieved by an optical layout (Rogers 1990[9]) configured such that the required athermalizing movement is reduced typically by a factor of four, but at the expense of somewhat greater optical complexity—Fig. 8.

Active Mechanical Athermalization

Active mechanical athermalization in its simplest form can be manual adjustment of a lens element or group for refocusing. For more complex optics, such as multi-field-of-view, a procedure can be specified for manual (or motorized) adjustment of several lens elements to maintain focus over a range of magnifications and temperatures (Thompson 1976,[21] Rogers and Andrews 1976[22]). Where automatic athermalization is required, a method can

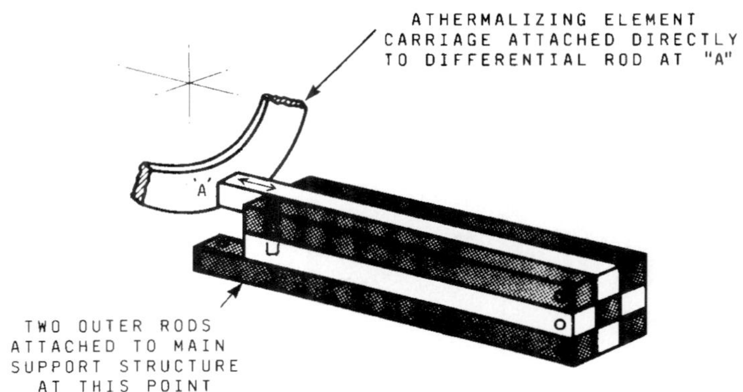

FIGURE 5 Passive mechanical thermal compensation using differential expansion rods *(from Povey 1986)*.

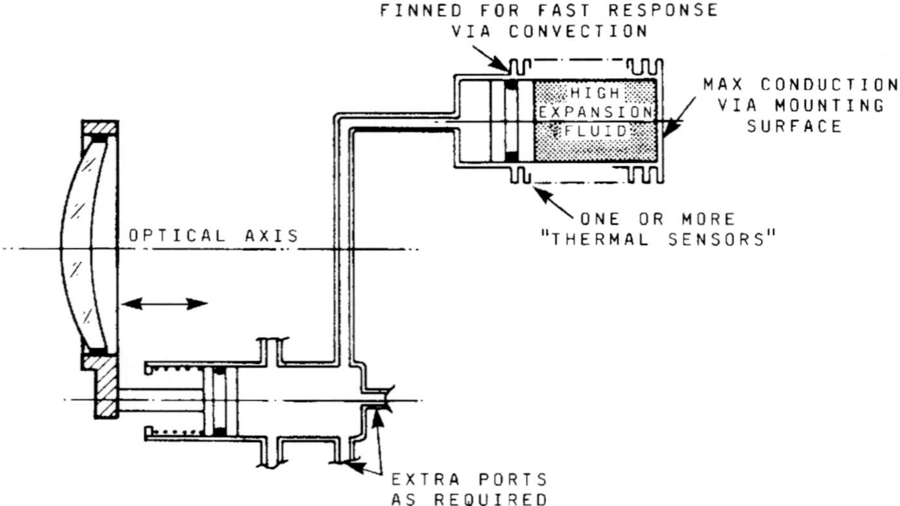

FIGURE 6 Passive mechanical thermal compensation using high-expansion-fluid thermal sensors *(from Povey 1986)*.

be employed that uses a combination of electronics and mechanics—Fig. 9. One or more temperature sensors located along the body of the optic feed their signals into an algorithm that calculates the required movement of a compensating lens and then initiates the motion. For simplicity, the compensating lens may be that which already provides close-distance focusing, thus requiring only an increase in the range of movement for athermalization. The location of sensors is especially important for infrared optics and should be dependent on the thermal sensitivity variations within the optical system.

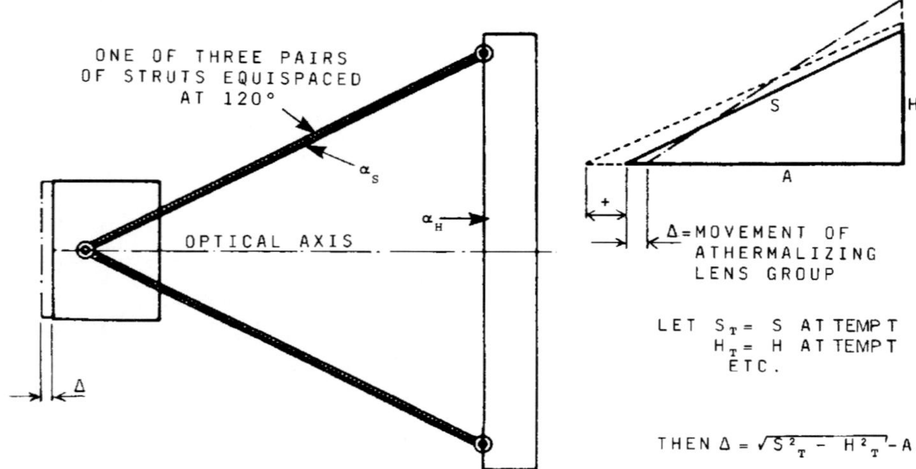

FIGURE 7 Geodetic support structure for positive or negative thermal compensation movement *(from Povey 1986.)*

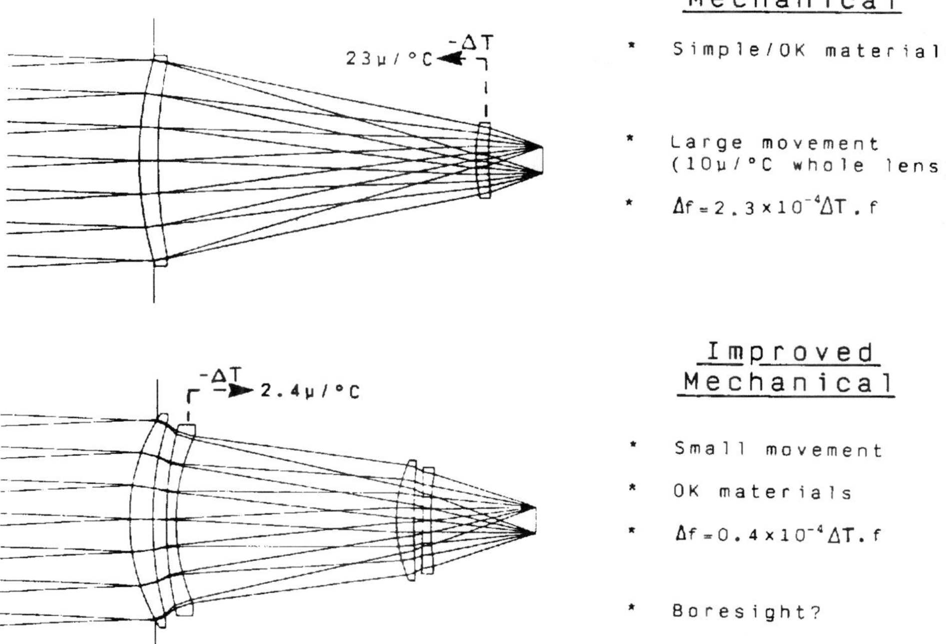

FIGURE 8 Alternative optical configurations for mechanically athermalized FLIR systems *(from Rogers 1990).*

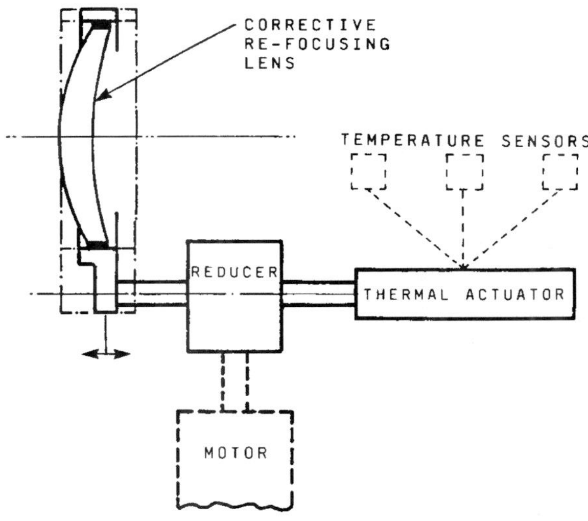

FIGURE 9 Active electromechanical athermalization—schematic.

Active electromechanical thermal compensation is particularly suitable where transient longitudinal temperature gradients are expected and for multi-field-of-view optics where thermal defocus is dependent on field-of-view setting. The algorithm required for elimination of the effects of a combination of both of the above is complex, but compensation may be accomplished by a single mechanical motion (Parr-Burman and Gardam 1985[23]).

A single motion does not, however, guarantee athermalization of image scale, in which case more than one compensatory movement may be required. Two motion athermalization in a zoom or dual-field-of-view infrared telescope can take advantage of the existing mechanisms required for field-of-view change. Also, by utilizing internal lens elements, problems associated with hermetically sealing an external focusing lens element can be avoided (Neil and McCreath 1983,[24] Neil 1984,[25] Neil and Turnbull 1985,[26] Nory 1985,[27] Parr-Burman and Madgwick 1988[28]). In order to maintain stability of aberration correction in infrared zoom telescopes, particularly those having a large zoom range, three-motion athermalization has been proposed (Simmons and Blaine 1988,[29] Shechterman, 1990[30]).

Active/passive Athermalization

An improvement over simple manual active athermalization is to include partial passive athermalization. This is best suited to systems that already contain axially moving lens components, for example a dual-field-of-view infrared telescope (Roberts and Crew 1987[31])—Fig. 10 (Roberts 1989[32]). Here the majority of the athermalization is provided by a mechanically passive device that adjusts the position of the rear lens group in the objective. The residual focus error is then corrected by small manual adjustments to the magnification change element. This technique can minimize the change of image scale and

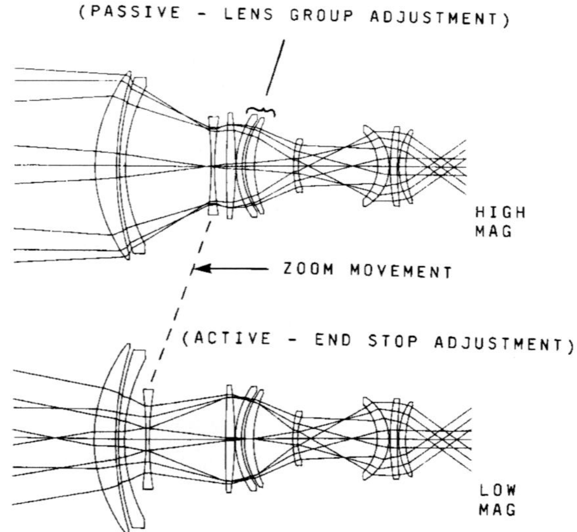

FIGURE 10 Part active, part passive mechanical athermalization *(from Roberts 1989.)*

aberrations with temperature. A potential problem, however, is the subjective nature of best-focus determination.

Athermalization by Image Processing

Athermalization by image processing is suitable for some applications. A range of automatic focusing techniques exists but, while this approach has the advantage of not requiring temperature sensors, it does suffer the potential disadvantage of misinterpretation of image information.

39.8 OPTICAL ATHERMALIZATION

General

Athermalization of the focus position of an optical system by choice of refractive materials has been described extensively in the literature, examples additional to those already referenced being Estes 1961,[33] Gibbons 1976,[34] Straw 1980,[35] Lidwell 1980,[36] Jamieson 1981,[37] Neil 1982,[38] Roberts 1984,[39] Rogers 1988,[40] Rayces and Lebich 1990,[41] and Yatsu et al. 1990.[42] The requirements of overall optical power, achromatism and athermalism demand that three conditions be satisfied for j thin lens elements in contact:

$$\text{Power:} \quad \sum_{i=1}^{j} \phi_i = \phi \tag{11a}$$

$$\text{Achromatism:} \quad \sum_{i=1}^{j} \frac{\phi_i}{V_i} = 0 \tag{11b}$$

$$\text{Athermalism:} \quad \sum_{i=1}^{j} (\gamma_i \phi_i) + \phi \alpha_h = 0 \tag{11c}$$

The presence of three conditions implies the need for three different materials in order to obtain an exact solution. It is possible, however, to find achromatic combinations of two materials that are also athermal, provided that a simple condition is satisfied (Rayces and Lebich 1990[41]):

$$V_1(\gamma_1 + \alpha_h) = V_2(\gamma_2 + \alpha_h) \tag{12}$$

Suitable combinations for thin-lens athermal achromats can be found by plotting a range of materials on a graph of γV against V, the slope of the line joining a chosen pair representing the required thermal expansion coefficient of the housing (Köhler and Strähle 1974[6]).

A number of approximately athermal optical glass achromats exist of which those listed in Table 4 (Rogers 1992b[43])—with the exception of the last entry—represent examples with low to moderate secondary spectrum over the visible to near infrared waveband. The data given for these achromats are: lens element total curvatures for unity focal length;

TABLE 4 Unity Focal Length Athermal Two-Material Achromatic Combinations

Material type	Material combination	Total curvatures	Secondary spectrum†	Petzval sum	Normalized mass
Optical glasses	BaLKN3 + KzFSN4	+7.24/−4.49	3.6×10^{-4}	0.77	2.1
	BaK2 + LaFN7	+4.43/−1.86	4.8×10^{-4}	0.76	1.4
	FK5 + LF5	+4.85/−2.34	4.8×10^{-4}	0.73	1.3
	PSK53A + SFN64	+3.06/−1.28	5.0×10^{-4}	0.65	1.0
	BaLKN3 + BaSF51	+5.21/−2.35	5.2×10^{-4}	0.79	1.5
Stabilized optical glasses	SK4 + KzFSN4	+7.30/−5.64	1.8×10^{-4}	0.62	2.0‡
	SK5 + SF5*	+3.97/−1.99	10.6×10^{-4}	0.67	1.0‡
3-to-5 μ Materials	As_2S_3 + MgO	+0.77/−0.12	8.6×10^{-4}	0.40	0.8§

* Thermally invariant housing, all others aluminum.
† Over wavebands of 480 to 644 nm, 546 to 852 nm, and 3 to 5μ respectively.
‡ Relative to SK5/SF5 solution.
§ Relative to lowest value in Table 5.

Source: From Rogers 1992*b*.

secondary spectrum (second-order color); thin-lens Petzval sum; and an approximate indication of mass, normalized to the lowest value. The pairing of radiation-stabilized versions of SK5 and SF5, both of which have a low value of γ makes a good choice for athermalized space optics in a temperature-invariant mount (Köhler and Strähle 1974[6]).

In the infrared wavebands the options are far more limited: at least one 3–5-micron waveband two-material athermal combination exists, namely, arsenic trisulfide and magnesium oxide, but there is currently no realistic pairing of materials in the 8–12-micron band.

Athermal Laser Beam Expanders

Many more two-material athermal combinations exist if the requirement for achromatism [Eq. (11*b*)] is removed. This is the situation that occurs with a (preferably) galilean laser beam expander, although here the two lens materials are separated (Palmer 1986[44]). From Eq. (4), making the thermal defocus Δf values equal and opposite for the two lenses leads to a value for magnification at which two given materials in a specific housing material will provide an athermal beam expander (for a homogeneous temperature distribution):

$$\text{Magnification} = \frac{\gamma_1 + \alpha_h}{\gamma_2 + \alpha_h} \tag{13}$$

Three-Material Athermal Solutions

Graphical methods have been described that allow investigation of preferred three-material athermalized achromatic solutions (Rayces and Lebich 1990[41]). An alternative method is the systematic evaluation of all possible combinations of three materials selected from a short list, each combination being allocated a risk factor dependent on material

TABLE 5 Unity Focal Length Athermal Three-Material Achromatic Combinations for the 3- to 5-Micron Waveband

Material combination	Total curvatures	Petzval sum	Normalized mass
Si + Ge + ZnS	+0.72/−0.36/+0.27	0.39	1.3
ZnSe + Ge + MgO	+1.16/−0.21/−0.06	0.51	1.8
[Si + Ge + KRS5]	+0.69/−0.26/+0.08	0.34	1.0
ZnS + MgO + Ge	+1.28/−0.17/−0.16	0.52	1.5
AMTIR1 + Ge + Si	+0.56/−0.32/+0.46	0.42	1.4
Si + MgO + KRS5	+0.31/−0.08/+0.22	0.31	1.1
ZnSe + ZnS + Ge	+1.80/−0.69/−0.23	0.50	2.8
Si + CaF_2 + KRS5	+0.32/−0.25/+0.24	0.29	1.1

[ ] Low residual high-order chromatic aberration.

characteristics and solution sensitivity (Rogers 1990[9]). The optical powers of the three in-contact thin-lens elements are determined by solving Eq. (11*a*–*c*) which give for a unity focal length:

$$\text{Given:} \quad a = \frac{V_1 V_2 - V_2 V_3}{V_1 V_3 - V_2 V_3} \qquad \phi_3 = \frac{(1-b)\gamma_1 + b\gamma_2 + \alpha_h}{(1-a)\gamma_1 + a\gamma_2 - \gamma_3} \tag{14a}$$

$$b = \frac{V_2}{V_2 - V_1} \qquad \phi_2 = b - a\phi_3 \tag{14b}$$

$$\phi_1 = 1 - (\phi_2 + \phi_3) \tag{14c}$$

Tables 5 and 6 give a selection of lower-risk three-material solutions, in approximate order of increasing risk, for 3–5- and 8–12-micron infrared combinations respectively: the data given are similar to those in Table 4, but the housing is assumed to be aluminum in all cases. Note that these tables are intended as a guide only and are based on currently available material data.

TABLE 6 Unity Focal Length Athermal Three-Material Achromatic Combinations for the 8- to 12-Micron Waveband

Material combination	Total curvatures	Petzval sum	Normalized mass
KRS5 + ZnSe + Ge	+0.34/−0.15/+0.25	0.30	1.0
ZnSe + ZnS + Ge	+2.05/−0.92/−0.26	0.50	2.5
GaAs + ZnS + KRS5	+0.38/−0.20/+0.26	0.31	1.0
AMTIR1 + ZnS + Ge	+1.42/−0.35/−0.24	0.48	1.3
{CdTe + ZnSe + KRS5}	+0.72/−0.37/+0.22	0.37	1.5
GaAs + ZnSe + KRS5	+0.68/−0.71/+0.33	0.25	1.8
[AMTIRI + ZnSe + KRS5]	+1.19/−0.72/+0.16	0.39	1.9
[CsI + NaCl + GaAs]	+0.68/−0.32/+0.29	0.38	1.1

{ } Very high transmission.

[ ] Low residual high-order chromatic aberration.

Athermalization of Separated Components

In many ways, thermal defocus and thermal change of focal length are analogous to longitudinal and lateral chromatic aberration, having the same first-order dependencies. For this reason it has been suggested that a thermal Abbe number, defined as γ^{-1} (Grey 1948[3]), be used to replace the chromatic Abbe number (V value) in the usual chromatic aberration equations. Thermal expansion of the housing—obviously not present in chromatic calculations—does, however, complicate the situation a little.

In equations thus far, Δf has meant both thermal defocus and focal length change, as numerically these are the same for a thin lens. For separated components, rules similar to those for chromatic aberration apply: for example, two separated thin-lens groups—such as those described by Fig. 1—must be individually athermalized if both types of thermal "aberration" are to be corrected simultaneously. More complex optics (for example, multistage) may have transfer of thermal aberration between constituent lens groups but may still be corrected simultaneously for thermal focus shift and focal length change as a whole. This procedure can, however, lead to one lens group requiring excessive optical powers in order to achieve full overall correction—transient longitudinal thermal gradients may also cause problems.

Use of Diffractive Optics in Optical Athermalization

The term "hybrid optic" is generally used to signify a combination of refractive and diffractive means in an optical element. The diffractive part of the hybrid is usually a transmission hologram which for high efficiency would be of surface relief form, the surface structure being machined or etched onto the refractive surface (Swanson and Veldkamp 1989[45]). The diffractive surface acts as a powered diffraction grating producing large amounts of chromatic aberration which could be employed in an optic where a lightweight optically athermalized combination of two materials could be chosen without regard to achromatism: residual chromatic aberration could then be corrected by the hologram (Rogers 1992*a*[46]).

39.9 REFERENCES

1. J. W. Perry, *Proc. Phys. Soc.,* vol. 55, 1943, pp. 257–285.
2. J. Johnson and J. H. Jeffree, U.K. Patent No. 561 503, 1942 U.K. priority.
3. D. S. Grey, *J. Opt. Soc. Am.,* vol. 38, 1948, pp. 542–546.
4. D. S. Volosov, *Opt. Spectrosc.,* U.S.S.R., vol. 4, pp. 663–669 and pp. 772–778, vol. 5, 1958, pp. 191–199.
5. R. Penndorf, *J. Opt. Soc. Am.,* vol. 47, 1957, pp. 176–182.
6. H. Köhler and F. Strähle, *Space Optics,* B. J. Thompson and R. R. Shannon (eds.), National Academy of Sciences, 1974, pp. 116–153.
7. P. J. Rogers, *SPIE Critical Reviews,* vol. CR38, 1991, pp. 69–94.
8. L. R. Estelle, *SPIE,* vol. 237, 1980, pp. 392–401.
9. P. J. Rogers, *SPIE,* vol. 1354, 1990, pp. 742–751.
10. M. Laikin, *Lens Design,* Marcel Dekker, New York, 1991, p. 28.
11. G. G. Slyusarev, *Opt. Spectrosc.,* U.S.S.R., vol. 6, 1959, pp. 134–138.
12. W. H. Turner, *Optical Sciences Center,* vol. 4, University of Arizona, Tucson, Arizona, 1970, pp. 123–125.

13. V. M. Mit'kin and O. S. Shchavelev, *Sov. J. Opt. Tech.,* vol. 40, 1973, pp. 558–561.
14. F. Reitmayer and H. Schroeder, *Appl. Opt.,* vol. 14, 1975, pp. 716–720.
15. P. J. Rogers, *SPIE,* vol. 147, 1978, pp. 141–148; and U.K. Patent No. 2,030,315.
16. D. G. Norrie, *Opt. Eng.,* vol. 25, 1986, pp. 319–322.
17. J. Angénieux et al., *SPIE,* vol. 399, 1983, pp. 446–448.
18. D. S. Garcia-Nuñez and D. Michika, *SPIE,* vol. 1049, 1989, pp. 82–85.
19. V. Povey, *SPIE,* vol. 655, 1986, pp. 142–153.
20. A. D. Michael and W. B. Hart, *Metallurgist and Materials Technologist,* August 1980, pp. 434–440.
21. G. V. Thompson, U.S. Patent No. 4,148,548, 1976 U.K. priority.
22. P. J. Rogers and G. N. Andrews, *SPIE,* vol. 99, 1976, pp. 163–175.
23. P. M. Parr-Burman and A. Gardam, *SPIE,* vol. 590, 1985, pp. 11–17.
24. I. A. Neil and W. McCreath, U.K. Patent No. 2,141,260, 1983 U.K. priority.
25. I. A. Neil, U.S. Patent No. 4,659,171, 1984 U.K. priority.
26. I. A. Neil and M. Y. Turnbull, *SPIE,* vol. 590, 1985, pp. 18–29.
27. P. Nory, *SPIE,* vol. 590, 1985, pp. 30–39.
28. P. M. Parr-Burman and P. Madgwick, *SPIE,* vol. 1013, 1988, pp. 92–99.
29. R. C. Simmons and P. A. Blaine, *SPIE,* vol. 916, 1988, pp. 19–26.
30. M. Shechterman, *SPIE,* vol. 1442, 1990, pp. 276–285.
31. M. Roberts and P. R. Crew, U.K. Patent No. 2,201,011, 1987 U.K. priority.
32. M. Roberts, *SPIE,* vol. 1049, 1989, pp. 72–81.
33. C. B. Estes, U.S. Patent No. 3,205,774, 1961 U.S. priority.
34. R. C. Gibbons, ERIM Report No. 120200-I-X, 1976, p. 71.
35. K. Straw, *SPIE,* vol. 237, 1980, pp. 386–391.
36. M. O. Lidwell, U.S. Patent No. 4,494,819, 1980 U.K. priority.
37. T. H. Jamieson, *Opt. Eng.,* vol. 20, 1981, pp. 156–160.
38. I. A. Neil, U.S. Patent No. 4,505,535, 1982 U.K. priority.
39. M. Roberts, U.S. Patent No. 4,679,891, 1984 U.K. priority.
40. P. J. Rogers, Thermal Imaging course notes, Institute of Optics, Summer School on Lens Design, Rochester, 1988 (unpublished).
41. J. L. Rayces and L. Lebich, *SPIE,* vol. 1354, 1990, pp. 752–759.
42. M. Yatsu et al., *SPIE,* vol. 1354, 1990, pp. 663–668.
43. P. J. Rogers, *SPIE,* vols. 1780 and 1781, 1992*b*, pp. 36–48.
44. J. M. Palmer, U.K. Patent No. 2,194,072, 1986 U.K. priority.
45. G. J. Swanson and W. B. Veldkamp, *Opt. Eng.,* vol. 28, 1989, pp. 605–608.
46. P. J. Rogers, *SPIE,* vol. 1573, 1992*a,* pp. 13–18.

P · A · R · T · 10

OPTICAL FABRICATION

CHAPTER 40
OPTICAL FABRICATION

Robert E. Parks
Optical Sciences Center
University of Arizona
Tucson, AZ 85721

40.1 INTRODUCTION

Traditional optical fabrication is the craft of producing spherical surfaces of about 0.1 micrometer peak-to-valley accuracy using common materials and relatively simple techniques applied in a consistent manner. Recently developed techniques can also produce nonspherical or steeply aspheric surfaces to similar accuracies. Many of the techniques used in optical fabrication tend to be unfamiliar because most optically worked materials are brittle (as opposed to metals which are ductile and easily worked with familiar tools) and the surfaces are truly spherical, or nearly so, as opposed to many other technical products that have rectilinear features.

Additional barriers to understanding optical working of materials include the statistical nature of producing geometrically perfect spheres and the actual mechanisms that produce specular surfaces during polishing. Even the most knowledgeable workers do not have a rigorous understanding of why two surfaces become more like each other as they are rubbed together using traditional methods as is evidenced by a lack of literature on that subject.

Polishing is likewise a rather miraculous process. It can improve the surface finish of ground glass by nearly three orders of magnitude, it uses unsophisticated materials, and it can be accomplished by persons with little training in polishing. On the other hand, the actual nature of the process by which this is accomplished has probably been debated since Newton's time and certainly since Lord Rayleigh's.[1] Works by Preston,[2] Silvernail,[3] Bach,[4] Golini,[5] and Cook[6] are now shedding some light on the nature of the glass-polishing mechanisms.

In this article, we introduce the various fabrication steps involved in producing accurately polished glass lenses and mirrors, touch on the somewhat different methods used to produce flats and prisms efficiently, and then address crystalline optics such as are used in the UV and IR. Then we address aspherics and the diamond-turning of optical surfaces. Finally, we say a few words about purchasing optics.

40.2 BASIC STEPS IN OPTICAL FABRICATION

The five basic steps in optical fabrication consist of[7–10b]

- *rough shaping* the raw glass into a blank about 1 mm oversize to the finished part

- *generating* the optical surfaces to the correct shape to better than 0.1 mm
- loose abrasive *lapping* the surfaces to remove damage due to generating and producing a curve good to a few micrometers of the desired shape
- *polishing* the ground surface to make it specular and accurate to about 0.1 μm
- *edging* the lens element to make the periphery coaxial with the optical axis as defined by the two lens surfaces.

Because glass is a brittle material, all material removal must be done by grinding rather than the more familiar machining processes used with metals that produce relatively large, curved chips. Grinding works by introducing a controlled network of small fractures into the surface such that glass particles fall out of the surface when the fractures intersect. The fractures must be kept below a critical fracture depth or the fractures will continue all the way through the brittle part.[11] On the other hand, the fractures must be deep enough to insure efficient material removal. In any case, material removal by grinding is a much slower process than machining because the brittle material removed must be disintegrated into micrometer-sized pieces. For this reason, the rough shaping is done by sawing or trepanning (core drilling) so that a minimum amount of material has to be disintegrated to achieve the rough blank shape. For production runs, the rough shaping is done at the glass manufacturer by pressing or molding the hot glass into correctly shaped blanks a few tenths of a millimeter oversize.

Generating is similar in purpose to the machining of raw metal parts. The generating imparts dimensionally true surfaces on the workpiece so that subsequent operations can be performed accurately and efficiently. With glass lenses, the generating is done with cup-shaped grinding wheels with a lip of diamond grit in a metal or resin matrix. When the rotating cup wheel touches the rotating glass blank, a spherical surface is produced, the radius of which depends on the angle between the two axes of rotation. The simple geometry in Fig. 1 shows that this arrangement will produce a spherical surface, either concave or convex, depending on the angle at which the grinding wheel is set.

Loose abrasive lapping is then used to remove sufficient glass to eliminate the deeper fractures caused by generating and to produce a more spherical surface by the action of two rigid bodies abrading each other. The abrasive material, typically between 30 and 5 μm in size, depending upon the stage of lapping, preferentially removes the high places on both the lap and the glass until both surfaces are true spheres to less than 0.1 μm and differ in radius by the several-micrometer grit size of the abrasive compound between the two surfaces.

Most lapping is done with diamond-impregnated pellets made of a resin matrix that are loose-abrasive lapped to the correct radius. This bound abrasive lapping with pellets is faster, less messy, and yields surfaces with a greater dimensional consistency and a more uniform finish. As with loose abrasive lapping, pellet lapping is done on conventional lapping and polishing machines where the work is rotated about its optical axis while the lap is moved over the workpiece by an oscillating pin. The lap takes its rotary motion due to friction between it and the workpiece. This combination of translation and rotation produces a random motion between work and lap that, in turn, produces truly spherical surfaces.

Once the lapping has removed all traces of damage due to generating and coarser lapping grits, and has produced the correct sagitta to within a few micrometers, the workpiece is ready for polishing. Polishing takes place on the same type of machine and uses a similar machine motion as with the lapping, but now the lap is made of pitch (or in many modern operations, a synthetic polishing pad) and the polishing compound used is a few percent mixture of cerium oxide in water.

Polishing converts the finely fractured, lapped surface with a roughness of about 1 μm rms into a specular surface with a roughness of typically 3 nm rms. With care, this finish

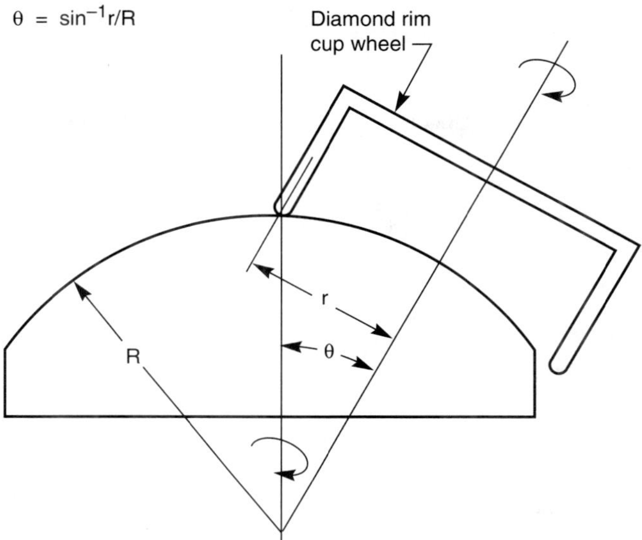

FIGURE 1 Geometry for generating a convex sphere where R is the radius of curvature of the sphere, r is the distance from the cup wheel axis of rotation to the point of contact with the sphere, and O is the angle between the axes of rotation of the sphere and the cup wheel. The geometry for a concave sphere is identical except the angle is negative.

can be reduced to 1 nm rms or less. Our best present understanding of polishing[12] is that the cerium oxide reacts chemically with the glass to soften the outer layer to a depth of a few nm. The relatively soft cerium oxide particles can then physically remove this softened layer and the process starts over again. Behind the lap, it appears that the silica-rich polishing slurry redeposits some of the silica on the surface, leaving a smooth but chemically slightly different outer surface a few nm thick on the bulk glass. Polishing is continued until the level of the surface has been driven below the last remaining fractures from the final lapping step. Typically 10 to 20 micrometers of glass are removed to reach this fully polished condition.

The last step of the fabrication cycle is edging the lens. While the two optical surfaces have been carefully applied to the glass blank, no precision alignment of the glass is usually made and thus the polished lens will have a wedge of some minutes of arc. Another way of stating this is that the optical axis as defined by the line joining the centers of curvature of the two spherical surfaces is decentered from the mechanical axis of the element as defined by its periphery by a few tenths of an mm. Prior to edging, the lens axis defined by the two optical surfaces is aligned to coincide with the axis of rotation of the edging machine spindle. Then a grinding wheel is brought in to reduce the diameter of the lens element to its final dimension while removing any runout in its periphery.

Research continues into methods to reduce the time taken during each step of fabrication and even methods to eliminate some steps. For example, small lenses with reasonably uniform cross sections can be directly molded into finished precision optics by the glass manufacturers and fabricators.[13] Also, a recent project has demonstrated that it is possible to generate spherical surfaces with sufficient accuracy and good enough finish that the surfaces can be polished without the usual lapping step.[14] This research was done on a numerically controlled machine that includes tooling for doing the edging, and, possibly, the polishing as well.

40.3 PLANO OPTICAL SURFACES

Plano optics and prisms are more difficult to produce than spherical optics in two ways. First, a plano surface is a spherical surface with a very specific radius, namely, infinity. Second, while a lens can be brought into correct mechanical alignment during the final edging step, this option is not possible in plano pieces. If a plane window is supposed to be made parallel to 2 seconds of arc, there is no last edging step to make this correction, so the wedge must be reduced to the tolerance level during polishing. Similarly, prism angles must be held within tolerance during polishing using precision tooling and fixtures.

For the polishers of plano work, there are two approaches that help reduce these problems. First, the generating and lapping steps can be more precisely controlled than in the case of spherical optics so the angles are very close to the finished tolerances before polishing begins. Second, much plano work is now done on continuous polishers (CPs) in which a large plano annular lap is "conditioned" to maintain lap flatness independent of the shape of the work put on the lap. Rather than the lap and work ending up at a "compromise" radius as the two work together, the lap on the CP machine is forced by a large glass "conditioner" to stay flat and then the lap forces the work flat in turn.[15] Flatwork fabricators must still keep everything adjusted correctly to maintain flatness, but much more of their time can be spent holding angles rather than fighting just to keep flatness.

The CP machines came into their own during the early days of making glass laser disks for the laser fusion programs and have proved their worth throughout the plano optics industry. These machines call for a large capital investment and need a sufficient volume of work to keep them running nearly full-time. Given those conditions, however, CPs are a cost-effective method of producing plano optics and prisms. Because the techniques for polishing plano work and lenses are different, it pays to find out what type of work potential vendors do best. The job will be less expensive and better when done at a firm that specializes in one type.

40.4 CRYSTALLINE OPTICS

As more and more optical work is being done in the UV and IR regions, there is more call for optics made of nonglass materials. Single and polycrystalline materials as well as some elemental materials are transparent far outside the usual spectral transmission range for glass. In most cases, the surfaces of these materials have differing hardnesses depending on the orientation of the crystal boundaries. Unless the lap is very hard (which leads to scratches) or the polishing material is very hard, the surfaces of the crystalline materials will have a light relief pattern as a function of the crystal structure and this produces wavefront errors and scattering. The use of finely graded diamond as the polishing compound eliminates the problem in many cases. Laps of tin, zinc, or cast iron and diamond polishing compound work well on the hardest crystalline materials. Beeswax-covered pitch or synthetic polishing material laps are often used with the softer crystalline materials and metals.[16,17]

40.5 ASPHERICS

Aspheric surfaces are surfaces that are not spherical and thus the convenient processes that nature has given us to make spherical surfaces no longer apply to making aspheric surfaces. This is why aspherics cost many times what spherical surfaces do if fabricated using traditional techniques. If the aspheric surface is very close (a few micrometers) to a spherical surface, then it may be treated rather similarly to a spherical surface and polished relatively simply. For large departure (fast) aspheres, however, the process is more like sculpturing and is quite time consuming.

In this case, the asphere is ground into the surface as accurately as possible, often with a numerically controlled lathe set up as a grinding machine. Then the surface is polished with a lap of sufficient flexibility to fit the changing curvatures of the asphere. The actual contour of the surface is controlled by dwelling longer on high places than in low spots. Frequent optical testing is required to determine the figure of the surface because convergence to the correct figure tends to be poor because the natural tendency is for the surface to return to a sphere. For larger-size aspheres, computer-controlled polishing or ion milling can be used to good advantage.

Several recent approaches to aspheric polishing are based on the principle of trying to imitate spherical polishing even though it is an asphere that is desired. This approach started with Schmidt, who used vacuum and simple edge support to deform a plane window into a concave shape that included the needed asphericity.[18] The window was polished spherical while held in this distorted condition. When the polishing was complete, the vacuum was released and the window relaxed into a plano shape with the desired corrector asphere imparted in the surface.

The polishing of the Keck telescope off-axis parabolic mirror segments is a logical continuation of the Schmidt method applied to off-axis aspheres, the so-called "bend and polish" technique.[19] For rigid optics where it is not possible to distort the workpiece, the lap shape may be actively controlled to fit the work as has recently been demonstrated.[20] For a broad treatment, Marioge[21] gives a nice review of more traditional methods for polishing aspheric surfaces.

40.6 DIAMOND TURNING

Another method of making both plano and aspheric optics is diamond turning. This precision engineering technology is an outgrowth of atomic weapons fabrication that requires the use of highly accurate, numerically controlled turning machines. Some of the materials used in the weapons were found to be difficult to machine using traditional tool bits so diamond-tipped tools were tried. It was found that some metals could be turned to specular finishes with these techniques employing diamond tooling and precision machines. This is the origin of diamond-turned optics as discussed in Vol. I, Chap. 41 of this Handbook.

From an optics viewpoint, some advantages of diamond turning are that on plano surfaces, the relationship of one surface to another can be controlled very accurately such as is required in a multifaceted scanning prism. Also, finished aspheric surfaces (including off-axis surfaces) can be turned in a few passes in metals such as aluminum and electroless nickel. This makes diamond turning a cost-effective method of producing types of optics that would be difficult and expensive using the traditional methods.

40.7 PURCHASING OPTICS

Purchasing optics to special order is a costly undertaking. If the needed optics can be found as catalog items, this is always the least costly approach. Failing this, most fabricators have a stock of optics that are too few in number to list in a catalog but are available at catalog prices just so they can be moved out of stock. Always inquire if there is something available that meets the requirements.

Often a small change in the design of an experiment or piece of hardware will permit the use of a catalog or standard item. Most optical shops maintain a list of test glasses for specific radii. If custom lenses can be designed using these radii, there will be smaller tooling costs. Along the same line, the Germans have a standard that lists "preferred" lens radii. Designs built around these preferred radii will be less expensive to manufacture than those using completely arbitrary radii.

To help protect the customer buying catalog optics, there is a new ISO standard that specifies minimum tolerances on lens parameters not specifically called out in a catalog. The standard is ISO 10110, Part 11—*Indications in optical drawings, non-toleranced dimensions and material indications.* If the catalog does not state that the standard applies to the listed lenses, this requirement can be given on a purchase order.

If custom optics are required, go to vendors that specialize in the type of optics required. Some are best at lenses, some at plano work and prisms, some at aspherics. Better prices and better quality are available from the specialist. When talking with potential vendors, ask if any of the requirements on the drawing unduly impact the cost of the optics. Some requirement that may seem trivial can cause the vendor substantial trouble and vice versa. Be sure too that the requested specifications and tolerance levels are really needed to make the hardware work. Many optics are overtoleranced just to be on the safe side or because the designer did not take the time to understand the true requirements.

When going out for bids, if one vendor comes in substantially lower than the others, there is probably some misunderstanding of the requirements. Rather than giving the order to the lowest bidder immediately, find out if they truly understand what is being asked for and if they have the capability to produce and test it. If it appears unlikely that the vendor can produce, it is better to excuse that vendor than try to hold them to the bid. That will just be a waste of time and money.

Finally, if the optical requirements are for something very specialized or at the cutting edge of the state of the art, the customer may have to supply the vendor with test equipment and supply a test method to demonstrate compliance. In this case, the customer and vendor will have to work together closely to ensure that the customer gets the optics that are needed.

40.8 CONCLUSIONS

We have shown that optical fabrication requires the serial application of a number of relatively simple, if not completely familiar, steps. If each of these steps is carefully carried out and good artisanship is observed, highly accurate optics can be made using traditional techniques by workers with reasonable skills. Nontraditional techniques are also being successfully applied to making complex optical surfaces and large volumes of similar surfaces. Finally, we have tried to indicate that purchasing optics is not a simple matter, but that the customer and potential vendors must work together to insure that what is required can be fabricated and tested in a cost-effective manner.

40.9 REFERENCES

1. Lord Rayleigh, *Scientific Papers,* vol. 14, Cambridge University Press, Cambridge, 1903, p. 542.
2. F. W. Preston, *Trans. Opt. Soc.* **23:**141 (1922).

3. W. L. Silvernail, "The Role of Cerium Oxide in Glass Polishing," in *The Science of Polishing,* OSA, Washington, D.C., 1984.
4. H. Bach, "Analysis of Subsurface Layers and Spots and the Reactivity of Glass Components," in *The Science of Polishing,* OSA, Washington, D.C., 1984.
5. D. Golini and S. D. Jacobs, "The Physics of Loose Abrasive Microgrinding," *Appl. Opt.* **30:**2761–2777 (1991).
6. L. M. Cook, "Chemical Processes in Glass Polishing," *J. Non-Cryst. Solids* **120:**152–171 (1990).
7. W. Zschommler, "Precision Optical Glassworking," *Proc. SPIE* **472.**
8. F. Twyman, *Prism and Lens Making,* 2d ed., Adam Hilger, New York, 1988.
9. G. Franke, "The Production of Optical Parts," in van Heel, A.C.S. (ed.), *Advanced Optical Techniques,* J. Wiley & Sons, 1967, pp. 253–308.

10*a*. D. F. Horne, *Optical Production Technology,* Crane, Russak & Co., New York, 1972.

10*b*. H. H. Karow, *Fabrication Methods for Precision Optics,* John Wiley & Sons, New York, 1993.

11. B. B. Lawn and M. V. Swain, "Microfracture Beneath Point Indentions in Brittle Solids," *J. Mat. Sci.* **10:** 113 (1975).
12. N. J. Brown, "Preparation of Ultrasmooth Surfaces," *Ann. Rev. Mater. Sci.* **16:**371–388 (1986).
13. H. M. Pollicove, "Precision Glass Molded Optics," *OF&T 1988 Technical Digest Series* **13:**121, OSA.
14. W. C. Holton, "Fast Microgrinding of High-Quality Surfaces," *Photonics Spectra* **26:**56–57 (1992).
15. N. J. Brown, Some Speculations on the Mechanisms of Abrasive Grinding and Polishing," *Precision Engineering* **9:**5 (1987).
16. G. W. Fynn and W. J. A. Powell, *Cutting and Polishing Optical and Electronic Materials,* 2d. (ed.), Adam Hilger, Philadelphia, 1988.
17. R. Sumner, "Polishing of IR Materials," in W. Wolfe and G. Zissis (eds.), *The Infrared Handbook,* 1978.
18. E. Everhart, "Making Corrector Plates by Schmidt's Vacuum Method," *Appl. Opt.* **5:**713–715 (1966) and *Errata* in *Appl. Opt.* **5:**1360 (1966).
19. J. Lubliner and J. Nelson, "Stressed Mirror Polishing," *Appl. Opt.* **19:**2332–2340 (1980).
20. H. M. Martin et al., "Progress in the Stressed Lap Polishing of a 1.8 m, f/1 Mirror," in *Advanced Technology Optical Telescopes, Proc. SPIE* **1236:**682 (1990).
21. J. P. Marioge, "Les Methodes de Fabrication de Surfaces Aspheriques (I) and (II), *J. Optics* (Paris) **15:**119–132 and 183–204 (1984).

CHAPTER 41
FABRICATION OF OPTICS BY DIAMOND TURNING

Richard L. Rhorer
Los Alamos National Laboratory
Los Alamos, New Mexico

Chris J. Evans
National Institute of Standards and Technology
Gaithersburg, Maryland

41.1 GLOSSARY

- f feed rate (mm/rev)
- h peak-to-valley height
- R tip radius
- R_a average absolute roughness

41.2 INTRODUCTION

The use of special machine tools with single-crystal diamond-cutting tools to produce metal optics is called *diamond turning*. The manufacture of optical surfaces by diamond turning is relatively new compared to the traditional optical-polishing methods. In terms of geometry and motions required, the diamond-turning process is much like the step of "generating the optical surface" in traditional optical fabrication. However, the diamond-turning machine is a more sophisticated piece of equipment that produces the final surface, which typically does not need the traditional polishing operation. But surface quality produced by the "best" diamond turning does not yet match the best conventional-polishing practice. Yet, the limits of diamond turning for both figure and surface finish accuracy have not yet been reached.

There are several important advantages of using diamond turning, including the ability to produce good optical surfaces to the edge of the element, to fabricate soft ductile materials difficult to polish, to eliminate alignment in some systems, and to fabricate shapes difficult to do by other methods. If the advantages of diamond turning suggest this fabrication method, then it is important to determine early in the design phase of a project whether the material specified is appropriate for diamond turning.

Sections in this chapter highlight the following:

- The diamond-turning process
- The advantages of diamond turning
- Diamond-turnable materials
- Comparison of diamond turning and traditional optical fabrication
- Machine tools for diamond turning
- Basic steps in diamond turning
- Surface finish in diamond-turned optics
- Measuring diamond-turned surfaces
- Conclusions

41.3 THE DIAMOND-TURNING PROCESS

The diamond-turning process produces finished surfaces by very accurately cutting away a thin chip or layer of the surface. Thus, it is generally applicable to ductile materials that machine well rather than to hard brittle materials traditionally used for optical elements. (However, by using a grinding head on a diamond-turning machine in place of the tool, hard brittle materials can be finished.) At very small effective depths of cut, brittle materials behave in an apparently ductile manner. This attribute allows fracture-free grinding of glasses and ceramics as well as diamond turning of optical surfaces on materials such as gemanium, zinc selenide, and potassium dihydrogen phosphate (KDP).

In diamond turning, both the figure and surface finish are largely determined by the machine tool and the cutting process. Note, however, that materials characteristics such as grain size, inclusion size, etc. limit the ultimate surface finish achievable. The tool has to be very accurately moved with respect to the optical element to generate a good optical surface, and the edge of the diamond tool has to be extremely sharp and free of defects.

41.4 THE ADVANTAGES OF DIAMOND TURNING

The advantages of diamond turning over the more traditional optical fabrication technique of lapping and polishing (see Chap. 40 by R. E. Parks) are

- It can produce good optical surfaces clear to the edge of the optical element. This is important, for example, in making scanners, polygons, special shaped flats, and when producing parts with interrupted cuts.
- It can turn soft ductile materials that are extremely difficult to polish.
- It can easily produce off-axis parabolas and other difficult-to-lap aspherical shapes.
- It can produce optical elements with a significant cost advantage over conventional lapping and polishing where the relationship of the mounting surface—or other feature—to the optical surface are very critical. Expressed differently, this feature of diamond turning offers the opportunity to eliminate alignment in some systems.
- The fabrication of some optical shapes, such as axicons and x-ray telescopes, would be extremely difficult by methods other than diamond turning.

Conflicts between optical requirements and diamond turnability on the one hand, and mechanical considerations on the other often lead to the use of platings. Plating

TABLE 1 General Guide to Optical Fabrication Methods

Size, m	Shape	Material	Preferred method
Less than 0.5	Flat or sphere	Glass/ceramic	Polish
		Ductile metal	Diamond turn
	Asphere	Glass/ceramic	Grind/polish*
		Ductile metal	Diamond turn
0.5 to 2	Any axisymmetric	Ductile metal	Diamond turn†
Greater than 2.0	Any	Any	Large polishing machines

* Can generate shape or figure on a diamond-turning machine with a grinding head replacing the diamond tool.
† Diamond-turning machines up to 2-m diameter have been made.

deficiencies, however, can cause as much trouble as poor bulk materials. For example, small changes in electroless nickel bulk composition may cause dramatic changes in tool wear.[1]

Residual stress introduced into the mirror blank, whether plated or not, can lead to changes in mirror shape with time. It is essential to pay careful attention to stress-relieving prior to final diamond turning. A decision to diamond turn an optical element, rather than fabricate it by the conventional polishing techniques, might be based on several different considerations such as type of element, size, and material. A general guide to different considerations in selecting diamond turning as a fabrication technique is presented in Table 1.

An axicon optical element produced by diamond turning is shown in Fig. 1. This type of element would be extremely difficult to make by means other than diamond turning.

41.5 DIAMOND-TURNABLE MATERIALS

It is not uncommon to hear materials described as either "diamond-turnable," or not, as if this were an inherent material property. In reality, some materials just wear out the diamond-turning tools at a much faster rate than other materials do. The reason for this fast wear rate is yet unknown. For example, it is commonly known that ferrous materials rapidly wear diamond tools. Some tests have been made—using different methods of diamond-turning ferrous materials to decrease wear rate—that show some encouraging results. These methods include operating at cryogenic temperatures,[2] in methane and acetylene environments,[3] and by ultrasonic vibration of the tool.[4] However, up to the present time, these tests have been done on only minuscule amounts of the material and will need to be demonstrated on realistic-sized components.

A number of listings, such as the one included in Table 2, have been published of diamond-turnable materials.

Such listings should be treated with caution. Typically, they are incomplete and do not provide sufficient information on the materials that are listed. For example, good optical surfaces are not generally produced on all aluminum alloys: 6061 is the most commonly used alloy, although certain 5000 series and 7000 series alloys have their proponents, and 2024 aluminum has been used but, in general, does not produce the best surfaces. Similarly, gold is considered diamond-turnable, but problems have been reported machining large gold-plated optics. Conventional electroplated nickels give rapid tool wear, but electroless nickel with phosphorous contents above about 10 percent, if

FIGURE 1 An axicon optical element being diamond turned. (*Courtesy of Rank Taylor Hobson, Keene, New Hampshire.*)

appropriately heat treated, can be machined effectively.[5] Recently, it has been shown that high-phosphorous electroplated nickel[6,7] also machines extremely well, although such platings are not widely available.

Silicon, although included in the listing given here, should be considered marginal as tool wear can be high. Reasonably large areas of amorphous silicon cladding are reported to have been successfully machined.

TABLE 2 Diamond-turnable Materials

Aluminum	Calcium fluoride	Polymethymethacrylate
Brass	Magnesium fluoride	Polycarbonates
Copper	Cadmium telluride	Polyimide
Beryllium copper	Zinc selenide	
Bronze	Zinc sulphide	
Gold	Gallium arsenide	
Silver	Sodium chloride	
Lead	Calcium chloride	
Platinum	Germanium	
Tin	Strontium fluoride	
Zinc	Sodium fluoride	
Electroless nickel	KDP	
	KTP	
	Silicon	

Diamond turnability of plastics varies, with some suggestion that parameters, such as surface speed, are more important than for metal and crystalline substrates. Some plastics are, however, diamond turned in volume production.

Therefore, it is important to involve experienced personnel early in the design phase[8] to ensure that the material specified is appropriate. In some projects, the part is so valuable and/or so difficult to produce by other techniques, it is worth consuming tools more rapidly than would normally be acceptable. However, such a decision should be taken consciously, not by default late in a project.

41.6 COMPARISON OF DIAMOND TURNING AND TRADITIONAL OPTICAL FABRICATION

In diamond turning, the final shape and surface of the optical produced depends on the machine tool accuracy, whereas, in traditional optical fabrication, the final shape and surface of the optical element are produced by lapping and polishing with an abrasive-loaded lap. The differences between diamond turning and traditional optical fabrication can be summarized by describing diamond turning as a *displacement-controlled process* versus a *force-controlled process* for traditional optical fabrication.[9] The goal in diamond turning is to have a machine tool that produces an extremely accurate path with the diamond tool, hence a displacement-controlled machine. A traditional polishing machine used for optical fabrication depends on the force being constant over the area where the abrasive-loaded lap—or tool—touches the surface being worked. Selective removal of material can be produced by increasing the lap pressure in selected areas or by use of a zone lap. The stiffness of a diamond-turning machine is important because, to control the displacement, it is important that cutting forces and other influences do not cause unwanted displacements. Feeds, speeds, and depth of cut are typically much lower in diamond turning than conventional machining, thus giving lower forces. However, the displacements of concern are also much lower. Thus the stiffness required is as much, or more, of a concern than conventional machining even though the total force capability may be lower for diamond turning.

41.7 MACHINE TOOLS FOR DIAMOND TURNING

In general, the machine tools used for diamond turning are very expensive compared to the equipment needed for traditional optical fabrication. The precision required for diamond turning is beyond the capability of conventional machine tools, thus some of the first diamond-turning machines for fabricating optics were built adapting Moore measuring machines.[10]

Although there are some records of machine tools being used to generate optical surfaces as early as the 17th century, most of the effort is modern, accelerated in the 1960s and 1970s with the advent of computer-based machine tool controls and laser interferometer systems used as positional feedback devices. Evans[11] has documented much of the history of diamond turning and provides an extensive reference list. Some of the research in metal cutting related to diamond turning and associated machine tools is summarized by Ikawa.[12]

Two commercial diamond-turning machines are shown in Fig. 2. Such machines may generally be configured to operate in a normal facing lathe mode (with the tool stationary and the part rotating), as a milling machine (part stationary, tool rotating), or, on occasion, as an optical generator—both part and tool rotating—with the addition of a second

(a)

FIGURE 2(*a*) Diamond-turning machine (*Courtesy of Rank Taylor Hobson, Keene, New Hampshire*).

(b)

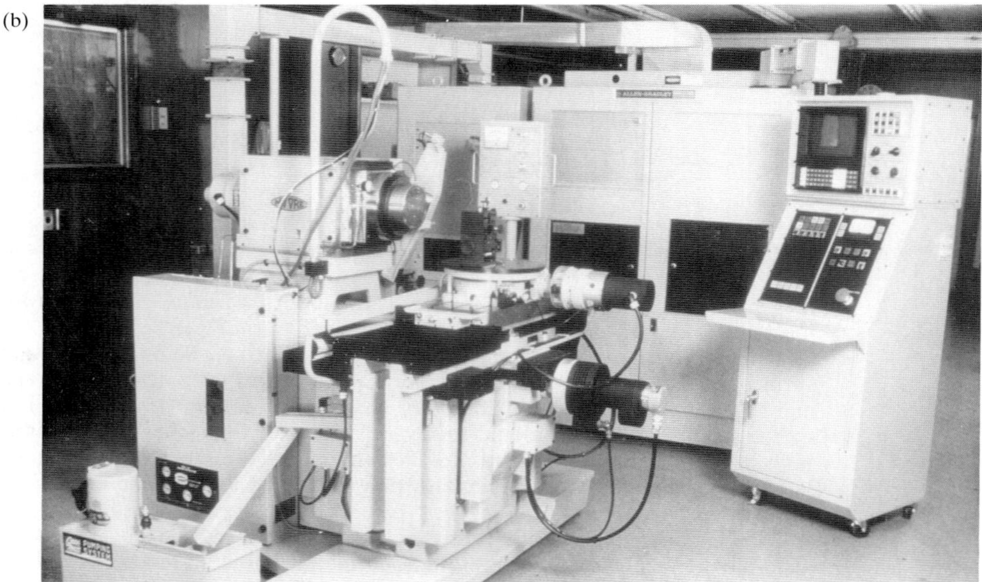

FIGURE 2(*b*) Diamond-turning machine (*Courtesy of Moore Special Tool Co., Bridgeport, Connecticut*).

spindle. The milling-type process is generally referred to as *flycutting*. Dedicated flycutting machines, also commercially available, are widely used in flat and polygon production. Also diamond-turning machines have been configured with a tailstock for special-purpose work.

41.8 BASIC STEPS IN DIAMOND TURNING

Much like the traditional optical-fabrication process, the diamond-turning process can be described as a series of steps used to make an optical element. The steps used in diamond turning are:

1. *Preparing the blank* with all the required features of the element with an extra thickness of material (generally 0.1 mm extra material or plating is adequate) on the surface to be diamond turned
2. *Mounting the blank* in an appropriate fixture or chuck on the diamond-turning machine
3. *Selecting the diamond tool* appropriate for material and shape of optical component
4. *Mounting and adjusting the diamond tool* on the machine
5. *Machining the optical surface* to final shape and surface quality
6. *Cleaning the optical surface* to remove cutting oils or solvents

Mounting the optical element blank on a diamond-turning machine is extremely important. If a blank is slightly distorted in the holding fixture, and then machined to a perfect shape on the machine, it will be a distorted mirror when released from the fixture. Therefore, fixtures and chucks to hold mirrors during diamond turning need to be carefully designed to prevent distortion. Often the best way to hold a mirror during machining is to use the same mounting method that will be used to hold the mirror in service.

It is advantageous in many applications to machine a substrate of an aluminum or copper and then plate on a surface to be diamond turned. The design and application of platings is part science and part art. Many aspects of the platings as related to diamond turning were covered at the ASPE Spring 1991 Topical Meeting.[13]

Tool setting—the mounting and adjusting of the diamond-tipped cutting tool—is often accomplished by cutting a test surface, either on the actual mirror blank to be later machined over, or by placing a test piece on the machine just for tool setting. If the cutting tool is too high, or too low, a defect at the center of a mirror is produced. It is possible, using reasonable care and patience, to set the tool height within about 0.1 μm of the exact center. Setting the tool in the feed direction after the height is correct is somewhat more difficult. For example, an error in setting will produce an ogive shape rather than a sphere. Gerchman[14] describes these types of defects.

The selection of the diamond tool for diamond turning is important. Large cutting tip radii (2 mm or more) are often used when producing flats and large, small-F-number focusing mirrors. However, small-radii diamond tools are available (in the range of 0.1 mm) for making small deep mirrors or molds. Tools with special geometries can be obtained for such applications as Fresnel lenses. In general, approximately zero degree rake tools, with about 5 or 6 degrees front clearance, are used for diamond turning ductile metals. Negative rake tools are often good for crystalline materials and positive rakes may be beneficial when machining some plastics. The cutting edge has to be chip-free to produce a good diamond-turned surface. A normal specification for edge quality is "chip free when examined at 1000×." The edge sharpness is also of concern in diamond turning, yet there is currently no convenient way to specify and inspect tools for edge sharpness.

The orientation of the diamond itself on the shank is of concern because the single-crystal diamond is anisotropic. The orientation of diamond tools has been studied, for example, by Wilks,[15] Decker,[16] and Hurt.[17] It is necessary for the tool manufacturer to mount the diamond so that it can be shaped to the required radius and produce a good cutting edge. The usual orientation for diamond tools is with the cleavage plane horizontal—parallel to the rake face.

The actual diamond turning, or machining to final size and surface finish, is often the fastest part of the process. The machine-tool controller has to be programmed to pass the tool through the correct path, the chip-removal system has to be positioned, and the cutting-fluid applicator needs to be adjusted to provide consistent clean cutting.

For machining of flats and spherical surfaces, the computer machine control programs are straightforward. But when cutting aspherical surfaces, caution has to be exercised so that the radius of the tool is properly handled in calculating the tool path. Modern CAD systems perform the necessary calculations, but test cases should be done prior to cutting a difficult or expensive component.

In general, the cutting speeds for diamond turning are similar to those used for conventional machining: less than one to over 100 m/min. However, the slower cutting speeds produced by facing to the center of a workpiece do not affect the surface finish in diamond turning as is often the case with nondiamond tools. Thus, varying the spindle speed to keep the cutting speed constant is not necessary in diamond turning. The upper speed for diamond turning is often limited by the distortion of the optical element due to centrifugal forces, expecially for larger elements. The upper spindle speed can also be limited due to any unbalance of the workpiece and fixture.

The feed rate in diamond turning is usually adjusted to give a good theoretical surface finish. (See the next section on the surface finish of diamond-turned optics.)

Cleaning of diamond-turned optics has a lot in common with cleaning conventionally polished optics. But because many of the diamond-turned elements are of soft metals, caution has to be exercised to prevent scratching. In general, a degreaser is used (soap or solvent), followed by a rinse in pure ethyl alcohol. The drag-wiping technique traditionally used on some glass optics can be used on some diamond-turned elements. Care must be taken to insure that the lens tissue is very clean and remains wet. Some work is being done to study the best solvents to use for cleaning diamond-turned optics from an environmental-impact standpoint.[18]

41.9 SURFACE FINISH IN DIAMOND-TURNED OPTICS

Surface structure is different for diamond-turned surfaces compared to conventionally polished surfaces. A diamond-turned surface is produced by moving a cutting tool across the surface of the turning component, as illustrated in Fig. 3. Therefore, diamond-turned elements always have some periodic surface roughness, which can produce a diffraction-grating effect, whereas polished optical surfaces have a random roughness pattern. The traditional "scratch and dig" approach to describing surfaces is not meaningful for diamond-turned surfaces.

The machining process produces a periodic surface structure directly related to the tool radius and feed rate. The theoretical diamond-turned surface is described in Fig. 4. The formula displayed in the figure for calculating the height of the cusps is

$$h = \frac{f^2}{8R} \tag{1}$$

where,

h is the peak-to-valley height of the periodic surface defect, usually expressed in

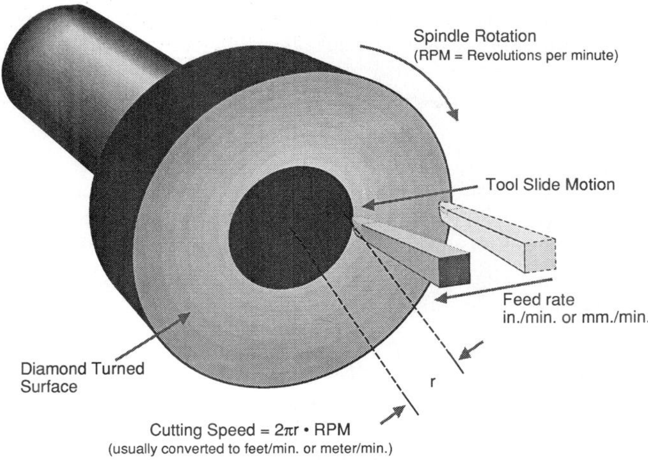

FIGURE 3 Diamond turning an optical element.

microinch, micrometer, nanometer, or angstrom units—the units have to be consistent with other parameters of the formula but any units will work.

f is the feed per revolution, expressed for example in mm per revolution, or thousandths of inch per revolution.

R is the tip radius of the tool, expressed for example in mm, fractions of an inch, or thousandths of an inch.

For example, if a surface is diamond turned using a spindle speed of 300 rpm, a feed of 7.5 mm/min, and a 5-mm tool nose radius:

$$h = \frac{(7.5/300)^2}{8 \times 5} = 1.56 \times 10^{-5}\ \text{mm}$$

$$h = 15.6\ \text{nm} \tag{2}$$

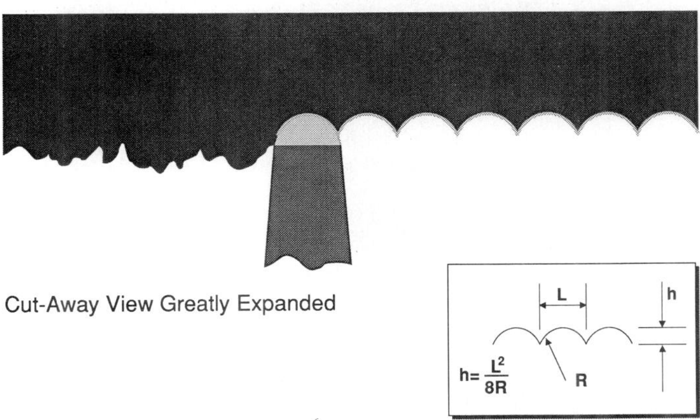

FIGURE 4 "Cusp" surface of diamond-turned optical element.

In addition to the "theoretical finish" based on cusp structure, the measured surface finish on diamond-turned parts is influenced by other factors such as

- Waviness within the long-wavelength cut-off for surface measurement. This type of waviness may be correlated, for example, with slide straightness errors.
- Asynchronous error motions. If, for a given angular spindle position, there is nonrepeatability in axial, radial, or tilt directions, these errors will transfer into surface structure. Details of spindle errors are important in diamond turning. Further information can be found in the Axis of Rotation Standard.[19]
- External and self-induced vibration, not at the spindle frequency nor at one of its harmonics, has the same effect on finish—measured across the lay—as asynchronous spindle motions.
- Materials effects. Differential elastic recovery of adjacent grains gives steps in the machined surface and an appearance commonly referred to as "orange peel." Impurities in the material can also degrade surface finish.
- Within each cusp, there is repeated structure related to "roughness" of the edge of the tool.

41.10 MEASURING DIAMOND-TURNED SURFACES

The measurement of diamond-turned surfaces presents some difficulties not encountered in conventional optics. In general, all methods used to measure optical surfaces are used on diamond-turned components. Bennett[20] presents a complete discussion of surface-roughness measurement. The Nomarski microscope is an excellent means of qualitatively evaluating diamond-turned surfaces. The Nomarski photos[21] in Fig. 5 illustrate the periodic machined structure of a diamond-turned surface. The feed rate used in producing the surface causes the periodic structure to be about 8 μm per revolution.

The measurement of the roughness of optical surfaces is performed by a number of different instruments, both stylus types and optical systems. If a surface is measured with an instrument that produces a profile, several different statistical methods can be used to describe the surface. Figure 6 illustrates a profile and some of the statistical parameters used in describing surfaces. The term peak-to-valley is often used in the diamond-turning shop to mean the difference between the highest and lowest points in any surface trace. The rms and Ra are also used to quantify a measured surface finish. These parameters are defined in Fig. 6 and as follows:[22]

> Although the *rms roughness* is generally used to describe the finish of *optical surfaces*, the *average roughness* R_a is normally used for roughness of machined surfaces. R_a is simply the average of the absolute values of the surface height variations z_i measured from the mean surface level (see Fig. 6). Expressed in equation form, this is
>
> $$R_a = \frac{1}{N} \sum_{i=1}^{N} |z_i|$$
>
> If a surface has a profile that contains no large deviations from the mean surface level, the values of δ and R_a will be similar. However, if there are appreciable numbers of large "bumps" or "holes," the largest values of the z_i's will dominate the surface statistics and δ will be larger than R_a.

(a)

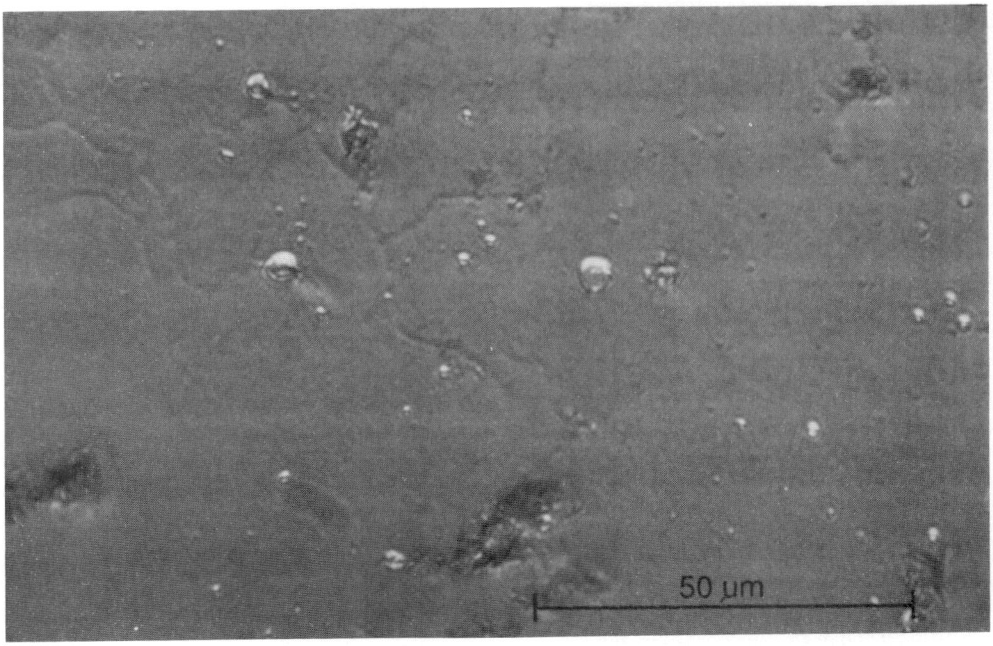

(b)

FIGURE 5 Nomarski micrograph of a diamond-turned aluminum alloy (*a*) aligned so that the grooves can be seen and (*b*) aligned so that the grooves are canceled. (*From Bennett, p. 84.*)

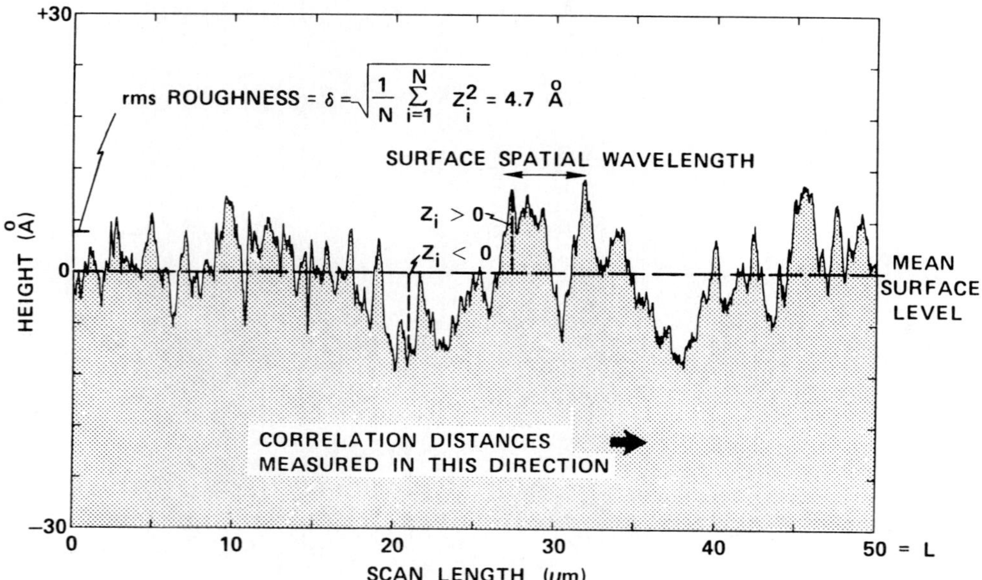

FIGURE 6 Schematic representation of a rough surface showing statistical parameters. (*From Bennett, p. 39.*)

41.11 CONCLUSIONS

Diamond turning has been used for many years to commercially produce infrared optics. Some visible and ultraviolet applications are now possible. Moreover, the limits of diamond turning for both figure and surface finish accuracy have not yet been reached. Taniguchi[23] and others have shown that precision in both conventional machining and ultraprecision machining, such as diamond turning, has steadily improved for many decades, with roughly a factor of three improvement-in-tolerances possible every ten years. If this trend continues, we could expect diamond-turning machines with accuracies below 10 nm by the year 2000. Yet, it is important to remember that it becomes increasingly difficult to push the capabilities in this regime. Research in the field of nanofabrication—working with dimensions on the order of molecules, or tenths of nm—may help extend diamond turning into the nanometer range.

The technology developed for diamond-turning optics in some industries is now beginning to impact the precision machining of nonoptical components. In the future, the improvement of all machine tools will likely be driven by both optical and nonoptical applications, with diamond-turning machines possibly reaching the accuracy level that will allow visible and ultraviolet optics to be fabricated by machining or grinding without postpolishing.

41.12 REFERENCES

1. C. K. Syn, J. S. Taylor, and R. R. Donaldson, "Diamond Tool Wear vs. Cutting Distance," *Proc. SPIE* **656**.
2. C. J. Evans, "Cryogenic Diamond Turning of Stainless Steel," *CIRP Annals* **40:**571–575 (1991).

3. J. Casstevens, "Diamond Turning of Steel in Carbon Saturated Environments," *Precision Engineering* **5:**9–15 (1983).

4. T. Moriwaki, "Ultraprecision Diamond Turning of Stainless Steel by Applying Ultrasonic Vibration," *CIRP Annals* **40:** 559–562 (1991).

5. J. S. Taylor, C. K. Syn, T. T. Saito, and R. R. Donaldson, "Surface Finish Measurements of Diamond Turned Electroless Nickel Plated Mirrors," *Optical Engineering* **25**(9)**:**1013–1020 (Sept 1986).

6. A. Mayer, et al., "Electrodeposited Coatings for Diamond Turning Applications," *Proceedings ASPE Spring Topical Meeting,* "Metal Platings for Precision Finishing Operations," 1991.

7. C. J. Evans, R. S., Polvani, and A. Mayer, "Diamond Turned Electrodeposited Nickel Alloys," *OSA Technical Digest Series* **9:**110 (1990).

8. J. S. Taylor and C. J. Evans, "Fabrication of a Metal Plated Mirror, Beginning from a Performance Specification," *Proceedings ASPE Spring Topical Meeting,* "Metal Platings for Precision Finishing Operations," 1991.

9. A. Gee, Cranfield Institute of Technology, in a private communication to the authors used the description of "displacement" controlled and "force" controlled.

10. W. R. Moore, *Foundations of Mechanical Accuracy,* Moore Special Tool Co., Bridgeport, Conn., 1970.

11. C. J. Evans, *Precision Engineering: An Evolutionary View,* Cranfield Press, Bedford, UK, 1989, pp. 135–155.

12. N. Ikawa, et al., "Ultraprecision Metal Cutting: the Past, the Present, and the Future," *CIRP Annals* **40:**587–594 (1991).

13. ASPE, "Metal Platings for Precision Finishing Operations," *Spring Topical Meeting,* Raleigh, N.C., 1991.

14. M. C. Gerchman, "Optical Tolerancing for Diamond Turning Ogive Error," *Proc. SPIE* (Reflective Optics II):224–229 (1989).

15. J. Wilks, "Performance of Diamonds as Cutting Tools for Precision Machining," *Precision Engineering* **2:**57–71 (1980).

16. D. C. Decker, H. H. Hurt, J. H. Dancy, and C. W. Fountain, "Preselection of Diamond Single Point Tools," *Proc. SPIE* **508** (1986).

17. H. H. Hurt and D. L. Decker, "Tribological Considerations of the Diamond Single Point Tool," *Proc. SPIE* **508** (1986).

18. L. A. Theye and R. D. Day, "Evaluation of Environmentally Safe Cleaning Agents for Diamond Turned Optics," *ASPE Proceedings,* ASPE, Raleigh, N.C., 1991.

19. ANSI/ASME B89.3.4M-1985 Standard: "Axes of Rotation, Methods for Specifying and Testing," ASME, New York, 1985.

20. J. M. Bennett and L. Mattsson, *Introduction to Surface Roughness and Scattering,* Optical Society of America, Washington, D.C., 1989.

21. Bennett, *ibid.,* p. 84.

22. Bennett, *ibid.,* p. 39.

23. N. Taniguchi, "Nanotechnology for Ultraprecision Instruments," *Precision Engineering* **16:**5–24 (1994).

P · A · R · T · 11

OPTICAL PROPERTIES OF FILMS AND COATINGS

CHAPTER 42
OPTICAL PROPERTIES OF FILMS AND COATINGS

J. A. Dobrowolski
Institute for Microstructural Sciences
National Research Council of Canada
Ottawa, Ontario, Canada

42.1 GLOSSARY

Quantities, Abbreviations

A	absorbtance of layer system
BW	base width of a bandpass filter
D	spectral detectivity of a detector
DW	Debye-Waller factor (used in XUV coatings)
d	metric thickness of a layer
E	electric vector of electromagnetic field
FP	Fabry-Perot (interference filter)
H	magnetic vector of electromagnetic field
H	quarterwave layer of high refractive index
HW	halfwidth of a bandpass filter
I	intensity of light
k	extinction coefficient of layer or medium
L	quarterwave layer of low refractive index
L	number of layers in system
LIDT	laser-induced damage threshold
M	layer matrix
n	refractive index of layer or medium
nd	optical thickness of a layer
$\tilde{n}$	complex refractive index
P	degree of polarization
R	reflectance of layer system
r	amplitude reflection coefficient

S	spectral intensity distribution
T	transmittance of layer system
t	amplitude transmission coefficient
δ	phase thickness of a layer
ε	phase change on reflection or transmission
η	effective refractive index
θ	angle of incidence
θ_o	angle of incidence in the incident medium
λ	wavelength
λ_o	reference or central wavelength
μ^*	effective index of a bandpass filter
σ	combined effect of interface layer and roughness (in XUV coatings)
σ	wavenumber
τ	internal transmittance of a substrate

Subscripts

ij	where $i,j = 1$ or 2, for elements of layer matrix
j	for j-th layer
m	for medium
p	for light polarized parallel to the plane of incidence
r	for reflected light
s	for light polarized perpendicular to the plane of incidence
s	for substrate
t	for transmitted light

42.2 INTRODUCTION

Scope of Chapter

In the broadest sense of the term, an optical filter is any device or material which is deliberately used to change the spectral intensity distribution or the state of polarization of the electromagnetic radiation incident upon it. The change in the spectral intensity distribution may or may not depend on the wavelength. The filter may act in transmission, in reflection, or both.

Filters can be based on many different physical phenomena, including absorption, refraction, interference, diffraction, scattering, and polarization. For a comprehensive review of this broader topic the interested reader is referred to the article entitled "Coatings and Filters" which appeared in the first edition of the *Handbook of Optics*.[1]

This chapter deals only with filters that are based on absorption and interference of electromagnetic radiation in thin films. Optical thin-film coatings have numerous applications in many branches of science and technology and there are also many consumer products that use them. The spectral region covered in this chapter extends from about 0.003 to 300 μm (3 to 3×10^5 nm), although the main emphasis is on filters for the visible and adjacent spectral regions.

The discussion in this chapter is largely confined to generic thin-film filters, such as antireflection coatings, cut-off filters, narrowband transmission or rejection filters, reflectors, beam splitters, and so forth. Filters for very specific applications, such as filters for colorimeters and other scientific instruments, color correction filters, and architectural coatings, are, as a rule, not treated. Of the filters described, many are available commercially while others are only research laboratory prototypes. This review does not cover thin-film filters whose properties can be changed by external electric or magnetic fields, temperature, or illumination level.

In this introductory section some general considerations on the use of optical filters are presented. In the following sections, the theory of optical multilayers and the methods for their deposition and characterization are briefly discussed. These sections are useful for gaining a proper understanding of the operation, advantages, and limitations of optical coatings. The remaining sections then describe the properties of various generic thin-film filters.

For further information on this subject, the interested reader is particularly encouraged to consult the books by Macleod[2] and Rancourt.[3]

General Theory of Filters

There are many different ways of describing the performance of optical coatings and filters. For example, transmission and reflection filters intended for visual applications are adequately described by a color name alone, or by reference to one of the several existing color systems (see Chap. 26). There also exist other specialized filter specifications for specific applications. However, the most complete information on the performance of a filter is provided by spectral transmittance, reflectance, absorptance, and optical density curves. This is the method adopted in this chapter.

Referring to Fig. 1, at a wavelength λ the normal incidence spectral transmittance $T(\lambda)$

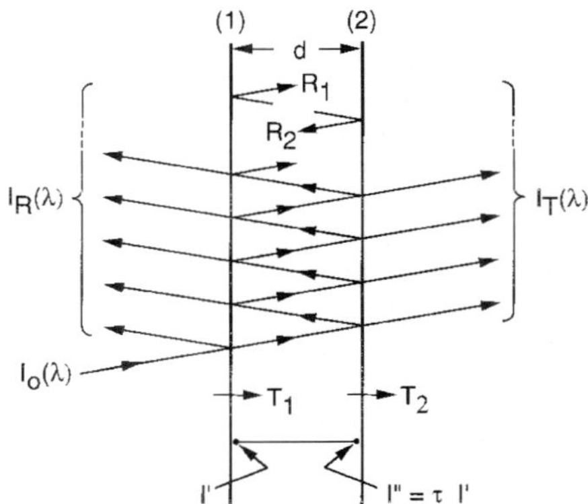

FIGURE 1 Specular transmission and reflection of light by a plane-parallel plate (see "General Theory of Filters").

of a filter placed between two semi-infinite media is equal to the ratio of the light intensity of that wavelength transmitted $I_T(\lambda)$ by the filter to that incident $I_o(\lambda)$ upon it,

$$T(\lambda) = \frac{I_T(\lambda)}{I_o(\lambda)} \tag{1}$$

At nonnormal incidence, the component of the intensity perpendicular to the interface must be used in the preceding equation.[2] The spectral reflectance $R(\lambda)$ of a filter is defined in a similar way,

$$R(\lambda) = \frac{I_R(\lambda)}{I_o(\lambda)} \tag{2}$$

The relationship between the transmittance $T(\lambda)$ and the density of a filter, sometimes also called the absorbance, is given by

$$D(\lambda) = \log \frac{1}{T(\lambda)} \tag{3}$$

In this chapter, transmittances and reflectances will be plotted using either linear or logarithmic scales. The logarithmic scale is particularly well suited whenever accurate information about the low-transmission or reflection region is to be conveyed. However, this is done at the expense of detail at the high end of the scale. Wavelengths are normally specified in micrometers (μm) because this is the most convenient unit for the spectral range covered in this chapter. In the following discussions, the dependence of the transmittance, reflectance, and absorptance on wavelength will be implicitly assumed.

Transmission and Reflection of Coatings on a Substrate

Many multilayer coatings are deposited onto a transparent or partially transparent substrate. Both the multilayer and the substrate contribute to the overall performance of the filter. For example, absorption in the substrate is frequently used to limit the transmission range of the filter. Reflectances at the filter interfaces need also to be considered. However, they can be reduced by antireflection coatings, or by cementing several components together.

In general, a filter can consist of multilayer coatings deposited onto one or both sides of a substrate. The overall transmittance T_{total} of a filter can be expressed in terms of the internal, or intrinsic transmittance τ of the substrate and the transmittances T_1, T_2 and internal reflectances R_1, R_2 of each surface of the substrate (Fig. 1).

The internal transmittance τ of a substrate is defined to be the ratio of the light intensity I'' just before reaching the second interface to the intensity I' just after entering the substrate (Fig. 1):

$$\tau = \frac{I''}{I'} \tag{4}$$

Expressions for the evaluation of the transmittance T and reflectance R of multilayer coatings are given in Sec. 42.3 under "Matrix Theory for the Analysis of Multilayer Systems."

Providing that the incident light is not coherent, there will be no interference between the beams reflected from the two surfaces of a substrate, even when the surfaces are plane-parallel. A summation of all the partial reflections leads to the following expression for the overall spectral transmittance T_{total} of a filter:

$$T_{\text{total}} = \frac{T_1 T_2 \tau}{1 - R_1 R_2 \tau^2} \tag{5}$$

The reflection coefficients of uncoated interfaces can be calculated from Eq. (80), providing that the complex refractive indices of the substrate and of the medium are known. If all the materials in the filter are nonabsorbing, then

$$T_{\text{total}} = \frac{T_1 T_2}{1 - R_1 R_2} \tag{6}$$

If R_1 is small, an appropriate expression for T_{total} is

$$T_{\text{total}} \approx [1 - R_1(1 - R_2)]T_2 \tag{7}$$

However, this last approximation is not valid in general; some infrared substrate materials have high reflection coefficients and in such cases Eq. (6) must be used.

Transmission Filters in Series and Parallel

To obtain a desired spectral transmittance, it is frequently necessary to combine several filters. One common approach is to place several filters in series (Fig. 2*a*).

Because of the many different different partial reflections that may take place between the various surfaces, precise formulas for the resulting transmittance are complicated.[4] Accurate calculations are best carried out using matrix methods.[5]

To a first approximation, the resultant transmittance T' of a filter system consisting of k individual filters placed in series is given by

$$T' \approx T'_1 T'_2 T'_3 \cdots T'_k \tag{8}$$

Here T'_i is the total transmittance of filter i. This expression is valid only if the reflectances

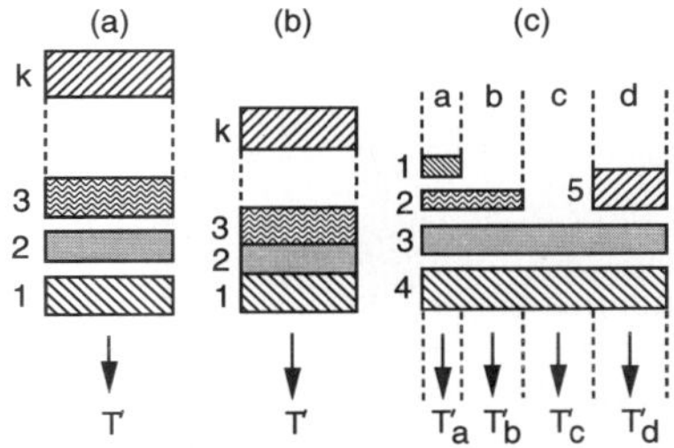

FIGURE 2 Transmission filters arranged in series and in parallel. The filters can be air-spaced (*a*) and (*c*) or cemented (*b*).

of the individual filters are small or if the interference filters are slightly inclined to one another and the optics are arranged in such a way that the detector sees only the direct beam.

Under other circumstances, the use of this expression with interference filters can lead to serious errors. Consider two separate filters placed in series and let T'_1, T'_2, R'_1, and R'_2 correspond to the transmittances and reflectances of the two filters. If $T'_1 = T'_2 = R'_1 = R'_2 = 0.5$, then according to Eq. (8), the resulting transmittance will be $T' = 0.25$. For this simple case the precise expression can be derived from Eq. (5) and is given by

$$T' = \frac{T'_1 T'_2}{1 - R'_1 R'_2} \tag{9}$$

Evaluating this expression one obtains $T' = 0.33$. This is significantly different from the result obtained from the application of Eq. (8).

Some spectral transmittance curves cannot be easily designed by placing filters in series alone. For certain applications it is quite acceptable to place filters not only in series, but also in parallel.[6] This introduces areas as additional design parameters. Thus, for example, the effective spectral transmittance T' of the filter shown in Fig. 2*c* would be given by

$$T' = \left(\frac{a}{A} T'_a + \frac{b}{A} T'_b + \frac{c}{A} T'_c + \frac{d}{A} T'_d\right) \tag{10}$$

where

$$A = \text{overall area of filter}$$

$$a, b, c, d = \text{areas of the four zones}$$

$$T'_a, T'_b, T'_c, T'_d = \text{transmittances of four zones}$$

The latter are given by

$$\begin{aligned} T'_a &= T'_1 \cdot T'_2 \cdot T'_3 \cdot T'_4 \\ T'_b &= T'_1 \cdot T'_2 \cdot T'_3 \\ T'_c &= T'_1 \cdot T'_2 \\ T'_d &= T'_1 \cdot T'_2 \cdot T'_3 \cdot T'_5 \end{aligned} \tag{11}$$

Great care must be exercised when using such filters. Because the spectral transmittance of each zone of the filter is different, errors will result unless the incident radiation illuminates the filter uniformly. Similar care must be used when employing the filtered radiation. One way proposed to alleviate these problems is to break the filter down into a large number of small, regular elements and to reassemble it in the form of a mosaic.[7]

Reflection Filters in Series

If radiation is reflected from k different filters, the resultant reflectance R' will be given by

$$R' = R'_1 \cdot R'_2 \cdot R'_3 \cdots R'_k \tag{12}$$

which is analogous to Eq. (8) for the resultant transmittance of filters placed in series. Many of the considerations of that section also apply here. For instance, R' will be significant only at those wavelengths at which every one of the reflectors has a significant reflectance. Metal layers (see "Metallic Reflectors" in Sec. 42.16) and thin-film interference coatings can be used exclusively or in combination.

For the sake of convenience the number of different reflecting surfaces used is

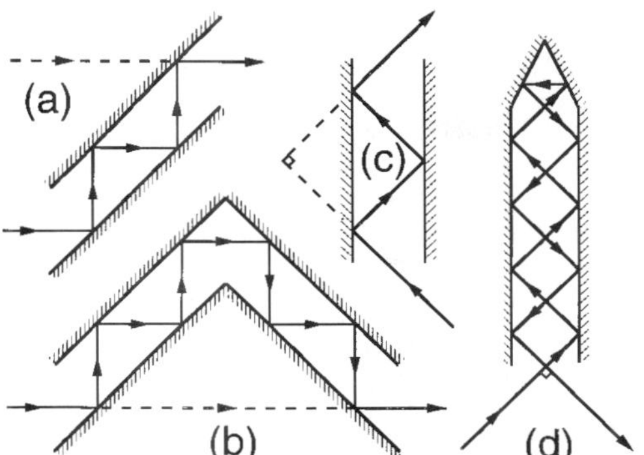

FIGURE 3 Various arrangements for multiple-reflection filters.

normally restricted. The outlines of some possible reflector arrangements given in Fig. 3 are self-explanatory. The arrangement shown in Fig. 3*b* does not deviate or displace an incident parallel beam. The number of reflections depends in each case on the lengths of the plates and on the angle of incidence of the beam. Other arrangements are possible.

Clearly, reflection filters placed in series require more space and are more complicated to use than transmission filters. But if in a given application these shortcomings can be accepted, multiple-reflection filters offer great advantages, which stem mainly from the nature of the reflectors available for their construction (see "Multiple-reflection Filters" later in this chapter).

42.3 THEORY AND DESIGN OF OPTICAL THIN-FILM COATINGS

Design Approaches

A thin-film designer may be asked to design a multilayer coating in which the transmittance, reflectance, and/or absorptance values are specified at a number of wavelengths, angles, and polarizations of the incident light. The designer may be required to provide a coating with many other more complicated properties, including integral quantities such as CIE color coordinates, solar absorptance, or emissivity.[8] The parameters that can be used to reach these goals are the number of layers in the multilayer, the layer thicknesses, and the refractive indices and extinction coefficients of the individual layers and of the surrounding media. Clearly, the more demanding the performance specifications, the more complex the resulting system. Many different methods have been developed for the design of multilayer coatings. For a good overview of this topic the interested reader is referred to the books by Macleod,[2] Knittl,[9] and by Kurman and Tikhonravov.[10] Here only the most important methods will be mentioned.

Graphical vector methods provide the most understanding of the problem, but the necessary approximations limit them to the solution of problems in which the final reflectance is not too high. Admittance diagrams and similar chart methods do not suffer from this limitation, but they are best applied to problems in which the specifications are relatively simple.[11] Many problems can be solved using the known properties of periodic multilayer systems.[12] Analytical synthesis methods yield solutions to problems in which quite complex spectral transmittace or reflectance curves are specified.[13,14] However, frequently solutions obtained in this way call for the use of inhomogeneous layers which are difficult to deposit, or for homogeneous layers with optical constants that are outside the range of known materials. Numerical design methods are the most flexible of all because they can be applied to problems with very complex specifications requiring a large number of layers for their solution.[15,16] They are usually based on the matrix theory of optical multilayer systems and are particularly powerful for the solution of complicated spectral problems when combined with analytical methods.

Matrix Theory for the Analysis of Multilayer Systems

If electromagnetic radiation falls onto a structure consisting of thin films of several different materials, multiple reflections will take place within the structure. Depending on the light source and the layer thicknesses, the reflected beams may be coherent and interfere with one another. This optical interference can be used to design optical multilayer filters with widely varying spectral characteristics. In this section the basic equations for thin-film calculations are presented and some general properties of interference filters are listed. For a thorough discussion of this topic the reader is referred to the work of Macleod[2,17] and Thelen.[18]

Consider the thin-film system consisting of L layers shown in Fig. 4. The construction parameters comprise not only the refractive indices n_j and the thicknesses d_j of the layers $j = 1, 2, \ldots, L$, but also the refractive indices n_s and n_m of the substrate and the incident medium. The angle of incidence θ, the wavelength λ, and the plane of polarization of the incident radiation are the external variables of the system.

The most general method of calculating the transmittance T and the reflectance R of a multilayer from these quantities is based on a matrix formulation[19,20] of the boundary conditions at the film surfaces derived from Maxwell's equations.[21]

It can be shown that the amplitude reflection r and transmission t coefficients of a multilayer coating consisting of L layers bounded by semi-infinite media are given by

$$r = \frac{\eta_m \mathbf{E}_m - \mathbf{H}_m}{\eta_m \mathbf{E}_m + \mathbf{H}_m} \tag{13}$$

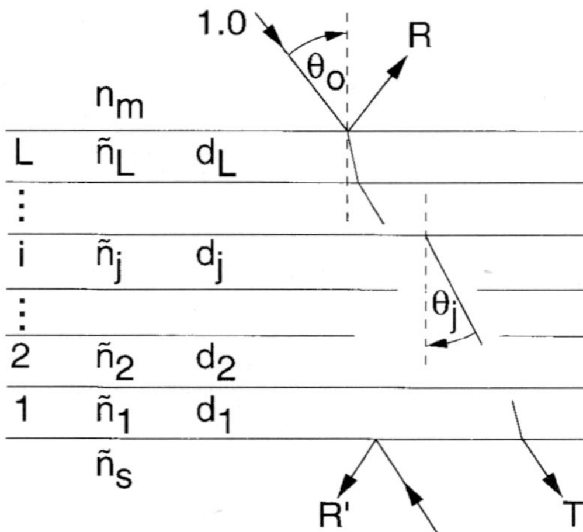

FIGURE 4 Construction parameters of a multilayer.

and

$$t = \frac{2\eta_m}{\eta_m \mathbf{E}_m + \mathbf{H}_m} \tag{14}$$

where

$$\begin{pmatrix} \mathbf{E}_m \\ \mathbf{H}_m \end{pmatrix} = \mathbf{M} \begin{pmatrix} 1 \\ \eta_s \end{pmatrix} \tag{15}$$

$\mathbf{E}_m$ and $\mathbf{H}_m$ are the electric and magnetic vectors, respectively, in the incident medium, and $\mathbf{M}$ is a product matrix given by

$$\mathbf{M} = \mathbf{M}_L \mathbf{M}_{L-1} \cdots \mathbf{M}_j \cdots \mathbf{M}_2 \mathbf{M}_1 \tag{16}$$

In the preceding equation, $\mathbf{M}_j$ is a 2×2 matrix which represents the jth film of the system:

$$\mathbf{M}_j = \begin{pmatrix} m_{11} & im_{12} \\ im_{21} & m_{22} \end{pmatrix} = \begin{pmatrix} \cos \delta_j & \dfrac{i}{\eta_j} \sin \delta_j \\ i\eta_j \sin \delta_j & \cos \delta_j \end{pmatrix} \tag{17}$$

where

$$\delta_j = \frac{2\pi}{\lambda} (n_j d_j \cos \theta_j) \tag{18}$$

The quantity $n_j d_j \cos \theta_j$ is the *effective optical thickness* of the layer j for an angle of refraction θ_j. In Eqs (13) to (17) η represents the *effective refractive index* of the medium, substrate, or layer, and is given by

$$\eta = \begin{cases} \dfrac{n}{\cos \theta} & p\text{-polarization} \\ n \cos \theta & s\text{-polarization} \end{cases} \tag{19}$$

depending on whether the incident radiation is polarized parallel (p) or perpendicular (s) to the plane of incidence. The angle θ_j is related to the angle of incidence θ_o by Snell's law

$$n_m \sin \theta_o = n_j \sin \theta_j \tag{20}$$

The intensity transmittance and reflectance are

$$T = \frac{\eta_s}{\eta_m} |t|^2 \tag{21}$$

$$R = |r|^2 \tag{22}$$

and the phase changes on transmission and reflection, ε_T and ε_R are given by

$$\varepsilon_T = \arg t \tag{23}$$

$$\varepsilon_R = \arg r \tag{24}$$

If the materials in a multilayer are all nonabsorbing, then $T + R = 1$. Should one or more materials absorb, then in the preceding equations the refractive indices of these materials must be replaced by their complex refractive indices $\tilde{n}$, defined by

$$\tilde{n} = n - ik \tag{25}$$

where k is the extinction coefficient of the material. Even though all the elements of the layer matrix for such a material are now complex, its determinant will still be unity. The absorptance of the multilayer is then calculated from

$$A = 1 - T - R \tag{26}$$

Certain important general conclusions about the properties of multilayer filters can be drawn from these equations.

1. The properties of thin-film systems vary with angle of incidence [Eqs. (18), (19)]. For some applications this is the major disadvantage of interference filters compared to absorption filters.
2. This variation depends on the polarization of the incident radiation [Eq. (19)]. The following equations define T and R for obliquely incident randomly polarized radiation:

$$T = \tfrac{1}{2}(T_p + T_s) \tag{27}$$

$$R = \tfrac{1}{2}(R_p + R_s) \tag{28}$$

The dependence of T and R on polarization has been used for the design of polarizers (see Sec. 42.11 and Chap. 3, Vol. II of this Handbook). However, like the angular variation, it is a disadvantage for most other applications. Many researchers have investigated ways of reducing these effects.[18,22–25]

3. Transmittance curves of nonabsorbing multilayers composed of layers whose optical thicknesses are all multiples of $\lambda/4$ show symmetry about λ_o when plotted on a relative wavenumber scale λ_o/λ [Eqs. (17) and (18)].
4. A proportional change of all the thicknesses of a nonabsorbing multilayer results merely in a displacement of the transmittance curve on a wavenumber scale [Eq. (18)]. Thus a thin-film design can be utilized in any part of the spectrum subject only to the limitations imposed by the dispersion of the optical constants of the materials used.
5. The reflectance and absorptance of a filter containing absorbing layers will depend, in general, on which side of the filter the radiation is incident (Fig. 5). However, the transmittance does not depend on the direction of the incident light.

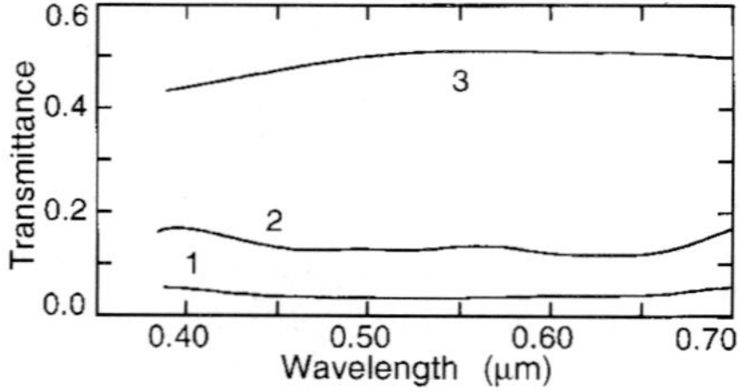

FIGURE 5 Spectral characteristics of a thin chromium alloy film on glass. Curve 1: transmittance; curves 2 and 3: reflectance from the glass and air sides, respectively. (*After Liberty Mirror.*[405])

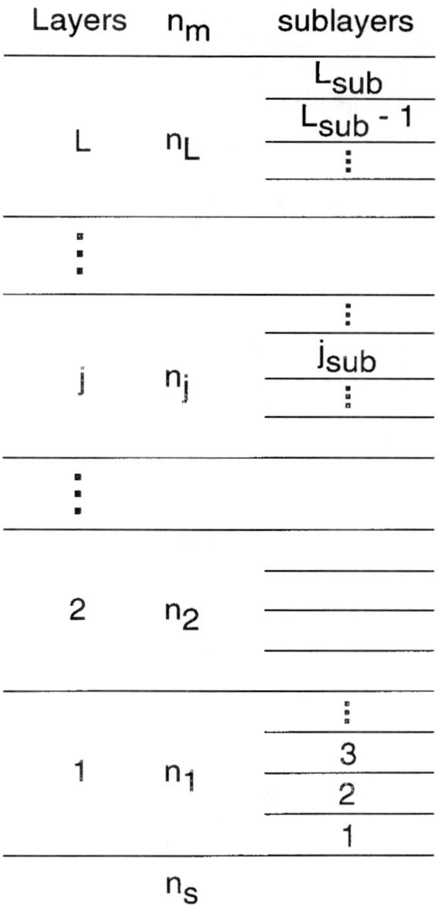

Layers	n_m	sublayers
		L_{sub}
L	n_L	$L_{sub} - 1$
		⋮
⋮		
		⋮
j	n_j	j_{sub}
		⋮
⋮		
2	n_2	
		⋮
1	n_1	3
		2
		1
	n_s	

FIGURE 6 Subdivision of layers for the evaluation of the electric field within a multilayer (see "Matrix Theory for the Analysis of Multilayer Systems").

An important parameter is the electric field amplitude squared, $|\mathbf{E}|^2$. The susceptibility of a multilayer to high-power laser damage is proportional to the highest value of this quantity within the multilayer. One way to evaluate $\mathbf{E}$ is to subdivide each layer of the system into L_{sub} sublayers (Fig. 6) and to evaluate the partial matrix products $\mathbf{P}_j$,

$$\mathbf{P}_j = \begin{pmatrix} p_{11} & ip_{12} \\ ip_{21} & p_{22} \end{pmatrix} = \prod_{k=1}^{j} (\mathbf{S}_k) \tag{29}$$

for all L_{sub} sublayers, where $\mathbf{S}_k$ is the matrix of the kth sublayer. Then, the electric field amplitude at a point j in the multilayer is given in terms of the elements of the total and partial product matrices by the following expression:[26]

$$\mathbf{E}_j = \frac{4[p_{11}^2 + (n_s p_{12})^2]}{[m_{11} + (n_s m_{22}/n_m)]^2 + [(m_{21}/n_m) + n_s m_{12}]^2} \tag{30}$$

The analysis of optical thin-film systems [with computer programs based on Eqs. (13) to (30)] is relatively simple. The design of filters with any but the simplest spectral characteristics remains a complicated problem and one of the methods listed under "Design Approaches" must be used.

42.4 THIN-FILM MANUFACTURING CONSIDERATIONS

The optical, mechanical, and environmental properties of multilayer coatings depend on the materials used, on the deposition process, and on the surface quality of the substrate.

There are many different methods for the deposition of thin films.[27–29] Some of the more common processes are reviewed below. The deposition methods and process parameters used affect the microstructure of the resulting layers. The films can be dense with an amorphous or a microcrystalline structure, or they may exhibit a columnar growth with considerable voids. The optical constants clearly depend on this microstructure and films of the same materials may sometimes have very different properties, depending on how they were deposited (see, for example, Ref. 30). The individual films in a multilayer may be under tensile or compressive stress and, unless materials and film thicknesses are selected to compensate for these stresses, the overall stress may be large enough to distort the substrate or cause the multilayer to break up. The mechanical properties of multilayer coatings also critically depend on the microstructure of the films. An excellent discussion of the effects of microstructure on the various properties of optical coatings has been given by Macleod.[17]

Optical Coating Materials

Many different materials have been used in the past for the construction of optical multilayer coatings. Some of the compounds used for the deposition of nonabsorbing layers for the ultraviolet, visible, and infrared parts of the spectrum are cryolite (1.35), LiF (1.37), MgF_2 (1.39), ThF_4 (1.52), CeF_3 (1.62), PbF_2 (1.73), ZnS (2.30), ZnSe (2.55), Si (3.5), Ge (4.20), Te (4.80), PbTe (5.50), SiO_2 (1.48), Al_2O_3 (1.60), MgO (1.72), Y_2O_3 (1.82), Sc_2O_3 (1.86), SiO (1.95), HfO_2 (1.98), ZrO_2 (2.10), CeO_2 (2.20), Nb_2O_5 (2.20), Ta_2O_5 (2.10), and TiO_2 (2.45). The numbers in parentheses represent the approximate refractive indices at the midpoints of the material transparency range. Some of the metals used in the same wavelength range for the deposition of reflecting or absorbing layers are Ag(0.12–i3.45), Al(1.02–i6.85), Au(0.31–i2.88), Cu(0.83–i2.60), Ni(1.80–i3.33), Cr(3.18–i4.41), Inconel(2.94–i2.92), and Rb(2.00–i5.11). The complex refractive indices of the metals given in the brackets correspond to a wavelength of 0.56 μm. These values are approximate and are intended as a rough guide only. More extensive listings of coating materials are given, for example, by Macleod[2] and Costich.[31]

As already mentioned, the properties of multilayers depend on the materials used for their construction. For example, layers made of oxides are, as a rule, harder than those made of fluorides, sulphides, or semiconductors. They are therefore preferred for use on exposed surfaces. Semiconductor materials should be avoided in filters that are to be used over a wide range of temperatures because their optical constants can change significantly. Some metals are soft and easily damaged while others tarnish when exposed to the atmosphere. Such coatings require further protective coatings, or should be cemented between two transparent plates. Other materials require the precoating with adhesion layers to ensure a good bond to the substrate. For example, frequently an Ni adhesive layer is deposited onto glass before coating with Au.

Evaporation

Conventional (*nonreactive*) *or reactive evaporation* from resistance, induction, or electron beam gun sources is a low-energy process (~0.1 eV) and the resulting films frequently have a porous structure. The porosity may vary with the material, the substrate temperature, the residual pressure in the deposition chamber, the deposition rate, and angle of incidence of the vapor on the substrate. Values of porosities ranging from 0 to 40 percent have been observed. On exposure to the atmosphere, some of the voids in the film may absorb water vapor. This increases the effective refractive index of the films and results in a shift of the spectral features of the multilayer towards longer wavelengths ("ageing"). This shift is partially reversible—by placing the filter in an inert atmosphere or in a vacuum, or on heating, some of the adsorbed water vapor can be removed. Unless it has been allowed for at the design stage, such ageing can render some filters useless.

The microstructure of the films can be significantly affected by bombarding the substrate during deposition with energetic ions from an auxiliary ion beam source.[32–34] The additional energy (~50 to 100 eV) results in denser films. Hence, coatings produced by *ion-assisted deposition* have higher refractive indices and exhibit less ageing on exposure to the atmosphere.

The *ion plating* process can result in even denser coatings.[35–37] In this high deposition rate process the starting material must be a good conductor and is usually a metal. Argon and a reactive gas species are introduced into the chamber and, together with the evaporant, are ionized. The ions are then accelerated to the substrate with energies of the order of 10 to 50 eV. Transparent films with near-bulklike densities and low temperature variation of refractive index can be obtained by this process. For most materials, the layers are glasslike and the interfaces remain smooth. This results in a lower scatter.

Conventionally evaporated thin films can be under compressive or tensile stresses. If not controlled, these stresses can distort the substrate or cause the multilayer to break up. The magnitude of the stresses depends on the material and on the deposition conditions. It is possible to select the materials and process parameters so that the stresses of the various layers counteract each other. In contrast, almost all ion plated layers are under compressive stress. It is therefore more difficult to produce stress-compensated multilayers by this process.

Sputtering

Reactive or nonreactive dc or RF magnetron sputtering is also used to deposit optical multilayer coatings. Many variants of this process exist.[38] Most are significantly slower than evaporation and the targets can be quite expensive. Filters produced by dc or RF sputtering may therefore also be more expensive. However, the process is stable, provides excellent control over the thicknesses of the layers, and can be readily scaled to provide uniform coatings over large areas. Both metal and metal oxide layers can be produced. Sputtering is an energetic process and results in dense, bulklike layers which exhibit virtually no ageing.

In *ion-beam sputtering,* an energetic beam of inert ions is aimed at a target made of the material that is to be deposited. Atoms or clusters of atoms of the material are dislodged from the target and land on the substrate with a high energy. This is the slowest physical vapor thin-film deposition method described here and it cannot be readily scaled for the coating of large components. However, it yields the highest quality coatings. Many of the high-reflectance coatings for laser gyros, in which no signficant losses can be tolerated, are produced in this way.

Deposition from Solutions

In this procedure the substrate is either dipped in an organo-metallic solution and withdrawn at a very steady rate from it, or the solution is applied from a pipette onto a spinning substrate. The substrate is then placed in an oven to drive off the solvent. The thickness of the film depends on the concentration of the solvent and on the rate of withdrawal or spinning. Other factors which influence the process are temperature and humidity, as well as the freshness of the solution. Although it yields quite porous films, this method is of interest because many of the layers produced in this way have a high laser damage threshold. The process has also been adapted for the coating of quite large area substrates with multilayer antireflection coatings for picture frame glass and for display windows.

Thickness Control during Deposition

The performance of many optical multilayers depends critically on the thicknesses of the individual layers. The control of the layer thicknesses during their formation is therefore very important. Many different methods exist for the monitoring of layer thicknesses. For very steady deposition processes, such as sputtering, simple timing can give good results. However, the most common techniques used are quartz crystal and optical monitoring. The former is very sensitive and can be used for thin and thick films, as well as transparent and opaque films. However, it is an indirect method and requires careful calibration. This is usually not a problem whenever layers of established coating materials are formed using a standard geometry and deposition conditions. Optical monitoring can be performed directly on the substrate, or indirectly on a witness glass. The quantities measured are usually T, R, or the ellipsometric parameters. One advantage of direct optical monitoring is that the parameters measured are usually closely related to the required performance. Furthermore, with optical monitoring, a real-time error determination and compensation is possible after the deposition of each layer, through the reoptimization of the remaining layers of the system.[39] With this method even quite complicated multilayer structures can be manufactured.

42.5 MEASUREMENTS ON OPTICAL COATINGS

Optical Properties

Transmission, Reflection, and Absorption. The most commonly used instrument for the measurement of the optical performance of thin-film coatings is the spectrophotometer. The wavelength dispersion of commercial instruments for the 0.185- to 8.00-μm spectral range is usually provided by prisms or by ruled or holographic gratings. Grazing incidence gratings, crystals, or multilayer coatings are used in the soft x-ray and extreme ultraviolet (XUV) spectral regions. Fourier transform spectrometers are capable of measurements from about 2 to 500 μm. A variety of attachments is available for the measurement of specular and diffuse reflectance. Absolute measurements of T and R that are accurate to within ± 0.1 percent are difficult even in the visible part of the electromagnetic spectrum.

Very small absorptions of single layers are normally measured with calorimeters.[40] The losses of high-performance laser reflectors are obtained from measurements of the decay times of Fabry-Perot interferometer cavities formed from these mirrors.[41]

Roughness of the substrate and irregularities occurring within individual films and the layer interfaces give rise to light scattering in all directions.[42–44] For many applications, it is important to minimize this scatter. Special instruments, called scatterometers, are used to measure the angular variation of the light scatter. Such data provides information about the substrate and multilayer.

The transmittance, reflectance, and absorptance of some optical coatings are affected by exposure to atomic oxygen and by electron, proton, and ultraviolet irradiation. They also depend critically on the cleanliness of the components measured. In space, contamination of optical components can also take place.[45,46]

Optical Constants. A reliable knowledge of the optical constants of all the materials used in the construction of optical multilayer coatings is essential. There exist many different methods for their determination.[47] These include methods that are based on refractometry, photometric and spectrophotometric measurements of R and/or T, polarimetry, single-wavelength or spectroscopic ellipsometry, various interferometric methods, attenuated total reflection, or on a combination of two or more of these methods. Excellent monographs on the various methods will be found in Palik's *Handbook of Optical Constants of Solids.*[48,49] Some are suitable for measurements on bulk materials and the results are valid only for films produced by the more energetic deposition processes described here. The optical constants of porous films must be measured directly. They will depend on the deposition parameters and on the layer thickness, and may differ significantly from those of bulk materials.[50] Special methods have to be used for the determination of the optical constants in the x-ray, XUV, and submillimeter regions. The accurate measurement of very small, residual extinction coefficients of transparent coating materials is difficult. Generally, it involves the use of laser calorimetry or the use of the film as a spacer layer in a bandpass filter. It is also very important to be able to measure the thickness of the film independently.

Laser Damage. A measure of the ability of an optical component to withstand high laser irradiations is its laser-induced damage threshold (LIDT). There are several ways of defining this quantity. One frequently used definition is based on a survival plot of the percentage of components that are damaged when they are exposed to different laser fluences. The value of the fluence corresponding to the intersection of a mean curve through the experimental points with the ordinate is nominally defined to be the LIDT. It is thus the maximum fluence at which no damage is expected to the component. The slope of the survival curve is also usually reported. Methods for the determination of the LIDT differ in the way in which the damage to the component is observed. This may, for example, be the observation of a catastrophic failure with a Nomarski microscope, or the detection of a specified change in scatter, or the relaxation of the material as seen with a photothermal deflection method, or an operational criterion, where, within specified limits, the component can no longer perform the function for which it was designed.

Absorption is the main cause for laser damage. The incident radiation that is absorbed in the optical component is usually converted into heat in the damage process. If the thermal conductance of the optical component is too low, the temperature of the local absorbing spot on the element will rise to a value at which damage ensues. The damage will therefore depend on the thermal conduction of the materials of which the element is made. For example, some high-reflectivity mirrors consist of coatings on Si, SiC, W, or Cu substrates that are water-cooled during use to increase heat removal.

Thermal conductivities of thin films are several orders of magnitude smaller than those of the corresponding bulk materials. (An exception to this are some fluoride layers.) This compounds the problem. To increase the LIDT, the deposition methods are optimized to obtain homogeneous thin films of low absorption and with more bulklike thermal conductivities.

Absorption damage usually initiates at defects and other imperfections. In the case of the substrate, it may occur at or below the surface, even when the substrate material is nonabsorbing. It is very important to avoid materials with color centers, subsurface damage, and inclusions. In point of fact, in most pulsed laser-induced damage to thin films, damage occurs at discrete locations such as nodules. The substrate surface must be very smooth and devoid of scratches, digs, and pores, otherwise polishing compounds and other contaminants can be trapped. It is imperative that the substrate be as perfectly clean as possible prior to coating. Electric fields associated with these imperfections can increase the absorption by an amount which is proportional to the refractive index of the material.

The coating materials used for the deposition of high LIDT multilayers must be very pure, with absorption edges far away from the wavelength of interest and, for many pulsed irradiations, from the harmonics as well. As a rule, materials in thin-film form have extinction coefficients that are orders of magnitude larger than those of the corresponding bulk materials. Currently, the processes used to produce high LIDT coatings include ion-assisted deposition, ion-beam sputtering, sol-gel deposition, and electron beam gun evaporation. To reduce the effects of the residual surface roughness, the thicknesses of the layers are often adjusted to shift the peaks of the electric field away from the layer boundaries where impurities tend to concentrate.

The form that the damage takes depends to a large extent on the materials. Pitting of the coatings is probably due to the explosion of nodules. Delamination may be due to poor adhesion of the layers to the substrate, to undue stresses in the films, and/or to a poor match between the expansion coefficients of the layers and the substrate.

The LIDT also depends on the laser pulse duration. For very short pulses (<10 ns) thermal conductance does not play as major a role in the process. For longer pulses, the LIDT is frequently seen to be proportional to the square root of the pulse duration, indicating a mechanism that depends mainly on diffusion. For high repetition rates the LIDT depends somewhat on the repetition rate. To achieve a long life ($>10^6$ pulses) in an industrial environment, the laser should be operated at a fraction (say, 1/4 or 1/10) of the nominal LIDT. In CW lasers it is not the LIDT, but the average power-handling capability (i.e., the energy) that is of essence. Long before damage takes place, the heating can cause a distortion of the surface which, in turn, can result in a loss of power, in mode and focusing problems, and even in material fracture.

The LIDT of an element can be increased through preconditioning. This consists of a special irradiation protocol applied to the component prior to its first use. On the other hand, some materials exhibit accumulation effects, e.g., through color centering, which reduce the LIDT on repeated exposure.

The development of laser coatings with high LIDT is so important that conferences have been held in Boulder, Colorado on this topic every year since 1969. For more detail the reader is referred to the proceedings of these conferences, as well as to a draft international standard on this topic. An increasing number of thin-film vendors include in their catalogues LIDT information on all or some of their products. Independent LIDT test services are now provided by several commercial companies and publicly funded institutions.

Mechanical Properties

Optical multilayer coatings are frequently required to operate under severe mechanical and environmental conditions. A number of standards deal with the substrate and coating quality (MIL-0-13830B), the adhesion of coatings (MIL-M-13508C; MIL C 48497), their abrasion resistance (MIL-C-675A, MIL C 675C, MIL-C-14806-A, MIL C 48497), hardness (MIL-M-13508C), and resistance to humidity (MIL-C-675A, MIL-C-14806-A, MIL-810-C,

MIL C 48497), salt solution (MIL-C-675-A), and salt spray (MIL-M-13508-C, MIL-C-14806-A). Most of these standards are reprinted in Rancourt's book.[3] Depending on the application, multilayer coatings may be required to meet one or more of these standards. An overview of the subject of stresses and hardness of thin films on a substrate has been recently published by Baker and Nix.[51]

Analytical Analysis Methods

In addition to optical and mechanical measurements, multilayer coatings can be subjected to a number of analytical measurements. These include Auger electron spectroscopy, energy dispersive x-ray analysis, Rutherford backscattering, secondary ion mass spectrometry, transmission electron microscopy, and x-ray photoelectron spectroscopy.[52,53] Some of these tests are destructive. However, when a multilayer coating is subjected to these tests, they yield fairly accurate information about the number of layers in a system, and on the composition, thickness, and structure of the individual layers.[54]

42.6 ANTIREFLECTION COATINGS

Effect of Surface Reflections on Performance of Optical Systems

The reflectance of an interface between two nonabsorbing media of refractive indices n_1 and n_2 is given by

$$R = \left[\frac{n_1 - n_2}{n_1 + n_2}\right]^2 \tag{31}$$

An expression for the total transmittance T_o of a nonabsorbing plane-parallel plate that takes into account the effect of multiple internal reflections within the plate can be obtained from Eq. (9):

$$T_o = \frac{1 - R}{1 + R} \tag{32}$$

Of this light only a fraction $(1 - R)^2$ passes through the plate without undergoing any reflections.

An expression for the transmission of a number of such plates placed in series is of interest. It helps to explain the effect of multiple reflections between the various plates on the performance of devices such as triple glazings and photographic objectives. It can be shown[5] that the total transmittance T_{total} of m plates placed in series is given by

$$T_{\text{total}} = \frac{T_o}{m - T_o(m - 1)} \tag{33}$$

The amount of light transmitted directly T_{direct}, is

$$T_{\text{direct}} = (1 - R)^{2m} \tag{34}$$

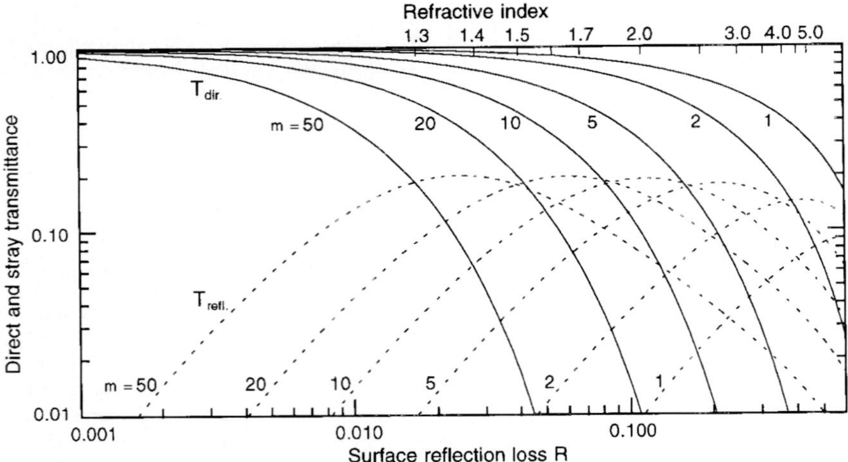

FIGURE 7 Transmittance of a system of m parallel plates of surface reflectances R for directly transmitted radiation T_{direct} and for radiation that suffers multiple reflections before transmission T_{stray}. The upper x axis is calibrated in terms of the refractive index of the plates for the special case when the plates are in air.

The light T_{stray} that undergoes multiple reflections before transmission is responsible for spurious images and stray light in the image plane. It is given by

$$T_{\text{stray}} = T_{\text{total}} - T_{\text{direct}} \tag{35}$$

The variation of T_{direct} and T_{stray} with R, for several values of m, is shown in Fig. 7. The refractive indices of the plate material plotted against the upper x axis assume that the plates are in air.

It will be seen that even for a relatively small number of low-refractive-index plates the ratio $T_{\text{stray}}/T_{\text{direct}}$ becomes significant. This means that, in an image-forming system under unfavorable conditions, the stray light can completely obscure the image.[55] Second, even in nonimaging optical systems, the loss of light $(1 - T_{\text{direct}} - T_{\text{stray}})$ can become quite prohibitive.

Both these problems can be overcome by reducing the surface reflection through the application of suitable antireflection coatings to the plate boundaries. Since antireflection coatings with zero reflectance across the whole spectrum cannot be constructed, the spectral reflectance $R(\lambda)$ of antireflection coatings is usually chosen to minimize the integral

$$\int_{\lambda} R(\lambda)I(\lambda)S(\lambda)\,d\lambda \tag{36}$$

where $I(\lambda)$ and $S(\lambda)$ are the spectral-intensity distribution of the incident radiation and the spectral sensitivity of the detector, respectively. A low reflectance is thus needed only in the spectral region in which $I(\lambda)S(\lambda)$ is significant. It should be emphasized here that in addition to this requirement, for most applications, antireflection coatings must be mechanically very tough, withstand drastic climatic and thermal variations, and stand up to the usual lens-cleaning procedures. Some examples of improvements in the performance of image-forming and non-image-forming optical systems obtained through the use of antireflection coatings are given by Mussett[55] and by Faber et al.[56]

Antireflection coatings can be based on homogeneous layers or on inhomogeneous

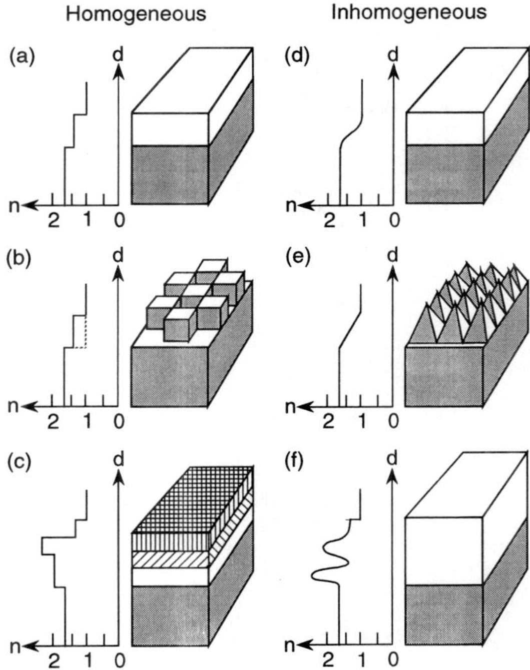

FIGURE 8 Structure and effective refractive index profiles of various types of antireflection coatings; (*a*), (*b*), (*c*) homogeneous single-layer, digital, and multilayer AR coatings; (*d*), (*e*), (*f*) simple inhomogeneous-layer, structured, and more complex inhomogeneous-layer AR coatings.

coatings. One can further classify them into single-layer, digital, or structured, and homogeneous multilayer or complex inhomogeneous layer coatings (Fig. 8*a* to *d*). Because of their industrial importance, antireflection coatings for the visible and infrared spectral regions have been the subject of much research and development. Two books have been written on this topic[57,58] and there exists a very extensive literature in scientific and technical journals. For a review of this literature and for a systematic discussion of antireflection coatings, the reader is referred to the excellent review articles by Cox and Hass[59] and by Mussett and Thelen.[55] In this section, only a brief summary will be given of the results obtained thus far, intended to aid in the selection of antireflection coatings for particular applications. The calculated data is presented on a logarithmic scale. The relative wave-number scale facilitates the calculations of the width of the effective region of coating in different parts of the spectrum.

Antireflection Coatings Made of Homogeneous Layers

The single homogeneous-layer antireflection coating (Fig. 8*a*) was the first antireflection coating and perhaps is still the most widely used. Theoretically, it should be possible to obtain a zero reflectance at one wavelength with single dense films. However, because of a lack of suitable low-index coating materials, this cannot be realized in practice for substrates with refractive indices less than about 1.9. Nevertheless, even with the available

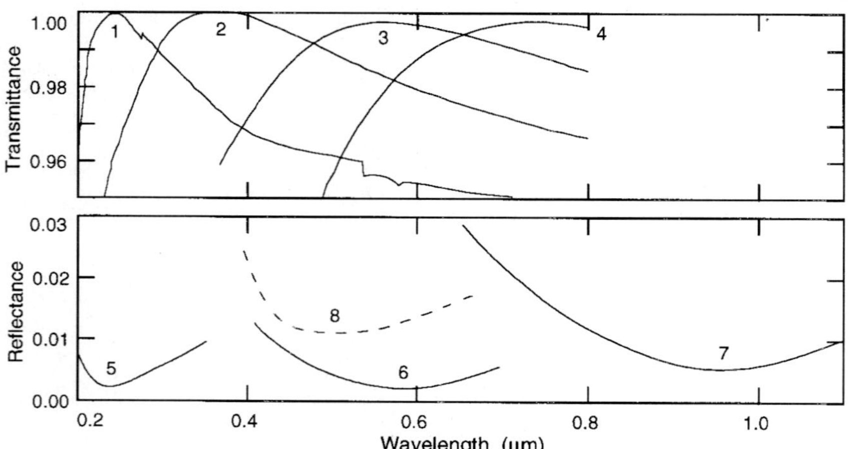

FIGURE 9 Performance of single-layer porous homogeneous antireflection coatings on various substrate surfaces. Porous silica on fused silica (1, 3, 5); on glass (6); on KDP (7); SF8 (4). Porous MgF_2 on fused silica (2). A conventional single-layer antireflection coating on glass is shown for comparison (8). *(Curve 1 after Wilder;*[60] *curve 2 after Thomas;*[62] *curves 3 and 4 after Thomas;*[63] *curves* 5–7 *after O'Neill;*[61] *curve 8 after Balzers.*[406]*)*

materials, a very useful reduction in reflection in a broad spectral region is obtained for all common glass types, the reflectance never rising above that of the uncoated surface (Fig. 9, curve 8). With the sol-gel method it is possible to produce homogeneous porous oxide and fluoride films with very low refractive indices.[60–63] Such films have excellent optical characteristics (Fig. 9) and have laser damage thresholds that are considerably higher than those produced by conventional means. However, these gains are at the expense of mechanical strength and long-term stability. Low-refractive-index coating materials can also be simulated by the deposition or etching of subwavelength structures (Fig. 8*b, e*). The effective refractive index depends on the volume fraction occupied by the structures.[64]

If more than one layer is used, all the degrees of freedom could be used to either (1) obtain a more complete antireflection in one particular spectral region; (2) increase the width of the spectral region over which the reflectance is generally low; or (3) obtain a coating in which the low reflectance is very uniform across the spectrum.[65,66] Vlasov has shown with the aid of a diagram of the type shown in Fig. 10 that even for a two-layer antireflection coating there exists a large number of refractive-index combinations which will yield zero reflectance at one wavelength.[57] As the number of layers and the overall thickness of the antireflection coating increases, it becomes possible to find solutions for a particular problem which not only fully meet the most important above desirata and almost satisfy the others, but are also based on the use of the mechanically most satisfactory coating materials.

The conditions that are satisfied by the refractive indexes and thicknesses of various types of antireflection coatings are given in Table 1. In a few cases where they are very complicated, reference is made to a paper in which they are set out in full. The calculated transmittance curves of antireflection-coated surfaces of glass (Fig. 11*a* to *c*), quartz (Fig. 11*d*), germanium (Fig. 11*e* and *f*), silicon (Fig. 11*g*), and other infrared materials (Fig. 11*h*) utilize refractive indices that for the most part correspond to real coating materials, and hence the curves represent realistic solutions rather than the theoretically best possible ones. The actual refractive indices used in the calculations (and optically thicknesses where they do not correspond to a multiple of $\lambda/4$) are given in Table 1.

Figure 11*a* shows the performance of several antireflection coatings on glass for

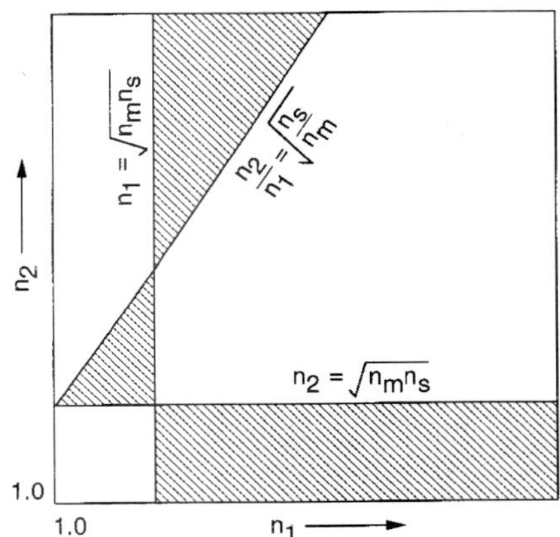

FIGURE 10 Refractive-index combinations (shaded areas) of two-layer antireflection coatings with which a zero reflectance at one wavelength can be achieved.

applications which require the highest possible efficiency for a limited wavelength region. Of these, the coating 2.1 has probably found the greatest acceptance. A typical measured-performance curve for such a coating is shown in Fig. 12. Antireflection coatings with a low reflectance in a broad spectral region are shown in Fig. 11*b* and 13. Coatings of the type 3.4 probably find the widest application in practice. The performance of commercial coatings for quartz substrates is shown in Fig. 14.

Solutions listed in Table 1 presume that coating materials with the required refractive indices exist. This is frequently not the case and the thin-film designer must seek solutions based on available coating materials. For example, Furman[67] gives a series of practical two-, three-, and four-layer solutions for ten different values of the substrate index ranging between 1.5 and 4.0, while Stolov[68] provides seven-layer solutions for ten values of n_s, $1.46 \le n_s \le 1.82$.

Willey has given a useful empirical expression which relates the average reflectance R_{avg}, in the visible and near-infrared spectral region $\lambda_{min} \le \lambda \le \lambda_{max}$ to the overall optical thickness $\Sigma(nd)$ of the coating of an antireflection coating composed of layers of refractive indices n_M, n_H and a single, outermost layer of index n_L:[69]

$$R_{avg} = 0.01 \cdot \left[\frac{\lambda_{max}}{\lambda_{min}}(n_L - 1)\right]^{3.4} \left(\frac{\lambda_{max}}{\Sigma(nd)}\right)^{0.63} [(1.2 - \Delta n)^2 + 0.42] \tag{37}$$

The preceding expression is valid for the following parameter values:

$$1.5 \le \frac{\lambda_{max}}{\lambda_{min}} \le 3.0 \qquad 1.17 \le n_L \le 1.46 \qquad 1.38 \le n_M, n_H \le 2.58$$
$$1.0 \le \frac{\Sigma(nd)}{\lambda_{max}} \le 3.0 \qquad 0.4 \le \Delta n = n_H - n_M \le 1.2 \tag{38}$$

A method for the design of optimum or near-optimum two-material antireflection coatings for a given substrate, coating materials, and overall thickness has recently been described.[70]

Since for some applications the color introduced into an optical system by antireflection coatings is of paramount importance, it has been the subject of many studies.[71–74] One way

TABLE 1 Some Antireflection Coatings

(The incident medium in all cases is air.)

Type	Conditions or reference	Substrate material	n_s	n_1 (n_1d_1/λ)	n_2 (n_2d_2/λ)	n_3 (n_3d_3/λ)	n_4 (n_4d_4/λ)	n_m
1.1	$n_1 = \sqrt{n_s n_m}; \quad n_1 d_1 = \frac{\lambda}{4}$	Glass Irtran II InSb InAs Ge Si	1.51 2.20 4.00 3.40 4.10 3.50	1.38 1.59 2.20 1.85 2.20 1.85				1.00 1.00 1.00 1.00 1.00 1.00
2.1	$\tan^2 \delta_1 = \frac{(n_1^2(n_m - n_s)(n_s n_m - n_2^2)}{(n_s n_2^2 - n_m n_1^2)(n_s n_m - n_1^2)}$ $\tan^2 \delta_2 = \frac{n_2^2(n_m - n_s)(n_s n_m - n_1^2)}{(n_s n_2^2 - n_m n_1^2)(n_s n_m - n_2^2)}$	Glass Quartz Ge	1.51 1.48 4.10	2.30 (0.0524) 2.09 (0.0947) 1.35 (0.0951)	1.38 (0.3250) 1.48 (0.3255) 4.10 (0.0586)			1.00 1.00 1.00
2.2	$n_2^2 n_s = n_1^2 n_m; \quad n_1 d_1 = n_2 d_2 = \frac{\lambda}{4}$	Glass Ge	1.51 4.10	1.70 3.30	1.38 1.57			1.00 1.00
2.3	$n_1 n_2 = n_m n_s; \quad n_1 d_1 = n_2 d_2 = \frac{\lambda}{4}$	Si	3.5	2.20	1.35			1.00
2.4	$n_1^2 - \frac{n_1 n_s}{2 n_2 n_m}(n_m^2 + n_2^2)(n_1 + n_2) + n_2 n_s^2 = 0$ $\frac{1}{2} n_1 d_1 = n_2 d_2 = \frac{\lambda}{4}$	Glass	1.51	1.70	1.38			1.00
2.5	$n_1 d_1 = n_2 d_2 = \frac{\lambda}{4}$ (see Kard, Ref. 529)	Glass	1.55	1.484	1.32			1.00
3.1	$n_3 n_s = n_2^2 = n_m n_s;$ $n_1 d_1 = n_2 d_2 = n_3 d_3 = \frac{\lambda}{4}$	Ge	4.1	3.30	2.20	1.35		1.00
3.2	$n_1 n_3 = n_2 \sqrt{n_m n_s};$ $n_1 d_1 = n_2 d_2 = n_3 d_3 = \frac{\lambda}{4}$	Glass	1.53	1.80	2.14	1.47		1.00
3.3	$n_2^2 = n_m n_s, \; n_m n_1^2 = n_3^2 n_s$ $n_1 d_1 = n_2 d_2 = n_3 d_3 = \frac{\lambda}{4}$	Si	3.45	2.56	1.86	1.38		1.00
3.4	$n_3^2 n_s = n_m n_1^2;$ $n_1 d_1 = \frac{1}{2} n_2 d_2 = n_3 d_3 = \frac{\lambda}{4}$	Glass Quartz	1.51 1.48	1.65 1.65	2.10 2.10	1.38 1.38		1.00
3.5	$n_3^2 n_s = n_m n_1^2,$ $\frac{1}{3} n_1 d_1 = \frac{1}{2} n_2 d_2 = n_3 d_3 = \frac{\lambda}{4}$	Glass	1.51	1.659	2.20	1.38		1.00
3.6	(see Thetford, Ref. 530.)	Glass	1.52	1.80 (0.1799)	2.20 (0.4005)	1.38 (0.2402)		1.00

TABLE 1 Some Antireflection Coatings (*Continued*)

(The incident medium in all cases is air.)

Type	Conditions or reference	Substrate material	n_s	n_1 (n_1d_1/λ)	n_2 (n_2d_2/λ)	n_3 (n_3d_3/λ)	n_4 (n_4d_4/λ)	n_m
3.7	$n_1d_1 = n_2d_2 = n_3d_3 = \frac{\lambda}{4}$ (see Kard, Ref. 529.)	Glass	1.55	1.53	1.454	1.32		1.00
4.1	$n_1n_4 = n_2\sqrt{n_mn_s}$; $n_1d_1 = n_2d_2 = \frac{1}{2}n_3d_3 = n_4d_4 = \frac{\lambda}{4}$	Glass	1.51	1.38	1.548	2.35	1.38	1.00
4.2	$n_1n_4 = n_2n_3 = n_mn_s$; $n_1d_1 = n_2d_2 = n_3d_3 = n_4d_4 = \frac{\lambda}{4}$	Ge	4.0	2.96	2.20	1.82	1.38	1.00
4.3	$n_1d_1 = n_2d_2 = n_3d_3 = n_4d_4 = \frac{\lambda}{4}$ (see Kard, Ref. 529.)	Glass	1.55	1.846	2.289	2.014	1.32	1.00
4.4	(Design derived from Kard, Ref. 529.)	Glass	1.55	1.656 (0.2417)	1.888 (0.2463)	1.832 (0.2390)	1.38 (0.2424)	1.00
10.1	$n_ld_s = n_s - (11 - l)\frac{n_s - n_9}{10}$, $l = 1, 2, \ldots, 10$; $n_1d_1 = n_2d_2 = \cdots = n_{10}d_{10} = \frac{\lambda}{4}$ [see Kard, Ref. 529.]	Ge	4.00	$n_1 = 3.735$	...	$n_9 = 3.735$	$n_{10} = 1.35$	1.00

to avoid the problem is to utilize coatings that are particularly achromatic (Fig. 11*c*). Thus, for example, 50 surfaces coated with antireflection coating 4.4 would have a transmittance of 78 ± 3 percent across the whole visible spectrum. Nevertheless, of the coatings shown in Fig. 11*c*, only the single-layer antireflection coating is being used extensively. Antireflection coatings can also be used to correct the residual color of lens systems.[75]

Often transmittance rather than reflectance measurements are performed to evaluate antireflection coatings. In general, it is incorrect to assume that $T = 1 - R$. For instance, the transmission of antireflection-coated infrared materials depends not only on the efficiency of the antireflection coating but also on the thickness and temperature of the material. This is because of the finite scatter and absorption in such materials and because of the dependence, in some cases, of the latter on temperature. The measured spectral transmittances of three common antireflection-coated infrared materials at room temperature are shown in Fig. 15. Curves for other materials are given by Cox and Hass.[76]

Inhomogeneous and Structured Antireflection Coatings

The interface between two media with refractive indices n_1 and n_2 can be antireflection-coated over a very broad spectral region by the application of a transition layer with an index that changes continuously from n_1 to n_2 (Fig. 8*d*). Many different refractive-index profiles have been investigated in the past.[77–80] Although some of these profiles are more

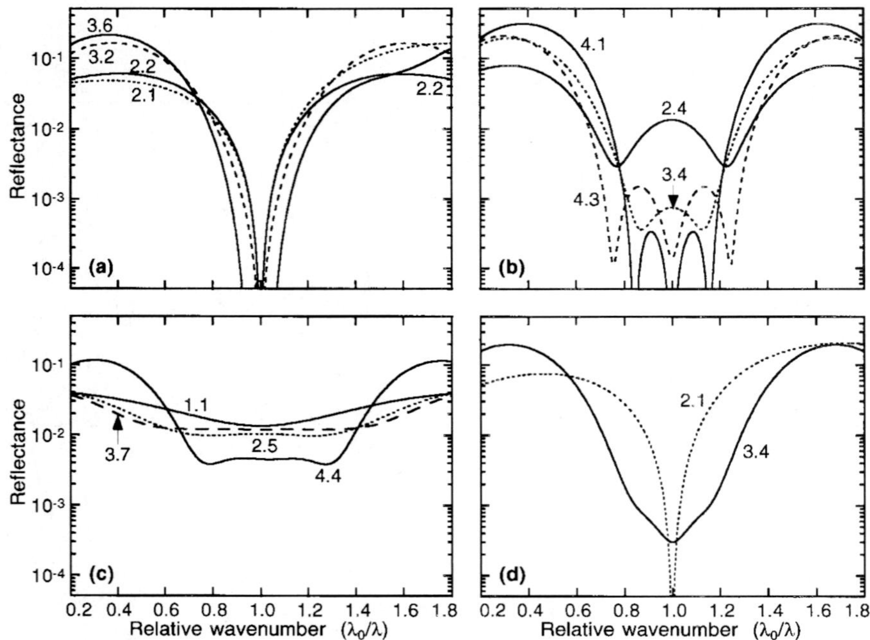

FIGURE 11 Calculated performance of various antireflection coatings. The numbers identifying the individual curves refer to Table 1. (*a*) High-efficiency antireflection coatings for glass; (*b*) broadband antireflection coatings for glass; (*c*) highly achromatic antireflection coatings for glass; (*d*) antireflection coatings for quartz.

effective than others, all reduce the reflectance to a fraction of a percent over the spectral region in which the coatings are transparent and do not scatter excessively, and for which the optical thickness of the layer is at least one half-wavelength (Fig. 16*a*). A further advantage of inhomogeneous antireflection coatings is that they are not sensitive to the angle of incidence.[81] Processes used for the production of inhomogeneous antireflection coatings include various additive, subtractive, additive/subtractive, and replication methods. Excellent reviews of this topic exist.[82,83]

In the additive method, relatively dense inhomogeneous layers of varying compositions of two or more compounds are formed on the substrate by physical or chemical deposition processes. Such coatings are mechanically more durable than any other described in this section. However, because of a lack of coating materials with refractive indices lower than about 1.35, solid inhomogeneous layers are not very suitable for the antireflection coating of air-glass interfaces. The several different inhomogeneous antireflection coatings of this type described in the past do not offer any special advantages over those composed of homogeneous layers.[84–86] The situation is different in the case of high-index materials (Fig. 17). An even lower reflectance can be achieved by ending the inhomogeneous layer when its index is equal to the square of that of the lowest-index coating material available. It is then possible to complete the coating by depositing an additional homogeneous quarter-wave-thick layer of that material (Fig. 17*a*).

A dense inhomogeneous layer can be approximated by a series of homogeneous layers of gradually decreasing refractive indices (curve 10.1, Fig. 11*f*).[87,88] Such layers can be prepared by evaporating a series of appropriate mixtures of two coating materials or, without mixing, by using the Herpin equivalent-index concept[89] to simulate intermediate

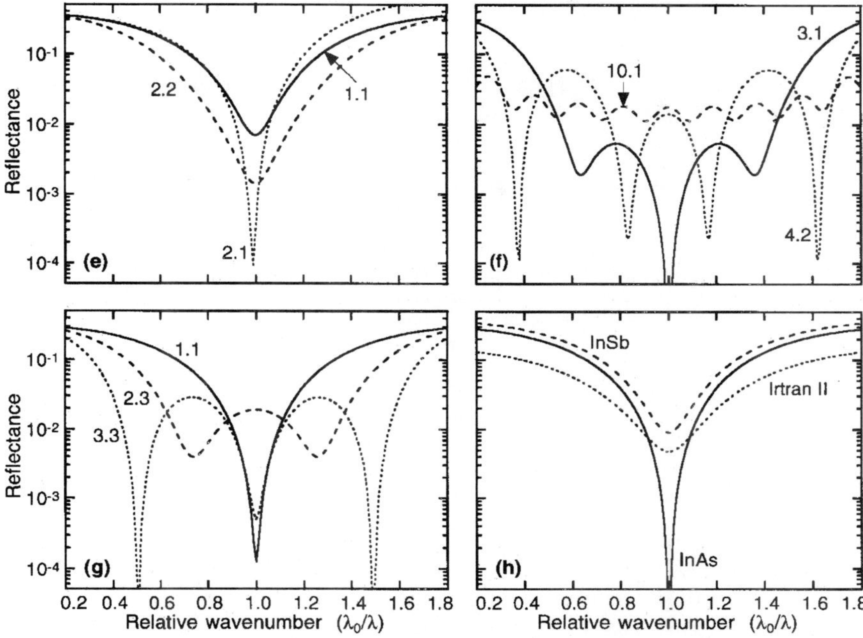

FIGURE 11 (*Continued*) (*e*) antireflection coatings for germanium; (*f*) broadband antireflection coatings for germanium, (*g*) antireflection coatings for silicon; (*h*) single-layer antireflection coatings for Irtran II, InAs, and InSb.

refractive indices. An even more practical solution is to replace the inhomogeneous layer by a series of thin homogeneous layers of two materials only (Fig. 18).[90]

In another additive process, a refractive index variation down to a value of 1.0 is achieved by depositing onto the substrate microspheres of transparent oxides or fluorides which form pyramidlike clumps (Fig. 8*e*).[91] If losses due to scattering are to be low, the average lateral size of the features of this structure must be a small fraction of the shortest wavelength for which the coating is to be effective. A reflectance of 0.3 percent can be achieved, but the films are fragile.

The subtractive methods are attractive because they do not require expensive deposition equipment. The surface to be antireflection-coated is leached and/or etched to

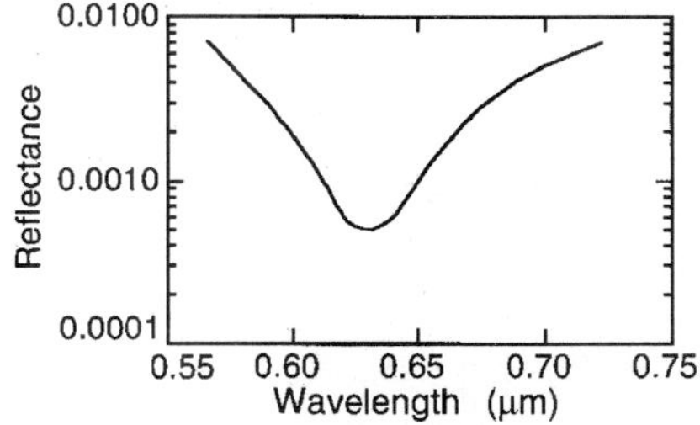

FIGURE 12 Reflectance of a high-efficiency antireflection coating on glass. *(After Costich.[407])*

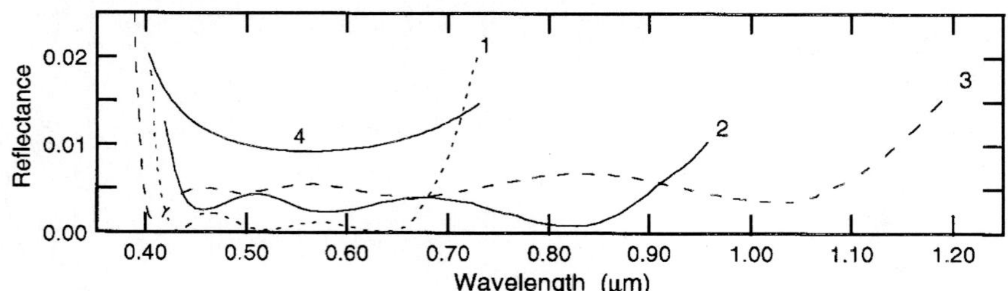

FIGURE 13 Reflectances of three broadband antireflection coatings. *(Curve 1, after Turner;*[408] *curve 2, Optical Coating Laboratory;*[409] *curve 3, Thin Film Lab.*[410]*)* Curve 4 *(after Balzers*[406]) corresponds to a single-layer AR coating on glass and is shown here for comparison purposes.

form a porous transition layer in which the index varies with thickness. Not all optical materials can be treated in this way. However, special phase-separable glasses have been developed that lend themselves well to this process.[92–95] Fairly durable antireflection coatings with a reflectance of less than 0.5 percent for the 0.35- to 2.5-μm spectral region have been produced in this way (Fig. 19*a*). Some other materials, such as Lexan, Mylar, and CR-39 plastic, require an ion implantation pretreatment before the etching can be applied.[78,96]

In the technologically important additive/subtractive method, a single glasslike film is first deposited by a sol-gel process onto the surface that is to be antireflection coated. The composition of the film is such that, after phase separation, it can be readily leached and/or etched to form a porous microstructure with a controlled refractive index gradient.[97,98] This eliminates the need for the use of expensive phase separation glass components. Such coatings have reflectances as low as 0.13 percent and a laser damage threshold that is four times higher than that of coatings produced by conventional physical vapor deposition techniques (Fig. 19*b*).[99] Variants of this process exist.

Microstructured surfaces can also be produced in polymeric and similar materials by a replication process from a suitable cast. Average reflectances in the visible of the order of 0.3 percent have been reported for surfaces treated in this way (Fig. 19*c*).[100]

Recently there has been a renewed interest in antireflection coatings in which the variable porosities of leached or etched layers are simulated by dense regular-shaped structures formed by photochemical or mechanical means. Clapham and coworkers appear to have been the first to demonstrate such devices.[101] They applied a photoresist to a surface, exposed it to two orthogonal sets of ultraviolet interference fringes, and then

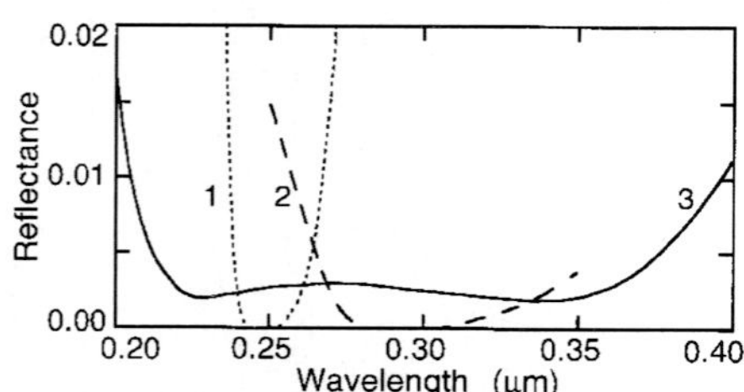

FIGURE 14 Ultraviolet antireflection coatings on fused silica. *(Curve 1, after TechOptics;*[411] *curve 2, after Reynard Corp.;*[412] *curve 3, after Spindler & Hoyer.*[413]*)*

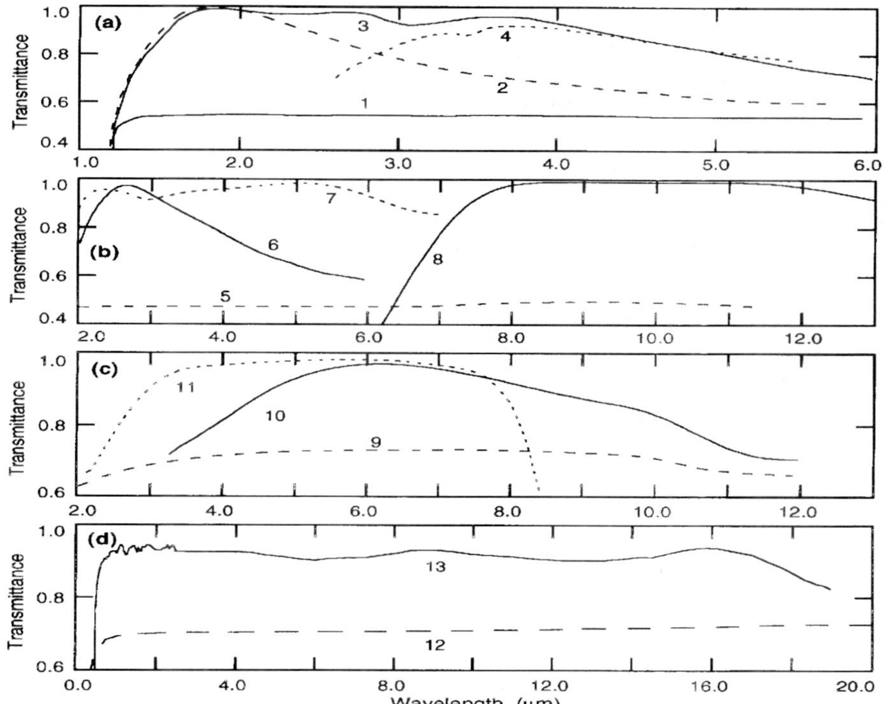

FIGURE 15 Transmittance of plates of infrared materials antireflection-coated on both sides: (*a*) Silicon—Curve 1: uncoated plate of 1.5-mm thickness; curve 2: single $\lambda_o/4$ layer of SiO_2 ($\lambda_o = 1.8$ μm); curve 3: $\lambda_o/4$ coatings of MgF_2, and CeO_2 ($\lambda_o = 2.2$ μm); curve 4: hard carbon layer; (*b*) Germanium. Curve 5: uncoated plate; curve 6: single $\lambda_o/4$ layer of SiO ($\lambda_o = 2.7$ μm); curve 7: $\lambda_o/4$ coatings of MgF_2, CeO_2, and Si ($\lambda_o = 3.5$ μm); curve 8: environmentally stable, high laser damage threshold three-layer design based on ThF_4 and Ge. (*c*) Irtran II. Curve 9: uncoated plate of 2-mm thickness; curve 10: single $\lambda_o/4$ coating of CeF_3; curve 11: $\lambda_o/4$ layers of MgF_2 and SiO ($\lambda_o = 4.2$ μm). (d) Zinc selenide. Curve 12: uncoated plate; curve 13: extremely broadband AR coating composed of 398 layers produced by molecular beam deposition. *(Curves 1–3, 5–7, 9–11 after Cox and Hass*[59]*; curve 4 after Balzers;*[406] *curve 8 after Oh;*[414] *curve* 13 *after Fisher.*[532]*)*

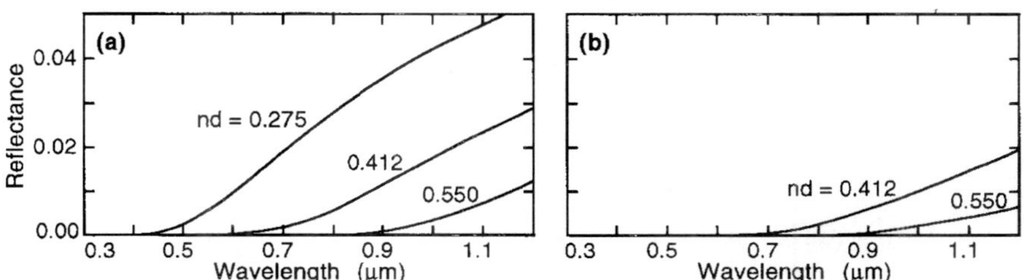

FIGURE 16 Calculated reflectances of the interfaces between two media antireflection-coated with inhomogeneous layers of indicated thickness and (complex) refractive indexes that vary smoothly, from the index of one medium to that of the other: (*a*) two nonabsorbing media of indexes 1.52 and 2.36; (*b*) glass-chromium interface with refractive indexes 1.52 and 2.26-i0.43, respectively. *(After Anders and Eichinger.*[111]*)*

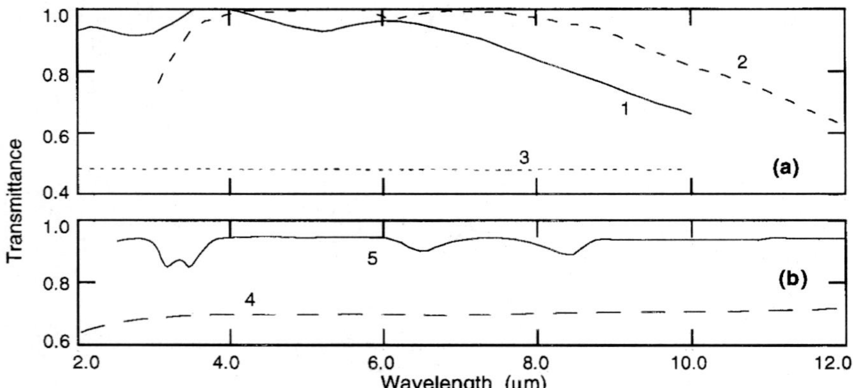

FIGURE 17 (*a*) Measured transmittance of a germanium plate coated on both sides with inhomogeneous antireflection coatings. Curve 1: 1.2-μm-thick film with an index that changes gradually from that of Ge to that of MgF_2 *(after Jacobsson*[415]*)*; curve 2: 1.76-μm-thick film with an index that varies from 4.0 to 1.5, overcoated with a 0.74-μm-thick homogeneous MgF_2 layer *(after Scheuerman*[188]*)*; curve 3: transmittance of an uncoated plate. (*b*) Measured transmittance of a thalium bromo-iodide plate. Curve 4: uncoated; curve 5: coated on both sides with ten 2.5-μm-thick homogeneous layers with refractive indices that vary between 1.3 and 2.29 and that are obtained through the evaporation of suitable NaF-CdTe mixtures *(after Kuznetsov and Perveyev*[416]*)*.

developed it to form a regular array of protuberances that could be optionally enlarged by additional ion-beam etching. Such surfaces can reduce the reflectance to less than 0.3 percent in the visible part of the spectrum (Fig. 19*d*).[102] The theory of such structures has been investigated by several workers[103,104] and devices for wavelengths extending into the mm and sub-mm region have been fabricated.[64,105]

Antireflection Coating of Absorbing and Amplifying Media

Antireflection coatings for glasses and semiconductors in regions of weak absorption, and for present-day laser materials in which k, the imaginary part of the complex refractive index [Eq. (25)], is small and negative, differ little from those described previously.[106]

The reduction of the reflectance of opaque materials for architectural, decorative, and

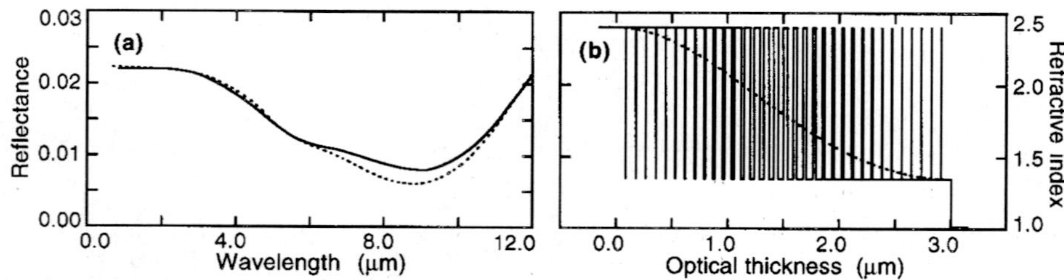

FIGURE 18 Calculated performances and refractive index profiles of an inhomogeneous-layer coating and its two-material equivalent. *(After Southwell.*[90]*)*

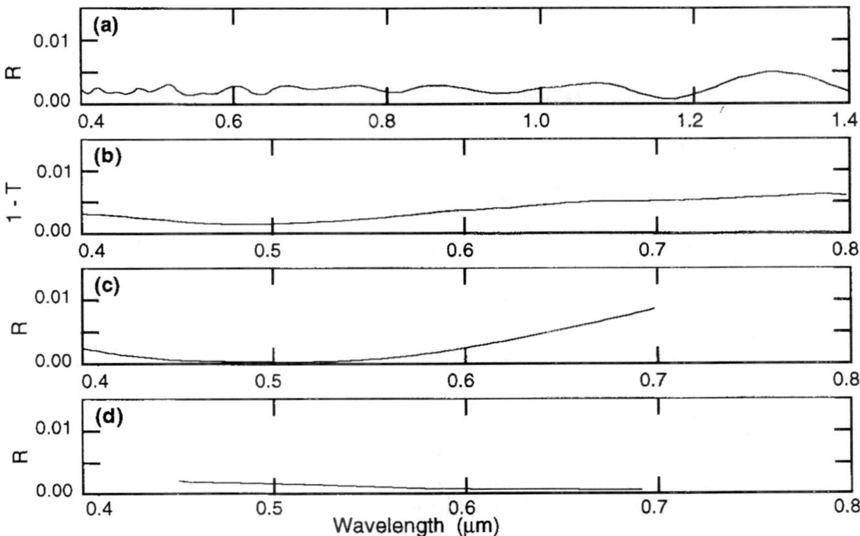

FIGURE 19 Measured performance of various broadband inhomogeneous antireflection coatings: (*a*) subtractive process—leached and etched phase-separation glass *(after Asahara*[94]*)*; (*b*) additive/subtractive process—leached and etched phase-separable film deposited onto fused silica *(after Yoldas*[98]*)*; (*c*) replication process on a cellulose acetate butyrate surface *(after Maffitt*[100]*)*; (*d*) structured antireflection coating in photoresist *(after Wilson*[102]*)*.

technical purposes leads to a corresponding increase in the absorption.[107,108] This can be utilized to improve the efficiency of radiation detectors and to control the solar-absorptance and thermal-emittance characteristics of surfaces (see, for example, Refs. 109 and 110). A measured example of the reduction in the reflectance of a metal surface attainable with a homogeneous nonabsorbing layer is shown in Fig. 20.

Inhomogeneous transition layers whose refractive index and extinction coefficient change gradually from the values of one of the media to those of the other are also very effective.[111] The calculated reflectance of a glass-chromium interface coated in this way, useful for blackening of prism faces, lens edges, scales, etc., is shown in Fig. 16*b*. Metal-air interfaces can be treated a little less effectively with single layers because of the lack of suitable low-index materials.

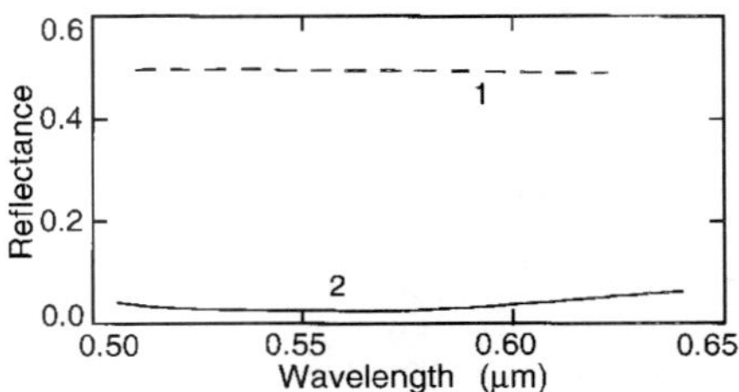

FIGURE 20 Antireflection coating of opaque metals. Curves 1 and 2: reflectance of chromium and of chromium with a ZnS layer. *(After Lupashko and Shkyarevskii.*[107]*)*

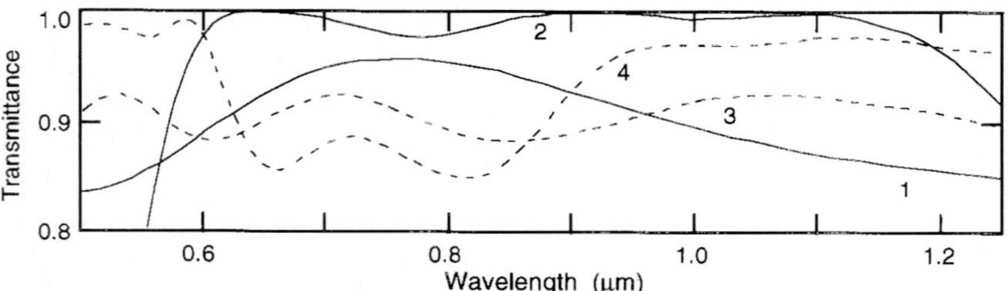

FIGURE 21 Calculated single-surface transmittance of antireflection coatings incorporating a conducting layer. Curves 1 and 2: glass of refractive index 1.755 coated with a 0.566-μm-thick SnO_2 layer before and after three dielectric layers were added to complete the antireflection coating; curves 3 and 4: as above except that the refractive index of the glass and the metric thickness of the SnO_2 film are 1.516 and 0.200 μm, respectively. *(After Veremey and Gorbunova.*[417]*)*

Antireflection Coating of Surfaces Carrying a Thin Film

For some applications it is necessary to deposit a certain film onto a glass surface. The objectionable reflectance that such a film would normally introduce can be avoided by incorporating it into an antireflection coating. Thus, for example, Fig. 21 shows the calculated transmittances of two conducting layers before and after inclusion in antireflection coatings. The use of homogeneous and of inhomogeneous antireflection coatings with absorbing layers for ophthalmic purposes is described by Katsube et al.[112] and Anders.[113]

Antireflection Coatings at Nonnormal Angle of Incidence

The calculated performances at 0, 45, and 60° incidence of the three commercially most important antireflection coatings and of a 10-layer coating for germanium are shown in Fig. 22. The deterioration with angle of incidence is particularly severe for a narrowband

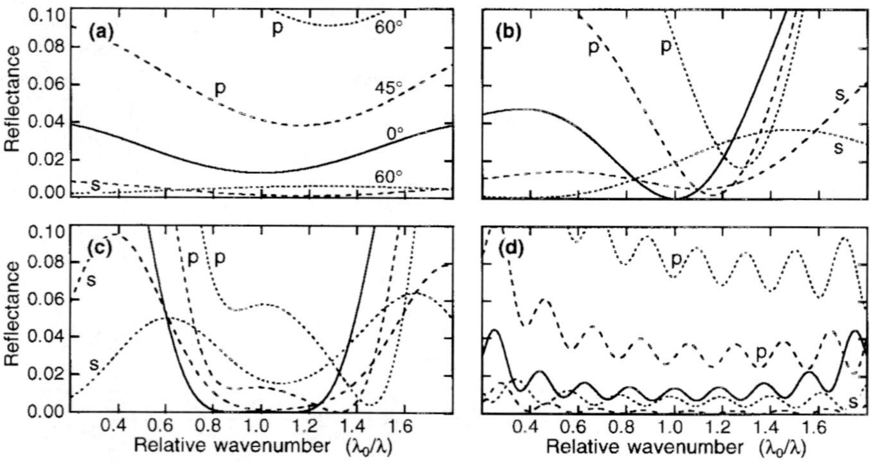

FIGURE 22 Calculated reflectance of (*a*) one-, (*b*) two-, (*c*) three-, and (*d*) ten-layer antireflection coatings (systems 1.1, 2.1, 3.4, and 10.1 in Table 1) at angles of incidence of 0° (solid curve), 45° (dashed curve), and 60° (dotted curve) for light polarized parallel and perpendicular to the plane of incidence.

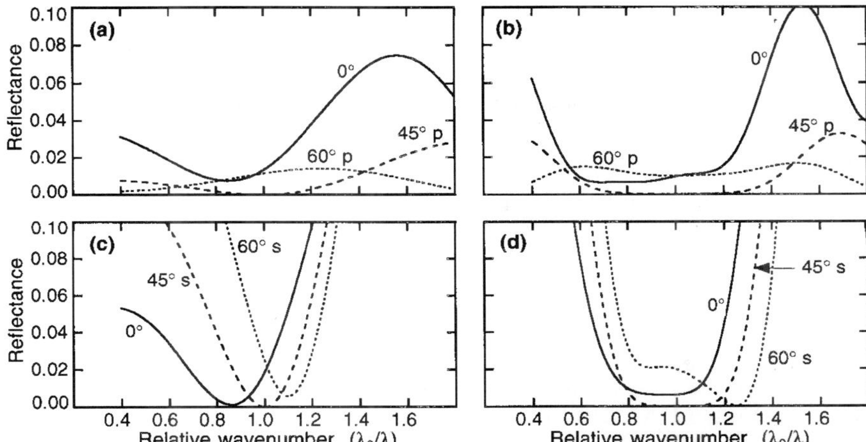

FIGURE 23 Calculated reflectance at 0° (solid curve), 45° (dashed curve), and 60° (dotted curve) of (*a*) and (*c*) two-layer, and (*b*) and (*d*) three-layer antireflection coatings designed for light incident at 45° and polarized parallel or perpendicular to the plane of incidence. *(After Turbadar.*[418,419]*)*

high-efficiency antireflection coating. Figure 22*d* suggests that the closer the design of an antireflection coating approximates an inhomogeneous transition layer (see "Inhomogeneous and Structured Antireflection Coatings" in Sec. 42.6), the less angle-dependent is its performance.

To design an antireflection coating for one angle of incidence and one plane of polarization, the effective thicknesses and refractive indices [Eqs. (18) and (19)] of its layers should satisfy the relations set out in Table 1. In practice, small departures from those conditions are required to optimize the performance with good coating materials. Calculated curves for two sets of two- and three-layer coatings designed for use at 45° are shown in Fig. 23.

If the obliquely incident radiation is unpolarized, a compromise is necessary. The effective thicknesses are matched for the design angle, but the refractive-index conditions set out in Table 1 are satisfied for normal incidence since they cannot be satisfied for both polarizations at the same time (Fig. 24).

Achromatic antireflection coatings can be designed that are optimized for both

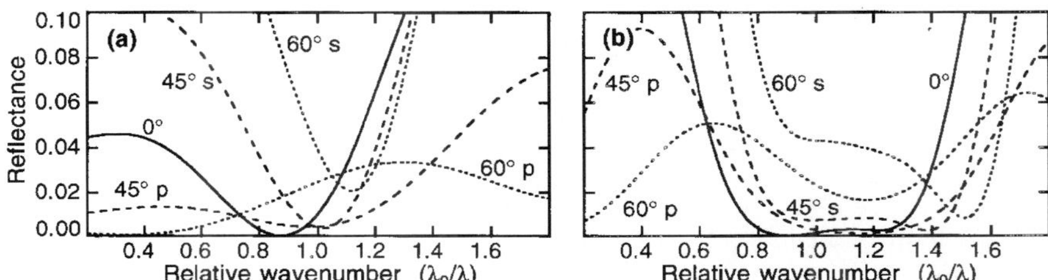

FIGURE 24 Calculated reflectance at 0° (solid curve), 45° (dashed curve), and 60° (dotted curve) of (*a*) two-layer and (*b*) three-layer antireflection coatings (systems 2.1 and 3.4 in Table 1) for use with unpolarized light, with the thicknesses of the layers optimized for an angle of incidence of 45°.

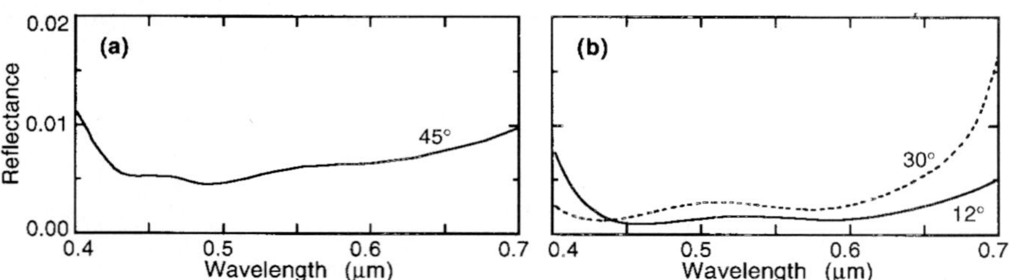

FIGURE 25 Average reflectances for unpolarized light of antireflection coatings designed (*a*) for 45° incidence and (*b*) for use with a convergent light cone of semiangle of 30°. *(Optical Coating Laboratory.*[409]*)*

polarizations at the same time or that are suitable for use over a wide range of angles of incidence (Fig. 25*b*).[22,114] But the problem becomes more difficult the greater the spectral and angular ranges required, especially if angles of incidence greater than 60° are involved. The performance of a single-wavelength antireflection coating that is effective for angles of incidence of up to 60° is shown in Fig. 26. The design of antireflection coatings for very high angles of incidence has been considered by Monga.[115,116]

42.7 TWO-MATERIAL PERIODIC MULTILAYERS—THEORY

Nonabsorbing $[AB]^N$ and $[AB]^N A$ Multilayer Types

Let a periodic multilayer be composed of N periods AB, where A, B represent layers of refractive indices n_A, n_B and optical thicknesses $n_A d_A$, $n_B d_B$. The most general representation of the complete multilayer is

$$\underset{1}{AB} \cdot \underset{2}{AB} \cdot \underset{3}{AB} \cdots \underset{N}{AB} = [AB]^N \tag{39}$$

In practice it is customary to write $[HL]^N$ or $[LH]^N$, depending on whether n_A is greater or less than n_B, respectively. The spectral-reflectance curve of the multilayer $[AB]^N$ will lie

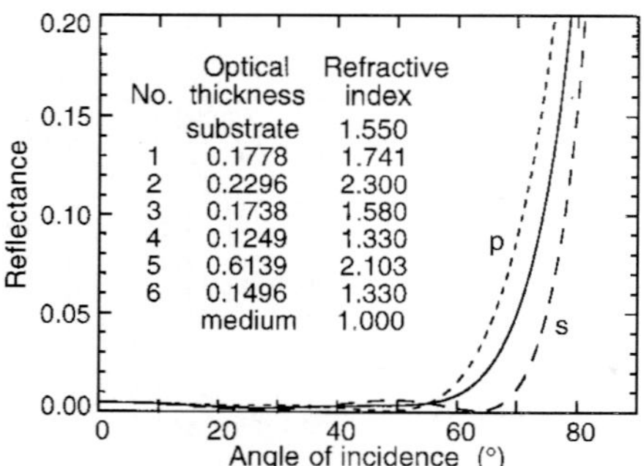

FIGURE 26 Wide-angle antireflection coating for $\lambda = 0.6328$ μm. *(After Dobrowolski and Piotrowski.*[420]*)*

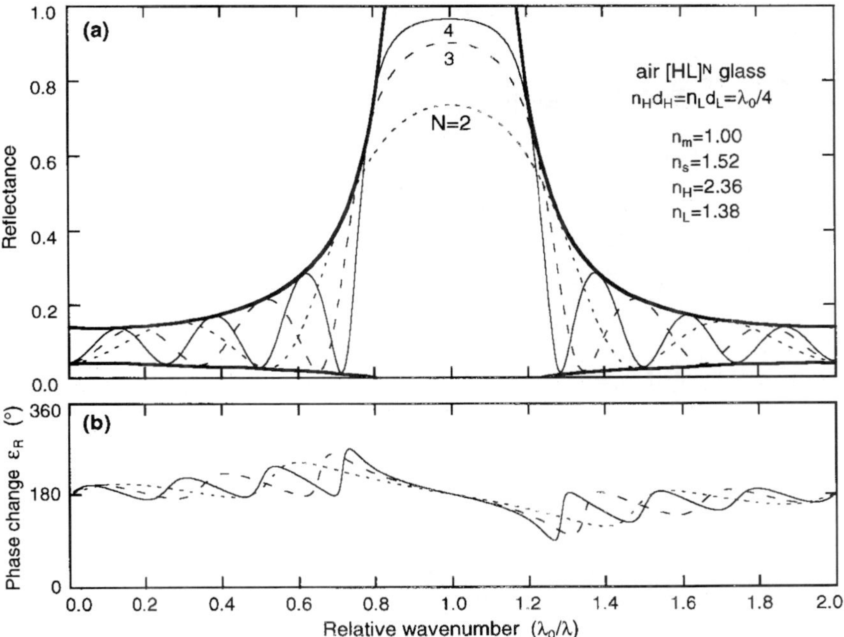

FIGURE 27 (*a*) Calculated reflectance and (*b*) phase change on reflection of periodic multilayers of the type $[HL]^N$. *H* and *L* stand for high- and low-refractive-index layers of quarter-wave thickness at $\lambda_o = 1.0\ \mu\text{m}$. The heavy lines are the envelopes of the reflectance curves.

within a pair of envelopes which, at normal incidence, depend only on $n_A d_A$, $n_B d_B$, n_A, n_B, n_s, and n_m.[117,118] For $n_s = n_m$ the lower of the envelopes becomes $R = 0$. The envelopes contain high-reflectance zones within which the reflectance at each wavelength increases monotonically with the number of periods, approaching 1.0 as *N* tends to infinity (Fig. 27). Outside these high-reflectance zones the curves exhibit subsidiary maxima and minima whose number depends on $n_A d_A / n_B d_B$ and which increases with *N*. The first-order high-reflectance zone occurs at a wavelength λ_1 given by

$$n_A d_A + n_B d_B = \frac{\lambda_1}{2} \tag{40}$$

and subsequent zones occur at wavelengths λ_q ($\lambda_1 > \lambda_2 > \lambda_3 \cdots$) given by

$$N(n_A d_A + n_B d_B) = q\frac{\lambda_q}{2} \qquad q = 2, 3, 4, \ldots \tag{41}$$

providing that at these wavelengths

$$n_A d_A,\ n_B d_B \neq p\frac{\lambda_q}{2} \qquad p = 1, 2, 3, \ldots \tag{42}$$

It is thus possible, by choosing suitable-thickness ratios, to arrange for or to suppress high-reflectance zones in several spectral regions at the same time. This is useful in the design of broadband reflectors, cut-off filters, hot and cold mirrors (Sec. 42.9) and laser

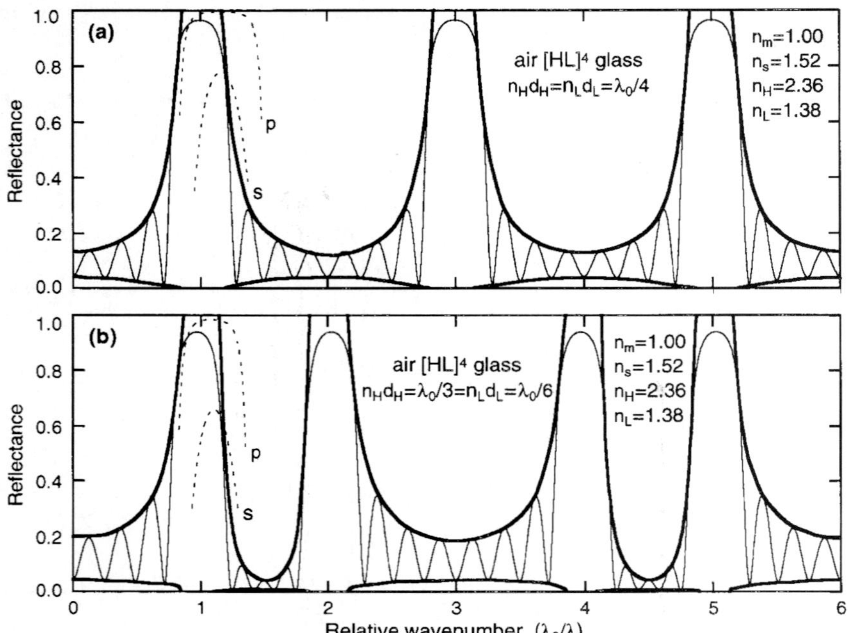

FIGURE 28 Calculated reflectance curves of periodic multilayers with (*a*) 1:1 and (*b*) 2:1 thickness ratios. The dotted curves represent the reflectance of polarized radiation incident at an angle of 60°.

reflectors, etc. Typical curves for thickness ratios 1:1 and 2:1 are given in Fig. 28. Plotted on a λ_o/λ scale, they show symmetry about all wavelengths λ for which $n_A d_A$ and $n_B d_B$ are both equal to some integral multiple of $\lambda/4$.

Maximum Reflectance. For a given refractive index ratio (n_A/n_B) and number of periods N, the highest reflectance occurs whenever $n_A d_A$, $n_B d_B$ are each equal to an odd multiple of $\lambda/4$. It is given by

$$R_{\max} = \left[\frac{n_m/n_s - (n_A/n_B)^{2N}}{n_m/n_s + (n_A/n_B)^{2N}}\right]^2 \tag{43}$$

Intermediate reflectances are obtained for the related symmetrical multilayers $[AB]^N A$:

$$R_{\max} = \left[\frac{n_m n_s/n_A^2 - (n_A/n_B)^{2N}}{n_m n_s/n_A^2 + (n_A/n_B)^{2N}}\right]^2 \tag{44}$$

Results for a number of such cases are given in Fig. 29. Intermediate reflectances can be obtained by changing the refractive index of any of the layers in the stack.

Explicit expressions for R for other thicknesses are complicated.

Phase Change on Reflection. The dispersion of the phase change on reflection from periodic all-dielectric multilayers is much greater than that of metal reflectors. Unless corrected for, it will lead to errors in some metrological and interferometric applications. Like reflectance, it varies rapidly outside the high-reflection zone (Fig. 27*b*). Within the high-reflectance zone, it changes almost linearly with wavenumber and is 180° at $\lambda_o/\lambda = 1$.

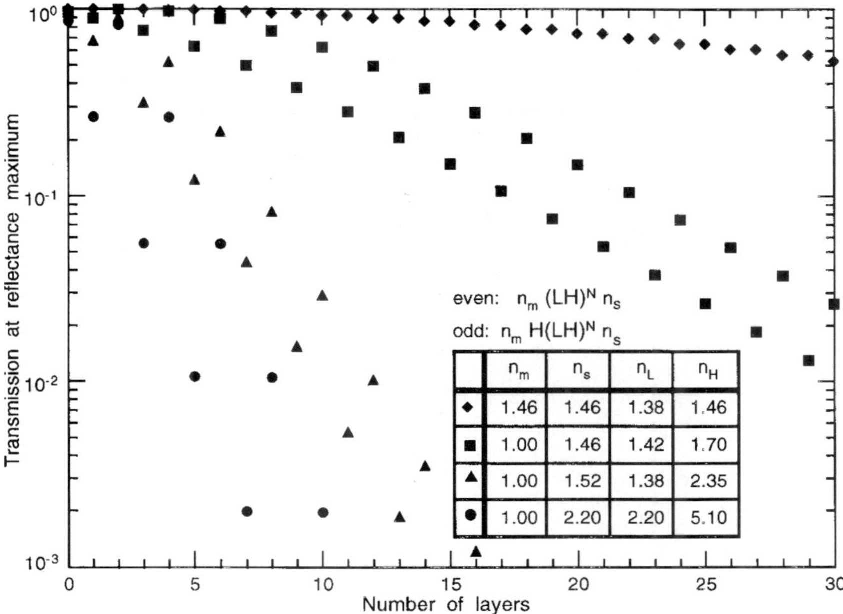

FIGURE 29 Transmittance at the reflection maximum of quarter-wave stacks composed of MgF_2 and SiO_2 in the visible region (◆), MgF_2 and MgO in the ultraviolet (■), MgF_2 and ZnS in the visible region (▲), and ZnS and PbTe in the infrared (●).

The slope of this portion of the graph increases and approaches a limiting value as N tends to infinity. The limiting values (in degrees per unit wavelength) is given by

$$\left[\frac{d\varepsilon}{d\lambda}\right]_\lambda = \begin{cases} \dfrac{180n_m}{\lambda_o(n_H - n_L)} & \text{for } n_H \\ \dfrac{180n_H n_L}{\lambda_o n_m(n_H - n_L)} & \text{for } n_L \end{cases} \tag{45}$$

depending on whether the light is first incident on a high-(n_H) or low-(n_L) refractive-index layer.[119] These values should be multiplied by 3, 5, . . . in stacks composed of $3\lambda/4$, $5\lambda/4$, . . . layers.

Böhme has shown that, by changing the refractive index of one of the layers in a quarter-wave stack, it is possible to obtain a zero value of the phase change on reflection ε at λ_o.[120]

Periodic Multilayers of the $[(0.5A)B(0.5A)]^N$ Type

The construction of such multilayers differs from that of the type $[AB]^N A$ discussed previously by having outermost layers of only half the thickness of the remaining layers.[89] The position and width of the high-reflection zones are the same in both cases, but in coatings of the $[(0.5A)B(0.5A)]^N$ type, it is possible to reduce the height of the subsidiary maxima on either one or the other side of the main reflectance maximum if $n_A d_A = n_B d_B$. Kard et al. describe the optimum choice of all the construction parameters (N, n_A, n_B, n_s,

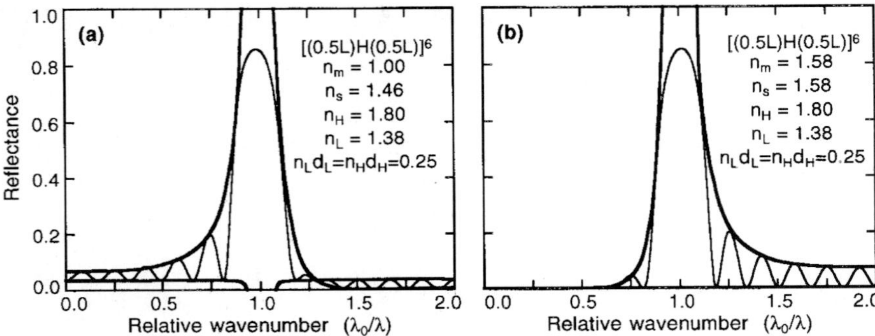

FIGURE 30 Calculated reflectance curves of two periodic multilayers with symmetrical periods of the type $[(0.5L)H(0.5L)]^6$ in which the subsidiary reflectance maxima on the short- and the long-wavelength side of the high-reflectance zone are reduced.

n_m) of such a coating.[121] If the refractive indices n_s and n_m of the surrounding media must be chosen on some other basis, then for maximum improvement of the short- and long-wavelength transmission, the refractive indexes n_A and n_B of the coating materials must satisfy

$$n_s n_m = \frac{n_A^3}{n_B} \tag{46}$$

or

$$n_s n_m = n_A n_B \tag{47}$$

respectively. Estimates of the maximum reflectance for larger values of N can be obtained from Fig. 29. Multilayers of this type are of importance in the design of cut-off filters (Sec. 42.9), and are illustrated in Fig. 30, which should be compared with the curves of Fig. 31.

Width of the High-Reflectance Zone

For a given value of (n_A/n_B), the width $\Delta\lambda_R/\lambda$ of the high-reflectance zone is greatest when $n_A d_A = n_B d_B = \lambda/4$, and it is then given by

$$\frac{\Delta\lambda_R}{\lambda} = \frac{4}{\pi}\sin^{-1}\left(\frac{1 - n_A/n_B}{1 + n_A/n_B}\right) \tag{48}$$

This width is reduced by a factor of $2N - 1$ if Nth-order quarter-wavelength layers are used. Graphs of the widths of the high-reflectance zones vs. refractive-index ratio for $\lambda/4$, $3\lambda/4$, and $5\lambda/4$ layer stacks are given in Fig. 32. Periodic multilayers with equal refractive-index ratios have high-reflectance zones of equal widths, but their reflectance curves will not be the same unless the refractive indices of the surrounding media are also increased by the same ratio (Fig. 31).

Periodic Multilayers of the $[xH \cdot (1-x)L]^N \cdot xH$ type

As already stated previously, it is not necessary for the optical thicknesses of the layers of a periodic multilayer reflector to be equal to a quarter-wave. A multilayer system of the type $[xH \cdot (1-x)L]^N \cdot xH$, where H, L correspond to quarter-wave layers of a high and low refractive index and where $0 < x < 1.0$, will have a high reflectance, providing that the

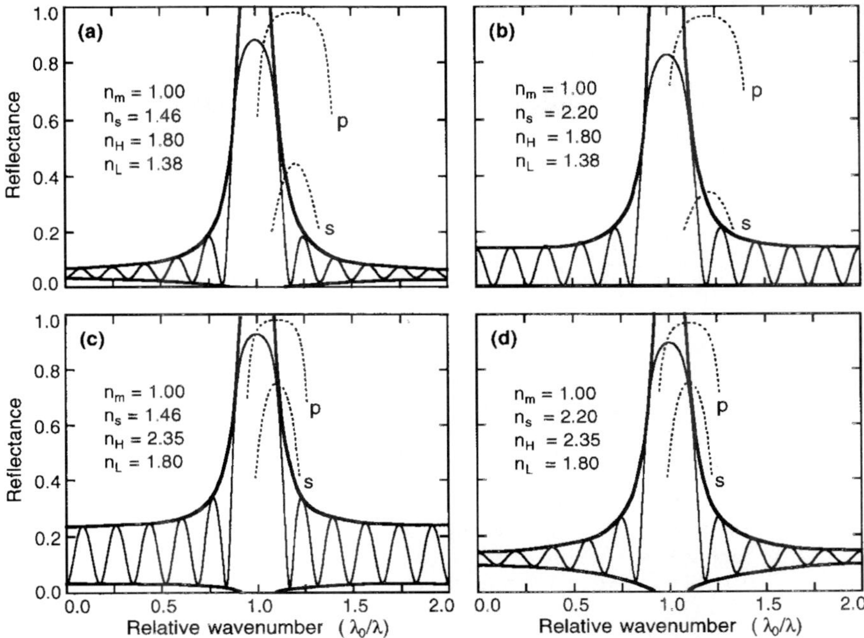

FIGURE 31 Calculated reflectance curves of two quarter-wave stacks of the type $H(LH)^N$ with different values of n_H and n_L but with the same ratio n_H/n_L, deposited onto two different substrate materials. The dotted curves represent the reflectances for light incident at 60°.

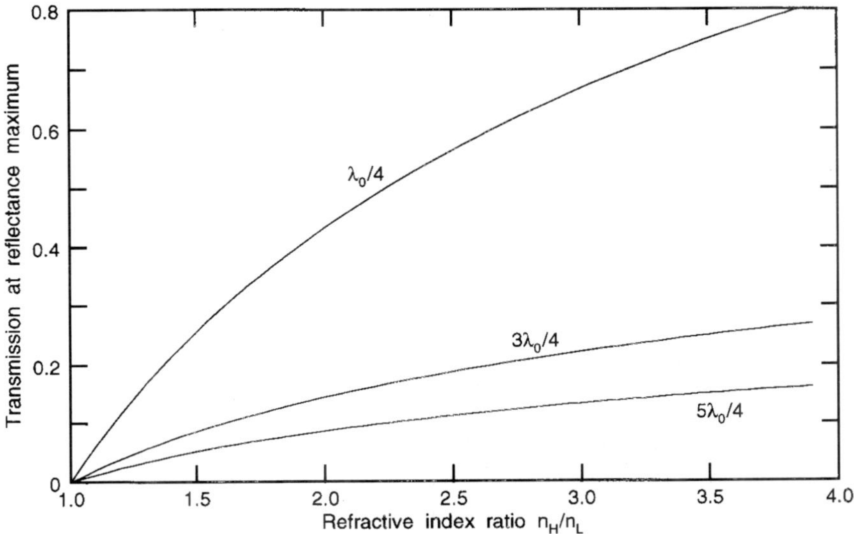

FIGURE 32 Calculated widths of high-reflectance zones of two-material periodic stacks of layers of optical thicknesses $\lambda_o/4$, $3\lambda_o/4$, and $5\lambda_o/4$ for different refractive-index ratios.

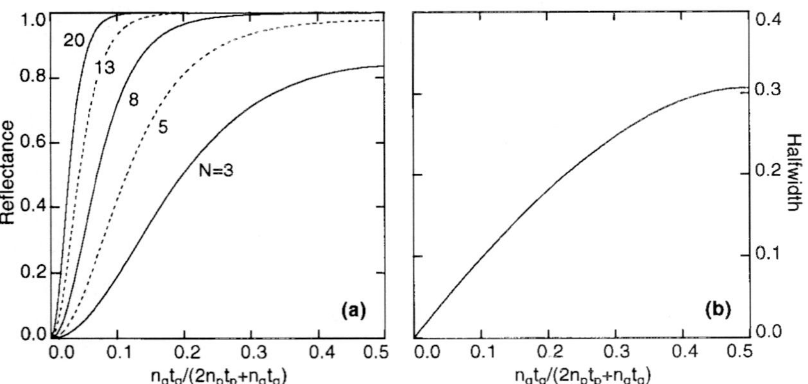

FIGURE 33 Maximum reflectance (*a*) and width of the high-reflectance zone (*b*) of periodic multilayers of the type $[xH \cdot (1-x)L]^N xH$. The curves were evaluated for refractive indices $n_H = 2.4$ and $n_L = 1.51$. (*After Shao.*[125])

number of periods N is high enough.[122–125] However, the smaller the value of x, the narrower the width of the high-reflectance zone. Given two coating materials, it is thus possible to independently select the reflectance and the width of the rejection region. Figures 33*a,b* show the relationship between $R_{\max}$, $\Delta\lambda_R/\lambda$, and x, N for values of n_H, n_L, corresponding to the refractive indices of ZnS and polyethylene at about 200 μm.

Angular Sensitivity

For some applications it is important to reduce the angular variation of the reflectance curve. This can be done by using materials with high refractive indices (Fig. 31) (see also "Angle-of-incidence Effects" in Sec. 42.9). Another way is to use periods in which the high-index film is thicker than the low-index film (Figs. 28*b* and 57*a*).

Multilayer Reflectors Made of Absorbing Materials

It is possible to achieve a high reflectance with multilayers of the $[AB]^N A$ type even when A, B correspond to absorbing materials. This is of particular interest for spectral regions for which no nonabsorbing coating materials exist. The optical thicknesses of the periods of such systems are still approximately equal to $\lambda/2$ but, for maximum reflectance, the individual thicknesses d_A, d_B may be quite different, depending on the number of periods N and on the optical constants of the materials used. Reflectors with $k_A > k_B$, have structures that are intermediate to those of quarter-wave stacks (with $k_A = k_B = 0$ and optical thicknesses of $\lambda/4$, in which constructive interference effects are maximized), and those of ideal Bragg crystals (with $k_A \gg k_B$, and in which $d_A \ll d_B$ to minimize absorption losses).

In the extreme ultraviolet (XUV) and in the soft x-ray regions, $(n-1)$ and k are much smaller than 1 for most coating materials.[126,127] To design a periodic multilayer with a high-normal-incidence reflectance for a given wavelength, it is first necessary to choose a material $(n_B - ik_B)$ with the lowest possible extinction coefficient.[128] Next, a second, chemically compatible material $(n_A - ik_A)$ is selected with the lowest extinction coefficient that will maximize the normal-incidence Fresnel reflection coefficient of the interface between the two materials given by:

$$r_{BA} = \frac{(n_B - n_A) - i(k_B - k_A)}{(n_B + n_A) + (k_B + k_A)} \tag{49}$$

Vinogradov and Zeldovich use a factor β_{opt} to relate the metric thicknesses of the layers A and B that yield a maximum reflectance to the overall thickness d_{opt} of the period:[129]

$$d_A = \beta_{\text{opt}} \cdot d_{\text{opt}} \quad \text{and} \quad d_B = (1 - \beta_{\text{opt}}) \cdot d_{\text{opt}} \tag{50}$$

where β_{opt} is obtained by the solution of the equation

$$\tan(\pi\beta_{\text{opt}}) = \pi\left[\beta_{\text{opt}} + \frac{n_B k_B}{n_A k_A - n_B k_B}\right] \tag{51}$$

d_{opt} is approximately equal to $\lambda/2$, but Vinogradov and Zeldovich give a more accurate expression for this quantity, as well as for the limiting reflectance R and the number of periods N required to reach that value.[129]

42.8 MULTILAYER REFLECTORS—EXPERIMENTAL RESULTS

In the calculations of the data for Figs. 27 to 33, the dispersion of the optical constants of the materials was ignored, and it was assumed that the films were absorption- and scatter-free and that their thicknesses had precisely the required values. In practice, none of these assumptions is strictly valid, and hence there are departures from the calculated values. In general, the agreement is better within the high-reflectance zone than outside.

Reflectors for Interferometers, Lasers, Etc.

The measured reflectances and transmittances of a number of quarter-wave reflectors suitable for use in Fabry-Perot interferometers are shown in Fig. 34. The transmission of a

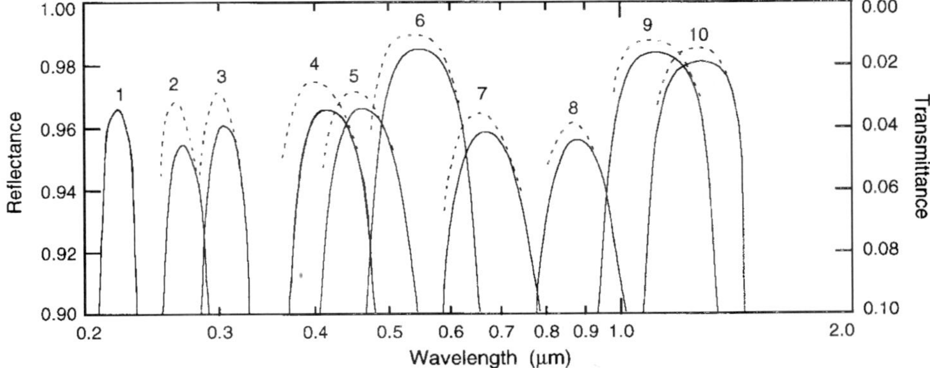

FIGURE 34 Measured spectral reflectance and transmittance curves of periodic all-dielectric reflectors of the type $[HL]^N H$ for the ultraviolet, visible, and near-infrared parts of the spectrum. Curve 1: 27 layers of MgO and MgF_2; curves 2 and 3: 11 and 13 layers of PbF_2 and cryolite, respectively; curves 4, 5, 7, and 8: 7 layers of ZnS and cryolite; curves 6, 9, and 10: 9 layers of ZnS and cryolite. *(Curve 1 after Apfel,*[421] *curves 2 to 10 after Hefft et al.*[422]*)*

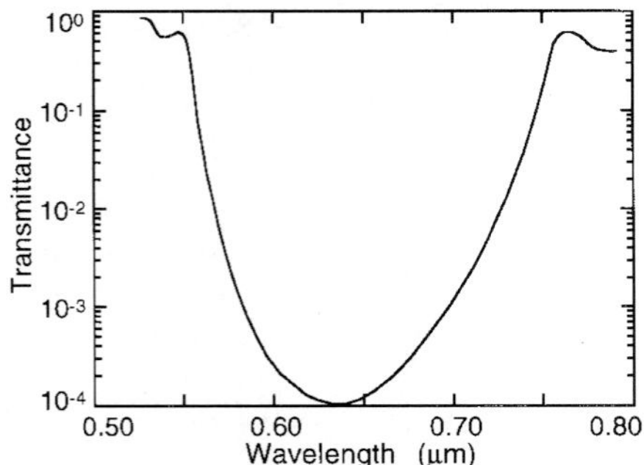

FIGURE 35 Measured spectral transmittance of a laser reflector. *(After Costich.[407])*

typical commercial laser reflector for $\lambda = 0.6328\ \mu m$ is shown in Fig. 35. The measured reflectances of a number of highly reflecting coatings for the infrared spectral region are shown in Fig. 36.

Both "soft" laser coatings that can be dissolved in weak acids and "hard" coatings that can be removed only through polishing are available commercially.

Effects of Imperfections

Thickness Errors. Small errors in the thickness of the layers of a quarter-wave stack have only a very small effect on the reflectance and on the phase change on reflection of the multilayer within the high-reflectance zone, but they do affect the performance outside the zone.[130,131] In fact, the effect may be quite serious: thickness variations can give rise to an apparent lack of flatness of the substrate surface.[132–134]

Dispersion. The most noticeable effects of dispersion in quarter-wave stacks of a given multilayer type are the increase in the peak reflectance with decrease in wavelength (see Fig. 34) and the asymmetry of the maxima on either side of the main reflectance maximum.

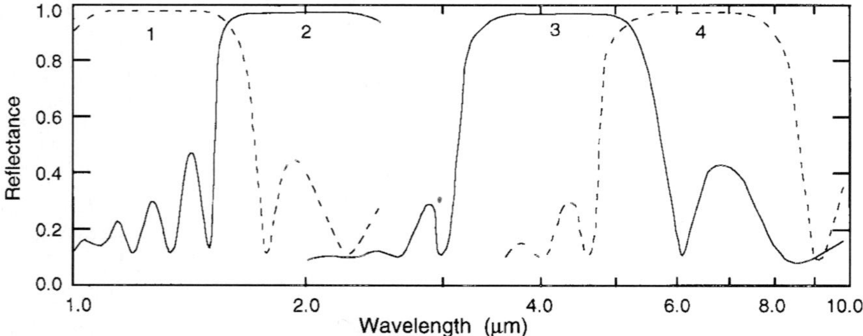

FIGURE 36 Measured spectral-reflectance curves of all-dielectric reflectors for the infrared made of stibnite and chiolite. Curves 1 and 2: multilayers of the type $[HL]^4H$ on glass; curves 3 and 4: type $[(0.5L)H(0.5L)]^4$ on barium fluoride. *(After Turner and Baumeister.[156])*

Absorption. The separation between the transmission and reflectance curves in Fig. 34 is mostly due to absorption. These losses can limit the usefulness of the reflectors for some applications. Thus, for example, in interference filters and Fabry-Perot interferometers they lead to a reduction in the peak transmissions and limit the attainable half-widths. In optical information-storage devices they set a limit to the highest reflectance attainable. In lasers, the losses counteract directly the gain in the laser medium. In addition, absorption within the layers is responsible for damage to laser reflectors (see also "Laser Damage" in Sec. 42.5).

If the two materials used for the construction of periodic quarter-wave stacks have small but finite extinction coefficients, the resulting absorption at first reduces both the transmission and reflection coefficients.[117,132,135,136] With an increase in the reflectance of a multilayer of the $[HL]^N H$ type, the absorption occurs more and more at the expense of the reflection coefficient, approaching a limiting value of

$$A = -\delta R = \frac{2\pi n_m}{n_H^2 - n_L^2}(k_H + k_L) \tag{52}$$

that is independent of the number of layers.[2] The corresponding expression for a multilayer $[HL]^N$ in which a low refractive index faces the incident medium is

$$A = -\delta R = \frac{2\pi(n_L^2 k_H + n_H^2 k_L)}{n_m(n_H^2 - n_L^2)} \tag{53}$$

Figure 37 shows the spectral variation of experimentally determined maximum reflectances and limiting losses of various quarter-wave multilayer reflectors.

The absorption in periodic multilayers composed of materials having finite extinction coefficients can be reduced below the values given by Eqs (52) and (53) if a structure of the $[xH \cdot (1-x)L]^N \cdot xH$ type is used.[124]

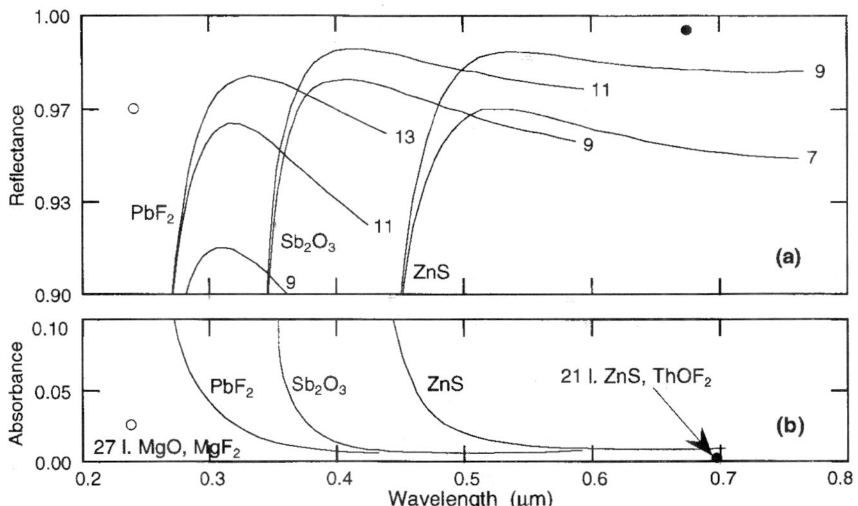

FIGURE 37 Variation with wavelength of the observed peak reflectance of 7-, 9-, 11- and 13-layer quarter-wave stacks made of PbF_2, Sb_2O_3, or ZnS with cryolite as the low-index material. The lower curves indicate the limiting value of the absorbance with these materials. *(After Honcia and Krebs.*[423] *Results obtained by Apfel*[421] (○) *and Behrndt and Doughty*[424] (●) *with other materials are included for comparison).*

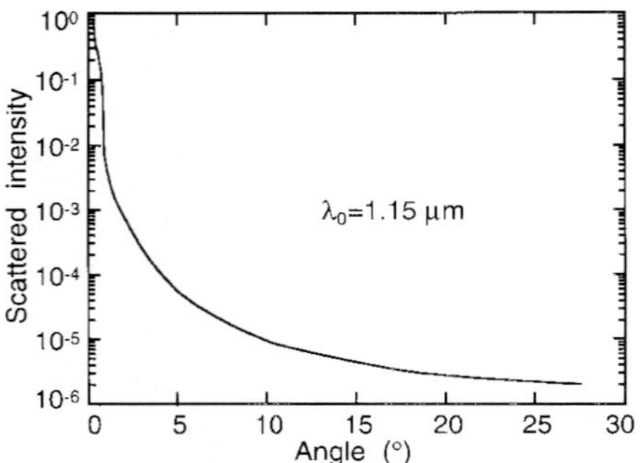

FIGURE 38 Measured intensity of the scattered radiation as a function of the angle away from the direction of specular reflection. *(After Blazey.[425])*

Surface and Interface Imperfections. In thin-film calculations, it is usually assumed that substrate surfaces and interfaces between the layers are smooth and that the layers are homogeneous. In practice, substrate surfaces and interfaces have a certain roughness and, at times, thin uniform or inhomogeneous interface layers formed between the boundaries of two layers as a result of oxidation, chemical interactions, or interdiffusion of the two coating materials. The interface layers, as a rule, are different at the AB and BA boundaries and are typically 0.0003 or 0.0010 μm thick. These and other imperfections of the layer system result in reduced reflection and/or in scatter. If ignored in the model used to represent the multilayer, they add to the discrepancies that are observed between the calculated and experimental data.

Scattering losses are particularly significant at shorter wavelengths. For this reason, there have been many theoretical and experimental investigations of scattering of surfaces and thin films. Investigations have shown that when scatter does occur, most of the light is scattered in directions that are close to that of the specularly reflected light.[117,137] The experimental results for a typical mirror are shown in Fig. 38.

Very Low Loss Reflectors. Mirrors with very low losses are required for use in laser cavities and in ring lasers. Mirrors with a combined loss L (=transmission + absorption + scatter) of the order of 5×10^{-5} are commercially available. With special manufacturing techniques, practically loss-free reflectors can now be made.

Recently a 41 quarter-wave stack made of Ta_2O_5 and SiO_2 layers with a combined loss of $L = 1.6 \cdot 10^{-6}$ corresponding to a reflectance of 0.9999984 at 0.633 μm, has been reported.[138] The absorption and scatter losses were estimated to be of the order of $1.1 \cdot 10^{-6}$. The essential starting point for the manufacture of such coatings are superpolished substrates with a surface roughness of 0.5-Å rms or less. The layers were deposited by reactive ion-beam sputtering from high-purity oxide targets in a cryogenically pumped, fully automated deposition system.[139]

Multilayers for the Soft X-ray and XUV Regions. The effect of roughness and of interface layers is particularly important in soft x-ray and XUV multilayers because the dimensions of these defects are comparable to the thicknesses of the individual layers. For this reason, much attention has been focused on the proper modeling of such coatings. Many workers use several very thin layers and the matrix method outlined in Sec. 42.3 to model roughness and interface layers.[140] Others make use of the following recursive formula for the amplitude reflectance r_j of the first j layers of the system:

$$r_j = \frac{r_{j-1} + r_{BA} \exp(2i\delta_j)}{1 + r_{j-1} r_{BA} \exp(2i\delta_j)} \tag{54}$$

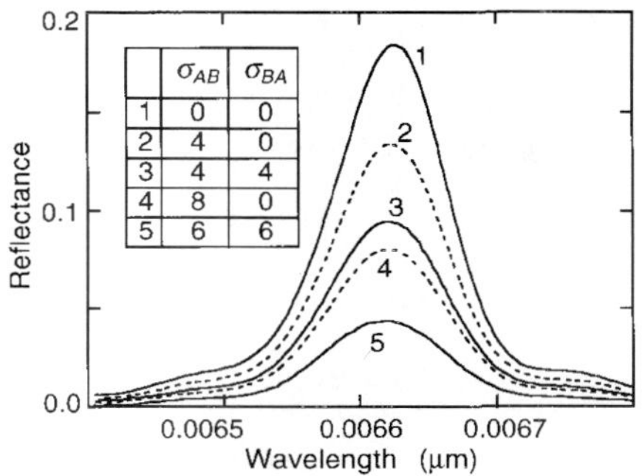

FIGURE 39 Calculated reflectances of 64 period x-ray mirrors with different imperfections σ (in Å) at the ReW-C and C-ReW interfaces. *(After Spiller.[128])*

Here δ_j is the effective optical thickness of the jth layer (Eq. 18), r_{j-1} is the amplitude reflectance of the first $(j-1)$ layers, and r_{BA} is the amplitude reflectance of the interface between the jth and $(j+1)$th layers. In this approach, if the Fresnel amplitude reflection coefficient r_{BA} is replaced by

$$r_{BA} \exp\left\{-\frac{1}{2}\left[\frac{4\pi}{\lambda_o}\sigma\mathcal{Re}(\tilde{n}\cos\tilde{\theta}_o)\right]^2\right\} \tag{55}$$

the combined effect of roughness and of interface layers σ can be allowed for.[141] In the hard x-ray region, where $n = 1$, $k = 0$ for all materials, the exponential term in the preceding expression reduces to the so-called Debye-Waller factor DW,

$$\mathrm{DW} = \exp\left[-\frac{1}{2}\left(\frac{4\pi}{\lambda_o}\sigma\cos\theta_j\right)^2\right] \tag{56}$$

The calculated reflectance of an XUV mirror, with and without the effect of surface imperfections, is shown in Fig. 39.

Narrowband Reflection Coatings

Narrowband rejection filters transmit freely all the radiation incident upon them except in one narrow spectral region in which the radiation is either wholly or partially reflected.[142] Lord Rayleigh observed a corresponding natural phenomenon in potassium chlorate crystals.[143] Subsequent experimentors reported crystals with rejection bands varying between 0.001 and 0.038 microns in width and reflectances between 33 and 99.9 percent.[144–146] But, unfortunately, at present the size of the crystals that can be grown is insufficient, and the position and width of the rejection region cannot be controlled. The same comments can be made about coextruded polymer films made of the thermoplastic materials polypropylene ($n = 1.49$) and polycarbonate ($n = 1.59$).[147,148] With thin films, these limitations can be overcome, but it is difficult to achieve the extremely narrow widths and high rejections observed in the crystals.

A quarter-wave stack of the type $[AB]^N A$ has been suggested as a model for the construction of such filters (see, for example, Refs. 145 and 149). It follows from Figs. 29 and 32 that the closer the refractive index ratio n_A/n_B is to unity, the narrower the width of the reflectance zone and the more layers required to achieve a certain rejection. Shown in

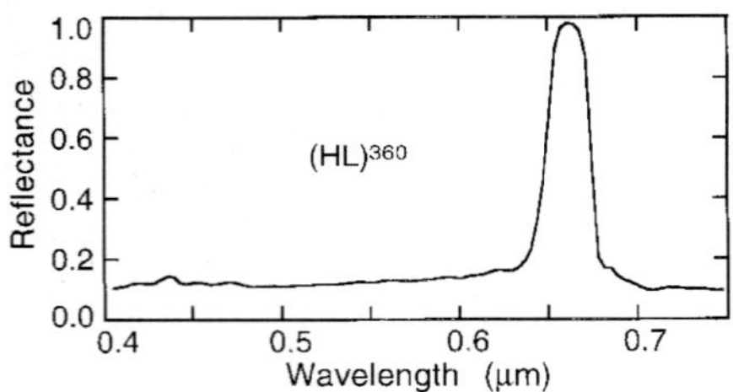

FIGURE 40 Measured reflectance of a 720-layer quarter-wave stack produced by chemical vapor deposition using materials with refractive indices of 1.575, 1.585. *(After Edmonds et al.*[150]*)*

Fig. 40 is the measured reflectance of a multilayer consisting of 720 layers that was produced by a plasma chemical deposition technique.[150] Unfortunately, the method could only deposit such coatings on the inside of a tube. To reduce the number of layers, films with higher n_A/n_B ratios could be utilized and the width reduced by using layers with thicknesses that are odd multiples of a quarter-wavelength. But this can be done only at the expense of bringing the adjacent higher- and lower-order reflection peaks closer. Resonant reflectors are an extreme example of this.

Resonant Reflectors

Even "hard" evaporated coatings cannot survive the power densities present in some high-power lasers. In the past, resonant reflectors have been used.[151–154] They consist of one or more accurately air-spaced plane-parallel plates of thicknesses of the order of millimeters made of a tough, high-optical-quality material. Because of the long coherence length of the laser radiation incident upon them, interference takes place in the same way as in thin films. Resonant reflectors may be regarded as being quarter-wave reflectors of enormously high order of interference, and all the equations given in Sec. 42.7 apply.

Resonant reflectors made of quartz and sapphire are commercially available. Since the refractive index of quartz is lower than that of sapphire, a larger number of plates is required to attain the same reflectance. On the other hand, quartz is much cheaper and is less temperature-sensitive. The calculated reflectance of one-, two-, three-, and four-plate sapphire resonant reflectors are shown in Fig. 41. In another development, diffusion-doped quartz plates are used in the construction of resonant reflectors.[155] The doping process causes the refractive index of the plate to increase smoothly in a 0.5-μm region from that of quartz to about 2.0 at the surface. As a consequence, fewer elements are needed to achieve a given reflectance.

All-dielectric Broadband Reflectors

There are several ways of obtaining a coating with a broad high-reflection region should the width attainable with quarter-wave stacks be inadequate. For a broad region with a very high-reflectance, one can superimpose two quarter-wave stacks tuned to two different wavelengths. The widest continuous high-reflectance region is attained when materials with the highest available refractive-index ratios are used and the thicknesses of the layers are so chosen that the two high-reflectance zones are contiguous (see Fig. 42*c*). For an even broader region, further quarter-wave stacks can be superimposed. If high-reflection regions overlap, special precautions must be taken to prevent the appearance of sharp reflectance

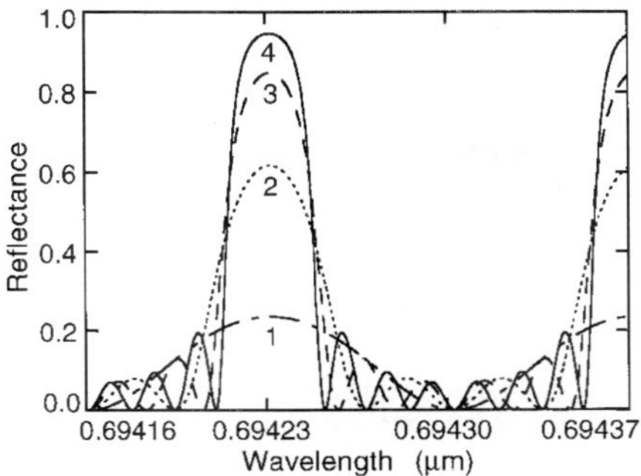

FIGURE 41 Calculated spectral reflectance of resonant reflectors consisting of one, two, three, and four sapphire plates ($n = 1.7$). The optical thicknesses of the plates and of the air spaces between them were assumed to be 1.7 mm.

minima in the high-reflectance region.[156] It is not easy to obtain a very uniform, moderately high reflectance in this way. Another approach is to deposit onto the substrate a series of alternating high- and low-refractive-index films of gradually increasing or decreasing thicknesses (Fig. 43). A broad high-reflection region can also be obtained with a multilayer in which the layers are different multiples of $\lambda/4$ of a selected wavelength (see Fig. 44). Finally, it has been suggested that multilayers with 10:1 high-reflectance regions might be produced by depositing hundreds of layers of random thicknesses made of two materials that are nonabsorbing throughout the spectral range of interest.[157] As yet, experimental data for this approach have not been presented.

If only a relatively small increase in the high-reflectance region is required, or if a uniform but only moderately high reflectance is required, the desired result can be achieved by modifying the thicknesses and refractive indices of a quarter-wave stack using a computer refinement program or by the addition of achromatizing $\lambda/2$ layers. The measured performances of several such reflectors are shown in Figs. 45 and 46. Broadband

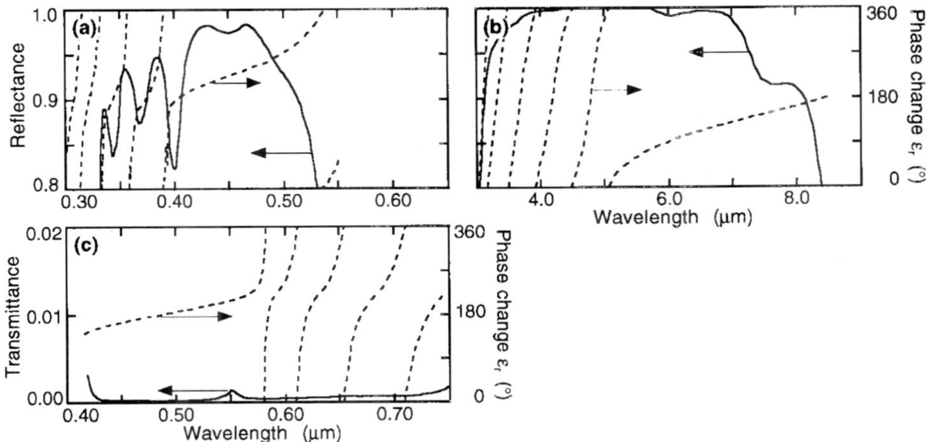

FIGURE 42 Measured reflectance of all-dielectric broadband reflectors consisting of two superimposed quarter-wave stacks with contiguous high-reflectance zones. The dotted curves in Figs. 42 to 46 represent the calculated phase changes on reflection. *[(a) and (b) after Turner and Baumeister;[156] (c) after Perry.[426]]*

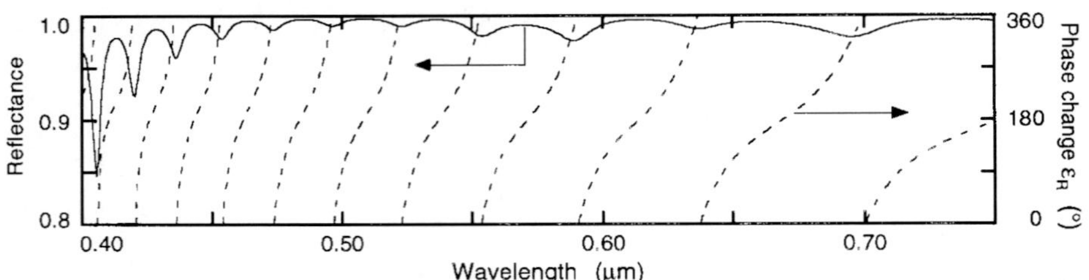

FIGURE 43 Calculated reflectance of an all-dielectric broadband reflector consisting of 35 layers made of a low- and a high-index material and having optical thicknesses that vary in a geometric progression. *(After Heavens and Liddell.*[427]*)*

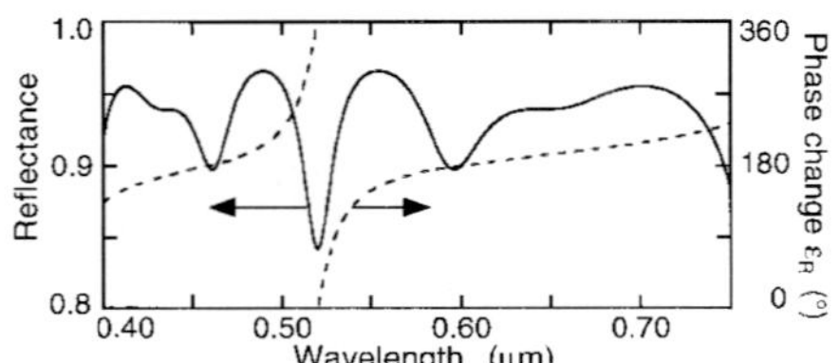

FIGURE 44 Calculated reflectance of an all-dielectric broadband reflector consisting of 11 layers all of which have optical thicknesses that are various multiples of 0.13 μm. *(After Elsner.*[428]*)*

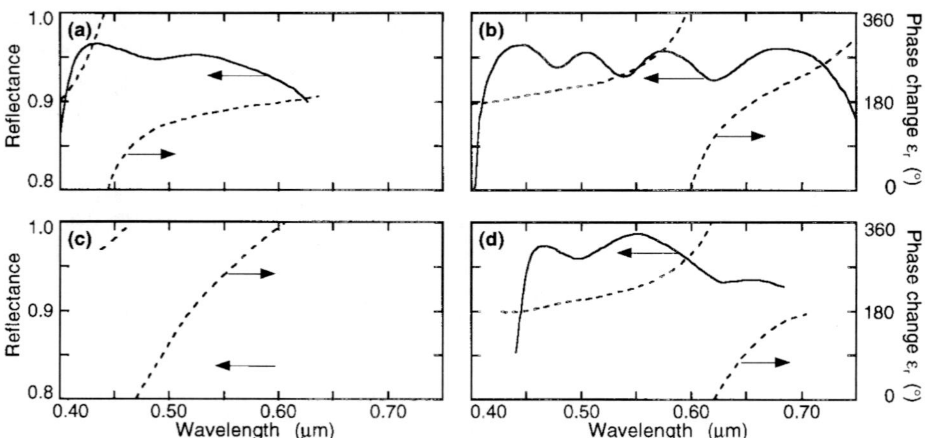

FIGURE 45 Measured reflectances of all-dielectric broadband reflectors designed with refinement programs. *[(a) after Penselin and Steudel,*[429] *(b) after Baumeister and Stone,*[430] *(c) after Ciddor,*[431] *(d) after Ramsay and Ciddor.*[134]*]*

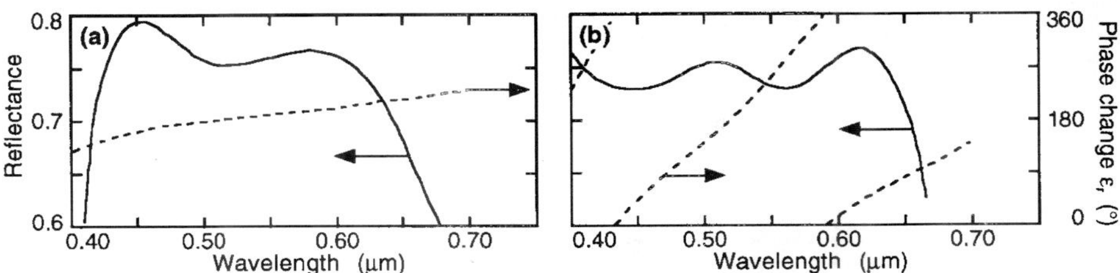

FIGURE 46 Measured reflectances of achromatic all-dielectric reflectors with an intermediate reflectance derived from a quarter-wave stack through (*a*) the addition of half-wavelength achromatizing layers *(after Turner*[220]*)* and (*b*) designed with a refinement program *(after Ciddor*[431]*)*.

reflectors with moderate and high reflectances for the ultraviolet spectral region have been reported by Korolev,[158] Sokolova,[159] and Stolov.[160]

Phase Change on Reflection from Broadband Reflectors. The phase change on reflection from broadband reflectors varies even more rapidly with wavelength (Figs. 42 to 46) than that from quarter-wave stacks (Fig. 27).[161,162] This can be a disadvantage in certain metrological applications. Another consequence of this rapid variation is that, in the presence of systematic thickness variations of the layers, it can give an impression of lack of flatness in the substrate.[163] Figures 45*d* and 46*a* represent designs of broadband reflectors in which an effort was made to reduce this effect.

Rejection Filters

Minus Filters. Minus filters are, in essence, multilayer reflectors in which the ripples in the transmission regions on either side of the high-reflectance zone have been reduced or eliminated. Filters of this kind with various widths and attenuations find applications as correction filters.[164] Narrow minus filters with high attenuations have various scientific and technological uses, including protection of equipment and personnel from harmful laser radiation. Figure 47 shows the measured transmittances of a number of rejection filters of various widths and attenuations.

Thelen has shown how to optimize the transmission on both sides of the rejection band of a minus filter simultaneously.[165] If the multilayer is surrounded on both sides by the same medium, it will be symmetrical and can be represented by $C[AB]^N AC$,

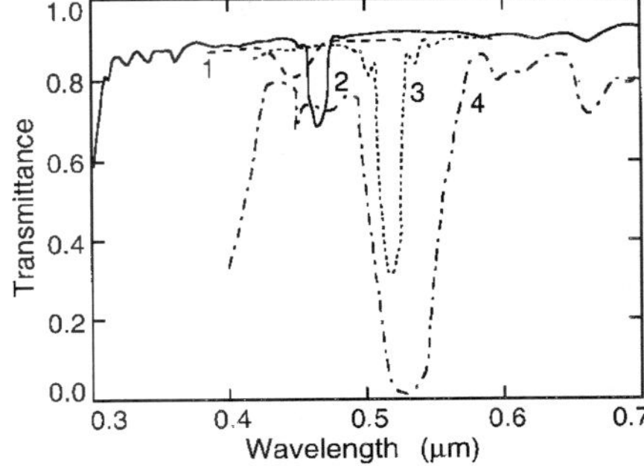

FIGURE 47 Measured spectral transmittances of several narrowband rejection filters. *(Curves 1 to 3 after Dobrowolski;*[164] *curve 4, Optical Coating Laboratory.*[340]*)*

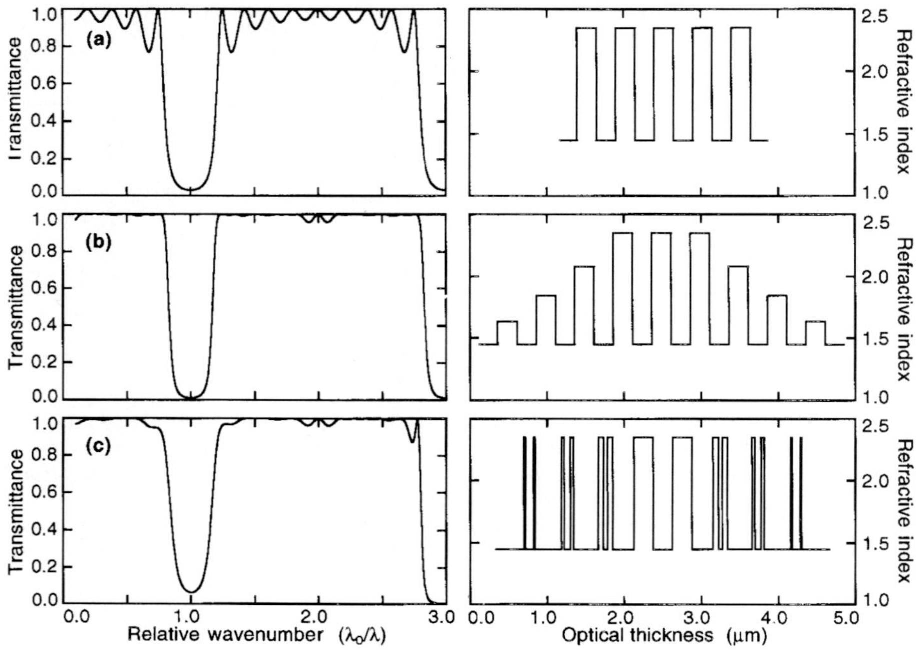

FIGURE 48 Suppression of ripples in the transmission region of rejection filters: (*a*) calculated spectral transmittance curves and refractive index profiles of a 9-layer quarter-wave stack; (*b*) a 17-layer, 5-material minus filter; and (*c*) a two-material equivalent of the minus filter.

$DAC/[AB]^N ACAD, \ldots$ Here $A, B, C, D, \ldots$ are layers of quarter-wave optical thickness at the design wavelength and

$$n_m = n_s = n_A \tag{57}$$

The refractive indices n_C, $n_D, \ldots$ depend on n_A and n_B in a more complicated way. The larger the number of different materials used in the construction of these multilayers, the better the performance in the transmission region. Should the use of more than two coating materials not be convenient, it is a simple matter to find a two-material version of this solution. These points are illustrated in Fig. 48.

Young[142] described two other design methods for narrowband rejection filters with improved transmission characteristics. The methods are based on analogies with antenna theory, and they yield nonperiodic equiripple designs in which all the layers either have equal thicknesses but different indices or are made of two materials only but have many different thicknesses.

Should the simultaneous rejection of several wavelengths be required, it is possible to achieve this by depositing several minus filters on top of one another (Fig. 49).

Rugate Filters. In inhomogeneous layers, the refractive index varies continuously in the direction of the thickness of the layer.[77,166,167] If the refractive index varies in a periodic manner between two extreme values, it is possible to design a minus filter with a high transmission on either side of the rejection band (Fig. 50). Such periodic inhomogeneous layers are sometimes called rugate filters. Some workers reserve this term for inhomogeneous layers in which the logarithm of the refractive index varies in a sinusoidal manner. The rejection wavelength corresponds to that wavelength for which the period of

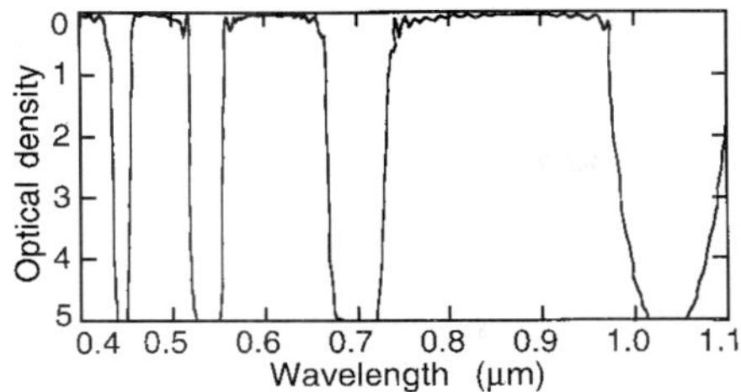

FIGURE 49 Transmission of laser goggles for the rejection of the ruby (0.694 μm) and the NdYag (0.532 and 1.064 μm) laser lines. *(After Omega Optical Inc.*[432]*)*

the index variation is equal to a half-wavelength. The attenuation depends on the ratio of the highest to lowest refractive index in the design and on the number of periods. As in the multilayer minus filters, the width of the rejection region also depends on the refractive index ratio. Rugate filters do not have the higher-order reflection peaks that are characteristic of periodic multilayers and this is one reason for their attractiveness. However, they are more difficult to produce. If necessary, they can be approximated by a homogeneous multilayer system consisting of a few materials.

As in the case of minus filters, it is possible to reject a number of wavelengths by depositing several rugate filters on top of each other. However, the combined overall

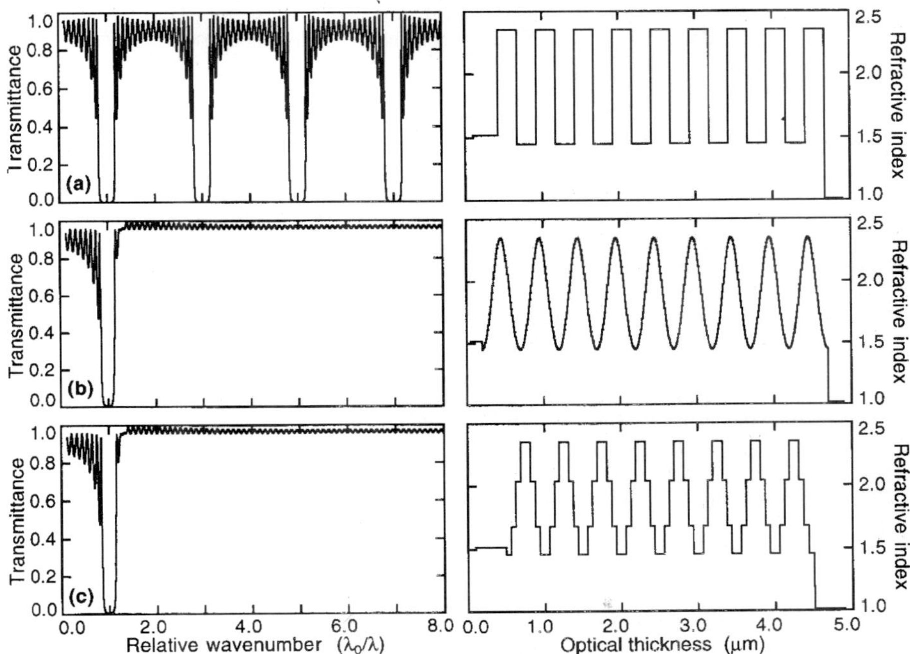

FIGURE 50 Suppression of higher-order reflectance peaks in rejection filters: calculated spectral transmittance curves and refractive index profiles of (*a*) a 17-layer quarter-wave stack; (*b*) a 9-period rugate filter; and (*c*) a 49-layer four-material design by Thelen.[18]

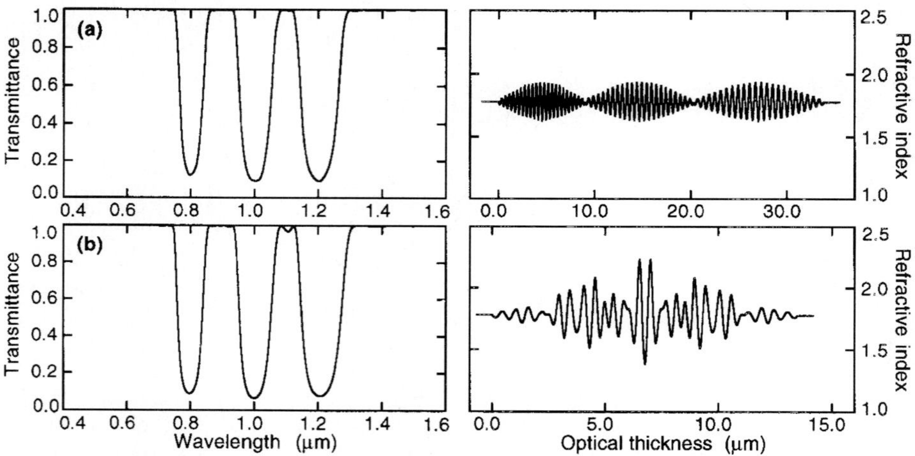

FIGURE 51 Simultaneous suppression of several laser wavelengths: (*a*) calculated spectral transmittance curves and refractive index profiles of a series and (*b*) a parallel solution to the problem. *(After Verly.*[433]*)*

thickness of the rugate filters will then be quite high. It is possible to find an inhomogeneous layer solution to this problem in which the refractive index profile is more complicated, but which requires a considerably thinner inhomogeneous layer (Fig. 51).[168]

In Lippmann-Bragg holographic mirrors, the refractive index varies continuously in a direction perpendicular to the plane of the substrate. These devices behave like thin-film systems and have properties similar to those of rugate filters. Holographic edge and narrowband rejection filters are available commercially.[169,170]

Graded Reflectivity Mirrors

Mirrors in which the absorption or reflection varies radially have been proposed in the past for the control of modes and of edge diffraction effects in laser resonators.[171,172] In addition to meeting the reflectance specifications, graded reflectivity mirrors must have a sufficiently high laser damage threshold for use with high-power lasers. Graded reflectivity mirrors are produced by depositing thin films through a suitable mask. The substrate and the mask can be stationary, or one or both can rotate. A single shaped layer suffices for a maximum reflectance of intermediate values. For higher values, the shaped layer can be inserted between a stack of layers of uniform thickness or, alternatively, all the layers can be deposited though the mask.[173] The experimentally measured reflections of two graded reflectivity mirrors are shown in Fig. 52.

Multilayer Reflectors for the Far-infrared Region

Thin-film filters cannot be produced by conventional deposition techniques for wavelengths greater than about 80 μm because of a lack of low-absorption coating materials that can be deposited in the form of thick, stable films. However, a hybrid process in which plastic sheets coated with relatively thin high-refractive-index materials are heat-bonded can be used to produce self-supporting optical multilayer filters.[125,174,175] As mentioned before, periodic multilayers of unequal optical thickness can have a high reflectance, providing that the number of periods is high enough. The measured reflectance curves of two typical hybrid multilayer filters are shown in Fig. 53.

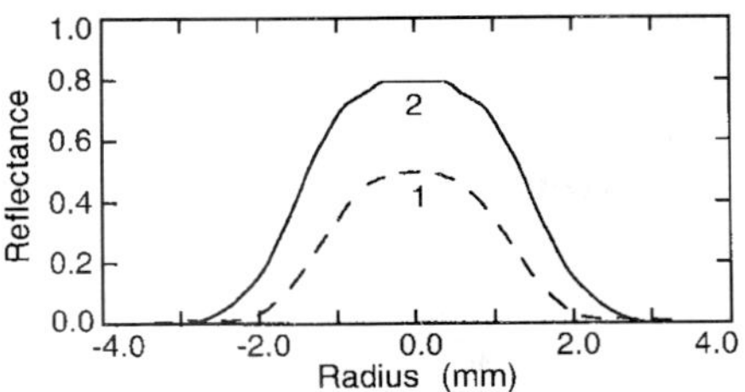

FIGURE 52 Experimental results for two super-gaussian graded reflectivity mirrors for $\lambda = 1.064\ \mu m$. Curve 1: three layer systems with one layer of varying thickness, with $R_{max} = 0.5$ and $\omega = 1.92$ mm *(after Piegari[434])*; curve 2: fully shaped mirror with $R_{max} = 0.85$ and $\omega = 1.90$ mm *(after Duplain[173])*.

Multilayer Reflectors for the Soft X-ray and XUV Regions

There are two main obstacles to obtaining multilayers with a high-normal-incidence reflectance in the soft x-ray and in the XUV regions. First, at these wavelengths all materials absorb; this limits the number of layers that can contribute to the overall reflectance. Second, roughness and the interdiffusion and alloying of the materials degrade the individual interfaces; this reduces their contribution to the overall reflectance.

XUV mirrors are normally produced from two chemically compatible materials by sputtering or by electron beam gun evaporation. As already discussed, the XUV multilayer mirrors have a period of optical thickness $\lambda/2$. Within this period, the thickness of the less-absorbing material is larger. This material usually has an absorption edge that is close to the design wavelength. Optical constants can vary widely in this region.[176] The second, more absorbing material is, therefore, selected to maximize the Fresnel reflection coefficient of the interface.

Theoretically, the best results are obtained with pure elements. However, sometimes alloys are used because they result in multilayers with better interfaces. Thus, $MgSi_2$ might be used in place of Mg, or B_4C instead of C or B. Examples of more commonly used material pairs are: (1) Mo/Si for the 130- to 250-Å region; (2) Mo/Y for the 90- to 130-Å region; (3) W/B_4C, Ru/B_4C, Mo/B_4C, etc. for the 70- to 130-Å region; and (4) Co/C, W/C, Re/C, ReW/C, Ni/C, etc., for the 45- to 70-Å. In these pairs, the second material in each pair has the lower extinction coefficient.

There are many reasons why there are differences beween the theoretical performance of a multilayer reflector and the reflectance measured on a synchrotron.[126,128] The highest

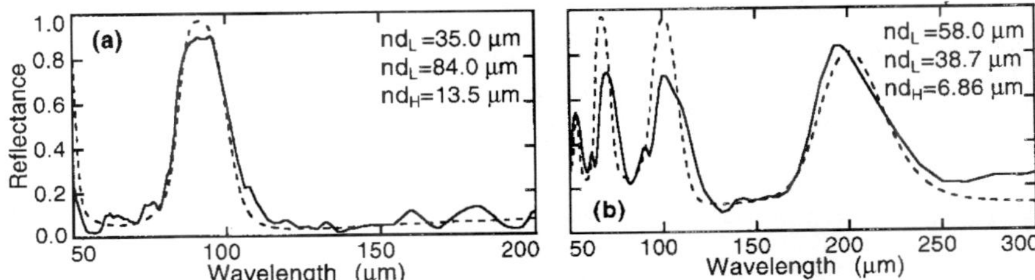

FIGURE 53 Measured and calculated performance of (*a*) $L'(HL)^{16}$ and (*b*) $L'(HL)^{16}$ reflectors made of ZnS and polyethylene of thicknesses indicated in the diagram. An instrumental resolution of 3 cm^{-1} was assumed in the calculations. *(After Shao.[125])*

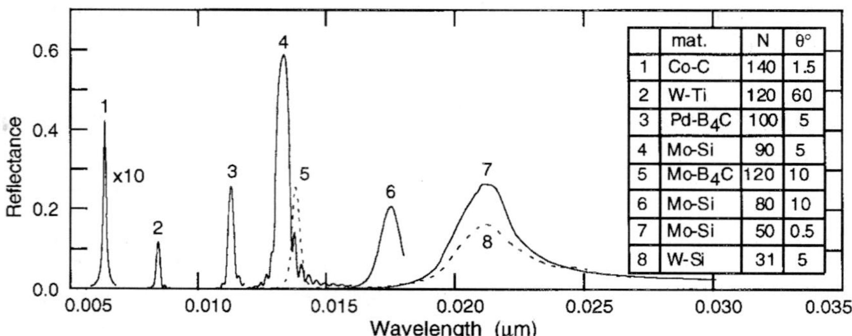

FIGURE 54 Measured spectral reflectance curves of some representative multilayer x-ray mirrors. The materials used, the number of layers N and the angles θ_0 at which the reflectance was measured are indicated in the table. *(Curve 1 after Spiller;*[435] *curves 2–4 after Montcalm et al.;*[436,437] *curves 5 and 6 after Zwicker;*[438] *curve 7 after Ceglio;*[439] *curve 8 after Falco.*[440]*)*

reflectance achieved thus far is $R \approx 0.65$ at 135 Å for a Mo/Si multilayer.[141] The measured reflectance of other experimentally produced x-ray and XUV mirrors is shown in Fig. 54.

42.9 CUT-OFF, HEAT-CONTROL, AND SOLAR-CELL COVER FILTERS

Ideal cut-off filters would reject all the radiation below, and transmit all that above a certain wavelength, or vice versa. Real cut-off filters, of course, are not perfect and so, in addition to the cut-off wavelength, the slope of the transition region and the extent and average transmission values of the transmission and rejection regions must be specified. The tolerable departures of these quantities from the ideal values depend greatly on the application. Most all-dielectric cut-off filters are based on periodic multilayers, whose basic properties were described in Sec. 42.7).

Transmission in the Passband Region

The usual way of avoiding the secondary transmission minima in the transmission band of a quarter-wave stack is through the use of eighth-wave layers on both sides of the stack (see "Periodic Multilayers of the $[(0.5A)B(0.5A)]^N$ Type" in Sec. 42.7).

Other, less frequently used methods of smoothing the transmission in the passband are the adjustment of the thicknesses of all the layers of a quarter-wave stack,[177] the use of homogeneous[178] and inhomogeneous[179] layers on either side of the stack; the choice of an optimum set of refractive-index values for the substrate and films,[180] and the use of an equiripple design[142] in which thicknesses are kept constant but the refractive indices are varied.

The Width of the Transmission Region

For short-wavelength cut-off filters of the type described here, the transmission region is limited only by the transmission characteristics of the materials used for their construction. In long-wavelength cut-off filters, the transmission regions can be limited by the appearance of higher-order reflectance maxima (see "Nonabsorbing $[AB]^N$ and $[AB]^NA$

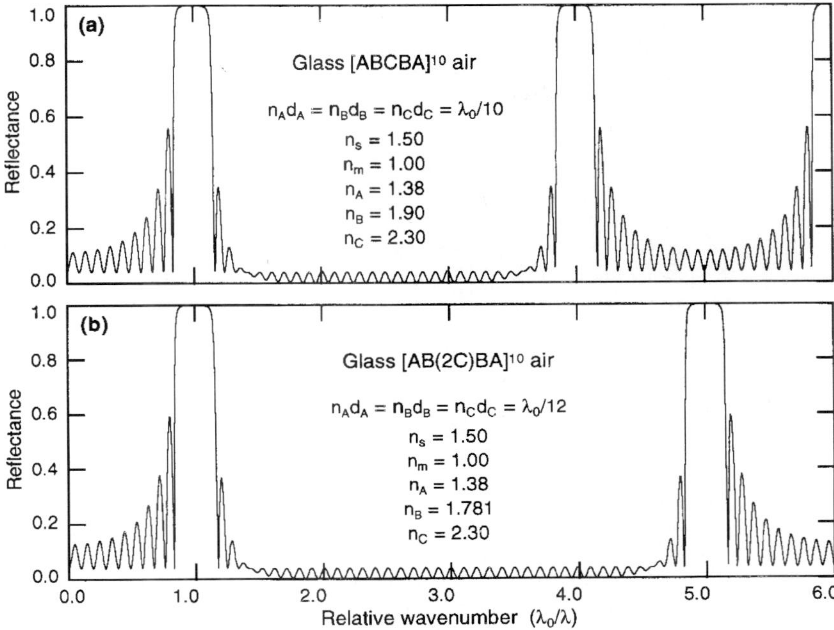

FIGURE 55 Calculated reflectance of three-material periodic multilayers with suppressed higher-order reflectance maxima. *(After Thelen.[441])*

Multilayer Types" in Sec. 42.7). Should this be a serious limitation, it is possible to suppress a number of adjacent reflectance maxima by using multilayers with periods composed of three or more different materials (Fig. 55). By using a period consisting of an inhomogeneous layer with a special refractive-index profile, an even larger number of consecutive reflectance maxima can be suppressed (see also Fig. 50).[77] Figure 56 shows the measured results for two such experimental coatings in which reflectance maxima are

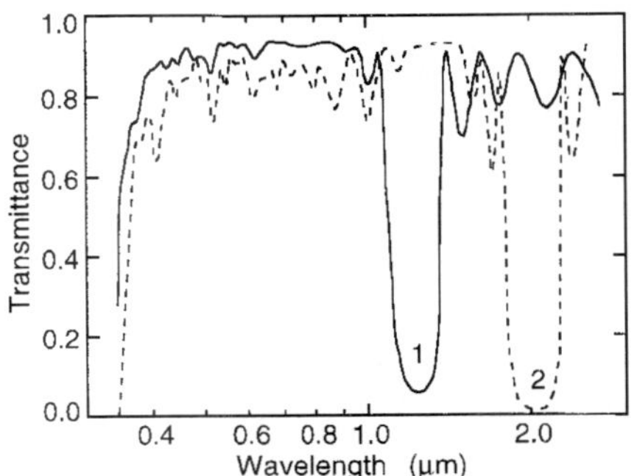

FIGURE 56 Measured transmittance of periodic multilayers in which higher-order reflectance maxima are suppressed through the use of periods that consist of an inhomogeneous layer. *(After Scheuerman.[188])*

suppressed at three and nine consecutive integer multiples of $1/\lambda_o$, a fact obscured in the case of the second filter by the absorption of the materials used.

Transmission in the Rejection Region

Figure 29 can be used for an estimate of the number of layers required to achieve a given transmission (see "Maximum Reflectance" in Sec. 42.7). Through the use of suitable substrate materials, the transmittances throughout the rejection region can typically be below 0.01 and 0.1 percent for short- and long-wavelength cut-off filters, respectively. Rejection filters with higher rejections can be provided or, alternatively, two or more filters in series can be used if they are placed at a small angle to one another.

The Width of the Rejection Region

The width of the high-reflection region of $[(0.5A)B(0.5A)]^N$ coatings can be estimated from Fig. 32. There is no shortage of absorbing materials should it be necessary to extend the rejection region of short-wave cut-off filters. The number of suitable absorbing materials for long-wavelength cut-off filters is more restricted and often it is necessary to use thin films to extend the cut-off region. In addition to the superposition of two or more cut-off filters tuned to different wavelengths (see "All-Dielectric Broadband Reflectors" in Sec. 42.8), one can deposit such coatings onto different substrates or onto the opposite sides of the same substrate.[181] The resulting transmission will be governed by the considerations of "Transmission Filters in Series and Parallel" in Sec. 42.2.

Slope of the Cut-off

This quantity is defined in a number of ways. One common definition is

$$\left|\frac{\lambda_{0.8} - \lambda_{0.05}}{\lambda_{0.5}}\right| \times 100\% \tag{58}$$

where $\lambda_{0.8}$, $\lambda_{0.5}$ and $\lambda_{0.05}$ refer to the wavelengths at which the transmittances are 0.8, 0.5, and 0.05 of the maximum transmittance of the filter. Explicit formulas for the slope are complicated.[121] The slope increases with the number of periods and with the refractive-index ratio. Slopes with values of the order of 5 percent are readily available in practice.

Angle-of-incidence Effects

The edges of cut-off filters move towards shorter wavelengths as the angle of incidence is increased. The use of higher-index materials reduces the effect. Measured results for a cut-off filter in which the shift was reduced by using high-refractive-index layers that have three times the thickness of the low-index layers are shown in Fig. 57*a*. See also "Nonpolarizing Edge and Bandpass Filters" in Sec. 42.10 on polarization-independent color-selective beam splitters.

Experimental Results

The spectral transmittance curves of a number of commercially available short- and long-wavelength cut-off filters are shown in Figs. 58 and 59. High performance long- and short-wavelength cut-off filters are depicted in Figs. 60 and 100. Similar filters for

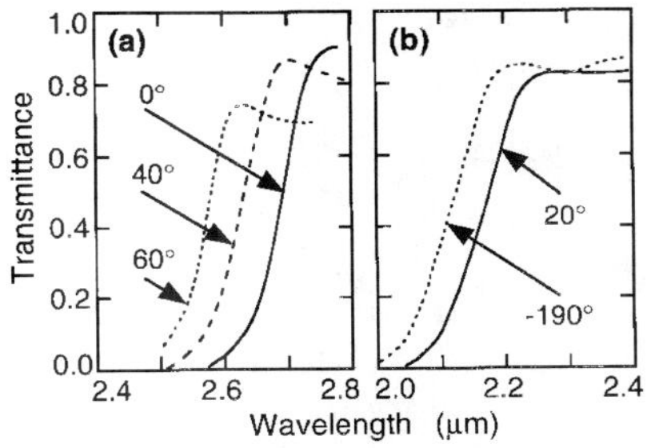

FIGURE 57 Effect of (*a*) incidence and (*b*) temperature on the performance of cut-off filters. *(Optical Coating Laboratory.[442])*

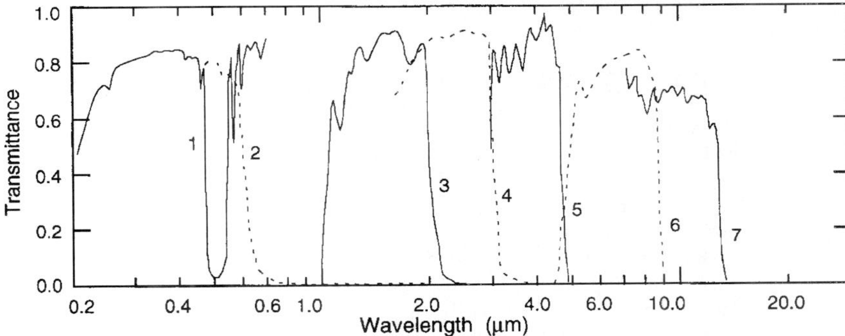

FIGURE 58 A series of commercial short-wavelength cut-off filters. *(Curves 1, 5, 8, 9, 10, Optical Coating Laboratory;[340,443] curve 2, Bausch & Lomb;[444] curves 3 and 6, after Turner;[408] curve 4, Eastman Kodak;[445] curve 5; Infrared Industries.[446])*

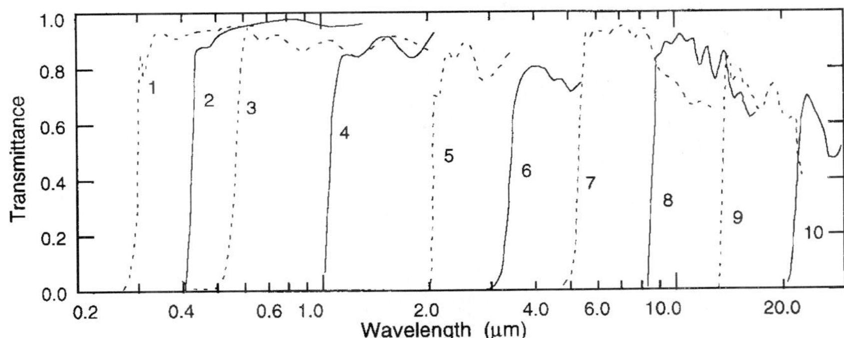

FIGURE 59 A series of commercial long-wavelength cut-off filters. *(Curve 1, after Apfel;[421] curve 2, Eastman Kodak;[445] curves 3, 4, 7, Optical Coating Laboratory;[443] curves 5 and 6, Infrared Industries.[446])*

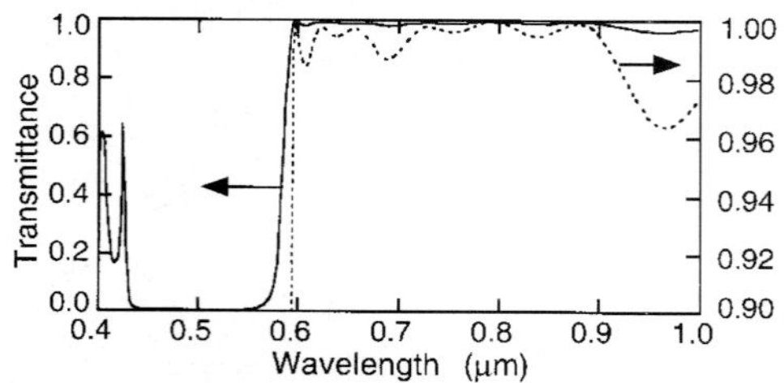

FIGURE 60 Measured transmittance of an unblocked short-wavelength cut-off filter. *(Afer Thin Film Lab.*[410]*)*

intermediate wavelengths can, of course, be constructed. It is also possible to construct edge filters in which the thicknesses of all the layers vary in proportion around the circumference of the substrate (see also "Linear and Circular Wedge Filters" in Sec. 42.12). A tuning of the cut-off wavelength is thus possible.

Heat Reflectors, Cold Mirrors, and Infrared-Suppressing Filters

Only 39 percent of the total radiation from carbon arcs and 13 percent from tungsten lamps operated at 3250 K represent visible light. Most of the remaining energy is infrared radiation, which is converted into heat on absorption. The use of heat reflectors and cold mirrors in film projectors,[182] in spot lamps for television and film studios,[183] and in other optical instruments can lead to a very significant reduction of this unwanted heat.

Heat reflectors (also called hot mirrors) are special long-wavelength cut-off filters with a cut-off at 0.7 μm which transmit the visible radiation from 0.4 to 0.7 μm without disturbing the color balance. The width of the rejection region depends on the light source to be used and on whether a heat-absorbing glass is also to be used. The spectral-transmittance curves of three typical commercial heat reflectors are shown in Fig. 61*a*. The measured spectral-transmittance and reflectance curves of two heat-reflecting coatings not based on periodic multilayers are shown in Fig. 62.

Cold mirrors reflect as much as possible of the visible light incident upon them and transmit the remaining radiation. The reflectance curves of two commercial cold mirrors are shown in Fig. 61*b*.

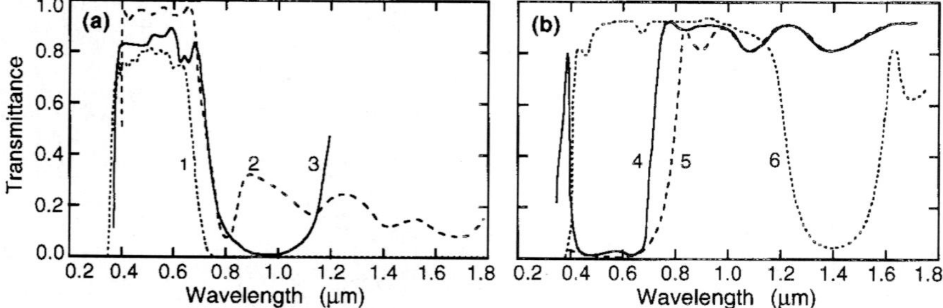

FIGURE 61 Measured performance of commercial multilayer coatings for heat control: (*a*) heat-reflecting coatings; (*b*) cold mirrors and blue-red solar-cell cover. (*Curve 1, Corion Corporation;*[447] *curve 2, Bausch & Lomb;*[444] *curve 3, Balzers;*[406] *curve 4, Optical Coating Laboratory;*[340] *curve 5, Heliotek;*[448] *curve 6, after Thelen.*[449])

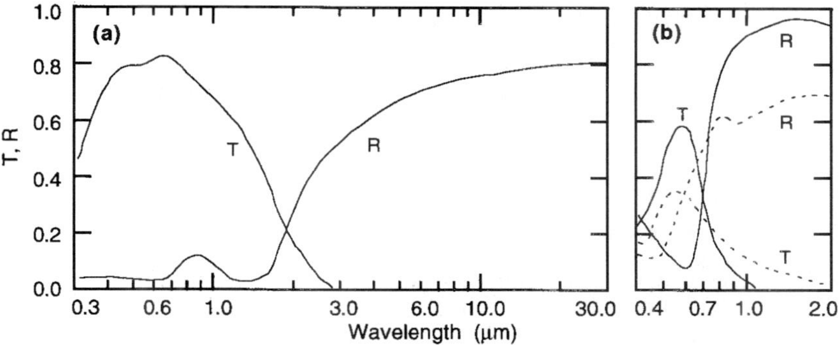

FIGURE 62 Two nonperiodic coatings with heat-reflecting properties: (*a*) spray-deposited conducting coating of tin oxide; (*b*) transmittance and reflectance of a gold film (dotted curves) and of a broadband metal dielectric filter (solid curves) with semitransparent gold films of the same total thickness. *(After Turner.*[450]*)*

Solar-cell Covers

Solar-cell covers remove the incident solar energy that does not contribute to the electrical output of the cell and protect it from possible deterioration of its performance through the action of ultraviolet radiation.[184,185] The spectral-transmittance of a blue-red solar-cell cover is shown in Fig. 61*b*. The earlier blue solar-cell covers (curve 2, Fig. 58) protected the cell only from the adverse effects of ultraviolet radiation.

Temperature Effects

Refractive indices of optical materials increase almost linearly with increase in temperature, thus causing cut-off edges to move towards longer wavelengths. In actual filters, the fractional-wavelength shift varies between 3×10^{-3} and 10^{-4}/°C. Ion-plated films have a smaller temperature shift than films prepared by conventional e-beam evaporated layers.[186] Higher-index materials tend to be more temperature sensitive, thus making it difficult to construct filters that are insensitive both to angle of incidence and temperature changes.[187] The measured performance of a cut-off filter at two temperatures is shown in Fig. 57*b*.

Metal-dielectric Reflection Cut-off Filters

It is possible to construct metal dielectric cut-off filters that act in reflected light (Fig. 63). Short-wavelength cut-off filters consist of an opaque metal layer and one or more

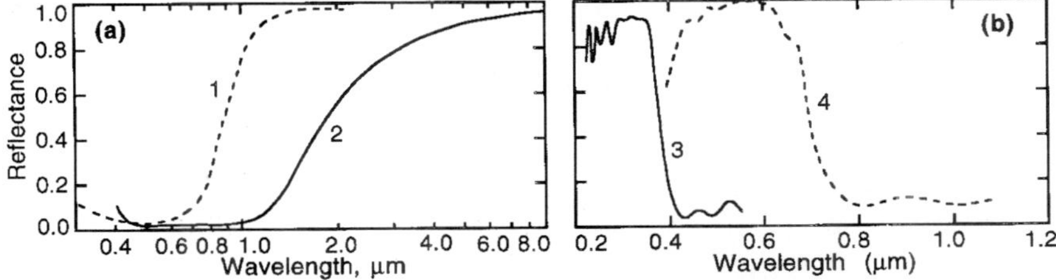

FIGURE 63 Short- and long-wavelength metal-dielectric cut-off filters based on reflection: (*a*) curves 1 and 2, three-layer coatings on aluminium *(after Drummeter and Hass*[109]*)*; (*b*) curves 3 and 4, multilayer reflecting coatings deposited onto black absorber coatings *(after Hoppert*[451]*)*.

additional layers. The light is removed through absorption within the absorbing layers of the system. The thicknesses of the individual layers are adjusted to maximize the absorption and width of the rejection region. Long-wavelength cut-off filters consist of all-dielectric multilayer reflectors superimposed onto a black absorbing coating. A high attenuation of the unwanted radiation can be achieved by placing identical filters of either type in a multiple-reflection arrangement of the kind depicted in Fig. 3.

Cut-off Filters Based on Absorption

All materials used in multilayer interference coatings possess short- and long-wavelength absorption edges. These can often be used to assist in the blocking of such filters. The admixture of small amounts of absorbing materials to an evaporant or organic coating solution is used at times to tune this absorption edge (Fig. 64*a* and *c*). For example, antireflection coatings containing such ultraviolet-absorbing materials can be used to protect works of art.

A series of commercial short-wavelength cut-off filters for the infrared spectral region consisting of chemically deposited silver sulfide coatings on silver chloride substrate substrates are also shown in Fig. 64*b*. These filters are quite delicate. When protected with a polystyrene layer, their transmittance is reduced and sharp absorption bands appear. The filters should not be used outdoors unless additionally protected, nor should they be exposed to ultraviolet radiation or temperatures in excess of 110°C.

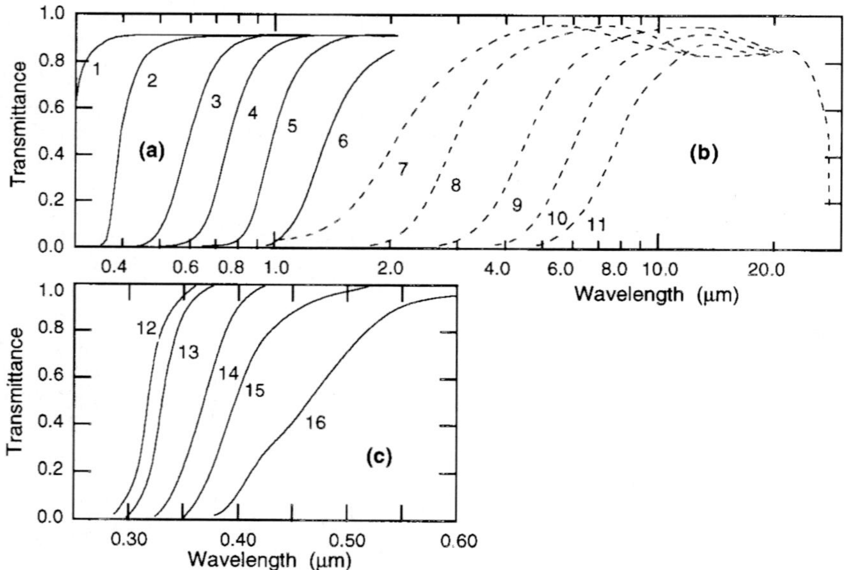

FIGURE 64 Spectral transmittance of absorbing films produced in various ways: (*a*) envelopes of the transmission maxima of thick evaporated films of ZnS (curve 2), Ge (curve 6), and various mixtures of ZnS and Ge (curves 3 to 5) on a glass substrate (curve 1). *(After Chang*[452]; (*b*) spectral transmittance of chemically deposited silver sulfide coating on silver chloride substrates (curves 7 to 11) *(Eastman Kodak*[445]*)*; (c) intrinsic transmission of thin films of titanium dioxide with admixtures of heavy metal oxides, deposited from organic solutions: curve 12, $TiO_2 + 1.5SiO_2$; curve 13, TiO_2; curve 14, $TiO_2 + 0.5PbO$; curve 15, $TiO_2 + 0.15Fe_2O_3$; curve 16, $TiO_2 + 5.7UO_3$ *(After Schröder*[453]*)*.

The greatest advantage of cut-off filters based on absorption in thin films is their much smaller angular dependance.

42.10 BEAM SPLITTERS AND NEUTRAL FILTERS

Geometrical Considerations

Several different physical forms of beam splitters are illustrated in Fig. 65. The simplest (Fig. 65*b*) consists of a coating on a transparent plane-parallel substrate. If the two derived beams are to traverse identical paths, a cemented beam splitter is used (Fig. 65*c*). The

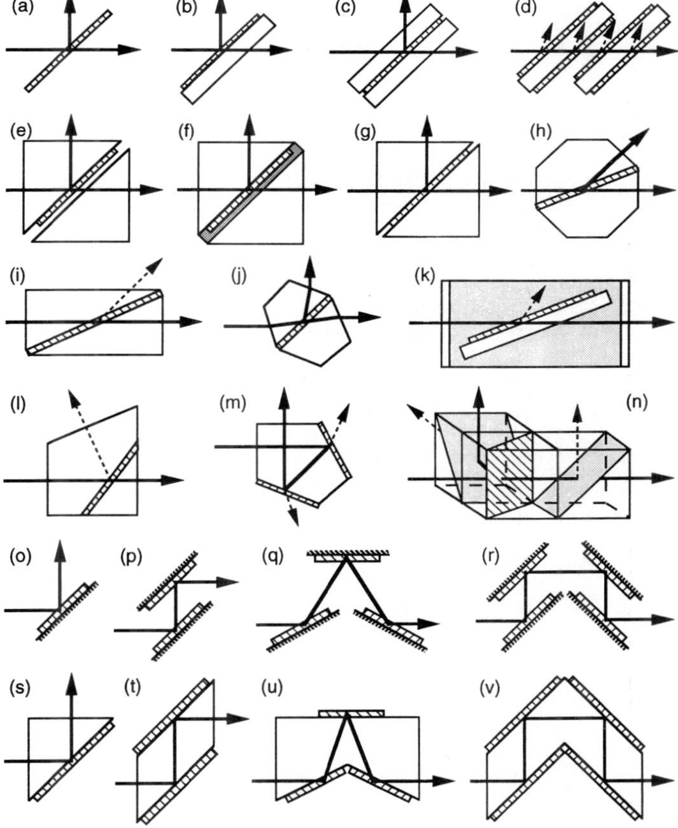

FIGURE 65 Schematic representation of some arrangements for beam splitters, polarizers, phase retarders, and multiple reflection devices. Thin films are represented in the diagrams by narrow shaded rectangles. Heavy lines ending in an arrow represent the path of the utilized radiation. Broken lines correspond to beams that are not used. The angles of incidence depend on the application.

lateral displacement of the transmitted beam introduced by these forms can be avoided with a beam-splitting cube (Fig. 65*f*). To reduce the stray reflected light in the system, the free surfaces of these beam splitters can be antireflection-coated. Alternatively, the coatings can be deposited onto an approximately 2-μm-thick nitrocellulose pellicle (Fig. 65*a*). The latter is an integral part of the multilayer and may introduce an interference pattern into the spectral reflectance and transmittance characteristics. Pellicle beam splitters are very light and yet quite sturdy.[188–190] They are, however, subject to vibrations caused by air currents and acoustical waves. The mechanical design of rugged, environmentally stable mounts for these types of beam splitters have been discussed by Heinrich et al., and Lipshutz.[191,192] Pellicles made of Mylar have been used at temperatures down to 4 K.[193]

In general, the transmission and reflection coefficients T and R will depend on the polarization of the incident radiation. The polarization effect can be reduced through the use of more complicated thin-film designs, but usually at the expense of other performance aspects. Achromatic or color-selective beam-splitting arrangements have been described in which the two derived beams have intensities that are completely polarization-independent over a very wide spectral region.[194] They consist of three identical beam splitters arranged in such a way that each beam undergoes identical reflections and transmissions on passing through the system (Fig. 65*n*).

Achromatic Beam Splitters

These devices are introduced into an incident beam of radiation when it is desired to divide it into two beams of approximately equal relative spectral composition but propagating in two different directions.[195] In neutral beam splitters, the quantity $0.5(R_p + R_s)_{\theta=45°}$ is always close to the reflectance at normal incidence, even though the individual R_p and R_s values may be quite different. The reflectance of absorbing, uncemented beam splitters depends also on the direction of incidence (see "Matrix Theory for the Analysis of Multilayer Systems in Sec. 42.3). The optimum values of T and R depend on the application. For example, for a binocular eyepiece on a nonpolarizing microscope, the most important requirement is $T_p + T_s = R_p + R_s$ (Fig. 66*a*).For a vertical illuminator $(R_pT_p + R_sT_s)$ should be a maximum (Fig. 66*b*). The condition for maximum fringe contrast in some interferometers requires that $R_{1,p}T_p = R_{2,p}T_p$ and $R_{1,s}T_s = R_{2,s}T_s$ (Fig. 66*c*). This is satisfied automatically by all nonabsorbing and by absorbing cemented beam splitters. The occasional requirement that the phase change on reflection be the same for radiation incident onto the beam splitter from opposite sides is automatically

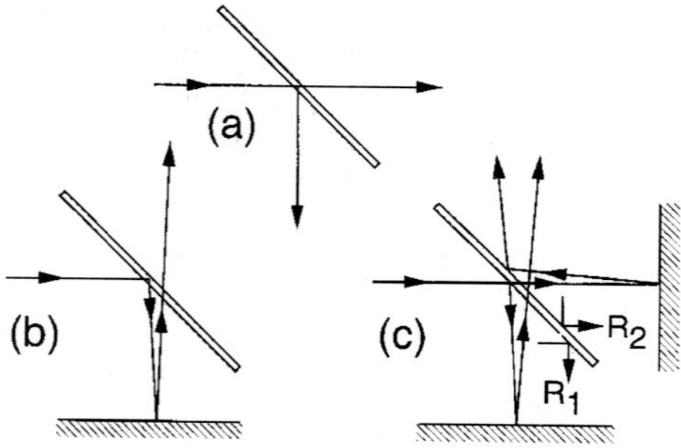

FIGURE 66 Three different ways of using beam splitters.

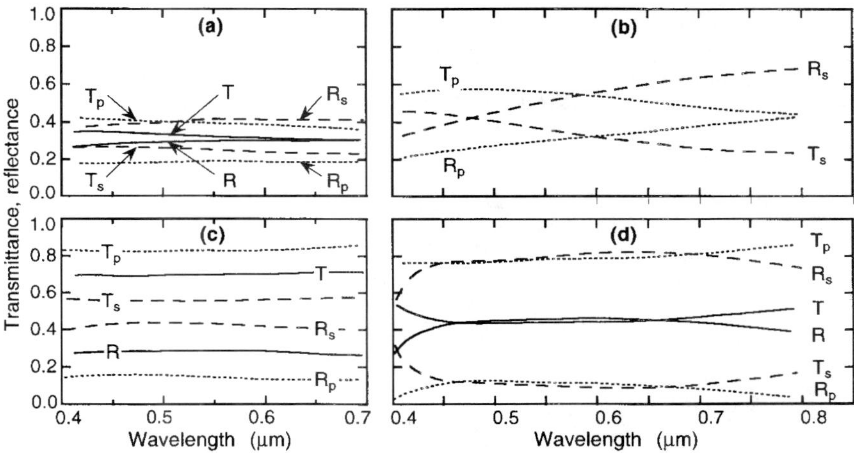

FIGURE 67 Measured spectral transmittance and reflectance for polarized and unpolarized light of (*a*) Inconel- and (*c*) dielectric-coated beam-splitting plates (*Oriel*[454]) and of (*b*) silver- and (*d*) dielectric-coated beam-splitting cubes (*after Anders*[195]).

satisfied at the design wavelength by all-dielectric coatings composed of $\lambda/4$ layers, but not by uncemented metal beam splitters.[196]

For maximum efficiency with unpolarized radiation, R_p, R_s, T_p and T_s should all approach 0.5. However, such a coating will not necessarily exhibit the best ratio of the intensities of the directly transmitted or reflected radiation to that which first undergoes multiple reflections.[195] Often, beam splitters are required that are uniform over a broad spectral region. Inconel films satisfy this requirement although about one-third of the incident radiation is lost through absorption (see also "Neutral Filters" in Sec. 42.10). The design of achromatic all-dielectric beam splitters has been discussed by many workers.[197–202] Knittl considered the design of beam splitters in which both the reflectance and the phase change on reflection are achromatised.[203] The measured performance of several beam splitters is shown in Figs. 67 and 68.

Beam splitters for the x-ray region described so far operate at close to normal incidence and are effective only over a very narrow range of wavelengths. They consist of multilayer reflecting stacks (see "Multilayer Reflectors Made of Absorbing Materials") in Sec. 42.7 and "Multilayer Reflectors for the Soft X-Ray and XUV Regions" in Sec. 42.8) deposited onto membranes or onto substrates that are thinned to enhance the transmitted component (Fig. 69).[204]

Nonpolarizing Beam Splitters. For some applications it is important that the beam splitter introduce no polarization effects. Azzam has shown that, with the appropriate single layer on the face of a suitable high-refractive index prism, it is possible to construct a polarization-independent beam splitter.[205] This device is quite achromatic and, in addition, by changing the angle of incidence of the beam on the prism, the beam-splitting ratio can be tuned over a wide range of values (Fig. 70*b*). The principle of frustrated total internal reflection can also be used to design beam splitters that have a very good performance.[206,207] In these devices, radiation is incident at a very oblique angle onto an air gap or low-refractive-index films at the interface between two prisms (Fig. 65*e,f*). Unfortunately, the performance of such systems is very sensitive to the angle of incidence (Fig. 70*c*).

In many applications, it is important that the beam splitter be relatively insensitive to

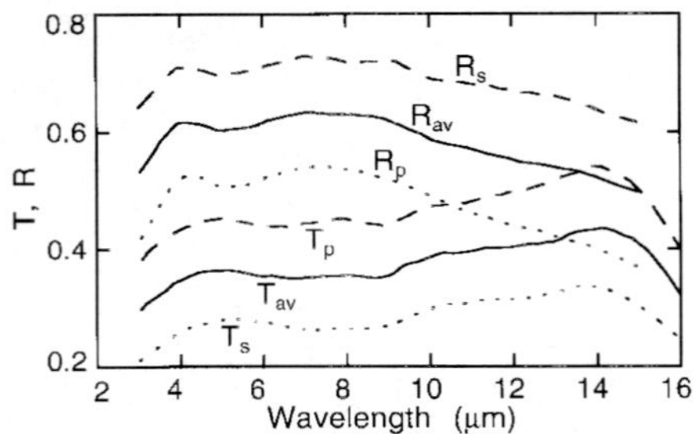

FIGURE 68 Measured performance of a 45° infrared beam splitter consisting of a suitably coated ZnSe plate. (*After Pellicori.*[455])

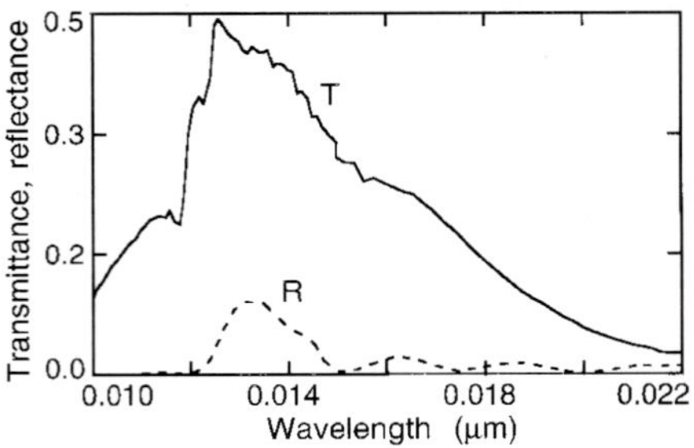

FIGURE 69 Measured performance of an x-ray beam splitter consisting of 11 pairs of Mo and Si layers on an 0.03-μm-thick Si_3N_4 membrane, operating at an angle of incidence of 0.5°. (*After Ceglio.*[439])

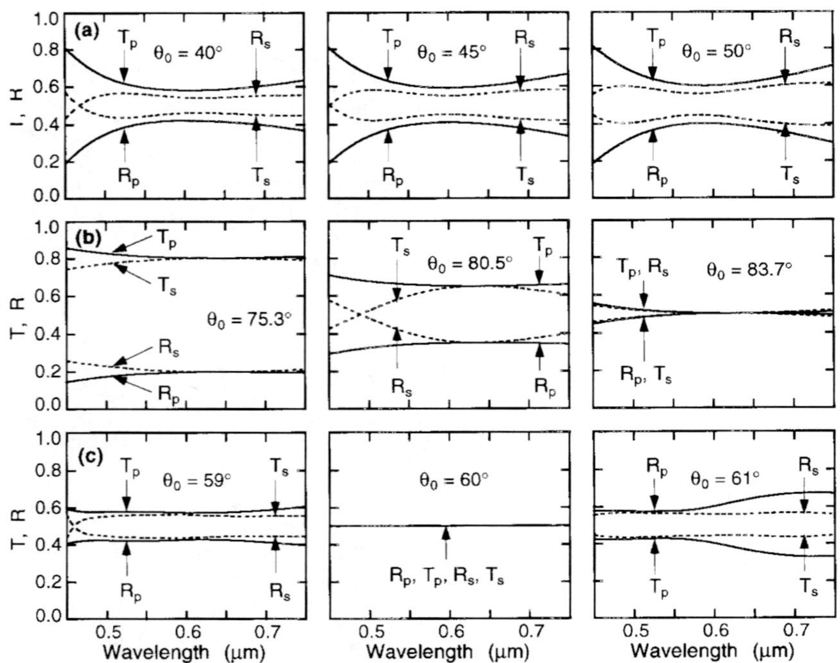

FIGURE 70 Calculated performance at three angles of incidence of beam splitters with an incident medium of air: (*a*) beam splitter of the type *glass* $(HL)^2$ *air,* where $n_H = 2.35$, $n_L = 1.38$; (*b*) single layer ($n = 1.533$, $d = 0.1356$ μm) on a prism ($n_s = 2.35$) (*after Azzam*[205]); (*c*) 15-layer frustrated-total-internal-reflection beam splitter (*after Macleod*[207]). The second diagram in each row corresponds to the performance at the design angle. The last two systems are fairly polarization-independent and yield different T/R ratios for different angles of incidence.

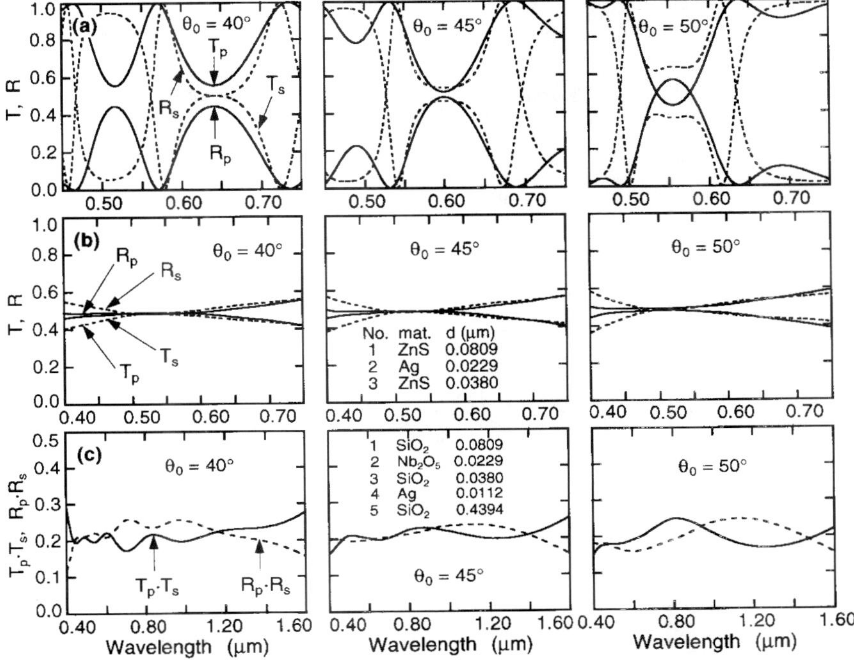

FIGURE 71 Calculated performance of polarization-insensitive achromatic beam splitters consisting of layer systems cemented between glass prisms: (*a*) all-dielectric system of the type n_s $(LMHMHML)^2$ n_s, where $n_s = 1.52$, $n_L = 1.38$, $n_M = 1.63$, $n_H = 2.35$. The quarter-wave layers are matched for 45° incidence *(After Thelen*[18]*)*; (*b*) three-layer metal/dielectric system *(after Chang*[456]*)*; (*c*) polarization-independent beam-splitting arrangement of the type of Fig. 65*n* composed of three identical prisms. *(After Ho.*[194]*)*

the angle of incidence. One way is to reduce the angle of incidence as much as possible (Fig. 65*h*).[208] However, in many cases a 45° angle of incidence is mandatory. Relatively polarization-insensitive beam splitters based on dielectric-metal-dielectric layer systems embedded between two prisms have been described.[209,210] Much work has been done to find solutions based on dielectric layers only.[23,211–218] However, the improvement is frequently at the expense of the width of the spectral region over which the beam splitter is effective. Some typical results are shown in Figs 71 and 72.

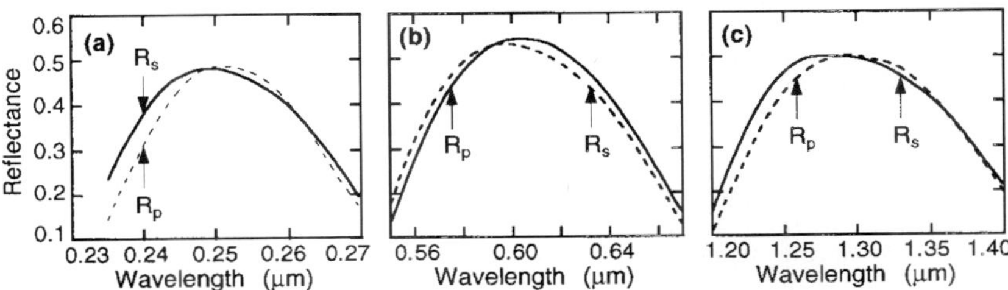

FIGURE 72 Measured performance of polarization-insensitive beam splitters of the type of Fig. 71*a* produced for three different spectral regions. *(After Konoplev.*[457]*)*

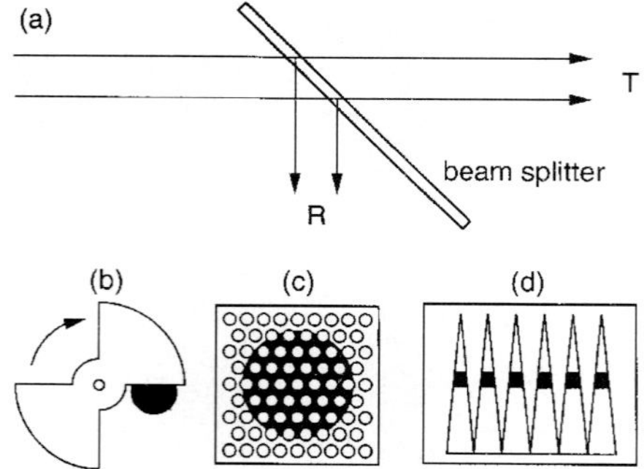

FIGURE 73 Achromatic and angle-insensitive mechanical beam-splitting arrangements: (*a*); (*b*) rotating aluminized blade which alternatively transmits and reflects the incident radiation; (*c*) stationary transparent plate with a polka-dot pattern in which either the circles or the background (each of equal total area) are aluminized; (*d*) aluminized sawtooth pattern on a transparent substrate that can be displaced to reflect different fractions of the incident radiation that is collimated into a narrow beam.

Simple angle- and polarization-insensitive mechanical solutions to the achromatic beam-splitting problem exist if the application can tolerate spatial or temporal beam sharing (Fig. 73).

Color-selective Beam Splitters

For various technological applications, a beam of light must be divided into several components of different color. All-dielectric color-selective beam splitters (*dichroics*) are used for this purpose because they are practically lossless and because their transition wavelengths can be selected at will. They are essentially cut-off filters (see Sec. 42.9) usually designed for use at 45° incidence. Their spectral characteristics normally depend on the polarization of the incident radiation. The effect of this and of the variations in the angle of incidence and thickness of the coatings on the chromaticity coordinates of dichroic beam splitters for television cameras was investigated by Pohlack.[219] If necessary, the polarization of the derived beams can be reduced through the use of auxiliary normal-incidence cut-off filters.[220] Typical transmittance curves of several color-selective beam splitters are shown in Fig. 74.

Nonpolarizing Edge and Bandpass Filters. For more exacting applications, such as for use in multiplexers and demultiplexers, or for the separation of emission or absorption lines in atmospheric physics or Raman spectroscopy, it is possible to design and construct short- and long-wavelength color-selective beam splitters in which the polarization splitting has been largely eliminated.[18,221] In the designs described, the polarization splitting is usually removed for all angles smaller than the design angle, but the cut-off wavelength

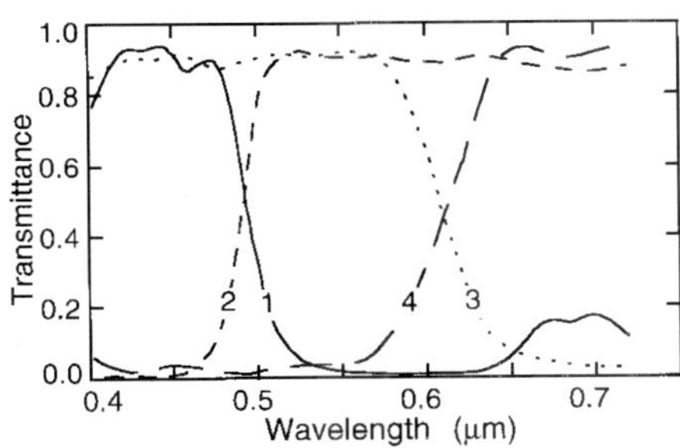

FIGURE 74 Measured spectral transmittance of four color-selective beam splitters. *(Optical Coating Laboratory.*[340]*)*

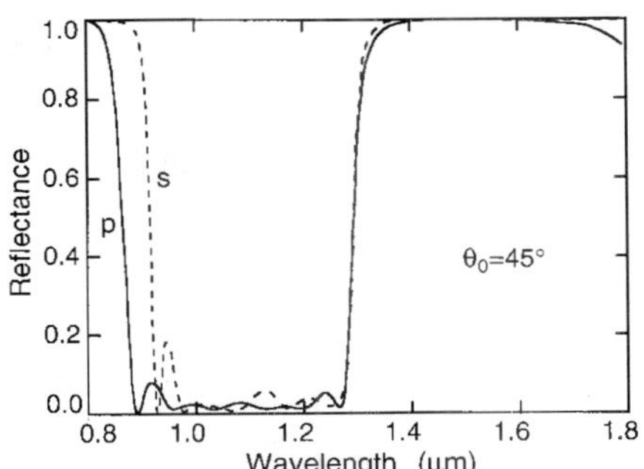

FIGURE 75 Calculated reflectance of a polarization-independent color-selective beam splitter. *(After Thelen.*[18]*)*

still shifts with the angle of incidence (Fig. 75). Some of the designs do not have a wide transmission region.

Neutral Filters

These devices are used whenever the intensity of the incident radiation is to be reduced uniformly throughout an extended part of the spectrum. The radiation usually traverses neutral filters at or near normal incidence. A number of absorbing glasses and gelatin filters are suitable for making neutral-density filters with densities of up to 5.0. However, their spectral transmission curves are not very uniform.

Evaporated films of metals such as aluminium, chromium, palladium, platinum, rhodium, tungsten, and alloys such as Chromel, Nichrome, and Inconel have been used for a long time to produce filters with densities of up to 6.0. A disadvantage of such filters is their high specular reflection. At present Inconel is commonly used for high-precision neutral-density filters. Chromium is favoured when tough, unprotected coatings are required (Fig. 76). The operating range and neutrality depends on the substrate materials.

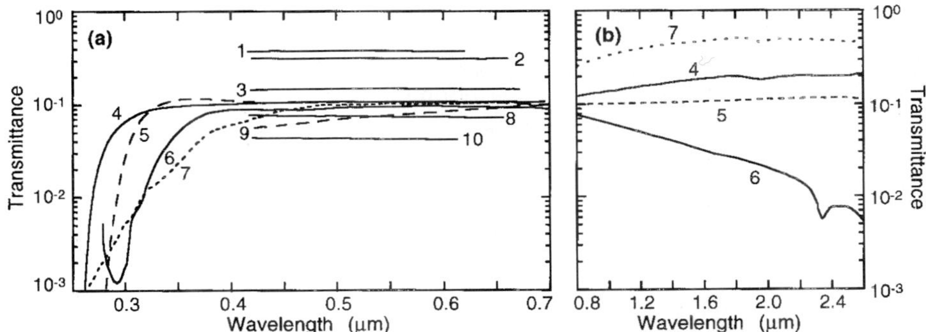

FIGURE 76 *(a)*, *(b)* Spectral transmittance of various neutral density materials. Curve 1: tungsten film on glass; curves 2 and 3: diffuse and specular transmittance of photographic emulsion; curve 4: M-type carbon suspension in gelatine; curve 5: Inconel film on glass; curve 6: photographic silver density; curve 7: Wratten 96 density filter; curve 8: chromium film; curves 9 and 10: Chromel A film on glass evaporated at pressures of 10^{-3} and 10^{-4} Torr, respectively. *(Curves 1 to 3, 9, and 10 after Banning;*[458] *curves 4 to 7, Eastman Kodak Company;*[459] *curve 8, Optical Coating Laboratory.*[340]*)*

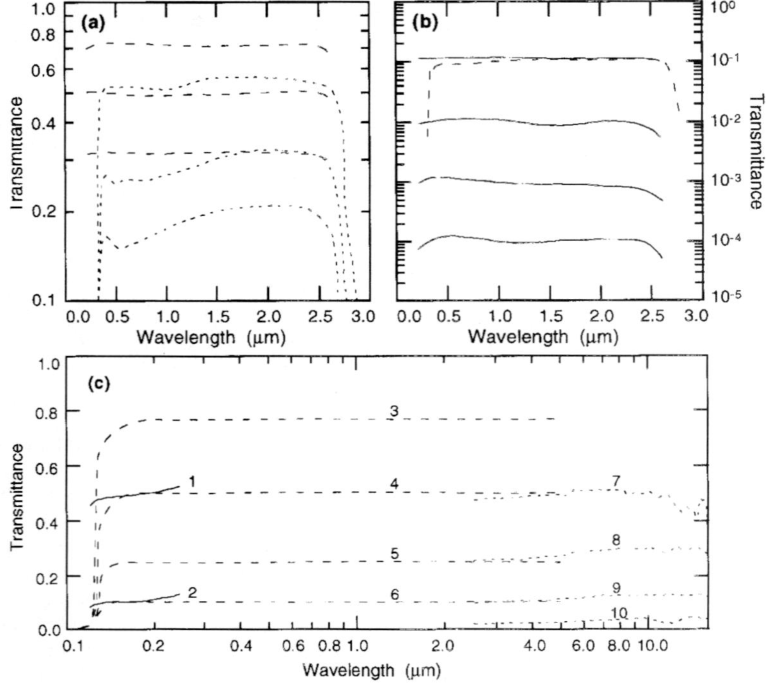

FIGURE 77 Neutral-density attenuators: (*a*), (*b*) Inconel films on glass (*dotted curves, Bausch & Lomb*[444]) and on quartz substrates (*full and broken curves, Corion*[460]); (*c*) Alloy films on MgF_2 (*solid curves, after Acton*[461]); on CaF_2 (*broken curves, after Spindler & Hoyer*[462]); and on Ge (*dotted curves, after Oriel*[454]).

The spectral-transmittance curves of neutral-density filters on magnesium fluoride, calcium fluoride, quartz, glass, sapphire, and germanium substrates are shown in Fig. 77. Linear and circular metal-film neutral-density wedges and step filters are also available commercially.

At times, there may be a need for a neutral attenuation that is not based on absorption. Sets of all-dielectric multilayer coating designs have been published with uniform transmission levels for the ultraviolet,[222] visible,[160,223] and near-infrared[224] parts of the spectrum (Fig. 78).

42.11 INTERFERENCE POLARIZERS AND POLARIZING BEAM SPLITTERS

The dependence of the optical properties of thin-film systems on the plane of polarization of obliquely incident radiation[22] (see "Matrix Theory for the Analysis of Multilayer Systems" in Sec. 42.3) can be exploited to design interference polarizers and polarizing beam splitters with properties that augment those attainable by other means (Chap. 3 in Vol. II of this Handbook). The main difference between a polarizer and a polarizing beam splitter is that in the former only one polarized beam is required, whereas in the latter both beams are to be utilized. A polarizing beam splitter can therefore also be used as a

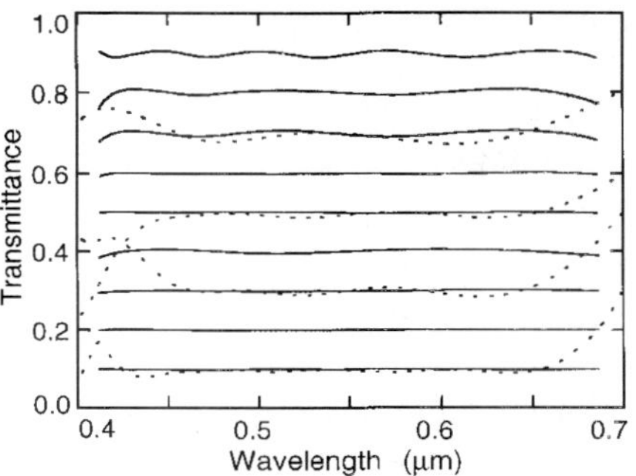

FIGURE 78 Achromatic all-dielectric attenuators. The solid curves are calculated results (*after Hodgkinson*[223]) and the dotted curves represent the performance of commercial coatings (*after TechOptics*[411]).

polarizer. The performance in transmission or reflection of both devices is usually characterized by their degree of polarization P,

$$P = \left| \frac{T_p - T_s}{T_p + T_s} \right| \quad \text{or} \quad P = \left| \frac{R_p - R_s}{R_p + R_s} \right| \tag{59}$$

In a polarizing beam splitter, a high degree of polarization is required in both beams and this is more difficult to achieve. Presently, efficient interference polarizers and polarizing beam splitters can be constructed for the ultraviolet, visible, and infrared spectral regions, and they are of particular interest whenever large areas and low losses are required. Schematic representations of the geometries of some of the devices are given in Fig. 65.

Multicomponent Polarizers

It is always possible to find an angle of incidence at which a substrate coated with a film of quarter-wave effective thickness will reflect only radiation polarized perpendicular to the plane of incidence.[225] This property can be used to construct efficient transmitting polarizers using far fewer plates than necessary in the conventional pile-of-plates polarizer of equal performance (Fig. 65*d*).[226] The calculated degree of polarization attainable with different numbers of plates is shown in Fig. 79 for a series of film indexes and polarizing angles. Experimental results agree closely with the calculations. In polarizers of this type, the variation of the degree of polarization is small over a wavelength span of one octave and for angular apertures of up to 10°.[225] Yet, because of its bulk, this type of polarizer is not frequently used. An exception are polarizers for the infrared, where it is more difficult to produce multilayers composed of many layers. Because of the high refractive indices available in that spectral region, it is possible to achieve a high degree of polarization even after a single transmission through a plate coated with one layer only.[227]

Several geometries of the reflection equivalent of the multiple-plate polarizer exist (Fig. 65). The reflectors can consist of one or more dielectric layers deposited onto nonabsorbing parallel plates or prisms (Fig. 65*s* to *v*). In other devices, the substrates are made of metal, or are coated with an opaque metallic film (Figs. 65*o* to *r*).[228] The angles of incidence on the various mirrors need not be the same. Polarizers of this type are particularly useful in the vacuum ultraviolet[229–231] and infrared[232] spectral regions. The measured performances in the XUV region of two polarizers that are based on three reflections are depicted in Fig. 80. Interesting variants are polarizers that are based on total internal reflection or frustrated total internal reflection (Figs. 65*l* and *m*).

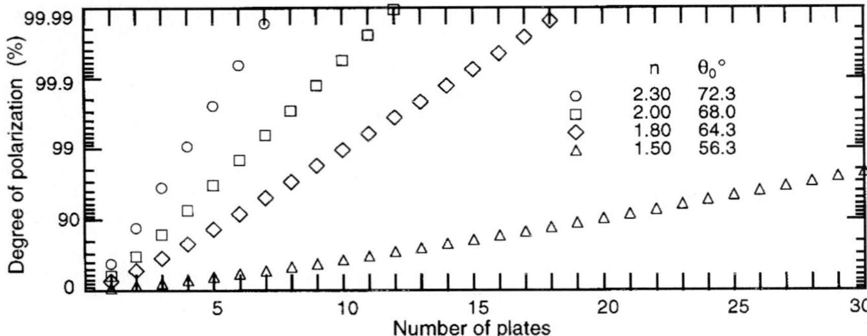

FIGURE 79 Calculated degree of polarization of different numbers of plates of refractive index 1.5 coated on both sides with films of different refractive indexes *n*. The films have effective optical thicknesses of a quarter-wavelength at the appropriate polarizing angle θ_0.

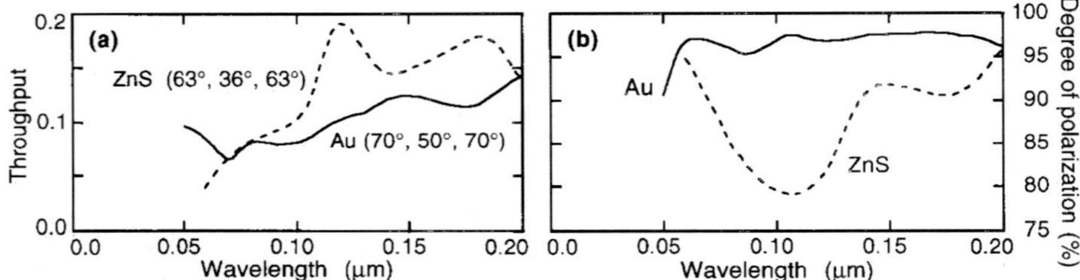

FIGURE 80 Measured performance of two extreme ultraviolet polarizers based on three reflections from Au (70°, 50°, 70°) and ZnS (63°, 36°, 63°) surfaces. The angles of incidence on the three mirrors (see Fig. 65*q*) are given in brackets. *(After Remneva et al.*[463]*)*

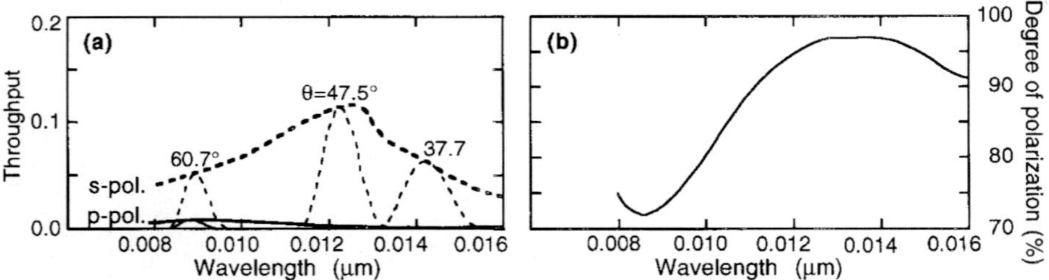

FIGURE 81 Measured performance of an x-ray polarizer consisting of two identical 21-layer Ru-C multilayers on Si substrates. The polarizer can be tuned to different wavelengths by changing the angle of incidence θ of the radiation on the two mirrors. *(After Yanagihara et al.*[464]*)*

Polarizers for the soft x-ray region can be based on the fact that the reflectance of an x-ray multilayer mirror at oblique angles is very different for radiation polarized parallel and perpendicular to the plane of incidence. Because the region of high reflection is very narrow, multilayer x-ray polarizers essentially operate at one wavelength only.[233] However, if the radiation is reflected from two identical mirrors (Fig. 65*p*), it is possible to construct a device with a reasonable throughput (>0.05) that can be tuned over a wide range of wavelengths without changing the direction of the emerging beam (Fig. 81).

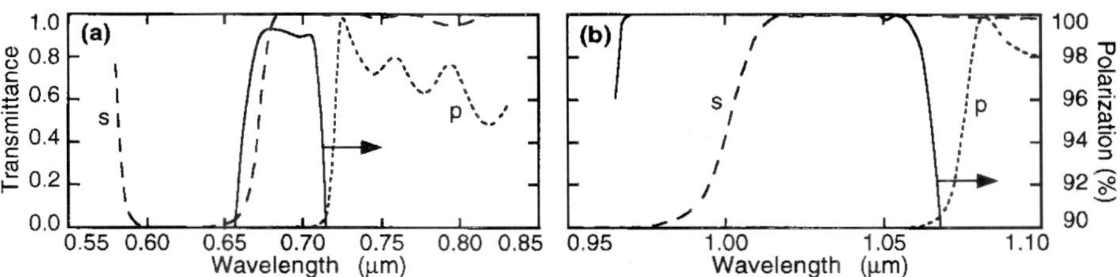

FIGURE 82 Measured performance of two plate polarizers. *(After TechOptics.[411])*

Plate Polarizers

The number of coated plates in the transmission polarizers described here can be reduced without compromising the performance by depositing more than one high-refractive-index layer onto the surface of a plate and by spacing them with low-index films. In particular, it is possible to minimize the surface scatter, plate absorption, and lateral beam displacement by combining all the layers into one coating.

The most common solution for high-power laser applications is plate polarizers that are based on the polarization splitting that occurs at higher angles, for example, at the edges of quarter-wave stack reflectors (Fig. 65*b*).[234] The plate on which the multilayer is deposited is held at Brewster's angle with respect to the incident light to avoid second surface reflections. Usually the long wavelength edge of the reflector is used and the design is somewhat modified to remove the ripples in the transmission band of the *p*-polarized radiation (Fig. 82). The use of other multilayer structures, such as bandpass filters, for the construction of plate polarizers has also been proposed.[235] The wavelength range over which plate polarizers are effective is much smaller than that of polarizers based on a series of coated plates, but this is acceptable for most laser applications.

Other methods for the design of narrowband plate polarizers based on two or three coating materials have been described by Minkov,[236] Mahlein,[211] and Thelen.[237] However, these solutions require many layers and sometimes very oblique angles of incidence.

Embedded Polarizers and Polarizing Beam Splitters

Polarizers and polarizing beam splitters effective over a wider spectral region are obtained when multilayer coatings of the type $[HL]^N H$ or $[(0.5H)L(0.5H)]^N$ are embedded between media of higher refractive index.[238] The higher the refractive-index ratio of the two coating materials used, the fewer the number of layers needed to achieve a certain degree of polarization and the wider the spectral region over which the polarizer will be effective. However, a certain relationship between the refractive indices of the materials and the angle of incidence must be satisfied.[239] The optical thicknesses of the quarter-wave layers should be matched for the angle of incidence.

A particularly convenient polarizer with no lateral beam displacement results when the multilayer is embedded between two right-angled prisms, as shown in Fig. 65*i*.[240,241] MacNeille polarizers operate over a very broad range of wavelengths (Figs. 83*a* and *b*). For best results, the following relationship between the angle of incidence θ_p and the refractive indices of the prism and the layers should be satisfied:

$$n_p \sin \theta_p = \frac{n_L n_H}{(n_L^2 + n_H^2)^{1/2}} \tag{60}$$

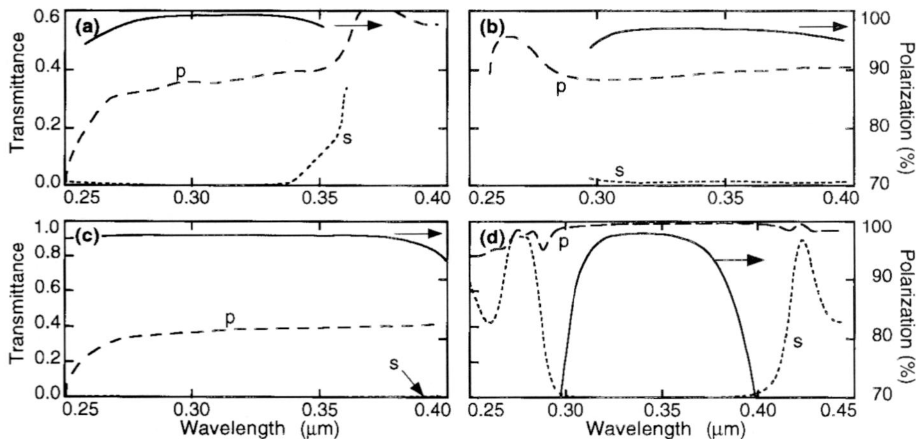

FIGURE 83 The measured degree of polarization P and transmittance of parallel and crossed McNeille interference polarizers for the ultraviolet spectral region: (c) represents the results obtained when the cemented polarizers shown in (a) and (b) are placed in series *(after Sokolova and Krylova*[465]*)*; (d) is the measured performance of an optically contacted polarizer with a high laser damage threshold *(after Wimperis*[466]*)*.

This expression is independent of the thicknesses of the layers and, were it not for the dispersion of the optical constants, the transmission for the p polarization would be 1.0 across the entire spectral region. However, the dispersion of the high-index material will tend to decrease the useful spectral range of the polarizer.[242] It is possible to select the V-number of the substrate material in order to decrease this disturbing effect. The rejection of the unwanted polarization will normally not be the same in both beams, although this can be achieved at the expense of the wavelength range.[243] If polarizers for an even wider spectral region are required, it is possible to extend the range by placing two polarizers in series (Fig. 83c), by using the technique of superposition of stacks with contiguous high-reflectance zones (see "All-Dielectric Broadband Reflectors" in Sec. 42.8) or of two periodic multilayers with thickness ratios of 1:1 and 1:2.[156] When both beams are used, the useful angular field of the MacNeille polarizer is of the order of ±2°. It is possible to increase it to ±10°, but again at the expense of the spectral range.[244]

When the MacNeille polarizer is to be used as a polarizing beam splitter, it is possible to design it for normal incidence of the beams onto the prism faces, or for a 90° deflection between the two beams (Fig. 65j).[243,245] However, it is more convenient to embed the multilayers between two 45° prisms. A higher-refractive-index prism material is required if Eq. (52) is to be satisfied[244] and the effective wavelength range will once again be reduced (Fig. 84).[246]

MacNeille polarizers and polarizing beam-splitter cubes suffer from several disadvantages. The aperture of the polarizers is limited by the size, cost, and availability of the prism material. The attainable degree of polarization is governed by the residual birefringence in the prisms. Cemented polarizers cannot be used with high-power lasers. To overcome these difficulties, a liquid prism polarizer has been proposed consisting of a multilayer on a quartz plate that is immersed in distilled water (Figs. 65k, 85).[247] Another way of avoiding the use of a cement is to optically contact the polarizer prisms (Figs. 65g, 83d).[248] More frequently, the layers are deposited onto the hypotenuse of one or two air-spaced prisms (Fig. 65e).[249] However, this arrangement is similar to a plate polarizer designed for use at 45° and so is its performance.

In Fig. 86a to h the calculated spectral and angular performances of a number of

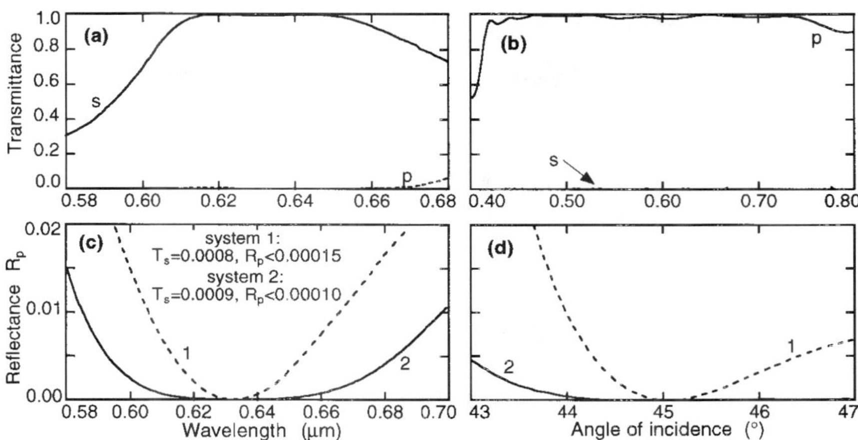

FIGURE 84 Measured performance of polarizing beam-splitting cubes: (*a*), (*b*) measured spectral transmittance curves of two commercial devices *(after TechOptics*[411]*)*; (*c*), (*d*) measured spectral and angular performance of two different systems *(after Netterfield*[246]*)*.

polarizers and polarizing beam splitters are compared. A theoretical comparison of the properties of the MacNeille, cube, and plate polarizers for one wavelength has been given by Cojocaru.[250]

42.12 BANDPASS FILTERS

An ideal bandpass filter transmits all the incident radiation in one spectral region and rejects all the other radiation. Such a filter is completely described by the width of the transmission region and the wavelength at which it is centered. Practical filters are not

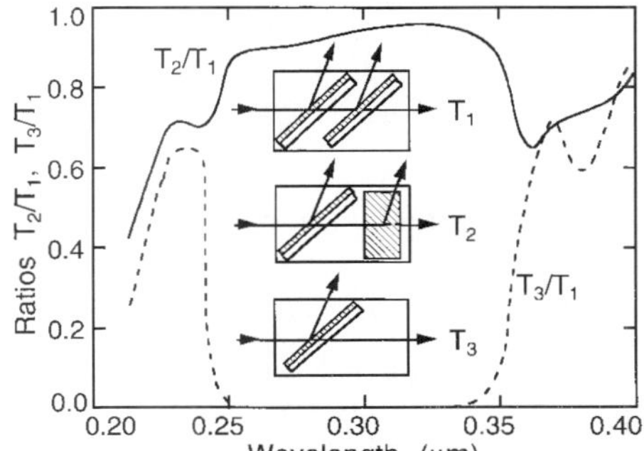

FIGURE 85 Measured spectral performance of two parallel and crossed coated plates forming a liquid prism McNeille polarizer for the ultraviolet part of the spectrum. The multilayers on both plates were identical and consisted of 13 alternate quarter-wave layers of HfO_2 and SiO_2. *(After Dobrowolski.*[247]*)*

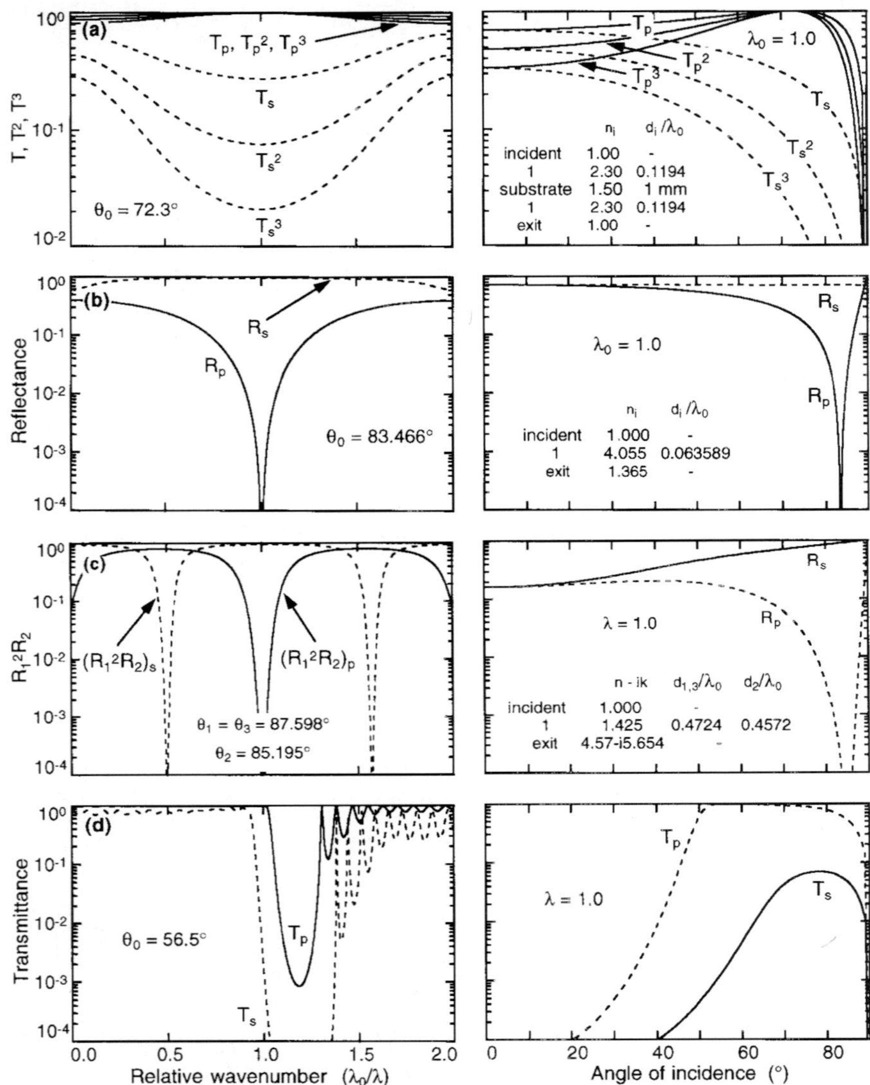

FIGURE 86 Calculated spectral and angular performance of several types of polarizers and polarizing beam splitters with an air incident medium. Calculations assume that all nonmetals are absorption- and dispersion-free. (*a*) Multiple plate polarizer; (*b*) single-reflection polarizer *(after Azzam*[227]*)*; (*c*) three-reflection polarizer *(after Thonn*[467]); (*d*) plate polarizer *(after Songer*[468]*)*.

perfect and require more parameters to adequately describe their performance. No uniform terminology has yet been developed for this purpose. Care should be taken when reading and writing specifications since often different terms are used to describe different types of filters, and sometimes quantities bearing the same name are defined differently.

The position of the transmission band is variously specified by the wavelength λ_{max} at which the maximum transmission occurs, the wavelength λ_o about which the filter

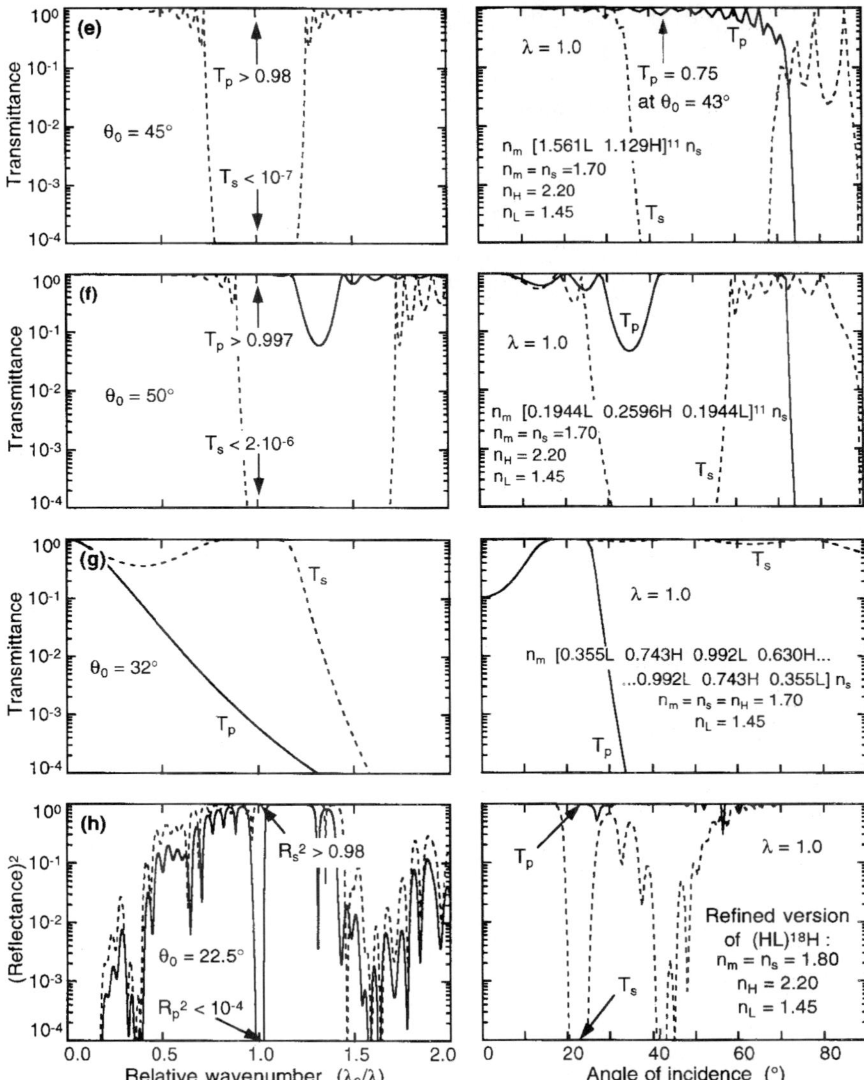

FIGURE 86 (*Continued*) Calculated spectral and angular performance of several types of polarizers and polarizing beam splitters with a nonair incident medium: (*e*) MacNeille polarizing beam splitter *(after Mouchart*[244]*)*; (*f*) wide-angle MacNeille polarizing beam splitter *(after Mouchart*[244]*)*; (*g*) frustrated-total-internal-reflection polarizing beam splitter *(after Lees and Baumeister*[469]*)*; (*h*) penta prism polarizer *(after Lotem and Rabinovich*[531]*)*.

passband is symmetrical, or the spectral center of gravity λ_c of the band. When specifying the tolerance on λ_o it should be remembered that the peak of interference filters can be moved only towards shorter wavelengths by tilting (see "Matrix Theory for the Analysis of Multilayer Systems" in Sec. 42.3).

The peak transmittance T_o may or may not take into account the absorption within the substrate and/or blocking filters used to remove the unwanted transmission of the interference filter away from the principal passband (Fig. 87).

The half-width (HW) $\Delta\lambda_{0.5}$ of the filter is the difference between the wavelengths at which the transmittance is a half of T_o. This quantity is also sometimes called the full-width half-maximum (FWHM). It is often expressed as a percentage of λ_o. The base-width (BW) $\Delta\lambda_{0.01}$ is similarly defined. The ratio $\Delta\lambda_{0.01}/\Delta\lambda_{0.5}$, sometimes called the *shape factor*, indicates how "square" the transmission band is. Sometimes widths corresponding to other fractions of the transmittance are used to define it.

The minimum transmittance $T_{\min}$ of the filter does not take into account the effect of blocking filters. The quantity $T_{\min}/T_o$ is called the *rejection ratio*.

In all-dielectric transmission-band filters, the transmittance rises at some distance on either side of the transmission band. The distance over which the transmittance is low is called the *rejection region*. The distance between the two transmission maxima adjacent to the principal transmission band is called *free spectral range*. Should either of these two quantities be inadequate, auxiliary blocking filters might have to be provided.

The ultimate measure of the suitability of a bandpass filter with a blocked spectral transmittance $T(\lambda)$ for a particular application is the signal-to-noise ratio SN, defined in terms of the spectral energy distribution $I(\lambda)$ of the source and the spectral detectivity $D(\lambda)$ of the detector,

$$\mathrm{SN} = \frac{\int_{\lambda_1}^{\lambda_2} I(\lambda)T(\lambda)D(\lambda)\,d\lambda}{\int_0^{\lambda_1} I(\lambda)T(\lambda)D(\lambda)\,d\lambda + \int_{\lambda_2}^{\infty} I(\lambda)T(\lambda)D(\lambda)\,d\lambda} \tag{61}$$

where λ_1, λ_2 are the lower and upper limits of the transmission region of the filter. The SN ratio is sometimes expressed in terms of optical density.

Useful general reviews of bandpass filters exist.[2,251,252] Interference filters, and especially bandpass filters, are increasingly deposited in complicated millimeter and submillimeter patterns for use with display devices and detectors.[253,254] Very fine masks or photolithographic processes are required to produce such structures (Fig. 88).

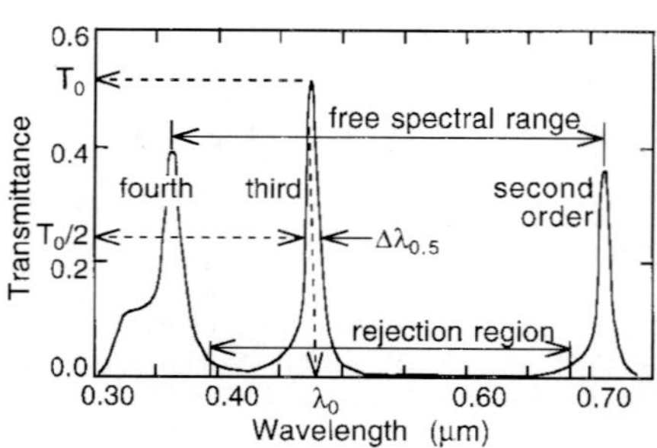

FIGURE 87 Definition of some of the terms used to describe the properties of narrow-bandpass filters (see section, "Bandpass Filters"). The curve represents the measured transmittance of an unblocked second-order metal-dielectric interference filter of the Fabry-Perot type (see "Filters with Metallic Reflecting Coatings"). *(After Bausch & Lomb.[444])*

FIGURE 88 Four-color checkerboard pattern for use with a 64 × 64 element focal plate HgCdTe detector array. The dimensions of each element are 100 × 100 μm. *(Reproduced with permission from Barr Associates.[470])*

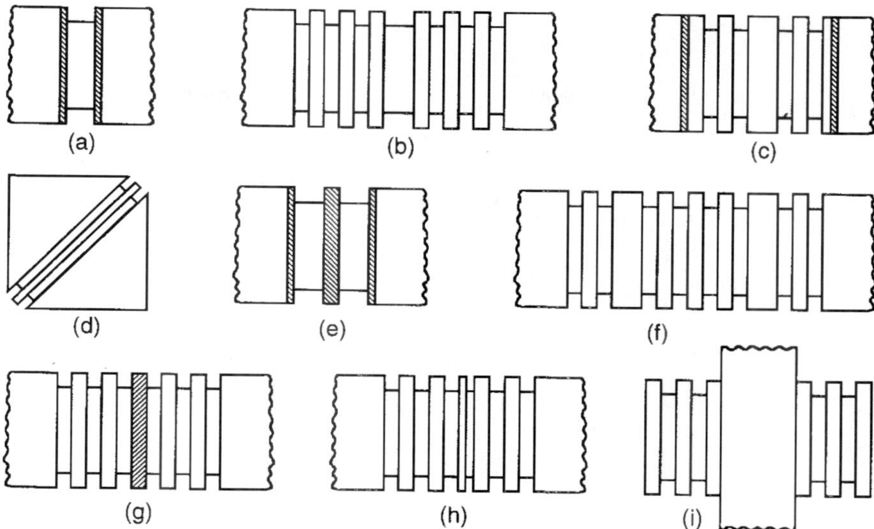

FIGURE 89 Schematic representations of various types of bandpass interference filters: (*a*) to (*c*) Fabry-Perot interference filters with metal, dielectric, and metal-dielectric reflectors; (*d*) frustrated-total-internal-reflection filter; (*e*) and (*f*) square-top multicavity filters with metal and dielectric reflectors; (*g*) induced-transmission filter; (*h*) phase-dispersion (spacerless) interference filter; (*i*) Fabry-Perot filter with a mica or quartz spacer.

Narrow- and Medium-bandpass filters (0.1 to 35 percent HW)

Even though the essential components of Fabry-Perot (FP) interference filters, i.e., the spacer and the two reflectors that surround it, can take on many different forms (see Fig. 89), the filters are essentially low-order FP interferometers and hence the theory developed for the latter (see, for example, Born and Wolf[21]) applies in full.

The transmittance of an FP-type filter, not allowing for absorption and multiple reflections within the substrate, is given by

$$T = \frac{T_R^2}{(1-R)^2 + 4R\sin^2\delta} \tag{62}$$

where

$$T_R = \sqrt{T_1 T_2}, \qquad R = \sqrt{R_1 R_2} \tag{63}$$

T_1, R_1 and T_2, R_2 are the transmittances and reflectances of the first and second reflectors, respectively, as seen from within the spacer medium, and δ is given by

$$\delta = \frac{2\pi}{\lambda} nd \cos\theta + \varepsilon \tag{64}$$

$$\varepsilon = \frac{\varepsilon_1 + \varepsilon_2}{2} \tag{65}$$

n, d, and ϕ are the refractive index, thickness, and angle of refraction of the spacer, respectively, and ε_1 and ε_2 are the phase changes on reflection from the spacer side of the first and second reflectors at a wavelength λ. Maxima of T occurs at wavelengths

$$\lambda_o = \frac{2nd \cos \theta}{k - \varepsilon/\pi} \qquad k = 0, 1, 2 \cdots \tag{66}$$

and are given by

$$T_o = \left(\frac{T_R}{1 - R}\right)^2 = \left(\frac{1}{1 + A/T_r}\right)^2 \tag{67}$$

$A(=1 - T_R - R)$ is the mean absorption of the reflectors. The minimum transmittance

$$T_{\min} = \left(\frac{T_R}{1 + R}\right)^2 \tag{68}$$

occurs at

$$\lambda_{\min} = \frac{2nd \cos \theta}{k - \varepsilon/\pi} \qquad k = \tfrac{1}{2}, \tfrac{3}{2}, \tfrac{5}{2} \cdots \tag{69}$$

If T_R and R are essentially the same at λ_o and $\lambda_{\min}$, the rejection ratio is given by

$$\frac{T_{\min}}{T_o} = \left(\frac{1 - R}{1 + R}\right)^2 \tag{70}$$

For $R > 0.7$ the half-width of the transmission band (expressed as a percentage of λ_o) is given by

$$\frac{\Delta\lambda_{0.5}}{\lambda_o} \times 100 \approx \frac{1 - R}{\sqrt{R}} \frac{100}{\dfrac{2\pi nd \cos \theta}{\lambda_o} - \lambda_o \dfrac{\partial \varepsilon}{\partial \lambda}} \tag{71}$$

For a given order of interference the half-width and rejection ratio cannot be varied independently. The formula

$$T \approx \frac{T_o}{1 + 4[(\lambda - \lambda_o)/\Delta\lambda_{0.5}]^2} \tag{72}$$

valid for FP filters with small values of $\partial\varepsilon/\partial\lambda$ in the neighborhood of λ_o, represents a lorentzian line shape with $\Delta\lambda_{0.1} = 3\Delta\lambda_{0.5}$ and $\Delta\lambda_{0.01} = 10\Delta\lambda_{0.5}$. The shape factor of all FP-type interference filters is therefore of the order of 10.

Southwell has recently shown that a spacer in an interference filter need not consist of a single layer only, and that there are some advantages when it is partitioned into a number of layers.[255]

Fabry-Perot Interference Filters (0.1 to 10 percent HW)

Filters with Metallic Reflecting Coatings. This is the first interference bandpass filter ever made, and the simplest.[256] It consists of two partially transmitting, highly reflecting metal layers separated by a dielectric film and is symbolically represented by MDM (Fig. 89*a*). The best metallic reflectors currently available are aluminium and silver for the 0.125- to 0.34-μm and 0.34- to 3.0-μm spectral ranges, respectively. The measured spectral-transmittance curves of a number of typical filters are shown in Fig. 90.

The phase changes on reflection at the spacer-metal-reflector surface are finite and hence they affect the position of the transmission maxima [Eq. (66)]. But the dispersion of the phase change on reflection can be neglected, and so the half-width depends only on the reflectance of the metal layers and the order of the spacer. Filters with half-widths of 1 to 8 percent are common. Because the ratio A/T is not very small for metals, the maximum transmission is limited [Eq. (67)]. Maximum transmittances of 40 percent are relatively common. For filters with the narrower half-widths, or for shorter wavelengths, transmittances of the order of 20 percent have to be accepted. This is much less than the transmittances of all-dielectric filters of comparable half-widths. Nevertheless, filters of this type are useful because their transmittances remain low except at wavelengths at which the first- and higher-order transmission maxima occur [Eq. (66)]. Blocking is thus easy. In particular, first-order filters usually do not require any blocking on the high-wavelength side—a difficult task at all times. The rejection of the filters is good, though not spectacular. If a better rejection is required and a lower transmittance can be tolerated, two identical filters may be cemented together (curve 6, Fig. 90). This is possible because of the finite absorption in the metal films. Alternatively, metal square-top filters or filters with even more complicated structures can be used (see Sec. 42.12).

Filters with All-dielectric Reflectors. Above 0.2 μm, the metallic reflectors can be replaced by all-dielectric quarter-wave stacks (*see* Sec. 42.7).[240] The symbolic representation of such a filter (Fig. 89*b*) is, for example, $[HL]^N 2mH[LH]^N$ or $H[LH]^N 2mL[HL]^N H$, H and L being quarter-wavelength layers of high and low refractive indices, respectively; m is the order of the spacer and N the number of full periods in the reflecting stacks. The phase change on reflection at the boundary between the spacer and such a reflector does not affect the position of λ_o (Eq. (66)] but the dispersion of the phase change on reflection is finite, depends on the materials used, and for lower-order spacers, contributes very significantly to the reduction of the half-width of the transmission band [Eq. (71)]. Expressions for the half-width

$$\frac{\Delta\lambda_{0.5}}{\lambda_o} \times 100 = \frac{4n_o n_L^{2N}(n_H - n_L) \times 100}{m\pi n_H^{2N+1}(n_H - n_L + n_L/m)}$$

$$= \frac{4n_o n_L^{2N-1}(n_H - n_L) \times 100}{m\pi n_H^{2N}(n_H - n_L + n_L/m)} \tag{73}$$

for high- and low-refractive index spacers, respectively have been given by Macleod.[2]

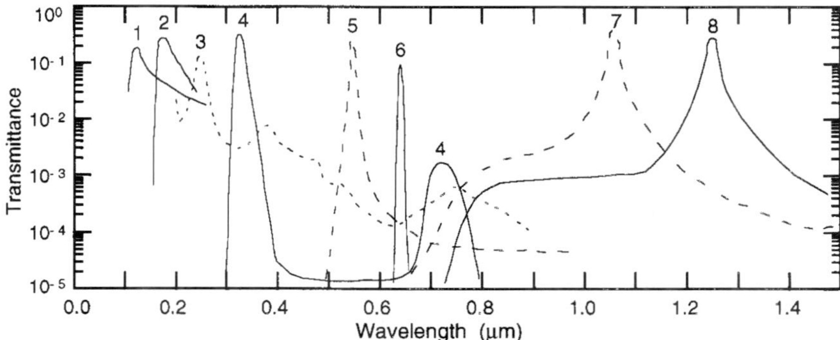

FIGURE 90 Measured transmittance of FP filters with metallic reflecting coatings. Curve 6 represents the transmission of two identical filters cemented together. *(Curve 1, after Harrison;[471] curve 2, after Bradley et al.;[472] curves 3 and 5 to 7, Balzers;[473] curves 4 and 8, Schott & Gen.[474])*

By choosing a suitable combination of the reflectance and order of the spacer almost any half-width between 0.1 and 5 percent can be achieved in the visible part of the spectrum while maintaining a useful rejection ratio.

The maximum transmittances of all-dielectric FP filters depart from unity because of the finite absorption, scattering, and errors in the thicknesses and refractive indexes of the films. In the central part of the visible spectrum, maximum transmittances of 0.8 are normal for unblocked filters with a half-width of 1 percent, although higher transmittances can be achieved. This figure is gradually reduced as λ_o approaches 0.2 or 20 μm, and filters with narrower half-widths become impracticable for lack of adequate transmittance.

The transmittance away from the transmission maximum is low only over the extent of the rejection region of the two materials used for the construction of the reflectors (Fig. 32), and additional blocking is often required on both the long- and the short-wavelength sides. This can result in a considerable lowering of the maximum transmittance of the blocked filter, a 30 to 40 percent loss being not uncommon for filters peaked in the ultraviolet or infrared spectral regions.

For those parts of the visible and infrared for which nonabsorbing mechanically robust coating materials abound, square-top interference filters (see "Square-Top Multicavity, Bandpass Filters" in Sec. 42.12) are often preferred because of their better shape factor and higher rejection ratio. All-dielectric FP filters are still attractive in the ultraviolet, where there is a lack of such materials and where thickness monitoring is difficult, and also in the far-infrared, where very thick layers are required.

The measured transmittance curves of a number of typical all-dielectric FP filters are shown later in Fig. 94. An additional curve on a smaller scale is given, whenever necessary, to show the transmittance away from the passband.

Filters with Metal-dielectric Reflectors. In these filters the reflectors consist of metal layers whose reflectance has been enhanced through the addition of several dielectric layers (see "Enhancement of Reflection" in Sec. 42.16) (Fig. 89*c*).[257] The properties of such filters are intermediate to those described in the two previous sections.

Frustrated-Total-Internal-Reflection Filters. These are essentially FP filters in which the spacer layer is surrounded by two frustrated-total-reflection surfaces (Fig. 89*d*).[257,258] They have not found wide applications as bandpass filters because the finite absorption and scattering within the layers have prevented the theoretically expected high transmittance and small half-widths from being realized and because the angular variation of the wavelength of the transmission peaks is very high.

Square-top Multicavity Bandpass Filters (0.1 to 35 percent HW). A filter with a "squarer" shape that does not suffer from some of the disadvantages of the FP filters results when the basic FP structure is repeated two or more times.[119,259,260] Such filters may be based on metal[261] or all-dielectric[262,263] reflectors (Fig. 89*e* and *f*). Thus, for example, $[HL]^N 2H[LH]^N C[HL]^N 2H[LH]^N$ represents an all-dielectric square-top filter in which the FP structure $[HL]^N 2H[LH]^N$ is repeated twice. The quarter-wavelength layer C is called a *coupling* or *tie layer,* and the half-wavelength-thick spacer layers $2H$ are often called *cavities.* Small departures from this model are made at times to improve the transmittance in the pass band or the angular properties of the filter.

The half-widths of the narrower multicavity filters of the preceding type do not differ very significantly from those of the basic FP structure [Eq. (71)]. The shape factors decrease with an increase in the number of the cavities and do not seem to depend on the materials.[259] They are approximately 11, 3.5, 2.0, and 1.5 for one, two, three, and four cavities, respectively. The minimum transmittance in the rejection region is roughly that which could be obtained if the filter were composed entirely of $\lambda/4$ layers [Eq. (44)]. Unlike in the FP filter, there is therefore some independent control of the half-width and rejection ratio. These various points are illustrated in Fig. 91. The peak transmittance of multicavity square-top filters is less affected by the residual absorption in the layers than

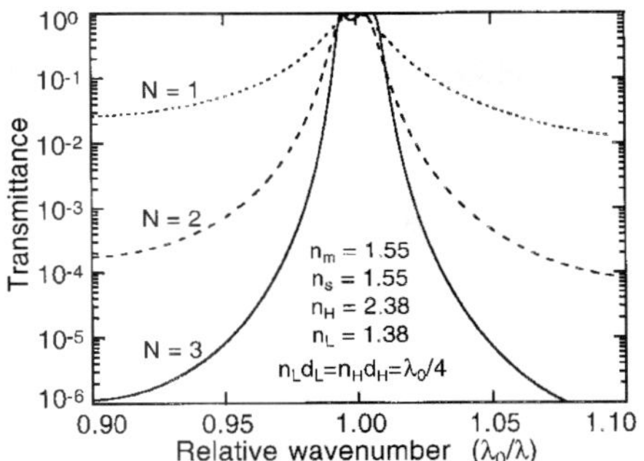

FIGURE 91 Calculated transmittance on a logarithmic scale of the bandpass filters: *air-*$[[(0.5H)L(0.5H)]^3H[(0.5H)L(0.5H)]^3]^N$*-glass*, $N =$ 1, 2, and 3.

that of FP-type filters. As in the case of their FP counterparts, metal-dielectric square-top filters can be cemented together to enhance the rejection (curve 5, Fig. 92).

The improvement in the performance of square-top bandpass filters over that of the FP type is such that, despite their more critical and expensive production, most manufacturers regard them as their standard line of filters. The spectral-transmittance characteristics of typical commercially produced metal-dielectric and all-dielectric bandpass interference filters of different half-widths are shown in Figs. 92 and 95 to 98, respectively. Filters with intermediate half-widths and peak wavelengths can readily be obtained.

For very critical or special applications, multicavity filters are designed and constructed with properties that exceed those shown in the preceding figures. For example, for the use in fiber-optic communications systems, multicavity filters are required in which the peak transmittance closely approaches unity. Various procedures for the design of such filters, including some that are based on the use of Chebyshev polynomials, have been described.[18,264,265] Special care has to be taken during the manufacture of the coatings to meet this requirement. Typical measured spectral transmittance curves are shown in Fig. 99. For other applications, such as fluorescence or Raman spectroscopy, the peak

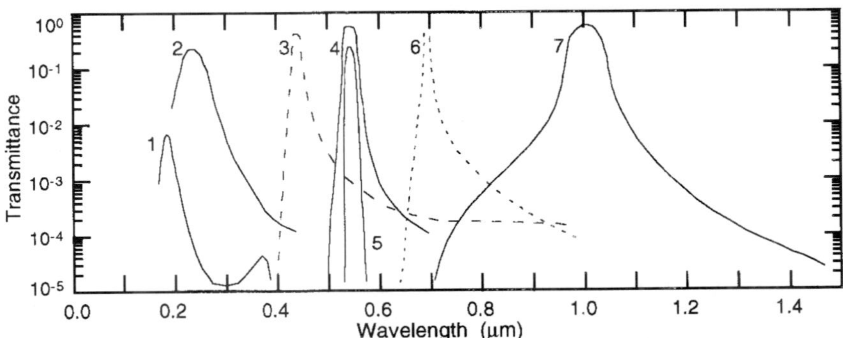

FIGURE 92 Measured transmittance of square-top bandpass filters with metallic reflecting coatings. Curve 5 corresponds to the transmission of two identical filters cemented together. *(Curves 1 and 2, Schröder;*[475] *curves 3 and 6, Balzers;*[473] *curves 4, 5, and 7, Schott & Gen.*[474]*)*

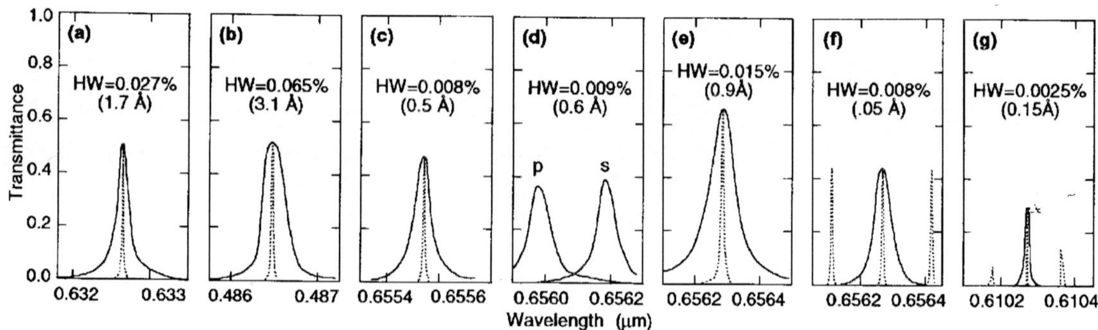

FIGURE 93 Measured transmittance of very narrow bandpass interference filters with half-widths less than 0.1 percent. Evaporated spacers: (*a*) *(after Meltzer*[476]*)*; (*b*) *(after Eather and Reasont*[276]*)*; (*c*) mica interference filter for H_α; (*d*) mica interference filter with transmission bands polarized at right angles to one another *(Heliotek*[448]*)*; (*e*) single and (*f*) and (*g*) double quartz-spacer interference filters *(after Austin*[287]*)*. The dotted curves in Figs. 93 and 94 represent the transmittances of the filters plotted over a ten-times-wider spectral region.

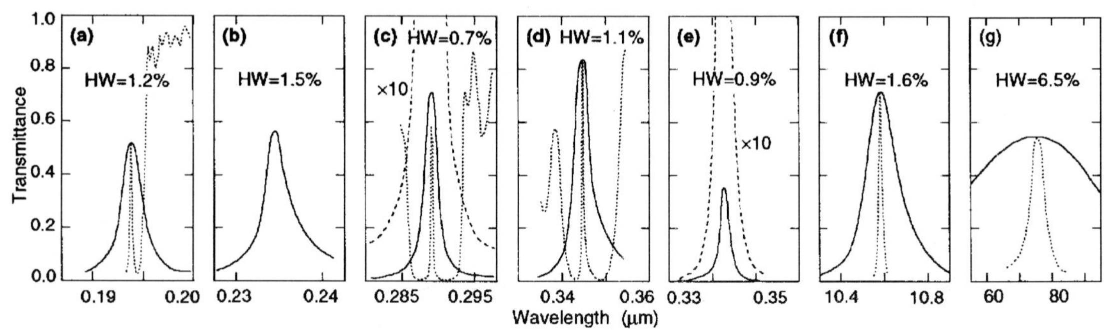

FIGURE 94 Measured transmittance of FP all-dielectric interference filters with narrow half-widths. Evaporated spacers: (*a*) *(after Cohendet and Saudreau*[477]*)*; (*b*) and (*d*) *(after Motovilov*[478]*)*; (*c*) and (*e*) *(after Neilson and Ring*[479]*)*; (*f*) *(after Turner and Walsh*[480]*)*; (*g*) *(after Smith and Seely*[481] *)*. The dashed curves in Figs. 94 and 95 correspond to a transmission range of 0.0 to 0.1.

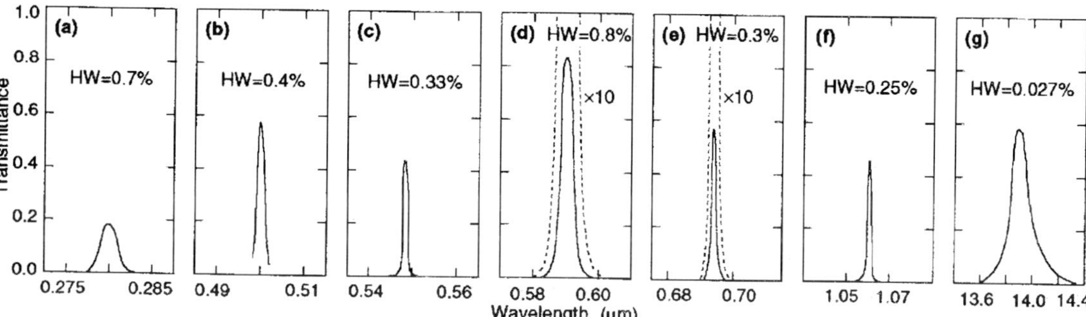

FIGURE 95 Measured transmittances of fully blocked square-top interference filters with half-widths between 0.25 and 1.4 percent. (*a*), (*c*), and (*f*) Corion;[460] (*b*) *(after Blifford*[339]*)*; (*d*) *(Heliotek*[448]*)*; (*e*) *(Baird Atomic*[482]*)*; (*g*) is the only filter in the series that is not blocked *(after Smith and Seely*[481]*)*.

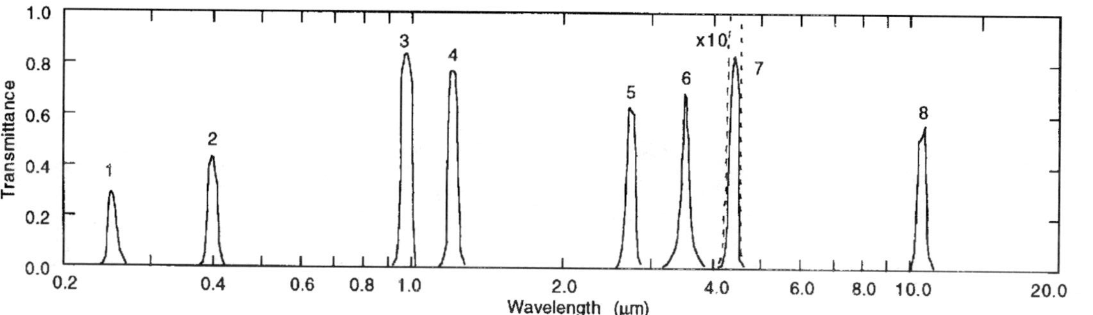

FIGURE 96 Measured transmittances of fully blocked square-top interference filters with half-widths of the order of 5 percent. *(Curves 1 and 2, Corion;*[460] *curve 3, Spectrum Systems;*[483] *curves 4 and 6, after Turner;*[408] *curve 5, Eastman Kodak;*[445] *curves 7 and 8, Optical Coating Laboratory.*[340,443]*)*

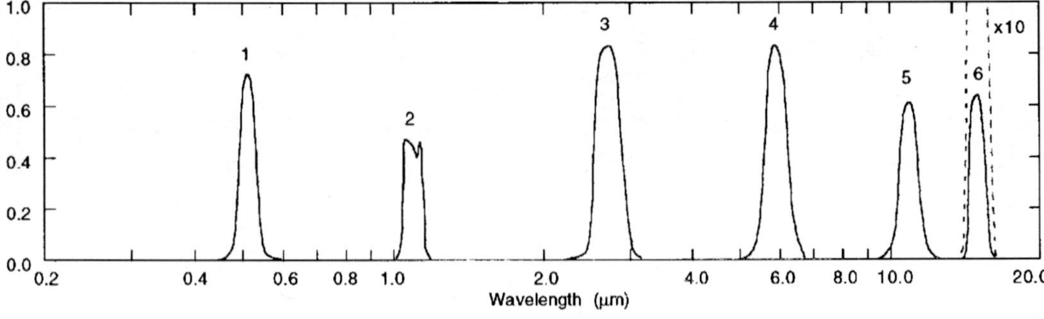

FIGURE 97 Measured transmittances of fully blocked square-top interference filters with half-widths of the order of 10 percent. *(Curve 1, Bausch and Lomb;*[444] *curve 2, Baird Atomic;*[482] *curve 3, Infrared Industries;*[484] *curve 4, after Turner;*[408] *curve 5, Eastman Kodak;*[445] *curve 6, Optical Coating Laboratory.*[443]*)*

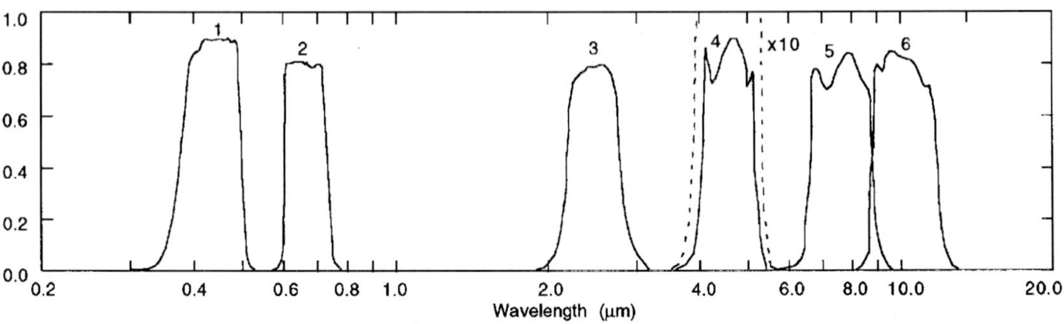

FIGURE 98 Measured transmittances of blocked square-top interference filters with half-widths of the order of 25 percent. *(Curve 1, Heliotek;*[448] *curves 2 and 4, Optical Coating Laboratory;*[340] *curves 3 and 6, after Turner and Walsh*[480] *and Turner;*[408] *curve 5, Infrared Industries.*[484]*)*

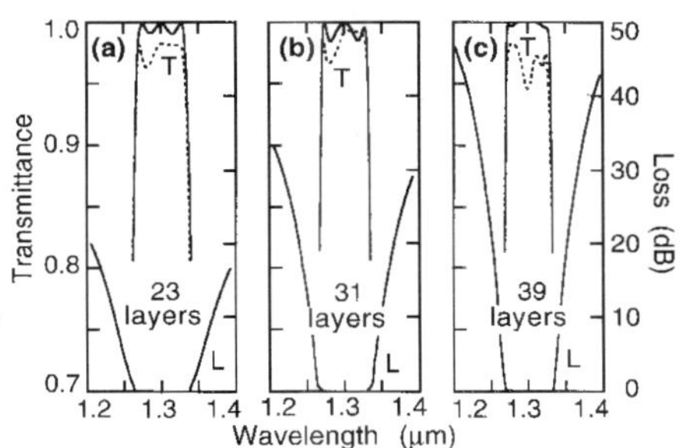

FIGURE 99 Calculated and experimental spectral transmittance and measured attenuation curves of (*a*) 3-, (*b*) 4-, and (*c*) 5-cavity bandpass filters. *(After Minowa.*[485]*)*

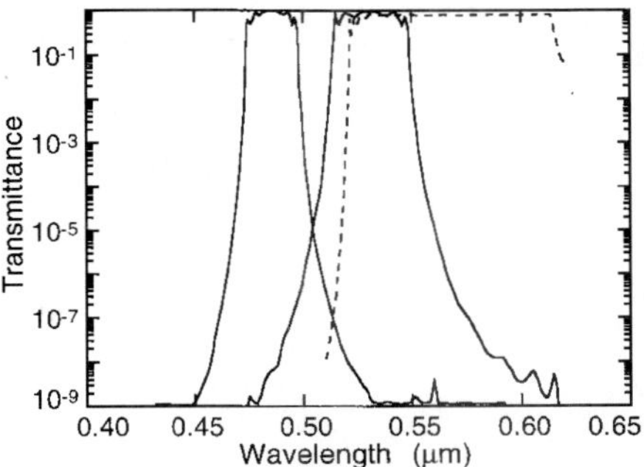

FIGURE 100 Measured spectral transmission characteristics of two bandpass filters for fluorescence applications and of a cut-off filter for Raman spectroscopy. *(After Omega Optical, Inc.*[432]*)*

transmittance is not important, but signal-to-noise ratios of the order of 10^{-8} are required. This necessitates the use of many cavities (Fig. 100).

Induced-transmission Filters. The transmittance of a metal layer can be considerably enhanced by surrounding it with suitable multilayer structures (Fig. 89*g*). Thus, for example, it is possible to induce a transmittance of 65 percent at $\lambda = 0.25\ \mu$m in a 0.03-μm-thick aluminum film which, when deposited directly onto a quartz substrate, would transmit only 2.5 percent of the same radiation.[266] The induced transmittance is highly wavelength-sensitive and can be used to construct bandpass filters containing one or more metal layers.[267–274] Induced-transmission filters combine the good long-wavelength attenuation properties of the more common types of metal/dielectric filters (earlier in this section) with peak transmittances that are closer to those of all-dielectric filters. The performances of some experimentally produced induced-transmission filters are shown in Fig. 101.

Very Narrow Bandpass Filters (HW—0.1 percent)

It follows from Eq. (71) that the half-widths of interference filters can be reduced by increasing the reflectance of the reflectors, the order of interference of the spacer, or the dispersion of the phase change on reflection. All these approaches have been tried in the past.

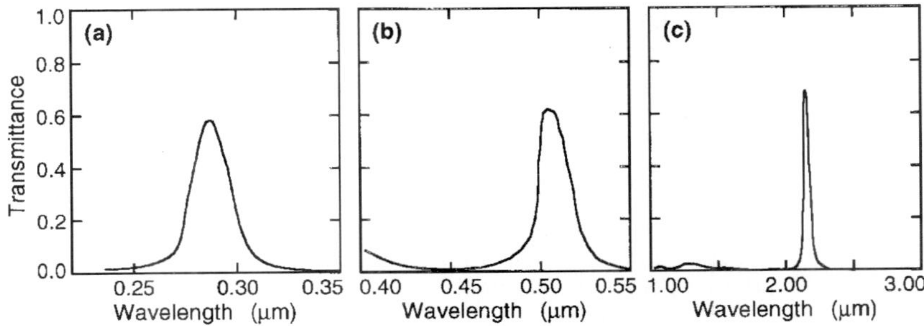

FIGURE 101 Measured transmittance of induced transmission filters for the ultraviolet, visible, and infrared spectral regions. (*a*) *(after Tsypin*[486]*)*; (*b*) *(after Berning and Turner*[267]*)*; (*c*) *(after Holloway and Lissberger*[270]*)*.

Filters with Evaporated Spacers (***HW*** $>$ **0.03 *percent***). In narrowband filters of conventional construction, both high-reflectance and higher-order spacers are used. The manufacturing process is quite critical, and attention must be paid to details. The films have to be very uniform over the filter area, and they must not absorb or scatter. They must not age, or, alternatively, their ageing must be capable of being accelerated or arrested. Monitoring must be precise, so that the peak occurs at or close to the desired wavelength.

Both FP and square-top filters of very narrow bandwidths can be made, the latter being preferable for most applications. The limit on the half-widths of this type of filter seems to be of the order of 0.03 percent.[275] The performance of two commercially produced filters of this type are shown in Fig. 93*a* and *b*.

Fabry-Perot Filters with Solid Spacers. In practice, the half-width of interference filters cannot be reduced indefinitely by increasing the optical thickness of an evaporated spacer [Eq. (71)] because when the latter exceeds about two wavelengths, it may become too rough to be useful.[276] A high-order filter can, however, be constructed by evaporating reflecting coatings on either side of a thin prefabricated spacer (Fig. 89*i*).

Mica Spacers (*HW* $>$ 0.01 *percent*). Transmission bands in silvered mica were probably first observed by Wood,[277] but the deliberate use of mica to construct filters came much later.[278–281] The construction of mica interference filters with transmittances of 30 to 80 percent per polarization for half-widths of the order of 0.01 to 0.1 percent in the 0.45- to 2.0-μm wavelength region is relatively straightforward.[280] The position of the transmission peak can be located within a fraction of an angstrom, does not change with time, and can be sufficiently uniform over areas of 2- to 5-cm diameter. Because of the very high order of interference (70 to 700 orders), the spectral free range is quite small, and auxiliary filtering is necessary for most applications. Unless the thickness of the mica is specially selected, the birefringence of mica will result in two mutually perpendicularly polarized sets of transmission bands, a fact that can be used to advantage in some applications. The spectral-transmittance curve of a fully blocked mica interference filter for H_α is shown in Fig. 93*c*.

Optically Polished Solid Spacers (*HW* $>$ 0.002 *percent*). It is possible to construct very narrow bandpass filters having thin fused-quartz spacers.[282–286] A good fused-quartz flat is coated with an all-dielectric reflector, and this coated surface is optically contacted to another flat.[287,288] The flat is then ground down and polished to form a spacer layer of the required thickness, and the second reflector layers are applied to complete the filter. As in the mica filters, the position of the transmission band is very stable, and auxiliary blocking filters are needed because of the small free spectral range. The transmittance of filters with silica spacers is higher than that of corresponding mica filters because fused quartz is highly transparent and is not birefringent. A typical unblocked filter with a clear aperture of 3.5 cm and a half-width of 0.007 percent had a transmittance of 45 percent for nonpolarized light. An important advantage of filters with fused-silica spacers is that it has been found possible, by repeating the process described, to construct square-top filters with rejection ratios of the order of 5×10^{-4} (Fig. 93*f*) and filters with half-widths as low as 0.002 percent (Fig. 93*g*). However, such filters are very expensive.

Other materials can also be used to produce solid spacers by optical polishing. Germanium was used by Smith and Pidgeon[289] and by Costich[290] to produce very narrow bandpass filters for the infrared. Roche and Title used a substrate made of a combination of yttrium and thorium oxides to produce a filter with a HW = 0.004 percent at 3.3 μm.[291]

Plastic Spacers (*HW* $>$ 0.15 *percent*). Mylar has very smooth surfaces and areas can be selected that have a sufficiently uniform thickness to permit the use of this material as a solid spacer. Candille and Saurel have used this material to produce narrow bandpass filters and obtained half-widths of the order of 0.0008 μm.[292,293] In some of their designs, a second, evaporated narrowband filter deposited onto one side of the solid spacer served to remove unwanted adjacent orders.

***Phase-dispersion Filters* (*HW* > *0.1 percent*).** The dispersion of the phase change on reflection enters into Eq. (71) for the half-width of FP filters. Typical values of this quantity at $\lambda = 0.5\ \mu\text{m}$ for a silver reflector, a nine-layer quarter-wave stack with $n_H/n_L = 1.75$, and for a broadband reflector are -0.5, -6.8, and -112.0, respectively.[161] As a result, the half-widths of FP filters constructed with such reflectors should be reduced by factors of about 1.05, 2, and 20. In the last case, the contribution of the spacer to the half-width is negligible, and a spacerless design is possible (Fig. 89*h*).[294,295] Unfortunately, the expected reduction in half-width has so far not been fully realized in practice, probably because of errors in the monitoring and lack of uniformity of the layers.[296]

***Tunable Filters* (*HW* > *0.001 percent*).** These are usually air-spaced FP interferometers, often provided with elaborate automatic plate-parallelism and spacing control, which are more akin to spectrometers than to filters.[297,298] The position of the passband can be tuned quite significantly by changing the separation between the reflector plates. This type of tuning, unlike the tuning of filters by tilting, does not affect the angular field or shape of the transmission band. Ramsey reviews the various problems associated with the construction and use of such instruments.[299,300]

An electrically tunable 0.005-percent half-width interference filter with a lithium niobate spacer sandwiched between two conducting reflecting coatings has also been described (Fig. 102).

Wide-bandpass Filters

Filters with half-widths ranging from 10 to 40 percent can be constructed using techniques described earlier in this section (Figs. 97 and 98). Filters with wider transmission bands are usually obtained by combining short- and long-pass filters with cutoffs at the desired wavelengths. The cut-off filters may be all-dielectric (Figs. 58 and 59), glass or gelatine cut-off filters, or antireflection-coated infrared materials. Some of the short- or long-pass filters may be regarded as being wide-bandpass filters in their own right. The cut-off filters may be combined into a single filter. Alternatively, it is possible to assemble a number of short- and long-wavelength cut-off filters into sets that make it possible to assemble wide-passband filters of different half-widths and peak wavelengths. In this latter arrangement, the cut-off positions of all-dielectric short- and long-pass filters can be tuned

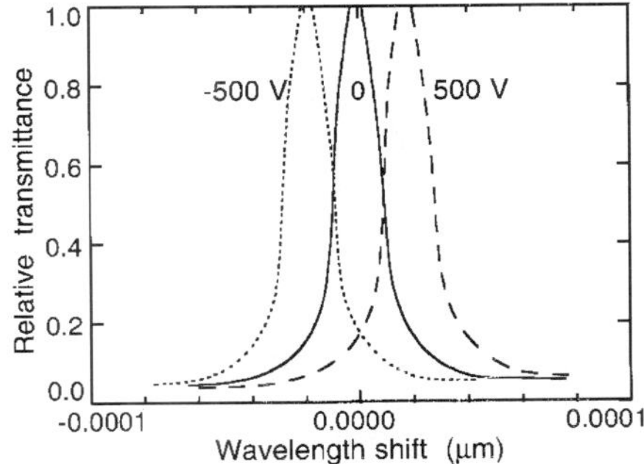

FIGURE 102 Performance of an electrically tunable narrowband filter with a lithium niobate spacer. *(After Burton.[487])*

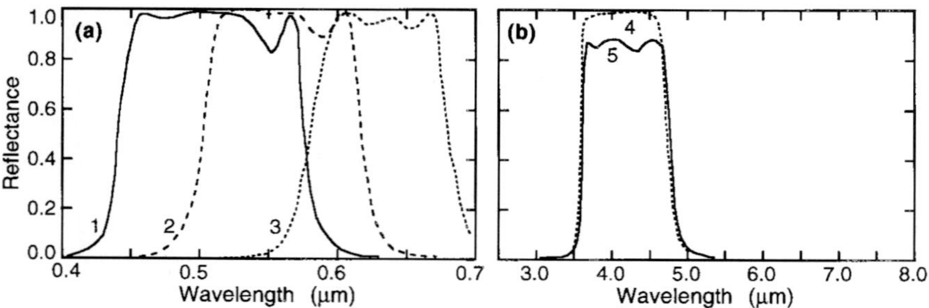

FIGURE 103 Broadband transmission filters: (*a*) calculated transmittance of three filters consisting essentially of two superimposed suitably tuned long-wavelength cut-off filter structures *(after McKenney and Turner*[488]*)*; (*b*) calculated (curve 4) and measured (curve 5) transmittances of a filter designed with an automatic synthesis program *(after Michael*[489]*)*.

individually by tilting to coincide with the desired wavelengths. Another way of tuning the edges of the transmission band is to pass the radiation through a pair of circular wedge short- and long-wavelength cut-off filters placed in series.

Filters with very broad transmission bands are also obtained when a multilayer is formed from two suitably displaced long-wavelength cut-off filters separated by an appropriate matching layer (Fig. 103*a*).[301] Automatic optimization programs can be used to design high-transmission broadband filters with a high rejection and a shape factor close to unity (Fig. 103*b*).[302]

Interference Filters with Multiple Peaks

For some applications, filters with multiple peaks in one particular spectral region are required. The design of such filters has been considered by Pelletier et al.[303] Typical results are shown in Fig. 104. Filters with different peak separations, rejections, and half-widths are possible.

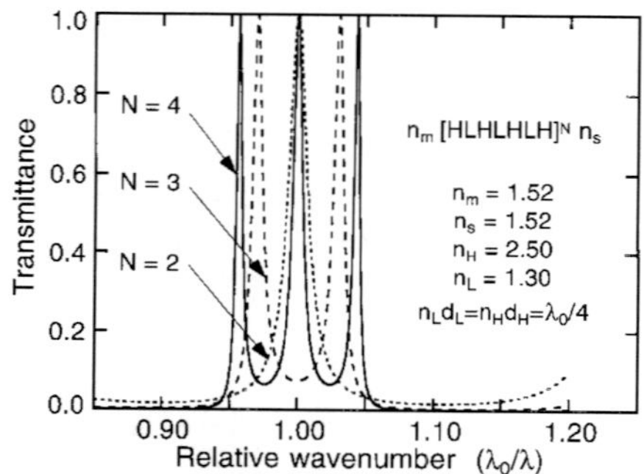

FIGURE 104 Calculated performance of interference filters with one and with two and three closely spaced peaks of the type glass $[HLHLHLH]^N$ glass for values of $N = 2, 3, 4$. *(After Pelletier and Macleod*[303]*)*.

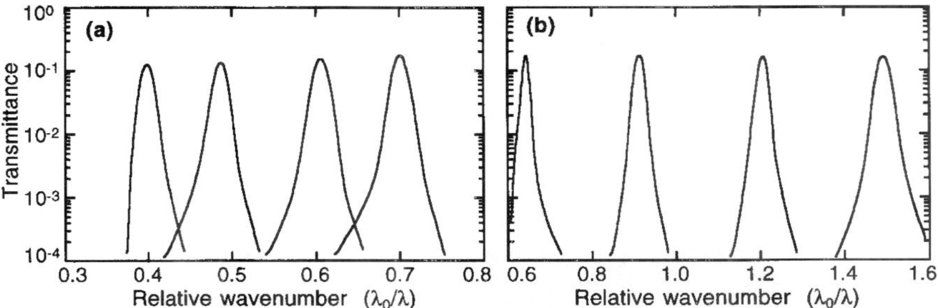

FIGURE 105 The transmission at different angular positions of two circular variable square-top filters for the (*a*) visible and (*b*) near-infrared spectral regions. *(After Mussett[490])*.

Linear and Circular Wedge Filters

If the thicknesses of all the layers of a bandpass filter vary in proportion across the surface of a substrate, the position of the transmission peak will vary in the same way (see "Matrix Theory for the Analysis of Multilayer Systems" in Sec. 42.3). Such wedge filters are as old as the interference filter itself and are available in versions in which the wavelength variation occurs along a straight line or a circle. The latter arrangement is particularly useful because it lends itself well to the construction of low-cost, small, and lightweight rapid-scan monochromators of moderate resolution that are robust and environmentally stable.[304] Methods for the production of circular variable filters with a linear dependence of wavelength on angle and references to some of the previous work on wedge filters are given in several papers.[305–309] Circular variable square-top filters for the 0.24- to 0.4- and for the 0.4- to 25-μm spectral regions are described by Avilov and by Yen.[310,311] The maximum transmittances for fully blocked filters vary between 15 and 75 percent, depending on the spectral region and the half-width of the filter. Rejection levels of 0.1 or 0.01 percent are possible. Typical transmission curves for several angular positions on two circular variable square-top filters are shown in Fig. 105. The ratio of the maximum to minimum wavelength available in one wheel varies between 1.11 and 16.[310] The angular width of the slit used in conjunction with a circular variable filter, expressed as a percentage of the angular size of the filter wedge, should not exceed the nominal half-width of the filter (expressed in percent) if it is not to cause a marked reduction in the resolution.

Angular Properties of Bandpass Interference Filters

With the gradual increase of the angle of incidence, the transmittance maximum of a typical bandpass filter moves towards shorter wavelengths. On further increase in the angle of incidence, the maximum transmittance and the half-width deteriorate; the transmission band becomes asymmetric and eventually splits up into p- and s-polarized components (Fig. 106*a*). The deterioration is more rapid for nonparallel radiation.

Properties of Bandpass Filters for Angles of Incidence Less than 20°. The behavior of bandpass filters for angles of incidence $\theta_o \leq 20°$ can be described quantitatively using the concept of an *effective index* μ^* of the filter. In terms of μ^* the transmittance T in the neighborhood of the transmission peak of any FP filter is given by Lissberger:[312]

$$T = \frac{T_o}{1 + \left[\frac{2(\lambda - \lambda_o)}{\Delta\lambda_{0.5}} + \frac{\lambda_o}{\Delta\lambda_{0.5}}\frac{\theta_o^2}{\mu^{*2}}\right]^2} \tag{74}$$

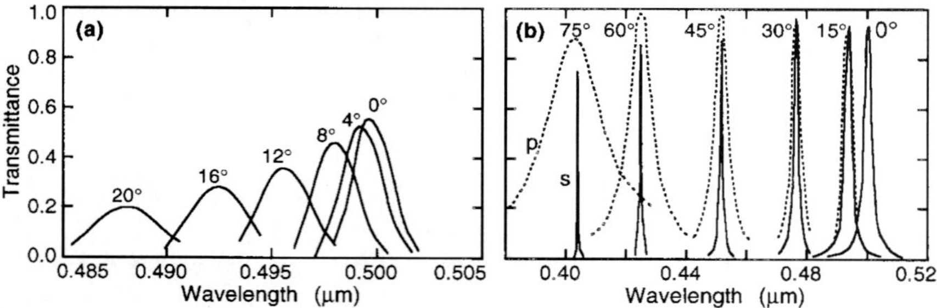

FIGURE 106 Angular properties of all-dielectric interference filters: (*a*) measured variation with angle of incidence of the spectral transmittance of a typical commercial interference filter (*after Blifford*[339]); (*b*) calculated transmittance of a filter in which the peaks of the two polarized transmission bands at nonnormal incidence coincide. The filter is of the type air-$[HL]^4(2A)[LH]^4$-glass, where $n_H d_H = n_L d_L = n_A d_A = \lambda/4$ and $n_s = 1.52$, $n_M = 1.00$, $n_H = 2.30$, $n_L = 1.38$, and $n_A = 1.825$.

$\Delta\lambda_{0.5}$ and T_o are the half-widths and the maximum transmittance (at λ_o) for normal incidence of the radiation. Formulas for μ^* in terms of the construction parameters have been found for the all-dielectric FP filter and the double-spacer filter,[313,314] and for the metal-dielectric FP and induced-transmission filters.[315] The change in position of the transmission peak $(\delta\lambda)_o$, and the half-width $(\Delta\lambda_{0.5})_\theta$ at angle θ are

$$\left(\frac{\delta\lambda}{\lambda_o}\right)_\theta = -\frac{\theta_o^2}{2\mu^{*2}} \tag{75}$$

and

$$\frac{(\Delta\lambda_{0.5})_\theta}{\Delta\lambda_{0.5}} = \left[1 + \left(\frac{\theta_o^2\lambda_o}{\mu^{*2}\,\Delta\lambda_{0.5}}\right)^2\right]^{1/2} \tag{76}$$

For convergent radiation of semiangle α, the corresponding expressions are

$$\left(\frac{\delta\lambda}{\lambda_o}\right)_\alpha = -\frac{\alpha^2}{4\mu^{*2}} \tag{77}$$

$$\frac{(\Delta\lambda_{0.5})_\alpha}{\Delta\lambda_{0.5}} = \left[1 + \left(\frac{\alpha^2\lambda_o}{2\mu^{*2}\,\Delta\lambda_{0.5}}\right)^2\right]^{1/2} \tag{78}$$

Linder and Lissberger discuss the requirements and design of filters for this case.[312,316]

Small tilts are commonly used to tune the peak of a filter to the desired wavelength even though they have an adverse effect on the angular field of the filter.

Bandpass Filters with Little or No Polarization Splitting. It has been shown numerically for the phase-dispersion filter,[317] for the frustrated-total-reflection filter,[318] and for the metal-dielectric[23] and all-dielectric[319] FP filters that it is possible to arrange for the two polarized transmission bands, which may have different widths, to coincide at high angles of incidence (Fig. 106*b*). The narrow, symmetrical high-transmittance bands that result may be useful for some applications even though the position of the maximum is still

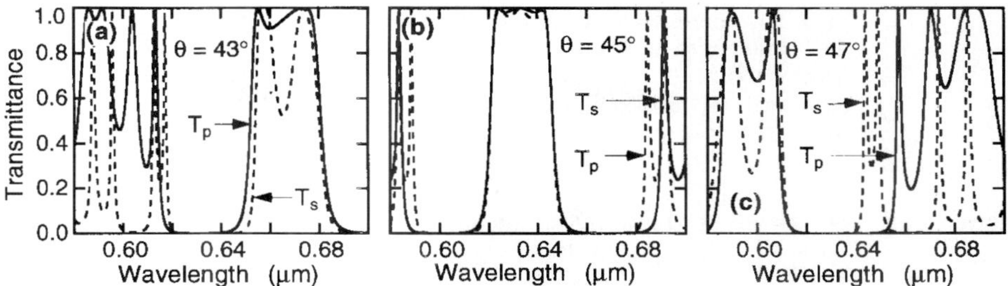

FIGURE 107 Multicavity bandpass filter with no polarization splitting for light incident at 45°. (*After Baumeister.*[320])

displaced with angle. Baumeister has shown how to design multicavity filters with no polarization splitting at one angle of incidence (Fig. 107).[320]

Wide-angle Bandpass Filters. Figure 108 shows the variation with effective index μ^* of the angular field of FP filters, defined as being twice the angle of tilt necessary to reduce to $0.8T_o$ the transmittance of the filter for radiation of wavelength λ_o. To increase the angular field, μ^* must be increased. Thus, for example, in all-dielectric FP filter, the expression for μ^* shows that $n_L < \mu^* < n_H$ and that with increasing spacer order, μ^* approaches the refractive index of the spacer (see also Ref. 321). The upper limits for μ^* for an all-dielectric FP filter in the ultraviolet, visible, and infrared parts of the spectrum are of the order of 2.0, 2.35, and 5.0, respectively. Little can be done about the angular field of solid spacer filters (see "Fabry-Perot Filters with Solid Spacers" earlier in this section).

For metal-dielectric FP and for induced-transmission filters, effective indexes μ^* of up to 3.2 and 2.0 have been reported.[315]

Wilmot and Schineller[322] and Schineller and Flam[323] have announced a filter consisting of a thin, plane-parallel fiber-optic face plate coated on both sides with all-dielectric mirrors. Since in such a filter the half-width is determined only by the thickness of the

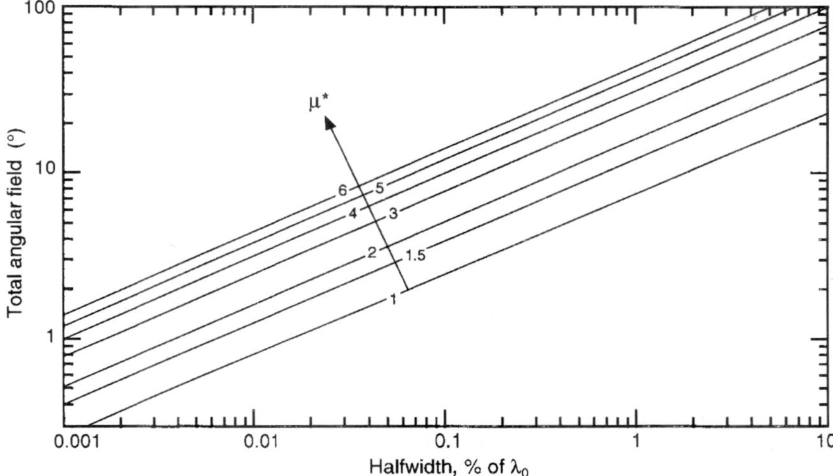

FIGURE 108 Angular field of FP filters as a function of the half-width for different effective indexes μ*.

plate and the reflectivity of the coatings, and since the field of view depends on the ratio of the wavelength to the fiber diameter, the two quantities are independent. The measured transmittance of a 15-Å-half-width, 6-mm-diameter filter composed of 1.5-μm-diameter fibers was 30 percent and the shift in wavelength with angle of incidence was one-eighth that of a conventional filter.

Stability and Temperature Dependence of Bandpass Filters

The stability of the position of the transmission peak has been studied by many workers.[259,276,324–332] The observed changes (up to 1 percent of λ_o) seem to depend greatly on the materials and manufacturing conditions. In filters with evaporated spacers both irreversible changes, probably due to changes in the structure of the films, and reversible changes due to the adsorption of water vapor, have been observed. Many manufacturers are now able to minimize these effects through the use of more stable materials, improved high-energy deposition methods (see Sec. 42.4) or accelerated artificial ageing processes. No changes were observed in solid spacer filters (see "Fabry–Perot Filters with Solid Spacers").

Changes in the operating temperature normally do not significantly affect the half-widths and peak transmittances of medium- and wide-bandpass interference filters (see, for example, Refs. 333–336). An exception are filters that contain semiconductors that start to absorb significantly on heating (e.g., germanium) or on cooling (PbTe).[337,338]

The position of the transmission peaks shift linearly towards longer wavelengths with an increase in temperature, the magnitude of the shift depending largely on the spacer material. The temperature coefficient, expressed as a percentage change in λ_o per degree Celsius change in temperature, lies between 2×10^{-4} and 3×10^{-3} for filters with evaporated spacers for the 0.3- to 1.0-μm spectral region[339] and between 2×10^{-3} and 2×10^{-2} for the infrared spectral region.[259,340] It is of the order of 1×10^{-3} for filters with mica[280] and quartz[287] spacers. Unless temperature control is provided, under adverse conditions all of these temperature coefficients could lead to serious shifts of the transmission peaks of very narrow band filters (see earlier in this section). The temperature control can take the form of an external constant-temperature enclosure, or it might be built right into the filter. Eather and Reasoner[276] and Mark et al.[341] describe arrangements of the latter type in which two sensors embedded in the filters are used to control the current flowing through two transparent conducting coatings that surround the filter.

Deliberate changes in the temperature can be used for a fine-tuning of the transmission wavelength without having an adverse effect on the angular field of the filter.

Bandpass Filters for the XUV and X-ray Regions

The construction of good bandpass filters for the extreme ultraviolet is hampered by the lack of coating materials with suitable optical constants. However, certain metals in thin-film form can be used as rudimentary bandpass filters in the extreme ultraviolet. The primary process in these filters is absorption, although at times interference within the film may have to be considered to explain the spectral transmission charateristics fully.

The measured spectral transmittance of some of these materials is shown in Figs. 109 and 110. By increasing the thicknesses of the films, higher rejection ratios could be obtained at the expense of peak transmissions, and vice versa. The transmittance of the most promising material, aluminum, would be higher if it were not for the formation of absorbing oxide layers.

Many of the layers are self-supporting (Fig. 109). Others must be deposited onto a suitable transparent substrate (Fig. 110). Thin aluminum films are sometimes used for this

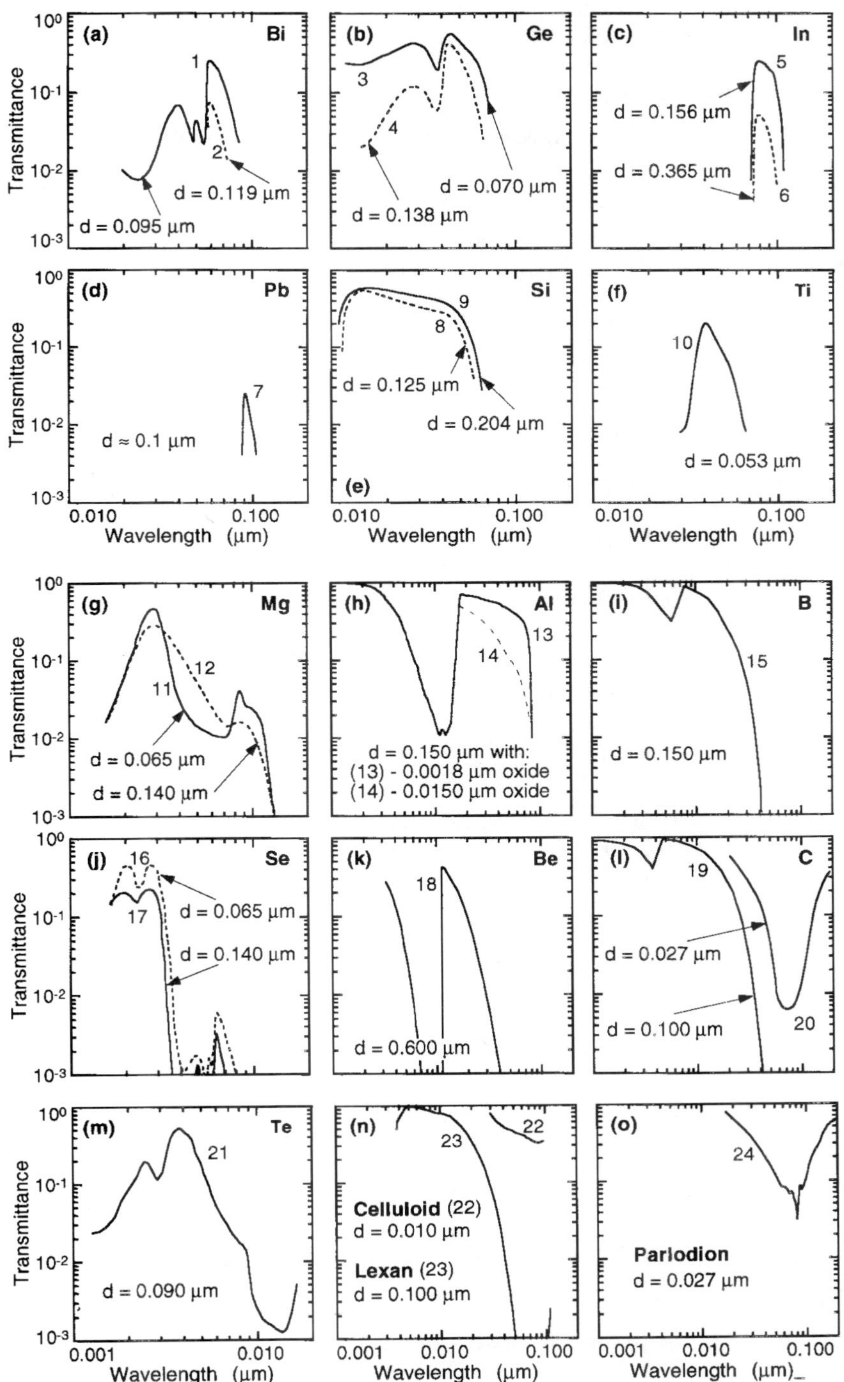

FIGURE 109 Measured extreme-ultraviolet and soft x-ray transmittance of several self-supporting metal films of indicated thicknesses.

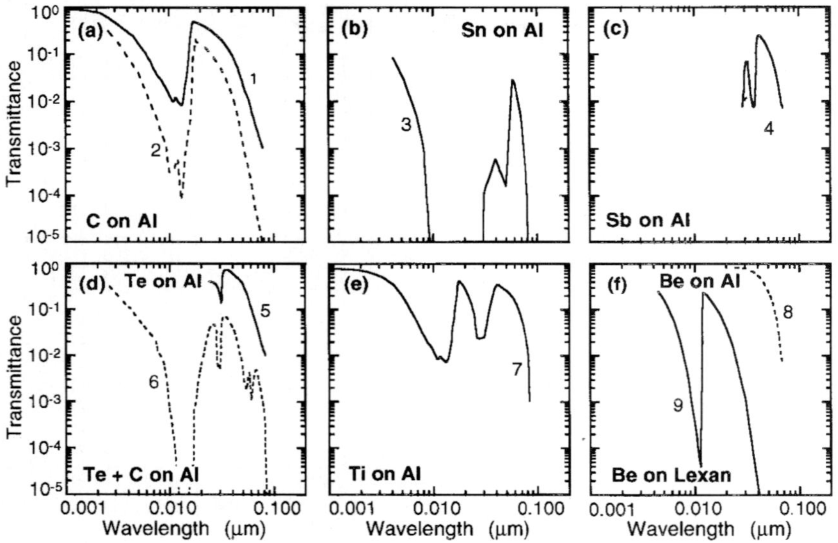

FIGURE 110 Measured extreme-ultraviolet and soft x-ray transmittance of several metal films of indicated thicknesses deposited onto thin aluminum or plastic films.

purpose. Other materials used in the past are Zapon (cellulose acetate); collodion, Parlodion, and Celluloid (cellulose nitrates); Mylar (polyethylene teraphtalate); and Formvar (polyvinyl formal) (Figs. 109*n,o*). Any residual absorption in the substrate contributes, of course, to the overall-transmission curve. The preparation of self-supporting thin films is described by Novikov and by Sorokin and Blank.[342,343] Because of their fragility, such films are usually mounted on a very fine mesh screen.

Multilayer Fabry-Perot interference filters for the soft x-ray region have also been constructed.[344–346] However, thus far the only spectral measurements reported are nonnormal incidence reflection.[347] The finesse of the filters is low and the modulation of the reflectance curve depends on the thickness of the spacer. The devices are useful for measurement purposes.

42.13 MULTILAYERS FOR TWO OR THREE SPECTRAL REGIONS

Increasingly, there are applications, especially in laser science, in which the spectral transmission and/or reflection has to be controlled at two or more wavelengths.[31] The design and construction of such coatings is more difficult than that of systems for one wavelength region only, especially when the ratio of the wavelengths of interest is very large.

Multilayers for Two Spectral Regions

Costich was the first to specify the construction parameters of coatings having all possible combinations of low- and high-reflection behavior at wavelength ratios of 1.5-, 2.0-, and 3.0:1.0.[348] His solutions were based on systems composed of quarter-wave layers or layers with other simple thickness relationships. Experimental results are in good agreement

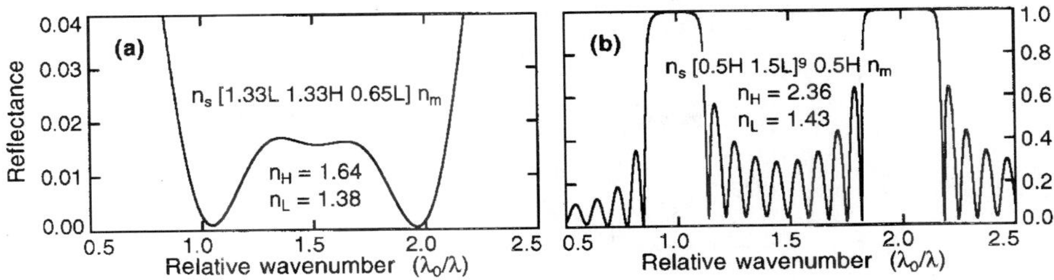

FIGURE 111 Calculated performance of (*a*) a commercial antireflection coating and (*b*) a high-reflectance coating for wave numbers 1.0 and 2.0 μm^{-1}. *(After Costich.[31])*

with the calculated values. The calculated performance of two coatings not shown before are given in Fig. 111. Figure 112 presents the performance of a number of commercially produced coatings of this type. Systems for other wavelength ratios and for combinations of other reflection values are possible.

Sometimes coatings are required in which the reflectance is controlled for wavelength ratios of 10:1 or more.[349,350] A systematic method for the design of such multilayers with different reflectance characteristics at the two wavelengths has been described.[351,352] The calculated performance of such coatings for the wavelengths of 0.6328 and 10.6 μm are shown in Fig. 113.

Multilayers for Three Spectral Regions

For some laser applications, the reflectance or transmittance has to be controlled at three or more wavelengths. Solutions to such problems can also be found. Costich has given designs for all possible combinations of low and high reflection for the important special case of a set of wavenumbers σ_o, $2\sigma_o$, and $3\sigma_o$.[348] The designs and performances of his solutions to this problem are shown in Fig. 114.

In principle, the method for the design of coating for two widely separated spectral regions mentioned here can be extended to the design of coatings for three or more

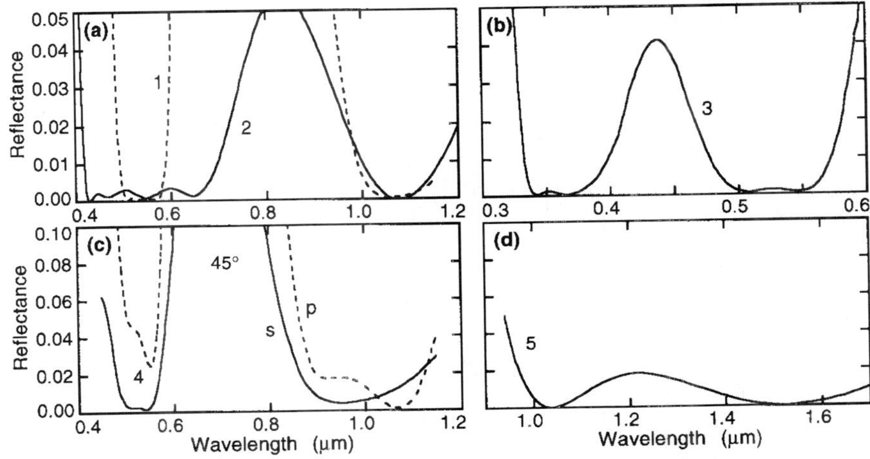

FIGURE 112 Measured performance of some commercial two-wavelength antireflection coatings. *(Curves 1, 3, and 4 after TechOptics;[411] curves 2, 5 after Thin Film Lab.[410])*

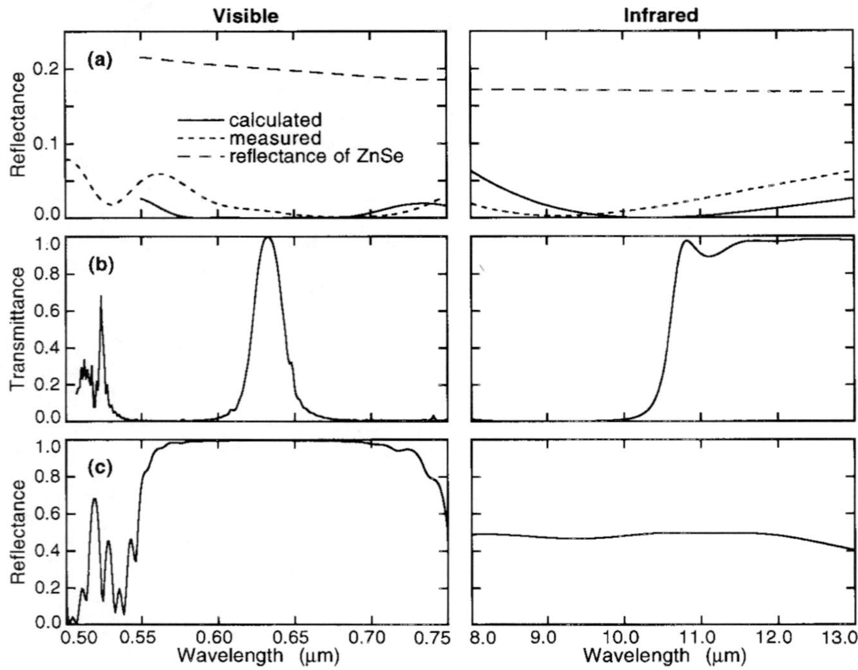

FIGURE 113 Calculated performance of three multilayer coatings (*a*), (*b*), and (*c*) designed for two widely separated spectral regions. Columns 1 and 2 represent the performance of the multilayers in the visible and in the infrared spectral regions, respectively. The experimental measurements for one coating are also shown. *(After Li.*[351,352]*)*

wavelengths. However, the number of layers required increases dramatically as the number of layers required for the longest wavelength region increases. Figure 115 shows the calculated performance of a coating that behaves like a high-reflection coating, a beam splitter, and an antireflection coating at 0.63, 2.52, and 10.6 μm, respectively.

42.14 PHASE COATING

In some applications, in addition to transmittance or reflectance requirements, special phase relationships have to be satisfied. These may be specific phase changes on reflection ε_R or transmission ε_T [Eqs. (23), (24)] for radiation incident at 0°. At other times it is required to displace or to deflect a beam without affecting its state of polarization. However, most frequently it is necessary to introduce a certain phase difference $(\varepsilon_p - \varepsilon_s)$ between p- and s-polarized light. Quarter-wave plates made of birefringent crystals are normally used to provide this phase difference.

Solutions to these and similar problems based on optical interference coatings can also be found. Porous films with an inclined columnar structure, formed in physical vapor deposition processes when the vapor is incident onto the substrate at an oblique angle, can also be birefringent.[353] Such films have been proposed for the construction of phase retardation plates for use with normal incidence of the radiation.[354] However, more frequently, solutions are based on the difference between the effective indices η_p, η_s of thin films for obliquely incident radiation [Eq. (19)]. Azzam has shown that, when

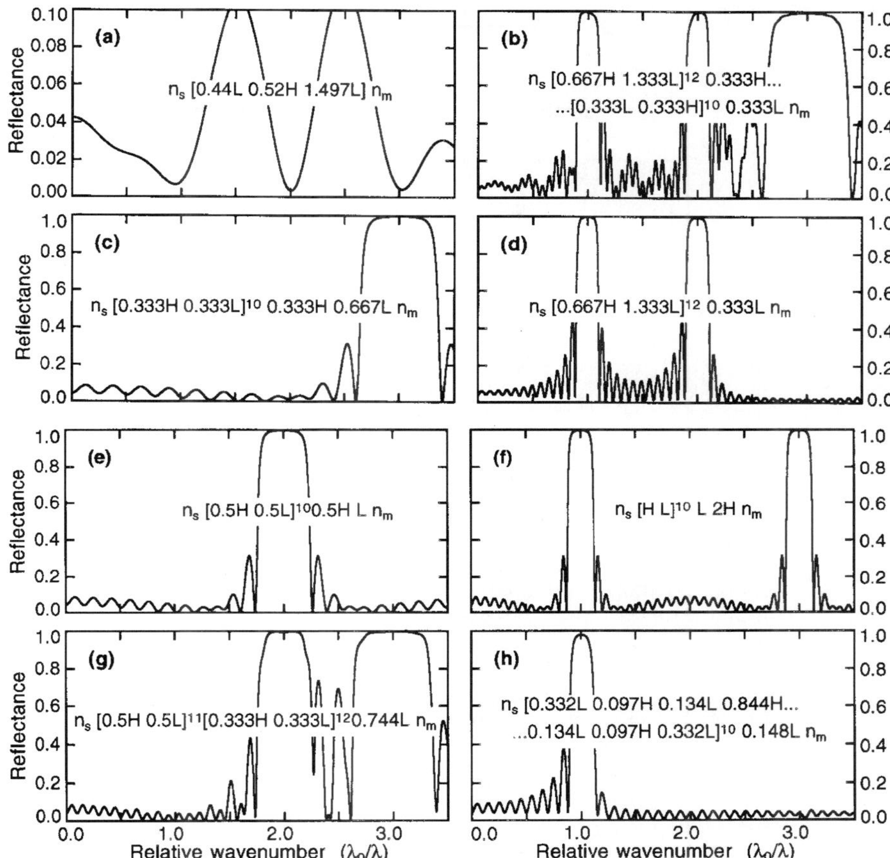

FIGURE 114 Calculated performance of multilayer coatings on glass with various combinations of high and low reflectance at relative wave numbers (λ_o/λ) 1.0, 2.0, and 3.0. In the designs H, L correspond to quarter-wave layers at $\lambda_o = 1.0$ μm. n_m, n_s, n_H, and n_L were assumed to be 1.00, 1.52, 1.95, and 1.43, except in (*a*) where n_H, n_L were 1.64, 1.38, respectively. *(After Costich.*[31]*)*

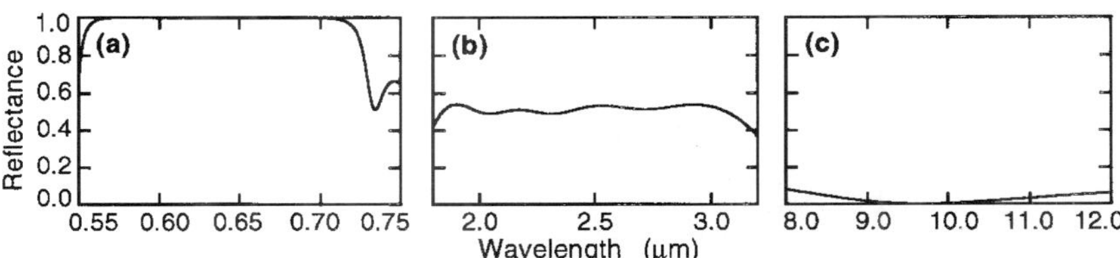

FIGURE 115 Calculated performance of a multilayer with different properties in three spectral regions. *(After Li.*[491]*)*

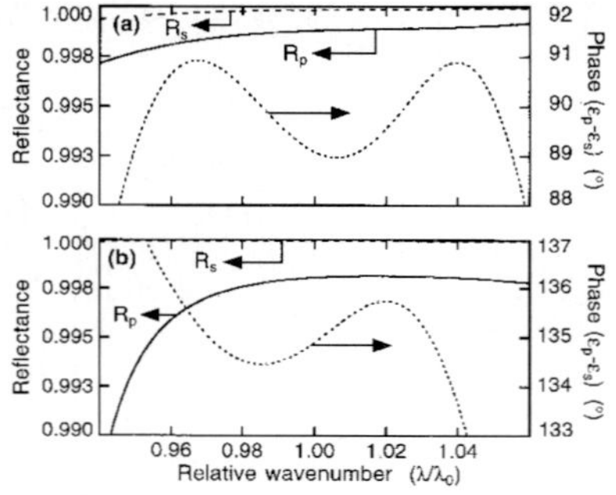

FIGURE 116 Calculated performance of two front surface 45° reflectors with different phase retardations: (*a*) 20 layers on an Ag substrate *(after Southwell*[492]*)*; (*b*) 22 layers on an Al substrate *(after Grishina*[493]*)*.

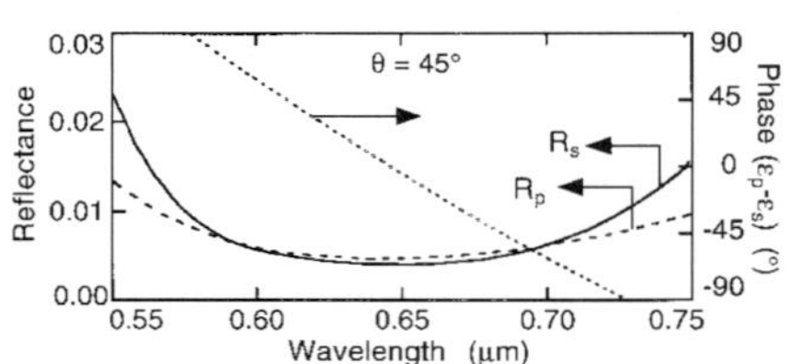

FIGURE 117 Antireflection coating on glass for use at 45° with zero phase retardation at $\lambda = 0.6471$ μm. *(After Thin Film Lab.*[410]*)*

performance at one wavelength only is specified and when oblique angles of incidence are acceptable, elegant solutions to many of the problems can be found that are based on a single layer only.[355]

Phase-retarding reflectors are commonly designed for use at 45°. Many layers are required when the radiation is incident from the air side. The performances of two multilayers of this type with different phase differences are shown in Fig. 116. The multilayers are optimized to maintain a constant phase difference in the vicinity of the design wavelength. Coatings with other phase differences and reflectances can also be constructed.[356,357] For example, an antireflection coating for 45° incidence in which the phase change is 180° is shown in Fig. 117.

When the radiation is incident on the layers from the substrate side, total internal reflection takes place.[358] The design of thin-film phase retarders based on this approach has also been examined by Apfel[359] and Azzam.[360] Total-internal-reflection phase retarders operate over broader spectral regions (Fig. 118) but their size is limited by the weight and homogeneity of the prism materials.

More complex phase-retardation devices have been constructed in which the radiation is allowed to undergo two, three, or even four internal reflections.[238,361,362] The performance of some typical total internal reflection devices that are based on the configurations of Figs. 65*t*, *u*, and *v* is shown in Fig. 119. For high-power laser beam delivery systems, front surface reflectors are usually employed (Fig. 120).

In Fig. 121 are shown the phase changes on reflection of a set of four metal/dielectric interferometer mirrors, all with $R > 0.97$, for which the differences in the phase changes on reflection for adjacent members in the set were approximately 90° over an extended spectral region. Other requirements for phase changes or phase-change differences can also be satisfied with thin films.

42.15 INTERFERENCE FILTERS WITH LOW REFLECTION

Reducing Reflection with a Thin Metal Film

A suitable thin metal film deposited onto a glass surface can act as a very efficient achromatic antireflection coating for light incident from the glass side (Fig. 122). The reflectance for light incident from the air side is not reduced and the transmittance suffers as a result of the absorption within the film.[363] By combining such films with additional layers, attractive colored sunglasses or architectural coatings are obtained.

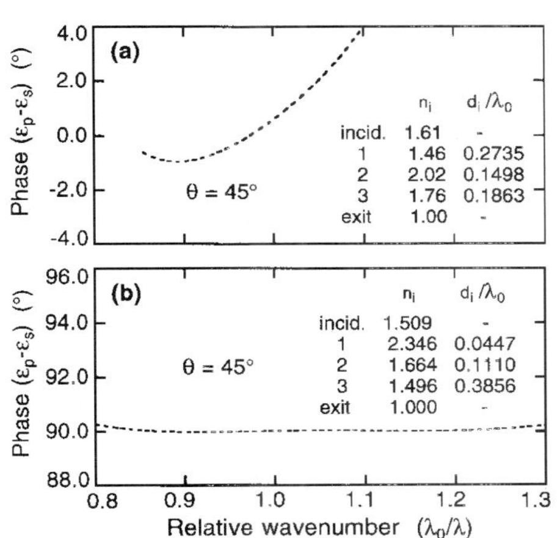

FIGURE 118 Phase retarders based on total internal reflection, consisting of three layers on glass and operating at an angle of incidence of 45°: (*a*) 0° phase retardation *(after Cojocaru*[494]*)*; (*b*) 90° phase retardation *(after Spiller*[495]*)*.

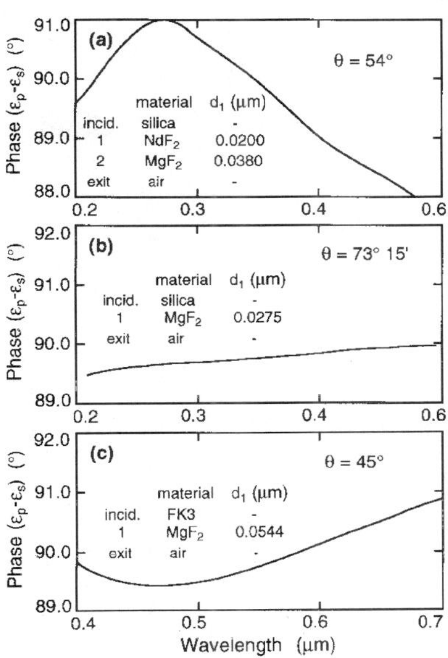

FIGURE 119 90° phase retardation devices based on (*a*) 2, (*b*) 3, and (*c*) 4 total internal reflections *(after King,*[496] *Clapham,*[202] *and Filinski*[362]*)*. The angle of incidence on the first reflecting surface is indicated in the diagrams.

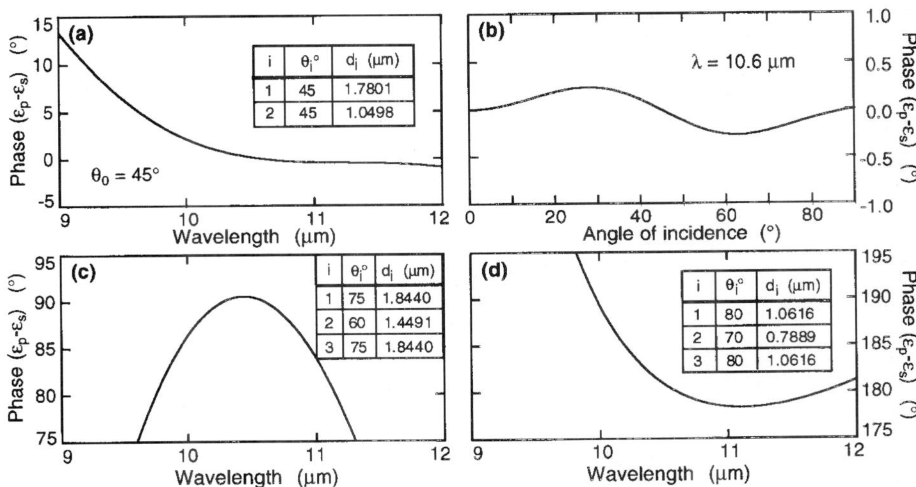

FIGURE 120 Phase retardation devices for $\lambda = 10.6\ \mu m$ based on multiple reflections from surfaces coated with opaque silver films and single layers of ZnS having specified thicknesses. (*a*), (*b*) 0° phase retardation device for all angles of incidence *(after Azzam*[497]*)*; (*c*), (*d*) 90°, 180° phase retardations *(after Thonn*[498]*)*.

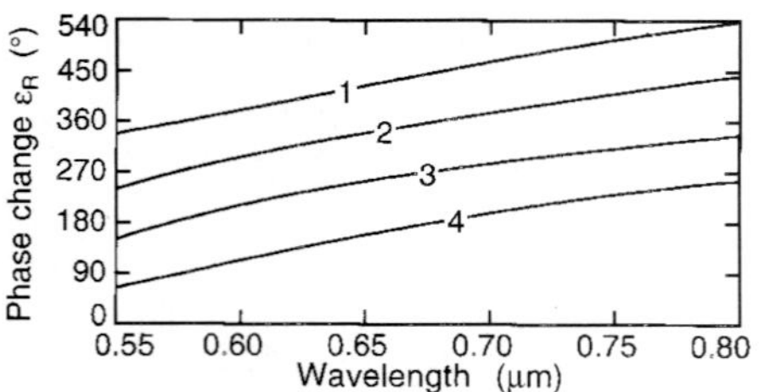

FIGURE 121 Normal incidence phase changes on reflection of a set of four highly reflecting mirrors for Michelson interferometers. *(After Piotrowski et al.*[499]*)*

Black Absorbers

Black absorbers efficiently absorb the radiation incident upon them in a specified spectral region. They are used, for example, to control radiant energy,[364] to remove stray light in optical systems, to enhance contrast in display devices,[365] and to increase the signal-to-noise ratio in multiplexers.[366] Black absorber coatings are based on interference in thin films and generally consist of an opaque metal layer and one or more dielectric layers interspersed with partially transparent metal layers (Figs. 20, 63*a*). They can be designed for first- and second-surface application (Fig. 123). Coatings of this type can also be designed for the ultraviolet and infrared spectral regions.[367]

Neutral Attenuators

Conventional metallic film attenuators described under "Neutral Filters" cannot readily be placed in series because multiple reflections between the components may result in unpredictable density values. However, by using metal and dielectric layer combinations of appropriate optical constants and thicknesses, it is possible to reduce the reflection of the

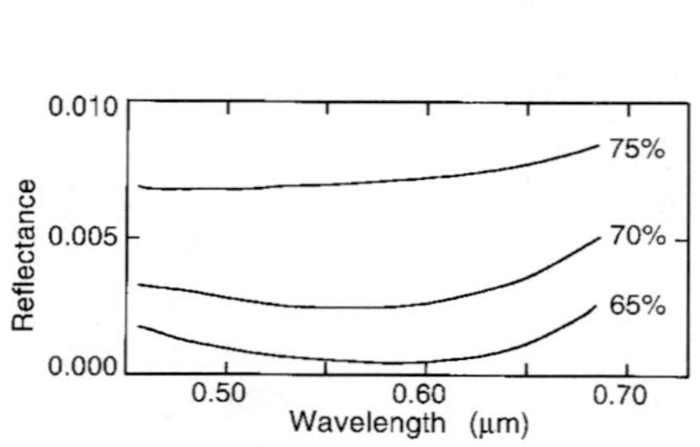

FIGURE 122 Spectral reflectance of thin chromium films on glass for light incident from the substrate side. The transmittance of the layers at $\lambda = 0.565$ μm is indicated. *(After Pohlack.*[363]*)*

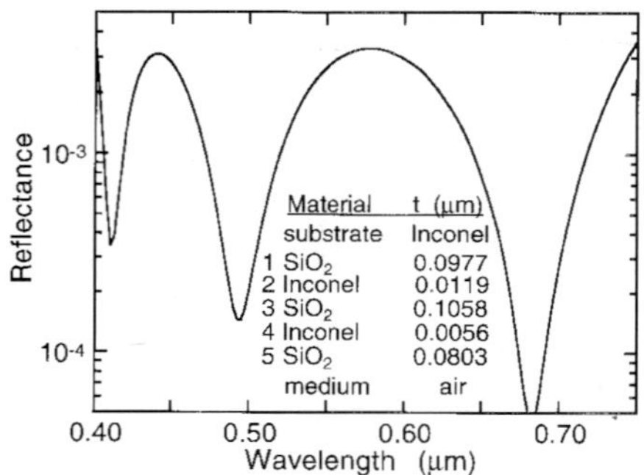

FIGURE 123 Calculated performance of a five-layer metal/dielectric black absorber. *(After Dobrowolski.*[500]*)*

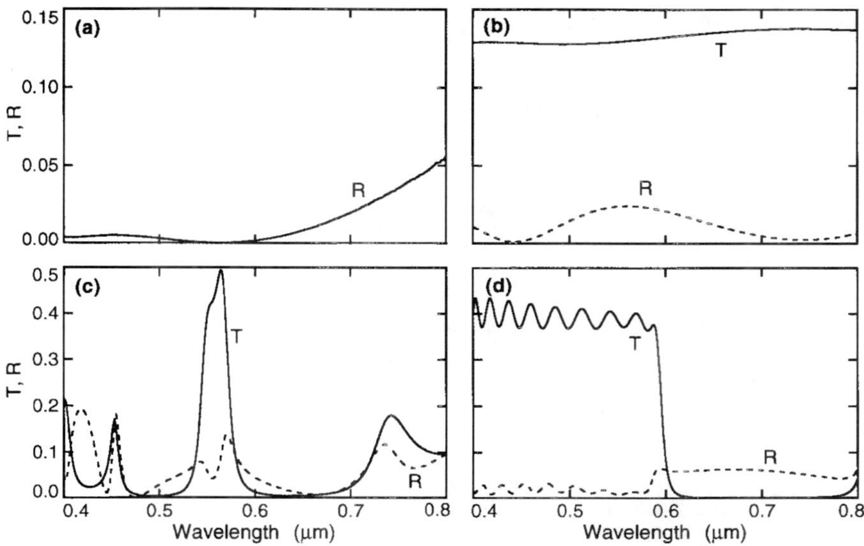

FIGURE 124 Measured performance of interference filters with reduced reflection: (*a*) black absorber; (*b*) neutral attenuator; (*c*) narrow bandpass filter; and (*d*) long-wavelength cut-off filter. (*a*) *(after Sullivan*[501]), (*b*) and (*d*) *(after Li*[502]), (*c*) *(after Dobrowolski*[503]).

metallic film from one or both sides of the substrate.[368–373] The experimental results for one such attenuator are given in Fig. 124*b*.

Other Interference Filters

It is possible, using a similar approach, to reduce the reflection of narrow-bandpass filters, cut-off filters, and other filter types. In particular, low-reflection narrowband interference filters for welding applications have been described by Jacobsson.[374] The experimental performance of a bandpass filter and of a long-wavelength cut-off filter are given in Figs. 124*c* and *d*. In both cases, the luminous reflectance has been reduced by an order of magnitude over that of a conventional design. However, this is at the expense of the transmittance.

42.16 REFLECTION FILTERS AND COATINGS

Metallic Reflectors

The Fresnel reflection coefficient of an interface between two semi-infinite media of complex refractive indices $\tilde{n}_m$, $\tilde{n}_s$ for polarized radiation incident at nonnormal angle is given by

$$R = \left| \frac{\eta_s - \eta_m}{\eta_s + \eta_m} \right|^{22} \tag{79}$$

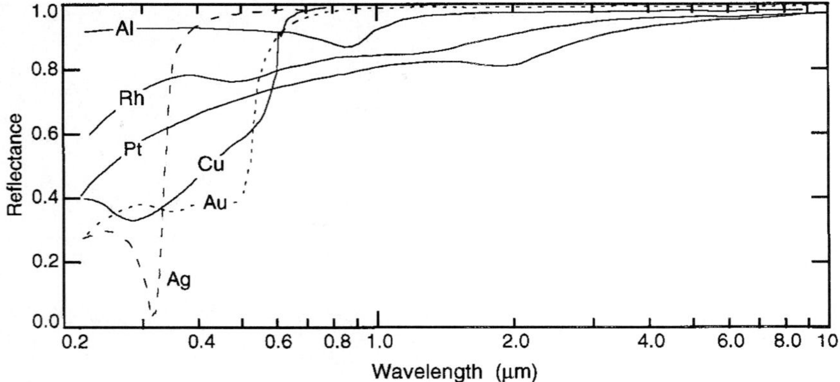

FIGURE 125 Reflectances of some metals. *(After Drummeter and Hass.*[109]*)*

where η_s, η_m are given by Eq. (19). When $\tilde{n}_m$, $\tilde{n}_s$ correspond to air and the metal, respectively, and when the angle of incidence is zero, the preceding expression reduces to:

$$R = \frac{(n_s - 1)^2 + k_s^2}{(n_s + 1)^2 + k_s^2} \tag{80}$$

If the substrate is opaque, this represents the total energy reflected, the remaining energy being absorbed within the material.

Metal reflectors are most commonly made by vacuum deposition of the material onto a suitable glass or quartz substrate. Before deposition, aluminium or beryllium mirror surfaces are sometimes first chemically plated with a nickel-phosphorus alloy (Kanigen process). Such deposits have excellent adhesion to the substrate and have a very hard surface that can be optically polished before coating.[375]

Visible, infrared, and ultraviolet spectral-reflectances of some of the more commonly used metals are shown in Figs. 125 and 126. Using Eq. (80) and the optical constants in Palik's handbook,[48,49] the spectral reflectances for many additional metals can be

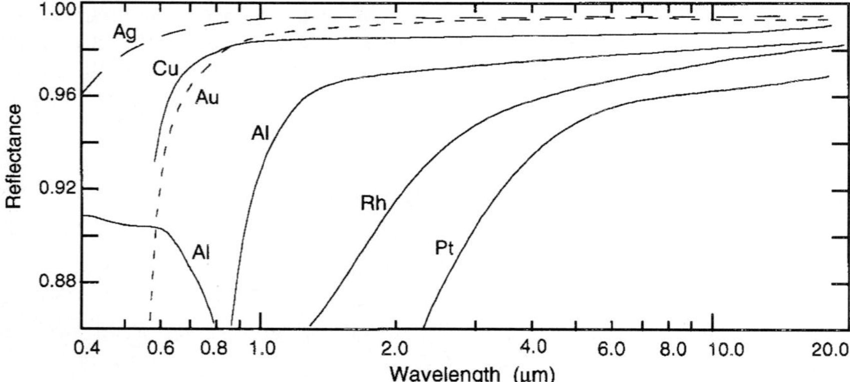

FIGURE 126 Visible and infrared reflectance of certain metals. *(Al, after Bennett et al.;*[504] *Ag and Au, after Bennett and Ashley;*[505] *Cu, after Hass and Hadley;*[506] *Rh and Pt, after Hass and Fowler, see Drummeter and Hass.*[109]*)*

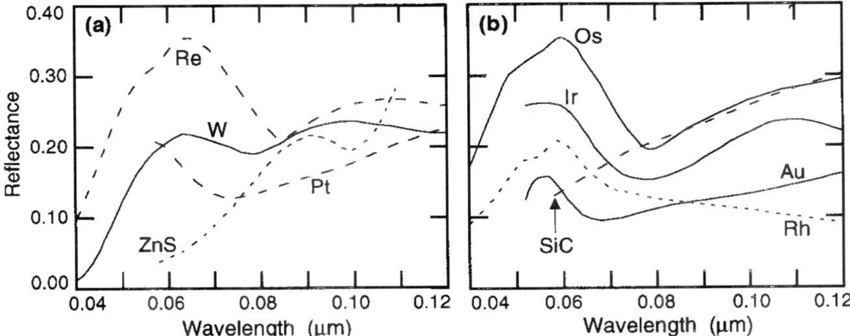

FIGURE 127 Measured ultraviolet reflectance of certain materials. *(Pt, Au, and ZnS, after Hunter;[507] Ir, after Hass et al.;[508] Rh, after Cox et al.;[509] Re, W, after Cox et al.;[510] Os, after Cox et al.;[509] and SiC, after Seely.[511])*

calculated. Silver has the highest visible and infrared reflectance, and hence is used for interferometer mirrors and interference filters. Exposed silver films tarnish readily. Aluminum has the broadest high-reflectance region of all metals and is commonly used in front-surface mirrors. It would reflect highly down to 0.1 μm were it not for the absorption below 0.18 μm of the thin oxide layer that starts to form seconds after deposition.[376,377]

Some of the highest known reflectances in the ultraviolet are shown in Fig. 127. At still shorter wavelengths, all materials have refractive indices that are close to unity and extinction coefficients that are rather small. It follows from Eq. (80) that normal-incidence reflectances in that part of the spectrum are small. However, for angles of incidence greater than the critical angle θ_c,

$$\theta_c = \cos^{-1}[\sqrt{2(1 - n_s)}] \tag{81}$$

total external reflection occurs resulting in high reflectances. Measured values of oblique angle reflection coefficients in the 0.0023- to 0.019-μm spectral region for a number of materials are given by Lukirskii et al.[378,379] Typical spectral reflectance curves are shown in Fig. 128. An Al reflectance of 0.987 for an angle of incidence of 80° at 0.0584 μm has been reported by Newnam.[380] Such coatings can be used in near-grazing incidence optics.

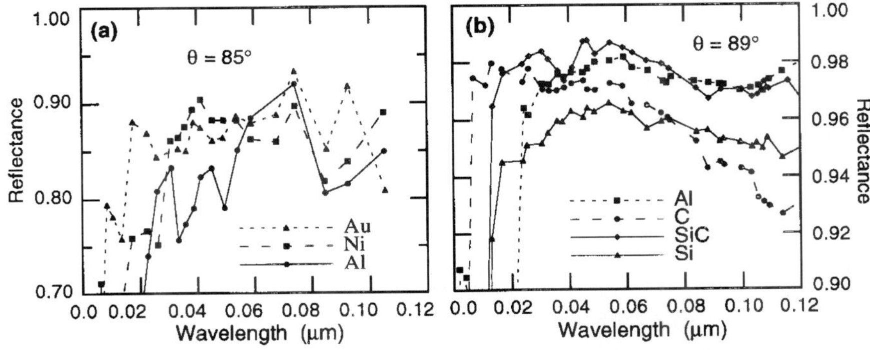

FIGURE 128 Measured x-ray reflectances of (*a*) Ni, Au, and Al films at 85° *(after Malina and Cash[512])*; and (*b*) Al, Si, C, and CVD SiC films at 89° angle of incidence *(after Windt et al., 1988[513])*.

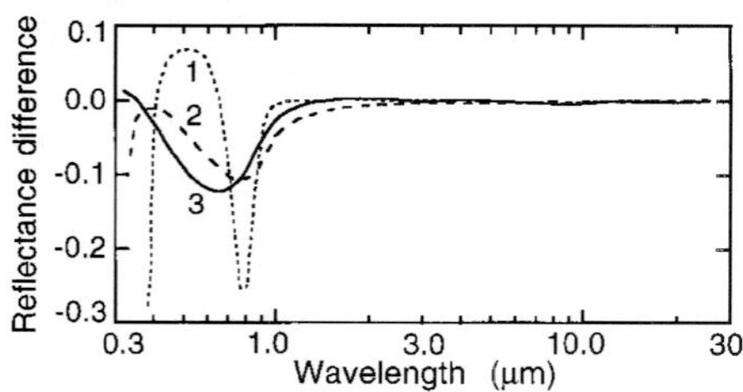

FIGURE 129 Measured difference between the reflectances of protected and unprotected aluminum mirrors. Curve 1: coating 4 of Fig. 1303*b*; curves 2 and 3: 0.1122 ± 0.002- and 0.0752 ± 0.001-μm-thick films of MgF_2 and SiO_2, respectively, on aluminum. *(After Bennett.*[514]*)*

Metal-dielectric Reflectors

Protective Coatings. For many applications, the thin aluminum oxide layer on an aluminum surface does not offer sufficient protection against abrasion and chemical attack, and therefore aluminum mirrors are often overcoated with single SiO_2 or MgF_2 protective layers. Such mirrors can be repeatedly cleaned with water and even withstand boiling in salt water.[381] Single protective layers reduce the reflectance (Fig. 129). If necessary, this problem can be overcome by using protective coatings of two or more layers (Fig. 130).

The deterioration of the ultraviolet reflectance of aluminum mirrors due to oxidation can be partially avoided by covering the freshly deposited aluminum layer immediately with a suitable coating of MgF_2[382–384] or LiF.[384–386] The reflectances of such overcoated aluminum reflectors are shown in Fig. 131. Their variation with angle of incidence in the 0.03- to 0.16-μm spectral region is discussed by Hunter.[387]

The reflectance of unprotected and protected silver mirrors has been investigated by Burge et al.[388] Highly adherent and chemically stable mirrors with a reflectance in excess of 0.95 for wavelengths greater than 0.5 μm have been reported.[389,390]

The reflectance of protected metal mirrors at oblique angles of incidence in the infrared part of the spectrum can be seriously reduced at the short-wavelength side of the Reststrahlen peak of the material used for its protection.[391–394]

Enhancement of Reflection. By depositing a quarter-wave stack (see "Nonabsorbing

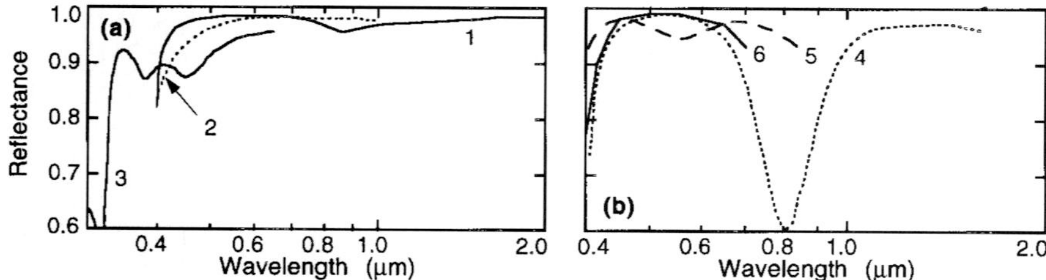

FIGURE 130 Reflectance of very durable overcoated metal mirrors: (*a*) silver mirrors—curve 1, protected front-surface mirror *(after Denton*[515]*)*; curve 2, enhanced reflection *(after Vvedenskii*[516]*)*; curve 3, extended reflection *(after Song et al.*[396]*)*; *(b)* aluminum mirrors—curve 4, with four layers of MgF_2 and CeO_2 *(after Hass*[381]*)*; curve 5, with four layers of SiO_2 and TiO_2 *(after AIRCO*[517]*)*; curve 6, with four layers MgF_2 and ZnS *(after Furman and Stolov*[518]*)*.

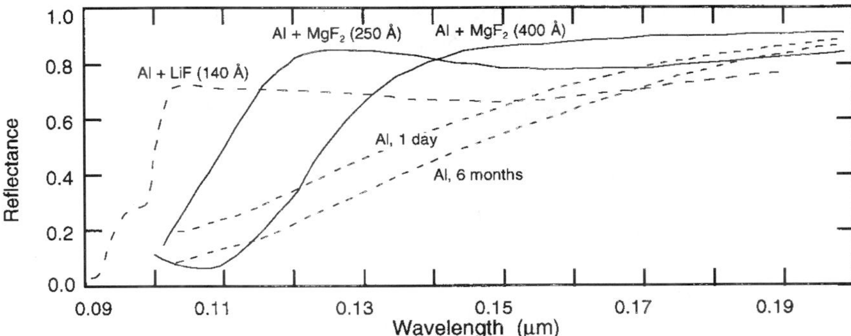

FIGURE 131 Measured spectral-reflectance curves of unprotected aluminum and aluminum overcoated with MgF_2 and LiF films of indicated thicknesses. *(Al + LiF coating, after Cox et al.,*[386] *all other curves after Canfield et al.*[383]*)*

$[AB]^N$ and $[AB]^NA$ Multilayers—Theory" and "Periodic Multilayers of the $[(0.5A)B(0.5A)]^N$ Type" in Sec. 42.7) onto the metal mirror, its reflectance can be enhanced considerably.[257] The thickness of the first layer should be adjusted to compensate for the phase change on reflection at the metal surface.[395] The spectral-reflectance curve dips on either side of the high-reflection region whose width is governed by the considerations under "Width of the High-Reflectance Zone" in Sec. 42.7 and which can be somewhat enhanced by the use of a half-wave outermost layer (Fig. 130). The measured spectral characteristics of three metal-dielectric reflectors for the ultraviolet region are shown in Fig. 132.

The reflectance of silver, although very high in the visible, falls off rapidly in the near-ultraviolet. Attempts to enhance the reflectance in that part of the spectrum and, at the same time, to protect the silver from tarnish, have been successful (Fig. 130*a*).[396]

A different kind of reflection enhancement has been reported for the extreme ultraviolet. By depositing semitransparent platinum films onto different substrates, the opaque-film reflectances of 19.3 and 12.8 percent at 0.0584 and 0.0736 μm were increased by up to 2.8 and 3.8 percent, respectively.[397] For space applications, suitably thick aluminum films on iridium are expected to yield reflectances as high as 40 and 52 percent at the same wavelengths (Fig. 133). Other proposed reflectance-increasing combinations can be found in Madden et al.[398]

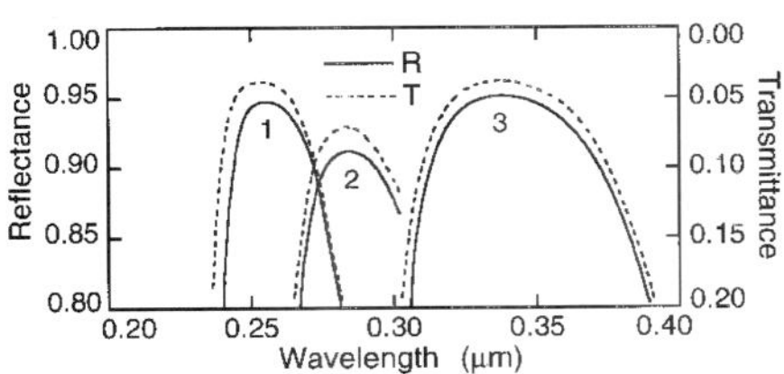

FIGURE 132 Enhanced ultraviolet reflectances of semitransparent aluminum films obtained through the addition of quarter-wave stacks. Curves 1 and 2: 11 layers of PbF and MgF_2 *(after Leś et al.*[519]*)*; curve 3: nine layers of Sb_2O_5 and MgF_2 *(after Leś and Leś*[520]*)*.

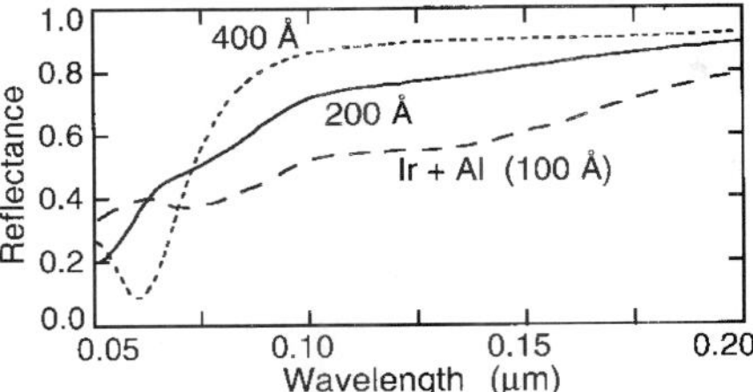

FIGURE 133 Calculated spectral reflectance curves of iridium overcoated with different thicknesses of unoxidized films of aluminum. *(After Hass and Hunter.*[521]*)*

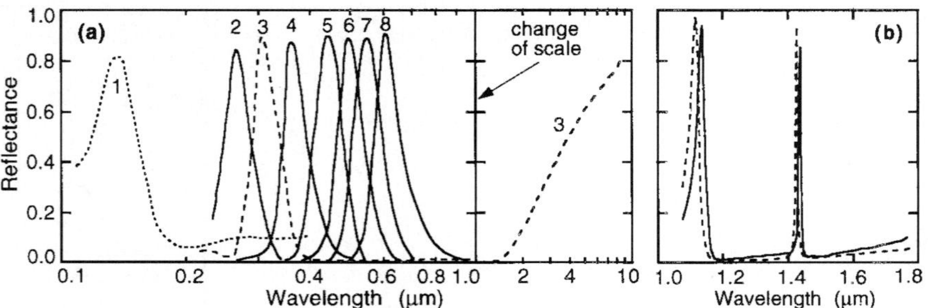

FIGURE 134 Measured performance of selective reflection filters: (*a*) reflecting filters for the ultraviolet and visible spectral regions *(curve 1 after Stelmack*[522]*);* curves 2–8 represent the performance of filters that consist of three half-wave cryolite or MgF_2 spacer layers separated by Inconel films of suitable transmission deposited on an opaque aluminum film *(after Turner and Hopkinson*[523]*)*; (*b*) calculated and experimental performance of a very narrowband near-infrared reflecting filter *(after Gamble*[524]*)*.

Selective Metal-Dielectric Reflectors. Several types of coatings that reflect highly in one spectral region, but not in another, have been developed in the past for different applications. Hadley and Dennison presented the theory and experimental results of reflection interference filters for the isolation of narrow spectral regions (Fig. 134*a*).[399,400] Very narrow reflection filters have been described by Zheng (Fig. 134*b*).[401] High-infrared and low-visible reflectance coatings (Fig. 63) are used to control the temperature of satellites.[109] These coatings could also be used to remove visible light from infrared optical systems. Several reflectors designed to reduce stray visible light in ultraviolet systems are shown in Fig. 135.

Multiple-Reflection Filters

Metal and Metal-dielectric Multiple-reflection Filters. Metals such as silver, copper, gold, and metal-dielectric coatings of the type shown in Fig. 63 used in a multiple-reflection arrangement should make cut-off filters with excellent rejection, sharp transition, and a long, unattenuated pass region far superior to those available with transmission filters.

Multiple-reflection Filters Made of Thin-film Interference Coatings. Interference coatings for use in a multiple-reflection filter need not be deposited onto substrates that

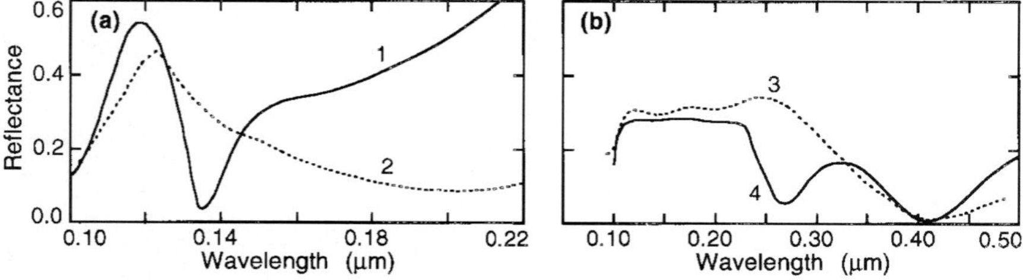

FIGURE 135 Measured performance of four multilayer selective ultraviolet reflectors for the control of stray radiation in the intermediate ultraviolet and visible spectral regions. *(Curve 1 after Hunter,*[507] *curve 2 after Berning et al.,*[525] *curves 3 and 4 after Hass and Tousey.*[382]*)*

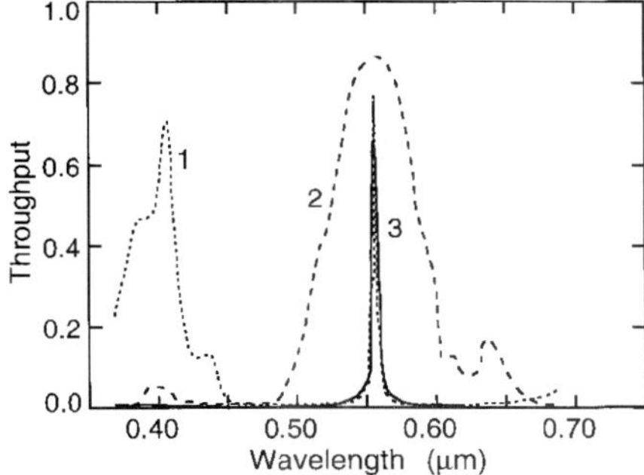

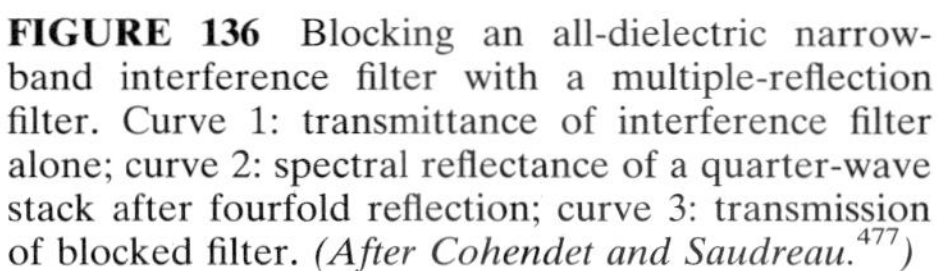

FIGURE 136 Blocking an all-dielectric narrowband interference filter with a multiple-reflection filter. Curve 1: transmittance of interference filter alone; curve 2: spectral reflectance of a quarter-wave stack after fourfold reflection; curve 3: transmission of blocked filter. *(After Cohendet and Saudreau.[477])*

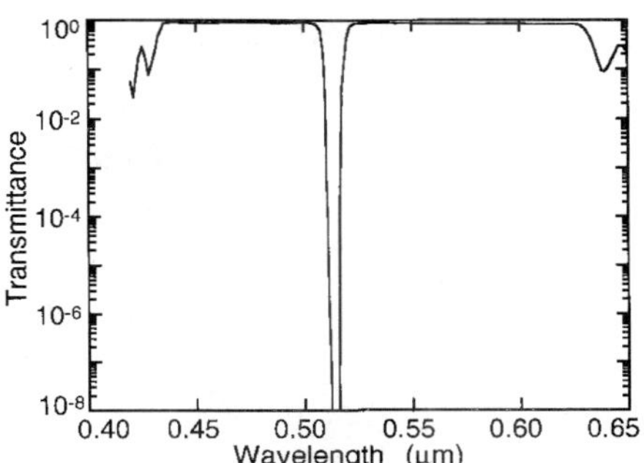

FIGURE 137 Measured transmission after four reflections from identical narrowband interference filters. *(After Omega Optical Inc.[432])*

transmit well in the spectral region of interest, but they should be used at small angles of incidence if disturbing effects due to polarization are not to occur.[402] The following are examples of some of the difficult filtering problems that can be easily solved with multiple-reflection filters composed of interference coatings, providing that there is space to use a multiple-reflection arrangement.

It is difficult to provide adequate blocking with transmission filters for narrow-bandpass filters of the type shown in Figs. 93 to 98 without considerably reducing the peak transmittance. This is done readily with a multiple-reflection filter composed of quarter-wave stacks of the same materials used for the construction of the bandpass filter and centered at the same wavelength (Fig. 136).

It should be possible to construct highly efficient short-pass filters with a very long and low rejection by using broadband reflectors consisting of several contiguous stacks (see "All-Dielectric Broadband Reflectors" in Sec. 42.8).

The use of a narrowband transmission filter in a multiple-reflection arrangement of the type shown in Fig. 3 results in a high-attenuation narrowband rejection filter surrounded by regions of high transmission (Fig. 137). However, such devices must be used with well-collimated light.

The transmittance curves of three multiple-reflection bandpass filters are shown in Figs. 138*a*–*c*. By using a number of multiple-reflection filters with sharp features, it is possible to separate signals transmitted by radiation of different, closely spaced wavelengths with a very low crosstalk and small insertion loss.[366]

Additional information on reflection coatings and filters will be found in the reviews by Hass et al. and by Lynch.[403,404]

42.17 SPECIAL-PURPOSE COATINGS

Space considerations limit the detailed description of coatings and filters for specific applications. Absorbing multilayer coatings on glass for enhancing the visual appearance, thermal and illumination control, and as "one-way mirrors" find applications in architecture and in the automotive industry. They are also used in solar-energy conversion and have been proposed for radiative cooling. Thin-film coatings are used in optical recording media and in optical multiplexers/demultiplexers. Bistable Fabry-Perot structures are

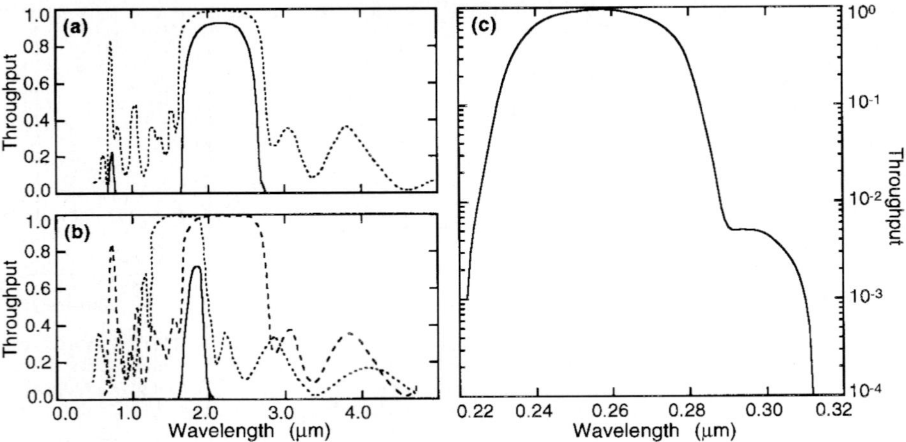

FIGURE 138 Measured spectral characteristics of broad- and narrowband multiple-reflection filters for the infrared and ultraviolet spectral regions: (*a*) eight reflections (solid curves) from identical quarter-wave stacks (dotted); (*b*) six reflections (solid) from each of two quarter-wave stacks (dotted, dashed) tuned to different wavelengths *(after Valeyev*[526]*); (c)* commercial multiple-reflection interference filter *(Schott and Gen.*[527]*)*

proposed for use as light switches in optical computers. Special filters and coatings are used in colorimetry, radiometry, detectors, and in high-contrast display devices. Consumer-oriented products include various kinds of decorative coatings, as well as coatings for the protection of documents and products from counterfeiting. There is no doubt that in the future even more applications will appear for optical multilayer coatings. Some of these will require very complex spectral characteristics. Methods for the design of such coatings have existed for some time now. At the time of writing, feasibility of the construction of such filters in the laboratory has also been demonstrated (Fig. 139).

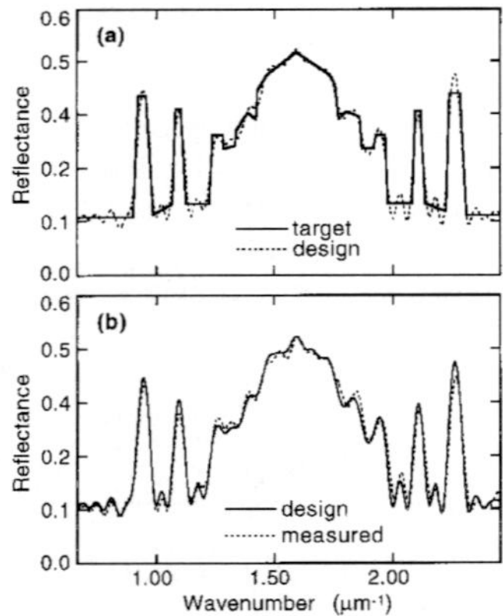

FIGURE 139 Calculated and experimentally measured reflectance of a filter that approximates the silhouette of the Taj Mahal. *(After Sullivan and Dobrowolski.*[528]*)*

42.18 ACKNOWLEDGMENTS

The author would like to acknowledge the help and encouragement of his colleagues, Brian T. Sullivan, Li Li, Claude Montcalm, Pierre Verly, Jeffrey Wong, and Allan Waldorf. He is also grateful to Philip Baumeister, Art Guenther, and Gaétan Duplain for their comments on the manuscript.

42.19 REFERENCES

1. J. A. Dobrowolski, "Coatings and Filters," in W. G. Driscoll and W. Vaughan (eds.), *Handbook of Optics,* McGraw-Hill, New York, 1978, pp. 8.1–8.124.
2. H. A. Macleod, *Thin Film Optical Filters,* McGraw-Hill, New York, 1986, 519 pp.
3. J. D. Rancourt, *Optical Thin Films Users' Handbook,* Macmillan, New York, 1987, 289 pp.
4. L. B. Tuckerman, "On the Intensity of the Light Reflected from or Transmitted through a Pile of Plates," *J. Opt. Soc. Am.* **37:**818–825 (1947).
5. A. M. Dioffo, "Treatment of General Case of *n* Dielectric Films with Different Optical Properties," *Rev. Opt. Theor. Instrum.* **47:**117–129 (1968).
6. A. Dresler, "Über eine neuartige Filterkombination zur genauen Angleichung der spektralen Empfindlichkeit von Photozellen an die Augenempfindlichkeitskurve," *Das Licht* **3:**41–43 (1933).
7. M. R. Nagel, "A Mosaic Filter Daylight Source for Aerophotographic Sensitometry in the Visible and Infrared Region," *J. Opt. Soc. Am.* **44:**621–624 (1954).
8. J. A. Dobrowolski, F. C. Ho, A. Belkind, and V. A. Koss, "Merit Functions for More Effective Thin Film Calculations," *Appl. Opt.* **28:**2824–2831 (1989).
9. Z. Knittl, *Optics of Thin Films,* Wiley & Sons, London, 1976, 548 pp.
10. S. A. Furman and A. V. Tikhonravov, *Optics of Multilayer Systems,* Editions Frontieres, Gif-sur-Yvette, 1992, 242 pp.
11. P. H. Berning, "Theory and Calculations of Optical Thin Films," in G. Hass (ed.) *Physics of Thin Films,* Academic Press, New York, 1963, vol. 1, pp. 69–121.
12. A. Thelen, *Design of Optical Interference Coatings,* McGraw-Hill, New York, 1988.
13. P. G. Kard, *Analysis and Synthesis of Multilayer Interference Coatings* (in Russian), Valrus, Tallin, 1971, 236 pp.
14. E. Delano and R. J. Pegis, "Methods of Synthesis for Dielectric Multilayer Filters," in E. Wolf (ed.), *Progress in Optics,* North-Holland Publishing Company, Amsterdam, 1969, vol. 7, pp. 68–137.
15. J. A. Dobrowolski and R. A. Kemp, "Refinement of Optical Multilayer Systems with Different Optimization Procedures," *Appl. Opt.* **29:**2876–2893 (1990).
16. L. Li and J. A. Dobrowolski, "Computation Speeds of Different Optical Thin Film Synthesis Methods," *Appl. Opt.* **31:**3790–3799 (1992).
17. H. A. Macleod, "Thin-Film Optical Coatings," in R. Kingslake, R. R. Shannon, and J. C. Wyant (eds.), *Applied Optics and Optical Engineering,* Academic Press, San Diego, 1987, vol. 10, pp. 1–69.
18. A. Thelen, *Design of Optical Interference Coatings,* McGraw-Hill, New York, 1988, 256 pp.
19. A. Herpin, "Calculation of the Reflecting Power of Any Stratified System," *Comptes Rendus* **225:**182–183 (1947).
20. W. Weinstein, "Computations in Thin Film Optics," in *Vacuum,* E. T. Heron & Co. Ltd., London, 1954, pp. 3–19.
21. M. Born and E. Wolf, *Principles of Optics,* Pergamon Press, New York, 1970.

22. P. Baumeister, "The Transmission and Degree of Polarization of Quarter-wave Stacks at Non-normal Incidence," *Opt. Acta* **8:**105–119 (1961).
23. V. R. Costich, "Reduction of Polarization Effects in Interference Coatings," *Appl. Opt.* **9:**866–870 (1970).
24. K. Rabinovitch and A. Pagis, "Polarization Effects in Multilayer Dielectric Thin Films," *Opt. Acta* **21:**963–980 (1974).
25. Z. Knittl, "Control of Polarization Effects by Internal Antireflection," *Appl. Opt.* **20:**105–110 (1981).
26. O. Arnon and P. Baumeister, "Electric Field Distribution and the Reduction of Laser Damage in the Multilayers," *Appl. Opt.* **19:**1853–1855 (1980).
27. J. L. Vossen and W. Kern, *Thin Film Processes,* Academic Press, New York, 1978.
28. J. L. Vossen and W. Kern, *Thin Film Processes II,* Academic Press, Boston, 1991.
29. M. R. Jacobson (ed.), *Selected Papers on Deposition of Optical Coatings,* SPIE Milestone Series vol. **MS 63,** SPIE Optical Engineering Press, Bellingham, Wash., 1990.
30. J. M. Bennett, E. Pelletier, G. Albrand, J. P. Borgogno, B. Lazarides, C. K. Carniglia, T. H. Allen, T. Tuttle-Hart, K. H. Guenther, and A. Saxer, "Comparison of the Properties of Titanium Dioxide Films Prepared by Various Techniques," *Appl. Opt.* **28:**3303–3317 (1989).
31. V. R. Costich, "Multilayer Dielectric Coatings," in M. J. Weber (ed.), *Handbook of Laser Science and Technology. Optical Materials*: *Part 3,* CRC Press, Inc., Boca Raton, Fla., 1987, pp. 389–430.
32. P. J. Martin, H. A. Maclean, R. P. Netterfield, C. G. Pacey, and W. G. Sainty, "Ion-assisted Deposition of Thin Films," *Appl. Opt.* **22:**178–184 (1983).
33. J. K. Hirvonen, "Ion Beam Assisted Thin Film Deposition," *Mater. Sci. Rep.* **6:**215–274 (1991).
34. F. A. Smidt, "Ion-beam-assisted Deposition Provides Control over Thin Film Properties," *NRL Publication 215-4670* issued May 1992.
35. H. Pulker, J. Edlinger, and M. Buehler, "Ion Plating Optical Films," in *Proceedings, 6th International Conference on Ion and Plasma Assisted Techniques,* Brighton, England, 1987, pp. 371–381.
36. K. H. Guenther, B. Loo, D. Burns, J. Edgell, D. Windham, and K.-H. Muller, "Microstructure Analysis of Thin Films Deposited by Reactive Evaporation and by Reactive Ion Plating," *J. Vac. Sci. Technol.* **A7:**1435–1445 (1989).
37. A. J. Waldorf, J. A. Dobrowolski, B. T. Sullivan, and L. M. Plante, "Optical Coatings by Reactive Ion Plating," *Appl. Opt.* **32:**5583–5593 (1993).
38. R. Parsons, "Sputter Deposition Processes," in J. L. Vossen and W. Kern (eds.), *Thin Film Processes II,* Academic Press, Boston, 1991, pp. 177–208.
39. B. T. Sullivan and J. A. Dobrowolski, "Deposition Error Compensation for Optical Multilayer Coatings: I. Theoretical Description," *Appl. Opt.* **31:**3821–3835 (1992).
40. P. A. Temple, "Thin-Film Absorptance Measurements Using Laser Calorimetry," in E. D. Palik (ed.), *Handbook of Optical Constants of Solids,* Academic Press, Orlando, 1985, pp. 135–153.
41. D. Z. Anderson, J. C. Friesch, and C. S. Masser, "Mirror Reflectometer Based on Optical Cavity Decay Time," *Appl. Opt.* **23:**1238–1245 (1984).
42. C. Amra, P. Roche, and E. Pelletier, "Interface Roughness Cross-correlation Laws Deduced from Scattering Diagram Measurements on Optical Multilayers: Effect of the Material Grain Size," *J. Opt. Soc. Am.* **B4:**1087–1093 (1987).
43. J. M. Bennett and L. Mattsson, *Introduction to Surface Roughness and Scattering,* Optical Society of America, Washington, D.C., 1989.
44. J. M. Bennett, "Recent Developments in Surface Roughness Characterization," *Measurement Science Technology* **3:**1119–1127 (1992).
45. G. Hass and W. R. Hunter, "Laboratory Experiments to Study Surface Contamination and Degradation of Optical Coatings and Materials in Simulated Space Environments," *Appl. Opt.* **9:**2101–2110 (1970).

46. W. R. Hunter, "Optical Contamination: Its Prevention in the XUV Spectrographs Flown by the U.S. Naval Research Laboratory in the Apollo Telescope Mount," *Appl. Opt.* **16:**909–916 (1977).
47. L. Ward, *The Optical Constants of Bulk Materials and Films,* Adam Hilger, Bristol, 1988, 244 pp.
48. E. D. Palik, *Handbook of Optical Constants of Solids,* Academic Press, Orlando, 1985.
49. E. D. Palik, *Handbook of Optical Constants of Solids II,* Academic Press, Boston, 1991.
50. D. P. Arndt, R. M. A. Azzam, J. M. Bennett, J. P. Borgogno, C. K. Carniglia, W. E. Case, J. A. Dobrowolski, U. J. Gibson, T. T. Hart, F. C. Ho, V. A. Hodgkin, W. P. Klapp, H. A. Macleod, E. Pelletier, M. K. Purvis, D. M. Quinn, D. H. Strome, R. Swenson, P. A. Temple, and T. F. Thonn, "Multiple Determination of the Optical Constants of Thin-Film Coating Materials," *Appl. Opt.* **23:**3571–3596 (1984).
51. S. P. Baker and W. D. Nix, "Mechanical Properties of Thin Films on Substrates," *Proc. Soc. Photo-Opt. Instrum. Eng.* **1323,** 263–276 (1990).
52. A. W. Czanderna, *Methods of Surface Analysis,* Elsevier Scientific Publishing Company, Amsterdam, 1975.
53. K. H. Guenther, "Non-optical Characterization Techniques for Optical Coatings," *Proc. Soc. Photo-Opt. Instrum. Eng.* **652:**210–214 (1986).
54. J. Bartella, P. H. Berning, B. Bovard, C. K. Carniglia, E. Casparis, V. R. Costich, J. A. Dobrowolski, U. J. Gibson, R. Herrmann, F. C. Ho, M. R. Jacobson, R. E. Klinger, J. A. Levitt, H.-G. Lotz, H. A. Macleod, M. J. Messerly, D. F. Mitchell, W.-D. Muenz, K. W. Nebesny, R. Pfefferkorn, S. G. Saxe, D. Y. Song, P. Swab, R. M. Swenson, W. Thoeni, F. V. Milligen, S. Vincent, and A. Waldorf, "Multiple Analysis of an Unknown Optical Multilayer Coating," *Appl. Opt.* **24:**2625–2646 (1985).
55. A. Mussett and A. Thelen, "Multilayer Antireflection Coatings," in E. Wolf (ed.), *Progress in Optics,* North-Holland Publishing Co., Amsterdam, 1970, vol. 8, pp. 203–237.
56. W. A. Faber, P. W. Kruse, and W. D. Saur, "Improvement in Infrared Detector Performance through Use of Antireflection Film," *J. Opt. Soc. Am.* **51:**115 (1961).
57. I. V. Grebenshchikov, *Prosvetlenie Optiki* (*Antireflection Coating of Optical Surfaces*), State Publishers of Technical and Theoretical Literature, Moscow-Leningrad, 1946.
58. T. Sawaki, "Studies on Anti-reflection Films," *Research Report No. 315,* Osaka Industrial Research Institute, issued September 1960.
59. J. T. Cox and G. Hass, "Antireflection Coatings for Optical and Infrared Optical Materials," in G. Hass and R. E. Thun (eds.), *Physics of Thin Films,* Academic Press, New York, 1964, vol. 2, pp. 239–304.
60. J. G. Wilder, "Porous silica AR coating for use at 248 nm or 266 nm," *Appl. Opt.* **23:**1448–1449 (1984).
61. F. O'Neill, I. N. Ross, D. Evans, J. U. D. Langridge, B. S. Bilan, and S. Bond, "Colloidal Silica Coatings for KrF and Nd:Glass Laser Applications," *Appl. Opt.* **26:**828–832 (1987).
62. I. M. Thomas, "Porous Fluoride Antireflective Coatings," *Appl. Opt.* **27:**3356–3358 (1988).
63. I. M. Thomas, "Method for the Preparation of Porous Silica Antireflection Coatings Varying in Refractive Index from 1.22 to 1.44," *Appl. Opt.* **31:**6145–6149 (1992).
64. M. E. Motamedi, W. H. Southwell, and W. J. Gunning, "Antireflection Surfaces in Silicon Using Binary Optics Technology," *Appl. Opt.* **31:**4371–4376 (1993).
65. P. Baumeister, "Antireflection Coatings with Chebyshev or Butterworth Response: Design," *Appl. Opt.* **25:**4568–4570 (1986).
66. J. Krepelka, "Maximally Flat Antireflection Coatings," *Jemna Mechanika a Optika* **37:**53–56 (1992).
67. S. A. Furman, "Broad-band Antireflection Coatings," *Sov. J. Opt. Technol.* **33:**559–564 (1966).
68. E. G. Stolov, "New Constructions of Interference Optical Antireflection Coatings," *Sov. J. Opt. Technol.* **58:**175–178 (1991).
69. R. R. Willey, "Broadband Antireflection Coating Design Performance Estimation," in *Proceedings, 34th Annual Technical Conference of the Society of Vacuum Coaters,* March 17–22, 1991, Philadelphia 1991, pp. 205–209.

70. A. V. Tikhonravov and J. A. Dobrowolski, "A New, Quasi-optimal Synthesis Method for Antireflection Coatings," *Appl. Opt.* **32:**4265–4275 (1993).
71. P. T. Scharf, "Transmission Colour in Camera Lenses," *J. SMPTE* **59:**191–194 (1952).
72. A. E. Murray, "Effect of Antireflection Films on Colour in Optical Instruments," *J. Opt. Soc. Am.* **46:**790–796 (1956).
73. H. Kubota, "Interference Color," in *Progress in Optics,* North-Holland Publishing Co., Amsterdam, 1961, vol. 1, pp. 211–251.
74. V. V. Obodovskiy, V. N. Rozhdestvenskiy, and I. Chernyy, "'Amber Coating' and Colour Photographic Objectives," *Sov. J. Opt. Technol.* **34:**324–330 (1967).
75. J. A. Dobrowolski and W. Mandler, "Color Correcting Coatings for Photographic Objectives," *Appl. Opt.* **18:**1879–1880 (1979).
76. J. T. Cox and G. Hass, "Antireflection Coatings for Optical and Infrared Optical Materials," in G. Hass and R. E. Thun (eds.), *Physics of Thin Films,* Academic Press, New York, 1964, pp. 239–304.
77. R. Jacobsson, "Light Reflection from Films of Continuously Varying Refractive Index" in E. Wolf (ed.), *Progress in Optics,* North-Holland, Amsterdam, 1966, vol. 5, pp. 247–286.
78. E. Spiller, I. Haller, R. Feder, J. E. E. Baglin, and W. N. Hammer, "Graded-index AR Surfaces Produced by Ion Implantation on Plastic Materials," *Appl. Opt.* **19:**3022–3026 (1980).
79. W. H. Southwell, "Gradient-index Antireflection Coatings," *Opt. Lett.* **8:**584–586 (1983).
80. B. Sheldon, J. S. Haggerty, and A. G. Emslie, "Exact Computation of the Reflectance of a Surface Layer of Arbitrary Refractive Index Profile and an Approximate Solution of the Inverse Problem," *J. Opt. Soc. Am.* **72:**1049–1055 (1982).
81. M. J. Minot, "The Angular Reflectance of Single-Layer Gradient Refractive-index Films," *J. Opt. Soc. Am.* **67:**1046–1050 (1977).
82. L. M. Cook and K.-H. Mader, "Integral Antireflective Surfaces on Optical Glass," *Opt. Eng.* **21:**SR-008–SR-012 (1982).
83. W. H. Lowdermilk, "Graded-Index Surfaces and Films," in M. J. Weber (ed.), *Handbook of Laser Science and Technology. Optical Materials: Part 3,* CRC Press, Inc., Boca Raton, Fla., 1987, pp. 431–458.
84. T. Sawaki, "Studies on Anti-reflection Films," *Research Report No. 315,* Osaka Industrial Research Institute, issued September 1960.
85. G. Lessman, "Optical Properties of Inhomogeneous Thin Films of Varying Modes of Gradation," *J. Opt. Soc. Am.* **56:**554 (1966).
86. R. Bertram, M. F. Ouellette, and P. Y. Tse, "Inhomogeneous Optical Coatings: an Experimental Study of a New Approach," *Appl. Opt.* **28:**2935–2939 (1989).
87. P. H. Berning, "Use of Equivalent Films in the Design of Infrared Multilayer Antireflection Coatings," *J. Opt. Soc. Am.* **52:**431–436 (1962).
88. J. A. Dobrowolski and F. C. Ho, "High Performance Step-Down AR Coatings for High Refractive-Index IR Materials," *Appl. Opt.* **21:**288–292 (1982).
89. L. I. Epstein, "The Design of Optical Filters," *J. Opt. Soc. Am.* **42:**806–810 (1952).
90. W. H. Southwell, "Coating Design Using Very Thin High- and Low-index Layers," *Appl. Opt.* **24:**457–460 (1985).
91. H. M. Moulton, "Method of Forming a Reflection Reducing Coating," U.S. patent 2536764, issued 2 January 1951.
92. S. M. Thomsen, "Skeletonizing Glass," U.S. patent 2490662, issued 6 December 1949.
93. M. J. Minot, "Single Layer, Gradient Refractive Index Antireflection Films—Effective 0.35 to 2.5 μ," *J. Opt. Soc. Am.* **66:**515–519 (1976).
94. Y. Asahara and T. Izumitani, "The Properties of Gradient Index Antireflection Layer on the Phase Separable Glass," *J. Non-Crystalline Solids* **42:**269–280 (1980).
95. W. H. Lowdermilk and D. Milam, "Graded-index Antireflection Surfaces for High-power Laser Applications," *Appl. Phys. Lett.* **36:**891–893 (1980).

96. S. Fujikawa, Y. Oguri, and E. Arai, "Antireflection Surface Modification of Plastic CR-39 by Means of ^{16}O- and ^{35}CI-ion Bombardment," *Optik* (Stuttgart) **84:**1–5 (1990).
97. S. P. Mukherjee, "Gel-derived Single-layer Antireflection Films with a Refractive Index Gradient," *Thin Solid Films* **81:**L89–L90 (1981).
98. B. E. Yoldas and D. P. Partlow, "Wide Spectrum Antireflective Coating for Fused Silica and Other Glasses," *Appl. Opt.* **23:**1418–1424 (1984).
99. S. P. Mukherjee and W. H. Lowdermilk, "Gel-derived Single layer Antireflection Films," *J. Non-Crystalline Solids* **48:**177–184 (1982).
100. K. N. Maffitt, H. U. Bruechner, and D. R. Lowrey, "Polymeric Optical Element Having Antireflecting Surface," U.S. patent 4114983, 19 September 1978.
101. P. B. Clapham and M. C. Hutley, "Reduction of Lens Reflexion by the 'Moth Eye' Principle," *Nature* **244:**281–282 (1973).
102. S. J. Wilson and M. C. Hutley, "The Optical Properties of 'Moth Eye' Antireflection Surfaces," *Opt. Acta* **29:**993–1009 (1982).
103. W. H. Southwell, "Pyramid-array Surface-relief Structures Producing Antireflection Index Matching on Optical Surfaces," *J. Opt. Soc. Am.* **A8:**549–553 (1991).
104. D. H. Raguin and G. M. Morris, "Antireflection Structured Surfaces for the Infrared Spectral Region," *Appl. Opt.* **32:**1154–1167 (1993).
105. J. Y. L. Ma and L. C. Robinson, "Night Moth Eye Window for the Millimetre and Sub-millimetre Wave Region," *Opt. Acta* **30:**1685–1695 (1983).
106. V. N. Smiley, "Conditions for Zero Reflectance of Thin Dielectric Films on Laser Materials," *J. Opt. Soc. Am.* **58:**1469–1475 (1968).
107. E. A. Lupashko and I. N. Shkyarevskii, "Multilayer Dielectric Antireflection Coatings," *Opt. Spectrosc.* (USSR) **16:**279–281 (1964).
108. K. C. Park, "The Extreme Values of the Reflectivity and the Conditions for Zero Reflection from Thin Dielectric Films on Metal," *Appl. Opt.* **3:**877–881 (1964).
109. L. F. Drummeter and G. Hass, "Solar Absorptance and Thermal Emittance of Evaporated Coatings," in G. Hass and R. E. Thun (eds.), *Physics of Thin Films,* Academic Press, New York, 1964, vol. 2, pp. 305–361.
110. R. N. Schmidt and K. C. Park, "High-temperature Space-stable Selective Solar Absorber Coatings," *Appl. Opt.* **4:**917–925 (1965).
111. H. Anders and R. Eichinger, "Die optische Wirkung und praktische Bedeutung inhomogener Schichten," *Appl. Opt.* **4:**899–905 (1965).
112. S. Katsube, Y. Katsube, K. Mitome, and S. Furuta, "Non-metallic Absorbing Films and the Anti-reflection Coating," *Oyo Buturi* **37:**225–230 (1968).
113. H. Anders, "Inhomogeneous, Absorbing Layers for Sunglasses," *Vakuum-Technik* **15:**123–126 (1966).
114. H. Pohlack, "Zum Problem der Reflexionsminderung optischer Gläser bei nichtsenkrechtem Lichteinfall," in P. Görlich (ed.), *Jenaer Jahrbuch 1952,* Fischer, Jena, 1952, pp. 103–118.
115. J. C. Monga, "Anti-reflection Coatings for Grazing Incidence Angles," *J. Modern Optics* **36:**381–387 (1989).
116. J. C. Monga, "Double-layer Broadband Antireflection Coatings for Grazing Incidence Angles," *Appl. Opt.* **31:**546–553 (1992).
117. P. Giacomo, "Les couches réfléchissantes multidiélectriques appliquées a l'interféromètre de Fabry-Perot etude théorique et expérimentale des couches rélles," *Rev. Opt. Theor. Instrum.* **35:**318–354 (1956).
118. J. Arndt and P. Baumeister, "Reflectance and Phase Envelopes of an Iterated Multilayer," *J. Opt. Soc. Am.* **56:**1760–1762 (1966).
119. J. S. Seeley, "Resolving Power of Multilayer Filters," *J. Opt. Soc. Am.* **54:**342–346 (1964).
120. H. Böhme, "Dielektrische Mehrfachschichtsysteme ohne Dispersion des Phasensprungs," *Optik* **69:**1–7 (1984).

121. P. Kard, E. Nesmelov, and G. Konyukhov, "Theory of Quarterwave Reflecting Filters," *Eesti NSV Tead. Akad. Toim., Fuus.-Mat.* **17:**314–323 (1968).

122. A. P. Ovcharenko and E. A. Lupashko, "Multilayer Dielectric Coatings of Unequal Thickness," *Opt. Spectrosc.* (USSR) **55:**316–318 (1983).

123. V. A. Smirnova and G. D. Pridatko, "Two-component Interference Coatings. Optical Properties and Applications," *Opt. Spectrosc.* (USSR) **55:**442–445 (1983).

124. M. Zukic and D. G. Toerr, "Multiple Reflectors as Narrow-band and Broadband Vacuum Ultraviolet Filters," *Appl. Opt.* **31:**1588–1596 (1992).

125. J. Shao and J. A. Dobrowolski, "Multilayer Interference Filters for the Far-infrared and Submillimeter Regions," *Appl. Opt.* **32:**2361–2370 (1993).

126. T. W. Barbee Jr., "Multilayers for X-ray Optics," *Opt. Eng.* **25:**898–915 (1986).

127. T. W. Barbee Jr., "Multilayer Optics for the Soft X-ray and Extreme Ultra-violet," *Physica Scripta* **T31:**147–153 (1990).

128. E. Spiller, "Multilayer Optics for X-rays," in P. D. a. C. Weisbuch (ed.), *Physics, Fabrication, and Applications of Multilayered Structures,* Plenum Publishing Corporation, New York, 1988, pp. 271–309.

129. A. V. Vinogradov and B. Y. Zeldovich, "X-ray and Far UV Multilayer Mirrors: Principles and Possibilities," *Appl. Opt.* **16:**89–93 (1977).

130. O. S. Heavens, "All-dielectric High-reflecting Layers," *J. Opt. Soc. Am.* **44:**371–373 (1954).

131. D. H. Rank and H. E. Bennett, "Problem of Phase Variation with Wavelength in Dielectric Films. Extension of Interferometric Standards into the Infrared," *J. Opt. Soc. Am.* **45:**69–73 (1955).

132. P. Giacomo, "Propriétés chromatiques des couches réfléchissantes multi-diélectrique," *J. Phys. Radium* **19:**307–311 (1958).

133. G. Bouwhuis, "A Dispersion Phenomenon Observable on Dielectric Multilayer Mirrors," *Philips Research Reports* **17:**130–132 (1962).

134. J. V. Ramsay and P. E. Ciddor, "Apparent Shape of Broad Band, Multilayer Reflecting Surfaces," *Appl. Opt.* **6:**2003–2004 (1967).

135. G. Koppelman, "Zur Theorie der Wechselschichten aus schwachabsorbierenden Substanzen und ihre Verwendung als Interferometerspiegel," *Annalen der Physik* **5:**387–396 (1960).

136. D. J. Hemingway and P. H. Lissberger, "Properties of Weakly Absorbing Multilayer Systems in Terms of the Concept of Potential Transmittance," *Opt. Acta* **20:**85–96 (1973).

137. D. Gloge, E. L. Chinnock, and H. E. Earl, "Scattering from Dielectric Mirrors," *Bell Syst. Tech. J.* **48:**511–526 (1969).

138. G. Rempe, R. J. Thompson, H. J. Kimble, and R. Lalezari, "Measurement of Ultralow Losses in an Optical Interferometer," *Opt. Lett.* **17:**363–365 (1993).

139. R. Lalezari, PMS Electro-Optics, 1855 South 57th Court, Boulder, CO 80301, U.S.A., private communication, 1993.

140. B. Vidal and P. Vincent, "Metallic Multilayers for X Rays Using Classical Thin-film Theory," *Appl. Opt.* **23:**1794–1801 (1984).

141. D. G. Stearns, "The Scattering of X Rays from Nonideal Multilayer Structures," *J. Appl. Phys.* **65:**491–506 (1989).

142. L. Young, "Multilayer Interference Filters with Narrow Stop Bands," *Appl. Opt.* **6:**297–315 (1967).

143. Lord Rayleigh, "On the Remarkable Phenomenon of Crystalline Reflexion Described by Prof. Stokes," *Philos. Mag.* **26** (Fifth Series):256–265 (1888).

144. R. W. Wood, *Physical Optics,* Macmillan, New York, 1940.

145. H. Schröder, "The Light Distribution Functions of Multiple-layers and Their Applications," *Z. Angew. Phys.* **2:**53–66 (1951).

146. J. Strong, "Iridescent $KClO_3$ Crystals and Infrared Reflection Filters," *J. Opt. Soc. Am.* **51:**853–855 (1961).

147. T. J. Alfrey, E. F. Gurnee, and W. J. Schrenk, "Physical Optics of Iridescent Multilayered Plastic Films," *Polymer Engineering and Science* **9:**400–404 (1969).

148. T. J. Alfrey, "Multilayer Thermoplastic Sheets and Films," in *Proceedings, 19th Sagainore Army Materials Research Conference,* 1973, pp. 195–200.

149. O. S. Heavens, J. Ring, and S. D. Smith, "Interference Filters for the Infra-red," *Spectrochimica Acta* **10:**179–194 (1957).

150. L. Edmonds, P. Baumeister, M. E. Krisl, and N. Boling, "Spectral Characteristics of a Narrowband Rejection Filter," *Appl. Opt.* **29:**3203–3204 (1990).

151. M. Hercher, "Single-mode Operation of a Q-switched Ruby Laser," *Appl. Phys. Lett.* **7:**39–41 (1965).

152. G. Magyar, "Simple Giant Pulse Ruby Laser of High Spectral Brightness," *Rev. Sci. Instrum.* **38:**517–519 (1967).

153. J. K. Watts, "Theory of Multiplate Resonant Reflectors," *Appl. Opt.* **7:**1621–1623 (1968).

154. H. F. Mahlein and G. Schollmeier, "Analysis and Synthesis of Periodic Optical Resonant Reflectors," *Appl. Opt.* **8:**1197–1202 (1969).

155. Anon., "Quartz Etalon Outperforms Sapphire," *Microwaves,* September 1969, p. 106.

156. A. F. Turner and P. W. Baumeister, "Multilayer Mirrors with High Reflectance over an Extended Spectral Region," *Appl. Opt.* **5:**69–76 (1966).

157. K. M. Yoo and R. R. Alfano, "Broad Bandwidth Mirror with Random Layer Thicknesses," *Appl. Opt.* **28:**2456–2458 (1989).

158. F. A. Korolev, A. Y. Klementeva, T. F. Meshcheryakova, and I. A. Ramazina, "Wide-band Reflectors with Multilayer Dielectric Coatings," *Opt. Spectrosc.* (USSR) **28:**416–419 (1970).

159. R. S. Sokolova, "Wide-band Reflectors for Ultraviolet Radiation," *Sov. J. Opt. Technol.* **38:**295–297 (1971).

160. E. G. Stolov, "Constructions of Neutral Lightsplitters and Wide-band Mirrors," *Sov. J. Opt. Technol.* **57:**632–634 (1990).

161. P. W. Baumeister and F. A. Jenkins, "Dispersion of the Phase Change for Dielectric Multilayers. Application to the Interference Filter," *J. Opt. Soc. Am.* **47:**57–61 (1957).

162. P. E. Ciddor, "Phase-dispersion in Interferometry. Interferometers with Solid Spacers and with Broadband-reflecting Surfaces," *Opt. Acta* **12:**177–183 (1965).

163. J. V. Ramsay and P. E. Ciddor, "Apparent Shape of Broad Band, Multilayer Reflecting Surfaces," *Appl. Opt.* **6:**2003–2004 (1967).

164. J. A. Dobrowolski, "Optical Interference Filters for the Adjustment of Spectral Response and Spectral Power Distributions," *Appl. Opt.* **9:**1396–1402 (1970).

165. A. Thelen, "Design of Optical Minus Filters," *J. Opt. Soc. Am.* **61:**365–369 (1971).

166. R. Jacobsson, "Inhomogeneous and Coevaporated Homogeneous Films for Optical Applications," in G. Hass, M. H. Francombe, and R. W. Hoffman (eds.), *Physics of Thin Films,* Academic Press, New York, 1975, vol. 8, pp. 51–98.

167. J. A. Dobrowolski and P. G. Verly (eds.), "Inhomogeneous and Quasi-inhomogeneous Optical Coatings," *Proceedings of the Society of Photo-Optical Instrumentation Engineers,* vol. **2046**, SPIE—The International Society for Optical Engineering, Bellingham, Wash., 1993.

168. W. J. Gunning, R. L. Hall, F. J. Woodberry, W. H. Southwell, and N. S. Gluck, "Codeposition of Continuous Composition Rugate Filters," *Appl. Opt.* **28:**2945–2948 (1989).

169. Physical Optics Corp., "Holographic Filters Aid Spectroscopy," *Photonics Spectra,* January 1992, pp. 113–114.

170. Kaiser Optical Systems Inc., "Holographic Notch Filters," 371 Parkland Plaza, P.O. Box 983, Ann Arbor, MI 48106, USA, 1993.

171. S. N. Vlasov and V. I. Talanov, "Selection of Axial Modes in Open Resonators," *Radio Eng. Electron. Phys.* **10:**469–470 (1965).

172. N. G. Vahitov, "Open Resonators with Mirrors Having Variable Reflection Coefficients," *Radio Eng. Electron. Phys.* **10:**1439–1446 (1965).

173. G. Duplain, P. G. Verly, J. A. Dobrowolski, A. Waldorf, and S. Bussière, "Graded Reflectance Mirrors for Beam Quality Control in Laser Resonators," *Appl. Opt.* **32:**1145–1153 (1993).

174. V. G. Vereshchagin and A. D. Zamkovets, "Polymer-crystalline Multilayer Systems for the Far IR Spectral Region," *Zh. Prikl. Spektrosk.* **47:**132–135 (1987).

175. V. R. Costich, "Study to Demonstrate a New Process to Produce Infrared Filters," *NASA Ames NAS2-12639,* issued 1 March 1990.

176. B. L. Henke, P. Lee, T. J. Tanaka, R. L. Shimabukuro, and B. K. Fujikawa, "Low-energy X-ray Interaction Coefficients: Photoabsorption, Scattering, and Reflection $-$ e $=$ 100–2000 EV. Z $= 1 - 94$," *Atom. Data and Nucl. Tables* **27:**1–144 (1982).

177. P. Baumeister, "Design of Multilayer Filters by Successive Approximations," *J. Opt. Soc. Am.* **48:**955–958 (1958).

178. M. A. Gisin, "Cutoff Interference Filters, Transparent to 25 μm," *Sov. J. Opt. Technol.* **36:**191–194 (1969).

179. R. Jacobsson, "Matching a Multilayer Stack to a High Refractive Index Substrate by Means of an Inhomogeneous Layer," *J. Opt. Soc. Am.* **54:**422–423 (1964).

180. E. A. Nesmelov and G. P. Konyukhov, "Theory of a Cutoff Interference filter," *Opt. Spectrosc.* (USSR) **31:**68–70 (1971).

181. M. Ploke, "Berechnung und Anwendung periodischer Mehrfachinterferenzschichten für ein Wärmereflexionsfilter mit breiter Reflexionsbande," *Zeiss-Mitteilungen* **4:**279–294 (1967).

182. H. H. Schroeder and A. F. Turner, "A Commercial Cold Reflector," *J. SMPTE* **69:**351–354 (1960).

183. F. E. Carlson, G. T. Howard, A. F. Turner, and H. H. Schroeder, "Temperature Reduction in Motion-picture and Television Studios Using Heat-control Coatings," *J. SMPTE* **65:**136–139 (1956).

184. A. Thelen, "The Use of Vacuum Deposited Coatings to Improve the Conversion Efficiency of Silicon Solar Cells in Space," *Progress in Astronautics and Rocketry* **3:**373–383 (1961).

185. D. L. Reynard and A. Andrew, "Improvement of Silicon Solar Cell Performance through the Use of Thin Film Coatings," *Appl. Opt.* **5:**23–28 (1966).

186. A. J. Waldorf, National Research Council of Canada, private communication, 1993.

187. F. Weiting and Z. Pengfei, "Determination of the Performance of Edge Filters Containing PbTe on Cooling," *Infrared Phys.* **33:**1–7 (1992).

188. R. J. Scheuerman, Perkin-Elmer Corporation, private communication, 1968.

189. L. J. Vande Kleft, E. A. Murray, W. M. Frey, and P. W. Yunker, "Construction and Evaluation of Thin Film Beam Dividers," AD 698025, issued October, 1969.

190. K. Hancock, "Membrane Optics," in *Proceedings, Electro-opt. Syst. Des. Conf.,* New York, Sept. 16–18, 1969, pp. 231–237 (1970).

191. P. L. Heinrich, R. C. Bastien, A. D. Santos, and M. Ostrelich, "Development of an All Dielectric Infrared Beamsplitter Operating in the 5 to 30 Micron Region," CR-703 issued February 1967.

192. M. L. Lipshutz, "Optomechanical Considerations for Optical Beam Splitters," *Appl. Opt.* **7:**2326–2328 (1968).

193. H. P. Larson, "Evaluation of Dielectric Film Beamsplitters at Cryogenic Temperature," *Appl. Opt.* **25:**1917–1921 (1986).

194. F. C. Ho and J. A. Dobrowolski, "Neutral and Color-Selective Beam Splitting Assemblies with Polarization-independent Intensities," *Appl. Opt.* **31:**3813–3820 (1992).

195. H. Anders, *Thin Films in Optics,* Focal Press, New York, 1967.

196. W. R. C. Rowley, "Some Aspects of Fringe Counting in Laser Interferometers," *I.E.E.E. Trans. Instrum. Meas.* **IM-15:**146–148 (1966).

197. H. Pohlack, "Zur Theorie der absortionsfreien achromatischen Lichtteilungsspiegel," in P. Görlich (ed.), *Jenaer Jahrbuch 1956,* Fischer, Jena, 1956, pp. 79–86.

198. L. A. Catalan and T. Putner, "Study of the Performance of Dielectric Thin Film Beam Dividing Systems," *Brit. J. Appl. Phys.* **12:**499–502 (1961).
199. K. P. Miyake, "Computation of Optical Characteristics of Dielectric Multilayers," *Journal de Physique* **25:**255–257 (1964).
200. R. S. Sokolova, "Beam-splitter Prisms with Dielectric Layers," *Sov. J. Opt. Technol.* **37:**318–320 (1970).
201. A. L. Sergeyeva, "Achromatic Nonabsorbing Beam Splitters," *Sov. J. Opt. Technol.* **38:**187–188 (1970).
202. P. B. Clapham, "The Design and Preparation of Achromatic Cemented Cube Beam-splitters," *Opt. Acta* **18:**563–575 (1971).
203. Z. Knittl, "Synthesis of Amplitude and Phase Achromatized Dielectric Mirrors," *Le Journal de Physique* **25:**245–249 (1964).
204. C. K. Malek, J. Susini, A. Madouri, M. Ouahabi, R. Rivoira, F. R. Ladan, Y. Lepêtre, and R. Barchewitz, "Semitransparent Soft X-ray Multilayer Mirrors," *Opt. Eng.* **29:**597–602 (1990).
205. R. M. A. Azzam, "Variable-reflectance Thin-Film Polarization-independent Beam Splitters for 0.6328- and 10.6-μm Laser Light," *Opt. Lett.* **10:**110–112 (1985).
206. V. V. Veremei, V. N. Rozhdestvenskii, and A. B. Khazanov, "Radiation Dividers for the Infrared," *Sov. J. Opt. Technol.* **48:**618–619 (1981).
207. H. A. Macleod and Z. Milanovic, "Immersed Beam Splitters—an Old Problem," in *Proceedings, Optical Interference Coatings,* Tucson, Arizona, 1992, pp. 28–30.
208. J. A. Dobrowolski, F. C. Ho, and A. Waldorf, "Beam Splitter for a Wide Angle Michelson Doppler Imaging Interferometer," *Appl. Opt.* **24:**1585–1588 (1985).
209. S. J. Refermat and A. F. Turner, "Polarization Free Beam Divider," U.S. patent 3559090, 26 January 1971.
210. S. Itoh and M. Sawamura, "Achromatized Beam Splitter of Low Polarization," U.S. patent 4415233, 15 November 1983.
211. H. F. Mahlein, "Non-polarizing Beam Splitters," *Opt. Acta* **21:**577–583 (1974).
212. A. A. M. Saleh, "Polarization-independent, Multilayer Dielectrics at Oblique Incidence," *The Bell System Tech. J.* **54:**1027–1049 (1975).
213. A. Thelen, "Nonpolarizing Interference Films Inside a Glass Cube," *Appl. Opt.* **15:**2983–2985 (1976).
214. A. R. Henderson, "The Design of Non-polarizing Beam Splitters," *Thin Solid Films* **51:**339–347 (1978).
215. Z. Knittl and H. Houserkova, "Equivalent Layers in Oblique Incidence: the Problem of Unsplit Admittances and Depolarization of Partial Reflectors," *Appl. Opt.* **21:**2055–2068 (1982).
216. C. M. d. Sterke, C. J. v. d. Laan, and H. F. Frakena, "Nonpolarizing Beam Splitter Design," *Appl. Opt.* **22:**595–601 (1983).
217. M. Zukic and K. H. Guenther, "Design of Nonpolarizing Achromatic Beamsplitters with Dielectric Multilayer Coatings," *Opt. Eng.* **28:**165–171 (1989).
218. M. Gilo, "Design of a Nonpolarizing Beam Splitter Inside a Glass Cube," *Appl. Opt.* **31:**5345–5349 (1992).
219. H. Pohlack, "Zur Theorie der absorptionsfreien achromatischen Lichtteilungsspiegel," in C. Zeiss (ed.), *Jenaer Jahrbuch,* Gustav Fischer Verlag, 1957, pp. 79–86.
220. A. F. Turner, "Design Principles for Interference Film Combinations," *J. Opt. Soc. Am.* **44:**352 (1954).
221. J. S. Seeley, "Simple Nonpolarizing High-pass Filter," *Appl. Opt.* **24:**742–744 (1985).
222. L. N. Kurochkina and N. M. Tulyakova, "Achromatic Lightsplitters for the Ultraviolet Spectrum," *Sov. J. Opt. Technol.* **52:**101–102 (1985).
223. I. J. Hodgkinson and R. G. Stuart, "Achromatic Reflector and Antireflection Coating Designs for Mixed Two-component Deposition," *Thin Solid Films* **87:**151–158 (1982).

224. J. Mouchart, J. Begel, and S. Chalot, "Dépôts achromatiques partiellement réfléchissants," *J. Mod. Opt.* **37:**875–888 (1990).
225. H. Schröder, "Die Erzeugung von linearpolarisiertem Licht durch dünne dielektrische Schichten," *Optik* **3:**499–503 (1948).
226. R. Messner, "Die theoretischen Grundlagen optischer Interferenzpolarisatoren," *Feinwerktechnik* **57:**142–147 (1953).
227. R. M. A. Azzam, "Efficient Infrared Reflection Polarizers Using Transparent High-index Films on Transparent Low-index Substrates," *Proc. Soc. Photo-Opt. Instrum. Eng.* **652:**326–332 (1986).
228. M. Ruiz-Urbieta and E. M. Sparrow, "Reflection Polarization by a Transparent-film-absorbing-substrate System," *J. Opt. Soc. Am.* **62:**1188–1094 (1972).
229. H. Winter, H. H. Bukow, and P. H. Heckmann, "A High Transmission Triple-reflection Polarizer for Lyman-α Radiation," *Opt. Commun.* **11:**299 (1974).
230. G. Hass and W. R. Hunter, "Reflection Polarizers for the Vacuum Ultraviolet Using $AI + MgF_2$ Mirrors and MgF_2 Plate," *Appl. Opt.* **17:**76–82 (1978).
231. W. R. Hunter, "Design Criteria for Reflection Polarizers and Analyzers in the Vacuum Ultraviolet," *Appl. Opt.* **17:**1259–1270 (1978).
232. J. T. Cox and G. Hass, "Highly Efficient Reflection-type Polarizers for 10.6-μm CO2 Laser Radiation Using Aluminum Oxide Coated Aluminum Mirrors," *Appl. Opt.* **17:**1657–1658 (1978).
233. P. Dhez, "Polarizers and Polarimeters for the X-UV Range," *Nucl. Instrum. Methods Phys. Res.* **A261:**66–71 (1987).
234. W. W. Buchman, S. J. Holmes, and F. J. Woodberry, "Single-wavelength Thin-film Polarizers," *J. Opt. Soc. Am.* **61:**1604–1606 (1971).
235. D. Blanc, P. H. Lissberger, and A. Roy, "The Design, Preparation and Optical Measurement of Thin Film Polarizers," *Thin Solid Films* **57:**191–198 (1979).
236. I. M. Minkov, "Theory of Dielectric Mirrors in Obliquely Incident Light," *Opt. Spectrosc.* (USSR) **33:**175–178 (1973).
237. A. Thelen, "Avoidance or Enhancement of Polarization in Multilayers," *J. Opt. Soc. Am.* **70:**118–121 (1980).
238. P. B. Clapham, M. J. Downs, and R. J. King, "Some Applications of Thin Films to Polarization Devices," *Appl. Opt.* **8:**1965–1974 (1969).
239. P. Kard, "Theory of Achromatic Multilayer Interference Polarizers," *Eesti NSV Tead. Akad. Toim., Fuus.-Mat.* **9:**26–32 (1960).
240. W. Geffcken, "Interference Light Filter," DB patent 913,005, issued 8 June 1954/44.
241. S. M. MacNeille, "Beam Splitter," U.S. patent 2403731, issued 6 July 1946.
242. M. Banning, "Practical Methods of Making and Using Multilayer Filters," *J. Opt. Soc. Am.* **37:**792–297 (1947).
243. W. B. Wetherell, "Polarization Matching Mixer in Coherent Optical Communications Systems," *Opt. Eng.* **28:**148–156 (1989).
244. J. Mouchart, J. Begel, and E. Duda, "Modified MacNeille Cube Polarizer for a Wide Angular Field," *Appl. Opt.* **28:**2847–2853 (1989).
245. R. Austin, "Thin Film Polarizing Devices," *Electro-Optical Systems Design* 30–35 (1974).
246. R. P. Netterfield, "Practical Thin-Film Polarizing Beam-splitters," *Opt. Acta* **24:**69–79 (1977).
247. J. A. Dobrowolski and A. J. Waldorf, "High-performance Thin Film Polarizer for the UV and Visible Spectral Regions," *Appl. Opt.* **20:**111–116 (1981).
248. V. P. Sobol', R. A. Petrenko, and A. S. Dimitreev, "Interference Polarizer Employing Optical Contact," *Sov. J. Opt. Technol.* **53:**692 (1986).
249. M. Gilo and K. Rabinovitch, "Design Parameters of Thin-film Cubic-type Polarizers for High Power Lasers," *Appl. Opt.* **26:**2518–2521 (1987).
250. E. Cojocaru, "Comparison of Theoretical Performances for Different Single-wavelength Thin-film Polarizers," *Appl. Opt.* **31:**4501–4504 (1992).

251. Anon., "Metal-dielectric Interference Filters," in G. Hass, M. H. Francombe, and R. W. Hoffman (eds.), *Physics of Thin Films,* Academic Press, New York, 1977, vol. 9, pp. 73–144.

252. E. E. Barr, "The Design and Construction of Evaporated Multilayer Filters for Use in Solar Radiation Technology," in A. J. Drummond (ed.), *Advances in Geophysics,* Academic Press, 1970, pp. 391–412.

253. R. Morf and R. E. Kunz, "Dielectric Filter Optimization by Simulated Thermal Annealing," *SPIE—Thin Film Technologies III* **1019:**211 (1988).

254. N. S. Gluck and W. J. Gunning, "Patterned Infrared Spectral Filter Directly Deposited onto Cooled Substrates," *Appl. Opt.* **28:**5110–5114 (1989).

255. W. H. Southwell, W. J. Gunning, and R. L. Hall, "Narrow-bandpass Filter Using Partitioned Cavities," *SPIE—Optical Thin Films II. New Developments* **678:**177–184 (1986).

256. W. Geffcken, "Interference Light Filter," DB patent 716,153, 8 December 1939.

257. A. F. Turner, "Some Current Developments in Multilayer Optical Films," *Le Journal de Physique et le Radium* **11:**444–460 (1950).

258. P. W. Baumeister, "Optical Tunneling and Its Applications to Optical Filters," *Appl. Opt.* **6:**897–905 (1967).

259. S. W. Warren, "Properties and Performance of Basic Designs of Infrared Interference Filters," *Infrared Phys.* **8:**65–78 (1968).

260. A. Thelen, "Design of Multilayer Interference Filters," in G. Hass and R. E. Thun (eds.), *Physics of Thin Films,* Academic Press, New York and London, 1969, vol. 5, pp. 47–86.

261. W. Geffcken, "The Wave-band Filter, an Interference Filter of Very High Efficiency," *Z. Angew. Phys.* **6:**249–250 (1954).

262. A. F. Turner, "Wide Passband Multilayer Filters," *J. Opt. Soc. Am.* **42:**878 (1952).

263. S. D. Smith, "Design of Multilayer Filters by Considering Two Effective Interfaces," *J. Opt. Soc. Am.* **48:**43–50 (1958).

264. P. Baumeister, "Use of Microwave Prototype Filters to Design Multilayer Dielectric Bandpass Filters," *Appl. Opt.* **21:**2965–2967 (1982).

265. A. Zheng, J. S. Seeley, R. Hunneman, and G. J. Hawkins, "Design of Narrowband Filters in the Infrared Region," *Infrared Phys.* **31:**237–244 (1991).

266. P. W. Baumeister, V. R. Costich, and S. C. Pieper, "Bandpass Filters for the Ultraviolet," *Appl. Opt.* **4:**911–914 (1965).

267. P. H. Berning and A. F. Turner, "Induced Transmission in Absorbing Films Applied to Band Pass Filter Design," *J. Opt. Soc. Am.* **47:**230–239 (1957).

268. R. L. Maier, "2M Interference Filters for the Ultraviolet," *Thin Solid Films* **1:**31–37 (1967).

269. P. W. Baumeister, "Radiant Power Flow and Absorptance in Thin Films," *Appl. Opt.* **8:**423–436 (1969).

270. R. J. Holloway and P. H. Lissberger, "The Design and Preparation of Induced Transmission Filters," *Appl. Opt.* **8:**653–660 (1969).

271. B. V. Landau and P. H. Lissberger, "Theory of Induced-transmission Filters in Terms of the Concept of Equivalent Layers," *J. Opt. Soc. Am.* **62:**1258–1264 (1972).

272. P. H. Lissberger, "Coatings with Induced Transmission," *Appl. Opt.* **20:**95–104 (1981).

273. N. P. Matshina, E. A. Nesmelov, I. K. Nagimov, R. M. Validov, and N. N. Soboleva, "Theory of Narrow-band Filters with Induced Transparency," *J. Appl. Spectrosc.* (USSR) **55:**1273–1278 (1991).

274. D. Gershenzon and L. Sossi, "Conditions for Maximum Transmission of Metal-dielectric Interference Filters," *Eesti NSV Tead. Akad. Toim., Fuus.-Mat.* **41:**142–149 (1992).

275. F. A. Korolev, A. Y. Klement'eva, and T. F. Meshcheryakova, "Interference Filters with a Transmission Band of 1.5 Å Half-width," *Opt. Spectrosc.* (USSR) **9:**341–343 (1960).

276. R. H. Eather and D. L. Reasoner, "Spectrophotometry of Faint Light Sources with a Tilting-filter Photometer," *Appl. Opt.* **8:**227–242 (1969).

277. R. W. Wood, "Some New Cases of Interference and Diffraction," in *The London, Edinburgh,*

and Dublin Philosophical Magazine and Journal of Science, Taylor and Francis, London, 1904, pp. 376–388.

278. A. I. Kartashev and N. M. Syromyatnikova, "Mica Interference Filters," in *Wavelength of Light as a Standard of Length, Proc. Mendeleev State Sci. Res. Inst.,* Leningrad, 1947, pp. 86–93.
279. J. Ring, R. Beer, and V. Hewison, "Reflectance Multilayers," *J. Phys. Radium* **19:**321–323 (1958).
280. J. A. Dobrowolski, "Mica Interference Filters with Transmission Bands of Very Narrow Half-Widths," *J. Opt. Soc. Am.* **49:**794–806 (1959).
281. G. S. Cheremukhin, V. P. Rozhnov, and G. I. Golubeva, "Narrow-band Interference Filters Made from Mica," *Sov. J. Opt. Technol.* **43:**312–315 (1976).
282. D. R. Herriott, J. R. Wimperis, and D. L. Perry, "Filters, Wave Plates and Protected Mirrors Made of Thin Polished Supported Layers," *J. Opt. Soc. Am.* **4:**546 (abstract) (1965).
283. J. D. Rehnberg, "Design Considerations for an Orbiting Solar Telescope," in *Proceedings, Electro-opt. Syst. Des. Conf.,* New York, Sept. 16–18, 1969, pp. 69–73 (1970).
284. A. Title, "Fabry-Perot Interferometers as Narrow-Band Optical Filters: Part One—Theoretical Considerations," TR-18 issued June 1970.
285. R. Fisher, "On the Use of a Solid Fabry-Perot Interferometer for Coronal Photography," *Solar Physics* **18:**253–257 (1971).
286. J. F. Markey and R. R. Austin, "High Resolution Solar Observations: the Hydrogen-alpha Telescopes on Skylab," *Appl. Opt.* **16:**917–921 (1977).
287. R. R. Austin, Perkin-Elmer Corporation, private communication, 1968.
288. R. R. Austin, "The Use of Solid Etalon Devices as Narrow Band Interference Filters," *Opt. Eng.* **11:**65–69 (1972).
289. S. D. Smith and C. R. Pidgeon, "Application of Multiple Beam Interferometric Methods to the Study of CO_2 Emissions at 15 μm," *Mem. Soc. R. Liege 5ieme serie* **9:**336–349 (1963).
290. V. R. Costich, "Study to Demonstrate a New Process to Produce Infrared Filters," *NASA Ames NAS2-12639,* issued 1 March 1990.
291. A. E. Roche and A. M. Title, "Tilt Tunable Ultra Narrow-band Filters for High Resolution Infrared Photometry," *Appl. Opt.* **14:**765–770 (1975).
292. J.-M. Saurel and M. Candille, "Réalisation de filtres multidiélectriques sur un substrat biréfringent, déformable et de faible épaisseur (mylar)," *Thin Solid Films* **16:**313–324 (1973).
293. M. Candille and J. M. Saurel, "Réalisation de filtres 'souble onde' à bandes passantes très étroites sur supports en matière plastique (mylar)," *Opt. Acta* **21:**947–962 (1974).
294. P. W. Baumeister, F. A. Jenkins, and M. A. Jeppesen, "Characteristics of the Phase-dispersion Interference Filter," *J. Opt. Soc. Am.* **49:**1188–1190 (1959).
295. R. R. Austin, "Narrow Band Interference Light Filter," U.S. patent 3528726, issued 15 September 1970.
296. P. Giacomo, P. W. Baumeister, and F. A. Jenkins, "On the Limiting Band Width of Interference Filters," *Proceedings of the Physical Society* **73:**480–489 (1959).
297. S. D. Smith and O. S. Heavens, "A Tunable Infra-red Interference Filter," *J. Sci. Instrum.* **34:**492–496 (1957).
298. P. D. Atherton, N. K. Reay, J. Ring, and T. R. Hicks, "Tunable Fabry-Perot Filters," *Opt. Eng.* **20:**806–814 (1981).
299. J. V. Ramsay, "Control of Fabry-Perot Interferometers and Some Unusual Applications," *The Australian Physicist* **5:**87–89 (1968).
300. J. V. Ramsay, H. Kobler, and E. G. V. Mugridge, "A New Tunable Filter with a Very Narrow Pass-band," *Solar Physics* **12:**492–501 (1970).
301. D. B. McKenney and P. N. Slater, "Design and Use of Interference Passband Filters with Wide-angle Lenses for Multispectral Photography," *Appl. Opt.* **9:**2435–2440 (1970).
302. J. A. Dobrowolski and R. C. Bastien, "Square-top Transmission Band Interference Filters for the Infra-red," *J. Opt. Soc. Am.* **56:**554 (1966).

303. E. Pelletier and H. A. Macleod, "Interference Filters with Multiple Peaks," *J. Opt. Soc. Am.* **72:**683–687 (1982).

304. W. A. Hovis Jr., W. A. Kley, and M. G. Strange, "Filter Wedge Spectrometer for Field Use," *Appl. Opt.* **6:**1057 (1967).

305. A. Thelen, "Circularly Wedged Optical Coatings. I. Theory," *Appl. Opt.* **4:**977–981 (1965).

306. J. H. Apfel, "Circularly Wedged Optical Coatings. II. Experimental," *Appl. Opt.* **4:**983–985 (1965).

307. V. A. Martsinovskii and F. K. Safiullin, "Determination of the Conditions of Formation of Annular Coatings with Linear Dependence of Thickness on Rotation Angle," *J. Appl. Spectrosc.* (USSR) **39:**976–980 (1983).

308. V. P. Avilov and A. Khosilov, "Apparatus for Producing Annular Variable Interference Filters," *Sov. J. Opt. Technol.* **55:**613–615 (1988).

309. I. M. Minkov, "Calculation of Narrow-band Circular Wedge Filter for 4–12 μm Spectral Region," *Sov. J. Opt. Technol.* **58:**491–492 (1991).

310. V. L. Yen, "Circular Variable Filters," *Optical Spectra* **3:**78–84 (1969).

311. V. P. Avilov, V. I. Tsypin, and E. M. Shipulin, "Circular-wedge Filter for the 0.24–0.40 μm Region," *Sov. J. Opt. Technol.* **54:**515 (1987).

312. P. H. Lissberger, "Effective Refractive Index as a Criterion of Performance of Interference Filters," *J. Opt. Soc. Am.* **58:**1586–1590 (1968).

313. P. H. Lissberger, "Properties of All-dielectric Interference Filters. I. A New Method of Calculation," *J. Opt. Soc. Am.* **49:**121–125 (1959).

314. C. R. Pidgeon and S. D. Smith, "Resolving Power of Multilayer Filters in Nonparallel Light," *J. Opt. Soc. Am.* **54:**1439–1466 (1964).

315. D. J. Hemingway and P. H. Lissberger, "Effective Refractive Indices of Metal-dielectric Interference Filters," *Appl. Opt.* **6:**471–476 (1967).

316. S. L. Linder, "Optimization of Narrow Optical Spectral Filters for Nonparallel Monochromatic Radiation," *Appl. Opt.* **6:**1201–1204 (1967).

317. P. W. Baumeister, F. A. Jenkins, and M. A. Jeppesen, "Characteristics of the Phase-dispersion Interference Filter," *J. Opt. Soc. Am.* **49:**1188–1190 (1959).

318. P. G. Kard, "On Elimination of the Doublet Structure of the Transmission Band in a Total-reflection Light-filter," *Opt. Spectrosc.* (USSR) **6:**244–246 (1959).

319. G. P. Konyukhov and E. A. Nesmelov, "On the Theory of a Dielectric Narrow-band Light Filter," *Zh. Prikl. Spektrosk.* **11:**468–474 (1969).

320. P. Baumeister, "Bandpass Design—Application to Nonnormal Incidence," *Appl. Opt.* **31:**504–512 (1992).

321. J. Ring, "The Fabry-Pérot Interferometer in Astronomy," in Z. Kopal (ed.), *Astronomical Optics,* North-Holland Publishing Co., Amsterdam, 1956, pp. 381–388.

322. W. Wilmot and E. R. Schineller, "A Wide-angle Narrow-band Optical Filter," *J. Opt. Soc. Am.* **56:**549 (1966).

323. E. R. Schineller and R. P. Flam, "Development of a Wide-angle Narrow-band Optical Filter," in *Proceedings, Spring meeting program—Optical Society of America,* pp. FF11-1–FF11-4 (1968).

324. J. Meaburn, "The Stability of Interference Filters," *Appl. Opt.* **5:**1757–1759 (1966).

325. J. Schild, A. Steudel, and H. Walter, "The Variation of the Transmission Wavelength of Interference Filters by the Influence of Water Vapor," *Journal de Physique* **28:**C2-276–C2-279 (1967).

326. L. D. Lazareva, "The Effect of Temperature on the Position of the Passband Maximum in Interference Filters," *Opt. Tech.* **36:**801–802 (1970).

327. S. A. Furman and M. D. Levina, "Effect of Moisture on the Optical Characteristics of Narrow Band Interference Filters," *Opt. Spectrosc.* (USSR) **30:**404–408 (1971).

328. S. A. Furman and M. D. Levina, "Strict Stabilization of the Optical Characteristics of Narrow-band Interference Filters," *Sov. J. Opt. Tech.* **38:**374–375 (1971).

329. S. A. Furman and M. D. Levina, "Stabilization of the Location of the Transmission Band of a Narrow-band Dielectric Interference Filter," *Sov. J. Opt. Tech.* **38:**272–275 (1971).

330. M. D. Levina and S. A. Furman, "Improving the Stability of the Optical Characteristics of Metal-dielectric Filters," *Sov. J. Opt. Technol.* **49:**128–129 (1982).

331. D. R. Gibson and P. H. Lissberger, "Optical Properties of Narrowband Spectral Filter Coatings Related to Layer Structure and Preparation," *Appl. Opt.* **22:**269–281 (1983).

332. A. Brunsting, M. A. Kheiri, D. F. Simonaitis, and A. J. Dosmann, "Environmental Effects on All-dielectric Bandpass Filters," *Appl. Opt.* **25:**3235–3241 (1986).

333. S. A. Furman and M. D. Levina, "Investigations of Temperature Dependence of Optical Properties of Dielectric Narrow-band Filters," *Opt. Spectrosc.* (USSR) **28:**412–416 (1970).

334. S. A. Furman, "Effect of Temperature, Angle of Incidence, and Dispersion of Index of Refraction of Layers on the Position of the Pass Bands of a Dielectric Narrow-band Filter," *Opt. Spectrosc.* (USSR) **28:**218–222 (1970).

335. A. S. Chaikin and V. V. Pukhonin, "An Investigation of the Temperature and Time Dependence of the Parameters of Narrow-band Interference Filters," *J. Appl. Spectrosc.* (USSR) **13:**1513–1515 (1970).

336. G. I. Golubeva, M. S. Abakumova, and A. M. Klochkov, "Results of Industrial Studies," *Opt. Tech.* **38:**228–229 (1971).

337. A. M. Zheng, J. S. Seeley, R. Hunneman, and G. J. Hawkins, "Ultranarrow Filters with Good Performance when Tilted and Cooled," *Appl. Opt.* **31:**4336–4338 (1992).

338. W. Feng and Y. Yan, "Shift in Infrared Interference Filters at Cryogenic Temperature," *Appl. Opt.* **31:**6591–6592 (1992).

339. I. H. Blifford Jr., "Factors Affecting the Performance of Commercial Interference Filters," *Appl. Opt.* **5:**105–111 (1966).

340. Optical Coating Laboratory, catalog, 2789 Northpoint Parkway, Santa Rosa, CA 95407-7397, USA, 1964.

341. R. Mark, D. Morand, and S. Waldstein, "Temperature Control of the Bandpass of an Interference Filter," *Appl. Opt.* **9:**2305–2310 (1970).

342. V. M. Novikov, "Vacuum Installation with a Manipulator," *Sov. J. Opt. Technol.* **35:**121–122 (1968).

343. O. M. Sorokin and V. A. Blank, "Thin-film Metal and Semiconductor Filters for the Vacuum Ultraviolet," *Sov. J. Opt. Tech.* **37:**343–346 (1970).

344. T. Barbee and J. H. Underwood, "Solid Fabry-Perot Etalons for X-rays," *Opt. Commun.* **48:**161–166 (1983).

345. Y. Lepetre, R. Rivoira, R. Philip, and G. Rasigni, "Fabry-Perot Etalons for X-rays: Construction and Characterization," *Opt. Commun.* **51:**127–130 (1984).

346. Y. Lepetre, I. K. Schuller, G. Rasigni, R. Rivoira, and R. Philip, "Novel Characterization of Thin Film Multilayered Structures: Microcleavage Transmission Electron Spectroscopy," *Proc. Soc. Photo-Opt. Instrum. Eng.* **563:**258–263 (1985).

347. R. J. Bartlett, W. J. Trela, D. R. Kania, M. P. Hockaday, T. W. Barbee, and P. Lee, "Soft X-ray Measurement of Solid Fabry-Perot Etalons," *Opt. Commun.* **55:**229–235 (1985).

348. V. R. Costich, "Coatings for 1, 2, even 3 Wavelengths," *Laser Focus Magazine* 41–45 (1969).

349. J. D. Rancourt and W. T. Beauchamp, "Articles Having Improved Reflectance Suppression," U.S. patent 4578527, 25 March 1986.

350. W. C. Herrmann and D. E. Morton, "A Design Technique for Multiband AR Coatings," in *Proceedings, 33rd Annual Technical Conference,* New Orleans, 1990, pp. 246–249.

351. L. Li, J. A. Dobrowolski, J. D. Sankey, and J. R. Wimperis, "Antireflection Coatings for Both Visible and Far Infrared Spectral Regions," *Appl. Opt.* **31:**6150–6156 (1992).

352. L. Li and J. A. Dobrowolski, "Design of Optical Coatings for Two Widely Separated Spectral Regions," *Appl. Opt.* **32:**2969–2975 (1993).

353. H. A. Macleod, "Structure-related Optical Properties of Thin Films," *J. Vac. Sci. Technol.* **A4:**418–422 (1986).

354. T. Motohiro and Y. Taga, "Thin Film Retardation Plate by Oblique Deposition," *Appl. Opt.* **28:**2466–2482 (1989).

355. R. M. A. Azzam, "Single-layer-coated Optical Devices for Polarized Light," *Thin Solid Films* **163:**33–41 (1988).

356. J. H. Apfel, "Graphical Method to Design Multilayer Phase Retarders," *Appl. Opt.* **20:**1024–1029 (1981).

357. J. H. Apfel, "Phase Retardance of Periodic Multilayer Mirrors," *Appl. Opt.* **21:**733–738 (1982).

358. P. Lostis, "Étude, réalisation et contrôle de lames minces introduisant une différence de marche déterminée entre deux vibrations rectangulaires," *Revue d'Optique, Theoretique et Instrumentale* **38:**1–28 (1959).

359. J. H. Apfel, "Graphical Method to Design Internal Reflection Phase Retarders," *Appl. Opt.* **23:**1178–1183 (1984).

360. R. M. A. Azzam and M. E. R. Khan, "Polarization-preserving Single-layer-coated Beam Displacers and Axicons," *Appl. Opt.* **21:**3314–3322 (1982).

361. J. M. Bennett, "A Critical Evaluation of Rhomb-type Quarterwave Retarders," *Appl. Opt.* **9:**2123–2129 (1970).

362. T. Filinski and T. Skettrup, "Achromatic Phase Retarders Constructed from Right-angle Prisms: Design," *Appl. Opt.* **23:**2747–2751 (1984).

363. H. Pohlack, "Über die reflexionsvermindernde Wirkung dünner Metallschichten auf Glas," in P. Görlich (ed.), *Jenaer Jahrbuch 1956,* Fischer, Jena, 1956, pp. 87–93.

364. G. Hass, H. H. Schroeder, and A. F. Turner, "Mirror Coatings for Low Visible and High Infrared Reflectance," *J. Opt. Soc. Am.* **46:**31–35 (1956).

365. J. A. Dobrowolski, B. T. Sullivan, and R. C. Bajcar, "Optical Interference, Contrast-enhanced Electroluminescent Device," *Appl. Opt.* **31:**5988–5996 (1992).

366. J. A. Dobrowolski, E. H. Hara, B. T. Sullivan, and A. J. Waldorf, "High Performance Optical Wavelength Multiplexer-demultiplexer," *Appl. Opt.* **31:**3800–3806 (1992).

367. J. A. Dobrowolski and R. A. Kemp, "Flip-flop Thin-film Design Program with Enhanced Capabilities," *Appl. Opt.* **31:**3807–3812 (1992).

368. J. H. Apfel and R. M. Gelber, "Multilayer Filter with Metal Dielectric Period," U.S. patent 3649359, 14 March 1972.

369. J. H. Apfel and R. M. Gelber, "Filter with Neutral Transmitting Multilayer Coating Having Asymmetric Reflectance," U.S. patent 3679291, 25 July 1972.

370. R. J. Fay and J. R. Cicotta, "Neutral Density Filter Element with Reduced Surface Reflection," U.S. patent 3781089, 25 December 1973.

371. K. Yamamoto and T. Koike, "Light Absorptive Film Provided with a Reflection Preventive Means," U.S. patent 4381883, 3 May 1983.

372. D. H. Cushing, "Neutral Density Filters for the Ultraviolet that Obey Beer's Law," *Proc. Soc. Photo-Opt. Instrum. Eng.* **1158:**189–193 (1989).

373. D. H. Cushing, "Broad Band Nonreflective Neutral Density Filter," U.S. patent 4,960,310, 2 October 1990.

374. J. R. Jacobsson, "Protective Device for Protection against Radiation During Welding," U.S. patent 4169655, 2 October 1979.

375. ASTM, "Symposium on Electroless Nickel Plating," *Spec. Tech. Publ.* **265:**(1959).

376. R. P. Madden, "Preparation and Measurement of Reflecting Coatings for the Vacuum Ultraviolet," in G. Hass (ed.), *Physics of Thin Films,* Academic Press, New York, 1963, vol. 1, pp. 123–186.

377. M. L. Scott, P. N. Arendt, B. J. Cameron, J. M. Saber, and B. E. Newnam, "Extreme Ultraviolet Reflectance Degradation of Aluminum and Silicon from Surface Oxidation," *Appl. Opt.* **27:**1503–1507 (1988).

378. A. P. Lukirskii, E. P. Savinov, O. A. Ershov, and Y. F. Shepelev, "Reflection Coefficients of Radiation on the Wavelength Range from 23.6 to 113 Å for a Number of Elements and Substances and the Determination of the Refractive Index and Absorption Coefficient," *Opt. Spectrosc.* (USSR) **16:**168–172 (1964).

379. A. P. Lukirskii, E. P. Savinov, O. A. Ershov, I. I. Zhukova, and V. A. Forichev, "Reflection of X Rays with Wavelengths from 23.6 to 190.3 Å. Some Remarks on the Performance of Diffraction Gratings," *Opt. Spectrosc.* (USSR) **19:**237–241 (1965).

380. J. Hecht, "A Novel Approach to High XUV Reflectivity," *Lasers and Optronics,* April 1989, p. 14.

381. G. Hass, "Filmed Surfaces for Reflecting Objects," *J. Opt. Soc. Am.* **45:**945–952 (1955).

382. G. Hass and R. Tousey, "Reflecting Coatings for the Extreme Ultraviolet," *J. Opt. Soc. Am.* **49:**593–602 (1959).

383. L. R. Canfield, G. Hass, and J. E. Waylonis, "Further Studies on MgF_2-overcoated Aluminum Mirrors with Highest Reflectance in the Vacuum Ultraviolet," *Appl. Opt.* **5:**45–50 (1966).

384. M. R. Adraens and B. Feuerbacher, "Improved LiF and MgF_2 Overcoated Aluminum Mirrors for Vacuum Ultraviolet Astronomy," *Appl. Opt.* **10:**958–959 (1971).

385. D. W. Angel, W. R. Hunter, and R. Tousey, "Extreme Ultraviolet Reflectance of LiF-coated Aluminum Mirrors," *J. Opt. Soc. Am.* **51:**913–914 (1961).

386. J. T. Cox, G. Hass, and J. E. Waylonis, "Further Studies on LiF-Overcoated Aluminum Mirrors with Highest Reflectance in the Vacuum Ultraviolet," *Appl. Opt.* **7:**1535–1539 (1968).

387. W. R. Hunter, J. F. Osantowski, and G. Hass, "Reflectance of Aluminum Overcoated with MgF_2 and LiF in the Wavelength Region from 1600 Å to 300 Å at Various Angles of Incidence," *Appl. Opt.* **10:**540–544 (1971).

388. D. K. Burge, H. E. Bennett, and E. J. Ashley, "Effect of Atmospheric Exposure on the Infrared Reflectance of Silvered Mirrors with and without Protective Coatings," *Appl. Opt.* **12:**42–47 (1973).

389. G. Hass, J. B. Heaney, J. F. Osantowski, and J. J. Triolo, "Reflectance and Durability of Ag Mirrors Coated with Thin Layers of Al_2O_3 plus Reactively Deposited Silicon Oxide," *Appl. Opt.* **14:**2639–2644 (1975).

390. E. A. Volgunova, G. I. Golubeva, and A. M. Klochkov, "Highly Stable Silver Mirrors," *Sov. J. Opt. Technol.* **50:**128–129 (1983).

391. W. R. Hunter, J. F. Osantowski, and G. Hass, "Reflectance of Aluminum Overcoated with MgF_2 and LiF in the Wavelength Region from 1600 Å to 300 Å at Various Angles of Incidence," *Appl. Opt.* **10:**540–544 (1971).

392. J. T. Cox, G. Hass, and W. R. Hunter, "Infrared Reflectance of Silicon Oxide and Magnesium Fluoride Protected Aluminum Mirrors at Various Angles of Incidence from 8 μm to 12 μm," *Appl. Opt.* **14:**1247–1250 (1975).

393. J. T. Cox and G. Hass, "Aluminum Mirrors Al_2O_3-protected with High Reflectance at Normal but Greatly Decreased Reflectance at Higher Angles of Incidence in the 8–12-μm Region," *Appl. Opt.* **17:**333–334 (1978).

394. S. F. Pellicori, "Infrared Reflectance of a Variety of Mirrors at 45° Incidence," *Appl. Opt.* **17:**3335–3336 (1978).

395. L. Young, "Multilayer Reflection Coatings on a Metal Mirror," *Appl. Opt.* **2:**445–447 (1963).

396. D.-Y. Song, R. W. Sprague, H. A. Macleod, and M. R. Jacobson, "Progress in the Development of a Durable Silver-based High-reflectance Coating for Astronomical Telescopes," *Appl. Opt.* **24:**1164–1170 (1985).

397. G. Hass, J. B. Ramsey, and W. R. Hunter, "Reflectance of Semitransparent Platinum Films on Various Substrates in the Vacuum Ultraviolet," *Appl. Opt.* **8:**2255–2259 (1969).

398. R. P. Madden, L. R. Canfied, and G. Hass, "On the Vacuum-ultraviolet Reflectance of Evaporated Aluminum Before and During Oxidation," *J. Opt. Soc. Am.* **53:**620–625 (1969).

399. L. N. Hadley and D. M. Dennison, "Reflection and Transmission Interference Filters. Part II. Experimental, Comparison with Theory, Results," *J. Opt. Soc. Am.* **38:**483–496 (1948).

400. L. N. Hadley and D. M. Dennison, "Reflection and Transmission Interference Filters. Part I. Theory," *J. Opt. Soc. Am.* **37:**451–465 (1947).

401. S.-y. Zheng and J. W. Y. Lit, "Design of a Narrow-band Reflection IR Multilayer," *Canadian Journal of Physics* **61:**361–368 (1983).

402. E. van Rooyen and E. Theron, "A Multiple Reflection Multilayer Reflection Filter," *Appl. Opt.* **8:**832–833 (1969).

403. G. Hass, J. B. Heaney, and W. R. Hunter, "Reflectance and Preparation of Front Surface Mirrors for Use at Various Angles of Incidence from the Ultraviolet to the Far Infrared," in *Physics of Thin Films,* Academic Press, 1982, vol. 12, pp. 1–51.

404. D. W. Lynch, "Mirror and Reflector Materials," in M. J. Weber (ed.), *Handbook of Laser Science and Technology. Optical Materials*: *Part 2,* CRC Press, Inc., Boca Raton, Fla., 1986, pp. 185–219.

405. Liberty Mirror—A Division of Libbey-Owens Ford Glass Co., "Coatings," 851 Third Avenue, Brackenridge, PA 15014, USA, 1966.

406. Balzers Aktiengesellschaft für Hochvakuumtechnik und Dünne Schichten, "Kurvenblätter und ihre Bezeichnungen," FL-9496 Balzers, Principality of Liechtenstein, 1962.

407. V. R. Costich, Spectra-Physics, private communication, 1968.

408. A. F. Turner, Bausch & Lomb, private communication, 1970.

409. Optical Coating Laboratory Inc., "Multilayer Antireflection Coatings," 2789 Northpoint Parkway, Santa Rosa, CA 95407-7397, USA, 1965.

410. Thin Film Lab, "Product Information," 501B Basin Rd., West Hurley, NY 12491, USA, 1992.

411. TechOptics Ltd., "Laser Optics and Instrumentation," Second Avenue, Onchan, Isle of Man, British Isles, 1992.

412. Reynard Corporation, "Optical Components: Thin Film Coatings; Optical Instruments," 1020 Calle Sombra, San Clemente, CA 92673–6227, USA, 1992.

413. Spindler & Hoyer Inc., "Precision Optics," 459 Fortune Boulevard, Milford, MA 01757-1757, U.S.A., 1990.

414. T. I. Oh, "Broadband AR Coatings on Germanium Substrates Using Ion-assisted Deposition," *Appl. Opt.* **27:**4255–4259 (1988).

415. R. Jacobsson, "Calculation and Deposition of Inhomogeneous Thin Films," *J. Opt. Soc. Am.* **56:**1435 (1966).

416. A. Y. Kuznetsov and A. F. Perveyev, "Multifilm Anti-reflection Coatings for Materials with High Refractive Index," *Sov. J. Opt. Technol.* **34:**593–595 (1967).

417. V. V. Veremey and T. A. Gorbunova, "Achromatic Coating of Glasses Covered by Conducting Films," *Sov. J. Opt. Technol.* **35:**519–521 (1968).

418. T. Turbadar, "Equi-reflectance Contours of Double-layer Anti-reflection Coatings," *Opt. Acta* **11:**159–205 (1964).

419. T. Turbadar, "Complete Absorption of Plane Polarized Light by Thin Metallic Films," *Opt. Acta* **11:**207–210 (1964).

420. J. A. Dobrowolski and S. H. C. Piotrowski, "Refractive Index as a Variable in the Numerical Design of Optical Thin Film Systems," *Appl. Opt.* **21:**1502–1510 (1982).

421. J. H. Apfel, "Multilayer Interference Coatings for the Ultraviolet," *J. Opt. Soc. Am.* **59:**553 (1966).

422. K. Hefft, R. Kern, G. Nöldeke, and A. Steudel, "Über Fabry-Perot-Verspiegelungen aus dielekrischen Vielfachschichten für den Spektralbereich von 2350 bis 20000 Å," *Z. Phys.* **175:**391–404 (1963).

423. G. Honcia and K. Krebs, "Highly Reflecting Substances for Dielectric Mirror Systems," *Z. Phys.* **165:**202–212 (1961).

424. K. H. Behrndt and D. W. Doughty, "High-reflectance Multilayer Dielectric Mirrors," *J. Vac. Sci. Technol.* **4:**199–202 (1967).

425. R. Blazey, "Light Scattering by Laser Mirrors," *Appl. Opt.* **6:**831–836 (1967).

426. D. L. Perry, "Low-loss Multilayer Dielectric Mirrors," *Appl. Opt.* **4:**987–991 (1965).
427. O. S. Heavens and H. M. Liddell, "Staggered Broad-band Reflecting Multilayers," *Appl. Opt.* **5:**373–376 (1966).
428. Z. N. Elsner, "On the Calculation of Multilayer Interference Coatings with Given Spectral Characteristics," *Opt. Spectrosc.* (USSR) **17:**238–240 (1964).
429. S. Penselin and A. Steudel, "Fabry-Perot-Interferometerverspiegelungen aus dielektrischen Vielfachschichten," *Z. Phys.* **142:**21–41 (1955).
430. P. W. Baumeister and J. M. Stone, "Broad-band Multilayer Film for Fabry-Perot Interferometers," *J. Opt. Soc. Am.* **46:**228–229 (1956).
431. P. E. Ciddor, "Minimization of the Apparent Curvature of Multilayer Reflecting Surfaces," *Appl. Opt.* **7:**2328–2329 (1968).
432. Omega Optical Inc., "Bandpass Filters," 3 Grove Street, P.O. Box 573, Brattleboro, VT 05302-0573, USA, 1992.
433. P. G. Verly, National Research Council of Canada, private communication, 1993.
434. A. Piegari, A. Tirabassi, and G. Emiliani, "Thin Films for Special Laser Mirrors with Radially Variable Reflectance: Production Techniques and Laser Testing," *Proc. Soc. Photo-Opt. Instrum. Eng.* **1125:**68–73 (1989).
435. E. Spiller and L. Golub, "Fabrication and Testing of Large Area Multilayer Coated X-ray Optics," *Appl. Opt.* **28:**2969–2974 (1989).
436. C. Montcalm, B. T. Sullivan, H. Pépin, J. A. Dobrowolski, and G. D. Enright, "Multilayer Mirrors for XUV Ge-laser Wavelengths," *SPIE—Multilayer Optics for Advanced X-Ray Applications* **1547:**127–133 (1991).
437. C. Montcalm, P. A. Kearney, J. M. Slaughter, C. M. Falco, and M. Chaker, "Preliminary Survey of Material Pairs for XUV Multilayer Mirrors for Wavelengths Below 130 Å," in *Physics of X-ray Multilayer Structures Technical Digest 1994,* Optical Society of America, Washington, D.C. (in press).
438. A. P. Zwicker, S. P. Regan, M. Finkenthal, H. W. Moos, E. B. Saloman, R. Watts, and J. R. Roberts, "Peak Reflectivity Measurements of W/C, Mo/Si, and Mo/B_4C Multilayer Mirrors in the 8-190-Å Range Using Both Ka Line and Synchrotron Radiation," *Appl. Opt.* **29:**3694–3698 (1990).
439. N. M. Ceglio, "Revolution in X-ray Optics," *J. X-Ray Sci. Technol.* **1:**7–78 (1989).
440. C. M. Falco, F. E. Fernández, P. Dhez, A. Khandar-Shahabad, L. Névot, B. Pardo, and J. Corno, "Normal Incidence X-UV Mirrors," *SPIE—Soft X-Ray Optics and Technology* **733:**343–352 (1986).
441. A. Thelen, "Multilayer Filters with Wide Transmittance Bands," *J. Opt. Soc. Am.* **53:**1266–1270 (1963).
442. Optical Coating Laboratory Inc., "Effect of the Variation of Angle of Incidence and Temperature on Infrared Filter Characteristics," 2789 Northpoint Parkway, Santa Rosa, CA 95407-7397, USA, 1967.
443. Optical Coating Laboratory Inc., *Infrared Handbook,* Santa Rosa, Calif., 1970.
444. Bausch & Lomb, "Bausch and Lomb Multi-Films," 1400 North Goodman Street, P.O. Box 450, Rochester, NY 14692-0450, USA, 1968.
445. Eastman Kodak Company, "Special Filters from Kodak for Technical Applications," 343 State Street, Bldg. 701, Rochester, NY 14650-3512, USA, 1968.
446. Infrared Industries Inc., "Infratron Interference Filters," Thin Film Products Division, 62 Fourth Avenue, Waltham, MA 02154, USA, 1966.
447. Corion Corporation, "Optical Filters and Coatings," 73 Jeffrey Ave., Holliston, MA 01746-2082, USA, 1992.
448. Heliotek—A Division of Textron Inc., "Data Sheets," 12500 Gladstone Avenue, Sylmar, CA 91342, USA, 1969.
449. A. Thelen and Optical Coating Laboratory Inc., "Optimum Heat Reflectors for Silicon Solar Cell Power Converters," issued 1961.

450. A. F. Turner, "Heat Reflecting Coatings," in H. H. Blau and H. Fischer (eds.), *Radiative Heat Transfer from Solid Materials,* Macmillan, New York, 1962.

451. B. Hoppert, "Heat Reducing Metal Mirrors," in *Proceedings, Annual Technical Conference,* 1983, pp. 131–143.

452. L. Chang, "Infrared Dispersion of ZnS-Metal Films," Ph.D. thesis, Colorado State University, 1964.

453. H. Schröder, "Oxide Layers Deposited from Organic Solutions," in G. Hass and R. E. Thun (eds.), *Physics of Thin Films,* Academic, New York, 1969, pp. 87–141.

454. Oriel Optics Corporation, "Catalog Section C: Optical Filters," 2789 Northpoint Parkway, Santa Rosa, CA 95407-7397, USA, 1968.

455. S. F. Pellicori, "Beam Splitter and Reflection Reducing Coatings on ZnSe for 3–14 μm," *Appl. Opt.* **18:**1966–1968 (1979).

456. L. Y. Chang and S. H. Mo, "Design of Non-polarizing Prism Beam Splitter," in *Proceedings, Topical Meeting on Optical Interference Coatings,* Tucson, Ariz., 1988, pp. 381–384.

457. Y. N. Konoplev, Y. A. Mamaev, V. N. Starostin, and A. A. Turkin, "Nonpolarizing 50 percent Beam Splitters," *Opt. Spectrosc.* (USSR) **71:**303–305 (1991).

458. M. Banning, "Neutral Density Filters of Chromel," *J. Opt. Soc. Am.* **37:**686–689 (1947).

459. Eastman Kodak Company, "Kodak Neutral Density Attenuators," 343 State Street, Bldg. 701, Rochester NY 14650-3512, USA, 1969.

460. Corion Instrument Corporation, "Thin Film Optical Filters," 73 Jeffrey Ave., Holliston, MA 01746-2082, USA, 1969.

461. Acton Research Corporation, "Optical Filters," P.O. Box 2215, 525 Main Street, Acton, MA 01720, USA, 1992.

462. Spindler & Hoyer, "Precision Optics," Königsalee 23, Postfach 3353, D-3400 Göttingen, Germany, 1990.

463. T. A. Remneva, A. V. Kozhevnikov, and M. M. Nikitin, "Polarizer of Radiation in the Vacuum Ultraviolet," *J. Appl. Spectrosc.* (USSR) **25:**1587–1590 (1967).

464. M. Yanagihara, T. Maehara, H. Normura, M. Yamamoto, T. Namioka, and H. Kimura, "Performance of a Wide-band Multilayer Polarizer for Soft X Rays," *Rev. Sci. Instrum.* **63:**1516–1518 (1992).

465. R. S. Sokolova and T. N. Krylova, "Interference Polarizers for the Ultraviolet Spectral Region," *Opt. Spectrosc.* (USSR) **14:**213–215 (1963).

466. J. Wimperis, Interoptics, A Division of Lumonics, Inc., 14 Capella Court, Nepean (Ottawa) Ontario, Canada K2E 7V6, private communication, 1993.

467. T. F. Thonn and R. M. A. Azzam, "Multiple-reflection Polarizers Using Dielectric-coated Metallic Mirrors," *Opt. Eng.* **24:**202–206 (1985).

468. L. Songer, "The Design and Fabrication of a Thin Film Polarizer," *Optical Spectra* 49–50 (1978).

469. D. Lees and P. Baumeister, "Versatile Frustrated-total-reflection Polarizer for the Infrared," *Opt. Lett.* **4:**66–67 (1979).

470. T. A. Mooney, private communication, 1993.

471. D. H. Harrison, "MDM Bandpass Filters for the Vacuum Ultraviolet," *Appl. Opt.* **7:**210 (1968).

472. D. J. Bradley, B. Bates, C. O. L. Juulman, and T. Kohno, "Recent Developments in the Application of the Fabry-Perot Interferometer to Space Research," *Journal de Physique* **28:**C2-280–C2-283 (1967).

473. Balzers Aktiengesellschaft für Hochvakuumtechnik und Dünne Schichten, "Interference Filters," FL-9496 Balzers, Principality of Liechtenstein, 1968.

474. Schott Glaswerke, "Monochromatic Interference Filters," Geschäftsbereich Optik, Hattenbergstrasse 10, D-6500 Mainz, Germany, 1965.

475. H. Schröder, "Properties and Applications of Oxide Layers Deposited on Glass from Organic Solutions," *Opt. Acta* **9:**249–254 (1962).

476. R. S. Meltzer, Industrial Optics Division, Yardney Razdow Laboratories, private communication, 1968.

477. M. A. Cohendet and B. Saudreau, DRME-66-34-443, issued 1967.

478. O. A. Motovilov, "Narrow-band Interference Filters for the Ultraviolet Region of the Spectrum," *Opt. Spectrosc.* (USSR) **22:**537–538 (1967).

479. R. G. T. Neilson and J. Ring, "Interference Filters for the Near Ultraviolet," *Journal de Physique* **28:**C2-270–C2-275 (1967).

480. A. F. Turner and R. Walsh, "Interference Filters for the 8–13 micron Atmospheric Window," *Bausch & Lomb Technical Report no. 2,* issued 1961.

481. S. D. Smith and J. S. Seely, "Multilayer Filters for the Region 0.8 to 100 Microns," AF61 (052)-833, issued May 7, 1968.

482. Baird Atomic Inc., "Optical Components," 125 Middlesex Turnpike, Bedford, MA 01730-1468, 1968.

483. Spectrum Systems Division—Barnes Engineering Co., "Visible and Near Infrared Filters, Bull. SS-14," 211 Second Avenue, Waltham, MA 02154, USA, 1967.

484. Infrared Industries Inc., "Thin Film Products at Infrared Industries Inc.," Thin Film Products Division, 62 Fourth Avenue, Waltham, MA 02154, USA, 1967.

485. J. Minowa and Y. Fujii, "High Performance Bandpass Filters for WDM Transmission," *Appl. Opt.* **23:**193–194 (1984).

486. V. I. Tsypin and E. A. Sukhanov, "Metal-dielectric Filters for the 200–350 nm Spectral Regions," *Sov. J. Opt. Technol.* **59:**438–439 (1992).

487. C. H. Burton, A. J. Leistner, and D. M. Rust, "Electrooptic Fabry-Perot Filter Development for the Study of Solar Oscillations," *Appl. Opt.* **26:**2637–2642 (1987).

488. D. McKenney and A. F. Turner, *Univ. Ariz. Opt. Cent. Newsl.* **1** (1967).

489. J. Michael, Perkin-Elmer Corporation, private communication, 1968.

490. A. Mussett, Optical Coating Laboratory, private communication, 1969.

491. L. Li, National Research Council of Canada, private communication, 1993.

492. W. H. Southwell, "Multilayer Coating Design Achieving a Broadband 90° Phase Shift," *Appl. Opt.* **19:**2688–2692 (1980).

493. N. V. Grishina, "Synthesis of Mirrors with Constant Phase Difference under Oblique Incidence of Light," *Opt. Spectrosc.* (USSR) **69:**262–265 (1990).

494. E. Cojocaru, "Polarization-preserving Totally Reflecting Prisms," *Appl. Opt.* **31:**4340–4342 (1992).

495. E. Spiller, "Totally Reflecting Thin-Film Phase Retarders," *Appl. Opt.* **23:**3544–3549 (1984).

496. R. J. King, "Quarter-wave Retardation Systems Based on the Fresnel Rhomb Principle," *J. Sci. Instrum.* **43:**627–622 (1966).

497. R. M. A. Azzam, "ZnS-Ag Film-substrate Parallel-mirror Beam Displacers that Maintain Polarization of 10.6 μm Radiation Over a Wide Range of Incidence Angles," *Infrared Phys.* **23:**195–197 (1983).

498. T. F. Thonn and R. M. A. Azzam, "Three-reflection Halfwave and Quarterwave Retarder Using Dielectric-coated Metallic Mirrors," *Appl. Opt.* **23:**2752–2759 (1985).

499. S. H. C. Piotrowski-McCall, J. A. Dobrowolski, and G. G. Shepherd, "Phase Shifting Thin Film Multilayers for Michelson Interferometers," *Appl. Opt.* **28:**2854–2859 (1989).

500. J. A. Dobrowolski, "Versatile Computer Program for Absorbing Optical Thin-film Systems," *Appl. Opt.* **20:**74–81 (1981).

501. B. T. Sullivan, National Research Council of Canada, private communication, 1993.

502. L. Li, "Low Reflection Interference Coatings with Arbitrary Transmittance Shapes," in *1993 Optical Society of America Annual Meeting Technical Digest,* October 3–8, 1993, Toronto, Ontario, Canada, p. 114 (abstract).

503. J. A. Dobrowolski, R. A. Kemp, and B. T. Sullivan, "Black Bandpass Interference Filters," in

1993 Optical Society of America Annual Meeting Technical Digest, October 3–8, 1993, Toronto, Ontario, Canada, p. 115 (abstract).

504. H. E. Bennett, M. Silver, and E. J. Ashley, "Infrared Reflectance of Aluminum Evaporated in Ultra-high Vacuum," *J. Opt. Soc. Am.* **53:**1089–1095 (1963).
505. J. M. Bennett and E. J. Ashley, "Infrared Reflectance and Emittance of Silver and Gold Evaporated in Ultrahigh Vacuum," *Appl. Opt.* **4:**221–224 (1965).
506. G. Hass and L. Hadley, in O. F. Gray (ed.), *American Institute of Physics Handbook,* McGraw-Hill, New York, 1972, pp. 6–118.
507. W. R. Hunter, "High Reflectance Coatings for the Extreme Ultraviolet," *Opt. Acta* **9:**255–268 (1962).
508. G. Hass and G. F. Jacobus, "Optical Properties of Evaporated Iridium in the Vacuum Ultraviolet from 500 to 2000 Å," *J. Opt. Soc. Am.* **57:**758–762 (1967).
509. J. T. Cox, G. Hass, and W. R. Hunter, "Optical Properties of Evaporated Rhodium Films Deposited at Various Substrate Temperatures in the Vacuum Ultraviolet from 150 to 2000 Å," *J. Opt. Soc. Am.* **61:**360–364 (1971).
510. J. T. Cox, G. Hass, and J. B. Ramsey, "Reflectance of Evaporated Rhenium and Tungsten Films in the Vacuum Ultraviolet from 300 to 2000 Å," *J. Opt. Soc. Am.* **62:**781–785 (1972).
511. J. F. Seely, G. E. Holland, W. R. Hunter, R. P. McCoy, K. F. Dymond, and M. Corson, "Effect of Oxygen Atom Bombardment on the Reflectance of Silicon Carbide Mirrors in the Extreme Ultraviolet Region," *Appl. Opt.* **1805:**1805–1810 (1993).
512. R. F. Malina and W. Cash, "Extreme Ultraviolet Reflection Efficiencies of Diamond-turned Aluminum, Polished Nickel, and Evaporated Gold Surfaces," *Appl. Opt.* **17:**3309–3313 (1978).
513. D. L. Windt, W. C. Cash Jr., M. Scott, P. Arendt, B. Newman, R. F. Fisher, A. B. Swartzlander, P. Z. Takacs, and J. M. Pinneo, "Optical Constants for Thin Films of C, Diamond, AI, Si and CVD SiC from 24 Å to 1216 Å," *Appl. Opt.* **27:**279–295 (1988).
514. H. E. Bennett, J. M. Bennett, and E. J. Ashley, "The Effect of Protective Coatings of MgF_2 and SiO on the Reflectance of Aluminized Mirrors," *Appl. Opt.* **2:**156 (1963).
515. R. Denton, Denton Vacuum Inc., private communication, 1970.
516. V. D. Vvedenskii, R. Y. Pinskaya, S. A. Furman, and T. V. Shestakova, "Wide-band Reflectors Based on Silver Films," *Sov. J. Opt. Technol.* **50:**781–782 (1983).
517. AIRCO Coating Technology, "Enhanced Front-Surface Aluminum Mirrors," 2700 Maxwell Way, P.O. Box 2529, Fairfield, CA 94533-252, USA, 1990.
518. S. A. Furman and E. G. Stolov, "Synthesis of Achromatic Metal-dielectric Mirrors," *Sov. J. Opt. Technol.* **44:**359–361 (1977).
519. M. Z. Leś, F. Leś, and L. Gabla, "Semitransparent Metallic-dielectric Mirrors with Low Absorption Coefficient in the Ultra-violet Region of the Spectrum," *Acta Physica Polonica* **23:**211–214 (1963).
520. F. Leś and M. Z. Leś, "Metallic-dielectric Mirrors with High Reflectivity in the Near Ultra-violet for the Fabry-Perot Interferometer," *Acta Physica Polonica* **21:**523–528 (1962).
521. G. Hass and W. R. Hunter, "Calculated Reflectance of Aluminum-overcoated Iridium in the Vacuum Ultraviolet from 500 Å to 2000 Å," *Appl. Opt.* **6:**2097–2100 (1967).
522. L. A. Stelmack, "Vacuum Ultraviolet Reflectance Filter," U.S. patent 4408825, 11 October 1983.
523. A. F. Turner and H. R. Hopkinson, "Reflection Filters for the Visible and Ultraviolet," *J. Opt. Soc. Am.* **43:**326 (1953).
524. R. Gamble and P. H. Lissberger, "Reflection Filter Multilayers of Metallic and Dielectric Thin Films," *Appl. Opt.* **28:**2838–2846 (1989).
525. P. H. Berning, G. Hass, and R. P. Maden, "Reflectance-increasing Coatings for the Vacuum Ultraviolet and Their Applications," *J. Opt. Soc. Am.* **50:**586–597 (1960).
526. A. S. Valeyev, "Multilayer Reflection-type Interference Filters," *Sov. J. Opt. Technol.* **34:**317–319 (1967).
527. Schott und Gen., "Interference Reflection Filter UV-R-250," Geschäftsbereich Optik, Hattenbergstrasse 10, D-6500 Mainz, Germany, 1967.

528. B. T. Sullivan and J. A. Dobrowolski, "Deposition of Optical Multilayer Coatings with Automatic Error Compensation: II. Magnetron Sputtering," *Appl. Opt.* **32:**2351–2360 (1993).

529. P. Kard, "Optical Theory of Anti-reflection Coatings," *Loodus ja matemaatika I* 67–85 (1959).

530. A. Thetford, "A Method of Designing Three-layer Anti-reflection Coatings," *Opt. Acta* **16:**37–43 (1969).

531. H. Lotem and K. Rabinovitch, "Penta Prism Laser Polarizer," *Appl. Opt.* **32:**2017–2020 (1993).

532. S. P. Fisher, J. F. Leonard, I. T. Muirhead, G. Buller, and P. Meredith, "The Fabrication of Optical Devices by Molecular Beam Deposition Technology," in *Optical Interference Coatings,* 1992 Technical Digest Series, Washington, D.C., Optical Society of America, vol. 15 pp. 131–133.

P · A · R · T · 12

TERRESTRIAL OPTICS

CHAPTER 43
THE OPTICAL PROPERTIES OF WATER

Curtis D. Mobley
Senior Research Engineer
Applied Electromagnetics and Optics Laboratory
SRI International
Menlo Park, California

43.1 INTRODUCTION

This article discusses the optical properties of three substances: pure water, pure sea water, and natural water. Pure water (i.e. water molecules only) without any dissolved substances, ions, bubbles, or other impurities, is exceptionally difficult to produce in the laboratory. For this and other reasons, definitive direct measurements of its optical properties at visible wavelengths have not yet been made. Pure sea water—pure water plus various dissolved salts—has optical properties close to those of pure water. Neither pure water nor pure sea water ever occur in nature. Natural waters, both fresh and saline, are a witch's brew of dissolved and particulate matter. These solutes and particulates are both optically significant and highly variable in kind and concentration. Consequently, the optical properties of natural waters show large temporal and spatial variations and seldom resemble those of pure water.

The great variability of the optical properties of natural water is the bane of those who desire precise and easily tabulated data. However, it is the connections between the optical properties and the biological, chemical, and geological constituents of natural water and the physical environment that define the critical role of optics in aquatic research. For just as optics utilizes results from the biological, chemical, geological, and physical subdisciplines of limnology and oceanography, so do those subdisciplines incorporate optics. This synergism is seen in such areas as bio-optical oceanography, marine photochemistry, mixed-layer dynamics, laser bathymetry, and remote sensing of biological productivity, sediment load, or pollutants.

43.2 TERMINOLOGY, NOTATION, AND DEFINITIONS

Hydrologic optics is the quantitative study of the interactions of radiant energy with the earth's oceans, estuaries, lakes, rivers, and other water bodies. Most past and current research within hydrologic optics has been within the subfield of oceanic optics, in particular the optics of deep ocean waters, as opposed to coastal or estuarine areas. This emphasis is reflected in our uneven understanding of the optical properties of various water types.

Although the optical properties of different water bodies can vary greatly, there is an overall similarity that is quite distinct from, say, the optical properties of the atmosphere. Therefore, hydrologic and atmospheric optics have developed considerably different theoretical formulations, experimental methodologies, and instrumentation as suited to each field's specific scientific issues. Chapter 44, by Killinger et al., in this *Handbook* discusses atmospheric optics. The text by Mobley[1] gives a comprehensive treatment of hydrologic optics.

Radiative transfer theory is the framework that connects the optical properties of water with the ambient light field. A rigorous mathematical formulation of radiative transfer theory as applicable to hydrologic optics has been developed by Preisendorfer[2] and others. Preisendorfer found it convenient to divide the optical properties of water into two classes: inherent and apparent. *Inherent optical properties* (IOPs) are those properties that depend only upon the medium and therefore are independent of the ambient light field within the medium. The two fundamental IOPs are the absorption coefficient and the volume scattering function. Other IOPs include the attenuation coefficient and the single-scattering albedo. *Apparent optical properties* (AOPs) are those properties that depend both on the medium (the IOPs) and on the geometric (directional) structure of the ambient light field and that display enough regular features and stability to be useful descriptors of the water body. Commonly used AOPs are the irradiance reflectance, the average cosines, and the various attenuation functions (K functions). (All of these quantities are defined below.) The radiative transfer equation provides the connection between the IOPs and the AOPs. The physical environment of a water body—waves on its surface, the character of its bottom, the incident radiance from the sky—enters the theory via the boundary conditions necessary for solution of the radiative transfer equation.

The IOPs are easily defined but they can be exceptionally difficult to measure, especially in situ. The AOPs are generally much easier to measure, but they are difficult to interpret because of the confounding environmental effects. (A change in the sea surface wave state or in the sun's position changes the radiance distribution, and hence the AOPs, even though the IOPs are unchanged.)

Hydrologic optics employs standard radiometric concepts and terminology, although the notation adopted by the International Association for Physical Sciences of the Ocean (IAPSO[3]) differs somewhat from that used in other fields. Table 1 summarizes the terms, units, and symbols for those quantities that have proven most useful in hydrologic optics. These quantities are defined and discussed in Secs. 42.3 to 43.5. Figure 1 summarizes the relationships among the various inherent and apparent optical properties. In the figure, note the central unifying role of radiative transfer theory. Note also that the spectral absorption coefficient and the spectral volume scattering function are the fundamental inherent optical properties in the sense that all inherent optical properties are derivable from those two. Likewise, spectral radiance is the parent of all radiometric quantities and apparent optical properties. The source term S in the radiative transfer equation accounts both for true internal sources such as bioluminescence and for radiance appearing at the wavelength of interest owing to inelastic scattering from other wavelengths.

Most radiative transfer theory assumes the radiant energy to be monochromatic. In this case the associated optical properties and radiometric quantities are termed *spectral* and carry a wavelength (λ) argument or subscript (e.g., the spectral absorption coefficient $a(\lambda)$ or a_λ, or the spectral downward irradiance $E_d(\lambda)$). Spectral radiometric quantities have the SI unit nm^{-1} added to the units shown in Table 1 (e.g., $E_d(\lambda)$ has units $\mathrm{W\,m^{-2}\,nm^{-1}}$). Many radiometric instruments, on the other hand, respond to a fairly wide bandwidth, which complicates the comparison of data and theory.

TABLE 1 Terms, Units, and Symbols for Quantities Commonly Used in Hydrologic Optics

Quantity	SI units	IAPSO recommended symbol*	Historic symbol† (if different)
Fundamental Quantities			
Most of the fundamental quantities are not defined by IAPSO, in which case common usage is given.			
geometric depth below water surface	m	z	
polar angle of photon travel	radian or degree	θ	
wavelength of light (in vacuo)	nm	λ	
cosine of polar angle	dimensionless	$\mu \equiv \cos\theta$	
optical depth below water surface	dimensionless	τ	
azimuthal angle of photon travel	radian or degree	ϕ	
scattering angle	radian or degree	ψ, γ or Θ	θ
solid angle	sr	Ω or ω	Ω
Radiometric Quantities			
The quantities as shown represent broadband measurements. For narrowband (monochromatic) measurements add the adjective "spectral" to the term, add nm^{-1} to the units, and add a wavelength index λ to the symbol (e.g., spectral radiance, L_λ or $L(\lambda)$) with units $W\,m^{-2}\,sr^{-1}\,nm^{-1}$. PAR is always broadband.			
(plane) irradiance	$W\,m^{-2}$	E	H
downward irradiance	$W\,m^{-2}$	E_d	$H(-)$
upward irradiance	$W\,m^{-2}$	E_v	$H(+)$
net (vertical) irradiance	$W\,m^{-2}$	$\vec{E}$	$\vec{H}$
scalar irradiance	$W\,m^{-2}$	E_o	h
downward scalar irradiance	$W\,m^{-2}$	E_{od}	$h(-)$
upward scalar irradiance	$W\,m^{-2}$	E_{ou}	$h(+)$
radiant intensity	$W\,sr^{-1}$	I	J
radiance	$W\,m^{-2}\,sr^{-1}$	L	N
radiant excitance	$W\,m^{-2}$	M	W
photosynthetically available radiation	photons $s^{-1}\,m^{-2}$	PAR or E_{PAR}	
quantity of radiant energy	J	Q	U
radiant power	W	Φ	P
Inherent Optical Properties			
absorptance	dimensionless	A	
absorption coefficient	m^{-1}	a	
scatterance	dimensionless	B	
scattering coefficient	m^{-1}	b	s
backward scattering coefficient	m^{-1}	b_b	b
forward scattering coefficient	m^{-1}	b_f	f
attenuance	dimensionless	C	
attenuation coefficient	m^{-1}	c	α
(real) index of refraction	dimensionless	n	
transmittance	dimensionless	T	
volume scattering function	$m^{-1}\,sr^{-1}$	β	σ
scattering phase function	sr^{-1}	$\tilde{\beta}$	p
single-scattering albedo	dimensionless	ω_o or $\tilde{\omega}$	

TABLE 1 Terms, Units, and Symbols for Quantities Commonly Used in Hydrological Optics (*Continued*)

Quantity	SI units	IAPSO recommended symbol*	Historic symbol† (if different)
	Apparent Optical Properties		
(vertical) attenuation coefficients:			
of downward irradiance $E_d(z)$	m^{-1}	K_d	$K(-)$
of total scalar irradiance $E_0(z)$	m^{-1}	K_o	k
of downward scalar irradiance $E_{0d}(z)$	m^{-1}	K_{od}	$k(-)$
of upward scalar irradiance $E_{0u}(z)$	m^{-1}	K_{ou}	$k(+)$
of PAR	m^{-1}	K_{PAR}	
of upward irradiance $E_u(z)$	m^{-1}	K_u	$K(+)$
of radiance $L(z, \theta, \phi)$	m^{-1}	$K(\theta, \phi)$	
irradiance reflectance (ratio)	dimensionless	R	$R(-)$
average cosine of light field	dimensionless	$\bar{\mu}$	
of downwelling light	dimensionless	$\bar{\mu}_d$	$D(-) = 1/\bar{\mu}_d$
of upwelling light	dimensionless	$\bar{\mu}_u$	$D(+) = 1/\bar{\mu}_u$
distribution function	dimensionless		$D = 1/\bar{\mu}$

* References 1 and 3.
† Reference 2.

43.3 RADIOMETRIC QUANTITIES USEFUL IN HYDROLOGIC OPTICS

Consider an amount ΔQ of radiant energy incident in a time interval Δt centered on time t, onto a surface of area ΔA located at (x, y, z). The energy arrives through a set of directions contained in a solid angle $\Delta\Omega$ about the direction (θ, ϕ) normal to the area ΔA and is produced by photons in a wavelength interval $\Delta\lambda$ centered on wavelength λ. Then an *operational* definition of the *spectral radiance* is

$$L(x, y, z, t, \theta, \phi, \lambda) \equiv \frac{\Delta Q}{\Delta t\, \Delta A\, \Delta\Omega\, \Delta\lambda} \qquad \mathrm{J\,s^{-1}\,m^{-2}\,sr^{-1}nm^{-1}}$$

In practice, one takes Δt, ΔA, $\Delta\Omega$, and $\Delta\lambda$ small enough to get a useful resolution of radiance over the four parameter domains but not so small as to encounter diffraction effects or fluctuations from photon shot noise at very low light levels. Typical values are $\Delta t \sim 10^{-3}$ to 10^3 s (depending on whether or not one wishes to average out sea surface wave effects), $\Delta A \sim 10^{-3}\,\mathrm{m}^2$, $\Delta\Omega \sim 10^{-2}$ sr, and $\Delta\lambda \sim 10$ nm. In the conceptual limit of infinitesimal parameter intervals, the spectral radiance is defined as

$$L(x, y, z, t, \theta, \phi, \lambda) \equiv \frac{\partial^4 Q}{\partial t\, \partial A\, \partial\Omega\, \partial\lambda} \qquad \mathrm{J\,s^{-1}\,m^{-2}\,sr^{-1}\,nm^{-1}}$$

Spectral radiance is the fundamental radiometric quantity of interest in hydrologic optics: it specifies the positional (x, y, z), temporal (t), directional (θ, ϕ), and spectral (λ) structure of the light field. For typical oceanic environments, horizontal variations (on a scale of tens to thousands of meters) of inherent and apparent optical properties are much less than variations with depth, and it is usually assumed that these properties vary only with depth z. Moreover, since the time scales for changes in IOPs or in the environment

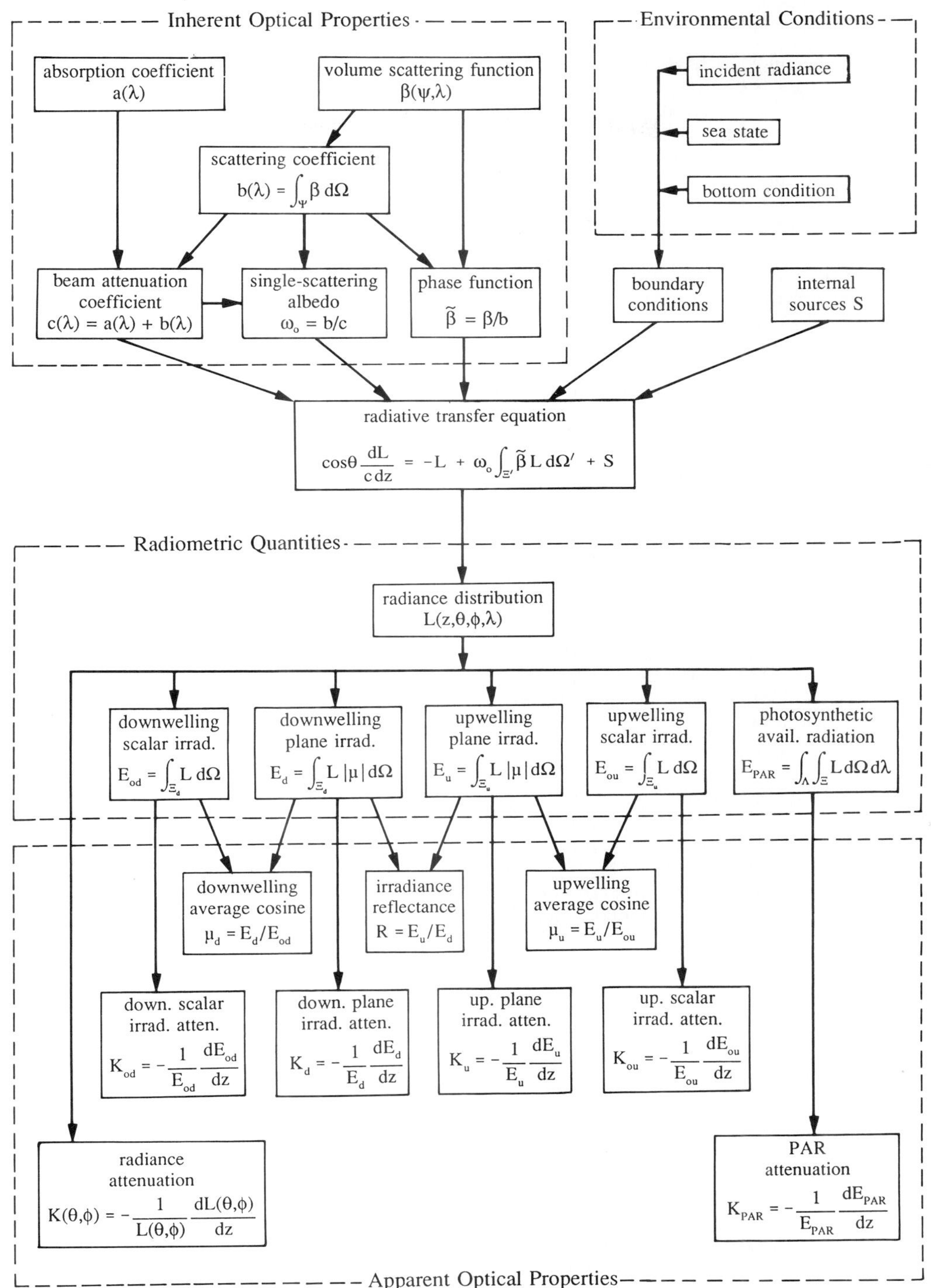

FIGURE 1 Relationships between the various quantities commonly used in hydrologic optics.

(seconds to seasons) are much greater than the time required for the radiance field to reach steady state (microseconds) after a change in IOPs or boundary conditions, time-independent radiative transfer theory is adequate for most hydrologic optics studies. The spectral radiance therefore usually is written as $L(z, \theta, \phi, \lambda)$. The exceptions are applications such as time-of-flight lidar.

There are few conventions on the choice of coordinate systems. Oceanographers usually measure the depth z positive downward from $z = 0$ at the mean water surface. In radiative transfer theory it is convenient to let (θ, ϕ) denote the direction of photon travel (especially when doing Monte Carlo simulations). When displaying data it is convenient to let (θ, ϕ) represent the direction in which the instrument was pointed (the viewing direction) in order to detect photons traveling in the opposite direction. Some authors measure the polar angle θ from the zenith (upward) direction, even when z is taken as positive downward; others measure θ from the $+z$ axis (nadir, or downward, direction). In the following discussion Ξ_u denotes the hemisphere of *upward* directions (i.e., the set of directions (θ, ϕ) such that $0 \le \theta \le \pi/2$ and $0 \le \phi \le 2\pi$, *if* θ is measured from the zenith direction) and Ξ_d denotes the hemisphere of *downward* directions. The element of solid angle is $d\Omega = \sin\theta\, d\theta\, d\phi$ with units of steradian. The solid angle measure of the set of directions Ξ_u or Ξ_d is $\Omega(\Xi_u) = \Omega(\Xi_d) = 2\pi$ sr.

Although the spectral radiance completely specifies the light field, it is seldom measured both because of instrumental difficulties and because such complete information often is not needed for specific applications. The most commonly measured radiometric quantities are various *irradiances.*

Consider a light detector constructed so as to be equally sensitive to photons of a given wavelength λ traveling in any direction (θ, ϕ) within a *hemisphere* of directions.[4] If the detector is located at depth z and is oriented facing *upward,* so as to collect photons traveling *downward,* then the detector output is a measure of the *spectral downward scalar irradiance* at depth z, $E_{0d}(z, \lambda)$. Such an instrument is summing radiance over all the directions (elements of solid angle) in the downward hemisphere Ξ_d. Thus $E_{0d}(z, \lambda)$ is related to $L(z, \theta, \phi, \lambda)$ by

$$E_{0d}(z, \lambda) = \int_{\Xi_d} L(z, \theta, \phi, \lambda)\, d\Omega \qquad \mathrm{W\, m^{-2}\, nm^{-1}}$$

The symbolic integral over Ξ_d can be evaluated as a double integral over θ and ϕ after a specific coordinate system is chosen.

If the same instrument is oriented facing *downward,* so as to detect photons traveling *upward,* then the quantity measured is the *spectral upward scalar irradiance* $E_{0u}(z, \lambda)$:

$$E_{0u}(z, \lambda) = \int_{\Xi_u} L(z, \theta, \phi, \lambda)\, d\Omega \qquad \mathrm{W\, m^{-2}\, nm^{-1}}$$

The *spectral scalar irradiance* $E_0(z, \lambda)$ is just the sum of the downward and upward components:

$$E_0(z, \lambda) \equiv E_{0d}(z, \lambda) + E_{0u}(z, \lambda) = \int_{\Xi} L(z, \theta, \phi, \lambda)\, d\Omega \qquad \mathrm{W\, m^{-2}\, nm^{-1}}$$

Here $\Xi = \Xi_d \cup \Xi_u$ is the set of all directions; $\Omega(\Xi) = 4\pi$ sr. $E_0(z, \lambda)$ is useful[2] because it is proportional to the spectral radiant energy density ($\mathrm{J\, m^{-3}\, nm^{-1}}$) at depth z.

Now consider a detector designed[4] so that its sensitivity is proportional to $|\cos\theta|$ where θ is the angle between the photon direction and the normal to the surface of the detector. This is the ideal response of a "flat plate" collector of area ΔA, which when viewed at an angle θ to its normal appears to have an area of $\Delta A\, |\cos\theta|$. If such a detector is located at depth z and is oriented facing *upward,* so as to detect photons traveling *downward,* then its output is proportional to the *spectral downward plane irradiance* $E_d(z, \lambda)$ (usually called *spectral downwelling irradiance*). This instrument is summing the downwelling radiance weighted by the cosine of the photon direction:

$$E_d(z, \lambda) = \int_{\Xi_d} L(z, \theta, \phi, \lambda)\, |\cos\theta|\, d\Omega \qquad \mathrm{W\, m^{-2}\, nm^{-1}}$$

Turning this instrument upside down gives the *spectral upward plane irradiance* (*spectral upwelling irradiance*) $E_u(z, \lambda)$:

$$E_u(z, \lambda) = \int_{\Xi_u} L(z, \theta, \phi, \lambda)\, |\cos\theta|\, d\Omega \qquad \mathrm{W\, m^{-2}\, nm^{-1}}$$

E_d and E_u are useful because they give the energy flux (power per unit area) across the horizontal surface at depth z owing to downwelling and upwelling photons, respectively.

The *spectral net irradiance* at depth z, $\bar{E}(z, \lambda)$, is the difference in the downwelling and upwelling plane irradiances:

$$\bar{E}(z, \lambda) = E_d(z, \lambda) - E_u(z, \lambda)$$

Photosynthesis is a quantum phenomenon (i.e., it is the *number* of available photons rather than the *amount* of radiant energy that is relevant to the chemical transformations). This is because if a photon of, say, $\lambda = 350$ nm, is absorbed by chlorophyll it induces the same chemical change as does a photon of $\lambda = 700$ nm, even though the 350-nm photon has twice the energy of the 700-nm photon. Only a part of the photon energy goes into photosynthesis; the excess is converted to heat or reradiated. Moreover, chlorophyll is equally able to absorb and utilize a photon regardless of the photon's direction of travel. Therefore, in studies of phytoplankton biology the relevant measure of the light field is the *photosynthetically available radiation,* PAR or E_{PAR}, defined by

$$\mathrm{PAR}(z) \equiv \int_{350\,\mathrm{nm}}^{700\,\mathrm{nm}} \frac{\lambda}{hc} E_0(z, \lambda)\, d\lambda \qquad \mathrm{photons\, s^{-1}\, m^{-2}}$$

where $h = 6.6255 \times 10^{-34}$ J s is Planck's constant and $c = 3.0 \times 10^{17}$ nm s^{-1} is the speed of light. The factor λ/hc converts the energy units of E_0 (watts) to quantum units (photons per second). Bio-optical literature often states PAR values in units of mol photons s^{-1} m^{-2} or einst s^{-1} m^{-2}. Morel and Smith[5] found that over a wide variety of water types from very clear to turbid, with corresponding variations in the spectral nature of the irradiance, the conversion factor for energy to quanta varied by only ± 10 percent about the value 2.5×10^{18} photons s^{-1} W^{-1} (4.2 μeinst s^{-1} W^{-1}).

For practical reasons related to instrument design, PAR is sometimes estimated using the spectral downwelling plane irradiance and the visible wavelengths only:

$$\mathrm{PAR}(z) \approx \int_{400\,\mathrm{nm}}^{700\,\mathrm{nm}} \frac{\lambda}{hc} E_d(z, \lambda)\, d\lambda \qquad \mathrm{photons\, s^{-1}\, m^{-2}}$$

However, it is now recognized[6,7] that the use of E_d rather than E_0 can lead to errors of 20 to 100 percent in computations of PAR. Omission of the 350–400-nm band is less troublesome since those wavelengths are rapidly absorbed near the water surface, except in very clear waters.

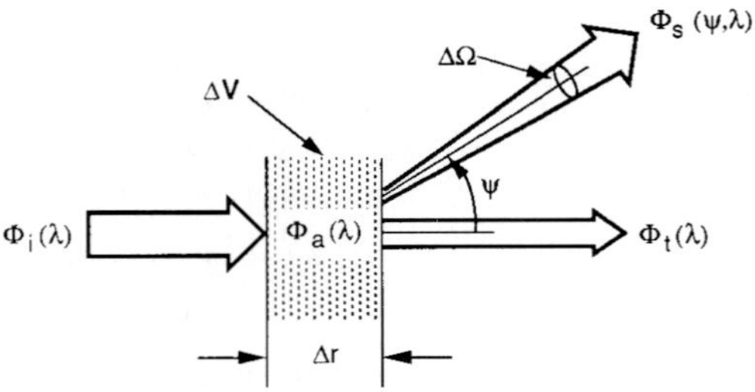

FIGURE 2 Geometry used to define inherent optical properties.

43.4 INHERENT OPTICAL PROPERTIES

Consider a small volume ΔV of water of thickness Δr as seen by a narrow collimated beam of monochromatic light of spectral radiant power $\Phi_i(\lambda)$ W nm^{-1} as schematically illustrated in Fig. 2. Some part $\Phi_a(\lambda)$ of the incident power $\Phi_i(\lambda)$ is absorbed within the volume of water. Some part $\Phi_s(\psi, \lambda)$ is scattered out of the beam at an angle ψ, and the remaining power $\Phi_t(\lambda)$ is transmitted through the volume with no change in direction. Let $\Phi_s(\lambda)$ be the total power that is scattered into all directions. Furthermore, assume that no inelastic scattering occurs (i.e., assume that no photons undergo a change in wavelength during the scattering process). Then by conservation of energy,

$$\Phi_i(\lambda) = \Phi_a(\lambda) + \Phi_s(\lambda) + \Phi_t(\lambda)$$

The *spectral absorptance* $A(\lambda)$ is the fraction of incident power that is absorbed within the volume:

$$A(\lambda) \equiv \frac{\Phi_a(\lambda)}{\Phi_i(\lambda)}$$

Likewise the *spectral scatterance* $B(\lambda)$ is the fractional part of the incident power that is scattered out of the beam,

$$B(\lambda) \equiv \frac{\Phi_s(\lambda)}{\Phi_i(\lambda)}$$

and the *spectral transmittance* $T(\lambda)$ is

$$T(\lambda) \equiv \frac{\Phi_t(\lambda)}{\Phi_i(\lambda)}$$

Clearly, $A(\lambda) + B(\lambda) + T(\lambda) = 1$. A quantity easily confused with the absorptance $A(\lambda)$ is the *absorbance* $D(\lambda)$ (also called *optical density*) defined as[8]

$$D(\lambda) \equiv \log_{10} \frac{\Phi_i(\lambda)}{\Phi_s(\lambda) + \Phi_t(\lambda)} = -\log_{10}[1 - A(\lambda)]$$

$D(\lambda)$ is the quantity actually measured in a spectrophotometer.

The inherent optical properties usually employed in hydrologic optics are the spectral absorption and scattering coefficients which are, respectively, the spectral absorptance and scatterance *per unit distance* in the medium. In the geometry of Fig. 2, the *spectral absorption coefficient* $a(\lambda)$ is defined as

$$a(\lambda) \equiv \lim_{\Delta r \to 0} \frac{A(\lambda)}{\Delta r} \qquad \mathrm{m}^{-1}$$

and the *spectral scattering coefficient* $b(\lambda)$ is

$$b(\lambda) \equiv \lim_{\Delta r \to 0} \frac{B(\lambda)}{\Delta r} \qquad \mathrm{m}^{-1}$$

The *spectral beam attenuation coefficient* $c(\lambda)$ is defined as

$$c(\lambda) \equiv a(\lambda) + b(\lambda)$$

Hydrologic optics uses the term attenuation rather than extinction.

Now take into account the angular distribution of the scattered power, with $B(\psi, \lambda)$ being the fraction of incident power scattered out of the beam through an angle ψ into a solid angle $\Delta\Omega$ centered on ψ as shown in Fig. 2. Then the angular scatterance per unit distance and unit solid angle, $\beta(\psi, \lambda)$, is

$$\beta(\psi, \lambda) \equiv \lim_{\Delta r \to 0} \lim_{\Delta\Omega \to 0} \frac{B(\psi, \lambda)}{\Delta r \, \Delta\Omega} = \lim_{\Delta r \to 0} \lim_{\Delta\Omega \to 0} \frac{\Phi_s(\psi, \lambda)}{\Phi_i(\lambda) \, \Delta r \, \Delta\Omega} \qquad \mathrm{m}^{-1}\,\mathrm{sr}^{-1}$$

The spectral power scattered into the given solid angle $\Delta\Omega$ is just the spectral radiant intensity scattered into direction ψ times the solid angle: $\Phi_s(\psi, \lambda) = I_s(\psi, \lambda)\,\Delta\Omega$. Moreover, if the incident power $\Phi_i(\lambda)$ falls on an area ΔA, then the corresponding incident irradiance is $E_i(\lambda) = \Phi_i(\lambda)/\Delta A$. Noting that $\Delta V = \Delta r\, \Delta A$ is the volume of water that is illuminated by the incident beam gives

$$\beta(\psi, \lambda) = \lim_{\Delta V \to 0} \frac{I_s(\psi, \lambda)}{E_i(\lambda)\, \Delta V}$$

This form of $\beta(\psi, \lambda)$ suggests the name *spectral volume scattering function* and the physical interpretation of scattered intensity per unit incident irradiance per unit volume of water; $\beta(\psi, \lambda)$ also can be interpreted as the differential scattering cross section per unit volume. Integrating $\beta(\psi, \lambda)$ over all directions (solid angles) gives the total scattered power per unit incident irradiance and unit volume of water or, in other words, the spectral scattering coefficient:

$$b(\lambda) = \int_{\Xi} \beta(\psi, \lambda)\, d\Omega = 2\pi \int_0^{\pi} \beta(\psi, \lambda) \sin\psi \, d\psi$$

The last equation follows because scattering in natural waters is azimuthally symmetric about the incident direction (for unpolarized sources and for randomly oriented scatterers). This integration is often divided into forward scattering, $0 \le \psi \le \pi/2$, and backward scattering, $\pi/2 \le \psi \le \pi$, parts. The corresponding spectral forward and backward scattering coefficients are, respectively,

$$b_f(\lambda) \equiv 2\pi \int_0^{\pi/2} \beta(\psi, \lambda) \sin\psi\, d\psi$$

$$b_b(\lambda) \equiv 2\pi \int_{\pi/2}^{\pi} \beta(\psi, \lambda) \sin\psi\, d\psi$$

The preceding discussion has assumed that no inelastic (transpectral) scattering processes are present. However, transpectral scattering does occur in natural waters attributable to fluorescence by dissolved matter or chlorophyll and to Raman or Brillouin scattering by the water molecules themselves (see Sec. 43.23). Power lost from wavelength λ by scattering into wavelength $\lambda' \neq \lambda$ appears in the above formalism as an increase in the spectral absorption.[9] In this case, $a(\lambda)$ accounts for "true" absorption (e.g., conversion of radiant energy into heat) as well as for the loss of power at wavelength λ by inelastic scattering to another wavelength. The gain in power at λ' appears as a source term in the radiative transfer formalism.

Two more inherent optical properties are commonly used in hydrologic optics. The *spectral single-scattering albedo* $\omega_0(\lambda)$ is

$$\omega_0(\lambda) \equiv \frac{b(\lambda)}{c(\lambda)}$$

The single-scattering albedo is the probability that a photon will be scattered (rather than absorbed) in any given interaction; hence, $\omega_0(\lambda)$ is also known as the *probability of photon survival.* The *spectral volume scattering phase function,* $\tilde{\beta}(\psi, \lambda)$, is defined by

$$\tilde{\beta}(\psi, \lambda) \equiv \frac{\beta(\psi, \lambda)}{b(\lambda)} \qquad \text{sr}^{-1}$$

Writing the volume scattering function $\beta(\psi, \lambda)$ as the product of the scattering coefficient $b(\lambda)$ and the phase function $\tilde{\beta}(\psi, \lambda)$ partitions $\beta(\psi, \lambda)$ into a factor giving the *strength* of the scattering, $b(\lambda)$ with units of m^{-1}, and a factor giving the *angular distribution* of the scattered photons, $\tilde{\beta}(\psi, \lambda)$ with units of sr^{-1}.

43.5 APPARENT OPTICAL PROPERTIES

The quantity

$$\bar{\mu}_d(z, \lambda) \equiv \frac{\int_{\Xi_d} L(z, \theta, \phi, \lambda)\, |\cos\theta|\, d\Omega}{\int_{\Xi_u} L(z, \theta, \phi, \lambda)\, d\Omega} \equiv \frac{E_d(z, \lambda)}{E_{od}(z, \lambda)}$$

is called the *spectral downwelling average cosine.* The definition shows that $\bar{\mu}_d(z, \lambda)$ is the average value of the cosine of the polar angle of all the photons contributing to the downwelling radiance at the given depth and wavelength. The *spectral upwelling average cosine* is defined analogously:

$$\bar{\mu}_u(z, \lambda) \equiv \frac{E_u(z, \lambda)}{E_{ou}(z, \lambda)}$$

The average cosines are useful one-parameter measures of the directional structures of the downwelling and upwelling light fields. For example, if the downwelling light field (radiance distribution) is collimated in direction (θ_0, ϕ_0) so that $L(\theta, \phi) = L_0\delta(\theta - \theta_0)\delta(\phi - \phi_0)$, where δ is the Dirac δ function, then $\bar{\mu}_d = |\cos\theta_0|$. If the downwelling radiance is completely diffuse (isotropic), $L(\theta, \phi) = L_0$ and $\bar{\mu}_d = \frac{1}{2}$. Typical values of the average cosines for waters illuminated by the sun and sky are $\bar{\mu}_d \approx \frac{3}{4}$ and $\bar{\mu}_u \approx \frac{3}{8}$. Older literature generally refers to *distribution functions,* D_d and D_u, rather than to average cosines. The distribution functions are just reciprocals of the average cosines:

$$D_d(z, \lambda) = \frac{1}{\bar{\mu}_d(z, \lambda)} \qquad \text{and} \qquad D_u(z, \lambda) = \frac{1}{\bar{\mu}_u(z, \lambda)}$$

The *spectral irradiance reflectance* (or *irradiance ratio*) $R(z, \lambda)$ is the ratio of spectral upwelling to downwelling plane irradiances:

$$R(z, \lambda) \equiv \frac{E_u(z, \lambda)}{E_d(z, \lambda)}$$

$R(z, \lambda)$ just beneath the sea surface is of great importance in remote sensing (see Sec. 43.22).

Under typical oceanic conditions for which the incident lighting is provided by the sun and sky, the various radiances and irradiances all decrease approximately exponentially with depth, at least when far enough below the surface (and far enough above the bottom, in shallow water) to be free of boundary effects. For example, it is convenient to write the depth dependence of $E_d(z, \lambda)$ as

$$E_d(z, \lambda) \equiv E_d(0, \lambda) \exp\left[-\int_0^z K_d(z', \lambda)\, dz'\right] \equiv E_d(0, \lambda) \exp\left[-\bar{K}_d(\lambda) z\right]$$

where $K_d(z, \lambda)$ is the *spectral diffuse attenuation coefficient* for spectral downwelling plane irradiance and $\bar{K}_d(\lambda)$ is the average value of $K_d(z, \lambda)$ over the depth interval 0 to z. Solving for $K_d(z, \lambda)$ gives

$$K_d(z, \lambda) = -\frac{d \ln E_d(z, \lambda)}{dz} = -\frac{1}{E_d(z, \lambda)} \frac{dE_d(z, \lambda)}{dz} \qquad \text{m}^{-1}$$

The distinction between *beam* and *diffuse* attenuation coefficients is important. The beam attenuation coefficient $c(\lambda)$ is defined in terms of the radiant power lost from a single, narrow, collimated beam of photons. The downwelling diffuse attenuation coefficient $K_d(z, \lambda)$ is defined in terms of the decrease with depth of the ambient downwelling irradiance $E_d(z, \lambda)$, which comprises photons heading in all downward directions (a diffuse or uncollimated light field). $K_d(z, \lambda)$ clearly depends on the directional structure of the ambient light field and so is classified as an apparent optical property. Other diffuse attenuation coefficients, e.g., K_u, K_{od}, K_{ou}, K_{PAR}, and $K(\theta, \phi)$, are defined in an analogous manner, using the corresponding radiometric quantities.

In *homogeneous* waters, these "K functions" depend only weakly on depth and therefore can serve as convenient, if imperfect, descriptors of the water body. Smith and Baker[10] have pointed out other reasons why K functions are useful:

1. The K's are defined as ratios and therefore do not require absolute radiometric measurements.
2. The K's are strongly correlated with chlorophyll concentration (i.e., they provide a connection between biology and optics).
3. About 90 percent of the diffusely reflected light from a water body comes from a layer of water of depth $1/K_d(0, \lambda)$ (i.e., K_d has implications for remote sensing).
4. Radiative transfer theory provides several useful relations between the K's and other quantities of interest, such as absorption and beam attenuation coefficients, the irradiance reflectance, and the average cosines.
5. Instruments are available for routine measurement of the K's.

It must be remembered, however, that in spite of their utility K functions are apparent optical properties—a change in the environment (e.g., solar angle or sea state) changes their value, sometimes by a negligible amount but sometimes greatly. However, numerical simulations by Gordon[11] show how with a few additional but easily made measurements measured values of $K_d(z, \lambda)$ and $\bar{K}_d(\lambda)$ can be "normalized" to remove the effects of solar angle and sea state. The normalized K_d and $\bar{K}_d$ are equal to the values that would be obtained if the sun were at the zenith and the sea surface were calm. *If* this normalization is performed, the resulting $K_d(z, \lambda)$ and $\bar{K}_d(\lambda)$ can be regarded as *inherent* optical properties for all practical purposes. It is strongly recommended that Gordon's procedure be routinely followed by experimentalists.

43.6 THE OPTICALLY SIGNIFICANT CONSTITUENTS OF NATURAL WATERS

Dissolved Substances

Pure sea water consists of pure water plus various dissolved salts, which average about 35 parts per thousand (%) by weight. These salts increase scattering above that of pure water by 30 percent (see Table 10 in Sec. 43.17). It is not well established what, if any, effect these salts have on absorption, but it is likely that they increase absorption somewhat at ultraviolet wavelengths.

Both fresh and saline waters contain varying concentrations of dissolved organic compounds. These compounds are produced during the decay of plant matter and consist mostly of various humic and fulvic acids.[8] These compounds are generally brown in color and in sufficient concentrations can color the water yellowish brown. For this reason the compounds are generically referred to as *yellow matter, Gelbstoff,* or *gilvin.* Yellow matter absorbs very little in the red, but absorption increases rapidly with decreasing wavelength. Since the main source of yellow matter is decayed terrestrial vegetation, concentrations are generally greatest in lakes, rivers, and coastal waters influenced by river runoff. In such waters yellow matter can be the dominant absorber at the blue end of the spectrum. In mid-ocean waters absorption by yellow matter is usually small compared to absorption by other constituents, but some yellow matter is likely to be present as the result of decaying phytoplankton, especially at the end of a bloom.

Particulate Matter

Particulate matter in the oceans has two distinct origins: biological and physical. The organic particles of optical importance are created as bacteria, phytoplankton, and zooplankton grow and reproduce by photosynthesis or by eating their neighbors. Particles of a given size are destroyed by breaking apart after death, by flocculation into larger aggregate particles, or by settling out of the water column. Inorganic particles are created primarily by weathering of terrestrial rocks and soils. These particles can enter the water as wind-blown dust settles on the sea surface, as rivers carry eroded soil to the sea, or as currents resuspend bottom sediments. Inorganic particles are removed from the water by settling, aggregating, or dissolving. This particulate matter usually is the major determiner of both the absorption and scattering properties of natural waters and is responsible for most of the temporal and spatial variability in these optical properties.

Organic Particles. These occur in many forms.

Viruses. Natural marine waters contain virus particles[12] in concentrations of 10^{12} to 10^{15} particles m^{-3}. These particles are generally much smaller (2–200 nm) than the wavelength of visible light, and it is not known what, if any, direct effect viruses have on the optical properties of sea water.

Colloids. Nonliving collodial particles in the size range 0.4–1.0 μm are found[13] in typical number concentrations of 10^{13} m^{-3} and colloids of size ≤0.1 μm are found[14] in abundances of 10^{15} m^{-3}. Some of the absorption traditionally attributed to dissolved matter may be due to colloids, some of which strongly resemble fulvic acids in electron micrographs.[14]

Bacteria. Living bacteria in the size range 0.2–1.0 μm occur in typical number concentrations of 10^{11}–10^{13} m^{-3}. It only recently has been recognized[15–17] that bacteria can be significant scatterers and absorbers of light, expecially at blue wavelengths and in clean oceanic waters where the larger phytoplankton are relatively scarce.

Phytoplankton. These ubiquitous microscopic plants occur with incredible diversity of species, size, shape, and concentration. They range in cell size from less than 1 μm to more than 200 μm, and some species form even larger chains of individual cells. It has long been recognized that phytoplankton are the particles primarily responsible for determining the optical properties of most oceanic waters. Their chlorophyll and related pigments strongly absorb light in the blue and red and thus when concentrations are high determine the spectral absorption of sea water. These particles are generally much larger than the wavelength of visible light and are efficient scatterers, especially via diffraction, thus influencing the scattering properties of sea water.

Organic Detritus. Nonliving organic particles of various sizes are produced, for example, when phytoplankton die and their cells break apart. They may also be formed when zooplankton graze on phytoplankton and leave behind cell fragments and fecal pellets. Even if these detrital particles contain pigments at the time of their production, they can be rapidly photo-oxidized and lose the characteristic absorption spectrum of living phytoplankton, leaving significant absorption only at blue wavelengths.

Large Particles. Particles larger than 100 μm include zooplankton (living animals with sizes from tens of micrometers to two centimeters) and fragile amorphous aggregates[18] of smaller particles ("marine snow," with sizes from 0.5 mm to tens of centimeters). Such particles occur in highly variable numbers from almost none to thousands per cubic meter. Even at relatively large concentrations these large particles tend to be missed by optical instruments that randomly sample only a few cubic centimeters of water or that mechanically break apart the aggregates. However, these large particles can be efficient diffuse scatterers of light and therefore may significantly affect the optical properties (especially backscatter) of large volumes of water, e.g., as seen by remote sensing instruments. Although such optical effects are recognized, they have not been quantified.

Inorganic Particles. These generally consist of finely ground quartz sand, clay minerals, or metal oxides in the size range from much less than 1 μm to several tens of micrometers. Insufficient attention has been paid to the optical effects of such particles in sea water, although it is recognized that inorganic particles are sometimes optically more important than organic particles. Such situations can occur both in turbid coastal waters carrying a heavy sediment load and in very clear oceanic waters which are receiving wind-blown dust.[19]

At certain stages of its life, the phytoplankton coccolithophore species *Emiliania huxleyi* is a most remarkable source of crystalline particles. During blooms *E. huxleyi* produces and sheds enormous numbers of small (2–4 μm) calcite plates; concentrations of 3×10^{11} plates m^{-3} have been observed.[20] Although they have a negligible effect on light absorption, these calcite plates are extremely efficient light scatterers: irradiance reflectances of $R = 0.39$ have been observed[20] at blue wavelengths during blooms (compared with $R = 0.02$ to 0.05 in the blue for typical ocean waters, discussed in Sec. 43.22). Such coccolithophore blooms give the ocean a milky white or turquoise appearance.

43.7 PARTICLE SIZE DISTRIBUTIONS

In spite of the diverse mechanisms for particle production and removal, observation shows that a single family of particle size distributions often suffices to describe oceanic particulate matter in the optically important size range from 0.1 to 100 μm. Let $N(x)$ be the number of particles per unit volume with size greater than x in a sample of particles; x usually represents equivalent spherical diameter computed from particle volume, but also can represent particle volume or surface area. The Junge (also called *hyperbolic*) cumulative size distribution[21] is then

$$N(x) = k\left(\frac{x}{x_0}\right)^{-m}$$

where k sets the scale, x_0 is a reference size, and $-m$ is the slope of the distribution when $\log N$ is plotted versus $\log x$; k, x_0, and m are positive constants.

Oceanic particle size distributions usually have m values between 2 and 5, with $m = 3$ to 4 being typical; such spectra can be seen in McCave,[22] fig. 7. It often occurs that oceanic particle size spectra are best described by a segmented distribution in which a smaller value of m is used for x less than a certain value and a larger value of m is used for x greater than that value. Such segmented spectra can be seen in Bader,[21] and in McCave,[22] fig. 8.

The quantity most relevant to optics, e.g., in Mie scattering computations for polydisperse systems, is not the *cumulative* size distribution $N(x)$, but rather the *number* size distribution $n(x)$. The number distribution is defined such that $n(x)\,dx$ is the number of particles in the size interval from x to $x + dx$. The number distribution is related to the cumulative distribution by $n(x) = |dN(x)/dx|$, so that for the Junge distribution

$$n(x) = kmx_0^{-m}x^{-m-1} \equiv Kx^{-s}$$

where $K \equiv kmx_0^{-m}$ and $s \equiv m + 1$; s is commonly referred to as the slope of the distribution. Figure 3 shows the number distribution of biological particles typical of open ocean waters; note that a value of $s = 4$ gives a reasonable fit to the plotted points.

It should be noted, however, that the Junge distribution sometimes fails to represent

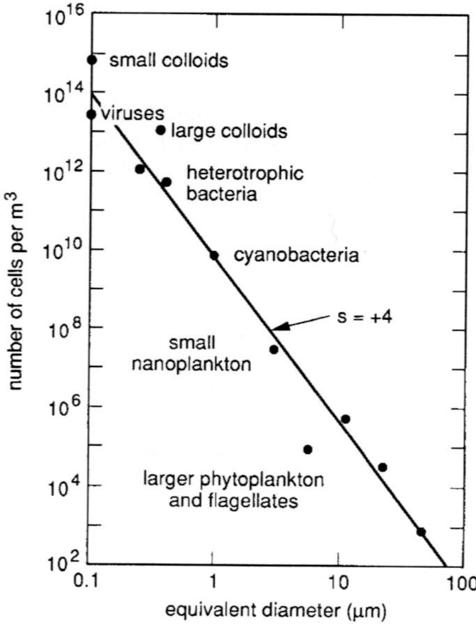

FIGURE 3 Number size distribution typical of biological particles in the open ocean. (*Based on Stramski and Kiefer,*[17] *with permission.*)

oceanic conditions. For example, during the growth phase of a phytoplankton bloom the rapid increase in population of a particular species may give abnormally large numbers of particles in a particular size range. Such bloom conditions therefore give a "bump" in $n(x)$ that is not well modeled by the simple Junge distribution. Moreover, Lambert et al.[23] found that a log-normal distribution sometimes better described the distributions of inorganic particles found in water samples taken from near the bottom at deep ocean locations. These particles were principally aluminosilicates in the 0.2–10.0-μm size range but included quartz grains, metal oxides, and phytoplankton skeletal parts such as coccolithophore plates. Based on the sampling location it was assumed that the inorganic particles were resuspended sediments. Lambert et al. found that the size distributions of the individual particle types (e.g., aluminosilicates or metal oxides) obeyed log-normal distributions which "flattened out" below 1 μm. For particles larger than ~1 μm, log-normal and Junge distributions gave nearly equivalent descriptions of the data. Biological particles were not as well described by the log-normal distribution, especially for sizes greater than 5 μm.

43.8 ELECTROMAGNETIC PROPERTIES OF WATER

In studies of electromagnetic wave propagation at the level of Maxwell's equations it is convenient to specify the bulk electromagnetic properties of the medium via the electrical permittivity ε, the magnetic permeability μ, and the elecrical conductivity σ. Since water displays no significant magnetic properties, the permeability can be taken equal to the free-space (in vacuo) value at all frequencies: $\mu = \mu_0 = 4\pi \times 10^{-7}\ \mathrm{N\,A^{-2}}$. Both ε and σ depend on the frequency ω of the propagating electromagnetic wave as well as on the water temperature, pressure, and salinity. Low-frequency ($\omega \rightarrow 0$) values for the permittivity are of order $\varepsilon \approx 80\varepsilon_0$, where $\varepsilon_0 = 8.85 \times 10^{-12}\ \mathrm{A^2\,s^2\,N^{-1}\,m^{-2}}$ is the free-space value. This value decreases to $\varepsilon \approx 1.8\varepsilon_0$ at optical frequencies. Extensive tabulations of $\varepsilon/\varepsilon_0$ as a function of temperature and pressure are given for pure water in Archer and Wang.[24] The low-frequency conductivity ranges from $\sigma \approx 4 \times 10^{-6}$ siemen m^{-1} for pure water to $\sigma \approx 4.4$ siemen m^{-1} for sea water.

The effects of ε, μ, and σ on electromagnetic wave propagation are compactly summarized in terms of the complex index of refraction, $m = n - ik$, where n is the real index of refraction, k is the dimensionless electrodynamic absorption coefficient, and $i = \sqrt{(-1)}$; n and k are collectively called the *optical constants* of water (a time dependence convention of $\exp(+i\omega t)$ is used in deriving wave equations from Maxwell's equations). The explicit dependence of m on ε, μ, and σ is given by[25]

$$m^2 = \mu\varepsilon c^2 - i\frac{\mu\sigma c^2}{\omega}$$
$$= (n - ik)^2 = n^2 - k^2 - i2nk$$

where $c = (\varepsilon_0\mu_0)^{-1/2}$ is the speed of light in vacuo. These equations can be used to relate n and k to the bulk electromagnetic properties. The optical constants are convenient because they are directly related to the scattering and absorbing properties of water. The real index of refraction $n(\lambda)$ governs scattering both at interfaces (via the laws of reflection and refraction) and within the medium (via thermal or other fluctuations of $n(\lambda)$ at molecular and larger scales). The spectral absorption coefficient $a(\lambda)$ is related to $k(\lambda)$ by[25]

$$a(\lambda) = \frac{4\pi k(\lambda)}{\lambda}$$

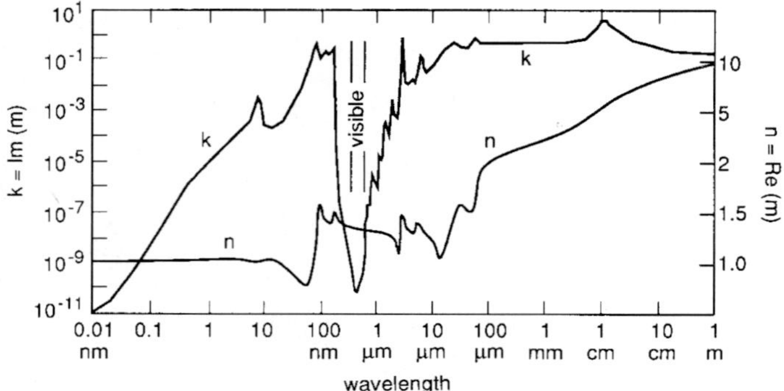

FIGURE 4 The optical constants of pure water. The left axis gives $k = \text{Im}(m)$ and the right axis gives $n = \text{Re}(m)$ where m is the complex index of refraction. (*Redrawn from Zoloratev and Demin,*[26] *with permission.*)

Here λ refers to the in vacuo wavelength of light corresponding to a given frequency ω of electromagnetic wave.

Figure 4 shows the wavelength dependence of the optical constants n and k for pure water. The extraordinary feature seen in this figure is the narrow "window" in $k(\lambda)$, where $k(\lambda)$ decreases by over nine orders of magnitude between the near ultraviolet and the visible and then quickly rises again in the near infrared. This behavior in $k(\lambda)$ gives a corresponding window in the spectral absorption coefficient $a(\lambda)$ as seen in Table 2. Because of the opaqueness of water outside the near-UV to near-IR wavelengths, hydrologic optics is concerned with only this small part of the electromagnetic spectrum. These wavelengths overlap nicely with the wavelengths of the sun's maximum energy output and with a corresponding window in atmospheric absorption, much to the benefit of life on earth.

43.9 INDEX OF REFRACTION

Seawater

Austin and Halikas[27] exhaustively reviewed the literature on measurements of the real index of refraction of sea water. Their report contains extensive tables and interpolation algorithms for the index of refraction (relative to air), $n(\lambda, S, T, p)$, as a function of wavelength ($\lambda = 400$–700 nm), salinity ($S = 0$–$43‰$), temperature ($T = 0$–$30°\text{C}$), and pressure ($p = 10^5$–10^8 Pa, or 1 to 1080 atm). Figure 5 illustrates the general dependence of n on these four parameters: n decreases with increasing wavelength or temperature and increases with increasing salinity or pressure. Table 3 gives the values of n for the extreme values of each parameter. The extreme values of n, 1.329128 and 1.366885, show that n varies by less than 3 percent over the entire parameter range relevant to hydrologic optics. Table 4 gives selected values of $n(\lambda, T)$ for fresh water ($S = 0$) and for typical sea water ($S = 35‰$) at atmospheric pressure ($p = 10^5$ Pa). The values in Table 4 can be multiplied by 1.000293 (the index of refraction of dry air at STP and $\lambda = 538$ nm) if values relative to vacuum are desired. Millard and Seaver[28] have developed a 27-term formula that gives the

TABLE 2 Absorption Coefficient a of Pure Water As a Function of Wavelength λ*

λ	a (m^{-1})	λ	a (m^{-1})
0.01 nm	1.3×10^1	700 nm	0.650
0.1	6.5×10^2	800	2.07
1	9.4×10^4	900 nm	7.0
10	3.5×10^6	1 μm	3.3×10^1
100	5.0×10^7	10	7.0×10^4
200	3.07	100 μm	6.5×10^4
300	0.141	0.001 m	1.3×10^4
400	0.0171	0.01	3.6×10^3
500	0.0257	0.1	5.0×10^1
600 nm	0.244	1 m	2.5

* Data for 200 nm $\leq \lambda \leq$ 800 nm taken from Table 6. Data for other wavelengths computed from Fig. 4.

index of refraction to part-per-million accuracy over most of the oceanographic parameter range.

Particles

Suspended particulate matter in sea water often has a bimodal index of refraction distribution. Living phytoplankton typically have "low" indices of refraction in the range 1.01 to 1.09 relative to the index of refraction of seawater. Detritus and inorganic particles generally have "high" indices in the range of 1.15 to 1.20 relative to seawater.[29] Typical values are 1.05 for phytoplankton and 1.16 for inorganic particles.

Table 5 gives the relative index of refraction of terrigenous minerals commonly found in river runoff and wind-blown dust. Only recently has it become possible to measure the refractive indices of individual phytoplankton cells.[30] Consequently, little is yet known

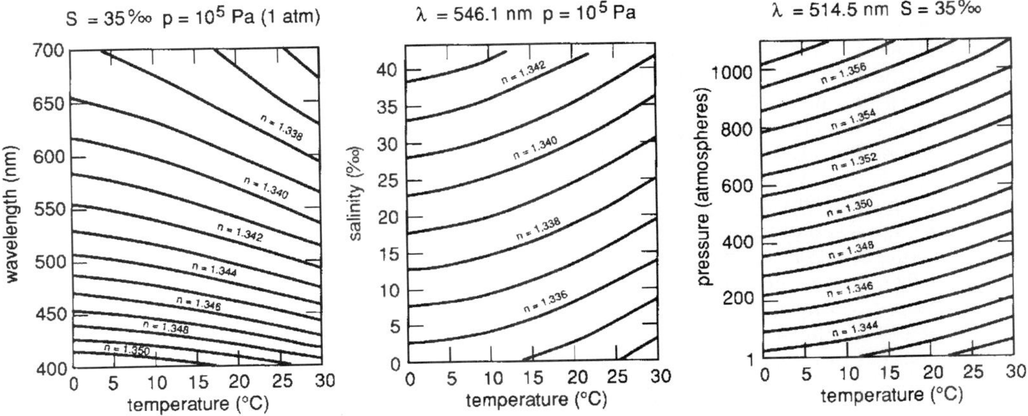

FIGURE 5 Real index of refraction of water for selected values of pressure, temperature, and salinity. (*Adapted from Austin and Halikas.*[27])

TABLE 3 Index of Refraction of Water n for the Extreme Values of Pressure p, Temperature T, Salinity S, and Wavelength λ Encountered in Hydrologic Optics*

p (Pa)	T (°C)	S (‰)	λ (nm)	n
1.01×10^5	0	0	400	1.344186
1.01	0	0	700	1.331084
1.01	0	35	400	1.351415
1.01	0	35	700	1.337906
1.01	30	0	400	1.342081
1.01	30	0	700	1.329128
1.01	30	35	400	1.348752
1.01	30	35	700	1.335316
1.08×10^8	0	0	400	1.360076
1.08	0	0	700	1.346604
1.08	0	35	400	1.366885
1.08	0	35	700	1.352956
1.08	30	0	400	1.356281
1.08	30	0	700	1.342958
1.08	30	35	400	1.362842
1.08	30	35	700	1.348986

* Reproduced from Austin and Halikas.[27]

about the dependence of refractive index on phytoplankton species, or on the physiological state of the plankton within a given species, although it appears that the dependence can be significant.[31]

43.10 MEASUREMENT OF ABSORPTION

Determination of the spectral absorption coefficient $a(\lambda)$ for natural waters is a difficult task for several reasons. First, water absorbs only weakly at near-UV to blue wavelengths so that very sensitive instruments are required. More importantly, scattering is never negligible so that careful consideration must be made of the possible aliasing of the absorption measurements by scattering effects. In pure water at wavelengths of $\lambda = 370$ to 450 nm, molecular scattering provides 20 to 25 percent (Table 10) of the total beam attenuation, $c(\lambda) = a(\lambda) + b(\lambda)$. Scattering effects can dominate absorption at all visible wavelengths in waters with high particulate loads. Additional complications arise in determining the absorption of pure water because of the difficulty of preparing uncontaminated samples.

Many techniques have been employed in attempts to determine the spectral absorption coefficient for pure water, $a_w(\lambda)$; these are reviewed in Smith and Baker.[32] The most commonly employed technique for routine determination of $a(\lambda)$ for oceanic waters consists of filtering a sample of water to retain the particulate matter on a filter pad. The spectral absorption of the particulate matter, $a_p(\lambda)$, is then determined in a spectrophotometer. The absorption of pure water, $a_w(\lambda)$, must be added to $a_p(\lambda)$ to obtain the total absorption of the oceanic water sample. Even though this technique for determining absorption has been in use for many years, the methodology is still evolving[33–35] because of the many types of errors inherent in the $a_p(\lambda)$ measurements (e.g., inability of filters to

TABLE 4 Index of Refraction of Fresh Water and of Sea Water at Atmospheric Pressure for Selected Temperatures and Wavelengths*

Fresh Water ($S = 0$)								
	Wavelength (nm)							
Temp (°C)	400	420	440	460	480	500	520	540
0	1.34419	1.34243	1.34092	1.33960	1.33844	1.33741	1.33649	1.33567
10	1.34390	1.34215	1.34064	1.33933	1.33817	1.33714	1.33623	1.33541
20	1.34317	1.34142	1.33992	1.33860	1.33745	1.33643	1.33551	1.33469
30	1.34208	1.34034	1.33884	1.33753	1.33638	1.33537	1.33445	1.33363
	Wavelength (nm)							
Temp (°C)	560	580	600	620	640	660	680	700
0	1.33494	1.33424	1.33362	1.33305	1.33251	1.33200	1.33153	1.33108
10	1.33466	1.33399	1.33336	1.33279	1.33225	1.33174	1.33127	1.33084
20	1.33397	1.33328	1.33267	1.33210	1.33156	1.33105	1.33059	1.33016
30	1.33292	1.33223	1.33162	1.33106	1.33052	1.33001	1.32955	1.32913
Sea Water ($S = 35‰$)								
	Wavelength (nm)							
Temp (°C)	400	420	440	460	480	500	520	540
0	1.35141	1.34961	1.34804	1.34667	1.34548	1.34442	1.34347	1.34263
10	1.35084	1.34903	1.34747	1.34612	1.34492	1.34385	1.34291	1.34207
20	1.34994	1.34814	1.34657	1.34519	1.34401	1.34295	1.34200	1.34116
30	1.34875	1.34694	1.34539	1.34404	1.34284	1.34179	1.34085	1.34000
	Wavelength (nm)							
Temp (°C)	560	580	600	620	640	660	680	700
0	1.34186	1.34115	1.34050	1.33992	1.33937	1.33885	1.33836	1.33791
10	1.34129	1.34061	1.33997	1.33938	1.33882	1.33830	1.33782	1.33738
20	1.34039	1.33969	1.33904	1.33845	1.33791	1.33739	1.33690	1.33644
30	1.33925	1.33855	1.33790	1.33731	1.33676	1.33624	1.33576	1.33532

* Data extracted from Austin and Halikas.[27]

TABLE 5 Index of Refraction Relative to Water, n, of Inorganic Particles Found in Sea Water

Substance	n
Quartz	1.16
Kaolinite	1.17
Montmorillonite	1.14
Hydrated mica	1.19
Calcite	1.11/1.24

retain all particulates, scattering effects within the sample cell, absorption by dissolved matter retained on the filter pad, and decomposition of pigments during the filtration process). Moreover, this methodology for determining total absorption assumes that absorption by dissolved organic matter (yellow substances) is negligible, which is not always the case. If the absorption by yellow matter, $a_y(\lambda)$, is desired, then the absorption of the *filtrate* is measured, and $a_y(\lambda)$ is taken to be $a_{\text{filtrate}}(\lambda) - a_w(\lambda)$. Several novel instruments under development[36–38] show promise for circumventing the problems inherent in the filter-pad technique as well as for making in situ measurements of total absorption which at present is difficult.[39]

43.11 ABSORPTION BY PURE SEA WATER

Table 2 showed the absorption for pure water over the wavelength range from 0.01 nm (x rays) to 1 m (radio waves). As is seen in the table, only the near-UV to near-IR wavelengths are of interest in hydrologic optics. Smith and Baker[32] made a careful but indirect determination of the *upper bound* of the spectral absorption coefficient of pure sea water, $a_w(\lambda)$, in the wavelength range of oceanographic interest, $200\ \text{nm} \leq \lambda \leq 800\ \text{nm}$. Their work assumed that for the clearest natural waters (1) absorption by salt or other dissolved substances was negligible, (2) the only scattering was by water molecules and salt ions, and (3) there was no inelastic scattering (i.e., no fluorescence or Raman scattering). With these assumptions the inequality (derived from radiative transfer theory)

$$a_w(\lambda) \leq K_d(\lambda) - \tfrac{1}{2} b_m^{sw}(\lambda)$$

holds. Here $b_m^{sw}(\lambda)$ is the spectral scattering coefficient for pure sea water; $b_m^{sw}(\lambda)$ was taken as known (Table 10). Smith and Baker then used measured values of the diffuse attenuation function $K_d(\lambda)$ from very clear waters (e.g., Crater Lake, Oregon, U.S.A., and the Sargasso Sea) to estimate $a_w(\lambda)$. Table 6 gives their self-consistent values of $a_w(\lambda)$, $K_d(\lambda)$, and $b_m^{sw}(\lambda)$.

The Smith and Baker absorption values are widely used. However, it must be remembered that the values of $a_w(\lambda)$ in Table 6 are upper bounds; the true absorption of pure water is likely to be somewhat lower, at least at violet and blue wavelengths.[40] Smith and Baker pointed out that there are uncertainties because K_d, an apparent optical property, is influenced by environmental conditions. They also commented that at wavelengths below 300 nm, their values are "merely an educated guess." They estimated the accuracy of $a_w(\lambda)$ to be within +25 and −5 percent between 300 and 480 nm and +10 to −15 percent between 480 and 800 nm. Numerical simulations by Gordon[11] indicate that a more restrictive inequality,

$$a_w(\lambda) \leq \frac{K_d(\lambda)}{D_0(\lambda)} - 0.62 b_m^{sw}(\lambda)$$

could be used. Here $D_0(\lambda)$ is a measurable distribution function $[D_0(\lambda) > 1]$ that corrects for the effects of sun angle and sea state on $K_d(\lambda)$ (discussed earlier). Use of the Gordon inequality could reduce the Smith and Baker absorption values by up to 20 percent at blue wavelengths. And finally, the Smith and Baker measurements were not made in optically pure water but rather in the "clearest natural waters." Even these waters contain a small amount of dissolved and particulate matter which will contribute something to both absorption and scattering.

There is evidence[41] that absorption is weakly dependent on temperature, at least in the red and near infrared ($\partial a/\partial T \sim 0.0015\ \text{m}^{-1}\,^\circ\text{C}^{-1}$ at $\lambda = 600$ nm and $\partial a/\partial T \sim 0.01\ \text{m}^{-1}\,^\circ\text{C}^{-1}$ at $\lambda = 750$ nm) and perhaps also slightly dependent on salinity; these matters are under investigation.

TABLE 6 Spectral Absorption Coefficient of Pure Sea Water, a_w, As Determined by Smith and Baker (Values of the molecular scattering coefficient of pure sea water, b_m^{sw}, and of the diffuse attenuation coefficient K_d used in their computation of a_w are also shown.*)

λ (nm)	a_w (m^{-1})	b_m^{sw} (m^{-1})	K_d (m^{-1})	λ (nm)	a_w (m^{-1})	b_m^{sw} (m^{-1})	K_d (m^{-1})
200	3.07	0.151	3.14	500	0.0257	0.0029	0.0271
210	1.99	0.119	2.05	510	0.0357	0.0026	0.0370
220	1.31	0.0995	1.36	520	0.0477	0.0024	0.0489
230	0.927	0.0820	0.968	530	0.0507	0.0022	0.0519
240	0.720	0.0685	0.754	540	0.0558	0.0021	0.0568
250	0.559	0.0575	0.588	550	0.0638	0.0019	0.0648
260	0.457	0.0485	0.481	560	0.0708	0.0018	0.0717
270	0.373	0.0415	0.394	570	0.0799	0.0017	0.0807
280	0.288	0.0353	0.306	580	0.108	0.0016	0.109
290	0.215	0.0305	0.230	590	0.157	0.0015	0.158
300	0.141	0.0262	0.154	600	0.244	0.0014	0.245
310	0.105	0.0229	0.116	610	0.289	0.0013	0.290
320	0.0844	0.0200	0.0944	620	0.309	0.0012	0.310
330	0.0678	0.0175	0.0765	630	0.319	0.0011	0.320
340	0.0561	0.0153	0.0637	640	0.329	0.0010	0.330
350	0.0463	0.0134	0.0530	650	0.349	0.0010	0.350
360	0.0379	0.0120	0.0439	660	0.400	0.0008	0.400
370	0.0300	0.0106	0.0353	670	0.430	0.0008	0.430
380	0.0220	0.0094	0.0267	680	0.450	0.0007	0.450
390	0.0191	0.0084	0.0233	690	0.500	0.0007	0.500
400	0.0171	0.0076	0.0209	700	0.650	0.0007	0.650
410	0.0162	0.0068	0.0196	710	0.839	0.0007	0.834
420	0.0153	0.0061	0.0184	720	1.169	0.0006	1.170
430	0.0144	0.0055	0.0172	730	1.799	0.0006	1.800
440	0.0145	0.0049	0.0170	740	2.38	0.0006	2.380
450	0.0145	0.0045	0.0168	750	2.47	0.0005	2.47
460	0.0156	0.0041	0.0176	760	2.55	0.0005	2.55
470	0.0156	0.0037	0.0175	770	2.51	0.0005	2.51
480	0.0176	0.0034	0.0194	780	2.36	0.0004	2.36
490	0.0196	0.0031	0.0212	790	2.16	0.0004	2.16
				800	2.07	0.0004	2.07

* Reproduced from Smith and Baker,[32] with permission.

43.12 ABSORPTION BY DISSOLVED ORGANIC MATTER

Absorption by yellow matter is reasonably well described by the model[42]

$$a_y(\lambda) = a_y(\lambda_0) \exp\left[-0.014(\lambda - \lambda_0)\right]$$

over the range $350\,\text{nm} \le \lambda \le 700\,\text{nm}$. Here λ_0 is a reference wavelength usually chosen to be $\lambda_0 = 440\,\text{nm}$ and $a_y(\lambda_0)$ is the absorption due to yellow matter at the reference wavelength. The value of $a_y(\lambda)$ of course depends on the concentration of yellow matter in the water. The exponential decay constant depends on the relative proportion of specific types of yellow matter; other studies have found exponents of -0.014 to -0.019 (Roesler et al.,[43] table 1). Both total concentration and proportions are highly variable. Table 7 gives measured values of $a_y(440)$ for selected waters. Because of the variability in yellow

TABLE 7 Measured Absorption Coefficient at $\lambda = 440$ nm Due to Yellow Matter, $a_y(440\text{ nm})$, for Selected Waters*

Water body	$a_y(440\text{ nm})$ (m^{-1})
Oceanic waters	
Sargasso Sea	≈0
off Bermuda	0.01
Gulf of Guinea	0.024–0.113
oligotrophic Indian Ocean	0.02
mesotrophic Indian Ocean	0.03
eutrophic Indian Ocean	0.09
Coastal and estuarine waters	
North Sea	0.07
Baltic Sea	0.24
Rhone River mouth, France	0.086–0.572
Clyde River estuary, Australia	0.64
Lakes and rivers	
Crystal Lake, Wisconsin, U.S.A.	0.16
Lake George, Australia	0.69–3.04
Lake George, Uganda	3.7
Carrao River, Venezuela	12.44
Lough Napeast, Ireland	19.1

* Condensed from Kirk,[8] with permission.

matter concentrations, the values found in Table 7 have little general validity even for the particular water bodies sampled, but they do serve to show representative values and the range of influence of yellow matter in determining the total absorption. Although the above model allows the determination of spectral absorption by yellow matter if the absorption is known at one wavelength, no model yet exists that allows for the direct determination of $a_y(\lambda)$ from given concentrations of yellow matter constituents.

43.13 ABSORPTION BY PHYTOPLANKTON

Phytoplankton cells are strong absorbers of visible light and therefore play a major role in determining the absorption properties of natural waters. Absorption by phytoplankton occurs in various photosynthetic pigments of which the chlorophylls are best known to nonspecialists. Absorption by chlorophyll itself is characterized by strong absorption bands in the blue and in the red (peaking at $\lambda \approx 430$ and 665 nm, respectively, for chlorophyll *a*), with very little absorption in the green. Chlorophyll occurs in all plants, and its concentration in milligrams of chlorophyll per cubic meter of water is commonly used as the relevant optical measure of phytoplankton abundance. (The term "chlorophyll concentration" usually refers to the sum of chlorophyll *a*, the main pigment in phytoplankton cells, and the related pigment pheophytin *a*.) Chlorophyll concentrations for various waters range from 0.01 mg m^{-3} in the clearest open ocean waters to 10 mg m^{-3} in productive coastal upwelling regions to 100 mg m^{-3} in eutrophic estuaries or lakes. The globally averaged, near-surface, open-ocean value is in the neighborhood of 0.5 mg m^{-3}.

The absorbing pigments are not evenly distributed within phytoplankton cells but are

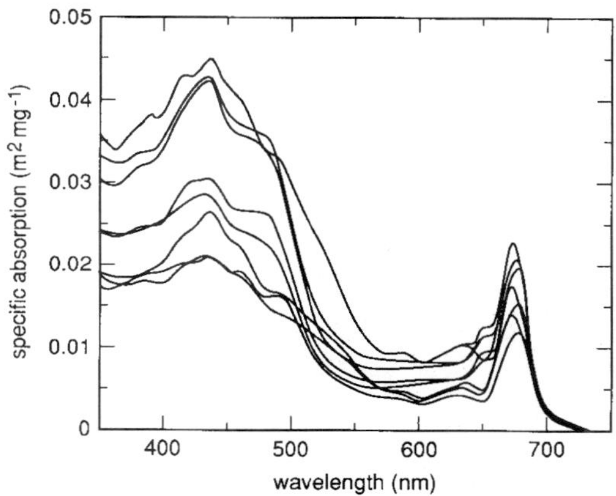

FIGURE 6 Chlorophyll-specific spectral absorption coefficients for eight species of phytoplankton. (*Redrawn from Sathyendranath et al.,*[44] *with permission.*)

localized into small "packages" (chloroplasts) which are distributed nonrandomly throughout the cell. This localized distribution of pigments means[8] that the spectral absorption by a phytoplankton cell or by a collection of cells in water is "flatter" (has less-pronounced peaks and reduced overall absorption) than if the pigments were uniformly distributed throughout the cell or throughout the water. This so-called "pigment packaging effect" is a major source of both inter- and intraspecies variability in spectral absorption by phytoplankton. This is because the details of the pigment packaging within cells depend not only on species but also on a cell's size and physiological state (which in turn depends on environmental factors such as ambient lighting and nutrient availability). Another source of variability in addition to chlorophyll *a* concentration and packaging is changes in pigment composition (the relative proportions of accessory pigments, namely chlorophylls *b* and *c*, pheopigments, biliproteins, and carotenoids) since each pigment displays a characteristic absorption curve.

A qualitative feel for the nature of phytoplankton absorption can be obtained from Fig. 6 which is based on absorption measurements from eight different single-species laboratory phytoplankton cultures.[44] Measured spectral absorption coefficients for the eight cultures, $a_i(\lambda)$, $i = 1$ to 8, were first reduced by subtracting $a_i(737)$ to remove the effects of absorption by detritus and cell constituents other than pigments: the assumption is that pigments do not absorb at $\lambda = 737$ nm and that the residual absorption is wavelength independent (which is a crude approximation). The resulting curves were then normalized by the chlorophyll concentrations of the respective cultures to generate the *chlorophyll-specific spectral absorption* curves for phytoplankton, $a_i^*(\lambda)$:

$$a_i^*(\lambda) = \frac{a_i(\lambda) - a_i(737)}{C_i} \qquad \frac{\text{m}^{-1}}{\text{mg m}^{-3}} = \text{m}^2\ \text{mg}^{-1}$$

which are plotted in Fig. 6.

Several general features of phytoplankton absorption are seen in Fig. 6:

1. There are distinct absorption peaks at $\lambda \approx 440$ and 675 nm.
2. The blue peak is one to three times as high as the red one (for a given species) due to the contribution of accessory pigments to absorption in the blue.
3. There is relatively little absorption between 550 and 650 nm, with the absorption minimum near 600 nm being 10 to 30 percent of the value at 440 nm.

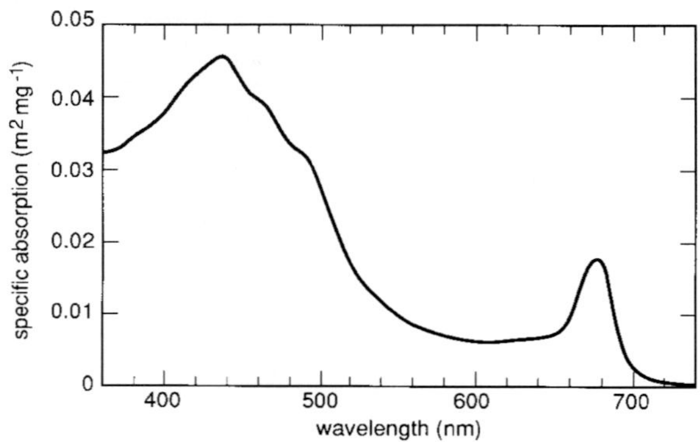

FIGURE 7 Average chlorophyll-specific spectral absorption coefficient for 14 species of phytoplankton. (*Redrawn from Morel,[45] with permission.*)

Similar analysis by Morel[45] yielded the average specific absorption curve shown in Fig. 7. Morel's curve is an average of spectra from 14 cultured phytoplankton species. The Morel curve is qualitatively the same as the curves of Fig. 6 and is as good a candidate as any for being called a "typical" phytoplankton specific absorption curve. The $a^*(\lambda)$ values of Fig. 7 are tabulated in Table 8 for reference.

43.14 ABSORPTION BY ORGANIC DETRITUS

Only recently has it become possible to determine the relative contributions of living phytoplankton and nonliving detritus to the total absorption by particulates. Iturriaga and Siegel[46] used microspectrophotometric techniques capable of measuring the spectral absorption of individual particles as small as 3 μm diameter to examine the absorption properties of particulates from Sargasso sea waters. Roesler et al.[43] employed a standard filter-pad technique with measurements made before and after pigments were chemically extracted to distinguish between absorption by pigmented and nonpigmented particles from fjord waters in the San Juan Islands, Washington, U.S.A. Each of these dissimilar techniques applied to particles from greatly different waters found very similar absorption spectra for nonpigmented organic particles derived from phytoplankton.

Figure 8 shows the microspectrophotometrically determined contributions of absorption by phytoplankton, $a_{\text{ph}}(\lambda)$, and of absorption by detritus, $a_{\text{det}}(\lambda)$, to the independently measured (by the filter-pad technique) total particulate absorption, $a_{\text{p}}(\lambda)$, for two depths at the same Atlantic location. The small residual, $\Delta a_{\text{p}}(\lambda) = a_{\text{p}}(\lambda) - a_{\text{ph}}(\lambda) - a_{\text{det}}(\lambda)$ shown in the figure is attributed either to errors in the determination of the phytoplankton and detrital parts (particles smaller than ~3 μm were not analyzed) or to contamination by dissolved organic matter of the filter-pad measurements of total particulate absorption. Note that at the shallow depth the phytoplankton are relatively more important at blue wavelengths whereas the detritus is slightly more important at the deeper depth. There is no generality in this result (other locations showed the reverse)—it merely illustrates the variability possible in water samples taken only 60 vertical meters apart.

The important feature to note in Fig. 8 is the general shape of the spectral absorption curve for detritus. Roesler et al. found essentially identical curves in their determination of $a_{\text{det}}(\lambda)$. These curves are reminiscent of the absorption curves for yellow matter and, indeed, Roesler et al. found that the model

$$a_{\text{det}}(\lambda) = a_{\text{det}}(400) \exp\left[-0.011(\lambda - 400)\right]$$

TABLE 8 Average Chlorophyll-Specific Spectral Absorption Coefficient a^* for 14 Species of Phytoplankton As Plotted in Fig. 7 (The standard deviation is ~30% of the mean except in the vicinity of 400 nm, where it is ~50%.*)

λ (nm)	a^* ($m^2\,mg^{-1}$	λ (nm)	a^* ($m^2\,mg^{-1}$)	λ (nm)	a^* ($m^2\,mg^{-1}$)
400	0.0394	500	0.0274	600	0.0053
405	0.0395	505	0.0246	605	0.0053
410	0.0403	510	0.0216	610	0.0054
415	0.0417	515	0.0190	615	0.0057
420	0.0429	520	0.0168	620	0.0059
425	0.0439	525	0.0151	625	0.0061
430	0.0448	530	0.0137	630	0.0063
435	0.0452	535	0.0125	635	0.0064
440	0.0448	540	0.0115	640	0.0064
445	0.0436	545	0.0106	645	0.0066
450	0.0419	550	0.0098	650	0.0071
455	0.0405	555	0.0090	655	0.0084
460	0.0392	560	0.0084	660	0.0106
465	0.0379	565	0.0078	665	0.0136
470	0.0363	570	0.0073	670	0.0161
475	0.0347	575	0.0068	675	0.0170
480	0.0333	580	0.0064	680	0.0154
485	0.0322	585	0.0061	685	0.0118
490	0.0312	590	0.0058	690	0.0077
495	0.0297	595	0.0055	695	0.0046
				700	0.0027

* Data courtesy of A. Morel[45].

provides a satisfactory fit to detrital absorption curves. Other studies have found coefficients of −0.006 to −0.014 (Roesler et al.,[43] table 1) instead of −0.011.

43.15 BIO-OPTICAL MODELS FOR ABSORPTION

Depending on the concentrations of dissolved substances, phytoplankton, and detritus, the total spectral absorption coefficient of a given water sample can range from almost identical to that of pure water to one which shows orders-of-magnitude greater absorption than pure water, especially at blue wavelengths. Figure 9 shows some $a(\lambda)$ profiles from various natural waters. Figure 9*a* shows absorption profiles measured in phytoplankton-dominated waters where chlorophyll concentrations ranged from $C = 0.2$ to $18.4\ mg\,m^{-3}$. In essence, the absorption is high in the blue because of absorption by phytoplankton pigments and high in the red because of absorption by the water. Figure 9*b* shows the absorption at three locations where $C \approx 2\ mg\,m^{-3}$ but where the scattering coefficient b varied from 1.55 to $3.6\ m^{-1}$ indicating that nonpigmented particles were playing an important role in determining the shape of $a(\lambda)$. Figure 9*c* shows curves from waters rich in yellow matter, which is causing the high absorption in the blue. One of the goals of bio-optics is to develop predictive models for absorption curves such as those seen in Fig. 9.

Case 1 waters are waters in which the concentration of phytoplankton is high compared to nonbiogenic particles.[47] Absorption by chlorophyll and related pigments therefore plays

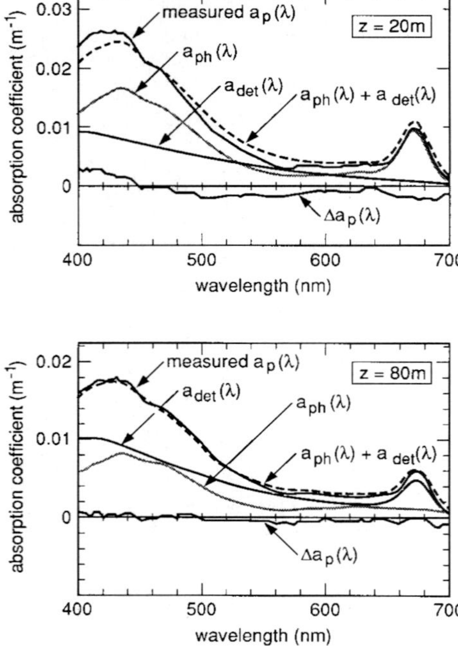

FIGURE 8 Examples of the relative contributions of absorption by phytoplankton $a_{ph}(\lambda)$, and by organic detritus $a_{det}(\lambda)$, to the total particulate absorption $a_p(\lambda)$, from Sargasso Sea waters. (*Redrawn from Iturriaga and Siegel,*[46] *with permission.*)

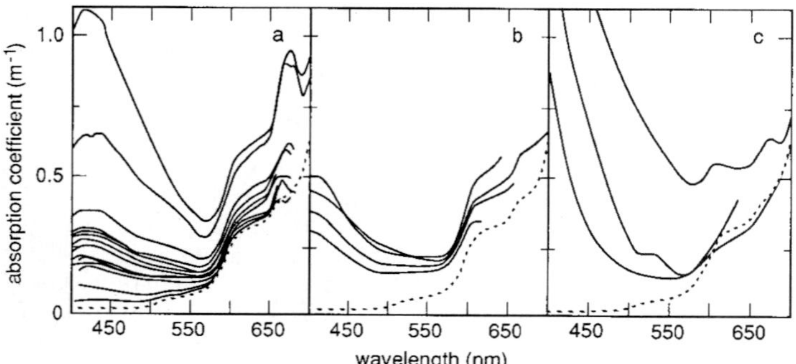

FIGURE 9 Examples of spectral absorption coefficients $a(\lambda)$ for various waters. Panel (a) shows $a(\lambda)$ for waters dominated by phytoplankton, panel (b) is for waters with a high concentration of nonpigmented particles, and panel (c) is for waters rich in yellow matter. (*Based on Prieur and Sathyendranath,*[48] *with permission.*)

a major role in determining the total absorption coefficient in such waters, although detritus and dissolved organic matter derived from the phytoplankton also contribute to absorption in case 1 waters. Case 1 water can range from very clear (oligotrophic) water to very turbid (eutrophic) water, depending on the phytoplankton concentration. *Case 2 waters* are "everything else," namely, waters where inorganic particles or dissolved organic matter from land drainage dominate so that absorption by pigments is relatively less important in determining the total absorption. (The case 1 and 2 classifications must not be confused with the Jerlov water *types* 1 and 2, discussed later.) Roughly 98 percent of the world's open ocean and coastal waters fall into the case 1 category, and therefore almost all bio-optical research has been directed toward these phytoplankton-dominated waters. However, near-shore and estuarine case 2 waters are disproportionately important to human interests such as recreation, fisheries, and military operations.

Prieur and Sathyendranath[48] developed a pioneering bio-optical model for the spectral absorption coefficient of case 1 waters. Their model was statistically derived from 90 sets of spectral absorption data taken in various case 1 waters and included absorption by phytoplankton pigments, by nonpigmented organic particles derived from deceased phytoplankton, and by yellow matter derived from decayed phytoplankton. The contribution of phytoplankton to the total absorption was parametrized in terms of the chlorophyll concentration C (i.e., chlorophyll a plus pheophytin a). The contributions of nonpigmented particles and of yellow matter were parametrized in terms of both the chlorophyll concentration and the total scattering coefficient at $\lambda = 550$ nm, $b(550)$. The essence of the Prieur-Sathyendranath model is contained in a more recent and simpler variant given by Morel[6]:

$$a(\lambda) = [a_w(\lambda) + 0.06a_c^{*\prime}(\lambda)C^{0.65}][1 + 0.2\exp(-0.014(\lambda - 440))] \tag{1}$$

Here $a_w(\lambda)$ is the absorption coefficient of pure water and $a_c^{*\prime}(\lambda)$ is a nondimensional, statistically derived chlorophyll-specific absorption coefficient; $a_w(\lambda)$ and $a_c^{*\prime}(\lambda)$ values are given in Table 9 [these $a_w(\lambda)$ values are slightly different than those of Table 6]. When C is expressed in mg m^{-3} and λ is in nm, the resulting $a(\lambda)$ is in m^{-1}.

Another simple bio-optical model for absorption has been developed independently by Kopelevich.[49] It has the form[50]

$$a(\lambda) = a_w(\lambda) + C[a_c^0(\lambda) + 0.1\exp[-0.015(\lambda - 400)]$$

TABLE 9 Absorption by Pure Sea Water, a_w, and the Nondimensional Chlorophyll-Specific Absorption Coefficient $a_c^{*\prime}$ Used in The Prieur-Sathyendranath-Morel Model for the Spectral Absorption Coefficient $a(\lambda)$*

λ (nm)	a_w (m^{-1})	$a_c^{*\prime}$	λ (nm)	a_w (m^{-1})	$a_c^{*\prime}$	λ (nm)	a_w (m^{-1})	$a_c^{*\prime}$
400	0.018	0.687	500	0.026	0.668	600	0.245	0.236
410	0.017	0.828	510	0.036	0.618	610	0.290	0.252
420	0.016	0.913	520	0.048	0.528	620	0.310	0.276
430	0.015	0.973	530	0.051	0.474	630	0.320	0.317
440	0.015	1.000	540	0.056	0.416	640	0.330	0.334
450	0.015	0.944	550	0.064	0.357	650	0.350	0.356
460	0.016	0.917	560	0.071	0.294	660	0.410	0.441
470	0.016	0.870	570	0.080	0.276	670	0.430	0.595
480	0.018	0.798	580	0.108	0.291	680	0.450	0.502
490	0.020	0.750	590	0.157	0.282	690	0.500	0.329
						700	0.650	0.215

* Condensed with permission from Prieur and Sathyendranath,[48] who give values every 5 mm.

where $a_c^0(\lambda)$ is the chlorophyll-specific absorption coefficient for phytoplankton ($m^2\,mg^{-1}$), and $a_w(\lambda)$ and C are defined as for Eq. (1). The Kopelevich model as presently used[49] takes $a_w(\lambda)$ from Smith and Baker[32] (Table 6) and takes $a_c^0(\lambda)$ from Yentsch.[51]

Although these and similar bio-optical models for absorption are frequently used, caution is advised in their application. Both models assume that the absorption by yellow matter covaries with that due to phytoplankton; i.e., each implies that a fixed percentage of the total absorption at a given wavelength always comes from yellow matter. The general validity of this assumption is doubtful even for open ocean waters: Bricaud et al.[42] show data (fig. 5) for which $a(375)$, used as an index for yellow matter concentration, is uncorrelated with chlorophyll concentration even in oceanic regions uninfluenced by freshwater runoff. Gordon[52] has developed a model that avoids assuming any relation between yellow matter and phytoplankton. However, his model becomes singular as $C \rightarrow 0.01\ mg\,m^{-3}$ and cannot be expected to work well for C much less than $0.1\ mg\,m^{-3}$. The Kopelevich model has the chlorophyll contribution proportional to C, whereas the Morel model has $C^{0.65}$. The exponent of 0.65 is probably closer to reality, since it reflects a change in the relative contributions to absorption by phytoplankton and by detritus as the chlorophyll concentration changes (absorption by detritus is relatively more important at low chlorophyll concentrations[52]). Moreover, the chlorophyll-specific absorption curve of Yentsch[51] used in the Kopelevich model is based on laboratory cultures of phytoplankton, whereas the later work by Prieur and Sathyendranath used in situ observations to derive the $a_c^{*\prime}(\lambda)$ values of Table 9—an additional point in favor of Eq. (1). Either of these bio-optical models is useful but clearly imperfect. They may (or may not) give correct *average* values, but they give no information about the *variability* of $a(\lambda)$. It can be anticipated that the simple models now available will be replaced, perhaps by models designed for specific regions and seasons, as better understanding of the variability inherent in spectral absorption is achieved.

43.16 MEASUREMENT OF SCATTERING

Scattering in natural waters is caused both by small scale ($\ll\lambda$) density fluctuations attributable to random molecular motions and by the ubiquitous large ($>\lambda$) organic and inorganic particles. Scattering by water molecules (and salt ions, in seawater) determines the minimum values for the scattering properties. However, as is the case for absorption, the scattering properties of natural waters are greatly modified by the particulate matter that is always present.

Scattering measurements are even more difficult than absorption measurements. The conceptual design of an instrument for measuring the volume scattering function $\beta(\psi, \lambda)$ is no more complicated than Fig. 2 and the defining equation $\beta(\psi, \lambda) = I_s(\psi, \lambda)/[E_i(\lambda)\,\Delta V]$: a collimated beam of known irradiance E_i illuminates a given volume of water ΔV and the scattered intensity I_s is measured as a function of scattering angle and wavelength. However, the engineering of instruments capable of the in situ determination of $\beta(\psi, \lambda)$ is quite difficult. The magnitude of the scattered intensity typically increases by five or six orders of magnitude in going from $\psi = 90°$ to $\psi = 0.1°$ for a given natural water sample, and scattering at a given angle ψ can vary by two orders of magnitude among water samples. The required dynamic range of an instrument is therefore great. Corrections must be made for absorption within the sample volume and also along the incident and scattered beam paths for in situ instruments. The rapid change in $\beta(\psi, \lambda)$ at small scattering angles requires very precise alignment of the optical elements, but rolling ships seem designed to knock things out of alignment. Because of these design difficulties only a few one-of-a-kind instruments have been built for in situ measurement of the volume scattering function, and measurements of $\beta(\psi, \lambda)$ are not routinely made. Petzold[53] gives the details of two such instruments, one for small scattering angles ($\psi = 0.085$, 0.17, and 0.34°) and one for larger angles ($10° \leq \psi \leq 170°$); these are the instruments used to obtain the data presented in Sec. 43.18. Other instruments are referenced in Kirk[8] and in Jerlov.[29]

Commercial instruments are available for laboratory measurement of $\beta(\psi, \lambda)$ at fixed scattering angles (e.g., ψ every 5° from ~20 to ~160°). These instruments are subject to their own problems, such as degradation of samples between the times of collection and measurement. Moreover, measurements of $\beta(\psi, \lambda)$ over a limited range of ψ are not sufficient for determination of $b(\lambda)$ by integration. In practice, the scattering coefficient $b(\lambda)$ is usually determined by the conservation of energy relation $b(\lambda) = c(\lambda) - a(\lambda)$ after measurements of beam attenuation and absorption have been made.

Both in situ and laboratory instruments sample (~cm^3) volumes of water and therefore may fail to detect the presence of optically significant large aggregates (marine snow) if such particles are too few in number to be reliably captured in the sample volume. However, such particles can affect the scattering properties of large volumes of water (e.g., as seen in remote sensing or underwater visibility studies).

Measurements at near forward ($\psi < 1°$) and near backward ($\psi > 179°$) angles are exceptionally difficult to make, yet the behavior of $\beta(\psi, \lambda)$ at these extreme angles is of considerable interest. Accurate determination of β at small angles is crucial to the determination of b by integration since typically one-half of all scattering takes place at angles of less than a few degrees. Scattering at small angles is important in underwater imaging and it is of theoretical interest for its connections to scattering theory, particle optical properties, and particle size distributions. The behavior of β very near $\psi = 180°$ is important in laser remote-sensing applications.

Spinrad et al.[54] and Padmabandu and Fry[55] have reported measurements at very small angles on suspensions of polystyrene spheres but no such measurements have been published for natural water samples. The Padmabandu and Fry technique is notable in that it allows the measurement of β at $\psi = 0°$ exactly by use of the coupling of two coherent beams in a photorefractive crystal to measure the phase shift that corresponds to 0° scattering. Measurement of $\beta(0, \lambda)$ is of theoretical interest because of its relation to attenuation via the optical theorem. Enhanced backscatter has been reported[56] in suspensions of latex spheres; a factor-of-two increase in scattered intensity between $\psi = 179.5$ and 180.0° is typical. Whether or not such backscattering enhancement ever occurs in natural waters is a subject of heated debate.

43.17 SCATTERING BY PURE WATER AND BY PURE SEA WATER

Morel[57] has reviewed in detail the theory and observations pertaining to scattering by pure water and by pure sea water. In pure water random molecular motions give rise to rapid fluctuations in the number of molecules in a given volume ΔV, where ΔV is small compared to the wavelength of light but large compared to atomic scales (so that the liquid within the volume is adequately described by statistical thermodynamics). The Einstein-Smoluchowski theory of scattering relates these fluctuations in molecular number density to associated fluctuations in the index of refraction, which give rise to scattering. In sea water the basic theory is the same but random fluctuations in the concentrations of the various ions (Cl^-, Na^+, etc.) give somewhat greater index of refraction fluctuations, and hence greater scattering. The net result of these considerations is that the volume scattering function for either pure water or for pure sea water has the form

$$\beta_w(\psi, \lambda) = \beta_w(90°, \lambda_0)\left(\frac{\lambda_0}{\lambda}\right)^{4.32}(1 + 0.835\cos^2\psi) \qquad \mathrm{m^{-1}\,sr^{-1}} \tag{2}$$

which is reminiscent of the form for Rayleigh scattering. The wavelength dependence of

TABLE 10 The Volume Scattering Function at $\psi = 90°$, $\beta_w(90°, \lambda)$, and the Scattering Coefficient $b_w(\lambda)$ for Pure Water and for Pure Sea Water ($S = 35$–$39‰$)*

	Pure water		Pure sea water	
λ (nm)	$\beta_w(90°)$ (10^{-4} m^{-1} sr^{-1})	b_w† (10^{-4} m^{-1})	$\beta(90°)$ (10^{-4} m^{-1} sr^{-1})	b_w† (10^{-4} m^{-1})
350	6.47	103.5	8.41	134.5
375	4.80	76.8	6.24	99.8
400	3.63	58.1	4.72	75.5
425	2.80	44.7	3.63	58.1
450	2.18	34.9	2.84	45.4
475	1.73	27.6	2.25	35.9
500	1.38	22.2	1.80	28.8
525	1.12	17.9	1.46	23.3
550	0.93	14.9	1.21	19.3
575	0.78	12.5	1.01	16.2
600	0.68	10.9	0.88	14.1

* Reproduced from Morel,[57] with permission.
† Computed from $b(\lambda) = 16.0\beta(90°, \lambda)$.

$\lambda^{-4.32}$ rather than λ^{-4} (for Rayleigh scattering) results from the wavelength dependence of the index of refraction. The 0.835 factor is attributable to the anisotropy of the water molecules. The corresponding phase function is

$$\tilde{\beta}_w(\psi) = 0.06225(1 + 0.835 \cos^2 \psi) \qquad \text{sr}^{-1}$$

and the total scattering coefficient $b_w(\lambda)$ is given by

$$b_w(\lambda) = 16.06\left(\frac{\lambda_0}{\lambda}\right)^{4.32} \beta_w(90°, \lambda_0) \qquad \text{m}^{-1} \tag{3}$$

Table 10 gives values of $\beta_w(90°, \lambda)$ and $b_w(\lambda)$ for selected wavelengths for both pure water and pure sea water ($S = 35$–$39‰$). Note that the pure sea water values are about 30 percent greater than the pure water values at all wavelengths. Table 11 shows the dependence of $b_w(546)$ on pressure, temperature, and salinity. Note that molecular

TABLE 11 Computed Scattering Coefficient b of Pure Water ($S = 0$) and of Pure Sea Water ($S = 35‰$) at $\lambda = 546$ nm As a Function of Temperature T and Pressure p (Numbers in the body of the table have units of m^{-1}.*)

	$p = 10^5$ Pa (1 atm)		$p = 10^7$ Pa (100 atm)		$p = 10^8$ Pa (1000 atm)	
T (°C)	$S = 0$	$S = 35‰$	$S = 0$	$S = 35‰$	$S = 0$	$S = 35‰$
0	0.00145	0.00195	0.00140	0.00192	0.00110	0.00167
10	0.00148	0.00203	0.00143	0.00200	0.00119	0.00176
20	0.00149	0.00207	0.00147	0.00204	0.00125	0.00183
40	0.00150	0.00213	0.00149	0.00212	0.00136	0.00197

* Data extracted from the more extensive table of Shifrin,[58] with permission.

scattering decreases as decreasing temperature or increasing pressure reduce the small-scale fluctuations.

43.18 SCATTERING BY PARTICLES

Heroic efforts are required to obtain water of sufficient purity that a Rayleigh-like volume scattering function is observed. As soon as there is a slight amount of particulate matter in the water—always the case for even the clearest natural water—the volume scattering function becomes highly peaked in the forward direction, and the scattering coefficient increases by at least a factor of ten.

Even for the most numerous oceanic particles (e.g., colloids at a concentration of $10^{15}\ \mathrm{m}^{-3}$) the average distance between particles is greater than ten wavelengths of visible light. For the optically most significant phytoplankton the average separation is thousands of wavelengths. Moreover, these particles usually are randomly distributed and oriented. Ocean water therefore can be treated as a very dilute suspension of random scatterers and consequently the intensity of light scattered by an ensemble of particles is given by the sum of the intensities due to the individual particles. Coherent scattering effects are negligible except perhaps at extremely small scattering angles.[58] An overview of scattering by particles is given in Chap. 6 by Bohren in this *Handbook*.

The conrtibution of the particulate matter to the total volume scattering function $\beta(\psi, \lambda)$ is obtained from

$$\beta_{\mathrm{p}}(\psi, \lambda) \equiv \beta(\psi, \lambda) - \beta_{\mathrm{w}}(\psi, \lambda)$$

Here the subscript p refers to particles, and w refers to pure water (if β is measured in fresh water) or pure sea water (for oceanic measurements). Figure 10 shows several particle volume scattering functions determined from in situ measurements of $\beta(\psi, \lambda)$ in a variety of waters ranging from very clear to very turbid. The particles cause at least a four-order-of-magnitude increase in scattering between $\psi \approx 90°$ and $\psi \approx 1°$. The contribution of molecular scattering to the total is therefore completely negligible except at backscattered directions ($\psi \gtrsim 90°$) in the clearest natural waters. The top curve in Fig. 10 is shown for small scattering angles in Fig. 11. The scattering function shows no indication of "flattening out" even at angles as small as 0.5°. Note that the scattering function increases by a factor of 100 over only a one-degree range of scattering angle.

Highly peaked forward scattering such as that seen in Figs. 10 and 11 is characteristic of diffraction-dominated scattering in a polydisperse system. Scattering by refraction and reflection from particle surfaces becomes important at large scattering angles ($\psi \gtrsim 15°$). Mie scattering calculations are well able to reproduce observed volume scattering functions given the proper optical properties and size distributions. Early efforts along these lines are seen in Kullenberg[59] and in Brown and Gordon.[60] Brown and Gordon were unable to reproduce observed backscattering values using measured particle size distributions. However, their instruments were unable to detect submicrometer particles. They found that the Mie theory properly predicted backscattering if they assumed the presence of numerous, submicrometer, low-index-of-refraction particles. It is reasonable to speculate that bacteria and the recently discovered colloidal particles are the particles whose existence was inferred by Brown and Gordon. Recent Mie scattering calculations[61] have used three-layered spheres to model the structure of phytoplankton (cell wall, chloroplasts, and cytoplasm core) and have used polydisperse mixtures of both organic and inorganic particles.

The most carefully made and widely cited scattering measurements are found in Petzold.[53] Figure 12 shows three of his $\beta(\psi, \lambda)$ curves displayed on a log-log plot to emphasize the forward scattering angles. The instruments had a spectral response

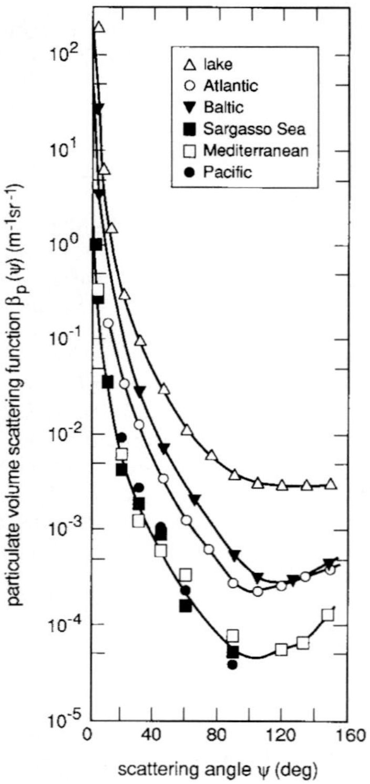

FIGURE 10 Particulate volume scattering functions $\beta_p(\psi, \lambda)$ determined from *in situ* measurements in various waters; wavelengths vary. (*Redrawn from Kullenberg,*[59] *with permission.*)

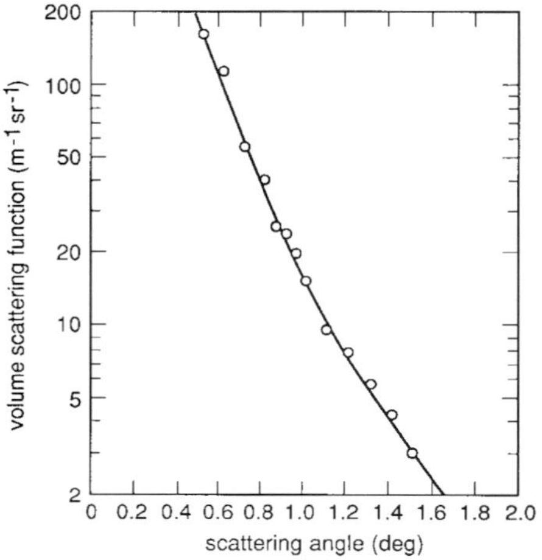

FIGURE 11 Detail of the forward scattering values of the "lake" volume scattering function seen in the top curve of Fig. 10. (*Redrawn from Preisendorfer.*[2])

centered at $\lambda = 514$ nm with an FWHM of 75 nm. The top curve was obtained in the very turbid water of San Diego Harbor, California; the center curve comes from near-shore coastal water in San Pedro Channel, California; and the bottom curve is from very clear water in the Tongue of the Ocean, Bahama Islands. The striking feature of these volume scattering functions (and those of Fig. 10) from very different waters is the similarity of their shapes. Although the scattering coefficients b of the curves in Fig. 12 vary by a factor of 50 (Table 13), the uniform shapes suggest that it is reasonable to define a "typical" particle phase function $\tilde{\beta}_p(\psi, \lambda)$. This has been done[62] with three sets of Petzold's data from waters with a high particulate load (one set being the top curve of Fig. 12), as follows: (1) subtract $\beta_w(\psi, \lambda)$ from each curve to get three particle volume scattering functions $\beta_p^i(\psi, \lambda)$, $i = 1, 2, 3$; (2) compute three particle phase functions via $\tilde{\beta}_p^i(\psi, \lambda) = \beta_p^i(\psi, \lambda)/b^i(\lambda)$; (3) average the three particle phase functions to define the typical particle phase function $\tilde{\beta}_p(\psi, \lambda)$. Table 12 displays the three Petzold volume scattering functions plotted in Fig. 12, the volume scattering function for pure sea water, and the average particle phase function computed as just described. This particle phase function satisfies the normalization $2\pi \int_0^\pi \tilde{\beta}_p(\psi, \lambda) \sin\psi \, d\psi = 1$ if a behavior of $\tilde{\beta}_p \sim \psi^{-m}$ is assumed for $\psi < 0.1°$ (m is a positive constant between zero and two, determined from $\tilde{\beta}_p$ at the smallest measured angles), and a trapezoidal rule integration is used for $\psi \geq 0.1°$, with linear interpolation used between the tabulated values. This average particle phase

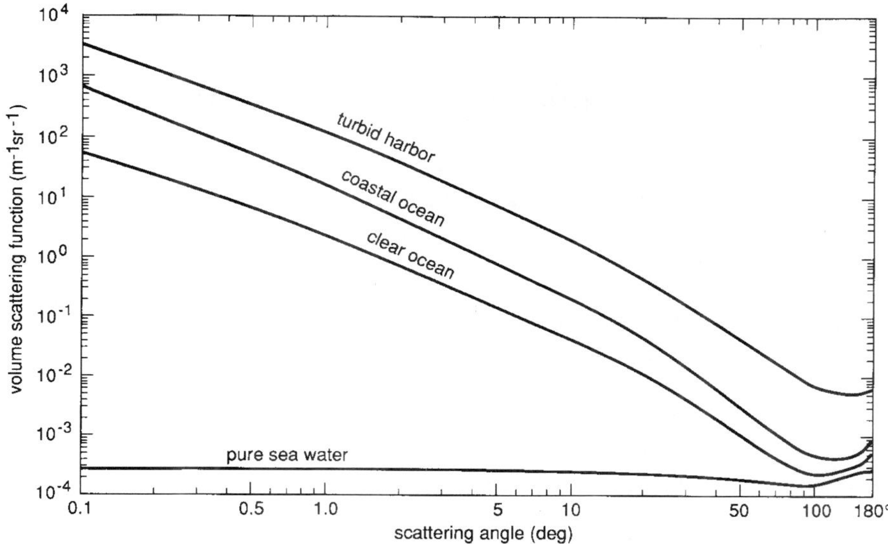

FIGURE 12 Measured volume scattering functions from three different natural waters and the computed volume scattering function for pure sea water, all at $\lambda = 514\,\text{nm}$. (*Redrawn from Petzold.*[53])

function is adequate for many radiative transfer calculations. However, the user must remember that significant deviations from the average can be expected in nature (e.g., in waters with abnormally high numbers of either large or small particles), although the details of such deviations have not been quantified.

Table 13 compares several inherent optical properties for pure sea water and for the three Petzold water samples of Fig. 12 and Table 12. These data show how greatly different even clear ocean water is from pure sea water. Note that natural water ranges from absorption-dominated ($\omega_0 = 0.247$) to scattering-dominated ($\omega_0 = 0.833$) at $\lambda = 514\,\text{nm}$. The ratio of backscattering to total scattering is typically a few percent in natural water. However, there is no clear relation between b_b/b and the water type, at least for the Petzold data of Table 13. This lack of an obvious relation is likely the result of differing particle types in the three waters. Since refraction and reflection are important processes at large scattering angles, the particle indices of refraction are important in determining b_b. Total scattering is dominated by diffraction and so particle composition has little effect on b values. The last row of Table 13 gives the angle ψ such that one-half of the total scattering occurs at angles between 0 and ψ. This angle is rarely greater than 10° in natural waters.

43.19 WAVELENGTH DEPENDENCE OF SCATTERING: BIO-OPTICAL MODELS

The strong $\lambda^{-4.32}$ wavelength dependence of pure-water scattering is not seen in natural waters. This is because scattering is dominated by diffraction from polydisperse particles that are usually much larger than the wavelength of visible light. Although diffraction depends on the particle size-to-wavelength ratio, the presence of particles of many sizes

TABLE 12 Volume Scattering Functions $\beta(\psi, \lambda)$ for Three Oceanic Waters and for Pure Sea Water and a Typical Particle Phase Function $\tilde{\beta}_p(\psi, \lambda)$, All at $\lambda = 514$ nm

Scattering angle (deg)	Volume scattering functions ($m^{-1}\,sr^{-1}$)				Particle phase function‡ (sr^{-1})
	Clear ocean*	Coastal ocean*	Turbid harbor*	Pure sea water†	
0.100	5.318×10^1	6.533×10^2	3.262×10^3	2.936×10^{-4}	1.767×10^3
0.126	4.042	4.577	2.397	2.936	1.296
0.158	3.073	3.206	1.757	2.936	9.502×10^2
0.200	2.374	2.252	1.275	2.936	6.991
0.251	1.814	1.579	9.260×10^2	2.936	5.140
0.316	1.360	1.104	6.764	2.936	3.764
0.398	9.954×10	7.731×10^1	5.027	2.936	2.763
0.501	7.179	5.371	3.705	2.936	2.012
0.631	5.110	3.675	2.676	2.936	1.444
0.794	3.591	2.481	1.897	2.936	1.022
1.000	2.498	1.662	1.329	2.936	7.161×10^1
1.259	1.719	1.106	9.191×10^1	2.935	4.958
1.585	1.171	7.306×10^0	6.280	2.935	3.395
1.995	7.758×10^{-1}	4.751	4.171	2.934	2.281
2.512	5.087	3.067	2.737	2.933	1.516
3.162	3.340	1.977	1.793	2.932	1.002
3.981	2.196	1.273	1.172	2.930	6.580×10^0
5.012	1.446	8.183×10^{-1}	7.655×10^0	2.926	4.295
6.310	9.522×10^{-2}	5.285	5.039	2.920	2.807
7.943	6.282	3.402	3.302	2.911	1.819
10.0	4.162	2.155	2.111	2.896	1.153
15.0	2.038	9.283×10^{-2}	9.041×10^{-1}	2.847	4.893×10^{-1}
20.0	1.099	4.427	4.452	2.780	2.444
25.0	6.166×10^{-3}	2.390	2.734	2.697	1.472
30.0	3.888	1.445	1.613	2.602	8.609×10^{-2}
35.0	2.680	9.063×10^{-3}	1.109	2.497	5.931
40.0	1.899	6.014	7.913×10^{-2}	2.384	4.210
45.0	1.372	4.144	5.858	2.268	3.067
50.0	1.020	2.993	4.388	2.152	2.275
55.0	7.683×10^{-4}	2.253	3.288	2.040	1.699
60.0	6.028	1.737	2.548	1.934	1.313
65.0	4.883	1.369	2.041	1.839	1.046
70.0	4.069	1.094	1.655	1.756	8.488×10^{-3}
75.0	3.457	8.782×10^{-4}	1.345	1.690	6.976
80.0	3.019	7.238	1.124	1.640	5.842
85.0	2.681	6.036	9.637×10^{-3}	1.610	4.953
90.0	2.459	5.241	8.411	1.600	4.292
95.0	2.315	4.703	7.396	1.610	3.782
100.0	2.239	4.363	6.694	1.640	3.404
105.0	2.225	4.189	6.220	1.690	3.116
110.0	2.239	4.073	5.891	1.756	2.912
115.0	2.265	3.994	5.729	1.839	2.797
120.0	2.339	3.972	5.549	1.934	2.686
125.0	2.505	3.984	5.343	2.040	2.571
130.0	2.629	4.071	5.154	2.152	2.476
135.0	2.662	4.219	4.967	2.268	2.377
140.0	2.749	4.458	4.822	2.384	2.329
145.0	2.896	4.775	4.635	2.497	2.313
150.0	3.088	5.232	4.634	2.602	2.365
155.0	3.304	5.824	4.900	2.697	2.506
160.0	3.627	6.665	5.142	2.780	2.662
165.0	4.073	7.823	5.359	2.847	2.835
170.0	4.671	9.393	5.550	2.896	3.031
175.0	4.845	9.847	5.618	2.926	3.092
180.0	5.109	1.030×10^{-3}	5.686	2.936	3.154

* Data reproduced from Petzold.[53]
† Computed from Eq. (2) and Table 10.
‡ Courtesy of H. R. Gordon; see also Ref. 62.

TABLE 13 Selected Inherent Optical Properties for the Waters Presented in Fig. 12 and in Table 12 (All values are for $\lambda = 514$ nm except as noted.)

Water	a (m^{-1})	b (m^{-1})	c (m^{-1})	ω_0	b_b/b	ψ for $\frac{1}{2}b$ (deg)
Pure sea water	0.0405*	0.0025†	0.043	0.058	0.500	90.00
Clear ocean	0.114‡	0.037	0.151§	0.247	0.044	6.25
Coastal ocean	0.179‡	0.219	0.398§	0.551	0.013	2.53
Turbid harbor	0.366‡	1.824	2.190§	0.833	0.020	4.68

* Value obtained by interpolation in Table 6.
† Value obtained by interpolation in Table 10.
‡ Estimated by Petzold[53] from $c(530\text{ nm}) - b(514\text{ nm})$.
§ Measured by Petzold[53] at $\lambda = 530$ nm.

diminishes the wavelength effects that are seen in diffraction by a single particle. Moreover, diffraction does not depend on particle composition. However, some wavelength dependence is to be expected, especially at backward scattering angles where refraction, and hence particle composition, is important. Molecular scattering also contributes something to the total scattering and can even dominate the particle contribution at backscatter angles in clear water.[63]

Morel[64] presents several useful observations on the wavelength dependence of scattering. Figure 13 shows two sets of volume scattering functions, one from the very clear waters of the Tyrrhenian Sea and one from the turbid English Channel. Each set displays $\beta(\psi, \lambda)/\beta(90°, \lambda)$ for $\lambda = 366$, 436, and 546 nm. The clear water shows a definite dependence of the shape of $\beta(\psi, \lambda)$ on λ whereas the particle-rich turbid water shows much less wavelength dependence. In each case the volume scattering function of shortest

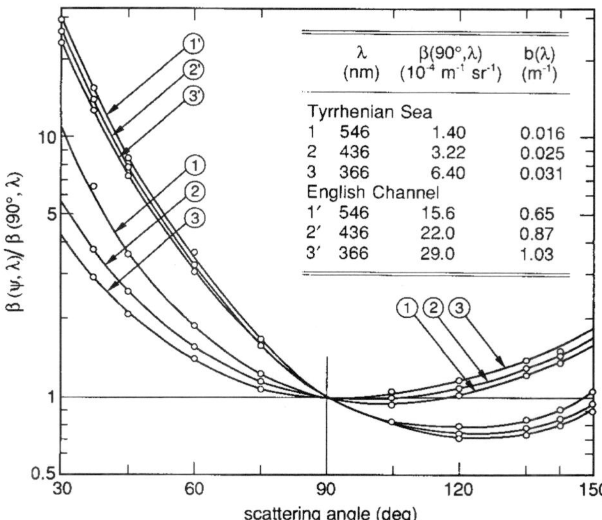

FIGURE 13 Wavelength dependence of total volume scattering functions measured in very clear (Tyrrhenian Sea) and in turbid (English Channel) waters. (*Redrawn from Morel.*[64])

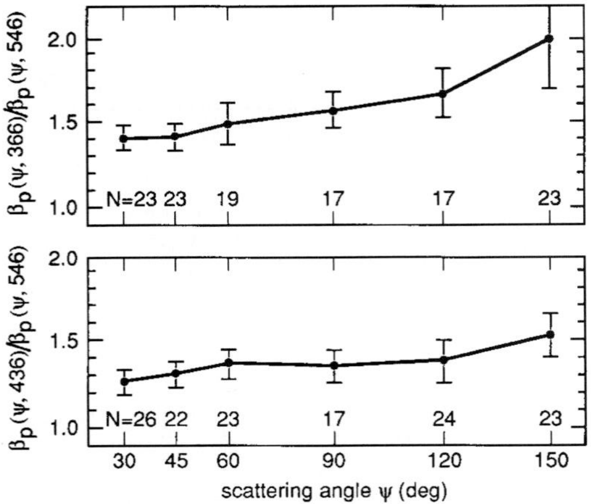

FIGURE 14 Wavelength dependence of particulate volume scattering functions. N is the number of samples. (*Redrawn from Morel.*[64])

wavelength is most nearly symmetric about $\psi = 90°$, presumably because symmetric molecular scattering is contributing relatively more to the total scattering at short wavelengths.

Figure 14 shows a systematic wavelength dependence of particle volume scattering functions. Figure 14*a* shows average values of $\beta_p(\psi, 366\text{ nm})/\beta_p(\psi, 546\text{ nm})$ for N samples as labeled in the figure. The vertical bars are one standard deviation of the observations. Figure 14*b* shows the ratio for $\lambda = 436$ to 546 nm. These ratios clearly depend both on wavelength and scattering angle. Assuming that $\beta_p(\psi, \lambda)$ has a wavelength dependence of

$$\beta_p(\psi, \lambda) = \beta_p(\psi, 546)\left(\frac{546\text{ nm}}{\lambda}\right)^n$$

the data of Fig. 14 imply values for n as seen in Table 14. As anticipated, the wavelength dependence is strongest for backscatter ($\psi = 150°$) and weakest for forward scatter ($\psi = 30°$).

Kopelevich[49,65] has statistically derived a two-parameter model for spectral volume scattering functions (VSFs). This model separates the contributions by "small" and "large" particles to the particulate scattering. Small particles are taken to be mineral particles less than 1 μm in size and having an index of refraction of $n = 1.15$; large particles are biologic

TABLE 14 Exponents n Required to Fit the Data of Fig. 14 Assuming That $\beta_p(\psi, \lambda) = \beta_p(\psi, 546)(546/\lambda)^n$

Wavelength λ (nm)	Scattering angle ψ		
	30°	90°	150°
366	0.84	1.13	1.73
436	0.99	1.33	1.89

particles larger than 1 μm in size and having an index refraction of $n = 1.03$. The model is defined by

$$\beta(\psi, \lambda) = \beta_w(\psi, \lambda) + v_s\beta_s^*(\psi)\left(\frac{550\text{ nm}}{\lambda}\right)^{1.7} + v_\ell\beta_\ell^*(\psi)\left(\frac{550\text{ nm}}{\lambda}\right)^{0.3} \tag{4}$$

with the following definitions:

- $\beta_w(\psi, \lambda)$ the VSF of pure sea water, given by Eq. (2) with $\lambda_0 = 550$ nm and an exponent of 4.30
- v_s the volume concentration of small particles, with units of cm^3 of particles per m^3 of water, i.e., parts per million (ppm)
- v_ℓ the analogous volume concentration of large particles
- $\beta_s^*(\psi)$ the small-particle VSF per unit volume concentration of small particles, with units of $m^{-1}\,sr^{-1}\,ppm^{-1}$
- $\beta_\ell^*(\psi)$ the analogous large-particle concentration-specific VSF

The concentration-specific VSFs for small and large particles are given in Table 15. Equation (4) can be evaluated as if the two parameters v_s and v_ℓ are known; the ranges of values for oceanic waters are $0.01 \le v_s \le 0.20$ ppm and $0.01 \le v_\ell \le 0.40$ ppm. However, these two parameters are themselves parametrized in terms of the total volume scattering function measured at $\lambda = 550$ nm for $\psi = 1$ and 45°:

$$\begin{aligned} v_s &= -1.4 \times 10^{-4}\beta(1°, 550\text{ nm}) + 10.2\beta(45°, 550\text{ nm}) - 0.002 \\ v_\ell &= 2.2 \times 10^{-2}\beta(1°, 550\text{ nm}) - 1.2\beta(45°, 550\text{ nm}) \end{aligned} \tag{5}$$

Thus $\beta(\psi, \lambda)$ can also be determined from two measurements of the total VSF.

The mathematical form of the Kopelevich model reveals its underlying physics. Large particles give diffractive scattering at very small angles; thus $\beta_\ell^*(\psi)$ is highly peaked for small ψ and the wavelength dependence of the large particle term is weak ($\lambda^{-0.3}$). Small

TABLE 15 The Concentration-Specific Volume Scattering Functions for Small (β_s^*) and Large (β_ℓ^*) Particles As a Function of the Scattering Angle ψ for Use in the Kopelevich Model for Spectral Volume Scattering Functions, Eq. (4)*

ψ (deg)	β_s^* $\left(\frac{m^{-1}\,sr^{-1}}{ppm}\right)$	β_ℓ^* $\left(\frac{n^{-1}\,sr^{-1}}{ppm}\right)$	ψ (deg)	β_s^* $\left(\frac{m^{-1}\,sr^{-1}}{ppm}\right)$	β_ℓ^* $\left(\frac{m^{-1}\,sr^{-1}}{ppm}\right)$
0	5.3	140	45	9.8×10^{-2}	6.2×10^{-4}
0.5	5.3	98	60	4.1	3.8
1	5.2	46	75	2.0	2.0
1.5	5.2	26	90	1.2	6.3×10^{-5}
2	5.1	15	105	8.6×10^{-3}	4.4
4	4.6	3.6	120	7.4	2.9
6	3.9	1.1	135	7.4	2.0
10	2.5	0.20	150	7.5	2.0
15	1.3	5.0×10^{-2}	180	8.1	7.0
30	0.29	2.8×10^{-3}	$b^* =$	$1.34\ m^{-1}/ppm$	$0.312\ m^{-1}/ppm$

* Reproduced from Kopelevich.[49]

particles contribute more to scattering at large angles and thus have a more symmetric VSF and a stronger wavelength dependence ($\lambda^{-1.7}$). This model gives a reasonably good description of VSFs observed in a variety of waters (Shifrin,[58] fig. 5.20).

Several simple models are available for the scattering coefficient $b(\lambda)$. A commonly employed bio-optical model for $b(\lambda)$ is that of Gordon and Morel[66] (see also Ref. 6):

$$b(\lambda) = b_w(\lambda) + \left(\frac{550\ \text{nm}}{\lambda}\right)0.30C^{0.62} \qquad \text{m}^{-1} \tag{6}$$

Here $b_w(\lambda)$ is given by Eq. (3) and Table 10. λ is in nm and C is the chlorophyll concentration in mg m^{-3}. A related bio-optical model for the backscatter coefficient $b_b(\lambda)$ is found in Morel[45]:

$$b_b(\lambda) = \tfrac{1}{2}b_w(\lambda) + \left[0.002 + 0.02(\tfrac{1}{2} - \tfrac{1}{4}\log C)\left(\frac{550\ \text{nm}}{\lambda}\right)\right]0.30C^{0.62}$$

The $(\frac{1}{2} - \frac{1}{4}\log C)$ factor gives $b_b(\lambda)$ a λ^{-1} wavelength dependence in very clear ($C = 0.01$ mg m^{-1}) water and no wavelength dependence in very turbid ($C = 100$ mg m^{-3}) water. These empirically derived models are intended for use only in case 1 waters.

A feeling for the accuracy of the $b(\lambda)$ model of Eq. (6) can be obtained from Fig. 15, which plots measured $b(550\ \text{nm})$ values versus chlorophyll concentration C in both case 1 and case 2 waters. Note that even when the model is applied to the case 1 waters from which it was derived, the predicated $b(550\ \text{nm})$ value easily can be wrong by a factor of 2. If the model is misapplied to case 2 waters, the error can be an order of magnitude. Note that for a given C value, $b(550\ \text{nm})$ is higher in case 2 waters than in case 1 waters, presumably because of the presence of additional particles that do not contain chlorophyll.

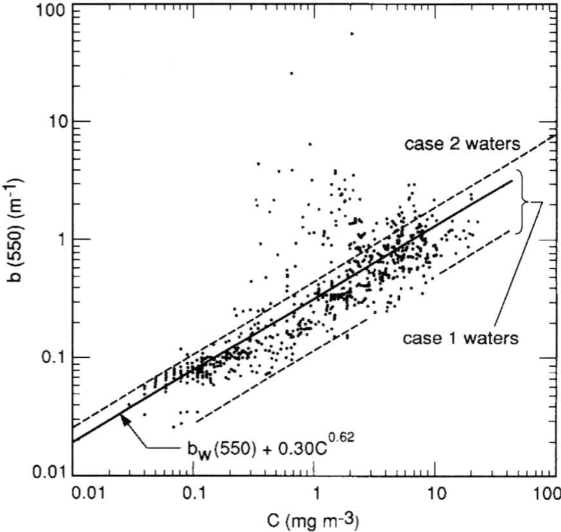

FIGURE 15 Measured scattering coefficients at $\lambda = 550$ nm, $b(550)$, as a function of chlorophyll concentration C. Case 1 waters lie between the dashed lines. Case 2 waters lie above the upper dashed line, which is defined by $b(550) = 0.45C^{0.62}$. The solid line is the model of Eq. (6). (*Redrawn from Gordon and Morel,*[66] *with permission.*)

Integration over ψ of the Kopelevich $b(\psi, \lambda)$ model of Eq. (4) yields another model for $b(\lambda)$:

$$b(\lambda) = 0.0017\left(\frac{550\text{ nm}}{\lambda}\right)^{4.3} + 1.34v_s\left(\frac{550\text{ nm}}{\lambda}\right)^{1.7} + 0.312v_\ell\left(\frac{550\text{ nm}}{\lambda}\right)^{0.3} \qquad \text{m}^{-1}$$

where v_s and v_ℓ are given by Eq. (5). Kopelevich claims that the accuracy of this model is ~30 percent.

A bio-optical model related to the Kopelevich model is found in Haltrin and Kattawar[50] (their notation):

$$b(\lambda) = b_w(\lambda) + b^0_{ps}(\lambda)P_s + b^0_{p\ell}(\lambda)P_\ell$$

Here $b_w(\lambda)$ is given by

$$b_w(\lambda) = 5.826 \times 10^{-3}\left(\frac{400}{\lambda}\right)^{4.322}$$

which is essentially the same as Eq. (3) and the data in Table 10. The terms $b^0_{ps}(\lambda)$ and $b^0_{p\ell}(\lambda)$ are the specific scattering coefficients for small and large particles, respectively, and are given by

$$b^0_{ps}(\lambda) = 1.1513\left(\frac{400}{\lambda}\right)^{1.7} \qquad \text{m}^2\text{ g}^{-1}$$

$$b^0_{p\ell}(\lambda) = 0.3411\left(\frac{400}{\lambda}\right)^{0.3} \qquad \text{m}^2\text{ g}^{-1}$$

P_s and P_ℓ are the concentrations in g m^{-3} of small and large particles, respectively. These quantities are parametrized in terms of the chlorophyll concentration C, as shown in Table 16. This work also presents a model for backscattering:

$$b_b(\lambda) = \tfrac{1}{2}b_w(\lambda) + B_s b^0_{ps}(\lambda)P_s + B_\ell b^0_{p\ell}(\lambda)\mathrm{P}_\ell$$

Here $B_s = 0.039$ is the backscattering probability for small particles and $B_\ell = 0.00064$ is the backscattering probability for large particles.

The bio-optical models for scattering just discussed are useful but very approximate.

TABLE 16 Parameterization of Small (P_s) and Large (P_ℓ) Particle Concentrations in Terms of the Chlorophyll Concentration C for use in the Kopelevich-Haltrin-Kattawar Models for $b(\lambda)$ and $b_b(\lambda)$*

C (mg m^{-3})	P_s (g m^{-3})	P_ℓ (g m^{-3})
0.00	0.000	0.000
0.03	0.001	0.035
0.05	0.002	0.051
0.12	0.004	0.098
0.30	0.009	0.194
0.60	0.016	0.325
1.00	0.024	0.476
3.00	0.062	1.078

* Reproduced from Haltrin and Kattawar,[50] with permission.

The reason for the frequent large discrepancies between model predictions and measured reality likely lies in the fact that scattering depends not just on particle concentration (as parameterized in terms of chlorophyll concentration), but also on the particle index of refraction and on the details of the particle size distribution which are not well parameterized in terms of the chlorophyll concentration alone. Whether or not the Kopelevich model or its derivative Haltrin-Kattawar form which partition the scattering into large and small particle components is in some sense better than the Gordon-Morel model is not known at present. Another consequence of the complexity of scattering is seen in the next section.

43.20 BEAM ATTENUATION

The spectral beam attenuation coefficient $c(\lambda)$ is just the sum of the spectral absorption and scattering coefficients: $c(\lambda) = a(\lambda) + b(\lambda)$. Since both $a(\lambda)$ and $b(\lambda)$ are highly variable functions of the nature and concentration of the constituents of natural waters so is $c(\lambda)$. Beam attenuation near $\lambda = 660$ nm is the only inherent optical property of water that is easily, accurately, and routinely measured. This wavelength is used both for engineering reasons (the availability of a stable LED light source) and because absorption by yellow matter is negligible in the red. Thus the quantity

$$c_p(660\text{ nm}) \equiv c(660\text{ nm}) - a_w(660\text{ nm}) - b_w(660\text{ nm}) \equiv c(660\text{ nm}) - c_w(660\text{ nm})$$

is determined by the nature of the suspended particulate matter. The particulate beam attenuation $c_p(660\text{ nm})$ is highly correlated with total particle volume concentration (usually expressed in parts per million), but it is much less well correlated with chlorophyll concentration.[67] The particulate beam attenuation can be used to estimate the total particulate load (often expressed as g m^{-3}).[68] However, the dependence of the particulate beam attenuation on particle properties is not simple. Spinrad[69] used Mie theory to calculate the dependence of the volume-specific particulate beam attenuation (particulate beam attenuation coefficient c_p in m^{-1} per unit suspended particulate volume in parts per million) on the relative refractive index and on the slope s of an assumed Junge size distribution for particles in the size range 1–80 μm; the result is shown in Fig. 16.

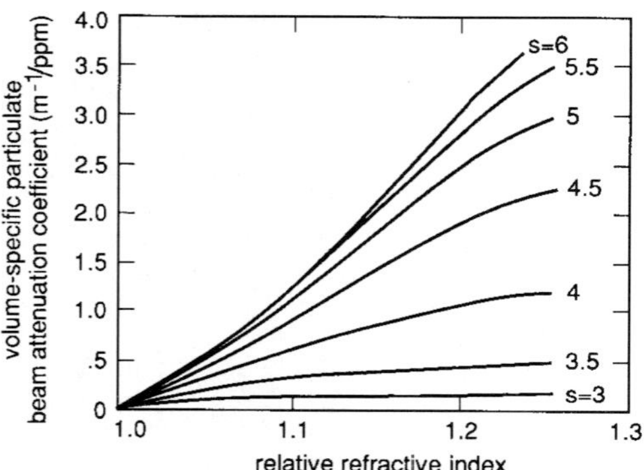

FIGURE 16 Computed relationship between volume-specific particulate beam attenuation coefficient, relative refractive index, and slope s of a Junge number size distribution. (*Reproduced from Spinrad,[69] with permission.*)

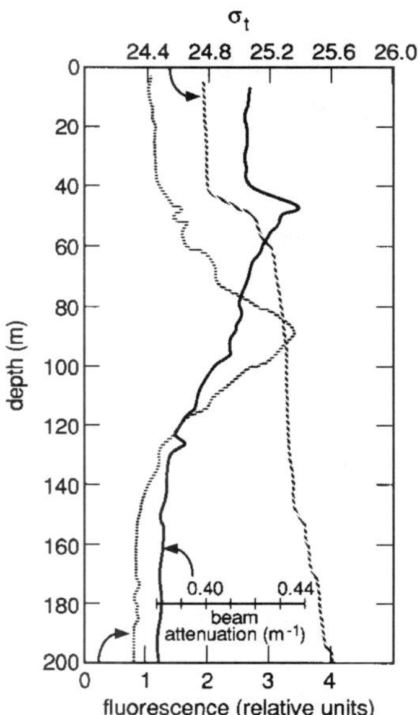

FIGURE 17 Example from the Pacific Ocean water of the depth dependence of beam attenuation (solid line), water density (σ_t, dashed line), and chlorophyll concentration (fluorescence, dotted line). (*Reproduced from Kitchen and Zaneveld,*[70] *with permission.*)

Although the details of the figure are sensitive to the choice of upper and lower size limits in the Mie calculations, the qualitative behavior of the curves is generally valid and supports the statements made in the closing paragraph of Sec. 43.19.

Because of the complicated dependence of scattering and hence of beam attenuation on particle properties, the construction of bio-optical models for $c(\lambda)$ is not easy. The reason is that chlorophyll concentration alone is not sufficient to parametrize scattering.[70] Figure 17 illustrates this insufficiency. The figure plots vertical profiles of $c(665\text{ nm})$, water density (proportional to the oceanographic variable σ_t), and chlorophyll concentration (proportional to fluorescence by chlorophyll and related pigments). Note that the maximum in beam attenuation at 46 m depth coincides with the interface (pycnocline) between less dense water above and more dense water below. Peaks in beam attenuation are commonly observed at density interfaces because particle concentrations are often greatest there. The maximum in chlorophyll concentration occurs at a depth of 87 m. The chlorophyll concentration depends not just on the number or volume of chlorophyll-bearing particles but also on their photoadaptive state, which depends on nutrient availability and ambient lighting. Thus chlorophyll concentration cannot be expected to correlate well with total scattering or with particulate beam attenuation $c_p(\lambda)$.

Voss[71] has developed an empirical model for $c(\lambda)$ given a measurement of c at $\lambda = 490$ nm:

$$c(\lambda) = c_w(\lambda) + [c(490\text{ nm}) - c_w(490\text{ nm})][1.563 - 1.149 \times 10^{-3}\lambda]$$

where λ is in nm and c is in m^{-1}. The attenuation coefficient for pure sea water, $c_w = a_w + b_w$, is given by the Smith-Baker data of Table 6. This model was statistically

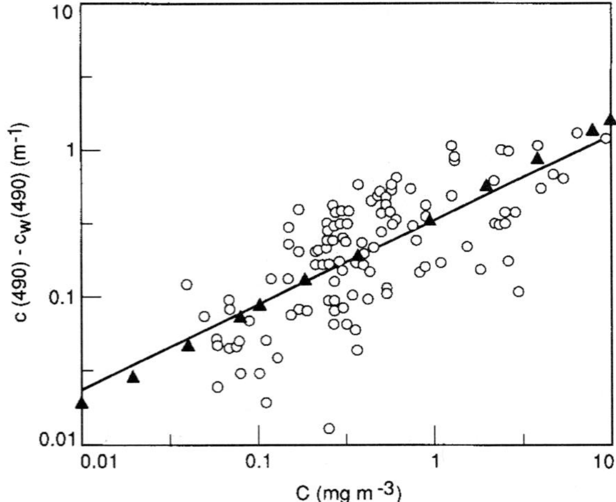

FIGURE 18 Particulate beam attenuation at 490 nm (open circles) as a function of chlorophyll concentration C as used to determine Eq. (7) which is given by the solid line. Solid triangles give values as predicted by the sum of Eqs. (1) and (6). (*Redrawn from Voss,*[71] *with permission.*)

derived from data of global extent. Testing of the model with independent data usually gave errors of less than 5 percent, although occasional errors of ~20 percent were found.

Voss also determined a least-squares fit of $c(490)$ to the chlorophyll concentration. The result

$$c(490\text{ nm}) = 0.39C^{0.57} \tag{7}$$

is similar in form to the chlorophyll dependence of the $a(\lambda)$ and $b(\lambda)$ models seen in Eqs. (1) and (6), respectively. Figure 18 shows the spread of the data points used to determine Eq. (7). Note that for a given value of C there is an order-of-magnitude spread in values of $c(490\text{ nm})$. The user of Eq. (7) or of the models for $b(\lambda)$ must always keep in mind that large deviations from the predicted values will be found in natural waters.

43.21 DIFFUSE ATTENUATION AND JERLOV WATER TYPES

As seen in Fig. 1 and in Table 1 there is a so-called diffuse attenuation coefficient for any radiometric variable. The most commonly used diffuse attenuation coefficients are those for downwelling plane irradiance, $K_d(z, \lambda)$, and for PAR, $K_{PAR}(z)$. Although the various diffuse attenuation coefficients are conceptually distinct, in practice they are often numerically similar and they all asymptotically approach a common value at great depths in homogeneous water.[2] The monograph by Tyler and Smith[72] gives tabulations and plots of $E_d(z, \lambda)$, $E_u(z, \lambda)$ and the associated $K_d(z, \lambda)$, $K_u(z, \lambda)$ and $R(z, \lambda)$ measured in a variety of waters.

Observation shows that $K_d(z, \lambda)$ varies systematically with wavelength over a wide range of waters from very clear to very turbid. Moreover, $K_d(z, \lambda)$ is often rather insensitive to environmental effects[73] except for extreme conditions[74] (such as the sun within 10° of the horizon) and in most cases correction can be made[11] for the environmental effects that are present in K_d. K_d therefore is regarded as a quasi-inherent optical property whose variability is governed primarily by changes in the inherent optical properties of the water body and not by changes in the external environment.

Jerlov[29] exploited this benign behavior of K_d to develop a frequently used classification scheme for oceanic waters based on the spectral shape of K_d. The *Jerlov water types* are in essence a classification based on water clarity as quantified by $K_d(z_s, \lambda)$ where z_s is a depth just below the sea surface. This classification scheme can be contrasted with the case 1 and case 2 classification described earlier, which is based on the nature of the suspended matter within the water. The Jerlov water types are numbered I, IA, IB, II, and III for open ocean waters, and 1 through 9 for coastal waters. Type I is the clearest and type III is the most turbid open ocean water. Likewise, for coastal waters type 1 is clearest and type 9 is most turbid. The Jerlov types I–III generally correspond to case 1 water since phytoplankton predominate in the open ocean. Types 1–9 correspond to case 2 waters where yellow matter and terrigenous particulates dominate the optical properties. A rough correspondence between chlorophyll concentration and Jerlov oceanic water type is given by[45]

$$\begin{array}{ll} C: & 0\text{–}0.01 \sim 0.05 \sim 0.1 \sim 0.5 \sim 1.5\text{–}2.0\,\text{mg m}^{-3} \\ \text{water type:} & \quad\text{I} \qquad \text{IA} \quad \text{IB} \quad \text{II} \qquad \text{III} \end{array}$$

Austin and Petzold[75] reevaluated the Jerlov classification using an expanded database and slightly revised the $K_d(\lambda)$ values used by Jerlov in his original definition of the water types. Table 17 gives the revised values for $K_d(\lambda)$ for the water types commonly encountered in oceanography. These values are recommended over those found in

TABLE 17 Downwelling Irradiance Diffuse Attenuation Coefficients $K_d(\lambda)$ Used to Define the Jerlov Water Types As Determined by Austin and Petzold* (All quantities in the body of the Table have units of m^{-1}.)

λ	Jerlov water type					
(nm)	I	IA	IB	II	III	1
350	0.0510	0.0632	0.0782	0.1325	0.2335	0.3345
375	0.0302	0.0412	0.0546	0.1031	0.1935	0.2839
400	0.0217	0.0316	0.0438	0.0878	0.1697	0.2516
425	0.0185	0.0280	0.0395	0.0814	0.1594	0.2374
450	0.0176	0.0257	0.0355	0.0714	0.1381	0.2048
475	0.0184	0.0250	0.0330	0.0620	0.1160	0.1700
500	0.0280	0.0332	0.0396	0.0627	0.1056	0.1486
525	0.0504	0.0545	0.0596	0.0779	0.1120	0.1461
550	0.0640	0.0674	0.0715	0.0863	0.1139	0.1415
575	0.0931	0.0960	0.0995	0.1122	0.1359	0.1596
600	0.2408	0.2437	0.2471	0.2595	0.2826	0.3057
625	0.3174	0.3206	0.3245	0.3389	0.3655	0.3922
650	0.3559	0.3601	0.3652	0.3837	0.4181	0.4525
675	0.4372	0.4410	0.4457	0.4626	0.4942	0.5257
700	0.6513	0.6530	0.6550	0.6623	0.6760	0.6896

* Reproduced from Austin and Petzold[75] with permission.

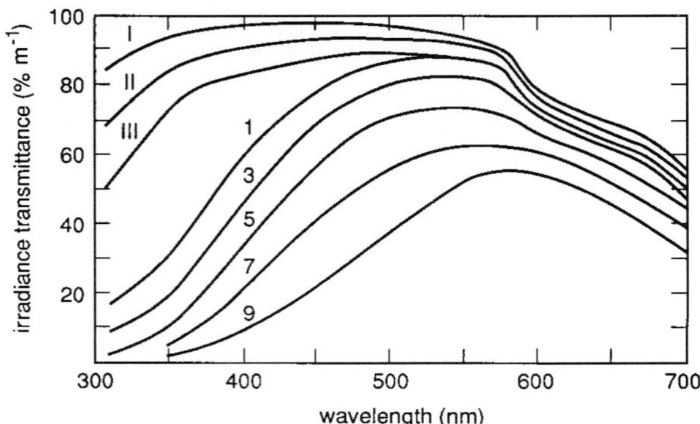

FIGURE 19 Percentage transmittance per meter of water of downwelling irradiance E_d as a function of wavelength for selected Jerlov water types. (*Reproduced from Jerlov,*[29] with permission.)

Jerlov.[29] Figure 19 shows the percent transmittance of $E_d(\lambda)$ per meter of water for selected Jerlov water types. Note how the wavelength of maximum transmittance shifts from blue in the clearest open ocean water (type I) to green (types III and 1) to yellow in the most turbid, yellow-matter-rich coastal water (type 9).

Austin and Petzold also presented a simple model that allows the determination of $K_d(\lambda)$ at all wavelengths from a value of K_d measured at any single wavelength. This model is defined by

$$K_d(\lambda) = \frac{M(\lambda)}{M(\lambda_0)} [K_d(\lambda_0) - K_{dw}(\lambda_0)] + K_{dw}(\lambda)$$

Here λ_0 is the wavelength at which K_d is measured and K_{dw} refers to values for pure sea water. $K_{dw}(\lambda)$ and the statistically derived coefficients $M(\lambda)$ are given in Table 18. (These

TABLE 18 Values of the Coefficient $M(\lambda)$ and of the Downwelling Diffuse Attenuation Coefficient for Pure Sea Water, $K_{dw}(\lambda)$, for Use in the Austin and Petzold Model for $K_d(\lambda)$*

λ (nm)	M (m^{-1})	K_{dw} (m^{-1})	λ (nm)	M (m^{-1})	K_{dw} (m^{-1})	λ (nm)	M (m^{-1})	K_{dw} (m^{-1})
350	2.1442	0.0510	470	1.1982	0.0179	590	0.4840	0.1578
360	2.0504	0.0405	480	1.0955	0.0193	600	0.4903	0.2409
370	1.9610	0.0331	490	1.0000	0.0224	610	0.5090	0.2892
380	1.8772	0.0278	500	0.9118	0.0280	620	0.5380	0.3124
390	1.8009	0.0242	510	0.8310	0.0369	630	0.6231	0.3296
400	1.7383	0.0217	520	0.7578	0.0498	640	0.7001	0.3290
410	1.7591	0.0200	530	0.6924	0.0526	540	0.7300	0.3559
420	1.6974	0.0189	540	0.6350	0.0577	660	0.7301	0.4105
430	1.6108	0.0182	550	0.5860	0.0640	670	0.7008	0.4278
440	1.5169	0.0178	560	0.5457	0.0723	680	0.6245	0.4521
450	1.4158	0.0176	570	0.5146	0.0842	690	0.4901	0.5116
460	1.3077	0.0176	580	0.4935	1.1065	700	0.2891	0.6514

* Condensed with permission from Austin and Petzold,[75] who give values every 5 nm.

K_{dw} values differ slightly from those seen in Table 6.) This model is valid in waters where $K_d(490) \leq 0.16\ \text{m}^{-1}$ which corresponds to a chlorophyll concentration of $C \leq 3\ \text{mg m}^{-3}$.

Unlike the beam attenuation coefficient $c(\lambda)$, the diffuse attenuation $K_d(z, \lambda)$ is highly correlated with chlorophyll concentration. The reason is seen in the approximate formula[11]

$$K_d(\lambda) \approx \frac{a(\lambda) + b_b(\lambda)}{\cos \theta_{sw}}$$

where θ_{sw} is the solar angle measured within the water. Since $a(\lambda) \gg b_b(\lambda)$ for most waters, $K_d(\lambda)$ is largely determined by the absorption properties of the water, which are fairly well parametrized by the chlorophyll concentration. Beam attenuation on the other hand is proportional to the total scattering which is not well parametrized by chlorophyll concentration. Observations show[76] that the beam attenuation at 660 nm is not in general correlated with diffuse attenuation.

A bio-optical model for $K_d(\lambda)$ is given by Morel[45]:

$$K_d(\lambda) = K_{dw}(\lambda) + \chi(\lambda)C^{e(\lambda)}$$

Here $K_{dw}(\lambda)$ is the diffuse attenuation for pure sea water and $\chi(\lambda)$ and $e(\lambda)$ are statistically derived functions that convert the chlorophyll concentration C in mg m^{-3} into K_d values in m^{-1}. Table 19 gives the K_{dw}, χ and e values used in the Morel model. This model is applicable to case 1 waters with $C \leq 30\ \text{mg m}^{-3}$, although the χ and e values are somewhat uncertain for $\lambda > 650$ nm because of sparse data available for their determination. Some feeling for the accuracy of the Morel $K_d(\lambda)$ model can be obtained from Fig. 20 which shows predicted (the line) and observed $K_d(450)$ values as a function of C. Errors can be as large as a factor of 2 in case 1 waters (dots) and can be much larger if the model is misapplied to case 2 waters (open circles). The Morel model allows the determination of $K_d(\lambda)$ if C is measured; the Austin and Petzold model determines $K_d(\lambda)$ from a measurement at one wavelength.

TABLE 19 Values of the Coefficients $\chi(\lambda)$ and $e(\lambda)$ and of the Downwelling Diffuse Attenuation Coefficient for Pure Sea Water, $K_{dw}(\lambda)$, for Use in the Morel Model for $K_d(\lambda)$*

λ (nm)	$\chi(\lambda)$	$e(\lambda)$	$K_{dw}(\lambda)$ (m^{-1})	λ (nm)	$\chi(\lambda)$	$e(\lambda)$	$K_{dw}(\lambda)$ (m^{-1})
400	0.1100	0.668	0.0209	550	0.0410	0.650	0.0640
410	0.1125	0.680	0.0196	560	0.0390	0.640	0.0717
420	0.1126	0.693	0.0183	570	0.0360	0.623	0.0807
430	0.1078	0.707	0.0171	580	0.0330	0.610	0.1070
440	0.1041	0.707	0.0168	590	0.0325	0.618	0.1570
450	0.0971	0.701	0.0168	600	0.0340	0.626	0.2530
460	0.0896	0.700	0.0173	610	0.0360	0.634	0.2960
470	0.0823	0.703	0.0175	620	0.0385	0.642	0.3100
480	0.0746	0.703	0.0194	630	0.0420	0.653	0.3200
490	0.0690	0.702	0.0217	640	0.0440	0.663	0.3300
500	0.0636	0.700	0.0271	650	0.0450	0.672	0.3500
510	0.0578	0.690	0.0384	660	0.0475	0.682	0.4050
520	0.0498	0.680	0.0490	670	0.0515	0.695	0.4300
530	0.0467	0.670	0.0518	680	0.0505	0.693	0.4500
540	0.0440	0.660	0.0568	690	0.0390	0.640	0.5000
				700	0.0300	0.600	0.6500

* Condensed with permission from Morel,[45] who gives values every 5 nm.

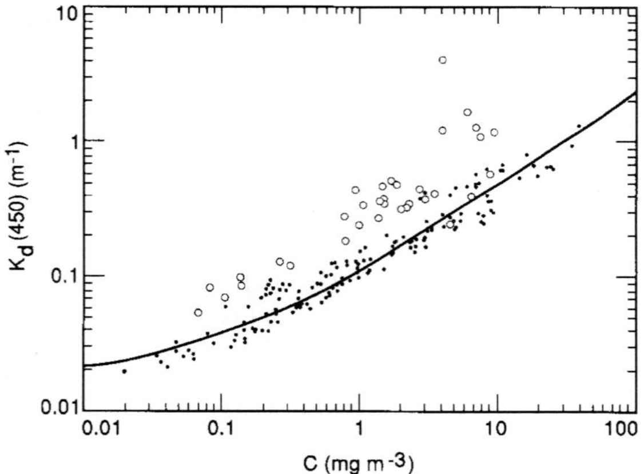

FIGURE 20 Measured K_d values at 450 nm as a function of chlorophyll concentration C. Dots are measurements from case 1 waters; open circles are from case 2 waters. The solid line gives $K_d(450)$ as predicted by the Morel bio-optical model. (*Redrawn from Morel,*[45] *with permission.*)

Morel[45] also presents a very simple bio-optical model for $\bar{K}_{\mathrm{PAR}}(0, z_{\mathrm{eu}})$ the value of $K_{\mathrm{PAR}}(z)$ averaged over the euphotic zone $0 \le z \le z_{\mathrm{eu}}$:

$$\bar{K}_{\mathrm{PAR}}(0, z_{\mathrm{eu}}) = 0.121 C^{0.428}$$

where C is the mean chlorophyll concentration in the euphotic zone in mg m^{-3} and $\bar{K}_{\mathrm{PAR}}$ is in m^{-1}. The euphotic zone is the region where there is sufficient light for photosynthesis to take place; it extends roughly to the depth where $E_{\mathrm{PAR}}(z)$ is 1 percent of its surface value (i.e., $E_{\mathrm{PAR}}(z_{\mathrm{eu}}) = 0.01\ E_{\mathrm{PAR}}(0)$). Table 20 gives z_{eu} as a function of C as determined by the Morel model.

43.22 IRRADIANCE REFLECTANCE AND REMOTE SENSING

The spectral irradiance reflectant $R(\lambda) \equiv E_u(\lambda)/E_d(\lambda)$ is an important apparent optical property. Measurements of $R(z, \lambda)$ within the water have been used[77] to estimate water quality parameters such as the chlorophyll concentration, the particle backscattering coefficient, and the absorption coefficient of yellow matter. More importantly, $R(\lambda)$ just below the water surface can be related to the radiance leaving the water[78]; this radiance is available for detection by aircraft- or satellite-borne instruments. Understanding the dependence of $R(\lambda)$ upon the constituents of natural waters is therefore one of the central problems in remote sensing of water bodies.

Figure 21 illustrates the variability of $R(\lambda)$ in natural waters. Figure 21*a* shows $R(\lambda)$ in percent for various case 1 waters. For low-chlorophyll concentrations $R(\lambda)$ is highest at blue wavelengths, hence the blue color of clean ocean water. As the chlorophyll concentration increases, the maximum in $R(\lambda)$ shifts to green wavelengths. The enhanced reflectance near $\lambda = 685$ nm is due to chlorophyll fluorescence. Also note the exceptionally

TABLE 20 Approximate Depth of the Euphotic Zone, z_{eu}, in Homogeneous Case 1 Water As a Function of Chlorophyll Concentration C.*

C ($mg\ m^{-3}$)	z_{eu} (m)	C ($mg\ m^{-3}$)	z_{eu} (m)
0.0	183	1	39
0.01	153	2	29
0.03	129	3	24
0.05	115	5	19
0.1	95	10	14
0.2	75	20	10
0.3	64	30	8
0.5	52		

* Data extracted from Morel,[45] with permission.

high values measured[20] within a coccolithophore bloom; $R(\lambda)$ is high there because of the strong scattering by the numerous calcite particles (see Sec. 43.6). Figure 21*b* shows $R(\lambda)$ from waters dominated by suspended sediments (i.e., by nonpigmented particles). For high-sediment concentrations $R(\lambda)$ is nearly flat from blue to yellow wavelengths, and therefore the water appears brown. Figure 21*c* is from waters with high concentrations of yellow substances; the peak in $R(\lambda)$ lies in the yellow. With good reason the term "ocean color" is often used as a synonym for $R(\lambda)$.

One of the main goals of oceanic remote sensing is the determination of chlorophyll concentrations in near-surface waters because of the fundamental role played by phytoplankton in the global ecosystem. Gordon et al.[78] define the *normalized water-leaving radiance* $[L_w(\lambda)]_N$ as the radiance that would leave the sea surface if the sun were at the zenith and the atmosphere were absent; this quantity is fundamental to remote sensing. They then show that $[L_w]_N$ is directly proportional to R and that R is proportional to $b_b/(a + b_b)$ (i.e., to b_b/K_d). Although a or K_d are reasonably well modeled in terms of

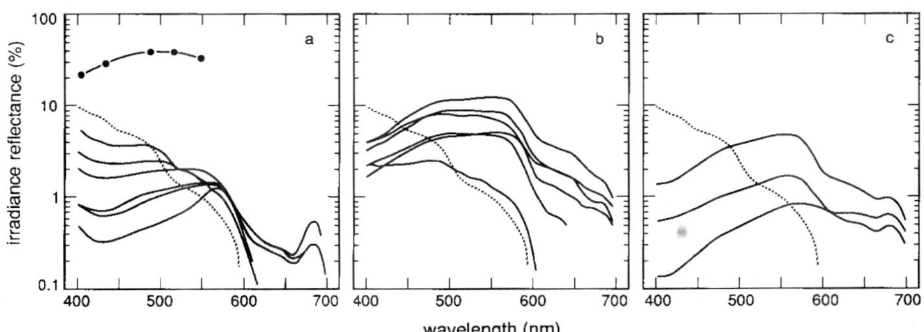

FIGURE 21 Measured spectral irradiance reflectances $R(\lambda)$ from various waters. Panel (*a*) is from case waters with different quantities of phytoplankton; the dotted line is $R(\lambda)$ for pure sea water. The heavy dots give values measured within a coccolithophore bloom.[20] Panel (*b*) is from case 2 waters dominated by suspended sediments and panel (*c*) is from case 2 waters dominated by yellow matter. (*Redrawn from Sathyendranath and Morel,*[79] *with permission.*)

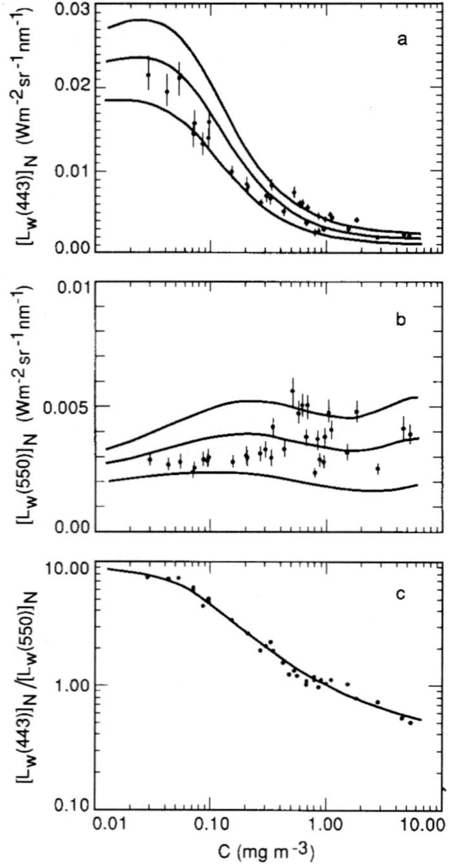

FIGURE 22 In panels (*a*) and (*b*) the solid lines are values of the normalized water-leaving radiances $[L_w(\lambda)]_N$ at $\lambda = 443$ and 550 nm, respectively, as predicted by various models that relate $[L_w(\lambda)]_N$ to the chlorophyll concentration *C*. The dots are measured values of $[L_w(\lambda)]_N$. Panel (*c*) shows the predicted (line) and observed ratio of the $[L_w(\lambda)]_N$ values of panels (*a*) and (*b*). (*Redrawn from Gordon, et al.*,[78] with permission.)

chlorophyll concentration *C* in case 1 waters, b_b is not well described in terms of *C*. Thus poor agreement is to be expected between observed values of $[L_w(\lambda)]_N$ and values predicted by a model parametrized in terms of *C*. This is indeed the case as is seen in Fig. 22*a* and *b* which shows observed and predicted $[L_w(443\,\text{nm})]_N$ and $[L_w(550\,\text{nm})]_N$ values as a function of chlorophyll concentration. Based on these figures there seems to be little hope of being able to reliably retrieve *C* from a remotely sensed $[L_w(\lambda)]_N$ value. However, in spite of the noise seen in Fig. 22*a* and 22*b*, the *ratios* of normalized water-leaving radiances for different wavelengths can be remarkably well-behaved functions of *C*. Figure 22*c* shows predicted (the line) and observed (dots) values of $[L_w(443\,\text{nm})]_N/[L_w(550\,\text{nm})]_N$; the agreement between prediction and observation is now rather good. Thus measurement of $[L_w(\lambda)]_N$ at two (carefully chosen) wavelengths along with application of a bio-optical model for their *ratio* can yield a useful estimate of chlorophyll concentration. Such models are the basis of much remote sensing.

43.23 INELASTIC SCATTERING AND POLARIZATION

Although the basic physics of inelastic scattering and polarization is well understood, only recently has it become computationally practicable to incorporate these effects into predictive numerical models of underwater radiance distributions. For this reason as well as because of the difficulty of making needed measurements, quantitative knowledge about the significance of inelastic scattering and polarization in the underwater environment is incomplete.

Inelastic scattering processes are often negligible in comparison to sunlight or artificial lights as sources of underwater light of a given wavelength. However, in certain circumstances transpectral scatter is the dominant source of underwater light at some wavelengths. Figure 23 illustrates just such a circumstance.

Figure 23 shows a measured depth profile from the Sargasso Sea of irradiance reflectance at $\lambda = 589$ nm (yellow-orange light); note that $R(589\text{ nm})$ increases with depth. Because of the fairly high absorption of water at this wavelength $[a_w(589\text{ nm}) = 0.152\text{ m}^{-1}]$ most of the yellow-orange component of the incident solar radiation is absorbed near the

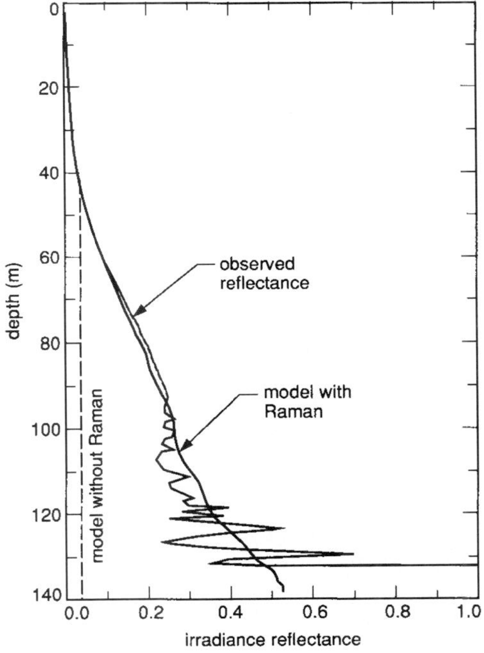

FIGURE 23 Observed irradiance reflectance R at 589 nm (light line) and values predicted by a model including Raman scattering (heavy line) and omitting Raman scattering (dashed line). (*Redrawn from Marshall and Smith,*[80] *with permission.*)

surface and monochromatic radiative transfer theory shows that the reflectance should approach a value of $R(589\,\text{nm}) \approx 0.04$ for the water body of Fig. 22. Calculations by Marshall and Smith[80] explain the paradox. Light of blue-green wavelengths ($\lambda \sim 500\,\text{nm}$) can penetrate to great depth in the clear Sargasso Sea water [$a_w(500\,\text{nm}) \approx 0.026\,\text{m}^{-1}$]. Some of this light is then Raman scattered from blue-green to yellow-orange wavelengths providing a source of yellow-orange light at depth. Moreover, since the phase function for Raman scattering is symmetric in the forward and backward hemispheres Raman scattered photons are equally likely to be heading upward [and thus contribute to $E_u(589\,\text{nm})$] or downward [and thus contribute to $E_d(589\,\text{nm})$] even though most of the blue-green light at depth is heading downward [e.g., $E_d(500\,\text{nm}) \gg E_u(500\,\text{nm})$]. Thus as depth increases and Raman scattering becomes increasingly important relative to transmitted sunlight as a source of ambient yellow-orange light, $E_u(589\,\text{nm})$ and $E_d(589\,\text{nm})$ become more nearly equal and the irradiance reflectance $R(589\,\text{nm})$ increases. Such an increase is not seen at blue-green wavelengths since E_d transmitted from the surface remains much greater than E_u at great depths. Since Raman scattering is by the water molecules themselves this process is present (and indeed relatively more important) even in the clearest waters. Another oceanographic effect of Raman scattering occurs in the filling of Fraunhofer lines in the solar spectrum as seen underwater; this matter is just now coming under detailed investigation.[81]

Fluorescence by chlorophyll or other substances can be significant if the fluorescing material is present in sufficient quantity. Chlorophyll fluoresces strongly near $\lambda = 685$ nm; this source of red light is responsible[82] for the enhanced reflectance near 685 nm noted in Fig. 21*a*. The spectral signature of fluorescence is a useful tool for analyzing many of the constituents of natural waters.[83]

Relatively little attention has been paid to the state of polarization of underwater light fields.[84] Some use of polarized light has been made in enhancing underwater visibility[85] and it is well established that many oceanic organisms sense polarized light when navigating.[86] Voss and Fry[87] measured the Mueller matrix for ocean water and Quinby-Hunt et al.[88] have studied the propensity of certain phytoplankton to induce circular polarization in unpolarized or linearly polarized light. Kattawar and Adams[89] have shown that errors of up to 15 percent can occur in calculations of underwater radiance if scalar (unpolarized) radiative transfer theory is used instead of vector (polarized) theory.

43.24 ACKNOWLEDGMENTS

This paper was written while the author held a National Research Council Resident Research Associateship (Senior Level) at the Jet Propulsion Laboratory, California Institute of Technology. This associateship was supported by the Ocean Biochemistry Program at NASA Headquarters. Final proofing was supported by SRI. Karen Baker, Annick Bricaud, Howard Gordon, Richard Honey, Rodolpho Iturriaga, George Kattawar, Scott Pegau, Mary Jane Perry, Collin Roesler, Shubha Sathyendranath, Richard Spinrad, and Kenneth Voss all made helpful comments on a draft of the paper; their efforts are greatly appreciated.

43.25 REFERENCES

1. C. D. Mobley, *Light and Water: Radiative Transfer in Natural Waters,* Academic Press, San Diego, 1994, 592 pp.
2. R. W. Preisendorfer, *Hydrologic Optics,* 6 volumes, U.S. Dept. of Commerce, NOAA, Pacific Marine Environmental Lab., Seattle, 1976, 1757 pp. Available from National Technical Information Service, 5285 Port Royal Road, Springfield, Virginia 22161.

3. A. Morel and R. C. Smith, "Terminology and Units in Optical Oceanography," *Marine Geodesy* **5**(4):335 (1982).

4. N. Højerslev, "A Spectral Light Absorption Meter for Measurements in the Sea," *Limnol. Oceanogr.* **20**(6):1024 (1975).

5. A. Morel and R. C. Smith, "Relation Between Total Quanta and Total Energy for Aquatic Photosynthesis," *Limnol. Oceanogr.* **19**(4):591 (1974).

6. A. Morel, "Light and Marine Photosynthesis: A Spectral Model with Geochemical and Climatological Implications," *Prog. Oceanogr.* **26:**263 (1991).

7. J. T. O. Kirk, "The Upwelling Light Stream in Natural Waters," *Limnol. Oceanogr.* **34**(8):1410 (1989).

8. J. T. O. Kirk, *Light and Photosynthesis in Aquatic Ecosystems,* Cambridge Univ. Press, New York, 1983, 410 pp.

9. R. W. Preisendorfer and C. D. Mobley, "Theory of Fluorescent Irradiance Fields in Natural Waters," *J. Geophys. Res.* **93**(D9):10831 (1988).

10. R. C. Smith and K. Baker, "The Bio-Optical State of Ocean Waters and Remote Sensing," *Limnol. Oceanogr.* **23**(2):247 (1978).

11. H. Gordon, "Can the Lambert-Beer Law be Applied to the Diffuse Attenuation Coefficient of Ocean Water?," *Limnol. Oceanogr.* **34**(8):1389 (1989).

12. C. A. Suttle, A. M. Chan, and M. T. Cottrell, "Infection of Phytoplankton by Viruses and Reduction of Primary Productivity," *Nature* **347:**467 (1990).

13. I. Koike, S. Hara, K. Terauchi, and K. Kogure, "Role of Submicrometer Particles in the Ocean," *Nature* **345:**242 (1990).

14. M. L. Wells and E. D. Goldberg, "Occurrence of Small Colloids in Sea Water," *Nature* **353:**342 (1991).

15. R. W. Spinrad, H. Glover, B. B. Ward, L. A. Codispoti, and G. Kullenberg, "Suspended Particle and Bacterial Maxima in Peruvian Coastal Water During a Cold Water Anomaly," *Deep-Sea Res.* **36**(5):715 (1989).

16. A. Morel and Y.-H. Ahn, "Optical Efficiency Factors of Free-Living Marine Bacteria: Influence of Bacterioplankton upon the Optical Properties and Particulate Organic Carbon in Oceanic Waters," *J. Marine Res.* **48:**145 (1990).

17. D. Stramski and D. A. Kiefer, "Light Scattering by Microorganisms in the Open Ocean," *Prog. Oceanogr.* **28:**343 (1991).

18. A. Alldredge and M. W. Silver, "Characteristics, Dynamics and Significance of Marine Snow," *Prog. Oceanogr.* **20:**41 (1988).

19. K. L. Carder, R. G. Steward, P. R. Betzer, D. L. Johnson, and J. M. Prospero, "Dynamics and Composition of Particles from an Aeolian Input Event to the Sargasso Sea," *J. Geophys. Res.* **91**(D1):1055 (1986).

20. W. M. Balch, P. M. Holligan, S. G. Ackleson, and K. J. Voss, "Biological and Optical Properties of Mesoscale Coccolithophore Blooms in the Gulf of Maine," *Limnol. Oceanogr.* **36**(4):629 (1991).

21. H. Bader, "The Hyperbolic Distribution of Particle Sizes," *J. Geophys. Res.* **75**(15):2822 (1970).

22. I. N. McCave, "Particulate Size Spectra, Behavior, and Origin of Nepheloid Layers over the Nova Scotian Continental Rise," *J. Geophys. Res.* **88**(C12):7647 (1983).

23. C. E. Lambert, C. Jehanno, N. Silverberg, J. C. Brun-Cottan, and R. Chesselet, "Log-Normal Distributions of Suspended Particles in the Open Ocean," *J. Marine Res.* **39**(1):77 (1981).

24. D. G. Archer and P. Wang, "The Dielectric Constant of Water and Debye-Hückel Limiting Law Slopes," *J. Phys. Chem. Ref. Data* **19:**371 (1990).

25. M. Kerker, *The Scattering of Light and Other Electromagnetic Radiation,* Academic Press, New York, 1969, 666 pp.

26. V. M. Zoloratev and A. V. Demin, "Optical Constants of Water over a Broad Range of Wavelengths, 0.1 Å–1 m," *Opt. Spectrosc.* (U.S.S.R.) **43**(2):157 (Aug. 1977).

27. R. W. Austin and G. Halikas, "The Index of Refraction of Seawater," SIO ref. no. 76-1, Scripps Inst. Oceanogr., San Diego, 1976, 121 pp.
28. R. C. Millard and G. Seaver, "An Index of Refraction Algorithm over Temperature, Pressure, Salinity, Density, and Wavelength," *Deep-Sea Res.* **37**(12):1909 (1990).
29. N. G. Jerlov, *Marine Optics,* Elsevier, Amsterdam, 1976, 231 pp.
30. R. W. Spinrad and J. F. Brown, "Relative Real Refractive Index of Marine Micro-Organisms: A Technique for Flow Cytometric Measurement," *Appl. Optics* **25**(2):1930 (1986).
31. S. G. Ackleson, R. W. Spinrad, C. M. Yentsch, J. Brown, and W. Korjeff-Bellows, "Phytoplankton Optical Properties: Flow Cytometric Examinations of Dilution-Induced Effects," *Appl. Optics* **27**(7): 1262 (1988).
32. R. C. Smith and K. S. Baker, "Optical Properties of the Clearest Natural Waters," *Appl. Optics* **20**(2):177 (1981).
33. T. T. Bannister, "Estimation of Absorption Coefficients of Scattering Suspensions Using Opal Glass," *Limnol. Oceanogr.* **33**(4, part 1):607 (1988).
34. B. G. Mitchell, "Algorithms for Determining the Absorption Coefficient of Aquatic Particles Using the Quantitative Filter Technique (QFT)," *Ocean Optics X,* R. W. Spinrad (ed.), *Proc. SPIE* **1302:**137 (1990).
35. D. Stramski, "Artifacts in Measuring Absorption Spectra of Phytoplankton Collected on a Filter," *Limnol. Oceanogr.* **35**(8):1804 (1990).
36. J. R. V. Zaneveld, R. Bartz, and J. C. Kitchen, "Reflective-Tube Absorption Meter," *Ocean Optics X,* R. W. Spinrad (ed.), *Proc. SPIE* **1302:**124 (1990).
37. E. S. Fry, G. W. Kattawar, and R. M. Pope, "Integrating Cavity Absorption Meter," *Appl. Optics* **31**(12):2025 (1992).
38. W. Doss and W. Wells, "Radiometer for Light in the Sea," *Ocean Optics X,* R. W. Spinrad (ed.), *Proc. SPIE* **1302:**363 (1990).
39. K. J. Voss, "Use of the Radiance Distribution to Measure the Optical Absorption Coefficient in the Ocean," *Limnol. Oceanogr.* **34**(8):1614 (1989).
40. F. Sogandares, Z.-F. Qi, and E. S. Fry, "Spectral Absorption of Water," presentation at the Optical Society of America Annual Meeting, San Jose, Calif., 1991.
41. W. S. Pegau and J. R. V. Zaneveld, "Temperature Dependent Absorption of Water in the Red and Near Infrared Portions of the Spectrum," *Limnol. Oceanogr.* **38**(1):188 (1993).
42. A. Bricaud, A. Morel, and L. Prieur, "Absorption by Dissolved Organic Matter of the Sea (Yellow Substance) in the UV and Visible Domains," *Limnol. Oceanogr.* **26**(1):43 (1981).
43. C. S. Roesler, M. J. Perry, and K. L. Carder, "Modeling in situ Phytoplankton Absorption from Total Absorption Spectra in Productive Inland Marine Waters," *Limnol. Oceanogr.* **34**(8):1510 (1989).
44. S. Sathyendranath, L. Lazzara, and L. Prieur, "Variations in the Spectral Values of Specific Absorption of Phytoplankton," *Limnol. Oceanogr.* **32**(2):403 (1987).
45. A Morel, "Optical Modeling of the Upper Ocean in Relation to Its Biogenous Matter Content (Case 1 Waters)," *J. Geophys. Res.* **93**(C9):10749 (1988).
46. R. Iturriaga and D. Siegel, "Microphotometric Characterization of Phytoplankton and Detrital Absorption Properties in the Sargasso Sea," *Limnol. Oceanogr.* **34**(8):1706 (1989).
47. A. Morel and L. Prieur, "Analysis of Variations in Ocean Color," *Limnol. Oceanogr.* **22**(4):709 (1977).
48. L. Prieur and S. Sathyendranath, "An Optical Classification of Coastal and Oceanic Waters Based on the Specific Spectral Absorption Curves of Phytoplankton Pigments, Dissolved Organic Matter, and Other Particulate Materials," *Limnol. Oceanogr.* **26**(4):671 (1981).
49. O. V. Kopelevich, "Small-Parameter Model of Optical Properties of Sea Water," *Ocean Optics,* vol 1, *Physical Ocean Optics,* A. S. Monin (ed.), Nauka Pub., Moscow, 1983, chap. 8 (in Russian).
50. V. I. Haltrin and G. Kattawar, "Light Fields with Raman Scattering and Fluorescence in Sea Water," Tech. Rept., Dept. of Physics, Texas A&M Univ., College Station, 1991, 74 pp.

51. C. S. Yentsch, "The Influence of Phytoplankton Pigments on the Color of Sea Water," *Deep-Sea Res.* **7:**1 (1960).

52. H. Gordon, "Diffuse Reflectance of the Ocean: Influence of Nonuniform Pigment Profile," *Appl. Optics* **31**(12):2116 (1992).

53. T. J. Petzold, "Volume Scattering Functions for Selected Ocean Waters," SIO Ref. 72–78, Scripps Inst. Oceanogr., La Jolla, 1972 (79 pp). Condensed in *Light in the Sea,* J. E. Tyler (ed.), Dowden, Hutchinson & Ross, Stroudsberg, 1977, chap. 12, pp. 150–174.

54. R. W. Spinrad, J. R. V. Zaneveld, and H. Pak, "Volume Scattering function of Suspended Particulate Matter at Near-Forward Angles: A Comparison of Experimental and Theoretical Values," *Appl. Optics* **17**(7):1125 (1978).

55. G. G. Padmabandu and E. S. Fry, "Measurement of Light Scattering at 0° by Small Particle Suspensions," *Ocean Optics X,* R. W. Spinrad (ed.), *Proc. SPIE* **1302:**191 (1990).

56. Y. Kuga and A. Ishimaru, "Backscattering Enhancement by Randomly Distributed Very Large Particles," *Appl. Optics* **28**(11):2165 (1989).

57. A. Morel, "Optical Properties of Pure Water and Pure Sea Water," *Optical Aspects of Oceanography,* N. G. Jerlov and E. S. Nielsen (eds.), Academic Press, New York, 1974, chap. 1, pp 1–24.

58. K. S. Shifrin, *Physical Optics of Ocean Water,* AIP Translation Series, Amer. Inst. Physics, New York, 1988, 285 pp.

59. G. Kullenberg, "Observed and Computed Scattering Functions," *Optical Aspects of Oceanography,* N. G. Jerlov and E. S. Nielsen (eds.), Academic Press, New York, 1974, chap. 2, pp 25–49.

60. O. B. Brown and H. R. Gordon, "Size-Refractive Index Distribution of Clear Coastal Water Particulates from Scattering," *Appl. Optics* **13:**2874 (1974).

61. J. C. Kitchen and J. R. V. Zaneveld, "A Three-Layer Sphere, Mie-Scattering Model of Oceanic Phytoplankton Populations," presented at Amer. Geophys. Union/Amer. Soc. Limnol. Oceanogr. Annual Meeting, New Orleans, 1990.

62. C. D. Mobley, B. Gentili, H. R. Gordon, Z. Jin, G. W. Kattawar, A. Morel, P. Reinersman, K. Stamnes, and R. H. Starn, "Comparison of Numerical Models for Computing Underwater Light Fields," *Appl. Optics* **32**(36):7484 (1993).

63. A. Morel and B. Gentili, "Diffuse Reflectance of Ocean Waters: Its Dependence on Sun Angle As Influenced by the Molecular Scattering Contribution," *Appl. Optics* **30**(30):4427 (1991).

64. A. Morel, "Diffusion de la lumière par les eaux de mer. Résultats experimentaux et approach théorique," NATO AGARD lecture series no. 61, *Optics of the Sea,* chap. 3.1, pp. 1–76 (1973), G. Halikas (trans.), Scripps Inst. Oceanogr., La Jolla, 1975, 161 pp.

65. O. V. Kopelevich and E. M. Mezhericher, "Calculation of Spectral Characteristics of Light Scattering by Sea Water," *Izvestiya, Atmos. Oceanic Phys.* **19**(2):144 (1983).

66. H. R. Gordon and A. Morel, "Remote Assessment of Ocean Color for Interpretation of Satellite Visible Imagery, A Review," *Lecture Notes on Coastal and Estuarine Studies,* vol. 4, Springer-Verlag, New York, 1983, 114 pp.

67. J. C. Kitchen, J. R. V. Zaneveld, and H. Pak, "Effect of Particle Size Distribution and Chlorophyll Content on Beam Attenuation Spectra," *Appl. Optics* **21**(21):3913 (1982).

68. J. K. Bishop, "The Correction and Suspended Particulate Matter Calibration of Sea Tech Transmissometer Data," *Deep-Sea Res.* **33:**121 (1986).

69. R. W. Spinrad, "A Calibration Diagram of Specific Beam Attenuation," *J. Geophys. Res.* **91**(C6):7761 (1986).

70. J. C. Kitchen and J. R. V. Zaneveld, "On the Noncorrelation of the Vertical Structure of Light Scattering and Chlorophyll *a* in Case 1 Water," *J. Geophys. Res.* **95**(C11):20237 (1990).

71. K. J. Voss, "A Spectral Model of the Beam Attenuation Coefficient in the Ocean and Coastal Areas," *Limnol. Oceanogr.* **37**(3):501 (1992).

72. J. E. Tyler and R. C. Smith, *Measurements of Spectral Irradiance Underwater,* Gordon and Breach, New York, 1970, 103 pp.

73. K. S. Baker and R. C. Smith, "Quasi-Inherent Characteristics of the Diffuse Attenuation Coefficient for Irradiance," *Ocean Optics VI,* S. Q. Duntley (ed.), *Proc. SPIE* **208:**60 (1979).
74. C. D. Mobley, "A Numerical Model for the Computation of Radiance Distributions in Natural Waters with Wind-Blown Surfaces," *Limnol. Oceanogr.* **34**(8):1473 (1989).
75. R. W. Austin and T. J. Petzold, "Spectral Dependence of the Diffuse Attenuation Coefficient of Light in Ocean Water," *Opt. Eng.* **25**(3):471 (1986).
76. D. A. Siegel and T. D. Dickey, "Observations of the Vertical Structure of the Diffuse Attenuation Coefficient Spectrum," *Deep-Sea Res.* **34**(4):547 (1987).
77. S. Sugihara and M. Kishino, "An Algorithm for Estimating the Water Quality Parameters from Irradiance Just Below the Sea Surface," *J. Geophys. Res.* **93**(D9):10857 (1988).
78. H. R. Gordon, O. B. Brown, R. E. Evans, J. W. Brown, R. C. Smith, K. S. Baker, and D. C. Clark, "A Semianalytic Model of Ocean Color," *J. Geophys. Res.* **93**(D9):10909 (1988).
79. S. Sathyendranath and A. Morel, "Light Emerging from the Sea—Interpretation and Uses in Remote Sensing," *Remote Sensing Applications in Marine Science and Technology,* A. P. Cracknell (ed.), D. Reidel, Dordrecht, 1983, chap. 16, pp. 323–357.
80. B. R. Marshall and R. C. Smith, "Raman Scattering and In-Water Ocean Optical Properties," *Appl. Optics* **29:**71 (1990).
81. G. W. Kattawar and X. Xu, "Filling-in of Fraunhofer Lines in the Ocean by Raman Scattering," *Appl. Optics* **31**(30):6491 (1992).
82. H. Gordon, "Diffuse Reflectance of the Ocean: The Theory of Its Augmentation by Chlorophyll *a* Fluorescence at 685 nm," *Appl. Optics* **18:**1161 (1979).
83. J. J. Cullen, C. M. Yentsch, T. L. Cucci, and H. L. MacIntyre, "Autofluorescence and Other Optical Properties As Tools in Biological Oceanography," *Ocean Optics IX,* M. A. Blizard (ed.), *Proc. SPIE* **925:**149 (1988).
84. A. Ivanoff, "Polarization Measurements in the Sea," *Optical Aspects of Oceanography,* N. G. Jerlov and E. S. Nielsen (eds.), Academic Press, New York, 1974, chap. 8, pp. 151–175.
85. G. D. Gilbert and J. C. Pernicka, "Improvement of Underwater Visibility by Reduction of Backscatter with a Circular Polarization Technique," *SPIE Underwater Photo-Optics Seminar Proc.,* Santa Barbara, Oct. 1966.
86. T. H. Waterman, "Polarization of Marine Light Fields and Animal Orientation," *Ocean Optics X,* M. A. Blizard (ed.), *Proc. SPIE* **925:**431 (1988).
87. K. J. Voss and E. S. Fry, "Measurement of the Mueller Matrix for Ocean Water," *Appl. Optics* **23:**4427 (1984).
88. M. S. Quinby-Hunt, A. J. Hunt, K. Lofftus, and D. Shapiro, "Polarized Light Studies of Marine *Chlorella,*" *Limnol. Oceanogr.* **34**(8):1589 (1989).
89. G. W. Kattawar and C. N. Adams, "Stokes Vector Calculations of the Submarine Light Field in an Atmosphere-Ocean with Scattering According to a Rayleigh Phase Matrix: Effect of Interface Refractive Index on Radiance and Polarization," *Limnol. Oceanogr.* **34**(8):1453 (1989).

CHAPTER 44
ATMOSPHERIC OPTICS

Dennis K. Killinger
Center for Laser Atmospheric Sensing
Department of Physics
University of South Florida
Tampa, Florida

James H. Churnside
National Oceanic and Atmospheric Administration
Environmental Technology Laboratory
Boulder, Colorado

Laurence S. Rothman
Air Force Geophysics Directorate/Phillips Lab
Optical Environment Division
Hanscom AFB, Massachusetts

44.1 GLOSSARY

c	speed of light
C_n^2	atmospheric turbulence strength parameter
F	hypergeometric function
$g(\nu)$	optical absorption lineshape function
h	height above sea level
h	Planck's constant
I	intensity of optical beam (watts/m^2)
J	Bessel function
L_o	outer scale size of atmospheric turbulence
l_o	inner scale size of atmospheric turbulence
N	density or concentration of molecules
P_ν	Planck radiation function
R	gas constant
S	molecular absorption line intensity
T	temperature

W	wind speed
β	backscatter coefficient of the atmosphere
γ_p	pressure-broadened half-width of absorption line
κ	optical attenuation
λ	wavelength
ν	optical frequency (wave numbers)
$\rho(I)$	probability density function of irradiance fluctuations
ρ_o	phase coherence length
σ_I^2	variance of irradiance fluctuations
σ_R	Rayleigh scattering cross section

44.2 INTRODUCTION

Atmospheric optics involves the transmission, absorption, emission, refraction, and reflection of light by the atmosphere and is probably one of the most widely observed of all optical phenomena.[1–5] The atmosphere interacts with light due to the composition of the atmosphere, which under normal conditions, consists of a variety of different molecular species and small particles or aerosols. This interaction of the atmosphere with light is observed to produce a wide variety of optical phenomena including the blue color of the sky, the red sunset, the optical absorption of specific wavelengths due to atmospheric molecules, the twinkling of stars at night, and the greenish tint sometimes observed during a severe wind storm due to the high density of dust and water-related aerosols in the atmosphere.

One of the most basic optical phenomena of the atmosphere is the absorption of light. This absorption process can be depicted as in Fig. 1 which shows the transmission spectrum of the atmosphere as a function of wavelength.[5] The transmission of the atmosphere is highly dependent upon the wavelength of the spectral radiation, and, as will be covered later in this chapter, upon the composition and specific optical properties of the

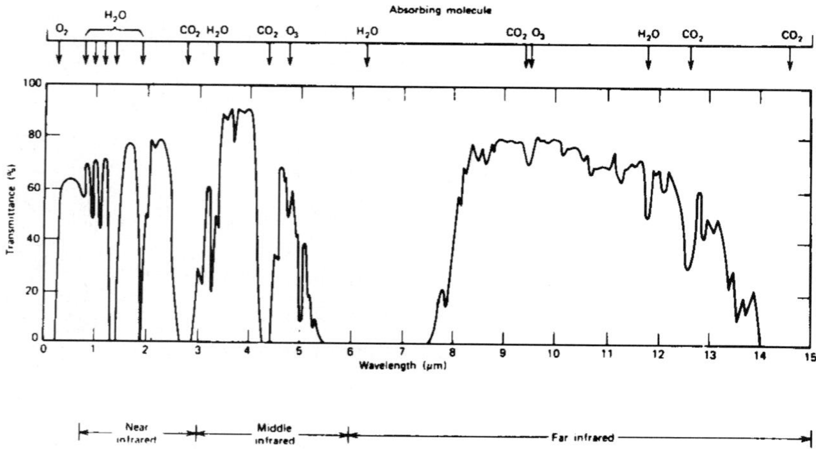

FIGURE 1 Transmittance through the earth's atmosphere as a function of wavelength taken with low spectral resolution (path length 1800 m). (*From Measures, Ref. 5.*)

constituents in the atmosphere. The prominent spectral features in the transmission spectrum in Fig. 1 are primarily due to absorption bands and individual absorption lines of the molecular gases in the atmosphere, while a portion of the slowly varying background transmission is due to aerosol extinction, continuum absorption, and particulates in the atmosphere.

This chapter presents a tutorial overview of some of the basic optical properties of the atmosphere, with an emphasis on those properties associated with optical propagation and transmission of light through the Earth's atmosphere. The physical phenomena of optical absorption, scattering, emission, and refractive properties of the atmosphere will be covered for optical wavelengths from the UV to the far-infrared. The primary focus of this chapter is on *linear* optical properties associated with the transmission of light through the atmosphere. Historically, the study of atmospheric optics has centered around the radiance transfer function of the atmosphere, and the linear transmission spectrum and blackbody emission spectrum of the atmosphere. This emphasis was due to the large body of research associated with passive, electro-optical sensors which primarily use the transmission of ambient optical light or light from selected emission sources. During the past decade, however, the use of lasers has added a new dimension to the study of atmospheric optics. In this case, not only is one interested in the transmission of light through the atmosphere, but also information regarding the optical properties of the backscattered optical radiation.

In this chapter, the standard linear optical interactions of an optical or laser beam with the atmosphere will be covered, with an emphasis placed on linear absorption and scattering interactions. It should be mentioned that the first edition of the OSA *Handbook of Optics* chapter on "Atmospheric Optics" had considerable nomographs and computational charts to aid the user in numerically calculating the transmission of the atmosphere.[2] Because of the present availability of a wide range of spectral databases and computer programs (such as the HITRAN Spectroscopy Database, and LOWTRAN, MODTRAN, and FASCODE atmospheric transmission computer programs) that model and calculate the transmission of light through the atmosphere, these nomographs, while still useful, are not as vital. As a result, the emphasis of this second edition of the "Atmospheric Optics" chapter is on the basic theory of the optical interactions, how this theory is used to model the optics of the atmosphere, the use of available computer programs and databases to calculate the optical properties of the atmosphere, and examples of instruments and meteorological phenomena related to optical or visual remote sensing of the atmosphere.

The overall organization of this chapter begins with a description of the natural, homogeneous atmosphere and the representation of its physical and chemical composition as a function of altitude. A brief survey is then made of the major linear optical interactions that can occur between a propagating optical beam and the naturally occurring constituents in the atmosphere. The next section covers several major computational programs (HITRAN, LOWTRAN, MODTRAN, and FASCODE) and U.S. Standard Atmospheric Models which are used to compute the optical transmission, scattering, and absorption properties of the atmosphere. The next major technical section presents an overview of the influence of atmospheric refractive turbulence on the statistical propagation of an optical beam or wavefront through the atmosphere. Finally, the last two sections of the chapter include a brief introduction to some optical and laser remote sensing experiments of the atmosphere and a brief introduction to the visually important field of meteorological optics.

It should be noted that the material contained within this chapter has been compiled from several recent overview/summary publications on the optical transmission and atmospheric composition of the atmosphere, as well as from a large number of technical reports and journal publications. These major overview references are: (1) the previous edition of the *OSA Handbook of Optics* (chapter on "Optical Properties of the Atmosphere"), (2) *Handbook of Geophysics and the Space Environment* (chapter on "Optical and Infrared Properties of the Atmosphere"), (3) *The Infrared Handbook,* (4) *Laser Remote Sensing,* and (5) *Atmospheric Radiation.*[1–5] The interested reader is directed toward these comprehensive treatments as well as to the listed references therein for detailed information concerning the topics covered in this brief overview of atmospheric optics.

44.3 PHYSICAL AND CHEMICAL COMPOSITION OF THE STANDARD ATMOSPHERE

The atmosphere is a fluid composed of gases, particulates, and aerosols whose physical and chemical properties vary as a function of time, altitude, and geographical location. Although these properties can be highly dependent upon local and regional conditions, many of the optical properties of the atmosphere can be described to an adequate level by looking at the compositon of what one normally calls a *standard atmosphere.* This section will describe the background, homogeneous standard composition of the atmosphere. This will serve as a basis for the determination of the quantitative interaction of the molecular gases, aerosols, and particulates in the atmosphere with a propagating optical wavefront.

Molecular Gas Concentration, Pressure, and Temperature

The majority of the atmosphere is composed of lightweight molecular gases. Table 1 lists the major gases and trace species of the terrestrial atmosphere, and their approximate concentration in units of atmospheres (atm) at standard room temperature (296 K), altitude at sea level, and total pressure of 1 atm.[6] The major optically active molecular constituents of the atmosphere are N_2, O_2, H_2O, and CO_2, with a secondary grouping of CH_4, N_2O, CO, and O_3. The other 22 species in the table are present in the atmosphere at trace-level concentrations (ppb, down to less than ppt by volume), however the concentration may be increased by many orders of magnitude due to local emission sources of these gases.

The temperature of the atmosphere varies both with seasonal changes and altitude. Figure 2 shows the average temperature profile of the atmosphere as a function of altitude presented for the U.S. Standard atmosphere.[7] The temperature decreases significantly with altitude until the level of the stratosphere is reached where the temperature profile has an inflection point. The U.S. Standard Atmosphere is one of six basic atmospheric models developed by the U.S. government; these different models furnish a good representation of the different atmospheric conditions which are often encountered. Figure 3 shows the temperature profile for the six atmospheric models.[7]

The pressure of the atmosphere decreases with altitude due to the gravitational pull of the earth and the hydrostatic equilibrium pressure of the atmospheric fluid. This is indicated in Fig. 4 which shows the total pressure of the atmosphere in millibars (1013 mb = 1 atm = 760 torr) as a function of altitude for the different atmospheric models.[7] The fractional or partial pressure of most of the major gases (N_2, O_2, CO_2, N_2O, CO, and CH_4) follows this profile and these gases are considered uniformly mixed. However, the concentration of water vapor is very temperature-dependent due to freezing and is not uniformly mixed in the atmosphere. Figure 5*a* shows the partial pressure (density) of water vapor as a function of altitude; the units of density are in molecules/cm^3 and are related to 1 atm by the appropriate value of Loschmidts number (the number of molecules in 1 cm^3 of air) at a temperature of 296 K, which is 2.479×10^{19} molecules/cm^3.[7]

The partial pressure of ozone (O_3) also varies significantly with altitude because it is generated in the upper altitudes and near ground level by solar radiation, and is in chemical equilibrium with other gases in the atmosphere which themselves vary with altitude and time of day. Figure 5*b* shows the typical concentration of ozone as a function

TABLE 1 List of Molecular Gases and Their Typical Concentration for the Ambient U.S. Standard Atmosphere (Note that the trace species have concentrations less than 1×10^{-9}, with a value that is variable and often dependent upon local emission sources.)

Molecule	p (atm)
N_2	0.781
O_2	0.209
H_2O	0.0775 (variable)
CO_2	3.3 E − 4
A (Argon)	0.0093
CH_4	1.7 E − 6
N_2O	3.2 E − 7
CO	1.5 E − 7
O_3	2.66 E − 8 (variable)
H_2CO	2.4 E − 9
C_2H_6	2 E − 9
HCl	1 E − 9
CH_3Cl	7 E − 10
OCS	6 E − 10
C_2H_2	3 E − 10
SO_2	3 E − 10
NO	3 E − 10
H_2O_2	2 E − 10
HCN	1.7 E − 10
HNO_3	5 E − 11
NH_3	5 E − 11
NO_2	2.3 E − 11
HOCl	7.7 E − 12
HI	3 E − 12
HBr	1.7 E − 12
OH	4.4 E − 14
HF	1 E − 14
ClO	1 E − 14
HCOOH	1 E − 14
COF_2	1 E − 14
SF_6	1 E − 14
H_2S	1 E − 14
PH_3	1 E − 20

of altitude.[7] The ozone concentration peaks at an altitude of approximately 20 km and is one of the principal molecular optical absorbers in the atmosphere at that altitude. Further details of these atmospheric models under different atmospheric conditions are contained within the listed references and the reader is encouraged to consult these references for more detailed information.[3,7]

Aerosols and Particulates

The atmospheric propagation of optical radiation is influenced by particulate matter suspended in the air such as dust, fog, haze, cloud droplets, and aerosols. Figure 6 shows

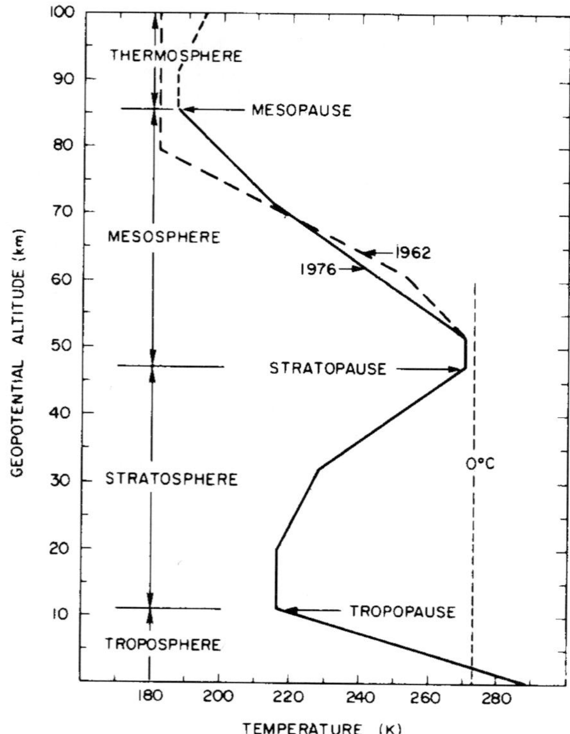

FIGURE 2 Temperature-height profile for U.S. Standard Atmosphere (0–86 km).

the basic characteristics of particulates in the amosphere as a function of altitude,[3] and Fig. 7 indicates the approximate size of common atmospheric particles.[5]

Aerosols in the boundary layer (surface to 1 to 2 km altitude) are locally emitted, wind-driven particulates, and have the greatest variability in composition and concentration. Over land, the aerosols are mostly soil particles, dust, and organic particles from vegetation. Over the oceans, they are mostly sea salt particles. At times, however, long-range global winds are capable of transporting land particulates vast distances across the oceans or continents, especially those particulates associated with dust storms or large biomass fires, so that substantial mixing of the different particulate types may occur.

In the troposphere above the boundary layer, the composition is less dependent upon local surface conditions and a more uniform, global distribution is observed. The aerosols observed in the troposphere are mostly due to the coagulation of gaseous compounds and fine dust. Above the troposphere, in the region of the stratosphere from 10 to 30 km, the background aerosols are mostly sulfate particles and are uniformly mixed globally. However, the concentration can be perturbed by several orders of magnitude due to the injection of dust and SO_2 by volcanic activity, such as the recent eruption of Mount Pinatubo.[8] Such increases in the aerosol concentration may persist for several years and significantly impact the global temperature of the earth.

Several models have been developed for the number density and size distribution of aerosols in the atmosphere.[7] Figures 8 and 9 show two aerosol distribution models appropriate for the rural environment and maritime environment, as a function of relative

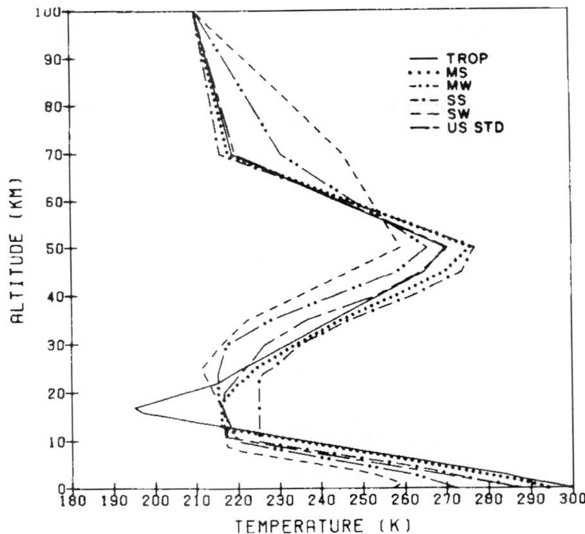

FIGURE 3 Temperature vs. altitude for the six model atmospheres: tropical (TROP), midlatitude summer (MS), midlatitude winter (MW), subarctic summer (SS), subarctic winter (SW), and U.S. standard (US STD).

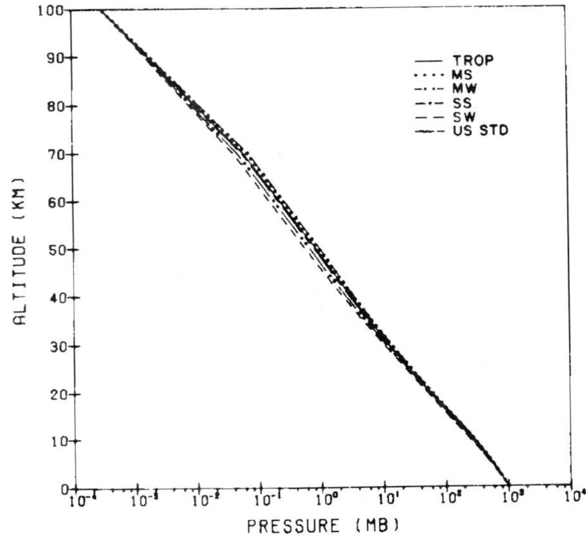

FIGURE 4 Pressure vs. altitude for the six model atmospheres.

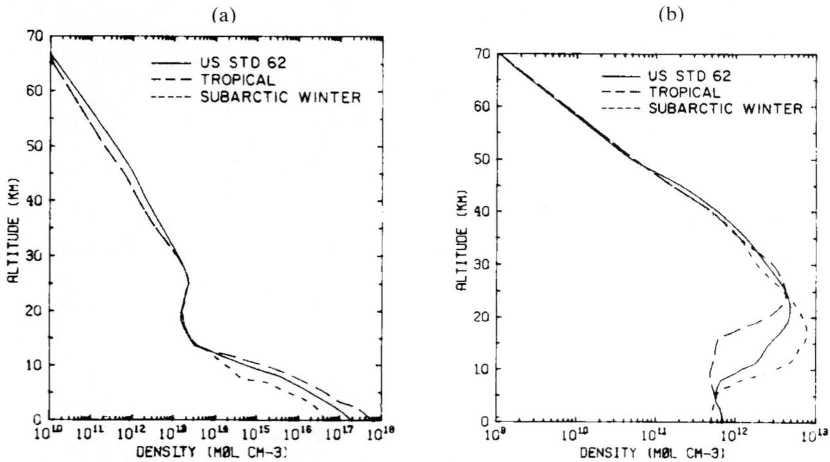

FIGURE 5 (*a*) Water vapor profile of several models; (*b*) ozone profile for several models; the U.S. standard model shown is the 1962 model.

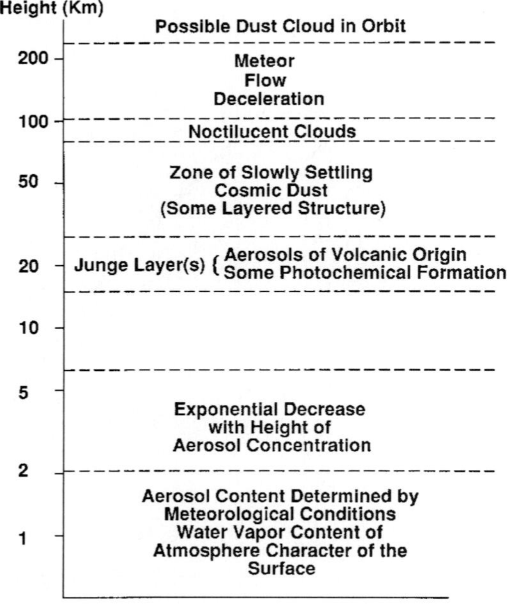

FIGURE 6 Physical characteristics of atmospheric aerosols.

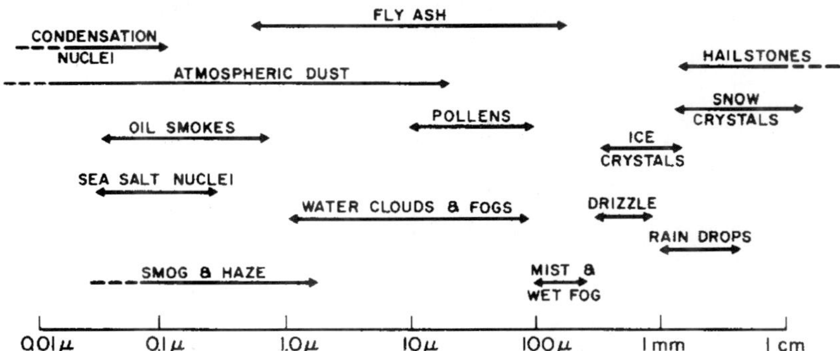

FIGURE 7 Representative diameters of common atmospheric particles. *(From Measures, Ref. 5.)*

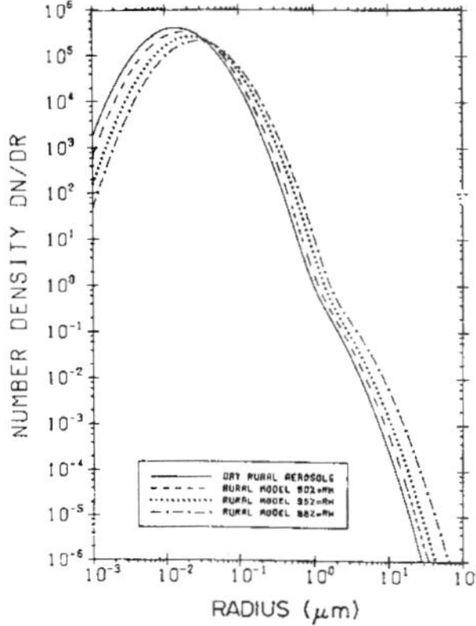

FIGURE 8 Aerosol number density distribution ($cm^{-3}\ \mu m^{-1}$) for the rural model at different relative humidities with total particle concentrations fixed at 15,000 cm^{-3}.

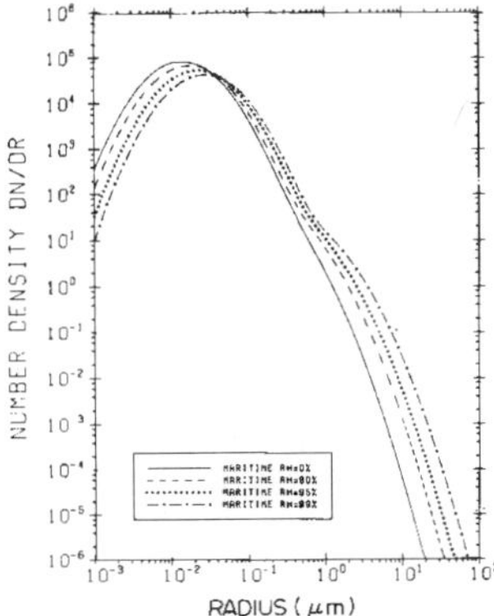

FIGURE 9 Aerosol number density distribution ($cm^{-3}\ \mu m^{-1}$) for the maritime model at different relative humidities with total particle concentrations fixed at 4000 cm^{-3}.

humidity[7]; the humidity influences the size distribution of the aerosol particles and their growth characteristics. The greatest number density (particles/cm^3) occurs near a size of 0.01 μm but a significant number of aerosols are still present even at the larger sizes near 1 to 2 μm. Finally, the optical characteristics of the aerosols can also be dependent upon water vapor concentration, with changes in surface, size, and growth characteristics of the aerosols sometimes observed to be dependent upon the relative humidity. Such humidity changes can also influence the concentration of some pollutant gases, if these gases have been absorbed onto the surface of the aerosol particles.[7]

44.4 FUNDAMENTAL THEORY OF INTERACTION OF LIGHT WITH THE ATMOSPHERE

The propagation of light through the atmosphere depends upon several optical interaction phenomena and the physical composition of the atmosphere. In this section, we consider some of the basic interactions involved in the transmission, absorption, emission, and scattering of light as it passes through the atmosphere. Although all of these interactions can be described as part of an overall radiative transfer process, it is common to separate the interactions into distinct optical phenomena of molecular absorption, Rayleigh scattering, Mie or aerosol scattering, and molecular emission. Each of these basic phenomena will be discussed in this section following a brief outline of the fundamental equations for the transmission of light in the atmosphere centered around the Beer-Lambert law.

The linear transmission (or absorption) of monochromatic light by species in the atmosphere may be expressed approximately by the Beer-Lambert law as

$$I(\lambda, t', x) = I(\lambda, t, 0)e^{-\int_0^x \kappa(\lambda)N(x', t)\,dx'} \tag{1}$$

where $I(\lambda, t', x)$ is the intensity of the optical beam after passing through a path length of x, $\kappa(\lambda)$ is the optical attenuation or extinction coefficient of the species per unit of species density and length, and $N(x, t)$ is the spatial and temporal distribution of the species density that is producing the absorption; λ is the wavelength of the monochromatic light, and the parameter time t' is inserted to remind one of the potential propagation delay. Equation (1) contains the term $N(x, t)$ which explicitly indicates the spatial and temporal variability of the concentration of the attenuating species since in many experimental cases such variability may be a dominant feature.

It is common to write the attenuation coefficient in terms of coefficients that can describe the different phenomena that can cause the extinction of the optical beam. The most dominant interactions in the natural atmosphere are those due to Rayleigh (elastic) scattering, linear absorption, and Mie (aerosol/particulate) scattering; elastic means that the scattered light does not change in wavelength from that which was transmitted while inelastic infers a shift in the wavelength. In this case, one can write $\kappa(\lambda)$ as

$$\kappa(\lambda) = \kappa_a(\lambda) + \kappa_R(\lambda) + \kappa_M(\lambda) \tag{2}$$

where these terms represent the individual contributions due to absorption, Rayleigh scattering, and Mie scattering, respectively. The values for each of these extinction coefficients will be described in the following sections along with the appropriate species density term $N(x, t)$. In some of these cases, the reemission of the optical radiation, possibly at a different wavelength, is also of importance. Rayleigh extinction will lead to Rayleigh backscatter, Raman extinction leads to spontaneous Raman scattering, absorption can lead to fluorescence emission or thermal heating of the molecule, and Mie extinction is defined primarily in terms of the scattering coefficient. Under idealized conditions, the scattering processes can be related directly to the value of the attenuation processes. However, if several complex optical processes occur simultaneously, such as in atmospheric propagation, the attenuation and scattering processes are not directly linked via a simple analytical equation. In this case, independent measurements of the scattering coefficient and the extinction coefficient have to be made, or approximation formulas are used to relate the two coefficients.[4,5]

Molecular Absorption

The absorption of optical radiation by molecules in the atmosphere is primarily associated with individual optical absorption transitions between the allowed quantized energy levels of the molecule. The energy levels of a molecule can usually be separated into those associated with rotational, vibrational, or electronic energy states. Absorption transitions between the rotational levels occur in the far-IR and microwave spectral region, transitions between vibrational levels occur in the near-IR (2 to 20 μm wavelength), and electronic transitions occur in the UV-visible region (0.3 to 0.7 μm). Transitions can occur which combine several of these categories, such as rotational-vibrational transitions or electronic-vibrational-rotational transitions.

Some of the most distinctive and identifiable absorption lines of many atmospheric molecules are the rotational-vibrational optical absorption lines in the infrared spectral region. These lines are often clustered together into vibrational bands according to the allowed quantum transitions of the molecule. In many cases, the individual lines are distinct and can be resolved if the spectral resolution of the measuring instrument is fine enough (i.e., $<0.1\ \mathrm{cm}^{-1}$). An example of such a region is the absorption feature near 2.04 μm in Fig. 1 which is actually composed of individual absorption lines if viewed under higher spectral resolution. Figure 10 is a computed high-resolution expansion of the

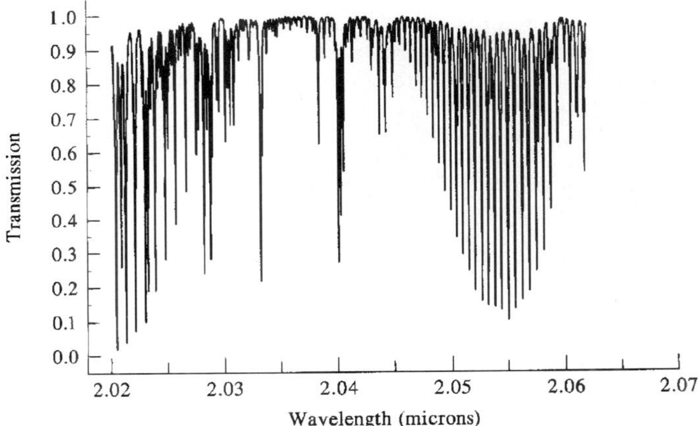

FIGURE 10 High-resolution transmission spectrum of the atmosphere for a horizontal path of 1800 m (similar to that in Fig. 1) for the spectral region near 2.04 μm. The individual rotational absorption lines due to CO_2 and H_2O are easily observed.

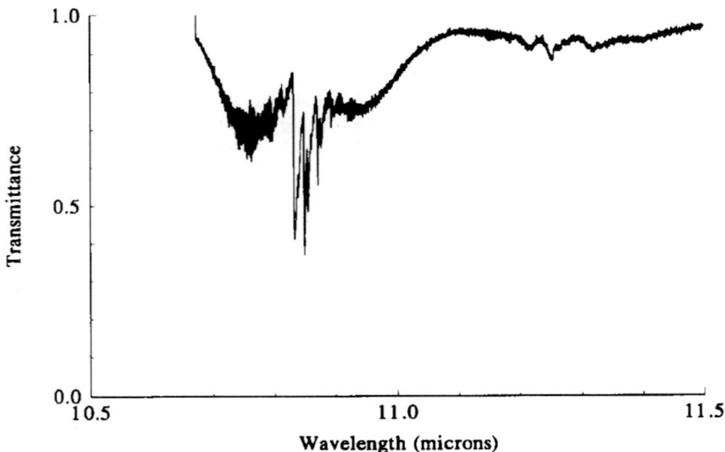

FIGURE 11 High-resolution absorption spectrum of Freon-12 gas showing complex band structure typical of heavy, complex molecules in the atmosphere (path of 1000 m with 1 ppm concentration).

atmospheric spectrum of Fig. 1 near 2.04 μm over a path length of 1800 m which shows these individual lines. In this case, the individual lines are well-separated and appear like a "picket fence" spectrum showing gaps between the absorption lines. Many of the atmospheric gaseous molecules listed earlier in Table 1 have similar spectral structure. These gases are relatively lightweight and have few (less than 5 or 6) atoms per molecule so that their moments of inertia are relatively small. The resulting energy spacing between the allowed rotational-vibrational absorption transitions is large and well-separated in wavelength.

In other spectral regions, however, the individual lines overlap or are so strong or saturated that the transmission spectrum displays only broad spectral features; an example of such a spectral region is the strong absorption seen near 2.7 μm, 5 to 8 μm, and beyond 13 μm in Fig. 1. Finally, the molecular absorption observed in the UV is often due to optical transitions to an electronic energy level, molecular energy continuum or a predissociation energy level. In some cases, such as that for O_3 or SO_2, this results in broad absorption bands that extend throughout the UV region (250 to 350 nm).

More complex, heavier molecules, such as benzene or chlorofluorocarbons, have absorption spectra which are blended or merged together into band spectra due to the complexity and overlap of the rotational-vibrational transitions of these molecules. Figure 11 shows a transmission spectrum of Freon-12 ($C\,Cl_2\,F_2$) which has a complex spectrum near 11 μm. As seen, the individual rotational lines are merged into a band spectrum. A band spectrum is often unique for each gas and can be used to identify the chemical composition of the gas. Heavy molecules in the atmosphere are not normally part of the natural atmosphere and are usually the result of pollution or gaseous plumes injected into the atmosphere.

The overall transmission or absorption of the atmosphere due to an individual molecular absorption line can be given quantitatively as

$$\kappa_a(\lambda)N(x,t) = Sg(\nu - \nu_o)NP_a \tag{3}$$

where S is the molecular transition line intensity (units of cm/molecule), $g(\nu - \nu_o)$ is the normalized lineshape function (units of cm or $1/\text{cm}^{-1}$), N is the number of molecules of absorbing species per cm^3 per atm, and P_a is the partial pressure of the absorbing gas in atm. The value of N is equal to the value of Loschmidt's number, which is 2.479 $\times 10^{19}$ molecules cm^{-3} atm^{-1} at a temperature of 296 K; The value of N is inversely proportional to temperature due to the change in gas concentration as a function of temperature for 1 atm of pressure. As will be seen later, the definition of S as given in Eq. (3) is that used in the U.S. Air Force Phillips Laboratory/Geophysics Directorate HITRAN database.[9] In this case, S contains the Boltzmann population factor and isotope fraction (natural abundance) as well as the stimulated emission term due to finite population in the upper energy states of the molecule.[9] In Eq. (3), $Sg(\nu - \nu_o)$ is the absorption cross section per molecule (cm^2/molecule) and NP_a is the number of absorbing molecules in units of molecules/cm^3.

The lineshape function can be described by several different models. The two most prevalent are the lorentzian lineshape associated with pressure broadening and the gaussian lineshape associated with Doppler broadening which becomes important at elevated temperatures or low pressures.

The lorentzian/pressure-broadened profile is given by

$$g_L(\nu - \nu_o) = (\gamma_p/\pi)/[(\nu - \nu_o)^2 + \gamma_p^2] \tag{4}$$

where γ_p is the pressure-broadened half-width at half-maximum (HWHM) in wave numbers (cm^{-1}). The pressure-broadened half-width is obtained from the air-broadened half-width parameter g as $\gamma_p = gP_t$, where P_t is the total background atmospheric pressure. Under ambient atmospheric conditions, g is approximately 0.05 cm^{-1} (i.e., 1.5 GHz) for many molecules in the atmosphere.

It should be noted that under very low pressure conditions, where the time between collisions with other molecules is relatively long, the intrinsic radiative lifetime of the molecule will determine the lineshape profile. Under these conditions, the linewidth is called the natural linewidth. The natural linewidth of many molecules is on the order of a few MHz (i.e., approximately 0.0001 cm^{-1}) or less.

The gaussian or Doppler line profile is expressed as

$$g_D(\nu - \nu_o) = (1/\gamma_D)(\ln 2/\pi) . 5 \exp[-\ln 2(\nu - \nu_o)^2/\gamma_D^2] \tag{5}$$

where γ_D is the Doppler linewidth (HWHM in cm^{-1}) given by

$$\gamma_D = (\nu_o/c)[2RT \ln 2/M]^{1/2} \tag{6}$$

where R is the gas constant, T is the temperature in Kelvin, and M is the molecular weight of the molecule.

The value for the lineshape at the peak (line center) is equal to $1/(\pi\gamma_p) = 0.318/\gamma_p$ for the pressure-broadened case. For the Doppler peak, the maximum value is $(\ln 2/\pi)^{1/2}/\gamma_D = 0.469/\gamma_D$. Under ambient atmospheric conditions, the Doppler linewidth is usually much smaller than the pressure-broadened linewidth.

For those cases where both lorentzian and Doppler broadening are present in approximately equal amounts, a convolution of the Doppler and lorentzian profile must be used. This convolution of a Doppler and lorentzian is called a Voigt profile and involves a double integral for an exact calculation. Fortunately, several numerical approximations are available for the computation of the Voigt profile and lineshape parameters.[10] The Voigt profile is important in the spectroscopy of molecules in the upper atmosphere where the ambient pressure is low and the Doppler and pressure-broadened linewidths are of the same order of magnitude.

Finally, the large number of transition lines of water vapor and other gases in the atmosphere can produce a significant level of background "quasi-continuum" absorption in the atmosphere. This phenomenon is primarily due to the additive contribution from the wings of the absorption lines even at wavelengths far removed (100 s of cm^{-1}) from the line centers. Such an effect has been studied by Burch and by Clough et al. for water

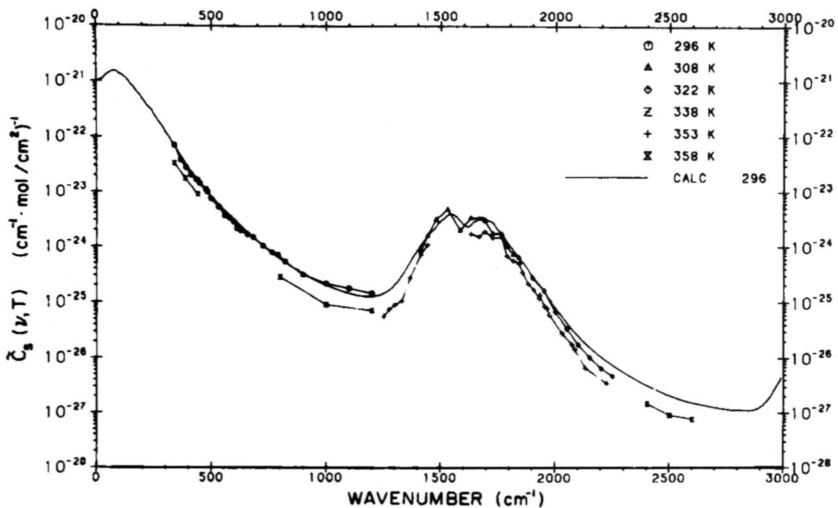

FIGURE 12 Self-density absorption continuum values C_s for water vapor as a function of wave number. The experimental values were measured by Burch. *(From Ref. 3.)*

vapor due to strong self-broadening interactions.[11] Figure 12 shows a plot of the relative continuum coefficient for water vapor as a function of wave number. Good agreement with the experimental data and model calculations is shown. Models for water vapor and nitrogen continuum absorption are contained within many of the major atmospheric transmission programs (such as FASCODE). The typical value for the continuum absorption is negligible in the visible to the near-IR, but can be significant at wavelengths in the range of 5 to 20 μm.

Molecular Rayleigh Scattering

Rayleigh scattering is elastic scattering of the optical radiation due to the displacement of the weakly bound electronic cloud surrounding the gaseous molecule which is perturbed by the incoming electro-magnetic (optical) field. This phenomenon is associated with optical scattering where the wavelength of light is much larger than the physical size of the scatterers (i.e., atmospheric molecules). Rayleigh scattering, which makes the color of the sky blue and the setting or rising sun red, was first described by Lord Rayleigh in 1871. The Rayleigh differential scattering cross section for polarized, monochromatic light is given by[5]

$$d\sigma_R/d\Omega = [\pi^2(n^2 - 1)^2/N^2\lambda^4][\cos^2\phi\cos^2\theta + \sin^2\phi] \tag{7}$$

where n is the index of refraction of the atmosphere, N is the density of molecules, λ is the wavelength of the optical radiation, and ϕ and θ are the spherical coordinate angles of the scattered polarized light referenced to the direction of the incident light. As seen from Eq. (7), shorter-wavelength light (i.e., blue) is more strongly scattered out from a propagating beam than the longer wavelengths (i.e., red), which is consistent with the preceding comments regarding the color of the sky or the sunset. A typical value for $d\sigma_R/d\Omega$ at a wavelength of 700 nm in the atmosphere (STP) is approximately 2×10^{-28} cm^2 sr^{-1}.[3] This value depends upon the molecule and has been tabulated for many of the major gases in the atmosphere.[10–12]

The total Rayleigh scattering cross section can be determined from Eq. (7) by integrating over 4π steradians to yield

$$\sigma_R\,(\text{total}) = [8/3][\pi^2(n^2 - 1)^2/N^2\lambda^4] \tag{8}$$

At sea level (and room temperature, $T = 296$ K) where $N = 2.5 \times 10^{19}$ molecules/cm^3, Eq. (8) can be multiplied by N to yield the total Rayleigh scattering extinction coefficient as

$$\kappa_R(\lambda)N(x, t) = N\sigma_R(\text{total}) = 1.18 \times 10^{-8}\,[550\text{ nm}/\lambda(\text{nm})]^4\text{ cm}^{-1} \tag{9}$$

The neglect of the effect of dispersion of the atmosphere (variation of the index of refraction n with wavelength) results in an error of less than 3 percent in Eq. (9) in the visible wavelength range.[5]

The molecular Rayleigh backscatter ($\theta = \pi$) cross section for the atmosphere has been given by Collins and Russell for polarized incident light (and received scattered light of the same polarization) as[12]

$$\sigma_R = 5.45 \times 10^{-28}\,[550\text{ nm}/\lambda(\text{nm})]^4\text{ cm}^2\text{ sr}^{-1} \tag{10}$$

At sea level where $N = 2.47 \times 10^{19}$ molecules/cm^3, the atmospheric volume backscatter coefficient, β_R, is thus given by

$$\beta_R = N\,\sigma_R = 1.39 \times 10^{-8}\,[550\text{ nm}/\lambda(\text{nm})]^4\text{ cm}^{-1}\text{ sr}^{-1} \tag{11}$$

The backscatter coefficient for the reflectivity of a laser beam due to Rayleigh backscatter is determined by multiplying β_R by the range resolution or length of the optical interaction being considered.

For unpolarized incident light, the Rayleigh scattered light has a depolarization factor δ which is the ratio of the two orthogonal polarized backscatter intensities. δ is usually defined as the ratio of the perpendicular and parallel polarization components measured relative to the direction of the incident polarization. Values of δ depend upon the anisotropy of the molecules or scatters, and typical values range from 0.02 to 0.11.[13] Depolarization also occurs for multiple scattering and is of considerable interest in laser or optical transmission through dense aerosol clouds.[14] The depolarization factor can sometimes be used to determine the physical and chemical composition of the cloud constituents, such as the relative ratio of water vapor or ice crystals in a cloud.

Mie Scattering: Aerosols, Particulates, and Clouds

Mie scattering is similar to Rayleigh scattering, except the size of the scattering sites is on the same order of magnitude as the wavelength of the incident light, and is, thus, due to aerosols and fine particulates in the atmosphere. The scattered radiation is the same wavelength as the incident light but experiences a more complex functional dependence upon the interplay of the optical wavelength and particle size distribution than that seen for Rayleigh scattering.

In 1908, Mie investigated the scattering of light by dielectric spheres of size comparable to the wavelength of the incident light.[15] His analysis indicated the clear asymmetry between the forward and backward directions, where for large particle sizes the forward-directed scattering dominates. Complete treatments of Mie scattering can be found in several excellent works by Deirmendjian and others, which take into account the complex index of refraction and size distribution of the particles.[16,17] These calculations are also influenced by the asymmetry of the aerosols or particulates which may not be spherical in shape.

The effect of Mie scattering in the atmosphere can be described as in the following figures. Figure 13 shows the aerosol Mie extinction coefficient as a function of wavelength for several atmospheric models, along with a typical Rayleigh scattering curve for

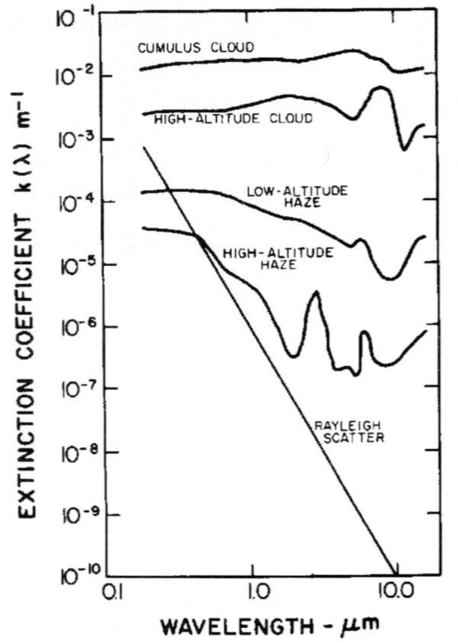

FIGURE 13 Aerosol extinction coefficient as a function of wavelength. *(From Measures, Ref. 5.)*

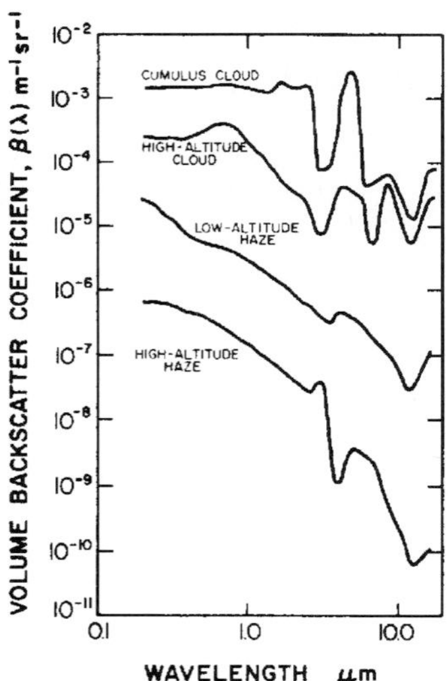

FIGURE 14 Aerosol volume backscattering coefficient as a function of wavelength. *(From Ref. 5.)*

comparison.[18] Figure 14 shows similar values for the volume Mie backscatter coefficient as a function of wavelength.[18] Extinction and backscatter coefficient values are highly dependent upon the wavelength and particulate composition.

Figures 15 and 16 show the calculated extinction coefficient for the rural and maritime aerosol models described in Sec. 44.3 as a function of relative humidity and wavelength.[3] Significant changes in the backscatter can be produced by relatively small changes in the humidity.

The extinction coefficient is also a function of altitude, following the dependence of the composition of the aerosol and particulates. Figure 17 shows an atmospheric aerosol extinction model as a function of altitude for a wavelength of 0.55 μm.[3,8] The influence of the visibility (in km) at ground level dominates the extinction value at the lower altitudes and the composition and density of volcanic particulate dominates the upper altitude regions. The dependence of the extinction on the volcanic composition at the upper altitudes is shown in Fig. 18 which shows these values as a function of wavelength and of composition.[3,8]

The variation of the backscatter coefficient as a function of altitude is shown in Fig. 19 which displays atmospheric backscatter data obtained by McCormick using a 1.06-μm Nd:YAG Lidar.[19] The boundary layer aerosols dominate at the lower levels and the decrease in the atmospheric particulate density determines the overall slope with altitude. Of interest is the increased value near 20 km due to the presence of volcanic aerosols in the atmosphere due to the eruption of Mt. Pinatubo in 1991.

Molecular Emission and Thermal Spectral Radiance

The same optical molecular transitions that cause absorption also emit light when they are thermally excited. Since the molecules have a finite temperature T they will act as blackbody radiators with optical emission given by the Planck radiation law. The allowed transitions of the molecules will modify the intensity distribution of the radiation due to emission of the radiation according to the thermal distribution of the population within

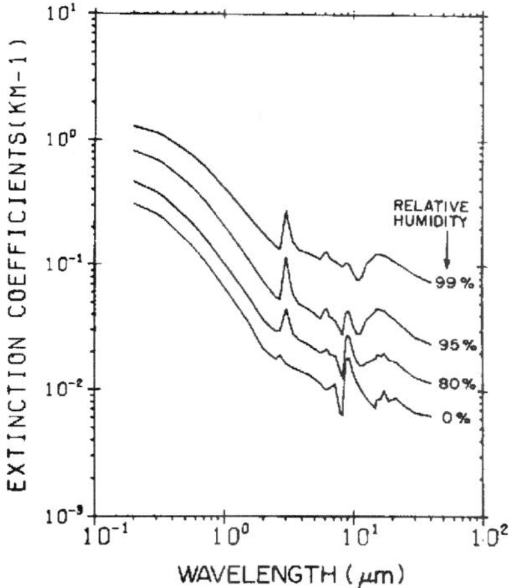

FIGURE 15 Extinction coefficients vs. wavelength for the rural aerosol model for different relative humidities and constant number density of particles.

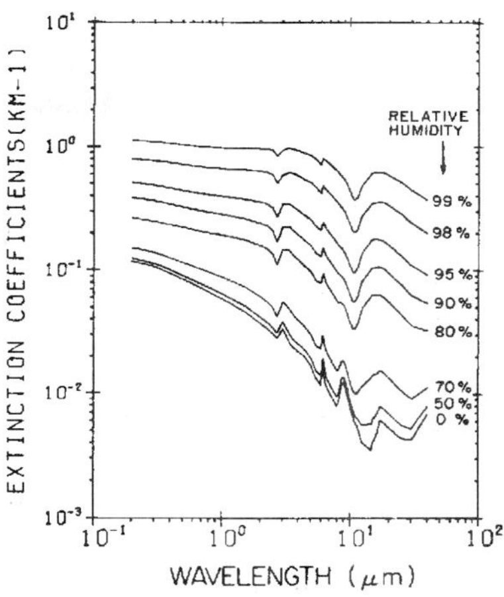

FIGURE 16 Extinction coefficients vs. wavelength for the maritime aerosol model for different relative humidities and constant number density of particles.

the energy levels of the molecule; it should be noted that the Boltzmann thermal population distribution is essentially the same as that which is described by the Planck radiation law for local thermodynamic equilibrium conditions. As such, the molecular emission spectrum of the radiation is similar to that for absorption. The thermal radiance from the clear atmosphere involves the calculation of the blackbody radiation emitted by each elemental volume of air multiplied by the absorption spectral distribution of the molecular absorption lines, $\kappa_a(s)$, and then this emission spectrum is attenuated by the rest of the atmosphere as the emission propagates toward the viewer. This may be expressed as

$$I_v = \int_0^S \kappa_a(s) P_v(s) \exp\left[-\int_0^S \kappa_a(s')\,ds'\right] ds \tag{12}$$

where the exponential term is Beer's law, and $P_v(s)$ is the Planck function given by

$$P_v(s) = 2hv^3/[c^2 \exp([hv/kT(s)] - 1)] \tag{13}$$

In these equations, s is the distance from the receiver along the optical propagation path, v is the optical frequency, h is Planck's constant, c is the speed of light, k is Boltzmann's constant, and $T(s)$ is the temperature at position s along the path. As seen in Eq. (12), each volume element emits thermal radiation of $\kappa_a(s)P_v(s)$, which is then attenuated by Beer's law. The total emission spectral density is obtained by summing or integrating over all the emission volume elements and calculating the appropriate absorption along the optical path for each element.

As an example, Fig. 20 shows a plot of the spectral radiance measured on a clear day with 1 cm^{-1} spectral resolution. Note that the regions of strong absorption produce more radiance as the foregoing equation suggests, and that regions of little absorption correspond to little radiance. In the 800 to 1200 wave number spectral region (i.e., 8.3- to 12.5-μm wavelength region), the radiance is relatively low. This is consistent with the fact that the spectral region from 8 to 12 μm is a transmission window of the atmosphere with relatively little absorption of radiation.

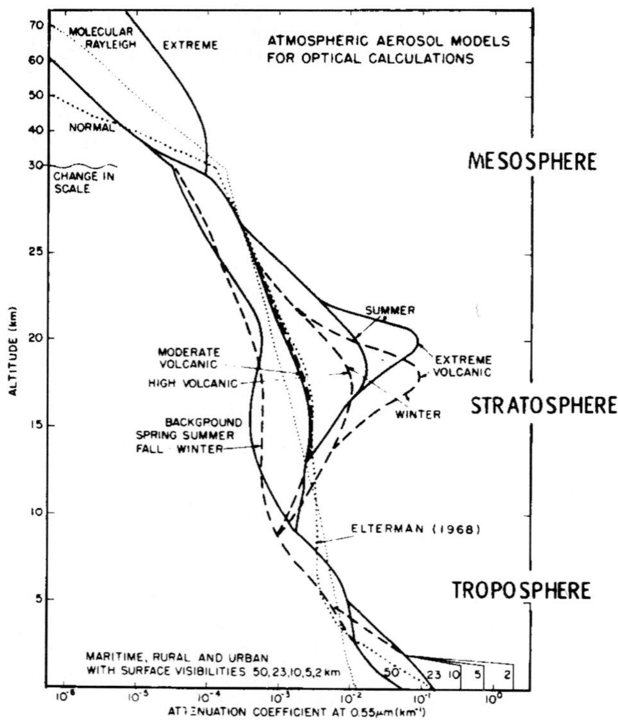

FIGURE 17 The vertical distribution of the aerosol extinction coefficient (at 0.55-μm wavelength) for the different atmospheric models. Also shown for comparison are the Rayleigh profile (dotted line). Between 2 and 30 km, where the distinction on a seasonal basis is made, the spring-summer conditions are indicated with a solid line and fall-winter conditions are indicated by a dashed line. *(From Ref. 3.)*

Surface Reflectivity and Multiple Scattering

The spectral intensity of naturally occurring light at the earth's surface is primarily due to the incident intensity from the sun in the visible to mid-IR wavelength range, and due to thermal emission from the atmosphere and background radiance in the mid-IR. In both cases, the optical radiation is affected by the reflectance characteristics of the clouds and surface layers. For instance, the fraction of light that falls on the earth's surface and is reflected back into the atmosphere is dependent upon the reflectivity of the surface, the incident solar radiation (polarization and spectral density), and the absorption of the atmosphere.

The reflectivity of a surface, such as the earth's surface, is often characterized using the bidirectional reflectance function (BDRF). This function accounts for the nonspecular reflection of light from common rough surfaces and describes the changes in the reflectivity of a surface as a function of the angle which the incident beam makes with the surface. In addition, the reflectivity of a surface is usually a function of wavelength. This latter effect can be seen in Fig. 21 which shows the reflectance of several common substances for normal incident radiation.[2] As seen in Fig. 21, the reflectivity of these surfaces is a strong function of wavelength.

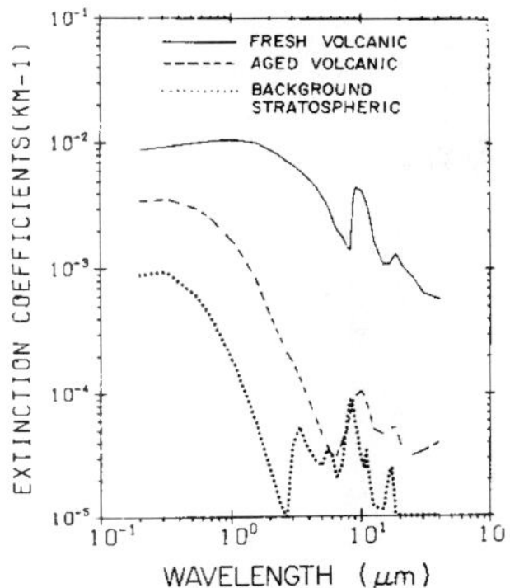

FIGURE 18 Extinction coefficients for the different stratospheric aerosol models (background, volcanic, and fresh volcanic). The extinction coefficients have been normalized to values around peak levels for these models.

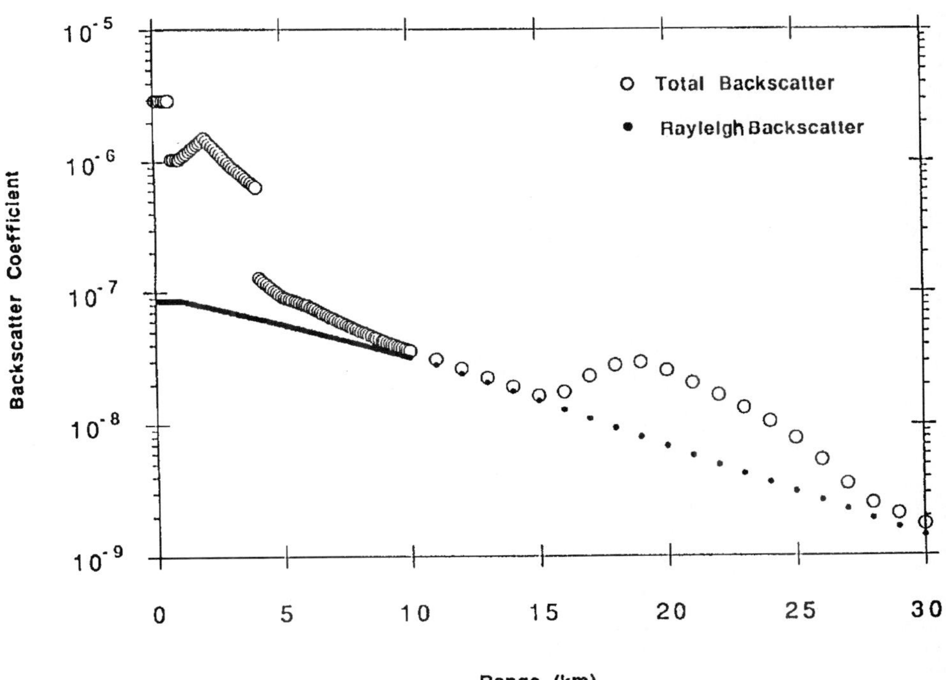

FIGURE 19 1.06-μm lidar backscatter coefficient measurements as a function of vertical altitude. *(From McCormick and Winker, Ref. 19.)*

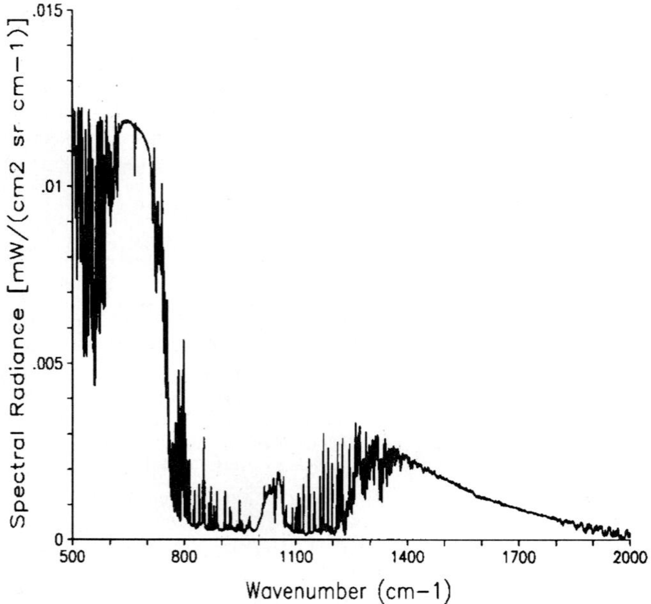

FIGURE 20 Spectral radiance (molecular thermal emission) measured on a clear day showing the relatively low value of radiance near 1000 cm^{-1} (i.e., 10-μm wavelength). *(Provided by Churnside.)*

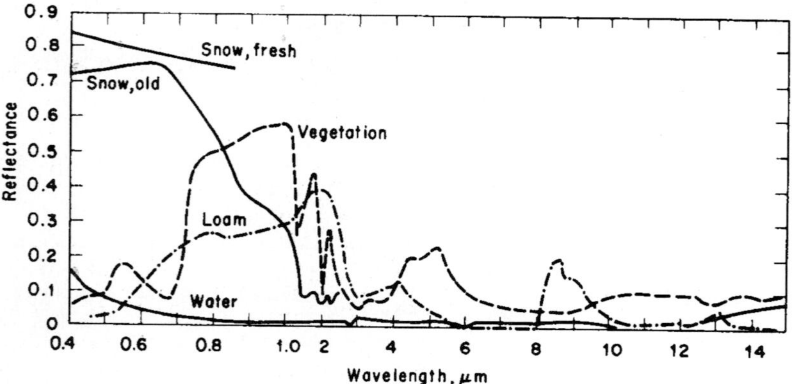

FIGURE 21 Typical reflectance of water surface, snow, dry soil, and vegetation. *(From Ref. 2.)*

The effect of multiple scattering sometimes must be considered when the scattered light undergoes more than one scatter event, and is rescattered on other particles or molecules. These multiple scattering events increase with increasing optical thickness and produce deviations from the Beer-Lambert law. Extensive analyses of the scattering processes for multiple scattering have been conducted and have shown some success in predicting the overall penetration of light through a thick dense cloud. Different computational techniques have been used including the Gauss-Seidel Iterative Method, Layer Adding Method, and Monte-Carlo Techniques.[3,5]

Additional Optical Interactions

In some optical experiments on the atmosphere, a laser beam is used to excite the molecules in the atmosphere to emit inelastic radiation. Two important inelastic optical processes for atmospheric remote sensing are fluorescence and Raman scattering.[5,20]

For the case of laser-induced fluorescence, the molecules are excited to an upper energy state and the reemitted photons are detected. In these experiments, the inelastic fluorescence emission is red-shifted in wavelength and can be distinguished in wavelength from the elastic scattered Rayleigh or Mie backscatter. Laser-induced fluorescence is mostly used in the UV to visible spectral region; collisional quenching is quite high in the infrared so that the fluorescence efficiency is higher in the UV-visible than in the IR. Laser-induced fluorescence is sometimes reduced by saturation effects due to stimulated emission from the upper energy levels. However, in those cases where laser-induced fluorescence can be successfully used, it is one of the most sensitive optical techniques for the detection of atomic or molecular species in the atmosphere.

Laser-induced Raman scattering of the atmosphere is a useful probe of the composition and temperature of concentrated species in the atmosphere. The Raman-shifted emitted light is often weak due to the relatively small cross section for Raman scattering. However, for those cases where the distance is short from the laser to the measurement cloud, or where the concentration of the species is high, it offers significant information concerning the composition of the gaseous atmosphere.

The use of an intense laser beam can also bring about nonlinear optical interactions as the laser beam propagates through the atmosphere. The most important of these are stimulated Raman scattering, thermal blooming, dielectric breakdown, and harmonic conversion. Each of these processes requires a tightly focused laser beam to initiate the nonlinear optical process. The reader is referred to Refs. 21 and 22 for information on these nonlinear optical processes.

44.5 PREDICTION OF ATMOSPHERIC OPTICAL TRANSMISSION: COMPUTER PROGRAMS AND DATABASES

During the past decade, several computer programs and databases have been developed which are very useful for the determination of the optical properties of the atmosphere. Many of these are based upon an ongoing program at the U.S. Air Force Geophysics Laboratory (AFGL), Phillips Laboratory, which has compiled extensive databases for the modeling of atmospheric optical phenomena. The latest versions of these programs and databases are the HITRAN database, FASCODE computer program, and the LOWTRAN or MODTRAN computer code. In addition, several PC (personal computer) versions of these database/computer programs have recently become available so that the user can easily use these computational aids.

The HITRAN database contains the spectroscopic optical parameters of over 30 different molecules in the atmosphere and lists these parameters for over 709,000 separate absorption lines from a wavelength of 0.44 μm to the microwave region.[9] The data contained in the database include the transition frequencies (or wave number ν), line intensity *S*, linewidth, transition probabilities, and energy levels of the transitions, along with 10 other parameters for each transition line. The HITRAN database has recently been updated in 1992, and a new version is expected in 1994 or 1995.

The FASCODE program calculates the high-resolution transmission spectrum of the atmosphere using the HITRAN database as input.[23,24] It uses computational algorithms to speed the computations of the line-by-line spectra, and contains information for calculating the continuum molecular extinction. FASCODE also calculates the radiance and transmittance of atmospheric slant paths, and can calculate the integrated transmittance through the atmosphere from the ground up to higher altitudes. Voigt lineshape profiles are also used to handle the transition from pressure-broadened lineshapes near ground level to the Doppler-dominated lineshapes at very high altitudes. Several representative models of the atmosphere are contained within FASCODE, so that the user can specify different seasonal and geographical models.

The LOWTRAN computer program calculates the low-resolution (broadband) transmission spectrum and background radiance of the atmosphere at moderate resolution $(20\ \text{cm})^{-1}$).[25,26] Molecular absorption, Rayleigh scattering, and aerosol extinction are computed. Slant path geometry can be specified. LOWTRAN contains many representative atmospheric models (geographical and seasonal), so that the user can see the differences between rural, maritime, winter, tropical, etc. models. Finally, it should be noted that a higher-resolution version of LOWTRAN, called MODTRAN, has recently been released, which calculates the transmission of the atmosphere with a resolution from 2 to 20 cm^{-1}.

Molecular Absorption Line Database: HITRAN

The HITRAN database contains optical spectral data on most of the major molecules in the atmosphere; details of HITRAN are covered in several recent journal articles.[9] The molecules contained in HITRAN are those given previously in Table 1, and cover the spectral range from 0.000001 cm^{-1} to 22,656 cm^{-1} (i.e., 0.4414 μm to 10^{10} μm). A copy of this database on magnetic tape can be obtained from the U.S. government, and has recently (1992) been put onto CD-ROM optical disks.[27] Each line in the database contains 14 molecular data items that consist of the molecule formula code, isotope type, transition frequency (cm^{-1}), line intensity *S* in cm/molecule, transition probability, air-broadened half-width (cm^{-1}/atm), self-broadened half-width (cm^{-1}/atm), lower state energy (cm^{-1}), temperature coefficient for air-broadened linewidth, upper-state global quanta index, lower-state global quanta index, upper- and lower-state quanta, error codes, and reference numbers. The density of the lines of the HITRAN database is shown in Fig. 22.

Figure 23 shows an output from a computer program that was used to search the HITRAN database and display some of the pertinent information. The data in HITRAN are in sequential order by transition frequency in wave numbers, and list the molecular name, isotope, absorption line strength *S*, transition probability *R*, air-pressure-broadened linewidth γ_g, lower energy state E'', and upper/lower quanta for the different molecules and isotopic species in the atmosphere.

Line-by-line Transmission Program: FASCODE

FASCODE is a large, sophisticated computer program that uses molecular absorption equations (similar to those under "Molecular Absorption") and the HITRAN database to calculate the high-resolution spectra of the atmosphere. The FASCODE program includes the effects of the continuum, Rayleigh and aerosol scattering, spectral radiance, slant

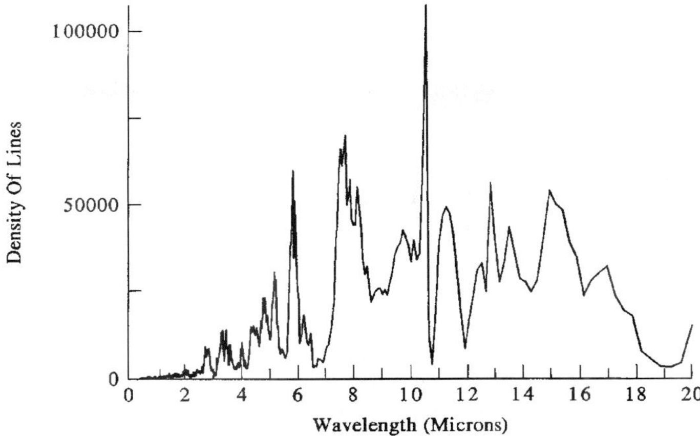

FIGURE 22 Density of absorption lines in HITRAN spectral database.

paths, and altitude density profiles. FASCODE generates the transmission, emission, and radiance of the atmosphere at a spectral resolution which is essentially infinite (i.e., $<0.0001\ \text{cm}^{-1}$).[24] Figure 24 shows a sample output generated from data produced from the FASCODE program and a comparison with experimental data obtained by J. Dowling at NRL.[23] As can be seen, the agreement is very good.

Broadband Transmission: LOWTRAN and MODTRAN

The LOWTRAN computer program does not use the HITRAN database directly, but uses absorption band models to calculate the moderate resolution ($20\ \text{cm}^{-1}$) transmission spectrum of the atmosphere. LOWTRAN uses extensive band-model calculations to speed up the computations, and provides an accurate and rapid means of estimating the transmittance and background radiance of the earth's atmosphere over the spectral

M	I	wn(cm-1)	S	R	g	E''	Q'	Q''
CO2	1	5000.05657	3.963E-27	6.206E	-6.0720	2783.34814		P 33
CO2	3	5000.06953	1.814E-26	1.495E	-6.0703	975.59039		P 51
CO2	1	5000.22347	1.477E-25	8.707E	-6.0674	2240.23755		R 63
H2O	1	5000.22500	2.160E-24	1.860E	-4.0088	2358.30396	14 014	15 015
CO2	2	5000.33473	1.649E-26	1.086E	-5.0894	1398.12122		R 8
CO2	1	5000.35108	4.283E-25	8.718E	-6.0686	2004.66040		R 58
N2O	1	5000.42800	2.990E-23	6.804E	-7.0720	316.67001		P 27
CO2	2	5000.47617	3.192E-25	4.751E	-6.0816	754.88672		P 16
CO2	1	5000.48083	4.913E-22	9.496E	-6.0719	464.17169		R 34
CO2	2	5000.48693	5.953E-27	4.687E	-6.0720	1710.65918		P 32
H2O	1	5000.54070	6.959E-27	1.096E	-8.0557	1411.61206	9 5 4	8 6 3
H2O	4	5000.75000	4.629E-27	3.108E	-6.0857	809.39502	8 3 5	8 4 4
CO2	1	5000.83037	9.346E-26	8.702E	-6.0669	2340.73218		R 65
H2O	4	5000.83600	4.469E-26	3.439E	-6.0923	308.61700	5 0 5	6 1 6
CO2	2	5000.85697	7.714E-24	5.195E	-6.0856	60.87380		R 12
CO2	3	5000.99115	2.146E-26	1.502E	-6.0704	938.09857		P 50

FIGURE 23 Example of data contained within the HITRAN database showing individual absorption lines, frequency, line intensity, and other spectroscopic parameters.

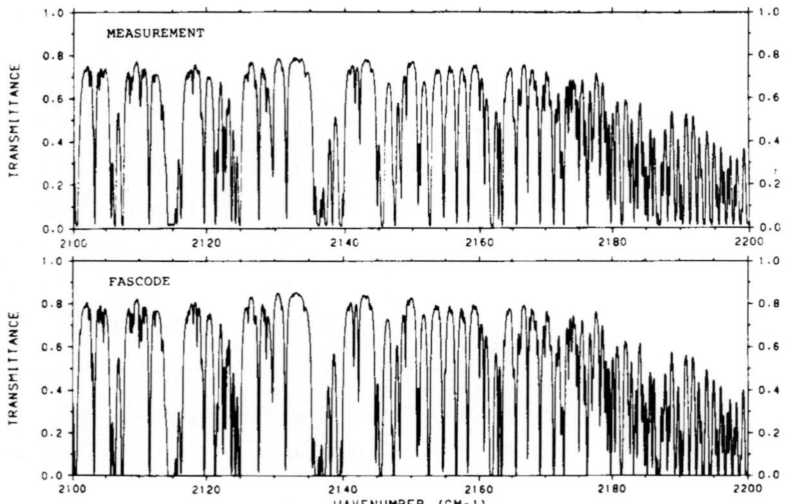

FIGURE 24 Comparison of an FASCOD2 transmittance calculation with an experimental atmospheric measurement (from NRL) over a 6.4-km path at the ground. *(Courtesy of S. Clough, Ref. 23.)*

interval of 350 cm^{-1} to 40,000 cm^{-1} (i.e., 250-nm to 28-μm wavelength). The spectral range of the LOWTRAN program extends into the UV. In the LOWTRAN program, the total transmittance at a given wavelength is given as the product of the transmittances due to molecular band absorption, molecular scattering, aerosol extinction, and molecular continuum absorption. The molecular band absorption is composed of four components of water vapor, ozone, nitric acid, and the uniformly mixed gases (CO_2, N_2O, CH_4, CO, O_2, and N_2).

The latest version of LOWTRAN (7) contains models treating solar and lunar scattered radiation, spherical refractive geometry, wind-dependent maritime aerosols, vertical structure aerosols, cirrus cloud model, and a rain model.[26] As an example, Fig. 25 shows a

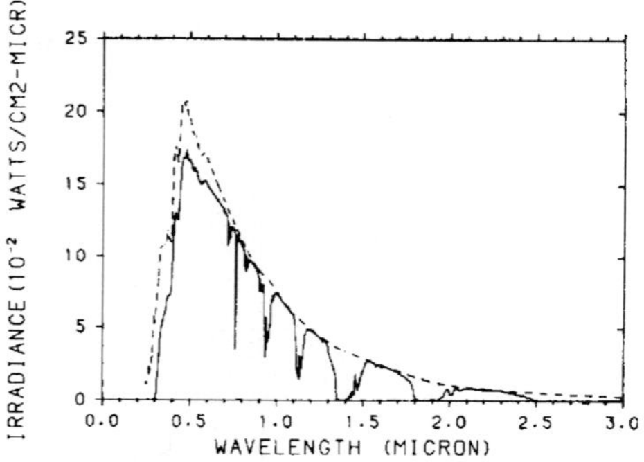

FIGURE 25 Solar radiance model (dashed line) and directly transmitted solar irradiance (solid line) for a vertical path, from the ground (U.S. standard 1962 model, no aerosol extinction) as used by the LOWTRAN program.

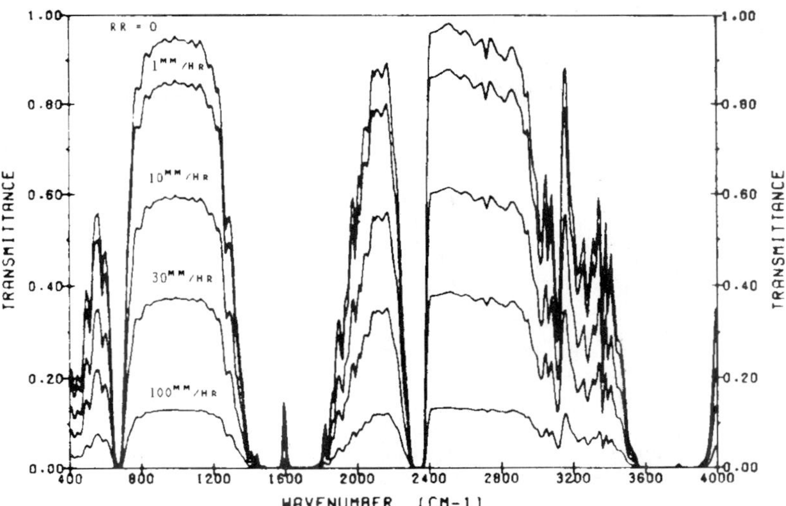

FIGURE 26 Atmospheric transmittance for different rain rates and for spectral frequencies from 400 to 4000 cm^{-1}. The measurement path is 300 m at the surface with $T = T_{\text{dew}} = 10°\text{C}$, with a meteorological range of 23 km in the absence of rain.

ground-level solar radiance model used by LOWTRAN, and Fig. 26 shows an example of a rain-rate model and its effect upon the transmission of the atmosphere as a function of rain rate in mm of water per hour.[26]

Extensive experimental measurements have been made to verify LOWTRAN calculations. Figure 27 shows a composite plot of the LOWTRAN-predicted transmittance and experimental data for a path length of 1.3 km at sea level.[3] As can be seen, the agreement is quite good. It is estimated that the LOWTRAN calculations are good to about

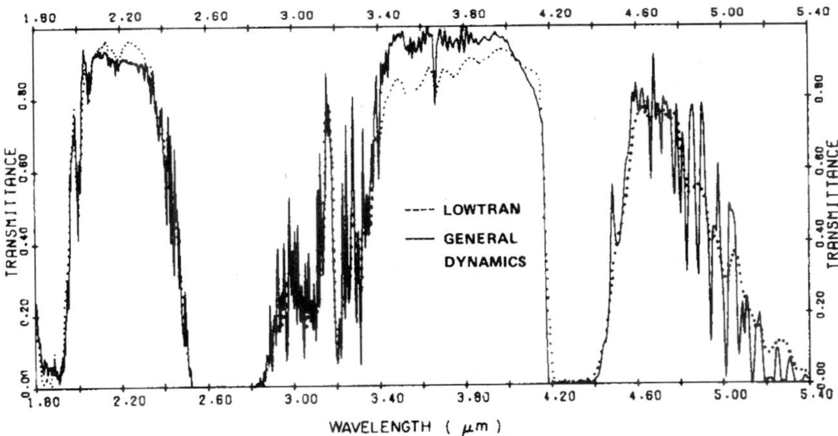

FIGURE 27 Comparison between LOWTRAN predicted spectrum and General Dynamics atmospheric measurements; range = 1.3 km at sea levels. *(From Ref. 3.)*

10 percent.[3] It should be added that the molecular absorption portion of the preceding LOWTRAN (moderate-resolution) spectra can also be generated using the high-resolution FASCODE/HITRAN program and then spectrally smoothing (i.e., degrading) the spectra to match that of the LOWTRAN spectra.

The most recent extension of the LOWTRAN program is the MODTRAN program. MODTRAN is similar to LOWTRAN but has increased spectral resolution. At present, the resolution for MODTRAN can be specified by the user between 2 and 20 cm^{-1}.

Programs and Databases for Use on Personal Computers

The preceding databases and computer programs have been converted or modified to run on different kinds of personal computers.[28,29] Currently, the HITRAN 1992 database can be obtained on CD-ROMs, and is also being offered as a random-access database which will fit on the hard disk drive of a personal computer (MS-DOS IBM-compatible computers). Several related programs are available, ranging from a complete copy of the FASCODE and LOWTRAN programs[28] to a simpler molecular transmission program of the atmosphere.[29] These programs calculate the transmission spectrum of the atmosphere and some show the overlay spectra of known laser lines. As an example, one version of these PC programs uses the HITRAN database and a transmission program to calculate the high-resolution spectrum of the atmosphere and displays the contribution of each individual molecule. Figure 28 shows the transmission spectrum produced by such a program.[29] The individual absorption lines are shown in the upper plot in Fig. 28 and the composite transmission is shown in the lower plot.

While these PC versions of the HITRAN database and transmission programs have become available only recently, they have already made a significant impact in the fields of atmospheric optics and optical remote sensing. They allow quick and easy access to atmospheric spectral data which was previously only available on a mainframe computer. It should be added that other computer programs are available which allow one to add or subtract different spectra generated by these HITRAN-based programs, from spectroscopic instrumentation such as FT-IR spectrometers or from other IR gas spectra databases. In the latter case, for example, the U.S. National Institute of Standards and Technology (NIST) has compiled a computer database of the qualitative IR absorption spectra of over 5200 different gases (toxic and other hydrocarbon compounds) with a spectral resolution of 4 cm^{-1}.[30] In addition, higher-resolution quantitative spectra for a limited group of gases can be obtained from several commercial companies.[30]

44.6 ATMOSPHERIC OPTICAL TURBULENCE

The most familiar effects of refractive turbulence in the atmosphere are the twinkling of stars and the shimmering of the horizon on a hot day. The first of these is a random fluctuation of the amplitude of light also known as scintillation. The second is a random fluctuation of the phase front that leads to a reduction in the resolution of an image. Other effects include the wander and break-up of an optical beam.

In the visible and near-IR region of the spectrum, the fluctuations of the refractive index in the atmosphere are determined by fluctuations of the temperature. These temperature fluctuations are caused by turbulent mixing of air of different temperatures. In the far-IR region, humidity fluctuations also contribute.

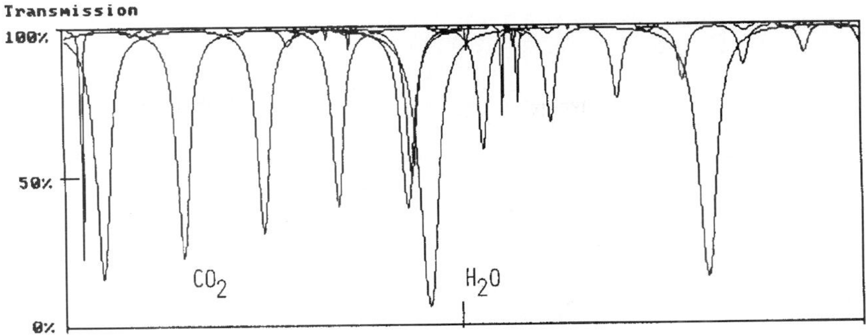

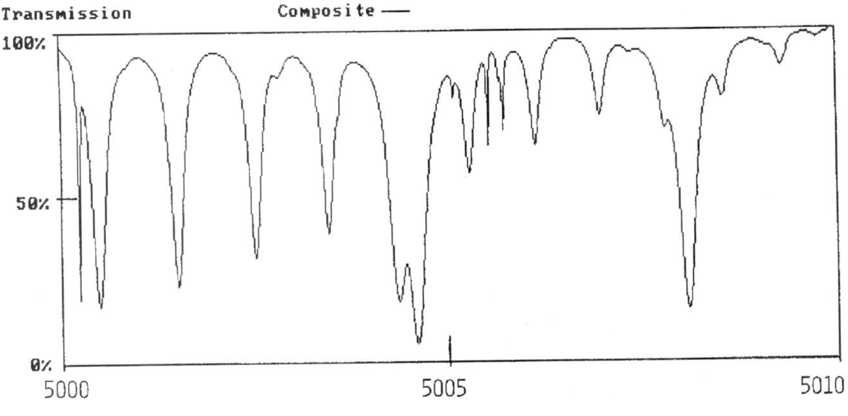

FIGURE 28 Example of generated atmospheric transmission spectrum obtained from HITRAN PC computer program showing individual molecular contribution and composite spectrum. *(From Ref. 29.)*

Turbulence Characteristics

Refractive turbulence in the atmosphere can be characterized by three parameters. The outer scale L_o is the length of the largest scales of turbulent eddies. The inner scale l_o is the length of the smallest scales. For eddies in the inertial subrange (sizes between the inner and outer scale), the refractive index fluctuations are best described by the structure function. This function is defined by

$$D_n(r_1, r_2) = \langle [n(r_1) - n(r_2)]^2 \rangle \tag{14}$$

where $n(r_1)$ is the index of refraction at point r_1 and the angle brackets denote an ensemble average. For homogeneous and isotropic turbulence it depends only on the distance between the two points r and is given by

$$D_n(r) = C_n^2 r^{2/3} \tag{15}$$

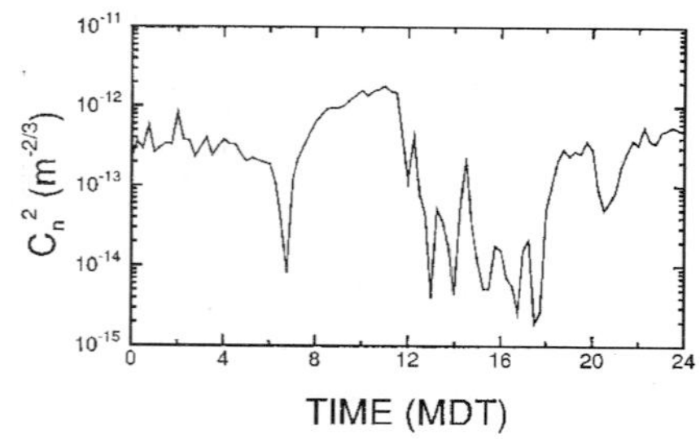

FIGURE 29 Plot of refractive-turbulence structure parameter C_n^2 for a typical summer day near Boulder, Colorado. *(Courtesy G. R. Ochs, NOAA WPL.)*

where C_n^2 is a measure of the strength of turbulence and is defined by this equation.

In the boundary layer (the lowest few hundred meters of the atmosphere), turbulence is generated by radiative heating and cooling of the ground. During the day, solar heating of the ground drives convective plumes. Refractive turbulence is generated by the mixing of these warm plumes with the cooler air surrounding them. At night, the ground is cooled by radiation and the cooler air near the ground is mixed with warmer air higher up by winds. A period of extremely low turbulence exists at dawn and at dusk when there is no temperature gradient in the lower atmosphere. Turbulence levels are also very low when the sky is overcast and solar heating and radiative cooling rates are low. Measured values of turbulence strength near the ground vary from less than 10^{-17} to greater than 10^{-12} $m^{-2/3}$ at heights of 2 to 2.5 m.[31,32]

Figure 29 illustrates typical summertime values near Boulder, Colorado. This is a 24-hour plot of 15-minute averages of C_n^2 measured at a height of about 1.5 m on August 22, 1991. At night, the sky was clear, and C_n^2 was a few parts times 10^{-13}. The dawn minimum is seen as a very short period of low turbulence just after 6:00. After sunrise, C_n^2 increases rapidly to over 10^{-12}. Just before noon, cumulus clouds developed, and C_n^2 became lower with large fluctuations. At about 18:00, the clouds, dissipated, and turbulence levels increased. The dusk minimum is evident just after 20:00, and then turbulence strength returns to typical nighttime levels.

From a theory introduced by Monin and Obukhov,[33] a theoretical dependence of turbulence strength on height in the boundary layer above flat ground can be derived.[34,35] During periods of convection C_n^2 decreases as the $-4/3$ power of height. At other times (night or overcast days), the power is more nearly $-2/3$. These height dependencies have been verified by a number of experiments over relatively flat terrain.[36–40] However, values measured in mountainous regions are closer to the $-1/3$ power of height day or night.[41] Under certain conditions, the turbulence strength can be predicted from meteorological parameters and characteristics of the underlying surface.[42–44]

Farther from the ground, no theory for the turbulence profile exists. Measurements have been made from aircraft[31,36] and with balloons.[45–47] Profiles of C_n^2 have also been measured remotely from the ground using acoustic sounders,[48–50] radar,[36,37,51–56] and optical techniques.[46,57–60] The measurements show large variations in refractive turbulence strength. They all exhibit a sharply layered structure in which the turbulence appears in layers of the order of 100 m thick with relatively calm air in between. In some cases these layers can be associated with orographic features; that is, the turbulence can be attributed to mountain lee waves. Generally, as height increases, the turbulence decreases to a minimum value that occurs at a height of about 3 to 5 km. The turbulence level then increases to a maximum at about the tropopause (10 km). Turbulence levels decrease rapidly above the tropopause.

Model turbulence profiles have evolved from this type of measurement. Perhaps the best available model for altitudes of 3 to 20 km is the Hufnagel model:[61,62]

$$C_n^2 = \left\{\left[(2.2 \times 10^{-53})h^{10}\left(\frac{W}{27}\right)^2\right] \exp\left(-\frac{h}{1000}\right) + 10^{-16} \exp\left(-\frac{h}{1500}\right)\right\} \exp\left[u(h, t)\right] \quad (16)$$

where h is the height above sea level in meters, W is the vertical average of the square of the wind speed, and u is a random variable that allows the random nature of the profiles to be modeled. W is defined by

$$W^2 = \frac{1}{1500} \int_{5000}^{20,000} v^2(h)\, dh \quad (17)$$

where $v(h)$ is the wind speed at height h. In data taken over Maryland, W was normally distributed with a mean value of 27 m/s and a standard deviation of 9 m/s. The random variable u is assumed to be a zero-mean, gaussian variable with a covariance function given by

$$\langle u(h, t)u(h + \delta h, t + \delta t)\rangle = A(\delta h/100) \exp(-\delta t/5) + A(\delta h/2000) \exp(-\delta t/80) \quad (18)$$

where

$$A(\delta h/L) = 1 - |\delta h/L|, \qquad \text{for} \qquad |h| < L \quad (19)$$

and equals 0 otherwise.

The time interval δt is measured in minutes. The average C_n^2 profile can be found by recognizing that $\langle \exp(u)\rangle = \exp(1)$. To extend the model to local ground level, one should add the surface layer dependence (e.g., $h^{-4/3}$ for daytime).

Another attempt to extend the model to ground level is the Hufnagel-Valley model.[63] This is given by

$$C_n^2 = 0.00594\left(\frac{W}{27}\right)^2 (h \times 10^{-5})^{10} \exp\left(-\frac{h}{1000}\right) + 2.7 \times 10^{-16} \exp\left(-\frac{h}{1500}\right) + A \exp\left(-\frac{h}{100}\right) \quad (20)$$

Where W is commonly set to 21 and A to 1.7×10^{-14}. This specific model is referred to as the $HV_{5/7}$ model because it produces a coherence diameter r_0 of about 5 cm and an isoplanatic angle of about 7 μrad for a wavelength of 0.5 μm. Although this is not as accurate for modeling turbulence near the ground, it has the advantage that the moments of the turbulence profile important to propagation can be evaluated analytically.[63]

The $HV_{5/7}$ model is plotted as a function of height in the dashed line in Fig. 30. The solid line in the figure is a balloon measurement taken in College Station, Pennsylvania. The data were reported with 20-m vertical resolution and smoothed with a gaussian filter with a 100-m $\exp(-1)$ full-width. This particular data set was chosen because it has a coherence diameter of about 5 cm and an isoplanatic angle of about 7 μrad. The layered structure of the real atmosphere is clear in the data. Note also the difference between the model atmosphere and the real atmosphere even when the coherence diameter and the isoplanatic angle are similar.

Less is known about the vertical profiles of inner and outer scales. Near the ground (1 to 2 m) we typically observe inner scales of 5 to 10 mm over flat grassland in Colorado. Calculations of inner scale from measured values of Kolmogorov microscale range from 0.5 to 9 mm at similar heights.[64] Aircraft measurements of dissipation rate were used along with a viscosity profile calculated from typical profiles of temperature and pressure to estimate a profile of microscale.[65] Values increase monotonically to about 4 cm at a height of 10 km and to about 8 cm at 20 km.

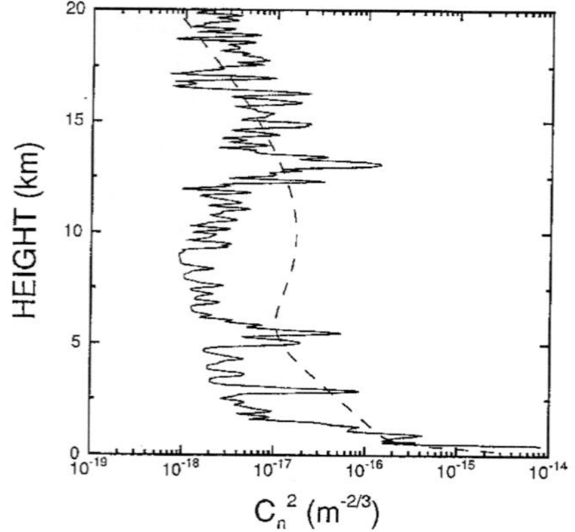

FIGURE 30 Turbulence strength C_n^2 as a function of height. The solid line is a balloon measurement made in College Station, Pennsylvania, and the dashed line is the $HV_{5/7}$ model. *(Courtesy R. R. Beland, Geophysics Directorate, Phillips Laboratory, U.S. Air Force.)*

Near the ground, the outer scale can be estimated using Monin-Obukhov similarity theory.[33] The outer scale can be defined as that separation at which the structure function of temperature fluctuations is equal to twice the variance. Using typical surface layer scaling relationships[66] we see that

$$L_0 = \begin{cases} 7.04h(1-7S_{\rm MO})(1-16S_{\rm MO})^{-3/2} & \text{for} \quad -2 < S_{\rm MO} < 0 \\ 7.04h(1+S_{\rm MO})^{-3}(1+2.75S_{\rm MO}^{2/3})^{-3/2} & \text{for} \quad 0 < S_{\rm MO} < 1 \end{cases} \tag{21}$$

where $S_{\rm MO}$ is the Monin-Obukhov stability parameter. For typical daytime conditions ($S_{\rm MO} < 0$), L_o is generally between $h/2$ and h.

Above the boundary layer, the situation is more complex. Weinstock[67] calculated that L_o should be about 330 m in moderate turbulence in the stratosphere. Barat and Bertin[68] measured outer scale values of 10 to 100 m in a turbulent layer using a balloon-borne instrument. Some recent optical data[69,70] suggest that the outer scale is generally smaller with a peak value of about 4 m at a height of 8.5 km. Other data[71] supply no evidence for an outer scale less than about 1 km. These results are still controversial and more measurements of outer scale are needed to resolve the issue.

Beam Wander

The first effect of refractive turbulence to consider is the wander of an optical beam in the atmosphere. The deviations of the centroid of a beam in each of the two orthogonal transverse axes will be independent gaussian random variables. This wander is generally characterized statistically by the variance of the angular displacement. In isotropic turbulence, the variances in the two axes are equal and the magnitude of the displacement is a Rayleigh random variable with a variance that is twice that of the displacement in a single axis. Both the single-axis and the magnitude variances are reported in the literature.

Three main approaches to the calculation of beam-wander variance have been used. (1) If diffraction effects are negligible and if the path-integrated turbulence is small enough, the geometric optics approximation[72–75] can be used. Diffraction effects are negligible when the aperture diameter is greater than the Fresnel zone size.[75] The other condition requires that the product of the transverse coherence of the field and the aperture diameter is greater than the square of the Fresnel zone size.[75] (2) If diffraction must be considered, the Huygens-Fresnel approximation[76–79] can be used. (3) The most complete theory uses the Markov random process approximation to the moment equation.[80–82] In the two types of calculation that include diffraction, beams with a gaussian irradiance profile are generally assumed.

In the geometric optics approximation, the variance of the angular displacement in a single axis is given by[75]

$$\sigma_d^2 = 2.92 D^{-1/3} \int_0^L dz C_n^2(z) \frac{\left(1 - \frac{z}{L}\right)^2}{\left|1 - \frac{z}{F}\right|^{1/3}} \tag{22}$$

where D is the diameter of the initial beam, z is position along the path, L is the propagation path length, and F is the geometric focal range of the initial beam (negative for a diverging beam). For homogeneous turbulence and a beam that does not come to a focus between the transmitter and observation plane, Eq. (22) reduces to

$$\sigma_d^2 = 0.97 C_n^2 D^{-1/3} L_2 F_1\left(\tfrac{1}{3}, 1; 4; \frac{L}{F}\right) \tag{23}$$

where ${}_2F_1$ is the hypergeometric function. The hypergeometric function is 1 for a collimated beam and 1.125 for a focused beam.

In the Markov approximation, the single-axis variance is given by[80]

$$\sigma_d^2 = 4\pi^2 \int_0^L dz \left(1 - \frac{z}{L}\right)^2 \int_0^\infty dK K^3 \Phi_n(K, z)$$
$$\times \exp\left\{-\frac{K^2 D^2}{8}\left[\left(1 - \frac{z}{F}\right)^2 + \frac{16 z^2}{K^2 D^4}\right] - \pi D_\Psi\left(\frac{Kz}{k}\right)\right\} \tag{24}$$

where $\Phi_n(K, z)$ is the path-dependent refractive index spectrum, K is the wave number of turbulence, D is the $\exp(-1)$ intensity diameter of the initial beam, k is the optical wave number, and $D_\Psi(\rho)$ is the wave structure function for separation ρ of a spherical wave. The structure function is given by

$$D_\Psi(\rho) = 8\pi^2 k^2 \int_0^z dz' \int_0^\infty dK' K' \Phi_n(K', z')[1 - J_0(K'\rho z'/z)], \tag{25}$$

where J_0 is the zero-order Bessel function of the first kind. The last term in the exponential of Eq. (24) is a correction term for strong turbulence. The middle term includes the effects of diffraction.

Beam Spreading

The next effect of refractive turbulence to consider is the spread of an optical beam as it propagates through the atmosphere. There are two types of beam spread denoted as long-term and short-term. The long-term beam spread is defined as the turbulence-induced beam spread observed over a long time average. It includes the effects of the slow wander of the entire beam. The short-term beam spread is defined as the beam spread observed at an instant of time. It does not include the effects of beam wander and is often approximated by the long-term beam spread with the effects of wander removed, although the two are not identical.

We can consider beam wander to be caused by turbulent eddies that are larger than the beam. Short-term beam spread is caused by turbulent eddies that are smaller than the beam. There are more small eddies than large in the beam at any time which implies that the beam spread at any instant is averaged over more eddies. As a result, the fluctuations of short-term beam spread are much smaller than those of beam wander and are typically neglected. The primary effect of short-term beam spread is to spread the average energy over a larger area. Thus, the average value of the on-axis irradiance is reduced, and the average value of the irradiance at large angles is increased.

The extended Huygens-Fresnel principle can be used to calculate the long-term spread of a gaussian beam in refractive turbulence.[83–87] The exp (−1) intensity radius of a gaussian beam is

$$p_1 = \left[\frac{4}{k^2D^2} + \frac{D^2}{4}\left(1 - \frac{L}{F}\right)^2 + \frac{4}{k^2\rho_0^2}\right]^{1/2} \tag{26}$$

where D is the exp (−1) intensity diameter of the source and ρ_o is the phase coherence length that would be observed for a point source propagating from the receiver to the transmitter. The first term in this equation is the diffraction beam spread, the second is the geometrical optics projection of the transmitter aperture, and the final term is the total turbulence-induced spread.

The phase coherence length is defined as the transverse separation at which the coherence of the field is reduced to exp (−1). If the coherence length ρ_o is much larger than the inner scale, the structure function is given by $D_\Psi(\rho) = 2(\rho/\rho_o)^{5/3}$ and

$$\rho_o = \left[1.46k^2 \int_0^L dz \left(\frac{z}{L}\right)^{5/3} C_n^2(z)\right]^{-3/5} \tag{27}$$

If the coherence length is much smaller than the inner scale, then $D_\Psi(\rho) = 2(\rho/\rho_o)^2$ and

$$\rho_o = \left[1.86k^2 \int_0^L dz \left(\frac{z}{L}\right)^2 C_n^2(z) l_o^{-1/3}(z)\right]^{-1/2} \tag{28}$$

Typical values of p_o for spherical-wave propagation through homogeneous turbulence are presented in Fig. 31. In these plots there is a slope change from $L^{-3/5}$ to $L^{-1/2}$ where p_o

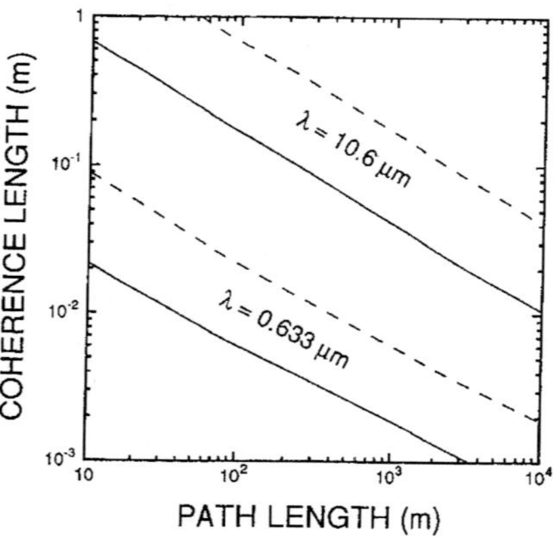

FIGURE 31 Coherence length ρ_o as a function of path length for two common laser wavelengths. An inner scale of 1 cm, and values of 10^{-12} (solid lines) and 10^{-13} (dashed lines) were used in the calculations.

is about equal to the inner scale of 1 cm. This small change is nearly imperceptible in the figure.

The extended Huygens-Fresnel principle has been used to calculate the short-term beam spread by explicitly subtacting the beam wander.[88–91] If p_o and l_o are much smaller than D, then the short-term beam spread is approximately given by

$$p_s = \left\{\frac{4}{k^2D^2} + \frac{D^2}{4L^2}\left(1 - \frac{L}{F}\right)^2 + \frac{4}{k^2\rho_0^2}\left[1 - 0.62\left(\frac{\rho_o}{D}\right)^{1/3}\right]^{6/5}\right\}^{1/2} \tag{29}$$

If ρ_o is much greater than D, the turbulence-induced component of beam spreading can be neglected.

If inner scale and outer scale effects are included, more complicated integral expressions result.[92] Numerical calculations were performed for truncated gaussian beams with central obscurations.[93] The following approximation was obtained by a curve fit to the results:

$$p_s = \left(1 + 0.182\frac{D_{\text{eff}}^2}{r_o^2}\right)^{1/2} p_d \qquad \text{for} \qquad \frac{D_{\text{eff}}}{r_o} < 3$$

$$p_s = \left[1 + \left(\frac{D_{\text{eff}}}{r_o}\right)^2 - 1.18\left(\frac{D_{\text{eff}}}{r_o}\right)^{-5/3}\right]^{1/2} p_d \qquad \text{for} \qquad 3 < \frac{D_{\text{eff}}}{r_o} < 7.5 \tag{30}$$

where D_{eff} is the effective diameter of the truncated aperture, $r_o = 2.099\ p_o$, and p_d is the diffraction-limited value. These expressions agree fairly well with available data.[78,94,95]

Imaging and Heterodyne Detection

The problem of imaging through the turbulent atmosphere is similar to the problem of beam propagation through the atmosphere. The dancing of an image in the focal plane of an imaging system is mathematically equivalent to the wander of a beam focused at the object by the same optical system. The resolution of the short-exposure image is equivalent to the short-term beam spread of the same focused beam. The resolution for a long exposure is related to the long-term beam spread.

Thus, the position of the image of a point object will drift in each axis in the focal plane. The variance of that drift is given by[96]

$$\sigma_i^2 = 1.10C_n^2D^{-1/3}LF^2 \tag{31}$$

where D is the aperture diameter of the imaging system, F is its focal length, and L is the distance to the object.

Fried[97] used the idea of tilt correction to calculate the average image resolution for a short-exposure image in the turbulent atmosphere. This problem is mathematically equivalent to the propagation of a beam in the opposite direction if the imaging aperture replaces the beam width. These results were refined by Lutomirski et al.[98] and applied to a space-to-ground path by Valley.[92]

Image resolution is also related to the signal-to-noise ratio of an optical heterodyne receiver. The long-exposure resolution is equivalent to a staring receiver. The short-exposure resolution is equivalent to a receiver that employs tilt-correction of the signal or of the local oscillator.[97,99–101]

Scintillation

The refractive index inhomogeneities that distort the optical phase front also produce amplitude scintillation at some distance. The first cases to be considered were plane-wave and spherical-wave propagation through weak path-integrated turbulence where the weak-turbulence condition requires that fluctuations of irradiance be much less than the mean value. Tatarskii[102] used a perturbation approach to the wave equation. Lee and Harp[103] used a physical approach to arrive at the same results. These results are summarized in a number of good reviews.[90,91,104,105]

We will first consider the weak-turbulence results. For propagation from space to the ground, the plane wave formula is generally valid. The variance of irradiance fluctuations (normalized by the mean irradiance value) is given by[104]

$$\sigma_r^2 = k^{7/6} \sec^{11/6} \theta \int_0^\infty dh h^{5/6} C_n^2(h) \tag{32}$$

where k is the optical wave number, θ is the zenith angle, and h is the height of the receiver above the ground. This expression is valid as long as the path-integrated turbulence is weak enough that the variance is much less than unity. This condition is usually met for near-zenith propagation.

For propagation of diverging waves near the ground, the spherical-wave approximation is often valid. Assuming constant turbulence along the path, the weak-turbulence variance in this case is given by[104]

$$\begin{aligned} \sigma_r^2 &= \exp\,[0.5k^{7/6}L^{11/6}C_n^2] - 1 \qquad \text{for} \qquad l_o < \sqrt{\lambda L} \\ \sigma_r^2 &= \exp\,[1.28L^3 l_o^{-7/3}C_n^2] - 1 \qquad \text{for} \qquad l_o > \sqrt{\lambda L} \end{aligned} \tag{33}$$

where L is the path length and l_o is the inner scale of turbulence.

For narrow-beam propagation, the effects of the finite beam must be considered. Kon and Tatarskii[106] calculated the amplitude fluctuations of a collimated beam using the perturbation technique. Schmeltzer[107] extended these results to include focused beams. These results were used to obtain numerical values for a variety of propagation conditions.[108–110] Ishimaru[111–113] used a spectral representation to obtain similar results. Under certain conditions, one sees a reduction in the variance on the optical axis[108,109,114,115] and an increase off of the optical axis.[116]

The spatial scale of weak scintillations is about equal to the larger of either the Fresnel zone size $(L/k)^{1/2}$ or the inner scale.[104] Scintillation will be reduced if an aperture larger than the scale size is used to collect the light. If the aperture diameter D is much larger than the scale size, the reduction factor can be expressed as[117,118]

$$A = C\left(\frac{D}{D_0}\right)^{-7/3} \tag{34}$$

where D_0 is $(\lambda L)^{1/2}$ and C is 4.71 for a plane wave and 23.5 for a spherical wave when the inner scale is much smaller than the Fresnel zone. If the inner scale is much larger than the Fresnel zone, D_0 is the inner scale and C is 0.45 for a plane wave and 9.17 for a spherical wave.

It is generally accepted that the probability density function for weak scintillation is log normal.[90,102,104,119] This density function has the form

$$p(I) = \frac{1}{\sqrt{2\pi}\,\sigma_I I} \exp\left[-\frac{1}{2\sigma_r^2}(\ln I + 0.5\sigma_r^2)^2\right] \tag{35}$$

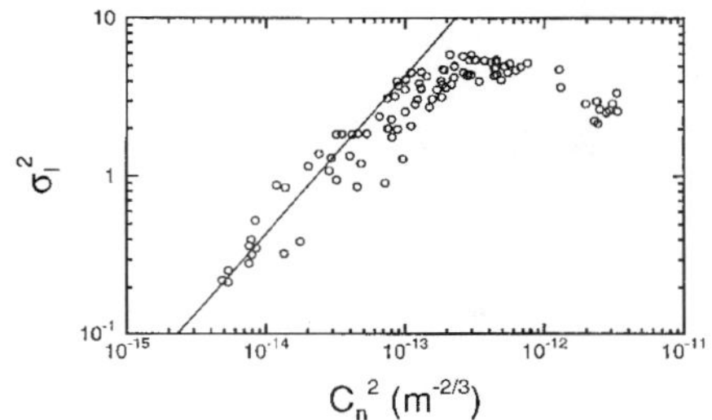

FIGURE 32 Plot of normalized variance of irradiance σ_I^2 as a function of turbulence strength C_n^2. Circles are data taken with a 488-nm wavelength laser over a 1200-m path, and the line is the corresponding weak-turbulence theory.

The weak-turbulence theory is, in essence, a single-scattering theory. As the path-integrated turbulence becomes larger, multiple-scattering effects become important, and this theory breaks down. Actual observed values of irradiance variance are smaller than predicted in this region as shown in Fig. 32. The circles in this figure are one-minute averages of irradiance variance for 488-nm laser light propagated across 1200 m of flat grassland. The solid line is the weak-turbulence approximation.

In these and other experiments, variance values reach a peak value of between 3 and 5 for a spherical wave and they begin to decrease with increasing turbulence strength.[120–123] In the limit of infinite path-integrated turbulence, the normalized variance of irradiance is predicted to approach unity.[124,125] In the intermediate region near the peak irradiance, numerical evaluation of currently available theories is impractical.

An asymptotic theory has been developed for very large path-integrated turbulence.[125–129] We assume that the inner scale is much less than the Fresnel zone in this regime because long paths are necessary to obtain large values of path-integrated turbulence. Then the irradiance variance can be approximated by

$$\begin{aligned} \sigma_r^2 &= 1 + 1.9\sigma_R^{-4/5} \qquad \text{for spherical waves} \\ \sigma_r^2 &= 1 + 0.86\sigma_R^{-4/5} \qquad \text{for plane waves} \end{aligned} \tag{36}$$

where σ_R^2 is the appropriate (spherical-wave or plane-wave) variance calculated from the Rytov approximation given by Eq. (32) or (33).

In very strong turbulence two distinct spatial scales are evident in the scintillation pattern.[125–129] The smaller scale is about the size of the coherence length ρ_o discussed under "Beam Spreading." The larger scale is the size of the scattering disk which is the ratio of the square of the Fresnel zone to the coherence length $(L/k\rho_o)$. As turbulence increases, the small scale becomes smaller and the large scale becomes larger. The strength of the small-scale fluctuations is constant in this regime and contributes a value of unity to the variance. The large-scale fluctuations contribute the rest of the variance and become weaker with increasing turbulence strength.

At high turbulence values, the density function is a log-normally modulated exponential.[123,130] This density function has the form

$$p(I) = \frac{1}{\sqrt{2\pi}\,\sigma_z} \int_0^\infty \frac{dz}{z^2} \exp\left[-\frac{I}{z} - \frac{(\ln z + \frac{1}{2}\sigma_z^2)^2}{2\sigma_z^2}\right] \tag{37}$$

where the parameter σ_z^2 is related to the irradiance variance by the relationship

$$\sigma_I^2 = 2\exp(\sigma_z^2) - 1 \tag{38}$$

An interesting feature of this density function is that the most likely value of the irradiance is 0. Although the observed signal is occasionally very bright, quite often there is no signal unless some sort of averaging is performed.

44.7 EXAMPLES OF ATMOSPHERIC OPTICAL REMOTE SENSING

One of the more important applications of atmospheric optics is optical remote sensing. Atmospheric optical remote sensing concerns the use of an optical or laser beam to remotely sense information about the atmosphere or a distant target. Optical remote sensing measurements are diverse in nature and include the use of a spectral radiometer aboard a satellite for the detection of trace species in the upper atmosphere, the use of spectral emission and absorption from the earth for the detection of the concentration of water vapor in the atmosphere, the use of lasers to measure the range-resolved distribution of several molecules including ozone in the atmosphere, and Doppler wind measurements. In this section, some typical optical remote sensing experiments will be presented in order to give a flavor of the wide variety of atmospheric optical measurements that are currently being conducted. More in-depth references can be found in several current journal papers and conference proceedings.[131–136]

The Upper Atmospheric Research Satellite (UARS) was placed into orbit in September 1991 as part of the Earth Observing System. One of the optical remote sensing instruments aboard UARS is the High Resolution Doppler Imager (HRDI) developed by P. Hays' and V. Abreu's group at the University of Michigan.[137] The HRDI is a triple etalon Fabry-Perot Interferometer designed to measure Doppler shifts of molecular absorption and emission lines in the earth's atmosphere in order to determine the wind velocity of the atmosphere. A wind velocity of 10 m/s causes a Doppler shift of 2×10^{-5} nm for the oxygen lines detected near a wavelength of 600–800 nm. A schematic of the instrument is given in Fig. 33*a* which shows the telescope, triple Fabry-Perots, and unique imaging Photo-Multiplier tubes to detect the Fabry-Perot patterns of the spectral absorption lines. The HRDI instrument is a passive remote sensing system and uses the reflected or scattered sunlight as its illumination source. Figure 33*b* shows the wind field measured by UARS (HRDI) for an altitude of 90 km.

Another kind of atmospheric remote sensing instrument is represented by an airborne laser radar (lidar) system operated by E. Browell's group at NASA/Langley.[138] Their system consists of two pulsed, visible-wavelength dye laser systems that emit short (10-ns) pulses of tunable optical radiation that can be directed toward aerosol clouds in the atmosphere. By the proper tuning of the wavelength of these lasers, the difference in the absorption due to ozone, water vapor, or oxygen in the atmosphere can be measured. Because the laser pulse is short, the timing out to the aerosol scatterers can be determined and range-resolved lidar measurements can be made. Figure 34 shows range-resolved lidar backscatter profiles obtained as a function of the lidar aircraft ground position. The variation in the atmospheric density and ozone distribution as a function of altitude and distance is readily observed.

A Coherent Doppler Lidar is one which is able to measure the Doppler shift of the backscattered lidar returns from the atmosphere. Several Doppler lidar systems have been developed which can determine wind speed with an accuracy of 0.1 m/s at ranges of up to 15 km. One such system is operated by M. Hardesty's group at NOAA/WPL for the mapping of winds near airports and for meteorological studies.[139] Figure 35 shows a two-dimensional plot of the measured wind velocity obtained during the approach of a wind gust front associated with colliding thunderstorms; the upper figure shows the real-time Doppler lidar display of the measured radial wind velocity, and the lower plot shows the computed wind velocity. As seen, a Doppler lidar system is able to remotely measure the wind speed with spatial resolution on the order of 100 m. A similar Doppler

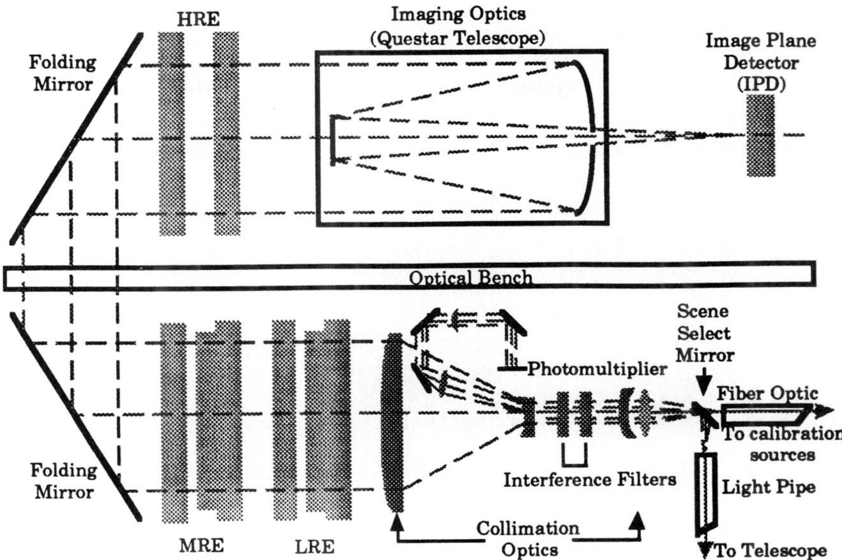

FIGURE 33 (*a*) Optical layout of the Upper Atmospheric Satellite (UARS) High Resolution Doppler Imager (HRDI) instrument. F.O. = fiber optic, LRE = low-resolution etalon, MRE = medium-resolution etalon, HRE = high-resolution etalon. *(From Hays, Ref. 137.)*

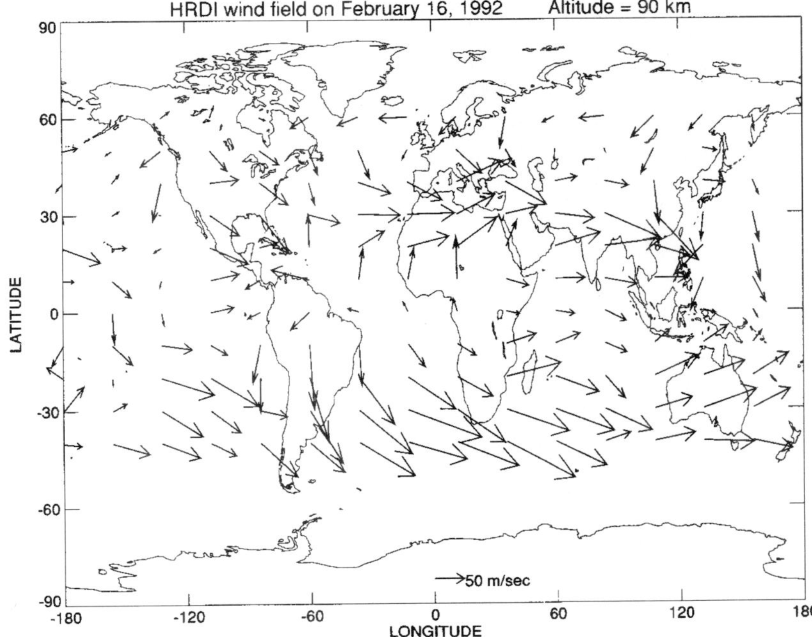

FIGURE 33 (*b*) Upper atmospheric wind field measured by UARS/HRDI satellite instrument. *(From Hays, Ref. 137.)*

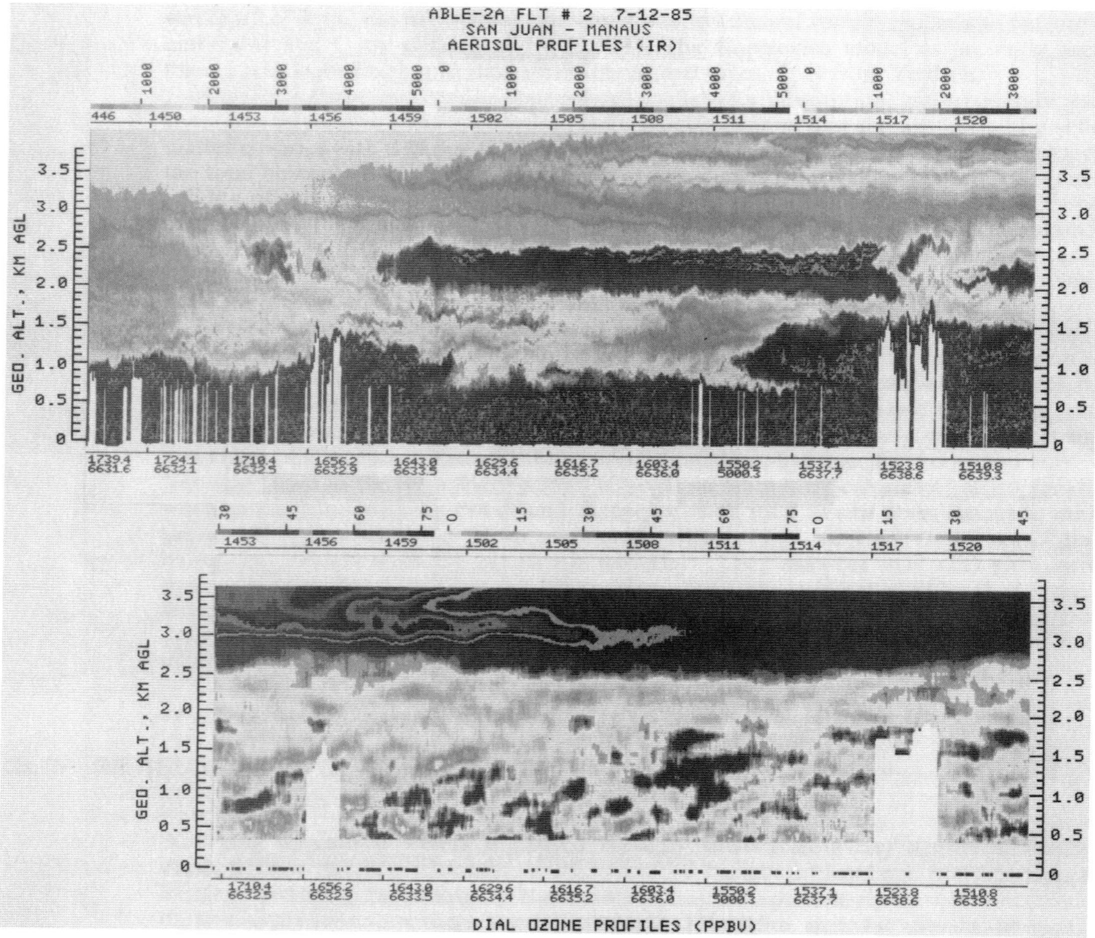

FIGURE 34 Range-resolved lidar measurements of atmospheric aerosols and ozone density. *(From E. Browell, Ref. 138.)*

lidar system is being considered for the early detection of windshear in front of commercial aircraft.

A further example of atmospheric optical remote sensing is that of the remote measurement of the global concentration and distribution of atmospheric aerosols and particulates. P. McCormick's group at NASA/Langley has developed the SAGE II satellite system which is part of a package of instruments to detect global aerosol and selected species concentrations in the atmosphere.[140] This system measures the difference in the optical radiation emitted from the earth's surface and the differential absorption due to known absorption lines or spectral bands of several species in the atmosphere, including ozone. The instrument also provides for the spatial mapping of the concentration of aerosols and particulates in the atmosphere, and an example of such a measurement is shown in Fig. 36. This figure shows the measured concentration of aerosols and particulates after the eruption of Mt. Pinatubo and demonstrates the global circulation and transport of the injected material into the earth's atmosphere.

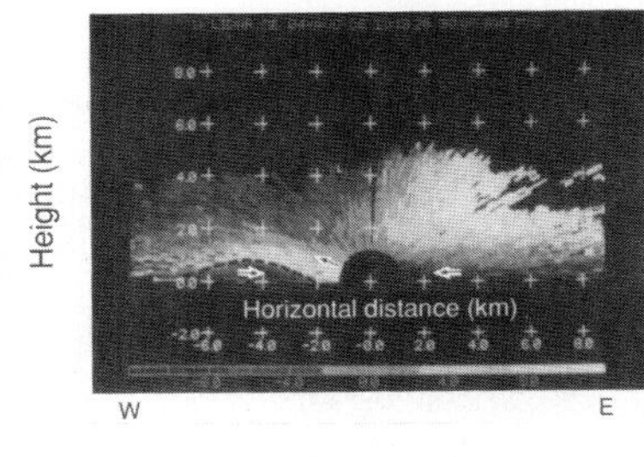

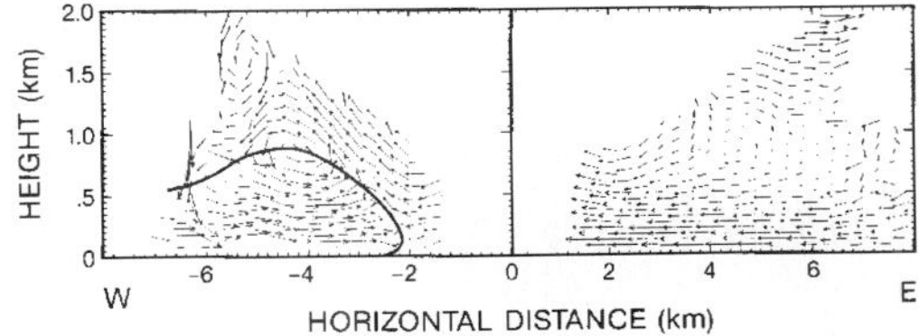

FIGURE 35 Coherent Doppler lidar measurements of atmospheric winds showing velocity profile of gust front. Upper plot is real-time display of Doppler signal and lower plot is range-resolved wind field. *(From M. Hardesty, Ref. 139.)*

The preceding examples are just a few of many different optical remote sensing instruments that are being used to measure the physical dynamics and chemical properties of the atmosphere. As is evident in these examples, an understanding of atmospheric optics plays an important and integral part in these measurements.

44.8 METEOROLOGICAL OPTICS

One of the most colorful aspects of atmospheric optics is that associated with meteorological optics. Meteorological optics involves the interplay of light with the atmosphere and the physical origin of the observed optical phenomena. Several excellent books have been written about this subject, and the reader should consult these and the contained references.[141,142] While it is beyond the scope of this chapter to present an overview of meteorological optics, some specific optical phenomena will be described to give the reader a sampling of some of the interesting effects involved in naturally occurring atmospheric and meteorological optics.

Some of the more common and interesting meteorological optical phenomena involve rainbows, ice-crystal halos, and mirages. The rainbow in the atmosphere is caused by internal reflection and refraction of sunlight by water droplets in the atmosphere. Figure 37 shows the geometry involved in the formation of a rainbow, including both the primary and larger secondary rainbow. Because of the dispersion of light within the water droplet, the colors or wavelengths are separated in the backscattered image. Although rainbows are commonly observed in the visible, such refraction also occurs in the infrared spectrum. As an example, Fig. 38 shows a natural rainbow in the atmosphere photographed with IR-sensitive film by R. Greenler.[142]

The phenomena of halos, arcs, and spots are due to the refraction of light by ice crystals suspended in the atmosphere. Figure 39 shows a photograph of collected ice crystals as

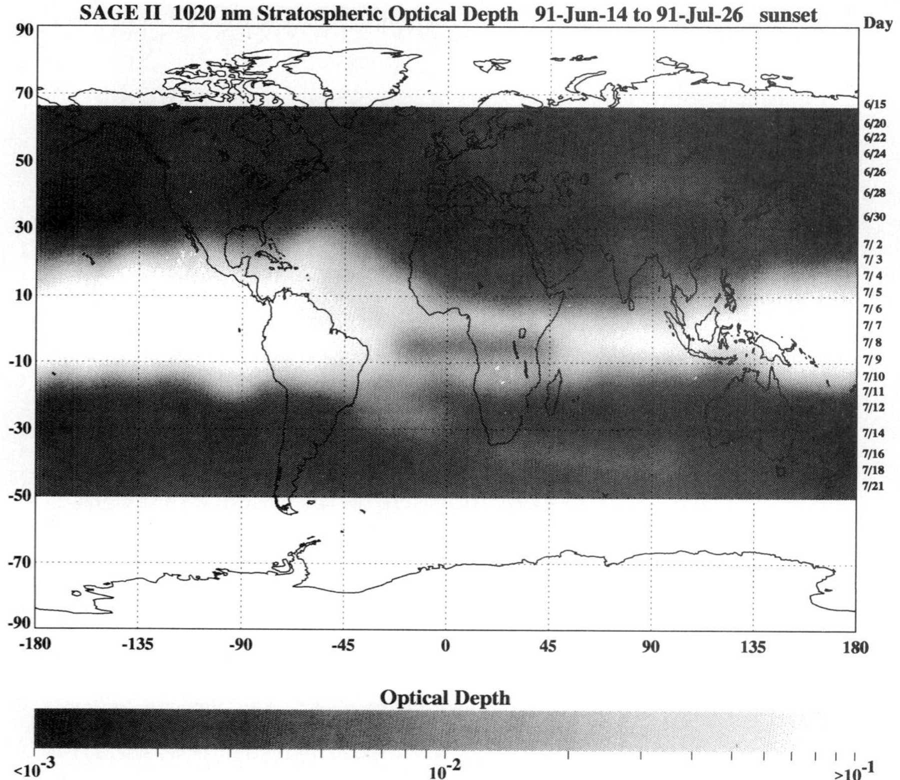

FIGURE 36 Measurement of global aerosol and particulate concentration using SAGE II satellite following eruption of Mt. Pinatubo. *(From P. McCormick, Ref. 140.)*

they fell from the sky. The geometrical shape, especially the hexagonal (six-sided) crystals, play an important role in the formation of halos and arcs in the atmosphere.

The common optical phenomenon of the mirage is caused by variation in the temperature and thus, the density of the air as a function of altitude or transverse geometrical distance. As an example, Fig. 40 shows the geometry of light-ray paths for a case where the air temperature decreases with height to a sufficient extent over the viewing angle that the difference in the index of refraction can cause a refraction of the image similar to total internal reflection. The heated air (less dense) near the ground can thus act like a mirror, and reflect the light upward toward the viewer. As an example, Fig. 41 shows a photograph taken by Greenler of motorcycles on a hot road surface. The reflected image of the motorcycles "within" the road surface is evident. There are many manifestations of mirages dependent upon the local temperature gradient and geometry of the situation. In many cases, partial and distorted images are observed leading to the almost surreal connotation often associated with mirages.

Finally, another atmospheric meteorological optical phenomenon is that of the green flash. A green flash is observed under certain conditions just as the sun is setting below the horizon. This phenomenon is easily understood as being due to the different relative displacement of each different wavelength or color in the sun's image due to spatially distributed refraction of the atmosphere.[142] As the sun sets, the last image to be observed

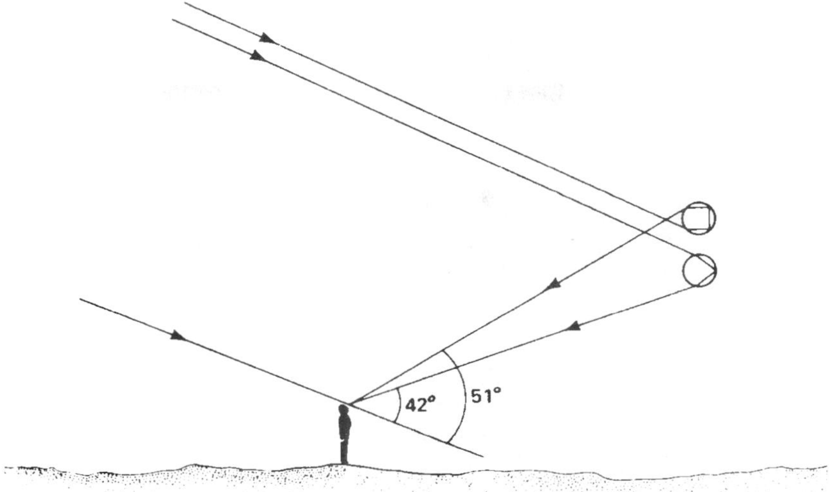

FIGURE 37 Different raindrops contribute to the primary and to the larger, secondary rainbow. *(From R. Greenler, Ref. 142.)*

FIGURE 38 A natural infrared rainbow. *(Photograph courtesy of R. Greenler, Ref. 142.)*

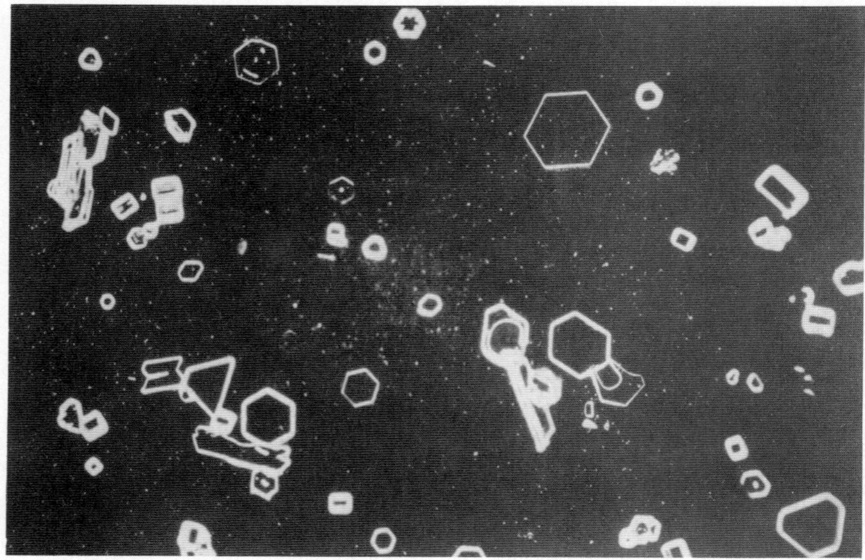

FIGURE 39 Photograph of magnified small ice crystals collected as they fell from the sky. *(From R. Greenler, Ref. 142.)*

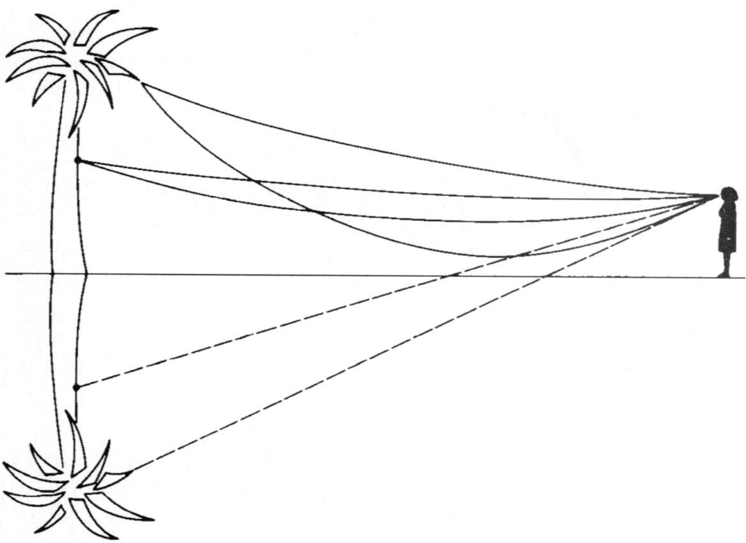

FIGURE 40 The origin of the inverted image in the desert mirage. *(From R. Greenler, Ref. 142.)*

FIGURE 41 The desert (or hot-road) mirage. In the inverted part of the image you can see the apparent reflection of motocycles, cars, painted stripes on the road, and the grassy road edge. *(From R. Greenler, Ref. 142.)*

is the shortest wavelength color, blue. However, most of the blue light has been Rayleigh scattered from the image seen by the observer so that the last image observed is closer to a green color. Under extremely clear atmospheric conditions when the Rayleigh scattering is not as preferential in scattering the blue light, the flash has been reported as blue in color.

44.9 ACKNOWLEDGMENTS

We would like to acknowledge the contributions and help received in the preparation of this chapter and in the delineation of the authors' work. The authors divided the writing of the chapter sections as follows: D. K. Killinger served as lead author and wrote Secs. 43.2 through 43.4 and Secs. 43.7 and 43.8. L. S. Rothman wrote and provided the extensive background information on HITRAN, FASCODE, and LOWTRAN in Sec. 43.5. The comprehensive Sec. 43.6 "Atmospheric Optical Turbulence" was written by J. H. Churnside. The data of Fig. 29 were provided by G. R. Ochs of NOAA/WPL and the data in Fig. 30 were provided by R. R. Beland of the Geophysics Directorate, Phillips Laboratory. We wish to thank Prof. Robert Greenler for providing original photographs of the meteorological optics phenomena; Paul Hays, Vincent Abreu, and Wilbert Skinner for information on the HRDI instrument; P. McCormick and D. Winker for SAGE II data; Mike Hardesty for Doppler lidar wind profiles; and Ed Browell for lidar ozone mapping data. We want to thank A. Jursa for providing a copy of the *Handbook of Geophysics.* R. Measures for permission to use diagrams from his book *Laser Remote Sensing,* and M. Thomas and D. Duncan for providing a prepublication copy of their chapter on atmospheric optics for *The Infrared Handbook.*

Finally, we wish to thank many of our colleagues who have suggested topics and technical items added to this work. We hope that the reader will gain an overall feeling of atmospheric optics from reading this chapter, and we encourage the reader to use the references cited for further in-depth study.

44.10 REFERENCES

1. R. M. Goody and Y. L. Young, *Atmospheric Radiation,* Oxford University Press, 1989.
2. W. G. Driscoll (ed.), Optical Society of America, *Handbook of Optics,* McGraw-Hill, 1978.
3. A. S. Jursa (sci. ed.), *Handbook of Geophysics and the Space Environment,* Air Force Geophysics Lab., NTIS Doc#ADA16700, 1985.
4. W. Wolfe and G. Zissis, *The Infrared Handbook,* Office of Naval Research, Wash. D.C., 1978.
5. R. Measures, *Laser Remote Sensing,* Wiley-Interscience, John Wiley & Sons, 1984.
6. "Major Concentration of Gases in the Atmosphere," NOAA S/T 76-1562, 1976; "AFGL Atmospheric Constituent Profiles (0–120 km)," AFGL-TR-86-0110, 1986; U.S. Standard Atmosphere, 1962 and 1976; Supplement 1966, U.S. Printing Office, Wash. D.C., 1976.
7. E. P. Shettle and R. W. Fenn, "Models of the Aerosols of the Lower Atmosphere and the Effects of Humidity Variations on Their Optical Properties," AFGL TR-79-0214; ADA 085951, 1979; A. Force, D. K. Killinger, W. DeFeo, and N. Menyuk, "Laser Remote Sensing of Atmospheric Ammonia Using a CO_2 Lidar System," *Appl. Opt.* **24:**2837 (1985); B. Nilsson, "Meteorological Influence on Aerosol Extinction in the 0.2–40 μm Range," *Appl. Opt.* **18:**3457 (1979).
8. J. F. Luhr, "Volcanic Shade Causes Cooling," *Nature* **354:**104 (1991).
9. L. S. Rothman, R. R. Gamache, A. Goldman, L. R. Brown, R. A. Toth, H. Pickett, R. Poynter, J. M. Flaud, C. Camy-Peyret, A. Barbe, N. Husson, C. P. Rinsland, and M. A. H. Smith, "The HITRAN Database: 1986 Edition," *Appl. Optics* **26:**4058 (1986); L. S. Rothman, R. R. Gamache, R. H. Tipping, C. P. Rinsland, M. A. H. Smith, D. C. Benner, V. Malathy Devi, J. M. Flaud, C. Camy-Peyret, A. Perrin, A. Goldman, S. T. Massie, L. R. Brown, and R. A. Toth, "The HITRAN Molecular Database: Editions of 1991 and 1992," *J. Quant. Spectrosc. Radiat. Transfer* **48:**469 (1992).
10. E. E. Whiting, "An Empirical Approximation to the Voigt Profile," *J. Quant. Spectrosc. Radiat. Transfer* **8:**1379 (1968); J. Olivero and R. Longbothum, "Empirical Fits to the Voigt Linewidth: A Brief Review," *J. Quant. Spectrosc. Radiat. Transfer* **17:**233 (1977); F. Schreir, "The Voigt and Complex Error Function—A Comparison of Computational Methods," *J. Quant. Spectrosc. Radiat. Transfer* **48:**734 (1992).
11. D. E. Burch, "Continuum absorption by H_2O," AFGL-TR-81-0300; ADA 112264, 1981; S. A. Clough, F. X. Kneizys, and R. W. Davies, "Lineshape and the Water Vapor Continuum," *Atmospheric Research* **23:**229 (1989).
12. Shardanand and A. D. Prasad Rao, "Absolute Rayleigh Scattering Cross Section of Gases and Freons of Stratospheric Interest in the Visible and Ultraviolet Region," NASA TN 0-8442 (1977); R. T. H. Collins and P. B. Russell, "Lidar Measurement of Particles and Gases by Elastic and Differential Absorption," in D. Hinkley (ed.), *Laser Monitoring of the Atmosphere,* Springer-Verlag, 1976.
13. E. U. Condon and H. Odishaw (eds.), *Handbook of Physics,* McGraw-Hill, 1967.
14. S. R. Pal and A. I. Carswell, "Polarization Properties of Lidar Backscattering from Clouds," *Appl. Opt.* **12:**1530 (1973).
15. G. Mie, "Bertrage Z. Phys. TruberMedien, Spezeziell kolloidaler Metallosungen," *Ann. Physik* **25:**377 (1908).
16. D. Deirnendjian, "Scattering and Polarization Properties of Water Clouds and Hazes in the Visible and Infrared," *Appl. Optics* **2:**187 (1964).

17. E. J. McCartney, *Optics of the Atmosphere,* Wiley, New York, 1976.

18. M. Wright, E. Proctor, L. Gasiorek, and E. Liston, "A Preliminary Study of Air Pollution Measurement by Active Remote Sensing Techniques," NASA CR-132724, 1975.

19. P. McCormick and D. Winker, NASA/LaRC: 1 μm lidar measurement of aerosol distribution, private communication, 1991.

20. D. K. Killinger and N. Menyuk, "Laser Remote Sensing of the Atmosphere," *Science* **235:**37 (1987).

21. R. W. Boyd, *Nonlinear Optics,* Academic Press, Orlando, Fla., 1992.

22. M. D. Levenson and S. Kano, *Introduction to Nonlinear Laser Spectroscopy,* Academic Press, Boston, 1988.

23. S. A. Clough, F. X. Kneizys, E. P. Shettle, and G. P. Anderson, "Atmospheric Radiance and Transmittance: FASCOD2," *Proc. of Sixth Conf. on Atmospheric Radiation,* Williamsburg, Va., published by Am. Meteorol. Soc., Boston, 1986; J. A. Dowling, W. O. Gallery, and S. G. O'Brian, "Analysis of Atmospheric Interferometer Data," AFGL-TR-84-0177, 1984.

24. R. Isaacs, S. Clough, R. Worsham, J. Moncet, B. Lindner, and L. Kaplan, "Path Characterization Algorithms for FASCODE," Tech Report GL-TR-90-0080, AFGL, 1990; ADA#231914.

25. F. X. Kneizys, E. Shettle, W. O. Gallery, J. Chetwynd, L. Abreu, J. Selby, S. Clough, and R. Fenn, "Atmospheric Transmittance/Radiance: Computer code LOWTRAN6," AFGL TR-83-0187, 1983; ADA#137786.

26. F. X. Kneizys, E. Shettle, L. Abreu, J. Chetwynd, G. Anderson, W. O. Gallery, J. E. A. Selby, and S. Clough, "Users guide to LOWTRAN7," AFGL TR-88-0177, 1988; ADA#206773.

27. The HITRAN, LOWTRAN, and FASCODE programs can be ordered in a magnetic tape format from the U.S. Department of Commerce, National Climatic Data Center of NOAA, Federal Building, Asheville, NC 28801 (USA); HITRAN in a CD-ROM format is available from L. Rothman, Phillips Laboratory, GPOS, Hanscom AFB, MA 01731-3013, or from ONTAR Corp., 9 Village Way, North Andover, MA 01845-2000.

28. PClnTRAN and PCTRAN computer programs, ONTAR Corp., 9 Village Way, North Andover, MA 01845-2000.

29. HITRAN-PC database and TRANS Program, USF Research Foundation, University of South Florida, Tampa 33620 (USA).

30. LAB_CALC, Galactic Industries, 395 Main St., Salem, NH, 03079 USA; NIST/EPA Gas Phase Infrared Database, U.S. Dept. of Commerce, NIST, Standard Ref. Data, Gaithersburg, MD 20899; Infrared Analytics, 1424 N. Central Park Ave, Anaheim, CA 92802; Aldrich Library of Spectra, Aldrich Co., Milwaukee, WS 53201; Sadtler Spectra Data, Philadelphia, PA 19104-2596; Coblentz Society, P.O. Box 9952, Kirkwood, MO 63122.

31. R. S. Lawrence, G. R. Ochs, and S. F. Clifford, "Measurements of Atmospheric Turbulence Relevant to Optical Propagation," *J. Opt. Soc. Am.* **60:**826–830 (1970).

32. M. A. Kallistratova and D. F. Timanovskiy, "The Distribution of the Structure Constant of Refractive Index Fluctuations in the Atmospheric Surface Layer," *Iz. Atmos. Ocean. Phys.* **7:**46–48 (1971).

33. A. S. Monin and A. M. Obukhov, "Basic Laws of Turbulent Mixing in the Ground Layer of the Atmosphere," *Trans. Geophys. Inst. Akad. Nauk. USSR* **151:**163–187 (1954).

34. A. S. Monin and A. M. Yaglom, *Statistical Fluid Mechanics: Mechanics of Turbulence,* MIT Press, Cambridge, 1971.

35. J. C. Wyngaard, Y. Izumi, and S. A. Collins, Jr., "Behavior of the Refractive-index-structure Parameter Near the Ground," *J. Opt. Soc. Am.* **61:**1646–1650 (1971).

36. L. R. Tsvang, "Microstructure of Temperature Fields in the Free Atmosphere," *Radio Sci.* **4:**1175–1177 (1969).

37. A. S. Frisch and G. R. Ochs, "A Note on the Behavior of the Temperature Structure Parameter in a Convective Layer Capped by a Marine Inversion," *J. Appl. Meteorol.* **14:**415–419 (1975).

38. K. L. Davidson, T. M. Houlihan, C. W. Fairall, and G. E. Schader, "Observation of the Temperature Structure Function Parameter, C_T^2, over the Ocean," *Boundary-Layer Meteorol.* **15:**507–523 (1978).

39. K. E. Kunkel, D. L. Walters, and G. A. Ely, "Behavior of the Temperature Structure Parameter in a Desert Basin," *J. Appl. Meteorol.* **15:**130–136 (1981).
40. W. Kohsiek, "Measuring C_T^2, C_Q^2, and C_{TQ} in the Unstable Surface Layer, and Relations to the Vertical Fluxes of Heat and Moisture," *Boundary-Layer Meteorol,* **24:**89–107 (1982).
41. M. S. Belen'kiy, V. V. Boronyev, N. Ts. Gomboyev, and V. L. Mironov, *Sounding of Atmospheric Turbulence,* Nauka, Novosibirsk, 1986, p. 114.
42. A. A. M. Holtslag and A. P. Van Ulden, "A Simple Scheme for Daytime Estimates of the Surface Fluxes from Routine Weather Data," *J. Clim. Appl. Meteorol.* **22:**517–529 (1983).
43. T. Thiermann and A. Kohnle, "A Simple Model for the Structure Constant of Temperature Fluctuations in the Lower Atmosphere," *J. Phys. D: Appl. Phys.* **21:**S37–S40 (1988).
44. E. A. Andreas, "Estimating C_n^2 over Snow and Sea Ice from Meteorological Data," *J. Opt. Soc. Am. A* **5:**481–494 (1988).
45. J. L. Bufton, P. O. Minott, and M. W. Fitzmaurice, "Measurements of Turbulence Profiles in the Troposphere," *J. Opt. Soc. Am.* **62:**1068–1070 (1972).
46. F. W. Eaton, W. A. Peterson, J. R. Hines, K. R. Peterman, R. E. Good, R. R. Beland, and J. W. Brown, "Comparisons of VHF Radar, Optical, and Temperature Fluctuation Measurements of C_n^2, r_0, and Θ_0," *Theor. Appl. Climatol.* **39:**17–29 (1988).
47. F. Dalaudier, M. Crochet, and C. Sidi, "Direct Comparison between in situ and Radar Measurements of Temperature Fluctuation Spectra: A Puzzling Result," *Radio Sci.* **24:**311–324 (1989).
48. D. W. Beran, W. H. Hooke, and S. F. Clifford, "Acoustic Echo-sounding Techniques and Their Application to Gravity-wave, Turbulence, and Stability Studies," *Boundary-Layer Meteorol.* **4:**133–153 (1973).
49. M. Fukushima, K. Akita, and H. Tanaka, "Night-time Profiles of Temperature Fluctuations Deduced from Two-year Solar Observation," *J. Meteorol. Soc. Jpn.* **53:**487–491 (1975).
50. D. N. Asimakopoulis, R. S. Cole, S. J. Caughey, and B. A. Crease, "A Quantitative Comparison between Acoustic Sounder Returns and the Direct Measurement of Atmospheric Temperature Fluctuations," *Boundary-Layer Meteorol.* **10:**137–147 (1976).
51. T. E. VanZandt, J. L. Green, K. S. Gage, and W. L. Clark, "Vertical Profiles of Refractivity Turbulence Structure Constant: Comparison of Observations by the Sunset Radar with a New Theoretical Model," *Radio Sci.* **13:**819–829 (1978).
52. K. S. Gage and B. B. Balsley, "Doppler Radar Probing of the Clear Atmosphere," *Bull. Am. Meteorol. Soc.* **59:**1074–1093 (1978).
53. R. B. Chadwick and K. P. Moran, "Long-term measurements of C_n^2 in the Boundary Layer," *Raio Sci.* **15:**355–361 (1980).
54. B. B. Balsley and V. L. Peterson, "Doppler-radar Measurements of Clear Air Turbulence at 1290 MHz," *J. Appl. Meteorol.* **20:**266–274 (1981).
55. E. E. Gossard, R. B. Chadwick, T. R. Detman, and J. Gaynor, "Capability of Surface-based Clear-air Doppler Radar for Monitoring Meteorological Structure of Elevated Layers," *J. Clim. Appl. Meteorol.* **23:**474 (1984).
56. G. D. Nastrom, W. L. Ecklund, K. S. Gage, and R. G. Strauch, "The Diurnal Variation of Backscattered Power from VHF Doppler Radar Measurements in Colorado and Alaska," *Radio Sci.* **20:**1509–1517 (1985).
57. D. L. Fried, "Remote Probing of the Optical Strength of Atmospheric Turbulence and of Wind Velocity," *Proc. IEEE* **57:**415–420 (1969).
58. J. W. Strohbehn, "Remote Sensing of Clear-air Turbulence," *J. Opt. Soc. Am.* **60:**948 (1970).
59. J. Vernin and F. Roddier, "Experimental Determination of Two-dimensional Power Spectra of Stellar Light Scintillation. Evidence for a Multilayer Structure of the Air Turbulence in the Upper Troposphere," *J. Opt. Soc. Am.* **63:**270–273 (1973).
60. G. R. Ochs, T. Wang, R. S. Lawrence, and S. F. Clifford, "Refractive Turbulence Profiles Measured by One-dimensional Spatial Filtering of Scintillations," *Appl. Opt.* **15:**2504–2510 (1976).

61. R. E. Hufnagel and N. R. Stanley, "Modulation Transfer Function Associated with Image Transmission through Turbulent Media," *J. Opt. Soc. Am.* **54:**52–61 (1964).

62. R. E. Hufnagel, "Variations of Atmospheric Turbulence," in *Technical Digest of Topical Meeting on Optical Propagation through Turbulence,* Optical Society of America, Washington, D.C., 1974.

63. R. J. Sasiela, *A Unified Approach to Electromagnetic Wave Propagation in Turbulence and the Evaluation of Multiparameter Integrals,* Technical Report 807, MIT Lincoln Laboratory, Lexington, 1988.

64. V. A. Banakh and V. L. Mironov, *Lidar in a Turbulent Atmosphere,* Artech House, Boston, 1987.

65. C. W. Fairall and R. Markson, "Aircraft Measurements of Temperature and Velocity Microturbulence in the Stably Stratified Free Troposphere," *Proceedings of the Seventh Symposium on Turbulence and Diffusion,* November 12–15, 1985, Boulder, Co.

66. J. C. Kaimal, *The Atmospheric Boundary Layer—Its Structure and Measurement,* Indian Institute of Tropical Meteorology, Pune, 1988.

67. J. Weinstock, "Vertical Turbulent Diffusion in a Stably Stratified Fluid," *J. Atmos. Sci.* **35:**1022–1027 (1978).

68. J. Barat and F. Bertin, "On the Contamination of Stratospheric Turbulence Measurements by Wind Shear," *J. Atmos. Sci.* **41:**819–827 (1984).

69. C. E. Coulman, J. Vernin, Y. Coqueugniot, and J. L. Caccia, "Outer Scale of Turbulence Appropriate to Modeling Refractive-index Structure Profiles," *Appl. Opt.* **27:**155–160 (1988).

70. C. E. Coulman and J. Vernin, "Significance of Anisotropy and the Outer Scale of Turbulence for Optical and Radio Seeing," *Appl. Opt.* **30:**118–126 (1991).

71. M. M. Colavita, "Design Considerations for Very Long Baseline Fringe-tracking Interferometers," *Proc. SPIE* **1237:**80–86 (1990).

72. L. A. Chernov, *Wave Propagation in a Random Medium,* Dover, New York, 1967, p. 26.

73. P. Beckmann, "Signal Degeneration in Laser Beams Propagated through a Turbulent Atmosphere," *Radio Sci.* **69D:**629–640 (1965).

74. T. Chiba, "Spot Dancing of the Laser Beam Propagated through the Atmosphere," *Appl. Opt.* **10:**2456–2461 (1971).

75. J. H. Churnside and R. J. Lataitis, "Wander of an Optical Beam in the Turbulent Atmosphere," *Appl. Opt.* **29:**926–930 (1990).

76. G. A. Andreev and E. I. Gelfer, "Angular Random Walks of the Center of Gravity of the Cross Section of a Diverging Light Beam," *Radiophys. Quantum Electron.* **14:**1145–1147 (1971).

77. M. A. Kallistratova and V. V. Pokasov, "Defocusing and Fluctuations of the Displacement of a Focused Laser Beam in the Atmosphere," *Radiophys. Quantum Electron.* **14:**940–945 (1971).

78. J. A. Dowling and P. M. Livingston, "Behavior of Focused Beams in Atmospheric Turbulence: Measurements and Comments on the Theory," *J. Opt. Soc. Am.* **63:**846–858 (1973).

79. J. R. Dunphy and J. R. Kerr, "Turbulence Effects on Target Illumination by Laser Sources: Phenomenological Analysis and Experimental Results," *Appl. Opt.* **16:**1345–1358 (1977).

80. V. I. Klyatskin and A. I. Kon, "On the Displacement of Spatially Bounded Light Beams in a Turbulent Medium in the Markovian-random-process Approximation," *Radiophys. Quantum Electron.* **15:**1056–1061 (1972).

81. A. I. Kon, V. L. Mironov, and V. V. Nosov, "Dispersion of Light Beam Displacements in the Atmosphere with Strong Intensity Fluctuations," *Radiophys. Quantum Electron.* **19:**722–725 (1976).

82. V. L. Mironov and V. V. Nosov, "On the Theory of Spatially Limited Light Beam Displacements in a Randomly Inhomogeneous Medium," *J. Opt. Soc. Am.* **67:**1073–1080 (1977).

83. R. F. Lutomirski and H. T. Yura, "Propagation of a Finite Optical Beam in an Inhomogeneous Medium" *Appl. Opt.* **10:**1652–1658 (1971).

84. R. F. Lutomirski and H. T. Yura, "Wave Structure Function and Mutual Coherence Function of an Optical Wave in a Turbulent Atmosphere," *J. Opt. Soc. Am.* **61:**482–487 (1971).

85. H. T. Yura, "Atmospheric Turbulence Induced Laser Beam Spread," *Appl. Opt.* **10:**2771–2773 (1971).

86. H. T. Yura, "Mutual Coherence Function of a Finite Cross Section Optical Beam Propagating in a Turbulent Medium," *Appl. Opt.* **11:**1399–1406 (1972).

87. H. T. Yura, "Optical Beam Spread in a Turbulent Medium: Effect of the Outer Scale of Turbulence," *J. Opt. Soc. Am.* **63:**107–109 (1973).

88. H. T. Yura, "Short-term Average Optical-beam Spread in a Turbulent Medium," *J. Opt. Soc. Am.* **63:**567–572 (1973).

89. M. T. Tavis and H. T. Yura, "Short-term Average Irradiance Profile of an Optical Beam in a Turbulent Medium," *Appl. Opt.* **15:**2922–2931 (1976).

90. R. L. Fante, "Electromagnetic Beam Propagation in Turbulent Media," *Proc. IEEE* **63:**1669–1692 (1975).

91. R. L. Fante, "Electromagnetic Beam Propagation in Turbulent Media: An Update," *Proc. IEEE* **68:**1424–1443 (1980).

92. G. C. Valley, "Isoplanatic Degradation of Tilt Correction and Short-term Imaging Systems," *Appl. Opt.* **19:**574–577 (1980).

93. H. J. Breaux, *Correlation of Extended Huygens-Fresnel Turbulence Calculations for a General Class of Tilt Corrected and Uncorrected Laser Apertures,* Interim Memorandum Report No. 600, U.S. Army Ballistic Research Laboratory, 1978.

94. D. M. Cordray, S. K. Searles, S. T. Hanley, J. A. Dowling, and C. O. Gott, "Experimental Measurements of Turbulence Induced Beam Spread and Wander at 1.06, 3.8, and 10.6 μm," *Proc. SPIE* **305:**273–280 (1981).

95. S. K. Searles, G. A. Hart, J. A. Dowling, and S. T. Hanley, "Laser Beam Propagation in Turbulent Conditions," *Appl. Opt.* **30:**401–406 (1991).

96. J. H. Churnside and R. J. Lataitis, "Angle-of-arrival Fluctuations of a Reflected Beam in Atmospheric Turbulence," *J. Opt. Soc. Am. A* **4:**1264–1272 (1987).

97. D. L. Fried, "Optical Resolution through a Randomly Inhomogeneous Medium for Very Long and Very Short Exposures," *J. Opt. Soc. Am.* **56:**1372–1379 (1966).

98. R. F. Lutomirski, W. L. Woodie, and R. G. Buser, "Turbulence-degraded Beam Quality: Improvement Obtained with a Tilt-correcting Aperture," *Appl. Opt.* **16:**665–673 (1977).

99. D. L. Fried, "Statistics of a Geometrical Representation of Wavefront Distortion," *J. Opt. Soc. Am.* **55:**1427–1435 (1965); **56:**410 (1966).

100. D. M. Chase, "Power Loss in Propagation through a Turbulent Medium for an Optical-heterodyne System with Angle Tracking," *J. Opt. Soc. Am.* **56:**33–44 (1966).

101. J. H. Churnside and C. M. McIntyre, "Partial Tracking Optical Heterodyne Receiver Arrays," *J. Opt. Soc. Am.* **68:**1672–1675 (1978).

102. V. I. Tatarskii, *The Effects of the Turbulent Atmosphere on Wave Propagation,* Israel Program for Scientific Translations, Jerusalem, 1971.

103. R. W. Lee and J. C. Harp, "Weak Scattering in Random Media, with Applications to Remote Probing," *Proc. IEEE* **57:**375–406 (1969).

104. R. S. Lawrence and J. W. Strohbehn, "A Survey of Clear-air Propagation Effects Relevant to Optical Communications," *Proc. IEEE* **58:**1523–1545 (1970).

105. S. F. Clifford, "The Classical Theory of Wave Propagation in a Turbulent Medium," in J. W. Strohbehn (ed.), *Laser Beam Propagation in the Atmosphere,* Springer-Verlag, New York, 1978, pp. 9–43.

106. A. I. Kon and V. I. Tatarskii, "Parameter Fluctuations of a Space-limited Light Beam in a Turbulent Atmosphere," *Izv. VUZ Radiofiz.* **8:**870–875 (1965).

107. R. A. Schmeltzer, "Means, Variances, and Covariances for Laser Beam Propagation through a Random Medium," *Quart. J. Appl. Math.* **24:**339–354 (1967).

108. D. L. Fried and J. B. Seidman, "Laser-beam Scintillation in the Atmosphere," *J. Opt. Soc. Am.* **57:**181–185 (1967).

109. D. L. Fried, "Scintillation of a Ground-to-Space Laser Illuminator," *J. Opt. Soc. Am.* **57:**980–983 (1967).
110. Y. Kinoshita, T. Asakura, and M. Suzuki, "Fluctuation Distribution of a Gaussian Beam Propagating through a Random Medium," *J. Opt. Soc. Am.* **58:**798–807 (1968).
111. A. Ishimaru, "Fluctuations of a Beam Wave Propagating through a Locally Homogeneous Medium," *Radio Sci.* **4:**295–305 (1969).
112. A. Ishimaru, "Fluctuations of a Focused Beam Wave for Atmospheric Turbulence Probing," *Proc. IEEE* **57:**407–414 (1969).
113. A. Ishimaru, "The Beam Wave Case and Remote Sensing," in J. W. Strohbehn (ed.), *Laser Beam Propagation in the Atmosphere,* Springer-Verlag, New York, 1978, pp. 129–170.
114. P. J. Titterton, "Scintillation and Transmitter-aperture Averaging over Vertical Paths," *J. Opt. Soc. Am.* **63:**439–444 (1973).
115. H. T. Yura and W. G. McKinley, "Optical Scintillation Statistics for IR Ground-to-Space Laser Communication Systems," *Appl. Opt.* **22:**3353–3358 (1983).
116. P. A. Lightsey, J. Anspach, and P. Sydney, "Observations of Uplink and Retroreflected Scintillation in the Relay Mirror Experiment," *Proc. SPIE* **1482** (in press) (1991).
117. D. L. Fried, "Aperture Averaging of Scintillation," *J. Opt. Soc. Am.* **57:**169–175 (1967).
118. J. H. Churnside, "Aperture Averaging of Optical Scintillations in the Turbulent Atmosphere," *Appl. Opt.* **30:**1982–1994 (1991).
119. J. H. Churnside and R. G. Frehlich, "Experimental Evaluation of Log-normally Modulated Rician and IK Models of Optical Scintillation in the Atmosphere," *J. Opt. Soc. Am. A* **6:**1760–1766 (1989).
120. G. Parry and P. N. Pusey, "K Distributions in Atmospheric Propagation of Laser Light," *J. Opt. Soc. Am.* **69:**796–798 (1979).
121. G. Parry, "Measurement of Atmospheric Turbulence Induced Intensity Fluctuations in a Laser Beam," *Opt. Acta* **28:**715–728 (1981).
122. W. A. Coles and R. G. Frehlich, "Simultaneous Measurements of Angular Scattering and Intensity Scintillation in the Atmosphere," *J. Opt. Soc. Am.* **72:**1042–1048 (1982).
123. J. H. Churnside and S. F. Clifford, "Lognormal Rician Probability-density Function of Optical Scintillations in the Turbulent Atmosphere," *J. Opt. Soc. Am. A* **4:**1923–1930 (1987).
124. R. Dashen, "Path Integrals for Waves in Random Media," *J. Math. Phys.* **20:**894–920 (1979).
125. K. S. Gochelashvily and V. I. Shishov, "Multiple Scattering of Light in a Turbulent Medium," *Opt. Acta* **18:**767–777 (1971).
126. K. S. Gochelashvily, V. G. Pevgov, and V. I. Shishov, "Saturation of Fluctuations of the Intensity of Laser Radiation at Large Distances in a Turbulent Atmosphere (Fraunhofer Zone of Transmitter)," *Sov. J. Quantum Electron.* **4:**632–637 (1974).
127. A. M. Prokhorov, F. V. Bunkin, K. S. Gochelashvily, and V. I. Shishov, "Laser Irradiance Propagation in Turbulent Media," *Proc. IEEE* **63:**790–810 (1975).
128. R. L. Fante, "Inner-scale Size Effect on the Scintillations of Light in the Turbulent Atmosphere," *J. Opt. Soc. Am.* **73:**277–281 (1983).
129. R. G. Frehlich, "Intensity Covariance of a Point Source in a Random Medium with a Kolmogorov Spectrum and an Inner Scale of Turbulence," *J. Opt. Soc. Am. A.* **4:**360–366 (1987).
130. J. H. Churnside and R. J. Hill, "Probability Density of Irradiance Scintillations for Strong Path-integrated Refractive Turbulence," *J. Opt. Soc. Am. A.* **4:**727–733 (1987).
131. D. K. Killinger and A. Mooradian (eds.), *Optical and Laser Remote Sensing,* Springer-Verlag, Optical Sciences, vol. 39, 1983.
132. L. J. Radziemski, R. W. Solarz, and J. A. Paisner (eds.), *Laser Spectroscopy and Its Applications,* Marcel Dekker, Optical Eng., vol. 11, 1987.
133. T. Kobayashi, "Techniques for Laser Remote Sensing of the Environment," *Remote Sensing Reviews,* **3:**1–56 (1987).
134. E. D. Hinkley (ed.), *Laser Monitoring of the Atmosphere,* Springer-Verlag, Berlin, 1976.

135. W. B. Grant and R. T. Menzies, "A Survey of Laser and Selected Optical Systems for Remote Measurement of Pollutant Gas Concentrations," *APCA Journal* **33:**187 (1983).

136. "Optical Remote Sensing of the Atmosphere," *Conf. Proceedings, OSA Topical Meeting,* Williamsburg, 1991.

137. P. B. Hays, V. J. Abreu, D. A. Gell, H. J. Grassl, W. R. Skinner, and M. E. Dobbs, "The High Resolution Doppler Imager on the Upper Atmospheric Research Satellite," *J. Geophys. Res. (Atmosphere)* **98:**10713 (1993); P. B. Hays, V. J. Abreu, M. D. Burrage, D. A. Gell, A. R. Marshall, Y. T. Morton, D. A. Ortland, W. R. Skinner, D. L. Wu, and J. H. Yee, "Remote Sensing of Mesopheric Winds with the High Resolution Imager," *Planet. Space Sci.* **40:**1599 (1992).

138. E. Browell, "Differential Absorption Lidar Sensing of Ozone," *Proc. IEEE* **77:**419 (1989).

139. J. M. Intrieri, A. J. Dedard, and R. M. Hardesty, "Details of Colliding Thunderstorm Outflow as Observed by Doppler Lidar," *J. Atmospheric Sciences* **47:**1081 (1990); R. M. Hardesty, K. Elmore, M. E. Jackson, in *21st Conf. on Radar Meteorology,* American Meteorology Society, Boston, 1983.

140. M. P. McCormick and R. E. Veiga, "Initial Assessment of the Stratospheric and Climatic Impact of the 1991 Mount Pinatubo Eruption—Prolog," *Geophysical Research Lett.* **19:**155 (1992).

141. R. A. R. Tricker, *Introduction to Meteorological Optics,* American Elsevier, New York, 1970.

142. R. Greenler, *Rainbows, Halos, and Glories,* Cambridge University Press, 1980.

INDEX

Index note: The *f.* after a page number refers to a figure, the *n.* to a note, and the *t.* to a table.

ABOUT THE EDITORS

Michael Bass is Professor of Physics and Electrical and Computer Engineering at the University of Central Florida and is on the faculty of the Center for Research and Education in Optics and Lasers (CREOL). He received his B.S. in Physics from Carnegie-Mellon and his M.S. and Ph.D. in Physics from the University of Michigan.

Eric Van Stryland is a Professor of Physics and Electrical Engineering at the University of Central Florida Center for Research and Education in Optics and Lasers. He received his Ph.D. from the University of Arizona.

David R. Williams is a Professor of Psychology and Visual Science and the Director of the Center of Visual Science at the University of Rochester. He received his B.S. in Psychology from Denison University, and his M.A. and Ph.D. in Psychology from the University of California, San Diego.

William L. Wolfe is a Professor at the Optical Sciences Center at the University of Arizona. He received his B.S. in Physics from Bucknell University, and his M.S. in Physics and M.S.E. in Electrical Engineering from the University of Michigan.